Scientific Results of the Viking Project

American Geophysical Union
Washington, D. C.

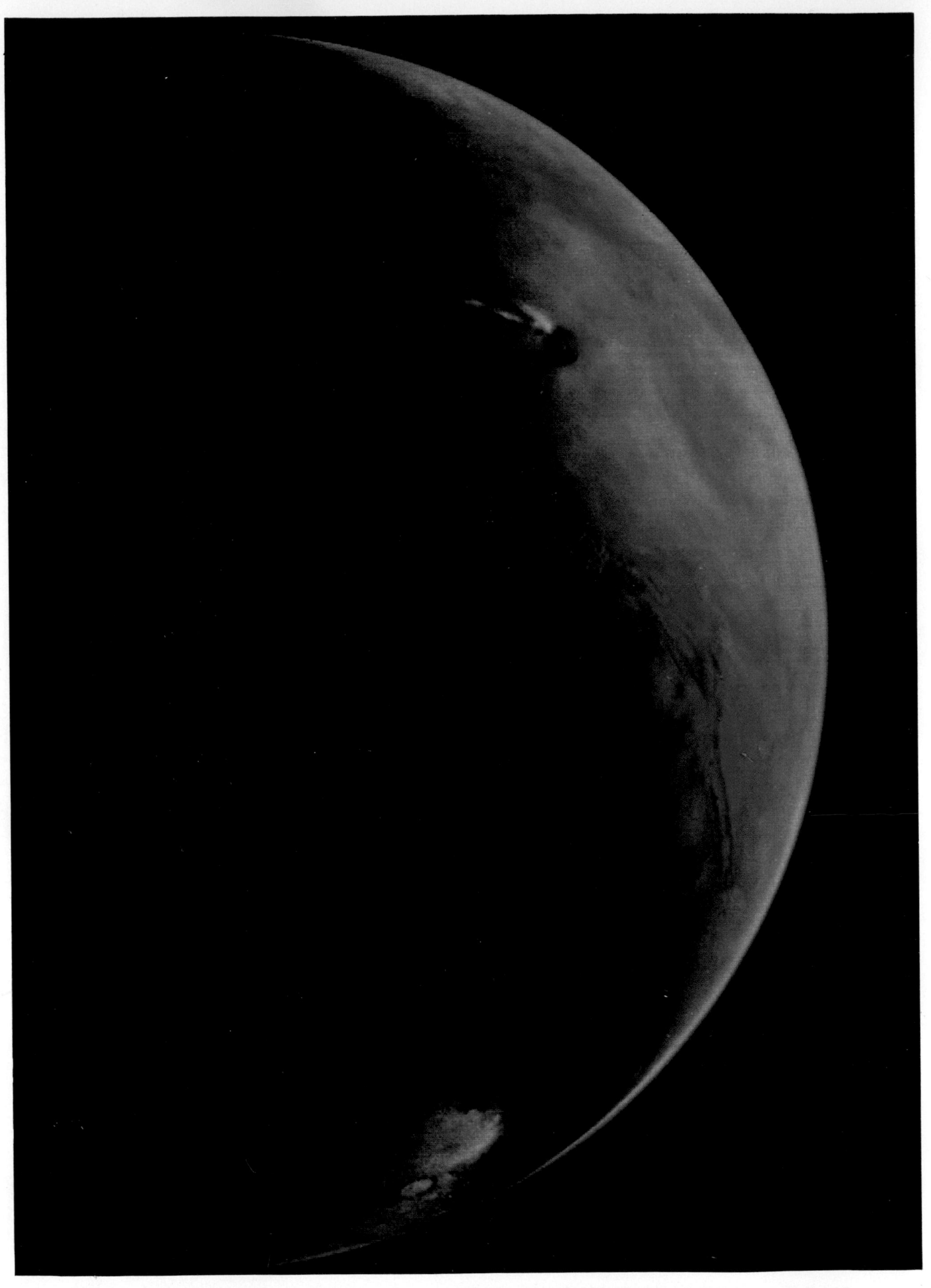

This view of Mars is a composite of pictures taken with different color filters by the second Viking spacecraft as it approached Mars in July of 1976. The morning terminator is seen, and the south pole is tilted toward the camera. In the south is the frost-filled Argyre basin with the crater Galle on its rim. The equatorial canyon system, Valles Marineris, is in the center of the picture, and to the north, clouds trail from the northernmost of the large Tharsis volcanos, Ascraeus Mons.

National Aeronautics and
Space Administration

Washington, D.C.
20546

Office of the Administrator

It has been scarcely a year since the Viking orbiters and landers started their intensive scientific investigations of the red planet. The preparations for the encounter with Mars started in earnest almost a decade earlier and at least that much time will probably be required to fully analyze and comprehend the enormous amount of information we are receiving. The ultimate scientific richness of the mission, however, is already clearly anticipated by over 50 papers in this special issue of the Journal of Geophysical Research. From seismology to biological sciences, the range is a record for a NASA planetary mission; more importantly, scientists, engineers and administrators had to and did work as a coherent team to make it all come together.

The Viking mission was clearly a high point of the U.S. Bicentennial year, and resulted in great national pride. It was perceived as an important technological and exploratory endeavor stemming from a vigorous, forward-looking national commitment made in 1969. The immediate benefits are a rich addition to the store of knowledge about one of our sister planets. Rapidly, as evidenced by some of the papers in the JGR, that knowledge will be used to give us a better perspective of and appreciation for our own planet Earth.

As the newly appointed Administrator of NASA, I congratulate all who participated in the Viking mission. I look forward to learning more about Mars and to working for a continuing, highly productive planetary exploration program.

Robert A. Frosch

Table of Contents

Radio Science

Entry Science

Lander Imaging

Physical Properties

Seismology

Magnetic Properties

Meteorology

Inorganic Chemistry

Molecular Analysis

Biology

VOL. 82, NO. 28 JOURNAL OF GEOPHYSICAL RESEARCH SEPTEMBER 30, 1977

The Viking Project

GERALD A. SOFFEN

NASA Langley Research Center, Hampton, Virginia 23665

The Viking project launched two unmanned spacecraft to Mars in 1975 for scientific exploration with special emphasis on the search for life. Each spacecraft consisted of an orbiter and a lander. The landing sites were finally selected after the spacecraft were in orbit. Thirteen investigations were performed: three mapping experiments from the orbiter, one atmospheric investigation during the lander entry phase, eight experiments on the surface of the planet, and one using the spacecraft radio and radar systems. The experiments on the surface dealt principally with biology, chemistry, geology, and meteorology. Seventy-eight scientists have participated in the 13 teams performing these experiments. This paper is a summary of the project and an introduction to the articles that follow.

PREFACE

The Viking project was initiated in 1968. The teams of scientists and their investigations were selected on a competitive basis by NASA headquarters in 1969. The Viking project is a highly complex organization consisting of these main elements: project management, teams of NASA-appointed scientists, the contractors and subcontractors, assisted by several NASA field centers, and the Jet Propulsion Laboratory (JPL).

Project management was assigned to NASA Langley Research Center by NASA Office of Space Science. Langley was also responsible for the lander system and the mission operations. The scientists interact with project management through the project scientist, who is a part of project management. The principal contractor for the lander was the Martin Marietta Aerospace Corporation, which was also responsible for its integration with the orbiter provided by JPL. JPL was also responsible for tracking and data acquisition and the mission control and computing center. The Lewis Research Center (LRC) was responsible for the launch vehicle, and the Kennedy Space Center for the launch facility. Other government centers such as NASA Ames Research Center, Goddard Space Flight Center, Johnson Space Flight Center, and the astrogeology branch of the U.S. Geological Survey (USGS) at Flagstaff, Arizona, provided assistance.

The main scientific policies were recommended by the Viking science steering group (SSG), which consisted of the project scientist, the orbiter and lander scientists, the program scientist from NASA headquarters, the science integration manager, and the 13 leaders of the science investigation teams. The SSG reported to the project manager. Issues such as science priority, payload weight allocation, data allocation, landing sites, strategy, finances, publication, etc., were the principal work of this group. Each science team handled its own investigation requirements (instrument specifications, operating parameters, data, etc.) by dealing directly with the responsible project engineers.

This elaborate organization proved to be eminently successful in conducting one of the most daring steps in man's exploration of space—the safe landing of two highly complex automated laboratories on the surface of Mars and the significant return of a wealth of scientific knowledge. During the flight operations a group of student interns were selected to aid in handling the heavy load of data returned by the four spacecraft. The interns were selected on the basis of a national competition and assigned to the various science and engineering teams. Their marvelous spirit and insatiable capacity for work was an inspiration to us all. Although the work was sometimes routine, their contributions were very important in maintaining the orderly flow of the daily bulk of data. The intern program illustrates the extraordinary nature of the Viking project. Besides the great scientific and remarkable technical accomplishments there was a spirit ignited from within by the participants and kindled by the outside world that gave the nation an opportunity to share in the thrill of exploration and discovery.

INTRODUCTION

The two Viking missions to Mars are NASA's most ambitious and memorable planetary missions to date. These highly successful ventures have yielded more data about Mars than all previous observations collectively. The scientific and public interest in planetary exploration reached an all-time high [*New York Times*, 1976*a*, *b*]. The effort was the collective activity of many thousands of people focused on a common dedicated goal: the scientific exploration of Mars by unmanned spacecraft.

The exploration of Mars is part of a general long-term scientific objective of understanding the formation and history of the solar system. Our knowledge of the planets has grown significantly over the past 2 decades and has led to some general as well as very specific questions. Some of the broader questions of Mars are:

What are the characteristics of the Martian surface, its chemistry, and its physical nature?

What are the constituents and the physical structure of the upper and lower atmosphere, and what events led to its present composition?

What is the nature of Martian climate, and what governs the daily and seasonal changes?

What is the internal structure of Mars and its history?

Has life ever started on Mars?

Most scientists agree on the kinds of investigations that were required for the first four questions. The differing arguments deal with priority.

The last question has provoked a good deal of controversy in the scientific community. While most scientists recognize the important consequence of discovering life on another planet, many believe that the likelihood of life on Mars is small. Another serious question was whether we were ready to perform an adequate experiment to test for life.

Biologists had recognized the improbability of any universal

Paper number 7S0535.

single test for life. Prior to Viking several laboratories had been developing life detection experiments based upon different premises. Most biologists believe that if there is life on Mars, it is likely to include a microbial form, although we have not overlooked the possibility of finding a larger form with the cameras. The various biology experiments under consideration were all tests for some biochemical response of a Martian soil sample innoculated with some kind of medium. No single test was considered an adequate first test for life.

All terrestrial life is related, with a common chemistry and likely a common origin. Since Mars today is quite different from the earth, the dilemma was to determine the conditions to be used in the life detection tests. In biological terms the major difference between the earth and Mars is in the state of water. The earth is a marine planet, Mars is not. The contemporary atmospheric pressure and temperature on Mars are too low to allow for the accumulation of large bodies of water as we have on the earth. Should the biological tests be done under aqueous conditions as would be done for terrestrial-type life, or should the tests be carried out under ambient Mars conditions with little or no water added? Also, what chemical agents should be added as nutrients? It was finally decided to send a set of biological tests that range in their environmental setting from a totally aqueous milieu, rich in organics, to a Marslike environment with no water or any other additives. Even so, only a very narrow set of all possibilities could be tested on the small samples acquired at our two landing sites.

In any case, because of the dramatic implications of the search for life it has been project policy to stress that not finding life does not prove the absence of life on Mars.

A positive biological result being recognized as a low probability (although of enormous scientific and social importance), there was a great deal of emphasis placed on the analysis of the organic material on Mars. The organic analysis is not a life detection test, but it obviously bears on the biological question and is of great interest to scientists studying chemical evolution, especially that related to biogenesis. It was generally agreed that besides biologically produced organics there were very likely two possible sources: meteoritic infall and de novo synthesis in the Mars atmosphere.

The carbonaceous chondrites are known to contain several percent of organics. The thin atmosphere of Mars should afford a barrier to infalling meteorites, but many should survive entry through the thin Martian atmosphere. The absence of liquid water on Mars suggested that many of these organics might be intact, weathered by the Martian dust storms and the UV radiation.

Numerous experiments have been done which indicate the possibility of organic synthesis in the Martian atmosphere. Horowitz and co-workers have shown that the simple known ingredients of the Mars atmosphere (CO, CO_2, and H_2O) in the presence of UV with appropriate inorganic substrates will form measurable quantities of organic compounds [*Hubbard et al.*, 1971].

Altogether, the organic analysis appeared to relate very strongly to the biological potential of the planet. Since nitrogen had not been detected, the question of nitrogen-bearing organics was of great interest. Since the project's stated goal was to 'improve our knowledge of Mars, with special emphasis on the biological question,' and since the biology was considered a long shot, the organic analysis was considered to be of high priority and an essential investigation for any subsequent life detection experiment.

Therefore the absence of organic compounds (at very low levels of sensitivity) at either of the landing sites is considered by many to be one of the more significant scientific surprises of the Viking results [*Biemann et al.*, 1976].

The Viking payload, consisting of 13 scientific investigations, is shown in Table 1. The investigators are listed in the section on the members of the Viking science teams.

TABLE 1. Science Investigations and Instruments

Investigations	Instruments
Orbiter Imaging	two vidicon cameras
Water Vapor Mapping	infrared spectrometer
Thermal Mapping	infrared radiometers
Entry Science	
Ionospheric properties	retarding potential analyzer
Atmospheric composition	mass spectrometer
Atmospheric structure	pressure, temperature, and acceleration sensors
Lander Imaging	two facsimile cameras
Biology	three analyses for metabolism, growth, or photosyntheses
Molecular Analysis	gas chromatograph mass spectrometer
Inorganic Analysis	X ray fluorescence spectrometer
Meteorology	pressure, temperature, wind velocity sensors
Seismology	three-axis seismometer
Magnetic Properties	magnet on sampler observed by cameras
Physical Properties	various engineering sensors
Radio Science	orbiter and lander radio and radar systems
Celestial mechanics, atmospheric properties, and test of general relativity	

PROJECT PROFILE

The planning phase which started in 1968 determined the major project goals, the vehicles to be used, the institutions involved, and the time line of events [*Soffen and Young*, 1972]. The implementation phase lasted the longest; from 1971 to 1975 the hardware and software were developed, built, tested, and assembled into two identical multisystem spacecraft. The lander portion of each spacecraft was heat sterilized within a bioshield which was not removed until the spacecraft was in space. Both spacecraft were launched in the summer of 1975.

After the launches, until the summer of 1976 when the spacecraft arrived at Mars, there was an intensive period of training the flight team (the scientists and engineers who would conduct the mission). Viking's unique adaptability, which was its hallmark, was in the hands of a newly assembled team, many of whose members came from the engineering and management elements of the previous phases. Additional personnel were added because of special skills in controlling the spacecraft, navigation, and data handling. One unique aspect of the flight team was its mosaic quality. Personnel were assembled on the basis of skill and experience rather than their institutional responsibility. In this sense the flight team was an ad hoc institution of its own, whose training was an essential ingredient to the success of the mission.

From the summer of 1976 the flight team has carried out the mission. During this period the two Viking spacecraft were injected into orbit around Mars, landing sites were found and certified by orbital reconnaissance and earth-based radar, and each lander was soft landed on the Martian surface [*Soffen and Snyder*, 1976; *Soffen*, 1976*a*]. The planned nominal mission

VIKING SCIENCE TEAMS

The project scientist is Gerald A. Soffen, the orbiter scientist is C. W. Snyder, and the program scientist is R. S. Young.

Orbiter Imaging

Michael H. Carr, USGS, Menlo Park
William A. Baum, Lowell Observatory
Karl R. Blasius, Science Applications
Geoffrey Briggs, JPL
James A. Cutts, Science Applications
Thomas C. Duxbury, JPL
Ronald Greeley, University of Santa Clara
John E. Guest, University of London, England
Keith A. Howard, USGS, Reston
Harold Masursky, USGS, Flagstaff
Bradford A. Smith, University of Arizona
Lawrence A. Soderblom, USGS, Flagstaff
John B. Wellman, JPL
Joseph Veverka, Cornell University

Lander Imaging

Thomas A. Mutch, Brown University
Alan B. Binder, Science Applications
Friedrich O. Huck, LRC
Elliott C. Levinthal, Stanford University
Sidney Liebes, Jr., Stanford University
Elliott C. Morris, USGS
James A. Pollack, Ames Research Center
Carl Sagan, Cornell University

Biology

Harold P. Klein, Ames Research Center
Norman H. Horowitz, Caltech
Joshua Lederberg, Stanford University
Gilbert V. Levin, Biospherics
Vance I. Oyama, Ames Research Center
Alexander Rich, MIT

Thermal Mapping

Hugh H. Kieffer, UCLA
Stillman C. Chase, Santa Barbara Research Center
Ellis D. Miner, JPL
Guido Munch, Caltech
Gerry Neugebauer, Caltech

Water Vapor Mapping

C. B. Farmer, JPL
Donald W. Davies, JPL
Dan LaPorte, Santa Barbara Research Center

Entry Science

Alfred O. C. Nier, University of Minnesota
William B. Hanson, University of Texas
Michael B. McElroy, Harvard University
Alfred Seiff, Ames Research Center
Nelson W. Spencer, Goddard Space Flight Center

Molecular Analysis

Klaus Biemann, MIT
Duwayne M. Anderson, Cold Regions Research and Engineering Laboratory, U.S. Army
Alfred O. C. Nier, University of Minnesota
Leslie E. Orgel, Salk Institute
John Oro, University of Houston
Tobias Owen, State University of New York
Priestley Toulmin III, USGS, Reston
Harold C. Urey, University of California, San Diego

Inorganic Chemical

Priestley Toulmin III, USGS, Reston
Alex K. Baird, Pomona College
Benton C. Clark, Martin Marietta Aerospace
Klaus Keil, University of New Mexico
Harry J. Rose, USGS, Reston

Meteorology

Seymour L. Hess, Florida State University
Robert M. Henry, LRC
Conway B. Leovy, University of Washington
Jack A. Ryan, California State University, Fullerton
James E. Tillman, University of Washington

Seismology

Don L. Anderson, Caltech
Fred Duennebier, University of Texas
Robert L. Kovach, Stanford University
Gary V. Latham, University of Texas
George Sutton, University of Hawaii
Nafi Toksöz, MIT

Physical Properties

Richard Shorthill, University of Utah
Robert E. Hutton, TRW
Henry J. Moore II, USGS, Menlo Park
Ronald F. Scott, Caltech

Magnetic Properties

Robert B. Hargraves, Princeton University

Radio Science

William H. Michael, LRC
George Born, JPL
Joseph P. Brenkle, JPL
Dan L. Cain, JPL
J. G. Davies, University of Manchester, England
Gunnar Fjeldbo, JPL
Mario D. Grossi, Raytheon
Robert Reasenberg, MIT
Irwin I. Shapiro, MIT
Charles T. Stelzried, JPL
Robert H. Tolson, LRC
G. Leonard Tyler, Stanford University

---

was completed in November with the onset of solar conjunction. An extended mission was begun at the end of conjunction and will continue through May 1978 or as long as useful data can be attained [*Soffen,* 1976*b*].

VIKING SPACECRAFT AND MISSION DESCRIPTION

Each spacecraft consisted of an orbiter and a heat-sterilized lander within its capsule (Plate 1). Together and with their fuel they weighed 3530 kg just after launch. After landing, the masses were approximately 900 kg for the orbiter and 600 kg for the lander. They were joined by a lander support structure that was to have been jettisoned by the orbiter after the landing. (In the case of orbiter 2 the support structure was not jettisoned because of an engineering problem that developed during separation. This had some consequence on the viewing of the instruments.) Each spacecraft was injected into orbit around Mars (Viking 1 on June 19, Viking 2 on August 7, 1976). During most of the primary mission (up to con-

Plate 1. The Viking spacecraft.

junction), each orbiter was in an elliptical orbit with a periapsis of about 1500 km, an apapsis of 33,000 km, and a period near 1 Mars day (24.6 hours).

The role of the orbiter was to transport the lander to Mars, to carry reconnaissance instruments for certifying the landing sites, to act as a relay station for the lander data, and to perform its own scientific investigations. The periapsis was placed over the candidate landing sites to allow for maximum viewing resolution and relay of the lander data.

A detailed description of the orbiter and its use appears in a subsequent report [*Snyder,* 1977]. Three reconnaissance instruments of the orbiter are mounted on a scanning platform: a pair of high-resolution television cameras, an infrared spectrometer for measuring atmospheric water vapor, and an infrared radiometer for making thermal measurements of the surface and atmosphere. These instruments are all sighted along a common axis to allow mapping of a common area of the planet. They were used during the first 4 weeks for selecting and certifying the landing sites and subsequently for investigations of the Martian surface and atmosphere over a considerable portion of the planet and of its satellites.

For several years prior to the Viking arrival, candidate landing sites were selected on the basis of all available Mariner and ground-based radar and telescopic data. They were periodically reviewed and adjusted as new data became available. The plan was to select sites and to use the Viking orbiters to certify that the sites were safe. The first photos of the Viking 1 site revealed a hazardous surface and forced a change in the site and a delay in the landing (from July 4 to July 20). For 4 weeks we were in a reconnaissance mode [*Masursky and Crabill,* 1976*a*].

Viking 1 arrived at the onset of summer in the northern hemisphere. Unlike the Mariner arrival, which was at Martian perihelion, Viking arrival occurred at aphelion, avoiding the problem of a severe planetary seasonal dust storm during the critical phase of landing.

The landing latitude is prescribed by the periapsis point and becomes essentially fixed at the time of the Mars orbit insertion. The longitude can be changed by changing the period (rate of rotation) of the spacecraft and allowing it to 'walk around the planet' and then stopping it with another maneuver. The selection of the original site (Chryse at 19°N, 34°W) was based on several criteria involving both 'safety' and 'science.' Clearly, safety was the main issue, but if there were several sites of equal safety, we selected the most scientifically interesting. Besides the prime candidate landing site there was a backup site selected at the same latitude (but not within the same geological domain). For lander 1 this was at 252°W longitude. The latitude band that was considered reasonable for Viking was 10°S–55°N based on the available data, the season on Mars, thermal constraints, and a communication link limit in the Martian southern hemisphere.

The safety issues were mainly altitude, winds, and local hazards. Since we did not know the atmospheric pressure at the surface to within several millibars, and since the atmosphere is a critical element in braking the descending lander, it was decided to avoid all regions where the surface is approximately 3 km above the mean altitude. This removed large young volcanic areas in the Tharsis region. The local hazard problem was more severe. The lander on the surface could have been damaged by any boulder larger than 22 cm. Since the best the orbiter cameras could resolve is 100 m, this hazard was dealt with by extrapolation of orbiter photos and interpretation of ground-based radar data [*Masursky and Crabill,* 1976*b*]. (Radar data are confined between 25°N and 25°S, the extremes of the subearth point on Mars.) Areas that had wind streaks or showed changes from Mariner data were avoided. avoided.

The landing ellipse was 100 × 300 km for targeting with 3σ accuracy. (The lander actually landed less than 20 km from the center of the ellipse.) The original Chryse site turned out to have extensive fluvial cratering activity and boulder fields at the 100-m scale (Figure 1), and common sense suggested that it would not be a safe site. The area to the south was known to be very rough with deep channel beds, while the region to the east of the candidate area revealed enormous ancient catastrophic flooding. To the west there was a vast area eroded by winds. A decision to go to the backup site was deferred in hope of finding an acceptable place in the Chryse region. The direction that appeared most promising was toward the northwest, and the spacecraft was maneuvered to photograph that region. Several weeks of intensive photography and interpretation led to a site at 22°N, 48°W for Viking 1 (Chryse).

The selection of the Viking 2 site caused a good deal more debate, since there was a strong desire by the biologists to land in the dampest region of Mars (water being thought to be the limiting ingredient to any indigenous Martian life). Several planetary models were considered [*Farmer,* 1976]. Eventually, the latitude of the candidate sites was fixed at 44°N, a compromise between the desire to go as close as possible to the area on the edge of the northern seasonal polar cap deposits while keeping temperature limits consistent with biologically available water and a requirement that any site must have some high-resolution Mariner photos in that region. The Viking 2 prime site was at 44°N, 10°W, and the backup at 44°N, 110°W.

In the case of Viking 2 the site selection process was even more arduous than for Viking 1 [*Masursky and Crabill,* 1976*b*]. The Viking 1 orbiter photographs of the candidate site revealed an extensive field of volcanic flow, and there was some consideration given to changing the latitude for the second lander to the equatorial belt or the southern hemisphere. Both orbiters continued photographing, and a large part of the belt at 45°N latitude was covered in a search for a region of deposited sand dunes that had migrated from the northern mantle. The final site for lander 2 was at 44°N, 226°W (Utopia).

Once the landing site was selected, the ground command was given to the spacecraft for separation and landing. The distance from the earth to Mars was such that the round trip time for the telemetry signal was about 40 min. This necessitated a completely automated system on board for carrying out the landing maneuver. Because of the thin Martian atmosphere it was necessary to use three sequential braking systems to assure a soft landing. At the time of separation the lander in its aeroshell was traveling at about 4 km/s. After separation it was oriented, and rockets fired to initiate the deorbit sequence. The lander coasted toward the Martian atmosphere for the next few hours, sending back its data to the orbiter, which in turn relayed it back to the earth (Figure 2).

Just prior to entering the Martian atmosphere, about 300 km above the Martian surface, the lander was reoriented for its aerodynamic entry. The aeroshell with an ablatable heat shield was the first braking system operating down to 6 km from the surface, dissipating most of the kinetic energy. It also provided some aerodynamic lift. Entry science instruments in the aeroshell made measurements of the ions and electrons in the upper atmosphere and the neutral species in the lower

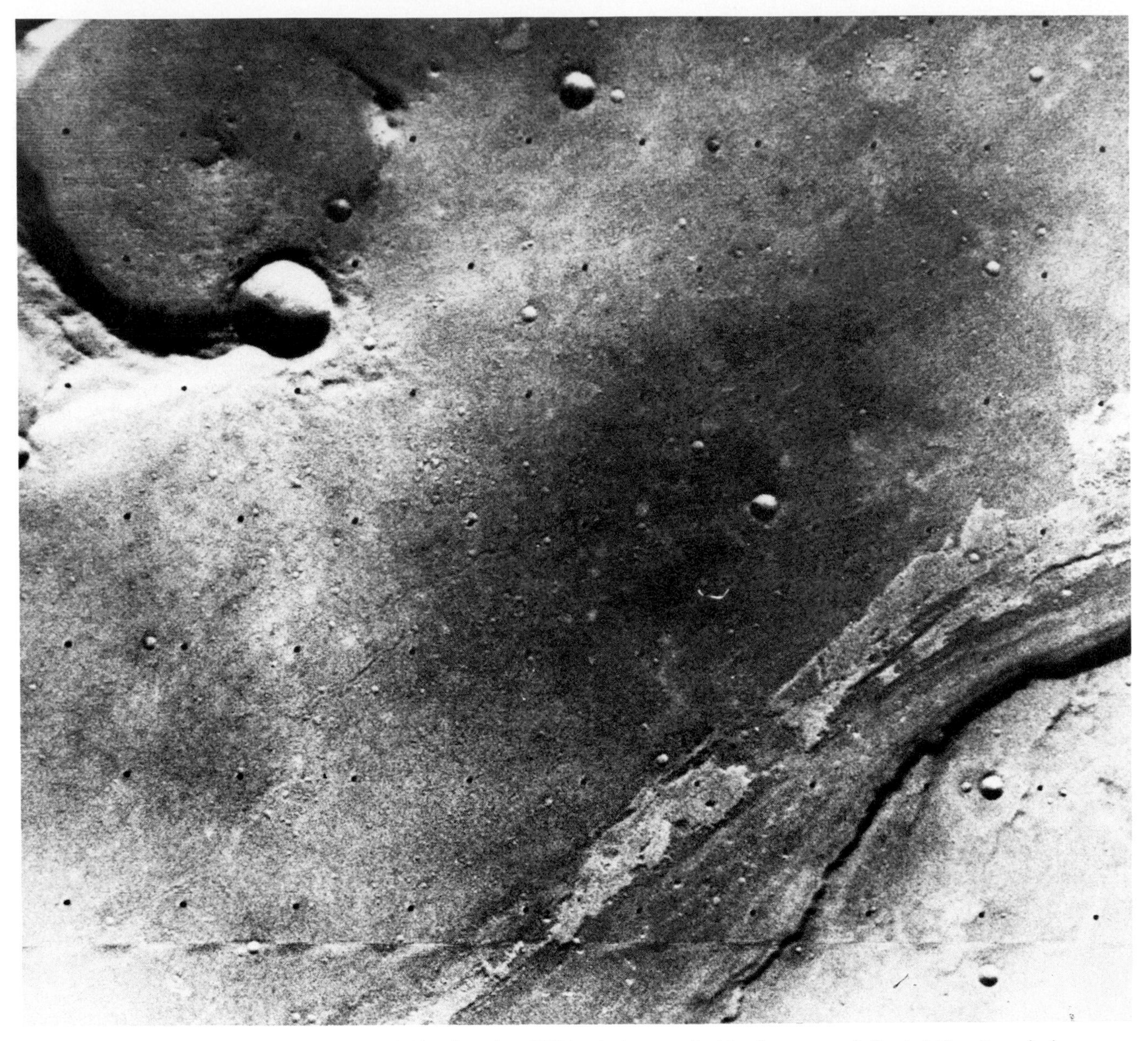

Fig. 1. A channel near the potential landing site of Viking 1 photographed by the spacecraft 3½ min before it reached the lowest point in its orbit around Mars on June 22. This frame was the 33rd of 58 taken during a 7-min period and the first of the sequence transmitted to the earth for scientific analysis. The slightly raised rim and floor markings of the channel are common in Mars channels. A highly degraded crater, about 13 km (8 mi) in diameter, and a fresher, smaller crater lie in the upper left corner. The channel bank cuts into the margin of the larger crater. The picture was taken from a range of 1562 km (963 mi).

atmosphere. Pressure, temperature, and acceleration measurements were also made during the descent of each of the landers [*Nier et al.*, 1976]. The descent of the lander was controlled by a landing radar system which located the surface and a computer system which commanded the events during descent. Accelerometer data were used to determine the rate of fall and subsequently to calculate the Martian atmospheric density profile. Gyros were used as an inertial reference of orientation.

At an altitude of 6 km (descending at 250 m/s), a 50-ft-diameter (15-m-diameter) parachute was deployed by a mortar, and 7 s later the aeroshell was jettisoned. Each lander has three retractable legs, each with a footpad; 8 s after parachute deployment these were extended. The parachute operated for 45 s in slowing the lander to 60 m/s. During this period of descent, measurements were made of the pressure and temperature.

At 1.5 km above the surface, three retro-engines were fired for the final descent to the surface, which was accomplished in 40 s. These engines were throttled by command from the computer, the signals of which came from the landing radar. The final landing was as gentle as falling from a low table (2 m/s). The last 30 m were a vertical path. A switch on the footpad shut off the descent engine. Viking 1 landed in the Martian afternoon on Mars on July 20, and Viking 2 in the morning on September 3, 1976.

During the design of the landing retro-rockets there was concern about the changes they would cause in the chemistry and physical character of the Martian surface in the area around the lander. This was a special problem, since the analytical instruments required samples from the upper layers of the surface. Much effort was spent in developing a retro-rocket engine that would have minimum mechanical and chemical

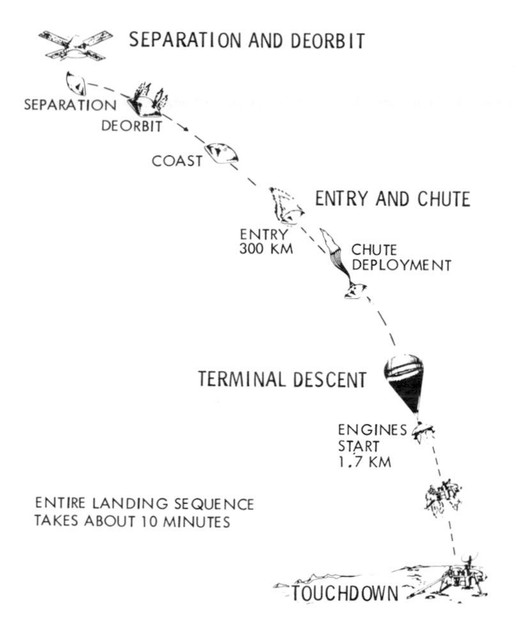

Fig. 2. The lander descent sequence.

effects on the surface. The retro-rocket selected used purified hydrazine as a propellant, which obtains its energy from the exergonic breakdown of ammonia into hydrogen and nitrogen, thus avoiding the contamination of the surface by uncombusted hydrocarbons. To avoid overheating and excessive erosion of the surface, the rockets used an 18-nozzle design to spread the impacting gas over a broad area. Many tests were conducted in Mars simulation chambers to examine the chemical, thermal, and mechanical influence on various models of a Marslike surface. It was determined that this design would cause the surface to be heated no more than 1°C at the hottest place and that no more than 1 mm of surface would be stripped away by the exhaust gases. Of particular concern was the injection of nitrogen (or ammonia) into the Martian soil, since one investigation was to analyze the atmosphere soon after arriving on the surface. Several tests of the residence time of these gases suggested that the analysis of the atmospheric constituents should be delayed for the first 3 days to allow the soil outgassing to come to equilibrium.

The first Viking lander landed within 1° of horizontal. The second lander apparently has one leg on a rock and is tipped about 8°.

The lander (see Figure 3) is a horizontal platform structure 0.5 m thick and 1.5 m across. It weighs about 600 kg and houses the scientific instruments and their attendant computer, tape recorder, data system, power system, transmitter, and receiver. The three analytical instruments, biology, X ray fluorescence spectrometer, and gas chromatograph mass spectrometer (GCMS), are mounted within the structure. Two identical facsimile cameras are mounted on top along with a three-axis seismometer and a meteorological boom with weather sensors. Electric power comes from two radioisotope thermoelectric generators using plutonium 238 that provide 70-W continuous power. Peak power loads are handled by rechargeable batteries. A sampling arm and scoop are mounted on the front side for returning samples to the analytical instruments.

The transmitting and receiving antennas for communication are also mounted on top. The lander can send information

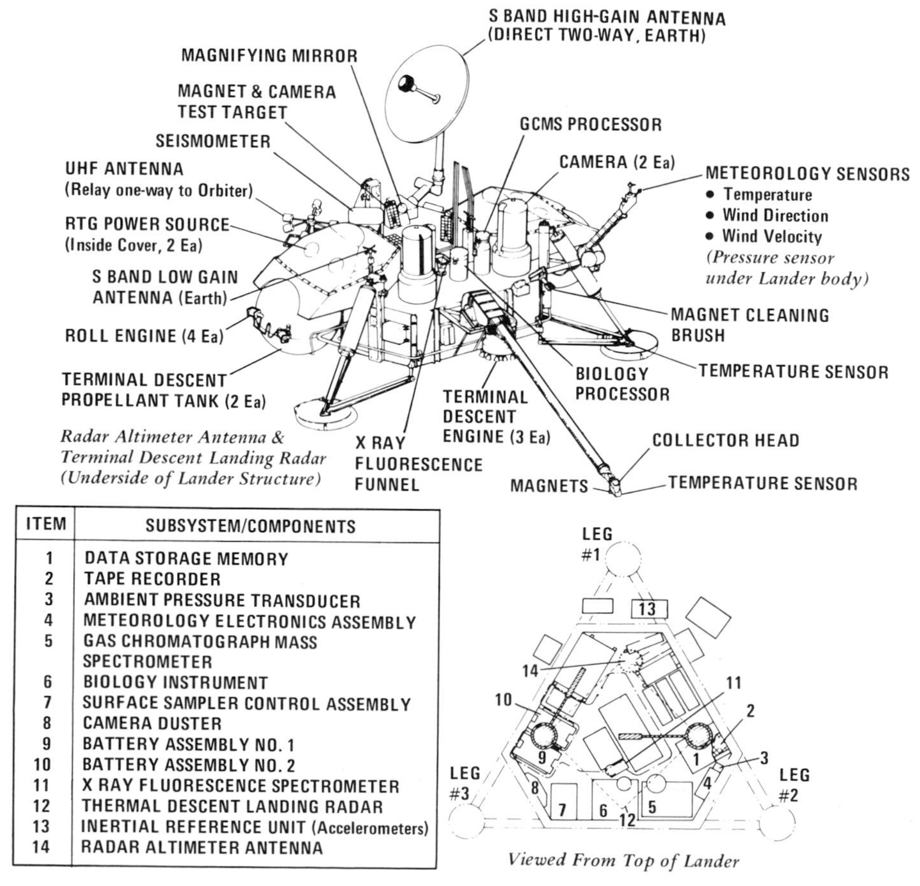

ITEM	SUBSYSTEM/COMPONENTS
1	DATA STORAGE MEMORY
2	TAPE RECORDER
3	AMBIENT PRESSURE TRANSDUCER
4	METEOROLOGY ELECTRONICS ASSEMBLY
5	GAS CHROMATOGRAPH MASS SPECTROMETER
6	BIOLOGY INSTRUMENT
7	SURFACE SAMPLER CONTROL ASSEMBLY
8	CAMERA DUSTER
9	BATTERY ASSEMBLY NO. 1
10	BATTERY ASSEMBLY NO. 2
11	X RAY FLUORESCENCE SPECTROMETER
12	THERMAL DESCENT LANDING RADAR
13	INERTIAL REFERENCE UNIT (Accelerometers)
14	RADAR ALTIMETER ANTENNA

Fig. 3. The Viking lander.

directly to the earth through the large movable parabolic *S* band antenna or through the orbiter over the fixed position UHF antenna. The lander also receives earth commands over an *S* band system. The length of this link changes with the geometry of the earth and Mars but characteristically is several hours each day.

Depending on the distance between lander and orbiter, the data rates between them were either 4000 or 16,000 bits/s, and the length of the link was from 15 min to slightly less than an hour. This means that on any one pass the total data that could be returned from the lander were between 10 and 50 million bits. For the photography, which requires 10^6–10^7 bits for every picture, this data rate was the major constraint. Other investigations that suffered for lack of data allocation were the seismometry and the meteorology. The analytical experiments were usually accommodated.

The landers were built so that they could operate completely autonomously. At the time of landing the on-board computers had instructions for performing many days of operations if we were not able to command the lander. In such an event it would have carried out a complete mission, taking pictures, getting samples from a preselected site, analyzing them, recording the weather and seismometry, and relaying the data back to the earth. Fortunately, both landers operated flawlessly, and this preprogrammed mission was overwritten beginning with the first command link.

Many of the elements that could cause a total loss such as power, communication, data, etc. were built with redundant units to avoid any single-point failure. This was fortunate in the case of Viking 2 when one of the battery chargers failed during cruise. Failure on each of the spacecraft of one of the command receivers has also borne out the wisdom of redundant systems.

It was early recognized that besides providing the scientific data the cameras would become one of the major tools in solving mechanical problems that occurred. Indeed, this proved to be the case time after time when we had trouble with mechanical parts such as the sampling arm and the antenna.

One of the most important requirements for performing the lander part of the mission was the ability to respond to the data. Since virtually nothing was known of the local surface topology, chemistry, or biology prior to the Viking landing, it was essential to be able to modify the experiments as they were progressing. This is the very nature of exploration and the place where the scientist plays a vital role. Our ability to receive data, interpret it, and make changes in our planned strategy normally required about 2 weeks. That was the time needed for preparing the software, for checking it to prevent errors that could be disastrous, and for sending and verifying the commands. (Of course, laboratory and field science on the earth is done differently, but one must remember that the automated laboratory is over 400,000,000 km away, operating by itself with our ability to command it only once a day at best!)

Under certain very special circumstances this reaction time was reduced to an absolute minimum; for example, a change in the strategy of using the GCMS for analysis of the atmosphere and an emergency problem of the sampling arm that failed to complete a sequence illustrate this. At the time this was less than 24 hours. During the descent through the atmosphere, repeated mass spectra were obtained, a prime objective being to determine the argon content of the atmosphere. This information was needed in order to decide upon the strategy for the initial use of the GCMS for performing the nitrogen analysis of the atmosphere prior to use of the GCMS for the soil sample. A high concentration of argon (30%) in the Martian atmosphere could have destroyed the ion pump of the mass spectrometer. The measurement of the atmosphere prior to soil analysis was highly desirable in order to obtain an accurate value. The required entry data were returned on the evening of the first day after landing. Within hours it was determined that the argon content was low enough (2%) to allow the GCMS to be used without jeopardy for the atmospheric analysis, and by the fourth day the command had been sent and the GCMS atmospheric analysis performed. The engineering problem with the sampling arm involved determining the exact location and position of the halted arm. The sequence for taking a set of pictures was written and verified, and commands sent, and the photographs were in the hands of the analysts in a little over 30 hours. These illustrations serve to establish that critical events could be dealt with heroically but only occasionally. In normal practice, making changes to the experiments required several days following the decision to do so.

Once they were on the surface, the landers began to perform the investigations, some sequentially and some simultaneously. The cameras systematically returned pictures of the entire panorama around the lander, only the landers own components partially obscuring some of the views. Both high- and low-resolution pictures were taken, in black and white, color, and infrared. One special feature of the cameras is a fixed scanning mode to detect motion. There have been almost 2000 pictures taken by the lander's cameras scanning from the nearby rock at the feet of the lander to the dunes on the horizon. Also photographed were the sky, the sun, the spacecraft components, the test targets, the small permanent magnets mounted on the sampling arm and affixed to the lander, the trenches dug by the sampler, and rocks that were moved (Figure 4).

The folded meteorology boom was deployed soon after landing and has made measurements at frequent intervals ever since. It is hoped that we can get a daily weather pattern of pressure, temperature, and wind speed and direction for a complete Martian year at both sites.

The seismometer on board Viking 1 failed to uncage, the only instrument that did not return usable data. The Viking 2 seismometer operated normally. The movements seen on the lander are largely due to wind moving the lander or to the mechanical motion of spacecraft components. The latter is easier to discount by knowing the spacecraft events. Accounting for motion due to winds requires that the meteorology data be collected during times of seismic monitoring.

The atmosphere was analyzed by the GCMS several times for the 2 days prior to the surface sample being analyzed and then many times afterward.

On the eighth sol after landing (sol is the coined word for a Mars day) the sampler delivered the first surface sample to the biology and the two chemical analytical instruments. During the course of the mission the sampler has operated many tens of times. Seventeen samples were delivered for the Viking 1 experiments and 26 for Viking 2 prior to the conjunction in November. This required the execution of about 2000 individual commands by each lander. After conjunction a trenching operation was performed in order to dig a 25-cm-deep trench. This also involved several hundred commands. This sampling device has proven to be a rugged machine, considering the intensive utilization on an unknown surface exposed to wide variations of temperature.

Fig. 4. The first photograph ever taken on the surface of the planet Mars, obtained by Viking 1 just minutes after the spacecraft landed. The center of the image is about 1.4 m (5 ft) from Viking lander camera 2. We see both rocks and finely granulated material (sand or dust). Many of the small foreground rocks are flat with angular facets. Several larger rocks exhibit irregular surfaces with pits, and the large rock at top left shows intersecting linear cracks. Extending from that rock toward the camera is a vertical linear dark band which may be due to a 1-min partial obscuration of the landscape due to clouds or dust intervening between the sun and the surface. Associated with several of the rocks are apparent signs of wind transport of granular material. The large rock in the center is about 10 cm (4 in.) across and shows three rough facets. To its lower right is a rock near a smooth portion of the Martian surface probably composed of very fine grained material. It is possible that the rock was moved during Viking 1 descent maneuvers, revealing the finer-grained basement substratum, or that the fine-grained material has accumulated adjacent to the rock. There are a number of other furrows and depressions and places with fine-grained material elsewhere in the picture. At right is a portion of footpad 3. Small quantities of fine-grained sand and dust are seen at the center of the footpad near the strut and were deposited at landing. The shadow to the left of the footpad clearly exhibits detail, owing to scattering of light either from the Martian atmosphere or from the spacecraft, observable because the Martian sky scatters light into shadowed areas.

The organic analysis has special requirements for organic cleanliness. Unusual precautions were taken in assembling the hardware and all components in the sample path to assure that the organic level was below the sensitivity of the detecting instrument. Cleaning reagents were specially purified. Lubricants and materials were substituted and extensive testing performed to establish the internal contamination level of the instrument.

The organic analyses were performed on two different samples at each of the sites. Each sample was heated at several different temperatures. The GCMS had three ovens on each lander, but one oven on each was found to be faulty.

The biology experiments were performed five to six times on each lander, depending on the specific test, and required from 5 days to several weeks for each test. Mostly, they shared a common sample, each sample taken from a different region. One sample was taken from underneath a rock on the possibility that protection from the UV radiation might offer another search opportunity.

Prior to the second landing the first lander was placed in a reduced mode of operation, only monitoring the meteorology, continuing X ray, and biology already started and taking a few pictures. Operation of the second lander lasted from September 7 until superior conjunction. For about a month around conjunction, November 25, 1976, all four vehicles were powered down, since no communication could be made. Following conjunction, all vehicles were reactivated and have functioned into the Martian winter taking pictures, continuing to monitor the weather, recording the seismometer, and performing additional analysis of the soil chemically and biologically. Extensive use of the sampler on the surface has provided a continuing source of data for determining the physical properties of the surface.

The communication systems have provided an opportunity to make measurements of the planet, its celestial mechanics, and its atmospheric and surface properties and, during the period of solar conjunction, to perform an experiment in general relativity.

THE RESULTS

The results given here are a brief summary of the highlights of each of the investigations. They are presented as a guide toward finding the material within this issue. The investigations will be treated in order of their appearance in the text.

Orbiter Imaging

Tens of thousands of pictures have been taken from orbit by the orbiter cameras. Prior to orbit some approach pictures were taken of a large part of the disc of Mars. The satellites of Mars have been photographed from as close as 100 km and reveal striking features. The surface of the planet has been photographed intensively during the site certification period, and a good deal of geological interpretation occurred during those times. Special targets of opportunity allowed photography of all major terrain types. Some color photographs have been taken, and some special pictures were taken to obtain stereoscopy for measuring elevation. Still, less than 10% of the surface has been photographed at 100-m resolution, and most areas have been seen at only one season.

The major conclusion of the orbiter imaging reported is that Mars has a very heterogeneous surface on the scale of tens of kilometers. Volcanism is extensive over vast areas of the planet. A large part of the northern hemisphere is covered by volcanic fields. There is widespread evidence of catastrophic flooding, but no collection basins such as lakes or oceans were found, and the source and sink of the water are still conjectural. The flooded regions do not appear to be recent. Some Martian craters have a unique morphology that suggests interactions with the underlying permafrost which causes lateral flows over the surface. The polar regions consist of extensive areas of terraced deposits and appear to be influenced by rapid erosion and deposition. Valles Marineris has large-scale vertical displacement and slumping to the floor of the canyon. The floor of the canyon has few craters and is swept by intense winds. Particles suspended in the atmosphere vary widely, both regionally and seasonally. The northern cratered highlands have undergone extensive erosion that appears to have taken place over several periods. On the basis of crater counts the major surface features of Mars are well over 1 billion years old.

The larger satellite, Phobos, has many linear surface features suggesting a strong influence by the tidal gravity of Mars.

Mars Atmospheric Water Detector (Mawd)

The water vapor in the Martian atmosphere is highly variable, changing with local time, elevation, latitude, and season. The amounts range from 0 ppm in the winter hemisphere to 85 ppm near the polar region of the summer hemisphere. The atmosphere above the north polar cap in midsummer is saturated, providing strong evidence that the permanent ice cap is made of water. The water in the atmosphere is concentrated near the surface and is believed to be moved from one hemisphere to the other during the changing seasons. The total global abundance of water vapor is about 1.3 km^3 and remained constant for the period of observation. An unusual kind of vapor boundary was discovered at 30°N latitude that is not correlated with any topographic or thermal features. Peak diurnal concentration appears to occur around noon, and it is postulated that a thin cloud or haze is present at dawn and dissipates by noon. Atmospheric aerosols complicate the problem of interpreting the data.

Infrared Thermal Mapping (IRTM)

Many maps of temperature and thermal inertia have been constructed, and two major discoveries made. One is a startling thermal characteristic of the southern winter pole. The temperature measured, which is well below the equilibrium temperature of dry ice at the pressure of the atmosphere, suggests a dynamic event during the winter solstice. The supercooling of the pole results in a lower vapor pressure of the volatile component, a concentration of the nonvolatile, and a movement of the new air mass to make up the atmospheric loss. It is known from the pressure measurements made on the lander that the atmospheric pressure drops steadily during the southern winter, but not to a level low enough to account for the polar cap temperature. The other important determination was the temperature of the permanent polar cap. Its temperature, measured at 200–215 K, indicates that the permanent cap is made of water ice, a result also consistent with the Mawd data. The amount of water deposited at the poles is many orders of magnitude greater than that in the atmosphere.

Radio Science

The telemetry and radar equipment was used very successfully in performing a variety of measurements. Prior to arrival at Mars several enhanced electron density events were detected in interplanetary space. The tracking data allowed location of the lander to about 11-km accuracy. The radius of Mars was determined at each landing site, and the orientation of the planetary axis and spin rate were established. The lander to orbiter link allowed for an estimate of the dielectric constant of the surface, which turned out to be similar to a pumice or tuff. Tracking data led to improved values of the gravitational field, the Mars to earth distance, and the ephemerides of the earth and Mars. About 50 occultation points were used to determine radio and atmospheric profiles. Small-scale structure of the solar corona was observed during conjunction and revealed a strong asymmetry of the corona. A special test of general relativity made by the time delay just before and after Mars conjunction confirmed the Einsteinian value of γ to 0.1%.

Entry Science

During the descent through the Martian atmosphere the physical structure and chemical composition were measured. The atmospheric pressure and temperature were measured, and a density profile has been calculated yielding a mean molecular weight of 43.34. The entry of Viking 1 in the afternoon and the entry of Viking 2 in the morning revealed a diurnal difference in temperature near the surface consistent with the IRTM data. The upper atmosphere consists mainly of CO_2 with small amounts of N_2, Ar, CO, O_2, O, and NO. Carbon monoxide and NO are enriched in relation to the composition of the lower atmosphere. The isotopic ratio of carbon and oxygen is similar to that of the earth, but the ratio of ^{15}N to ^{14}N is enriched, a result suggesting that Mars has had a denser atmosphere in its past. This denser atmosphere might account for the appearance of ancient rivers seen in the photos. (The current pressure would not support the accumulation of liquid water.) The atmosphere is well mixed to heights above 120 km. The ionosphere has also been measured, and the major constituent is O_2^+ at 130-km altitude with CO^+ less abundant by about an order of magnitude. A principal reaction in the ionosphere is probably $CO_2^+ + O \rightarrow CO + O_2^+$.

Lander Imaging

Both sites are dominated by a variety of rocks among fine-grained material, and both have the brownish to orange color of the surface and the sky. Beyond this superficial similarity the sites are quite different in their appearance. The Chryse site topography is undulating and has a great range of rock size and type from rocks a few centimeters across to one nearby large rock almost 2 m across and others in the distance much larger. Many multilayered sand drifts in the lee of the large boulders suggest periodic deposition and deflation. The fine material has a great deal of cohesion. The rocks are of a basaltic igneous type, angular and with a coarsely pitted surface of grain size 3 mm to 1 cm. In addition to the rocks several outcrops are visible. The terrain is formed by the chemical and mechanical destruction of the upper layer of a volcanic flow of basalt. This exposed the jointed and fractured outcrop and left behind the angular rocks. To this was added debris from impact, and the fine material was deposited by the winds which also faceted some of the blocks. In contrast, the Utopia site is a flat plain with numerous evenly distributed boulders that are vesicular and appear to be ejecta of a nearby crater. The fine-grained material dispersed among the rocks has been swept from the more northerly polar regions. No evidence of life has been found in the pictures at either site. The aerosols of about 1-μm diameter result in a brown to pink colored sky.

Physical and Magnetic Properties

The surface material of Mars is more firm than the lunar regolith. Rocky material occupies a large part of the sample field at both sites. During the landing the surface beneath the lander was eroded slightly. At the Chryse site, one of the footpads was buried several centimeters into the soil. Rocks that were immovable by the sampling arm are buried among the fine-grained material. This material is very adhesive (it sticks to the lander components) but weakly cohesive, with a bulk density of about 1.2 g/cm^3. The rocky portion has a density of about 2.9 g/cm^3, and the frequency of rocks at the Utopia site is about twice that at the Chryse site. Repeated failure to collect rocks in the size range of 0.2–1.2 cm suggests that they are very scarce. Many small weakly cohesive clods (probably duricrusts) can be seen that were fragmented by the sample collector. Magnetic material has been detected in both the aerosol fraction and the surface material and probably comprises 1–7%. Some particles several millimeters across have been observed clinging to the permanent magnets. While several candidate materials have been suggested, the most likely is maghemite, γ Fe_2O_3, which could account for the reddish pigment.

Seismology

No major seismic events have been detected, a result which indicates that Mars is less seismically active than the earth. The winds are a major source of seismic background noise in the instrument. One local seismic event was detected having a magnitude of 2.8 (on the Richter scale) at a distance of 110 km from Viking lander 2. The shear wave reflections indicate that the crustal thickness in this region is 15 km. The natural background of seismicity, discounting noise due to wind, is low. The seismic signal was significantly damped within a few minutes, very likely because of the water and trapped atmosphere in the crust (unlike the moon, which rings for long periods).

Meteorology

The pressure of the Martian atmosphere varies seasonally by about 30% owing to condensation at the polar caps. Winds measured at the landing sites in the summer were relatively mild, generally less than 20 m/s. Diurnal and semidiurnal pressure oscillations probably due to solar tides have been observed. Wind patterns in the summer were highly repetitious on a daily basis, generated by the global circulation and modified by the local terrain. Diurnal temperature during the period reported varied between 150 K at night and 240 K in the summer midafternoon. (More recently, with the approach of winter, the CO_2 condensation temperature (149 K) has been observed.) The phase retardation of heating suggests that both convection and absorption of solar radiation play significant roles, affecting the global atmospheric tides.

Inorganic Chemistry

The fine-grained material at Chryse and that at Utopia have similar elemental abundances. A variety of surface and subsurface samples shows a surprisingly uniform composition indicating a homogenization of the fine-grained material, probably well mixed by the winds. No rocks were sampled. The material is not similar to any simple terrestrial sample but can be matched by a mixture. There is a low level of alkali and alumina, indicating that most of the material is mafic in origin (less differentiated than the salic terrestrial rocks). The material is largely comprised of an iron-rich clay, oxides of silica and iron making up more than half of the components. Sulfur, probably as sulfate, accounts for another 8–10%; water and carbonates make up about 5%; magnesium about 9%, and calcium 5%. A plausible model of the material suggests that most of it resembles nontronite, a material formed under hydrothermal conditions. On the earth this is a common weathering product of basaltic lava flows. This would suggest the importance of the volcanoes and the permafrost in forming this material. The duricrust may be formed by evaporative precipitation.

Molecular Analysis

The atmospheric composition has been measured with the GCMS and found to be in agreement with that found by the upper-atmosphere mass spectrometer. The discovery of 2.5% nitrogen is one of the most important Viking accomplishments (this bears directly on the chemical history of Mars). The argon 36 to argon 40 ratio is about 10% of the terrestrial value, and the abundance of argon 36 per unit mass of the entire planet is only 1% of that of the earth. The abundance of the isotopes of carbon, oxygen, and nitrogen were similarly measured. All of this indicated that Mars has not outgassed as much as the earth and that the present atmosphere is only a small fraction of the atmosphere in the past. Estimates of ancient atmospheric pressure range from 50 to 500 mbar.

The analysis of soil samples for organic molecules was performed at both sites. No organics have been found, although the instruments are known to have obtained samples and worked perfectly. The upper limits for organics at these two sites are at the parts per million for small one- or two-carbon compounds and at the parts per billion for the larger molecules. The destruction of organics by UV radiation or oxidation is strongly suggested by the data (and consistent with biology results). A sample taken from underneath a rock revealed a higher water content than the exposed surface material. This is important in interpreting the biology experiments.

Biology

The biological results were by far the most complex of all investigations. There was no unambiguous discovery of life by the Viking landers, and three of the results appear to indicate the absence of biology in the samples tested. Nevertheless, the experiment gave significant results revealing the chemical nature of the Martian surface and at least one result that could still be consistent with a biological interpretation. One experiment indicates that the Martian soil has an agent capable of rapidly decomposing organic chemicals used in the medium or that life is present. This oxidizing agent is heat labile to temperatures as low as 45°C for 3 hours and may disappear when held at spacecraft temperature (6°–20°C) for several months. In another experiment the addition of water vapor to the Martian sample caused a vigorous release of oxygen for a few hours. This oxygen release is heat stable. Heating the dry sample generates large amounts of CO and CO_2. In one experiment a small amount of carbon monoxide (or carbon dioxide) was incorporated into the organic fraction (or made organic de novo). This process does not appear to be stimulated by light or the addition of water vapor. The surface of Mars is obviously highly reactive and contains at least one and probably several highly oxidizing substances. While inorganic chemical reactions may be sufficient to explain the data seen, biological processes cannot be ruled out at this time.

REPORTING OF VIKING RESULTS

The results of Viking have been released through nearly every known medium. During the early part of the mission, when public interest was highest, there was extensive television and press coverage. Over 100 reporters covered the mission, many staying in Pasadena throughout the summer of 1976, writing daily.

Articles in scientific journals were released as rapidly as possible (*Science, 193,* August 27, 1976, *194,* October 1, 1976, and December 17, 1976). *Eos* had an early release in October 1976. NASA prepared a special publication, 'Viking 1 Early Results' (SP 408). The journal of the AIAA, *American Scientist* (November, December 1976), and *National Geographic* (January 1977) had special articles.

Many major professional societies had special sessions on Viking (American Astronomical Society, American Association for the Advancement of Science, American Physical Society, American Meteorology Society, American Microbiology Society, American Chemical Society, American Geophysical Union, COSPAR, and the International Society for the Study of the Origin of Life). Many investigators who were abroad for other reasons gave lectures in England, Canada, Australia, Mexico, France, Germany, Italy, Holland, Denmark, Finland, USSR, Hungary, Japan, Korea, Israel, India, and a number of South American, African, Southeastern, and Asian countries.

I believe that the exploration of Mars came at an important time when mankind must be made aware of the earth as a planet. Comparative planetology was conceived with Mariner and born with Viking. Our rendezvous with history was a major milestone in human affairs.

REFERENCES

Biemann, K., et al., Search for organic and volatile inorganic compounds in two surface samples from the Chryse Planitia region of Mars, *Science, 194,* 72–75, 1976.
Farmer, C. B., Liquid water on Mars, *Icarus, 28,* 279, 1976.
Hubbard, J. S., J. P. Hardy, and N. H. Horowitz, Photocatalytic production of organic compounds from CO and H_2O in a simulated Martian atmosphere, *Proc. Nat. Acad. Sci. U. S., 68,* 574–578, 1971.
Masursky, H., and N. L. Crabill, The Viking landing sites: Selection and certification, *Science, 193,* 809–811, 1976*a*.
Masursky, H., and N. L. Crabill, Search for the Viking 2 landing site, *Science, 194,* 62–65, 1976*b*.
New York Times, editorial, August 31, 1976*a*.
New York Times, editorial, November 18, 1976*b*.
Nier, A. O. C., M. B. McElroy, and Y. L. Yung, Isotopic composition of the Martian atmosphere, *Science, 194,* 68–69, 1976.
Snyder, C. W., The missions of the Viking orbiters, *J. Geophys. Res., 82,* this issue, 1977.
Soffen, G. A., Status of the Viking missions, *Science, 194,* 57–58, 1976*a*.
Soffen, G. A., Scientific results of the Viking mission, *Science, 194,* 1274–1276, 1976*b*.
Soffen, G. A., and C. W. Snyder, The first Viking mission to Mars, *Science, 193,* 759–765, 1976.
Soffen, G. A., and A. T. Young, The Viking missions to Mars, *Icarus, 16,* 1–16, 1972.

(Received May 25, 1977;
revised June 10, 1977;
accepted June 10, 1977.)

The Missions of the Viking Orbiters

CONWAY W. SNYDER

Jet Propulsion Laboratory, California Institute of Technology
Pasadena, California 91103

The two Viking orbiters carried the two landers into orbit around Mars, observed the planet to certify the landing sites, released the landers for the landings, and subsequently served as telemetry relays for the lander data. In addition, they conducted scientific investigations using two cameras, an infrared radiometer for temperature measurements, an infrared spectrometer for water vapor measurements, and the radio communication system. The nature and extent of the orbiter observations have been influenced by the requirements for lander support, the capabilities of various orbiter subsystems, and the visibility of the planet from the orbits. All the orbiter scientific experiments are continuing.

The prime goal of the Viking missions was to land and make observations on the surface of Mars, and in support of this goal the orbiters were conceived as buses to get the landers to their destinations and as tenders to service them after they arrived. Indeed in the earliest planning it was uncertain whether the orbiters would have any scientific mission of their own or whether the bus would even go into orbit. The fundamental consideration that tipped the decision in favor of orbiter science was the belief that the probability of landing successfully would be augmented by having better information on the basis of which to choose the landing sites. At the same time it was appreciated that remote sensing instruments on the orbiters that could provide site certification information could also acquire valuable scientific data independent of the landers and that the complementarity of data from the landers and orbiters would be expected to increase the value of both. These considerations were paramount in the selection by NASA of the orbiter experiments.

The papers that follow this introduction discuss the results so far obtained by the scientific instruments on the orbiters (by 'orbiter science' in the jargon of the project). It will be seen that the orbiter science missions are by no means concluded, even though we are long past the end of the primary mission and some of the major lander experiments have been terminated. On the contrary, we are still looking forward to the achievement of some of the important goals of orbiter science.

THE ORBITER SCIENCE INSTRUMENTS

The orbiter science instruments are (1) a pair of cameras, incorporated in a visual imaging subsystem (VIS), which provide a larger format and higher spatial resolution than the Mariner 9 cameras and can lay down swaths of pictures that overlap to cover large areas, (2) a multiple-sensor, multiple-wavelength infrared radiometer, called the infrared thermal mapper (IRTM), which was evolved from a Mariner 9 instrument but has more wavelengths and more spatial coverage capability, and (3) an infrared spectrometer of a type not previously flown, the Mars atmospheric water detector (affectionately called Mawd), which can map the total abundance of water vapor in the atmosphere over approximately the same areas that are being observed by the other two instruments.

In the mission operations, each of these investigations is under the jurisdiction of a science team which was partially or completely responsible for the design, construction, and preflight testing of the instrument. A fourth team, the Radio Science Team, is included in the orbiter science group for administrative convenience, even though it is concerned with data from both landers and orbiters. The membership of the teams during the primary mission is given by *Soffen* [1977].

The VIS consists of two high-resolution, slow-scan television framing cameras, each with a telescope having a 475-mm focal length and a 37-mm-diameter vidicon, the central region of which is scanned with a raster format of 1056 lines by 1182 samples. Each field of view is 1.54° × 1.69°, and each picture element subtends 16 μrad. The optic axes are offset by 1.38° so that the fields overlap slightly as is shown in Figure 1. The cameras are shuttered and read out alternately, providing a single continuous digital data stream, the interval between pictures (one frame time) being 4.48 s. This interval determines the basic timing of the other two science instruments as well. Each camera has five color filters and one clear 'filter.'

Among the objectives of the imaging investigation were the following: to characterize potential landing sites in sufficient detail to support the site choice; to study the topographic, photometric, and colorimetric characteristics of the surface; to study in greater detail the various interesting geologic features (volcanos, canyons, channels, faults, polar cap formations, etc.) that had been discovered by Mariner 9 to improve our knowledge of the geologic evolution of Mars; to follow the changes in cloudiness, in the polar caps, in the polar hoods, and in the contrast of surface features; and to observe the vertical structure of the atmosphere. It will be seen in the papers that follow that these objectives have been achieved to a considerable degree. The clarity of the pictures has exceeded expectations, and many previously undetected and often puzzling types of terrain have been observed. The volume of data has been so great that the scientific interpretations and conclusions from these pictures will not be completed for several years.

The IRTM has four small telescopes, each focusing incident light on an array of seven small thermopile detectors to measure the thermal emission of the Martian surface and atmosphere and the total reflected sunlight. During planet observations using the 'planet port' the set of seven 5-mrad circular fields of view in a chevron pattern for each telescope is superimposed on the VIS and Mawd fields of view as shown in Figure 1 so that simultaneously recorded data from the three instruments are directly correlatable. Various filters divide the 28 detectors among six wavelength bands; one of these is designed particularly to measure temperature in the upper atmosphere, and the other five are for surface observations. Making a full set of measurements every 1.12 s (i.e., four times

Paper number 7S0504.

in each VIS frame time) and utilizing the motion of the spacecraft or of the scan platform to move the line of sight over the surface, the IRTM can map the variations of radiation intensity at each wavelength.

A very significant feature of the IRTM is its so-called 'scan mirror,' which is pivoted. In its normal position it directs the line of sight out through the planet port and parallel to the lines of sight of the other instruments. As part of every observational sequence the mirror is rotated at the beginning and end or periodically throughout the sequence to provide absolute calibration information. A rotation by 90° directs the line of sight away from the planet to look at space (through the 'space port'), and a 180° rotation directs it inward toward a radiator plate at a controlled temperature. Measurement of detector response to these two known temperatures provides the data required to derive absolute radiation intensities. The necessity for the IRTM to have unobstructed views in two perpendicular directions has caused considerable difficulty in the mission.

Objectives of the thermal investigation included the following: to contribute to the site selection decision by determining the temperature and the thermal inertia of the proposed sites and relating them to other parts of the planet; to measure the local surface kinetic temperature over the whole planet and its variations as functions of time of day and of season; to look for regions with anomalous cooling during the Martian night and for regions that depart from thermal balance; to determine the global atmospheric temperature; and to measure the temperature and hence the composition of haze, clouds, and ground frosts, including the polar caps.

Some very significant discoveries have been made by this investigation, and a large number of data have been accumulated, but the constraints on the times and places of observations (as explained in a following section) have so far prevented the full potential of the experiment from being realized. Only a small fraction of the data have been intensively analyzed and interpreted. It is hoped that the operations in the coming year of the extended mission can effectively surmount many of these difficulties.

The Mawd is a grating spectrometer operating in the 1.4-μm region of the infrared. In its principal operating mode it measures, every 280 ms (i.e., $\frac{1}{16}$ of a VIS frame time), the intensity of reflected sunlight in five narrow spectral bands, from which the quantity of water vapor along the line of sight can be inferred. Thus it can measure the abundance of atmospheric water vapor with a spatial resolution considerably surpassing that of any earlier measurements. It can also map the albedo of the surface at this one wavelength.

The instantaneous field of view of the Mawd sensors is a narrow rectangle, approximately 2 × 17 mrad, and a small stepping mirror just inside the instrument aperture moves the line of sight so as to give 15 such adjacent fields in measurement cycle, which is one VIS frame time (4.48 s). During the 16th step time the mirror is reset, and engineering data are inserted into the data stream. The Mawd acquires data simultaneously with the IRTM, and their fields of view overlap as shown in Figure 1.

Objectives of the water vapor investigation were to measure the abundance of water vapor at and near the landing sites to contribute to the site certification process and to map the abundance of water vapor over the daylight side of the planet, determining its variation with latitude, surface elevation, nature of the surface, time of day, and season with the objective of discovering the sources, sinks, and transport of water. Achievement of this general objective requires that fairly complete water vapor maps be generated covering essentially the whole lighted planet at intervals of approximately a month. Observational constraints have so far made it impossible to approach this ideal, and the Mawd is farther from fulfilling its major objectives than either of the other two orbiter investigations. The temporal and spatial coverage has been much too spotty. Much valid and interesting information has been obtained, however, and some important features of the water transport are beginning to be understood.

The radio science investigation utilizes the radio signals to and from the four spacecraft to obtain a variety of information. Transponders on the spacecraft send back signals that are coherent with those received from earth so that precision Doppler and ranging measurements can be made. The downlinks from the orbiters include two coherent frequencies (*S* band and *X* band) so that dispersion in the interplanetary medium can be measured and corrected for.

From analysis of the radio signals a surprising diversity of information can potentially be obtained, and the Radio Science Team has identified a large number of scientific objectives, which fall into three categories: dynamical, surface, and internal properties of Mars, atmospheric and ionospheric properties of Mars, and miscellaneous solar system properties. Several of these objectives can only be achieved if the spacecraft orbits are appropriately controlled, which has usually not been possible because of the requirements of other experiments. In addition to their scientific and telemetric uses the radio signals also provide precise information on the locations of the orbiters and landers.

CHRONOLOGY OF VIKING MISSION 1

The first spacecraft was launched on August 20, 1975, and arrived at Mars on June 19, 1976. En route the orbiter science instruments were usually dormant, but pictures of earth, Jupiter, Mars, and several star fields were taken at various times for use in calibrating the cameras and in refining the accuracy of knowledge of the pointing of the camera optic axes and the scan platform. Pictures of Phobos and Deimos were also taken, which were analyzed to provide more precise information on the relative positions of the spacecraft and Mars than could be derived from the radio tracking. These observations contributed materially to the outstanding precision that was achieved for the initial orbit. The Mawd was turned on approximately monthly to monitor its performance, and the IRTM made some observations of Mars to acquire data for evaluating the off-axis response of the sensors. Between 120 and 30 hours before the Mars orbit insertion (MOI) maneuver, many observations were made to get an overall view of the planet surface with all three instruments and especially to obtain some three-color pictures.

A few days before Mars encounter the engineering telemetry disclosed a problem in the propulsion subsystem, and the maneuvers prior to MOI were redesigned. As a result, Viking arrived at Mars about 6 hours later than had been planned. Since all the orbiter observations for the first 10 days after MOI had already been scheduled in exhaustive detail, it was decided to put the spacecraft into a more eccentric orbit than had been planned. This stratagem enabled us to hold to our preplanned time line, since the new orbit, having a period of 42.35 hours instead of 24.66 hours, brought the spacecraft at its second periapsis passage to the same space-time point that had been planned for the third periapsis passage. Near this time a short propulsive maneuver achieved very nearly the preplanned orbit, which had a periapsis altitude approximating 1500 km and passed very close to the proposed landing site just before each periapsis passage.

For each spacecraft the Viking 'primary' mission began with orbit insertion and ended early in November, when daily radio communication for commanding and data transmission was terminated because of solar conjunction. During more than 5 weeks around conjunction, the orbiter science instruments were turned off, but two-way radio links continued to be used intermittently to both landers and orbiters to acquire data on the solar corona and on the relativistic time delay. In early December, commanding and data transmission resumed with all four spacecraft, and the Viking 'extended' mission commenced. This phase of the mission is currently expected to continue through May 1978. The flight team is considerably smaller than it was in the primary mission, so the frequency of command sessions to the orbiters is less, and the coverage by the telemetry receiving stations of the Deep Space Net is also reduced. Nevertheless, the rate of orbiter data acquisition is still a substantial fraction of that in the primary mission.

The chronology of mission events of significance for Orbiter 1 science is given in Table 1. Time is reckoned by date and by orbit revolution (rev) number. The convention is that rev n begins at apapsis An and is centered on periapsis Pn. The other time variable of interest for orbiter science is the areocentric longitude of the sun L_s, which was 90° at the northern summer solstice and 180° at the equinox beginning northern autumn.

The primary mission for orbiter science included four phases: (1) site certification for 29 revs, when all observations were devoted to looking for and characterizing potential landing sites [*Masursky and Crabill,* 1976*a*], (2) 52 revs synchronized over Lander 1, (3) 14 revs in an asynchronous 'walking' orbit around the planet, and (4) 41 revs synchronized over Lander 2. The subsequent conjunction period with only radio science active was 38 revs in length.

The periapsis altitude remained close to 1500 km through the primary mission, although the orbit was changed six more times. The orbital elements that are significant for the science observations are shown in Table 2. The dates included in the tabulation are only those on which changes in orbit occurred that had a major effect on the observations. Prior to the initial orbital observations on June 22, 1976, Viking 1 was placed in what we call a 'synchronous orbit,' which means that it has a period approximately equal to 1 Mars sidereal day (24.623 hours) so that every periapsis point is approximately over the same spot on the planet surface. Synchronous orbits were used during most of the primary mission because they provided the capability for daily relay transmissions of lander data to the orbiter. For the whole planet surveillance that is of interest to the orbiter scientists, a synchronous orbit is a disadvantage. Accordingly, on September 11, after Lander 2 was in operation and the intensity of Lander 1 operations had been cut back, the period was reduced by 2.8 hours so that each successive periapsis point occurred about 40° farther east and all longitudes could be observed in a period of 9 days. A nonsynchronous orbit was maintained for 13 days, permitting a so-called walk 1.5 times around the planet, and Orbiter 1 was then synchronized to provide relay support for Lander 2 while Orbiter 2 went on a similar walk. No further orbit changes were made during the primary mission.

In the extended mission a set of three orbit trims was executed beginning on January 21, 1977, to bring the spacecraft orbit into synchronism with that of Phobos (period, 23.05 hours) so that closeup pictures of it could be acquired for several days. It had been hoped to get as close as 50 km, but because of the uncertainty of our knowledge of the mass of Phobos the closest approach was 90 km. Then on March 11 the periapsis altitude was lowered to 300 km to obtain pictures with significantly better spatial resolution than any earlier ones. In this maneuver the period was reduced to 21.9 hours, which gave too rapid a walk around the planet, and the March 24 maneuver brought it back up to 23.5 hours. In this orbit the spacecraft walks once around Mars in approximately 22 revolutions. On May 15 a small maneuver was executed to assure against a possible collision with Phobos. No further maneuvers are definitely planned at present.

Gravitational anomalies cause the orbital elements to change with time. In the primary mission the periapsis altitude of Orbiter 1 was usually decreasing by about 0.3 km/d. Both the inclination and the latitude of periapsis increased slowly from the beginning, and by April 1977, both were at about 39°.

The orbits that are optimum for the radio science experiments have an annoying tendency to be quite different from those that are suitable for various other experiments. A project specification established at the beginning stated that no earth occultations were to occur until after the completion of the primary missions of the landers. On Orbiter 1, occultations

TABLE 1. VO-1 Chronology

Date	Rev	L_s	Event
June 19, 1976	0	83	Mars orbit insertion
June 21, 1976	2	84	Trim to planned site certification orbit
July 4, 1976	15	90	Northern summer solstice
July 9, 1976	19	92	Orbit trim to move westward
July 14, 1976	24	94	Synchronous orbit over landing site
July 20, 1976	30	97	VL-1 landing at 1153:06 UT
Aug. 3, 1976	43	103	Minor orbit trim to maintain synchronization over VL-1
Sept. 3, 1976	75	117	VL-2 landing
Sept. 11, 1976	82	121	Decrease of orbit period to begin eastward walk
Sept. 20, 1976	92	125	Orbit trim to permit synchronization over VL-2
Sept. 24, 1976	96	127	Synchronous orbit over VL-2
Oct. 6, 1976	107	133	First earth occultation
Nov. 1, 1976	133	146	End of earth occultations
Nov. 5, 1976	137	148	Final data transmission in primary mission
Nov. 25, 1976	156	158	Solar conjunction
Dec. 14, 1976	175	168	First command load in extended mission
Jan. 4, 1977	195	180	Northern autumnal equinox
Jan. 22, 1977	213	190	Period change to approach Phobos
Feb. 4, 1977	227	198	Orbit synchronization with Phobos period
Feb. 12, 1977	235	203	Precise correction to Phobos synchronization
March 11, 1977	263	219	Reduction of periapsis to 300 km
March 24, 1977	278	227	Adjustment of orbit period to 23.5 hours
May 15, 1977	331	261	Small Phobos avoidance maneuver
May 30, 1977	347	270	Northern winter solstice

TABLE 2. Orbits of VO-1

	Date of Maneuver					
	June 21, 1976	Sept. 11, 1976	Sept. 24, 1976	Jan. 22, 1977	March 11, 1977	March 24, 1977
Rev number	2	82	96	213	263	278
Mean period, hours	24.66	21.88	24.65	23.05	21.92	23.50
Periapsis altitude,* km	1513	1491	1515	1480	299	303
Inclination, deg	37.9	38.1	38.2	39.1	39.2	39.3
Periapsis latitude, deg	23.2	30.8	32.3	38.1	39.2	39.2

* The difference between the range from the planet center of mass to the orbiter and the equatorial radius (3394 km). The actual periapsis altitude depends upon the elevation of the surface at the subperiapsis point.

took place only between October 6 and November 1, and the requirements of other experiments dictated that the orbiter should be in a synchronous orbit throughout this period, a condition not in accord with radio science desires. The extended mission occultations of Orbiter 1 occur daily for more than a year and a half, commencing on March 22, and for this period the radio scientists would like the spacecraft to be in a 24-hour orbit so that all occultations would occur when the orbiter is in sight of two receiving stations. Current plans provide for transferring to this orbit about July 1. About 50 good occultation measurements were made in the primary mission, yielding information on the topography and the atmosphere, and more are being made in the extended mission. The 300-km periapsis of Orbiter 1 in the extended mission does make available data that will characterize the gravitational anomalies with resolution and precision not hitherto achieved. It is hoped that the other orbiter can be brought down close to that altitude in the autumn.

TABLE 3. VO-2 Chronology

Date	Rev	L_s	Event
Aug. 7, 1976	0	105	Mars orbit insertion
Aug. 9, 1976	2	106	Period and altitude adjustment; walking westward
Aug. 14, 1976	6	108	Increase of period to increase walk rate
Aug. 25, 1976	16	113	Decrease of walk rate to proceed to landing site
Aug. 27, 1976	18	114	Synchronous orbit over landing site
Sept. 3, 1976	25	117	VL-2 landing at 2237:50 UT
Sept. 29, 1976	50	130	Propulsion system test to prepare for orbit plane change
Sept. 30, 1976	51	131	Change of orbit plane to 75° inclination and beginning westward walk
Nov. 8, 1976	85	149	Final data transmission in primary mission
Nov. 25, 1976	101	158	Solar conjunction
Dec. 15, 1976	119	169	First command load in extended mission
Dec. 20, 1976	123	172	Lowering of periapsis to 800 km and increase of inclination to 80°
Jan. 4, 1977	137	180	Northern autumnal equinox
Jan. 14, 1977	146	186	First earth occultation
March 2, 1977	189	214	Synchronous over VL-2
April 18, 1977	235	244	Period change: 13 revs equals 12 Mars days
May 29, 1977	279	269	End of earth occultations
May 30, 1977	280	270	Northern winter solstice

CHRONOLOGY OF VIKING MISSION 2

Viking 2 was launched on September 9, 1975, and arrived at Mars on August 7, 1976. The operations of the orbiter science instruments en route and during the last 5 days before encounter were similar to those of Orbiter 1. The primary mission included three phases: (1) site certification for 25 revs, (2) 26 revs synchronized over Lander 2, and (3) 34 revs in a walking orbit before the cessation of telemetry for solar conjunction. The conjunction blackout period was 34 revs long, before the extended mission began. Table 3 summarizes the chronology, and Table 4 the orbital characteristics.

In accordance with the plan the insertion orbit was nonsynchronous so that a fairly thorough survey of the surface between 40° and 50°N could be made. Approximately 40% of this area was examined before the landing site was chosen [*Masursky and Crabill,* 1976*b*], and the orbiter was synchronized over it. After Orbiter 1 had completed its walk and stopped over Lander 2, Orbiter 2 commenced a walk on September 30 at a rate of 31.7°/rev, which continued into the extended mission. The maneuver that established this new orbit was a very large one, requiring a change in the orbiter velocity vector by 343 m/s to increase the orbit inclination from 55° to 75°. This change was made to permit the very important north polar observations to be made before the formation of the polar hood.

At the beginning of the extended mission, another fairly large maneuver further increased the inclination to 80° and lowered the periapsis altitude to 778 km, which is near the optimum for all three orbiter science instruments. On March 2, Orbiter 2 was synchronized over Lander 2 to provide more frequent relay links during the approach of winter at the northern landing site. On April 14 a maneuver placed the orbiter in a walking orbit that overflies the lander every 13th revolution to provide two consecutive days of relay transmission while the lander is in its cyclic winter survival mode. There will probably be no other orbit changes until next autumn.

Earth occultations of Orbiter 2, which are of interest to radio science, occurred between January 14 and May 29, 1977. As in the case of Orbiter 1, the orbits were not optimum for acquiring the necessary Doppler and ranging data very frequently. Additional occultations will occur near the end of the year.

THE VIKING ORBITER SPACECRAFT

The Viking orbiters (designated VO-1 and VO-2) were designed and built by the Jet Propulsion Laboratory, and they are similar in basic design to their predecessors, the Mariner spacecraft, especially Mariner 9. They are much larger, how-

TABLE 4. Orbits of VO-2

	Date of Maneuver					
	Aug. 9, 1976	Aug. 27, 1976	Sept. 30, 1976	Dec. 20, 1976	March 2, 1977	April 18, 1977
Rev number	2	18	51	123	189	235
Mean period, hours	27.32	24.62	26.78	26.48	24.73	22.73
Periapsis altitude,* km	1499	1489	1518	778	751	723
Inclination, deg	55.2	55.4	75.1	80.1	80.2	80.5
Periapsis latitude, deg	50.4	52.2	63.6	60.6	54.4	51.9

* The difference between the range from the planet center of mass to the orbiter and the equatorial radius (3394 km). The actual periapsis altitude depends upon the elevation of the surface at the subperiapsis point.

ever, since they are the first planetary spacecraft to carry a passenger, the Viking landing capsule. The complete Viking spacecraft (lander and orbiter) is shown in Plate 1 of *Soffen* [1977]. When the lander separates from the orbiter, the inboard half of the lander capsule (termed the bioshield) and the supporting struts holding it to the orbiter remain in place. They were subsequently jettisoned by VO-1, but because of problems with the separation of the second spacecraft it was deemed unwise to attempt this maneuver.

The orbiters can be characterized as solar-powered, three-axis stabilized vehicles with propulsive capability. This capability was very large, since each orbiter had to put both itself and its passenger into Mars orbit and still be able to change that orbit several times. In normal flight its axis of symmetry is along the orbiter-sun line. Directions relative to an orbiter are specified by a spherical coordinate system in which the azimuth angle is called 'clock angle' and is measured from the pointing direction of the star sensor, and the colatitude is called 'cone angle'; the direction of zero cone angle is normally toward the sun.

A Viking orbiter spacecraft is a highly integrated system, and virtually every one of its subsystems contributes to and constrains the observations that can be made by the science instruments. In this section the characteristics and capabilities of the various subsystems are discussed to the extent that they influence the observations that have been or may yet be made.

Propulsion

The propulsion subsystem has at its heart a bipropellant liquid-fueled rocket engine which provides the thrust as re-

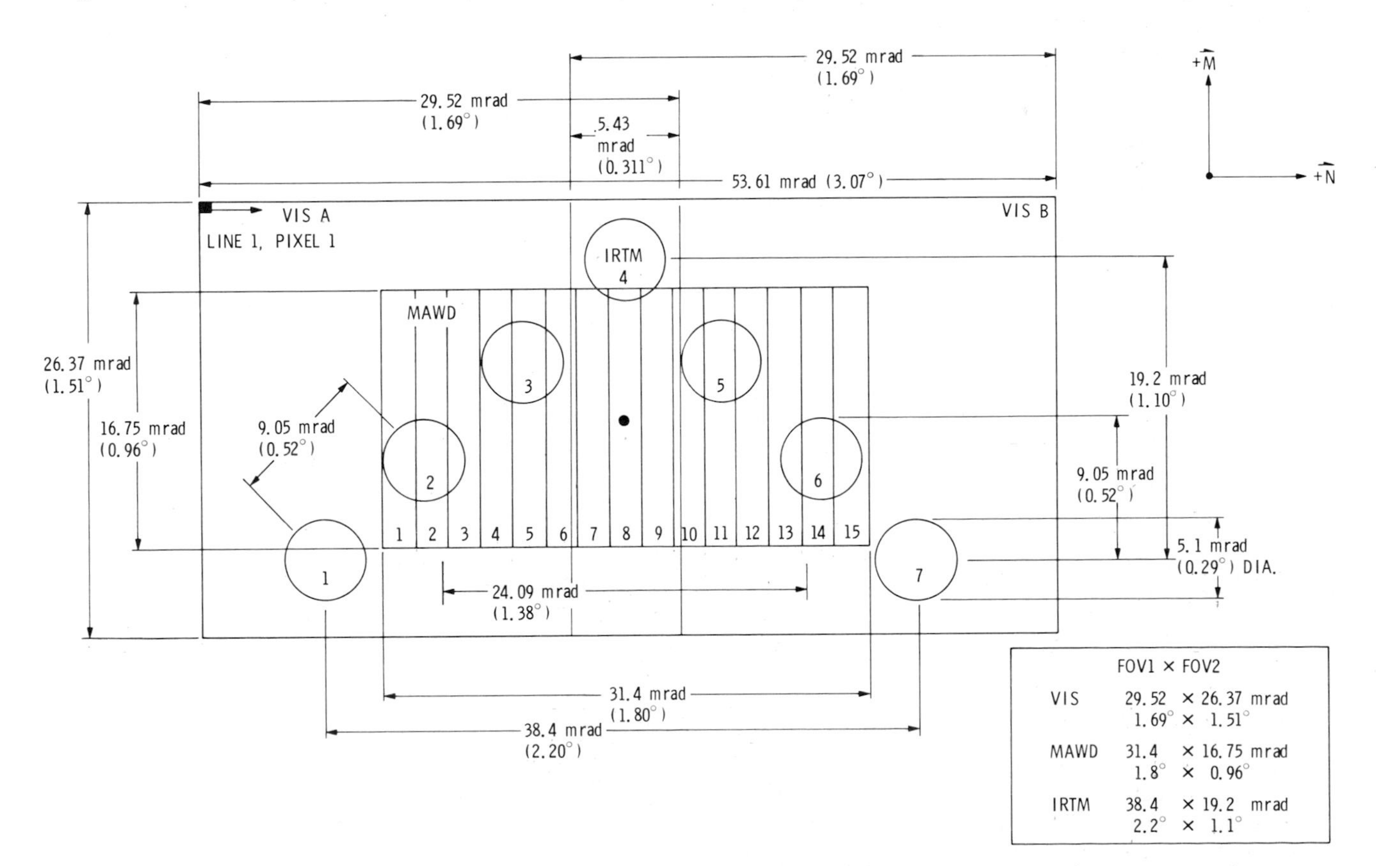

Fig. 1. Orbiter science instrument fields of view. Depicted are the fields of the two VIS cameras, the seven sensors in each of the four IRTM telescopes, and the 15 individual positions of the Mawd stepping mirror as they are superimposed for high-altitude observations for which orbiter-planet motion is negligible over a few seconds. Since the various observations are not all simultaneous but spread over one frame time (4.48 s), the pattern is greatly distorted in low-altitude sequences.

quired to correct the interplanetary trajectory, place the spacecraft into Mars orbit, and subsequently alter this orbit a number of times. The propellant capacity was designed to assure that orbit could be achieved under the '3σ' worst conditions and to provide the number of orbit trims that was deemed necessary to carry out the primary mission as it was defined several years before launch. This capacity is most conveniently expressed in terms of the velocity increment ΔV in meters per second that can be imparted to the spacecraft. Since the ΔV required before and during orbit insertion proved to be moderate on both spacecraft, the capacity for subsequent trims has been fully adequate. On VO-1 the trims during the primary mission changed only the orbit period, requiring very little propellant, and we entered the extended mission with 241 m/s still available. Lowering the periapsis altitude from 1500 to 300 km used up 70 m/s of this, and at present (June 1977) there is 104 m/s left. Because the two orbiters were identical and the orbit insertion of the second spacecraft required less fuel for orbit insertion than the first, it was possible to rotate the orbital plane of VO-2, a very energy costly maneuver. The change in inclination from 57° to 80° in two steps and the lowering of periapsis to 778 km consumed 432 m/s of ΔV, and the remaining capacity is 75 m/s as of June 1, 1977.

Attitude Control

The attitude control subsystem includes a sun sensor that provides a signal indicating orientation in pitch and yaw, a star sensor that similarly indicates orientation in roll, and a set of small compressed-gas jets that intermittently provides the small torque couples that are required to maintain the images of the two celestial bodies near the center of the fields of view of the sensors. It also includes a gyroscopic inertial reference unit that can be activated whenever the orientation provided by the celestial references is not appropriate (e.g., during propulsive maneuvers). The gyros are normally turned off, both because they have a limited lifetime and because the attitude control gas is expended at a more rapid rate when the spacecraft is on inertial reference. The supply of this gas (nitrogen) controls the lifetime of the orbiters, on the assumption of the absence of catastrophic failure of some component. When the spacecraft is quiescent, the rate of gas expenditure is very low (it averaged 0.003 kg/d on VO-1 during the solar conjunction period), but propulsive maneuvers and large motions of the planetary scan platform can markedly increase the rate. At launch the two spacecraft held 14.7 and 14.0 kg of gas, respectively, and the residual amount on April 5, 1977, was 10.2 and 10.1 kg, respectively. At the rate of consumption in the first 3 months of 1977 the supply would be exhausted in June 1978. The actual time will depend upon how often we move the scan platform to observe the planet, since this motion accounts for about half of the total usage. Before that time arrives, a valve can be opened to connect the nitrogen tanks to the helium tanks that pressurize the propulsion system. This helium will provide additional attitude control capability equivalent to about 3.2 kg of nitrogen. With some exercise of frugality in the scan platform use we should be able to continue acquiring valuable data with both orbiters until the end of 1978.

The characteristics of the attitude control subsystem constrain the science observations in several ways. The gas jets are actuated only momentarily when the image of the celestial object moves beyond the edge of the 'dead band' in the sensor, which is normally 0.5° in width. Thus the direction of pointing of the instrument line of sight cannot be predicted more precisely than this amount. After the fact, however, with the use of engineering telemetry from the star sensor, the direction can be determined to an accuracy of 0.2°. The orbiter orientation is constrained by the choice of a reference star. The star must be sufficiently bright (at least about one fifth as bright as Canopus), it must not disappear behind the planet at any point in the orbit, and it must be in a direction that allows the high-gain antenna to point toward the earth. Various stars have been used, including Canopus, Vega, Capella, Arcturus, Rigel, Deneb (the dimmest that is acceptable), and Jupiter, but during the extended mission, when the relative positions of the sun, Mars, and the earth are very different from those for which the orbiters were designed, it has frequently been the case that no star was accessible that permitted the desired regions of the planet to be observed without going onto inertial reference and executing a roll maneuver. Restrictions on the frequency with which these maneuvers could be designed and executed has imposed a serious constraint on the observations that could be made.

Planetary Scan Platform

The three orbiter science instruments are enclosed in a structure called the scan platform (Figure 2), which provides temperature control for them and can be rotated in both cone and clock angles to point them as is desired. After the lander capsule and its mounting structure have been separated from the orbiter, the physical constraints on the platform permit the line of sight of the instruments to be pointed anywhere within the region bounded by cone angles of 90° and 175° and clock angles of 80° and 310° as shown in Figure 3. In the narrow 'keyhole' between 80° and 130° clock angles, cone angles down to 45° are also accessible. Portions of the physically accessible pointing region are not usable by one or another of the instruments because various parts of the orbiter come close enough to the line of sight to cause problems. This situation is depicted in Figure 3, and it will be seen that the constraints are particularly severe for the IRTM because of its sensitivity to diffuse thermal radiation from the spacecraft and because both its planet port and its space port must be unobstructed.

The restricted viewing region has been a major constraint on the orbiter observations. The design and placement of the scan platform were fixed in the early years of the project with the goal of providing the best viewing of the planet (and especially of the landing sites) during the primary mission. At this time the landings were specified to occur within 30° of the equator, and the platform placement was not optimum for the orbits of the second spacecraft required to land at 48° north. With the retention on Orbiter 2 of the bioshield the accessible region has been reduced by more than half. The bioshield also makes it impossible to point the Mawd at its diffuser plate, which is a small diffusely reflecting metal plate placed in the sunlight. By observing it the Mawd can record a portion of the solar spectrum for calibrating the instrument. Thus no postlaunch calibrations have been available on VO-2.

The scan platform can be rotated smoothly in either clock angle, cone angle, or both at rates of either 0.25 or 1.0 deg/s. The motions are quantized in steps that are approximately $\frac{1}{4}$°. The higher slew rate is used infrequently because it is considerably more wasteful of attitude control gas. Subroutines stored in the memory of the on-board computer can be called up to move the platform in specified patterns during the observations.

Command Processing

The computer command subsystem (CCS) consists of two identical and independent data processors which receive and store all commands from earth and control everything that the

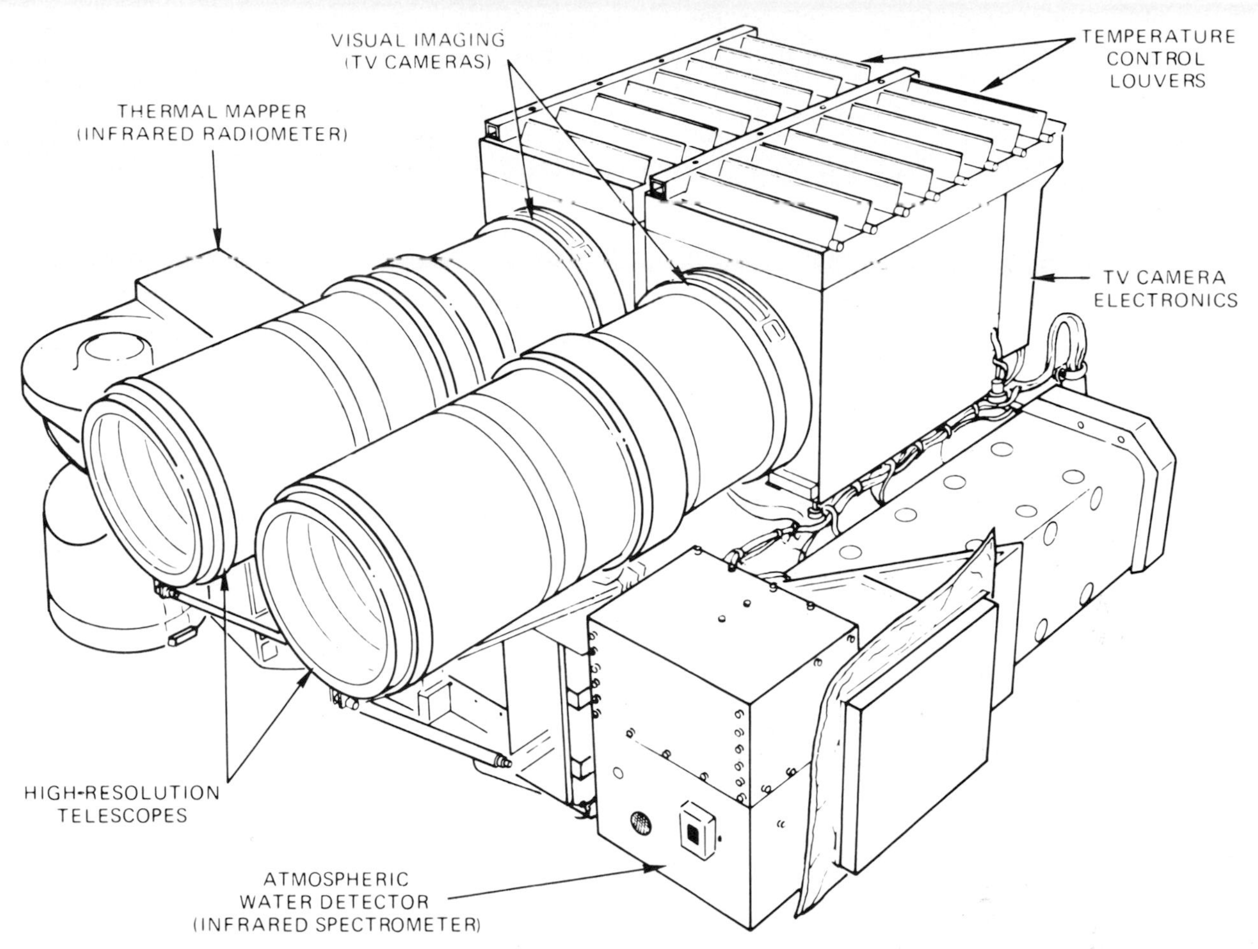

Fig. 2. Orbiter science instruments mounted on the scan platform depicted without their covering of thermal insulation. Through the large aperture of the thermal mapper (IRTM) can be seen the edge of the movable mirror. The rectangular protuberance on the side of the water detector (Mawd) is the radiator plate that keeps its sensors cold.

spacecraft does. Each processor contains a 4096-word memory, and for data acquisition sequences, only one processor is used. This relatively large memory and the capability of reprograming it in flight give to Viking the capability for much greater complexity and flexibility of observation sequences than any of its Mariner antecedents had. The memory is partitioned such that a maximum of about 1500 words are available for each load of ground commands to operate the orbiter, and during the primary mission it was the rule that the initial planning would use no more than 1200 of these. Since it was not feasible to predict in advance precisely how many command words would be required, this limitation assured that the command generation process could proceed expeditiously and that planned observations would not have to be dropped at the last minute because they would not fit.

The command words not only must specify how, when, and for what duration each of the science instruments operates but also must control the motions of the scan platform, any spacecraft maneuvers that occur, the tape recorders during both recording and playback of the data, and any switching between celestial and inertial attitude reference that may be required by the intermittent presence in the star sensor of stray light from Mars, Phobos, or Deimos. The command word limitation is usually the factor that constrains the quantity of observations that can be made. This constraint was fairly comfortable when command loads were 2 days apart at times during the primary mission but quite uncomfortable for 4- or 5-day loads; when loads are separated by a week or more, very large gaps in the desirable coverage of the planet must be accepted.

A fairly typical example of a 2-day primary mission command load is shown below. Since each orbiter roll maneuver requires about twice as many words as an average science observation and since there is a certain amount of daily overhead, a load that runs for several days or includes rolls may provide considerably fewer observations than this one.

A typical VO-1 command load, for example, for the period 1800 hours on August 17 to 2100 hours on August 19 (revs 58 and 59) was nine VIS sequences for a total of 70 frames; three 2.5-min 'headers' preceding VIS sequences for both Mawd and IRTM; eight box scans primarily for IRTM, 92 min total; two other IRTM sequences, 16 min total; eight box scans primarily for Mawd, 102 min total; 2 hours of simultaneous lander and orbiter ranging for radio science; and 2 hours with the Canberra tracking station recording signals alternately from the orbiter and from a quasar.

Instrument Operation and Data Processing

The detailed sequencing of the orbiter science instruments is accomplished by the flight data subsystem (FDS) in response

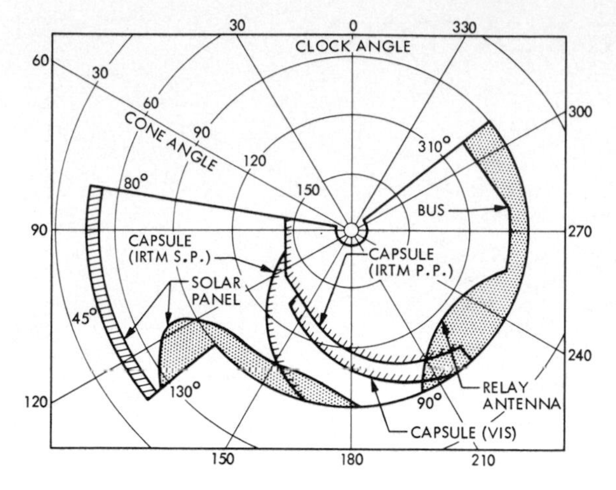

Fig. 3. Region in clock and cone angle space that is available for viewing by the orbiter science instruments. The outer boundary represents the physical constraint to prevent touching any part of the structure. The interior lines represent (somewhat simplified) boundaries of regions which must be avoided to prevent the reflection of excessive stray light into an instrument from some spacecraft component. The capsule (bioshield) constraint is no longer present on VO-1; on VO-2 it is more severe for the space port (SP) of the IRTM than for its planet port (PP). The striped area is a VIS constraint, and the dotted areas are IRTM space port constraints. The bus is the orbiter itself.

to the general instructions of the CCS. The FDS performs all the data handling for all orbiter subsystems. In response to command pulses from it, Mawd steps and resets its raster mirror, IRTM moves its mirror and informs FDS of the mirror's position, and VIS shutters its vidicons and subsequently erases each picture from the photosurface. The FDS controls data acquisition modes, data rates and formats, gain settings, etc. From every subsystem it receives data, converts them from analog to digital, combines them with other data in the proper format, and routes them either to storage or to immediate transmission. Data from the two IR instruments, which come in at 1000 bits/s, are interleaved bit by bit with engineering data at the same rate which contain, among many other measurements, the position of the spacecraft in its orientation limit cycles and the position of the scan platform. This scheme assures that the pointing and position information that are essential to the analysis and interpretation of the IR data are received at earth simultaneously with the scientific data themselves. For VIS the FDS separates the picture data at 2.2 Mbit/s into seven streams at 314 kbit/s (one for each recording track) and inserts (during the flyback of the picture readout raster between scan lines) the engineering data on camera performance, the IR data that are acquired simultaneously, and the entire engineering data stream of the orbiter. The observing conditions of each picture (filter, exposure, light flood, and amplifier gain and offset) are specified by a word in a table in the FDS memory. The 256-word capacity of this table has only occasionally limited the number of pictures that could be acquired in one command load.

Data Quantity Limitations

Many factors influence the quantity of data return from the orbiters, and any one of them may be the limiting factor in particular circumstances. For the IR instruments the limit is usually the number of command words available in the command load, since their data rate is not high and the tape recording track that is used for their data and the lander relay data has adequate capacity. The number of VIS picture se-

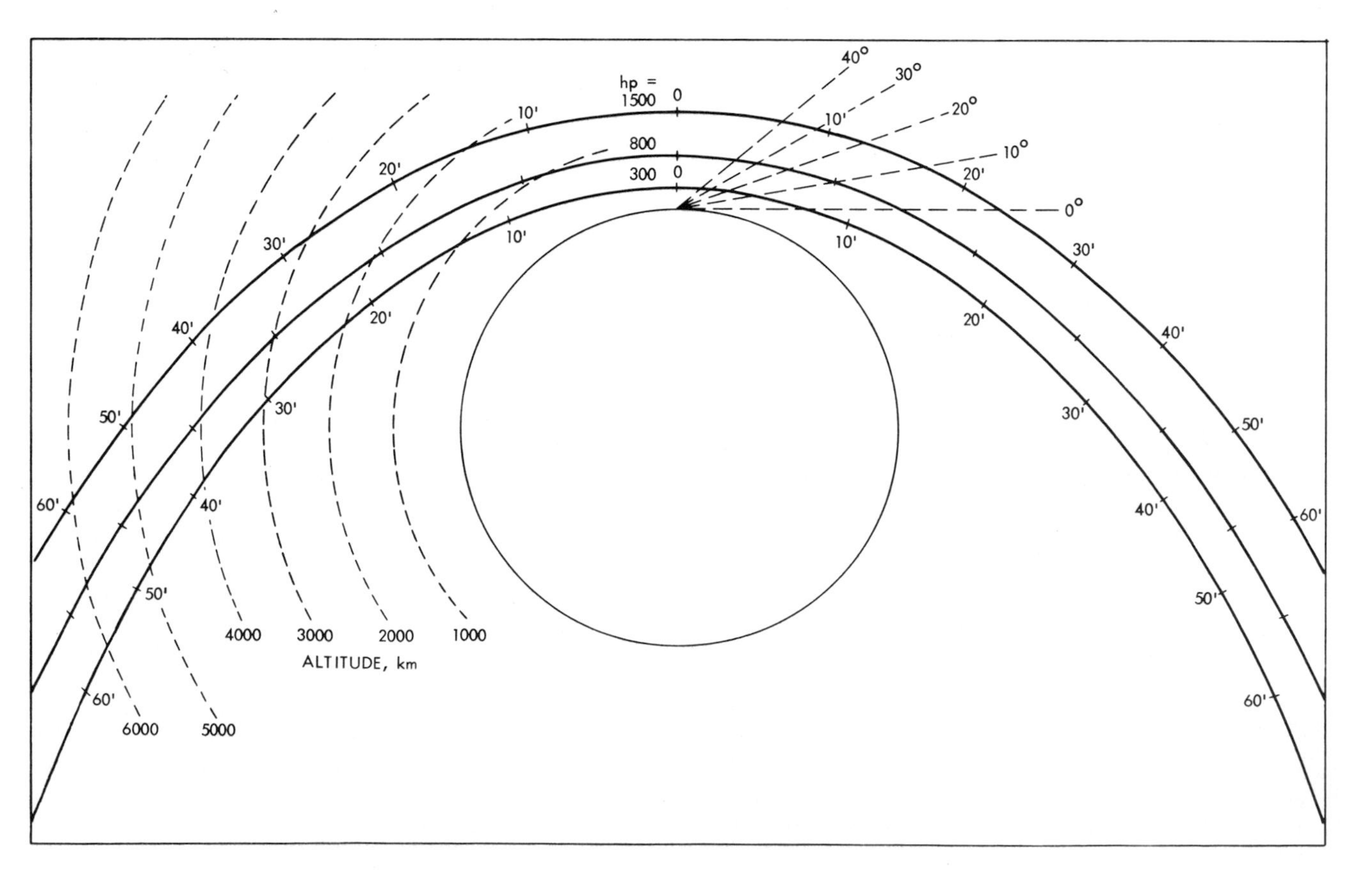

Fig. 4. Near-periapsis portion of the three standard orbits with periapsis altitudes of 1500, 800, and 300 km. The tick marks show times from periapsis in minutes.

quences is likewise limited by the command words, but the number of pictures is not, since a sequence of any number of pictures up to 64 requires the same number of command words as does a single frame. The limitation on picture quantity is the tape capacity, the telemetry link capacity, or (rarely) the FDS table capacity. Each of the two tape recorders can hold approximately 60 pictures acquired in one sequence, but since it is not feasible to pack the sequences tightly, the picture acquisition and playback are planned so that only rarely do we exceed 100 unretrieved pictures on the two recorders. The link capacity is determined by the acceptable data rate and the time available on the Deep Space Net. During the primary mission, Viking had nearly 100% coverage by 64-m antennas at the three stations, and data playback was scheduled approximately 20 hours/d. In the extended mission, more than one station is rarely available to the project. The telemetry data rate is limited by the increase in noise and bit errors with increasing rates. A maximum acceptable bit error rate was set by the science teams: one error in 50 bits for imaging data and one in 800 for IR data. The telemetry rates available are 1, 2, 4, 8, and 16 kbit/s. At the commencement of each mission planning cycle the total link capability for each command load is predicted by considering the factors of station availability, distance to the spacecraft, and elevation above the horizon and using the assumption that the highest data rate consistent with the specified bit error rate will be used. This predicted capacity usually proves to be the limiting factor on the number of VIS pictures that can be planned to be acquired. It requires about 10 min to play back one picture at 16 kbit/s and about 80 min at 2 kbit/s. The 1-kbit/s rate has never been used. Unfortunately, the 16-kbit/s capability ran out just about the time that Viking 1 reached the planet. With Mars now again approaching the earth the highest data rate is available part of the time, and its use compensates in part for the reduction in receiving station coverage.

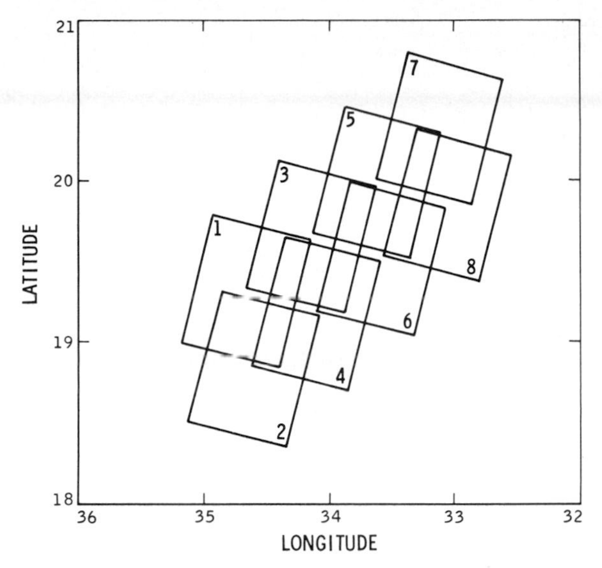

Fig. 5. A typical swath of VIS pictures near periapsis. The numbers give the order in which the pictures were taken, odd numbered frames by camera A and even by camera B. The orientation of the frames relative to the orbit motion depends upon the position of the sun relative to the orbit plane. With the sun in the plane the alignment is perfect, and the edges of the swath are not jagged. This particular swath was taken by VO-1 on rev 1 between 10 and 42 s after periapsis passage.

VIKING ORBITS AND VISIBILITY OF THE PLANET

The geometric relations between the orbiter and the planet determine what kind of observations can be made. Throughout both missions the orbital periods have never been much different from the synchronous value of 24.623 hours; hence by Kepler's law the major axis of the orbital ellipses has always approximated 40,850 km, and the apapsis altitudes have been much higher than the periapsis altitudes. The periapsis region of the three standard orbits (1500, 800, and 300 km) that have been used is shown in Figure 4. In all three cases the elapsed time between periapsis and the end of the latus rectum, where the altitude is greater than 3000 km, is less than 51 min, and the time spent at the lowest altitude, where the spatial resolution is the best, is very short.

All three orbiter science instruments were designed to make observations most efficiently at periapsis in the 1500-km orbit that was standard for the primary mission. In this orbit the angular velocity relative to the surface near periapsis is 2.616 mrad/s; thus the motion during one Mawd raster is about 12 mrad, so successive fields of view overlap by about one quarter of their 17-mrad width; similarly, the motion in two VIS frame times (23.4 mrad) provides overlap between successive frames from either camera, which have a minimum width of 26.9 mrad. A typical swath of pictures taken under these conditions with no scan platform motion is shown in Figure 5, and since the two IR instruments are boresighted with the cameras, their coverage of the surface contains the area pictured. At higher altitudes the orbital angular velocity is reduced, and since it is not feasible to increase the interval between exposures to control the overlap, the scan platform is moved between pictures in a regular pattern that is appropriate for the speed and direction of relative ground motion at the time. Platform slews of $\frac{1}{4}°$ or $\frac{1}{2}°$ in clock angle, cone angle, or both can be made, and the resulting vibration will be damped out sufficiently so that the next picture is not smeared. An example of a strip of pictures taken in this way is shown in Figure 8*a* of *Carr et al.* [1977]. This intermittent platform motion does not provide good coverage for the two IR instruments, and so these VIS sequences are frequently preceded by a standard 'box scan' pattern of about a 2-min duration that covers the same area. At the lower periapsis altitudes of the extended mission orbits, continuous scan platform motion is commanded during VIS sequences partially to compensate for the ground motion and reduce the image smear. This technique has also made it feasible to take pictures of Phobos from as close as 100 km without getting excessive smear.

A majority of the data for the IR instruments are acquired by using the so-called box scan, which provides a boustrophedonic pattern of alternate clock and cone motions that efficiently covers a fairly large area. The platform motion is controlled by a subroutine in the CCS, and the size and shape of the coverage are specified by two cone slews (in the positive and negative directions) and one clock slew. The shape can be rectangular, if both cone slews are equal, or that of a skewed parallelogram, if alternate cone slews are unequal. Figures 6 and 7 show examples of box scans. They may last as long as an hour and provide efficient coverage with a minimum requirement of command words. Unfortunately, they have a high usage of attitude control gas, and we are now having to cut back on their size and frequency to conserve it. More complex or nonrepetitive platform sequences can be executed by specifying each motion individually, but many more command words are required than are required for a box scan.

At periapsis in the standard 1500-km orbit the visible surface of the planet is a circle of a 2730-km radius, but a large fraction of that circle is, of course, seen at near-grazing in-

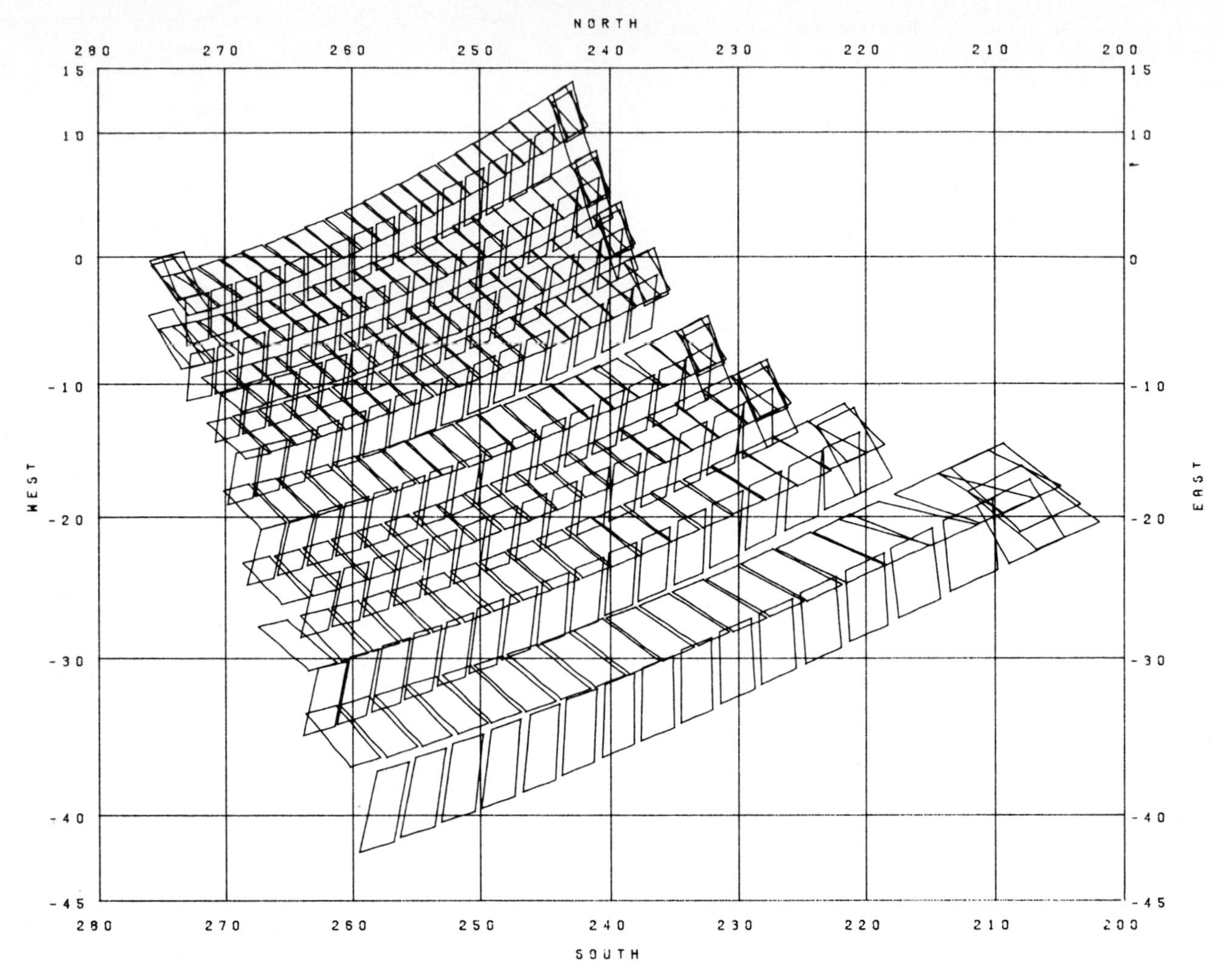

Fig. 6. Surface coverage in a typical Mawd box scan at intermediate altitude, acquired by VO-1 on rev 137. The sequence began at the lower left, 74.8 min before periapsis with the orbiter 7660 km above 23°S, 272°W; it ended at the upper left 48.8 min before periapsis, 5170 km above 14°S, 263°W. It consisted of 16 cone angle slews, alternately 18.75° toward the east and 21.0° toward the west, with northward clock angle slews of 1.0° between cone slews. Each parallelogram depicts the coverage in one 15-step scan of the mirror lasting 4.48 s. This was an early morning observation; approximately the first field of view at the left of each swath was beyond the terminator in the dark.

cidence. A swath of VIS frames is approximately 80 km wide, the exact value depending upon the angle between the orbiter's symmetry axis (normally pointed at the sun) and the plane of the orbit, and the swath covered by the IR instruments is just slightly larger. Figure 8 depicts the surface as seen from the orbiter at periapsis and outlines the composite field of view in a typical 8-frame swath. A typical box scan observation by the Mawd at an intermediate altitude is shown in Figure 6 in Mercator projection. At high altitudes it is possible for the IRTM (which can see in the dark) to view nearly half the planet. A typical box scan of this type is shown in Figure 7; the chevrons represent every eighth observation by the IRTM.

The latitude of periapsis for VO-1, given in Table 2, started at 23.2°N and moved slowly northward to 39.2°N during the first 9 months. For VO-2 the periapsis latitude started at 50.4°N, jumped to 63.6°N with the first plane change maneuver, and subsequently has moved down to about 50°N. As a result, all the low-altitude high-resolution observations have been in the northern hemisphere, the observations at intermediate altitudes have been in the equatorial regions or the northern hemisphere, and the broad area low-resolution coverage has been in the south. One reason for hoping to be able to continue observations into 1978 is the chance that is offered to move high resolution southward and broad coverage northward at the same Martian season that was observed in the primary mission.

The ground tracks for the two orbiters shortly after the landings are shown in Figures 9 and 10. In synchronous orbits the tracks are approximately closed curves as shown. In walking orbits the shape of the ground track is not greatly different, since the change in longitude in successive orbits has never been more than 40°. After the VO-1 walk in September the new orbit was about 160° west of the position shown in Figure 9. That position was chosen to give the maximum relay link from VL-2; the orbiter did not go far enough north to fly over the lander. Figure 10 shows the ground track of VO-2 at the time of landing and the track after the second inclination change; the track for the 75° inclination orbit would be similar to the latter. When this orbiter was again synchronized over its lander in March, the orbiter flew over the lander on the eastern, southward moving portion of the orbit; this arrangement was chosen because by that time the northward pass over the latitude of the lander was in the dark, a situation precluding photography of the lander's vicinity from that point in the

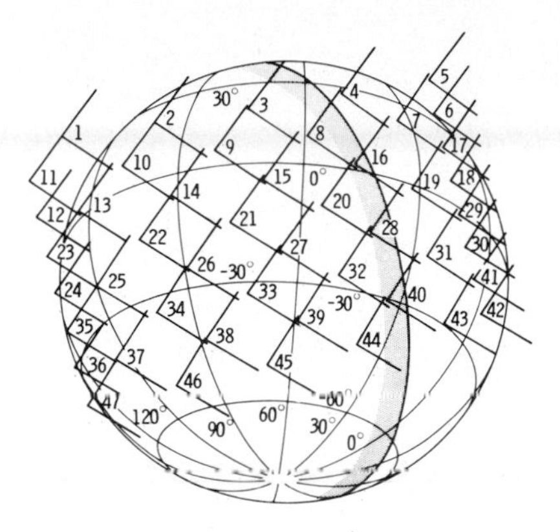

Fig. 7. Perspective view of the planet and surface coverage in a typical IRTM box scan at high altitude, acquired by VO-1 on rev 40. As shown by the numbering of the chevrons, the sequence began at the upper left and ended at the lower left. The alternating cone angle slews were 10° in length, separated by 1° southward clock angle slews. The sequence lasted for 7 min beginning 10.5 hours before periapsis. During this time the orbiter moved about 240 km, approximately 32,000 km above 29°S, 63°W. The chevrons, representing the fields of view of each set of seven sensors, are shown for every eighth observation (8.96-s intervals). This was a predawn observation with most of the planet in the dark; note the terminator.

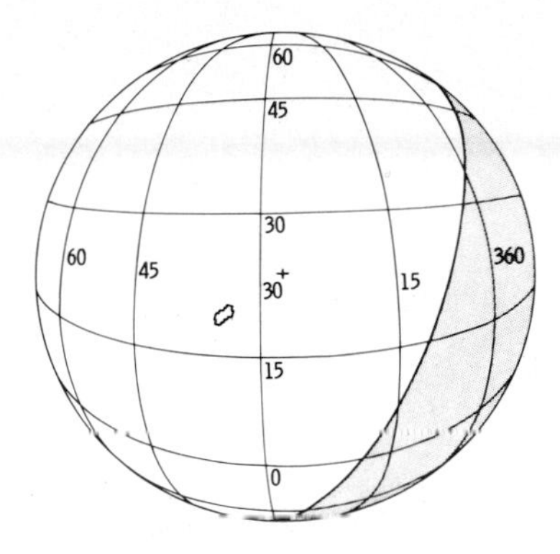

Fig. 8. Perspective view of the planet near periapsis and surface coverage of a typical VIS swath. At the exposure of the first picture the orbiter was over the cross at the center, and it moved 126 km northeastward in the 32 s before the eighth picture. An enlarged view of the coverage is given in Figure 5.

orbit, and because the higher altitude of overfly increased the length of time that data could be transmitted from lander to orbiter.

To assure optimum illumination for the site certification photography, the initial orbit of VO-1 was chosen to place the subperiapsis point and the landing site near the terminator at the time of overfly. A sun elevation of about 20° was considered optimum. Periapsis passage for VO-1 on rev 2 occurred in the late afternoon with the sun elevation 18°. The elevation increased about ½°/d, first rising above 45° on rev 65 (August 25), and remained above that value throughout the primary mission. Thus the high-resolution pictures became progressively less useful.

On rev 108 (October 7) the orbit plane of VO-1 crossed the sun so that the elevation angle at periapsis reached a maximum (89°), and thereafter it decreased, falling below 45° again on rev 217 (January 25) in the extended mission. In March and April, as the sun at periapsis approached the evening terminator, conditions were again very good for near-terminator photography. On rev 329 (May 13) the subperiapsis point went into the dark, from which it will not emerge until July or August 1978.

On the first rev after MOI, VO-1 crossed the morning terminator nearly 5 hours before periapsis, about 26,500 km above 35°S latitude, and it remained over the lighted portion of the planet until 11 min after periapsis, when it crossed the evening terminator at 35°N. Thus a large portion of the planet was visible in the morning hours, and the IRTM was able to observe large areas in the predawn, but very little coverage could be obtained late in the day because of the low altitude at this time. As time went by, the time of morning terminator crossing became rapidly later, while the evening terminator crossing time changed more slowly, and at the end of the primary mission the times for the two crossings were 1.5 hours before periapsis and 28 min after periapsis, respectively, so the area and duration of visibility on the planet were reduced (except for IRTM, which can see in the dark), but the duration of high-resolution coverage in the north was increased.

At the start of the extended mission (rev 175) the morning terminator crossing took place 67 min before periapsis over 14°S, and the evening crossing 35 min after periapsis over 14°N. Thus a considerable area near and north of the equator was visible with good resolution, but the far southern latitudes were in the dark. As the periapsis point (at the time of periapsis passage) approached the morning terminator, the time of visibility late in the day increased rapidly, and the entire southern hemisphere came into view. On rev 329 (May 13), when the periapsis point crossed the terminator into the dark, the orbiter was over the lighted portion of the planet for more than 11.5 hours. Subsequently, this time will become much longer.

Orbiter 2, in contrast to its mate, initially passed through its periapsis in the early morning, and it was not possible to provide optimum lighting for the site certification photography. The sun elevation angle at the first periapsis passage was 63°, and it had decreased to 54° by the time of landing. It dropped below 45° at the time of the first plane change (rev 51) and continued decreasing, reaching 22° at the end of the primary mission. Thus lighting for high-resolution photography improved continuously. At the time of the second plane change early in the extended mission (rev 123) the periapsis point crossed the morning terminator, and it will reemerge into the light in early September 1977.

In the initial orbit, VO-2 crossed the morning terminator onto the lighted side of Mars 48 min before periapsis at 17°S and crossed the evening terminator 54 min after periapsis at 17°N. Prior to the plane change to 75° inclination on rev 51, the terminator crossing times moved later to 28 min before periapsis at 10°N and 97 min after periapsis at 10°S, respectively. In the new orbit the orbiter was over the lighted hemisphere from 61 min before periapsis at 31°S to 41 min after at 32°N. Making its closest approach to the north pole about 9 min after periapsis at an altitude of 1700 km and rising rapidly, the orbiter was able to look a few degrees beyond the pole and get excellent though oblique coverage of the entire summer polar cap. By the end of the primary mission the terminator crossings had moved to 19 min before periapsis at 17°N and 140 min after periapsis at 17°S.

By the time of the change of inclination to 80° and periapsis altitude to 800 km at the start of the extended mission, the periapsis passage was barely beyond the morning terminator at 62°N, and the orbiter remained over the lighted hemisphere for more than 14 hours of the 26.5 hours in one revolution, crossing the evening terminator at 63°S. Thus the north pole

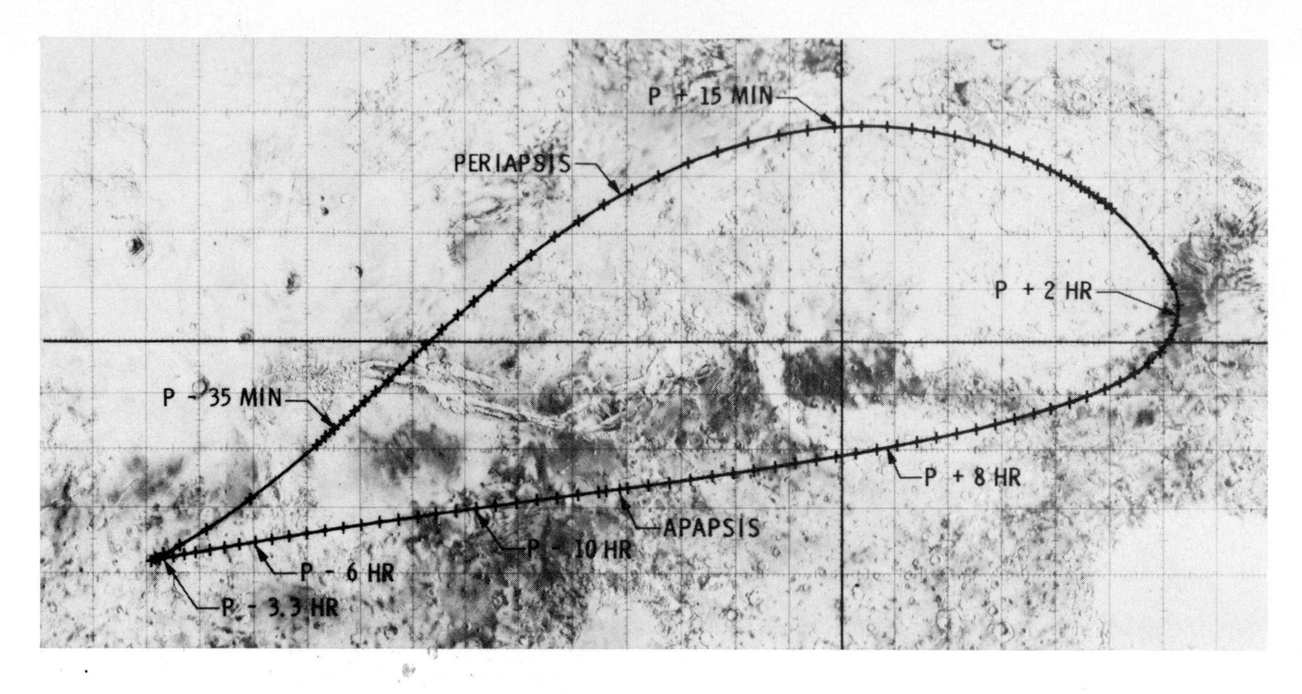

Fig. 9. Suborbital ground track of VO-1 in synchronous orbit just after the landing. Motion is clockwise. Tick marks are spaced 2 min apart near periapsis and 20 min apart near apapsis. Note that the track passes over the lander at 22°N, 48°W, about 3 min before periapsis.

could still be observed at relatively high resolution, and most of the southern hemisphere was visible at low resolution. Since the orbit was nonsynchronous at the time, periapsis moving westward 27°/d, excellent coverage of the entire planet would have been possible at this time except for the pointing constraints imposed by the platform. Two months later (rev 181) the suborbiter point was in the light for 24 hours from 77°N to 77°S, but the close approach to the north pole was now in the dark. In the following 100 orbits (up to June 1, 1977), as the periapsis moved farther into the dark and the sun moved southward, the northern limit of visibility has moved southward (morning terminator crossing at 52°N on the southward portion of the orbit), but the time spent over the lighted hemisphere was still more than 21 hours, and excellent coverage of the southern hemisphere was possible.

FUTURE OF THE VIKING ORBITER MISSIONS

At the time of writing (June 1, 1977) we are approaching the first anniversary of the first Viking orbiter observations of Mars with four healthy spacecraft still continuously acquiring

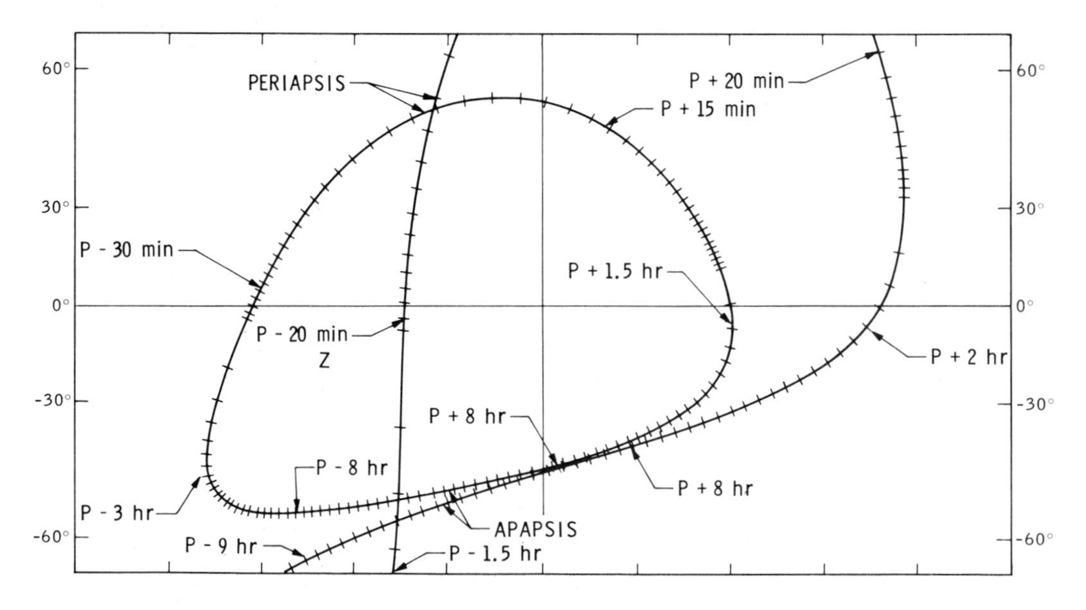

Fig. 10. Suborbital ground tracks for two orbits of VO-2. The closed track with inclination 55.4° is rev 18 on August 27, synchronized over the landing site at 48°N, 226°W. Moving clockwise, the orbiter passes over the lander about 3 min before periapsis. The open track with inclination 80.2° is rev 191 on March 4; moving southward, the orbiter passes near the lander about 29 min after periapsis. Tick marks are spaced 2 min apart near periapsis and 20 min apart near apapsis. The horizontal (longitude) position of the tracks is arbitrary.

scientific data. Detailed plans are now being generated for 1 more full year of operations (to the currently authorized termination date of May 31, 1978), and consideration is being given to the desirability and feasibility of continuing beyond that date, perhaps through the next solar conjunction in January 1979. For several lander experiments and all the orbiter experiments, there is great scientific value in extended observations, especially over a full Martian year or longer.

Since Viking was principally a lander mission, the orbiter investigations have generally had to accommodate to the lander requirements. In particular, the choice of orbits has usually been dictated by the necessities of lander relays. Hereafter the emphasis will shift more and more toward the orbiter investigations so that they will be more nearly free to pursue their own major scientific goals. Thus relays will be less frequent, and orbits will be chosen to suit the requirements of orbiter science.

Current plans provide for maneuvering both orbiters into orbits with periapsis altitudes of about 300 km and periods of 24.0 hours, in July for VO-1 and in September for VO-2. Far from optimum for lander relays, these orbits have several advantages for orbiter science. The low altitude provides VIS with an increase in resolution of a factor of between 3 and 5 and affords radio science the opportunity for much increased precision and resolution in measuring gravitational anomalies. The 24-hour period will simplify the operations to the advantage of everyone, and it will place all the earth occultations at a time when they can be observed simultaneously by two tracking stations. The time for one walk around the planet will now be 40 days, and the motion of 9°/rev is slow enough so that good overlapping coverage of the planet is attainable by the two infrared experiments.

Thus if the orbiters continue to perform well, all the orbiter science investigations will continue to acquire data for another year and perhaps somewhat longer. VIS will be systematically photographing the planet at moderate resolution; improving on the mapping done by Mariner 9 (which was especially poor in the northern hemisphere); observing the change of seasons in both polar regions; making high-resolution observations of small regions that are accessible near periapsis, especially those that are candidate landing sites for the next Mars mission; and considerably augmenting the collection of stereo and color pictures. IRTM will be observing both polar areas more systematically and with better resolution than was possible in the primary mission, studying the thermal changes in the surface and atmosphere with the changing seasons, and attempting to improve the characterization of surface properties by better coverage and resolution than has been achieved. Mawd will be attempting to acqu're complete global coverage of the atmospheric water vapor every 40 days to study the seasonal changes, observing the diurnal variation of water vapor at various places, and determining the phase function at various places. Radio science will be acquiring occultation data at many places on the planet to provide atmospheric profiles and topographic information, making Doppler measurements at low altitude to determine the gravity field, perhaps carrying out a bistatic radar experiment to measure surface properties, and (possibly at the very end) repeating the relativity and solar corona experiments in the next solar conjunction. All the investigators will be continuing to analyze and interpret the great quantity of data that has been obtained by the remarkably successful Viking missions.

Acknowledgments. In such a highly organized and cooperative endeavor as the Viking project, every scientist can think of scores of associates whose competence and devotion to the goal of doing good science on Mars has contributed in a major way to his investigation. As orbiter scientists we feel especially indebted to the Mission Planning Group, which supplied the detailed information necessary to plan all the observations and transformed our ideas into specific computerized instructions, to the Orbiter Performance Analysis Group, which turned these instructions into commands to the orbiters and tirelessly monitored every single action that the orbiters took in response to those commands, and to the Data Support Group, which processed all the data into formats that were usable to us. Personally, I want to acknowledge the important contribution of the Orbiter Science Group staff, Neil L. Nickle, Stephen Z. Gunter, and Nicholas K. Simon, in coordinating and assisting the activities of the Orbiter Science Team.

REFERENCES

Masursky, H., and N. L. Crabill, The Viking landing sites: Selection and certification, *Science*, *193*, 809–812, 1976*a*.

Masursky, H., and N. L. Crabill, Search for the Viking 2 landing site, *Science*, *194*, 62–68, 1976*b*.

Carr, M. H., et al., Some Martian volcanic features as viewed from the Viking orbiters, *J. Geophys. Res.*, *82*, this issue, 1977.

Soffen, G. A., The Viking project, *J. Geophys. Res.*, *82*, this issue, 1977.

(Received May 23, 1977;
revised June 6, 1977;
accepted June 6, 1977.)

Some Martian Volcanic Features as Viewed From the Viking Orbiters

M. H. CARR,[1] R. GREELEY,[2] K. R. BLASIUS,[3] J. E. GUEST,[4] AND J. B. MURRAY[4]

The plains to the south and southwest of Arsia Mons are composed of elongate lava flows, many of which appear to originate at a reentrant in the Arsia Mons edifice. Individual flows can be traced for up to 300 km. Their widths range from 5–6 km close to Arsia Mons to 40–60 km at distances greater than 400 km from the volcano. The age of the flows, as indicated by the number of superposed impact craters, increases systematically with increasing distance from Arsia Mons. The length of the flows varies with altitude of the vent: the higher the vent, the shorter the flow. The distribution of vents on Arsia Mons and Pavonis Mons suggests the presence of deep-seated fractures trending NW–SE and NE–SW. Flows on the flanks of Olympus Mons extend far beyond the basal scarp, an indication of a much larger size for the volcano than was formerly believed. Around the basal scarp are numerous lobate features interpreted as landslides. Several types of flows are recognized on the flanks of Alba Patera. These include tube-fed flows, sheet flows, and tube-channel flows. The different types of flows are believed to indicate different eruption rates of a low-viscosity lava similar to basalt in its rheological properties.

INTRODUCTION

One of the most exciting results of the Mariner 9 mission was the discovery of huge volcanos on Mars. They range widely in size and age. The largest and probably the youngest are in the Tharsis region, where four volcanos achieve elevations of about 27 km above the Mars reference level. In addition to such discrete volcanic edifices, there are extensive volcanic plains. In many places, particularly in the Tharsis region, the plains have numerous low lobate escarpments that resemble flow features on the lunar maria, suggesting that the plains in these areas are formed by accumulation of very fluid lavas. This paper summarizes the results of some new observations on Martian volcanic features made from the Viking orbiters. The new coverage is rather uneven, comprising what was obtained as part of site certification, what was visible while the orbiters were synchronized over the landers, and what was observed during the short periods in which the orbiters were desynchronized from the landers and allowed to make wider planetary observations. We have at the time of writing good coverage of Alba and most of Tharsis but rather poor coverage of other areas including the prominent volcanic province of Elysium. This has largely controlled the choice of topics to be discussed. Most of the paper concerns the Tharsis and Alba regions (Figure 1). Volcanic features in other areas are only mentioned briefly in the concluding statement.

ARSIA MONS AND THE SOUTH THARSIS PLAINS

On revolutions 52 and 56, Viking Orbiter 1 acquired moderate-resolution (~450 m) coverage of Arsia Mons and the plains to the south and west. Mariner 9 B frames of the region show numerous lobate flow fronts that resemble those of Mare Imbrium on the moon. The Viking pictures reveal a systematic pattern of flows which suggests that many of the lavas making up these plains originate in a reentrant on the southwest flank of Arsia Mons. Individual flows can be traced for distances of up to 300 km, but from the general pattern it may be inferred that some flows may have traveled as far as 800 km from the Arsia Mons source. The morphology of the flows tends to change with increasing distance from Arsia Mons. Close in, the flows tend to be narrower and thinner and to have well-defined central channels; farther from Arsia Mons the flows are generally broader, lack central channels, and appear to be thicker. This difference may be caused by the steeper regional slopes close to Arsia, by changes in lava properties, such as viscosity and volatile content, with increasing distance from the vent, or by different eruption rates from vents at different altitudes. The characteristics of the flows are consistent with lavas of basaltic composition or at least lavas with low viscosities and yield strengths. The ages of the flows, as suggested by the number of superimposed impact craters, range widely. The flows close to Arsia Mons have more than a factor of 10 fewer superimposed craters than those in the peripheral areas. The general impression is of a continuous accumulation of basaltic lavas, primarily from one source, over a long period of time with a higher proportion of young flows close to the vent.

The general pattern of flows is seen in Figure 2, a mosaic made of frames from revolutions 52 and 56. The pictures were taken at 15°–30° from the morning terminator at a range of 8500–9500 km and have a resolution of 200–250 m per picture element. The surface is partly obscured by clouds, which are evident as wavelike brightenings or as a cellular pattern. The clouds are particularly visible just to the east of Arsia Mons and to the southwest. The surface morphology is also obscured by albedo variations of the surface, especially to the SW and WSW of Arsia Mons. The albedo patterns in places run transverse to the flow patterns, making the flows very difficult to discern. In other places they highlight the flow pattern, as they do to the south of Arsia Mons. Despite these problems the pattern of flows is clear. Close to Arsia Mons the flows are relatively narrow and appear to originate from an irregular alcove in the south flank of the main Arsia Mons edifice. Flows that trend eastward from the source cut across the radial fabric of Arsia Mons, an indication that the surface of Arsia is older than the uppermost flows of the adjacent plains.

The lava plains south of Arsia Mons can be divided into four zones. The inner zone extends from Arsia Mons down to approximately the 9-km contour (Figure 2), a distance of

[1] U.S. Geological Survey, Branch of Astrogeology, 345 Middlefield Road, Menlo Park, California 94025.
[2] NASA Ames Research Center, Mail Stop 204A-1, Moffett Field, California 94035.
[3] Planetary Science Institute, 283 S. Lake Avenue, Pasadena, California 91101.
[4] University of London Observatory, Mill Hill Park, London, NW7 2QS, England.

Paper number 7S0470.

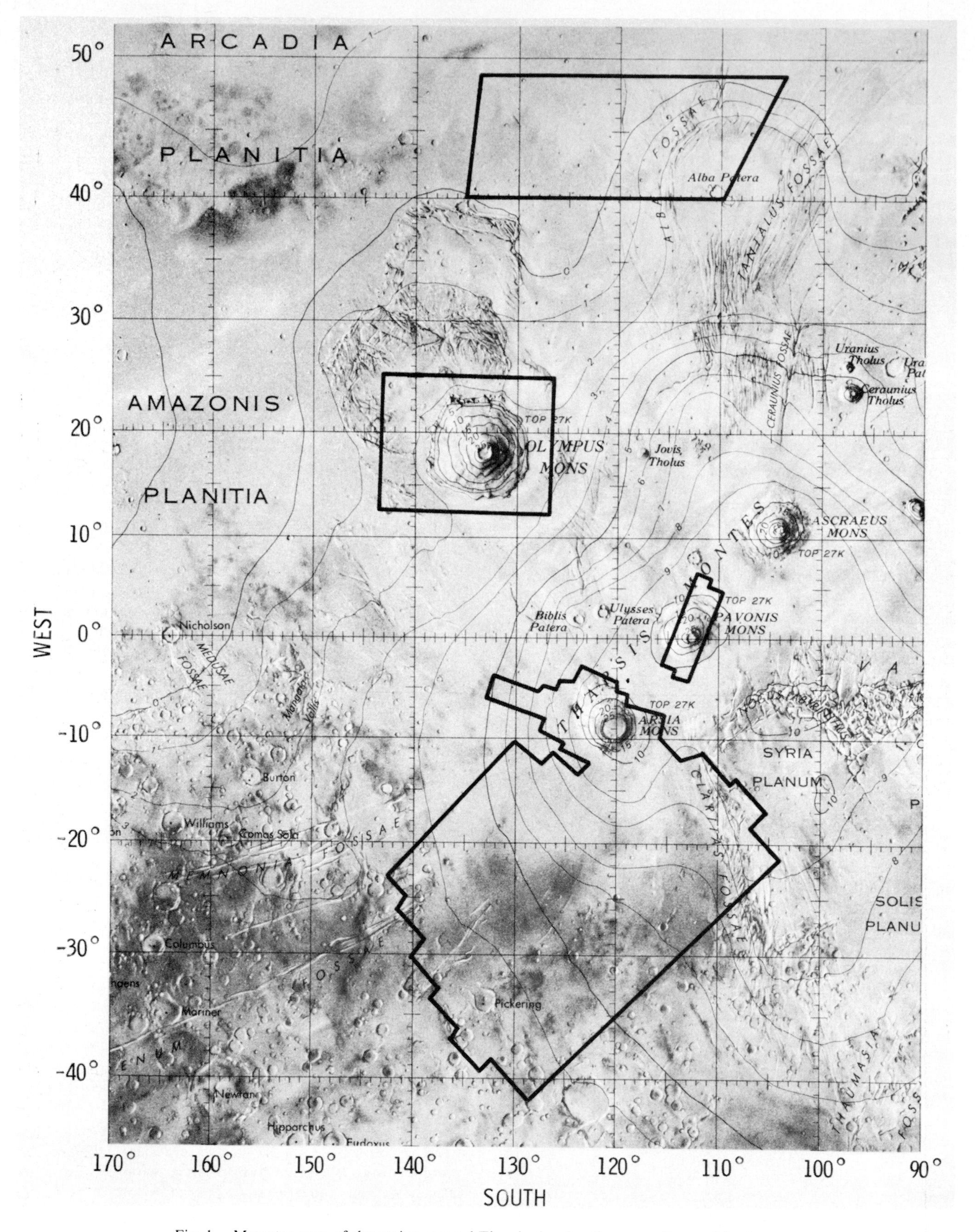

Fig. 1. Mercator map of the region around Tharsis showing the areas discussed in the text.

roughly 400 km. Within this zone the flows are relatively narrow, being mostly less than 3 km wide, and commonly have a central channel (Figure 3). In some cases the central channel has levees, although in general they are absent. The second zone, also approximately 400 km wide, extends from the outer edge of the first zone down to the 6.5-km elevation. In this zone the flows are somewhat wider, being generally 4–7 km across, and central channels are visible in many of the more

Fig. 2. Mosaic of the volcanic plains south of Arsia Mons. Individual flows are outlined where they are identifiable, fissures radial to Arsia Mons are shown with solid lines, and a crosshatched line indicates the edge of the Arsia Mons-related flow field. Dashed lines are 1-km contours starting at 4 km above datum at the bottom of the picture and increasing to 9 km close to Arsia Mons (frames 56A01-56A58 and 62A41-62A48).

Fig. 3. Details of a flow 150 km east of the vent area on the southwest flank of Arsia Mons. The two flows that form a 'V' are each approximately 2 km across, and each has a clearly identifiable central channel (frame 56A27).

TABLE 1. Dimensions of Leveed Flows

w, km	w_b, km	F
1.7 ± 0.4	0.44 ± 0.01	0.0376 ± 0.012
1.7 ± 0.4	0.35 ± 0.08	0.122 ± 0.040
1.2 ± 0.3	0.26 ± 0.06	0.0898 ± 0.03
1.2 ± 0.3	0.26 ± 0.06	0.0898 ± 0.03

Parameter w is the flow width, w_b is the width of the levee, and F is a dimensionless parameter [*Hulme*, 1974].

fresh appearing flows. At distances greater than 800 km from the edge of Arsia Mons the flows broaden substantially, so that the direction of flow is no longer clear. Flows in this third zone achieve widths in excess of 40 km, and channels are no longer visible. A fourth zone can be identified (in the bottom right of the Figure 1 mosaic) which seemingly has been unaffected by the Arsia flows. In this zone, flow fronts are rare. The surface is characterized by numerous features resembling lunar mare ridges and by a relatively high number of superimposed impact craters. It appears to be an older volcanic plain not covered by the Arsia-related volcanics.

The changes in morphology of the Arsia Mons flows appear to be related, at least in part, to changes in slope. The available data on slopes in this area are very general [*U.S. Geological Survey*, 1976]. These data give gradients of approximately 0.008 in the innermost zone, 0.007 in the middle zone, and 0.003 in the outermost zone of flows. The geometry of the contact between the plains and the main edifice of Arsia Mons suggests that the generalization of the contours around Arsia Mons is in error and that the true slopes close to Arsia are higher than those indicated on the map. The range in slope is probably from in excess of 0.008 down to at least as low as 0.003. The differences in flow morphology in the different zones are consistent with the known behavior of lava (or any Bingham fluid) which results in narrower flows on steeper sloping ground [*Hulme*, 1974].

The morphology of a lava flow provides clues concerning its composition. Over the last several years, considerable attention has been given to the morphology of lava flows and what it indicates about the rheological properties of the lava and the eruption rates [*Walker*, 1971, 1973; *Hulme*, 1974; *Sparks et al.*, 1976], and the results have been applied to the moon [*H. J. Moore and G. G. Schaber*, 1975; *Schaber*, 1973] and Mars [*Hulme*, 1976]. Both *Hulme* [1974] and *H. J. Moore and G. G. Schaber* [1975] treat a lava flow as a Bingham fluid and derive the yield strength from the width of channel levees and the flow thickness. Measurement of the latter two parameters from photographs presents considerable difficulties, but estimates are possible. The edge of a levee is difficult to identify and easy to confuse with the edge of the flow in cases in which the flow is narrow. Measurements were made on four flows where levees were clearly distinguishable from flow boundaries (Table 1).

If Hulme's model of levee formation is assumed, the shear strength of the lava can be calculated. From *Hulme* [1974, p. 366],

$$Sy = 2g\rho\alpha^2 w_b$$

If the acceleration due to gravity $g = 371$ cm/s², the lava density $\rho = 2.7$ g/cc, and the slope $\alpha = 0.007$, then the shear strength $Sy = 1.4 \times 10^2$ N/m². This is clearly only a crude estimate of the yield strength and is dependent on a specific model for the formation of a levee. This model assumes that the levee forms by lateral viscous flow and the flow is arrested by the finite yield strength of the lava. Levees can form in several other ways such as by accretion and overflow [*Greeley*, 1971; *Sparks et al.*, 1976]. The above calculations are not valid for the latter cases.

The shear strength as calculated above can be checked against an alternative method derived by *H. J. Moore and G. G. Schaber* [1975]. Their method requires knowledge of the thickness of the flows, which is not easily derived from photographs. The thickness cannot be calculated directly from width of the 'shadow' at the edge of the flow for the following reasons. First, the dark line marking the edge of the flows is almost certainly not a shadow. Computer enhancement of the images accentuates the contrast such that areas in an image only slightly less bright than the surroundings are made to look much darker. Second, camera artifacts tend to broaden features at the limiting resolution of the camera. A better way of estimating flow heights is by comparison with the rim heights of adjacent craters. An estimate of the rim height can be derived from the diameter by using the rim height/diameter relation for fresh lunar craters [*Pike*, 1976]. This is almost certainly an upper limit because the relation is for fresh craters, and rim height on Mars may be somewhat less than it is on the moon. Flow fronts in frames 56A01, 56A03, and 56A017 were examined to find cases where a flow front had inundated part of the ejecta of a crater and the crater rim height was comparable in height to the thickness of the flow.

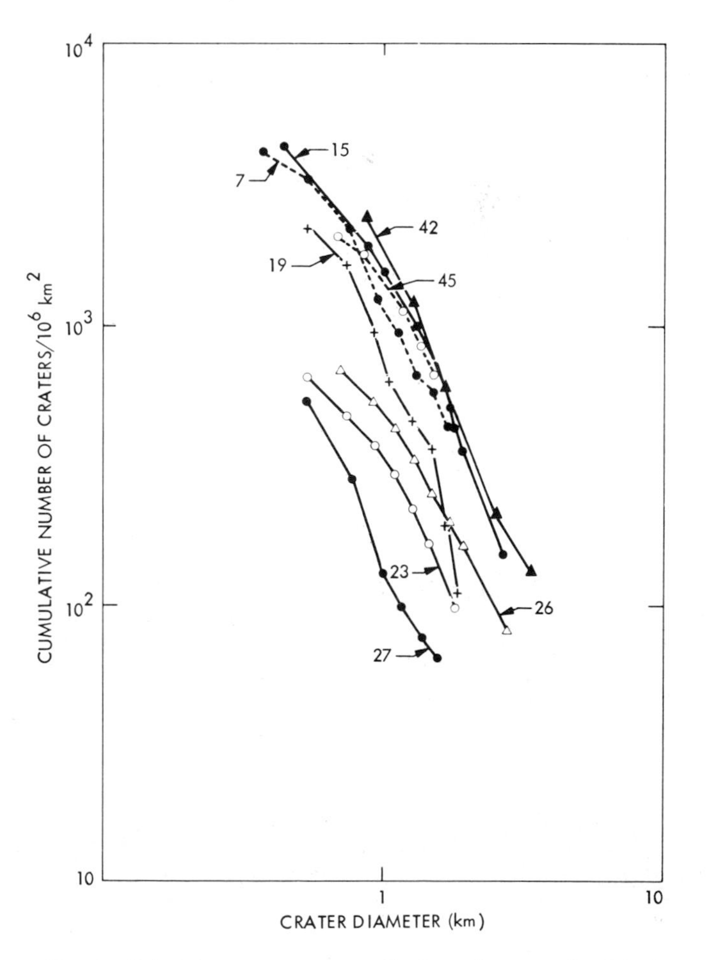

Fig. 4. Selected crater counts on the lava plains south of Arsia Mons. The numbers refer to individual frames taken on revolution 56 from orbiter 1. The location of the frames is indicated in Figure 5.

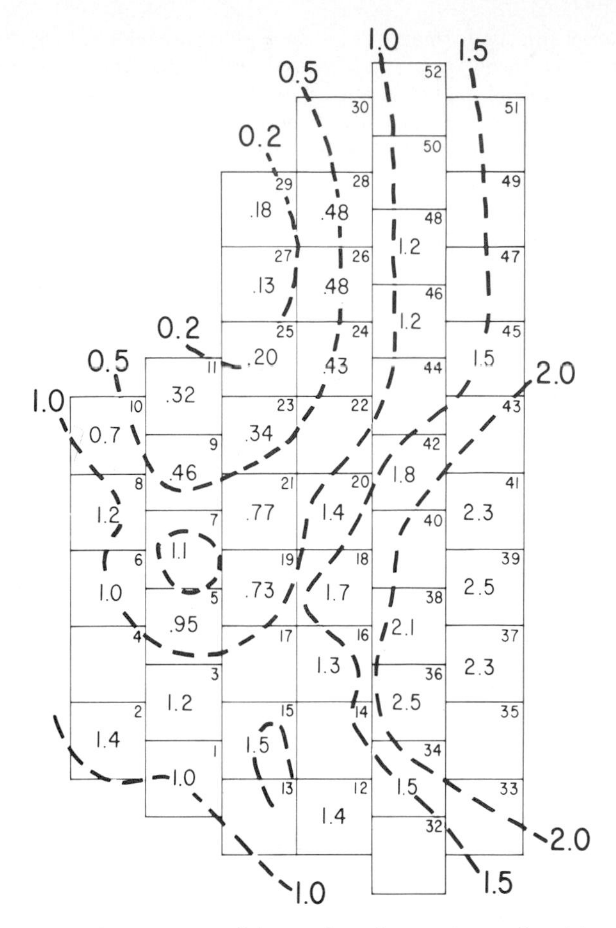

Fig. 5. Contour map of the number of craters larger than 1 km in diameter, per 10^3 km², for the region south of Arsia Mons. Rectangles represent frames in the Figure 2 mosaic. The number in the right-hand corner in each frame is the frame number. The number in the center is the number of craters larger than 1 km in diameter, per 10^3 km², in that frame. A progressive decrease in the number of superposed craters with increasing proximity to Arsia Mons (in the top left corner) is clearly evident.

Several such craters were found in the 700-m- to 1-km-diameter size range. Applying *Pike*'s [1976] relation for crater rim heights and diameter gives a maximum depth of 30 m for the flows. Attempts to do the same for flows closer to Arsia Mons failed because of the lack of craters. The flows closer in do, however, appear to be thinner on the basis of visibility of the flow edge in the image.

From *H. J. Moore and G. G. Schaber* [1975],

$$dP/dL = \rho g \alpha = 3.01 \quad \text{dyn/cm}$$

where P is the pressure at the base of the flow acting along length L, and

$$Sy = (dP/dL)h \leq 9 \times 10^2 \quad \text{N/m}^2$$

This result is in reasonable agreement with that using the Hulme method. A yield stress in the 10^3–10^2 range is comparable with that of mafic magma [*Hulme,* 1974, p. 378].

The length of the individual flows may give some indication of effusion rates. *Walker* [1973] found that for terrestrial lavas the most significant factor affecting flow length is not composition but eruption rate. He found a fairly regular relation between flow length and eruption rate that holds irrespective of composition. The flows that he studied are, however, all less than 100 km in length, and all those in excess of 10 km are basaltic, a situation suggesting that dacitic and andesitic lava flows on earth are rarely more than 10 km in length. Individual flows around Arsia Mons can be traced for 300 km, but Arsia-derived lavas may travel as much as 800 km from the source. Flows of comparable length occur on the moon in Mare Imbrium [*Schaber,* 1973]. If Walker's relation is valid for flows of this size, then very high eruption rates are implied. A 300-km flow gives rates of 2×10^4 to 2×10^6 m³/s; an 800-km flow gives values of 5×10^5 to 5×10^7 m³/s. Each eruption may thus have involved the availability of large volumes of lava in a short time span. These values for eruption rate can be compared with theoretical values. According to *Hulme* [1974, p. 366],

$$F = E\eta(g\rho)^3(\alpha/Sy)^4$$

where E is the eruption rate and η is the viscosity. If $g = 371$ cm/s², $\rho = 2.7$ g/cc, $\alpha = 0.008$, $Sy = 2 \times 10^3$ dyn/cm, and $F = 0.0376$ (from Table 1) are assumed, then

$$E = (1.46 \times 10^{11})/\eta \quad \text{cm}^3/\text{s}$$

Viscosities comparable to those for the Mare Imbrium flows (10^1–10^2 P) give eruption rates of 10^3–10^4 m³/s, which are lower than the values derived above for a 300-km lava flow from Walker's curve.

The eruptions appear from the crater count data to have been extremely intermittent and spread out over a long period of time. To demonstrate the wide spread of crater ages, complete crater counts were obtained on 18 frames within the Figure 2 mosaic. In addition, the number of craters larger than 1 km in diameter was counted for nearly every frame in the mosaic, and then the entire mosaic was contoured. Some of the results are shown in Figures 4 and 5. Figure 5 shows that at the southern end of the mosaic, outside the region of prominent flow features, there are more than 2×10^3 craters larger than 1 km in diameter, per 10^6 km². Close to Arsia Mons, there are less than 2×10^2. The total spread is more than a factor of 20. What these numbers mean in terms of absolute ages is uncertain; interpretation of crater ages is a subject of considerable controversy [*Soderblom et al.,* 1974; *Neukum and Wise,* 1976]. The crater counts in the region peripheral to the Arsia-derived flows are comparable to those of the lunar maria. If the ages are also comparable, say, 2–4 b.y., and the meteorite influx has been approximately constant since the surfaces formed, ages of 100–200 m.y. are derived for the plains next to Arsia Mons. Such a model or one similar would imply that the flow fields being discussed here had accumulated over much of Mars's history and continued to accumulate into the relatively recent geologic past. The preponderance of young ages close to Arsia Mons may merely be an indication that most eruptions did not have a high enough eruption rate to carry the lavas far from the vent. Resurfacing thus takes place at a much higher rate close to Arsia than it does in the peripheral regions, and the average age of the surface is correspondingly younger.

Although Viking pictures show individual flows clearly, it is not always possible to determine the exact location of the vent area from which individual flows erupted. Where the vent area is clear, we have measured the length of flow and noted the altitude of the vent (Figure 6). There appears to be a correlation between flow length and altitude of the vent: the flows that originate higher on the mountain tend to be shorter. *Walker* [1973] has noted a similar correlation on Mount Etna, where lavas erupted at higher altitude did so at a lower rate and thus gave shorter flows. A corollary of this is that the maximum

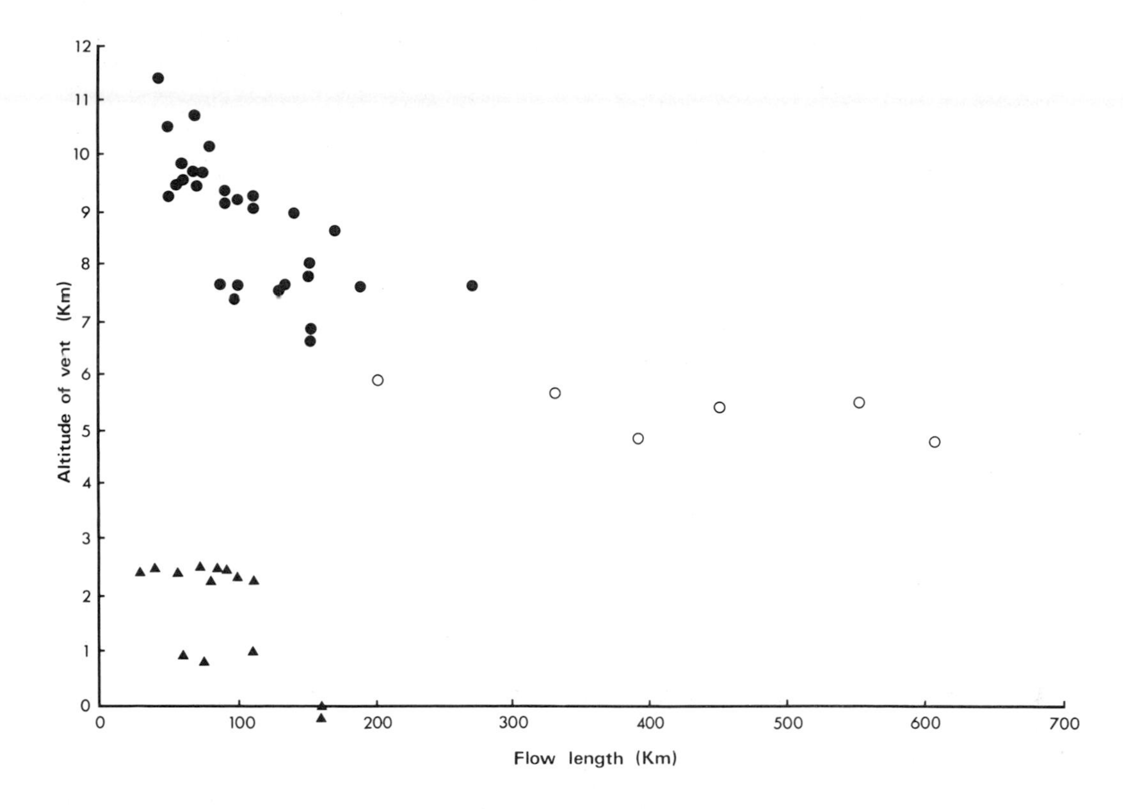

Fig. 6. Lava flow length as a function of altitude of vent for Arsia Mons and Alba Patera. Circles are for Arsia Mons. Open circles indicate that the vent areas are obscured and flows may be longer. Triangles are for Alba Patera.

height of a volcano is that at which the lava eruption rate approaches zero. Plots for Alba Patera show a tendency for flow length to increase with decreasing altitude of the vent; however, the lengths of flows compared with altitude of the vent are considerably less than they are for Arsia Mons. This might suggest different physical properties of the erupting magma, a different crustal thickness in the region of Alba Patera, or that the area is at a lower altitude than it was at the time when the volcano was forming.

A final feature of the flow field south of Arsia Mons that is of interest is a set of fractures radial to Arsia Mons. In the peripheral areas the fractures both cut and are covered by flows, an indication that the fractures and flows were forming penecontemporaneously. In some cases, as in frames 56A14 and 56A16, the fractures appear to have been a source of lava, but the extent to which the lavas in the peripheral areas are fissure fed or derived from Arsia Mons is uncertain.

A faint trace of a fissure is visible within the caldera of Arsia Mons itself (Figure 7). A line of very subdued mounds, some of which have summit pits, connects the reentrants on the northeast and southwest sides of the volcano, from which most of the lavas around Arsia Mons appear to be derived. The very low slopes in the flanks of these mounds suggest that they are secondary shield volcanos and not cinder cones or spatter cones.

One of the most puzzling features of Arsia Mons occurs to the WNW, where a lobe-shaped feature extends approximately 350 km from the base of the volcano (Figure 8). The terrain within the lobe consists of closely spaced equidimensional hills mostly ranging from 100 to 500 m across. At the outer edge of the lobe and paralleling it are numerous closely spaced ridges which give the surface a striated appearance. Individual ridges can be traced several hundred kilometers around the edge of the lobe, often cutting across flow features and craters without deflection. Even albedo features continue undisturbed through the striated zone. Although comparable Viking photographs were not available at the time of writing, a similar feature seems to occur on the northwest side of Pavonis Mons.

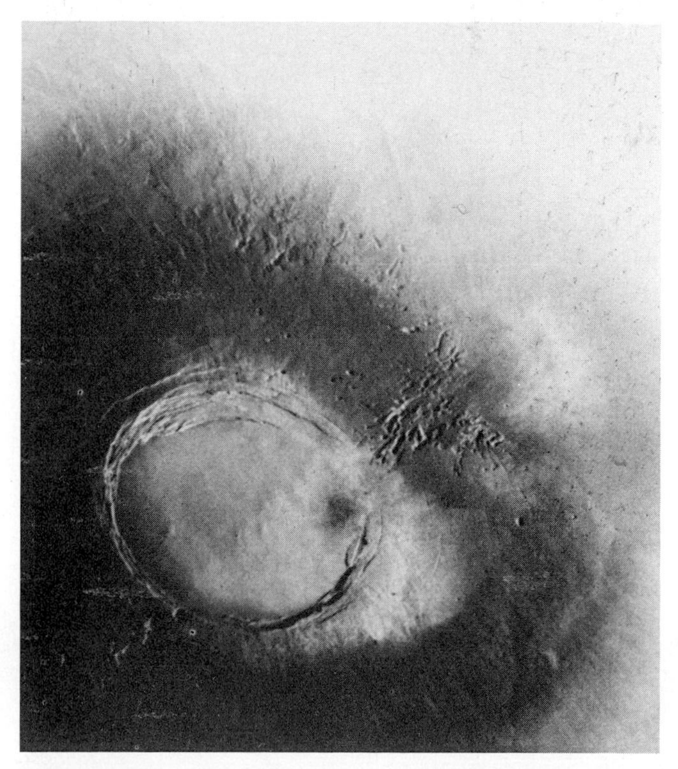

Fig. 7. Image Processing Laboratory enhancement of frame 62A36 showing a line of low hills on the floor of the Arsia Mons caldera. The line connects the two reentrants on the flanks of the volcano. The hills are probably secondary shield volcanos.

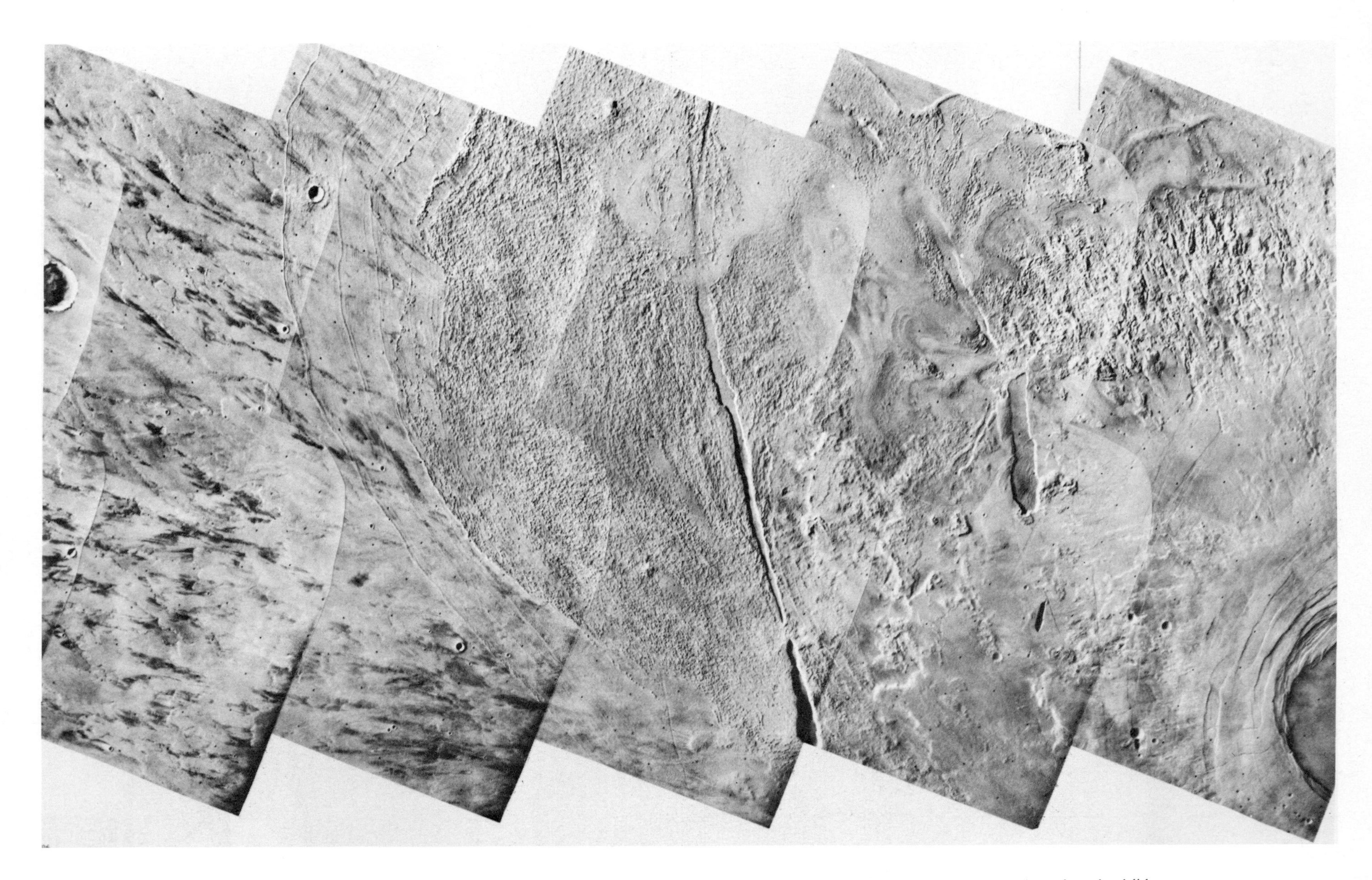

Fig. 8*a*. Mosaic of the area west of Arsia Mons. The granular textured terrain in the center of the mosaic is interpreted as a huge landslide that resulted from collapse of the flanks of Arsia Mons. The irregular outward facing scarps on the west flanks of Arsia Mons may be detachment features. Each frame in the mosaic is 210 km across (Jet Propulsion Laboratory (JPL) mosaic 211-5317).

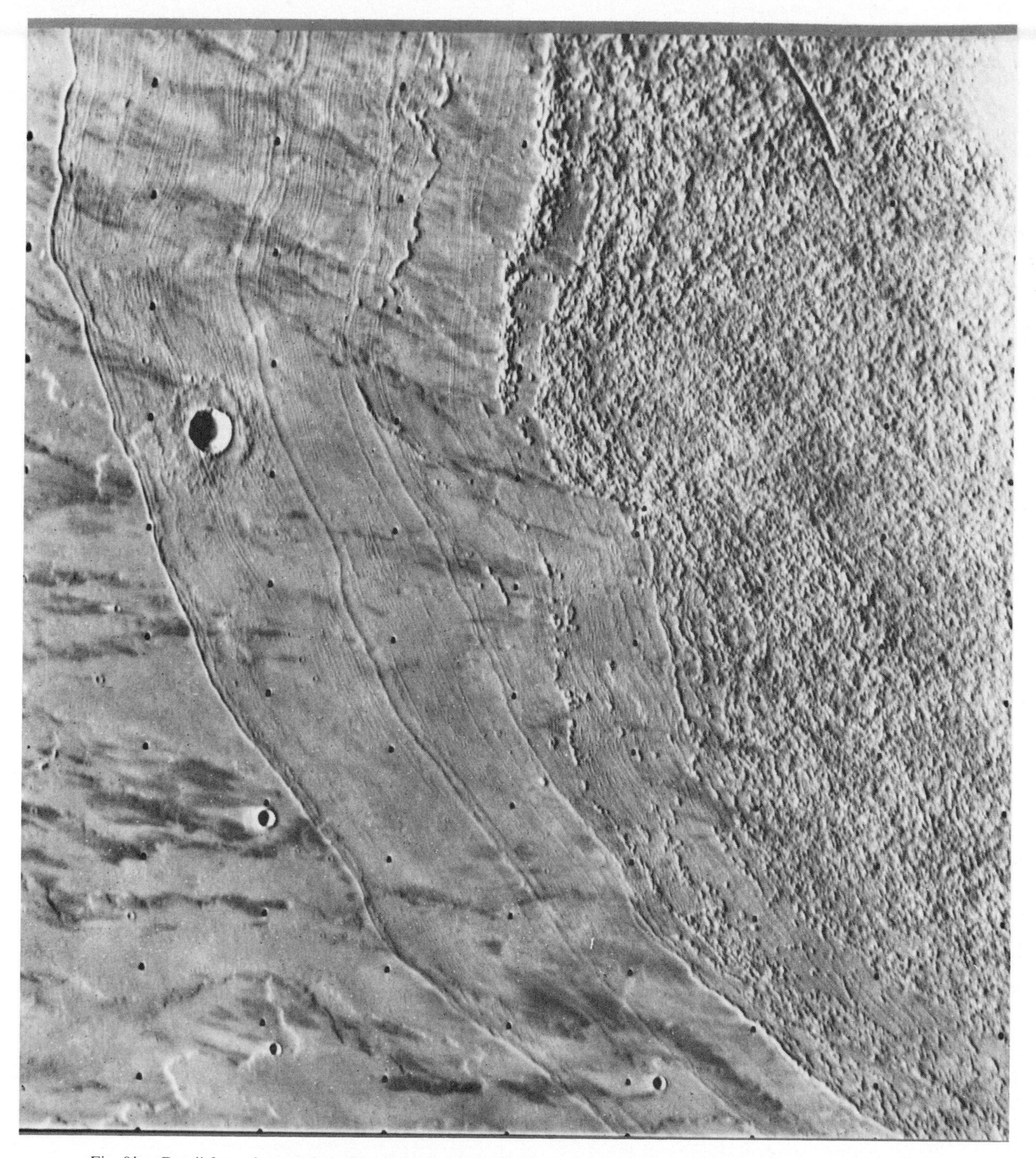

Fig. 8*b*. Detail from the mosaic in Figure 8*a* showing striated terrain around the periphery of the postulated slide. Note that the striations pass undeflected across a crater and its ejecta.

The origin of these features is not known, but a good possibility is that they are formed by huge landslides shed from the lower flanks of the volcanos. Large landslidelike features have been observed on the flanks of terrestrial volcanos [*J. G. Moore*, 1964]. The presence of the features to the northwest of the Martian volcanos suggests gravity assistance, since the plain around both Arsia Mons and Pavonis Mons slopes at approximately $\frac{1}{2}°$ in that direction. The flank of Arsia Mons, adjacent to the supposed slide, appears extremely dissected. This appearance contrasts with the smoothly sloping flanks that occur elsewhere around the volcano. The peculiar appearance was formerly attributed to 'stubby flows' [*Masursky*, 1973] or etching by the wind [*Carr*, 1973]. A detachment zone for a large slide now appears more likely.

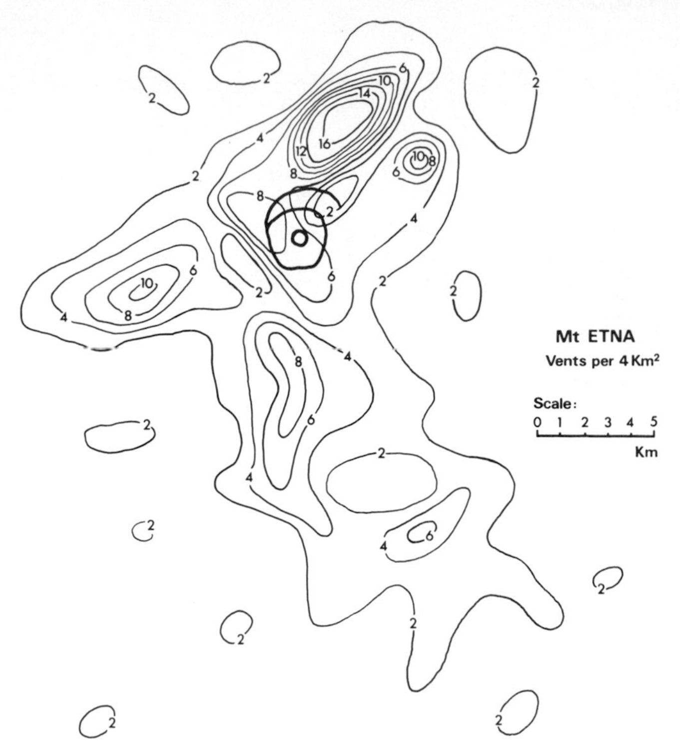

Fig. 9. Contour map of the distribution of vents on Mount Etna, Sicily. Contours represent number of vents per 4 km².

Similar but smaller features also occur at the base of Olympus Mons (see below). The main body of the lobate feature, that which has a granular appearance, is thus envisaged as a chaotic mass of debris that has detached from the volcano and slid over the adjacent plains downslope to the northwest. An alternative is that the feature is an ash flow deposit. This we find less convincing because of the lack of pyroclastic features on the volcano flanks and because such an origin fails to explain the proximity to the 'detachment zone': the position on the downslope side of the volcano and the outer striated margin.

The nature of the striated margin is unclear. There are two main possibilities, first, that the ridges represent material that has moved outward with the slide and second, that the ridges are deformational in nature. The first possibility seems unlikely on several grounds. Any material on the surface would be expected to cover features of the underlying terrain, yet flow features of the surrounding plains can be readily traced through the striated zone. Furthermore, the individual ridges are undeflected by obstacles such as craters. If flow was involved, then the ridges should wrap around such obstacles, but the lineaments cross the ejecta and inner bowl of a prominent crater seemingly undisturbed by it (Figure 8*b*). These observations strongly imply a deformational origin.

The ridges could be the surface expression of a series of closely spaced, shallow, very low angle thrust faults that parallel the base of the slide. These would be produced by the same deformational forces that produce the slide. This possibility seems unlikely from the smooth traces of the ridges and from the fact that they are undeflected by surface features. Low-angle thrusts would be more sinuous and deflected by obstacles. A more likely explanation is that the ridges are folds or reverse faults caused by drag of the slide over the underlying terrain. Such a process would tend to form deformational features that run roughly parallel to the edge of the slide. A somewhat similar process has been suggested by *Harris* [1977] for the formation of the aureole around Olympus Mons. He suggests that the aureole deposits are vast thrust sheets emplaced by the aid of gravity. The aureole deposits, like the lobate features west of Arsia Mons and Pavonis Mons, are concentrated in the direction of the regional slope. He suggests that melting of ground ice, perhaps as a result of volcanic action, could have lubricated the slides. It is not certain that a slide such as that proposed here for Arsia Mons requires a lubricant, but the possibilities include melted ice, aeolian surface deposits, or lava. Separation of the slide from the flanks of Arsia Mons may have been triggered by flank eruptions.

DISTRIBUTION OF VENTS

A common feature of major basaltic composite volcanos on earth is a central conduit with associated rift zones marked by clusters of cones and fissures. In volcanos such as Kilauea or Mount Etna, lava may erupt from the central conduit directly or migrate laterally from the central conduit to cause eruptions along a rift zone lower on the mountain. Figure 9 shows the distribution of vents on Mount Etna illustrating the prominent NE rift, the S rift, and a less extensive rift to the SW. Activity is more common at specific nodes on the rifts. Arsia Mons on Mars clearly forms a NE trending alignment with Pavonis Mons and Ascraeus Mons, possibly extending to Uranius Patera. Qualitatively, groups of pits are concentrated on the NE and SW sides of the summit calderas of these volcanos, the existence of a major NE trending rift through the Tharsis volcanos thus being suggested. To test this quan-

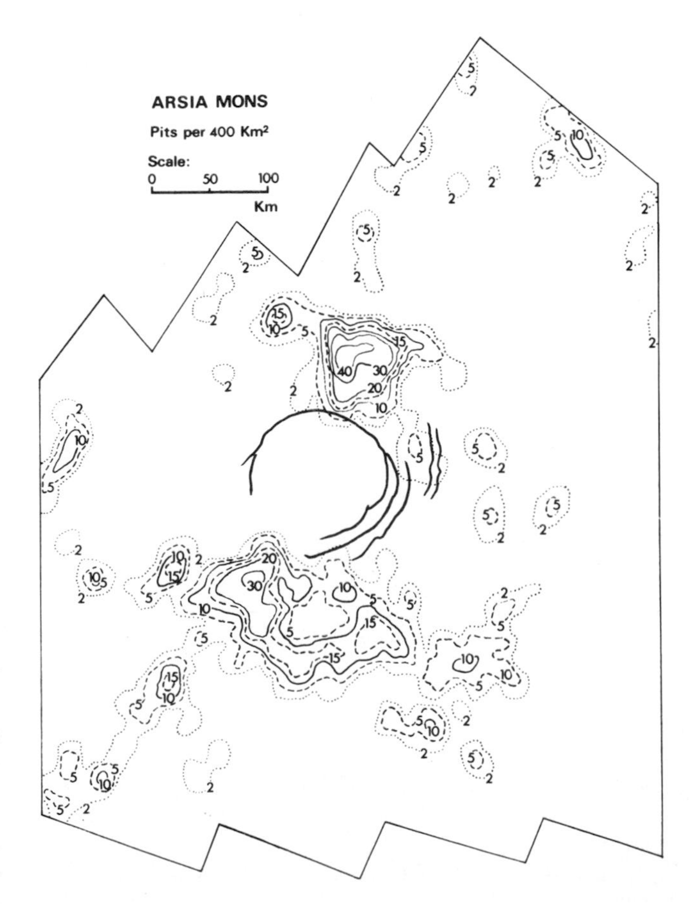

Fig. 10. Contour map of the distribution of pits on Arsia Mons. Contours represent number of pits per 400 km². The distribution suggests deep-seated fractures trending NE-SW and NW-SE.

titatively, the distribution of pits and vents was plotted for Arsia Mons and Pavonis Mons. A contoured plot of vent density for these two volcanos (Figures 10 and 11) shows pit and vent densities to be as many as 3 or 4 times higher in linear zones to the NE and SW of the volcanos, values indicating a preference for activity on the rift zones through the mountains. Other concentrations of vents and pits occur elsewhere, presumably associated with other structures. On both Arsia and Pavonis, there appears to be a crosscutting rift trending WNW on the southern side of each volcano.

Olympus Mons and Vicinity

On orbits 45 to 48, Viking Orbiter 2 acquired 145 images, including a large block of stereo coverage of the shield volcano Olympus Mons and its immediate vicinity (Figure 12). Although there was considerable obscuration by clouds over the western flank and summit of the shield, the general clarity of the atmosphere elsewhere allowed recognition of many features not visible in Mariner 9 images. These include (1) long narrow flows that mantle and extend beyond the volcano's basal scarp, (2) lava flows on the plains to the northeast, east, and south of Olympus Mons which appear smooth on Mariner 9 images, and (3) landslides of enormous size and varied morphology which occur along the entire foot of the west basal scarp.

Olympus Mons is an enormous shield volcano situated on the northwest flank of the Tharsis-Syria upland. It is approximately 500 km across, stands about 25 km above the surrounding plains, and has a summit caldera approximately 80 km across. Around much of its margin is an escarpment up to 6 km high, along which have developed a variety of mass movement features: slumps, landslides tens of kilometers long, and broad regions of apparent mass flow [*Blasius*, 1976*a*]. The basal scarp is subdued or absent along the northeast and southwest margins of the shield. Beyond the scarp is a ring of plains, mostly 100–200 km wide, which on Mariner 9 images appears nearly featureless, the only exceptions being an occasional crater and some mass movement deposits primarily at the base of the western scarp. Outside the ring of plains are areas of grooved terrain, 200–700 km across, which form an incomplete ring around the volcano [*McCauley et al.*, 1972]. The origin of the grooved terrain is obscure; deep erosion of earlier shield volcanos, enormous ignimbrite flows [*King and Riehle*, 1974], erosion of the prevolcano surface [*Head et al.*, 1976], and vast gravity-assisted thrust sheets [*Harris*, 1977] have been proposed.

The Shield of Olympus Mons

During the early Viking Lander 2 operations, Viking Orbiter 2 was held in an orbit with a period synchronous with Mars's rotation so as to communicate daily with the lander. Once each orbit in the late afternoon the spacecraft also passed over the great shields of the Tharsis–Olympus Mons region. This daily opportunity to photograph Mars's youngest geologic province was used to acquire contiguous coverage of about 4.5 million km². The uniquely large local relief of Olympus Mons and the puzzling character of the basal scarp and associated grooved terrain led to the decision to acquire a large block of stereo imaging in this region (Figure 12). Because Mars's atmosphere was far clearer than it was during the Mariner 9 mission, these images show many new features, particularly at the lowest altitudes.

Qualitatively, the Viking images confirm the impression of

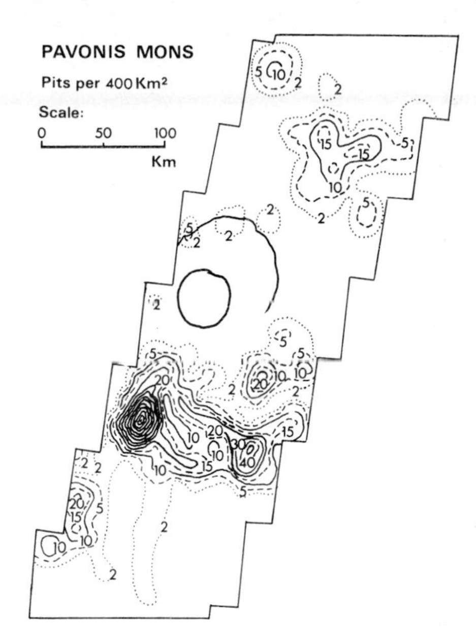

Fig. 11. Contour map of the distribution of pits on Pavonis Mons. Contours represent number of pits per 400 km². The distribution suggests deep-seated fractures trending NE–SW and NW–SE.

Olympus Mons derived from Mariner 9 B frames. However, the lateral extent of the shield is considerably greater than was previously supposed. The shield has three physiographic subdivisions: the summit caldera, the terraced upper flanks, and the lower flanks. The summit caldera is composed of at least five coalesced collapse craters (Figure 12). Superposition relationships indicate that the last craters to form are on the northeast and southwest margins of the present caldera. Crater floors are characterized by concentric fractures and networks of ridges not unlike 'wrinkle ridges' of lunar maria. The fractures suggest central subsidence, while the ridges may be formed by compression or upwelling of material from below. If the activity of Olympus Mons is characterized by cyclical swelling with magma and emptying, as is the case for larger terrestrial basaltic shields [*Eaton and Murata*, 1960], then uplift and subsidence of the summit caldera floor are easily understood.

The upper flanks of Olympus Mons are terraced in a roughly concentric pattern. Terrace spacing is typically 15–50 km. Viking images show the terrace surfaces to be made up primarily of thin lava flows which cannot be traced more than typical terrace spacing (Figure 13). Also present are perhaps a dozen clusters and chains of small collapse craters individually up to 1 km in diameter (example A in Figure 13). A population of scattered circular craters of probable impact origin is also present. Most are smaller than 1 km in diameter, but a larger 14-km-diameter crater immediately east of the summit caldera (Figure 12) possesses all the major characteristics of an impact origin: a raised rim, a halo of hummocky terrain, and radial chains of shallow craters and grooves.

Lower on the flanks of Olympus Mons, terraces disappear, and individual lava flows are more visible in the images, often because of the presence of levees (example A in Figure 14). Relatively smooth radial ridges up to 10 km wide and 60 km long, often having crater chains or sinuous channels along their crests, are also common (examples B–E in Figure 14). These ridges are similar to radial ridges extending outward from the Alba volcanic center (this paper) and are interpreted as leveed lava channels which have become roofed-over lava

Fig. 12. Mosaic of images of Olympus Mons and vicinity taken by Viking Orbiter 2 on orbits 45–48 from ranges of 4800–6100 km. The mosaic is one half of a stereo pair. The main volcanic edifice is 550 km across.

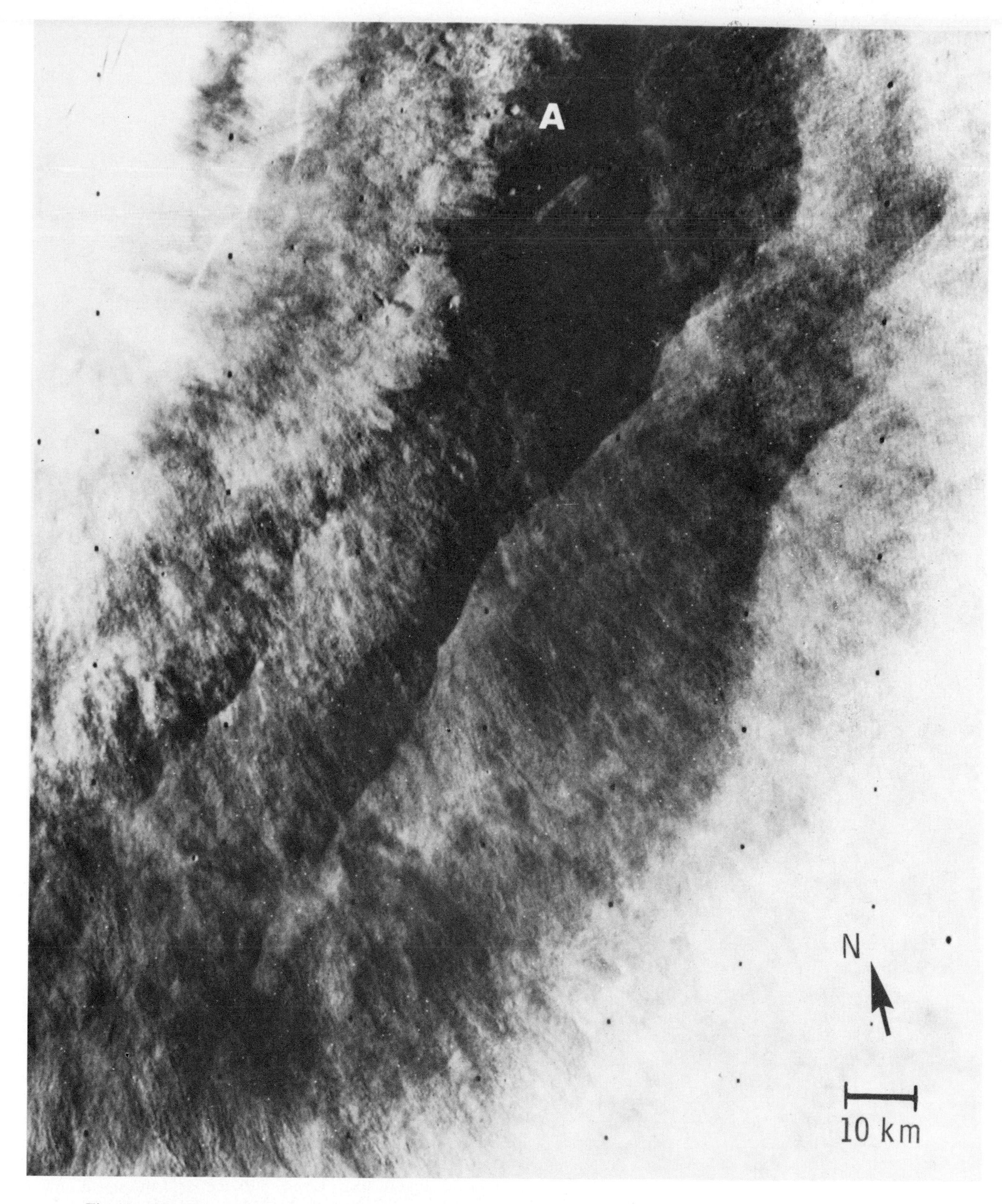

Fig. 13. The terraced upper southeast flanks of Olympus Mons (61.5°N, 132.3°W). Note a corner of the summit caldera at the upper left, the clusters and chains of collapse pits (as at example A), and the very thin, barely distinguishable flows which form the terrace surfaces. Image height is approximately 160 km (portion of frame 46B32, filtered version).

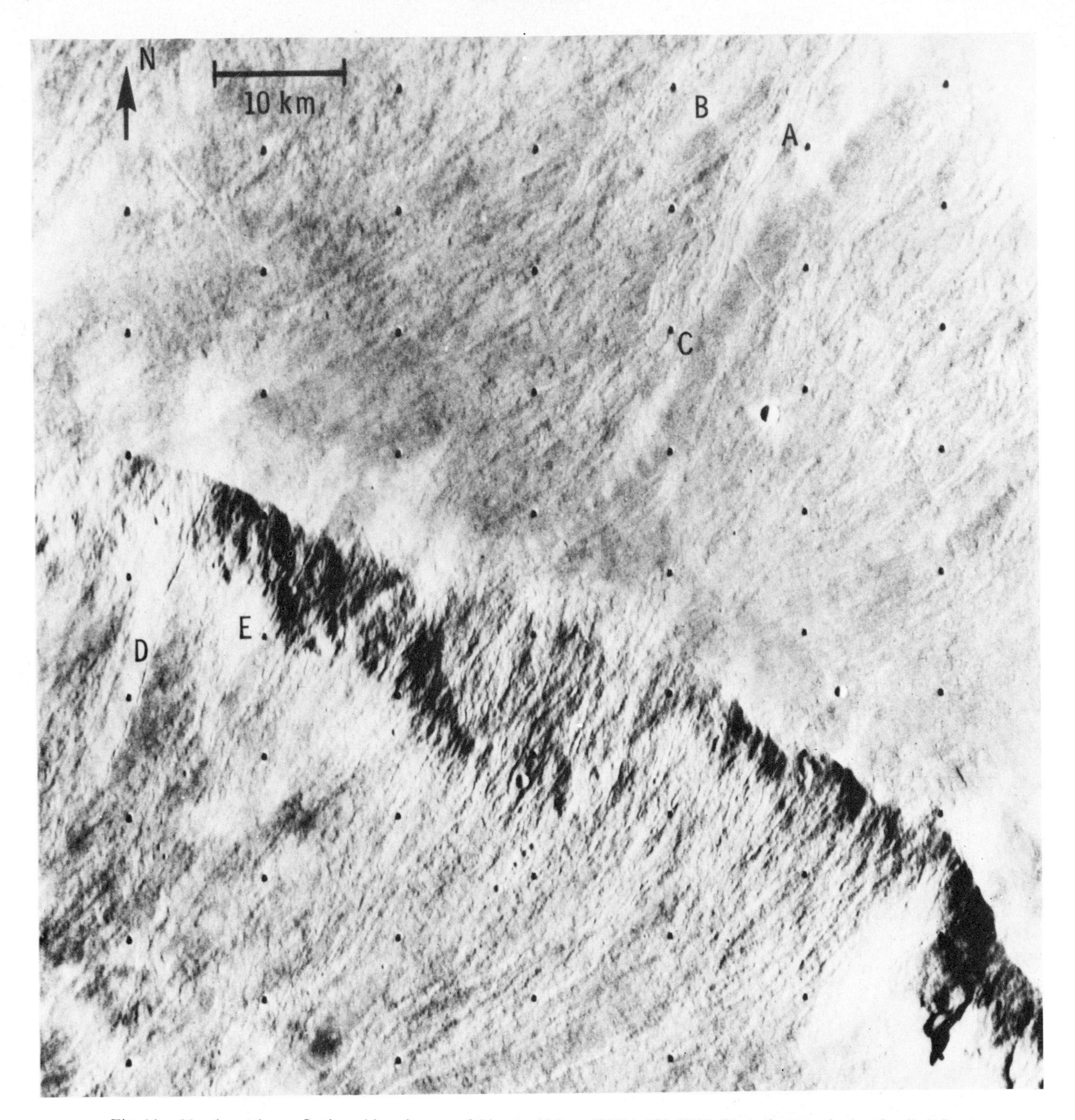

Fig. 14. Northeast lower flank and basal scarp of Olympus Mons (22°N, 130.5°W). Note the prevalence of well-defined leveed flows (as at example A) and smooth ridges (examples B–E), some of which have crater chains or rilles along their crests. Flows completely mantle and greatly subdue basal scarp here. Image height is approximately 80 km (portion of frame 47B25, filtered version).

tubes during prolonged eruptions [*Greeley*, 1970, 1971, 1972]. Another common landform on the lower flanks is the lava fan, apparently built by a point source of lava, as at the terminus of a lava tube (examples A–C in Figure 15).

Obscuration of surface features by dust during the Mariner 9 mission was most severe for regions of lowest altitude, so the lowest levels of Olympus Mons flanks are seen clearly for the first time in Viking images. Except to the west, lava flows have poured over the basal scarp (Figures 14 and 15), and to the northeast and southwest, where the scarp is particularly subdued, large areas formerly thought to be part of the basal plains are now clearly seen as just the lowest flanks of the shield (Figure 12). The mean diameter of the shield now appears to be approximately 700 km rather than 500 km estimated from Mariner 9 images.

The differences in the character of flows on the upper and lower flanks of Olympus Mons suggest that most volcanic vents are located on the terraced upper flanks, while the lower

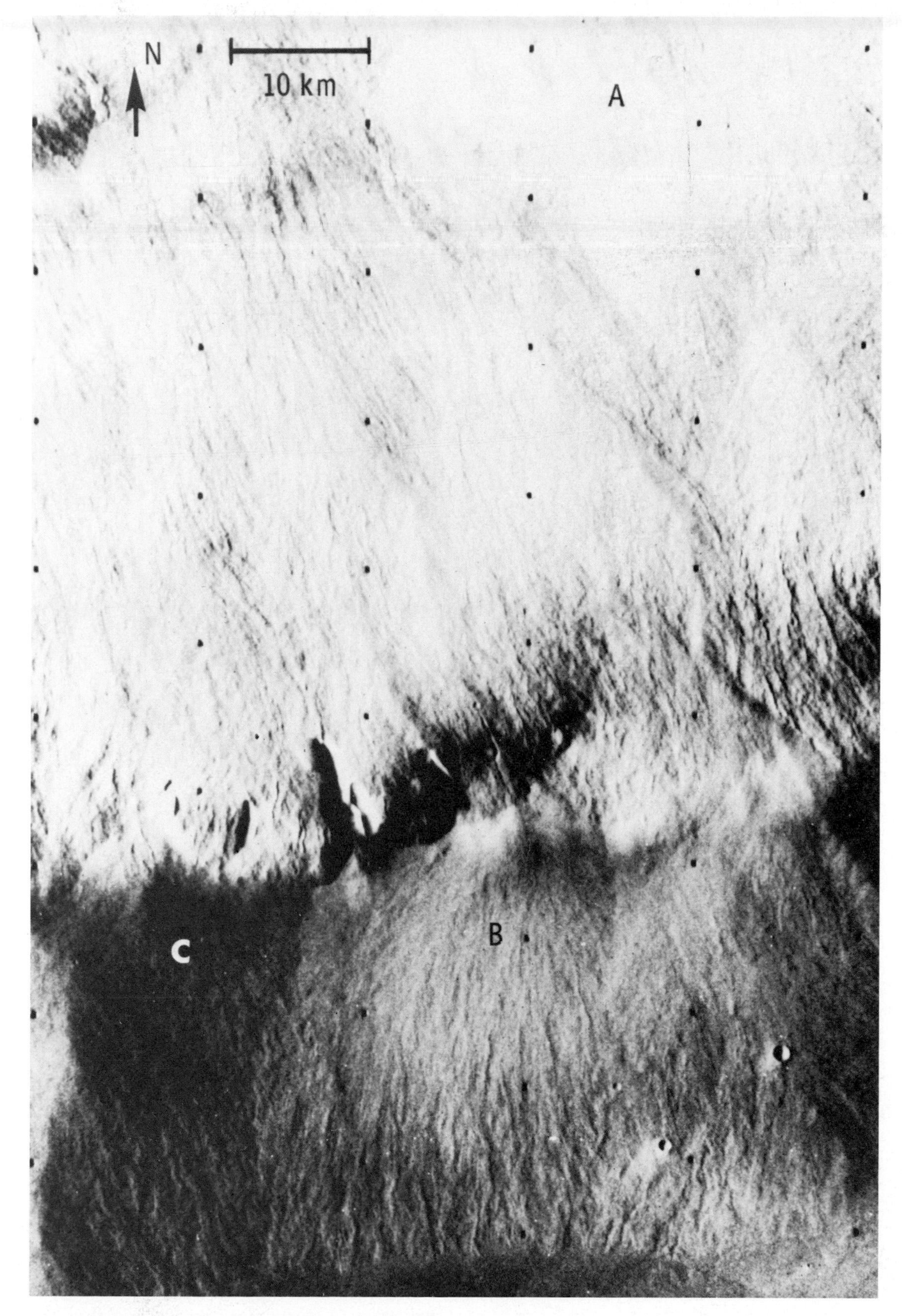

Fig. 15. Part of the southern lower flank and basal scarp of Olympus Mons (14.5°N, 132.5°W). Note the leveed flows and lava fans (examples A–C) and the chain of collapse pits along the ridge crest at the head of fan B. Image height is approximately 100 km (portion of frame 45B42, filtered version).

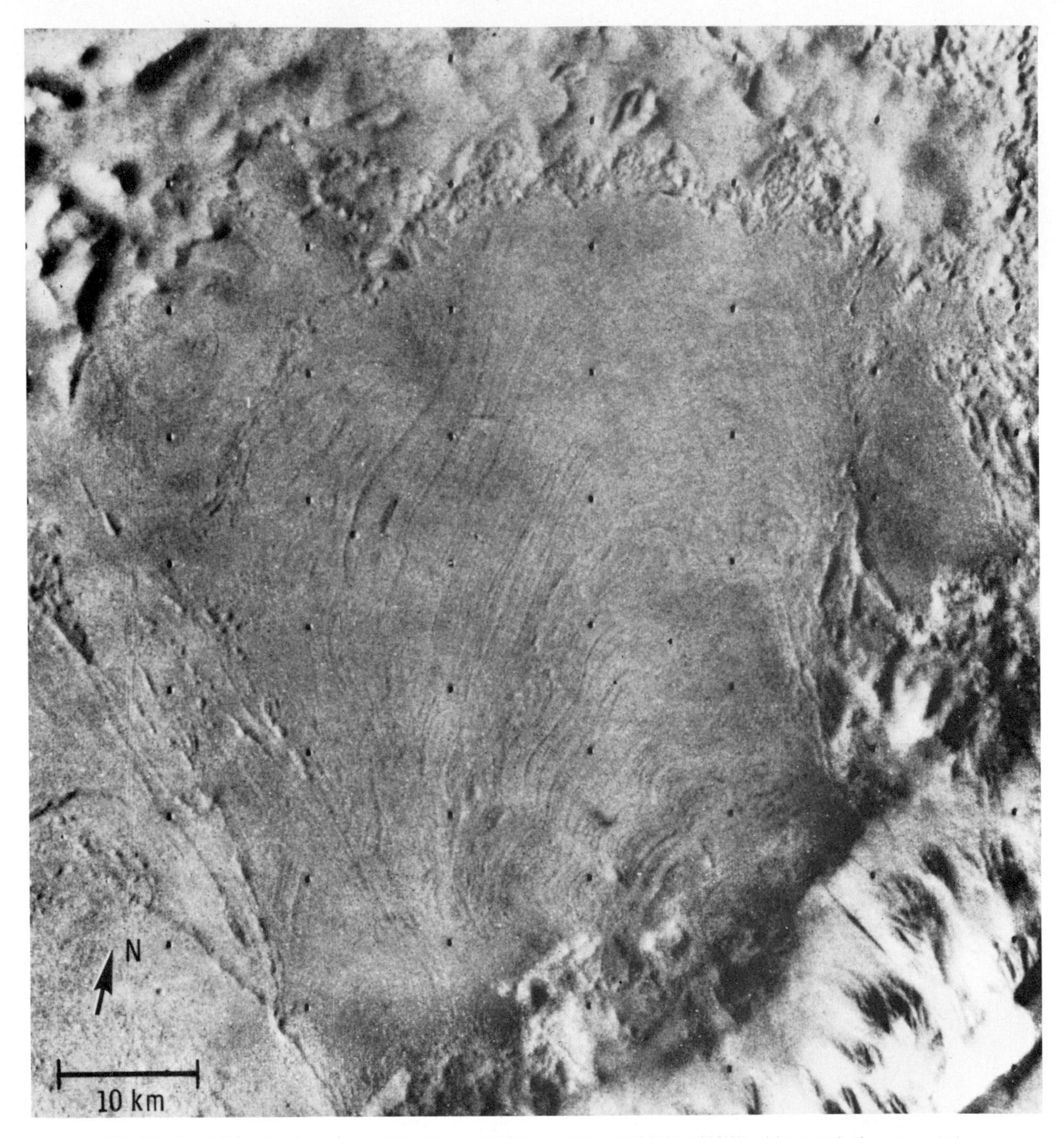

Fig. 16. Landslide lobe along the west basal scarp of Olympus Mons (22.5°N, 138°W) with a terminal zone consisting of tens of conformal ridges and troughs. The slide apparently originated from a very high and steep reach of basal scarp at the lower right. Image height is approximately 90 km (frame 48B04, filtered version).

flanks are fed primarily by long-range distributaries consisting of leveed channels and lava tubes. This hypothesis is also consistent with the relative scarcity of clusters and chains (except over lava tubes) of collapse craters, which would be expected at vent sites, on the lower flanks. The concentric pattern of terraces on the upper flanks of Olympus Mons suggests control by some form of faulting concentric about a central vent or magma chamber.

Mass Movement Features Along the Basal Scarp

A wide variety of apparent slumps, slides, and flows were noted from Mariner 9 images of the Olympus Mons basal scarp [*Blasius*, 1976*a*]. Large slump blocks are prominent all along the southeast basal scarp and locally to the north (Figure 12), but little can be added to their description from Viking images. Lobate deposits from slides or flows along the west basal scarp (Figure 12) are, however, much more diverse and complex in character than was previously realized. Eight enormous overlapping lobes measuring 18 × 18 km in length and 100 × 72 km in breadth have been identified. Each lobe has two major morphological subdivisions: (1) a marginal zone consisting of a terminal ridge or a series of conformal ridges and troughs and (2) a hummocky interior zone. The marginal zone varies greatly in size from as much as 95% of the area of a lobe, containing tens of conformal arcuate features

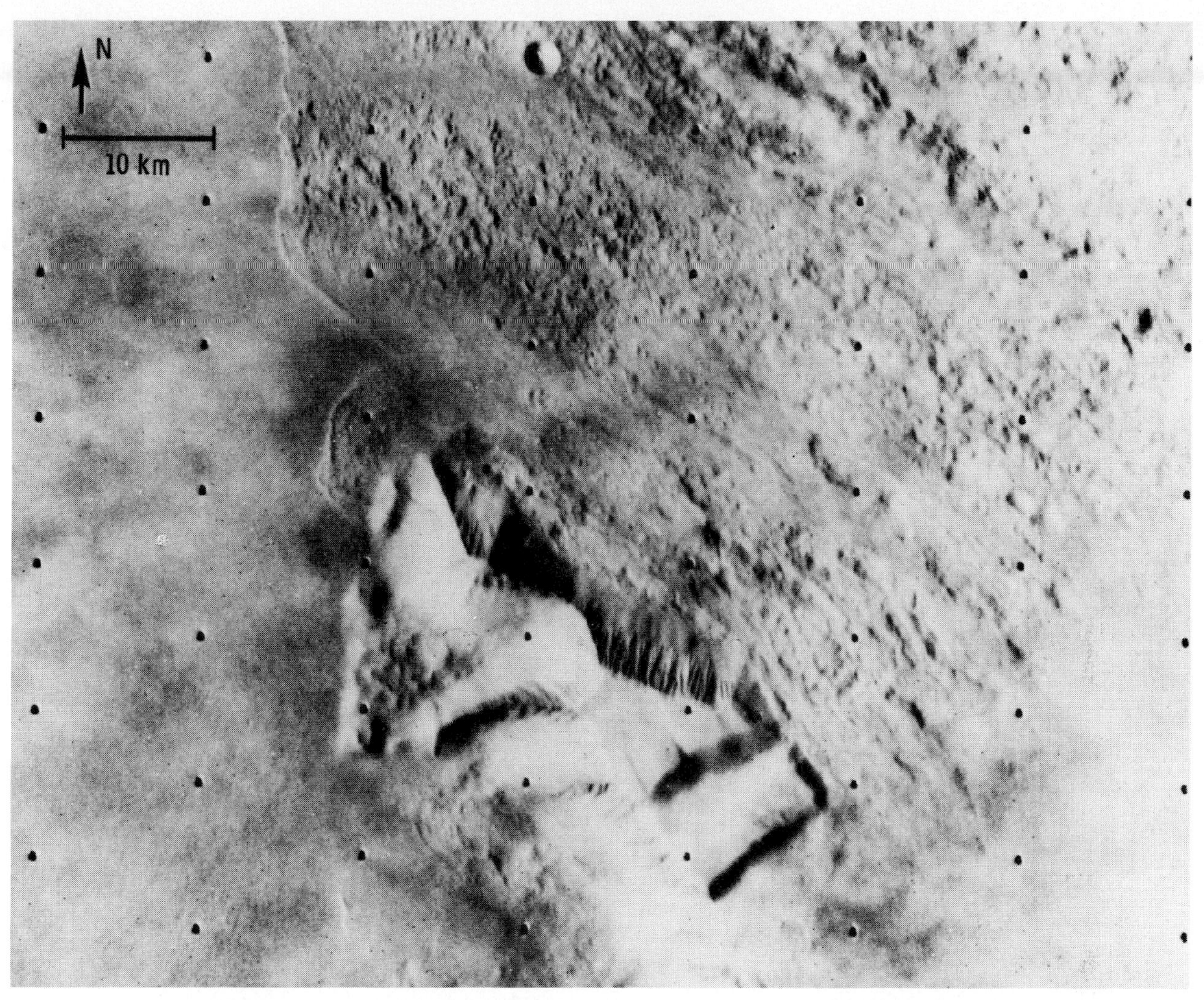

Fig. 17. Landslide or flow lobes along the west basal scarp of Olympus Mons (18°N, 139°W) with extremely large hummocky interior zones and only a small marginal zone of conformal ridges and troughs. Material apparently originated by massive slope failure of a subdued reach of basal scarp. Image height is approximately 60 km (portion of frame 46B01, filtered version).

(Figure 16), to a single terminal ridge (Figure 17). The character of the scarp from which a landslide originates appears to correlate with lobe morphology, the steeper reaches of basal scarp giving rise to lobes with larger marginal zones of conformal ridges and troughs.

The morphology of these slide deposits, by analogy with smaller terrestrial slides [*Shreve*, 1968], suggests lubrication of their movements by cushions of compressed gas. Gas trapped from the present rarefied atmosphere is probably insufficient, but a denser atmosphere in the past or the release of gases from surface materials may provide adequate lubrication [*Blasius*, 1976*a*]. After the quantitative reduction of the stereo data for Olympus Mons and the Valles Marineris, where enormous slides of quite different morphology are seen [see *Blasius et al.*, 1977], the energetics of Martian landslides will be studied in greater detail.

Basal Plains

Beyond the flanks of Olympus Mons and inside and between the lobes of grooved terrain, Viking images show a plain surfaced by broad lava flow lobes (Figures 12 and 18). To the southeast and northwest (Figure 18), flows of the basal plains clearly embay the shield flanks, but elsewhere the relationship is unclear at present Viking image resolution. Aside from scattered small craters of probable impact origin the plains display only local concentrations of fractures. The latter may indicate locations of volcanic vents, but higher-resolution images will be necessary to trace individual lava flows back to their sources.

Relative Ages of Olympus Mons and the Basal Plains

Preliminary counts of apparent impact craters were made to investigate the range of surface ages for Olympus Mons and the basal plains (Table 2 and Figure 19). The cumulative number of craters larger than 1 km in diameter, per 10^6 km^2, was estimated to provide a single number indicative of relative age [*Blasius*, 1976*b*]. Ages seen thus far cover only a range of a factor of 3, considerably smaller than the age range of terrains associated with Arsia Mons (this paper); however, many data for Olympus Mons and its vicinity remain to be studied, and a great deal of the shield is obscured by clouds in currently available Viking images.

Though statistical significance is only marginal, the general pattern of age appears to be, from youngest to oldest, basal plains, lower flank surfaces, and upper flank surfaces. A similar pattern was derived from a smaller sample of Mariner 9 data on the shield alone [*Blasius*, 1976*b*].

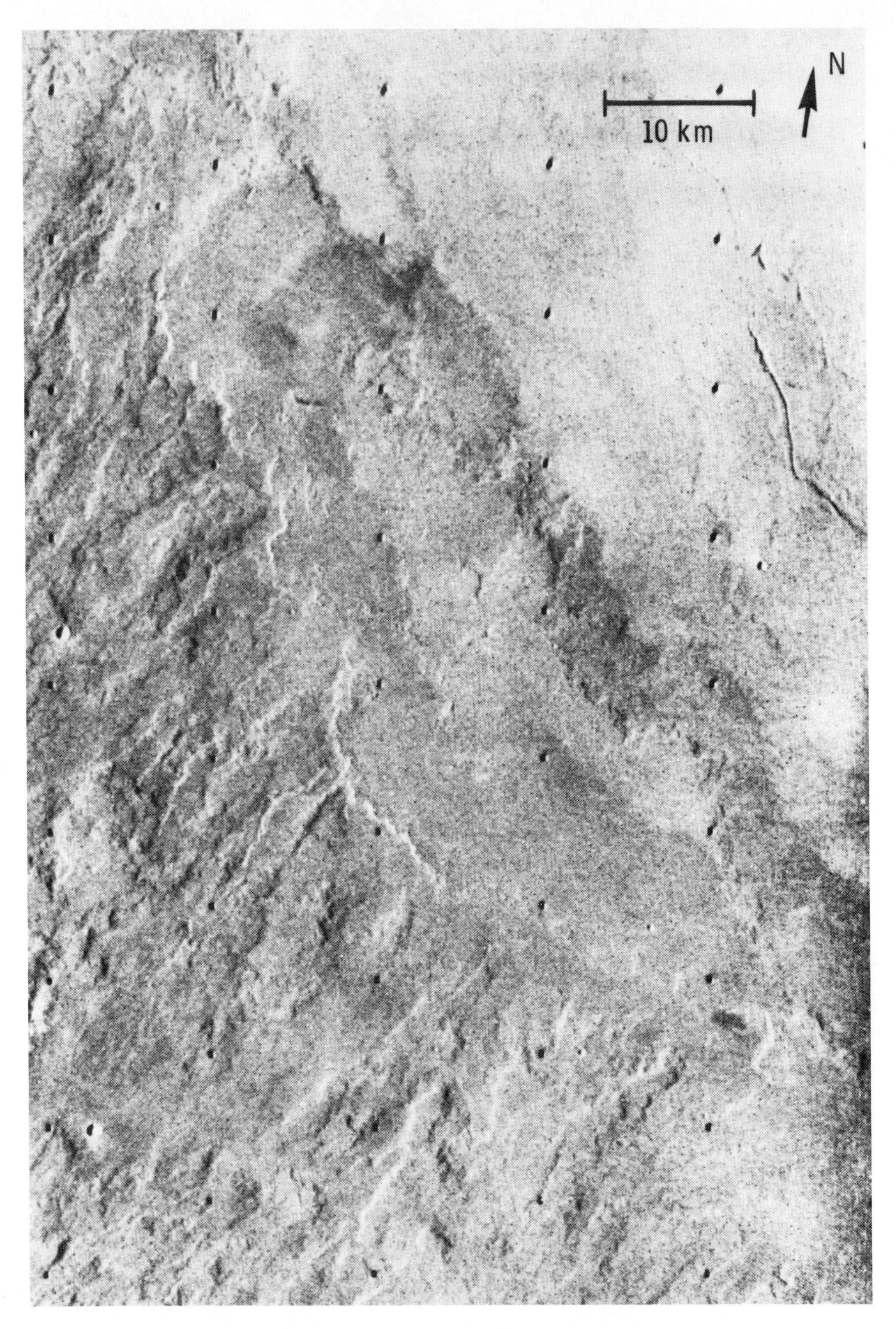

Fig. 18. Flows of the basal plains type in contact with and embaying the northeast flank of Olympus Mons (23.5°N, 129.5°W). Image height is approximately 85 km (portion of frame 48B18, filtered version).

TABLE 2. Preliminary Impact Crater Densities on Olympus Mons

Image Picture Number	Probable Impact Crater Count	Terrain	F*
45B34	13	shield flank	49 ± 13
45B36	13	shield flank	25 ± 7
45B38	16	shield flank	32 ± 8
45B39	23	shiled flank	47 ± 10
45B40	17	shield flank	35 ± 8
45B41	25	shield flank	52 ± 10
45B42	22	shield flank	57 ± 12
45B43	23	shield flank	48 ± 7
45B44	12	shield flank	56 ± 16
45B44	8	basal plain	29 ± 10
45B45	20	shield flank	48 ± 11
45B46	17	basal plain	40 ± 10
45B47	16	basal plain	37 ± 9
46B12	26	shield flank	47 ± 9
46B13	38	shield flank	72 ± 12
46B14	30	shield flank	60 ± 11
46B15	27	shield flank	54 ± 10
46B16	13	shield flank	43 ± 12
46B17	9	shield flank	23 ± 8
46B18	9	basal plain	17 ± 6
47B22	29	shield flank	53 ± 10
47B23	9	shield flank	15 ± 5
47B24	29	shield flank	49 ± 9
47B25	26	shield flank	45 ± 9
47B26	15	shield flank	26 ± 7
47B27	22	shield flank	40 ± 9
47B28	30	shield flank	53 ± 10
47B29	16	basal plain	48 ± 12
48B16	10	shield flank	30 ± 10
48B17, 19	13	basal plain	18 ± 5
48B18	15	shield flank	39 ± 10
48B20, 22	22	basal plain	27 ± 6

* F is the estimated cumulative impact crater density of craters with diameters of ≥ 1 km, per 10^6 km^2.

Grooved Terrain

Perhaps the most puzzling feature of the Olympus Mons region is the aureole of grooved terrain. The young flows of the basal plains appear to embay grooved terrain wherever they contact. One interesting contact relationship observed in Mariner 9 images but not widely discussed is now seen duplicated at the margins of lobes of grooved terrain separated by over 1000 km. Lobes of grooved terrain far NW and SE of Olympus Mons (Figures 20 and 21) are delimited by the toe of a broad lobate flow. This suggests that grooved terrain is of volcanic origin or possibly is the site of flow type mass movements which merge at lobe margins. In general, the origin of grooved terrain remains uncertain, but a great deal of Viking imaging beyond the area covered by the mosaic of Figure 16 remains to be studied in detail.

VOLCANICS OF ALBA PATERA

General Statement

Alba Patera (Figure 1) is probably the largest central vent volcanic structure on Mars, having a possible maximum diameter of nearly 1600 km [*Carr*, 1975]. It belongs to a type of volcanic structure named 'patera' (meaning saucer) that appears to have no counterpart on the earth, the Moon, or the half of Mercury that has been photographed. Alba Patera, as is typical for paterae in general, has a very low profile. Its volcanic origin can be inferred from the presence of a central caldera complex made up of numerous coalescing crater vents, with lava flows and flow structures, such as lava tubes and lava channels, radiating from the central vent region.

The volcano has fractures encircling the central caldera which are part of a regional set of fractures which in this area trend NE–SW or N–S. Originally called the Arcadia ring [*Carr*, 1973], the fractures are formally named Alba Fossae on the west side and Tantalus Fossae on the east side. The ring of fractures has a diameter of about 600 km, nearly equal to the diameter of Olympus Mons (Figure 22).

Mariner 9 images of Alba Patera show a faint radial texture which is interpreted as indicating radial flows. Because of the general lack of detail in the Mariner 9 pictures and the fact that most of the fractures appear to cut the radial flows, Alba Patera was interpreted to be a relatively old volcanic feature that had been degraded and tectonically modified. It was suggested that it might have been an enormous shield volcano that had collapsed to its present low profile. One high-resolution frame showed numerous channellike features arrayed in a dendritic pattern interpreted to be part of a fluvial system. The central region appeared so smooth on Mariner 9 images that it was considered a prime backup landing site for the Viking 2 lander.

During Viking landing site certification activities the Viking landing site originally selected in the Cydonia region was rejected, and alternatives such as Alba Patera were examined with the Viking orbiter cameras [*Masursky and Crabill*, 1976]. Although the pictures are oblique, they reveal a rugged terrain made up of numerous relatively fresh appearing lava flows, and the area was abandoned as a potential landing site.

The swath of pictures acquired confirmed some of the interpretations based on Mariner 9 and caused rejection of others. Well-preserved flow features (Figure 22) are traceable for more than 1000 km westward from the central caldera complex. Most of the flows in the northwest quadrant are cut by the ring fractures; some flows, however, are superimposed over the fractures (Figure 23), and this suggests that subsidence of the volcano took place concurrently with its construction. Many of the supposed fluvial drainage features appear in fact to be volcanic (Figure 24). The discussion that follows applies only to that part of Alba Patera imaged by the Viking orbiter, primarily the northwest quadrant of the structure as shown in Figure 1.

Alba Patera Flow Types

One of the most striking aspects of the Viking pictures of Alba Patera is the diversity and relative crispness of the lava flows that make up the structure. Four main types of lava flows are mapped (Figure 22): (1) tube-fed flows, (2) sheet flows, (3) tube-channel flows, and (4) undifferentiated flows.

Tube-fed flows consist of ridges that tend to be radial to the center of Alba Patera. Lava channels and partly collapsed lava tubes (Figure 25) can be traced down the axes of the flow ridges and were the main conduits feeding the advancing flow front. From terrestrial and lunar experience the only commonly erupted lavas that are fluid enough for lava tubes to develop are basalts, and it is inferred that either the Alba lavas are basaltic or they have the rheology of lavas of basaltic composition. The flows extend very long distances, the longest tube-fed flow so far identified being more than 340 km long. The average width for this flow and other tube-fed flows is about 8 km and is remarkably uniform. This width, however,

Fig. 19. Map of estimated cumulative impact crater densities (number of craters larger than 1 km in diameter, per 10^6 km^2) for images of the eastern flank and basal plain of Olympus Mons (from the mosaic of Figure 2). See Table 2 for tabulation of actual crater counts and statistical uncertainties in crater densities. For the most part the basal plain flows are younger than the lower flank surfaces, which in turn may be slightly younger on the average than the upper flank surfaces.

represents only the exposed part of the flow. From experience with terrestrial volcanos, lava channels and tubes (Figure 26) frequently form constructional ridges along their axes, principally by overflow. These act as topographic barriers [*Greeley*, 1971]. Subsequent flows may bury the lower flanks of the ridge but may not be able to breach it. Similar relationships appear on Alba Patera, so the actual width (exposed buried parts) of the tube-fed flows cannot be determined.

Some of the tube-fed flows have domes or series of domes at the end of flows (Figure 27). These may result from the hydrostatic pressure of lava within the tube system. At the lower end of the flow the lava may rupture the roof of the tube to form a pseudovent, local outpourings of flows to build the domes thus resulting. Similar features have been described for lava systems associated with volcanos on earth [e.g., *Greeley and Hyde*, 1972].

Sheet flows (Figures 24 and 25) constitute the most conspicuous type of flow on Alba Patera. Although most appear to originate near the ring fracture, the actual vents for the sheet flows are rarely visible. As is typical of similar appearing flows on the earth and the moon, the vents for the sheet flows may have been buried by their own products. Typically, the sheet flows lack flow surface structures such as lava tubes, lava channels, and flow festoons but form multiple overlapping lobes that have fairly level surfaces. Their closest analogies are the Imbrium flows on the moon and possibly the flood type basalts of the Columbia River Plateau.

Sheet flows on Alba Patera are best developed northwest of the ring fracture (Figure 22), although they also occur within the ring fracture, where some appear to have originated from the central caldera complex. A particularly long (more than 270 km) flow is visible in the southwest part of the Viking

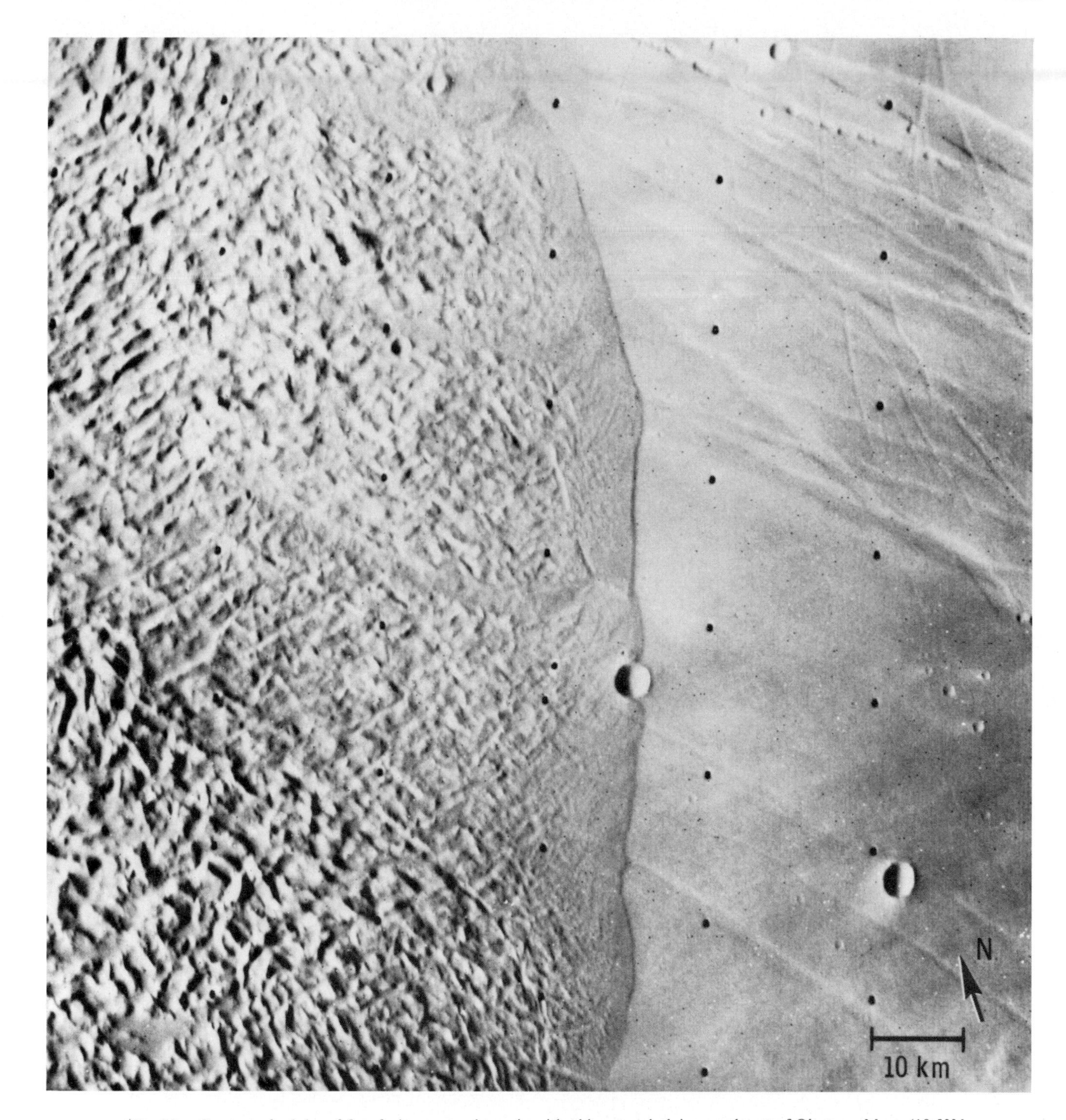

Fig. 20. Contact of a lobe of fine-facies grooved terrain with old cratered plains southeast of Olympus Mons (10.5°N, 125.5°W). The margin of the lobe seems to consist of a broad flow front which has overridden the cratered plains. Height of image is approximately 110 km (frame 44B27).

orbiter coverage, although its morphology is slightly different from that of the more typical sheet flows and it may be a variant.

Tube-channel flows cover above 4.7×10^4 km^2 west of the ring fracture and a smaller area north of the ring fractures. These flows consist of a complex series of channel-fed and tube-fed anastomosing flows which represent multiple flow units (Figure 24). The presence of tubes again probably indicates lavas of basaltic composition. In some respects these flows are similar to those that occur on certain terrestrial volcanos. For example, some of the flows on Mount Etna (Figure 6) show a similar morphology made up of anastomosing channel-fed flows, although the area covered is only a fraction as large as that of the Alba flows. One particular flow on Mount Etna, the historic 1614–1624 sequence, shows a terraced aspect which is similar in morphology to that of some

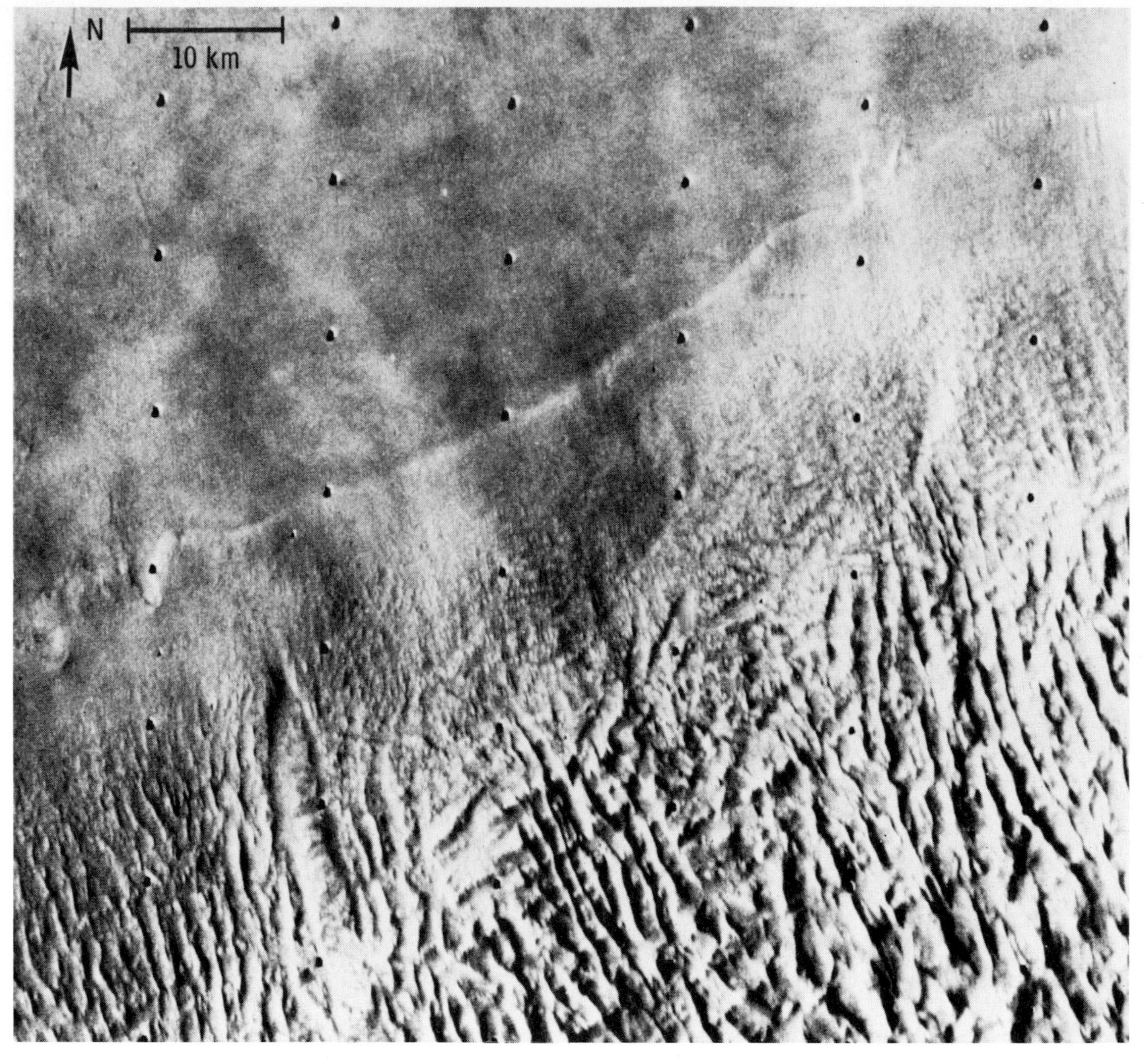

Fig. 21. Contact of a lobe of grooved terrain with plains to the northwest of Olympus Mons (33.5°N, 144°W). The margin of the lobe has flow front morphology similar to that seen in Figure 11. Height of image is approximately 60 km (frame 34B39, filtered version).

of the Alba Patera flows. The Mount Etna flows were emplaced through lava tubes which fed numerous 'toes,' building up an imbricate sequence of overlapping flow units.

The remainder of the flows that make up the part of Alba Patera photographed by Viking are shown as undifferentiated on Figure 22. Many of these flows are also fed by tubes and channels but differ by having surfaces of rugged relief and flow margins that are indistinct. From superposition relations and a general impression of greater degradation the undifferentiated flows at the distal ends of Alba Patera may be the oldest flows. Channel and tube segments, suggestions of festoon flow structure, and lineations can be used to indicate flow direction, as depicted on the map. The flow festoons, by terrestrial analogy, may represent more viscous lavas than those in which lava tubes form.

The undifferentiated flows within the ring fractures are fresh appearing and have well-preserved channels with levees. Many of the flows originate from the central caldera complex, and some of the flows are superimposed on the rim of the caldera; other flows are cut by the caldera wall, an indication of an essentially contemporaneous eruption–caldera formation sequence. The same relationship can be seen for some of the flows and the ring fractures (Figure 23).

Except for the obvious flows that originate from the caldera, it is not possible to determine the source vents for most of the flows and flow units. For such an enormous structure it is

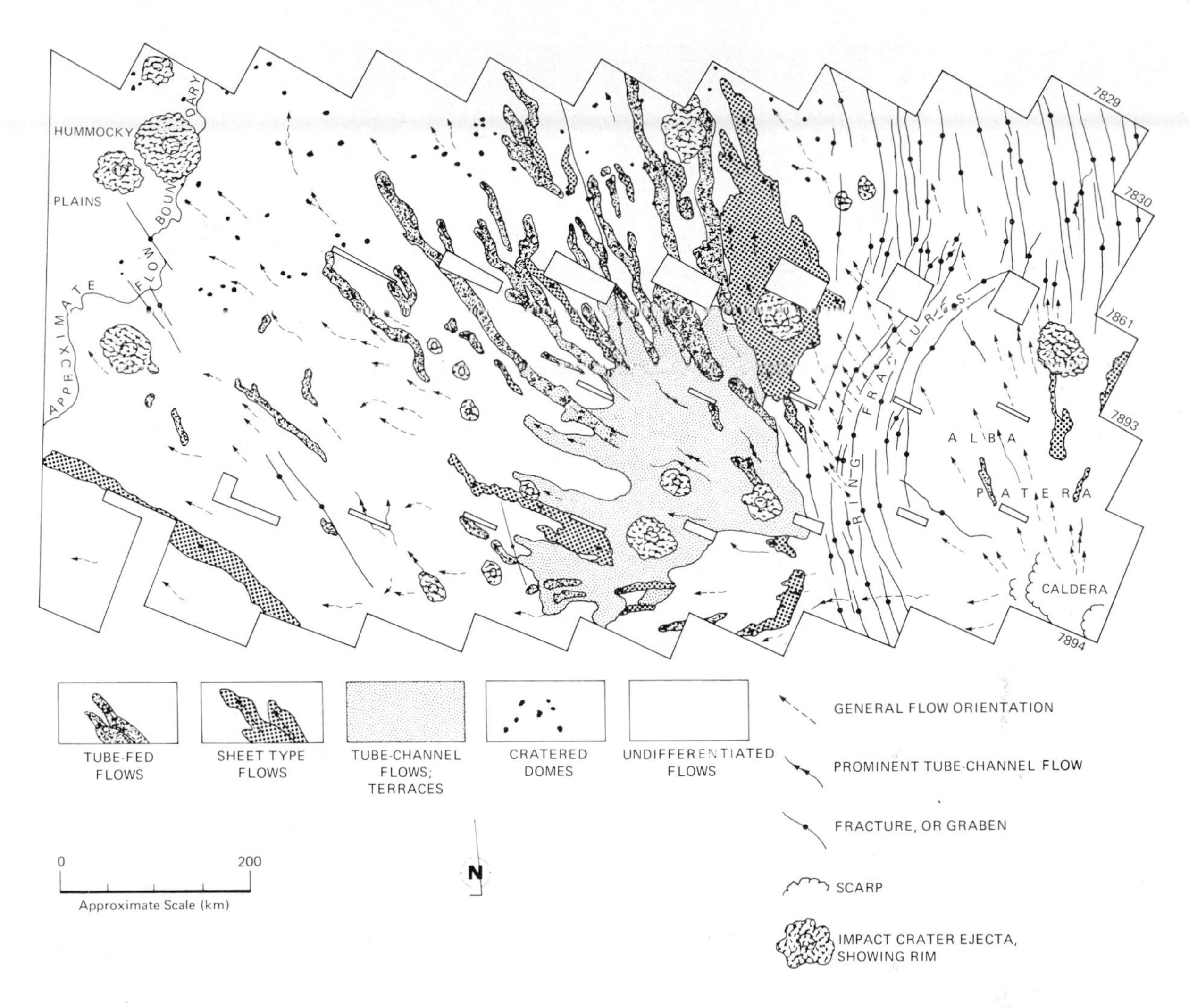

Fig. 22. Geologic sketch map showing the distribution of various flow textures, flow directions, structural features, and maximum extent of flows identified with Alba Patera based on images acquired by Viking Orbiter 2 on revolution 7. Images and this diagram are oblique; therefore the scale bar is approximate.

rather surprising that subsidiary vents are not present or more obvious. Either the subsidiary vents are buried by flows or they do not exist.

Cratered Domes

At the outer margins of Alba Patera are numerous small domes, many of which have summit craters whose floors are at a higher elevation than are the surrounding plains (Figure 25). These structures are tentatively interpreted as volcanos, possibly cinder cones. However, these features could be impact craters modified to form pedestal type craters [*Arvidson et al.*, 1976] in which the ejecta deposits have been eroded and the floors have been filled or partly filled. The distribution of this class of structure on the outer margins of Alba could be a consequence of lower topographic elevation, where aeolian processes are more effective.

On the other hand, the cratered domes occur adjacent to and superimposed on lava flows that have primary structures such as channel levees and flow features that are the same scale and smaller than the features that would have to be eroded if the craters were impact. Thus in order for the craters to be of impact origin some explanation must be derived to explain the preservation of some features and the erosion of others at the same scale. Moreover, the cratered domes are remarkably similar in size, and one would expect a range of sizes for eroded impact craters.

Styles of Volcanism

The flow textures and other possible volcanic features identified on Alba Patera provide some indication of the styles of eruption that were involved in the construction of the volcano. From the interpretations that follow, Alba Patera appears to have had a complex eruptive history involving several different forms of volcanic activity.

Tube-fed flows probably represent effusion of modest volumes of lava from point source vents erupted over a long period of time, probably years or dozens of years. The single-tube system characteristic of these flows acted as a pipeline from the vent to the flow front. Although by terrestrial analogy most of the flow probably occurred through roofed tubes, there were probably sections that were unroofed. As eruption continued, the flow slowly extended by addition of lava at the

Fig. 23. Mosaic of oblique photographs of the northwest quadrant of Alba Patera showing the western set of ring fractures (Alba Fossae) and part of the central caldera complex (lower right corner). Note the intricately patterned flows (undifferentiated flows on the map) associated with the caldera. Some flows are superimposed on the rim. Although the ring fractures (primarily graben) transect most of the flows, the western set (bottom of the photograph) is superimposed by flows. Differences in the amount of detail in individual frames result partly from differences in processing. Area covered is about 294 × 377 km (JPL mosaic 211-5065).

end of the tube, resulting in the extremely long (by terrestrial standards) 'noodles' of flows reaching out from the ring fractures of Alba Patera.

In a recent study of the various factors involved in the ultimate length of lava flows, *Walker* [1973] demonstrated that the rate of effusion was the single most important parameter, with very high rates required to produce long flows. This study, however, did not compare flows that were emplaced by lava tubes with those that were not, and it has been proposed [*Greeley*, 1976] that lava tubes provide efficient conduits resulting in longer flows at lower rates of eruption than result for flows not emplaced through tubes. The primary reason is that heat and perhaps some volatiles are better retained in tube systems, so lava remains fluid for a longer period of time.

In contrast to the tube-fed flows the sheet flows represent a relatively high rate of effusion and therefore also produce long flows. The effusion rate and volumes of lava were probably so great that individual flow structures (tubes, channels, and festoons) were not able to form. These flows were probably erupted from fissures; many of the flows can be traced back to the zone of ring fractures, and although vents are not obvious, some of the fractures may have served as fissure vents which were subsequently buried by flows and modified by further tectonic processes. Sheet flows are probably analogous to flood type flows on earth.

Tube-channel flows as mapped on Alba Patera are interpreted to represent multiple eruptions, probably occurring over a long period of time. The complex anastomising network of individual flows is characteristic of near-vent flows; some of the flows having this texture grade into tube-fed flows which may represent active lavas within the tube-channel system that eventually concentrated into a narrow zone to produce a tube-fed flow. Tube-channel flows on Alba Patera may be very thick, made up of countless hundreds of flows and flow units, some of which are probably interconnected both vertically and horizontally through a maze of lava tubes. The undifferentiated flows represent a range of eruptive styles, including tube-fed, channel-fed, and some sheet flows, plus some that

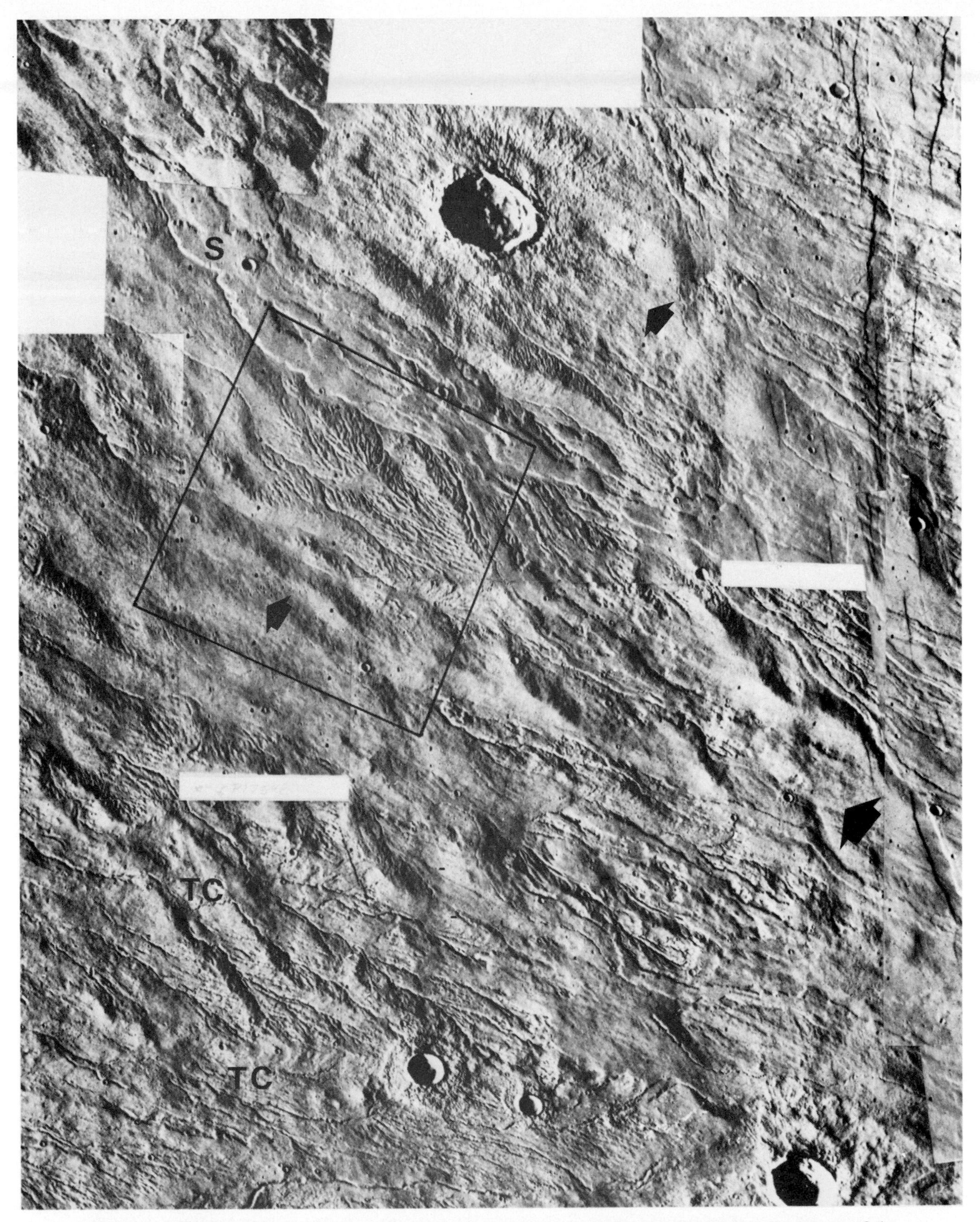

Fig. 24. Oblique view of the flank of Alba Patera northwest of the ring fractures (visible at the bottom of the photograph), showing sheet flows (S) and tube-channel flows (TC). Flow direction is indicated by arrows. The region formerly considered to be fluvial dendritic drainage patterns on the basis of a Mariner 9 B frame is outlined. Note that the pattern is oriented 180° out of phase with slope for a dendritic drainage system; instead the pattern is interpreted to be a primary volcanic structure resulting from multiple flows through lava tubes which 'bud' to produce distributary lava tubes; similar systems, but on a smaller scale, are observed on orbiter frames 7B24 and 7B53–7B58 (JPL negative 211-5065B).

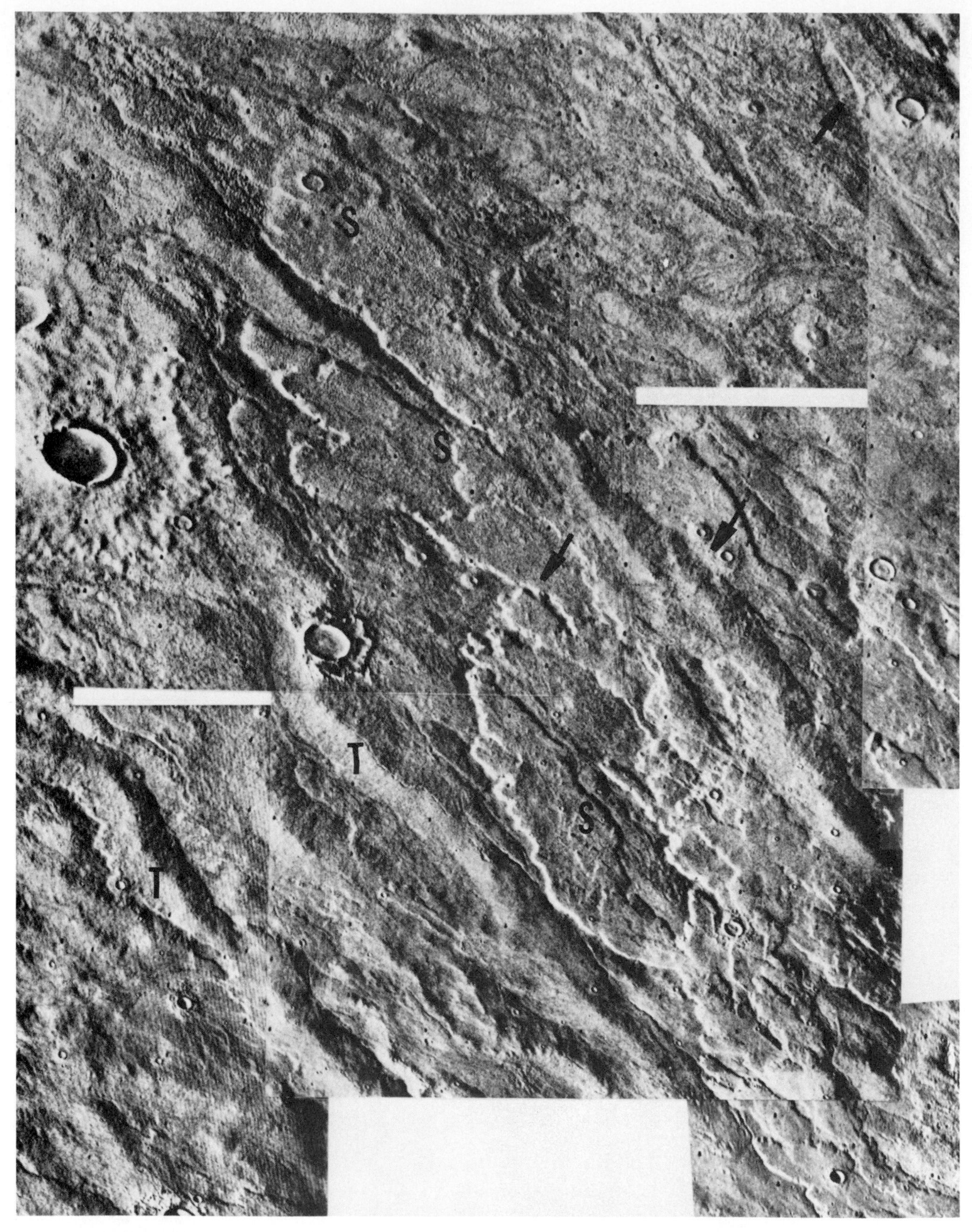

Fig. 25. Oblique view of sheet flows (S) and tube-fed flows (T) northwest of Alba Patera. Sheet flows typically lack flow surface structures such as lava tubes, lava channels, and festoons and have relatively level surfaces. Tube-fed flows form sharp-crested ridges of relatively uniform average width. Partly collapsed lava tubes and lava channels are visible along the crest of the flows. Possible volcanic domes (with summit craters) in the area are identified with arrows. Alternatively, these structures may be modified impact craters, although it seems unlikely that the craters would be so highly modified yet preserve adjacent flow features of comparable scale. Area of the mosaic is about 146 × 187 km, centered at 48°N, 115°W (Viking orbiter frames 7B22–7B27; JPL negative 211-5065B).

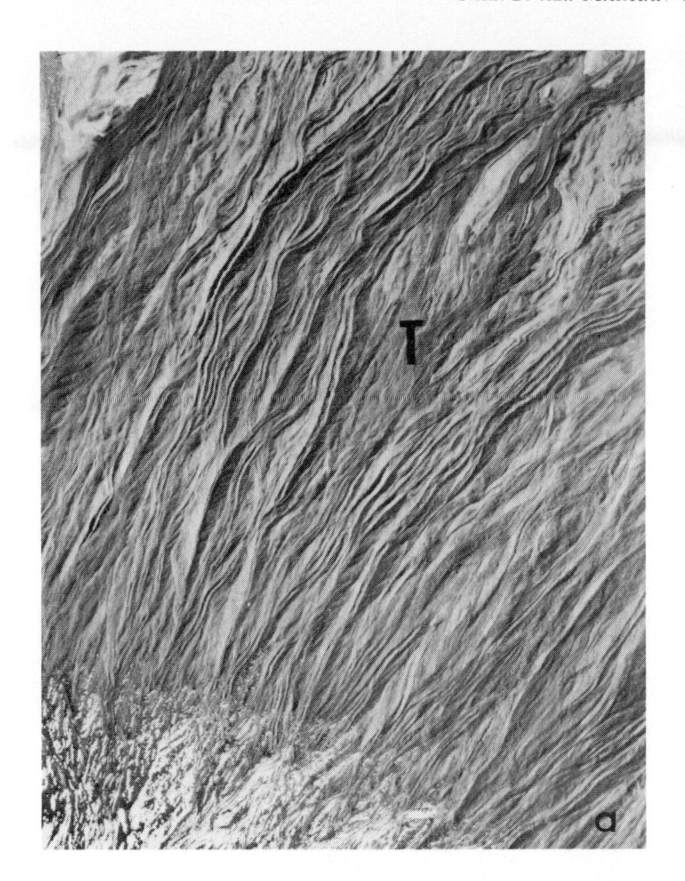

Fig. 26. Aerial photograph (*a*) Mount Etna showing near-vent summit flows composed of numerous flow units fed through tubes and leveed channels (compare the ridge-tube structure at point T with the tube-fed flows on Alba Patera in Figure 25) and (*b*) flank flows on Mount Etna showing finely textured tube- and channel-fed flows arrayed in an anastomosing and distributary system network (compare with Figure 24). Each photograph of Mount Etna covers about 3.2 × 4 km, an area many times smaller than the areas shown on Alba Patera. Flow direction is toward the top of the photographs; illumination is from the left.

display festoons which may indicate viscous flow. However, in the reconnaissance mapping presented here, these individual flows are too limited in areal extent and generally too poorly defined to map individually.

Regardless of the precise styles of volcanism involved with the flows on Alba Patera, the total volumes of lava erupted to produce single flows are orders of magnitude greater than they are in terrestrial lava flows, and the total volumes of lava erupted from essentially a single-vent volcano are enormous. This implies a model of magma generation rather different from any current models for earth.

General Discussion

The discussion above focused on the regions of Tharsis and Alba. The youth, size, and style of the volcanic features in these areas render them particularly susceptible to analysis. Volcanic features also were observed by the Viking orbiters in many other areas. Excellent coverage was acquired of Apollinaris Patera (Figure 28). It has many features of the Tharsis shields, with a cliff around its circumference and a large central caldera. To the south a series of flows emanate from a fissure and extend up to 250 km southward, much as in the case of Arsia Mons. The degree of degradation and the number of superposed craters, many of which could be impact, suggest, however, a much older age. On revolution 87, Orbiter 1 acquired comparable coverage of Tyrrhenum Patera (Figure 29) which showed it to be extremely degraded and embayed by the surrounding plains. Similar coverage was obtained of Hadriaca Patera (revolution 87) and Amphitrites Patera (revolution 95).

Some of the most puzzling features observed during the Viking mission occur in the 40°–50°N latitude band. Throughout this region, but particularly between longitudes 240° and 270°, many craters are on raised platforms or pedestals. In the case of small craters the pedestals appear dome shaped, so the features look like cratered domes. These were the subject of considerable debate during the mission. The issue was whether they are volcanic or whether they are impact and the result of a complex burial and erosional history. Although little systematic work has been done, the crater frequency curves in some of these areas have unusual slopes, and a volcanic component is suspected. Similar features were also observed in parts of Chryse Planitia [*Greeley et al.*, 1977]. In several areas therefore, some evidence exists for central volcanic constructs distinctively different from those of Tharsis and Alba discussed in this paper. We have restricted ourselves to the obvious and unambiguous. Some of the more controversial aspects of Martian volcanism remain to be addressed.

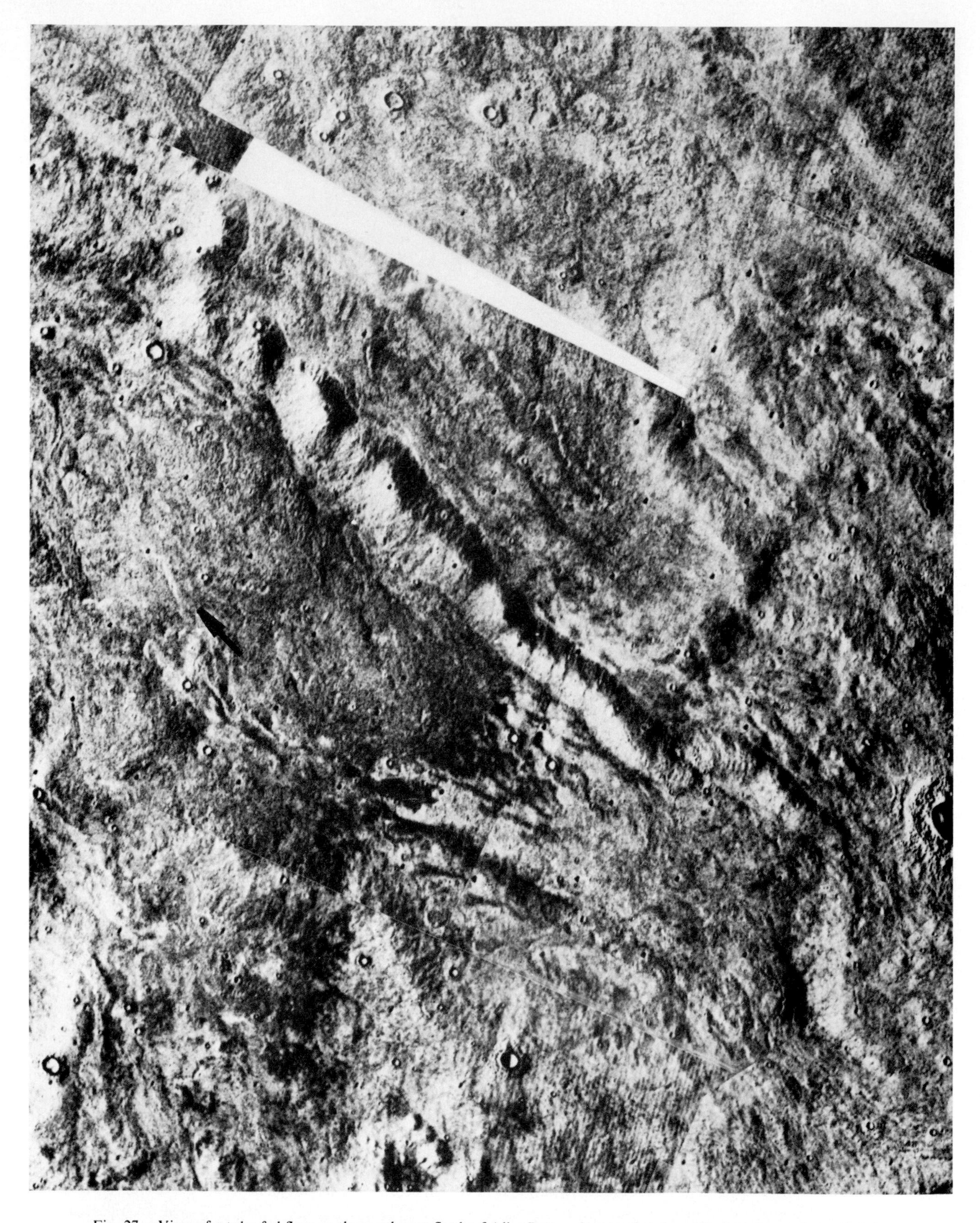

Fig. 27. View of a tube-fed flow on the northwest flank of Alba Patera showing five domelike features at the end of the flow; some of the domes have summit craters. These structures may result from extrusion of lava from the feeding closed-system lava tube at the end of the flow. Similar dome-tube relationships have been noted on earth on a smaller scale [*Greeley and Hyde*, 1972]. Flows surrounding the tube-fed flow are mapped as undifferentiated on the flow map; note the general rough texture and the discontinuous flow structure enhanced by albedo patterns. Area covered by the mosaic is about 110 × 140 km, centered at 46.5°N, 126°W; the arrow indicates general flow direction (Viking orbiter frames 7B18, 7B47, and 7B49; JPL negative 211-5065).

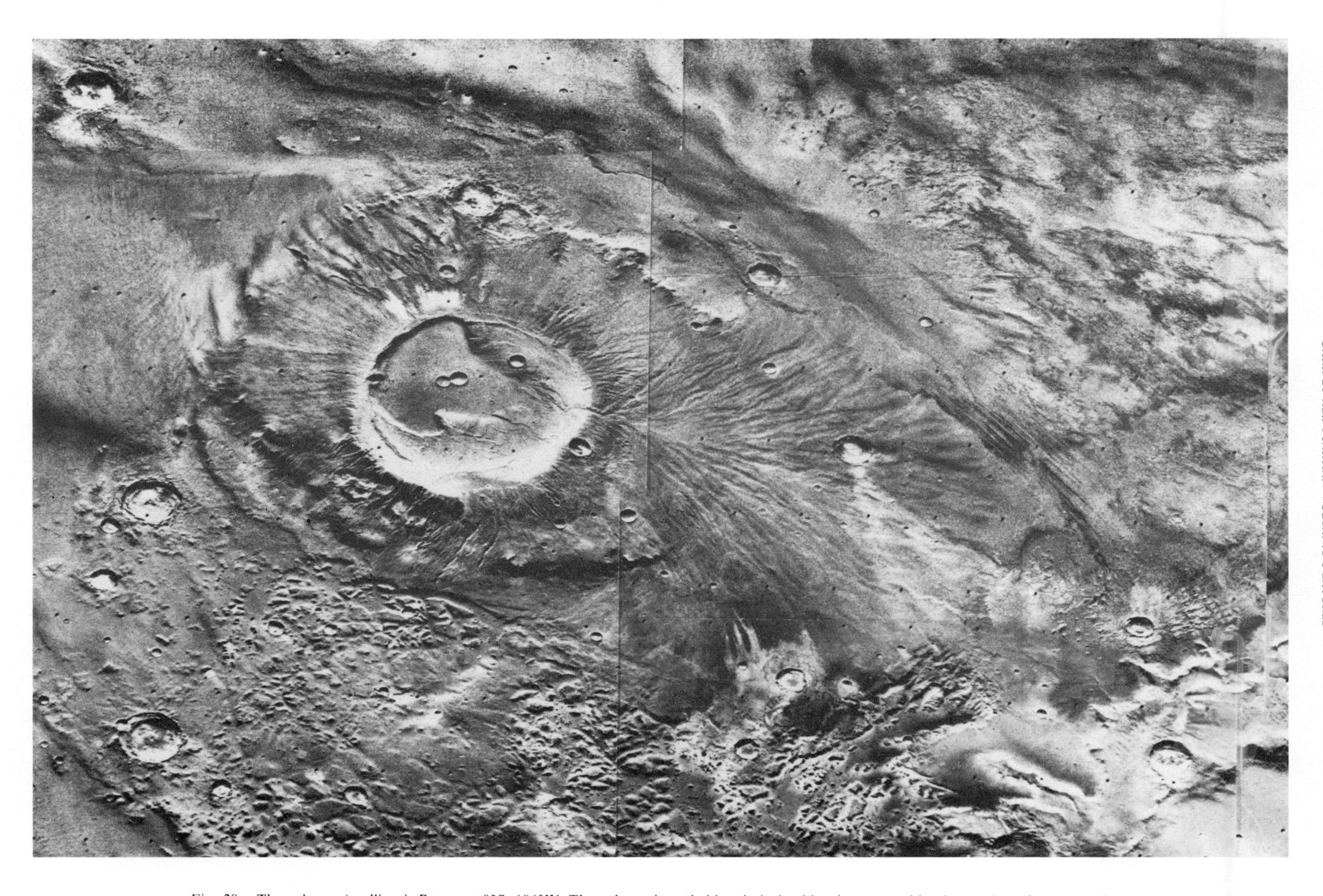

Fig. 28. The volcano Apollinaris Patera at 8°S, 186°W. The volcano is probably relatively old as is suggested by the number of superposed craters, some of which are probably impact. This picture was taken from Viking Orbiter 1 on revolution 88.

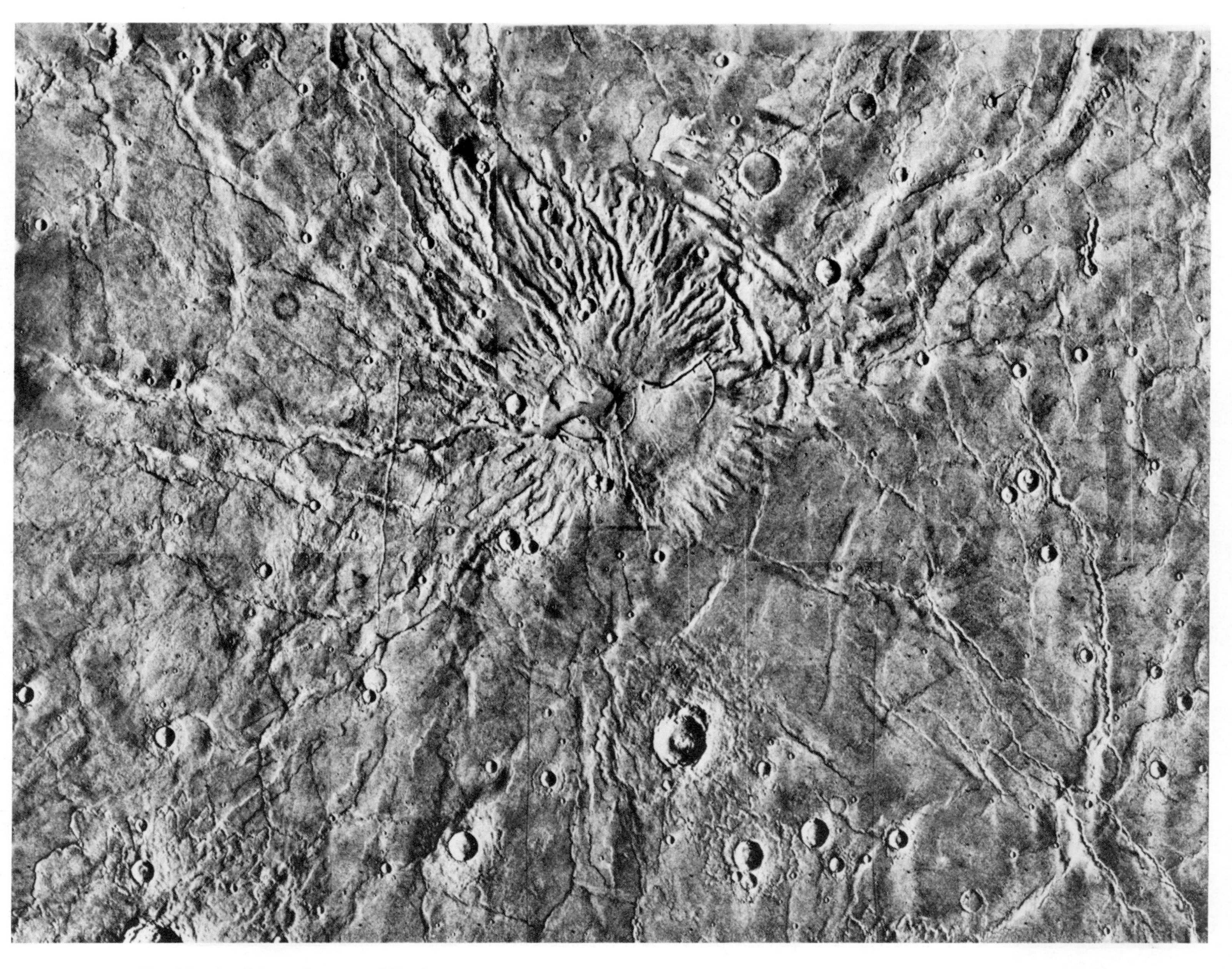

Fig. 29. Mosaic of the region around Tyrrhenum Patera at 22°S, 253°W made of frames taken from Viking Orbiter 1 on revolution 87. The flanks of the volcano appear to be partly covered by the surrounding lava plains (section of mosaic 211-5213).

References

Arvidson, R. E., M. Coradini, A. Carusi, A. Coradini, M. Fulchignoni, R. Funiciello, and M. Salomone, Latitudinal variation of wind erosion of crater ejecta deposits on Mars, *Icarus, 27*, 503–516, 1976.

Blasius, K. R., Topical studies of the geology of the Tharsis region of Mars, Ph.D. thesis, Calif. Inst. of Technol., 85 pp., 1976*a*.

Blasius, K. R., The record of impact cratering on the great volcanic shields of the Tharsis region of Mars, *Icarus, 29*, 343–361, 1976*b*.

Blasius, K. R., J. A. Cutts, J. E. Guest, and H. Masursky, Geology of the Valles Marineris: First analysis of imaging from the Viking 1 orbiter primary mission, *J. Geophys. Res., 82*, this issue, 1977.

Carr, M. H., Volcanism on Mars, *J. Geophys. Res., 78*, 4049–4062, 1973.

Carr, M. H., The volcanoes of Mars, *Sci. Amer., 234*, 32–43, 1975.

Eaton, J. P., and K. J. Murata, How volcanoes grow, *Science, 132*, 925, 1960.

Greeley, R., Observations of actively forming lava tubes and associated structures, Hawaii, *Mod. Geol., 2*, 207–223, 1970.

Greeley, R., Lava tubes and channels in the lunar marius hills, *Moon, 3*, 289–314, 1971.

Greeley, R., Mariner 9 photographs of small-scale volcanic structures of Mars, *NASA Tech. Memo., X-62*(222), 22, 1972.

Greeley, R., Modes of emplacement of basalt terrains and an analysis of mare volcanism in the Orientale Basin, *Proc. Lunar Sci. Conf. 7th*, 2747–2759, 1976.

Greeley, R., and J. H. Hyde, Lava tubes of the cave basalt, Mount St. Helens, Washington, *Geol. Soc. Amer. Bull., 83*, 2397–2418, 1972.

Greeley, R., E. Theilig, J. E. Guest, M. H. Carr, H. Masursky, and J. A. Cutts, Geology of Chryse Planitia, *J. Geophys. Res., 82*, this issue, 1977.

Harris, S. A., The aureole of Olympus Mons, Mars, *J. Geophys. Res., 82*, in press, 1977.

Head, J. W., M. Settle, and C. A. Wood, Origin of Olympus Mons escarpment by erosion of pre-volcano substrate, *Nature, 263*, 667–668, 1976.

Hulme, G., The interpretation of lava flow morphology, *Geophys. Jour. Roy. Astron. Soc., 39*, 361–383, 1974.

Hulme, G. The determination of the rheological properties and effusion rate of an Olympus Mons lava, *Icarus, 27*, 207–213, 1976.

King, J. S., and J. R. Riehle, A proposed origin of the Olympus Mons escarpment, *Icarus, 23*, 300–317, 1974.

Masursky, H., An overview of geologic results from Mariner 9, *J. Geophys. Res., 78*, 4000–4030, 1973.

Masursky, H., and H. L. Crabill, Search for the Viking 2 landing site, *Science, 194*(4260), 62–68, 1976.

McCauley, J. F., M. H. Carr, J. A. Cutts, W. K. Hartmann, R. P. Sharp, and D. E. Wilhelms, Preliminary Mariner 9 report on the geology of Mars, *Icarus, 17*, 289–327, 1972.

Moore, H. J., and G. G. Schaber, An estimate of the yield strength of the Imbrium flows, *Proc. Lunar Sci. Conf. 6th*, 101–118, 1975.

Moore, J. G., Giant submarine landslides on the Hawaiian ridge, *U.S. Geol. Surv. Prof. Pap., 501D*, 95–98, 1964.

Neukum, G., and D. V. Wise, Mars: A standard crater curve and possible new time scale, *Science, 194*, 1381–1387, 1976.

Pike, R. J., Simple to complex craters: The transition on the moon, *Lunar Science VII*, part II, pp. 700–702, Lunar Science Institute, Houston, Tex., 1976.

Schaber, G. G., Lava flows in Mare Imbrium: Geologic evaluation from Apollo orbital photography, *Proc. Lunar Sci. Conf. 4th*, 73–82, 1973.

Shreve, R. L., The Blackhawk landslide, *Geol. Soc. Amer. Spec. Pap., 108*, 47, 1968.

Soderblom, L. A., C. D. Condit, R. A. West, B. M. Herman, and T. J. Kreidler, Martian planetwide crater distributions: Implications for geologic history and surface processes, *Icarus, 22*, 239–263, 1974.

Sparks, R. S. J., H. Pinkerton, G. Hulme, Classification and formation of lava levees on Mount Etna, Sicily, *Geology, 4*, 269–270, 1976.

U.S. Geological Survey, Topographic map of Mars, *I-961*, U.S. Geol. Surv., Reston, Va., 1976.

Walker, G. P. L., Compound and simple lava flows and flood basalts, *Bull. Volcanol., 35*(3), 579–590, 1971.

Walker, G. P. L., Lengths of lava flows, *Phil. Trans. Roy. Soc. London, Ser. A, 274*, 107–118, 1973.

(Received April 1, 1977;
revised May 23, 1977;
accepted May 23, 1977.)

Classification and Time of Formation of Martian Channels Based on Viking Data

HAROLD MASURSKY, J. M. BOYCE, A. L. DIAL, G. G. SCHABER AND M. E. STROBELL

U.S. Geological Survey, Flagstaff, Arizona 86001

Fluviatile and volcanic Martian channels, first discovered on Mariner 9 pictures, have been reexamined by using Viking orbital photography. The superior discrimination of the Viking photographs, resulting from clearer atmospheric conditions and an improved camera system, has permitted us to map additional channels and to estimate their relative ages, using a technique based on crater counting. Broad channels like the Ares and Tiu/Simud Valles are situated along the margin of the southern highlands near Chryse Planitia, the landing site of Viking 1. They originate in areas of collapsed terrain that may have been formed when subsurface water-ice (permafrost) was melted by geothermal heat from deep-seated volcanic centers. When permafrost melting reached an abrupt topographic slope, the interstitially stored meltwater 'lakes' were breached suddenly, releasing the great floods that modified the channels. The volume of material involved in the collapsed terrain is large enough to furnish the water calculated to have filled the broad channels. Conditions are reviewed for persistence of liquid water on Mars under present and more favorable pressures and temperatures. Sinuous channels of intermediate size, like the Ma'adim and Hrad Valles and other shorter, stubby channels, have multiple tributaries; in the limited coverage available, they appear to result from 'spring sapping,' with the underground permafrost meltwater emerging in box canyons at their heads. The widespread distribution of this type of channel makes their origin by local geothermal heating less likely; climatic warming may be required to explain their formation. The final fluviatile type, dendritic channel networks, has the widest areal distribution and appears to have been formed during at least two episodes. The filamentous channels in their source areas (often the rims of craters) seem to resemble terrestrial river systems; rainfall would seem to be required to form these features. All these channel types debouch onto lowland plains or crater floors, where they disappear in short distances; these abrupt terminations may have resulted from percolation and/or evaporation. Simple and complex lava channels are common; they originate at volcanic centers and are usually morphologically distinct from the aqueous channels. Three types of lava channels are recognized. The wide variation in crater densities implies varying channel ages. Water must have flowed on the Martian surface at many different times in the past, although this would be possible only with great difficulty under the present Martian thermal conditions. Based on a crater flux curve derived by Soderblom et al. (1974) the fluviatile channel ages vary from 3.5 to 0.5 Gy. Lava channel ages range from 3.5 Gy to an age too young to date by the crater counting technique (perhaps 200 m.y.). Methods for dating the channels and volcanic episodes are still insufficiently developed to determine whether episodes of volcanic heating and climatic change are coincident. It is possible that the large floods and volcanic eruptions might trigger a short 'interglacial' interval. Alternatively, the floods may be related to episodic volcanic activity, and the dendritic channels to rainfall that was associated with independent interglacial climatic episodes resulting from variations in solar output or other causes.

INTRODUCTION

Possible fluviatile channels on Mars were first recognized from Mariner 9 photographs (H. Masursky, in the paper by *Driscoll* [1972]). Channel descriptions, classification, and possible genesis, based on Mariner imaging, have been reported by many authors [*Masursky*, 1973; *Milton*, 1973; *Baker and Milton*, 1974; *Hartmann*, 1974; *Schumm*, 1974; *Sharp and Malin*, 1975; *Malin*, 1976; *Pieri*, 1976; *Masursky*, 1976; *Nummedal*, 1976; *Baker*, 1977]. This paper reconsiders the classification, genesis, and age relations of several channel systems based on Viking photographs taken from July 1976 to February 1977 (Figure 1 and Table 1).

Because it is difficult to retain water on the Martian surface under the present climatic conditions, many observers are still convinced that the channels photographed were eroded by the wind [*Cutts et al.*, 1976; *Cutts*, 1977; D. Anderson, personal communication, 1976] rather than by flowing water. Three types of terrain features indicate that the channels were carved by liquid flow: (1) Martian channels having many tributaries form integrated systems (Figure 2*a*), like the integrated terrestrial stream systems first noted by *Playfair* [1802] (see also *Hack* [1956] and *Leopold and Maddock* [1953]); (2) the configuration of the complex anastomosing channels (Figure 2*b*) and the shapes of islands within the channels (Figure 2*c*) are unlike wind-eroded shapes (Figure 2*d*) [*Milton*, 1973; *Baker and Milton*, 1974; *Nummedal*, 1976; *Carr et al.*, 1976]; and (3) features typical of fluid erosion are seen within the Ares channel near its shoreline, whereas the terrain outside (Figure 2*c*) is unmodified.

Martian wind-eroded forms occur at differing elevations, such as near the base of Olympus Mons (Figure 2*d*) and near the base and summit of Arsia Mons, but fluvial erosional forms appear to be controlled by elevation and local topography.

AGE DETERMINATION METHODOLOGY

The density of impact craters superposed on a surface is generally assumed to be proportional to the relative age of the surface. Viking photographs were used to make counts of craters occurring within Martian channels and on adjacent surfaces in order to obtain relative ages for the channels. Craters larger than ~0.3 mm in diameter were counted; on 1500-km photographs, 0.3 mm equals ~0.4 km. In most cases the counts were done by the same observer to ensure internal consistency. Obvious secondary craters and endogenic craters

Paper number 7S0566.

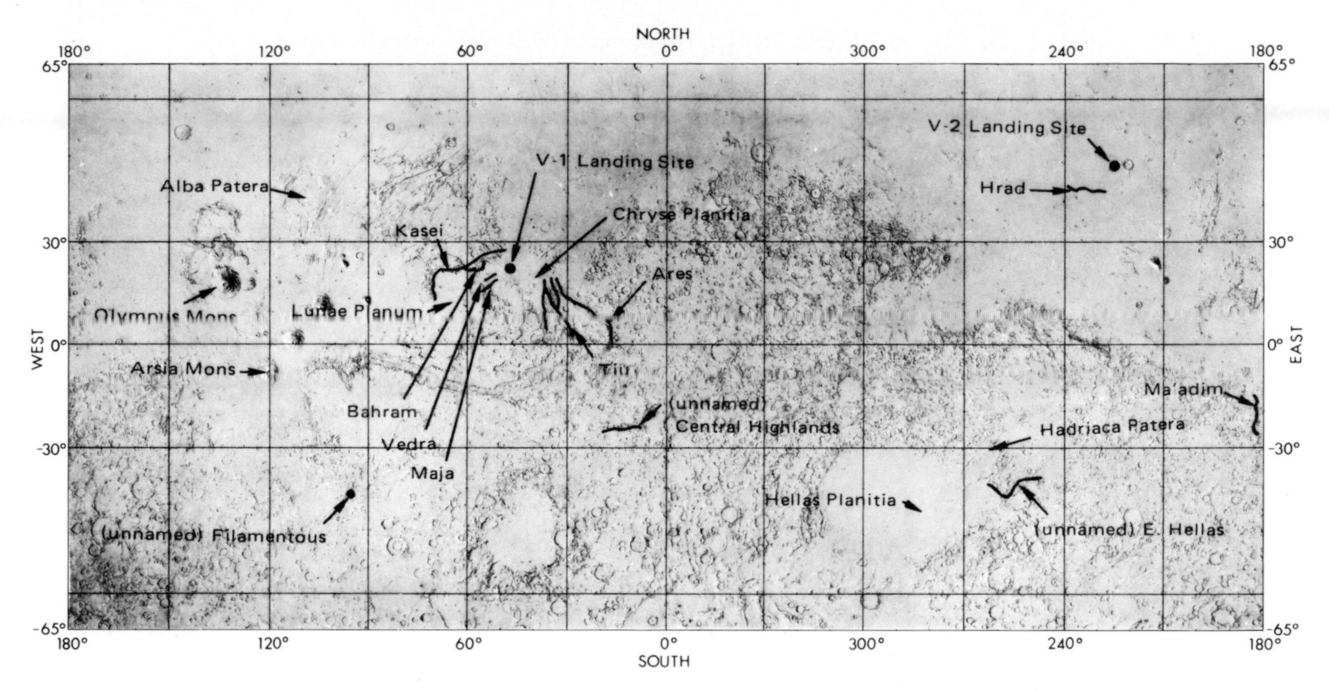

Fig. 1. Index map showing location of Martian fluviatile and volcanic channels described in this report. All names are approved by the International Astronomical Union. The following list gives the figures associated with each channel or other feature: Ares, Figures 2*b*, 2*c*, 5*c*, 5*d*, and 7*c*; Bahram, Figure 4*a*; Hrad, Figure 8*c*; Kasei, Figure 6*c*; Ma'adim, Figure 3*c*; Maja, Figures 6*a*, 7*a*, and 7*b*; Tiu, Figure 8*a*; Vedra, Figures 2*a* and 4*c*; (unnamed) East Hellas, Figure 9*a*; (unnamed) southern highlands, Figures 10*a* and 10*b*; (unnamed) 'filamentous,' Figure 9*c*; Alba Patera, Figures 10*c*, 11*d*, and 12*a*; Arsia Mons, Figure 11*b*; Chryse Planitia, Figures 2*b*, 2*c*, 3*a*, 6*a*, 6*c*, 7*a*, and 7*b*; central highlands, Figures 3*a*, 9*a*, 9*c*, 10*a*, and 10*b*; Hadriaca Patera, Figure 12*c*; Lunae Planum, Figures 4*a* and 4*c*; and Olympus Mons, Figure 2*d*.

were excluded from the counts. The crater count data were used to obtain cumulative crater size-frequency curves, normalized to an area of 1 km², for each channel or surface area counted. All crater curves have been plotted against an idealized average lunar curve [*Neukum et al.*, 1974]. The crater size-frequency distribution curve for the Viking 1 landing site is also included for comparison (see Figures 3*a* and 3*b*). Error bars are standard error ($\pm(N/A)^{1/2}$, where N is the cumulative number of craters in a class interval/unit area A).

The crater size-frequency distribution curves for the channels can then be compared by noting their relative displacements; areas having the lowest crater density determinations are assumed to be the youngest in age [*Masursky and Crabill*, 1976*a*, *b*]. This assumption has been tested by earlier workers [*Shoemaker*, 1962; *Shoemaker and Hackman*, 1962; *Baldwin*, 1964] and found to be valid. Early workers also discovered that the form and shape of the lunar curves for areas of widely differing locations and ages are similar and that the Martian curves derived from Mariner 9 data are also similar. These curves usually have regular slopes of $-2°$ to $-4°$. However, some of our Martian curves derived from Viking data show unusually low slopes or anomalous bends in the slope. We infer that a surface, such as that shown in Figure 3*c* (Ma'adim Valles), has been modified by continuous erosion or deposition by either wind or water because the surface curve has a lower than normal slope (Figure 3*d*). A bend in the curve, connecting two parallel but offset segments, is usually interpreted to mean that a particular size range of craters has been destroyed or buried without affecting craters of other sizes [*Hartmann*, 1973, 1977]. This situation occurs when lava flows or other types of mantles are thick enough to obliterate the larger craters [*Neukum and Horn*, 1976]. We interpret the surface cut by Bahram Valles (Figure 4*a*) to be resurfaced because of the obvious bend seen in the Bahram surface curve (Figure 4*b*). Photographic resolution is another factor that can affect the shape of the curve, causing an apparent reduction in crater density at small diameters, so that the curve acquires a progressively lower slope or 'rollover.' Resolution of photograph 22A76 of Vedra Valles (Figure 4*c*) is shown to be ~300 m by the curve rollover at that position (Figure 4*d*). Figure 5*a* shows the relation between the effective photographic resolution, as determined by the rollover position on the crater curve, and changes in photographic resolution with variations in distance between the orbital camera and the surface photographed by the camera (the 'range'). Figure 5*b* is the highest-resolution picture of the surface of Mars yet taken. Abundant craters down to the limit of resolution are clearly shown.

It is conceivable that these variations in curve slope and shape could all result from changes in the impact flux, but the data do not seem to support such an hypothesis. No regular variation in slope angle or slope discontinuity has been observed on a planet-wide basis. The observed variations in crater density distribution curves are therefore interpreted as reflecting variations in the age and erosional state of the terrain rather than changes in the impact flux. The relative ages of channels and surfaces show clearly in the several crater density plots discussed.

Several attempts have been made to correlate Martian flux curves with lunar curves and radiometric ages of returned lunar samples [*Soderblom et al.*, 1974; *Chapman*, 1976; *Neukum and Wise*, 1976; *Hartmann*, 1977] so that absolute ages for the Martian events can be determined. The shape and position of both the lunar and the Martian flux curves, from which absolute ages are ascertained, are still subjects of intense de-

TABLE 1. Data Used to Prepare Crater Size-Frequency Distribution Curves for Figures in Text

Figure	Feature	Area Counted, km²	Number of Craters Counted		Median Number of 1-km Craters per 1 km²	Range, km	Coordinates		Viking Photo Numbers
			Per 500 m	Per 1 km			Latitude	Longitude	
2*a*	Vedra Valles, other dendritic channels					1630–1640	18°–19°N	54°–56°W	Mosaic: 46A53–46A60, 47A52, 47A54, 47A56, and 47A58
2*b*	Ares channel, NE of A-1 landing site					1920–1950	24°N	31°W	Mosaic: 3A13 and 3A14
2*c*	Ares channel, NE of A-1 landing site					1590	20°–22°N	30°–32°W	Mosaic: 4A48–4A54
2*d*	Surface, NE of Olympus Mons					3140	25°N	132°W	48B15
3*a*	V-1 landing site					1655	22°N	48°W	20A71
3*b*	V-1 landing site surface curve	11,000	120	20	2.1×10^{-3}	1650–1810	22.5°N	48.0°W	20A71, 20A46, 20A48, 20A50, 20A52, 20A69, 20A71, and 20A73
3*c*	Ma'adim Vallis					7320	18°S	182°W	88A69
3*d*	Ma'adim surface curve	69,000	0	200	$2.9 \times 10^{+1}$	7320	18°S	182°W	88A69
4*a*	Bahram Vallis					1750–1780	22°N	55°–57°W	Mosaic: 22A33, 22A34, 22A36, 22A38, and 22A40
4*b*	Bahram surface curve	10,000	200	40	1.2×10^{-2}	1750–1780	22°N	55°–57°W	22A34, 22A36, 22A38, and 22A40
	Bahram channel curve	680	6	2	2.9×10^{-3}	1750–1780	22°N	55°–57°W	22A34, 22A36, 22A38, and 22A40
4*c*	Vedra Valles					1710	20°N	55°W	22A76
4*d*	Vedra surface curve	9,200	130	35	3.7×10^{-3}	1680–1690	19°–24°N	52°–54°W	22A56, 22A76, 22A77, and 22A78
	Vedra channel curve	400	2	0	6.0×10^{-4}	1680–1690	19°–24°N	52°–54°W	22A56, 22A76, 22A77, and 22A78
5*a*	High-resolution at Alba Patera curve	80	2	1		317	39°N	107°W	268A22
	A-1 landing site	2,700	32	8	3.3×10^{-3}	1920–1950	23°–24°N	31°W	03A13, 03A14, and 4A25
	B-3 landing site	19,900	154	29	1.4×10^{-3}	3540	47°N	231°W	09B10
5*b*	High resolution at Alba Patera photo					317	39°N	107°W	268A22
5*c*	Hydaspis Chaos					2700	3°N	27°W	83A37
5*d*	Tiu Valles					2710	40°N	27°W	83A38
6*a*	Maja Vallis					1659	19°N	51°W	20A56
6*b*	Maja Vallis surface	11,000	120	15	2.0×10^{-3}	1659–1700	50°–52°N	19°–20°W	20A55, 20A56, 20A58, 44A43, and 44A44
6*c*	Kasei Vallis mouth					2080	25°N	50°W	20A26
6*d*	Kasei Vallis surface at mouth	8,130	155	18	2.2×10^{-3}	2080	25°N	50°W	20A26 and 20A28
7*a*	Maja Vallis mouth					1660	21.3°N	47.8°W	20A62
7*b*	Maja Vallis in Chryse Planitia					1820	22.2°N	44.0°W	20A40
7*c*	Ares Vallis					2700–3050	7°–12°N	21°–27°W	Mosaic: 83A18, 83A19, 83A20, 83A21, 83A47, 83A48, 83A49, and 83A50
7*d*	Ares surface curve	31,500	230	100	6.0×10^{-1}	2700–3060	8°–10°N	24°–25°W	83A18, 83A19, 83A48, 83A49, and 83A50
	Ares channel curve	4,000	35	6	1.5×10^{-3}	2700–3060	8°–10°N	24°–25°W	83A18, 83A19, 83A48, 83A49, and 83A50
8*a*	Tiu Valles					2700–2710	4°N	27°W	Mosaic: 83A06, 83A08, 83A0[illegible], 83A37, and 83A38
8*b*	Tiu surface curve	28,100	185	65	2.0×10^{-1}	2710–3050	1°–6°N	16°–29°W	Mosaic: 83A02, 83A04, 83A06, 83A08, 83A10, 83A34, 83A36, 83A37, 83A38, and 83A39
	Tiu channel curve	7,000	22	2	2.8×10^{-4}	2710–3050	1°–6°N	16°–29°W	Mosaic: 83A02, 83A04, 83A06, 83A08, 83A10, 83A34, 83A36, 83A37, 83A38, and 83A39
8*c*	Hrad Valles					3960	42°N	226°W	9B54
8*d*	Hrad surface curve	13,600	40	16	1.0×10^{-2}	3960–4060	41°–44°N	222°–239°W	9B22, 9B43, 9B45, 9B47, 9B48, 9B50, 9B52, 9B54, and 9B56
	Hrad channel curve	1,640	21	3	1.5×10^{-4}	3960–4060	41°–44°N	222°–239°W	9B22, 9B43, 9B45, 9B47, 9B48, 9B50, 9B52, 9B54, and 9B56
9*a*	East Hellas unnamed channel					8660–8840	39°–48°S	260°–270°W	Mosaic: 97A48–97A54 and 97A60–97A68
9*b*	East Hellas surface curve	500,000	0	670	2.4×10^{-2}	8660–8840	39°–48°S	260°–270°W	97A48–97A54 and 97A60–97A68

TABLE 1. (continued)

Figure	Feature	Area Counted, km²	Number of Craters Counted: Per 500 m	Number of Craters Counted: Per 1 km	Median Number of 1-km Craters per 1 km²	Range, km	Coordinates: Latitude	Coordinates: Longitude	Viking Photo Numbers
	East Hellas channel curve	22,800	0	6	2.6×10^{-4}	8660–8840	39°–48°S	260°–270°W	97A48–97A54 and 97A60–97A68
9*c*	Unnamed filamentous channel					8890	43°S	94°W	63A09 (partial frame)
9*d*	Unnamed filamentous channel surface	40,000	0	62	6.0×10^{-1}	8890	43°S	94°W	63A09 (partial frame)
10*a*	Unnamed, central highlands					8550	27°S	14°W	84A47
10*b*	Unnamed, central highlands					8590	24°S	9°W	84A43
10*c*	Alba Patera dendritic channels					1780	44°N	104°W	04B57
10*d*	Alba (dendritic) surface curve	2,700	0	3	1.1×10^{-3}	1780	44°N	104°W	04B57
11*b*	Arsia Mons flank					6950	12°S	120°W	52A04 (partial frame)
11*c*	Arsia flank surface curve	7,400	0	3	2.7×10^{-4}	6530	9°S	121°W	39A09
11*d*	West Alba Patera					4200	42°N	118°W	07B88
12*a*	Alba Patera summit					4120–4220	43°–44°N	108°–111°W	Mosaic: 7B60 and 7B62
12*b*	Alba summit surface curve	5,500	0	123	2.2×10^{-3}	4120–4220	43°–44°N	108°–111°W	Mosaic: 7B60 and 7B62
12*c*	Hadriaca Patera					8950–8975	28°–34°S	265°–274°W	Mosaic: 97A40–97A42
12*d*	Hadriaca surface curve	64,000	0	110	4.6×10^{-1}	8957	30°S	269°W	97A42

bate. We have used the *Soderblom et al.* [1974] calibration curve, rather than the *Neukum and Wise* [1976] curve, to obtain the estimates of absolute ages for Martian surfaces that we quote here. We favor the *Soderblom et al.* [1974] curve because it is consistent with the data recently presented by *Shoemaker* [1977] for asteroid populations; his data show that the impact flux at Mars is higher by a factor of ~2, in relative production of craters 4–10 km in diameter, than the lunar flux. Based on the *Soderblom et al.* [1974] curve the estimated ages of the Martian fluviatile channels vary from 3.5 Gy to 0.5 Gy; lava channels vary from 3.5 to an age too young to determine by crater counts (less than 200 m.y.). These estimates were made by extrapolating the curves to the diameter range used by *Soderblom et al.* [1974]. These age estimates may change following additional analysis of the Viking orbiter imaging data.

DESCRIPTION OF CHANNELS AND IMPLICATIONS FOR CHANNEL FORMATION

Some of the broad channels, such as the Ares and Tiu/Simud valles (Figure 1), rise in the highlands area south of Chryse Planitia, the landing site of Viking 1. These channels head in areas of collapsed terrain, like Hydaspis Chaos (Figure 5*c*), and then flow into the Chryse lowland. The spacing and size of the collapsed areas resemble those of volcanic centers in the Tharsis plateau and the Elysium region. A possible mechanism for the release of water is subterranean emplacement of local volcanic complexes which melt the overlying permafrost layer. When the outward spreading front of meltwater (from ice wedges and interstitial ice) intersects a cliff face, the underground 'lake' rapidly drains over the lowered ice dam in the subsurface; as the water emerges from the base of the chaotic terrain, it modifies preexisting fault troughs or other topographic lows and erodes channel floors (Figure 5*d*).

Other broad channels, like the Maja and Kasei Valles (Figure 1), have less obvious areas of collapsed terrain in their source areas (on Lunae Planum) than do the broad channels that rise in the highlands south of Chryse Planitia. Abundant evidence for permafrost activity does occur, however, at Lunae Planum and elsewhere on the planet [*Carr and Schaber*, 1977]. Photographs of the Maja and Kasei Valles near their mouths (Figures 6*a* and 6*c*) show crater densities that are similar to those seen at the Viking 1 landing site (Figure 3*a*), implying similar ages for the three areas; curves derived from crater counts of the three areas (Figures 3*b*, 6*b*, and 6*d*) also imply similar ages.

Viking temperature and water vapor measurements support the theory that abundant water is present and probably was also present on Mars in the past [*Kieffer*, 1976; *Kieffer et al.*, 1976*a*; *Farmer et al.*, 1976]. These data show that the residual polar ice caps are water-ice and that the atmosphere is saturated with water vapor during the Martian night.

When the proposed channel flow reaches the lowland plain, it spreads widely and forms complex braided networks. In the region directly west of the Viking 1 landing site the floodwater apparently was dammed behind marelike ridges until it overflowed low points and cut gaps in the ridge crests (Figure 7*a*). Evidence of stream erosion occurs far out into the Chryse basin, where grooves have been cut on the downslope sides of mare ridges (Figure 7*b*). Evidence of stream erosion is seen as far as 350 km from the canyon mouths of the Ares, Tiu/Simud, and Maja Valles. Evidence of ponded water is not visible in photographs of the basin, nor do shorelines, beach ridges, or bars appear to be present; photographs of terrestrial

Fig. 2*a*. Photomosaic of two dendritic channel systems that drain Lunae Planum west of Chryse Planitia. Vedra Valles is the northern channel system shown. The integrated system and accordant tributary junctions are similar to terrestrial drainage systems. The tributaries start at points, not in collapsed terrain or box canyons. Collected rainfall seems necessary to erode this type of channel system.

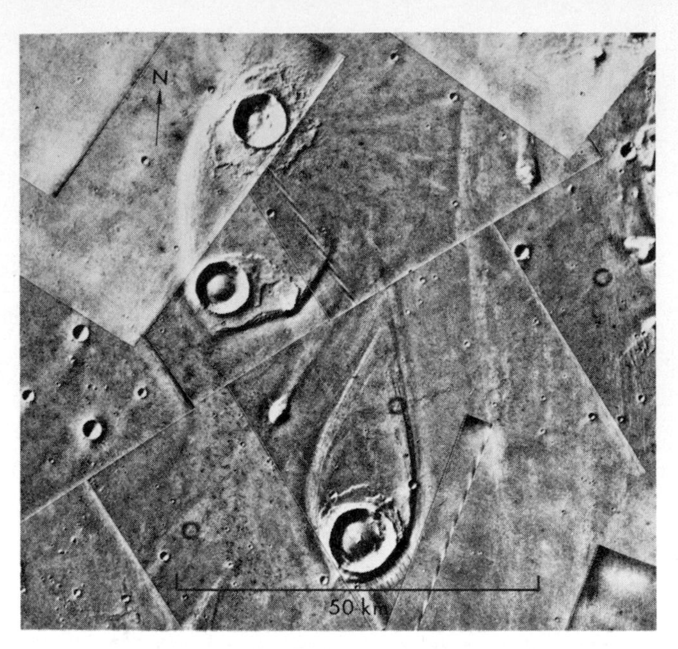

Fig. 2*c*. Photomosaic of part of Ares Valles in Chryse Planitia. The islands have impact craters at the south (upstream) ends and long streamlined forms extending downstream. In the central island a prow points upstream—a feature commonly observed in terrestrial streams that have rapid turbulent flow. The rock layers, probably basaltic lava flows, show clearly where they have been eroded by the flowing water. Ejecta blankets from some impact craters have been partially cut away by the channel, indicating that the craters preceded the channel. The ejecta blanket at the north end of the northern island lies across the channel, showing that it was later. Smaller obstacles at the south end of the picture have deep moats at the upstream end, confirming the turbulent flow. In the north east corner is a shoreline about 100 m high. The terrain to the west has surface grooves cut by the water, while the terrain to the east is unmodified. The slope of the channel floor, determined photogrammetrically, is 10 m/km to the north.

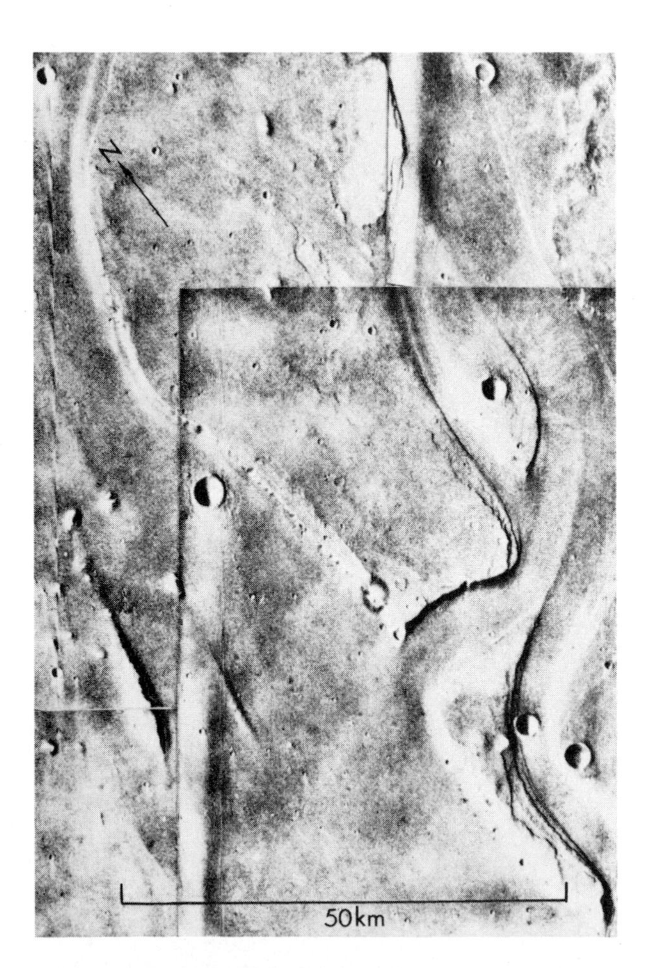

Fig. 2*b*. Photomosaic showing anastomosing channels at the north end of Ares Vallis in Chryse Planitia near the originally designated A-1 landing site. The streamlined forms and intersecting channels are characteristic of water erosion. Discrete layers, possibly basaltic lava flows, show along the edges of the islands. A number of small impact craters are seen that formed after the channel was eroded. The islands are about 100 m high.

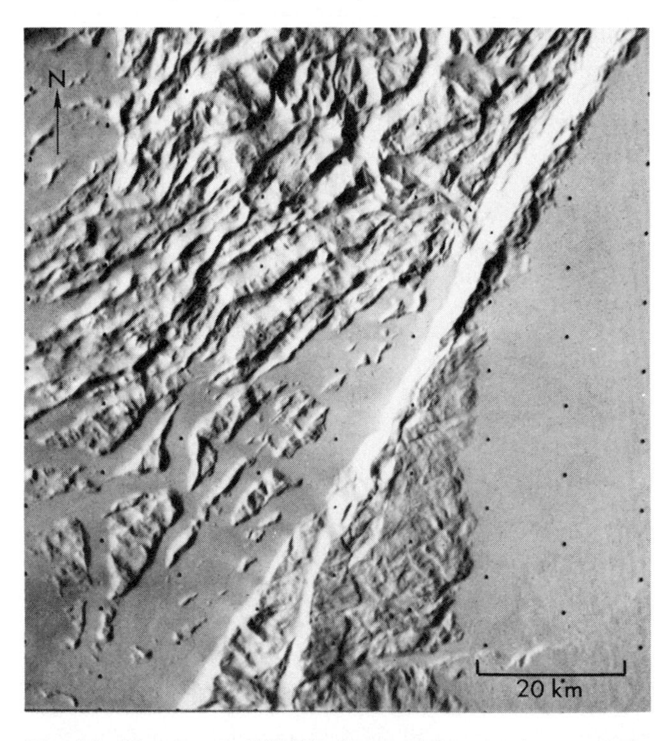

Fig. 2*d*. Photograph (48B15) of part of the aureole around the base of Olympus Mons 600 km north of the volcanic summit. The picture is about 100 km wide. The ridges probably are yardangs—positive features formed by wind erosion of old lava flows. The ridges are shaped like inverted boat hulls as are terrestrial yardangs; *McCauley* [1973] first pointed out the morphologic similarity of the Martian and terrestrial features. The shapes are distinct from the shapes produced by flowing water in the Mars channels.

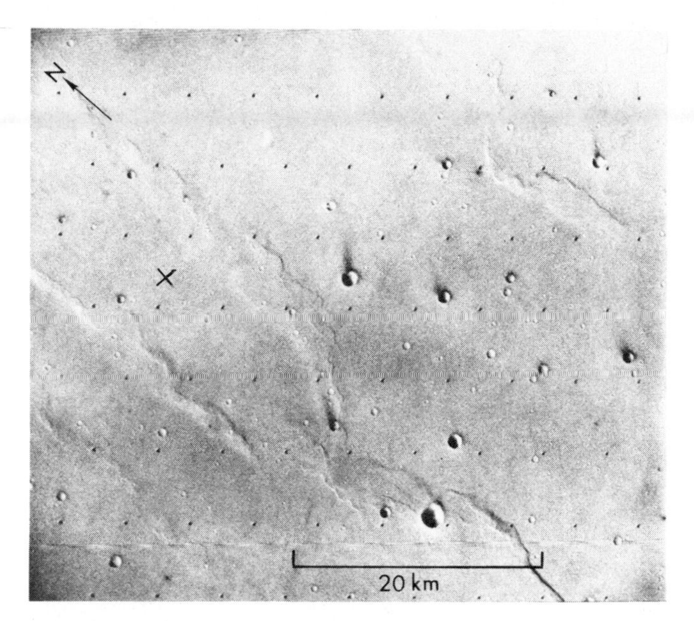

Fig. 3*a*. Photograph (20A71) of the Viking 1 landing site (marked by a cross) in Chryse Planitia. The ridges resemble mare ridges on the moon that most commonly form in basaltic lava flows. Also visible are many impact craters ranging up to 1.5 km in diameter and crater streaks which are characteristic of wind erosion and deposition. Wind dunes and drifts are shown in the lander pictures. Grooves that probably were formed by water are visible 60 km southwest of the largest crater near the bottom of the picture.

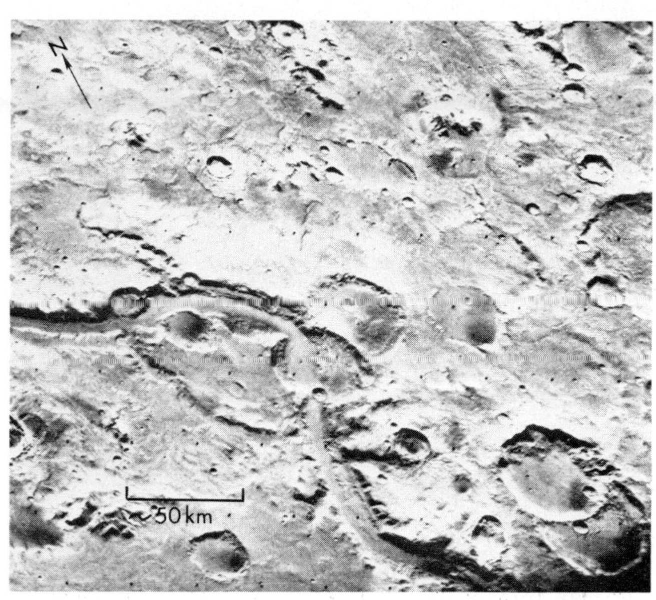

Fig. 3*c*. Oblique photograph (88A69) of Ma'adim Vallis. This sinuous intermediate size valley south of Elysium Planitia has many short, stubby tributaries that join the main channel. The channel cuts across an ancient crater in the middle of the picture; two sizable young craters lie in the channel.

dry lakes, taken at comparable resolution, show such features clearly. Either the Martian water disappeared by percolation, evaporation, or freezing before it ponded in the Martian playas, or these features are covered by younger deposits of either volcanic or eolian origin. However, only a small portion of the water calculated to flow through the channels could percolate into the subsurface [*Kieffer*, 1976; *Kieffer et al.*, 1976*b*, *c*; D. Davies, personal communication, 1977] because the permafrost layer that lies at a shallow depth would prevent deep percolation; the remainder of the water would have to freeze and then sublimate. However, there is no geomorphic evidence in the photographs taken to date for either ponded or frozen water within or adjacent to the Martian channels.

Crater counts show that the heavily cratered Martian terrain is quite ancient, whereas the less cratered channel floors are younger. Figure 7*c* shows Ares Vallis where it dissects part of the ancient cratered highlands; Figure 7*d* shows relative age relations of Ares Vallis and the surface it cuts. The Tiu/Simud

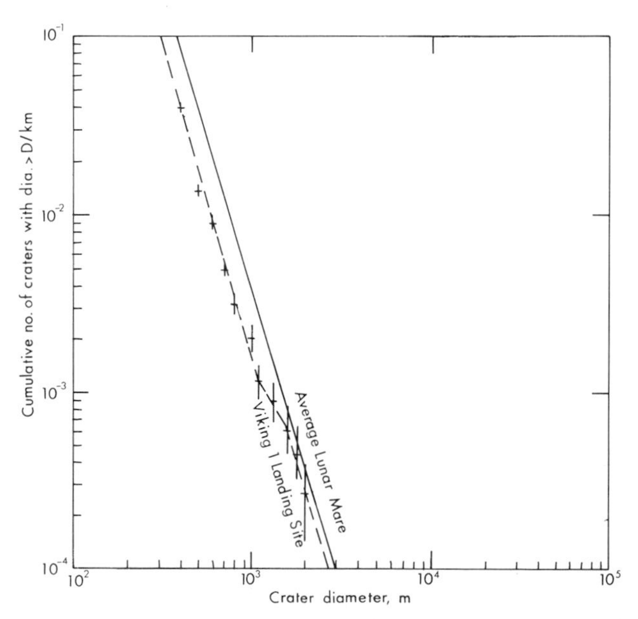

Fig. 3*b*. Crater size-frequency distribution curve of the Viking 1 landing site. It follows the lunar curve with one offset, possibly indicating a modifying event partway through the development of the surface.

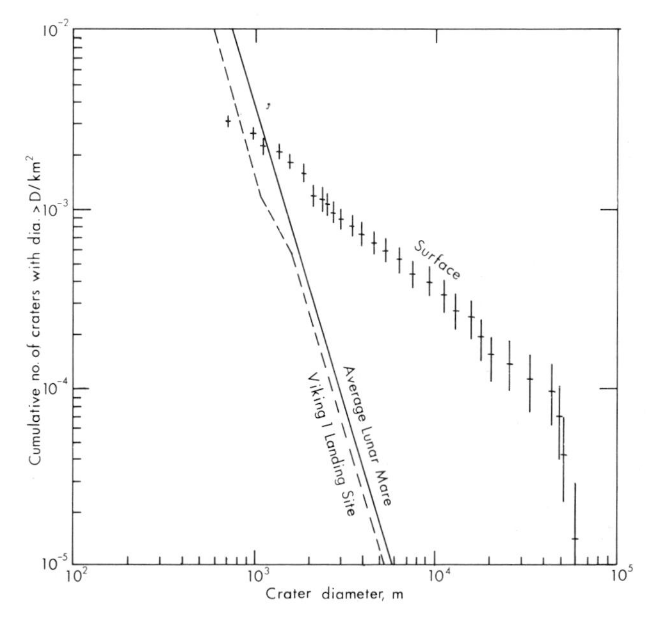

Fig. 3*d*. Crater size-frequency distribution curves for Ma'adim Vallis and the adjacent terrain. The upland is heavily cratered; the low slope on the crater curve may indicate continual degradation of craters smaller than 14 km in diameter. The obliquity of the photograph may also have biased the data for the curves.

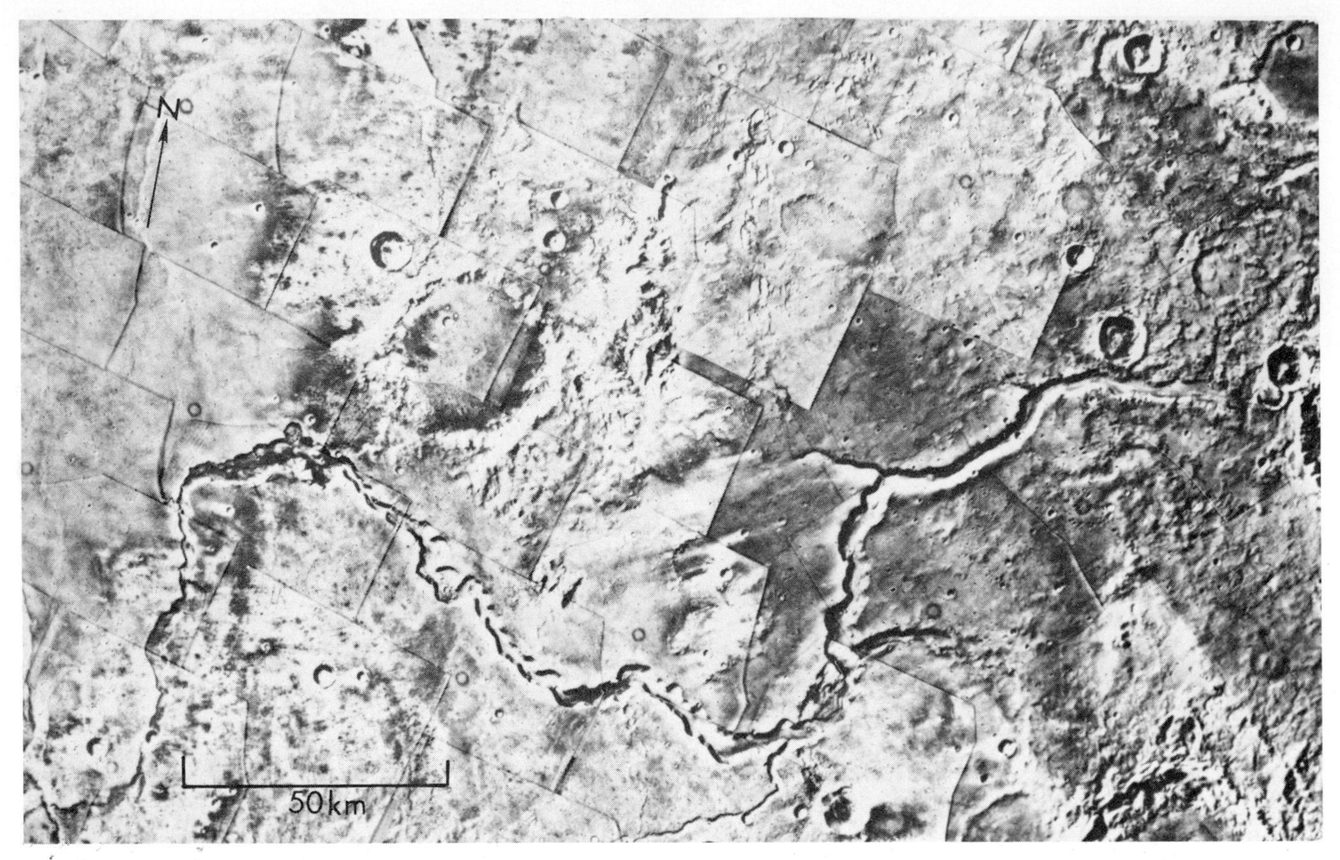

Fig. 4*a*. Viking mosaic of Bahram Vallis due west of the Viking 1 landing site. The sinuous, intermediate size channel is cut into Lunae Planum.

channel, shown in Figure 8*a*, and the crater curve for this channel (Figure 8*b*) show a lower density of impact craters than do the Ares or Kasei channels; Tiu is therefore thought to be younger. These crater counts indicate that volcanic or geothermal heating may have occurred during a number of distinct episodes. Additional high-resolution Viking photography of Martian channels will allow us to establish better relative age relations for the broad channels.

Sinuous channels like the Ma'adim (Figure 3*c*), Bahram (Figure 4*a*), and Hrad Valles (Figure 8*c*) are intermediate in size and have multiple tributaries; they are recognized as far north as 44°N and as far south as 45°S. Sinuosities and braids on the channel floors closely resemble features seen in terrestrial stream channels; their courses are smoothly integrated, and their tributaries are accordant. Their tributaries commonly head in box canyons and resemble terrestrial desert streams that are fed by springs rather than by collected surface precipitation; some tributaries are dendritic (see following discussion). We infer that the Martian box canyons are formed by 'spring sapping,' that is, erosion by water escaping from a permeable layer, with subsequent collapse of the overlying resistant rock. On Mars, such streams are probably fed by escaping ice meltwater. The ice may have been melted by volcanic heat, as was the case with the broad channels; however, their wide areal distribution suggests that an episode of climatic warming is a more likely cause for their formation.

Many short, stubby channels that head in box canyons occur either along the continental margin, where the ancient cratered highlands meet the northern lowlands, or along the rim of Valles Marineris. These streams closely resemble the tributaries of the intermediate size sinuous valleys and probably have a similar origin.

The floors of sinuous channels are wide enough to allow valid crater counts to be made of the younger channel floors, as well as of the unchanneled adjacent terrain (Figure 8*c*). Figure 8*d* shows the variation in crater densities between the Hrad channel and the surrounding surface; a similar variation between an unnamed channel northeast of Hellas Planitia and the surface it cuts (Figure 9*a*) is seen in the crater curves for that area (Figure 9*b*). The variation in crater densities that exists between counts made of the intermediate channels is as wide as that which exists for the broad streams; they probably were formed over a similar span of time.

Dendritic channel networks on Mars closely resemble terrestrial stream patterns (Figures 9*c*, 10*a*, and 10*b*). They occur ubiquitously in the southern highlands [*Masursky*, 1973; *Sagan et al.*, 1973; *Pieri*, 1976] and are found as far north as 40°N and as far south as 45°S. The networks commonly head on the outside rims of ancient degraded craters in the central highlands, originating just below the rim crest. No trace of collapsed terrain or box canyons in the headwater areas of these channels is seen at the resolution of pictures obtained thus far; the filamentous character of the stream network probably persists beyond the present limit of resolution. These networks resemble terrestrial stream networks that are the result of collected rainfall. The channels do not appear to be deeply incised, and, in the ancient cratered terrain, they dissect the outer rims of the craters, as do the intermediate channels. The

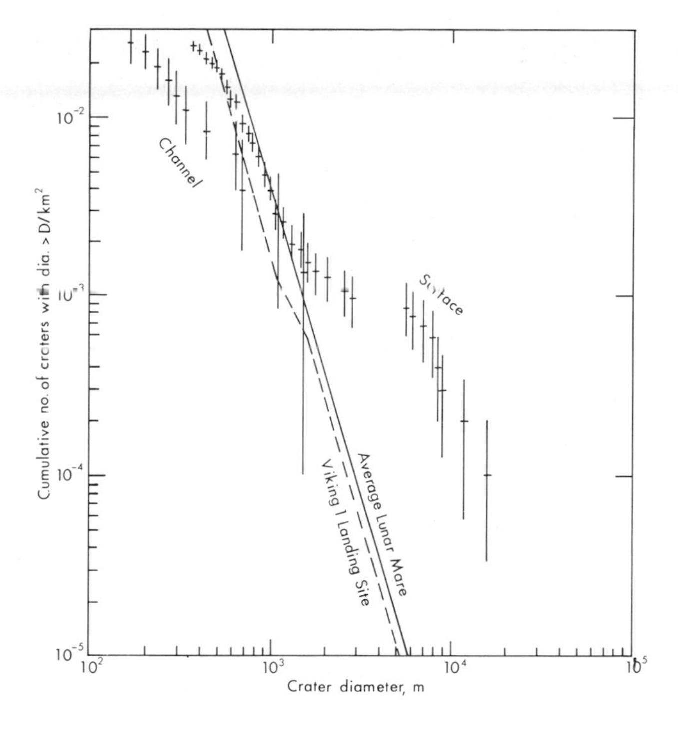

Fig. 4*b*. Crater size-frequency distribution curves for Bahram Vallis and the prechannel surface. The crater curve for the older surface has a sharp inflection indicating an erosional or depositional interruption in the crater distribution. The surface curve lies well above the average lunar mare curve, indicating greater antiquity. The channel crater curve lies along the Viking 1 landing site curve as do the curves for the Maja and Kasei valles; all of these channels emerge from Lunae Planum and flow across the western part of Chryse Planitia toward the Viking 1 landing site.

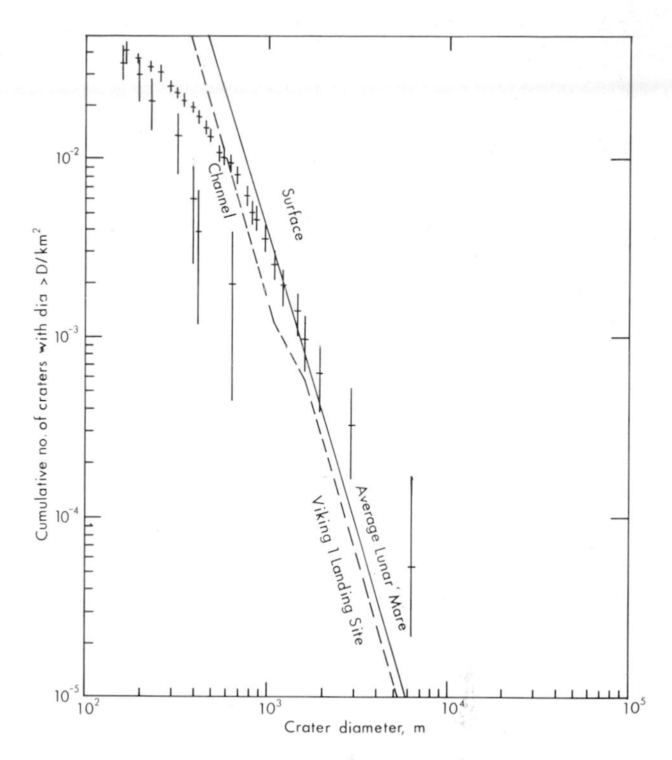

Fig. 4*d*. Crater size-frequency distribution curves based on photos 22A56, 22A76, 22A77, and 22A78 of the prechannel and channel surfaces showing the relatively younger age of the channel. The relatively steep slope of the crater curve for the old surface suggests that degradation is less than that at the other broad channels (Figures 5*d*, 6*d*, and 7*b*). The apparent age of the channel is younger than that of the Viking 1 landing site, but the area counted is not statistically significant.

Fig. 4*c*. Viking photograph (22A76) showing part of the branched headwaters of Vedra Valles. The channel system drains the eastern flank of Lunae Planum. Note that the channel has cut the degraded rim of a crater 14 km in diameter.

total time needed to form each channel must have been relatively short; limited and sporadic rainfall is thus inferred. Only a few of the dendritic channel networks have been photographed by Viking to date, but crater density determinations have been obtained for several of these areas; one (Figure 9*d*) is shown here. The crater densities obtained for the surfaces these channels cut indicate that the surfaces are very old. The channels are clearly younger than the surface, but a crater count for the channels themselves cannot be obtained because of the restricted channel width and area. The crater size-frequency curve for the surface cut by the fluvial channel network on the flank of Alba Patera (Figures 10*c* and 10*d*) yields a much younger age than do the curves obtained for areas in the Martian highlands. At least two ages for dendritic channel formation are proposed, based on these counts.

Many channel networks in the southern highlands that were recognized in Mariner 9 pictures were assigned informal relative ages, based on measurements of the degree of degradation of the channel walls ('degradation index'). Some channels have very irregular edges; others have smooth margins. One hypothesis for this variation in channel morphology is that the irregularity of the walls, caused by cratering and slumping, is caused by an increase in age, the youngest channels having the smoothest edges. However, a similar effect would be produced by variations in physical properties or atmospheric conditions due to differences in elevation of the terrain: less cohesive material would produce irregular channels and highly cohesive material would produce smooth margins, even though the

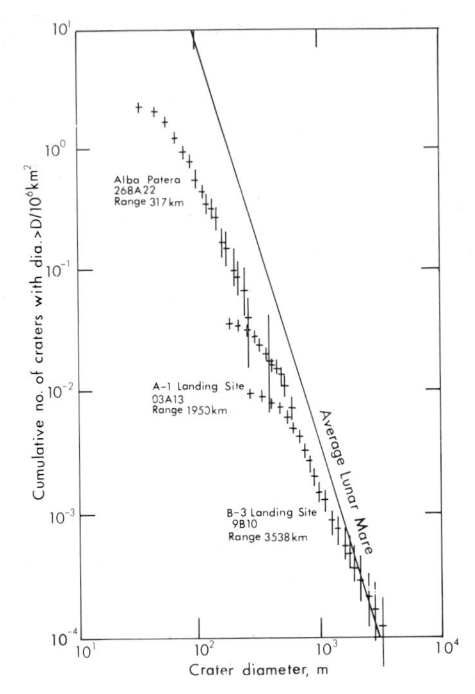

Fig. 5*a*. Crater size-frequency curves showing the effect of resolution. The slant range is indicated and varies from 3830 to 317 km. The rollover in the crater curves shows the true identification resolution, which is 5–6 times as great as the pixel resolution. The pixel resolution varies from 400 m at the B-3 site, to 40 m at A-1, to 8 m at Alba Patera.

Fig. 5*c*. Photograph (83A37) of Hydaspis Chaos and Tiu Vallis extending north from it. Hydaspis Chaos is an elongate area of collapsed terrain, nearly 100 km wide, where permafrost ice may have melted and the water escaped, modifying the valley floor.

channels might have been formed at the same time. Similarly, channels at a lower elevation, with the corresponding higher atmospheric pressures, would be subject to greater erosion. We are therefore concluding that variations in age of the channels and surfaces are best measured by variations in crater densities.

The small-scale dendritic channels may have formed during an interglacial period when warmer climatic conditions on the planet resulted in a denser atmosphere; under such conditions, rain could form, fall, drain, and dissect the ancient terrain. Some workers [*McElroy et al.*, 1976; *Owen and Biemann*, 1976] have postulated that this could have happened only during the early history of Mars, when the atmosphere was denser. The younger ages determined for some channel networks suggest that the history of the Martian atmosphere and climate may be too complex to have been the result of a monotonic pressure decrease since its formation. Warm periods may have been caused by variations in solar output [*Hartmann*, 1974], variations in obliquity of the planet [*Ward*, 1973, 1974], or critical variations in the CO_2 and other gas abundances [*Sagan et al.*, 1973]. Additional photography and crater counts may allow

Fig. 5*b*. Photograph (268A22) of an area near Alba Patera taken at a slant range of 317 km with a pixel resolution of 8 m. Degraded microchannels are shown; small craters down to the resolution limit are common.

Fig. 5*d*. Photograph (83A38) of Tiu Vallis. In this area the channel is 30 km wide.

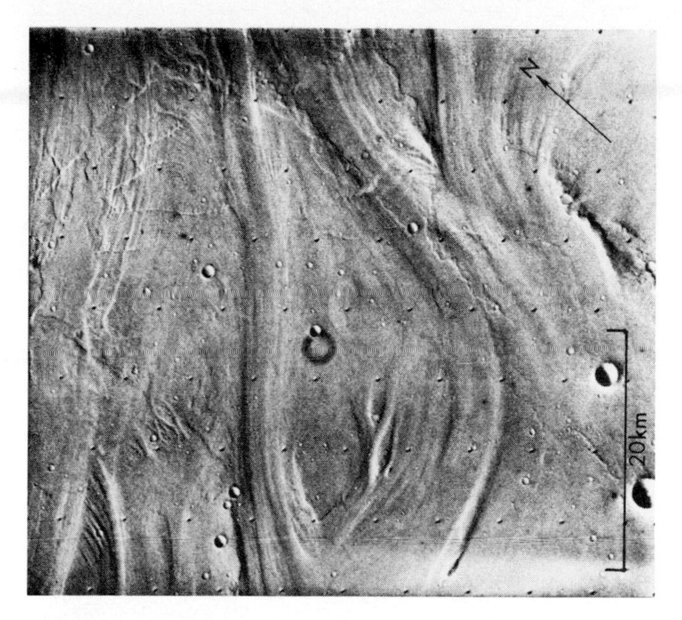

Fig. 6*a*. Photograph (20A56) showing braided channel pattern formed by water erosion at the mouth of Maja Vallis. This channel rises in Lunae Planum, spreads out in Chryse Planitia, and runs east, disappearing before it reaches the area of the Viking 1 landing site.

us to determine whether periods of intensive volcanic eruptions and channel formation coincide.

PERSISTENCE OF WATER ON THE MARTIAN SURFACE

In this section we will first consider a possible method by which running water might exist on the Martian surface under present atmospheric conditions, and then we will present the evidence given to date that more favorable conditions existed for surface running water during various episodes of Martian history.

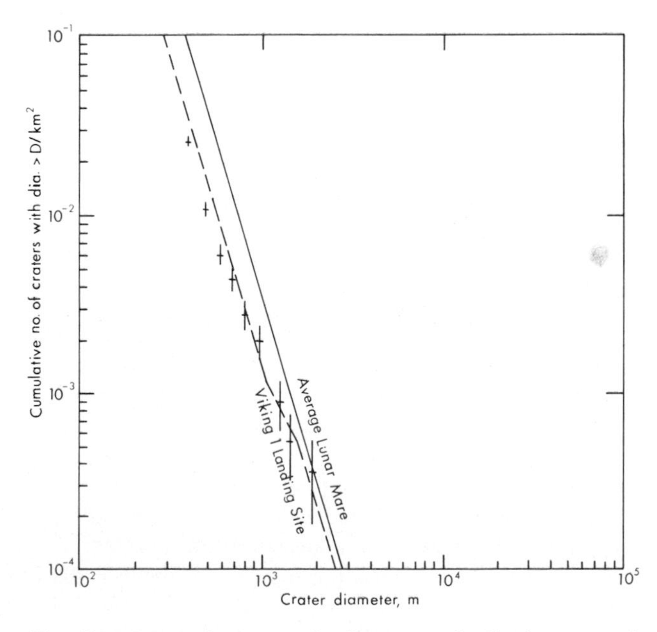

Fig. 6*b*. Composite crater size-frequency distribution curve for photographs 20A55, 20A56, 20A58, 44A43, and 44A44 of the Maja channel; the similarity of the curve to that for the Viking 1 landing site implies a similar age. It is also similar to the curve at Kasei Vallis.

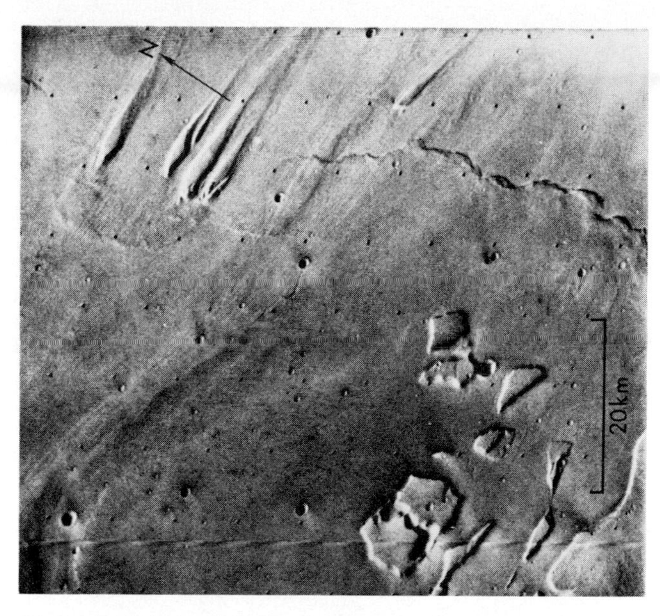

Fig. 6*c*. Photograph (20A26) of mouth of Kasei Vallis showing grooves on the floor of the channel and streamlined islands, which were probably shaped by flowing water. The southern shore of the channel is seen in the lower part of the picture. A mare ridge is crossed by the channel. Abundant small impact craters pepper the floor of the channel.

Surface Water Under Present Martian Conditions

Wallace and Sagan [1977] have calculated the rate of evaporation for water and ice under present Martian atmospheric conditions. They showed that although liquid water would evaporate very rapidly, evaporative cooling would cause a layer of ice (1-m maximum thickness) to form on top of the water; further evaporation would thus be reduced to a level of about 10^{-6} g cm^2 s. It is difficult to envision subice flow as the mechanism of formation for any of the three types of Martian

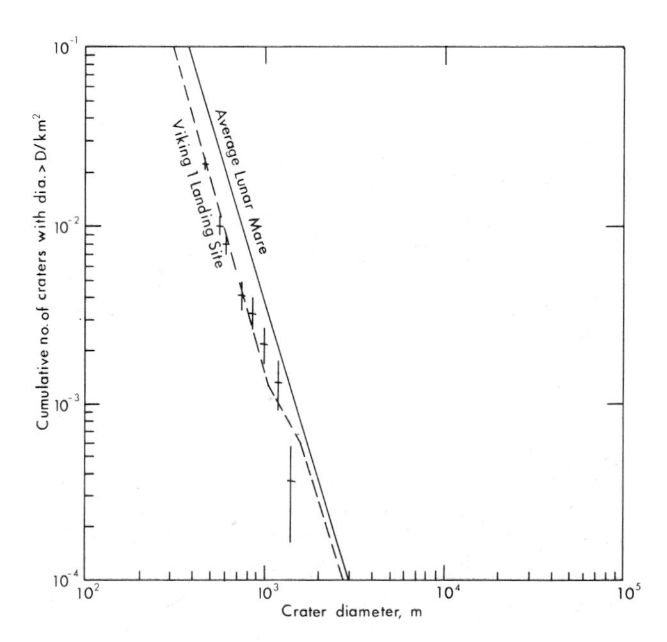

Fig. 6*d*. Crater size-frequency curve of the Kasei Vallis channel floor. The curve follows that of the surface at the Viking 1 landing site, southeast of this area.

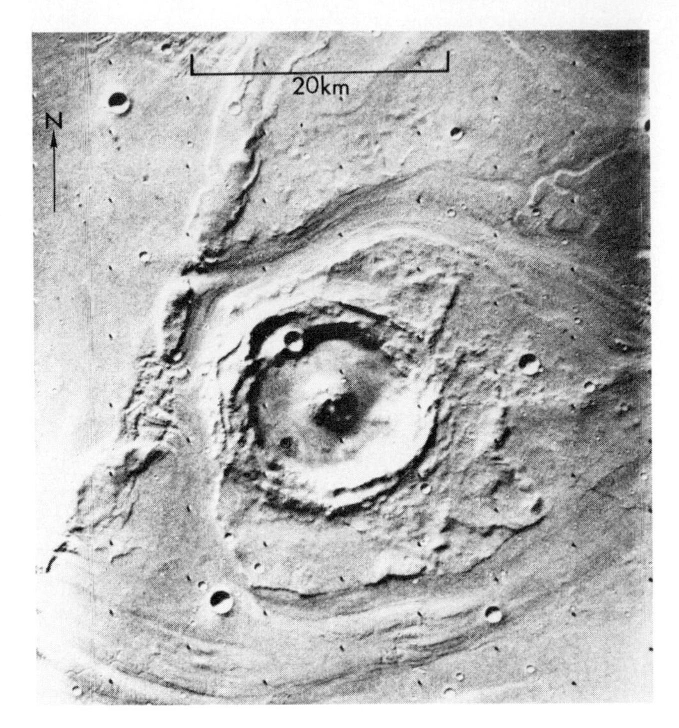

Fig. 7*a*. Photograph (20A62) of Maja Vallis where it enters Chryse Planitia, 150 km southwest of the Viking 1 landing site. Terrain slopes to the right (east). Water ponded west of the mare ridge, and then cut gaps as it overflowed the low points on the ridges; similar terrestrial features are seen in the channeled scabland area in Washington state. The impact crater in the center is 15 km in diameter; the gap is about 2 km wide. After crossing the ridge the water spread widely and cut many grooves in the channel floor. The high-water mark can be seen clearly where it cuts the ejecta blanket on both sides of the large crater. The channel erodes the ejecta blanket of a large crater and, in turn, is pockmarked by younger craters.

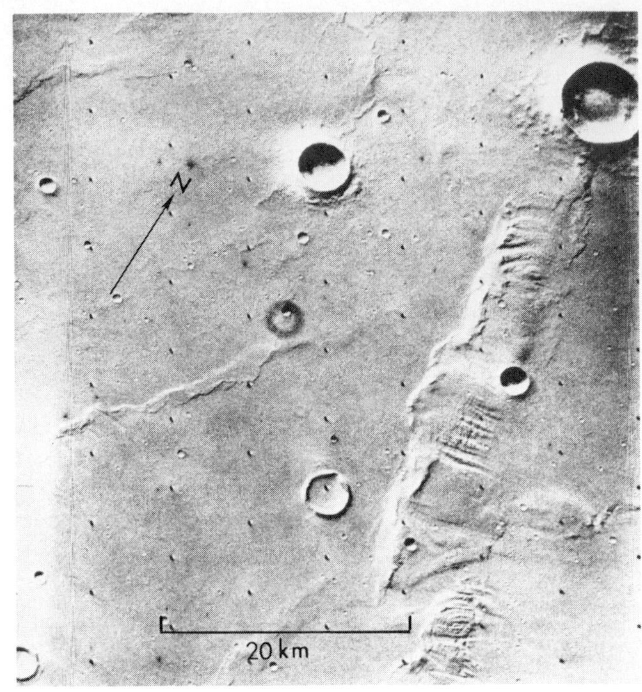

Fig. 7*b*. Photograph (20A40) of Maja Vallis in Chryse Planitia, 130 km west southwest of the Viking 1 landing site. The mare ridges are part of the Xanthe Dorsa system. The largest impact crater is 7 km in diameter and displays a hummocky ejecta blanket. Two smaller craters 2.5–3.5 km in diameter have been flooded by younger lavas. The mare ridges have been incised on the downhill side; grooves 200–400 m across apparently were cut where the water accelerated in crossing the ridge, like a dam spillway; cavitation may have helped cut the grooves. Obscure signs of water erosion are present 80 km east of this area. Perhaps higher-resolution pictures at lower sun elevation angles could bring out more subtle erosion features on the channel floor.

channels described here; in particular, this mechanism would not seem to apply to the broad channels. In order to maintain an ice layer formed above streams with large volumes of flow, such as Ares and Tiu/Simud, smooth, laminar flow conditions would be required rather than the turbulent flow that would result from the calculated discharge rates proposed in this report. Laminar flow might occur if the restricted channel walls were iced over to form large-scale ice 'tubes,' similar to conventional lava tubes. The initial turbulence of the flood events might create, by rapid evaporative cooling, an ice crystal fog above the flowing water and the overlying ice cover; this phenomenon is seen above arctic ice leads and in terrestrial subglacial streams. It is clear that the water and its protective ice cover could not have been in contact; continued disruption of the surface ice layer by the turbulence of the flowing water would have caused freezing of the entire water mass due to evaporative cooling and freezing of the water exposed when the ice cover was breached by the turbulent flow of the stream. When the channel flow reached the northern lowland plains, the water might have spread laterally in distributaries, become thinned, and finally frozen (the edges of the flows would form natural ice levees) and subsequently sublimated.

Moisture could also be added to the atmosphere by creating fog banks of ice crystals above the floods (S. Hess and C. Leovy, personal communication, 1977), which if blown southward against the highlands might precipitate as snow; it is difficult to envision transport of the moisture in an unfrozen state so that rain would fall. The fallen snow might later melt and erode the channel network if it were first covered by a layer of volcanic or eolian dust that would darken it and cause melting. Only a small part of the water involved in one of the great floods would have to be evaporated and precipitated during the warmest part of each day to engender enough flow to carve the small channels.

A more logical, less complicated, and thus more appealing mechanism for forming the Martian channels of intermediate and small size is to assume that temperature and pressure conditions of the Martian atmosphere in the past were such that liquid water could flow uninhibited for at least short periods of time. What, then, is the evidence that varying atmospheric and thermal conditions have existed in the past?

Surface Water Under More Favorable Martian Conditions in the Past

McElroy et al. [1976] have reported that the observed ratio of ^{15}N to ^{14}N in the Martian upper atmosphere suggests that the partial pressure of this gas (presently 2.5% of the Martian atmosphere) may have reached 2–30 mbars. *Nier et al.* [1976] suggested that the initial abundance of oxygen, present as CO_2 or H_2O, must have been equivalent to an interchangeable atmospheric pressure of at least 2 bars in order to have inhibited escape-related enrichment of ^{18}O. *Pollack* [1977], using equations for greenhouse heating, has calculated that a Martian atmospheric pressure in the range of 1–2 bars (CO_2 plus a trace of water vapor) is required to reach atmospheric temper-

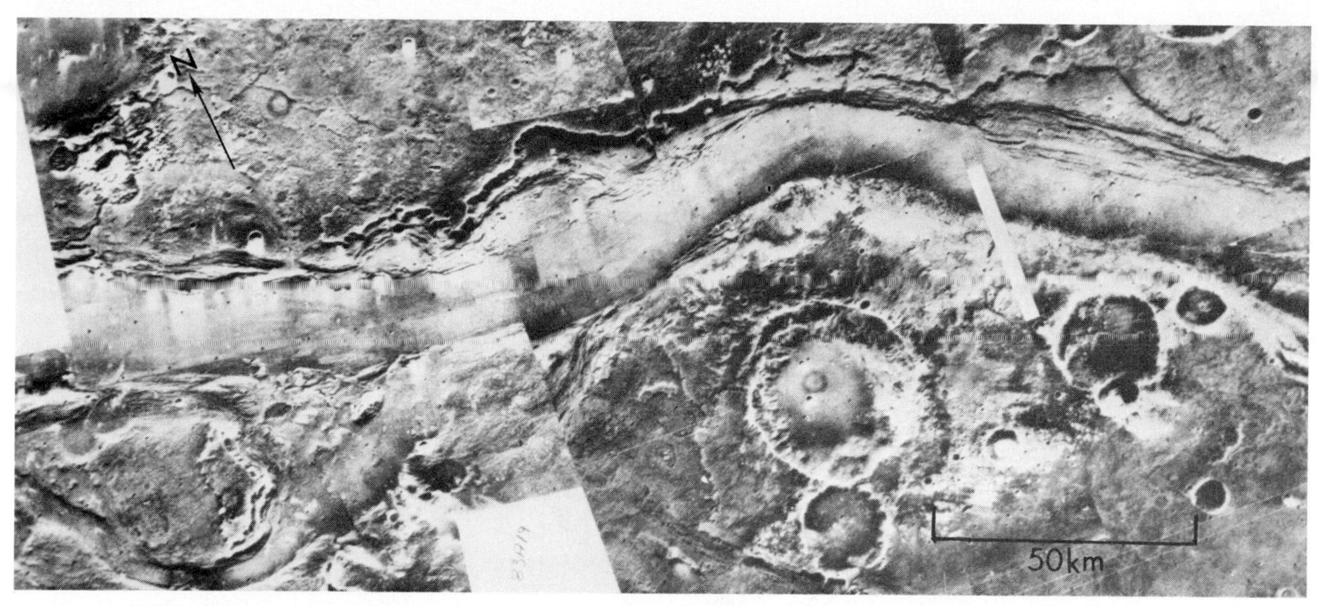

Fig. 7*c*. Photomosaic of part of Ares Vallis, where it is incised in the heavily cratered upland. The channel is 25 km wide and about a kilometer deep; stereoscopic photographs that would provide more accurate measurements are not yet available. Four to eight layers, probably lava flows, are exposed in the walls of the canyon where they have been eroded by flowing water and mass wasting processes. Abundant impact craters up to 50 km in diameter mark the upland; the younger channel floor is sparsely cratered.

atures above 273°K and permit liquid water to exist on the surface for at least short periods of time.

The large amounts of CO_2 reported from initial Viking results must represent the total amount of carbon gases outgassed over the lifetime of Mars [*Pollack*, 1977]. *Fanale* [1977] has suggested that if the total amount of CO_2 in the atmosphere-regolith system has remained constant and the temperature of any regolith column (or of Mars as a whole) has increased by 40°C over a period of $\sim 10^6$ years or more, equilibrium pressures of only 25–35 mbars are likely to result. It is clear that additional feedback mechanisms, such as global heat transport, greenhouse effects, additional atmospheric water vapor, and presence of reducing compounds in the atmosphere, are required. *Pollack* [1977] has shown that liquid water would flow on Mars under substantially lower atmospheric pressure conditions (~ 150 mbars) if the Martian atmosphere were enriched in reducing compounds such as ammonia and methane. Pollack points out that this type of atmosphere, formed during the early history of earth, provided the chemical basis from which life originated. The atmospheric reducing agents probably resulted from extensive volcanism on both planets.

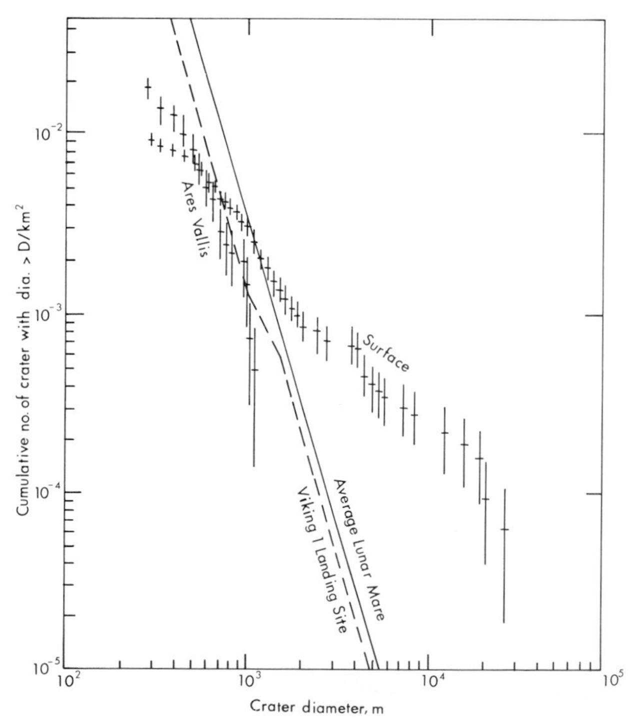

Fig. 7*d*. Crater size-frequency distribution curves for the old cratered upland surface adjacent to Ares Vallis and its channel floor. The surface crater curve has a low slope, indicating a steady degradation of essentially the entire crater population. The channel curve closely follows the Viking 1 landing site curve and probably has a similar age.

We have concluded from photogeologic observations that both the dendritic and the large channel system may have formed during relatively short periods of time at several different times; the larger channel systems have few tributary streams, and the smaller, older dendritic channels that are found primarily in heavily cratered or sloping terrain are not deeply incised. Flowing water on the Martian surface caused by the above or other conditions may have eventually frozen and then sublimated after a short period of time, may have percolated into the ground, or may have evaporated rapidly under conditions of elevated wind velocities and low water vapor pressures in the atmosphere. Depending on the initial temperature of the released water the flowing streams may also have melted the near-surface permafrost layers, so that additional volumes of meltwater were dispersed on the channel floors.

In the following discussion of Martian channel formation by flowing water, we will assume that one or more of the above criteria for increased atmospheric pressure and temperature were achieved during short periods of Martian history. The rate of evaporation of water under such conditions can be at

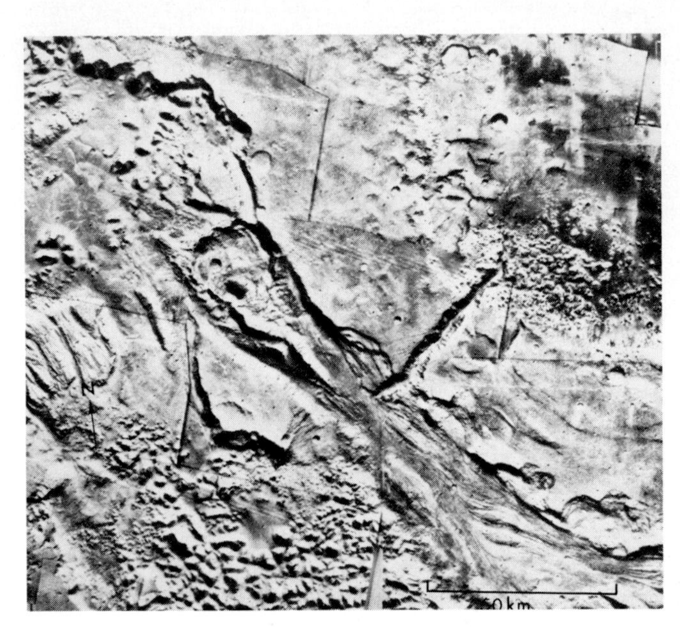

Fig. 8*a*. Photomosaic of Tiu Vallis near its source area, Hydaspis Chaos. Relative youthfulness of this channel is inferred from the detailed structure seen within the channel.

Fig. 8*c*. Photograph (9B54) of Hrad Vallis area (~350 km) south of the Viking 2 landing site. The tributaries head in box canyons that probably formed by spring sapping of nonresistant rock materials by meltwater derived from subsurface ice. Melting may have been caused by geothermal heating or climatic warming.

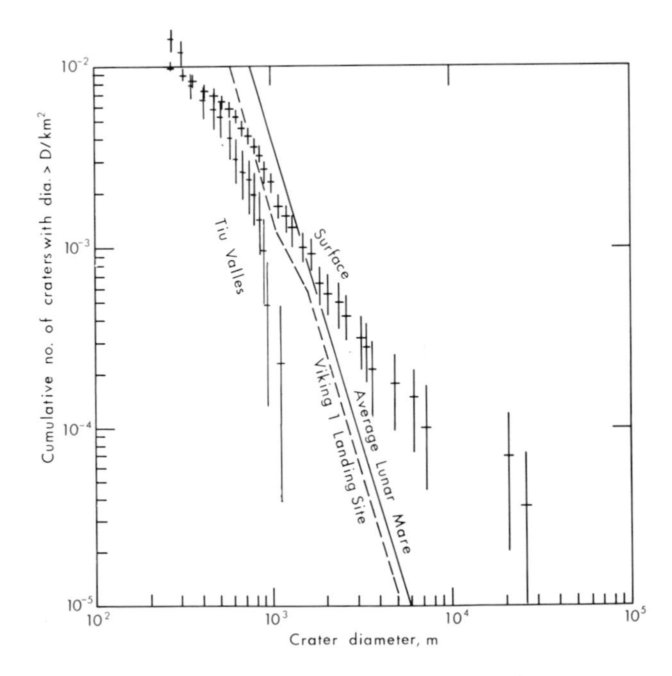

Fig. 8*b*. Crater size-frequency distribution curves for the Tiu channel and the cratered upland surface it cuts. Crater curve of the cratered upland surface is similar to that at Ares Vallis. The position of the channel curve, well below that of the Viking 1 landing site curve, may indicate that the feature is younger than any of the other large channels on Mars.

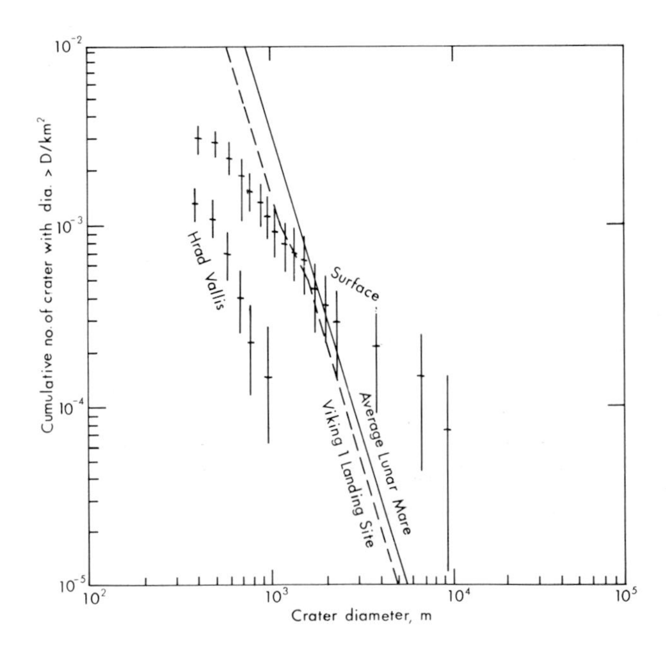

Fig. 8*d*. Crater size-frequency distribution curves of Hrad Vallis and adjacent surface based on counts of frames 9B22, 9B43, 9B45, 9B47, 9B48, 9B50, 9B52, 9B54, and 9B56. The calculated age for the heavily cratered surface is about 3.5 Gy. There are fewer craters on the channel; the area has been mantled by windblown debris that subsequently has been partially stripped. Position of the crater curve may be biased by this blanket of younger sediments.

Fig. 9*a*. Photomosaic of an unnamed channel in Hellas. The sinuous channel coming in from the north cuts through only one layer. The deeper east-west part of the channel heads in a box canyon where at least three layers are exposed. Many small tributaries occur along the channel; most head in box canyons, indicating spring sapping. However, small dendritic networks at the headwaters of two tributaries may indicate rainfall. The channel flows westward into Hellas Planitia. A rock layer etched by the wind crops out on the plain shown; all of these layers may be lava flows. The debris apron around the hill and ridge in the center of the mosaic was probably caused by mass wasting. (Photographs 97A48–97A54 and 97A60–97A68; 39°–48°S, 260°–270°W.)

least estimated using a basic equation [*Dalton*, 1798–1802] of the form

$$E = C(P_w - P_a) \quad (1)$$

where

E rate of evaporation;
P_w vapor pressure just above the water surface;
P_a vapor pressure in the lower atmosphere;
C coefficient dependent on barometric pressure, wind velocity, and other variables.

Dalton's original equation was modified by many workers, including *Fitzgerald* [1886], *Carpenter* [1891], and *Magin and Randall* [1960], by introducing an empirical calculation of the constant and the actual wind velocity:

$$E = (0.6)(W)(P_{H_2O} - P_{atm}) \quad (2)$$

where

E evaporation rate, mm/day;
0.6 empirically derived constant;
W wind velocity, m/s;
P_{H_2O} vapor pressure immediately in contact with water, mm Hg;
P_{atm} vapor pressure of the atmosphere near the surface.

Figure 11*a* shows the relation between wind velocity and water temperature (the latter being the controlling factor for P_{H_2O}) for an arbitrarily high atmospheric vapor pressure of 5 mm Hg and on the assumption of a near-surface atmospheric temperature of slightly over 0°C (273°K). Note that for water temperature ranging from 5°C to 80°C (due to volcanic heating at depth) and wind velocities between 2 and 20 m/s, evaporation rates from 2 mm/day to over 4 m/day can be achieved. The magnitude of the atmospheric vapor pressure P_{atm} is insignificant as long as it does not approach the P_{H_2O} value, indicating a saturated condition. These preliminary calculations do not address the rather important effects of evaporative cooling on the water and the presence of solute or sediment in the water that might alter its freezing temperature or evaporation rate.

Another problem that must be addressed is concerned with the dispersal of part of the floodwaters so that the channel

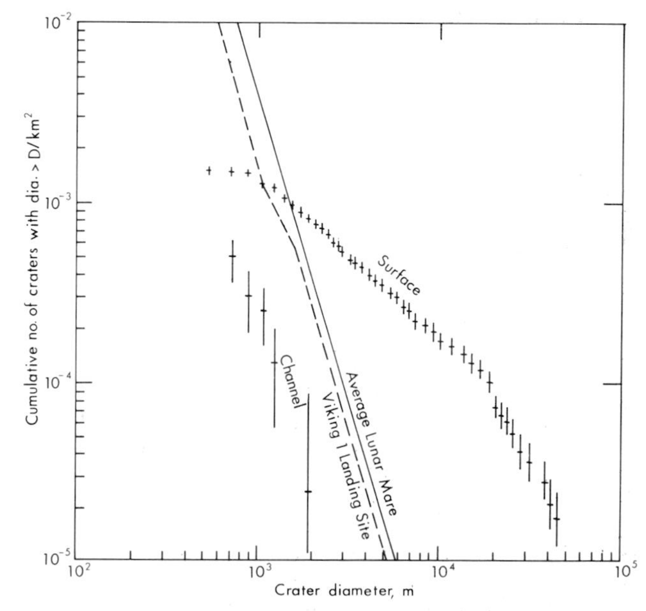

Fig. 9*b*. Crater size-frequency distribution curves for the floor of a channel east of Hellas Planitia and the older surface which it cuts. The curve for the surface has a low slope—possibly due to degradation of the smaller craters by wind erosion. The curve for the channel floor lies below the Viking 1 site curve, indicating that it may be much younger than the Bahram and Vedra valles. However, the channel floor may have been covered by subsequent windblown or other volcanic basalt deposits; if so, the curve shows the age of these deposits rather than the time of channel formation. (Photographs 97A48–97A54 and 97A60–97A68; 39°–48°S, 260°– 270°W.)

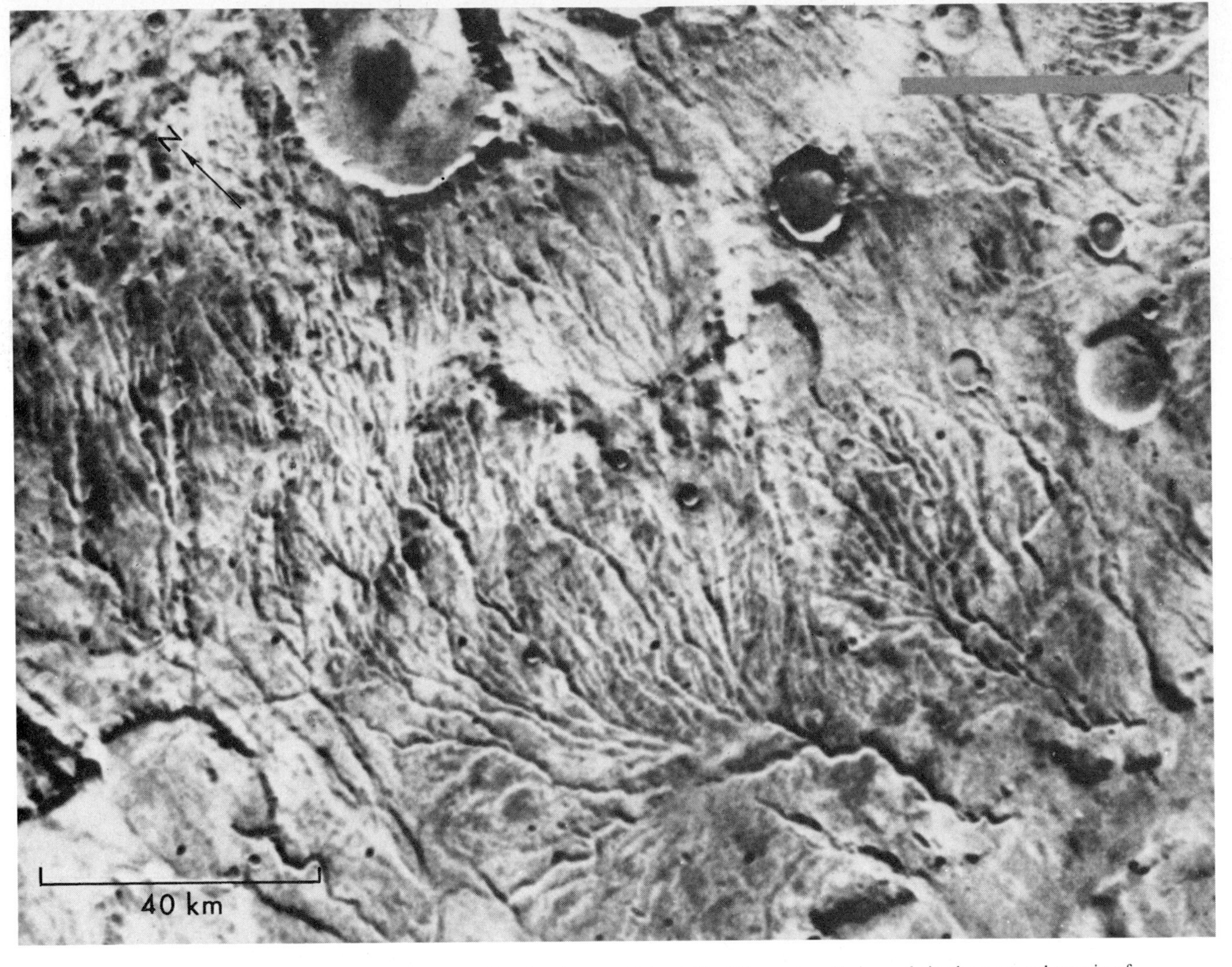

Fig. 9*c*. Part of Viking photograph (63A08) showing small dendritic ('filamentous') channels in the cratered terrain of the southern Martian highlands. Channels intersect and erode some large craters. The trunk stream of the channel system is deeply incised.

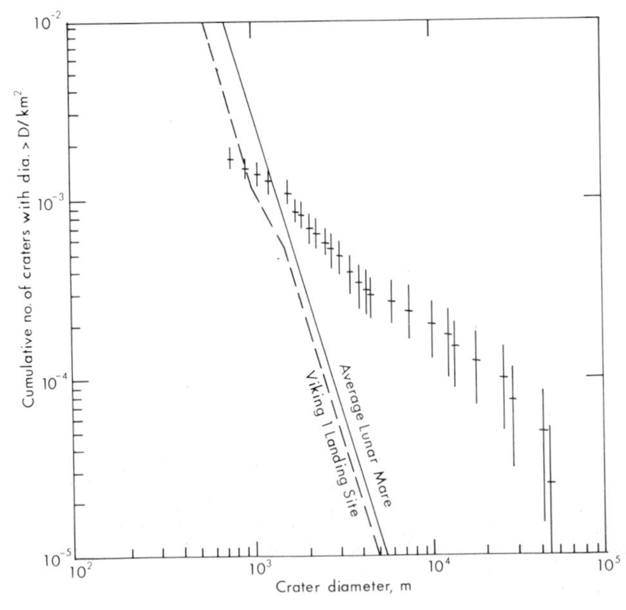

Fig. 9*d*. Crater size-frequency distribution curve of an old surface on which the dendritic channel was superposed. No crater counts were attempted for the channel because of its small size and the restricted number of postchannel craters. The low slope of the curve for the heavily cratered surface suggests degradation and aggradation of the surface; photogeologic interpretation confirms this interpretation. (Part of photograph 63A09; 43°S, 94°W.)

erodes the plains surface for only about 350 km beyond the end of the channel, as is the case with the larger Martian features. The peak discharge in the flood would last from 2 to 4 days, and the flood could continue for 2 weeks if it were similar to the Missoula flood [*Bretz*, 1923; *Malde*, 1968; *Baker*, 1973*a*]. If the atmosphere were only slightly above 0°C and the water in the channel were warm owing to the volcanic heat, 10–70% could be lost to the atmosphere with sufficient winds and the rest could be lost by percolation into the regolith or plains surface. Much of the water that spread thinly over the plains might have frozen temporarily, owing to increased area available for evaporative cooling, but could then be lost to the atmosphere by slow ice sublimation.

Channel Parameter Estimates

Preliminary estimates of the open channel flow in the Ares channel where it spreads out in Chryse Planitia (21°N, 31°–34°W) have been computed by Lawrence Mann and C. H. Bell (personal communication, 1977) of the U.S. Geological Survey, using the Manning formula [*Chow*, 1959] adjusted for Mars gravity. Ares channel measurements obtained photographically from Viking pictures by S. S. C. Wu, R. Jordan, and F. Schafer, also of the U.S. Geological Survey, indicate that the channel is 300 km wide and 100 m deep. The calculated slope of the channel floor is 10 m/km to the north, and the slope of the channel banks is ~20°. Boulders as large

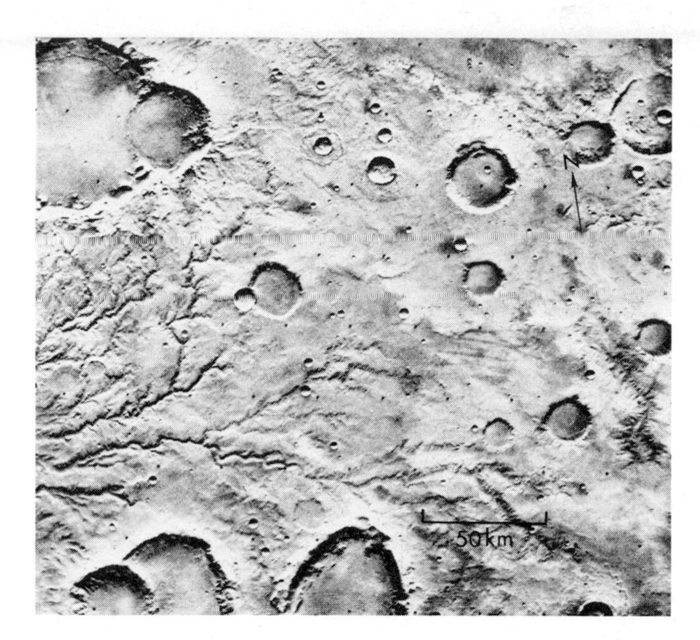

Fig. 10*a*. Photograph (84A47) showing dendritic channel network in southern highlands 2000 km northeast of Argyre Planitia. These channels are part of a network of streams that extend for more than 200 km; the channel network is almost as long as channels in the intermediate size range. The trunk channels are about 2 km wide.

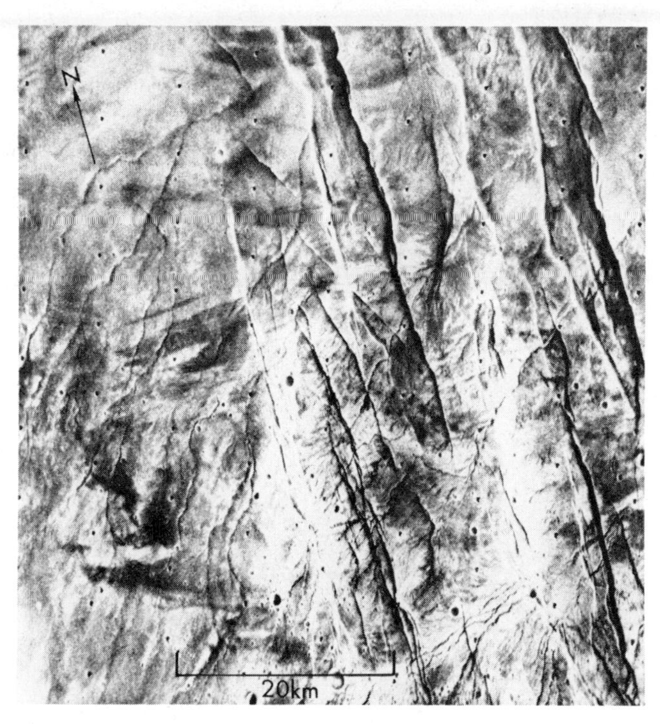

Fig. 10*c*. Photograph (4B57) of fluviatile channels southeast of Alba Patera. These form a dendritic channel network; many filamentous channels join the trunk stream. The headwater area of the channels does not begin as a box canyon or in collapsed terrain. The channels are offset by en echelon faults that are the most sharply defined, least modified ones on Mars; they are therefore thought to be quite young in age. These dendritic channels, at 44°N, 104°W, are the most northerly ones observed.

Fig. 10*b*. Photograph (84A43) of several channel networks; some are tributaries to an intermediate size sinuous channel that is nearly 5 km wide. This part of the channel is located about 1400 km northeast of Argyre Planitia in the southern highlands. Old craters ranging up to 50 km in diameter pockmark the surface. There may have been two channeling episodes in this area.

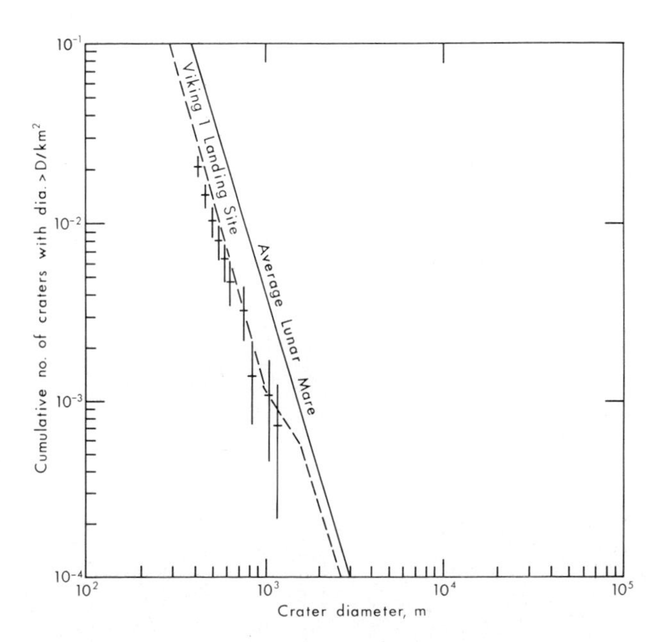

Fig. 10*d*. Crater size-frequency distribution curve for the northeast side of Alba Patera. The crater count curve for this volcanic surface coincides with that for the Viking 1 landing site, and this surface is probably the same age or slightly younger. These dendritic channels are superimposed on a younger surface than those that occur in the southern highlands where the dendritic channels cut ancient cratered terrain. (Photograph 4B57; 44°N, 104°W.)

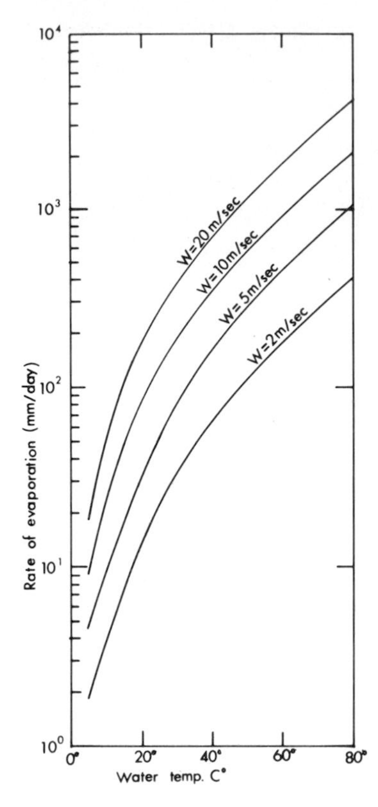

Fig. 11*a*. Graph relating water temperature and wind velocity to evaporation rate given an atmospheric vapor pressure of 5 mm Hg (see equation (2) in text).

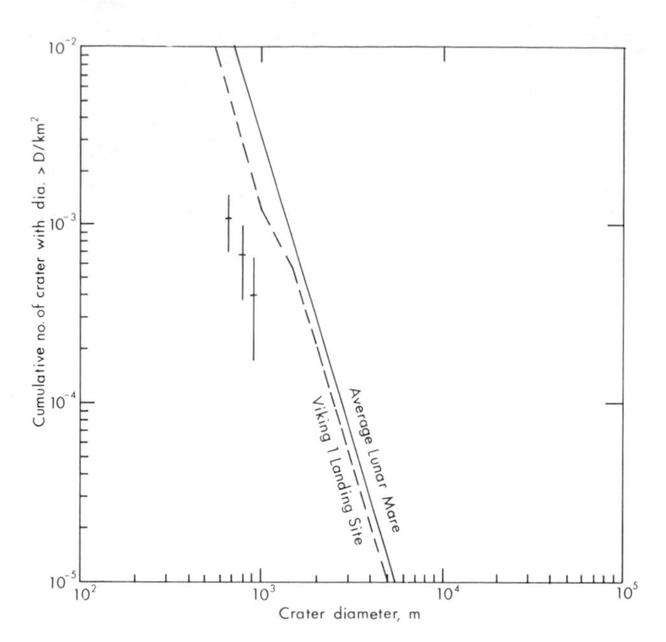

Fig. 11*c*. Crater size-frequency distribution curve for photograph (39A09) near the summit caldera of Arsia Mons. Few impact craters are visible; the lava channels and flows therefore are younger than those on Alba Patera.

Fig. 11*b*. Part of photograph (52A04) of lava channels near the summit caldera of Arsia Mons. Two types are shown: those that originate in source craters and become narrower and shallower down the shield slope and those with marginal natural levees formed when the lava freezes along the borders of the long, thin, narrow flow.

Fig. 11*d*. Photograph of the west flank of Alba Patera showing lava flows and many lava channels. Several channels, particularly those seen in the lower left corner of the picture, lie along the crests of the ridges. The ridges apparently were formed by the cooling flows in which the center top remained mobile longest and drained the crest of the flow. (Photograph 39A09; 42°N, 118°W.)

Fig. 12*a*. Viking photographs (7B60 and 7B62) of lava flows west of the summit of Alba Patera. The lava flows exhibit many lava channels with marginal natural levees. Similarities between these flows and terrestrial lava flows in Hawaii and the Galapagos Islands suggest that they are basaltic in composition. Many small impact craters are visible; the surface is little modified by wind or water.

as 2 m in diameter, like the boulder observed in the Viking 1 landing site [*Mutch et al.*, 1976], are inferred to lie on the channel floor. From these parameters the velocity is estimated to be about 3 m/s, and peak discharge is estimated to be 8.6 × 10^7 m^3/s. The velocity is comparable to derived velocities of 11–30 m/s and peak discharge of 2.1 × 10^7 m^3/s for the Pleistocene Missoula flood [*Baker*, 1971, 1973*a*, *b*] and the Bonneville flood [*Malde*, 1968]. Total flood volume for the Missoula flood is 7 × 10^{13} m^3/s, and for the Bonneville flood, 3 × 10^{12} m^3. Volume of release for recent Icelandic glacial bursts is 7 × 10^9 m^3/s; *Thorarinsson* [1957] quotes a peak discharge of 3–4 × 10^5 m/s for the Katla Jökulhlaup. The volume for Ares is 1.5 × 10^{13} m^3, a flood duration on Mars of 4 days with lower stages for about 2 weeks being assumed. These values appear to be comparable to those inferred for the terrestrial floods, but these estimates are generalized. There is considerable variation in the infrared durations of Pleistocene and recent floods.

We calculate the volume of water that could have been derived from chaotic terrain in the headwaters of Ares Valles to be 2.6 × 10^{14} m^3 and that for the Simud/Tiu valles to be 4.4 × 10^{14} m^3. Because stereoscopic measurements from Viking photographs are not yet available for the collapsed terrain, our calculated depth is based on Mariner 9 ultraviolet spectrometer and Goldstone radar measurements. Three ultraviolet spec-

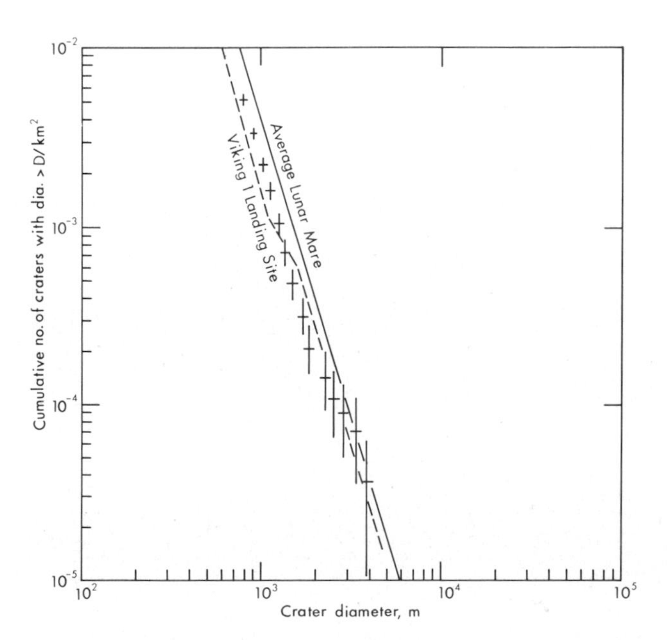

Fig. 12*b*. Crater size-frequency distribution curve of the volcanic surface west of the summit of Alba Patera. The curve is coincident with the one for the Viking 1 landing site; the two surfaces are probably similar in age. (Photographs 7B60 and 7B62; 43°–44°N, 108°–111°W.)

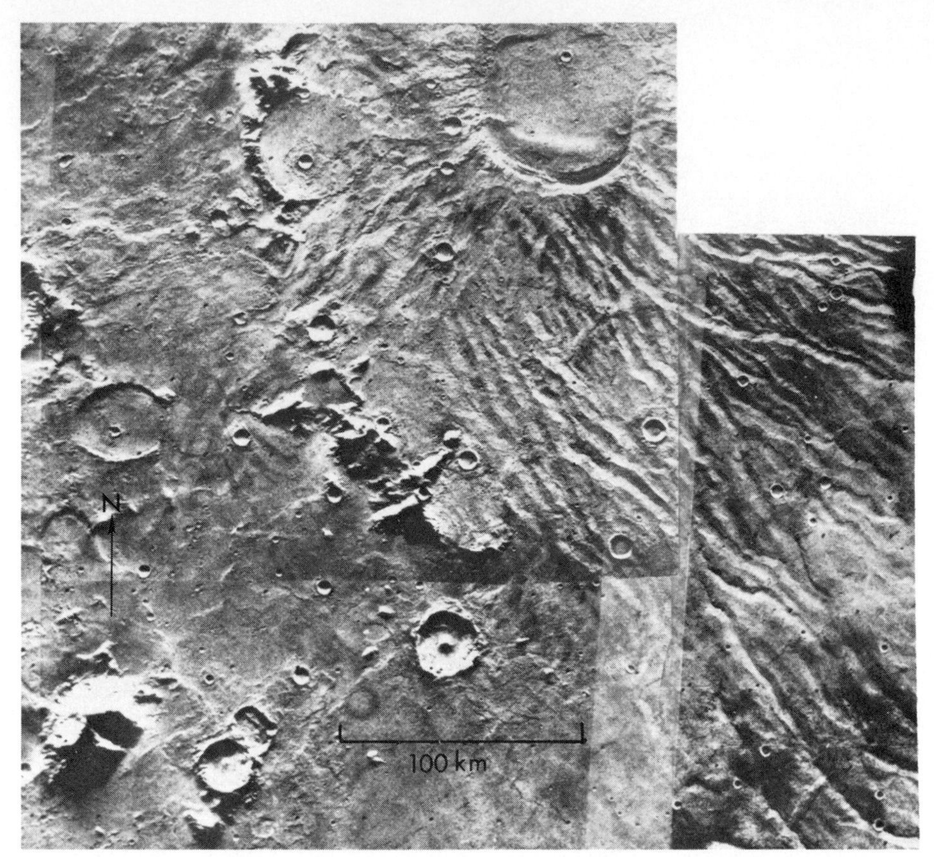

Fig. 12*c*. Photomosaic of Hadriaca Patera on the northeast rim of Hellas Planitia. This volcanic center has a central caldera 60 km in diameter and radiating lava flows and channels that are degraded and impact cratered. The western of two unnamed channels originates in the collapsed area in the upper right part of the picture. These are intermediate between the spring sapped box canyons, where most intermediate size sinuous channels start, and the chaotic terrain, where the large channels originate. The streams may be fed by water coming from melting permafrost. Because these channels originate in a volcanic area they may be activated by geothermal heating. Alternatively, they may be due to climatic warming. (Photographs 97A40 and 97A42; 28°–34°S, 265°–274°W.)

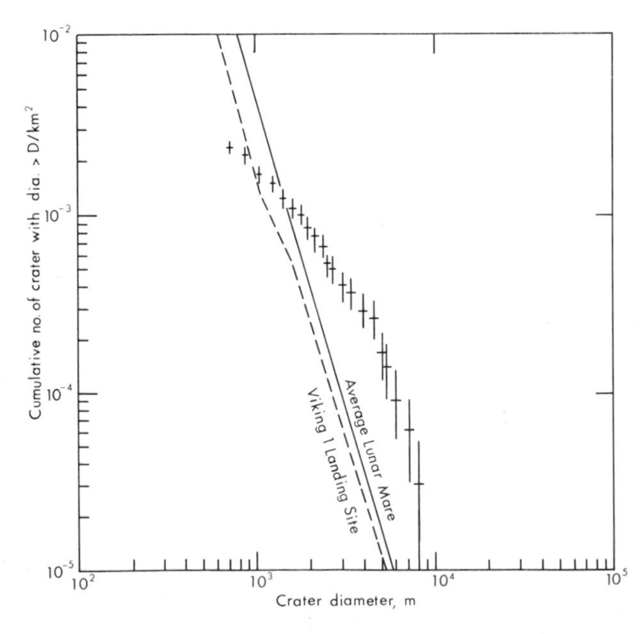

Fig. 12*d*. Crater size-frequency distribution curve of Hadriaca Patera (photograph 97A42). The crater densities on this volcanic center are much higher than are those on Alba Patera. This curve indicates an age of 3.5 Gy. A less-cratered fluviatile channel crosses the flows that emanate from Hadriaca Patera; it is therefore thought to be younger in age.

trometer transects of the collapsed terrain indicate a depth of about 1 km [*Barth and Hord*, 1971; *Hord et al*., 1974]. Two Goldstone radar (R. Goldstein, personal communication, 1973) traverses indicate depths of 1 km and 0.5 km, respectively. We therefore have used a 1.0-km nominal depth and assumed that the total maximum volume represents ice that was melted by the volcanic heat to form the water released in the floods. There are three separate areas of collapsed terrain in the headwaters of Ares Valles; collapse in these areas probably did not occur at the same time, and there may have been three separate episodes of flooding. It is apparent that the volume of water contained as permafrost was large enough to form floods on Mars comparable to terrestrial late Pleistocene and recent floods on the earth.

VOLCANIC CHANNELS

Three types of volcanic channels have been observed. The first type of channel originates in volcanic craters and becomes narrower and shallower as it flows downslope; it resembles the lava channels and collapsed lava tubes seen on the earth and the moon. Channels of this type are seen near the summit of Arsia Mons, a great shield volcano (Figure 11*b*). Both the gross shape of the volcanic shield and the digitate shape of the flows suggest that the flows may be basaltic in composition. The number of impact craters in Figure 11*b* is not sufficient to obtain a good size-frequency curve (Figure 11*c*) using the crater counting technique; however, the small number of impact craters clearly indicates a very young surface—perhaps

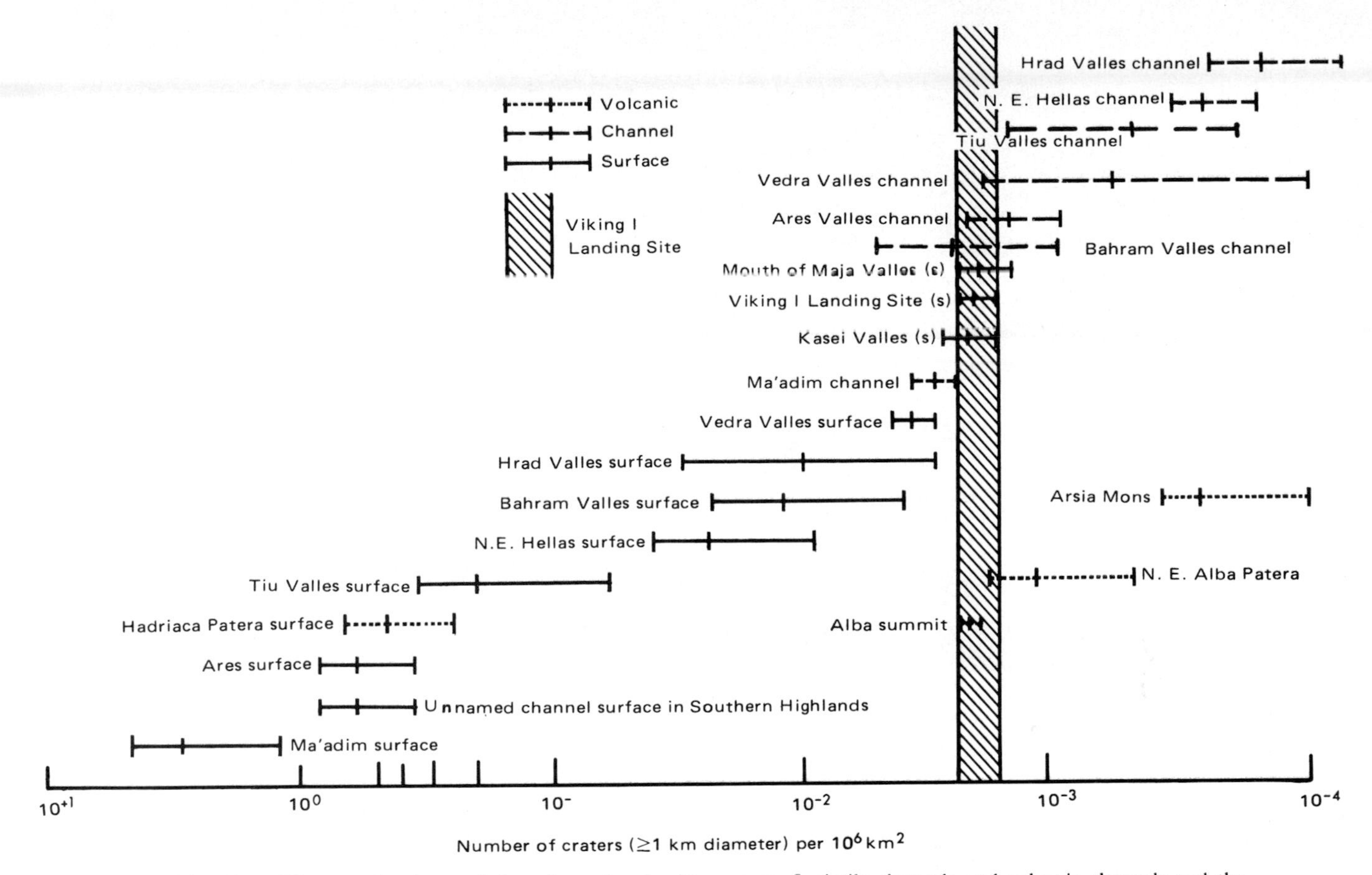

Fig. 13*a*. Diagram showing variations in crater densities among fluviatile channels and volcanic channels and the surfaces they cut. The wide variation in crater densities implies widely varying ages. The fluviatile channels have substantially lower crater densities than the surfaces into which they are cut. Both fluvial and volcanic channels vary greatly in age, but those plotted here are only a small subset of the Mars occurrences. The position of the Viking 1 landing site is plotted as a reference.

about 200 m.y. Examples of this channel type were also identified in Mariner 9 photographs of Elysium Mons [*Masursky*, 1973]. The second type of volcanic channel is characterized by natural levees that occur along the edges of the channels. This type was observed in Mariner 9 pictures of Olympus Mons [*Carr*, 1973] and is also seen in Figure 11*b*. The third type of lava channel occurs on the crests of ridges that were built up by the cooling lava. This type also was recognized in Mariner photographs of Olympus Mons and in Viking photographs of the western flank of Alba Patera (Figure 11*d*).

Photographs of Hadriaca Patera (Figure 12*a*) show a volcanic center with radiating flows and channels that are heavily impact cratered (Figure 12*b*). Relative age determinations indicate that this feature may be very old. The crater densities for the channels on Alba Patera are lower, and those for Arsia Mons are even lower, implying very different ages for these volcanic units.

The channels that have natural levees and those that lie along the crests of ridges are clearly volcanic in origin; only in the complex channels with distributaries seen near Alba (Figure 10*d*) is there a possibility of confusion with the fluviatile channels.

FUTURE WORK

Viking orbital photography during the nominal mission covers only a small part of the area where channels have been incised. Stereoscopic photographs that allow measurements to be made of channel cross sections and slopes cover only a few of these channel areas. In a future report we will discuss additional channels and channel measurements elucidated by extended mission photography. In addition, the increased photographic resolution resulting from the lowered periapsis (from 1500 km to 300 km) should clarify the details of the tributary network source areas and the distributaries where the channels spread out in the plains. Finally, we hope to construct a new Martian crater flux curve so that we can estimate more accurately the relative ages of the fluviatile and lava channels, using the crater counting technique described above. These counts should demonstrate whether the water and lava channels are related in time and genesis or whether water episodes and volcanism have independent histories, one controlled by internal dynamics, the other possibly by varying solar output.

SUMMARY

Crater densities, both for fluvial channel surfaces and lava channels and for the surfaces incised by the channels, are summarized in Figure 13*a*. The wide variation in crater densities shown by the Martian channels discussed in this paper strongly implies widely differing ages for both fluviatile and lava channels. Plots of fluvial channel occurrences based on Mariner 9 photographs are latitude dependent and peak south of the equator [*Sagan et al.*, 1973; G. Granata and H. Masursky, personal communication, 1974]. It is unclear

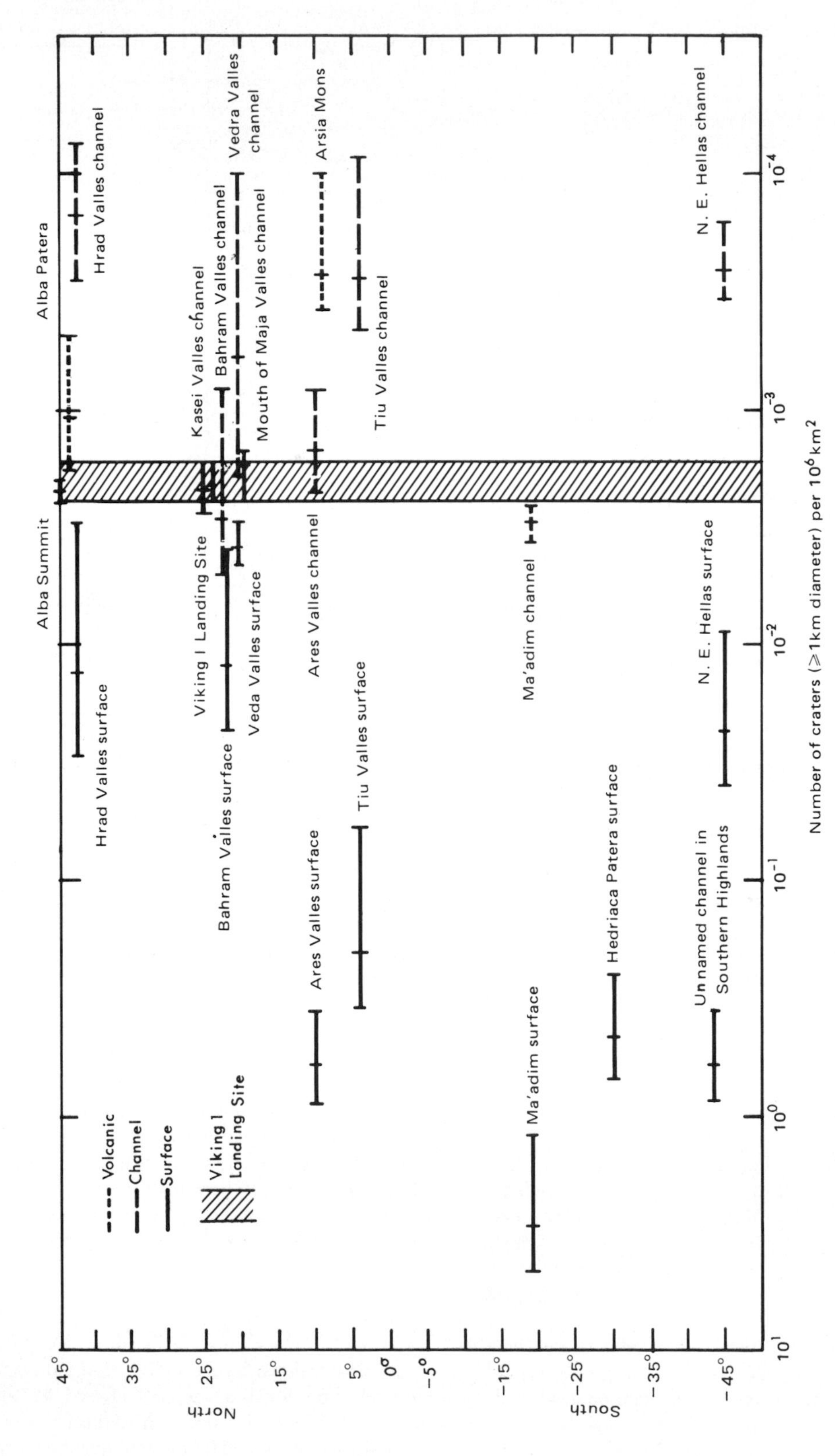

Fig. 13*b*. Crater densities plotted against latitude of occurrence. The data set is too small to show whether channels were never formed at higher latitudes or whether they are buried in the higher latitudes by younger eolian deposits.

whether this distribution is a true reflection of original channel locations or whether channels north and south in the equatorial region have been eroded and/or covered by eolian material [*Soderblom et al.*, 1974]. In Figure 13*b* the channels discussed in this paper are plotted by latitude. No distributional pattern is recognized, but our data set is too small to confirm or refute the bias toward channel formation or location in the equatorial latitudes that was shown in the Mariner photographs. Additional Viking photography should help clarify this issue.

Even the small number of channels imaged to date show that a considerable variety of crater densities are observed, implying a variety of ages.

Using the *Soderblom et al.* [1974] curve and age dating technique, it appears that the summit of Arsia Mons and the Hrad Vallis channel surface may be about 0.5 Gy old. Ares Vallis and the Viking 1 landing site may be 1–2.5 Gy old. The heavily cratered upland surfaces, like Hadriaca Patera and the surface cut by the southern highland channels, may be 3.5–4.5 Gy old. If these latter counts, where channel networks have modified the surface considerably, are accurate representations of channel ages, as well as of surface ages, then we have documented channel ages (by crater counts) that span the entire time interval that can be analyzed. The important results of this study are that the greatly varying crater densities imply variations in time for channel formation and that water and volcanic episodes must have spanned much of the decipherable Mars history.

Acknowledgments. Thanks are due to Lawrence Soderblom, David Scott, and Victor Baker for helpful review of the manuscript. The work was carried out under NASA contract WO-8259.

REFERENCES

Baker, V. R., Paleohydrology of catastrophic Pleistocene flooding in eastern Washington, *Geol. Soc. Amer. Abstr. Programs, 3,* 497, 1971.

Baker, V. R., Paleohydrology and sedimentology of Lake Missoula flooding in eastern Washington, *Geol. Soc. Amer. Spec. Pap., 144,* 79, 1973*a*.

Baker, V. R., Erosional forms and processes for the catastrophic Pleistocene Missoula floods in eastern Washington, in *Fluvial Geomorphology, A Proceedings Volume of the 4th Annual Geomorphology Symposium Series,* edited by M. Morisawa, State Univ. of N. Y., Binghamton, 1973*b*.

Baker, V. R., Viking—Slashing at the Martian scabland problem (abstract), Reports of the Planetary Geology Program, 1976–1977, *NASA Tech. Memo., X-3511,* 169, 1977.

Baker, V. R., and D. J. Milton, Erosion by catastrophic floods on Mars and earth, *Icarus, 23,* 27–41, 1974.

Baldwin, R. B., Lunar crater counts, *Astron. J., 69,* 377–392, 1964.

Barth, C. A., and C. W. Hord, Mariner ultraviolet spectrometer—Topography and polar cap, *Science, 173,* 197–201, 1971.

Bretz, J. H., The channeled scablands of the Columbia Plateau, *J. Geol., 31,* 617–649, 1923.

Carpenter, L. D., Section of meteorology and irrigation engineering, *Colo. Agr. Exp. Sta. Annu. Rep., 4,* 1891.

Carr, M. H., Volcanism on Mars, *J. Geophys. Res., 78,* 4049–4062, 1973.

Carr, M. H., and G. G. Schaber, Mars permafrost features, *J. Geophys. Res., 82,* this issue, 1977.

Carr, M. H., H. Masursky, W. A. Baum, K. R. Blasius, G. A. Briggs, J. A. Cutts, T. Duxbury, R. Greeley, J. E. Guest, B. A. Smith, L. A. Soderblom, J. Veverka, and J. B. Wellman, Preliminary results from Viking orbiter imaging experiment, *Science, 193,* 766–776, 1976.

Chapman, C. R., Asteroids as meteorite parent bodies: An astronomical perspective, *Geochim. Cosmochim. Acta, 40,* 701–719, 1976.

Chow, V. T., *Open-Channel Hydraulics,* p. 680, McGraw-Hill, New York, 1959.

Cutts, J. A., The origin of the channels of the Chryse basin region, paper presented at the 7th Lunar Science Conference, Lunar Sci. Inst., Houston, Tex., March 15, 1977.

Cutts, J. A., K. Blasius, and K. W. Farrell, Mars: New data on Chryse basin land forms (abstract), *Bull. Amer. Astron. Soc., 8,* 480, 1976.

Dalton, J., Experimental essay on the constitution of mixed gases; on the force of steam or vapor from water and other liquids in different temperatures, both in a Torricellian vacuum and in air; on evaporation; and on the expansion of gases by heat, *Mem. Proc. Manchester Lit. Phil. Soc., 5,* 535–602, 1798–1802.

Driscoll, E., Mariner views a dynamic, volcanic Mars, *Sci. News, 101,* 106–107, 1972.

Fanale, F. P., Volatile evolution, Reports of the Planetary Geology Program, 1976–1977, *NASA Tech. Memo., X-3511,* 183–186, 1977.

Farmer, C. B., D. W. Davies, and D. D. LaPorte, Mars: Northern summer ice cap–water vapor observations from Viking 2, *Science, 194,* 1339–1341, 1976.

Fitzgerald, D., Evaporation, *Trans. Amer. Soc. Civil Eng., 15,* 581–646, 1886.

Hack, J. T., Studies of longitudinal stream profiles in Virginia and Maryland, shorter contributions to general geology, *U.S. Geol. Surv. Prof. Pap., 294B,* 45–94, 1956.

Hartmann, W. K., Martian cratering, 4, Mariner 9 initial analysis of cratering chronology, *J. Geophys. Res., 78,* 4096–4116, 1973.

Hartmann, W. K., Geological observations of Martian arroyos, *J. Geophys. Res., 79,* 3951–3957, 1974.

Hartmann, W. K., Cratering in the solar system, *Sci. Amer., 236,* 84–99, 1977.

Hord, C. W., K. E. Simmons, and L. K. McLaughlin, Mariner ultraviolet-spectrometer experiment pressure-altitude measurements on Mars, *Icarus, 21,* 293–302, 1974.

Kieffer, H. H., Soil and surface temperatures at the Viking landing sites, *Science, 194,* 1344–1346, 1976.

Kieffer, H. H., S. C. Chase, T. Z. Martin, E. D. Miner, and F. D. Palluconi, Martian north polar summer temperature: Dirty water ice, *Science, 194,* 1341–1343, 1976*a*.

Kieffer, H. H., S. C. Chase, Jr., E. D. Miner, F. D. Palluconi, G. Munch, G. Neugebauer, and T. Z. Martin, Infrared thermal mapping of the Martian surface and atmosphere: First results, *Science, 193,* 780–785, 1976*b*.

Kieffer, H. H., P. R. Christensen, T. Z. Martin, E. D. Miner, and F. D. Palluconi, Temperatures of the Martian surface and atmosphere: Viking observations of diurnal and geometric variations, *Science, 194,* 1346–1351, 1976*c*.

Leopold, L. B., and T. Maddock, The hydraulic geometry of stream channels and some physiographic implications, *U.S. Geol. Surv. Prof. Pap., 252,* 57 pp., 1953.

Magin, G. B., Jr., and L. E. Randall, Review of literature on evaporation suppression, Studies of Evaporation, *U.S. Geol. Surv. Prof. Pap., 272C,* 69 pp., 1960.

Malde, H. E., The catastrophic late Pleistocene Bonneville flood in the Snake River plain, Idaho, *U.S. Geol. Surv. Prof. Pap., 596,* 52 pp., 1968.

Malin, M. C., Age of Martian channels, *J. Geophys. Res., 81,* 4825–4845, 1976.

Masursky, H., An overview of geological results from Mariner 9, *J. Geophys. Res., 78,* 4009–4030, 1973.

Masursky, H., Martian channels, *NASA Tech. Memo., X-3364,* 169–171, 1976.

Masursky, H., and N. L. Crabill, Search for the Viking 1 landing site, *Science, 194,* 62–68, 1976*a*.

Masursky, H., and N. L. Crabill, Search for the Viking 2 landing site, *Science, 194,* 809–812, 1976*b*.

McCauley, J. F., Mariner 9 evidence for wind erosion in the equatorial and mid-latitude regions of Mars, *J. Geophys. Res., 78,* 4123–4137, 1973.

McElroy, M. B., Y. L. Yung, and A. O. Nier, Isotopic composition of nitrogen: Implications for the past history of Mars atmosphere, *Science, 194,* 70–72, 1976.

Milton, D. J., Water and processes of degradation in the Martian landscape, *J. Geophys. Res., 78,* 4037–4047, 1973.

Mutch, T. A., R. E. Arvidson, A. B. Binder, F. O. Huck, E. C. Levinthal, S. Liebes, Jr., E. C. Morris, D. Nummedal, J. B. Pollack, and C. Sagan, Fine particles on Mars: Observations with the Viking 1 lander cameras, *Science, 194,* 87–97, 1976.

Neukum, G., and P. Horn, Effects of lava flows on lunar crater populations, *Moon, 15,* 205–222, 1976.

Neukum, G., and D. U. Wise, A standard crater curve and possible new time scale, *Science, 194,* 1381–1387, 1976.

Neukum, G., B. Konig, H. Fechtig, and D. Storzer, Cratering in the earth-moon system: Consequences for age determination by crater counting, *Proc. Lunar Sci. Conf. 6th,* 239–263, 1974.
Nier, A. O., M. B. McElroy, and Y. L. Yung, Isotopic composition of the Martian atmosphere, *Science, 194,* 68–70, 1976.
Nummedal, D., Fluvial erosion on Mars: A review, paper presented at Colloquium on Water in Planetary Regoliths, NASA, Hanover, N. H., Oct. 5–7, 1976.
Owen, T., and K. Biemann, Composition of the atmosphere at the surface of Mars: Detection of argon-36 and preliminary analysis, *Science, 193,* 801–803, 1976.
Pieri, D., Martian channels: Distribution of small channels on the Martian surface, *Icarus, 27,* 25–50, 1976.
Playfair, J., Illustrations of the Huttonian theory of the earth, pp. 102, 116, 124, and 354, Edinburgh, 1802. (Reproduced in *Geology From Original Sources,* edited by W. Agar, R. P. Flint, and C. R. Longwell, Henry Holt, New York, 1925.)
Pollack, J. B., Climatic changes on Mars: Inferences based on Viking and Mariner data, Reports of the Planetary Geology Program, 1976–1977, *NASA Tech. Memo., X-3511,* 187–188, 1977.
Sagan, C., O. B. Toon, and P. J. Gierasch, Climatic change on Mars, *Science, 181,* 1048–1049, 1973.
Schumm, S. A., Structural origin of large Martian channels, *Icarus, 22,* 371–384, 1974.
Sharp, R. P., and M. C. Malin, Channels on Mars, *Geol. Soc. Amer. Bull., 86,* 593–609, 1975.
Shoemaker, E. M., Interpretation of lunar craters, in *Physics and Astronomy of the Moon,* edited by Z. Kopal, pp. 283–360, Academic, New York, 1962.
Shoemaker, E. M., Present cratering rates on the terrestrial planets and the moon, Reports of the Planetary Geology Program, 1976–1977, *NASA Tech. Memo., X-3511,* May 1977.
Shoemaker, E. M., and R. J. Hackman, Stratigraphic basis for lunar time scale, in *The Moon: Symposium of the International Astronomical Union,* edited by Z. Kopal and Z. K. Mikhailov, pp. 289–300, Academic, New York, 1962.
Soderblom, L. A., R. A. West, B. M. Herman, T. J. Kreidler, and C. D. Condit, Martian planet-wide crater distribution: Implications for geologic history and surface processes, *Icarus, 22,* 239–263, 1974.
Thorarinsson, S., The jökulhlaup from the Katla area in 1955 compared with other jökulhlaups in Iceland, *Reykjavik Mus. Natur. Hist. Misc. Pap., 18,* 21–25, 1957.
Wallace, D., and C. Sagan, Evaporation of ice-choked rivers: Application to Martian channels, Reports of the Planetary Geology Program, 1976–1977, *NASA Tech. Memo., X-3511,* 161, 1977.
Ward, W. R., Large scale obliquity variations on Mars, *Science, 181,* 260–262, 1973.
Ward, W. R., Climatic variations on Mars: Astronomical theory of insolation, *J. Geophys. Res., 79,* 3375–3395, 1974.

(Received April 13, 1977;
revised June 3, 1977;
accepted June 3, 1977.)

Martian Permafrost Features

MICHAEL H. CARR

U.S. Geological Survey, Menlo Park, California 94025

GERALD G. SCHABER

U.S. Geological Survey, Flagstaff, Arizona 86001

The outgassing history of Mars and the prevailing temperature conditions suggest that ground ice may occur to depths of kilometers over large areas of the planet. The presence of permafrost is also indicated by several topographic features that resemble those found in periglacial regions of the earth. East of Hellas and in the Protonilus and Nilosyrtis regions there are features that resemble those formed on earth by gelifluction, the slow creep of near-surface materials aided by freeze-thaw of ground ice. In the south part of Chryse Planitia there are irregular depressions that resemble thermokarst features, and the pattern of tributaries to the equatorial canyons is suggestive of a sapping process that would result from the melting of ground ice. The morphology of ejecta around fresh Martian impact craters is distinctively different from that around lunar and Mercurian craters. Such differences could be ascribed to the presence of ground ice in the target materials. The convergence of these different observations supports permafrost conditions not only at present but also for much of the planet's history.

The presence of permafrost on Mars has long been suspected [*D. M. Anderson et al.*, 1967; *Lederberg and Sagan*, 1962; *Salisbury*, 1966; *Wade and de Wys*, 1968] simply on the basis of the observed annual variations in surface temperatures and an assumed planetary outgassing. Acquisition of close-up photographs of the planet by the three Mariner Mars missions revealed many surface features that could reasonably be attributed to permafrost action. *L. W. Gatto and D. M. Anderson* [1975], *D. M. Anderson et al.* [1973], and *Belcher et al.* [1971] pointed to several terrestrial permafrost features that could be analogs for features seen on Mars, and *Sharp et al.* [1971] and *Sharp* [1973] suggested that ground ice could have played a significant role in the formation of the etch-pitted, chaotic, and fretted terrains and some of the channel features. Acquisition of additional high-resolution photographs of the planet from the Viking orbiters has reinforced the suspicion that permafrost has played a significant role in landscape development and may be continuing to do so. Much of the evidence is indirect. The convergence of several different lines of evidence is, however, strongly suggestive. The purpose of this paper is to indicate the kind of evidence that is available in the orbiter pictures. A more definitive summary of the evidence will be presented in a later paper.

A first condition for permafrost is appropriate temperature conditions. The present mean annual temperature, with a maximum of approximately −60°C close to the equator [*Leighton et al.*, 1967; *Opik*, 1966; *Sinton and Strong*, 1960], favors development of frozen ground over the entire surface. Only between latitudes 70°S and 30°N do near-surface temperatures exceed 0°C, and here only near midday during the summer [*D. M. Anderson et al.*, 1973]. The top few centimeters in these regions are probably devoid of ground ice, since any available water will tend to be lost to the atmosphere when temperatures rise above freezing. Below this dehydrated zone, however, ice-cemented conditions probably persist down to some unknown depth which depends on the internal heat flow. For the low heat flow values expected on Mars the mean annual temperature may remain below 0°C to depths of several kilometers. This does not necessarily mean, however, that ice is stable throughout this entire region. At latitudes lower than approximately 40°, water ice is unstable with respect to the present atmosphere and could sublime and dissipate if exposed to it [*Fanale*, 1976]. Ground ice can exist in these latitudes only at depths at which it is effectively blocked from atmospheric exchange by the overlying materials. At latitudes higher than approximately 40°, ice can exist in equilibrium with the present atmosphere.

Conditions may have been different in the past. The presence of seemingly water eroded channels suggests higher surface temperatures in the past and so greater depths to the top of the permafrost layer. Also, models of the thermal evolution of Mars [e.g., *Johnston et al.*, 1974] and evidence from the large volcanoes [*Carr*, 1976] suggest higher heat flows in the past, hence shallower depths for the bottom of the permafrost. Despite possible warmer climates and higher heat flows in the past, it appears that conditions favoring permafrost prevailed for extended periods, since much of the evidence for permafrost comes from terrain of different ages including relatively old features.

A second condition for a permafrost layer is the availability of water. Many estimates have been made of the amount of water outgassed from the planet [*Owen*, 1976]. Some of the pre-Viking estimates are based on an erroneously high value (35 ± 10%) for argon in the Martian atmosphere inferred from anomalous ion pump readings on Mars 6 [*Istomin and Grechnev*, 1976]. Viking results indicate that the Martian atmosphere contains only 1–2% argon and that the $^{36}Ar/^{40}Ar$ ratio is a factor of 10 lower than in the earth's atmosphere [*Owen et al.*, 1976; *Owen and Biemann*, 1976]. The interpretation that these authors place on these data is that (1) Mars has not outgassed as much as the earth, (2) the present Martian atmosphere represents about one tenth of the total outgassed volatiles, excluding water, and (3) the corresponding amount of water is equivalent to a layer a few tens of meters deep. These conclusions are consistent with the relatively high $^{15}N/^{14}N$ ratio in the Martian atmosphere which suggests that the atmosphere was formerly more dense than at present and subsequent exospheric loss of nitrogen has effectively enriched the atmosphere in ^{15}N [*McElroy et al.*, 1976].

Not all the implied few tens of meters of water is available to

Paper number 7S0450.

Fig. 1. Mosaic of the Nilosyrtis region at 34°N, 228°W. The mosaic is approximately 180 km across. High-standing areas are believed to be remnants of old cratered terrain, which is much more extensive to the south of the mosaic area. In many of the low-lying areas there are subparallel ridges and grooves which are suggestive of creep of near-surface materials such as might result from freeze-thaw of interstitial ice. North is at the top. (JPL mosaic 211–5207.)

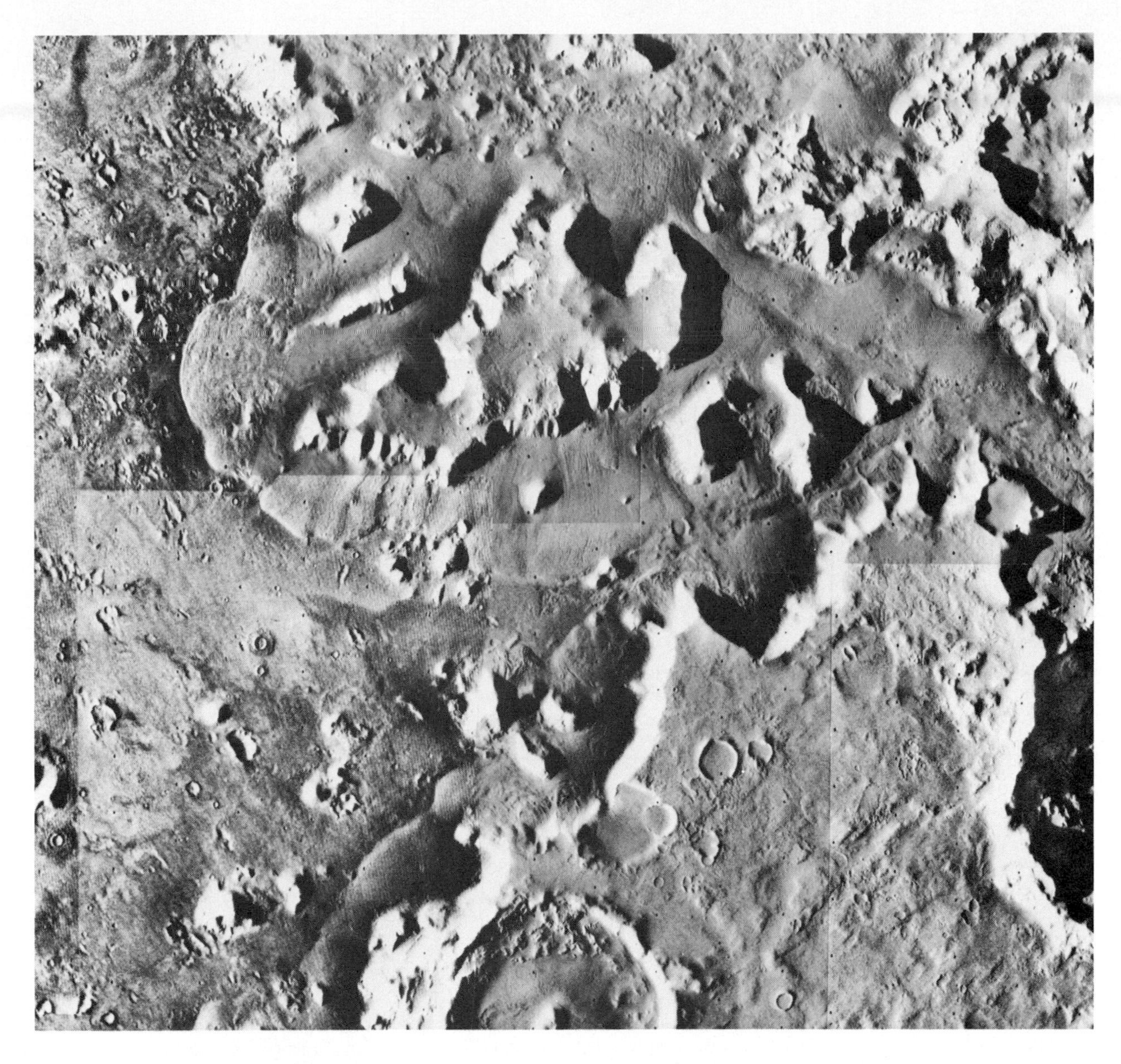

Fig. 2. Debris flows around remnants of old cratered terrain in the Protonilus region, 44°N, 315°W. The debris flows commonly have a curvilinear texture at right angles to the source scarp and extending up to 30 km from it. (JPL mosaic 211–5266.)

form ground ice. Some is lost by photolysis of water in the upper atmosphere and exospheric escape of hydrogen and oxygen [*McElroy,* 1972; *McElroy and Donahue,* 1972]. Some may be irreversibly removed from the atmosphere by photochemical weathering [*Huguenin,* 1976]. Considerable uncertainty exists as to the efficacy of these two mechanisms. On one extreme the entire estimated 5×10^3 g/cm^2 H_2O outgassed from the planet could be removed by either process, so that little water would be left over as permafrost for landscape development. On the other extreme, very little water would be lost by either process. Ground ice may thus have existed in the past but not survived to the present because of exospheric escape or a loss of water by weathering. Identification of permafrost features and determination of their ages are therefore important in understanding the near-surface history of water on the planet. In the following sections, different kinds of permafrost phenomena are described, and possible Martian examples are discussed.

MASS-WASTING

Mass-wasting is 'the gravitative movement of rock debris downslope without the aid of the flowing medium such as air at ordinary pressure, water or glacier ice' [*Longwell et al.,* 1969, p. 162]. It is generally more effective in periglacial than in temperate regions and includes frost creep, gelifluction, and slumping. Frost creep is 'the ratchet-like downslope movement of particles as the result of frost heaving of the ground and subsequent settling and thawing, the heaving being predominantly normal to the slope and the settling more nearly vertical' [*Washburn,* 1967]. The term solifluction was defined by *J. G. Anderson* [1906] to cover slow flowing from higher to lower ground of masses of waste saturated with water. Gelifluction is one type, the term having been coined by *Baulig* [1956] to describe flow which results from the presence of ground ice. In this case a permafrost table prevents drainage of the near-surface materials, so that during the thaw season they

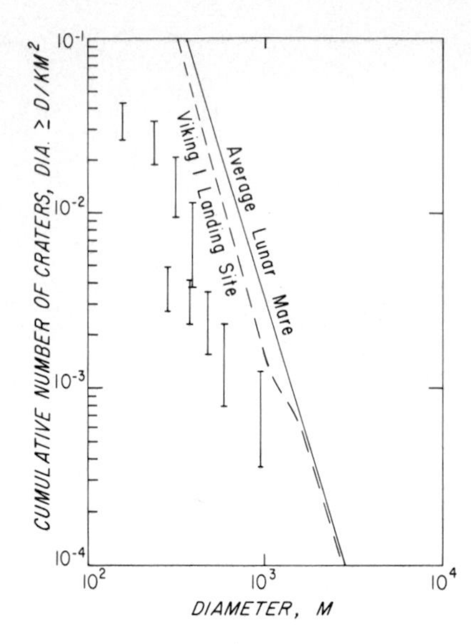

Fig. 3. Crater counts on the postulated gelifluction deposits in Nilosyrtis (upper curve) and Protonilus (lower curve), indicating a relatively young age for the deposits.

become saturated and are able to flow. Frost creep and gelifluction frequently operate under the same conditions and form common deposits by their joint action [*Washburn*, 1973]. The deposits may exist as sheets, lobes, or streams and can form on slopes as low as 1°.

In the Nilosyrtis and Protonilus regions of Mars there are features that plausibly could be interpreted as the result of frost creep and gelifluction (Figure 1). Both areas are at the boundary between the old cratered terrain to the south and the northern plains. The old cratered terrain is dissected by numerous flat-floored valleys, some of which have semicircular terminations at their upper ends. Within most of the valleys there are numerous longitudinal ridges. Where valleys meet, the longitudinal ridges merge much in the manner of median glacial moraines. The general curvilinear patterns, particularly at valley junctions, and the continuity of the ridges along the length of the valley are strongly suggestive of flow down the valleys. Cross flow, away from the valley walls toward the valley center, also appears to have occurred and formed ridges that follow local indentations of the valley walls and in places override the longitudinal ridges. The extent to which the longitudinal ridges are due to flow from the valley sides or flow downstream is uncertain, but both directions of flow imply mass-wasting on a grand scale. Where isolated positive features occur (Figure 2), flow of debris is not constrained by an opposing valley wall and continues outward away from the source. The flow lines are generally radial but are deflected where obstacles are present. The debris aprons can extend more than 50 km from the source. We believe that gelifluction and frost creep are more probable causes of such large-scale mass flow than mechanisms that operate in more temperate climates and require massive amounts of liquid water.

The rates of flow could be very slow. Rates of flow on terrestrial gelifluction and frost creep deposits depend on slope and other factors such as grain size and water content. The slopes within the Nilosyrtis and Protonilus areas are not known, but downstream slopes within the valleys and slopes on the extensive debris aprons are unlikely to exceed 1°–2° because of the lengths over which they are sustained. Transverse slopes could be much larger. *Washburn* [1973] summarizes various measures of rates of the combined effects of frost creep and gelifluction, and values range from 0.9 to 12 cm/yr on slopes of 3°–15°. The relative rates of gelifluction and frost creep vary from site to site and year to year, but in general, they are within a factor of 3 of each other [*Washburn*, 1967]. Although the lowest rates measured are 0.9 cm/yr, slower rates are surely possible. The yield strength of the flowing material may impose a lower limit on the rate of gelifluction flow, but frost creep could occur at extremely low rates and be detectable only after sustained action over many years.

Crater counts on the deposit may give some indication of the rates involved and the time since the deposits formed. Unfortunately, at the time of writing, only limited areas had been photographed. Although the statistics are poor, both counts in Figure 3 suggest relatively young ages. There are fewer craters on these deposits than on almost any Martian features other than the volcanics of the Tharsis ridge and some of the youngest channels [see *Carr et al.*, 1977*b*; *Masursky et al.*, 1977]. The counts may still, however, imply ages in excess of a billion years [*Neukum and Wise*, 1976; L. S. Soderblom, personal communication, 1977]. If the flows had been active during this time, the small craters would have been preferentially destroyed, and the slope of the size-frequency curve reduced. The Nilosyrtis and Protonilus data in Figure 3 give, however, no indication of an anomalously low slope; the slopes are identical to those for the Tharsis volcanics. The tentative conclusions are that the deposits were formed in the second half of the history of Mars, partly contemporaneous with accumulation of the volcanics of the Tharsis ridge. These particular deposits have not, however, been actively moving in the recent geologic past.

In the region west of Hellas there are similar mass-wasting features that could be interpreted as resulting from gelifluction or frost creep (Figure 4). In this region, almost all positive features are surrounded by aprons of debris that extend as far as 20 km from their source. The debris aprons are not made up of discrete lobate flows as would be expected if they were formed by landslides, nor are they talus deposits close to the angle of repose. The surface slopes are probably less than 10°, although we have no direct measures. An alluvial origin also appears unlikely because the deposits are not fan shaped, nor is there any evidence of gullying either in the source area or on the apron surface. The features appear to have been formed by slow creep of debris away from the source area. As was discussed above, such slow creep, while not necessarily requiring ground ice, is certainly facilitated by the freeze-thaw of interstitial water.

The presence of possible permafrost features in the Protonilus and Nilosyrtis regions may have profound implications regarding the evolution of the Martian surface. Both regions are at the boundary between the old cratered terrain and the younger plains. Some of the most puzzling questions regarding Mars concern the nature of this boundary and the process whereby the old cratered terrain has been destroyed. Fretted terrain occurs at the boundary along much of its length. *Sharp* [1973] suggested that ice may have played a significant role in the formation of fretted terrain and the destruction of the old cratered areas. In the two fretted areas so far photographed on Viking (Nilosyrtis and Protonilus) we found convincing evidence to support Sharp. It appears that disaggregation of the old terrain has occurred along exposed escarpments and provided debris to be carried away from the high areas by mass-wasting processes. Freeze-thaw of ground ice is a plausible mechanism both for the mechanical breakup of the old terrain

Fig. 4. Debris flows around remnants of old cratered terrain east of Hellas at 41°S, 257°W. The frame is approximately 280 km across. Debris flows around positive features extend up to 20 km from the source. (Frame 97A62.)

and for the transport of the resulting debris. These observations raise the following question. Is ice action the primary agent responsible for destruction of old cratered terrain, or does it merely modify features formed by some more fundamental process? If the first alternative is correct, then substantial amounts of ice within the old cratered terrain are implied. This theme will be discussed again at the end of the paper after all the evidence for ice action has been reviewed.

CHAOTIC TERRAIN

Chaotic terrain was initially recognized from Mariner 6 pictures [*Sharp et al.*, 1971] and later observed in detail from Mariner 9 [*Sharp*, 1973], and it appears to have formed at the expense of the surrounding cratered upland. It consists largely of a jumble of angular-shaped blocks at a lower elevation than the surrounding terrain. The general impression is of undermining and collapse of the preexisting upland surface. Many large channels start in chaotic areas, indicating a genetic relation between the two features. The mode of formation, although still a puzzle, has been tentatively attributed to melting of ground ice [*Sharp*, 1973] or dissociation of clathrates [*Milton*, 1974]. Because chaotic terrain has been discussed in detail elsewhere, it will not be discussed further here. It is mentioned as additional supporting evidence for extensive ground ice in the cratered uplands.

An area that resembles chaotic terrain but has significant differences appears in Figure 5. The picture shows a section of plains north of Elysium which appears to have been partly destroyed. To the south the uppermost layer that forms the plains is intact; to the north there are only disconnected remnants. In the transition zone between, the upper layer is broken into angular blocks to form terrain that resembles chaotic areas in the cratered uplands. The general impression is of a process of planation that has destroyed surface materials down to a specific depth and created a new planar surface at that depth. The process is somewhat analogous, although at a smaller scale, to the formation of fretted terrain along the boundary between the cratered highlands and the northern

Fig. 5. Chaotic terrain in the plains north of Elysium at 33°N, 213°W. The plains to the south appear to have partly collapsed and been eroded away such that only isolated remnants remain in the northern part of the mosaic. Collapse by removal of subsurface ice is a plausible possibility. The crater in the northwest is 13 km in diameter. (JPL mosaic 211–5274.)

Fig. 6. Possible thermokarst terrain in the southern part of Chryse Planitia. Small depressions in the plateau area appear to enlarge and merge to form valleys as the plateau is progressively destroyed. Formation of the depressions and their subsequent enlargements may result from removal of ground ice. (Frames 8A70 and 8A74.)

plains where remnants of a former surface occur to the north, the surface is intact to the south, and a transition zone occurs between. In both cases there is a problem as to where the missing materials went and what the erosive process was. The presence of large fractions of ground ice close to the surface and its subsequent thawing and sublimation would provide a mechanism for dissolution of the near-surface materials. Their dispersal by wind would readily follow.

THERMOKARST AND FORMATION OF ALASES

Thermokarst is the process whereby melting of ground ice caused local collapse and formation of rimless depressions, or alases, which generally are roughly circular and flat floored. Alases are well developed in such regions as Siberia and reach diameters of 15 km or more [*Washburn,* 1973]. Alas valleys form as individual alases coalesce. Ultimately, extensive areas of lower terrain can result. Possible alases were seen in Chryse Planitia during the early part of the mission (Figure 6). Here a tableland comprised of layered deposits has been partly destroyed by processes which involve the formation of isolated depressions within the tableland, the merging of depressions to form valleys, and scarp retreat. A scalloped cliff marks the edge of the tableland. Individual scallops appear to have been formed partly by the walls of depressions that have been breached by the escarpment and partly by the process of scarp retreat. A plausible explanation of what is observed is that we are seeing a combination of two processes at work, both of which depend on the presence of ground ice. The first process is collapse to form alaslike depressions within the tableland, and the second is thermal erosion at exposed cliffs involving melting of ground ice and the collapse and disaggregation of the host rock materials. The process may be analogous to the formation of thermocirques [*Czudek and Demek,* 1970]. An albedo feature that follows the outline of the cliff is an in-

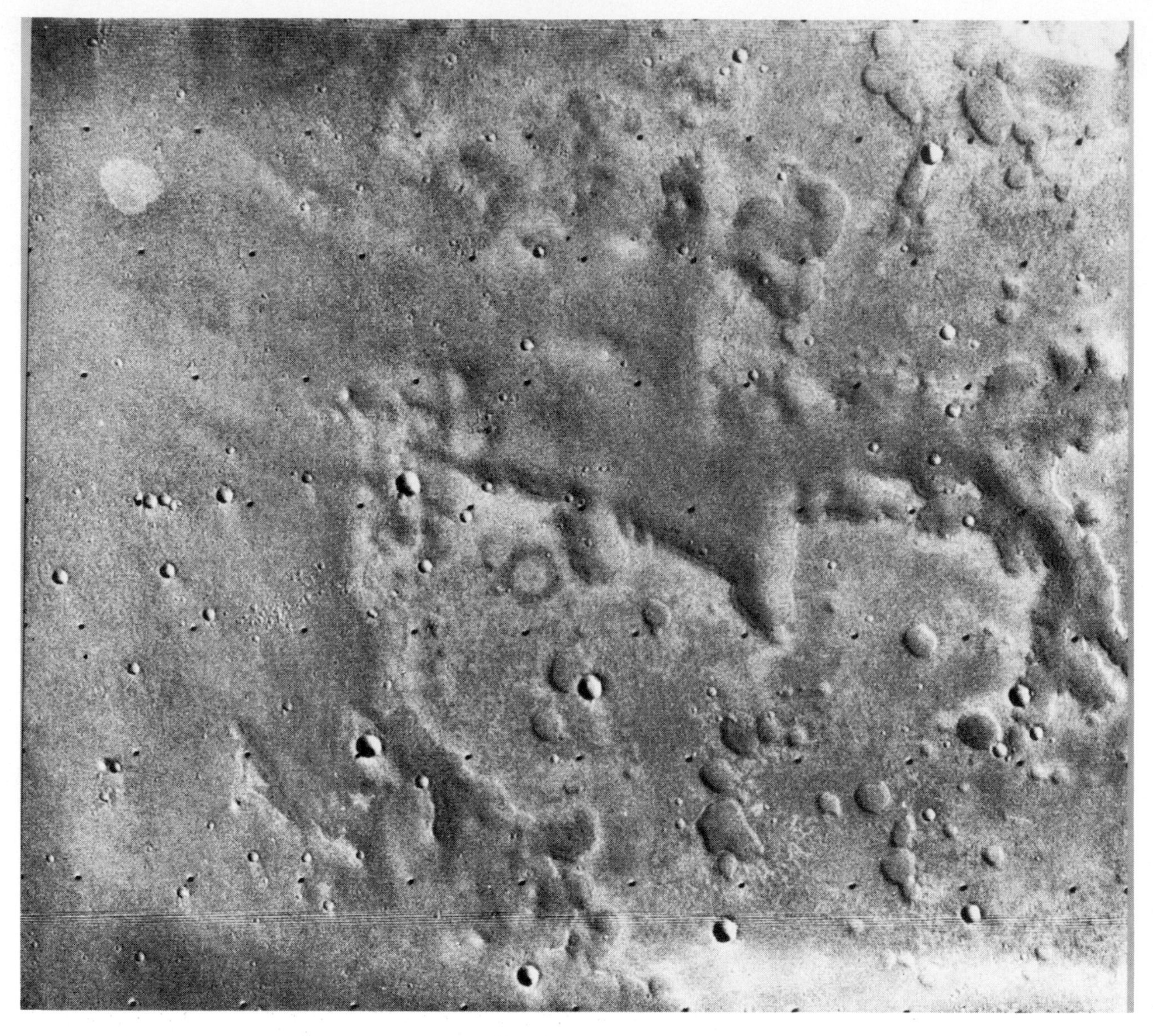

Fig. 6. (continued)

dication of some change in the tableland materials caused by proximity to the cliff.

If the isolated hollows within the tableland are alases, then large amounts of ground ice are implied. The quantity of ground ice in thermokarst regions of the earth is commonly 80–90% of the total volume of the deposits [*Washburn,* 1973]. It exists in two basic forms: (1) segregated ice dispersed in the soil and (2) ice veins and ice wedges. One of the primary differences between the Siberia and Martian alases is that on Mars the alas development continues (in concert with scarp retreat) until the whole surface is lowered. In terrestrial cases, alas development usually is arrested before extensive planation occurs. This may indicate more uniform distribution of ice within the surface materials in the case of Mars. Such would be the case if the ice were syngenetic, that is, deposited contemporaneously with the host materials.

In other regions, particularly in the 30°–50°N latitude belt, there are extensive areas of complex disordered terrain. One such area is to the south and west of the VL-2 landing site between latitudes 40° and 45°N and longitudes 220° and 240°W (Figure 7). Here are numerous extremely irregular depressions at a variety of scales. Hrad Vallis rises in the region and extends southeast toward Elysium. It does not have a regular set of tributaries but appears instead to emerge from this complex terrain by the coalescing of irregular depressions. These features are consistent with supply of water by the thawing of ground ice and the consequent collapse of the local surface.

Canyon Features

Sharp [1973] suggested that wherever scarps develop on Mars, they will tend to be undermined as a result of sapping, either by groundwater or water derived from the melting of ground ice. One style of sapping appears to have occurred along the boundary between the plains and old cratered terrain, as discussed above; other styles are illustrated by the walls of Valles Marineris. Massive landslides occur along much of the length of the canyon, particularly on the north side. In only rare cases are massive blocks of country rock involved in the slides; generally, the slumped material appears to be largely disaggregated, with only the former uppermost layer retaining its coherence [*Carr et al.*, 1976]. The landslides indicate that some process has been active whereby the slopes become unstable and collapse. On earth, slumping can occur in most any environment, but it is especially common in periglacial regions, where escarpments tend to be undercut at their base by meltwater or to fail because the strength of the materials is lowered by melting of ground ice. In extreme cases the permafrost-bearing strata may liquify and form voluminous mudflows. Such a process may be the cause of the numerous lobate features in Valles Marineris.

Most of the canyon walls have been modified by gullies of some kind. Their general morphology indicates that they formed by some sapping process rather than surface runoff.

Fig. 7. Possible seepage area at 41°N, 229°W, just south of the VL-2 landing site. A large channel Hrad Vallis arises in this region of complex and highly irregular terrain. The almost complete lack of tributaries feeding the main channel suggests subsurface drainage such as could result from melting of ground ice. The picture is 120 km across. (Frame 9B50.)

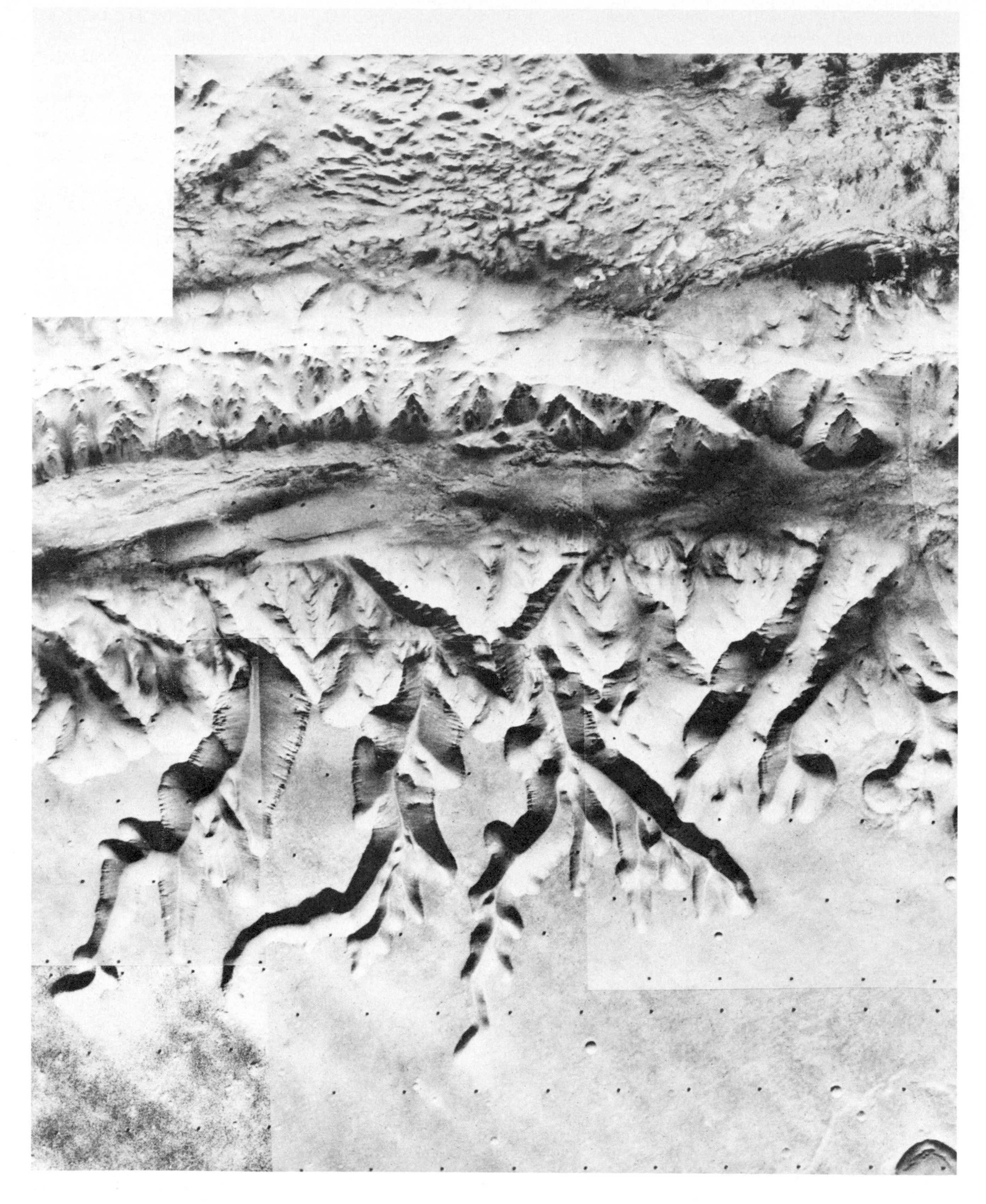

Fig. 8. Section of Valles Marineris at 80°S, 85°W. The tributaries on the southern wall each head in a cirquelike depression and lack a fine-scale drainage net. The morphology suggests that they form by groundwater sapping, not surface runoff. Ground ice is a possible source for the water. (JPL mosaic 211–5158.)

Fig. 9. Patterned ground at 44°N, 18°W. The polygonal pattern of fractures seen here is typical of extensive areas of plains in the 40°–50°N latitude belt. The pattern resembles that which develops by ice wedging in terrestrial periglacial regions except that the scale is totally different. Terrestrial ice polygons rarely achieve dimensions in excess of 100 m. These polygons are up to 20 km across. (Frame 32A18.)

Fig. 10*a*

Fig. 10. (*a*) Striped ground at 45°N, 354°W. The faint parallel lines are believed to mark successive positions of the retreat of an escarpment during removal of a former mantle. The picture is 60 km across. (Frame 52A35.) (*b*) Striped ground at 50°N, 289°W. Here the parallel markings are caused by low ridges and, less commonly, shallow depressions. The frame is approximately 110 km across. (Frame 11B01.)

The south wall of Ius Chasma, for example, is dissected by a series of branching valleys. Most of the valleys have no identifiable catchment area. Each tributary heads in a semicircular cirquelike feature rather than a net of tributaries as would be expected if they were formed by surface runoff (Figure 8). The valleys are mostly V shaped, with steep sides probably formed by talus. A narrow stripe down the center of each valley is believed to have been formed by channel deposits, although rock glaciers are possible alternatives. Most valley junctions are concordant, and the valleys tend to line up along two

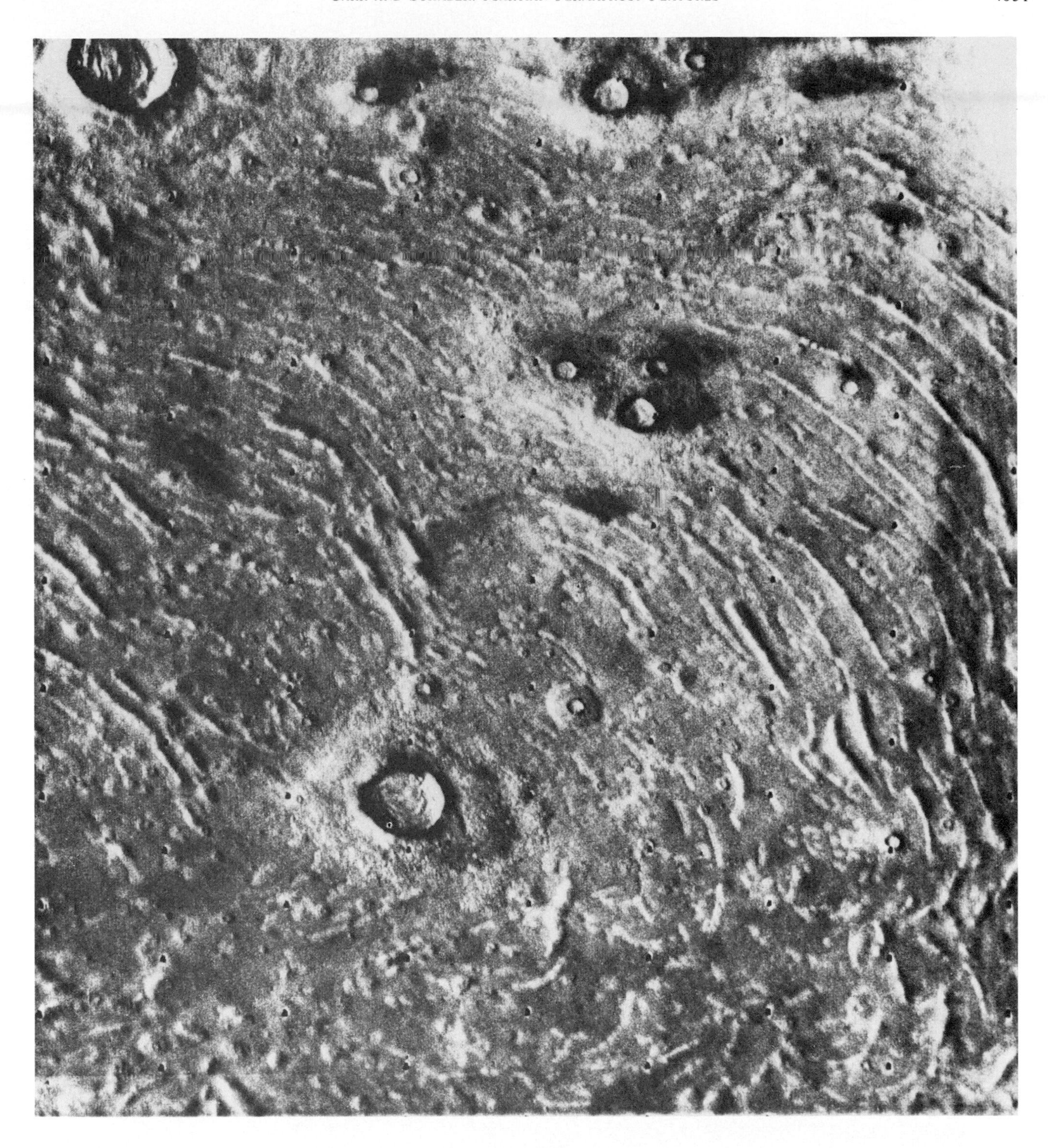

Fig. 10*b*

mutually orthogonal directions. These features are all consistent with headward erosion by sapping and with preferential erosion along structural lines of weakness.

PATTERNED GROUND AND MANTLED TERRAIN

Patterned ground is one of the most unique features of terrestrial periglacial regions. Many types have been recognized, such as circles, polygons, nets, and stripes [*Washburn,* 1973]. The Viking orbiter pictures reveal several types of patterned ground which resemble terrestrial examples except for their scale. In the region of 40°–50°N there are extensive plains believed to have been formed by lava. Most of these plains have a striking polygonal pattern (Figure 9), with individual polygons ranging in size up to 20 km across. The polygonal pattern of cracks suggests a contraction mechanism of some kind. Ice wedging is a possibility, although the pattern could equally well have formed by contraction of the lava as it cooled after deposition. The mechanics of the two processes are almost identical [*Lachenbruch,* 1961, 1962]. The crack pattern is therefore not very diagnostic, and the scale of the

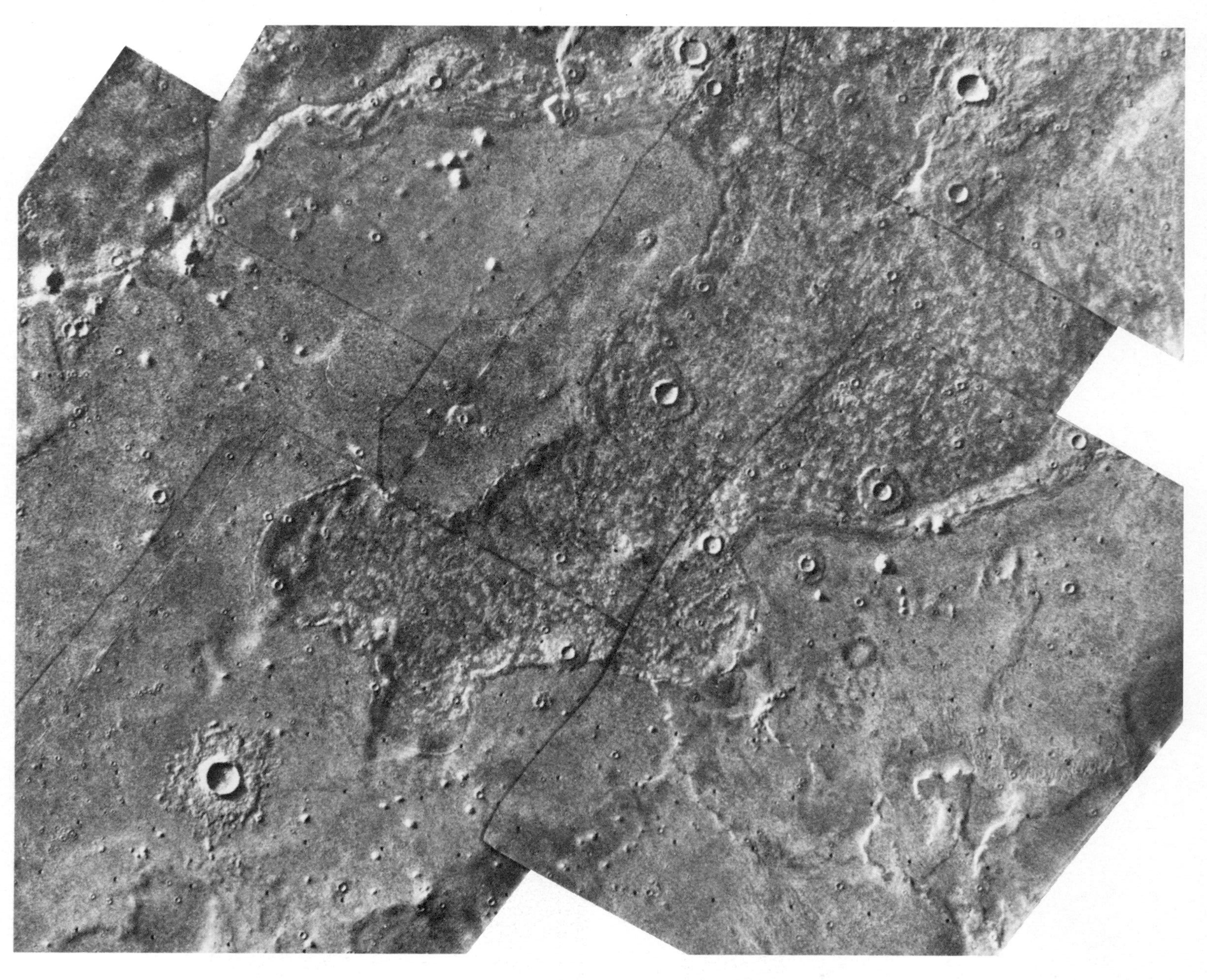

Fig. 11. Contrasting terrain at 44°N, 352°W. The smooth areas may be either debris mantles or remnants of older terrain. Note faint outlines of crater in the smooth terrain. In the textured areas the irregular or linear markings are believed to mark the former position of the escarpment that bands the smooth terrain. (JPL mosaic 211–5066.)

phenomena is equally difficult to reconcile with either process. The occurrence of widespread polygonal ground cannot therefore itself be taken as indicating permafrost. Its restriction to a specific latitude belt is, however, suggestive, particularly when combined with the other phenomena here described.

Peculiar parallel linear patterns occur on the surface in several places in the 30°–50°N latitude band of Mars. The linear features may be formed by low, sinuous ridges, by lines of troughs, or by lines of disconnected hollows. In places the different kinds occur together (Figure 10). Their origin is uncertain. They are probably not directly analogous to striped ground found in terrestrial periglacial areas. Terrestrial striped ground appears to be closely related in origin to sorted polygons. It forms as a result of downslope creep of sorted polygonal ground and rarely occurs on slopes less than 3°. The stripes are usually only a few meters across. In contrast, the Martian features occur on seemingly level plains with stripes up to several hundred meters across.

The Martian stripes appear to be related to the former presence of 'mantles' covering the area. *Soderblom et al.* [1973] postulated the existence of partly eroded debris mantles in many areas poleward of 30°N and 30°S. Smooth areas in the 40°–50° latitude belt have been tentatively suggested as remnants of these mantles (Figure 11). At the junction between the smooth areas and the more textured terrain there is commonly a low escarpment, with the smooth terrain on the higher ground. Stripes are occasionally found adjacent to and parallel to the outline of this escarpment, suggesting that the stripes are in some way related to the escarpments, perhaps reflecting former positions of the escarpment.

The postulated mantles were a subject of debate during the Viking mission because of their implication regarding potential landing sites. The possibility that they may be ice cemented and contain significant amounts of ground ice was proposed, principally by R. Hargraves (personal communication, 1976). Removal of the mantles would then not simply be by deflation, as originally proposed by Soderblom et al., but by escarpment retreat with progressive melting of ground ice, disaggregation of the debris, and then deflation. Repeated mantling and removal of the mantles would then reflect changes in the temperature regime. Such a proposal, while very speculative, is consistent with the observed relation between stripes and mantles. The mechanism is similar to but apparently more efficient than the mechanism proposed above for erosion of the old cratered terrain to form fretted terrain and alases.

CRATER MORPHOLOGY

The morphology of ejecta around most fresh-appearing Martian impact craters is distinctively different from that around lunar and Mercurian craters. The ejecta commonly consist of discrete layers, each with a low rampart marking its outer margin. Radial 'flow lines' are commonly within the ejecta, diverging around obstacles or 'pressure ridges' at the outer margin [*Carr et al.*, 1977*a*]. Continuous ejecta also extends much further from the crater than in the lunar and Mercurian cases. These characteristics were interpreted by Carr et al. as indicating flow of ejecta radially outward along the ground after ballistic deposition. Lubrication of the ejecta by water or entrained gases as a result of impact into ice-laden ground was suggested by Carr et al. as a possible mechanism to explain the unique Martian features. Such a mechanism is consistent with the presence of other possible permafrost features described in this paper.

CONCLUSION

We have seen that a wide variety of evidence supports the presence or former presence of permafrost over much of the surface of the planet. Some features, particularly the alases, the gelifluction features, and chaotic terrain, suggest large proportions of ground ice, perhaps as much as 50%, dispersed throughout the host materials. We have also suggested that the debris mantles at high latitudes may contain large proportions of ice and that their deposition and removal were largely processes of thermal erosion. If these suggestions are correct, then there are wide implications for the evolution of the Martian surface and particularly for the origin of fluvial features. Several lines of evidence (scarp retreat to form fretted terrain, gelifluction deposits, collapse of canyon walls, chaotic terrain) suggest that permafrost is particularly widespread in the old cratered terrain. The concentration of fine channels in this terrain may result because they form not by surface runoff or rainfall but by thawing of ground ice. The presence of extensive ground ice in the old terrain also might explain why most of the chaotic terrain is restricted to the cratered uplands and why the large flood channels arise in these regions.

REFERENCES

Anderson, D. M., E. S. Gaffney, and P. F. Law, Frost phenomena on Mars, *Science, 155,* 319–322, 1967.

Anderson, D. M., L. W. Gatto, and F. Ugolini, An examination of Mariner 6 and 7 imagery for evidence of permafrost terrain on Mars, in *Permafrost: The North American Contribution to the Second International Conference,* pp. 449–508, National Academy of Sciences, Washington, D. C., 1973.

Anderson, J. G., Solifluction, a component of subaerial denudation, *J. Geol., 14,* 91–112, 1906.

Baulig, H., Peneplaines et pediplaines, *Soc. Belg. Etud. Geog., 25,* 25–58, 1956.

Belcher, D., J. Veverka, and C. Sagan, Mariner photography of Mars and aerial photography of earth: Some analogies, *Icarus, 15,* 241–252, 1971.

Carr, M. H., H. Masursky, W. A. Baum, K. R. Blasius, G. A. Briggs, J. A. Cutts, T. Duxbury, R. Greeley, J. E. Guest, B. A. Smith, L. S. Soderblom, J. Veverka, and J. B. Wellman, Preliminary results from the Viking orbiter imaging experiment, *Science, 193,* 766–776, 1976.

Carr, M. H., L. A. Crumpler, J. A. Cutts, R. Greeley, J. E. Guest, and H. Masursky, Martian craters and emplacement of ejecta by surface flow, *J. Geophys. Res., 82,* this issue, 1977*a*.

Carr, M. H., R. Greeley, K. R. Blasius, J. E. Guest, and J. B. Murray, Some Martian volcanic features as viewed from the Viking orbiters, *J. Geophys. Res., 82,* this issue, 1977*b*.

Czudek, T., and J. Demek, Thermokarst in Siberia and its influence on the development of lowland relief, *Quaternary Res., 1,* 103–120, 1970.

Fanale, F. R., Martian volatiles: The degassing history and geochemical fate, *Icarus, 28,* 179–202, 1976.

Gatto, L. W., and D. M. Anderson, Alaskan thermokarst terrain and possible Martian analog, *Science, 188*(4185), 255–257, 1975.

Huguenin, R. L., Mars chemical weathering as a massive volatile sink, *Icarus, 28,* 203–212, 1976.

Istomin, V. G., and R. V. Grechnev, Argon in the Martian atmosphere: Evidence from the Mars 6 descent module, *Icarus, 28,* 155–158, 1976.

Johnston, D. H., T. R. McGetchin, and M. N. Toksoz, The thermal state and internal structure of Mars, *J. Geophys. Res., 79,* 3959–3971, 1974.

Lachenbruch, A. H., Depth and spacing of tension cracks, *J. Geophys. Res., 66,* 4273–4292, 1961.

Lachenbruch, A. H., Mechanics of thermal contraction cracks and ice wedge polygons in permafrost, *Geol. Soc. Amer. Spec. Pap., 70,* 69, 1962.

Lederberg, J., and C. Sagan, Microenvironments for life on Mars, *Proc. Nat. Acad. Sci. U.S., 48,* 1473–1475, 1962.

Leighton, R. B., B. C. Murray, R. P. Sharp, J. D. Allen, and R. K. Sloan, Mariner Mars 1964 project report: Television experiment, I,

Investigators' report, Mariner IV pictures of Mars, *Tech. Rep. 32-884*, Jet Propul. Lab., Pasadena, Calif., 1967.
Longwell, C. R., R. F. Flint, and J. E. Sanders, *Physical Geology*, p. 685, John Wiley, New York, 1969.
Masursky, H., J. M. Boyce, A. L. Dial, G. G. Schaber, and M. E. Strobell, Classification and time of formation of Martian channels based on Viking data, *J. Geophys. Res., 82*, this issue, 1977.
McElroy, M. B., Mars: An evolving atmosphere, *Science, 175*, 443, 1972.
McElroy, M. B., and T. M. Donahue, Stability of the Martian atmosphere, *Science, 177*, 986–988, 1972.
McElroy, M. B., Y. L. Yung, A. O. Nier, Isotopic composition of nitrogen: Implications for the past history of Mars' atmosphere, *Science, 194*, 70–72, 1976.
Milton, D. J., Carbon dioxide hydrate and floods on Mars, *Science, 183*, 654–656, 1974.
Neukum, G., and D. V. Wise, Mars: A standard crater curve and possible new timescale, *Science, 194*, 1381–1387, 1976.
Opik, E. J., The Martian surface, *Science, 153*(3733), 255–265, 1966.
Owen, T., Volatile inventories on Mars, *Icarus, 28*, 171–177, 1976.
Owen, T., and K. Biemann, Composition of the atmosphere at the surface of Mars: Detection of argon-36 and preliminary analysis, *Science, 193*, 801–803, 1976.
Owen, T., K. Biemann, D. R. Rushneck, J. E. Biller, D. W. Howarth, and A. L. LaFleur, The atmosphere of Mars: Detection of krypton and xenon, *Science, 194*, 1293–1295, 1976.
Salisbury, J. W., The light and dark areas of Mars, *Icarus, 5*, 291–298, 1966.
Sharp, R. P. Mars: Fretted and chaotic terrain, *J. Geophys. Res., 78*, 4073–4083, 1973.
Sharp, R. P., L. A. Soderblom, B. C. Murray, and J. A. Cutts, The surface of Mars, 2, Uncratered terrains, *J. Geophys. Res., 76*, 331, 1971.
Sinton, W. M., and J. Strong, Radiometric observations of Mars, *Astrophys. J., 131*(2), 459–469, 1960.
Soderblom, L. S., T. J. Kreidler, and H. Masursky, Latitudinal distribution of a debris mantle on the Martian surface, *J. Geophys. Res., 78*, 4117–4122, 1973.
Wade, F. A., and J. N. de Wys, Permafrost features in the Martian surface, *Icarus, 9*, 175–185, 1968.
Washburn, A. L., Instrumental observations of mass wasting in the Mesters Vig district, northeast Greenland, *Medd. Gronland, 166*, 303, 1967.
Washburn, A. L., *Periglacial Processes and Environments*, p. 320, St. Martin's, New York, 1973.

(Received April 1, 1977;
revised May 23, 1977;
accepted May 24, 1977.)

Martian Impact Craters and Emplacement of Ejecta by Surface Flow

M. H. CARR,[1] L. S. CRUMPLER,[2] J. A. CUTTS,[3] R. GREELEY,[4] J. E. GUEST,[5] AND H. MASURSKY[6]

Several types of Martian impact craters have been recognized. The most common type, the rampart crater, is distinctively different from lunar and Mercurian craters. It is typically surrounded by several layers of ejecta, each having a low ridge or escarpment at its outer edge. Outward flow of ejecta along the ground after ballistic deposition is suggested by flow lines around obstacles, the absence of ejecta on top and on the lee side of obstacles, and the large radial distance to which continuous ejecta is found. The peculiar flow characteristics of the ejecta around these craters are tentatively attributed to entrained gases or to contained water, either liquid or vapor, in the ejecta as a result of impact melting of ground ice. Ejecta of other craters lacks flow features but has a marked radial pattern; ejecta of still other craters has patterns that resemble those around lunar and Mercurian craters. The internal features of Martian craters, in general, resemble their lunar and Mercurian counterparts except that the transition from bowl shaped to flat floored takes place at about 5-km diameter, a smaller size than is true for Mercury or the moon.

INTRODUCTION

The first Viking spacecraft was injected into orbit around Mars on June 30, 1976, and was followed by a second on August 7, 1976. Each orbiter carries two cameras. To this time they have taken several thousand pictures of the planet, many of which reveal surface detail that was largely unsuspected from previously acquired data [*Carr et al.*, 1976]. This paper describes the variety of crater forms seen in the Viking pictures and makes some tentative suggestions about their origin. Emphasis is on a particular type of Martian crater for which the final mode of ejecta emplacement appears to be flow. The intent here is not to present a definitive treatise on Martian craters but is rather to bring to the attention of the scientific community some of the more novel aspects of Martian craters as seen in the Viking orbiter photographs.

The two identical vidicon cameras on each spacecraft have fields of view of $1.5° \times 1.7°$, which are offset just enough to provide slight overlap between frames. The cameras take pictures alternately, every 4.5 s, so that when they are looking down at the planet close to periapsis, they record a two-frame-wide strip, the length of which is controlled by the number of frames taken. At the nominal periapsis altitude of 1500 km a resolution element (pixel) is 38 m, and an individual frame is 44×40 km. All pictures referred to in this paper were taken at ranges from 1500 to 2200 km. Although the size of a pixel on the ground is close to that of the Mariner 9 B cameras (50 m), the Viking pictures contain much more detail. This stems partly from the greater ability of the Viking cameras to preserve high-frequency information and partly from the greater clarity of the atmosphere at the time of observation.

Several types of Martian craters have been recognized. Some resemble those on the moon and Mercury; others are distinctively different. Lunar and Mercurian craters typically have a coarse, disordered texture close to the rim. Farther out, the texture becomes finer and grades imperceptibly into dense fields of secondary craters, grading finally into discrete secondary craters and rays [*Shoemaker*, 1962]. The most distinctive Martian craters have a quite different pattern. The ejecta commonly appears to consist of several layers, the outer edge of each being marked by a low ridge or escarpment. On the ejecta surface are a variety of features, including closely spaced radial striae, discontinuous concentric ridges, grooves and scarps, particularly toward the outer margin, and low, rounded hills. This type of crater was recognized from the Mariner 9 images and termed a rampart crater [*McCauley*, 1973]. The peculiar morphology was attributed to modification of the ejecta by the action of wind [*McCauley*, 1973; *Arvidson*, 1976], and this is clearly true in many cases. It appears now, however, that some of the features formerly attributed to wind action are primary emplacement features. The more fresh appearing is the crater, the better preserved are the striae, the ramparts, and the concentric features. We suggest that some of the unique features of Martian craters are primary and that they are the result of flow of the ejecta along the ground after ballistic deposition and are not caused by secondary modification. This in no way denies the presence of wind-modified forms. We have narrowed our attention to fresh-appearing craters and only to those in the 5- to 50-km size range. The wide range of crater types that result from modification subsequent to impact is not discussed.

Three main types of fresh-appearing craters have been recognized: (1) rampart craters, (2) craters with ejecta that has a very marked radial texture, and (3) craters that resemble those on the moon and Mercury. The proportions of the different types appear to vary regionally, although this cannot be fully documented at the time of writing because of the restricted Viking coverage. At the south end of the Tharsis ridge, for example, there appears to be a larger proportion of the lunar-Mercurian type than occurs in Chryse Planitia. Documentation of the regional variations, if any, must await acquisition of additional photographic coverage.

RAMPART CRATERS

Rampart craters are the most common type in most of the areas so far observed at Viking resolution. They occur in a wide range of morphologies, from those that seemingly have only one layer of ejecta with a simple, almost circular, outline to those having several ejecta layers, each being complexly lobed. The simple rampart crater of Figure 1 shows most of

[1] U.S. Geological Survey, Menlo Park, California 94025.
[2] University of New Mexico, Albuquerque, New Mexico 87131.
[3] Science Applications, Inc., Pasadena, California 91101.
[4] University of Santa Clara, Santa Clara, California 95053.
[5] University of London, London, England.
[6] U.S. Geological Survey, Flagstaff, Arizona 86001.

Paper number 7S0488.

Fig. 1. A 9-km-diameter rampart crater in Chryse Planitia. An inner ejecta layer is clearly demarcated by a low ridge or escarpment. Beyond the edge of the continuous ejecta, clusters of secondary craters, low hills, and faint radial striae are visible (Viking orbiter frames 10A56 and 10A54).

their general characteristics. This crater has many features typical of lunar craters, such as terraced walls and a central peak partly surrounded by a flat floor. The ejecta pattern, however, is distinctly nonlunar. A layer of ejecta appears to extend from the crater rim out to approximately one crater radius. The outer edge of the layer is marked either by a low ridge or by an escarpment that gives the rampart crater its name. Most of the ejecta surface has a concentric pattern of low ridges and grooves, but part has a fine radial pattern. Beyond the edge of the continuous ejecta are low hills, lines of shallow craters, and faint radial striae, all being reminiscent of the distal edges of ejecta from lunar and Mercurian craters. These features appear to be transected by the edge of the continuous ejecta as though they were overridden.

The description given above applies in general to smaller craters (5–20 km). Larger craters generally have more ejecta layers. Of the craters examined, approximately half in the 10- to 40-km size range and all those larger than 40 km have more than one layer. The typical multilayered ejecta crater shown in Figure 2 has an upper layer clearly overlying a lower, more lobate layer, indicating successive deposition. At the arrow the upper layer appears to have flowed around but not over a protrusion. When it was mobile, the ejecta was therefore no thicker than the protrusion is high. Since the present ejecta thickness is comparable to the height of the protrusion, there has been very little volume reduction of the ejecta after emplacement, as would be expected if the ejecta were deposited from a low-density solid-gas mixture like a ground surge [*Sparks and Walker,* 1973]. In general, the edges of the ejecta layers are more lobate for larger craters. The crater Yuty (Figure 3), for example, has several layers of ejecta with complex lobate outer margins. A faint radial pattern is visible in places. A small crater close to the rim of Yuty is partly covered with ejecta, yet is clearly visible, indicating that the ejecta is thin. In another, more complex example the ejecta cannot as easily be separated into discrete layers as it can at Yuty, although the mode of emplacement (near-surface outward flow) of the ejecta appears to be similar (Figure 4). Northwest of the crater a low mesa forms an obstacle to radial flow. Very little ejecta is visible on the top of the mesa, whereas ejecta is

Fig. 2. A 15-km-diameter crater in Chryse Planitia, having several layers of ejecta. Each ejecta layer has a distal ridge or escarpment. At the arrow the ejecta appears to have flowed around a low obstacle (Viking orbiter frames 10A66 and 10A95).

present on either side, indicating that at this distance, ballistic deposition is negligible and radial flow is the dominant emplacement mechanism.

The crater Arandas (Figure 5) is one of the least modified large craters yet observed. Several tiers of ejecta are clearly visible. The most distant lobes appear to have ridden over the polygonal fractures of the surrounding plains, whereas the inner lobes appear to have ridden over just deposited ejecta. To the south and southwest the flow of ejecta has been diverted by small craters where it is so indicated by arrows on the photograph (Figure 5). The ejecta extended almost to the rim crest of the south crater but did not flow over and into the crater, again suggesting little volume change after deposition. On the Arandas side of the southwest crater is a series of ridges that could be pressure ridges caused because the crater impeded the flow. Each of the layers has a strong radial pattern. In places, small ridges parallel the edge of individual ejecta blankets. Lines of secondary craters are visible to the west and southwest.

MODE OF EMPLACEMENT OF RAMPART CRATER EJECTA

The mode of emplacement of ejecta around impact craters has been a subject of intense study over the last few years [*Chao*, 1977; *Gault et al.*, 1968; *Moore et al.*, 1974; *Morrison and Oberbeck*, 1975; *Oberbeck*, 1975; *Oberbeck et al.*, 1975*a*, *b*]. Attention has focused mainly on the extent of continuous ejecta and the degree of admixing with the surface materials on which it is deposited. Of particular interest with respect to Martian rampart craters is the extent to which the ejecta (together with any admixed substrate materials) continues its motion outward after deposition from ballistic trajectories. In the case of the moon it appears that only limited outward movement follows ballistic deposition [*Howard*, 1972; *Moore et al.*, 1974]. The avalanches around larger craters such as

Fig. 3. The 18-km-diameter crater Yuty (22°N, 34°W), having several ejecta layers, each being complexly lobed. A pre-Yuty crater, close to the rim of Yuty, is buried by ejecta, yet is still visible, indicating a thin ejecta blanket (Viking orbiter frame 3A07).

Tsiolkovsky [*Guest and Murray,* 1969], Tycho [*Shoemaker et al.,* 1968], and Aristarchus [*Guest,* 1973] have been interpreted as part of the impact process, although this is not universally accepted. The final distribution of ejecta around lunar craters predominantly reflects ballistic emplacement; only in rare cases do modifications result from radial transport of material after initial deposition.

In contrast, the configuration of the ejecta around most Martian rampart craters appears to be produced by flow. The pattern of ejecta around obstacles to radial flow, such as older craters and hills, strongly supports the flow mechanism. The ejecta is deformed into ridges which wrap around the obstacles as though the ejecta flowed around them. 'Shadow zones' appear on the lee side of many obstacles. The lack of ejecta on the obstacle itself and in the lee zones rules out subaerial deposition. Additional support for the flow mechanism is the strong radial pattern and transection of older features. The mode of formation of the ridges at the distal ends of each flow

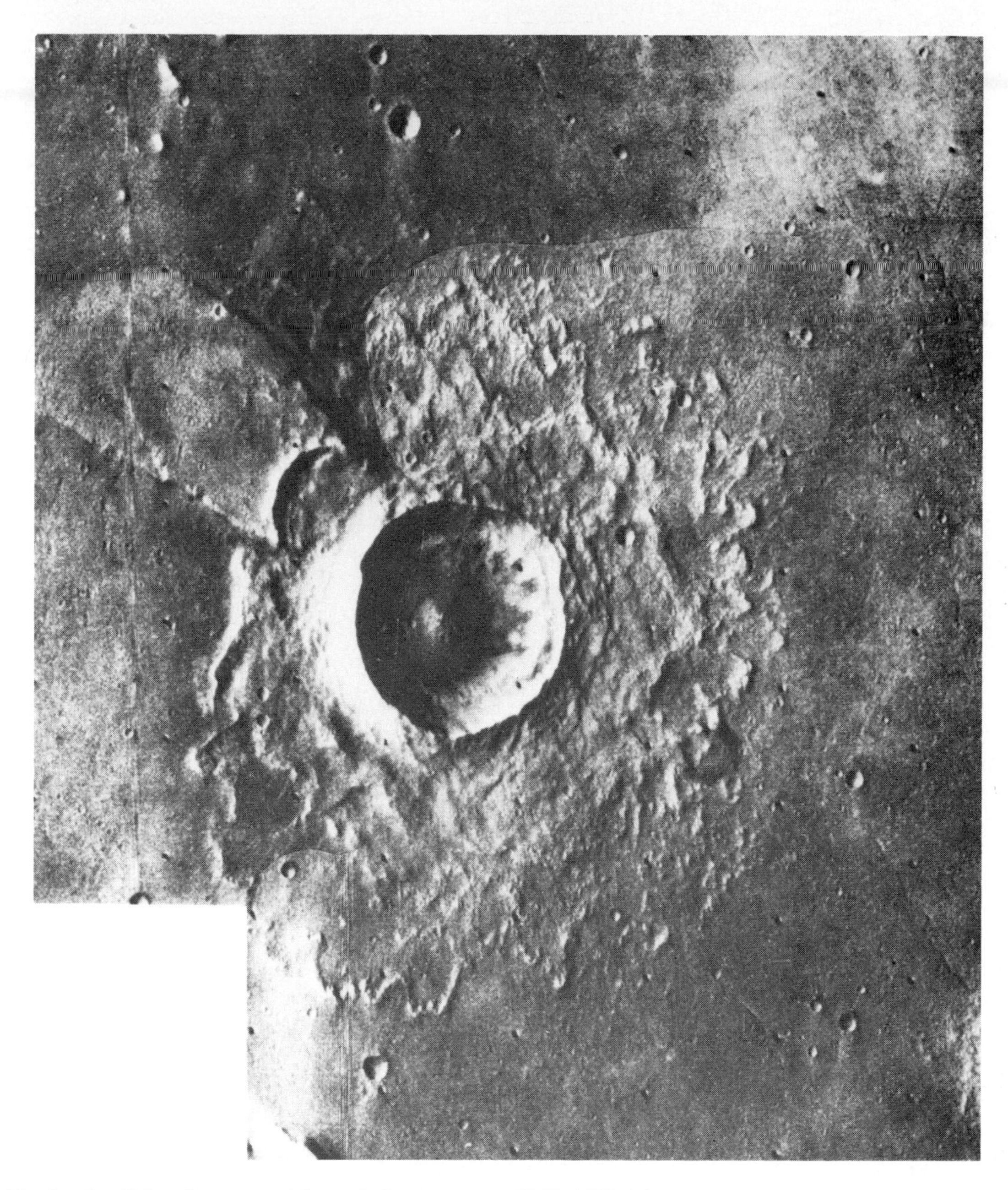

Fig. 4. An 11-km-diameter multilayered ejecta crater at 21°N, 36°W. The crater formed just to the southeast of a preexisting mesa. Little ejecta is visible on the mesa, although to the north the ejecta is present at comparable radial distances, suggesting that final emplacement of ejecta is by surface flow and is not ballistic (Viking orbiter frames 34A77 and 34A78).

is not known, but they could be caused by buildup of ejecta at the edge of the flow as the flow front slows and finally stops. Some ridges have been previously attributed to selective erosion [*McCauley*, 1973], but the preservation of very fine primary textures, such as shallow impact craters and rays immediately adjacent to the ridge (Figure 1), indicates that the ridge is a primary texture.

Additional evidence supporting flow is the larger radial extent of continuous ejecta. Because the gravity fields on Mercury and Mars are similar, the radial extents of continuous ejecta on both planets should be approximately the same (≤ 0.4 crater diameter from the rim) if the ejecta configuration is controlled by primarily ballistics. Continuous ejecta may, however, extend more than two crater diameters from the rim of Martian rampart craters (Figure 6), an observation which suggests that flow has been not merely a minor modification but the dominant mechanism of ejecta emplacement.

A critical aspect is the nature of the ejecta flows: were they gas-supported flows similar to ground surges, or were they more akin to landslides and mud flows? The evidence suggests that they were ground-hugging, relatively thin, dense flows rather than low-density ground surges. In several examples the flows appear to have been deflected by obstacles comparable in thickness to the height of the present ejecta surface. If the flows were of low density, such as those in a ground surge, collapse after flow ceased would cause a significant contrast in elevation between obstacles and the ejecta surface. The original flow must therefore have had a relatively high density, and emplacement by ground surge is unlikely. A relatively small thickness of the flows is suggested by the visibility, in several

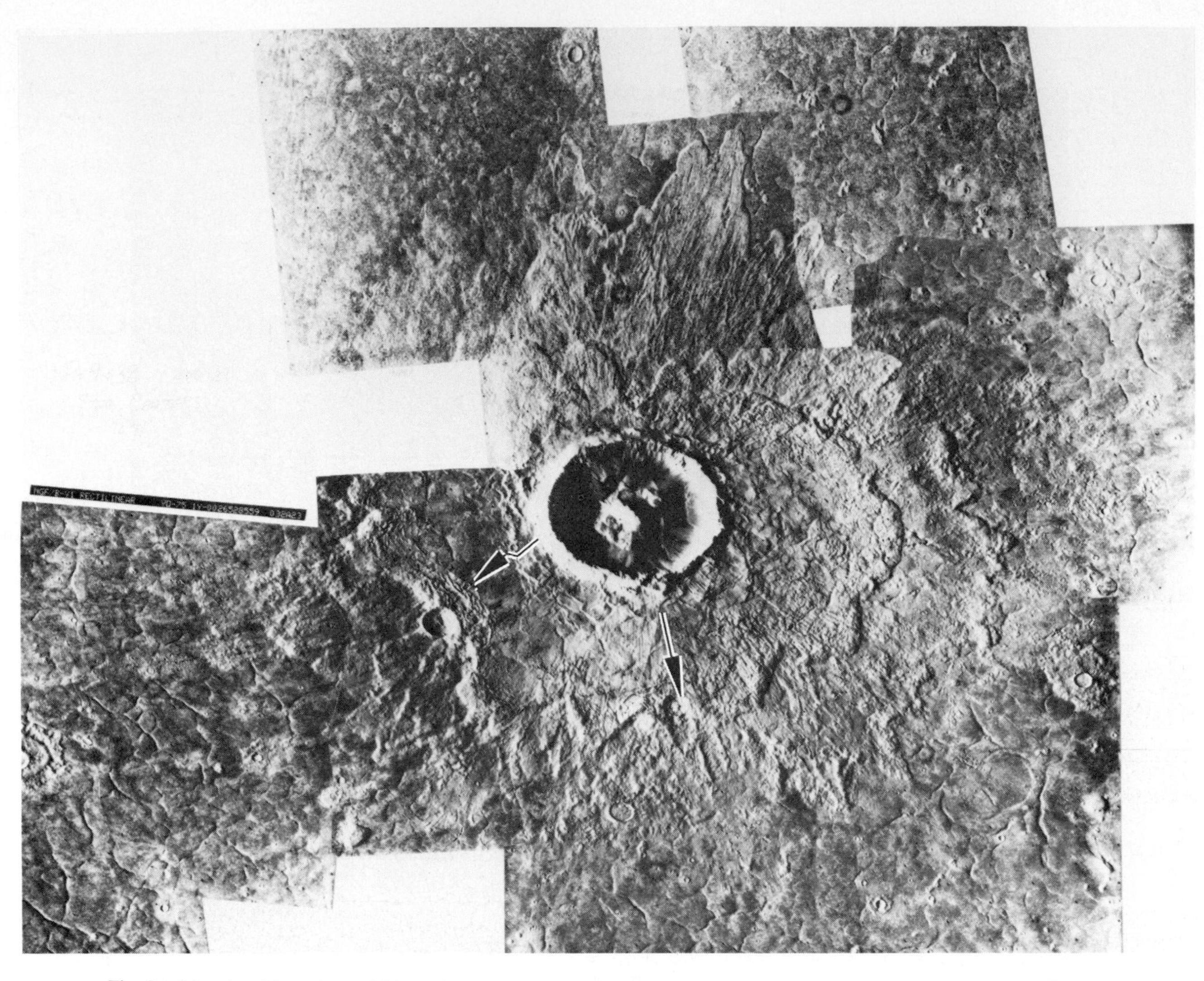

Fig. 5. Mosaic of Arandas, a 28-km-diameter crater at 43°N, 14°W. The outer layer of ejecta appears to have flowed over the surrounding fractured plains. At the two arrows an inner ejecta layer has flowed around preexisting craters, and massive ridges have formed on the ejecta surface 'upstream' from the obstacles (Viking orbiter mosaic P-17871).

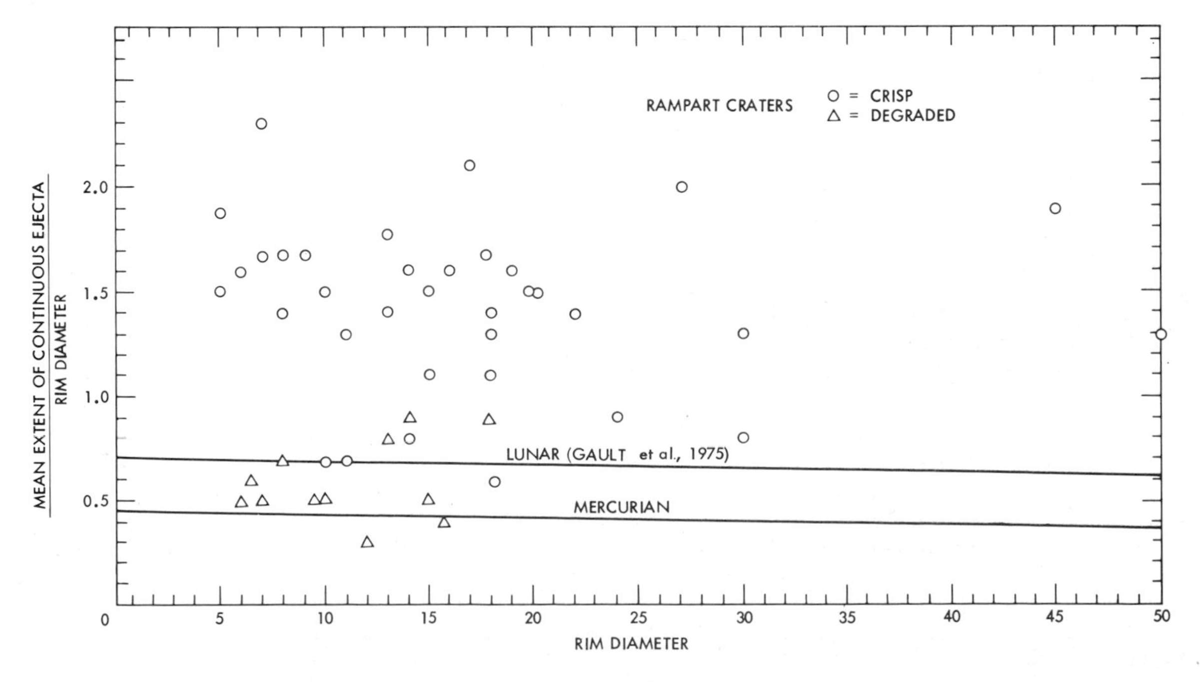

Fig. 6. Comparison of extent of continuous ejecta around Martian rampart craters with extent around lunar and Mercurian craters [after *Gault et al.*, 1975].

cases, of subjacent craters within the ejecta blanket area. These craters are clearly only thinly mantled, for their shapes show well through the overlying blanket.

Arcuate ridges at the front of some flows were probably produced by 'buckling' as the flow was obstructed by the slower moving distal parts. Similar ridges are commonly present around obstacles in the path of the flow (Figures 2 and 5). Farther away from the flow edge, concentric cracks and low, outward facing scarps are common on the surface. These may be caused by tension in the uncompressed surface layers as they are carried forward on the flow or may be a result of tensional stresses in the waning phases of flow. Striations parallel to the flow direction are typical of certain types of landslides (for example, the Sherman Glacier landslide [*Shreve,* 1966]) where they are formed by vertical sheer planes parallel to the direction of flow.

We therefore envisage that after emplacement from ballistic trajectories the fluid consistency of the ejecta allowed the outward motion to continue by flow along the surface. The whole ejecta sheet apparently moved outward en masse, substantially enlarging the area covered by continuous ejecta over that expected from predominantly ballistic deposition. A similar process may take place in some large lunar craters [*Howard,* 1972; *Moore et al.,* 1974] but to a very limited extent, and there is considerable uncertainty as to whether the observed landslides were part of the crater-forming process. Post-ballistic flow may also take place in some terrestrial impact craters. The overturned flap of Meteor Crater extends more than twice the distance [*Roddy et al.,* 1975] expected from lunar and experimental studies [*Gault et al.,* 1975], and *Chao* [1976, 1977] finds ample evidence of lateral surface flow in the ejecta from Ries Crater.

If our conclusion that surface flow of ejecta took place in the late stages of crater formation commonly on Mars but rarely on the moon and Mercury is right, then the question arises, Why is such a flow more common on Mars? Mars differs from

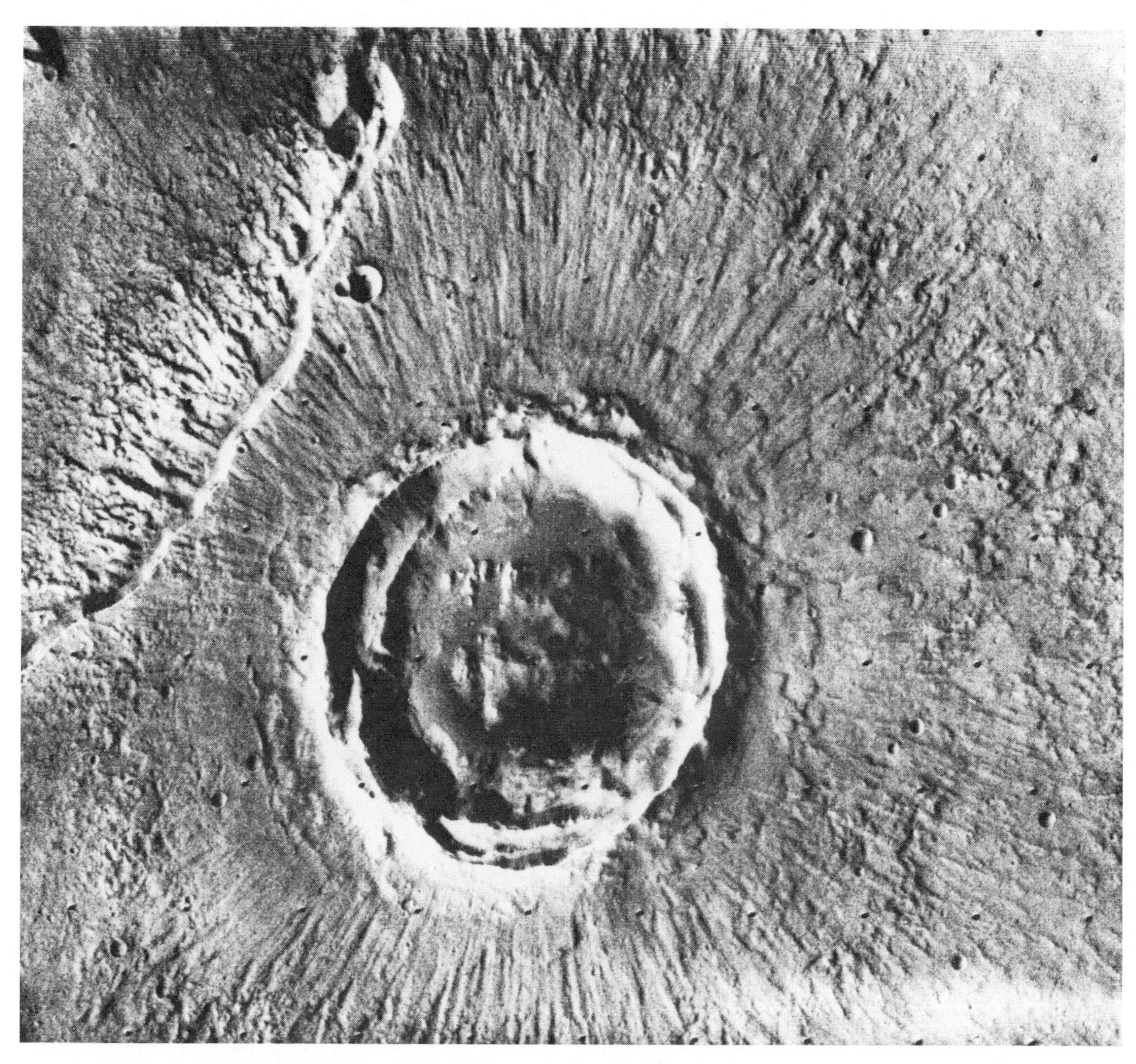

Fig. 7. An 18-km-diameter crater at 24°N, 52°W, close to the edge of Kasei Vallis. The ejecta has a marked radial pattern and no outer rampart (Viking orbiter frame 22A54).

Fig. 8. Region south of the Tharsis ridge centered at 29°S, 121°W, showing several craters that are transitional between the three main types of craters discussed in the text: rampart craters, radially patterned ejecta craters, and lunar-Mercurian type craters (Viking orbiter frame 56A40).

the moon and Mercury in having an atmosphere, and the properties of the surface materials may be different. The peculiar form of Martian craters could be explained by entrainment of atmospheric gases in the ejecta or by incorporation into the ejecta of volatiles such as water that were formerly in the ground. Studies of the regional variations of the different crater types are required to determine which of these two factors is more important. Properties of the surface that could contribute are permafrost or interstitial water present at the time of impact. Such conditions have been inferred from other evidence [*Carr et al.*, 1976]. Water as either vapor or liquid would then be the lubricant that permitted extensive flow. Also, *Mutch et al.* [1976] show that at the Viking Lander 2 site the surface is broken into patterned ground. A likely explanation is that the pattern is caused by freeze-thaw of ground ice.

Craters of the Lunar-Mercurian Type

Craters of the lunar-Mercurian class typically are smaller than about 10 km in diameter, although the maximum size appears to be at least partly a function of terrain type. The ejecta pattern of craters of this class is similar to that typical of lunar and Mercurian craters. Continuous ejecta with coarse, disordered topography close to the rim grades outward into discontinuous ejecta with finer-scale topography. In some cases, beyond the continuous ejecta are concentric bands of ridges and hollows. The greatest frequency of this type of crater from the limited observations to date is on the lava flows south of Arsia Mons, where ejecta-flow craters are less frequent.

Craters With Pronounced Radially Patterned Ejecta

Relatively few craters of this type have been found thus far. Their strong radial pattern makes them distinctively different from typical lunar and Mercurian craters. The most spectacular occurs in Chryse Planitia, superimposed on the bank of Kasei Vallis (Figure 7). The ejecta of this 20-km-diameter crater can be subdivided into three zones based on surface morphology. The inner ejecta facies occurs as a discontinuous concentric band extending from the base of the crater rim outward for a maximum distance of 5 km. This band consists of sheets and small 'plates' of ejecta, some being imbricate

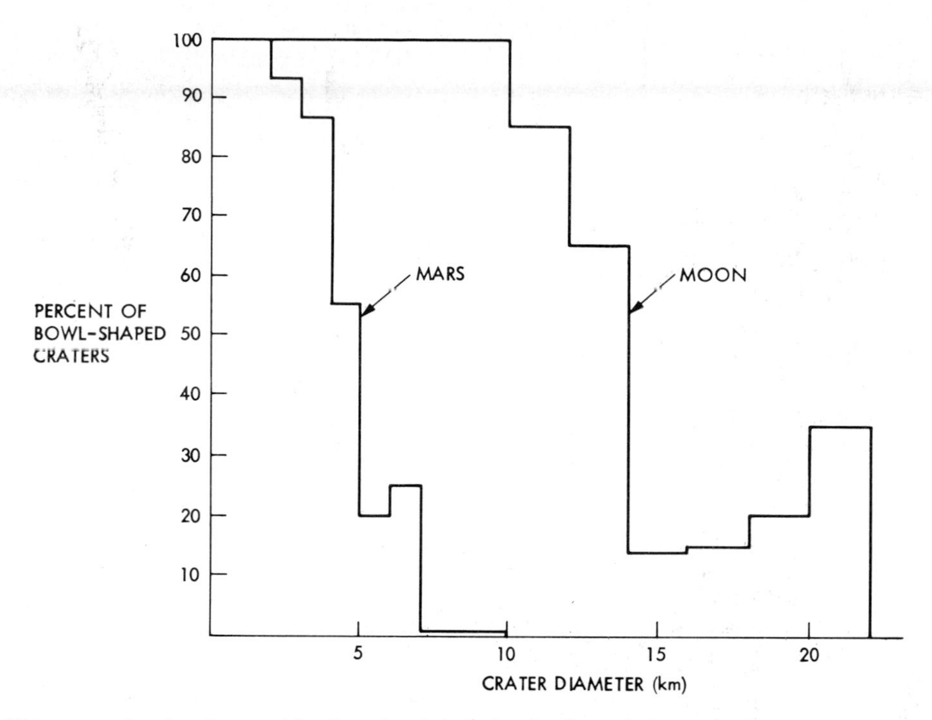

Fig. 9. Histograms showing the transition from bowl-shaped to flat-floored crater morphology for the moon and Mars as a function of rim diameter. The Martian data are based on 59 craters analyzed in this study; the 'break' on the 4- to 5-km histogram denotes that the value was indeterminate in this diameter range (no craters in this size range were measured). The lunar data are an analysis of data tabulated by *Smith and Sanchez* [1973] and are based on 47 craters with diameters less than 22 km. Smith and Sanchez observed no bowl-shaped craters on the moon with diameters exceeding 72 km.

with abrupt termini, forming a scarp estimated to be approximately 20 m high. Although some parts of the inner ejecta surface are radially grooved, most of the surface is smooth.

The intermediate ejecta facies extends outward as a continuous deposit to about 15 km from the rim, or about three fourths of a crater diameter, where it forms a ragged terminus and breaks up into the discontinuous radial clots of the outer ejecta facies. The surface of the intermediate facies has both radial grooves and radial ridges, the widths of which are relatively constant or increase slightly outward.

The outer ejecta facies of Kasei Vallis extends to a maximum distance of 30 km from the crater rim. It consists of discontinuous deposits of ejecta, some forming discontinuous ridges roughly radial to the crater. There are no well-defined radial depressions. In some cases, radial ridges can be traced roughly from the continuous intermediate ejecta zone outward into the discontinuous ejecta zone.

Several craters with radially patterned ejecta appear to be transitional with ejecta-flow craters, having characteristics of both forms. These transitional forms occur primarily in Chryse Planitia, but some examples do occur elsewhere. Although each is unique in some respects, all have zones of radial ridges (Figure 8). The nonridged part may have concentric bands of hummocky material or may be smooth or consist of flow type lobes. Many of the radial ridges and depressions at radially patterned ejecta craters extend to the base of the crater rim, but at no place do they extend up onto or over the rim. This relation suggests an origin within the ejecta deposits.

Interior Morphology

Although the ejecta around Martian craters commonly is distinctively different from that around lunar and Mercurian type craters, the differences in morphology of the crater interiors are more subtle. The final form of an impact crater depends on several factors, including properties of the target medium and gravity. Surface gravity on Mars and Mercury is nearly the same and thus should not cause differences in crater morphology. However, differences in surface materials may affect crater shapes, just as they are suspected of being the primary cause for differences in the ejecta patterns.

On Mars, as on the moon and Mercury, craters have a considerable variation in the planimetric form and morphology of rims, walls, and floors. Some have smooth, convex, bowl-shaped interiors; others have flat floors with sharp breaks in slope at the walls. Craters are also seen with single and multiple central peaks and complex central mountainous

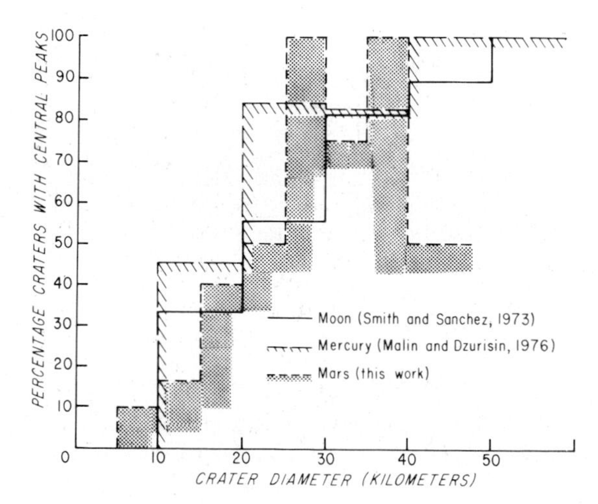

Fig. 10. Histograms showing the occurrence of central peaks and terraces as a function of rim diameter. The drop-off in central peaks at large diameters for Martian craters is a consequence of the inclusion of large degraded craters in the population analyzed.

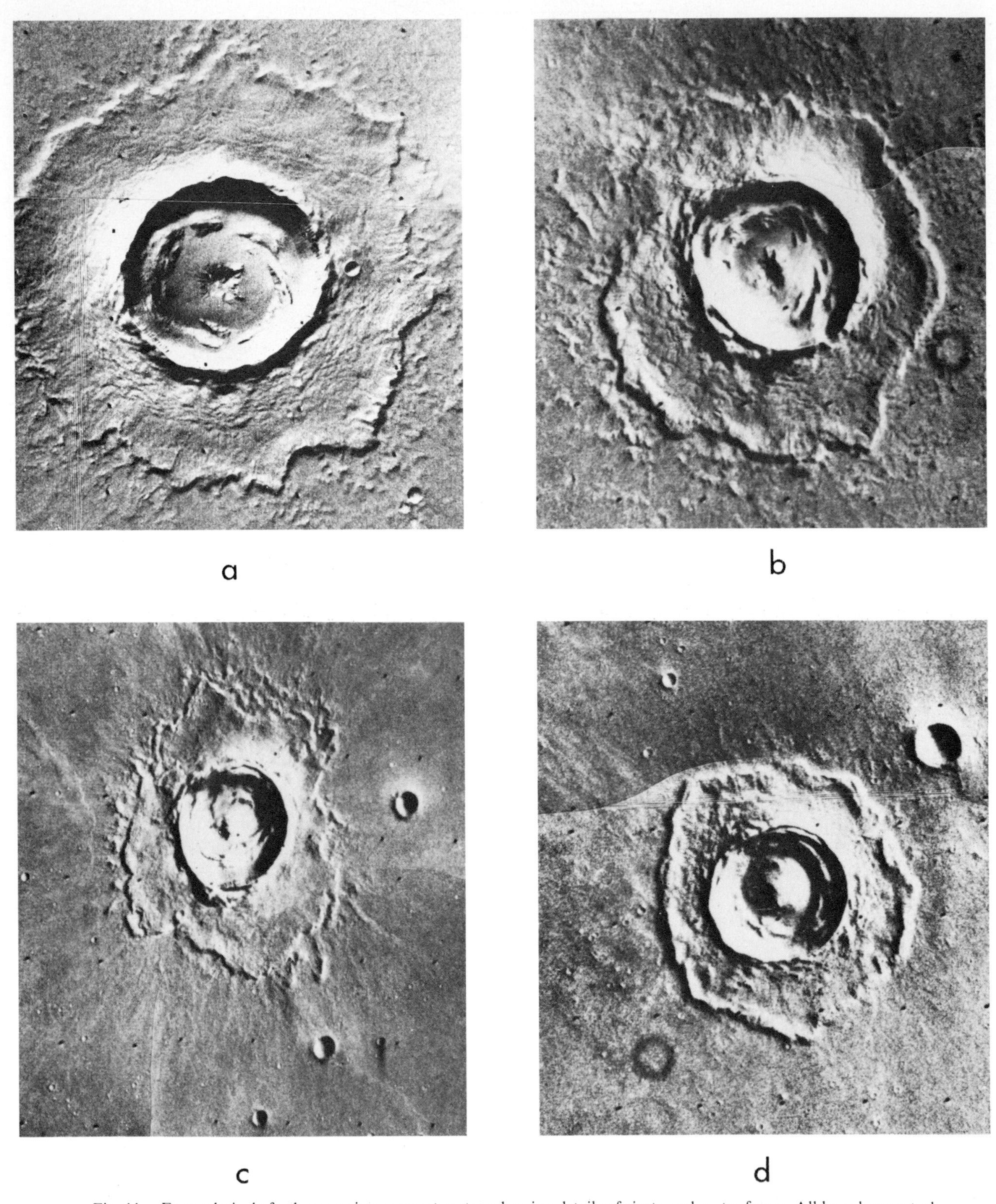

Fig. 11. Four relatively fresh appearing rampart craters showing details of ejecta and crater forms. All have low central peaks, arcuate terraces on the walls, and arcuate ridges on the floors. (*a*) 12.8-km crater at 23°N, 42°W (10A61, 10A62). (*b*) 9.5-km crater at 21°N, 43°W (10A56). (*c*) 10.2-km crater at 23°N, 45°W (10A21). (*d*) 7.4-km crater at 22°N, 31°W (6A52).

zones. Concentric ridge features occur in crater floors, and complex terracing is observed on some crater walls.

Students of cratering on the moon and Mercury have sought to systematize the morphological characteristics of craters in terms of crater dimensions. Many variations in crater morphology appear to be a function of scale [*Pike,* 1976]. For example, as the diameter increases, craters change from bowl shaped to flat floored, a change that results in a break in slope on a plot of crater depth relative to diameter. *Gault et al.* [1975] and, more recently, *Malin and Dzurisin* [1976] have compared morphological characteristics of craters as a function of diameter for the moon and Mercury. Their data for these two bodies, whose surface gravities differ by approximately a factor of 2, show that the crater diameters corresponding to the onset of central peaks, the onset of terracing, and the appearance of central peaks are not greatly different. Malin and Dzurisin have concluded that gravity controls neither the permanent nor the transient and postcratering modifications of the simple bowl-shaped crater form. Another interpretation of the same data is that differences in the other factors (e.g., substrate properties, density of impacting body) governing crater formation on the moon and Mercury have largely offset the effects of gravity.

From the Viking images, data on Martian craters have been obtained that are similar to data collected from Mariner 9 reported by *Cintala et al.* [1976, 1977]. Comparable data were compiled for the moon by *Smith and Sanchez* [1973]. The data strongly suggest (Figure 9) that the transition from bowl-shaped to flat-floored morphology occurs at much smaller sizes on Mars (4–5 km) than on the moon (10–12 km). We note parenthetically that Malin and Dzurisin found that the 'break' in the depth/diameter curve for both Mercury and the moon is very near 10 km, which corresponds to the onset diameter for flat floor.

A plot of the percentage of craters with central peaks as a function of crater diameter (Figure 10) shows that on the moon [*Pike,* 1976] and Mercury [*Gault et al.,* 1975] the development of central peaks is transitional over a diameter range of several tens of kilometers. The occurrence of terracing as a function of crater diameter (Figure 10) also is transitional. The statistics at this time are not good enough to permit exact comparisons of these curves with those for the moon and Mercury, but it appears that central peaks are common in Martian craters larger than 25 km in diameter.

The origin of arcuate ridges that appear on the flat floors of many Martian craters (Figure 11) is unclear. Their morphology and association with scalloped rims and crater terraces suggest mass movements from the crater rim. On the other hand, their location closer to the center of the crater than to the rim suggests a more deep-seated origin, such as that for central peaks and concentric ring structures. Hummocky floors, marelike fill, and concentric rilles, which occur in some large fresh lunar craters such as Tycho, are not seen in the Martian craters, but it is not clear whether such features ever formed on Mars, whether they did but were rapidly covered up by eolian materials, or whether their apparent absence is simply a resolution effect.

CONCLUSIONS

Only a small fraction of the Martian surface has, at the time of writing, been examined at the resolution of the Viking cameras. The ejecta of most of the craters in these areas has a distinctive morphology. It generally appears to consist of several layers, each of which terminates abruptly at its margin with a low ridge or escarpment. Deflection of the ejecta layers by low obstacles suggests that after ballistic ejection from the crater the ejecta continued its motion outward from the crater as a debris flow. Ground ice and entrained atmospheric gases are suggested as possible causes for this unique mechanism of ejecta emplacement as compared with the mechanisms on the moon and Mercury.

REFERENCES

Arvidson, R. H., A. Carusi, A. Coradini, M. Coradini, M. Fulchignoni, C. Frederico, R. Funicello, and M. Salomone, Latitudinal variation of wind erosion of crater ejecta deposits on Mars, *Icarus, 27,* 503–516, 1976.

Carr, M. H., H. Masursky, W. A. Blaum, K. R. Blasius, G. A. Briggs, J. A. Cutts, T. Duxbury, R. Greeley, J. E. Guest, B. A. Smith, L. A. Soderblom, J. Veverka, and J. B. Wellman, Preliminary results from the Viking orbiter imaging experiment, *Science, 193,* 766–776, 1976.

Chao, E. C. T., The Ries crater, a model for the interpretation of the source areas of lunar breccia samples (abstract), in *Lunar Science VII,* pp. 126–128, Lunar Science Institute, Houston, Tex., 1976.

Chao, E. C. T., The Ries crater of southern Germany—A model for large basins on planetary surfaces, *Geol. Jahrb.,* in press, 1977.

Cintala, M. J., J. W. Head, and T. A. Mutch, Characteristics of fresh Martian craters as a function of diameter: Comparison with the moon and Mercury, *Geophys. Res. Lett., 3,* 117–120, 1976.

Cintala, M. J., J. W. Head, and T. A. Mutch, Martian crater depth/diameter relationships: Comparison with the moon and Mercury, *Proc. Lunar Sci. Conf. 7th,* 3575–3587, 1977.

Gault, D. E., W. L. Quaide, and V. R. Oberbeck, Impact cratering mechanics and structures, in *Shock Metamorphism of Natural Materials,* edited by B. M. French and N. M. Short, pp. 87–99, Mono, Baltimore, Md., 1968.

Gault, D. E., J. E. Guest, J. B. Murray, D. Dzurisin, and M. C. Malin, Some comparisons of impact craters on Mercury and the moon, *J. Geophys. Res., 80,* 2444–2460, 1975.

Guest, J. E., Stratigraphy of ejecta from the lunar crater Aristarchus, *Geol. Soc. Amer. Bull., 84,* 2873–2893, 1973.

Guest, J. E., and J. B. Murray, Nature and origin of Tsiolkovsky Crater, lunar farside, *Planet. Space Sci., 17,* 121–141, 1969.

Howard, K. A., Ejecta blankets of large craters exemplified by King Crater, Apollo 16 Preliminary Science Report, *NASA Spec. Publ., SP-315,* 29-70–29-79, 1972.

Malin, M. C., and D. Dzurisin, Modification of fresh crater landforms: Evidence from Mercury and the moon (abstract), in *Papers Presented to the Conference on Comparisons of Mercury and the Moon,* pp. 25–26, Lunar Science Institute, Houston, Tex., 1976.

McCauley, J. F., Mariner 9 evidence for wind erosion in the equatorial and mid-latitude regions of Mars, *J. Geophys. Res., 78,* 4123–4137, 1973.

Moore, H. J., C. A. Hodges, and D. H. Scott, Multiringed basins—Illustrated by Orientale and associated features, *Proc. Lunar Sci. Conf. 5th,* 71–100, 1974.

Morrison, R., and V. Oberbeck, Geomorphology of crater and basin deposits—Emplacement of the Fra Mauro formation, *Proc. Lunar Sci. Conf. 6th,* 2503, 1975.

Mutch, T. A., S. V. Grenander, K. L. Jones, W. Patterson, R. A. Arvidson, E. A. Guiness, P. Arvin, C. E. Carlston, A. B. Binder, C. Sagan, E. W. Dunham, P. L. Fox, D. C. Pieri, F. O. Huck, C. W. Rowland, G. R. Taylor, S. D. Wall, E. C. Levinthal, S. Liebes, R. B. Tucker, E. C. Morris, J. B. Pollack, R. S. Saunders, and M. R. Wolf, The surface of Mars: The view from the Viking 2 lander, *Science, 194,* 1277–1283, 1976.

Oberbeck, V., The role of ballistic erosion and sedimentation in lunar stratigraphy, *Rev. Geophys. Space Phys., 13,* 337, 1975.

Oberbeck, V., R. Morrison, and F. Horz, Transport and emplacement of crater and basin deposits, *Moon, 13,* 9, 1975*a*.

Oberbeck, V., F. Horz, R. Morrison, W. Quaide, and D. Gault, On the origin of lunar smooth plains, *Moon, 12,* 19, 1975*b*.

Pike, R. J., Simple to complex impact craters: The transition on the moon (abstract), in *Lunar Science VII,* pp. 700–702, Lunar Science Institute, Houston, Tex., 1976.

Roddy, D. J., J. M. Boyce, G. W. Colton, and A. L. Dial, Jr., Meteor Crater, Arizona, rim drilling with thickness, structural uplift, diameter, depth, volume, and mass-balance calculations, *Proc. Lunar Sci. Conf. 6th,* 2621–2644, 1975.

Shoemaker, E. M., Interpretation of lunar craters, in *Physics and Astronomy of the Moon,* edited by Z. Kopal, p. 538, Academic, New York, 1962.

Shoemaker, E. M., R. M. Batson, H. E. Holt, E. C. Morris, J. J. Remilson, and E. A. Whitaker, Television observations from Surveyor VII, Surveyor VII: A Preliminary Report, *NASA Spec. Publ., SP-173,* 13–81, 1968.

Shreve, R. L., Sherman landslide, Alaska, *Science, 154,* 1639–1643, 1966.

Smith, E. E., and A. G. Sanchez, Fresh lunar craters: Morphology as a function of diameter, a possible criterion for crater origin, *Mod. Geol., 4,* 51–59, 1973.

Sparks, R. S. J., and G. P. L. Walker, The ground surge deposit: A third type of pyroclastic rock, *Nature, 241,* 62–64, 1973.

(Received April 1, 1977;
revised June 2, 1977;
accepted June 3, 1977.)

VOL. 82, NO. 28 JOURNAL OF GEOPHYSICAL RESEARCH SEPTEMBER 30, 1977

Geology of the Valles Marineris: First Analysis of Imaging From the Viking 1 Orbiter Primary Mission

KARL R. BLASIUS AND JAMES A. CUTTS

Planetary Science Institute, Pasadena, California 91101

JOHN E. GUEST

University of London Observatory, London, England

HAROLD MASURSKY

Astrogeology Center, U.S. Geological Survey, Flagstaff, Arizona 96001

The Valles Marineris, an enormous canyon system spanning more than one quarter of the equatorial girth of Mars, exhibits in its landforms the consequences of uniquely Martian extensional tectonics and a variety of erosional and depositional processes. Reported here are new insights into the evolution of the canyon system and possible evidence for cyclical climate change from the equatorial region. Tectonic control appears to be the fundamental influence on canyon form and evolution, but the style or intensity of tectonism appears to be regionally variable. In the region of the west and central canyons, chains of elongate pits, graben, and the actual troughs are all inferred to be manifestations of pronounced north-south and secondary east-west crustal extension. It is proposed that this region of the Valles Marineris is made up of a large number of discrete elongate blocks which have shifted vertically and tilted in relation to one another. Depending on the geometry of this adjustment the surface layer subsides as a coherent block or collapses to form a chain of pits. In the eastern canyons and chaotic terrain the tectonics appear to follow a different pattern, as the crustal layer breaks up into large patches of equant blocks, seemingly reflecting a lesser amount of crustal extension. Diverse wall features and varied landslide morphologies in the canyons indicate that material has been transported from the walls of the troughs to the floors, contributing to their widening. The paucity of small impact craters inside the canyons suggests that this process spans a large part of Martian history, perhaps even up to the present day. Small, sharply defined scarps cut erosional features on many canyon walls, suggesting that the driving force for canyon enlargement continues to be downfaulting. This conclusion is supported by the absence of evidence for excavation by exogenic process. Many troughs are segmented into closed basins, precluding lateral transport of debris except by aeolian deflation, for which there is little evidence. Substantial accumulations of regularly layered sediments have been recognized on the floors of at least two unconnected canyons. Cyclical variations in sedimentation rates or conditions are implied. The stratigraphic relationships among various layered and unlayered canyon interior materials as well as dune fields require that substantial deposition and reerosion have taken place within some canyons since their formation. The pervasive influence of extensional tectonics in the Valles Marineris leads us to doubt that the decay of ground ice or volcanism needs to be invoked to explain collapse in the eastern canyons or chaotic terrain. We find no persuasive evidence for former fluvial episodes in the trough system proper, where most erosional features can be accounted for by mass wasting. There are suggestions, however, that cyclical climate change has played a role in the development of layered materials in the canyons in the same way that it has apparently controlled layering of deposits in the polar regions. Prolonged tectonic activity in the Valles Marineris, on the east flank of the Tharsis-Syria Rise, may correlate with prolonged volcanism and tectonism in evidence on the west flank. The long-lived dynamic process in the deep Martian interior suggested by this relationship may also be responsible for large gravity anomalies associated with the rise.

Within the Valles Marineris, an enormous group of steep-walled equatorial canyons extending over an area more than 4000 km long and up to 700 km wide (Figure 1), the most complex geological relationships seen at the surface of Mars are magnificently displayed through variations in terrain morphology and coloration. The Viking 1 orbiter imaging experiment has added greatly to our knowledge of these features (Figure 2), which were first seen in images returned by the Mariner 9 orbiter during 1971 and 1972. Mariner 9 mapping images, with surface resolution of 1–1.5 km, were adequate to define the gross physiography of the Valles Marineris system, and a small sample of higher-resolution images (about 3% of the surface at 100- to 150-m resolution) led to the identification of spectacular landslide features and layered rocks on the canyon walls and floors. However, these data were inadequate to seriously pursue the problem of the prime mode of origin and the detailed subsequent evolution of the Valles Marineris. The Viking images show new features in the canyons and clarify our perceptions of others; so major outstanding questions can be addressed anew.

This report is an initial survey and interpretation of the images of the Valles Marineris region acquired during the Viking primary mission (July to November 1976). These data provide an increase of approximately a factor of 10 in resolution (to 100–150 m) over most of the canyon system. Moreover, a substantial number of data have been acquired in stereo, further enhancing the visibility of critical geologic relationships. These new data have allowed us to discern vertical displacements along faults at the bases of canyon walls, giving support to the hypothesis that crustal subsidence was important in initiating canyon formation and furthermore has been a continuing influence in canyon evolution. We have also seen new details of canyon floor morphology—scarps, closed depressions, fractured surfaces, dunes, and great lobes of hum-

Paper number 7S0524.

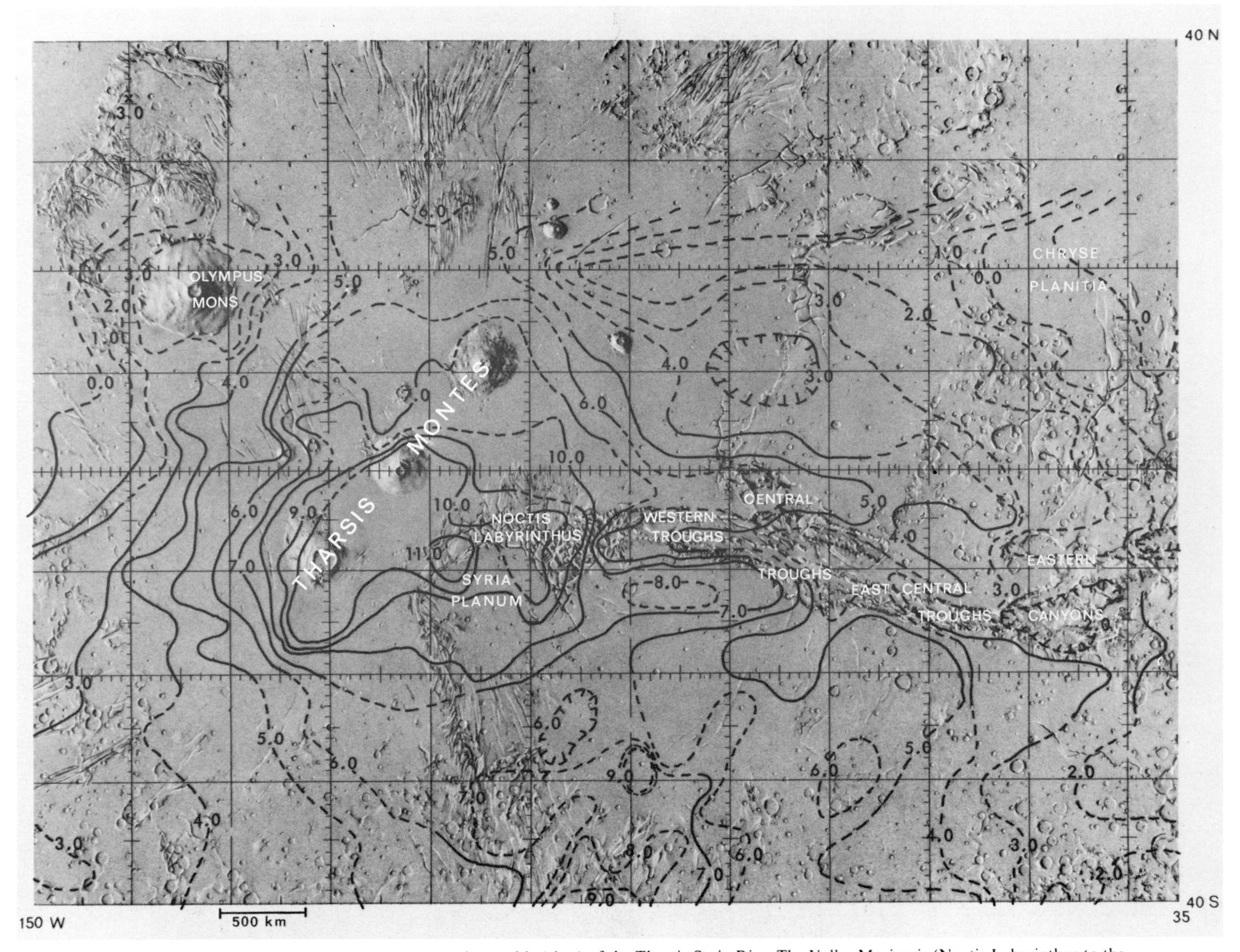

Fig. 1. Contoured topographic map (contour interval is 1 km) of the Tharsis-Syria Rise. The Valles Marineris (Noctis Labyrinthus to the Eastern Canyons) extend from the summit of the rise down the crest of a ridgelike swell on its eastern flank. Young volcanic plains occur to the west around the shields Olympus Mons and Tharsis Montes. The mercator shaded relief map base was provided by the U.S. Geological Survey. Topographic data were compiled by *Christensen* [1975].

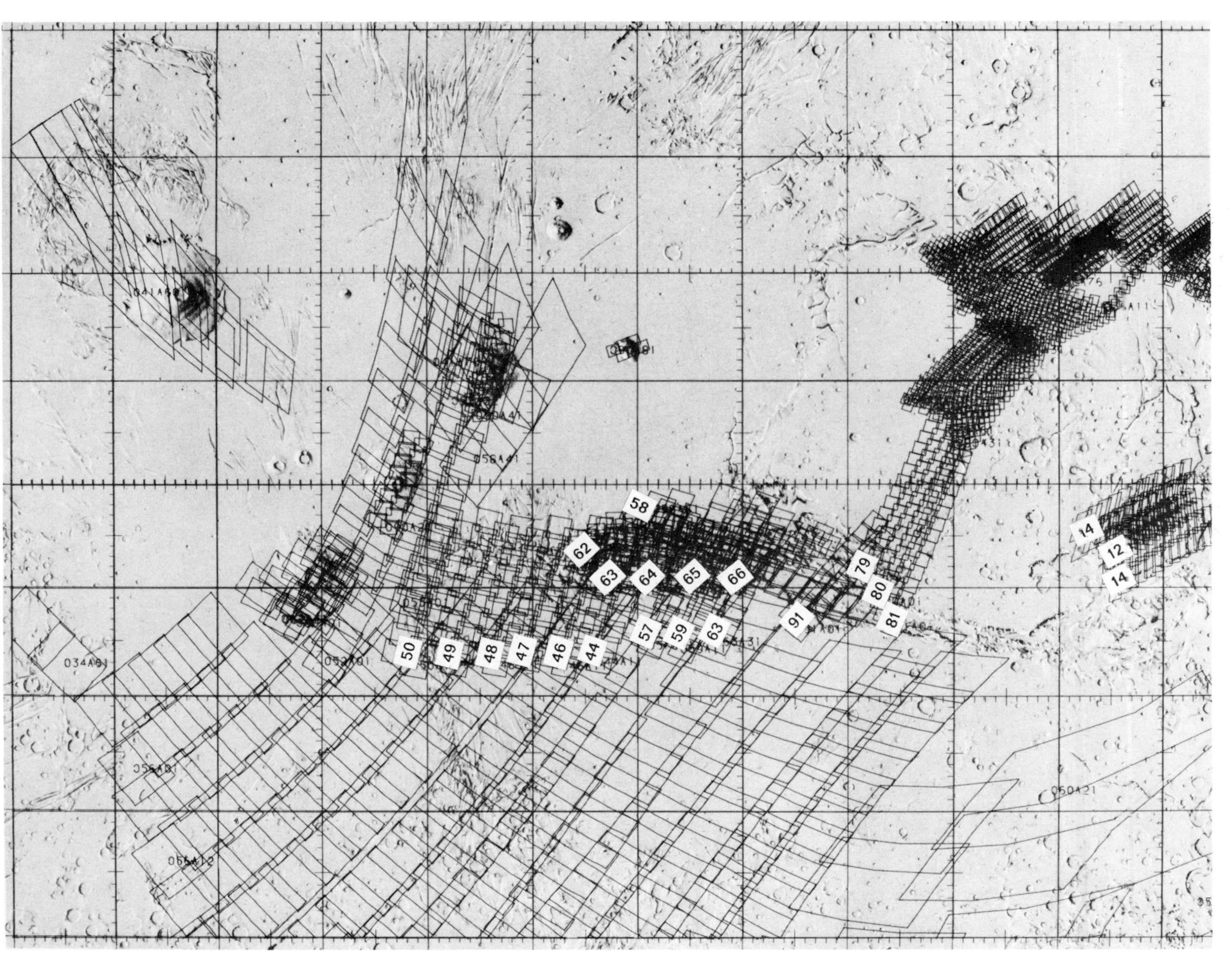

Fig. 2. Footprints of high- and moderate-resolution images of the Tharsis-Syria Rise, including the Valles Marineris, acquired by the Viking 1 orbiter during its primary mission (July to December 1976). Strips of frames covering parts of the canyons are numbered at their beginnings according to the orbits on which they were taken (see Table 1).

mocky and striated terrain where canyons have widened catastrophically during great landslides. Clear evidence has also been found for widespread cyclical sedimentation in the past, producing extremely regularly layered materials exposed within unconnected canyons hundreds of kilometers apart. Substantial sedimentation may postdate, at least locally, episodes of tectonism and erosion which produced portions of the canyons.

In this paper we first survey the gross physiography of the Valles Marineris, as defined by Mariner 9 and now clarified by Viking 1. We then describe the variety of floor and wall morphologies observed in the canyon system and discuss their interpretation in terms of tectonic, depositional, and erosional processes. Finally, we attempt a synthesis of the regional tectonic and erosional patterns and discuss the key scientific issues presented by the Valles Marineris system.

The terminology used in this paper to describe the Valles Marineris (literally, the Mariner Valleys, in honor of the achievements of the Mariner spacecraft) is drawn from official International Astronomical Union approved map nomenclature, published papers in descriptive geology, and *Webster's Collegiate Dictionary* (fifth edition): vallis, valles (valley, valleys), an elongate depression; chasma (canyon), an elongate steep-sided depression; labyrinthus (valley labyrinth), a labyrinthine complex of valleys; trough, an elongate depression with steep and generally parallel walls; planum (plateau), smooth elevated area; and catena (crater chain), chain or line of craters.

AN OVERVIEW OF VALLES MARINERIS

Occupying a region measuring 4500 km east-west and 150–700 km north-south, the great equatorial canyons of Mars, the Valles Marineris, lie along the crest of a broad, ridgelike swell on the eastern flank of the Tharsis-Syria Rise (Figure 1). For convenience in description we have divided the canyons into five sections based on physiographic differences. From west to east these subdivisions are Noctis Labyrinthus, the Western Troughs, the Central Troughs, the East Central Troughs, and the Eastern Canyons.

Canyon System Subdivisions

Farthest to the west is Noctis Labyrinthus (Figure 1), a canyon network 1000 km across, located at the summit of the Tharsis-Syria Rise 10 km above the standard Mars areoid datum (a reference ellipsoid determined from Mariner 9 radio occultation and gravity tracking data [*Christensen,* 1975]). These canyons for the most part show less evidence of erosion and are rather narrower and shallower than the other Valles Marineris.

Three subdivisions of the Valles Marineris (Western, Central, and East Central troughs) comprise the 'troughed terrain' of *Sharp* [1973*a*], a group of linear steep-walled valleys, typically 300–1000 km long and 50–150 km wide and trending approximately E15°S. The troughs are aligned parallel to each other, two to four wide, separated by ridges or plateaus similar in dimensions to the troughs themselves. Some troughs are joined laterally to form composite features up to 300 km wide or lengthwise to form features up to 2500 km long. Troughs also terminate in places, either with broad blunt ends or by gradually pinching out.

Extending eastward from the labyrinth for 700–900 km are the Western Troughs (Figure 1). These two parallel canyons range from less than 50 km to 150 km in width and are bounded by steep walls indented by numerous alcoves tens of kilometers across. The upland surface to the south is dissected by prominent tributary canyons extending up to 150 km back from the trough wall.

The Central Troughs are four parallel valleys, generally broader than the Western Troughs and only partially integrated with each other and the troughs to the east and west. The Valles Marineris system broadens considerably here, extending from the equator to 14°S, and individual troughs are up to 850 km long. The northernmost troughs are unconnected to other Valles Marineris, one being a completely closed depression and the other opening to plains to the north.

The two parallel East Central Troughs extend out of the south member of the Central Troughs for about 900 km to the ESE. Together these troughs are about 200 km wide and have the most strikingly linear wall trends in the troughed terrain.

The Eastern Canyons, connected at their southwest limit to the East Central Troughs, are generally broader and shallower than the other Valles Marineris. They are intermingled with and they open into chaotic terrain, areas of collapsed cratered upland which appear to evolve into hilly terrain [*Sharp,* 1973*b*]. These canyons form the east end of the Valles Marineris, cutting across cratered terrain as low as about 1-km elevation. In this region the dominant ESE trend of the canyons is lost, and a NE-SW trend emerges. Like the Central Troughs the broad Eastern Canyons extend across an enormous area measuring approximately 1000 km E-W and 900 km N-S.

The Valles Marineris cut across upland surfaces of moderately to heavily cratered plains and (east of 55° longitude) heavily cratered terrain [*Carr et al.,* 1973]. The ages of these surfaces, as evidenced by the numbers of superimposed impact craters, contrast sharply with the sparsely cratered, relatively fresh appearing volcanic plains on the north and west flanks of the Tharsis-Syria Rise. Immediately adjacent to the canyons the plains display prominent to subdued fault scarps aligned mostly parallel to canyon wall segments.

The floors of the Valles Marineris are also sparsely cratered or uncratered, implying either a young age for their excavation or an active erosional or depositional regime renewing floor surfaces. The canyons need not be entirely very young, however, as great lobes of hummocky terrain have blanketed large areas. These features have been interpreted as landslides of tremendous proportions, up to 100 km across [*Sharp,* 1973*a*]. Other areas of canyon floor are relatively smooth or characterized by ridges or plateaus rising up to at least 2 or 3 km. Within some broader segments of canyons, plateaus are seen which have a distinctly layered structure [*Masursky,* 1973; *Sharp,* 1973*a*].

Ultraviolet spectrometer elevation profiles [*Hord et al.,* 1974] across the canyons at six locations (Figure 3 [from *Malin,* 1976]) show relief ranging from 2 to 7 km. Slopes at or near the angle of repose for unconsolidated materials are implied and confirmed by limited photogrammetric measurements [*Blasius,* 1973; *Wu et al.,* 1973]. The dominant scarp morphology within the canyons is fluted spurs separated by smooth-floored gullies, similar to steep terrestrial scarps in desert or alpine environments where the dominant process of landscape evolution is dry mass wasting [*Sharp,* 1973*a*; *Lucchitta,* 1977]. The limited high-resolution imaging of Mariner 9 showed evidence of layering also near the tops of canyon walls—terraces, lines of knobs, and albedo bands [*Malin,* 1976].

Other common patterns of trough wall morphology are formed by dissection by branching tributary canyons and mas-

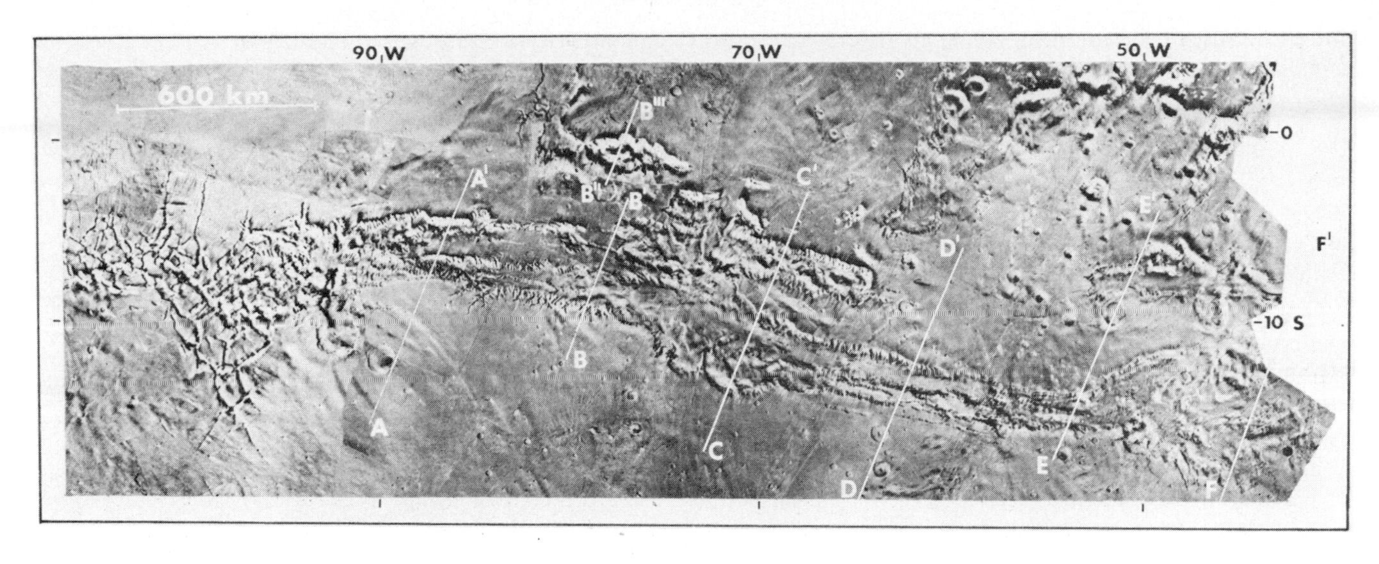

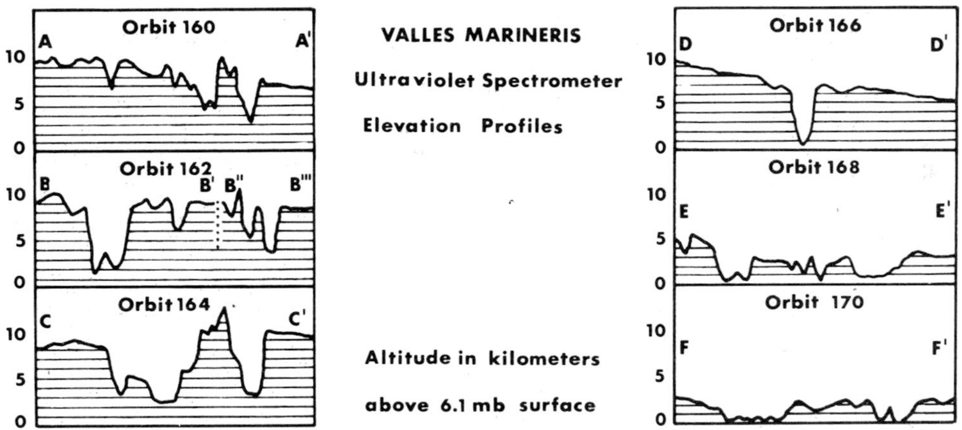

Fig. 3. Elevation profiles across the Valles Marineris derived from surface pressure measurements by the Mariner 9 ultraviolet spectrometer [*Hord et al.*, 1974]. Figure from *Malin* [1976].

sive landsliding, the latter producing broad alcoves bounded by high scarps which lack spurs and gullies.

Imaging and Topographic Data

In 1971 and 1972, Mariner 9 photographed the Valles Marineris at a resolution of about 500 m per picture element and acquired a small sample, approximately 50 frames showing a total of 8×10^4 km^2, at a resolution better by a factor of 10. Topographic data were also acquired by the radio occultation and ultraviolet and infrared spectral experiments. Combined with ground-based radar altimetry these data form a topographic data set [*Christensen,* 1975], of somewhat uneven spatial resolution, for the entire Valles Marineris system. A contoured map of these data is shown in Figure 1.

During the period of the Viking 1 orbiter primary mission, June to November 1976, 315 images were acquired, showing parts of the Valles Marineris at resolutions ranging from 75 to 130 m per picture element (Table 1). Footprints of individual photographs of these and neighboring Viking imaging data are shown in Figure 2. Most of these data were acquired primarily for geological interpretation as monoscopic coverage viewing the surface no more than 30° from vertical through a clear filter. Lighting conditions for these images are generally more nearly vertical than is desirable for geologic imaging of other parts of Mars (sun elevation angles were mostly 50°–60°) but appear to be near optimum for photographing the steep scarps

TABLE 1. High-Resolution Imaging of the Valles Marineris by Viking Orbiter 1 During the Primary Mission

Image*	Target Areas	Comment
12A51–12A70	Gangis	
14A29–14A56	Gangis, Capri	
14A57–14A80	Gangis	
44A14–44A28	Labyrinth, Ius, Tithonium	Stereo 1
46A13–46A28	Labyrinth	Stereo 1
47A17–47A28	Labyrinth	
48A21–48A28	Labyrinth	
49A22–49A28	Labyrinth	
50A14–50A28	Labyrinth	
57A38–57A48	Ius, Tithonium	Stereo 1
58A61–58A92	Ius, Tithonium, Melas, Coprates	Violet
59A16–59A28	Ius, Tithonium, Echus	Stereo 1
62A61–62A71	Labyrinth	Stereo 2
63A38–63A48	Ius, Tithonium, Hebes, Candor	Stereo 1
63A51–63A71	Labyrinth, Ius, Tithonium	Stereo 2
64A01–64A27	Labyrinth, Ius, Tithonium	Stereo 2
65A01–65A27	Ius, Tithonium, Candor	Stereo 2
66A01–66A28	Ius, Candor, Ophir	Stereo 2
79A31–79A38	Ophir	Stereo 2
80A01–80A14	Coprates, Ophir	
81A01–81A21	Coprates, Ophir, Juventae	
91A01–91A18	Melas, Ophir	Stereo 1

*The first two digits of the image number indicate the orbit, the letter indicates the spacecraft, and the last two digits indicate the frame.

and prominent albedo contrasts within the Valles Marineris. A sample of this coverage, showing Noctis Labyrinthus and some adjacent troughed terrain to the east, is shown in Figure 4.

Additional imaging coverage was acquired to study atmospheric phenomena and to view surface features stereoscopically. For example, high-resolution images taken through violet filters were acquired on orbit 58 (shown in part in Figure 5) to search for condensate clouds within the canyons. No discrete clouds were observed, and the overall impression of all the imaging coverage is that the Martian atmosphere was remarkably clear during the Viking primary mission in comparison with much of the period of Mariner 9 observation.

On orbits 62–66, 114 frames of near-vertical contiguous coverage were taken of parts of the Western and Central troughs and Noctis Labyrinthus (Figures 2 and 6). These frames make a large block of stereo coverage when paired with slightly oblique images acquired on orbits 44, 46, 57, 59, and 63 (Table 1). The visibility of surface relief and the enhanced areal resolution achieved when viewing a stereo pair make these data extremely valuable for geological interpretation.

The Valles Marineris are also shown in many low-resolution or highly oblique images taken primarily to observe color differences or atmospheric phenomena [*Briggs et al.,* 1977]. Examples of such coverage which show the canyons in a regional setting are two mosaics taken from oblique high-altitude perspectives (Figures 7 and 8).

Canyon Physiography

We now examine in more detail the major physiographic provinces of Valles Marineris and their interrelationships. This description is based in large part on selected low-resolution Viking images, which show large-scale features more clearly than did the Mariner 9 mapping images.

Noctis Labyrinthus. This canyon complex consists of an intricate pattern of intersecting valleys, some clearly graben, dividing the cratered upland into a mosaic of blocks. In Figure 7, part of a group of images taken through green filters from a range of about 24,000 km, the labyrinth is partially filled with morning clouds; however, one can discern that scarps bounding individual canyons there are very straight to strongly cuspate in plan, the latter pattern seeming to be transitional into strings of discrete pits. The surrounding cratered plains of the summit of the Tharsis-Syria Rise are crossed by prominent parallel fractures which extend out from the labyrinth; north trending fractures occur to the north, and southeast trending fractures to the southeast. Higher-resolution images (Figure 4) show clearly the structural origin of most canyons in the west labyrinth. Floors of straight-walled canyons are for the most part faulted and cratered plains like the surrounding upland surface. In the eastern labyrinth the more irregular canyon walls and rough floor terrains suggest that erosion and deposition have been more significant.

Western Troughs. Out of the labyrinth to the east extend several parallel linear troughs, of which the two largest are, to the north, Tithonium Chasma, and to the south, Ius Chasma (Figure 7). These troughs are generally narrower than the others to the east. Ius is choked by landslide deposits beneath alcoves which interrupt the otherwise linear trend of its north wall (Figure 6), and tributary canyons extend up to 150 km into the cratered upland plain from its south wall. Scattered shorter side canyons occur along most walls of the Tithonium and Ius chasmas. From west to east the two largest troughs change in width and depth in a complementary fashion. Tithonium Chasma begins as a trough about 100 km across and 200 km long and narrows to about 50 km in width for the next 150 km, but farther to the east it ends in a string of slightly elongate closed depressions 10–50 km across. Ius Chasma, on the other hand, is not segmented and broadens gradually and deepens from west to east (AA′ and BB′ in Figure 3). For about 300 km the trough is only about 50 km wide, but beyond that distance, where a catena on the upland intersects the south trough wall (C in Figure 6), Ius Chasma gradually broadens to 100–150 km before opening into Melas Chasma. East of the point where the catena intersects Ius Chasma, the canyon is composed of two parallel troughs separated by a prominent ridge, which at places appears to crest near the level of the cratered upland plain.

The Ius and Tithonium chasmas cut across moderately cratered plains to the west and heavily cratered plains to the east. There is no strong change in large-scale canyon morphology across this boundary, but the plains themselves are distinct in character. South of Ius Chasma the heavily cratered plains are crossed by prominent NNE trending ridges (Figure 7), similar to those seen on lunar maria, while the moderately cratered plains appear relatively smooth. Catena aligned parallel with the troughs occur on the upland plains between them and north of westernmost Tithonium Chasma.

Central Troughs. The midsection of the troughed terrain is made of of four extremely broad canyons. From south to north they are the Melas, Candor, Ophir, and Hebes chasmas (Figure 7). West of Hebes but separated from it by a narrow ridge is Echus Chasma, a small trough which opens to smooth plains to the north by way of a broad, shallow, steep-walled canyon. The floor of Echus Chasma is uniquely smooth among the Valles Marineris, but it is similar to and contiguous with the smooth floor of the broad canyon and the plains to the north. The floor of Melas Chasma, linking Ius in the west to Coprates in the east, appears relatively smooth in its central part, but to the south where the chasma wall is broadly arcuate in plan, several plateaus, locally as high as 2 km where the trough is 5 km deep (CC′ in Figure 3), rise above the canyon floor. To the north a gap in the trough walls links Melas Chasma to Candor Chasma. Across Candor to the north, another gap links Candor Chasma to west Ophir Chasma. Two ridges or plateaus, elongate generally north-south, stand above the floor of eastern Candor, between the gaps in the canyon walls. Plateaus and a shelf rising 2–3 km above the trough floor are also present in the south half of east Ophir Chasma (CC′ in Figure 3). The foot of the north bounding scarp of east Ophir Chasma is remarkably linear, whereas the brink is complexly arcuate and the scarp face is deeply gullied.

The two broadest troughs, Melas and Candor–east Ophir, are both joined to the east and west to pairs of parallel structures (Figure 7). Melas is open to the two parallel troughs of Ius Chasma to the west and the similar parallel troughs of Coprates to the east. The Candor–east Ophir trough has two parallel smaller troughs or catena extending beyond its broad blunt ends, both to the east and to the west. The extremely broad troughs thus give the impression of having formed by the lateral merger of two parallel structures.

Farther north the troughs of west Ophir Chasma and Hebes Chasma (the latter a completely closed depression) have less of the strongly linear character typical of other troughs. Each is distinctly elongate in the direction of the overall trend of the Valles Marineris system, but their walls are more irregular in plan, being formed of numerous arcuate segments. There is a prominent plateau about 3 km high within the 3- to 4-km-deep

Fig. 4. Mosaic of images of Noctis Labyrinthus taken by Viking 1 orbiter on orbits 44, 46, 47, 48, 49, and 50 (see Table 1). Many canyons have a classic graben form with the upland plain surface preserved on the valley floor (E and F). Other canyons are more irregular in form and have smooth to rough floor textures (H). Canyon trends are continued in places by strings of discrete pits (A and B).

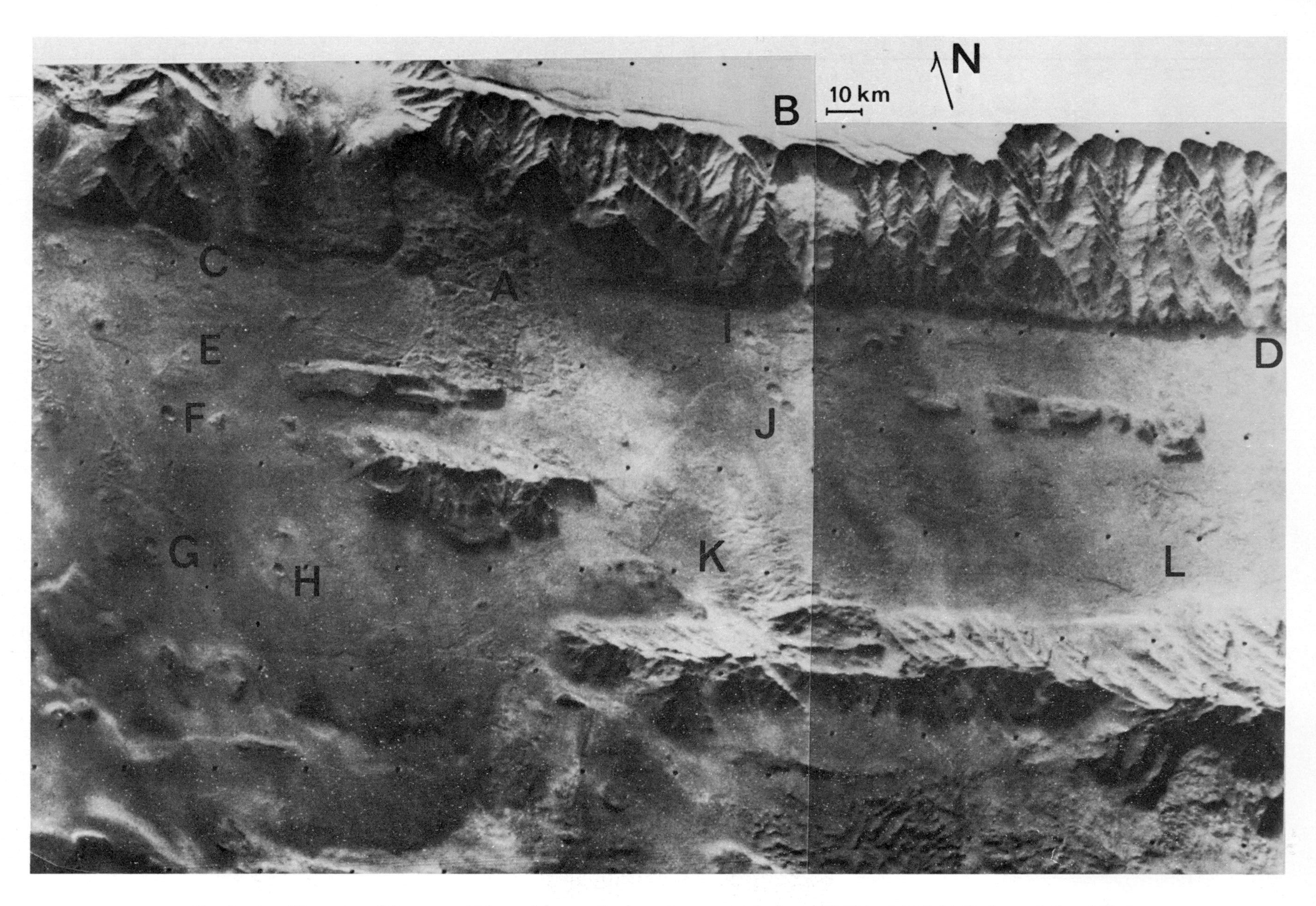

Fig. 5. An oblique view of the eastern Melas and western Coprates chasmas acquired on orbit 58 in violet light. To the north the relatively flat canyon floor is bounded by a low steep scarp (ID) which truncates spurs and gullies on the canyon walls. Landslide lobes and scars (A, K, and J) show that considerable erosion has occurred after gullying and the formation of the low steep scarp. Craters (E–I) indicate a surface of considerable age, perhaps downfaulted upland. Coprates Chasma is interpreted as an erosionally broadened but still tectonically active rift valley.

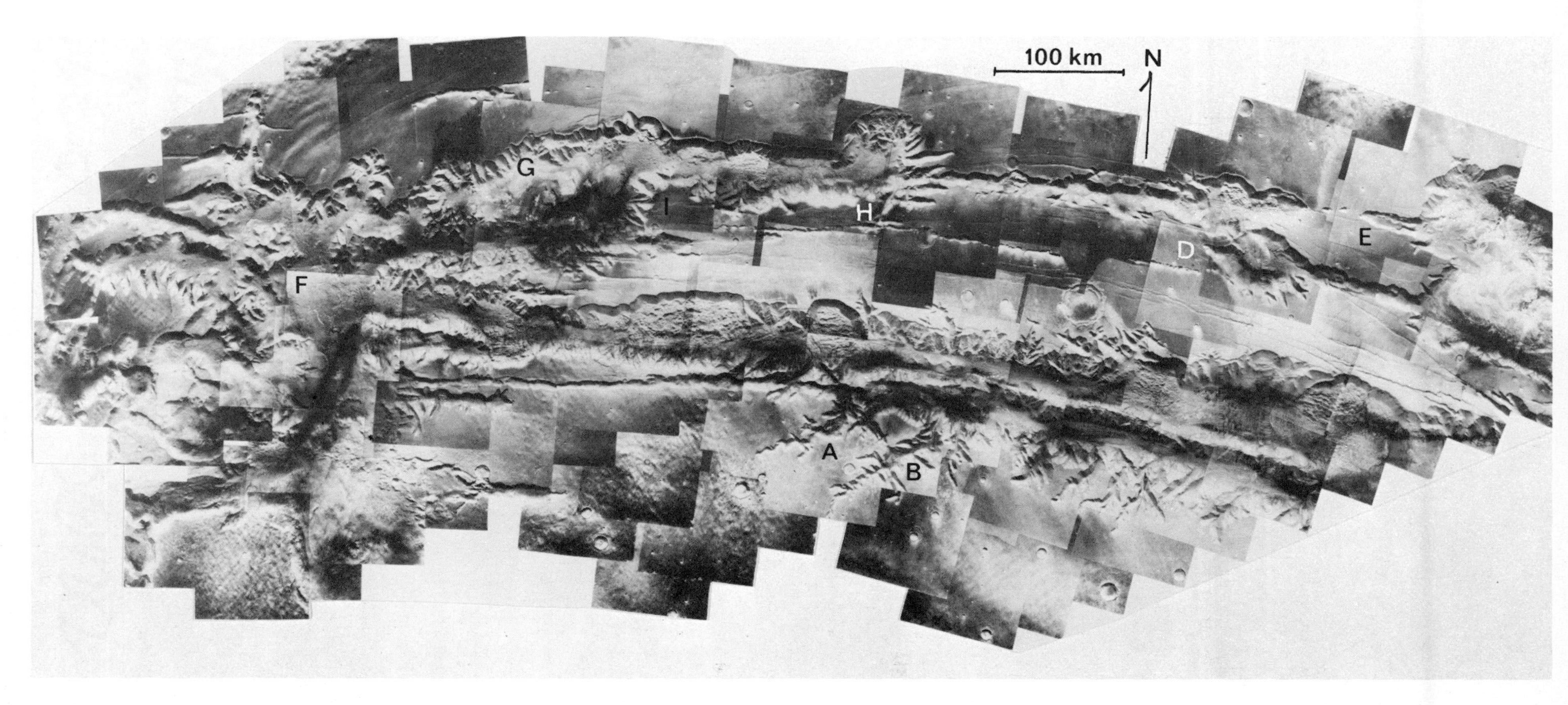

Fig. 6. Mosaic of images of the eastern labyrinth, the Western Troughs, and part of the Central Troughs, Candor Chasma, taken on orbits 62–66 to form contiguous stereo coverage of an extended region. Note the lobes of rugged terrain beneath broad canyon wall alcoves. Layered bright and dark materials occur on the canyon floor (at G), and a similar (possibly the same) sequence is seen in the canyon wall (west of H). Branching tributary canyons are best developed on the south wall of Ius Chasma (A, B, etc.).

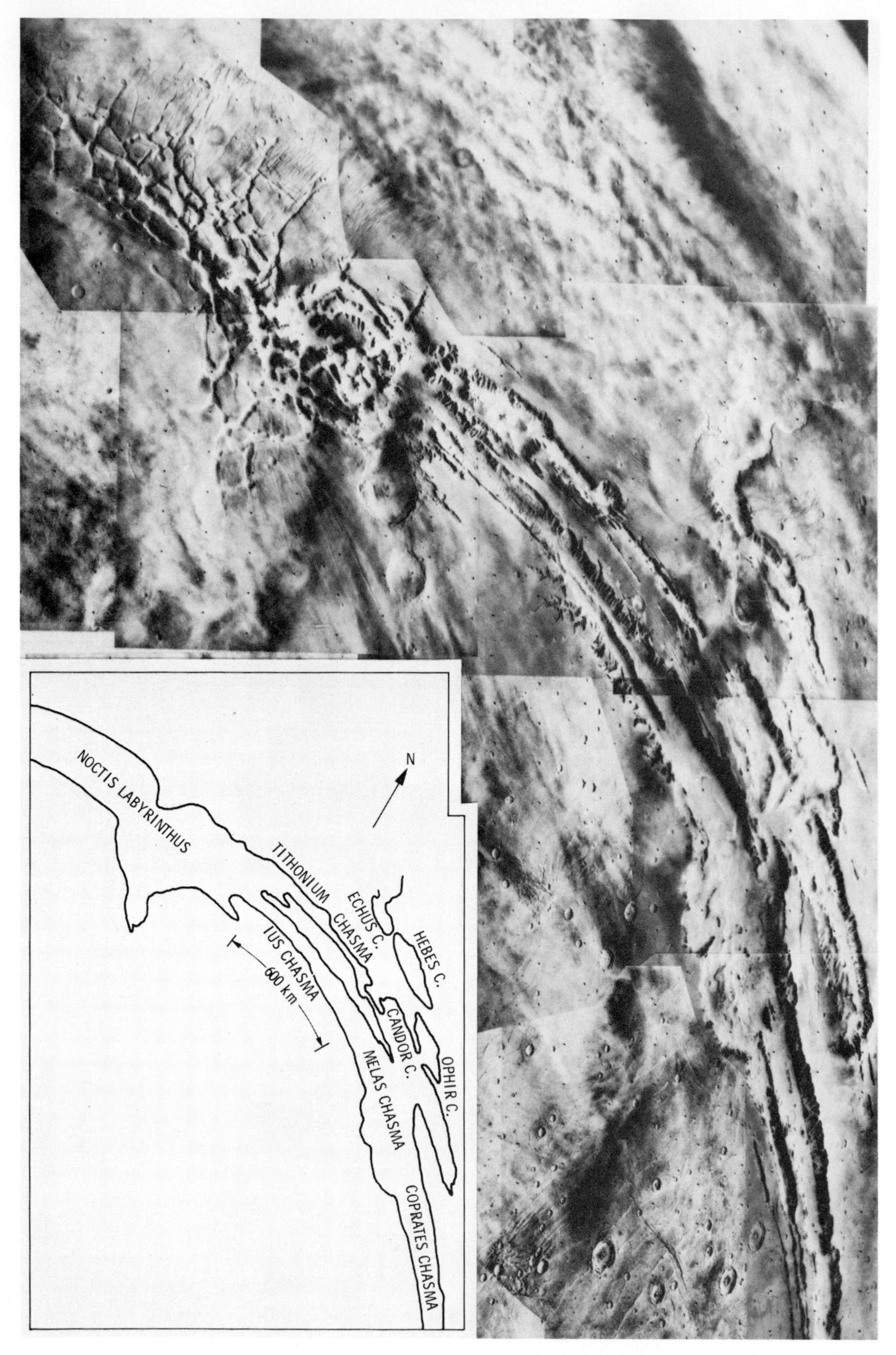

Fig. 7. Mosaic of images of Noctis Labyrinthus and the troughed terrain taken from high altitude in green light on orbit 40. Names of the individual canyons (chasmas) are shown in the key. Note the poor integration of the canyon system, including large variations in apparent depth along Tithonium Chasma.

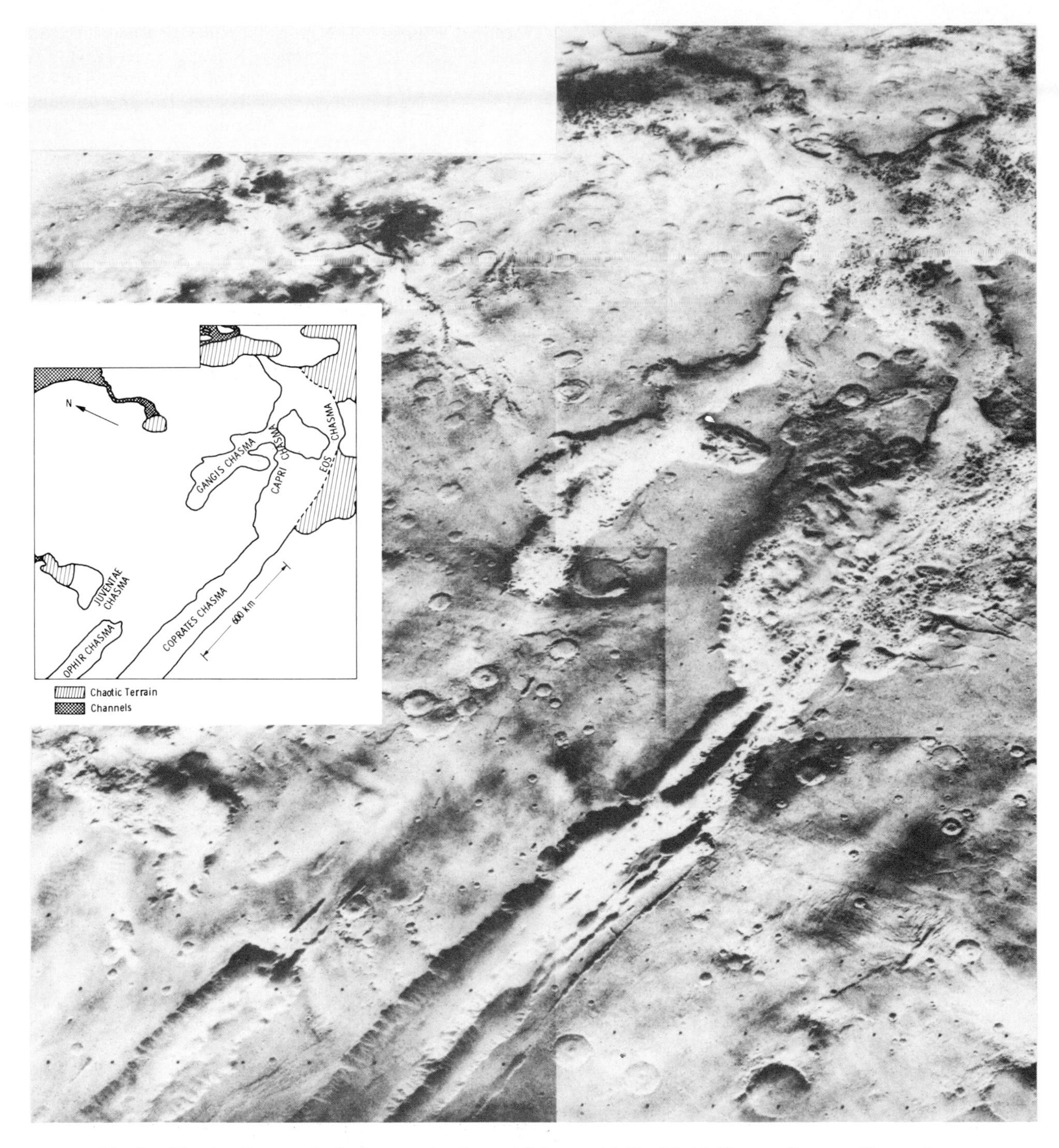

Fig. 8. Mosaic of low-resolution images taken in green light on orbit 32 of Ophir Chasma, Coprates Chasma, Juventae Chasma, and the broad shallow Eastern Canyons Eos, Capri, and Gangis. Juventae Chasma and the Eastern Canyons have associated with them patches of chaotic terrain, apparently produced by collapse and erosion of the cratered upland. Channels beginning at the chaotic terrain extend down into the Chryse Basin, where the Viking 1 lander set down. In contrast to the irregular forms of the Eastern Canyons, Ophir and Coprates are troughlike in form and have remarkably linear structures associated with them. Long catena (crater chains) parallel Coprates and extend from the blunt end of Ophir.

Hebes Chasma (B″ B″ in Figure 3), and a ridge or plateau rises from the southwest floor of Ophir. To the east of these troughs but unconnected to them, catena continue the general ESE regional structural trend.

East Central Troughs. To the east of Melas Chasma extend two parallel troughs comprising Coprates Chasma. Aside from a few mountains elongate parallel to the trend of the canyon walls the floor of the larger Coprates trough appears very smooth. The gullied north wall of that trough has an extremely straight foot but a more irregular brink, similar to the north wall of east Ophir Chasma.

The east end of Coprates Chasma, shown in a mosaic of pictures acquired through green filters on orbit 32 (Figure 8), crosses from heavily cratered plains into heavily cratered terrain with greater local relief. Within 200 km of that transitional terrain boundary the canyons become shallower and more irregular in plan and merge with large patches of chaotic terrain.

Eastern Canyons. Coprates Chasma opens into a broad, relatively shallow depression which includes the west end of Capri Chasma and Eos Chasma. This depression is partly bounded by low steep scarps, but to the south and northeast there is a transitional boundary with large patches of chaotic terrain. The floors of Capri and Eos are also irregularly hummocky, except in central Eos and eastern Capri. North of Capri Chasma and joining it to the east is Gangis Chasma, a complex of several tongue-shaped depressions which is elongate generally east-west. Like Capri and Eos, Gangis Chasma is shallow in comparison with portions of the troughs to the west (EE′ and FF′ in Figure 3), and some of its floor has the hummocky character of chaotic terrain. Plateaus rise from the floors of both the Capri-Eos depression and Gangis Chasma. Between the central sections of the Capri and Eos chasmas, heavily cratered terrain is preserved. The area previously mapped as central Capri Chasma (Figure 8) resembles heavily cratered terrain partly modified to chaos more closely than it does the flat-floored and steep-walled canyons of the remainder of the Valles Marineris, suggesting that Capri Chasma is not a single physiographic entity; so some redefinition of the surface feature nomenclature in this part of Mars is needed.

North of Coprates Chasma a small trough, Juventae Chasma, opens to the north in a manner similar to Echus Chasma (Figure 8). The north end of Juventae Chasma is floored by a patch of chaotic terrain. Because of this association we have grouped it with the eastern canyons.

To the north, broad channels extend from the chaos at the mouths of the Juventae, Capri, and Eos chasmas and run down into the Chryse Basin, in which the Viking 1 lander set down. The Valles Marineris bridge the gap geographically between the youngest volcanic province on Mars, on the summit and western flank of the Tharsis-Syria Rise, and the chaotic terrain and channels, possibly of fluvial origin [*Masursky,* 1973; *Milton,* 1973; *Baker and Milton,* 1974; *Sharp and Malin,* 1975], associated with the Chryse Basin. It is not surprising therefore that both volcanic and fluvial processes have been assigned major roles in some recent interpretations of the Valles Marineris [*McCauley,* 1977; *Masursky et al.,* 1977]. In this respect the most striking aspect of the Viking high-resolution images of the canyons is the absence of obvious analogs for terrestrial volcanic landforms and the sparse evidence for fluvial activity; however, high-resolution images of canyon walls and floors do show a wealth of new evidence that tectonic activity, sedimentation, and mass movements all played major roles in shaping the Valles Marineris.

Canyon Wall Morphology

The walls of the Valles Marineris canyons display features of both erosional and tectonic origin. We consider here examples of several types of erosional features and evidence for recent tectonic activity. We also describe instances of layering visible in the canyon wall.

Erosional Features

The three basic types of canyon wall morphology recognized by previous investigators—gullied, landslide scarred, and dissected by tributary canyons—span the range of wall types seen in Viking images. However, composite wall types and secondary structural features are seen much more clearly than in Mariner 9 images. The spur-and-gulley and landslide alcove morphologies are shown side by side in a stereogram of west Ius Chasma (Figure 9). About 90 km to the east (C in Figure 6) the similarly gullied south wall of this trough was found to be 4 km high from analyses of Mariner 9 stereo imaging [*Wu et al.,* 1973; *Blasius,* 1973] and ultraviolet spectroscopy (see Figure 3). Fluted spurs commonly descend to the floors of canyons, suggesting fairly uniform erosional characteristics through the entire exposed section. The gullied canyon walls resemble some high terrestrial scarps which form in fairly resistant homogeneous rocks [*Lucchitta,* 1977]. On most scarps the average trend of spurs and gullies is downslope, but some reaches of gullied scarp show other strong alignments, suggesting structural control. For example, spurs on the north wall of eastern Ophir Chasma (Figure 10) are aligned northeast, approximately 60° from the local strike of the trough wall. Low-relief lineations (B, C, D, and E in Figure 10), including a scarp and a trough, on the upland plain are similarly aligned, suggesting structural control of gullying by joints or faults.

Landslide alcove scarps differ from gullied scarps in having only short, stubby, and very irregular spurs. These are mostly limited to the upper third of the scarp face; the lower face is smooth to complexly conical, presumably owing to accumulation of talus in aprons and cones.

Dissection by tributary canyons produces another widespread type of trough wall [*Sharp,* 1973*a*] best developed on the south side of Ius Chasma (Figure 11). The heads of the canyons have blunt terminations sharply defined by steep scarps. Where the floor and wall features of larger canyon reaches are resolved, the cross section is typically V shaped [*Ruiz et al.,* 1977] or shows only a narrow flat floor. Lower slopes of canyon walls generally appear smooth or shallowly striated, while the upper one third to one fourth of the scarp face is fluted like the crests of spurs along the walls of the major canyons. Elongate light and dark markings up to 300 m across are seen on some canyon floors and suggest flow along their length. No sizeable accumulations of material have been recognized at the mouths of tributary canyons, and neither canyons nor their floors broaden systematically downstream. In a few cases, tributary canyons broaden considerably for short stretches (A and B of Figure 6). The canyon floors in those places are irregularly hummocky, similar in character to trough floors near landslide alcoves. *Lucchitta* [1977] notes that in places, secondary branches of these canyons do not join the main valley concordantly. Although apparently not common, these relationships indicate that some tributaries are inactive for extended periods while others continue to deepen. A sporadic locally intensive process of erosion seems indicated. *Sharp* [1973*a*] and *Sharp and Malin* [1975] hypothesized that a sapping process, either seepage of liquid water or sublimation of ground ice, has been the principal cause of canyon wall recession in general and the development of dendritic tributary canyons in particular. In this view, tributary canyons either are enlarged by intermittent catastrophic mass movements, between which most evidence of material transport along the canyons is buried by slow mass wasting, or are fossils of a period in Mars history when surface conditions were more conducive to liquid water sapping and runoff.

Sharp and Malin [1975] point out strong trends toward parallel alignments and large junction angles among major segments of the Ius tributary canyons. Northeast and northwest trends seem to predominate (Figures 6 and 11). These characteristics suggest strong structural control, perhaps by

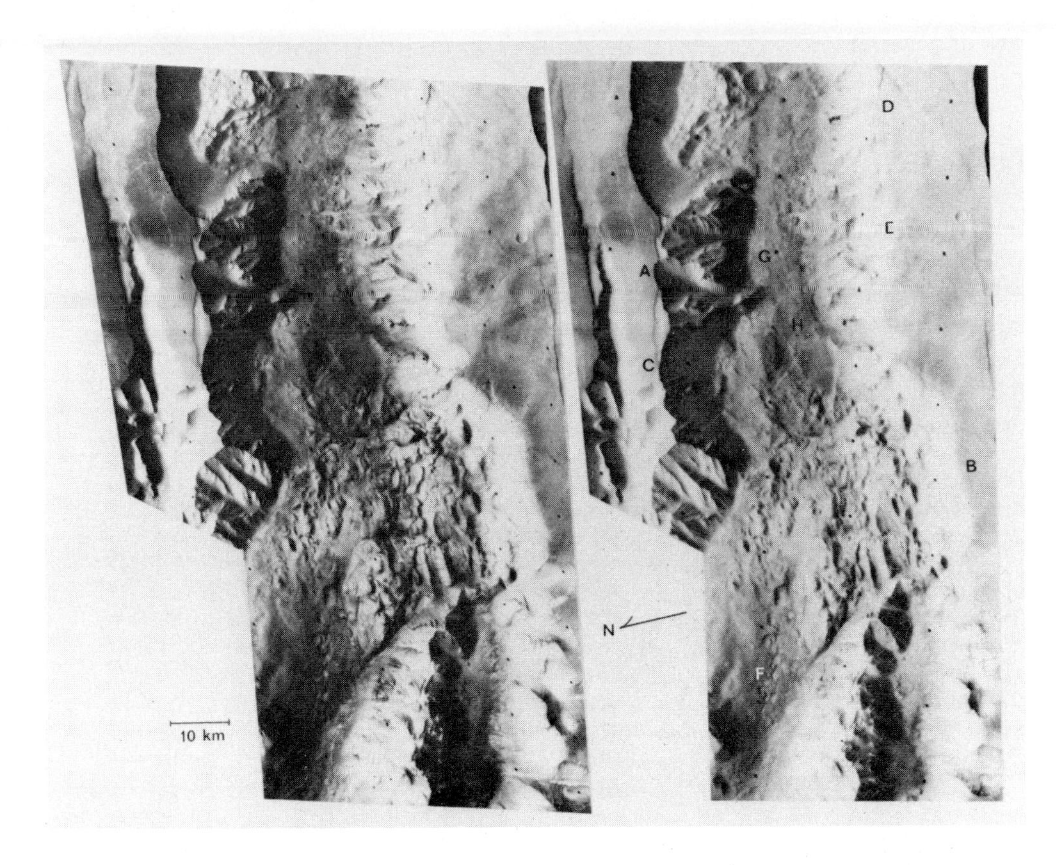

Fig. 9. Stereogram of a portion of west Ius Chasma (left, 64A12 and 64A14; right, 44A25) showing a large landslide scar (B) and slide debris on the canyon floor. Smaller landslide lobes can be seen at G and H. Also illustrated is the spur-and-gully canyon wall morphology (below E). North of A, a tributary canyon appears to have advanced headward along a graben valley on the upland.

fractures. *Sharp and Malin* [1975] also suggest an explanation for the much better development of tributary canyons on the south wall as compared to the north wall of Ius Chasma. They point out that regional slope or dip of bedding, into or away from a large canyon, can influence the rates of erosion by sapping. Tributaries of the Grand Canyon in Arizona are much better developed to the north, up the dip of bedding and the regional slope, than to the south. There tributary canyons work headward by groundwater sapping, the water migrating downslope along the surface and bedding planes. A similar contrast in structure may also influence the tributaries of Ius Chasma. The elevation profile AA′ in Figure 3 indicates a decided northward slope of the north rim of the canyon about 100 km west of the area of best tributary canyon development.

The three classes of trough wall morphology rarely occur separately. Tributary canyons are always mingled with gullies and spurs where they enter a large canyon, and individual secondary valleys intermediate in morphology between gullies and tributary canyons are common (A in Figure 9). Landslide alcoves can occur on the same scarp above or below spurs and gullies (A and B in Figure 5). Combinations of scarp morphology which have not yet been recognized are landslide alcoves which have been significantly modified by subsequent formation of gullies or tributary canyons. This may be an indication that massive landsliding has generally occurred in a later period than gulley or tributary canyon formation, or it may be that older landslide alcoves are not easily recognized.

Layered Rocks in Canyon Walls

The rocks which form the bounding walls of the Valles Marineris probably have a layered structure in the topmost one third to one fourth of the scarp face everywhere, and in places, features can be seen lower on the canyon walls that are similar to those on the toes of spurs. In high-resolution Viking images of west Ius Chasma, layering is expressed as an alternately light and dark sequence of lines below the brink of the sunward facing slope (at B in Figure 9) and as a line of knobs or a groove cutting across the fluted upper face of the opposite scarp (C in Figure 9). Albedo and patterns of fluting vary similarly between layers exposed in the walls of tributary canyons of Ius Chasma (Figure 11, C and D).

Tectonic Features

The secondary wall features which probably are most diagnostic of the processes responsible for the formation of the Valles Marineris are the remarkably straight low scarps which occur at the bases of canyon walls and often cut across spurs and gullies or leave them hanging above the main canyon floor. One of the best examples imaged by Viking runs along the base of the north wall of west Coprates Chasma (CD in

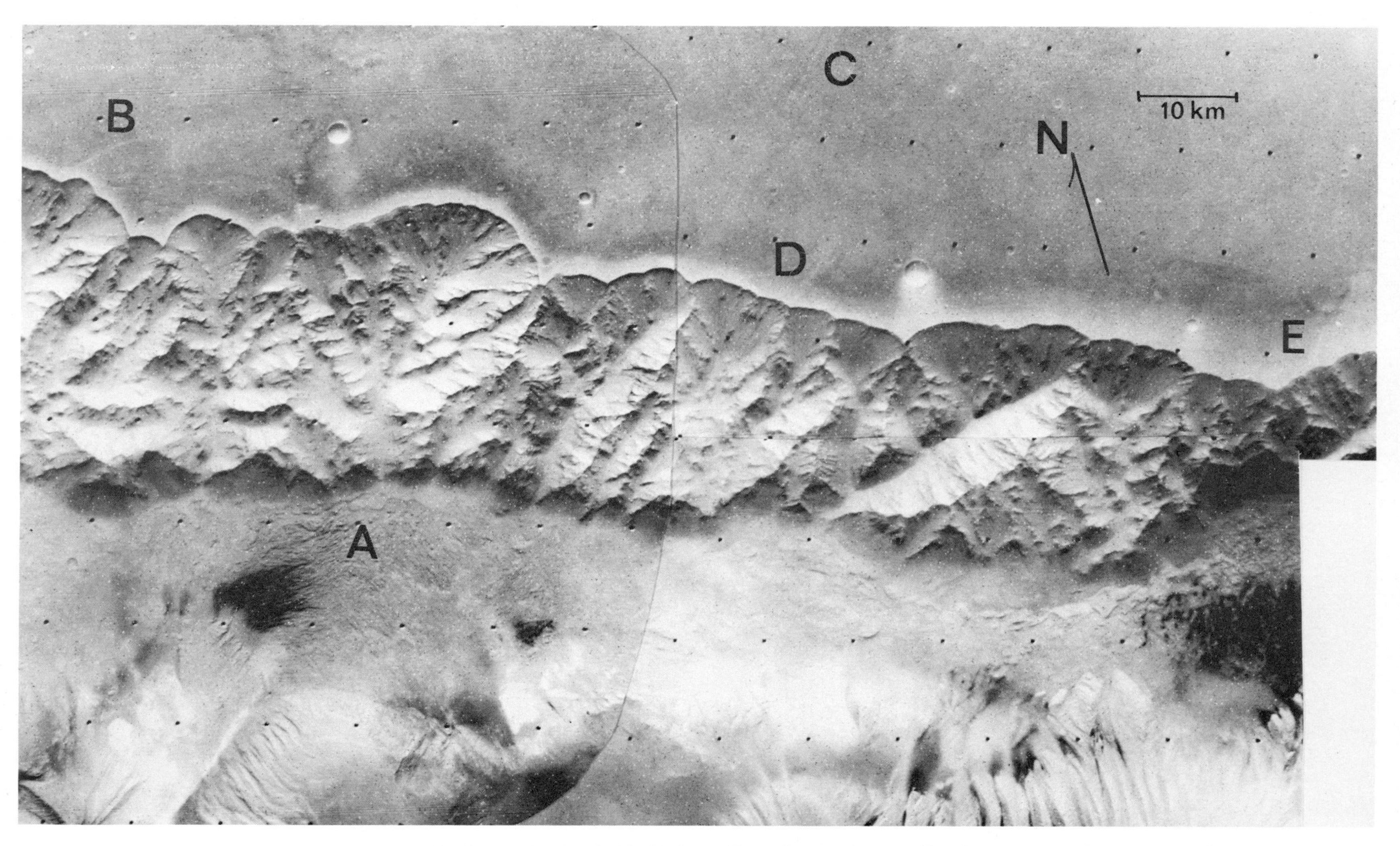

Fig. 10. The north wall of east Ophir Chasma (80A08–80A09) displays gullies and spurs whose peculiar alignment, cross slope, appears to be structurally controlled. Low scarps and shallow troughs (at BE) on the upland surface have this same trend.

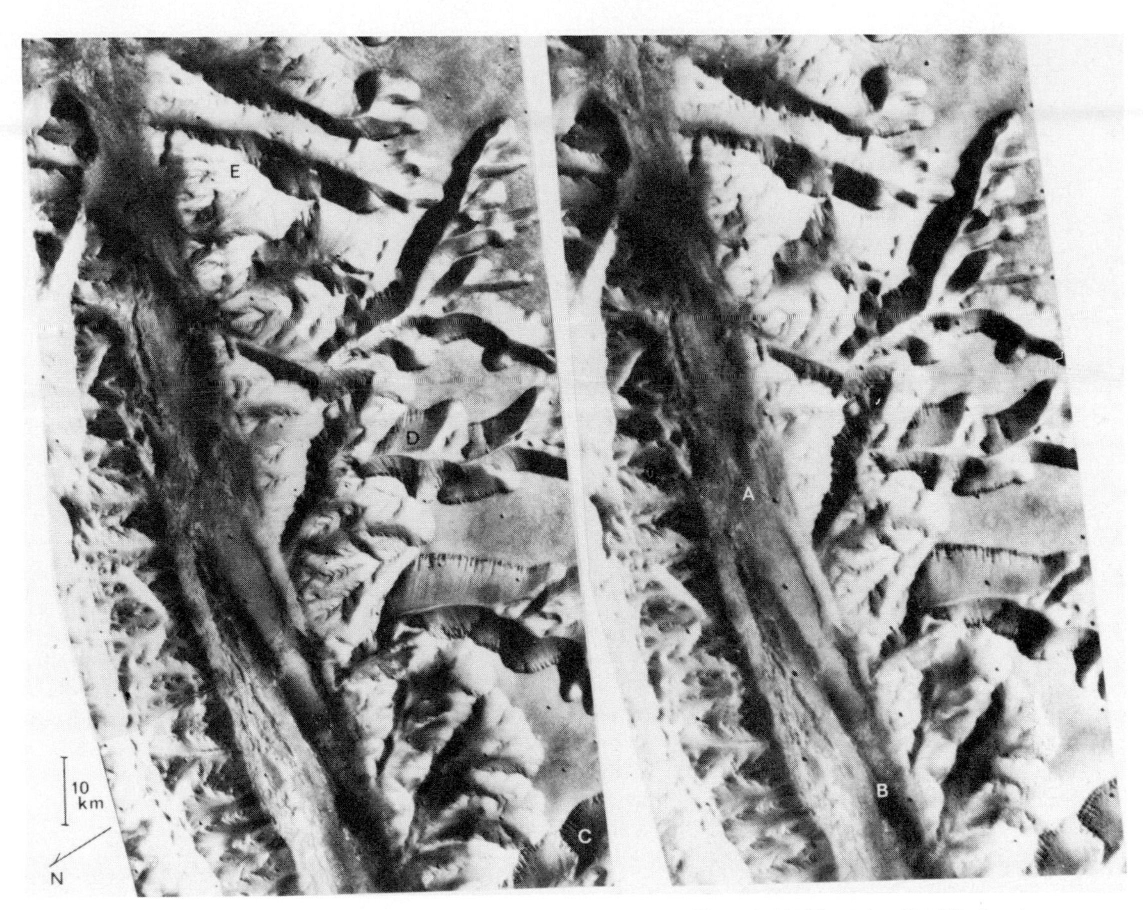

Fig. 11. Stereogram of part of central Ius Chasma (left, 66A07, 66A08, and 66A09; right, 63A40) showing a canyon wall incised by several large branching tributary canyons. The canyon floor (AB) appears to have dropped down below the erosional base level of the tributary canyons, being fractured in the process. Layering, expressed as albedo and erosional morphology differences, can be seen in canyon walls (C and D).

Figure 5). The formation of this small scarp clearly postdates the gullying of the main scarp face, but it preceded some major landslides which overran it (at A in Figure 5). In east Ius Chasma a similar scarp (AB in Figure 11) leaves gullies and tributary canyons hanging above the main canyon floor. The structure may continue on east (A to E, Figure 11) as a groove cutting across spurs on the canyon wall. At the foot of the opposite wall of Ius Chasma, spurs terminate abruptly in crude facets along a gently arcing line. Similarly faceted spurs occur along the north wall of east Ophir Chasma (Figure 10). In Ius Chasma a scarp cutting across spurs and gullies on a canyon wall angles up onto the cratered upland where it becomes part of a set of scarps defining shallow graben which parallel the major troughs (DE in Figure 9).

A similar small scarp is seen at the base of the west wall of Candor Chasma (EF in Figure 12). It separates a narrow ledge, onto which a short tributary canyon opens, from the distinctly lower main canyon floor. Farther north along another scarp at the west end of Candor there is a suggestion of a similar vertical offset of an erosional wall feature. A spur extending approximately 20 km down into the canyon (G in Figure 12) has a sharp break in the slope and change in direction along its crest approximately one third of the way up from the canyon floor. The upper segment of the spur is narrow and has a rugged fluted appearance, while lower down, the spur is much broader and its crest smoother. The lower spur segment appears to be in a much more advanced state of erosion than the remainder of the canyon wall. On the upland surface to the west, two alignments of shallow graben are seen—the usual trend ESE, parallel to the overall trend of the Valles Marineris, and NNE, parallel to the large scarps which form the west wall of Candor.

The linearity of small scarps which offset varied erosional features of canyon walls and their relations to obvious faulting on the cratered upland indicate that the scarps are traces of faults along which vertical displacements have occurred. It is these displacements which mark the boundaries of the floors of several major troughs and, in many cases, postdate gullying of trough walls. The spurs and gullies are adequately explained as the products of dry mass wasting, certainly an active process on Mars today, so it seems probable that in recent times, major segments of the Valles Marineris have been tectonically active as deepening graben valleys. The conclusion of *Sharp* [1973*a*] that materials removed to form the Valles Marineris must have in large part been removed downward is therefore strongly supported by new Viking data. The first analyses of Mariner 9 data suggested that the Valles Marineris may have formed initially as graben [*McCauley et al.*, 1972], but the disposition of material subsequently eroded from canyon walls could not be satisfactorily explained. It now appears that canyon floors continued to subside, removing the need to invoke aeolian, fluvial, or other surface processes to transport debris out of the canyons.

CANYON FLOOR MORPHOLOGY

For the most part, the floors of the Valles Marineris are rougher, at scales of 200 m to 50 km, than the surrounding

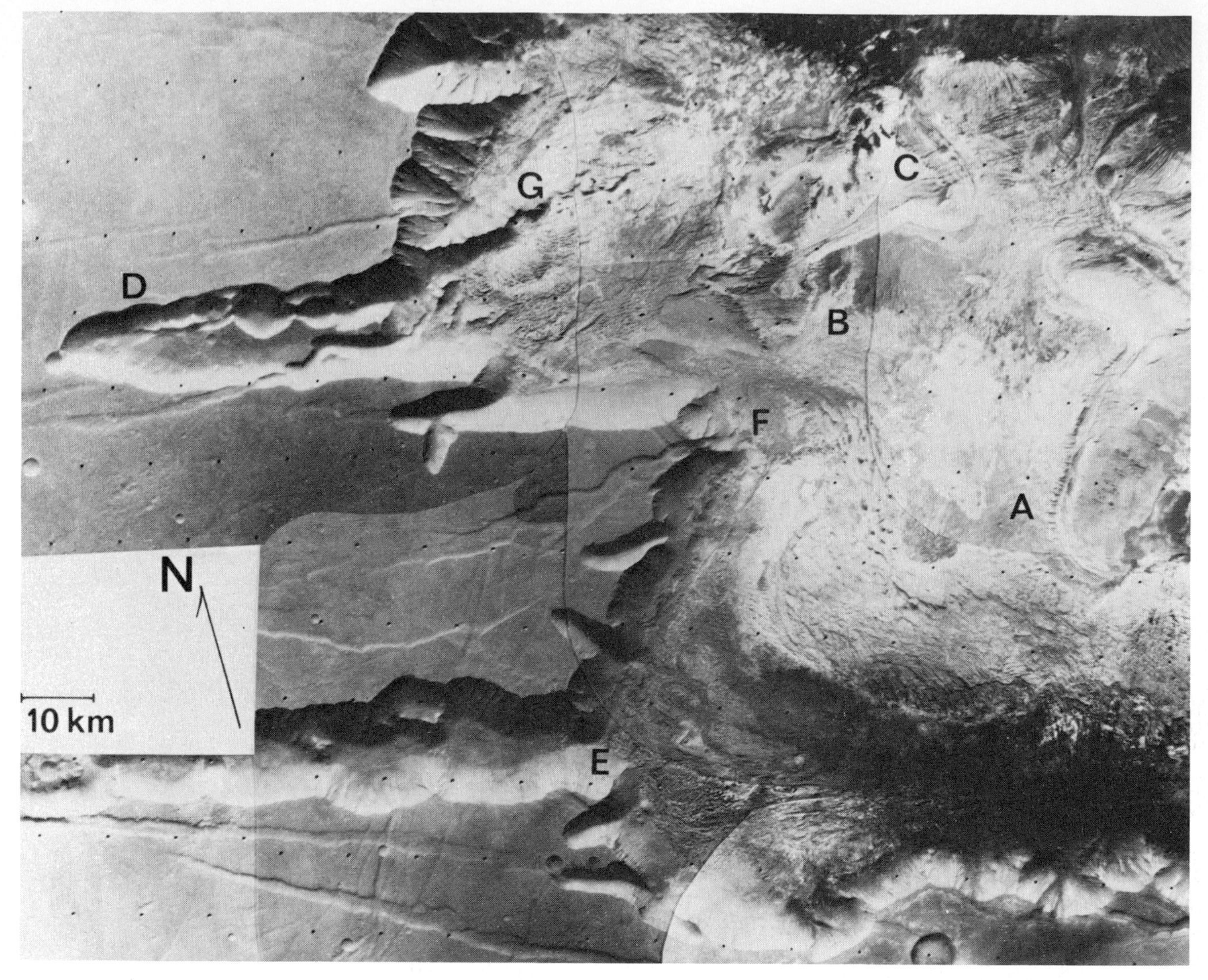

Fig. 12. Mosaic of images of west Candor Chasma (65A25, 65A57, and 66A17–66A27) showing eroded layered materials covering nearly the entire canyon floor. Layering is most prominent at A, B, and C. A vertically offset spur (G) and a low steep scarp (EF) along the west walls of the canyon suggest control by faulting. The tributary canyon at D appears to have originated by two processes: subsidence of a coherent block and collapse into a string of pits. South of E, a tributary canyon appears to have been eroded headward along a graben.

cratered upland terrains. The canyon floors are also extremely varied: plains, hills, plateaus, and ridges show fresh evidence of mass movements, tectonic activity, and wind action. Below, a number of broad classes of floor morphology are discussed, beginning first with those most closely related to features of the canyon walls and the uplands.

Plains

Cratered or smooth planar surfaces similar in character to those of the surrounding uplands occur widely within the Valles Marineris, most commonly in the western half of Noctis Labyrinthus (C, D, and E in Figure 4) but also farther east in Tithonium Chasma (at D and E in Figure 6) and Candor Chasma (B, F, and G in Figure 13). The locally elevated abundance of rather subdued craters on the broad flat floors of the Melas and Coprates chasmas (E–J, Figure 5) also suggests a cratered plain like the local upland, perhaps degraded by mantling or erosion. In the labyrinth, most occurrences of cratered plains on canyon floors are obvious instances of downfaulting within graben. The similar tectonic setting of cratered plains in Coprates and the general pattern of vertical tectonism within the Valles Marineris suggest that most cratered plains on the canyon floors are remnants of the formerly continuous upland surface. Alternatively, *Malin* [1976] has pointed out evidence of exhumation of ancient cratered surfaces from within the Martian intercrater plains. Thus erosion of downfaulted blocks of upland or deposition, cratering, and reerosion in canyon interiors may also expose cratered surfaces. In most places, downfaulting of upland surfaces was followed by considerable deposition and/or erosion, so the present canyon floor does not at all resemble the cratered upland.

Landslide Deposits

Landslide deposits, most easily recognized when they possess a sharply defined lobate form, mantle a large fraction of present canyon floors. Within and extending from many of the broad canyon wall alcoves described above there are large lobes of blocky terrain, and in the broader canyons these

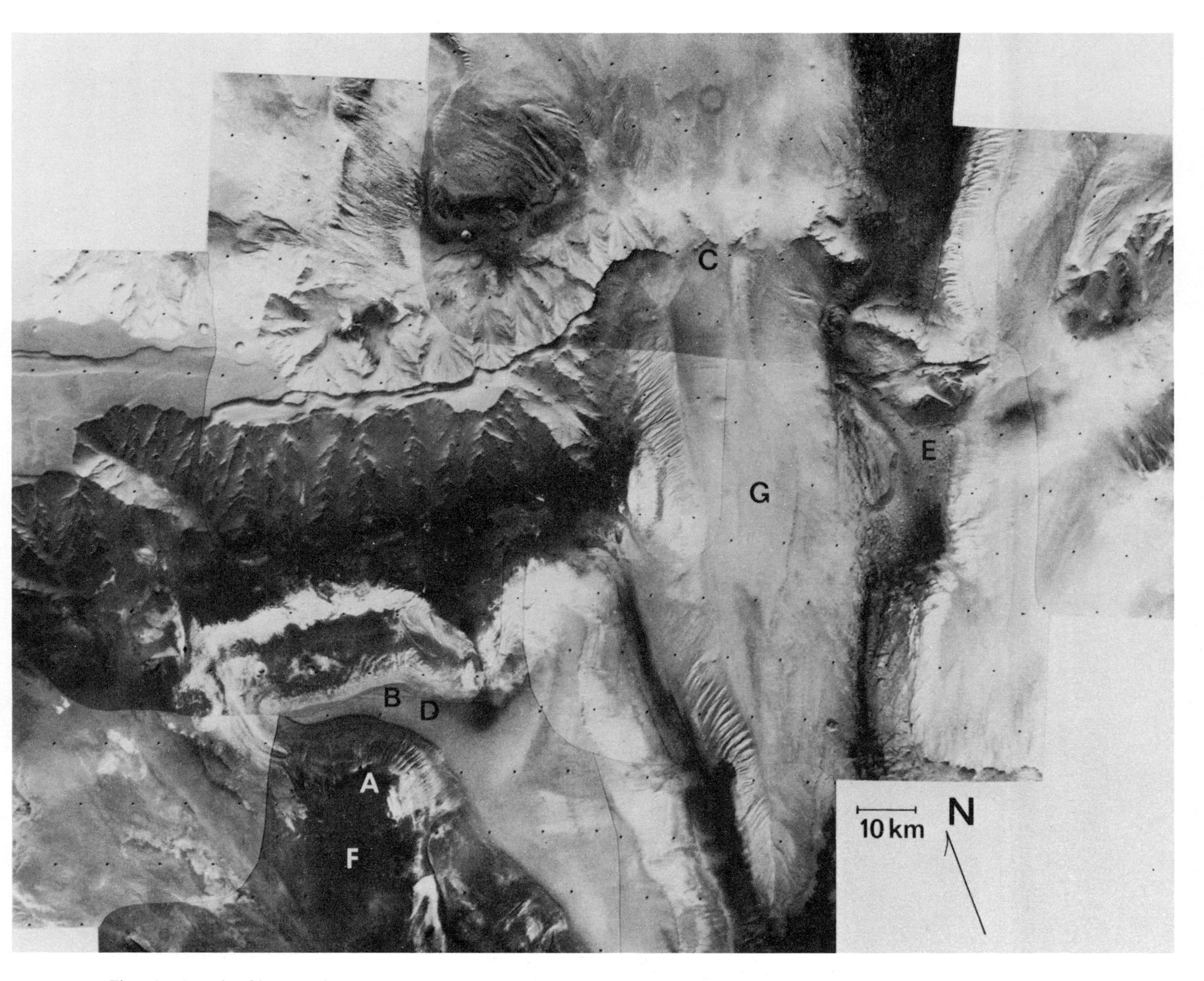

Fig. 13. Mosaic of images of the east Candor and Ophir chasmas (66A23–66A30). Large plateaus (D and G) in part formed of regularly layered materials (D) rise from the floor of Candor and extend across the gap between the two canyons. At C, plateau materials apparently were deposited upon an eroded spur and are now themselves being eroded away.

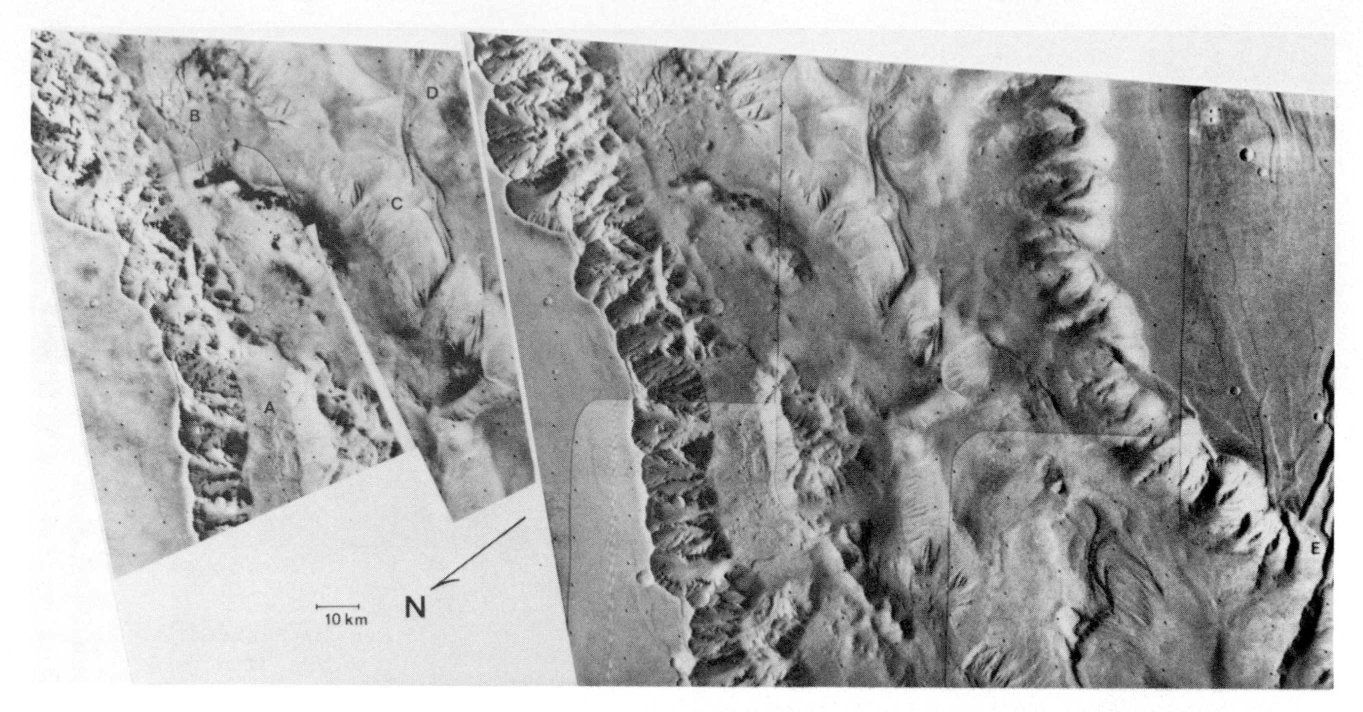

Fig. 14. Stereogram of part of east Ophir Chasma (left, 79A34–79A37; right, 91A11–91A15). Canyon floor morphology is highly varied, with small landslide lobes on a shelf at A, low-elevation fractured terrain at B, and ridges possibly formed by inclined bedding at C and D. At E, a large spur appears to reflect the same structure as several graben on the upland.

deposits may be surrounded by a thinner and smoother lobe several times broader. Typically, the blocky inner deposits are characterized by short ridges and valleys, aligned subparallel to the trend of the canyon wall (Figure 5, A and K; Figure 9), while surfaces of outer lobes are finely striated in a pattern which diverges from the alcove in the canyon wall (Figure 15).

Smaller thin lobes, of the order of 10 km across or less, also occur on the floors of the Valles Marineris but not exclusively in association with wall alcoves (Figure 9, G and H; Figure 14, A; Figure 15, A). They are seen at the mouths of gullies and tributary canyons or below steep scarp faces on which shallow chutes are sometimes visible. The surfaces of these lobes vary greatly in character from smooth, perhaps with lateral levee-like ridges, to hummocky or longitudinally striated.

B. K. Lucchitta (personal communication, 1976) suggests that small leveed lobes seen in Mariner 9 photos originate as debris or mudflows. The melting or sublimation of ground ice has been suggested by previous workers [*Sharp,* 1973*a, b*; *Milton,* 1974] as the possible cause of subsidence and disintegration of crustal materials in the chaotic terrain and wall recession in the Valles Marineris.

Longitudinal striations on landslide lobes are a new discovery on Mars. The largest lobes are enormous, tens of kilometers across, comparable to other large Martian mass movement deposits [*Sharp,* 1973*c*; *Carr et al.,* 1977], but their defining characterisitcs, low marginal scarps, and fine surface striations are too small to have been visible in Mariner 9 mapping images. Slide deposits with similar surface morphologies are known on earth. *Shreve* [1966] and *Post* [1965] discuss lobate landslide deposits with distinct longitudinal striations on top of valley glaciers in Alaska. They offer no specific hypothesis for the origin of the striations, however, only the suggestion that patches of snow or a slurry of snow and air overriden by the slide may have caused adjacent parts of the mass to move at differing velocities, thus creating longitudinal zones of shear. The larger problem of the general physical mechanisms by which massive energetic slides move on the earth is unresolved in the literature. *Shreve* [1968] proposes that large rockfall avalanches may trap and slide upon compressed air, while *Hsu* [1975] suggests that a large energetic slide carries its load by the buoyant force of colliding grains in an internal dust cloud.

Hilly Terrain

Another type of canyon floor, important in terms of total area occupied but geographically confined, is the hilly terrain on the floors of the broad shallow Eastern Canyons (Figure 8). Hilly terrain is composed of closely spaced, more or less equant, sometimes conical, mountains, individually less than 10 km and more typically about 1 km across. This terrain is transitional into the chaotic terrain, which intermingles with the eastern Valles Marineris. In high-resolution images (B in Figure 15) the individual hills are remarkably symmetric and appear to have uniform slopes, so the hills with smaller basal diameters have proportionately less relief.

Fractured Floor

Widely separated areas of canyon floor are characterized by patterns of small scarps or troughs which give the impression that a formerly continuous surface has been fractured. The largest such area is in the broad depression where the west Ius and Tithonium troughs merge (F in Figure 6). The canyon floor is divided into polygons up to 10 km across by a network of shallow troughs, each up to 1 km across. Clusters and chains of pits, all smaller than approximately 1 km, are also abundant. Pitting extends eastward into Ius Chasma proper, where it appears to have modified the surface of a large landslide deposit (F in Figure 9). The network of troughs and the chains of pits suggest an analogy with the much larger Noctis Labyrinthus (Figure 4). Many troughs there are obvious gra-

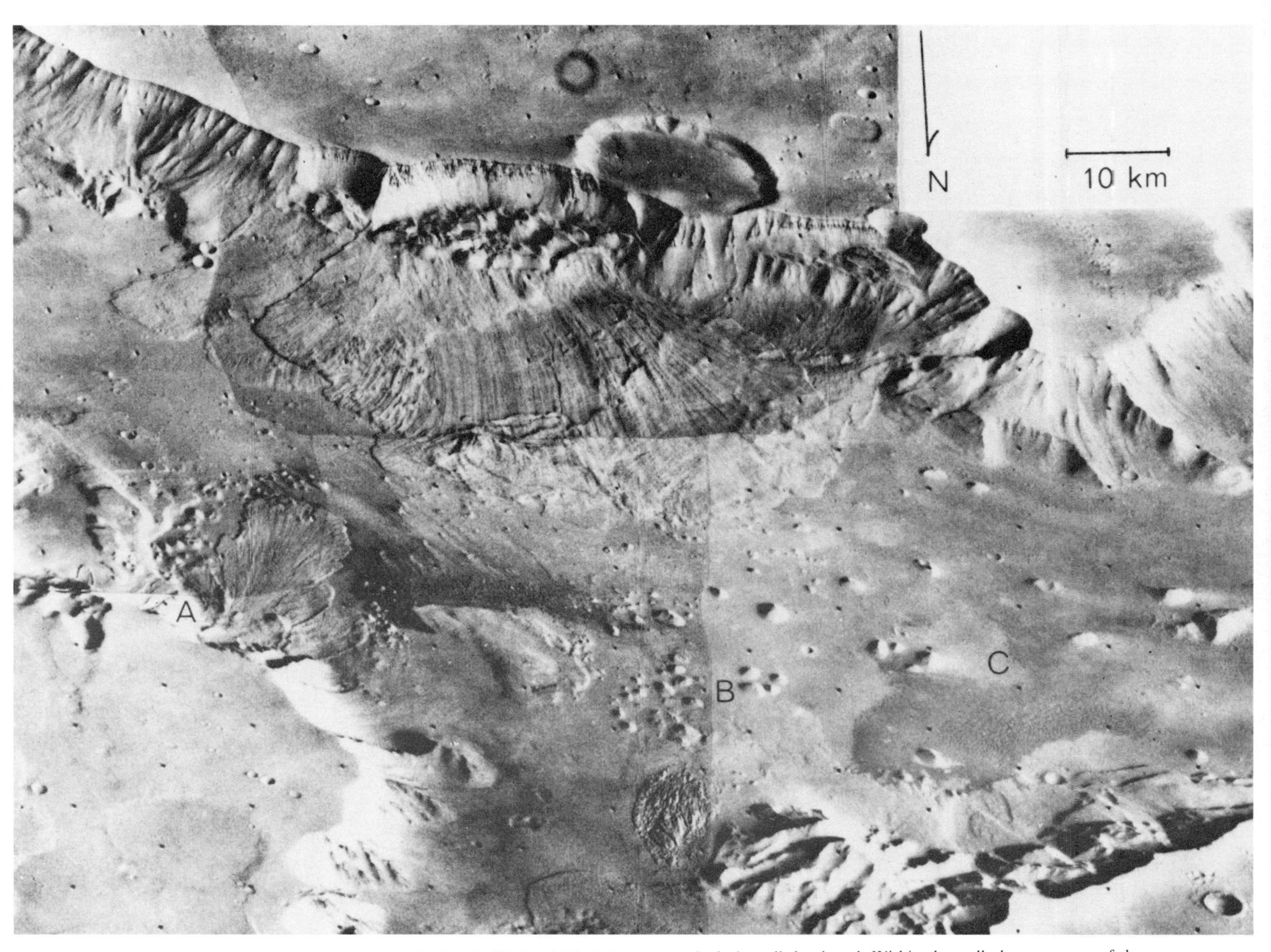

Fig. 15. In broad Gangis chasma (14A29–14A32), landslide lobes are particularly well developed. Within the wall alcove or scar of the largest is a lobe of blocky terrain, and extending outward across the flat canyon floor is a larger thin lobe with a diverging pattern of surface striations. Smaller striated slide lobes (A) are also present. A field of dark dunes (C) and hilly terrain (B) contribute to a varied canyon floor landscape. Note also the layers near the brink of the south canyon wall; they are seen particularly well from this oblique perspective looking south.

ben, apparently formed by extensional stresses in the center of the continental size Tharsis-Syria Rise. Perhaps this area of canyon floor has been similarly stressed and fractured.

Two other areas of canyon floor disruption are bounded in part by small canyon wall scarps interpreted above as evidence of recent faulting. In central Ius Chasma (AB in Figure 11) the 10-km-wide canyon floor, when viewed stereoscopically, is seen to consist of a trough about 40 km long. The central portion of the trough is broadly V shaped in profile, with slopes made up of smooth planar surfaces separated by low scarps. Flanking the central trough are two shelves about 1 km wide and steep low scarps rising up to the erosional base level of spurs and gullies on the main canyon wall. In east Ophir Chasma (B in Figure 14) a reticulate pattern of canyon floor fractures is bounded to the north by a low steep scarp which also truncates spurs and gullies on the trough wall.

In Ius Chasma the canyon floor was apparently formerly smooth and surfaced by a layer of material with some coherency. The canyon was then deepened by faulting, as evidenced by the steep scarps truncating gully floors, during which the canyon floor deposit was stretched across a newly enlarged canyon interior. Differential movement of multiple blocks beneath the canyon floor might also have contributed to the warping and breaking of the surface crust. In east Ophir Chasma, stresses near the surface of the downfaulted block may have been more uniform with azimuth, leading to the opening of fissures oriented in several directions. Downfaulting of an equant block, as in evidence in the labyrinth (C, R, and G in Figure 4), might produce relatively uniform horizontal tensile stresses by drag on the canyon walls.

Floor Plateaus and Ridges

The most striking landforms of the canyon floors, and at the same time the most enigmatic, are ridges and plateaus which have substantial relief compared to the depths of the canyons in which they are located [*Masursky*, 1973; *Sharp*, 1973*a*; *Malin*, 1976]. One plateau in Gangis Chasma, photographed at high resolution by Mariner 9, has a distinctly layered structure and total relief approximately equal to the depth of the canyon in which it is located (D. Dzurisin, personal communication, 1977). Where floor plateaus and ridges have been rephotographed by Viking at improved resolution, in broad sections of the Tithonium, Candor, Ophir, Melas, and Juventae chasmas, many are seen to have distinct layering. In some locations the layering is made apparent by differential erosion, more competent layers forming ledges on scarp faces or preferentially developing flutes (A and B in Figure 12, A in Figure 13, and C and D in Figure 14). Layers may also appear as striped albedo variations on slopes (Figure 16; C in Figure 12). In most cases, layers appear to be flat lying, but a significant dip is suggested in east Ophir Chasma (C in Figure 14) where ledges curve around and angle downslope through a gap in a ridge. One slope face apparently formed along or at a shallow angle to the dipping beds and was dissected by broad gullies, in contrast to the fluted ledges on the opposite side of the ridge.

Layered plateaus may be topped by plains surfaces that are similar to the upland outside the canyon (B in Figure 13) except for an apparent lack of the fault traces common on the upland.

The most remarkable aspect of the canyon floor layered materials is the extreme regularity of bedding in several locations. In Candor Chasma (D in Figure 13; C in Figure 12) and Juventae Chasma (Figure 16), 10–25 alternately light and dark layers of apparently uniform thickness can be seen. The cyclical recurrence of similar depositional events implied by these sequences is taken as strong evidence that they formed by sedimentation, perhaps controlled by cyclical changes in climate. Variations in solar isolation caused by oscillations of Mars obliquity and orbital eccentricity [*Ward*, 1974] produce climatic changes which may modulate sedimentation of, for example, windblown materials [*Ward et al.*, 1974]. Other processes which might form layered deposits, such as volcanism and mass wasting, would not be expected to act so uniformly with each recurrence.

Speculation on the origin of canyon floor plateaus and ridges has produced two very different hypotheses. *Malin* [1976] suggested that the layering is inherent to parts of the ancient crust of Mars which was faulted and eroded to form the Valles Marineris. *Sharp* [1973*a*] and *McCauley* [1977] suggested that deposition may have occurred after canyon formation. Sharp pointed out that recession of the canyon walls could have provided the source of materials. Mass movement processes would probably not act with sufficient regularity over time, however, to produce the uniform layering we have now observed in several locations. *McCauley* [1977] suggested that lakes have existed within the broader reaches of the Valles Marineris and are responsible for deposition of layered floor materials. Sedimentation within lakes, perhaps modulated by dust storm activity, could certainly explain some features of regularly layered materials, but conclusive evidence for the implied stratigraphic relationships has yet to be found. Lake bed sediments would be expected to thin at their margins, and erosional terraces might be preserved on the canyon walls. *Malin* [1976] also points out the seeming improbability of a geologic history for the canyons that requires that they be refilled, in one case to the rim, with sediment and then reexcavated. Some less regularly layered light and dark materials can be traced into the main walls of the canyons (as in Tithonium Chasma, Figure 6, between G and I), so not all layered materials were emplaced after canyon formation. On the other hand, relationships in the gap between the Ophir and Candor chasmas indicate that substantial deposition has occurred within the canyons subsequent to the erosion of some canyon walls. There (C in Figure 13) plateau-forming, though apparently unlayered, materials seem to embay the eroded ridge separating the two canyons. This plateau is so large, at least 20 by 70 km, that its existence suggests that large parts of the Candor and Ophir chasmas have been partially filled and then reeroded.

Other Types of Canyon Floor

In addition to the major classes of canyon floor morphology described above, dunes, subkilometer hummocks, and pits are also seen. Dune fields are rare and generally small in comparison with canyon dimensions. Both light albedo (west Tithonium Chasma) and dark albedo (C in Figure 15) dunes have been recognized. Hummocky terrain is seen in the labyrinth (H in Figure 4), Candor Chasma (E in Figure 13), and Ophir Chasma (A in Figure 10). In some cases these hills are probably eroded remnants of more coarsely textured landslide deposits. The floor of western Coprates Chasma (L in Figure 5) is pocked with large numbers of irregular pits up to 5 km but mostly less than 1 km across. There is no apparent pattern to the locations of the pits, but the larger ones seem similar in depth and have flat floors. Aeolian deflation of a thin surface layer is a possible explanation for their origin.

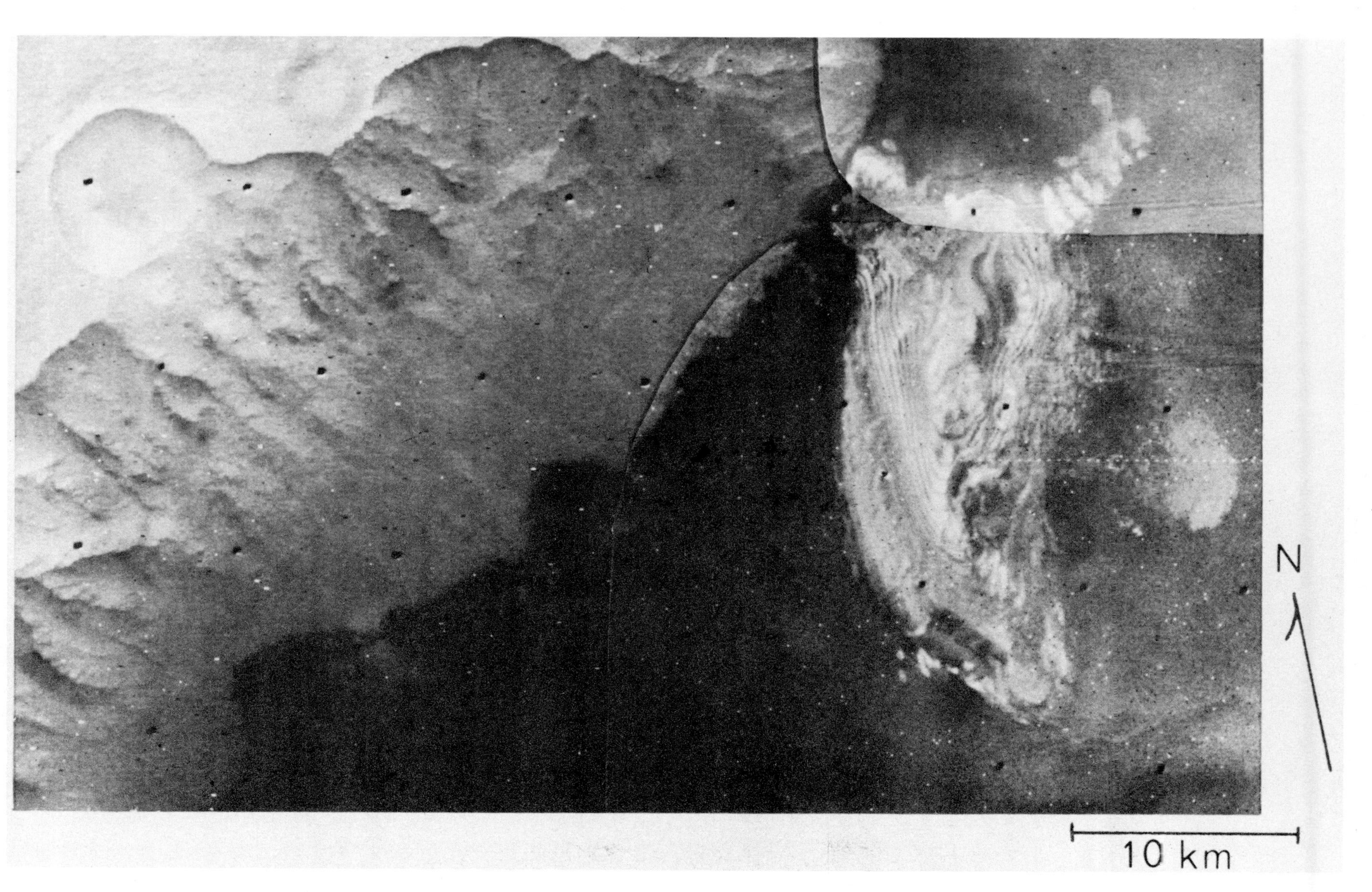

Fig. 16. Ridge of extremely regularly layered light and dark materials on the floor of Juventae Chasma (81A15–81A17). Qualitatively, this layered sequence resembles strata seen on a scarp bounding a plateau in Candor Chasma (D in Figure 13).

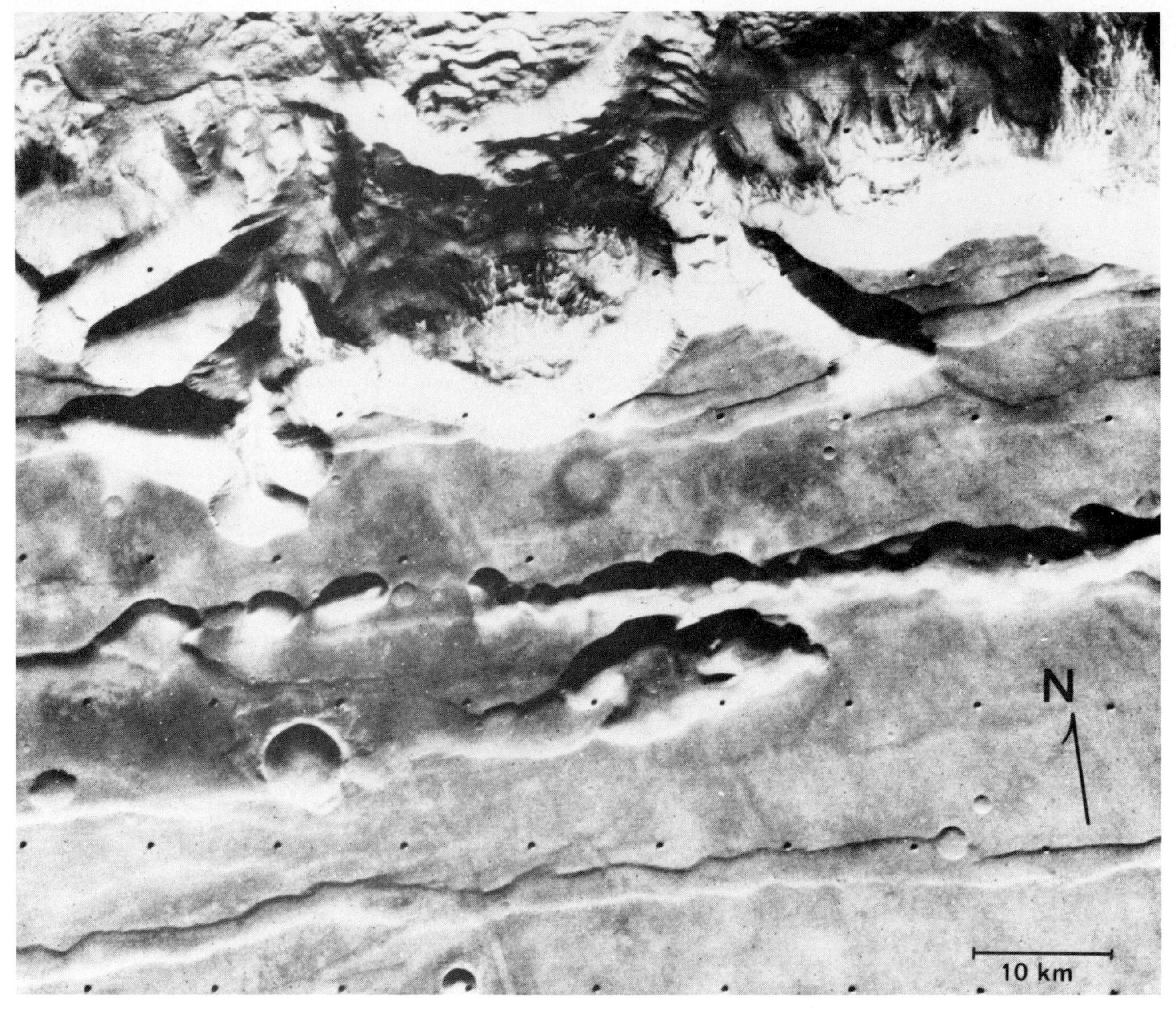

Fig. 17. Chain of craters (catena) developed along a graben on the upland between the Ius and Tithonium chasmas (62A20). Short tributary canyons along north facing canyon wall also seem to follow closely upland graben.

Upland Pits and Closed Depressions in the Canyons

The upland surfaces outside the Valles Marineris are marked by elongate, closed depressions aligned parallel to major canyon segments. Most of these features are clearly small graben (less than about 5 km across), some of which are simple trenches and some of which have associated with them pits that are circular or elongate in plan. The pits form discontinuous chains along graben floors (A and B in Figure 4), or they may have diameters greater than the width of the graben itself (Figure 17). In other cases the pits are not directly related to visible graben, but alignments roughly parallel graben directions. Some pits occur nested within larger pits (D in Figure 12). In the labyrinth, chains of pits of varied sizes suggest that canyons may develop from relatively small pits to large depressions by coalescence of growing pits. This relation between pits and development of the canyons is not so clearly seen elsewhere in the Valles Marineris system, where the closed depressions tend to be much larger and associated with extremely long rifts, such as Coprates (Figure 5).

The pits themselves are always deeper than the graben along which they form. We assume that the graben are a result of tensional fracturing of surface materials. However, the pits, being discrete closed depressions, cannot have such a simple origin. Similar pits are common in volcanic terrain, for example on the east rift of Kilauea, Hawaii, where they are explained by withdrawal of magma at depth, leaving cavities into which the surface rocks fall, and *Sharp* [1973*a*] has suggested this origin for the Valles Marineris pits. However, no evidence has been found for volcanism contemporary with the rifting or erosion of the canyons, and to invoke a volcanic mechanism for pit formation would involve an improbable style of volcanism where intrusion occurred repeatedly close to the surface but eruption was absent.

Two other mechanisms might cause the collapse; one is the removal of some other material such as subsurface ice, also suggested by *Sharp* [1973*a*]. While this is possible, it would necessitate the removal of extremely large volumes of segregated ground ice. A second possibility is that the faults bounding the graben are not simple normal faults but that beneath some graben the faults splay out at depth to form reverse faults, thereby allowing substantial collapse at the surface outside the intersection of the fault plane with the ground surface.

There are a number of examples in the Valles Marineris system of almost closed, large depressions with blunt ends associated with, but wider than, graben. Extrapolating from the smaller pits, one finds that it is at least possible that the larger valleys have been broadened not only by normal erosive processes but also by collapse similar in character to that which has produced the smaller pits. This process of pit formation on graben is important to an understanding of the tectonics of Mars and also of the moon, where, for example, the pits on the Hyginus Rille are again not directly related to any observed volcanic activity.

REGIONAL TECTONICS

A detailed examination of the tectonic history of the Valles Marineris is beyond the scope of this paper, but some important clues to the evolution of these features may have come from the recognition that graben deepening, and probably also associated collapse pit formation, are probably contemporary processes equal in significance to erosion in defining the present canyons. The Valles Marineris as rift valleys are accounted for as the products of extensional stresses concentric about the summit of the domed Tharsis-Syria Rise [*Carr,* 1974]. The Valles Marineris graben on the east flank of the rise are of exceptional size compared to individual rifts seen on other flanks. This is possibly due to burial by younger volcanics to the north and west, but development of a secondary ridgelike swell, reflected in present topography (Figure 1), and concentration of extensional stresses along its crest account more simply for the exceptional width and depth of individual rift valleys. Radial tensional stresses are also expected on the flanks of a dome and may account for the small upland graben aligned transverse to the primary alignment of the canyons. These graben appear to associate with the blunt ends of the Candor and Ophir chasmas, so even those mostly closed depressions may be fundamentally fault-controlled valleys.

Simple doming and extensional tectonics do not, however, account for the collapse pits which form some sections of the canyons and run parallel to others on the cratered upland. Stresses along the axis of the Valles Marineris system may very well be far more complex than simple extension along and transverse to the crest of a ridge. The ultraviolet spectrometer elevation profiles (Figure 3) show that upland surfaces adjacent to and between the Valles Marineris are far from level in several locations. Canyon rims locally slope sharply toward or away from the canyons (B, C, and E in Figure 3), and an intratrough plateau slopes decidedly to the south (CC′ in Figure 3). If these departures of upland surfaces from horizontal are due to the tectonics of canyon formation, both extension and compression may occur at depth at the boundaries of lithospheric blocks which have been tilted in relation to one another.

REGIONAL EROSIONAL STYLES

Walls of the Valles Marineris

The distribution of erosional styles in the Valles Marineris can be attributed to a number of factors. Dominant among these factors through most of the canyon system is the changing tectonic setting, although material variations, and possibly climatic conditions (temperature, pressure, winds), may account for some important distinctions in erosional style.

Tributary canyons and, more generally, valleys intermediate in scale between upland graben and great canyons are often colinear and contiguous with the shallow upland graben (Figure 9, north of A; Figure 12; Figure 17).

Tectonic activity also appears to have strongly influenced the development of spurs and gullies locally. At E in Figure 14, for example, the western slope of the ridge in Ophir Chasma can be interpreted as the trace of a fault that defines a graben on the adjacent upland.

The occurrence of landslides appears to be related to faulting parallel to the canyon orientation. Clearly, smaller mass movements have occurred throughout the canyons, but the most massive landslides, located on the north wall of Ius Chasma (Figure 6), have wall alcove scars with lineaments parallel to local upland graben and tributary canyons. Conversely, where there is a strong structural trend perpendicular or at high angle to canyon walls, as in west Candor (Figure 12), Ophir (Figure 10), or west Tithonium (I in Figure 6), massive landsliding is suppressed. It seems that the shape and size of individual landslide blocks which detach from canyon walls may be determined by the local pattern of faulting on the upland.

All styles of canyon wall erosion thus appear to be influenced by local tectonic patterns. Where local patterns are weakly developed, one might expect less regular patterns of wall recession. The Eastern Canyons and the chaotic terrain may have developed under such circumstances.

Trough and Chaos Formation

In the Eastern Canyons a rather gradual and uneven transition occurs from the characteristic trough morphology which is dominant west of 50°W to the morphology of chaotic terrain, which is the dominant terrain type between 40°W and 10°W. Chaotic terrain occurs in basins which are quite shallow (1–2 km deep) in comparison with the troughs to the west and appears to involve collapsing or sagging due to removal of material support from beneath. Chaotic terrain evolution appears to involve several stages: the development of a mosaic of upland blocks in direct response to removal of support, the erosion of these blocks into a landscape of rounded hummocks, and transport of debris, partly in channels, over considerable horizontal distances away from the original location of the collapse [*Sharp,* 1973*b*]. The channels in many cases descend into Chryse Planitia. We now consider whether it is plausible that the Eastern Canyons and chaotic terrain can be attributed to the same basic processes inferred to have formed troughed terrain, that is, extensional block tectonics acting in concert with mass wasting, or whether other processes such as volcanism, thermokarst activity, or aeolian activity are necessary to explain the observed features.

In the Eastern Canyons, where chaos is developed, the geological setting is quite different from that in most of the Valles Marineris. Structural influences are much less obvious; small graben and catena seen on the upland around troughs are absent. Moreover, the cover of plains-forming deposits appears discontinuous. Ridged plains and ghost craters do occur, however, and suggest partial volcanic flooding in this area. Most areas of chaos are developed in cratered terrain which lies below the 2-km level. However, at Juventae Chasma the adjoining upland is at the 5-km level. Chaotic terrain develops locally in areas immediately adjacent to canyons, such as the south margin of Eos Chasma (Figure 8), or in apparently random locations in the middle of the cratered terrain.

We hypothesize that the primary cause of collapse in the Eastern Canyons and chaotic terrains is tectonic. We attribute much of the erosion of these terrains to mass wasting, perhaps driven by ground ice sapping [*Sharp,* 1973*b*], followed by lateral transport by either aeolian or fluvial processes. Lateral transport is strongly evident, in the form of channels, only on the margins of some chaotic terrain. We attribute the other contrasts in erosional style between chaotic and troughed terrain to differences in geological setting, including tectonic patterns and near-surface materials in the region.

Our present lack of understanding of the apparent poor structural and topographic integration of Eastern Canyons and patches of chaotic terrain is comparable to the puzzle posed by the broad blunt terminations of some troughs (such as Candor) and the topographic isolation of others (such as Hebes). In discussions of both troughed and chaotic terrains, *Sharp* [1973*a, b*] proposed mechanisms for substantial collapse of large patches of preexisting terrain, intervening areas being left undisturbed. These mechanisms, withdrawal of intruded magma and local sublimation or melting of segregated ground ice, presumably also due to igneous activity, both have the advantage of conceivably operating without forming new regional tectonic patterns. The absence of recognized volcanic constructional forms in the canyons and chaos makes these hypotheses less attractive. Improved imaging from Viking now shows upland faulting, which suggests a tectonic origin for closed depressions found within troughed terrain. It may be that upland fault traces connecting loci of chaotic terrain formation, comparable to the shallow graben running between major chasma, are just too subtle in character to have been recognized in Mariner 9 mapping images or the present relatively small sample of Viking imaging. Alternatively, differences in response of surface materials to like stresses may prevent the development of sharply defined low-relief scarps in the heavily cratered terrain but allow them in cratered plains. An analogy with the formation of wrinkle ridges on the moon may be appropriate. A structure producing prominent wrinkle ridges on the lunar maria may be expressed as a simple scarp after it crosses into highland terrain [*Lucchitta,* 1976]. Though improved imaging data may provide support for the above hypothesis for the initiation of chaos formation, a major distinction between chaotic and troughed terrains will remain in that fluvial or aeolian processes or both have apparently transported material out of chaotic terrain.

Conclusions

After Mariner 9 the relative importance of tectonic, volcanic, fluvial, and mass movement processes which formed the Valles Marineris was poorly understood. Initial examination of Viking imaging clarifies matters somewhat in that volcanism is nowhere apparent and evidence for fluvial activity is only indirect, relating to chaotic terrain, which in turn is associated with the easternmost Valles. On the other hand, tectonic activity now appears to have been protracted in time, acting to deepen canyons and competing with the effects of erosion and deposition which cause canyons to broaden and fill. The primary process appears to have been vertical adjustment of crustal blocks under the influence of N-S and E-W extensional stresses. Some tilting of blocks in relation to one another may also have taken place, giving rise to peculiar slopes near canyon rims and on an intratrough plateau and possibly causing the formation of strings of collapse pits.

The past erosional and depositional history of the canyons is also more complex than previously realized. Layered materials, including some very regularly bedded sediments first recognized in Viking images, are highly diverse in character and widespread in occurrence.

An underlying issue which motivates study of the Valles Marineris is the problem of the relationship through time of the development of the canyons to the tectonic and volcanic activity in the Tharsis–Olympus Mons region to the west (Figure 1). Both phenomena are relatively recent developments in Martian history, and associated patterns of faults suggest control by the doming of the Tharsis-Syria Rise [*Carr,* 1974]. Volcanism has been found to be protracted in time down, perhaps, to the present [*Blasius,* 1976*a*; *Carr et al.,* 1977] and strongly influenced by faulting at the largest scale [*Blasius,* 1976*b*]. Erosion and deposition within the canyons will make comparisons of age with volcanic terrains, through impact crater densities, extremely difficult, if not impossible. This first qualitative evaluation of Viking images suggests, however, that the tectonic activity in the Valles Marineris may also have continued for a long period up to the recent past. Though the doming of the Tharsis-Syria Rise may have been largely accomplished at the very beginning of the extended period of volcanism [*Carr,* 1974], we speculate that tectonic activity persisted at a lower level over the entire region to control volcanism in the west and the continued deepening of the canyons in the east. This implies a long-lived dynamic process in the Martian lithosphere or mantle, which also seems to be required by the large gravity anomalies of the Tharsis-Syria Rise [*Phillips and Saunders,* 1975; *Blasius and Cutts,* 1976].

Acknowledgments. The authors thank M. Malin and R. Stockman for constructive reviews of this paper which improved considerably its final form. M. Malin generously allowed the use of a figure (Figure 3) from his Ph.D. thesis. We are also indebted to the members of the Viking Primary Mission Flight Team, without whose extraordinary efforts we would not have the pleasure and satisfaction of studying the detailed geology of great Martian canyons. Jet Propulsion Laboratory contract 954149. Planetary Science Institute contribution 88.

References

Baker, V. R., and D. J. Milton, Erosion by catastrophic floods on Mars and earth, *Icarus, 23,* 27–41, 1974.

Blasius, K. R., A study of Martian topography by analytic photogrammetry, *J. Geophys. Res., 78,* 4411–4423, 1973.

Blasius, K. R., The record of impact cratering on the great volcanic shields of the Tharsis region of Mars, *Icarus, 29,* 343–361, 1976*a*.

Blasius, K. R., Architecture of great shield volcanoes of the Tharsis region of Mars (abstract), *Bull. Amer. Astron. Soc., 8,* 480, 1976*b*.

Blasius, K. R., and J. A. Cutts, Shield volcanism and lithospheric structure beneath the Tharsis Plateau, Mars, *Proc. Lunar Sci. Conf. 7th,* 3561–3573, 1976.

Briggs, G., K. Klaasen, T. Thorpe, J. Wellman, and W. Baum, Martian dynamical phenomena during June–November 1976: Viking orbiter imaging results, *J. Geophys. Res., 82,* this issue, 1977.

Carr, M. H., Tectonism and volcanism of the Tharsis region of Mars, *J. Geophys. Res., 79,* 3943–3949, 1974.

Carr, M. H., H. Masursky, and R. S. Saunders, A generalized geologic map of Mars, *J. Geophys. Res., 78*(20), 4031–4036, 1973.

Carr, M. H., K. R. Blasius, R. Greeley, J. E. Guest, and J. B. Murray, Some Martian volcanic features as viewed from the Viking orbiters, *J. Geophys. Res., 82,* this issue, 1977.

Christensen, E. J., Martian topography derived from occultation, radar, spectral, and optical measurements, *J. Geophys. Res., 80,* 2909–2913, 1975.

Hord, C. W., K. E. Simmons, and L. K. McLaughlin, Mariner 9 ultraviolet spectrometer experiment: Pressure-altitude measurements on Mars, *Icarus, 21,* 292–302, 1974.

Hsu, K. J., Catastrophic debris streams (Sturzstroms) generated by rockfalls, *Geol. Soc. Amer. Bull., 86,* 129–140, 1975.

Lucchitta, B. K., Mare ridges and related highland scarps—Result of vertical tectonism?, *Proc. Lunar Sci. Conf. 7th,* 2761–2782, 1976.

Lucchitta, B. K., Morphology of chasma walls, Mars, *Interagency Rep. Astrogeol. 83*, 51 pp., U.S. Geol. Surv., Reston, Va., in press, 1977. (Also *J. Res. U.S. Geol. Surv.*, in press, 1977.)
Malin, M. C., Nature and origin of intercrater plains on Mars, Part 3 of Ph.D. thesis, 176 pp., Calif. Inst. of Technol., Pasadena, 1976.
Masursky, H., An overview of geological results from Mariner 9, *J. Geophys. Res., 78*, 4009–4030, 1973.
Masursky, H., A. L. Dial, and M. H. Strobell, Geologic map of the Phoenicus Lacus quadrangle of Mars, U.S. Geol. Surv., Reston, Va., in press, 1977.
McCauley, J. F., Geologic map of the Coprates quadrangle of Mars, U.S. Geol. Surv., Reston, Va., in press, 1977.
McCauley, J. F., M. H. Carr, J. A. Cutts, W. K. Hartmann, R. P. Sharp, and D. E. Wilhelms, Preliminary Mariner 9 report on the geology of Mars, *Icarus, 17*, 289–327, 1972.
Milton, D. J., Water and processes of degradation in the Martian landscape, *J. Geophys. Res., 78*, 4037–4048, 1973.
Milton, D. J., Carbon dioxide hydrate and floods on Mars, *Science, 183*, 654–656, 1974.
Phillips, R. J., and R. S. Saunders, The isostatic state of Martian topography, *J. Geophys. Res., 80*, 2893–2898, 1975.
Post, A., Effects of the March, 1964 Alaska earthquake on glaciers, *U.S. Geol. Surv. Prof. Pap.*, 544-D, 1965.
Ruiz, R. M., D. A. Elliott, G. M. Yagi, R. B. Pomphrey, M. A. Power, K. W. Farrell, Jr., J. J. Lorre, W. D. Benton, R. E. Dewar, and L. E. Cullen, IPL processing of the Viking orbiter images of Mars, *J. Geophys. Res., 82*, this issue, 1977.
Sharp, R. P., Mars: Troughed terrain, *J. Geophys. Res., 78*, 4063–4072, 1973*a*.
Sharp, R. P., Mars: Fretted and chaotic terrains, *J. Geophys. Res., 78*, 4073–4083, 1973*b*.
Sharp, R. P., Mass movements on Mars, in *Geology, Seismicity, and Environmental Impact*, pp. 115–122, Association of Engineering Geologists, Dallas, Tex., 1973*c*.
Sharp, R. P., and M. C. Malin, Channels on Mars, *Geol. Soc. Amer. Bull., 86*, 593–609, 1975.
Shreve, R. L., Sherman landslide, Alaska, *Science, 154*, 1639–1643, 1966.
Shreve, R. L., The Blackhawk landslide, *Geol. Soc. Amer. Spec. Pap., 108*, 47 pp., 1968.
Ward, W. R., Climatic variations on Mars, 1, Astronomical theory of insolation, *J. Geophys. Res., 79*, 3375–3386, 1974.
Ward, W. R., B. C. Murray, and M. C. Malin, Climatic variations on Mars, 2, Evolution of carbon dioxide atmosphere and polar caps, *J. Geophys. Res., 79*, 3389–3395, 1974.
Wu, S. S. C., F. J. Schafer, G. M. Nakata, R. Jordan, and K. R. Blasius, Photogrammetric evaluation of Mariner 9 photography, *J. Geophys. Res., 78*, 4405–4423, 1973.

(Received April 5, 1977;
revised June 9, 1977;
accepted June 9, 1977.)

Geology of Chryse Planitia

RONALD GREELEY,[1] EILENE THEILIG,[1] JOHN E. GUEST,[2] MICHAEL H. CARR,[3]
HAROLD MASURSKY,[4] AND JAMES A. CUTTS[5]

On July 20, 1976, Viking 1 made the first successful landing on Mars in Chryse Planitia, a plains-covered basin in the northern hemisphere. Viking orbiter pictures reveal more surface detail of the area and show the basin to be more complex than was seen on Mariner 9 images. The plains consist of areas with smooth and relatively uniform surfaces with prominent lunarlike mare ridges, mesas and plateaus, surfaces that appear to be 'etched,' fields of knobs, low shields that may be volcanic, and vast areas that have been subjected to channel-forming processes. At least two sets of channels, originating from distant sources, terminate in Chryse Planitia. The four major units of the basin are basal hilly and cratered terrain, plateau material which can be divided into upper and lower units, lower smooth plains, and upper smooth plains. There is no evidence for the origin of the basin. Deposition of the plateau-forming material to the east followed the period of bombardment, and since that time the history of the basin includes eruptions of flood lavas, channel formation, and deposition with at least two channel-forming events. The later history of the basin includes possible local volcanic events, etching, and aeolian activity.

INTRODUCTION

The objective of this paper is to present the general geology of the Chryse Planitia region on Mars, as derived primarily from Viking orbiter photographs. Chryse Planitia, the site of the first successful landing on Mars by Viking 1, is an asymmetrical basin, centered at 45°W and 24°N, about 2000 km northeast of Valles Marineris (Figure 1). Previous Mariner 9 based geologic studies of the region include the 1:5,000,000 scale geologic maps of *Milton* [1974] and *Wilhelms* [1976].

Viking orbiter pictures reveal more surface detail of the area than has been previously available [*Carr et al.*, 1976]. During site certification of the Viking 1 lander [*Masursky and Crabill*, 1976] it became evident that the site chosen from Mariner 9 images was potentially unsafe for landing, and sites were systematically examined in Chryse Planitia until a site suitable for both engineering considerations and scientific objectives was identified. Blocks of orbiter photographs were mapped geologically within a few hours after receipt of the data, ultimately resulting in a series of geological maps forming a swath across the basin. In the half-year following the landing the area has been reanalyzed, and the preliminary geological maps that were prepared during site certification have been remapped and correlated to derive a more unified picture of the geology of Chryse Planitia (Plate 1). Only part of the basin has been photographed by Viking orbiter, and the discussion which follows is restricted to these areas.

GEOLOGY AND GEOMORPHOLOGY

Regional Setting

Chryse Planitia is bounded to the east and south by moderately cratered terrain; to the west and southwest a narrow strip of cratered terrain separates Chryse Planitia from Lunae Planum, an elevated plain characterized by low albedo. North and northeastward, Chryse Planitia grades into another low-albedo plains area, Acidalia Planitia, which is part of the large northern hemisphere plains (Figure 1).

Chryse Planitia is 2–3 km below the mean Mars elevation, one of the lowest regions on the planet. There is no indication that it is an impact-generated basin (radial grooves, furrows, and secondary craters typical of impact basin geology are lacking); however, this may be a function of age. There is also no evidence of crustal downwarp; therefore its structural origin remains unknown.

Many major channel systems terminate in the basin. One group originates in chaotic terrain associated with the eastern end of Valles Marineris and trends northward more than 1000 km to the southern end of the basin. This group includes Ares Vallis, Tiu Vallis, Simud Vallis, and Shalbatana Vallis. Another system of channels, dominated by Kasei Vallis, originates from both chaotic terrain and Echus Chasma and enters the basin from the west. Thus Chryse Planitia may have received materials via the channels from two distinctive source areas separated by more than 2500 km.

Smooth Plains

Low-resolution approach views of Chryse Planitia show that it is a moderate to high albedo area of little topographic relief. The A frames (~2-km resolution) from Mariner 9 show the plains to be smooth and relatively uniform with suggestions of marelike ridges and lobate forms [*Wilhelms*, 1976]. Viking orbiter pictures reveal a heterogeneous surface with prominent lunarlike mare ridges, mesas and plateaus, areas of the plains that appear to be 'etched,' fields of knobs, vast areas that have been subjected to channel-forming processes, and low shields that may be volcanic. The term shield as used here indicates a positive relief feature with low slopes.

In places the plains have a mottled appearance, particularly in the central and eastern parts. Many crater-associated streaks are present, indicating wind action; the mottling may also reflect aeolian activity. Two lobate albedo patterns occur in the eastern part of the basin (Figure 2), and although they could indicate flow deposits of some sort (volcanic?), there appears to be little topographic relief, and their occurrence with aeolian features suggests that they may also be aeolian, perhaps 'sand sheets.' Viking lander pictures [*Mutch et al.*, 1976*a*, *b*] show that aeolian deposits are variable in both thickness and albedo, at least in the vicinity of the lander.

The plains are pock-marked with impact craters ranging in size from the limit of detection (~100 m) up to the largest

[1] University of Santa Clara at NASA Ames Research Center, Moffett Field, California 94035.
[2] University of London Observatory, London NW7 2QS, England.
[3] Branch of Astrogeologic Studies, U.S. Geological Survey, Menlo Park, California 94025.
[4] Center of Astrogeology, U.S. Geological Survey, Flagstaff, Arizona 86001.
[5] Planetary Science Institute, Pasadena, California 91101.

Paper number 7S0509.

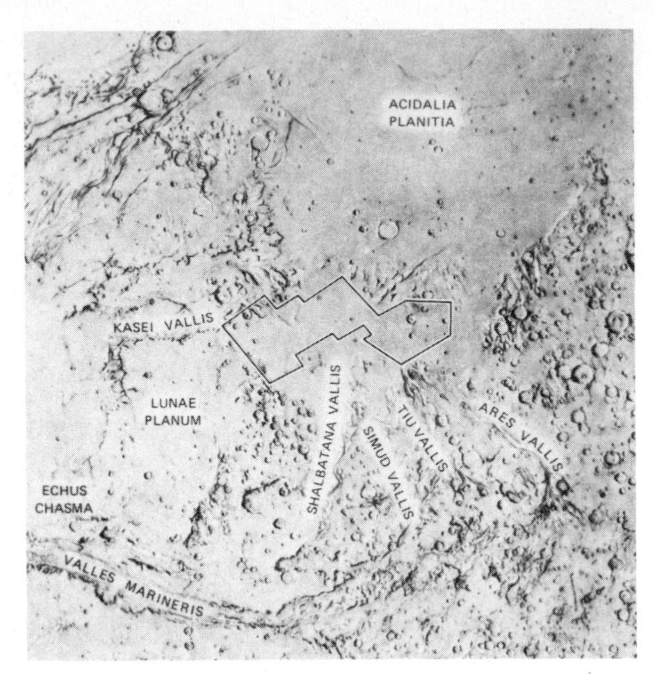

Fig. 1. Airbrush-shaded relief map showing the regional physiographic setting of Chryse Planitia and the major channels. Area outlined corresponds to available Viking orbiter photography across Chryse Planitia and to the geologic map of Plate 1. Area shown here is 3140 × 4000 km, centered at 45°W, 24°N (base map courtesy of U.S. Geological Survey).

crater in the basin, the 60-km-diameter crater Wahoo. The general morphology of some of the large craters in Chryse Planitia and elsewhere on Mars is distinctly different from the moon and Mercury [*Carr et al.*, 1977]. The unusual ejecta patterns (Figure 3) may be attributed primarily to emplacement involving flow of ejecta close to the ground after ballistic deposition, perhaps as a result of ejecta entrainment with water or released gases or of lateral sorting of ballistic ejecta in an atmosphere [*Schultz and Gault*, 1976]. Craters in Chryse Planitia smaller than about 4 km in diameter do not exhibit this ejecta flow pattern but rather are bowl shaped and have ejecta deposits similar to typical impact craters on the lunar mare.

Several clusters of craters are found within the basin, primarily in the south. Although these may be secondary crater clusters, they cannot be readily identified with any large primary crater, and adjacent doublet and triplet craters within the clusters lack the herringbone patterns of ejecta typically associated with secondary craters [*Oberbeck and Morrison*, 1973]. Moreover, one such cluster occurs in an area of suspected volcanic features (described below), and the possibility of their being endogenic cannot be discounted. Alternatively, they may represent impacts formed by the breakup of a larger, incoming meteoroid, although even in this case, herringbone ejecta patterns would be expected.

Dark halo craters smaller than 1 km occur locally in Chryse Planitia, notably in the southeast and west. Except for the dark ejecta the craters have the same morphology as normal small impact craters. The dark halo craters probably represent impact into a multilayered surface in which darker subsurface material was excavated.

Chryse Planitia Plateau

A plateau mapped by *Wilhelms* [1976] as a possible remnant of cratered plateau material, the main body of which lies to the southeast, occurs in the central southeast part of the basin (Plate 1). The plateau, estimated to rise 300 m above the surrounding plains, is about 270 km long by 118 km wide. Unlike the plain the surface of the plateau has not been eroded by channels, although some of the scarps around the plateau are channeled; therefore the processes that formed the channels in Chryse Planitia must have been a discontinuous surface phenomenon that did not affect the topographically higher areas.

Where they have not been eroded by the channels, the margins of the plateau are prominently scalloped. Numerous irregular pits and hollows occur on the surface of the plateau (Figure 4), particularly near the margins. Their irregular form and distribution, lack of ejecta, and lack of raised rims suggest a nonimpact origin. Rather they appear to be the result of collapse, perhaps by a ground-sapping process. Alternatively, they could be wind deflation hollows, although the lack of streaks and other aeolian features argues against this origin. Regardless of their origin the irregular pits contribute to the scalloped form of the plateau margin in that the pits appear to enlarge and eventually connect with the retreating plateau scarp (Figure 4). However, on other plateaus within this area the irregular pits are sparse, and the margins are scalloped nevertheless; therefore the scalloped edges are not necessarily formed by the same process as that which formed the pits.

Mare-Type Ridges

Ridges resembling mare ridges on the moon are common in parts of Chryse Planitia, being most prominent in the west and nearly absent in the central and southeastern parts of the basin. Their distribution is somewhat similar to that on the moon, where mare ridges typically form zones several hundred kilometers basinward from the mare-highlands boundary. Their morphology is also similar to lunar mare ridges in both form and size. Typically, they have a two-part profile consisting of a lower, broad arch which may be many kilometers wide, and an upper, steeper, narrower element that is sinuous and typically several hundred meters wide (Figure 3). Many ridges can be traced more than 150 km as continuous struc-

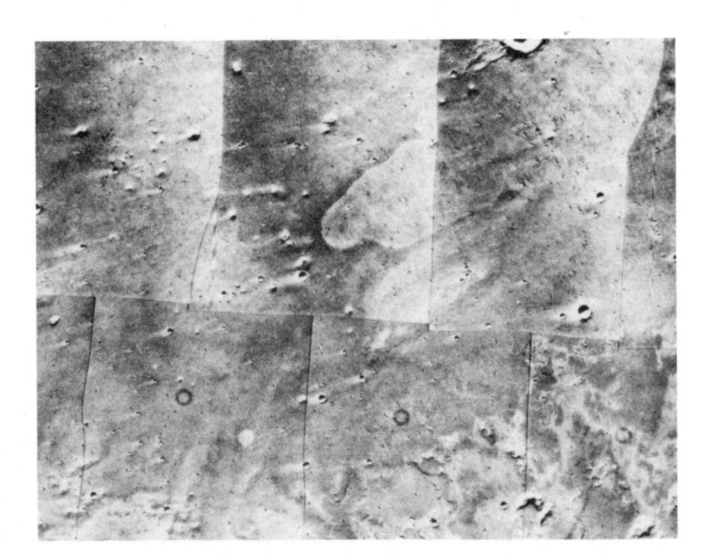

Fig. 2. Lobate albedo pattern in Chryse Planitia, centered at 24.5°N, 35.3°W, immediately north of the Chryse Planitia 'plateau.' Lack of topographic relief on the lobe and its association and approximate orientation with other aeolian features (crater streaks) suggest that the lobes may also be aeolian, perhaps 'sand' sheets. Area shown is about 65.5 × 90 km. (Viking orbiter frames 8A 47, 49, 51, 53 and 8A 75, 77, 79.)

Plate 1. Geologic map of Chryse Planitia.

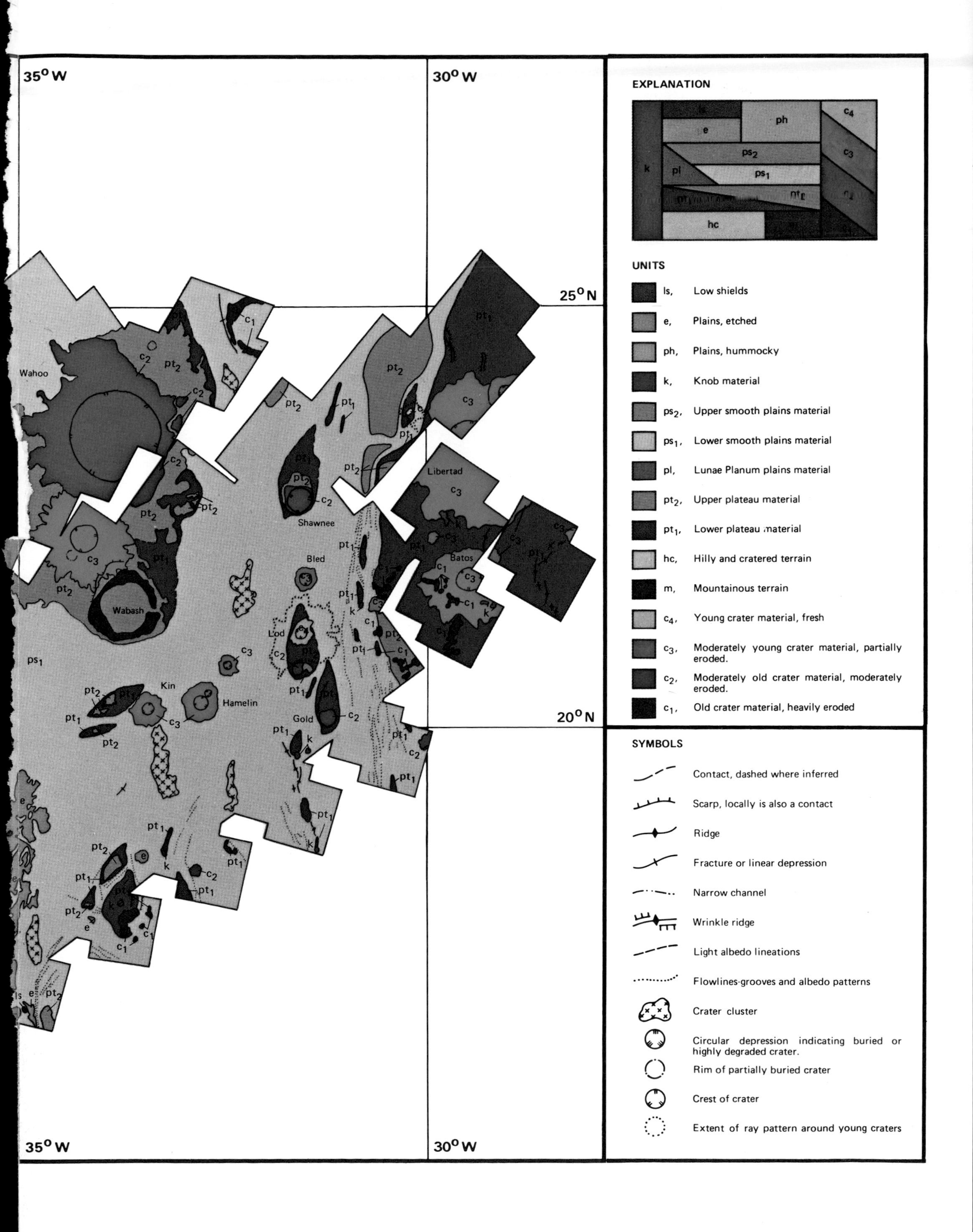
35° W
30° W
25° N
20° N
35° W
30° W
Wahoo
Wabash
Shawnee
Bled
Lod
Kin
Hamelin
Gold
Libertad
Batos
EXPLANATION
UNITS
ls, Low shields
e, Plains, etched
ph, Plains, hummocky
k, Knob material
ps2, Upper smooth plains material
ps1, Lower smooth plains material
pl, Lunae Planum plains material
pt2, Upper plateau material
pt1, Lower plateau material
hc, Hilly and cratered terrain
m, Mountainous terrain
c4, Young crater material, fresh
c3, Moderately young crater material, partially eroded.
c2, Moderately old crater material, moderately eroded.
c1, Old crater material, heavily eroded
SYMBOLS
Contact, dashed where inferred
Scarp, locally is also a contact
Ridge
Fracture or linear depression
Narrow channel
Wrinkle ridge
Light albedo lineations
Flowlines-grooves and albedo patterns
Crater cluster
Circular depression indicating buried or highly degraded crater.
Rim of partially buried crater
Crest of crater
Extent of ray pattern around young craters

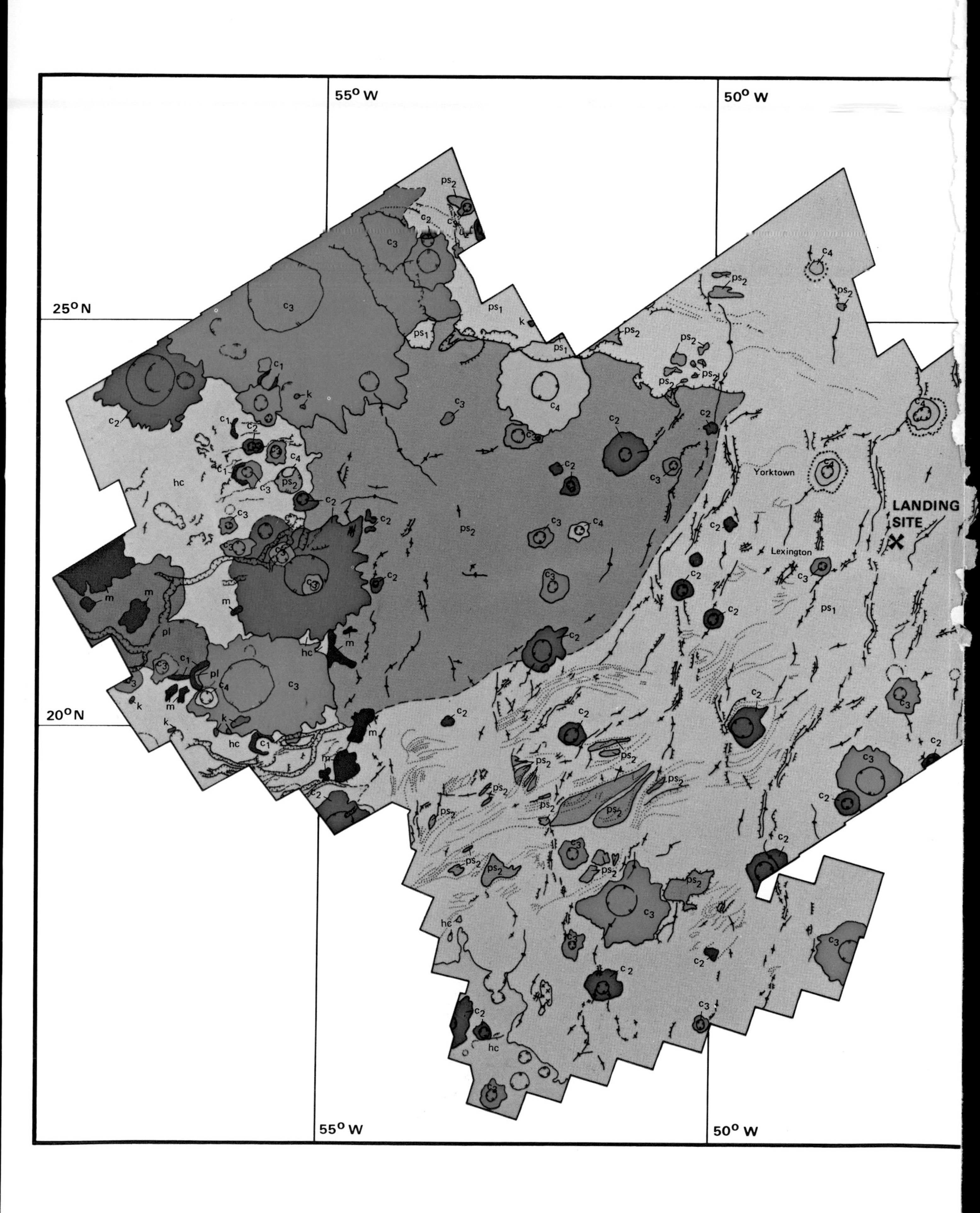
55° W
50° W
25° N
20° N
LANDING SITE
Yorktown
Lexington
55° W
50° W

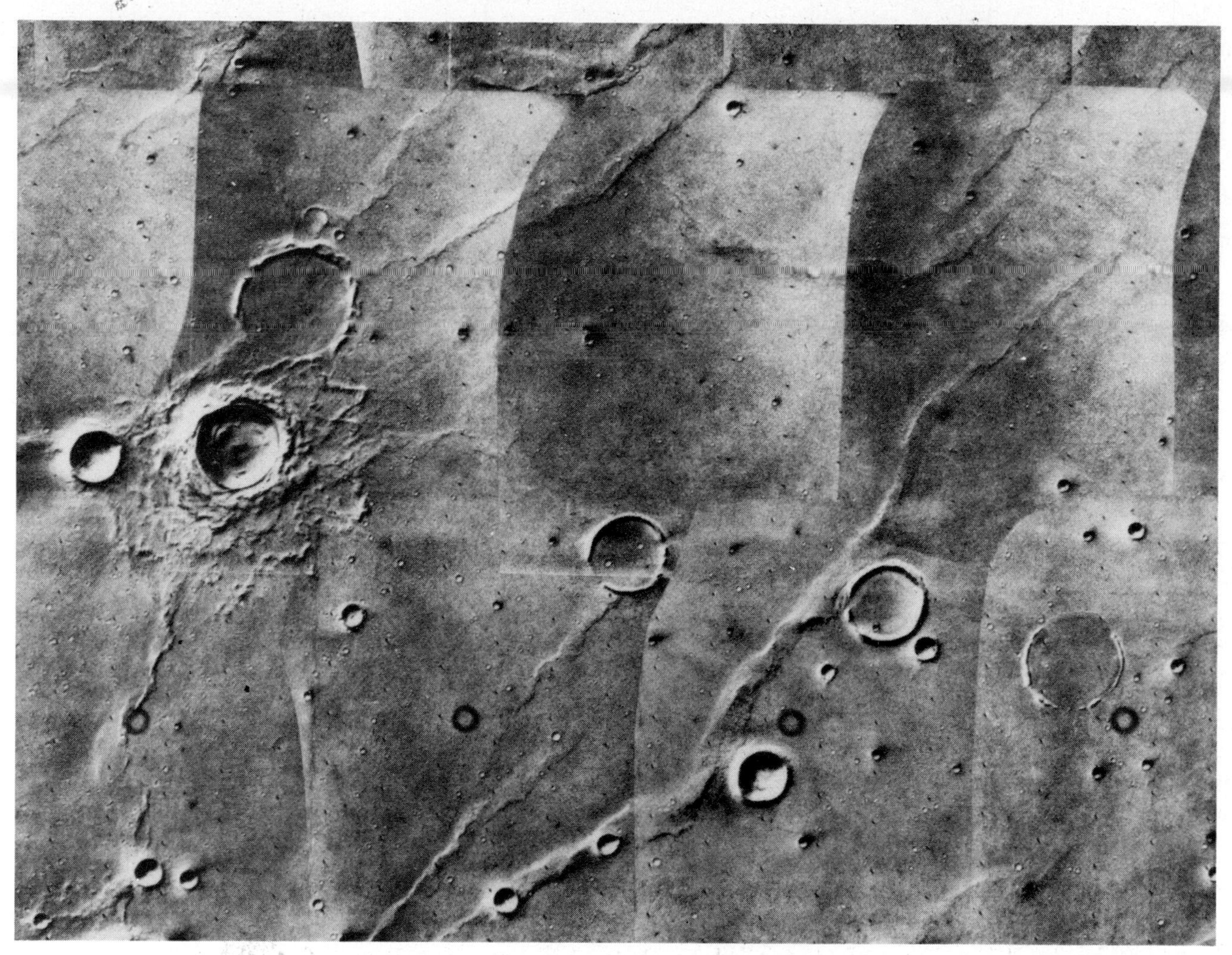

Fig. 3. Photomosaic covering an 84 × 105 km area of southwestern Chryse Planitia (centered at 21°N, 47°W) showing mare-type ridges and impact craters in various stages of burial and degradation. The morphology of the ridges is identical to that of lunar mare ridges. The 8-km-diameter impact crater on the left displays a 'flow ejecta' pattern typical of many fresh Martian craters; the pattern is believed to result from entrainment of atmospheres and other volatiles and perhaps liquid water with the ejecta [see *Carr et al.*, 1977]. Also shown are four large crater rings interpreted to be impact craters that have been buried to varying depths by the mare basalts of Chryse Planitia. These craters may represent a 'basement' of heavily cratered terrain beneath the plains units. The cratered terrain–plain boundary lies about 380 km west (to the left) of this area; the possible progressive burial of the craters from left to right suggests that the basement surface dips to the right, or basinward, on the present surface of the plains. (Viking orbiter frames 20A 84–92.)

tures; others occur as en echelon sets that total several hundred kilometers.

The origin of lunar mare ridges is a matter of controversy, and because of their morphological similarities the controversy similarly concerns the Martian equivalents. The two general hypotheses involve igneous and tectonic origins. First, the ridges may be volcanic features, either pressure ridges that result from buckling of a solidified crust on top of a vast lava lake, in which case they are primary landforms and the plains in which they occur are volcanic, or combinations of intrusive features (the broad arches) and extrusive features (the narrow upper level), in which case the plains may or may not be volcanic. Second, the ridges may result from tectonic processes, such as faulting. Several authors have demonstrated that some lunar mare ridges are continuous across mare-highland boundaries and therefore must be unrelated to the basalt flows of the maria. Other ridges truncate impact craters, indicating that the mare lavas must have been solid enough to have preserved the crater, which argues against a plastic volcanic crust [e.g., *Howard and Muehlberger*, 1973].

It is possible that both volcanic and tectonic mechanisms have operated on the moon to produce landforms that have essentially the same appearance, and the same may be true on Mars; however, we favor a volcanic origin for the mare-type ridges and some of the plains in Chryse Planitia for the following reasons.

1. None of the mare-type ridges are traceable from the plains into the adjacent cratered uplands, as is common for lunar ridges cited as faults.

2. None of the ridges truncate impact craters, although many craters are superimposed on top of the ridges. Were the ridges tectonic, it seems reasonable to expect at least some transected craters.

3. Several ring ridges, similar to mare ridges, are clearly associated with partly buried crater rims and appear to represent surface deformation of plains units laid down on top of the craters (Figure 5). Such deformation would be expected for basalt flows emplaced on top of a cratered surface; with cooling and degassing, differential settling of the lavas over the crater rims would account for the formation of the ring ridges.

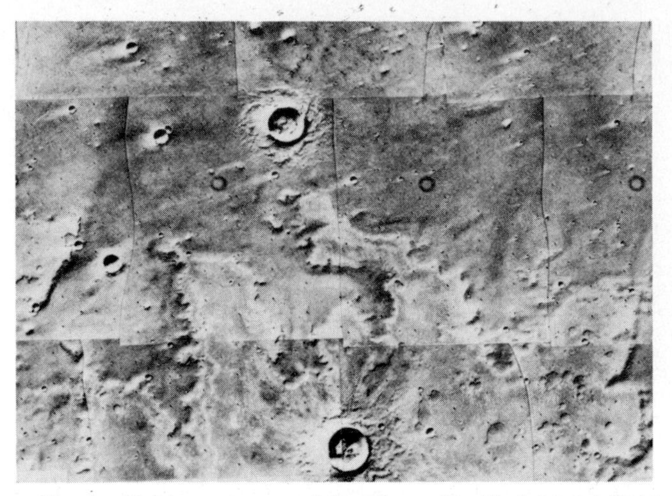

Fig. 4. Northern contact of the Chryse Planitia 'plateau' (light mottled surface, lower part of photomosaic), and the darker albedo plains. Note the irregular scalloped scarp of the plateau and the irregular depressions on the plateau (right side). As the plateau scarp retreats, it produces a hummocky, mass wasted deposit that grades into the plains. (Viking orbiter frames 8A 68–74.)

Similar adjustments in other parts of the basin not controlled by craters may account for the mare-type ridges.

4. The margins of the plains in some areas show an irregular, hummocky surface (Figure 6) resembling thin basalt flows on earth, suggesting that some of the plains in Chryse Planitia are also lava flows.

Channels

Since their discovery, Martian channels have generated considerable interest. Viking orbiter pictures show them in considerable detail, particularly in Chryse Planitia and environs, permit more refined interpretations of their geology, and place constraints on various models of their origin. Although a more detailed study of Martian channels in light of Viking data is currently in preparation, it is necessary to consider the channels here because they are an integral part of the geology of Chryse Planitia.

At least six major channel systems terminate in the basin (Figure 1). They appear to arise from at least two widely separated source areas. Assuming that channel formation involves erosion and transportation, then transport to Chryse Planitia of mixtures of materials of different compositions and/or ages is likely. The headward regions of the channels either have dendritic tributary systems, like most fluvial systems on earth, or originate from a 'point' source, as might be the case when sapping processes have taken place.

The channeled surfaces in Chryse Planitia are characterized as zones of braided and anastomosing channels and depressions that become progressively diffuse toward the basin center and disappear before reaching the lowest part of the basin. The morphology of the channels appears to be at least partly a function of the terrain in which they occur. In moderately to heavily cratered regions (Figure 7) the channels are relatively uniform in width, comparatively deep, and fairly well defined. In plains units the same channel becomes very broad and difficult to trace (Figure 8). The unnamed channel south of Kasei Vallis, for example, alternates from the diffuse form in Lunae Planum to a well-defined channel as it crosses the cratered terrain eastward, then again becomes diffuse as it terminates in Chryse Planitia. The cause of the correlation of channel morphology and terrain is unknown, but differences in material units and in local topography are likely contributing factors. Several channels in the cratered terrain, however, truncate crater rims and evidently were little influenced by the topographic barrier presented by the rims, but these appear to be exceptions (Figure 9). Nonetheless, the hypothesis [*Schonfeld*, 1976] that these channels may be endogenic cannot be ruled out. Channeling controlled by material differences and topography within the basin is illustrated by many teardrop-shaped 'islands' with impact craters on the upslope end that apparently acted as a barrier to the flow (Figure 10). In many cases, crater ejecta deposits are still visible behind the crater but have been removed 'upstream.' More deeply scoured channels on the flanks of such islands apparently resulted from vortical flow and turbulence related to a horseshoe vortex pattern similar to that described for aeolian flow [*Greeley et al.*, 1974].

In some places, long streamlined islands occur within the channel which do not appear to be related to any form of topographic obstruction. Figure 11 illustrates several examples in Kasei Vallis where the upper surfaces of the islands appear to be at about the same level as the nonchanneled part of the plains to the immediate southwest.

The mare-type ridges also appear to have influenced the channeling process. In some cases the ridges appear to have diverted the channels (Figure 12), causing a venturi-type flow; in other cases, where the channels cross ridges, deep channel segments develop on the downflow side of the ridge, perhaps in response to turbulence generated by the ridge or by an increase in flow velocity associated with an increase in slope 'downstream' of the ridge.

On the plains the channels are best developed in areas where the flow appears to have been restricted, as between large craters, between ridges, or where confined by large topographic features such as the plateau in the southeast part of the basin. In these areas the sides of the channels show distinctive layering (Figure 10), which could be erosional terraces representing different levels of channel flow or differential erosion of rock strata. Also present are lineations on the channel floors parallel to the sides; the widths of the lineations, however, are at the limit of resolution (~35 m), and it is not possible to determine whether they are topographic features or albedo units.

Although there are no obvious large deltalike accumulations of materials where the channels pass from the cratered areas onto the plains, there are albedo patterns that appear to be related to the channels and fine-scale albedo patterns in the western transition zone from the hilly and cratered terrain to the plains that could represent deposits (Figure 8).

Fracture systems are found in some areas of well-defined channeling on the plains (Figure 12). The fractures are as wide as several hundred meters and are several kilometers long. Typically, they occur in two crudely arranged sets, one set approximately parallel to the channel and a second, weakly developed set approximately perpendicular to the first. The origin of the fractures, particularly in relation to the channels, is not clear, but the fractures may be related to the etching process described below.

One of the main issues in the general controversy of the Martian channels is whether there has been more than one episode of channel formation. This issue is also crucial to the general geology of the Chryse basin, since the sequence of the plains units and their modification is at least partly a function of the channeling process. Although Mariner 9 images show several regions where channels truncate other channels, it is

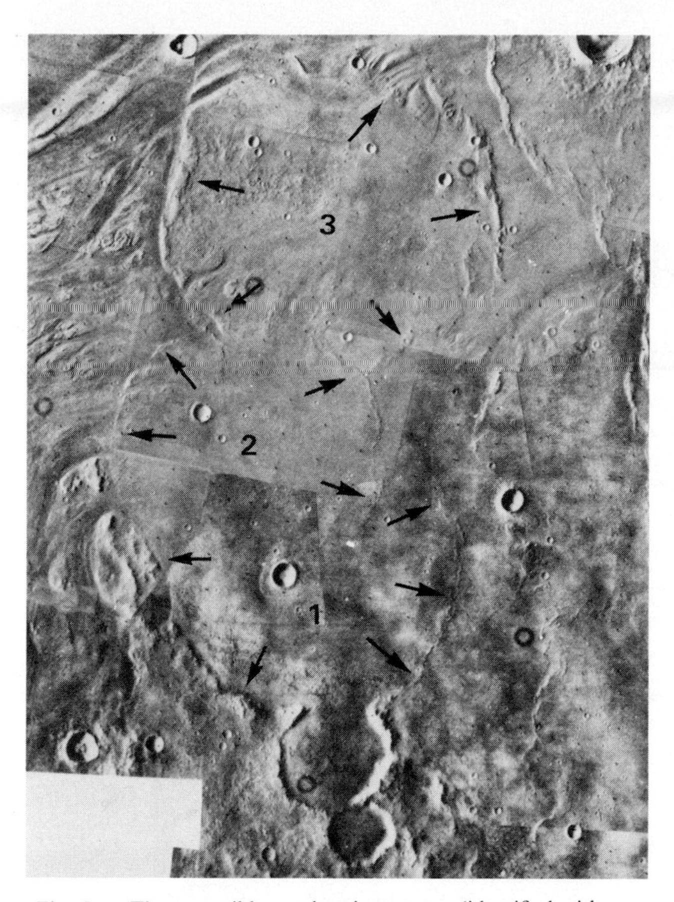

Fig. 5*a*. Three possible overlapping craters (identified with numbers and outlined with arrows) that have been buried by plains units on the western side of Chryse Planitia. Crater 1 is fairly well defined by its southern rim in the cratered terrain; the eastern part of the rim apparently caused a small mare-type ridge to form in the mantling plains unit. Craters 2 and 3 are suggested by the presence of similar mare-type ridge rings. The area of the photomosaic is about 84 × 107 km. (Viking orbiter frames 44A 57–62 and 54A 27–30.)

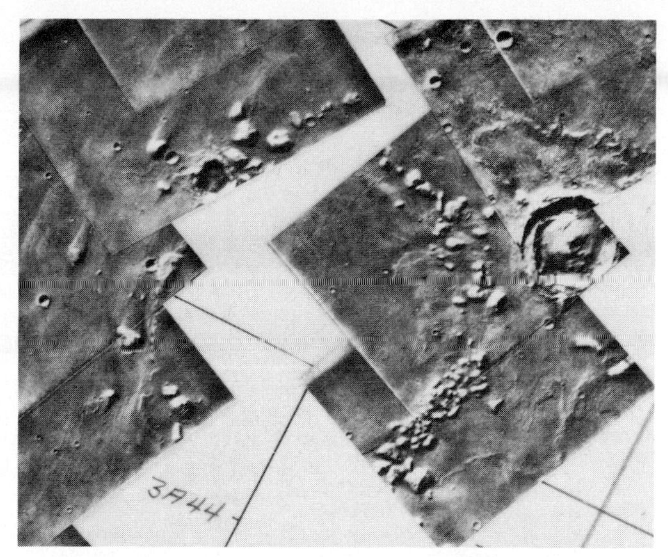

Fig. 5*b*. Photomosaic showing ring of knobs in southeastern Chryse Planitia interpreted to be remnants of a crater rim (ancient heavily cratered terrain) that has been nearly buried by plains units.

not possible to determine how much older the other channels were at the time of truncation, and it can be argued that such relations could occur during a single episode of channel formation.

Viking orbiter images show evidence that either there was more than one period of channel formation or the process was of very long duration. For example, Figure 9 shows a region in the cratered terrain west of Chryse Planitia in which there are two sets of channels, one having a relatively fresh appearance with sharp channel banks and the other having a degraded appearance and truncated by the younger set. Moreover, the 28-km-diameter impact crater in the figure has ejecta deposits that are superimposed on the older channel set and eroded by the younger set of channels on the opposite side of the crater. A similar relationship between two sets of channels and an impact crater is found on the plains within the Chryse basin (Figure 13). In this region the older set of channels occurs as isolated, more degraded channel segments and small remnant 'islands.' It could be argued that impact craters formed contemporaneously with the channeling, thus explaining the superposition relationships observed. However, it seems to be too much of a coincidence for two impact events forming craters more than 21 km in diameter in virtually the same region of Mars to have occurred other than over a long period of time. Moreover, the existence of two sets of channels in the area distinguished by their state of preservation suggests two distinctive episodes of channel formation rather than simply a prolonged period of channeling.

The mode of formation of Martian channels has also been a matter of debate. While most investigators agree that fluids were responsible for their formation, the kind of fluid is in question, water, lava, and wind being the primary candidates. Consideration of the size and morphology of the channels in Chryse Planitia has led some investigators to the conclusion that volcanic processes [*Schonfeld,* 1976] and/or aeolian processes dominate in the formation of the channels. However, we favor a fluvial (water) origin for the channels in Chryse Planitia, perhaps in a manner similar to that described for the formation of the 'scablands' of the Pacific northwest [*Baker*

Fig. 6. Photomosaic showing the southern boundary of the plains (top), the cratered terrain (bottom left), and the hummocky irregular edge of the plains. This irregular terrain could be the result of erosion, or it may represent the flow margin of the mare-type basalts that fill the basin. This surface texture is typical of certain thin (less than 10 m thick) basalt flows on earth, such as the Amboy lava field in California [*Greeley and Bunch,* 1976]. Note the bench on the floor of the crater at the bottom center, below the dark ring (photo artifact); this bench may represent a high lava mark, left by pooled lava intruded into the crater. Area of photograph is 45 × 59 km, centered at 16.5°N, 51.5°W. (Viking orbiter frames 65A 53–58.)

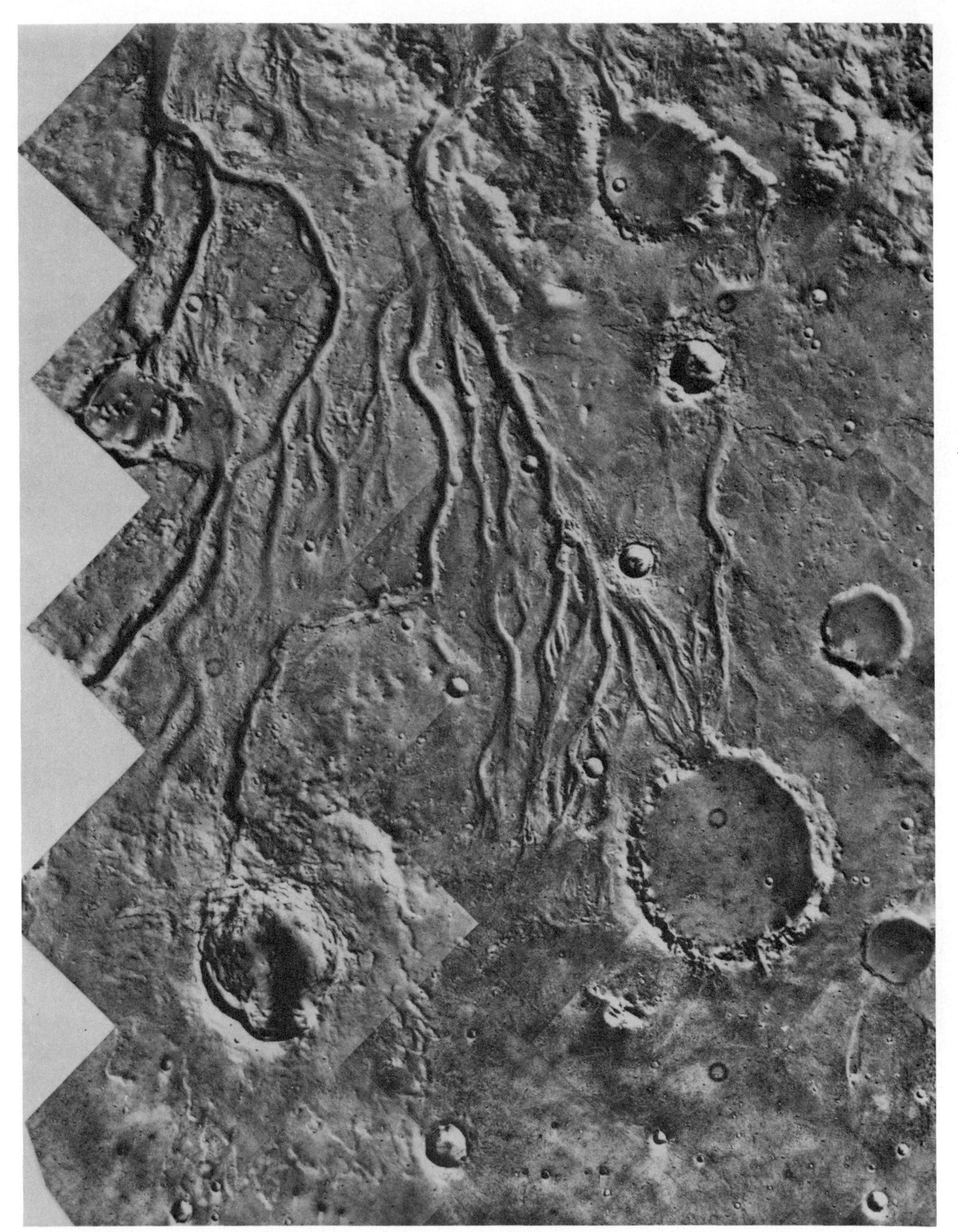

Fig. 7. Photomosaic of cratered terrain (partly mantled) between Chryse Planitia (east, to the right) and Lunae Planum (west, to the left), showing typical channels in cratered terrain having well-defined channel banks. Area of mosaic is 99 × 134 km, centered at 55.5°N, 19.5°W. (Viking orbiter frames 47A 48, 50, 52, 54, 58 and 46A 51–60.)

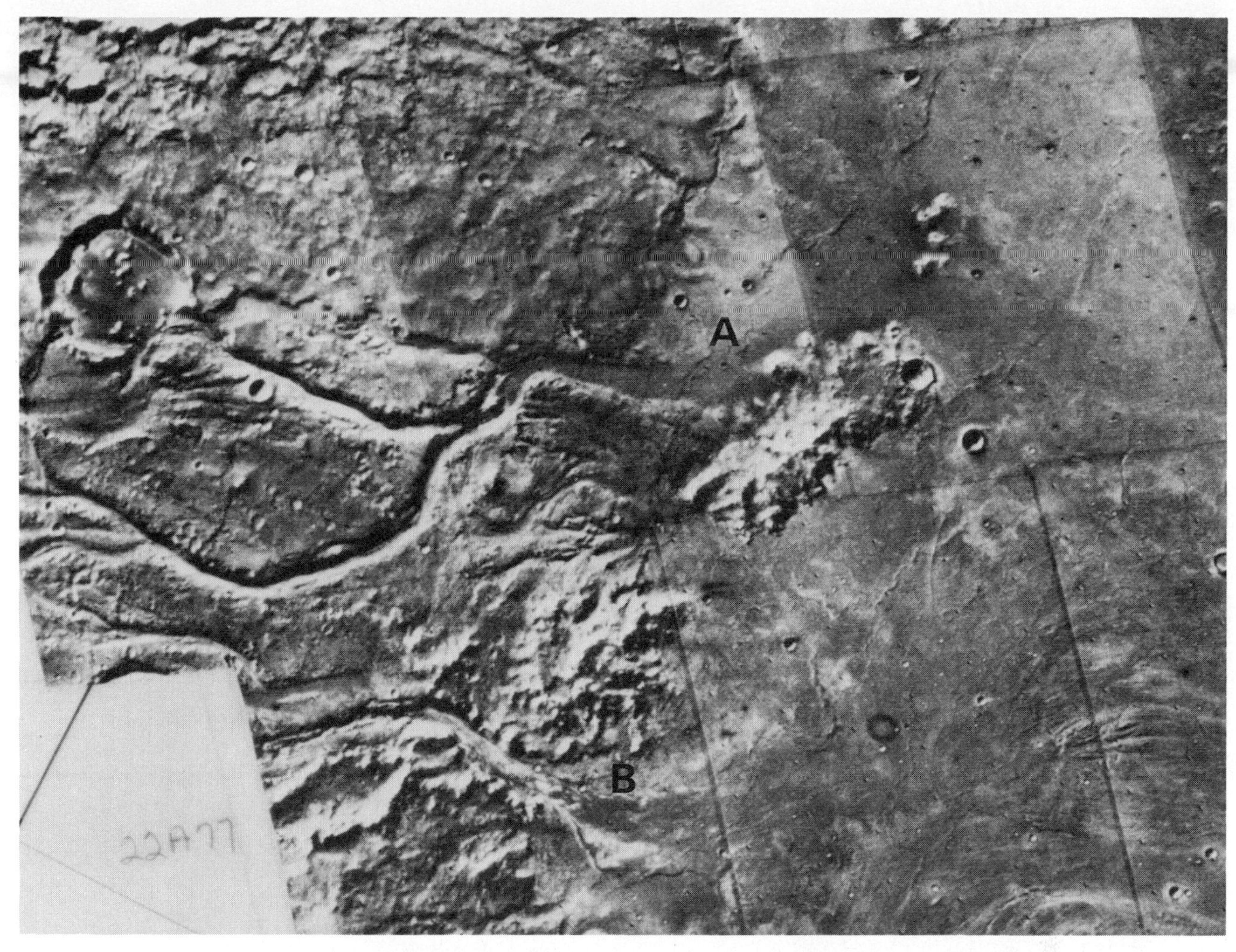

Fig. 8. The 'mouths' of two unnamed channel systems at the cratered terrain (left) and plains boundary of Chryse Planitia showing albedo markings (A, B) that may represent fluvial deposits. Area of photomosaic is 83 × 101 km, centered at 20°N, 54°W. (Viking orbiter frames 22A 76–81.)

and Milton, 1974]. The channels originate from many sources in diverse terrains (such as plains, cratered units, and canyonlands). The lack of volcanic features, the presence of tributary systems, and the lack of 'roofed' sections that are characteristic of lava channels and tubes all argue against a volcanic origin. Furthermore, the orientation of the channels is totally unrelated to the present wind patterns indicated by crater-associated streaks [*Greeley et al.,* 1977]. In addition, the channels occur in topographically low regions, without any indication of channeling on elevated surfaces such as the plateau. Thus if wind was the agent, extreme local topographic control is implied. In other regions, where there is abundant evidence of wind action such as crater streaks, erosion appears limited; we consider that wind is of secondary importance in carving channels in Chryse Planitia.

Etched Terrain

Part of the plains in southeastern Chryse Planitia has been subjected to a process in which the upper ~25–75 m of the plains has been stripped away. Informally named 'etched' terrain (Figure 14), the area covers more than 16,000 km². The alignment of the general zone within a major channelway and its association with the fracture sets to the northwest (described above) suggest a genetic relationship. Also the etched depressions are rectilinear, suggesting structural control. It is proposed that the plains were initially fractured and that etching resulted either from fluvial plucking of fractured plains units or from aeolian deflation after the channel formed.

The floors of the etched region are light toned, which may be caused by either exposure of a subsurface unit or a mantling of light-colored debris (perhaps wind transported), trapped in the depressions. The former case is preferred here because there is a general lack of aeolian features such as streaks that would indicate aeolian deposits.

Volcanic Features

In addition to the probable volcanic origin for some of the plains units and the associated mare-type ridges, two other types of possible volcanic features occur in Chryse Planitia: low shields and knobs. Low shields occur in the same general area as the etched terrain in the southeast part of the basin. They consist of low-profile rises, many of which have summit craters and some of which have steeply profiled summit knobs. About 60 shields have been identified over a region of some 10,000 km². They are typically circular to elongate in plan view and are uniform in size, having an average minimum width of 1.9 km and an average maximum of 3.0 km, for an average diameter of 2.4 km. Of the 60 low shields, 55% have summit craters, 6% have summit domes, and 5% have both craters and

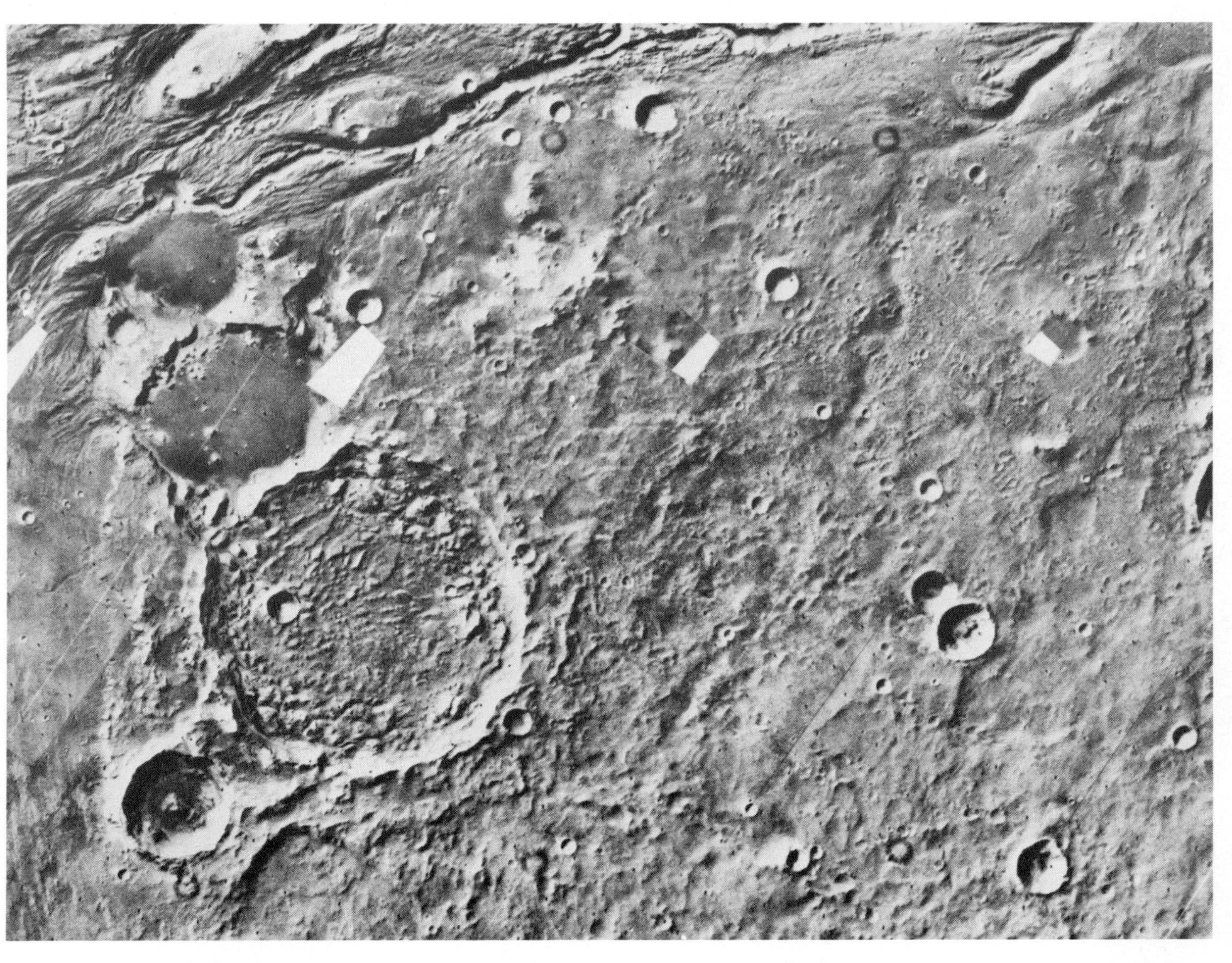

Fig. 9. Stratigraphic relationship of early-age channel, intermediate-age impact crater, and late-age channel. The early channel (right half of photograph) is overlain by ejecta from the impact crater, whereas the late channel (left and top) erodes ejecta of the same crater.

Fig. 10. Part of channel in southeastern Chryse Planitia showing distinctive layering in the channel banks of teardrop-shaped islands. Layering may represent stream-cut terraces formed by different flow levels, or it may represent differential erosion of layered rock units. A field of possible secondary craters is visible on the left. Area covered by photomosaic is 31 × 43 km, centered at 20.5°N, 31°W. (Viking orbiter frames 4A 50–52.)

domes. The craters average 310 m across, and the domes average 390 m across.

Typically, the low shields are of higher albedo than the plains on which they occur, although they are variable, and one low shield within etched terrain has about the same albedo as the plains. The age relationship of the shield, plains, and etched terrain is suggested in at least one case in which the shield is superimposed on both the plains and the etched terrain. Alternatively, the appearance of superposition of the low shields may be an effect of differential etching of material not covered by low shields.

The low shields are considered to be small volcanic constructs, probably built from low-volume eruptions of relatively fluid (basaltic?) lavas that spread out as thin sheets. The pit-type craters on the summits of more than half the shields suggest that the shields are centered on vents; the domes on the summits of some may represent spatter and pyroclastic activity, typical of some shields on earth. Several long (up to 15 km) narrow (50 m) high-albedo lineations are associated with some of the low shields, and these features may be dikes. Origins other than volcanic which could account for the constructional nature of the low shields are not discounted.

Numerous steep-sided knobs with summit craters, interpreted to be volcanic pyroclastic cones, are found in the same general region as the low shields but are of wider areal extent. They are usually less than 1 km in diameter. Uncratered knobs occur in the same region and elsewhere. Not all knobs are considered volcanic; in many cases, they appear to be remnants of preplains units that remain as steptoes or kiupkas, some of which are almost certainly the remains of impact crater rims (Figure 5*b*). However, in some cases the uncratered knobs may be volcanic features which lack craters or on which the crater may be smaller than the resolution of the photography.

STRATIGRAPHY

In this section the units on the geologic map (Plate 1) are described, and their possible origins discussed.

Rugged Terrain

Hilly and cratered terrain (hc). The material of this terrain forms a rough, heavily cratered, highly modified unit that separates Chryse Planitia from Lunae Planum. Most of the craters are highly degraded, and it is difficult to determine the age and extent of crater ejecta. Topographic roughness decreases westward, which may indicate a mantling of the hilly and cratered terrain by either endogenic material or material transported into the area. However, it may also indicate that the eastern area was also mantled but the material has subsequently been stripped away, exhuming the original surface. Channels through the hilly and cratered terrain are incised (Figure 7) and are shown on the map (Plate 1) as scarps. The hilly and cratered terrain is interpreted as the original surface

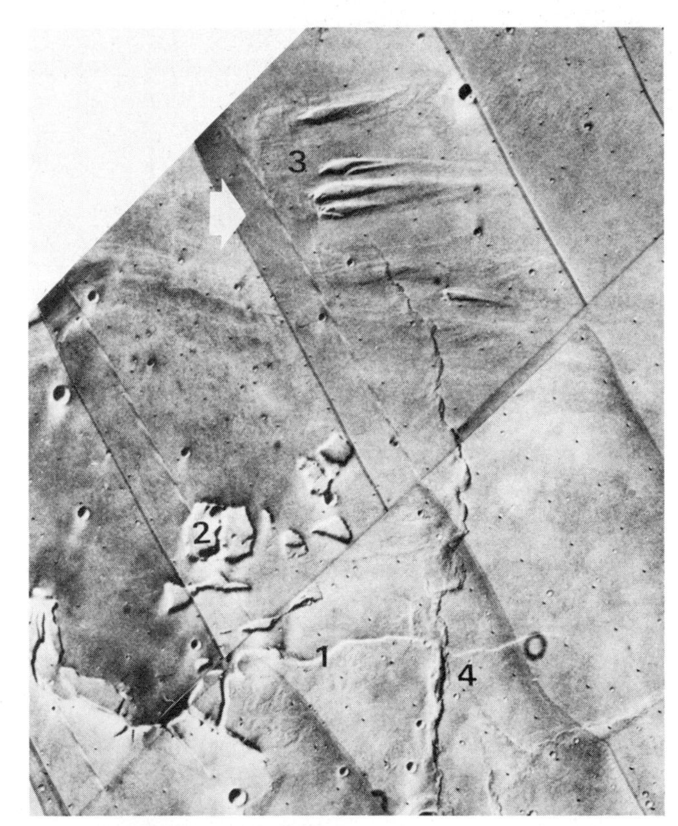

Fig. 11. Photomosaic of part of Kasei Vallis (upper three fourths of figure) where the channel terminates in Chryse Planitia; number 1 marks the channel bank; the arrow indicates direction; number 2 locates a series of mesas of remnant plains units not eroded; number 3 shows several fusiform islands of presumed plains material that were shaped by the channeling process; and number 4 marks a mare-type ridge that appears to have been partly exhumed by the channel. Note that within the channel (north of the channel bank) the broad arch element of the ridge is exposed but ends abruptly at the channel bank; presumably it continues beneath the plains units. Area of photomosaic is 77 × 102 km, centered at 25°N, 50°W.

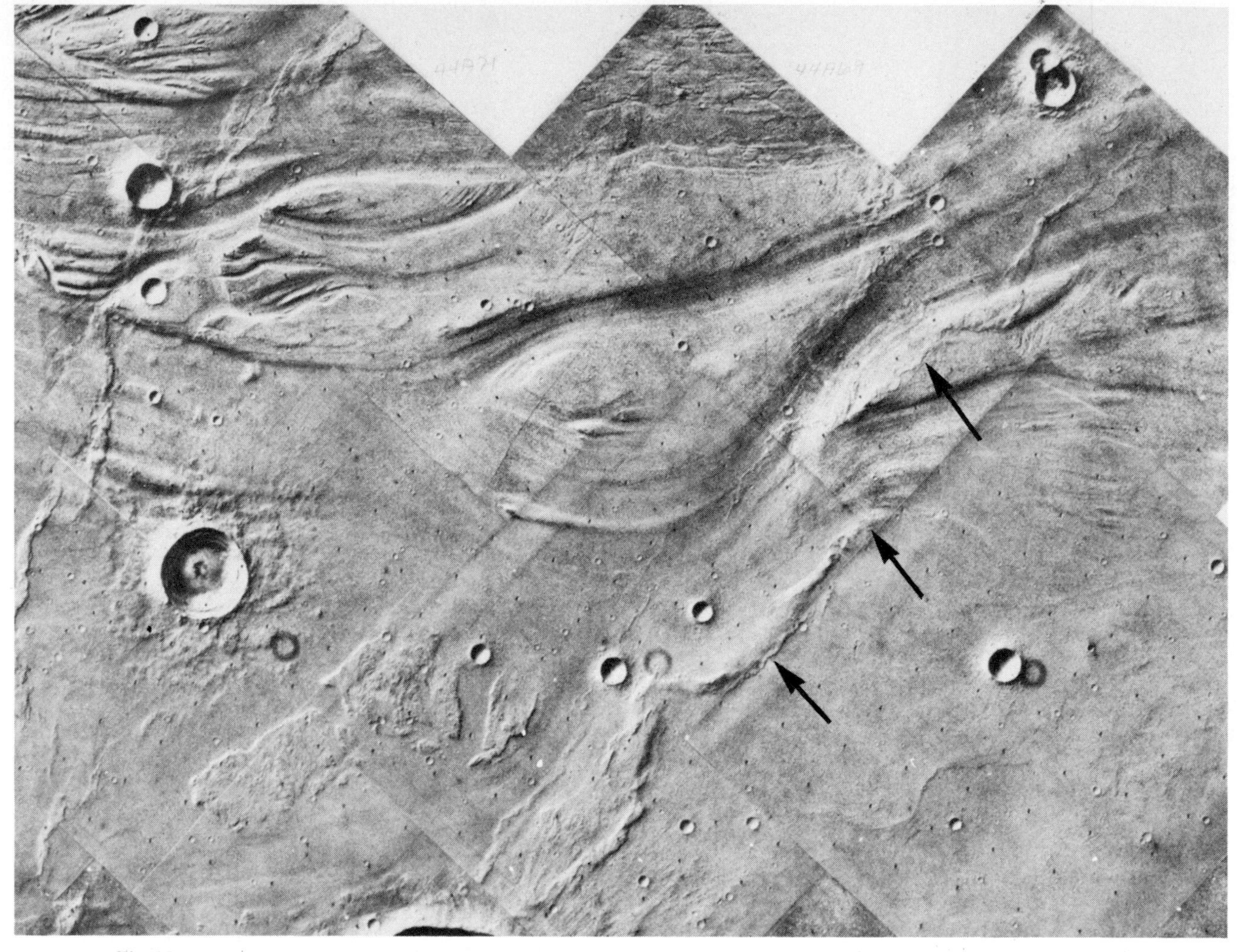

Fig. 12*a*. Part of the channel system in western Chryse Planitia showing the partial diversion of channels by mare-type ridges (arrows) and fracture systems on the floor of a main channelway (top central part of photomosaic). Degraded segments of an older channel system are visible in the lower left quadrant of the photomosaic. Area shown is 79 × 96 km, centered at 19°N, 51°W. (Viking orbiter frames 44A 68–72.)

of Chryse Planitia and may form the basement beneath the basin.

Mountainous terrain (m). Material of this terrain within and along the edge of the hilly and cratered terrain forms a high-albedo rugged surface that rises above the surrounding surface. We consider the mountainous material to be genetically associated with the hilly and cratered terrain. This material may represent large craters that were highly modified before the deposition of the Lunae Planum units, the Chryse Planitia plains, and the mantling material of the hilly and cratered terrain.

Plateau Materials

The plateau materials of southeastern Chryse Planitia consist of at least two different lithologies, both of which may show a facies change. These units are defined by their erosional characteristics, the upper plateau being more resistant and forming a scarp and the lower plateau being easily eroded. Irregular pits occur on the northern upper plateau material but are lacking on the unit in the south, which may be a function of lithology or geomorphic processes. The lower plateau thins northward and may indicate a change to a more resistant lithology, a thinning out of the entire unit, or possibly the burial of more lower plateau material by the subsequent flooding of the basin by lava. The plateau units are interpreted as deposits on an originally cratered surface that were then eroded to form topographic highs before the flooding of the basin by lavas.

Lower plateau (pt_1) material. This material forms a moderately to gently dipping slope and is best seen in the teardrop-shaped islands, where the upper plateau material has been partly or entirely stripped away. This material is easily streamlined by channeling and often shows stair steps and lineations that are interpreted as either differential erosion or stream-cut terraces and flow lines (Figure 10). Along the downflow ends of the teardrop-shaped islands the lower plateau grades into the lower plains, and the boundary is difficult to determine.

The lower plateau material is not always visible in the northern inlier and the eastern plateau. Where it can be seen, it forms a narrow moderately sloping apron around the plateau and may be defined as the material that crops out between the scarp and the lower smooth plains material. However, this narrow band of material may be talus.

Upper plateau (pt_2) material. This material forms a smooth surface with variable albedo and is distinguished by a scalloped erosional scarp (Figure 4). In many cases the large-scale

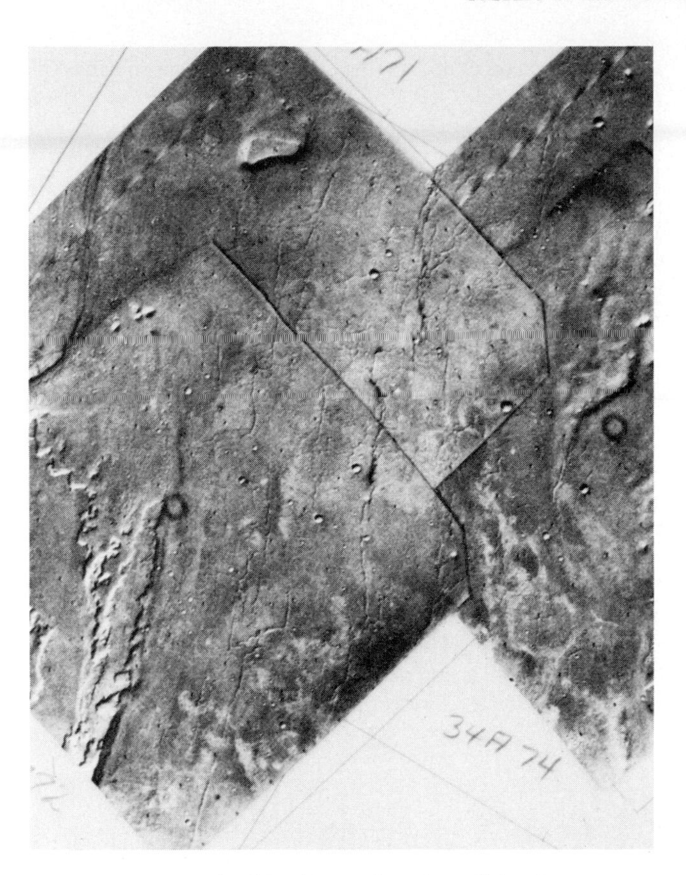

Fig. 12*b*. Network of fractures oriented parallel and transverse to a channelway in southern Chryse Planitia; flow direction was apparently from the south (bottom of mosaic) toward the north.

scallops have been smoothed by channeling processes, although small scallops are preserved. The surface of the material is similar to the lower smooth plains but has a higher average albedo. The surface is mottled, and mare ridges are sparse and subdued. Partially buried crater rims are also visible in the upper plateau material. Shallow irregular depressions, or pits, are unique to the upper plateau material, as was discussed above.

Plains Units

Lunae Planum plains (lp) material. This material forms a smooth surface west of the hilly and cratered terrain. The contact between the two is gradational, suggesting that the plains material at least partly mantles the hilly and cratered terrain. Where Lunae Planum extends into the map area, channels appear incised; to the west the channels are unconfined. The age relationship between Lunae Planum plains and Chryse Planitia plains cannot be established because they are not in contact in the available images. The two units are considered approximately contemporaneous because both overlie the ancient cratered surface, are modified by the same channel systems, and are similar in appearance.

Chryse Planitia plains. The plains-forming materials of Chryse Planitia consist of basal lower smooth plains materials, overlain in the west by upper smooth plains material. The upper smooth plains have been separated from the lower smooth plains because (1) a scarp in a mantling material crosses a mare ridge of the lower smooth plains, dividing it into a partly buried ridge to the south where the broad portion of the ridge is not visible and a northern portion where the complete ridge is exposed (it is proposed that the upper smooth plain is the mantling material which partially covers the mare ridge of the lower smooth plain), (2) in the western part of the basin, dark halo craters occur only in the upper smooth plains area, and (3) the mare ridges in the upper smooth plains can barely be seen, only the sharp narrow portion being visible (it is proposed that the broad arch element is buried), whereas immediately east of the area the mare ridges are distinct. Subdued mare ridges also occur locally on the lower smooth plains, and it is our belief that these ridges may also be partially buried by a thin and patchy veneer of sediments. However, there is no evidence of the extent of this veneer, and therefore it cannot be mapped as a separate unit. Estimates of the thickness of the upper plains based on the diameters of the dark halo craters range from about 50 to 75 m, a figure roughly approximated by the depth of burial for the broad arch element of the mare ridges. However, this material may thicken to the northwest.

Lower smooth plains (ps_1): These plains gently slope toward the center of the basin and form the major unit of Chryse Planitia. The plains are subtly to distinctly mottled, showing no topographic relationship between the dark and light areas. In the area of greatest mottling, dark halo craters are found. The lighter areas may be depositional features with dark halo craters bringing the darker material up from the lower mantled surface. Alternatively, since this area is near the etched zone, it may be highly modified etched terrain. Two lobate albedo patterns (Figure 2) in the eastern part of the plains are discussed above.

Very gently sloping mounds occur throughout the lower plains unit. Elongated mounds in the central plains may be buried mare ridges. Mare ridges in this plains unit form a band about 180 km wide in the west and appear randomly throughout the rest of the plains. Many of the ridges in the band appear unmodified, whereas others are subdued or have been eroded (Figure 3). Ridges in other areas of the lower plains are moderately subdued.

Crater-associated streaks (both dark and light) occur in certain areas of the lower and upper plains, suggesting that surface material is available for wind transport.

We consider the lower smooth plains to consist of a marelike basalt (or comparably fluid lava) throughout the entire area with a thin discontinuous veneer of sediment of variable thickness. The marelike lavas flooded an older cratered surface, as is shown by the presence of ring structures in the western part of the basin (Figure 5*a*). The mottled appearance of the plains and the gentle sloping mounds may be indications of variable thicknesses of sedimentary deposits overlying the basalts. In areas where mare ridges are subdued, the depositional veneer is thin, whereas in areas where the mare ridges are distinct, the veneer may be absent. The veneer covering the lower plains may have formed in association with one or more channeling events.

Upper smooth plains (ps_2): These plains on the western slope of the Chryse Basin are very similar to the lower smooth plains. The upper plains have a mottled appearance, but the gently sloping mounds found on the lower smooth plains are absent, and where mare ridges occur, they are barely visible. Dark halo craters are found in the south central upper smooth plains.

The erosional scarp of Kasei Vallis marks the approximate northern extent of the upper smooth plains units in the area mapped. The scarp decreases eastward and cuts across a mare ridge, dividing it into a northern part where the broad arch part is visible and a southern part where only the narrow ridge element is exposed. Figure 11 indicates that the upper smooth plains material partly mantles the mare ridge south of the scarp but has been stripped away north of the scarp. Lineations visible in portions of the scarp may be different lithological units of the upper smooth plains. Irregularly shaped mesas and steep-sided streamlined islands north of the scarp form outliers of the upper smooth plains (Figure 11).

Islands of upper smooth plains occur in the channeled area to the south. The islands are streamlined, and some show lineations that could be flow terraces or differential erosion of strata, possibly corresponding to the lineations in the Kasei Vallis scarp wall. The different erosional characteristics of the northern and southern upper smooth plains may be explained by different erosional processes.

Lobelike and irregular positive relief features in the eastern area of the upper plains may be flow fronts of volcanic origin, or, alternatively, they may represent aeolian etching around more resistant parts of the plains.

We consider the upper smooth plains to be a mantle formed from an early depositional event associated with channeling. The mare ridges to the west formed an incomplete barrier resulting in a greater accumulation of deposits upstream of the ridges. This deposit was later eroded to the north and south by subsequent channeling, leaving the area between the channels relatively unmodified (Plate 1).

Hummocky plains (ph): These plains are associated with the upper plateau and are generally found on the north or northwest sides of the upper plateau remnants, although the unit also is found elsewhere. The materials of this unit form subdued hummocks which may be erosional products of the upper plateau or may represent the lower plateau after the upper plateau has been removed. In the latter case the hummocks would show the erosional characteristics of the lower plateau where not influenced by channel action.

Etched plains (pe): These plains are in the southeast part of Chryse Basin, where the overlying plains materials have been stripped away leaving shallow steep-sided depressions. These depressions are rectilinear, and the floor material has a high albedo (Figure 14). The rectilinear appearance may reflect structural control in the formation of this unit. Partly etched plains and intricate networks of etching are also included in this unit. The origins and implications of the etched plains are discussed above.

Low shields (ls): These are circular to elongate low-profile flat features that occur near the etched terrain in the southeastern area of the Chryse Basin (Figure 14). Typically, they

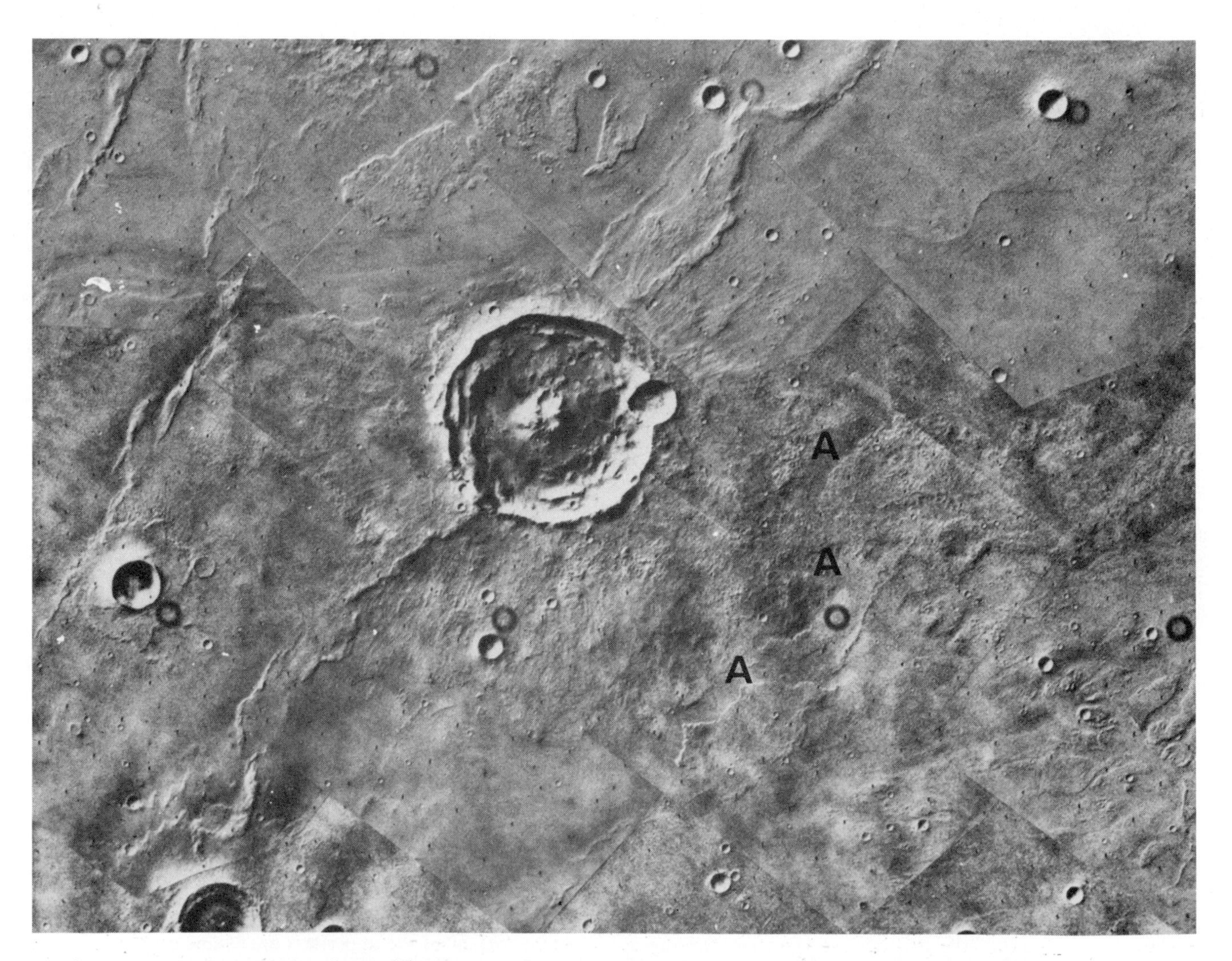

Fig. 13. Large (21-km diameter) impact crater in western Chryse Planitia showing relationships of channel episodes to crater ejecta similar to those shown in Figure 9. A set of degraded (older) channels (A) is overlain by ejecta deposits from the crater. On the north (upper half of figure) side of the crater the ejecta deposits have been eroded by a later episode of channeling. (Viking orbiter frames 44A 62, 64, 66, 68, and 54A 31–36.)

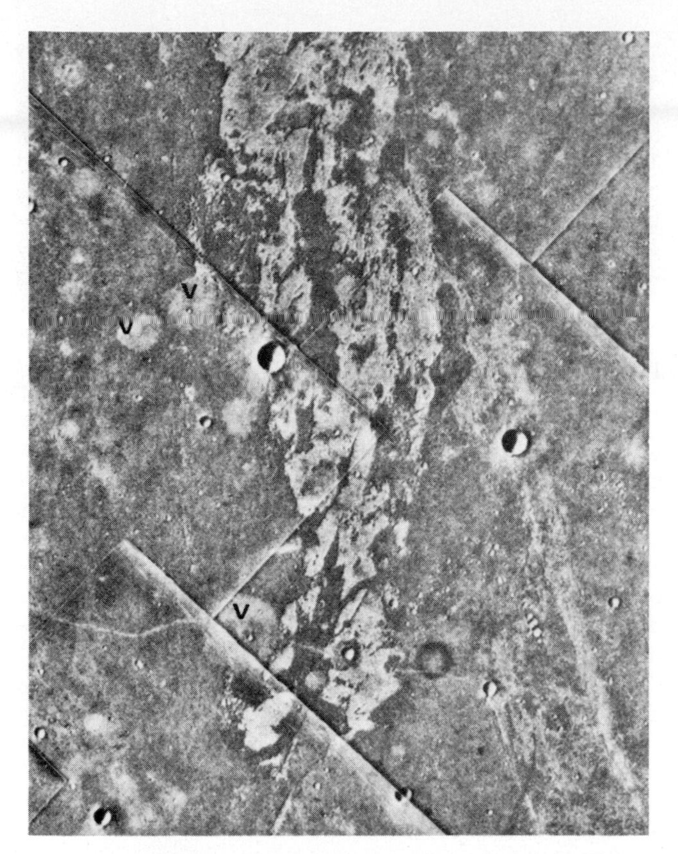

Fig. 14. Etched terrain showing light-toned angular depressions where plains material has been removed and circular features (three of which are labeled with a V) interpreted to be low-profile constructional shield volcanoes, many of which have summit craters. The white stripe associated with the low shield at the bottom of the mosaic may be a dike. Etching processes may be the result of aeolian deflation, channeling processes, sapping processes, or a combination. Photomosaic width is about 60 km.

have a higher albedo than the surrounding plains. Craters and/or central domes are found on many shields, although some low shields are featureless. Low shields are superimposed on both the lower smooth plains and the etched terrain and are considered to be volcanic constructs built by relatively fluid lava erupted late in the history of the Chryse Basin. The craters may indicate possible vent areas, and the central domes may represent spatter and pyroclastic activity.

Knob material (k): This material forms smooth steep-sided mounds that occur throughout Chryse Planitia and are considered to be formed by several processes. The knobs are typically circular and less than 1 km across. Knobs near plateau material or hilly and cratered terrain are thought to be remnants of older, more widespread units. Some knobs are arranged in a circular pattern and clearly represent old crater rims. Several knobs have summit craters and are interpreted as volcanic features. In eastern Chryse Planitia there is a knob cluster which may be either a volcanic field or the remnants of plateau material.

Craters

Craters have been mapped by relative age on the basis of state of preservation. In this report the convention of assigning cratering to a fourfold relative age sequence follows that adapted for the Mars Geologic Mapping Program of the U.S. Geological Survey, typified by the map of *Wilhelms* [1976]. It is emphasized that rate of degradation is probably not uniform geographically on a global scale for Mars but is considered fairly uniform over the limited area of Chryse Planitia, so that the scheme is consistent within the area mapped here.

Old crater material (c_1). This material represents highly degraded craters with only parts of the rim or central peak remaining. The rim material may consist of a rugged or partly buried ring or a circular group of knobs (Figure 5*b*).

Moderately eroded crater material (c_2). This material represents craters that are moderately eroded; the ejecta is partly stripped away or lacks texture. Throughout Chryse Planitia there are many small impact craters that are mapped as moderately eroded crater material because ejecta cannot be seen clearly around them. Although this may result from poor image resolution, it must be recognized that smaller craters will erode faster than larger craters.

Partly eroded crater material (c_3). These craters are relatively fresh with sharp features and textured ejecta but lack the rayed ejecta pattern of young craters. Lobate ejecta deposits are best seen on these craters.

Young crater material (c_4). These craters are crisp and typically have ray patterns in the form of needlelike features radiating from the rim or a high-albedo ray pattern superimposed on the surrounding material.

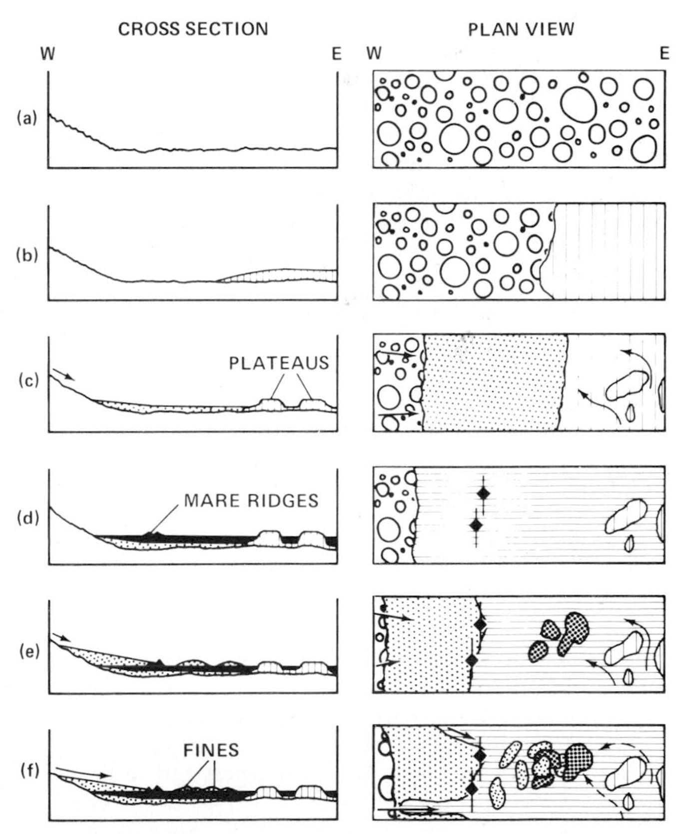

Fig. 15. Schematic diagram of geologic history of Chryse Planitia (thicknesses of units are greatly exaggerated). (*a*) Original heavily cratered surface; basin probably formed before or during the period of early heavy bombardment. (*b*) Deposition of material from east. (*c*) Formation of plateaus by erosion and possible erosion of western highlands. (*d*) Flooding of basin by lava (possibly multiple eruptions) surrounding the plateaus and burying sediments; formation of mare ridges. (*e*) Deposition in the west behind mare ridges; some sediments (heavy dot pattern) may have also been carried into the basin from the southeast. (*f*) Channeling of deposits in west, exhuming parts of the lava plains; fine material carried out to center of basin and deposited as patchy thin veneer. Throughout the sequence there may have been redistribution of surficial material by aeolian processes and impact cratering, not shown in this sequence.

Geologic History of Chryse Planitia

Although the origin and the earliest history of the Chryse Basin cannot be determined from the foregoing descriptions of the general geology and stratigraphy of Chryse Planitia, it is possible to derive a tentative geologic history beginning with the termination of a period of heavy bombardment and the formation of heavily cratered terrain and associated deposits (Figure 15). This unit crops out east, south, and southwest of the basin and appears to underlie the plains, forming a basement, as is indicated by partly buried craters on the margins of the basin and by buried craters (mare-type ring ridges).

Within the cratered terrain around the margins of Chryse Planitia are possible mantling deposits [*Wilhelms,* 1976] which partly bury the floors and flanks of some of the craters. This mantle may be wind-transported material, volcanic deposits, or (less likely) deposits of an early stage channeling event. Although the age relationships of the mantling are not clear, the unit appears to be older than the plains in the Chryse Basin. Following the period of heavy impact bombardment and possibly the episode(s) of mantling, material was deposited in the eastern part of the basin. This material may be a combination of both volcanic and sedimentary deposits, which may explain the lineations visible where channels have cut through this material. There is no evidence within the photographic coverage as to whether the sedimentary deposits are alluvial or aeolian. After deposition this material was channeled, forming the plateaus and mesas that are seen today. This period may represent the earliest channeling event in Chryse Planitia in which the eroded material was carried westward and deposited in the deepest part of the basin. At the same time, material may have been removed from the westward highlands and also deposited in the basin.

Following the plateau formation, low-viscosity lavas were extruded in the basin. The lavas were probably of basaltic composition, resembling those of the lunar maria, which formed the mare-type ridges. Emplacement of these lavas was probably of the flood basalt form, involving great volumes of lava erupted at very high rates of effusion, creating vast lava lakes and destroying all vestiges of the vents from which the lavas were erupted.

The emplacement of the lava plains was followed by the formation of channels in the west. This set of channels eroded parts of the mantling deposits and the craters in the hilly and cratered terrain, as well as some of the lava plains within the basin. Most traces of this episode of channeling have been obliterated, primarily by subsequent channel processes and by degradation from impact cratering. Segments of channels of this stage are present in the cratered terrain west of Chryse Planitia and on the plains in the southwest part of the basin. However, there is no evidence of this period of channeling in the southeast part of the basin or in the adjacent cratered terrain, although the Viking orbital coverage is limited in this area. We propose that this episode of channeling resulted in the deposition of fluvial sediments in some parts of the basin. The mare ridges that form a zone within the basin on the western side may have served as a discontinuous dam behind which the sediments accumulated. The thickness of the sediments was probably rather thin and perhaps even absent in some places, since they did not completely bury the ridges. Deposition probably occurred beyond the ridges as well, since the ridges do not form a continuous dam.

Evidence for deposits between the boundary of the plains and cratered terrain and the mare-type ridge system is offered, as discussed above, by the character of the mare-type ridges and by the distribution of dark halo craters.

The latest channeling formed the prominent channels visible today in western Chryse Planitia and the adjacent regions. These channels represent high-energy systems that eroded both the cratered terrain and associated units and the plains units within the basin. In many places, formerly buried topographic surfaces appear to have been exhumed, a possibility noted by *Milton* [1974] on Mariner 9 images for the cratered terrain immediately west of Chryse Planitia.

Exhumation also appears to have occurred within the plains units in the western part of the basin. Deposits of the earlier episode of channels have been stripped away along the course of the later channels, as indicated perhaps by the general lack of dark halo craters. Moreover, some of the mare ridges appear to have been partly exhumed by the channeling. For example, the southern bank of Kasei Vallis has cut plains units along one mare-type ridge, exposing both the broad arch element of the ridge and the upper narrow element, whereas immediately south of the channel, only the narrow element is exposed; presumably the broad arch remains buried where the channeling was not active (Figure 11).

Thus the main effects of the second episode of channeling in the west were to redistribute sediment of the early channeling and to form the prominent channels visible now. Some of the deposits must have been transported basinward to lower topographic areas. The center of the basin lacks mare-type ridges, which may indicate burial, or they may never have formed in the first place.

As was discussed in the section on the channels, either the interval of time for the fluvial activity was very long, indicated by the frequency of large impact craters interspersed within the channeling events, or the formation of channels was sporadic, with intervening periods when channels were not formed. The presence of at least two distinctive systems of channels suggests the latter to be the case. Although expressed here in a simplified form, the actual sequence may have involved repeated volcanic and fluvial episodes resulting in a complex sequence of volcanic and fluvial sediments. In this case the channeling after the lava flooding as discussed above would represent the last two events. From the relatively high frequency of craters superimposed on even the youngest channels it would appear that the channel-forming events are rather old. One of us (J.A.C.) believes that the two episodes of channeling arise from two distinct mechanisms: lava erosion and aeolian processes, as discussed elsewhere (J. A. Cutts and K. R. Blasius, unpublished manuscript, 1977).

One of the surprises from the Viking lander was the presence of numerous large angular rocks on the plains in an area considered to be primarily a depositional regime of presumably fluvial sediments. Within the framework presented here the materials observed at the Viking 1 landing site are considered to be part of the impact-generated regolith (accounting for some of their angularity) derived both from in situ and reworked channel deposits and from the underlying mare-type basalt flows. The multiple source (transported plus local) for the rocks may account for the wide variability in textures and colors, noted by *Mutch et al.* [1976*a*].

One of the principal arguments against a fluvial origin for the channels associated with Chryse Planitia is the lack of easily identifiable sedimentary deposits in the basin. Such deposits may be difficult to recognize at the scale and resolution of the Viking pictures. Moreover, deposition may not have occurred everywhere within the basin but rather as scattered patches with the margins of the deposits thinning to feather edges; they may also have been rather thin in comparison to the size of the basin. In addition, some of the fluvial deposits may have been removed from the basin by aeolian activity. The latter may be expected because only fine material may have traveled the distance to the basin. Partial removal of fine particles by the wind could account in part for the blocky appearance of the surface at the landing site. The laminated terrain and the vast dune fields of the north polar region [*Cutts et al.*, 1976] may be one of the depositional sites for the aeolian material removed from Chryse Planitia.

Following and perhaps even during the last episode of channeling, limited volcanic activity took place in the southeast part of the basin to form the low shields and perhaps some of the knobs. Impact cratering continued through the period of volcanism and to a minor scale undoubtedly continues today.

The most active process currently modifying the surface of Chryse Planitia appears to be aeolian activity, evidenced by the streaks associated with many of the craters and the drift deposits viewed by the Viking lander.

SUMMARY AND CONCLUSIONS

High-resolution Viking orbiter images show Chryse Planitia to be much more complex than had been suspected from Mariner 9 images. Ancient heavily cratered terrain appears to form the basement for the basin. Much of the heavily cratered terrain is mantled with deposits that may be of aeolian, fluvial, or volcanic origin. The basin is partly filled with a sequence of plains units, some of which resemble mare basalt flows on the moon and hence are considered volcanic; other units are probably sedimentary deposits resulting from both fluvial and aeolian processes.

The surface of Chryse Planitia has been modified by several processes, the most dominant of which formed channels. Several mechanisms have been proposed to explain channels in the area, including erosion by water, wind, and lava. Fluvial, or water, origins are preferred here. Regardless of their origin, Viking orbiter pictures demonstrate that there was more than one period of channel formation in Chryse Planitia and the surrounding terrain (Figures 9 and 13). Regions of fracturing and etched terrain, where the upper plains units have been stripped away, appear to be related to the channels.

The youngest activity in the basin includes suspected volcanism, which formed features interpreted to be low constructional shields, and impact cratering. Aeolian activity is indicated by albedo patterns associated with craters and undoubtedly constitutes an active process.

In conclusion, the geological history of Chryse Planitia involves a complex sequence of impact cratering, mantling by extensive deposits of unknown origin, redistribution of mantling and crater materials by erosion and deposition with concurrent eruptions of flood-type basalts, and aeolian activity.

Acknowledgments. We wish to thank the Viking Flight Team for their part in making the mission a success, particularly in regard to the acquisition and processing of the orbiter images. During site certification activities when the preliminary geologic maps were derived, many individuals made substantial contributions, and we wish to acknowledge with thanks E. A. Flinn, B. Lucchitta, and K. W. Farrell. The Viking Interns played a significant role during these activities and contributed to the geologic analyses; we thank in particular A. Spruck, P. Spudis, R. Papson, and J. MacQueen. B. Lucchitta and P. Spudis provided valuable reviews for both the text and the geologic map. Support for the work of the team was provided by the NASA Viking Project Office, NASA Office of Planetary Geology (R.G.), and the United Kingdom Natural Environment Research Council (J.E.G.).

REFERENCES

Baker, V. R., and D. J. Milton, Erosion of catastrophic floods on Mars and earth, *Icarus, 23,* 27–41, 1974.

Carr, M. H., H. Masursky, W. A. Baum, K. R. Blasius, G. A. Briggs, J. A. Cutts, T. Duxbury, R. Greeley, J. E. Guest, B. A. Smith, L. A. Soderblom, J. Veverka, and J. B. Wellman, Preliminary results from the Viking Orbiter Imaging Experiment, *Science, 193,* 766–776, 1976.

Carr, M. H., L. A. Crumpler, J. A. Cutts, R. Greeley, J. E. Guest, and H. Masursky, Martian craters and emplacement of ejecta by surface flow, *J. Geophys. Res., 82,* this issue, 1977.

Cutts, J. A., K. R. Blasius, G. A. Briggs, M. H. Carr, R. Greeley, and H. Masursky, North polar region of Mars: Imaging results from Viking 2, *Science, 194,* 1329–1337, 1976.

Greeley, R., and T. E. Bunch, Basalt models for the Mars penetrator mission: Geology of the Amboy lava field, *NASA Tech. Memo. X-73,* 125, 1976.

Greeley, R., J. D. Iversen, J. B. Pollack, N. Udovich, and B. White, Wind tunnel simulations of light and dark streaks on Mars, *Science, 183,* 847–849, 1974.

Greeley, R., R. Papson, and J. Veverka, Crater streaks in the Chryse Planitia region of Mars—Early Viking results, submitted to *Icarus,* 1977.

Howard, K. A., and W. R. Muehlberger, Lunar thrust faults in the Taurus-Littrow region, Apollo 17 Preliminary Science Report, *NASA Spec. Rep. SP-330,* 31-22 to 31-25, 1973.

Masursky, H., and N. L. Crabill, The Viking landing sites: Selection and certification, *Science, 193,* 809–812, 1976.

Milton, D. J., Geologic map of the Lunae Palus quadrangle of Mars, *Misc. Inv. Map I-894,* U.S. Geol. Surv., Reston, Va., 1974.

Mutch, T. A., A. B. Binder, F. O. Huck, E. C. Levinthal, S. Liebes, Jr., E. C. Morris, W. R. Patterson, J. B. Pollack, C. Sagan, and G. R. Taylor, The surface of Mars: The view from the Viking 1 lander, *Science, 193,* 791–800, 1976*a*.

Mutch, T. A., R. E. Arvidson, A. B. Binder, F. O. Huck, E. C. Levinthal, S. Liebes, Jr., E. C. Morris, D. Nummedal, J. B. Pollack, and C. Sagan, Fine particles on Mars: Observations with the Viking 1 lander cameras, *Science, 194,* 87–91, 1976*b*.

Oberbeck, V. R., and R. H. Morrison, On the formation of the lunar herringbone pattern, *Proc. Lunar Sci. Conf. 4th, 1,* 107–123, 1973.

Schonfeld, E., On the origin of the Martian channels (abstract), *Eos Trans. AGU, 57,* 948, 1976.

Schultz, P. H., and D. E. Gault, Atmospheric effects on ballistic impact ejecta: Implications for Martian craters (abstract), *Eos Trans. AGU, 57,* 948, 1976.

Wilhelms, D. E., Geologic map of the Oxia Palus quadrangle of Mars, *Misc. Inv. Map I-895,* U.S. Geol. Surv., Reston, Va., 1976.

(Received April 1, 1977;
revised June 8, 1977;
accepted June 8, 1977.)

VOL. 82, NO. 28 JOURNAL OF GEOPHYSICAL RESEARCH SEPTEMBER 30, 1977

Geological Observations in the Cydonia Region of Mars From Viking

J. E. GUEST AND P. S. BUTTERWORTH

University of London Observatory, Mill Hill Park, London NW7, England

R. GREELEY

University of Santa Clara, Santa Clara, California 95053

During the Viking mission, three broad areas of the northern plains were investigated as possible landing sites for Viking lander 2. We present a geological map of the area designated the B1 landing site in Cydonia centered on 45°N latitude, 4°W longitude. Viking imagery of this area has given detailed coverage, allowing the northern plains to be examined in more detail over wider areas than was possible from Mariner 9. Some plains areas with polygonal fracture patterns are considered to be pediments in ancient southern hemisphere rocks. The fracture patterns predate at least some of the younger northern plains material. Several northern plains units are distinguished as well as surface textures of unknown origin. Viking pictures provide good data for future studies of stratigraphy and surface process in this region and others like it in northern latitudes.

INTRODUCTION

During the search for a second Viking landing site on Mars [*Masursky and Crabill*, 1976], three areas in the northern latitudes were imaged by Viking orbiters: the Cydonia region (B1 site) centered at 44°N, 12°W, the Alba Patera Volcano (B2 site) [*Carr et al.*, 1977], and the Utopia Planitia region (B3) at 45°N, 250°W. A site in the B3 region was eventually chosen for the landing. The geology of these regions was studied briefly but intensely in an attempt to construct geological models for the sites and thus to allow the prediction of small-scale structure, critical to the lander's safety.

These areas were selected as potential landing sites on the basis of Mariner 9 imagery in which at A frame resolution the surfaces appeared to be relatively smooth plains [*Underwood and Trask*, 1977], although B frames showed surface irregularities [*Scott*, 1976]. However, Viking pictures show the plains to have an intricate topographic relief and to have had a much more complex history than was expected. Most of the elements that make up the terrain can be seen on Mariner 9 B frames of a few small areas [*U.S. Geological Survey*, 1976]. The advantage of Viking pictures is that the resolution is sensibly better and provides extensive contiguous cover. The purpose of this paper is to describe the B1 area from Viking pictures to illustrate the style of geology and geomorphology of the plains in a region close to the boundary between the southern cratered hemisphere of Mars and the northern plains. This study is based on mapping done during the site certification phase of the Viking mission and shows that material is now available to reach a closer understanding, by future work, of surface history and process in the northern plains.

The coverage of this area was acquired during numerous revolutions (revolutions 9, 26, 32, 35, 37, 38, 39, 43, and 52) of the A spacecraft. The total area covered is more extensive than the area described here, which extends from 15° to 350°W longitude between 40°N and 50°N latitudes (Figure 1). The area includes Cydonia Mensae to the south and extends north into the Acidalia Planitia [*U.S. Geological Survey*, 1975].

On a global scale, Mariner 9 pictures demonstrate that the ancient cratered surface of Mars is largely restricted to the southern hemisphere. In the northern hemisphere, which is topographically lower, the cratered terrain appears to be covered in most places by younger units. The cratered terrain consisting of hilly and cratered material is partly buried by a flat-lying unit, known as cratered plateau material [*Milton*, 1975; *Wilhelms*, 1976], which lies stratigraphically between the hilly and cratered material and the northern plains material. From Mariner 9 it appears that the boundary between the northern plains and the southern cratered terrain is characterized by scarp retreat mainly in cratered plateau material. Outliers of cratered plateau material occur as far as 400 km north of the main scarp, and ancient craters below cratered plateau material often appear as rings of knobs.

The area described here contains the northern end of Cydonia Mensae, where it consists of mesalike and hilly outliers of cratered plateau material as well as underlying crater material. To the north of the Cydonia Mensae, and making up most of the area described here, are the northern plains.

GEOLOGICAL UNITS AT B1

Knob and Mesa Materials

These units crop out at Cydonia Mensae and consist of flat-topped mesas and conical hills or knobs. The mesas may be as wide as 10 km, although most are less than 5 km. Knobs are usually about 2 km across; they may be isolated hills with a shallow apron around the base, or they may rise above mesas (Figures 2 and 3).

Mesas are interpreted as remnants of cratered plateau material which probably comprises flat-lying beds of eolian sediment and lava. Cratered plateau materials form a continuous outcrop to the south of Cydonia, and it appears that erosion has stripped back this unit, leaving mesalike outliers. This explanation is consistent with that derived from Mariner 9 pictures.

Knobs, on the other hand, are more likely to be polygenetic. They may be erosional remnants of the underlying heavily cratered terrain as parts of crater rims and central peaks that were once overlain by cratered plateau material and then exhumed to give the present form; some may be igneous intrusions, while many are simply remnants of cratered plateau material itself.

Paper number 7S0486.

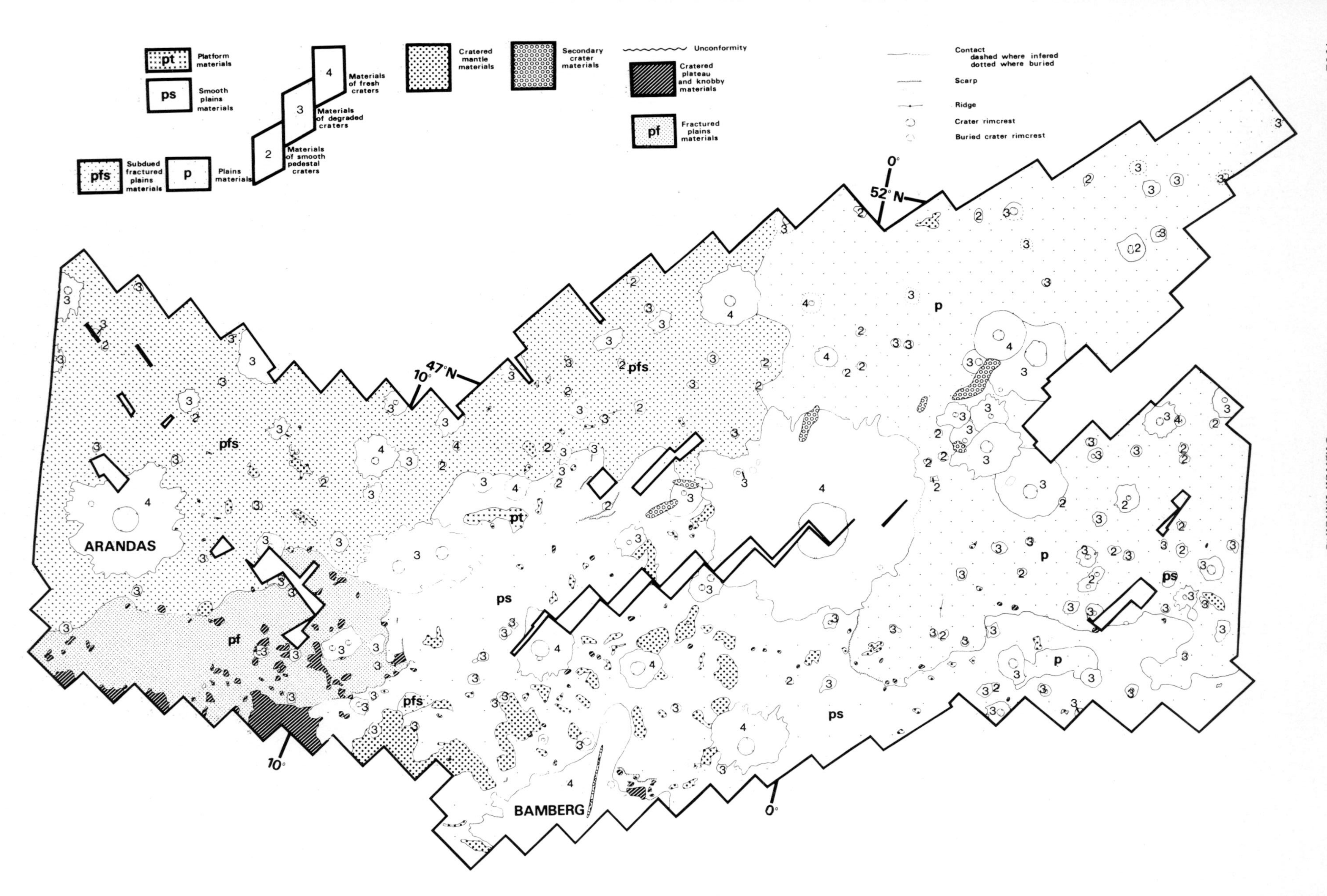

Fig. 1. Geological map of part of the Cydonia region.

Fractured Plains Materials

This unit occurs in the southwestern part of the area (Figure 1) in the region of mesas and knobs (Figure 3), although some mesas lie outside the outcrop of this unit. At Viking resolution the surface is smooth to rolling and is characterized by fractures forming polygonal patterns. The present surface expression of the fractures may be up to about 1 km wide, but it is likely that they have been enlarged by erosion. The rectilinear pattern may be controlled in part by tectonic trends, and the fractures may also have controlled the planimetric form of the mesas.

The stratigraphic position of the fractured plains material cannot be determined with complete confidence. It may postdate the mesas and knobs, being sedimentary or volcanic material that was emplaced around the mesas. However, it is more probable that it is a stratigraphic unit below the mesas, the mesa material having been stripped off, leaving the fractured plains surface bare. If the latter explanation is correct, the cover of fractured plains material must be thick toward the north, since there are so few old craters of the heavily cratered terrain showing through.

The origin of the polygonal fracture pattern has been discussed by *Carr et al.* [1976]. They conclude that it is most likely to be a permafrost feature analogous to terrestrial patterned ground and that the large size of the polygons indicates a deep layer of permafrost at the time of formation of the patterned ground; the polygons may be in part controlled by buried crater forms, there being deep lenses of ice in the material filling the craters.

Subdued Fractured Plains Materials

The surface of this unit superficially resembles that of fractured plains, but the fractures are subdued, and the pattern includes more circular forms. This surface appears to represent a mantle of variable thickness giving a small-scale mottled appearance (Figure 4). Many small craters on the surface of this unit appear to have been wind deflated.

Plains Materials

Toward the east of the map area (Figure 1), the subdued fractured plains surface grades into one with few fractures, although there is still a small-scale mottled appearance with numerous small hills (Figure 5). These surface characteristics may represent a thickening of subdued fractured plains material such that the underlying fractured surface no longer shows through. To the southeast the plains have peculiar textures of arcuate grooves and ridges often forming a 'thumbprint' texture (Figure 5). This texture may be a surface expression of breakdown of the surface material by subsurface ice.

Smooth Plains Materials

This material overlies all the previously described units. It has a relatively level surface, and the outcrop is well defined to the east and west by a scarp some tens of meters high. On the western side the scarp appears to cut across mesas in the older terrain, indicating that smooth plains material was emplaced over an already eroded terrain (Figures 2 and 6). In some places, subdued mesa forms within the smooth plains outcrop suggest that the eroded terrain below is only thinly mantled by smooth plains material.

To the east the scarp is winding. Immediately in front of the scarp are ridgelike forms and close-spaced hummocks, apparently formed by slumping of the scarp front (Figure 5). We

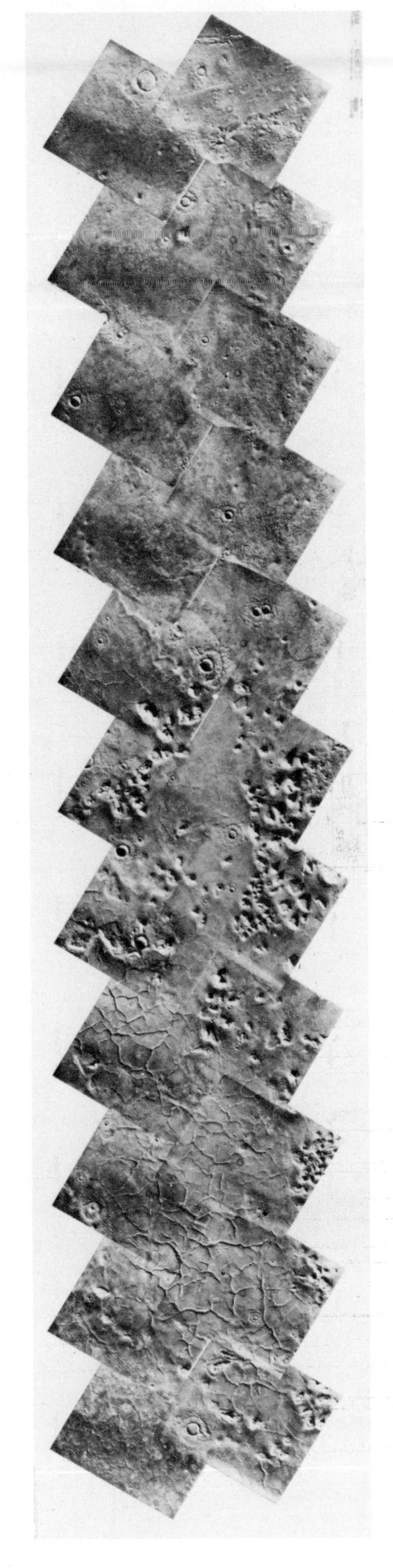

Fig. 2. Photomosaic across the northern end of Cydonia Mensae showing mesas and knobs. Mesa and knob materials occur in the middle of the mosaic. To the right are smooth plains materials, and to the left are fractured plains materials. At top left is a small area of subdued fractured plains materials. The strip of pictures runs east-west and forms the lower left corner of Figure 1 centered on 10°W longitude (revolution 35).

Fig. 3. Mesa materials (M) and knob materials (K). The mesa materials are considered to be remnants of a unit overlying fractured plains (pf). The two craters (c) are mapped in Figure 1 as class 3. The figure shows an area about 60 km across near 42°N latitude, 10°W longitude. North is toward top left (picture number 9A62).

consider that scarp retreat occurred by breakdown of the rocks as a result of sapping and other ground ice activity and that as the scarp retreated, it left piles of debris, forming patterns on the surface of rocks stratigraphically below the smooth plains material.

Cratered Mantle Materials

Superimposed on the plains are patches of material usually associated with isolated craters or crater clusters (Figures 4 and 6). In some places the crater clusters are distinct but may grade into areas where craters sit on top of small hills. This unit is interpreted as a thin mantle of eolian sediments with impact craters of primary and possibly secondary origin superimposed. It appears that in places the unit has been deflated. Craters and crater clusters tend to anchor the mantle, but in the intercrater areas the mantle has been deflated, leaving craters on hills. Materials of similar character occur on the subdued fractured plains, suggesting that this mantle extended over much of the area.

Crater Materials

The large variety of Martian crater forms suspected from Mariner 9 results has been confirmed by Viking [*Carr et al.*, 1976]. The B1 site contains examples of the three principal fresh crater classes recognized by *Carr et al.* [1977]: (1) rampart craters, bounded by low ridges or scarps, (2) craters with strong radially textured ejecta blankets, and (3) lunarlike craters. There are also several other types representing different stages and styles of degradation.

In the most common type of medium-sized crater (1–15 km in diameter) the outer edge of the ejecta terminates in a scarp at several crater diameters from the crater rim. In plan view the scarp may be circular or lobate. Those with a lobate plan tend to be smaller and have lower scarps. Within the ejecta outcrop the majority have a subcircular shallow scarp at about 1 crater diameter from the rim. Where a crater of this type is superimposed on another of the same type, the older crater usually has a more pronounced scarp at 1 crater diameter and a lower outer bounding scarp.

Fig. 4. Subdued fractured plains materials. The fracture pattern is patchily subdued and mottled. Craters on small hills are remnants of cratered mantle materials (m). A pedestal crater of class 2 is marked with an arrow. The figure shows an area about 60 km across at 45°N latitude, 15°W longitude. North is between top left and top (picture number 26A40).

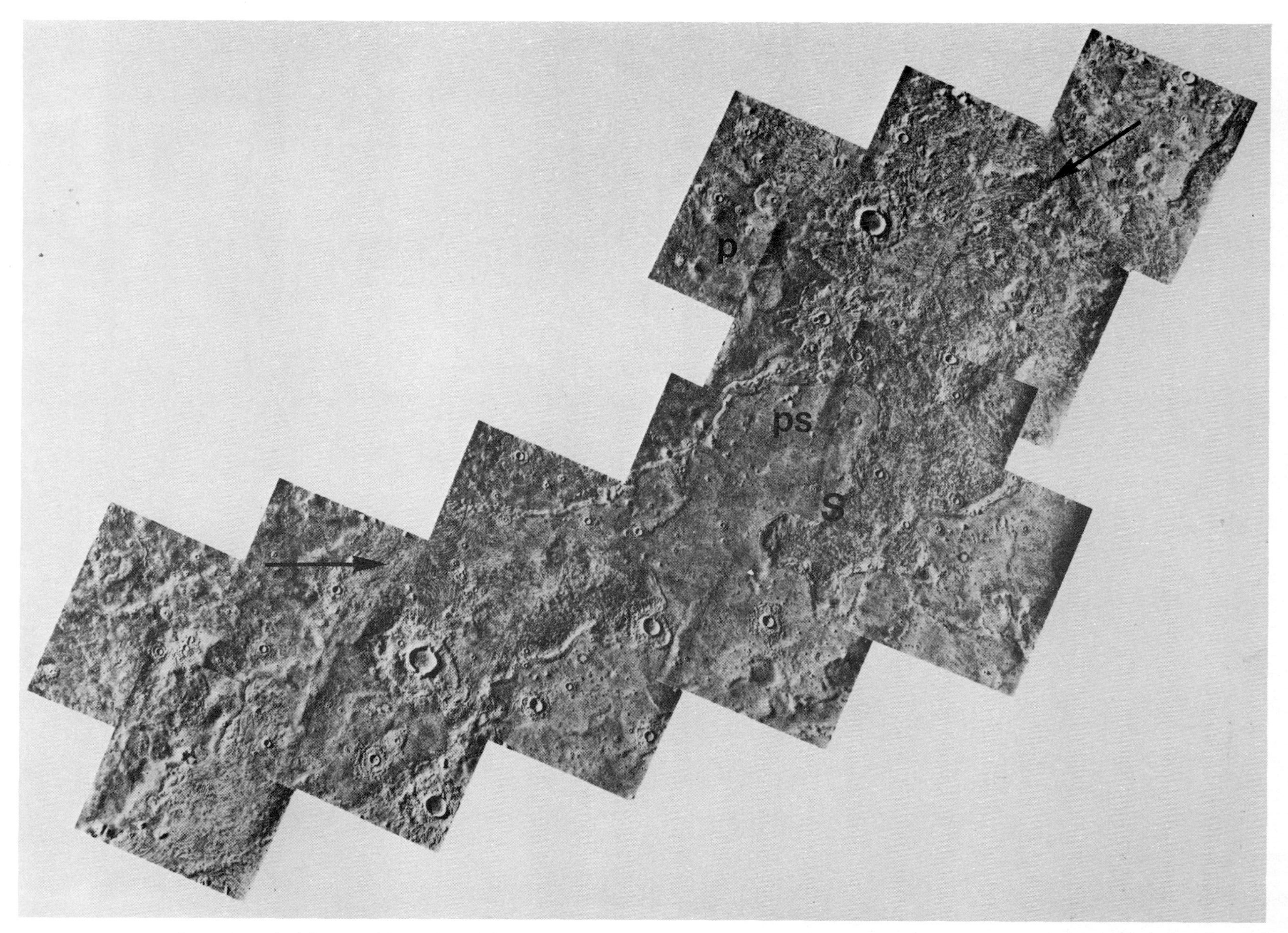

Fig. 5. Smooth plains materials (ps) bounded by scarp (S) on the eastern side of an outcrop, overlying plains materials (p). Thumbprint texture (arrows) is well displayed in this area. The photomosaic is about 350 km across. The bottom right corner of the mosaic is the bottom right corner of Figure 1. The extreme right of the mosaic lies outside Figure 1 (revolution 52).

Fig. 6. Western scarp (S) of smooth plains materials (ps) overlying exhumed mesas (M) and fractured plains (pf). An area of cratered mantle materials (m) overlies smooth plains materials. The figure shows an area about 60 km across at 41°N latitude, 8.5°W longitude (picture number 9A64).

Several of the largest craters in the B1 site are not bounded by a continuous scarp. Instead, their outer ejecta consists of several overlapping flaps with lobate edges. The best example is Arandas (Figure 7).

Several small craters have strikingly radial extensive ejecta blankets with serrated edges. These craters appear to be among the youngest features in the area and commonly show faint bright ray patterns. Other small craters, particularly on the western fractured plains, resemble similar-sized craters on lunar maria.

In some parts of the site, pedestal craters preponderate. In others they occur together with craters of similar ejecta texture and extent but without bounding scarps. Superposition relations show that the latter are younger than many pedestal craters.

In the mantled eastern part of the site, craters commonly

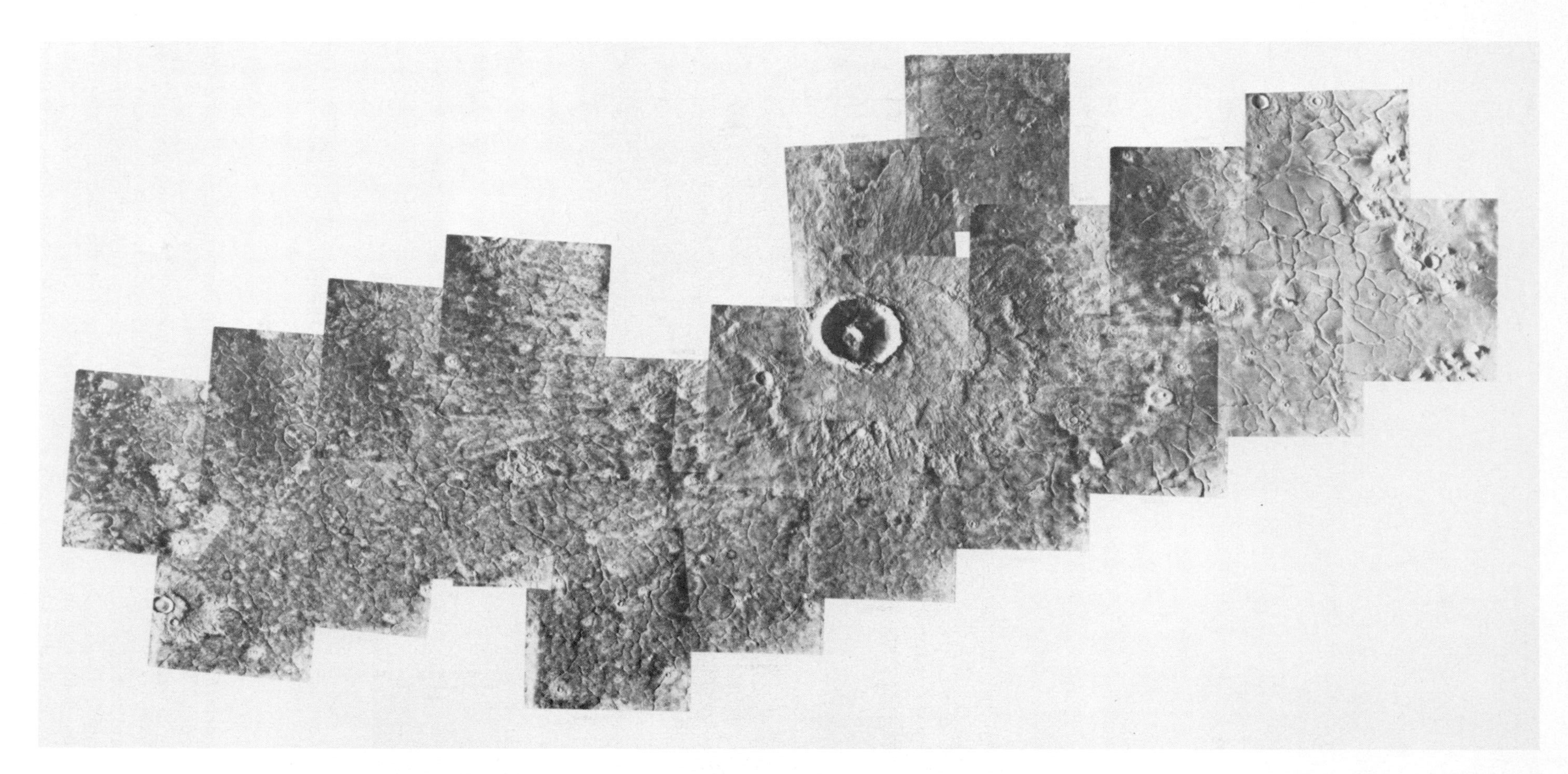

Fig. 7. Photomosaic showing the 30-km-diameter crater Arandas (center). This is a relatively fresh crater (class 4) with well-developed flow lobes and textures. At the far right of the mosaic the surface consists of fractured plains materials, while to the far left there are subdued fractured plains materials. North is towards the top (revolution 32).

only have a circular bounding scarp at about 1 crater diameter from the rim. The apparent lack of ejecta outside the scarp could be the result of burial of the outer parts of the ejecta blanket.

For the geological map (Figure 1) we have attempted to classify craters on the basis of their state of preservation. Class 4 craters (Figure 7) appear to be relatively fresh with crisp surface textures and well-defined ejecta edges. These craters also tend to show faint bright ray patterns and have associated with them clusters of secondary craters and secondary crater chains [see *Carr et al.*, 1976, Figure 8]. In general, class 4 craters are younger than the smooth plains, although in some places they may be thinly overlain by smooth plains material at the outer margins of their ejecta blankets.

Class 3 craters (Figure 3) also have textures on their ejecta surfaces, but the outer edges of the continuous ejecta tend to be ragged. Bright rays and secondary craters are absent, a greater degree of degradation being indicated than that seen on class 4 craters. Class 3 craters are post-fractured plains in age; some are partly buried by smooth plains material, and others appear to be post-smooth plains.

The most degraded are designated class 2 (Figure 4). It will be noted, however, that those craters mapped in this class tend to be relatively small, and it is likely that they owe their greater degradation to size rather than to age. Many class 2 craters may therefore be comparable in age at least to class 3 craters and possibly to craters of class 4 age.

It is interesting to note that a high percentage of the craters in the B1 area are relatively young, suggesting that either cratering was spasmodic after a period of heavy bombardment (i.e., pre-fractured plains material) or there was a period of erosion that removed many craters. A further possibility is that deposition of rock units was rapid and spasmodic while the impact flux remained steady.

Numerous crater clusters exist. Some of the most dense and most irregular can be related to large craters and are probably their secondaries. Several extensive clusters, consisting of circular bowl-shaped craters, could be the result of atmospheric fragmentation of a single incident body.

GEOLOGIC HISTORY

The following summary assumes that (1) the mesa materials are equivalent to the highland cratered plateau material, identified by mapping from Mariner 9 frames, and (2) the fractured plains materials represent a rock unit below the mesa materials.

On this basis the geological history of the area opened with the emplacement of the fractured plains and mesa materials over an irregular and probably eroded cratered terrain, partly represented by knobby materials. This was followed by erosion that stripped back the mesa materials, leaving a bare surface of fractured plains materials and isolated patches of mesa materials above. It is not clear whether the fractures formed at this stage or were preexisting fractures exhumed by the removal of mesa materials. Whichever is correct, it is clear that the fractures are old in relation to the other rock units in the area. Part of the area was then covered by thin deposits of subdued fractured plains and plains materials.

The next unit to be emplaced was smooth plains material, which was deposited over an eroded surface and was relatively thin. This unit probably thinned to the northwest and was overlain over the whole area by a thin deposit of cratered mantle materials.

There followed another erosional episode stripping back the smooth plains materials to the east and west of their outcrop by scarp retreat, and wind erosion removed part of the upper mantle, leaving patchy areas on top of both the plains materials and the fractured plains materials to give the subdued fractured plains.

It is clear from these studies that the area developed by a series of depositional and erosional episodes. If this area is characteristic of other regions between the southern cratered highlands and the northern plains of Mars, it can be expected that future studies based on Viking pictures of other areas of this type on Mars will lead to a complex history of deposition and pediplanation. As *Milton* [1975] has pointed out, there is some difficulty in distinguishing between the age of an erosional surface and the age of the rock units exposed on that surface from Mariner 9 pictures. The B1 area illustrates this point. Fractured plains, for example, are relatively young surfaces cut in much older bedrock. From Mariner 9 pictures, such surfaces would be mapped as relatively young on geological maps, and it is likely that many areas in the northern plains are young pediments cut in much older rocks.

CONCLUSIONS

We have attempted to describe briefly the surface morphology and geological relations in a part of the northern plains photographed by Viking during the nominal mission. Several conclusions can be drawn from these observations.

1. The erosional history of the northern boundary of the cratered highlands and cratered plateau material of the southern hemisphere is complicated. There have been distinct periods of erosion and deposition, giving a number of rock units separated by unconformities.

2. As on earth, there is a distinct difference between the ages of rocks exposed and the ages of the surfaces.

3. The northern 'plains' are composed of an intricate mixture of young rock units, young surfaces cut in old rocks that have been modified by surface processes such as wind, permafrost, etc., and old exhumed rock surfaces probably equivalent in age to units exposed in the southern cratered highlands of Mars.

Hence the old Mariner 9 view of Mars in which the older cratered terrain of the southern hemisphere is buried by younger units in the northern hemisphere may be simplistic. From Viking photography it is suggested that not only is the northern hemisphere geology much more complicated than was expected, but as *Milton* [1975] predicted, although the present surfaces are young, some of the rocks exposed at the surface may be old.

Acknowledgments. The authors thank other members of the Orbiter Imaging Team for discussions about this area; also, J. Boyce, D. Scott, and J. Underwood constructively reviewed the manuscript. We thank the U.K. Natural Environmental Research Council (J.E.G. and P.S.B.), and the NASA Planetary Geology Office (R.G.) for financial support.

REFERENCES

Carr, M. H., H. Masursky, W. A. Baum, K. R. Blasius, G. A. Briggs, J. A. Cutts, T. Duxbury, R. Greeley, J. E. Guest, B. A. Smith, L. A. Soderblom, J. Veverka, and J. B. Wellman, Preliminary results from the Viking orbiter imaging experiment, *Science, 193*, 766–776, 1976.

Carr, M. H., L. A. Crumpler, J. A. Cutts, R. Greeley, J. E. Guest, and H. Masursky, Martian craters and emplacement of ejecta by surface flow, *J. Geophys. Res., 82*, this issue, 1977.

Masursky, H., and N. L. Crabill, Search for Viking 2 landing site, *Science, 194*, 62–68, 1976.

Milton, D., Geologic map of the Lunae Palus quadrangle of Mars, *Map I-894*, U.S. Geol. Surv., Reston, Va., 1975.

Scott, D. H., Geologic maps of the Cydonia region of Mars Prime B Viking landing site, *Open File Rep. 76-500*, U.S. Geol. Surv., Reston, Va., 1976.

Underwood, J. R., Jr., and N. J. Trask, Geologic map of the Mare Acidalium quadrangle of Mars, *Misc. Inv. Map I-1048*, U.S. Geol. Surv., Reston, Va., 1977.

U.S. Geological Survey, Topographic map of the Mare Acidalium quadrangle of Mars, *Map I-979*, Reston, Va., 1975.

U.S. Geological Survey, High resolution Mariner 9 pictures in the Cydonia region of Mars, *Misc. Inv. Ser. Map I-990*, Reston, Va., 1976.

Wilhelms, D. E., Geologic map of the Oxia Palus quadrangle of Mars, *Map I-895*, U.S. Geol. Surv., Reston, Va., 1976.

(Received April 1, 1977;
revised June 1, 1977;
accepted June 3, 1977.)

Martian Dynamical Phenomena During June–November 1976: Viking Orbiter Imaging Results

GEOFFREY BRIGGS, KENNETH KLAASEN, THOMAS THORPE, AND JOHN WELLMAN

Jet Propulsion Laboratory, Pasadena, California 91103

WILLIAM BAUM

Lowell Observatory, Flagstaff, Arizona 86001

The Viking primary mission, principally covering the northern summer, has provided observations of a season previously unstudied from close range. The morning formation stage of the clouds, apparently orographic, associated with the Tharsis Montes and Olympus Mons, shows northwest slope clouds developing in late morning, at which time, discrete clouds, indicative of convection in a ~6-km layer near the surface, are formed over much of the elevated terrain. Low-level morning condensate clouds, or fogs, are associated with the canyons of Labyrinthus Noctis at this season. In Memnonia, at only 15°S, CO_2 frost condenses on the surface at night over a large area. The southern winter polar cap reaches 40°S, having an irregular margin about 10° in width. Frost deposits within the cap are nonuniform and patchy. Discrete clouds have been observed at mid-latitudes in the south reaching 50-km altitude. One example of a local dust storm has been detected, but generally, the season has been one of comparative calm, having a relatively clear atmosphere in the south and dust and condensate hazes in the north.

INTRODUCTION

This report, which summarizes the orbital imaging observations of dynamical phenomena on Mars during the Viking primary mission, covers a range of areocentric longitudes of the sun, L_s, from 87° to 145°, corresponding to the last days of northern spring through late summer. Because of the season and because of restrictions imposed both by the orbital geometries and by a variety of operational considerations, synoptic imaging observations have been limited to the local morning portion of the globe between latitudes ~25°N and ~65°S for Viking Orbiter 1 (VO 1) and to high southern latitudes for Viking Orbiter 2 (VO 2). The results discussed here are mostly concerned with such observations rather than with imaging data taken at the highest resolution for geological purposes. However, the high-resolution data have been analyzed to obtain statistical information about the opacity of the atmosphere [*Thorpe*, 1977], and the high-resolution pictures of the north polar region have provided important information about dynamical Martian phenomena [*Cutts et al.*, 1976]. *Veverka et al.* [1977] have described the extent of aeolian activity during the period under consideration here. The reader should be aware that all images have been computer-processed to improve scene contrast. Where it is indicated, certain images have been high-pass-filtered to provide improved high-frequency discrimination.

During the period of the observations, Mars has been near aphelion, far removed from the season of perihelic dust storms that apparently characterize Mars. On approach, however, some longitudes in the northern hemisphere of Mars appeared quite hazy in comparison to the south, where the atmosphere was sufficiently clear for the growing CO_2 polar cap to be seen (Figure 1). In the south, surface morphology is readily seen even at resolutions of ~20 km, whereas little surface detail can be seen in many images at latitudes north of about 10°S. In part, this difference is due to the smoother topography of the more northerly terrains and also to the increased contrast that the southern polar cap provides. Imaging observations made by both Viking landers (VL 1 and VL 2) show substantial sky brightness, which has been attributed mainly to dust particle aerosols [*Mutch et al.*, 1976; *Pollack et al.*, 1977].

CLOUDS IN THE THARSIS/AMAZONIS REGIONS

One of the most notable Martian cloud phenomena observed telescopically is the 'W cloud,' a group of white clouds that forms seasonally and is most prominent in the afternoon. The Mariner 9 observations (Figure 2) showed that the cloud elements comprising the W cloud are located at the sites of the four large volcanos (Olympus Mons, Ascraeus Mons, Pavonis Mons, and Arsia Mons) and at the western end of the equatorial canyon Valles Marineris [*Leovy et al.*, 1973]. This cloud activity is generally most intense beginning in northern spring and continues into summer [*Smith and Smith*, 1972]. Although Mariner 9 was able to observe such activity at the beginning of summer, viewing was limited to the afternoon hours. On the basis of these data it was found that the clouds are formed mainly on the western flanks of the volcanos, from where they stretch over many hundreds of kilometers. The clouds near Arsia Mons, the most southerly volcano, were relatively weakly developed. Clouds in the area of Valles Marineris were only poorly observed by Mariner 9. Those associated with the volcanos showed very little variation during the afternoon. The height of these clouds above the local surface has not been well determined, since shadows were not clearly evident. C. B. Leovy (personal communication, 1977) has measured apparent shadows that suggest heights of about 20 km above the mean reference level.

The Viking orbiter observations of the Amazonis/Tharsis clouds are, on approach, at a slightly earlier season and, in orbit, at a slightly later season than those at the end of the Mariner 9 mission. They are largely limited to the morning hours, thereby providing an opportunity to see the cloud development phase. The initial appearance of the volcanos (Figure 1) as they rotate over the morning terminator is reminiscent of that observed at the beginning of the Mariner 9 mission, when only the upper slopes of the mountains could be seen above the dust. In the present case the surrounding obscu-

Paper number 7S0480.

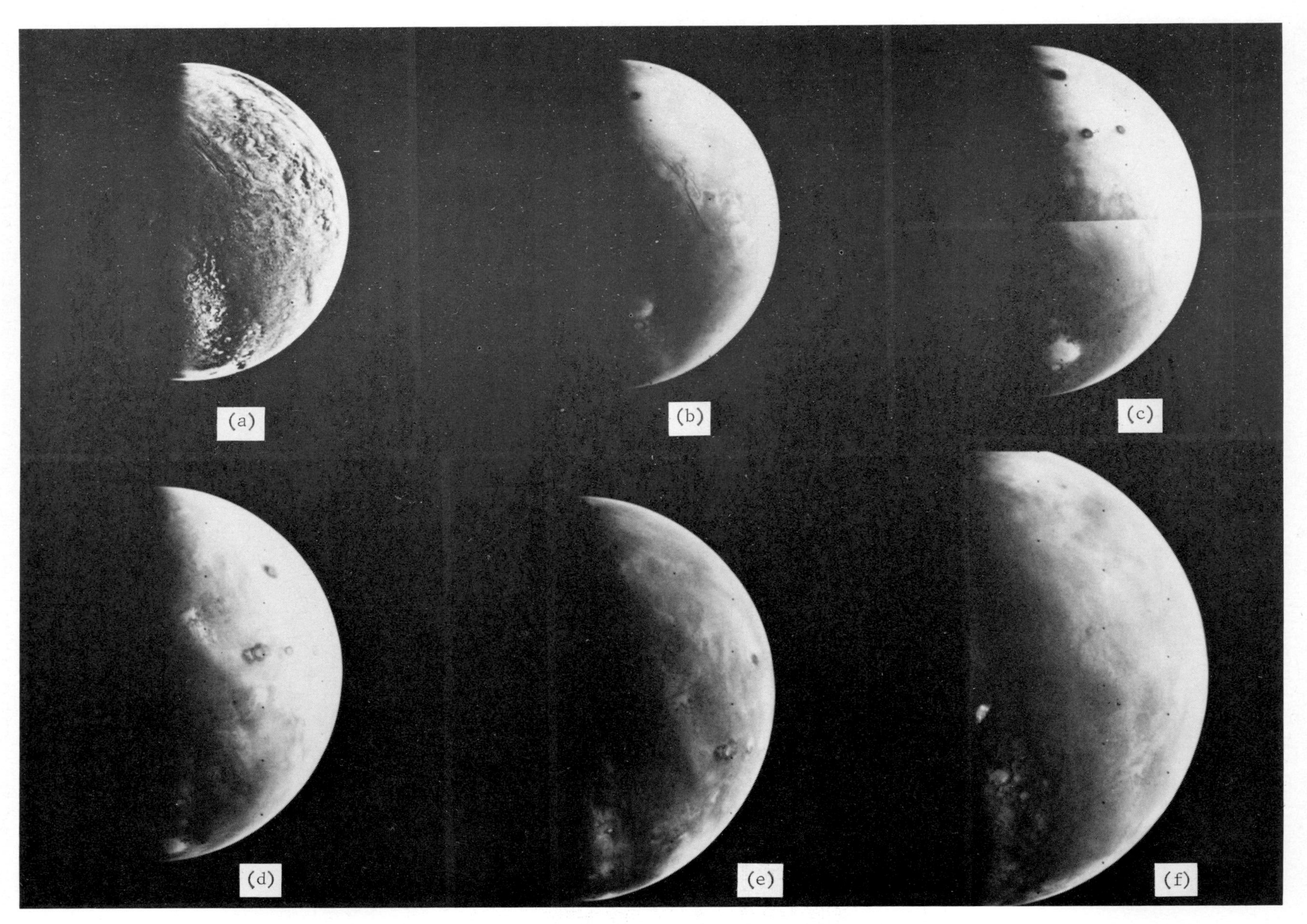

Fig. 1. Successive views of Mars at 30° longitude intervals taken by VO 1 on approach in mid-June 1976 (L_s = 83°). Part *a* has been filtered to improve visibility of predawn cloud. Part *d* is saturated near limb.

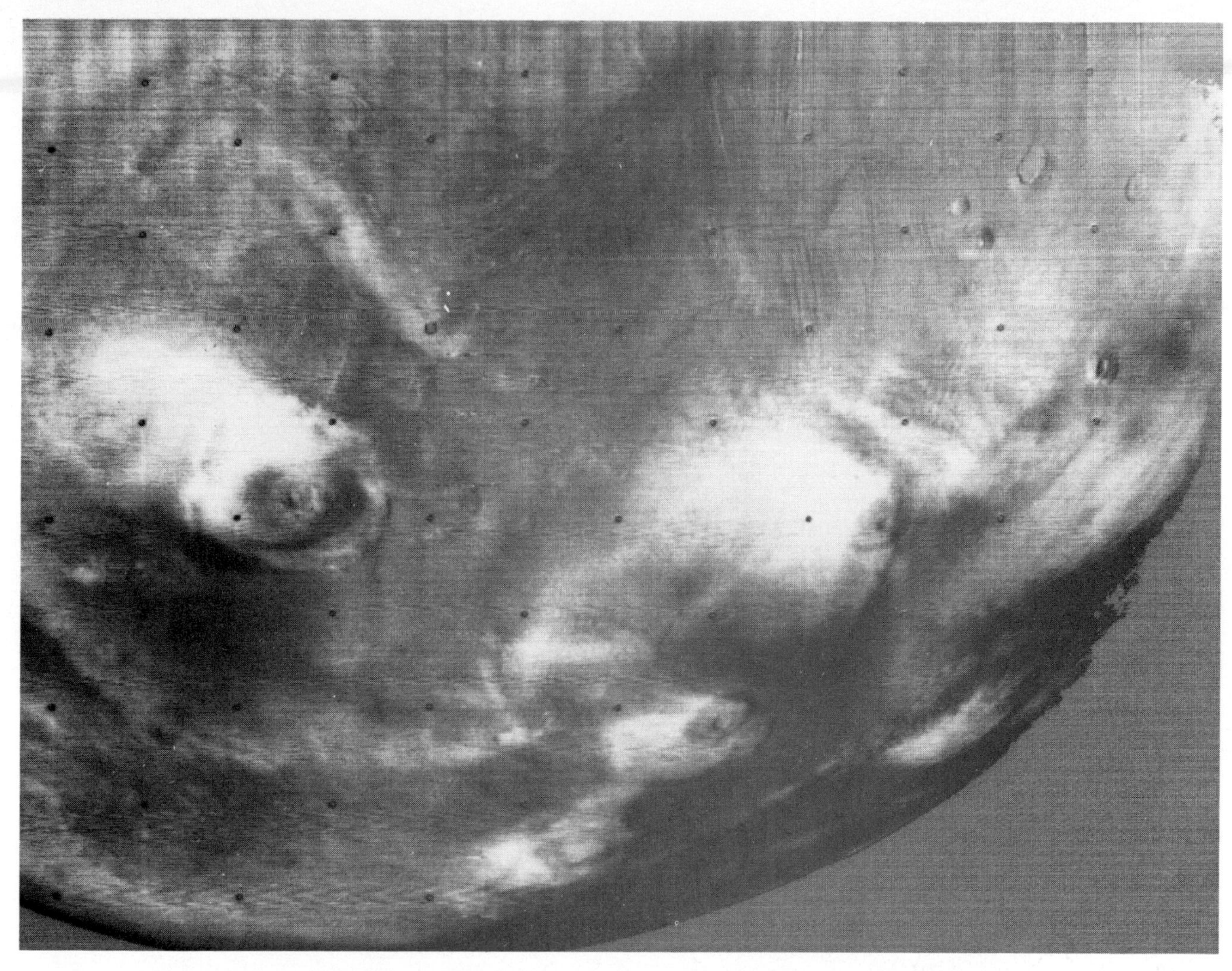

Fig. 2. Mariner 9 afternoon view of the Amazonis/Tharsis region in early northern summer (L_s = 96°).

ration (Figures 1, 3, and 4) is inferred to be due primarily to water ice mists rather than to dust, since the visibility of the region in violet light increases as the volcanos move further on to the dayside. This is consistent with Viking lander imaging determinations of diurnal changes in atmospheric optical depth [*Pollack et al.*, 1977]. Waves, at apparently low levels, can be clearly seen in the Figure 4 view of Arsia Mons, which has just crossed the morning terminator, and imply a stable lapse rate near the surface. The implications of the wind directions that can be inferred from the wave orientations remain to be analyzed. The remarkable extended curvilinear cloud feature to the southwest of Pavonis Mons in Figure 4 was observed only on this occasion, revolution 62. The clouds appear to mark an unusual extended gravity wave that resembles a bore. For the more northerly volcanos (i.e., all we have named except Arsia Mons), discrete, optically thick clouds form over the northwestern flanks in the second half of the morning (Figures 1 and 5). Olympus Mons appears to show the least clearing of the early morning hazes before more well defined clouds develop in the middle of the day. This is presumed to be the result of the greater amounts of water vapor at the more northerly latitudes. In the afternoon there are relatively few observations. Those available suggest that the cloud development at this early summer season differs somewhat from that in the Mariner 9 observations. Figure 6 shows Olympus Mons in the afternoon (a mosaic of pictures taken by VO 2 between revolutions 45 and 47). The region around the giant volcano is generally clear of clouds except for some to the northwest. The summit and flanks of the volcano, however, are substantially obscured by clouds, which are mainly of a diffuse nature. The southeastern slopes are the least haze covered. The prevailing wind direction is uncertain but may be from the southeast. At the slightly earlier season when the Mariner 9 images were acquired the inferred wind direction is more nearly from the west, and the wind strength at that time was apparently greater and thereby able to carry most of the water ice clouds to the lee of the volcanos, as is shown in Figure 2.

Measurements of the shadows of clouds above the summit of Olympus Mons indicate cloud heights of over 10 km above the 27-km-high mountaintop. An orographic origin for the clouds [*Leovy et al.*, 1973; *Webster*, 1977] seems most likely: upslope winds produced by the thermal low created by the mountain lead to adiabatic cooling of the air and condensation of the water vapor. The distribution of the clouds depends upon the basic wind flow onto which the upslope winds are superimposed. In the case of Arsia Mons the insolation at 10°S may be insufficient to create a major effective heat source and thus may not cause sufficiently strong upslope winds. The relative lack of water vapor at this latitude is also presumed to be a factor.

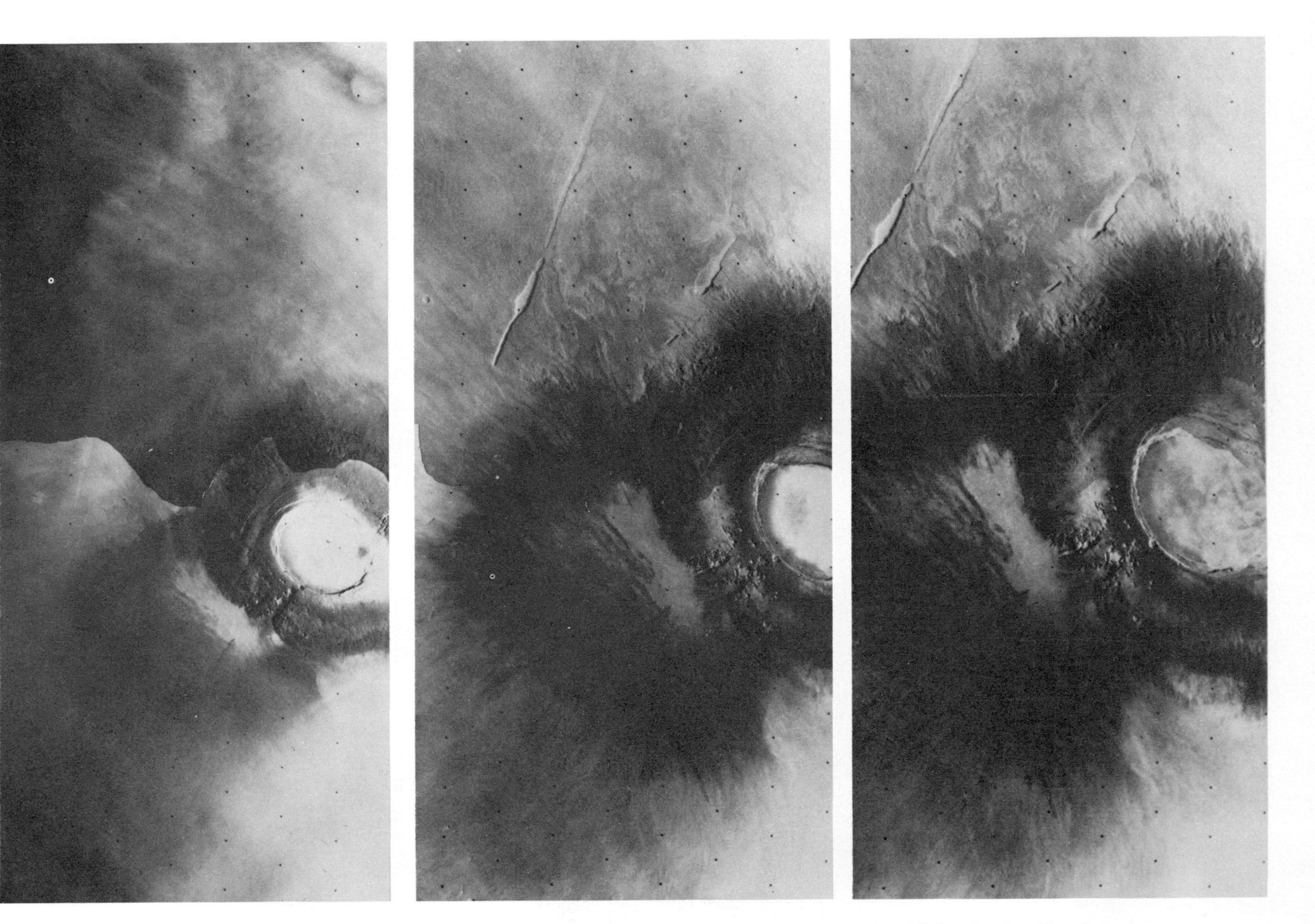

Fig. 3. Arsia Mons (VO 1, revolution 39, $L_s = 101°$) at three early morning times of day showing increasing visibility of surface. The volcano was ~10° from the dawn terminator in the first view (left), the other observations being 90 (middle) and 113 (right) min later than the first. The mosaic join shows as an irregular boundary.

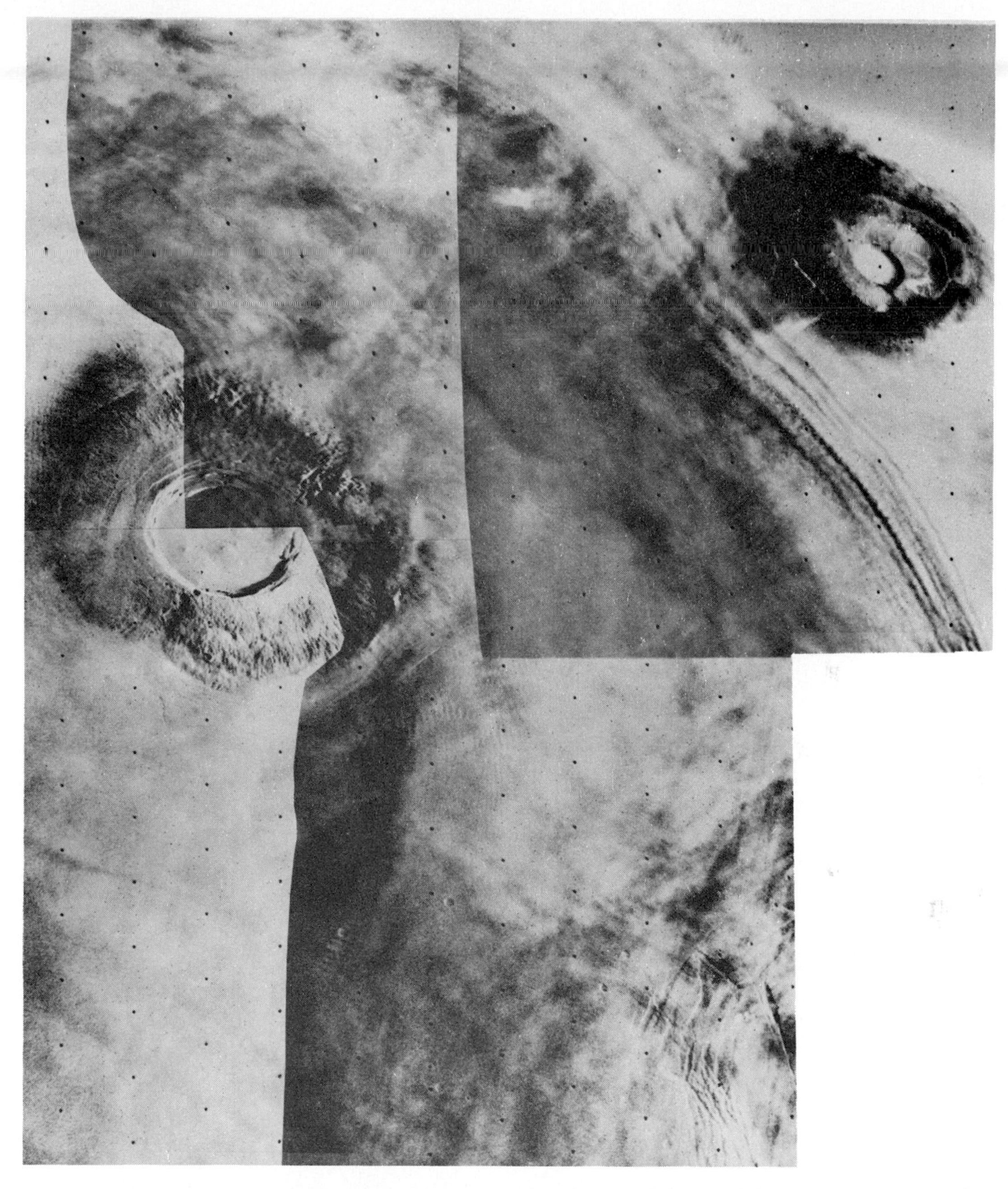

Fig. 4. Pavonis Mons and Arsia Mons near the dawn terminator on revolution 62 (VO 1, L_s = 112°). The region is generally shrouded in haze, and numerous waves are visible, including one near Pavonis Mons that stretches for several hundred kilometers. The western slopes of Arsia Mons lie at the terminator.

The most northerly of the Tharsis Montes, Ascraeus Mons, exhibits a unique plumelike cloud to its west when it is first seen near the morning terminator at this time of year (Figures 1 and 7). This extended cloud, which stretches for several hundred kilometers, was observed by both spacecraft during their approaches and, in orbit, by VO 1 on revolutions 40 and 58 when images were acquired at the appropriate time of day. An interesting aspect of the cloud is its highly turbulent character. The repeatability of the phenomenon indicates that the tropical circulation is very stable at this time of year. No similar clouds have been observed to be associated with Olympus Mons, 8° further north, or with Pavonis Mons, 10° further south. The closeup view of the cloud in Figure 7 somewhat resembles one seen within the north polar hood by Mariner 9 (Figure 8). *Pirraglia* [1976] analyzed the form of this cloud and concluded that it was a variety of lee wave cloud, commonly seen within the hood, formed when the vertical wind shear was exceptionally strong. The resemblance is, however, by no means complete, even though both are characterized by two diverging streamers. The Ascraeus Mons cloud displays a morphology suggestive of the vortical flow which *Greeley et al.* [1974] observed in wind tunnel experiments aimed at understanding the streak patterns of aeolian material behind Martian craters. It is possible that the presence of the cloud to the lee of this volcano, and not of the others, is the result of particularly strong winds at 10°N. The composition of the cloud is uncertain. If the cloud marks a vortex, it may be that water vapor or carbon dioxide is con-

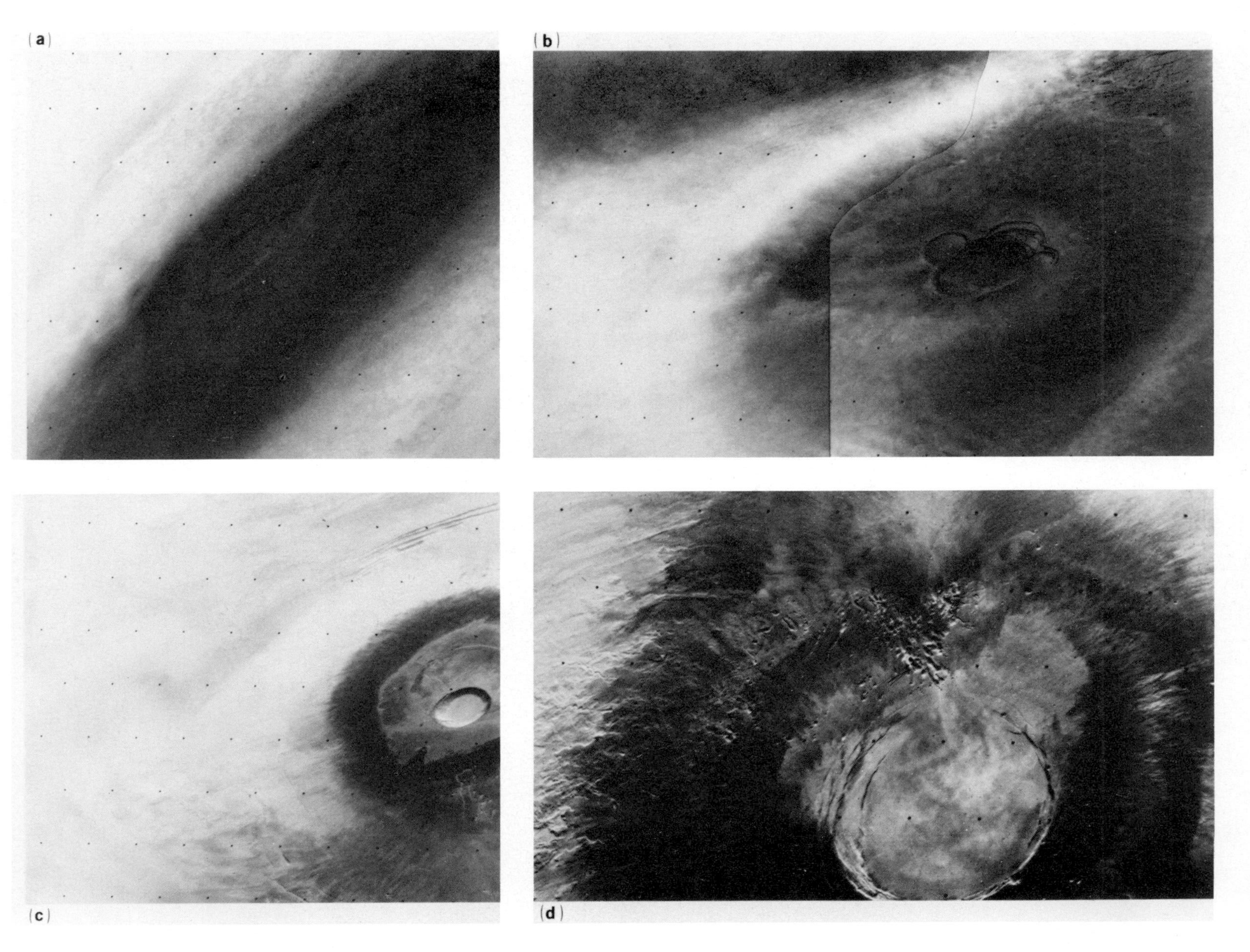

Fig. 5. Olympus Mons and Tharsis Montes in middle to late morning at $L_s = 102°$ (VO 1, revolution 41). All of the mountains except Arsia Mons have developed optically thick clouds on their northwestern flanks.

Fig. 6. Olympus Mons in late afternoon as seen in a mosaic put together from data gathered on several successive days (VO 2, revolutions 45–47, L_s = 128°).

densed by expansive cooling of the air due to the upward motion of the air on one side of the vortex and/or to the low-pressure center. At the ~25-km altitude (above the mean) in question, simultaneously acquired infrared radiometer thermal mapping (IRTM) data imply temperatures around 170°K (T. Z. Martin, private communication, 1977). The relative paucity of water vapor at such temperatures and the high optical thickness of the clouds tend to argue against a water ice composition. On the other hand, the CO_2 condensation temperature is around 135°K at this altitude, and if the inferred temperature is approximately correct, it is unclear whether a sufficient pressure drop could be achieved by a vortex to cause CO_2 condensation.

There is additional evidence for very high velocity easterlies, which are predicted at this season by *Leovy and Mintz* [1969] and by *Pollack et al.* [1976], in this region near dawn. Discrete clouds were photographed by VO 1 in the region between Pavonis Mons and Olympus Mons at two separate times 12 min apart on revolution 40. The clouds (Figure 9) lie partly over the dawn terminator and are seen clearly because of their evidently high altitude and the lack of illumination of the underlying surface. The individual clouds exhibit delicate cirrus forms and are up to 60 km in length. There appears to be a distinct boundary to the cloud field, marked by a highly linear feature oriented approximately SW-NE. In the time between observations this boundary appeared to remain fixed, while

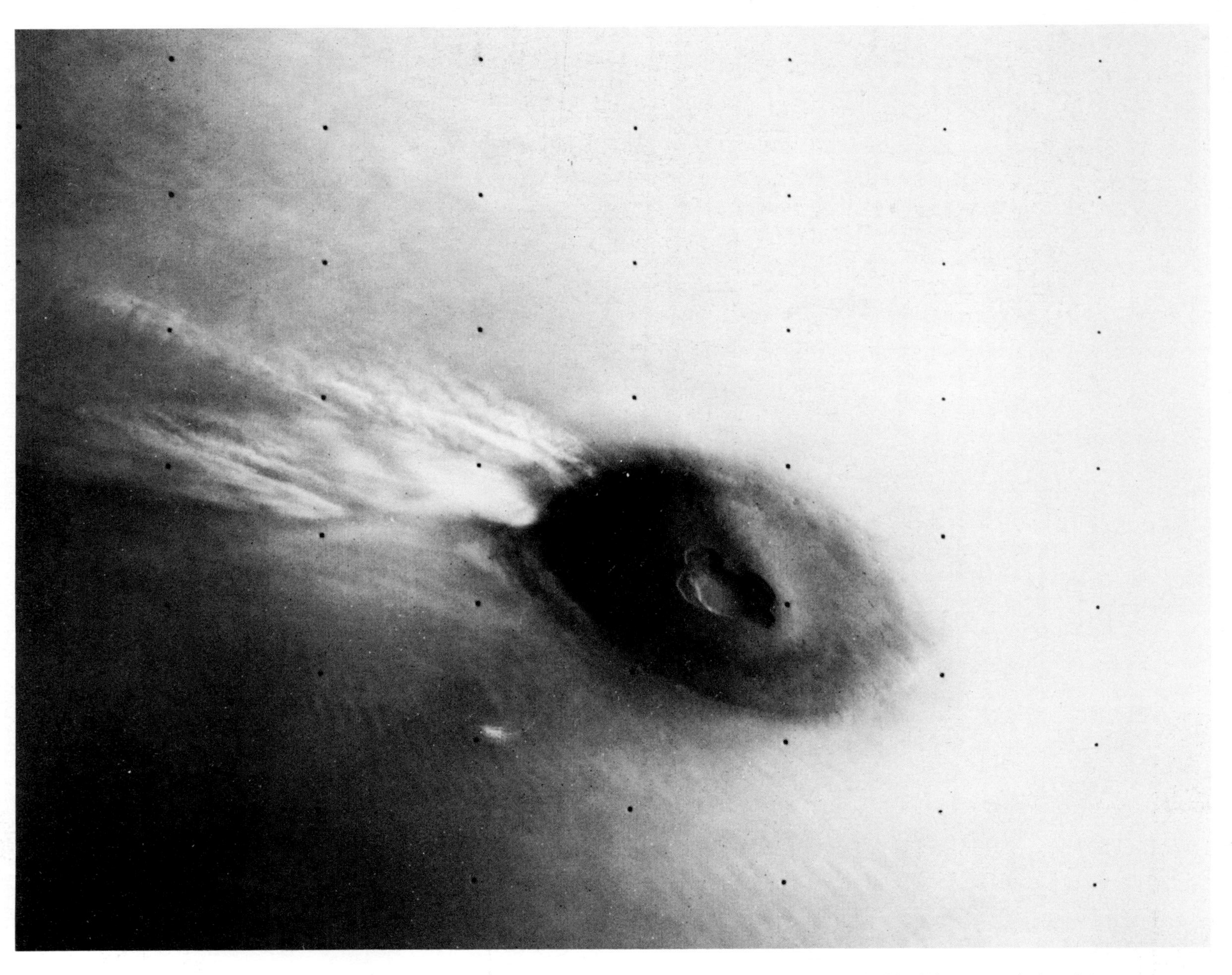

Fig. 7. High-resolution view of an early morning plumelike cloud to the west of Ascraeus Mons, acquired by VO 1 on revolution 58 (L_s = 110°).

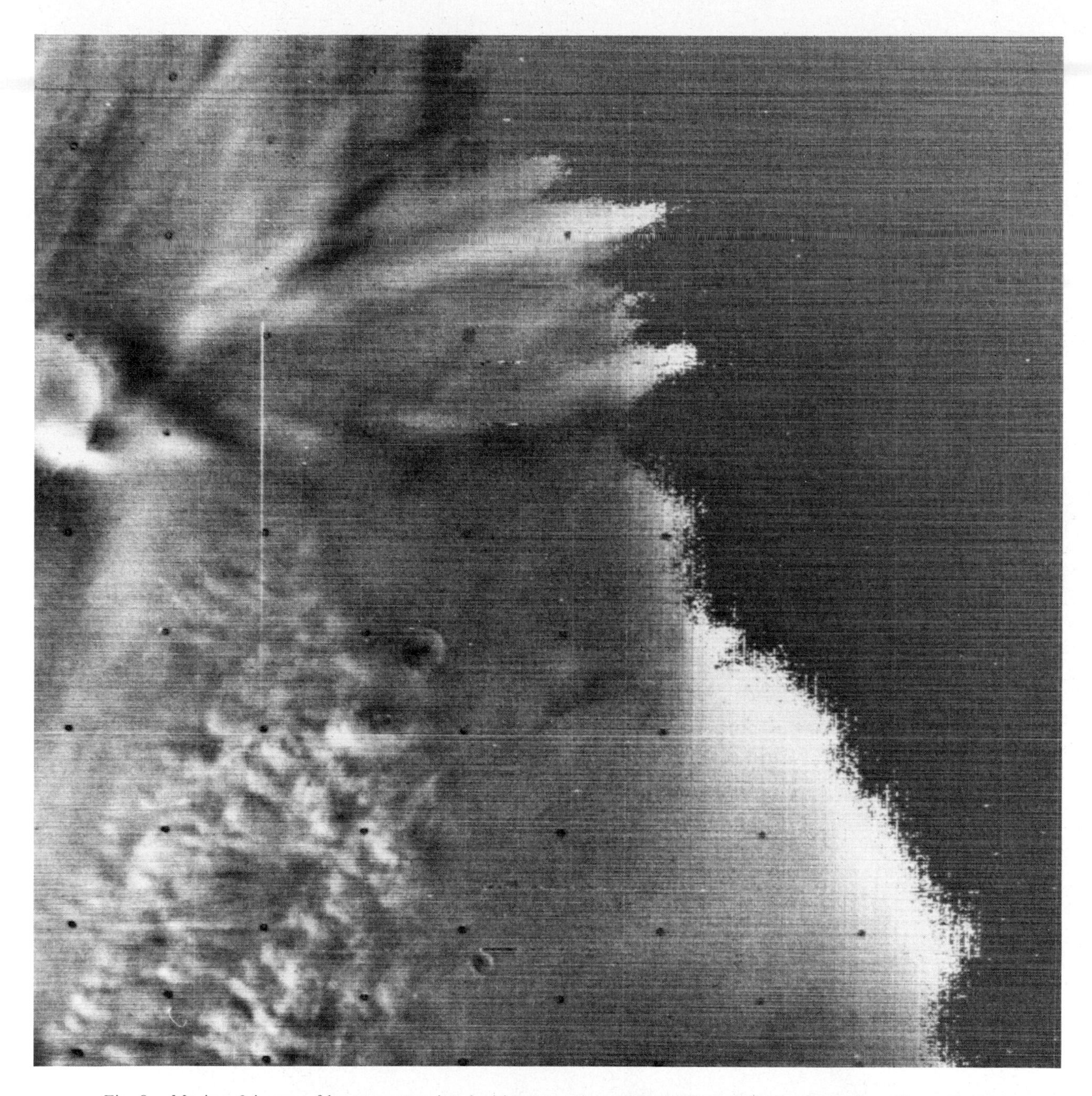

Fig. 8. Mariner 9 image of lee wave associated with a crater in middle northern latitudes during late winter. Ragged boundary to right is evening terminator. Turbulent cloud at bottom of picture is thought to be a dust storm. Image is filtered and contrast-enhanced.

individual cloud elements were determined to approach the boundary. Typical displacements are 30 km, corresponding to a velocity of about 40 m/s. The maximum displacement, however, was twice this amount, indicating that the wind field was not uniform and that the wind speeds reached very high values. It is presumed that the stationary cloud boundary marked a front between two distinct cloud masses where downward flow led to the sublimation of condensed material. The location of the supposed front in the region between the Tharsis Ridge and Olympus Mons may be associated with the slopes in this region or, perhaps, with the unusual temperature contrasts measured in this area by the Viking orbiter infrared radiometer experiment [*Kieffer et al.*, 1976*c*]. The cloud composition is uncertain. Clouds with similar cirrus appearance have also been observed near dawn over Valles Marineris, but in that case also, no altitude information could be clearly established.

Observations of the Labyrinthus Noctis region, where an element of the classical W cloud occurs, reveal diurnal changes also. Figure 10 shows red and violet images acquired by VO 1 in early morning on revolution 40. Diffuse patches of haze are seen throughout the area, and in a number of places within the confines of the canyon there is distinctively greater brightness, particularly in the violet image. This increased brightness indicates that some of the haze, presumed to be condensate from its color, lies within the canyon itself. Data acquired on the following day indicate that the haze within the canyon was still present in the middle of the morning. Later in the day, toward noon (Figures 11 and 12), the haziness in the canyon dis-

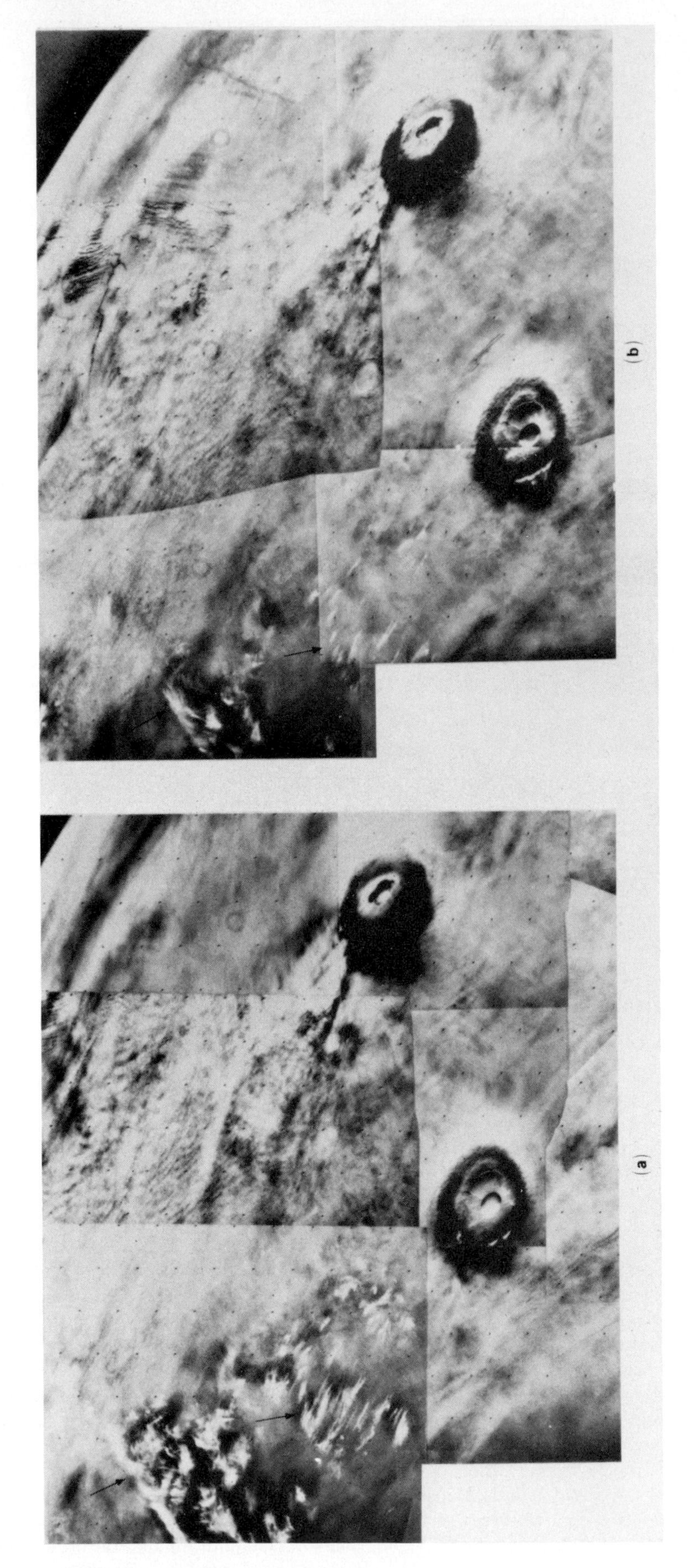

Fig. 9. (*a*) Field of extensive cirruslike clouds northwest of Pavonis Mons close to morning terminator, acquired by VO 1 on revolution 40 ($L_s \sim 101°$). (*b*) Second observation 12 min later. Pictures in mosaic have been filtered and contrast-enhanced.

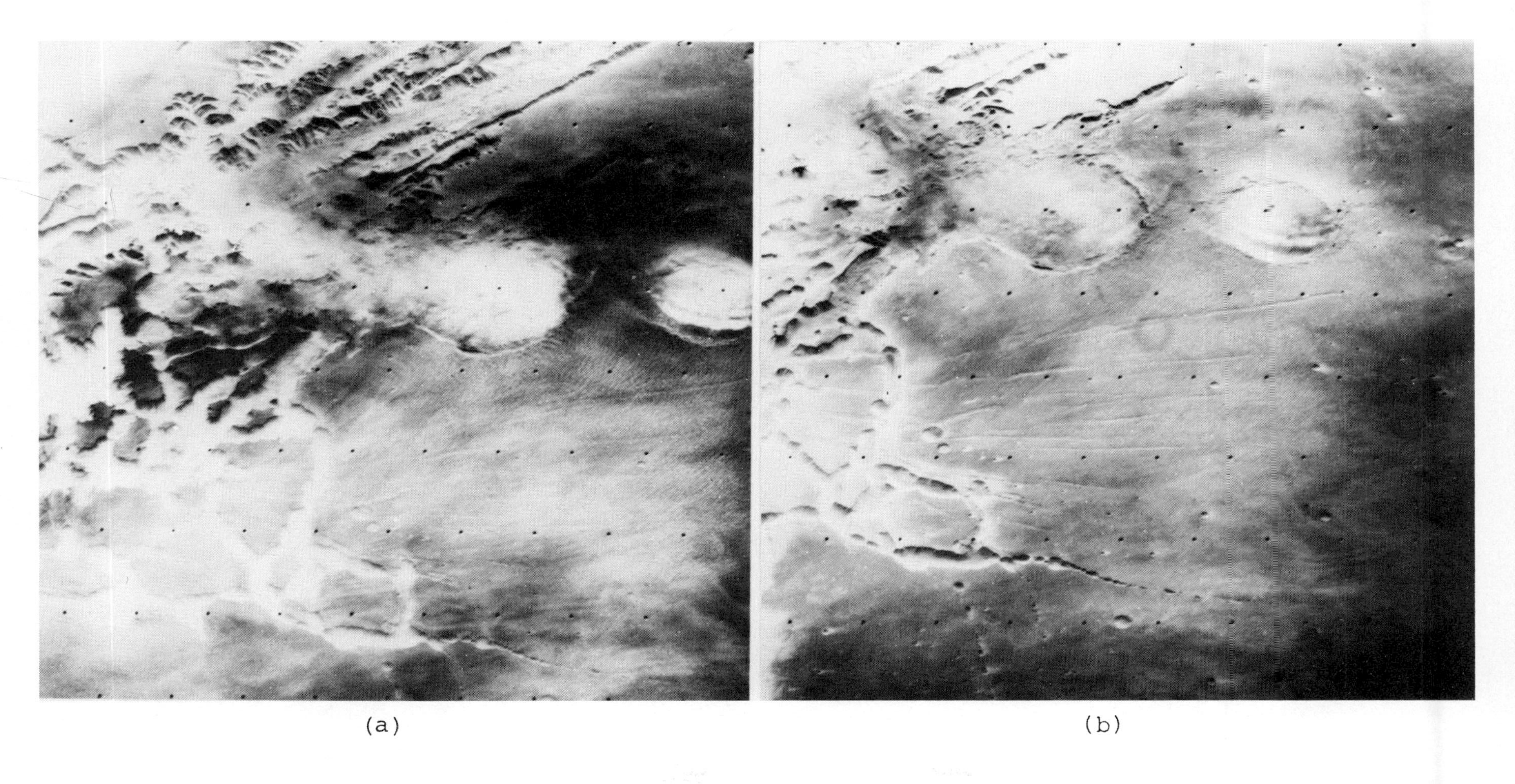

Fig. 10. *(a)* View of Labyrinthus Noctis acquired by VO 1 on revolution 40 shortly after dawn using violet filter. *(b)* View in red light, same revolution.

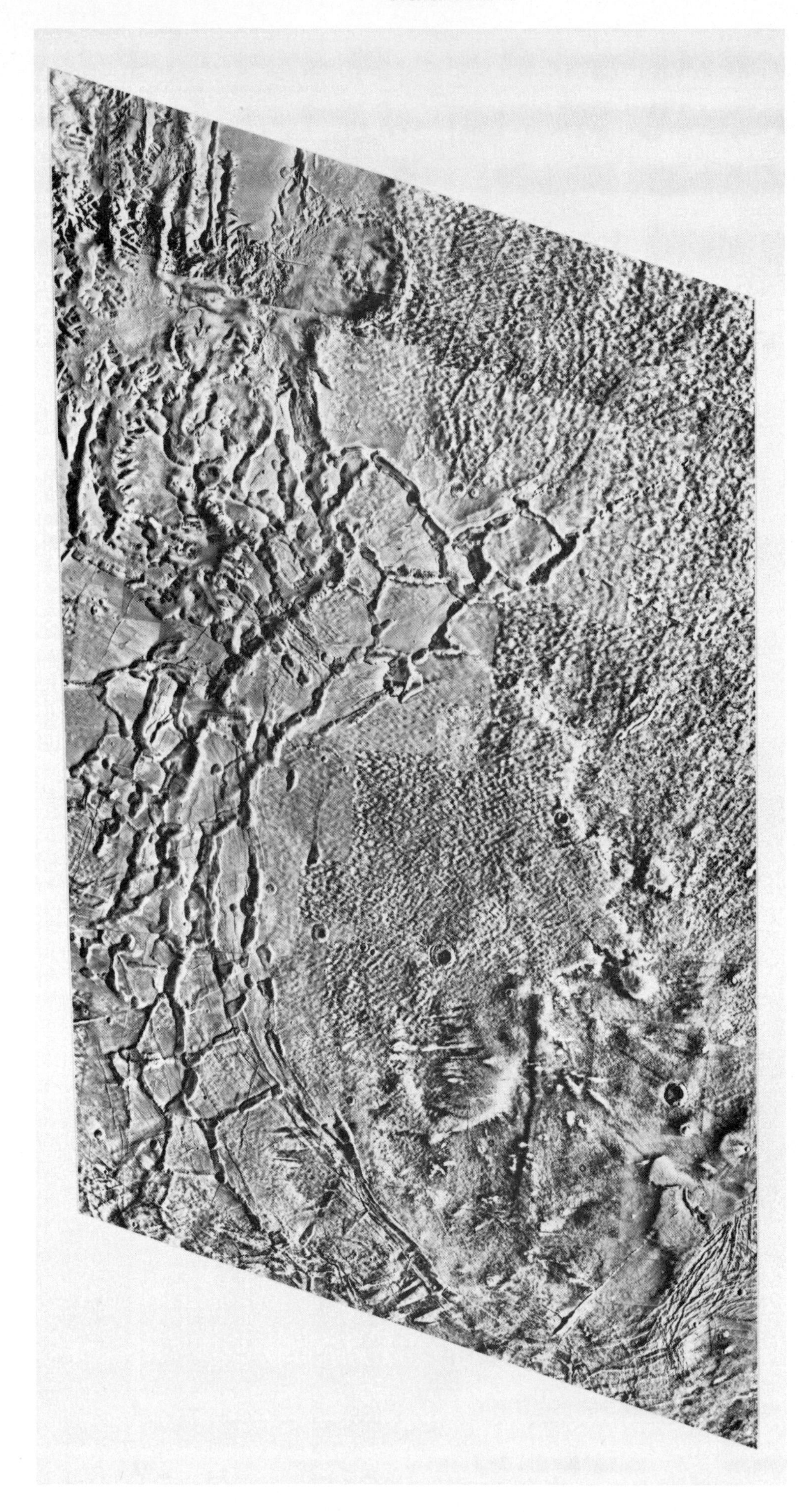

Fig. 11. Mosaic of pictures acquired by VO 1 on revolutions 44–49 showing field of cellular clouds in the vicinity of Labyrinthus Noctis near midday ($L_s \sim 104°$).

Fig. 12. View of Labyrinthus Noctis (VO 1, revolutions 56–57, $L_s \sim 109°$) showing alignment of cellular clouds and indicating general boundaries of cloud field to the east and to the south.

sipates. This behavior seems consistent with the overnight condensation of water vapor in the bottom few kilometers of the atmosphere, as has been discussed by a number of authors.

Toward noon and in the afternoon the Valles Marineris appear largely cloud free, but over the plateau region into which the canyons are cut an extensive field of cellular clouds is evidently a daily occurrence in this season. The mosaics of frames shown in Figures 11 and 12 were acquired over several days, and imaging of the Tharsis Montes near this time of day (Figure 13) indicates that the cloud field extends into this region also. The individual clouds have sufficient optical thickness to cast shadows, from measurements of which the cloud heights are calculated to be 4–6 km. Cloud size and intercloud spacing are of comparable dimensions, suggesting that this is the height of the convective cells. There is a clear indication, especially in Figures 10 and 12, that the clouds are aligned into streets. More than one direction of alignment can be seen in Figure 12, a finding which may indicate that the clouds are at more than one elevation or that the wind direction at the cloud base differs from that higher up. A clear indication of cloud heights from shadows is not available for the clouds aligned in the north-south direction to the west of the more southerly crater. A crescent-shaped wave cloud over the more southerly crater in both figures suggests that the wind here is from the southwest. The streets of clouds to the crater's west are aligned in the wind direction in Figure 10 but apparently at a large angle in Figure 12. In the latter case the streets may mark a lower-level wind direction more from the south.

In Figure 12 the convective cloud field is seen to be a relatively localized phenomenon that disappears to the south and east. The localization may be due to time of day, to the availability of water vapor, or to the elevation of the region. Insofar as the region is a dome of several kilometers' elevation over the Mars average, it is unlikely that the water vapor is especially abundant here. The time of day when ground temperatures are maximum lags behind the maximum (noon) insolation, so that the lack of such clouds to the west, where surface temperatures are probably slightly higher, suggests that time of day is not the only consideration. The high elevation of this region (about 9 km) may be important in promoting a higher than average lapse rate, since the tropopause over this area may not be displaced upward by a comparable

Fig. 13. Cellular cloud field near Ascraeus Mons (VO 1, revolution 55, $L_s \sim 109°$). Pictures have been filtered and contrast-enhanced.

Fig. 14. Memnonia region observed by VO 1 on revolution 34 ($L_s \sim 99°$) about 45 min after dawn. Picture is taken in red light. Line drawn across crater indicates source of photometric data displayed in Figure 15.

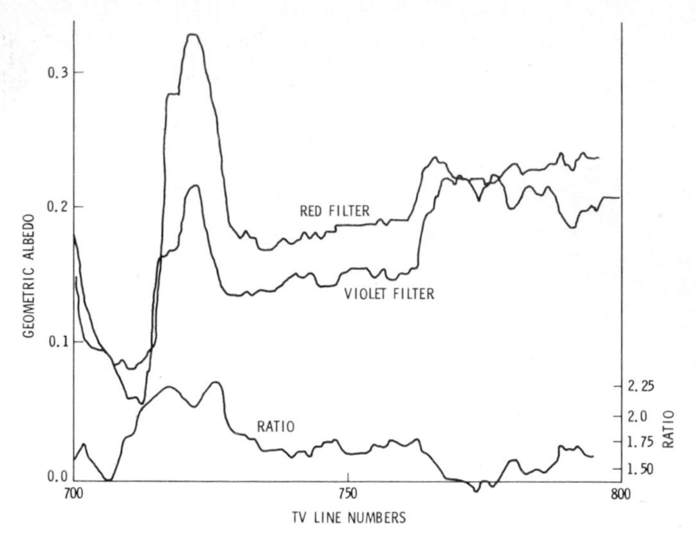

Fig. 15. Geometric albedo in red light (594 nm) and violet light (450 nm) for the trace within the large crater shown in Figure 14.

amount. Additionally, the dust content of the atmosphere may be less over regions of high elevation, so that direct heating is reduced and the lapse rate approaches the adiabatic rate more closely. C. B. Leovy (personal communication, 1977) has suggested that large-scale uplift over the high elevation area at this time of day would favor convection. The uplift would be balanced by large-scale subsidence over surrounding lowland regions [*Webster*, 1977].

Memnonia Brightening

An area of unusually high brightness (Figure 1*d*) in Memnonia to the southwest of Arsia Mons was noticed in the Viking approach imaging. Images acquired 2 hours later (Figure 1*e*) indicated that the brightening had disappeared, and the region was noted as one of potential interest for the study of condensate phenomena. Operational considerations for VO 1 were such that only one opportunity to observe this region in the early morning was presented. Figure 14 shows the Memnonia area, centered at about 15°S, 145°W, as it appeared in red light to VO 1 on revolution 34 at about 45 min after sunrise. The region is seen to be one of extensive volcanic flooding, flows of lava being superimposed on one another. To the west of the area shown is a sinuous channel cut into the plain. The region of brightening observed in the approach pictures can be readily identified by comparing the two views. It consists primarily of the embayed area in the center of the mosaic which is bounded by the scarp to the west and by a number of craters to the north and east. In the orbiter view the principal areas of brightness, which are less extensive than those in the approach observation, are within these craters and within the northern part of the embayment. The relative sharpness of the bright regions suggests a surface rather than an atmospheric phenomenon.

The area was photographed at the same time in violet light, and one area of brightening has been studied to determine albedo and color. A photometric trace has been made through the large crater (as indicated by the line) by using data from both red (594 nm) and violet (450 nm) images. The result is shown in Figure 15, and it is concluded that the bright area is substantially closer to grey in color than the Mars average and that the albedo is only slightly higher than average. A thin or incomplete frost layer would most readily explain a diurnally varying phenomenon with these characteristics.

In view of the position of Memnonia within the southern tropics, water ice condensed overnight might seem, a priori, to be the most likely explanation of such a phenomenon. However, the VO 1 infrared radiometer data show that the region in question attains temperatures of less than 150°K several hours before dawn and that at the time that the imaging data were acquired the surface temperatures in the area were still depressed by about 20°K from the values expected (F. D. Palluconi, private communication, 1977). Thus the explanation for the Memnonia brightening appears to be that CO_2 frost forms in the area for a few hours before dawn (perhaps about 0.1 g cm^{-2}) and takes 1 hour or more after dawn to dissipate. The region in question exhibits an exceptionally low radar reflectivity [*Simpson et al.*, 1974] and evidently has a remarkably low thermal inertia to allow the surface temperature to plunge sufficiently at night for CO_2 to condense.

The channel region to the west of the scarp was observed about 30 min later, in violet light. In Figure 16 the earlier and later views are compared, and it is evident that a number of features (arrows) have become brighter in this interval. The brightening inferred from visual inspection is confirmed by a study of the numerical imaging data. In this case the bright areas are relatively diffuse, and the temperatures believed to be too high to allow CO_2 condensation. The most likely explanation is considered to be that the brightening is caused by a fog of water ice particles, presumably condensed on the surface at night and later released into the atmosphere in the morning, subsequent recondensation occurring in the cooler air above the low-level inversion layer [*Hess*, 1976].

In view of the extremely cold night temperatures of this region the Memnonia area may be an exceptional area for atmosphere surface exchange of water vapor and of potential interest for future lander missions. Surprisingly, the area is not one that has been singled out by telescopic observers.

South Polar Region

Clouds

During the first half of the southern winter the middle latitudes in this hemisphere were notable for the clarity of the atmosphere. Figure 1 shows Mars as it appeared on the last day of the approach of the first spacecraft and indicates the relatively greater visibility at these latitudes than at equatorial and northern latitudes. Some of this better visibility may also be related to the increased surface contrast that results from the deposited frost of the southern polar cap. Figure 17 is an oblique view of the region around the Argyre basin. Craters can be seen to within a few degrees of the limb, indicating further the low optical depth, estimated here to be 0.05. The clarity of the middle latitude atmosphere was not expected because most interpretations of telescopic observations of Mars during the winter have argued for the presence of a cloudy 'hood,' although *Tombaugh* [1968] has reported that such a hood does not form in the south. In addition, the Mariner 9 imaging observations in the north [*Briggs and Leovy*, 1974], which were taken on the afternoon side of Mars during the second half of the northern winter, had shown substantial obscuration due to diffuse haze and clouds. These clouds were generally wave clouds and were interpreted to be composed of water ice on the basis of shadow heights and IR spectrometer temperature profiles.

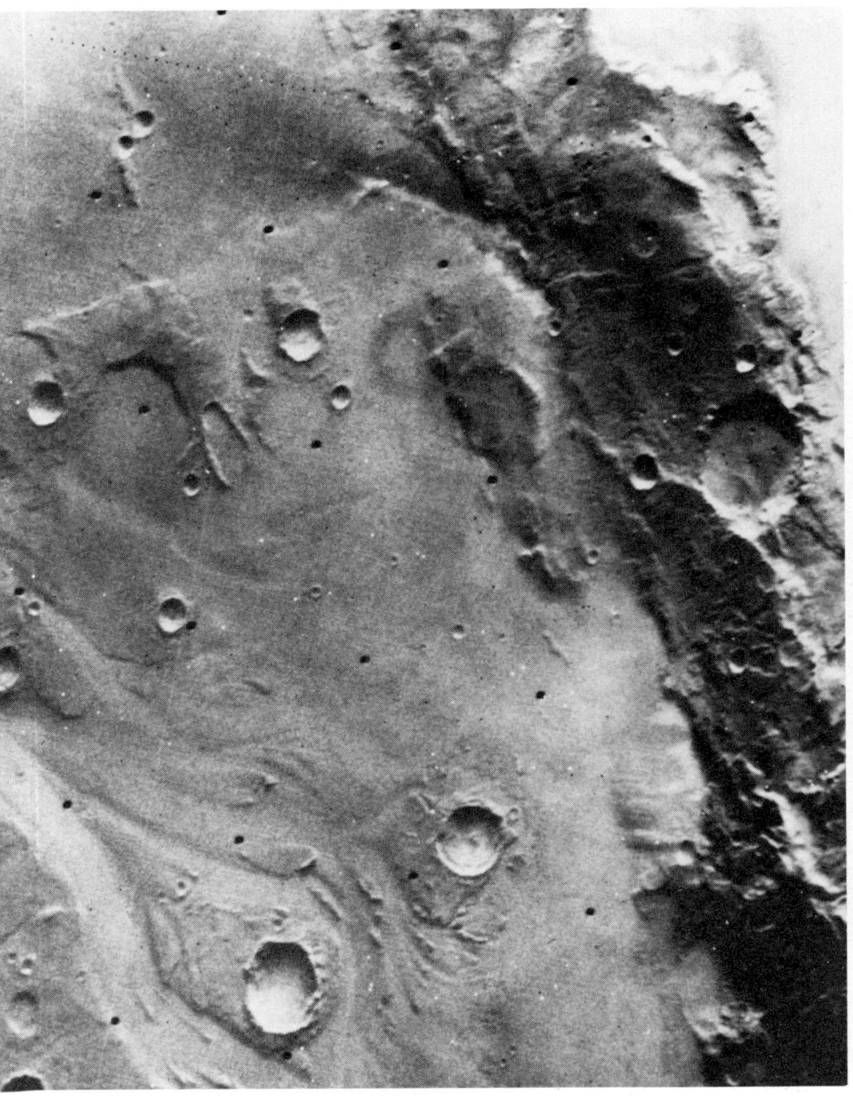

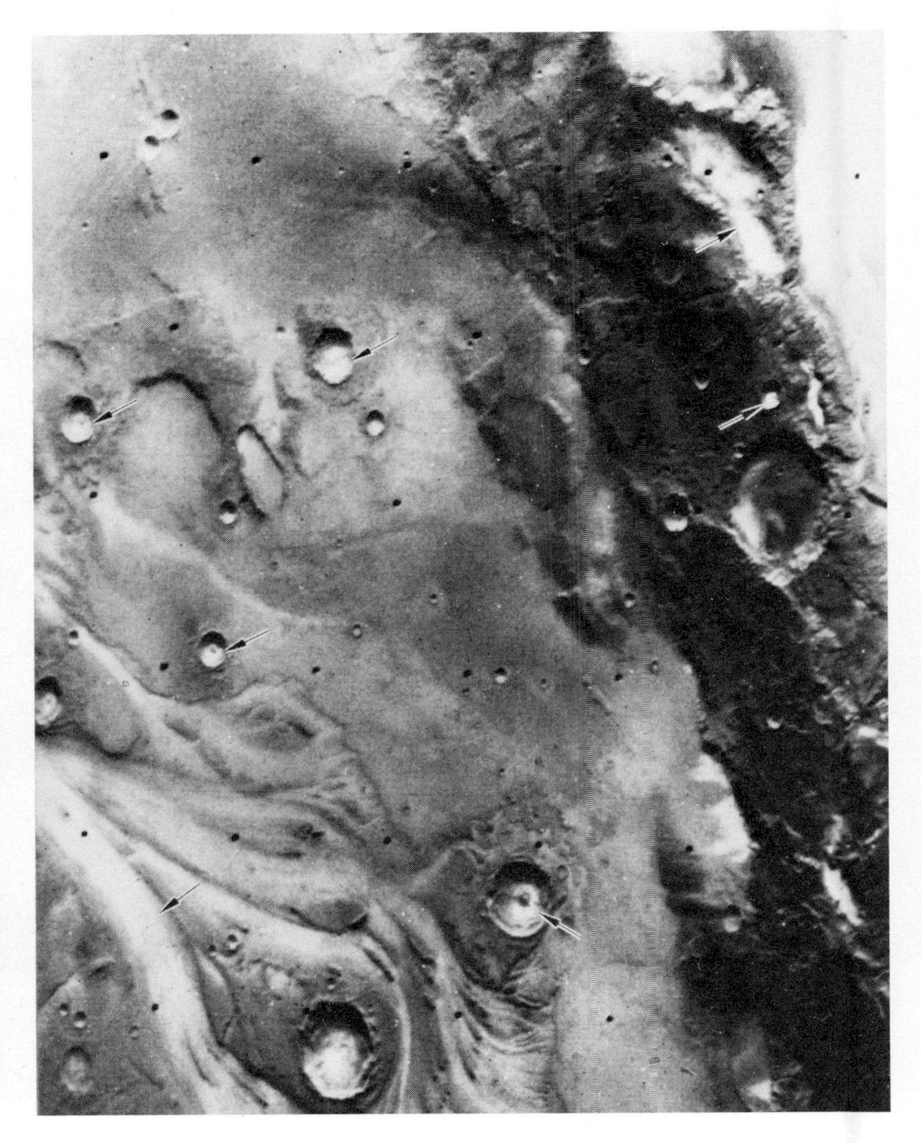

Fig. 16. Comparison of channel region in Memnonia (VO 1, revolution 34) at (left) dawn plus 45 min and (right) 30 min later. Arrows indicate locations of brightenings observed at later times.

Fig. 17. Oblique view of frost-filled Argyre basin on revolution 34 (VO 1), in violet light ($L_s \sim 99°$).

Although, in general terms, there was a notable absence of an early winter polar hood, some discrete clouds and streaky haze can be seen in the southern coverage (Figure 18). From shadow measurements the wave clouds seen are determined to be at an altitude of about 2 km. While no atmospheric temperature profiles are available, surface infrared radiometer measurements [*Kieffer et al.*, 1976*a*] indicate temperatures of about 150°K at the latitudes and times in question. Infrared determinations of total column abundance of water vapor [*Farmer et al.*, 1976] show values close to zero, as might be expected at such low temperatures. The observed wave clouds at this season may be composed of carbon dioxide rather than water ice. The orientation of the wave clouds indicates prevailing westerlies, in common with the opposite hemisphere during winter, as is expected from general circulation models [*Leovy and Mintz*, 1969].

Synoptic pictures taken in the middle and toward the end of southern winter indicate that there was a major change in the clarity of the atmosphere in the southern hemisphere. Figure 19 is a mosaic of pictures taken in violet light by VO 2 on revolution 81 (L_s = 145°) looking north across the Argyre basin, and a comparison of this view with Figure 17 indicates the extent of the change. An earlier observation at L_s = 135° showed a similar state of cloudiness, but many of the images were accidentally overexposed and saturated. The coldest time of year at these latitudes is centered about the beginning of

Fig. 18. Synoptic view of the middle southern latitudes in early winter (VO 1, revolution 40, L_s = 101°) showing streaky haze and wave clouds.

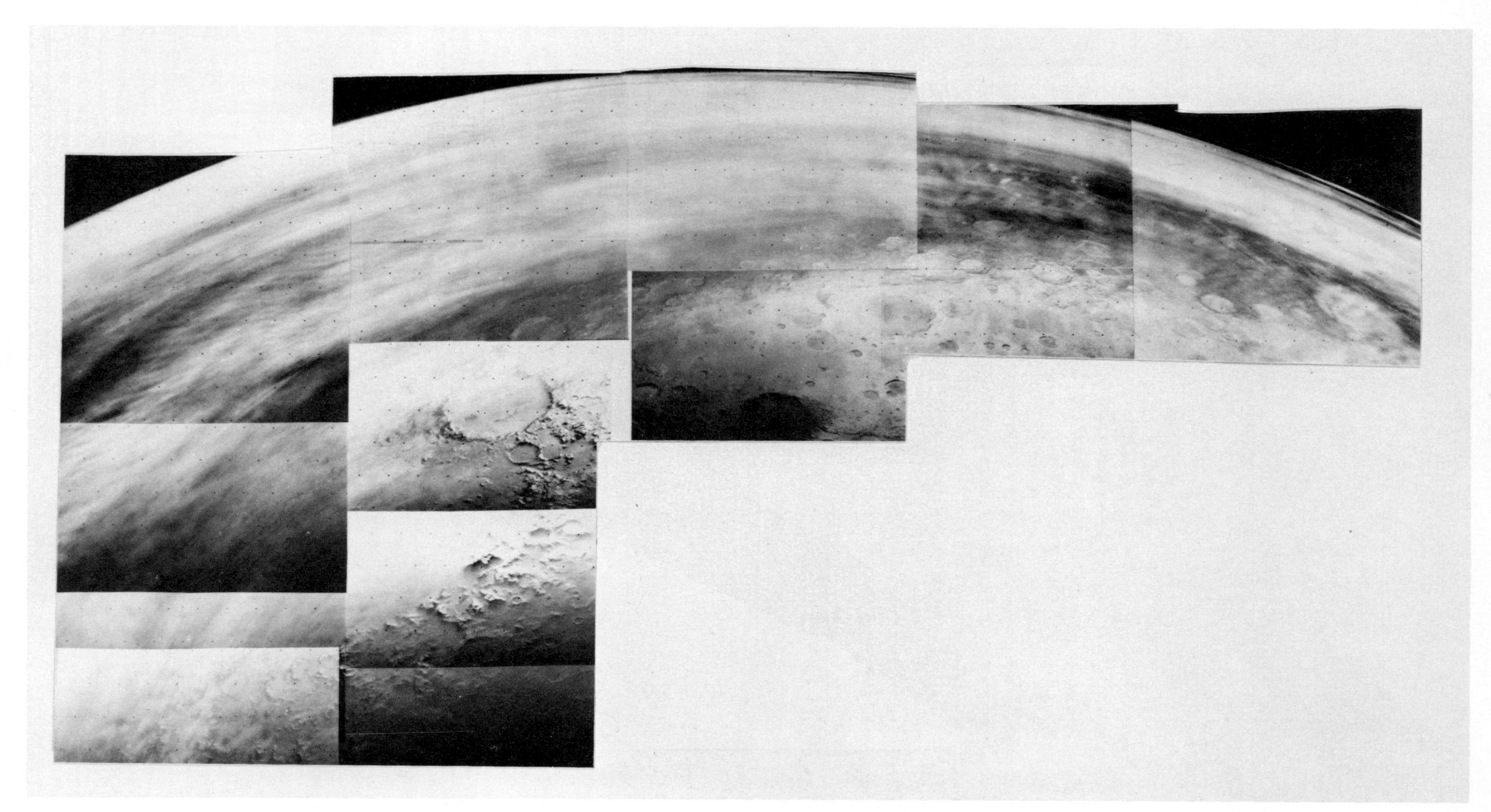

Fig. 19. Argyre region in the second half of winter (VO 2, revolution 81, $L_s = 145°$) showing extensive cloudiness. Picture was taken by using violet filter.

Fig. 20. Synoptic view of cratered terrain east of Hellas basin showing high-altitude clouds and associated shadows (VO 1, revolution 97, L_s = 128°).

Fig. 21. VO 2 approach image showing limb cloud at a height estimated to be 50 km. Also seen is a dawn cloud in the Electris region.

winter, $L_s = 90°$, so that the onset of cloudiness in midwinter occurs at a time when temperatures are rising. The cloudiness may be the result of the seasonal migration of water vapor from the northern hemisphere and/or may be due to the release of water ice from the margin of the south polar cap.

Some other unexpected cloud observations have also been made in the south. Figure 20 shows a region east of the Hellas basin where several irregular bright patches (arrowed) obscure underlying topography. These patches are believed to be clouds. The cloud shadows (also arrowed) are about 200 km away from the clouds, and after corrections for parallax and for planetary curvature are made, all cloud heights are determined to be close to 50 km. Such optically thick clouds at this elevation are presumed to be composed of CO_2, whose condensation temperature at this altitude (pressure of ~0.03 mbar) would be around 120°K. Inspection of some of the approach pictures shows the presence of a few comparably high altitude hazes or clouds at the limb (Figure 21). A meridional alignment of the clouds is evident in Figure 20, and given the expectation of prevailing westerlies in the southern hemisphere, this suggests the possibility that the formation of these clouds may be influenced by waves set up by the Hellas basin. The formation of clouds at temperatures of about 120°K does not necessarily imply that the atmospheric temperature at 50-km altitude is this cold, since the cloud temperature may refer only to a localized wave crest. The Viking entry data [*Nier et al.,* 1976; *Seiff and Kirk,* 1976] indicate temperatures at 50 km of about 140°K for the northern summer latitudes in question. If the atmospheric temperatures at similar altitudes in the winter mid-latitudes are in fact anywhere near 120°K, then moderate- to high-velocity westerly winds would be implied.

If the clouds to the east of Hellas and at other locations as indicated by Figure 21 are caused by waves, then they may be a manifestation of the same wave phenomena inferred to be present in the upper atmosphere from the temperature profiles deduced from the entry data [*Seiff and Kirk,* 1976].

Another southern mid-latitude area of interest as a locale of persistent cloud cover is in the Electris region (~45°S, 190°W), where a discrete cloud was seen predawn and just postdawn in the approach pictures from both spacecraft. The cloud is not seen in pictures taken on approach when the region in question is about 2 hours beyond dawn. These observations (Figures 1*f* and 21) cover the seasonal longitudes L_s = 82° and 102°. From the predawn observation the cloud is determined to be at least 30 km in altitude and is presumed to be composed of CO_2. In-orbit observations (L_s = 122° and L_s = 135°) taken near the morning terminator allowed the cloud feature to be seen in some detail (Figure 22). In Figure 22*a*, at the very edge of the coverage, close to the terminator, is seen another cloud that is presumed to be of a similar nature. The surface region in question is unexceptional, being part of an extensive ancient cratered unit, no particularly large craters or ridges being in the vicinity. The Hellas basin is some thousands of kilometers to the west. The clouds do not resemble wave clouds seen at relatively low altitude elsewhere on the planet, but there is an indication of an alignment of the cells making up the patch. If this alignment is real, it is along a direction approximately NW-SE. The persistence of the dawn cloud over many months of southern winter and its unvarying

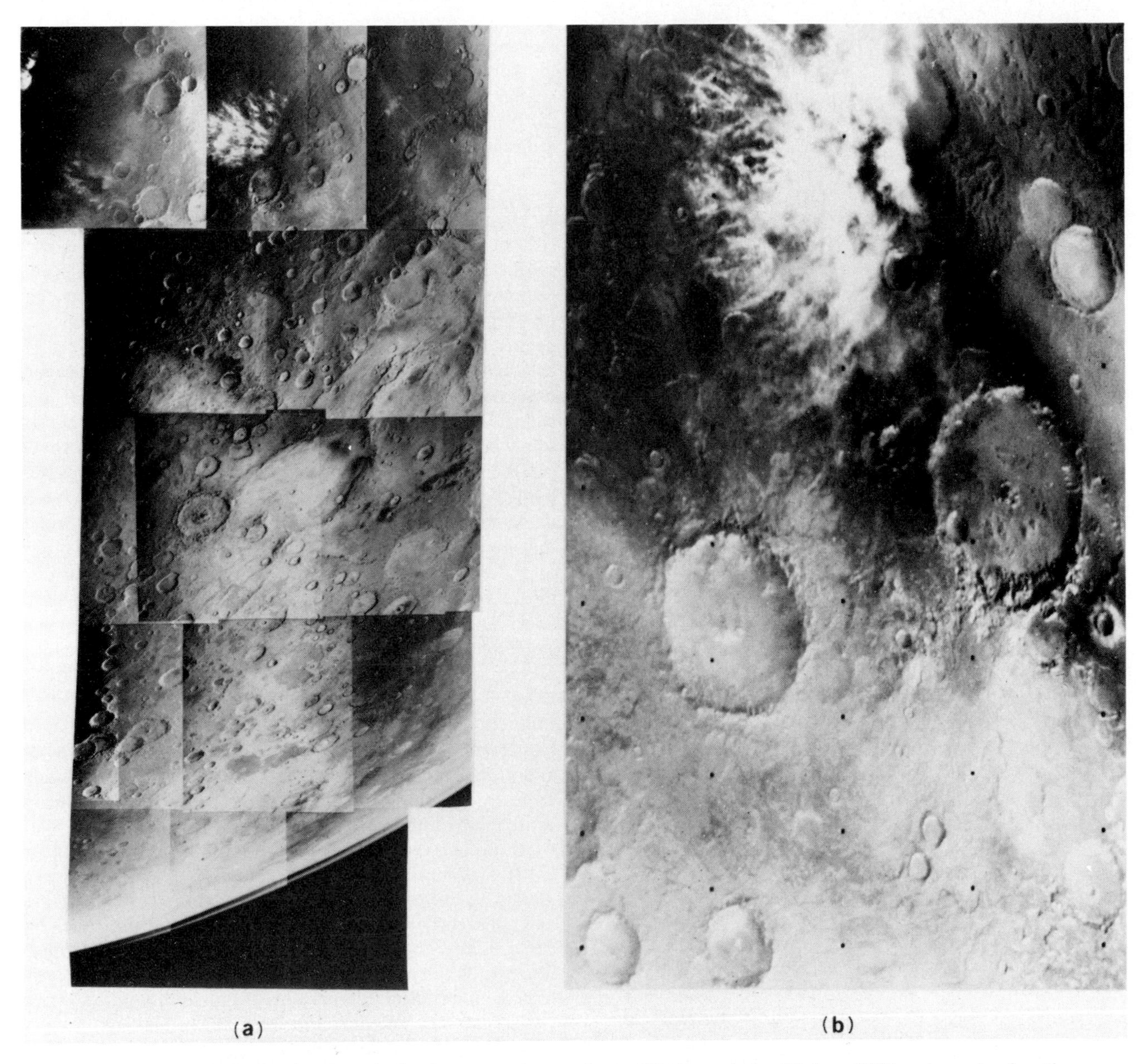

Fig. 22. (*a*) Mosaic showing Electris cloud near morning terminator (VO 1, revolution 88, L_s = 122°) and margin of south polar cap. (*b*) View of Electris cloud, again near morning terminator (VO 2, revolution 62, L_s = 136°).

location are remarkable. An association of the cloud with a lee wave connected with the Hellas basin, perhaps reinforced by local topography, is a possible explanation. No IRTM signature has been identified for the cloud feature (H. Kieffer, private communication, 1977).

Very high altitude clouds are a common feature of the middle latitudes in the winter hemisphere. The temperatures of the observed clouds must be in the vicinity of 120°K, and the orbiter pictures suggest that the clouds are optically thick. Generally, the clouds do not persist for more than a short time in daylight, although an instance (the clouds east of Hellas) has been found involving a daytime life of at least 1 hour. Presumably, the combined heating from surface irradiation and insolation causes the CO_2 particles to sublime. *Kieffer et al.* [1976*a*] have reported temperatures, on the nightside of the planet in the vicinity of the south pole, that reach 130°K. These are well below the condensation temperature of CO_2 at 5 or 6 mbar, and one explanation that has been proposed is that the surface condensation of CO_2 in winter leads to a depletion of CO_2 in favor of nitrogen and argon, so that the condensation temperature falls to the levels measured. The imaging observations of persistent localized clouds at high altitudes in the middle winter latitudes suggest that the presence of similar clouds in the polar region may, in part, be responsible for the very low temperatures observed by the infrared radiometer.

Fig. 23. Closeup views of margin of south polar cap. Frames from (*a*) Figure 17 and (*b*, *c*) Figure 22.

Surface Frost

The observed mid- and high-latitude brightening in the southern hemisphere during autumn and winter has been interpreted in different ways by ground-based observers. Some interpretations have favored a largely atmospheric condensation, the polar hood, on account of the relatively weak brightening compared to the retreating cap and because of observed variability of the brightening. *Tombaugh* [1968], in particular, has expressed a different point of view, i.e., that the south polar cap seldom develops a hood. He reports that the cap extends to 40°S and has an irregular margin which varies by 10° in latitude at different longitudes. The observations made by the Viking orbiter cameras described above indicate that there is relatively little polar cloud cover until the middle of winter, at which time the Viking orbiter observations indicate the abrupt onset of thick cloud cover. Thus the discrepancy between interpretations of telescopic data is understandable. The surface condensate coverage seen in the images corresponds exactly with Tombaugh's description: the cap margin is irregular and reaches about 40°S. The two large southern hemisphere basins, Argyre and Hellas, both have frost cover. Within the margin of the cap the influence of slope and elevation, and probably of albedo and thermal inertia, critically determines where frost is deposited. The effects of wind in drifting the frost into low-lying areas and against windward slopes may also be important. In Figure 23, individual frames from the mosaics of Figures 17 and 22 show the margin in more detail. The floors of some craters are bright and apparently frost covered, while others show no such lightening. Crater bottoms may be favored for frost deposition because of their lower elevation (which leads to a relatively high condensation temperature) and ability to trap windborne material. The northern inner rim of these craters is also favored by its antisolar slope, and at the cap margin in a number of places an apparent reverse illumination is observed where the northern wall is brighter than the more directly

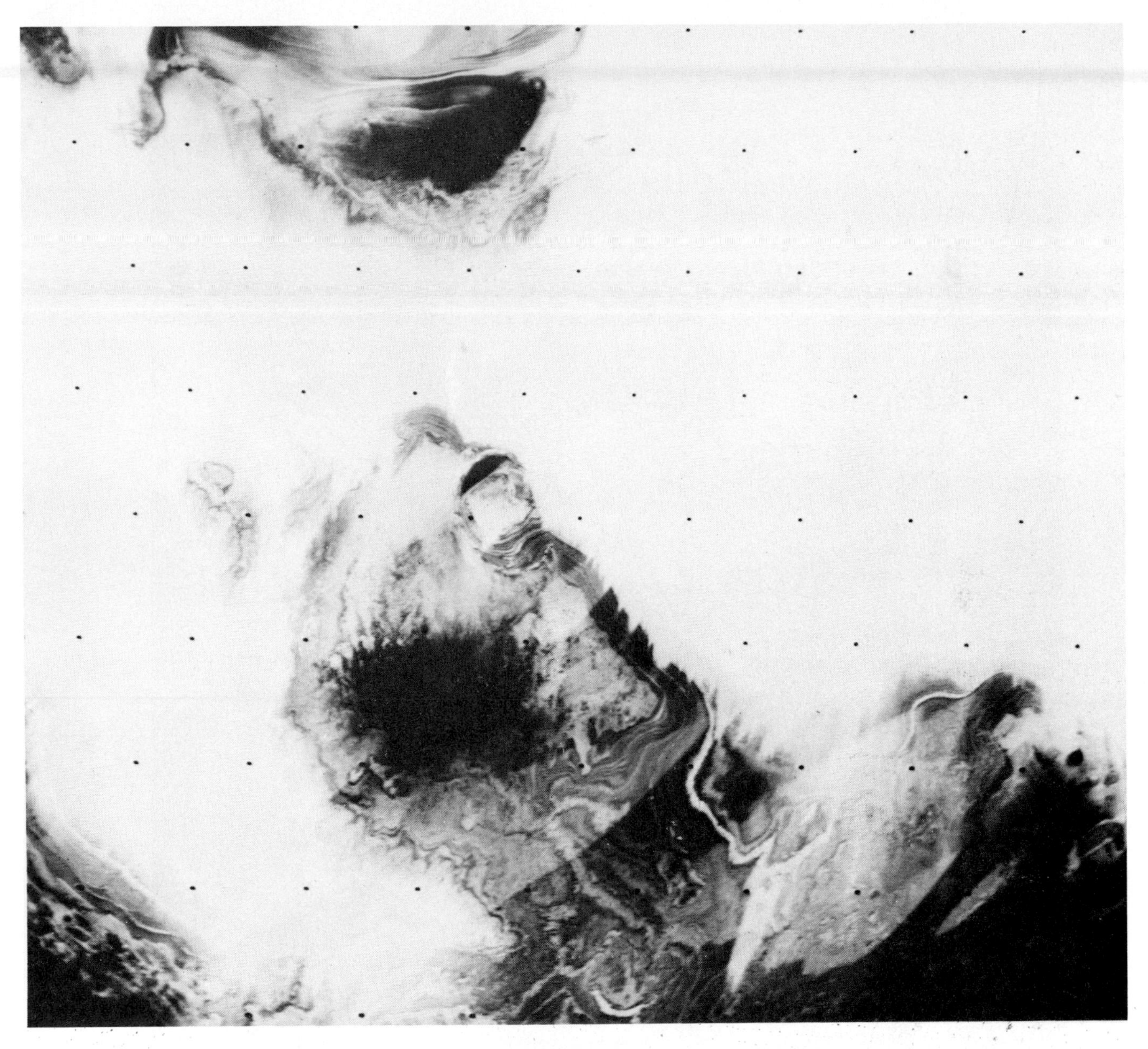

Fig. 24. Margin of north polar residual ice cap at 78°N, 350°W, showing complex distribution of ice (VO 2, revolution 56, L_s = 133°).

illuminated southern wall. The adverse effect of high elevations for frost formation is indicated by the prominent dark peaks around the Argyre basin. The low albedo of these regions is also probably a factor. Within the cap proper, preliminary photometric studies suggest that nowhere examined to date is the frost cover complete. For the areas where the ratio of red to violet reflectances is closest to 1 (i.e., closest to white) the normal albedo in red light reaches values of up to 0.32 only. This may be compared with values of up to 0.51 for the north polar remnant cap, which has been identified as 'dirty water ice' [*Kieffer et al.*, 1976*b*]. Thus it appears that the southern seasonal CO_2 cap in regions that are observable in winter provides only patchy cover. The atmosphere appears relatively clear above the polar cap, and there has been little evidence of dust storms to provide material to reduce the albedo. Moreover, the seasonal polar cap is being constantly resurfaced during early winter, so that a substantial dust contribution, as opposed to a contribution by exposed surface, seems unlikely. By using the observed reflectances of the seasonal cap region in red and violet light, R_r^1 and R_v^1, and by making certain assumptions about the albedo of the surface material, R_r and R_v, estimates of the fractional frost cover and the albedo of the (white) frost may be made. In this preliminary study a clear atmosphere is assumed. The fractional cover is thus described by

$$f = 1 - [(R_r^1 - R_v^1)/(R_r - R_v)]$$

and the albedo of the frost, A, by

$$A = [R_r^1 - (1 - f)R_r]/f$$

By using this simple model for the whitest area in the southern cap to the east of Hellas and assuming that a dark, frost-free spot observed nearby (R_r = 0.180 and R_v = 0.083) is representative of the underlying bare ground, a value of 0.25 is found for the fractional cover and of 0.75 for the frost albedo. Such a

Fig. 25. Appearance of region within south polar residual ice cap (85°S, 355°W) as seen by Mariner 9 on revolution 231 near the end of summer.

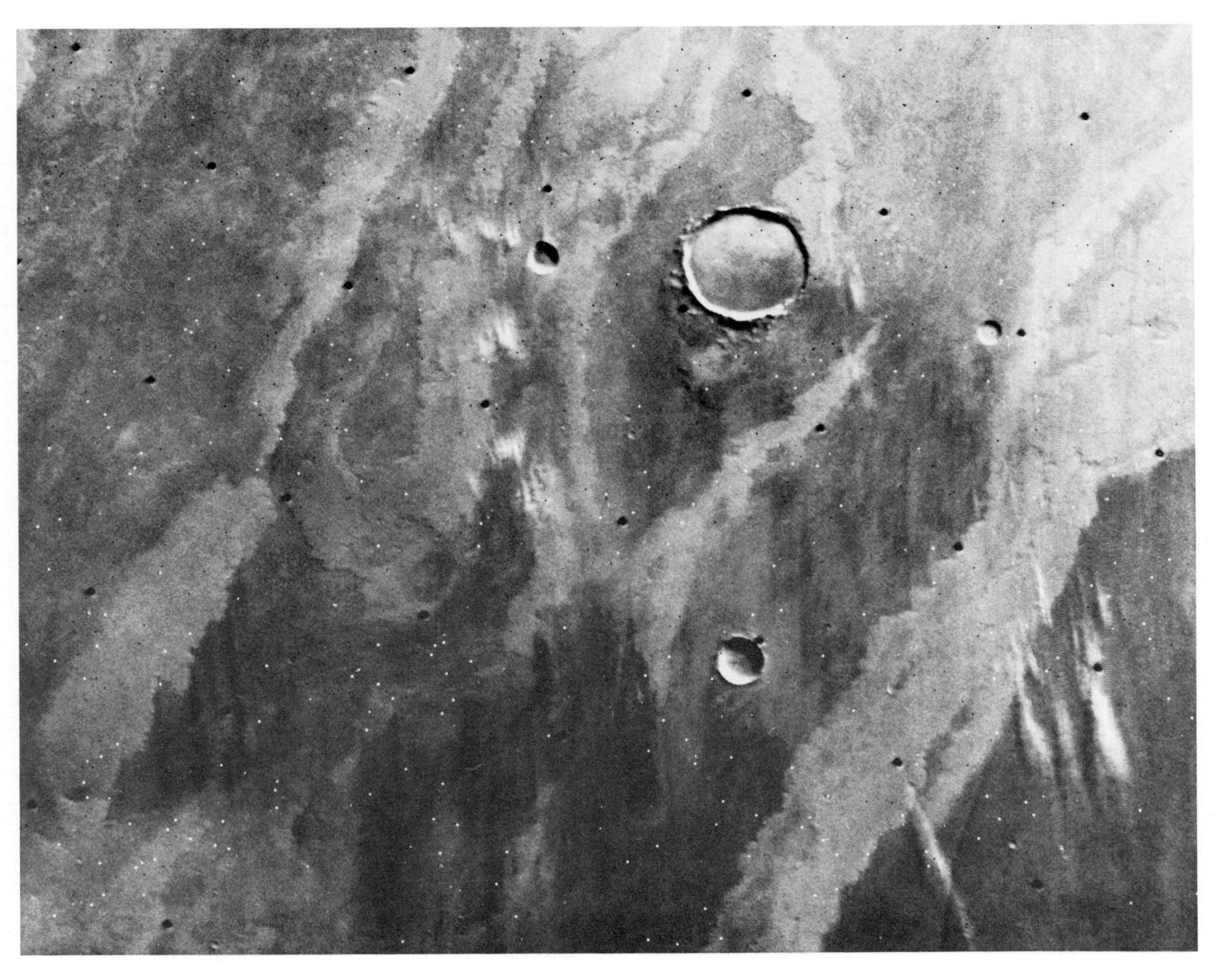

Fig. 26. Plumelike clouds in a region to the southeast of Arsia Mons (23°S, 117°W) that may be dust clouds produced by winds from the north. The large crater is about 20 km in diameter (VO 1, revolution 56, L_s = 109°).

low value for the cover suggests that surface roughness as such is unlikely to be a sufficient reason for the low albedo. Forested terrestrial areas in winter exhibit low albedos and low apparent frost cover, but a comparable surface geometry is hard to conceive of on Mars. Roughness due to blocks is unlikely to reduce fractional coverage below 0.5. Possibly, wind drifting of frost may be responsible for piling up material in restricted areas. In this event the sublimation of such drifted deposits may show a significant delay relative to simple thermal models when the cap retreat takes place. At that time it should be possible to measure the actual albedo of the exposed bare ground to improve the above estimates, which certainly are susceptible to errors due to the assumption made about the underlying terrain and the neglect of atmospheric contributions.

North Polar Cap

From infrared radiance measurements [*Kieffer et al.*, 1976*b*] it has been inferred that the north polar cap remnant is composed principally of water ice. It is also inferred on the basis of thermal balance considerations that the ice albedo must be quite low, about 0.4, so that an admixture of dirt is suggested. Imaging data also provide a similar low measurement of the albedo. Data are available in both red and violet light, and by means of the same simple model applied to the seasonal southern cap the possibility has been examined that the remnant is patchy, high-albedo ice rather than uniform, dirty ice. A frost-free area of layered terrain adjacent to the whitest area (R_r^1 = 0.598 and R_v^1 = 0.478) of the polar cap was used as a reference surface (R_r = 0.292 and R_v = 0.099). The fractional coverage, f, was calculated to be 0.38, and the ice albedo to be an impossible value of 1.10. The uniform, dirty (and therefore reddish) ice model seems to be more reasonable for the brightest parts of the remnant cap.

The margin of the northern remnant displays a highly complex, nonuniform ice distribution [*Cutts et al.*, 1976], as is illustrated in Figure 24. Local slopes and winds appear to play important controlling roles. There is a notable difference in the appearance of the northern and southern remnants late in summer, as can be seen by examining a typical Mariner 9 image of the southern cap (Figure 25). The relatively uniform ice distribution in the north and the very patchy distribution in the south may be related to the present phase of the ~180,000-year equinoctial precessional cycle: southern summer occurs near perihelion. The ice of the southern remnant, presumed also to be water ice [*Murray et al.*, 1972; *Briggs*, 1974], may be in the process of being transferred to the north, where the remnant is not only more uniform but also of larger diameter. If so and if a basic north-south symmetry is assumed, then neither remnant may be very thick, since the total amount of water that could reasonably be transferred in the time available is likely to be no more than a few meters. If water is irreversibly transferred during the time that the seasonal CO_2 is condensing in the same mixing ratio as the water vapor to CO_2 in the atmosphere ($\sim 10^{-4}$), then about 10^{-2} g cm^{-2} might be transferred in 1 year. Over half of the precessional cycle such a rate could transfer less than 10^3 g cm^{-2} of water ice. To contain an appreciable fraction of the total water (some tens of meters for the whole planet [*Owen and Biemann*, 1976; *McElroy et al.*, 1976]) that has been outgassed by Mars, then the residual caps would have to be kilometers thick. There is little evidence to suggest that they are. A number of other sinks (physically and chemically bound water, permafrost, weathering products, exospheric escape) could readily account for the 'missing' water, as has been suggested by various investigators. It seems unlikely that the present remnant caps can play an important role in climate change, and perhaps the most interesting question concerns their survival when competitive sinks have apparently accounted for the bulk of the outgassed water. Perhaps the basic balance is achieved between subsurface water (permafrost and adsorbed water) and exospheric escape, a saturated atmosphere and thin polar caps serving a buffering role.

Dust Storm Activity

Only one candidate dust storm has been identified to date in the Viking orbiter primary mission imaging. The apparent dust clouds (Figure 26) were observed in a region about 600 km southeast of Arsia Mons. The bright plumelike features range in size from a few kilometers to about 25 km and appear to have associated shadows that imply heights of about 1 km. The clouds are optically thick and are relatively well defined. They are oriented approximately north-south and broaden to the south. These characteristics suggest that the clouds are dust rather than water ice, the other potential constituent of clouds in this location, and also indicate northerly winds. The available elevation data indicate that the region is at about 7-km altitude and that the wind is aligned approximately downslope.

In view of the limited orbiter coverage it seems highly likely that the observed dust-raising activity was not unique during the season in question. Such local dust activity, though relatively limited by Martian standards, helps to maintain an aerosol burden in the atmosphere. These data and other orbiter and lander data, acquired at this aphelion season, imply that the Martian atmosphere is never entirely clear.

Acknowledgments. Only through the dedicated efforts of innumerable engineers and project personnel have the data described herein been acquired. We are particularly indebted to the engineers at Ball Brothers, the camera manufacturers, and in the photoscience and image-processing sections at the Jet Propulsion Laboratory. Conway Leovy has been a greatly appreciated source of advice. We acknowledge his valuable comments on a draft of this manuscript. This paper presents the results of one phase of research conducted at the Jet Propulsion Laboratory, California Institute of Technology, under contract NAS 7-100, sponsored by the National Aeronautics and Space Administration.

References

Briggs, G. A., The nature of the residual martian polar caps, *Icarus*, *23*, 167–191, 1974.

Briggs, G. A., and C. B. Leovy, Mariner 9 observations of the Mars north polar hood, *Bull. Amer. Meteorol. Soc.*, *55*, 278–296, 1974.

Cutts, J. A., K. R. Blasius, G. A. Briggs, M. H. Carr, R. Greeley, and H. Masursky, North polar region of Mars: Imaging results from Viking 2, *Science*, *194*, 1329–1337, 1976.

Farmer, C. B., D. W. Davies, and D. L. LaPorte, Mars: Northern summer ice cap—Water vapor observations from Viking 2, *Science*, *194*, 1339–1341, 1976.

Greeley, R., J. D. Iverson, J. B. Pollack, N. Udovich, and B. White, Wind tunnel simulations of light and dark streaks on Mars, *Science*, *183*, 847–849, 1974.

Hess, S. L., The vertical distribution of water vapor in the atmosphere of Mars, *Icarus*, *28*, 269–278, 1976.

Kieffer, H. H., S. C. Chase, E. D. Miner, F. D. Palluconi, G. Munch, G. Neugebauer, and T. Z. Martin, Infrared thermal mapping of the Martian surface and atmosphere: First results, *Science*, *193*, 780–786, 1976*a*.

Kieffer, H. H., S. C. Chase, T. Z. Martin, E. D. Miner, and F. D. Palluconi, Martian north pole summer temperatures: Dirty water ice, *Science*, *194*, 1341–1344, 1976*b*.

Kieffer, H. H., P. R. Christensen, T. Z. Martin, E. D. Miner, and F. D. Palluconi, Temperatures of the Martian surface and atmosphere: Viking observations of diurnal and geometric variations, *Science*, *194*, 1346–1351, 1976*c*.

Leovy, C. B., and Y. Mintz, *J. Atmos. Sci.*, *26*, 1167–1190, 1969.

Leovy, C. B., G. A. Briggs, and B. A. Smith, Mars atmosphere during the Mariner 9 extended mission: Television results, *J. Geophys. Res.*, *78*, 4252–4266, 1973.

McElroy, M. B., Y. L. Young, and A. O. Nier, Isotopic composition of nitrogen: Implications for the past history of Mars' atmosphere, *Science*, *194*, 70–72, 1976.

Murray, B. C., L. A. Soderblom, J. A. Cutts, R. P. Sharp, D. J. Milton, and L. D. Leighton, Geological framework of the south polar region of Mars, *Icarus*, *17*, 328–345, 1972.

Mutch, T. A., R. E. Arvidson, A. B. Binder, F. O. Huck, E. C. Levinthal, S. Liebes, E. C. Morris, D. Nummedal, J. B. Pollack, and C. Sagan, Fine particles on Mars: Observations with two Viking 1 lander cameras, *Science*, *194*, 87–91, 1976.

Nier, A. O., W. B. Hanson, A. Seiff, M. B. McElroy, N. W. Spencer, R. J. Duckett, T. C. D. Knight, and W. S. Cook, Composition and structure of the Martian atmosphere: Preliminary results from Viking 1, *Science*, *193*, 786–788, 1976.

Owen, T., and K. Biemann, Composition of the atmosphere at the surface of Mars: Detection of argon-36 and preliminary analysis, *Science*, *193*, 801–803, 1976.

Pirraglia, J. A., Martian atmospheric lee waves, *Icarus*, *27*, 517–530, 1976.

Pollack, J. B., C. B. Leovy, Y. H. Mintz, and W. Van Camp, Winds on Mars during the Viking season: Predictions based on a general circulation model with topography, *Geophys. Res. Lett.*, *3*, 479, 1976.

Pollack, J. B., D. Colburn, R. Kahn, J. Hunter, W. Van Camp, C. E. Carlston, and M. R. Wolfe, Properties of aerosols in the Martian atmosphere, *J. Geophys. Res.*, *82*, this issue, 1977.

Seiff, A., and S. B. Kirk, Structure of Mars' atmosphere up to 100 kilometers from entry measurements of Viking 2, *Science*, *194*, 1300–1302, 1976.

Simpson, R. A., G. L. Tyler, and B. J. Lipa, Analysis of radar data from Mars, *Sci. Rep. 3276-1, 66*, Stanford Electron. Lab., Stanford Univ., Stanford, Calif., 1974.

Smith, S. A., and B. A. Smith, Diurnal and seasonal behavior of discrete white clouds on Mars, *Icarus*, *16*, 509–521, 1972.

Thorpe, T. E., Viking orbiter observations of atmospheric opacity during July–November 1976, *J. Geophys. Res.*, *82*, this issue, 1977.

Tombaugh, D. W., A survey of long-term observational behavior and various Martian features that affect some recently proposed interpretations, *Icarus*, *8*, 227–258, 1968.

Veverka, J., P. Thomas, and R. Greeley, A study of variable features on Mars during the Viking primary mission, *J. Geophys. Res.*, *82*, this issue, 1977.

Webster, P. J., The low latitude circulation of Mars, *Icarus*, *30*, 626–649, 1977.

(Received April 1, 1977;
revised June 1, 1977;
accepted June 1, 1977.)

Viking Orbiter Observations of Atmospheric Opacity During July–November 1976

THOMAS E. THORPE

Jet Propulsion Laboratory, Pasadena, California 91103

Viking orbiter photography during the primary mission combined with lander indications of surface properties have permitted the estimation of atmospheric optical depths and phase functions. Highly variable time of day opacities ranging from 0.05 to 0.6 are seen to occur in three principle regions. A wavelength-dependent particulate component plus a time variable grey aerosol of higher density may explain these opacities versus time of day. These data should serve as a basis for extended mission comparisons.

INTRODUCTION

Among the most important atmospheric properties to be modeled from a photometric analysis of Viking orbiter (VO) pictures of Mars are the optical depth and phase function. The thin hazes may be characterized by using single-scattering theory; these parameters are found to vary as a function of season, time of day, wavelength, and latitude. The primary orbiter mission has provided over 9000 television images during the Martian northern summer (aerocentric solar longitude, 83°–157°) versus 7300 Mariner 9 late winter pictures (aerocentric solar longitude, 330°–0°). Reconnaissance of lander sites and selected geologic targets at a variety of luminance geometries has produced time of day observations at several wavelengths (0.44–0.59 μ) in several regions: the Viking lander 1 (VL-1) site (10°–30°N, 30°–60°W), 'Smooth Plains' (10°–50°N, 150°–270°W), and 'Old Terrain' (20°–60°S, 60°–260°W) (see Figure 1).

VARIOUS APPROACHES

Terrestrial observations of Mars, limited to coarse resolution and small phase angles, have nevertheless provided wavelength and seasonal discrimination of integrated planetary photometry as yet unmatched via spacecraft. Unfortunately, descriptions of optical depth τ and atmospheric phase functions $f(g)$ are not readily separable from these data without assumptions for surface photometry, atmospheric physical properties, and phase angle extrapolation. The Mariner 9 and Viking pictures offer the potential to complete the terrestrial model and, by subtracting this image contribution, to permit better discrimination of Martian surface properties.

Evidence for the measurement of τ and $f(g)$ in orbiter pictures is manifest by limb hazes, the reduction of scene contrast as a function of viewing geometry, and by image brightness contributions in shadows and at the terminator. Despite the existence of these data both a generalized planetary model and a limited local photometric description have been limited by the inability to remove instrumental effects adequately by the preference for certain observational geometries and by lack of knowledge of the underlying surface behavior. Limb profiles provide large air mass data at high signal-to-noise levels, but these large emission angle calculations are often complicated by (1) failure to locate properly the planet's limb without an obvious brightness discontinuity, (2) crude nonspherical air mass corrections, and (3) large extrapolation from limited viewing geometry requiring aerosol-gas ratio assumptions.

Paper number 7S0457.

Because the bulk of VO pictures consist of on-planet images at moderate emission angles, measurement of the surface contrast reduction at various spatial frequencies versus emission angle offers a statistical advantage. Optical depths determined from contrast changes as pictures approach the limb or terminator implicitly require knowledge of the underlying surface contrast. Although topography may provide such a reference, slopes present in a specific terrain (and the corresponding phase function difference) are usually deduced from craters and, in the absence of stereo, are often precariously inferred from lunar diameter-to-depth ratios. Brightness estimates made from obvious shadows and terminator hazes are subject to scattered light contributions from surface reflection and possible meteorological changes in atmospheric transparency characteristic of large incident angles.

Observations of the sun upward through the atmosphere have been made by the Viking landers [*Mutch et al.,* 1976*a*]. The large optical depths measured (0.45) are considerably higher (3 times) than would be deduced by using the above methods on the high-resolution orbiter images of the VL-1 site. However, the lander data were taken at low sun elevation ($\leq$20°), and the estimates could be attributable to the time of day, local obscuration (orbiter pictures showed an apparent cloud at the VL-1 site during the prelanded operations), and instrumental effects. The avenue of depth calculation from surface pressure and composition leading to number densities and scale height measurement yields only the Rayleigh brightness component ($\tau \sim 0.02$) and neglects the apparently dominant aerosol extinction.

VL-1 SITE ANALYSIS

The Viking landers have provided an indication of 'ground truth' surface reflectance properties. Preliminary findings [*Mutch et al.,* 1976*b*] indicate a surface similar to that of the moon, i.e., '. . . complex microrelief, porosity, and fine grain size distribution not very dissimilar from those of lunar regolith' with 'lunar-like backscattering.' Prelanded predictions of signal versus phase dependence for the appropriate azimuths, compared with observations, showed a strong backscattering tendency (Figure 2). Consequently, modeling of the Martian atmosphere above the VL-1 site was performed by subtracting a lunar phase function, normalized to the Mars average desert albedo [*Adams and McCord,* 1969], from orbiter observations. Although the average albedo contains atmospheric reflectance, the initial values for $f(g, \lambda)$ and $\tau(\lambda)$ are not strongly dependent on this number. The average data numbers for the central 200 × 200 sample area of over 6000 VO television

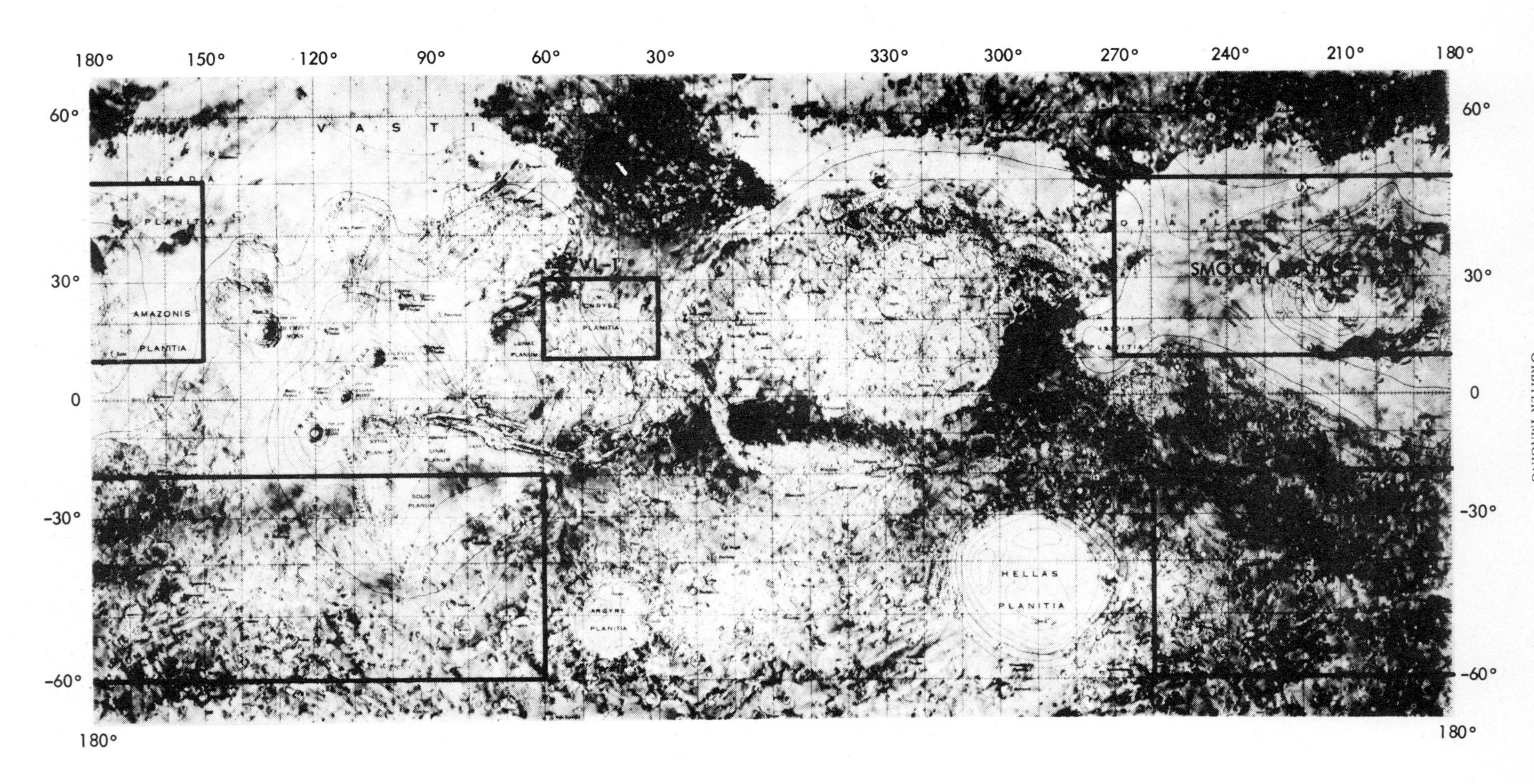

Fig. 1. Areas with extensive time of day coverage.

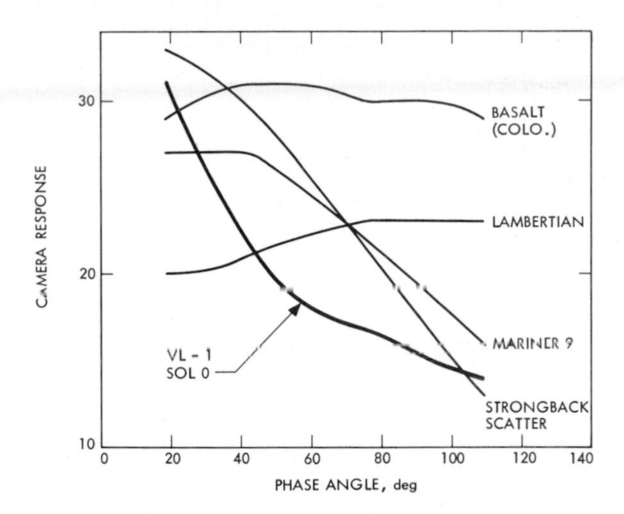

Fig. 2. Predicted versus observed camera response at VL-1 site.

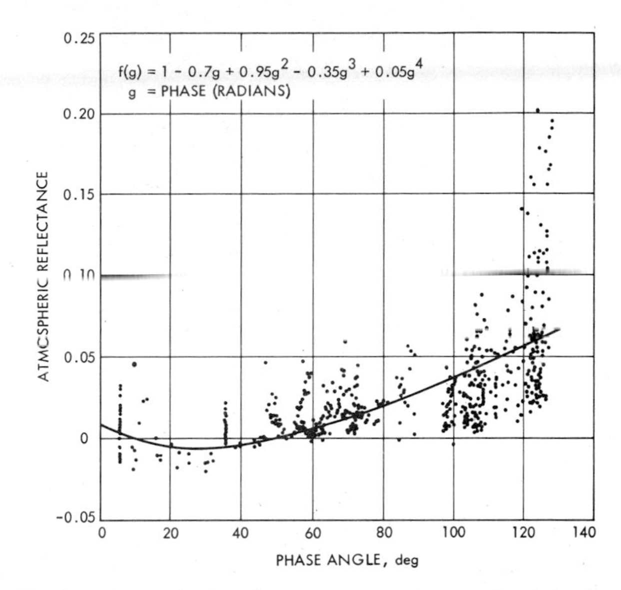

Fig. 4*a*. Atmospheric reflectance versus phase angle; violet filter.

frames were converted to Mars reflectance by using calibration relations [*Klaasen et al.*, 1977]. Plotting the Mars minus normalized moon reflectance ($M - nM$) versus phase angle shows the increasing difference between these two photometric functions at large phase angles (Figure 3). It may be argued that these differences are attributable solely to surface differences despite the lander observations. However, the increased data scatter at large phase angles (ranging from typical moon to typical Mars) for a specific Martian area (i.e., the VL-1 site) suggests temporal variations difficult to characterize in terms of surface properties. Atmospheric phase functions, then, were calculated for three orbiter filter wavelengths (violet, 0.443 μ; clear, 0.532 μ; red, 0.592 μ) by measuring the ($M - nM$) reflectance versus phase angle for a constant incidence angle (to avoid time of day effects). At moderate incidence angles both large and small phase angle observations were made at nearly constant air mass, allowing the calculation of $f(g)$ independent of τ. The phase functions determined take the form

$$f(g, \lambda) = 1 + A(\lambda)g + B(\lambda)g^2 + C(\lambda)g^3 + D(\lambda)g^4$$

where g is phase angle and λ is filter wavelength. Figures 4*a* and 4*b* show the differences in the coefficients versus filter position. The more elongated scattering indicatrix in violet shown in Figure 5*a* is consistent with limb profiles seen during approach (Figure 5*b*) and suggests smaller particles than are observed in red light.

An atmospheric phase function having been determined, the optical depth was calculated as follows:

$$\tau(\lambda) = \frac{-1}{m} \log \left(1 - \frac{R(M - nM)}{sf(g)}\right)$$

where

m air mass;
$s = [(\cos i)/4(\cos i + \cos e)]$;
$R(\ )$ atmosphere reflectance.

A time-of-day-dependent depth was then plotted, both isotropic single scattering (Figures 6*a*, 6*c*) and the above $f(g)$

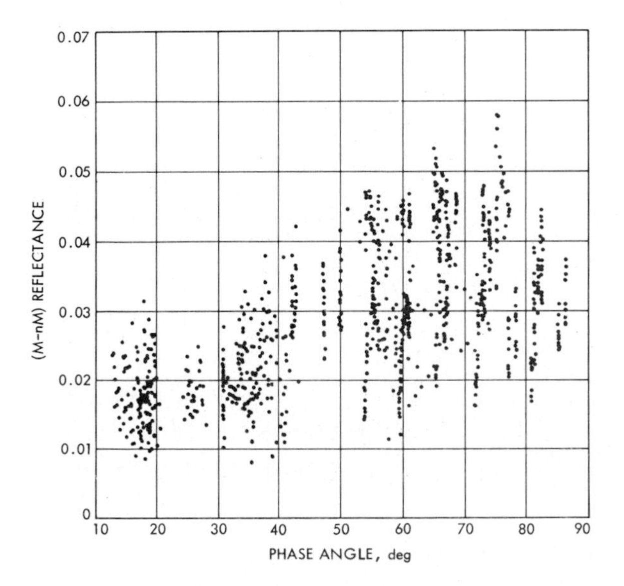

Fig. 3. ($M - nM$) reflectance versus phase angle; clear filter, VL-1 site.

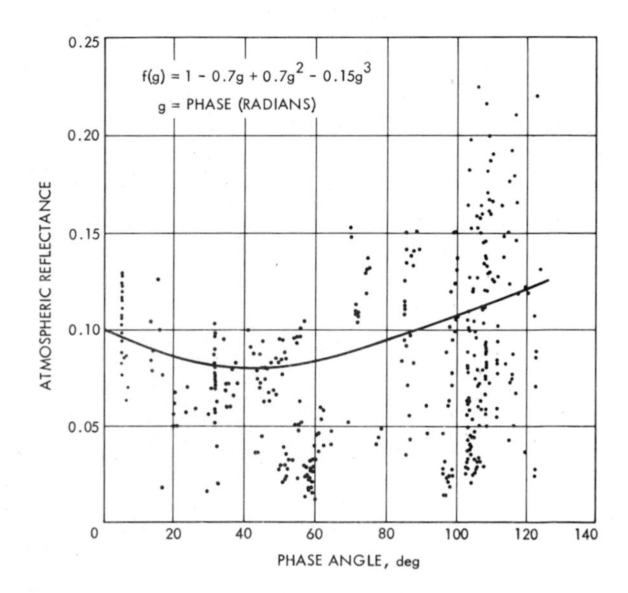

Fig. 4*b*. Atmospheric reflectance versus phase angle; red filter.

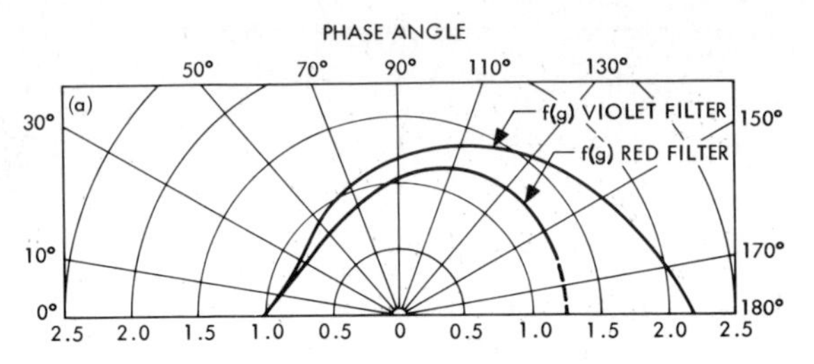

Fig. 5*a*. Red and violet filter atmospheric phase functions.

function (Figures 6*b*, 6*d*) being assumed. Less data scatter occurs with the latter nonisotropic relation for planet-wide observations and less depth versus incidence dependence exists. However, the VL-1 site data still exhibit somewhat higher early morning and late afternoon opacity (Figure 7). These data were closely examined for possible inclusion of secular or longitudinal effects. The trends appear to be mainly time of day dependent. For example, on VO-1 revolution 20, observations at +21° ± 1° latitude and 50° ± 1° longitude cover a range of incidence i = 40°–80°. Thirty-five days later (revolution 54 and VO-2 revolution 10), pictures at this location were recorded at i = 35°–77°. The clear filtered pictures show a maximum opacity reached near sunrise (6–8 A.M.) and midafternoon (3–4 P.M.) local time with depths slightly higher in the morning (0.4) than in the evening (0.20). These values are consistent with the lander measurements, and the opacity decrease to 0.1 near midday is in agreement with a significant fraction of the orbiter pictures reduced by the methods previously described. Comparing violet versus red filter data (Figures 8*a*, 8*b*) indicates a greater atmospheric reflectance increase at the longer wavelength through midafternoon. However, contrast measurements (Figure 9) suggest that surface resolution is higher in red light near midmorning. Some of this resolution difference may be due to surface limb darkening changes. Figure 10 illustrates optical depths calculated from atmospheric brightness versus depths obtained from an assumed slope modulation reduction, i.e.,

$$\tau_c = \frac{-1}{m} \log_e \left(\frac{Mosf(g)}{MsnM + Mo[sf(g) - nM]} \right) \quad (1)$$

where Mo is observed modulation and Ms is modulation of 5° slope (lunar limb darkening). One possible model for this behavior might be an atmospheric mixture of a small amount of red material (e.g., particles of desert albedo), which produces the larger red filter brightness, combined with a greater amount of material with neutral spectral reflectivity which reduces the violet filter contrasts. For example, if we assume the presence of dust particles with red to violet brightness ratios of B_R/B_V = 4 and condensates with B_R'/B_V' = 1, where

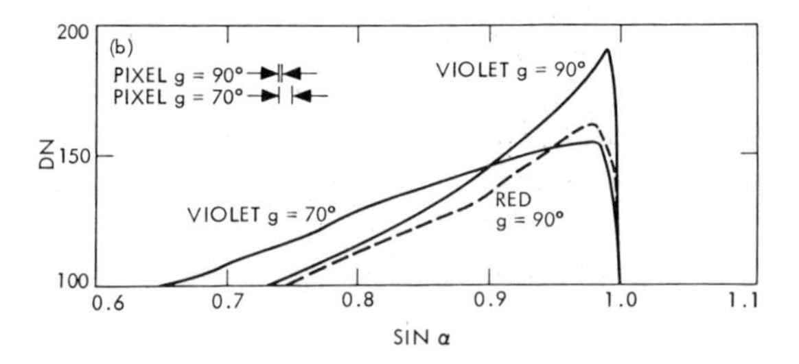

Fig. 5*b*. Limb profiles across approach images (violet shutter speed is 2 times red filter shutter speed).

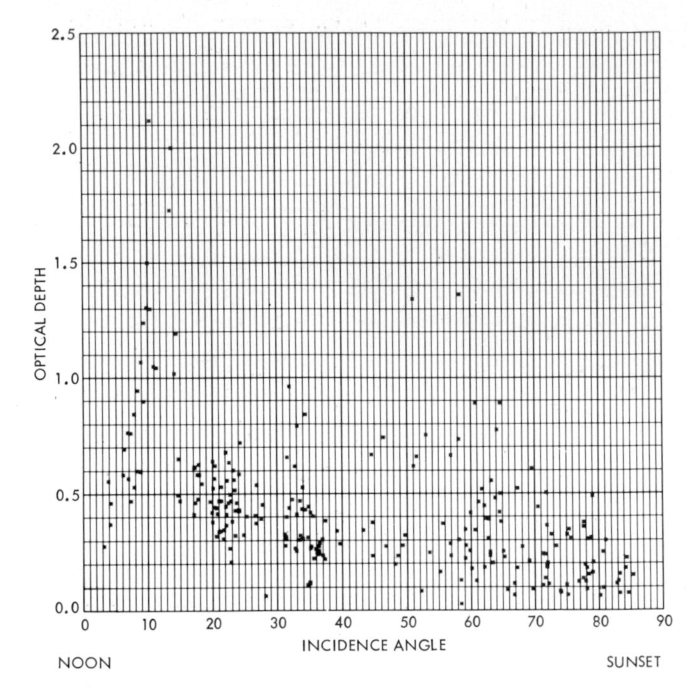

Fig. 6*a*. Optical depth versus incidence; red filter, $f(g)$ = 1.

$B_R' = \frac{1}{2}B_R$, then by using (1) with equal surface modulation in red and violet a 30% increase in red versus violet filter total modulation is observed for τ_R = 0.4 and τ_V = 0.35, i.e.,

$$\frac{Mo_R}{Mo_V} = 3 \cdot \frac{\exp(-\tau_R m)}{\exp(-\tau_V m)} \cdot \frac{\frac{2}{3}N \exp(-\tau_V m) + \frac{3}{2}B_R}{2N \exp(-\tau_R m) + 3B_R}$$

where

Mo_R observed modulation in red;
$N = nM$;
$B_R = 1 - Ne^{-\tau m}N \leq 1$.

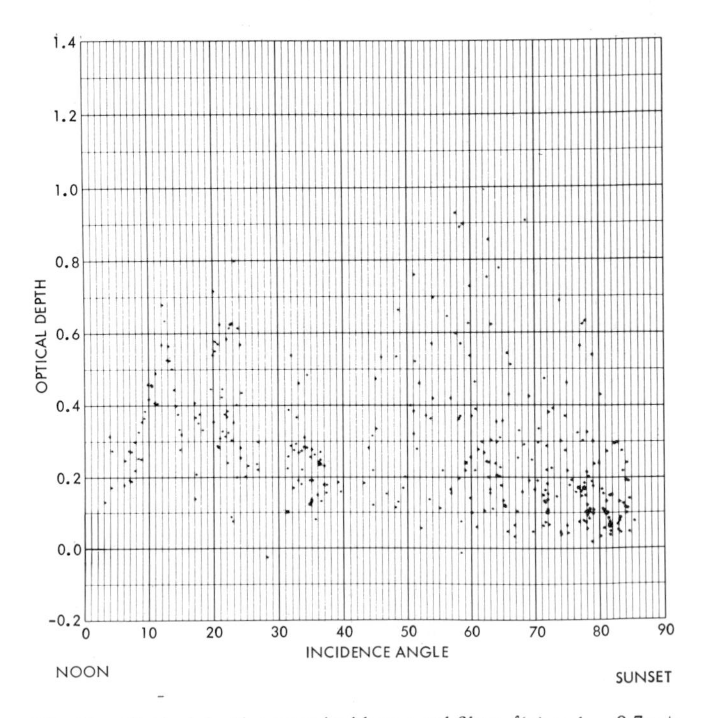

Fig. 6*b*. Optical depth versus incidence; red filter, $f(g) = 1 - 0.7g + 0.7g^2 - 0.16g^3$.

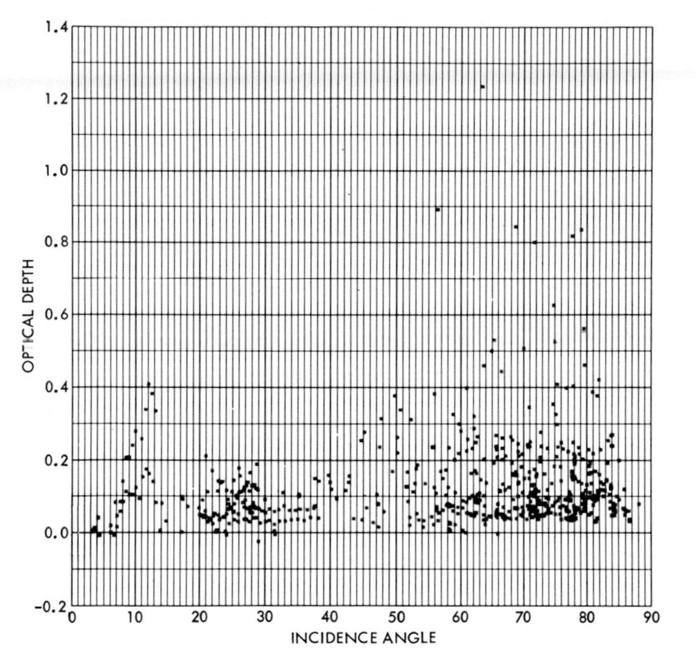

Fig. 6*c*. Optical depth versus incidence; violet filter, $f(g) = 1$.

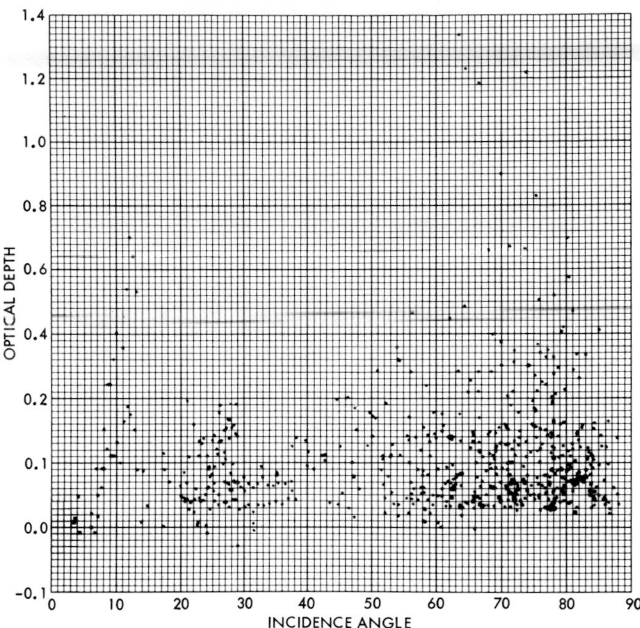

Fig. 6*d*. Optical depth versus incidence; violet filter, $f(g) = 1 - 0.7g + 0.95g^2 - 0.35g^3 + 0.05g^4$.

Hence a mixing ratio $\tau_V/\tau_R = n\sigma h_V/n\sigma h_R$ of condensates with 10% additional opacity provided by dust could account for the modulation differences seen in Figure 9, the same particle cross section and scale height being assumed. Although these assumptions are unlikely, it is apparent that only a small amount of dust is necessary to produce the results given in Figure 9.

Near midday the atmosphere is most transparent despite the lack of contrast resolution (apparently due to the surface limb darkening constancy at low phase angles). But the reduction of condensates has greatly diminished the violet reflectance (Figure 8). By midafternoon, however, the increasing haze is present again in reverse order from the morning trend. One might argue, of course, that the lunar phase function is more sharply peaked than the Mars surface function and hence masks the near-noon opacity. However, observations at small incidence angles at the VL-1 site were generally made at phase angles of 20°–40°, consistent with data taken at higher incidence angles.

OTHER REGIONS

Similar comparisons were made between the Smooth Plains (30°N) and the Old Terrain (40°S) (see Figure 1). It was assumed that the surface functions were similar to those at the VL-1 site and that the atmospheric equation was equally valid. Care was taken to separate obvious surface albedo features

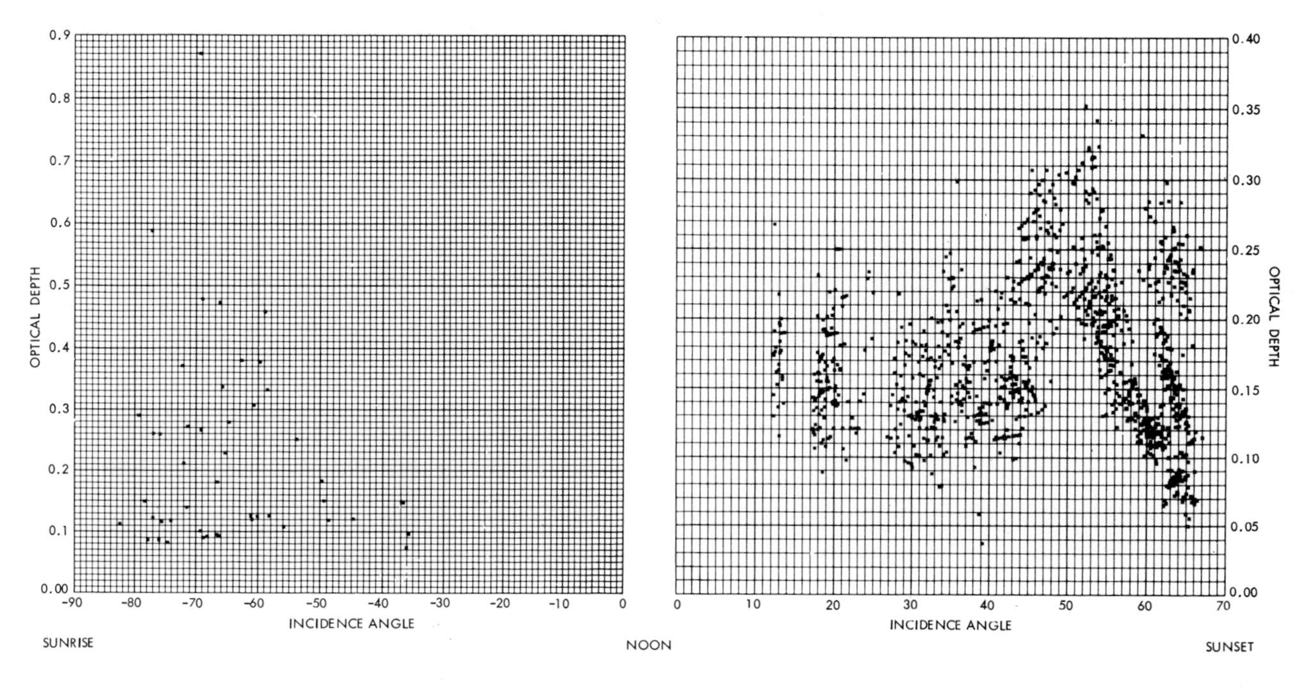

Fig. 7. Optical depth versus incidence; clear filter, VL-1 site.

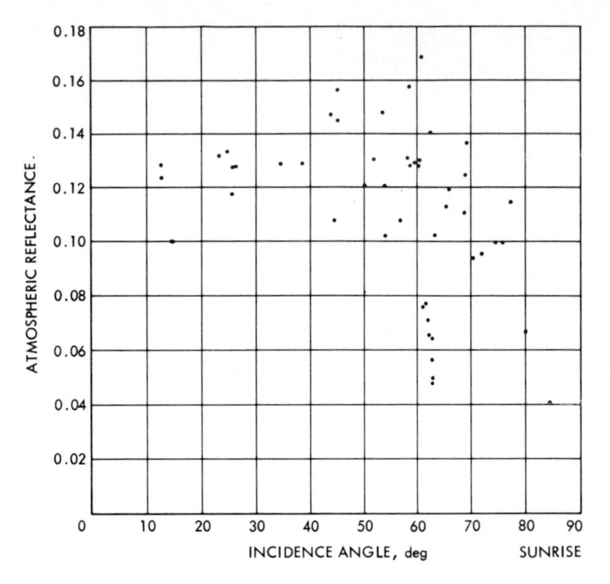

Fig. 8*a*. Atmospheric reflectance versus incidence; red filter, VL-1 site.

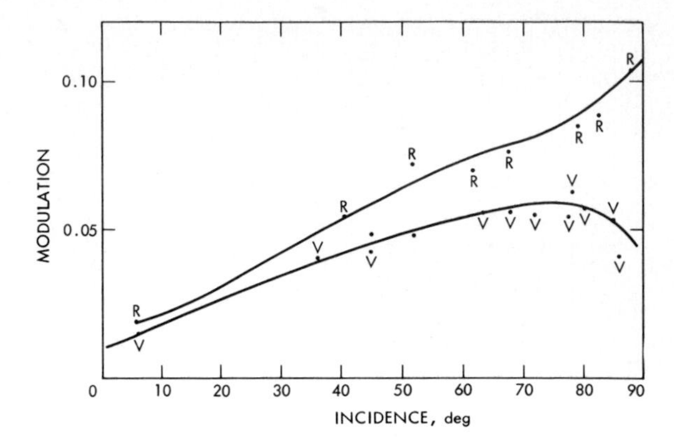

Fig. 9. Image modulation of surface topography at VL-1 site versus incidence.

from geometric tendencies in the observations. Figure 11 shows that the late summer northern observations exhibit the greatest clear filter optical depth of the three regions. A maximum above 0.6 in some pictures is reached at sunrise in a fashion similar to that at the VL-1 site; however, an afternoon counterpart is not as apparent. A near-sunset maximum is suggested that is not as large as the early morning optical depth. Violet filter near-noon backscatter is visible in the Smooth Plains data (Figure 12), producing a large opacity (>0.15) even at high sun. These pictures were taken at low phase angle (5°), suggesting that the Mars surface backscattering is at least as large as that of the moon. Red values are more or less constant over the range of observations with some decrease in the afternoon opacity. Again, aerosols (ice or dust) account for most of the atmospheric reflectance.

More data scatter is present in the Old Terrain observations (Figure 13), perhaps owing to greater surface roughness or frost patches. In this region during late winter an early morning maximum (0.4) occurs near sunrise without an obvious midmorning decrease, and late afternoon measurements show nearly the same opacity, reached only at the end of the day. Red filter opacities are also larger than those in violet. However, the lack of high sun pictures in this area prevents time of day red-violet comparisons.

Mariner 9 Comparison

The Viking reflectance measurements were also compared with the Mariner 9 B-camera data [*Thorpe*, 1973] corrected to the Viking clear filter effective wavelength ($\Delta\lambda_{eff} \sim 0.04\ \mu$) by using the Adams and McCord spectral curve for the Martian desert. Figure 14 shows that little difference exists (~0.008 rms) between the planet-wide reflectances recorded by Viking and the mean Mariner 9 phase function. If we assume that most changes are attributable to variations in the atmosphere and calculate τ by using the phase function described earlier, a slight planet-wide increase of 0.03 is seen to occur mainly at high sun (Figure 15). Considering the dispersion in the earlier curve, this difference may not be significant.

At the VL-1 site (Figure 16) the primary difference from a late winter 1972 versus early summer 1976 comparison is slightly higher optical depth at high sun ($\Delta\tau = 0.04$) or in the

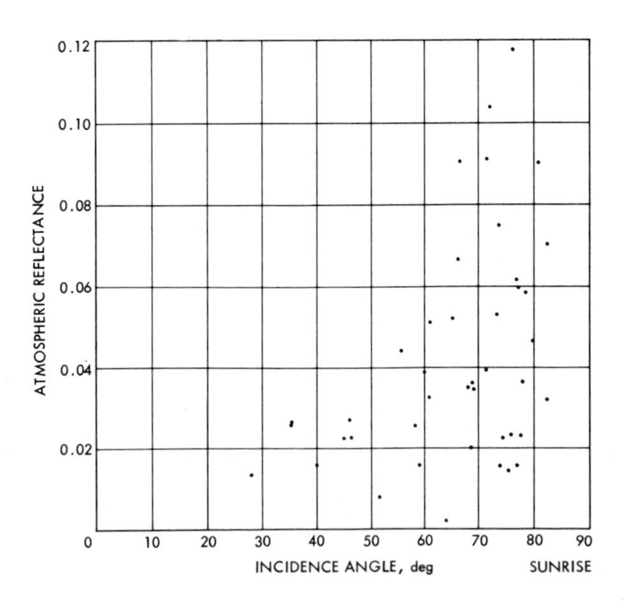

Fig. 8*b*. Atmospheric reflectance versus incidence; violet filter, VL-1 site.

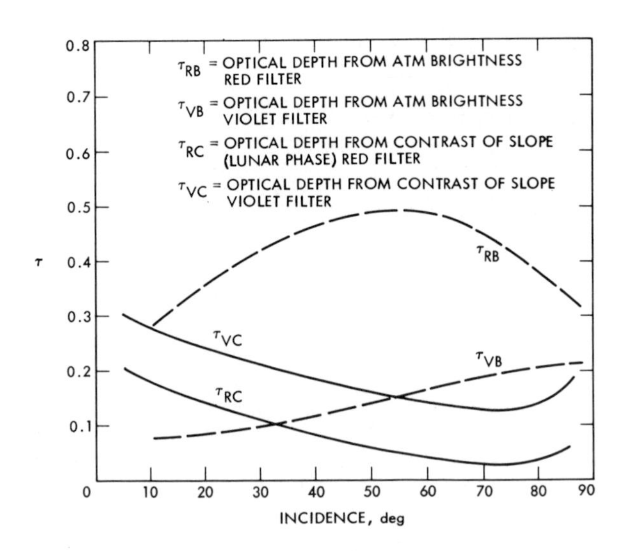

Fig. 10. Optical depth from reflectance versus contrast.

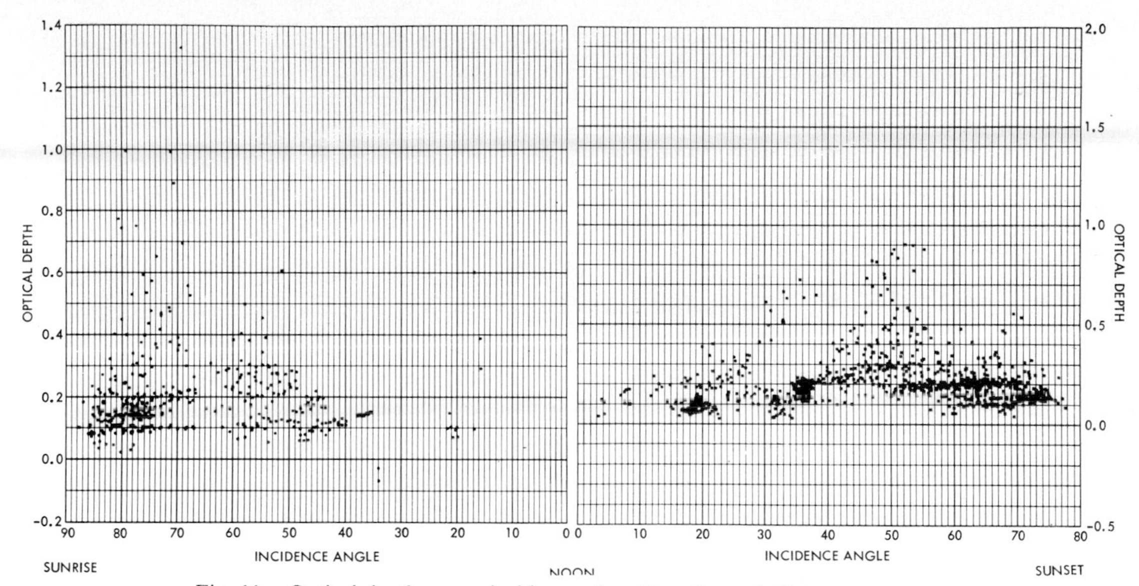

Fig. 11. Optical depth versus incidence; clear filter, Smooth Plains.

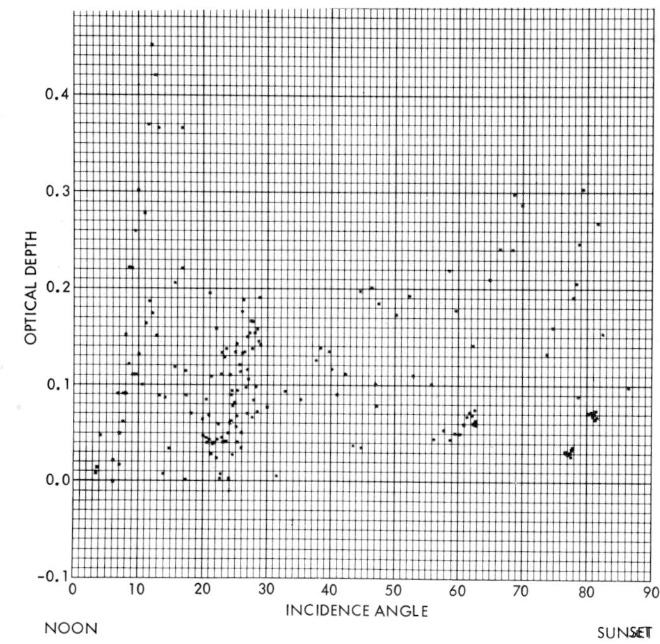

Fig. 12. Optical depth versus incidence; violet filter, Smooth Plains.

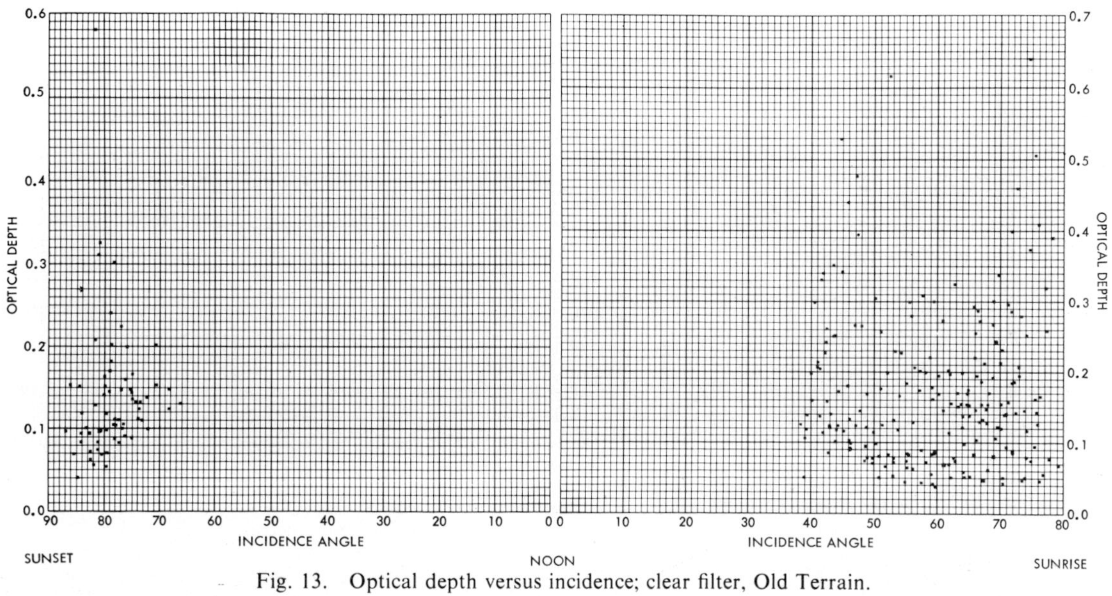

Fig. 13. Optical depth versus incidence; clear filter, Old Terrain.

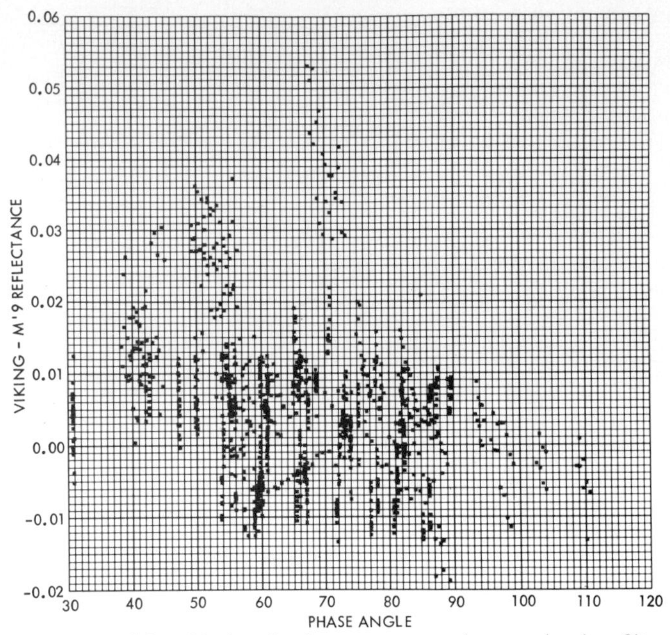

Fig. 14. Viking–Mariner 9 reflectance versus phase angle; clear filter, all planet.

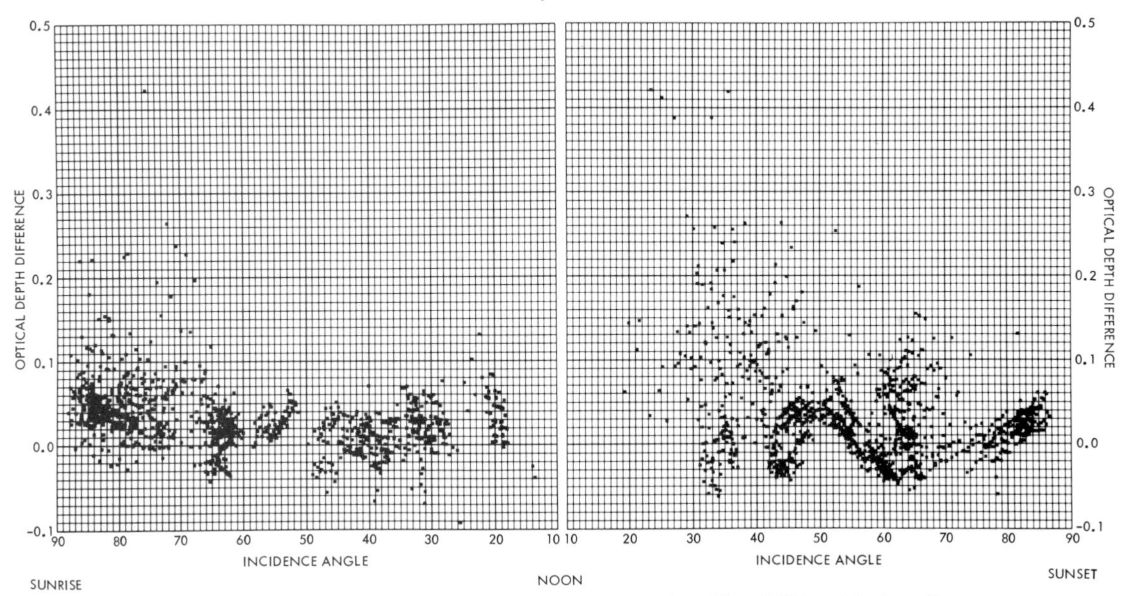

Fig. 15. Optical depth difference versus incidence; clear filter (Viking–Mariner 9).

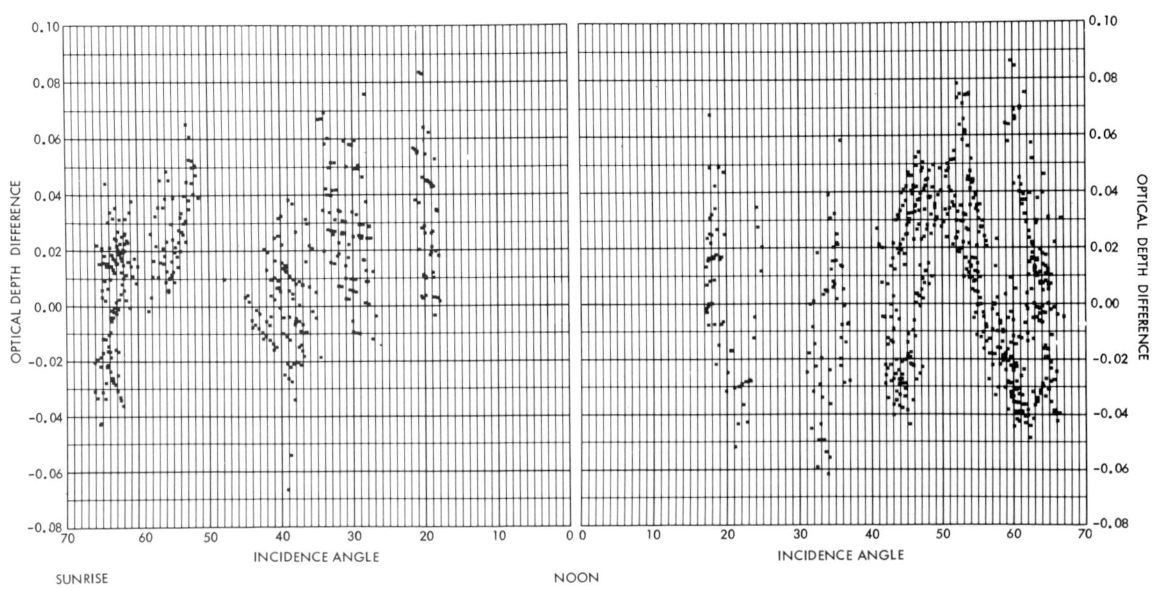

Fig. 16. Optical depth difference versus incidence; clear filter, VL-1 site.

early afternoon, a result suggesting increasing aerosols with increasing daily temperature. In the Smooth Plains region most of the Mariner 9 difference occurs near sunset ($\Delta\tau$ = 0.006, i = 80°). This may also be true in the Old Terrain; however, a larger data scatter (0.05 rms) exists about $\Delta\tau$ = 0 throughout the day.

CONCLUSIONS

Viking observations combined with lander 1 images obtained during the primary mission have provided a data base for subsequent seasonal comparisons in the extended mission. These data should allow further modeling of the aerosol production versus time of day and the separation of dust versus condensate reflectances. Additional correlations with lander measurements should permit seasonal opacity mapping and application to other regions. Identification of typical atmospheric trends has helped to separate surface versus isolated atmospheric phenomena and should also benefit VO Mars atmospheric water detector and infrared radiometer thermal mapping data interpretation. More detailed comparisons, however, await improved lander surface photometry.

Acknowledgment. This paper presents the results of one phase of research carried out at the Jet Propulsion Laboratory, California Institute of Technology, under contract NAS 7-100, sponsored by the National Aeronautics and Space Administration.

REFERENCES

Adams, J. B., and T. B. McCord, Mars: Interpretation of spectral reflectivity of light and dark regions, *J. Geophys. Res., 74*(20), 4851, 1969.

Klaasen, K., L. Morabito, and T. E. Thorpe, Inflight performance of the Viking visual imaging subsystem, *Appl. Opt., 16,* in press, 1977.

Mutch, T. A., et al., The surface of Mars: The view from the Viking 1 lander, *Science, 193*(4255), 799, 1976*a*.

Mutch, T. A., et al., Fine particles on Mars: Observations with the Viking 1 lander cameras, *Science, 194*(4260), 89, 1976*b*.

Thorpe, T. E., Mariner 9 photometric observations of Mars from November 1971 through March 1972, *Icarus, 20,* 482, 1973.

(Received April 1, 1977;
revised May 25, 1977;
accepted May 25, 1977.)

Viking Orbiter Photometric Observations of the Mars Phase Function July Through November 1976

THOMAS E. THORPE

Jet Propulsion Laboratory, Pasadena, California 91103

Over 7200 Viking Orbiter pictures have provided phase function information over a large range in viewing geometry. Comparison with the earlier Mariner 9 data reveals possibly significant changes. A two-component limb darkening characterization is shown to fit the data better at large phase angles than the traditional Minnaert or Lommel-Seeliger approach. The phase integral is 15% larger than the Mariner 9 observations owing in part to data obtained at larger phase angles revealing apparent condensate phenomena.

INTRODUCTION

The Viking Orbiter (VO) television experiment has provided over 9000 images of Mars during a period (July–November 1976) comparable to that of the earlier Mariner 9 reconnaissance (November–March 1972). In several ways, however, the Viking data set differs from the Mariner 9 observations.

1. Pictures were taken during the Martian northern summer (L_s = 83°–157°) versus the Mariner record of late winter (L_s = 330°–0°) 2 years prior following a global dust storm (L_s is the aerocentric solar and longitude measured from the equatorial ascending node).

2. The Viking images are mainly confined to areas of uniform albedo (Chryse Planitia, Utopia Planitia) at a wide range of viewing geometries (5°–135° phase).

3. The VO cameras provided improved photometric operation at several wavelengths.

4. This analysis utilized only the central 200 × 200 element picture area (~8 × 8 km surface footprint) of 7200 frames versus the full frame modal values of Mariner 9 reported earlier [*Thorpe,* 1973].

Significant optical parameters of the Viking Orbiter cameras are described elsewhere [*Thorpe,* 1976]. In-flight performance checks [*Klaasen et al.,* 1977] have produced minor updates to these parameters, so that reflectances may be calculated as follows:

$$R = (DNP * N * FF)/(A * E * S) \qquad (1)$$

where the notation is defined as follows:

- *DNP* data number (camera response to Mars) above the light flood residual for the high-gain state [see *Thorpe,* 1976];
- *N* ratio of sunlight to calibration source energy within the camera's filter band pass;
- *FF* filter factor for sunlight (camera response without filter divided by camera response with filter);
- *A* sensitivity of camera to calibration source irradiance, e.g., *DNP*/(ergs/cm²);
- *E* shutter speed, s;
- *S* solar illuminance at Mars for the date of observation, ergs/cm² s.

For example, on revolution 118, picture A43, recorded by Viking Orbiter 1 camera A, had a mean data number of 50 (violet filter; 0.05-s exposure; pedestal = −1*DN*). The corresponding reflectance is

$$R = (51 * 0.882 * 5.381)/(0.588 * 0.05 * 80100.) = 0.103$$

Paper number 7S0456.

PHASE FUNCTION

The observed reflectance computed from the mean *DNP* of the center 200 × 200 elements of each picture using (1) is plotted versus phase angle in Figures 1*a*–1*c* for three filter positions: violet (0.443 μ), clear (0.532 μ), and red (0.592 μ). The families of curves obtained at a variety of luminance longitudes α (emission angle projected into the phase plane) illustrate the range of viewing geometries represented by the Viking data. Figure 2 illustrates the data scatter at α = 10° with the clear filter and α = 30° with the violet filter. At large negative luminance longitudes and large phase angles (incidence and emission angles on opposite sides of the surface normal), increased reflectance and data scatter are observed at all wavelengths. The data scatter suggests variations in atmospheric opacity. In violet at these geometries (e.g., α = −30°; phase > 100) the atmospheric forward scattering characterisitc is especially prominent (Figure 2*b*). At low phase angle (left portions of the curves shown in Figures 1*a*–1*c*), reflectances obtained primarily in the Elysium Planitia region tend to suggest a backscatter peak (especially in red) nearly as pronounced as is observed for the moon. Figure 3 shows additional lunar similarities at low phase angle: bright crater rims and rays.

The Mariner 9 post-dust storm observations were confined to a single filter (0.558 Å) between 30° and 80° phase. If a mean fit to the earlier observations [*Thorpe,* 1973] is subtracted from the clear filter values given in Figure 1*a*, a mean reflectance difference of only 0.02 (10%) is observed at moderate α values (Figure 4).

Significant differences, however, would occur at both large and small phase angles if the Mariner data were extrapolated to larger and smaller phase angles. These data may simply indicate the error in extrapolation of the Mariner 9 observations or real atmosphere-surface changes. Figure 5 gives the full disk-integrated functions for the three wavelengths using 16-point cubature. A more pronounced backscatter peak, especially in the red filter, together with increased brightness at large phase angles, is again indicated when a comparison is made with the prior data. These curves lead to the following phase integral values:

Clear filter

$$q = 2\int_0^{\pi} \phi \sin g \, dg = 1.19 \pm 0.1 \qquad (2a)$$

Violet filter

$$q = 2\int_0^{\pi} \phi \sin g \, dg = 1.30 \pm 0.1 \qquad (2b)$$

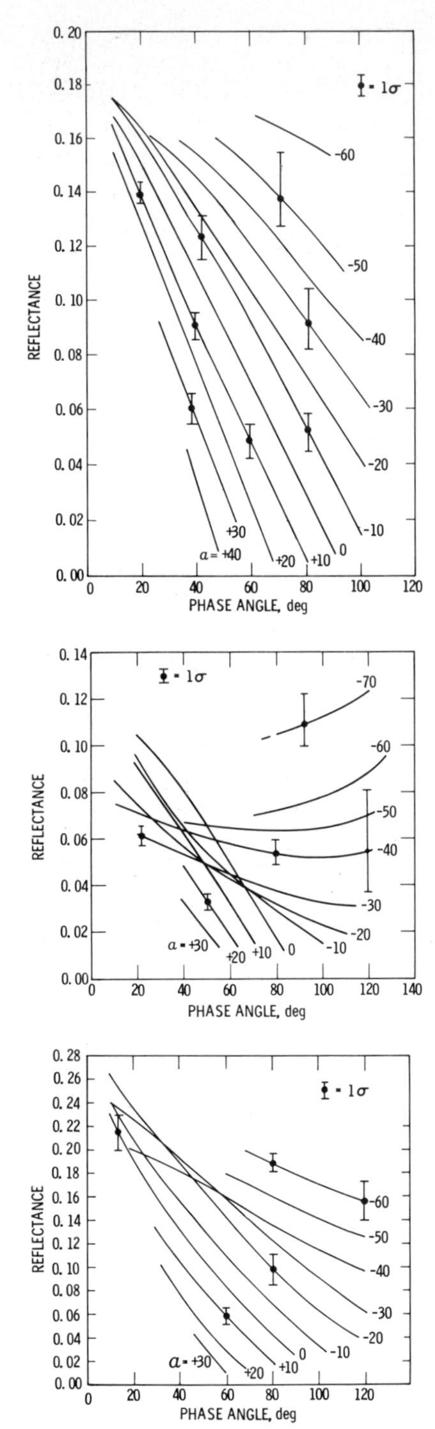

Fig. 1. Reflectance versus phase angle (α is luminous longitude) for (top) clear filter, (middle) violet filter, and (bottom) red filter.

Red filter

$$q = 2 \int_0^{\pi} \phi \sin g \, dg = 1.22 \pm 0.1 \qquad (2c)$$

By means of the albedos of *Adams and McCord* [1969] for the appropriate effective wavelengths, Bond albedos of 0.25, 0.16, and 0.31 are calculated for the three filters, respectively. These values are 15% higher than both terrestrial [*de Vaucouleurs,* 1964] and Mariner 9 observations and indicate the importance of large phase angle data, which appear to be more susceptible to local atmospheric effects.

LIMB DARKENING

Reflectance at points of equal separation across the Martian disk ($\sin \alpha$) is plotted in Figures 6*a* and 6*b*. At moderate phase angles the location of maximum brightness occurs close to the subsolar point (arrows), suggesting Lambertian behavior and limb darkening. At large phase, however, limb brightening is evident, especially in violet, emphasizing a forward scattering brightness contribution. Figure 7 shows this difference in a limb profile across two VO-1 approach pictures taken at 90° phase (shutter speed twice as long in violet filter as in red). The atmospheric origin of this scattering is indicated by the larger data scatter encountered in looking toward the limb (subsolar point) over regions of uniform surface albedo and topography (Figure 8). A typical Minnaert plot of log ($B \cos e$) versus log ($\cos i \cos e$) reveals two distinct populations of data, corresponding to different values of slope (Figures 9*a* and 9*b*). At small incidence and emission angles ($\cos i \cos e > 0.4$) the reflectance variation is consistent with strong backscatter

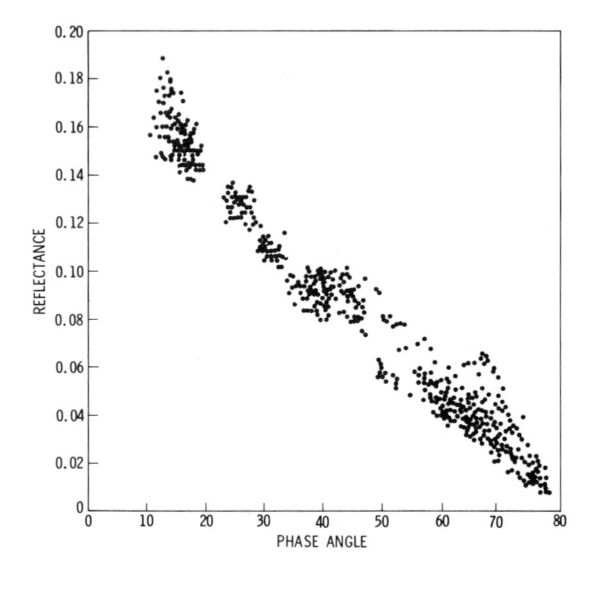

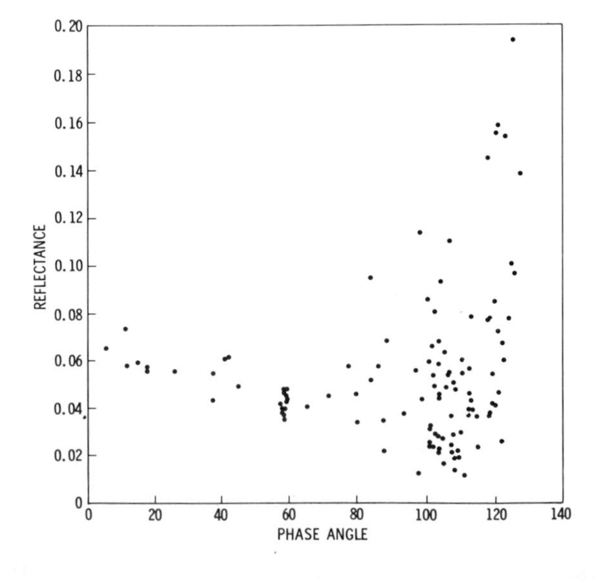

Fig. 2. Phase function (dots are Viking observed reflectances) for (top) clear filter, α = 10°, and (bottom) violet filter, α = 30°.

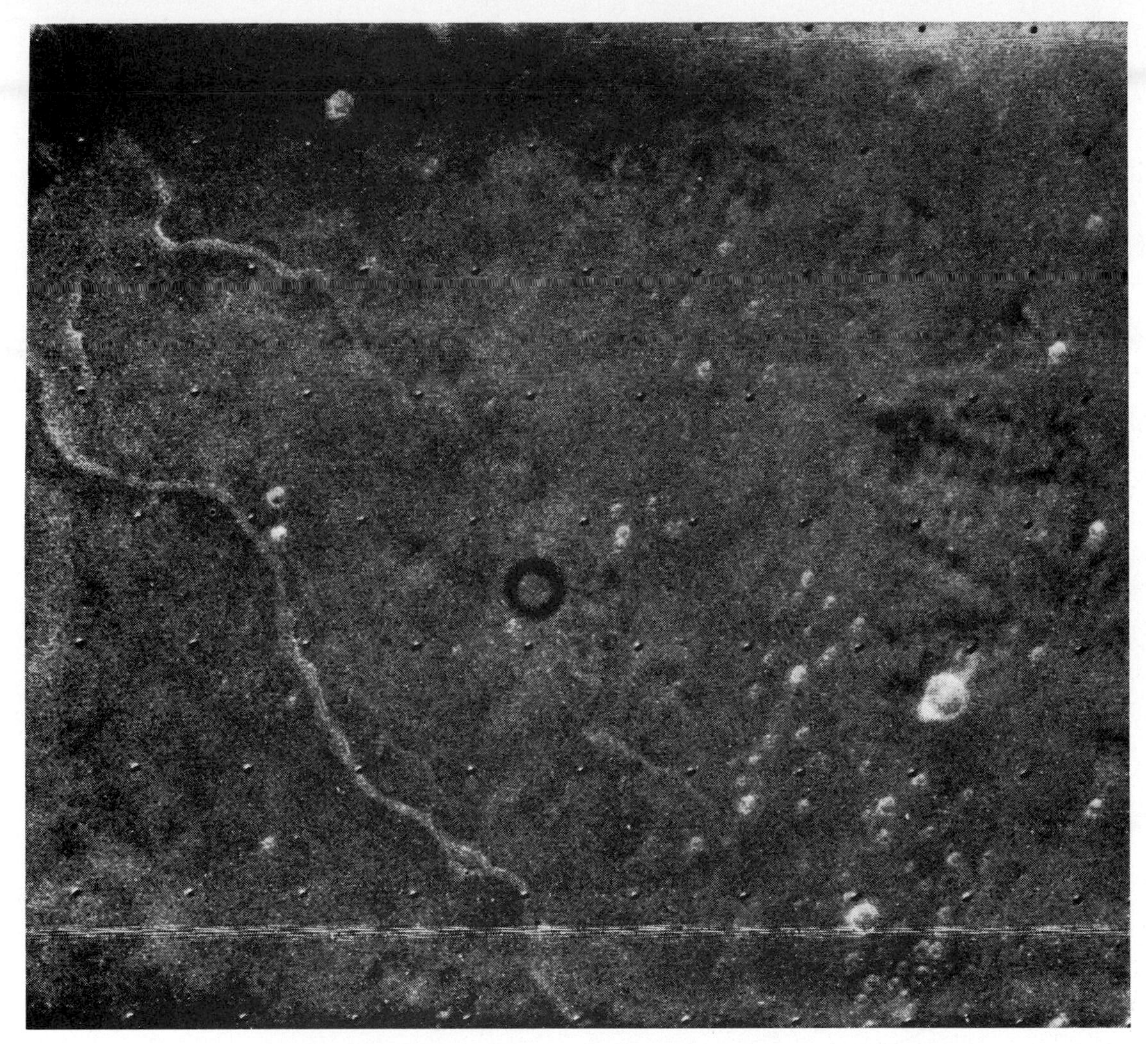

Fig. 3. The 5° phase observation: red filter (Elysium Planitia).

(slope exceeds 1.0). At laıge incidence angles, corresponding to $i > 70°$, the signal is well represented by small values of K (slope), consistent with limb brightening. Attempts to model this behavior by using Lommel-Seeliger and Minnaert functions are shown in Figure 10, which illustrates the inability of a normalized Lommel-Seeliger (L-S) function to characterize the observed behavior fully, i.e., where

$$B = \frac{2\mu_0 B_0}{\mu_0 + u}\left(\frac{[1 - \exp(-\tau M)]f(g)}{8} + \phi(g)\exp(-\tau M)\right) \quad (3)$$

and where

- μ_0, μ cosines of incidence and emission angles ($\cos i$, $\cos e$);
- B_0 normalization constant;
- M air mass, equal to $\mu_0^{-1} + \mu^{-1}$;
- τ optical depth;
- $\phi(g)$ normalized surface phase function ($\phi(0) = 1$ at g = phase = 0°);
- $f(g)$ normalized atmospheric phase function.

The L-S function exhibits a decidedly nonlinear behavior not displayed by the data ($f(g)$ and $\phi(g)$ do not affect the shape of the curve at a constant phase angle). The plots can be fitted by two Minnaert functions, the values for K and B_0 being given in Figures 11 and 12. The 60° phase plot suggests limb darkening (slope > 1) from atmospheric constituents observed at high sun (large values of $\cos i \cos e$) and limb brightening from phenomena seen near the terminator of $K_1 = 1.2$ for $i < 70°$ and $K_2 = 0.6$ for $i > 70°$. The linearity of the K_2 population up to 88° incidence angle would require increasing optical depth near the terminator with a Lommel-Seeliger representation which would be consistent with condensate formation at low sun. Where Mars is observed at large phase angles (>90°), e.g., approach images, this scattering source dominates the limb profile; i.e., a terminator near the limb produces limb brightening. At low phase angles, however, backscattering produced perhaps by larger-diameter aerosols (or surface porosity), especially at high sun, results in limb darkening. The Minnaert characterization is then

$$B = B_0 \mu_0^K \mu^{K-1} \quad (4)$$

where

$$B_0 = B_1 \quad K = K_1 \quad i < 70°$$
$$B_0 = B_2 \quad K = K_2 \quad i > 70°$$

which avoids a singularity at the limb for a large range of sun

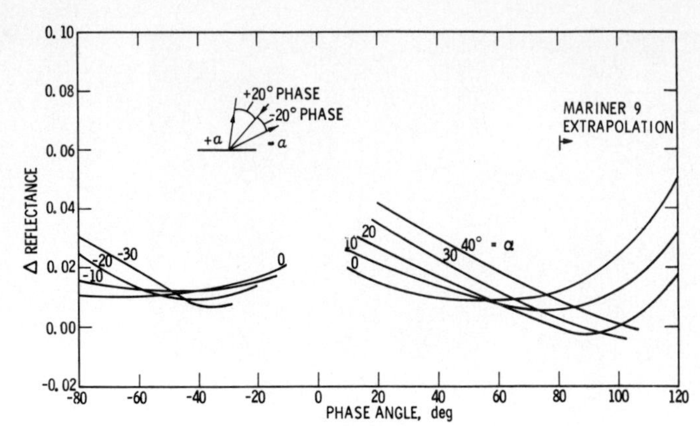

Fig. 4. Mean reflectance difference between Viking and Mariner 9: clear filter.

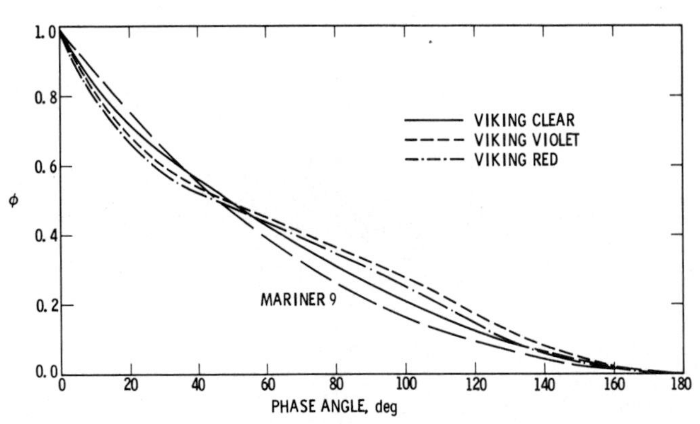

Fig. 5. Integrated phase function.

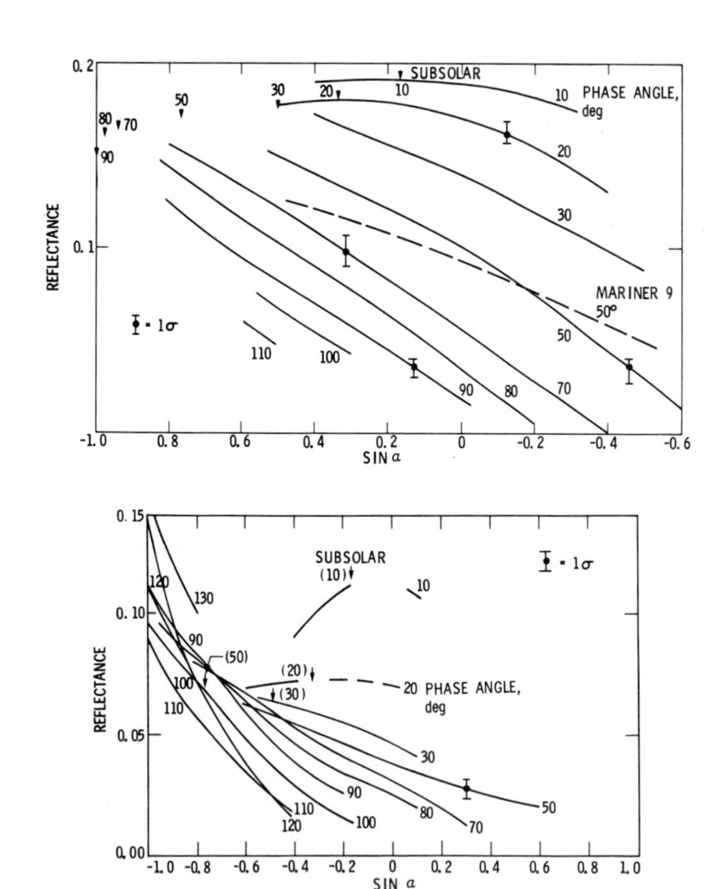

Fig. 6. Limb darkening (subsolar points are indicated by arrows) for (top) clear filter and (bottom) violet filter.

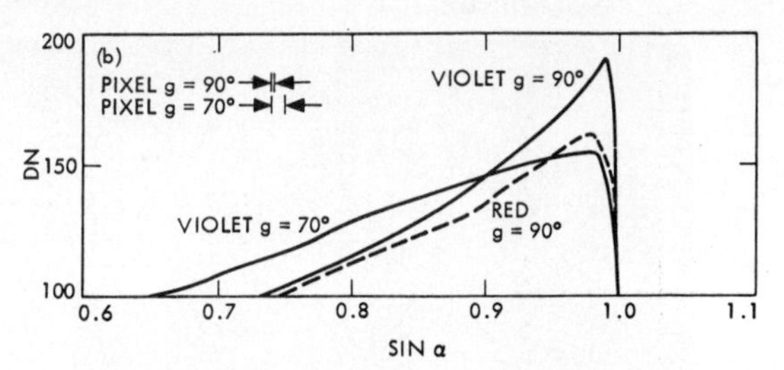

Fig. 7. Limb profile: VO-1 approach images.

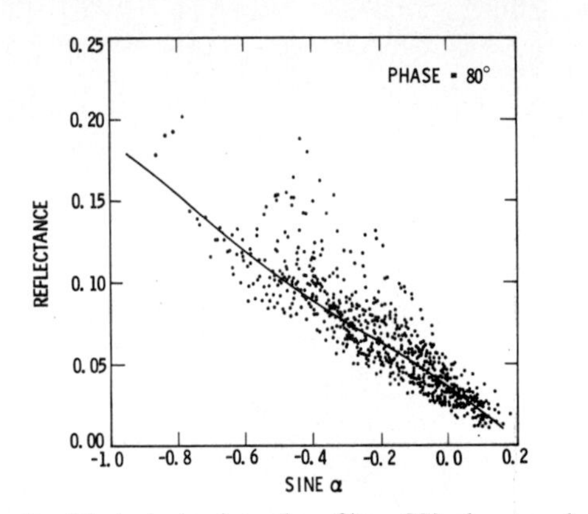

Fig. 8. Limb darkening: clear filter, 80° phase angle.

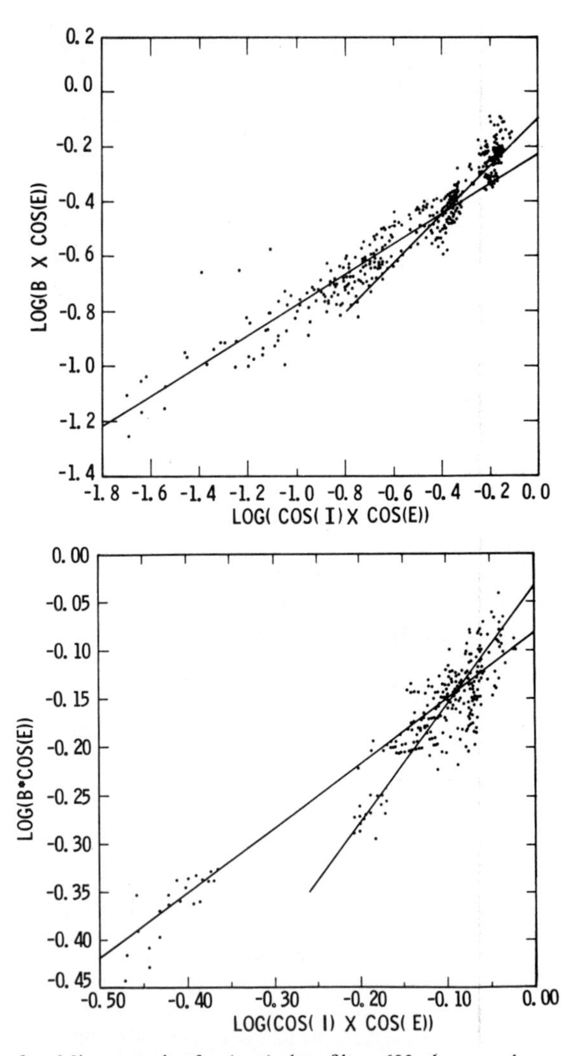

Fig. 9. Minnaert plot for (top) clear filter, 60° phase angle, and VL-1 site and (bottom) clear filter, 30° phase angle, and VL-1 site.

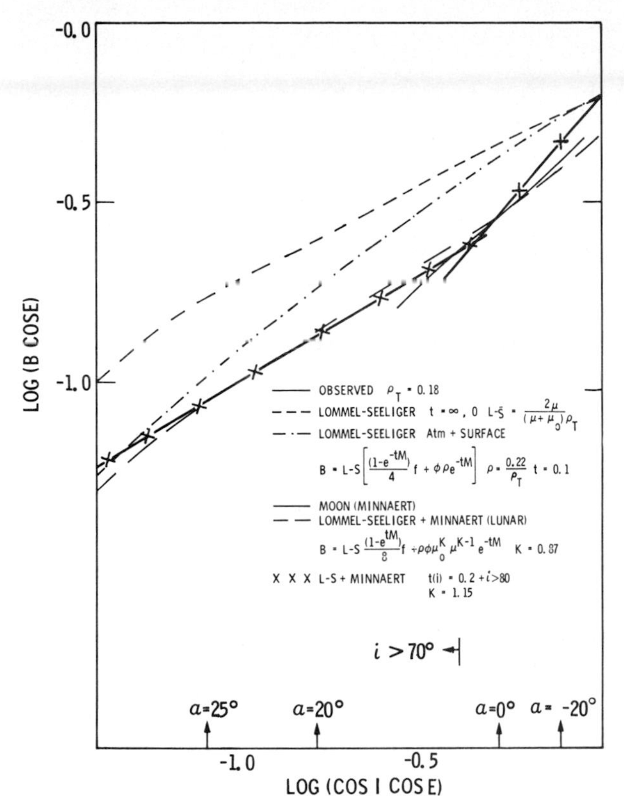

Fig. 10. Observation versus model fit at 60° phase angle, clear filter.

elevations for $K_1(>1)$ given in Figures 11 and 12. Although several brightness populations (B_2) at large incidence angles with similar K have been recorded, at moderate phase and incidence angles, roughly 14% of the total light reflected is from the B_2 populations. Assuming that this is all due to atmospheric phenomena leads to an L-S optical depth of τ = 0.12, consistent with observations made at large phase angles.

A satisfactory fit to the data in Figure 9 may also be obtained by a combination of L-S and Minnaert functions. In this case the latter function is similar to the lunar surface phase characterization [*Kuiper and Middlehurst,* 1961] with somewhat larger values of K, a finding which agrees with preliminary Viking lander results [*Thorpe,* 1977]. When the values for ϕ(moon) and $K = K_1$ and a decreasing optical depth with increasing terminator distance are used, the 60° phase data fit is given by

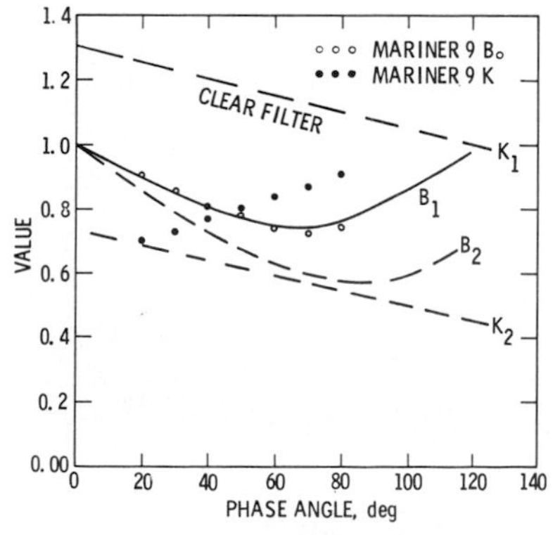

Fig. 11. Two-component Minnaert parameters: clear filter.

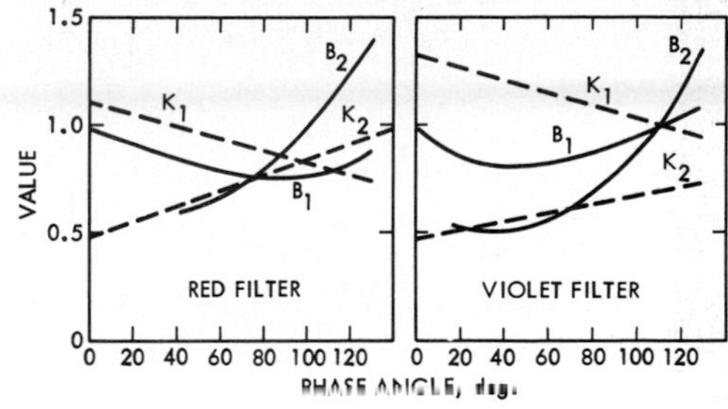

Fig. 12. Two-component Minnaert parameters: red and violet filters.

$$r = \frac{\mu_0}{\mu + \mu_0} \frac{1 - \exp(-\tau M)}{4} f(g) + \rho\phi(g)\mu_0{}^K\mu^{K-1} \exp(-\tau M) \qquad (5)$$

where

$f(g)$ defined as by *Thorpe* [1977];
ρ surface albedo, equal to 0.22 for clear filter;
$\tau(i) = (i - 1)_{\text{rad}}$ for $i > 63°$;
$\tau \,|= 0.1$ for $i < 63°$;
$B(\text{Fig. 9}) = r/r_0$, where $r_0 = 0.18$ (0° phase reflectance).

CONCLUSIONS

The phase function information obtained by the Viking Orbiter imaging experiment over a large range of viewing geometries suggests that the earlier Mariner 9 data cannot be directly extrapolated to large and small phase angles. A brightness increase is observed in both cases which may be attributable to the areas chosen for this analysis (desert). The limb darkening at low phase angles and the limb brightening at large phase angles may, however, be a general atmospheric and possibly seasonal effect. These data are well represented by a two-component, two-phase dependent Minnaert function, which for all its physically relatable shortcomings is an improvement over the typical Lommel-Seeliger approaches, especially at large phase angles. Regional differences and secular changes will continue to be investigated during the Viking extended mission.

REFERENCES

Adams, J. B., and T. B. McCord, Mars: Interpretation of spectral reflectivity of light and dark regions, *J. Geophys. Res., 74,* 4851, 1969.

de Vaucouleurs, G., Geometric and photometric parameters of the terrestrial planets, *Icarus, 3,* 187–235, 1964.

Klaasen, K., T. Thorpe, and L. Morabito, In flight performance of the Viking visual imaging subsystem, *Appl. Opt., 16,* 1977.

Kuiper, G., and B. Middlehurst, Planets and satellites, in *The Solar System,* vol. 3, p. 322, University of Chicago Press, Chicago, Ill., 1961.

Thorpe, T. E., Mariner 9 photometric observations of Mars from November 1971 through March 1972, *Icarus, 20,* 482, 1973.

Thorpe, T. E., The Viking Orbiter cameras' potential for photometric measurement, *Icarus, 27,* 229, 1976.

Thorpe, T. E., Viking Orbiter observations of atmospheric opacity during July–November 1976, *J. Geophys. Res., 82,* this issue, 1977.

(Received April 1, 1977;
revised May 25, 1977;
accepted May 25, 1977.)

A Study of Variable Features on Mars During the Viking Primary Mission

J. VEVERKA AND P. THOMAS

Laboratory for Planetary Studies, Cornell University, Ithaca, New York 14853

RONALD GREELEY

University of Santa Clara, Santa Clara, California 95053

Very few surface changes were seen during the Viking primary mission in 1976, a result consistent with predictions of relatively low wind velocities during northern summer. No eolian activity was detected from orbit in the vicinity of either landing site. This lack of eolian activity was representative of Mars as a whole at this season with two exceptions: (1) Persistent eolian activity was observed on the flanks of the three large Tharsis volcanos (Arsia Mons, Pavonis Mons, and Ascraeus Mons) but most markedly on Arsia Mons. The observed changes are best explained in terms of the erosion of bright albedo material by downslope winds. (2) The formation of a number of dark crater-associated streaks by E to W winds was documented on the eastern edge of Hellas. Comparison of specific albedo boundaries in the 1972 Mariner 9 coverage and in the 1976 Viking coverage revealed that in many cases, subtle changes in outline and/or contrast have occurred during the past 4 years. In a few areas (Syria Planum, for example) the albedo patterns in 1976 are dramatically different from those in 1972. As in 1972, the most conspicuous wind markers on the planet in 1976 are light crater-associated streaks whose pattern still defines the wind flow expected during southern summer. While some new light streaks have formed since 1972 and a few old ones have disappeared, by and large, most light streaks appear essentially unchanged in outline and in direction since 1972. The Viking observations confirm that the mean lifetime of dark streaks is short in comparison to that of light streaks. Many dark streaks have changed conspicuously in both outline and direction since 1972. It appears that contemporary albedo variations on the surface of Mars are due solely to eolian effects or to the deposition or sublimation of volatiles in the polar regions. The combined Mariner 9 and Viking observations contain no convincing evidence to support the once traditional concept of a seasonal wave of darkening.

1. INTRODUCTION

Since the Mariner 9 mission in 1971-1972 it has been known that a variety of albedo markings which can be attributed to eolian activity exist on the surface of Mars [*Sagan et al.*, 1972, 1973]. None of these markings is permanent on geological time scales; many are now known to have lifetimes ranging from several days to a few Martian years. Collectively, they have been labeled variable features [*Sagan et al.*, 1972]. The most important of the variable features are the crater-associated wind streaks. These are abundant in the planet's equatorial zone and serve as wind markers to map out the wind regimes which determine the patterns of eolian redistribution of surface materials.

The crater-associated streaks can be classified into two major categories, light and dark, on the basis of their albedo relative to their surroundings. These two types appear to be fundamentally different in morphology, mean lifetime, and origin [*Veverka et al.*, 1977*a*, *b*]. The light streaks have been interpreted as accumulations, in the lee of craters, of high-albedo fallout from the major global dust storms which usually occur during southern summer [*Arvidson*, 1974; *Gierasch*, 1974; *Veverka et al.*, 1977*a*]. Mariner 9 data indicate that the mean lifetime of a light streak may be at least 1 Martian year, since no changes in light streaks were documented during the Mariner 9 mission [*Sagan et al.*, 1973]. However, *Veverka et al.* [1974] found evidence that some light streaks in Sinus Meridiani had changed in the interval between the Mariner 6 and 7 coverage in 1969 and the Mariner 9 coverage in 1971-1972. With a few exceptions [cf. *Veverka et al.*, 1976], dark streaks have been interpreted as erosion scars: regions from which a thin veneer of high-albedo material has been removed, a darker underlying surface [*Veverka*, 1975] thus being revealed. Statistically, dark streaks differ in morphology from light streaks [*Veverka et al.*, 1977*a*, *b*] and tend to have short lifetimes [*Veverka*, 1975]. The above suggestions for different origins for light and dark streaks are consistent with the results of wind tunnel simulations by *Greeley et al.* [1974*a*, *b*].

Mariner 9 observed Mars during summer and fall in the southern hemisphere and witnessed the waning stages of a major global dust storm. Eolian activity was high at this time, and rapid variations in dark streaks and other more diffuse dark albedo markings were observed [*Sagan et al.*, 1972, 1973, 1974; *Veverka*, 1975]. As was already noted, no variations in light streaks were detected during the Mariner 9 mission. At the time of Mariner 9 the light streaks defined a coherent wind flow pattern in the equatorial zone of Mars (Figure 1), indicating a near-surface flow from the northern to the southern hemisphere. This flow had a characteristic curved pattern veering from a NE to SW trend north of the equator to a NW to SE trend south of the equator, a wind flow pattern expected during southern summer, the season of strongest winds on Mars [*Gierasch*, 1974]. At other seasons, surface winds are predicted to be too weak to alter the pattern of light streaks laid down during global dust storms [*Pollack et al.*, 1976; *Gierasch*, 1974].

Except in the latitude belt 30°-40°S the wind flow pattern recorded by dark streaks in 1972 was more erratic (Figure 1); near 30°S the dark streaks recorded a consistent E to W flow [*Veverka et al.*, 1977*b*].

Since the Mariner 9 systematic coverage extended over only half a Martian year, only inferences could be made about the nature and extent of eolian activity during the other half of the year (summer and fall in the north). The expectation was that

Paper number 7S0571.

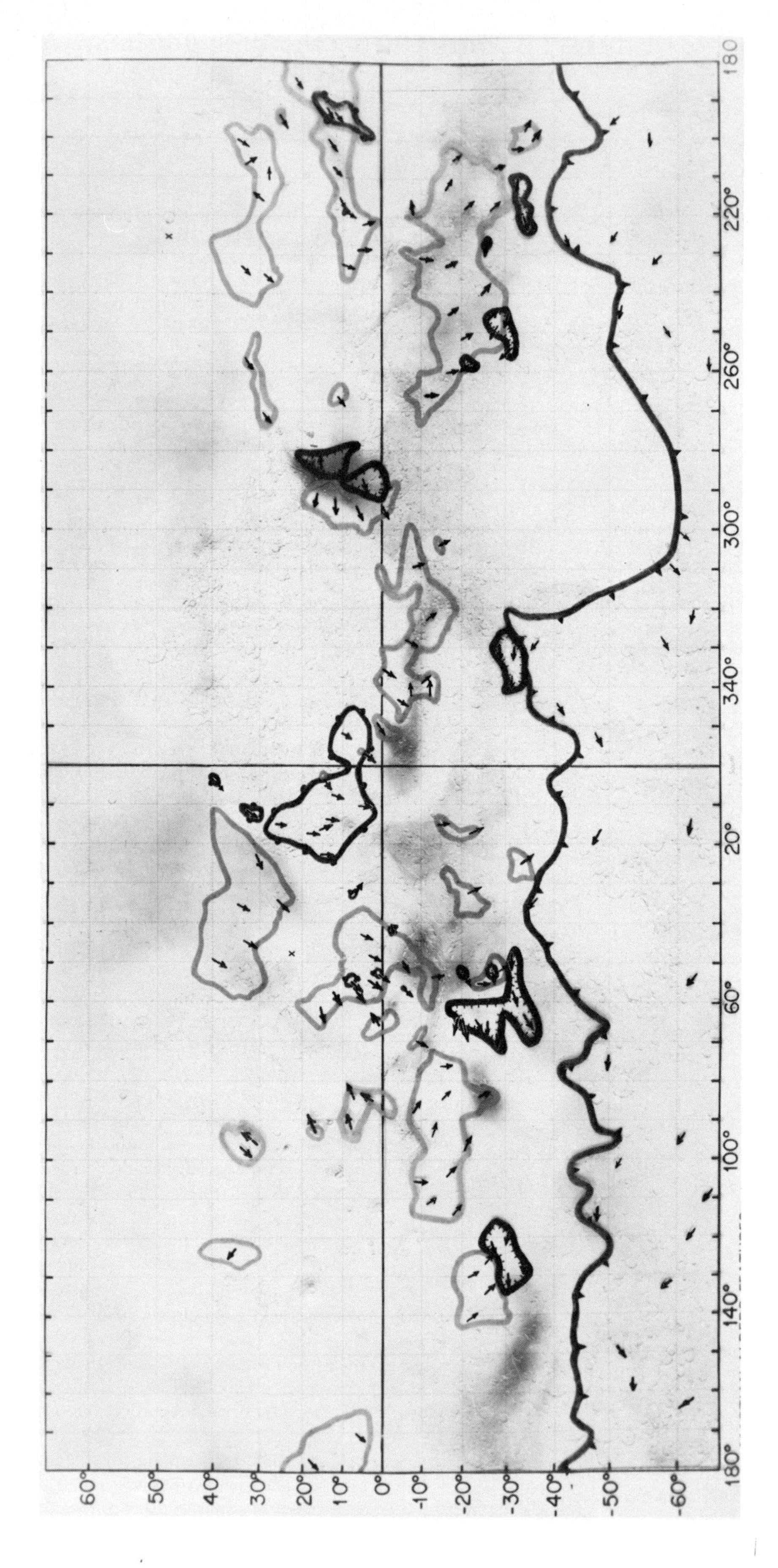

Fig. 1. Sketch map of streak directions in 1971–1972, based on Mariner 9 data. Only the major concentrations of streaks have been mapped. The grey boundaries outline concentrations of light streaks; the black boundaries outline concentrations of various dark streaks. The crosses indicate the two Viking landing sites at 22°N, 48°W and 48°N, 226°W.

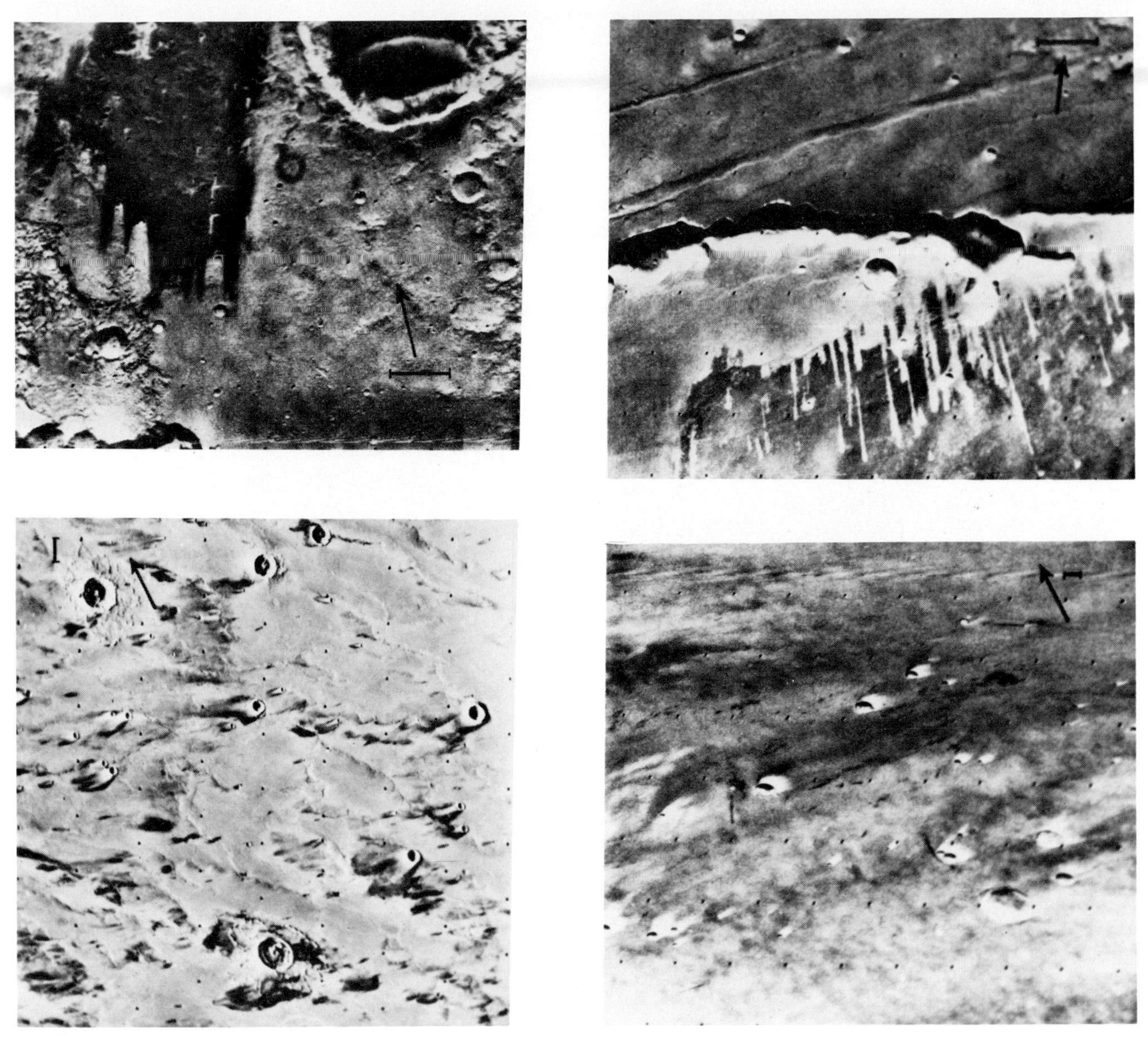

Fig. 2. A sample of some of the more unusual wind streaks in the Viking coverage. (Top left) At 6°N, 28°W on September 12, 1976; L_s = 121° (083A12; NGF/B). (Top right) At 3°N, 90°W on August 22, 1976; L_s = 112° (062A71; NGF/B). (Bottom left) At 13°S, 139°W on July 31, 1976; L_s = 102° (041B51; NGF/B). (Bottom right) At 5°N, 86°W on July 2, 1976; L_s = 88° (013A31; NGF/B). In all figures the scale bar represents 10 km, and the arrow points north.

during the Viking primary mission, there should be little if any eolian activity on Mars. (The Viking primary mission extended from June to November 1976, corresponding to the last days of northern spring through late summer. The range of aerocentric longitudes of the sun (L_s) covered is 87°–165°.) Wind markers were expected to be common on the surface, but winds were predicted to be too weak to produce variations in the albedo patterns [*Pollack et al.,* 1976]. The observations made by the two Viking orbiters have confirmed this prediction and have shed light on some other related questions: (1) How much had major albedo boundaries changed during the 4-year interval between 1972 and 1976? (2) How much had small-scale albedo boundaries changed?

This paper contains a summary of the answers to these and related questions obtained from a study of the images obtained by the two Viking orbiters during the primary mission. With some minor exceptions (section 10) the Viking observations generally confirm the major predictions made on the basis of the Mariner 9 data.

2. VIKING OBSERVATIONS OF WIND STREAKS

As was expected, the Viking coverage revealed numerous wind streaks on the surface of Mars. A few of the more interesting examples are shown in Figure 2. Since in many cases the effective resolution of the Viking images is superior to that of the Mariner 9 images (better cameras and clearer atmosphere), important new information about the morphology of streaks and of their associated craters has been obtained. This information will prove useful in refining various models of streak formation. For instance, the Viking images have provided the first good examples of very distinctive mixed tone streaks which appear to consist of a tapered light streak

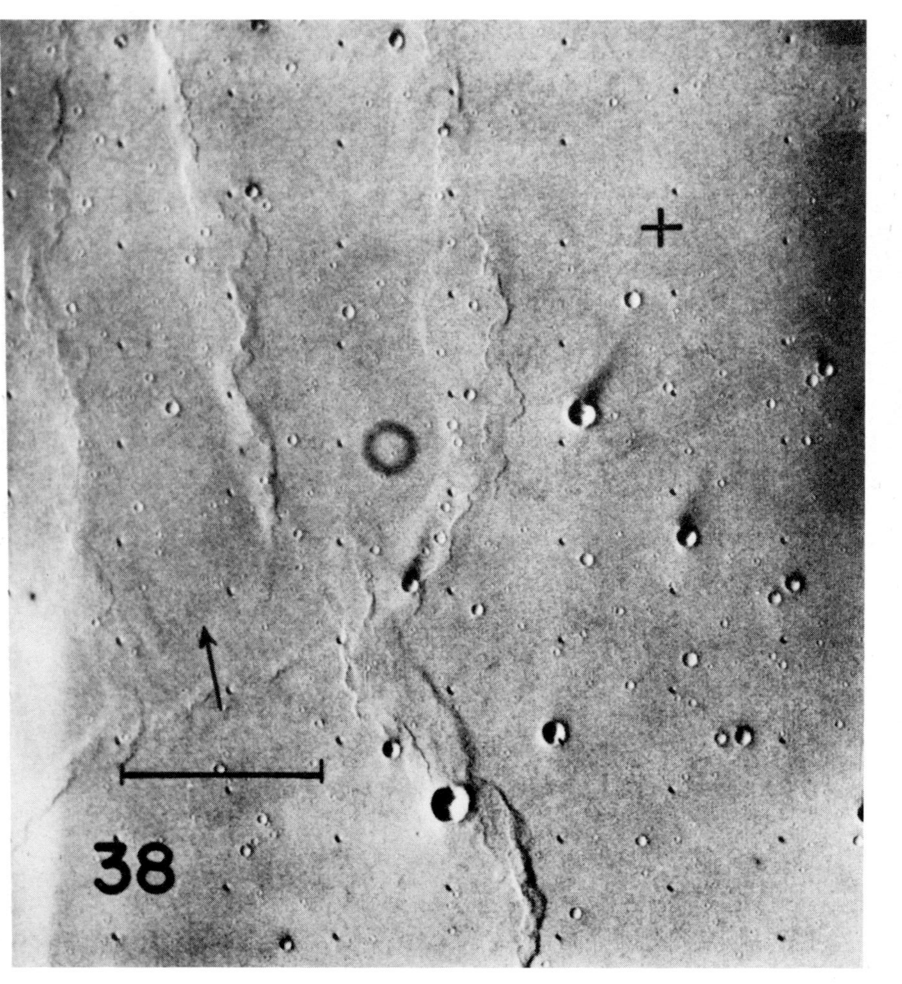

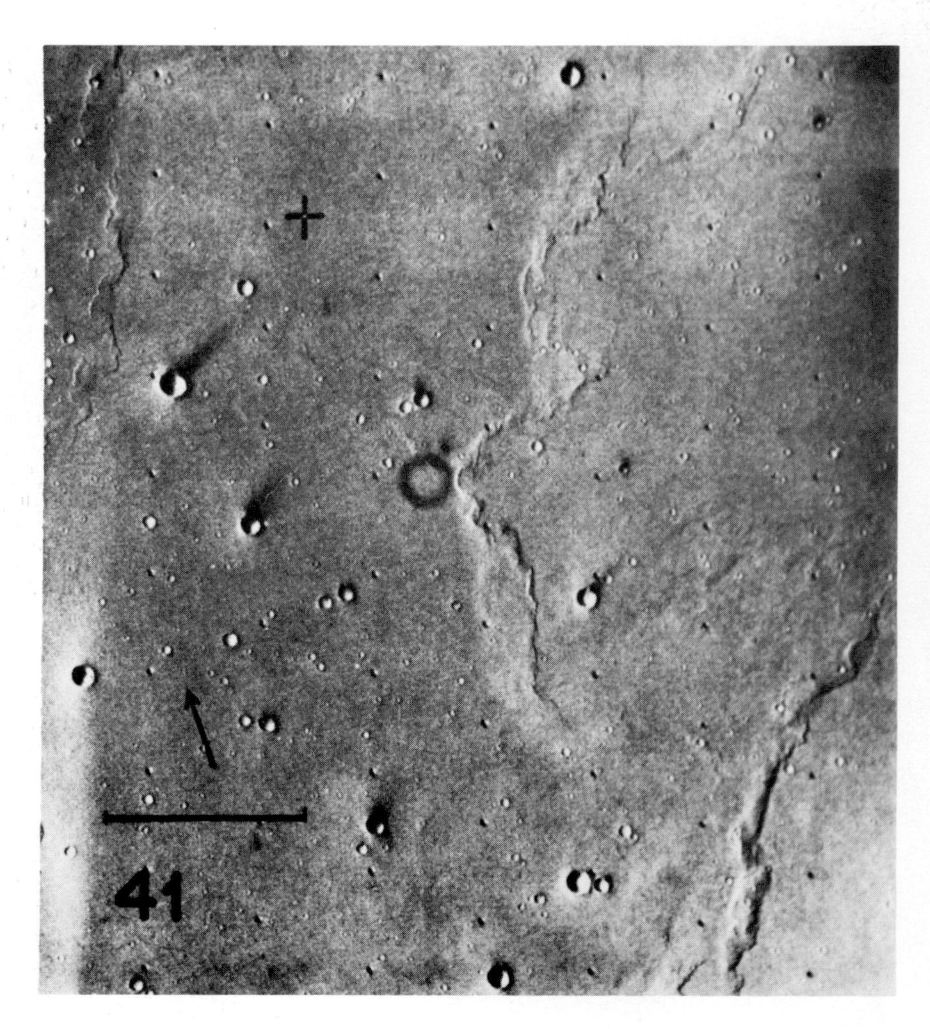

Fig. 3. Two views of the VL-1 landing site (cross) 4 days apart. No changes in the surface albedo patterns have occurred. (Left) At 22°N, 47°W on July 28, 1976; L_s = 100° (038A02; NGF/B). (Right) At 22°N, 47°W on July 31, 1976; L_s = 102° (041A91; NGF/B).

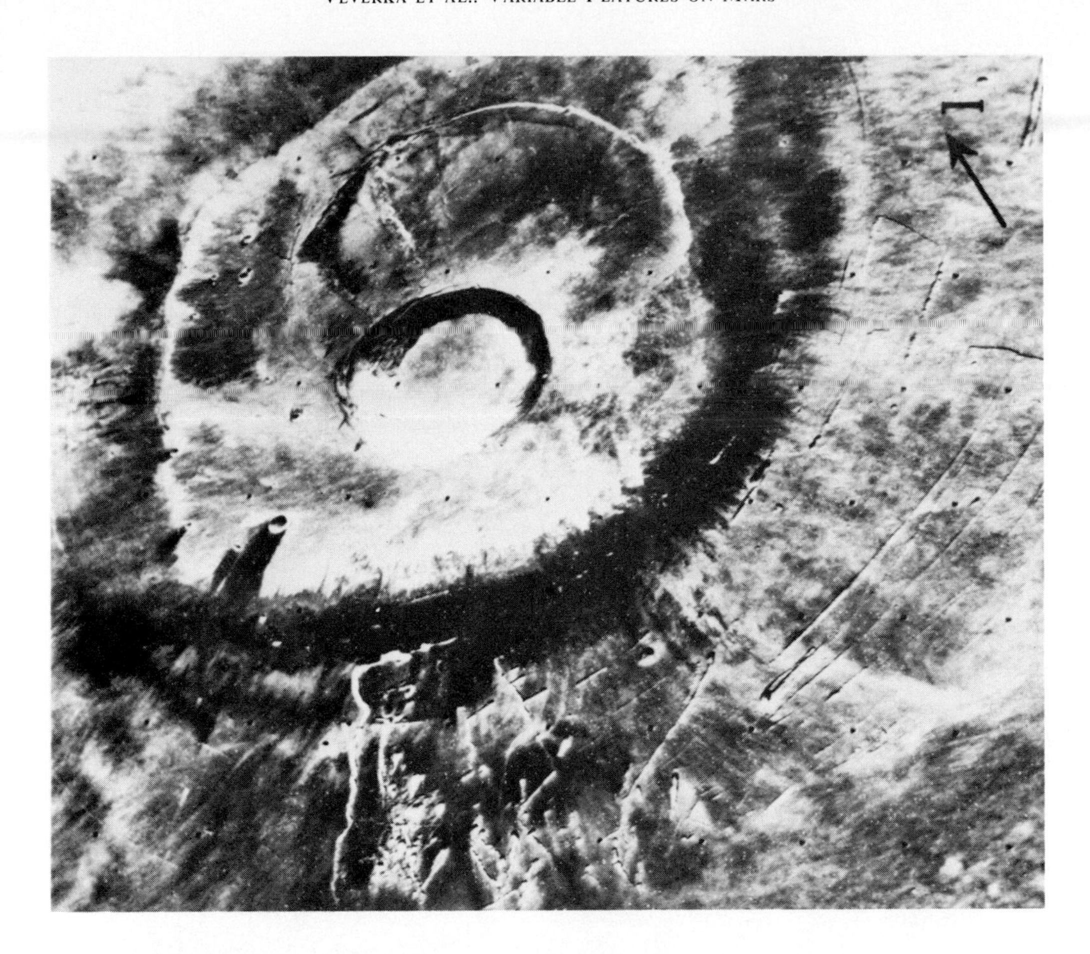

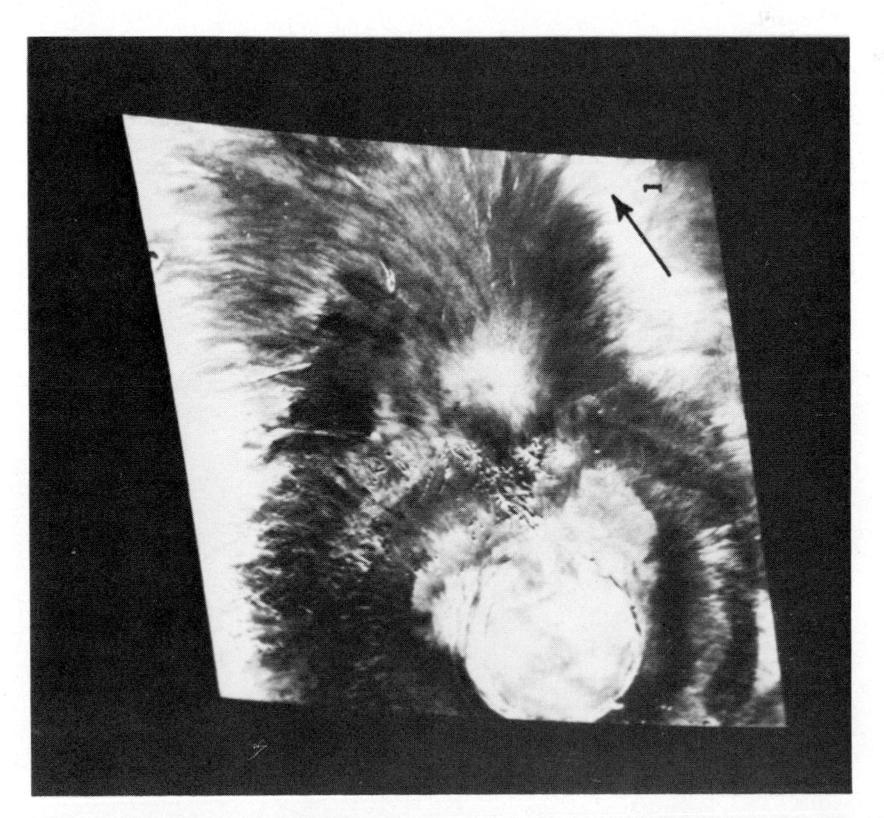

Fig. 4. Ringlike dark albedo patterns on the flanks of (top) Pavonis Mons at 0°N, 112°W on August 12, 1976, for L_s = 107° (052A15; NGF/B) and (bottom) Arsia Mons at 6°S, 119°W on July 31, 1976, for L_s = 102° (041A56; NGF/B).

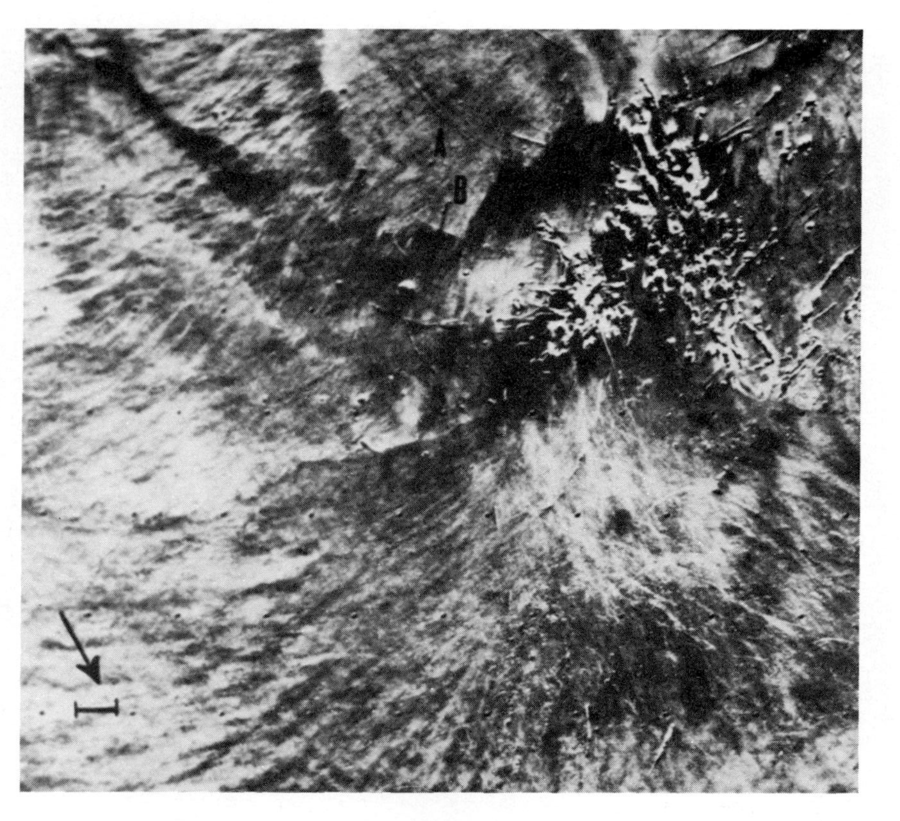

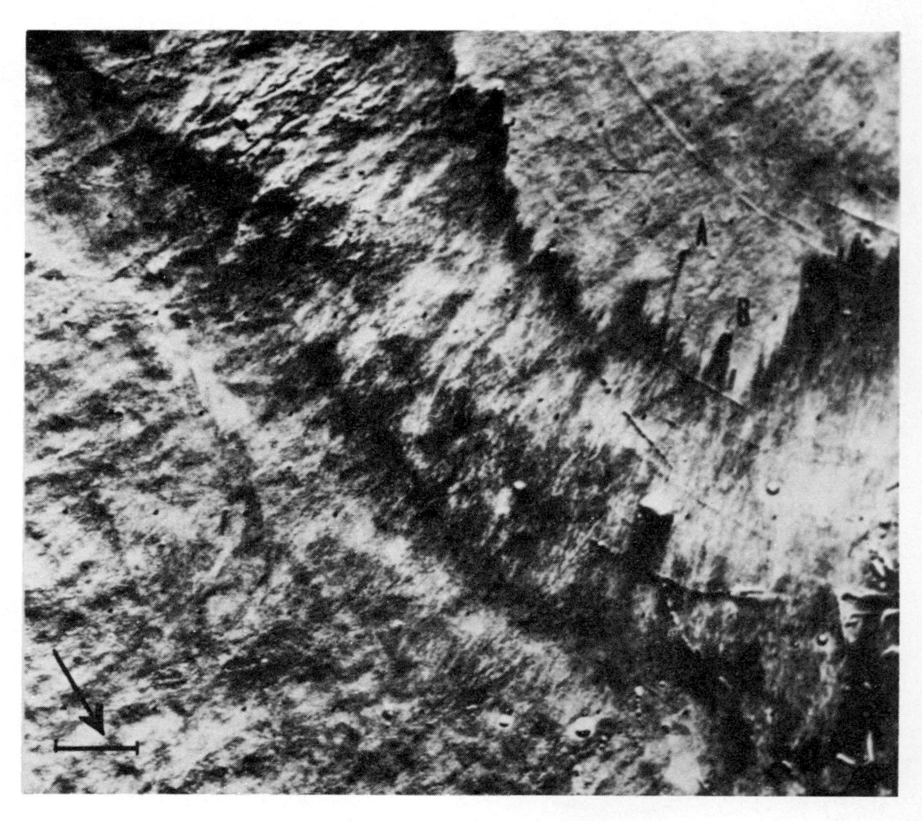

Fig. 5. Changes on the flanks of Arsia Mons. (Left) At 7°S, 117°W on August 12, 1976; L_s = 107° (052A08; NGF/B). (Right) At 8°S, 118°W on September 19, 1976; L_s = 125° (090A09; NGF/B). Prominent albedo changes have occurred near points A and B during the 39-day interval separating the two views. The summit caldera lies to the upper right; the upslope direction is approximately from lower left to upper right.

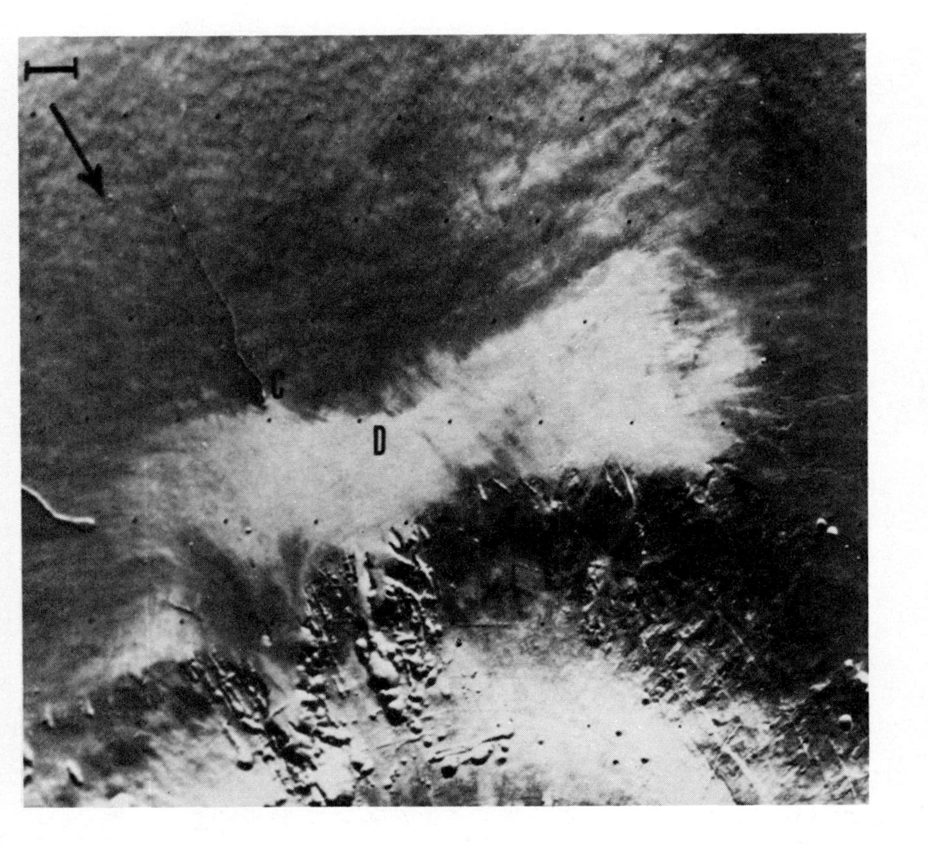

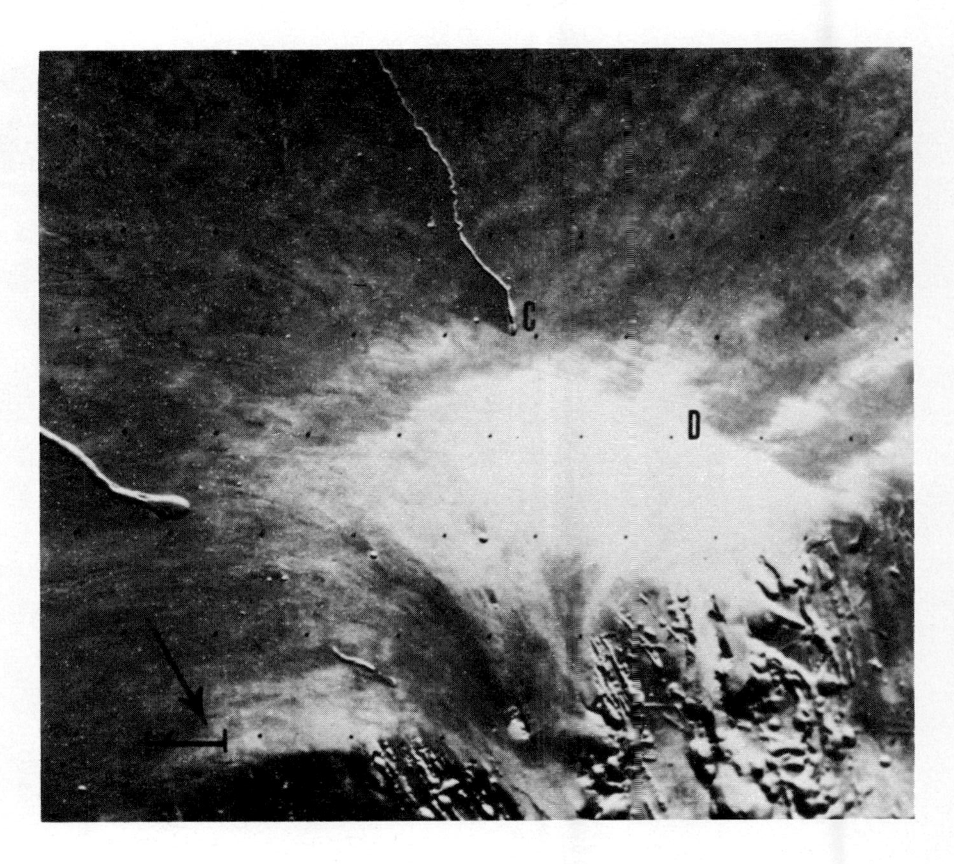

Fig. 6. Changes on the flanks of Arsia Mons. (Left) At 11°S, 121°W on August 22, 1976; L_s = 112° (062A43; NGF/B). (Right) At 12°S, 121°W on September 19, 1976; L_s = 125° (090A03; NGF/B). Prominent albedo changes have occurred throughout the area during the 29-day interval but especially near points C and D. The summit caldera lies to the bottom center of the left-hand frame.

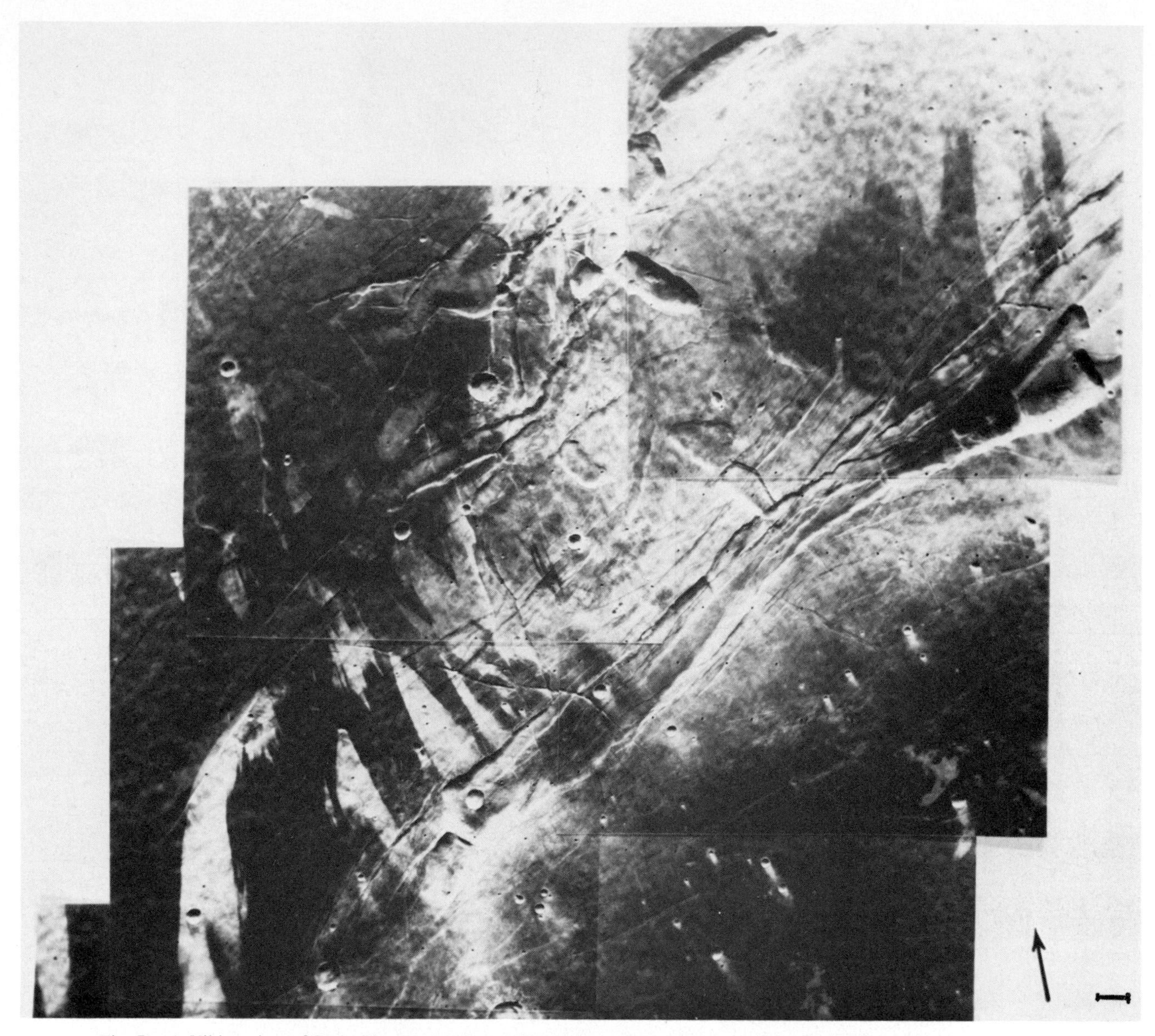

Fig. 7. A Viking view of Syria Planum (10°S, 110°W) on August 9, 1976; L_s = 105°. The albedo patterns in 1972 were very different, but the wind markers indicated a similar wind direction.

bordered by two dark side lobes. Very similar streaks were produced earlier by *Greeley et al.* [1974*a*, *b*] in wind tunnel simulations. If the wind tunnel analogy is valid, the dark lobes represent regions of erosion.

With few exceptions, no variations in wind streaks or in other variable features were documented during the Viking primary mission. The few exceptions are discussed in more detail in sections 4 and 7.

3. General Lack of Eolian Activity

The general lack of eolian activity during the Viking season can be illustrated by considering the observations of the areas surrounding the two landing sites. No eolian activity was detected from orbit in either locale, a result consistent with the observations made by the lander cameras [*Mutch et al.*, 1976*a*, *b*] and with the low wind velocities reported by the meteorology experiments [*Hess et al.*, 1976*a*, *b*]. Although both light and dark streaks occur in the general vicinity of the VL-1 landing site in Chryse (22°N, 48°W), no changes in either were documented (Figure 3). The bright streaks in the area displayed the same general NE to SW trend that they had in 1972. The mean azimuth (measured clockwise from north) of the light streaks in 1976 was 218°, very close to the 1972 value of about 225° and consistent with the azimuth of 197° ± 14° of the drifts of wind-blown material in the landing site [*Mutch et al.*, 1976*a*]. The dark streak trend in the VL-1 area is almost directly opposite to that of the light streaks. The mean azimuth of dark streaks is 42°, close to the value of 32° for the dominant winds recorded during the primary mission [*Hess et al.*, 1976*a*].

The vicinity of the VL-2 landing site (48°N, 226°W) is notable for its lack of recognizable variable features at orbital resolution. The closest recognizable wind streaks lie about 600 km to the south. No surface changes in the vicinity of the VL-2 site were observed during the primary mission.

4. Eolian Activity on the Slopes of the Tharsis Volcanos

In only one locale was persistent eolian activity documented during the Viking primary mission: on the flanks and in the

Fig. 8*a*. Changes during 1972–1976 in the appearance of Syria Planum (10°S, 108°W). Mariner 9 image (DAS 07182738) on January 31, 1972; L_s = 340°.

Fig. 8*b*. Changes during 1972–1976 in the appearance of Syria Planum (10°S, 108°W). Viking mosaic on August 8, 1976; L_s = 105° (049A16–049A22; NGF/B).

immediate vicinity of the three large Tharsis volcanos, most conspicuously on the flanks of Arsia Mons. The three Tharsis volcanos and Olympus Mons were conspicuous as prominent dark spots in the approach imagery on both missions. Subsequent orbital imagery revealed that since 1972, each of the three Tharsis volcanos had developed a more or less complete dark albedo ring on its flanks. The dark ring was especially well developed on Pavonis Mons (Figure 4), where it was typically 20 km wide and situated at altitudes between 20 and 25 km. The boundaries of the ring had a ragged appearance, the upper boundary being clearly made up in part of coalescing dark crater-associated wind streaks trending downslope.

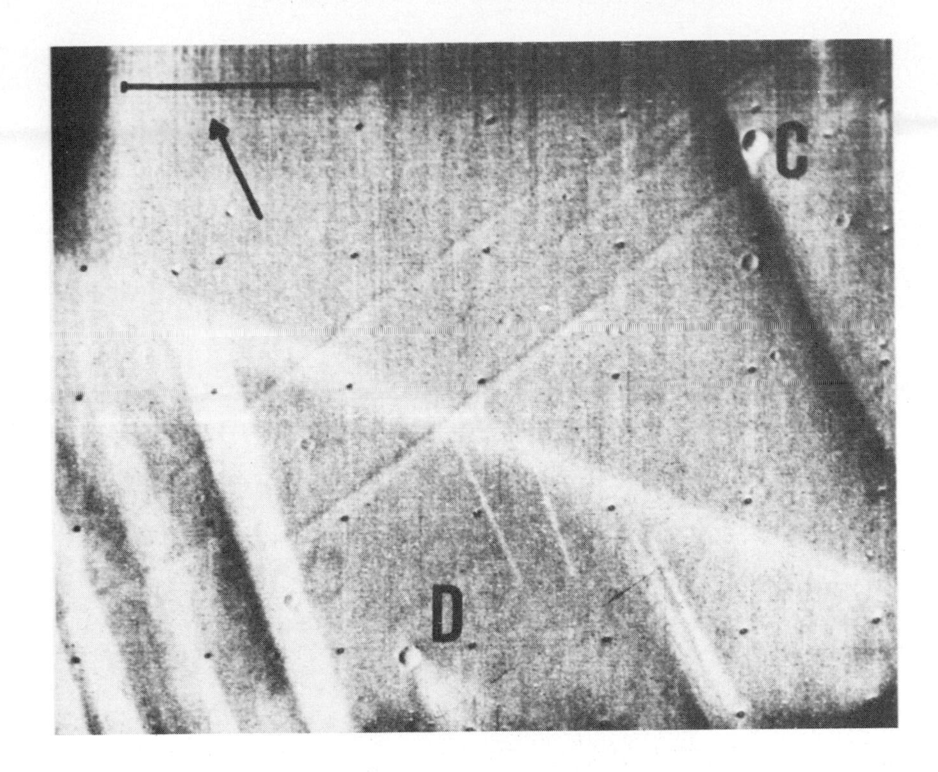

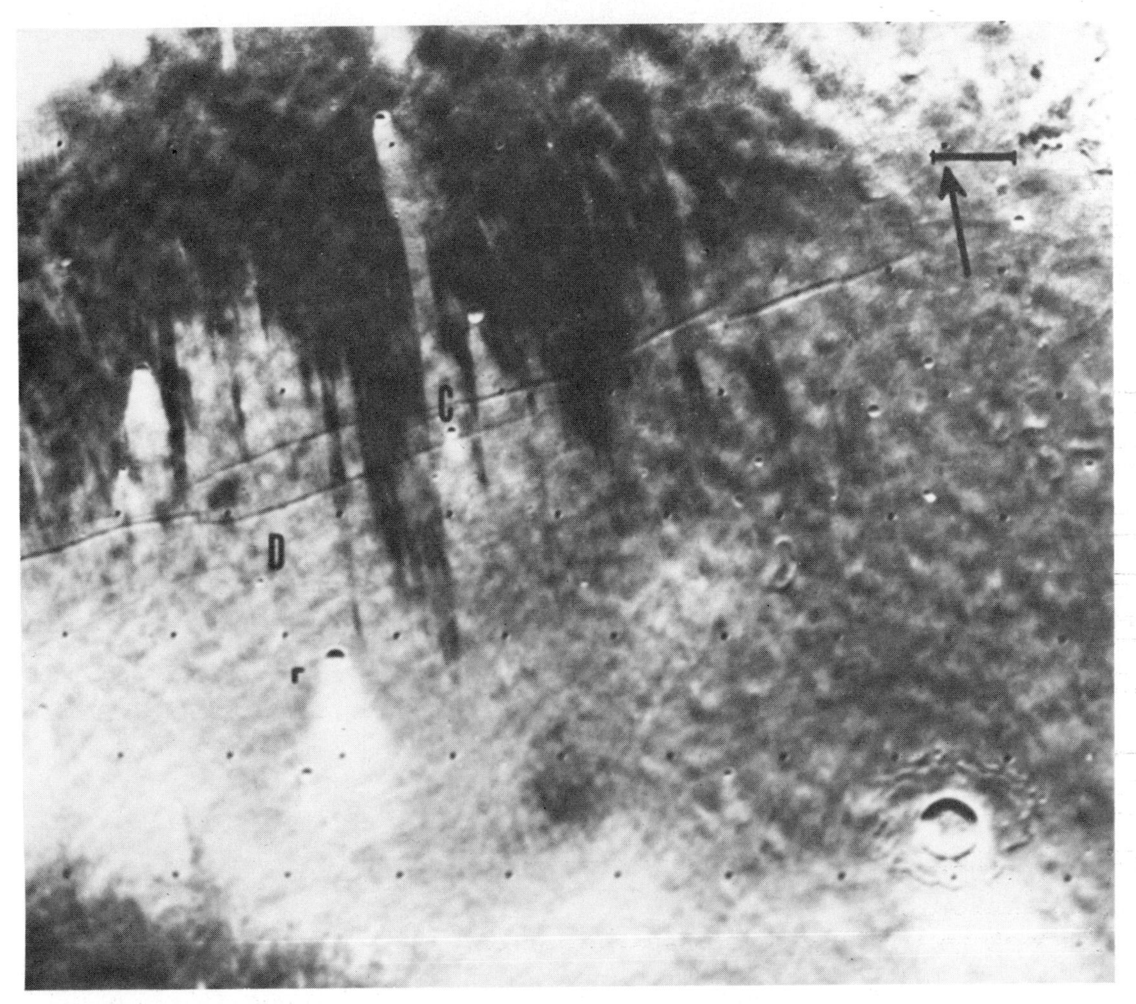

Fig. 9. Changes during 1972–1976 in Syria Planum. (Top) Mariner 9 image (DAS 09988659) on March 10, 1972; L_s = 355°. (Bottom) Viking image (050A21; SCR2) on August 9, 1976; L_s = 105° at 9°S, 107°W.

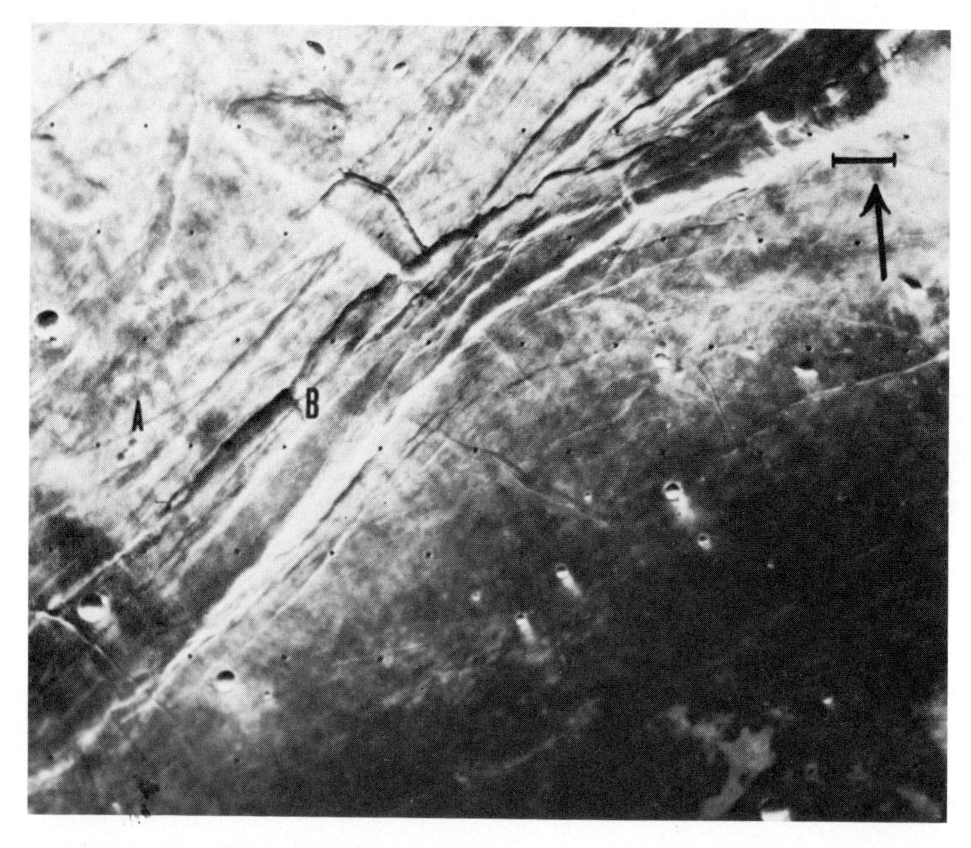

Fig. 10. Changes during 1972–1976 in Syria Planum. (Top) Mariner 9 image (DAS 10060549) on March 11, 1972; L_s = 359°. (Bottom) Viking image (049A19; NGF/B) on August 8, 1976; L_s = 105° at 10°S, 104°W.

Fig. 11. Viking mosaic of the Mesogaea area (8°N, 192°W) on September 17, 1976; L_s = 124° (88A; NGF/B). Subtle albedo changes have occurred in the prominent dark streak between 1972 and 1976.

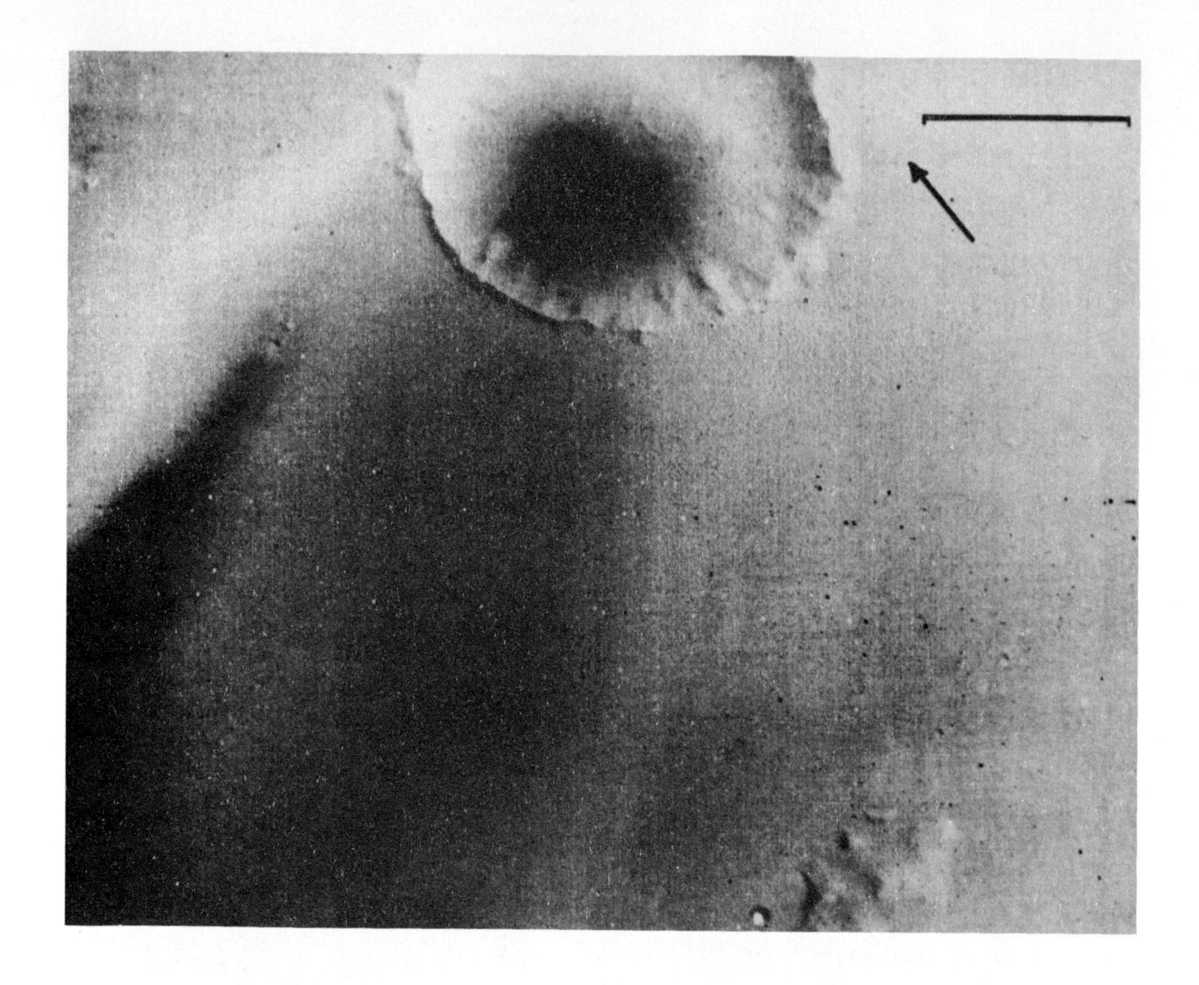

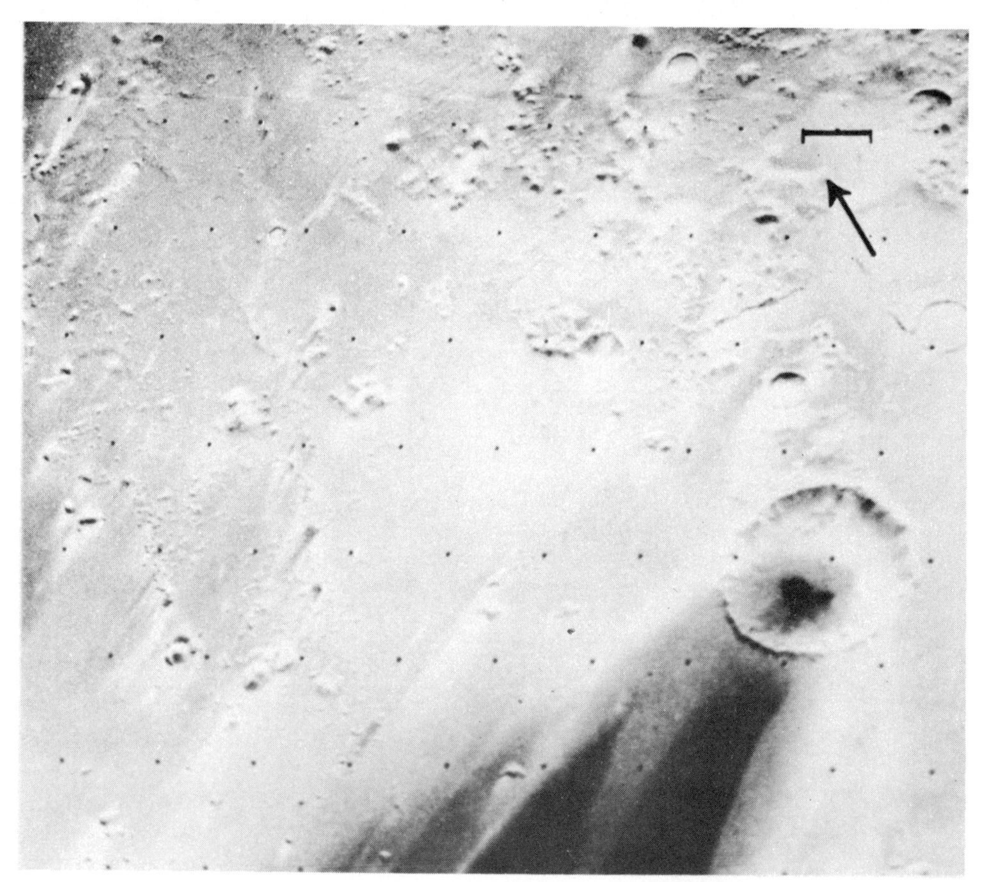

Fig. 12. An example of an albedo change within the dark streak shown in Figure 11. (Top) Mariner 9 image (DAS 12326691) on June 26, 1972; L_s = 2°. (Bottom) Viking image (088A87; NGF/B) on September 17, 1976; L_s = 124° at 10°N, 191°W.

Fig. 13. A Viking view of a small portion of Hellas near 39°S, 283°W. A number of dark ragged crater-associated streaks can be seen near the top. The picture was taken on September 23, 1976; L_s = 127° (095A53; NGF/B).

The dark ring around Arsia Mons (Figure 4) was much more extensive, in places reaching to the edge of the summit caldera. As in the case of its counterpart on Pavonis Mons, the dark ring around Arsia Mons had ragged boundaries, and the upper boundary was clearly made up in part of coalescing dark crater-associated wind streaks pointing downhill (Figure 4). Significant changes in the boundaries of the dark ring on Arsia Mons were observed throughout the primary mission. A conspicuous change observed between orbits 62 and 90 on VO-1 is shown in Figure 5. The change involves the uphill advance of the dark boundary by the development of crater-associated dark streaks trending downslope. These changes occurred near an altitude of 10 km and are definitely true surface changes; they are not caused by variations in atmospheric transparency or in the amount of cloud cover.

Figure 6 shows another example of surface changes on the flanks of Arsia Mons, this time near an altitude of 14 km. The pattern of the observed changes, that it is the tendency for the dark albedo boundaries to move uphill, is best explained in terms of the scouring of bright material from the volcano flanks by downslope winds. Similar but more subtle changes were observed on the flanks of Pavonis Mons and Ascraeus Mons. Eolian activity at comparable elevations on Pavonis Mons was observed in 1972 [*Sagan et al.*, 1974].

While the flanks of Arsia Mons provided the best example of persistent eolian activity during the primary mission, a few isolated eolian events were observed in some other locales. One especially interesting case, the development of a number of dark streaks on the eastern edge of the Hellas basin, is described in section 7.

5. SPECTACULAR ALBEDO CHANGES SINCE 1972

A few areas have experienced spectacular albedo changes since 1972. A good example is the northern section of Syria Planum (Figure 7), which lies just south of the western end of the Valles Marineris. Between 1972 and 1976, albedo boundaries have changed drastically (Figures 8*a* and 8*b*); thus dramatic evidence has been provided of a significant redistribu-

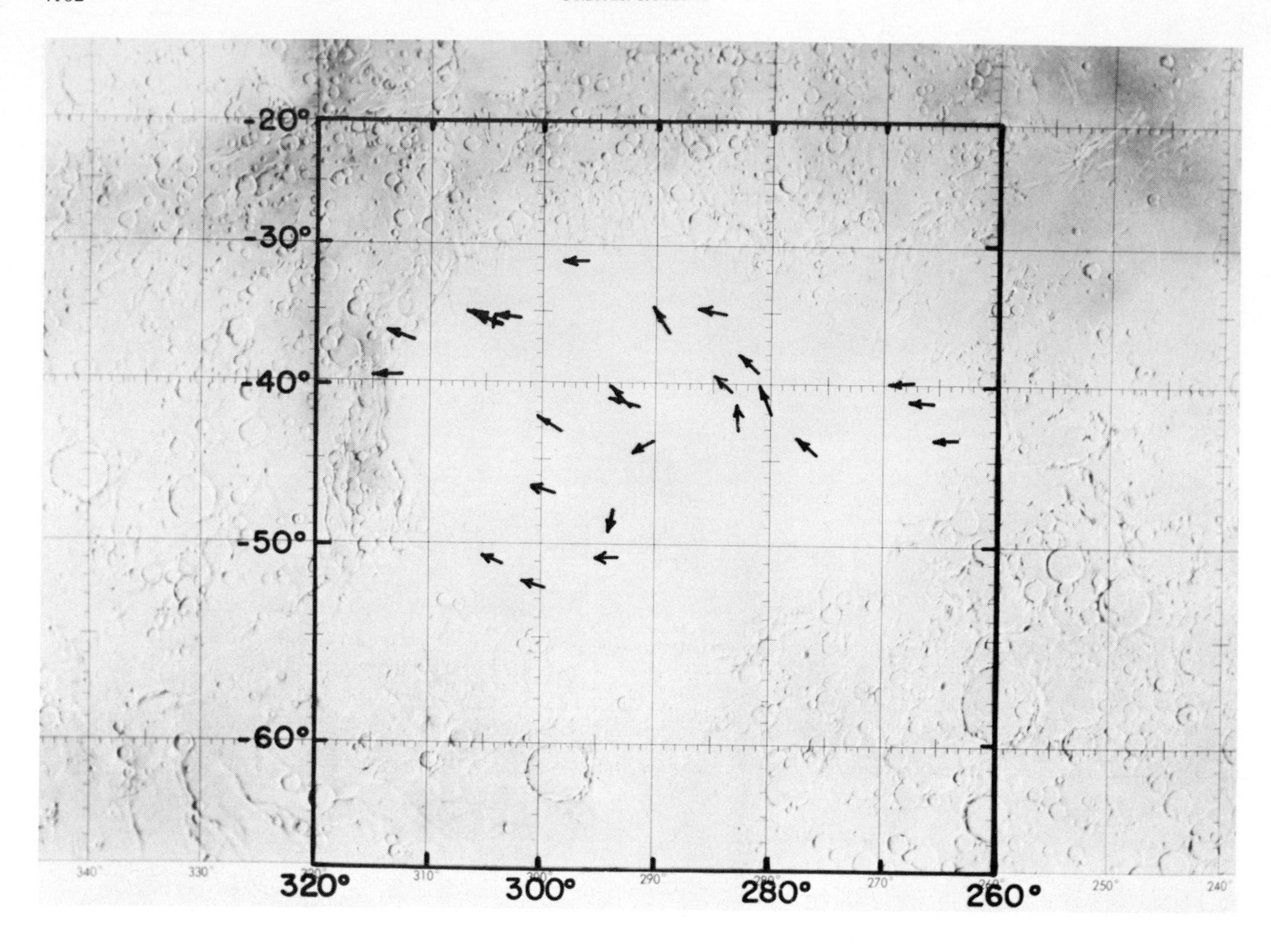

Fig. 14. Map of the locations and directions of prominent wind streaks in Hellas. The single crossed arrow represents a concentration of light streaks. All other arrows represent dark streaks such as those in Figure 13.

tion of surface materials by Martian winds. The surface area involved in these changes is several times 10^5 km^2, but the thickness of the layer redistributed may be quite small; i.e., about 1 mm would seem to be sufficient to produce the observed changes in contrast.

A high-resolution comparison showing a striking change between 1972 and 1976 near the center of the area of Figure 8 is shown in Figure 9. Another comparison involving the region near the center of Figure 7 is shown in Figure 10.

Even though the albedo patterns seen in this area in 1976 bear little resemblance to those that prevailed in 1972, the dominant wind flow direction is the same at both times: roughly N to S, consistent with the direction of strong winds in this equatorial region during southern summer. No changes in these particular markings were detected during the Viking primary mission.

6. More Subtle Albedo Changes

In many areas where distinctive variable feature markings occur, some subtle changes in the albedo boundaries between 1972 and 1976 can be detected. For example, consider the region in Mesogaea shown in Figure 11, which was also thoroughly imaged by Mariner 9 [*Veverka et al.*, 1976]. While the albedo patterns seen in 1976 are very similar to those obtained in 1972, a number of subtle changes have taken place within and along the boundaries of the prominent dark streak (Figure 12). The dominant wind direction indicated in the 1976 images (Figure 11) is identical to that in the 1972 images [*Veverka et al.*, 1976], essentially NE to SW. The subtle changes in the albedo markings documented in this area are typical of many cases on the planet where a detailed comparison between high-quality Mariner 9 and Viking images is possible.

7. Surface Detail and Wind Flow Patterns in Hellas

During the primary mission the atmosphere within the Hellas basin was unusually clear, so surface detail was readily visible. Among the several types of albedo markings discovered were crater-associated streaks (Figure 13). Most of the streaks within Hellas are of the dark, ragged variety [*Veverka*, 1975], although a few light streaks are also present.

The streaks indicate a dominant wind flow E to W across the basin (Figure 14) consistent with the direction inferred from dark streaks at similar latitudes to the east and to the west of Hellas. It is interesting that this E to W flow coincides with the direction of motion of major dust storms which develop to the

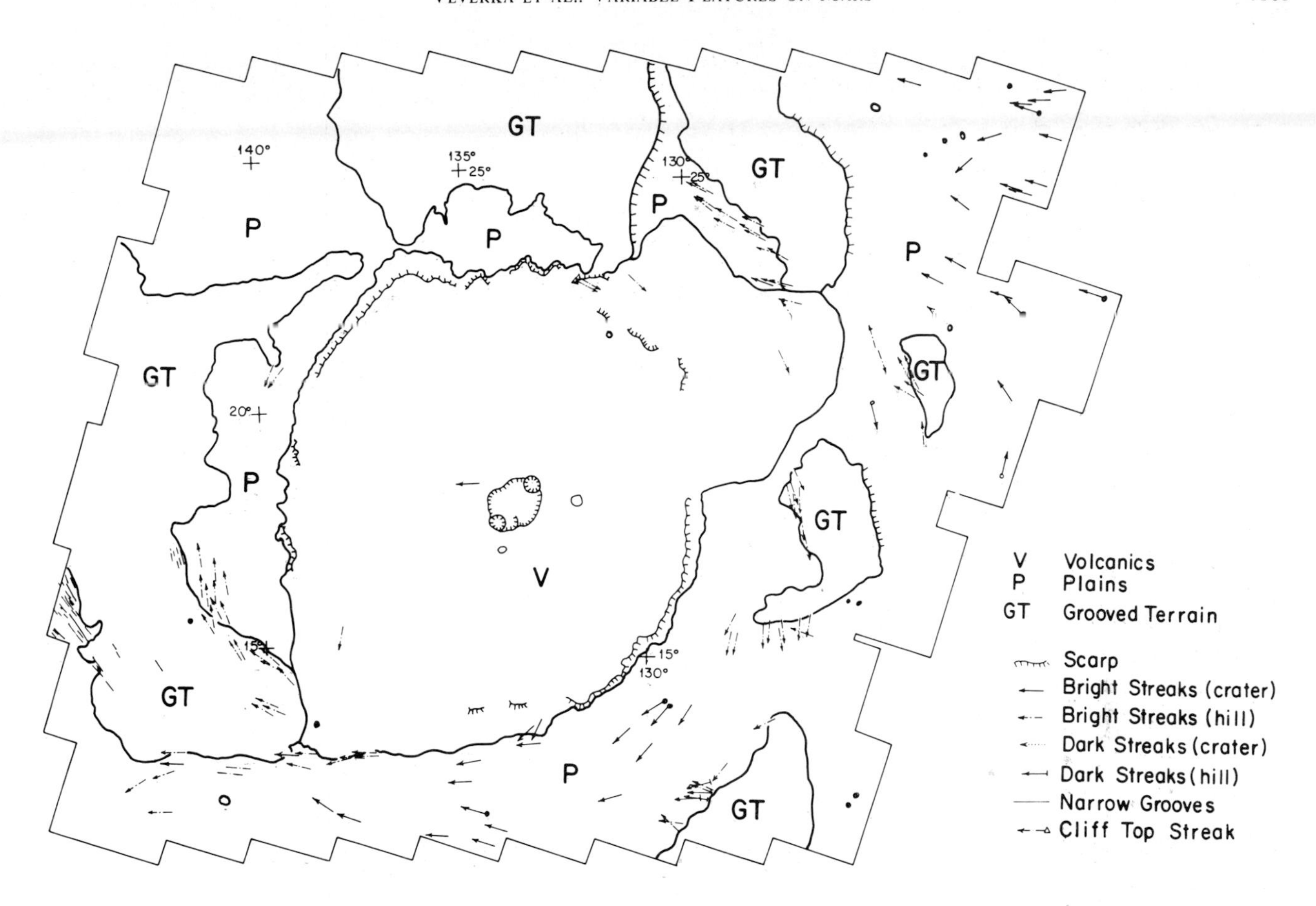

Fig. 15. Sketch map of the Olympus Mons region showing prominent wind markers and major terrain types.

northwest of Hellas [*Martin*, 1974]. The streak pattern within Hellas does not indicate a strong circumbasin wind flow within Hellas of the sort proposed by *Gierasch and Sagan* [1971].

Localized eolian activity near Hellas was observed late in the Viking primary mission. Between VO-1 orbits 106 and 124 a number of dark streaks developed near 43°S, 260°W. Their direction indicated an E to W wind flow.

The remarkable clarity of the atmosphere within Hellas at the Viking season is consistent with a trend observed during the Mariner 9 mission. Although the floor of Hellas was obscured by clouds during the bulk of the Mariner 9 mission, a clearing trend was observed near the end of the Mariner 9 mission (end of southern summer) [*Leovy et al.*, 1972].

8. Wind Flow in the Vicinity of Olympus Mons

The Viking data reveal an interesting wind flow pattern near the base of Olympus Mons (Figure 15). The pattern is well defined by numerous crater- and hill-associated streaks (Figure 16) which are mostly light in tone. No variations in these features were detected during the Viking primary mission.

The streak pattern shown in Figure 15 suggests the deflection of near-surface winds by the base of Olympus Mons. It is noteworthy that the three Tharsis volcanos, Arsia Mons, Pavonis Mons, and Ascraeus Mons, do not have comparable deflection patterns associated with them. The pattern may be especially well developed around the base of Olympus Mons because this construct is an isolated topographic high bounded by an abrupt scarp.

Incidentally, the very prominent dark rings so conspicuous on the flanks of the three Tharsis volcanos have no evident counterpart on Olympus Mons, although a few wind streaks which do point downhill can be found near the summit caldera of Olympus Mons.

9. Comparison of the 1972 and 1976 Wind Streak Patterns

Although the coverage of Mars obtained during the Viking primary mission is not comprehensive enough to construct a streak map for 1976 as complete as that shown for the Mariner 9 season in Figure 1, nevertheless, it is possible to compare streak patterns in a number of key areas and reach the following conclusions.

As in 1972, light streaks in 1976 occur only in a belt some 40° wide about the equator. In agreement with predictions the 1976 pattern of light streaks is essentially the same as that seen by Mariner 9 in 1972. The light streaks outline the wind flow pattern expected in the equatorial zone of Mars during southern summer, the season in which the strongest surface winds are expected [*Gierasch*, 1974]. The light streaks appear to be unaffected by the wind regimes prevalent at the Viking season.

In most cases where a detailed comparison of individual light streaks in 1972 and 1976 is possible, no change in direction, and little if any change in appearance, is seen. This observation confirms that the lifetimes of most light streaks are long (many Martian years). Yet, some light streaks have disappeared, and some new ones have formed during the past 4 years (Figures 17 and 18). This observation suggests that the thickness of bright material in some light streaks is not very large, since the accumulated bright material can be removed by Martian winds within the span of a few seasons.

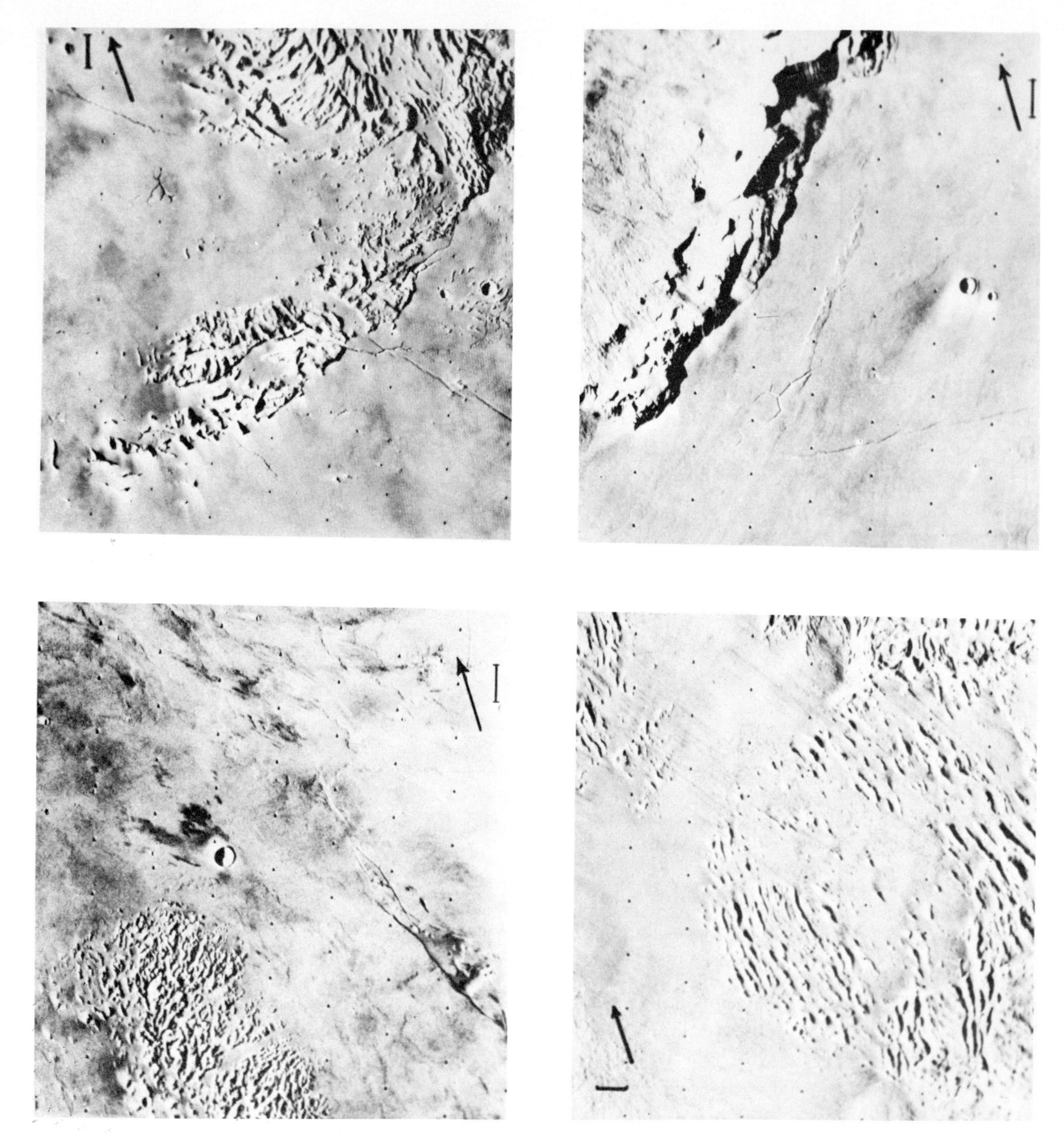

Fig. 16. A sample of wind markers around the base of Olympus Mons. Most common are the hill- and crater-associated light streaks (top left and top right). Dark streaks which include some mixed tone streaks (bottom left) are rarer. Narrow grooves (bottom right), possibly of eolian origin, also occur near Olympus Mons (cf. Figure 15). (Top left) At 17°N, 126°W on August 5, 1976; L_s = 104° (046B38; NGF/B). (Top right) At 15°N, 130°W on August 4, 1976; L_s = 103° (045B63; NGF/B). (Bottom left) At 22°N, 123°W on August 6, 1976; L_s = 123° (047B49; NGF/B). (Bottom right) At 14°N, 144°W on August 4, 1976; L_s = 103° (045B51; NGF/B).

As in 1972, dark streaks in 1976 were much less common on Mars than light streaks and were much less extensively distributed over the planet [cf. *Veverka et al.,* 1977*a*]. In several areas, individual dark streaks had changed since 1972, the short lifetimes predicted for them thus being confirmed [*Sagan et al.,* 1973; *Veverka,* 1975]. In some areas, for example, Mare Erythraeum, many dark streaks visible in 1972 are apparently no longer there. In the Daedalia region the dark streaks are still there but have changed both in direction and in outline

Fig. 17. (Opposite) Comparison of the 1972 and 1976 appearance of Daedalia. (Top) Photomosaic from 1972 Mariner 9. $L_s \simeq$ 335°–345°. (Bottom) Photomosaic from 1976 Viking. $L_s \simeq$ 109°. Numerous changes have occurred in the 4-year interval. Although most craters which had dark streaks in 1972 have dark streaks in 1976, the dark streaks have changed in outline and in mean direction (Figures 18 and 19). Some light streaks which were prominent in 1972 have disappeared. The mottled appearance of the upper half of the Viking mosaic suggests the presence of clouds, but significant surface albedo changes have definitely taken place since 1972.

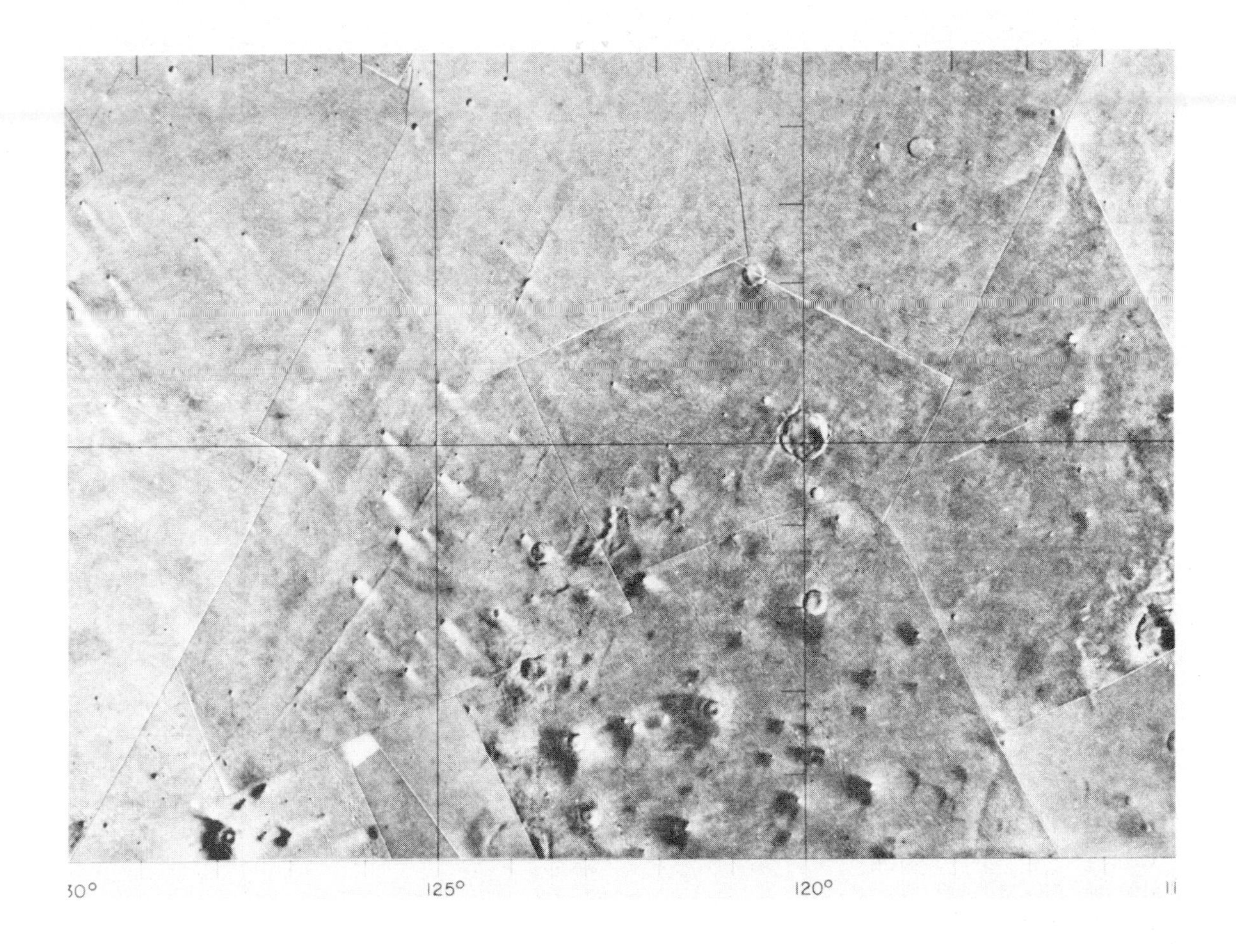

125°
120°

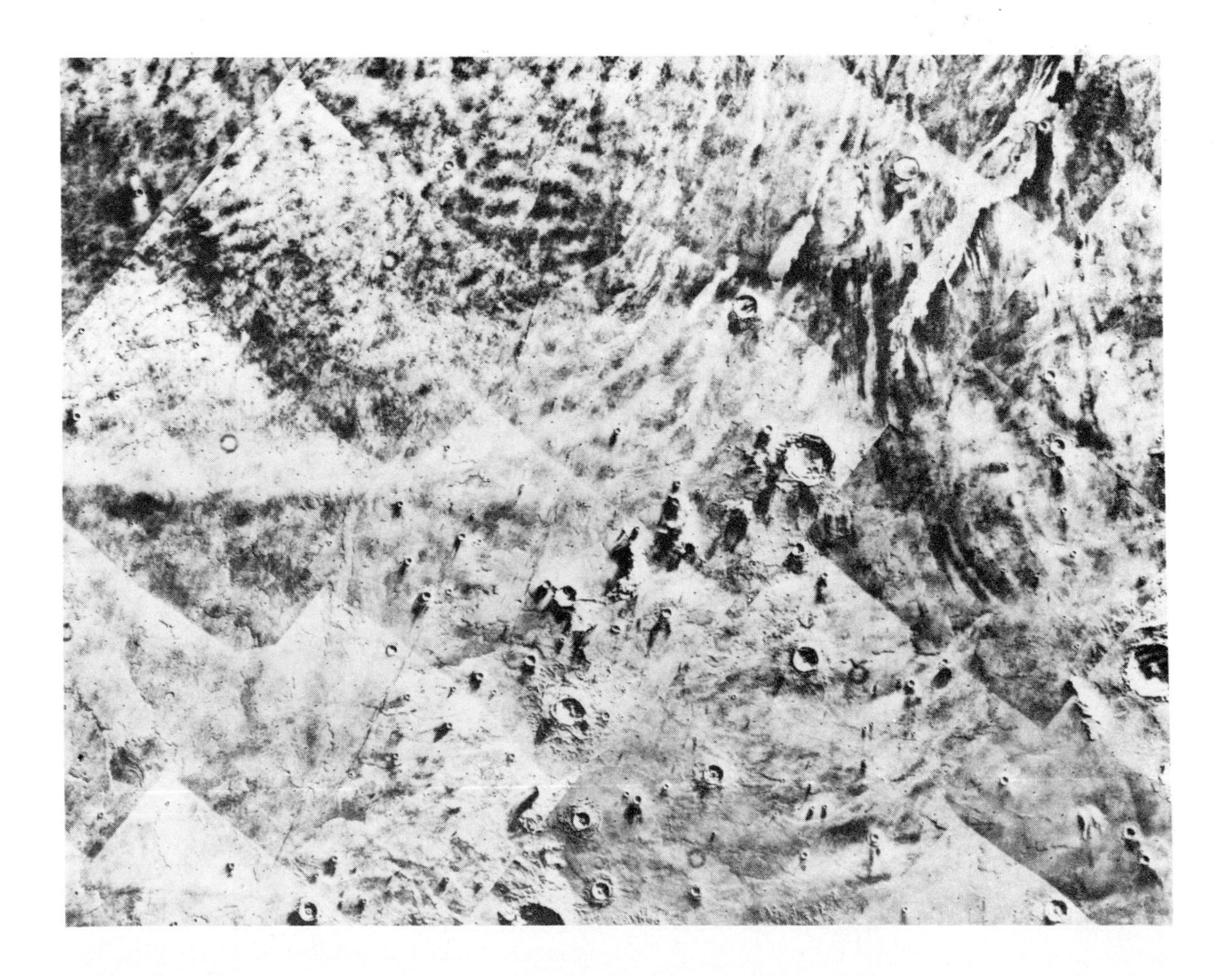

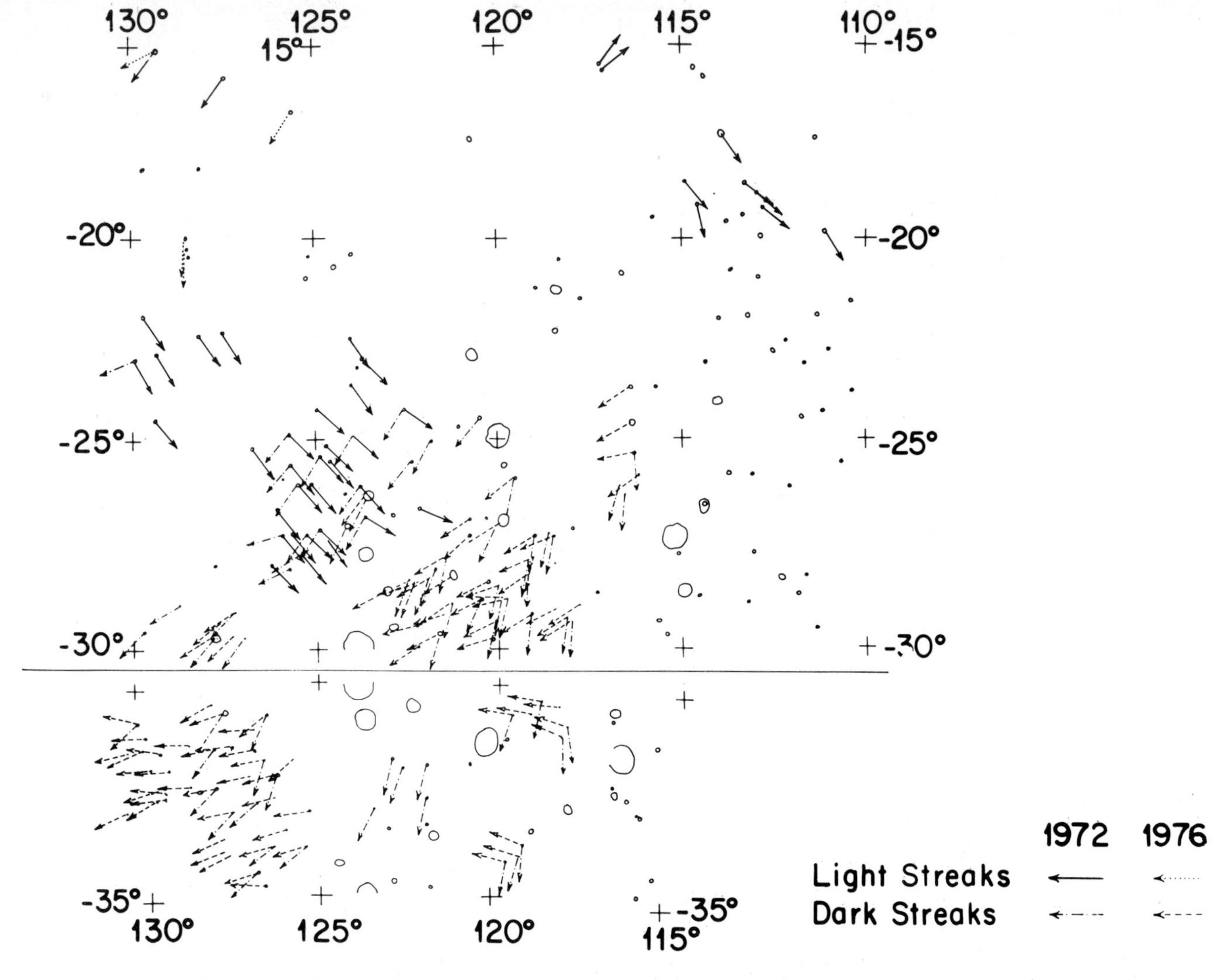

Fig. 18. Sketch map of the locations and mean directions of crater-associated streaks in Daedalia in 1972 and 1976, based on Figure 17.

(Figures 17 and 18). Note that many other albedo changes have occurred in this region of the planet during the past 4 years: for example, some light streaks have disappeared, a few new ones have appeared, some new albedo boundaries have formed, etc.

The manner in which the outlines of the dark streaks in Daedalia differ in 1972 and 1976 is especially interesting (Figure 19). In 1972 the streaks were conspicuously fan shaped with a general E to W trend, while the 1976 streaks are much narrower and have a NE to SW trend. The patterns sketched in Figure 19 suggest that wide fan-shaped dark streaks such as those which occurred in this area in 1972 are formed by a succession of erosion events, each involving winds from a slightly different direction. Each strong wind adds a small section to the fan until the entire pattern is erased by fallout from a global dust storm.

10. Summary and Conclusions

Although numerous instances of eolian activity were documented during the Mariner 9 mission [*Sagan et al.*, 1972, 1973] in 1971–1972 (southern summer and fall), very few changes were seen during the Viking primary mission in 1976 (northern summer). No eolian activity was detected from orbit in the vicinity of either landing site. This lack of eolian activity at this season was typical of Mars as a whole with two important exceptions:

1. Persistent eolian activity was observed on the flanks of the three large Tharsis volcanos, especially on Arsia Mons. The observed changes are best explained in terms of the erosion of bright albedo material by downslope winds.
2. The formation of a number of dark crater-associated streaks by E to W winds on the eastern edge of Hellas was documented.

Comparison of specific albedo boundaries reveals that in many cases, subtle changes in outline and/or contrast have occurred between 1972 and 1976. There are also many specific albedo boundaries in which no variations can be detected; this is especially true of many bright streaks. There are a few areas where the albedo patterns in 1976 are dramatically different from those seen in 1972, for example, the region of Syria Planum near 10°S, 105°W, where a large-scale redistribution of surface materials has occurred since 1972. Significantly, although the specific wind markers in this area in 1976 differ from those in 1972, the dominant wind direction that they indicate remains unchanged: generally N to S, the dominant wind direction expected in this equatorial region during southern summer.

As in 1972, the most conspicuous wind markers on the planet in 1976 are the light crater-associated streaks. While some new light streaks have appeared since 1972 and a few old ones have disappeared, by and large, most light streaks appear unchanged in outline and direction since 1972. The wind flow

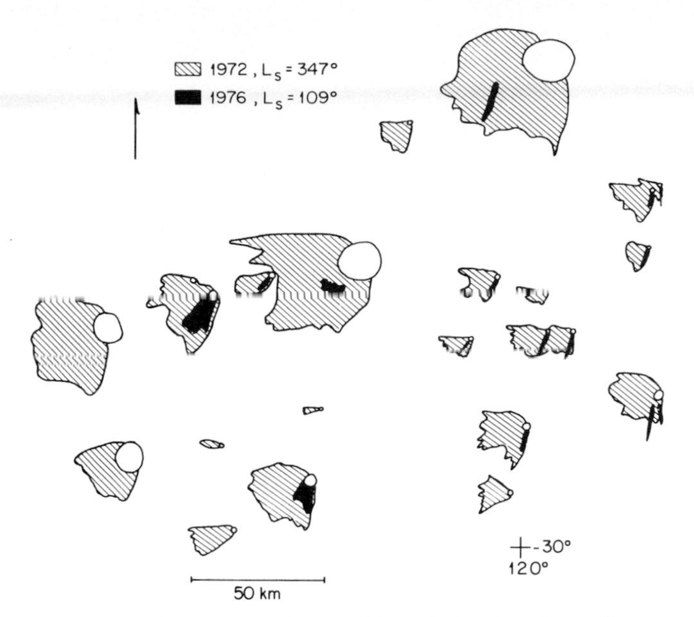

Fig. 19. Sketch map of the 1972 and 1976 outlines of some prominent dark streaks near 30°S, 120°W in Daedalia (cf. Figure 17).

defined is that expected for strong winds during southern summer [*Sagan et al.,* 1973; *Gierasch,* 1974]. As in 1972, the light streaks in 1976 occur only within a belt of about ±40° of the equator.

The Viking observations confirm that the mean lifetime of dark streaks is short in comparison with that of light streaks. Many dark streaks have changed conspicuously in both outline and direction since 1972.

Owing to the unusual atmospheric clarity in the Hellas basin at this season, it was possible for the first time to map wind streaks on the floor of Hellas. They indicate an E to W flow across the basin consistent with that outlined by dark streaks in adjacent regions at the same latitude to the east and to the west.

With few exceptions the Viking data generally support most of the hypotheses developed about variable features on the basis of Mariner 9 data. The most significant exception is that the dark band around the permanent polar cap in the north is not a region of eolian erosion as was suggested by *Sagan et al.* [1973] but is clearly a region of eolian deposition, since it turns out to be a vast dune field [*Cutts et al.,* 1976].

The Mariner 9 data combined with the Viking observations obtained during the primary and extended missions will span the whole cycle of Martian seasons and provide some overlap. By comparing the Viking extended mission observations with those obtained by Mariner 9 at a comparable season it should become possible to assess the relative amplitudes of seasonal and secular changes in the albedo patterns on Mars.

The combined Mariner 9–Viking primary mission experience indicates that albedo variations on the surface of Mars are to be attributed solely to eolian effects or to the deposition or sublimation of volatiles in the polar regions. There is no convincing evidence in the spacecraft data to support the concept of a seasonal wave of darkening once advanced by a number of earth-based observers [e.g., *Focas,* 1962]. If the phenomenon is nonillusory, then it is unlikely to be a surface effect and could represent a systematic seasonal change in atmospheric clarity.

Acknowledgments. We are grateful to the referees and to Conway Snyder for helpful and detailed comments. This research was supported by the Viking Project Office and by the NASA Office of Planetary Geology under grant NSG 7156.

REFERENCES

Arvidson, R. E., Wind-blown streaks, splotches, and associated craters on Mars: Statistical analysis of Mariner 9 photographs, *Icarus, 21,* 12–27, 1974.

Cutts, J. A., K. R. Blasius, G. A. Briggs, M. H. Carr, R. Greeley, and H. Masursky, North polar region of Mars: Imaging results from Viking 2, *Science, 194,* 1329–1337, 1976.

Focas, J. H., Seasonal evolution of the fine structure of the dark areas of Mars, *Planet. Space Sci., 9,* 371–381, 1962.

Gierasch, P. J., Martian dust storms, *Rev. Geophys. Space Phys., 12,* 730–734, 1974.

Gierasch, P., and C. Sagan, A preliminary assessment of Martian wind regimes, *Icarus, 14,* 312–318, 1971.

Greeley, R., J. D. Iversen, J. B. Pollack, N. Udovich, and B. White, Wind tunnel studies of Martian aeolian processes, *Proc. Roy. Soc. London, Ser. A, 341,* 331–360, 1974*a*.

Greeley, R., J. D. Iversen, J. B. Pollack, N. Udovich, and B. White, Wind tunnel simulations of light and dark streaks on Mars, *Science, 183,* 847–849, 1974*b*.

Hess, S. L., R. M. Henry, C. B. Leovy, J. A. Ryan, J. E. Tillman, T. E. Chamberlain, H. L. Cole, R. G. Dutton, G. C. Greene, W. E. Simon, and J. L. Mitchell, Preliminary meteorological results on Mars from the Viking 1 lander, *Science, 193,* 788–791, 1976*a*.

Hess, S. L., R. M. Henry, C. B. Leovy, J. L. Mitchell, J. A. Ryan, and J. E. Tillman, Early meteorological results from the Viking 2 lander, *Science, 194,* 1352–1353, 1976*b*.

Leovy, C. B., G. A. Briggs, A. T. Young, B. A. Smith, J. B. Pollack, E. N. Shipley, and R. L. Wildey, The Martian atmosphere: Mariner 9 television experiment progress report, *Icarus, 17,* 373–394, 1972.

Martin, L. J., The major Martian yellow storm of 1971, *Icarus, 22,* 175–188, 1974.

Mutch, T. A., T. A. Binder, F. O. Huck, E. C. Levinthal, S. Liebes, E. C. Morris, W. R. Patterson, J. B. Pollack, C. Sagan, and G. R. Taylor, The surface of Mars: The view from the Viking 1 lander, *Science, 193,* 791–801, 1976*a*.

Mutch, T. A., S. U. Grenander, K. L. Jones, W. Patterson, R. E. Arvidson, E. A. Guinness, P. Avrin, C. E. Carlston, A. B. Binder, C. Sagan, E. W. Dunham, P. L. Fox, D. C. Pieri, F. O. Huck, C. W. Rowland, G. R. Taylor, S. D. Wall, R. Kahn, E. C. Levinthal, S. Liebes, R. B. Tucker, E. C. Morris, J. B. Pollack, R. S. Saunders, and M. R. Wolf, The surface of Mars: The view from the Viking 2 lander, *Science, 194,* 1277–1283, 1976*b*.

Pollack, J. B., C. B. Leovy, Y. H. Mintz, and W. Van Camp, Winds on Mars during the Viking season: Predictions based on a general circulation model with topography, *Geophys. Res. Lett., 3,* 479–482, 1976.

Sagan, C., J. Veverka, P. Fox, R. Dubisch, J. Lederberg, E. Levinthal, L. Quam, R. Tucker, J. B. Pollack, and B. A. Smith, Variable features on Mars: Preliminary Mariner 9 television results, *Icarus, 17,* 346–372, 1972.

Sagan, C., J. Veverka, P. Fox, R. Dubisch, R. French, P. Gierasch, L. Quam, J. Lederberg, E. Levinthal, R. Tucker, B. Eross, and J. B. Pollack, Variable features on Mars, 2: Mariner 9 global results, *J. Geophys. Res., 78,* 4163–4196, 1973.

Sagan, C., J. Veverka, R. Steinbacher, L. Quam, R. Tucker, and B. Eross, Variable features on Mars, IV, Pavonis Mons, *Icarus, 22,* 24–47, 1974.

Veverka, J., Variable features on Mars, V, Evidence for crater streaks produced by wind erosion, *Icarus, 25,* 595–601, 1975.

Veverka, J., C. Sagan, L. Quam, R. Tucker, and B. Eross, Variable features on Mars, III, Comparison of Mariner 1969 and Mariner 1971 photography, *Icarus, 21,* 317–368, 1974.

Veverka, J., C. Sagan, and R. Greeley, Variable features on Mars, VI, An unusual crater streak in Mesogaea, *Icarus, 27,* 241–253, 1976.

Veverka, J., K. Cook, and J. Goguen, A statistical study of crater-associated wind streaks in the north equatorial zone of Mars, *Icarus,* in press, 1977*a*.

Veverka, J., K. Cook, and J. Goguen, A statistical study of dark crater-associated wind streaks in the southern hemisphere of Mars, *Icarus,* in press, 1977*b*.

(Received April 1, 1977;
revised June 22, 1977;
accepted June 22, 1977.)

VOL. 82, NO. 28 JOURNAL OF GEOPHYSICAL RESEARCH SEPTEMBER 30, 1977

IPL Processing of the Viking Orbiter Images of Mars

REUBEN M. RUIZ, DENIS A. ELLIOTT, GARY M. YAGI, RICHARD B. POMPHREY, MARGARET A. POWER, K. WINSLOW FARRELL, JR., JEAN J. LORRE, WILLIAM D. BENTON, ROBERT E. DEWAR, AND LOUISE E. CULLEN

Jet Propulsion Laboratory, Pasadena, California 91103

The Viking orbiter cameras returned over 9000 images of Mars during the 6-month nominal mission. Digital image processing was required to produce products suitable for quantitative and qualitative scientific interpretation. Processing included the production of surface elevation data using computer stereophotogrammetric techniques, crater classification based on geomorphological characteristics, and the generation of color products using multiple black-and-white images recorded through spectral filters. The Image Processing Laboratory of the Jet Propulsion Laboratory was responsible for the design, development, and application of the software required to produce these 'second-order' products.

INTRODUCTION

Digital computer processing of Martian images is not a new technology at the Image Processing Laboratory (IPL). Historically, the scene distortions introduced by the recording instrument and the low-contrast nature of Mars have imposed computer processing prerequisites to scientific interpretation [*Rindfleisch et al.*, 1971; *Green et al.*, 1975]. The processing which evolved from the support of Mariner 4, 6, 7, 9, and 10 provided a legacy of inherited software and hardware expertise and a better understanding of Mars as an imaging target. This expertise was used to build a hardware and software system to support the Viking lander and orbiter mission objectives by providing image-processing products of unprecedented scientific usefulness.

History

Computer image processing has been applied to data from previous missions to remove distortions introduced by the recording imaging system [*Green et al.*, 1975; *Soha et al.*, 1975] and to provide versions of images whose contents display maximum scene discriminability [*Dunne et al.*, 1971]. Removal of distortions introduced by the camera system is called decalibration, and improving scene visibility is called enhancement. Decalibration is generally a quantitative process which uses distortion correction data files constructed from instrument characterization data recorded during prelaunch and inflight calibration of the cameras. Image enhancement is the general name given to any process which makes no attempt to maintain absolute data relationships while exaggerating relative differences. Decalibration and enhancement techniques developed for previous missions were routinely applied to Viking orbiter (VO) images as preliminary processing. Table 1 lists the principal decalibration and enhancement software technology contributions of the previous Mars missions. This table shows the pattern of an evolving image-processing technology. Techniques developed for one mission were refined and systematically applied to data from the next. The Mariner Mars 1969 (MM'69) mission introduced techniques for the removal of geometric, photometric, and residual image distortions introduced by the camera system. These methods were refined for the Mariner Mars 1971 (MM'71) mission and used to produce over 7000 reduced data records (RDR) [*Seidman et al.*, 1973].

The MM'69 mission also introduced the concept of geometric transformation of the perspective view of the imaging instrument to one of the common map projections [*Gillespie and Soha*, 1972]. This process was used during MM'71 to produce over 4000 rectified and scaled (R&S) products [*Green et al.*, 1975]. An examination of the utilization of these products establishes the foundation for the Viking orbiter image-processing objectives. The RDR products were used by the U.S. Geological Survey (USGS) at Flagstaff, Arizona, for the production of elevation contour and shaded relief maps. Processing at USGS included computer and photographic preparation of the RDR for stereo plotter production of the elevation maps. Jet Propulsion Laboratory (JPL) color processing used the RDR products as input to the preparation of gray level color separation transparencies for color reconstruction in a photographic laboratory. The MM'71 R&S products were combined into global and regional mosaics used for geologic analysis.

These examples demonstrate that previous IPL computer processing of Martian images resulted in the production of preliminary products which required subsequent processing prior to the majority of the scientific interpretation. The Viking orbiter contribution to the evolution of computer image processing and the IPL objectives were to bridge the gap between preliminary products and scientific analysis by producing surface elevation, crater morphology classification, and color data suitable for direct analysis.

Viking Orbiter Camera

The dual Landing Site Certification requirements of 100-m resolution at periapsis and contiguous coverage of candidate landing sites governed the design of the VO visual imaging subsystem (VIS). The JPL Space Instrument and Photography Section responded to these requirements by designing and building a slow scan vidicon camera system with a 475-mm focal length, an 8.96-s frame time, and an 1182 active element by 1056 line image format [*Wellman et al.*, 1976]. The camera system included a filter wheel with violet, blue, clear, minus blue, green, and red spectral filters. The system also included light flooding to minimize residual image problems and two gain states to provide an increased dynamic range capability. The camera system characteristics are documented in the VO calibration report [*Benesh and Thorpe*, 1976].

IPL Hardware Configuration

The IPL Viking mission hardware configuration was designed to provide a flexible batch and interactive processing environment that satisfied the overall lander and orbiter requirements. An overall system configuration is shown in Fig-

Paper number 7S0505.

TABLE 1. Principal Decalibration and Enhancement Technology Contributions of Previous Missions to Mars

Project	Year	Images Returned	Processing Technology Contribution
Mariner Mars 4	1964	20	Utilized qualitative methods to restore missing data lost in RF link to spacecraft and produced contrast-enhanced version of each image.
Mariner Mars 6 and 7	1969	201	Developed electronic coherent noise removal and elementary photometric, geometric, and residual image correction procedures. Applied convolutional filtering to reduce degradation to fine scene detail resulting from camera system modulation transfer function. Introduced multispectral color reconstruction and map projection techniques.
Mariner Mars 9	1971	7300	Systematically applied refined photometry, geometry, and residual image correction techniques to produce digital archival record of each picture (reduced data record). Applied improved map projection software to over 4000 MM'71 images producing rectified and scaled products.

ure 1. The central processing unit is an IBM 360/65 digital computer with one megabyte of core memory operating under the IBM operating system (OS) with the time-sharing option (TSO). Principal peripherals include Ramtek and Comtal black-and-white and color image display devices, remote terminals, and a Bendix *X/Y* coordinate digitizing table. The displays, coordinate digitizer, and remote terminals are interfaced to the 360 through a PDP 11/40 minicomputer [*Jepsen,* 1976].

The Ramtek device consists of a 640 element by 512 line display which has six-bit gray level resolution, two independent character and graphics overlays, and two track ball controlled cursors.

The Comtal system consists of a 1024 × 1024 element and three 512 × 512 element gray level displays. The three 512 channels can be combined into one 512 × 512 color display. This system also has a graphics overlay and track ball cursor capability. A video switching network allows routing of any

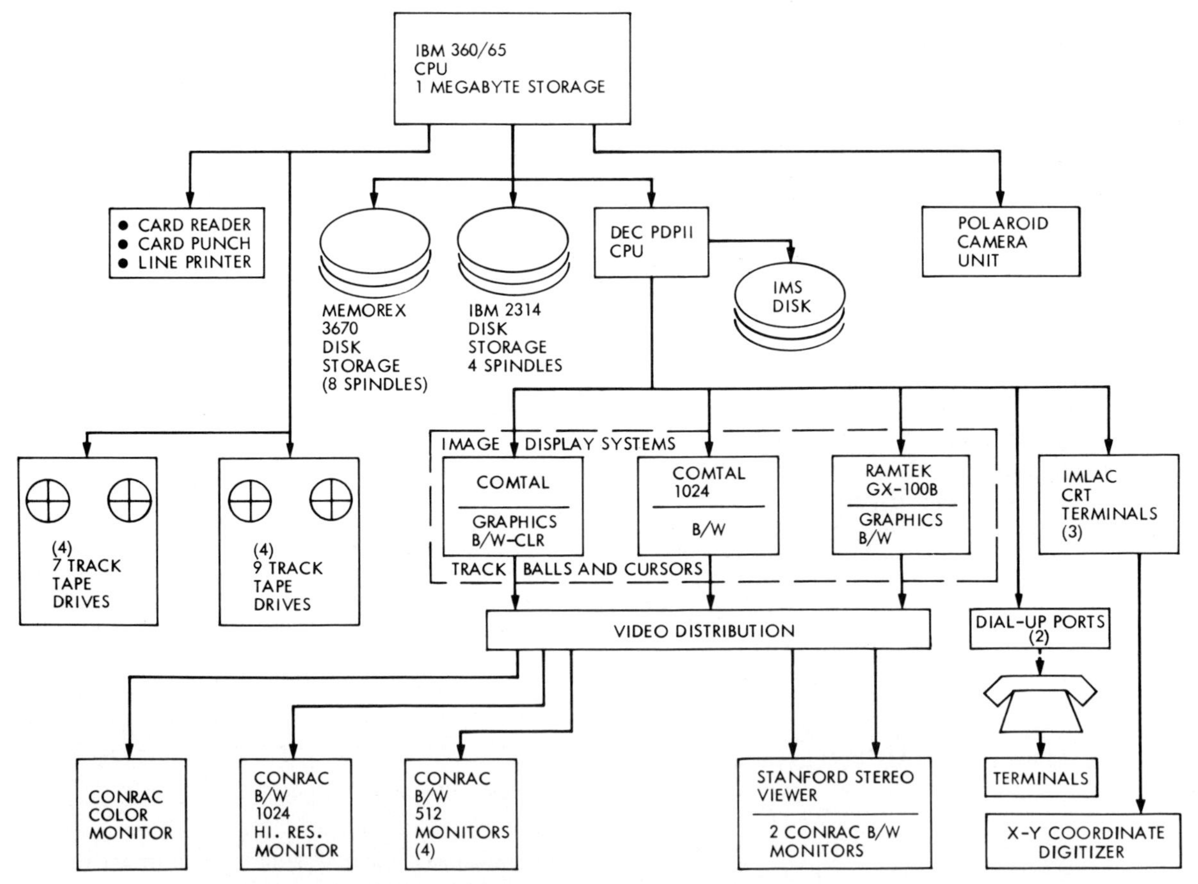

Fig. 1. Schematic representation of the IPL hardware configuration used for Viking orbiter and Viking lander image processing.

512 square gray level display to any one of four high-resolution monitors. Color images can be displayed only on the Comtal color monitor.

Three separate user areas were configured to support interactive image processing, computer stereophotogrammetry, color reconstruction, and crater morphology classification.

COMPUTER STEREOPHOTOGRAMMETRY

Topographic data in the form of elevation shaded relief maps, contours, profiles, and point statistics are critical tools in the analysis of Martian geologic phenomena. Generating these products for previous Mars missions required serial processing of stereo pairs of images at the IPL and the USGS at Flagstaff, Arizona. Processing consisted of decalibration and enhancement at IPL followed by photolab preparation and stereo plotter operation at the USGS. This process was slow, was susceptible to subjective errors, and required large amounts of time of highly trained personnel. The development of a semiautomated topographic mapping system that would directly produce elevation data for selected areas was one of the objectives of the IPL support of the Viking mission.

The method implemented at IPL consists of digitally computer-processing stereo pairs of images to (1) decalibrate and enhance each image, (2) transform them geometrically to a common map projection and scale, (3) characterize elevation by determining relief at selected points (tie points), (4) refine these tie point data interactively, and (5) produce topographic relief products. This sequence of processing and a description of each step are illustrated in Table 2.

Data Preparation

Each input frame is first corrected for telemetry bit errors. The image data resulting from the reseau marks inscribed on the vidicon are replaced by four-point bilinear interpolation. Camera blemishes caused by dust spots are similarly replaced. Camera-induced geometric distortions are corrected. At this point the image pair may be enhanced to improve the discriminability of fine detail by high-pass filtering and contrast expansion. These decalibration and enhancement techniques are described by *Green et al.* [1975]. Finally, the stereo overlap area is identified, and the two images are projected to a common scale, oriented so that the stereopsis line is nearly horizontal. The pair need not be registered; that is, there does not need to be any point whose line and sample locations on both images are the same, since all the programs utilized can compensate for an overall translation between the two images.

A typically processed stereo pair showing a portion of Ius Chasma, part of the Valles Marineris canyon system, is shown in Figure 2.

Tie Point Generation

The interactive program Stergen is used to locate features accurately in each image of the stereo pair. This is accomplished as follows. The analyst displays similar portions of the two images on one of the monitors. Corresponding points in each image are then located by using the two track ball controlled cursors. When a pair of points has been selected, the user initiates computer correlation. Stergen then computes, to subpixel accuracy, the position of the right image which yields the best correlation with the area identified in the left image. The tie point locations and a correlation quality index are then stored in a direct access data set. In addition, Stergen draws a numbered cross to identify the point in the left image and displays an exaggerated vector showing the difference in position of the corresponding point in the right image. The analyst then moves the cursors to a new tie point location and begins the process again. Figure 3 shows the left and right images as they appear in a typical session. This illustration displays 15 tie points whose locations and magnitudes are shown by the vectors. The number of tie points required to map a stereo pair depends on the accuracy desired and the amount of information in the images. Typical tie point data sets range from 250 for a polar pair up to a maximum of 550 for a topographically complex area.

The cross-correlation algorithm used to refine the accuracy of tie point locations uses the Fourier transforms of small areas surrounding the initial points. Let the two areas selected be A_{mn} and B_{mn} of dimension $M \times M$. The Fourier transforms of each area (FA_{kj} and FB_{kj}) are computed by using the standard formula for the discrete transform:

$$FA_{kj} = \sum_{m=0}^{M-1} \sum_{n=0}^{M-1} A_{mn} e^{-2\pi j} \left[\frac{km}{M} - \frac{jn}{M} \right] \tag{1}$$

where M is the size of the square correlation matrix and A is the digital value of the pixel located at coordinates m and n. Then the product of the two transforms is taken:

$$FC_{kj} = FA_{kj} \cdot FB_{jk} \quad {\scriptstyle 0 \le k \le M-1 \atop 0 \le j \le M-1} \tag{2}$$

TABLE 2. Computer Stereophotogrammetry Processing Steps

Step No.	Processing Step	Description
1	Decalibration and cosmetic enhancement	Fill missing lines, remove bit errors, correct camera geometric and photometric distortions, stretch contrast, and/or filter for maximum discriminability of details.
2	Map projection	Geometrically transform both frames to the same standard map projection, such as Mercator, at same scale and orientation.
3	Tie point generation	Examine both frames on an interactive monitor, indicating common points with movable track balls and allowing the computer to optimize the correlation.
4	Tie point editing	Make small-scale elevation maps and vector maps, checking for consistency of vector orientations and agreement of elevation surface with visual impression of input frames, delete bad points, and add new ones as needed.
5	Final product preparation	Choose appropriate relief interval and contour elevation map, and superimpose contours on image; plot intensity of elevation map over selected lines for terrain profiles; superimpose point elevations on image.

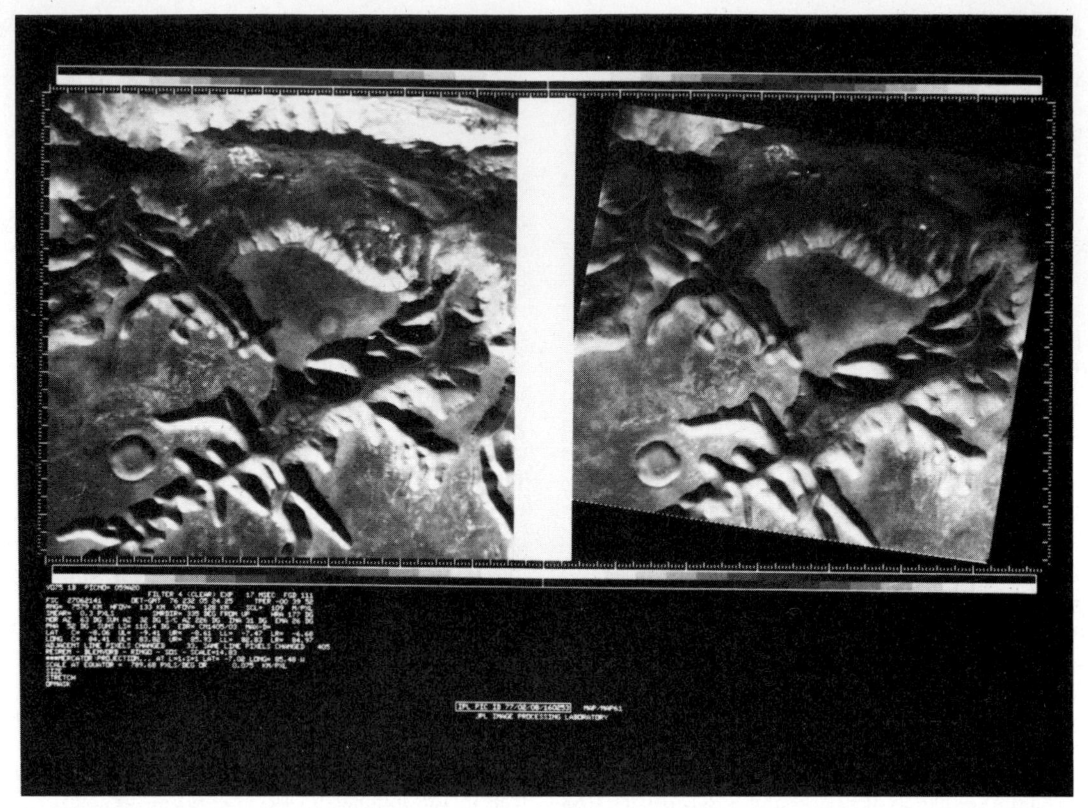

Fig. 2. Stereo pair of images showing the Ius Chasma region of Valles Marineris after map projection to a common scale.

A high-pass filter is performed to remove the low-frequency brightness variations so that differences in overall image brightness do not affect the correlation quality. The filter suppresses all components having zero horizontal ($k = 0$) or vertical ($j = 0$) spatial frequencies.

The inverse transform is then computed to give the cross correlation:

$$C_{mn} = \sum_{k=0}^{M-1} \sum_{j=0}^{M-1} FC_{kj} e^{+2\pi i} \left[\frac{km}{M} - \frac{jn}{M} \right] \quad (3)$$

The displacement from the origin of the correlation maximum determines the error in the initial estimate of the common feature locations. If this displacement is not within a user-specified tolerance, the location of the right cursor position is automatically moved toward the maximum, and a new correlation is computed. This process is repeated up to three times or until the correlation maximum is within the tolerance.

Generation of Point Elevations

Point elevations are generated for each tie point collected by using either the relative elevation program (Stermap) or the absolute elevation program (PHOTOGM). Relative elevations within an image pair are computed directly from the vector displacements between tie point members, while absolute elevations are triangulated from the vectors connecting each spacecraft position with the feature location in the image recorded at that position.

Relative elevation. For relative elevations, Stermap provides a choice of several algorithms for removing vector components that are displacements caused by image distortions such as projection errors rather than true parallax displacements. These biases are removed by modifying the location of the left image relative to the right image by using translation only, translation plus uniform scaling, translation plus rotation, translation plus rotation plus uniform scaling, or an unconstrained first-order least squares fit.

By means of these options, several small vector maps can be drawn, and a detrending performed which produces vectors oriented along the known parallax line (parallel to the direction of spacecraft travel). Usually, that option is chosen which requires the fewest degrees of freedom.

Stermap then computes relative elevations from the vector length at each tie point. For points in the image between tie points, elevations are computed as a distance-weighted average of the elevations of the nearest four tie points. The output from this stage is a gray level elevation map with a range of 250 levels, in which zero (black) represents the lowest elevation within the image. Figure 4 shows the gray level elevation map for the Ius Chasma pair.

In order to convert the gray level map to a contoured topographic map it is necessary to know the pixels of displacement represented by a single brightness level and the amount of relief represented by a pixel of displacement. The former quantity is printed out by Stermap, but the latter must be computed. The simple stereo base equation used for vertical terrestrial aerial photography is not used because the spacecraft is generally at different altitudes when the two images are recorded and the plane formed by these two positions and the image center is generally not perpendicular to the image plane. Figure 5 shows the camera positions for the pair in Figure 2. Meters of relief per pixel of displacement may be calculated by using the following equation derived by *Blasius and Cutts* [1975]:

$$H/H_x \approx SF/|SB| \quad (4)$$

where

H height, m;

H_x parallax displacement, pixels;

SF projection scale, m/pixel;
$SB = \tan EMA_1 \cos w + \tan EMA_2 \cos (180° - ASC - w)$;
EMA_1 emission angle for image 1 (left image), the angle at the center of the image between the local surface normal and the line to the spacecraft;
EMA_2 emission angle for image 2 (right image);
ASC angle subtended by the two spacecraft at the center of the overlap area;
w 'base angle' chosen to maximize SB.

$$\tan w = \frac{\sin ASC \tan EMA_2}{\tan EMA_1 - \cos ASC \tan EMA_2} \qquad (5)$$

$$0° \leq w < 180°$$

Absolute elevations. Absolute elevation is expressed in terms of the absolute radius of Mars from its center of mass to the feature. The program PHOTOGM uses the camera-pointing and spacecraft location parameters at the times when the images were recorded to compute the vectors from spacecraft to feature; these vectors converge at an elevation above the

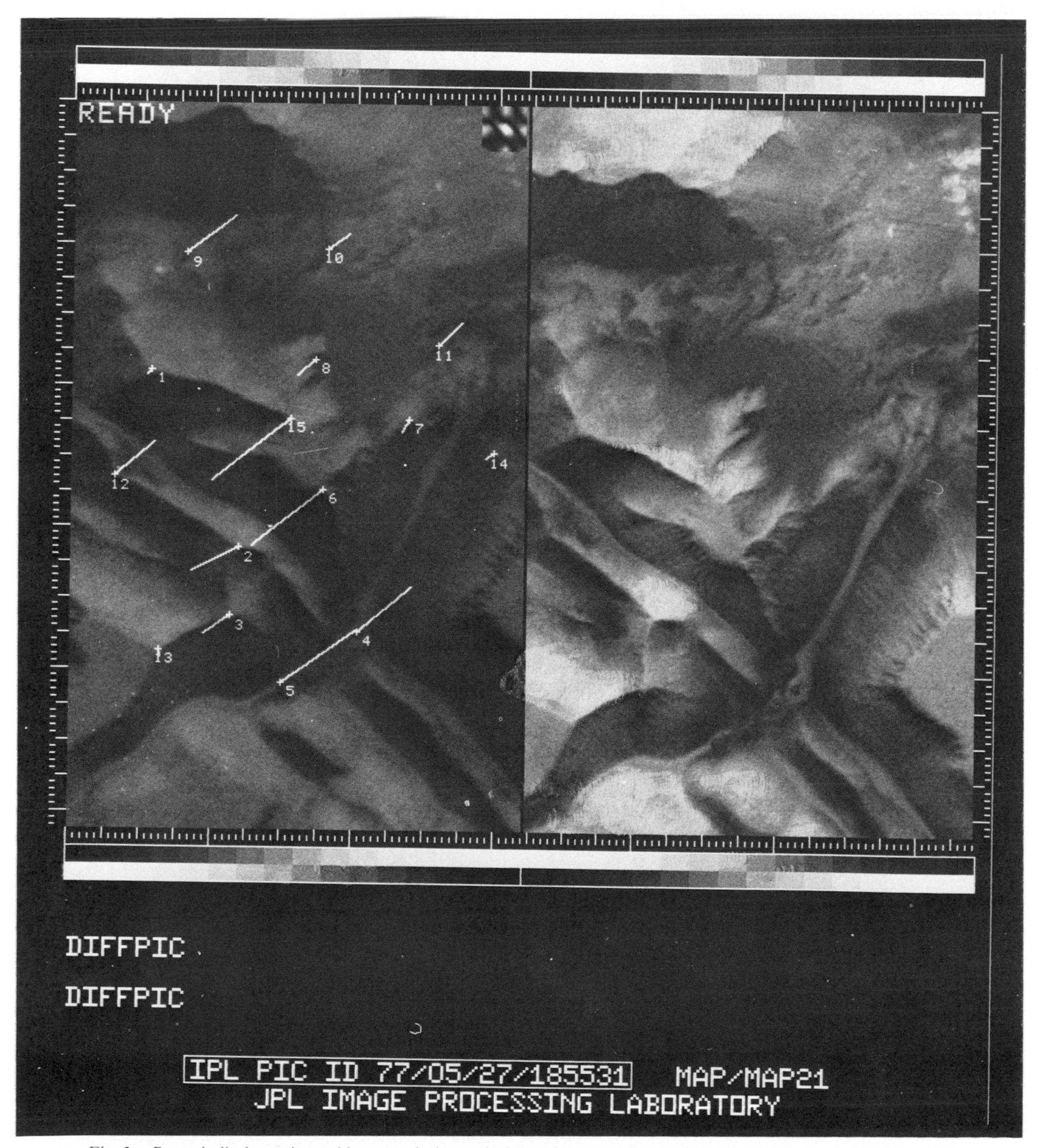

Fig. 3. Ramtek display as it would appear during typical tie point generation session. Higher elevations are indicated by vectors pointing to the lower left, and lower elevations by vectors pointing to the upper right.

Plate 1*a*

Plate 1*b*

Plate 1. (*a*) Decalibrated, unenhanced multispectral composite picture consisting of three images exposed during revolution 40 of Viking Orbiter 1. Image is centered at approximately −29° latitude and 69°W longitude. (*b*) The scene illustrated in part *a* after a color coordinate transformation has been employed. A contrast stretch of the intensity component has been applied before execution of the inverse color coordinate transformation.

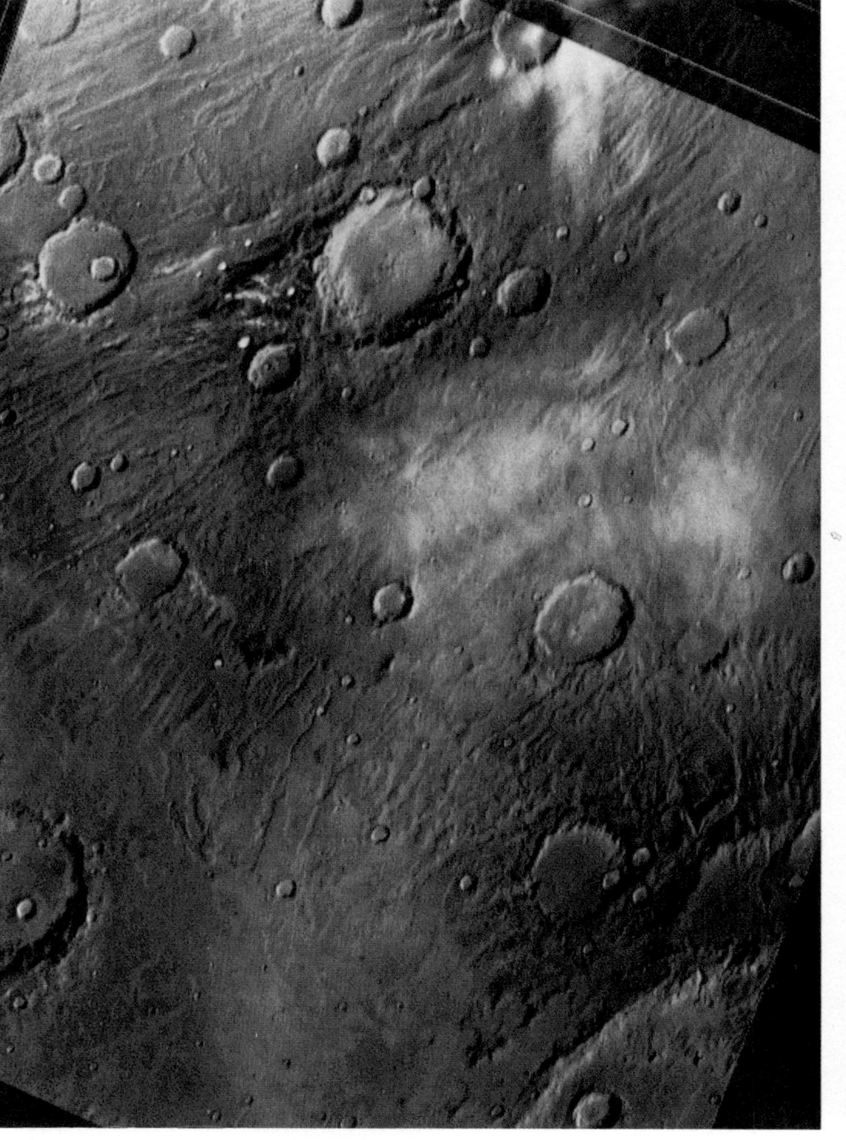

Plate 2*a*

Plate 2*b*

Plate 2. (*a*) The scene illustrated in Plate 1*a* after color coordinate transformation has been employed. The intensity component has been corrected for variations due to sun angle differences, has been spatially filtered to correct for the VIS modulation transfer function (MTF), and has been contrast-stretched, before the application of the inverse color coordinate transformation. (*b*) The scene illustrated in Plate 1*a* after color coordinate transformation has been employed. The intensity component is the same as that used in Plate 2*a*. In addition, the hue and saturation components have been contrast-stretched before the application of the inverse color transformation.

Fig. 4. Gray-level elevation map for the Ius Chasma pair shown in Figure 2. Lighter values represent relatively higher areas.

center of Mars. A lengthy series of matrix rotations is required to translate and rotate the camera coordinate system of the right image into that of the left. This series of rotations has been fully described by *Benesh* [1973].

Point elevations are calculated in this manner for all tie point pairs and passed to Stermap for interpolation as described above. The elevation for each tie point is computed from the tie point location and the camera parameters only. No detrending algorithm is used.

There are advantages and disadvantages to both the absolute and the relative mapping algorithms. Relative mapping has the following advantages.

1. A fit is computed from the relationship of all tie points to each other so that errors in camera-pointing and spacecraft location parameters, which cause projection errors, can be averaged out.

2. The method is directly transferable to mapping planets other than Mars, since elevations are computed directly from parallax by an algorithm which does not require data formats specific to Viking.

Absolute mapping has the following advantages.

1. Planetary curvature is removed. This error source can produce erroneous elevation artifacts in topographic data compiled from high-altitude or highly oblique images.

2. Maps of adjacent areas are referenced to a common origin, the center of the planet, and can be compared and perhaps mosaicked.

3. Large perturbations in the shape of the planet, such as the Tharsis bulge on Mars, may be measured and compared with measurements made by gravitational and radar mapping.

Interpreting Elevation Maps

A variety of products can be made from the gray-level elevation map illustrated in Figure 4. It may be contoured at a convenient interval, and the contours removed and superimposed on the image, as is shown in Figure 6. This map may be used with a reference map of all tie points or with a map of elevations at selected points. Elevation profiles across desired features may also be plotted.

Accuracy

The most important piece of work yet to be completed is accuracy testing. Relative elevation maps produced thus far compare favorably with maps made from Mariner Mars 1971 images and with Viking maps produced with a stereo plotter by the USGS in Flagstaff. Absolute elevation maps generally contain a systematic error primarily due to inaccuracies in camera-pointing data; for example, the Martian radius within the pair in Figure 2 is computed as 3413 ± 5 km, whereas the true value would lie in the range 3400 ± 4 km. Further accuracy analysis will be performed when more precise pointing information becomes available as a result of an expanded control net with increased resolution.

CRATER STRUCTURE AND MORPHOLOGY

Impact craters are the dominant landform on the surface of Mars. They are valuable tools for the study of the effects of obliteration, or resurfacing, of Mars, since they appear in nearly every type of geologic terrain.

A hardware/software system is under development at the Image Processing Laboratory whose goal is to allow the rapid accumulation, in a consistent manner, of a large quantity of crater descriptive data from the Viking orbiter images of Mars. The data can be used to study the physical properties and geologic history of the surface. The system is an extensive modification of the Brown University crater classification system, which was used to analyze Mariner 9 wide-angle imagery [*Arvidson et al.,* 1974].

The IPL crater analysis system allows crater size and morphological characteristics to be gathered for all craters visible on a given photographic print. Data gathering and reduction are highly automated so that the large volume of data required for statistical analyses can be obtained in a reasonable amount of time. Yet considerable flexibility has been built into the system so that special circumstances, such as unusual craters, can be handled as they arise.

The data acquisition and reduction phases of the system are now operational. Programs to aid in the interpretation of the data are also available; more may be added as users gain experience with the system and recognize additional requirements not currently met by the software.

The generation of a master data file is also now in progress. This data set will contain one record for each crater analyzed. Every record will contain the diameter of the rim crest and the latitude and longitude of the crater center. In addition, each record will contain morphological information pertaining to crater shape, floor structure, ejecta, and the relation of the crater to its surroundings. Records will contain size information for the ejecta blanket, features on the crater floor, streaks emanating from the crater, etc., when they are applicable. The 'morphological information' referred to above consists of the values of 15 key words, each of which has four or five possible values. These key words are given in Table 3.

Preliminary analysis of some data collected by using the system and a comparison with a sample of representative lunar and Mercurian craters, also digitized with the IPL system, have been undertaken (J. A. Cutts, unpublished data, 1977).

The hardware used by the system consists of a Bendix

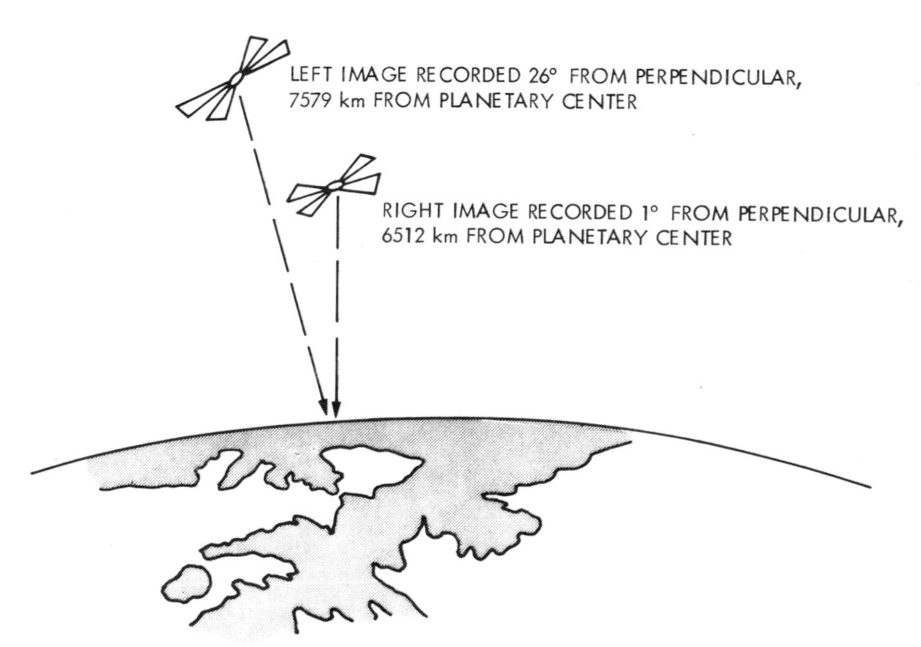

Fig. 5. Approximate spacecraft positions for the Ius Chasma stereo pair showing the difference in altitude and viewing angle.

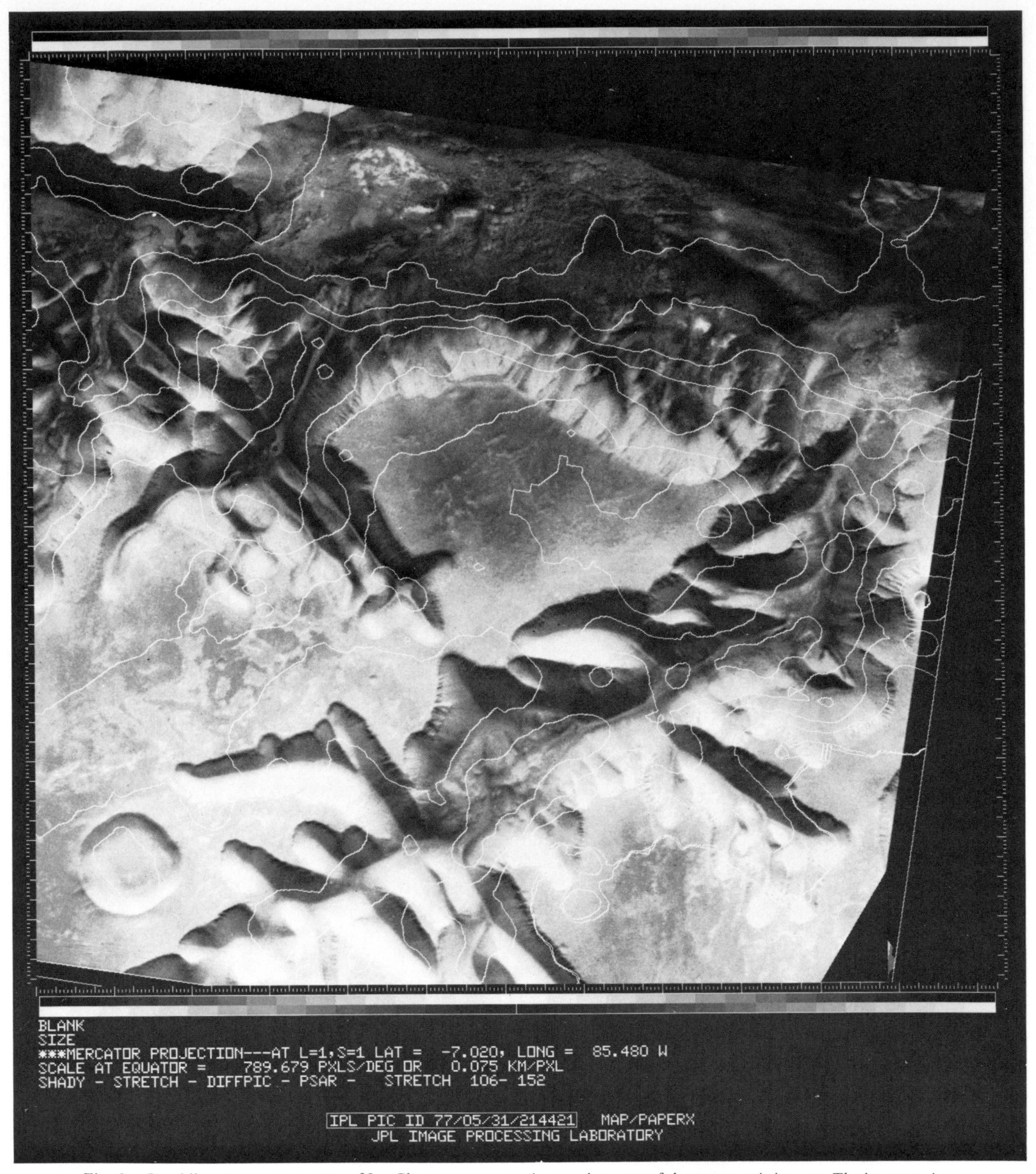

Fig. 6. One-kilometer contour map of Ius Chasma area superimposed on one of the stereo pair images. The lowest points occur within the canyon in the upper part of the image.

coordinate digitizer, an Imlac minicomputer and graphics display device, a PDP 11/40 minicomputer, and an IBM 360/65 computer. Figure 1 shows the relationship of the devices. Photographic prints are taped to the digitizing table. An Imlac program displays a series of options called 'menus' on its screen to prompt the user as he digitizes and morphologically classifies craters. Digitized coordinates are sent from the Imlac to the IBM computer via the PDP interface and are immediately written to disk for later processing. Some data editing is possible before records are transmitted from the Imlac.

A series of interactive programs is available on the 360 which permits extensive editing of the raw data and which converts the digitizer coordinate values to crater size, shape, and position information before storing the data in the master file. Any subset of the records in this file lying within a specified region can be quickly extracted for analysis. The region chosen may be irregularly shaped as long as it is bounded by a simple polygon. This capability is important because it permits craters lying in some specific region to be isolated for separate study. If the area covered by a crater field is of importance in a given analysis, undigitized portions of the surface can be excluded so as not to bias the results.

The display and analysis software can generate crater maps and diameter-frequency plots similar to those produced by the Brown system [*Jones*, 1974]. In addition, a morphological classification display program has been written which can be a powerful tool in the analysis of adjacent or overlapping zones with different ages or histories. Any combination of the 15 morphological parameters from Table 3, plus diameter, can be used to define a class. Up to 10 classes may be defined per execution of the program. All craters within a specified polygonal area are read from the master file and classified by using the given criteria. An output picture is produced which is an orthographic projection of the polygon. The brightness of each picture element is determined by that class to which the nearest crater belongs. The user is free to combine as many variables and their respective values as he wishes in defining a class. Any of the logical operators 'and,' 'or,' 'equals,' 'not,' 'less than,' or 'greater than' may be used as connectives. Thus one class might be 'all craters with scalloped or polygonal rims which are deep or moderately deep.' Figure 7 is an example of the output of this program. Floor shape was the sole criterion. Dark areas in the figure contain craters with flat floors; bright areas contain craters with bowl-shaped floors. The figure contains an area approximately 3° × 5° centered near 29°N, 130°W (Tharsis). It is a complex area containing craters with various floor shapes scattered randomly. When larger areas have been digitized, such a map may clearly delineate different geologic zones. Using the interactive image-processing system at IPL, a user can define a new set of classes and generate and display a map in 5 min. In this way he can experiment with a wide variety of classification schemes to see which ones are the most informative.

IPL PROCESSING OF VIKING ORBITER MULTISPECTRAL IMAGES

The Viking orbiter mission has presented the first opportunity to conduct high-resolution multispectral investigations of the Martian environment. Multispectral images can contribute significantly to the discrimination and evaluation of features in the observed scene because they display interrelationships among large quantities of multispectral data in a single presentation. For example, on the basis of color differences, color images allow easy discrimination between atmospheric phenomena and surface features which are sometimes barely distinguishable in a black-and-white image.

All color composites to date have used black-and-white images exposed through the violet and green interference filters and the red absorption filter. These filters have effective wavelengths of approximately 440, 529, and 591 nm, respectively. Unfortunately, while vidicon spectral sensitivity is similar to the photopic curve, it declines markedly at wavelengths greater than 600 nm and is less than 5% of its maximum sensitivity at 650 nm [*Benesh and Thorpe*, 1976]. Thus although the VIS is imaging the 'red planet,' we are unable to reproduce accurately the red spectral range that would be perceived by the human eye, because we do not have adequate data for wavelengths greater than 650 nm. Therefore all color reconstruction produced from Viking orbiter images must be characterized as being approximate.

Approximate Color Reconstruction

Synthesis of a color composite requires digital processing of three images of the same scene with similar viewing and camera conditions. In addition to the decalibration and map projection techniques described in the stereophotogrammetry section of this paper this processing includes normalization for exposure time and camera sensitivity differences and registration of the three black-and-white images. Normalization is accomplished through the application of filter factors as described by *Klaasen et al.* [1977] to compensate for the different spectral response of each camera system, including telescope and filter transmissions and vidicon sensitivity. Ideally, when these filter factors are used to decalibrate a triplet of images of a gray scene under achromatic illumination, the reconstructed color is gray. Image registration is accomplished by using cross correlation to determine the direction and magnitude of error and geometrically transforming two of the images with respect to the third.

Color Enhancement

Generally, the processing described above yields individual black-and-white images and color composites which exhibit low contrast. This is a consequence of the generally low contrast photometric nature of Mars. Plate 1*a* displays an example of a low-contrast color composite. These versions do not provide the relative multispectral information required for complete analysis. Thus two enhancement techniques are em-

TABLE 3. Crater Morphology Menu

Key Word	Value 1	Value 2	Value 3	Value 4	Value 5
Floor shape	bowl	intermediate	flat	convex	hummocky
Rim shape	circular	elongated	scalloped	polygonal	concentric rings
Floor texture	smooth	textured	fractured	filled	inundated
Central feature	none	peak	cluster	ring	pit
Depth	deep	moderately deep	shallow	very shallow	ghost
Walls	smooth	slumped	subdued slump	ringed, plain	dissected
Raised rim	sharp	subdued	absent	hummocky	breached
Ejecta texture	absent	radial	concentric rings	hummocky	smooth
Secondary	none	secondary	dark halo	bright rays	
Ejecta rings	none	1	2	3	
Clustering	singlet	doublet, no septa	doublet septa	doublet wide	multi
Ejecta lobes	none	1–20	20–40	40–60	>60
Streak	none	bright	dark	bright bifurcated	dark bifurcated
Terraces	none	1	2	3–5	>5
Ejecta margin	none	sharp scarp	subdued scarp	smooth	hummocky

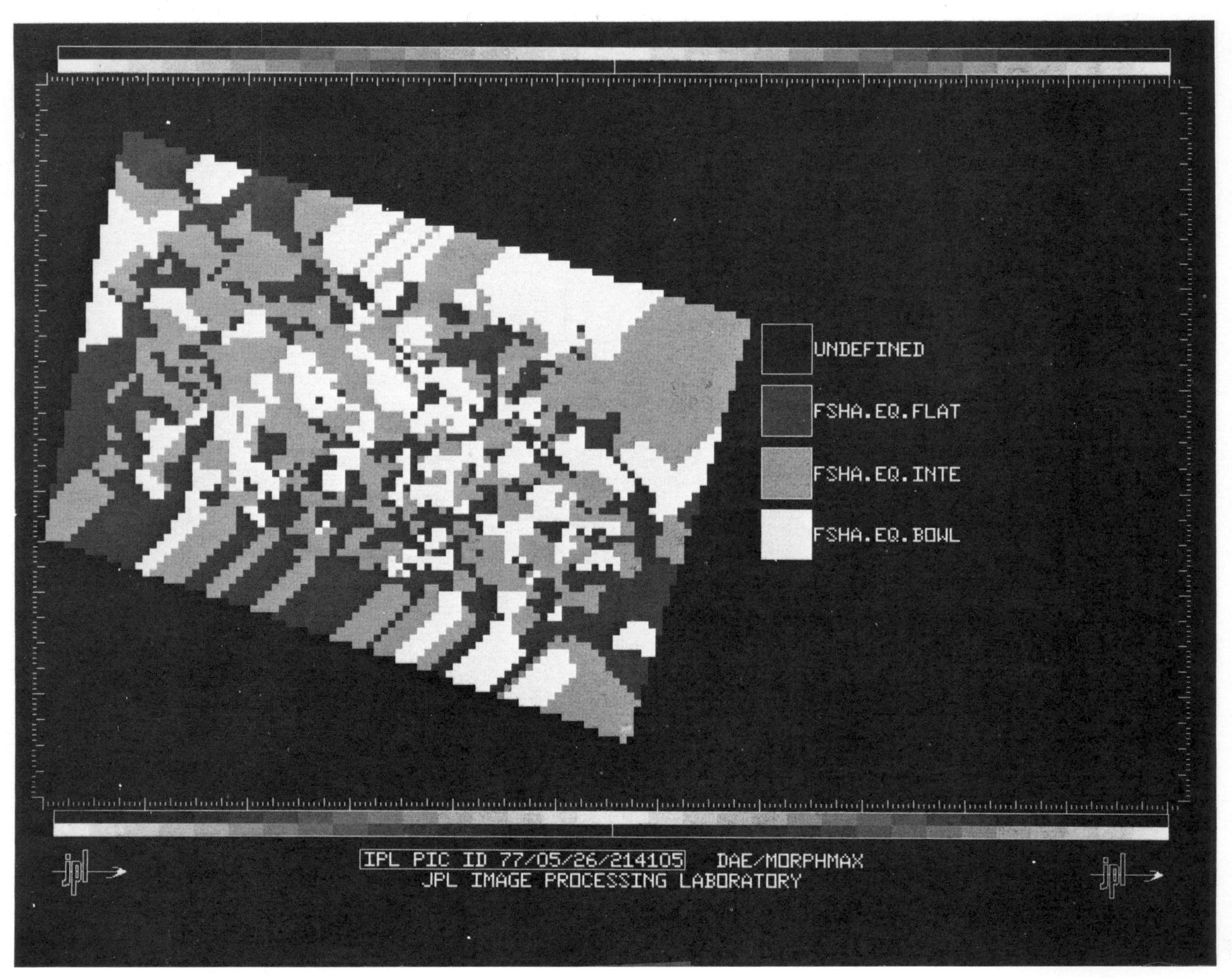

Fig. 7. Crater classification map of a 3° × 5° area centered near 29°N, 130°W (Tharsis). Dark areas contain craters with flat floors; bright areas contain craters with bowl-shaped floors.

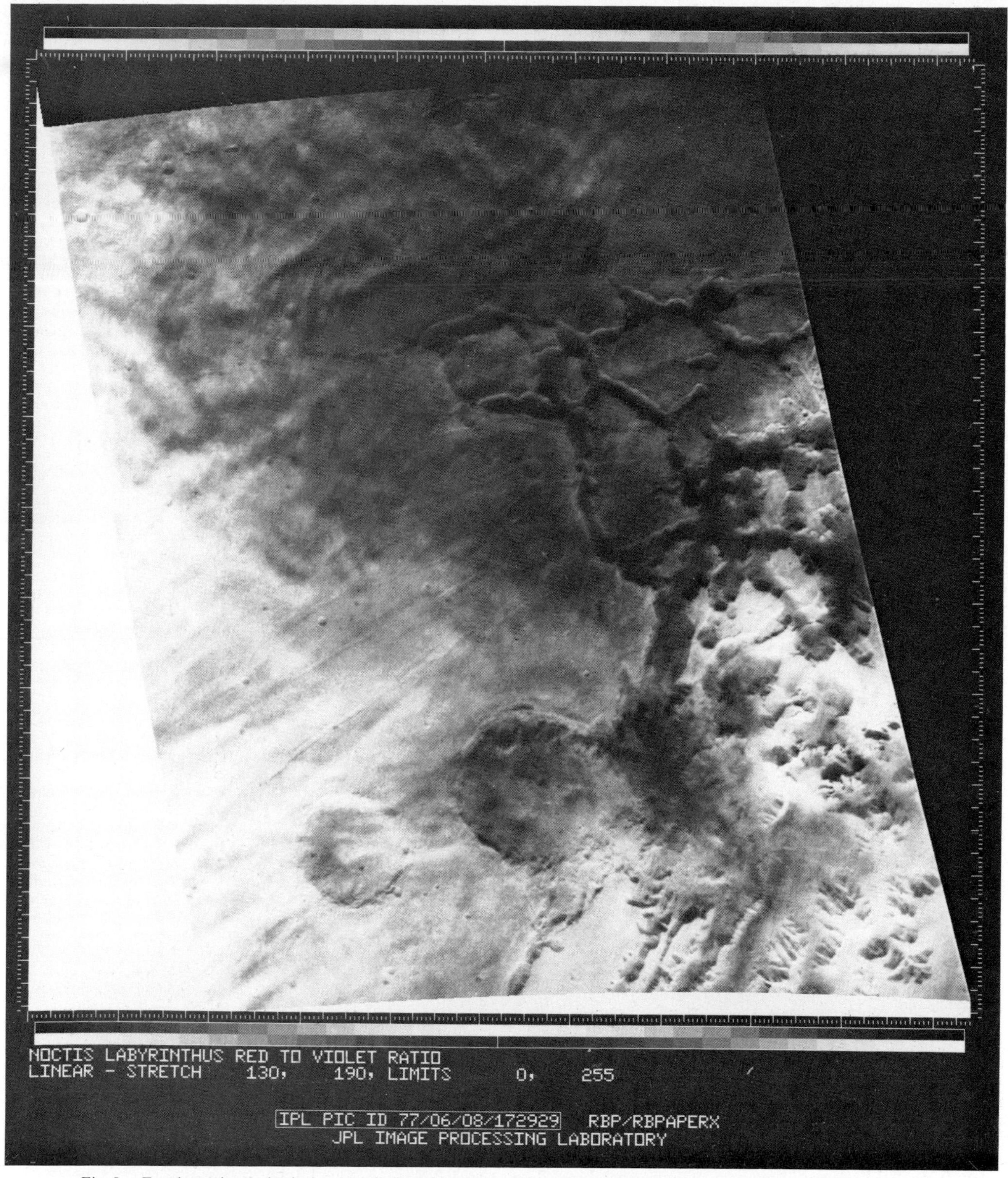

Fig. 8. For the region Labyrinthus Noctis the red to violet ratio was calculated, and the contrast increased to emphasize differences. The darker areas correspond to a more violet color, while the brighter areas correspond to a relatively redder surface.

ployed to produce data with highly visible scene content. Ratio pictures are generated to display relative color differences, and color coordinate transformation [*Soha et al.,* 1976; A. R. Gillespie, unpublished data, 1977] is used to separate hue, saturation, and intensity components of a color composite. Once they are separated, these components can be independently enhanced.

Color ratios. Taking the ratio of two spectral images is a quantitative approach to color discrimination. The resultant black-and-white 'color ratio' image may then be enhanced by using standard algorithms, such as a contrast stretch, to aid in the identification and investigation of subtle color variations. The red to violet ratio picture of the Labyrinthus Noctis region in Figure 8 is a typical example of how a color ratio image can be used to discriminate between areas which are relatively 'bluer' (shown as being dark in Figure 8) and areas which are relatively 'redder' (shown as being bright in Figure 8).

Color coordinate transformation. The introduction to this paper describes the loss of absolute data relationships which generally occurs when qualitative image enhancement techniques are used to improve scene visibility. The independent application of the standard high-frequency and contrast enhancement techniques to the violet, green, and orange images invalidates the relative color balance and renders a color composite scientifically uninterpretable. Thus it was necessary to utilize a technique which allows exaggeration of atmospheric and surface detail in a color composite image while maintaining relative color information. Processing of Landsat multispectral images has demonstrated that a coordinate transformation of input violet, green, and orange images to an orthogonal system of hue, saturation, and intensity images allows enhancement of scene detail while retaining relative color balance. Isolation of all intensity information in only one of the three images allows standard enhancements such as spatial filtering and contrast stretching to be performed without introducing artifacts into the color information, which is contained in the other two images. Furthermore, this transformation allows the exaggeration of color differences by contrast enhancement of the hue and saturation images to discriminate among atmospheric and geologic features. The reverse transformation is then employed to obtain blue, green, and red images for the production of an enhanced color composite picture.

Plates 1*b*, 2*a*, and 2*b* display the results of applying a color coordinate transformation to Viking orbiter multispectral images. Plate 1*b* displays the same scene as Plate 1*a* after color coordinate transformation followed by contrast enhancement has been applied to the intensity component. Plate 2*a* illustrates the effect of applying a filter and a contrast enhancement to the intensity component to, respectively, correct for the high-frequency loss resulting from the VIS modulation transfer function and improve the visibility of the corrected version. Plate 2*b* is a color-enhanced version of Plate 2*a* which resulted from contrast enhancement of both the hue and the saturation components. It can be seen that Plate 2*a* improves the discrimination of fine ground detail, whereas Plate 2*b* readily allows the differentiation between surface and atmospheric features.

In summary, a number of specialized techniques have been employed at the IPL for the production and enhancement of Viking orbiter multispectral imagery. These techniques include methods by which the VIS distortions have been minimized, while scene detail and color discrimination have been optimized. The successful utilization of these techniques has provided Viking scientists with previously unavailable high-resolution multispectral information on the Martian environment.

Summary

IPL processing of the Viking orbiter images produced data that were directly suitable for qualitative and quantitative scientific interpretation. Crater morphology data were used to differentiate the characteristics and the relative ages of the Martian surface. Computer-generated topographic relief data were used to measure relative and absolute surface elevations, and color ratios and reconstructions were used to isolate the spectral characteristics of the atmosphere and the surface.

This processing and these products represent a logical evolution of IPL image-processing techniques inherited from previous missions. Refinements to this processing will continue through the Viking orbiter extended mission, providing an even greater legacy of technology for future missions.

Acknowledgments. The task of defining, developing, and applying the software and procedures required to process the Viking orbiter images spanned several years and involved the help of many individuals. The authors gratefully acknowledge the camera calibration and in-flight mission data processing contributions of R. J. Quiros, F. G. Staudhammer, N. J. Constantinides, and the JPL Space Instrumentation and Photography Section, supervised by D. D. Norris. J. E. Kreznar, M. A. Girard, A. R. Gillespie, and J. D. Addington contributed essential new software. M. Benesh provided the technique and the software used to derive absolute topographic elevations, and K. Pang and D. Petrie converted this software for IPL. The IPL operations group supervised by J. B. Seidman provided computer and photolab support. Library services were provided by the Viking Data Library team, guided by M. D. Martin and D. L. Miller. Recent processing was provided by L. Shigg and C. Y. Han. The authors would particularly like to acknowledge W. B. Green and K. S. Watkins for their overall guidance, D. J. Lynn for his critical review of this paper, M. H. Carr and the Orbiter Imaging team for their patience and cooperation, J. A. Cutts and K. R. Blasius for their contribution and guidance to the computer stereo process, and J. B. Wellman, K. P. Klaasen, and T. E. Thorpe for their relentless search for processing that would yield improved products. This paper presents the results of one phase of research conducted at the Jet Propulsion Laboratory, California Institute of Technology, under National Aeronautics and Space Administration contract NAS 7-100.

References

Arvidson, R. E., T. A. Mutch, and K. L. Jones, Craters and associated aeolian features on Mariner 9 photographs: An automated data gathering and handling system and some preliminary results, *Moon*, *9*, 105, 1974.

Benesh, M., Mariner 9 stereophotogrammetry, *Photogramm. Eng.*, *XXXIX*(11), 1171–1179, 1973.

Benesh, M., and T. Thorpe, Viking orbiter 1975 visual imaging subsystem calibration report, *JPL Intern. Doc. 611-125*, Jet Propul. Lab., Pasadena, Calif., 1976.

Blasius, K. R., and J. A. Cutts, A test digital topographic map compilation using the Vicar image processing language, *Sci. Appl. Intern. Doc. SA15-420-393-E1*, Sci. Appl., Inc., Pasadena, Calif., 1975.

Dunne, J. A., W. D. Stromberg, R. M. Ruiz, A. Collins, and T. E. Thorpe, Maximum discriminability versions of the near-encounter frames, *J. Geophys. Res.*, *76*, 438–472, 1971.

Gillespie, A., and J. Soha, An orthographic photomap of the south pole of Mars from Mariner 7, *Icarus*, *16*, 522–527, 1972.

Green, W. B., P. L. Jepsen, J. E. Kreznar, R. M. Ruiz, A. A. Schwartz, and J. B. Seidman, Removal of instrument signature from Mariner 9 television images of Mars, *Appl. Opt.*, *14*, 105–114, 1975.

Jepsen, P. L., The software/hardware interface for interactive image processing at the Image Processing Laboratory of the Jet Propulsion Laboratory, paper presented at 1976 Fall Decus Symposium, Digital Equip. Comput. Users Soc., Las Vegas, Nev., Dec. 6–9, 1976.

Jones, K. L., Evidence for an episode of crater obliteration intermediate in Martian history, *J. Geophys. Res.*, *79*, 3917–3931, 1974.

Klaasen, K. P., T. E. Thorpe, and L. A. Morabito, In-flight performance of Viking visual imaging subsystem, *J. Geophys. Res.*, *82*, this issue, 1977.

Rindfleisch, T. C., J. A. Dunne, H. J. Frieden, W. D. Stromberg, and R. M. Ruiz, Digital processing of the Mariner 6 and 7 pictures, *J. Geophys. Res.*, *76*, 394–417, 1971.

Seidman, J. B., W. B. Green, P. L. Jepsen, R. M. Ruiz, and T. E. Thorpe, User guide to the Mariner 9 reduced data record, *JPL Tech. Memo. 33-628*, Jet Propul. Lab., Pasadena, Calif., 1973.

Soha, J. M., D. J. Lynn, J. J. Lorre, J. A. Mosher, N. N. Thayer, D. A. Elliott, W. D. Benton, and R. E. Dewar, IPL processing of the Mariner 10 images of Mercury, *J. Geophys. Res.*, *80*, 2394–2417, 1975.

Soha, J. M., A. R. Gillespie, M. J. Abrams, and D. P. Madura, Computer techniques for geological applications, in *Proceedings of the Caltech/JPL Conference on Image Processing Technology, Data Sources and Software for Commercial and Scientific Applications*, California Institute of Technology, Pasadena, Calif., 1976.

Wellman, J. B., F. P. Landauer, D. D. Norris, and T. E. Thorpe, The Viking orbiter visual imaging subsystem, *J. Spacecr. Rockets*, *13*, 660–666, 1976.

(Received April 1, 1977;
revised June 6, 1977;
accepted June 6, 1977.)

Viking Imaging of Phobos and Deimos: An Overview of the Primary Mission

T. C. DUXBURY

Jet Propulsion Laboratory, California Institute of Technology, Pasadena, California 91103

J. VEVERKA

Laboratory for Planetary Studies, Cornell University, Ithaca, New York 14853

The Viking primary mission (June 20 to November 15, 1976) yielded approximately 50 images of Phobos and Deimos. These pictures completed the surface coverage of Phobos begun by Mariner 9 in 1971–1972 and extended the coverage of Deimos. The effective resolution of the Viking images exceeds that obtained by Mariner 9 for two main reasons: improved camera performance and closer encounter distances. The improved resolutions revealed a number of unexpected surface features such as linear chains of irregular craters and elongated grooves and striations. The Viking coverage has provided the first high-resolution color information on the surfaces of the two satellites and has extended the phase angle coverage to 125°. Additionally, a number of images of the satellites were obtained against star backgrounds to refine further the ephemerides.

INTRODUCTION

The first satellite picture was taken of Phobos on July 24, 1976, by Viking orbiter 1 (VO-1). During the Viking primary mission the cameras on the two orbiters acquired about 50 pictures of the two Martian moons. Viking imaging completed the high-resolution (<500 m) surface coverage of Phobos and extended the surface coverage of Deimos obtained by Mariner 9. Like the close encounters of Mariner 9, VO-1 and Viking orbiter 2 (VO-2) close encounters with Deimos were inside the orbit of Deimos. Such encounters did not permit high-resolution viewing of the backside of Deimos facing away from Mars.

The Viking images of the satellites have a higher surface resolution than those obtained by Mariner 9. Improved Viking camera modulation transfer functions increased camera response to high-frequency (small scale) scene variations. Also, the orbits of the two spacecraft were such that many satellite encounters at ranges within 5000 km occurred, while all Mariner 9 pictures were taken at ranges greater than 5000 km. The Viking coverage produced the first high-resolution color information on the satellite surfaces and extended the phase angle coverage up to 125°. Additionally, some star pictures were taken in conjunction with the satellite pictures to give precision fixes for satellite ephemeris improvement. The following sections give an overview of these various pictures.

HIGH-RESOLUTION SURFACE COVERAGE

The increment in high-resolution coverage of the satellite surfaces produced by Viking is shown in Figure 1. Viking essentially filled in the gap left by Mariner 9 in our high-resolution coverage of the northern hemisphere of Phobos. The coverage of Deimos was not extended significantly, since Viking could only image the Mars-facing side of Deimos. The typical surface resolution achieved was 100–200 m, although detail as small as 40 m was imaged on Phobos during a particularly close passage (see Figure 6).

Figure 2 shows a comparison of Viking and Mariner 9 surface resolution. Both images were taken at a range of about 7000 km and show the southern hemisphere of Phobos under similar lighting conditions. The intriguing grooves and crater chains visible in the Viking image are at best marginally detectable (with hindsight, that is) in the Mariner 9 frame.

Figure 3 shows a pair of Deimos pictures in which the

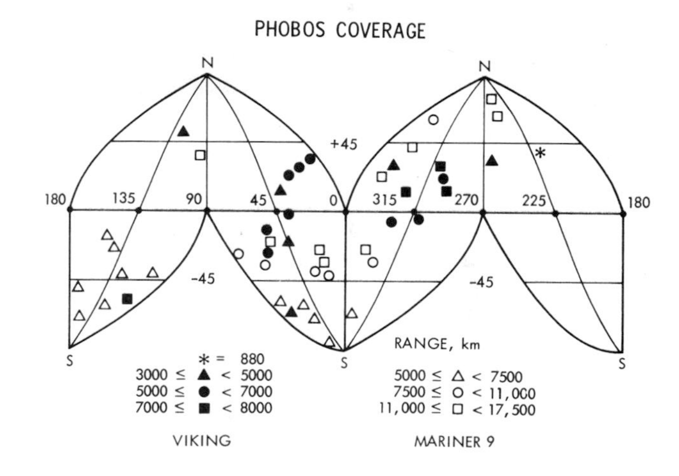

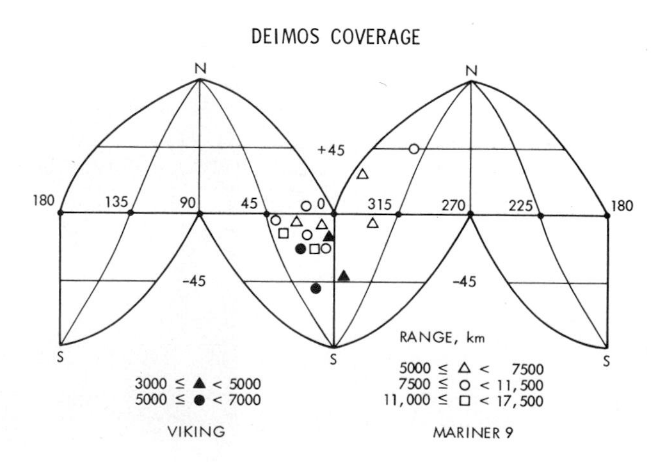

Fig. 1. The subspacecraft points of Mariner 9, VO-1, and VO-2 on Phobos and Deimos indicated at times when high-resolution pictures were taken. The Viking pictures were chosen to complement the Mariner 9 pictures by filling in mission coverage and increasing resolution.

Paper number 7S0510.

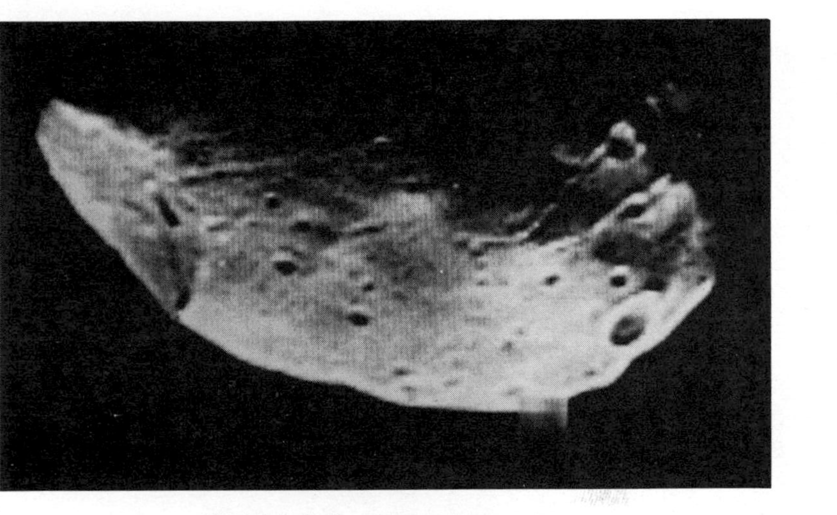

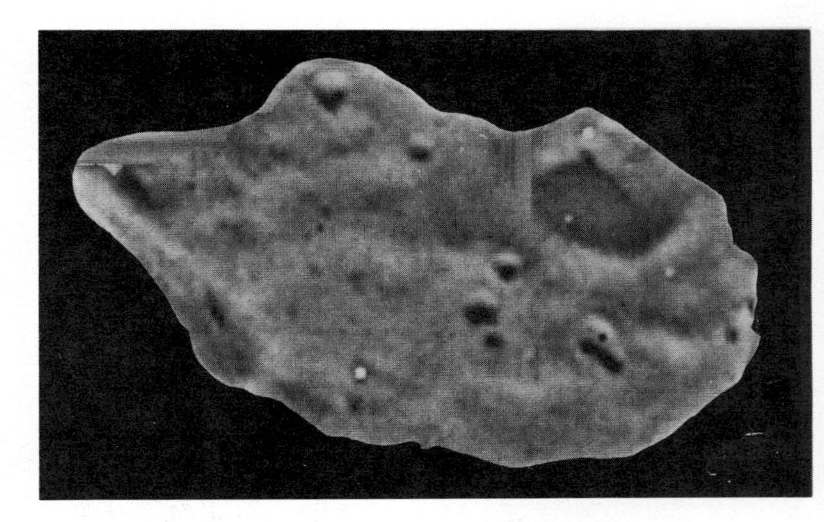

Fig. 2. A comparison between the resolving capability of the Mariner 9 and Viking cameras. Both pictures are of the southern portion of Phobos taken under similar lighting conditions. The Viking picture (055A32), even though it was taken at a greater range, clearly shows significantly more surface detail than the Mariner 9 picture (DAS 04470630). The detection of the crater chains was a surprising discovery by Viking.

Fig. 3. The highest-resolution pictures (056A93 and 056A96) ever taken of Deimos. The picture taken at 3000 km is the southernmost picture taken during the Viking primary mission. The highly irregular surface was found to be heavily cratered and had small areas of higher albedo near the limb. Also, a long linear depression running from lower left to upper right is seen in the upper half of each picture.

surface resolution approaches 100 m, the highest resolution for Deimos obtained during the nominal mission and higher than any achieved by Mariner 9 [*Veverka et al.*, 1974].

Color Sequences

Color sequences (covering the range of ~0.4–0.6 μ) were obtained for each satellite. On orbit 056A, two color images of Deimos were obtained at a range of 3000 km and at a phase angle of ~80°. On orbit 115A, violet, clear, green, and red images of Phobos were obtained at a range of 3700 km and a phase angle of ~90°. Analysis of such color sequences will provide color difference information at scales as small as 200 m on the satellite surfaces. Preliminary analysis indicates that such color differences are, at most, very small.

Phase Angle Coverage

The Viking cameras have more pointing flexibility than did those on Mariner 9. In principle the range of phase angles observable by Viking is 0°–135° compared with about 20°–90° for Mariner 9. The actual range achieved during the primary mission was 25°–125° (See Figure 4). The extension of the coverage to large phase angles is particularly important for improving our knowledge of the photometric functions of the satellites [*Noland and Veverka*, 1976].

Pictures for Ephemeris Improvement

The fact that the orbiters have twin imaging systems [*Wellman et al.*, 1976] can be used to advantage for determining satellite positions relative to star backgrounds. A satellite can be centered in one camera and imaged with the appropriate exposure while a maximum exposure of 2.7 s is taken with the second camera recording background stars. Such pairs of satellite and star images can be used to determine the spacecraft-satellite vector in inertial space. The accuracy achievable is of the order of several tenths of a kilometer (in fixing the position of the satellite in its orbit) compared to accuracies of 5–10 km obtainable from Mariner 9 images, which have no star backgrounds. The Viking cameras are ideally suited for imaging stars, having large fields of view (>2.0 deg^2) and high sensitivity (stars as faint as ninth visual magnitude can be recorded). Over half of the normal satellite exposure pictures also had an accompanying star picture from the other camera.

On a few occasions, when picture playback was limited, images of Phobos were obtained against a star background in a single frame. Owing to the long exposure needed to record the stars, the image of Phobos becomes saturated, no surface detail being visible. An example of a satellite-star ephemeris picture is illustrated in Figure 5. The image of Phobos is saturated, and no surface detail is visible. Additionally, the picture is smeared (~15 pixels (picture elements)) owing to the motion of Phobos relative to the spacecraft during the long 2.7-s exposure.

Unique Pictures

In this section we comment briefly on several unique satellite pictures obtained during the nominal mission. Outstanding among these is the image of Phobos obtained from a range of only 880 km on orbit 38B (see Figure 6). This picture shows surface features as small as 40 m and is the highest-resolution

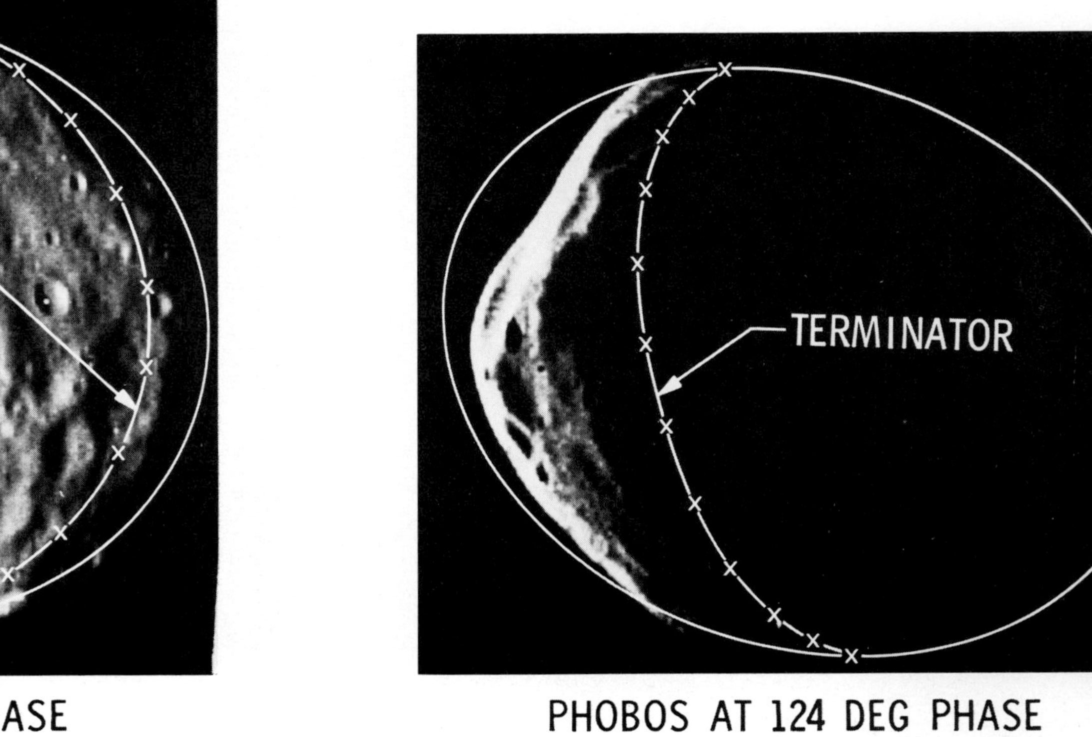

Fig. 4. Viking provided a greater range in phase angle coverage than was obtained by Mariner 9. The extremes in coverage obtained during the primary mission are shown, Phobos being observed at 35° (079B41) and 124° (111A03) phase angles. The overlays indicate the nominal limb and terminator for an ellipsoid. The 35° phase angle picture gives an excellent view of the largest crater, Stickney, with Kepler Dorsum at the bottom. Also, a crater chain in the middle of the picture is seen running from Stickney to the right.

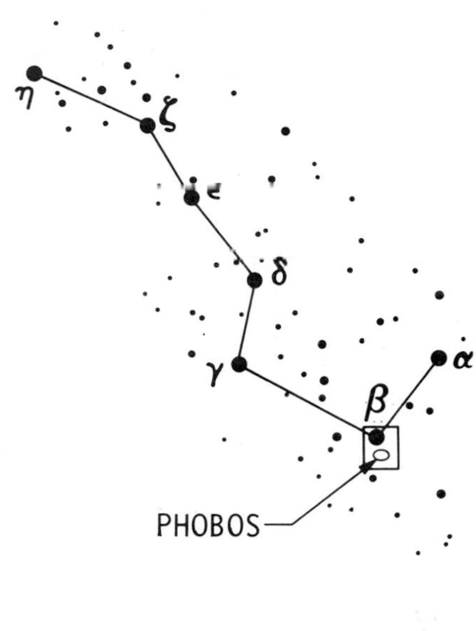

Fig. 5. Periodically, Phobos and stars were imaged in the same picture to obtain an accurate position fix without requiring two pictures to be played back to earth. Here Phobos was viewed as it passed by one of the pointer stars in the Big Dipper (β Ursa Major). The star image shows up as a little white dot near the top of the figure, while the overexposed Phobos image shows no surface detail (118A15).

picture taken by Viking during the primary mission. To achieve this remarkable resolution, it was necessary to compensate for the motion of Phobos relative to the spacecraft during the exposure by slewing the camera at 0.25°/s. In the absence of slewing the image would have shown a smear of about 15 pixels and an effective resolution of only ~250 m. This sequence marked the first time that the camera-slewing technique was attempted. Now that the technique has been demonstrated, it will become an integral part of the Viking extended mission and will permit high-resolution imaging of the planet at distances of 300 km and of the satellites at distances as small as 90 km.

In Figure 7 we see an example of a large phase angle view of Phobos in which the dark side of the satellite is illuminated by Mars light. Such images not only are useful in extending our knowledge of the photometric function of Phobos to large phase angles [*Noland and Veverka*, 1976] but are important in determining the size and shape of the satellite from its outline.

In Figure 8 we see an image of Phobos in the penumbra of Mars. This long-exposure picture is typical of the 'satellite with star background' pictures taken for ephemeris improvement. A sequence of such pictures as the satellite progressively goes through the penumbra into the umbra would be very valuable for analyzing the scattering properties of the Martian upper atmosphere.

VIKING EXTENDED MISSION

The satellite pictures obtained during the primary mission have given us a direction for taking additional pictures during the extended mission. High-resolution coverage of the crater chains and striations on Phobos as well as backside coverage of Deimos is needed. Additional high-resolution surface coverage giving overlapping stereo coverage will be needed for global cartography and geodesy. Photo coverage throughout the satellite orbits will be needed to determine amplitudes and periods of rotational librations. More complete phase angle and color coverage is needed for detailed surface photometry and composition studies. Also, satellite-star ephemeris pictures will be important to maintain kilometer or subkilometer position computations for studying a possible secular acceleration in the longitude of Phobos.

Flybys within 100 km are possible in February and May of 1977 for Phobos and in October and December of 1977 for Deimos. These close flybys could not only provide the additional picture coverage described above but also be used to determine the satellite masses. The extended mission data may provide the answer to the questioned origins of Phobos and Deimos. Since the submission of this article, multiple flybys of Phobos within 300 km have been obtained in February and March of 1977. The closest flyby of 88 km occurred on February 20, 1977.

Fig. 6. Phobos observed at 880 km to reveal the presence of striations and crater chains (039B84). The striations appear to follow the surface topography and are seen inside of the larger craters. The crater chains appear to be similar to secondary craters observed on the moon and Mercury. This picture was taken while the camera was slewing at 0.25°/s to compensate for the relative velocity between Phobos and VO-2.

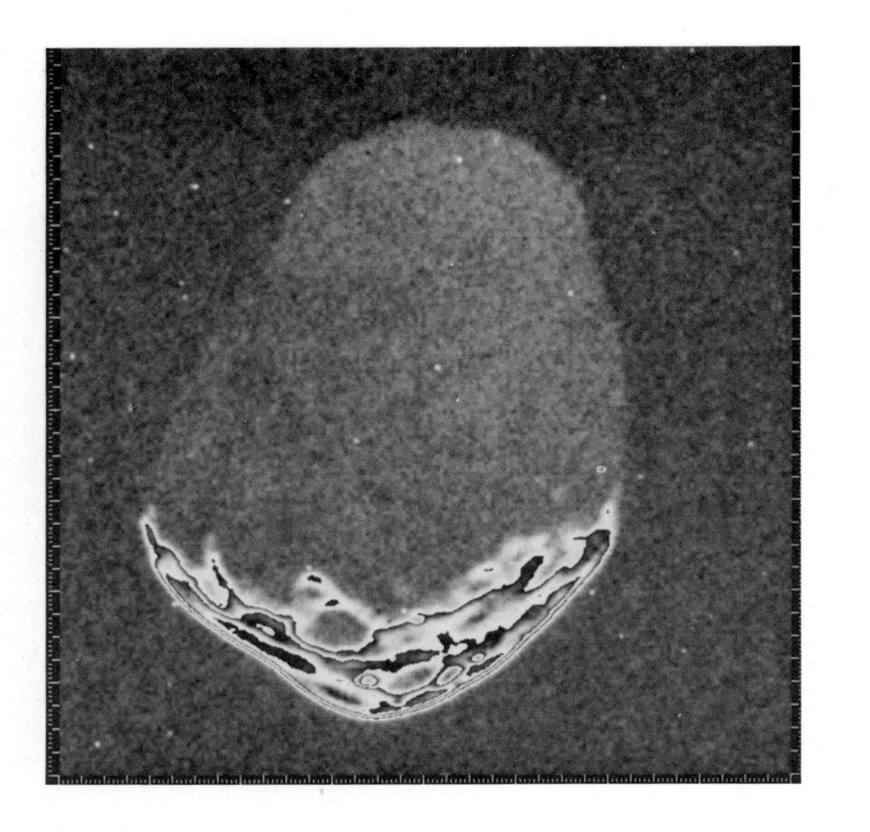

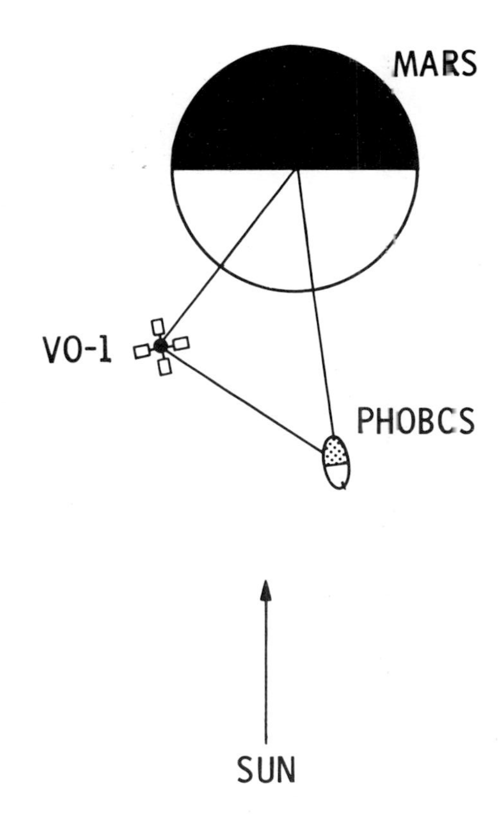

Fig. 7. Phobos observed at a high phase angle, where a portion was illuminated by reflected sunlight from Mars (130A13). The geometry associated with this picture is shown where both VO-1 and Phobos are on the sunlit side of Mars. Special processing of the picture was performed to enhance the reflected light.

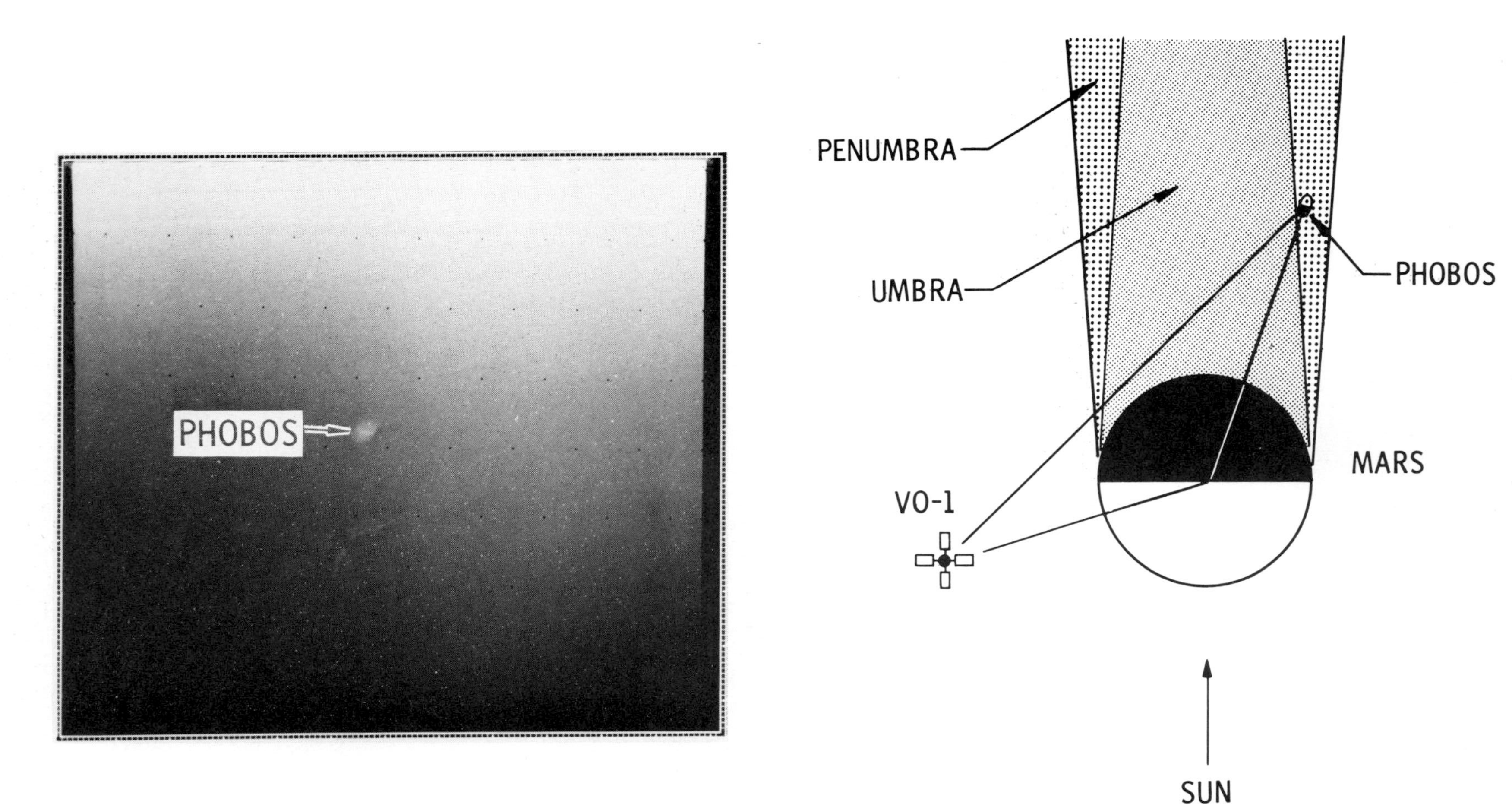

Fig. 8. Phobos appearing barely detectable in the center of this picture (118A71), which was a 2.7-s exposure of Phobos in the penumbra of Mars. Star images were also detected in this picture. As indicated, VO-1 was on one side of the Mars shadow while we looked behind Mars to see Phobos in the opposite side penumbra.

Acknowledgment. This paper presents the results of one phase of research conducted at the Jet Propulsion Laboratory, California Institute of Technology, under National Aeronautics and Space Administration contract NAS 7-100.

REFERENCES

Noland, M., and J. Veverka, The photometric functions of Phobos and Deimos, I, Disc-integrated photometry, *Icarus, 28,* 405–414, 1976.

Veverka, J., M. Noland, C. Sagan, J. Pollack, L. Quam, R. Tucker, B. Eross, T. Duxbury, and W. Green, A Mariner 9 atlas of the moons of Mars, *Icarus, 23*(2), 206–289, 1974.

Wellman, J. B., F. P. Landauer, D. D. Norris, and T. E. Thorpe, The Viking orbiter visual imaging subsystem, *J. Spacecr. Rockets, 13,* 660–666, 1976.

(Received April 1, 1977;
revised June 7, 1977;
accepted June 7, 1977.)

Viking Observations of Phobos and Deimos: Preliminary Results

J. VEVERKA

Laboratory for Planetary Studies, Cornell University, Ithaca, New York 14853

T. C. DUXBURY

Jet Propulsion Laboratory, California Institute of Technology, Pasadena, California 91103

The improved resolution of the Viking orbiter images has led to the discovery of a number of unusual features on the surface of Phobos: (1) elongated rill-like depressions associated with the crater Stickney (possibly surface fractures), (2) chains of irregular craters which sometimes show a 'herringbone' pattern (possibly secondaries), and (3) sets of almost parallel linear striations of uncertain origin. The crater chains are not randomly oriented but tend to lie parallel to the orbital plane of Phobos. The striations, on the other hand, appear to form arcs of small circles which are normal to the Mars-Phobos direction. With the possible exception of feature (2), similar features have not been recognized on Deimos, possibly because of the coarser resolution of available imagery. The Viking data demonstrate that the surfaces of both satellites are definitely saturated with craters ≥300 m across.

1. INTRODUCTION

During the Viking primary mission (June–November 1976), about three dozen high-resolution images of Phobos and Deimos were obtained by the Viking orbiter cameras [*Duxbury and Veverka*, 1977]. The prime objective was to extend the high-resolution coverage of the satellite surfaces obtained by Mariner 9 in 1971–1972. For example, in the case of Phobos the Mariner 9 coverage of the north polar regions and of the potentially interesting area around 230°W, 10°N, which is antipodal to Stickney (the largest crater on Phobos), was of mediocre quality. In the case of Deimos the quality of the Mariner 9 coverage was generally poor; in fact, so few craters were visible on the images that a statistically meaningful crater density curve could not be constructed for the outer satellite [*Thomas and Veverka*, 1977].

The Viking coverage of the satellites is superior to that obtained by Mariner 9 in several respects but primarily in terms of effective resolution. While Viking obtained images of Phobos and Deimos from distances as small as 880 and 3000 km, respectively, the corresponding minimum ranges in the case of Mariner 9 were 5710 and 5490 km [*Veverka et al.*, 1974]. In addition, the performance of the Viking cameras is superior to that of the Mariner 9 B camera [*Carr et al.*, 1976]. For example, whereas in the Mariner 9 satellite images, limbs are typically 3–4 pixels (picture elements) wide [*Noland and Veverka*, 1976*a*], in the Viking images their width does not exceed 1–2 pixels.

The improved resolution has led to better crater density counts on Phobos and to the first statistically valid measurement of the crater density on Deimos. More important, the improved resolution has led to the discovery of a suite of unusual surface features on Phobos. It is the purpose of this paper to give a preliminary account of these new results. An overview of the satellite coverage during the Viking primary mission is given in a companion paper [*Duxbury and Veverka*, 1977].

2. DISCOVERY OF UNUSUAL SURFACE FEATURES ON PHOBOS

The improved resolution of the Viking images has led to the discovery of a number of unusual surface features on Phobos. On the basis of their morphology these can be divided into three major categories: (1) elongated rill-like depressions associated with the crater Stickney, possibly surface fractures or rows of coalescing secondaries, (2) chains and clusters of irregular elongated craters, possibly secondaries, and (3) parallel linear striations or grooves of enigmatic origin.

With the possible exception of category (2), similar features have not been recognized on the surface of Deimos, possibly because of insufficient resolution. On Phobos, such features become prominent when the surface resolution exceeds 100 m and approaches 50 m.

a. Elongated Rill-Like Depressions Associated With Stickney

A number of elongated rill-like depressions occur in the vicinity of Stickney, the largest crater on Phobos. These features are clearly visible in Figure 1.

The most prominent of the elongated depressions seems to originate at the rim of Stickney and extends some 10 km in almost a straight line. At the resolution of Figure 1 (~100 m) it is difficult to be sure whether this depression is a trough or a string of coalescing irregular pits, although it does look more like a trough which may taper somewhat distally from Stickney. It is conceivable that this feature represents a fracture in the surface of Phobos produced by the severe impact which formed Stickney.

Note that one somewhat similar feature can be followed more or less continuously from the rim of Stickney to the far limb. For most of its trend the feature appears to consist of almost coalescing elongated depressions and terminates in a puzzling arcuate trough near the limb. Clearly, still higher resolution images of these features are needed before their origin can be interpreted more definitively. Nevertheless, the association of at least some of them with Stickney appears certain.

Unfortunately, it was not possible during the nominal mission to examine the neighborhood of Hall, the large 6-km crater near the south pole, at high resolution for similar features. None have been identified in the vicinity of Roche, the 5-km crater near the north pole (Figure 2).

Paper number 7S0567.

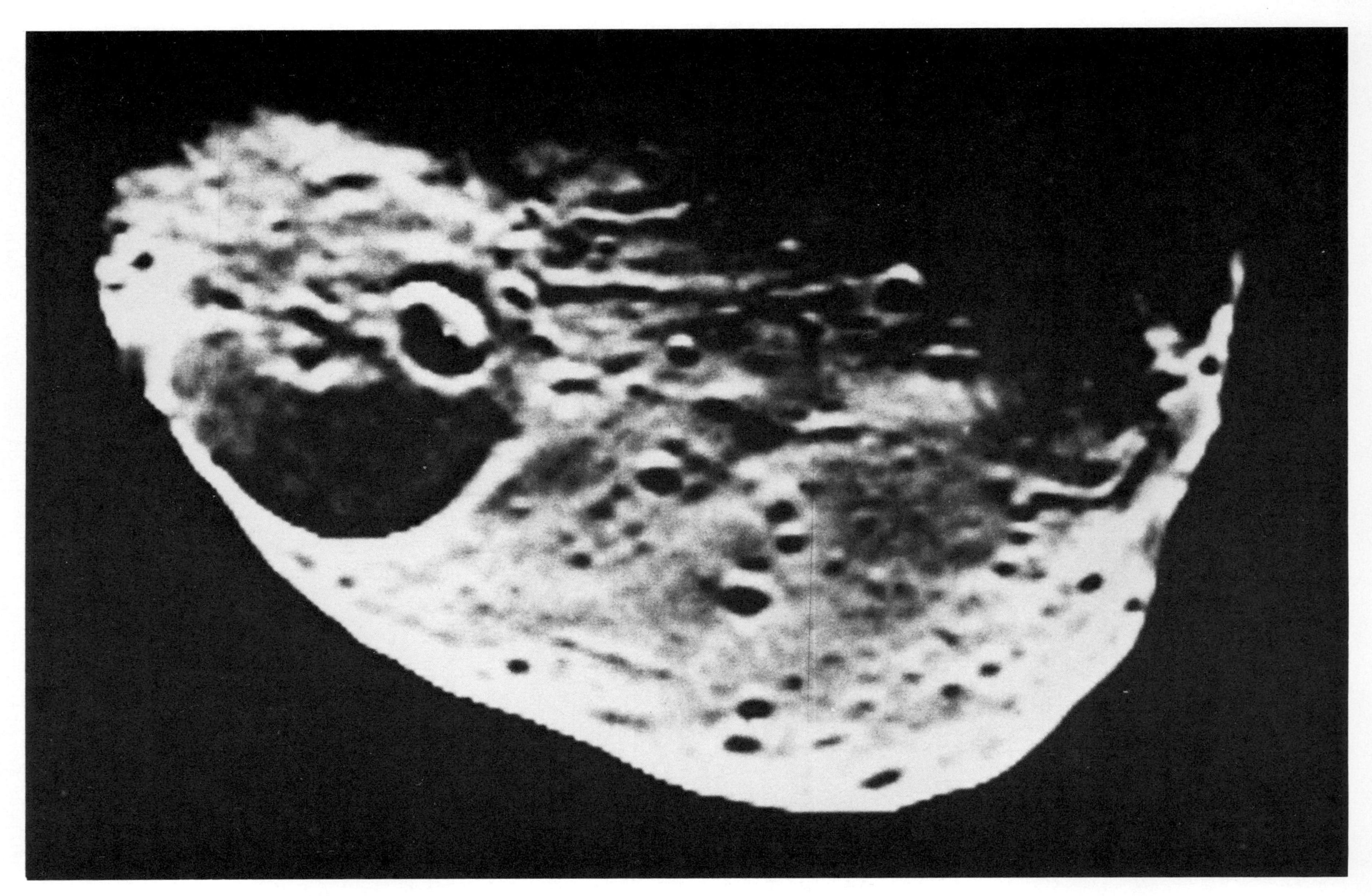

Fig. 1. View of Phobos from Viking Orbiter 1 (87A52). Stickney, the largest crater on Phobos, is seen at left. Note the rill-like depressions trending from left to right which appear to emanate from the rim of this 10-km crater.

Fig. 2. View of Phobos from Viking Orbiter 2 (039B84). The large crater at top (Roche) lies close to the north pole of Phobos and is about 5 km across. Enlarged segments of this frame are shown in Figures 3, 7, and 8.

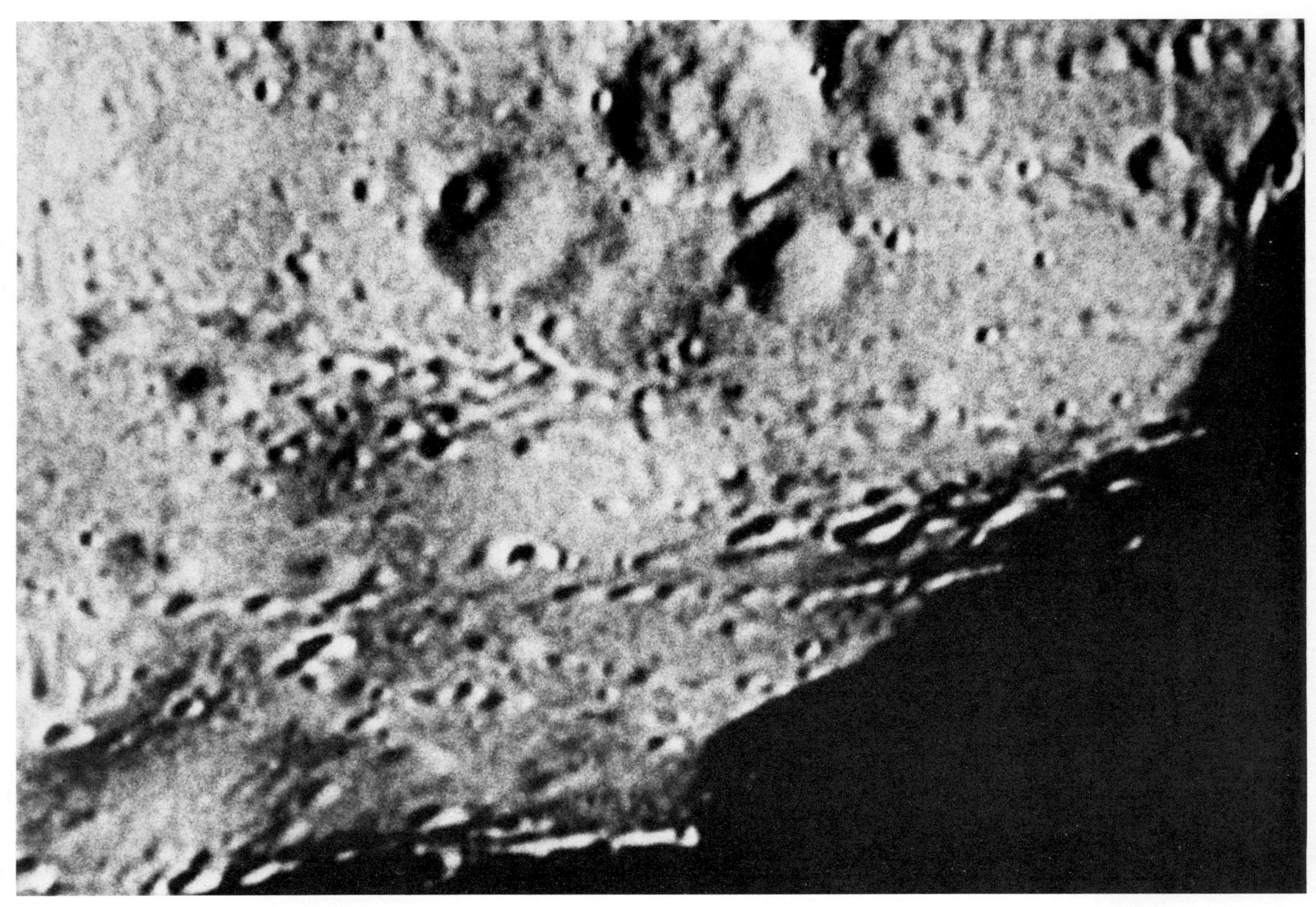

Fig. 3. Enlarged segment of frame 039B84 (Figure 2) showing clusters of irregular craters arranged in herringbone patterns characteristic of secondary impacts. The largest crater in this view (top center) is about 1.5 km across.

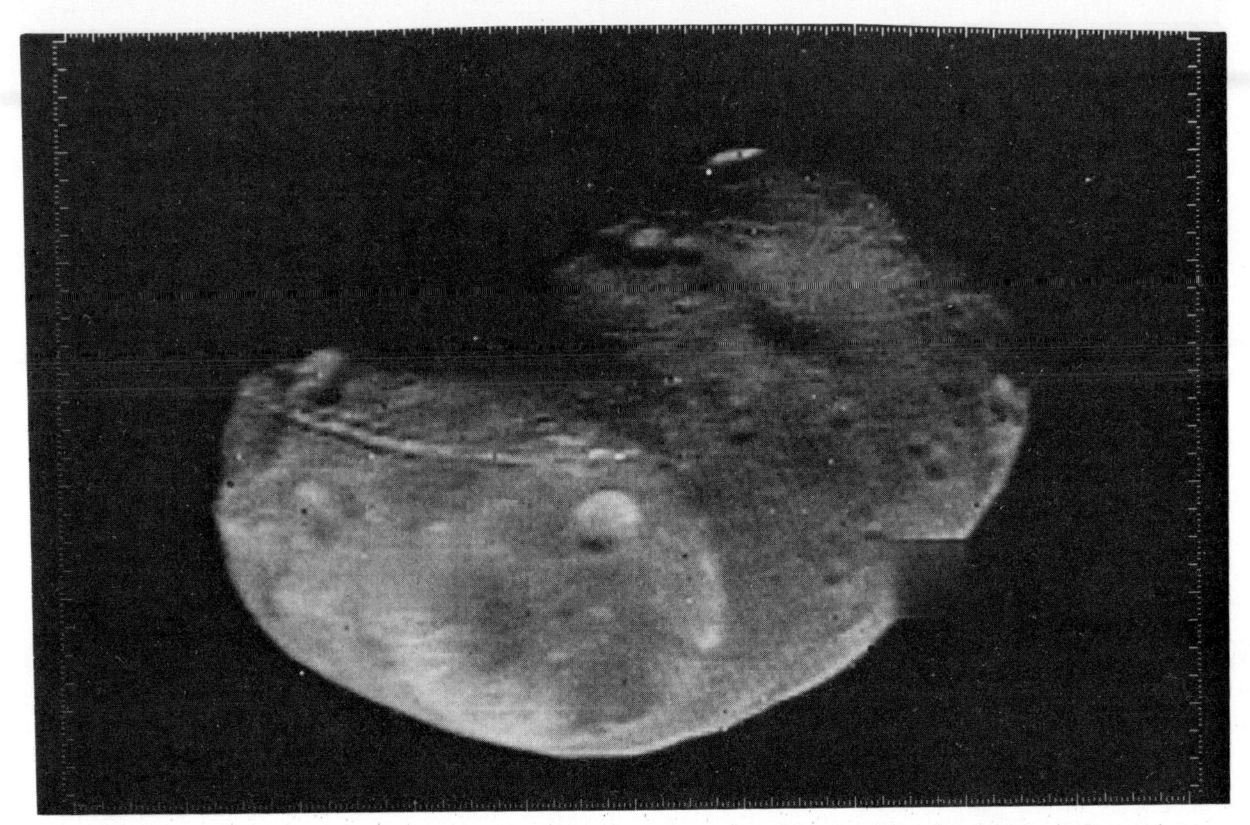

Fig. 4. View of Phobos from Viking Orbiter 2 (081B33) showing a prominent chain of irregular craters superimposed on the crater Stickney.

b. Chains and Clusters of Irregular Elongated Craters

A number of prominent crater chains are seen in Figure 2. Typically, the chains consist of irregular craters ~50–200 m across, which sometimes cluster into the 'herringbone' pattern (Figure 3) characteristic of secondary ejecta [*Oberbeck and Morrison*, 1974]. Such crater chains appear to be common on Phobos (Figure 4) but do not appear to be randomly oriented. Rather, they all seem to be closely parallel to the orbital (equatorial) plane of Phobos (Figure 5).

Traditionally, there has existed the feeling that since the escape velocity from Phobos is so low (~12 m/s), very little ejecta would reimpact the surface and that secondary craters should be absent. Thus when *Pollack et al.* [1973] described a row of several 200-m craters visible in some Mariner 9 images of Phobos, they did not consider the possibility that these were secondaries. Instead, they suggested that such rows of small craters could be produced by localized outgassing along fractures following a severe impact such as that which formed Stickney. However, *Soter* [1971] had noted that some ejecta from Phobos would remain in Mars orbit and would have an opportunity to reimpact Phobos at fairly low velocities.

The evidence that clusters of low-velocity objects do impact the surface of Phobos seems clear. But until detailed calculations are carried out, it is not clear whether any secondaries can be produced on Phobos in single-hop trajectories. It is possible that many of the crater chains are produced by clumps of ejecta which originally were thrown out at slightly more than the escape velocity of Phobos, ended up in Mars orbit, and subsequently reimpacted the surface of the satellite. Although such a mechanism appears to be feasible, the details remain to be worked out. It is also conceivable that some of the crater chains on Phobos were produced by ejecta originating from the outer satellite Deimos.

The nonrandom orientation of the crater chains on Phobos (Figure 5) seems to be more consistent with the 'escape-into-Mars-orbit and reimpact mechanism' than with the idea of a single-hop process. The crater chains on Phobos definitely do not radiate from any particular crater.

Incidentally, at least some of the crater chains on Phobos are younger than the large crater Stickney, as the superposition of the prominent crater chain across the crater seen in Figure 4 demonstrates.

It appears that there may be clusters of irregular craters ('secondaries') on Deimos, although in the case of this satellite we still lack images with adequate resolution to settle the issue. For example, in the upper left of Figure 6 one sees a cluster of irregular craters which may be secondaries. This particular group of craters had been noticed in the Mariner 9 coverage [*Veverka et al.*, 1974, orbit 149] and found to be associated with an area of higher albedo, by about 30% relative to its surroundings [*Noland and Veverka*, 1976*a*].

c. Linear Striations or Grooves

The most enigmatic features discovered on the surface of Phobos are the swarms of almost parallel striations or grooves best seen on frame 039B84 (Figure 2). Typically, the striations are 150–200 m wide, appear to have rather shallow cross-sectional profiles, and are separated from their neighbors by distances comparable to their widths. Individual striations can be followed for distances in excess of 5 km (Figure 2).

The striations seem to occur in at least two sets which are not precisely parallel (Figure 7) but inclined at about 10° in relation to each other. It can be seen from Figures 7 and 8 that elements of the two sets do not cross each other.

The striations do cross the large craters near the north pole

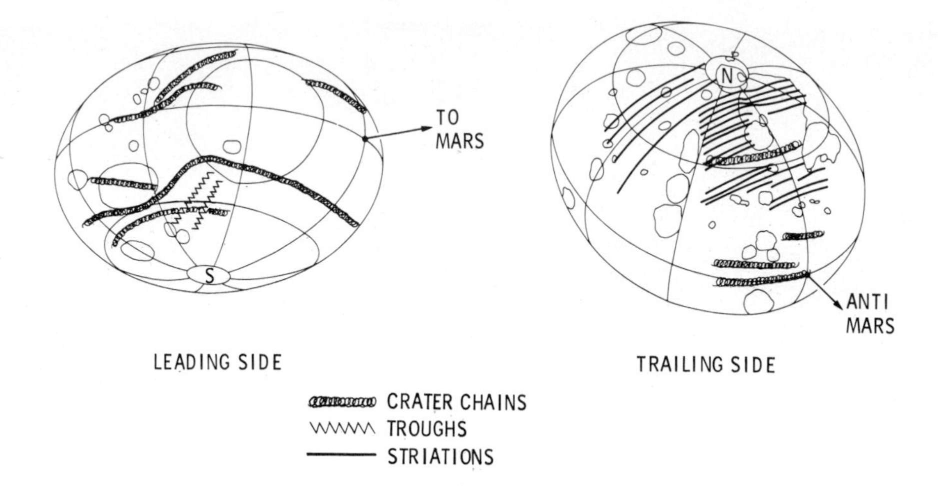

Fig. 5. Sketch map of the distribution of various linear features on the surface of Phobos based on data available from the Viking primary mission. Features such as those shown in Figures 3 and 4 are mapped as crater chains. Rill-like depressions associated with Stickney (Figure 1) are mapped as troughs, while the linear features shown in Figures 7 and 8 are mapped as striations.

visible in Figure 2 but are overlain by a number of smaller craters (Figure 7). Thus they are neither the oldest nor the youngest features visible on the surface of Phobos.

A fundamental question concerns the longitudinal profile of a typical striation. Is a typical striation made up of a chain of coalescing craters, or is it more properly a crack or a gouge? Unfortunately, the evidence obtained during the primary mission is far from conclusive. A few of the striations visible in Figure 7 have segments which have a vaguely pitted appearance and could be interpreted as modified rows of pits. However, many other striations (Figures 7 and 8) look more like grooves at the resolution currently available.

Coverage of Phobos at a resolution adequate to show striations was of very limited extent during the Viking primary mission. The only region of Phobos covered adequately is that shown in Figure 2. The south polar region and the area around Stickney were not imaged with resolution adequate to show striations. The distribution of striations on the surface of Phobos based on the limited standard mission coverage is shown in Figure 5. The striations appear to lie along small circles perpendicular to the Mars-Phobos direction (parallel to the satellite's orbital velocity vector) and show an apparent concentration at high northern latitudes near longitude 180°W.

The origin of the striations remains unclear. Although the striations qualitatively resemble gouges, no reasonable mechanism for gouging the surface of Phobos has been suggested.

The theories proposed to date to account for the striations can be divided into three categories: first, the striations represent layering within Phobos; second, the striations are rows of impact craters; and third, the striations are cracks resulting from tensional stresses.

According to the first hypothesis the striations represent layering within Phobos. The satellite is considered to be a fragment of a much larger parent body which had a radially layered structure. According to this view the layering should pervade Phobos, and expressions of it might be found in other areas of the surface. It is not expected that the striations should be resolvable into rows of craters.

One possible difficulty with this view is that craters should disturb the layering pattern locally, while they evidently do not. Specifically, the formation of the large crater Roche might be expected to have disturbed the layering in some radially symmetric fashion, which is clearly not the case (Figure 8).

According to the second view the striations are actually rows of coalescing impact craters. It is assumed that it is possible (in some unspecified manner) to accumulate strings of debris in Mars orbit with which Phobos subsequently collides. Although a detailed mechanism has not been worked out, the scheme may be possible. It predicts that all striations should be resolvable into rows of small craters (possibly modified by subsequent erosion by micrometeoroids).

It should be noted that this mechanism must be different from that which produces the crater chains discussed in section *2b*, since the distributions on the surface of Phobos of the two types of features are markedly different (Figure 5).

According to the third hypothesis the striations represent cracks produced by tensional stresses. Specifically, A. W. Harris and S. Soter (private communication, 1977) have shown that tidal forces tend to distort an object at the distance of Phobos from Mars into a triaxial ellipsoid and that tensional stresses would be concentrated in a narrow belt near the poles at right angles to the Mars-Phobos direction. If the striations are simple stress cracks or stretch marks in a regolith, they are not resolvable into rows of craters, and another swarm of striations should be located near the south pole of Phobos.

It should be noted that large stresses in the surface of Phobos can also be produced during the formation of large craters. It is conceivable that the formation of a crater such as Stickney may have initiated failure in the polar regions, which could have been under tension due to the Harris-Soter mechanism. In this event, Stickney would be younger than Roche (compare Figure 8).

It is possible that the striations may have a more complex origin. They could be primarily faults which have, in part, been modified by local outgassing of the sort suggested by *Pollack et al.* [1973]. If we assume that Phobos was originally made of a volatile-rich material similar to that found in type 1 carbonaceous chondrites, severe impacts such as the impact responsible for Stickney would have heated a significant fraction of the satellite's interior to the point where outgassing could have occurred. It is reasonable to expect that such outgassing would have been concentrated along faults which themselves could be the result of impacts. Such a mechanism

Fig. 6. View of Deimos from Viking Orbiter 1 (056A98). The largest crater is about 1.3 km across. A cluster of irregular depressions occurs near top left.

Fig. 7. Enlargement of a segment of frame 039B84 (Figure 2) showing striations. The largest crater in this view is about 900 m across. At least two sets of striations inclined at about 10° to each other can be seen. Some craters are clearly superimposed on the striations.

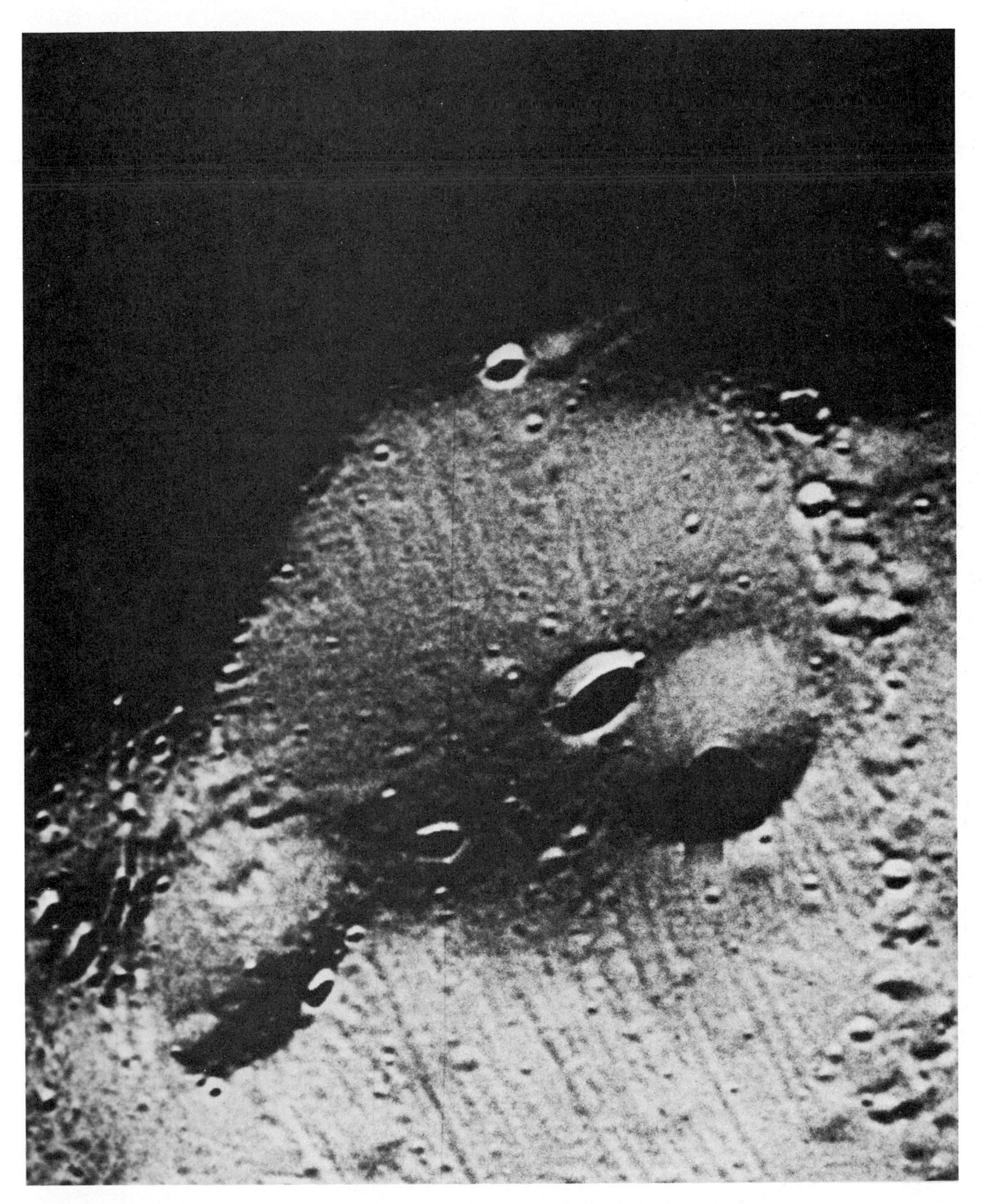

Fig. 8. Enlargement of a segment of frame 039B84 (Figure 2) showing striations crossing the floor of the large crater Roche, which is about 5 km across.

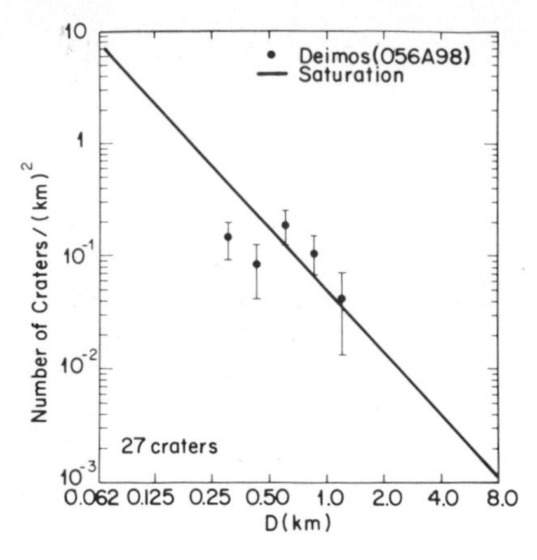

Fig. 9. Crater density on Deimos measured from frame 056A98 (Figure 6).

would account for the pitted appearance of some of the striations.

In order to choose among the above theories one needs additional information. Specifically two important questions must be answered: Do all striations resolve into crater chains given sufficient resolution? What is the distribution of striations on the surface of Phobos? Hopefully, these key questions will be answered during the close encounters with Phobos which will take place during the Viking extended mission. (Preliminary data from the Viking extended mission show that many of the striations are pitted in a manner consistent with the faulting plus outgassing hypothesis outlined above.) The examination of Deimos at very high resolution during the extended mission to see whether similar features exist on the surface of the outer satellite should further limit the scope of speculation.

3. Crater Counts

Crater counts were carried out on the best Deimos frame available (frame 056A98, Figure 6). A total of 27 craters ranging in size from 0.3 to 1.4 km are visible. In Figure 9, the crater counts on Deimos are compared with the 'saturation curve,' an empirical curve based on the most heavily cratered regions of the lunar uplands taken from *Hartmann* [1973]. Hartmann and some other workers hypothesize that the saturation curve represents the limiting crater density that can be reached before crater production and crater obliteration processes balance each other.

From the data in Figure 6 we conclude that the surface of Deimos is saturated with craters in the diameter range of 0.3–1.4 km. Since the flux of impacting bodies over the last 4.6 b.y. at the orbit of Mars is not known accurately, it is not certain how long it takes a fresh surface at the orbit of Mars to reach 'saturation.' *Pollack* [1977] estimated a minimum time of 1.5 b.y. We emphasize that our observation that the satellite surfaces are saturated with small craters provides only a minimum exposure age.

Crater counts were also carried out on frame 039B84 (Figure 2), the best of the Phobos images obtained during the nominal mission. Craters ranging from 50 m to 4.9 km in diameter are visible, and clusters of irregular secondary craters are conspicuous. Only circular craters, not obviously associated with clusters, were included in the crater count. The data for 237 craters are plotted in Figure 10; for crater diameters of ≥0.3 km the data points lie close to the saturation curve. The falloff at smaller sizes probably indicates a progressive loss of resolution as well as obliteration of small primary craters by secondaries.

Over 60 small craters can be counted within the large 4.9-km crater (Roche) near the north pole of Phobos (Figure 2). None of these show any of the characteristics of secondaries. The crater counts for the inside of Roche (Figure 11) do not differ significantly from those for the surrounding terrain or from the saturation curve. The fact that the density of impact craters within Roche is identical to that on the surrounding terrain suggests that erosion mechanisms specific to the interior of large craters (e.g., slumping of walls) are not effective, a fact which is understandable in view of the negligible surface gravity on Phobos ($g \sim 1$ cm/s^2).

4. Other Topics

Three sequences of color coverage, using filter passbands from 0.42 to 0.62 μm, were obtained during the nominal mission. This represents the first time that color measurements have been made of individual surface features on Phobos and Deimos. Preliminary analysis has not revealed any color variations on the surfaces. Between 0.4 and 0.6 μm the satellites appear to be uniformly gray.

The Viking nominal mission has also extended the phase angle coverage of the satellites. Mariner 9 observations were limited to the range 18°–83° [*Veverka et al.*, 1974]. For Phobos, Viking has extended this range to phase angles as large as 125°. Such data are being used to refine the satellite's photometric function and in particular its phase integral [*Noland and Veverka*, 1976*b*].

5. Conclusions

The Viking images revealed a number of unusual features on the surface of Phobos. Prominent chains of noncircular craters suggest that clusters of objects impact the surface at relatively low velocity. It is possible that these clusters are made up of ejecta from either Phobos or Deimos which do not have sufficient velocity to escape Mars [*Soter*, 1971]. The apparent common occurrence of low-velocity impacts is consistent with the observations that both Phobos and Deimos have regoliths [e.g., *Gatley et al.*, 1974; *Noland and Veverka*, 1976*a*, *b*].

The elongated depressions which seem to be associated with the large crater Stickney may be evidence of the local fracturing of a small body by an almost catastrophic impact. This would be the first time that such evidence has been observed in the solar system.

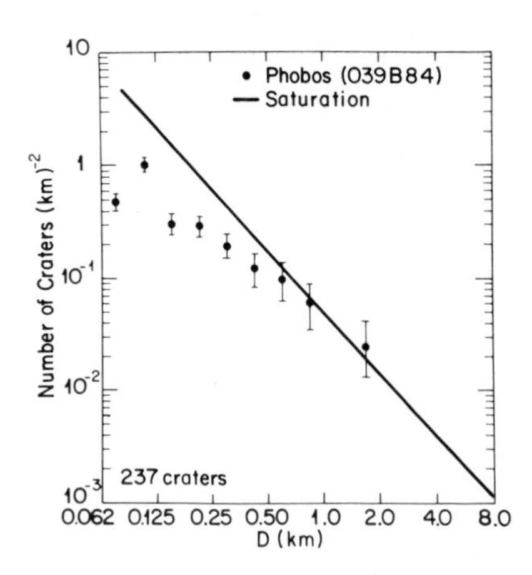

Fig. 10. Crater density on Phobos measured from frame 039B84 (Figure 2).

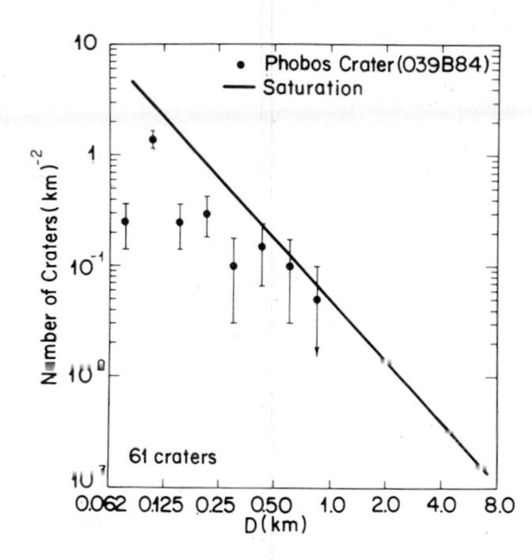

Fig. 11. Crater density within the crater Roche on Phobos measured from frame 039B84 (Figure 8).

Although a definitive explanation of the enigmatic striations is still being worked out, it can be expected that the solution of this puzzle will expand our understanding of small objects in the solar system. This is a case where more coverage at the highest resolution of large extents of Phobos is needed. Such coverage is being obtained during the Viking extended mission.

At the present time we lack adequate imagery of Deimos to say whether similar features occur on the surface of the outer satellite. Clearly, this is another important question which should be answered during the Viking extended mission.

The crater densities on both Phobos and Deimos have now been established reliably. Both surfaces appear to be saturated. Since the flux of impacting objects at the orbit of Mars throughout the history of the solar system is most uncertain, it is difficult to determine how long it takes a surface at the orbit of Mars to become saturated with craters. *Pollack* [1977] estimated at least 1.5 b.y., with time scales comparable to the age of the solar system being quite probable. Thus the surfaces of the satellites are probably very old. If Phobos and Deimos are fragments of larger objects, the fragmentation occurred a long time ago.

Acknowledgments. We thank our colleagues on the Viking Orbiter Imaging Team for helpful discussions, W. Quaide for specific comments, and R. Ruiz and W. Farrel of the Jet Propulsion Laboratory's Image Processing Laboratory for enhancing the satellite images used in this paper. This work was supported by the Viking Project and by NASA's Planetary Geology Program. NASA contract NSG 7156.

REFERENCES

Carr, M. H., et al., Preliminary results from the Viking orbiter imaging experiment, *Science, 193*, 766–776, 1976.

Duxbury, T. C., and J. Veverka, Viking imaging of Phobos and Deimos: An overview of the primary mission, *J. Geophys. Res., 82*, this issue, 1977.

Gatley, I., H. Kieffer, E. Miner, and G. Neugebauer, Infrared observations of Phobos from Mariner 9, *Astrophys. J., 190*, 497–503, 1974.

Hartmann, W. K., Martian cratering, 4, Mariner 9 initial analysis of cratering chronology, *J. Geophys. Res., 78*, 4096–4116, 1973.

Noland, M., and J. Veverka, The photometric functions of Phobos and Deimos, II, Surface photometry of Deimos, *Icarus, 29*, 200–211, 1976*a*.

Noland, M., and J. Veverka, The photometric functions of Phobos and Deimos, I, Disc-integrated photometry, *Icarus, 28*, 405–414, 1976*b*.

Oberbeck, V. R., and R. H. Morrison, Laboratory simulations of the herringbone pattern, associated with lunar secondary crater chains, *Moon, 9*, 415–455, 1974.

Pollack, J. B., Phobos and Deimos: A review, in *Planetary Satellites*, edited by J. Burns, University of Arizona Press, Tucson, 1977.

Pollack, J. B., et al., Mariner 9 television observations of Phobos and Deimos, 2, *J. Geophys. Res., 78*, 4313–4326, 1973.

Soter, S., Studies of the terrestrial planets, Ph.D. thesis, Cornell Univ., Ithaca, N. Y., 1971.

Thomas, P., and J. Veverka, Phobos: Surface density of impact craters, *Icarus, 30*, 595–597, 1977.

Veverka, J., et al., A Mariner 9 atlas of the moons of Mars, *Icarus, 23*, 206–289, 1974.

(Received April 29, 1977;
revised June 21, 1977;
accepted June 21, 1977.)

VOL. 82, NO. 28 JOURNAL OF GEOPHYSICAL RESEARCH SEPTEMBER 30, 1977

Mars: Water Vapor Observations From the Viking Orbiters

C. B. FARMER, D. W. DAVIES, AND A. L. HOLLAND

Jet Propulsion Laboratory, California Institute of Technology, Pasadena, California 91103

D. D. LAPORTE

Santa Barbara Research Center, Goleta, California 93017

P. E. DOMS

Department of Earth and Space Sciences, University of California at Los Angeles
Los Angeles, California 90024

The results of observations of the spatial and temporal variation of water vapor during the Viking primary mission are reported. The instrument, the Mars atmospheric water detector (Mawd), is a five-channel grating spectrometer operating in the 1.4-μm water vapor bands. The seasonal period covered here is the northern summer solstice to the following equinox. The global water vapor, mapped at low resolution at approximately 1-month intervals, has been observed to undergo a gradual redistribution, the latitude of maximum column abundance moving from the northern polar area to the equatorial latitudes, and the integrated global atmospheric vapor content remaining constant. The peak abundances ($\sim$100 precipitable microns) occurred over the dark material of the circumpolar region. The summer residual cap is dirty water ice; at the season of maximum vapor the atmosphere above it is saturated and has a stable lapse rate of temperature. High-resolution maps show local structure controlled by abrupt changes of surface elevation, suggesting that large variations at a given latitude are orographic in nature and only occur in association with features whose horizontal scale is small in comparison to the product of the atmospheric relaxation time and the local mean wind speed. These results are at variance with the low-resolution global maps, however, which seem to show topographic control even at the regional scale. Attempts to isolate the diurnal variation of the vapor have shown a variety of effects at different latitudes and locations; scattering by dust and condensate particles obscures the intrinsic diurnal variation of the vapor phase. The large diurnal variation reported from earth-based measurements may be largely an observational effect.

INTRODUCTION

Prior to the Viking mission, knowledge of the behavior of water vapor in the Martian atmosphere was based almost entirely on the results of earth-based observations, with correspondingly limited information related to the details of its spatial and seasonal dependences. It was known that the atmospheric vapor was seasonally variable, apparently reaching a maximum in each hemisphere at some time after the summer solstice, and that there appeared to be a diurnal variation of the water vapor column abundance at mid-latitudes during a part, at least, of the season of maximum vapor content. The inaccessibility of the extreme polar regions from earth, coupled with the relatively poor spatial resolution, the limitation in sensitivity, and the extreme difficulty of the measurements, are factors which prevented any further light being thrown on a number of fundamental planetological and meteorological questions involving the present-day behavior of Mars's condensates and the extent to which the planet may have retained its primitive fraction of water.

The water vapor mapping instruments on the two Viking orbiters have provided the first opportunity to make measurements with sufficient spatial and temporal resolution and coverage to enable some insight to be gained into these questions. The vapor has been seen to exhibit variability with local time, elevation, and latitude, each of these in turn varying with season. It is intended that the observations should continue for a long enough period to encompass the complete cycle of Martian seasons so that the individual effects can be isolated and the variability on both long and short time scales tested against models which seek to describe the spatial and temporal variations in terms of plausible planetary mechanisms. Some trends, however, are already beginning to emerge from the data; this report presents the results of the observations made during the Viking primary mission period, i.e., from June to mid-November, 1977 (a time span equal to a quarter of the Mars year, commencing at the northern summer solstice), and such interpretation of them as has been attempted at this interim stage.

INSTRUMENT

The observations are made by measuring the absorption of solar radiation by the strong lines at the center of the 1.4-μm (7200 cm^{-1}) combination vibration-rotation bands of water vapor. The instrument is a grating spectrometer (Figure 1) having five channels which, in the normal ('locked') mode of operation, are centered on three absorption lines and two nonabsorbing, or continuum, regions between the lines. The PbS detectors, which define the exit apertures of the spectrometer, are operated at 200°K and are cooled to this temperature by being connected via a flexible copper strap to a plate radiating to space. The spectral resolution of each detector channel is $\sim$1.2 cm^{-1}; the grating (1200 lines/mm) is operated in first order for the five water vapor channels, the second order of diffraction being rejected by a silicon cut-on filter. The precise frequency location of the water vapor channels is maintained against possible shifts due to thermal distortion of the instrument optical alignment by a neon reference system operating in the grating second order, using a set of silicon photo-

Paper number 7S0500.

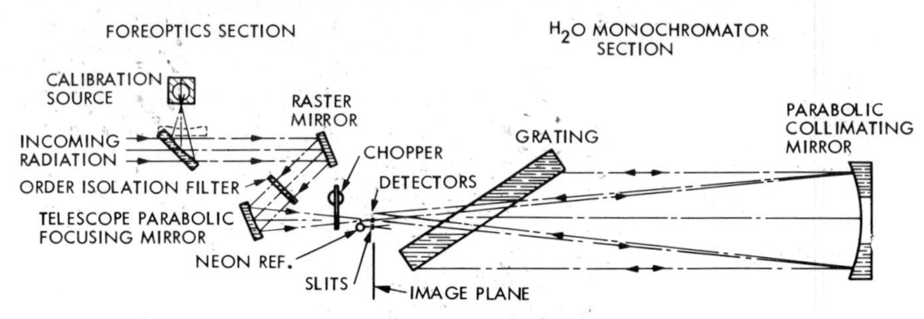

Fig. 1. Mawd optical configuration.

diodes as detectors. Motion of the neon lamp exit plane image with respect to the silicon reference detectors is sensed and used to drive a cam which repositions the grating to preserve the correct wavelength alignment.

The instantaneous field of view (Ifov) of the detectors projected onto the surface of the planet at periapsis (1500 km) is 3 $\times$ 24 km; the Ifov can be step-scanned sideways (i.e., perpendicular to its long dimension) through 15 positions by a raster mirror which forms part of the input optical train, or it can be held in its central position. The time spent at each step (approximately the signal integration time) is 280 ms, and the duration of the raster is 4.48 ms. The combination of the motions of the raster mirror, the orbiter scan platform, and the orbital ground track is used to generate sets of box scans and slew patterns over the planet's surface; these are designed to meet the experimental needs for high- and low-resolution mapping or the more localized pointing required for the investigation of the diurnal characteristics of selected locations. The instantaneous and raster fields of view are shown, including the effects of spacecraft ground track motion, in Figure 2.

Although the instrument is used as a five-channel monochromator in its normal data gathering mode, the ability provided by the wavelength servo system to scan the grating through a small angular range is used occasionally to record the continuous spectrum of the atmosphere. A range of 8 cm^{-1} on either side of the nominal (locked) position of each channel is covered, giving a total composite spectrum covering $\sim$40 cm^{-1}. The wavelength-scanning mode was included for several reasons: (1) to verify that the Martian atmospheric spectrum in the region of water vapor absorption is not contaminated by absorptions due to other constituents (particularly weak transitions of CO_2), (2) to validate the theoretical spectrum generated for Martian conditions, based on extensive laboratory measurements of the 1.4-μm region water vapor transition

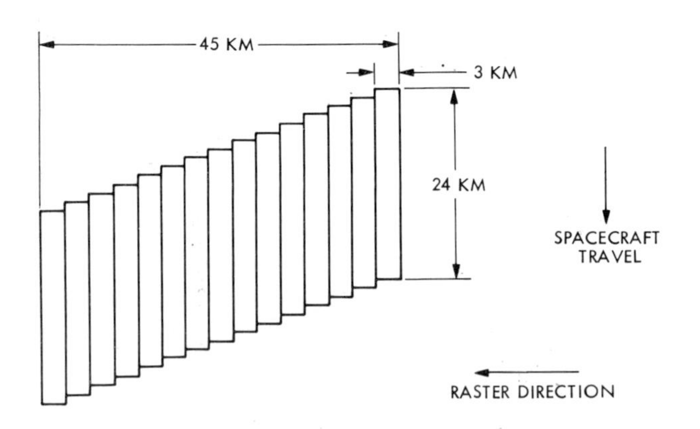

Fig. 2. Projection of the instrument field of view onto the surface of the planet from an altitude of 1500 km.

strengths, frequencies, and pressure-broadened half widths, on which the quantitative analysis of the received data is based, and (3) to determine to what extent, if any, absorption of lines of solar origin may affect the distribution of continuum radiation at the nominal 'line' and 'continuum' frequencies. The latter (solar) spectrum is obtained by orienting the scan platform to view the sun's radiation falling on a small, diffusely reflecting plate mounted on a structural member of the spacecraft. Details of the characteristics of the spectra obtained in the wavelength-scanning mode are discussed in a separate section.

When it is viewing the planet, the instrument obtains a measurement of the radiance, in each of the five channels, of reflected solar radiation which has passed twice through the Martian atmosphere. To determine these radiance values in absolute terms, the channel responses are corrected by means of periodic calibrations made by inserting a mirror into the input beam so that the detectors view an illuminated cavity. The ratios of the radiances in the line channels to the continuum radiance give three values of absorptance A_n, from which the line of sight water vapor content can be derived. It should be pointed out that the contribution of surface thermal emission to the received radiation is negligible at the chosen wavelength of operation. The optimum choice of spectral band and the extent to which the spectrum in this region is expected to be contaminated by absorptions due to other gaseous species of the Martian atmosphere are discussed in an earlier paper dealing with the instrument and experiment design [*Farmer and LaPorte*, 1972].

DATA ANALYSIS

In the absence of scattering by atmospheric particulates the vertical column abundance of water vapor W is obtained directly from the measured line of sight abundance along the optical path W' and a knowledge of the sun-planet-spacecraft geometry of the observation. Specifically, $W = W'/\eta$, where the air mass factor of the observation is $\eta = \sec i + \sec e$, i and e being the incidence and emission angles, respectively. As will be noted later, particulate scattering in the Mars atmosphere does affect the formation of the absorption spectrum at the wavelength used here, the effect being to underestimate the vapor content, particularly at the larger values of η. In general, however, for most regions of the planet the effects of scattering can be ignored in observations made at $\eta < 4$, and the bulk of the data presented here have been taken under, or selected to meet, this condition. The subject of scattering is considered in more detail in appropriate later sections of this report.

The line of sight water vapor content of the Martian atmosphere could be determined directly and simply from the oberved channel absorptions by the use of an empirically derived calibration made with the flight instruments prior to launch if the absorption were dependent only on the abundance of water vapor along the optical path. The difficulty with this approach arises from the fact that the absorption is, of course, dependent on pressure (and to some extent temperature) and that the laboratory simulation of Martian water vapor quantities under representative conditions of partial pressure and temperature requires prohibitively long absorption paths.

The method adopted here therefore was to compare the measured channel absorption values with theoretical values generated for a wide range of conditions of pressure and temperature. The set of theoretical 'curves of growth' was stored in the form of a three-dimensional table (i.e., A_n as a function of W, P, and T) as part of the routine Mawd data processing software. Briefly, the table was generated by using standard algorithms for the spectral extinction coefficients of lines with mixed collision and Doppler profiles and the known (measured) instrument spectral response function. The molecular parameters (transition strengths, collision-broadened widths, ground state energies, and foreign and self-broadening coefficients) were determined from extensive laboratory measurements; a total of 81 transitions arising from the five vibration-rotation bands that occur in the 1.38-μm water vapor spectrum were included in the final computations. The resultant theoretical spectra and the corresponding channel absorption values at the specific spectral locations of the detectors were verified over the restricted range of partial pressures and temperatures that could be achieved in the laboratory. Figure 3 illustrates computed spectra, at the instrument resolution, for water vapor amounts from 3 to 100 precipitable microns (pr μm) at a total pressure of 6.9 mbar and at a temperature of 225°K. In a later section the comparison between an observed Martian atmospheric spectrum and a corresponding theoretical spectrum is shown. The fit that can be achieved between computed and observed data provides strong evidence, albeit not proof, of the validity of the application of the spectral calculations to the physical conditions of the Mars atmosphere.

The dependence of absorption on water vapor amount, total pressure, and temperature is shown in Figure 4 for each of the three line channels. In principle, the three independent values of absorption, made at frequencies having sensibly different absorption coefficients, can be used to determine the effective temperature and pressure of the atmosphere at the level of line formation, as well as the water vapor amount itself. However, the accuracy with which temperature and pressure can be found is poor, since the pressure and temperature dependence of the curves of growth under Martian conditions is similar for all three channels, notwithstanding their intrinsic absorption differences. Thus satisfactory solutions for temperature and pressure can only be extracted when the data from a large number of individual rasters can be averaged to reduce the uncertainty due to noise. This requires also that all of the sources of systematic error be reduced to a very small level (i.e., less than 0.1%). In the case of the measurement of water vapor alone, however, the signal-to-noise ratio under typical conditions, for observations at the single-raster integration level (120:1), is high enough to give a minimum detectable vapor content of 1.0 pr μm and an average accuracy (dependent on surface albedo and observation geometry) of ±20%. When an improvement in signal-to-noise ratio is needed (for example, in areas with very low radiance) or when higher precision in the water content is required (for mapping purposes), measurements within a selected surface resolution element are combined. This 'binning' has been done at a resolution of 10° in latitude and longitude for global maps and 2.5° for the fine-scale results discussed later.

For the first-order analysis of the data taken in the fixed grating mode (constituting 98% of the experimental records) the water vapor abundances were determined by using fixed input pressure and temperature values of 6.0 mbar and 200°K. At the present stage of the data analysis this assumption is felt to be justified, since the dependence of W on the assumed value of P is not sufficiently strong to change qualitatively the major conclusions that can be drawn from the data. Nevertheless, in those circumstances in which the relationship between water vapor content and local topography is of interest, we have investigated the effects which variation of the assumed input pressure has on the results and their interpretation (as, for example, in the section on fine-scale structure).

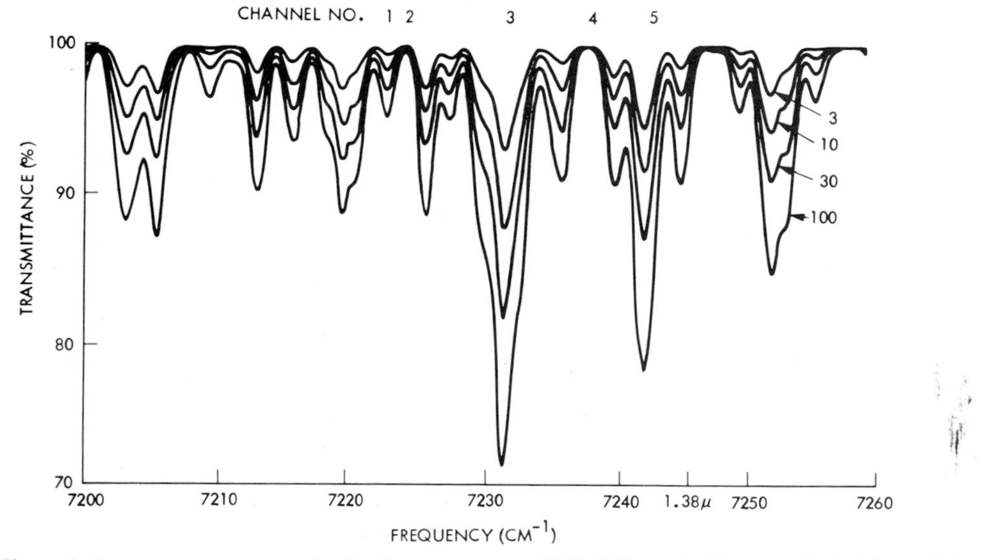

Fig. 3. Theoretical water vapor spectra in the frequency range 7200–7260 cm^{-1}. The spectral positions of the five detectors are indicated by the numbers 1–5. The curves are for vapor of 3, 10, 30, and 100 pr μm at a total pressure of 6.9 mbar and a temperature of 225°K.

OBSERVATIONS AND COVERAGE

Because of the variability of the water vapor over a large range of spatial and temporal scales the observations must be made with sufficient coverage and resolution to allow the dependence of vapor abundance on each of the controlling parameters to be extracted separately. Ideally, this requires that complete global coverage be obtained at several local times of day and with sufficient spatial resolution to reveal the effects of local topography (and perhaps the existence of localized sources of vapor). This global coverage must be repeated at intervals throughout the Martian year in order to obtain the seasonally changing variation with latitude. However, such extensive coverage would require almost continuous observation from a vehicle in a near-polar, asynchronous orbit. Within the constraints imposed by the demands to provide adequate relay links to the landers, by the observational needs of the other orbiter instruments, and by the conservation of spacecraft consumables, the observational coverage which in practice can be realized for the water vapor mapping studies must of necessity be a compromise which falls somewhat short of ideal.

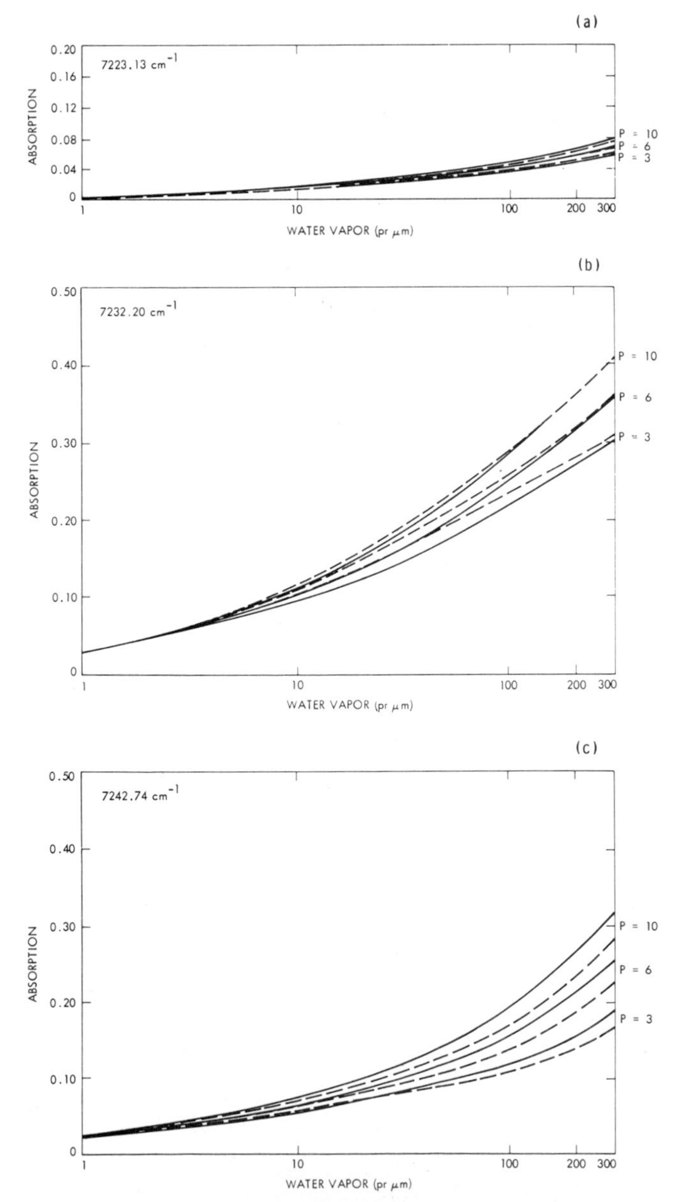

Fig. 4. Curves of growth for the three absorption channels, used for the calculation of vapor abundance from the locked mode data, for (*a*) 7223.13 cm^{-1}, (*b*) 7232.20 cm^{-1}, and (*c*) 7242.74 cm^{-1}. T = 200°K (solid lines) and 250°K (dashed lines) for total pressures of 3, 6, and 10 mbar.

To this end, the aim during the postlanded period (of both the primary mission and the extended mission) has been to achieve complete global coverage at low resolution once per earth month, higher-resolution mapping and diurnal sequences being inserted for selected regions of interest on the basis of opportunity. Low-resolution mapping, in the context of the acquisition of data related to the seasonal variation of the global distribution, refers to the measurement of the average column abundance over an area of 10° × 10° in latitude and longitude at one or more fixed times of day. The high-resolution mapping sequences have yielded regional maps at resolution down to 2½° × 2½° and have covered areas such as the northern polar region, the Tharsis Ridge and terrain to the east, the Valles Marineris region, and the area around the two landing sites.

Table 1 summarizes the history of those characteristics of the orbits of the two spacecraft which are relevant to observational coverage; Figure 5 shows the coverage in terms of the density of data acquired during the course of the mission. The figure illustrates graphically the dense but spatially restricted coverage obtainable during those periods when the orbits were synchronous.

The observation sequences, which involve a series of more or less complex motions of the orbiter scan platform, are defined and designed in the orbiter-centered space coordinate system of cone and clock angles. For details of the relationships between the angles defining the movement of the scan platform and the celestial and spacecraft coordinate systems, see *Snyder* [1977]. The sequences are designed as 'box scans' and 'slews,' which involve motions of the platform coupled with the spacecraft orbital motion, or as 'fixed cone-clock' scans, in which no platform slews are used.

Box scans are essentially mapping sequences which employ a series of reciprocating cone slews stepped by a small clock angle increment at each reversal of the cone slew direction. Box scans can be designed to cover all or part of the accessible area of the planetary surface as densely as necessary; they are seldom used in the orbital time period within ~15 min of

TABLE 1. Orbital Characteristics During the Primary Mission

Revolution	L_s, deg	Orbit	Periapsis Latitude, deg	Inclination, deg
		Viking Orbiter 1		
3–81	84–123	Essentially synchronous, periapsis over Chryse	25	38
83–95	124–130	Asynchronous, 38° rev^{-1}*	31	38
97–137	131–151	Synchronous, over VL-2	32	38
		Viking Orbiter 2		
4–15	109–115	Asynchronous, VL-2 site certification, −38° rev^{-1}	50	55
20–50	118–133	Synchronous, over VL-2	52	55
52–85	134–152	Asynchronous, −32° rev^{-1}	64	75

The periapsis altitude of both orbiters throughout the period covered by the table was ~1500 km.

*Expression indicates drift or walk rate in terms of the longitude separation of the periapses on successive revolutions; negative values signify a supersynchronous or westward drift.

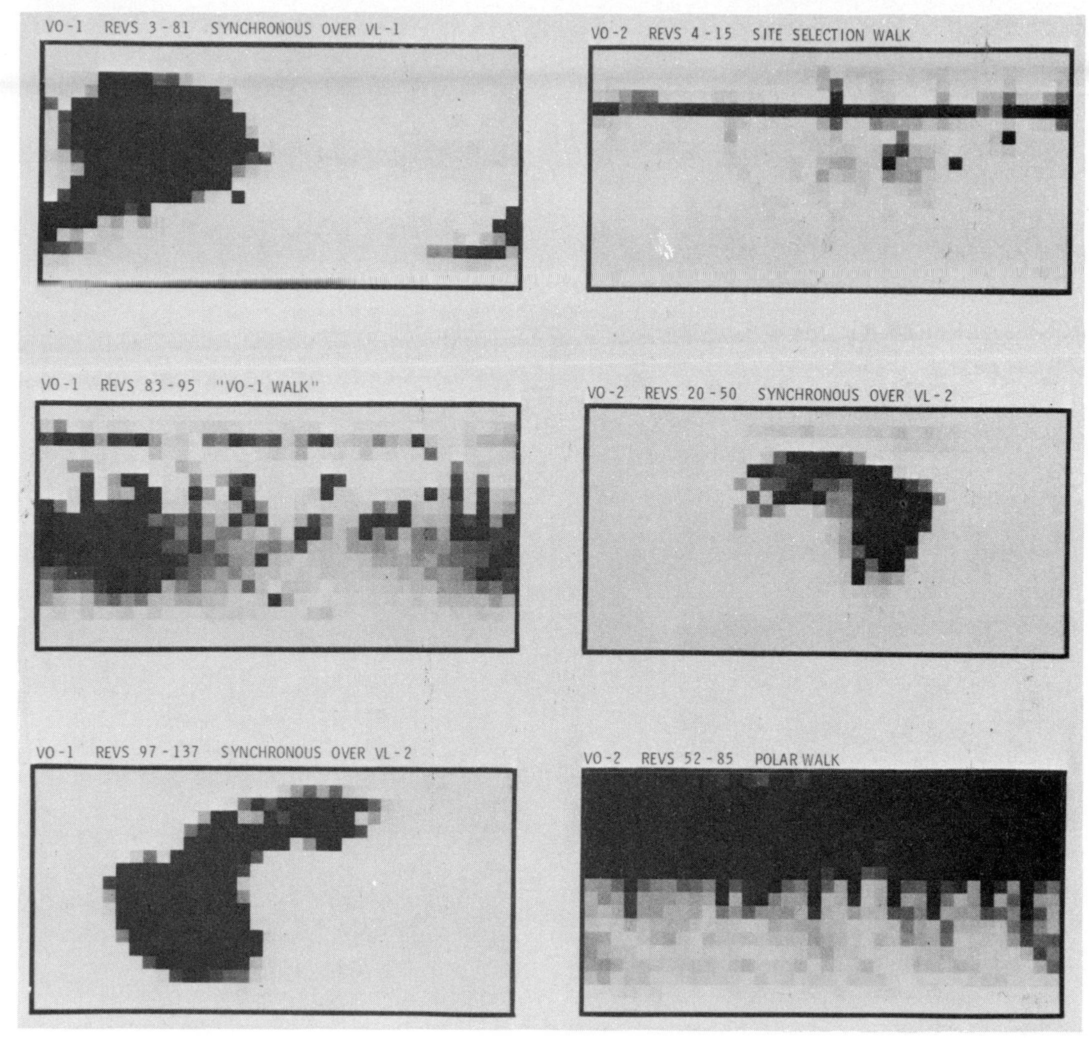

Fig. 5. Global coverage obtained during the primary mission water vapor observations. The diagrams indicate the density of observations within 10° square elements (bins) of latitude and longitude on a simple Cartesian projection of the planet with 0° west longitude at the right-hand end of each grid. The gray scale saturates at black when at least 50 raster centers lie within a given area element; white indicates no observational coverage.

periapsis, because the motion of the planet in the coordinate system of the spacecraft is too rapid to allow useful sequences to be designed. Consequently, box scans are used primarily for medium- and high-altitude mapping purposes. Examples of typical global and medium-resolution box scans are shown in Figures 6*a* and 6*b*. For observations close to periapsis the spacecraft ground track motion is used with the scan platform in a fixed cone-clock position to provide a one-dimensional scan across a region of interest. Several of these scans can be juxtaposed (from consecutive synchronous orbits) to build up the highest-resolution area coverage. In addition to these, 'fixed target' sequences are occasionally employed in which the scan platform motion is designed to cancel the orbital ground track velocity and thus maintain the position of the instrument field of view on the surface with little or no smear. These are used, for example, for the relatively long duration wavelength scans, which require upward of 15 min of observation with minimal smear.

RESULTS

Solar Spectra

Wavelength scans of the solar diffuser plate were recorded on VO-1 revolutions (revs) 34, 55, 89, and 107; additional scans were made (during the extended mission) on revs 124 and 210. No observations of the diffuser plate on VO-2 have been possible to date, because the view of the plate from the instrument remains obstructed by the bioshield. While the solar spectra were recorded for the main purpose of monitoring the spectral distribution of the continuum radiation over the wavelength range encompassed by the water vapor measurements, they revealed a number of solar absorption features which to our knowledge have not previously been observed. The spectra are thus of some interest in their own right and warrant inclusion with the experimental results as a whole.

Each of the observations on revs 55, 107, 124, and 210 comprised four repeated scans. These have been combined and averaged to produce the solar spectrum reproduced in Figure 7. The signal-to-noise ratio of this spectrum is estimated to be 10^3: 1. In combining the individual scans to produce this average, the relative channel gain levels and frequency offsets were normalized to channel 3 (chosen arbitrarily), and the detector radiance level variations from observation to observation normalized to the rev 55 data. The detector radiances were then grouped in frequency intervals of 0.2 cm^{-1} over the total range 7215–7250 cm^{-1}.

The frequencies, intensities, and identifications for the solar

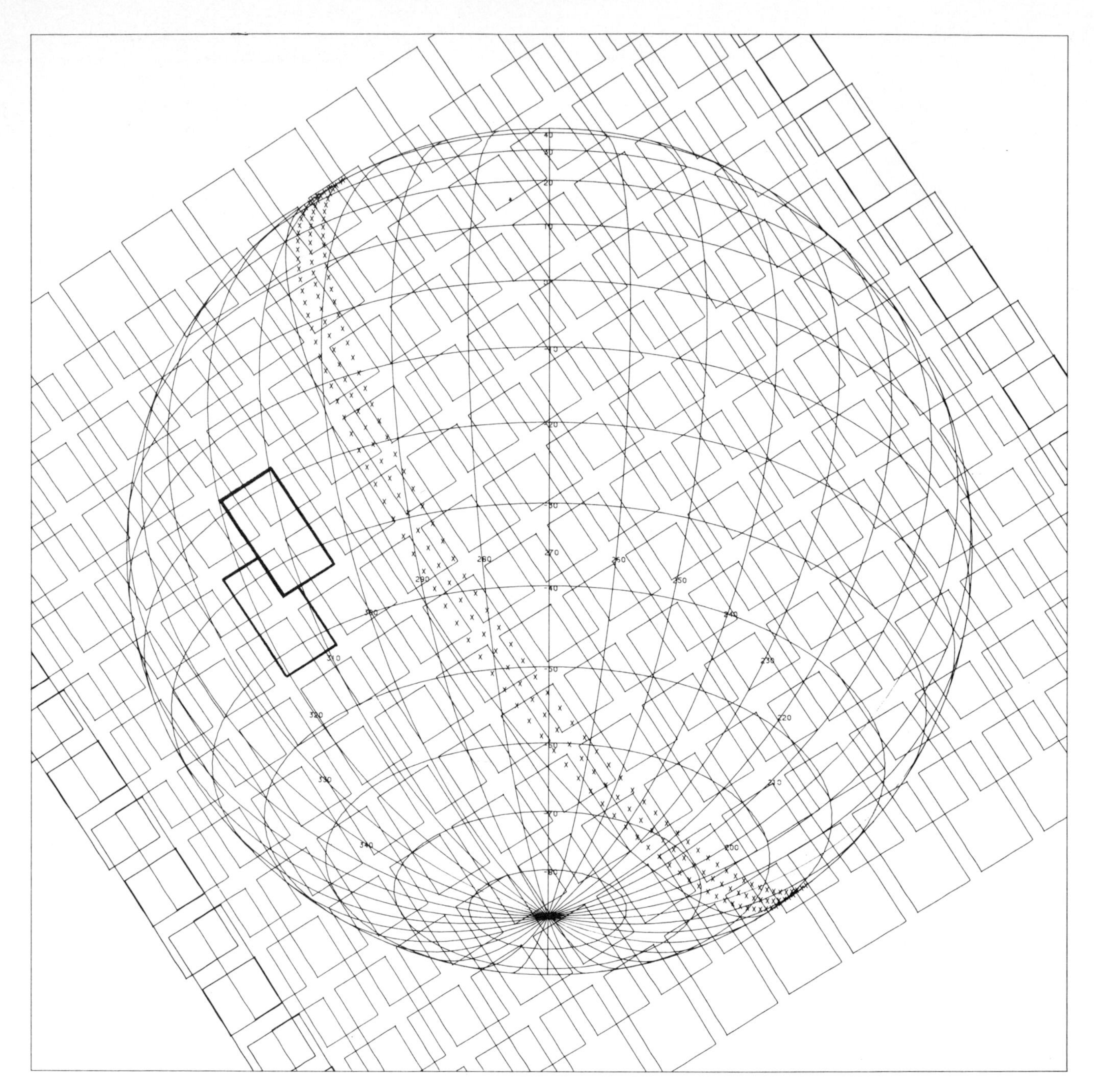

Fig. 6*a*. High-altitude global box scan. The rectangles represent the raster field of view.

absorptions appearing in the spectra are listed in Table 2. The estimated accuracy of the frequencies of the individual lines that combine to form unresolved pairs or triplets (at the Mawd resolution, ~1.2 cm^{-1}) is ±0.5 cm^{-1}, and of single lines is ±0.1 cm^{-1}. The intensities of the lines are believed to be accurate to ±25% for blended components, and ±10% for the stronger single lines.

With the exception of the line at 7231.58 cm^{-1}, the presence of the solar absorptions superimposed on the continuum radiance distribution does not significantly affect the radiance ratios between continuum and water vapor absorption channels at their locked frequencies. Although the line at 7231.58 cm^{-1} does contribute to the apparent absorption of water vapor as measured by channel 3, its effect on the calculated water content derived from the locked grating mode data is small. In the analysis of the spectral measurements of the Martian water vapor, however, corrections must be made for all of the solar features, since their intensities are comparable with the intensities of many of the water vapor absorptions for average conditions of vapor abundance (see the following section on Martian spectra).

Mars Water Vapor Spectrum

A number of wavelength scans of the planet were made during the primary mission, the majority of these being taken with the VO-1 instrument during the first period of synchronous orbits (see Table 1). A typical spectrum is shown in Figure 8 (second curve from the top); this observation was made from VO-1 (rev 45) viewing an area centered at 10°W, 83°N, at approximately 2 hours before periapsis, when the projected size of the raster field of view was about 5° in latitude and longitude.

The planet wavelength scan data were treated in a manner similar to that for the solar spectra, i.e., normalizing the frequencies and radiances to channel 3 and grouping them in 0.2-cm^{-1} frequency intervals. In this case the wavelength scan is of a 12-min total duration, so the resulting spectrum is noisier than the solar spectrum.

Before the observed spectrum can be used to estimate water content, pressure, and temperature, the solar spectrum must be subtracted. Figure 8 shows the solar spectrum and the corrected atmospheric spectrum. A spectrum-fitting program was then used to determine the values of W, P, and T which give the best fit to the corrected Mars spectrum. The results for the rev 45 data shown in the figure are $W' = 27 \pm 2$ pr μm ($W = 7.3 \pm 0.5$ pr μm), $P = 5.0 \pm 2$ mbar, and $T = 200 \pm 15$°K. The theoretical spectrum calculated for these conditions is shown as the bottom curve in Figure 8.

The differences between the theoretical and corrected observed spectra are 0.5% or less. No absorption lines, other than those expected for water vapor, appear in the Martian atmospheric spectrum over this frequency range at this spectral resolution. As was pointed out in the section on data analysis, the close agreement between observed and computed spectra, under Martian atmospheric conditions, is an important verification of the algorithms and molecular parameters used to generate the transmittance tables on which the analysis of the Mawd locked grating mode data is based.

Global Distribution

Global maps of the distribution of water vapor are constructed from data accumulated during the asynchronous orbital periods. As has been mentioned previously, since the global distribution itself changes with season (and in some regions, very rapidly), coverage of all latitudes and longitudes must be obtained within a period of time which is short in comparison with the time in which significant changes due to seasonal variation of the global distribution occur. For the present analysis and presentation of the global-seasonal behavior the data have been divided into seasonal 'periods' of equal (15°) increments of planetocentric longitude L_s; during the primary mission each period corresponds to approximately 1 earth month, and the results included here cover the first 7 of the 24 periods into which the Mars year is divided. Period 1, which includes the initial site certification phase of VO-1, covers the interval L_s = 80°–95°. The first opportunity to acquire data over an extensive range of longitudes occurred with the asynchronous 'walk' of VO-2, during the VL-2 site selection phase (in period 3); the spatial density of data acquired throughout the orbital operations to date is varied, as a result of the observational opportunities which have occurred. In order to preserve a maximum uncertainty in the derived

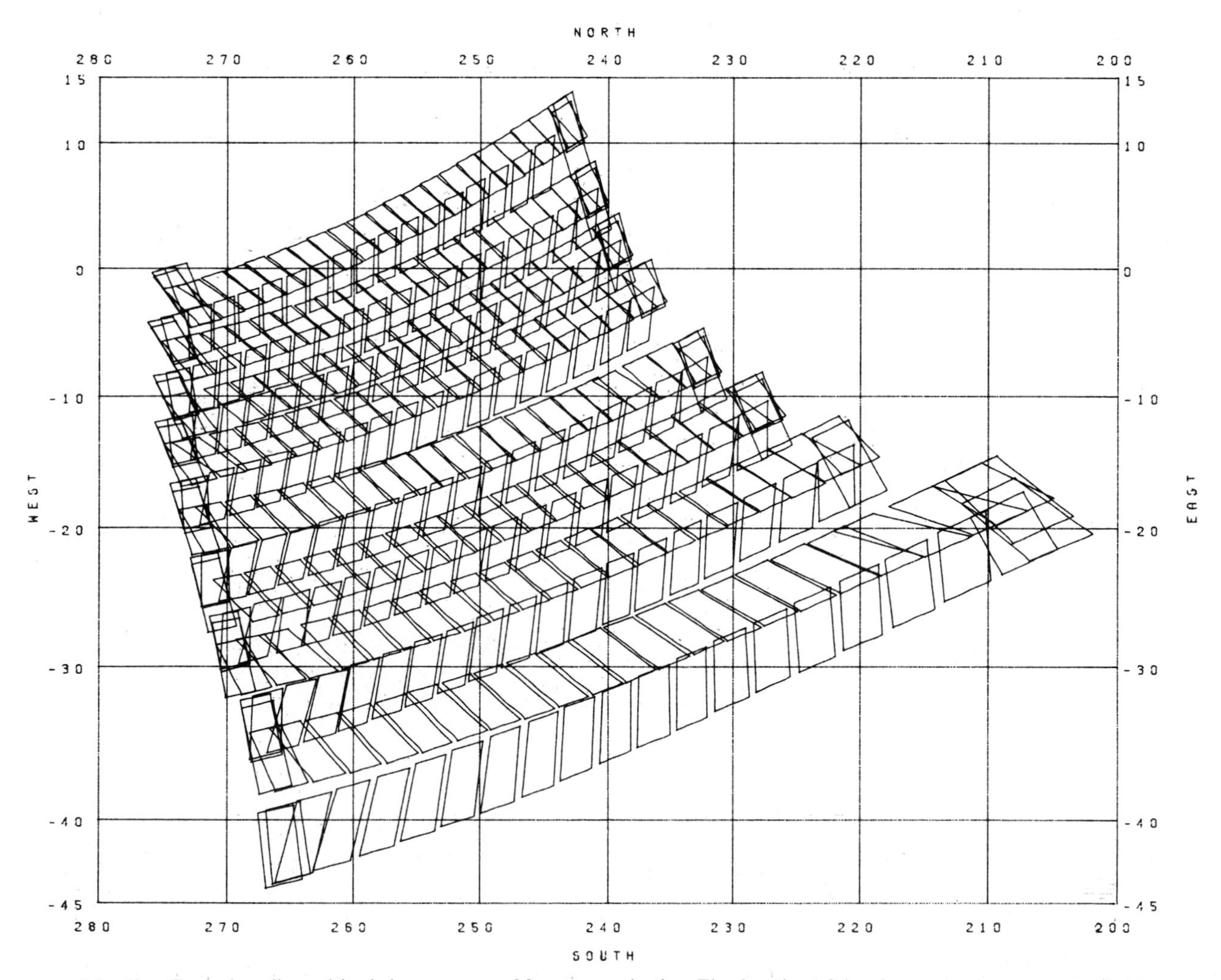

Fig. 6*b*. Typical medium-altitude box scan on a Mercator projection. The duration of the observation is approximately 20 min, commencing 1¼ hours before periapsis.

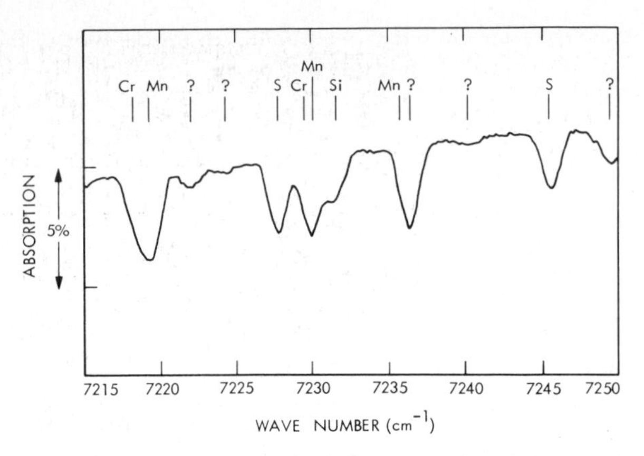

Fig. 7. Solar spectrum obtained from wavelength scans of the diffuser plate on VO-1 revs 55, 107, 124, and 210.

TABLE 2. Table of Solar Lines in the Mawd Frequency Range

Line	Identification*	Frequency, cm^{-1}	Intensity, cm^{-1} × 10^3
1	Cr	7218.38	30.
2	Mn	7219.46	53.
3	...	7222.08	14.
4	...	7224.57	5.
5	S	7227.88	45.
6	Cr	7230.05	51.
7	Mn		
8	Si	7231.58	23.
9	Mn	7236.42	47.
10	...		
11	...	7240.35	4.
12	S	7245.66	34.
13	...	7249.57	30.

*We are indebted to M. Geller for these identifications.

water vapor abundance values of ±10%, over regions where the observations are sparsely distributed, the global data are averaged over bins of 10° × 10° in latitude and longitude.

The total coverage obtained to date from both VO-1 and VO-2 is represented at this reduced spatial resolution in Figure 5. In this figure, and in the global maps which follow, the planet is represented as a simple Cartesian grid of latitude and longitude.

Plates 1*a*–1*d* show global contour maps of water vapor distribution for each of the L_s periods; the contours of equal vertical column abundance are derived from averages of all data obtained within each 10° × 10° area bin, within a selected local time period (1000–1600 hours Martian local time) and with incidence and emission angles less than 70°. The maps have been generated by using both a constant (6 mbar) pressure and a varying pressure derived from elevations on a topographic map of the *U.S. Geological Survey* [1976]. The resulting water vapor distributions are very similar; the variable pressure version is reproduced here. As is indicated in Figure 5, the coverage during periods 1 and 2 is insufficient in global extent to warrant presentation in this form, although the apparent latitude averages obtained from the partial coverage during these periods are included in the graphical summaries which follow. Period 6 (L_s = 155°–170°) corresponds to the period of solar conjunction, during which no orbital data were taken. For reference the map of surface elevation, also at a resolution of 10° in latitude and longitude, is shown in Figure 9.

A number of general comments can be made from the global maps. Period 3 (Plate 1*a*) corresponds to a mean L_s value of 117°, the mean subsolar latitude being 21°N, i.e., the sun moving toward the equator 1 Martian month after the northern summer solstice. The extreme southern latitudes are virtually devoid of atmospheric water vapor (<0.1 pr μm); the contours show a gradual increase from the south toward the equator into the northern equatorial latitudes, with a marked increase through the northern mid-latitudes, reaching a maximum in the circumpolar region. Maximum column abundances in excess of 90 pr μm in the dark area surrounding the residual summer ice cap were observed during this period [*Farmer et al.*, 1976*b*]; the maximum value for the 10° latitude band average was 76 pr μm at 70°–80°N. As the subsequent seasonal periods progress (toward the northern autumnal equinox, period 7), the latitude distribution is seen to be modified by a rapid decrease in the northern polar abundances and a gradual increase in the equatorial and southern values.

In addition to this first-order behavior the vapor distribution can be seen in a general sense to follow the gross regional scale topography, the coincidence of the minima and the locations of the principal volcanic summits being preserved even at the 10° × 10° averaging scale. No anomalous 'wet' spots have been seen, with the possible exception of the region of relatively high concentration to the northwest of the Hellas Basin. This feature was observed in periods 3 and 5.

Figure 10 shows the 10° latitude band averages (i.e., taken over all longitudes) plotted as a function of L_s. As was noted, the longitude coverage during L_s periods 1 and 2 was incomplete, so the corresponding latitude averages derived for these periods may be incorrect. No polar coverage was obtained until period 3, but there is strong reason to believe that the abundances observed for the extreme northern latitudes were, in fact, close to their seasonal maxima at the time that they were first observed from VO-2 (see the discussion of polar data below).

During the seasonal period represented in Figure 10 the subsolar point moves from its most northerly position (25°) at L_s = 90° to the equator at L_s = 180° (equinox); thus the maximum value of the average abundance at a given latitude occurs at a time which appears to lag behind the solstice by an increasing amount, the closer that latitude is to the equator.

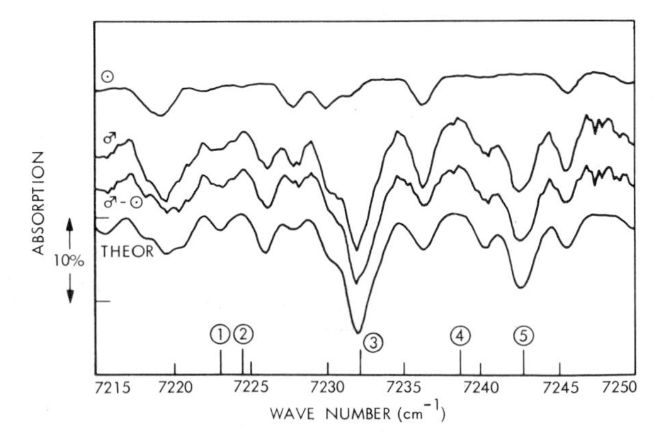

Fig. 8. Spectra of the sun (top curve), Mars (at 10°N, 83°W) (second curve), and Mars corrected for the solar features (third curve). The bottom curve is the theoretical spectrum for the conditions giving the best fit to the corrected atmospheric spectrum (see the text). The curves have been offset for clarity.

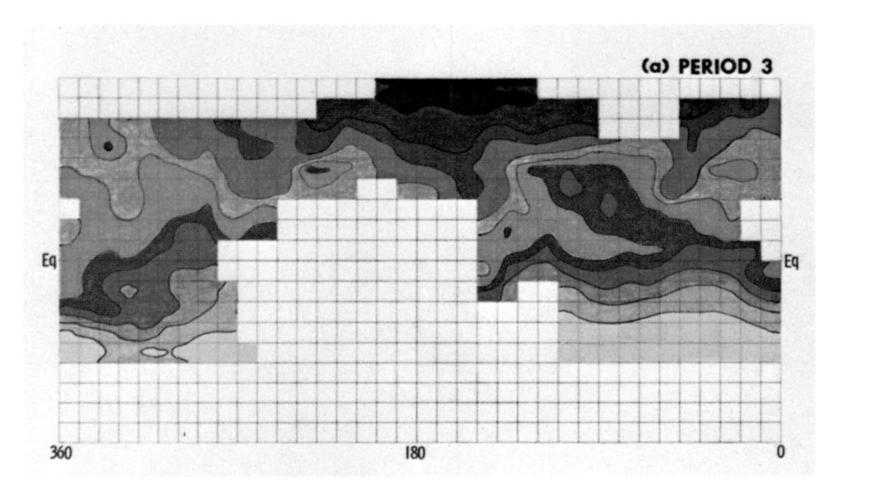

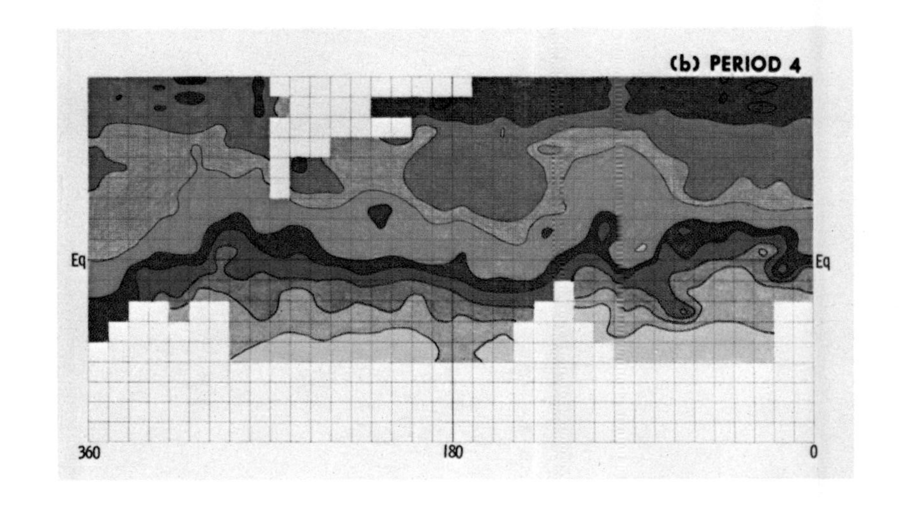

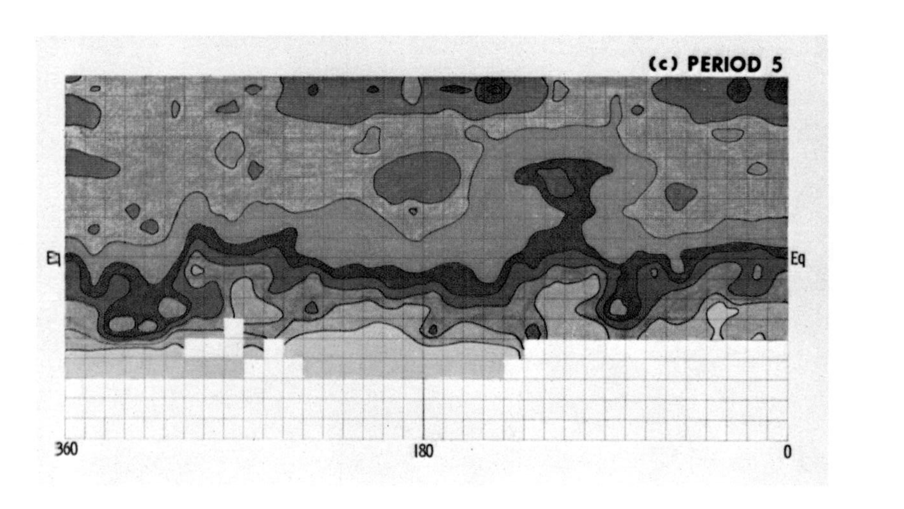

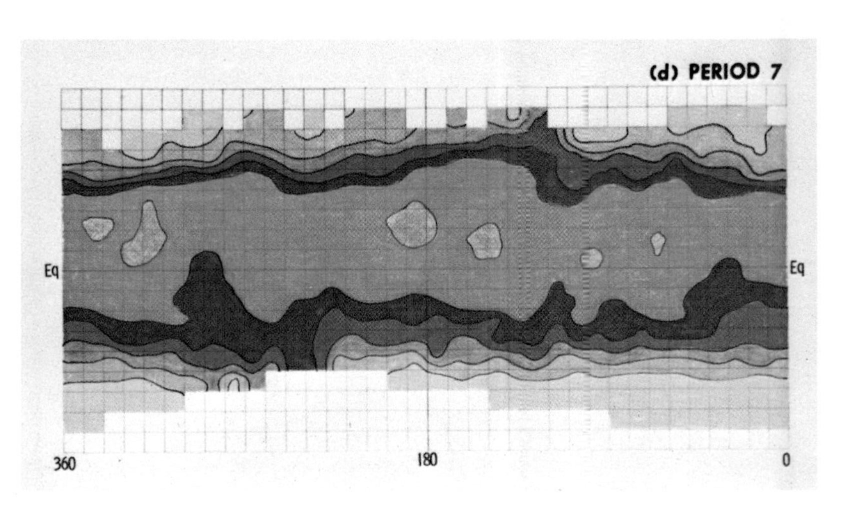

Plate 1. Seasonal variation of the global distribution of water vapor at planetocentric longitude intervals of 15°, commencing with period 3 (L_s = 110°–125°). The contoured values represent average column abundances between 1000 and 1600 hours local time. The white areas indicate regions with no observational coverage. Solar conjunction occurred in period 6, during which no data were taken.

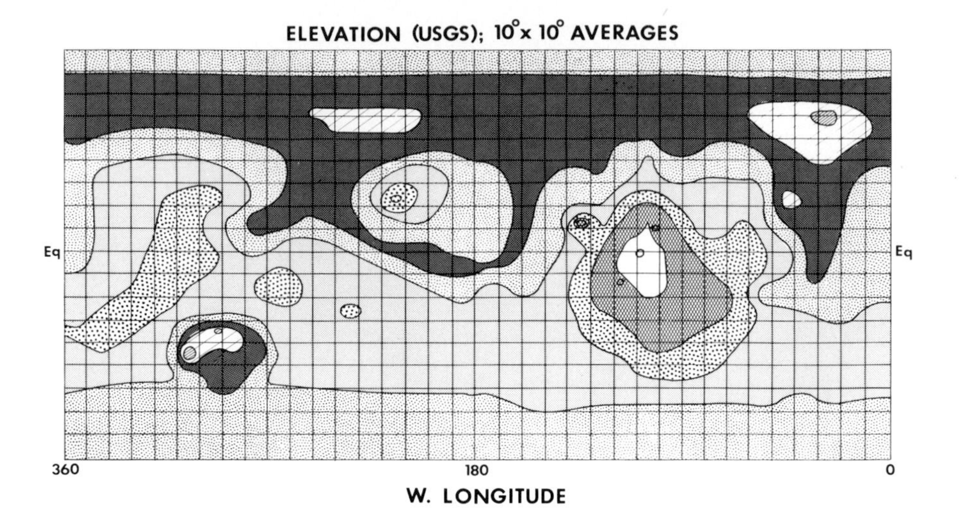

Fig. 9. Topographic map of Mars showing contours of average elevation over areas of 10° latitude and longitude to match Plate 1. (Data are taken from the *U.S. Geological Survey* [1976] map.)

This trend can be seen for latitudes from the pole to 20°N, but the seasonal coverage must be extended in order to determine whether it holds for the equatorial latitudes and, if so, whether a twice-yearly maximum occurs for these latitudes as would be expected if the hemispheric symmetry is preserved.

The variation of column abundance with latitude is shown in Figure 11, and the redistribution of the total global vapor abundance, shown in terms of the latitude-normalized values (i.e., the latitude average column abundance times the cosine (latitude)), in Figure 12. The total global water vapor abundance is the integral of the latitude-normalized distribution; throughout the period of the primary mission during which global coverage was possible (periods 3–7), the global total remained approximately constant, the amount being equivalent to 1.3 km³ of ice.

It is interesting to note in this context that the earlier earth-based telescopic observations had led to the belief that the atmospheric vapor was at a minimum during the equinoxes. It is easy to see how, with a detection limit of 10–15 pr μm and

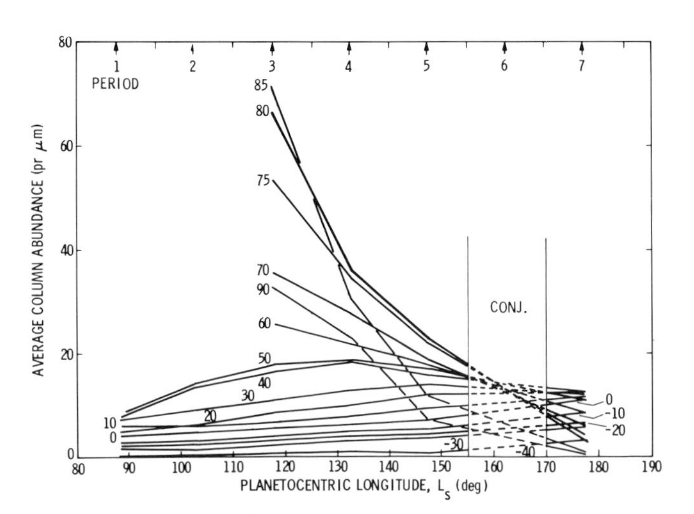

Fig. 10. Average (over all longitudes) of the column abundance of vapor within 10° latitude bands (5° bands from 70°N to the pole) as a function of L_s.

the inability to see the polar areas, the redistribution of the vapor could give the impression of 'dry' seasons when in fact the total atmospheric vapor may not have varied significantly. Nevertheless, it should be added that during the extended mission, at the time when major dust activity occurred ($L_s \sim 210°$), the vapor abundance apparently decreased by a significant factor. This may be the direct result of obscuration of the true column abundance by atmospheric scattering at levels above the bulk of the vapor.

Fine-Scale Structure

When the spacecraft are in Mars synchronous orbits and are therefore overflying the same region of the planet on each orbit, it is possible to design a series of observations to map systematically a restricted area at much higher spatial resolution than can be achieved from an asynchronous orbit. (It will be recalled that the instrument is limited more by signal-to-noise ratio than by field of view, so the resolution is determined primarily by how much observation time is obtained per unit area. By concentrating the observations a small area can be mapped at higher resolution at the expense of area coverage.) During the primary mission, there were two areas that could be mapped systematically at high resolution. For an extended time after the VL-1 landing, VO-1 was in a synchronous orbit passing over VL-1; after VL-2 landed, VO-1 was moved over to the VL-2 site, where it again stayed in a synchronous orbit for an extended period of time. The large number of data accumulated from these two areas have provided maps of water vapor distribution with a resolution of 2.5° in latitude and longitude (i.e., about 150 km). This resolution is comparable to that of the raster field of view at a range of 5000 km; observations obtained at ranges greater than 5000 km were not used for these maps. Figure 13 displays the maps generated from VO-1 observations taken on revs 34–67 (VL-1 area; $100° < L_s < 116°$) and revs 97–137 (VL-2 area; $131° < L_s < 151°$). Contours of equal vertical column water abundance at intervals of 2 pr μm are shown; bins (area elements) for which the data did not allow determination of the water amount to an accuracy better than 10% (or 0.7 pr μm for amounts less than 7 pr μm) are not included in the map.

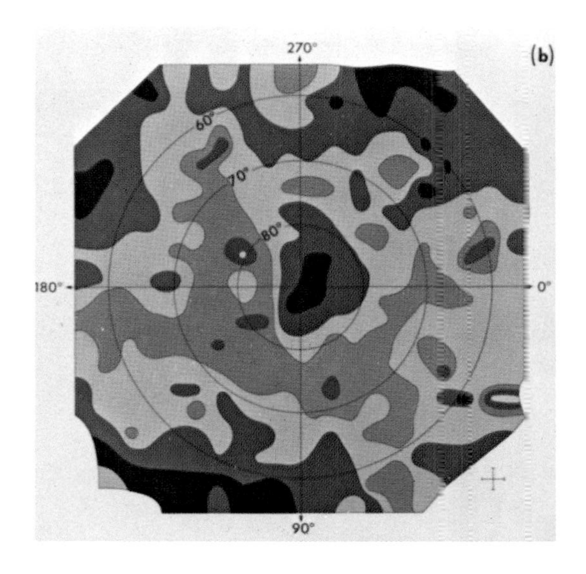

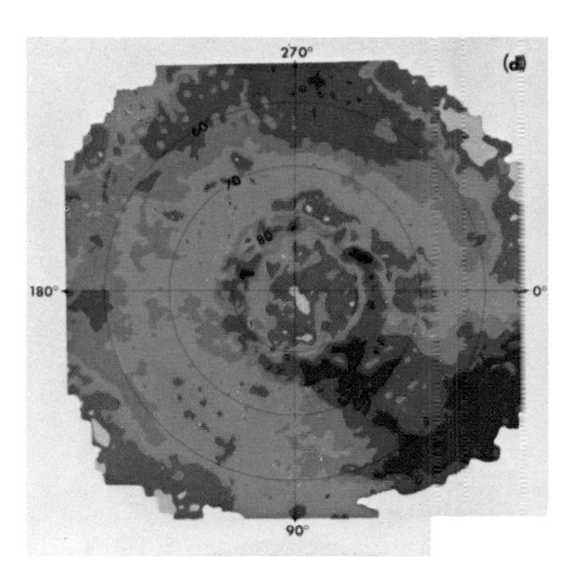

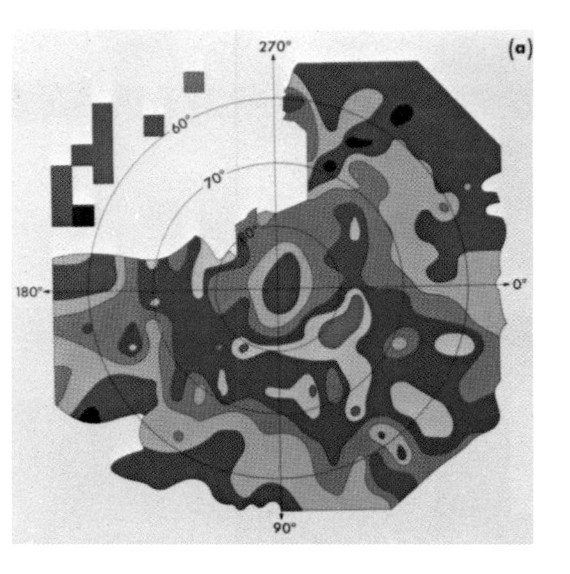

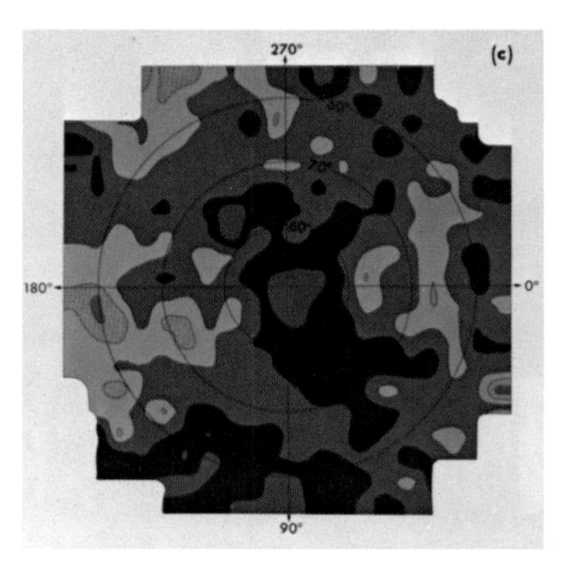

Plate 2. Distribution of water vapor over the north polar region. Data from VO-2; (*a*) revs 52–64, (*b*) revs 63–75, and (*c* revs 74–85. (*d*) The surface brightness map corresponding to the water vapor distribution in (*a*) is included for comparisor (see the text). The spatial resolution of the water vapor data is indicated at the lower right-hand corner of (*b*).

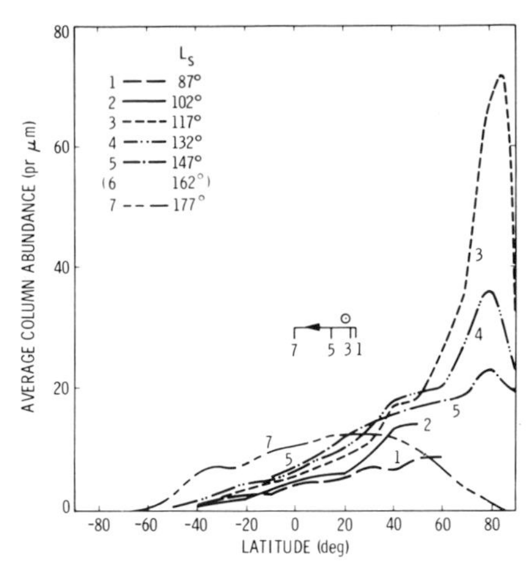

Fig. 11. Latitude distribution of water vapor for L_s periods 1 through 7.

In addition to these two areas which were mapped in two dimensions, observations were also made over the three Tharsis volcanos and Nix Olympica by using fixed cone-clock sequences, the spacecraft motion slowly sweeping the instrument line of sight across the calderas. This series of observations, obtained from VO-1 on rev 71, yielded a one-dimensional water abundance profile across each of the four volcanos; parts of these profiles cross into the area covered by the eastern portion of the high-resolution map. Radiances were averaged over groups of 10 rasters to improve the signal-to-noise ratio; the water abundances computed from these averages are shown in Figures 14*a*–14*d*. The location of each of the measurements shown in Figure 14 is given in Figure 13*a*; also included is the cross track size of the field of view. Where the rev 71 observations overlap the high-resolution map, the agreement in calculated water amounts is very good. Figure 14 also includes the surface elevation of each point, computed from a table of elevations obtained from the *U.S. Geological Survey* [1976] topographic map of Mars.

As was mentioned in the section on data analysis, the water vapor abundances (unless otherwise noted) have been calculated under the assumption that the vapor is at a pressure of 6 mbar. In the vicinity of the volcanos the effective pressure of the water vapor might be expected to be changing rapidly with location, and the assumption of constant pressure could result in significant errors. To demonstrate the effect of input pressure on calculated vapor abundance in this region, the water vapor amounts shown in Figure 14*a* have been recalculated by using pressures appropriate to the elevations shown in that figure, a 'zero-elevation' pressure of 6 mbar and a scale height of 10 km for the bulk atmosphere being assumed. These 'pressure-corrected' results are represented by the dashed curve in Figure 14*a*. It can be seen that decreasing the assumed pressure increases the calculated amount of water, the magnitude of the increase depending on the amount (for small amounts of water the calculated values tend to be independent of pressure; see Figure 4). Thus, although the distribution is modified, the use of a varying pressure has little effect on the qualitative dependence of water vapor on elevation in the vicinity of the volcanos.

In the following paragraphs we discuss several of the interesting properties displayed in Figures 13 and 14 and then offer some possible explanations.

Dependence of water amount on elevation. The most striking feature of the rev 71 data and the high-resolution maps is the decrease in water abundance over the summits of the volcanos. This occurs not only for the four volcanos covered by the rev 71 data but also over Tharsis Tholus (13°N, 91°W) and Elysium Mons (25°N, 214°W) in the high-resolution map. Conspicuously absent is any indication of a decrease of vapor over Hecates Tholus, to the northeast of Elysium Mons. Also absent is any pronounced change in water abundance associated with the 4-km drop in elevation between the Tharsis Ridge area and Lunae Planum and the 6-km drop between Lunae Planum and Chryse Planitia. The detailed map displays a small (~3 pr μm) increase in going from the Tharsis Ridge (at a 9-km elevation) to the Chryse Basin (at −1 km); this small increase in vapor content disappears if a surface pressure appropriate to the higher elevation in the Tharsis area is used. The same situation is encountered in the region of the gradual slopes surrounding Elysium Mons—the map shows a weak anticorrelation of elevation and water amount; most of this is removed when appropriate pressures are used.

There is one other topography-related change in the amount of water vapor shown on the high-resolution maps which occurs at 19°N, 71°W and coincides with the top edge of a 3-km-high cliff marking the edge of Lunae Planum. The effect of this rapid change in elevation persists for about 200 km east of the cliff edge.

Vapor boundary at 30°N. Another striking feature of the high-resolution map is the boundary or 'front' marked by the very sharp increase (from 10 to more than 20 pr μm) in crossing 30°N latitude in the region northwest of the Chryse Basin. There is no corresponding rapid change in elevation shown on the elevation map [*U.S. Geological Survey,* 1976]. To determine whether a change in the local surface temperature could be responsible, the same set of observations used to construct the high-resolution maps was used to produce maps of 1.4-μm surface brightness. The 1.4-μm brightness (corrected by dividing by the cosine of the incidence angle) is a good indication of the short-wavelength albedo, and variations in albedo should indicate whether an area is relatively warm or cool. Figure 15 shows the brightness map at the same (2.5°) spatial resolution as that of the corresponding water vapor distribution map of Figure 13. There appears to be no correlation of water amount with surface albedo in the region of the vapor boundary or in any other area included in the two high-resolution maps.

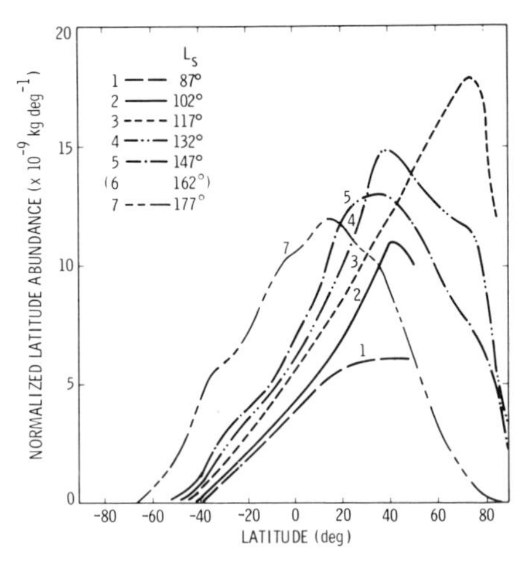

Fig. 12. Seasonal redistribution of the water vapor shown in terms of the latitude-normalized abundances versus latitude.

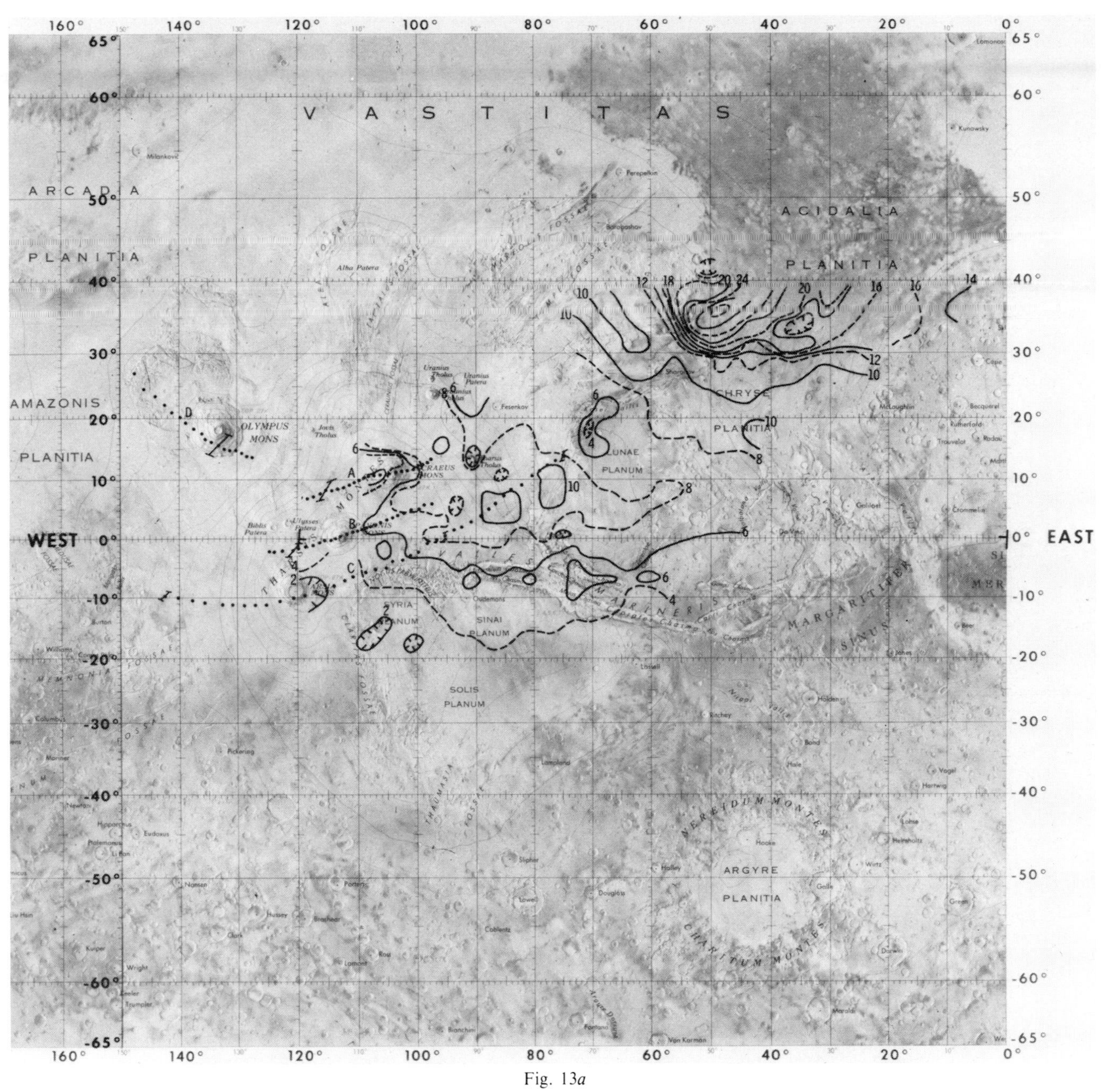

Fig. 13a

Fig. 13. (*a*) High-resolution map of the Tharsis-Chryse region, near the VL-1 site. The spatial resolution is 2.5°; the four dotted tracks, labeled A–D, indicate the locations of the measurements shown in Figure 14. (*b*) High-resolution map of the Elysium region.

Interpretation of fine-scale structure. It appears that areas characterized by rapid increases in elevation have associated with them a reduction in vertical water vapor content; however, gradual changes in elevation do not appear to affect the amount of water (for example, the one atmospheric scale height difference in elevation between the Tharsis Ridge and the Chryse Basin is not associated with any corresponding large change in the column abundance of vapor). A clue to the possible explanation of this behavior is seen in the dry area at the top of a 3-km cliff marking the edge of Lunae Planum (at 19°N, 71°W). The atmosphere moving from west to east would be adiabatically cooled by about 15°K in rising rapidly to the level of Lunae Planum; this will reduce the saturation vapor pressure of the lower layers of the atmosphere by about a factor of 6 (independent of the initial temperature, to first order) and cause a reduction in water vapor if the atmosphere initially contained a substantial fraction of its maximum capacity. On earth the atmosphere would remain at this lower temperature for an extended period of time; however, because of the much shorter radiative relaxation time for Mars [see *Goody and Belton,* 1967], the atmosphere will lose its 'memory' of the cooling and will be warmed back to its initial temperature profile relative to the elevation of the local surface. This implies that effects of topographic relief will only persist for as far as the atmosphere is transported in a time comparable to the radiative relaxation time. If we assign a value of 1 day to this characteristic time, a wind velocity of 2 m s^{-1} is required to give a corresponding 'relaxation distance' of 200 km, whereas a wind velocity of 40 m s^{-1} would be required to span the 4000-km distance between the Tharsis Ridge and the Chryse Basin. Since during most seasons of the year the expected velocities are of the order of 2 m s^{-1} rather than 40 m

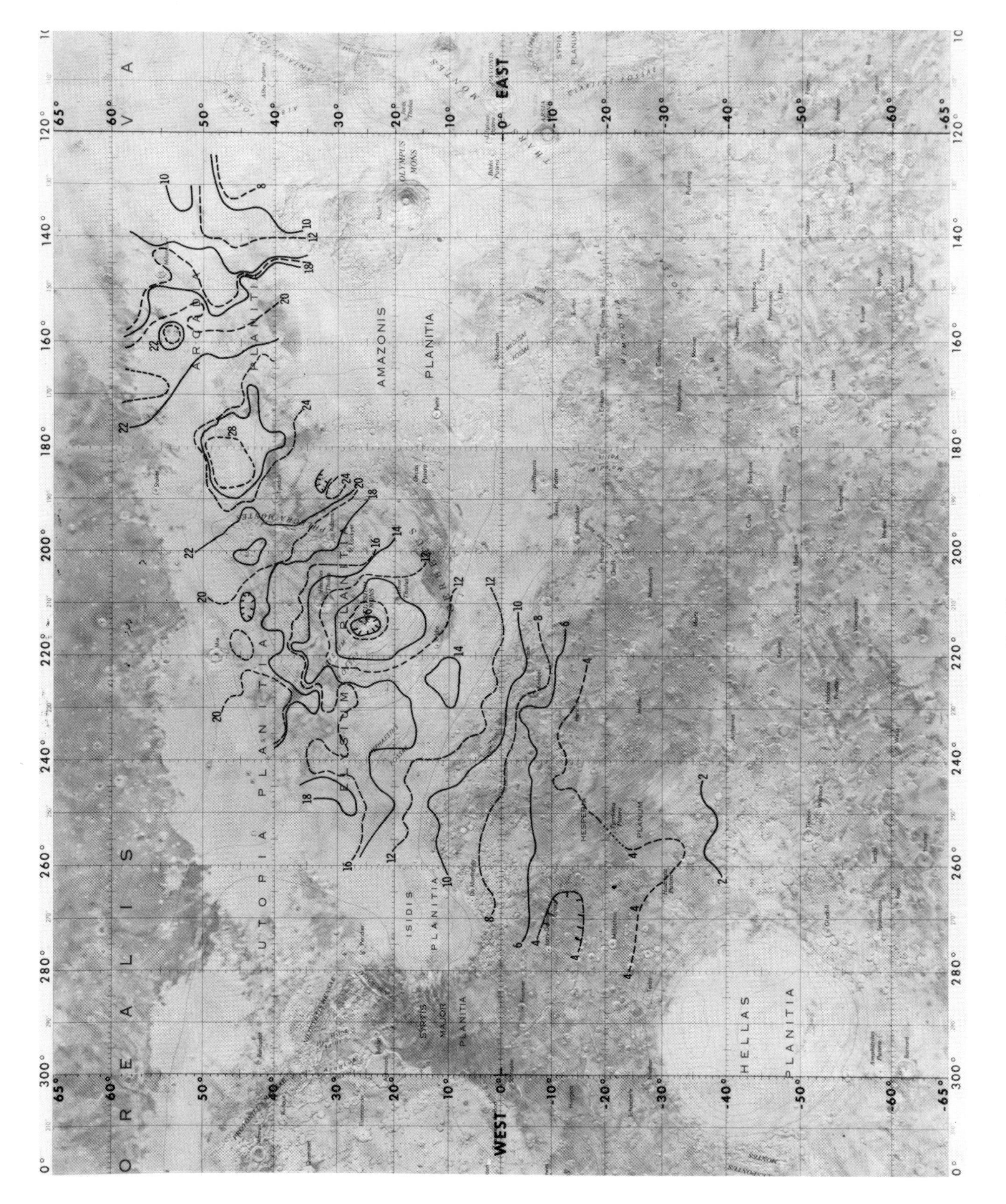

WEST
EAST
BOREALIS
VASTITAS
UTOPIA PLANITIA
ARCADIA PLANITIA
AMAZONIS PLANITIA
ELYSIUM PLANITIA
ISIDIS PLANITIA
SYRTIS MAJOR PLANITIA
HESPERIA PLANUM
HELLAS PLANITIA
OLYMPUS MONS
SYRIA PLANUM

Fig. 13b

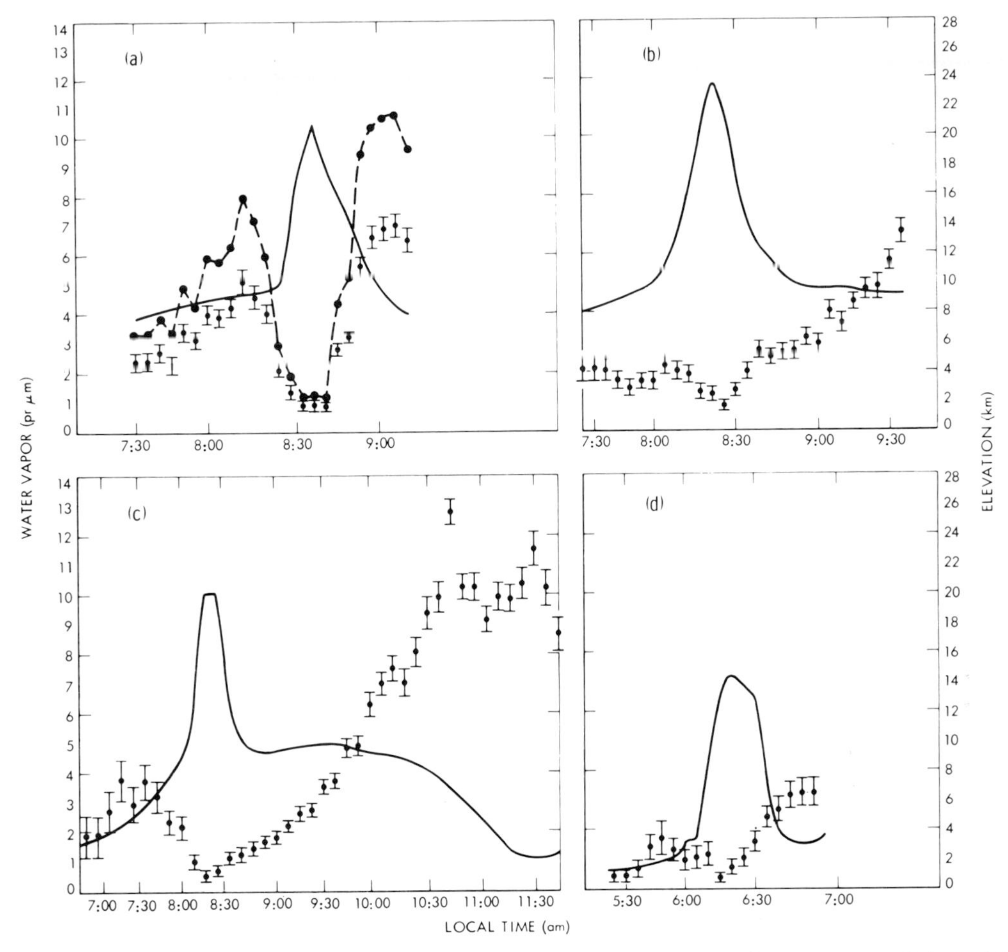

Fig. 14. Water vapor and elevation profiles across the three Tharsis volcanos and Olympus Mons. Points with error bars represent water vapor amounts (left-hand scale), the variation of elevation along the ground track is given by the curve (right-hand scale), and the local times corresponding to the observations can be read from the horizontal scale.

s^{-1} [*Leovy and Mintz*, 1969], the behavior shown by the water vapor in Figures 13 and 14 is consistent with an explanation based on the radiative response of the atmosphere.

An alternate explanation for the vapor distributions shown in Figure 13 involves particulates in the atmosphere. Measurements of optical opacity made from the Viking Lander 1 show a surprisingly large amount of suspended material in the Martian atmosphere over the VL-1 site. At the time of the measurements included in Figure 13 an atmospheric opacity τ of 0.45 (at a wavelength of 0.67 μm) was reported [*Mutch et al.*, 1976]. At about the same time (VO-1 rev 42), orbiter measurements were made of the VL-1 site over a large range of emission angles in order to determine the emission angle dependence of thermal and reflected radiation from the surface. From these observations it is apparent that at least for the VL-1 site the computed vertical column abundance of water is strongly dependent on the emission angle. The rev 42 observations indicate that at large values of the air mass factor η the amount of water in the optical path tends toward a constant value instead of being proportional to η. Figure 16 illustrates this behavior. Also shown in Figure 16 is the dependence of the calculated water amount on phase angle ϕ. The observations were made in an essentially continuous fashion from the time that the landing site appeared on one limb and passed directly below the spacecraft to when it crossed to the other limb. On the inbound leg the phase angle was much smaller (backscatter direction) than it was on the outbound leg. The difference in the computed amounts of water vapor for the same air mass factor can again be explained by a strong component of the radiation arising from scattering in the atmosphere, if the particulate matter in the atmosphere is more strongly forward scattering than the surface. Figure 16 strongly suggests that at least at the VL-1 site we do not measure the true total vertical column abundance, even when we are observing at $\eta = 3$. It is therefore possible that the apparent uniformity of water vapor over the area from the Tharsis Ridge to the Chryse Basin is due to a fortuitous combination of dust opacity and water content that gives the appearance of a constant water vapor abundance. We do not consider this possibility very likely, however, and the explanation in terms of the (short) radiative relaxation time is simpler and more natural, particularly so, bearing in mind the existence of the dry area on the edge of Lunae Planum. The data obtained on rev 42 are, however, an indication that there may be a systematic underestimation of the vertical column abundance and that measurements made with large air masses are especially suspect.

The remaining anomaly is the boundary, or front, at 30°N, which does not appear to be associated with any local change in surface elevation or any change in albedo that might produce a local thermal anomaly. This conclusion is supported by preliminary analysis of the thermal mapping data, which do not indicate any significant variation in surface temperature at the location of the boundary (H. H. Kieffer, private communi-

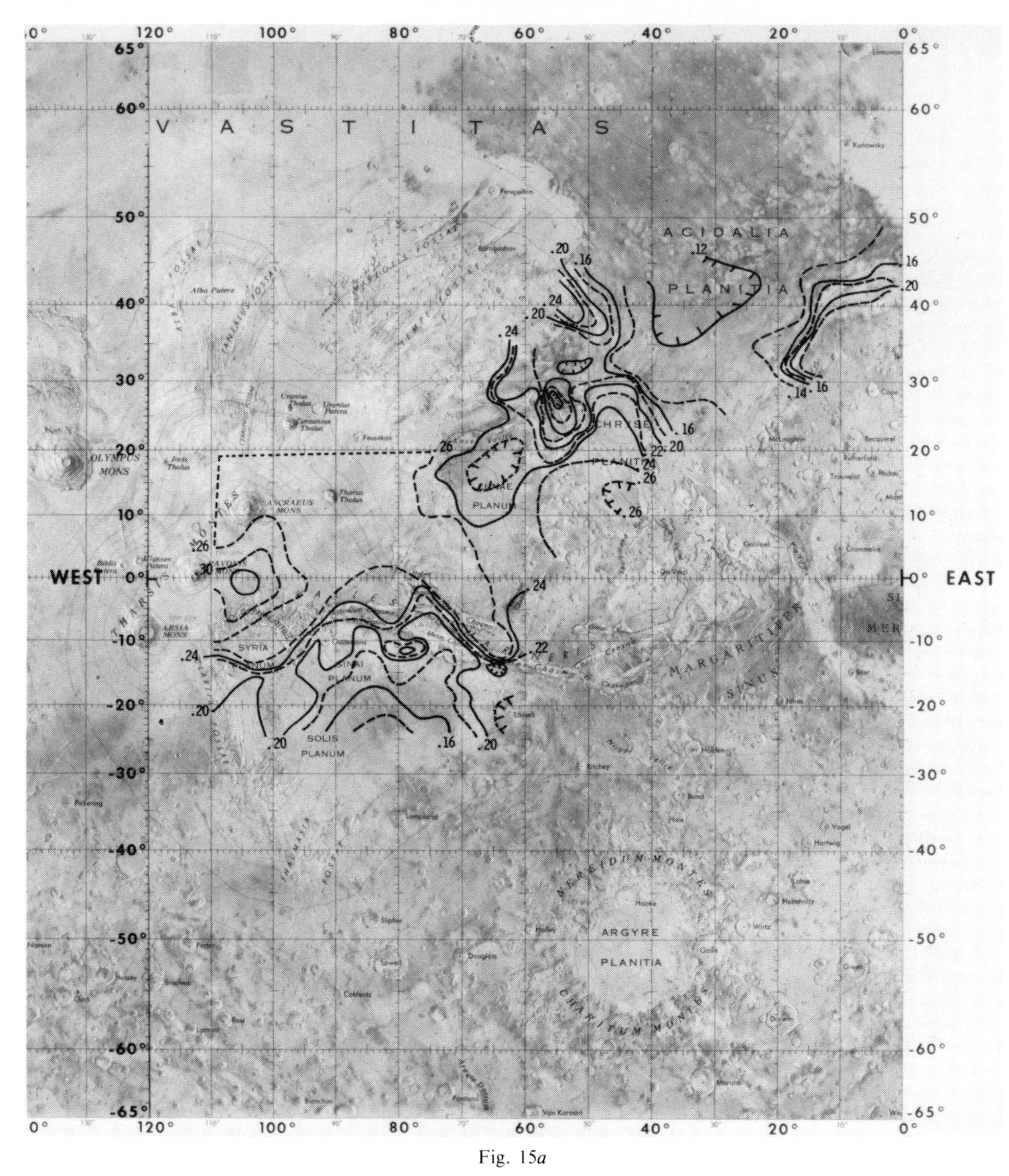

Fig. 15*a*

Fig. 15. (*a*) Observed brightness, at 1.4 μm, of the areas included in Figure 13*a*. The brightnesses have been calculated by dividing the radiances by the cosine of the incidence angle. (*b*) Same as Figure 15*a* for the area included in Figure 13*b*.

cation, 1977). On rev 91 an observation was obtained which covered the area from the equator to 45°N at 45°W longitude. The results from these data are identical to the results shown in Figure 13 for that longitude. Since the most recent observations contained in the mapping data were made on rev 52, we may conclude that the feature remained stationary for a period of at least 6 weeks.

We can only speculate as to its cause: The feature could be the result of a local change in surface properties causing a larger than normal release of water to the atmosphere at this season; alternatively, it could perhaps be the result of the fact that the transport of vapor from more northern latitudes is interrupted by strong winds or by a persistent high-pressure region at 30°N. It is even possible that the front marks the boundary of the region of large atmospheric opacity seen over the VL-1 site. The choice between these or other possible explanations must await a much more complete understanding of the surface properties and the atmospheric circulation.

Diurnal Variation

A number of observations made during the primary mission were designed specifically for monitoring the diurnal behavior of the water vapor. Most of these were made from VO-1 while it was in synchronous orbit, which enabled sites south of 10°N latitude, within a restricted longitude range, to be observed from dawn until noon. Several sites in this accessible region were chosen for repeated observation, the most frequent being at 10°N, 83°W on the lower slopes of the Tharsis Ridge close to Tharsis Tholus.

As was reported earlier [*Farmer et al.*, 1976*a*], the sites that were monitored exhibited a characteristic pattern of variation with local time, in which the apparent vertical column abun-

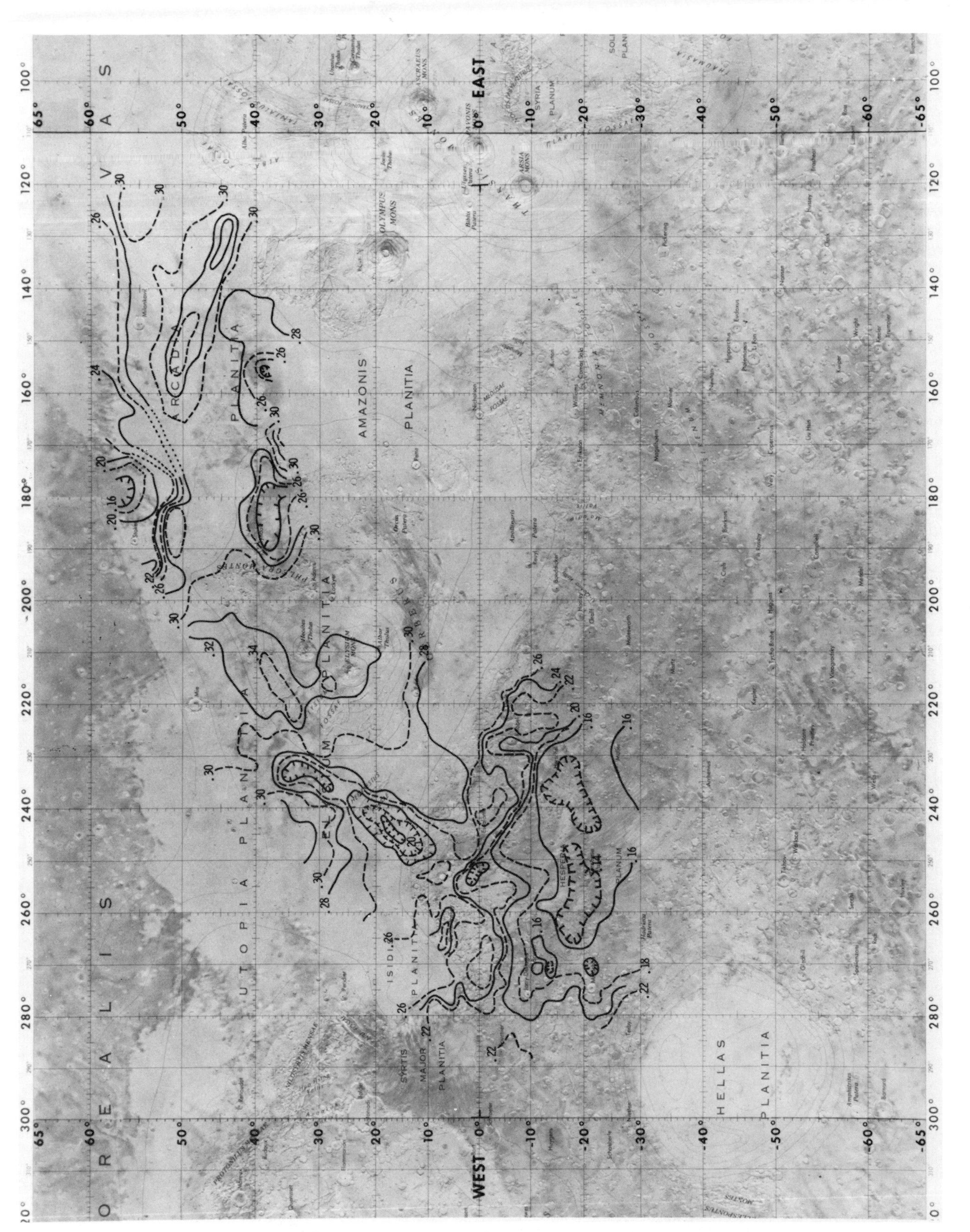
WEST
EAST
UTOPIA PLANITIA
AMAZONIS PLANITIA
ARCADIA PLANITIA
ISIDIS PLANITIA
SYRTIS MAJOR PLANITIA
HELLAS PLANITIA
HESPERIA PLANUM
SYRIA PLANUM
OLYMPUS MONS
ARSIA MONS
ASCRAEUS MONS

Fig. 15b

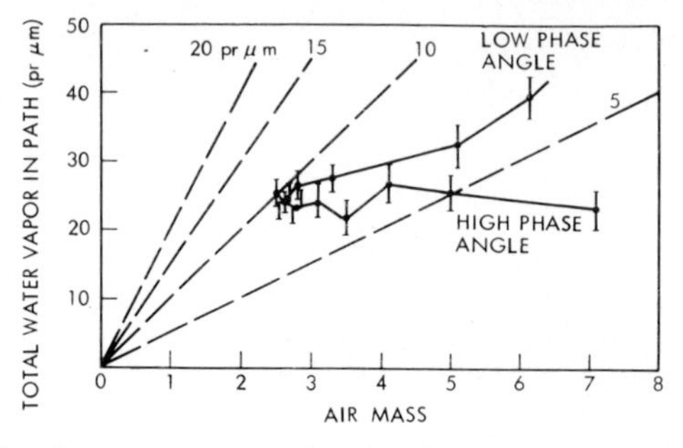

Fig. 16. Apparent water vapor abundances along the line of sight for observations of VL-1 made on VO-1 rev 42. If the atmosphere were free of particulates, the calculated vapor contents would be proportional to air mass.

dance of vapor increased linearly from zero at dawn to some maximum value at or soon after mid-day. In the case of the site at 10°N, 83°W, this behavior was repeated over a period of more than 110 days (see Figure 17). The diurnal variation was also investigated at other sites farther to the north (from VO-2), where the vapor quantities during this period of the mission were higher. For these sites the orbital conditions necessitated reconstructing the apparent diurnal variation by selecting observations made during asynchronous periods (in particular, the VL-2 site selection period). These gave data for several local times of day from dawn until dusk at several chosen locations; however, since the diurnal curves were compiled from data taken on successive orbits, they did not have the continuity of the equatorial observations from the synchronous (VO-1) orbit. The results for the 40°–50°N latitude band are shown in Figure 18 and suggest far more variability in the apparent diurnal behavior.

Before studying in detail the plausibility of mechanisms which might explain the diurnal behavior, it is certainly necessary to demonstrate that the observed diurnal variation is real rather than apparent. As was discussed previously, at large air masses, particulates in the atmosphere obscure the surface at the wavelength of the observations; large air mass factors are, of course, unavoidable with observations made near sunrise or sunset. When it became apparent from the rev 42 phase function observations that the atmospheric opacity over the VL-1 site was unexpectedly high, further phase function measurements were planned at several additional sites. Four such observations were made during the primary mission, and four (including one more of the VL-1 site) have been made during the extended mission to date. While it has not yet been possible to analyze the results of these in detail, it is clear that none of the other sites exhibit particulate scattering effects as severe as those at VL-1. For all of the other sites (which include the site at 10°N, 83°W) the observations to date indicate that for η less than 3.5, atmospheric scattering does not affect the measured quantities of water vapor, at least at the local time of the phase function measurements (noon to midafternoon).

It is entirely possible therefore that at least part of the observed diurnal variation at sites other than VL-1 is due to real changes in local water vapor content. However, it is also possible that a thin cloud or haze (located above the bulk of the water vapor), which is present at dawn but which dissipates by noon, could cause the observed variation. Such a haze layer, containing a very small amount of water ice (less than 1 pr μm), would have a large enough optical depth, when the layer is viewed at large incidence or emission angles, to obscure most of the water vapor. At present we cannot distinguish between these possibilities; until the present analysis to isolate the effects of particulate scattering from the data is

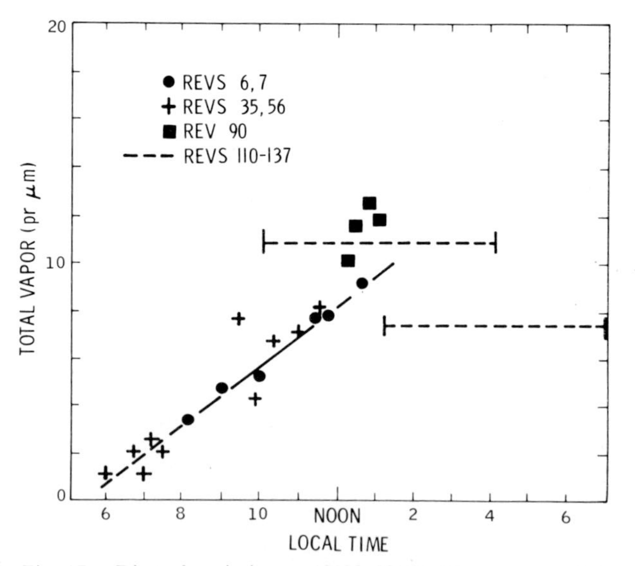

Fig. 17. Diurnal variation at 10°N, 83°W. The data are all from VO-1; the values indicated for revs 110–137 (dashed lines) are taken from low-resolution global observations and represent averages over the indicated local time intervals.

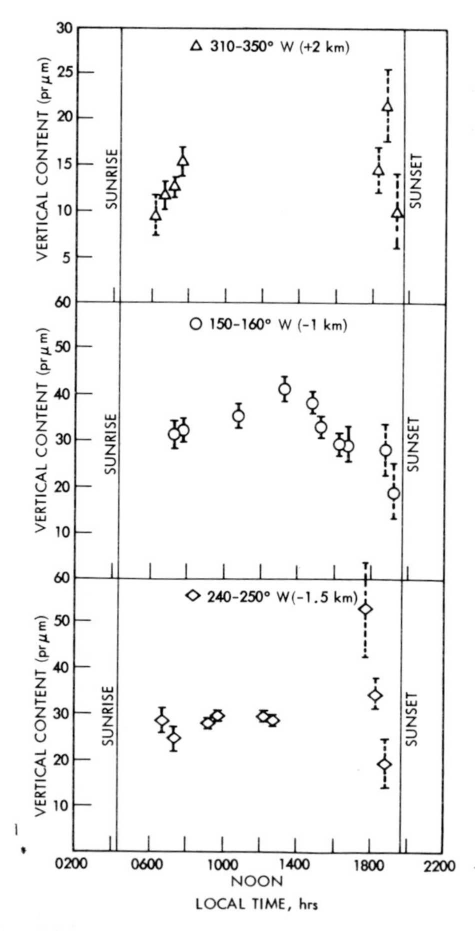

Fig. 18. Diurnal variation at three areas in the 40°–50°N latitude band during the period covered by VO-2 revs 7–15. The data used for this figure were restricted to slant range <5000 km, $i < 85°$, and $e < 75°$. Air mass factors η are less than 4.0 for all the plotted data points with the exception of those shown with dashed error bars.

completed, no firm conclusion regarding the magnitude of the diurnal variations can be drawn. Finally, it should be pointed out that the results obtained here raise the possibility that the diurnal variation observed by *Barker* [1974*a*, *b*] from ground-based observations of the 8200-Å band may also be a manifestation of scattering at large air masses. In the case of earth-based observations the effect is more severe, since both emission and incidence angles corresponding to the morning and evening limbs (or terminators) are always large.

Polar Regions

Before describing the observed behavior of water vapor in the polar regions, it will be useful to restate briefly the seasonal conditions which applied during the period covered by the primary mission. The commencement of the mission (VO-1 orbit insertion) to the end of the orbital observations prior to conjunction covered the seasonal interval corresponding to the planetocentric longitude range from L_s = 83°–152°. The subsolar point during this period moved from the maximum northern latitude (25°N) southward to 12°N. (The fall equinox, L_s = 180°, occurred about 1 month after conjunction, i.e., January 3, 1977.) Thus the beginning of the period corresponded with midwinter in the southern hemisphere, and all latitudes poleward of 65°S were in constant darkness. By the end of the period the sun was illuminating areas as far south as 78°S, but latitudes south of 45°S were still covered with CO_2 frost deposits [*Briggs et al.*, 1977]. It might be noted here that if the atmosphere over a CO_2 deposit were in thermal equilibrium with the frost, it would be too cold to hold a detectable quantity of water vapor. (For example, the saturation vapor pressure at 150°K corresponds to a vapor content equivalent to $<10^{-2}$ pr μm/km.) The sensitivity limit of the instrument for the observations in the south polar region was 0.1 pr μm; during the primary mission, no water vapor was detected poleward of 60°S.

Measurements of the north polar region (i.e., latitudes above 60°N) were not possible from VO-1 because of the low inclination and high eccentricity of its orbit. The initial inclination of the VO-2 orbit was 55°, and this enabled observations to be made at all northern latitudes up to the pole, although the pole itself was seen under rather poor viewing conditions. Prior to the separation of VL-2, all VO-2 observations were directed toward the search for a landing site in the 40°–50°N latitude band; the site selection and certification observations were made from an asynchronous orbit which allowed a few widely spaced scans from the pole to equatorial latitudes to be made at different longitudes on successive orbits. These observations were included in order to be able to characterize the 40°–50°N band in terms of the latitude distribution of water vapor in the northern hemisphere as a whole. From these early VO-2 observations, very large column abundances of water vapor (approaching 100 pr μm in some cases) were found to occur over the Arctic region, especially in the vicinity of the dark material surrounding the residual polar ice cap. These results [*Farmer et al.*, 1976*b*], as well as measurements of the surface temperature [*Kieffer et al.*, 1976], led to the conclusion that the remnant cap was composed of water ice (rather than being a CO_2 ice reservoir) and that the vapor in the atmosphere above the ice cap was close to the saturation limit.

As a result of the difficulties encountered in finding a suitable landing site for VL-2 and of damage to the orbiter during VL-2 separation, further observations of the north polar area were not obtained until rev 52, one revolution after the inclination of the orbit of VO-2 was increased to 75° and its period desynchronized again (to an 11-day global walk rate). This high-inclination asynchronous orbit was ideally suited to mapping the north polar area, and mapping observations were made essentially continuously from rev 58 to rev 85 (137° < L_s < 152°), that is, to the end of the primary mission. From the data obtained on these VO-2 orbits, three maps of the north polar area were constructed (Plate 2). In addition, since a strong anticorrelation between water amount and surface albedo had been noted from the earlier data, maps of 1.4-μm surface brightness were also made. Plate 2*d* shows the brightness distribution corresponding to the water vapor map of Plate 2*a*; all three of the brightness maps are essentially identical.

It is apparent from comparison of Plates 2*a* and 2*d* that the anticorrelation between atmospheric water vapor amount and albedo noted in the earlier data is still present; note particularly the minimum in water vapor over the high-albedo polar ice and the maximum in the 0°–90° longitude quadrant corresponding to the low-albedo material in this area. This anticorrelation would be expected if the atmospheric temperature were controlled by the surface temperature and if the atmosphere were near saturation.

In Plates 2*b* and 2*c* the anticorrelation is no longer evident except for the minimum over the polar ice. At this stage, if the atmospheric water content is still controlled by the surface temperature in the areas not covered by the remnant cap, the thermal inertias of the light and dark areas must differ so that the temperature contrast between high- and low-albedo material is significantly reduced. At the present time of writing, thermal maps, made from data taken by the infrared thermal mapper during the period under consideration, have not yet been constructed. When these become available, it should be possible to determine whether any such reduction in thermal

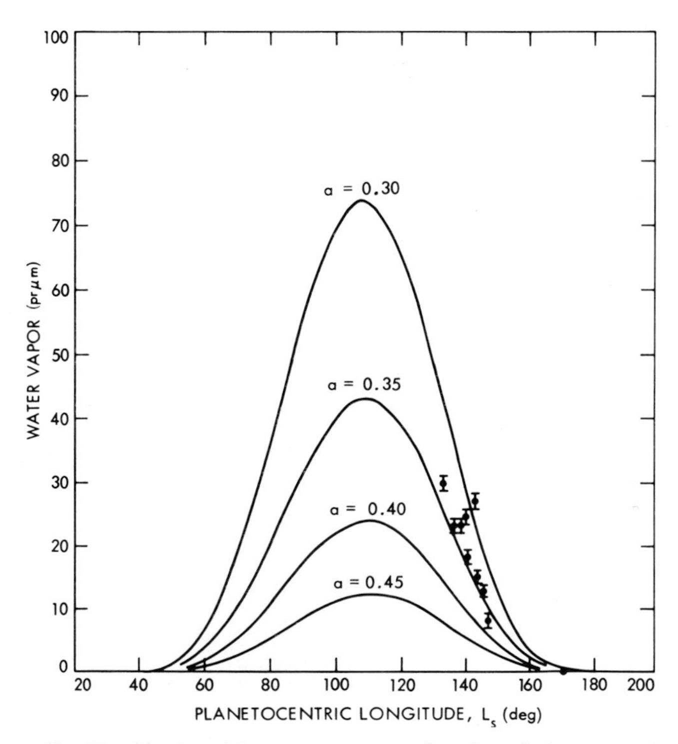

Fig. 19. North polar water vapor as a function of planetocentric longitude. A comparison of observations and model calculations for various assumed albedos [from *Davies et al.*, 1977] is included.

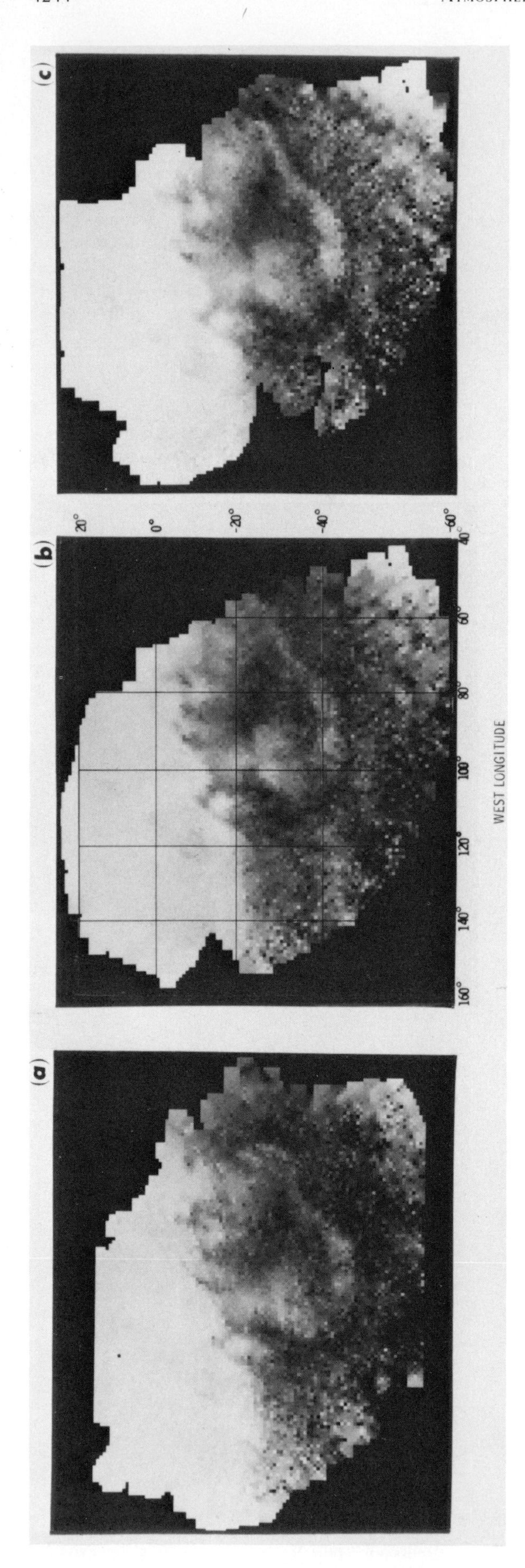

Fig. 20. Images at 1.4 μm of the planet from VO-1 on a Mercator projection: (*a*) revs 6–13, (*b*) revs 38–43, and (*c*) revs 51–63.

contrast did occur and to test the correlation between surface temperature and atmospheric vapor content.

As far as the vapor over the residual cap (which we shall define as the area above 85°N) is concerned, the seasonal variation can be examined with finer time resolution than is possible for other areas, as a result of the high density of data taken over the pole. These data have been averaged over 3-rev intervals and are shown in Figure 19, together with one point taken during the early part of the extended mission. It should be stressed that these results refer entirely to the atmosphere over the ice of the remnant cap only.

The observed variation with time over the seasonal interval covered so far has been compared to a thermal model of the polar surface and atmosphere by *Davies et al.* [1977]. The polar water vapor content is extremely sensitive to the albedo of the ice of the remnant cap, as a consequence of the strong dependence of the vapor pressure on temperature. (As the albedo decreases, the surface temperature rises, the atmospheric temperature and with it the saturation vapor pressure of the water thus being increased.) From the model studies it was found that the best fit to the north polar water vapor quantities was obtained with an ice albedo of 0.34 and a value for the thermal inertia of the surface material appropriate for water ice (Figure 19).

An albedo of 0.34 is low, an indication that the ice is dirty; this is not surprising, since condensation of the small amount of vapor in the atmosphere over the pole each year would not produce a layer of ice thick enough to cover up dust deposited that year. Agreement between observed and theoretical quantities of vapor could not be achieved if a value for the inertia of the surface material consistent with that of typical Martian soil was used. The high inertia of the polar material indicates that the ice must be at least as thick as the depth penetrated by the seasonal thermal wave, which is approximately 1 m. The model results also suggest the following:

1. The peak water content at the north pole occurred just before the first polar observations were made from VO-2.
2. Because of the sensitivity of water vapor to surface albedo the amounts observed in the polar areas could be strongly influenced by the previous few years' dust storm activity.
3. If the albedo of the south remnant cap is the same as that of the cap in the north and if the cap were able to saturate the atmosphere above it, the peak amount of vapor in the south would be approximately 5 times higher than that in the north.

Other results derived from the application of the model include an estimate of the thickness and extent of the polar CO_2 deposits and lead to a possible explanation of the asymmetric location of the south remnant water ice cap. It is inappropriate to discuss these at length here, and the reader is referred to *Davies et al.* [1977] for details of the model itself and the results obtained from it.

Appearance of Mars at 1.4 μm

In association with the water vapor measurements the Mawd observations provide a direct measure of the brightness of the surface and atmosphere at 1.4 μm from the radiance values recorded in the two continuum channels. Since most of the data are obtained by using the scan platform to provide large-area scans of the planetary surface, the instrument can be considered to operate as a very wide angle low-resolution facsimile camera, which generates 1.4-μm images of large areas

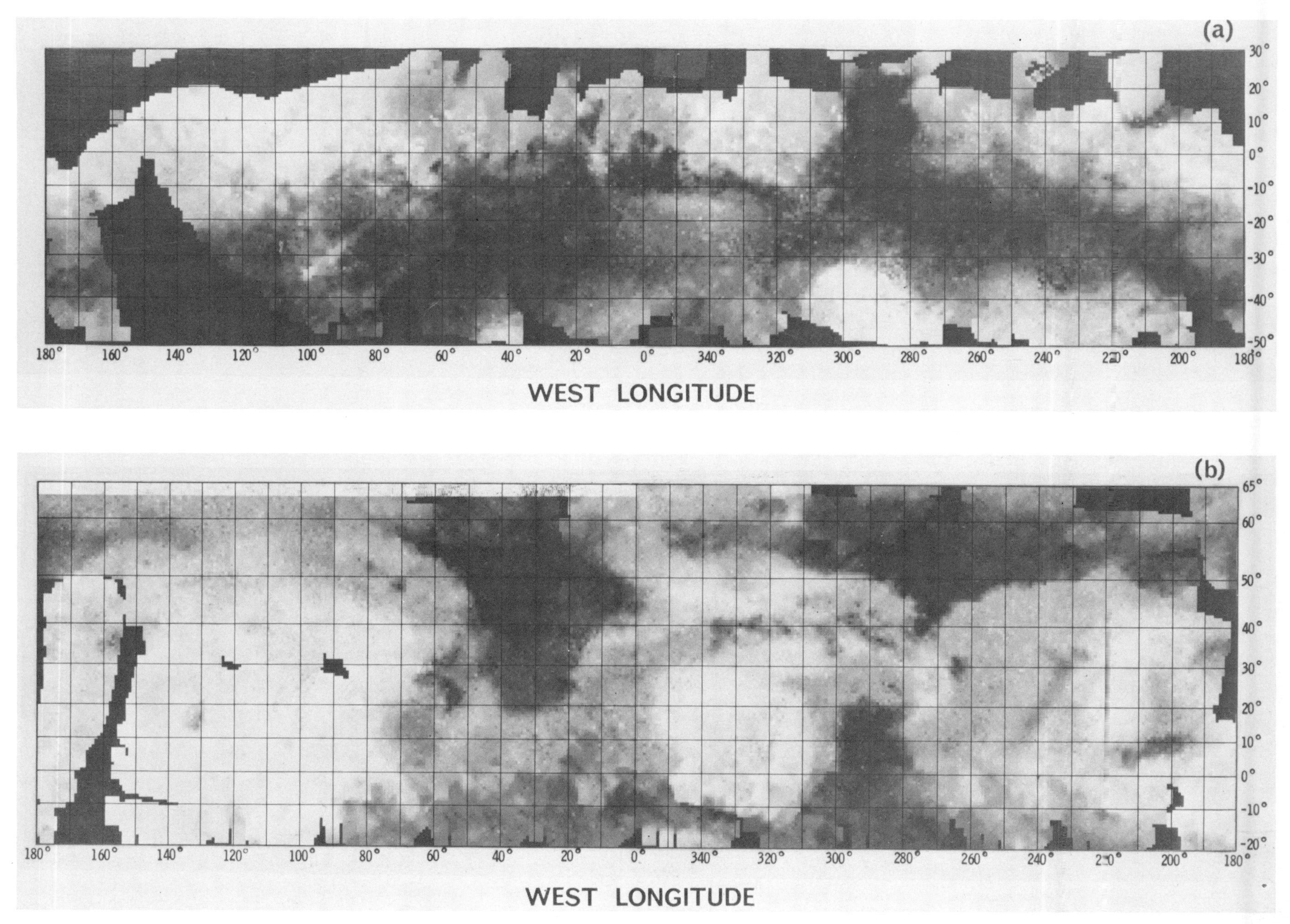

Fig. 21. Global image of brightness at 1.4 μm from (*a*) VO-1 walk and (*b*) VO-2 walk (revs 73–84). Mercator projection, with 0°W longitude at the center.

of the planet. Several of these images have been constructed in order to monitor changes in the global and regional scale appearance (e.g., clouds in the Tharsis area) and to map the 1.4-μm albedo.

In general, these images have been produced by combining many box scans taken under similar viewing conditions, binning the instantaneous field of view radiances, and filling empty bins by interpolating between adjacent values. The images produced in this way are typically made up from some 3×10^4 randomly placed individual radiance measurements and have approximately the same number of picture elements. The spatial resolution is somewhat better than 1° of latitude and longitude. In order to correct for variations in solar incidence angle, the images are usually displayed in terms of 'Lambert brightness,' the radiance divided by the cosine of the incidence angle, where the incidence angle is referred to a spherical planet (i.e., ignoring local slopes). It should be noted that the shapes of the spectral reflectance curves for dark and light material are essentially the same [*McCord and Westphal,* 1971]; measurements made at 1.4 μm are therefore a good indication of the visual appearance of the planet.

The first images were constructed from data taken by the VO-1 instrument while VO-1 was in a synchronous orbit passing over the VL-1 site. Figures 20*a*–20*c* show these images, obtained on revs 6–13, 38–43, and 51–63, respectively (corresponding to mean values of L_s of 88°, 103°, and 111°). The Tharsis area can be seen to be considerably brighter than the region to the south, the three Tharsis volcanos appearing as dark spots. The southern border of the bright area is sharp, correlating well with Valles Marineris between 60° and 90°W in longitude and then running south of Noctus Labyrinthus. The northern brightening is probably caused by diffuse cloudiness in that area. This pattern appears to have been remarkably stable over the time period from rev 6 to rev 85.

During revs 83–95 the orbit of VO-1 was asynchronous; this provided observations of different longitudes on each orbit and a view of the entire planet between 50°S and 20°N latitude. Near the end of the primary mission, VO-2 was also in an asynchronous orbit; thus maps of 1.4-μm Lambert brightness from the north pole to 10°S could be constructed. These are shown in Figures 21*a* and 21*b*, covering the region to 65°N. An attempt has been made to relate the observed brightnesses to the true 1.4-μm albedo by scaling to the earth-based measurements of near-infrared albedo made by *Binder and Jones* [1972]. The resulting contours of 1.4-μm albedo are shown in Figures 22*a* and 22*b*.

The albedo maps (Figure 22) agree fairly well with the appearance of the planet shown on the *U.S. Geological Survey* [1976] topographic map, which combines Mariner 9 and earth-based National Planetary Patrol Program data. There are differences in specific areas, e.g., the dark 'Y' shaped feature to the west of Elysium Mons in Figure 22*b*.

The albedo maps can also be used to locate areas of partial cloud cover. If we assume a maximum albedo for Martian soil, then regions brighter than this maximum must have at least partial cloud or condensate (CO_2 frost) cover. (Obviously, the converse is not true: areas darker than this maximum could still have partial cloud cover, especially if the surface material is dark.)

If we assume that the maximum surface albedo is 0.35, then Figures 22*a* and 22*b* indicate several areas of cloud cover. The most extensive area includes Tharsis and Olympus Mons, the area northward to a latitude of 45°, and Amazonis Planitia. This bright region also has a broad finger running west around the south of Cerberus and then north to cover a broad area over Elysium Mons and Hecates Tholus. An extensive bright area between Syrtis Major Planitia and Acidalia Planitia is also apparent. These high-albedo areas are consistent with earth-based measurements of individual locations reported by Binder and Jones. High-albedo regions in the southern hemisphere include Argyre, Hellas, and the region to the east of Hellas; presumably, the high brightness of these is due to CO_2 frost. As data for other seasons become available during the extended mission, it will be possible to monitor changes in the extent and thickness of the cloud cover and frost deposits.

Summary

It is appropriate to conclude by summarizing the observations of water vapor made during the primary mission and the main conclusions that can be drawn from the results at this stage.

The global distribution of the vapor has been mapped at low resolution throughout the period from the northern summer solstice to the following equinox. During this seasonal period the water vapor underwent a gradual redistribution, the latitude of maximum column abundance moving from the northern polar area to the equatorial latitudes. The total global vapor content remained approximately constant at the equivalent of ~1.3 km^3 of ice.

The peak vapor abundances observed were found to occur over the dark circumpolar region, an area that was inaccessible to earth-based observers. Maximum vapor column abundances of ~100 pr μm were measured. These results, together with measurements of the corresponding local surface temperatures, indicate that the residual polar caps are composed of water ice; at the season of maximum polar vapor content the atmosphere above the ice is at or close to saturation.

Observations of the diurnal variation of the vapor have shown a variety of effects at different latitudes and locations. The apparent vertical column abundance of the vapor phase is influenced by scattering from dust and/or condensate particles, especially at large incidence and emission angles. The large diurnal variation observed from earth may be, in part at least, an observational effect.

High-resolution observations have been made covering limited areas in the equatorial region (Tharsis–Lunae Planum–Chryse and Isidis–Elysium–Arcadia). The results of these measurements suggest that local variations in the vertical column abundance of vapor occur only where there are known topographic features characterized by abrupt elevation changes. The vapor remains fairly constant over regions where the slopes are gradual, even though the total elevation change may be large. This behavior may result from the short radiative relaxation time of the Martian atmosphere and the expected low wind velocities at this season. However, during the course of the primary mission, only two such areas were mapped at the higher resolution, and this apparent dependence of the vapor abundance on the magnitude of the local slopes may not occur everywhere. Indeed, the low-resolution global maps show, superimposed on the general (seasonal) variation with latitude, large spatial variations which seem to exhibit some topographic control even on the regional scale.

The outstanding questions concerning the specific mechanisms controlling the seasonal redistribution of water vapor and its possible hemispheric migration, and the existence of a major subsurface reservoir of water ice at other than polar latitudes, remain to be answered. Further insight should be

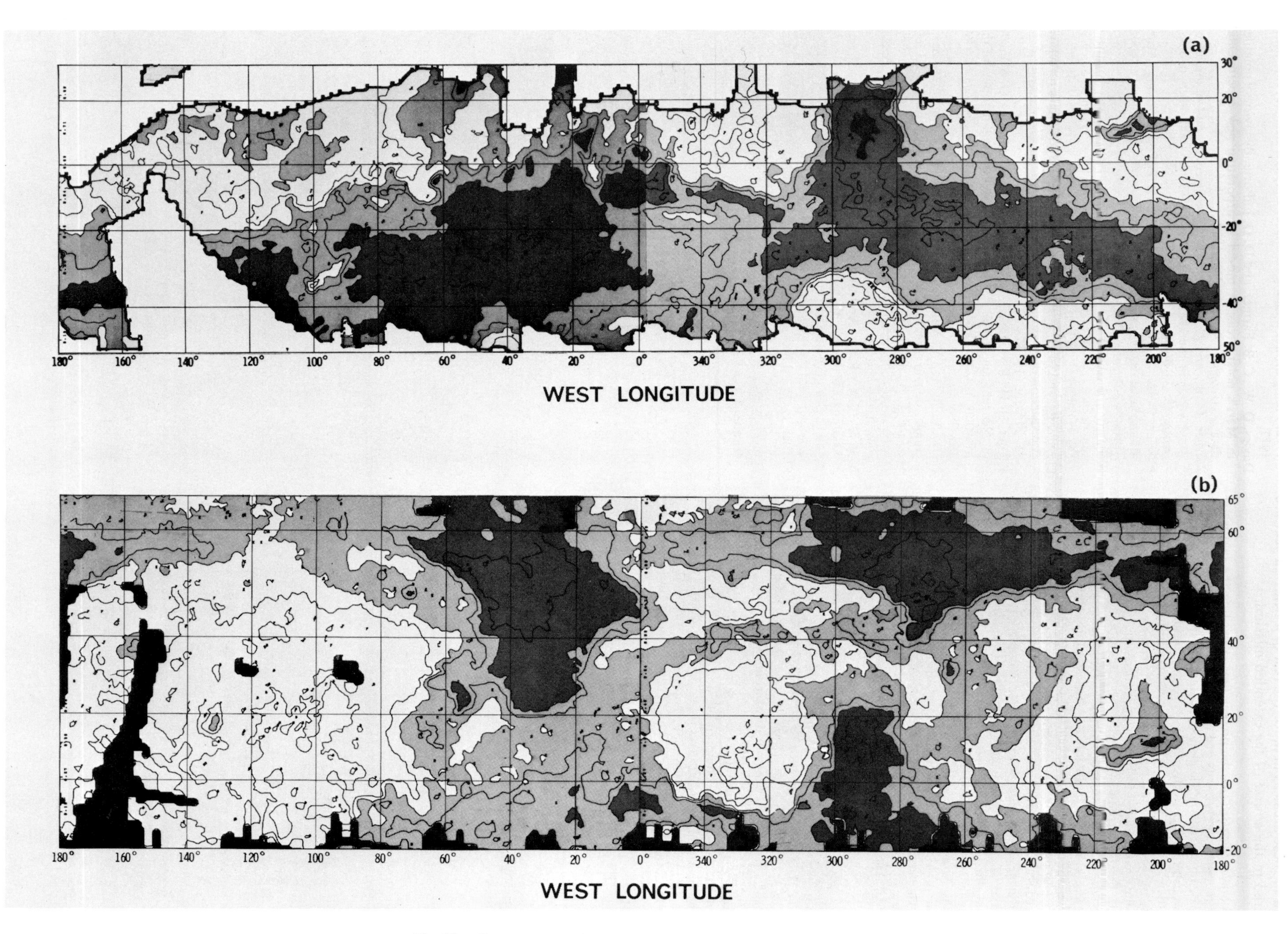

Fig. 22. Contour map of 1.4-μm albedo corresponding to Figure 21.

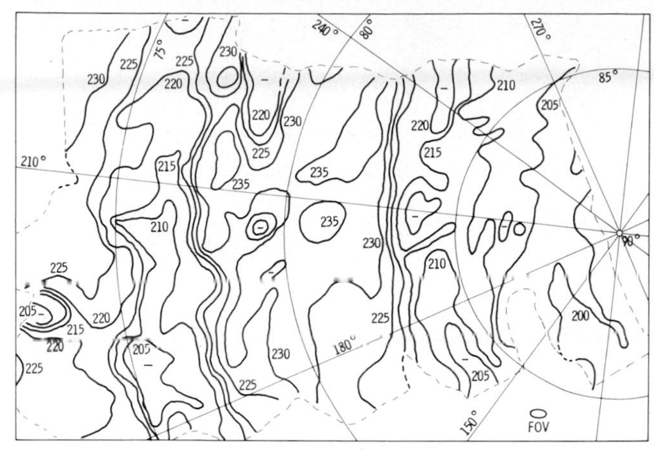

Fig. 4*a*

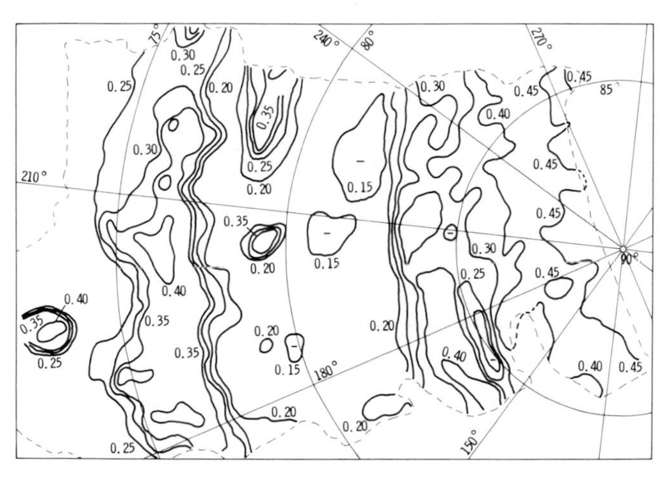

Fig. 4*b*

Fig. 4. IRTM observations of a portion of the north polar region at L_S = 141°. Absolute values are dependent on the use of a smooth atmospheric model to remove instrument drift (see text). The dashed line indicates the extent of IRTM data. The local time was 16 H at 210°W longitude. (*a*) 20-μm brightness temperatures. The field of view is indicated in the lower right-hand corner. (*b*) Lambert albedos A_L.

temperatures show very strong thermal boundaries, with temperatures of 198–210 K corresponding to frosted areas and temperatures in the range 220–237 K over unfrosted regions, similar to those for observations made 45 days earlier (Figure 4*a*). The crater Korolev (+73°, 193°W) still appeared to be filled with frost, T_{20} being near 205 K. The ice north of +83° becomes colder toward the pole, where the sun was 16° above the horizon.

The region is composed of three materials of quite different albedo (Figure 4*b*). The area north of +83°, the band along +75° to +77°, and other smaller areas have $A_L \geq 0.27$. Most of the area south of +75° has $A_L \sim 0.22$. The band between +77° and +83° has $A_L \leq 0.20$. There is little area with albedo of 0.23–0.30. The brightest material appears to represent frost, or ground ice, and the darkest material is seen in the imaging data to be sand dunes [*Cutts et al.*, 1976].

Although the temperatures are anticorrelated with albedo in the manner expected for particulate material, the dark material is about 5 K cooler than was expected for these albedos. However, analysis of the behavior of absolute temperature with season will have to await more detailed reduction of the observations made after the VO 2 plane change.

There is a strong anticorrelation between T_{20} and A_L throughout this region. The distribution of bright areas generally agrees with the late summer pattern observed 4 years earlier (Figure 5*a*), although individual bright areas, not present in 1972, are observed, such as at +79°, 205°W and at +83°, 245°W. Viking imaging suggests ground frosts in these areas (Figure 5*b*). Both surface and atmospheric temperatures remained well above the CO_2 frost point, and both thermal and albedo contrasts continued undiminished throughout the Viking primary mission, indicating that obscuration by the polar hood [*Briggs and Leovy*, 1974] had not begun by L_S = 151°.

B. *South Polar Region*

Observations of the South Polar Region, the first of any kind of the Martian polar night, revealed brightness temperatures well below those expected for the planet and suggested that major changes of atmospheric composition might be occurring in the polar night (paper 1). Important questions are whether the coldest areas would stay at the same locations and how they would behave as the spring sublimation commenced.

1. *Surface behavior.* Throughout the primary mission, temperatures over the winter polar cap have shown a decrease toward the pole rather than the anticipated nearly constant value associated with CO_2 condensation at the surface pressure. The basic temperature pattern observed from synchronous orbit early in the mission has persisted, the foremost feature being a distinct minimum near the South Pole. There are several localized features with a few degrees thermal contrast.

At the time of the VO 1 walk, much of the south polar cap had T_{20} near 143 K (Figure 6). There were several prominent lower temperature features between −68° and −82° latitudes. The thermal trough extending out along 175°W then curving east toward 140°W was, apart from the polar minimum, the most constant feature persisting throughout the primary Viking mission. A similar but broader cold region existed 90° further east. A third cold arm extended outward along 350°W but was not clearly connected to the thermal low near −76°, 300°W. There was not a well-established cold arc in the 225°W quadrant at this time that would have completed a crude fourfold symmetric pattern. The trend of these apparent low-temperature arcs eastward away from the pole is the opposite of that expected to be associated with coriolis force deflection of the atmospheric flow toward or away from the pole.

With the oblique viewing available from VO 1, the polar minimum generally appeared about 2° of latitude beyond the geometric pole but varied to either side rather than appear directly opposite the spacecraft at all times. This general behavior indicates that there is an emission angle effect which offsets the apparent position of the polar minimum, but the deviation away from the direction of increasing emission angle indicates that there is about a 200-km motion of the minimum temperature position. To minimize the emission angle effect, a composite map was made. This map uses observations with emission angles less than 60° from the VO 1 walk for the circumpolar region, and the first observation sequence following the plane change of VO 2 is used to fill in the region within 5° of the pole (Figure 6). The VO 2 observation shows the polar minimum to be centered on the geometric pole at that time, 2130 UT, October 3, 1976. In all VO 2 observations after the plane change, with emission angles less than 50° at the pole, the minimum temperature has appeared within 100 km of the South Pole.

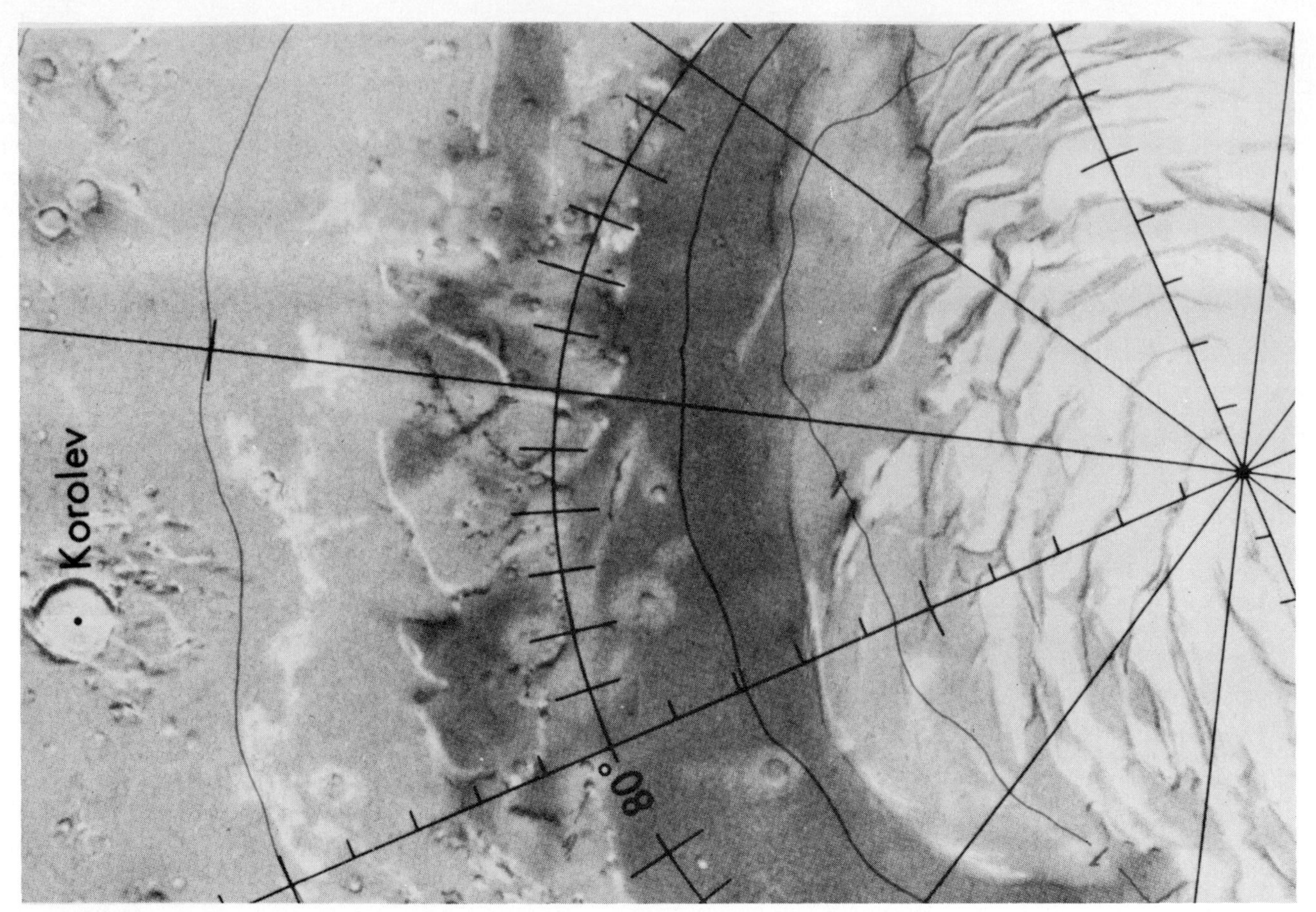

Fig. 5a

Fig. 5b

Fig. 5. Imaging appearance of the same region as that in Figure 4. (*a*) Detail of the topographical map of Mars as it appeared at the end of the Mariner 9 mission in late 1972 at $L_S = 65°$. (*b*) Mosaic of high-resolution Viking imaging obtained at the same season as that in Figure 4.

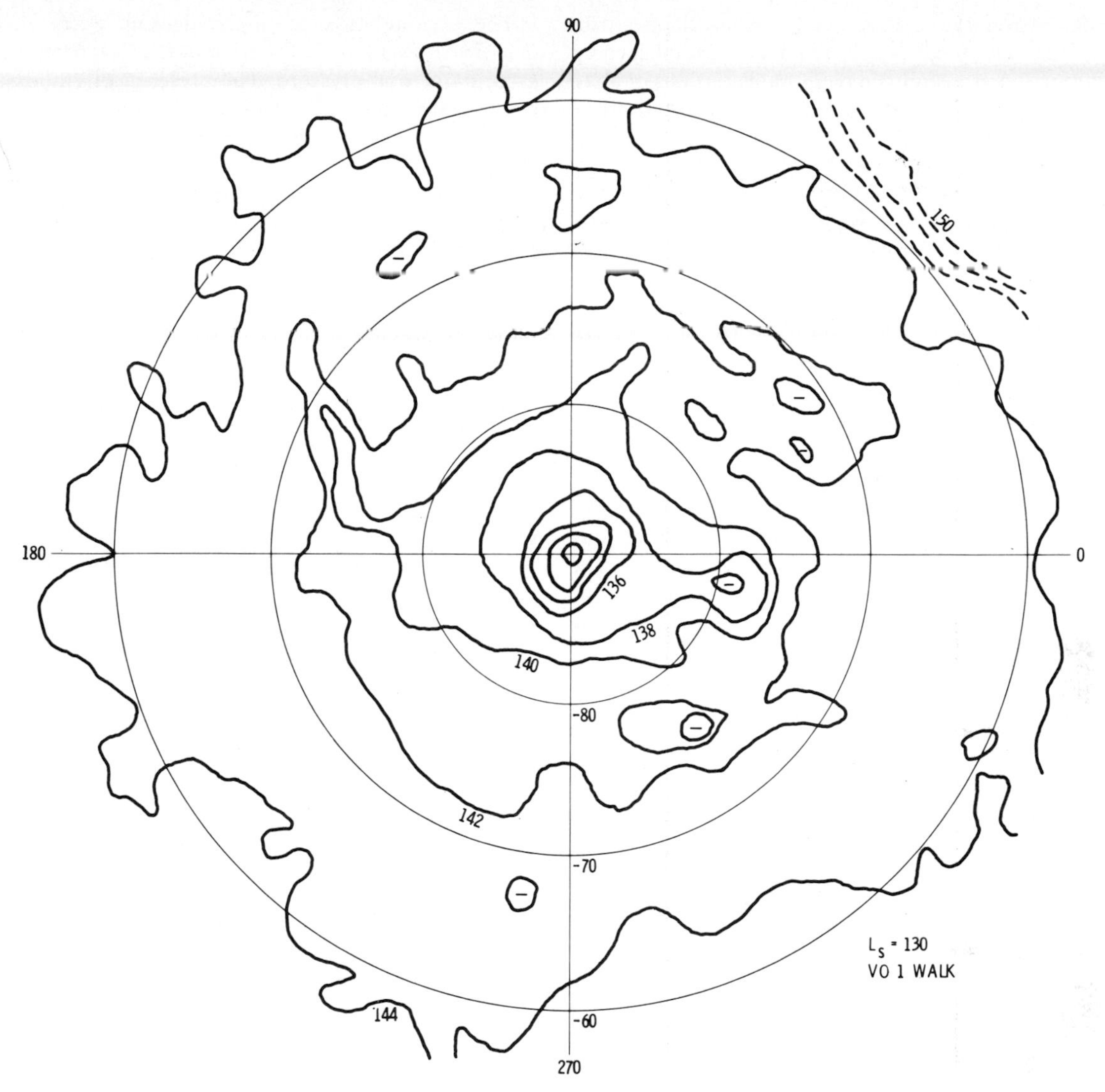

Fig. 6. T_{20} observed in the south polar region near L_S = 130°. This thermal map is a composite; north of −84° it uses seven slices, extending up to 45° of longitude from the subspacecraft longitude (local time 6 H), obtained during the VO 1 walk (L_S = 124°–129°). Poleward of −84° the data are from VO 2 rev 54 (L_S = 135°; see text). The contour interval is 2 K. The region mapped extends to −55°. The dashed contours at the upper right-hand side are examples of the rapid temperature rises toward noon at −60°. The IRTM resolution is about 80 km throughout this composite.

After resynchronization of VO 1 the South Polar Region was monitored at about a 3-day interval, and some temporal behavior was observed. The subspacecraft longitude for these observations was about 260°W; five observations, on revs 97, 101, 104, 107, and 110, have been examined, and two prominent changes have been observed. On rev 101 a cold region at −82°, 340°W had a minimum T_{20} of 133 K and was 6–8 K colder than areas about 400 km away except where a narrow saddle with T_{20} = 135 K separated this low from the polar minimum. The temperature contrast of this feature was only 3 K both 3 days earlier and 3 days later. On rev 107 the south polar minimum was 127 K, 4 K colder than for the other four sequences. Rather than a uniform cooling, the central area previously below 140 K became cooler and the surrounding adjacent areas previously above 140 K became warmer during a 3-day period. This behavior reversed to reestablish the prevailing pattern within the next 3 days.

Several thermal features with contrast of about 2 K and dimensions of a few hundred kilometers were seen in the region −65° to −85°, 270°W to 30°W. However, more detailed analysis will be required before the possible motion of individual features may be established.

Observations at the end of the primary mission, L_S = 151°, still showed thermal structure in the South Polar Region similar to the first Viking measurements. The minimum temperature was 133 K (Figure 7). The cold area at 160°W had a maximum amplitude of 7 K. A large amount of thermal structure with 1 or 2 K amplitude was present within 30° of the pole. There was no noticeable change in the average T_{20} variation with latitude during the Viking primary mission (Figure 8).

The low temperatures may be explained in part by emissivity of less than 1.0 and by elevations above the Martian average. The T_{20} of 143 K, representative of a large circumpolar area, could be explained by an emissivity of 0.86 or by surface pressures of 3 mbar, corresponding to an elevation of 5 km. As perfect emissivity is not expected and Viking radio occultations have shown elevations higher than those predicted at −63° to −74° latitudes [*Fjeldbo et al.*, 1977], both mechanisms probably contribute to the low brightness temperatures.

If a 5 K depression due to elevation and emissivity is attributed to all of the south polar cap, there remains unexplained up to 13 K additional depression and temporal variations of up to 4 K. At least these amounts of the low temperatures must be due to atmospheric causes.

The two atmospheric mechanisms discussed in paper 1 are clouds and CO_2 depletion. To result in appreciable brightness temperature depressions, clouds would have to be high above the polar cap. For that reason, low-lying fogs are not differentiated from ground frost in this discussion.

Because of the strong absorption of both gaseous and solid CO_2 at 15 μm the brightness temperature at 15 μm cannot be greater than it is at 20 μm when CO_2 clouds are present (unless both warmer air and an appreciable part of the 15-μm weighting function are above the cloud top). If CO_2 clouds are present, they must be at higher altitude than the pressure level implied by saturation at T_{20} and beneath the pressure level implied by T_{15} (e.g., T_{20} = 140 K implies 1.9 mbar or an altitude of 8.6 km). H_2O clouds cannot contribute to the low-temperature observations, since the atmosphere around the polar cap was quite clear, and there can be no appreciable source of H_2O to form clouds in the polar night, where the saturated mixing ratio is below 10^{-6}.

The accumulation of noncondensing gases as CO_2 condenses will result in the lowering of the CO_2 partial pressure at a constant total pressure. The presence of clouds is difficult to separate from CO_2 depletion by thermal measurements alone, since they have a similar effect on the origin of the 15- and 20-μm radiation. Cloud formation raises the altitude at which 20- and 15-μm radiation originates; high clouds can cut off much of the 15-μm weighting function as derived from CO_2 gaseous absorption. CO_2 depletion does not affect T_{20} emission but moves the 15-μm weighting function to lower altitudes, so that it becomes terminated by the surface. Both processes move the effective surface to lower CO_2 partial pressure.

Since direct positive identification of CO_2 depletion has not been made, the demonstration that clouds with appreciable thermal opacity are present or absent would be very valuable. The absence of clouds can be shown by either warm atmospheric temperatures or crisp surface imaging. Because of the small size of the coldest regions, the fact that T_{15} is measured in only one of seven IRTM fields of view, and the large noise equivalent delta temperature of T_{15} at these temperatures, to date it is only certain that T_{15} was near T_{20} south of $-75°$ latitude at the start, and south of $-85°$ latitude at the end of the primary mission (see section IIB2).

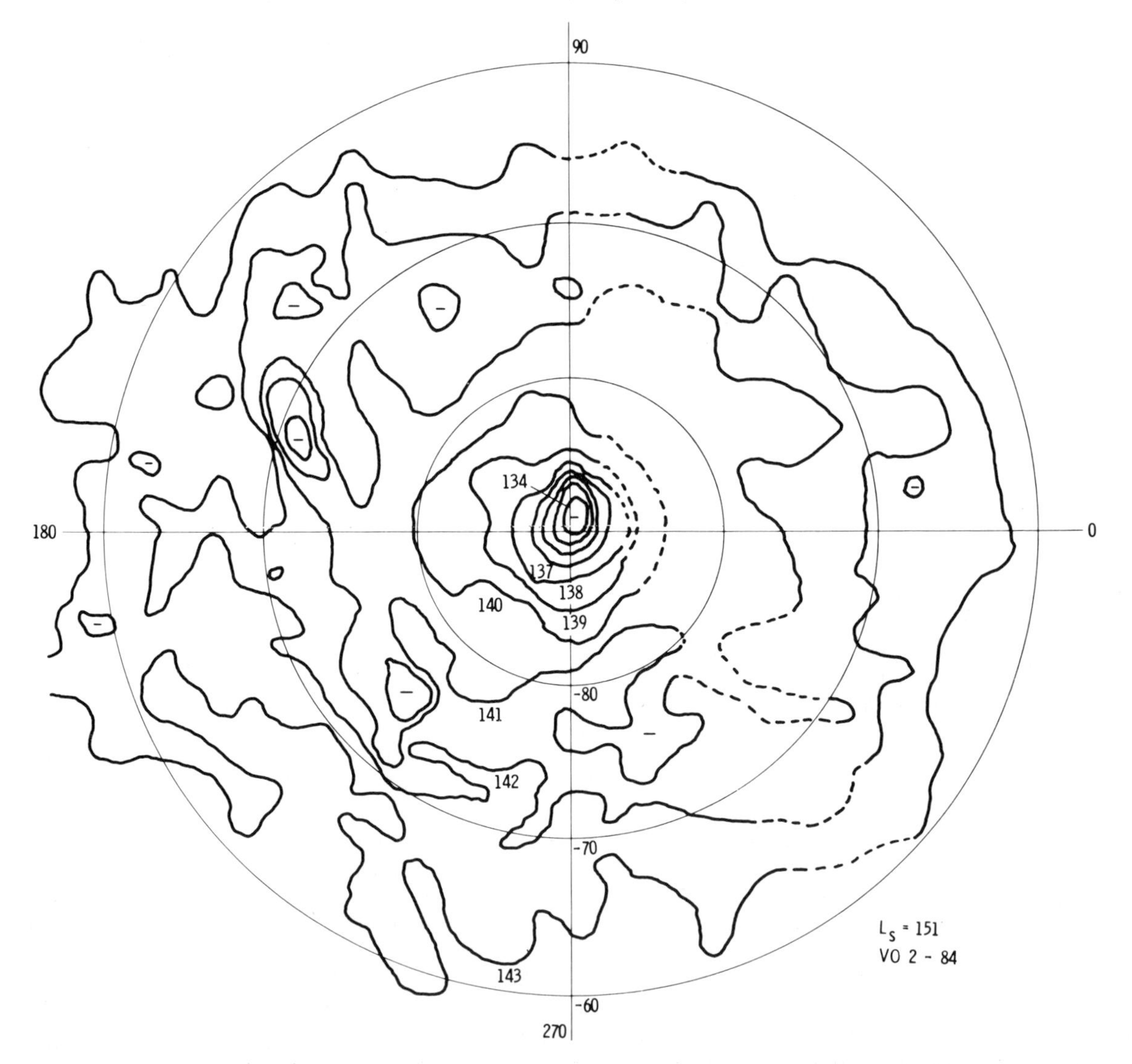

Fig. 7. T_{20} observed near the South Pole at L_S = 151° during a single sequence on VO 2 rev 84. The region outside the 144 K line is not contoured. The dashed lines indicate one strip where error caused a temperature offset in data reduction. The contour interval is 1 K.

South of −78°, the region including most of the area below 140 K, no imaging was possible in the primary mission. However, the absence of clouds at one of the cold areas can be established. Near the end of the primary mission, imaging of colder areas became possible as the noon terminator moved south. In particular, two orbital imaging sequences covered the northern portion of the cold area near −72°, 160°W. On VO 2 rev 72, images extending to −75°, 149°W, within 1° of the geometrical terminator, showed distinct ground features; no clouds were discernible south of 60°. Violet and red images were nearly identical in contrast. Clouds do appear north of −50°. The area of minimum temperature appears as a generally smooth plain with few craters, no marked relief, and no indication of structure suggestive of major regional elevation variations. Nothing in the imaging distinguishes the coldest area from the surrounding terrain. There were no IRTM data obtained simultaneously with this sequence, but a minimum T_{20} of 137 K at this location was measured both on rev 62 and on rev 84 of VO 2. VO 2 rev 62 imaging, obtained 40 min earlier than a complete south polar thermal map, extended south to −70° along 150°W; there are no clouds apparent near the cold area. At the edge of the images, 200 km from the coldest point, T_{20} was 140 K. Although simultaneous imaging and temperatures are not available for the coldest part of the thermal low around −72°, 160°W, this low persisted in the IRTM sequences covering this area throughout the primary mission. The conclusion is that at least this one polar low-temperature feature is not readily attributable to unusual elevation, emissivity, or clouds; the low temperature is therefore attributed to CO_2 depletion at the Martian surface. Why this particular area became enriched in noncondensing gases remains unexplained.

The thickness of the low–molecular weight layer or whether it extends high into the atmosphere remains undetermined. The processes which would operate against local depletion are diffusion, turbulence, and convection due to the density variation. In the absence of information on the near-surface turbulence above the polar cap and in ignorance of any theory of convection including loss of the medium into the lower boundary, we consider diffusion alone to establish an upper limit on the CO_2 depletion rate. The diffusion constant of CO_2 in air under standard conditions (1.013 bars and 273 K) is 0.139 cm² s⁻¹ [*Dushman,* 1962]. The theoretical variation with temperature and pressure is $D = D_0(T/T_0)^{3/2}(P_0/P)$, whereas measured values follow a temperature exponent of 1.7. If the measured temperature dependence is used, D would be 7.54 $cm^2\ s^{-1}$ at 6 mbar and 140 K. The CO_2 condensation time scale of 44 days (paper 1), with a scale height of 7.4 km, corresponds to a net downward velocity V of 0.195 $cm\ s^{-1}$. The characteristic diffusion scale $(Dt)^{1/2}$ and motion scale Vt would be equal for t = 200 s and a length of 0.4 m. Thus if diffusion were the only competing process, it would be possible for a thin depleted layer to form on a very short time scale. The shape of CO_2 partial pressure profiles which might be maintained against static instability by the net downward velocity remains to be studied.

CO_2 depletion must occur to some extent throughout the polar night. The main uncertainty in the degree to which this lowers polar temperatures is the magnitude of atmospheric mixing that results from vertical and lateral dynamic instability associated with the polar thermal field and the variation of atmospheric mean molecular weight. About 5% of the Martian atmosphere, composed of nitrogen, oxygen, and argon with a mean molecular weight of 31 [*Nier and McElroy,* 1976], will not condense. The relation between temperature and density of the saturated, depleted polar air differs drastically from the familiar form. The density falls rapidly with decreasing temperature to a minimum at 134 K (10% CO_2), where the density is 82% of that at 147.7 K, when CO_2 condensation begins. This temperature dependence is opposite that for a gas of uniform composition, is much steeper, and might result in atmospheric motions not accounted for by normal circulation models.

A numerical model of polar cap formation which allows the noncondensing portion of the Martian atmosphere to accumulate while CO_2 condenses has at least confirmed that CO_2 depletion in poleward-moving parcels will result in the low temperatures observed if mixing between latitudes is small. This model has a one-layer atmosphere, so that the entire atmospheric column must become depleted for surface temperatures to decrease. This dynamically naive axisymmetric model predicts a catastrophic temperature drop near the pole in the late winter as the CO_2 content rapidly approaches zero. Temperatures below the minimum observed can be avoided only by increasing the poleward flow of the atmosphere above that required for uniform surface pressure and by recirculating the CO_2-depleted atmosphere back toward the equator.

The ability of a simple model accounting for the noncondensing gases to produce the low temperatures observed suggests that the CO_2 depletion is their major cause. Radio occultation determination of the atmospheric density scale height over the winter pole would allow, in combination with the temperature observations, the determination of the atmospheric molecular weight and could resolve this question. Unfortunately, earth-Mars-spacecraft geometry severely limits polar occultation at the seasonal extremes. Study of the atmospheric motion adjacent to the polar night may provide useful estimates of the density scale height within the polar night by establishing the extent of 'molecular weight' winds of direction

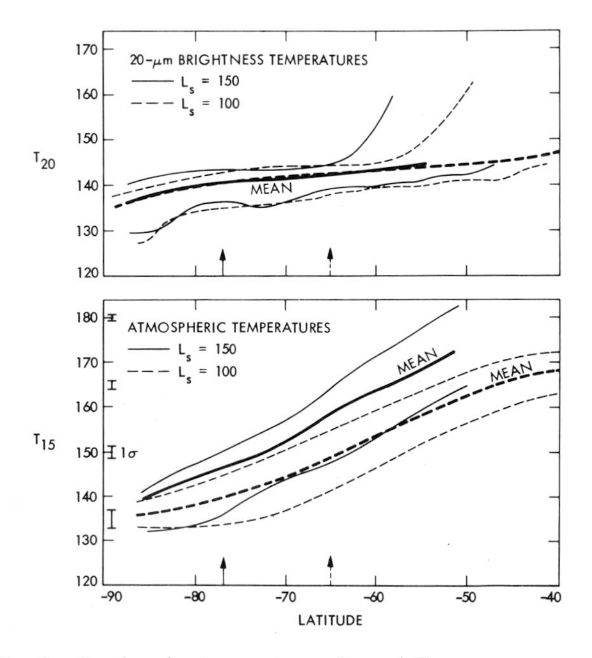

Fig. 8. South polar temperatures. T_{20} and T_{15} measurements are shown for early and late southern winter. The arrows at the bottom indicate the southernmost sunlit latitude at the two times. The total range of observed temperatures is indicated by the lighter dashed and solid lines. Noise equivalent temperature difference, which is dependent on radiance, is shown for T_{15} on the left margin by error bars.

opposite those expected for thermal winds. Whatever explanation is proposed, it must be compatible with both the general latitude dependence and the local structure of the temperatures observed.

2. *Atmospheric temperatures.* During the Viking primary mission the southern hemisphere of Mars was repeatedly mapped by the IRTM to monitor seasonal changes in surface and atmospheric conditions. The strong latitudinal gradient in T_{15} identified in early global mapping sequences (paper 1) is the most pronounced feature of southern hemisphere atmospheric behavior during the Viking primary mission. Changes in temperature with season and with respect to surface temperature have been especially interesting, since these trends were incompletely established by earlier studies [*Hanel et al.*, 1972*a*, *b*; *Conrath et al.*, 1973].

Study of the atmospheric temperatures is complicated in general by the interplay of air mass effects with diurnal and latitudinal variation. In a single observation sequence the sampling of higher altitudes with increasing emission angle may obscure the dependence of T_{15} on time of day or on latitude, which also changes with emission angle. If vertical temperature profiles are known or can be inferred, the air mass effect can be estimated and removed from T_{15} data. That procedure has been employed in the study of diurnal behavior at +48° latitude (paper 4). In southern latitudes, no Viking temperature profiles are yet available, although radio occultation measurements have been made [*Fjeldbo et al.*, 1977]. Lander entry profiles, obtained at +22° and +48° [*Seiff and Kirk*, 1976, 1977], are not applicable in the south polar cap region, where a strong inversion clearly exists (see below). However, Mariner 9 winter profiles above the north polar cap [*Conrath et al.*, 1973] are nearly isothermal in the range of IRTM-sampled altitudes above the 1-mbar level; similar structure in the southern winter would yield small air mass dependence. In the South Polar Region, T_{15} does not show a strong variation normal to the limb. Therefore we have not used an air mass correction in presenting the T_{15} data below. The behavior of T_{15} with latitude and season will not be affected substantially by ignorance of air mass effects, since no systematic bias would be applied and since the discussed variations are larger than any conceivable emission angle dependent error.

Diurnal variation can partly be determined by plotting T_{15} data versus time of day in narrow latitude bands. It is assumed that no longitude dependent variations occur; global maps in fact have shown no pronounced local features except where the surface elevation exceeds about 15 km and where the surface radiance becomes important in the atmospheric channel. Estimation of total diurnal amplitude is hampered north of the polar cap edge near −40° by the lack of complete coverage and the possible onset of important air mass effects. Comparison of data from observations with different subspacecraft local times permits some checking of data consistency. In general, it seems that a diurnal amplitude between 5 and 10 K exists between −10° latitude and the edge of the polar night, little change occurring in the Viking primary mission.

The mean diurnal atmospheric temperatures are presented as functions of latitude for L_S = 100° and 150° (early and late southern winter) in Figure 8. At any given latitude the total observed range of temperatures is typically 12 K. Much of this is due, especially at the lower temperatures, to electronic noise associated with this narrowband high-gain channel. The remainder of the variation is attributed primarily to diurnal effects. Three important features are evident: there is a strong poleward cooling in the atmosphere, the atmosphere is warmer than the surface over virtually the entire extent of the polar cap even into areas that receive no insolation, and the atmosphere warms as winter progresses even where surface temperatures remain constant. The poleward temperature gradient may be compared to the model prediction of *Pollack et al.* [1976*a*] for temperatures at the 1.87-mbar level at L_S = 110°. There appears to be a significant gradient south of −60° in the higher-altitude T_{15} data, indicating atmospheric heat transport to internal regions of the cap. There may be heat transport by baroclinic waves at moderate latitudes. Unfortunately, thermal differences across low-temperature moving fronts are difficult to detect in T_{15} data because of the noise level; furthermore, the major differences are expected at lower altitudes than are effectively sampled at 15 μm. Certainly, some sensible heat is transported into the cap by the net poleward flow. The part of the vertical column that radiates to space is roughly equivalent to the part sensed by the IRTM; the weighting function (paper 4) expresses the importance of different layers. The radiative time constant for such a column is approximately 1 day, when the calculation scheme of *Goody and Belton* [1967] is used. The poleward flow is such that parcels of gas may move considerable distances before radiative losses bring them to equilibrium with the surface.

The general warming in the southern atmosphere between L_S = 100° and 150° is attributed to two causes: (1) the continued transport of heat in the atmosphere from sunlit zones, either by baroclinic waves or by mass advection, and (2) the onset of direct solar heating of the atmosphere. The trend is clearly not due to the warming of the surface; T_{20} measurements are essentially unchanged south of −60°. High surface albedos of the polar cap will tend to promote atmospheric heating above areas covered by frost. Dusty northern air brought south over the cap during winter would readily absorb sunlight. The observed increase in atmospheric opacity in the southern hemisphere [*Briggs et al.*, 1977] may be due in part to the presence of dust though water vapor released at the cap edge will certainly enter the atmosphere and condense.

III. GLOBAL ALBEDOS

The surface markings on Mars have been the object of study by earth-based astronomers for many centuries. Improved telescopic resolution led observers early in the century to draw charts of the markings they saw; these constituted the first crude albedo maps of Mars (for a review of early maps see *de Vaucouleurs* [1954]). The advent of photoelectric and photographic photometry enabled observers to assign relative quantitative albedos to the major (≥500-km extent) features, but absolute calibration of the albedos was difficult [*de Vaucouleurs*, 1967]. The situation improved considerably when spacecraft carried high-resolution imaging systems to the red planet; absolute calibration of the albedos was difficult [*de Vaucou-* topographic mapping of the surface was done mainly near the terminators, where albedo determination depends on accurate knowledge of the photometric function. The most detailed relative albedo map presently available was produced on the basis of Mariner 9 pictures and ground-based observations [*USGS*, 1976]. The most useful albedo for thermal calculations is the radiometric (equal to the bolometric bond) albedo. No radiometric albedo maps of the surface of Mars have ever been produced, though such a map is likely to be qualitatively similar to nonbolometric albedo maps. *Kieffer et al.* [1973] concluded on the basis of Mariner 9 thermal data that the mean radiometric albedo of Mars was near 0.25. The Viking IRTM experiment has provided absolutely calibrated bolometric reflectivities for Mars which are interpreted here in terms of a bolometric normal albedo map of the near-equatorial region of Mars (Plate 2*a*). Values of A_g ($\simeq p^*$) are found to vary from 0.089 to 0.429; the mean is 0.214 ± 0.063 (1σ). The distribution frequency is depicted in the histogram associated with Plate 2*a*. A marked bimodality of albedo values is evident.

Several restrictions were placed on the data accepted for the albedo map. Data taken from ranges in excess of 15,000 km were discarded because of their poorer resolution and because small errors in instrument-pointing knowledge can result in large errors in incidence and emission angles at that range. Data within 1° of the limb of the planet were discarded to be certain that the field of view was entirely on the planet and that atmospheric limb brightening effects were avoided. Data at incidence angles in excess of 60° were discarded to minimize atmospheric effects near the terminator. A total of 84,480 points that passed the above tests were sorted into 60 latitude bins in each of 360 longitude bins, thus providing an effective spatial resolution of 1° latitude by 1° longitude.

For phase angles less than about 50° the solar band brightness was found to vary as the cosine of the incidence angle. However, at higher phase angles the calculated Lambert albedos A_L were consistently higher than those at low phase angles. It was found that by dividing the calculated A_L values by an empirical phase angle correction, $\Phi(g) = 1 + 0.000146e^{0.075g}$, where g is the phase angle in degrees, the resulting A_g were generally self-consistent. The functional form of this correction was suggested by the relative constancy of the calculated A_L values at low phase angle and by the exponential increase in A_L values observed for high phase angles.

It is apparent that for certain areas of the planet a fair amount of mottling exists at the scale of the spatial resolution. Because, in general, only about 70% of the latitude/longitude bins in a particular area are filled by a single sequence, construction of a complete map necessitates the combination of data from several sequences. Occasionally, selected sequences are separated by periods ranging from several hours to many days, and adjacent bins in latitude/longitude space may have A_g values determined from those time-separated sequences. Mottling is therefore an indication of disagreement between the A_g measured from these different data sets and is probably due to changes in the atmospheric opacity during the intervening time. It is not yet clear whether those changes are primarily diurnal in nature or whether they are the signature of secular atmospheric variations on Mars or some combination of the two. The mottling would be even more accentuated by the inclusion of data from incidence angles of 60°–80°, where A_g values to the west of the Tharsis Montes sometimes exceed 0.60.

In addition to the separation in time between sequences covering the same area, their incidence, emission, and phase angles can also differ markedly. Hence it is also possible that the corrections applied to the solar band brightnesses in order to obtain A_g values, though apparently valid in many areas, may not be universally applicable. This is especially true if, as is suggested in section VI, the phase correction is due entirely or in large part to dust or condensates suspended in the Martian atmosphere, since the amount and distribution of such material is unlikely to be both spatially and temporally constant.

Aside from these differences the general appearance of the map is similar to earth-based albedo maps of Mars (see, for example, *Lowell Observatory* [1976]) and to the Mariner 9 albedo map [*USGS*, 1976]. The minimum albedo of 0.089 was found in the middle of Syrtis Major Planitia (+12°, 289°W); the maximum A_g was 0.429 at +20°, 105°W. This latter area lies to the north of Ascraeus Mons, where all of the points on the map with A_g values in excess of 0.40 lie. It is suspected that this area was covered with clouds or fog at the time of one of the observations. This suspicion is strengthened by the presence of adjacent bins with A_g values near 0.310 obtained from observations on a different day. The mean A_g, 0.214 ± 0.063 (1σ), is lower than the 0.25 found from Mariner 9 data [*Kieffer et al.*, 1973], but the area covered by Plate 2*a* has a disproportionately large percentage of dark areas, and the mean A_g from the map may therefore be an underestimate of the Martian mean.

The marked bimodal distribution of albedos is separately characteristic of each of the four quadrants of the albedo map (Plate 2*a*). The peaks generally occur near A_g values of 0.15 and 0.26. The quadrant including the Tharsis Montes and Olympus Mons, however, is characterized by greater reflectivity, and the positions of the two maxima and intervening minimum are at higher A_g values than those for the other three quadrants. This bimodal behavior is apparently the consequence of relatively sharp boundaries between low- and high-albedo areas. Histograms of the latitude ranges −30° to −5° and −5° to +30° (each covering all longitudes) have single maxima centered near 0.15 and 0.26, respectively. These two albedos may be characteristic of two major material types at the surface of Mars, as is borne out by the corresponding bimodality of surface thermal inertias (section V).

The five large volcanoes, Olympus (+19°, 233°W), Arsia (−9°, 120°W), Pavonis (0°, 113°W), Ascraeus (+11°, 104°W), and Elysium (+25°, 213°W) montes, each appear darker in Plate 2*a* than their surroundings. This may in fact be a result of their extreme elevations and correspondingly diminished atmospheric obscuration rather than an intrinsically lower surface albedo. Claritas Fossae (extending from near −15°, 110°W to near −25°, 105°W) is somewhat brighter than its surroundings, possibly a consequence of morning ground fog along this feature. Noctis Labyrinthus (−7°, 110°W), at the west end of the Valles Marineris complex, is also bright, though it does not stand out against the large presumably cloudy area to the northwest. Unfortunately, Memnonia Fossae, which appeared bright in approach photographs of the planet [*Briggs et al.*, 1977], falls mostly in the gap caused by the loss of data on rev 89.

The most prominent dark features charted by ground-based observers are all seen in Plate 2*a*. Solis Planum (near −27°, 75°W) is very dark, with A_g = 0.11. The southern tip of Acidalium Planitia is visible between 20°W and 25°W at the north edge of the map. The Viking Lander (VL) 1 site (+22°, 48°W) is also covered and has an A_g of 0.244 ± 0.002. The northern edge of Hellas Planitia is just visible near 290°W.

The craters Huygens (−14°, 304°W), Schroeter (−2°, 304°W), Schiaparelli (−2°, 343°W), and Herschel (−15°, 230°W) all have rims that are bright in relation to their surroundings and darker centers. Antoniadi (+22°, 299°W) and Cassini (+24°, 327°W), on the other hand, are both within a relatively bright area which extends about 50° in longitude and is hereafter referred to as Arabia, and neither crater is apparent in Plate 2*a*. It is conceivable that atmospheric obscuration is responsible both for the brightness of the area and for the lack of contrast of Antoniadi and Cassini. Alternatively, a thick mantle of dust may have obscured the characteristic crater features.

The albedo maps presented here must be considered a preliminary attempt at deriving bolometric normal albedos. Atmospheric effects obviously remain in some portions of the maps. No time independent photometric function will be successful in eliminating those effects. Only by comparison with albedo maps constructed from a different data set is it likely to be possible to separate surface and atmospheric effects.

IV. PREDAWN TEMPERATURES

Temperatures just before dawn are of particular interest because that is the coldest time of day and because the dependence of temperature on albedo or slope is then a minimum, so that individual thermal observations can most directly be related to surface physical properties. The magnitude of observed predawn thermal variations was unexpected; many of the areographic relationships were not anticipated, most are not well understood, and some are simply baffling. Clouds cannot account for most of the predawn temperature variations observed; these variations are attributed to differences in surface thermal inertia.

Earth-based infrared observations of Mars have necessarily pertained to the warmest parts of the planet. Prior to spacecraft measurements the poor spatial resolution available with reasonable signal levels confined Martian thermal observations to the periods near opposition, thereby precluding observation of the nighttime portion of the planet. Nonetheless, knowledge of surface physical properties, probable nighttime temperatures, and regional variations of temperature has steadily improved.

Thermal observations of Mars made half a century ago [*Coblentz*, 1922; *Coblentz and Lampland*, 1927; *Pettit and Nicholson*, 1924] established that midday equatorial temperatures rose above the melting point of water; this had been a major question at that time. Although subject to fairly large absolute temperature uncertainty, due to the broad wavelength regions used and the poorly determined amount of terrestrial water vapor attenuation, the observations with resolution about $\frac{1}{10}$ the diameter of the planetary disk revealed the dawn-to-dusk variation in the equatorial zone to be comparable to the equator-to-pole variation near midday [*Coblentz and Lampland*, 1927], the sunrise limb generally being cooler than the sunset limb. Dark areas were shown to be warmer than bright areas, and bright clouds were associated with temperature depressions. A synthesis of the early observations indicated that there was a clear seasonal variation of the dependence of midday temperature on latitude [*Gifford*, 1956].

The first observations of Mars to provide an estimate of the surface thermal inertia, and hence to allow calculation of nighttime temperatures, were made by *Sinton and Strong* [1960]. Their observations could be satisfactorily fit with I = 4–6 and yielded predicted minimum equatorial temperatures between 170 and 180 K [*Morrison et al.*, 1969].

The first direct observations of Martian predawn temperatures were made by the Mariner 9 spacecraft [*Kieffer et al.*, 1973]. These late southern summer observations were largely confined to 0° to −60° latitude and had a total range of temperature deviation from a homogeneous model of 38 K; the prominent features were a cool region around the south Tharsis Ridge, a warm area along the northern edge of Hellas Planitia, and local high temperatures where the radiometer path crossed Valles Marineris.

The VO 1 walk provided the first reasonably uniform coverage of large areas of Mars. The standard set of VO 1 asynchronous observations (Figures 3*a*–3*c*) was spaced such that over successive revolutions the predawn terminator was observed at increments of 5° of longitude. For the predawn residual temperature map (Plate 2*b* and Figures 9*a*–9*d*) the sequence at 2.2 hours before periapsis was used for local times from 0.5 H to 2 H before dawn where these data were available. Where coverage gaps occurred, the observations further from periapsis, with the commensurate lower resolution, were utilized. Good predawn coverage is available from the South Polar Region to approximately +20°. However, south of −55° the presence of the winter polar cap inhibits appreciable diurnal temperature variation.

The predawn difference T_r between the observed 20-μm brightness temperature and the standard model (A^* = 0.25 and I = 6.5) temperature was mapped from −60° to +25° latitude. The resolution varies from 60 km between 0° and −30° to 200 km locally along the northern border. The model temperature minimum is a constant 149 K south of −45°; model temperature increases smoothly northward to a broad peak at 180 K between +40° and +60° latitude.

Predawn temperature behavior generally repeated on the several revolutions of coverage available for each surface location. The major exception is the region between the Tharsis Ridge and Olympus Mons and an extension several hundred kilometers to the west. Here, variations of 10 K in T_r through the VO 1 walk imply that the atmosphere is having a major effect, directly or indirectly, on T_{20}. High and variable albedos are observed in this area (section III). T_r typically varied by 1–3 K on successive days. On one day a region of about 400-km extent centered at +7°, 127°W warmed by 4 K in 1½ hours rather than cooling by 2 K as was predicted for the surface. Predawn warming has been seen at other locations and at other times (section VIIIA).

There are three major low-temperature areas: a very large region extending east and west from the four major volcanoes, a small area around Elysium Mons, and another large area corresponding to the classic bright feature Arabia. There are two major regions of high temperature: along Valles Marineris continuing into the Chryse Basin and along the southern border of Isidis Planitia.

The strong thermal contact outlining the Tharsis area crosses topographic features, physiographic terrain types, and modest albedo boundaries. The thermal boundary trends south across Lunae Planum, skirts west and northwest around Valles Marineris, cuts directly across the chaotic terrain of Noctis Labyrinthus onto Sinai Planum, meanders around and across the graben of Claritas Fossae, then trends primarily west downslope across Memnonia Fossae and across cratered terrain, crosses Apollinaris Patera, generally includes the smooth plains which constitute the southwest extension of Amazonis Planitia, and loops northeast along the southern edge of the dark feature Cerberus. The lowest temperatures west of Arsia Mons generally follow the downslope direction into Amazonis Planitia. Near −15°, 152°W, adjacent to Memnonia Fossae, T_{20} is 149 K, and the surface elevation is near 2 km, yielding temperatures near the CO_2 frost point. Bright features have been observed in the orbiter imaging of this region [*Briggs et al.*, 1977], and it is possible that CO_2 frosts are forming before dawn. The lowest temperatures are approximately centered on the summits of the major volcanoes. T_r values for Olympus Mons and Arsia Mons are both −34 K, corresponding to temperatures of 144 and 138 K, respec-

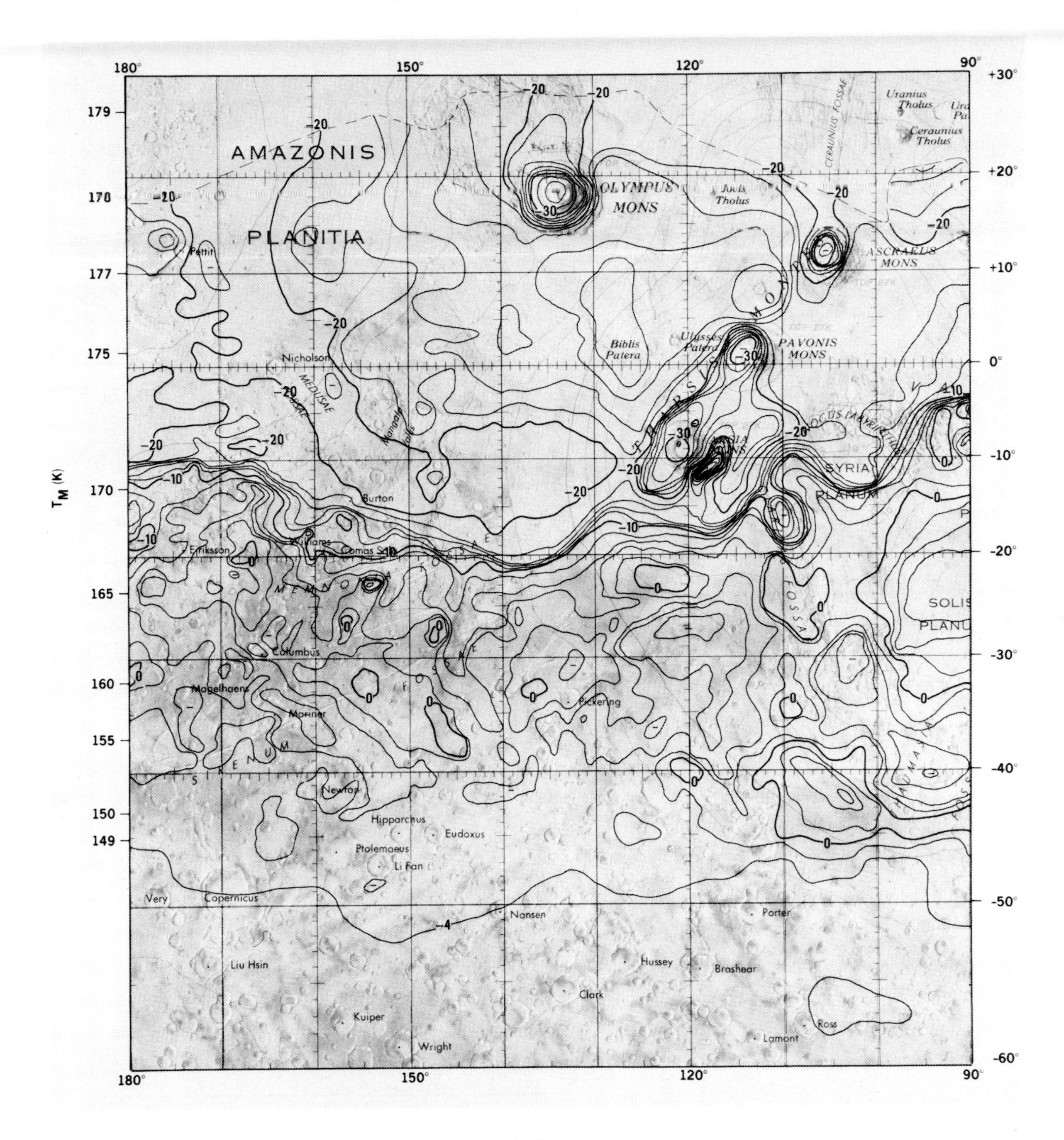

Fig. 9a

Fig. 9. Predawn residual temperatures observed during the VO 1 walk (L_S = 124°–129°). The contour interval is 2 K, with 10 K lines accentuated. The model temperature at dawn at the middle of the VO 1 walk period is indicated at the map margin. The northern coverage limit resulted from the details of the slew pattern on individual sequences. Dashed contours indicate the few places where good coverage is not available. The same data are shown in Plate 2*b* as a color contour map. (*a*) 90°W–180°W longitudes. (*b*) 0°W–90°W longitudes. (*c*) 270°W–360°W longitudes. (*d*) 180°W–270°W longitudes.

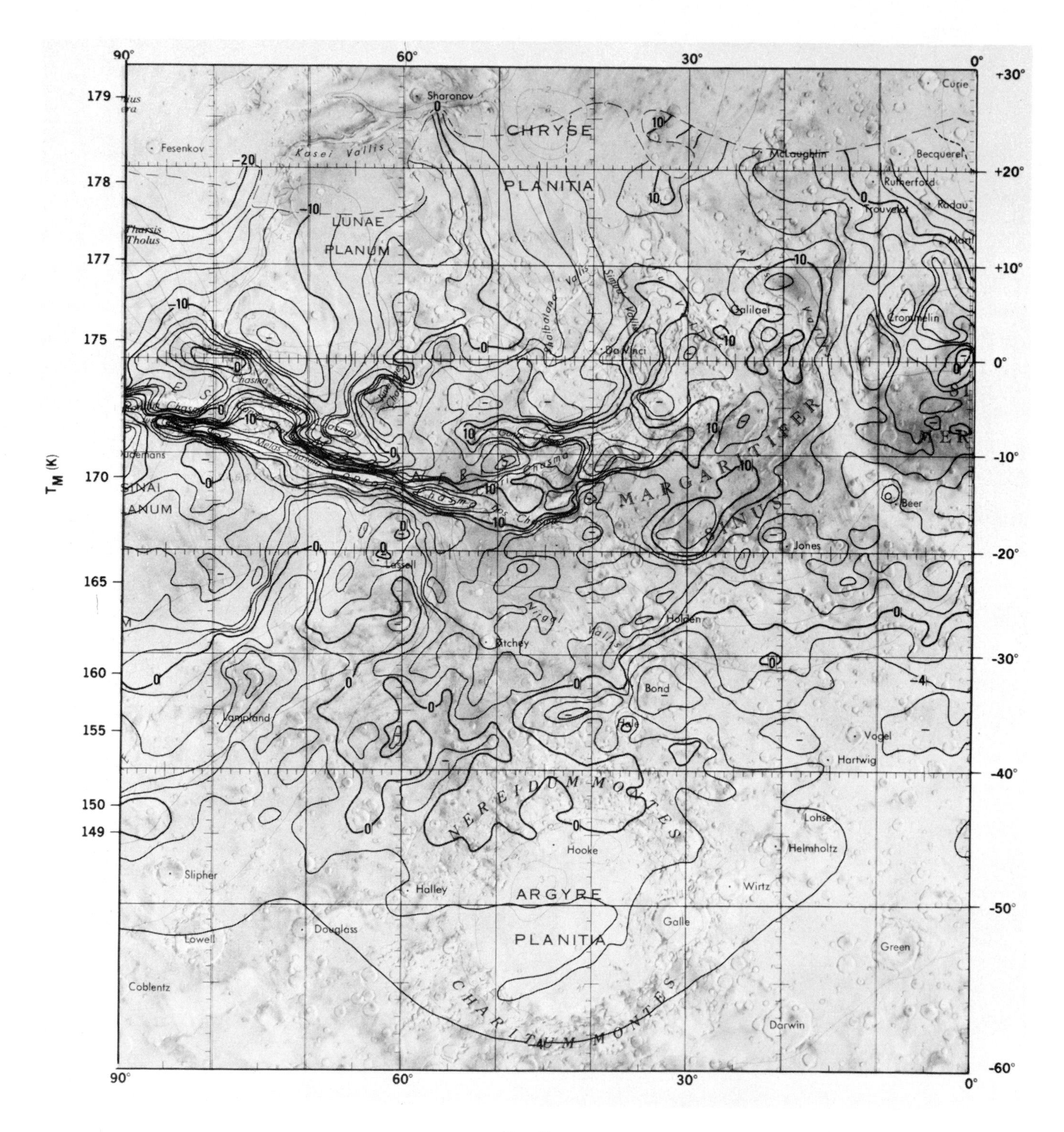
90°
60°
30°
0°
+30°
+20°
+10°
0°
-10°
-20°
-30°
-40°
-50°
-60°
179
178
177
175
170
165
160
155
150
149
T_M (K)
Sharonov
Curie
CHRYSE
PLANITIA
Fesenkov
Kasei Vallis
McLaughlin
Becquerel
Rutherford
Trouvelot
Radau
LUNAE
PLANUM
Tharsis
Tholus
Mart
Galilaei
Cronmelin
Da Vinci
Chasma
MARGARITIFER
SINUS
Beer
Jones
SINAI
PLANUM
Lassell
Holden
Ritchey
Bond
Hale
Vogel
Hartwig
Kampland
NEREIDUM MONTES
Lohse
Helmholtz
Hooke
Slipher
Halley
ARGYRE
PLANITIA
Wirtz
Galle
Douglass
Lowell
Green
Coblentz
CHARITUM MONTES
Darwin
-20
-10
0
10
-4

Fig. 9b

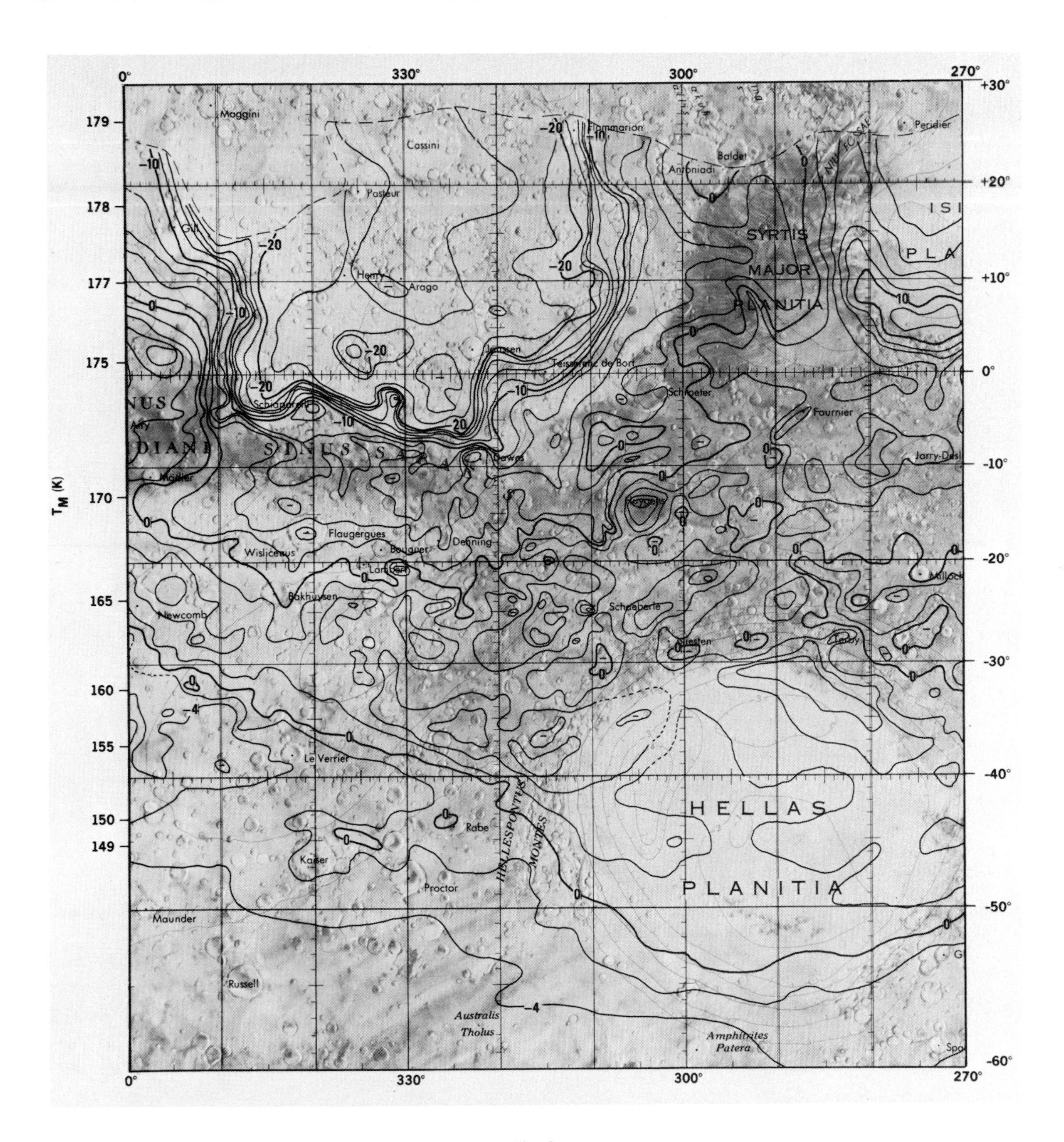

0°
330°
300°
270°
+30°
+20°
+10°
0°
-10°
-20°
-30°
-40°
-50°
-60°
T_M (K)
179
178
177
175
170
165
160
155
150
149
Maggini
Cassini
Flammarion
Peridier
Baldet
Antoniadi
Pasteur
Gill
Henry
Arago
SYRTIS
MAJOR
PLANITIA
ISI
PLA
Janssen
Teisserenc de Bort
Schiaparelli
Schroeter
Fournier
Airy
SINUS SA
Dawes
Jarry-Desl
Huygens
Flaugergues
Bouguer
Denning
Wislicenus
Lambert
Millochau
Bakhuysen
Newcomb
Schaeberle
Niesten
Le Verrier
HELLAS
PLANITIA
HELLESPONTUS MONTES
Rabe
Kaiser
Proctor
Maunder
Russell
Australis
Tholus
Amphitrites
Patera

Fig. 9c

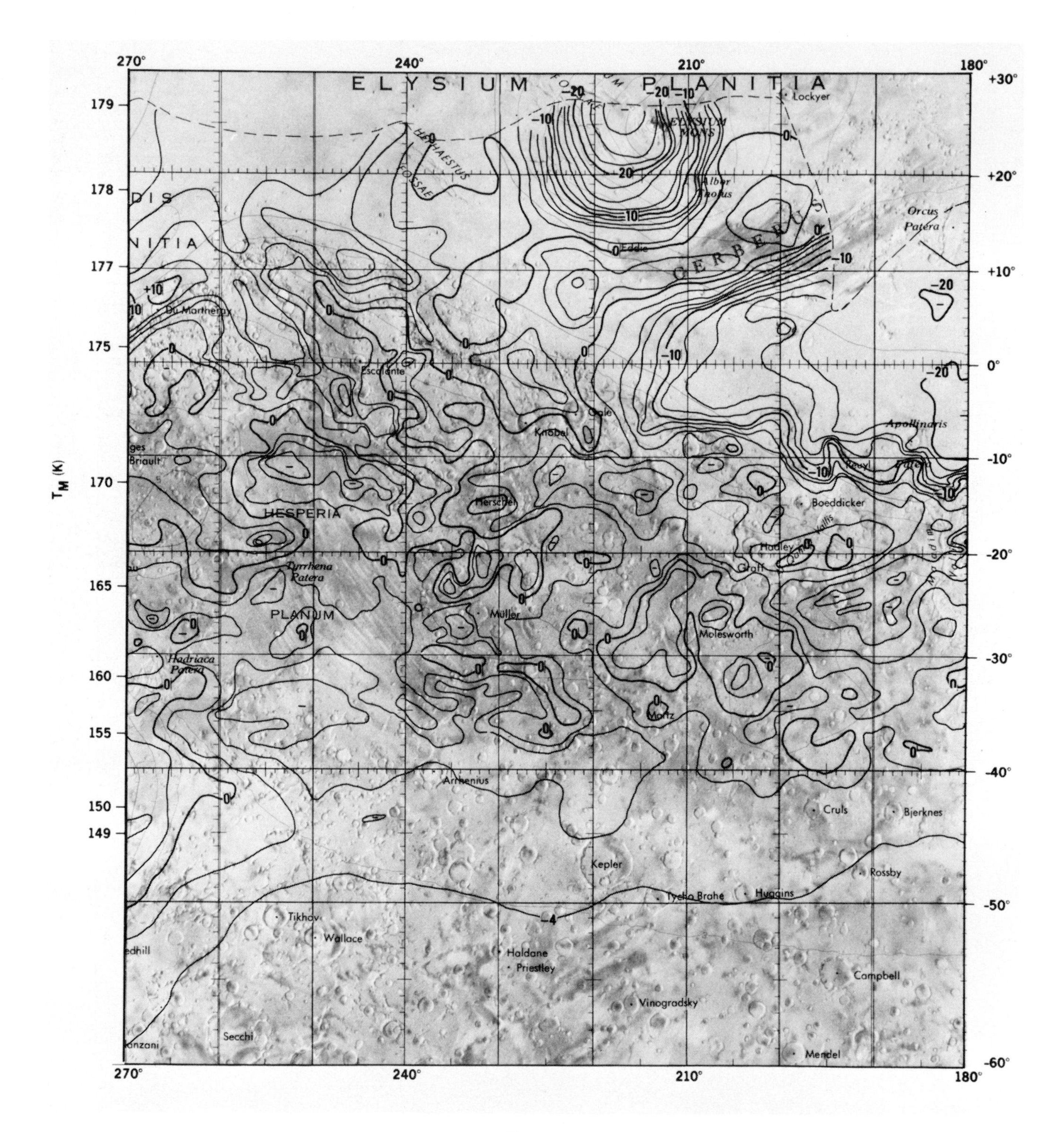

270°
240°
210°
180°
+30°
+20°
+10°
0°
-10°
-20°
-30°
-40°
-50°
-60°
179
178
177
175
170
165
160
155
150
149
T_M (K)
ELYSIUM PLANITIA
Lockyer
ELYSIUM MONS
HEPHAESTUS FOSSAE
Albor Tholus
CERBERUS
Orcus Patera
Eddie
Du Martheray
Escalante
Gale
Knobel
Apollinaris
Briault
HESPERIA
Herschel
Boeddicker
Hadley
Graff
Tyrrhena Patera
PLANUM
Müller
Molesworth
Hadriaca Patera
Martz
Arrhenius
Cruls
Bjerknes
Kepler
Rossby
Tycho Brahe
Huygens
Tikhov
Wallace
Haldane
Priestley
Campbell
Vinogradsky
Secchi
Mendel

Fig. 9*d*

tively. Although the lowest temperatures are at the highest elevations, there is considerable thermal structure in regions of approximately uniform elevation around Arsia Mons (paper 4).

In contrast, the low temperatures surrounding Elysium Mons are centered approximately 200 km west of that volcano, and much of the low-temperature region is at elevations below 3 km. The lateral extent of this low-temperature feature is considerably larger in comparison with the size of the volcano than that which occurs for the Tharsis and Olympus volcanoes.

The cool region in Arabia is approximately symmetric around the large crater Cassini (+24°, 327°W). Although the major portion of the area has T_r below -20 K, the region has no major topography other than a smooth rise in elevation from 2 to 5 km toward the southeast. The boundary of the low-temperature region corresponds approximately with the extent of a high-albedo feature, but the albedo contrast is completely inadequate to explain predawn thermal contrasts of this magnitude (see below). Local convolutions of this sharp thermal boundary have no obvious relation to the surface morphology; for example, the thermal contact cuts directly across the crater Schiaparelli.

Two of these cool areas are obviously associated with topographic features extending one to three scale heights into the Martian atmosphere; these features are also volcanic. The third area has neither obvious topographic relief nor volcanic features. It is not known how, or if, topography and/or volcanic history are related to the cause of the large cold areas. Although there are not yet predawn measurements to determine directly whether these cool areas connect at higher latitudes, these associations would suggest that they do not.

An increase in albedo of 0.1 would decrease the predawn temperature only 2.5 ± 0.2 K for all latitudes mapped. Thus the direct effects of possible albedo variations are much smaller than the range of T_r observed. Because the atmosphere is expected to be warmer than the surface in the predawn, atmospheric thermal opacity cannot readily account for the low temperature observed. The fact that some visual opacity is present over much of this cold area is indicated by greater than average albedo over these low-T_r areas and enhanced visual brightness near the dawn terminator. The major cause of these low predawn temperatures must be low thermal inertia. The low surface inertias may result in enhanced condensation of atmospheric H_2O on the surface; any albedo increase would have positive feedback, further contributing to the low temperatures. Day-to-day variations, and perhaps 5 K average temperature depressions (corresponding to an albedo increase of 0.2), are caused by atmosphere effects.

The highest T_r is observed in the main chasm of Valles Marineris, where values up to 16 K are found. The warm features are at the limit of resolution for these predawn observations, and there are probably warmer areas at scales below 30 km. The highest T_r values are observed along areas of reasonably well-defined escarpments. Bordering the chasm eastward of 40°W the thermal field is more complex, but there is a general continuation of high temperatures curving to the north and northwest along the inferred path of flow associated with the formation of the 'fluvial' morphology observed at the borders of the Chryse Basin. A warmer region centered at $-17°$, 28°W is not obviously associated with the canyons and may result largely from the low albedo occurring in Margaritifer Sinus. Both Ophir Chasma ($-6°$, 70°W) and Ganges Chasma ($-8°$, 146°W) have T_r comparable to the main chasm of Valles Marineris. The canyons themselves are not especially dark, so that it is likely that the high T_r is associated with high-inertia material.

The warm area near +5°, 270°W is associated with the boundary between Isidis Planitia, Syrtis Major Planitia to the west, and the heavily cratered uplands to the south. The regions with T_r greater than +10 K are at the base of steep (>1%) slopes. The two adjacent warm regions at +30°, 255°W and +10°, 220°W, both trending northwest-southeast, occur in regions of moderate slopes and are in moderately cratered plains. The general morphology shown by Mariner 9 and Viking imagery does not indicate any obvious cause for these high predawn temperatures. Where T_r is greatest, the minimum nighttime temperature is 189 K. It is very unlikely that condensate clouds would form in this region, and there is no evidence of unusual dust clouds. The high inertia is probably associated with the surface material (see section VC3). This region does not have the low albedo expected from the general relation between low albedo and high inertia determined for the Martian surface in this latitude range.

In addition to these five specific regions with large deviations from the thermal model, there are a multitude of local features with T_r contrasts from 2 to 8 K. The most extensive feature includes the Hellas Basin, with an average T_r of approximately 4 K and localized warm areas at the west and northeast margin of the basin. This last area corresponds to the location of the highest thermal inertia inferred from Mariner 9 data [*Kieffer et al.*, 1973], although T_r was much larger then, in southern autumn.

The predawn T_r pattern at Hellas is repeated with somewhat smaller magnitude at Argyre Planitia. The Nereidum Montes, along the north edge of the Argyre Basin, appear warmer than the model, and the -4 K contour, which otherwise circles the globe at approximately $-55°$, is displaced south at both Hellas and Argyre. The floors of Hellas and Argyre are both warmer and brighter than average for this latitude. A_L for the major portion of the floors of the Hellas and Argyre basins was greater than 0.45 and 0.35, respectively, whereas the remainder of the $-40°$ to $-50°$ latitude band had an average A_L near 0.28. As the predawn temperatures in both regions are near 148 K, the increased brightness of these basins is probably due to CO_2 frosts. The higher temperatures in the basins could in part be due to the lower altitudes, as the CO_2 frost point increases about 1 K for a 1-km decrease in altitude. However, the temperature pattern and the topography do not agree in detail. Other possible contributing effects such as heterogeneity of albedo and temperature below the IRTM resolution or warming associated with wind patterns related to these two large basins have not been examined.

Most of the small thermal features at $-10°$ to $-40°$ latitudes do not correspond to specific features depicted on the topographic map of Mars. A few examples of correlation are the warm interior of the crater Huygens at $-14°$, 304°W (paper 4) and a broad warm region at $-20°$, 90°W corresponding to the classical dark feature Solis Lacus. A general study of the correlation between thermal inertia and geomorphic features is presented in section VC3.

One useful aspect of residual temperature mapping is that, to the extent that albedo, slopes, and atmospheric effects can be ignored, residual temperatures can easily be related to apparent thermal inertia. The variation of model temperature with inertia, $\partial T_M/\partial I$, is not strongly dependent on albedo, season, time within a few hours before dawn, or latitude in the equatorial region. Variation of apparent thermal inertia could reflect either a uniform change of the effective soil particle size or mixtures of different inertia material at any scale below the IRTM spatial resolution; both effects are probably present. The set of possible solutions can be studied graphically by displaying residual temperature and inertia on a scale which is linear with in-band radiance (Figure 10). On such a graph the effect of mixing materials of different inertia is linear with areal abundance. For example, $T_r = 0$ could be due to 100% material of $I = 6.5$ (the nominal value) or to a soil of $I = 4$ with a 22% admixture of $I = 40$ exposed bedrock. More complex graphic solutions would be possible if more components were felt to be warranted.

V. THERMAL INERTIA

Thermal inertias derived from brightness temperature measurements are useful in understanding the physical properties of Martian surface materials. They may also be used to provide predictions of surface temperature for use in global circulation models and radiometric calibration. Thermal inertia estimates have been derived for a number of solar system objects including the moon [*Winter and Saari,* 1969], Phobos [*Gatley et al.,* 1974], earth [*Gillespie and Kahle,* 1977; *Pohn et al.,* 1974], and Mercury [*Chase et al.,* 1976]. These estimates range from a low near 1 (found on the moon) to over 80 (found on the earth). For Mars, measurements with the 8- to 40-μm radiometers on Mars 3 and Mars 5 [*Moroz and Ksanfomaliti,* 1972; *Ksanfomaliti and Moroz,* 1975; *Moroz et al.,* 1976] lead to estimates of Martian thermal properties within the range inferred from the 8- to 12-μm and 18- to 24-μm Mariner 6, Mariner 7, and Mariner 9 results [*Neugebauer et al.,* 1971; *Kieffer et al.,* 1973]. In the Mars survey reported here the thermal inertia range is 1.6 to ~12. There are several large well-defined areas of low thermal inertia ($I \leq 4$) where the surface is likely to be covered by fine material (diameter ≤ 100 μm) with very few rocks exposed. The large volcanoes of the Tharsis region are included in one of the low-inertia areas, indicating that the extrusive material around these features is either blanketed by fine material or that the upper flow layer is highly porous. The upper limit of $I = 12$ implies the absence of large exposures of bare rock comparable to our 2° latitude by 2° longitude sample size. Where rocks are exposed at the surface they are likely to be associated with the fine material. The low-inertia areas correspond to the brightest regions in our survey. Except at the extremes in elevation, no correlation between elevation and inertia was found. Correlation of thermal inertia with surficial geology provides additional insight into the processes which have shaped the surface of Mars.

A. *Global Map*

A thermal inertia contour map (Plate 2*c*) was constructed on the basis of a grid of inertia values computed for 2° latitude by 2° longitude bins. The data used were collected from VO 1 during its asynchronous walk period (revs 81–91) and supplemented by data collected a month earlier over a more limited area (the Tharsis region).

Without a priori knowledge of radiometric albedo, thermal inertia is most convincingly determined by establishing the diurnal temperature behavior of a surface. Observations made during the VO 1 walk run from 2 H through 10 H, and in small areas through 12 H, provided the requisite diurnal variation. The predawn sequences were taken far from periapsis and were consequently of low resolution. The bin size was chosen on the basis of the lowest-resolution data. The latitude range was selected so as to avoid large emission angles in the north and the complicating effects of nighttime CO_2 condensation in the south, where it was winter.

Inertias were calculated with both a one-parameter least squares fit to the thermal data by using measured albedos (A^* was assumed equal to A_g) and a two-parameter fit in which both the radiometric albedo and inertia were determined by a least squares fitting of the brightness temperatures alone. In order to provide reliable brightness temperatures at low surface temperatures (below 170 K), T_{20} values were used exclusively. Data (T_{20} and A_g) were not used if the emission angle was greater than 60° or if the incidence angle was in the range 60°–90°. The rationale for the emission angle constraint is discussed in section VIA. The incidence angle constraint is used to provide a better estimate of surface albedo, since early morning brightenings were known to occur in some areas. In both the one- and the two-parameter fit the thermal data were weighted in such a way that each consecutive interval of 2 H was given equal weight. The morning data were of higher spatial resolution than the predawn data, and consequently, there were generally more morning points in a given bin than predawn points. The weighting procedure reduced this resolution-related bias in the data set.

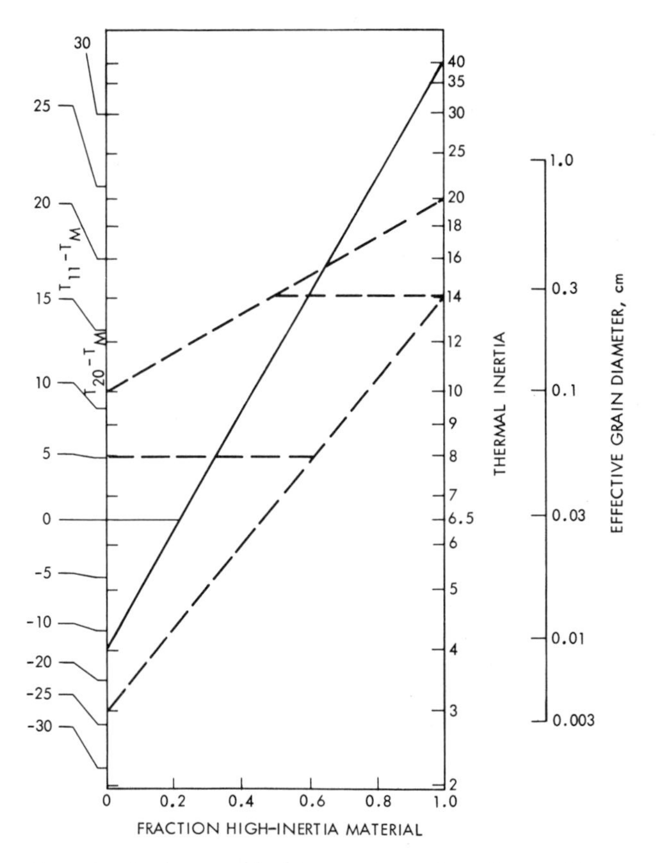

Fig. 10. Predawn residual temperatures $T_{20} - T_M$ and $T_{11} - T_M$ related to the thermal inertia of surfaces with one or more components. The temperature scales are linear with in-band radiance; nonlinearity in temperature increases at shorter wavelengths. A straight line between a fine soil of $I = 4$ ($\langle d \rangle = 0.009$ cm) and rock of $I = 40$ relates the residual temperature to the areal fraction of the two materials; 22% would yield $T_r = 0$, as would a homogeneous surface of material with $I = 6.5$ (solid line). As an example of a three-component surface, $T_r = 5$ could result from 38% of the material with $I = 3$ and 62% of the surface having equal exposure of $I = 10$ and $I = 20$ material (dashed lines). This figure is for a latitude of + 10° at $L_S = 126°$. The effective grain diameter scale is for 6-mbar surface pressure [*Kieffer et al.,* 1973, Figure 11].

In the one-parameter fit, each observed brightness temperature is reduced to a temperature difference ΔT as follows:

$$\Delta T = T_{20} - T_M - (\bar{A}_g - 0.25)\ \partial T/\partial A^* \qquad (1)$$

where $\bar{A}_g$ is the mean A_g for the bin. ΔT is then paired with $\partial T/\partial I$ weighted by $1/N$, where N is the number of points in the 2 H time interval, and a least squares fit is carried out. The two-parameter fit is carried out in a similar fashion by using $\Delta T = T_{20} - T_M$, $\partial T/\partial A^*$, and $\partial T/\partial I$. The partial derivatives of temperature with albedo and inertia are approximated by differences between the three Viking thermal models (Appendix 1).

A single pass through this procedure is adequate if the derived inertia falls in the range of the three Viking models. Outside this range the procedure misestimates the inertia due to the nonlinearity of $\partial T/\partial I$. This difficulty is overcome by an iterative procedure in which the derived inertia and an estimate of $\partial^2 T/\partial I^2$ at the local time of each observation are used to apply a correction to $\partial T/\partial I$ for each observation. Convergence is rapid.

For the more than 3000 bins which contained at least one morning and one predawn point the agreement between inertia values determined by the one- and the two-parameter fit was good, the largest disagreement being less than 25%. The uncertainty in the fit, measured by the rms deviation of ΔT, was significantly better for the two-parameter method. These two results are in accord with the weak dependence in the region surveyed of surface temperature on albedo for the range $0.2 \leq A^* \leq 0.3$. Because the fit was better and because of uncertainties in the translation of IRTM solar band measurements into A^*, we use only inertias derived from the two-parameter method.

As an overall check on the two-parameter method, inertias were computed for an area in the Tharsis region (285 bins) by using data obtained from VO 1 on revs 60 and 62; T_{20} was plotted versus local time, and each diurnal plot was compared with a set of model diurnal curves representing a range of A^* and I. For the inertia range found ($1.6 \leq I \leq 8$), the agreement between methods was excellent except at the lowest inertias ($I < 3$), where the machine-calculated inertias were lower. The difference probably lies in the correction factor used to modify $\partial T/\partial I$ at low inertia. Since the model diurnal curves use the correct value of $\partial T/\partial I$, all two-parameter fit inertias below $I = 3$ were increased according to the results of the comparison between the two methods.

The differences between the inertia values derived by the two-parameter method and the actual area average values for Mars lie in the following factors: the formal uncertainty in the fit; uncertainties in T_{20}; and processes not modeled, including surface roughness and general slopes, thermal opacity in the atmosphere, surface CO_2 condensation (paper 4), and conductive coupling between the surface and atmosphere (Appendix 1). While the effect of these processes cannot be evaluated quantitatively for the region surveyed, the emission, incidence, and latitude limits selected mitigate their being a major factor. The uncertainty resulting from the fit and noise in T_{20} is about 20% of I.

The thermal models used and discussed in Appendix 1 assume thermophysical properties which are homogeneous with depth. This condition is satisfied if the material is homogeneous down to several thermal skin depths. The thermal skin depth varies from about 1.4 cm for a thermal inertia of 2 to 11 cm for an inertia of 16. If less than a thermal skin depth of material overlies higher thermal inertia material, such as dust over rock, the inertia deduced by the technique described here will be intermediate between the upper- and lower-layer thermal inertias. For example, if an $I = 3$ layer one skin depth in thickness overlies $I = 18$ material, the surface temperature will be similar to that for a single layer with $I = 3.3$. If the upper layer is only a half thermal skin depth in thickness, the homogeneous analog would have an $I = 4.7$. To the extent that such multilayer models are appropriate for Mars, the thermal inertia of the upper layer is at least as low as the values derived when homogeneity with depth is assumed.

Interestingly, the subsurface diurnal temperature behavior for vertically inhomogeneous materials can be significantly different from that of the homogeneous case while their surface temperatures are similar. Remote subsurface temperature sensing would nicely complement the surface inertia deductions presented here.

The predawn residual temperature and thermal inertia maps (Plates 2*b* and 2*c*) show that large regions of Mars will have temperatures substantially different from those predicted by any thermal model which assumes physical properties to be uniform over the entire planet. The use of Mars as an infrared or radio calibration source presuming such uniform properties [see *Becklin et al.*, 1973; *Logan et al.*, 1973; *Wright*, 1976; *Loewenstein et al.*, 1977] will result in error to the extent that the area-averaged brightness temperature of the subearth hemisphere for a homogeneous model differs from that for the actual surface. This practice should be pursued with caution until the error resulting from the neglect of areographic variations and seasonal weather has been evaluated.

By using the two-parameter method, inertias were computed for a range of $-30°$ to $+20°$. Coverage north of $+10°$ was too sparse to permit contouring.

B. *Relation to Surface Properties*

The surface physical properties implied by thermal inertia have been discussed by *Kieffer et al.* [1973] and *Neugebauer et al.* [1971]. For Martian conditions the principal thermophysical property determining inertia is the conductivity, which is closely related to the particle size (from *Kieffer et al.* [1973] on the basis of measurements and a literature survey of *Wechsler and Glaser* [1965]).

For the lower range of observed inertias ($I \lesssim 4$), two general descriptions of the surface are possible: a surface covered with fine particles with diameters less than about 0.1 mm or a surface covered with highly vesicular volcanic material. Where the surface is completely covered with fine material, only a few centimeters are necessary to produce a low thermal inertia signature. Thicker layers are possible and may be required to provide complete blanketing of the surface. The boundary in particle diameter between saltating and suspendable particles is near 0.1 mm. This diameter is also near that at which the threshold wind speed curve has its minimum [*Pollack et al.*, 1976*b*]. Thus particles in the size range consistent with low inertia are those which can be suspended. The upturn in threshold wind speed for particles less than 0.1 mm in diameter provides a mechanism that leads to net deposition where the suspended particles return to the surface in regions of low peak wind speed.

Volcanism could also provide a low-inertia surface. The expansion of volatile rich lavas in a low-pressure environment would be expected to produce a highly porous material, as has been demonstrated by *Dobar et al.* [1961]. The material produced by the expansion of simulated magma in vacuum is

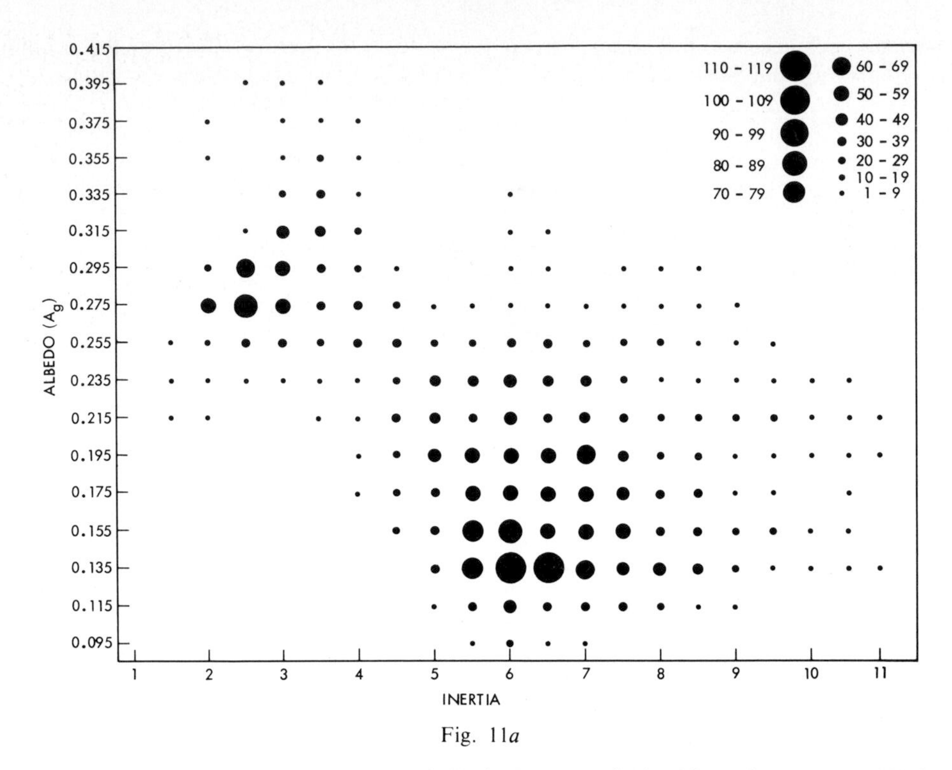

Fig. 11*a*

Fig. 11. (*a*) Thermal inertia versus phase-corrected albedo. Data are obtained from the −30° to +20° data set. Total sample size is 3345 A_g-I pairs. (*b*) Population of thermal inertia. The data set is the same as that used for part *a*.

similar to that of a naturally occurring product of terrestrial volcanism, reticulite. Both are highly porous, with interconnecting voids separated by a glassy netlike structure, possess low density (0.12 g cm^{-3} for the simulated magma), and are likely to be of low thermal inertia. Martian analogs of other terrestrial volcanic products such as vesicular basalt, tuff, ash, scoria, and pumice may have very low inertias.

Thermal inertia is also related to elevation through the dependence of thermal conductivity on gas pressure when the mean free path is comparable to the dimensions of the soil pores. For pressures between 1 and 10 mbar the thermal conductivity of powdered materials increases approximately as the square root of the gas pressure [*Kieffer et al.*, 1973]. At pressures of 3 and 1 mbar, corresponding roughly to the elevation of the Tharsis upland and the upper area of the large volcanoes, the inertia of powdered materials would be lower than their values at 6 mbar by factors of 1.2 and 1.6, respectively. If material from the summit region of Arsia Mons, where the inertia is ~1.6, were brought to the 6-mbar level, the inertia would increase to 2.6.

Inertias greater than 4 can be described in terms of one or more of the following: (1) particle distributions with mean particle diameter greater than 0.1 mm, (2) vesicular volcanic products, (3) distributions of small particles, larger blocks, and exposed bedrock, and (4) surfaces consisting of bonded fine particles. The two Viking landing sites illustrate the combination of several of these possibilities. The thermal inertia derived from thermal measurements of Viking landing sites 1 and 2 are 9 ± 0.5 and 8 ± 1.5, respectively (paper 3). If we assume that these values are appropriate to the immediate environs of the landers, they result from the combined effect of exposed blocks (and bedrock at VL 1), a fine soil component, and crustal bonding (see *Mutch et al.* [1976*a*, *b*, *c*] and *Shorthill et al.* [1976*a*, *b*] for a description of landing sites). This exemplifies the caution that should be exercised in interpreting higher values of thermal inertia because of the great variety of surface types which can result in the same area average thermal inertia.

No areas were found where the thermal inertia exceeded 12. This can be contrasted with the earth, where remote determination of thermal inertia has led to values above 50 [*Pohn et al.*, 1974; *Gillespie and Kahle*, 1977]. Such large values of thermal inertia are associated with massive exposure of rock at the surface. If such material were present on the Martian surface, similar values of inertia would be obtained. The fact that we have not found inertia values anywhere near 50 implies that such material is not exposed on the Martian surface on a scale comparable to our sample size (120 × 120 km). In the terrestrial examples cited, the resolution was 8 km or better.

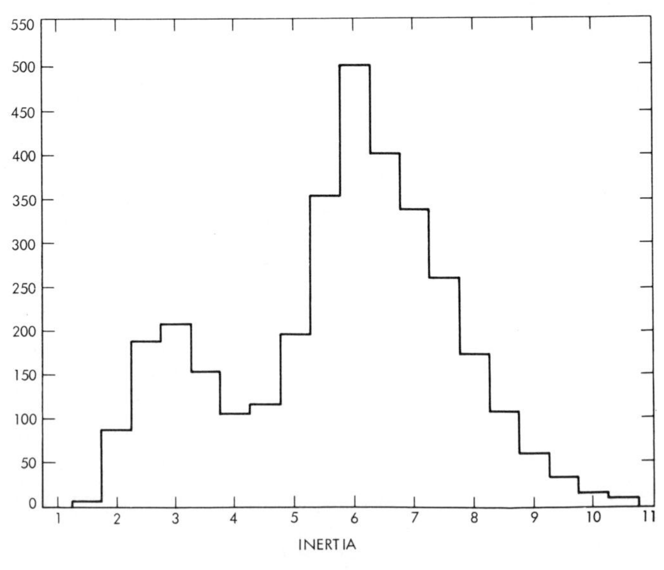

Fig. 11*b*

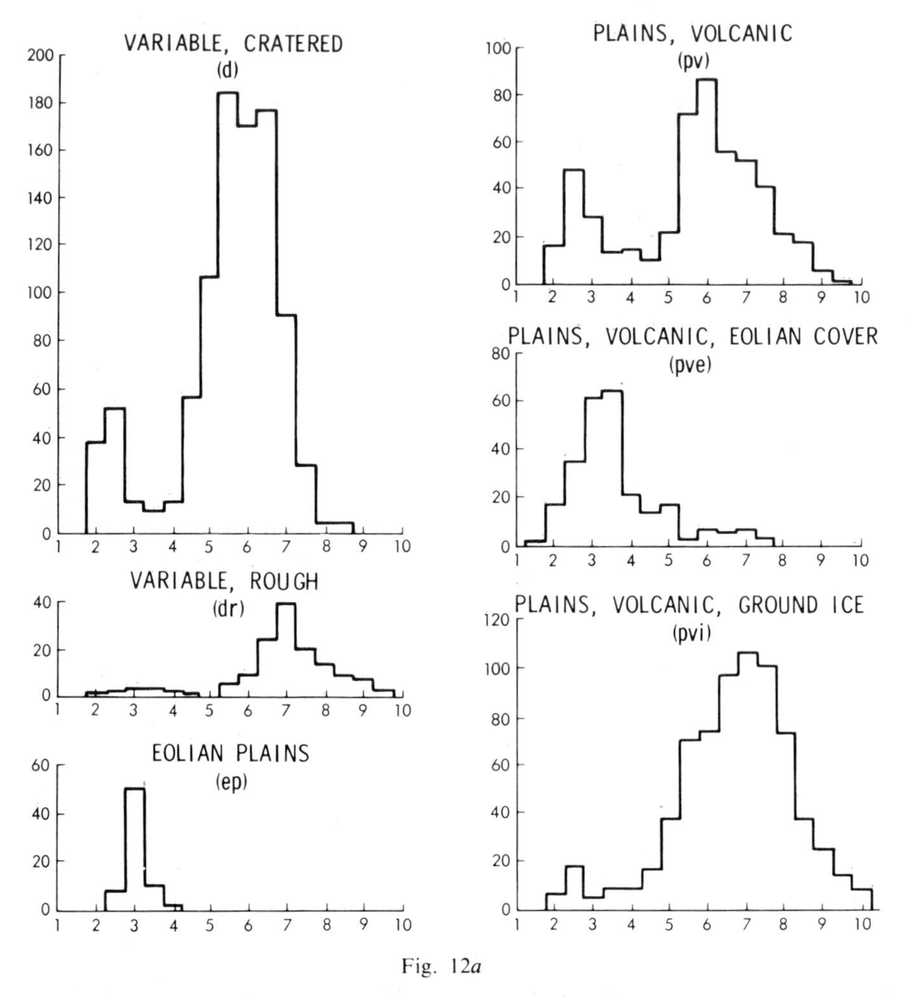

Fig. 12*a*

Fig. 12. Distribution of inertias for surficial geology units. A single sample size is equal to a 2° latitude-longitude square. Each inertia sample was paired with a surficial geologic unit as assigned by *Spudis and Greeley* [1976]. The data set is the same as that used for Figure 11.

(dvr). These two units both have low inertia and relatively low crater counts [*Blasius*, 1976]. The low inertia might imply a highly vesicular nature for the material labeled bv, suggesting that the most recent endogenic material is exposed on the surface. Hadriaca Patera (−30°, 265°W) and Tyrrhena Patera (−22°, 253°W) are identified from morphologic characteristics and crater counts as being the oldest volcanic structures on Mars [*Carr*, 1976]. These regions have the highest inertia in the bv units. It appears that the older volcanic structures have become thermally indistinct from the surrounding terrain.

The eolian plains (ep) also have low thermal inertia. In the region near 0°, 180°W this unit is generally featureless, with a very low crater frequency [*Scott and Allingham*, 1976]. The low inertia supports a depositional history for this terrain involving grain sizes of the order of 100 μm or less. The area in which VL 2 landed was identified as the unit ep from Mariner 9 pictures. However, the thermal inertia is 8 ± 1.5 (paper 3), which is much higher than any of the inertias seen for this unit in the latitude band studied here. The inertia and other surface characteristics of the ep unit may vary with latitude.

VL 1 is in an area that fits the assigned characteristics of the unit defined by Spudis and Greeley as plains, eolian cover (pe). The unit is assigned a variety of features such as crater interiors and low plains. The large range in inertia seen in this unit suggests that at the scale of resolution available to the photogeologist the extent of eolian cover can be difficult to determine.

At 2° resolution, only six craters were large enough to be sampled properly. Of these craters, only for Huygens (−14°, 304°W) did the thermal inertia deviate from that of the surrounding terrain. Four of the craters, Newcomb (−24°, 359°W), Flaugergues (−17°, 341°W), Schroeter (−2°, 304°W), and Herschel (−15°, 230°W), showed essentially no variation in inertia, and Schiaparelli (−2°, 343°W) exhibited a lower inertia which appears to be related to the large low-inertia region of Arabia.

Within Isidis Planitia, which is mapped as pe, exists an area of $I > 11$ (+5°, 273°W). A tentative explanation is that localized eolian transport, associated with the regional slopes to the south and west, may have resulted in an accumulation of coarse material near the edge of the basin.

Most of the Valles Marineris is mapped as variable, mass wasted, ground ice (dmi). The unit ground ice, variable (id) is associated with the chaotic terrain that extends northeast from the canyon. Generally, the floors of the chasms appear to contain high-inertia material, and the surrounding intercrater plains have low to moderate inertia. The terrain feature associated with the highest inertias is the steep ridge in the center of Coprates Chasma (−15°, 55°W). In this region, major differences in surficial units exist within the 2° bin size used here. High-inertia material might be expected if the chaotic terrain

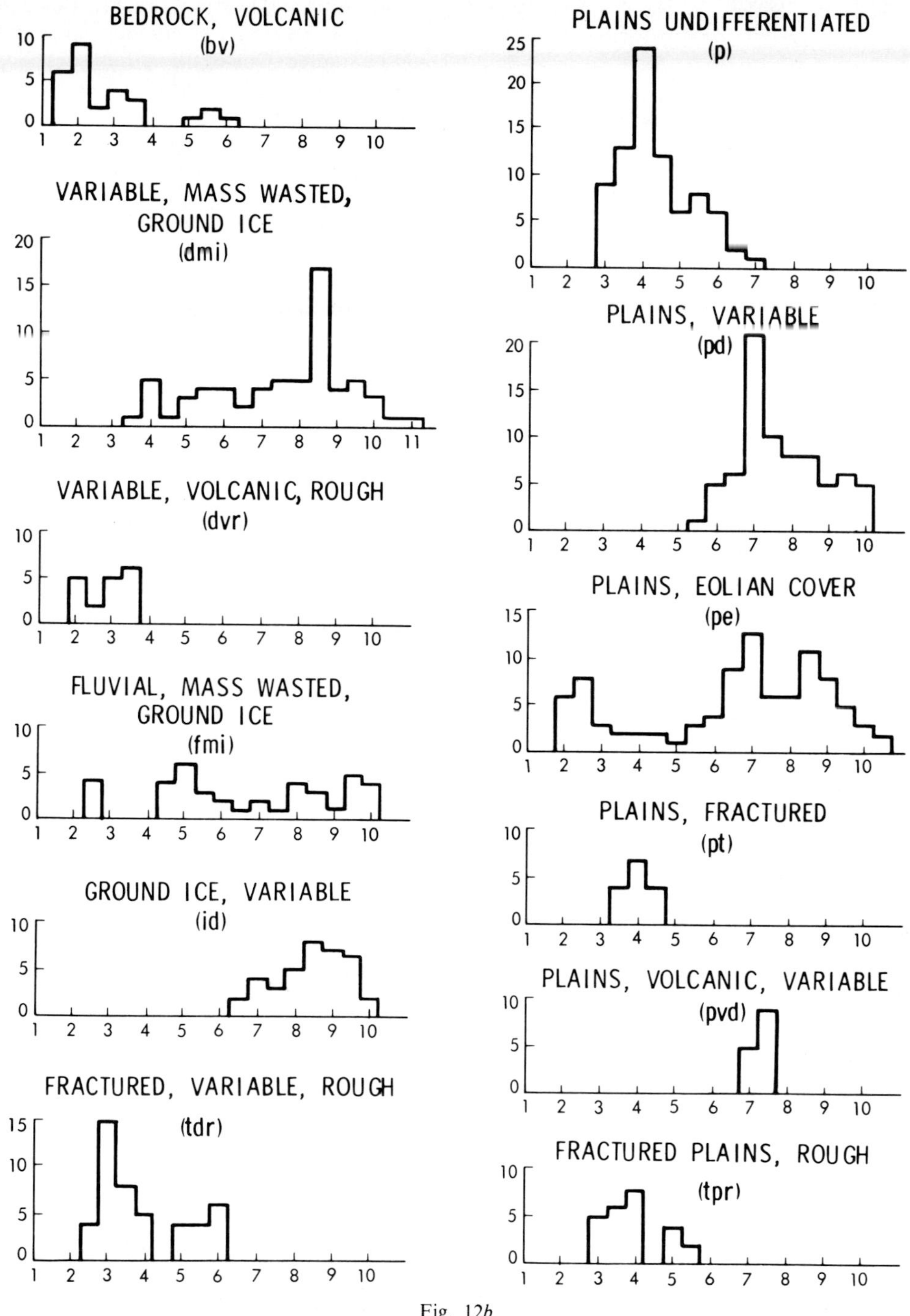

Fig. 12*b*

and areas of apparent headward erosion are still active sites of surface degradation and if blocky debris resulting from this breakup of the cratered plains surface is locally abundant.

The unit fluvial, mass wasted, ground ice (fmi) is primarily assigned to the channel features. Seven isolated examples of fmi were identified within the latitude of this study. The inertias of each of the fmi examples, which varied in size, were generally those of the surrounding terrain, which accounts for the large variation in inertia. The highest inertia observed in this unit was south of Chryse (+2°, 34°W), and the lowest was at Locras Valles (+8°, 312°W), in the Arabia region.

The fractured terrain, represented by fractured, variable, rough (tdr); fractured, plains, rough (tpr); and plains, fractured (pt), appears to be fairly heterogeneous with regard to inertia, the median value being near 4. Geographically, these features are located mainly near Claritas Fossae, Noctis Labyrinthus, and the escarpment on the western edge of Lunae Planum. The relatively low inertia of this unit indicates that surface conditions, on the centimeter scale sensed by the IRTM, can differ radically from those suggested by the large-scale features of an area.

VI. Geometry Considerations

Observed reflected and emitted radiances from the Martian surface and atmosphere depend on the geometry of the observations. The large range of incidence, emission, and phase angles involved in the Viking observations makes possible an investigation into the effects of these geometry variations on the deduced temperatures and albedos. This investigation is necessary to establish the validity of conclusions reached by

TABLE 2. Letter Classification System for Major Surficial Units

Letter	Designation	Character
b	bedrock	consolidated lithology
d	variable (diverse)	heterogeneous surficial nature
e	eolian	wind-related deposits
f	fluvial	water-related (river) deposits
i	ground ice	permafrost related deposits
m	mass wasted	colluvium
p	plains	relatively smooth, low to moderate relief areas
r	rough	high topographic relief
t	tectonically modified	debris derived from fracturing and faulting
v	volcanic	characteristic landforms of volcanic activity

Data are from *Spudis and Greeley* [1976].

using uncorrected data, but it also allows some inferences to be made about the sources of the geometric effects. Both objectives are addressed in this section. The magnitudes of the geometric effects were used to establish constraints on the data used in other sections, so that the resulting conclusions would not be strongly dependent on the specific observational geometries involved.

Planetary surface brightness temperatures are known to decrease with increasing emission angles even for atmosphereless bodies. The presence of an appreciable atmosphere can alter both the amount of solar insolation reaching the surface and the levels from which the emitted thermal radiation emanates. Mariner 9 results for Mars (reviewed in paper 4, where preliminary Viking results are also discussed) showed little difference in angular emissivity between Mars and the moon and Mercury. Viking IRTM results extend the range of useful emission angles from 60° to 80°, show a dependence on phase angle not previously observed, and provide evidence for wavelength dependent atmospheric thermal opacity. In contrast to the results from Mariner 9, the Viking thermal angular emission function is not represented well by a simple cosine power law.

The Martian surface photometric function and atmospheric scattering function have been discussed by several authors [*Harris*, 1961; *Young*, 1969; *Young and Collins*, 1971; *Thorpe*, 1973; *Pleskot and Kieffer*, 1977]. The first four papers deal with observations through relatively narrow visual passbands, and all of the authors conclude that the Martian photometric function is adequately represented by a lunarlike Minnaert formulation with exponent $K \sim 0.5$. Pleskot and Kieffer find from a synthesis of visual and infrared data from Mariner 6 and Mariner 7 that a bolometric passband yields a photometric function with Minnaert exponents near unity. Viking IRTM solar passband data are consistent with this near-Lambertian ($K \sim 1.0$) behavior but reveal an additional strong forward-scattering component which is apparently due to small particles of airborne dust or condensates even in areas which appear to be cloud free in Viking pictures. Photometric results of the Viking visual imaging system are presented in two other papers in this issue [*Thorpe*, 1977*a*, *b*].

A. *Thermal Measurements*

On several occasions during the Viking primary mission, observations were made of individual small areas on the surface of the planet as the spacecraft passed over them from horizon to horizon. To keep the solar incidence angle and time of day relatively constant, these observations were made near periapsis, so that they could occur in a short period of time. In this manner the effects of varying emission and phase angles could be studied. One of the surface sites chosen for study was the VL 1 site at +22°, 48°W; observations of this site provided a certain degree of tie in with 'ground truth.'

Prior data sets from the moon (ground-based observations of *Buhl et al.* [1968]), Mercury (data from Mariner 10 of *Chase et al.* [1976]), and Mars (Mariner 9) have shown that emitted surface thermal radiance could be expressed as a vertical (e = 0) radiance multiplied by the cosine of the emission angle raised to a power β, typical values of β being near 0.2 for each of the three bodies. A practical limit to the range of emission angles of about 60° has existed due to a combination of large target ranges and extended radiometer field of view responses.

Viking observations have extended the range of usable data to an emission angle of nearly 80° due to shorter (<2000 km) ranges and less extended fields of view [*Chase et al.*, 1977]. The resulting data show two effects not apparent in prior data from Mars: (1) a sharper falloff of thermal radiation at emission angles in excess of 60° than is predicted by a simple cosine power law and (2) a dependence of thermal emission on phase angle such that measurements of thermal emission at phase angles in excess of 90° yield significantly reduced brightness temperatures.

For the VL 1 site measurements a cosine power law fit to the data at emission angles less than 60° yields β values of 0.11, 0.17, 0.20, and 0.16 for 20-, 11-, 9-, and 7-μm data, respectively. The corresponding β values for the entire range of emission angles are 0.18, 0.27, 0.32, and 0.25, respectively. Although the range of emission angles at other sites measured during the primary mission is somewhat more limited, it is nevertheless apparent that the derived β values vary both spatially and secularly. It may not be possible to represent the entire Martian surface accurately with a single thermal angular emission function. The spatial variation of surface thermal inertia, depicted in Plate 2c and discussed in section V, provides further evidence that a corresponding spatial variability of β is likely. The observed secular variations lead to the conclusion that airborne dust and condensates play a much larger role in the control of measured brightness temperatures than was previously anticipated.

Lander imaging measurements of the atmospheric optical depth above VL 1 give values near 0.5 at the time of our VL 1 site thermal angular emission function observations [*Mutch et al.*, 1976*a*, *b*]. This visual opacity has been attributed to atmospheric dust by these authors. *Conrath et al.* [1973] have shown that airborne dust can cause significant infrared opacity and that this opacity is greatest near 9 μm (1100 cm^{-1}). It is interesting to note that the β values are largest at 9 μm. Following the line of reasoning that the measurements might be explained by airborne dust opacity at infrared wavelengths, we have fit our VL 1 site observations with a function of the form

$$R_\lambda(T) = R_\lambda(T_s)e^{-\tau\lambda M} + R_\lambda(T_a)(1 - e^{-\tau\lambda M}) \quad (2)$$

where $R_\lambda(T)$ is the observed thermal radiance at wavelength λ, $R_\lambda(T_s)$ is the radiance that would be observed for an unobscured surface, $R_\lambda(T_a)$ is the radiance corresponding to the lower atmosphere, τ_λ is the infrared optical depth, and M is the air mass along the line of sight to the surface (approximately equal to sec e). Better fits to the VL 1 site observations were obtained with this formulation than with the cosine power law (Figure 13). The derived surface temperature T_s was 3–4 K warmer than the observed vertical viewing bright-

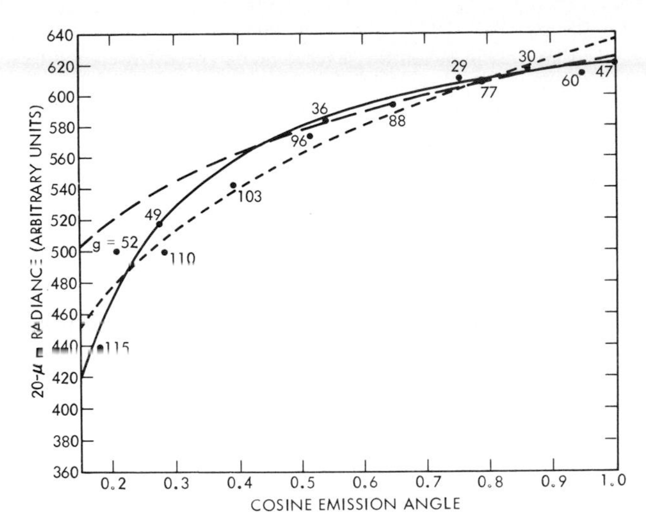

Fig. 13. Variation of 20-μm radiance with emission angle for the VL 1 site. The individual data points are marked with their associated mean phase angles g for both the inbound (lower phase) and outbound (higher phase) measurements. The short dashes represent the best fit cosine power law ($\beta = 0.18$) for the whole data set, and the long dashes represent the best fit cosine power law ($\beta = 0.11$) for emission angles less than 60°. The solid line is for an atmosphere with 20-μm infrared opacity represented by optical depth $\tau_{20} = 0.08$ (see text).

ness temperatures, and the derived infrared optical depths ranged from 0.08 at 20 μm to 0.14 at 9 μm.

It should be noted that the VL 1 site observations may not be representative of the whole planet. The landing site was specifically chosen to be at low elevation, and atmospheric effects are consequently accentuated in relation to Martian averages. Direct observational evidence exists for higher than average opacity at 1.4 μm at the VL 1 site [*Farmer et al.*, 1977].

The observational results discussed elsewhere in this paper use only data taken at emission angles less than 60° unless otherwise specified. Effects of the dependency of emissivity on emission angle are therefore negligible, and no explicit correction has been made for them. Atmospheric thermal opacity has also been neglected. To the first order, thermal opacity would cause derived thermal inertias to be slightly higher than actual surface thermal inertias.

The dependence of thermal emission on phase angle (Figure 13) is easily explained. The brightness temperature is independent of phase angle only in the idealized case of a homogeneous smooth locally flat surface devoid of blockiness. Observations of the two landing sites by the lander imaging systems have shown terrain that is neither locally flat nor homogeneous [*Mutch et al.*, 1976*b*, *c*]. For a soil strewn with rocks of higher thermal inertia than the surface the brightness temperatures are expected to be highest at small phase angles, where sunlit faces are preferentially observed. At larger phase angles the observations include larger and larger percentages of surfaces that are either shadowed or obliquely illuminated; the corresponding brightness temperatures would be lower. At incoming and outgoing emission angles of 74° the VL 1 site phase angles were 49° and 110°, respectively. The corresponding temperature difference was between 1 and 3 K, depending on wavelength. The observations are well represented by a model having 10% surface coverage of 15-cm cubical blocks. The details of the calculations are given in Appendix 2.

B. *Photometric Function*

More than 10^6 individual measurements of the reflected solar band brightness of the surface of Mars were obtained during the Viking primary mission. However, only a small fraction has been analyzed in any detail at this time. The observations cover a range of phase angles from 5° to 135°, with additional measurements of atmospheric forward scattering during solar occultation at phase angles of about 173°. The lower limit of 5°, due to spacecraft constraints, precludes easy observation of the Martian opposition effect, which is most noticeable at phase angles between 0° and 4° [*de Vaucouleurs*, 1968].

The special observations discussed in section VIA also provide useful data for solar band studies. These measurements were made at five different surface locations during the primary mission. One of the measurements was made of a relatively dark area near the northern end of Syrtis Major Planitia (+20°, 285°W). The data have been interpreted for each area in terms of best fit coefficients in the standard Minnaert formulation

$$B^* = p^*\Phi(g) \cos^K i \cos^{K-1} e \tag{3}$$

Because of the continual variation of phase angle during the observations, accurate determination of the Minnaert coefficients was only possible after the addition of observations from other times during the primary mission.

For all of the areas studied the Minnaert exponents fell within the range of $K = 0.95 \pm 0.10$ with only a very weak dependence on phase angle. The apparent brightness at normal incidence and emission as derived from the Minnaert formulation for each of the areas was found to be strongly dependent on phase angle, particularly for large phase angles. A good fit to the data for the five areas (Table 3) can be obtained from

$$B^*(i, e, g) = p^*(1 + 0.000146e^{0.075g}) \cos^{0.94} i \cos^{-0.06} e \qquad i, e < 70° \quad 5° < g < 130° \tag{4}$$

where g is the phase angle in degrees and p^* is the normal albedo instrumentally averaged over all wavelengths. Again it should be emphasized that the p^* derived in this manner takes no account of an opposition effect and may therefore be subject to a scale error. However, because our prime interest in thermal balance calculations is the radiometric albedo and because any opposition effect scale error in B^* is counteracted by an offsetting error in the phase integral q^* (see, for example, *Pleskot and Kieffer* [1977]), the radiometric albedo $A^* = p^*q^*$ should be affected very little by the presence or absence of an opposition effect. (During the extended mission of Viking, one of the Viking orbiters will come close to crossing above the subsolar point on the planet. At that time in the mission it will be possible to maneuver the spacecraft and make some direct measurements of the opposition effect at several points on the planetary surface.)

The departure of the Minnaert exponent from Lambertian ($K = 1$) behavior is likely to be due to atmospheric limb brightening. The brightening is apparent in all of the data near the limb and is approximated by a small negative exponent in the cosine of the emission angle. The variation of measured brightness with phase, incidence, and emission angles is shown in Figure 14.

The solar band data show a strong dependence on atmospheric conditions. Even in those circumstances where the surface albedo is high, the brightness of the limb haze exceeds the underlying surface brightness. The brightness near the terminator is also higher than was expected from surface re-

TABLE 3. Normal Albedo and Minnaert Exponent Values for Five Areas

Observation Date, 1976	Latitude, deg	Longitude, deg W	Local Time, H	p^*	K
Aug. 1	+22	48	15.5	0.24	0.94
Sept. 16	+30	145	14.5	0.31	1.02
Sept. 18	+10	83	12.5	0.30	1.01
Sept. 22	+20	285	13.2	0.15	0.93
Sept. 23	+25	245	13.5	0.24	1.01

flectance alone. Furthermore, although a certain amount of near-terminator brightening is always present, the total amount is definitely spatially variable and probably temporally variable as well. The secular variations may be due to thin clouds or ground fogs, but the persistence of brightening and the associated thermal behavior point to airborne dust. Localized brightenings have been found to coincide with discrete clouds observed by orbital imaging where simultaneous imaging data are available, but they are uncommon.

The phase angle dependence of brightness given in (4) is likely to be due to airborne particles with strong forward-scattering characteristics. Very high phase angle observations ($g \simeq 173°$) were obtained on September 24, 1976, while the spacecraft was within the shadow of Mars. From the increase in brightness with increasing phase angle and from the assumption that the brightness was due to single scattering in the Martian atmosphere a scattering scale height of approximately 5.7 km was deduced. The absolute brightness is estimated to exceed 3 times the brightness of a vertically illuminated white Lambert surface.

Previous observations of dark surface areas on Mars seemed to indicate a departure from Minnaert behavior [*Pleskot and Kieffer*, 1977]. No such departure is seen in the data from the northern portion of Syrtis Major Planitia mentioned earlier. It is likely that previous observations included surface areas of varying normal albedo in the data samples analyzed. The present analysis was limited to a surface area 0.8° in latitude by 0.8° in longitude. IRTM solar band observations do not seem to require surface scattering functions more complex than the Minnaert formulation.

For most of the planet, the atmospheric contribution to the solar band brightness at small incidence and emission angles may probably be neglected. If the presence of an opposition effect is ignored and departures from Lambertian behavior in the photometric function represented by (4) are entirely attributed to atmospheric effects, the phase integral q^* for the surface of Mars is unity, and $A_g = p^* = A^*$. *Pleskot and Kieffer* [1977] determined the bolometric phase integral for Mars and found $q^* \sim 1.05$. A_g is therefore likely to be a good approximation for A^*, in accord with the general agreement found between A_g values derived from solar band measurements and A^* values derived for best fit surface thermal models (section VA).

VII. CLOUDS

Martian clouds are known to be composed of water ice, carbon dioxide ice, or dust. Condensates and dust storms have been noted in earth-based observations, and the presence of all three types has been inferred from extensive spacecraft observations, especially those of Mariner 9 and Viking [*Leovy et al.*, 1972; *Briggs et al.*, 1977]. The usefulness of the Viking IRTM in cloud studies is based on several facts: the different clouds have spectral signatures that may be discerned in IRTM data, clouds seen in images may be interpreted with the aid of IRTM temperatures and albedos, and synoptic information relating to clouds can affect the modeling of surface thermal behavior.

During the Viking primary mission, although examples of extensive cloud cover were observed in orbital images, many of these clouds were not discernible in corresponding infrared data, presumably because of the small infrared optical depths involved. The few discrete dust clouds noted in pictures were too small to have recognizable thermal signatures. No independent search of the infrared data has been made to discover dust storms, except during the limited time period of landing site selection. Historically, large dust events tend to occur in the opposite season, southern spring to summer, when Mars is near perihelion and surface temperatures are conducive to the onset of strong dynamic activity. Likewise, carbon dioxide clouds, which would be difficult to distinguish from water ice clouds in images alone, have not been sought thoroughly in the vast quantity of IRTM data. Expected thermal signatures of optically thick clouds and the necessarily limited observations of dust and water ice clouds are presented below.

Dust clouds have received wide attention. The infrared spectrometer on Mariner 9 determined that suspended dust produces a broad spectral feature centered at 1075 cm^{-1} (9.3 μm) due to stretching vibrations of silicates in the dust particles [*Hanel et al.*, 1972*a*]. The same behavior is expected in dust clouds observable by Viking, with differences in absorption band shape and depth arising from the degree of thermal contrast between cloud and surface, the particle sizes involved, the optical depth, and possibly local compositional variations. The IRTM 9- and 11-μm bands each fall partially within the spectral range of the dust feature; however, because individual bandwidths are nearly equal to the width of the dust feature, no accurate band shape information can be derived from Viking data. If the band location varies, the band shape may be deduced by the ratio of T_9 to T_{11}. The band depth may be inferred by comparing T_9, which should be most strongly affected, with T_7, T_{11}, and T_{20}, which are less altered by dust. A dust cloud extending in height to more than several kilometers will begin to be sensed in the atmospheric channel at 15 μm if the local air temperatures are affected. The presence of dust with an infrared optical depth of 0.1 in the atmosphere above VL 1 has been inferred from IRTM observations of that region designed to determine emission angle dependence (section VIA). The spectral variation of the derived optical depths is consistent with the dust band wavelength location given above, but the observations do not permit analysis of the cloud beyond estimation of its optical thickness. General atmospheric dustiness at the VL 2 latitude (+48°), optical depths being near 0.2, was implied by an observed strong diurnal variation of atmospheric temperatures (paper 4). The lander imaging experiments have also measured extinction attributed to airborne dust [*Mutch et al.*, 1976*a*; *Pollack et al.*, 1977]. Additionally, the strong forward-scattering characteristic of IRTM solar band measurements (part B of section 6) implies that atmospheric dust is present over much of the planet.

The spectral characteristics of CO_2 ice are poorly known, but there should be strong emission within the IRTM 15-μm band. Carbon dioxide clouds can form only at temperatures below 148 K (152 K for the lowest elevations). For clouds forming at high altitudes the expected thermal signature is a marked reduction in brightness temperature in the atmo-

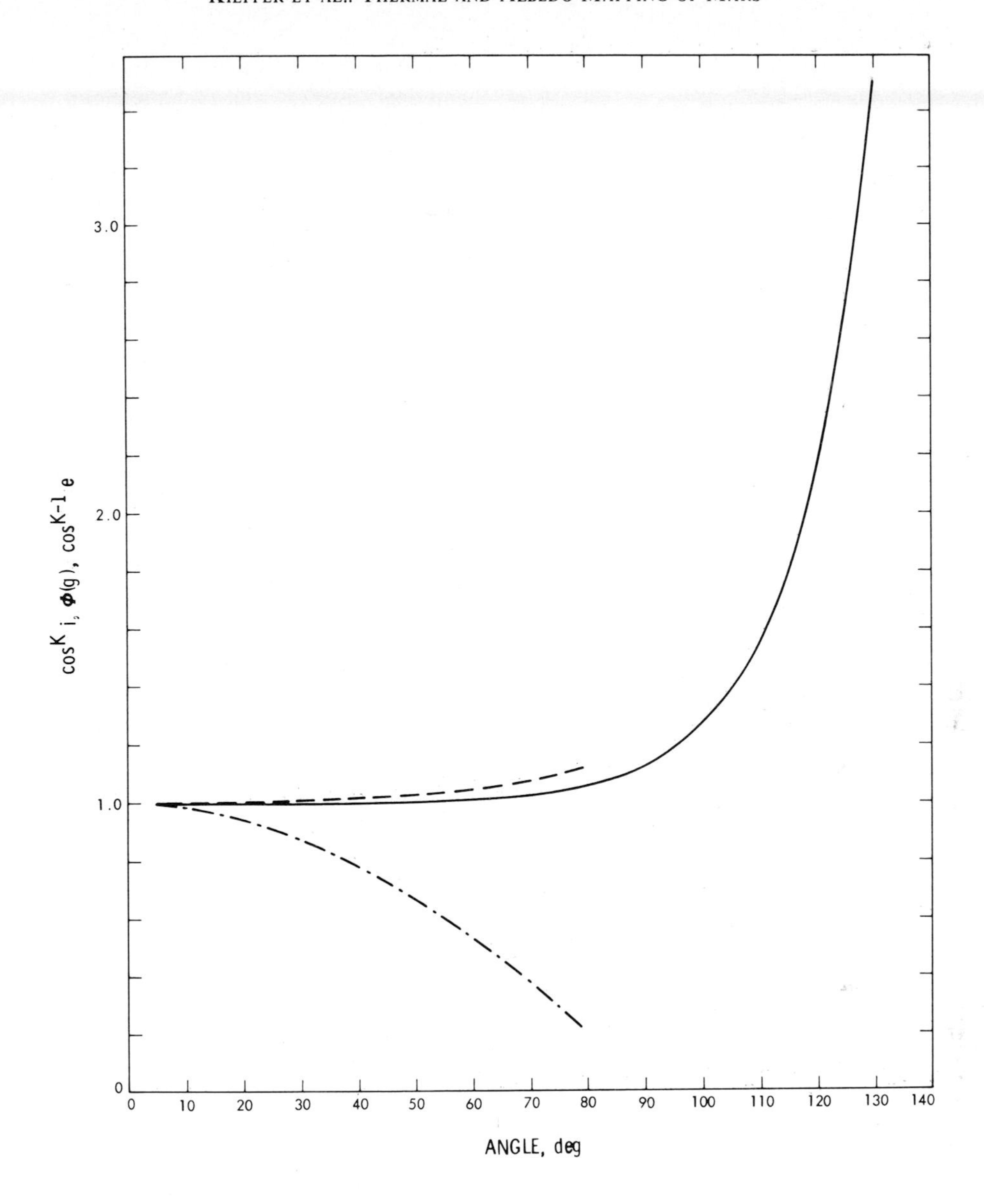

Fig. 14. Empirical variation of solar band brightness B^* with phase (solid line), incidence (lower dashed line), and emission (upper dashed line) angles. The variation depicted for the incidence and emission angles is for a phase independent Minnaert exponent of $K = 0.94$.

spheric channel. Although CO_2 condensation might be expected to occur in the atmosphere during polar cap formation, there is scant evidence for this phenomenon in Viking coverage of the South Polar Regions commencing at $L_S = 100°$. Atmospheric temperatures over most of the polar cap are well above surface temperatures (Figure 8); inferred temperature profiles are incompatible with CO_2 condensation except near the surface (below about 5 km). Condensation in the broad regions where T_{15} exceeds T_{20} may be occurring strictly at the surface; images of southerly areas in early winter have shown a clear atmosphere [*Carr et al.*, 1976; *Briggs et al.*, 1977]. Within 15° of the South Pole, T_{15} measurements are essentially equal to T_{20} at $L_S = 100°$; in this area, CO_2 clouds could form, but there is no direct evidence for them (section IIB1).

Water clouds are common in the Martian atmosphere. They occur as diffuse hazes, morning ground fogs, various orographically controlled structures, and other forms [*National Aeronautics and Space Administration*, 1974; *Briggs et al.*, 1977]. One recurrent cloud forms with regularity to the west of the volcanic shield Ascraeus Mons (+11°, 104°W). Similar clouds often appear to the west of the summits of the other Tharsis volcanoes and Olympus Mons [*Carr et al.*, 1976; *Briggs et al.*, 1977]. Analysis of spectra of a region partially covered by the cloud west of Ascraeus Mons led *Curran et al.* [1973] to conclude that it was composed of water ice with a particle size near 2 μm.

During rev 90 of VO 1 a series of IRTM observations were made in the Tharsis area to document the behavior of this area in the dawn to late morning time period. This data set was found to be particularly valuable for determining cloud effects. The time dependent behavior of several parameters of interest for a cloudy area bounded by +11° to +14° latitude, 107°W to 109°W, and a control area with the same latitude limits between 103°W and 105°W is shown in Figure 15. The areas

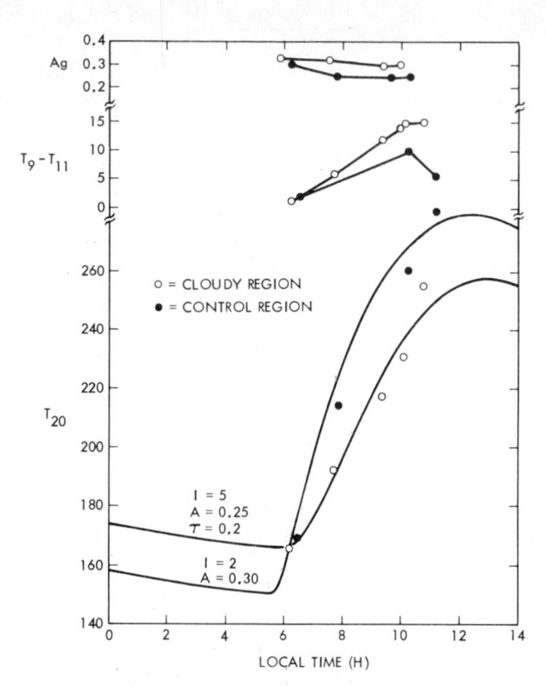

Fig. 15. Cloud effects on several IRTM-measured parameters. The top graph shows phase-corrected albedos for the cloudy area (+11° to +14°, 107°W to 109°W) and for the control area (+11° to +14°, 103°W to 105°W). The middle graph shows the difference between T_9 and T_{11} for the two regions. The bottom graph shows the behavior of T_{20} for the two regions. Model curves are labeled with inertia and radiometric albedo. The model represented by the curve for the cloudy area includes a modification for diminution of sunlight through an atmosphere with extinction optical depth of 0.2.

were selected by using maps of T_r that show a well-defined cold region in the late morning northwest of Ascraeus Mons and by using albedo maps that show a general brightening around the slopes of the volcano. There were no images of the Ascraeus Mons region taken on this day. The control area is located on the north-facing flank of the volcano in a region which possesses lower albedo and has the 20-μm thermal behavior characteristic of the Tharsis region.

Albedos for the two areas (Figure 15) have been corrected for phase angle dependence. Because of the large phase angles, the range of variation (123°–89°), and the consequent magnitude of the corrections, it is not possible to deduce from the data that the albedo decline between 6 H and 10 H is due to cloud dissipation. We emphasize only that the cloudy area albedo is consistently greater than that of the control area. Near 6 H at the highest phase angle the Lambert albedos for the two areas are about 0.80, indicating strong forward scattering, probably from cloud particles, since this value is higher than that for the general northern atmosphere. Both cloud and dust particles have been assessed in past investigations to be somewhat greater than 1 μm in size [*Curran et al.*, 1973; *Conrath et al.*, 1973]; forward scattering at visible wavelengths is characteristic of such particles.

The behavior of the difference $T_9 - T_{11}$ (Figure 15) supports the conclusion of *Curran et al.* [1973] that the cloud seen in this region was water ice. Near dawn, when little temperature contrast exists between the surface and an overlying cloud, T_9 and T_{11} are in good agreement. By 10 H a large depression of T_{11} relative to T_9 has occurred. The 9-μm channel presumably samples the surface, and T_{11} is lowered by the ice absorption within its passband. For the cloudy area the difference is as great at 15 K. Since half of the T_{11} passband is in the ice absorption feature, we may estimate roughly that the true band depth is 30 K and that the cloud is at least that much cooler than the surface. The rise in $T_9 - T_{11}$ for the control area implies cloud absorption there also though to a lesser extent. The 5° slope of the north flank of Ascraeus Mons will not sensibly affect albedo at this time and season. If the control area surface is intrinsically dark, the cloud above it may explain that region's Mars nominal albedo of 0.25. Cloud dissipation, or clearing, is suggested by the decline of $T_9 - T_{11}$ after 10.5 H; corroborative evidence may be present in the T_{20} data.

Measurements of T_{20} and the behavior of two thermal model diurnal curves are also presented in Figure 15. The control area data show a morning temperature rise consistent with the behavior of a low-inertia soil at this latitude, although a slight delay, probably caused by the cloud cover responsible for the $T_9 - T_{11}$ behavior noted above, is evident. The cloudier area exhibits a strong delay in the temperature rise. This delay is not shown by thermal models free of atmospheric opacity nor is the delay characteristic of models incorporating frost formation. The cloud probably inhibits radiative heating in the first few daylight hours more effectively than would a surface of the same normal incidence albedo because of the large scattering losses at high incidence angles. The best-fitting model curve employs an inertia higher than is likely in this province and also an atmospheric optical depth of 0.2; in the absence of predawn data and modeling of nighttime cloud effects, only approximate parametric values can be determined. Delays of this size have been predicted by *Flasar and Goody* [1976] for surface ice fogs (section VIIIA). A morning delay is anticipated whenever clouds intercept sunlight near dawn. It is uncertain if the cloud studied here is in contact with the surface; the $T_9 - T_{11}$ difference carries no implication for the cloud bottom. Both of the areas observed here are at high elevations: the control area extends from 9.5 to 27 km, the mean being near 17 km, and the cloudy region is nearly flat at 9.5 km. Thus cloud behavior here is not necessarily applicable to Mars in general, although the thermal signatures are likely to be characteristic: T_{11} will be altered by water ice clouds relative to the other IRTM bands, permitting cloud identification when optical thickness and time of day are adequate to establish radiance contrasts.

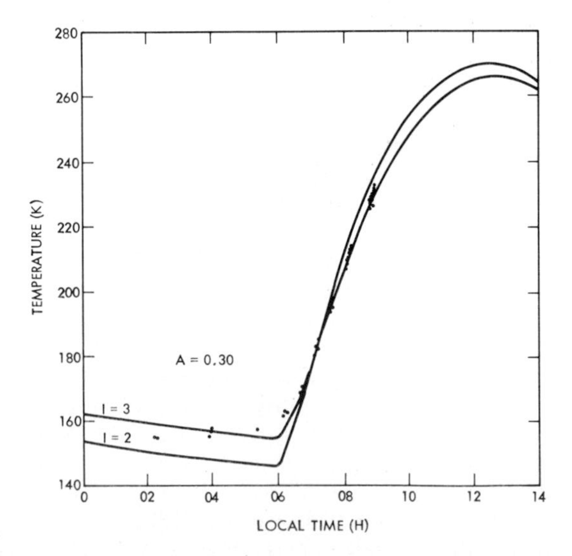

Fig. 16. Unusual predawn behavior of T_{20}. Measurements are collected from the area 0° to −2°, 124°W to 126°W. The model curves are characterized by $A^* = 0.30$.

VIII. MISCELLANEOUS

A. *Predawn Warming*

During the study of the detailed thermal behavior of the region near Arsia Mons, several instances were found in which the temperature rises during the 4 hours preceding dawn. T_{20} measurements for the area 0° to −2°, 124°W to 126°W between 2 and 9 H from VO 1 revs 60 and 62 are shown in Figure 16. Model curves are presented for $A = 0.30$ and $I = 2$ and 3. The anomalous early rise is similar to the behavior predicted by *Flasar and Goody* [1976] for water ice ground fogs with particle sizes in the range 5–10 μm, infrared optical depths near 0.5, and visual extinction optical depths between 2 and 3. The brightness temperature rise is attributed to radiative emission from the cloud top, which rises in the predawn along the inverted atmospheric temperature profile as the fog grows. In contradistinction to the Flasar and Goody curves, however, there is no strong postdawn delay in temperature rise in our data; the delay is expected because the fog dissipates during the first few hours after dawn, during which it contributes to observed radiance and inhibits surface warming. The extremely low surface inertias may play some role in reducing the magnitude of the fog-induced delay. A surface rapidly heated by the weak sunlight penetrating the fog will more effectively dissipate the fog than a surface with high inertia. Temperatures approach those of the standard model about 7 H (Figure 16). Predawn warming also occurs in some areas with inertias of 6. Typically, for these areas, modellike behavior is not resumed until 8 H. The difference in delay between low- and higher-inertia areas is difficult to establish with precision because uncertainty in albedo affects the late morning model temperatures. In all cases, however, the size of the delay is much less than the amount predicted by Flasar and Goody. It should be noted that surface elevations in the area under discussion are 8–9 km and that atmospheric pressures are near 3 mbar. The low pressure may promote early fog dissipation and thus explain the discrepancy with behavior of a fog model having a surface pressure of 6.1 mbar.

Simultaneous measurements by the atmospheric water detection instrument of the region discussed above showed a vertical column abundance of 3.1 ± 0.5 precipitable μm at 8.5 H (C. B. Farmer, personal communication, 1977). Although this value may represent water vapor abundance above the fog rather than including it, the number is different from the 10-μm abundance employed by Flasar and Goody. Possibly, modeling based on the smaller amount would yield better agreement with the data shown in Figure 16.

B. *Observations of Earth*

During the earth-Mars cruise period of the second Viking spacecraft the first positive thermal detection of the earth from deep space was made. The observation occurred at 1650 UT on October 16, 1975, from a range of 1.15×10^7 km. The apparent diameter of the earth was 1.11 mrad, and the phase angle was 127.5°. The disk center was near 0° latitude, 300°W longitude at about 8:30 P.M. local time.

The deduced temperatures are $T_{20} = 234^{+9}_{-2}$ K, $T_{11} = 275^{+7}_{-1}$ K, and $T_9 = 270^{+7}_{-3}$ K. The larger positive uncertainties are due to the uncertainty in the apparent position of the earth relative to the center of the instrument field of view. The reflected solar brightness is equivalent to $0.43^{+0.09}_{-0.05}$ of the brightness of a Lambert sphere at the same phase angle. The 20-μm temperature is lower than the 11- and 9-μm temperatures because of atmospheric water vapor opacity at 20 μm. For a mean tem-

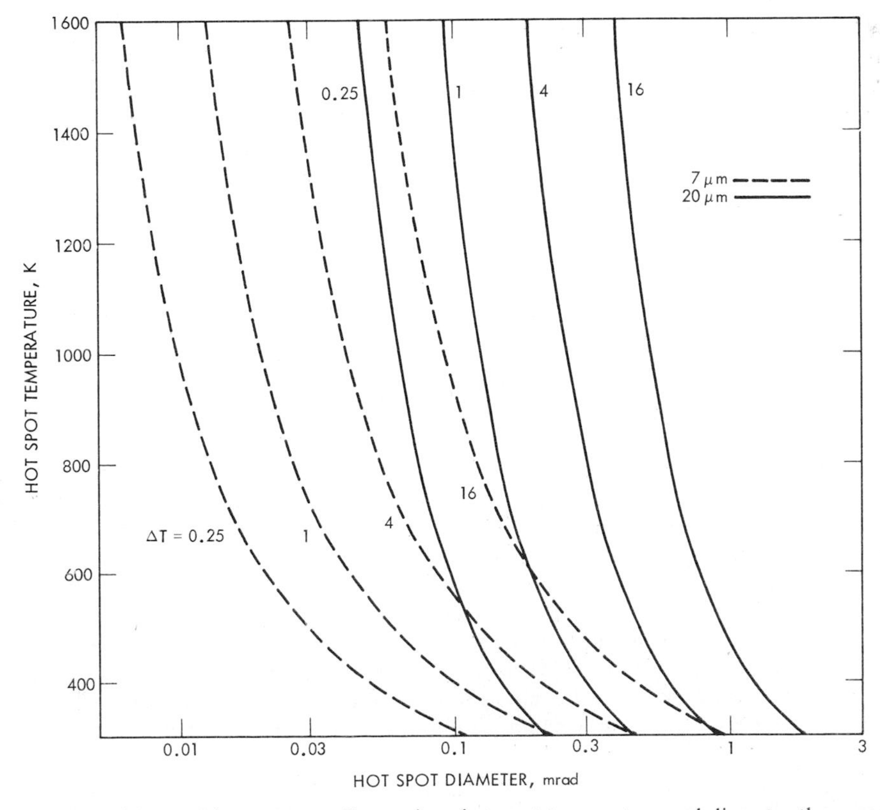

Fig. 17. Detectability of internal heat sources. For a given hot spot temperature and diameter the curves show the amount by which T_7 and T_{20} will exceed the background temperature which is taken as 200 K. The milliradian scale is equivalent to kilometers at an altitude of 1000 km. Digitization is 1 K for T_7 and 0.2 K for T_{20} at 200 K.

perature gradient of -6.5 K km^{-1} (and if we assume T_{11} and T_9 refer to the surface temperature of the earth) the altitude sampled by T_{20} is near 6 km.

C. *Detectability of Internal Heat Source*

Although detection of active Martian volcanism is unlikely, the geologic value of such detection justifies careful searching. The IRTM has several advantages over the Mariner 9 infrared radiometer in regard to the ability to detect sites of exceptionally high heat flow. The IRTM field of view area is smaller by a factor of 10, a fact which enhances the detectability of small sources. The areal coverage of the planet is greatly increased by the larger number of detectors available and by the scanning capability of the Viking spacecraft. Finally, the number of spectral bands available and their wavelength placement enhance the distinction between small areas with large temperature ranges and those for which the temperature is uniform across the field of view.

The detectability of thermal anomalies as a function of source temperature and angular size is shown in Figure 17. The IRTM field of view is taken to be a circle of 5.2-mrad diameter. At 200 K the digitization is 1 K in the 7-μm band and 0.2 K in the 20-μm band. Lava fields that are several hours old or quiescent lava lakes would probably have brightness temperatures in the range of 400–500 K due to crust formation; they would need to be at least 200 m in diameter to produce a 4 K change in T_7 at periapsis with a background temperature of 200 K. There are many variables. Hot spots would be more obvious in winter or at nighttime with a much lower background brightness temperature. Although thermally bright small areas appear in IRTM data, no systematic search has yet been made incorporating the spectral information in order to exclude regions that heat easily during daylight hours because of low thermal inertia or regions that retain heat at night because of high inertia. In the area covered by the predawn map and at the resolution associated with these observations, areas attributed to internal heat have not been found. Individual measurements with temperature differences of 4 K from adjacent samples would have been noticed.

APPENDIX 1: VIKING THERMAL MODEL

The large variation of surface temperature with time of day, latitude, and season make it desirable to remove the average Martian response to the temporal variation of insolation before studying spatial variations of the thermal behavior. This is particularly true when obserations taken at widely different local times or seasons are involved. The utility of this technique is somewhat tempered by the fact that the variations of Martian thermal behavior are large, in many instances larger than those anticipated, and an 'average' Martian model is in some respects not as well defined nor as satisfactory as it appeared to be when much less data were available.

The Viking thermal models are intentionally simple and do not include a variety of geophysical processes which certainly occur but whose quantitative behavior is not well known. These include the total radiative effect of clouds, non-Lambertian emission, the latent heat of water ice, subsurface inhomogeneity, and variable conductance at the air-surface boundary. The objective of the models was to account for the best-understood and largest terms in the heat equation: the diurnal and annual variation of insolation and the conduction into the ground.

The models used assume surface physical properties homogeneous with depth and uniform over the planet. The atmospheric properties are constant, with no cloud formation of any kind at any time. A CO_2 polar cap forms, and the surface radiometric albedo changes discontinuously to 0.65 when any frost is present, but the condensation temperature is fixed at 149 K rather than following the predicted surface pressure variation. There is no latitudinal transport of heat; the downgoing atmospheric radiation F_a is constant through the day at 0.02 of the greater of the noontime insolation or the surface frost emission (in the winter night). No slopes are included, the absorption of sunlight is Lambertian, ideal blackbody surface emission is assumed ($\epsilon = 1$), and no heat flows across the lower boundary.

The surface boundary condition and heat diffusion equation are solved in the normalized forms

$$\frac{S_S}{R^2}(1 - A^*) \cos i + I \left.\frac{\partial T}{\partial Z'}\right|_{Z'=0} + F_a + L\frac{dm}{dt} = \epsilon\sigma T^4$$

and

$$\partial T/\partial t = (P/\pi)(\partial^2 T/\partial Z^2)$$

where

S_S solar constant;
R Mars heliocentric distance;
P period of the Martian mean solar day;
$Z' = Z/[(K/PC)(\rho/\pi)]^{1/2}$, the depth normalized to the diurnal skin depth;
L latent heat of CO_2 condensation;
m mass per unit area of CO_2 frost;
ϵ surface emissivity;
σ Stephan-Boltzmann constant.

The term $L\, dm/dt$, representing CO_2 phase change, controls the minimum temperature and establishes the seasonal extent of the polar caps.

The program uses an explicit finite-difference scheme with layer thickness increasing exponentially downward; the time step size progressively doubles at various lower layers where allowed by convergence criteria, maintaining a time step smaller, by a conservative factor, than that required for convergence of the diffusion equation [*Richtmeyer and Morton*, 1967, p. 189]. By using the mean orbital elements for 1976.0 and the spin-axis orientation determined by Mariner 9 [*de Vaucouleurs et al.*, 1973] the heliocentric distance and subsolar declination are determined at uniform time intervals through the Martian year. The corresponding insolation function is used for detailed calculations for several days until the surface temperature behavior has converged and subsurface midnight temperatures are steady or progressing smoothly. Then the surface temperatures through the last day are stored, and the midnight temperatures at all depths are predicted for the end of this seasonal time step.

For the Viking mission, three separate major computer runs were made; in each case, computations covered three Martian years, the last year being stored and utilized for data analysis. Temperatures at 37 latitudes, at 5° increments from pole to pole, were evaluated at seasonal intervals of $\frac{1}{40}$ of a Martian year and a fine time step size of $\frac{1}{384}$ of a day. The top layer has a thickness of 0.18 of the diurnal thermal skin depth; 25 layers progressively thicker by a ratio of 1.2 extend to 2.24 times the annual thermal skin depth. Surface temperatures at each $\frac{1}{48}$ of a day ($\frac{1}{2}$ H) and the minimum and maximum temperatures of each subsurface layer were stored for each season.

The three base models have radiometric albedo and thermal inertia of 0.2 and 6.0; 0.3 and 6.0; and 0.2 and 8.0. By assuming a volume specific heat of 0.24 cal cm^{-3} K^{-1} (paper 3) the surface layer thicknesses and total depths are 0.756 and 247 cm

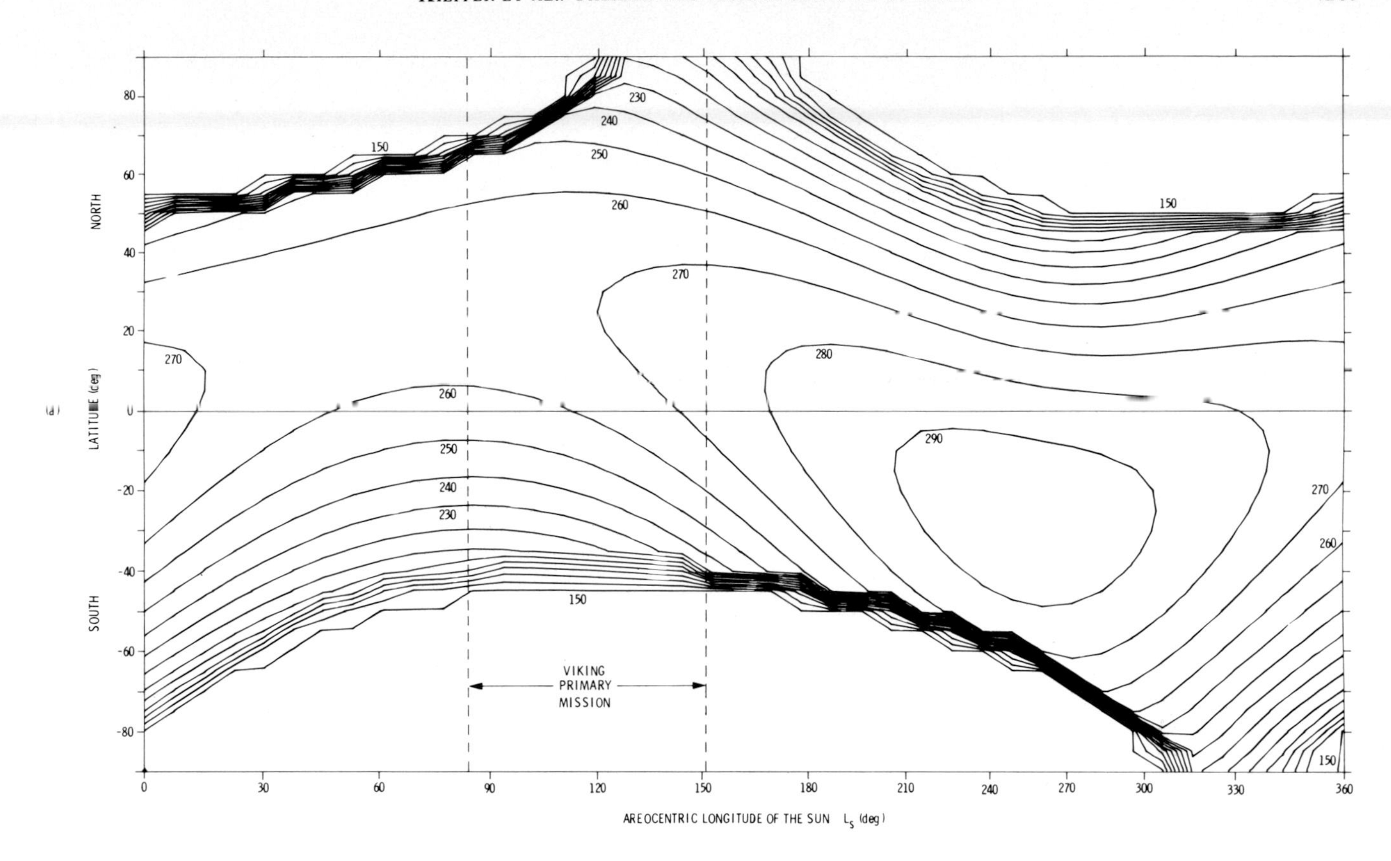

Fig. 18*a*

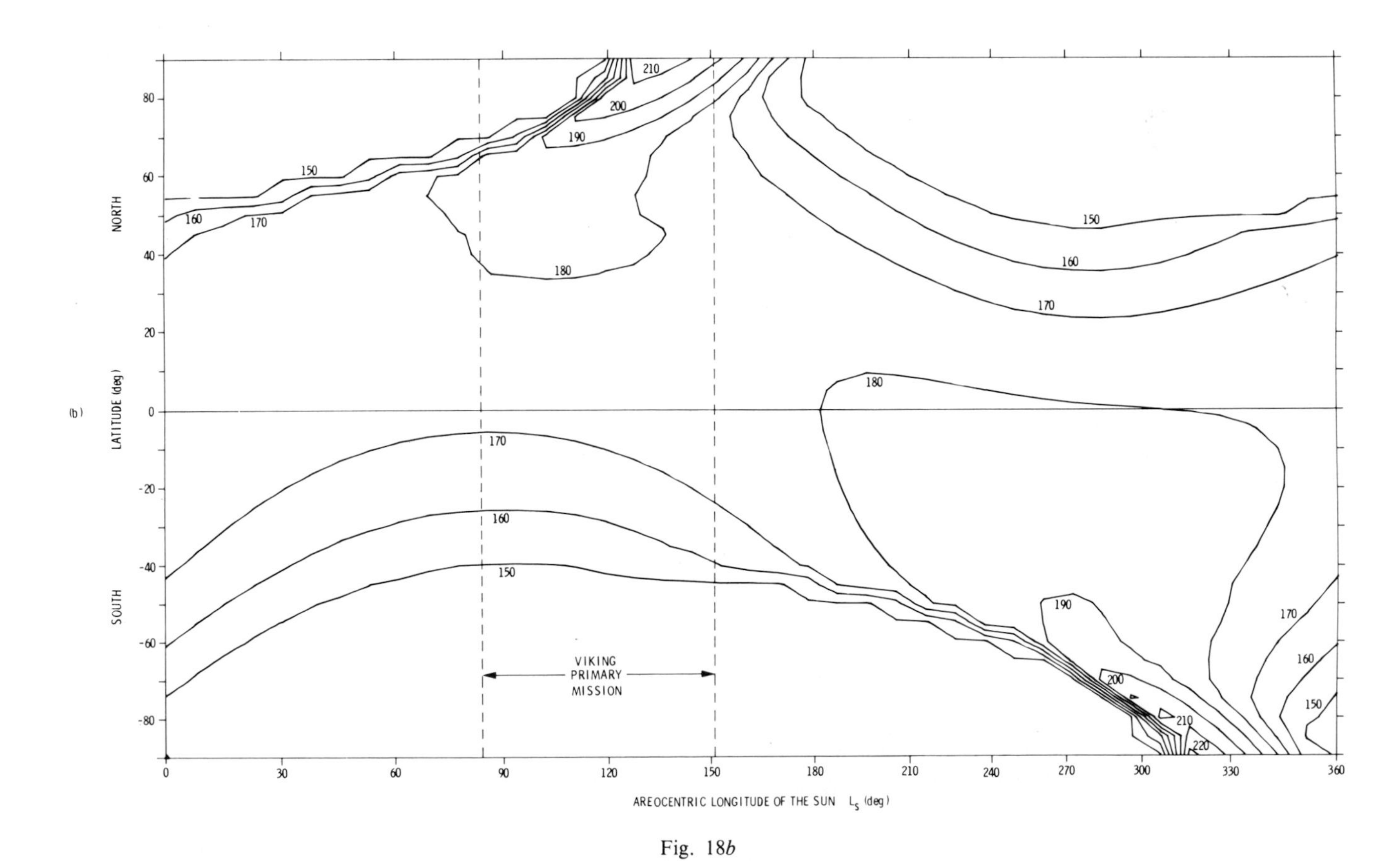

Fig. 18*b*

Fig. 18. Diurnal surface temperature mean and extremes for the primary Viking thermal model ($A^* = 0.25$ and $I = 6.5$). The dashed lines indicate the seasonal range of the primary mission. (*a*) Maximum temperatures. (*b*) Minimum temperatures. (*c*) Mean temperatures.

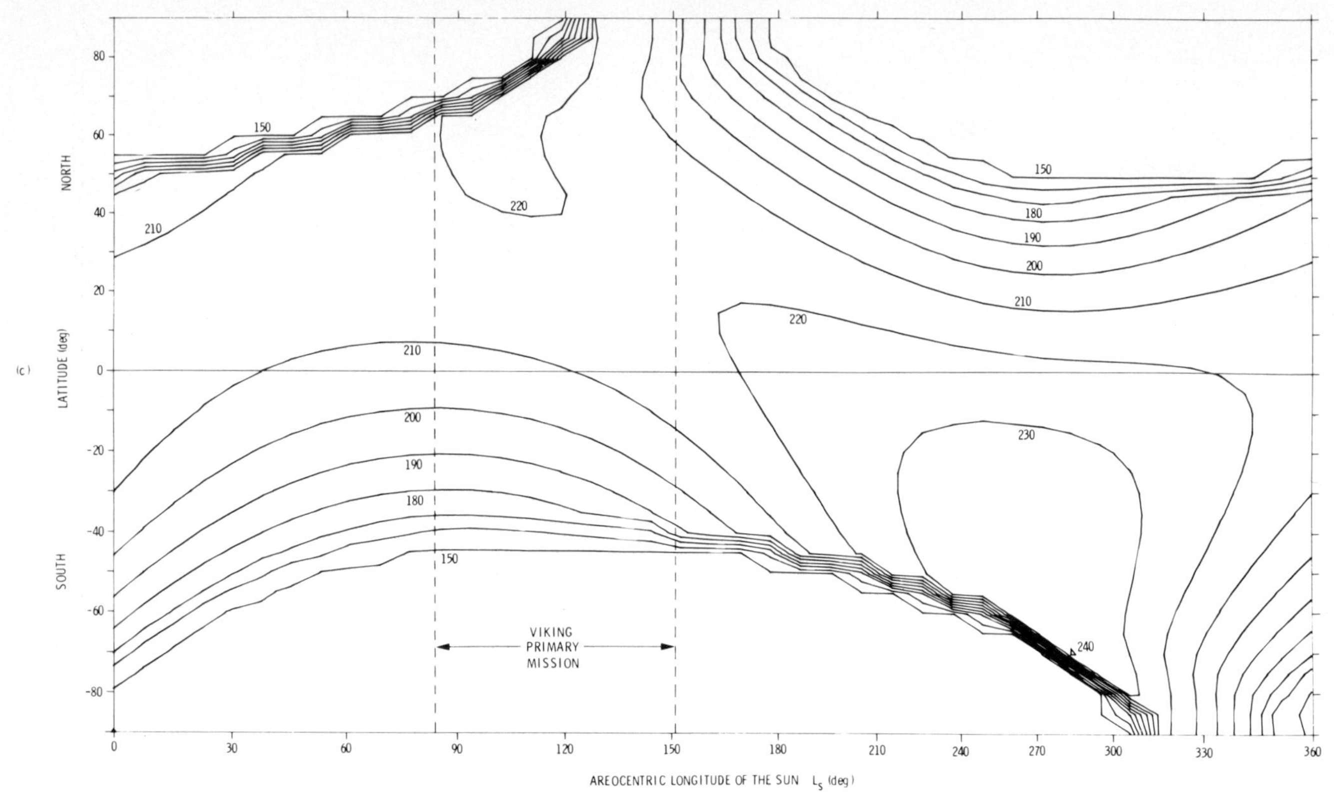

Fig. 18*c*

for 6.0 inertia models and 1.01 and 329 cm for the 8.0 inertia model. These three models surround, in A^*-I space, the average Martian values derived from Mariner 9 observations after clearing of the 1971 global dust storm [*Kieffer et al.*, 1973]. The partial derivatives of surface temperature with respect to albedo or inertia are approximated by differences between these three models.

A standard thermal model with $A^* = 0.25$ and $I = 6.5$ is used for comparison with the Viking data. The annual variation of daily extreme and daily average temperatures for the standard model is shown in Figure 18. For each revolution of the Viking orbiters the three base models are linearly interpolated in A^* and I to form the standard model at the nearest two model seasons (spaced by 17.17 days). These two results and the partial derivatives are linearly interpolated to the time of periapsis of the orbiter. The resulting model temperature T_M, $\partial T/\partial A^*$, and $\partial T/\partial I$ on a 5° latitude and $\frac{1}{2}$ H grid are each bilinearly interpolated for all seven IRTM spots for each observation.

The resulting thermal model is continuous and piecewise linear in albedo, inertia, latitude, and time. It is believed not to differ significantly from the exact solution in these four variables. Detailed tests with finer step sizes in these four parameters and thinner subsurface layers indicate that model errors are less than 1 K except very near dawn or near the edge of the polar cap. The error near dawn results from the finite size of the final grid in H in comparison with the rapid rise in temperature at dawn; it is, however, in the worst case, equivalent to a local slope of about 2°. Errors near the edge of the polar cap can be large, particularly in the partial derivatives, since only one or two of the base models may have frost at some latitudes and seasons. This difficulty is inherent in models which do not include heat transport across latitude. There is, in fact, currently no numerical model which approximates very well the behavior of the polar cap throughout the year. In any event, the large thermal contrasts near the edge of the polar cap cannot be modeled well with a 5° latitude grid.

The use of unit emissivity allows observed brightness temperatures to be compared directly to the model surface temperature. In fact, the broadband brightness temperature is essentially independent of surface emissivity ϵ if it is spectrally gray, since the basic thermal process is a balance between incident and emitted radiation. The surface physical temperature will, however, be higher than the model temperatures with a scaling factor of $\epsilon^{-1/4}$; and the appropriate thermal conductivity is decreased by a scale $\epsilon^{1/4}$ from that used in the Viking thermal model. Thus if the actual broadband emissivity were 0.95, the

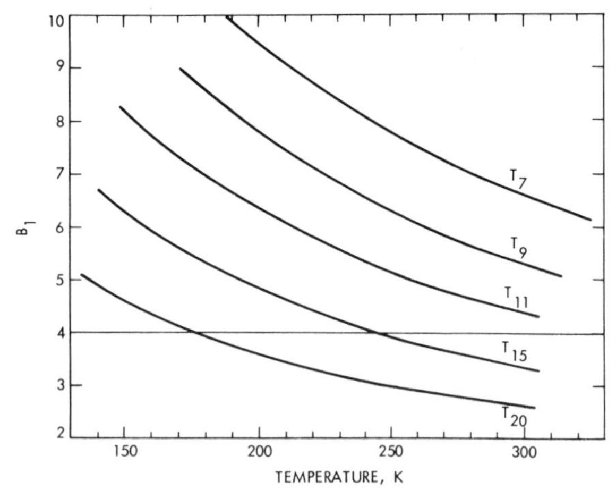

Fig. 19. Exponent of the dependence of in-band radiance on temperature. $B_1 = (\partial E_j/\partial T)(T/E_j)$. Lines are shown only where they are within the IRTM dynamic range and where the signal exceeds to 10 DN.

surface kinetic temperatures would be 1.25%, or typically 2.5 K, greater than T_M, and the appropriate conductivity would be 1.25% less than that used in the model which best fits the observations.

The above relation, however, depends on the Stephan-Boltzmann law, and for finite bandwidth the difference between brightness temperature and the computed T_M is approximately

$$(T_B - T_M)/T_M = (1 - \epsilon_T)(\tfrac{1}{4} - B_1^{-1})$$

where

$$B_1 = (\partial E_j/\partial T)(T/E_j)$$

is the power dependence upon temperature of E_j, the Planck function integrated over the band spectral response. The behavior of B_1 for the IRTM bands is shown in Figure 19. The spectral variation of brightness temperature resulting from gray nonunit emissivity is quite small and probably smaller than the effects of spectral emissivity of the surface. Comparison of this relation with the response of the IRTM bands as a function of temperature (Figure 2) shows that this effect is near or below the instrument resolution.

Inclusion of diffusion of heat into the atmosphere by conductive coupling with the surface will have a form similar to heat diffusion into the subsurface and will not generate a significantly different class of model solutions. If we assume that convection is unimportant in the lower 1.4 m of the atmosphere, the Viking meteorology temperature measurements [*Hess et al.*, 1976] can be compared with the surface temperature measurements (paper 1) to provide a lower limit to the heat exchange with the atmosphere. By using an average temperature gradient of 10 K in 1.5 m for one half of a Martian day the conductive heat transfer is 0.08 cal cm^{-2}, only 10^{-3} of the heat transfer into the subsurface. Near-surface convection or turbulence will increase the total heat transfer, perhaps as much as 5% of the subsurface value, as was suggested by early modeling of the Martian atmosphere [*Gierasch and Goody*, 1968].

As is discussed herein, the range of thermal behavior observed on Mars far exceeds the corrections associated with terms neglected in the thermal model.

APPENDIX 2: BLOCKY SURFACE MODEL

The thermal angular emission function, described in section VIA, cannot be explained solely by a homogeneous surface and/or the presence of an atmosphere; the phase angle dependence of the observed radiance must have another explanation. On the basis of the large number of rocks present on the Martian surface [*Mutch et al.*, 1976*b, c*] a blocky surface model has been constructed which will predict the apparent temperature of the surface and attempt to reproduce the observed phase and emission angle dependencies.

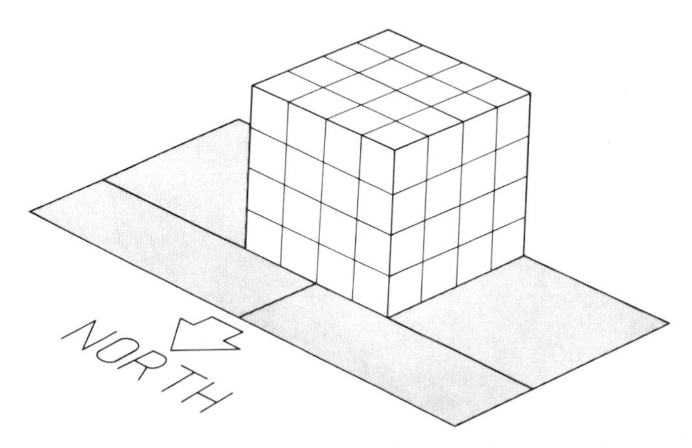

Fig. 20. Relationship between a block and the surrounding surface for the block model. The individual elements of the block and the regions adjacent to the block which can be shadowed at various times of day are shown.

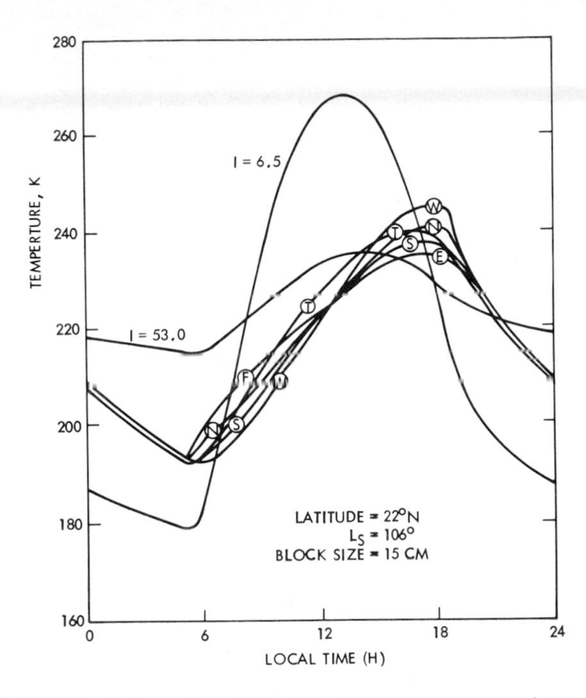

Fig. 21. Typical block face diurnal temperature curves along with curves for a horizontal homogeneous surface of Martian average thermal inertia (I = 6.5) and of the same thermal inertia as the block (I = 53). The block face curves are labeled for the top (T), north (N), south (S), east (E), and west (W) faces.

Surface roughness models have been used before to explain observational characteristics of planetary surfaces in the visible and infrared. *Gear and Bastin* [1962] suggested that a corrugated model of the lunar surface, the peaks and troughs having different temperatures, might explain thermal anomalies caused by emission angle effects during lunar eclipses. *Buhl* [1967] determined the effects of spherical indentations on radiation patterns on the moon and then the effects of rocks and exposed bedrock to explain thermal anomalies in the Tycho crater [*Buhl*, 1971]. *Roelof*'s [1968] extensive study of the thermal behavior of rocks on the lunar surface concludes that infrared observations from spacecraft will strongly depend on the viewing geometry, while earth-based observations will not. *Bastin and Gough* [1969] constructed a three-dimensional version of the peak and trough model which accounted for several thermal anomalies on the moon during lunar eclipses. *Veverka and Wasserman* [1972] calculated Minnaert coefficients appropriate for a Mars with macroscopic surface roughness characterized by paraboloidal craters and found them to be different from the coefficients for a flat surface.

The possible sources of radiative flux emitted at the surface in our model are the blocks, the surface shadowed by the blocks, and the unperturbed surface, generally all at different temperatures. The block, represented here by a cube (Figure 20), has five faces which can be observed, the sixth face resting on the surface, with the average temperature of each block face determined by the block model. Parts of the soil surface immediately adjacent to the block will be shadowed by the block at various times of day. For simplicity, these shadowed areas are divided into four separate regions with different daily temper-

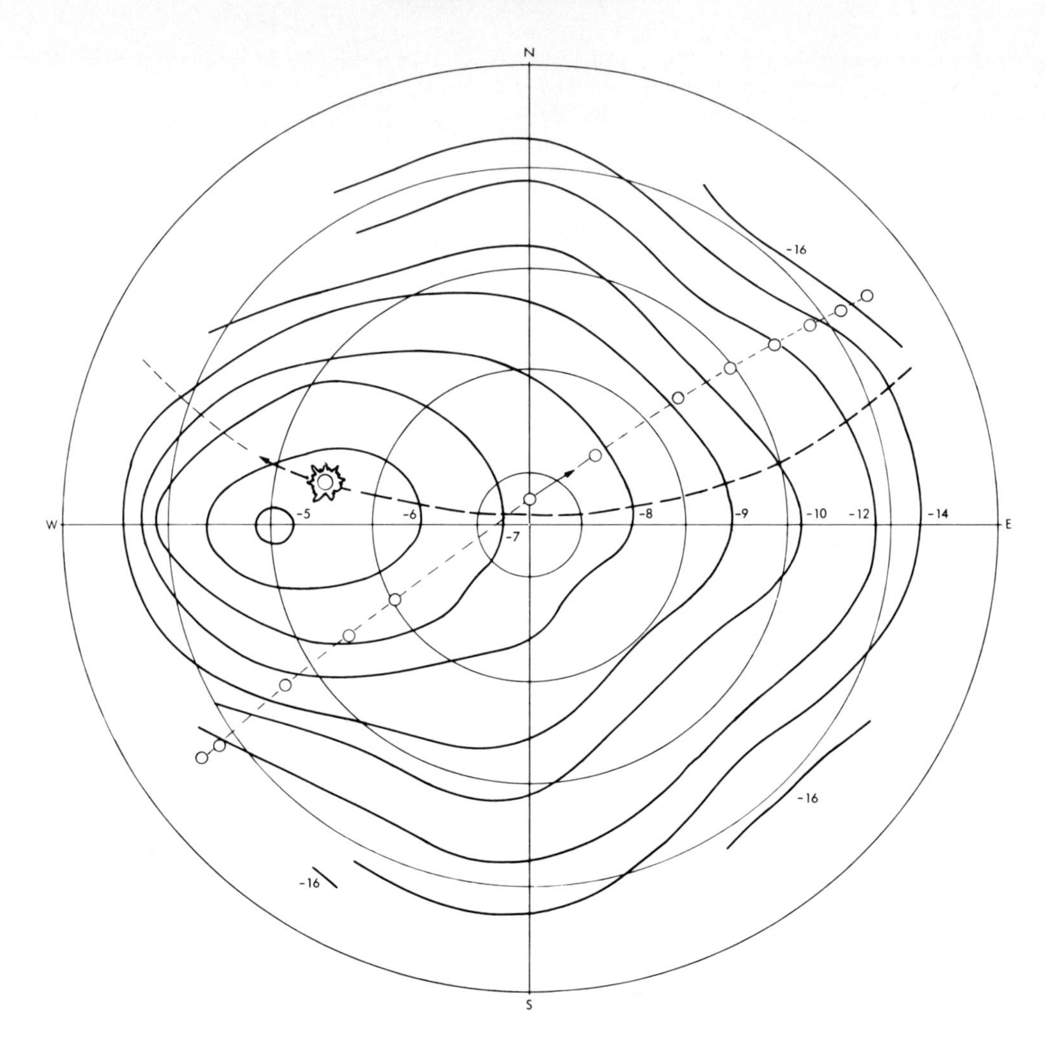

Fig. 22. Contours of predicted 20-μm brightness temperature variation $T_{\text{blocky model}} - T_M$ shown as a function of emission and azimuth angles at the time of IRTM observations of the VL 1 site (15.5 H on August 1, 1976). The location of the sun and its daily path across the sky are shown. Small circles show the location of the spacecraft at each of the observation times. Atmospheric opacity was assumed to be zero. Lines of constant emission angle are at intervals of 20°, beginning at the horizon ($e = 90°$). The apparent distortion of the contours and the offset of the maximum observable temperature from the solar incidence direction are due to the choice of cubes to represent the surface rocks.

ature histories (Figure 20): one on either side of the block, to the east and to the west of the block, each having the same dimensions as a block face and being shadowed in the early morning or late afternoon, respectively; and the other two on that side of the block which is shadowed in the midday, shadow times being midmorning to about noon and noon to midafternoon. The remainder of the observed flux comes from that part of the surface which feels no effects of the presence of blocks and is determined from the standard model for soil at the appropriate latitude.

An iterative one-dimensional finite-difference solution to the thermal diffusion equation already exists for calculating planetary surface temperatures as a function of season, time of day, latitude, and surface physical properties (Appendix 1). This standard model solution was modified to include the presence of blocks and shadows cast by the blocks and their effects on surface temperatures.

The block model consists of a cubical rock sitting on the surface of the planet, with the faces along the cardinal directions. The rock is composed of $n \times n \times n$ independent elements, each of which can communicate thermally with those around it (Figure 20). In iterating the block temperatures, each element can exchange energy with the adjacent elements, the top layer of surface (if the element is on the bottom side of the block), or the surroundings (if the element is on the outer layer of the block). Energy transfer is by conduction, except for exchange with the surroundings for which it is by radiation, with the energy gain in the element per unit time per unit face area being

$$dE/dt = (S_S/R^2)(1 - A_f)\cos(i_f) + 0.02\cos(i_{s,\text{noon}}) + \epsilon\sigma T_s^4/2 - \epsilon\sigma T_f^4$$

or

$$dE/dt = (S_S/R_2)(1 - A_f)\cos(i_f) + 0.02\cos(i_{s,\text{noon}})/2 + \cos(i_s)(A_s/2) + \epsilon\sigma T_s^4/2 - \epsilon\sigma T_f^4$$

where

- A_f radiometric albedo of the block face;
- A_s radiometric albedo of the surface;
- i_f solar incidence angle on the block face;
- i_s solar incidence angle on the surface;
- $i_{s,\text{noon}}$ solar incidence angle on the surface at local noon;
- T_s soil surface temperature;
- T_f block element face temperature;
- ϵ emissivity of the soil and block.

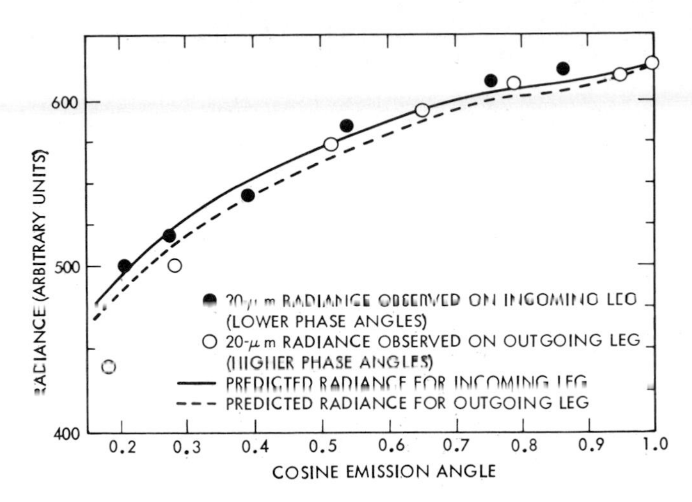

Fig. 23. 20-μm predicted and observed thermal angular emission functions. The predicted curve, done from a blocky surface model, has a phase and emission angle dependence similar to that of the observed curve. The observations of the VL 1 site occurred on VO 1 rev 42 at about 15.5 H Mars local time.

The first equation is appropriate if the element is on the horizontal top face of the block; the second equation is appropriate if the element is on a vertical side face, in which case the element receives only half as much atmospheric radiation and also receives solar radiation reflected off the surface. The first term on the right side of each equation is incident solar radiation, and the second term is background radiation from the sky, taken as 2% of the noontime flux. The next to last term is the thermal energy emitted from the surface and absorbed by the block, the $\frac{1}{2}$ being included since the ground covers only half of the sky visible from a vertical block face. The final term is the thermal energy emitted by the block face. Where the blocks cover a small fraction of the surface (less than about 20%), thermal radiation emitted by adjacent blocks will be negligible. Temperatures of the soil layers directly beneath the block are computed simultaneously with those layers not beneath the block, and thermal conduction can occur both vertically and horizontally in the soil. Iterations of the energy transfer processes are carried out until the surface layer temperatures converge.

When performing the calculations the emissivity is set to 1.0, the block albedo is set equal to the soil albedo, the block specific heat, thermal conductivity, and density used are values typical of terrestrial basalts (C = 0.2 cal g^{-1} K^{-1}, k = 0.005 cal s^{-1} cm^{-1} K^{-1}, and ρ = 2.80 g cm^{-3}) and the surface soil parameters used are the Martian average (I = 6.5 and A_s = 0.25). The number of elements n along a block face is generally about four (values as large as n = 10 have been used for large blocks, the results differing by less than about 1.5 K).

An example of the block solution is shown in Figure 21, which gives the average temperatures of the five observable faces of a 15-cm block at the VL 1 site as a function of time of day at L_S = 104°. Shown for comparison are the diurnal temperature curves of the Martian average surface and of a planar horizontal surface of the same thermal inertia as the block (I = 53).

Average temperatures for the shadowed regions, similar to typical eclipse curves, are calculated by setting the insolation to zero at the shadowed times of day while retaining the atmospheric emission term. To set the scale of the shadow, the soil layer temperatures beneath a specified depth are set equal to the corresponding layer temperatures of the standard model at each iteration of the program. The shadow times are determined empirically by a geometry routine which determines the shadow location on the surface as a function of time of day.

To get the apparent brightness temperature for a particular geometry relative to the surface, the area-weighted thermal fluxes in one of the IRTM wavelength bands from each of the visible surfaces (block face, soil surface, or shadowed soil surface) are added to get an apparent flux, which is converted to brightness temperature. Because the surface surrounding the block will be obscured partially by the block, the surface is divided into a large array of elements, each of which is considered separately; the block faces will not be obscured by other blocks except at very high surface emission angles ($e \gtrsim 75°$ for a 10% block covering).

To compare the apparent brightness temperature with the observed thermal angular emission function, brightness temperature is determined for the incidence, phase, and emission angles of the actual observations. As is discussed in section VIA, a nonzero infrared opacity of the atmosphere is expected such that the actual observed radiance is given by (2), T_s being the apparent brightness temperature of the blocky surface in the absence of an atmosphere. The atmospheric emission $R_\lambda(T_a)$ was obtained for an atmospheric temperature of 250 K and an emissivity of 0.3; the final result has only a small dependence on these numbers.

A model was computed for 20-μm observations taken of the VL 1 site in the late afternoon by VO 1 on rev 42 (August 1, 1976). The ground truth of block populations is known, and we approximate it with a 10% covering of 15-cm blocks, which give about the same total surface coverage of rocks as is observed (paper 3). Figure 22 shows the results, giving the predicted brightness temperature variation, $T_{\text{blocky model}} - T_M$, as a function of viewing geometry for the case τ = 0. The maximum observable temperature is at a phase angle of about zero, and the result is almost symmetric about the east-west axis, since the sun never ventures very far away from the axis (the solar declination was 24° at the time of the model, and the latitude of the observation was 22°). Calculations were then done for several values of the infrared opacity, and the opacity which best fit the observations was determined. For a surface with no blocks, τ = 0.08 fits the data, but there is no phase angle dependence of observed brightness temperature. For a blocky surface with no atmosphere (τ = 0) a phase angle dependence is present, but the decrease of radiance with emission angle is not strong enough. For 10% blocks a best fit occurs with τ = 0.05 (estimated uncertainty of $\pm$0.015), and the phase angle dependence is adequately reproduced through emission angles of about 65° (Figure 23); at larger emission angles the phase angle dependence is not strong enough. Had we used either a block albedo different from that of the surface, lower by about 0.10, or a larger block size, the temperature difference across the block would have tended to be enhanced by several degrees, and at large emission angles the phase angle dependence would have increased.

The observed thermal angular emission function can satisfactorily be explained in terms of both the emission angle and the phase angle dependencies by the observed presence of a large number of rocks on the Martian surface and an infrared opacity of the atmosphere at 20 μm of about 0.05.

Acknowledgments. The success of the IRTM investigation has required prolonged effort on the part of many individuals. We would like to thank Michael Agabra, Jack Engel, Howard Eyerly, Claude Michaux, Richard Ruiz, and Donald Schofield for their diligence in design, fabrication, and operation of the instrument; Robert Mehlman, John Gieselman, and Elliot Goldyn for design of the infrared data processing system; Philip Christensen and Judy Bennett for their data processing and program development help; Viking interns Mark Jennings, Wayne Roberge, Judy Burt, and Glen Stewart for assistance at critical times; I. E. Sedor for graphics assistance; and Bonnie Long for her professional skill and patience. We were the beneficiaries of a superbly designed spacecraft and a well-managed operational system. The accuracy and precision with which observational desires were translated into observations and returned as data were frequently amazing; the groups responsible are known as MPG, FPAG, OPAG, and DSG. Plates 1 and 2*a* were produced through the facilities of the Viking lander imaging ground reconstruction equipment. The Santa Barbara Research Center built the IRTM's, and the care taken in their design and calibration has amply been justified. The understanding of our close relatives and friends through the long Viking project and its remarkable summer was deeply appreciated. Discussions with members of all the Viking science teams have been beneficial and enjoyable. Coinvestigators Stillman Chase, Guido Münch, and Gerry Neugebauer have made many valuable contributions of all kinds throughout the duration of this investigation. Finally, we would like to thank Conway W. Snyder for his encouragement and judgment. This work was supported by NASA contract NAS 7-100 at the Jet Propulsion Laboratory, Pasadena, California, and contract 952988 to the University of California at Los Angeles. Additional support was received from the NASA Viking Project Office.

REFERENCES

Bastin, J. A., and D. O. Gough, Intermediate scale lunar roughness, *Icarus, 11,* 289–319, 1969.

Becklin, E. E., O. Hansen, H. Kieffer, and G. Neugebauer, Stellar flux calibration at 10 and 20 μm using Mariner 6, 7, and 9 results, *Astron. J., 78,* 1063–1066, 1973.

Blasius, K. R., The record of impact cratering on the great volcanic shields of the Tharsis region of Mars, *Icarus, 29,* 343–361, 1976.

Briggs, G. A., The nature of the residual martian polar caps, *Icarus, 23,* 167–191, 1974.

Briggs, G. A., and C. B. Leovy, Mariner 9 observations of the Mars north polar hood, *Bull. Amer. Meteorol. Soc., 55,* 278–296, 1974.

Briggs, G. A., K. Klaasen, T. Thorpe, J. Wellman, W. Baum, and J. Veverka, Martian dynamical phenomena during June–November 1976: Viking orbiter imaging results, *J. Geophys. Res., 82,* this issue, 1977.

Buhl, D., Radiation anomalies on the lunar surface, *Space Sci. Lab. Rep.,* Ser. 8, Univ. of Calif., Berkeley, 1967.

Buhl, D., Lunar rocks and thermal anomalies, *J. Geophys. Res., 76,* 3384–3390, 1971.

Buhl, D., J. Welch, and D. G. Rea, Radiation and thermal emission from illuminated craters on the lunar surface, *J. Geophys. Res., 73,* 5281–5295, 1968.

Carr, M. H., The volcanoes of Mars, *Sci. Amer., 234,* 33–43, 1976.

Carr, M. H., H. Masursky, W. A. Baum, K. R. Blasius, G. A. Briggs, J. A. Cutts, T. Duxbury, R. Greeley, J. E. Guest, B. A. Smith, L. A. Soderblom, J. Veverka, and J. A. Wellman, Preliminary results from the Viking orbiter imaging experiment, *Science, 193,* 766–776, 1976.

Chase, S. C., E. D. Miner, D. Morrison, G. Münch, and G. Neugebauer, Mariner 10 infrared radiometer results: Temperatures and thermal properties of the surface of Mercury, *Icarus, 28,* 565–578, 1976.

Chase, S. C., J. L. Engel, and H. W. Eyerly, H. H. Kieffer, and F. D. Palluconi, Design and performance of the thermal mapper, submitted to *Appl. Opt.,* 1977.

Coblentz, W. W., Further measurements of stellar temperatures and planetary radiation, *Proc. Nat. Acad. Sci. U.S., 8,* 330–333, 1922.

Coblentz, W. W., and C. O. Lampland, Further radiometric measurements and temperature estimates of the planet Mars, 1926, *Sci. Pap. Bur. Stand., 22,* 237–276, 1927.

Conrath, B. J., Thermal structure of the martian atmosphere during the dissipation of the dust storm of 1971, *Icarus, 24,* 36–46, 1975.

Conrath, B., R. Curran, R. Hanel, V. Kunde, W. Maguire, J. Pearl, J. Pirraglia, and J. Welker, Atmospheric and surface properties of Mars obtained by infrared spectroscopy on Mariner 9, *J. Geophys. Res., 78,* 4267–4278, 1973.

Cross, C. A., The heat balance of the martian polar caps, *Icarus, 15,* 110–114, 1971.

Curran, R. J., B. J. Conrath, R. A. Hanel, V. G. Kunde, and J. C. Pearl, Mars: Mariner 9 spectroscopic evidence for H_2O ice clouds, *Science, 182.* 381–383, 1973.

Cutts, J. A., K. R. Blasius, G. A. Briggs, M. H. Carr, R. Greeley, and H. Masursky, North polar region of Mars: Imaging results from Viking 2, *Science, 194,* 1329–1337, 1976.

de Vaucouleurs, G., *Physics of the Planet Mars,* Faber and Faber, London, 1954.

de Vaucouleurs, G., A low-resolution photometric map of Mars, *Icarus, 7,* 310–349, 1967.

de Vaucouleurs, G., On the opposition effect of Mars, *Icarus, 9,* 598–599, 1968.

de Vaucouleurs, G., M. E. Davies, and F. M. Sturms, Jr., Mariner 9 areographic coordinate system, *J. Geophys. Res., 78,* 4395–4404, 1973.

Dobar, W. I., O. L. Tiffany, and J. P. Gnaedinger, Simulated extrusive magma solidification in vacuum, *Icarus, 3,* 323–331, 1961.

Dushman, S., *Scientific Foundation of Vacuum Technology,* 2nd ed., pp. 66–69, John Wiley, New York, 1962.

Farmer, C. B., D. W. Davies, A. L. Holland, D. D. LaPorte, and P. E. Doms, Mars: Water vapor observations from the Viking orbiters, *J. Geophys. Res., 82,* this issue, 1977.

Fjeldbo, G., D. Sweetnam, J. Brenkle, E. Christensen, D. Farless, J. Mehta, B. Seidel, W. Michael, Jr., A. Wallio, and M. Grossi, Viking radio occultation measurements of the Martian atmosphere and topography, Primary mission coverage, *J. Geophys. Res., 82,* this issue, 1977.

Flasar, F. M., and R. M. Goody, Diurnal behavior of water on Mars, *Planet. Space Sci., 24,* 161–181, 1976.

Gatley, I., H. H. Kieffer, E. Miner, and G. Neugebauer, Infrared observations of Phobos from Mariner 9, *Astrophys. J., 190,* 497–503, 1974.

Gear, A. E., and J. A. Bastin, A corrugated model for the lunar surface, *Nature, 196,* 1305, 1962.

Gierasch, P. J., and R. M. Goody, A study of the thermal and dynamical structure of the martian lower atmosphere, *Planet. Space Sci., 16,* 615–646, 1968.

Gifford, F., Jr., The surface-temperature climate of Mars, *Astrophys. J., 123,* 154–161, 1956.

Gillespie, A. R., and A. B. Kahle, The construction and interpretation of a digital thermal inertia image, *Photogramm. Eng. Remote Sensing,* in press, 1977.

Goody, R., and M. J. S. Belton, Radiative relaxation times for Mars: A discussion of martian atmospheric dynamics, *Planet. Space Sci., 15,* 247–256, 1967.

Hanel, R., B. Conrath, W. Hovis, V. Kunde, P. Lowman, W. Maguire, J. Pearl, J. Pirraglia, C. Prabhakara, B. Schlachman, G. Levin, P. Straat, and T. Burke, Investigation of the martian environment by infrared spectroscopy on Mariner 9, *Icarus, 17,* 423–442, 1972*a*.

Hanel, R. A., B. J. Conrath, W. A. Hovis, V. G. Kunde, P. D. Lowman, J. C. Pearl, C. Prabhakara, B. Schlachman, and G. V. Levin, Infrared spectroscopy experiment on the Mariner 9 mission: Preliminary results, *Science, 175,* 305–308, 1972*b*.

Harris, D. L., Photometry and colorimetry of planets and satellites, in *Planets and Satellites,* edited by G. P. Kuiper and B. M. Middlehurst, pp. 272–342, University of Chicago Press, Chicago, Ill., 1961.

Hartmann, W. K., Martian surface and crust: Review and synthesis, *Icarus, 19,* 550–575, 1973.

Herr, K. C., and G. C. Pimentel, Infrared absorptions near three microns recorded over the polar cap of Mars, *Science, 166,* 496–499, 1969.

Hess, S. L., R. M. Henry, C. B. Leovy, J. A. Ryan, J. E. Tillman, T. E. Chamberlain, H. L. Cole, R. G. Dutton, G. C. Greene, W. E. Simon, and J. L. Mitchell, Preliminary meteorological results on Mars from the Viking 1 lander, *Science, 193,* 788–791, 1976.

Ingersoll, A. P., Mars: The case against permanent CO_2 frost caps, *J. Geophys. Res., 79,* 3403–3410, 1974.

Kieffer, H. H., Soil and surface temperatures at the Viking landing sites, *Science, 194,* 1344–1346, 1976.

Kieffer, H. H., G. Neugebauer, G. Münch, S. C. Chase, Jr., and E. Miner, Infrared thermal mapping experiment: The Viking Mars orbiter, *Icarus, 16,* 47–56, 1972.

Kieffer, H. H., S. C. Chase, Jr., E. Miner, G. Münch, and G. Neugebauer, Preliminary report on infrared radiometric measurements from the Mariner 9 spacecraft, *J. Geophys. Res., 78,* 4291–4312, 1973.

Kieffer, H. H., S. C. Chase, Jr., E. D. Miner, F. D. Palluconi, G. Münch, G. Neugebauer, and T. Martin, Infrared thermal mapping of the martian surface and atmosphere: First results, *Science, 193,* 780–786, 1976*a*.

Kieffer, H. H., S. C. Chase, Jr., T. Z. Martin, E. D. Miner, and F. D. Palluconi, Martian north pole summer temperatures: Dirty water ice, *Science, 194,* 1341–1344, 1976*b*.

Kieffer, H. H., P. R. Christensen, T. Z. Martin, E. D. Miner, and F. D. Palluconi, Temperatures of the martian surface and atmosphere: Viking observation of diurnal and geometric variations, *Science, 194,* 1346–1351, 1976*c*.

Ksanfomaliti, L. V., and V. I. Moroz, Infrared radiometry on board Mars-5, *Cosmic Res., 13,* 65–67, 1975.

Leighton, R. B., and B. C. Murray, Behavior of carbon dioxide and other volatiles on Mars, *Science, 153,* 136–144, 1966.

Leovy, C., Note on thermal properties of Mars, *Icarus, 5,* 1–6, 1966.

Leovy, C. B., G. A. Briggs, A. T. Young, B. A. Smith, J. B. Pollack, E. N. Shipley, and R. L. Wildey, The martian atmosphere: Mariner 9 television experiment progress report, *Icarus, 17,* 373–393, 1972.

Loewenstein, R. F., D. A. Harper, S. H. Moseley, C. M. Telesco, H. A. Thronson, Jr., R. H. Hildebrand, S. E. Whitcomb, R. Winston, and R. F. Stiening, Far-infrared and submillimeter observations of the planets, *Icarus, 31*(3), 315–324, 1977.

Logan, L., S. R. Balsamo, and G. B. Hunt, Absolute measurements and computed values for martian irradiance between 10.5 and 12.5 μm, *Icarus, 18,* 451–458, 1973.

Lowell Observatory, Mars 1975–76, 1:25M mercator projection, ground-based albedo map of Mars, Flagstaff, Ariz., 1976.

MacDonald, T. L., The origins of martian nomenclature, *Icarus, 15,* 233–240, 1971.

Moroz, V. I., and L. V. Ksanfomaliti, Preliminary results of astrophysical observations of Mars from Mars-3, *Icarus, 17,* 408–422, 1972.

Moroz, V. I., L. V. Ksanfomaliti, G. N. Krasovskii, V. D. Davydov, N. A. Parfent'ev, V. S. Zhenulëv, and G. S. Filippov, Infrared temperatures and thermal properties of the martian surface measured by the Mars-3 orbiter, *Cosmic Res., 13,* 346–358, 1976.

Morrison, D., C. Sagan, and J. B. Pollack, Martian temperatures and thermal properties, *Icarus, 11,* 36–45, 1969.

Murray, B. C., and M. C. Malin, Polar wandering on Mars?, *Science, 179,* 997–1000, 1973.

Mutch, T. A., R. E. Arvidson, A. B. Binder, F. O. Huck, E. C. Levinthal, S. Liebes, Jr., E. C. Morris, D. Nummedal, J. B. Pollack, and C. Sagan, Fine particles on Mars: Observations with the Viking 1 lander cameras, *Science, 194,* 87–91, 1976*a*.

Mutch, T. A., A. B. Binder, F. O. Huck, E. C. Levinthal, S. Liebes, Jr., E. C. Morris, W. R. Patterson, J. B. Pollack, C. Sagan, and G. R. Taylor, The surface of Mars: The view from the Viking 1 lander, *Science, 193,* 791–801, 1976*b*.

Mutch, T. A., S. U. Grenander, K. L. Jones, W. Patterson, R. E. Arvidson, E. A. Guinness, P. Avrin, C. E. Carlston, A. B. Binder, C. Sagan, E. W. Dunham, P. L. Fox, D. C. Pieri, F. O. Huck, C. W. Rowland, G. R. Taylor, S. D. Wall, R. Kahn, E. C. Levinthal, S. Liebes, Jr., R. B. Tucker, E. C. Morris, J. B. Pollack, R. S. Saunders, and M. R. Wolf, The surface of Mars: The view from the Viking 2 lander, *Science, 194,* 1277–1283, 1976*c*.

National Aeronautics and Space Administration, Mars as viewed by Mariner 9, *NASA Spec. Publ. 329,* 149–161, 1974.

Neugebauer, G., G. Münch, S. C. Chase, Jr., H. Hatzenbeler, E. Miner, and D. Schofield, Mariner 1969: Preliminary results of the infrared radiometer experiment, *Science, 166,* 98–99, 1969.

Neugebauer, G., G. Münch, H. H. Kieffer, S. C. Chase, Jr., and E. Miner, Mariner 1969 infrared radiometer results: Temperatures and thermal properties of the martian surface, *Astron. J., 76,* 719–728, 1971.

Nier, A. O., and M. B. McElroy, Structure of the neutral upper atmosphere of Mars: Results from Viking 1 and Viking 2, *Science, 194,* 1298–1300, 1976.

Peterfreund, A. R., H. H. Kieffer, and F. D. Palluconi, Thermal inertia of the Elysium region of Mars, in *Lunar Science VIII,* pp. 765–767, Lunar Science Institute, Houston, Tex., 1977.

Pettit, E., and S. B. Nicholson, Measurements of the radiation from the planet Mars, *Pop. Astron., 32,* 601–608, 1924.

Pleskot, L. K., and H. H. Kieffer, The infrared photometric function of Mars and its bolometric albedo, *Icarus, 30,* 341–359, 1977.

Pohn, H. A., T. W. Offield, and K. Watson, Thermal inertia mapping from satellite-discrimination of geologic units in Oman, *J. Res. U.S. Geol. Surv., 2*(2), 147–158, 1974.

Pollack, J., and C. Sagan, An analysis of martian photometry and polarimetry, *Space Sci. Rev., 9,* 234–299, 1969.

Pollack, J. B., C. B. Leovy, Y. H. Mintz, and W. Van Camp, Winds on Mars during the Viking season: Predictions based on a general circulation model with topography, *Geophys. Res. Lett., 3,* 479–482, 1976*a*.

Pollack, J. B., R. Haberle, R. Greeley, and J. Iverson, Estimates of the wind speeds required for particle motion on Mars, *Icarus, 29,* 395–417, 1976*b*.

Pollack, J. B., D. Colburn, R. Kahn, J. Hunter, W. Van Kamp, C. E. Carlston, and M. R. Wolf, Properties of aerosols in the Martian atmosphere as inferred from Viking lander imaging data, *J. Geophys. Res., 82,* this issue, 1977.

Richtmeyer, R. P., and K. W. Morton, *Difference Methods for Initial-Value Problems*, 2nd ed., Wiley-Interscience, New York, 1967.

Roelof, E. C., Thermal behavior of rocks on the lunar surface, *Icarus, 8,* 138–159, 1968.

Sagan, C., O. B. Toon, and P. J. Gierasch, Climatic change on Mars, *Science, 181,* 1045–1049, 1973.

Scott, D. H., and J. W. Allingham, Geologic map of the Elysium quadrangle of Mars, Map I-935, U.S. Geol. Surv., Washington, D. C., 1976.

Scott, D. H., and M. H. Carr, Geologic map of Mars, U.S. Geol. Surv., Washington, D. C., 1976.

Seiff, A., and D. B. Kirk, Structure of Mars' atmosphere up to 100 km altitude from the entry measurements of Viking 2, *Science, 194,* 1300, 1976.

Seiff, A., and D. B. Kirk, Structure of the atmosphere of Mars in summer at mid-latitudes, *J. Geophys. Res.,* this issue, 1977.

Shorthill, R. W., R. E. Hutton, H. J. Moore II, R. F. Scott, and C. R. Spitzer, Physical properties of the martian surface from the Viking 1 lander: Preliminary results, *Science, 193,* 805–809, 1976*a*.

Shorthill, R. W., H. J. Moore II, R. E. Hutton, R. F. Scott, and C. R. Spitzer, The environs of Viking Lander 2, *Science, 194,* 1309–1318, 1976*b*.

Sinton, W., and J. Strong, Radiometric observations of Mars, *Astrophys. J., 131,* 459–469, 1960.

Snyder, C. W., The missions of the Viking orbiters, *J. Geophys. Res., 82,* this issue, 1977.

Soffen, G. A., and C. W. Snyder, The first Viking mission to Mars, *Science, 193,* 759–766, 1976.

Spudis, P., and R. Greeley, Surficial geology of Mars: A study in support of a penetrator mission to Mars, *Tech. Mem. TM X-73, 184,* NASA, Washington, D. C., 1976.

Thorpe, T. E., Mariner 9 photometric observations of Mars from November 1971–March 1972, *Icarus, 20,* 482–489, 1973.

Thorpe, T. E., Viking orbiter observations of atmospheric opacity during July–November 1976, *J. Geophys. Res., 82,* this issue, 1977*a*.

Thorpe, T. E., Viking orbiter photometric observations of the Mars phase function July–November 1976, *J. Geophys. Res., 82,* this issue, 1977*b*.

U.S. Geological Survey, Topographic map of Mars, Map I-961, U.S. Geol. Surv., Washington, D. C., 1976.

Veverka, J., and L. Wasserman, Effects of surface roughness on the photometric properties of Mars, *Icarus, 16,* 281–290, 1972.

Ward, W. R., B. C. Murray, and M. C. Malin, Climatic variations on Mars: Evolution of carbon dioxide atmosphere and polar caps, *J. Geophys. Res., 79,* 3387–3395, 1974.

Wechsler, A. E., and P. E. Glaser, Pressure effects on postulated lunar materials, *Icarus, 4,* 335–352, 1965.

Winter, O. F., and J. M. Saari, A particulate thermophysical model of the lunar soil, *Astrophys. J., 156,* 1135–1151, 1969.

Wright, E. L., Recalibration of the far-infrared brightness temperatures of the planets, *Astrophys. J., 210,* 250–253, 1976.

Young, A. T., High-resolution photometry of a thin planetary atmosphere, *Icarus, 11,* 1–23, 1969.

Young, A. T., and S. A. Collins, Photometric properties of the Mariner cameras and of selected regions on Mars, *J. Geophys. Res., 76,* 432–437, 1971.

(Received April 11, 1977;
revised June 7, 1977;
accepted June 7, 1977.)

VOL. 82, NO. 28 JOURNAL OF GEOPHYSICAL RESEARCH SEPTEMBER 30, 1977

The Viking Radio Science Investigations

W. H. MICHAEL, JR.,[1] R. H. TOLSON,[1] J. P. BRENKLE,[2] D. L. CAIN,[2] G. FJELDBO,[2] C. T. STELZRIED,[2] M. D. GROSSI,[3] I. I. SHAPIRO,[4] AND G. L. TYLER[5]

The Viking radio science investigations use the highly accurate radio tracking and communications systems data from the Viking orbiters and landers to perform a number of analyses concerning properties of Mars and its environment. This paper gives a general description of the investigations and of the instrumentation used; detailed results are presented in the companion papers.

The Viking Radio Science Team employs the radio tracking and communications systems data from the orbiters and landers to perform a number of investigations relating to properties of Mars and its environment. Specific objectives of the investigations can be categorized in three general areas, dynamical, surface, and internal properties of Mars; atmospheric and ionospheric properties of Mars; and solar system properties, as is illustrated in the appendix. Investigations that involve the locations of the landers and the dynamical properties of Mars primarily use radio tracking of the landers but also rely on radio tracking of the orbiters for calibration and auxiliary information. Determinations of the gravity field, figure, and atmospheric and ionospheric properties of Mars primarily use radio tracking of the orbiters, whereas the solar system and surface properties investigations use various combinations of both orbiter and lander radio tracking data. Here we provide general descriptions of these investigations and of the instrumentation used. Detailed results from several of the investigations are included in other papers in this issue. Additional results from analyses of the data obtained in the primary and extended mission will be reported as they become available.

The orbiter, lander, and ground-based tracking and communications instrumentation used for the radio science investigations has been described in some detail by *Michael et al.* [1972] and *Johnston et al.* [1977]. Each lander and orbiter is equipped with transponder circuitry which provides two-way Doppler and range data at *S* band (~2.3 GHz) frequencies. Each orbiter is also equipped with an *X* band (~8.4 GHz) system which provides *X* band signals on the downlink, coherent with the *S* band signals, and which also yields both Doppler and range data. These dual-frequency downlinks from the orbiters are especially valuable for making dispersive measurements and for providing calibration data. Signal amplitude data on the lander-to-orbiter data relay link, measured at the orbiter, are used for the surface properties investigation. Ground-based instrumentation includes the transmitting, receiving, and data collection facilities of the NASA Deep Space Network at the 64- and 26-m stations in California, Australia, and Spain. Use is also made of the NASA ground communications and data processing facilities.

In general, the radio science investigations make use of instrumentation and facilities which are required by the Viking project for trajectory and orbit determination, spacecraft control, data transmission, etc., and which are thus available for scientific purposes. The major exception for Viking is the *X* band downlink on the orbiters, which has been included to enhance the radio science capabilities and to conduct communications experiments.

Viking radio science operations began during the cruise phase of the mission, when the orbiter high-gain antenna was activated and *S* and *X* band tracking data became available. Activities during this period were mostly related to check-out of the systems and of data processing procedures, with some effort devoted to data and systems calibration. The dual-frequency dispersive data collected during the cruise phase were used to monitor the electron content between the spacecraft and the earth. Several enhanced electron density events were detected [*Winn et al.,* 1977], and these have been catalogued for use in more detailed analyses of the interplanetary medium.

Shortly after the successful touchdown of the first Viking lander on Mars, Doppler and range data became available for the first time between the earth and a spacecraft on a planetary surface. With use of the first few days of tracking data it was possible to determine the location of the lander, the radius of Mars at the landing site, and the orientation of the spin axis of Mars. Additional data from both landers improved the accuracy of such determinations and also led to a first determination from Viking data of the spin rate of Mars [*Michael et al.,* 1976*a*, *b*]. The most recent results from analyses of lander tracking data are given by *Mayo et al.* [1977], who also discuss other parameters which can be determined from these data. Of particular interest is the possibility of the determination of spin axis precession and nutation from long arcs of lander tracking data. These spin axis motion constants, combined with low-degree gravitational field parameters, will provide useful constraints on models for the variations of density in the interior of Mars, which bear on its origin and evolution.

For some time after each landing the period of the corresponding orbiter was synchronized such that at periapse the orbiter passed over the lander and could receive lander data for transmission to earth. Signal amplitude data from the Lander 1 to Orbiter 1 relay link, measured at the orbiter when it rose and set as viewed from the lander, have been analyzed to estimate the dielectric constant of the surface. Preliminary results of these analyses [*Tang et al.,* 1977] indicate that the surface material around Lander 1 has a dielectric constant similar to that for pumice or tuff.

Several times during the primary Viking mission, maneuvers were made to change the orbital period so that it was nonsynchronous with the Mars rotation period; thus the orbiters could make longitudinal surveys of the planet. Tracking data

[1] NASA Langley Research Center, Hampton, Virginia 23665.
[2] Jet Propulsion Laboratory, California Institute of Technology, Pasadena, California 91103.
[3] Raytheon Company, Sudbury, Massachusetts 01776.
[4] Massachusetts Institute of Technology, Cambridge, Massachusetts 02139.
[5] Stanford University, Stanford, California 94305.

Paper number 7S0383.

obtained at such times, particularly near periapsis passages, are analyzed to determine the gravitational field of Mars. The Viking orbiters, unlike Mariner 9, have periapses in the northern hemisphere. A new determination of the gravitational field, based on a combination of Viking data and Mariner 9 data, has been made by *Gapcynski et al.* [1977]. The results indicate improved estimates of the coefficients in the harmonic expansion of the gravitational potential of Mars, particularly those that relate to the potential for the northern hemisphere, where the Viking data are most pertinent.

As lander and orbiter tracking data have accumulated, improvements have been made in our knowledge of the precise distance between Mars and the earth and thus of the ephemerides of the earth and Mars and of the value of astrodynamic constants.

Another type of measurement, the very long base line interferometry (VLBI) measurement of the angular separation of the Viking orbiters with respect to distant quasar sources, serves to determine the orientation of the orbits of Mars and the earth with respect to an inertial reference frame. VLBI measurements are being made periodically during the Viking mission, and when they are available over a period of many years, they will be used to determine the relativistic advance of the perihelion of Mars' orbit as a further test of the theory of general relativity. The measurements of the range between the earth and the landers will also be most useful in the determination of this perihelion advance.

Starting in early October 1976 and continuing for about a month, the orientation of the orbit of Orbiter 1 was such that the spacecraft, during a portion of each orbit, passed behind Mars as seen from the earth, and earth occultations resulted. *S* and *X* band measurements were made of the variations in frequencies of the signals immediately preceding and following occultations for about 50 locations on Mars with latitudes from about 80° south to 70° north; such measurements have been used to determine the radii to the surface at the occultation points and to determine the profiles of atmospheric and ionospheric properties at these points [*Fjeldbo et al.,* 1977]. The mean atmospheric pressure at the surface agrees well with the pressures measured at the landing sites. The radii determined for these additional locations will also allow a better estimate to be made of the figure of Mars.

For a period of several weeks before and after Mars was in conjunction with the sun (November 25, 1976), a sustained effort was made to obtain measurements of the range to each lander and simultaneous measurements of the range to an orbiter at both *S* and *X* band frequencies to perform a time delay test of the theory of general relativity. The unique combination of ranging to the landers fixed on the surface of Mars, and thus at a precisely known position, and of dual-frequency ranging to the orbiters to provide a calibration of the disturbing effects introduced by the solar corona should lead to the most precise test yet achieved [*Shapiro et al.,* 1977]. Preliminary analysis of the data has confirmed the delay predicted by Einstein's theory of general relativity to an accuracy of better than 0.5%. If plasma-calibrated ranging to the landers is continued throughout the planned extended mission, the predictions can be tested to an accuracy of about 0.1%.

During the conjunction period, excellent data were also obtained for studies of the solar corona, including data on small-scale structure within at least 0.8 and 0.3 R_S of the photosphere at *S* and *X* band wavelengths, respectively. These observations are the first of this type involving direct solar conjunction. Since conjunction occurred at solar minimum, the data are free from the usual confusion caused by active regions of the sun. Preliminary analysis of the dispersive range data shows a strong asymmetry, with respect to conjunction, in the plasma density integrated along the propagation path. This asymmetry is evidently the result of a latitudinal variation in the plasma density [*Tyler et al.,* 1977], which can be more fully evaluated with further analysis of the existing Viking data.

In February 1977 the relative orbital orientations of Viking Orbiter 1 and Phobos, the inner satellite of Mars, were such that it was possible to perform maneuvers resulting in small changes in the orbital period and orientation of the orbiter to effect a series of close flybys of Phobos. One objective was to determine the mass of Phobos by analysis of the perturbations that it introduced in the orbit of the Viking orbiter; other objectives were to obtain high-resolution photography and thermal measurements. From analysis of the orbiter tracking data from nine passes of the orbiter within about 200 km of Phobos, a preliminary determination has been made that the mass of Phobos is about 1×10^{19} g (W. T. Blackshear and G. M. Kelly, private communication, 1977). The present uncertainty in our knowledge of the volume of Phobos prevents the determination of a useful value for the density of Phobos, but it is expected that additional Viking orbiter photographs of Phobos during future encounters will improve this situation.

Most of the Viking radio science experiments, as indicated, are being continued during the extended mission. The additional data which are becoming available will increase the precision of many of the results, and some of the investigations will benefit considerably from the extended mission data. One reason for this is that the signal paths are moving away from the sun as Mars proceeds toward opposition, the disturbing effects introduced by the near-solar environment thus being decreased. Other reasons are that many additional occultations are being obtained for both orbiters, which could provide seasonal variation atmospheric data; the periapses of the orbiters have been lowered from about 1500 to 300 and 800 km for Orbiters 1 and 2, respectively, the data for both global and local gravitational analyses thus being enhanced; flybys of Deimos, the outer satellite of Mars, are feasible and could lead to determination of its mass; and in some cases the decreased earth to Mars distance is an enhancement to or is necessary for an investigation, such as bistatic radar studies of the surface properties of Mars. Thus it is expected that a number of additional results will be obtained from these investigations during the remaining part of the Viking mission.

APPENDIX: VIKING RADIO SCIENCE INVESTIGATIONS

Dynamical, surface, and internal properties of Mars
- Spin axis orientation and motion
- Spin rate
- Gravity field
- Figure
- Surface dielectric constant

Atmospheric and ionospheric properties of Mars
- Pressure, temperature, and density altitude profiles
- Electron number density altitude profiles

Solar system properties
- Ephemerides of Mars and earth
- Masses of Martian satellites
- Interplanetary medium
- Solar corona
- Tests of general relativity

Acknowledgments. We thank all those members of the Viking Flight Team, the Deep Space Network, the Viking Navigation Team, and other elements of the Viking project who have been involved in the radio science investigations for their dedicated contributions, assistance, and support. This work was supported by the National Aeronautics and Space Administration.

REFERENCES

Fjeldbo, G., D. N. Sweetnam, J. P. Brenkle, E. J. Christensen, D. L. Farless, J. S. Mehta, B. Seidel, W. H. Michael, Jr., H. A. Wallio, and M. D. Grossi, Viking radio occultation measurements of the Martian atmosphere and topography: Primary mission coverage, *J. Geophys. Res., 82,* this issue, 1977.

Gapcynski, J. P., R. H. Tolson, and W. H. Michael, Jr., Mars gravity field: Combined Viking and Mariner 9 results, *J. Geophys. Res., 82,* this issue, 1977.

Johnston, D. W., T. W. Howe, and G. M. Rockwell, Viking mission support, Deep Space Network, *Progr. Rep. 42-37,* pp. 12–25, Jet Propul. Lab., Pasadena, Calif., 1977.

Mayo, A. P., W. T. Blackshear, R. H. Tolson, W. H. Michael, Jr., G. M. Kelly, J. P. Brenkle, and T. A. Komarek, Lander locations, Mars physical ephemeris, and solar system parameters; Determination from Viking lander tracking data, *J. Geophys. Res., 82,* this issue, 1977.

Michael, W. H., Jr., D. L. Cain, G. Fjeldbo, G. S. Levy, J. G. Davies, M. D. Grossi, I. I. Shapiro, and G. L. Tyler, Radio science experiments: The Viking Mars orbiter and lander, *Icarus, 16,* 57–63, 1972.

Michael, W. H., Jr., R. H. Tolson, A. P. Mayo, W. T. Blackshear, G. M. Kelly, D. L. Cain, J. P. Brenkle, I. I. Shapiro, and R. D. Reasenberg, Viking lander location and spin axis of Mars: Determination from radio tracking data, *Science, 193,* 803, 1976*a*.

Michael, W. H., Jr., A. P. Mayo, W. T. Blackshear, R. H. Tolson, G. M. Kelly, J. P. Brenkle, D. L. Cain, G. Fjeldbo, D. N. Sweetnam, R. B. Goldstein, P. E. MacNeil, R. D. Reasenberg, I. I. Shapiro, T. I. S. Boak III, M. D. Grossi, and C. H. Tang, Mars dynamics, atmospheric and surface properties: Determination from Viking tracking data, *Science, 194,* 1337, 1976*b*.

Shapiro, I. I., R. D. Reasenberg, P. E. MacNeil, R. B. Goldstein, J. P. Brenkle, D. L. Cain, T. Komarek, A. Zygielbaum, W. F. Cuddihy, and W. H. Michael, Jr., The Viking relativity experiment, *J. Geophys. Res., 82,* this issue, 1977.

Tang, C. H., T. I. S. Boak III, and M. D. Grossi, Measurements of Mars surface electrical properties by Viking lander-to-orbiter relay link, *J. Geophys. Res., 82,* this issue, 1977.

Tyler, G. L., J. P. Brenkle, T. A. Komarek, and A. I. Zygielbaum, Viking solar corona experiment, *J. Geophys. Res., 82,* this issue, 1977.

Winn, F. B., S. C. Wu, T. A. Komarek, V. W. Lam, H. N. Royden, and K. B. W. Yip, A solar plasma stream measured by DRVID and dual-frequency range and Doppler radio metric data, Deep Space Network, *Progr. Rep. 42-37,* pp. 43–54, Jet Propul. Lab., Pasadena, Calif., 1977.

(Received April 1, 1977;
accepted May 6, 1977.)

Lander Locations, Mars Physical Ephemeris, and Solar System Parameters: Determination From Viking Lander Tracking Data

A. P. MAYO, W. T. BLACKSHEAR, R. H. TOLSON, AND W. H. MICHAEL, JR.

NASA Langley Research Center, Hampton, Virginia 23665

G. M. KELLY

Analytical Mechanics Associates, Inc., Hampton, Virginia 23665

J. P. BRENKLE AND T. A. KOMAREK

Jet Propulsion Laboratory, Pasadena, California 91103

Radio tracking data from the Viking landers have been analyzed to determine the parameters of the Mars physical ephemeris, the radii of Mars at the landing sites, and the lander locations. The orientation of the Mars rotation axis, referred to the 1950.0 earth mean equator, equinox, and epoch, was determined to be 317.340 ± 0.003° right ascension and 52.710 ± 0.002° declination. The planet's rotation period was determined to be 24 h, 37 min, 22.663 ± 0.002 s. Analyses indicate that the determination of the motions of the Mars rotation axis will require additional tracking data. The Mars radii at the sites of landers 1 and 2 are 3389.38 ± 0.06 km and 3381.91 ± 0.08 km, respectively. The areocentric location of lander 1 is 22.272 ± 0.002°N, 47.94 ± 0.2°W. The lander 2 location is 47.670 ± 0.002°N, 225.71 ± 0.2°W. The areocentric right ascensions of the landers are determined to be 277.314 ± 0.002° for lander 1 and 99.546 ± 0.002° for lander 2 at 0000 hours, January 1, 1977 (Julian date 2443144.5). Possible determinations of relativity parameters, solar oblateness, asteroid mass, and variations of the universal gravitational constant, from their effects on the planetary motions, will require the additional tracking data of the Viking extended mission.

INTRODUCTION

The lander tracking data contain information on the physical ephemeris of Mars (rotation axis orientation, rotation rate, precession, nutation), orbits of Mars and the earth, and parameters affecting the orbital motions. Approximately 6 months of data have been analyzed to obtain the results presented in this paper. The results obtained to date are the lander locations, the Mars rotation axis orientation, and the rotation rate. The analyses indicate that the additional tracking data of the Viking extended mission are required before significant improvements in the Mars rotation axis motion and parameters affecting the orbital motion can be obtained.

DATA SUMMARY

The lander tracking data analyses presented have used approximately the first 6 months of S band range and counted Doppler measurements made to both landers by the NASA Deep Space Network tracking stations at Goldstone, California (Deep Space Station (DSS) 14), Canberra, Australia (DSS 43), and Madrid, Spain (DSS 63). Summaries of the data from each station, as used in the analysis, are given in Table 1. Over the approximately 6-month total data interval the data consisted of 2387 Doppler points for Viking lander 1 (VL 1) and 704 points for lander 2 (VL 2). Eighty-two range measurements were obtained for VL 1 and VL 2. The first 3 months of data contained an average of about 25 Doppler points per sol for VL 1 and about 15 points per sol for VL 2. In this interval, 16 range measurements were made to VL 1 and 28 to VL 2. On November 25, 1976, about 4 months after VL 1 touchdown, Mars was in conjunction with the sun. Near conjunction the lander ranging data were affected by the charged particles of the interplanetary medium, the solar corona, and relativity. These effects on the tracking data near conjunction (mid-October 1976 to mid-January 1977) will require further analysis. Lander data during this interval were deleted in the data processing. Approximately 3 weeks of data subsequent to conjunction were combined with the preconjunction data in the analyses.

SOFTWARE DESCRIPTION

The lander tracking data software has the capability to estimate the parameters defining the Mars physical ephemeris, the planetary ephemeris of the earth and Mars, and parameters affecting the latter. The statistics are generated by considering significant error sources. For the lander data analysis the estimated parameters are the latitude, longitude, and radius of Mars at the lander locations; the right ascension α_0 and declination δ_0 of the Mars rotation axis; and the Mars sidereal rotation rate $\dot{V}$. The axis orientation angles are referenced to the earth equator and equinox of 1950.0 and the epoch of 0000 hours, January 1, 1950. The orientation at any time is expressed [*de Vaucouleurs*, 1964] as

$$\alpha = \alpha_0 + \mu \sin I \cos \Delta \,(\sec \delta) T \quad (1)$$

$$\delta = \delta_0 + \mu \sin I \,(\sin \Delta) T \quad (2)$$

where μ is the precession constant in degrees per century, I is the Mars mean obliquity, Δ is the angle along the Mars mean equator from its ascending node on the earth mean equator of 1950.0 to the Mars autumn equinox, and T is Julian centuries since epoch. The Mars hour angle V at any time is given by

$$V = V_0 + \dot{V} d \quad (3)$$

where d is days since the epoch of V_0. The values of μ and V_0 were −708 are sec/century [*Lowell*, 1914] and 4.376° on 0000 hours, November 27, 1971 (Julian date (J.D.) 2441282.5) [*de*

Paper number 7S0426.

TABLE 1. Summary of Lander Tracking Data Used in the Analysis

Station	Lander	No. Points Two-Way Doppler*	No. Points Mu 2 Range*	No. Points Plop Range*
14	1	576	0	NA
	2	264	9	NA
43	1	991	NA	19
	2	230	NA	21
63	1	820	NA	24
	2	210	NA	9

The data intervals are July 20–October 11, 1976, and January 9–30, 1977. The solar conjunction region is deleted.

*Mathematical definitions of these measurements are given by *Khatib et al.* [1972].

Vaucouleurs et al., 1973], respectively. The Mars mean orbit was that of *Sturms* [1971], and the earth and Mars planetary ephemerides were obtained from Jet Propulsion Laboratory ephemeris tape DE 96 [*Standish et al.*, 1976]. The angles I and Δ are given in terms of the axis orientation angles and the mean orbit by *Her Majesty's Nautical Almanac Office* [1961]. The range and Doppler observables are expressed by *Moyer* [1971] and *Khatib et al.* [1972]. Batch processing, a Bayes estimation technique, and statistics involving considered parameters [*Moyer*, 1971] were used. Any solution parameter could be included in the considered as well as the solve-for category.

The statistics on the estimated parameters considered the uncertainties of the tracking station locations, precession constant, nutations, earth and Mars orbits, and location of crater Airy-0 defining the prime meridian, as well as range, Doppler noise, and bias. The data from each lander could be processed separately or in combination to obtain estimates of the solution parameters.

The software also has the additional capabilities to solve for the tracking station locations, astronomical unit, corrections to the earth and Mars orbits, Mars precession constant, Mars nutation amplitudes, Mars pole wander, solar oblateness, variations in the universal gravitational constant, asteroid mass, and relativity parameters affecting the tracking signal and the orbital motions. These capabilities are applicable mainly to the additional data that will be obtained during the extended mission.

DATA PROCESSING TECHNIQUE

The lander tracking data were processed at the end of the first 3 days [*Michael et al.*, 1976*a*], 3 months [*Michael et al.*, 1976*b*], and 6 months. The first 3 days of data obtained from VL 1 were processed to solve for the Mars spin axis orientation and the lander location. Processing the longer arcs of VL 1 data and the VL 2 data in a simultaneous solution yielded more accurate values of the lander locations, the rotation axis orientation, and an improved spin rate of Mars. Six months of lander tracking data alone did not contain sufficient orbit information to allow accurate simultaneous evaluation of the earth and Mars orbits and the Mars rotation axis orientation, the rotation rate, and the lander locations. In order to obtain the Mars physical ephemeris parameters with high accuracy the orbiter range data were used to evaluate the effects of ephemeris errors on the tracking data, and the corrections were applied to the lander data as suggested by *Blackshear et al.* [1973]. This technique permitted the solution parameters of the lander data analysis to be limited to lander locations, Mars rotation axis orientation, and rotation rate.

The ephemeris corrections to the lander data were made in the following manner. By using the sixth-order Mariner 9 gravity field of Mars [*Daniels and Tolson*, 1976] some 18 individual orbits of Viking 1, distributed over the data interval, were determined from the orbiter Doppler data, and the orbiter range residuals were established. After calibration for hardware delays [*Komarek*, 1976] and charged particle effects the residuals were attributed to errors in the pre-Viking Mars and earth ephemerides. Since these effects are similar for both the orbiter and the lander, the orbiter range residuals were used to calibrate the lander range data for ephemeris errors. Specifically, the orbiter ranging residuals, relative to ephemeris DE 96, were least squares fitted with a quartic time polynomial as

$$\Delta\rho = -956.1 + 34.0738t - 0.01335t^2 - 0.0006421t^3 + 0.0000006357t^4 \quad (4)$$

where $\Delta\rho$ is the increment, in meters, to be added to the range computed from the pre-Viking ephemeris to obtain measured range and t is the number of days from 0000 hours, day 180, 1976, ephemeris time. The root mean square of the fit to the range residuals was 9 m. The polynomial was evaluated at the lander range and Doppler observation times in order to calibrate the lander tracking data for ephemeris errors.

The lander tracking data are corrected for the effect of the earth's troposphere and the small motions of the tracking stations due to the wandering of the earth's pole. Differences between measurements of station time, universal time, and atomic time are also considered. The earth pole wander values and polynomials relating the types of time were obtained by the Tracking System Analytic Calibration (TSAC) group [*Fliegel and O'Neil*, 1973] from data obtained from the Bureau International de l'Heure in Paris. The troposphere zenith range corrections were determined by the TSAC group from tracking station troposphere data. Elevation effects were applied, and range and Doppler corrected, for the effects of the earth's troposphere [*Madrid et al.*, 1974].

The total charged particle effects on the range data cannot be determined directly from the lander data, since the landers have only S band radio systems. The orbiter ranging signals, however, traversed almost the same charged particle environment. The charged particle calibrations were determined for the orbiter from its dual frequency S and X band ranging data [*Madrid*, 1974] and were then interpolated to the lander ranging times and used to calibrate the lander range data. When the data deleted near solar conjunction are excluded, the accuracy of the charged particle calibrations for the lander range

TABLE 2. A Priori Values and Prelander Standard Deviations of Solution Parameters Used in the Analysis of 3 Days of Data

Parameter	A Priori Value	A Priori Standard Deviation
Mars spin axis		
α_0, deg	317.32	0.2
δ_0, deg	52.68	0.2
Lander 1 location*		
u_1, km	3138.0017	25.0
v_1, km	1293.3910	60.0
λ_1, deg W	47.5	2.5

*Here u and v are components perpendicular and parallel, respectively, to the Mars rotation axis and λ is the longitude.

TABLE 3. A Priori Values and Standard Deviations of Considered Parameters Used in the Data Analysis

Parameter	A Priori Value	A Priori Standard Deviation
Mars precession constant, ″/century	−708.0	50.0
Mars dynamical ellipticity $(C - A)/C$*	0.0052	0.0005
Tracking station locations†		
u, km		0.0015
v, km		0.0150
λ, deg		3.0×10^{-5}
Mu 2 range noise, range unit‡		12800.0
Plop range noise, range unit§		125.0
Plop range bias, range unit‖		300.0
Counted Doppler noise, Hz		0.015
Counted Doppler bias, Hz		0.01

*The nutations depend on the dynamical ellipticity $(C - A)/C$, where C is the Mars polar moment of inertia and A is the equatorial moment.

†Here u and v are components perpendicular and parallel, respectively, to the earth's 1903.0 pole and λ is the longitude. See *Jet Propulsion Laboratory* [1976] for their a priori values.

‡One range unit approximately equals 0.0011 m for Mu 2.

§One range unit approximately equals 0.15 m for Plop.

‖ Includes data bias and bias equivalent of errors in ephemeris and charged particle calibrations using the orbiter data.

measurement is estimated at about 30 m. The effect of the mean charged particles on the Doppler data was small, and Doppler calibrations for charged particle effects were not made. The Doppler variations due to the highly dynamic portion of the charged particle environment were also probably small as evidenced by small scatter in the Doppler residuals.

The a priori estimates and statistics for the analysis of 3 days of data are given in Table 2. A priori statistics on the solution parameters were not used in the analysis of 3 months and 6 months of data. The statistics of the considered error sources are given in Table 3. The residuals for the approximately 6-month data arc are shown in Figures 1 and 2. The Doppler residuals are generally less than 0.03 Hz, and the range residuals are less than 60 m.

ROTATION AXIS ORIENTATION

Numerous determinations of the orientation of the Mars axis of rotation have been made since Schiaparelli obtained an estimate in 1886 from a decade of observing the position angles of the polar caps. Many determinations have been made from earth-based observations of surface markings and the natural satellites, Phobos and Deimos. The Mariner 9 photographs and tracking data of 1971 have also yielded estimates. The rotation axis orientation, obtained from the lander track-

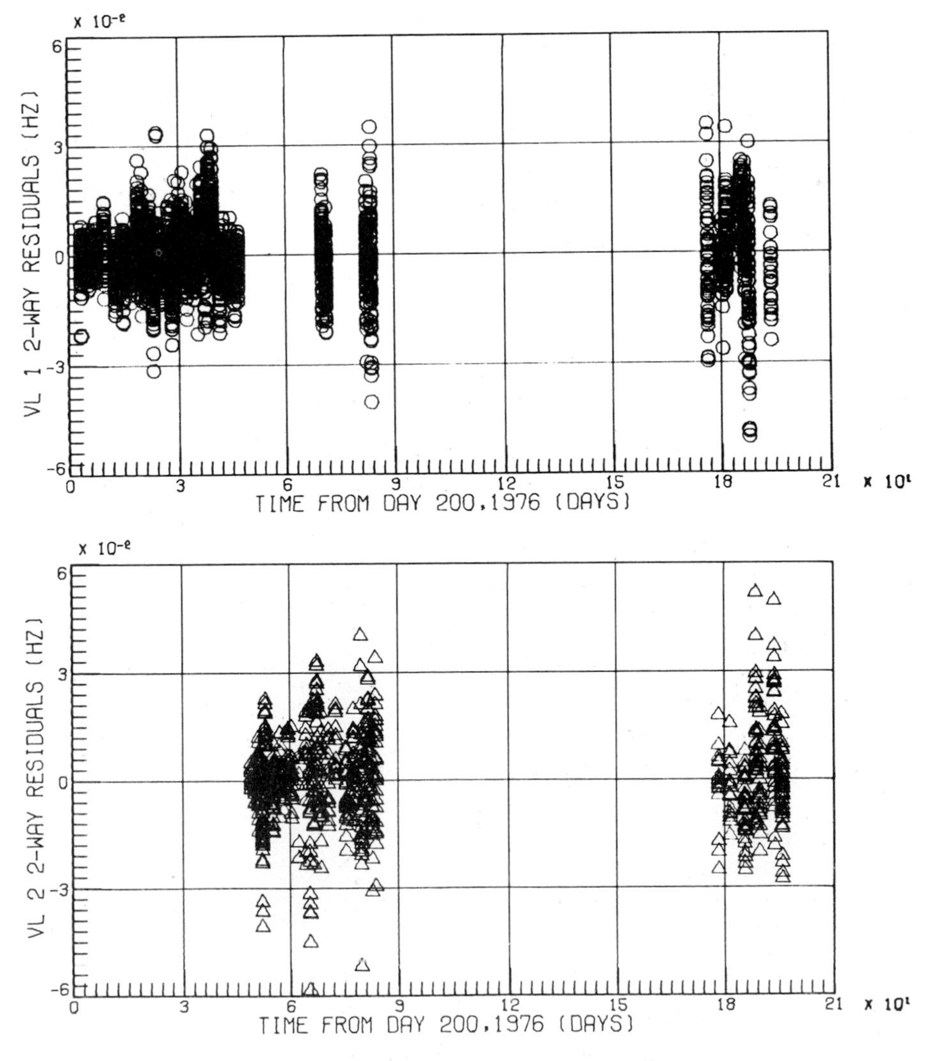

Fig. 1. Doppler residuals for processed lander data.

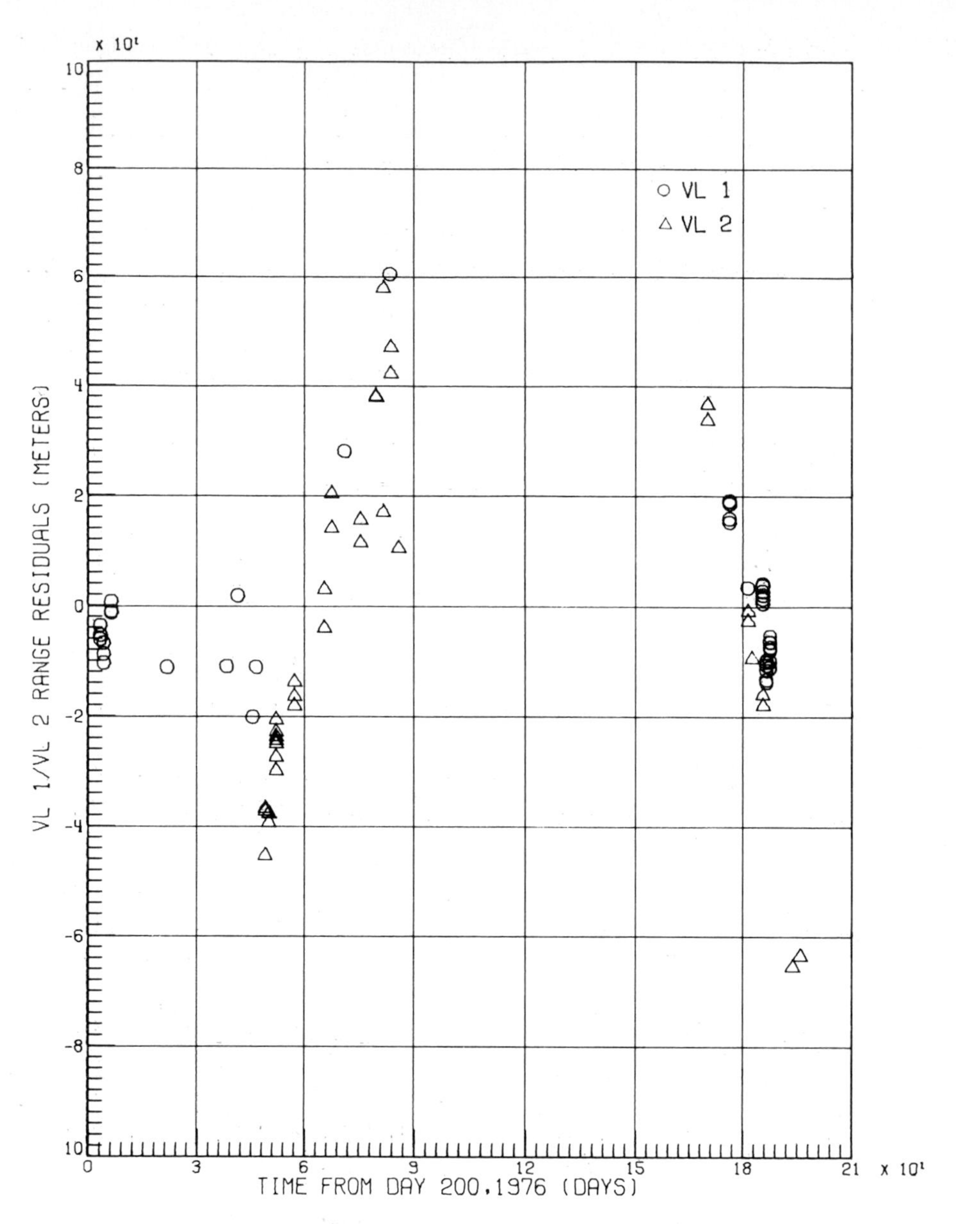

Fig. 2. Range residuals for processed lander data.

ing data, is compared in Figure 3 with the axis orientations obtained from Mariner 9 radio and landmark tracking [*Born et al.*, 1972], earth-based observations of Phobos and Deimos from 1877 to 1969 [*Sinclair*, 1972], and post-Mariner 9 analyses [*de Vaucouleurs et al.*, 1973]. With the exception of the results of de Vaucouleurs et al., for which no specific uncertainty is given, the circles and ellipses shown in the figure are the uncertainties and demonstrate the high accuracy of the lander results. This accuracy of axis orientation is equivalent at the surface of Mars to a 200-m uncertainty in the position of the rotation axis. The axis orientation estimates shown in Table 4 for analyzing approximately 3 days, 3 months, and 6 months of lander data demonstrate the ability of the data to give consistent values of spin axis orientation. The values shown in the table are for the 0000 hours, January 1, 1950, epoch. The Mars rotation axis orientation, at any epoch, was determined to be

$$\alpha = 317.340° - 0.10106T \quad (5)$$

$$\delta = 52.710° - 0.05706T \quad (6)$$

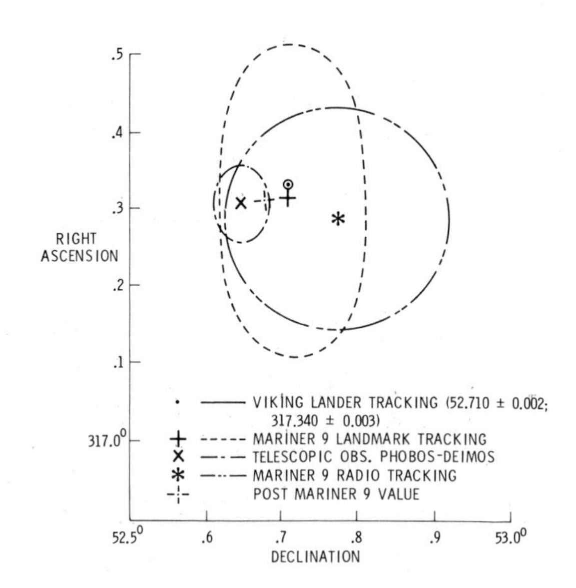

Fig. 3. Mars rotation axis orientation for the 1950.0 epoch.

TABLE 4. Mars Rotation Axis Orientation, Rotation Rate, and Lander Location Solutions for Various Amounts of Data

Estimated Parameter	Data Arc Length		
	3 sols at VL 1*	81 sols at VL 1, 35 sols at VL 2	188 sols at VL 1, 144 sols at VL 2
α_0, deg	317.35 ± 0.06	317.34 ± 0.006	317.340 ± 0.003
δ_0, deg	52.71 ± 0.01	52.710 ± 0.004	52.710 ± 0.002
$\dot{V}$, deg/d		350.891986 ± 12 × 10^{-6}	350.891986 ± 7 × 10^{-6}
r_1, km	3389.5 ± 0.3	3389.38 ± 0.08	3389.38 ± 0.06
φ_1, deg N†	22.27 ± 0.02	22.272 ± 0.006	22.272 ± 0.002
λ_1, deg W	48.0 ± 0.2‡	47.94 ± 0.2	47.94 ± 0.2
r_2, km		3381.88 ± 0.22	3381.91 ± 0.08
φ_2, deg N†		47.669 ± 0.006	47.670 ± 0.002
λ_2, deg W		225.71 ± 0.2	225.71 ± 0.2

*A sol is the interval from Martian midnight to the next midnight.
†Areocentric.
‡This uncertainty is with respect to the location of crater Airy-0. The longitude uncertainty is 0.07° when referenced to a mathematically defined prime meridian located 148.24° right ascension at 0000 hours, January 1, 1950.

where T is measured in Julian centuries from 0000 hours, January 1, 1950, J.D. 2433282.5. The right ascension and declination rates were determined by using the *Lowell* [1914] precession constant.

MOTION OF THE AXIS OF ROTATION

The total motion of the Mars axis of rotation is composed of both precession and nutation components. The Mars equinox precesses westward about 7 arc sec/earth year, amounting to an equivalent equinox motion at the Mars radius of 115 m/yr. This motion converts to a yearly change of 0.001° in right ascension of the Mars axis of rotation and 0.0006° in declination as measured in the earth equator and equinox of the 1950.0 system. These yearly motions are smaller than the present accuracies of the right ascension and declination of the axis shown in Table 4. However, the major source of the 1950 axis location error is the uncertainty in the precession constant, used to map the axis to the 1950 epoch. The uncertainty in the precession constant was assumed to be 50″/century, which is conservatively larger than the 30″/century difference between the precession constants of *Lowell* [1914] and *Lorell et al.* [1973]. The nutations of Mars have a miximum equivalent surface motion amplitude of about 30 m in longitude and 10 m in obliquity, with a period slightly less than 1 earth year. The additional tracking data of the extended mission will improve the analyses and may yield measurements of the axis motion.

MARS ROTATIONAL PERIOD

The rotational period of Mars is determined very effectively from tracking landed spacecrafts. The rotation period has also been determined from earth-based observations spanning some 3 centuries. Wislicenus analyzed the earth-based observations made from 1659 to 1881 and established the sidereal rotational period as 24 h, 37 min, 22.655 ± 0.013 s. The results of the lander analysis, along with those of Wislicenus, Bakhuyzen, and Ashbrook, as obtained from earth-based observations [*Michaux,* 1967], and those of *de Vaucouleurs et al.* [1973], obtained from Mariner 9 data, are compared in Figure 4. The dots shown in the figure are the values of the rotational period, and the vertical bars are the uncertainties.

The Viking lander results give the sidereal rotational period as 24 h, 37 min, 22.663 s, some 8 ms longer than the period given by de Vaucouleurs et al. and 6 ms shorter than the estimate of Ashbrook. The consistency of the Viking lander to give this rotation period with 3 months of data and 6 months of data is evident in the results shown in Table 4.

LANDER LOCATIONS

The lander locations as determined from different arc lengths of lander tracking data are shown in Table 4. The lander latitudes, longitudes, and radii are shown in the table, although the lander perpendicular distance from the Mars spin axis, distance from the equator parallel to the rotation axis, and longitude were actually estimated from the data. The distance of the landers from the Mars rotation axis and the right ascension of the landers are determined by both the Doppler and the range data. The Doppler data are relatively insensitive to the component of the lander position parallel to the Mars spin axis. This component is determined by the range data, and its accuracy is influenced by the accuracy with which the ephemeris and charged particle calibrations were made.

The lander tracking data contain no information on the

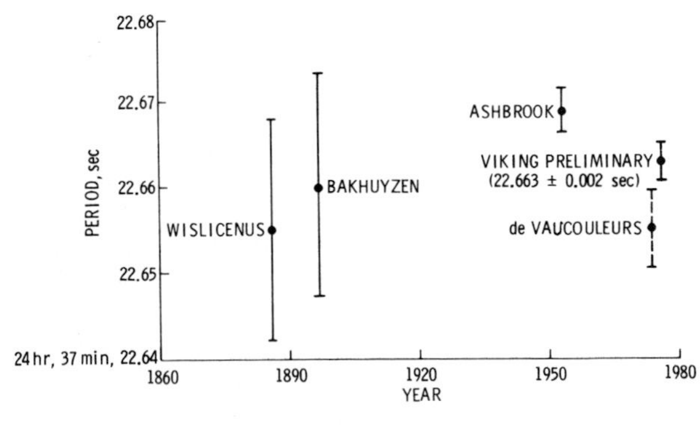

Fig. 4. Mars sidereal rotation period.

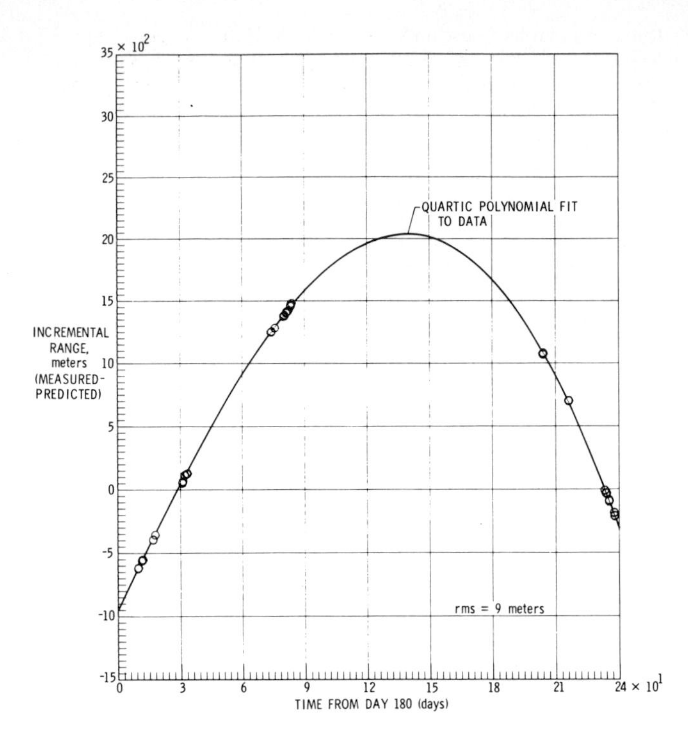

Fig. 5. Comparisons of measured range to Mars and ephemeris predictions.

location of the Mars prime meridian. The right ascension of the Mars prime meridian, passing through crater Airy-0, was determined from the Mariner 9 photographs to be at right ascension 4.376° at 0000 hours, November 27, 1971 (J.D. 2441282.5) [*de Vaucouleurs et al.*, 1973]. The uncertainty in the right ascension of the crater at the 1971 epoch was 10 km [*Davies and Arthur*, 1973]. Owing to the uncertainty of the Mars spin rate since 1971, the prime meridian uncertainty at the 1976 epoch is about 11 km. Until improved knowledge is obtained on the right ascension of crater Airy-0, the longitudes of the landers are limited to an accuracy of about 11 km or 0.2°. The right ascensions of the landers, in contrast, are determined to an accuracy of 0.002° over the data arc and are 277.314° (VL 1) and 99.546° (VL 2) at 0000 hours, January 1, 1977 (J.D. 2443144.5).

Planetary Ephemeris and Heliocentric Parameters

The lander tracking data contain information on the earth and Mars planetary ephemerides as well as on the Mars physical ephemeris. The influence of planetary ephemeris errors was removed from the analyses through the use of the orbiter range data, as was previously discussed, and the remaining information on the Mars physical ephemeris was analyzed. The differences between the measured range to Mars, as reduced from the orbiter data, and the range to Mars, as predicted by the pre-Viking development ephemeris DE 96, are shown in Figure 5. The additional data of the extended Viking mission may permit a combined simultaneous solution for both the physical and the planetary ephemeris. In this case the analyses would investigate the parameters affecting the planetary ephemeris such as the solar oblateness, variation of the universal gravitational constant, asteroid mass, and relativistic effect on planetary motion, as well as the determination of the elements of the earth and Mars orbits.

Conclusion

Approximately 6 months of lander data have been analyzed. The lander locations, Mars rotation axis orientation, and rotation period were accurately determined. The analyses indicate that the additional tracking data of the Viking extended mission are required before significant improvements may be obtained in the Mars rotation axis motion and parameters affecting the earth and Mars orbits.

References

Blackshear, W. T., R. H. Tolson, and G. M. Day, Mars lander position estimation in the presence of ephemeris biases, *J. Spacecr. Rockets, 10,* 284, 1973.

Born, G. H., E. J. Christensen, S. N. Mohan, J. F. Jordan, and T. C. Duxbury, Determination of the Mars spin axis direction from Mariner 9, paper presented at meeting of the AAS Planetary Science Division, Amer. Astron. Soc., Kona, Hawaii, March 20–24, 1972.

Daniels, E. F., and R. H. Tolson, Spherical harmonic representation of the gravity field of Mars using a short arc technique, paper presented at the AIAA/AAS Astrodynamics Conference, Amer. Inst. of Aeronaut. and Astronaut., Amer. Astron. Soc., San Diego, Calif., Aug. 18–20, 1976.

Davies, M. E., and D. W. G. Arthur, Martian surface coordinates, *J. Geophys. Res., 78,* 4355, 1973.

de Vaucouleurs, G., The physical ephemeris of Mars, *Icarus, 3,* 236, 1964.

de Vaucouleurs, G., M. E. Davies, and F. M. Sturms, Jr., Mariner 9 areographic coordinate system, *J. Geophys. Res., 78,* 4395, 1973.

Fliegel, H. J., and W. J. O'Neil, Viking 75 project software requirement document, The Platform Observable Calibration Software Assembly (Plato) for the Support of Viking 75, report p. 1-1, Jet Propul. Lab., Pasadena, Calif., April 16, 1973.

Her Majesty's Nautical Almanac Office, *Explanatory Supplement to the Astronomical Ephemeris and to the American Ephemeris and Nautical Almanac,* p. 333, Her Majesty's Stationery Office, London, 1961.

Jet Propulsion Laboratory, *Deep Space Network Progr. Rep. 42-35, July–August 1976,* Pasadena, Calif., Oct. 15, 1976.

Khatib, A. R., G. W. Null, and J. W. Zielenbach, The alphabet system, *Eng. Plann. Doc. 900-578,* Jet Propul. Lab., Pasadena, Calif., Oct. 15, 1972.

Komarek, T. A., Ranging calibration data for June (reported monthly). *Rep. IOM 3395-76-139,* Jet Propul. Lab., Pasadena, Calif., July 7, 1976.

Lorell, J., J. D. Anderson, J. F. Jordan, R. D. Reasenberg, and I. I. Shapiro, Celestial mechanics experiment, Mariner Mars 1971 project final report, vol. 5, *Tech. Rep. 32-1550,* p. 22, Jet Propul. Lab., Pasadena, Calif., Aug. 20, 1973.

Lowell, P., Precession of the Martian equinoxes, *Astron. J., 28*(21), 169, 1914.

Madrid, G. A., The measurement of dispersive effects using the Mariner 10 *S* and *X* band spacecraft to station link, *Deep Space Network Progr. Rep. 42-22, May–June 1974,* p. 22, Jet Propul. Lab., Pasadena, Calif., 1974.

Madrid, G. A., C. C. Chao, H. F. Fliegal, R. K. Leavitt, N. A. Mottinger, F. B. Winn, R. N. Wimberly, K. B. Yip, and J. W. Zielenbach, Tracking system analytic calibration activities for the Mariner Mars 1971 mission, internal document, Jet Propul. Lab., March 1, 1974. (For update of constants, see Critical planetary constants, *Viking Flight Team Memo. FPAG-14969-WJO KRW FPAG,* Jet Propul. Lab., Pasadena, Calif., April 30, 1976.)

Michael, W. H., Jr., R. H. Tolson, A. P. Mayo, W. T. Blackshear, G. M. Kelly, D. L. Cain, J. P. Brenkle, I. I. Shapiro, and R. D. Reasenberg, Viking lander location and spin axis of Mars: Determination from radio tracking data, *Science, 193,* 803, 1976*a*.

Michael, W. H., Jr., A. P. Mayo, W. T. Blackshear, R. H. Tolson, G. M. Kelly, J. P. Brenkle, D. L. Cain, G. Fjelbo, D. N. Sweetnam, R. B. Goldstein, P. E. MacNeil, R. D. Reasenberg, I. I. Shapiro, T. I. S. Boak III, M. D. Grossi, and C. H. Tang, Mars dynamics, atmospheric and surface properties: Determination from Viking tracking data, *Science, 194,* 1337, 1976*b*.

Michaux, C. M., Handbook of the physical properties of the planet Mars, *NASA Spec. Publ. 3030,* 31, 1967.

Moyer, T. D., Mathematical formulation of the double precision orbit

determination program (DPODP), *Tech. Rep. 32-1527,* Jet Propul. Lab., Pasadena, Calif., May 15, 1971.
Sinclair, A. T., The motions of the satellites of Mars, *Mon. Not. Roy. Astron. Soc., 155,* 249, 1972.
Standish, E. M., M. S. W. Keesey, and XX Newhall, Jet Propulsion Laboratory development ephemeris number 96, *Tech. Rep. 32-1603,* Jet Propul. Lab., Pasadena, Calif., 1976.
Sturms, F. M., Jr., Polynomial expressions for the planetary equators and orbits with respect to the mean 1950.0 coordinate system, *Tech. Rep. 32-1508,* p. 8, Jet Propul. Lab., Pasadena, Calif., Jan. 15, 1971.

(Received April 7, 1977;
revised May 13, 1977;
accepted May 13, 1977.)

Bistatic Radar Measurements of Electrical Properties of the Martian Surface

C. H. TANG, T. I. S. BOAK III, AND M. D. GROSSI

Raytheon Company, Wayland, Massachusetts 01778

The Viking lander-to-orbiter relay links make it possible to perform measurements of the electrical properties of the Martian surface by the bistatic technique. The electromagnetic signals radiated by the lander antenna at 381 MHz, in fact, reach the orbiter both directly and after reflection from the Martian terrain as the orbiter rises and sets with respect to the lander. The fading pattern of the signal intensity received at the orbiter therefore contains information on the reflection coefficient of the terrain and hence on the relative dielectric constant ϵ_r and the conductivity σ in the vicinity of the lander. The signal amplitude's fading patterns collected with the Lander 1 to Orbiter 1 relay link were of good quality and led to the determination of $\epsilon_r = 3.3 \pm 0.7$ in the vicinity of Lander 1 (when the quasi-specular theory was used and σ was assumed to be between 10^{-3} and 10^{-5} mho/m). These electrical properties are similar to those of pumice and tuff. The dielectric constant of the surface near the Lander 2 site is estimated to be ϵ_r = 2.8–12.5.

INTRODUCTION

The electrical properties of the Martian surface have been investigated previously by many authors through earth-based measurements performed by monostatic radar [e.g., *Pollack and Sagan,* 1970; *Tyler et al.,* 1976]. Several bistatic radar experiments have also been conducted since the inception of the space age aimed at the measurement of the electrical properties of the surface of the moon and planets, including Mars. The scheme consisted of receiving on earth the resultant of the direct wave and the surface reflection from a transmitter installed on an orbiter or a flyby.

Fjeldbo [1964] established the theoretical foundation of the bistatic method and used it to detect reflections from the Martian surface on the occasion of the Mariner 1969 mission to Mars [*Fjeldbo,* 1972]. Separation between direct and reflected waves was achieved by exploiting the differences in their Doppler spectrum. *Tyler et al.* [1967], *Tyler* [1968], and *Tyler and Howard* [1973] then applied this method to Lunar Orbiter 1 and 3, to Explorer 35, and to Apollo 14 and 15 moon flights, respectively.

A variant in system configuration, consisting of locating both the transmitting and the receiving terminal of the bistatic radar on the surface to be measured, was then used by *Simmons et al.* [1973] for the measurements of the electrical properties of the surface of the moon.

In the bistatic radar experiment performed by us on Mars with the 381-MHz link between lander and orbiter of the Viking mission, we used an approach intermediate between the two configurations mentioned above, inasmuch as the transmitter was located on the surface of the planet and the receiver was in orbit around the planet.

The goal of the experiment was to measure the mean value of the reflection coefficient ρ of the Martian surface in the vicinity of each lander at the frequency of 381 MHz. From these measurements, through a suitable data inversion procedure, we derived the value of the relative dielectric constant ϵ_r of the Martian terrain, again in the vicinity of each lander.

This inversion procedure applied to the raw data was complemented with (1) additional information concerning the geometry of the lander-to-orbiter links (and the consequent geometry of terrain probing) and the roughness of the terrain (as observed by stereo TV) and (2) a set of hypotheses that reduced in a justifiable fashion the unknowns of the problem to a manageable number (these hypotheses will be listed one by one in the sections that follow). The 'observables' in our experiment were the amplitude 'fading pattern' data collected at the receiving end (orbiter) of the lander-to-orbiter link. Finally, the ultimate goal of the experiment was to infer from the measurements the mean density and the mineralogical nature of the Martian surface, on the basis of the relationship with ϵ_r of these parameters.

THEORETICAL FOUNDATIONS OF THE MEASUREMENT METHOD

A highly idealized representation of the link geometry involved in the measurements is given in Figure 1, which for simplicity is plotted with a disuniform scale throughout. In fact, the height of the lander antenna above the surface and the orbital altitude of the orbiter are approximately 1 m and 1500 km (at periapsis), respectively.

Figure 2 gives a more detailed representation of the geometrical quantities at the lander terminal of the link. These will be repeatedly called out in our analysis.

Figure 3 is a qualitative illustration of the amplitude fading pattern that we would expect to observe, again in highly idealized conditions. The diagram in the figure is based on a two-ray model (the direct and the reflected, or indirect, ray). The footnote to Table 2 defines the various quantities that appear in the figure. Subsurface reflections can in fact be excluded on three accounts: (1) at the frequency of operation (381 MHz), skin depth is as short as 2.5 m, with a layer conductivity as low as 10^{-4} mho/m, (2) owing to the almost grazing conditions for the rays used in the measurements (elevation angle ψ always less than 30°), the electromagnetic energy that penetrates below the surface is absorbed at very shallow depths; and (3) 'impedance contrasts' that would cause significant subsurface reflections do not appear physically plausible at that shallow depth.

The equation of the curve in Figure 3 (the curve of the so-called 'residuals') is

$$y(\psi)\ (\text{dB}) = 10 \log_{10} (P_c/P_d) = 10 \log_{10} [1 + (P_i/P_d) - 2(P_i/P_d)^{1/2} \cos (K\Delta R + \Delta\alpha)] \quad (1)$$

where P_i is the power of a reflected ray; P_d is the power of a direct ray; P_c is the total received power (the observable); $K = 2\pi/\lambda$ is the phase constant (λ is the free space wavelength); and

Paper number 7S0534.

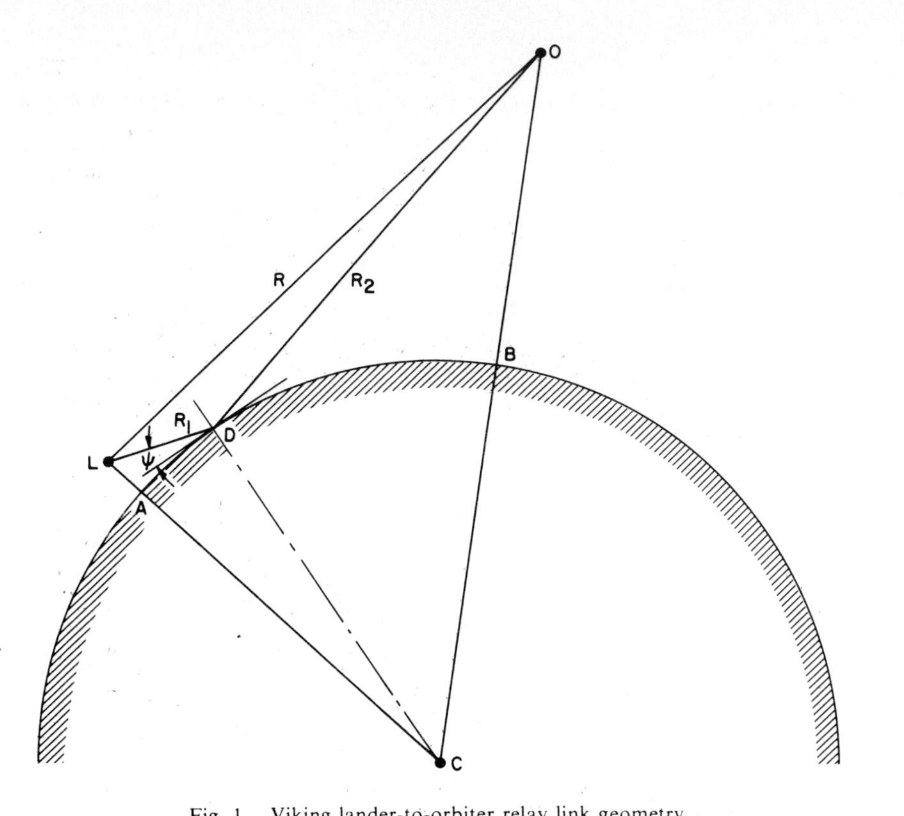

Fig. 1. Viking lander-to-orbiter relay link geometry.

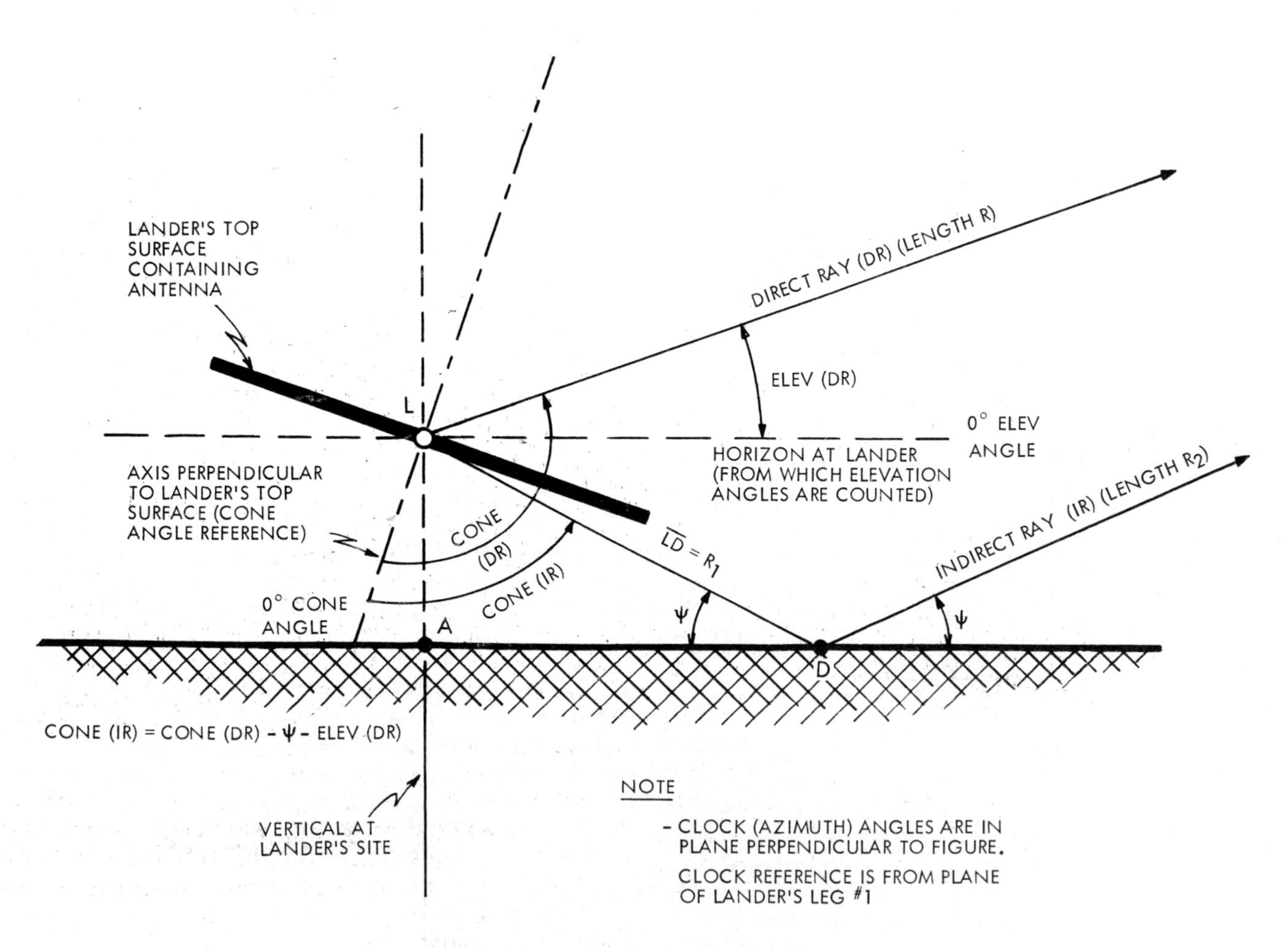

Fig. 2. Geometrical quantities at the lander terminal of the link.

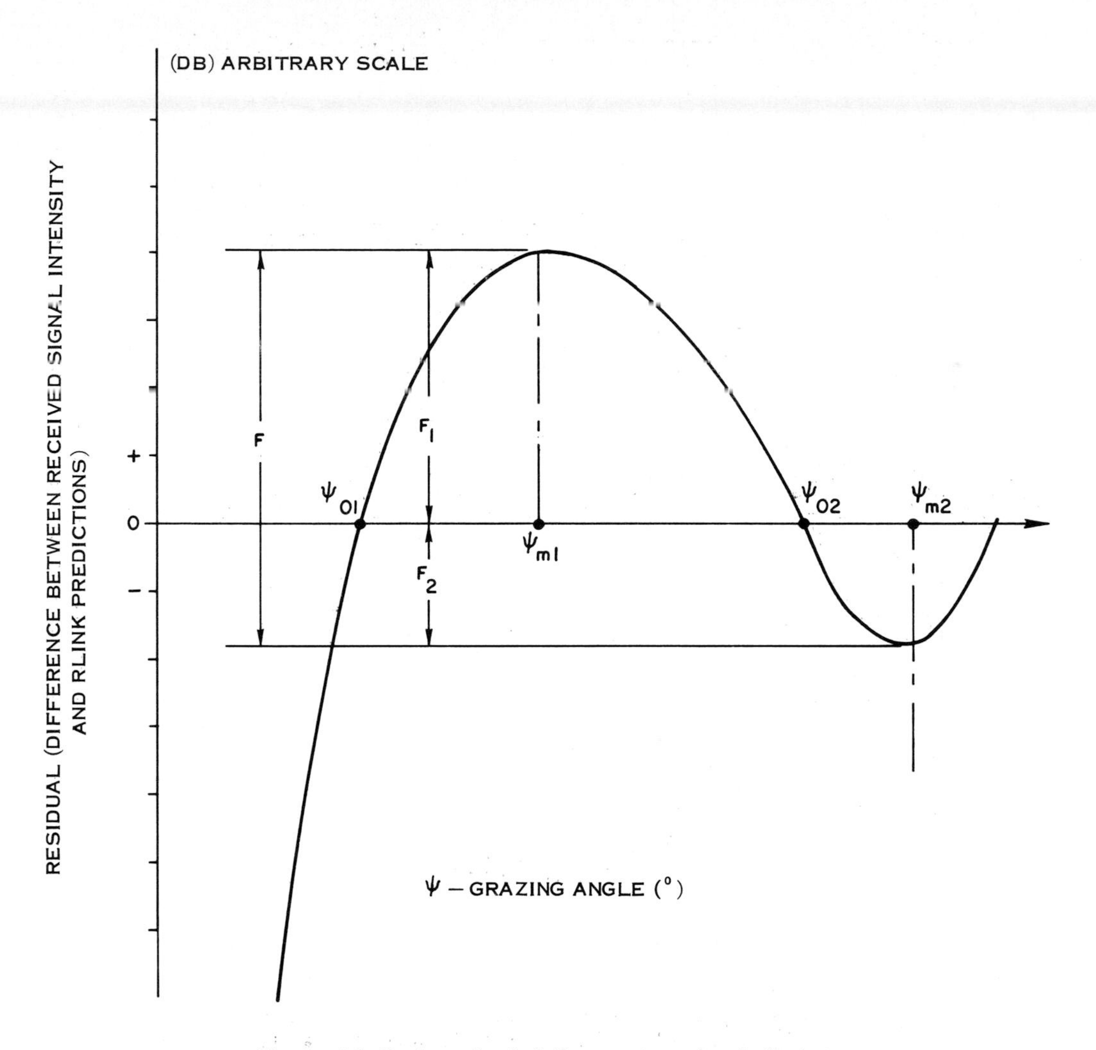

Fig. 3. Idealized amplitude fading pattern and main features.

$\Delta R = (R_1 + R_2) - R$ is the path length difference. Because $R_2 \gg R_1$, we have

$$(R_1 + R_2) - R = 2R_1 \sin^2 \psi = 2h_l \sin \psi \quad (2)$$

Also in (1), $\Delta\alpha$ is the relative antenna pattern phase difference between the direct and the indirect path. Since the curve of the residuals is a function of the P_i/P_d ratio, it is independent of the RF circuit losses and the propagation medium loss for the Viking link geometry.

The expression that we have used for the total received power is

$$P_c = P_d + P_i - 2(P_d P_i)^{1/2} \cos (K\Delta R + \Delta\alpha) \quad (3)$$

where all the symbols have already been introduced. This formula is based on the theory of quasi-specular scattering, defined as a surface scattering process that is dominated by single-ray specular reflections and in which the surface roughness intervenes only in affecting the amplitude of the reflected ray [*Barton and Ward,* 1969].

The planet's surface, within the illuminated region of the transmitter, is assumed to be rough with a Gaussian distribution of surface heights. Since the elevation angle of interest is small, the rough terrain is assumed to have some shadowing effect in addition to the modification of the reflection amplitude.

The longest ground range which scatters energy toward the receiver would be 2.6 km if a $\psi = 0$ incidence could be used in collecting data. However, $\psi = 0$ is never used in the experiment. This limit corresponds, in fact, to the case in which the receiver is at the radio horizon of the transmitter, where the Fresnel coefficients approach -1. There the direct and the surface-reflected fields cancel each other, and only the surface wave contribution is left. As the orbiter elevation increases beyond certain limits, the surface wave contribution can be neglected in comparison to both the direct wave and the surface-reflected wave contributions. At 381 MHz, this limit is given by ψ (the grazing angle) greater than 3°. Figure 4 shows the corresponding specular point ranges for various grazing angles. Thus for most orbiter elevation angles of interest (5°–30°) the indirect path contributions are reflections from the immediate vicinity of the lander (1.7 to ~10 m).

Figure 4 shows the separation between the specular and the diffuse scattering regions. These are defined as a function of the surface roughness and the grazing angle. A surface may be considered specular when

$$(\sigma_h/\lambda) \sin \psi \leq 0.065 \quad (4)$$

where σ_h is the rms deviation in surface height and λ is the wavelength. This is approximately the same as the Rayleigh criterion for surface roughness [*Beckmann and Spizzichino,* 1963; *Barton and Ward,* 1969]. Because, as has already been indicated, in our study we have used the quasi-specular scatter-

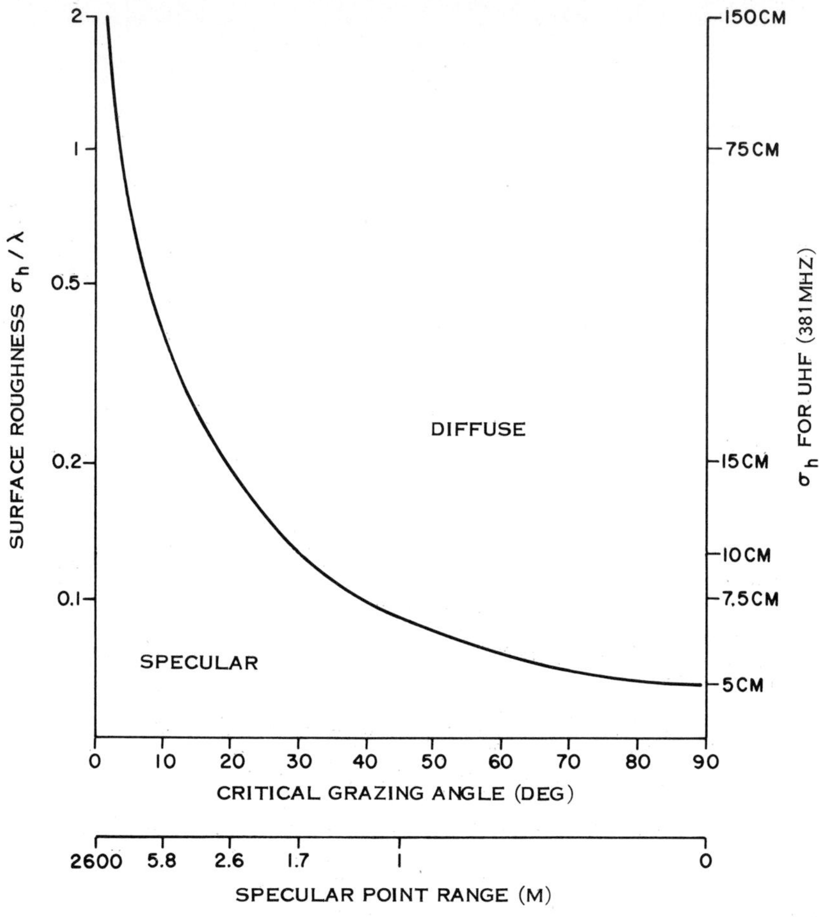

Fig. 4. Separation between specular and diffuse scattering regions and specular point ranges.

ing theory, the surface reflection for each orbiter elevation angle is represented by a single ray with the reflection amplitude modified by the scattering factor $\langle \rho_s^2 \rangle$.

In addition, a shadowing function $S(\psi)$ is employed to account for the screening of parts of the surface by other parts interrupting the illuminating radiation [*Beckmann and Spizzichino*, 1963; *Beckmann*, 1965]. Figure 5 plots the shadowing function for three surface correlation parameter values (C = 5, 5.6, and 6), using Beckmann's approximate expression.

With G as the antenna gain, $\mathcal{R}$ as the reflection factor, $\mathcal{P}$ as the polarization efficiency, $\langle \rho_s^2 \rangle$ as the roughness scattering factor, and $S(\psi)$ as the shadowing function, P_d and P_i can be written as

$$P_d = \frac{P_t G_{dt} G_{dr} \lambda^2}{(4\pi R)^2} \mathcal{P}_d \tag{5}$$

$$P_i = \frac{P_t G_{it} G_{ir} \lambda^2}{[4\pi(R_1 + R_2)]^2} \mathcal{R} \cdot \mathcal{P}_i \cdot \langle \rho_s^2 \rangle \cdot S^2(\psi) \tag{6}$$

where P_t is the power of the lander's transmitter. The ratio P_i/P_d can then be computed as follows:

$$\frac{P_i}{P_d} = \frac{G_{it}}{G_{dt}} \left(\frac{R}{R_1 + R_2} \right)^2 \left(\frac{\mathcal{P}_i}{\mathcal{P}_d} \right) \cdot \mathcal{R} \cdot \langle \rho_s^2 \rangle \cdot S^2(\psi) \tag{7}$$

The various quantities involved in the above formulas are introduced hereafter.

Antenna gain G. The subscripts t and r denote, respectively, the transmitter and receiver. The subscripts d and i denote, respectively, the direct and indirect ray.

Polarization efficiency $\mathcal{P}$. The polarization efficiency $\mathcal{P}$ of an antenna with the polarization factor r receiving an incident wave with polarization factor t is given by [*Beckmann*, 1968]

$$\mathcal{P} = \frac{|1 + rt^*|^2}{(1 + |t|^2)(1 + |r|^2)} \tag{8}$$

where the complex polarization factor is the ratio of two mutually perpendicular linearly polarized components. For example, the transmitting antenna's polarization characteristic can be described by

$$t = E_V / E_H \tag{9}$$

which can also be expressed in terms of the axial ratio a and the polarization ellipse tilt β as

$$t = \frac{a(1 + e^{j2\beta}) + (1 - e^{j2\beta})}{a(1 - e^{j2\beta}) + (1 + e^{j2\beta})} e^{-j\pi/2} \tag{10}$$

Thus we have the polarization efficiency for the direct path d

$$\mathcal{P}_d = \frac{|1 + r_d t_d^*|^2}{(1 + |t_d|^2)(1 + |r_d|^2)} \tag{11}$$

while

$$\frac{\mathcal{P}_i}{\mathcal{P}_d} = \frac{|1 + rt_i'^*|^2}{|1 + rt_d^*|^2} \frac{(1 + |t_d|^2)}{(1 + |t_i'|^2)} \tag{12}$$

where t_i' is the indirect path polarization factor. (We note that $G_{dr} \approx G_{ir}$ and $r_d \approx r_i$, because the direct and indirect paths are almost parallel.)

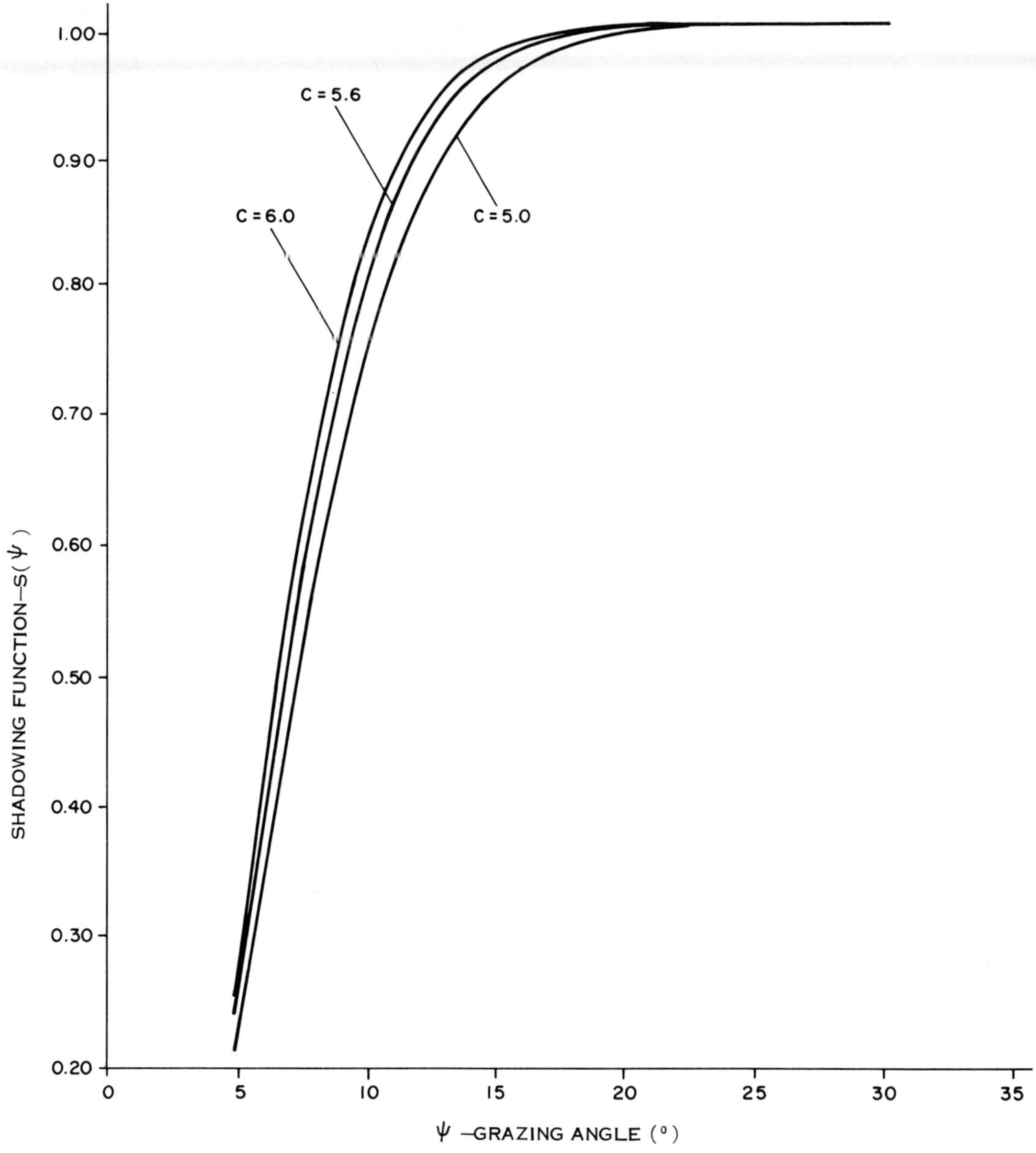

Fig. 5. Shadowing function generated from Beckmann's equation.

Reflection factor $\mathcal{R}$. The indirect path component i contains the surface reflection coefficient, which affects both the reflection amplitude and the polarization efficiency. Thus when the received indirect path power is normalized with respect to the total indirect path incident power, it is proportional to

$$\mathcal{R}\cdot\mathcal{P}_i = \frac{|1 + r_i t_i'^*|^2}{(1 + |t_i|^2)(1 + |r_i|^2)} = \frac{(1 + |t_i'|^2)}{(1 + |t_i|^2)} \cdot \frac{|1 + r_i t_i'^*|^2}{(1 + |t_i'|^2)(1 + |r_i|^2)} \quad (13)$$

where t_i and t_i' are the indirect path polarization factors before and after the surface reflection, respectively. Hence

$$t_i' = \frac{E_V'}{E_H'} = \frac{R_V}{R_H}\frac{E_V}{E_H} = \frac{R_V}{R_H} t_i \quad (14)$$

where R_V and R_H are the familiar Fresnel reflection coefficients given by

$$R_V = \frac{n^2 \sin\psi - (n^2 - \cos^2\psi)^{1/2}}{n^2 \sin\psi + (n^2 - \cos^2\psi)^{1/2}} \quad (15)$$

$$R_H = \frac{\sin\psi - (n^2 - \cos^2\psi)^{1/2}}{\sin\psi + (n^2 - \cos^2\psi)^{1/2}} \quad (16)$$

with

$$n^2 = (\text{index of refraction})^2 = \epsilon_r - j18 \times 10^{-3}\sigma f \quad (17)$$

ϵ_r as the relative dielectric constant of the reflecting surface, σ as the conductivity of the reflecting surface in mhos per meter, and f as the frequency in megahertz.

Shadowing function $S(\psi)$. Shadowing effects due to terrain undulation become important for low elevation angles. An approximate shadowing function (as previously defined) was derived from geometric considerations by *Beckmann* [1965]

$$S(\psi) = \exp\left[-\tfrac{1}{4}\cot\psi \operatorname{erfc}(C\tan\psi)\right] \quad (18)$$

where C is a parameter related to the surface correlation function. For the bistatic case, from the theorem of joint probability,

$$S(\psi, \psi_S) = P(|\psi_S| \psi)S(\psi) = S(|\psi_S|)S(\psi) \quad (19)$$

where ψ_S is the scattering angle.

Roughness scattering factor ρ_s. The 'specular scattering factor' ρ_s gives the rms value of the field which is reflected without being perturbed (apart from the amplitude) by the surface irregularities and can be expressed as

$$\langle \rho_s^2 \rangle = \exp\left[-\left(\frac{4\pi\sigma_h \sin\psi}{\lambda}\right)^2\right] \quad (20)$$

where σ_h is the rms deviation in surface height.

From the analytical formulations outlined above, it clearly emerges that we are required to add a substantial amount of information to the actual experimental data; these data, in fact, are limited to the quantity P_c. Also, as we mentioned already in the introduction, we need to formulate several assumptions and introduce them in the reduction process in order to finally obtain the value of ϵ_r from the observables. A substantial part of the required supplemental information is contained in a tape (called RLINK) and in tables that we, as experimenters, routinely receive from mission operations together with the raw data collected by Viking.

As a function of orbiter event time (OET), the following data are available to us as auxiliary information: the modulus of the lander-to-orbiter range vector, the cone angle of the lander-to-orbiter range vector, the clock angle of the lander-to-orbiter range vector, the gain of the lander antenna in the direction of the lander-to-orbiter range vector, the axial ratio of the lander antenna in the direction of the lander-to-orbiter range vector, the gain of the orbiter antenna in the direction of the lander-to-orbiter range vector, the axial ratio of the orbiter antenna in the direction of the lander-to-orbiter range vector, the output power of the lander transmitter, and the estimate of the power of the direct ray P_d. In these data the direction of the lander-to-orbiter range vector is assumed to be coincident with the direct ray (DR). We are therefore in the position of computing the cone and clock angles of the indirect ray (IR)

$$\begin{aligned} \text{cone(IR)} &= \text{cone(DR)} - \psi - \text{elev(DR)} \\ \text{clock(IR)} &= \text{clock(DR)} \end{aligned} \quad (21)$$

as is indicated in Figure 2.

Once we have these angles of the indirect path, we can enter into the table of the antenna pattern and read the gain of the lander antenna in the direction of the indirect path and the axial ratio of the lander antenna in the direction of the indirect path. At the orbiter end of the link, indirect path gain and axial ratio are identical to those of the direct path, because the two directions (direct and indirect) are virtually coincident at that end of the link.

The data on the antennas were obtained from measurements (scale model and otherwise) performed before the Viking launch. We assume that these are representative of the actual patterns; also, since the shortest range to the reflecting region in the vicinity of the lander is approximately 2 m, we can conclude that surface reflections take place in the far field of the lander antenna, which implies that the measured far field patterns are applicable to our reduction process. The link geometry is quite reliable, since it is readjusted after every orbiter's revolution on the basis of tracking data.

Another substantial set of supplemental information is represented by the stereo TV observations of the surface around each of the landers [*Mutch et al.*, 1976] that yield the rms surface roughness σ_h from which $\langle \rho_s^2 \rangle$ can be determined. From stereo TV data it was found that $\sigma_h = 0.15\lambda$ for the terrain near Lander 1 and $\sigma_h = 0.25\lambda$ for the terrain near Lander 2. These determinations are rather reliable because although the observations made by the stereo TV system of the lander pertain to an angular sector that is not involved in the lander-to-orbiter indirect ray propagation, around-the-horizon observations made with a single TV camera have confirmed the uniformity of the terrain roughness all around each lander.

Now, to illustrate the set of assumptions that we need to formulate, we have the following parameters.

Terrain conductivity σ. On the basis of plausible estimates that have appeared in the literature, we assumed a value $\sigma = 10^{-4}$ mho/m. We have also verified that variations of the conductivity between 10^{-3} and 10^{-5} mho/m scarcely influence our determination of the value of ϵ_r.

Tilt angle β of the polarization ellipse for lander and orbiter antenna. These antenna radiation quantities are difficult to measure, and no measured data were available to the experimenters. We have assumed $\beta_r = 45°$ for the receiver antenna on both orbiters and $\beta_t = 0°$ for the transmitting antenna in both landers, following the estimates of β_r and β_t formulated by the antenna designers. We have verified that the values assumed for β_r and β_t have scarce influence on the slope of the curve of the residuals and ultimately on our determination of ϵ_r.

Finally, the following quantities are derived from particular features of the observed curve of the residuals.

Lander's antenna height above the terrain h_l for Lander 1 and Lander 2. This quantity is derived from the pair of points (ψ_{m1}, ψ_{m2}) (or (ψ_{01}, ψ_{02})) in the curve of the residual through the formula

$$h_l(\sin\psi_{m2} - \sin\psi_{m1}) = \lambda/4 \quad (22)$$

(assuming that $\Delta\alpha$ is a slowly varying function of ψ). Usually, the pair (ψ_{01}, ψ_{02}) is less convenient than the pair (ψ_{m1}, ψ_{m2}) in this derivation. We found that $h_l = 0.9$ m for Lander 1 and $h_l = 1.4$ m for Lander 2.

Pattern-related phase difference $\Delta\alpha$ between direct and indirect rays. This determination was obtained from the position in the curve of the residuals of the points ψ_{01} and ψ_{02}. We obtained $\Delta\alpha = -45°$ in the link from Lander 1 to Orbiter 1, $\Delta\alpha = +170°$ in the link from Lander 2 to Orbiter 2, and $\Delta\alpha = -90°$ in the link from Lander 2 to Orbiter 1. In all cases, $\Delta\alpha$ is assumed to be constant in the range $5° \leq \psi \leq 30°$.

Local terrain undulation (parameter C of shadowing function). Again with reference to Figure 3, the value of C was obtained from the ratio F_1/F_2. We found $C = 5.6$, and we verified that the value of ϵ_r (as determined by our experiment) is insensitive to the value of C for $C > 5.6$.

When all the above is done, we can proceed to the determination of ϵ_r. This is performed with a model matching approach based on simulating the experimental results with the aid of (1) and searching for their best match to the experimental data by using ϵ_r as the parameter. Once ϵ_r is determined, surface density and mineralogical composition can be inferred from it [*Tyler et al.*, 1976; *Campbell and Ulrichs*, 1969].

DATA COLLECTION, REDUCTION, AND PROCESSING

The raw data of the measurement of the electrical properties of the Martian surface consisted of the signal amplitude of the lander-to-orbiter links. This amplitude was recorded on board each orbiter (Table 1).

TABLE 1. Summary of Observations by the Viking Relay Links

Link	Sols (1976)	Data Quality*	Sols Used for Averaging†
Lander 1 to Orbiter 1	1–43 July 21– Sept. 2	Consistent, excellent data quality but only rising side data usable.	16 sols (1, 4, 6, 7, 8, 9, 10, 15, 16, 17, 21, 22, 25, 26, 27, 32)
Lander 2 to Orbiter 2	1–26 Sept. 4–30	Fluctuating, not too consistent and only rising side data usable.	10 sols (11, 12, 13, 14, 15, 16, 17, 18, 19, 20)
Lander 2 to Orbiter 1	21–61 Sept. 25– Nov. 5	Fluctuating, not too consistent and only rising side data usable.	4 sols (27, 28, 29, 30)

*With respect to residuals for the multipath analysis, because of the blockage caused by the lander structures, the setting side multipath data of all links are complicated and not useful.

†Only those residuals which have all features described in Figure 3 are selected for averaging.

These data were available to the experimenters as part of the engineering telemetry data stream, which consists of a collection of 6-bit data numbers (DN's) which sample and measure selected housekeeping functions in the time domain and are recorded at the earth stations of the Viking mission. The two specific DN's which were used in the experiment are called 'RRS RCVD SIG STRENGTH' and '16 KBPS SIG MEAN' and are available as channel numbers E-626 and E-653, respectively. Both of them are related in a known fashion to the received signal strength. The 'SIG MEAN' provides better resolution in the dynamic range from about −100 dBm (dBm represents decibels with respect to 1 mW) to −120 dBm. The 'predictions' of the so-called RLINK computer program are used in conjunction with these engineering telemetry data and are correlated with them according to their OET. These RLINK predictions include information on the orbital parameters, on the lander-to-orbiter relative geometry, and on the characteristics of the patterns of the lander and orbiter relay link antennas. The RLINK software is updated on the basis of the actual passes of the orbiter above the lander; it is therefore more than the prediction that the name implies.

The signal amplitude observations and the RLINK predictions are used to generate 'residual curves,' obtained by subtracting the latter from the former. Observations of good quality were selected for averaging and for generating residuals. Examples of residuals are shown in Figure 6, and Table 2 summarizes some of their relevant characteristics.

As Table 1 indicates, the data collected with the VL-1 to VO-1 link are of particularly good quality and show that surface reflections are well above threshold of detectability.

The curve of the residuals for slowly varying antenna char-

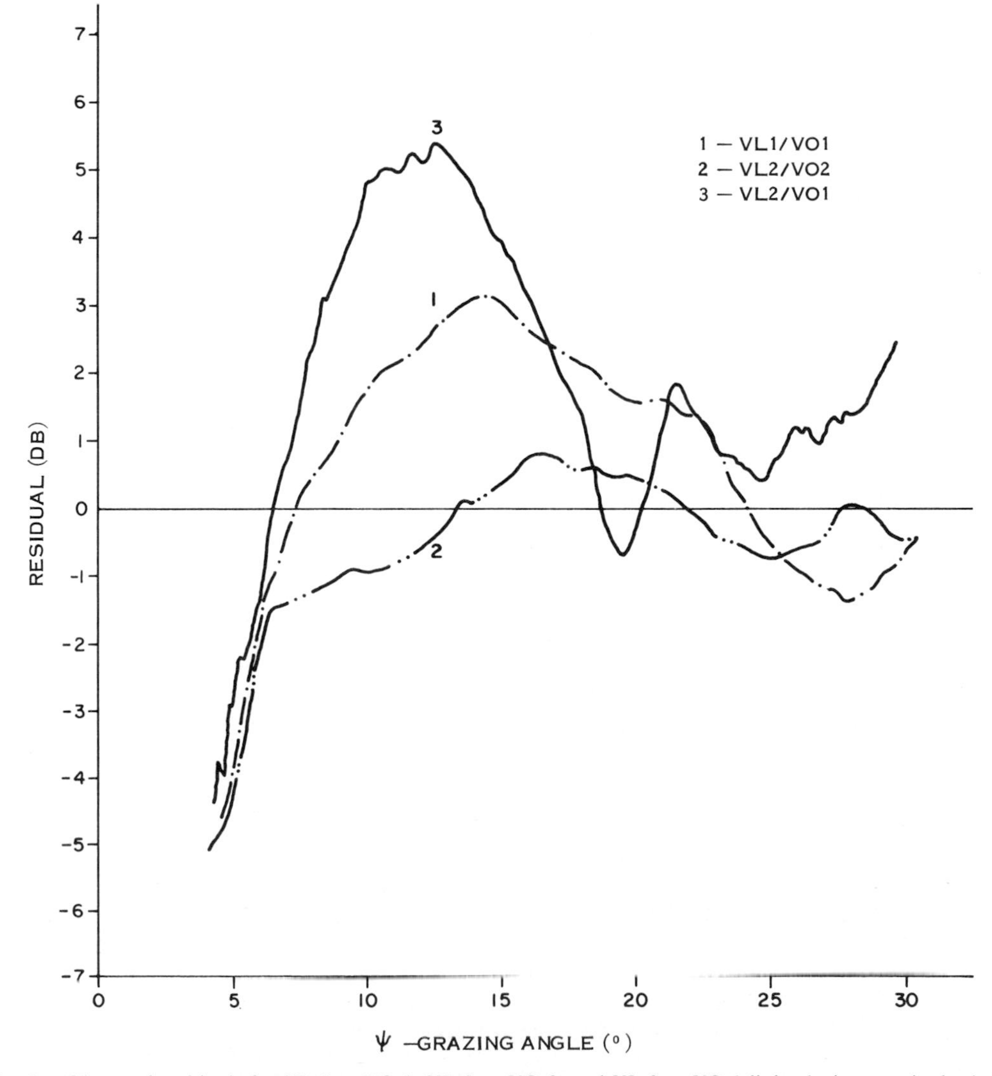

Fig. 6. Observed residuals for VL-1 to VO-1, VL-2 to VO-2, and VL-2 to VO-1 links during nominal mission.

TABLE 2. Characteristics of Averaged Residual Data

Link	ψ_{01}, deg	ψ_{02}, deg	ψ_{m1}, deg	ψ_{m2}, deg	F_1, dB	F_2, dB	F, dB
VL-1 to VO-1	7.3	24.0	14.8	28.0	3.15	1.35	4.5
VL-2 to VO-2	13.3	22.0	16.7	25.0	0.8	0.7	1.5
VL-2 to VO-1	6.5	18.7	12.5	19.5	5.4	0.7	6.1

The symbols are (see also Figure 3): ψ_{01}, first zero crossing of the residual curve (in the first upgoing branch): ψ_{02}, second zero crossing of the residual curve (in the first downgoing branch); ψ_{m1}, the abscissa of the first maximum of the residual curve (between ψ_{01} and ψ_{02}); ψ_{m2}, the abscissa of the first relative minimum of the residual curve (beyond ψ_{02}); F_1, the amplitude of the first maximum of the residual curve (at abscissa ψ_{m1}) with respect to reference level; F_2, the amplitude of the first minimum of the residual curve (at abscissa ψ_{m2}) with respect to the reference level; and $F = |F_1| + |F_2|$.

acteristics can be defined by means of the various parameters shown in Figure 3. From (1), (7), and (12) it is seen that the fading amplitude F is a function of ϵ_r, the surface dielectric constant, while (2) and (3) indicate that the lander antenna height h_l can be determined by either the pair (ψ_{m1}, ψ_{m2}) or the pair (ψ_{01}, ψ_{02}) through expression (22). Here it is assumed that $\Delta\alpha$ (see equation (1)) is a slowly varying function of ψ. Usually, it can be expected that the pair (ψ_{m1}, ψ_{m2}) is more convenient to use than the pair (ψ_{01}, ψ_{02}).

When (22) is applied, lander antenna heights are estimated to be Lander 1, from VL-1 to VO-1 data, 88.7 cm; Lander 2, from VL-2 to VO-2 data, 138 cm; and Lander 2, from VL-2 to VO-1 data, 165 cm. (We have been informed that the Lander 1 antenna height determined by the optical technique is 90 ± 2 cm, but no other estimate of the Lander 2 antenna height is known.) On the basis of these results, 90 cm and 140 cm were used as antenna heights for Lander 1 and Lander 2, respectively, for the surface dielectric constant data reduction.

DATA ANALYSIS

The Martian surface electrical properties are reconstructed by model matching, based on the comparison of the observed multipath residuals with those obtained from the computer simulation model. As was described, the simulation model was developed on the basis of a quasi-specular scattering theory. The model provides sufficient detail to determine the Martian terrain's electrical properties from the information contained in the residual data derived from lander-to-orbiter links. Thus

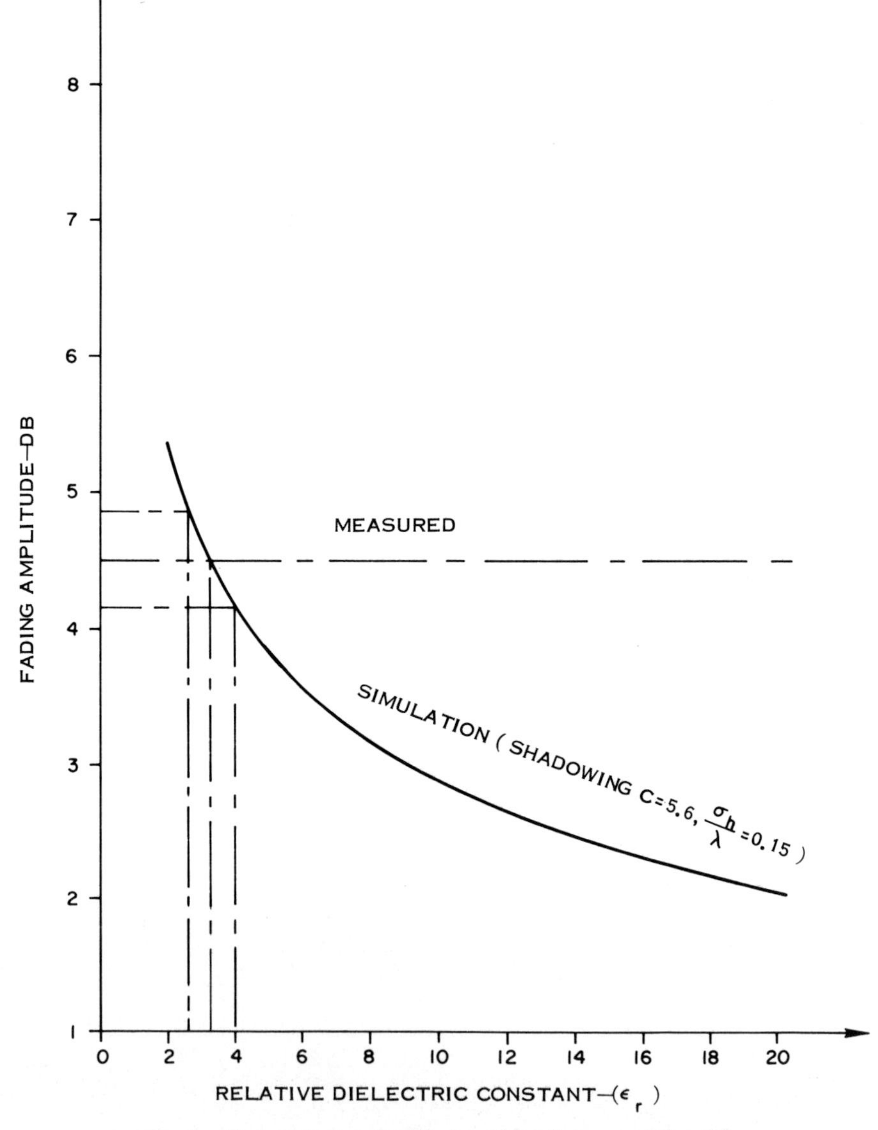

Fig. 7. Data reduction for VL-1 to VO-1 link: ϵ_r = 3.3 ± 0.7.

the period of fading oscillation is related to the lander's antenna height, and the fading amplitude is a function of the surface dielectric constant.

Clearly, the ultimate accuracy of the data reduction procedure will depend on the knowledge of link parameters and the quality of measured channel data. The link parameters, as was mentioned previously, include the orbiter parameters and the antenna characteristics—the antenna gain and phase patterns, polarization axial ratios, and ellipse tilts. As was previously indicated, accurate orbiter parameters and nominal values of antenna parameters are available, except for the ellipse tilts and the antenna phase patterns which were not measured before mission launch.

An error analysis was performed on the simulation procedure by using the Monte Carlo method. The expected deviations of the antenna characteristics as well as the other parameters of the lander and orbiter intervening in the measurements are found to produce only 0.5-dB total standard deviation of the fading amplitude (see Table 3). This is much less than the fluctuation that existed in the measured data. The combined effect of all these assumed errors is found to be 0.494 dB for a 100-sample test (the result for a 1000-sample test is 0.502 dB).

In the data collected with the VL-1 to VO-1 relay link, fading amplitudes with peak-to-peak relative intensities of 4–5 dB were recorded in most of the 43 sols (Table 2 shows, in fact, that the average value for this quantity was 4.5 dB).

The derivation of the value of the dielectric constant ϵ_r of the terrain in the vicintiy of VL-1 and VL-2 was obtained with a procedure that, generally speaking, consisted of performing a multiparameter computer simulation of the relay link residuals and adjusting the parameters involved in order to 'match' this residual curve to the averaged observations.

TABLE 3. Results of Error Analysis

Assumed Errors	Standard Deviation of Fading Amplitude, dB
Antenna gain error (0.5 dB)	0.249
Axial ratio error (1.0 dB)	0.328
Antenna height error (2 cm)	0.114
Surface roughness error (2 cm)	0.227
Shadowing parameter error (0.5)	0.085
Ellipse tilt error (10°)	0.059
Antenna phase error (10°)	0.160
Orbital position error (3 km)	1.8×10^{-6}
Orbital angular error (0.25°)	0.072

For the VL-1 to VO-1 link the procedure of model matching (the estimated channel data accuracy of ±0.7 dB also being taken into account) yielded a value of $\epsilon_r = 3.3 \pm 0.7$ for the terrain in the vicinity of VL-1 (Figure 7). The reasonably good accuracy in ϵ_r determination is due to the fact that the VL-1 to VO-1 observations were of good quality. Table 1, in fact, shows that the average peak-to-peak amplitude F in the residuals, as was already remarked, is 4.5 dB, while the sol-to-sol spread is only 1 dB (from 4 to 5 dB). In the VL-2 to VO-2 link, on the contrary, the average value of F is only 1.5 dB, while the sol-to-sol spread of values is as high as 2 dB.

The data collected with the VL-2 to VO-1 link were also of a less than desirable quality.

It was found that the antenna pattern characteristics for the

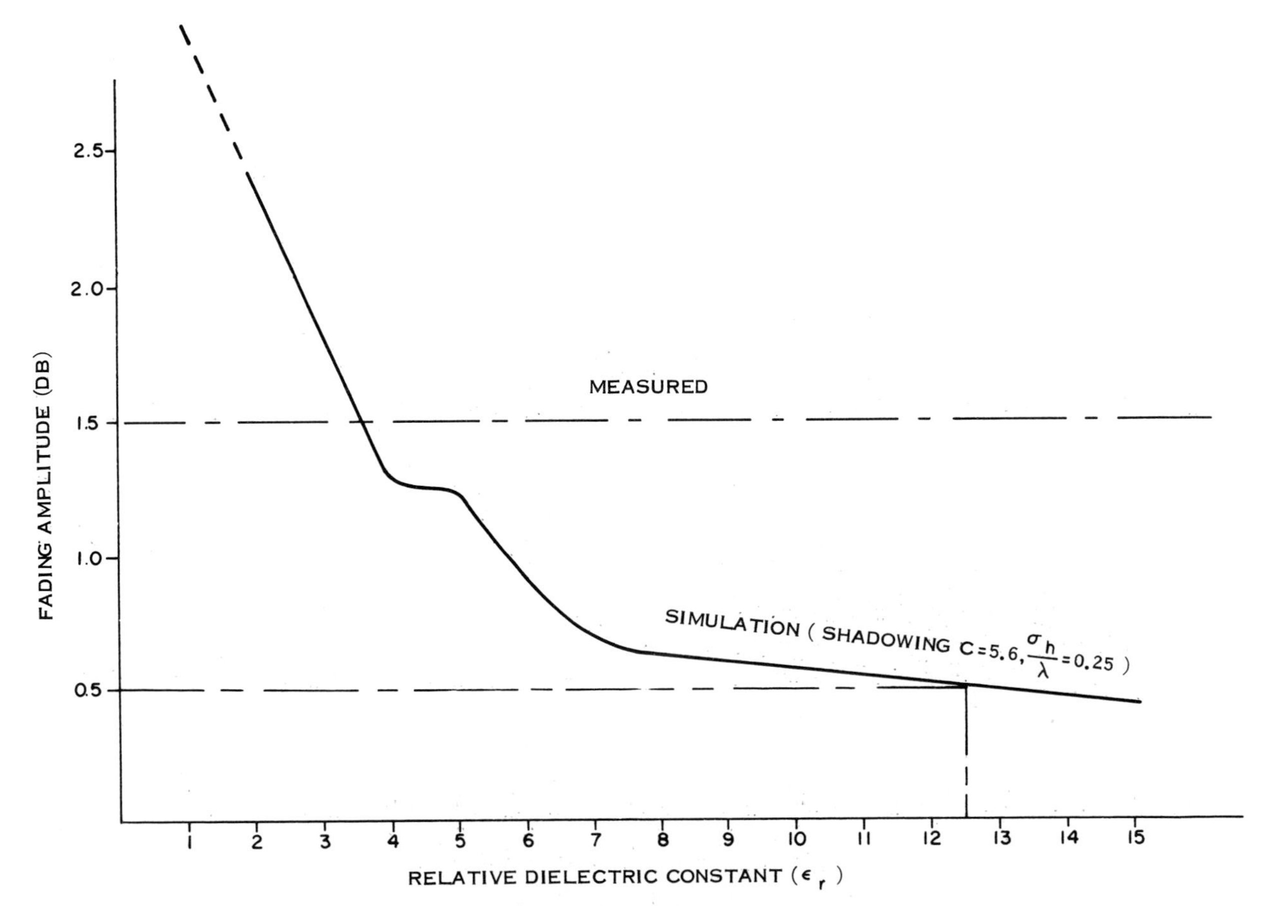

Fig. 8. Data reduction for VL-2 to VO-2 link: $\epsilon_r < 12.5$.

two links that use Lander 2 as the surface terminal fluctuate greatly (for example, the polarization axial ratios vary from approximately 2 to more than 100). This makes it difficult to deduce a precise value of surface dielectric constant at the Lander 2 site. Nevertheless, attempts were made to arrive at best estimates by using the VL-2 to VO-2 and VL-2 to VO-1 links. Procedures similar to those used for the VL-1 to VO-1 link were used in data reduction.

Results for the VL-2 to VO-2 link data reduction are summarized in Figure 8. With an estimated channel data accuracy of ±1.0 dB, the best estimate for the surface dielectric constant for the Lander 2 site is $\epsilon_r < 12.5$.

Although the VL-2 to VO-1 link appears to contain more than 6 dB of multipath residual, recognizable multipath features exist only for a few sols. Also, difficulties were experienced in this case in matching the observed results with the simulation, because accurate antenna side lobe patterns were not available for data reduction. Extrapolated gain patterns were then used.

Here, for the VL-2 to VO-1 link, simulated results are sensitive to the shadowing function and the surface roughness (Figure 9). The fading amplitude for this link is affected by a large measurement error in the data channel (±1 dB) and is also characterized by a wider statistical spread from sol to sol (2 dB). Thus from the VL-2 to VO-1 observations the most probable range for the surface dielectric constant around VL-2 appears to be $2.8 < \epsilon_r < 11$. As a result of the data from the VL-2 to VO-2 and VL-2 to VO-1 links the dielectric constant at the Lander 2 site is $2.8 < \epsilon_r < 12.5$.

SUMMARY AND CONCLUSIONS

The Viking lander-to-orbiter relay links at 381 MHz made it possible to perform measurements of the electrical properties of the Martian surface by the bistatic technique. The fading pattern of the signal intensity received on board the Viking orbiters via the lander-to-orbiter links contains information on the reflection coefficient of the Martian terrain in the vicinity of the lander site. While the lander-to-orbiter communication link is designed, in fact, primarily for an optimum direct path propagation, there are two major reasons why the received

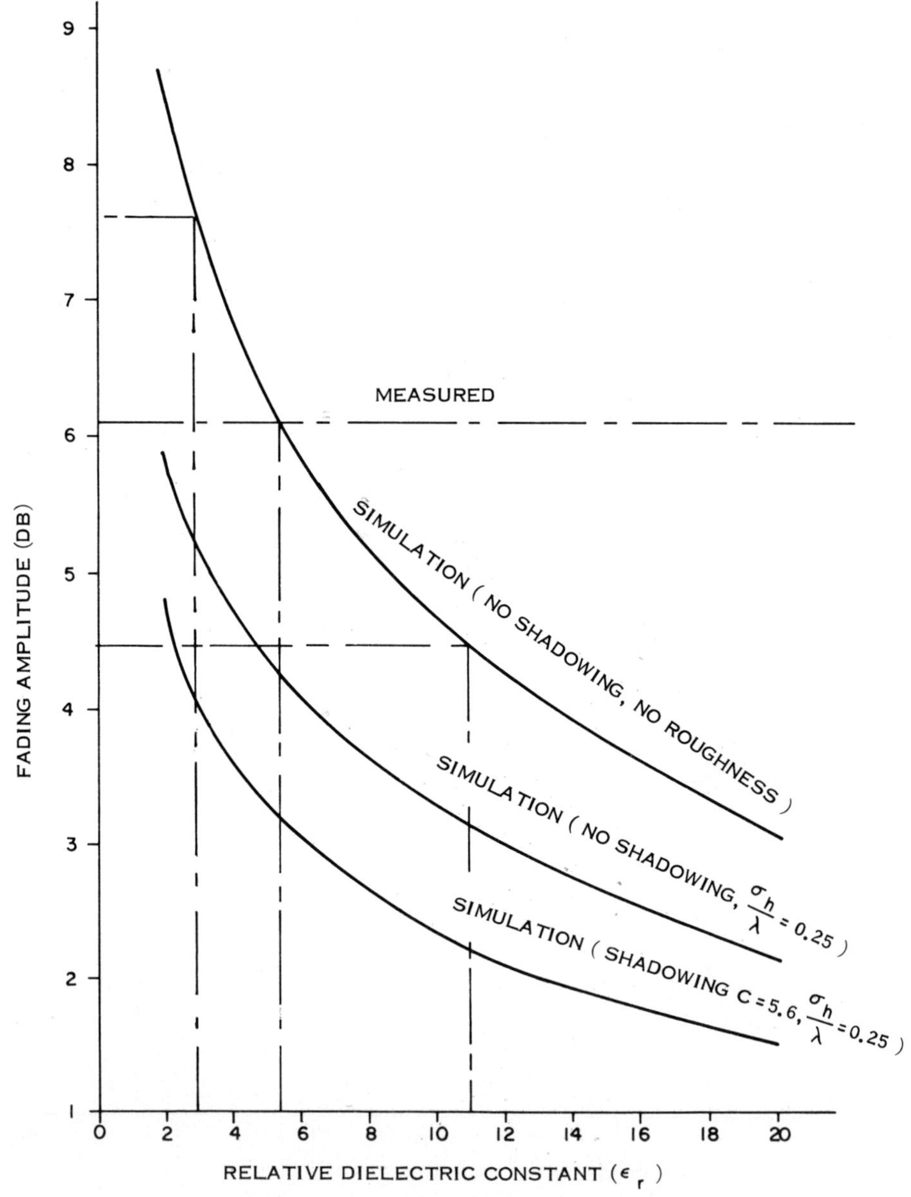

Fig. 9. Data reduction for VL-2 to VO-1 link: $2.8 < \epsilon_r < 11$.

signal intensity contains information on the reflectivity of the surface. First, in order to maintain a maximum lander's radio field of view, the lander has a low-directivity UHF antenna; the reception on the orbiter of lander signals will thus take place via a direct and unavoidable surface-reflected path. Second, while the technique of polarization rejection can be used to maximize the direct path signal, no unique polarization can be selected to reject the surface-reflected signal for all orbiter elevation angles.

In order to determine the Martian surface electrical properties, the observed multipath residuals were analyzed on the basis of a quasi-specular scattering theory. The surface reflection for each orbiter elevation angle was represented by a single ray. The surface irregularities were accounted for by a roughness scattering factor, and the terrain undulation was modeled by using a shadowing function. Antenna characteristics, such as gain patterns and relative phases, polarization axial ratios, and ellipse tilt, were also included in the simulation.

The data collected with the VL-1 to VO-1 link were of good quality and have shown that surface reflections are well above the threshold of detectability. The observed residuals give a Lander 1 site ϵ_r of 3.3 ± 0.7 and a Lander 1 site σ of 10^{-3}–10^{-5} mho/m. The electrical properties are similar to those of pumice and tuff.

The fading pattern data for the VL-2 to VO-2 and VL-2 to VO-1 links are not as self-consistent, and their quality is not as good as that of the VL-1 to VO-2 link. Hence the observed data only lead to a range of estimated values: the Lander 2 site ϵ_r is 2.8–12.5 and the Lander 2 site σ is 10^{-3}–10^{-5} mho/m.

REFERENCES

Barton, D. K., and H. R. Ward, *Handbook of Radar Measurement*, Prentice-Hall, Englewood Cliffs, N. J., 1969.

Beckmann, P., Shadowing of random rough surfaces, *IEEE Trans. Antennas Propagat.*, *13*, 384–388, 1965.

Beckmann, P., *The Depolarization of Electromagnetic Waves*, Golem Press, Boulder, Colo., 1968.

Beckmann, P., and A. Spizzichino, *The Scattering of Electromagnetic Waves From Rough Surfaces*, Pergamon, New York, 1963.

Campbell, M. J., and J. Ulrichs, Electrical properties of rocks and their significance for lunar radar observations, *J. Geophys. Res.*, *74*(25), 5867–5881, 1969.

Fjeldbo, G., Bistatic radar methods for studying planetary ionospheres and surfaces, Ph.D. dissertation, Stanford Univ., pp. 64–82, Stanford, Calif., April 1964.

Fjeldbo, G., A. Kliore, and B. Seidel, Bistatic radar measurements of the surface of Mars with Mariner 1969, *Icarus*, *16*, 502–508, 1972.

Mutch, T. A., A. B. Binder, F. O. Huck, E. C. Levinthal, S. Liebes, Jr., E. C. Morris, W. R. Patterson, J. B. Pollack, C. Sagan, and G. R. Taylor, The surface of Mars: The view from the Viking 1 lander, *Science*, *193*(4255), 791–801, 1976.

Pollack, J. B., and C. Sagan, Studies of the surface of Mars, *Radio Sci.*, *5*(2), 443, 1970.

Simmons, G., D. W. Strangeway, L. Bannister, D. Cubley, and G. LaTorraca, The surface electron properties experiments, *Moon*, *6*, 258, 1973.

Tyler, G. L., Oblique-scattering radar reflectivity of the lunar surface: Preliminary results from Explorer 35, *J. Geophys. Res.*, *73*(24), 7609–7620, 1968.

Tyler, G. L., and H. T. Howard, Dual-frequency bistatic radar investigations of the moon with Apollo 14 and 15, *J. Geophys. Res.*, *78*(23), 4852–4874, 1973.

Tyler, G. L., V. R. Eshleman, G. Fjeldbo, H. T. Howard, and A. M. Pitman, Bistatic-radar detection of lunar scattering centers with Lunar Orbiter 1, *Science*, *157*, 193–195, 1967.

Tyler, G. L., D. B. Campbell, G. S. Donn, R. R. Green, and H. J. Moore, Radar characteristics of Viking 1 landing site, *Science*, *193*, 812–815, Aug. 27, 1976.

(Received April 8, 1977;
revised June 6, 1977;
accepted June 6, 1977.)

Viking Radio Occultation Measurements of the Martian Atmosphere and Topography: Primary Mission Coverage

GUNNAR FJELDBO, DONALD SWEETNAM, JOSEPH BRENKLE, EDWARD CHRISTENSEN, DAVID FARLESS, JITENDRA MEHTA, AND BORIS SEIDEL

Jet Propulsion Laboratory, California Institute of Technology, Pasadena, California 91103

WILLIAM MICHAEL, JR., AND ANDREW WALLIO

Langley Research Center, Hampton, Virginia 23365

MARIO GROSSI

Raytheon Company, Sudbury, Massachusetts 02139

Radio occultation measurements were made at approximately 50 locations on Mars with the Viking Orbiter 1 *S* (2.3 GHz) and *X* (8.4 GHz) band tracking links during October 1976. The measurements have been used to study the topography and atmosphere of Mars at latitudes ranging from about 75°S to 70°N. By using the ingress and egress times obtained from the observed limb diffraction effects together with the best ephemerides available for the orbiter and the planet we have determined the surface elevations at the occultation points relative to the reference areoid. The observations agree with Mariner 9 and radar data to within 2 km. The mean atmospheric pressure at the areoid level was found to be 5.9 mbar during the northern midsummer season, a value which agrees quite well with data obtained at the landing sites. By comparing the new electron density measurements with earlier Mariner data we have determined that the temperature and the plasma scale height of the upper atmosphere appear to be functions of solar activity.

INTRODUCTION

Two spacecraft were placed in orbit around Mars by the NASA Viking Project, one on June 19 and the other on August 7, 1976 [*Martin and Young*, 1976; *Soffen and Snyder*, 1976; *Soffen*, 1976]. One of these spacecraft, denoted Viking Orbiter 1 (VO-1), was occulted daily by Mars during the October 6 to November 1, 1976, time interval. This orbital configuration allowed a number of occultation measurements to be made of the Martian atmosphere and topography by utilizing the radio tracking links with the orbiter.

The occultation experiment was conducted a few weeks prior to superior conjunction, and the sun-earth-Mars angle was decreasing from about 14° at the start of the measurements to 7° on the last day of the experiment. The measurements were therefore affected by radio phase scintillations in the solar corona. Only the better portions of the occultation data are discussed here; studies of the solar corona are reported elsewhere in this issue [*Tyler et al.*, 1977].

In describing the measurements we have organized the material into three parts. The first section of the report, entitled 'Instrumentation,' briefly outlines how the experiment was conducted. The results inferred from the observations, including the atmospheric pressure near the surface, the vertical temperature profiles, and the depth of the surface below the reference areoid, are summarized in the second section, entitled 'Data Analysis.' In the last portion of the report we compare our results with data obtained from other experiments.

INSTRUMENTATION

The tracking system utilized in the Viking occultation experiment consists of an *S* band (2.1 GHz) uplink (earth to orbiter) carrier frequency and coherently related *S* (2.3 GHz) and *X* (8.4 GHz) band downlink (orbiter to earth) carriers. The exact ratio between the *S* and *X* band downlink carrier frequencies transmitted from the spacecraft is 3/11.

In the so-called two-way mode a stable earth-based oscillator is used to generate the uplink carrier, and the orbiter radio transponder coherently retransmits the new RF carriers back to the tracking station. The ratio between the *S* band uplink and downlink frequencies is, in this mode, 221/240. All ingress measurements were conducted in the two-way mode.

An on-board crystal oscillator is automatically switched in to provide the reference frequency for the downlinks when the spacecraft receiver loses lock. This operating mode is denoted one-way tracking and was used during egress measurements. The two-way mode provides the highest frequency stability, since all link carrier frequencies in that case are derived from a stable earth-based reference oscillator.

The tracking signals from the orbiter were transmitted from a steerable paraboloidal reflector with a diameter of approximately 1.5 m. Prime tracking stations involved in the reception of these signals were the NASA Deep Space Stations 14 (Mars), 43 (Ballima), and 63 (Robledo), located in California, Australia, and Spain, respectively. All these stations are equipped with 64-m-diameter paraboloidal antennas and phase-locked loop receivers. In addition, stations 14 and 43 are instrumented with wide-band receiver channels and tape recorders which allow recordings to be made of the downlink signals for later digitization and computer analysis. The recording bandwidths are approximately 5 and 16 kHz at the *S* and *X* bands, respectively. For the interested reader, more information about the radio tracking instrumentation is available in a previous publication [*Michael et al.*, 1972]. A similar radio system was utilized in the Mariner 10 mission to Venus and Mercury [*Howard et al.*, 1974].

Paper number 7S0429.

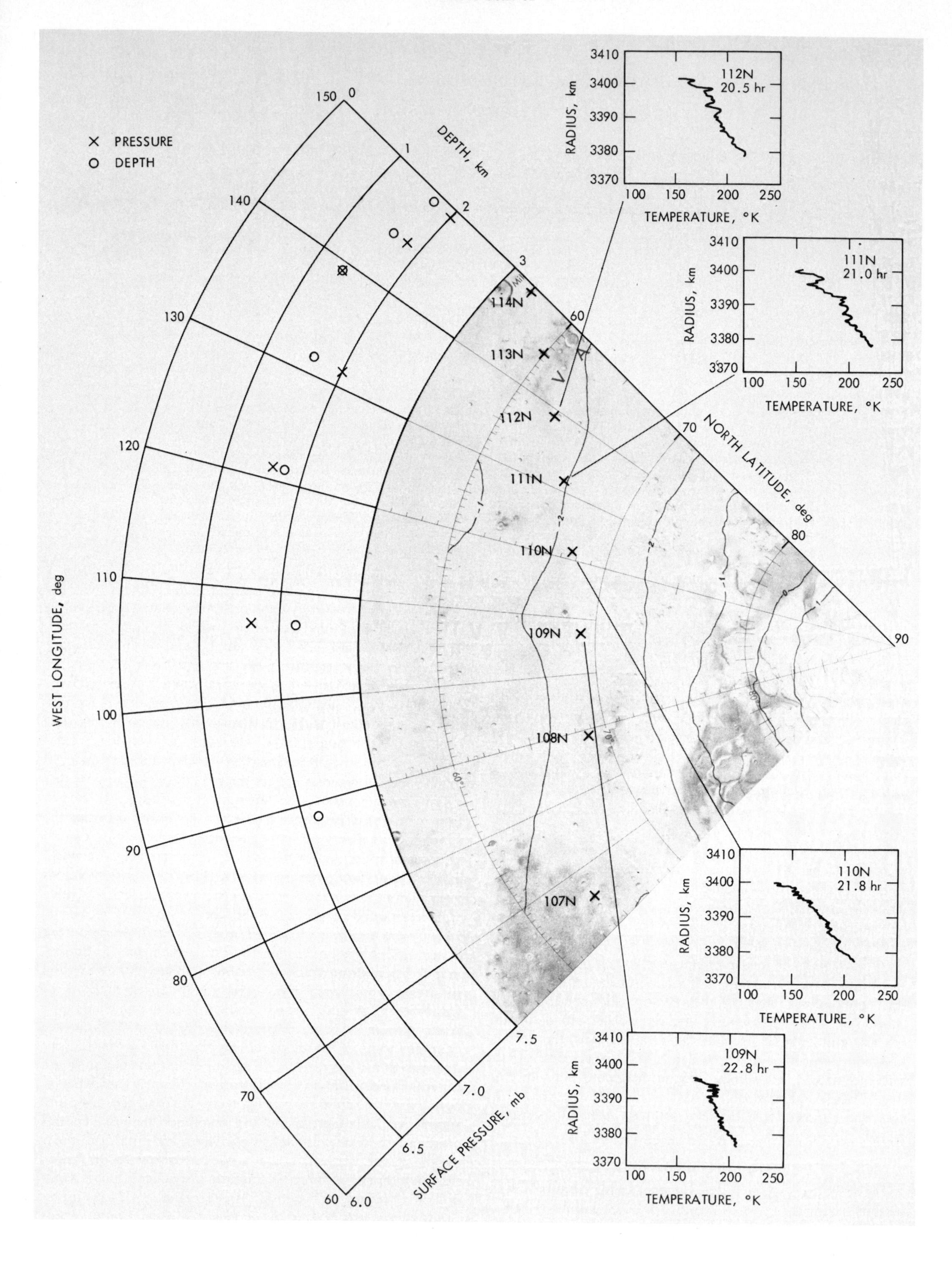

× PRESSURE
○ DEPTH
DEPTH, km
0
1
2
3
150
140
130
120
110
100
90
80
70
60
WEST LONGITUDE, deg
SURFACE PRESSURE, mb
6.0
6.5
7.0
7.5
NORTH LATITUDE, deg
60
70
80
90
114N
113N
112N
111N
110N
109N
108N
107N
112N
20.5 hr
111N
21.0 hr
110N
21.8 hr
109N
22.8 hr
RADIUS, km
3410
3400
3390
3380
3370
TEMPERATURE, °K
100
150
200
250

DATA ANALYSIS

The first step in the analysis of the radio frequency data consists of computing the Doppler perturbations imposed on the links by the Martian atmosphere. The differential dispersive downlink Doppler, defined as the received *S* band frequency minus 3/11 of the *X* band frequency, is used to study ionization in the atmosphere of Mars. The effects of plasma changes in the solar corona and the terrestrial ionosphere are approximated by subtracting a linear frequency term from the data. This frequency drift term is determined by fitting a straight line in a least squares sense to the differential dispersive Doppler data obtained above the Martian atmosphere.

The frequency perturbations imposed on the *S* band links by the atmosphere of Mars are determined by taking the received *S* band carrier frequency and subtracting the corresponding frequency computed from changes in the geometrical distance between the spacecraft and the tracking station. This procedure yields what is commonly referred to as atmospheric Doppler residuals. A correction for non-Martian propagation effects is again approximated by subtracting a linear frequency drift term from the Doppler residuals.

Frequency perturbations imposed on the radio links during atmospheric occultation measurements can be employed to determine the vertical distributions of the tropospheric gas refractivity and the ionospheric electron density [e.g., *Fjeldbo and Eshleman,* 1968]. This computation is done by inverting the integral equations relating the gas refractivity and electron density profiles to the atmospheric Doppler frequency perturbations. The location of the solid surface is obtained from limb diffraction effects observed at ingress and egress.

Figure 1 shows the results obtained from the first ingress measurements made with Viking Orbiter 1 in the Vastitas Borealis region of Mars. The points marked by crosses on the map indicate the locations where the radio links grazed the surface. The integer to the left of each occultation point gives the number of orbital revolutions since orbit insertion, and the capital letter N indicates that the measurements were made during ingress or entry into occultation. Egress or exit from occultation is denoted by an X after the revolution number.

In the Martian troposphere the gas refractivity is proportional to the molecular number density. By employing information on atmospheric composition obtained with the molecular analysis instrumentation on the Viking lander modules [*Owen and Biemann,* 1976] we have been able to calculate from the gas refractivity data the number density as a function of altitude above each occultation point. On the assumption of hydrostatic atmospheric equilibrium these data on the density distribution have, in turn, been utilized to compute pressure and temperature profiles above the occultation points. Examples of temperature profiles determined in this manner are shown in Figure 1. The vertical scales in these temperature plots give the distance from the planetary center of mass. The smallest radius for which a temperature is plotted corresponds to the solid surface. As is expected, the planetary radius is seen to increase with decreasing latitude. The values obtained for the radii appear to agree quite well with previous occultation measurements conducted with Mariner 9 [*Kliore et al.,* 1972].

Data shown on the left-hand side of Figure 1 give surface pressure (marked by crosses) and depth (marked by circles) at the occultation points. The depth of the surface is measured from the same reference gravity equipotential surface or areoid that was utilized by the U.S. Geological Survey in making the map [*Wu,* 1975]. This reference areoid was determined from Mariner 9 tracking data [*Christensen,* 1975]. The numerical values marked on the elevation contours on the map give the height in kilometers above the reference areoid. The map is based on Mariner 9 and radar data, and it is labeled M 25M 3 RMC.

In order to separate the effects of electrons and neutral molecules experimentally, we used both *S* band and *X* band Doppler data in the calculation of the gas refractivity distribution in the Martian troposphere. However, since the *X* band was only available on the downlink, we could not correct for nonlinear dispersive Doppler noise produced on the *S* band uplink by the solar corona. For this reason, some of the data are too noisy to yield reliable atmospheric temperatures and pressures. Only the better data sets are shown in Figure 1 and the following illustrations.

Subsequent to the Vastitas Borealis measurements, data were acquired on the atmosphere and topography of the Arcadia Planitia and Amazonis Planitia regions of Mars. These results are summarized in Figure 2. Here again, we have only shown the better data sets, since the noisier ones yield very unreliable temperature and pressure results. In fact, some of the tracking data were even too noisy to be utilized to calculate the depth of the surface at the occultation points because the spacecraft orbit could not be determined with sufficient accuracy.

Figures 1 and 2 show results obtained from ingress measurements made at Deep Space Stations 14 (Mars), 43 (Ballima), and 63 (Robledo). In Figure 3 we show topographic data obtained from egress measurements in the Acidalia Planitia, Chryse Planitia, and Margaritifer Sinus regions of Mars. These measurements were made at station 43 during orbital revolutions 108–116. As was described in the previous section on instrumentation, the egress data were acquired in the one-way tracking mode. The on-board crystal oscillator utilized as a frequency reference during the one-way measurements was stable enough to allow us to study the topography of Mars. However, the frequency stability was not adequate for tropospheric studies.

Poor crystal oscillator stability has also affected the accuracy of a number of past occultation experiments. The problem now appears to have been solved, however. Future spacecraft may carry a radiation-hardened crystal placed in a thermally controlled oven. The new oscillator was designed by Frequency Electronics, Inc., and it has a short-term stability of a few parts in 10^{12} after linear drift has been removed (G. E. Wood, unpublished data, 1977). The first mission scheduled to employ this oscillator is the Voyager mission to Jupiter, Saturn, and Uranus [*Eshleman et al.,* 1977].

Subsequent to orbital revolution 116 the egress events occurred over station 63, which was only instrumented to pro-

Fig. 1. (Opposite) Radio occultation data obtained with Viking Orbiter 1 in the north polar region of Mars during October 1976. The crosses on the map show the locations where the radio links grazed the Martian surface. The integer to the left of each occultation point gives the number of revolutions that the spacecraft had completed around Mars since insertion into a planetary orbit. The capital letter N following the revolution number denotes ingress. Plots giving the temperature in the atmosphere above the occultation points as a function of the distance from the planetary center of mass are shown in the right-hand portion of the figure. The lowest point at which a temperature is given corresponds to the radius of the solid surface. Local time in hours past midnight is indicated in the upper right-hand corner of each temperature plot. Data on the left-hand side of the map give surface pressure (marked by crosses) and surface depth below the reference areoid (marked by circles) at the occultation points.

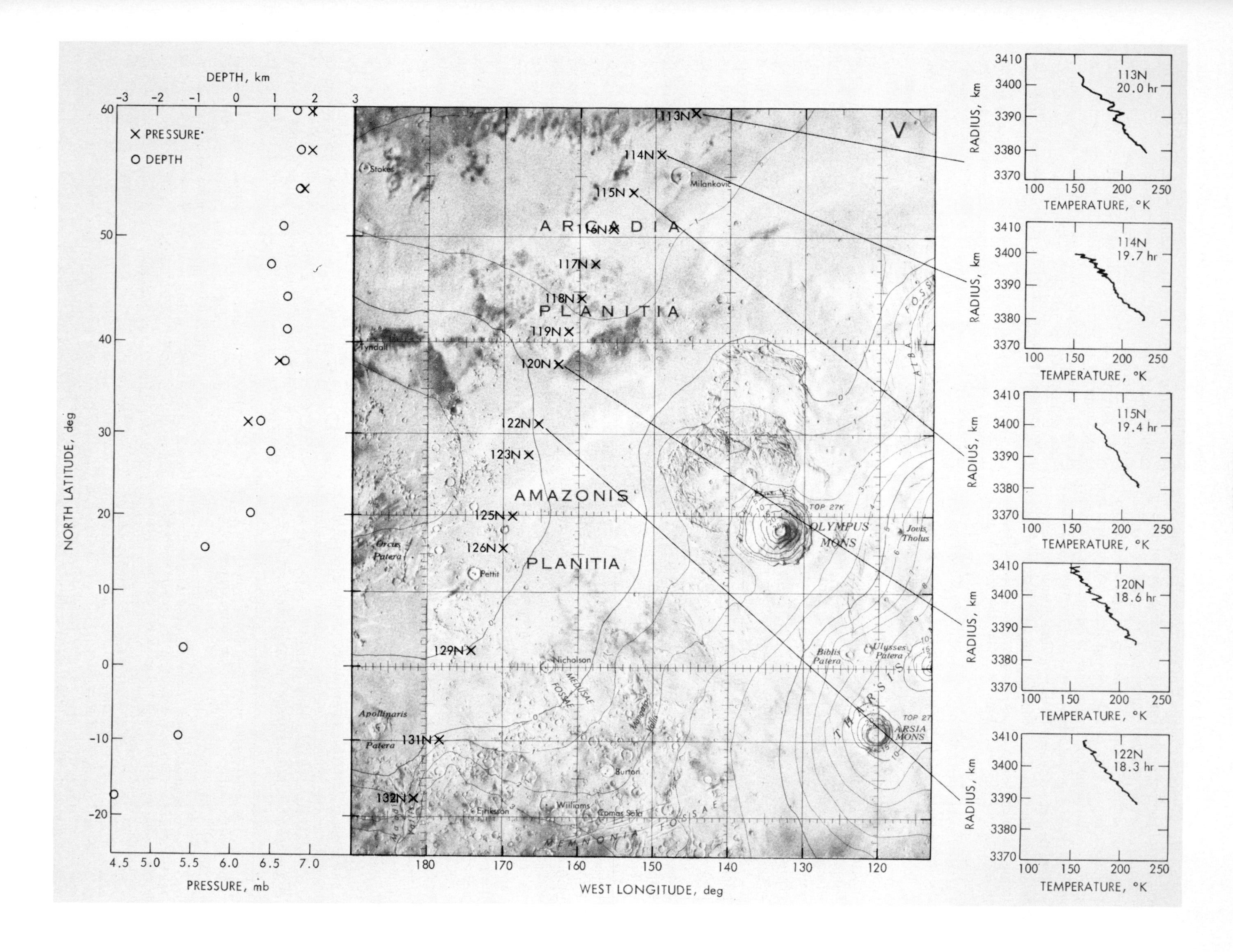
DEPTH, km
-3
-2
-1
0
1
2
3
PRESSURE
DEPTH
NORTH LATITUDE, deg
60
50
40
30
20
10
0
-10
-20
4.5
5.0
5.5
6.0
6.5
7.0
PRESSURE, mb
V
ARCADIA
PLANITIA
AMAZONIS
PLANITIA
113N
114N
115N
116N
117N
118N
119N
120N
122N
123N
125N
126N
129N
131N
132N
Stokes
Milankovič
Tyndall
Orcus
Patera
Pettit
Nicholson
MEDUSAE
FOSSAE
Apollinaris
Patera
Burton
Williams
Comas Sola
Ericsson
MEMNONIA
TOP 27K
OLYMPUS
MONS
Jovis
Tholus
Biblis
Patera
Ulysses
Patera
THARSIS
ARSIA
MONS
180
170
160
150
140
130
120
WEST LONGITUDE, deg
113N
20.0 hr
114N
19.7 hr
115N
19.4 hr
120N
18.6 hr
122N
18.3 hr
RADIUS, km
3410
3400
3390
3380
3370
100
150
200
250
TEMPERATURE, °K

vide Doppler data from phase-locked loop receiver channels. No amplitude data were obtained at station 63. The egress data from this station are not suitable for topographic studies because there is no way that one can reliably distinguish limb diffraction effects from the Doppler transients that occur when the station's phase-locked loop receivers lock on to the reappearing spacecraft signals. Also, the phase-locked loop receivers will, in many cases, simply not lock on to the downlink signals until well after egress has occurred.

In addition to the data displayed in Figures 1–3 we acquired egress data in the south polar region of Mars at station 14 during orbital revolutions 127–134. These measurements were made between approximately 225° and 350° west longitude and 63° and 75° south latitude. While all the other topographic data that we acquired agree very well with the elevation contours indicated on the map prepared by the U.S. Geological Survey, the data for the south polar region show surface elevations 2–6 km larger than those given on the map, in approximate agreement with earlier occultation measurements made with Mariner 9. A summary of the new measurements of the shape of Mars is given in Figure 4.

Finally, we have compared the new data on the electron density distribution in the Martian upper atmosphere with results deduced from previous radio occultation experiments conducted with the Mariner series of spacecraft. As a result of this work, we have determined that the topside plasma scale height changed by about a factor of 2 during the current solar cycle. These results are illustrated in Figure 5.

If one assumes that the topside of the Martian ionosphere is in photochemical equilibrium (plasma being formed through the process of photo-ionization of carbon dioxide by solar extreme ultraviolet) and, furthermore, that the effective electron-ion recombination rate coefficient is independent of altitude, one finds that the plasma scale height must be twice as large as the scale height of the neutral CO_2 gas. Data obtained during the Mariner 1969 mission to Mars may be used to check these assumptions. The scale height of the neutral CO_2 above the ionospheric peak was at that time observed to be approximately 23 km on the basis of UV measurements [*Barth et al.*, 1969]. The radio occultation measurements yielded a plasma scale height of 45 km [*Fjeldbo et al.*, 1970], i.e., about twice the neutral scale height.

By using the assumptions outlined above, one can infer the temperature of the neutral gas from the plasma scale height measurements. The calculation is done by noting that the temperature (T) and the scale height (H) of CO_2 are related by $H = kT/mg$. Here k, m, and g denote Boltzmann's constant, the molecular mass of CO_2, and the acceleration of gravity, respectively. The resulting gas temperatures can also be read off the chart shown in Figure 5.

In order to allow the reader to correlate the observed changes in the temperature and scale height with the solar activity we have also plotted the 10.7-cm solar flux in Figure 5. These data were obtained by taking the monthly mean flux values published by the Solar-Terrestrial Data Services Division of the National Oceanic and Atmospheric Administration and adjusting them for the appropriate range to Mars. The resulting flux data were subsequently filtered by computing yearly mean values.

Fig. 2. (Opposite) Radio occultation data obtained with Viking Orbiter 1 in the Arcadia Planitia and Amazonis Planitia regions of Mars during October 1976. The measurements were made in the evening atmosphere during the midsummer season.

SUMMARY AND CONCLUSIONS

Since Mars has no seas and therefore no sea level that may be used to define a reference gravity equipotential surface or reference areoid, one must employ a somewhat arbitrary mathematical surface instead. The shape of the reference areoid that currently is being used in topographic studies of Mars is defined by a gravity field described by an expansion in spherical harmonics to fourth order and fourth degree. The coefficients in this expansion have been determined by utilizing Mariner 9 tracking data [*Lorell et al.*, 1973; *Jordan and Lorell*, 1975]. The size of the reference areoid was determined by picking the areoid which in a least squares sense best fits the 6.1 mbar isobaric surface deduced from the Mariner 9 occultation experiment [*Christensen*, 1975]. Of course, we now know that the atmospheric pressure changes with the Martian seasons, but we still use the same reference areoid. This reference areoid was also utilized by the U.S. Geological Survey in making contour map M 25M 3 RMC.

On the basis of the observed ingress and egress times and the best ephemerides available for the spacecraft and the planet we have determined the radius of the Martian surface at the occultation points to an accuracy of approximately ± 1 km. The altitude differences between the solid surface and the reference areoid represent topographic data that readily can be compared with the elevation contours that are shown on U.S. Geological Survey map M 25M 3 RMC. Except for the measurements made in the south polar region of Mars the topographic observations appear to agree very well with the altitude contours on the map (see Figures 1–3). The data obtained between 63° and 75° south latitude give surface elevations 2–6 km larger than those indicated on the map, in approximate agreement with previous occultation measurements conducted with the Mariner 9 spacecraft [*Kliore et al.*, 1973]. The question of surface elevation in the south polar region could assume considerable practical importance, since this may be the landing site for a mobile vehicle, if current discussions within NASA bear fruit. The new data on the shape of the Martian surface are summarized in Figure 4.

The atmospheric occultation data consist of temperature and pressure profiles near the surface and electron density profiles in the upper atmosphere of Mars. The tropospheric data, which were obtained at latitudes ranging from 30° to 70° N, are displayed in Figures 1 and 2.

Because of the small sun-earth-Mars angle, all the Primary Mission occultation measurements were conducted near the terminator. As is indicated by the local Martian time given in the upper right-hand corner of each temperature plot, the atmospheric measurements were all made in the evening. The 3σ uncertainty in the temperature computed near the surface is approximately 10°K. The vertical resolution, as measured by the diameter of the first S band Fresnel zone, is about 3 km. This resolution appears to be inadequate for resolving the nighttime inversion layer expected near the surface. However, new occultation data acquired during the winter of 1977 have provided better altitude resolution and clearly show an inversion layer near the morning terminator. The temperature inversion extends to an altitude of 1–2 km and resembles theoretical profiles published by *Gierasch and Goody* [1968].

Of particular interest are the atmospheric temperature profiles obtained in the north polar region of Mars, which show near-surface nighttime temperatures in excess of 200°K (see Figure 1). These measurements tend to lend credence to the hypothesis that the summer residual ice patches found between about 70° and 85° north latitude consist of water ice, since

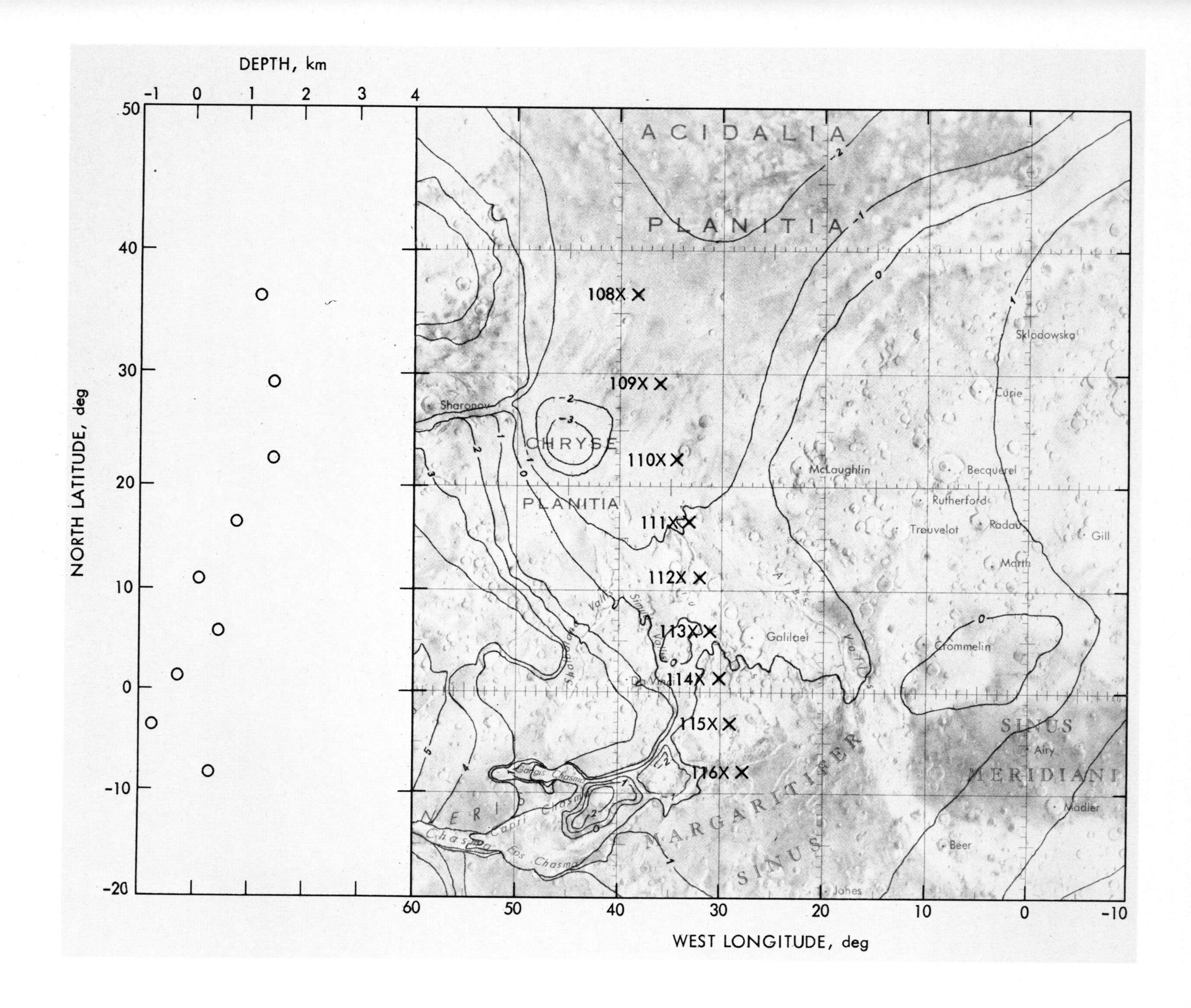

DEPTH, km
NORTH LATITUDE, deg
WEST LONGITUDE, deg
ACIDALIA PLANITIA
CHRYSE PLANITIA
SINUS MERIDIANI
MARGARITIFER SINUS
Sklodowska
Curie
Becquerel
Rutherford
Radau
Gill
Marth
Trouvelot
McLaughlin
Galilaei
Crommelin
Airy
Mädler
Beer
Jones
Sharonov
Ares Vallis
Simud Vallis
Shalbatana Vallis
Ganges Chasma
Capri Chasma
Eos Chasma
108X
109X
110X
111X
112X
113X
114X
115X
116X

Fig. 3. (Opposite) Radio occultation data obtained with Viking Orbiter 1 in the Acidalia Planitia, Chryse Planitia, and Margaritifer Sinus regions of Mars. The depth of the surface is measured in relation to the reference areoid currently in use [*Christensen*, 1975]. The maps shown in this report were prepared for NASA by the U.S. Geological Survey on the basis of Mariner 9 and radar data. The numerical values marked on the elevation contours on the maps give the height of the surface above the reference areoid in kilometers. The estimated vertical accuracy of each source of data indicates a probable error of 1–2 km for the contour lines [*Wu*, 1975].

CO_2 ice would require a temperature of 150°K or less. Similar conclusions have been reached on the basis of infrared measurements [*Farmer et al.*, 1976; *Kieffer et al.*, 1976].

Temperature profiles similar to those given in this report have been derived from infrared and entry science measurements conducted in conjunction with the second Viking landing [*Kieffer et al.*, 1976; *Seiff and Kirk*, 1976]. The entry science measurements were made at approximately 0900 hours local time in the Utopia Planitia region of Mars on September 3, 1976. Similar temperature profiles have also been published by *McElroy et al.* [1976]. The latter report contains results from both Viking landings.

In addition to temperature profiles, Figures 1 and 2 give the atmospheric pressure near the surface. The 3σ uncertainty in each pressure measurement is approximately 0.3 mbar. To within the accuracy of the data the observed surface pressures are consistent with the topographic measurements.

The Viking lander modules were instrumented to monitor the atmospheric pressure at the landing sites [*Hess et al.*, 1976]. During October 1976, mean evening pressures of approximately 6.7 and 7.4 mbar were observed at the Viking Lander 1 and 2 locations, respectively. In order to compare our pressure measurements with data obtained from the meteorology instrumentation on the landers it is convenient to use the areoid as a reference level. The average areoid pressure deduced from the occultation measurements described in this report is 5.9 ± 0.1 mbar. The data utilized to compute this value were obtained in the evening atmosphere during the Martian midsummer season ($L_s \approx 137°$) at northern latitudes ranging from 30° to 70°. The corresponding areoid pressure deduced from lander meteorology data agrees very well with the value reported here.

Finally, the differential dispersive Doppler measurements have yielded new data on the electron density distribution in the Martian dayside atmosphere. Because of phase scintillations in the solar corona the noise level was considerably

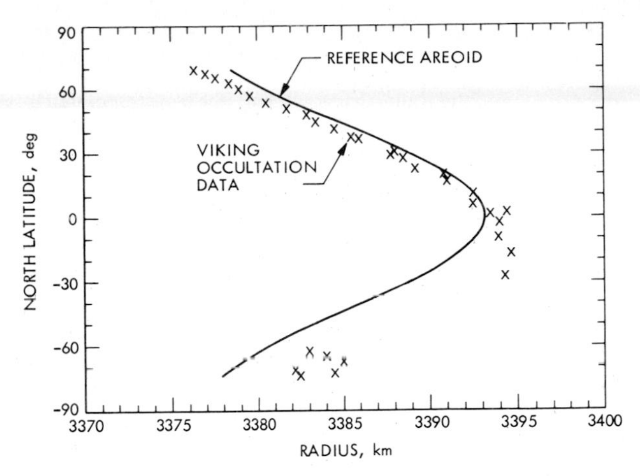

Fig. 4. Radius of Mars, measured from the planetary center of mass, versus areographic latitude. Also shown is the radius of the reference areoid at the longitudes sampled by the tracking links.

higher than that seen in previous experiments. It was nevertheless possible to detect layers similar to those reported earlier. There was one important difference from previous results, however; the upper atmosphere appears to have been considerably colder during the Viking measurements than was reported for the years 1965 [*Fjeldbo and Eshleman*, 1968], 1969 [*Fjeldbo et al.*, 1970], and 1971 [*Kliore et al.*, 1972]. As is illustrated in Figure 5, this difference appears to be due to changes in solar activity. The new occultation data yielded a temperature of about 200°K in the upper atmosphere of Mars. By comparison, the measurements conducted during the entry of Viking Lander 1 indicated an average temperature of 180° ± 20°K between 140- and 200-km altitude [*Nier et al.*, 1976].

The results discussed above were obtained from data acquired on the sunlit side of the planet. Mars is also known to have a nighttime ionosphere. It was first detected by *Kolosov et al.* [1975], who reported observing a nighttime peak density of 7×10^3 el cm^{-3}. This nighttime ionization level is too low, however, to be detectable in the Viking data acquired near superior conjunction.

The measurements discussed in this report were conducted in October 1976. More data were acquired during the winter and spring of 1977, and preliminary mission-planning data indicate that occultations will continue to occur through 1978. Continued Viking measurements may therefore allow us to study the atmosphere of Mars through a complete Martian year.

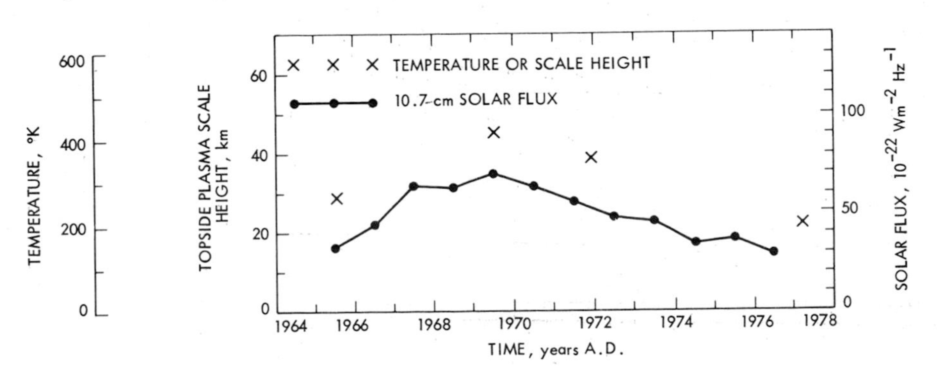

Fig. 5. Changes in the temperature and the topside plasma scale height (marked by crosses) of the Martian upper atmosphere during the current solar cycle. The data, which apply to the 150- to 200-km altitude region of the sunlit side of Mars, were acquired by conducting radio occultation measurements with Mariner 4, 6, 7, and 9 and with Viking. Also shown are data on the 10.7-cm solar flux adjusted to the range of Mars.

Acknowledgments. The radio occultation measurements described here were accomplished through the combined efforts of the Viking Project staff, the Deep Space Net personnel, the Martin Marietta Corporation, and the Radio Science Team. The team appointed by the National Aeronautics and Space Administration to conduct radio science experiments with the Viking spacecraft consisted of J. Brenkle, D. Cain, G. Fjeldbo, and C. Stelzried from the Jet Propulsion Laboratory; J. Davies from the University of Manchester; M. Grossi from the Raytheon Company; W. Michael, Jr. (team leader), and R. Tolson from the Langley Research Center; I. Shapiro from the Massachusetts Institute of Technology; and G. Tyler from Stanford University. This paper constitutes one portion of the Radio Team's final report. G. Cowdery, L. Dicken, R. Diehl, T. Komarek, D. Pradhan, Z. Shippony, and J. Wackley provided invaluable support in the acquisition and analysis of the data. Management support was given generously by D. Johnston and D. Mudgway. Stimulating discussions with D. Cain, V. Eshleman, R. Henry, T. Howard, A. Kliore, H. Masursky, C. Snyder, and P. Woiceshyn are gratefully acknowledged. The work reported here was supported by the National Aeronautics and Space Administration.

REFERENCES

Barth, C. A., W. G. Fastie, C. W. Hord, J. B. Pearce, K. K. Kelly, A. I. Stewart, G. E. Thomas, G. P. Anderson, and O. F. Raper, Ultraviolet spectroscopy, Mariner Mars 1969, a Preliminary Report, *NASA Spec. Publ., SP-225,* 97–104, 1969.

Christensen, E. J., Martian topography derived from occultation, radar, spectral, and optical measurements, *J. Geophys. Res., 80,* 2909–2913, 1975.

Eshleman, V. R., G. L. Tyler, J. D. Anderson, G. Fjeldbo, G. S. Levy, G. E. Wood, and T. A. Croft, Radio science investigations with Voyager, *Space Sci. Rev.,* in press, 1977.

Farmer, C. B., D. W. Davies, and D. D. LaPorte, Mars: Northern summer ice cap—Water vapor observations from Viking 2, *Science, 194,* 1339–1341, 1976.

Fjeldbo, G., and V. R. Eshleman, The atmosphere of Mars analyzed by integral inversion of the Mariner 4 occultation data, *Planet. Space Sci., 16,* 1035–1059, 1968.

Fjeldbo, G., A. Kliore, and B. Seidel, The Mariner 1969 occultation measurements of the upper atmosphere of Mars, *Radio Sci., 5,* 381–386, 1970.

Gierasch, P., and R. Goody, A study of the thermal and dynamical structure of the Martian lower atmosphere, *Planet. Space Sci., 16,* 615–646, 1968.

Hess, S. L., R. M. Henry, C. B. Leovy, J. A. Ryan, J. E. Tillman, T. E. Chamberlain, H. L. Cole, R. G. Dutton, G. C. Greene, W. E. Simon, and J. L. Mitchell, Mars climatology from Viking 1 after 20 sols, *Science, 194,* 78–81, 1976.

Howard, H. T., G. L. Tyler, G. Fjeldbo, A. J. Kliore, G. S. Levy, D. L. Brunn, R. Dickinson, R. E. Edelson, W. L. Martin, R. B. Postal, B. L. Seidel, T. T. Sesplaukis, D. L. Shirley, C. T. Stelzried, D. N. Sweetnam, A. I. Zygielbaum, P. B. Esposito, J. D. Anderson, I. I. Shapiro, and R. D. Reasenberg, Venus: Mass, gravity field, atmosphere, and ionosphere as measured by the Mariner 10 dual-frequency radio system, *Science, 183,* 1297–1301, 1974.

Jordan, J. F., and J. Lorell, Mariner 9: An instrument of dynamical science, *Icarus, 25,* 146–165, 1975.

Kieffer, H. H., P. R. Christensen, T. Z. Martin, E. D. Miner, and F. D. Palluconi, Temperatures of the Martian surface and atmosphere: Viking observations of diurnal and geometric variations, *Science, 194,* 1346–1351, 1976.

Kliore, A. J., D. L. Cain, G. Fjeldbo, B. L. Seidel, M. J. Sykes, and S. I. Rasool, The atmosphere of Mars from Mariner 9 radio occultation measurements, *Icarus, 17,* 484–515, 1972.

Kliore, A. J., G. Fjeldbo, B. L. Seidel, M. J. Sykes, and P. M. Woiceshyn, *S*-band radio occultation measurements of the atmosphere and topography of Mars with Mariner 9: Extended mission coverage of polar and intermediate latitudes, *J. Geophys. Res., 78,* 4331–4351, 1973.

Kolosov, M. A., O. I. Yakovlev, G. D. Yakovleva, A. I. Efimov, B. P. Trusov, T. S. Timofeeva, Yu. M. Kruglov, V. A. Vinogradov, and V. P. Oreshkin, Results of investigations of the atmosphere of Mars by the method of radio transillumination by means of the automatic interplanetary stations Mars 2, Mars 4, and Mars 6, *Cosmic Res., 13,* 54–59, 1975.

Lorell, J., G. H. Born, E. J. Christensen, P. B. Esposito, J. F. Jordan, P. A. Laing, W. L. Sjogren, S. K. Wong, R. D. Reasenberg, I. I. Shapiro, and G. L. Slater, Gravity field of Mars from Mariner 9 tracking data, *Icarus, 18,* 304–316, 1973.

Martin, J. S., and A. T. Young, Viking to Mars: Profile of a space expedition, *Astronaut. Aeronaut., 14,* 22–47, 1976.

McElroy, M. B., T. Y. Kong, Y. L. Yung, and A. O. Nier, Composition and structure of the Martian upper atmosphere: Analysis of results from Viking, *Science, 194,* 1295–1298, 1976.

Michael, W. H., D. L. Cain, G. Fjeldbo, G. S. Levy, J. G. Davies, M. D. Grossi, I. I. Shapiro, and G. L. Tyler, Radio science experiments: The Viking Mars orbiter and lander, *Icarus, 16,* 57–73, 1972.

Nier, A. O., W. B. Hanson, A. Seiff, M. B. McElroy, N. W. Spencer, R. J. Duckett, T. C. D. Knight, and W. S. Cook, Composition and structure of the Martian atmosphere: Preliminary results from Viking 1, *Science, 193,* 786–788, 1976.

Owen, T., and K. Biemann, Composition of the atmosphere at the surface of Mars: Detection of argon-36 and preliminary analysis, *Science, 193,* 801–803, 1976.

Seiff, A., and D. B. Kirk, Structure of Mars' atmosphere up to 100 kilometers from the entry measurements of Viking 2, *Science, 194,* 1300–1303, 1976.

Soffen, G. A., Status of the Viking missions, *Science, 194,* 57–59, 1976.

Soffen, G. A., and C. W. Snyder, The first Viking mission to Mars, *Science, 193,* 759–766, 1976.

Tyler, G. L., J. P. Brenkle, T. A. Komarek, and A. I. Zygielbaum, The Viking solar corona experiment, *J. Geophys. Res., 82,* this issue, 1977.

Wu, S. S. C., Topographic mapping of Mars, *Interagency Rep. Astrogeology 63,* U.S. Geol. Surv., Flagstaff, Ariz., 1975.

(Received April 4, 1977;
revised May 17, 1977;
accepted May 17, 1977.)

Mars Gravity Field: Combined Viking and Mariner 9 Results

J. P. GAPCYNSKI, R. H. TOLSON, AND W. H. MICHAEL, JR.

NASA Langley Research Center, Hampton, Virginia 23665

A Martian gravity field of sixth degree and order has been determined from an analysis of a combination of Viking and Mariner 9 spacecraft Doppler tracking data. A short-arc technique utilizing approximately 4 hours of data centered at periapsis was used, and the data covered 16 arcs from Mariner 9 and 17 arcs from the Viking orbiters. The data were selected so as to obtain a uniform distribution of periapsis longitudes over the surface of Mars, and both S band and X band data were used where possible to eliminate charged particle effects. Inclusion of the Viking data arcs altered the Martian geoid features, as defined by previous short-arc analysis techniques of Mariner 9 data, by about 80 m in the southern hemisphere and about 140 m in the northern hemisphere.

INTRODUCTION

Prior to the Viking missions the gravitational field of Mars had been investigated extensively from analyses of the Mariner 9 mission spacecraft tracking data, and different techniques were utilized to obtain results. For example, the analyses performed by *Reasenberg et al.* [1975] and *Lorell et al.* [1973] utilized data over several continuous orbital periods, whereas *Sjogren et al.* [1975] and *Daniels and Tolson* [1976] utilized short data arcs centered at spacecraft periapsis for each of several orbits. For the high-eccentricity orbits which characterize the Mars missions the higher-degree gravitational perturbations are primarily effective when the spacecraft is in the vicinity of periapsis. During the Mariner mission the spacecraft periapsis was restricted to a southern latitude of about 25°, and it may be anticipated that the gravitational results in each case would overemphasize the effect of the southern latitude regions of Mars. The Viking missions, on the other hand, have periapsis locations in the northern hemisphere, and an analysis of these tracking data should complement the Mariner data and result in a more accurate definition of the Martian gravity field.

This paper presents the results from an investigation of the gravity field of Mars utilizing both Mariner 9 and Viking 1 and 2 spacecraft data.

ANALYSIS

The analysis technique is an extension of the multiple short-arc method used by *Daniels and Tolson* [1976] for the reduction of Mariner 9 tracking data. This procedure involves the utilization of short periods of tracking data centered about spacecraft periapsis, and data arcs are selected so as to establish a satisfactory distribution of periapsis longitudes over the surface of Mars. For high-eccentricity orbits of the type established for both Mariner 9 and Viking spacecraft, it was shown [*Daniels and Tolson,* 1976] that the use of data arcs covering a period of about 4 hours and having a ground track periapsis longitude separation approximately equal to periapsis altitude was sufficient to establish a valid spherical harmonic representation of the gravity field of Mars to sixth, and possibly seventh, degree. By use of the short-arc technique, aliasing by unmodeled gravitational and nongravitational forces can be reduced to an acceptable level, computational time can be reduced over that required for multiple orbital period analysis, and greater flexibility can be obtained in data selection.

Standard weighted least squares and differential correction concepts were utilized. The equations of motion and the variational equations were integrated with a general purpose orbit integration program using power series methods [*Hartwell,* 1967]. The estimation accuracy of the coefficients was obtained from the 'consider' parameter technique outlined by *Pfeiffer et al.* [1969]. The considered parameters for this report were the seventh-degree and seventh-order terms, tracking station locations, Mars' pole location and spin rate, and ephemeris errors. The Jet Propulsion Laboratory DE96 ephemeris was used.

The availability of ranging data for the Viking missions permitted the use of a calibration technique for Doppler biases due to ephemeris errors. The technique consisted of determining the ephemeris bias for the ranging data, as outlined by *Blackshear et al.* [1973], and then using the rate of change of the ranging bias as a calibration for the Doppler data. These corrections were not available for the Mariner data but were applied to all of the Viking data [*Mayo et al.,* 1977]. The magnitude of the corrections was of the order of 0.004 Hz. For the ephemeris uncertainty in the 'consider' analysis, a Doppler bias of 0.006 Hz, the maximum encountered for the Viking data, was used for the Mariner data, and no bias was assumed for the Viking data, since this was accounted for in the analysis.

DATA SELECTION

For this investigation a combination of Mariner 9 and Viking 1 and 2 data was utilized. The Mariner 9 spacecraft orbits had nominal values of inclination to the Mars equator of 64°, periods of 12 hours, and eccentricities of 0.6. During the 4-hour data time span the altitude of the spacecraft varied from approximately 1660 km to over 9800 km. The Viking orbits had nominal inclinations of 38° for Viking 1 and 75° for Viking 2, periods ranging from 22 to 27 hours, and eccentricities of approximately 0.75. The nominal spacecraft altitudes varied from 1,500 km to 11,700 km during the 4-hour time span.

The data consisted of coherent S band Doppler observations, sampled once per minute and, where possible, extending over a 4-hour time period centered at spacecraft periapsis for each arc. For the majority of the Viking arcs the data were corrected for charged particle effects with the use of simultaneous X band Doppler information. This technique is derived by *Madrid* [1974] and assumes identical uplink and downlink charged particle conditions. This assumption is required because X band Doppler information is only available on the downlink between spacecraft and tracking station, and no attempt has been made yet to model the interplanetary media to provide independent uplink calibrations.

Paper number 7S0487.

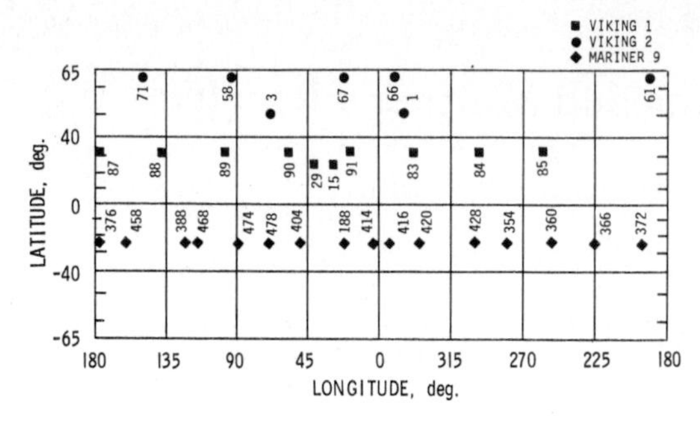

Fig. 1. Data arc periapsis locations. Identification numbers refer to orbit numbers.

A total of 33 data arcs was used in the analysis. The Mariner 9 data (16 arcs) are identical to those used by *Daniels and Tolson* [1976]. The Viking data consisted of 10 arcs from the Viking 1 spacecraft and 7 arcs from the Viking 2 spacecraft. The distribution of periapsis locations and the arc identification numbers are shown in Figure 1. The longitudinal distribution of Viking arcs is not ideal, particularly for Viking 2. The data were restricted by the requirements of nonsynchronous orbital motion with respect to the rotational period of Mars, simultaneous *S* and *X* band Doppler, and maneuver free periapsis passage. The data arcs shown in Figure 1 are the only ones available for this analysis prior to the November 1976 solar conjunction.

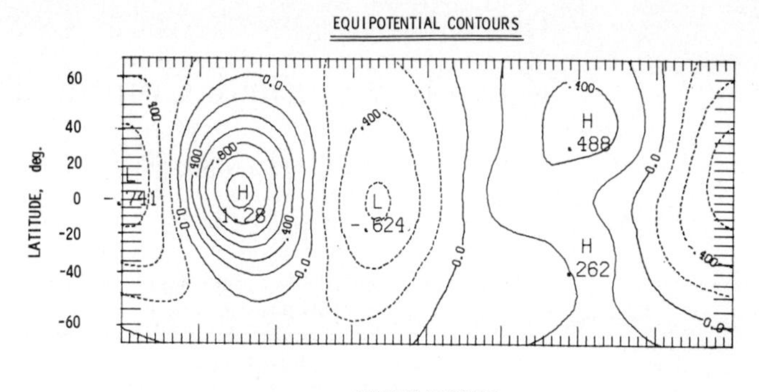

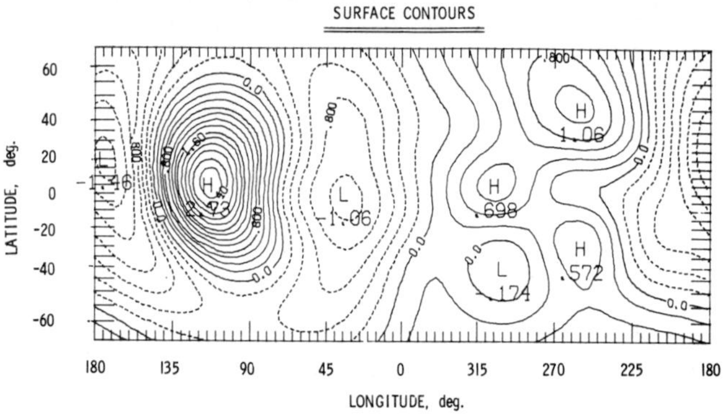

Fig. 2. Equipotential and surface height contours associated with the gravity field of Table 1.

RESULTS AND DISCUSSION

The Martian gravity field coefficients obtained from the short-arc analysis of a combination of Viking and Mariner 9 data are presented in a normalized form in Table 1. Comparison of these coefficients with the results obtained by *Daniels and Tolson* [1976] indicates changes larger than the quoted uncertainties in a number of the coefficients. These differences are due not only to the addition of the Viking data but also to the use of a different planetary ephemeris (DE96, as opposed to DE79). The changes in the values of the low-degree zonals are due primarily to the change in the planetary ephemeris.

Contours of the equipotential and surface undulations associated with the field given in Table 1 are presented in Figure 2. The C_{20} effect is deleted in both cases, and the surface undulations assume a homogeneous body. Both sets of contours are referenced to a spherical body with an average radius of 3397.5 km. The contours are remarkably similar to those obtained by *Daniels and Tolson* [1976] and *Sjogren et al.* [1975]. Between

TABLE 1. Mars Gravity Field—Combined Viking and Mariner Data

n	m	$C_{nm} \times 10^7$	$S_{nm} \times 10^7$
2	0	-8746 ± 2*	...
2	2	-850 ± 3	497 ± 2
3	0	-126 ± 9	...
3	1	34 ± 3	243 ± 6
3	2	-157 ± 3	80 ± 3
3	3	352 ± 4	250 ± 4
4	0	61 ± 3	...
4	1	44 ± 3	33 ± 4
4	2	-9 ± 2	-88 ± 3
4	3	68 ± 2	-1 ± 3
4	4	-2 ± 4	-121 ± 5
5	0	-23 ± 4	...
5	1	-2 ± 4	25 ± 3
5	2	-44 ± 3	-7 ± 4
5	3	30 ± 3	0 ± 3
5	4	-48 ± 6	-36 ± 4
5	5	-50 ± 6	31 ± 5
6	0	20 ± 6	...
6	1	24 ± 4	-1 ± 4
6	2	14 ± 5	20 ± 3
6	3	9 ± 4	-7 ± 4
6	4	23 ± 4	31 ± 5
6	5	21 ± 5	2 ± 6
6	6	28 ± 5	2 ± 5

*$C_{20} = (-1.9557 \pm 0.0004) \times 10^{-3}$ unnormalized.

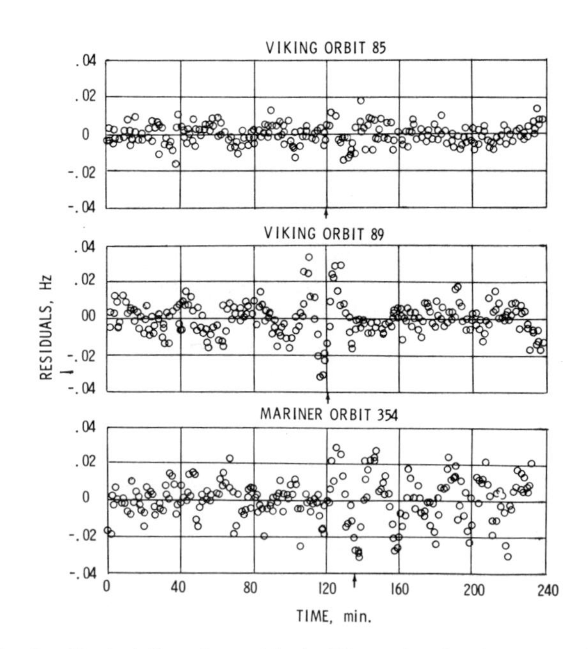

Fig. 3. Typical Doppler residuals. Time of periapsis passage is indicated by an arrow.

latitudes of 65° south and 30° north the three equipotential surfaces are consistent to less than 80 m. Up to 65° north latitude the current field is consistent with that of *Sjogren et al.* [1975] to about 110 m and that of *Daniels and Tolson* [1976] to about 140 m. Comparisons with the results of *Born* [1974], *Jordan and Lowell* [1973], and *Reasenberg et al.* [1975] range from 150 m in the southern hemisphere and equatorial regions to 350 m in the northern regions. The short-arc methods yield sufficiently consistent results that it is felt that the new gravity field defines the equipotential surface to an accuracy of about 75 m between ±65° latitude.

Residual patterns for three representative data arcs are shown in Figure 3. These arcs were chosen to illustrate the fact that the pattern may vary considerably, depending upon the longitude of periapsis. For orbit 85 the sixth-degree field has removed most of the signal from the data. This is typical for the majority of the arcs, but in some cases there is evidence of a signal at periapsis which cannot be accounted for by the sixth-degree field. Orbit 89, for example, has a periapsis approximately over the Tharsis region, and the residuals exhibit a periapsis signal 3-4 times the noise level of the data. All of the orbits which pass over the Tharsis region exhibit this characteristic. The residual pattern for Mariner orbit 354 not only shows a similar periapsis signal but also exhibits an increased signal level after periapsis passage. This orbit passes over the Hellas region, approximately 180° of longitude away from the Tharsis region. Thus, to fully account for the Tharsis and Hellas regions, higher-degree solutions, or techniques such as simultaneous solutions for spherical harmonics and mass points, will be required. This requirement will also be true when data from the extended mission are analyzed, since these orbits have lower periapsis altitudes.

CONCLUDING REMARKS

A sixth-degree and sixth-order Martian gravitational field has been obtained from a short-arc analysis of Viking and Mariner 9 spacecraft tracking data. The equipotential contours obtained from this field are expected to be accurate to about 75 m within the ±65° latitude band. Analysis of the residual patterns indicates that a higher-order gravity field will be necessary to adequately represent the Tharsis and Hellas regions of Mars.

REFERENCES

Blackshear, W. T., R. H. Tolson, and G. M. Day, Mars lander position estimation in the presence of ephemeris biases, *J. Spacecr. Rockets, 10,* 284, 1973.

Born, G. H., Mars physical parameters as determined from Mariner 9 observations of the natural satellites and Doppler tracking, *J. Geophys. Res., 79*(32), 4837-4844, 1974.

Daniels, E. F., and R. H. Tolson, Spherical harmonic representation of the gravity field of Mars using a short-arc technique, paper presented at the Astrodynamics Conference, Amer. Inst. of Aeronaut. and Astronaut./Amer. Astronaut. Soc., San Diego, Calif., Aug. 18-20, 1976.

Hartwell, J. G., Simultaneous integration of N-bodies by analytical continuation with recursively formed derivatives, *J. Astronaut. Sci., 14*(4), 173-177, 1967.

Jordan, J. F., and J. Lowell, Mariner 9, an instrument of dynamical science, paper presented at the Astrodynamics Specialists' Conference, Amer. Astronaut. Soc./Amer. Inst. of Aeronaut. and Astronaut., Vail, Colo., July 16-18, 1973.

Lorell, J., G. H. Born, J. E. Christensen, P. B. Esposito, J. F. Jordan, P. A. Laing, W. L. Sjogren, S. K. Wong, R. D. Reasenberg, I. I. Shapiro, and G. L. Slater, Gravity field of Mars from Mariner 9 tracking data, *Icarus, 18,* 304-316, 1973.

Madrid, G. A., The measurement of dispersive effects using the Mariner 10 *S*-band and *X*-band spacecraft to station link, *Deep Space Network Progr. Rep. 42-22,* pp. 22-27, Jet Propul. Lab., Pasadena, Calif., 1974.

Mayo, A. P., W. T. Blackshear, R. H. Tolson, W. H. Michael, Jr., G. M. Kelly, J. P. Brenkle, and T. A. Komarek, Lander locations, Mars physical ephemeris, and solar system parameters: Determination from Viking lander tracking data, *J. Geophys. Res., 82,* this issue, 1977.

Pfeiffer, C. G., D. D. Morrison, R. E. Mortensen, J. V. Breakwell, W. H. Berry, and M. H. Merel, Sequential processing techniques for trajectory estimation, *NASA Contract. Rep. 1360,* Oct. 1969.

Reasenberg, R. D., I. I. Shapiro, and R. D. White, The gravity field of Mars, *Geophys. Res. Lett., 2*(3), 89-92, 1975.

Sjogren, W. L., J. Lorell, L. Wong, and W. Downs, Mars gravity field based on a short-arc technique, *J. Geophys. Res., 80*(20), 2899-2908, 1975.

(Received April 7, 1977;
revised June 2, 1977;
accepted June 3, 1977.)

The Viking Relativity Experiment

I. I. SHAPIRO, R. D. REASENBERG, P. E. MACNEIL, AND R. B. GOLDSTEIN

Massachusetts Institute of Technology, Cambridge, Massachusetts 02139

J. P. BRENKLE, D. L. CAIN, T. KOMAREK, AND A. I. ZYGIELBAUM

Jet Propulsion Laboratory, Pasadena, California 91103

W. F. CUDDIHY AND W. H. MICHAEL, JR.

NASA Langley Research Center, Hampton, Virginia 23665

Measurements of the round-trip time of flight of radio signals transmitted from the earth to the Viking spacecraft are being analyzed to test the predictions of Einstein's theory of general relativity. According to this theory the signals will be delayed by up to ~250 μs owing to the direct effect of solar gravity on the propagation. A very preliminary qualitative analysis of the Viking data obtained near the 1976 superior conjunction of Mars indicates agreement with the predictions to within the estimated uncertainty of 0.5%.

1. INTRODUCTION

The theory of general relativity predicts that the propagation time of electromagnetic signals traveling between two points will be increased by the presence of a massive object near the signal path [*Shapiro*, 1964]. For signals propagating between the earth and a planet near superior conjunction the predicted increase in round-trip delay can reach about 250 μs. This prediction has been tested several times, first by means of radar signals reflected from the planets Mercury and Venus [*Shapiro et al.*, 1968, 1971] and later via radio tracking of the Mariner 6, 7, and 9 spacecraft [*Anderson et al.*, 1975, 1976; *Reasenberg and Shapiro*, 1976]. The results from these tests were consistent with general relativity to within the estimated errors, which ranged from 5% for the radar experiment to 2% for Mariner 9.

The Viking spacecraft offer the opportunity to increase the accuracy of this time delay test about twentyfold. The limitations on prior experiments that Viking can help to overcome fall into two categories: accuracy of measurement and accuracy of interpretation. The 'ranging' transponders on all four of the Viking spacecraft, and the associated ground equipment, permit measurements of the round-trip group delays of signals propagating between the earth and the spacecraft to be made with an uncertainty of about 10 ns under most circumstances. This level of accuracy, by itself, represents a severalfold improvement over that attained during prior interplanetary ranging experiments. To interpret these accurate delay measurements fully in terms of a test of general relativity, we must be able to determine, with at least comparable accuracy, all of the other, nonrelativistic, contributions to the delays. These other contributions stem primarily from two sources: (1) the orbits of the spacecraft and of the 'tracking' stations on the earth and (2) the solar corona, which increases group delays significantly for signal paths that pass near the sun. The Viking spacecraft, collectively, enable the contributions from these two sources to be determined accurately because (1) the two landers are firmly emplaced on another planet and (2) the two orbiters possess ranging transponders that can each transmit at both S and X band frequencies. In the most accurate previous time delay test of general relativity both of these features were lacking.

How do these features lead to a more accurate determination of the orbital and coronal contributions to the measured delays? The Viking landers allow us to take full advantage of the very low area-to-mass ratios of the earth and Mars and the consequent immunity of their orbits to the effects of nongravitational forces which usually plague spacecraft. The landers also provide a point target fixed to a planetary surface, an advantage not achievable with radar measurements which are affected severely by planetary topography. For these reasons the orbits of the landers (and of the tracking stations) can be represented accurately by parameterized theoretical models, the orbital contributions to the measured delays thereby being determined with corresponding accuracy.

Each lander, unfortunately, possesses a transponder that can transmit only at S band frequencies, and so delay measurements between a tracking station and a lander are affected significantly by the solar corona, as were the measurements from prior spacecraft tests. The increase in group delay from this cause can reach about 100 μs for signals passing close to the solar limb. The erratic variations of the electron density in the corona prevent us from determining the plasma contribution with the desired accuracy via a parameterized theoretical model. The plasma contribution must be measured directly.

Because of the dispersive nature of the corona, delay measurements between a tracking station and a Viking orbiter, whose transponder has a dual-band transmitter, can be used to obtain useful estimates of the coronal contributions to the delay measurements made between a tracking station and a lander. These dual-band plasma calibration measurements must be made simultaneously with the lander measurements to minimize the effect of variations in the corona. Nonetheless, the calibration is considerably less than perfect: the orbiters can receive only S band signals from the earth, although, as is explained in section 2, they retransmit these signals, suitably altered, along with phase-coherent X band signals [*Michael et al.*, 1972]. Thus dual-band delay information can be obtained only for the downlink portion of the signal path between a tracking station and an orbiter, whereas this information is desired for both the uplink and the downlink portions.

This difference between the space-time path for which calibration is desired and that for which it is available is expected to constitute the main limitation on the accuracy achievable in the Viking relativity experiment. The calibration problem is

Paper number 7S0564.

most severe where it is most important: near superior conjunction. For larger sun-earth-Mars angles the density of plasma is lower, the scale of plasma inhomogeneities larger, and the signal path shorter.

In the remainder of this paper we discuss the method used to measure group delays, the procedures employed in data collection and data analysis, and the results obtained.

2. METHOD OF DELAY MEASUREMENT

We describe here briefly the equipment and the technique used to obtain the group delay measurements upon which the relativity experiment is based.

Three tracking stations were used for these measurements, one each in Australia, California, and Spain. Each of these stations has a 64-m-diameter antenna equipped with a transmitter-receiver system and various calibration devices. The local oscillator signals used in each transmitter-receiver system are derived from an atomic frequency standard, or clock, whose instability is about 1 part in 10^{12}, or less, over time scales corresponding to the round-trip delay of signals propagating between station and spacecraft. Time at each station is maintained to within about 20 μs of universal time, coordinated (UTC).

How is this equipment in conjunction with the spacecraft transponders used to measure group delays? The nearly monochromatic, approximately 2.2-GHz *S* band 'carrier' signal to be transmitted toward a spacecraft is first modulated by the introduction of approximately 65° phase changes, of alternating sign, at evenly spaced intervals. This 'square wave' modulation results in about a 9-dB carrier suppression, the remaining 90% of the signal power being in the side bands. At the spacecraft this phase modulation is first separated from the carrier, the carrier frequency is then multiplied by 240/221, the modulation is reapplied with 0.9-dB carrier suppression, and the resultant signal is transmitted. The *X* band downlink signal is similarly generated except that the frequency multiplier is 880/221. The frequency change of the *S* band carrier signals is necessary to enable simultaneous transmission and reception at both the spacecraft and the tracking station.

The group delay is determined by cross correlation of the modulation on the signals received on the earth with a delayed 'stretched' replica of the transmitted modulation. The delay used for the cross correlation is based on prior knowledge, whereas the stretching, needed to accommodate the effect of the Doppler shift, is controlled directly by the frequency change detected in the received carrier signal. Both the 'in-phase' and the 'quadrature' (90° out of phase) cross-correlation coefficients are determined. The expected normalized values of these coefficients for a true square wave modulation in the absence of noise are illustrated in Figure 1 as a function of the difference $\delta\tau$ between the value of the measured delay and the value of the delay assumed in forming the replica. For example, for $\delta\tau = 0$ the coefficient I for the correlation of the received modulation with the in-phase replica will be unity, whereas the coefficient Q for the corresponding correlation with the quadrature replica will vanish. The 'triangle' patterns traced by I and Q as functions of $\delta\tau$ are periodic with the period T of the modulation. It is easy to show that within the interval T the value of $\delta\tau$ can be expressed as

$$\delta\tau = \frac{T}{4}\left(1 - \frac{I}{|I| + |Q|}\right)\text{sign } Q \tag{1}$$

In practice, modifications to this formula are used in the estimation of $\delta\tau$ to account for the deviations of the modulation from a true square wave.

As can be seen from Figure 1, $\delta\tau$ can be determined only to within an integral multiple of T. How can this ambiguity in $\delta\tau$ be removed? Our a priori knowledge of the delay is of insufficient accuracy for this purpose when the modulation of shortest period, ~2 μs, is used. Therefore modulations with periods longer by successive multiples of two are transmitted sequentially until a period is encountered that is large in comparison with the a priori uncertainty in the delay and that therefore insures the proper removal of the ambiguity. Only the signs of the values of I for the longer-period modulations are needed to remove the ambiguity. Thus its removal does not add appreciably to the time required to make a delay measurement.

Despite the ~2-μs extent of the shortest period available with the present equipment, the signal-to-noise ratios usually attained in a few minutes of integration are so high that the standard deviation in the estimate of $\delta\tau$ due to random noise alone is often only about 2–3 ns. The uncertainties in the epoch and the rate of each station's clock usually introduce a comparable error in delay measurement. More important, however, is the degradation in measurement accuracy caused by uncertainties in the calibrations of the delays through the spacecraft and through the ground equipment: transmitter, antenna, and receiver [*Komarek and Otoshi,* 1976]. These instrumental delays are not constant and for the spacecraft, for example, vary by several tens of nanoseconds over the ranges of signal level and temperature encountered during the mission. The overall uncertainty in these calibrations is estimated to be about 10 ns and generally provides the limit on achievable measurement accuracy. Many of these calibration problems would be ameliorated if shorter-period modulations and higher-bandwidth transponders and receivers were available.

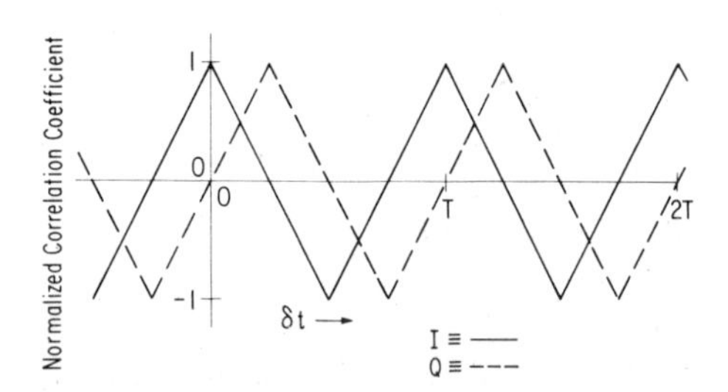

Fig. 1. Estimation of signal delay via cross correlation of the modulation on the received signal with a suitably modified (see text) replica of the modulation imparted to the transmitted signal. The in-phase I and quadrature Q normalized correlation coefficients are each shown as a function of δt, the error in the assumed value of the round-trip group delay of the radio signals propagating between the earth and a Viking spacecraft.

3. DATA COLLECTION

In this preliminary report we discuss only those delay measurements obtained during an approximately 1-month period centered on November 25, 1976, the date of the superior conjunction of Mars. Here we describe the procedures that, with only a few exceptions, were used to obtain these data.

To estimate with useful accuracy the coronal contribution to the delay measurements made between tracking station and lander during this period, it was necessary, as stated earlier, to make simultaneous dual-band delay measurements between tracking station and orbiter. However, because a given ground station could track only one spacecraft at a time, such simultaneous tracking had to involve those two of the three available 64-m-diameter antennas that could view Mars at the same time:

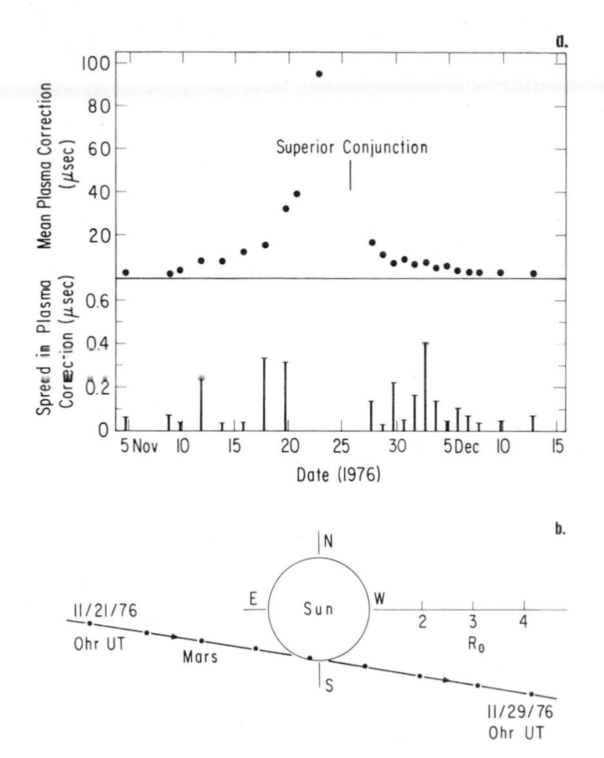

Fig. 2. (*a*) The upper portion shows the daily means of the inferred effect of plasma on delay measurements. The lower portion shows the total spread (maximum minus minimum) of the plasma corrections applied to the individual delay measurements made on a given day. (*b*) The relative positions on the plane of the sky of the sun and Mars as seen from the earth on the days surrounding the superior conjunction of Mars on November 25, 1976. Solar equatorial coordinates are shown.

deep-space stations (DSS) 14 and 43 in Goldstone, California, and Canberra, Australia, respectively. The two landers are separated from one another by nearly 180° in longitude on Mars, so at most one lander was visible during the entire 3-hour period that Mars could be observed simultaneously from both stations. On each of those days on which lander and orbiter tracking was scheduled a series of separate measurements was made of the round-trip group delays between a tracking station and that lander which was in view. Technical constraints on the landers actually limited such measurements to a period of about 100 min and 40 min for lander 1 and lander 2, respectively.

Because of certain of its unique features that increase the probability for successful measurement the Mu-II sequential ranging system [*Martin and Zygielbaum,* 1977], available only at DSS 14, was almost always used to make the delay measurements between tracking station and lander. Simultaneously, dual-band delay measurements were made between DSS 43 and a conveniently placed orbiter, i.e., one near its apoapsis orbital position, where tracking is least difficult. These latter delay measurements utilized the planetary ranging assembly (PRA) [*Osborn,* 1974], which unfortunately has an inherent ambiguity interval of ~ 2 μs for X band group delay measurements, since the PRA is not configured to accommodate longer-period modulations at X band. To try to ensure the accurate removal of this ambiguity, dual-band ranging to an orbiter was first carried out from DSS 14 on each day of scheduled observations until the time Mars 'rose' at DSS 43. Since the Mu-II ranging system has no such ambiguity problem, a comparison of the dual-band data from the two tracking stations enabled the ambiguity in the DSS 43 delay data to be removed reliably with a few possible exceptions: the coronal effect on the (downlink) measurements of group delay at X band never exceeded about 5 μs, and the differences in the coronal effects on the DSS 14 and DSS 43 measurements of delay were usually under 1 μs, since the two sets of measurements were rarely separated in time by more than ~ 1 hour and never by more than ~ 2 hours. (For sun-earth-Mars angles under about 0.5°, where the effects would have been larger, the severe gradients and turbulence in the corona prevented any useful measurements of delays from being obtained.)

4. DATA ANALYSIS

Our preliminary analysis of the delay data proceeded in two steps. First, we corrected the relevant delay measurements for the effects of the solar corona. Second, we compared the corrected data with theoretical predictions based on the theory of general relativity.

Plasma Corrections

In the determination of plasma corrections we have not yet attempted to account for the differences in spatial locations between lander and orbiter and between DSS 14 and DSS 43. However, to account for the more important fact that the dual-band delay measurements were available only for the downlink paths, we did employ the 'thin-screen' model to estimate the uplink plasma delay: we assumed that the entire plasma effect was confined to the planar region perpendicular to the earth-Mars line and passing through the center of the sun. Thus for the plasma correction for the uplink path of a given measurement we took that value of the plasma delay measured on the downlink path at a time earlier by the round-trip travel time of a radio signal propagating between the thin screen and Mars.

The mean plasma corrections are shown in Figure 2*a* for each day on which delay measurements were obtained near superior conjunction. The erratic behavior of these daily means demonstrates vividly that a simple parameterized model of the corona could not yield an adequate representation of the plasma delays. The scatter of the daily means about the predicted plasma delays obtained from such a model [*Tyler et al.,* 1977] is approximately a hundredfold greater than the delay measurement error. Thus without direct plasma calibration the accuracy achievable in the Viking relativity experiment would be drastically reduced.

On the lower portion of Figure 2*a* we show for each appropriate day the difference between the maximum and the minimum of the plasma corrections applied to the individual delay measurements obtained on that day. These differences varied from about 20 to 400 ns but by no means varied monotonically with the sun-earth-Mars angle. The ratios of these differences to their corresponding daily means were also far from constant, even when they were corrected for the differences in the time spans of the data obtained on different days. These characteristics provide further evidence of the erratic time variability of the corona.

One systematic effect is clearly discernible in Figure 2*a*: the asymmetry of the daily means with respect to reflection about the date of superior conjunction; i.e., the daily means are substantially smaller after superior conjunction. This qualitative behavior was anticipated because (1) the solar latitude of the intersection of the earth-Mars line with the thin screen just before conjunction was nearly equatorial, whereas at the corresponding time after conjunction the solar latitude of this

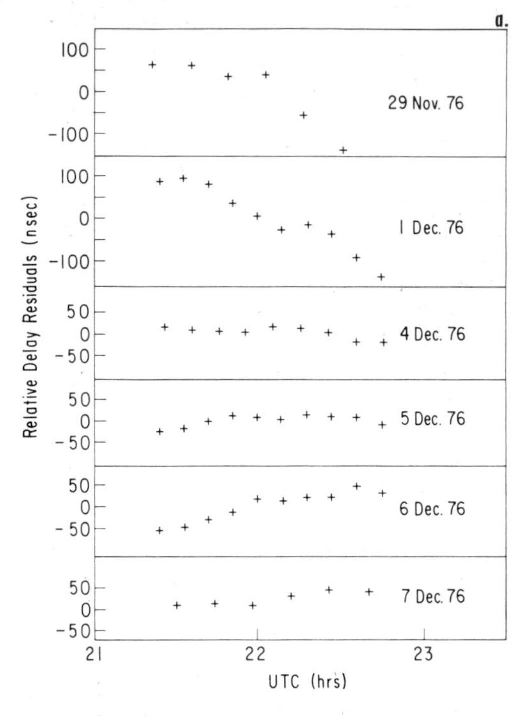

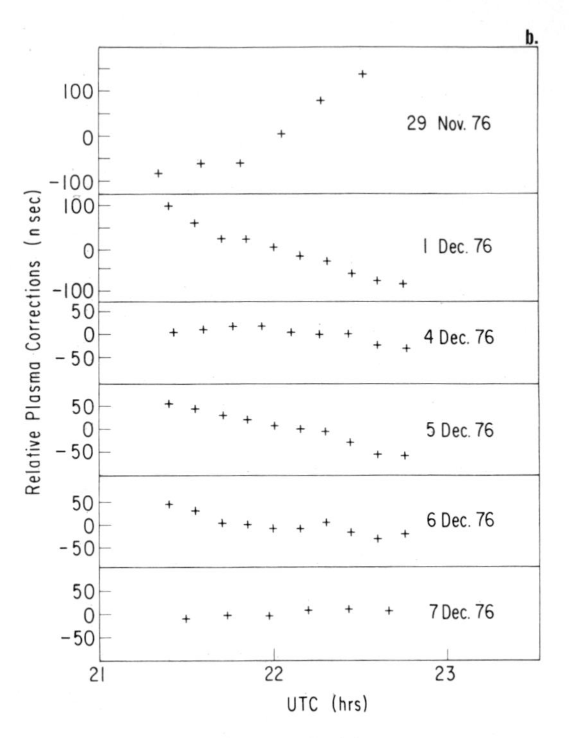

Fig. 3. (*a*) Relative delay residuals from lander 1 for six representative days on which six or more delay measurements were analyzed. (*b*) Relative values of the plasma corrections applied to the delay measurements for each given day.

point was much higher (see Figure 2*b*), and (2) at minimum sunspot activity, such as in 1976, the mean coronal electron density at a given radial distance from the sun is as much as 10 times lower for polar than for equatorial latitudes (see, for example, *Counselman and Rankin* [1972] and *Berman et al.* [1976]). The asymmetry in Figure 2*a* does not fully exhibit this latitude variation of the mean electron density partly because the coronal effect on delays is proportional to the integral of the electron density along the ray path of the radio signals which sample only low solar latitudes far from the sun.

Finally, we remark that the dual-band technique for plasma calibration is appropriate for this test of gravitation theory because at the level of accuracy of concern to us here, general relativity and all other well-known theories of gravitation are color blind: the same vacuum delays are predicted for all signal frequencies.

Comparison of Corrected Data With Theory

The delays measured between tracking stations and landers, after correction for plasma contributions, were compared with theoretical predictions of the corresponding round-trip vacuum delays. These predictions were based on (1) pre-Viking knowledge of the orbits of the earth and Mars, based primarily on radar observations of the inner planets, radio tracking of the Mariner series of spacecraft, laser ranging to the moon, and optical observations of the asteroids and the outer planets; (2) pre-Viking determinations of the locations of the tracking stations on the earth from radio tracking of Mariner spacecraft; (3) a standard model of the earth's rotation, including the effects of polar motion and variations in the earth's rate of rotation; (4) preliminary determinations of the locations of the landers on Mars and of the rotation vector of Mars from radio tracking of the landers in the weeks immediately following their respective arrivals on Mars (the results were in good agreement with those of *Michael et al.* [1976] and *Mayo et al.* [1977]); and (5) the predicted relativistic contribution to the delay, given by the generalized equation

$$\Delta\tau = \frac{2r_0}{c}(1+\gamma)\ln\left(\frac{r_e + r_m + R}{r_e + r_m - R}\right) \quad (2)$$

where $r_0 \equiv GM_S/c^2 \simeq 1.5$ km is the length equivalent of the sun's mass; G is the universal constant of gravitation; M_S is the mass of the sun; c is the speed of light; r_e, r_m, and R are the distances between the sun and the earth, the sun and Mars, and the earth and Mars, respectively; and γ is a parameter that appears in the generalized Schwarzschild metric (see, for example, *Weinberg* [1972] and *Misner et al.* [1973]). In Einstein's theory of general relativity, γ is replaced by unity. In other metric theories of gravitation, γ can take on values different from unity (see, e.g., *Will* [1974]). With $\gamma = 1$, (2) shows that $\Delta\tau$ can reach as high as $\sim$250 μs for signals that just graze the limb of the sun.

The delay measurements are also affected by the atmospheres and ionospheres of the earth and Mars. Their combined contribution to a measured delay was as large as 100 ns. But only the variations of these contributions over the set of delay measurements affect the relativity experiment. Since these variations about the mean were no greater than one quarter of the maximum contribution, we could safely ignore the atmospheres and ionospheres in our theoretical model for this preliminary analysis. (Of course, partial correction for the earth's ionosphere is incorporated automatically with the coronal plasma correction and is incomplete mainly insofar as the ionosphere over DSS 14 differs from that over DSS 43.)

We also ignored the precession, nutation, and possible polar motion of Mars in our theoretical model, again justifiable because the changes introduced into the delay measurements by such motions are not expected to exceed 10 ns for the period of interest.

As an illustration of the consistency of the results from separate delay measurements made during a single day we

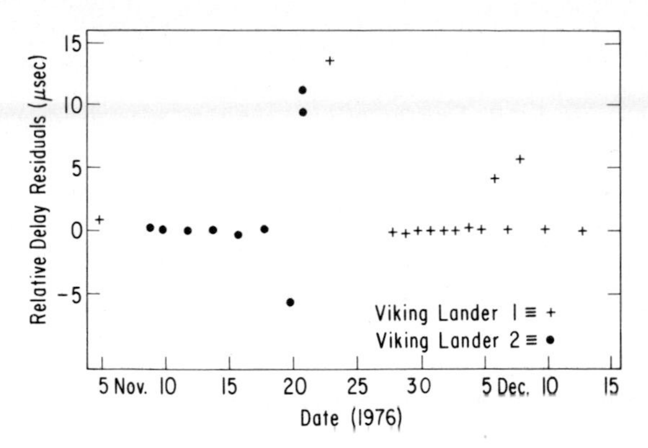

Fig. 4. Daily means of the relative residuals for the delay measurements made during the approximately 1-month period centered on November 25, 1976, the date of the superior conjunction of Mars. Anomalous residuals appear on five different days. On November 20, 1976, delay measurements were obtained from both DSS 14 and DSS 43; the means from the measurements at the two stations are shown separately. Note that useful data were obtained for signal paths passing as close as ~3 solar radii from the center of the sun (see Figure 2*b*).

show in Figure 3*a* the differences ('residuals') between the corrected measurements and the corresponding theoretical predictions for typical days on which six or more individual delay measurements are available. (For some of these days there are additional delay measurements which have not yet been analyzed.) In Figure 3*b* we show the corresponding corrections that had been applied for coronal plasma. Each residual and each plasma correction is given relative to its corresponding daily mean. These examples are all from lander 1 measurements for the period following superior conjunction, since with one exception the earlier samples were all from lander 2 measurements for which the daily tracking time was more severely limited (see section 3).

Systematic trends are apparent in each set of residuals, especially those obtained during the first week following superior conjunction. Systematically changing errors in the plasma calibration may well be the cause. It is possible, but hardly assured, that these trends will be lessened after we implement and apply a more sophisticated algorithm for the determination of plasma corrections.

What about the consistency between the daily means of the delay residuals? Here the story is entirely different. To illustrate, we plot the daily means of the residuals in Figure 4, after adding a single constant to all of the data for a given lander. These constants serve mostly to remove in an ad hoc fashion an apparent average error of about 5 μs in our theoretical model of the earth-Mars distance during this period. The difference between the constants used for the two landers is slight and reflects the difference in the errors in the estimates of their locations.

Returning our attention to Figure 4, we note that there are five days for which the residuals are far removed from the mean of the others. On one of those days, delay measurements were obtained sequentially from DSS 14 and DSS 43, and the results from each tracking station were significantly different from each other, as well as from the mean. On a sixth day, the first day chronologically, the residual is about 1 μs larger than the majority of the others obtained from observations of the same lander; however, its epoch is sufficiently far removed from the epochs of the others that orbital errors cannot be ruled out as an explanation. What of the other anomalous results? How can they be explained? Plausible errors in orbits, in tracking station locations, and in lander locations cannot be the cause; such errors would introduce not erratic behavior but only a slowly varying drift in the residuals. Errors in the software used to analyze the delay measurements can also be eliminated as a cause: although the residuals shown in the figures were based on theoretical values obtained with a computer program developed at the Massachusetts Institute of Technology, substantially the same results were obtained from the use of a program developed wholly independently at the Jet Propulsion Laboratory. The consistency of the separate measurements on a given day and other evidence make hardware errors seem unlikely culprits. It is possible that one or two, but no more, of the anomalous results could be attributable to an incorrect removal of the ambiguity in the X band delay measurements for days in which there was a large gap in time between the DSS 14 and the DSS 43 plasma calibration measurements. Such an error, when it is converted to its equivalent in round-trip S band delay, would be a (small) multiple of 4.6 μs. The only other possibility that we can conceive as an explanation for the anomalous results involves intermittent procedural errors in the setup for ranging, but no such error has been positively identified.

Let us assume that the relative residuals in Figure 4 that are far removed from the overall mean, and only these, are in fact affected by procedural or other errors. If we compare all other residuals, as in Figure 5, with the 'excess' delays predicted by general relativity ($\gamma \equiv 1$), we may infer by inspection that if γ were to differ from unity by more than about 1%, the rest of (2) remaining valid, then there would be a noticeable systematic 'dip' or 'bump' in the relative residuals, centered at the time of superior conjunction. (Recall that the theoretical predictions used in forming these residuals are completely independent of the measurements; i.e., the residuals are prefit, not postfit.)

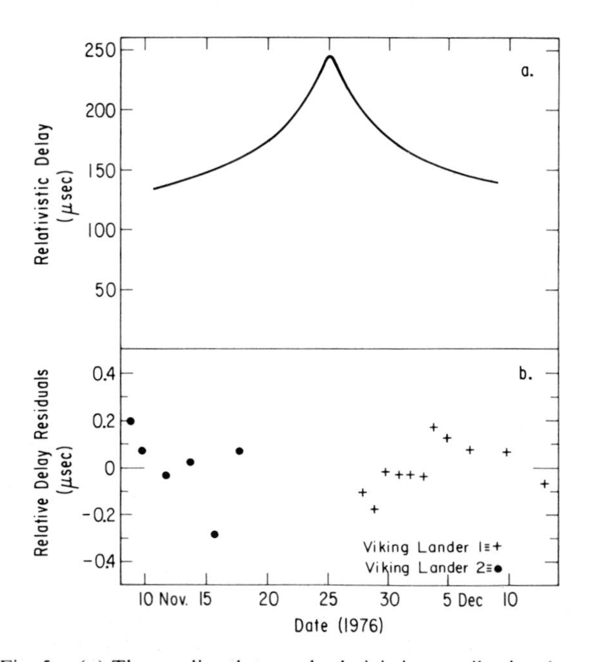

Fig. 5. (*a*) The predicted general relativistic contribution (see (2)) to delay measurements made near superior conjunction. (*b*) The relative residuals (see text) of delay measurements based on the assumption that general relativity is the correct theory of gravitation. Note that the ordinate scale is expanded by a factor of 200 relative to that in Figure 5*a*. Note also that the theoretical predictions used in the construction of the residuals are completely independent of the corresponding measurements; i.e., the residuals are prefit, not postfit.

Since no such systematic deviation is in evidence, we conclude tentatively that γ differs from unity by no more than 1% and equivalently that the Viking data agree with the predictions of general relativity to within 0.5%.

5. DISCUSSION AND CONCLUSIONS

The best prior determination of γ from analysis of a time delay experiment was based on the radio tracking of Mariner 9. This experiment, as indicated earlier, showed that γ was unity to within the estimated uncertainty of 4% [*Anderson et al.*, 1976; *Reasenberg and Shapiro*, 1976]. A different experiment based on the gravitational deflection of radio waves [*Fomalont and Sramek*, 1976] yielded a more stringent limit of 2% on the uncertainty in any possible deviation of γ from unity, a result that is twofold lower than that of Mariner 9 and twofold higher than our present preliminary result.

What are the prospects for improvement of this Viking result? If calibrated delay measurements are accumulated throughout the Viking mission, without large gaps, we anticipate that our final analysis will lower the uncertainty in γ to about 0.2%. In this analysis we will also include all other useful data and will estimate γ simultaneously with all of the other relevant parameters that govern the dynamics of the solar system and hence the inertial trajectories of the deep-space stations and the Viking landers.

What alternative theories of gravitation could be proven invalid by, say, a confirmation that γ was unity to within 0.2%? Unfortunately, none. All other widely discussed and analyzed theories of gravitation, such as the theory of *Brans and Dicke* [1961] and that of *Rosen* [1974], have a special property: their predictions either agree or can be made to agree with those of general relativity in the post-Newtonian regime that applies to the Viking relativity experiment. In many of these theories the agreement is secured by the selection of an appropriate value of an adjustable parameter, such as γ (no such parameters exist in general relativity). For example, the scalar-tensor theory of Brans and Dicke contains only one additional parameter, a dimensionless constant ω (>0) related to γ by

$$\gamma = (\omega + 1)/(\omega + 2) \tag{3}$$

A limit of 0.2% on the deviation of γ from unity would restrict ω to being greater than 500, since $\omega \geq (1 - \gamma)^{-1}$ when $0 < 1 - \gamma \ll 1$. But no matter how accurate the experiment, it could not prove such a theory invalid, provided that general relativity is a correct description of gravitation to that level. Only if general relativity is proven wrong, or if experiments can be made sensitive to 'post-post-Newtonian' effects, will there be a substantial winnowing of competing theories.

Acknowledgments. We thank the Viking project staff, the personnel at the deep-space stations, and our colleagues on the Radio Science Team for their indispensable aid in this experiment. We wish particularly to thank D. L. Brunn and W. L. Martin for their contributions to the radio-tracking system and J. S. Martin, G. A. Soffen, and A. T. Young for their generous support. The Massachusetts Institute of Technology experimenters were sponsored in part by the National Aeronautics and Space Administration (NASA) under contract NAS1-9702 and in part by the National Science Foundation under grant PHY72-05104A05. The portion of the research carried out at the Jet Propulsion Laboratory, California Institute of Technology, was under contract NAS7-100 sponsored by NASA.

REFERENCES

Anderson, J. D., P. B. Esposito, W. Martin, C. L. Thornton, and D. O. Muhleman, Experimental test of general relativity using time-delay data from Mariner 6 and Mariner 7, *Astrophys. J., 200,* 221–233, 1975.

Anderson, J. D., M. S. W. Keesey, E. L. Lau, E. M. Standish, Jr., and X. X. Newhall, Tests of general relativity using astrometric and radiometric observations of the planets, paper presented at the Third International Space Relativity Symposium, 27th Congress, Int. Astronaut. Fed., Anaheim, Calif., 1976.

Berman, A. L., J. A. Wackley, S. T. Rockwell, and J. G. Yee, The Pioneer 11 1976 solar conjunction: A unique opportunity to explore the heliographic latitudinal variations of the solar corona, *Deep Space Network Progr. Rep. 42-35,* pp. 136–147, Jet Propul. Lab., Pasadena, Calif., 1976.

Brans, C., and R. H. Dicke, Mach's principle and a relativistic theory of gravitation, *Phys. Rev., 124,* 925–935, 1961.

Counselman, C. C., III, and J. M. Rankin, Density of the solar corona from occultations of NP 0532, *Astrophys. J., 175,* 843–856, 1972.

Fomalont, E. B., and R. A. Sramek, Measurements of the solar gravitational deflection of radio waves in agreement with general relativity, *Phys. Rev. Lett., 36,* 1475–1478, 1976.

Komarek, T., and T. Otoshi, Terminology of ranging measurements and DSS calibrations, *Deep Space Network Progr. Rep. 42-36,* pp. 35–40, Jet Propul. Lab., Pasadena, Calif., 1976.

Martin, W. L., and A. I. Zygielbaum, Mu-II ranging, *Tech. Memo. 33-768,* pp. 1–70, Jet Propul. Lab., Pasadena, Calif., 1977.

Mayo, A. P., W. T. Blackshear, R. H. Tolson, W. H. Michael, Jr., G. M. Kelly, J. P. Brenkle, and T. Komarek, Lander locations, Mars physical ephemeris, and solar system parameters: Determination from Viking lander tracking data, *J. Geophys. Res., 82,* this issue, 1977.

Michael, W. H., Jr., D. L. Cain, G. Fjeldbo, G. S. Levy, J. G. Davies, M. D. Grossi, I. I. Shapiro, and G. L. Tyler, Radio science experiments: The Viking Mars orbiter and lander, *Icarus, 16,* 57–74, 1972.

Michael, W. H., A. P. Mayo, W. T. Blackshear, R. H. Tolson, G. M. Kelly, J. P. Brenkle, D. L. Cain, G. Fjeldbo, D. N. Sweetnam, R. B. Goldstein, P. E. MacNeil, R. D. Reasenberg, I. I. Shapiro, T. I. S. Boak III, M. D. Grossi, and C. H. Tang, Mars dynamics, atmospheric and surface properties: Determination from Viking tracking data, *Science, 194,* 1337–1339, 1976.

Misner, C., K. S. Thorne, and J. A. Wheeler, *Gravitation,* W. H. Freeman, San Francisco, Calif., 1973.

Osborn, G. R., Planetary ranging operational software, *Deep Space Network Progr. Rep. 42-21,* pp. 87–91, Jet Propul. Lab., Pasadena, Calif., 1974.

Reasenberg, R. D., and I. I. Shapiro, Solar system tests of general relativity, paper presented at the International Meeting on Experimental Gravitation, Acad. Naz. dei Lincei, Coll. Ghislieri, Pavia, Italy, 1976.

Rosen, N., A theory of gravitation, *Ann. Phys., 84,* 455–473, 1974.

Shapiro, I. I., Fourth test of general relativity, *Phys. Rev. Lett., 13,* 789–791, 1964.

Shapiro, I. I., G. H. Pettengill, M. E. Ash, M. L. Stone, W. B. Smith, R. P. Ingalls, and R. A. Brockelman, Fourth test of general relativity: Preliminary results, *Phys. Rev. Lett., 20,* 1265–1269, 1968.

Shapiro, I. I., M. E. Ash, R. P. Ingalls, W. B. Smith, D. B. Campbell, R. B. Dyce, R. F. Jurgens, and G. H. Pettengill, Fourth test of general relativity: New radar result, *Phys. Rev. Lett., 26,* 1132–1135, 1971.

Tyler, G. L., J. P. Brenkle, T. A. Komarek, A. I. Zygielbaum, The Viking solar corona experiment, *J. Geophys. Res., 82,* this issue, 1977.

Weinberg, S., *Gravitation and Cosmology: Principles and Applications of the General Theory of Relativity,* John Wiley, New York, 1972.

Will, C. M., The theoretical tools of experimental gravitation, in *Experimental Gravitation,* edited by B. Bertotti, pp. 1–110, John Wiley, New York, 1974.

(Received May 31, 1977;
revised June 16, 1977;
accepted June 16, 1977.)

The Viking Solar Corona Experiment

G. LEONARD TYLER

Stanford University, Stanford, California 94305

JOSEPH P. BRENKLE, THOMAS A. KOMAREK, AND ARTHUR I. ZYGIELBAUM

Jet Propulsion Laboratory, Pasadena, California 91103

The 1976 Mars solar conjunction resulted in complete occultations of the Viking spacecraft by the sun at solar minimum. During the conjunction period, coherent 3.5- and 13-cm wavelength radio waves from the orbiters passed through the solar corona and were received with the 64-m antennas of the NASA Deep Space Network. Data were obtained within at least 0.3 and 0.8 R_S of the photosphere at the 3.5- and 13-cm wavelengths, respectively. The data can be used to determine the plasma density integrated along the radio path, the velocity of density irregularities in the coronal plasma, and the spectrum of the density fluctuations in the plasma. Observations of integrated plasma density near the south pole of the sun generally agree with a model of the corona which has an 8:1 decrease in plasma density from the equator to the pole. Power spectra of the 3.5- and 13-cm signals at a heliocentric radial distance of about 2 R_S have a $\frac{1}{2}$-power width of several hundred hertz and vary sharply with proximate geometric miss distance. Spectral broadening indicates a marked progressive increase in plasma irregularities with decreasing ray altitude at scales between about 1 and 100 km.

INTRODUCTION

The Viking mission and solar conjunction of Mars on November 25, 1976, provided a unique opportunity for studies of the K corona very near solar minimum. As Mars moved into and away from conjunction, the radio paths between the earth-based tracking station and Mars sliced through a region of the corona that included both equatorial and south polar solar latitudes. The observations extended from a heliocentric radial distance of 60 R_S (solar radii) to signal loss very near the photosphere. The coherent 3.5- and 13-cm (called X and S band) radio transmissions from the orbiters have been detected within at least 0.3 and 0.8 R_S of the photosphere, respectively. The regions through which the ray paths passed are particularly interesting, since they are thought to contain both the critical point (3–5 R_S), where coronal expansion goes from subsonic to supersonic flow, and the Alfvén point (5–10 R_S), where the magnetic and dynamic forces in the plasma are equal. These radio signals were received by the 64-m antennas of the NASA Deep Space Network in California and Australia. This paper discusses the observational factors involved and presents some very preliminary results.

Radio waves traversing the corona are affected in several ways: (1) they are deflected away from the sun by refraction, (2) the phase observed at the receiver is less, because the phase velocity is increased, but the time delay for a wave packet is greater, because group velocity is decreased, than would have been the case for signals propagating over the same path in the absence of the coronal electrons, and (3) the waves are scattered by turbulence in the corona so that there is partial randomization of amplitude and phase. All these effects are important in the Viking solar corona experiment, and since they are wavelength dependent, observation at two wavelengths permits a determination of the properties of the medium integrated along the radio path. The modulation time delay introduced by the corona is of the same sign as that of the time delay of general relativity, but the refraction is of opposite sign to that of the relativistic bending of light. Both are additionally of interest as corrections to the general relativity experiment utilizing the Viking landers [*Shapiro et al.*, 1977].

Observations of the radio propagation through the corona can be used to infer several fundamental quantities including the mean variation of coronal density with radial distance and the strength and structure of turbulence and/or waves in the coronal plasma. It may be possible to separate latitudinal and longitudinal effects on the basis of the variations in the geometry during the conjunction period.

The coronal medium has refractivity

$$\nu = -4.48 \times 10^{-16}\lambda^2 N_e \qquad (1)$$

where λ is the wavelength in meters, N_e is the electron number density per cubic meter, and the magnetic field and collisions are neglected. Absorption is negligible at the wavelengths of interest here. The refractive index $n = 1 + \nu$ is less than unity.

We have used the expression

$$N_e \left[\left(\frac{2.99}{\rho^{16}} + \frac{1.55}{\rho^{6}}\right) \times 10^{14} + \frac{3.44}{\rho^2} \times 10^{11}\right] \cdot (\cos^2\theta + \tfrac{1}{64}\sin^2\theta)^{1/2} \text{el}^-/\text{m}^3 \qquad (2)$$

where ρ is the distance (in solar radii) from the solar center of mass and θ is solar latitude, as a model of the corona. This model is the same as the Baumbach-Allen model [*Allen*, 1947; *Pottasch*, 1960] at the equator, where $\theta = 0$, except that we have added a latitudinal factor and a ρ^{-2} term corresponding to 7.5 el$^-$/cm^3 in the solar wind at the orbit of the earth. The latitudinal factor is heuristic; it approximates results from work on solar effects on radio observations of pulsars [*Weisberg et al.*, 1976]. This factor is an ellipse of axial ratio 8:1 and is of the same order as latitudinal variations determined from white light coronagraphs and eclipse observations near solar minimum. The value for the solar wind term was determined (by one of us, T.A.K.) from differential group delay measurements using Viking radio tracking prior to conjunction. It is in essential agreement with other similar determinations [*Brandt*, 1970; *Croft*, 1972].

The ρ^{-6} and ρ^{-2} terms are equal at just under 5 solar radii.

Paper number 7S0455.

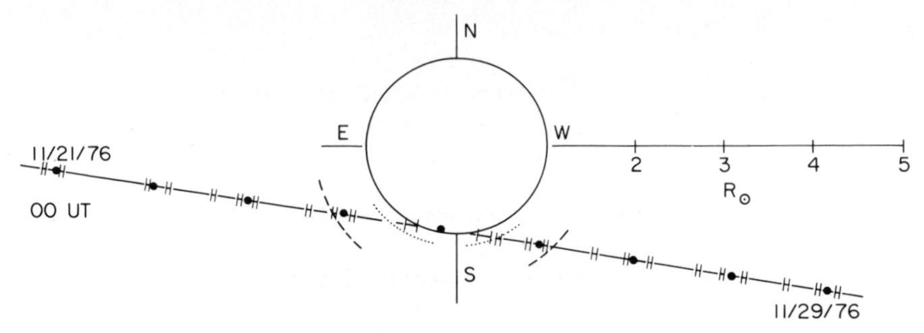

Fig. 1. Geometry of Viking solar corona observations and the geometric path of Mars in the plane of the sky for November 21–29, 1976. Bold points are positions at 0000 UT. The large circle represents the photosphere. Dotted and dashed segments are the geometric sizes of the 3.5- and 13-cm occulting disks, the limiting closest approach of the rays due to refraction, based on the same radial variation in plasma density as (2) in text, but with the latitudinal factor omitted. E-W directions are astronomic convention. Short hatch marks indicate positions where scintillation data were obtained. Farther from the sun the path of Mars is a smooth continuation of the curves shown. Data in Figure 2 are from E and W of this figure; data in Figures 5 and 6 are included in this figure.

This limit can be considered as defining a radio corona in terms of a rapid onset of propagation effects, but it should be noted that the solar plasma has a significant, often highly detrimental effect on radio propagation at the 13-cm wavelength as far out as 60–80 R_S.

OBSERVATIONS

Large- and Intermediate-Scale Coronal Structure

Differential dual-frequency measurements of apparent spacecraft range (time delay) and velocity (Doppler frequency) were used to determine the value and the time derivative, respectively, of the columnar electron content integrated along the propagation path. These values apply to a column with a cross-sectional area of Fresnel zone size, or about a 100-km diameter in the vicinity of the sun.

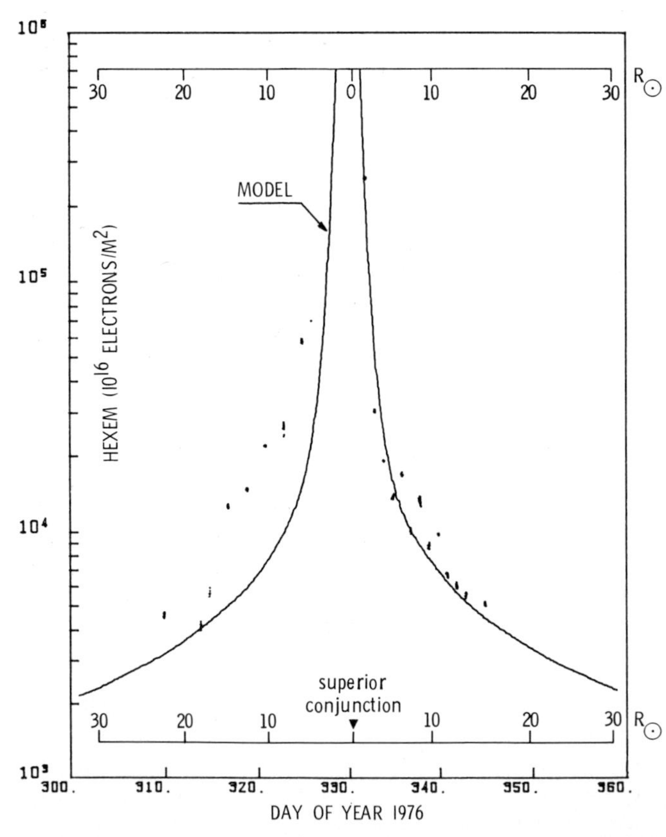

Fig. 2. Variation of measured columnar electron content of corona. The solid curve is a result of numerical integration of (2) in the text along geometric ray with closest approach A (see Figure 3) given by the inner scale. A total of 312 separate observations are included, many of which are indistinguishable at this scale. Note the strong asymmetry with respect to conjunction in both model and data. Day of year 330 is November 25, 1976.

The techniques employed have been discussed extensively elsewhere (for a discussion of the Viking system and operations, see *Michael et al.* [1972] and *Johnston et al.* [1977]). We note, however, that small changes in the total content are best determined when it is possible from the integral of the measured derivative with time and that the constant of integration is determined from the differential range. For observations within about 20 R_S, only the differential range was reliable because of the difficulties in accurate Doppler measurements under dynamic signal conditions.

Figure 1 depicts the geometric position of Mars with respect to the sun. The view is in the plane of the sky. The short intervals between small marks along the path indicate scintillation data coverage as explained in the caption. At distances greater than 5 R_S the path is a smooth continuation of the one shown. The coverage depicted here is only a fraction of the total but is all within 5 R_S of the sun's center.

Total electron content as determined from differential range measurements is given in Figure 2. The smooth curve in the figure corresponds to the modified Baumbach-Allen model of (2) above with the proper pole geometry included. The small groups of points in the figure are the result of independent range measurements on a single day. The standard error in these measurements is about 70×10^{16} el$^-$/m^2, so standard error bars in integrated content would not be visible in the presentation of Figure 2; the spread among points in a single day is the result of variations in the solar plasma with time. Figure 2 contains 312 separate observations, many of which overlie one another.

Most of the data presented for a single day in Figure 2 are based on differential range measurements using a device called the planetary ranging assembly (PRA) at the Australian 64-m antenna station. Unfortunately, this device had an ambiguity in the resolution of differential range of only 2 μs. This corresponds to an ambiguity in the integrated electron content of 8.456×10^{19} el$^-$/m^2. A different device, called the Mu-2 machine, with an ambiguity resolution of about 1 s, was in use at the California station. Mu-2 observations obtained just prior to the PRA observations were used to resolve the ambiguity associated with the use of the PRA alone. When the range residuals in tracking the Viking landers on the surface of Mars are corrected for the effects of the plasma shown in Figure 2, they are typically much less than one-half the ambiguity in the PRA results, i.e., less than 1 μs; thus confidence in the results is increased. However, on isolated days, there are larger problems [see *Shapiro et al.,* 1977].

The model curve in Figure 2 is not symmetrical about the time of closest approach of the ray path to the center of the sun but is skewed leftward by the latitudinal factor in (2). Clearly, the measured values during the period just prior to conjunction exceed those predicted by the model by up to a factor of about 4. During the period immediately following conjunction the measured values and the model agree within a factor of less than 2 and are less than the values at the same distance from the sun prior to conjunction by a factor of about 5. Evidently, this rather strong change in integrated plasma density is indicative of pronounced variations along the ray path in the vicinity of the sun. The variations may be purely latitudinal, as is presumed in (2), or may represent a combination of latitudinal, temporal, and longitudinal effects.

Note that the increased plasma density observed prior to conjunction persisted for a time period of approximately 15 days. The measurements near the sun were effectively terminated by the effects of the corona prior to conjunction at about 5 R_S. Just after conjunction, differential range measurements were obtained as close as about 2 R_S, giving indirect evidence of asymmetry. In time these measurements corresponded to about one-half solar rotation following the maximum deviation from the model at about 10 R_S during occultation entry. Thus a purely longitudinal variation observed just before conjunction would have been carried across the path after conjunction by solar rotation. Further, no sunspots were observed in the southern hemisphere of the sun for 27 days prior to and for 18 days after occultation, although some calcium plagues were present. During all of November, only one small sunspot group appeared in the northern hemisphere near midmonth and disappeared over the western limb near the end of November [*National Oceanic and Atmospheric Administration,* 1976, 1977*a, b*]. While additional work on this point is required, the conditions of geometry and the lack of features on the photosphere are good preliminary evidence that the strong asymmetry observed is due to latitudinal variations similar to those in (2).

Velocity in the Coronal Plasma

For this discussion we need to distinguish between the motions of irregularities in the refractive index of the corona due to wave propagation through a particular region and those due to convective transport or mass flow. Radio methods are sensitive only to the motions of the refractive index fluctuations themselves, but these different events may have distinct signatures.

Under the frozen-in hypothesis for plasma it is expected that small fluctuations in the refractive index that occur outside the Alfvén point are carried along at the bulk velocity of the medium over some distance. Waves propagating through the medium are similarly translated by the large-scale motion. Near the sun the wave velocities and mass flow are thought to be of the same order, but at sufficiently small heliocentric distances the wave velocity is greater. The Viking measurements can be used to determine the velocities of coronal irregularities from cross correlation of the induced phase and intensity scintillations on two well-separated radio paths and from a characteristic feature in intensity scintillation spectra (see, for example, *Hewish* [1972], *Young* [1971], and *Jokipii* [1973]).

Velocity can be obtained rather directly by cross correlation of signals at the same wavelength over several Viking spacecraft-to-earth paths. These include signals transmitted from a single spacecraft to two earth stations and signals from two or more Viking spacecraft, usually an orbiter and a lander, received at the same ground station. The vector separation of the rays at the point of their closest approach to the sun is approximately 0.4 times the vector separation of the spacecraft in the plane of the sky. This separation is typically of the order of 10,000 km. If the lifetimes of smaller-scale irregularities are greater than the time to traverse this distance, about 30 s, then direct cross correlation of the signals will yield accurate velocities. Large-scale (100,000 km) irregularities persist much longer than 30 s [*Callahan,* 1975], but evidently, not much is known about lifetimes of irregularities at smaller scales. Radio astronomy observations of the solar wind suggest that the lifetime of a 100-km structure is greater than about 1 s, but there are suggestions that correlation over very large path separations may not be useful [*Dennison and Hewish,* 1967; *Hewish,* 1972]. If the lifetime of small-scale irregularities is too short to permit their use, we would hope to determine bulk velocity from such widely separated paths by searching for events associated with the passage of density gradients of material. The method of spaced receivers has been applied to observations of intensity scintillations of radio sources and tested by comparison with spacecraft data [*Jokipii,* 1973]. Simple theory gives agreement to within about 20%. More refined calculations improve the results.

We also expect to be able to use paths with much smaller separations to determine the radial component of velocity near the sun. Figure 3 schematically illustrates the geometry. Signals at the two wavelengths follow paths that are determined by the mean refractivity of the medium. The 3.5-cm wave undergoes the least bending, relative to which the 13-cm wave is always displaced toward the sun. The resulting ray path separation at ray perihelion is given in Figure 4. A disturbance moving radially outward from the sun will first perturb the 13-cm and then the 3.5-cm signal. The velocity would be determined by cross correlating simultaneous 3.5- and 13-cm wavelength observations of a single spacecraft by a single station. In the corona, where wave velocities may be typically 150 km/s at 2 R_S and the bulk velocities considerably less, the time delay will be of the order of seconds to tens of seconds, depending on the event and the geometry. Of additional interest, the magnitude of the correlation as a function of the vector path spacing is expected to be useful in weighting the plasma corrections for the Viking General Relativity experiment. Estimates of the coherency between the 3.5- and the 13-cm wavelength intensity, if no path separation is assumed, give values near 0.5

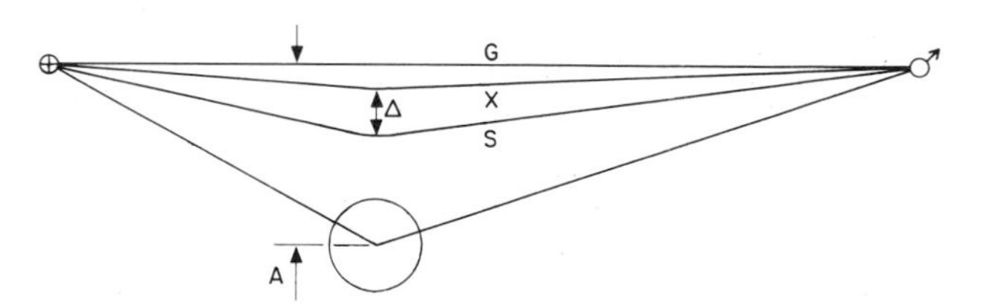

Fig. 3. Refraction of 3.5- and 13-cm waves in the corona and a schematic illustration of ray path bending due to coronal plasma. Paths are G, geometric ($\lambda = 0$); X, 3.5-cm; and S, 13-cm wavelengths. Distance A is closest approach of geometric ray, and distance Δ is ray path separation.

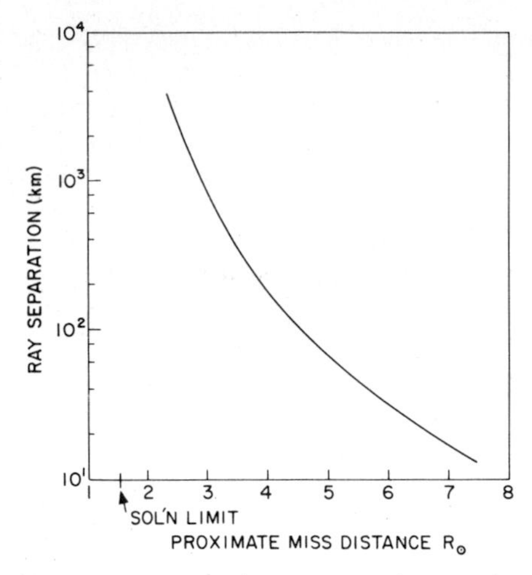

Fig. 4. Separation of 3.5- and 13-cm paths due to differential refraction. Ray path separation corresponds to distance Δ plotted versus distance *A* in Figure 3.

for this quantity [*Woo*, 1975]. This would be the approximate value of the cross correlation squared if the medium is perfectly rigid over the times required to traverse Δ in Figure 4.

The detectability of these effects based on the phase observable depends, as do the dual spacecraft observations, primarily on the short-term stabilities of the Viking spacecraft oscillators (about 1 part in 10^{10}). The passage of a plasma cloud of content equal to the earth's daytime ionosphere (10^{17} el^-/m^2) would correspond to a 10% fluctuation in plasma density at 5 R_S over a path length of about a 1/6 solar radius.

Finally, intensity scintillation spectra display a characteristic knee at a temporal frequency whose precise value depends on the geometry and distribution of the turbulence but is approximately the velocity divided by the Fresnel zone size. More specifically, this frequency is

$$f_F = ku/(\pi\lambda z)^{1/2}$$

where u is the velocity, z is approximately 1 AU, λ is the wavelength as before, the constant k is approximately unity, and the fluctuations of the medium are assumed to be isotropic [*Young*, 1971; *Jokipii*, 1973]. For plasma waves in the corona this frequency is of the order of 1 Hz and is easily observable in frequency spectra.

Because the Viking data offer several independent methods for determination of velocity, it will be possible to test the results for internal consistency and to test inferences based on the theory of scintillation against cross correlations of signals following different paths. Hopefully, it will be possible to separate events associated with wave phenomena from mass flow on the basis of theoretical considerations, ancillary observations, and event characteristics such as spectral shape.

Small-Scale Coronal Structure

The Viking coherent 3.5- and 13-cm radio systems are the first to provide direct occultation of the sun by coherent dual-frequency signals. Several previous spacecraft have provided observations at a single wavelength or, in the case of Mariner 10, which carried a similar coherent dual-frequency system, have not approached closer than about 7 R_S.

Most past radio astronomy and spacecraft studies of rapid (fluctuations faster than about 1 s) interplanetary or coronal scintillation have been restricted to fluctuations in signal intensity or apparent source size. These fluctuations are strongly coupled to variations in the refractive index of the medium that are approximately Fresnel zone size and smaller, or about 70 and 140 km for the 3.5- and 13-cm waves at solar distance. Viking dual-frequency spacecraft signals make it possible to observe rapid coronal scintillations in the phase of the signals as well. (Differential phase measurements or their equivalent have been made by numerous others; they have been specifically interpreted as large-scale scintillations by *Callahan* [1975] and *Woo et al.* [1976].) Differential phase scintillations are sensitive to refractive index fluctuations over all scales significantly greater than the tracking station antenna size, a lower bound that is shared with the effects of the medium on intensity fluctuations. Thus conservatively, the Viking data contain information on all scales in the corona greater than about 1 km without the need to consider the effects of finite antenna beam width. The problem of extracting scale size information from the observed intensity and phase scintillations is straightforward, since the relationship between the turbulence spectrum and the scintillations is known (see, for example, *Jokipii* [1973]).

Here we present examples of spectral broadening observed during the Viking conjunction period. Figure 5 shows the power spectra of the received signals at three different times during the experiment. The data shown were all taken by using

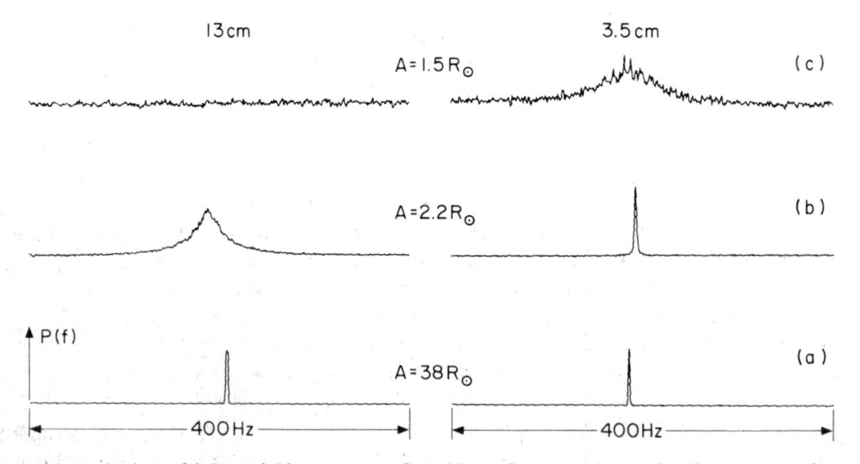

Fig. 5. Spectral broadening of 3.5- and 13-cm waves. Sequence of power spectra for three geometric positions of earth, sun, and Mars are characterized by closest approach of geometric ray path *A* (see Figure 3). Absence of the 13-cm signal in 5*c* is due to corona. A spurious signal, 0.2 intensity of the 3.5-cm wave, about 50 Hz higher in frequency than that of the 3.5-cm wave, has been deleted from the Figure 5*b* 3.5-cm data.

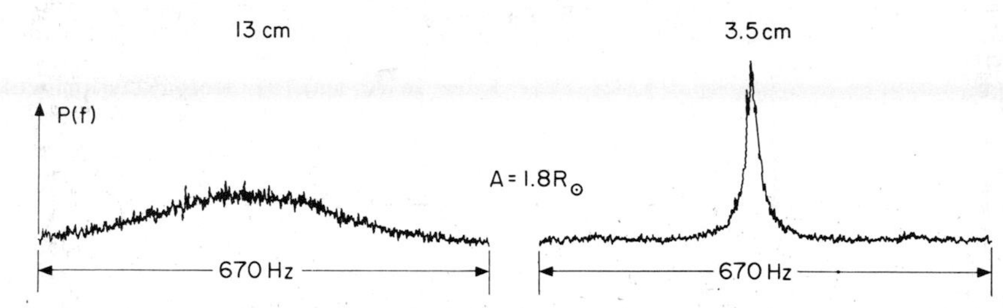

Fig. 6. Simultaneous 3.5 and 13 cm spectral broadening at 1.8 R_S. There is well developed spectral broadening near the limit of the 13-cm penetration. A spurious signal has been deleted from the center of the 13-cm spectrum; otherwise spectra are as computed. Leftward skew of the 13-cm signal is apparently real but needs further study. These spectra were obtained 5 hours before those of Figure 5*c*.

the spacecraft auxiliary oscillator as a frequency reference for the spacecraft. On the time scale of these observations, about 200 s, the spectral width of the transmitted signal is near 1 and 4 Hz at the 13- and 3.5-cm wavelengths, respectively. Figure 5*a* is typical of ray paths that miss the coronal region but are still influenced by the interplanetary medium. Power spectral density is plotted linearly, and the horizontal axes of both the 3.5- and the 13-cm wavelength signal are the same, 400 Hz full width. The signals, which appear near the center of the plots in Figure 5*a*, show slight spectral broadening at 13 cm and less at 3.5 cm. Figures 5*b* and 5*c* show progressive effects of propagation in deeper portions of the corona. In Figure 5*b* the 13-cm wavelength is significantly affected, but on the scales shown, the 3.5-cm signal shows no significant effects. In Figure 5*c*, which corresponds to a ray path near the theoretical 13-cm occulting disk (the limiting distance of closest approach based on refraction, see Figure 1), the 13-cm signal has degraded to the point where it is no longer apparent. Note that the transition from Figure 5*b* to Figure 5*c* is very rapid owing to the ρ^{-6} dependence in coronal structure. The 3.5-cm signal is still readily detectable in spectra such as these when the ray path is 1.3 R_S from the center of the sun.

An intermediate condition to Figures 5*b* and 5*c* is illustrated in Figure 6, in which there is marked spectral broadening at both wavelengths. Note that the horizontal scale in Figure 6 has been expanded. The conditions of Figure 6, which occurred at about 1.8 R_S, or only slightly outside the conditions of Figure 5*c*, lasted only a few hours.

In the broadened scintillation spectra the shift in frequency from the undisturbed (by the irregularities) position is directly related to the velocity of irregularities u and the vector component of wave number of the plasma irregularity in the direction of the velocity k_u. The frequency offset is then $uk_u/2\pi$, independent of the radio wavelength. If 200 km/s represents a typical plasma bulk velocity high in the corona, then the frequency displacements easily visible in Figure 6 represent the effects of structure down to about 2 km in size.

APPLICATIONS

Independent measures of coronal plasma density, velocity, and structure larger than about 1 km in scale make it possible to determine (1) the distribution of coronal plasma, (2) the velocity of irregularities in the corona and probably the mass flow, and (3) the spectrum of the turbulence down to scales as small as about 1 km. The geometry of the Viking experiment permits comparison of equatorial and south polar regions as given by the example of the integrated plasma content above.

One problem of particular interest is that of regions of relatively low X ray, UV, and radio emissivity (and therefore relatively low temperature and/or density) in the corona, called 'coronal holes.' Coronal holes are observed to be the source of high-speed solar wind streams having relatively high particle fluxes [*Nolte et al.*, 1976]. To account for the apparent contradiction between this source's behavior and its being a region of low density and/or temperature, *Kopp and Holzer* [1976] and others have suggested that the critical point in coronal expansion is located much closer to the sun ($\lesssim 2\ R_S$) in coronal holes than it is elsewhere. Thus the flow becomes supersonic low in the corona and remains supersonic at larger heliocentric distances. We hope to test this hypothesis experimentally. Because the ray path trajectory passes through the southern polar corona before and after occultation, it provides an opportunity to study one such region continuously located above the sun's polar latitudes at solar minimum. It has also been suggested [*Jackson*, 1977] that there is significant energy deposition beyond 3 R_S, far above the base of the corona. Since the radio link data are a measure of wave and shock disturbances, on a time scale greater than that required to form a reliable spectrum, about 1 s, energy deposition by these means can be investigated.

Disturbances in the solar wind observed near 1 AU such as shocks and magnetohydrodynamic waves are often assumed to originate in the low corona [*Hundhausen*, 1972; *Wu et al.*, 1976]. Such disturbances in the low corona are presumably detectable by the Viking radio link experiment, although the ray path trajectory made near-ecliptic measurements possible only at relatively large distances from the sun ($\gtrsim 3\ R_S$). The measurement of events both in the low corona and in the interplanetary medium is expected to be useful in evaluating theoretical work on disturbance propagation in the solar wind (for example, that of *Hundhausen and Gentry* [1969] and *Wu et al.* [1976]).

The fact that the occultation took place very near solar minimum provides a considerable advantage. First, the flow structure of the corona was in its most simple state, relatively undisturbed by active regions. Second, studies of the corona above active regions will be relatively simple. That is, there were very few active regions on the sun at this time, and thus studies of the corona above an individual active region (for example, using the radio link data in conjunction with ground-based and spacecraft data from Oso 8 and Helios 1 and 2) are not subject to confusing effects from nearby active regions as would likely be the case at times of more solar activity. Further, the data will be extremely interesting for comparison with similar observations at the same wavelengths from Mariner 10, which passed about 7 R_S over the north pole of the sun in 1974, and from Voyager during the next solar maximum.

CONCLUSIONS

Viking 3.5- and 13-cm radio data penetrated the solar corona to very near the limits expected on the basis of previous optical studies. The measured columnar electron content is strongly asymmetric in terms of radial dependence with respect to conjunction. This asymmetry reflects either a strong latitudinal dependence of coronal plasma density or a latitudinal dependence in combination with other variations. Spectral broadening of the signals penetrating the corona indicates a marked progressive increase in plasma irregularities with decreasing distance from the sun at scales greater than about 1 km. The data are not yet completely reduced; these results are based on the first preliminary analyses of the observations.

Acknowledgments. The authors gratefully acknowledge the assistance and support of the Viking Project staff and especially their colleagues on the Viking Radio Science Team. We also note many useful discussions with J. F. Vesecky and the efforts of B. Seidel in the collection of the scintillation data. This work at Stanford and at the Jet Propulsion Laboratory was carried out as a part of the NASA Viking mission by using the facilities of the NASA Deep Space Network but also depended in part on ongoing NASA programs in the use of radio propagation data at both institutions. NASA contract NAS1-9701.

REFERENCES

Allen, C. W., Interpretation of electron densities from corona brightness, *Mon. Notic. Roy. Astron. Soc., 107,* 426–432, 1947.

Brandt, J. C., *Introduction to Solar Wind,* W. H. Freeman, San Francisco, Calif., 1970.

Callahan, P. S., Columnar content measurements of the solar-wind turbulence near the sun, *Astrophys. J., 199,* 227–236, 1975.

Croft, T. A., Solar wind concentration averaged in time and space (abstract), *Eos Trans. AGU, 53,* 505, 1972.

Dennison, P. A., and A. Hewish, The solar wind outside the plane of the ecliptic, *Nature, 213*(5074), 343–346, 1967.

Hewish, H., Observations of solar plasma using radio scattering and scintillation methods, Solar Wind, *NASA Spec. Publ., 308,* 477–493, 1972.

Hundhausen, A. J., *Solar Wind and Coronal Expansion,* Springer, New York, 1972.

Hundhausen, A. J., and R. A. Gentry, Numerical simulation of flare-generated disturbances in the solar wind, *J. Geophys. Res., 74*(11), 2908, 1969.

Jackson, B. V., A coronal hole equatorial extension and its relation to a high speed solar wind stream, paper presented at Topical Conference on Solar and Interplanetary Physics, AGU and Amer. Astron. Soc., Tucson, Ariz., Jan. 12–15, 1977.

Johnston, D. W., T. W. Howe, G. M. Rockwell, Viking mission support, *Progr. Rep. 42-37,* pp. 12–25, Deep Space Network, Jet Propul. Lab., Pasadena, Calif., 1977.

Jokipii, J. R., Turbulence and scintillations in the interplanetary plasma, in *Annu. Rev. Astron. Astrophys., 11,* 1–28, 1973.

Kopp, R. A., and T. E. Holzer, Dynamics of coronal hole regions, I, Steady polytropic flows with multiple critical points, *Solar Phys., 49,* 43–56, 1976.

Michael, W. H., Jr., D. L. Cain, G. Fjeldbo, G. S. Levy, J. G. Davies, M. D. Grossi, I. I. Shapiro, and G. L. Tyler, Radio science experiments: The Viking Mars orbiter and lander, *Icarus, 16,* 57–73, 1972.

National Oceanic and Atmospheric Administration, Solar geophysical data prompt report, I, *Rep. 388,* U.S. Dep. of Commer., Boulder, Colo., Dec. 1976.

National Oceanic and Atmospheric Administration, Solar geophysical data prompt report, I, *Rep. 389,* U.S. Dep. of Commer., Boulder, Colo., Jan. 1977*a*.

National Oceanic and Atmospheric Administration, Solar geophysical data prompt report, I, *Rep. 390,* U.S. Dep. of Commer., Boulder, Colo., Feb. 1977*b*.

Nolte, J. T., A. S. Krieger, A. F. Timothy, R. E. Gold, E. C. Roelof, G. Viana, A. J. Lazarus, J. D. Sullivan, and P. S. McIntosh, Coronal holes as sources of the solar wind, *Rep. ASE-3817,* Amer. Sci. and Eng., Cambridge, Mass., 1976.

Pottasch, S. R., Use of the equation of hydrostatic equilibrium in determining the temperature distribution in the outer solar atmosphere, *Astrophys. J., 131,* 68–74, 1960.

Shapiro, I. I., R. D. Reasenberg, P. E. MacNeil, R. B. Goldstein, J. P. Brenkle, D. L. Cain, T. A. Komarek, A. Zygielbaum, W. F. Cuddihy, and W. H. Michael, Jr., The Viking relativity experiment, *J. Geophys. Res., 82,* this issue, 1977.

Weisberg, J. M., J. M. Rankin, R. R. Payne, and C. C. Counselman III, Further changes in the distribution of density and radio scattering in the solar corona in 1973, *Astrophys. J., 209,* 252–258, 1976.

Woo, R., Multifrequency techniques for studying interplanetary scintillations, *Astrophys. J., 201,* 238–248, 1975.

Woo, R., F. C. Yang, K. W. Yip, and W. B. Kendall, Measurements of large scale density fluctuations in the solar wind using dual-frequency phase scintillations, *Astrophys. J., 210*(2), part I, 568–574, 1976.

Wu, S. T., M. Dryer, and S. M. Han, Interplanetary disturbances in the solar wind produced by density, temperature or velocity pulses at 0.08 A.U., *Solar Phys., 49,* 187–204, 1976.

Young, A. T., Interpretation of interplanetary scintillations, *Astrophys. J., 168,* 543–562, 1971.

(Received March 31, 1977;
revised May 23, 1977;
accepted May 25, 1977.)

Composition and Structure of Mars' Upper Atmosphere: Results From the Neutral Mass Spectrometers on Viking 1 and 2

A. O. NIER

School of Physics and Astronomy, University of Minnesota, Minneapolis, Minnesota 55455

M. B. MCELROY

Center for Earth and Planetary Physics, Harvard University, Cambridge, Massachusetts 02138

The upper atmospheric mass spectrometers flown on Viking 1 and 2 are described, and results obtained for the composition and structure of Mars' upper atmosphere are summarized. Carbon dioxide is the major constituent of the atmosphere at all heights below 180 km. The thermal structure of the upper atmosphere is complex and variable with average temperatures below 200°K for both Viking 1 and 2. The atmosphere is mixed to heights in excess of 120 km. The isotopic composition of carbon and oxygen in the Martian atmosphere is similar to that in the terrestrial atmosphere: ^{15}N is enriched in Mars' atmosphere by a factor of 1.62 ± 0.16.

INTRODUCTION

The scientific payload of Viking landers 1 and 2 included mass spectrometers designed to measure the composition of Mars' atmosphere at altitudes above about 100 km. The instruments were mounted on the spacecraft aeroshell, and measurements were taken during the entry phase of the missions [*Nier et al.*, 1972].

Viking 1 landed on Mars on July 20, 1976, at a latitude of 22.5°N and a longitude of 48°W about 4 hours after local noon with a solar zenith angle of approximately 44°. Viking 2 landed at 48°N, 226°W on September 3, 1976, about 10 hours after local midnight with a solar zenith angle similar to that for Viking 1. Preliminary accounts of results obtained by using the upper atmospheric mass spectrometers have appeared elsewhere [*Nier et al.*, 1976*a*, *b*; *McElroy et al.*, 1976*a*, *b*; *Nier and McElroy*, 1976]. This paper gives an updated account of results from both missions.

EXPERIMENTAL DETAILS

The compositional measurements were carried out by using double-focusing magnetic deflection type spectrometers described in more detail elsewhere [*Nier and Hayden*, 1971; *Nier et al.*, 1972]. Mass analyses were performed by using Mattauch-Herzog geometry [*Mattauch and Herzog*, 1934], an arrangement with special advantages for space investigations in that it allows efficient use of available space, weight, and power without serious compromise to performance characteristics of the instrument. The electron bombardment ion source adopted for Viking is similar in design to quasi-open sources employed successfully to study properties of the earth's atmosphere by using rockets and satellites and was used most recently in the Atmosphere Explorer aeronomy satellites AE-C, AE-D, and AE-E [*Nier et al.*, 1973].

Figure 1 shows a schematic illustration of the instrument as mounted on the Viking aeroshell. Ambient gas enters the ionizing region through three high-transmission grids, labeled 1, 2, and 3. Ions are formed by a tightly collimated electron beam indicated by the black dot between two small bar magnets designated by the letter M. The magnets are viewed end on in the figure. Ions formed in the source region are directed to the right by an electric field produced in part by a repelling voltage on grid 3 and in part by field penetration through the slit to the right of the electron beam. Field penetration arises because of the relatively low potential of the ion-focusing electrodes J_1 and J_2.

Grid 1 is maintained at the same potential as the instrument housing. Grid 2 is held at a negative potential, chosen to cancel fields produced outside by the positive potentials of the ion box sh and the ion repeller grid 3. Potentials on grids 2 and 3 vary in a linear fashion in response to a change in the potential of the ion box sh, and a similar linear relationship applies for the potentials on the focusing plates J_1 and J_2 and for the electric analyzer plates positive and negative. Mass scans are obtained by varying the potential applied to the ion box exponentially with time between limits selected to ensure that ions reaching the high-mass collector have mass numbers in the approximate range 7–49 amu. A permanent magnet was employed to provide the field in the magnetic analyzer. The mass of ions reaching the collector varies inversely as a function of the voltage V applied to the ion box, and in order to restrict the range of values required for V a second collector was used to capture ions in the mass range 1–7 amu. Thus low and high mass numbers were sampled simultaneously. A complete spectrum for the range 1–49 amu was accumulated in an interval of 5 s.

Ions were detected by using conventional electrometer type amplifiers with the input stage employing field effect transistors. The sensitivity of the instruments could have been enhanced significantly if electron multipliers had been used instead of electrometer amplifiers, and indeed multipliers have been used with high reliability on Atmosphere Explorer [*Nier et al.*, 1973]. Use of electrometers was dictated, however, by weight restrictions imposed by the Viking project during the instrument design phase of the mission. The sensitivity of the flight instruments was such that for nitrogen the peak analyzed at mass number 28 gave a signal of somewhat better than 2×10^{-5} A per torr of pressure in the ion source. Background fluctuations of the amplifiers were in the range $1\text{-}5 \times 10^{-14}$ A. The ionizing electron beam had a nominal value of 100 μA. Mass scans were made with both 75-eV elec-

Paper number 7S0485.

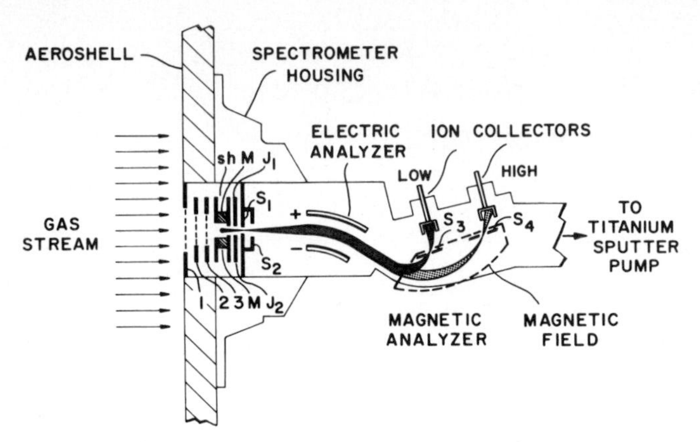

Fig. 1. Schematic diagram of the Viking upper atmosphere mass spectrometer mounted on the aeroshell of the lander. Until descent to planet was started, the instrument was sealed by a cap covering the ion source and pumped continuously. The radius of curvature of ions reaching the high-mass collector is 2.54 cm, and the nominal strength of the magnetic field 5000 G. Accordingly, the ion accelerating potential was 177 V for ions of mass 44 amu.

trons and 25-eV electrons. The high-energy source was used in six out of eight consecutive mass scans, and all of the results given here were derived from data obtained during the 75-eV mode.

Calibration procedures were similar to those employed for the open source mass spectrometers on Atmosphere Explorer [*Nier et al.*, 1973]. The instrument, with exposed ion source, was attached to a vacuum chamber pumped by a liquid helium pump separated from the chamber by an orifice of accurately known dimensions. Gas was admitted to the chamber through a molecular leak of known conductance from a reservoir at a high and known pressure. The equilibrium pressure of the gas in the chamber surrounding the mass spectrometer ion source could be computed accurately by using the known rate for flow into the chamber and the known rate at which gas was removed through the orifice to the liquid helium pumped vacuum chamber. Calibration runs were made for CO_2, N_2, Ar, CO, O_2, He, and H_2. The CO_2 runs covered the pressure range 2×10^{-7}–3×10^{-4} torr; reduced pressure ranges were employed for the other less abundant planetary gases.

The instrument was removed from the calibration stand following completion of the calibration procedure. A cap was installed over the ion source, and the mass spectrometer was evacuated through a small tube attached to the cap. The tube was pinched off after the instrument had been evacuated and baked. A small titanium sputter ion pump, permanently attached to the analyzer part of the instrument vacuum housing, was operated almost continuously for the remaining life of the instrument. This procedure ensured that the pressure in the instrument due to outgassing of surfaces or undetected leaks would remain below 10^{-8} torr during the long period (almost 2 years) between calibration and entry into the Martian atmosphere. The pump, operating during calibration and entry, had the additional benefit that it allowed measurements at higher pressures than would otherwise have been the case and permitted therefore an extension of the range of useful measurements to lower altitudes in the Martian atmosphere.

The cap covering the ion source was cut off flush with the surface of the aeroshell, exposing the ion source as shown in Figure 1, immediately following separation of the lander from the orbiter. The cutting mechanism employed an improved version of a device described elsewhere [*Thorness and Nier*, 1962]. The instrument received additional pumping from an essentially zero-pressure environment during the long 2½-hour descent to the sensible atmosphere. The mass spectrometer was turned on 60 min prior to the anticipated encounter with sensible planetary gas in order to allow it to warm up. Useful measurements were taken over the approximate height range of 100–200 km. During this time the lander was traveling at a speed of 4.5 km s^{-1}, the normal to the instrument ion source lying within 10° of the velocity vector. For all practical purposes the instrument was exposed to a high-speed gas beam from the normal direction as illustrated in Figure 1.

Problems and complications associated with the measurement of ambient gas densities by using instrumentation on high-speed spacecraft have been discussed extensively elsewhere [*Wiener*, 1949; *Horowitz and La Gow*, 1957; *Nier et al.*, 1964; *Hedin et al.*, 1964]. A stagnation effect resulting in a substantial increase in particle density is known to occur in the ion source for instruments looking forward as in Figure 1. The stagnation ratio can be accurately calculated in terms of the ambient and instrument temperatures and spacecraft velocity for instruments with ideal pinhole openings. A correction factor must be employed to allow for the more open source used here. The correction factor depends on the geometry of the ion source, and because of the complexity of the electrode and grid structures of the Viking instrument, the correction factor must be determined experimentally by using a molecular beam simulation [*Hayden et al.*, 1974; *French et al.*, 1975]. The correction factor was found to have a value of 1.17 in a laboratory test carried out with an instrument identical to the flight hardware; i.e., the ambient gas number densities as found by using the idealized formula are too high by 17%. The data presented below were corrected to allow for this effect.

Density Profiles for Major Constituents: Implications for Thermospheric Temperature

Useful data were obtained over the height range 120–200 km for Viking 1. Data from Viking 2 extend somewhat lower in altitude, to about 115 km. Some 12 spectra were analyzed for each entry. We shall describe here results derived for CO_2, N_2, Ar, CO, and O_2. Carbon dioxide is the major constituent at all altitudes below 180 km. The variation with altitude for the density of CO_2 may be used therefore to provide relatively direct information on the temperature of the upper atmosphere over the height range studied by Viking.

Figure 2 shows a typical mass spectrum taken at an altitude of about 140 km during the descent of Viking Lander 1. The

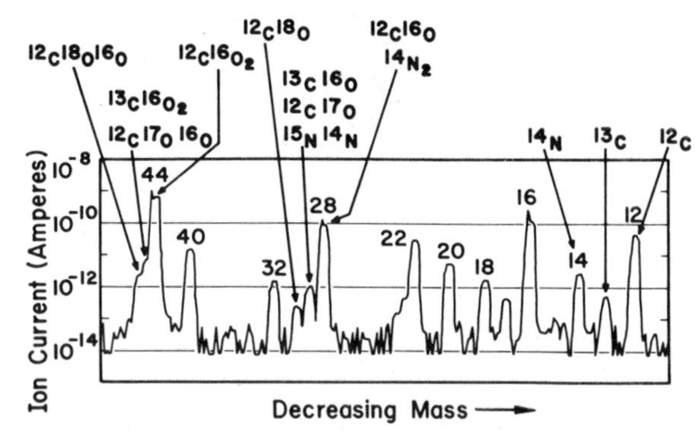

Fig. 2. Mass spectrum found at the 140-km altitude during the descent of Viking 1. Since the deflection, and hence separation, of ions in instruments of this type depends upon their ratio of mass to charge, doubly charged ions such as CO_2^{2+} and Ar^{2+} appear at the mass 22 and 20 amu positions in the spectrum, respectively. Time increases to the right of the figure.

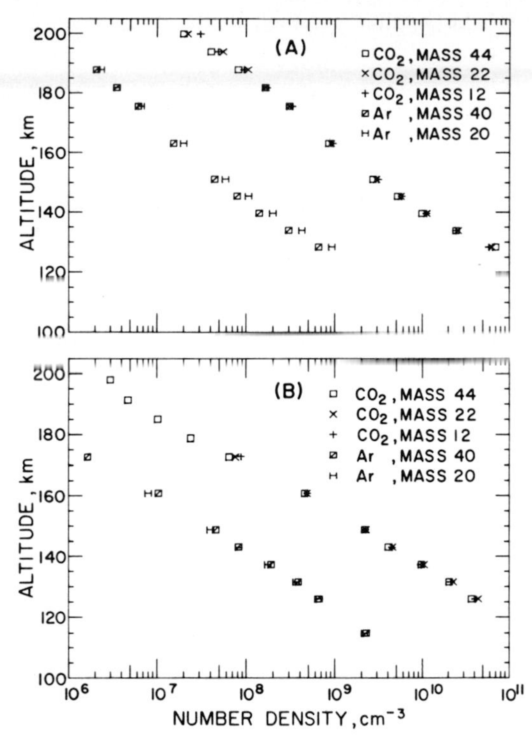

Fig. 3. Number densities of CO_2 and Ar computed from analysis of mass peaks at 44, 22, and 12 amu in case of CO_2 and mass peaks at 40 and 20 amu in case of Ar for (*a*) Viking 1 and (*b*) Viking 2.

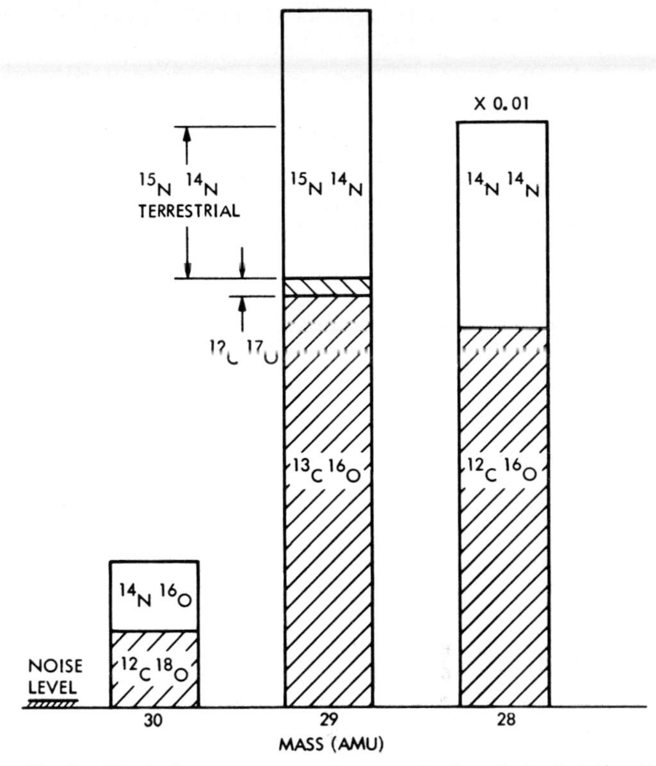

Fig. 5. Block diagram representing to scale the relative heights of the mass peaks at 30, 29, and 28 amu in a spectral scan found near 140 km during the descent of Viking 1. The contributions of the various isotopic combinations are shown. For reference the noise level of the amplifier which measured the ion currents is also shown.

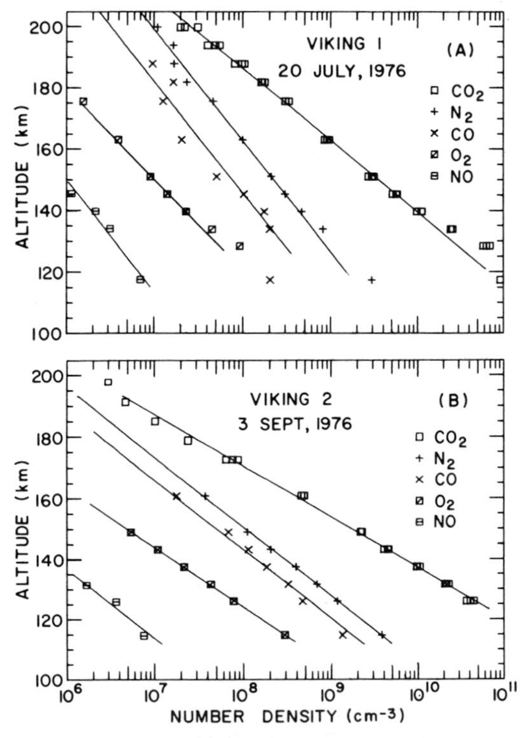

Fig. 4. Number densities of CO_2, N_2, CO, O_2, and NO for (*a*) Viking 1 and (*b*) Viking 2. The lines drawn are the least squares fit of a straight line to the data above 140 km. The NO points were obtained by measuring the heights of the 30-amu peaks and subtracting the contributions due to $^{12}C^{18}O$, assuming the difference is due to the presence of NO. Since laboratory calibrations of the instruments were not made for NO in computing the NO particle densities, the cross section for ionization by 75-eV electrons was assumed to be 0.8 of that for N_2 in accordance with generally accepted laboratory calibration data.

spectrum includes all of the important fragments which result from ionization and dissociation of CO_2 as indicated by mass peaks at 44, 28, 22, 16, and 12 amu associated with CO_2^+, CO^+, CO_2^{++}, O^+, and C^+. The spectrum includes in addition peaks at 40 and 20 amu associated with Ar^+ and Ar^{++}. The peak at 14 amu may be attributed to N^+ formed by dissociative ionization of N_2. Molecular nitrogen makes a significant contribution also to the peak at 28 amu. The peak at 32 amu is due to O_2. Peaks at 18 and 17 amu are due most probably to terrestrial H_2O released from the surfaces of the ion source by interactions with the incident gas stream. This phenomenon is well known and is seen invariably by mass spectrometers carried on sounding rockets and earth orbital satellites.

The spacecraft moves appreciably in the vertical direction during the 5-s interval required to complete a spectral scan. The vertical component of the spacecraft velocity has a magnitude of approximately 1.2 km s^{-1}. Thus the atmosphere sampled during the later stages of a spectral scan is denser than that seen at the beginning of the scan, and the data must be corrected accordingly. All ion peak magnitudes in a particular spectrum were corrected to reflect values which should apply at the beginning of the scan. The correction amounts to as much as 40% for the C^+ peak at mass number 12 amu.

The heights of peaks at 44, 22, and 12 amu, in combination with laboratory calibrations, provide independent information on the density of CO_2. In a similar manner the peaks at 40 and 20 provide redundant information for Ar. The degree to which the data are internally consistent may be assessed from the results shown in Figure 3.

Carbon dioxide, CO, and N_2 all contribute to the peak at 28 amu. Similarly, CO and N_2 contribute to the signal observed at

TABLE 1. Constants Deduced for (2) From Number Densities Plotted in Figures 3 and 4

	b_i	a_i
	Viking 1	
CO_2	−0.042696	15.930
O_2	−0.033945	12.074
N_2	−0.027514	12.452
CO	−0.027514	11.984
NO	−0.027514	10.111
Ar	−0.027514	13.681
	Viking 2	
CO_2	−0.059714	18.164
O_2	−0.050729	14.272
N_2	−0.043866	14.580
CO	−0.043866	14.254
NO	−0.043866	11.995
Ar	−0.053521	15.567

14 amu, and CO and CO_2 contribute at mass number 12. Densities for N_2 and CO may be derived by solving an appropriate set of simultaneous equations which incorporate relevant information obtained during the preflight calibration procedure. Results are shown in Figure 4, which includes also densities derived for O_2 and NO. The presence of NO may be inferred from an analysis of the peak measured at mass number 30, as illustrated in Figure 5. Only about half of this peak can be attributed to $^{12}C^{18}O^+$ formed by dissociative ionization of $^{12}C^{16}O^{18}O$, as is discussed in more detail elsewhere [*Nier and McElroy*, 1976]. The remainder is due to NO formed by reaction of $N(^2D)$ with CO_2 [*Yung et al.*, 1977; *McElroy et al.*, 1976*b*].

Figure 4 includes a number of simple parametric fits to the data as measured for CO_2, N_2, CO, O_2, and NO. We assumed that the density n_i of a particular species i might be taken to vary with height according to the relation

$$n_i(z) = n_i(z_0) \exp [-(z - z_0)m_i g/kT] \quad (1)$$

where z_0 denotes an appropriate reference altitude and m_i, g, k, and T indicate the mass of constituent i, the acceleration of gravity, Boltzmann's constant, and an effective temperature, respectively. The parameter T was adjusted to obtain a best least squares fit for each constituent. Temperatures obtained in this manner for CO_2, N_2, Ar, and O_2 on Viking 1 had values of 179.0°, 174.7°, 176.0°, and 163.7°K, respectively. Corresponding temperatures for these gases on Viking 2 had values of 128.0°, 110.9°, 129.8°, and 109.6°K, respectively. For purposes of the fits shown in Figure 4 we assumed that the appropriate values for T to describe CO and NO would be identical to those found for N_2. With our assumptions the variation with altitude for the density of gas i may be described by the empirical relation

$$\log_{10} n_i(z) = a_i + b_i z \quad (2)$$

where z denotes height above the mean surface of the planet. Values of the parameters a_i and b_i are listed for both Viking 1 and 2 in Table 1.

More careful examination of the CO_2 data in Figure 3 reveals an interesting and complex thermal structure. The range of temperatures found by fitting individual gases to the simple relation (1) hints indeed that this should be the case, though part of the discrepancy must be attributed to complications introduced by chemistry and by mixing processes at lower altitude. Mass density ρ and pressure p are constrained to satisfy the familiar barometric relation

$$dp/dz = -\rho g \quad (3)$$

which may be recast in the form

$$n(z_2) = \frac{T_1 n(z_1)}{T_2} \exp\left(-\int_{z_1}^{z_2} \frac{dz}{H'}\right) \quad (4)$$

where $n(z_1)$ and $n(z_2)$ denote densities of CO_2 at heights z_1 and z_2, T_1 and T_2 denote corresponding temperatures, and H' is the scale height given by

$$H' = kT/mg \quad (5)$$

Equation (4) is essentially exact at high altitudes. It should be adjusted to reflect a somewhat smaller value for m at lower altitudes, where dynamical effects may dominate. Equation (4) may be employed therefore in an iterative scheme to determine temperature as a function of altitude, using measured values for n as given in Figure 3. The results are shown in Figure 6. The error bars reflect the range of temperatures derived by using independent analyses of the data at mass numbers 44, 22, and 12. The figure includes for comparison results derived for the lower atmosphere by *Seiff and Kirk* [1977]. Temperatures indicated by the results in Figure 6 are colder than was anticipated. The variety of vertical structure in T was also unexpected and most probably reflects the influence of wave motion generated in the lower atmosphere [*McElroy et al.*, 1976*b*; *Nier and McElroy*, 1976; *Seiff and Kirk*, 1976, 1977].

The density of NO in the Martian upper atmosphere is determined mainly by in situ chemical processes. The gas is formed by

$$N(^2D) + CO_2 \rightarrow NO + CO \quad (6)$$

and removed by

$$N(^4S) + NO \rightarrow N_2 + O \quad (7)$$

Nitrogen atoms are produced by

$$e + N_2 \rightarrow N + N \quad (8)$$

where e denotes energetic electrons formed mainly by photoionization of CO_2, and the product atoms may be released in either ground or excited states. Extrapolation of the upper atmospheric NO measurements to lower altitudes requires a complex chemical-dynamical model, and the concentration of NO in ground level air is uncertain within at least 4 orders of

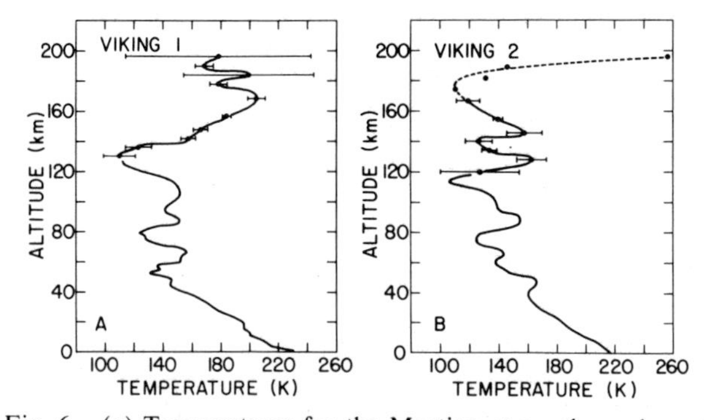

Fig. 6. (*a*) Temperatures for the Martian atmosphere above 120 km obtained from an analysis of ion peaks at mass numbers 44, 22, and 12, as measured by Viking 1. Uncertainties implied by the spread in values obtained from the individual mass peaks are indicated by the error bars. Temperatures obtained by *Seiff and Kirk* [1977] for the lower atmosphere are shown for comparison. (*b*) Same as (*a*) but for Viking 2. The dashed lines above 170 km indicate altitudes for which there are data only at mass number 44.

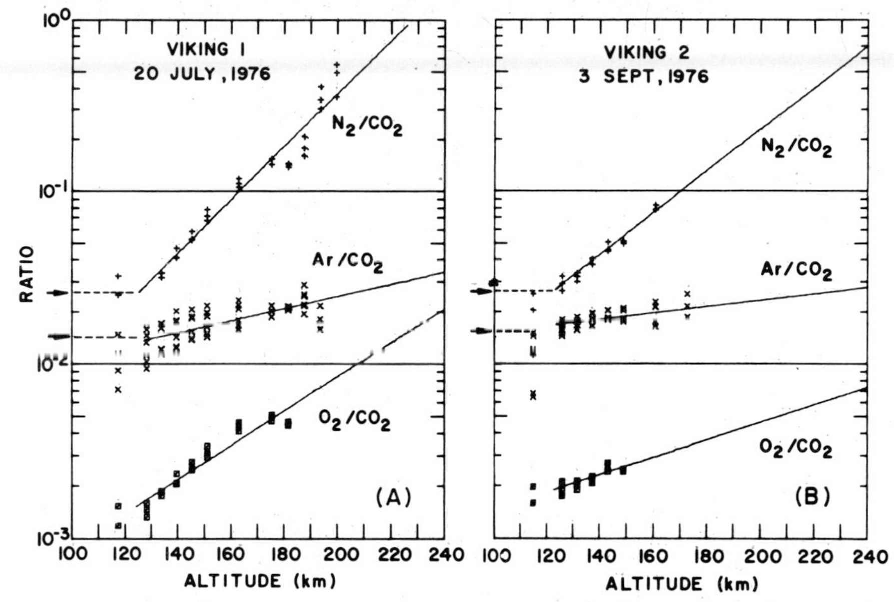

Fig. 7. N_2/CO_2, Ar/CO_2, and O_2/CO_2 mixing ratios as a function of altitude for (*a*) Viking 1 and (*b*) Viking 2.

magnitude. *Yung et al.* [1977] calculate NO concentrations for the troposphere in the range 10^5–10^9 cm^{-3} with the uncertainty attributed mainly to lack of information on the chemical reactivity of odd nitrogen at the Martian surface.

The density of CO in the upper atmosphere is set by a balance of production due to photodissociation of CO_2 with removal by diffusive and eddy transport of the gas to lower altitudes. Mixing ratios for CO in the upper atmosphere are significantly higher than values expected and indeed observed [*Kaplan et al.*, 1969] at lower altitudes. Present results are consistent with a mixing ratio for CO in the bulk atmosphere of magnitude 8×10^{-4}, in excellent agreement with ground-based spectroscopic measurements [*Kaplan et al.*, 1969].

Molecular oxygen is formed mainly at lower altitudes, below 40 km [*Kong and McElroy*, 1977*a*, *b*; *Liu and Donahue*, 1976; *McElroy and Donahue*, 1972]. The density of O_2 above 100 km is dynamically determined, and present results are in excellent agreement with theoretical models based on an assumed mixing ratio for O_2 in the bulk atmosphere of magnitude 1.6×10^{-3}, a result in excellent accord with ground-based spectroscopic determinations of O_2 [*Carleton and Traub*, 1972; *Barker*, 1972].

COMPOSITIONAL MEASUREMENTS: IMPLICATIONS FOR ATMOSPHERIC MIXING

As may be readily seen from the data in Figure 4, the relative abundance of light gases grows steadily as a function of increasing altitude. This phenomenon arises because of gravitational separation and is a dominant characteristic of the terrestrial thermosphere at heights above about 120 km. The tendency toward gravitational separation is opposed either by turbulence or by large-scale dynamical processes, and it is customary in the terrestrial context to idealize the mixing process by introducing the notion of a well-defined level, the turbopause, above which gravitational separation may be taken to dominate. The actual situation is of course more

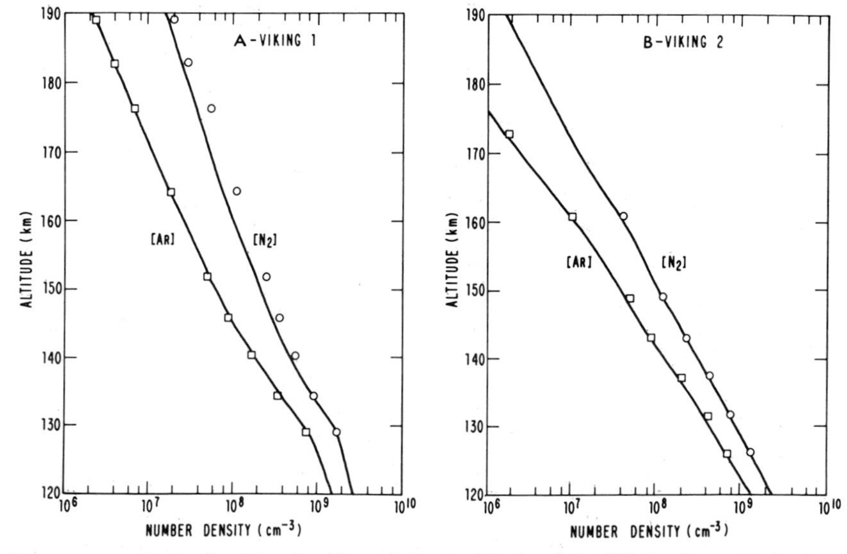

Fig. 8. (*a*) Upper atmospheric densities for N_2 and Ar as calculated for Viking 1 on the basis of eddy diffusion coefficients tabulated in Table 2. Volume mixing ratios for N_2 and Ar near the Martian surface are taken as 2.4×10^{-2} and 1.5×10^{-2}. Data obtained by Viking 1 are indicated by circles (N_2) and squares (Ar). (*b*) Same as (*a*) except for Viking 2.

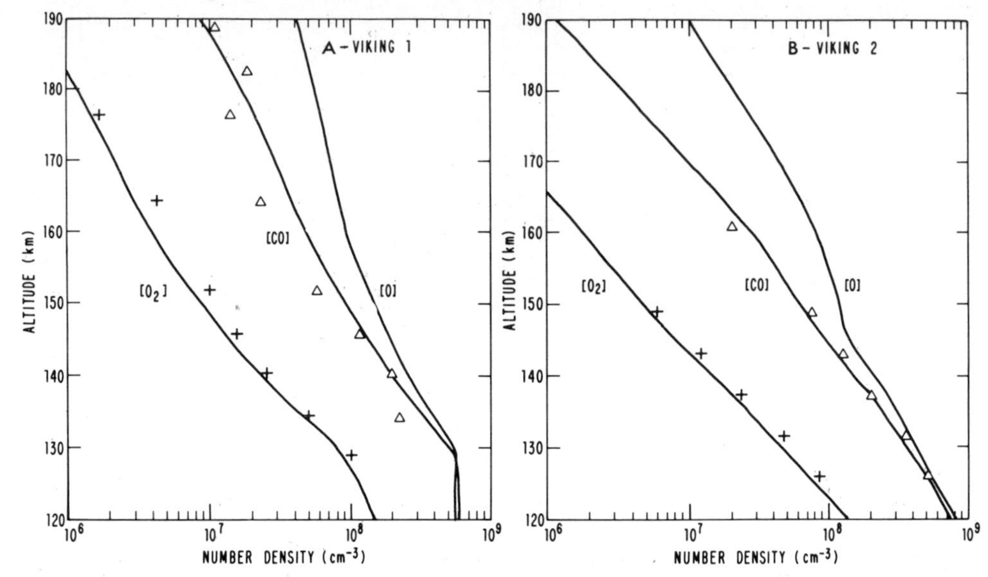

Fig. 9. (*a*) Upper atmospheric densities for CO, O_2, and O calculated for Viking 1 on the basis of eddy diffusion coefficients tabulated in Table 2. Volume mixing ratios for CO and O_2 near the Martian surface are taken as 8×10^{-4} and 1.6×10^{-3}, respectively. Data obtained by Viking 1 are indicated by triangles (CO) and pluses (O_2). (*b*) Same as (*a*) except for Viking 2.

complex. An alternate approach seeks to parameterize dynamical mixing in terms of an effective diffusion coefficient. The vertical flux (molecules cm^{-2} s^{-1}) for species i may be written then in the form

$$\phi_i = -K\left(\frac{dn_i}{dz} + \frac{n_i}{H_{av}} + \frac{n_i}{T}\frac{dT}{dz}\right) - D_i\left(\frac{dn_i}{dz} + \frac{n_i}{H_i} + \frac{n_i}{T}\frac{dT}{dz}\right) \quad (9)$$

where K represents the eddy diffusion coefficient (cm^2 s^{-1}), D_i is the coefficient for molecular diffusion (cm^2 s^{-1}), H_i is the scale height for i in the gravitationally separated regime ($kT/m_i g$), and H_{av} is the scale height which should apply if the gas were homogeneously mixed ($kT/m_{av}g$ with $m_{av} = \sum m_i n_i / \sum n_i$). For inert gases (Ar and N_2) the flux should average to zero. The density n_i should vary then with altitude according to the simple relation (4) with

$$H' = (K + D_i)\left(\frac{K}{H_{av}} + \frac{D_i}{H_i}\right)^{-1} \quad (10)$$

The flux for a chemically active species is given by solution of an appropriate continuity equation:

$$d\phi_i/dz = P_i - L_i \quad (11)$$

where P_i and L_i denote volume production and loss rates for species i.

Mixing ratios, by volume relative to CO_2, for N_2, Ar, and O_2 are shown in Figures 7*a* and 7*b* for Viking 1 and 2. The straight lines indicate least square fits to the mixing ratios, subject to the constraint that ratios should vary exponentially with altitude. The arrows to the left of the ordinate scale denote ratios determined by *Owen and Biemann* [1976; also K. Biemann, private communication, 1976] for ground level N_2 (0.028 ± 0.003) and Ar (0.017 ± 0.003). Their value for the mixing ratio of O_2 is rather more uncertain, between 0.001 and 0.004. The results in Figure 7 indicate that if one were to attempt to define a turbopause, it should be located between about 120 and 130 km.

A more complete analysis by *McElroy et al.* [1976*b*] using (9)–(11) gave the results shown in Figures 8 and 9 for N_2, Ar, CO, O_2, and O. Corresponding values for K are listed in Table 2. Results for K are subject to largest uncertainty at the highest altitude. Measurements of CO provide the most sensitive diagnostic of K, and though these data are subject to considerable uncertainty, there can be little doubt that atmospheres probed by both Viking 1 and 2 are mixed to surprisingly high levels.

The profiles for K shown in Table 2 were designed to provide an optimal fit for N_2, Ar, CO, and O_2. Figure 10 shows results which would apply for CO if K were held fixed at a value of 5×10^7 cm^2 s^{-1} for all altitudes above 110 km. The agreement with the observational data is satisfactory perhaps for Viking 1, but a serious discrepancy emerges for Viking 2. There can be little doubt that the atmosphere probed by Viking 2 is mixed to higher levels than that seen by Viking 1. The effective eddy coefficient is higher by at least a factor of 3 above 120 km. The trend in K agrees with what one might have anticipated given the temperature profiles shown in Figure 6. The Viking 1 atmosphere is statically more stable than Viking 2 over the height range 120–160 km.

Isotopic Composition

The mass spectrometric data may be used to draw a number of important conclusions with regard to the isotopic composition of carbon, oxygen, and nitrogen in the Martian atmosphere. Some 13 spectra are useful in this regard, eight from Viking 1 taken between 120 and 180 km and five from Viking 2

TABLE 2. Eddy Diffusion Coefficients as Derived From an Analysis of Height Profiles for Ar and N_2

Altitude, km	Coefficient, cm^2 s^{-1}	
	Viking Lander 1	Viking Lander 2
170	1.2×10^9	4.2×10^9
160	5.9×10^8	2.0×10^9
150	2.8×10^8	9.3×10^8
140	1.3×10^8	4.4×10^8
130	6.2×10^7	2.1×10^8
120	5.0×10^7	9.8×10^7
110	5.0×10^7	4.6×10^7
100	5.0×10^7	2.1×10^7

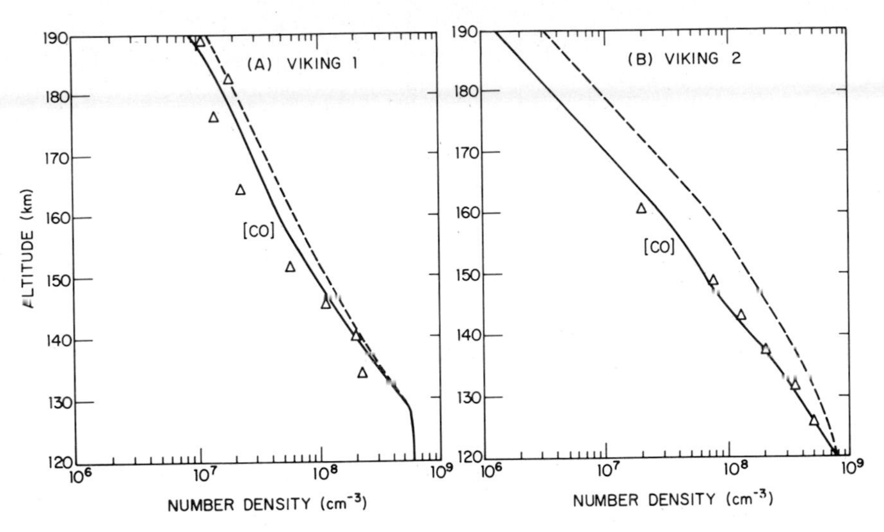

Fig. 10. Calculated profiles for CO compared to observational data for both Viking 1 and 2. The solid curves were computed by using the K profile in Table 2, and the dashed lines with $K = 5 \times 10^7$ cm^2 s^{-1} at altitudes above 110 km.

taken between 125 and 150 km. A number of isotopic features are illustrated in Figure 2.

In principle, the abundance ratio $^{13}C/^{12}C$ may be obtained from an analysis of the relative magnitude of peaks at 45/44, 22.5/22, and 13/12 amu. Likewise, the abundance ratio $^{18}O/^{16}O$ may be determined from the relative magnitude of peaks at 46/44 and 23/22 amu. In practice, the observed peaks at 23, 22.5, and 13 amu are too small to allow useful analysis. Information on the isotopic composition of carbon and oxygen was derived therefore primarily on the basis of data at mass numbers 46, 45, and 44 amu.

The results are shown in Figure 11. The ratios 45/44 and 46/44 were fitted to simple exponential functions of the form

$$\frac{[45]}{[44]} = \frac{[45]_0}{[44]_0} \exp\left(-\frac{\Delta mg(z - z_0)}{kT}\right) \qquad (12)$$

and

$$\frac{[46]}{[44]} = \frac{[46]_0}{[44]_0} \exp\left(-\frac{\Delta mg(z - z_0)}{kT}\right) \qquad (13)$$

where z_0 defines a reference level and Δm denotes the relevant mass difference, 1 amu for (12) and 2 amu for (13). The effective temperature was set equal to the value derived earlier in Figure 4.

If we assume a sharp turbopause at 125 km, results in Figure 11a may be interpreted to indicate concentration ratios $^{13}C^{16}O^{16}O/^{12}C^{16}O^{16}O$ and $^{12}C^{18}O^{16}O/^{12}C^{16}O^{16}O$ of magnitude 0.0124 ± 0.0005 and 0.0041 ± 0.0002 for the lower atmosphere of Mars. A similar analysis for the Viking 2 data gives values of 0.0128 ± 0.0003 and 0.0042 ± 0.0002, respectively. The errors quoted here reflect standard deviations for the averages computed in the manner described above. A change in turbopause altitude of 5 km would alter the Viking 1 results by 1 and 2% for 45/44 and 46/44, respectively. Temperatures are lower for Viking 2, and results are correspondingly more sensitive to the choice of level for the turbopause. A shift of turbopause altitude by 5 km for Viking 2 would alter the abundance ratios derived for the lower atmosphere by 1.6 and 3.2% for 45/44 and 46/44, respectively.

Average terrestrial values for the concentration ratios $^{18}O/^{16}O$, $^{17}O/^{16}O$, and $^{13}C/^{12}C$ are 0.00204, 0.000375, and 0.0112, respectively [*Johnson and Nier,* 1967]. The terrestrial ratio 46/44 in CO_2 is thus 0.00408, a value which falls within the errors of the Viking results obtained above. The average value for the terrestrial ratio 45/44 is 0.0120. Here one must allow for the contribution due to $^{12}C^{16}O^{17}O$. The average of the 45/44 ratios derived for Mars falls 5% higher than this result. It is doubtful, however, that the difference is numerically significant. A detailed examination of Figure 2 indicates that although the instrumental resolution was sufficient to provide reasonably flat-topped peaks in the 45–46 amu mass region, the peaks are not perfectly flat and a small error in evaluation of the mass 44 contribution at 45 could account for the apparent difference in terrestrial and Martian abundance ratios. We may conclude that the ratio $^{13}C/^{12}C$ in the Martian atmosphere should lie within 5% of the average terrestrial value. We assume here that $^{17}O/^{16}O$ is similar for both planets, a reasonable assumption in light of the results discussed earlier for $^{18}O/^{16}O$.

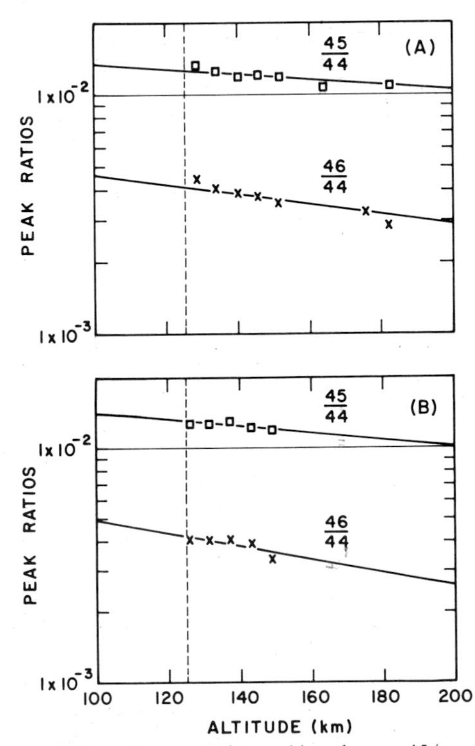

Fig. 11. Variation of mass 46/mass 44 and mass 45/mass 44 abundance ratios as a function of altitude for (a) Viking 1 and (b) Viking 2.

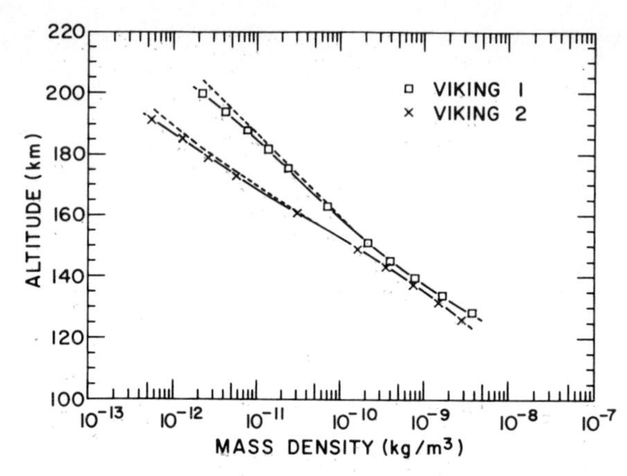

Fig. 12. Mass densities computed from particle number densities measured for the atmospheric constituents. Dashed curves include atomic oxygen densities deduced from the retarding potential analyzer ion density measurements.

The abundance ratio $^{15}N/^{14}N$ may be determined from an analysis of peaks at 28 and 29 amu. Peaks at 15 and 14 amu are too small to allow a meaningful check on this result. As may be seen from Figure 5, $^{13}C^{16}O^+$ and $^{12}C^{16}O^+$ from CO_2 and CO account for a little more than half the observed peaks at 29 and 28 amu. The residuals are due to N_2. We find, with an analysis similar to that described above for carbon and oxygen, a ratio $^{15}N^{14}N/^{14}N^{14}N$ of magnitude 0.0119 ± 0.0015 for Viking 1 and 0.0119 ± 0.0005 for Viking 2; both values refer to an altitude of 125 km. The errors quoted reflect standard deviations of the averages of results from seven spectra on Viking 1 and four spectra on Viking 2. The terrestrial ratio is 0.00736, and the results indicate therefore that Mars' atmosphere is enriched in $^{15}N/^{14}N$ in relation to the terrestrial value by a factor of 1.62 with an estimated uncertainty of about 10%. This result agrees favorably with the enrichment, 1.75, quoted earlier based on a preliminary analysis of Viking 1 data and is consistent also with an independent measurement by *Biemann et al.* [1976] which suggests an enrichment factor in the range 1.4–1.7. The enrichment in ^{15}N may be attributed to selective escape of ^{14}N from an atmosphere initially rich in N_2 [*McElroy et al.*, 1976*a*].

A search in the mass 36 region of the spectrum failed to provide a quantitative measure for the abundance of ^{36}Ar. On the basis of our data we can conclude only that the ratio $^{36}Ar/^{40}Ar$ must be less than 10^{-3}. Our analysis is consistent therefore with *Owen and Biemann* [1976; also K. Biemann, private communication, 1976] who reported a value of about 3×10^{-4} for this ratio.

An examination of the mass 2 and 4 amu spectral regions indicates that the mixing ratios for H_2 and He at lower altitudes must not exceed about 10^{-4}.

The mass 16 peak includes contributions from CO_2, CO, O_2, and H_2O. After these contributions are subtracted, an appreciable residual remains, attributable to atomic oxygen. A small peak is seen also in the mass 8 portion of the spectrum. This can only be due to doubly charged atomic oxygen formed in the ionization process. A quantitative evaluation of atomic oxygen has not been possible as yet owing to uncertainties in the extent to which gas may be removed on surfaces of the instrument. It is hoped that laboratory simulation experiments now under way may provide useful information on this important topic.

Knowledge of the number densities of the various atmospheric constituents allows one to compute profiles for mass density as a function of altitude. Results of such a computation are plotted in Figure 12. The experimental points shown in Figures 3 and 4 were used in making these calculations. The wavelike structure of the particle density curves exhibited in Figures 3 and 4 and discussed earlier by *McElroy et al.* [1976*b*] is readily apparent, as is the appreciably lower density found by the Viking 2 instrument at the highest altitudes.

The solid curves of Figure 12 do not include atomic oxygen, whose abundance could not be measured quantitatively with the mass spectrometer as discussed above. Because of its low mass, atomic oxygen might be expected to contribute in substantial part to the mass density at highest altitudes. An evaluation of the magnitude of the effect was made by employing the $n(O)$ value, 1.5% of $n(CO_2)$ at 130 km, computed from the retarding potential analyzer ion measurements (W. B. Hanson, private communication, 1977). Concentrations at other altitudes were calculated by assuming diffusive equilibrium with an average temperature of 169.2°K for Viking 1 and 110°K for Viking 2. The effect on the total mass densities is shown by the dashed curves of Figure 12. It is seen that atomic oxygen makes an appreciable contribution at altitudes above 160 km and indeed may be the major constituent of the Martian atmosphere at heights above 200 km, similar to the situation which prevails in the terrestrial atmosphere.

Acknowledgments. Work at the University of Minnesota and Harvard University was supported under NASA contracts NAS-1-9697 and NAS-1-10492, respectively. A.O.N. is indebted to Michael Wade and Ward Johnson for assistance in the computations and John Ballenthin for a determination of the stagnation characteristics of the mass spectrometer ion sources. M.B.M. thanks T. Y. Kong for valuable discussions. We are also indebted to Kai Hsi, James Rice, Donald Bianco, and their colleagues at the Bendix Aerospace Corporation who were responsible for the successful production of the mass spectrometers which made the present work possible. Special thanks are due Jeffrey L. Hayden of the University of Minnesota, Roy Duckett of Langley Research Center, and Tony C. D. Knight, Walter Chamberlin, and Roland Younger of the Martin Marietta Aerospace Corporation for their untiring efforts in guiding the development and construction of the mass spectrometers throughout the duration of the project.

REFERENCES

Barker, E. S., Detection of molecular oxygen in the Martian atmosphere, *Nature, 238,* 447–448, 1972.

Biemann, K., T. Owen, D. R. Rushneck, A. La Fleur, and D. W. Howarth, The atmosphere of Mars near the surface: Isotopic ratios and upper limits on noble gases, *Science, 194,* 76–77, 1976.

Carleton, N. P., and W. A. Traub, Detection of molecular oxygen on Mars, *Science, 177,* 988–992, 1972.

French, J. B., N. M. Reid, A. O. Nier, and J. L. Hayden, Rarefied gas dynamic effects on mass spectrometric studies of upper planetary atmospheres, *AIAA J., 13,* 1641–1646, 1975.

Hayden, J. L., A. O. Nier, J. B. French, N. M. Reid, and R. J. Duckett, The characteristics of an open source mass spectrometer under conditions simulating upper atmosphere flight, *Int. J. Mass Spectrom. Ion Phys., 15,* 37–47, 1974.

Hedin, A. E., C. P. Avery, and C. D. Tschetter, An analysis of spin modulation effects on data obtained with a rocket-borne mass spectrometer, *J. Geophys. Res., 69,* 4637–4648, 1964.

Horowitz, R., and H. E. La Gow, Upper air pressure and density measurements from 90 to 220 kilometers with the Viking 7 rocket, *J. Geophys. Res., 62,* 57–78, 1957.

Johnson, W. H., Jr., and A. O. Nier, Nuclear masses, in *Handbook of Physics,* 2nd ed., edited by E. U. Condon and H. Odishaw, chap. 2, part 9, p. 63, McGraw-Hill, New York, 1967.

Kaplan, L. D., J. Connes, and P. Connes, Carbon monoxide in the Martian atmosphere, *Astrophys. J., 157,* L187–L192, 1969.

Kong, T. Y., and M. B. McElroy, Photochemistry of the Martian atmosphere, *Icarus,* in press, 1977*a*.

Kong, T. Y., and M. B. McElroy, The global distribution of O_3 on Mars, *Planet. Space Sci.,* in press, 1977*b*.

Liu, S., and T. M. Donahue, The regulation of hydrogen and oxygen escape from Mars, *Icarus, 28,* 231–246, 1976.

Mattauch, J., and R. Herzog, Über einen neuen Massenspektrographen, *Z. Phys., 89,* 786–795, 1934.

McElroy, M. B., and T. M. Donahue, Stability of the Martian atmosphere, *Science, 177,* 986–988, 1972.

McElroy, M. B., Y. L. Yung, and A. O. Nier, Isotopic composition of nitrogen: Implications for the past history of Mars' atmosphere, *Science, 194,* 70–72, 1976*a*.

McElroy, M. B., T. Y. Kong, Y. L. Yung, and A. O. Nier, Composition and structure of the Martian upper atmosphere: Analysis of results from Viking, *Science, 194,* 1295–1298, 1976*b*.

Nier, A. O., and J. L. Hayden, A miniature Mattauch-Herzog mass spectrometer for the investigation of planetary atmospheres, *Int. J. Mass Spectrom. Ion Phys., 6,* 339–346, 1971.

Nier, A. O., and M. B. McElroy, Structure of the neutral upper atmosphere of Mars: Results from Viking 1 and Viking 2, *Science, 194,* 1298–1300, 1976.

Nier, A. O., J. H. Hoffman, C. Y. Johnson, and J. C. Holmes, Neutral composition of the atmosphere in the 100- to 200-kilometer range, *J. Geophys. Res., 69,* 979–989, 1964.

Nier, A. O., W. B. Hanson, M. B. McElroy, A. Seiff, and N. W. Spencer, Entry science experiments for Viking 1975, *Icarus, 16,* 74–91, 1972.

Nier, A. O., W. E. Potter, D. R. Hickman, and K. Mauersberger, The open-source neutral-mass spectrometer on the Atmosphere Explorer-C, -D and -E, *Radio Sci., 8,* 271–276, 1973.

Nier, A. O., W. B. Hanson, A. Seiff, M. B. McElroy, N. W. Spencer, R. J. Duckett, T. C. D. Knight, and W. S. Cook, Composition and structure of the Martian atmosphere: Preliminary results from Viking 1, *Science, 193,* 786–788, 1976*a*.

Nier, A. O., M. B. McElroy, and Y. L. Yung, Isotopic composition of the Martian atmosphere, *Science, 194,* 68–70, 1976*b*.

Owen, T., and K. Biemann, Composition of the atmosphere at the surface of Mars: Detection of argon-36 and preliminary analysis, *Science, 193,* 801–803, 1976.

Seiff, A., and D. B. Kirk, Structure of Mars' atmosphere up to 100 kilometers from the entry measurements of Viking 2, *Science, 194,* 1300–1303, 1976.

Seiff, A., and D. B. Kirk, Structure of the atmosphere of Mars in summer at mid-latitudes, *J. Geophys. Res., 82,* this issue, 1977.

Thorness, R. B., and A. O. Nier, Device for remote opening of a vacuum system, *Rev. Sci. Instrum., 33,* 1005–1007, 1962.

Wiener, B., Ambient pressure determination at high altitudes by use of free-molecule theory, *Nat. Adv. Comm. Aeronaut. Annu. Rep., 1821,* 1949.

Yung, Y. L., D. F. Strobel, T. Y. Kong, and M. B. McElroy, Photochemistry of nitrogen in the Martian atmosphere, *Icarus, 30,* 26–41, 1977.

(Received April 21, 1977;
revised May 6, 1977;
accepted May 25, 1977.)

VOL. 82, NO. 28 JOURNAL OF GEOPHYSICAL RESEARCH SEPTEMBER 30, 1977

The Martian Ionosphere as Observed by the Viking Retarding Potential Analyzers

W. B. HANSON, S. SANATANI, AND D. R. ZUCCARO

University of Texas at Dallas, Richardson, Texas 75080

The retarding potential analyzers on the Viking landers obtained the first in situ measurements of ions from another planetary ionosphere. Mars has an F_1 ionospheric layer with a peak ion concentration of approximately 10^5 cm^{-3} just below 130-km altitude, of which ~90% are O_2^+ and 10% CO_2^+. At higher altitudes, O^+ ions were detected with peak concentration near 225 km of less than 10^3 cm^{-3}. Viking 1 measured ion temperatures of approximately 150°K near the F_1 peak increasing to an apparent exospheric temperature of 210°K near 175 km. Above this altitude, departures from thermal equilibrium with the neutral gas occur, and T_i increases rapidly to >1000°K at 250 km. An equatorward horizontal ion velocity of the order of 100–200 m/s was observed near 200 km and near the F_1 peak, with a minimum velocity at intermediate heights. Both landers entered the F_1 layer at a solar zenith angle near 44°, though the local times of the Viking 1 and 2 entries were 16:13 and 9:49 LT, respectively. On Viking 2, considerably more structure was observed in the height profiles of ionospheric quantities, although they were similar in shape to the Viking 1 profiles.

INTRODUCTION

The Viking landers have provided the first in situ measurements of the ionosphere of another planet. These observations, together with a brief discussion of their significance, are presented herein. A recent review by *Whitten and Colin* [1974] has summarized the previous experimental data bearing on the ionospheres of Mars and Venus. It is probably fair to say that a quantitative knowledge of the upper atmosphere of Mars began with the Mariner 4 occultation data in 1965 [*Kliore et al.*, 1965*a*]. Just prior to the acquisition of these data the ionosphere had been predicted by *Kliore et al.* [1965*b*] to have its ionization maximum at 3 times the observed altitude and with 7 times the observed concentration; similarly, a model upper atmosphere with no CO_2 and several times the observed exospheric temperature was postulated by *McElroy et al.* [1965].

Immediately following the Mariner 4 encounter, many ionospheric models were put forward, but the F_2 region possibilities of *Johnson* [1965] and *Fjeldbo et al.* [1966] were convincingly eliminated on the basis that the reaction

$$O^+ + CO_2 \rightarrow O_2^+ + CO \tag{1}$$

is simply too rapid to permit the development of a significant layer of O^+ ions [*Norton et al.*, 1966; *Donahue*, 1966]. Further, it was very difficult to accept the extremely low temperatures required by these models [*Chamberlain and McElroy*, 1966; *McElroy*, 1967]. *Norton et al.* [1966] argued that in addition to (1) the reaction

$$CO_2^+ + O_2 \rightarrow O_2^+ + CO_2 \tag{2}$$

would also produce O_2^+ ions and that O_2^+ was probably the dominant ion in the principal layer.

An E layer model had also been suggested by *Chamberlain and McElroy* [1966] and *McElroy* [1967]. It was based on an atmospheric model that had a neutral particle concentration 2 orders of magnitude larger near the ionization peak than could be accommodated with an F_1 model. The E model required an extremely large recombination coefficient for CO_2^+ ions that was quickly shown by *Weller and Biondi* [1967] to be inconsistent with laboratory measurements. Thus even before Mariner 6 and 7 reached Mars, it seemed most likely that the dayside ionization maximum was an F_1 peak formed near unit optical depth for solar ultraviolet radiation in the wavelength range from 200 to 900 Å.

Fehsenfeld et al. [1970] showed that the atomic oxygen reaction

$$\begin{aligned} CO_2^+ + O &\rightarrow O_2^+ + CO \\ CO_2^+ + O &\rightarrow O^+ + CO_2 \text{ (followed by (1))} \end{aligned} \tag{3}$$

was even faster than (2). So although it was expected that there would be more O than O_2 at high altitude, it was likely that reactions (3) and (1) would cause O_2^+ to be the principal ion even in the region where O became dominant. Model calculations of *McElroy and McConnell* [1971*a*] showed that a mixing ratio of 5×10^{-3} for [O]/[CO_2] at the F_1 peak was sufficient to make the ionosphere mostly O_2^+. *Stewart* [1972] found that 2% O at the F_1 peak would bring coincidence with the Mariner 6 and 7 data for resonance scattering from CO_2^+ ions but that the layer was only 30% CO_2^+ ions. Direct calculations of the O mixing ratio at the ionization maximum from the atomic oxygen emissions observed by Mariner 6 and 7 varied from a fraction of 1% [*McElroy and McConnell*, 1971*b*] to 3% [*Thomas*, 1971] with a most probable value [*Strickland et al.*, 1972] near 1%.

Our understanding of the thermal structure of the upper atmosphere has had a colorful history since Mariner 4. As was previously mentioned, the F_2 ionosphere models required exospheric temperatures of approximately 100°K, which were too low to satisfy any of the model calculations. On the other hand, *Chamberlain and McElroy* [1966] and *McElroy* [1967] concluded that a lower limit for T_∞ was 400°K for the Mariner 4 (low sunspot number) case. Such high temperatures could not be simply reconciled with an F_1 region interpretation of the plasma scale heights from Mariner 4. However, agreement with this scale height data was achieved [*Cloutier et al.*, 1969; *McElroy*, 1969] by postulating an interaction with the solar wind that depressed the ionospheric scale height to essentially the neutral CO_2 scale height. (In photochemical equilibrium, $n_e \propto \{n(CO_2)\}^{1/2}$.) On the other hand, *Hogan and Stewart* [1969] obtained agreement with the lower inferred temperatures (~270° for Mariner 4 and 350° for Mariner 6 and 7) by assuming a lower effective heating efficiency for the UV energy absorbed in the upper atmosphere of Mars, though they used

Paper number 7S0523.

the same effective heating efficiency as *McElroy* [1969] for Venus. *Johnson* [1967] and *Shimizu* [1968] suggested that the lower temperatures might be caused by eddy transport of heat to lower altitudes. Large eddy coefficients were also suggested to achieve the small apparent dissociation of CO_2 in the thermosphere [*Shimizu*, 1968]. The Mariner 6 and 7 airglow measurements made during high sunspot activity yielded a temperature of only 350°K [*Stewart*, 1972] and strongly supported the low-temperature models, i.e., those with large eddy coefficients and/or small heating efficiencies. The need for the solar wind depression of the F_1 topside scale height was no longer apparent.

While the Mariner 9 mission supplied a great deal more data on the Martian neutral atmosphere, the basic conclusions about the ionosphere remained unchanged; the main ionospheric maximum on the dayside is an F_1 layer where most of the initial ions formed are subsequently converted to O_2^+ before recombining with electrons. In addition, however, the long observation period provided by Mariner 9 permitted some morphological studies. A good correlation was found by *Stewart et al.* [1972] between UV airglow brightness and $F_{10.7}$ (flux of solar 10.7-cm radio emission), but none between $F_{10.7}$ and airglow scale heights, which had a large variance. The authors concluded that at least during the dust storm, thermospheric temperatures were strongly influenced by atmospheric disturbances. The large body of 1304-Å emission data was found to contain evidence for larger atomic oxygen concentrations in the afternoon than in the morning, perhaps by as much as a factor of 3 [*Strickland et al.*, 1973]. Occultation measurements of the altitude of the main ionospheric layer [*Kliore et al.*, 1973] as a function of solar zenith angle showed good agreement with theory during the standard mission, but there was a very large discontinuity between these data and those recorded during the extended mission. The drastically lower peak heights during the extended mission were attributed to the fact that Mars was much further from the sun and thus probably had a colder more contracted atmosphere (even though the plasma scale height was essentially unchanged). These data, taken at rather large solar zenith angles, deserve further consideration.

As we shall see, the direct ionospheric measurements from the Viking landers are reasonably consistent with the deductions inferred from the remote Mariner data, although the large-amplitude, small-scale structure observed with Viking 2 is rather surprising.

Experimental Details

A brief preliminary description of the Viking retarding potential analyzer (RPA) was given by *Nier et al.* [1972], but a more complete and more accurate description of the devices actually flown will be provided here. The sensor was mounted in the aeroshell of the lander at 1.5 m from the axis of symmetry, where the aeroshell surface was coned back 20° from perpendicularity to this axis. During the ionospheric measurement phase the lander was pitched over at least 10°, so that the RPA angle of attack was actually less than 10° for all the measurements made below 350-km altitude. To avoid danger of burn-through to the aeroshell, some compromises were made to the usual RPA sensor design. As can be seen in Figure 1, the entrance aperture area is defined by blunt surfaces rather than a knife edge, and the spacing between the grounded entry grids is relatively large. The remaining grid structure is quite standard, however, and all exposed surfaces were gold plated to reduce extraneous electric fields within the sensor. Externally, the entire aeroshell was covered with aluminum-coated mylar, which was grounded to the sensor.

A complete cycle of instrument operations consists of four consecutive main frames, two of which are illustrated in Figure 2. Each 4-s main frame is divided into a 1-s energetic electron mode, a 1-s thermal electron mode, and a 2-s ion mode. In each mode the suppressor grid is held at a fixed voltage while the retarding grids are stepped through nearly equal voltage segments every 10 ms.

The collector is held at essentially ground potential, and the collector current is measured by an automatic range-changing linear electrometer with eight different sensitivity ranges. Gain on the most sensitive range (range 1) is 8×10^{-14} A/bit, and the sensitivity is decreased by 4 each succeeding range. The electrometer output is set to zero before each major frame, while the retarding grid is held at −75 V and the suppressor grid at +15 V. This procedure was adopted to avoid any zero offsets that might occur in a dc electrometer during sterilization and the subsequent long journey to Mars. It was also done as a hedge to protect against large currents of energetic particles or photoemission currents that might tend to mask the presence of particles in the energy ranges being monitored. As a further precaution the zero adjustment was carried out on range 5 during the first main frame of each cycle in the event that the spurious currents were too large to be nullified on range 1. This conservative philosophy was also followed in the choice of the positive ion retarding potential range, which extended to +15 V, even though sensible ion currents were expected out to only approximately 8 V. This was done to allow for the possibility of a very negatively biased lander or the presence of ions of unexpectedly large mass or bulk velocity.

As was stated above, half the telemetry data were devoted to electron measurements, even though the geometry of the sensor was optimized for making thermal ion measurements. Substantial currents were measured during the electron modes, but their interpretation does not appear to be straightforward, since there is contamination by photoelectrons from the lander and by secondary electrons from the various elements of the sensor itself. The data are being studied further but will not be treated in this paper.

The raw data recorded for the ion characteristic curve at 130 km on lander 1 are shown in Figure 3. The finite beginning current at the left arises because of the grid potential differences between the start of the ion sweep and the time when the electrometer output was zeroed (retarding grids from −75 V to +15 V and suppressor grid from +15 V to −15 V). The data would have been improved had we rezeroed the electrometer for each individual sweep with its own starting grid potentials. The crosses that appear in the plot are retarding grid potential measurements to which the RPA telemetry words are assigned every 160 ms. At the end of each sweep the electrometer range is measured. In this case the telemeter voltage at the start is positive and on range 2; the positive ions can first be observed near word 300 (7.5-V retarding potential) where the net current starts to increase rapidly. Several changes in electrometer sensitivity take place before the ion current saturates on range 7 near word 360 (3 V). The inflection in the data points on the range just prior to saturation (near word 345) results from the presence of two species of ions with substantially different masses.

The electrometer noise level as seen at the beginning of the sweep in Figure 3 is nearly 2 decades higher than was measured during cruise tests. There appear to be spurious mono-

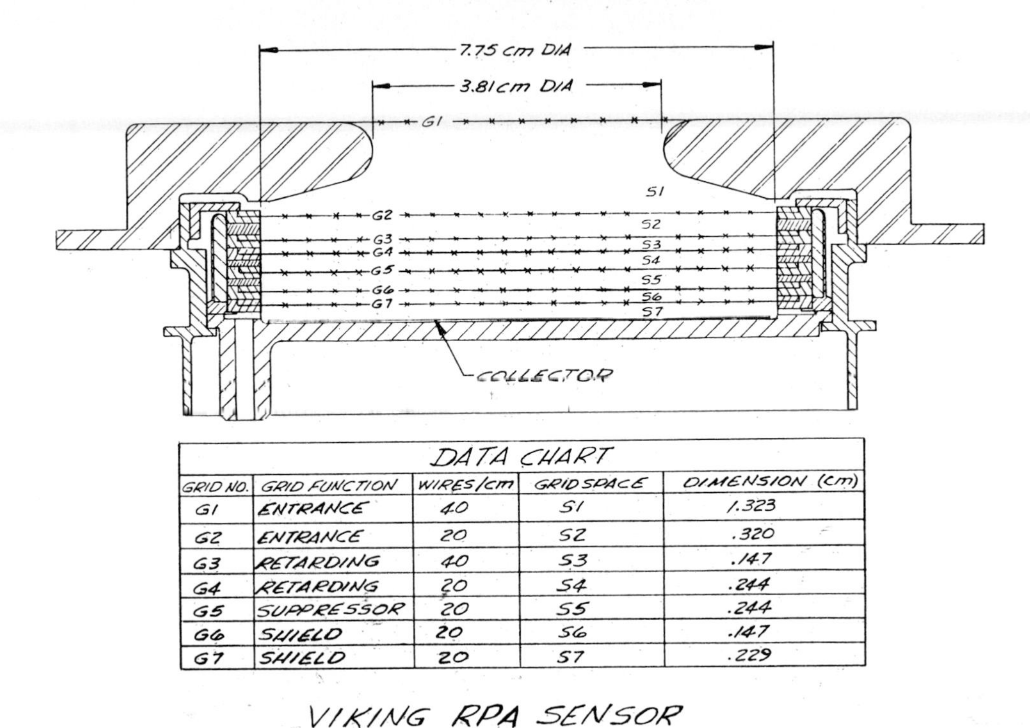

DATA CHART

GRID NO.	GRID FUNCTION	WIRES/cm	GRID SPACE	DIMENSION (cm)
G1	ENTRANCE	40	S1	1.323
G2	ENTRANCE	20	S2	.320
G3	RETARDING	40	S3	.147
G4	RETARDING	20	S4	.244
G5	SUPPRESSOR	20	S5	.244
G6	SHIELD	20	S6	.147
G7	SHIELD	20	S7	.229

Fig. 1. Cross section of the Viking RPA sensor. The entrance plane was mounted flush with the aeroshell surface. Transmission of the grid stack is 0.392.

chromatic signals in the data, the amplitude and frequency of which vary throughout both flights. In addition, several curves were irretrievable owing to severe and sporadic noise, particularly on Viking 2.

Figure 4*a* shows the same data as Figure 3 but plotted on a semilogarithmic scale after it has been appropriately adjusted for the different range sensitivities. It is remarkably like the pre-Viking sample curve shown by *Nier et al.* [1972]. The solid line is a theoretical curve fitted to the data points by a least squares technique [*Patterson*, 1969; *Hanson and Heelis*, 1975]. The parameters determined from this curve fitting are ion temperatures T_i; vehicle potential ψ; component of the ion bulk velocity normal to the sensor face, V_d; and the concentrations n_i of the individual ions selected for analysis. Ion current regimes of the two ion species separated by the large inflection in the curve are attributed to O_2^+ and CO_2^+ ions. The CO_2^+ ions dominate the curve at higher retarding potentials (higher translational energies in the lander coordinate frame) because of their greater ram energy (greater mass). The energy separation between the O_2^+ and CO_2^+ ions allows the determination of the total bulk ion velocity component normal to the sensor face. Since the lander velocity and Mars' rotational velocity are both known, the bulk velocity component of the ions in Mars' rotating frame can be deduced (see *Hanson and Heelis* [1975] for further details). In this instance it is 112 m s^{-1} toward the lander. If the ions had zero temperature, i.e., no random thermal motion, the CO_2^+ and O_2^+ current regimes would be simple step functions, since all ions of a given species

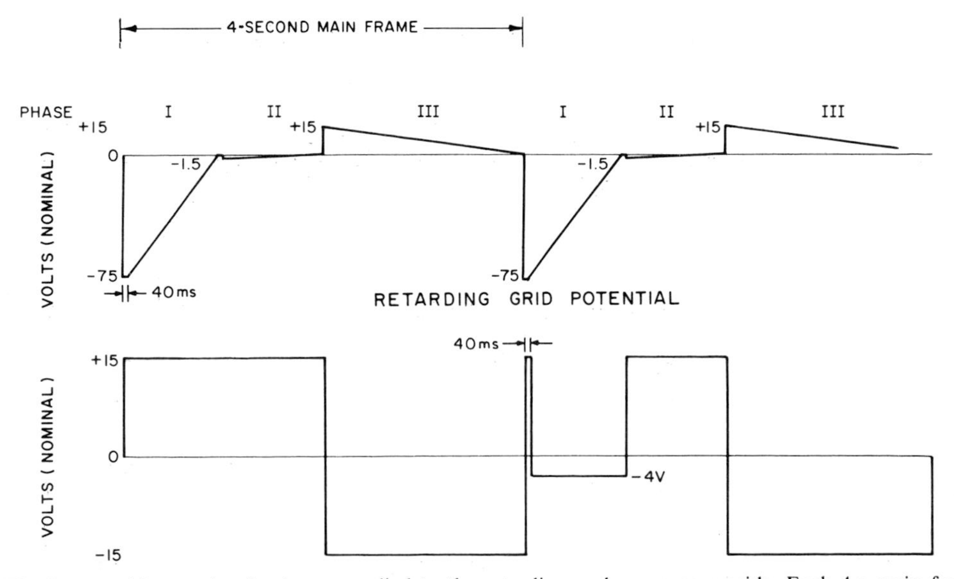

Fig. 2. The 8-s repetition cycle of voltages applied to the retarding and suppressor grids. Each 4-s main frame has a suprathermal electron phase (I), a thermal electron phase (II), and an ion phase (III).

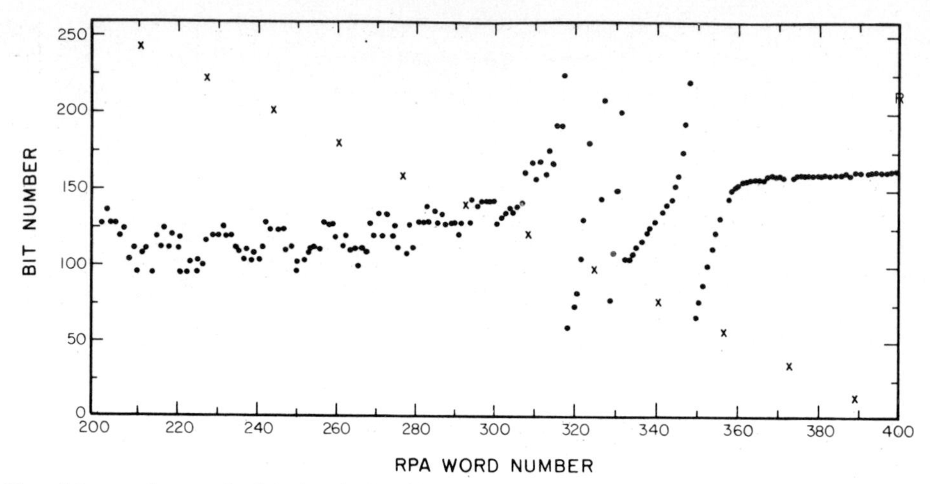

Fig. 3. Plot of the raw data received during the 2-s ion sweep recorded near 130-km altitude. Eight bit words are allotted to the RPA every 10 ms, and every 16th word (crosses) is assigned to monitor the retarding grid potential. The last word on the right (R) gives the sensitivity range of the collector.

would have the same translational energy. It is assumed in the least squares program that a common temperature exists for all ions, and indeed the rounded shoulders of both ion components are fitted quite well at 130 km with T_i = 152°K. The quality of the least squares fit is measured by σ, the root mean square fractional deviation of the experimental points from the theoretical curve, which is 1.58%.

Figure 4*b* shows graphically the best fit that can be obtained if it is assumed that the ions are at rest in Mars' corotating frame. The visibly poor fit corresponds to σ of 5% and rather convincingly demonstrates the need for including the bulk velocity parameter in the least squares program.

The analysis program requires specification of ion species as input, and the choice of O_2^+ and CO_2^+ is dictated by our prior knowledge of the probable photochemistry on Mars. Most of the curves obtained below 200 km are fit better with a few tens per cubic centimeter of ions with mass greater than 44 amu. In this low-signal (small S/N) region, however, no single mass ion current fits the shape of the measured curve. The currents may arise from suprathermal ions or may be due to an instrumental artifact. We have arbitrarily used a mass of 76 for our solutions, but the results are not sensitive to this value.

Reasonable fits to the data in Figures 4*a* and 4*b* can also be achieved by assuming that the lighter ions are of 30-amu (NO^+) or 28-amu (CO^+ or N_2^+) mass. In these cases, of course, different temperatures, vehicle potentials, and drift velocities are also obtained. Results from least squares tests using these alternate ion masses are summarized in Table 1, where it can be seen that both T_i and V_d are sensitive functions of the ions chosen for the least squares fitting procedure. Rows 4 and 5 show that the minimum σ values associated with O_2^+ and NO^+ together correspond very nearly to

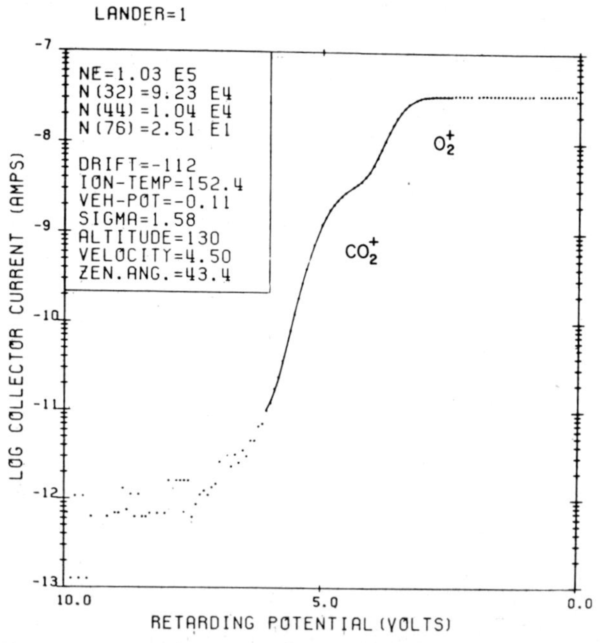

Fig. 4*a*. Logarithmic plot of data shown in Figure 3 together with a least squares theoretical fit (solid line). Parameters of fit are listed in the legend. Ion (electron) concentrations are in units per cubic centimeter (E5 indicates 10^5), ion drift is in meters per second, ion temperature is in degrees Kelvin, and vehicle potential is in volts.

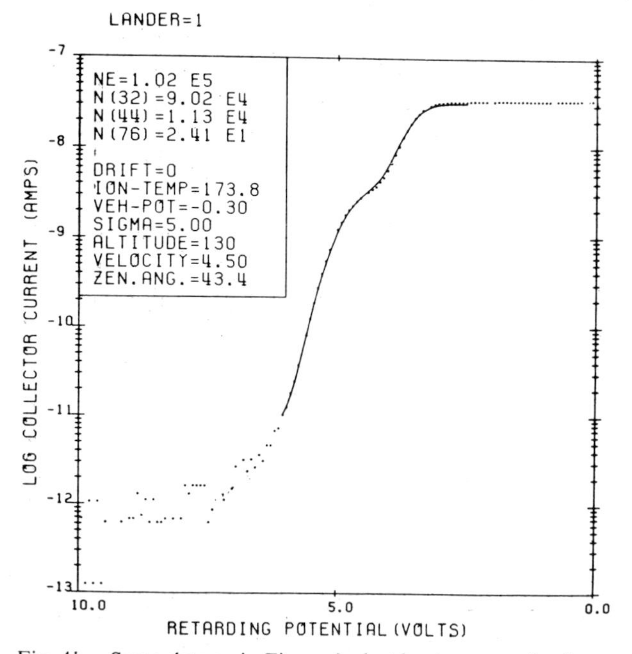

Fig. 4*b*. Same data as in Figure 4*a*, but least squares fit allows no ion drift. The value of σ, the rms fractional deviation $(x_i - x)/x$ of the points x_i from the theoretical (x) curve, has increased from 1.58% to 5%, and the points no longer fit the curve shape.

TABLE 1. Results From Least Squares Tests Using Alternate Ion Masses

No.	Composition	Vehicle Potential, V	Bulk Velocity, m/s	Ion Temperature, °K	σ, %	$n(O_2^+)/n(NO^+)$
1	$O_2^+ + CO_2^+$	−0.11	−112	152	1.58	
2	$NO^+ + CO_2^+$	−0.82	+245	180	1.57	
3	$N_2^+ + CO_2^+$	−1.35	+536	209	1.76	
4	$O_2^+ + NO^+ + CO_2^+$	−0.08	−125	152	1.58	−19.3*
5	$NO^+ + O_2^+ + CO_2^+$	−0.75	+208	175	1.55	0.090

*See text.

solutions 1 and 2. (For 4 and 5 the starting conditions were in the σ minima found for 1 and 2, respectively, and the search capability of the program was restricted.) Thus when a mixture of O_2^+ and NO^+ is allowed, there are two minima in σ, one near pure NO^+ and one near pure O_2^+. This tells us that whatever the identities of the low mass ions are, their concentration ratio is probably not very close to unity. The fraction of the minor molecular ion is, in fact, so small that it cannot be reliably determined from the RPA data (e.g., −5% NO^+ yields a best fit for 4). Model calculations based on the observed neutral atmosphere composition suggest that O_2^+ is the dominant molecular ion, and this assumption is made throughout the RPA data analysis presented here, even though for the data in Figures 4*a* and 4*b* the smallest value of σ is obtained with NO^+, not O_2^+.

At higher altitudes the increasing ion temperature tends to blend the O_2^+ and CO_2^+ curves together, as shown in Figure 5*a* for the data from 207 km. The inflection point can still be clearly seen, however, so that the technique used at 130 km is still applicable. At this height a small quantity of a lighter ion (identified as O^+) is also detectable at low retarding potentials. At still higher altitudes the ability to distinguish O_2^+ from CO_2^+ is lost, the data from 237 km shown in Figure 5*b* still permit determination of the bulk velocity, since O^+ and O_2^+ can be clearly distinguished. But at 231 km the Viking 2 data (Figure 5*c*) show that the separation into two or more ions is not obvious, and the usual quantities cannot be obtained unambiguously. In this case we have arbitrarily set the ion velocity to be zero and solved for only T_i, $n(O_2^+)$, $n(O^+)$, and ψ. A reasonably small σ is obtained in this manner which cannot be achieved with either O^+ or O_2^+ ions alone, although the uncertainty in the concentration ratio of these two ions is

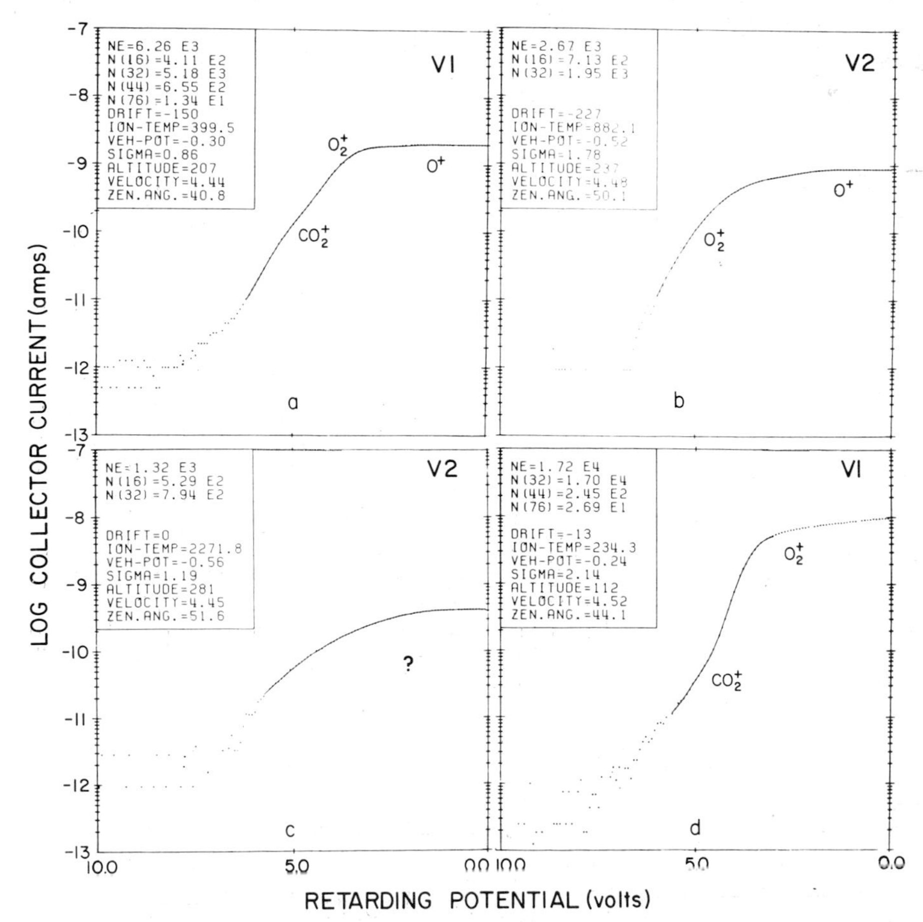

Fig. 5. A selection of RPA characteristic curves from both landers that illustrate how the data change with altitude. The interpretation of the curves is discussed in the text.

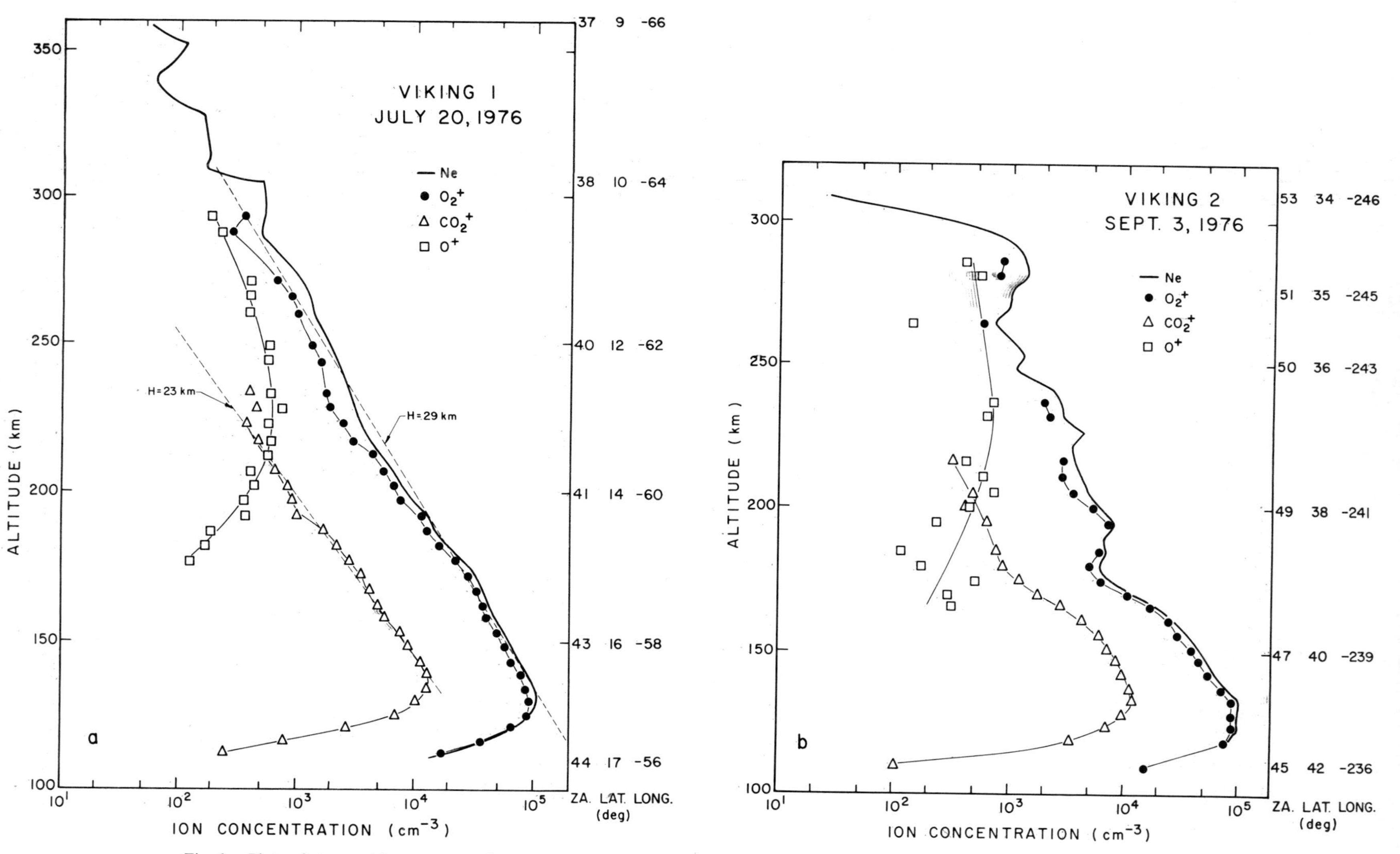

Fig. 6. Plots of observed ion concentrations versus altitude. The solid lines labeled N_e represent the sum of the individual ion concentrations. The dashed lines in Figure 6a are eyeball fits to the CO_2^+ and O_2^+ (N_e) data and correspond to the scale heights shown. At the right of each plot the solar zenith angle and sublander Mars coordinates are shown at several altitudes.

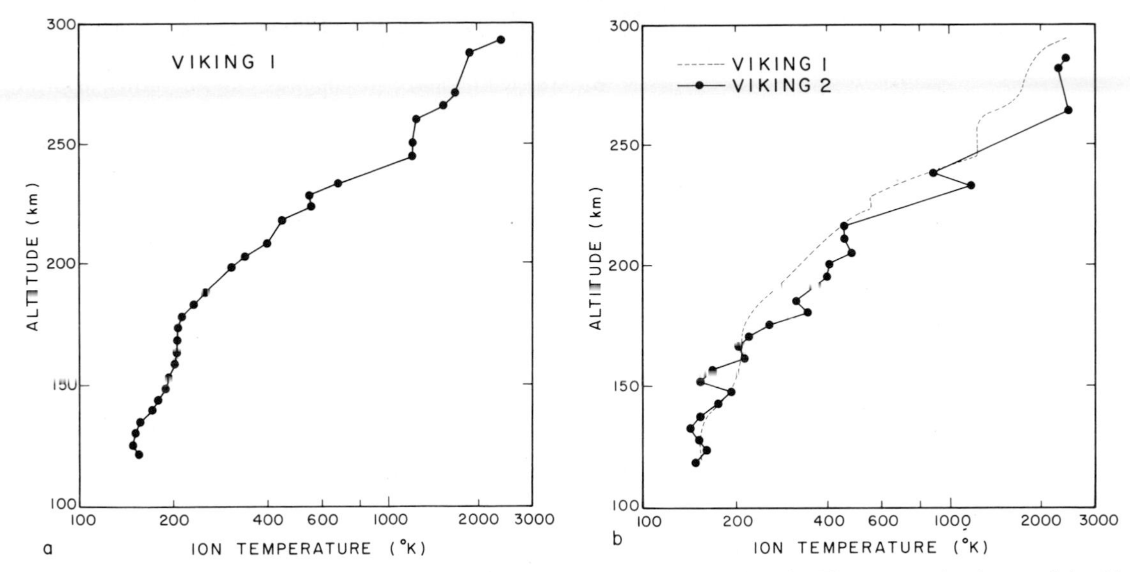

Fig. 7. Plots of ion temperature versus altitude. T_i is plotted with a logarithmic scale. The measured points are joined by straight lines, and the curve from (*a*) is lightly dashed in (*b*) for easy comparison of the two profiles.

much larger than that for the case shown in Figure 5*b*. The uncertainty in T_i is also much greater at the higher heights; only the total ion concentration determination can be accorded much confidence, and even this is so only if the ion bulk velocity remains small in comparison to the lander velocity (~4.5 km/s).

At low altitudes another kind of difficulty arises that is related to the short mean free path encountered there. These effects distort the characteristic curves so badly that the theoretical curve-fitting technique is no longer valid. Figure 5*d* shows an attempt to fit the data at 112 km on Viking 1; both the derived temperature and the total ion concentration are clearly unreliable, but the O_2^+ to CO_2^+ concentration ratio still retains some validity. Consequently, only data obtained above 120 km are suitable for analysis.

RESULTS

Ion concentration profiles determined from the two landers are shown in Figure 6. For various reasons related to data quality there are regions, mostly at higher altitudes, where values for the individual constituents are missing. In many instances it is still possible to determine N_i, the total ion (electron) concentration. The absolute magnitude of N_i has an uncertainty of approximately ±15%. For Viking 2 the plotted N_i values stop just above 300 km, where they decrease rapidly by over a decade. The Viking 1 data extend above 350 km and decrease more gradually at the top. At higher altitudes the currents that are modulated by the grid potential in the ion mode become very small or do not increase monotonically with decreasing grid potential, and they cannot be interpreted simply in terms of ion concentration. At the low-altitude end the N_i values cease because of the mean free path problems alluded to earlier.

The measured ion temperature profiles are shown in the semilogarithmic plots of Figure 7. In both cases, T_i increases with altitude by more than an order of magnitude. Considerably more structure in T_i is observed for lander 2, as was the case for N_i, yet the general shapes of the two altitude profiles are similar.

Ion velocity profiles are plotted in Figure 8, where negative velocity refers to a bulk ion motion toward the lander. The measured velocity component is very nearly horizontal (within approximately 5°), so that very little information is obtained about the vertical ion velocity. In both cases we find a bulk ion motion toward the equator above and below a minimum near 150 km.

DISCUSSION

The RPA data shown for Viking 1 in the previous figures are generally consistent with the occultation data from Mariner 4 [*Kliore et al.*, 1965*a*], which also reached Mars during minimum sunspot activity. Not only are the peak electron concentrations equal within experimental error, but if one accepts the dashed line in Figure 6*a* as being parallel to the N_i profile out to 250 km, then the average plasma scale heights are also indistinguishable. The RPA data do not extend to low enough altitudes to observe the ledge in N_e detected by the occultation measurements near 100 km.

Since no ion composition information is available from the occultation measurements, our comparisons here must be made with inferences from the Mariner airglow data or with theory. *Stewart* [1972] estimated that during the Mariner 6 and 7 encounters the topside ionosphere was approximately 30% CO_2^+, although the data were consistent with even smaller amounts of CO_2^+. Dissociative ionization of CO_2 by either photons or fast electrons can produce C^+, O^+, CO^+, and O_2^+; the first three of these react rapidly with CO_2 to produce CO^+, O_2^+, and CO_2^+, respectively. According to *Golden and Rapp* [1965], dissociative ionization accounts for ~28% of the total ions formed by energetic electrons in CO_2, but they give no data on the distribution among these four ions. Laboratory data of *Nier et al.* [1976] show a branching ratio ($C^+:O^+:CO^+:O_2^+$) of approximately 1:2:1:0 for 75-eV electrons. *McElroy and McConnell* [1971*a*] and *McConnell* [1973] have calculated model ionospheres using these dissociative product ions (with unspecified branching ratios) together with reaction (3) [*Fehsenfeld et al.*, 1970] and obtained $[O_2^+]/[CO_2^+]$ ratios greater than unity. *McConnell* [1973] produced a graph of the $[O_2^+]/[CO_2^+]$ ratio at 150 km versus the mixing ratio of atomic oxygen at 135 km. In Figure 9 are plotted the $[O_2^+]/[CO_2^+]$ ratios determined from the RPA data for Viking 1 and 2. The minimum values (~5–6) are

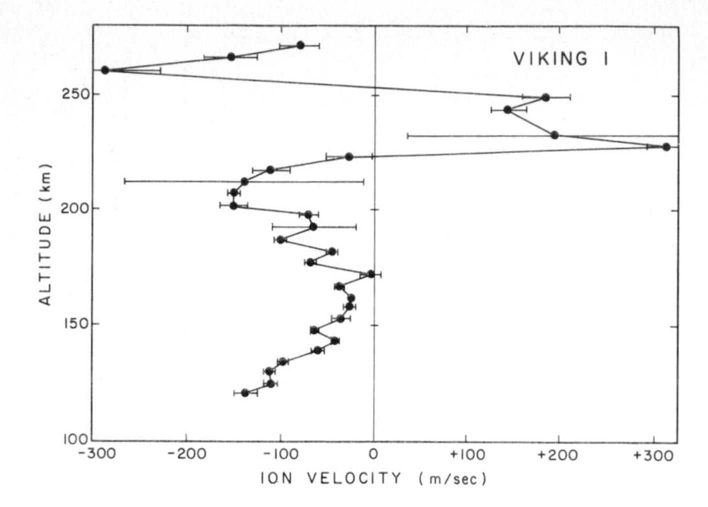

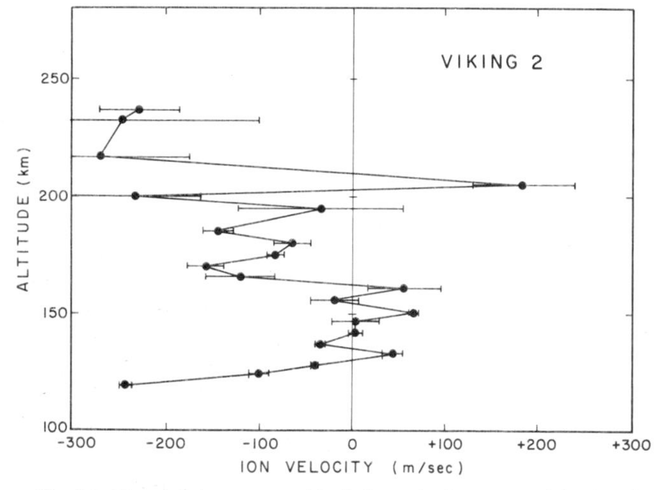

Fig. 8. Plots of the measured bulk ion velocity versus altitude. The velocity component normal to the sensor face is determined, and it is nominally horizontal. Negative velocities are toward the landers and are approximately in a south-southwest direction.

reached near 150 km, and from *McConnell*'s [1973] Figure 10 this should correspond to an O mixing ratio of approximately 3% at 135 km. This value is slightly large but certainly within the range of pre-Viking expectations. It should be kept in mind, however, that the neutral atmosphere used by McConnell is more appropriate to sunspot maximum conditions and is quite unlike that measured by the Viking neutral spectrometers [*Nier and McElroy*, 1976, 1977].

The determination of the atomic oxygen concentration from the intensities of 1304-Å emission measured by Mariner 6, 7, and 9 has received considerable attention, and the derived values for the O mixing ratio at the F_1 peak are of the order of 1% [*Thomas*, 1971; *McElroy and McConnell*, 1971*b*; *Strickland et al.*, 1972, 1973]. Because atomic oxygen adsorbs and recombines on surfaces, it is difficult to obtain a reliable estimate of the ambient atomic oxygen concentration from the upper-atmosphere mass spectrometer (UAMS) on Viking. From the ion composition profile measured by the RPA, however, it is possible to deduce an atomic oxygen mixing ratio by chemical modeling. A fit to the Viking 1 data is shown in Figure 11, where we have used the neutral model atmosphere plotted in Figure 10 and Hinteregger's solar flux of 1970 [*Hinteregger*, 1970]. The solar flux values were decreased slightly on the basis of data from Atmosphere Explorer on July 20, 1976 (H. Hinteregger, private communication, 1976). Our neutral model is in diffusive equilibrium above 115 km, and the neutral temperature profile was chosen rather arbitrarily to be similar to the ion temperature profile but subject to the further constraint that it should produce an atmosphere that is reasonably consistent with the measured density data of *Nier and McElroy* [1976, 1977] above 125 km. At lower altitudes the atmosphere probably becomes well mixed for the permanent gases, while the behavior of the dissociation products, O and CO, is more complicated. In any case our simple diffusive equilibrium model is poor below 125 km. The electron temperature profile (right-hand side of Figure 10) used in the model is completely fabricated and is without experimental or theoretical justification. The chemical modeling was carried out in some detail, but only the ions with concentrations greater than 200 cm^{-3} are shown in Figure 11, and only the dominant chemical reactions are listed in Table 2. A similar truncated set of reactions was used by *McElroy et al.* [1976] to construct an ionospheric model, but they did not specify an electron temperature profile. Also, their value for the ionization rate coefficient of atomic oxygen is more than 3 times as large as ours, though there is good agreement for $J(CO_2)$. We have not found a reason for this discrepancy.

The only free parameter used in constructing the ionosphere model is the atomic oxygen mixing ratio at the F_1 peak near 130 km (where the slant column density of CO_2 to the sun is 5.6×10^{16} cm^{-2}). It can be seen that the absolute concentrations of both O_2^+ and CO_2^+ (and hence their ratio) are matched very well with 1.25% atomic oxygen at 130 km. This value is relatively insensitive to everything except the quantity (k_1/k_{10}), and it is likely that the uncertainty in this ratio may be as large as 40% when sensitivity of k_{10} to electron temperature is considered. Certainly the modeling should be reexamined when a good thermal calculation is completed taking into account the new Viking data.

At higher altitudes the agreement between the model and the observations deteriorates, though there is fair accord with the total $N_i(N_e)$ profile to above 250 km. The underestimate of CO_2^+ ions suggests that we have too much atomic oxygen in the model at the higher altitudes. This could be attributed to the fact that eddy mixing has been neglected in our model; its inclusion would have resulted in a smaller scale height for O above the ionization peak and hence a smaller conversion rate of CO_2^+ to O_2^+. Suitable adjustment of T_e could be made to achieve a better fit to the O_2^+ data (even with the decrease in the source of O_2^+ ions brought about by the reduced O concentrations), but it will be better to await a proper calculation of T_e before attempting to deduce a mixing strength from the O profile inferred from the data in Figure 9.

The chemical equilibrium values for O^+ are in modest harmony with the data up to approximately 250 km, well into the exosphere. This may not be too surprising, since the O^+ lifetime at 200 km is only 40 s. If anything, the O^+ discrepancy seems to indicate that the model O concentration is too small rather than too large. We do not know how to place a proper upper-boundary condition on the ionosphere because of the uncertainty in the strength and nature of the interaction between the solar wind and Mars. However, the ion temperature profile and the concentrations of O_2^+ and O^+ at 280 km (coupled with the absence of H^+ ions) place certain constraints on the nature of this interaction.

The Viking 2 profiles in Figure 6*b* have an overall resemblance to the Viking 1 data, but there are very large gradients in the ion concentrations and even places where N_i increases

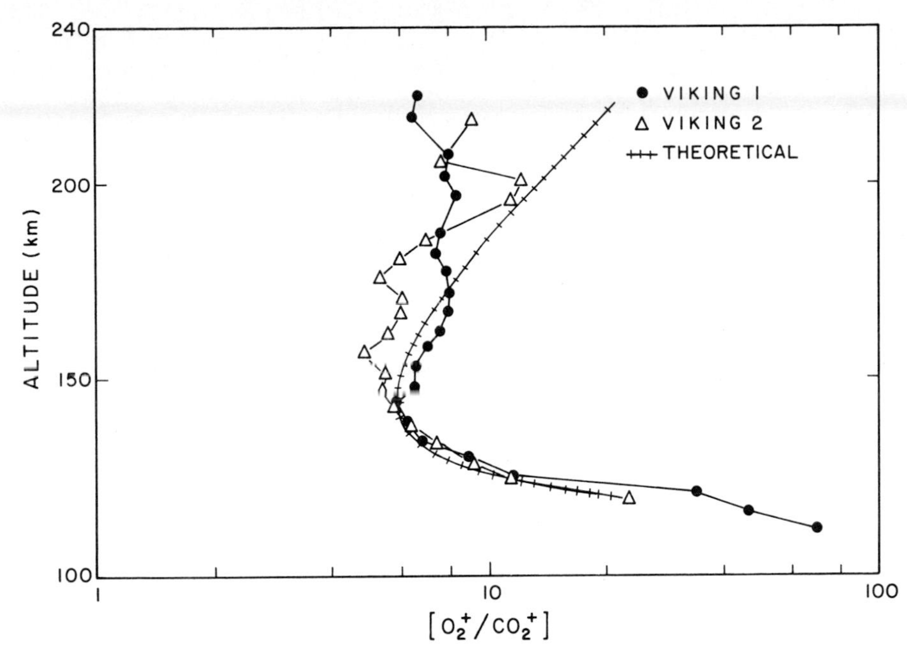

Fig. 9. Plots of the concentration ratio of O_2^+ to CO_2^+ measured from the two landers and determined from the model calculations presented in Figure 11.

with altitude. Without knowledge of the electron temperature profile, one cannot say for certain that the total plasma pressure increases with altitude, but it may well do so. The observed plasma gradients imply either some very large amplitude wave motions or the presence of a magnetic field or both. The total ion pressure never exceeds 2×10^{-9} erg cm^{-3}, which corresponds to the pressure of a 23-γ magnetic field, a value well below that measured by *Dolginov et al.* [1975] at higher altitudes.

The concentrations of O_2^+ and CO_2^+ near the peak are nearly the same as for Viking 1, though the peak altitude appears to be a few kilometers lower on Viking 2. This difference is quantitatively consistent with the lower CO_2 concentrations found by *Nier and McElroy* [1976, 1977] on Viking 2. It would appear that any attempt to deduce plasma scale heights from the Viking 2 profiles is a high-risk venture. We shall refrain.

The measured ion temperatures below 175 km from both Viking 1 and 2 are lower than the calculated temperatures for all pre-Viking neutral atmospheric models except those of *Fjeldbo et al.* [1966], and this work has been seriously criticized by both *Hunten* [1968] and *McElroy* [1967]. Above 175 km, thermal contact with the neutral gas becomes progressively weaker, and the increased ion temperature reflects the appreciable plasma heat sources that exist at high altitudes (including an indeterminate component from the solar wind). It seems reasonable to expect that at low altitudes the ion and neutral temperatures would be nearly equal and that the ion temperatures below 165 km shown in Figures 7*a* and 7*b* should be in good agreement with the values inferred from neutral particle scale heights. There is, however, a discrepancy of approximately 10% between the measured ion temperatures and the neutral temperatures derived from the UAMS data by *McElroy et al.* [1976] and *Nier and McElroy* [1977].

It is possible to argue that the Viking 'altitude' profiles are not really that, since the descent rate was approximately one fourth the horizontal velocity, and in view of the dynamic state of the atmosphere, perhaps the ion temperature need not exactly follow the horizontal vagaries in the atmospheric density. The argument is not particularly appealing. One could also argue with more vigor that large neutral winds (and wind shears) are associated with the observed wave structure and that these velocities are not taken into account in the analysis of the UAMS data. The source pressure of the UAMS is quite insensitive to the ambient neutral temperature but is essentially proportional to the ram velocity, so that an unaccounted-for wind component of 135 m s^{-1} would correspond to a 3% error in the inferred neutral particle concentration.

These effects could lead to significant errors in the UAMS density profiles and temperatures as determined by the technique of *McElroy et al.* [1976], but there would have to be a continuous increase in the velocity away from the spacecraft to explain the net mean difference in T_i and T_n. There are two other obvious possibilities to explain the temperature difference: (1) the RPA data overestimate the true ion temperature by ~10% or (2) the ions are actually hotter than the neutral gas. It is certainly possible that the RPA is at fault. T_i data from earth-orbiting RPA's that we have flown in the past have been compared extensively with backscatter measurements of T_i, and the two techniques have consistently shown excellent agreement [*Benson et al.*, 1977; *McClure et al.*, 1973]. In the case of Viking, however, the sensor geometry and the lower vehicle ram velocity are both less favorable toward producing good ion characteristic curves.

It seems unlikely that the normal UV energy input to the electrons and ions would raise the ion temperature appreciably above the neutral temperatures below 160 km. It may be, however, that ion currents flow in this region, perhaps induced by the solar wind interaction, and that Joule heating of the ions is occurring. A relative velocity of only 100 m s^{-1} between the ions and the neutrals would be sufficient to raise the O_2^+ temperature by 7%, and the effect varies with the square of the relative velocity. Thus if the observed ion velocities are not also associated with the neutral gas, a relative velocity of sufficient magnitude to explain the temperature difference would exist. This phenomenon is commonly observed in the

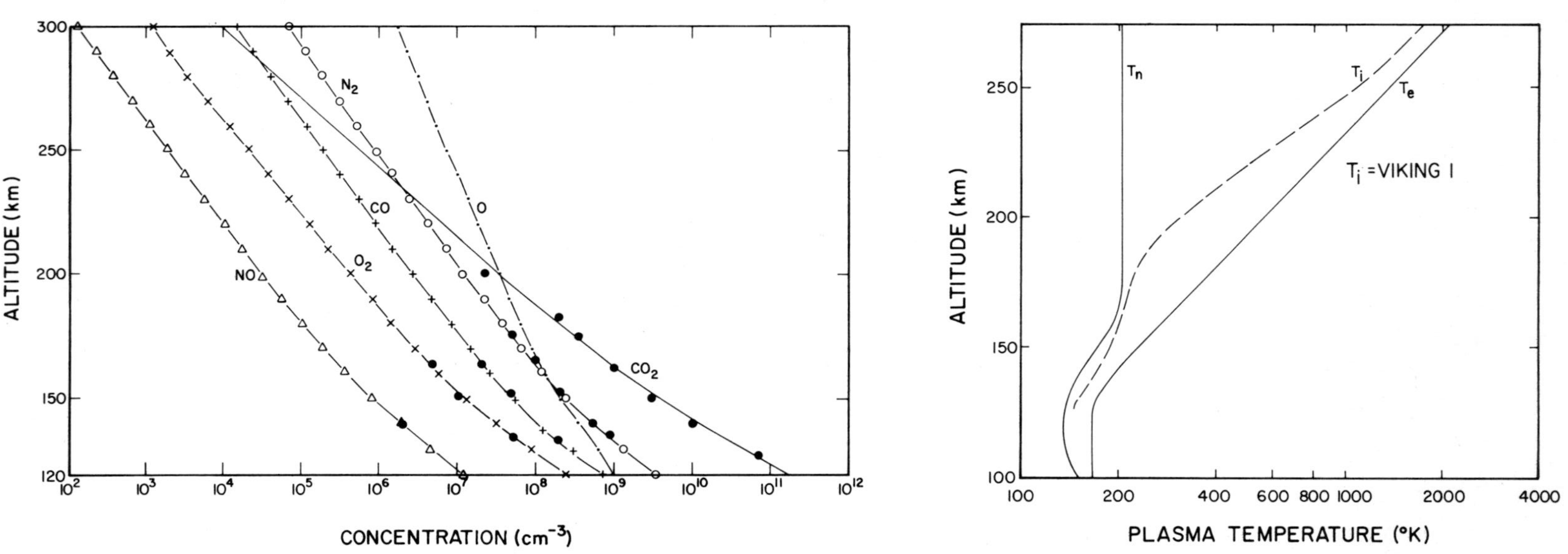

Fig. 10. A model Martian atmosphere constructed to be consistent with the data from Viking 1. (Left) In addition to the model curves, some of the UAMS data for Viking 1 from *Nier and McElroy* [1977] are shown as solid dots. Argon and hydrogen are also known quite well but were left off the figure because they are of minor importance to the ionosphere. (Right) The neutral temperature profile used to calculate the concentrations in the left-hand side, the measured ion temperature profile, and an assumed T_e distribution needed to evaluate critical rate coefficients.

earth's ionosphere [e.g., *Hanson*, 1975], but there it is intimately related to the existence of a large magnetic field.

The Viking 1 data below 175 km appear to be similar to the terrestrial ionosphere except for the low exospheric temperature. The positive vertical temperature gradient is associated with the downward conduction of heat absorbed in the upper atmosphere. In contrast, the Viking 2 ion temperatures have large vertical gradients of both signs and appear to show the presence of a relatively large amplitude wave structure with a vertical wavelength of approximately 15–20 km. Large-amplitude waves are not unexpected in Mars' upper atmosphere. The damping of these waves is probably associated with the very large vertical mixing that was postulated by *Shimizu* [1968] and *Johnson* [1968] to explain an undissociated and cold thermosphere. The fact that the Viking 2 measurements of both ion and neutral particle properties [*Nier and McElroy*, 1977] show more structure and also lower temperatures than Viking 1 may be evidence that the energy convected out of the thermosphere is greater than the energy deposited there by wave attenuation.

The Viking 2 ion concentration and drift velocity profiles also have very structured appearances, but there does not appear to be a close correlation of features among the N_i, T_i, and V_d profiles. As was previously discussed, the ion velocity determinations shown in Figure 8 require the identification of two separate ions in the characteristic curves. At the higher altitudes (>200 km) this identification becomes suspect; even though O^+ appears to be a solid candidate for the light ion, the ratio of NO^+ to O_2^+ is rather uncertain and may increase with altitude more rapidly than is shown in Figure 11. The statistical error bars of Figure 8 do not take into account this uncertainty in ion identification.

It is perhaps of historical interest to examine at least the Viking 1 ion concentration profile in terms of 'scale height' temperatures, since the ionization profiles from occultation data have been so interpreted in the past. The line shown to the right of the total ion concentration in Figure 6*a* has a scale height of 29 km, and it is in reasonable accord with N_e from 130 to 280 km. If we assume that this value is twice the CO_2 scale height, as would be the case for an ideal F_1 region, the (isothermal) temperature would be 250°K, approximately 40% higher than the neutral temperature inferred from UAMS data. The scale height of the CO_2^+ ions is approximately 23 km, and with a similar interpretation a CO_2 temperature of

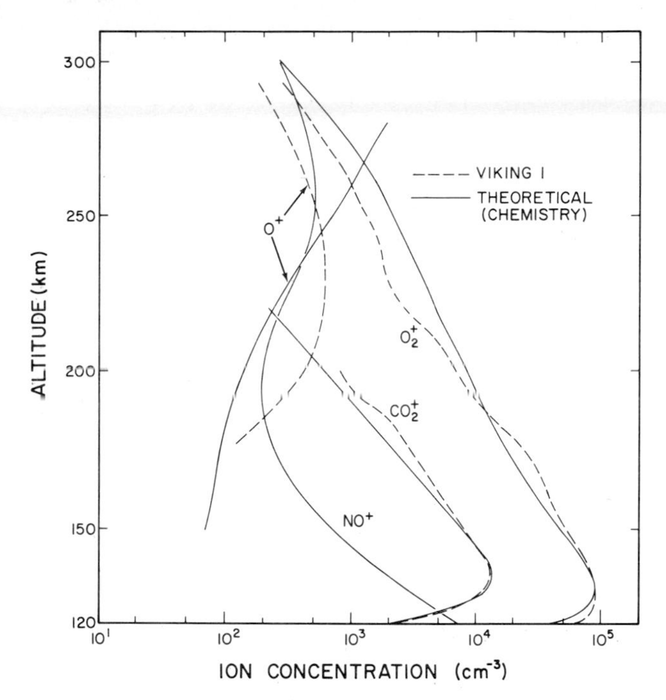

Fig. 11. A comparison of the measured ion concentration from Viking 1 profiles and a theoretical model ionosphere in chemical equilibrium. Only ions with concentrations greater than 200 ions cm^{-3} are plotted. NO^+ could not be uniquely identified from the RPA data.

200°K would result, a value fortuitously close to the measured ion temperature.

The exospheric temperature measured by Viking is lower than had been inferred from any of the previous Martian encounters. This is probably understandable on the basis of the data presented in Table 3, where Mars-sun distances and solar activities are summarized. Mariner 9 took data over an extended time period, but a single example was arbitrarily chosen for the table. Not only are the Viking Mars-sun distances greatest, but also the solar activity is least. Thus if solar UV intensity is the principal determinant of exospheric temperature, the relatively low Viking numbers should not be unexpected. Dust storms, such as the one associated with Mariner 9, tend to occur for small Mars-sun distances. It is possible that the amount of atmospheric mixing and even the chemical composition of the atmosphere may be affected by

TABLE 2. Model Ionosphere Chemistry

Reaction	Rate Coefficient*	Reference
$CO_2 + h\nu \rightarrow CO_2^+ + e$	$J_1 = 2.18 \times 10^{-7}$	*Bortner and Baurer* [1972]
$CO_2 + h\nu \rightarrow O^+ + CO + e$	$J_2 = 3.3 \times 10^{-8}$	*Bortner and Baurer* [1972]
$CO_2 + h\nu \rightarrow CO^+ + O + e$	$J_3 = 1.6 \times 10^{-8}$	*Bortner and Baurer* [1972]
$CO_2 + h\nu \rightarrow C^+ + CO + e$	$J_4 = 1.6 \times 10^{-8}$	*Bortner and Baurer* [1972]
$O + h\nu \rightarrow O^+ + e$	$J_5 = 7.5 \times 10^{-8}$	*Bortner and Baurer* [1972]
$N_2 + h\nu \rightarrow N_2^+ + e$	$J_6 = 1.26 \times 10^{-7}$	*Bortner and Baurer* [1972]
$CO_2^+ + O \rightarrow O_2^+ + CO$	$k_1 = 1.6 \times 10^{-10}$	*Fehsenfeld et al.* [1970]
$CO_2^+ + O \rightarrow O^+ + CO_2$	$k_2 = 1 \times 10^{-10}$	*Fehsenfeld et al.* [1970]
$CO_2^+ + NO \rightarrow NO^+ + CO_2$	$k_3 = 1.2 \times 10^{-10}$	*Fehsenfeld et al.* [1970]
$O^+ + CO_2 \rightarrow O_2^+ + CO$	$k_4 = 1.2 \times 10^{-9}$	*Norton et al.* [1966]
$O^+ + N_2 \rightarrow NO^+ + N$	$k_5 = 1.2 \times 10^{-12} (300/T_n)$	*Ferguson* [1967]
$O^+ + NO \rightarrow NO^+ + O$	$k_6 = 1.0 \times 10^{-11}$	*McFarland et al.* [1974]
$O_2^+ + NO \rightarrow NO^+ + O_2$	$k_7 = 6.3 \times 10^{-10}$	*Fehsenfeld et al.* [1970]
$N_2 + O \rightarrow NO^+ + N$	$k_8 = 1.4 \times 10^{-10}$	*Fehsenfeld et al.* [1970]
$CO_2^+ + e \rightarrow CO + O$	$k_9 = 3.8 \times 10^{-7}$	*Biondi* [1972]
$O_2^+ + e \rightarrow O + O$	$k_{10} = 8.5 \times 10^{-8} (1000/T_e)^{0.63}$	*Biondi* [1972]
$NO^+ + e \rightarrow N + O$	$k_{11} = 1.08 \times 10^{-7} (1000/T_e)^{1.2}$	*Biondi* [1972]

*The J_i are in units per second and the k_i are in units of cubic centrimeters per second.

TABLE 3. Mars-Sun Distances and Solar Activities

Vehicle	Date	R_m, AU	$(R_m/R_{m7})^2$	$F_{10.7}$
Viking 1	July 20, 1976	1.648	1.343	69.4
Viking 2	Sept. 3, 1976	1.613	1.287	75.7
Mariner 4	July 15, 1965	1.554	1.194	76.9
Mariner 6	July 31, 1969	1.427	1.007	167.0
Mariner 7	Aug. 5, 1969	1.422	1.00	187.7
Mariner 9	Nov. 14, 1971	1.412	0.986	103.1

these storms. Changes in the mixing strength, in particular, could have an important influence on the exospheric temperature as suggested by the Mariner 9 data [*Stewart et al.*, 1972].

Acknowledgments. We express our appreciation to the great many people who contributed so much to making the retarding potential analyzer data finally become available to us. In particular, we thank Roy Duckett of Langley Research Center; Don Bianco and Steve Smith of Bendix Aerospace Corporation; Lee Antes, Tony Knight, and Sid Cook of Martin Marietta Corporation; and C. R. Lippincott, B. J. Holt, and L. A. Swaim from the University of Texas at Dallas. This work was supported by the National Aeronautics and Space Administration under contract NAS-1-9699.

REFERENCES

Benson, R. F., P. Bauer, L. H. Brace, H. C. Carlson, J. Hagen, W. B. Hanson, W. R. Hoegy, M. R. Torr, R. H. Wand, and V. B. Wickwar, Electron and ion temperatures—A comparison of ground-based incoherent scatter and AE-C satellite measurements, *J. Geophys. Res.*, *82*, 36–42, 1977.

Biondi, M. A., Charged-particle recombination processes, Reaction Rate Handbook, *Rep. DNA 1948H*, p. 16-1, Defense Nuclear Agency, Dep. of Def. Inform. and Anal. Center, Santa Barbara, Calif., 1972.

Bortner, M. H., and T. Baurer, Reaction Rate Handbook, *Rep. DNA 1948H*, Defense Nuclear Agency, Dep. of Def. Inform. and Anal. Center, Santa Barbara, Calif., 1972.

Chamberlain, J. W., and M. B. McElroy, Martian atmosphere: The Mariner occultation experiment, *Science*, *152*, 21–25, 1966.

Cloutier, P. A., M. B. McElroy, and F. C. Michel, Modification of the Martian ionosphere by the solar wind, *J. Geophys. Res.*, *74*, 6215–6228, 1969.

Dolginov, Sh. Sh., Ye. G. Yeroshenko, L. N. Thuzzov, V. A. Sharova, K. I. Gringauz, V. V. Bezrukikh, T. K. Breus, M. I. Verigin, and A. P. Remizov, Magnetic field and plasma inside and outside of the martian magnetosphere, Solar Wind Interaction With the Planets Mercury, Venus, and Mars, *NASA Spec. Publ.*, *397*, 1, 1975.

Donahue, T. M., Upper atmosphere and ionosphere of Mars, *Science*, *152*, 763–764, 1966.

Fehsenfeld, F. C., D. B. Dunkin, and E. E. Ferguson, Rate constants for the reaction of CO_2^+ with O, O_2 and NO; N_2^+ with O and NO; and O_2^+ with NO, *Planet. Space Sci.*, *18*, 1267–1269, 1970.

Ferguson, E. E., Ionospheric ion-molecule reaction rates, *Rev. Geophys. Space Phys.*, *5*, 305, 1967.

Fjeldbo, G., W. Fjeldbo, and V. R. Eshleman, Models for the atmosphere of Mars based on the Mariner 4 occultation experiment, *J. Geophys. Res.*, *71*, 2307–2316, 1966.

Golden, P. E., and D. Rapp, Total cross sections for ionization of gases by electron impact, *J. Chem. Phys.*, *43*, 1464, 1965.

Hanson, W. B., Earth's dynamic thermosphere, *Astronaut. Aeronaut.*, *13*, 16–22, 1975.

Hanson, W. B., and R. A. Heelis, Techniques for measuring bulk gas-motions from satellites, *Space Sci. Instrum.*, *1*, 493–524, 1975.

Hinteregger, H. E., The extreme ultraviolet solar spectrum and its variation during solar cycle, *Ann. Geophys.*, *26*, 547, 1970.

Hogan, J. S., and R. W. Stewart, Exospheric temperatures on Mars and Venus, *J. Atmos. Sci.*, *26*, 332–333, 1969.

Hunten, D. M., The ionosphere and upper atmosphere of Mars, in *Atmospheres of Venus and Mars*, edited by J. C. Brandt and M. B. McElroy, pp. 147–180, Gordon and Breach, New York, 1968.

Johnson, F. S., Atmosphere of Mars, *Science*, *150*, 1445–1448, 1965.

Johnson, F. S., The atmosphere of Mars, in *Moons and Planets*, edited by A. Dolfus, pp. 240–245, North-Holland, Amsterdam, 1967.

Johnson, F. S., Mariner IV and the atmosphere of Mars, in *Atmospheres of Venus and Mars*, edited by J. C. Brandt and M. B. McElroy, pp. 181–187, Gordon and Breach, New York, 1968.

Kliore, A. J., D. L. Cain, G. S. Levy, V. R. Eshleman, G. Fjeldbo, and F. D. Drake, Occultation experiment: Results of the first direct measurements of Mars' atmosphere and ionosphere, *Science*, *149*, 1243–1248, 1965*a*.

Kliore, A., D. L. Cain, G. S. Levy, V. R. Eshleman, G. Fjeldbo, and F. D. Drake, The Mariner IV occultation experiment, *Astronaut. Aeronaut.*, *7*, 72–80, 1965*b*.

Kliore, A. J., G. Fjeldbo, B. L. Seidel, M. J. Sykes, and P. M. Woiceshyn, *S* band radio occultation measurements of the atmosphere and topography of Mars with Mariner 9: Extended mission coverage of polar and intermediate latitudes, *J. Geophys. Res.*, *78*, 4331–4351, 1973.

McClure, J. P., W. B. Hanson, A. F. Nagy, R. J. Cicerone, L. H. Brace, M. Baron, P. Bauer, H. C. Carlson, J. V. Evans, G. N. Taylor, and R. F. Woodman, Comparison of T_e and T_i from Ogo 6 and various incoherent scatter radars, *J. Geophys. Res.*, *78*, 197–205, 1973.

McConnell, J. C., The atmosphere of Mars, in *Physics and Chemistry of the Upper Atmospheres*, edited by B. M. McCormac, pp. 309–334, D. Reidel, Hingham, Mass., 1973.

McElroy, M. B., The upper atmosphere of Mars, *Astrophys. J.*, *150*, 1125–1138, 1967.

McElroy, M. B., Structure of the Venus and Mars atmospheres, *J. Geophys. Res.*, *74*, 29–42, 1969.

McElroy, M. B., and J. C. McConnell, Atomic carbon in the atmosphere of Mars and Venus, *J. Geophys. Res.*, *76*, 6674–6690, 1971*a*.

McElroy, M. B., and J. C. McConnell, Dissociation of CO_2 in the martian atmosphere, *J. Atmos. Sci.*, *28*, 879, 1971*b*.

McElroy, M. B., J. L'Ecuyer, and J. W. Chamberlain, Structure of the martian upper atmosphere, *Astrophys. J.*, *141*, 1523–1535, 1965.

McElroy, M. B., T. Y. Kong, Y. L. Yung, and A. O. Nier, Composition and structure of the martian upper atmosphere: Analysis of results from Viking, *Science*, *194*, 1295–1298, 1976.

McFarland, M., D. L. Albritton, F. C. Fehsenfeld, A. L. Schmeltekopf, and E. E. Ferguson, Energy dependence of rate constant for the reaction $O^+ + NO \rightarrow NO^+ + O$, *J. Geophys. Res.*, *79*, 2005–2006, 1974.

Nier, A. O., and M. B. McElroy, Structure of the neutral upper atmosphere of Mars: Results from Viking 1 and Viking 2, *Science*, *194*, 1298–1300, 1976.

Nier, A. O., and M. B. McElroy, Composition and structure of Mars' upper atmosphere: Results from the neutral mass spectrometers on Viking 1 and 2, *J. Geophys. Res.*, *82*, this issue, 1977.

Nier, A. O., W. B. Hanson, M. B. McElroy, A. Seiff, and N. W. Spencer, Entry science experiments for Viking 1975, *Icarus*, *16*, 74–91, 1972.

Nier, A. O., W. B. Hanson, A. Seiff, M. B. McElroy, N. W. Spencer, R. J. Duckett, T. C. D. Knight, and W. S. Cook, Composition and structure of the martian atmosphere: Preliminary results from Viking 1, *Science*, *193*, 786–788, 1976.

Norton, R. B., E. E. Ferguson, F. C. Fehsenfeld, and A. L. Schmeltekopf, Ion-neutral reactions in the martian ionosphere, *Planet. Space Sci.*, *14*, 969–978, 1966.

Patterson, T. N. L., Deduction of ionospheric parameters from retarding potential analyzers, *J. Geophys. Res.*, *74*, 4799–4801, 1969.

Shimizu, M., The recombination mechanism of CO and O in the upper atmospheres of Venus and Mars, *Icarus*, *9*, 593–597, 1968.

Stewart, A. I., Mariner 6 and 7 ultraviolet spectrometer experiment implications of CO_2^+, CO, and O airglow, *J. Geophys. Res.*, *77*, 54–68, 1972.

Stewart, A. I., C. A. Barth, C. W. Hord, and A. L. Lane, Mariner 9 ultraviolet spectrometer experiment: Structure of Mars' upper atmosphere, *Icarus*, *17*, 469–474, 1972.

Strickland, D. J., G. E. Thomas, and P. R. Sparks, Mariner 6 and 7 ultraviolet spectrometer experiment: Analysis of the O I 1304- and 1356-Å emissions, *J. Geophys. Res.*, *77*, 4052–4068, 1972.

Strickland, D. J., A. I. Stewart, C. A. Barth, C. W. Hord, and A. L. Lane, Mariner 9 ultraviolet spectrometer experiment: Mars atomic

oxygen 1304-Å emission, *J. Geophys. Res.*, *78*, 4547–4559, 1973.
Thomas, G. E., Neutral composition of the upper atmosphere of Mars as determined from the Mariner UV spectrometer experiments, *J. Atmos. Sci.*, *28*, 859–868, 1971.
Weller, C. S., and M. A. Biondi, Measurements of dissociative recombination of CO_2^+ ions with electrons, *Phys. Rev. Lett.*, *19*, 59, 1967.
Whitten, R. C., and L. Colin, Ionosphere of Mars and Venus, *Rev. Geophys. Space Phys.*, *12*, 155–191, 1974.

(Received April 22, 1977;
revised June 9, 1977;
accepted June 9, 1977.)

Structure of the Atmosphere of Mars in Summer at Mid-Latitudes

ALVIN SEIFF AND DONN B. KIRK

Ames Research Center, NASA, Moffett Field, California 94035

The structure of Mars' atmosphere was measured in situ by instruments on board the two Viking landers from an altitude of 120 km to near the surface. The two entries were separated by 178° in longitude, 25° in latitude, 45 days elapsed time, and 6 hours in Mars local time. Atmosphere structure was very well defined by the measurements and was generally similar at the two sites. Viking 1 and 2 surface pressures and temperatures were 7.62 and 7.81 mbar and 238°K and 226°K, respectively, while pressures at the elevation of the reference ellipsoid were 6.74 and 6.30 mbar. Mean temperature decreased with a lapse rate of about 1.6°K/km, significantly subadiabatic, from above the boundary layer to about 40 km, then was near isothermal but with a large-amplitude wave superimposed, attributed to the diurnal thermal tide. The mean profile appears to be governed by radiative equilibrium. Differences between the two temperature profiles are due to diurnal effects in the boundary layer, a small cooling of the Viking 2 profile up to 40 km due to latitude and season, and effects of time of day, latitude, terrain, and season on the wave structure. The density data merge well with those of the upper-atmosphere mass spectrometer to define a continuous profile to 200 km. The temperature wave continues above 100 km, increasing in amplitude and wavelength.

INTRODUCTION

Knowledge of the structure of the neutral atmosphere of Mars has advanced rapidly in the age of space astronomy. Remote sensing experiments from flyby and orbiting spacecraft have indicated the mean surface pressure to be near 5 mbar, much lower than was previously accepted (see, for example, *Kliore et al.* [1972]), and in conjunction with ground-based spectroscopy have indicated that the atmosphere is predominantly CO_2. Remote sensing has also indicated that the temperature structure up to 45-km altitude is highly variable, dependent on latitude, season, and the dust content of the atmosphere [*Kliore et al.*, 1972; *Hanel et al.*, 1972].

The temperature data have been limited to the lowest 35 km (occultation) and 45 km (IR sounding) and have been somewhat puzzling and hard to assess because of their diversity. It has not been possible to say conclusively what accuracy, temperature resolution, and altitude resolution should be assigned to these results.

The USSR spacecraft Mars 6 made some measurements of the atmosphere structure during its entry into Mars in 1974 [*Kerzhanovich*, 1977]. Very useful information was obtained, although it was limited by the lack of direct temperature sensing, an extensive radio blackout, and acceleration sensing confined to four points. It confirmed the magnitude of the surface pressure obtained from remote sensing, measuring 5.45 ± 0.3 mbar, and put bounds on the temperature structure, indicating a lapse rate of 2.9°K/km up to 33 km and an isothermal middle atmosphere at 149° ± 8°K to 90 km in early spring in the southern hemisphere (−24° latitude). There was significant uncertainty in temperatures below 30 km, with a maximum of ±18° at 29 km.

The Viking mission provided an opportunity for in situ measurements of the atmosphere, and experiments to make use of that opportunity were described by *Nier et al.* [1972]. The atmosphere structure measurement approach, initially proposed in 1962 [*Seiff*, 1963], was the subject of intensive study and development in the ensuing 9 years, culminating in a test flight of the experiment in the earth's atmosphere in 1971. This experiment showed that the techniques for measuring the atmosphere during high-velocity entry were capable of providing data of comparable quality to that normally obtained from meteorological sounding techniques [*Seiff et al.*, 1973]. A description and discussion of the techniques and instruments which have been applied to the Viking mission has been given [*Seiff*, 1976].

This paper reports the results of the atmosphere structure measurements from Viking 1 and 2 as of May 1977. The analysis of the data is not yet complete in all respects, but we expect that the material we present herein will be substantially unchanged by further analysis. Two topics not yet ready for reporting are omitted—data on the winds encountered during entry and descent of the two landers and data on the terrain under the entry trajectories. These will be the subject of later reports. Preliminary accounts of the atmospheric data have been given [*Nier et al.*, 1976; *Seiff and Kirk*, 1976].

INSTRUMENT DESCRIPTION

The overall definition of the structure of Mars' atmosphere to be described herein was a result of synthesis of data from several instruments. In the upper regions of the experiment, altitudes from 120 to 26 km, a set of three-axis accelerometers measured the atmospheric density as a function of altitude from the vehicle deceleration. To calibrate this measurement, a thorough and extensive ground test program was conducted to define the drag coefficient as a function of velocity and Reynolds number in an atmosphere of CO_2. These tests were made with models of the Viking entry configuration in free flight through a ballistic range.

During parachute descent, which began nominally at 6 km, pressure and temperature were directly measured. Also throughout the high-speed entry the flow stagnation pressure was measured, and below about 25 km the flow recovery temperature (a temperature closely related to stagnation temperature) was directly sensed.

The lander sytems provided three kinds of data which were important to the determination of atmosphere structure: (1) altitudes, from a radar altimeter; (2) attitude change data, from gyroscopes, which were used in the determination of the trajectories and for analysis of wind effects; and (3) velocity data during parachute descent, from a three-axis Doppler radar.

The nominal measurement altitudes, telemetry resolution,

Paper number 7S0499.

TABLE 1. Viking Atmosphere Structure Instruments

Sensors	Measurement Altitudes, km	Telemetry Resolution	Sample Interval, s	Altitude Resolution, km
Accelerometer	120–0	ΔV = 0.0127 m/s	0.1	0.006–0.1
Pressure				
Aeroshell*	90–6	0.16 and 0.74 mbar	0.2	0.01–0.5
Parachute†	4.5–1.5	0.085 mbar	0.5	0.03
Temperature				
Aeroshell	27–6	1.2°K	1.0	0.01–0.1
Parachute	3.8–1.5	1.2°K	0.5	0.03
Radar altimeter	132–0	5 m	0.2	0.01–0.2
Doppler radar (TDLR)	5–0	0.06 m/s	1.0	0.012
Gyros	250–0	0.0008°	0.1	0.006–0.1

*The aeroshell phase of the experiment is from entry to nominally 6-km altitude.
†The parachute phase is from 6 km to descent engine ignition at 1.5 km.

sampling intervals, and altitude resolution of all these measurements are shown in Table 1. These were generally more than adequate for atmospheric definition. The accuracy of definition of temperature and pressure gradients with altitude in the parachute descent was constrained by telemetry resolution, as discussed below, but it was possible to define these gradients to the order of 1% or 2% on pressure and to within 0.1°K/km on temperature, or better.

The locations of the sensors on the landers are shown in Figure 1. The accelerometers and gyros were located within the inertial reference unit, external to the lander body, along the z axis, which was in the nominally vertical plane, pointing obliquely upward. These sensors were retained with the lander through the landing and provided necessary guidance data as well as scientific data on the atmosphere.

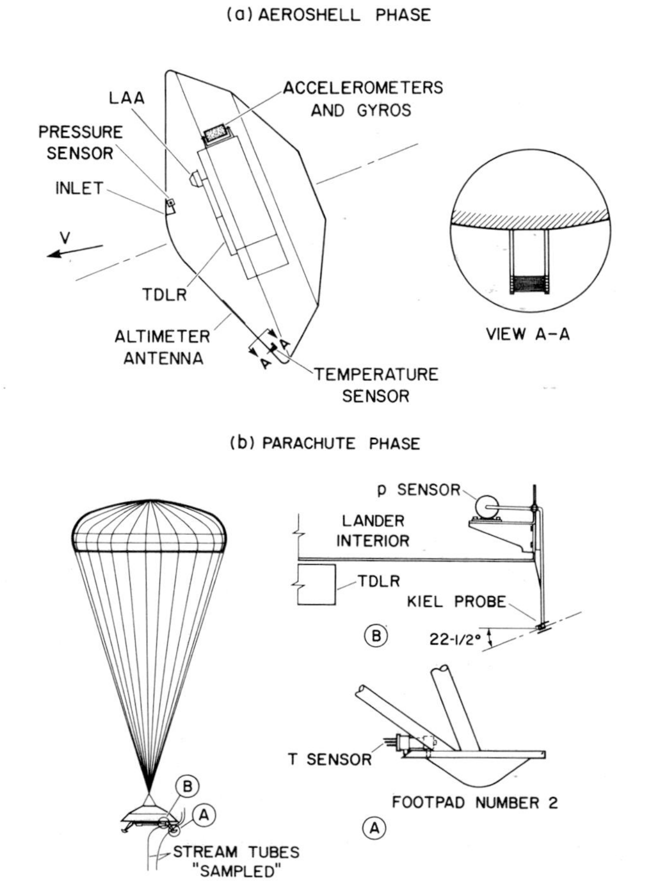

Fig. 1. Location of atmosphere structure instruments on the Viking landers. (a) The configuration of the spacecraft at atmosphere entry. (b) The instruments on the landers after heat shields were jettisoned and parachutes were deployed at a nominal altitude of 6 km.

The pressure inlet during high-speed entry was at the nominal flow stagnation point on the heat shield at the entry attitude. The aeroshell temperature sensor was deployed at a velocity of 1.1 km/s through the surface of the conical heat shield to a position safely outside the aeroshell boundary layer, where it could sense the atmospheric recovery temperature, without convective influence by the heat shield.

The pressure inlet during parachute descent was mounted on an edge of the lander body, with a Kiel probe geometry facing into the theoretical flow direction at that location (Figure 1b). This insured that the stagnation pressure was sensed. The temperature sensor in this phase was mounted on the inboard edge of the footpad on landing leg 2, where it sampled the temperature of oncoming stream tubes well away from thermal contact with the lander. This location also ensured a vigorous flow over the sensing elements at essentially the descent velocity.

The location of the Doppler radar (terminal descent and landing radar (TDLR)) on the lander bottom is also indicated. Two altimeter antennas were provided, one on the surface of the heat shield, used during high-speed entry, and the other on the lander bottom (LAA), used during parachute descent.

The acceleration sensors were derived from guidance quality accelerometers manufactured by Bell Aerospace Company (Bell model IX). They sense acceleration by electromagnetically constraining a test mass to a precise null position. The restoring force is provided by a current flowing in a coil mounted in the test mass, which reacts against the field of a permanent magnet in the sensor. The nulling current is the measure of the acceleration. The scale factor accuracy achieved is believed to have been better than 0.02%, with bias uncertainties <100 μg.

The temperature sensors were multiple fine wire (0.0127-cm diameter) thermocouples, directly exposed to the atmospheric flow. They were designed to minimize errors due to spurious inputs (radiation, conduction, etc.) by maximizing thermal coupling to the atmosphere. Response times were typically ~0.3 s in the aeroshell phase and 0.8 s in parachute descent. Reference junction temperatures were measured with platinum resistance thermometers, within the sensor housing.

The pressure sensors were thin stretched stainless steel diaphragms referenced to vacuum. Diaphragm displacement is the measure of the applied pressure. Unsupported diameters of the diaphragms were ~2.5 cm. Sensor characteristics are given in Table 2. In a thorough preflight test evaluation the sensors

TABLE 2. Pressure Sensors

Mission Phase	Range, mbar	Repeatability	Displacement Sensing
Aeroshell	0–150	±0.5% rdg	capacitive
	0–20	±0.5% rdg	capacitive
Parachute	0–18	±0.01 mbar	inductive

were found to be stable through the flight environments including prelaunch sterilization and launch and entry vibration. The sensor temperatures were nearly constant during entry and descent. Data were corrected for the effects of temperature on calibration where appropriate.

The design philosophy, development, and accuracy evaluation programs for all these sensors have been described in detail elsewhere [*Seiff*, 1976]. Some remarks on the accuracies achieved by the actual sensors during Mars entry will be incorporated into the discussions of results below.

Atmospheric Measurements During Parachute Descent

The two Viking landers deployed parachutes at altitudes of 5.80 and 5.92 km above the terrain. Parachutes were fully open at 5.60 and 5.63 km, and the heat shields were jettisoned at 5.03 and 4.96 km. This exposed the parachute phase pressure sensors to the atmosphere, but there was some interference with both pressure and altitude data by the departing heat shield, down to an altitude of about 4.5 km. The temperature sensors, on the footpad of lander leg 2, were deployed at altitudes near 3.9 km. From there until the terminal descent rocket engines were ignited, at about 1.45 km, both landers transmitted good-quality data on pressure and temperature of the atmosphere (Figures 2 and 3).

Only a small fraction of the data points received are shown, because many repetitions occurred at every digital level. Corrections made for dynamic pressure and temperature effects are indicated and will be discussed. These were of the order of 0.2 mbar and 1.5°K.

Below 1.45 km the descent rockets clearly affected the measurements, and no quantitative use was made of the data. The effect on measured pressures was, surprisingly, to lower them to a level very near ambient. The jets aspirated the region beneath the landers. In further detail the atmospheric pressure provides a boundary condition for the rocket jets, and the jets block atmospheric flow from impinging on the lander bottom and hence suppress dynamic effects on the pressure measurement. It is evident that pressures measured then were within 0.1 or 0.2 mbar of ambient. The effect on temperatures sensed was to raise them abruptly, presumably a result of mixing of the rocket effluent with the local atmosphere.

Pressure Measurements and Landing Site Elevations

Pressure was sampled during descent at 0.5-s intervals, corresponding to nominally 30 m in altitude. The telemetry resolution was 0.085 mbar. There were typically 6–8 repetitions of the reading at every digital level, of which only the first are plotted in Figure 2. These points have just surmounted, and hence are very close to, the reading level. It can be shown that the effective resolution under these circumstances is the nominal resolution/n, where n is the number of reading repetitions at the given level. Thus the effective resolution was about 0.014 mbar. The curves are put through the highest data points, since with pressure rising the resolution error is always negative. By this combination of practices the uncertainty in the relative pressure levels due to resolution was reduced to the order of 0.01 mbar. However, the uncertainty in the absolute pressure level is up to 0.085 mbar because of resolution uncertainty in the zero reading.

The zero readings of the sensors taken just before atmosphere entry established the zero readings for the measurement period. For both landers the sensors read a consistent zero level without deviation over a period of several hundred seconds prior to the actual pressure rise, and these readings differed by 0.10-mbar equivalent for Viking 1 and 0.027 mbar

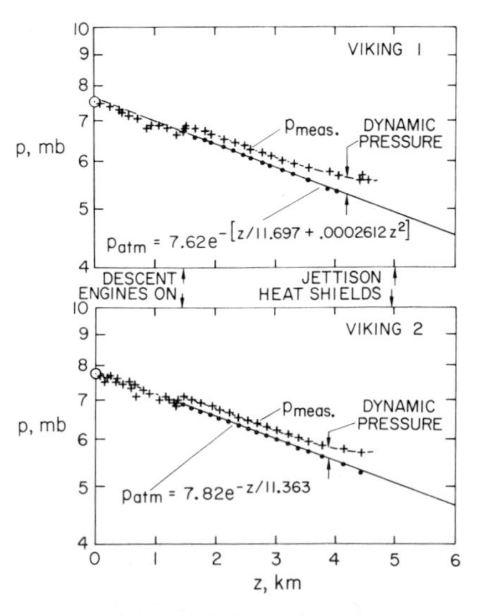

Fig. 2. Pressures measured during parachute descent. Dynamic corrections are indicated. The equations define the variation of atmospheric pressure with altitude in the parachute phase.

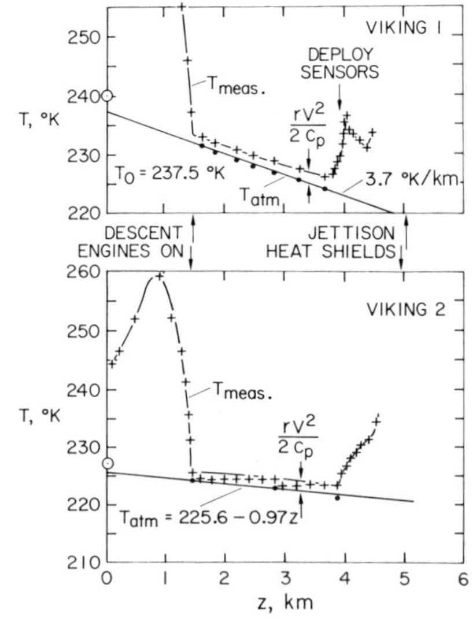

Fig. 3. Atmospheric temperatures; direct sensing in the lowest 5 km. Dynamic corrections are of the order of 2°K. The altitude of full deployment of lander leg number 2, on which the sensor was mounted, can be readily identified.

for Viking 2 from those obtained in calibration in July–November 1974. The repeatability of sensor scale factors, from data taken over the course of 4 months during the preflight test program in which the sensors were exposed to vibration, thermal cycling, etc., was better than 0.2% or 0.01 mbar at 5 mbar. The Viking 2 sensor calibrations were independent of sensor temperature within the data scatter. For Viking 1, scale factor changed by 0.018%/°C, which, for 10° uncertainty in sensor temperature during the measurement period, would lead to 0.01-mbar error at 5 mbar. Overall, we believe the errors due to calibration uncertainties to be within 0.01–0.02 mbar.

The dynamic pressure correction $\frac{1}{2}\rho V^2$ proceeds iteratively with the density initially defined from the uncorrected pressure. The correction accuracy is estimated to be 0.004 mbar. Lander descent velocities on the parachute were between 50 and 90 m/s. Dynamic pressure corrections were applied to the individual data points, prior to the defining of the curve $p_{atm}(z)$, and the corrected data are shown by the heavy dots in Figure 2.

For Viking 2 the plot of log p versus z was highly linear, consistent with the isothermal character of the lowest 4 km of the atmosphere. For Viking 1 a linear fairing also represented the pressure data very well. For both sets of data, however, a curve of the form

$$p/p_0 = \exp(-z/H_0 - k_2 z^2)$$

was defined to represent the curvature in the log $p(z)$ plot and to merge smoothly with the lowest few pressures obtained from the accelerometry near 30-km altitude. The values of surface pressure p_0 defined by simple logarithmic extension of the data between 1.5 and 4.0 km were 7.64 mbar for Viking 1 and 7.81 mbar for Viking 2. The values defined by non-isothermal fits were 7.62 and 7.80 mbar, respectively. The recommended values are 7.62 and 7.81 mbar, since Viking 2 was indeed very close to isothermal below 4 km.

The fitted equations had rms deviations from the data points of the order of 0.003–0.007 mbar, implying a high relative accuracy in the measured pressures. Since relative accuracy was important for defining the slopes dp/dz used in determining the mean molecular weight, pressures are stated in Table 3 to the nearest 0.001 mbar. It must be understood that only relative accuracy is implied, since absolute accuracy is limited by telemetry resolution of the zero reading. The maximum absolute uncertainty is 0.085 mbar, and the expected average uncertainty is 0.04 mbar in a large population of measurements.

Pressures measured after landing on sol 1 at the same time of day by the same sensor were reported by the Viking Meteorology Team [*Hess et al.*, 1976*a*]. These points are shown by circular symbols at $z = 0$ in Figure 2. They are in satisfactory agreement, within the telemetry resolution, with those projected downward from the parachute phase data. The scale heights at ground level were determined from the pressure data to be 11.70 km (Viking 1) and 11.36 km (Viking 2). Equations for $p(z)$ suitable for use in the lowest 30 km (Viking 1) and the lowest 5 km (Viking 2) are given in Figure 2.

The pressure difference recorded between the two landing sites is the combined result of differences in elevation, season, and time of day. The elevations were best given by the radio determinations of the landed radii 3389.38 ± 0.08 and 3381.88 ± 0.22 km [*Michael et al.*, 1976] in relation to the equipotential reference ellipsoid radii at the landing coordinates 3390.87 ± 0.23 and 3384.33 ± 0.23 km, respectively [*Christensen*, 1975]. These data indicate landing site elevations relative to the reference ellipsoid of −1.49 ± 0.24 km (Viking 1) and −2.45 ± 0.32 km (Viking 2). (The relative accuracy of these two determinations should be significantly better than the absolute accuracy.) Thus the second lander came to rest 0.96 km lower than the first, relative to the equipotential surface. Pressures at the equipotential surface at the times of landing are determined from Figure 2 to be 6.74 mbar (Viking 1) and 6.30 mbar (Viking 2). The difference, 0.44 mbar, reflects primarily the seasonal reduction in surface pressure between the landings. The seasonal decrease reported (Viking Meteorology Team, private communication, 1977) was 0.56 mbar at the Viking 1 site. The effect of diurnal variations was significant at the Viking 1 site, the pressure at the time of landing being 0.12 mbar below the daily mean [*Hess et al.*, 1976*a*]. Subtracting this from the seasonal effect, 0.56 − 0.12, gives 0.44 mbar. Hence the measurements of pressure and of landing site elevation are consistent with the seasonal and diurnal variations seen in the Viking meteorology experiment.

Temperature

Selected points from the temperature data from Viking 1 and 2 are plotted in Figure 3. Since the sensors were read every 0.5 s and readings were repeated at every digital level (7–15 times on Viking 1), the effective resolution was ~0.1°K, compared to the telemetry resolution of 1.1°K. On Viking 2, which

TABLE 3. Lower Atmospheric State Properties From Direct Sensing

	Viking 1			Viking 2		
z, km	p, mbar	T, °K	ρ,* kg/m³	p, mbar	T, °K	ρ,* kg/m³
0	7.620†	237.3†	0.01680†	7.820†	225.6†	0.01813†
0.5	7.301†	235.5†	0.01622†	7.48†	225.1†	0.01738†
1.0	6.994†	233.7†	0.01565†	7.16†	224.6†	0.01667†
1.5	6.707	231.8	0.01513	6.853	224.0	0.01600
2.0	6.427	230.1	0.01461	6.564	223.6	0.01536
2.5	6.150	228.3	0.01409	6.282	223.1	0.01473
3.0	5.885	226.5	0.01359	6.015	222.6	0.01413
3.5	5.635	224.7	0.01312	5.747	222.1	0.01353
4.0	5.39	222.8	0.01265	5.483	221.6	0.01294
4.5	5.16	222.1†	0.01221	5.222	221.2†	0.01235

For Viking 1, H_0 = 11.697 km, μ = 44.36 ± 0.41; for Viking 2, H_0 = 11.363 km, μ = 43.36 ± 0.35.
*Calculated for μ = 43.49, R = 191.18 J/kg°K.
†Extrapolated.

encountered a nearly isothermal atmosphere below 4 km, the first 36 readings were at a common digital level, followed by 52 readings at the next level, and one reading at a third level, just before the retro-rockets were ignited. Every eighth one of these data points is shown in Figure 3, but only the first at each level was used to establish the lapse rate with an effective resolution of 0.02°K.

The sensor reference junctions were at 265° on Viking 1 and 263.7° on Viking 2. The sensors were automatically compensated for cold junction temperature by means of a resistance network employing a platinum resistance element, but a small residual correction was applied, as defined by the preflight calibration data. The calibration data were repeatable within ±0.1°C, and interpolation errors between the widely spaced calibration points are estimated to be less than 0.2°C.

The accuracy expected from these data had been extensively analyzed in advance of flight [*Seiff*, 1976] and was concluded to be ~1°K, after consideration of errors due to the electronics, response lag, conduction and radiation, calibration uncertainties, and error in the dynamic temperature corrections. The data presented are still subject to small refinements, of the order of 1°K overall, by application of corrections for small perturbations due to conduction and radiation. Lag and dynamic correction errors are presently ~0.1° or less.

The dynamic correction is given by $rV^2/2c_p$, where c_p is the specific heat at constant pressure of the atmosphere, $V^2/2c_p$ is the temperature increment from conversion of kinetic energy to thermal energy in the gas flow approaching the sensor, and r, the recovery factor, is the fraction of this energy actually experienced by sensors in laboratory calibrations in 6-mbar CO_2 flows. The fraction $r = 0.80$ for the nominal wire diameter Reynolds number (8.4) and Mach number (0.23) during parachute descent. The corrections to the individual data points range from 1.3° to 2.3°.

Straight lines put through these data points so as to pass through the highest corrected points and within the resolution uncertainty of all lower points indicate lapse rates of 3.7°K/km and 0.97°K/km and are shown in Figure 3. The difference is a diurnal effect, as will be discussed below. Surface temperatures are indicated.

The linear lapse rate defined by the lower two points from Viking 2 passes slightly above the third point. This indicates that the initial reading at the 223.3°K level would have been observed to a somewhat higher altitude with earlier leg deployment.

Ground level atmospheric temperatures obtained on the first day after landing, at the same Mars local time, from the meteorology temperature sensor are plotted as circular symbols on the T axis. Agreement is satisfactory but suggests a region of slightly increased lapse rate near the surface.

The steadiness of the temperature readings on both landers, evidenced by a complete lack of fluctuation in the data returned, indicates a thermally homogeneous atmosphere. Because the sensor response lag was 0.77 s (at 2.5 km), small fluctuations (~1°) of ~20-m scale could have occurred without detection.

Mean Molecular Weight

Mean molecular weights were determined from data given in Figures 2 and 3. The defining relationship is

$$\mu = -(R_u T/pg)(dp/dz) \tag{1}$$

which is a generalization to a nonisothermal atmosphere of the expression

$$\mu = R_u T/gH \tag{2}$$

where R_u is the universal gas constant, g is local acceleration due to gravity, and H is scale height. All planetary properties on the right side of these expressions were determined by measurements. The acceleration due to gravity determined from the accelerometers after landing was extended upward as $(R_0 + z)^{-2}$. The accuracy-determining factor is dp/dz, which could be defined only to within about 1% in the presence of the finite telemetry resolution.

The mean molecular weight from Viking 2 data was 43.34 ± 0.8%. This is the average of six values at 0.5-km altitude increments from 1.5 to 4.0 km. The largest deviations from the mean occurred at the end points, where dp/dz is less well determined. Without the two end points the mean value is $43.36^{+0.2\%}_{-0.4\%}$. The derivative was determined from the equations fitted to the pressure data, of the form $p/p_0 = \exp(-k_1 z - k_2 z^2)$, to accommodate pressures measured near 30 km (see below). The fits were both within 0.007-mbar rms, well within the accuracy of determination of pressure.

Since the Viking 2 parachute phase data were essentially isothermal, μ was also calculated from the isothermal relationship to obtain 43.82 at $z = 2.5$ km, $T = 223.1$°K, and $g = 3.7252$ m/s². This is slightly outside the scatter band of the more precise measurement.

The Viking 2 mean molecular weight agrees well with that derived from atmospheric composition data. The landed mass spectrometer has reported a mixture of CO_2 with 0.027 ± 0.003 mol fraction of N_2, 0.016 ± 0.003 mol fraction of Ar, and 0.0015 ± 0.0005 mol fraction of O_2 [*Owen and Biemann*, 1976], for which the mean molecular weight is 43.486 ± 0.066. The present measurement corresponds to a nitrogen fraction of 0.035 for argon fractions of 0.013–0.019.

Similar analysis of the Viking 1 data gave $\mu = 44.36 \pm 1.0\%$, and a simple scale height analysis gave 44.21, about 2% higher than the Viking 2 and the mass analysis values. A considerable effort was made to understand this discrepancy. It could be due to accumulation of maximum possible errors in pressure and temperature. The measurement of T is accurate to ~0.4%, and while the molecular weight is insensitive to scale factor errors in pressure, since $\mu \sim p^{-1}\,dp/dz$, it is sensitive to bias errors in pressure, such as that due to the resolution uncertainty in the zero reading, which could be as large as 1.5%. Hence an accumulation of the maximum expectable errors could add up to 1.9%. Two other possible explanations for the discrepancy are (1) that the small dynamic corrections require further refinement for the effects of wind (a first-order correction for wind is already incorporated) and (2) that the definition of dp/dz has been affected by terrain slopes under the lander. During the descent from 3.5 to 1.5 km the lander was transported about 0.7 km horizontally by winds, and if terrain elevation changed by 40 m along this track, the mean molecular weight would be close to that expected from the other measurements. The necessary ground slope is 3.27°. It has been reported that the lander came to rest at an angle of 2.99°–3.6° [*Mutch et al.*, 1976], and the same source reports, 'The topography (around Viking 1) is gently rolling . . . . The nominal horizon is 3 km away; however, nearby hills obscure our view of large segments of the far horizon.' We may have detected this topography in our molecular weight analysis.

The molecular weight analysis has another significance. It shows the internal consistency of the measurements. Even including the extraneous effects of winds and terrain, the entire set of measurements is consistent to within 1% or 2% in mean molecular weight.

Atmospheric Densities and Tabulated State Properties

The atmospheric densities computed from the pressure and temperature profiles for $\mu = 43.49$ are plotted in Figure 4 and listed in Table 3, together with the pressures and temperatures at 0.5-km intervals, scale heights, and mean molecular weights determined from the experiment.

ATMOSPHERE RECONSTRUCTED FROM ACCELEROMETER DATA

The determination of the density, pressure, and temperature of the Martian atmosphere from an altitude of about 120 km to an altitude of about 26 km depended primarily on data from accelerometers carried on board the entry vehicles. In its simplest sense the density of the atmosphere, ρ, is proportional to the acceleration along the flight path, $-a_s$, through the equation

$$\rho = -2ma_s/C_D A V_r^2 \qquad (3)$$

where m is the vehicle mass, C_D the aerodynamic drag coefficient, A the vehicle cross-sectional area, and V_r the vehicle velocity relative to the atmosphere. The mass and reference area are known constants, and the velocity can be continually tracked from the on-board deceleration measurements.

The analytical approach used to define the vehicle trajectory from the measured accelerations by use of the equations of motion is described in Appendix A. This analysis yields the vehicle velocity and altitude as functions of time and uses the altimeter data as an input. The measured densities are coupled with the reconstructed altitudes, which differ somewhat from the measured altitudes because of variable terrain elevation and roughness, to define $\rho(z)$, from which $p(z)$ is derived under the assumption of hydrostatic equilibrium. From p, ρ, and μ the temperature profile is obtained through the equation of state.

In (3) the drag coefficient must be known precisely, since fractional errors in C_D are the negative equivalent of fractional errors in density. The drag coefficient was measured in a series of simulation flight tests in the laboratory, by the use of a ballistic range with an atmosphere of CO_2. Parameters simulated were model velocity, Reynolds number, gas composition, and, of course, model geometry [*Intrieri et al.*, 1977; P. F. Intrieri, personal communication, 1976]. These tests led to the

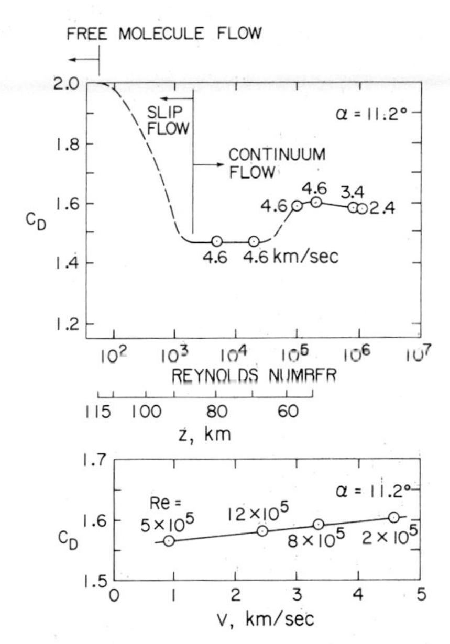

Fig. 5. Drag coefficients used to reconstruct the atmosphere. Test points are at velocity–Reynold number combinations which lie along the nominal entry trajectory. The angle of attack is near the equilibrium flight attitude. The jump in C_D just below $Re = 10^5$ is due to laminar turbulent transition in the vehicle wake and is also seen in the flight accelerometer records.

variations of C_D with Reynolds number and velocity shown in Figure 5 for continuum flow. The accuracy with which the drag coefficient was measured was within 1% (Appendix B).

For altitudes above 90 km the entry vehicle Reynolds number drops below 1000, and continuum flow gives way to slip flow. Near 115 km, free molecule flow begins. For this region we have not yet collected calibrating test data, although work is currently in progress to do so, and one preliminary measurement confirms the slip flow curve to within 5% at $Re = 460$. We have used literature values for the transitional regime from continuum to free molecule drag of spheres [*Masson et al.*, 1961] to model similarly the transition from laminar continuum $C_D = 1.47$ to an assumed value of 2.0 in free molecule flow, with Knudsen number as the governing parameter. Accordingly, the density results above 90 km are less accurate than those below 90 km.

Acceleration Data

The axial acceleration data from the two entries are shown in Figure 6. The accelerations were integrated on board the

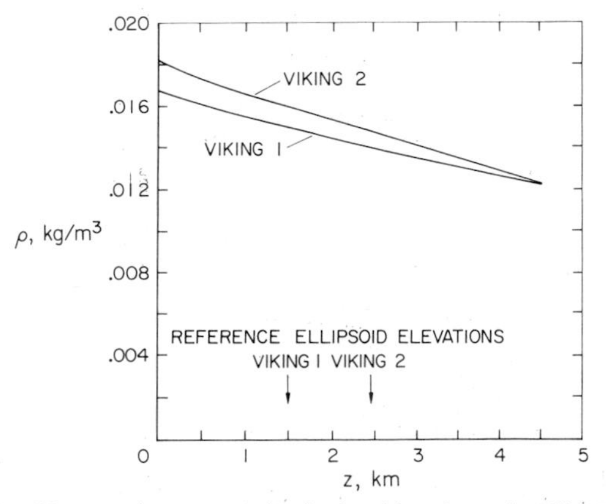

Fig. 4. Near-surface atmospheric densities above the Viking landing sites. Density at the reference ellipsoid was 0.015 kg/m³ at both landings.

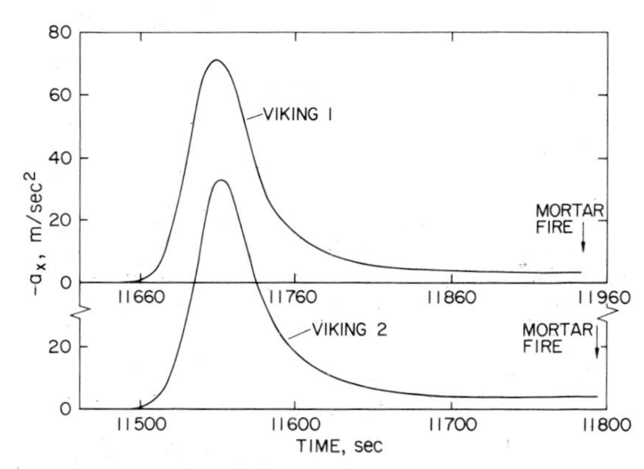

Fig. 6. Axial deceleration data. Time is measured from the instant of separation of the lander from the orbiter.

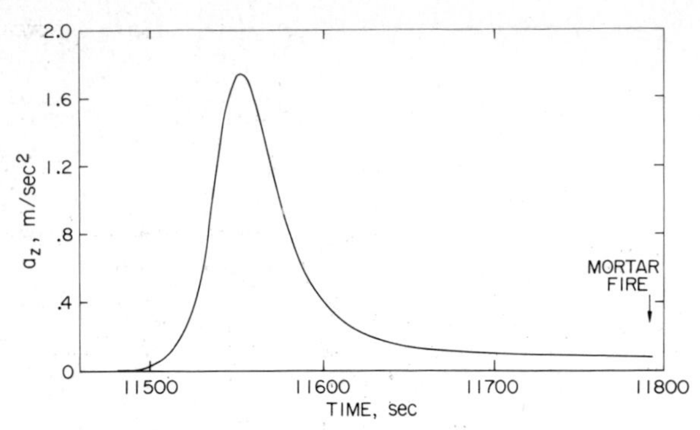

Fig. 7. Viking 2 z axis acceleration data.

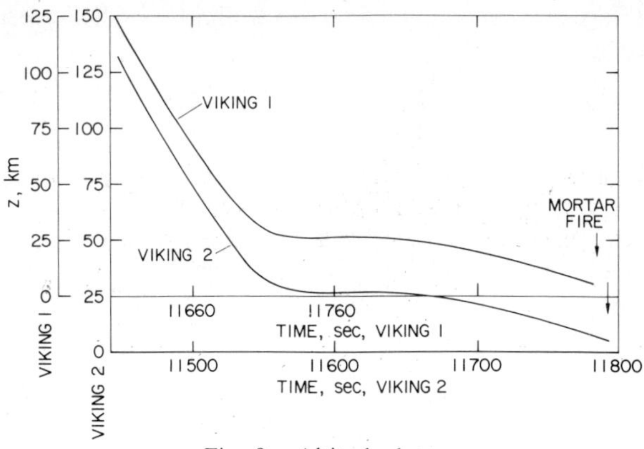

Fig. 9. Altitude data.

Viking landers and transmitted to earth as velocity pulse counts. Each axial pulse represented a velocity change of about 1.27 cm/s, and each lateral axis acceleration pulse about 0.32 cm/s. Atmosphere-relative velocities at entry (arbitrarily defined as an altitude of 244 km above the reference ellipsoid) were 4.418 and 4.477 km/s, respectively, for Viking 1 and 2, and so the axial velocity resolution provided was about 3 ppm of the entry velocity. At 10 samples/s the plotted data appear almost continuous. The peak axial deceleration, about 73 m/s² on Viking 2 and 71 m/s² on Viking 1, occurred at altitudes just above 30 km.

The z axis acceleration data, normal to x in the plane of the aerodynamic lift, are shown for Viking 2 in Figure 7. They essentially duplicate the shape of the axial deceleration profile but at a much lower level, peaking at 1.74 m/s². The ratio a_z/a_x is a measure of the aerodynamic lift-to-drag ratio and of the lander instantaneous angle of attack. The third component of acceleration, a_y, remained essentially zero.

From Figure 6 it can be seen that the axial acceleration threshold extends to at least 11490 s (Viking 2, z = 83 km), but at higher resolution it is seen to extend to appreciably higher altitude (Figure 8). At early times, less than one count/s is being measured, and the number alternates between adjoining levels, reflecting the fact that fractional values are being interpreted by a data system dealing only in integers. An 11-point running average of these data, representing the mean deceleration over an 11-s interval and plotted at the central point, produces the smoothed variation shown by the dots, which exhibit small scatter about the fitted line. The line, in turn, merges nicely with the values of instrument bias determined over long integration periods just prior to atmosphere entry. It is apparent that significant atmospheric decelerations are defined to an altitude near 125 km.

Altitude data from the altimeters on Viking 1 and 2 are shown in Figure 9. These data, at five samples/s, are also near continuous. They are uncorrected for variable terrain elevation and local roughness. These terrain characteristics can be extracted by detailed comparison of the altimeter data with the altitudes determined from the equations of motion analysis.

Both Viking entries, by virtue of their near glancing incidence relative to the atmosphere and the use of aerodynamic lift, experienced periods of essentially horizontal flight extending over about 50 km at altitudes near 28 km. It was found to be difficult to carry the atmosphere definition from the acceleration data through this region successfully, because of extreme sensitivity of the results to very small variations in the near-zero flight path angle and possibly also because of wind effects. We believe that we will be able to accomplish this, but it has not been done at this writing. Hence the accelerometric definition of the atmosphere presently terminates at the region where horizontal flight begins.

Atmospheric Profiles

The density data for the atmosphere of Mars derived from the accelerometer data of Viking 1 and 2 are given in Figure 10. Over the altitude range from about 28 to 120 km, the density ranges from 10^{-3} to 10^{-8} kg/m³ with a mean scale height of about 8 km but with significant curvature or waviness on the log ρ (z) plot, evidence of temperature variation

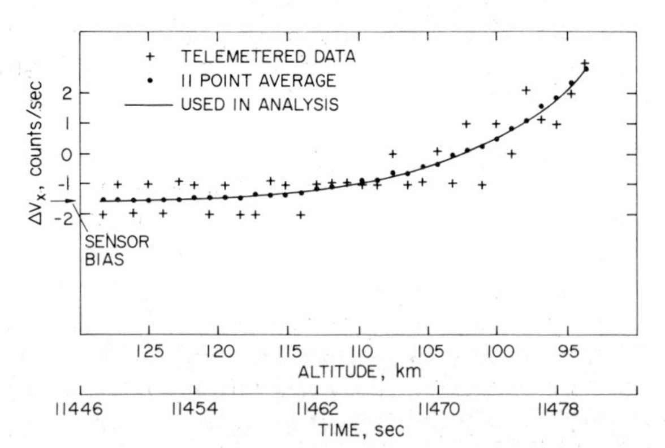

Fig. 8. Viking 2 axial velocity pulse counts near threshold. The telemetered data have been corrected for the small-velocity increments due to firing of the attitude control jets (note the occasional deviations from integral numbers of counts).

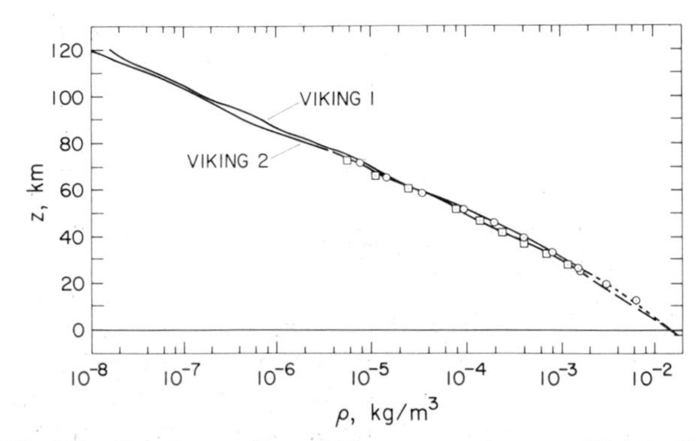

Fig. 10. Density profiles of Mars atmosphere to 120 km. Data shown by curved lines above 28 km are derived from accelerations. Points are from stagnation pressures (circles, Viking 1; squares, Viking 2). Densities from measured pressure and temperature during parachute descent are shown in the lowest 5 km.

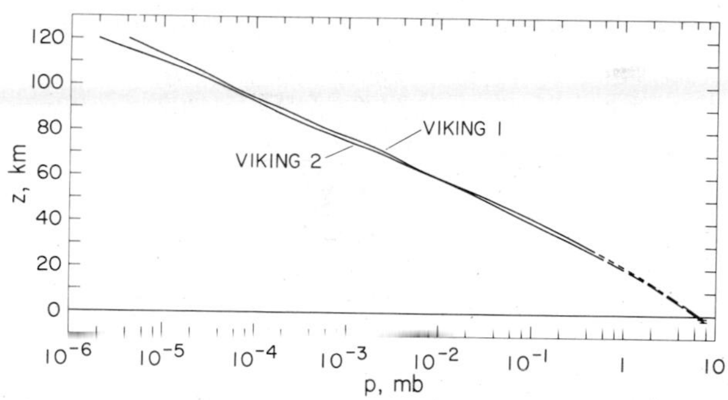

Fig. 11. Variation of pressure with altitude in Mars atmosphere. Above 28 km, pressure is derived from the density data, hydrostatic equilibrium being assumed; below 5 km, pressure is derived from direct sensing during parachute descent.

with altitude. Accuracy analysis indicates that the absolute accuracy of the densities should be within ~1% for the region of continuum aerodynamics, $z < 90$ km. At higher altitudes, present uncertainties in the drag coefficient in slip flow and free molecule flow could lead to errors of the order of 10%.

The pressure structure is plotted in Figure 11, and the temperature profiles in Figure 12. Since the data are essentially continuous, they are represented by lines rather than discrete points. Table 4 is a listing of the atmospheric state properties at 4-km intervals. The parachute phase direct-sensing data are included on these figures to show their relationship to the state properties determined by accelerometry and to define the atmosphere to ground level. In addition, the data obtained from aeroshell phase pressure and temperature sensing are included as symbols. These are described below.

The wave structure in the temperature profiles was identified in early discussions as evidence for strong thermal tidal oscillations in the atmosphere of Mars. The implications of these waves in relation to theory will be discussed in a later section.

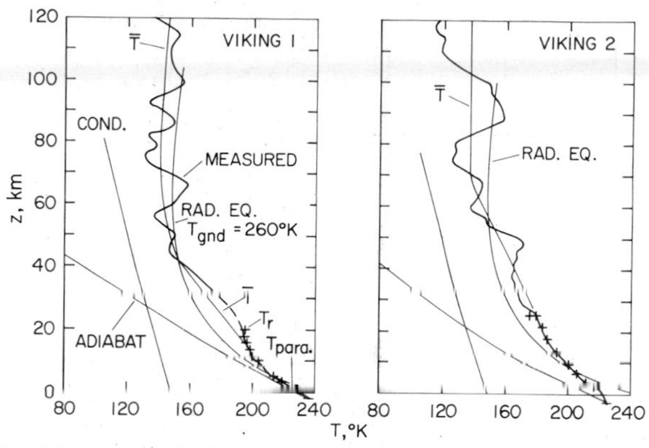

Fig. 12. Atmospheric temperature profiles below 120 km from Viking 1 and 2. The wave structure is the dominant characteristic. Comparisons are shown with the radiative equilibrium prediction of Gierasch and Goody and with an adiabatic profile. The condensation boundary lies well below the atmospheric temperatures. A curve which seeks to define the mean temperature $\bar{T}$ about which the oscillations are centered is indicated.

Supporting Data From Aeroshell Phase Pressure and Temperature Sensing

The stagnation pressure readings taken throughout the high-speed entry are convertible to measurements of atmospheric density through the relation

$$C_{p_s} = (p_s - p_{\text{atm}})/(\tfrac{1}{2}\rho V_r^2) \qquad (4)$$

wherein C_{p_s}, the stagnation pressure coefficient, is a theoretically known function of velocity (and weakly of gas composition), $p_s/p_{\text{atm}} \approx 500$ for $z > 35$ km (so that p_{atm} can be neglected in first approximation), and V_r is available from the trajectory reconstruction. The atmospheric densities derived from these independent data are plotted as symbols on Figure

TABLE 4. Atmosphere State Properties from Accelerometer Data

Altitude, km	Viking 1 ρ, kg/m³	Viking 1 p, mbar	Viking 1 T, °K	Viking 2 ρ, kg/m³	Viking 2 p, mbar	Viking 2 T, °K
120	1.60(−8)	4.14(−6)	136.3	8.86(−9)	1.99(−6)	116.0
116	2.42(−8)	6.91(−6)	149.2	1.69(−8)	3.76(−6)	115.5
112	3.95(−8)	1.12(−5)	148.6	3.08(−8)	6.98(−6)	118.2
108	6.59(−8)	1.84(−5)	146.4	5.62(−8)	1.30(−5)	121.1
104	1.06(−7)	3.03(−5)	149.4	9.36(−8)	2.35(−5)	131.8
100	1.67(−7)	4.94(−5)	154.8	1.42(−7)	4.01(−5)	147.9
96	2.88(−7)	8.02(−5)	145.9	2.30(−7)	6.60(−5)	150.2
92	5.39(−7)	1.38(−4)	133.6	3.63(−7)	1.08(−4)	155.1
88	8.33(−7)	2.33(−4)	146.7	5.79(−7)	1.74(−4)	157.5
84	1.40(−6)	3.87(−4)	144.2	1.06(−6)	2.88(−4)	141.4
80	2.57(−6)	6.70(−4)	136.6	2.11(−6)	5.08(−4)	126.1
76	4.66(−6)	1.16(−3)	130.5	3.83(−6)	9.22(−4)	125.9
72	7.70(−6)	2.05(−3)	139.1	6.71(−6)	1.68(−3)	130.9
68	1.17(−5)	3.43(−3)	152.9	1.07(−5)	2.91(−3)	143.0
64	1.88(−5)	5.55(−3)	154.6	1.82(−5)	4.94(−3)	142.3
60	3.19(−5)	9.11(−3)	149.5	3.26(−5)	8.54(−3)	137.3
56	5.96(−5)	1.56(−2)	136.8	5.27(−5)	1.47(−2)	146.4
52	9.56(−5)	2.67(−2)	146.3	8.23(−5)	2.46(−2)	156.5
48	1.57(−4)	4.45(−2)	148.6	1.20(−4)	3.92(−2)	170.7
44	2.65(−4)	7.46(−2)	147.5	1.94(−4)	6.19(−2)	166.8
40	4.10(−4)	1.23(−1)	157.4	3.14(−4)	9.87(−2)	164.5
36	6.25(−4)	1.98(−1)	166.1	5.04(−4)	1.58(−1)	164.4
32	9.32(−4)	3.12(−1)	175.1	7.92(−4)	2.54(−1)	167.6
28	1.38(−3)	4.83(−1)	183.8	1.22(−3)	4.04(−1)	173.2

Read 1.60(−8) as 1.60×10^{-8}.

10 and show very satisfactory agreement with those given by the accelerometers.

Although intuitively one would like to obtain atmospheric pressure from stagnation pressure data, (4) cannot be directly solved for p_{atm} without knowledge of ρ and hence of T_{atm}. That is, in the absence of independently obtained temperatures the equation cannot give directly more than a first approximation of the atmospheric pressure. It is evident, though, that p_{atm} can be obtained from ρ through integration of (14) in Appendix A and that if $\rho(z)$ from the stagnation pressure measurement agrees with that from the accelerometers, atmospheric pressures and temperatures will likewise agree.

The recovery temperature data taken during the aeroshell phase below a velocity of 1.1 km/s were used to define the temperature of the atmosphere in the altitude gap between the accelerometer and the parachute phase measurements. Since atmospheric specific heat is not constant over a broad range of temperatures, the solution proceeds through the enthalpies. The relation between recovery and ambient enthalpies is

$$h_{atm} = h_r + (V_r^2/2)[(1 - r)(V_l/V_r)^2 - 1] \qquad (5)$$

where h_r is the gas enthalpy at the recovery temperature, V_l is the local gas velocity incident on the sensor within the lander shock layer, and r is the recovery factor defined earlier. The temperatures from this analysis are shown in Figure 12 as the pluses between 3 and 25 km. The continuity of the data with the parachute phase direct sensing is good, and the data suggest that the wave structure of the temperature profile begins in this low-lying region, at initially small amplitudes.

Merging of the Accelerometer and Upper-Atmosphere Mass Spectrometer Data

Densities of the atmosphere above 120 km also were measured by the Viking Entry Science Team, by use of the upper-atmosphere mass spectrometer (UAMS) [*Nier et al.*, 1976; *Nier and McElroy*, 1976; *Nier and McElroy*, 1977]. The possibility was recognized well in advance of the Viking entries of joining these data with the accelerometer data to define the structure of the atmosphere from ground level to ~200 km, and it was with some eagerness that we compared the data to see the degree of compatibility that existed. This is shown in Figure 13.

To prepare this figure, the UAMS atmospheric number densities were converted to mass densities through the relation $\rho = \sum N_i\mu_i/N_A$, where N_i is the number density of species i, μ_i is its molecular weight, and N_A is Avogadro's number. (A. O. C. Nier has pointed out that we have a definition of Avogadro's number implicit in the data in regions of overlap of the UAMS and accelerometers.) To extend the data to the highest altitudes, it was necessary to extrapolate number densities of some of the minor species. Furthermore, an allowance was made for the presence of atomic oxygen, based on the observation by *Hanson et al.* [1977] that a number density of 5×10^8 was indicated for an altitude of 135 km on Viking 1. This observation was extended upward by assuming that the O was in hydrostatic equilibrium, at the same temperature as the other neutral constituents. The implied assumption is that there are no chemical reactions acting as a major source or sink for O in this altitude range, which may not be the case. With this assumption, atomic oxygen becomes an important bulk constituent above about 150 km, rising to a mole fraction of 0.4 at 170 km. Neglecting the O fraction would not, however, change the essential character of the log ρ (z) plot. The densities would be shifted downward by about 15% at 170 km, a small shift on the logarithmic plot. (The small effect of O on mass density is due to its low molecular weight relative to those of other species present.)

The UAMS density points, although more widely spaced, exhibit a curvature with altitude on the semilogarithmic plot similar to that shown by the accelerometer-derived densities. (Because of the relatively wide point spacing, some of the fine structure may be lost.) Furthermore, it is possible to join data from the two instruments without any discontinuities in density or its gradient if the curve is not forced through the last data point from each instrument. There is excellent reason to give these last-obtained points low weight. For the accelerometry the reason is that the measurement is near threshold, where accuracy is diminishing, and is in free molecule flow, where C_D is more uncertain. For the UAMS the reason is that the instrument is becoming saturated by the high inlet flux, and the calibration is less certain; also, if the transition to continuum flow has begun, the ground test data obtained in the molecular beam cannot give the ratio between measured and free stream densities reliably. It is evident that the joining of these two sets of data was very satisfactory, and in combination they define the density profile essentially from ground level to nearly 200 km.

The density profiles were used to derive pressures and temperatures as described earlier, by means of (14) and the equation of state, with the results shown in Figures 14 and 15. Values of the state properties derived from the UAMS data are

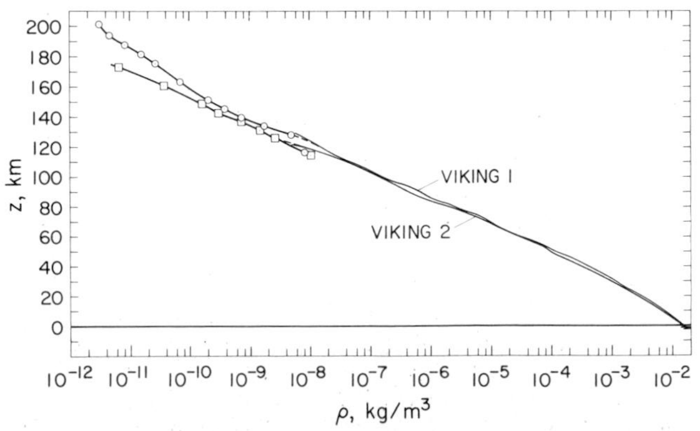

Fig. 13. Density profiles from ground level to 200 km obtained by merging the mass spectrometer and accelerometer density data. The merging altitude is around 120 km. Differences between the Viking 1 and Viking 2 soundings are the combined result of differences in latitude, season, and time of day. Viking 1 landed at 22.3°N at 4:13 P.M. MLT on July 20, 1976, Viking 2 at 47.7°N at 9:49 A.M. MLT on September 3, 1976.

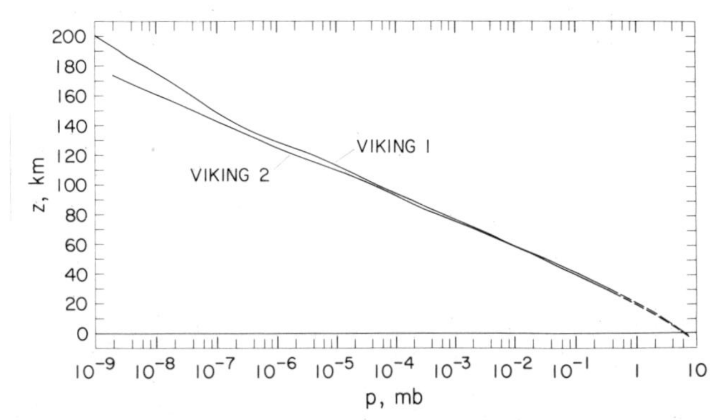

Fig. 14. Pressure structure of Mars atmosphere to 200 km derived from measured density profiles of Figure 13.

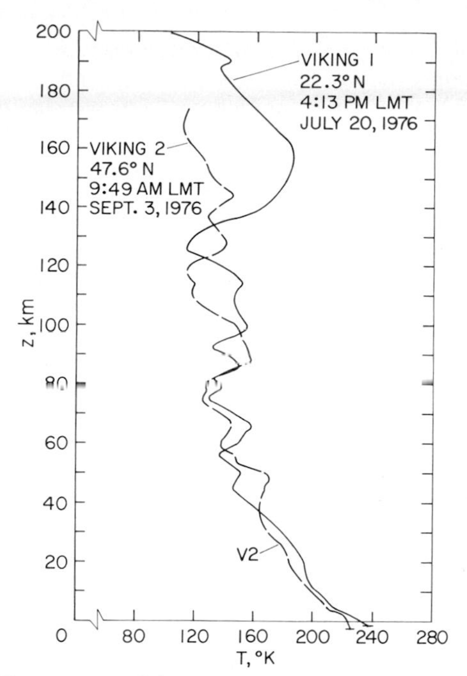

Fig. 15. Temperature of the neutral atmosphere from synthesis of all available entry science data.

listed in Table 5. The mean molecular weight was allowed to vary with altitude in the fashion shown by the UAMS data. The wave structure determined below 120 km continues upward in a continuous fashion, increasing in both wavelength and amplitude with increasing altitude.

Our procedure for deriving the temperatures differs from that used by *Nier and McElroy* [1977] in several respects. First, it is applied to the smoothed density variation indicated by the total set of data, including the merging of the accelerometer densities with the UAMS densities, rather than being a linear, point-to-point type of calculation. (We make no assumption of local linearity.) Second, it uses the total atmospheric density as a basis rather than the number density of CO_2 alone. Nevertheless, the temperatures from the two analyses agree within 10° or 15° below 160 km but diverge at higher altitudes, although our values are generally within the uncertainty bars which they have indicated.

It should be explained that the starting temperature at the highest altitude depends on the starting pressure, which in turn depends on the column density above that altitude. Since this was not experimentally determined, the starting temperature was fixed by fitting a straight line to the uppermost 10 km of the density curve to define the scale height and hence the temperature at that level, which was used with the measured density and composition to compute the starting pressure.

Comparison of Atmosphere Structure From Viking 1 and 2 Entries

The atmosphere density data from Viking 1 and 2 are compared in Figure 13. The first-order conclusion is that they are closely similar. In the lower atmosphere, below 50 km, densities seen by Viking 2 were as much as 28% lower than those seen by Viking 1. The two profiles come together around 60 km and approach closely near 100 km before diverging strongly above 100 km. Near the surface, Figure 4 showed larger densities for Viking 2 than for Viking 1, but if the difference in landed elevation is taken into account, then at the reference ellipsoid, $\rho_{0_{V1}}$ (0.0150 kg/m³) is essentially equal to $\rho_{0_{V2}}$ (0.0148 kg/m³).

It is natural to inquire whether these density profiles satisfy the observed surface pressure differences through the equation of hydrostatic equilibrium. The answer is yes, since hydrostatic equilibrium has been assumed in interpreting the accelerometer data, and it has been shown that for both sets of data an equation of the form $p/p_0 = \exp(-z/H)$ (representing isothermal hydrostatic equilibrium) fits the near-surface data well. Between 5 and 30 km, temperature variations with altitude require only that the scale height vary slightly with altitude.

Since 95% of the pressure at any altitude is contributed by the mass of gas within 3 scale heights upward, one might expect to find differences in pressure between Viking 1 and 2 that are greater at higher altitudes than those at the reference surface, where the difference is 6.7%. This proves to be the case, as shown in Figure 14. A pressure difference of ~15% of the mean occurs, for example, at 35 km. At 58 km, however, pressures from the two soundings were within 4%.

If these data were simultaneous, they would indicate a strong driving potential for atmosphere circulation at some altitudes. However, they are not simultaneous, and so a

TABLE 5. Mass Densities, Pressures, and Temperatures Derived from UAMS Data

Viking 1				Viking 2			
Altitude, km	ρ, kg/m³	p, mbar	T, °K	Altitude, km	ρ, kg/m³	p, mbar	T, °K
200	3.30(−12)	1.05(−9)	102	174	5.80(−12)	1.93(−9)	115
195	4.70(−12)	1.67(−9)	125	170	1.07(−11)	3.06(−9)	112
190	6.62(−12)	2.53(−9)	143	166	1.90(−11)	5.04(−9)	112
185	1.13(−11)	4.02(−9)	139	162	3.23(−11)	8.46(−9)	116
180	1.80(−11)	6.48(−9)	148	160	4.15(−11)	1.10(−8)	119
175	2.77(−11)	1.03(−8)	158	156	6.82(−11)	1.83(−8)	125
170	4.21(−11)	1.61(−8)	167	152	1.13(−10)	3.05(−8)	129
165	6.27(−11)	2.48(−8)	177	148	1.81(−10)	5.04(−8)	135
160	9.35(−11)	3.78(−8)	186	144	2.75(−10)	8.15(−8)	145
155	1.48(−10)	5.80(−8)	185	140	4.90(−10)	1.33(−7)	135
150	2.41(−10)	9.05(−8)	182	136	8.65(−10)	2.24(−7)	130
145	4.10(−10)	1.45(−7)	175	132	1.43(−9)	3.81(−7)	136
140	7.25(−10)	2.39(−7)	166	128	2.31(−9)	6.36(−7)	141
135	1.59(−9)	4.21(−7)	136				
130	3.80(−9)	8.77(−7)	120				

Read 3.30(−12) as 3.30 × 10^{-12}.

primary difference to be considered is the effect of seasonal condensation of the atmosphere [*Hess et al.*, 1976*b*]. Approximately 7.35% of the atmosphere of Mars condensed between the landings of Viking 1 and 2. Thus the pressure difference between the two soundings oscillates about this mean reduction.

Very sizable pressure differences are indicated between the two entries at altitudes above 100 km, e.g., a factor >2 at 150 km and a factor >5 at 170 km. Pressure differences of this magnitude are suggestive of very strong circulations in the upper atmosphere of Mars. These differences are a consequence of the lower temperatures, hence smaller scale heights, of the Viking 2 upper atmosphere, which lead to increasing differences in density with altitude. Thus the number densities of CO_2 at 170 km are 4.9×10^8 (Viking 1) and 1.04×10^8 (Viking 2) [*Nier and McElroy*, 1977]. Pressure is then lowered because of both density and temperature, but ultimately, the lower temperature is the cause.

A comparison of the temperature profiles is given in Figure 15. The remarkable thing that emerges is the very close correspondence of the two profiles. We were fully expecting to find significant effects of the change in season and latitude, both of which should tend to lower Viking 2 temperatures relative to those of Viking 1. The data show that their combined effect is to lower temperatures a few degrees at altitudes from 5 to 35 km and to lower the mean temperature 4°–10° up to 120 km. A larger effect on mean temperature is seen above 120 km, where the mean appears to be about 126° for Viking 2, compared to 150° for Viking 1. In the intermediate altitudes, 35–100 km, the major differences appear to be due primarily to the phase and modal makeup of the diurnal oscillation and not to differences in the mean.

The upper-atmospheric temperature difference appears to be a result of one large dynamic half-wave extending from 135 km to upward of 185 km. Thus the difference in mean temperature may be a consequence of more vigorous dynamic oscillations at the near subsolar Viking 1 site. However, the direct solar heating of the upper atmosphere should also be greater at this location, and a diurnal effect (late afternoon versus early morning) may also contribute to the difference in upper-atmospheric temperatures.

In the lowest 5 km there is the diurnal difference referred to earlier. Thus we can summarize three differences between the temperature profiles at the two landing sites: (1) diurnal differences near the surface, (2) a small temperature difference (~3°–10°) below 35 km and in mean temperature up to 120 km due to season and latitude, and (3) a 25° temperature difference above 120 km, possibly controlled by dynamics.

It will now be of interest to compare these findings with pre-Viking theoretical expectations.

Comparisons With Theory, Theoretical Implications

Mean Temperature Profiles

Processes governing the temperature profiles in the atmosphere of Mars were studied theoretically by *Gierasch and Goody* [1967, 1968]. In the 1967 paper (paper A) they calculate radiative equilibrium profiles, and in the 1968 paper (paper B) they incorporate effects of free convection and the diurnal radiative wave emanating from the surface, while arguing that planetary-scale circulation is not a first-order consideration in determining temperatures. Although dynamical oscillations were not treated, they were alluded to as a possibly significant phenomenon for Mars.

We will not attempt a quantitative comparison with this theory, simply because the cases calculated and presented do not coincide closely with conditions of season, latitude, and ground temperature actually encountered by the Viking landers. However, it is possible to examine the theory relative to the observations and to draw some conclusions. The first concerns the important role of radiative equilibrium.

The radiation equilibrium temperature profile for a clear (dust free) atmosphere (interpolated from paper A for a ground temperature of 260°) gives a first-order representation of the mean profiles observed (Figure 12). The differences between the calculated and observed (mean) profiles could possibly be explained by the presence of dust in the lower atmosphere, which would be expected to warm the dust-carrying levels while shielding the higher levels from ground radiation to some degree. They could also be explained by an underestimate of atmospheric radiative absorption.

The major effect of a dust storm on the temperature profiles was well displayed during the early stages of the Mariner 9 mission [*Hanel et al.*, 1972; *Kliore et al.*, 1972]. *Gierasch and Goody* [1971] showed that an absorption of 0.1 of the solar heating by dust in the atmosphere would modify their results to agree with the Mariner 9 data during the height of the storm. During the entries of the Viking landers the atmosphere was quite transparent, but there was evidence for a small dust fraction, seen in the scattering of sunlight to produce a pink sky [*Mutch et al.*, 1976].

The Viking 1 entry at 4:15 P.M. was at the time of day for which paper B predicts a sharply defined tropopause at altitudes near 15 km, with essentially adiabatic lapse below that. This type of structure, predicted for winter and summer mid-latitudes and for the equatorial region at equinox, was not encountered by the Viking landers.

Below about 5 km the radiative equilibrium temperature profiles are unstable, for T_{gnd} = 260°K, and should give way to a region of adiabatic temperature lapse. Experimentally, the lapse rate remains stable, in both the late afternoon and morning profiles, down to the lowest altitude of observation, 1.5 km. Linear extension of the measured temperature profiles to the ground agrees with the landed temperatures, sensed on sol 1 at the same time of day, to within 2.5° on Viking 1 and 1.5° on Viking 2. If these comparisons are error free, they would admit an adiabatic lapse region in the lowest 1.5 km of Viking 1 and the lowest 0.4 km of Viking 2.

However, there is a clear change in slope of the profiles a few kilometers above the surface, especially in the Viking 1 profile at 4 km, which suggests that a convective region occurs near the surface. Since the temperature profile does not admit natural (thermal) convection, the data suggest the presence of turbulent forced convection due to winds and a boundary layer thickness of 6 km, extending to 4 km above the reference ellipsoid. In the boundary layer the temperatures are near adiabatic, indicating the occurrence of vertical mixing, while above it they veer sharply away from the adiabat. Since potential temperature increases with height in the boundary layer, the forced convection transports heat downward.

Forced convection can explain an adiabatic profile, but some further explanation is needed for the subadiabatic profiles observed. *Stone* [1972] has indicated that stable lapse rates will occur in winds driven by baroclinic instability. *Blumsack et al.* [1973] incorporate an empirical dynamic vertical heat flux term in their model and present an example in which the lapse rate becomes subadiabatic roughly 1 scale height above the surface. Perhaps these kinds of processes are responsible

TABLE 6. Measured Characteristics of the Temperature Wave Structure

Peak No.	Viking 1				Viking 2			
	z, km	ΔT, °K	Δz, km	λ, km	z, km	ΔT, °K	Δz, km	λ, km
1	44.5	−3.5	+0.75	11	33	−5	+1.2	31
2	50	+4	−1.0	12	48.5	+14.5	−3.4	22
3	56	−7	+1.7	20	59.5	−8	+1.8	16
4	66	+17	−4.2	20	67.5	+6.5	−1.5	13
5	76	−10	+2.2	6	74	−10.5	+2.4	8
6	79	−2.5	+0.7	6	78	−9	+2.0	4
7	82	−8	+1.8	9	80	−10	+2.1	17
8	86.5	+6	−1.6	12	88.5	+21.5	−5.3	13
9	92.5	−9	+2.1	14	96	+14	−3.6	8
10	99.5	+12	−3.2	17	100	+12	−2.8	20
11	108	+2.5	−0.9	13	110	−18.5	+4.2	8
12	114.5	+7.5	−1.8	23	114	−17	+3.7	8
13	126	−32	+7.5	68	118	−22	+4.9	19
14	160	+40	−12.4		127.5	+5	−1.3	18
15					136.5	−7.5	+1.8	15
16					144	+9	−2.4	46
17					167	−25	+7.0	

for stable lapse rates in the lowest 5 or 6 km of Mars atmosphere.

The effect of the ground temperature cycle in cooling the atmosphere during the night and warming it during the day was predicted in paper B to be limited to the lowest levels of the atmosphere, having little effect above 5 km and large effect only below 2 km. This phenomenon is seen in the comparison of Viking 1 and 2 near-surface data to extend up to ~4 km above the surface, and it is apparently very much as anticipated. The heat exchange is radiative, so that dust in the atmosphere would perhaps be expected to enhance and extend the region of diurnal variation in temperature.

The CO_2 condensation boundary has been included in Figure 12. In both the morning and late afternoon profiles there is a minimum margin of ~20°K between atmospheric temperature and condensation. In view of the lack of diurnal variation of the mean profile above the boundary layer it appears that only a large negative amplitude of the thermal tide could condense CO_2 and that hazes and clouds seen in the Mars atmosphere in the summer season below polar latitudes are most probably water ice.

Wave Structure

We have previously discussed the relationship of the wave structure in the Viking 2 temperature profile to the tidal model of *Zurek* [1976] [*Seiff and Kirk,* 1976]. The Viking 1 wave structure is generally similar to that of Viking 2, with differences in phase and amplitude associated with latitude and time of day (Figure 15). The waves appear to be a complex superposition of modes. Table 6 lists approximately the temperature amplitudes and wavelengths obtained relative to the mean temperature curve drawn through the data (Figure 12). The wavelengths are measured over half a cycle, from one peak to the next higher peak. The significance of the Δz column will be explained below.

The wavelength varies from 4 to 31 km below 100 km and grows to a maximum in the uppermost half cycle of each sounding. For comparison the theoretical model has indicated wavelengths of 22–24 km. The wavelength values depend, of course, on just how small a disturbance is considered to define a peak. For example, the Viking 2 'peaks' at 78 km and 96 km, which lead to two of the smaller wavelengths, are of very small amplitude. Their omission from Table 6 would increase λ at 74 km to 12 km and at 88.5 km to 23 km. They are suggestive of a tendency for a minor peak to appear from a mode locally out of phase with the fundamental mode. The reality of these details is not clear to us, but we are interpreting them literally, as measured. The most prominent of these out-of-phase peaks is at 79 km on Viking 1.

There appear to be two or three peaks (not included in the table) in the wave structure of the Viking 1 sounding below the 45-km peak. These suggest that the wave structure persists down to the level of the boundary layer edge. The temperature amplitude is initially small (a few degrees), then increases with altitude to the order of 15° or 20°, then is damped somewhat before increasing to very large values above 100 km.

Zurek has calculated that horizontal velocities excited by the diurnal heating cycle are as large as 150 m/s, above 70 km, under clear air conditions. Vertical flow velocities are about 2 orders less, up to 1.5 m/s. If adiabatic compression and expansion resulting from the vertical flow component are assumed to be responsible for the temperature waves, the amount of vertical movement thus implied by the observed peaks is given in Table 6 under the heading Δz. Vertical excursions of a few kilometers are sufficient to produce the observed temperature structure. Vertical motion at 0.1 m/s can account for the smaller displacements in 4 hours (1.44 km), while 1 m/s for 4 hours (14.4 km) would account for the largest displacement. These velocity magnitudes are compatible with the theoretical model.

Horizontal winds of the magnitude described, in the presence of wind shear, can be expected to break down into turbulence and to contribute to mixing of the atmosphere. The larger velocities and greater range of motion observed at the higher altitudes suggest a more vigorous turbulent mixing there. The possible relationship of this dynamical turbulent mixing process to the observed photochemical stability of the atmosphere has been pointed out by Zurek and others.

Finally, Zurek has discussed the effect of terrain on tidal oscillations and concludes that it will have first-order effects, adding modes and varying the contributing modes with longitude. We suspect that some of the modal complexity observed in our data is the effect of the terrain, which for Viking 1 was a rapidly descending rolling surface, near the center of a deep

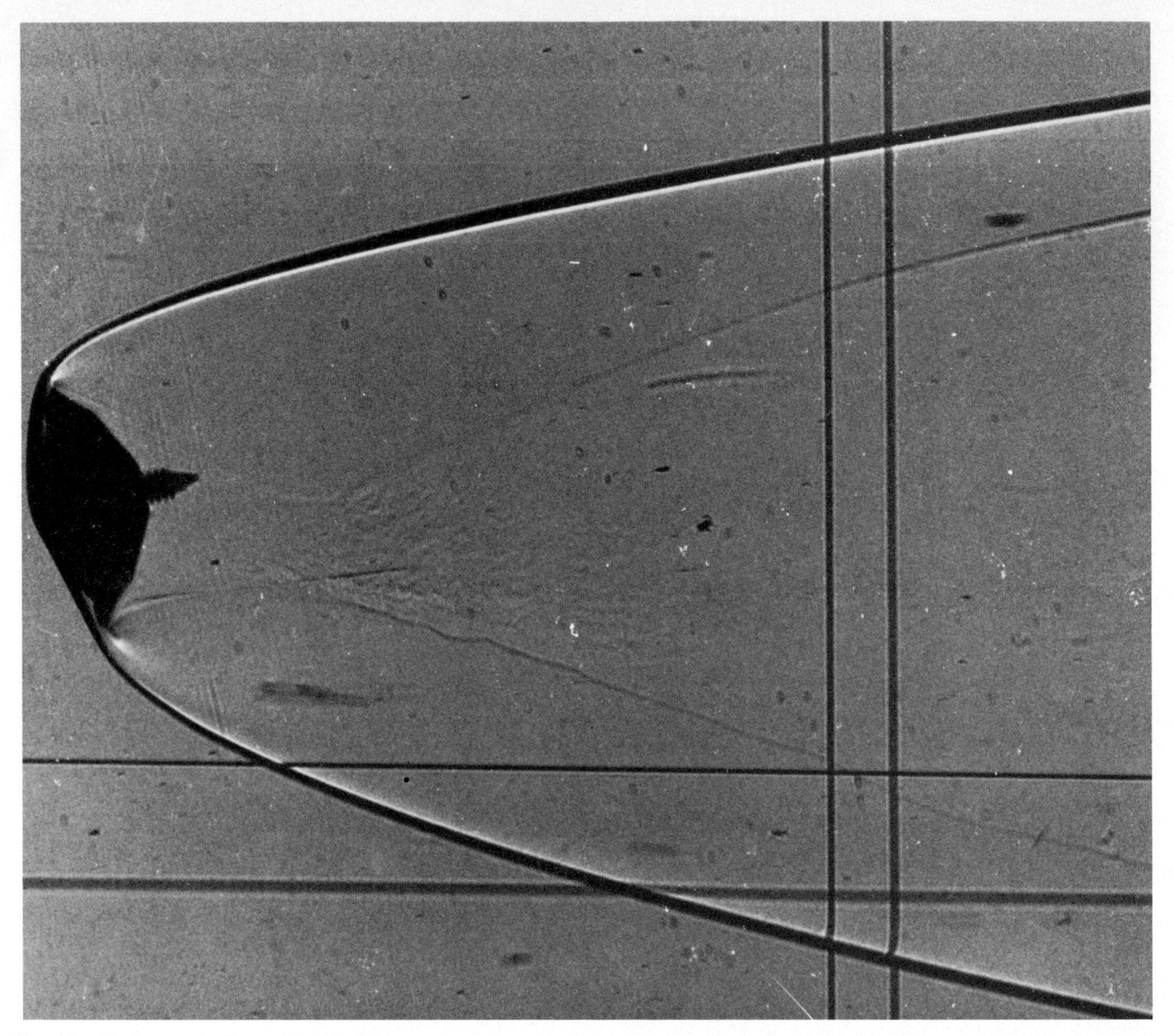

Fig. 16. Shadowgraph of a model of the Viking entry configuration in a laboratory test flight. The atmosphere is CO_2, the model's velocity is 3.4 km/s, and the Reynolds number is 8×10^5. In this figure the model is at the Viking equilibrium flight attitude, $\alpha \approx -11°$.

basin. The theory shows that low basins enhance the velocities and the temperature amplitude (see, for example, the results for 120°W longitude, at the equator and at 19°N, in *Zurek*'s [1976] paper).

Appendix A: Data Analysis

The equations of motion used to reconstruct the trajectory were as follows:

Inertial velocity

$$dV_I/dt = g \sin \gamma_I - a_{s_I} \quad (6)$$

Resolution of accelerations

$$a_{s_I} = (a_x \cos \alpha + a_z \sin \alpha) \cos \epsilon_1 - (a_x \sin \alpha - a_z \cos \alpha) \cos \epsilon_2 \quad (7)$$

$$a_{L_I} = -(a_x \sin \alpha - a_z \cos \alpha) \cos \epsilon_3 + (a_x \cos \alpha + a_z \sin \alpha) \cos \epsilon_4$$

where ϵ_1 is the angle between V_I and V_r; ϵ_2, the angle between V_I and L_r; ϵ_3, the angle between L_I and L_r; and ϵ_4, the angle between L_I and V_r. L_I and L_r are the lift vectors perpendicular to V_I and V_r, respectively. Subscript I denotes an inertial reference frame, and subscript r denotes a reference frame anchored in the atmosphere rotating with the planet. Note that ϵ_1 and $\epsilon_3 \approx 0°$, while ϵ_2 and $\epsilon_4 \approx 90°$.

Inertial flight path angle

$$d\gamma_I/dt = -\{[V_I/(R + z)] - (g/V_I)\} \cos \gamma_I + (a_{L_I}/V_I) \quad (8)$$

Gravitational acceleration

$$g = g_0 R^2/(R + z)^2 \quad (9)$$

Altitude

$$dz/dt = -V_I \sin \gamma_I \quad (10)$$

Downrange angle

$$d\delta/dt = (V_I \cos \gamma_I)/(R + z) \quad (11)$$

Relative velocity

$$V_r^2 = V_I^2 + V_a^2 - 2V_I V_a \cos \gamma_I \sin \lambda_I \quad (12)$$

Velocity of atmosphere rotating with planet

$$V_a = (R + z)\, \omega \cos \Phi \tag{13}$$

Hydrostatic equilibrium

$$dp/dt = g\rho V_I \sin \gamma_I \tag{14}$$

Here, a_{s_I} is the acceleration in the V_I direction, R is the local radius of the planet (measured to the reference ellipsoid), γ_I is the flight path angle below horizontal, α is the angle of attack, δ is the range angle subtended at the planet center from entry to time t, λ_I is the trajectory heading angle ($0°$ is due north), ω is planetary rotation rate ($7.08821765 \times 10^{-5}$ rad/s), and Φ is latitude.

The data analysis was programed for computer solution. The acceleration and altitude data were processed by the program to yield the trajectory and the properties of the atmosphere. The iterative procedure used follows.

1. First approximations of the initial conditions (inertial velocity, flight path angle, altitude, latitude, and heading angle) are made on the basis of the expected, or mission nominal, trajectory. These conditions are for a selected arbitrary time prior to encountering the sensible atmosphere.

2. A table of altitudes measured by the radar altimeter is stored in memory as a function of time.

3. Differential equations (6) and (8) are solved simultaneously, supported by auxiliary equations (7) and (9). Altitudes are calculated for each of the times specified in step 2. (Note that in (7), a_x and a_z are the measured accelerations and that the angle of attack, α, is calculated from their ratio through the experimentally determined lift and drag coefficients.)

4. When a specified final altitude is reached, a differential corrections process [*Chapman and Kirk*, 1970] is applied to yield corrections to the initial conditions. This involves solving another set of differential equations in which partial derivatives of velocity, flight path angle, etc., with respect to the initial conditions, are determined. Given a new set of initial conditions, the process is repeated until a best fit in a least squares sense is obtained to the altitude data. Normally three or four iterations lead to convergence.

5. Relative velocity is determined as a function of time from (12), and the density is then determined from (3). Pressure and temperature are computed from the assumption of hydrostatic equilibrium (equation (14)) and from the gas law. The mean molecular weight used for the mixed lower atmosphere in this step was 43.49, corresponding to 2.7 mol % N_2, 1.6 mol % Ar, and 0.15 mol % O_2.

APPENDIX B: DRAG COEFFICIENT DETERMINATION

Continuum flow drag coefficients were measured in a series of experiments conducted at Ames Research Center in a ballistic range with CO_2 as the test medium [*Intrieri et al.*, 1977; P. F. Intrieri, personal communication, 1976]. The tests were designed to define the effects of gas composition, angle of attack, Reynolds number, and velocity on drag coefficient. Accurately made scale models of the Viking entry configuration were fired from light-gas guns through an atmosphere of carbon dioxide in an enclosed test range, at velocities and Reynolds numbers selected to lie along the Viking entry trajectory. Reynolds number was controlled by selection of gas pressure within the test range. Drag coefficients were determined from precise measurements of the model retardation in flight (see, for example, *Canning et al.* [1970]). In addition, the lift, static and dynamic stability, and trim angle of attack were determined.

A shadowgraph picture of one of the models in flight is reproduced in Figure 16 to show the flow configuration typical of the high entry speeds and attitude. An example of the drag data is given in Figure 17. It shows the accuracy of the data to be within 1%. The effect of angle of attack is well defined. The comparison with air data shows a 4% effect on C_D of the change in gas composition at the flight attitude. Test data for CO_2, slightly contaminated by air leakage, were used in reducing the Viking flight data. A more complete report of these tests is planned.

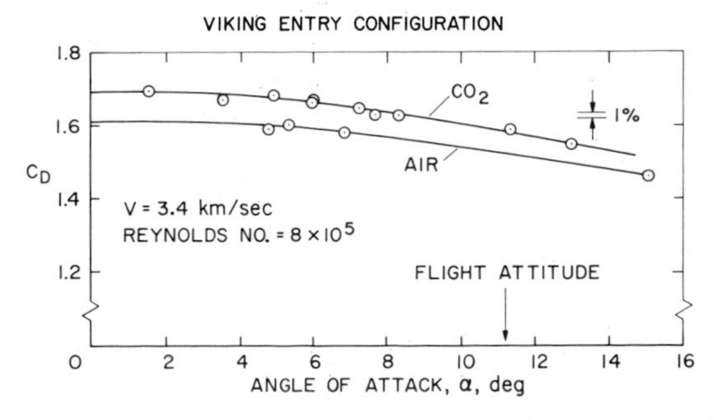

Fig. 17. Representative data from laboratory measurements of drag coefficient.

Acknowledgments. The authors are indebted to many people who, over the years of preparation of this experiment, contributed significantly to its development, implementation, and analysis. Unfortunately, we cannot acknowledge all who contributed here, but in particular we wish to thank Robert C. Blanchard of the Langley Research Center Viking Project Office (VPO) and Fred Hopper of Martin Marietta Corporation (MMC) for invaluable contributions to the trajectory reconstruction effort; Simon C. Sommer and David E. Reese of Ames Research Center (ARC) for major contributions to the demonstration of the concept; Robert Corridan and Carlton S. James (ARC) for laboratory studies of the instruments; and Roy Duckett and Brooks Drew (VPO) and Sid Cook, Tony Knight, and Ernie Carlston (MMC) for contributions to the experiment integration in the spacecraft and operations planning; and finally, Roland Dupree (Kennedy Space Center) and Sally Rogallo (ARC) for invaluable assistance in the computer analysis of the flight data.

REFERENCES

Blumsack, S. L., P. J. Gierasch, and W. R. Wessel, An analytical and numerical study of the Martian planetary boundary layer over slopes, *J. Atmos. Sci.*, *30*, 66–82, 1973.

Canning, T. N., A. Seiff, and C. S. James (Eds.), Ballistic range technology, *Rep. 138*, Adv. Group for Aeronaut. Res. and Develop., Brussels, 1970.

Chapman, G. T., and D. B. Kirk, A method for extracting aerodynamic coefficients from free flight data, *AIAA J.*, *8*, 753–758, 1970.

Christensen, E. J., Martian topography derived from occultation, radar, spectral, and optical measurements, *J. Geophys. Res.*, *80*, 2909–2913, 1975.

Gierasch, P. J., and R. M. Goody, An approximate calculation of radiative heating and radiative equilibrium in the Martian atmosphere, *Planet. Space Sci.*, *15*, 1465–1477, 1967.

Gierasch, P. J., and R. M. Goody, A study of the thermal and dynamical structure of the Martian lower atmosphere, *Planet. Space Sci.*, *16*, 615–646, 1968.

Gierasch, P. J., and R. M. Goody, The effect of dust on the temperature of the Martian atmosphere, *J. Atmos. Sci.*, *29*, 400–402, 1971.

Hanel, R., B. Conrath, W. Hovis, V. Kunde, P. Lowman, W. Maguire, J. Pearl, J. Pirraglia, C. Prabhakara, and B. Schlachman, Investigation of the Martian environment by infrared spectroscopy on Mariner 9, *Icarus*, *17*, 423–442, 1972.

Hanson, W. B., S. Sanatani, and D. R. Zuccaro, The Martian ionosphere as observed by the Viking retarding potential analyzers, *J. Geophys. Res.*, *82*, this issue, 1977.

Hess, S. L., R. M. Henry, C. B. Leovy, J. A. Ryan, J. E. Tillman, T. E. Chamberlain, H. L. Cole, R. G. Dutton, G. C. Greene, W. E. Simon, and J. L. Mitchell, Preliminary meteorological results on Mars from the Viking 1 lander, *Science*, *193*, 788–791, 1976*a*.

Hess, S. L., R. M. Henry, C. B. Leovy, J. A. Ryan, J. E. Tillman, T. E. Chamberlain, H. L. Cole, R. G. Dutton, G. C. Greene, W. E. Simon, and J. L. Mitchell, Mars climatology from Viking 1 after 20 sols, *Science, 194,* 78–81, 1976*b*.

Intrieri, P. F., C. E. De Rose, and D. B. Kirk, Flight characteristics of probes in the atmospheres of Mars, Venus, and the outer planets, *Acta Astronaut.*, in press, 1977.

Kerzhanovich, V. V., Mars 6: Improved analysis of the descent module measurements, *Icarus, 30,* 1–25, 1977.

Kliore, A. J., D. L. Cain, G. Fjeldbo, B. L. Seidel, and M. J. Sykes, The atmosphere of Mars from Mariner 9 radio occultation measurements, *Icarus, 17,* 484–516, 1972.

Masson, D. J., D. N. Morris, and D. E. Bloxsom, Measurements of sphere drag from hypersonic continuum to free-molecule flows, in *Rarified Gas Dynamics,* edited by L. Talbot, p. 643, Academic, New York, 1961.

Michael, W. H., Jr., A. P. Mayo, W. T. Blackshear, R. H. Tolson, G. M. Kelly, J. P. Brenkle, D. L. Cain, G. Fjeldbo, D. N. Sweetnam, R. B. Goldstein, P. E. MacNeil, R. D. Reasenberg, I. I. Shapiro, T. I. S. Boak, M. D. Grossi, and C. H. Tang, Mars dynamics, atmospheric, and surface properties: Determination from Viking tracking data, *Science, 194,* 1337–1338, 1976.

Mutch, T. A., A. B. Binder, F. O. Huck, E. C. Levinthal, S. Liebes, Jr., E. C. Morris, W. R. Patterson, J. B. Pollack, C. Sagan, and G. R. Taylor, The surface of Mars: The view from the Viking 1 lander, *Science, 193,* 791–801, 1976.

Nier, A. O., and M. B. McElroy, Structure of the neutral upper atmosphere of Mars: Results from Viking 1 and Viking 2, *Science, 194,* 1298–1300, 1976.

Nier, A. O., and M. B. McElroy, Composition and structure of Mars upper atmosphere: Results from the neutral mass spectrometers on Vikings 1 and 2, *J. Geophys. Res., 82,* this issue, 1977.

Nier, A. O., W. B. Hanson, M. B. McElroy, A. Seiff, and N. W. Spencer, Entry science experiments for Viking, 1975, *Icarus, 16,* 74–91, 1972.

Nier, A. O., W. B. Hanson, A. Seiff, M. B. McElroy, N. W. Spencer, R. J. Duckett, T. C. D. Knight, and W. S. Cook, Composition and structure of the Martian atmosphere: Preliminary results from Viking 1, *Science, 193,* 786–788, 1976.

Owen, T., and K. Biemann, Composition of the atmosphere at the surface of Mars: Detection of argon 36 and preliminary analysis, *Science, 193,* 801–803, 1976.

Seiff, A., Some possibilities for determining the characteristics of the atmospheres of Mars and Venus from gas dynamic behavior of a probe vehicle, *NASA Tech. Note, D-1770,* 1963.

Seiff, A., The Viking atmosphere structure experiment—Techniques, instruments, and expected accuracies, *Space Sci. Instrum., 2,* 381–423, 1976.

Seiff, A., and D. B. Kirk, Structure of Mars' atmosphere up to 100 km from the entry measurements of Viking 2, *Science, 194,* 1300–1303, 1976.

Seiff, A., D. E. Reese, S. C. Sommer, D. B. Kirk, E. E. Whiting, and H. B. Niemann, Paet, An entry probe experiment in the earth's atmosphere, *Icarus, 18,* 525–563, 1973.

Stone, P. H., A simplified radiative-dynamical model for the static stability of rotating atmospheres, *J. Atmos. Sci., 29,* 405–418, 1972.

Zurek, R. W., Diurnal tide in the Martian atmosphere, *J. Atmos. Sci., 33,* 321–337, 1976.

(Received April 13, 1977;
revised June 6, 1977;
accepted June 6, 1977.)

VOL. 82, NO. 28 JOURNAL OF GEOPHYSICAL RESEARCH SEPTEMBER 30, 1977

Photochemistry and Evolution of Mars' Atmosphere: A Viking Perspective

MICHAEL B. MCELROY, TEN YING KONG, AND YUK LING YUNG

Center for Earth and Planetary Physics, Harvard University
Cambridge, Massachusetts 02138

Viking measurements of the Martian upper atmosphere indicate thermospheric temperatures below 200°K, temperatures much colder than those implied by remote sensing experiments on Mariner 6, 7, and 9 and Mars 3. The variability in thermospheric temperature may reflect an important dynamical coupling of upper and lower regions of the Martian atmosphere. Absorption of extreme ultraviolet solar radiation can account for observed features of the ionosphere and provides an important source of fast N and O atoms which may escape the planet's gravitational field. Isotopic measurements of oxygen and nitrogen impose useful constraints on models for planetary evolution. It appears that the abundance of N_2 in Mars' past atmosphere may have exceeded the abundance of CO_2 in the present atmosphere and that the planet also has copious sources of H_2O. The planet acquired its nitrogen atmosphere early in its history. The degassing rate for nitrogen in the present epoch must be less than the time-averaged degassing rate by at least a factor of 20.

1. INTRODUCTION

The entry science experiments on Viking provide a wealth of new information on the structure and composition of Mars' upper atmosphere. They may be used, in combination with remote sensing data from earlier spacecraft, to develop a reasonably consistent model for Martian aeronomy.

It is clear that escape processes have played a major role in the evolution of Mars' atmosphere. Recombination of O_2^+ in the planetary exosphere provides a significant source for fast atoms which can escape the planet's gravitational field. Compositional data inferred from the retarding potential analyzer experiment [*Nier et al.*, 1976*b*; W. B. Hanson, private communication, 1977] indicate that O_2^+ is the major constituent of the Martian ionosphere and suggest an average escape rate of about 6×10^7 oxygen atoms cm^{-2} s^{-1}. The chemistry of the bulk atmosphere is regulated by oxygen escape on a time scale of the order of 10^5 years in such a manner as to ensure an escape rate for H atoms of the magnitude of 1.2×10^8 atoms cm^{-2} s^{-1}. Hydrogen molecules are formed in the lower atmosphere by reaction of H with HO_2. Escaping H atoms are released by ionospheric reactions involving H_2 and CO_2^+.

Models for the Martian ionosphere are developed in section 2 and are shown to agree satisfactorily with in situ measurements by Viking. The upper atmosphere measured by Viking is unusually cold. The scale height of CO_2 is about 8 km, which may be compared with scale heights in the range 15–22 km as inferred from the ultraviolet spectrometer experiment on Mariner 9 [*Stewart et al.*, 1972]. It is clear that the temperature of Mars' upper atmosphere is quite variable, ranging from as low as 120°K to perhaps as high as 400°K. This variability may be seen also in the topside plasma scale heights as measured by Mariner 4 [*Kliore et al.*, 1965], Mariner 6 and 7 [*Kliore et al.*, 1969; *Fjeldbo et al.*, 1970], and Mariner 9 [*Kliore*, 1974]. The general characteristics of the ionosphere may be reproduced by a relatively simple photochemical model if proper account is taken of the variability of the extreme ultraviolet solar flux. It is unlikely, however, that such a simple model can account for the observed variation in atmospheric temperature.

The photochemistry of Mars' atmosphere is discussed in section 3. In accord with earlier models [*Parkinson and Hunten*, 1972; *McElroy and Donahue*, 1972; *Liu and Donahue*, 1976; *Kong and McElroy*, 1977*a*, *b*] we assume that photolysis of CO_2 is balanced mainly by gas phase reaction of CO with OH. Models are constrained to agree with recent measurements (B. A. Thrush, private communication to R. T. Watson, 1977) of the rate constant for reaction of OH with HO_2 and are adjusted to provide a hydrogen escape flux of 1.2×10^8 atoms cm^{-2} s^{-1} in agreement with fluxes measured by Mariner 9 [*Barth et al.*, 1972]. Results are developed to illustrate possible variations in upper atmospheric composition over a Martian year.

Measurements of the isotopic composition of oxygen and nitrogen may be used to place important constraints on the evolution of Mars' atmosphere. The observed enrichment of ^{15}N relative to ^{14}N for Mars' atmosphere as compared to that of the earth implies that Mars had a much larger nitrogen atmosphere in the past. In a similar manner the lack of a detectable enrichment of the isotopic ratio $^{18}O/^{16}O$ in Mars' atmosphere may be taken to imply significant exchange between the atmosphere and an extensive surface or subsurface reservoir containing a volatile form of oxygen, most probably H_2O. Measurements of noble gases in Mars' atmosphere [*Owen and Biemann*, 1976; *Owen et al.*, 1976] pose additional constraints on planetary evolution, and the implications of these data are explored in section 4.

2. IONOSPHERIC CHEMISTRY AND STRUCTURE

Carbon dioxide is the major constituent of Mars' atmosphere over the ionospherically important height range 120–180 km [*Nier et al.*, 1976*b*; *Nier and McElroy*, 1976, 1977]. Electrons are produced mainly by photo-ionization of CO_2:

$$h\nu + CO_2 \rightarrow CO_2^+ + e \quad (1)$$

The primary photo-ion, CO_2^+, may be removed either by dissociative recombination,

$$CO_2^+ + e \rightarrow CO + O \quad (2)$$

or by reaction with O [*Stewart*, 1972; *McElroy and McConnell*, 1971; *Fehsenfeld et al.*, 1970],

$$CO_2^+ + O \rightarrow O_2^+ + CO \quad (3)$$

Paper number 7S0558.

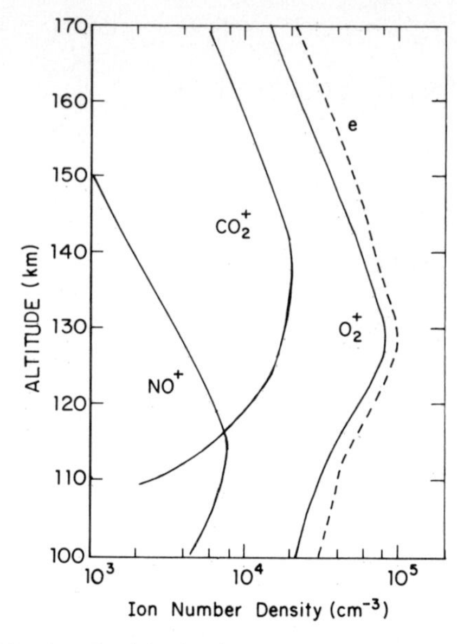

Fig. 1. Number densities for the Martian ionosphere. Results were obtained by using densities for neutral species measured by Viking 1 [*Nier and McElroy*, 1976, 1977]. Corresponding ionospheric reactions are given in Table 1.

Reaction (3) ensures that O_2^+ should be the dominant component of Mars' ionosphere at all heights below about 300 km.

A model ionosphere derived by using densities for neutral species measured by Viking 1 [*Nier and McElroy*, 1976, 1977] is shown in Figure 1. A summary of important reactions and relevant rate constants is given in Table 1. Values for the flux of sunlight at extreme ultraviolet wavelengths were taken from a paper by *Hinteregger* [1976] and reflect recent measurements by the Atmosphere Explorer aeronomy satellites.

The Viking measurements were taken during a period of exceptionally low solar activity. The solar flux at 10.7 cm had a value of 69.4×10^{-22} W m^{-2} Hz^{-1} at 1 AU on July 20, 1976, which may be compared with a flux of 75.7×10^{-22} W m^{-2} Hz^{-1} at 1 AU on September 3, 1976, during the entry of Viking 2, and fluxes in the range (110–145) $\times 10^{-22}$ W m^{-2} Hz^{-1} for the Mariner 9 standard mission (November 14 to December 23, 1971) or (110–169) $\times 10^{-22}$ W m^{-2} Hz^{-1} during the extended mission I (May 7 to June 25, 1972). A limited number of computations were carried out under conditions appropriate for Mariner 9. For the extended mission I in 1972 a Vikinglike thermal structure was used for the middle and lower portions of Mars' atmosphere. For the standard mission a considerably warmer atmosphere was assumed (for altitudes below 100 km) in order to account for the large amount of dust present in the atmosphere at that time [*Kliore et al.*, 1972]. If one defines an effective temperature T_{eff} given by

$$T_{eff} = Z_m \left(\int_0^{Z_m} \frac{dz}{T} \right)^{-1}$$

for the atmosphere below the ionospheric peak Z_m, then the effective temperature (T_{eff}) for the standard mission is about 20°K warmer than the value assumed for the first extended mission.

TABLE 1. Important Ionospheric Reactions in the Martian Ionosphere

Reaction	Rate Coefficient
$CO_2 + h\nu \rightarrow CO_2^+ + e$	$J_1 = 2.4 \times 10^{-7}$
$N_2 + h\nu \rightarrow N_2^+ + e$	$J_2 = 8.7 \times 10^{-8}$
$O + h\nu \rightarrow O^+ + e$	$J_3 = 1.2 \times 10^{-7}$
$CO + h\nu \rightarrow CO^+ + e$	$J_4 = 4.4 \times 10^{-7}$
$CO_2^+ + O \rightarrow CO + O_2^+$	$k_1 = 1.6 \times 10^{-10}$
$CO_2^+ + O \rightarrow CO_2 + O^+$	$k_2 = 1.0 \times 10^{-10}$
$N_2^+ + CO_2 \rightarrow N_2 + CO_2^+$	$k_3 = 9 \times 10^{-10}$
$O^+ + CO_2 \rightarrow CO + O_2^+$	$k_5 = 1 \times 10^{-9}$
$CO^+ + CO_2 \rightarrow CO + CO_2^+$	$k_6 = 1 \times 10^{-9}$
$CO_2^+ + NO \rightarrow CO_2 + NO^+$	$k_7 = 1.2 \times 10^{-10}$
$O_2^+ + NO \rightarrow O_2 + NO^+$	$k_8 = 6.3 \times 10^{-10}$
$CO_2^+ + e \rightarrow CO + O$	$k_9 = 3.8 \times 10^{-7}$
$O_2^+ + e \rightarrow O + O$	$k_{10} = 2.2 \times 10^{-7}(300/T_e)$
$NO^+ + e \rightarrow N + O$	$k_{11} = 4.3 \times 10^{-7}(300/T_e)^{0.37}$

T_e denotes electron temperature, taken to be equal to the value for neutral temperature in the present study. Photodissociation rates (J) at the top of the atmosphere are given in reciprocal seconds. Two-body reaction rates (k) are in cubic centimeters per second. This table is taken from *McElroy et al.* [1976].

An excellent fit to the Mariner data may be obtained if the EUV flux is assumed to vary according to the relation

$$F_{EUV} \quad 0.0222 F_{EUV}^0 (F_{10.7} - 25)/R^2 \tag{4}$$

(F_{EUV}^0 is the solar EUV flux taken from *Hinteregger* [1976], and R (in astronomical units) denotes the planet's radial distance from the sun) and if upper atmospheric temperatures are fixed by using plasma scale heights reported by *Kliore et al.* [1973]. Calculated values for peak electron density are compared with observations in Figure 2. Estimated values for the height of the ionospheric maximum are compared with observations in Figure 3. The variation of EUV flux with solar activity implied by (4) is somewhat larger than the range of intensities observed for dayglow emission in the Cameron bands of CO [*Stewart et al.*, 1972] and is also larger than variations observed by *Hinteregger* [1970] for the chromospheric EUV emission lines at 1025, 977, 630, 584, and 304 Å. The discrepancy may not be serious, however. It may reflect the role of high-excitation solar lines (for example, lines from Fe XV and Fe XVI) and differences in spectral regions important for photo-ionization and excitation of Cameron bands in the Martian atmosphere. We may note that the trend with zenith angle for both peak electron density and peak ionospheric height during the extended mission I is satisfactorily described by the model. Interpretation of the variation of peak

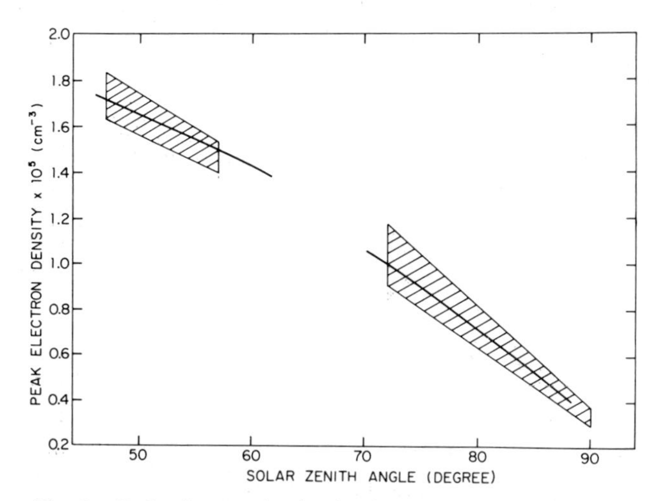

Fig. 2. Peak electron density in the Martian ionosphere versus solar zenith angle. Shaded regions represent data obtained by the Mariner 9 radio occultation experiment during its standard mission (47°–57°) and extended mission I (>72°) [*Kliore et al.*, 1973]. Solid curves are densities from theoretical computations as described in the text. Solar EUV flux was assumed to vary with $F_{10.7}$, and R according to (4).

altitude with zenith angle during the standard mission is complicated because of temporal variations in the structure of the lower atmosphere which occurred during this period. Mariner 9 arrived at Mars when the planet was enveloped by a global dust storm [*Kliore et al.*, 1972]. The solid curve for the standard mission in Figure 3 assumes that the effective temperature was 20°K warmer than values applicable for the extended mission. The trend of peak altitude with zenith angle differs significantly from the observed trend. We believe, however, that the discrepancy may be attributed to a gradual clearing of dust accompanied by steady cooling of the lower atmosphere. If we assume that the effective temperature declined by 8°K over the first 6 weeks of the Mariner mission, we obtain the trend indicated by the dashed line in Figure 3, a result evidently consistent with observation [*Conrath*, 1975]. The comparisons shown in Figures 2 and 3 indicate that solar EUV radiation may provide the dominant source of ionization for Mars.

It is more difficult to account for the range of values inferred for thermospheric temperature. The intensity of the CO Cameron band emission is nicely correlated with observed variations in the 10.7-cm solar flux. In contrast, the scale height of the Cameron bands exhibits no such correlation [*Stewart et al.*, 1972]. There can be little doubt, however, that the thermospheric temperature varies considerably with time. Figure 4 summarizes the available information, including data from Mariner 4, 6, 7, and 9 [*Kliore et al.*, 1965, 1969; *Fjeldbo et al.*, 1970; *D. E. Anderson and C. W. Hord*, 1971; *Kliore*, 1974; *Stewart et al.*, 1972], Mars 3 [*Izakov*, 1973], and Viking 1 and 2 [*Nier and McElroy*, 1976; *McElroy et al.*, 1976]. The lowest temperatures were observed for Viking 1 and 2 and Mariner 4. Noting the absence of a correlation of airglow scale height with solar activity, at least over short time periods on Mariner 9, one is tempted to postulate that the temperature of Mars' upper atmosphere may be affected to a considerable extent by processes originating in the lower atmosphere [*Stewart et al.*, 1972]. The Viking data indicate that gravity waves excited in the lower atmosphere may propagate to high altitudes on Mars [*McElroy et al.*, 1976]. These waves may deliver significant amounts of energy to the upper atmosphere and may be responsible, at least in part, for the high-altitude mixing detected by Viking [*McElroy et al.*, 1976; *Nier and McElroy*, 1977]. Excitation of these waves should be favored when Mars is closest to the sun and may be enhanced further by the additional aerosol burden known to be present in the air at that time [*Gierasch and Goody*, 1972]. Further data are clearly required in order to resolve this issue. It must be pointed out, however, that the data in Figure 4 could also be used to argue a gross correlation of thermospheric temperature with solar activity. It would be difficult, though, to account for the range of temperatures exhibited in Figure 4 if EUV solar radiation should be the only important thermospheric heat source [see *Stewart et al.*, 1972].

3. Photochemistry

The chemistry of the bulk Martian atmosphere should be relatively insensitive to short-period changes in the temperature of the upper atmosphere. Carbon monoxide, formed by photodissociation of CO_2, has a time constant of about 3 years, and a somewhat longer time constant applies for O_2, formed mainly by

$$O + OH \rightarrow O_2 + H \tag{5}$$

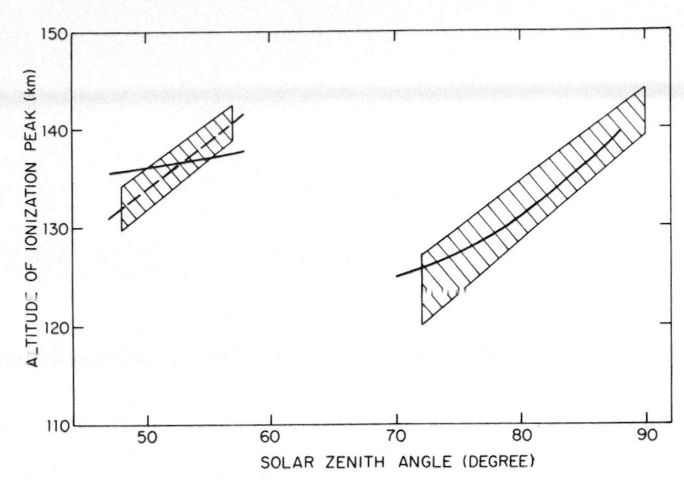

Fig. 3. Altitude of the ionospheric maximum (above a 3-mbar pressure level) versus solar zenith angle. Notation and conditions are the same as those in Figure 2. The solid curve for the standard mission assumes that the effective temperature T_{eff} was 20°K warmer than the value applicable for the extended mission. The dashed curve assumes that the effective temperature declined by 8°K over the first 6 weeks of the Mariner 9 mission.

Molecular hydrogen, produced by

$$H + HO_2 \rightarrow H_2 + O_2 \tag{6}$$

has a time constant of approximately 10^3 years. We shall assume that the chemistry of the bulk atmosphere may be adequately described by using the average model given in Figure 5.

Reactions important for a carbon-hydrogen-oxygen atmosphere are summarized in Table 2. Odd hydrogen compounds are supplied by photochemical decomposition of H_2O, either by photolysis,

$$h\nu + H_2O \rightarrow OH + H \tag{7}$$

or by reaction with $O(^1D)$,

$$O(^1D) + H_2O \rightarrow OH + OH \tag{8}$$

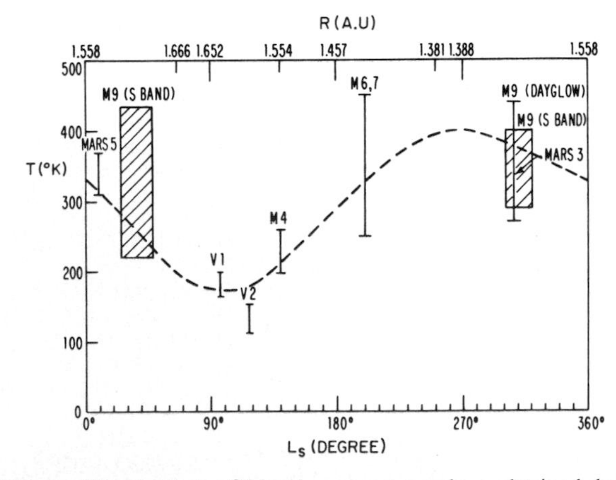

Fig. 4. Temperature of Mars' upper atmosphere obtained during the missions of Mariner 4, 6, 7, and 9, Mars 3 and 5, and Viking 1 and 2. Data (each uncertainty being indicated by an error bar) are plotted against the planetocentric solar longitude of Mars, L_s, and the radial distance from the sun, R, during the period of each mission. Shaded regions represent temperatures derived from topside plasma scale heights obtained by the Mariner 9 radio occultation experiment [*Kliore et al.*, 1973]. Dashed curve indicates a possible seasonal variation for temperature in the Martian upper atmosphere.

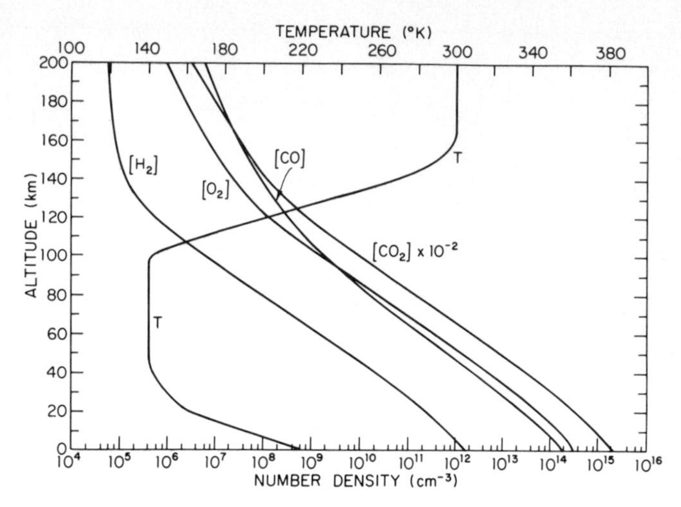

Fig. 5. Time-averaged model for the Martian atmosphere.

Water may be reformed by

$$OH + HO_2 \rightarrow H_2O + O_2 \quad (9)$$

There is, however, a small net sink for H_2O at low altitudes associated with formation of H_2 by reaction (6). Hydrogen is transported upward and escapes mainly in atomic form, atoms being released by reaction of CO_2^+ with H_2:

$$CO_2^+ + H_2 \rightarrow CO_2H^+ + H \quad (10)$$

followed by

$$CO_2H^+ + e \rightarrow CO_2 + H \quad (11)$$

Hydrogen escape is limited by the supply of H_2 from below. Production of H_2 is determined by (6) and is limited by the abundance of tropospheric H atoms. The abundance of tropospheric H is set by balance between

$$CO + OH \rightarrow CO_2 + H \quad (12)$$

and

$$H + O_2 + CO_2 \rightarrow HO_2 + CO_2 \quad (13)$$

and is therefore sensitive to the net oxidation state of the atmosphere. The oxidation state of the atmosphere is determined by the relative magnitude of H and O escape rates. The escape rates are regulated at the present epoch [*McElroy*, 1972; *McElroy and Donahue*, 1972; *McElroy and Kong*, 1976] to ensure a relatively steady oxidation state, H atoms being transported upward as H_2 and O atoms being supplied to the exosphere mainly as components of upward flowing CO_2. The hydrogen escape rate is set ultimately by the rate at which oxygen escapes. The escape rate for O is set by the rate at which CO_2 is photo-ionized in the exosphere and should maintain a relatively steady value over large intervals of geologic time. At the present epoch, water evaporates from the surface, is processed photochemically by the atmosphere, and escapes to space at a steady rate of 6×10^7 molecules $cm^{-2}\ s^{-1}$. If this rate had applied over the past 4.5×10^9 years, Mars would have lost to space an amount of H_2O sufficient to coat the surface of the planet with ice to an average depth of about 2.5 m. It should be emphasized, however, that the net quantity of H_2O processed by the planet could significantly exceed this figure if heterogeneous reactions at the planetary surface might be shown to represent a sink for O of a magnitude comparable to or larger than that due to escape [e.g., *Huguenin*, 1973*a*, *b*, 1974, 1976].

Densities for H, OH, HO_2, H_2O_2, O, and O_3 as calculated for Mars' atmosphere are given in Figure 6. Computational details are described by *Kong and McElroy* [1977*a*]. Rates for

TABLE 2. Chemical Reactions in the Neutral Martian Atmosphere

Reaction No.	Reaction	Rate Expression
(1)	$h\nu + CO_2 \rightarrow CO + O$	
(2)	$CO + O(^3P) + CO_2 \rightarrow CO_2 + CO_2$	2×10^{-37}
(3)	$CO + OH \rightarrow CO_2 + H$	$9 \times 10^{-13} \exp(-500/T)$
(4)	$H + O_2 + CO_2 \rightarrow HO_2 + CO_2$	$2 \times 10^{-31}(T/273)^{-1.3}$
(5)	$O + HO_2 \rightarrow OH + O_2$	7×10^{-11}
(6)	$O + O + CO_2 \rightarrow O_2 + CO_2$	$3 \times 10^{-33}(T/300)^{-2.9}$
(7)	$HO_2 + HO_2 \rightarrow H_2O_2 + O_2$	5.5×10^{-12}
(8)	$h\nu + H_2O_2 \rightarrow OH + OH$	
(9)	$O + OH \rightarrow O_2 + H$	5×10^{-11}
(10)	$h\nu + H_2O \rightarrow OH + H$	
(11)	$O(^1D) + H_2O \rightarrow OH + OH$	3×10^{-10}
(12)	$O(^1D) + H_2 \rightarrow OH + H$	1.9×10^{-10}
(13*a*)	$h\nu + O_3 \rightarrow O_2(^1\Delta_g) + O(^1D)$	
(13*b*)	$h\nu + O_3 \rightarrow O_2 + O(^3P)$	
(14)	$OH + HO_2 \rightarrow H_2O + O_2$	$8.24 \times 10^{-11} \exp(-150/T)$
(15)	$H + HO_2 \rightarrow H_2 + O_2$	$9 \times 10^{-12} \exp(-333/T)$
(16)	$h\nu + O_2 \rightarrow O + O$	
(17)	$O + O_2 + CO_2 \rightarrow O_3 + CO_2$	$1.4 \times 10^{-33}(T/300)^{-2.5}$
(18)	$O + O_3 \rightarrow O_2 + O_2$	$1.32 \times 10^{-11} \exp(-2140/T)$
(19)	$H + O_3 \rightarrow OH + O_2$	2.6×10^{-11}
(20)	$O(^1D) + CO_2 \rightarrow O(^3P) + CO_2$	1.8×10^{-10}
(21)	$O_2(^1\Delta_g) \rightarrow O_2(^3\Sigma_g^-) + h\nu$	2.6×10^{-4}
(22)	$O_2(^1\Delta_g) + CO_2 \rightarrow O_2(^3\Sigma_g^-) + CO_2$	$\{4 \times 10^{-18};\ \leq 8 \times 10^{-20};\ \leq 1.5 \times 10^{-20}\}$

Units for rate constants are s^{-1} for unimolecular reactions, $cm^3\ s^{-1}$ for bimolecular reactions, and $cm^6\ s^{-1}$ for termolecular reactions. Photodissociation rates (s^{-1}) are calculated from solar flux and relevant cross-section data. Rate for reaction (14) is taken from the recent measurement by *Thrush* [1977], and the references for the rest of the reactions are given by *Kong and McElroy* [1977*a*].

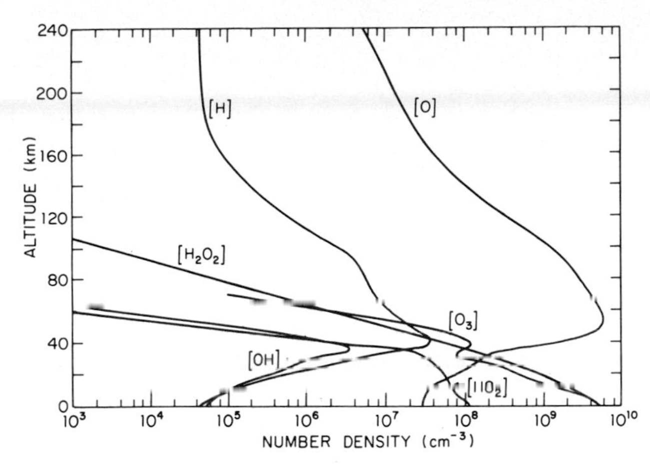

Fig. 6. Density distributions of H, OH, HO_2, H_2O_2, O, and O_3 in the Mars atmosphere. Results were obtained by using the averaged model atmosphere given in Figure 5. Surface was assumed to be inert, and mean water vapor abundance in the atmosphere was taken to be 10 precipitable μm.

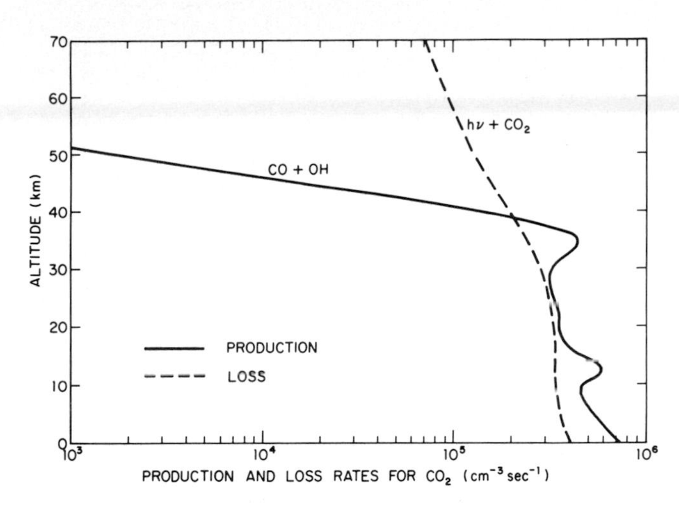

Fig. 8. Production and loss rates for CO in the Martian atmosphere (conditions are the same as those in Figure 6).

reactions contributing to the budget of odd hydrogen are illustrated as functions of altitude in Figure 7. Sources and sinks for CO_2 and O_2 are summarized in Figures 8 and 9, and the variation of upper atmospheric H and H_2 with exospheric temperature is illustrated by Figure 10.

It may be of interest to consider the response of the atmosphere to a large transient source for H_2O, which might arise, for example, during periods of active volcanism. An injection of H_2O at high altitudes could lead to a significant source for H_2, formed directly by photodissociation at wavelengths near Lyman α. The source strength could be as large as 5×10^{10} molecules cm^{-2} s^{-1}. Hydrogen formed in this manner could escape readily to space, leaving oxygen in the atmosphere in concentrations sufficient perhaps even to perturb the total atmospheric pressure. Mars could acquire a relatively long lived transient atmosphere with O_2 as a major constituent. This atmosphere would relax by escape. The associated relaxation time could be relatively long, however, since the O escape rate cannot exceed the value noted earlier, 6×10^7 atoms cm^{-2} s^{-1}

The chemistry of a nitrogen-hydrogen-oxygen-carbon system was investigated recently by *Yung et al.* [1977]. Nitric oxide is formed in the upper atmosphere by reaction of $N(^2D)$ with CO_2,

$$N(^2D) + CO_2 \rightarrow NO + CO \quad (14)$$

and is removed mainly by

$$N(^4S) + NO \rightarrow N_2 + O \quad (15)$$

Odd nitrogen atoms are released by ionospheric reactions and by electron impact dissociation of N_2. A relatively simple model gives results in satisfactory accord with measurements of upper atmospheric NO reported by *Nier et al.* [1976*a*]. Results are shown in Figure 11, which includes several theoretical curves which differ mainly in assumptions made with regard to the yield of $N(^2D)$ in electron impact dissociation of N_2 [*McElroy et al.*, 1976]. Densities for major forms of odd nitrogen in the lower atmosphere are shown [after *Yung et al.*, 1977] in Figure 12. Densities computed for the lower atmosphere are sensitive to assumptions made with regard to the role of surface chemistry. We assumed that HNO_2 and HNO_3 could be incorporated in surface minerals at rates (molecules cm^{-2} s^{-1}) given by γnv, where γ denotes an activity coefficient, n indicates the density (cm^{-3}) of $HNO_2 + HNO_3$, and v is an appropriate thermal velocity (cm s^{-1}). The results in Figure 12 were derived with γ set equal to 10^{-2}.

4. MODELS FOR PLANETARY EVOLUTION

Measurements of isotopic composition may be used to impose important constraints on the range of permissible models

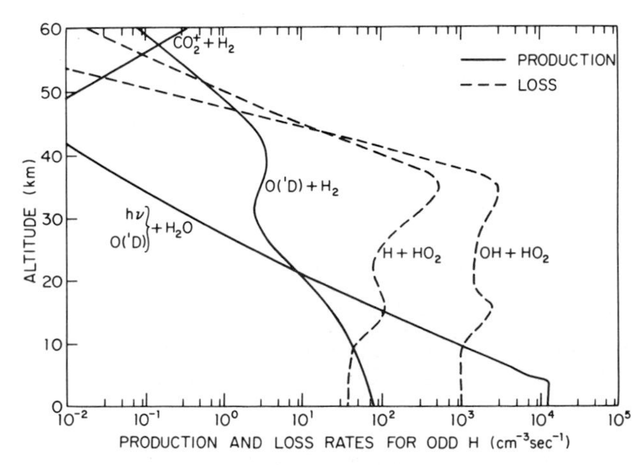

Fig. 7. Production and loss rates for odd H in the lower Martian atmosphere (conditions are the same as those in Figure 6).

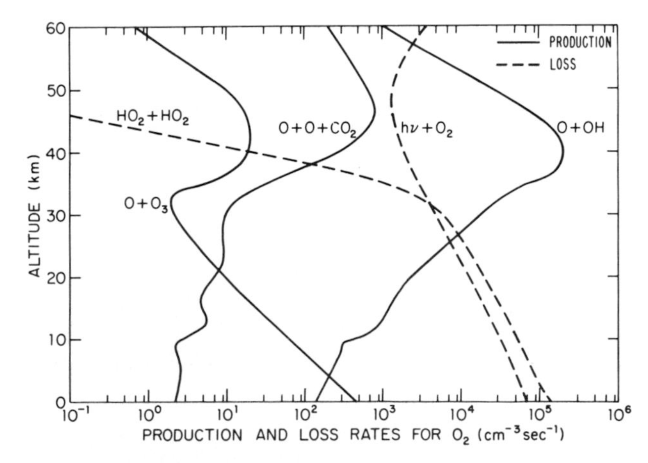

Fig. 9. Production and loss rates for O_2 in the Martian atmosphere (conditions are the same as those in Figure 6).

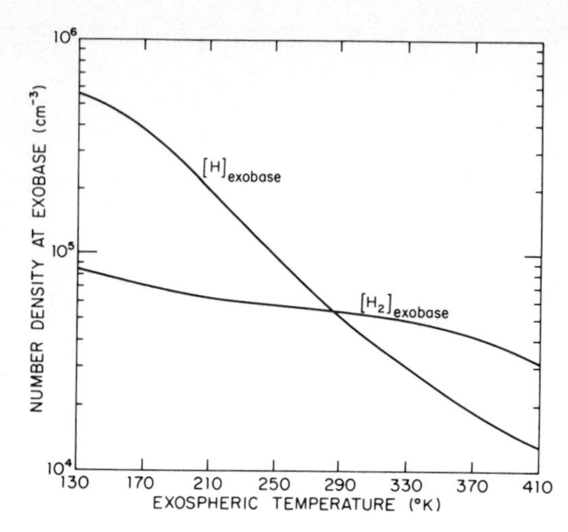

Fig. 10. Density variations of H and H_2 at critical level versus exospheric temperature. Ionospheric structure was assumed to be fixed in the present model.

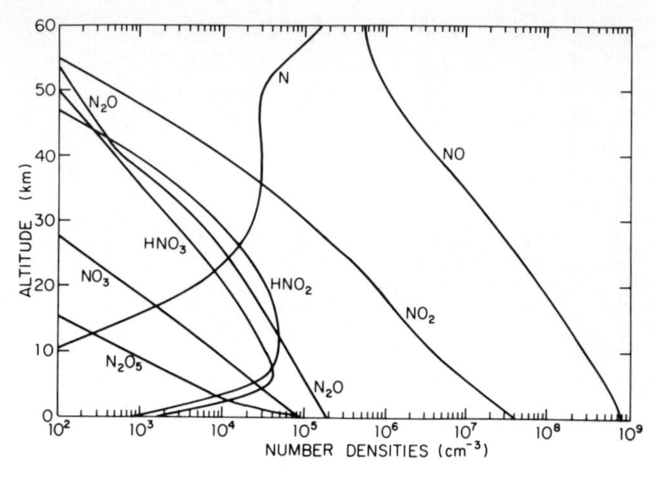

Fig. 12. Number densities of N, NO, NO_2, NO_3, N_2O_5, HNO_2, HNO_3, and N_2O in the lower atmosphere of Mars. The surface reactivity coefficient γ for HNO_2 and HNO_3 equals 1×10^{-2}. Dissociation of N_2 is assumed to proceed through $e + N_2 \rightarrow e + N(^2D) + N(^4S)$. The model atmosphere used is described by *Yung et al.* [1977].

for planetary evolution. Mars has lost significant amounts of oxygen and nitrogen over geologic time. Oxygen escape proceeds by production of fast atoms in the exosphere, atoms being formed primarily by the sequence, reaction (3), or

$$O^+ + CO_2 \rightarrow O_2^+ + CO \quad (16)$$

followed by

$$O_2^+ + e \rightarrow O + O \quad (17)$$

Nitrogen atoms with sufficient velocity for escape may be formed by

$$e + N_2 \rightarrow e + N + N \quad (18)$$

and

$$e + N_2^+ \rightarrow N + N \quad (19)$$

and we may note that these reactions are sufficiently energetic that they should proceed with equal efficiency for all oxygen and nitrogen isotopes present in Mars' exosphere. Carbon atoms may also escape. Fast atoms in this case could be formed by

$$e + CO \rightarrow e + C + O \quad (20)$$

$$e + CO_2 \rightarrow e + C + O_2$$
$$\rightarrow e + C + O + O \quad (21)$$

$$e + CO_2^+ \rightarrow C + O_2 \quad (22)$$

or

$$e + CO^+ \rightarrow C + O \quad (23)$$

Diffusive separation in the Martian thermosphere will result in a preferential supply of light isotopes to the exosphere. The deficiency of the heavier isotopes at the critical level may be characterized by a parameter R defined by the relation [*McElroy and Yung*, 1976]

$$R = f_C/f_0 \quad (24)$$

where f_C denotes the abundance of the heavy relative to the light isotopes at the critical level and f_0 denotes the analogous quantity for the bulk atmosphere. The deficiency parameters R may be readily derived as a function of the eddy diffusion coefficient K, taken to model effects of mixing near the turbopause. Values for R as functions of K are shown in Figure 13. Curve A gives the values of R for ^{18}O relative to ^{16}O; curve B gives similar information for $^{15}N/^{14}N$ and $^{13}C/^{12}C$. Analysis of the Viking data [*McElroy et al.*, 1976; *Nier and McElroy*, 1977] suggests a value for K of about 10^8 cm^2 s^{-1}.

Consider a reservoir which contains an initial concentration of gas $b(0)$ atoms cm^{-2}, of known isotopic composition $f_0(0)$. The concentration of gas in the reservoir at time t, $b(t)$, will vary with time according to the equation

$$db/dt = -(\phi_1 + \phi_2) \quad (25)$$

where ϕ_1 denotes the rate at which gas escapes to space (cm^{-2} s^{-1}), an isotopically dependent quantity as was noted above; ϕ_2 defines the loss rate (cm^{-2} s^{-1}) for all isotopically insensitive removal processes. The time evolution of the bulk isotopic composition is given then by

$$b\, df_0/dt = (1 - R)\phi_1 f_0 \quad (26)$$

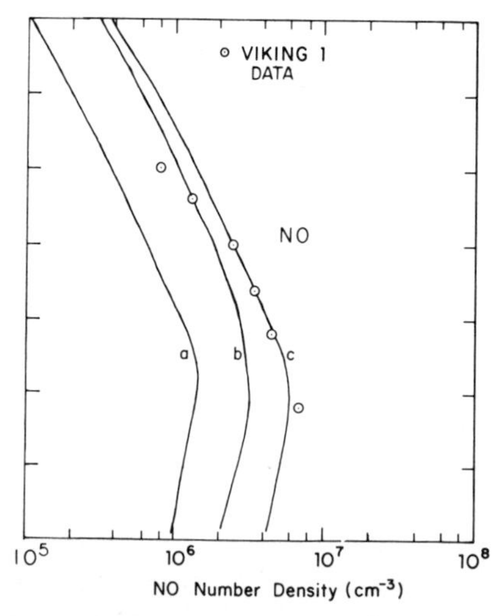

Fig. 11. Comparison of computed and measured number densities of NO in the upper atmosphere of Mars. Curve *a* is obtained by using cross sections for electron impact dissociation of N_2 as measured by Winters with a quantum yield for $N(^2D)$ set equal to 50%. Curves *b* and *c* allow for uncertainties in Winters' cross sections at low energy and in the quantum yield for $N(^2D)$, as described in more detail by *McElroy et al.* [1976].

Consider the application of this simple model to study the time evolution of ^{16}O and ^{18}O. The escape rate for O has a magnitude of 6×10^7 atoms cm^{-2} s^{-1}, as discussed above. The additional flux ϕ_2 may be used to model loss of O due to oxidation of surface rocks. We shall assume that $^{18}O/^{16}O$ is enriched in the present Martian atmosphere by less than 5% [*Nier and McElroy*, 1977] with respect to initial conditions. The manner in which the enrichment of ^{18}O should vary with respect to the initial reservoir size is illustrated in Figure 14. It is clear that Mars' atmosphere must be in contact with a reservoir containing a source of oxygen at least as large as 4.5×10^{25} atoms cm^{-2}. It is probable that this reservoir reflects the presence of a relatively large concentration of atmospherically exchangeable subsurface H_2O [*McElroy and Yung*, 1976].

Viking's measurement of isotopically enriched ^{15}N may be used in a similar manner to place a lower bound on the initial concentration of volatile nitrogen. An enrichment of 1.62 [*Nier and McElroy*, 1977] requires an initial N_2 concentration of no less than 7.8×10^{22} molecules cm^{-2}, equivalent to a partial pressure of 1.3 mbar. The time evolution of Martian nitrogen will be sensitive to surface reactions involving HNO_2 and HNO_3 as described earlier. The enrichment as predicted for ^{15}N will depend on assumptions made with regard to the magnitude of the initial source of volatile N and its isotopic composition, surface reactivity, and escape efficiency. Two possible models are illustrated in Figure 15. The surface reactivity γ (for HNO_2 and HNO_3) was taken to be 3×10^{-2} (case A) and 1×10^{-2} (case B). More detailed discussion of the possible range of values for γ is given by *Yung et al.* [1977]. The present calculations imply an initial N_2 abundance of about 1.7×10^{24} molecules cm^{-2}, equivalent to a partial pressure of 30 mbar. Considerable uncertainty is attached to this value, arising in part from our assumption that the eddy diffusion coefficient K is constant in time and in part from the lack of precision in our estimate for the escape efficiency of N. Our analysis nonetheless provides a reasonable estimate (factor of 3) for the initial nitrogen abundance. One might argue that K is set mainly by dynamical processes controlled by insolation and topography. Uncertainties in the escape rate (and consequently in γ) may be removed by suitable laboratory experimentation.

The observed enrichment of ^{15}N may be used to place limits

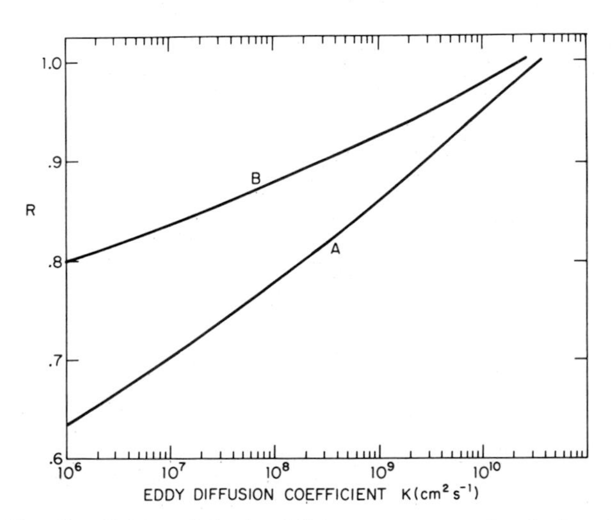

Fig. 13. Values of R, the deficiency parameter for the heavier isotope in the exosphere, as a function of the eddy diffusion coefficient K. Curve A is appropriate for $^{18}O/^{16}O$. Curve B applies to $^{15}N/^{14}N$ and $^{13}C/^{12}C$.

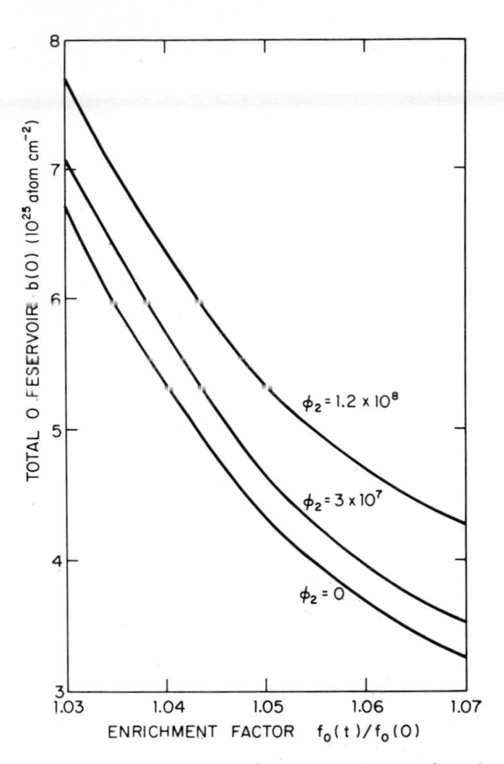

Fig. 14. Total oxygen reservoir at $t = 0$ as a function of the enrichment factor $f_0(t)/f_0(0)$ at $t = 4.5$ b.y. for $\phi_2 = 1.2 \times 10^8$, 3×10^7, and 0 atoms cm^{-2} s^{-1}, respectively.

on the rate at which the atmosphere may gain nitrogen because of slow, steady degassing from the interior. Here, as before, we denote the initial atmospheric abundance of nitrogen by a quantity $b(0)$ atoms cm^{-2}. The time independent steady source

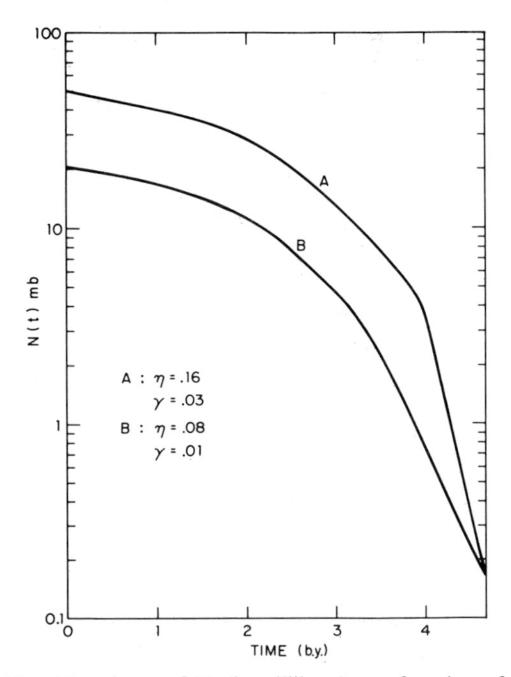

Fig. 15. Abundance of N_2 (in millibars) as a function of time (in units of 10^9 years). The enrichment factor at $t = 4.5$ b.y. is 1.62 for both curves, in accord with measurements. For case A the escape efficiency η for the reaction $e + N_2 \rightarrow e + N + N$ to produce an escaping atom is 0.16, the surface reactivity coefficient γ (for HNO_2 and HNO_3) is 0.03. The corresponding values for η and γ in case B are 0.08 and 0.01, respectively. The eddy diffusion coefficient K used in these calculations equals 1×10^8 cm^2 s^{-1}. This figure is adapted from *McElroy et al.* [1976].

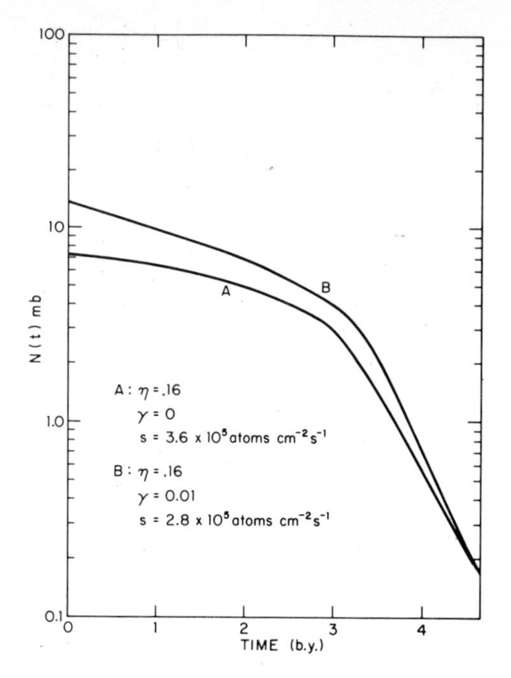

Fig. 16. Abundance of N_2 as a function of time. Enrichment at t = 4.5 b.y. is 1.62; escape efficiency η = 0.16 for both curves. Case A assumes $\gamma = 0$ and $S = 3.6 \times 10^5$ atoms cm^{-2} s^{-1}. Case B assumes γ = 0.01 and $S = 2.8 \times 10^5$ atoms cm^{-2} s^{-1}.

of nitrogen is defined by a quantity S atoms cm^{-2} s^{-1}, and the abundance of atmospheric nitrogen at time t, $b(t)$, satisfies the relation

$$db/dt = -(\phi_1 + \phi_2) + S \tag{27}$$

where the various symbols have the significance discussed earlier. The time evolution of the isotopic composition is given now by

$$b\, df_0/dt = (1 - R)\phi_1 f_0 - (f_0 - f_0^*)S \tag{28}$$

where f_0^* defines the enrichment associated with the source S, assumed to be equal to $f_0(0)$. Results for several combinations of the parameters ϕ_2 and S are shown in Figure 16. It is clear that S cannot exceed about 4×10^5 atoms cm^{-2} s^{-1}, which may be compared with the escape rate in the present atmosphere, estimated at 10^6 atoms cm^{-2} s^{-1}. The escape rate in the past depends on the instantaneous abundance of N_2 and was calculated as described above. It follows that Mars must have acquired its nitrogen atmosphere early in the evolutionary history of the planet. The degassing rate for nitrogen at the present epoch must be less than the average degassing rate over the planet's history by a factor of at least 20. One might suppose that a similar conversion should hold for other volatiles with the exception of radiogenic gases such as ^{40}Ar.

It may be of interest now to examine the possibility of a common origin for the volatile budgets of Mars and earth. A number of papers have appeared in the recent literature attributing planetary volatiles to different classes of chondrites [*Turekian and Clark*, 1975; *Owen et al.*, 1976; *Rasool and Le Sergeant*, 1977]. Information for earth is summarized in Table 3. An acceptable model for Mars requires either selective degassing of nitrogen relative to noble gases or a volatile composition for Mars significantly different from that indicated for earth in Table 3.

Suppose that the initial volatile compositions of earth and Mars were similar. A difference in the formation temperatures of the two planets could lead to differential rates for release of H_2O, CO_2, N_2, ^{36}Ar, Kr, and Xe. Laboratory experiments [*Zahringer*, 1962; *Heymann*, 1971] suggest that noble gases in meteorites are released at relatively elevated temperatures, above about 900°K for ^{36}Ar, Kr, and Xe. Suppose that nitrogen were present in more volatile forms, either as interstitial atoms or molecules or as components of the organic complexes identified in carbonaceous chondrites [*Moore*, 1971]. In this case, Mars could have acquired an early atmosphere rich in H_2O, CO_2, and N_2. Thermal constraints would have limited the gas phase concentrations of H_2O and CO_2. Molecular nitrogen might have accumulated as the major constituent of the atmosphere, while H_2O and CO_2 would have been stored in condensed form in near-surface regions of the planet. An illustrative model for this situation, case A, is included in Table 3. We assume here that most of the noble gases remain trapped by the bulk planet. ^{40}Ar is released at a rate (g/g s^{-1}) less than that appropriate for earth by a factor of 20. The smaller source for ^{40}Ar may reflect either a smaller concentration of crustal ^{40}K on Mars or a slower degassing rate or both. Note that with this model the atmosphere acquires its argon and nitrogen at quite distinct phases of planetary evolution.

TABLE 3. Models for the Evolution of Martian Volatiles

	H_2O	C	N	Ne	^{36}Ar	^{40}Ar	Kr	Xe
Earth*	2.8×10^{-4}	1.7×10^{-5}	7.7×10^{-7}	1.1×10^{-11}	3.5×10^{-11}	1.1×10^{-8}	2.6×10^{-12}	3.6×10^{-13}
Martian atmosphere, present†	$\sim 2.2 \times 10^{-12}$	1.1×10^{-8}	6.2×10^{-10}	$<1.8 \times 10^{-13}$	1.9×10^{-13}	5.7×10^{-10}	2.2×10^{-14}	3.5×10^{-15}
Martian atmosphere and crust‡								
Case A								
Initial	3.0×10^{-5}	2.9×10^{-6}	1.3×10^{-7}	small	small	small	small	small
Present	2.9×10^{-5}	2.9×10^{-6}	2.4×10^{-8}	6.0×10^{-14}	1.9×10^{-13}	5.7×10^{-10}	2.7×10^{-14}	3.5×10^{-15}
Case B								
Initial	3.0×10^{-5}	2.9×10^{-6}	small	6.0×10^{-14}	5.9×10^{-12}	small	2.9×10^{-13}	1.7×10^{-13}
Intermediate	3.0×10^{-5}	2.9×10^{-6}	1.3×10^{-7}	small	small	small	small	small
Present	2.9×10^{-5}	2.9×10^{-6}	2.4×10^{-8}	6.0×10^{-14}	1.9×10^{-13}	5.7×10^{-10}	2.7×10^{-14}	3.5×10^{-15}
Case C								
Initial	3.0×10^{-5}	2.9×10^{-6}	1.3×10^{-7}	6.0×10^{-14}	1.9×10^{-13}	small	2.7×10^{-14}	3.5×10^{-15}
Present	2.9×10^{-5}	2.9×10^{-6}	2.4×10^{-8}	6.0×10^{-14}	1.9×10^{-13}	5.7×10^{-10}	2.7×10^{-14}	3.5×10^{-15}

Units are grams of volatile per gram of total planetary material.
*From *McElroy* [1976] and *Turekian and Clark* [1975].
†From *Nier and McElroy* [1977] and *Owen et al.* [1976].
‡Discussed in this paper.

The nitrogen concentration included in the table reflects our best estimate based on studies of escape as constrained to satisfy the Viking ^{15}N observation. The H_2O and CO_2 concentrations are defined by scaling the terrestrial abundances listed in the first row of the table. An H_2O concentration of magnitude 4.7×10^{-5} g/g would require a planet-wide ice layer of thickness 200 m. Alternatively, it could be accommodated by a crustal material containing 3% H_2O if this crust had an average thickness of 3 km. If one were to adopt the lower bound as discussed earlier for the initial nitrogen concentration, the abundances of H_2O, CO_2, and N_2 could be reduced by a factor of 15.

Case B considers a rather different model in which we assume that nitrogen in the early Martian system may in fact have been less volatile than ^{36}Ar, Kr, and Xe. This situation could have arisen if nitrogen had been present mainly in stable compounds such as sinoite (Si_2N_2O) or osbornite (TiN), as appears to be the case for enstatite chondrites and achondrites [*C. A. Anderson et al.*, 1964; *K. Keil and C. A. Anderson*, 1965; *Bannister*, 1941]. Then devolatilization of the early planet might have favored release of H_2O, CO_2, and noble gases, nitrogen being released subsequently during the period when the planet underwent major differentiation. The concentrations of H_2O and CO_2 in the primitive atmosphere would be limited for case B, as for case A, by thermal constraints, and the planet would have developed an initial atmosphere rich in ^{36}Ar, Kr, and Xe. Noble gases in this system would be exposed directly to the solar wind. Under present solar wind conditions a magnetic field of $\sim 20\ \gamma$ would be required to shield the upper atmosphere of Mars from such an interaction. It is clear that Mars does not possess a magnetic field of this magnitude today, nor is it likely that it ever did. The number of particles that can be swept away by the solar wind is limited by the rate of photo-ionization and by mass loading of the solar wind itself. It may be shown [*Michel*, 1971] that the latter factor determines the maximum escape rate. *Michel* [1971] gives a largely model independent formula for the mass loss rate,

$$dM/dt = 0.86(T/10^3)(40/m) \text{ g/s} \quad (29)$$

where dM/dt denotes the total mass loss rate expressed in grams per second, T denotes the temperature of the exosphere in degrees Kelvin, and m denotes the molecular weight of the escaping gas. Applying Michel's simple formula, we can show that the time required for about 1×10^{20} atoms cm^{-2} of ^{36}Ar to escape is less than 0.5 b.y., and similarly short times would be associated with the removal of Kr and Xe. Light noble gases would be stripped first, followed by the heavier components. There would be an associated loss of CO_2 and H_2O, whose extent would depend fairly critically on the duration of this hypothetical early evolutionary phase. The time interval associated with this phase would be relatively brief if the early sun were more active than is assumed here. Properties of case B are summarized in Table 3.

Case C in Table 3 summarizes a model in which we relax the requirement that volatiles on Mars and earth should have similar compositions. We assume here that all volatile compounds may be released with equal efficiency. The nitrogen model requires that the planet undergo a period of rapid initial degassing. The early atmosphere should be rich in N_2 and noble gases and would be protected from the scavenging effects of the solar wind. Thus we must assume that Mars was assembled from material rich in N_2 relative to noble gases, and the various parameters in the table reflect this view. As was noted earlier, the H_2O concentration for all models must exceed 5×10^{-6} g/g. The concentrations for CO_2 and H_2O in case C were obtained by scaling from N_2 by using terrestrial ratios for $H_2O:CO_2:N_2$.

Model C seems the most plausible. It avoids the ad hoc assumption of time differential degassing for N_2 and noble gases implicit in models A and B. Differences between Mars and earth might be attributed to relatively more extensive degassing in the latter case, degassing extending to greater depths where the material may be deficient in low-temperature condensates. It is interesting to note in this context that the nitrogen to noble gas ratios inferred here for the primitive Mars are similar to values found for a wide class of meteorites [*Gibson*, 1969; *Eugster et al.*, 1969; *Mazor et al.*, 1970; *Heymann*, 1971; *Van Schmus*, 1974]. It suggests that scaling of noble gas abundances from earth to other planets may be hazardous. Measurements of noble gas abundances in the atmosphere of Venus, scheduled for the upcoming Pioneer probe, should provide additional insights. Major uncertainties remain in the interim.

5. Summary

Viking results, in combination with earlier data from Mariner 4, 6, 7, and 9, have been used to develop a relatively comprehensive model of Martian aeronomy. Escape of H, O, and N played an important role in the evolution of Mars' atmosphere. It is not possible, however, to identify a unique combination of parameters to define the initial inventory of Martian volatiles, though the planet appears to have undergone a period of rapid early degassing, at least for N_2. Measurements of noble gases in the present Martian environment are especially puzzling. The possibility that the composition of Mars' atmosphere may have been influenced by interactions with the solar wind over the early stages of planetary evolution should not be ignored. The absence of a significant Martian magnetic field may have contributed to differences in the evolutionary paths of Mars and earth, a possibility which could be illuminated further by mass spectrometric measurements on the scheduled Pioneer mission to Venus.

Acknowledgments. This work was supported by the National Aeronautics and Space Administration under grant NAS-1-10492 and by the National Science Foundation under grant NSF-ATM75-22723, both to Harvard University. We are indebted to E. Anders, T. Owen, and S. Wofsy for illuminating discussions.

References

Anderson, C. A., K. Keil, and B. Mason, Silicon oxynitride: A meteoritic mineral, *Science*, *146*, 256–257, 1964.

Anderson, D. E., and C. W. Hord, Mariner 6 and 7 ultraviolet spectrometer experiment: Analysis of hydrogen Lyman α data, *J. Geophys. Res.*, *76*, 6666, 1971.

Bannister, F. A., Osbornite, meteoritic titanium nitride, *Mineral. Mag.*, *26*, 36–44, 1941.

Barth, C. A., A. I. Stewart, C. W. Hord, and A. L. Lane, Mariner 9 ultraviolet spectrometer experiment: Mars airglow spectroscopy and variations in Lyman α, *Icarus*, *17*, 457, 1972.

Conrath, B. J., Thermal structure of the Martian atmosphere during the dissipation of the dust storm of 1971, *Icarus*, *24*, 36, 1975.

Eugster, O., P. Eberhardt, and J. Geiss, Isotopic analyses of krypton and xenon in fourteen stone meteorites, *J. Geophys. Res.*, *74*, 3874–3896, 1969.

Fehsenfeld, F. C., D. B. Dunkin, and E. E. Ferguson, Rate constants for the reaction of CO_2^+ with O, O_2 and NO^-, N_2^+ with O and NO, and O_2^+ with NO, *Planet. Space Sci.*, *18*, 1267, 1970.

Fjeldbo, G. A., A. Kliore, and B. Seidel, The Mariner 1969 occultation measurements of the upper atmosphere of Mars, *Radio Sci.*, *5*, 381–386, 1970.

Gibson, E. K., Ph.D. thesis, Ariz. State Univ., Tempe, 1969.

Gierasch, P. J., and R. M. Goody, The effect of dust on the temperature of the Martian atmosphere, *J. Atmos. Sci.*, *29*, 400, 1972.

Heymann, D., The inert gases: $He^{(2)}$, $Ne^{(10)}$, $Ar^{(18)}$, $Kr^{(36)}$ and $Xe^{(54)}$, in *Handbook of Elemental Abundances in Meteorites*, edited by B. Mason, chap. 2, pp. 29–66, Gordon and Breach, New York, 1971.

Hinteregger, H. E., The extreme ultraviolet solar spectrum and its variation during a solar cycle, *Ann. Geophys.*, *26*, 547–554, 1970.

Hinteregger, H. E., EUV fluxes in the solar spectrum below 2000Å, *J. Atmos. Terr. Phys.*, *38*, 791–806, 1976.

Huguenin, R. L., Photostimulated oxidation of magnetite, 1, Kinetics and alteration phase identification, *J. Geophys. Res.*, *78*, 8481, 1973*a*.

Huguenin, R. L., Photostimulated oxidation of magnetite, 2, Mechanism, *J. Geophys. Res.*, *78*, 8495, 1973*b*.

Huguenin, R. L., The formation of goethite and hydrated clay minerals on Mars, *J. Geophys. Res.*, *79*, 3895, 1974.

Huguenin, R. L., Surface oxidation: A major sink for water on Mars, *Science*, *192*, 138, 1976.

Izakov, M. N., On the temperature of the Martian upper atmosphere (in Russian), *Kosm. Issled.*, *11*, 761, 1973.

Keil, K., and C. A. Anderson, Occurrences of sinoite in meteorites, *Nature*, *207*, 745, 1965.

Kliore, A. J., Radio occultation exploration of Mars, Exploration of the Planetary System, *IAU Rep. 65*, p. 295, Int. Astron. Union, 1974.

Kliore, A. J., D. L. Cain, G. S. Levy, R. Eshleman, G. A. Fjeldbo, and D. Drake, Occultation experiment: Results of the first direct measurement of Mars' atmosphere and ionosphere, *Science*, *149*, 1243, 1965.

Kliore, A. J., G. Fjeldbo, B. L. Seidel, and S. I. Rasool, Mariner 6 and 7: Radio occultation measurements of the atmosphere of Mars, *Science*, *166*, 1393, 1969.

Kliore, A. J., D. L. Cain, G. Fjeldbo, B. L. Seidel, and M. J. Sykes, The atmosphere of Mars from Mariner 9 radio occultation measurements, *Icarus*, *17*, 484, 1972.

Kliore, A. J., G. Fjeldbo, B. L. Seidel, M. J. Sykes, and P. M. Woiceshyn, *S* band radio occultation measurements of the atmosphere and topography of Mars with Mariner 9: Extended mission coverage of polar and intermediate latitudes, *J. Geophys. Res.*, *78*, 4331–4351, 1973.

Kong, T. Y., and M. B. McElroy, Photochemistry of the Martian atmosphere, *Icarus*, in press, 1977*a*.

Kong, T. Y., and M. B. McElroy, The global distribution of O_3 on Mars, *Planet. Space Sci.*, in press, 1977*b*.

Liu, S., and T. M. Donahue, The regulation of hydrogen and oxygen escape from Mars, *Icarus*, *28*, 231–246, 1976.

Mazor, E., D. Heymann, and E. Anders, Noble gases in carbonaceous chondrites, *Geochim. Cosmochim. Acta*, *34*, 781–824, 1970.

McElroy, M. B., Mars: An evolving atmosphere, *Science*, *175*, 443, 1972.

McElroy, M. B., Chemical processes in the solar system: A kinetic perspective, in *International Review of Science*, vol. 9, *Chemical Kinetics*, edited by D. R. Herschbach, pp. 127–211, Butterworths, Boston, 1976.

McElroy, M. B., and T. M. Donahue, Stability of the Martian atmosphere, *Science*, *177*, 986–988, 1972.

McElroy, M. B., and T. Y. Kong, Oxidation of the Martian surface: Constraints due to chemical processes in the atmosphere, *Geophys. Res. Lett.*, *3*, 569, 1976.

McElroy, M. B., and J. C. McConnell, Dissociation of CO_2 in the Martian atmosphere, *J. Atmos. Sci.*, *28*, 879, 1971.

McElroy, M. B., and Y. L. Yung, Oxygen isotopes in the Martian atmosphere: Implications for the evolution of volatiles, *Planet. Space Sci.*, *24*, 1107–1113, 1976.

McElroy, M. B., T. Y. Kong, Y. L. Yung, and A. O. Nier, Composition and structure of the Martian upper atmosphere: Analysis of results from Viking, *Science*, *194*, 1295–1298, 1976.

Michel, F. C., Solar-wind-induced mass loss from magnetic field free planets, *Planet. Space Sci.*, *19*, 1580, 1971.

Moore, C. B., Nitrogen, in *Handbook of Elemental Abundances in Meteorites*, edited by B. Mason, chap. 7, pp. 93–98, Gordon and Breach, New York, 1971.

Nier, A. O., and M. B. McElroy, Structure of the neutral upper atmosphere of Mars: Results from Viking 1 and Viking 2, *Science*, *194*, 1298–1300, 1976.

Nier, A. O., and M. B. McElroy, Composition and structure of Mars' upper atmosphere: Results from the neutral mass spectrometers on Viking 1 and 2, *J. Geophys. Res.*, *82*, this issue, 1977.

Nier, A. O., M. B. McElroy, and Y. L. Yung, Isotopic composition of the Martian atmosphere, *Science*, *194*, 68–70, 1976*a*.

Nier, A. O., W. B. Hanson, A. Seiff, M. B. McElroy, N. W. Spencer, R. J. Duckett, T. C. D. Knight, and W. S. Cook, Composition and structure of the Martian atmosphere: Preliminary results from Viking 1, *Science*, *193*, 786–788, 1976*b*.

Owen, T., and K. Biemann, Composition of the atmosphere at the surface of Mars: Detection of Ar^{36} and preliminary analysis, *Science*, *193*, 801–803, 1976.

Owen, T., K. Biemann, D. R. Rushneck, J. E. Biller, D. W. Howarth, and A. L. La Fleur, The atmosphere of Mars: Detection of krypton and xenon, *Science*, *194*, 1293, 1976.

Parkinson, T. M., and D. M. Hunten, Spectroscopy and aeronomy of O_2 on Mars, *J. Atmos. Sci.*, *29*, 1380, 1972.

Rasool, S. I., and L. Le Sergeant, Volatile outgassing from earth and Mars: Implications of Viking results, *Nature*, in press, 1977.

Stewart, A. I., Mariner 6 and 7 ultraviolet spectrometer experiment: Implications of the CO_2^+, CO, and O airglow, *J. Geophys. Res.*, *77*, 54, 1972.

Stewart, A. I., C. A. Barth, C. W. Hord, and A. L. Lane, Mariner 9 ultraviolet spectrometer experiment: Structure of Mars' upper atmosphere, *Icarus*, *17*, 469–474, 1972.

Turekian, K. K., and S. P. Clark, The non-homogeneous accumulation model for terrestrial planet formation and the consequence for the atmosphere of Venus, *J. Atmos. Sci.*, *32*, 1257, 1975.

Van Schmus, H., Chemical and detrographic correlations among carbonaceous chondrites, *Geochim. Cosmochim. Acta*, *38*, 47–64, 1974.

Yung, Y. L., D. F. Strobel, T. Y. Kong, and M. B. McElroy, Photochemistry of nitrogen in the Martian atmosphere, *Icarus*, *30*, 26, 1977.

Zahringer, J., Ueber die Uredelgase in den Achondriten kapoeta und staroe pesjanoe, *Geochim. Cosmochim. Acta*, *26*, 665–680, 1962.

(Received April 21, 1977;
revised June 8, 1977;
accepted June 8, 1977.)

The Viking Lander Imaging Investigation: An Introduction

THOMAS A. MUTCH

Department of Geological Sciences, Brown University, Providence, Rhode Island 02912

The nine articles that summarize the results of the Lander Imaging Investigation are briefly reviewed.

The nine articles in this section describe the results of the Viking Lander Imaging Investigation. In the lead article by *Patterson et al.* [1977] the performance characteristics of the cameras are summarized. Calibration procedures are reviewed. The information contained is important for any analyst who anticipates quantitative study of imaging data. The second article by *Levinthal et al.* [1977*a*] reviews the picture inventory. At the time of this writing, more than 1500 pictures have been acquired. Levinthal and his coauthors present essential information regarding the content and processing history for these many images. In the third article, *Huck et al.* [1977] summarize spectrophotometric and color data acquired by lander cameras. Even though some color pictures have been released in accordance with public information requirements, this is the first definitive discussion of the colorimetric properties of the Martian surface. The conclusion of the authors that the soil color is predominantly yellowish brown will be of interest to layman and scientist alike. The fourth article by *Liebes and Schwartz* [1977] describes a little appreciated capability of the lander camera system. Two cameras provide stereoscopic capability. Sophisticated data processing techniques developed by the authors capitalize on these capabilities. Three-dimensional products are visually impressive but also provide important data for morphological interpretation of the Martian scene. The fifth and sixth articles by *Binder et al.* [1977] and *Mutch et al.* [1977] review the geologic characteristics of the two landing sites. The more likely models for the evolution of the observed landscape are critiqued. In the seventh article by *Pollack et al.* [1977] the atmospheric characteristics as determined by the lander cameras are summarized. Diurnal and seasonal ice-crystal fogs have been detected, in addition to a constant haze of fine-grained dust. The mode of data collection, whereby the camera is used as a scanning photometer, underlines the unusual capabilities of this particular imaging device. In the eighth article, *Sagan et al.* [1977] review the conditions required for initiation of particle motion on Mars. In that context they interpret the significance of the generally static landscape monitored by the Viking cameras. Finally, *Levinthal et al.* [1977*b*] summarize the several experimental and analytical techniques employed in an unsuccessful search for biologic forms.

As a convenience to the reader we have indexed in Figure 1 the positions of pictures cited in all nine articles.

During the 6 years of preparation and planning prior to the 1976 landings we discussed at length a strategy for the first few days of the mission. We spent many hours debating the priorities for the first two pictures. Implicit in these deliberations was a concern that the landers might not survive for many days on the surface of Mars. Viewed against this background of cautious planning, the actual results are astounding. According to current plans we will be acquiring pictures through May 1978—perhaps longer. The total number of pictures probably will be more than 2000.

This bonanza of data has proved something of an embarrassment. Confronted with continuing mission operational requirements, we have fallen behind in areas of picture processing and analysis. To give only one example, many picture pairs have yet to be compared in the search for temporal variations.

Because so many data have been received, there are many opportunities for image analysis by scientists not associated with Viking. Partly for this reason we have included here articles which spell out in some detail the way that the camera works and the way that the data are formatted. Our intention is to provide the potential analyst with background information that will permit him to evaluate the raw data in his own research efforts.

Any project such as Viking is a group effort. The contributions of many persons are intricately entwined, and it is virtually impossible to compile a list, however long, of those who contributed to a particular investigation. The pictures taken by lander cameras reflect contributions by many individuals at the Langley Research Center, Martin Marietta Aerospace, the ITEK Corporation, the Image Processing Laboratory of the Jet Propulsion Laboratory, and other private and federal institutions. To all who helped, we are grateful.

REFERENCES

Binder, A. B., R. E. Arvidson, E. A. Guinness, K. L. Jones, E. C. Morris, T. A. Mutch, D. C. Pieri, and C. Sagan, The geology of the Viking Lander 1 site, *J. Geophys. Res.*, *82*(28), this issue, 1977.

Huck, F. O., D. J. Jobson, S. K. Park, S. D. Wall, R. E. Arvidson, W. R. Patterson, and W. B. Benton, Spectrophotometric and color estimates of the Viking lander sites, *J. Geophys. Res.*, *82*(28), this issue, 1977.

Levinthal, E. C., W. Green, K. L. Jones, and R. Tucker, Processing the Viking lander camera data, *J. Geophys. Res.*, *82*(28), this issue, 1977*a*.

Levinthal, E. C., K. L. Jones, P. Fox, and C. Sagan, Lander imaging as a detector of life on Mars, *J. Geophys. Res.*, *82*(28), this issue, 1977*b*.

Liebes, S., Jr., and A. A. Schwartz, Viking '75 Mars lander interactive computerized video stereophotogrammetry, *J. Geophys. Res.*, *82*(28), this issue, 1977.

Mutch, T. A., R. E. Arvidson, A. B. Binder, E. A. Guinness, and E. C. Morris, The geology of the Viking Lander 2 site, *J. Geophys. Res.*, *82*(28), this issue, 1977.

Patterson, W. R., F. O. Huck, S. D. Wall, and M. R. Wolf, Calibration and performance of the Viking lander cameras, *J. Geophys. Res.*, *82*(28), this issue, 1977.

Pollack, J. B., D. Colburn, R. Kahn, J. Hunter, W. Van Camp, C. E. Carlston, and M. R. Wolf, Properties of aerosols in the Martian atmosphere as inferred from Viking lander imaging data, *J. Geophys. Res.*, *82*(28), this issue, 1977.

Sagan, C., P. Fox, R. E. Arvidson, and E. A. Guinness, Particle motion on Mars inferred from the Viking landers, *J. Geophys. Res.*, *82*(28), this issue, 1977.

(Received June 9, 1977;
accepted June 9, 1977.)

Paper number 7S0557.

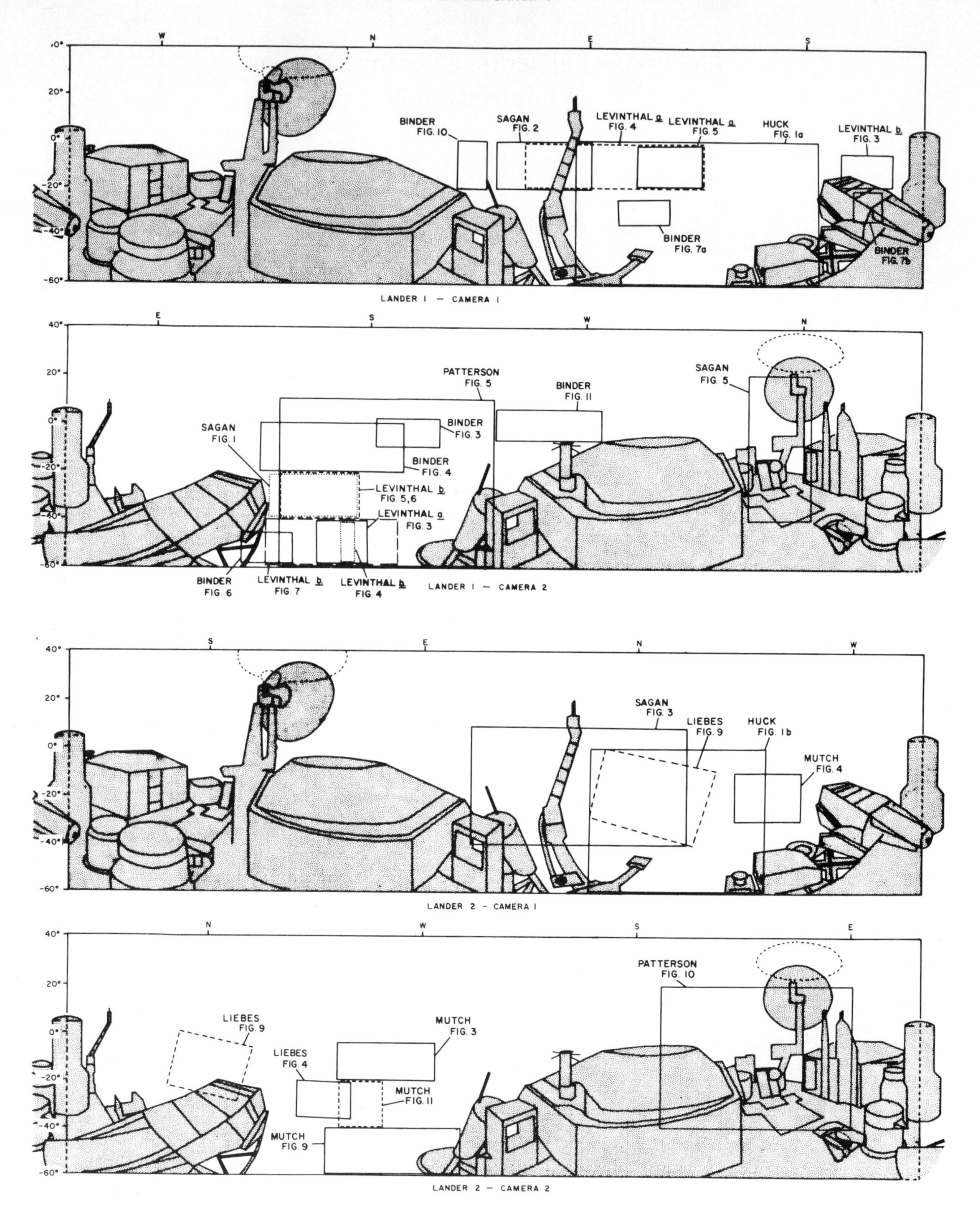

Fig. 1. Skyline drawings for both cameras on both Viking landers showing the locations of pictures used in the nine articles of this issue that describe the Lander Imaging Investigation. A depression angle of 60° corresponds to a slant range of 1.5 m, 40° corresponds to 2.0 m, and 20° corresponds to 4 m.

Calibration and Performance of the Viking Lander Cameras

WILLIAM R. PATTERSON III

Division of Engineering, Brown University, Providence, Rhode Island 02912

F. O. HUCK AND S. D. WALL

Flight Instrument Division, Langley Research Center, Hampton, Virginia 23665

M. R. WOLF

Image Processing Laboratory, Jet Propulsion Laboratory, Pasadena, California 91103

The Viking lander cameras acquire data in six spectral bands for color and near-IR imaging with an angular resolution of 0.12° and for broadband imaging with a resolution of 0.04°. Accurate determination of spatial and spectral brightness variations from image data depends both on calibration and on evidence of calibration stability. A 'rubber sheet transformation' is used to correct camera-dependent geometric errors. Edge matching between pictures taken with different photodiodes and parallax errors less than three picture elements for distant objects indicate that the calibration for this transformation is still valid. The absence of aliasing effects despite undersampling indicates that most surface particles are less than 0.3 mm in diameter. Radiometric measurements have been corrected for significant photodiode degradation due to neutrons from the radioisotopic thermoelectric generators. Vignetting occurs between +25° and +40° elevation, and false 'clouds' caused by light scattered from the camera's outer housing may appear between +12° and +20°. Contouring, most noticeable in uniformly bright areas, is caused by quantization of the photodiode signal.

INTRODUCTION

The cameras on board the two Viking landers are, in effect, radiometers with an optical-mechanical scanning mechanism which provides the line-scan raster and which selects the field of view for a detector array of 12 silicon photodiodes. Because the small number of detectors is particularly stable, the cameras have the potential for radiometric measurements of high relative accuracy and even of fairly good absolute accuracy. Because of the precision servo aiming the scanning mirror the cameras also have the potential to make highly accurate measurements of direction. Since there are two cameras at known positions on each lander, their combined direction measurements make possible stereoscopic mapping of the near field, that is, high-accuracy photogrammetry.

The reasons for the choice of this type of instrument have been discussed by *Mutch et al.* [1972], and the camera design has been described by *Huck et al.* [1975]. The intention of this paper is to describe how closely the cameras fulfilled their potential as quantitative measuring instruments by considering what calibrations were performed, what evidence shows that the calibration has been maintained, and what the consequent limits on the accuracy of each type of measurement are. The first section briefly reviews the camera design. The second section describes the photogrammetric calibration efforts, while the third qualitatively assesses the effects of sampling on the achievable photogrammetric precision. The cameras have six spectrally narrow band detectors having approximately 100-mm resolution spaced over the 400- to 1100-nm wavelength region. The fourth section discusses the preflight spectral calibration and the required corrections for degradation in the infrared response of the detectors due to neutron radiation damage from the radioisotopic thermoelectric power sources on the lander. A fifth section discusses the effects of the known internal reflections on the qualitative images (e.g., the appearance of artifact clouds) and their effects on sky radiometry. The final section deals briefly with the qualitative and quantitative effects of the signal quantization.

Paper number 7S0560.

1. CAMERA DESCRIPTION

Figure 1 shows a cutaway view of the Viking lander camera, while Figure 2 represents schematically the relative placement of the window, scanning mirror, lens, and detector. Light from the scene is reflected off the mirror through an objective lens onto an array of 12 photodiodes, each masked by a small aperture which defines the individual field of view. The mirror rotates about an axis normal to the plane of Figure 2 providing the elevation scan of the camera. Only one of the photodiodes is electrically active during a given scan, and that diode converts the light to an electrical signal which is amplified, sampled, and quantized 512 times during each scan line for digital transmission. After each elevation scan the entire upper elevation assembly rotates a small step about a vertical axis through the center of the camera to provide the azimuthal scan.

The 12 photodiodes are separately selectable and make possible pictures with a variety of characteristics. Four of them have small apertures subtending a solid angle of 0.04° at the lens and are customarily used with 0.04° steps of elevation and azimuth for high-resolution black and white pictures. Since these four are located at different distances from the lens, it is possible to set the focal distance by selecting the appropriate diode electrically; no mechanical focal adjustment is necessary. The other eight diodes have apertures subtending 0.12°. Six of these have interference filters with approximately 100-nm-wide passbands positioned over them. Three filters were selected in the visible color range, and three are in the near infrared. Their relative spectral sensitivity is given in Figure 3 of *Huck et al.* [1977]. Color and infrared pictures are taken by making three successive elevation scans, one with each diode, between each azimuth step. The eleventh diode has no filter and is used for rapid large-area survey scans. The twelfth diode has no ampli-

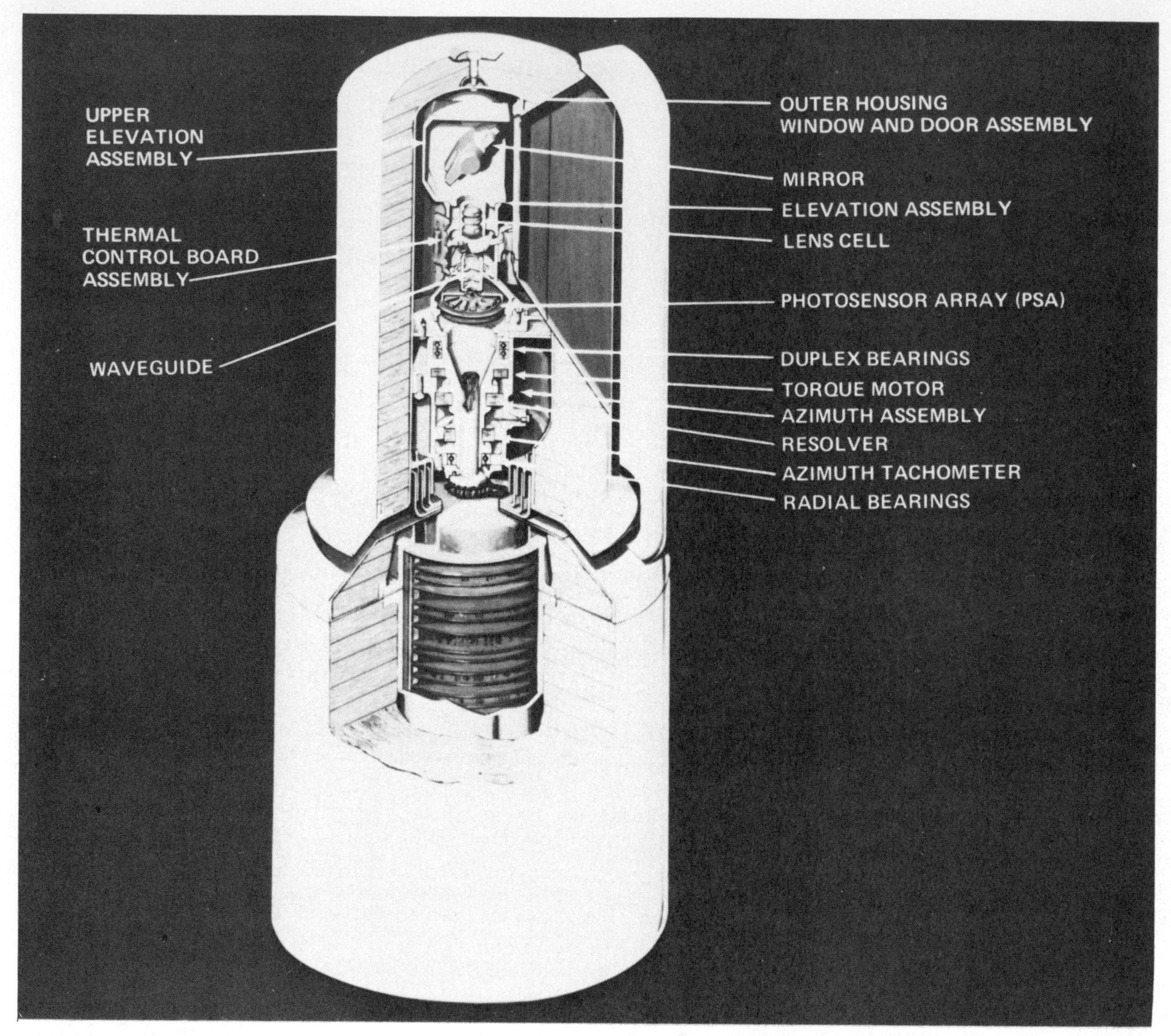

Fig. 1. Cutaway view of the camera.

fication associated with it and is used to view the sun directly through a red filter for atmospheric extinction measurements.

The selection of picture type, time, pointing directions, and scan width can be programed through the guidance control and sequencing computer on the lander. The resulting data are either sent directly by S band microwave link to the earth, stored on tape until a more convenient time for transmission, or sent to a tape recorder on one of the orbiters for relay to the earth over the orbiter's faster radio link. Tables 1 and 2, reproduced from *Huck et al.* [1975], summarize the camera design capabilities.

2. Geometric Calibration

To use Viking lander images for photogrammetric purposes (e.g., isoelevation contouring of the Martian surface, distance determination to discrete object points, etc.), we must be able to infer object space pointing directions from image coordinates. The elevation sampling is discrete and is synchronized to the mirror position through the servo resolver on the mirror shaft. Together with the discrete azimuth steps this results in a digital image of individual picture elements labeled by azimuth line number L and elevation sample number S. If the scanning geometry were ideal, that is, if the mirror rotation axis lay in the reflection plane and if the ray from the diode center through the lens principal points intercepted the mirror center on the axis of rotation, then with thin windows the relation of the object space coordinates to the image coordinates would be:

$$\phi = \phi_0 + R(1 - L) \tag{1}$$

$$\theta = \theta_0 + R(S - 1) \tag{2}$$

where ϕ and θ are the object space azimuth and elevation, respectively, of the object of interest relative to the camera coordinate system origin, ϕ_0 and θ_0 are the object space azimuth and elevation corresponding to the origin of the image coordinate system, R is the angular picture element (pixel) spacing (equal in azimuth and elevation), and L and S are the horizontal and vertical image coordinates, respectively.

The actual camera system geometry shown in Figure 2 differs significantly from the ideal. Note that none of the 12 diodes are on the optical axis of the lens, the rotation axis for

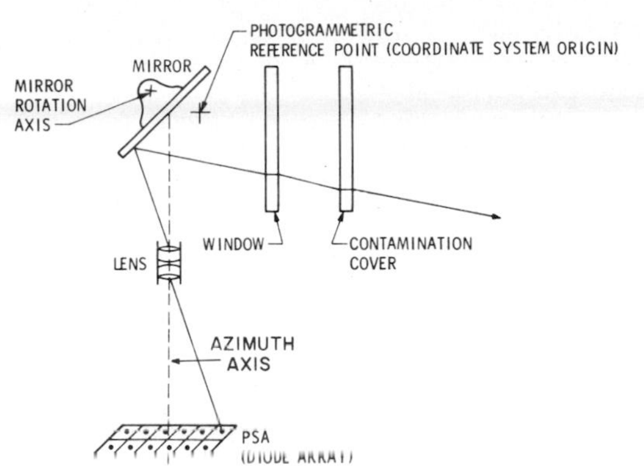

Fig. 2. Schematic representation of the actual relative placement of the elevation scanning components.

the mirror does not lie in the plane of its reflecting surface, and there are one or optionally two windows in the optical path.

The nonideal camera geometry is handled best by a ray trace approach which associates an object space viewing vector with each mirror position and each diode. This is done by tracing a ray backward, starting at the diode aperture and ultimately finding the ray direction and position as it passes through the last optical surface in the optical system. In contrast to ray trace problems which involve systems symmetric about the optical axis the system we are dealing with must be traced in three dimensions instead of two. This is because there is no single plane which contains all segments of the ray as it winds its way through the optical system. To find the new ray direction from the old ray direction as the ray is reflected off a mirror or refracted by an optical element, the appropriate three-dimensional vector equations must be solved at each reflecting surface and at each interface between different indexes of refraction. Thus it must be done either three times (one window in optical path) or five times (two windows in optical path).

The result of the ray trace program is a correction to be added to (1) for azimuth and to (2) for elevation. These corrections are a function of the positions of all the optical elements in the camera, the selected diode, and the elevation that the camera is viewing. The positions of the optical elements were estimated by means of an iterative process, starting from the nominal engineering positions and using the convergence of the pointing directions of the four diodes at the extreme corners of the diode array as the success criterion.

Before launch but after the cameras had been mounted onto the landers, each one was used to image a large (1.2 × 2.4 m) grid target mounted on a magnesium tooling plate. The target intersections were carefully surveyed with respect to the camera coordinate system origin [*Wolf et al.*, 1977]. Although other pointing angle data were taken at the camera assembly level to prove the camera servo performance, these target images were the principal photogrammetric calibration actually applied to the final data. They served as input data for the iterative calculations of corrections to (1) and (2) and for the determination of the constants ϕ_0 and θ_0 in those equations.

To test the ray trace algorithm, a series of pictures of a small survey target was taken at the Jet Propulsion Laboratory (JPL) using all four high-resolution diodes with the target at various elevation angles. Figure 3*a* is one such picture with the target at −58° elevation. In an ideal camera the target would appear in the same place in the image, regardless of the diode used to take the picture. Because of the actual camera geometry the scatter in apparent target position among the four high-resolution images was 0.95° in azimuth and 0.16° in elevation. Figure 3*b*, which is a superposition of images of the same target from all four high-resolution diodes with no ray trace correction, shows this misalignment. After a geometric transformation to align the four images based on a ray trace correction generated from another set of calibration target measurements, the four pictures were again superposed to form Figure 3*c*. Although the lineup is not perfect, it is a substantial improvement over the raw images.

With the ray trace approach we can accurately calculate the object space viewing direction of any image point. Alternatively, we can apply a geometric transform to the data of a type referred to as a 'rubber sheet transformation' so that (1) and (2) hold nearly exactly. In other words, we can transform the image so that it will look as though it had been taken by an ideal camera. Figure 3*c* was generated with this transform. Figures 4*a* and 4*b* show the average of four scans of the survey grid target, taken by the four corner diodes, which were used to iteratively determine the positions of the optical elements in a typical flight camera. Figure 4*a* is the average before the rubber sheet transformation, and Figure 4*b* is the average after the transformation had been applied to all four images. The residual average angular misalignment error was about 0.04° rms over the 300 intersections in a 20° × 60° field of view.

Stereo images of these grid targets were used to determine the absolute *x*, *y*, *z* coordinates to selected intersections. (See also *Liebes and Schwartz* [1977].) Since these intersections had also been surveyed, this gave us an indication of the accuracy of our geometric correction algorithm (ray trace). Errors were generally a few millimeters on intersections that ranged from 2 to 4 m in distance from the cameras.

It is difficult to quantitatively assess the accuracy of the geometric correction algorithm for image data returned from Mars. There is a grid target on top of the lander which was to have been used to compare the prelanded and postlanded performance of the cameras, but it has buckled badly on both landers and thus been made useless for this purpose. However,

TABLE 1. Spatial Characteristics

Characteristics	Survey	Color and IR	High Resolution
Instantaneous field of view, deg	0.12	0.12	0.04
Picture element registration error, deg	±0.036	±0.013	±0.006
Absolute angle error, deg			
Elevation	±0.3	±0.2	±0.2
Azimuth	±0.15	±0.1	±0.1
Frame width, deg			
Elevation	61.44	61.44	20.48
Azimuth	2.5–342.5	2.5–342.5	2.5–342.5
Field of view, deg			
Elevation	100*	100*	100*
Azimuth	342.5†	342.5†	342.5†
Geometric depth of field, m	1.7–∞	1.7–∞	1.7–∞
In-focus distance, m	3.7	3.7	1.9, 2.7, 4.5, 13.3
Picture elements per line	512	512	512
Bits per picture element	6	6	6
Bits per degree azimuth	2.84 × 10⁴	8.53 × 10⁴	8.53 × 10⁴
Time per degree azimuth			
Rapid scan, s	1.84	5.52	5.52
Slow scan, min	2.0	6.0	6.0

*From 40° above to 60° below horizon in 10° steps.
†In multiples of 2.5° steps.

TABLE 2. Radiometric Characteristics

Characteristics	Survey	Color: Blue	Color: Green	Color: Red	Infrared: IR-1	Infrared: IR-2	Infrared: IR-3	High Resolution
Noise-equivalent radiance, W m^{-2} sr^{-1}								
Rapid scan	0.02	0.08	0.08	0.04	0.06	0.07	0.07	0.07
Slow scan	0.008	0.015	0.015	0.01	0.012	0.014	0.014	0.014
Signal-to-noise ratio K/N*	420	200	200	320	270	220	220	220

*For average Mars radiance with rapid scan.

there are two pieces of information that are strong evidence that our geometric camera model is still valid. First, large mosaics have been made of many smaller lander images, taken with several different diodes, to produce large high-resolution panoramas. It is difficult to detect where the edges of the component images are in the mosaics (see Figure 5). In other words, the convergence of pointing direction for the different diodes must still be good, for otherwise there would be discontinuities in the mosaics. Second, the geometrically corrected object space azimuths and elevations of a single very distant point derived independently first from the left camera image and then the right camera image agree to about three pixels or less.

3. Resolution Limitations

Although the results of the photogrammetric calibration and analysis as discussed above suggest that the angle between two pointing vectors can be reliably determined by interpolation to within about 0.06°, the resolution of the size of small objects is further limited by defocus blur and aliasing. Figure 6 shows the effects of distance and focus selection on the estimated smallest resolvable object [*Huck and Wall*, 1976]. Defocus blur has the dual effect of increasing the apparent size of an object while decreasing its contrast to surrounding material.

A more subtle effect to which the Viking cameras are susceptible is aliasing. This is generally less severe in vidicon cameras but may be more severe in solid state array cameras. Aliasing is a phenomenon in sampled data systems which occurs when the scene imaged on the sensors contains frequencies which are higher than one half the sampling rate. If the scene has periodic or quasi-periodic detail smaller than the sampling interval, then the reconstructed image contains false low spatial frequencies which represent interference between the sampling rate and the scene frequencies. This effect is most easily recognized for a single periodic component, as is shown in Figure 7. Increased sampling windows in this example would decrease the magnitude of the false signal but would not eliminate it; the lander cameras have just contiguous sample windows.

Aliasing is not generally a serious problem for an orbiter camera because the Wiener spectrum of natural scenes almost always decreases rapidly with spatial frequency to detail comparable with the grain size, which is far below orbital resolution [*Nill*, 1976]. However, this is not the case at the resolution that can be obtained with the Viking lander cameras. Fine-grained material, either sand or vesicles in rocks, of fairly uniform size distribution near the sampling limit has sufficient high-frequency detail to show serious aliasing effects. For example, test pictures were taken of sand with a size distribution around 1 mm, but in the reconstructed image the distribution appeared centered around 3 mm. Also, during prelaunch testing, pictures were taken at the Great Sand Dunes National Monument, Alamosa, Colorado, where the sand had a particle distribution from about 0.1 to 0.3 mm. The pictures clearly showed structure that is due to aliasing. That the lander pictures from Mars do not show such structure is a probable indication that the Martian surface powder does not have an appreciable size fraction as large as 0.2–0.3 mm.

4. Radiometric Calibration

Radiometric calibration data were acquired for the cameras in three distinct ways. First, the relative radiometric response of the camera as a function of wavelength, temperature, gain, and offset was determined by separate measurements of the photosensor array and of the optical and electronic components. Second, laboratory measurements were made on the

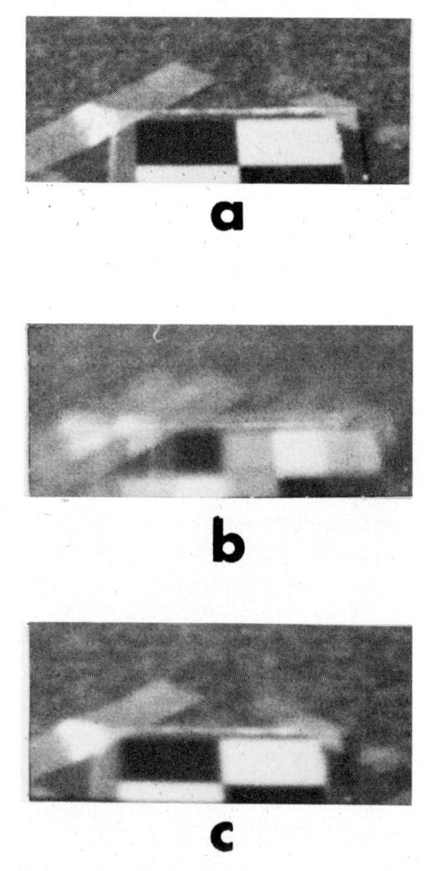

Fig. 3. (*a*) Survey target used in JPL testing of ray trace corrections. (*b*) Superposed images of the target of Figure 3*a* from the four high-resolution diodes without ray trace corrections. (*c*) Same as Figure 3*b* except with ray trace corrections.

absolute radiometric response of the cameras with calibrated light sources and reference reflectances. Third, periodic measurements of the stability of this response were made by using a light source internal to the camera during the year-long journey to Mars and after landing. To account for the small but significant degradation in the infrared sensitivity caused by radiation damage to the detectors, an analytic model of the diode response was developed and fitted to this internal calibration data. As prelaunch calibration activities have been reviewed at length elsewhere [*Huck et al.*, 1974; *Wall et al.*, 1974; *Fleming*, 1974; *Wolf et al.*, 1977], we shall only discuss them briefly.

The relative spectral response of the photosensor array was determined by treating the entire array including window filter, detectors, and amplifiers as a single unit. A monochromator equipped with an $f/22$ exit lens and set for a full width at half maximum bandwidth of 10 nm was used to illuminate alternately one diode in the photosensor array and one reference silicon diode. The reference diode responsivity multiplied by the ratio of the two readings at each wavelength yields the channel responsivity. A second reference diode illuminated by a fraction of the monochromatic beam served to check the stability of the light source between the measurement on the array and the one on the calibration diode. Run to run reproducibility was about $\pm 1\%$. The reference photodiode and preamplifier were calibrated at two laboratories: the Optics Laboratory of the Flight Instrumentation Division, NASA Langley Research Center, Hampton, Virginia, and the Air Force Cambridge Research Laboratory, Bedford, Massachusetts. The two absolute radiometric measurements fell within $\pm 3\%$ of their average value. One problem with this measurement was the use of an $f/22$ converging lens instead of an $f/5.6$ lens similar to that in the actual camera. Because of light scattering effects inside the array [*Huck et al.*, 1975, p. 214] which are dependent on the angle at which the light enters the array this difference in lenses introduces an additional error. This error probably amounts to no more than $\pm 10\%$ at any given wavelength. An experiment was performed in which the photosensor array was exposed to a diffuse large-area broadband light source at different distances, simulating different f number sources. The changes in channel to channel signal ratios between $f/22$ and $f/5.6$ conditions are a measure of the probable integrated error, which was about 6%.

The other optical components which affect the relative spectral response include the outer (contamination cover) and inner windows, the scanning mirror (Al with SiO overcoat), and the lens. Standard measurements of transmission or reflection were performed on all components, but since variations were found to be negligible between samples, a single standard table of these properties was used for all cameras.

A pair of pinlight bulbs in the post on the outside of the camera allowed periodic measurement of the relative transmission of the optical train. Although this measurement could only be performed with one of the unfiltered diodes, it seemed to indicate no appreciable dust accumulation on the contamination cover during the primary mission. Consequently, the contamination cover was retained throughout that time.

There is a small metal frame on the contamination cover edge which causes a diode-dependent vignetting effect at the higher elevation angles. The onset of this effect is at approximately +25°. Since the camera can view up to +40° above the horizontal, this effect affects a significant portion of the camera's field of view. It is especially important in the images of the Martian sky from which scattering models are to be inferred. This problem was not fully appreciated during the camera design, at least in part because the contamination cover design was added after the camera construction had already begun. A preliminary analysis using a flight camera which is now at JPL has been completed using two images per diode, one taken with the contamination cover open and one with it closed, of a white paper backdrop which subtended the entire camera field of view in elevation. The 'window closed' and 'window open' images, after correcting for the dc offset in the camera's electronic system and for the cover transmission, were ratioed. A plot of this vignetting ratio versus elevation is shown in Figure 8 for the blue diode.

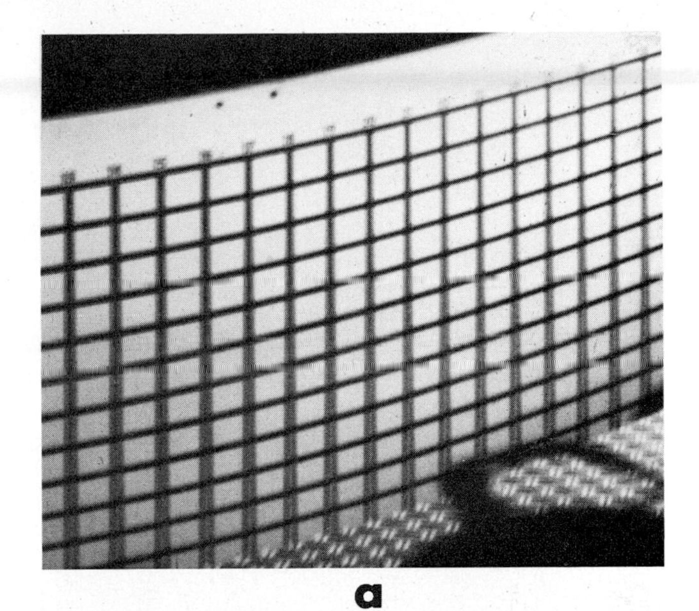

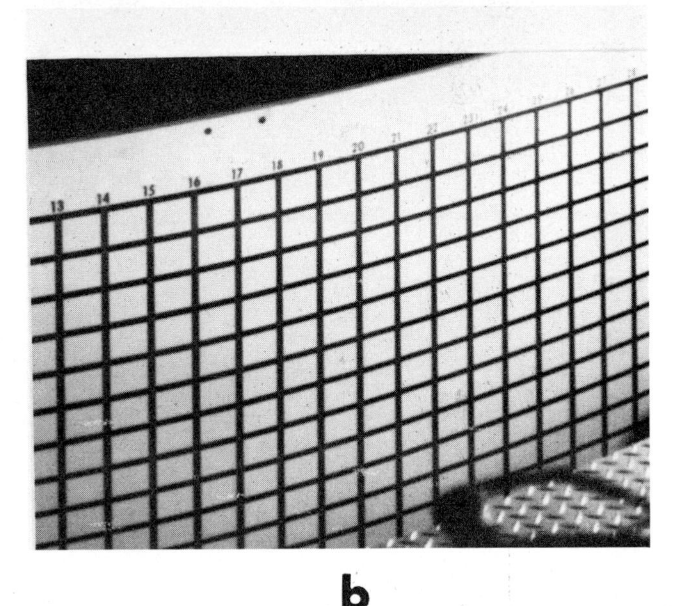

Fig. 4. (*a*) Flight lander picture of the grid target used for principal calibration data. This picture is a superposition of four high-resolution pictures without ray trace corrections. (*b*) Same as Figure 4*a* except with ray trace corrections.

The gain constants k_g, commandable offset constants k_{co} and fixed offset constants k_0 relate the output of the photosensor to the encoded digital output signal as:

$$D_s = (k_g/2^g)(V_p + k_0 - nk_{co})$$

where D_s is the digital output, g is the gain number (0, 1, $\cdots$,

Fig. 5. High-resolution mosaic assembled from separate images taken with the four high-resolution diodes corrected by the ray trace algorithm. The picture joins are nearly invisible, indication of the probable stability of the geometric calibration.

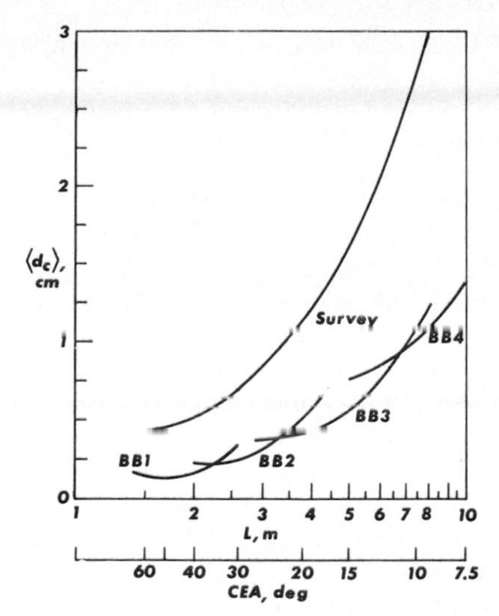

Fig. 6. Smallest linear dimension resolvable by the camera as a function of distance and focus step. Camera elevation angle is shown along the distance scale based on nominal lander orientation parallel to the planar landing surface.

5), V_p is the photosensor voltage, and n is the commanded offset number (ranging from 0 to 31). These constants were determined for each gain and each camera as a function of temperature by introducing a known voltage at the array to A/D interface through a test connector after initial camera tests. The total error in the analog to digital conversion process due to uncertainty or drift in these constants is estimated to be less than ±2% for the small offset ($n \leq 2$) and low gains most commonly used for multispectral imaging.

To improve the absolute radiometric calibration over that determined by component measurements, the camera responsivity was assumed to be that determined by the component tests multiplied by a set of channel- and camera-dependent constants, near unity, determined by end to end system tests in which the camera viewed a broadband source of known radiance. Table 3 gives these constants for the flight cameras. Although the reasons that all the constants are somewhat greater than unity are not well understood, there are three known contributing factors. The first is the choice of lens in the apparatus used to measure the photosensor array as discussed above. The second is that, on the basis of comparing a calculation of the theoretical diode response from a calculation made for estimating radiation damage effects with the measured response, we believe that the sensitivity of the reference photodiode used to calibrate the photosensor array may have been underestimated by a factor of 0.92. This would increase the constants by 8%. The third factor is a small error in the calculation of the expected performance in which the range of wavelength integration in the infrared did not extend quite as far as the array response. This error is probably of the order of 5%.

The radiance source for this test was a reference test chart [see *Huck et al.*, 1977, Figure 5; *Wall et al.*, 1974], identical to the ones carried on the landers for in situ reflectance calibration, which was illuminated by a lamp certified by the National Bureau of Standards.

A special calibration fixture was used to insure that the lamp-to-chart distance and the lighting and viewing angles remain constant for all measurements. The fixture itself is also calibrated to account for peculiarities of lighting geometry and for internal reflections which occurred despite careful baffling. Four major error sources are as follows: lamp irradiance, ±3%; reference test chart reflectances, ±6%; fixture, ±3%; and camera gains and offsets, ±3%. The root sum square error is ±8%.

System-level calibrations were made at the Itek Corporation during the spring of 1974, at the Martin Marietta Corporation (MMC) during the fall of 1974, and at Cape Kennedy during the spring of 1975. Two different calibration fixtures were used, each with its own reference test chart and lamp. Results of all three measurements were in close agreement, well within the estimated ±8% measurement error.

A tungsten filament pinlight mounted inside the camera just above the photosensor array is used as the principal source for verifying the stability of the camera's responsivity. A metal flag moves into the optical path just below the lens to block external light when the lamp is turned on.

The lamp operates at a relatively low absolute temperature (2100°K) to maximize its life. A current-regulated power sup-

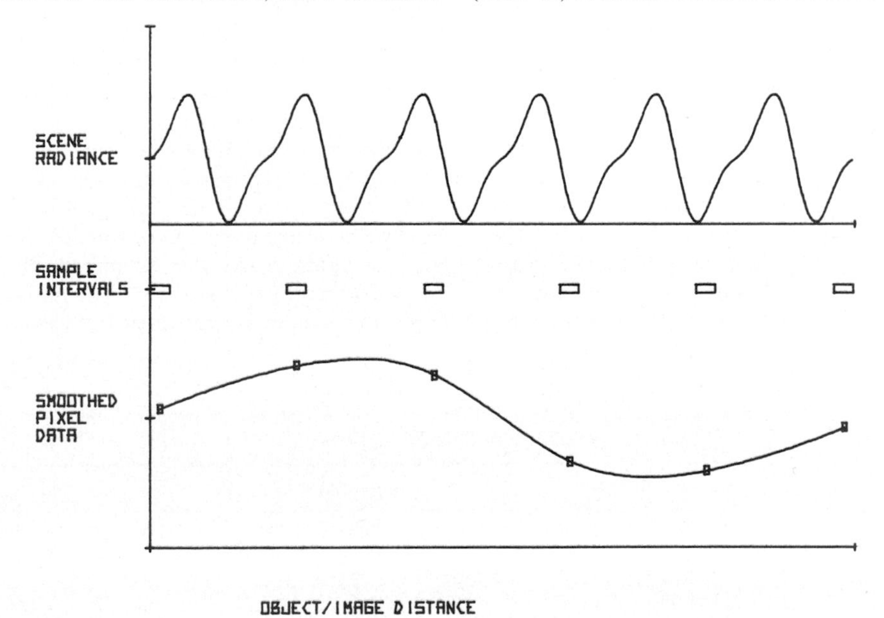

Fig. 7. Origin of aliasing effects. Note the low-frequency component in the smoothed pixel data which is not present in the scene radiance.

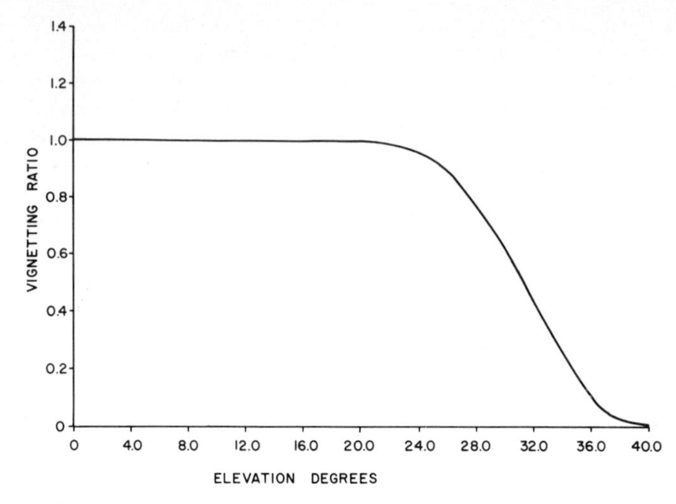

Fig. 8. Vignetting of the blue channel above the horizon caused by the metal frame of the contamination cover. This is typical of an effect present in all channels.

ply sets the lamp current with a short-term stability and reproducibility of about ±0.3%; the constancy of response of those channels which are relatively insensitive to temperature and radiation indicates that the long-term stability has probably been of nearly the same order.

For internal calibration the camera is commanded to 'image' nearly 1 s (four line-scans) with each diode. About 256 pixels are acquired before the lamp is turned on, and another 256 pixels are acquired after the pinlight has reached steady state illumination. Then the lamp is turned off, the next diode selected, and so on, until the integrated response of 11 of the 12 diodes has been measured. (The sun diode, having no preamplifier, shows no appreciable response to the lamp.)

Data are reduced by subtracting the average value of the 256 pixels acquired without illumination from the average value of those acquired with steady state illumination. A first-order estimate of temperature correction was made to all postlanded data by applying a least squares linear fit of diode response versus temperature as measured by a thermistor located on the photosensor array housing. The corrected data are plotted versus time and examined for any remaining trend which might indicate drift or degradation.

Internal calibration data have been recorded for the flight cameras since their manufacture in early 1974. The reduced data showed a tendency toward long-term drift of about 3% upward in all channels during the first 5 months. Thereafter the data showed only random variations of about 1–2% until the landers' heat sterilization in June of 1975. Marked changes (mostly downward) in internal calibration levels were noted after sterilization, and therefore only data acquired after this time were used in preparing a base line for operations on Mars.

A total of 14 internal calibrations were made between sterilization and landing on Mars: 12 were taken before launch, 1 during the voyage, and 1 just prior to separation of lander from orbiter. From 8 to 14 calibrations were performed on each camera during the primary mission period. The reduced data show that while the cameras on both landers have remained radiometrically stable since landing, their integrated responsivity was significantly reduced between launch and landing in the IR-1, -2, and -3 channels. Table 4, part *a*, shows the total variation in the internal calibration data for the blue, green, and red channels. Since these channels are independent of temperature and radiation and since there is no consistent trend to the data, it must be assumed that these numbers represent the long-term camera stability. Part *b* of this table shows the fractional degradation of the internal calibration response attributed to radiation damage from the steady flux of neutron radiation from the radioisotopic thermoelectric generators which were installed just prior to sterilization. Finally, part *c* shows the temperature coefficients of the broadband and infrared diodes as inferred from the postlanded data.

Although the reduction in integrated sensitivity in the infrared channels is small, it was known through tests performed during the camera design activities that this degradation had a marked wavelength dependence. The semiconductor parameter which affects photodiode operation and which is most sensitive to changes due to radiation is the minority carrier diffusion length [*Loferski and Rappaport,* 1958]. Neutrons are known to introduce simple deep-level defects in silicon [*Corbett,* 1966] which may act as recombination centers. Our tests indicated that for the dosage levels expected all the effects might be attributed to the change in diffusion length. Infrared radiation is absorbed in the bulk material behind the actual junction, and the infrared responsivity depends upon the ability of the photogenerated carriers to diffuse back toward the junction. This is why the infrared response and not the visible response is altered by the radiation. Because the dosage level was uncertain and because the net damage depends on the particular fabrication and annealing history of the device, it was difficult to predict the exact effect.

Instead, a simple theoretical model of the photodiode response curve was developed in which the bulk minority carrier diffusion length appeared as an explicit parameter. The other parameters were junction depth, depletion width, chip thickness, surface recombination velocity, and diffused layer carrier diffusion length. Those parameters which were fixed accurately in fabrication (junction depth, depletion width, and chip thickness) were treated as being known. The other parameters were adjusted about reasonable estimated values so as to fit the theoretical response curve to the measured curve for the survey diode. (This diode is fabricated in the same monolithic block

TABLE 3. Ratio of Sensitivity as Measured End to End in Absolute Calibration to Sensitivity Estimated by Subsystem Tests

		Channel						
Lander	Camera	Blue	Green	Red	IR-1	IR-2	IR-3	Survey
1	1	1.24	1.31	1.08	1.28	1.37	1.36	1.18
1	2	1.27	1.37	1.09	1.26	1.35	1.41	1.18
2	1	1.40	1.39	1.27	1.44	1.43	1.42	1.28
2	2	1.20	1.26	1.04	1.22	1.32	1.37	1.14

TABLE 4. Internal Calibration Data

Channel	Viking 1 Camera 1	Viking 1 Camera 2	Viking 2 Camera 1	Viking 2 Camera 2
	*a. Color Channel Variation With Time**			
Blue	10†	2.3	2.2	1.7
Green	4	1.9	1.3	1.8
Red	1.9	1.7	1.6	1.6
	b. Radiation Degradation‡			
BB-1	1.6	2.7	2.6	3.1
BB-2	0	3.1	2.4	2.9
BB-3	0	2.7	2.6	2.0
BB-4	0	2.8	3.2	3.4
IR-1	5.0	7.7	6.7	7.9
IR-2	10.4	13.5	12.5	13.4
IR-3	9.0	11.8	11.1	11.8
	c. Temperature Sensitivity§			
BB-1	0.052	0.066	0.045	0.065
BB-2	0.062	0.052	0.054	0.060
BB-3	0.059	0.055	0.050	0.058
BB-4	0.061	0.064	0.027	0.068
IR-1	0.168	0.187	0.150	0.166
IR-2	0.323	0.280	0.260	0.366
IR-3	0.218	0.228	0.180	0.221

*The difference between the highest and lowest of all poststerilization internal calibrations is shown as a percent of the average level.

†This channel was defective in manufacture. It exhibits noise greater by a factor of 3 than the corresponding channel in other units.

‡In broadband channels. Shown are decreases in internal calibration levels attributable to radiation damage between lander assembly and the middle of the prime mission.

§As determined by postlanding internal calibration. First-order temperature coefficients are shown as percent decrease per Fahrenheit degree. Internal calibration levels generally decrease with increasing temperature.

as the IR diodes.) The details of the model are given in the appendix.[1] Figure 9 shows the resulting fit. The closeness of the fit (no worse than ±3%) increases confidence in both the model and the photosensor calibration. Then the minority carrier diffusion length was adjusted to simulate the degradation of the IR-1 channel, the results of which are also shown in Figure 9. Determining the ratio of the two theoretical curves wavelength by wavelength gave a correction curve which could be applied to the relative responsivity of all the narrowband diodes.

5. RADIOMETRIC ARTIFACTS—CLOUDS

A significant source of radiometric errors can be camera-internal reflections. The only such effect of any significance found to date first appeared in the first panoramic view returned from Mars. This picture showed what appeared to be a toroidal shaped cloud with a break in it in the antisolar direction. Another example of this phenomenon appears as a cloud extending horizontally across even a portion of the microwave antenna in Figure 10. These clouds are artifacts caused by direct sunlight incident on a portion of the camera's outer housing around the dust seal just below the window recess. The light scatters upward off the housing through the window onto the scanning mirror, which reflects it back almost perpendicularly to the window. The window then reflects a portion of the light onto the scanning mirror in such a way that it is reflected down into the lens and onto the detector. The apparent gap in the cloud occurs where the camera shades the strategic part of its housing. Engineering analysis shows that this path becomes significant abruptly as the elevation mirror rises to a pointing angle of 12° above the horizon and decreases slowly to a negligible effect at 22°. Under worst-case conditions the effect amounts to about 3% of the brightness of the camera housing or about 5% of the dynamic range of the camera.

[1]The supplementary appendix and the entire article are available on microfiche. Order from American Geophysical Union, 1909 K Street, N. W., Washington, D. C. 20006. Document J77-006; $1.00. Payment must accompany order.

6. QUANTIZATION EFFECTS

The analog electrical signal from the photosensor array is digitized to 64 linearly spaced gray levels by an A/D converter circuit. In this conversion, six gains can be commanded in steps of powers of 2, and 32 offsets can be commanded in uniform small steps up to nearly the full dynamic range of the camera. This capability provides for adjusting the dynamic range and quantization interval of the digital signal to the radiance variations of the scene. In practice, however, the dynamic range of scene radiance has proven too large to allow effective use of the offsets except for a few special studies of exceptionally dark or bright spots. At high light levels the signal to noise ratio is high, and quantization errors are not reducible below ±0.5 levels by averaging over a number of pixels. At low light levels when the detector noise is comparable with or larger than a change of plus and minus one gray level, it is possible to average over several pixels and reduce errors due to both noise and quantization. The practical limit to this process is set by the size of the uniform areas in the scene and by the converter linearity and does not seem to reduce the uncertainty below ±0.3 gray levels.

Quantization of the signal also causes marked qualitative effects in the pictures in the form of banding or isointensity contours over all relatively bright uniform areas, such as the sky or the lander structure. This is apparent, for example, in Figure 10. The contour lines occur where the light level changes through the critical value for the A/D converter to make a step of one full gray level. The contours are especially noticeable because of the way the eye enhances edge effects. Where two uniform fields of different brightness meet a sharp

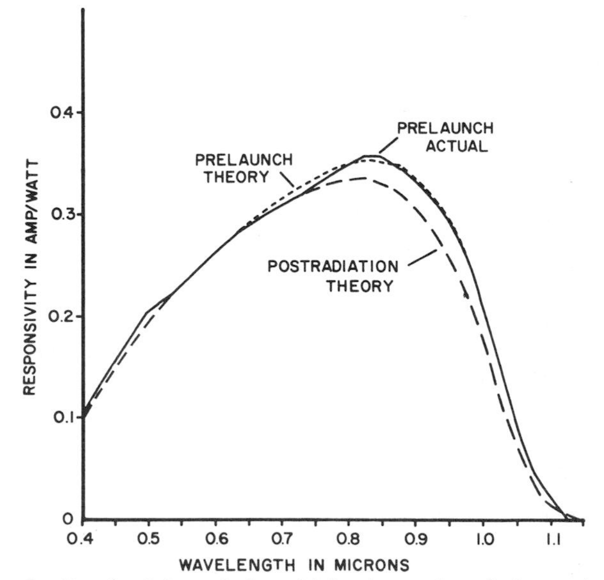

Fig. 9. Result of theoretical model fitted to prelaunch data and the estimated change thereto caused by long-term neutron damage.

Fig. 10. Portion of a survey picture from day of landing on lander 2 showing isointensity contouring due to quantization effects both in the uppermost part of the sky and over the lander body. Also note the apparent 'cloud' just above the horizon running parallel to the camera scan and extending even over the microwave antenna. This is an artifact of internal reflections.

line, the eye perceives a band on the brighter side of the line which is brighter than the true brightness. There is a similar band on the darker side which is darker than the true brightness. These are the Mach bands due to lateral inhibition effects in the retina [*Ratliff,* 1965, 1972].

Acknowledgments. The authors would like to thank S. Grenander of Brown University for the analysis of cloud artifacts and Joe Berry and Steven Albers of the Jet Propulsion Laboratory for the mosaic image production. NASA contract NAS-1-9680.

REFERENCES

Corbett, J. W., *Electron Radiation Damage in Semiconductors and Metals, Solid State Physics,* vol. 7, p. 79, Academic, New York, 1966.

Fleming, J., A review of the accuracy of the absolute calibration of the Viking photosensor array, *Contract Rep. MCR 74-165,* Martin Marietta Aerospace Corp., Denver, Colo., 1974.

Huck, F. O., and S. D. Wall, Image quality prediction: An aid to the Viking lander imaging investigation on Mars, *Appl. Opt., 15,* 1748, 1976.

Huck, F. O., E. E. Burcher, E. J. Taylor, and S. D. Wall, Radiometric performance of the Viking Mars lander cameras, *NASA Tech. Memo. TMX-72692,* 1974.

Huck, F. O., H. F. McCall, W. R. Patterson, and G. R. Taylor, The Viking Mars lander camera, *Space Sci. Instrum., 1,* 189–241, 1975.

Huck, F. O., R. E. Arvidson, D. J. Jobson, S. K. Park, W. R. Patterson III, and S. D. Wall, Spectrophotometric and color estimates of the Viking lander sites, *J. Geophys. Res., 82,* this issue, 1977.

Liebes, S., and A. A. Schwartz, Viking 1975 Mars lander interactive computerized video photogrammetry system, *J. Geophys. Res., 82,* this issue, 1977.

Loferski, J. J., and P. Rappaport, Radiation damage in Ge and Si detected by carrier lifetime changes, *Phys. Rev., 111,* 432, 1958.

Mutch, T. A., et al., The Viking lander imaging investigation, *Icarus, 16,* 92–110, 1972.

Nill, N. B., Scene power spectra: The moment as an image quality merit factor, *Appl. Opt., 15,* 2846, 1976.

Ratliff, F., *Mach Bands: Quantitative Studies on Neural Networks in the Retina,* Holden Day, New York, 1965.

Ratliff, F., Contour and contrast, *Sci. Amer., 226,* 90–103, 1972.

Wall, S. D., E. E. Burcher, and D. J. Jobson, Reflectance characteristics of the Viking lander camera reference test charts, *NASA Tech. Memo. TMX-72762,* 1974.

Wolf, M. R., et al., Viking lander camera radiometric calibration report, special report, Jet Propul. Lab., Pasadena, Calif., 1977.

(Received April 13, 1977;
revised June 17, 1977;
accepted June 17, 1977.)

Spectrophotometric and Color Estimates of the Viking Lander Sites

FRIEDRICH O. HUCK,[1] DANIEL J. JOBSON,[1] STEPHEN K. PARK,[1] STEPHEN D. WALL,[1]
RAYMOND E. ARVIDSON,[2] WILLIAM R. PATTERSON,[3] AND WILLIAM D. BENTON[4]

The spectral radiance and color of the Martian sky and soil and the spectral reflectance of soil features are estimated from six-channel (0.4–1.0 μm) spectral data obtained with the Viking lander cameras. Images taken near local noon from the two landers reveal a sky that is brighter near the horizon than the soil but with a similar spectral radiance shape and color. The scenes are predominantly moderate yellowish brown in color with only subtle variations except for some dark grey rocks. Most spectral reflectance estimates are similar: they rise rapidly with increasing wavelength between 0.4 and 0.8 μm and with only a few exceptions exhibit a pronounced minimum centered about 0.93 μm. These characteristics are consistent with an abundance of Fe^{+3}-rich weathering products, notably nontronite. However, the delineation of the number and abundances of total mineral phases requires further analyses and laboratory comparisons. Reflectance estimates for rocks have not been repeatable, probably because most rocks have irregular pitted surfaces that introduce significant shadowing components.

INTRODUCTION

In this paper we present spectral radiance and color estimates of the Martian soil and sky and spectral reflectance estimates of soil features. The estimates are based on six-channel spectral data obtained with the Viking lander cameras [*Huck et al.*, 1975; *Patterson et al.*, 1977]. The spectral range extends from 0.4 to 1.0 μm, and the spatial resolution ranges from several millimeters near the lander to several meters at the horizon. A special analysis technique is required to account properly for the camera responsivity shapes, which are irregular and exhibit appreciable out-of-band sensitivity. Atmospheric attenuation is approximately accounted for with the aid of a set of reference grey and color patches (referred to as the reference test chart) mounted on each lander. However, our estimates must be regarded as preliminary results until atmospheric attenuation has been more rigorously included in the analysis.

Our estimates are based mostly on data from two multispectral images taken near local noon with one of the two cameras on each lander. Estimates based on image data taken with the other cameras at nearly the same time are shown to be in agreement with these results. Table 1 identifies the data, and Figure 1 shows the locations included in the analysis. We selected these locations with the general objective of characterizing those features that appear to be most prevalent in each scene and a few others that appear to be distinctly different. Rocks are not included because of our inability to obtain reliable and reproducible measurements, probably because most rocks have irregular pitted surfaces that introduce significant shadowing components.

The predominantly yellowish brown color of the two pictures shown in Figure 1 departs appreciably from the reddish brown color of pictures that were released for many months after the landings on Mars. The incorrect reddish renditions resulted from initial attempts to balance pictures by reproducing the color of the reference test charts and of the grey paint and orange cables of the lander without, however, also properly accounting for the camera responsivities (as described in this paper) and the exposure characteristics of a newly developed laser film reproducer.

SPECTRAL RADIANCE

Multispectral samples. The photosensor array output signal for each of the $i = 1, 2, \cdots, 6$ spectral channels is [*Huck and Wall*, 1976]

$$V_i = \pi c_i \int_0^\infty N(\lambda) T_i(\lambda)\, d\lambda \qquad (1)$$

where $N(\lambda)$ is the spectral radiance of the object and c_i and $T_i(\lambda)$ are the camera calibration constants and transfer functions, respectively. The constants c_i are determined from preflight and internal camera calibrations with an absolute accuracy of about ±10% [*Patterson et al.*, 1977]. The functions $T_i(\lambda)$ are given by

$$T_i(\lambda) = t_i^{-1} \tau_c(\lambda) R_i(\lambda) \qquad (2)$$

where $\tau_c(\lambda)$ is the camera optical throughput and $R_i(\lambda)$ the photosensor array responsivity. Typical camera transfer functions are shown in Figure 2 with the scaling constant t_i chosen so that

$$\int_0^\infty T_i(\lambda)\, d\lambda = 1$$

The voltage V_i is amplified, sampled, and quantized to six-bit digital numbers that are transmitted to earth. In the analog-to-digital conversion, six gains can be commanded in steps of powers of 2, and 32 offsets can be commanded in small steps up to nearly the full dynamic range of the camera. This capability provides a means for matching the dynamic range and quantization interval of the digital signal to the predicted radiance variations of the scene.

Formulation of estimate. The camera transfer functions $T_i(\lambda)$ shown in Figure 2 are generally irregular, and several exhibit appreciable out-of-band sensitivity. It is thus inappropriate to generate a data point for each spectral channel by associating a spectral radiance value with a distinct wavelength. Instead, we express the output of each of the six spectral channels as a linear (integral) function of the spectral radiance and camera transfer functions (equation (1)), producing six equations that are used to determine the coefficients in a representation of the spectral radiance as a linear combination of basis (i.e., curve fitting) functions. The spectral radiance

---

[1] Flight Electronics Division and Analysis and Computation Division, NASA Langley Research Center, Hampton, Virginia 23665.

[2] Department of Earth and Planetary Sciences, McDonnell Center for the Space Sciences, Washington University, Saint Louis, Missouri 63160.

[3] Division of Engineering, Brown University, Providence, Rhode Island 92912.

[4] Jet Propulsion Laboratory, Pasadena, California 91103.

Paper number 7S0502.

TABLE 1. Identification of Multispectral Data

Camera Event*	Type of Picture	Camera Gain, Offset	Local Lander Time	Sun Elevation Angle, deg
11A147/026†	color	4,1	11:44	80
11A149/026†	IR	4,1	12:01	80
12A168/028†	color	4,1	11:43	80
12A170/028†	IR	4,1	12:01	80
21A187/023‡	color	4,1	12:15	60
21A188/023‡	IR	4,1	12:22	60
22A190/023‡	color	4,1	12:34	60
22A191/023‡	IR	4,1	12:42	60

*The first number designates the lander; the second, the camera; the third, the frame count; and the fourth, the Martian day after landing.

†Six-channel multispectral data of surface and sky used for the estimates presented in Figures 3*a* and 7*a*; $\tau_a(0.55) = 0.84$. ($\tau_a(0.55)$ is the factor by which the atmosphere attenuates the solar irradiance that reaches the surface at a wavelength of 0.55 μm [from *Pollack et al.*, 1977].)

‡Six-channel multispectral data of surface and sky used for the estimates presented in Figures 3*b* and 7*b*; $\tau_a(0.55) = 0.88$.

estimate $\langle N(\lambda)\rangle$ becomes then [*Park and Huck*, 1976]

$$\langle N(\lambda)\rangle = \sum_{i=1}^{6} b_i f_i(\lambda) \qquad (3)$$

where the constants b_i are the spectral samples

$$b_i = V_i/\pi c_i \qquad (4)$$

and the functions $f_i(\lambda)$ are camera characteristic functions which satisfy the equation

$$\int_0^\infty T_j(\lambda) f_i(\lambda)\, d\lambda = 1 \qquad j = i$$
$$\int_0^\infty T_j(\lambda) f_i(\lambda)\, d\lambda = 0 \qquad j \neq i \qquad (5)$$

In essence, we require the estimated radiance $\langle N(\lambda)\rangle$ to yield the same values b_i as are obtained from the actual spectral radiance $N(\lambda)$. Natural cubic splines with six equally spaced interior knots are used as a basis function for computing $f_i(\lambda)$. The radiance estimates suppress and smooth short-period small-amplitude variations that cannot be reliably estimated because of undersampling inherent in the limited number of spectral channels, without adversely affecting the longer-period variations that are sufficiently sampled. Numerical details of the formulation and general theoretical aspects of the estimation technique have been presented by *Park and Huck* [1976, 1977].

Estimates obtained with this approach are generally as accurate as those that could be obtained with idealized camera transfer functions represented by impulse functions that yield six spectral samples b_i at distinct wavelengths. However, the estimates are inevitably and primarily limited by the few spectral channels. It would therefore be misleading to assign confidence intervals about the estimate $\langle N(\lambda)\rangle$; such intervals can have meaning only if the accuracy of the estimate is limited by the uncertainty of the spectral samples b_i (due to electronic noise and calibration errors) rather than by undersampling.

Results. Figure 3 presents spectral radiance estimates for a patch of sky and soil. The estimates $\langle N(\lambda)\rangle$ vary smoothly with wavelength, without reproducing any of the short-period variations between 0.4 and 0.55 μm that one could expect from the solar irradiance $S(\lambda)$. The small part of the sky that can be observed near the horizon is brighter than the soil at both landing sites, contrary to commonly held expectations before the Viking landings. Otherwise, the shape of the spectral radiances are remarkably similar. Both sky and soil radiances rise rapidly with increasing wavelength from 0.4 μm to a peak between 0.68 and 0.75 μm and then drop again above 0.75 μm. The surface radiances exhibit a slightly more pronounced change of slope around 0.5–0.6 μm and drop more slowly in the near IR.

We will use these and similar results later to compute the color of the sky and soil. It would also be possible to obtain approximate surface spectral reflectance estimates from these results if the surface illumination were accurately known. However, it is more reliable and accurate to estimate surface reflectance properties directly by (1) including the known variations of solar irradiance into the estimation process and (2) using calibration data from a reference test chart to balance the six spectral samples b_i relative to each other.

SPECTRAL REFLECTANCE

Constraints. Atmospheric transmittance and scattering as well as surface scattering are not known accurately at this time, so it is possible only to estimate relative rather than absolute spectral reflectances. Furthermore, we must rely on reference test chart data to adjust the integrated response of camera transfer functions and atmospheric attenuation for the six channels relative to each other. This approach is feasible because the atmospheric attenuation of solar irradiance varies slowly with wavelength in comparison to the spectral sampling intervals [*Pollack et al.*, 1977].

The multispectral images analyzed in this paper were taken near local noon. High sun elevation angles are desirable because (1) the shortest light path through the atmosphere obviously minimizes atmospheric scattering and (2) near-vertical light incidence on the surface minimizes the dependence of surface reflectance on lighting and viewing geometry.

Formulation of estimate. Analogous to (3), the relative spectral reflectance estimate can be formulated as

$$\langle \tilde{\rho}(\lambda)\rangle = \sum_{i=1}^{6} \tilde{b}_i \tilde{f}_i(\lambda) \qquad (6)$$

The spectral samples $\tilde{b}_i$ are obtained as the ratio

$$\tilde{b}_i = V_i/\tilde{c}_i \qquad (7)$$

where the illumination-camera calibration constants $\tilde{c}_i$ are intended to account for the spectral variation of atmospheric attenuation. That is, if the atmosphere attenuates the solar irradiance that reaches the surface in channel i by $\tau_{a,i}$, then $\tilde{c}_i = \tau_{a,i} c_i$. The corresponding illumination-camera characteristic functions $\tilde{f}_i(\lambda)$ must satisfy the equation

$$\int_0^\infty S(\lambda) T_j(\lambda) \tilde{f}_i(\lambda)\, d\lambda = 1 \qquad j = i$$
$$\int_0^\infty S(\lambda) T_j(\lambda) \tilde{f}_i(\lambda)\, d\lambda = 0 \qquad j \neq i \qquad (8)$$

where $S(\lambda)$ is the solar irradiance of Mars.

To test the estimation technique, the spectral samples corresponding to a known reflectance $\tilde{\rho}(\lambda)$ can be computed as

$$\tilde{b}_i = \int_0^\infty S(\lambda) \tilde{\rho}(\lambda) T_i(\lambda)\, d\lambda \qquad (9)$$

in which case the illumination $S(\lambda)$ and transfer functions $T_i(\lambda)$ are assumed to be exactly known. Figure 4 presents several estimates for this case. The reflectances in Figures 4*a*

(a) Lander 1

(b) Lander 2

Fig. 1. View of Lander 1 and 2 scene (camera events 11A147/026 and 21A187/023). The color pictures for this print were processed to represent our estimated hue and chroma but not value (darkness/lightness). The reflectance of the two scenes is generally less than 0.15, which would result in a very dark print with little dynamic range available for contrast differences. The value has been enhanced therefore to take full advantage of the available dynamic range for color prints, which generally allows reflectances up to about 0.7. The color of the two lander sites is virtually the same; the color difference between the two pictures is mostly the result of unintentional variations in film processing. Table 3 presents CIE, Munsell, and NBS-ISCC designations for the estimated colors.

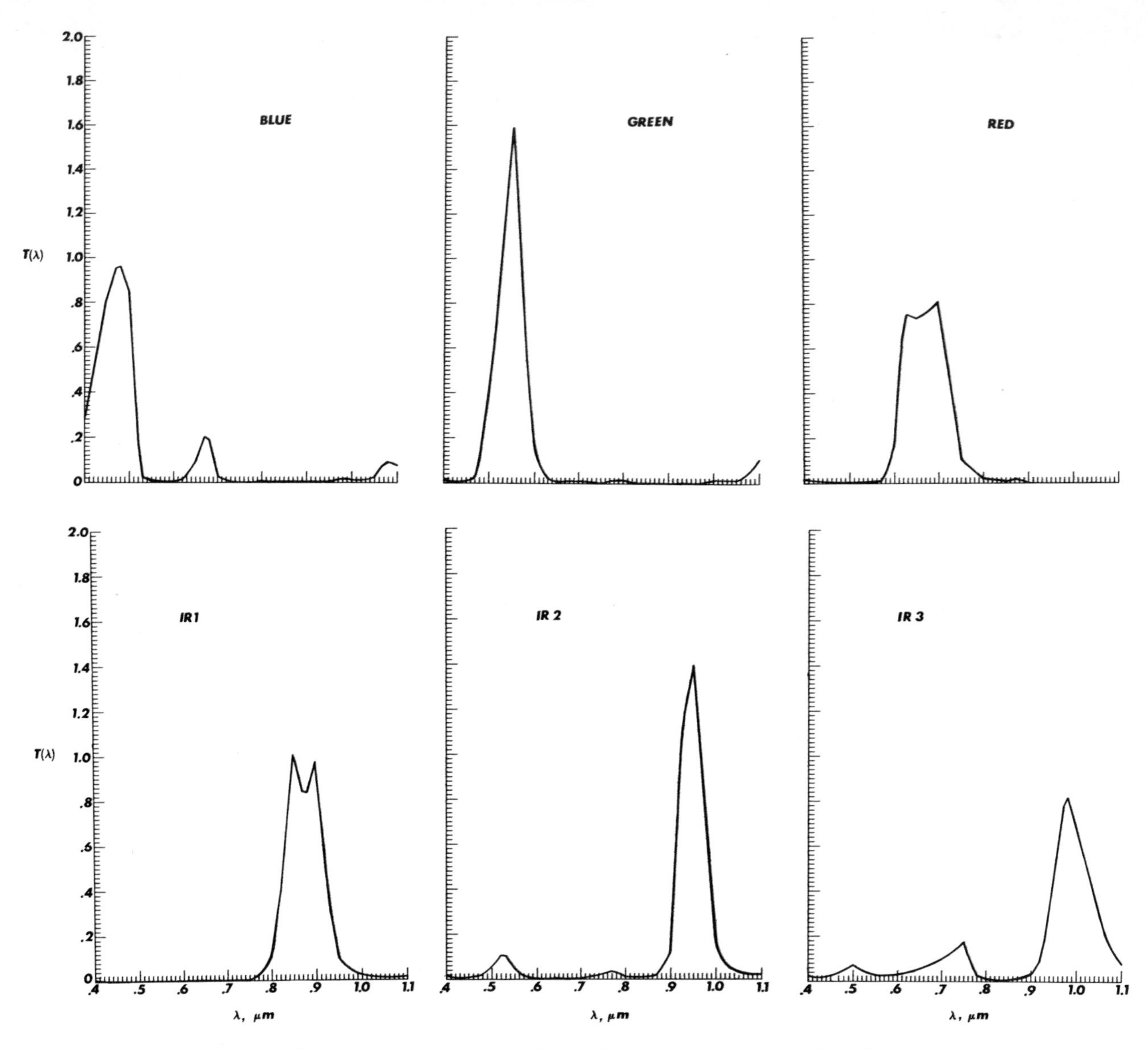

Fig. 2. Typical camera transfer functions for the six spectral channels.

and 4*b* are representative of dark and bright areas on Mars, respectively, as obtained by *McCord et al.* [1977] in 1973 with a 25-channel (0.33–1.1 μm) spectrophotometer attached to a 230-cm telescope, and the reflectances in Figures 4*c* and 4*d* are representative of hypersthene and limonite, respectively, which are thought to be among the possible candidates for mineral phases present on Mars. The dominant reflectance features are correctly reproduced; only the small-period features are lost. In particular, the very subtle near-IR reflectance minima in McCord et al.'s data are either missed completely (Figure 4*a*) or slightly misplaced (Figure 4*b*). But the 0.92-μm Fe^{+2} band in hypersthene (Figure 4*c*) is quite accurately estimated.

Calibration. Each camera can view two of the three identical reference test charts that are mounted on each lander. The chart (see Figure 5) consists of 11 grey and three color patches for radiometric calibration and three tribars for checking the camera frequency response. The geometric reflectance properties are approximately Lambertian (within ±3%) for incident angles ι between 10° and 70°; that is, the reflectance of each patch is $\rho_r(\lambda)$ cos ι for 10° $\leq \iota \leq$ 70°, where $\rho_r(\lambda)$ is the spectral reflectance of the patch measured in relation to the standard white diffuse surface of magnesium oxide [*Wall et al.*, 1975].

The reference test charts cannot be used for absolute calibrations because the exact amount of light falling onto the charts is difficult to determine. Total lighting includes contributions from direct sunlight as well as from light scattered by the atmosphere and reflected off the lander structure. The latter two contributions may be especially significant near noon, when the direct sunlight is grazingly incident on the charts, which are mounted nearly vertical to face directly toward the cameras. Furthermore, the incidence angle of the direct sunlight onto the charts can be determined with an accuracy of only a few degrees, so that the corresponding error in estimating the direct sunlight contribution may be quite large.

However, even though we cannot determine the absolute magnitude of the reference test chart illumination, we can assume that the spectral variation of the illumination is approximately the same as that for the surface, since the lander structure is grey. The calibration constants $\tilde{c}_i$ and multispectral signals V_i contain therefore approximately the same

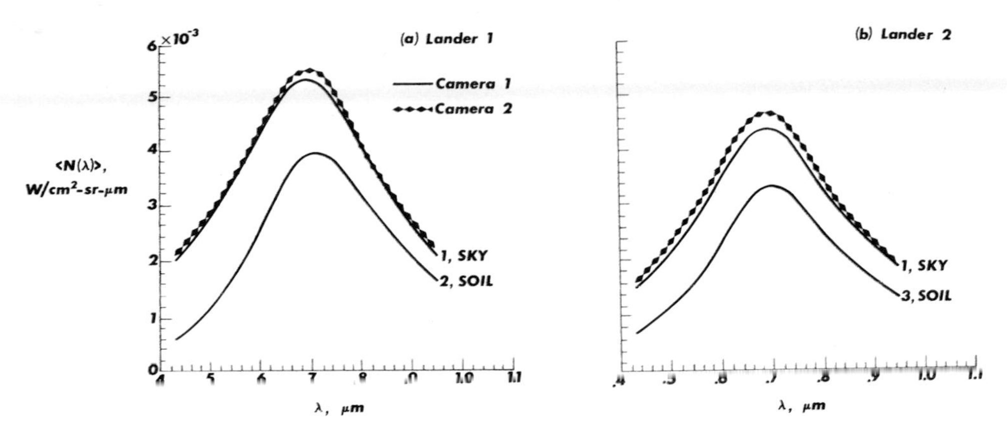

Fig. 3. Spectral radiance estimates of a patch of sky and soil. The sky (rather than soil) is used to compare estimates from two cameras because it provides a common lighting and viewing geometry and a uniform radiance over a wide angular field.

relative spectral contribution of atmospheric transmittance and skylight. One difference is unavoidable: the illumination of the reference test charts contains a larger fraction of skylight than that of the surface. This difference might introduce a subtle effect on the overall shape of the estimated reflectance curves, but it would not change their major features.

Those reference test chart reflectances $\rho_r(\lambda)$ with corresponding photosensor array output voltages $V_1^r, V_2^r, \cdots, V_6^r$ that fall within the selected camera dynamic range satisfy the equation

$$V_i^r = \tilde{c}_i \cos \iota \int_0^\infty S(\lambda)\rho_r(\lambda)T_i(\lambda)\, d\lambda \qquad i = 1, 2, \cdots, 6 \tag{10}$$

Consequently, the constant $\tilde{c}_i$ can be determined (for each spectral channel) as the slope of the linear least square error fit to the data points of the measured values V_i^r and the computed values $\cos \iota \int_0^\infty S(\lambda)\rho_r(\lambda)T_i(\lambda)\, d\lambda$.

To show that relative spectral reflectance estimates based on data returned from Mars are not appreciably degraded either by electronic and quantization noise or by uncertain atmospheric scattering, we present in Figure 6 two estimates for the known reflectances of one of the grey patches (patch 5) and the three color patches of the reference test chart. One estimate is determined from spectral samples $\tilde{b}_i$ computed directly from (9), and the other from the actual data.

The fact that the estimates cannot accurately reproduce the color reflectances simply demonstrates a basic limitation of estimates based on only six spectral channels. However, since the two sets of estimates are nearly the same, we can conclude that the estimates based on actual data from Mars should be nearly as accurate as those shown in Figure 4. Furthermore, we can conclude that the reference test charts used for calibra-

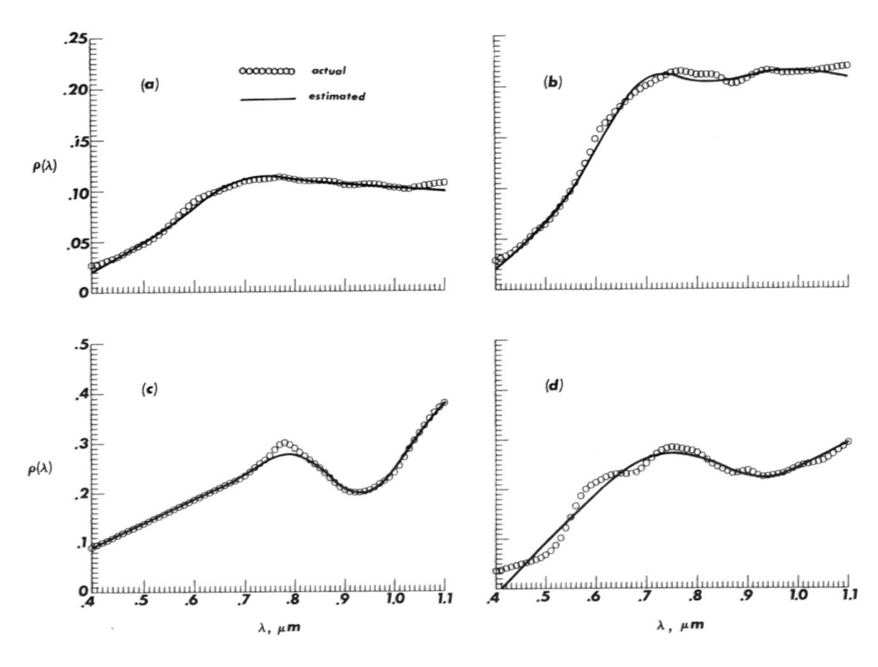

Fig. 4. Spectral reflectance estimates for the ideal case in which the spectral samples are exactly known. The spectral reflectances are representatie of (*a*) dark areas on Mars, (*b*) bright areas on Mars, (*c*) hypersthene, and (*d*) limonite. The Mars data measurements by *McCord et al.* [1977] are adjusted somewhat arbitrarily to visual brightnesses typical for dark and bright areas.

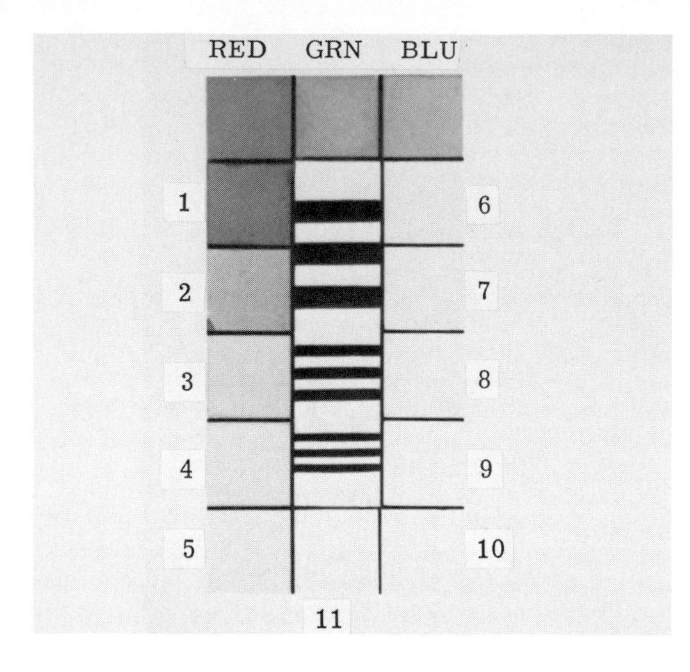

Fig. 5. Reference test chart.

tion have not been covered by windblown dust. This has also been confirmed by visual inspections of reconstructed images. The two reference test charts that have been used for calibration are mounted well above the lander deck on the cover of the radioisotope thermoelectric generators, whereas a third chart is mounted toward the rear of and close to the deck. An appreciable fraction of the material delivered to the entrance ports for the biological and organic and inorganic chemical analysis instruments has missed the ports and accumulated on the lander deck and the rear chart.

Results. Figure 7 presents relative spectral reflectance estimates $\langle\tilde{\rho}(\lambda)\rangle$. Their scale is based on internal camera calibrations of the green channel (which has a well-defined peak responsivity at $\lambda = 0.55\ \mu$m). The absolute spectral reflectance $\langle\rho(\lambda)\rangle$ can be determined approximately from these estimates as

$$\langle\rho(\lambda)\rangle = \langle\tilde{\rho}(\lambda)\rangle/\tau_a(0.55) \tag{11}$$

where $\tau_a(0.55)$ is the factor (listed in Table 1) by which the atmosphere attenuates the solar irradiance that reaches the surface at $\lambda = 0.55\ \mu$m. However, the results are strictly limited to a specific lighting and viewing geometry; it is still necessary to determine surface scattering properties before the albedo of the soil can be estimated. Nevertheless, the adjustment suggested by (11) is necessary to determine the color of surface features.

There is little significant difference between most of the spectral reflectance estimates of the two landing sites. All estimates rise rapidly with increasing wavelength between 0.4 and 0.8 μm and with one or two exceptions exhibit a pronounced minima centered about 0.93 μm. The exception of curve 7 for Lander 1 and to a lesser extent curve 6 for Lander 1 appears to be real: it was obtained consistently with several estimates from those two areas. These two curves correspond to a drift of soil located among dark rocks (Figure 1*a*). It is conceivable that this soil may be less chemically altered than typical soils at the landing sites. The reflectances of the trench material (curve 12 for Lander 1 and curve 8 for Lander 2) are similar in shape to the other curves, an implication that little difference exists between surface and near-surface materials.

The major difference between our reflectance estimates and telescopic data (such as shown in Figures 4*a* and 4*b*) is that the latter exhibit only a very subdued absorption band in the near IR. For bright areas the band occurs between 0.83 and 0.90 μm for data obtained in 1973 and between 0.93 and 0.97 μm for data obtained in 1969 [*McCord et al.,* 1977]. For dark areas the positions of the band are more variable. The difference between in situ and remote observations could be due to the fact that the latter include dust-laden atmospheric contributions which might reduce spectral contrast. Furthermore, the integrated albedo of the material of both landing sites would include reflectance components of rocks and their shadows as well as of the soil at telescopic resolutions.

The shape of our spectral reflectance estimates is most consistent with the presence of Fe^{+3}-rich weathering products, such as limonite (shown in Figure 4*d*) and nontronite [*Gillespie et al.*, 1974]. Limonite has an Fe^{+3} band centered at 0.89 μm [*Adam,* 1975], and nontronite at 0.93 μm [*Bragg,* 1977; *Gillespie et al.*, 1974]. The Fe^{+3} band of nontronite agrees most closely with our estimates. This interpretation is also consistent with the results of the Viking lander X ray fluorescence investigation [*Baird et al.,* 1976]. It is conceivable from the curves that some part of the reflectance could be dominated by pyroxenes, such as hypersthene (shown in Figure 4*c*), which have Fe^{+2} bands located at 0.85–0.95 μm [*Adam,* 1975]. However, this interpretation would not be consistent with the X ray fluorescence results. Delineation of the number and abundance of total mineral phases must await further analyses and comparisons to laboratory spectral reflectance measurements of mixtures of various candidate minerals. The extent to which the spectral reflectance estimates can be related to mineral components is, however, severely limited by the few spectral channels of the cameras.

COLOR

Color estimate from spectral reflectance. The color of an object viewed under standard illumination can be computed uniquely from its spectral reflectance. According to the Commission Internationale de l'Eclairage (CIE) system the tristimulus color values that correspond to the spectral reflectance estimate $\langle\rho(\lambda)\rangle$ are [*Wright,* 1964]

$$\langle X\rangle = \frac{1}{K}\int_{0.38}^{0.77} \langle\rho(\lambda)\rangle S_c(\lambda)\bar{x}(\lambda)\,d\lambda \tag{12a}$$

$$\langle Y\rangle = \frac{1}{K}\int_{0.38}^{0.77} \langle\rho(\lambda)\rangle S_c(\lambda)\bar{y}(\lambda)\,d\lambda \tag{12b}$$

$$\langle Z\rangle = \frac{1}{K}\int_{0.38}^{0.77} \langle\rho(\lambda)\rangle S_c(\lambda)\bar{z}(\lambda)\,d\lambda \tag{12c}$$

The normalization constant K is defined as

$$K = \int_{0.38}^{0.77} \rho_s(\lambda)S_c(\lambda)\bar{y}(\lambda)\,d\lambda \tag{13}$$

where $\rho_s(\lambda) \approx 1$ is the reflectance of a standard white diffuse surface such as magnesium oxide. Of three illuminants that have been established as standards for matching color, we use the illuminant C, or $S_c(\lambda)$, because it most closely approximates diffuse sunlight on a clear day on earth. The distribution coefficients $\bar{x}(\lambda)$, $\bar{y}(\lambda)$, and $\bar{z}(\lambda)$ account for average color perception for foveal vision with a field subtending 2°. The

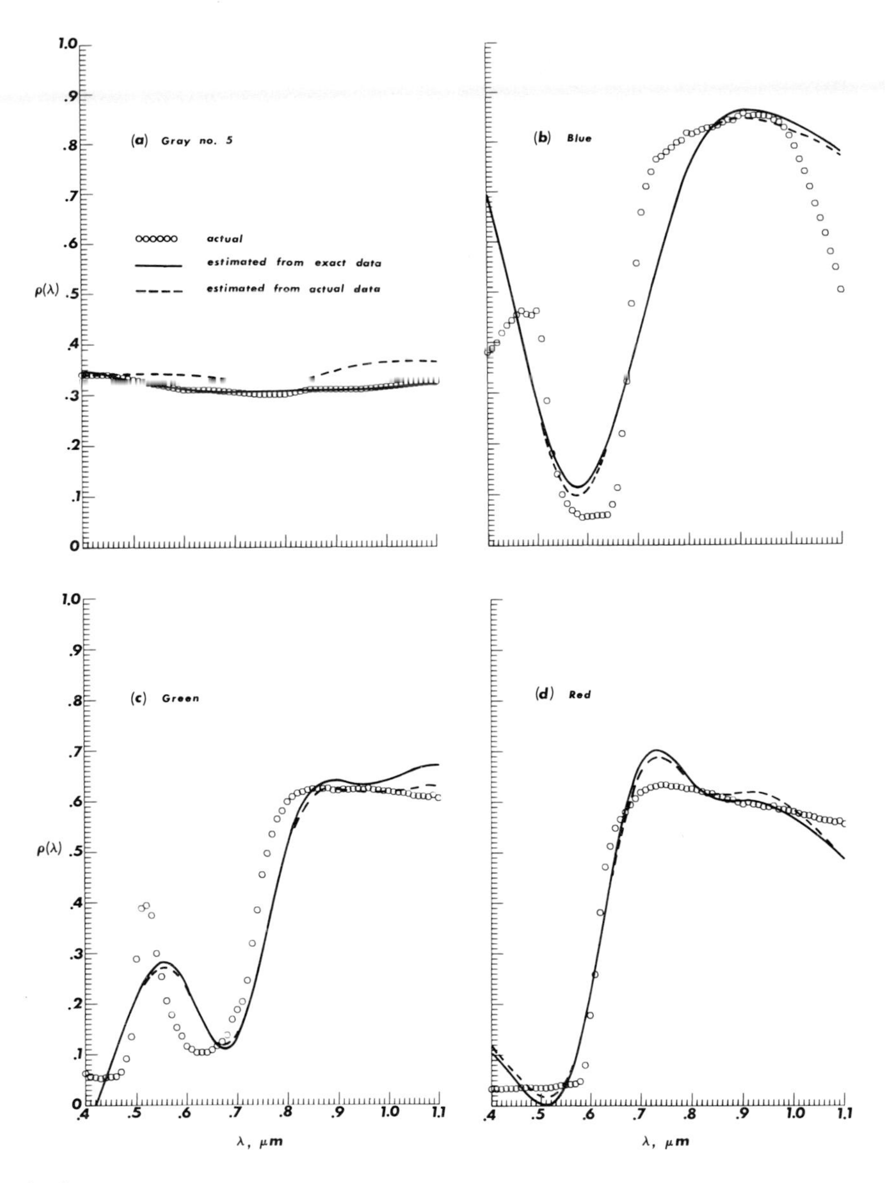

Fig. 6. Spectral reflectance estimates of reference test chart reflectances for the ideal case in which the spectral samples are exactly known and for data returned from Mars.

products $S_c(\lambda)\bar{x}(\lambda)$, $S_c(\lambda)\bar{y}(\lambda)$, and $S_c(\lambda)\bar{z}(\lambda)$ are referred to as illuminant C weighted distribution coefficients.

The tristimulus value Y is generally regarded as the visual reflectance factor. It completely specifies color in the CIE system together with the chromaticity factors:

$$x = X/(X + Y + Z) \qquad (14a)$$

$$y = Y/(X + Y + Z) \qquad (14b)$$

By definition, for a white diffuse surface viewed with normally incident illuminant C, $Y = 1.0$, $x = 0.310$, and $y = 0.316$.

Color estimates from spectral radiance. The color of the Martian surface and sky, as it would appear to a human observer on Mars, can be calculated directly from the spectral radiance estimates $\langle N(\lambda)\rangle$ by substituting $\pi\langle N(\lambda)\rangle$ for $\langle\rho(\lambda)\rangle S_c(\lambda)$ in (12*a*)–(12*c*) and $\tau_a(0.55)S(\lambda)$ for $\rho_s(\lambda)S_c(\lambda)$ in (13). $S(\lambda)$ is the solar irradiance of Mars. The approximation $\tau_a(\lambda) \approx \tau_a(0.55)$ affects primarily the visual brightness Y and not the chromaticity factors x and y.

Color representations. Once the tristimulus values have been estimated, there still remains the problem of determining the appropriate color chart and illumination that together provide the same color. *Newhall et al.* [1943] present a procedure for converting the CIE quantities to the Munsell system which describes color in terms of hue (color), value (darkness/lightness), and chroma (departure from greyness). The *Munsell Book of Color* [*Kollmorgen Corporation,* 1976] contains color patches which when viewed with illuminant C or, as should be usually satisfactory, with diffuse sunlight represent the estimated color.

Whereas the CIE and Munsell systems provide a means of numerically or symbolically defining a color, the National Bureau of Standards–Inter-Society Color Council (NBS-ISCC) system [*Kelly and Judd,* 1968] provides the more familiar but less precise designation of a color by name. This system standardizes color designations by the use of nouns with adjective modifiers. The noun describes the primary characteristics of the color: the basic hue of coloration or lightness of grey. The modifiers describe secondary characteristics: secondary hues, saturation, and lightness.

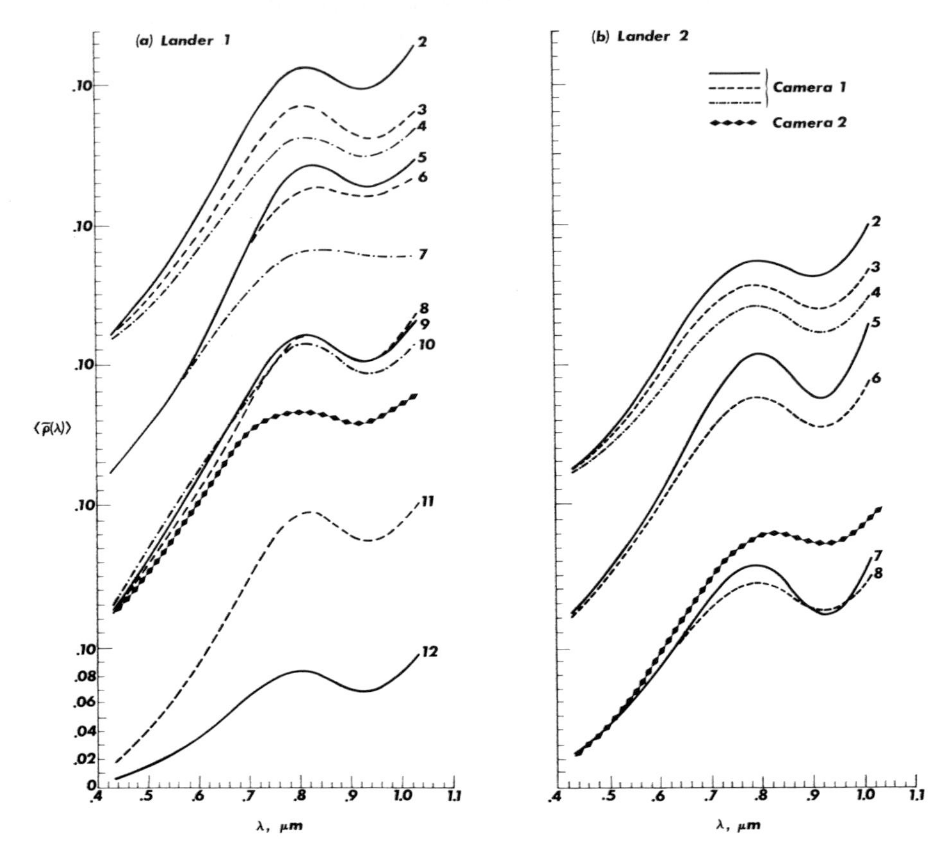

Fig. 7. Relative spectral reflectance estimates. The locations for the camera 1 curves are identified in Figure 1; the locations for the camera 2 curves are visually similar to patches 2, 3, and 4 of Lander 1 and to patches 3 and 4 of Lander 2. The differences between the shape of the near-IR reflectance minima from camera 1 and 2 estimates result from uncertainties in calibration and possibly also from differences in lighting and viewing geometry. The estimates tend to agree that the reflectance minima are centered about 0.93 μm but are inconclusive about the depth of the minima.

Since our spectrophotometric estimates contain almost inevitably some errors, it is interesting to investigate briefly the sensitivity of color estimates to such errors. To do so, we list in Table 2 the colors that correspond to the spectral reflectances and their estimates shown in Figures 4 and 6. The colors are specified in the CIE, Munsell, and NBS-ISCC systems.

There are no significant differences between the actual and estimated colors for six out of the eight reflectances (Figures 4, 6*a*, and 6*d*). It should be noted in particular that the small-period reflectance variations of limonite and its smooth reflectance estimate (Figure 4*d*) yield very similar CIE values which can be barely distinguished in the Munsell and NBS-ISCC systems. Only the estimated color of the blue and green color patches of the reference test chart (Figures 6*b* and 6*c*) are significantly different. It is for this reason insufficient to simply balance color pictures obtained with the Viking lander cameras so that the color of these patches agrees approximately with the actual data. Initial attempts to do so without proper treatment of camera responsivities and film exposure led to the construction of pictures in which the scene was shown as predominantly reddish brown. Although data analyses and other color reconstructions gave early indications that this color might be incorrect, several technical difficulties in data processing and film exposure had to be overcome before an accurate color rendition could be produced. Proper color reconstruction on film requires not only the treatment of camera responsivities as described in this paper but also the linearization of exposure, prevention of color channel cross talk, and other related problems.

Results. Table 3 presents color estimates of the sky and soil. The colors computed from spectral radiance estimates represent colors as seen on Mars and from spectral reflectance estimates as seen on earth in diffuse sunlight. The shift in color 'as seen on Mars' to 'as seen on earth' is consistent with the contribution of skylight due to atmospheric scattering that increases toward long wavelengths on Mars [*Pollack et al.*, 1977] and short wavelengths on earth. However, the difference is within the range of uncertainty of our estimates and perhaps cannot be regarded as significant. It should be recalled that the two sets of estimates rely on independent calibrations: radiance estimates on internal camera calibrations and reflectance estimates on external calibrations with a reference test chart.

The color of the soil of both landing sites is predominantly moderate yellowish brown with only subtle variations, especially in the Viking 2 site, where it is almost uniform. Colors appear to range mostly from dark brown to strong yellowish brown and even moderate olive brown in the Viking 1 site. Some of the rocks in the Viking 1 site appear dark grey in carefully balanced color pictures.

These results are in close agreement with the colors computed for the telescopic reflectance curves of *McCord et al.* [1977] which, as listed in Table 2, range from dark to moderate yellowish brown in diffuse sunlight. Lighting and viewing conditions for visual telescopic observations are, of course, different from diffuse sunlight. For the situation in which the planet is directly illuminated by the sun and viewed through the earth's atmosphere [*Carpenter and Chapman*, 1965], our color computations indicate a slight shift further away from red

TABLE 2. Color Corresponding to Actual and Estimated Spectral Reflectance

Identification of Spectral Reflectance		Color Computed From Actual Reflectance					Color Computed From Estimated Reflectance*				
Figure	Description	Y	x	y	Munsell Notation†	NBS-ISCC‡	$\langle Y \rangle$	$\langle x \rangle$	$\langle y \rangle$	Munsell Notation†	NBS-ISCC‡
4*a*	dark area of Mars§	0.07	0.393	0.375	(9.7YR)3/3	dark yellowish brown (78)	0.07	0.387	0.377	(1.2Y)3/2	moderate olive brown (95)
4*b*	bright area of Mars§	0.11	0.420	0.385	(8.3YR)4/4	moderate yellowish brown (77)	0.11	0.414	0.388	(9.6YR)4/4	moderate yellowish brown (77)
4*c*	hypersthene	0.17	0.357	0.355	(0.6Y)4.5/2	greyish yellowish brown (80)	0.17	0.354	0.353	(1.2Y)4.5/2	greyish yellowish brown (80)
4*d*	limonite	0.15	0.438	0.401	(9.0YR)4.5/5	moderate to strong yellowish brown (77–74)	0.15	0.432	0.419	(2.0Y)5/3	light olive brown (94)
6*a*	RTC grey patch 5‖	0.32	0.303	0.310	(4.6PB)6/1	bluish grey (191)	0.32	0.303	0.310	(4.6PB)6/1	bluish grey (191)
6*b*	RTC blue	0.15	0.177	0.185	(2.5PB)4.5/12	strong blue (178)	0.16	0.227	0.179	(9.0PB)4.5/11	strong to brilliant purplish blue (196–195)
6*c*	RTC green	0.20	0.295	0.514	(0.4G)5/10	strong yellowish green (131)	0.25	0.354	0.443	(5.7GY)5.5/6	moderate yellowish green (120)
6*d*	RTC red	0.11	0.519	0.308	(4.4R)4/11	moderate to strong red (15–12)	0.10	0.534	0.306	(5.4R)3.5/11	moderate to strong red (15–12)

For standard illuminant C.

*The estimates are based on computed spectral samples b_i, using (9).

†Read as (hue) value/chroma.

‡Description (numerical notation).

§The relative spectral reflectance measurements of *McCord et al.* [1977] have been somewhat arbitrarily adjusted to absolute reflectances (see Figure 4); an error in this adjustment would effect the CIE visual reflectance Y and Munsell value but not the CIE chromaticity factors and Munsell hue and chroma.

‖RTC stands for reference test chart.

TABLE 3. Color Estimates of Soil and Sky

	Color Computed From Spectral Radiance Estimates*					Color Computed From Spectral Reflectance Estimates†				
Location‡	⟨Y⟩	⟨x⟩	⟨y⟩	Munsell Notation	NBS-ISCC§	⟨Y⟩	⟨x⟩	⟨y⟩	Munsell Notation	NBS-ISCC§
					Lander 1					
Sky	0.18	0.40	0.38	(9YR)5/4	light to moderate yellowish brown (76–77)					
Soil (range)	0.09–0.12	0.43–0.46	0.39–0.40	(8–10YR)3–4/4–5	55, 58, 59, 74, 75, 77, 78, 95	0.08–0.11	0.40–0.42	0.39–0.40	(9YR–2Y)3–4/3–4	77, 78, 95
Soil (most common)	0.10	0.44	0.40	(9YR)4/5	moderate yellowish brown (77)	0.10	0.41	0.39	(10YR)3.5/4	moderate to dark yellowish brown (77–78)
Trench	0.036	0.46	0.39	(7.5YR)2/4	dark brown (59)	0.033	0.43	0.39	(10YR)2/3	dark yellowish brown (78)
					Lander 2					
Sky	0.14	0.40	0.38	(9YR)4/4	moderate yellowish brown (77)					
Soil (range)	0.08–0.11	0.43–0.44	0.39	(8–9YR)3–4/4–5	55, 58, 74, 77, 78	0.07–0.10	0.41–0.42	0.39–0.40	(10YR–1Y)3–4/3–4	77, 78
Soil (most common)	0.09	0.44	0.40	(8YR)4/5	moderate yellowish brown (77)	0.09	0.42	0.39	(10YR)3.5/4	moderate to dark yellowish brown (77–78)
Trench	0.084	0.44	0.39	(7.5YR)3/4	moderate brown (58)	0.077	0.41	0.39	(10YR)3/3	dark yellowish brown (78)

*As seen on Mars.

†As seen on earth in diffuse sunlight (illuminant C).

‡See Figure 1.

§The descriptions of the numerical notations are as follows: 55, strong brown; 58, moderate brown; 59, dark brown; 74, strong yellowish brown; 75, deep yellowish brown; 76, light yellowish brown; 77, moderate yellowish brown; 78, dark yellowish brown; and 95, moderate olive brown.

toward more yellowish to olive brown. These results are also in general agreement with color pictures that have been reconstructed from telescopic photographic observations [*Baum,* 1973; *Jones and Cook,* 1974]. The comparison of numerical color estimates and color pictures via the *Munsell Book of Colors* should be made with original color prints. But even then the comparison is usually restricted by film constraints (e.g., dynamic range of linear reflectance) and visual objectives (e.g., enhanced contrast). Nevertheless, these results agree that contrary to a commonly held belief, the red planet does not appear to be red, as seen either from its surface or from earth.

Acknowledgments. We are indebted to S. J. Katzberg of the Langley Research Center for major contributions in the effort to construct accurate color pictures from the Viking lander camera data. He (together with D. Jobson) discovered that the early color renditions were in error and produced color pictures that subsequently proved to be in close agreement with the results presented in this paper. He also helped to identify film exposure problems with the Viking image reconstruction facility. We thank C. W. Rowland of the Langley Research Center for the calibration and operation of the laser film recorder that made the color prints shown in Figure 1. One of us (R.E.A.) was partially supported by the Planetary Geology Program Office of the National Aeronautics and Space Administration. We thank S. Bragg of Washington University for the independent analysis of reference test chart image data that supported our use of this data for camera calibrations.

REFERENCES

Adam, J., Interpretation of visible and near-IR diffuse reflectance spectra of pyroxenes and other rock-forming minerals, in *IR and Raman Spectroscopy of Lunar and Terrestrial Minerals,* edited by C. Karr, p. 375, Academic, New York, 1975.

Baird, A. K., P. Toulmin III, B. C. Clark, H. J. Rose, Jr., K. Keil, R. Christian, and J. Gooding, Mineralogic and petrologic implications of Viking geochemical results from Mars: Interim report, *Science, 194,* 1288, 1976.

Baum, W. A., The international planetary patrol program: An assessment of the first three years, *Planet. Space Sci., 21,* 1511, 1973.

Bragg, S., Characteristics of Martian soil on Chryse Planitia as inferred by reflectance properties, magnetic properties, and dust accumulation of Viking Lander I, M.A. thesis, Washington Univ., Saint Louis, Mo., 1977.

Carpenter, R. O., and R. M. Chapman, in *Applied Optics and Optical Engineering,* edited by R. Kingslake, p. 139, Academic, New York, 1965.

Gillespie, J. B., J. D. Lindberg, and M. S. Smith, Visible and near-infrared absorption coefficients of montmorillonite and related clays, *Amer. Mineral., 59,* 1113, 1974.

Huck, F. O., and S. D. Wall, Image quality prediction: An aid to the Viking lander imaging investigation on Mars, *Appl. Opt., 15,* 1748, 1976.

Huck, F. O., H. F. McCall, W. R. Patterson, and G. R. Taylor, The Viking Mars lander camera, *Space Sci. Instrum., 1,* 189, 1975.

Jones, S. E., and N. O. Cook, Color pictures of planets from black-and-white images, *Sky Telesc., 47*(1), 57, 1974.

Kelly, K. L., and D. B. Judd, ISCC-NBS centroid color system, in *Manual of Color Aerial Photography,* 1st ed., edited by J. T. Smith, Jr., p. 523, American Society of Photogrammetry, Falls Church, Va., 1968.

Kollmorgen Corporation, *Munsell Book of Color,* New York, 1976.

McCord, T. B., R. L. Huguenin, D. Mink, and C. Pieters, Spectral reflectance of Martian areas during the 1973 opposition: Photoelectric filter photometry 0.33–1.0 μm, *Icarus,* in press, 1977.

Newhall, S. M., D. Nickerson, and D. B. Judd, Final report of the O. S. A. subcommittee on the spacing of the Munsell colors, *J. Opt. Soc. Amer., 33,* 305, 1943.

Park, S. K., and F. O. Huck, Spectral reflectance estimation technique using multispectral data from the Viking lander camera, *NASA Tech. Note, D-8292,* 1976.

Park, S. K., and F. O. Huck, Estimation of spectral reflectance curves from multispectral image data, *Appl. Opt.,* in press, 1977.

Patterson, W. R., III, F. O. Huck, S. D. Wall, and M. R. Wolf, Calibration and performance of the Viking lander cameras, *J. Geophys. Res., 82,* this issue, 1977.

Pollack, J. B., D. Colburn, R. Kahn, J. Hunter, W. Van Camp, C. E. Carlston, and M. R. Wolf, Properties of aerosols in the Martian atmosphere, as inferred from Viking lander imaging data, *J. Geophys. Res., 82,* this issue, 1977.

Wall, S. D., E. E. Burcher, and D. J. Jobson, Reflectance characteristics of the Viking lander camera reference test charts, *NASA Tech. Memo., X-72762,* 1975.

Wright, W. D., *The Measurement of Colors,* 3rd ed., p. 118, D. Van Nostrand, Princeton, N. J., 1964.

(Received April 11, 1977;
revised June 7, 1977;
accepted June 7, 1977.)

Processing the Viking Lander Camera Data

ELLIOTT C. LEVINTHAL,[1] WILLIAM GREEN,[2] KENNETH L. JONES,[3] AND ROBERT TUCKER[1]

Over 1000 camera events were returned from the two Viking landers during the Primary Mission. A system was devised for processing camera data as they were received, in real time, from the Deep Space Network. This system provided a flexible choice of parameters for three computer-enhanced versions of the data for display or hard-copy generation. Software systems allowed all but 0.3% of the imagery scan lines received on earth to be placed correctly in the camera data record. A second-order processing system was developed which allowed extensive interactive image processing including computer-assisted photogrammetry, a variety of geometric and photometric transformations, mosaicking, and color balancing using six different filtered images of a common scene. These results have been completely cataloged and documented to produce an Experiment Data Record.

INTRODUCTION

During the nominal Viking mission the two landers acquired 1025 different camera events: 451 from the VL-1 site in Chryse Planitia and 574 from the VL-2 site in Utopia Planitia. Details of image acquisition, on-board data tape storage, and transmission to earth are presented elsewhere [*Huck et al.*, 1975]. Table 1 summarizes the camera events. For many camera events, multiple replays from the lander or orbiter tape recorders were received and later merged. Figure 1 illustrates the frequency of imaging data acquired as a function of day of mission; Figure 2 diagrams the ground system flow of camera data and the processing at the Jet Propulsion Laboratory's Viking Mission Control and Computing Center (VMCCC) and Image Processing Laboratory (IPL).

Each 3413-bit camera scan line, acquired as the facsimile camera mirror scans upward, is returned to earth as a 'frame' of data. Each data frame starts with a 31-bit pseudonoise word followed by a 5-bit instrument identification (ID) word, which are used to synchronize the beginning of each imaging line and to identify the source of data as the camera, respectively. The remainder of each imaging frame contains 512 6-bit picture elements as well as engineering data bits.

As lander telemetry is received in near real time at the Deep Space Network (DSN) from the tracking stations, it is relayed to the VMCCC for processing. By means of the pseudonoise and instrument ID words, imaging frames are decoded by the Telemetry Processor Program (TLMP). Imaging data are then passed via in-core transfer to the first-order image-processing program, described below, which resides in the same IBM 360/75 computer and which operates as part of the same real time task as the program TLMP. Imaging is the only lander science experiment allowed to process data in real time. Optionally, the System Data Record (SDR) produced by TLMP (see the appendix, glossary of acronyms) may be accessed at a later time from tape by the first-order imaging program.

After the decoding by TLMP the actual reduction of imaging data is divided into two parts: a real time first-order processing system (Fovlip) for rapid display of the data and production of hard-copy products and a second-order processing system at the Image Processing Laboratory that made available a wider range of analytical techniques and which produced the final versions of the digital data. The preparation of the raw data sets for the latter activity was called Experiment Data Record (EDR) generation and involved the use of both processing facilities. The EDR, in turn, was used to generate the final photoproducts, which are the Team Data Record (TDR).

[1] Department of Genetics, Stanford University, Stanford, California 94305.

[2] Jet Propulsion Laboratory, California Institute of Technology, Pasadena, California 91103.

[3] Department of Geological Sciences, Brown University, Providence, Rhode Island 02912.

Paper number 7S0440.

FOVLIP (REAL TIME FIRST-ORDER PROCESSING)

The first-order Viking Lander image-processing software consists of a number of distinct subprograms, each of which performs a single function on a complete lander picture. The subprograms are of three types: those which perform input processing of data, those which create enhanced versions of the images, and those which format the raw and enhanced images for volatile displays and tape output.

When receipt of an incoming real time image is completed, up to three enhanced versions can be created and stored on a data base for later display or hard-copy generation. Each version is the result of one to five serially applied enhancement subprograms. Fovlip has the capability for either operator-directed enhancement processing or automatic processing of incoming images using a library of 100 predetermined enhancement sequences.

Within Fovlip, enhancement processing subprograms consist of despiking (Adespike), contrast stretching (Aconalt), box filtering (Ahipass), and radiometric corrections (Aradcam). Adespike is used to remove random noise. The digital value of each picture element is compared to its adjacent neighbors in the vertical and horizontal directions and, depending upon user-specified tolerances, may be replaced by the average of its neighbors. Aconalt allows considerable freedom to map input pixel values to output values. This mapping can be specified directly by the creation by the user of a look-up table with an arbitary set of values, or Aconalt can calculate the table automatically by specifying the input and output values of two different brightness levels with other values linearly interpolated or extrapolated. Optionally, one can define a linear stretch so that a specified percentage of the input images pixel values are above and below specified *DN* values. A table can be generated from a cumulative distribution function, in which case the relative frequencies of different *DN* levels are examined, more frequently occurring levels being more highly stretched. Ahipass uses a box filter algorithm to boost high spatial frequencies in an image. Aradcam corrects video data for errors in relative and absolute radiometry due to known distortion introduced by the camera by using tables of correction factors based on preflight and inflight calibration

TABLE 1. Summary of Camera Events

		VL-1				VL-2			
		PRIMARY		EXTENDED (1)		PRIMARY		EXTENDED (1)	
	STEP SIZE	CE's	LINES × 10^3	CE's	LINES × 10^3	CE's	LINES × 10^3	CE's	LINES × 10^3
CAMERA EVENTS									
BY EVENT TYPE									
HIGH RESOLUTION	0.04	188	175.6	74	49.7	240	210.9	84	36.3
SURVEY (2)	0.12	44	17.3	0	0	63	23.5	0	0
COLOR (3)	0.12	74	41.9	31	17.7	100	91.4	21	22.0
IR (3)	0.12	48	24.6	19	13.5	50	41.7	1	1.5
COLOR/IR SINGLETS (4)	0.12	33	16.2	7	0.4	70	59.5	29	9.4
SUN (5)	0.12	20	3.9	32	2.0	29	3.5	33	2.0
CALIBRATION	——	25	1.5	13	0.7	22	1.3	11	0.6
SCAN VER. (6)	0.12	19	0.8	7	0.3	0	0	2	0.1
TOTAL		451	281.8	183	84.3	574	431.8	181	71.9
REAL TIME RESCAN		CE's	LINES			CE's	LINES		
< 25 LINES		11	65			9	66		
≥ 25 LINES		17	1349			31	3184		
TOTAL		28	1414			40	3250		

(1) EXTENDED MISSION REFERS TO THE PERIOD FROM CONJUNCTION TO MARCH 1, 1977.

(2) FIVE SURVEY CE's WERE RECORDED AT 0.04° STEP SIZE, SEE NOTE (7) BELOW.

(3) THE LINE TOTALS FOR COLOR AND IR TRIPLET CE's ARE GIVEN IN THE NUMBER OF "SINGLE DIODE" LINES SCANNED.

(4) MANY OF THESE COLOR AND IR, SINGLET MODE, CE's WERE RECORDED AT 0.04° STEP SIZE, SEE NOTE (7) BELOW.

(5) ONLY TWO OF THE SUN CE's WERE RECORDED AT 0.12° STEP SIZE, ALL OTHERS WERE RECORDED AT 0.04° (NONNOMINAL) SEE NOTE (7).

(6) THE SCAN VERIFICATION CE's ARE IMAGES OF A LIGHT SOURCE IN THE CAMERA POST ASSEMBLY USING AN HR DIODE AT 0.12° STEP SIZE.

(7) HIGH RESOLUTION DIODES SAMPLED AT 0.12° STEP SIZE RESULTS IN A +5.6° CAMERA ELEVATION POINTING SHIFT, LOW-RESOLUTION DIODES SAMPLED AT 0.04° STEP SIZE RESULTS IN A -5.6° ELEVATION POINTING SHIFT.

(8) THE CAMERA CAN RESCAN AT THE STOP AZIMUTH OF A CAMERA EVENT IN BOTH REAL-TIME AND RECORDED MODES. IN RECORDED IMAGES RESCAN WAS DONE IN 14 CE's WHICH WERE ENTIRELY RESCANNED LINES USING COLOR MODE AT THE SLOW (250 bps) SCAN RATE. MANY OTHER RECORDED IMAGES HAD SMALL AMOUNTS (< 15 LINES) OF RESCAN RESULTING FROM A STRATEGY RELATED TO TAPE RECORDER PERFORMANCE.

data. It calculates and presents to the user values of radiance associated with pixel digital values of 0–63.

In addition to providing real time display of processed images, Fovlip produces a digital tape of the processed data which is sent to the Mission and Test Imaging System (MTIS), where a Dicomed cathode-ray tube (CRT) scanning device produces master negatives. This in turn is delivered to the Mission and Test Photographic System (MTPS) to produce photocopies for general distribution. The copies were generally available within 24 hours of the receipt of data. A digital facsimile printing device (Digifax) provides prompt prints of Fovlip-processed data within minutes of data receipt. The MTIS product is a negative transparency on a 5-inch-wide roll of film generated by the Dicomed raster of 4096 × 4096 elements. Each lander camera pixel is reproduced on film by using a 6 × 6 array of recorder elements giving a crisp square just barely perceptible to the unaided eye but permitting the user to establish precise line and sample values. Figure 3 illustrates a processed version of the first 512 lines of the first lander camera image received from Mars. The annotation, histograms, and marginal scales of the MTIS photoproduct provided the information needed for analysis of these data.

Fovlip also produces digital tapes of the raw images which are forwarded to IPL as input to the second-order image-processing system.

EDR GENERATION

In addition to the real time data stream a second stream, referred to as the Data Record Stream, has the objective of providing the best possible data on a somewhat slower time line. It takes data from the Deep Space Station selected as having the best data and records the data on a tape called the Network Data List. A list is created of all missing telemetry frames along with a System Performance Record. These two are used to generate a request to a second station, which also received similar data; the data from the second source are merged with the data in hand to create an Intermediate Data Record, which, after being read into TLMP, yields the Intermediate System Data Record (ISDR).

One problem encountered late during simulation testing was the severity of the effects of data corruption (noisy data) on telemetry processing. Single bit errors in the 5-bit instrument ID words or 8-bit frame count indicators associated with each 3413-bit scan line resulted in the loss of the entire line. Additionally, bit errors in the pseudonoise word used for frame synchronization resulted in the inability of TLMP to identify imaging frames for Fovlip properly. For a bit error rate of 3/1000, 4% of all imaging lines were lost; for a rate of 20/1000, over half of all lines were lost. Since the latter bit error rate was typical of real time imaging and since despiking can produce

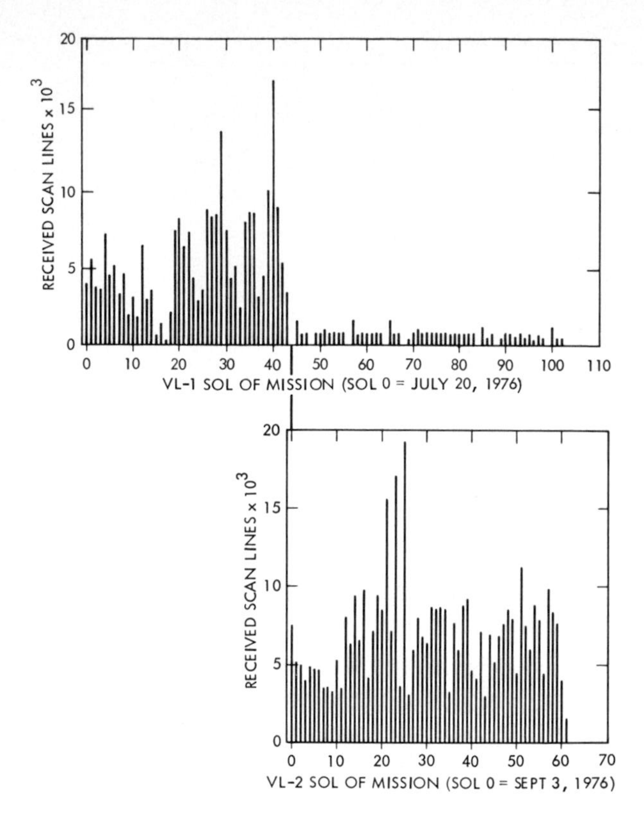

Fig. 1. Histograms showing the amount of data returned as a function of the sol (solar day) of each mission. Note that sol 0 for VL-2 corresponds to sol 44 for VL-1. At this point the data return from VL-1 was reduced to accommodate the second lander. This reduced data flow from VL-1 was transmitted directly from the lander to earth by using the *S* band communication system.

useful results for such a bit error rate, a determined effort was made to recover all such corrupted data. The program RESTIMG (Restore Imaging Data) corrected the problems listed above. Since the techniques required for successful data recovery are not feasible for real time telemetry processing, the input for program RESTIMG was either an SDR or an ISDR tape, and the output was a similarly formatted tape suitable for input to Fovlip. By means of RESTIMG, several hundred degrees of real time imaging, as well as hundreds of missing lines within images, were recovered and became part of the EDR.

Of approximately 723,000 imagery scan lines commanded from earth associated with 1036 camera events during the Primary Mission, all but about 9000 were received from 1025 camera events and form part of the Lander Camera Experiment Data Record of this part of the mission. Three camera events from Viking Lander 1 were not received owing to shortened real time relay communication links, and eight were lost owing to a direct communication link problem late in the operation of Viking Lander 2, accounting for 6886 of the missing lines.

At this point, many images still contained bit errors. In some cases, multiple data transmission provided the opportunity to select pixels from two or more transmissions and remove this source of bit error noise. Programs Despike and VLMERGE, operating at IPL, were used for this function. One of the playbacks is despiked, and, rather than replace the pixels by an average of neighboring pixels, they are replaced by the value of the corresponding pixels in the second playback. Figure 4 illustrates the results of this processing, which produced, as the EDR, an ordered, cataloged, verified digital record of the imagery data received from Mars. This will be the basis of the Experiment Data Record Picture Catalog to be published.

SECOND-ORDER IMAGE-PROCESSING SYSTEM

The Image Processing Laboratory provided computer resources required to support both batch and interactive image processing. Non–real time processing of imagery in support of science objectives had been performed by the Jet Propulsion Laboratory (JPL) Image Processing Laboratory on prior missions [*Rindfleisch et al.*, 1971; *Green et al.*, 1975; *Levinthal et al.*, 1973; *Soha et al.*, 1975]. The Viking requirements represented an increased emphasis on flexible, adaptive, and interactive image processing in comparison to previous missions, a new requirement for color processing, and a considerably greater throughput capacity.

In order to meet the above demands of the Viking Lander mission the computer facilities at IPL were substantially upgraded. The IBM 360/44 central processing unit (CPU), used on prior missions [*Levinthal et al.*, 1973], was replaced with an IBM 360/65 CPU, and 800 Mbyte of on-line disk storage was added to the configuration. A Digital Equipment Corporation PDP 11/40 minicomputer was interfaced to a channel on the 360/65 and was used to support a variety of interactive terminals and image display systems [*Jepsen*, 1976]. The available terminals included both dial-up terminals and direct line Imlac CRT terminals. The image display systems include a Ramtek GX-100B system (black-and-white system with graphics overlay and track ball/cursor unit with resolution of 640 × 512 picture elements), a Comtal 8003 system (a system that displays either three separate black-and-white images or a single color image, with track ball/cursor unit with resolution of 512 × 512 picture elements), and a Comtal 1024 system (a black-and-white system with graphics overlay and track ball/cursor unit with resolution of 1024 × 1024 picture elements). The overall configuration of the IPL computer facility is shown in Figure 1 of *Ruiz et al.* [1977]. The details of the hardware and software design of the interactive terminal and display system controlled by the PDP 11/40 are described by *Jepsen* [1976].

A specially designed video system supports stereo viewing and computer-assisted stereo mapping. It was developed at Stanford University to interface to the Ramtek GX-100B system at IPL for use with the application program Ranger. The details of its design and functioning are described by *Liebes and Schwartz* [1977].

The IBM 360/65 operates under the IBM OS/MVT operating system, and the Time Sharing Option (TSO) is used to support interactive applications. The PDP 11/40 operating system is Comtex, and all terminals and image display systems emulate standard IBM-supported devices when controlled via the Comtex system.

The Image Processing Laboratory began development for Viking operations with a large inheritance of existing applications software that had been successfully utilized to support previous JPL flight projects. The major developments required for Viking support included the following classes of software: (1) an executive system for control of interactive image-proc-

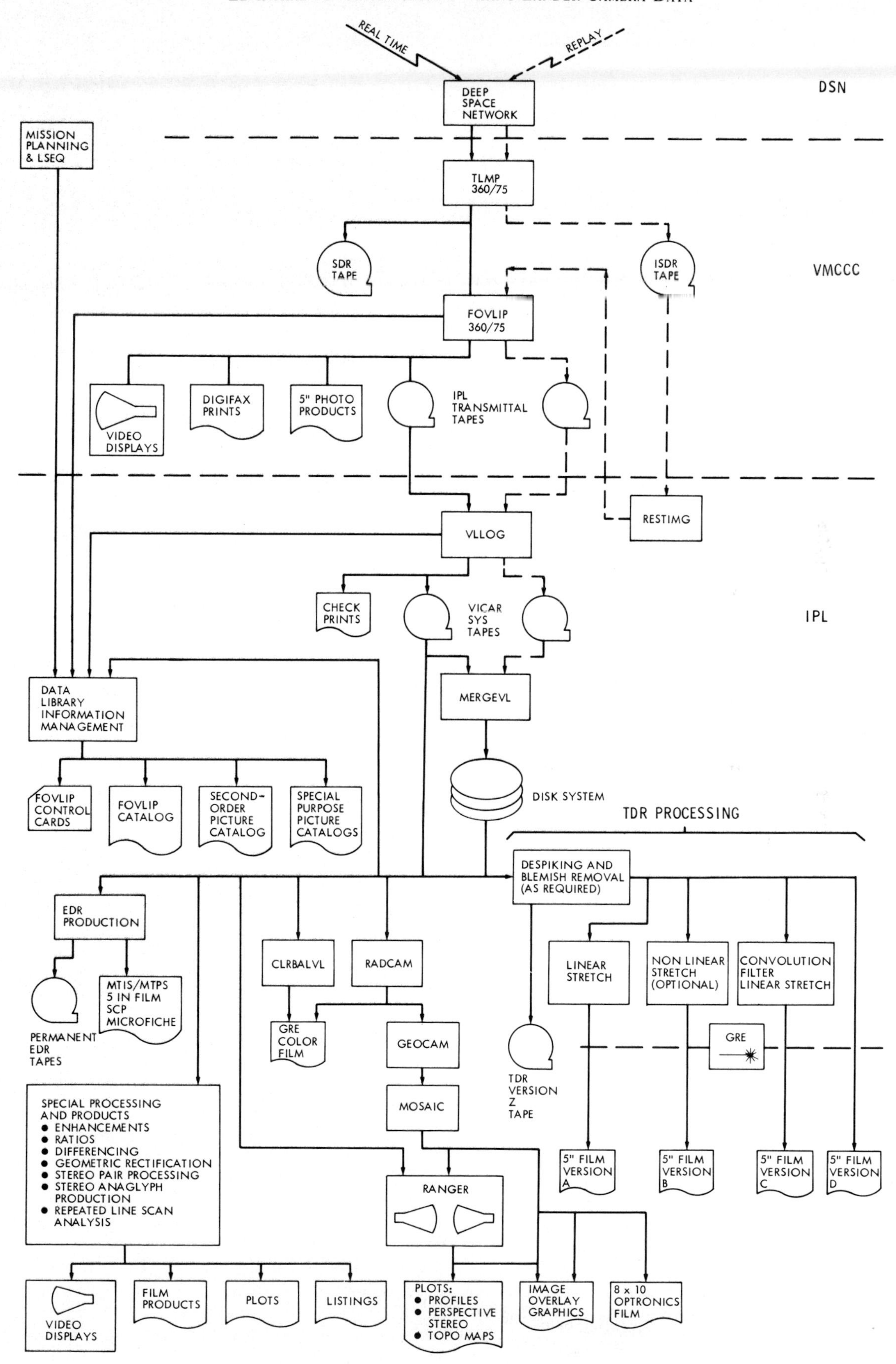

Fig. 2. Flow chart indicating the steps involved in the generation of the lander camera data base and lander image processing. The dashed path indicates the flow of the non-real time data prior to their merging with the real time data stream. The data base is generated on the Disk System, from which it is accessed for the subsequent processing steps.

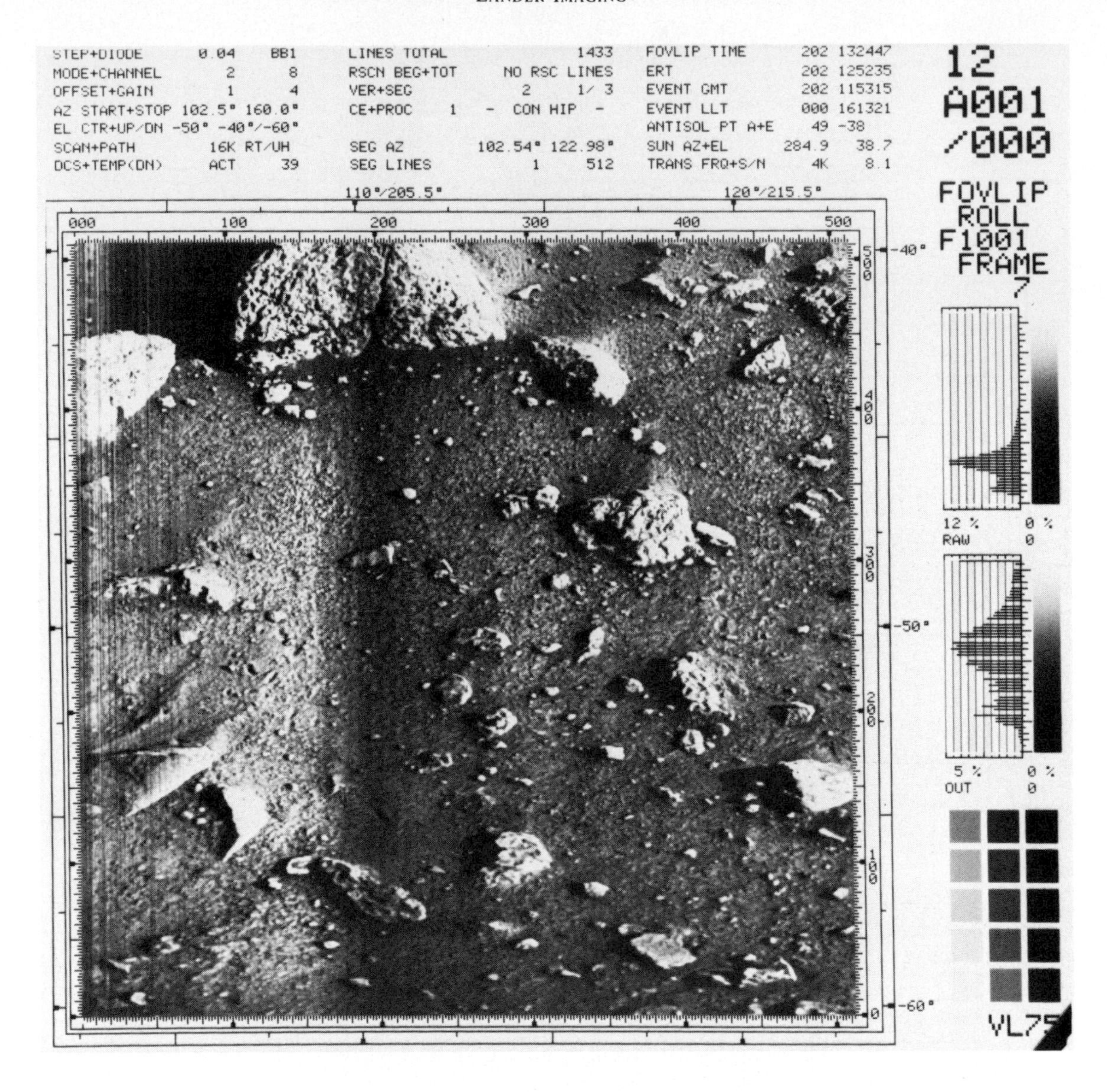

Fig. 3. Example of a photoproduct output of the real time first-order processing program (Fovlip) showing the first 512 lines of the first image returned by VL-1. The vertical brightness variations from left to right are the effects of dust settling immediately after touchdown. The image has been box-filtered to emphasize high spatial frequencies followed by a contrast enhancement algorithm (see text). The histograms at the right represent the (top) unprocessed data and (bottom) processed data.

essing sequences, (2) an automated method for cataloging auxiliary data relating to each camera event, the sequence of programs used to process a particular camera event, and the numerical parameters used, (3) a capability for interrogating the results of the cataloging, and (4) new applications programs required to accommodate the unique features of the Viking Lander facsimile camera and particular mission objectives.

The new software executive developed for interactive processing, Libexec, represents an outgrowth of the Vicar image-processing system used at JPL for 9 years in batch-processing applications. The existing Vicar system, operational on the 360/44 computer, was modified to operate under the OS/MVT operating system and to incorporate the automated cataloging functions of Libexec. The Mark IV data base management system (acquired from Informatics Inc.) was used to maintain and search the picture catalogs created under Libexec and Vicar. An interactive software package, OQL, was also acquired from Informatics Inc. which enables interactive interrogation of the picture catalogs by using the TSO.

An extensive set of applications programs were developed to support second-order Viking Lander image-processing activities. Software development began during lander camera calibration and system test activities. Several programs were written to evaluate lander camera radiometric response and geometric distortion characteristics by using test data recorded by the camera manufacturer (Itek Corporation) and system level calibration data acquired with the cameras mounted on the flight spacecraft prior to launch. Additional programs were written to generate camera calibration data regarding camera performance characteristics. These data were later used by a variety of programs to support flight science analysis.

Although the Viking Lander cameras are less susceptible to the types of distortions returned by vidicon systems utilized on

Fig. 4. Three images illustrating the results of image restoration and noise removal heuristics used on the Viking Lander images. (Top) The data as they were output by the telemetry processor program (TLMP) to program Fovlip. (Middle) The results of RESTIMG, which recovers missing lines but performs no cosmetic enhancement. (Bottom) The image after having been passed through a despiking program, which removed many noise errors. All three images have been contrast-enhanced by different amounts.

previous missions, it is still necessary to remove the small amounts of radiometric and geometric distortions for many applications. The program Geocam removed camera system induced geometric distortion from the imagery, and the program Radcam converted raw camera output into standard radiometric units.

Several other geometric transformations were required for lander applications. Because the facsimile camera samples at uniform intervals in solid angle space, objects in the scene appear in the image with their true shapes distorted. The Geotran program reprojected lander images so that shape distortion was removed and objects would appear as if they were acquired by a film plane camera. The same program was used to project images so that hard-copy imagery could be mosaicked on a cylindrical surface for viewing with correct shape representation. Geotran could also provide an overhead view of a scene imaged with the lander camera by using topographic information as input.

After the spacecraft landed, it was apparent from the shape of the horizons that the landers were tilted (this was a much larger effect for the second lander than for the first). By means of horizon information a geometric transformation was performed that removed the scene distortion caused by the tilt of the landers.

A series of programs (CLRBALVL and GREFIX) were written to color-balance imagery from sets of six images of the scene: three in visible color and three in infrared. The programs accounted for the quantitative radiometric performance of the camera systems and for the fact that the three visible color diodes also responded to some extent to infrared wavelengths. An additional correction was required to remove nonlinearities in the color film and print paper. The details of the spectrophotometric analysis are described by *Huck et al.* [1977].

Another major activity within IPL was the production of computer-generated mosaics of the high-resolution imagery recorded over several months. Several mosaics were constructed by using imagery from both cameras on both landers.

Fig. 5. Portion of the same image as is shown in Figure 4 (11A077/011). The data have been subjected to a linear contrast stretch to maximize the Martian surface detail. This is an example of the output of the TDR processing (version A, see text).

Each mosaic is an ensemble of images taken at about the same time of day: 0700 and 1400 for Viking Lander 1 and 0700, 1200, and 1730 for Viking Lander 2.

FILM RECORDING FOR SECOND-ORDER PROCESSING

Four film recorders were used in recording second-order versions of the lander imagery onto film. Two 70-mm black-and-white film recorders utilizing a flying spot cathode-ray tube to expose imagery onto film spot by spot were employed.

An Optronics black-and-white recorder was used for playback whenever large format and/or high geometric precision was required. This recorder records onto 8 × 10 inch sheet film and was used for mosaics and stereo pair production.

The Viking Ground Reconstruction Equipment (GRE) recorder was used for both black-and-white and color imagery. This recorder was developed by the Itek Corporation. By means of a laser as the light source, three spectral lines are separated, individually modulated, and then recombined to produce a single beam for color recording. For black-and-white recording, only the green band is used.

TDR PROCESSING

The EDR (described earlier) was the data used to generate the final photoproducts, which are the Team Data Record and which are furnished to others through the National Space Science Data Center (NSSDC). Different IPL software programs were used to produce computer-processed versions of the raw data, appropriate to the peculiarities of each camera event. After much experimentation, three enhancement algorithms were chosen that provided images suitable for the detailed geological interpretation of lander images. Version A consists of a simple linear contrast stretch, chosen for each image to maximize Martian surface detail. Version B is a logarithmic contrast stretch ($DN_{\text{out}} = A + B \log DN_{\text{in}}$) which 'stretches' the darker levels more than the lighter, the result being more representative of one's visual impression of the scene than a linear contrast stretch. Version C is a convolutional filter algorithm which applies a truncated modulation transfer function correction followed by a linear contrast stretch identical to that used in version A. Program MASKVL generated the scales, annotation, histograms, and appropriate format for the GRE. The GRE was used with an 8-mil spot to produce the negatives. For camera events of more than 900 lines the image was broken into multiple overlapping negatives. These negatives were used, at MTPS, to produce a master positive from which all the photoproducts for final distribution were derived. Figure 5 illustrates these products.

DATA CATALOGING AND MANAGEMENT

The inventory of camera events and the experiment parameters associated with them were cataloged and managed with the Mark IV system described above.

The system has three major inputs. A summary of the commanded images was accumulated from the Lander Sequencing Software (LSEQ). This summary provided Fovlip and IPL with an expected camera event list from which control information could be generated for processing the images. A record of the first-order image processing was acquired via an interface with Fovlip. It provided a list of the transmitted images, the engineering parameters, and the first-order photoproducts. A second-order image-processing record was received from the Vicar/IPL system supplying tape storage information, processing parameters, and photoproduct identifiers.

The major catalogs generated are as follows: (1) a complete description of all camera events comprising the EDR sorted by camera event number, (2) camera events sorted by type (see Table 1), (3) high-resolution events sorted by time of day on Mars, and (4) events sorted by elevation pointing angle and starting azimuth.

For the Primary Mission the above listings together with an offset reproduction of each camera event suitable for identification purposes will be published as the Viking Lander Camera System Experiment Data Record Picture Catalogue and will be available through the Langley Research Center, Hampton, Virginia 23665, or the NSSDC, Code 601, Goddard Space Flight Center, Greenbelt, Maryland 20771 (NASA document RP 1007).

APPENDIX: GLOSSARY OF ACRONYMS

Digifax	Digital Facsimile Printing Device
DSN	Deep Space Network
EDR	Experimental Data Record
Fovlip	First-Order Viking Lander Image Processing
GRE	Ground Reconstruction Equipment
IPL	Image Processing Laboratory
ISDR	Intermediate System Data Record
JPL	Jet Propulsion Laboratory
LSEQ	Lander Sequencing Software
MTIS	Mission and Test Imaging System
MTPS	Mission and Test Photographic System
NSSDC	National Space Science Data Center
RESTIMG	Restore Image Data
SDR	System Data Record
TDR	Team Data Record
TLMP	Telemetry Processing Program
VMCCC	Viking Mission Control and Computing Center

Acknowledgments. The first-order processing system (Fovlip) was written by L. G. Green, H. F. Lesh, and E. Morita at JPL. Many individuals at the JPL Image Processing Laboratory developed and implemented the second-order image-processing programs and capabilities. The following identifies areas of major responsibilities: Libexec, Michael Girard; library and catalog programs, Michael Martin, Edward Y. S. Lee, and Ted Sesplaukis; calibration and distortion removal, Michael Wolf; ranging, Arnold Schwartz; color reconstruction, William Benton; mosaic production, Joseph Berry and Steven Albers; data logging and graphics, Rodger Philips; TDR formatting and production, Susan LaVoie and Deborah Spurlock; PDP 11/40 software, Paul Jepsen; IPL facility upgrade, Joel Seidman; and IPL management and administration, Don Lynn. The Viking Image Processing System, as a whole, was under the management responsibility, at JPL, of Kermit Watkins of the Viking Project Office and the direction of Elliott Levinthal and Sidney Liebes, representing the Lander Imaging Team. This paper presents the results of one phase of the tasks carried out at the Jet Propulsion Laboratory, California Institute of Technology, under contract NAS-7-100; by the Genetics Department, Stanford University, under contract NAS-1-9682; and by the Department of Geological Sciences, Brown University, under contract NAS-1-9680. All of these were sponsored by the National Aeronautics and Space Administration.

REFERENCES

Green, W. B., P. L. Jepsen, J. E. Krezner, R. M. Ruiz, A. A. Schwartz, and J. B. Seidman, Removal of instrument signature from Mariner 9 television images of Mars, *Appl. Opt.*, *14*, 105, 1975.

Huck, F. O., H. F. McCall, W. R. Patterson, and G. R. Taylor, The Viking Mars lander camera, *Space Sci. Instrum.*, *1*, 189, 1975.

Huck, F. O., D. J. Jobson, S. K. Park, S. D. Wall, R. E. Arvidson, W. R. Patterson, and W. D. Benton, Spectrophotometric and color estimates of the Viking lander sites, *J. Geophys. Res.*, *82*, this issue, 1977.

Jepsen, P. L., The software/hardware interface for interactive image

Laboratory, in *Proceedings of the Digital Equipment Computer Users Society*, 1976.

Levinthal, E. C., W. B. Green, J. A. Cutts, E. D. Jahelka, R. A. Johansen, M. J. Sander, J. B. Seidman, A. T. Young, and L. A. Soderblom, Mariner 9 image processing and products, *Icarus, 18*, 75, 1973.

Liebes, S. L., and A. A. Schwartz, Viking 1975 Mars lander interactive computerized video photogrammetry system, *J. Geophys. Res., 82*, this issue, 1977.

Rindfleisch, T. C., J. A. Dunne, H. J. Frieden, W. D. Stromberg, and R. M. Ruiz, Digital processing of the Mariner 6 and 7 pictures, *J. Geophys. Res., 76*, 394, 1971.

Ruiz, R. M., D. A. Elliott, G. M. Yagi, R. B. Pomphrey, M. A. Power, K. W. Farrell, Jr., J. J. Lorre, W. D. Benton, R. E. Dewar, and L. E. Cullenn, IPL processing of the Viking orbiter images of Mars, *J. Geophys. Res., 82*, this issue, 1977.

Soha, J. M., D. J. Lynn, J. J. Lorre, J. A. Mosher, N. N. Thayer, D. A. Elliott, W. D. Benton, and R. E. Dewar, IPL processing of the Mariner 10 images of Mercury, *J. Geophys. Res., 80*, 2394, 1975.

(Received April 11, 1977;
revised May 20, 1977;
accepted May 20, 1977.)

Viking 1975 Mars Lander Interactive Computerized Video Stereophotogrammetry

SIDNEY LIEBES, JR.

Department of Genetics, Stanford University Medical Center, Stanford, California 94305

ARNOLD A. SCHWARTZ

Image Processing Laboratory, Jet Propulsion Laboratory, California Institute of Technology Pasadena, California 91103

A novel computerized interactive video stereophotogrammetry system has been developed for analysis of Viking 1975 lander imaging data. Prompt, accurate, and versatile performance is achieved. Earth-returned digital imagery data are driven from a computer to a pair of video monitors. Powerful computer support enables a photogrammetrist, stereoscopically viewing the video displays, to create diverse topographic products. Profiles, representing the intersection of any definable surface with the Martian relief, are readily generated. Vertical profiles and elevation contour maps, including stereo versions, are produced. Computer overlays of map products on stereo images aid map interpretation and permit independent quality evaluation. Slaved monitors enable parallel viewing. Maps span from the immediate foreground to the remote limits of ranging capability. Surface sampler arm specific vertical profiles enable direct reading of arm commands required for sample acquisition, rock rolling, and trenching. The ranging accuracy of ±2 cm throughout the sample area degrades to ±20 m at 100-m range.

INTRODUCTION

This paper discusses the interactive computer-based stereo ranging system developed jointly by Stanford University and the Jet Propulsion Laboratory to analyze the topography imaged by the Viking 1975 Mars lander cameras. The presentation includes descriptions of the stereo analysis system and its operational characteristics and examples of the output products.

The system has been utilized to develop vertical profiles and elevation contour maps systematically for the entire stereoscopically imaged surroundings of both landers, extending from the immediate foreground to the remote limits of ranging capability, several hundred meters from the spacecraft. The system has been used in support of all surface sample acquisitions, trenchings, and rock rollings to quantify the topography in a sampler arm specific coordinate system that enables a direct reading of the arm commands required to execute the desired sequences.

LANDER CAMERA CONFIGURATION

Figure 1 indicates the relative locations of the lander cameras, the surface sampler arm, and the Viking project defined Lander Aligned Coordinate System (LACS), adopted as the reference coordinate system for the representation of all range data. The unlikely placement of this system derives from its introduction for lander-orbiter integration purposes.

The lander facsimile cameras have been described in detail elsewhere [*Huck et al.*, 1975; *Mutch et al.*, 1972]. It will suffice for the purpose of this discussion to consider each of the identical cameras to be an image-scanning device that samples scene brightness in a raster of viewing directions relative to a photogrammetric reference point. Vertical scans are accomplished by the upward nod of a mirror located in the object space of the camera. Following the acquisition of 512 samples at equal angular increments along a vertical scan the camera rotates clockwise in azimuth about a vertical axis, and the process is repeated. Each of the azimuthal and elevation steps is equal to the angular resolution of the selected detector in the photosensor array, either 0.04° or 0.12°. In the vertical direction the camera can be commanded to scan about image centers at 0°, ±10°, ±20°, ···, elevation. In the azimuthal direction the camera can be commanded to a set of start and stop azimuth values at 2.5° intervals. In this fashion, each camera can accumulate data from 40° above the horizon to 60° below, through nearly 360° in azimuth. The photogrammetric reference points of the two cameras are located 0.822 m apart and 1.300 m above the nominal landing plane.

IMAGE RECONSTRUCTION

The straightforward and standard format of image reconstruction is illustrated in Figures 2*a*–2*c*. Picture elements, acquired by sampling in equal increments in elevation angle and azimuth angle, are reconstructed 1:1 on a square raster, elevation displacements being aligned along the ordinate and azimuth displacements along the abscissa. It is a characteristic of this acquisition/reconstruction sequence that most straight lines in the scene will appear curved in the reconstruction. Specifically, consider the infinitely long straight line shown in Figure 2*c*, oriented parallel to the intercamera base line and lying immediately in front of the lander on the nominal landing plane. Each camera will record the line to vanish at horizon points 180° apart, at 0° elevation (Figures 2*a* and 2*b*). Since all intermediate points will be below the horizon, the line will reconstruct as being curved. Scene points will generally not be perceived at equal elevation angles by the two cameras, to wit, point *P*, located along the line directly in front of camera 2. This point will appear at a higher elevation angle for camera 1 than for camera 2.

ANALYTICAL STEREO ANALYSIS SYSTEM

General

The analytical stereo analysis system capitalized on the introduction for Viking of a time-shared interactive computing

Paper number 7S0468.

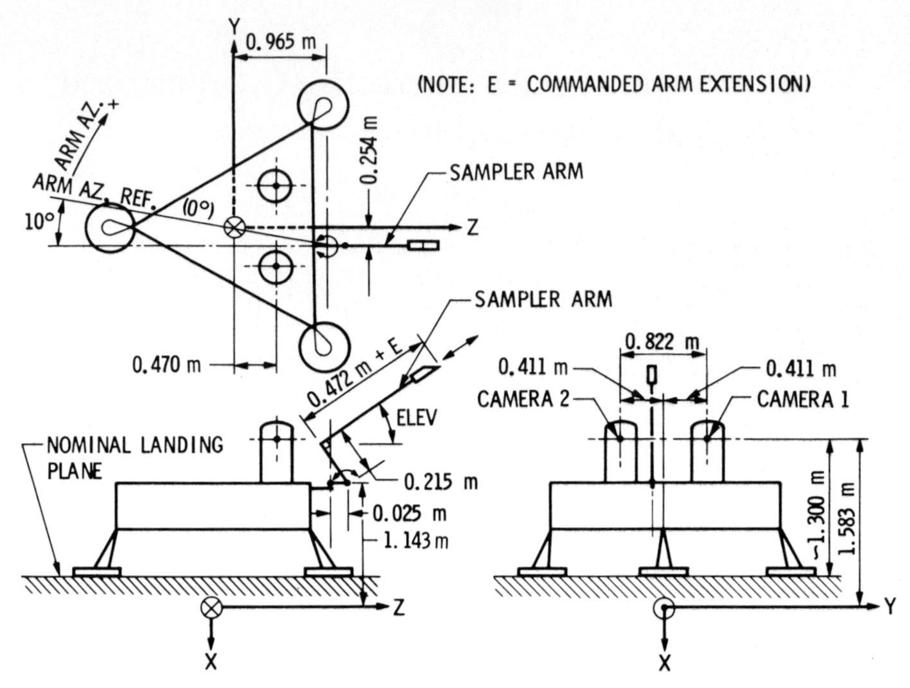

Fig. 1. Schematic illustration of the relative location of the lander cameras, the surface sampler arm, and the Lander Aligned Coordinate System (LACS).

facility at the Jet Propulsion Laboratory's (JPL) Image Processing Laboratory (IPL).

The stereo system was developed to enable prompt, accurate, and versatile analysis of the topography recorded by the lander cameras. Time limitations and budgetary considerations precluded development of an automated mapping system that could operate adequately on near-ground-based stereo over distances spanning from the very near field to the very far field. Consequently, three basic design decisions were made: (1) the system would rely on a photogrammetrist, employing the traditional skills of visual stereo correlation and manual control of a three-space mark; (2) powerful analytical assistance would be provided by the computer; and (3) all imagery and visual aids required by the photogrammetrist would be formatted by the computer to stereo video monitors for his viewing. The adoption of a computer-based video system offered great analytical power and flexibility to the designers and, furthermore, enabled topographic analysis to commence essentially as soon as the images had been logged into the computer.

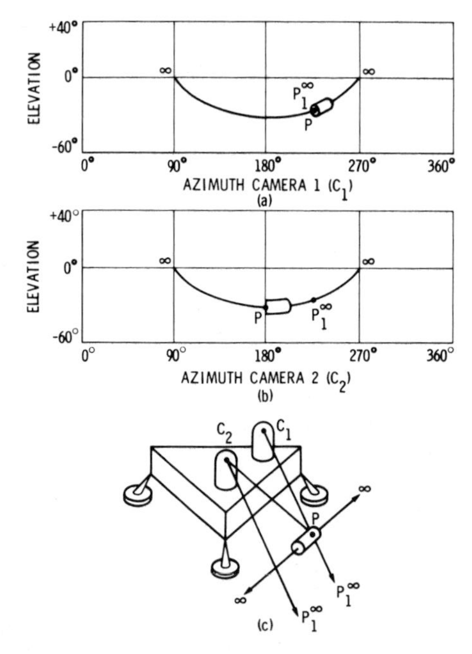

Fig. 2. Standard image reconstruction format characteristics for (*a*) camera 1 and (*b*) camera 2. (*c*) Diagram showing an infinite straight line resting on the landing surface parallel to the intercamera base line. One end P of a 'tennis ball can,' coaxial with the line, lies directly in front of camera 2. The reconstructions of the camera 1 and camera 2 recordings of the straight line appear curved. The tennis ball can appears rotated, displaced, and magnified in one image relative to the other. The vanishing point P_1^{∞} is at the infinite extension of the ray passing from camera 1 through point P.

Computer System

The basic hardware configuration of the Interactive Image Processing Facility (IIPF) has been described elsewhere [*Levinthal et al.*, 1977]. We need only note here that the main frame computer, an IBM 360/65, interfaces with the complex of on-line peripherals through a DEC PDP-11 minicomputer. Imagery and graphics data directed from the 360/65 to the video monitors of the stereo station, described below, are routed via a Ramtek random access solid-state memory device. The Ramtek memory consists of eight 512 × 640 bit planes, six of which support the imagery and two of which are utilized for graphics overlay. Operator control of the software program Ranger, which supports the stereo analysis system, is generally accomplished through an Imlac keyboard and volatile alphanumeric display.

Stereo Station

The stereo station, to which the video images are driven and at which the stereophotogrammetry is accomplished, is illustrated schematically in Figure 3. It consists of an 8-foot-long (2.438-m-long) table with a pair of modified Conrac QQA-17C video monitors mounted at opposite ends and swiveled to face

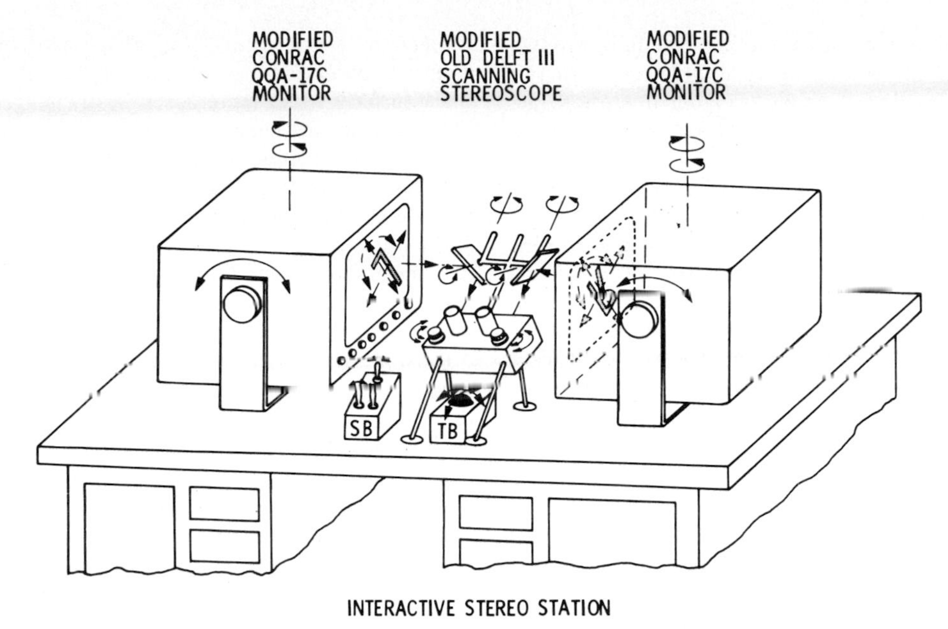

Fig. 3. The stereo station consists of a pair of 17-in. (43-cm) video monitors facing one another from opposite ends of an 8-foot-long (2.438-m-long) table, a scanning stereoscope, a switch box (SB) that controls both the image routing and the rotational state of the motorized beam deflection yokes in the monitors, and a track ball (TB) that controls the location of the three-space mark overlaid on the stereo images.

one another for stereo operations. The P-4 white phosphor commonly employed in video monitors is composed of yellow and blue components that tend to aggregate into colored patches characterized by differing decay constants. We avoided potential multicomponent phosphor issues by utilizing a single-component white P-45 phosphor.

Left and right camera imagery data are driven to the left and right monitors, respectively. An Old Delft III scanning stereoscope, modified for lateral viewing, is mounted midway between the monitors. Slaved video monitors enable parallel viewing of all stereo operations.

During the primary mission, quantitative topographic analysis was generally accomplished by using raw returned imagery data formatted in the standard manner illustrated in Figures 2*a* and 2*b*. Ranger can additionally operate on images rectified to plane camera perspective as well as on those transformed to remove horizon distortion due to spacecraft tilt. However, geometric transformation of the imagery generally results in some degradation of spatial resolution and some sacrifice of geometric fidelity.

The stereo station video image alignment requirements can be seen by reference, again, to Figure 2. Consider the 'tennis ball can,' shown coaxial with the previously discussed straight line in Figure 2*c* and located closer to camera 2 than to camera 1. The following observations can be made by reference to Figures 2*a* and 2*b*: (1) the can will appear at different elevation angles for the two cameras, (2) it will project to different image sizes because of the unequal elevation angles and distances from the cameras, and (3) the can axis will appear rotated in one image relative to the other.

It follows that there is a necessity, especially in the near field, to translate, magnify, and rotate one image in relation to the other in order to obtain a suitable stereo model. Coarse translational displacements are quickly achieved by command to the computer, but fine vertical and horizontal adjustments are more conveniently accomplished by analog vernier controls installed on the monitors, and by parallax and panning controls on the stereoscope. Analog controls on the monitors also enable independent substantial variation of image size in mutually orthogonal directions. Finely controlled image rotation is accomplished by a motorized rotary drive on the beam-deflecting yokes of the video monitors.

A track ball constitutes the prime means by which the photogrammetrist communicates with the computer, generally to control placement of one point cursor or a stereo pair of point cursors on the video displays. The track ball is a baseball-sized sphere protruding from the top of a retaining box and capable of being rotated freely and indefinitely about its center. The ball rests on two orthogonally oriented angular sensors that enable the input to the 360/65 computer of two independently variable parameters that may be employed for any desired program control purpose.

Software

All analytical stereo analysis is controlled by the applications program Ranger performing under the IBM time-sharing operating system, TSO. Ranger commands can fetch images from disc storage, extract and forward to the stereo video monitors selected image subwindows, alter the video stretch tables of the Ramtek, command the read-back from the Ramtek into the 360/65 of any video and graphics data displayed on the stereo station monitors, magnify images, incorporate camera calibration files, respond to keyboard and track ball commands, and generally accomplish all of the required calculations and input/output file management associated with the suite of operating modes. It will be recalled that the Ramtek video refresh device has storage space for images measuring 512 × 640 pixels. Ranger excises subwindows 512 pixels high by 320 pixels wide from the left and right images and concatenates these 512 × 640 pixels to the Ramtek. The concatenated pair can be routed directly to any number of video monitors, in addition to those of the stereo station. The stereo station also incorporates video processing circuitry that can decompose the concatenated pair into the constituent 512 × 320 pixel images and route these separately to the two monitors for stereo work.

Quantitative topographic analysis of the scene falls into two classes of operations: (1) point ranging and (2) contouring or profiling. We describe first the point-ranging operation, designed to establish the lander body Cartesian coordinates, in the LACS, of points in the relief.

Ranger is commanded to forward to the stereo station monitors a suitable image pair. The operator then uses the track ball to move a point cursor to a feature of interest in one of the images, say, the point P in the camera 1 image of Figure 2*a*. Ranger is able, by virtue of incorporated lander camera geometry and calibration files, to associate with the cursor a viewing vector extending from camera 1 through the selected scene point P to the vanishing point P_1^∞ at infinity. Knowledge of the camera geometry allows this vector to be projected into the camera 2 image, as is indicated by a portion of the solid segment of the curved line in Figure 2*b*. The operator uses the track ball to move a point cursor, constrained to the projection of the view vector in the camera 2 image, to the location of the conjugate scene point. The values of three independent degrees of freedom required to locate the feature of interest are thereby established, and Ranger evaluates the LACS values.

If residual geometrical discrepancies, resulting, for example, from camera model deficiencies, calibration file limitations, or lander settling, cause the projection of the view vector to miss the field point in the conjugate image, the track ball may be used to define an independent viewing direction for the second camera. The field point is then taken to be at the midpoint of the shortest three-space path connecting the view vectors from the two cameras.

It is a critical feature of the system that since the signal path from the track ball to the video screen is via the 360/65 and a graphics plane of the Ramtek, there is a precise and unambiguous registration of all graphics overlay relative to the displayed imagery. Since the graphics and video planes of the Ramtek are summed prior to output to the displays, the registration is independent of the performance of the subsequent circuitry and video monitors.

Though it is desirable, it is not essential that point ranging involve the fusion of the stereo images into a three-dimensional model. The perception of a stereo model is mandatory, however, for the continuous profiling or contouring mode.

Profiling and contouring are functionally identical. They represent particular applications of the general capability of Ranger to support generation of a set of three-dimensional intersections of families of any analytically definable three-space surfaces with the stereoscopically imaged Martian relief. We illustrate this capability by outlining the procedure for generating a gravity-normal contour map.

Initial postlanding inertial reference data are utilized to provide Ranger with the transformation between the LACS and a north-oriented gravity-normal coordinate system. Suitable images are identified and windowed to the stereo station monitors. The photogrammetrist uses the combination of previously described video monitor and stereoscope analog controls of position, size, and rotation, as well as brightness and contrast, to establish a satisfactory stereo mode. Ranger is then given an elevation value for a contouring plane. The location of a three-space point, common to the two images and on the plane, is established, and Ranger projects this point into each of the images via a graphics overlay plane of the Ramtek. The operator will observe these two points to fuse into a three-space cursor (TSC) resting upon the contouring plane though generally not yet coincident with his perception of the Martian relief.

Ranger incorporates a choice of scale-variable trigonometric parameterizations relating quantized track ball angular coordinates to the location of the TSC on the surface of constraint, in this case the elevation contouring plane. If mapping is to span from the near to the far field, a parameterization is selected that enables the operator to move the TSC from proximate to remote features in a reasonable number of track ball rotations.

Upon completion of initialization procedures the photogrammetrist moves the TSC along the contouring plane to an intersection of the plane with the perceived relief and initiates contour generation. As it is necessary in this mode that there be no perceivable delay between displacement of the track ball and tracking response of the TSC, approximately 30 incremental TSC updates per second are processed through the time-sharing system.

As the contour is traced, the succession of quantized LACS three-space coordinates is recorded into a range data set (RDS) computer file. The operator generally exercises the valuable option to display visually the projections of the contours as image overlays on the stereo monitors as they are generated. Function button control on the track ball housing

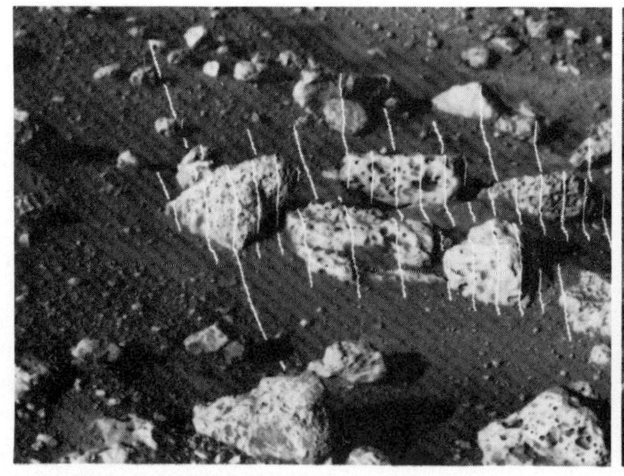

Fig. 4. Family of arm specific vertical profiles overlaid on a stereo pair of images of VL 2 rock displacement candidates. The profiles are 1° apart. The rocks are approximately 2 m from the azimuth gimbal axis of the sampler arm. The intensity modulating bands running diagonally across the images are a printing artifact resulting from a beating between the halftone screen and the camera pixels.

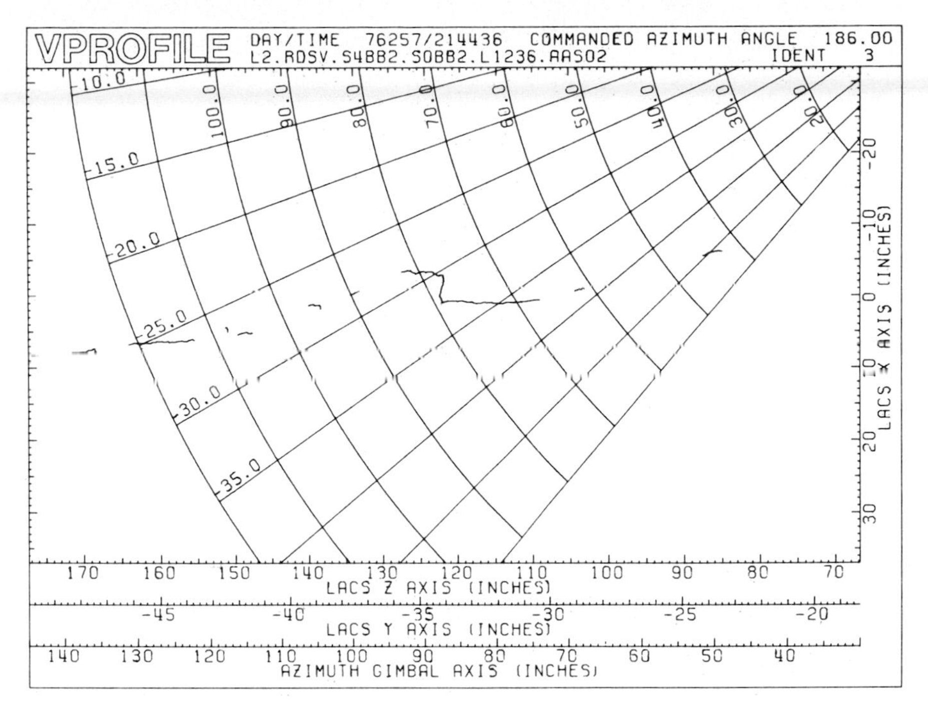

Fig. 5. Plot of the third vertical profile from the left in Figure 4. The LACS x, y, and z coordinate scales appear in the margins, the y and z being azimuth dependent. The distance from the sampler arm azimuth gimbal axis is indicated at the bottom. The concentric curves denote sampler arm collector head tip isoextension loci, labeled with command values of arm extension (in inches). The diagonal fan of rays indicates tip position isoelevation loci for indicated arm elevation angle commands. The Julian day and time of profile generation appear at top center, immediately beneath which is the RDS name. The arm command value of azimuth angle appears in the upper right corner. The identification number, immediately beneath the azimuth, designates the profile member of the RDS.

enables the photogrammetrist to make and break the contour as conditions require. Erasure can be accomplished when it is necessary. Contour generation continues over the range of selected elevations and from window to window until mapping is completed.

The contouring principle may be applied to any family of analytically definable surfaces. In practice, contour maps have been generated for both gravity-normal and lander-aligned coordinate systems. Vertical profiles of the relief have been generated in both of the above systems. Cylindrical profiles concentric with the lander have also been generated.

One of the important mission services provided by the stereo system has been the creation of map products specifically generated and formatted to support the development of surface sampler arm commands for sample acquisition, trenching, and rock rolling. It will be noted from Figure 1 that the sampler arm articulates in azimuth about an axis normal to the deck of the lander. Arm contact with the Martian surface is generally accomplished by commanding the arm to deelevate to surface contact microswitch cutoff at specified values of extension and azimuth. Consequently, there was incorporated into Ranger a mode that enables the generation of arm specific

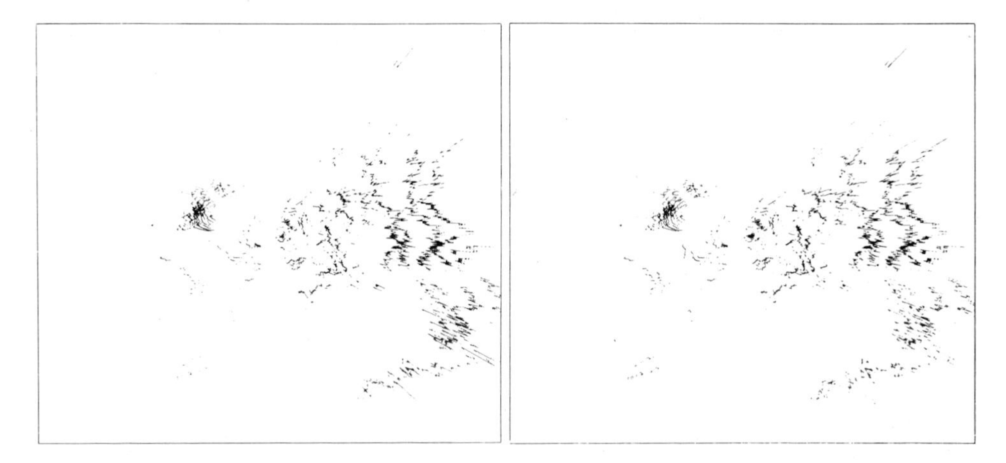

Fig. 6. Stereo representation of a portion of the gravity-normal VL 1 site contour map extending out approximately 100 m in all directions from the centrally located lander. The front of the lander faces to the right. The view is obliquely downward from the right side of the lander; i.e., the surface of Mars recedes as one moves from the bottom to the top of the figure. Absolute accuracy of the primary data varies from less than 1 cm in the near field to upward of 20 m at 100-m range. The contour interval, which increases with range, is 1 m on the conspicuous 5-m-high ridge to the right. The construction combines a convergence angle of 10° and a ×2 vertical exaggeration.

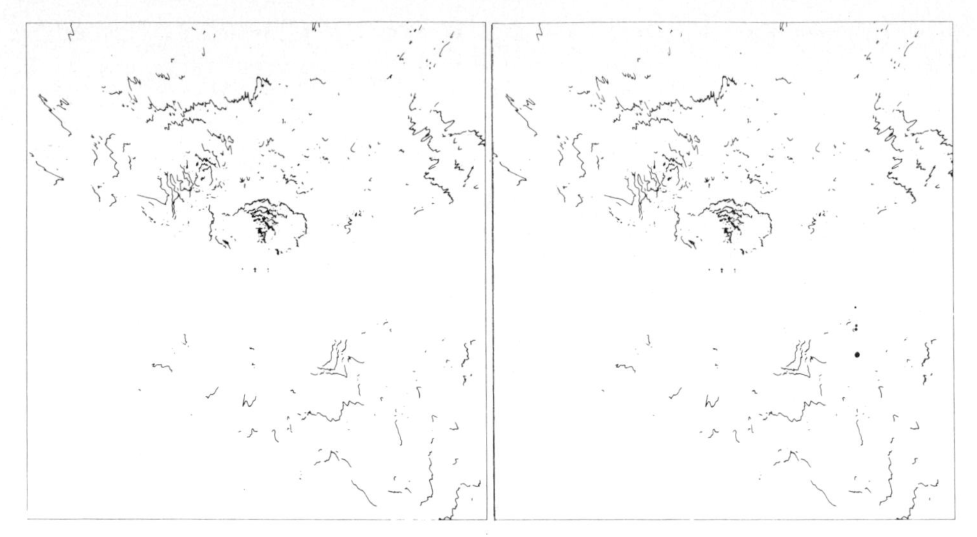

Fig. 7. Stereo representation of a portion of the VL 1 site map shown in Figure 6. The view here is downward, perpendicular to the lander deck. The front of the lander faces toward the top of the figure. The map extends out approximately 20 m from the centrally located lander. Two depressed areas are conspicuous. One, several meters across, is immediately in front of the lander (at 12 o'clock), and another of comparable size is further out (at 10–11 o'clock). The central plus at the lander location denotes the LACS origin. One-meter scale marks appear to each side of the plus. The contour interval is variable. The total relief is about 1 m. A much larger scale would be required to reveal the substantial map detail close to the lander.

vertical profiles of the relief on planes passing through the sampler arm azimuth axis. The profile plane is established either by commanding it to pass through a cursor-identified three-space point or by specifying an arm command value of azimuth.

Output Products

A map and an overlay of contours of the image of a dune field at the VL 1 (Viking Lander 1) site, generated by the present system, have been published [*Mutch et al.*, 1976*a*], as have an overlay and plots of vertical profiles of the first trench dug at the VL 1 site [*Shorthill et al.*, 1976]. Figure 4 is an illustration of the stereo image products that result from the generation of a family of arm specific vertical profiles. Such images, with overlaid profiles or contours, are synthesized by either of two means. Either the Ramtek memory is read back to the 360/65 in real time, or alternatively, at some subsequent time the overlays may be constructed from the magnetic tape or disc-stored primary images and RDS's. Prints of profiles and contours overlaid on imagery serve two particularly valuable functions: (1) they enable a ready association of topographic map or profile plot details with features in the scene, and (2) they enable an independent assessment at any future time of the quality of the stereophotogrammetry.

Prompt hard copy of any displayed video data, in particular, the imagery overlaid with map graphics, was produced by an on-line Tektronix 4632 video hard copy unit. Film products were generated by off-line film recorders.

Figure 5 illustrates a plot of the third vertical profile from the left in Figure 4. Arm command values of azimuth, extension, and elevation required to accomplish any desired sequence can be read directly from such vertical profile plots. Plots can be generated at any desired scale, including full size, with flexibility regarding fiducial scales, overlay grids, and system of units. A summary of primary mission surface sampler operations based on overlay and profile data has been reported [*Clark et al.*, 1976]. The utilization of vertical profile data in support of rock pushing and sampling under rocks on Mars has also been described [*Moore et al.*, 1977].

Computer-synthesized mosaics of high-resolution (0.04°) images were generated during the early phase of the extended mission. The component images were acquired with four different high-resolution diodes, each requiring independent geometric calibration treatment in the process of mosaic generation. Thus Ranger need not access the diode calibration files during mapping with mosiacs. For both landers the mosaics have been used for the systematic generation of gravity-normal contour maps and vertical profiles covering from the immedi-

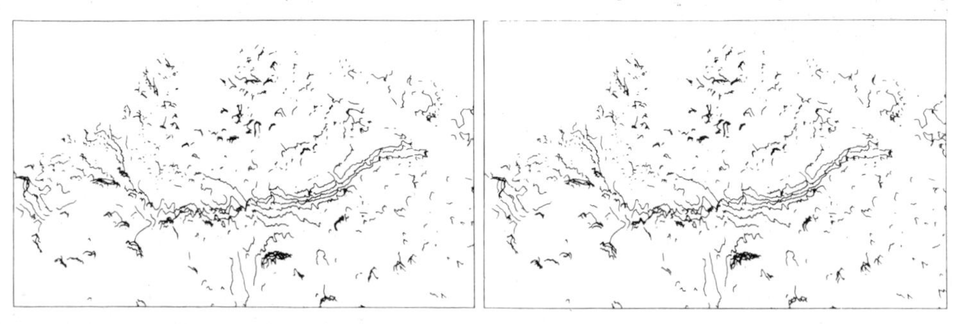

Fig. 8. Stereo representation of a contour map of a portion of the troughlike feature present in front of VL 2. The view is normal to the tilted deck of the lander. The front of the lander faces upward in the figure from a location near bottom center. The mapped region is approximately 15 m wide. The convergence angle is again 10 ° with ×2 vertical exaggeration. The contour interval within the trough is 5 cm.

Fig. 9. Contours spanning the left central portion of the trough in Figure 8 overlaid upon the mosaic stereo image pair. Spacecraft tilt has been removed by computer processing, revealing a flat horizon. The images are not, however, geometrically rectified to planar camera perspective.

ate vicinities of the landers to the remote limits of ranging capability, in excess of 100 m from each craft. These maps are characterized by range dependent accuracies and contour spacings.

Landers 1 and 2 were tilted 3° [*Mutch et al.*, 1976*b*] and 8° [*Mutch et al.*, 1976*c*], respectively. By the time that the VL 1 systematic mapping had been completed, computer processing had removed the tilt from the VL 2 mosaics. The VL 2 mapping was accomplished with these detilted images. This paper cannot include the full resolution contour maps, characterized by 1-cm accuracy in the near field. However, as companions to the contour maps, stereo representations of the map products are also generated by using a computer program that accomplishes perspective stereo projections of the computer resident map data. Orthographically projected pairs enable arbitrarily large mapped areas to be viewed stereoscopically. Even small size, limited resolution, reconstructions of these stereo map materials can indicate the topography. Examples of such orthographic stereo maps are presented in Figures 6–8. Each of these figures employs a stereo convergence angle of 10° and ×2 vertical exaggeration. Figure 6 shows a computer-generated oblique stereo view of the portion of the VL 1 landing site extending out from the centrally located lander to approximately 100 m in all directions. The view is similar in perspective to that seen by an airline passenger looking down obliquely at the terrain beneath him. The most conspicuous relief is indicated by 1-m-interval contours on the 5-m-high ridge in front of the lander (to the right in the figure). The associated contour map products along with such stereo pairs served as the basis for the VL 1 site map presented and discussed in a companion article [*Binder et al.*, 1977].

Figure 7 is an overhead stereo view (normal to the deck of the lander) of the portion of Figure 6 extending out approximately 20 m from the centrally located lander. Two depressions can be seen, one immediately in front of the lander (up in the figure) and another further out, at 10–11 o'clock.

Figure 8 is a stereo view, normal to the tilted deck of VL 2, showing details of the conspicuous troughlike structure discussed in an earlier publication [*Mutch et al.*, 1976*c*.] The contour interval within the trough is 5 cm. The width of the projection window is approximately 15 m. The contour maps and stereo map representations of this area served as the basis for the detailed map of this troughlike feature presented in a companion article [*Mutch et al.*, 1977]. The contours overlaid on the stereo pair of mosaics in Figure 9 are those of the left central region of Figure 8.

Calibration and Ranging Accuracy

The use of preflight survey data and preflight test target images for the generation of geometric calibration files for the flight configuration hardware has been described elsewhere [*Patterson et al.*, 1977; *Wolf*, 1977]. These files accounted for the geometrical idiosyncrasies of each of the 12 photosensitive diodes in each camera. Ranger utilizes these files in either of two distinct fashions. Ranger either incorporates the calibration data into its calculations in order to operate on geometrically untransformed images or operates on geometrically corrected images, in which latter case, Ranger only indirectly uses the calibration data.

Consideration must be given to the applicability of the preflight calibration data to the landed hardware on Mars. Geometrical perturbations could result from a variety of causes, e.g., irreversible landing-induced strains, thermal gradients, wind buffeting, settling of the lander, and anomalous camera behavior. Several procedures were employed to check the geometrical performance of the landed system. Postlanding internal elevation and azimuth scan checks were performed. Images were subjectively inspected for clues to scan misbehavior. Images of hard points on the lander were compared with preflight data. Stereo images enabled two internal consistency checks. (1) The view vector point-ranging mode in Ranger was utilized to ascertain the precision of the intersection of the projection of the vector in the conjugate image with the conjugate field point; TSC projection into the image pair in the profiling and contouring modes constituted a similar common check. (2) Intercamera parallax checks of remote points in the scene were made. Image comparison and differencing of images taken of the same hard points in the natural scene at different times enabled lander displacement to be detected.

The following findings were made. VL 2 experienced a slight displacement manifesting itself in the most extreme case, in high-resolution (0.04°) images, as $1\frac{1}{2}$ pixels of image displacement. The cause has not been established, but the displacement may have resulted either from firing of the sampler arm protective shroud into the surface near one of the lander footpads, or from forces resulting from sampler arm interactions with the surface. For both landers, parallax checks on

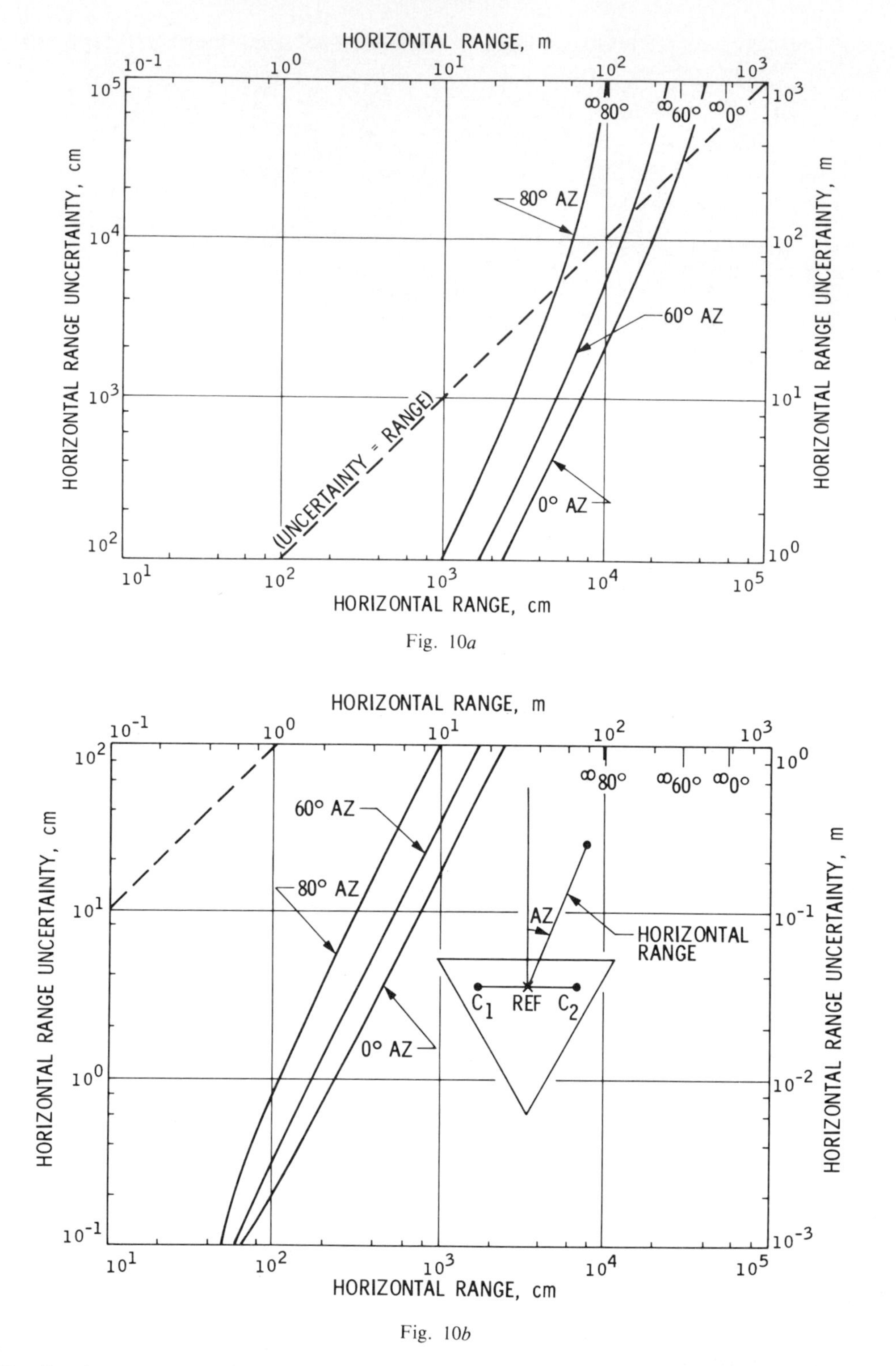

Fig. 10. Ranging accuracy as a function of range. Highest accuracy is in the direction perpendicular to the intercamera base line. The dashed line indicates accuracy equal to range. Infinite asymptotes for the curves appear in the upper right corner. The practical interpretation of these theoretical curves is that uncertainties are typically plus or minus the values indicated.

remote features and stereo pair disparities manifested via the mapping checks described above revealed typical relative angular displacements of 1–2 high-resolution pixels, extreme cases in the elevation direction being as high as 4 or, on occasion, 5 pixels. When such deviations were identified, the corrective difference was arbitrarily split between the cameras.

Ranging errors more nearly reflect disparities in azimuth than those in elevation and tend to be disproportionately greater at larger distances. However, high-resolution images spanning remote distances often incorporate what are effectively infinitely distant features that enable the relative parallax error to be removed. Thus it is concluded that ranging errors may be taken to be typically those associated with 1 or 2 pixels of relative pointing direction uncertainty.

Figures 10*a* and 10*b* are plots of the theoretical relative ranging accuracy as a function of range, parameterized by azimuth angle. The zero of azimuth of these plots is taken to be perpendicular to the intercamera base line. The curves are calculated for camera resolutions of 0.04° each and are constructed as follows. In plan view, a pair of wedges, each being of apex angle equal to the camera resolution, is radiated out from each camera to the region of wedge crossover at the test point in question. The length of the diamondlike overlap region in the direction away from the lander is taken for the purposes of the plot to be the ranging uncertainty. Uncertainties will be 3 times as large at the same range for paired 0.12° resolution diodes and roughly twice as large for crossed 0.04° × 0.12° diodes. In consideration of the above remarks regarding camera pointing direction uncertainties we conclude that one will obtain a fair measure of Viking lander absolute ranging accuracy by taking the uncertainty to be typically plus or minus the values indicated on the ordinates of Figures 10*a* and 10*b*. Transverse positional errors will generally be less than the plotted radial errors.

Comments

The fact that the stereo analysis system described here is so substantially computer oriented enabled great flexibility in design and ease of modification. The storage of all system products in computer accessible form permits ready manipulation of data and arbitrary formatting of output. The utilization of video monitors with unambiguous registration of graphics and imagery offers advantages in both start-up time and convenience over photographic input, as well as freedom from photographic projective errors. The digital and analog controls on image brightness and contrast and, indeed, the unlimited capacity to process the images enable effective utilization of the full dynamic range of the data. This range can be extreme with solid-state light detectors such as those employed in the Viking lander cameras.

The quantization of the digitally controlled graphics overlay raster is inherently identical to that of the underlying imagery. In using the stereo system, one sometimes wishes that one had the capacity for interpolative placement of the point-ranging and three-dimensional cursors. This could have been achieved by digitally magnifying the images, but this was not desirable without concurrently incorporating corresponding unavailable reducing optics. Time-shared operation posed certain limitations of convenience and speed that could be resolved by use of a stand-alone minicomputer.

Acknowledgments. Many individuals contributed to the successful development of the present system. Given that the stand-alone computer approach was not a viable option at the time, this development could not have been undertaken without the achievement of the IIPF, promoted with foresight and perseverance by E. C. Levinthal. W. B. Green, D. Lynn, and J. B. Seidman of IPL assumed major administrative responsibilities in implementing and managing the IIPF. We are indebted to many analysts at IPL. We wish to express our appreciation to M. A. Girard for implementing the time-sharing executive software and for much of the geometric image transformation code and to P. Jepsen for programing the TSO interface software for the PDP-11. We are grateful to M. R. Wolf for conducting the laborious calibration tasks and for creating the camera calibration files as well as for accomplishing the transformation routines that removed lander tilt from the images. J. W. Berry assumed major responsibility for the generation of the high-resolution computer-formatted mosaics and spent long hours commanding Ranger in support of the photogrammetrists. R. N. Philips wrote the computer code for the arm specific vertical profile format as well as that for the stereo projection of the map products. He also wrote the logging routines for reading the raw image tapes into the IIPF. J. J. Lorre wrote backup software to enable monoscopic ranging along the sampler arm shadow in the event of failure of one camera. We are indebted to D. L. Atwood for extensive and careful work in logging images. We are grateful to S. K. LaVoie for extensive formatting of images for film recorders and for creating files of RDS's and to W. A. Parkyn for executing many profile and map plots. W. B. Harlow of Stanford University expertly did virtually all of the machining required in the construction of the stereo station. L. H. Quam designed and constructed the image-splitting and formatting circuitry for the stereo station. Much of the sample support photogrammetry was performed by R. Jordan of the U.S. Geological Survey in Flagstaff, Arizona. We wish to acknowledge valuable discussions with J. E. Wurtz and F. J. Marshall of Litton Industries Electron Tube Division, San Carlos, California, and with R. Dorr of Dorr Electronics, Mt. View, California, regarding video hardware. C. Bientemma and C. Odom of Conrac Division, Conrac Corporation, West Covina, California, rendered great assistance regarding selection and modification of the Conrac monitors. The work of S. Liebes was supported by contract NAS-1-9682, and that of A. A. Schwartz by contract NAS-7-100, sponsored by NASA.

References

Binder, A. B., R. E. Arvidson, K. L. Jones, E. C. Morris, T. A. Mutch, and D. C. Pieri, The geology of the Viking Lander 1 site, *J. Geophys. Res., 82,* this issue, 1977.

Clark, L. V., D. S. Crouch, and R. D. Grossart, Viking '75 project summary of primary mission surface sampler operations, *Viking Flight Team Doc. VFT-019,* NASA, 1976.

Huck, F. O., H. F. McCall, W. R. Patterson, and G. R. Taylor, The Viking Mars lander camera, *Space Sci. Instrum., 1,* 189–241, 1975.

Levinthal, E. C., W. B. Green, K. L. Jones, and R. B. Tucker, Processing the Viking lander camera data, *J. Geophys. Res., 82,* this issue, 1977.

Moore, H. J., S. Liebes, Jr., D. S. Crouch, and L. V. Clark, Rock pushing and sampling under rocks on Mars, paper presented at 20th Plenary Meeting of COSPAR and Associated Symposia, Comm. on Space Res., Tel Aviv, Israel, June 7–18, 1977.

Mutch, T. A., A. B. Binder, F. O. Huck, E. C. Levinthal, E. C. Morris, C. Sagan, and A. T. Young, Imaging experiment: The Viking lander, *Icarus, 16,* 92–110, 1972.

Mutch, T. A., R. E. Arvidson, A. B. Binder, F. O. Huck, E. C. Levinthal, S. Liebes, Jr., E. C. Morris, D. Nummedal, J. B. Pollack, and C. Sagan, Fine particles on Mars: Observations with the Viking 1 lander cameras, *Science, 194,* 87–91, 1976*a*.

Mutch, T. A., A. B. Binder, F. O. Huck, E. C. Levinthal, S. Liebes, Jr., E. C. Morris, W. R. Patterson, J. B. Pollack, C. Sagan, and G. R. Taylor, The surface of Mars: The view from the Viking 1 lander, *Science, 193,* 791–810, 1976*b*.

Mutch, T. A., S. U. Grenander, K. L. Jones, W. R. Patterson, R. E. Arvidson, E. A. Guinness, P. Avrin, C. E. Carlston, A. B. Binder, C. Sagan, E. W. Dunham, P. L. Fox, D. C. Pieri, F. O. Huck, C. W. Roland, G. R. Taylor, S. D. Wall, R. Kahn, E. C. Levinthal, S. Liebes, Jr., R. B. Tucker, E. C. Morris, J. B. Pollack, R. S. Saunders, and M. R. Wolf, The surface of Mars: The view from the Viking 2 lander, *Science, 194,* 1277–1283, 1976*c*.

Mutch, T. A., R. E. Arvidson, A. B. Binder, E. A. Guinness, and E. C. Morris, The geology of the Viking Lander 2 site, *J. Geophys. Res., 82,* this issue, 1977.

Patterson, W. R., F. O. Huck, S. D. Wall, and M. R. Wolf, Calibration and performance of the Viking lander cameras, *J. Geophys. Res., 82,* this issue, 1977.

Shorthill, R. W., H. J. Moore, R. F. Scott, R. E. Hutton, S. Liebes, Jr., and C. R. Spitzer, The 'soil' of Mars (Viking 1), *Science, 194,* 91–97, 1976.

Wolf, M. R., Viking lander camera geometric calibration report, Jet Propul. Lab., Pasadena, Calif., 1977.

(Received April 13, 1977;
revised May 31, 1977;
accepted June 1, 1977.)

Particle Motion on Mars Inferred From the Viking Lander Cameras

CARL SAGAN, DAVID PIERI, AND PAUL FOX

Laboratory for Planetary Studies, Cornell University, Ithaca, New York 14853

R. E. ARVIDSON AND E. A. GUINNESS

Department of Earth and Planetary Sciences, Washington University, St. Louis, Missouri 63160

The cameras of the Viking landers have uncovered several lines of evidence for fine particle mobility on the Martian surface, including particulate drifts, rock-associated raised streaks, and probable ventifacts. Inferred peak wind directions in both Chryse and Utopia are roughly the same and are consistent with peak winds inferred by orbiter photography. A 24° systematic offset between the direction of rock-associated streaks in the Viking 1 landing site and Mariner 9 and Viking observations of crater-associated streaks is consistent in both sign and magnitude with a Coriolis acceleration of particles entrained by high-velocity winds in the course of the production of crater-associated streaks. If a significant fraction of the impact energy upon collision goes into deformation, strain, and rupture, there should be a preferential destruction of the most easily saltated grains, which are here called kamikaze particles, and a depletion of 150-μm-diameter grains. Observations of fine particulates dumped on the VL-1 grid indicate that major saltation events occurred between sols 96 and 207 and were caused by winds of >50 m s^{-1}, normalized to the top of the velocity boundary layer. This is the first observation of saltation on another planet and a rough confirmation of the usual Bagnold saltation theory applied to another planet.

As long ago as the late nineteenth century, ground-based observations of transient yellow clouds on Mars were interpreted as evidence for fine mobile dust on the planet's surface. Mariner 9 observations provided compelling evidence that both local and global sand and dust storms occur at least once each Martian year, that substantial alterations of the local distribution of albedo result, and that the classical seasonal and secular changes of bright and dark markings on the Martian surface are probably caused by wind-blown particulates [*Sagan et al.*, 1972, 1973, 1974; *Veverka et al.*, 1974]. A range of calculations and low-pressure wind tunnel experiments all indicate that the threshold friction velocity for saltation for fine particles on the Martian surface is about 2 m s^{-1} at the grain surface level and many tens of meters per second above the surface velocity boundary layer [*Sagan and Pollack*, 1969; *Sagan et al.*, 1972; *Greeley et al.*, 1976]. The basic physics behind these estimates for Mars was first derived by *Bagnold* [1941]. Such high winds are very likely provided by the steep latitudinal temperature gradient and by major elevation differences on the planet [*Gierasch and Sagan*, 1971; *Mass and Sagan*, 1976; *Pollack et al.*, 1976*a*]. The Viking landers provided the first opportunity for in situ study of Martian mobile particulates. Dune forms had been detected by Mariner 9 with relatively poor spatial resolution [*Cutts and Smith*, 1973; *Sagan et al.*, 1972], and there was some reason to believe that smaller-scale dune and drift fields might be very abundant on Mars and might with luck be within the range of the Viking lander cameras. The Viking 1 landing site, on the western slopes of the Chryse Planitia basin, and the Viking 2 landing site, in Utopia Planitia west of the crater Mie, are both very bland terrains, as viewed from orbit, consciously so chosen for lander safety reasons. Sand and dust accumulations might be expected preferentially in such locales. Pretouchdown study of the two landing sites suggested to some analysts that extensive dune fields might be found in the Viking 2 but not in the Viking 1 landing sites; but the data on which such interpretations were based were always recognized to be marginal. The actual situation at the surface was the reverse, with extensive (as well as small scale) accumulations apparent in the Viking 1 landing site but almost absent in the Viking 2 landing site [*Mutch et al.*, 1976*a*, *b*, *c*]. Views of these drifts can be seen in Figures 1–3. Because aeolian abrasion by mobile particles is proportional to at least the cube of the transport velocity, because the saltation threshold is about an order of magnitude larger on Mars than on earth, and because the high-energy tail of the velocity distribution function is not known to be more truncated on Mars than on earth, it was expected that signs of wind erosion would be common on Mars [*Sagan*, 1973]. In both landing sites, there are a number of rocks which seem to be clear ventifacts, including some with rounded 'dreikanter' shapes, and other features which are possible but by no means certain indicators of such abrasion, for example, the rock vesicles apparent in the Viking 2 landing site which might possibly be preferential aeolian pitting of the more friable mineral inclusions. In the case of the environment of Venus the recent surprise was that in the absence of so many erosion mechanisms which are effective on earth, the Venus rocks were so highly eroded [*Sagan*, 1976]. In the case of Mars the situation seems to be the opposite: in the presence of so potent an abrasion agent as aeolian abrasion, it is striking how many fresh looking rocks there are. One possible resolution of this dilemma is that both landing sites have been buried, for substantial periods, beneath dust deposits at least meters in thickness until comparatively recent times. An alternative explanation, discussed in more detail below, is that the grains most readily saltated have been preferentially depleted by collision. Finally, the relatively highest velocity winds may be less common on Mars than on earth. The Viking 1 landing site, on the slopes of the great Chryse basin, may be a region of enhanced slope winds [*Gierasch and Sagan*, 1971; *Sagan et al.*, 1971], a possibility which might conceivably have some bearing on the preferential presence of particulate drifts there.

Paper number 7S0590.

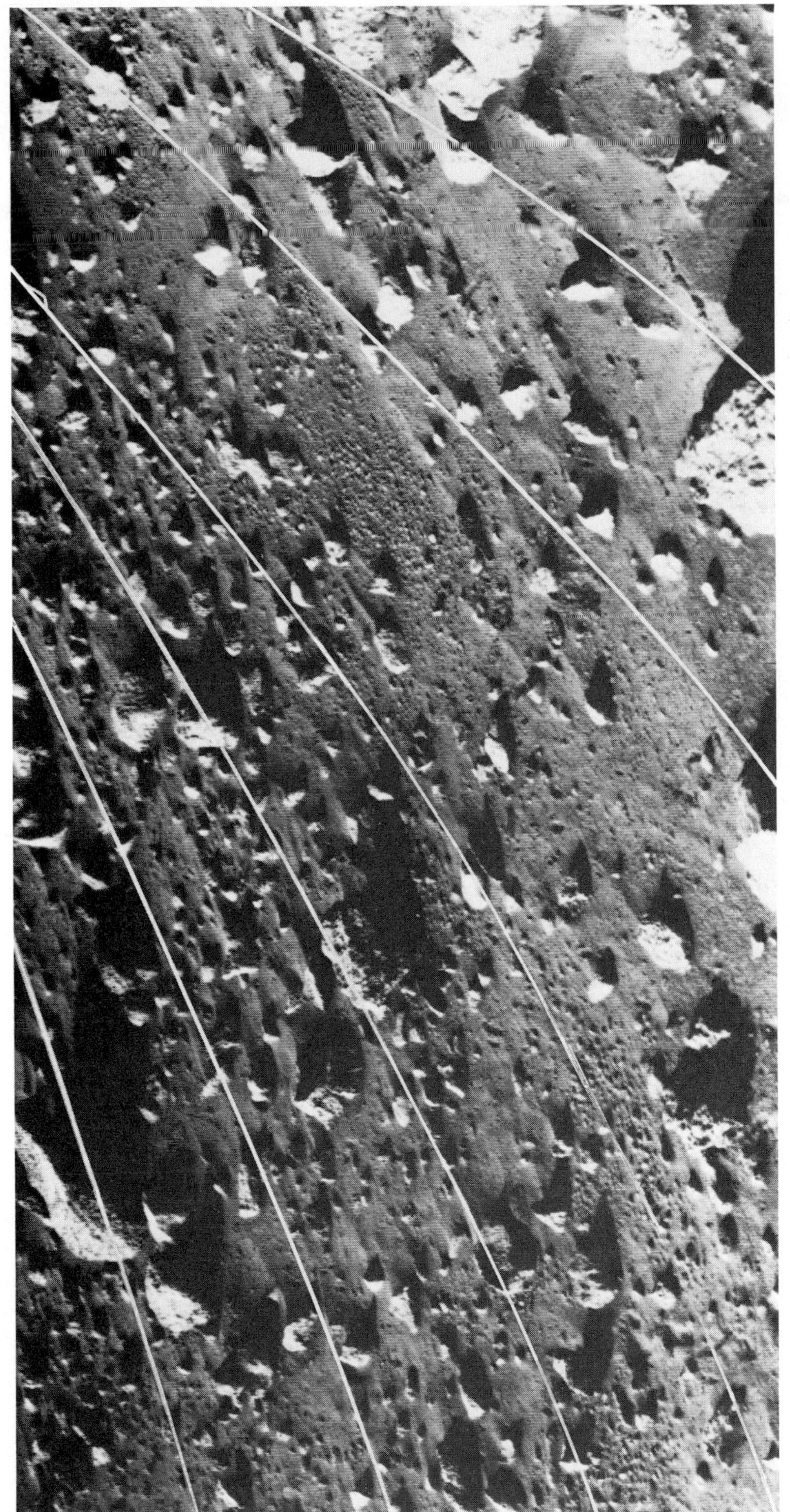

Fig. 1. This Viking Lander 1 frame shows a number of wind streaks extending from rocks. The white lines trend in the mean direction derived from wind streaks in this frame and other frames. CE label 12A003-C, segment 1.

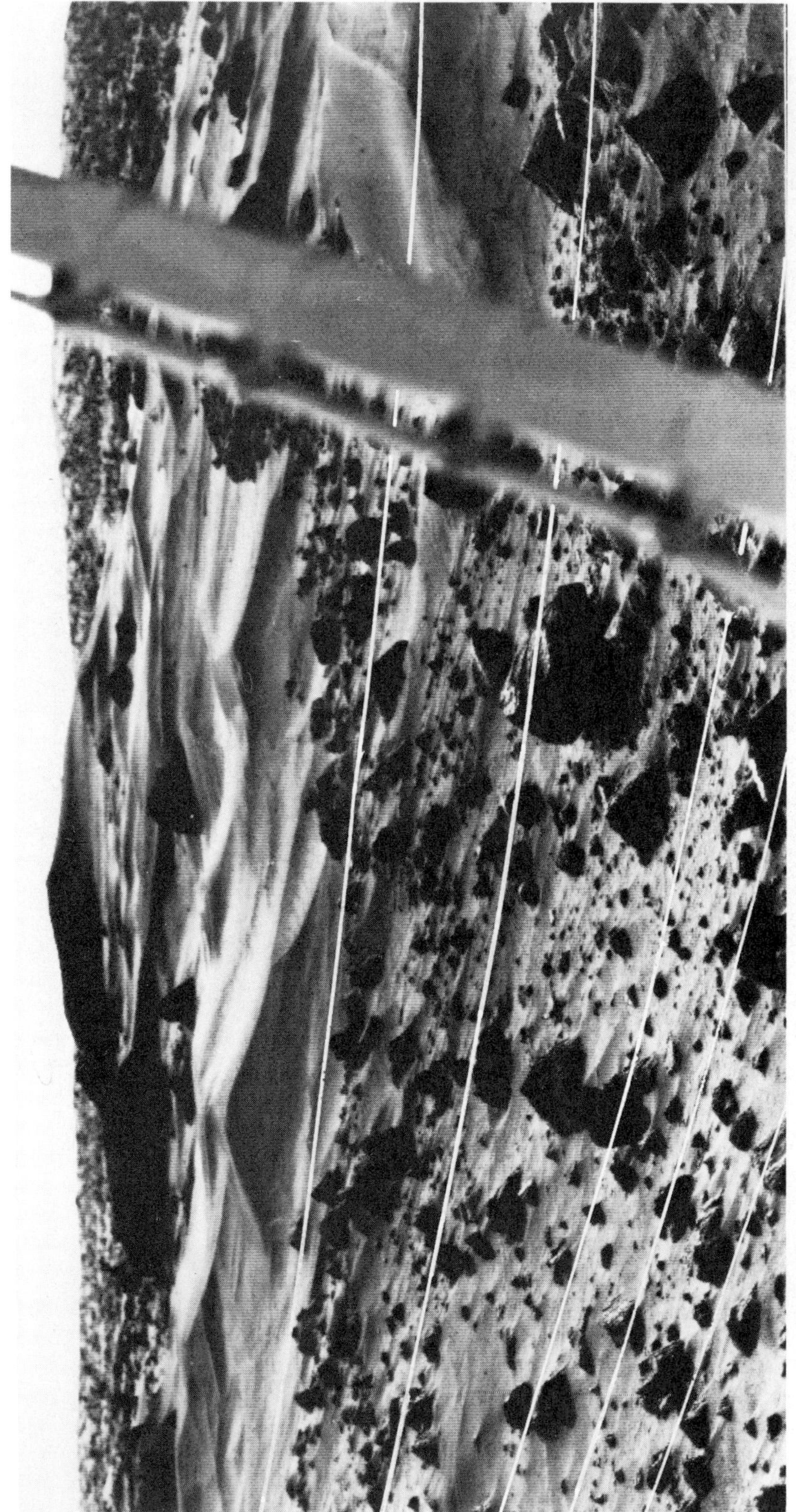

Fig. 2. This frame shows the drift field at the Lander 1 site. White lines trend in the same direction as the mean vector direction of wind streaks. The drift field components are aligned either parallel or perpendicular to the wind streaks, indicating that both the drift and the streaks were produced by the same wind system. CE label 11A097, segment 2.

Fig. 3. Mosaic of the camera 1 frames at Utopia Planitia. The white lines trend in a direction 215° clockwise from north, which is the mean trend of sediment streaks extending from rocks in the left-hand portion of the frame. Note that the drifts located to the right midfield part of the frame have their long axes roughly aligned with the wind streak directions, a situation implying that both features were produced by the same or very similar wind systems.

Both orbital and lander imagery suggest that the Viking 2 landing site is within the boundaries of a debris blanket mapped from Mariner 9 data [*Soderblom et al.*, 1973; *Mutch et al.*, 1976*c*].

One striking feature of the rock-strewn area of the Viking 1 landing site, away from the fields of particulate drifts, is the presence of bright raised streaks associated with rocks and other positive relief obstacles (Figure 1). Such streaks continue typically for many obstacle diameters and are remarkably parallel. On a scale larger by a factor of $\sim 10^5$, similar features (although not necessarily raised) were uncovered by Mariner 9 orbital photography, most characteristically associated with crater ramparts [*Sagan et al.*, 1972; *Veverka et al.*, 1976]. In both cases it is likely that the streaks are aeolian in origin and oriented leeward of the associated obstacle. While there may be several different mechanisms contributing to such streaks and splotches at large scale, the interpretation of the Viking rock-associated streaks at small scale seems straightforward. As was mentioned, these streaks all appear to have positive relief and to compose an elongated pile of sediment in the lee of the obstacle. In many cases a small depression can be made out in the regolith on the windward side of the obstacle. (On a slightly larger scale this difference between leeward and windward sediment accumulation becomes quite striking; cf. the large boulder in the background in Figure 5 below.) It seems clear that the streaks represent wind shadows: locales protected from wind scouring by the rock or other positive relief obstacle. Since such elongated sediment piles seem particularly vulnerable to crosswinds, it follows that we are seeing the results of particulate transport during the last major wind storm above saltation threshold. The close parallelism of the streaks requires a prevailing wind with small angular deviation; since Martian winds are known to change directions with large angular excursions [*Hess et al.*, 1976, 1977], the circumstances seem to imply that small-scale streaks are produced by the last brief episode of high winds at the Chryse landing site.

If the Viking 1 cameras survive the next major wind storm, it will be interesting to see whether the previous streaks have been obliterated and replaced with a new array whose direction is consistent with that of the highest winds detected by the Viking meteorology and seismology packages. The prevailing preconjunction winds in the Viking 1 landing site have been approximately in a direction opposite to that required to produce the streaks; but the wind velocities have, with an exception discussed below, less than saltation threshold, and there is no inconsistency in the interpretation of streaks as aeolian. The relative absence of such small streaks in the Viking 2 landing site may be due to the past aeolian removal or destruction of fine mobile particulates, to the rarity of prevailing winds well above saltation threshold, or to the relative sparsity of rough terrain like that in the streak fields of the Chryse site. Terrain roughness lowers saltation thresholds through the roughness scaling number in the Prandtl equation for logarithmic wind shear. The observed variation of both dark and bright crater-associated streaks on Mars [*Veverka et al.*, 1974] suggests that time variation of the smaller rock-associated Viking 1 streaks can be anticipated on a several year time scale.

Figure 4 shows a rosette diagram of streak directions, measured clockwise from north, in the Viking 1 landing site as observed in small scale in 1976 from the Viking lander and in large scale in 1971–1972 for the same terrain by the Mariner 9 orbiter. The vector mean direction for small-scale streaks is 191.5°, and for large-scale streaks it is 216°. Viking orbiter de-

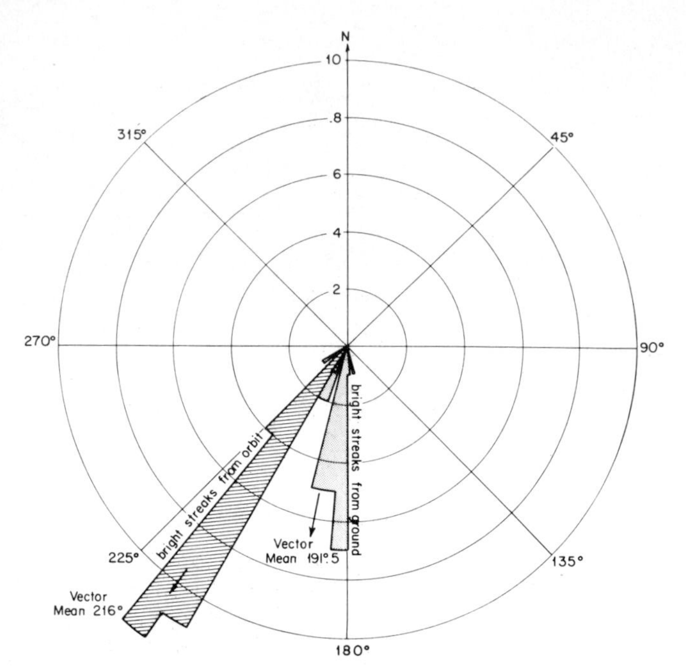

Fig. 4. Rosette diagram of frequency as a function of direction of wind streaks in the region of the Viking 1 lander. Twenty-five obstacle-associated streaks seen from the lander were measured in stereo by using frames 11A023/004 and 12A003/001. Azimuth is measured clockwise from north. Twenty-nine streaks, most associated with craters, were also measured on Mariner 9 frames 205A13/23 and DAS 08946674 with central latitude and longitude 29.3°N and 45.6°W, respectively. The vector mean direction is calculated as arctan $(\sum_i \sin_i \theta / \sum_i \cos_i \theta)$.

termination of 1976 Chryse bright streaks gives a direction of 218° with a spread of $\simeq 10°$ (R. Greeley et al., manuscript in preparation, 1977). Thus there is a 24° offset to the west of large-scale from small-scale streaks, which are statistically perfectly separated. The Coriolis acceleration in the northern hemisphere corresponds to a deflection to the right of the direction of motion. It seems plausible that the short rock streaks are produced locally by particles on short trajectories and that the long crater streaks involve entrained particles on much longer trajectories. The latter but not the former should experience a significant Coriolis acceleration and a variation of direction with altitude. *Pollack et al.* [1976*b*] have calculated this turning angle as a function of H/z_0, where H is the thickness of the Martian boundary layer and z_0 is the local roughness scale at the surface. Typical numbers for the thickness of the boundary layer are several hundred meters. The roughness in the rock-strewn area of the Viking 1 landing site which exhibits short streaks is in the range of several centimeters to several tens of centimeters. A turning angle of 20° requires [*Pollack et al.*, 1976*b*] $H/z_0 \sim 10^3$. Thus within the errors of measurement and the uncertainty of theory, both the sign and the magnitude of the offset between large- and small-scale streaks can be understood in the above models for streak generation and turning angle. The agreement in direction between 1972 and 1976 large-scale streaks in Chryse implies that (1) the highest velocity winds often have the same prevailing direction and/or (2) most streaks are formed or reformed less frequently than once each Martian year.

It might be argued that both short and long streaks are produced by particles which remain very close to the surface and which therefore are never affected by the Coriolis acceleration. The offset between short and long streaks would then have to be attributed to local meteorological conditions in the rocky area of the Viking 1 landing site. This is, of course, entirely possible, but it is not a very parsimonious explanation; we would have to believe that in the one case where local high-velocity wind directions can be determined, local meteorological conditions precisely simulate the offset in the direction to be expected from the Coriolis turning angle.

The drift features in the Chryse site (Figure 2) range from a few centimeters across to more than 100 m in the long dimension. Mariner 9 and Viking orbital photography of this region shows no sign of dunes, even at a resolution of 80 m/line pair. Barchan (crescentiform), longitudinal, and other dune forms have been identified in other areas of Mars. The drifts seen by Viking Lander 1 are therefore unlikely to be part of a large and topography-tied dune field, as may be, for example, in the interior of a crater with breached ramparts. Instead, the rocky terrain may have been inundated with a moving sand drift. Dunes only rarely form in such a stony surface on the earth; therefore the dune field is likely to be in motion. Alternatively, locales with large numbers of blocks may have trapped and accumulated material, a situation analogous to the trapping of sand in pebble fields on earth [*Bagnold*, 1941]. Leeward accumulation of particulates behind boulders in Chryse is consistent with this interpretation. The small angle of dispersion in the orientation of the small-scale streaks, described above, is evidence for the unidirectionality of the most recent high-velocity winds. In Figures 1 and 2 we see that the small-scale streak direction is roughly parallel to the long dimension of most of the larger sand drifts seen in Chryse. The large foreground drift in Figure 2 seems to have undergone deflation recently, and cross-bedded strata are thus exposed [*Mutch et al.*, 1976*b*]. We note that the strata, where they are exposed along their steepest dip, are parallel to the streaks, while strata exposed along their strike are aligned perpendicular to streak direction. These observations imply that the drifts were formed by winds identical in direction to the prevailing winds during streak formation and possibly by the same winds.

A comparison of Figures 1–3 shows that the highest prevailing winds observed by Viking lander cameras in Chryse (from streaks and drift orientation) seem to have the same orientation as the comparable winds in Utopia (from streaks). Mariner 9 orbital data comparing the direction of large streaks in these two regions suggest such a parallelism as well [*Sagan et al.*, 1973, Figures 2 and 3], although the streaks near Utopia are not as plentiful as those near Chryse. Thus the VL-1 and VL-2 observations are consistent with Mariner 9 and Viking orbiter evidence on the directions of peak winds in the two sites, which are in turn roughly consistent with the directions expected from crude models of Martian general circulation [*Sagan et al.*, 1973].

The most readily moved particle diameter under average Martian conditions is calculated to be about 150 μm [*Sagan and Pollack*, 1969; *Greeley et al.*, 1976]. Both larger and smaller particles require higher saltation threshold velocities. Particles both enter and leave the diameter interval corresponding to the most easily moved particles by weathering, which may involve chemical, aeolian, or other erosional processes. On the earth the principal agents of such weathering are weak acids, chiefly carbonic acid. If interstitial liquid water forms on Mars [*Farmer*, 1976], the high atmospheric carbon dioxide content makes it very likely that carbonic acid weathering would occur. The high ultraviolet flux on the Martian surface should make a very important contribution to such weathering, both directly and through the production of chemical intermediaries of high oxidation state, including peroxides and superoxides. Weathering in place is clearly occurring in both the Chryse and Utopia sites, which exhibit dark drifts near dark rocks [e.g., *Binder et al.*, 1977] and many cases of broken rocks with the likely fragments nearby. Mariner 9 infrared interferometer spectrometer observations of the great 1971 dust storm [*Toon et al.*, 1977], Viking X ray fluorescence spectrometer data [*Toulmin et al.*, 1976; *Baird et al.*, 1976], and analyses of Viking lander camera multispectral data [*Huck et al.*, 1977] all suggest that there is a significant component of fine grained clays, perhaps something like montmorillonite, in great abundance all over Mars. Montmorillonite typically has 90% of its particles by weight with radii smaller than 1 μm, even clayey sandy silt has 90% of its particle population smaller than 100 μm in radius [*Scott*, 1963]. Thus it is very likely that the topmost portion of the Martian surface is composed mainly of very fine particles too small to be moved by saltation under any plausible velocity regime. Additional evidence for very fine particles comes from Viking physical properties investigations and from orbital infrared and Viking optical investigations of the fine particles suspended in the Martian atmosphere [*Shorthill et al.*, 1976; *Toon et al.*, 1977; *Pollack et al.*, 1977]. All these analyses are consistent with the observation that the bulk of the particles are smaller than 1 μm. Such a particle population is consistent with the estimated cohesion strength [*Shorthill et al.*, 1976] of the Martian soil near the Viking landers of 10^3–10^4 dyn cm^{-2}. This is, of course, not necessarily the cohesion strength of individual particles; but if, for example, a 150-μm particle were an agglomerate of particles $\lesssim 1$ μm in diameter, the strength estimate would not be far off. Many apparent 'rocks' in the Chryse site are very likely clods or agglomerates of much smaller particles (B. Clark, private communication, 1977).

When a saltating particle strikes some target, energy is dissipated, both in the saltating particle and in the target, in many forms, including heat, plastic deformation, permanent strain, catastrophic rupture, and the kinetic energy of postcollision motion. The energy density dissipated in such a collision is the same as the instantaneous dynamic pressure of the collision $\rho v^2/2$, where ρ is the particle density and v is the relative velocity at impact. At a typical Martian threshold frictional velocity of 2 m s^{-1}, this dynamic pressure corresponds to 4×10^4 dyn cm^{-2}, or more than the cohesion strength. (By comparison, threshold frictional velocities on earth correspond to dynamic pressures around 400 dyn cm^{-2}, values substantially less than even the Martian cohesion strength.) At 5, 20, and 50 m s^{-1} the corresponding pressures are 2.5×10^5, 4×10^6, and 2.5×10^7 dyn cm^{-2}, all for particles of density 2 g cm^{-3}. Thus if a significant fraction of the collision energy goes into creating and propagating fractures, at the saltation threshold the cohesion will be exceeded by a factor of between 4 and 40. But a simple scaling analysis suggests that saltating particles on Mars will be entrained by the wind [*Sagan*, 1973] and will have much higher velocities, between 5 and 50 m s^{-1}. For such entrained particles the fraction of the energy which goes into deformation, strain, and rupture can be as small as 10^{-3}–10^{-4}, and the particle cohesion will still be exceeded. As a result, there is a kind of collisional natural selection which seems to be occurring on the Martian surface, in which the most readily moved particles are preferentially destroyed by collision. Because of their self-destructive tendencies we call such grains kamikaze particles, a name which is additionally apt because its literal meaning in Japanese is 'divine wind.' We have already mentioned evidence that the bulk of the particles on the outer layers of the Martian surface are significantly smaller

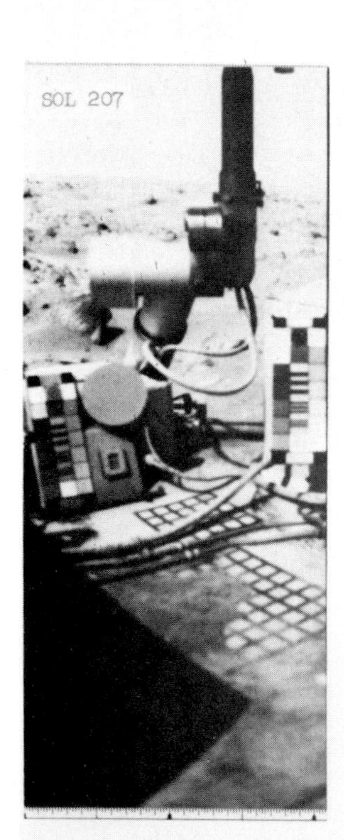

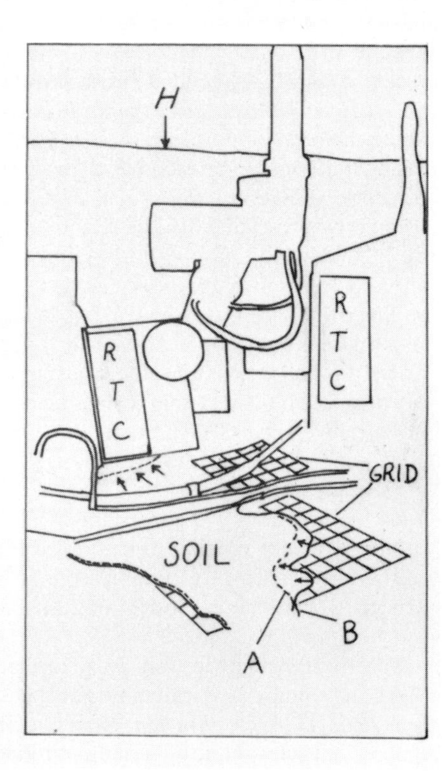

Fig. 5. This sequence of five low-resolution, blue diode Viking lander images illustrates the movement of soil across the deck of Lander 1 at Chryse Planitia. On sol 41, soil was dumped on the central deck by the sampler arm as an experiment to detect winds of sufficient magnitude to initiate grain movement. The next image (sol 76) shows a better view of the same area and provides the reference base for the sketch at the bottom right. Notice the small outlier pile between the main pile and the left reference test chart (RTC). On sol 96 the small outlier pile has progressed toward the RTC and has become disaggregated. The sol 207 image shows the outlier pile now abutting the base of the RTC, having moved about 7 cm away from the main pile. By sol 207 the lower right edge of the large pile has retreated in relation to sol 76. There appears to be little difference between the soil positions on sol 207 and sol 268, although there may be evidence of debris movement just below the left RTC. The sketch map at the lower right shows the change in relative position of the soil pile between sol 76 (solid line labeled B) and sol 268 (dashed line labeled A). Small arrows indicate roughly inferred directions of motion (wind blowing from the southeast). For scale, each grid square is 6.45 cm² in area. In all five photographs a boulder can be seen beyond the high-gain *S* band antenna support strut with a marked accumulation of sediment on its probable windward side.

than those particles at the optimum radius for saltation. We expect that the bulk of the population of particles near the 150-μm diameter will be grains with cohesions significantly higher than the Martian average or particles recently weathered into this size range. If the rate of aeolian destruction of kamikaze particles could be estimated, the steady state population would provide evidence on the rate of formation of kamikaze particles and therefore on the net erosion rate on the Martian surface.

One of the principal objectives of the Viking lander imaging experiments was to search directly for wind-initiated motion of fine particulates on Mars. Partly for this reason, several grids with 2.54-cm spacing were painted on the lander's horizontal surface to provide a clear reference for particle motion. On sol 41 a largish sample of particulates was deposited by the sample arm on the grid surface. The initial appearance of the pile and its subsequent motion are displayed in Figure 5. It is clear that some small motion occurred between sols 76 and 96, and major particle transport between sols 96 and 207.

Although wind seems the most obvious explanation for the observed movement, some other possibilities were investigated. With the help of William De Sazor and Robert Grossart of the Surface Sampler Team, sample retrieval and dump sequences were simulated on the full-scale Viking Science Test Lander to determine whether vibration due to arm maneuvers would be strong enough to induce movement on the lander deck. Biology instrument and gas chromatograph–mass spectrometer comminution sequences were carried out, and the high-gain antenna was stepped in both azimuth and elevation to check for vibrations from those sources. The strongest vibrations in the deck were induced by upward movement of the high-gain antenna; however, in no case was there any movement of soil greater than one or two grain diameters. It is highly unlikely that such vibrations would be strong enough to cause the gross rearrangement of material observed. Although there is a 3° slope to the deck at the VL-1 site, the soil movement seems to be across slope and is more probably slightly uphill.

Thus it seems likely from imaging data alone that local wind velocities above saltation threshold were experienced on the grid surface in the interval of 111 Martian days following sol 96 and that the wind was southeasterly. This is not the wind direction inferred for the production of major streaks and drifts (Figure 4). The Meteorology Team reports (R. Henry, private communication, 1977) winds from virtually every cardinal direction during this time interval with some winds of quite high speed. A continuous monitoring of Martian winds of >3 m s^{-1}, normalized to the meteorology boom, is provided by the Viking 2 seismometer [*Anderson,* 1977]. While winds are generally found to have been <10 m s^{-1} as determined by both instruments, unusual activity was recorded on sols 118–131, with winds greatly in excess of 22 m s^{-1} measured on sol 131 [*Anderson*, 1977]. This corresponds to winds of $\gg$40–50 m s^{-1} above the velocity boundary layer and therefore in the probable range of the saltation threshold [*Sagan and Pollack*, 1969]. We have already mentioned the correlation in peak wind direction between the VL-1 and the VL-2 site. If the seismometer-detected winds are large scale, the measurements at VL-2 may apply also to the winds at VL-1.

Thus the combined Viking imaging/meteorology/seismology data imply that saltation threshold was achieved probably more than once in the interval between sols 96 and 207 and that the corresponding velocity above the boundary layer is >50 m s^{-1}. This represents the first in situ evidence on the threshold velocity for saltation ever made on another planet and tends, at least roughly, to confirm the standard theory which is essentially due to *Bagnold* [1941].

It is possible that venturi effects induced by the wind blowing between obstacles may enhance the average wind velocity across the VL-1 deck, although wind tunnel experiments for the top of the lander have not as yet been carried out. There is, however, a relatively clear path between the sampler arm assembly, camera, and radioactive thermoelectric generator housing. It is also probable that whatever small-grain movement is induced by vibration may have the result of lowering effective particle cohesion and thereby altering wind transportability. However, both these effects are probably rather small.

Martian general circulation models suggest highest winds in local winter. The Viking camera systems are designed to survive the inclemencies of Martian dust and sand storms. There are commandable CO_2 jets to remove fine particles adhering to exterior optical surfaces. If excessive aeolian pitting occurs, the exterior optics can be jettisoned on command. In a severe sandstorm the exterior optics of the cameras can be rotated so they abut firmly against a vertical post. If observations during or after such storms can be made, it will be of interest to determine, among other matters, (1) motion of the sand dunes, (2) possible dissipation, elongation, or direction change of the rock-associated streaks, (3) new production of depressions on the windward side of rocks, (4) albedo changes which may be attributed to the preferential transport of particles in a given size range, and (5) further evolution of the grid sediment pile. In addition, the Viking cameras are equipped with a single-line scan mode called rescan which is under some circumstances capable of detecting the motion of individual particles. By April 1976 no such particle motion had been detected. Laboratory and field calibration of rescan mode and a summary of Rescan searches for particle motion on Mars will be the subject of a future publication.

Acknowledgments. This research was supported by the Viking Project Office and by the Planetary Geology Office of NASA headquarters under grants NGR 33-010-22 and NSG-7087. We acknowledge helpful discussions with J. Veverka, H. Moore, R. Greeley, J. B. Pollack, and B. White and the laboratory assistance of W. De Sazor and R. Grossart and appreciate a stimulating exchange of letters with R. A. Bagnold, who initiated comparable studies of fine particle motion on the earth more than 4 decades ago.

REFERENCES

Anderson, D. L., Development of a seismometer for flight missions, Reports of Planetary Geology Program, 1976–1977, *NASA Tech. Memo., TMX-3511*, 1977.

Bagnold, R. A., *The Physics of Blown Sand and Desert Dunes*, Methuen, London, 1941.

Baird, A. K., P. Toulmin III, B. C. Clark, H. J. Rose, K. Keil, R. P. Christian, and J. L. Gooding, Mineralogic and petrologic implications of Viking geochemical results from Mars: Interim report, *Science, 194*, 1288–1293, 1976.

Binder, A. B., R. E. Arvidson, K. L. Jones, E. C. Morris, T. A. Mutch, D. C. Pieri, and C. Sagan, The geology of the Viking Lander 1 site, *J. Geophys. Res., 82*, this issue, 1977.

Cutts, J. A., and R. S. U. Smith, Eolian deposits and dunes on Mars, *J. Geophys. Res., 78*, 4139–4154, 1973.

Farmer, C. B., Liquid water on Mars, *Icarus, 28*, 279–289, 1976.

Gierasch, P., and C. Sagan, A preliminary assessment of Martian wind regimes, *Icarus, 14*, 312–318, 1971.

Greeley, R., B. White, R. Leach, J. Iversen, and J. Pollack, Mars: Wind friction speeds for particle movement, *Geophys. Res. Lett., 3*, 417–420, 1976.

Hess, S. L., R. M. Henry, C. B. Leovy, J. A. Ryan, J. E. Tillman, T. E. Chamberlain, H. L. Cole, R. G. Dutton, G. C. Greene, W. E.

Simon, and J. L. Mitchell, Mars climatology from Viking 1 after 20 sols, *Science*, *194*, 78–81, 1976.

Hess, S. L., et al., Meteorological results from the surface of Mars: Viking 1 and 2, *J. Geophys. Res.*, *82*, this issue, 1977.

Huck, F. O., R. Arvidson, D. Jobson, S. Park, W. Patterson, and S. Wall, Spectrophotometric and color estimates of the Viking landing sites, *J. Geophys. Res.*, *82*, this issue, 1977.

Mass, C., and C. Sagan, A numerical circulation model with topography for the Martian southern hemisphere, *J. Atmos. Sci.*, *33*, 1418–1430, 1976.

Mutch, T. A., A. B. Binder, F. O. Huck, E. C. Levinthal, S. Liebes, E. C. Morris, W. R. Patterson, J. B. Pollack, C. Sagan, and G. R. Taylor, The surface of Mars: The view from the Viking 1 lander, *Science*, *194*, 791–801, 1976*a*.

Mutch, T. A., R. E. Arvidson, A. B. Binder, F. O. Huck, E. C. Levinthal, S. Liebes, E. C. Morris, D. Nummedal, J. B. Pollack, and C. Sagan, Fine particles on Mars: Observations with the Viking 1 lander cameras, *Science*, *194*, 87–91, 1976*b*.

Mutch, T. A., S. U. Grenander, K. L. Jones, W. Patterson, R. E. Arvidson, E. A. Guinness, P. Avrin, C. E. Carlston, A. B. Binder, C. Sagan, E. W. Dunham, P. L. Fox, D. C. Pieri, F. O. Huck, C. W. Rowland, G. R. Taylor, S. D. Wall, R. Kahn, E. C. Levinthal, S. Liebes, R. B. Tucker, E. C. Morris, J. B. Pollack, R. S. Saunders, and M. R. Wolf, The surface of Mars: The view from the Viking 2 lander, *Science*, *194*, 1277–1283, 1976*c*.

Pollack, J. B., C. B. Leovy, Y. H. Mintz, and W. Van Camp, Winds on Mars during the Viking season: Predictions based on a general circulation model with topography, *Geophys. Res. Lett.*, *3*, 479–482, 1976*a*.

Pollack, J. B., R. Haberle, R. Greeley, and J. Iversen, Estimates of the wind speeds required for particle motion on Mars, *Icarus*, *29*, 395–417, 1976*b*.

Pollack, J. B., D. Colburn, R. Kahn, J. Hunter, W. Van Camp, C. E. Carlston, and M. R. Wolf, Properties of aerosols in the Martian atmosphere, as inferred from Viking lander imaging data, *J. Geophys. Res.*, *82*, this issue, 1977.

Sagan, C., Sandstorms and eolian erosion on Mars, *J. Geophys. Res.*, *78*, 4155–4162, 1973.

Sagan, C., Erosion and the rocks of Venus, *Nature*, *261*, 31, 1976.

Sagan, C., and J. B. Pollack, Windblown dust on Mars, *Nature*, *223*, 791–794, 1969.

Sagan, C., J. Veverka, and P. Gierasch, Observational consequences of Martian wind regimes, *Icarus*, *15*, 253–278, 1971.

Sagan, C., J. Veverka, P. Fox, R. Dubisch, J. Lederberg, E. Levinthal, L. Quam, R. Tucker, J. B. Pollack, and B. A. Smith, Variable features on Mars: Preliminary Mariner 9 television results, *Icarus*, *17*, 346–372, 1972.

Sagan, C., J. Veverka, P. Fox, R. Dubisch, R. French, P. Gierasch, L. Quam, J. Lederberg, E. Levinthal, R. Tucker, and B. Eross, Variable features on Mars, 2, Mariner 9 global results, *J. Geophys. Res.*, *78*, 4163–4196, 1973.

Sagan, C., J. Veverka, R. Steinbacher, L. Quam, R. Tucker, and B. Eross, Variable features on Mars, IV, Pavonis Mons, *Icarus*, *22*, 24–47, 1974.

Scott, R. R., *Principles of Soil Mechanics*, Addison-Wesley, Reading, Mass., 1963.

Shorthill, R. W., H. J. Moore, R. F. Scott, R. E. Hutton, S. Liebes, and C. R. Spitzer, The 'soil' of Mars (Viking 1), *Science*, *194*, 91–97, 1976.

Soderblom, L. A., T. J. Kreidler, and H. Masursky, Latitudinal distribution of a debris mantle on the Martian surface, *J. Geophys. Res.*, *78*, 4117–4122, 1973.

Toon, O. B., J. B. Pollack, and C. Sagan, Physical properties of the particles composing the Martian dust storm of 1971–1972, *Icarus*, *30*, 663–696, 1977.

Toulmin, P., III, B. C. Clark, A. K. Baird, K. Keil, and H. J. Rose, Preliminary results from the Viking X-ray fluorescence experiment: The first sample from Chryse Planitia, Mars, *Science*, *194*, 81–83, 1976.

Veverka, J., C. Sagan, L. Quam, R. Tucker, and B. Eross, Variable features on Mars, III, Comparison of Mariner 1969 and Mariner 1971 photography, *Icarus*, *21*, 317–368, 1974.

Veverka, J., C. Sagan, and R. Greeley, Variable features on Mars, VI, An unusual crater streak in Mesogaea, *Icarus*, *27*, 241–253, 1976.

(Received May 18, 1977;
revised June 24, 1977;
accepted June 24, 1977.)

The Geology of the Viking Lander 1 Site

ALAN B. BINDER,[1] RAYMOND E. ARVIDSON,[2] EDWARD A. GUINNESS,[2] KENNETH L. JONES,[3] ELLIOT C. MORRIS,[4] THOMAS A. MUTCH,[3] DAVID C. PIERI,[5] AND CARL SAGAN[5]

Viking 1 landed on volcanic terrain in the plains of Chryse. Stereo pictures reveal an undulating topography. Bedrock is exposed along several ridge crests. Blocks are more numerous than can be attributed to impact ejecta. The presence of an apparent variety of rock types suggests in situ weathering of extrusive and near-surface basaltic igneous rocks along a linear volcanic vent. Fine-grained sediment is present in drift complexes and isolated drifts. During the course of the Viking mission a small patch of fine-grained sediment slumped down one of the drift faces. Otherwise, no morphological changes unrelated to spacecraft activity have been observed.

GENERAL GEOLOGIC SETTING

On July 20, 1976, Viking 1 set down on a rock-strewn rise in Chryse Planitia, thereby becoming the first spacecraft to successfully land on the red planet. The lander is located at 22.483°N and 47.82°W (aerographic coordinates), as determined by Doppler tracking of spacecraft radio signals [*Michael et al.*, 1976; M. E. Davies, unpublished data, 1977]. The maximum error in position of the orbiter photobase, considering both accuracy of radio tracking and accuracy in fitting measurements to the cartographic base, is less than 120 m in latitude and less than 11 km in longitude [*Mayo et al.*, 1977] (Figure 1). Orbiter pictures reveal a lunarlike mare surface, complete with wrinkle ridges and fresh-appearing craters (Figure 2). Lava flows are inferred to be present, but lava flow fronts are not observed. This may be because the lavas were of such low viscosity that fringing scarps did not form or because original topographic relief has been degraded.

Large scoured channels, 2–3 km wide, emerge from Lunae Planum, a high plateau 500 km to the west, and trend eastward toward the landing site. Although there is substantial evidence that erosion by flowing water has produced these channels [*Greeley et al.*, 1977], orbiter pictures do not show recognizable indications of fluvial erosion in the immediate vicinity of the lander.

Some of the marelike ridges, approximately 60 km to the southwest of the landing site, are breached by channels, but other ridges transect channels and therefore appear to have formed after incision of channels. This suggests several episodes of volcanism, interleaved with channeling events. Channels show interior terracing, and some relatively fresh-appearing channels are superimposed on older depressions. This suggests multiple episodes of fluvial activity.

Relying exclusively on analysis of orbiter pictures, at least three models of geological evolution can be presented: (1) The Chryse plains are predominantly volcanic, but fluvial erosion to the west is coupled with emplacement of a sedimentary deposit in the landing area. (2) The Chryse plains are volcanic and in part are younger than the channel deposits, particularly in the landing area. (3) The channel deposits are younger than the plains but do not reach the landing site area. The absence of any recognizable fluvial landforms or deposits in the lander pictures adds support to the latter two models.

[1]Institut für Geophysik, University of Kiel, Federal Republic of Germany.

[2]Department of Earth and Planetary Sciences, McDonnell Center for the Space Sciences, Washington University, Saint Louis, Missouri 63160.

[3]Department of Geological Sciences, Brown University, Providence, Rhode Island 02912.

[4]Branch of Astrogeologic Studies, U.S. Geological Survey, Flagstaff, Arizona 86001.

[5]Laboratory for Planetary Studies, Cornell University, Ithaca, New York 14853.

Paper number 7S0541.

THE LANDING SITE

The lander pictures, the first of which was initiated only 25 s after touchdown, show a surface which only superficially resembles the lunar mare scenes viewed by Apollo astronauts and Surveyor and Luna spacecraft. The surface is strewn with rocks in the centimeter- to meter-size range (Figure 3). Several small areas interpreted as bedrock are present (Figures 3 and 4). In addition, the site contains abundant fine-grained (<100 μm) material [*Moore et al.*, 1977] which is present in large-scale (~10 m) drift complexes, smaller (~1–3 m) individual drifts, and wind tails (~10 cm) behind rocks [*Mutch et al.*, 1976*a*]. There is an apparent lack of sand-sized particles, intermediate between the fines and blocks [*Moore et al.*, 1977].

TOPOGRAPHY

A topographic map of the landing site (Figure 5), derived from stereoscopic data [*Liebes and Schwartz*, 1977], shows prominent relief (~5 m). The dominant elements are ridges and troughs which have two approximately orthogonal trends, northwest and northeast. There is a general slope of about 1.5° toward the northwest. The major northeast-trending ridge at the site parallels the crest of a wrinkle ridge viewed in orbiter pictures. The second set of ridges and troughs is parallel to transverse structural elements at the junction of wrinkle ridge crests (Figure 2). As such, the topographic texture in the immediate vicinity of the lander is similar to that of the larger-scale wrinkle ridge viewed in orbiter pictures. The observed ridge and trough topography is consistent with the speculation that the lander is situated on a wrinkle ridge.

The topography may be partly controlled by bedrock fabric. Several crests are underlain by jointed bedrock. The joints in exposed or near-surface rocks may influence the pattern of erosional relief. Eolian influences cannot be discounted. The dominant wind direction, from a geomorphological point of view, is from north-northeast. This is indicated by both crater lee streaks in orbiter pictures and wind tails in lander pictures [*Sagan et al.*, 1977].

As a final speculation it is noted that the northeast-trending depressions are parallel to channels located about 60 km to the southwest of the landing site. Although it seems unlikely that there is actually a genetic relationship between the two sets of features, this possibility cannot be definitively excluded.

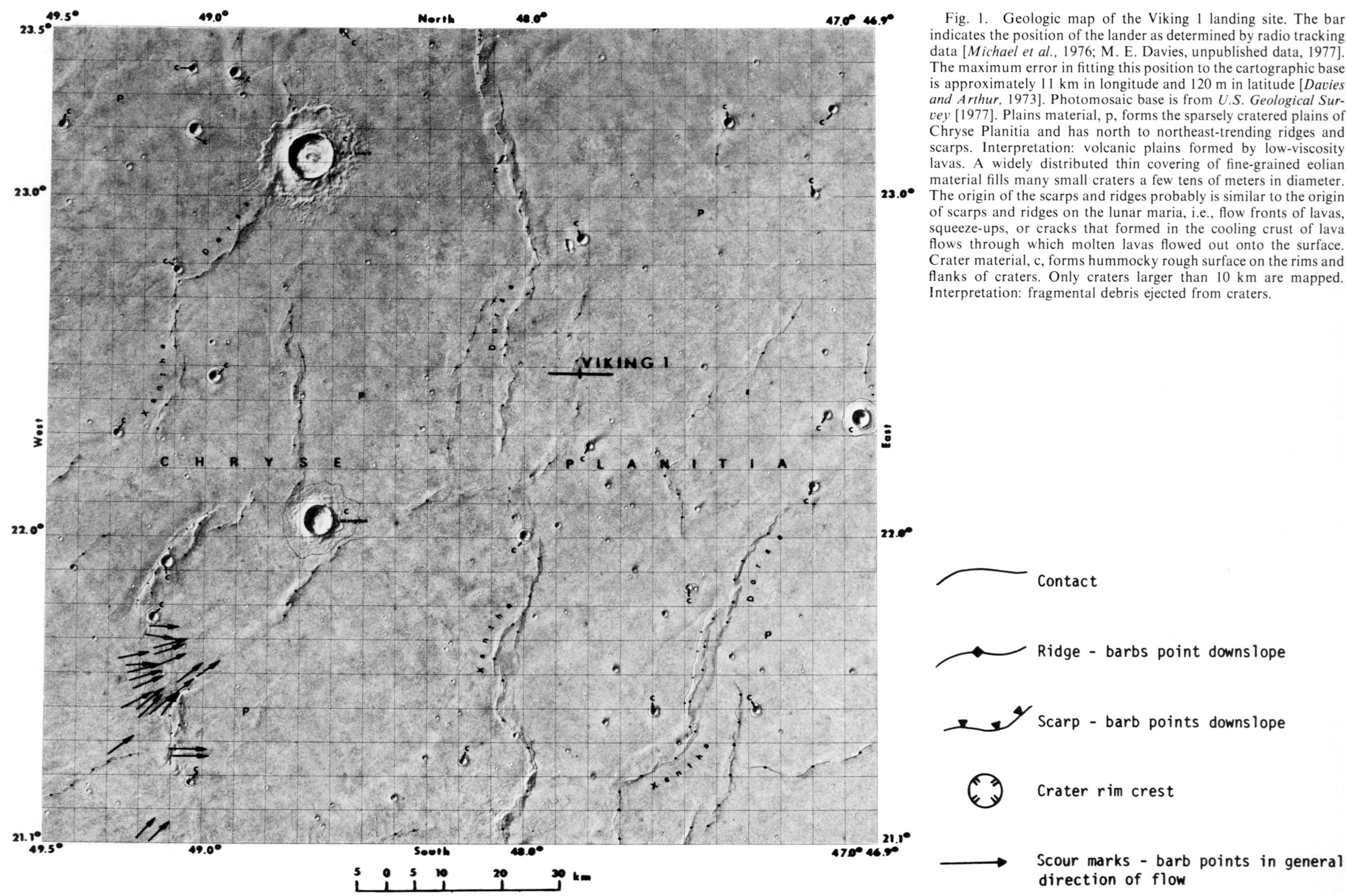

Fig. 1. Geologic map of the Viking 1 landing site. The bar indicates the position of the lander as determined by radio tracking data [*Michael et al.*, 1976; M. E. Davies, unpublished data, 1977]. The maximum error in fitting this position to the cartographic base is approximately 11 km in longitude and 120 m in latitude [*Davies and Arthur*, 1973]. Photomosaic base is from *U.S. Geological Survey* [1977]. Plains material, p, forms the sparsely cratered plains of Chryse Planitia and has north to northeast-trending ridges and scarps. Interpretation: volcanic plains formed by low-viscosity lavas. A widely distributed thin covering of fine-grained eolian material fills many small craters a few tens of meters in diameter. The origin of the scarps and ridges probably is similar to the origin of scarps and ridges on the lunar maria, i.e., flow fronts of lavas, squeeze-ups, or cracks that formed in the cooling crust of lava flows through which molten lavas flowed out onto the surface. Crater material, c, forms hummocky rough surface on the rims and flanks of craters. Only craters larger than 10 km are mapped. Interpretation: fragmental debris ejected from craters.

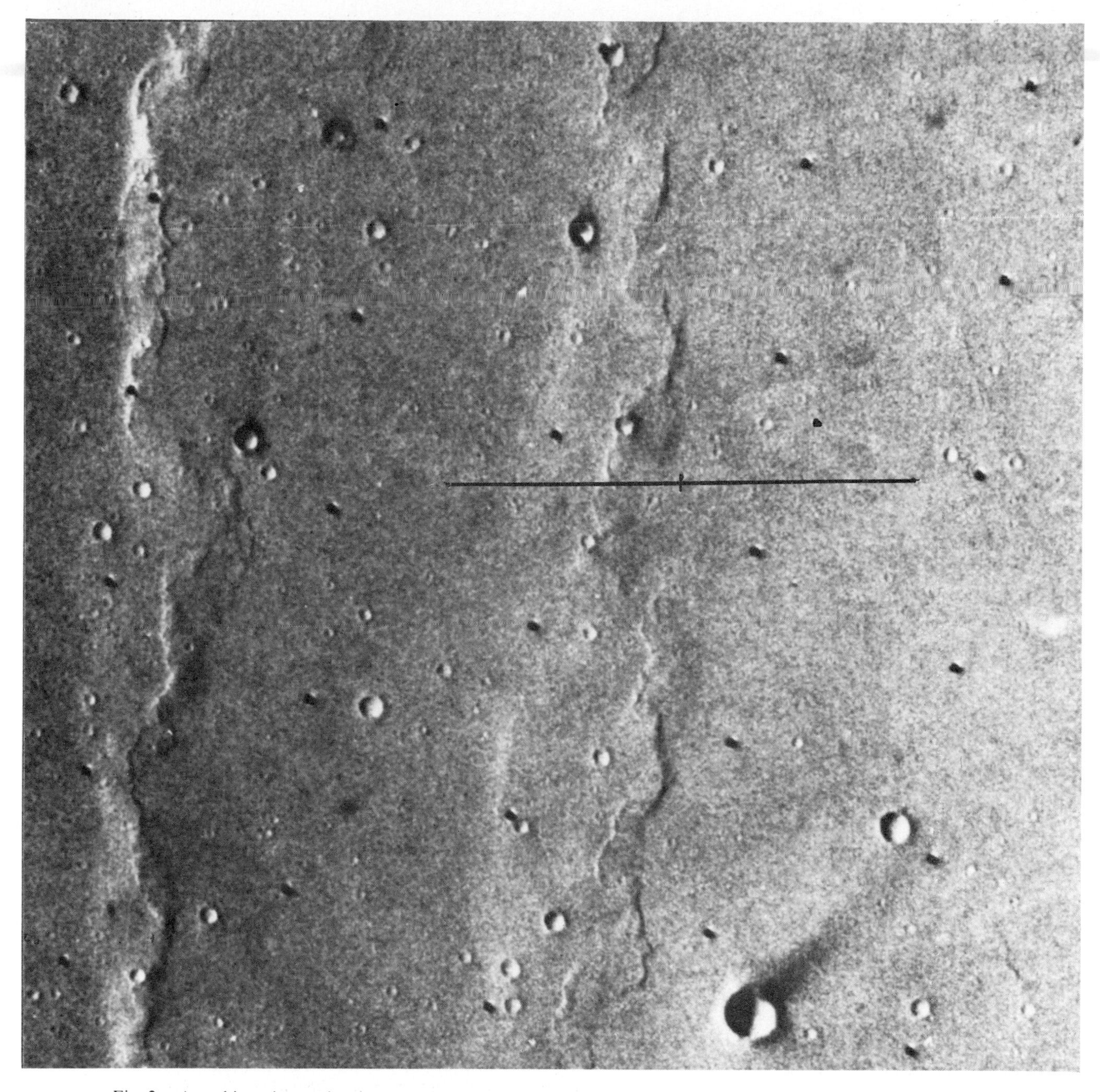

Fig. 2. An orbiter picture showing the Viking 1 landing site. Scene is approximately 25 km on a side. Possible positions of the Viking lander, as determined by radio tracking, are shown by a bar. The nominal location is astride a gentle ridge. A 400-m crater, located about 2.5 km to the southwest, may be the same one visible on the horizon with Viking lander cameras.

CHEMICAL WEATHERING

Color pictures [*Huck et al.*, 1977] show that most of the materials exposed at the surface, both fines and blocks, are a yellowish brown color. The observed characteristics are consistent with a model proposed independently by *Binder and Cruikshank* [1964] and *Von Tassel and Salisbury* [1964]. The model, as proposed by Binder and Cruikshank, was based on earth-based telescopic colorimetric measurements of Mars, similar colorimetric measurements for terrestrial samples, and analog studies in the Sonoran desert of Arizona. These several lines of inquiry suggested that the yellowish color of Mars was due to a limonite ($FeO \cdot OH \cdot nH_2O$) stain, or patina, which coats blocks and fines and which was formed by chemical weathering of mafic minerals in igneous rocks. By analogy with terrestrial deserts it was proposed that 'the Mars surface is composed partly of [limonite] stained outcrops, rock fragments, and finer material' [*Binder and Cruikshank*, 1964].

Binder and Cruikshank originally suggested that the limonite stain on the Martian blocks and fines was due to chemical weathering in an earlier epoch when water and oxygen were more plentiful in the Martian atmosphere. However, *Huguenin* [1974] has argued that UV photooxidation of mafic minerals under present Martian atmospheric conditions also could be responsible for producing the observed limonite. Viking imaging data do not permit one to distinguish between these two hypotheses. Since there is independent evidence [e.g., *Nier et al.*, 1976; *McElroy et al.*, 1976] in support of an earlier chem-

Fig. 3. A rock-strewn surface in front of Lander 1. The nearest part of the scene is 4 m from the spacecraft. The larger midfield are approximately 15 m from the spacecraft, and the horizon is approximately 80 m distant. A ridge of knobby occur in the vicinity of the dunes but are absent in the near field. For exact location of the picture, refer to *Mutch* [1977,

ically active atmosphere and since there is a strong UV flux reaching the surface at present, it seems likely that both modes of formation of limonite have been active. In any event, the apparent limonite stain indicates that there has been some chemical weathering and resulting degradation of surface materials during some or all of Martian history. This conclusion is strengthened by two other sets of observations.

First, the Viking X ray fluorescence data are interpreted to indicate the dominance of iron-rich montmorillonitic clay (nontronite) with small amounts of sulfates and carbonates [*Baird et al.*, 1976]. Spectral reflectance measurements of fine material made with Viking lander cameras [*Huck et al.*, 1977] resemble spectral values for nontronite. Previously, presence of montmorillonite [*Hunt et al.*, 1973] or a mixture of basalt and clay particles [*Toon et al.*, 1977] was suggested by Mariner 9 infrared spectra. Thus several different measurements appear to document the chemical alteration of iron-rich basic igneous rocks.

Second, as is shown in Figure 6, the area under one of the retro-engines was swept clean of fines during the landing. The material which is exposed forms a fractured crust. This crust is widely exposed at both Viking landing sites and is believed to be formed by accumulation of soluble minerals deposited by mineral-bearing waters that move upward by capillary action and then evaporate. We call this distinctive layer duricrust. The term was coined by *Woolnough* [1928] to describe a rock-like weathering zone overlying softer decomposed sediment, widely exposed in western Australia. Subsequently, it has been used more generally to describe a hard crust in the upper horizon of soils in a semiarid environment. Although genetic identity between terrestrial and Martian duricrust has not been established, a general equivalence is nonetheless suggested. For example, X ray fluorescence data indicate that carbonates and sulfates are probable constituents of the fines and that sulfur is relatively enriched in the duricrust [*Clark et al.*, 1977].

Thus the pictures, as well as the X ray data and earlier spacecraft and astronomical observations, suggest that chemical weathering in the presence of relatively large amounts of water (by present Martian standards) played an important role in developing the landscape viewed by Viking 1.

Rock Types

The blocks at the Lander 1 site, because they are so numerous and diverse, convey a great deal of information. As has been noted in the very first pictures received from Viking [*Mutch et al.*, 1976*a*], many types of blocks are identifiable in the hectare (10^4 m^2) available for study with the lander cameras. One of the authors (A.B.B.) has recorded 30 block groups based on color, shape, and texture. Although it is unlikely that each of these 30 groups represents a distinct petrologic type, as opposed to variations in age and exposure to weathering, it appears likely that at least several distinctly different lithologies are present. The majority of rocks are angular, with coarsely pitted surfaces (Figures 6 and 7). Terrestrial field studies [e.g., *Morris et al.*, 1972] show that aeolian erosion produces pitting more commonly on coarse-grained

blocks in the near field are approximately 0.4 m across. The bright dune in the left midfield and the dark dune in the center bedrock extends from center to far right, between the dark dune and the foreground. Distinctively dark angular blocks Figure 1]. Camera event is 12A119.

rocks than on fine-grained rocks. Making the intuitively reasonable inference that the scale of the pits is the same as that of the constituent grains in the rock, we believe that these rocks contain 3-mm to 1-cm grains. Most probably they were derived from depths of 10 m or less, either from shallow intrusions or from the bottom of lava flows.

Many lunar basalts are coarse grained, even though they were derived from flow units which cooled rapidly [*Weill et al.*, 1970]. This is because the viscosities of lunar magmas were so low (~20 P) that crystals grew to large sizes even though the cooling time was short [*Weill et al.*, 1970]. Since the Viking X ray data suggest that parent rocks were mafic and iron-rich, it is reasonable to expect that viscosities of the Martian magmas were similar to those of the iron-rich lunar basalts. The viscosity might have been even further decreased by the inclusion of volatiles, apparently more common on Mars than on the moon. These several lines of argument negate any requirement to couple the coarse-grained character of blocks with original deep-seated intrusions.

Rocks at the Lander 1 which are demonstrably from the tops of lava flows are rare. A concentration of dark blocks in the midfield (Figure 3) have flat angular faces which, in some terrestrial desert situations, are characteristic for fine-grained basalt [*Morris et al.*, 1972]. On some of these rocks a sharply shadowed structure suggests vesiculation. A few blocks in the near field show layering suggestive of flow banding and vesiculation and texture characteristic of aa or pahoehoe lavas. Additional rare blocks found near the lander include one fragmental rock which has the appearance of welded tuff, one fluted mottled rock which looks like a lunar breccia (Figure 7*a*), and a few small bright fine-grained faceted blocks.

In addition to the blocks, several areas of possible outcrop are visible in the scene. Most of these outcrops are found at ridge crests, and two of them show definite layering. One of the exposures is relatively bright with steeply dipping joints (Figure 3). Another outcrop is darker and displays subhorizontal layering (Figure 4).

CRATER DENSITY

Craters at the Viking Lander 1 site are less abundant than would be expected by extrapolation from orbital observations. We can test the probability that craters of a given size below the orbital resolution limit have been produced but that they cannot be seen from the lander because of some random spatial clustering. The lander cameras stand 1.3 m above the Martian surface. At such a height, assuming a spherical Mars, the horizon is 3 km away. The most likely number of craters of a given size range that would fall within the camera field of view is

$$E = \left[\int_d^{d+\Delta d} A d^\gamma \, d(d) \right] \pi r^2$$

in which E is the number of craters of size d to $d + \Delta d$ to be expected in a circle of radius r, where r is the effective range of the cameras. A and γ are the intercept and slope of the crater

Fig. 4. Two views of a bedrock exposure extending from left to right in midfield. (*a*) The outcrop area appears bright and can be easily delineated (camera event 12A221). (*b*) The topographic details of the bedrock are revealed (camera event 12A116). Near-horizontal surfaces are cut by steeply dipping fractures. For exact location of the pictures, refer to *Mutch* [1977, Figure 1].

size-frequency distribution determined in orbiter pictures. Assuming that the impacting population forms in a random manner with respect to location is the same as assuming that the population obeys a binomial probability distribution:

$$\frac{E}{N} = \frac{\pi r^2}{\text{count area}} = P$$

where N is the total number of craters of size d to $d + \Delta d$ within the count area (about 30,000 km^2) and P represents the fraction of the total number of craters to be expected within a circle of radius r.

Admittedly, E is subject to variations due to random spatial clustering. However, the number of observed craters should be within $E \pm 3\sigma$, where σ^2 represents the variance about E. For a binomial distribution:

$$\sigma^2 = NP(1 - P) = 1 - \frac{\pi r^2}{\text{count area}} E$$

For the analysis we have chosen r = 500 m, the distance at

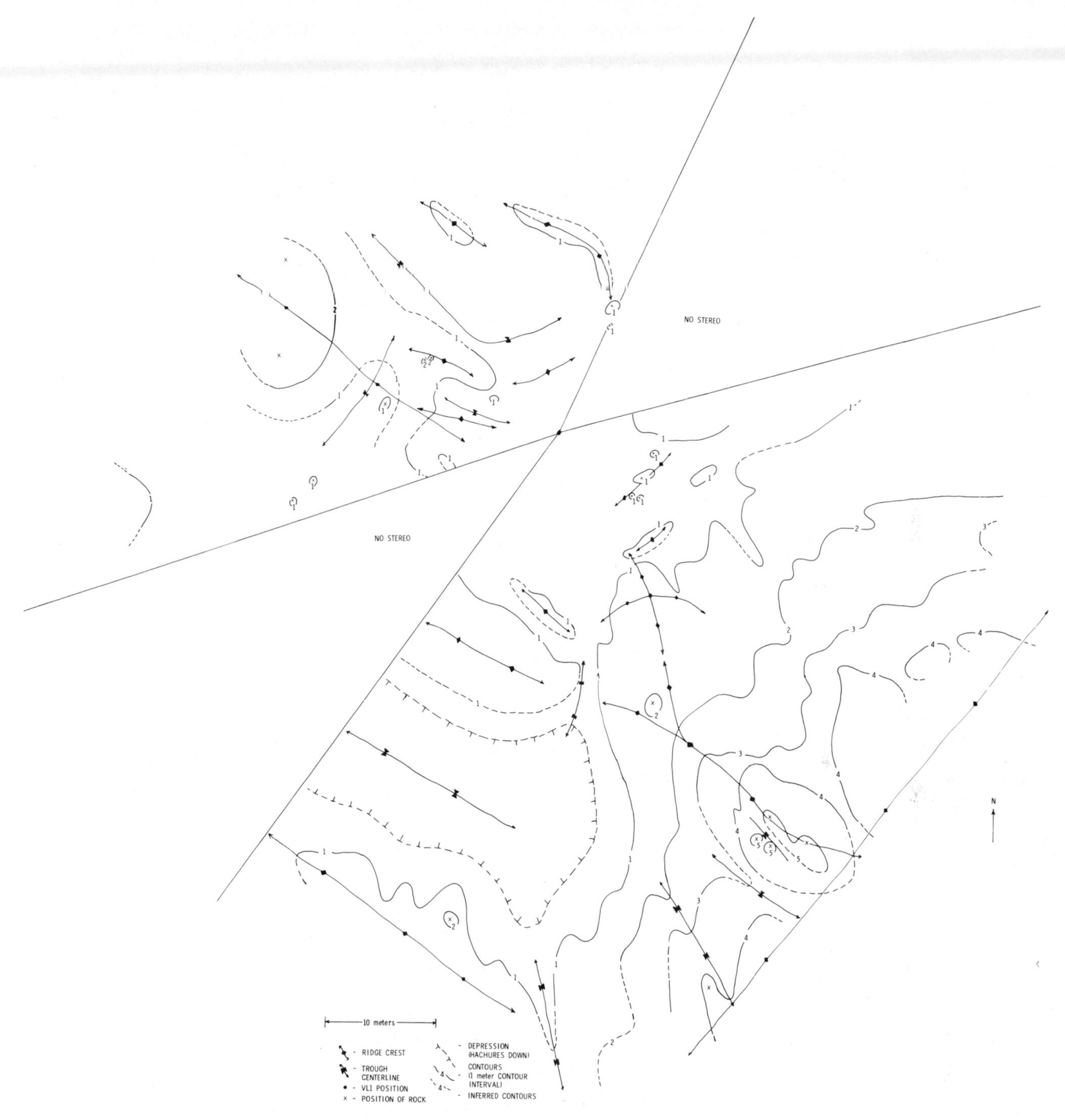

Fig. 5. A topographic map for the Viking Lander 1 site, extending to a distance of approximately 50 m from the spacecraft. The map was constructed by analysis of stereoscopic picture pairs. For further details, refer to *Liebes and Schwartz* [1977].

which a 25-m diameter crater is just resolvable. For craters between 100 and 300 m, $E \pm 3\sigma$ is 0.75 ± 2.6; for 50–100 m, $E \pm 3\sigma$ is 4.9 ± 6.6; and for 25–50 m, $E \pm 3\sigma$ is 35.4 ± 17.9. Since craters larger than about 50 m are rare, there is a reasonable chance that they are not in the field of view. However, for craters smaller than 50 m the probability of seeing them is high enough that they would be in the lander field of view if they existed. But only one crater can be identified with certainty at the site. That crater is located on the far horizon (Figure 11), and it is certainly larger than 50 m in diameter. Topographic maps constructed out to 100-m range show that there are no discernible craters of any size (Figure 5). *Gault and Baldwin* [1970] showed that craters $\lesssim 50$ m should be absent because of ablation and breakup of meteoroids by the Martian atmosphere. If former atmospheric pressure was greater than the present value, then the lower limit to the crater-size distribution would be proportionally larger. Our results are consistent with such theoretical treatment (Figure 8). The decay of the primary crater population can be explained without recourse to obliteration of craters by surface processes.

Fig. 6. In the lower left an area of fractured duricrust, which has been swept clear of overlying sediment by retro-rocket exhaust, is revealed. The prominently pitted rock is approximately 10 cm across. It appears more elongated than it actually is because at large depression angles the azimuthal dimension of the scene is exaggerated. For further details, refer to *Huck et al.* [1977]. A pin that dropped from the surface sampler assembly is in the lower right. For exact location of the picture, refer to *Mutch* [1977, Figure 1]. Camera event is 12A136.

ORIGIN OF THE STREWN FIELD

The diversity of lithologies in the small area viewed by the lander cameras is surprising, since the orbiter pictures reveal a surface resembling the lunar maria. Pursuing this analogy, one might expect to find rocks from basaltic flow units in addition to blocks of breccia formed by impact.

Since the surface of Mars is cratered, an obvious suggestion is that the blocks are emplaced as ejecta from craters which have penetrated several different types of bedrock. However, the distribution of blocks is more regular than might be expected for ejecta, which is characteristically arranged in rays and lobes. In addition, the concentration of craters as determined in both orbiter and lander pictures appears insufficient to produce the large number of observed blocks (Figure 9). Admittedly, this is a qualitative argument, especially because it is difficult to determine the exact position of the lander relative to several craters $\gtrsim 500$ m in diameter that are observed in orbiter pictures.

As was mentioned above, observational and theoretical ar-

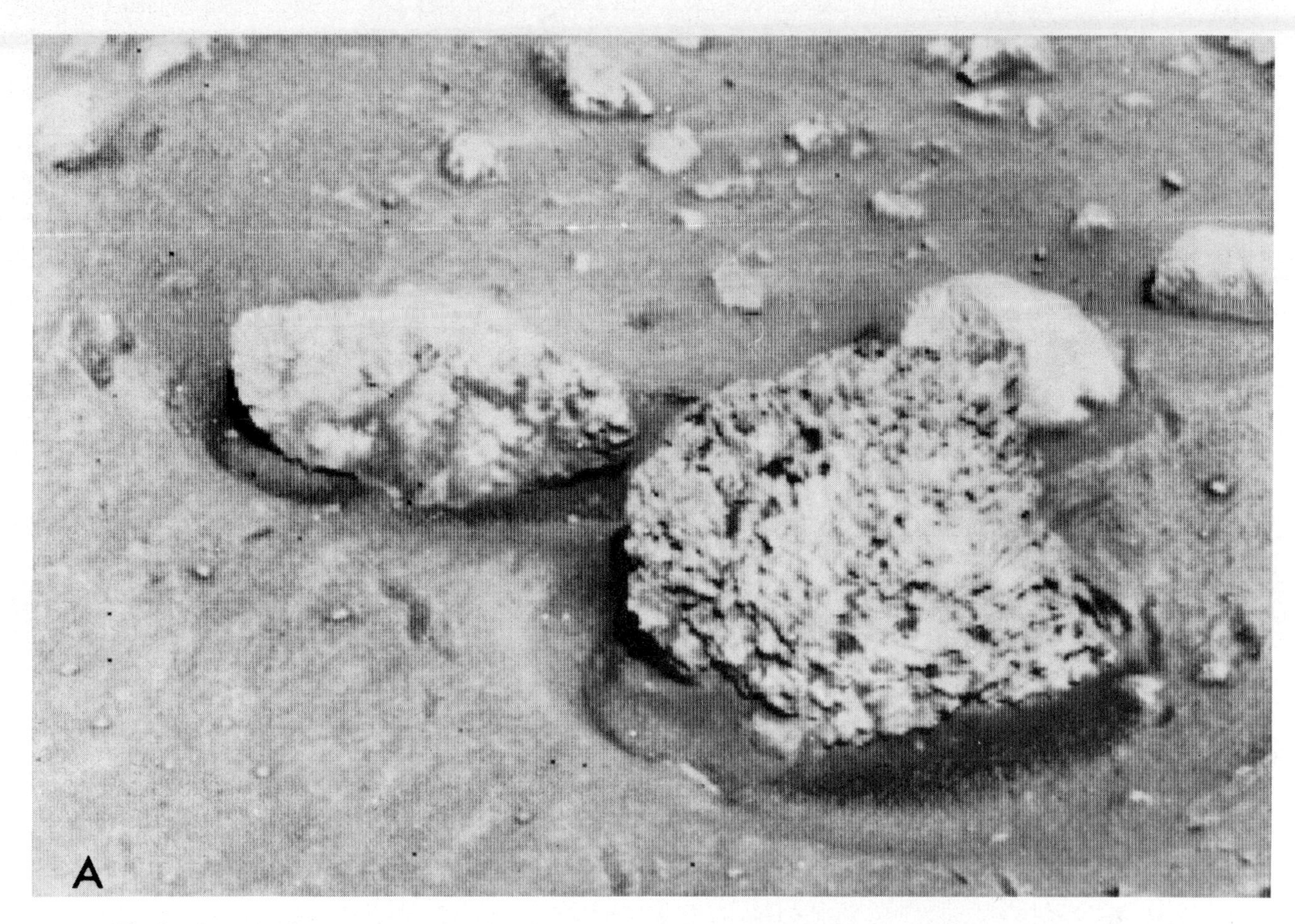

Fig. 7. Several different types of blocks at the Viking Lander 1 site. (*a*) A bright rock with darker spots, possibly an impact breccia with dark clasts. The block is about 20 cm across. Camera event is 11A037. (*b*) A coarsely pitted, conchoidally fractured block which may be a coarse-grained 'gabbroic' rock. Camera event is 11A141. For exact locations of the pictures, refer to *Mutch* [1977, Figure 1].

guments suggest absence of craters smaller than 50 m in diameter. Thus the numerous small-body impacts which are so important in developing the lunar regolith were not correspondingly important on Mars. For the same reason, breccias, formed on the moon by successive impact fragmentations and impact lithification, may be a relatively subordinate rock type on Mars.

EOLIAN FEATURES

Drifts of fine-grained sediment present in the vicinity of the Viking Lander 1 have been described previously [*Mutch et al.*, 1976*b*]. They occur both as isolated drifts (Figure 3) and as larger complexes (Figure 10). There is no evidence that any of the drifts are presently growing in size or migrating downwind. Instead, the exposure of internal stratification suggests that the most recent event was erosion of sedimentary deposits.

EROSION

Orbiter pictures indicate a geomorphologically fresh surface. Wrinkle ridges and craters are preserved in pristine form. Pedestal craters, which are abundant in other regions on Mars where intercrater deposits have been stripped, are not present here. These observations argue against deep erosion of the order of tens of meters. However, features visible in lander pictures argue for a lesser but significant amount of erosion. The several possible exposures of bedrock indicate that the Martian 'soil' or regolith is not uniformly distributed, as it is on the moon. One bedrock unit is topped by a 2-m boulder with texture and brightness similar to those of underlying bedrock. Assuming that the boulder has weathered in situ, erosion of the order of several meters is suggested. The general absence of vesiculated fine-grained rocks from the tops of lava flows also suggests erosion of several meters, assuming, of course, that extrusive flows were originally present.

A small-scale but highly significant erosional event has been observed in the course of the Viking mission. Between September 5, 1976, and March 22, 1977, some fine-grained material in the region of drifts in front of a large boulder called Big Joe slumped (Figure 10). A patch of fine-grained material approximately 15 cm across moved downslope. Apparently, slippage occurred along a boundary between two layers with different internal cohesion. A distinct scarp is visible along the upper edge of the cavity. The specific cause of this erosional event is not known, but it clearly is in the general category of gravity-induced mass movement and is not related to wind-induced saltation.

Statistics of one are of doubtful value. Nonetheless, the observation of this erosional event over a period of 8 months

Fig. 7. (continued)

suggests that the drifts have been sculpted relatively recently. For example, assuming one such event every year, it is obvious that the landforms would be dramatically changed in the course of, say, 1 m.y.

LANDER LOCATION

Accepting the most probable position of the lander, as indicated in Figure 2, two nearby craters are visible in orbiter pictures. One crater, approximately 400 m in diameter, is located about 2.5 km to the southwest. Another crater, about 800 m in diameter, is located about 2 km to the northwest. The first crater may be the same one viewed on the horizon from the lander (Figure 11). It intercepts 12° in azimuth on a lander picture, which is consistent with the presence of a 400-m crater at a distance of 2.5 km. The other crater, though larger and closer, may be obscured by a closer ridge situated to the northwest of the spacecraft, prominently visible above the radiothermal generator windscreen.

The proposed correlation of features between orbiter and lander pictures is strengthened by the fact that, moving to the east or west along the 11-km longitude uncertainty bar, there are no other craters appropriately situated to the southwest.

SUMMARY

The following interpretive geological history for the Viking Lander 1 site is suggested:

1. Sometime early in the history of Mars a series of basic flood basalts filled the Chryse basin.

2. During the waning stages of volcanism, differentiation produced a diverse series of magmas intruded as dike complexes or extruded on the surface as stubby flows. A mare wrinkle ridge was formed.

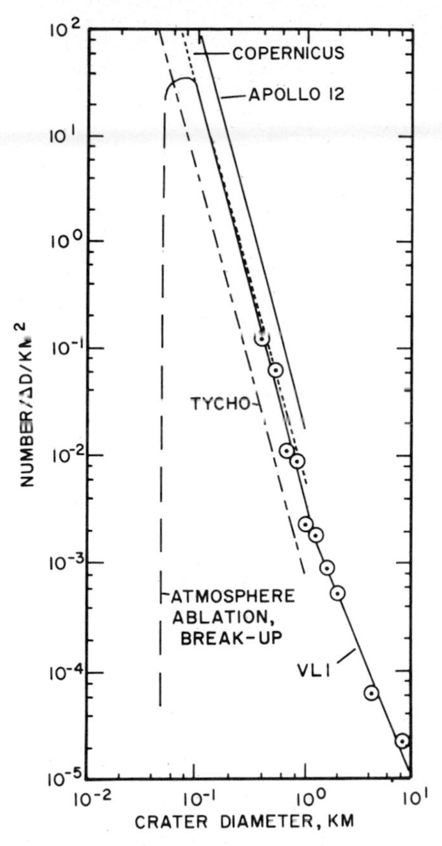

Fig. 8. Crater distributions at the Viking Lander 1 site compared to measurements on the moon in the vicinity of Tycho, Copernicus, and Apollo 12. Lunar data are from *Arvidson et al.* [1977].

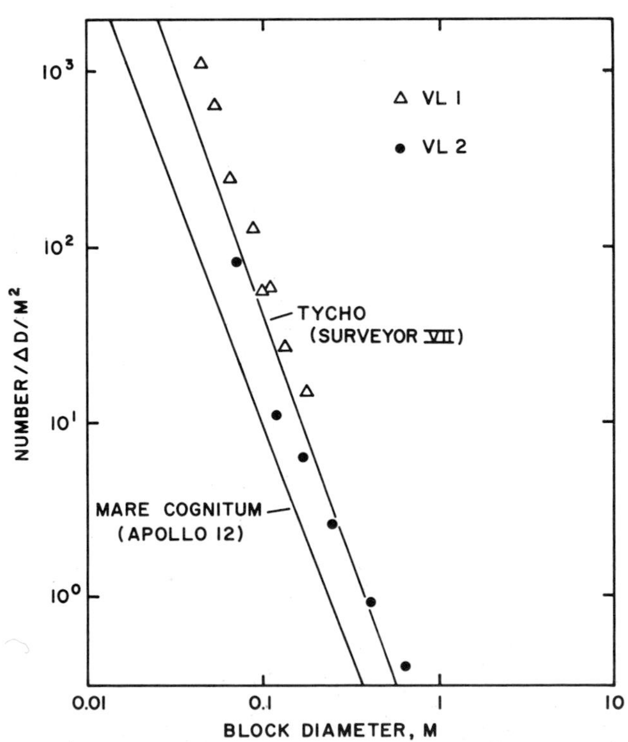

Fig. 9. Graph showing block distribution at Viking Lander 1 and 2 sites. Distributions at the Apollo 12 and Surveyor 7 sites are shown for comparison. Note that block abundances at the Viking 1 site are higher than those for the rim of Tycho, even though no large craters are nearby.

Fig. 10. (opposite) Two pictures of a distinctive boulder, approximately 2 m across, called Big Joe. (*a*) Taken on September 5, 1976 (camera event 11B126). (*b*) Taken on March 22, 1977 (camera event 11C162). In the time period between the two pictures a small amount of sediment (shown by an arrow in (*b*)) slumped. For exact locations of the pictures, refer to *Mutch* [1977, Figure 1].

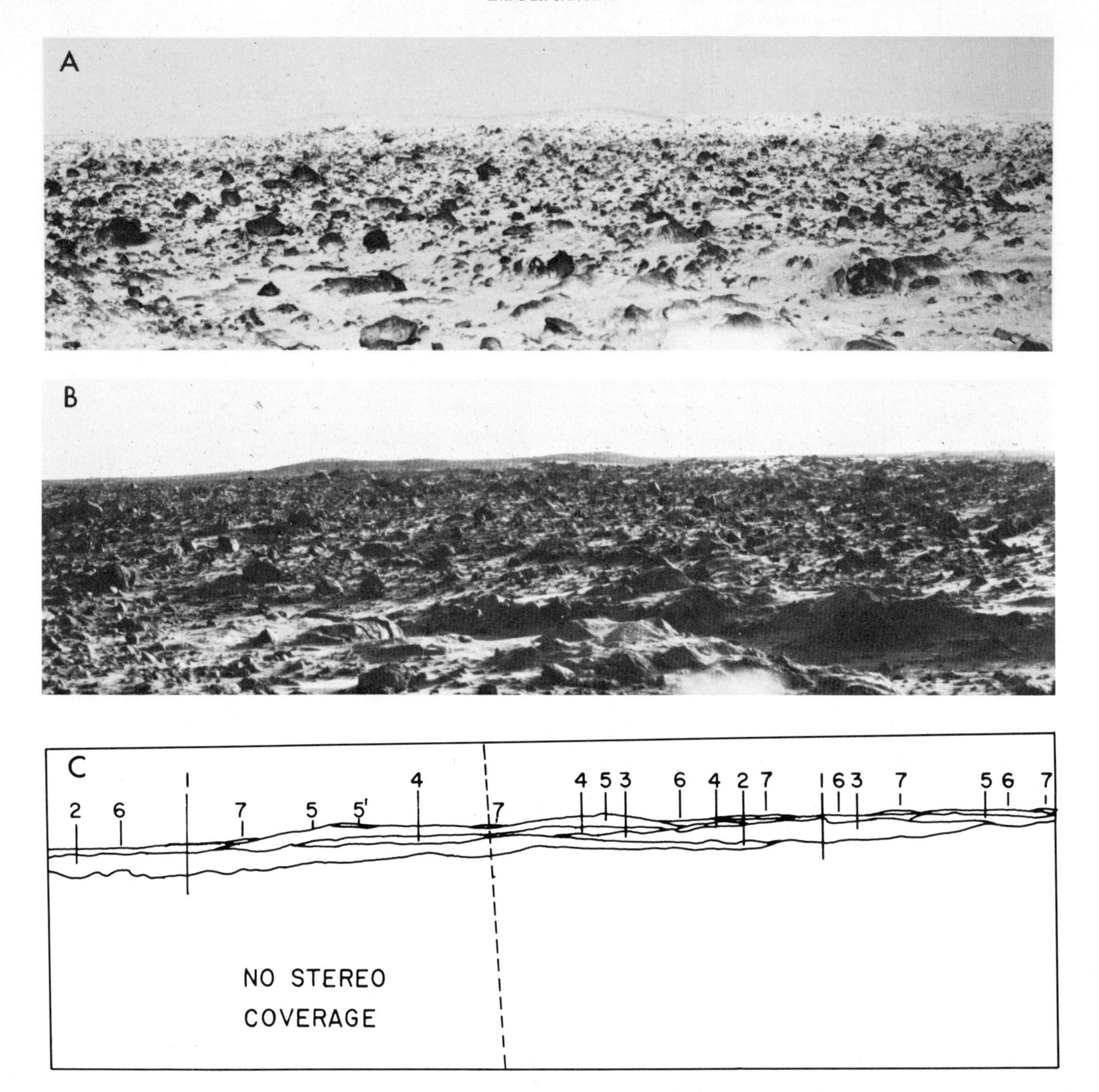

Fig. 11. Two pictures (camera events 12A116 and 12A235) and a line drawing showing horizon details. Though both pictures are of the same area, different lighting conditions and picture enhancement techniques emphasize different details. A prominent depression is visible in the foreground of (*b*) (12A235). A sequence of ridge crests is indicated on the line drawing. Positions of a probable crater rim (ridge 5), a small segment of the far crater wall (point 5′), and bluffs behind the crater (ridges 6 and 7) are shown along the horizon. For exact location of the pictures, refer to *Mutch* [1977, Figure 1].

3. Chemical and mechanical erosion broke down the upper few meters of the ridge, exposing jointed and fractured outcrops and leaving behind angular blocks. Fluvial activity may have occurred at this time.

4. Some impact debris was ballistically transported to the ridge, increasing the number and diversity of blocks.

5. Wind action formed fine-grained drifts, swept clear some outcrops, and faceted many of the blocks.

Acknowledgments. We are indebted to the many people within the Viking project who contributed to the acquisition and interpretation of the lander pictures. Our special thanks go to P. Avrin, S. H. Bosworth, C. E. Carlston, E. W. Dunham, P. L. Fox, S. U. Grenander, J. W. Head, F. O. Huck, R. A. Kahn, E. C. Levinthal, S. Liebes, Jr., B. K. Luchitta, D. Nummedal, W. R. Patterson, J. B. Pollack, C. W. Rowland, G. R. Taylor, R. B. Tucker, S. D. Wall, and M. R. Wolf, all of whom were closely involved with the Lander Imaging Investigation. R. P. Sharp reviewed the manuscript and made valuable suggestions. This paper presents results of tasks carried out under the following NASA contracts: contract NAS1-9680 to Department of Geological Sciences, Brown University; order L-9718 to Branch of Astrogeologic Studies, U.S. Geological Survey; contract NAS1-11851 to Science Applications; contract NAS1-9683 to Center for Radio Physics and Space Research, Cornell University; and contract NAS1-13889 to Department of Earth and Planetary Sciences, Washington University.

REFERENCES

Arvidson, R. E., E. Guinness, and C. Hohenberg, On the constancy of the lunar cratering flux over the past 3.3 billion years (abstract), *Proc. Lunar Sci. Conf.*, *8*, 50–52, 1977.

Baird, A. K., P. Toulmin, B. C. Clark, H. Rose, Jr., K. Keil, R. P. Christian, and J. L. Gooding, Mineralogic and petrologic implications of Viking geochemical results from Mars: Interim report, *Science*, *194*, 1288–1293, 1976.

Binder, A. B., and D. P. Cruikshank, Comparison of the infrared spectrum of Mars with the spectrum of selected terrestrial rocks and minerals, *Commun. Lunar Planet. Lab.*, *2*, 193–196, 1964.

Clark, B. C., A. K. Baird, K. Keil, H. Rose, Jr., and P. Toulmin, Element concentrations in Martian surface materials, paper presented at Special Mars Session of 8th Lunar Science Conference, Johnson Space Center, Houston, Tex., March 15, 1977.

Davies, M. E., and D. W. G. Arthur, Martian surface coordinates, *J. Geophys. Res.*, *78*, 4355–4394, 1973.

Gault, D. E., and B. S. Baldwin, Impact cratering on Mars—Some effects of the atmosphere, *Eos Trans. AGU*, *51*, 343, 1970.

Greeley, R., E. E. Theilig, J. E. Guest, M. H. Carr, H. Masursky, and J. A. Cutts, Geology of Chryse Planitia, *J. Geophys. Res.*, *82*, this issue, 1977.

Huck, F. O., D. J. Jobson, S. K. Park, S. D. Wall, R. E. Arvidson, W. R. Patterson, and W. B. Benton, Spectrophotometric and color estimates of the Viking lander sites, *J. Geophys. Res.*, *82*, this issue, 1977.

Huguenin, R. L., The formation of goethite and hydrated clay minerals on Mars, *J. Geophys. Res.*, *79*, 3895–3905, 1974.

Hunt, G. R., L. M. Logan, and J. W. Salisbury, Mars: Components of infrared spectra and the composition of the dust cloud, *Icarus*, *18*, 459–469, 1973.

Liebes, S., Jr., and A. A. Schwartz, Viking 1975 Mars lander interactive computerized video stereophotogrammetry, *J. Geophys. Res.*, *82*, this issue, 1977.

Mayo, A. P., W. T. Blackshear, R. H. Tolson, W. H. Michael, Jr., G. M. Kelly, J. P. Brenkle, and T. A. Komarek, Lander locations, Mars physical ephemeris, and solar system parameters: Determination from Viking lander tracking data, *J. Geophys. Res.*, *82*, this issue, 1977.

McElroy, M. B., Y. L. Yung, and A. O. Nier, Isotopic composition of nitrogen: Implications for the past history of Mars' atmosphere, *Science*, *194*, 70–72, 1976.

Michael, W. H., Jr., A. P. Mayo, W. T. Blackshear, R. H. Tolson, G. M. Kelly, J. P. Brenkle, D. L. Cain, G. Fjeldbo, D. N. Sweetnam, R. B. Goldstein, P. E. MacNeil, R. D. Reasenberg, I. I. Shapiro, T. I. S. Boak III, M. D. Grossi, and C. H. Tang, Mars dynamics, atmospheric and surface properties: Determination from Viking tracking data, *Science*, *194*, 1337–1339, 1976.

Moore, H. J., R. E. Hutton, R. F. Scott, C. R. Spitzer, and R. W. Shorthill, Surface materials of the Viking landing sites, Mars, *J. Geophys. Res.*, *82*, this issue, 1977.

Morris, E. C., T. A. Mutch, and H. E. Holt, Atlas of geologic features in the dry valleys of South Victoria Land, Antarctica, *Interagency Rep. Astrogeology 52*, 156 pp., U.S. Geol. Surv., Flagstaff, Ariz., 1972.

Mutch, T. A., The Viking Lander Imaging Investigation: An introduction, *J. Geophys. Res.*, *82*, this issue, 1977.

Mutch, T. A., A. B. Binder, F. O. Huck, E. C. Levinthal, S. Liebes, E. C. Morris, W. R. Patterson, J. B. Pollack, C. Sagan, and G. R. Taylor, The surface of Mars, The view from the Viking 1 lander, *Science*, *193*, 791–801, 1976*a*.

Mutch, T. A., R. E. Arvidson, A. B. Binder, F. O. Huck, E. C. Levinthal, S. Liebes, E. C. Morris, D. Nummedal, J. C. Pollack, and C. Sagan, Fine particles on Mars, Observations with the Viking 1 lander cameras, *Science*, *194*, 87–91, 1976*b*.

Nier, A. O., M. B. McElroy, and Y. L. Yung, Isotopic composition of the Martian atmosphere, *Science*, *194*, 68–70, 1976.

Sagan, C., P. Fox, R. E. Arvidson, and E. A. Guinness, Particle motion on Mars inferred from the Viking landers, *J. Geophys. Res.*, *82*, this issue, 1977.

Toon, O. B., J. B. Pollack, and C. Sagan, Physical properties of the particles comprising the Martian dust storm of 1971–72, *Icarus*, *3*, 663–696, 1977.

U.S. Geological Survey, Atlas of Mars, Yorktown region, M250K 22/48 CmC, scale 1:250,000, *Topogr. Ser.*, 1977.

Von Tassel, R. A., and J. W. Salisbury, The composition of the Martian surface, *Icarus*, *3*, 264–269, 1964.

Weill, O. F., I. S. McCallum, Y. Bottinga, M. J. Drake, and G. A. McKay, Mineralogy and petrology of some Apollo II igneous rocks, *Proc. Lunar Sci. Conf.*, *1*, 937–955, 1970.

Woolnough, W. G., Presidential address, II, The duricrust of Australia, *J. Roy. Soc. N. S. W.*, *61*, 1–53, 1928.

(Received April 13, 1977;
revised June 9, 1977;
accepted June 9, 1977.)

The Geology of the Viking Lander 2 Site

THOMAS A. MUTCH,[1] RAYMOND E. ARVIDSON,[2] ALAN B. BINDER,[3] EDWARD A. GUINNESS,[2] AND ELLIOT C. MORRIS[4]

Viking Lander 2 landed on a flat plain of fine-grained sediment overlain by dispersed, evenly distributed boulders. The fine-grained material is probably part of a high-latitude mantle comprising material swept south from the polar regions. The boulders, which have distinctive deep pits, or vesicles, may be the residue of an ejecta deposit from the crater Mie. Alternatively, they may be the remnants of lava flows which formerly covered the region. Polygonal sediment-filled cracks may have been formed by ice wedging, similar to the process that occurs in terrestrial permafrost regions. Alternatively, they may be desiccation polygons.

GENERAL GEOLOGIC SETTING

Viking Lander 2 landed on the Utopia Planitia at 47.968°N, 225.59°W (aerographic coordinates), as determined by Doppler tracking of radio signals [*Michael et al.*, 1976; M. E. Davies, unpublished data, 1977]. The maximum error in position on the orbiter photobase, considering both accuracy of radio tracking and accuracy in fitting measurements to the cartographic base, is less than 120 m in latitude and less than 11 km in longitude [*Mayo et al.*, 1977] (Figure 1).

This region is part of the vast plains that occupy much of the northern hemisphere of Mars. Although Chryse Planitia, the site for the first landing, also lies within the broad morphological province of northern plains, Viking orbiter pictures reveal that the two regions are much different. The surface at Utopia Planitia has relatively few craters. Dominant topographic elements are hummocks, knobs, and irregular pits. Some protuberances in the vicinity of the lander are interpreted as volcanic domes (Figure 1).

Approximately 150 km to the south and west of the site, sharply incised fissures form a polygonal network (Figure 1). The fissures characteristically are 0.5–1.5 km wide and 50–100 m deep, values based on shadow measurements. The polygons are approximately 5–10 km in diameter. The cracks have been interpreted as cooling fractures in lava or, alternatively, as patterned ground produced by thermal contraction and expansion in a permafrost environment [*Masursky and Crabill,* 1976]. They also might have been formed by tectonic forces. The fissures become muted and subdued where they are traced toward the landing site, suggesting superposition of a layer of fine-grained sediment. In the immediate vicinity of the landing site the surface presents a diffuse, mottled appearance, interpreted to be caused by mantling material (Figure 2). Regularly spaced ridges, a few hundred meters apart and a few hundred meters to 1 km long, are superimposed on a larger-scale topography of hummocks and knobs. These have been interpreted as dunes [*Masursky and Crabill,* 1976]. They impart an irregularly reticulate pattern to the landscape as viewed from orbit (Figure 2).

On the basis of Mariner 9 pictures, *Soderblom et al.* [1973] suggested that a mantle of eolian debris extends equatorward from the pole to approximately 35°N. The relationships seen in Viking pictures confirm that there have been successive periods of stripping and deposition along the margins of the mantle. For example, many crater interiors are filled with sediment, suggesting that they are the remnants of a sedimentary layer that once mantled the entire region. In both the fissured terrain and the mantled terrain, small craters and their associated annuli of ejecta stand higher than the surrounding terrain. Apparently, the substrate has been deflated by eolian erosion. Ejecta deposits are more resistant to erosion, probably because of an abundance of coarse-grained rubble that serves as protective armor.

TOPOGRAPHY

The Viking Lander 2 site is remarkably flat, as viewed from the surface. Quantitative stereoscopic mapping [*Liebes and Schwartz,* 1977] indicates relief of less than 1 m to a radial distance of approximately 100 m. The maximum apparent slopes along the horizon are only 1° or 2°. A gently sloping ridge of bright material is visible to the east of the spacecraft, beyond the middle-ground horizon defined by the block-littered plains (Figure 3).

A network of shallow troughs is visible in the vicinity of the spacecraft (Figures 4 and 5). The distinctive morphology of these depressions suggests a specific geomorphological process, one which is not similarly expressed at the Viking Lander 1 site.

The best displayed trough is situated approximately 8 m north of the spacecraft and trends approximately east-west. It is 1 m across and 10 cm deep. Small drifts of fine-grained sediment occur in the trough. By comparison with the adjacent terrain, blocks are rare. Some of those rocks that are present are partly buried, as opposed to the majority of rocks outside the trough. Topographic cross sections through the trough indicate a lack of symmetry. The northern wall is relatively steep and high (Figure 6). A prominent upraised lip is present along the northern edge. Subtle differences in elevation also suggest the presence of an upraised rim along the southern margin of the trough.

Troughs trending in other directions are present at the landing site. Although a comprehensive map is precluded both by obscuration and by the oblique viewing, the presence of several trough junctions suggests a polygonal network. Orbiter pictures of this region also reveal distinctive polygonally arranged depressions, but a change in scale of approximately 3 orders of magnitude renders any hypothesis of common origin highly speculative.

There are at least four processes that might account for trough formation. The first is cooling of lava to form con-

[1]Department of Geological Sciences, Brown University, Providence, Rhode Island 02912.

[2]McDonnell Center for the Space Sciences, Department of Earth and Planetary Sciences, Washington University, St. Louis, Missouri 63160.

[3]Institut für Geophysik, University of Kiel, Federal Republic of Germany.

[4]Branch of Astrogeologic Studies, U.S. Geological Survey, Flagstaff, Arizona 86001.

Paper number 7S0537.

traction fractures, which typically are polygonal. This is a common phenomenon in terrestrial basalt flows. Since lava flows are not revealed at the surface, it is necessary to postulate that polygonally fractured lava is present at shallow depth and that overlying sediment reflects subjacent topography. Were the thickness of the sedimentary cover more than a few meters, it would be unlikely that the bedrock cracks would be mimicked at the surface. However, the general flatness of the landscape, the absence of bedrock exposures, and the pervasive exposure of sediments argue against near-surface presence of volcanic bedrock.

A second mechanism for polygonal fracturing is desiccation of water-saturated clay minerals. X ray fluorescence results [*Baird et al.*, 1976] and spectral reflectance estimates from lander camera data [*Huck et al.*, 1977] from both Viking Lander 1 and Viking Lander 2 suggest the presence of clay minerals. The pedestal craters and generally muted appearance of the Viking Lander 2 area, as viewed from orbit, suggest a considerable thickness of fine-grained material. If the fines are dominated by clays and if they were deposited with substantial water content, then subsequent dehydration could have led to contraction and fracturing. Large desiccation polygons have been identified in terrestrial playas, where they reach diameters of up to 300 m [*Neal et al.*, 1968]. Shrinkage of about 20% in linear dimension is observed when water-saturated montmorillonite is dehydrated [*Grim*, 1962].

It is conceivable that the troughs have a fluvial origin. However, the absence of a unidirectional pattern and also the absence of recognizable fluvial bed forms (sand or gravel bars, channel bottom ripples, and so forth) argue against fluvial processes.

A fourth mechanism is thermal expansion and contraction of frozen ground, possibly accompanied by some melting. Thermal stresses, generated in frozen ground because of low temperature and rapid cooling, can produce contraction fractures [*Lachenbruch*, 1962]. In the terrestrial case, melting of ice during spring and summer leads to the accumulation of water in contraction cracks. The water then freezes to form an ice wedge. Repetition of this annual cycle leads to growth of the fractured zone and formation of polygonally patterned ground (Figure 7). Expansion of permafrost during summer months results in upbulging on either side of the crack. Where wind-driven sediment is available, the crack is filled with fine-grained material, exhibiting small-scale ripples. The total appearance of a terrestrial sand-filled crack closely resembles that of the feature observed on Mars (Figure 8).

At the latitude of the landing site, current ground temperatures never rise high enough to permit melting of ice [*Wade and DeWys*, 1968]. Conceivably, the cracks could have been nourished by frost crystals, formed by condensation of atmospheric water vapor [*Wade and DeWys*, 1968]. Alternatively, small amounts of metastable liquid water might be able to form and migrate through the soil, even though the atmospheric pressure is below the triple point. Yet another possibility is that the liquid is not pure water but is rather a saline solution. *Malin* [1974] has demonstrated that typical salt solutions might remain liquid for part of the Martian day, at least in the more temperate latitudes. Finally, it can be argued that the patterned ground is a fossil landform, produced at a time when the atmosphere was denser and liquid water was stable. For example, preservation of patterned ground formed during the Pleistocene maximum ice advance and now observed in temperate climatic zones is documented on earth [*Washburn*, 1973].

BLOCKS

The blocks at the Viking Lander 2 site are remarkable for their similarity to one another. They stand in marked contrast to the rocks at the Viking Lander 1 site, which display a wide range of shape, texture, and brightness. Blocks at the Viking Lander 2 site are generally subangular and equidimensional. Although the cumulative distributions are approximately the same as those at the Viking Lander 1 site, large blocks are more numerous at the Viking Lander 2 site [*Binder et al.*, 1977, Figure 9].

It is apparent that there is a paucity of particles in the range from 2 mm to 1 cm, especially when one considers that many of the fragments in this size range are cemented soil fragments. This leads to a bimodal distribution, one mode being in the vicinity of 100 μm or less and a second mode in the vicinity of 10–20 cm. In terrestrial situations, some bimodal size distributions indicate two depositional mechanisms or events. The Martian situation is probably similar. The fine-grained sediment is interpreted to be wind transported and part of an extensive mantle that covers much of the high-latitude regions [*Soderblom et al.*, 1973]. The boulders are interpreted to be formed during a later, unrelated event: the emplacement and subsequent erosion of dominantly coarse-grained material on top of the fine-grained eolian deposits.

The most provocative features of the rocks at the Viking Lander 2 site are the numerous pits a few millimeters to a few centimeters across, which impart a spongelike appearance (Figure 9). More than 90% of the rocks close to the spacecraft display such pits. The majority of depressions are equidimensional, but some have elliptical cross sections.

The most obvious interpretation is that these pits are volcanic vesicles, formed by solidification of a frothy gas-charged lava. One particularly instructive block is partly massive and smooth and partly pitted. At least four bands, defined in part by variable pitting, can be identified (Figure 4). These observations suggest that this block may have come from the top of a volcanic flow, the vesicular part of the rock solidifying at the top. Analogous textures are common in terrestrial basaltic lavas. Following their initial formation, the vesicles probably have been enlarged and modified by eolian abrasion.

Although a volcanic explanation for the vesicular character of the rocks is suggested by terrestrial analogy, the extreme prominence and ubiquity of the pits are disquieting. Alternate mechanisms, perhaps peculiar to the Martian environment, are hinted at. Sandblasting of terrestrial rocks commonly produces pitting. Although pitting can occur on relatively homogeneous rocks, the process is enhanced by internal heterogeneity. If the original rocks contained large crystals or clasts in a finer-grained matrix, differential erosion may have etched out the softer component. Porphyritic basalts or polymict impact breccias would yield the requisite large crystals/clasts. However, it is unlikely that the erosion rate of, say, plagioclase phenocrysts would be that different from that of fine-grained groundmass. Indeed, in many terrestrial situations the phenocrysts are more resistant than the groundmass, resulting in a knobby appearance, the reverse of what is observed. Finally, the presence of large phenocrysts should be indicated by characteristic specular reflection on some of the less altered surfaces. The presence of impact breccias can be discounted for similar reasons. Polymict breccias characteristically have a mottled appearance not observed in Martian blocks.

More intriguing is the possibility that the pits were origi-

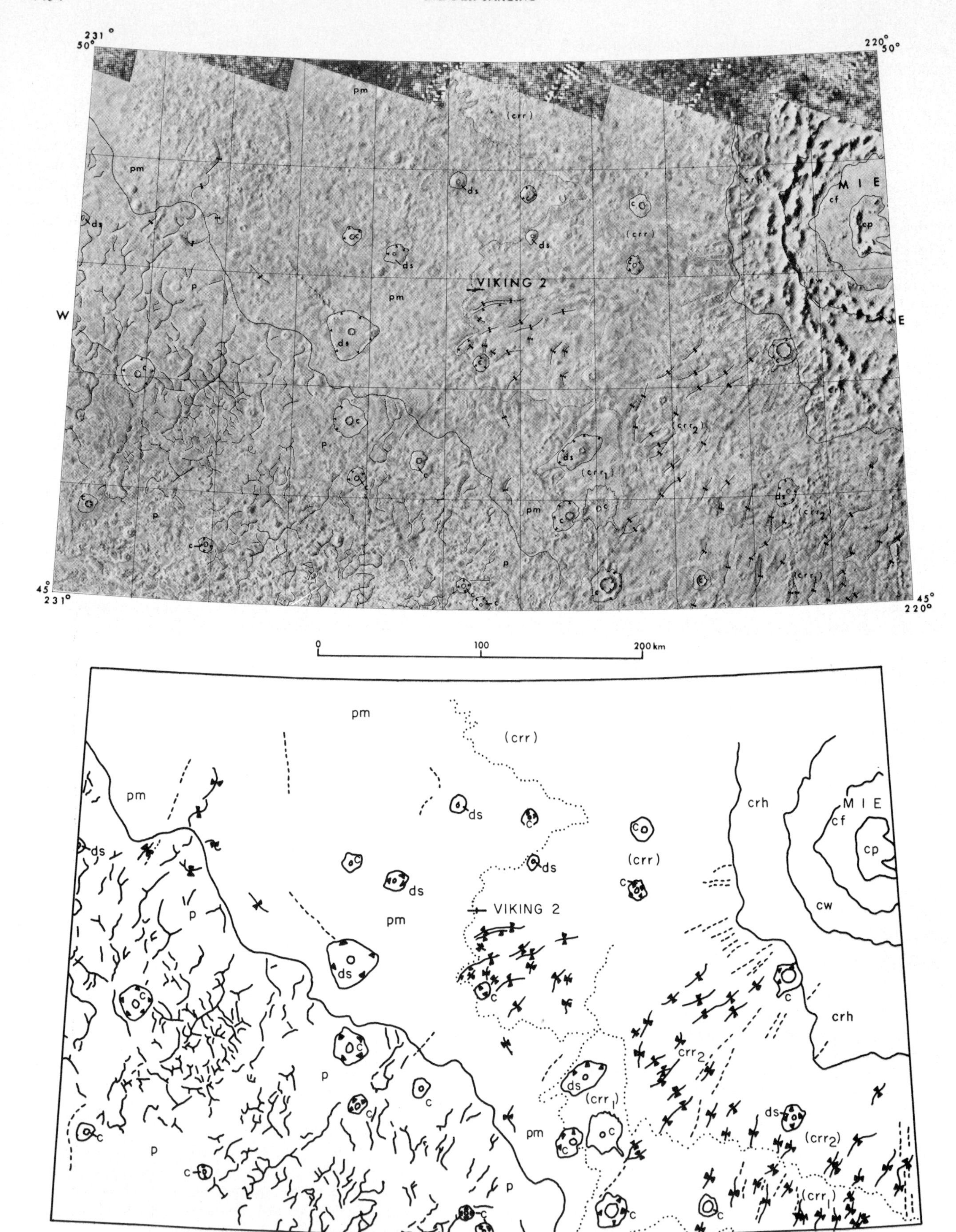

Fig. 1. Photomap (top), geologic map (bottom), and legend (opposite page) of Viking 2 landing site. Bar indicates position of lander as determined by radio tracking data [*Michael et al.*, 1976; M. E. Davies, unpublished data, 1977]. The maximum error in fitting this position to the cartographic base is approximately 11 km in longitude and 120 m in latitude [*Davies and Arthur*, 1973]. Photomosaic base is by the U.S. Geological Survey.

EXPLANATION

PLAINS MATERIAL

pm MANTLED PLAINS Surface composed of subdued rolling topography hummocks, ridges and knobs. Small craters a few kilometers in diameter and their aprons of ejecta stand above the surface on knobs and mesas. Interiors of the craters are often filled with smooth material. A fine reticulate pattern of ridges a few hundred meters to a kilometer long are superimposed on the larger pattern of hummocks and knobs. Interpretation: Volcanic plains blanketed by a mantle of fine-grained aeolian deposited material. Aprons of ejecta around craters that have formed on the blanket of aeolian material form an armor or protection to subsequent deflation or stripping of the surface. Fine reticulate patterns of ridges are interpreted to be dunes of fine-grained material.

p PLAINS MATERIAL Surface composed of rolling topography of hummocks, irregular pits, hollows, ridges and knobs. The surface is broken by sharply incised curvilinear cracks that form roughly polygonal patterns 5 to 10 kilometers wide. Interpretation: Volcanic plains formed by very fluid lavas. Cracks probably are cooling fractures in the lavas that subsequently have been etched and enlarged by aeolian action.

DOME AND SHIELD MATERIAL

ds DOME AND SHIELD MATERIALS Forms irregular shaped broad, low convex upward structures or conical shaped structures with small pit craters at their summits. Surfaces are rough with small knobs and grooves. The terminal edges of the broad low structures extend up to 5 crater diameters from summit craters. Some structures are bounded by steep scarps. Interpretation: Volcanic domes and small shields formed from low viscosity lavas and pyroclastic deposits. Volcanic materials provide a resistant cover to aeolian stripping. Some structures are on mesas with bounding scarps.

CRATER MATERIAL

c CRATER MATERIALS Forms hummocky, blocky surfaces around small craters. Steep scarps form the terminal edges of most deposits. Only craters with rim deposits greater than 10 km diameter were mapped. Interpretation: Fragmental debris ejected from craters. Material forms a protective armor to deflation of the surface by aeolian action. The crater and its apron of ejecta remain on knobs or mesas after surrounding material is eroded.

crr_{1-2} CRATER RIM RADIAL FACIES Hummocky, knobby surface similar in appearance to mantled plains but has gentle ridges and intervening shallow open grooves radial to the crater Mie. Two units of the radial facies are evident south of Mie but are not differentiable east of the crater. Unit crr_2 has sharper appearing hummocks and radial grooves than unit crr_1. The boundary between the two units is a very subdued gentle scarp. Interpretation: Fragmental debris ejected from the crater Mie, emplaced as fluid flow and subsequently buried under a mantle of fine-grained aeolian debris. The radial grooves and ridges and scarps that form its distal boundaries are muted by the overlying mantle. Craters on pedestal and mesas indicate the surface has had several episodes of burying and stripping.

crh CRATER RIM HUMMOCKY FACIES Surface composed of large hummocks and knobs several km across and ridges up to 10km in length arranged in a roughly concentric pattern around the flanks of the crater Mie. Surface between the ridges and knobs is smooth and in places has a fine reticulate pattern of ridges a few hundred meters in length. Interpretation: Fragmental debris ballistically ejected from the crater Mie and subsequently partially buried by an aeolian mantle. Between the ridges and hummocks are dunes of fine-grained aeolian deposited material.

cw CRATER WALL Concentric steep blocky ridges and benches form the rim and sloping inner wall of the crater Mie. Surface smooth between blocks and ridges. Interpretation: Fractured material and slumped blocks emplaced during the late stages of the formation of the crater Mie. The walls of the crater subsequently have been mantled by aeolian material. Fields of large dunes of fine-grained material are prominent amid the ridges and blocks.

cf CRATER FLOOR Smooth nearly level surface with clumps of rounded knobs projecting above the surface. A fine reticulate pattern of ridges and grooves cover the north and west part of the unit. Interpretation: Floor of the crater Mie that is buried by a blanket of aeolian material. Blocks of brecciated material project through the aeolian cover.

cp CENTRAL PEAK Cluster of knobs and ridges in the center of the crater Mie. Interpretation: Brecciated crater floor uplifted by rebound during shock decompression following the formation of the crater by an impact.

Contact

Concealed contact

Lineament - aligned ridges and knobs

Trace of grooves and cracks on plains material

Trough or depression - line marks center of depression, barbs point down slope

Scarp - barbs point down slope, line at bottom of slope

Crater rim crest

Location of Viking 2 Lander. Ellipse shows approximate precision of location determined from radar tracking data.

Fig. 2. Mosaic of three orbiter pictures showing features in the vicinity of the Viking Lander 2 landing site. The scene is approximately 75 × 100 km. The possible locations of the lander are shown by a bar 11 km long.

Fig. 3. Illustrations showing different aspects of the same horizon features at the Viking Lander 2 site. (Top) Camera event 22A252. (Middle) Camera event 22A148. (Bottom) Line drawing. Several bluffs (numbered 2, 3, and 4) are visible in the far distance, beyond the closer rock-studded ridge (numbered 1). The apparent tilt of the horizon results from spacecraft tilt of about 8°. The horizon is, in fact, horizontal. For the exact location of the pictures, refer to the paper by *Mutch* [1977, Figure 1].

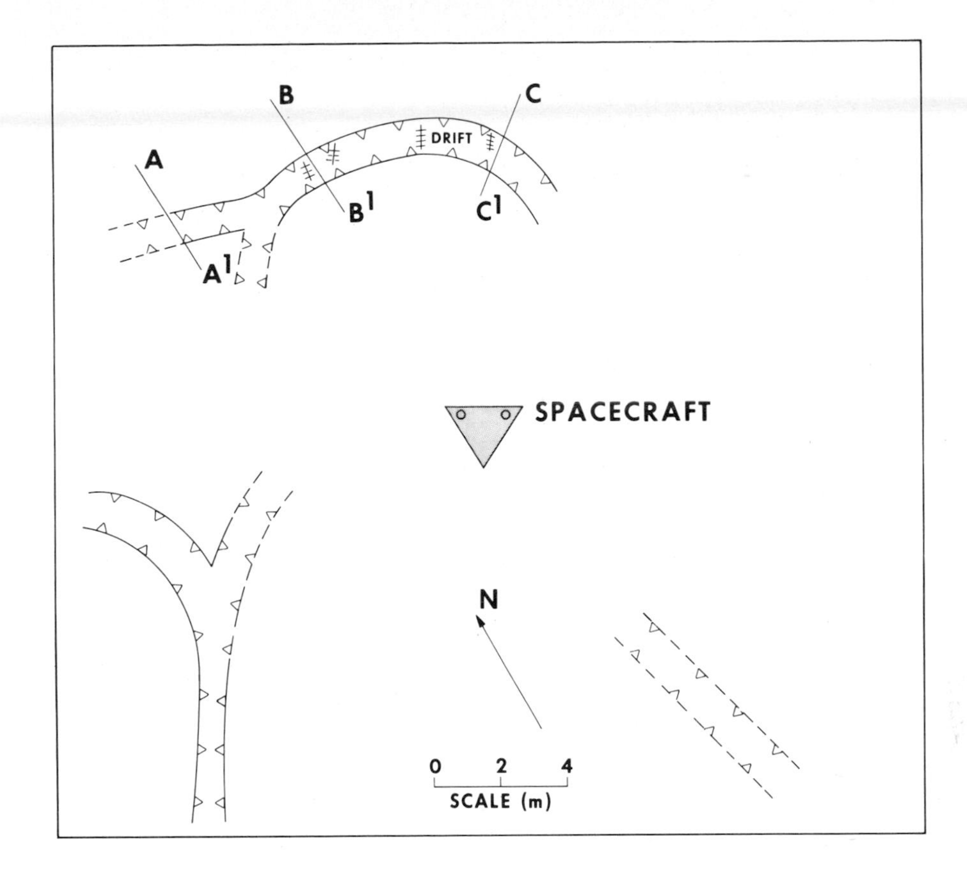

Fig. 5. The general distribution of troughs in the vicinity of Viking Lander 2. The trough at the top of the figure (in front of the spacecraft) is shown in Figure 4 and is mapped in more detail in Figure 6.

nally occupied by evaporite crystals. The occurrence of large single crystals of halides, sulfates, and carbonates is common in terrigenous sedimentary accumulations in highly saline basins. Subsequent leaching by freshwater produces a cavity, essentially a cast of the original evaporite crystal. Similar precipitation and solution of sulfates and carbonates might have taken place as saline waters migrated through the Martian soil. The presence of sulfates and carbonates, probably as secondary cementing material, is suggested by Viking X ray fluorescence results.

When the precipitation model is carried to an extreme, it can be argued that the pits were originally occupied by ice crystals that were dispersed throughout frozen soil. Subsequent sublimation of ice has led to the present vesicular appearance (A. A. Mills, personal communication, 1976). A serious objection is that sublimation of ice should have resulted in disintegration of surrounding soil.

Still another possibility is that spongelike structures are formed as a characteristic weathering product of Martian near-surface materials. Accumulation of iron oxides and hydroxides in concretional layers in terrestrial lateritic soils produces a 'ferricrete' rock which superficially resembles vesicular volcanic rock (Figure 10) [*Pullen,* 1967].

Those several arguments just mentioned which involve formation of the pitted blocks by chemical interactions in the soil are weakened by the observation that the Martian rocks appear to be relatively durable. Pushing and nudging of several rocks, accomplished with the sampler head, provided no evidence of spalling, chipping, or fracturing.

After considering all ways in which the deep pits might have formed we conclude that volcanic vesiculation is the most likely explanation.

DURICRUST

An assemblage of observations suggests the presence of a zone of lightly cemented fine-grained sediment. This layer, of the order of 1 or 2 cm thick, occurs at both landing sites. We have proposed [*Binder et al.,* 1977] that it be called duricrust. Its presence was first suggested by a picture showing a polygonally fractured cohesive stratum where a thin veneer of dust had been swept away by the retro-rocket exhaust on Viking Lander 1. Subsequently, a larger area of fractured crust was mapped at the Viking Lander 2 site [*Mutch et al.,* 1976]. The duricrust fractures are characteristically much smaller and more closely spaced than those previously discussed as possible permafrost troughs.

The presence of duricrust was further indicated during trench operations. Where the shovel penetrated the soil and pushed forward in order to obtain a soil sample, the upper

Fig. 4. (Opposite) The near field at the Viking Lander 2 site. A trough filled with fine-grained sediment trends from upper left to lower right. Just beyond the trough, in the right half of the picture, there is a large boulder approximately 1 m wide. Layering in the block runs from upper left to lower right. The left end of the boulder is much more pitted than the central part. Throughout the scene, most rocks are deeply pitted on the scale of ~1 cm. The camera event is 21A024. For the exact location of the picture, refer to the paper by *Mutch* [1977, Figure 1].

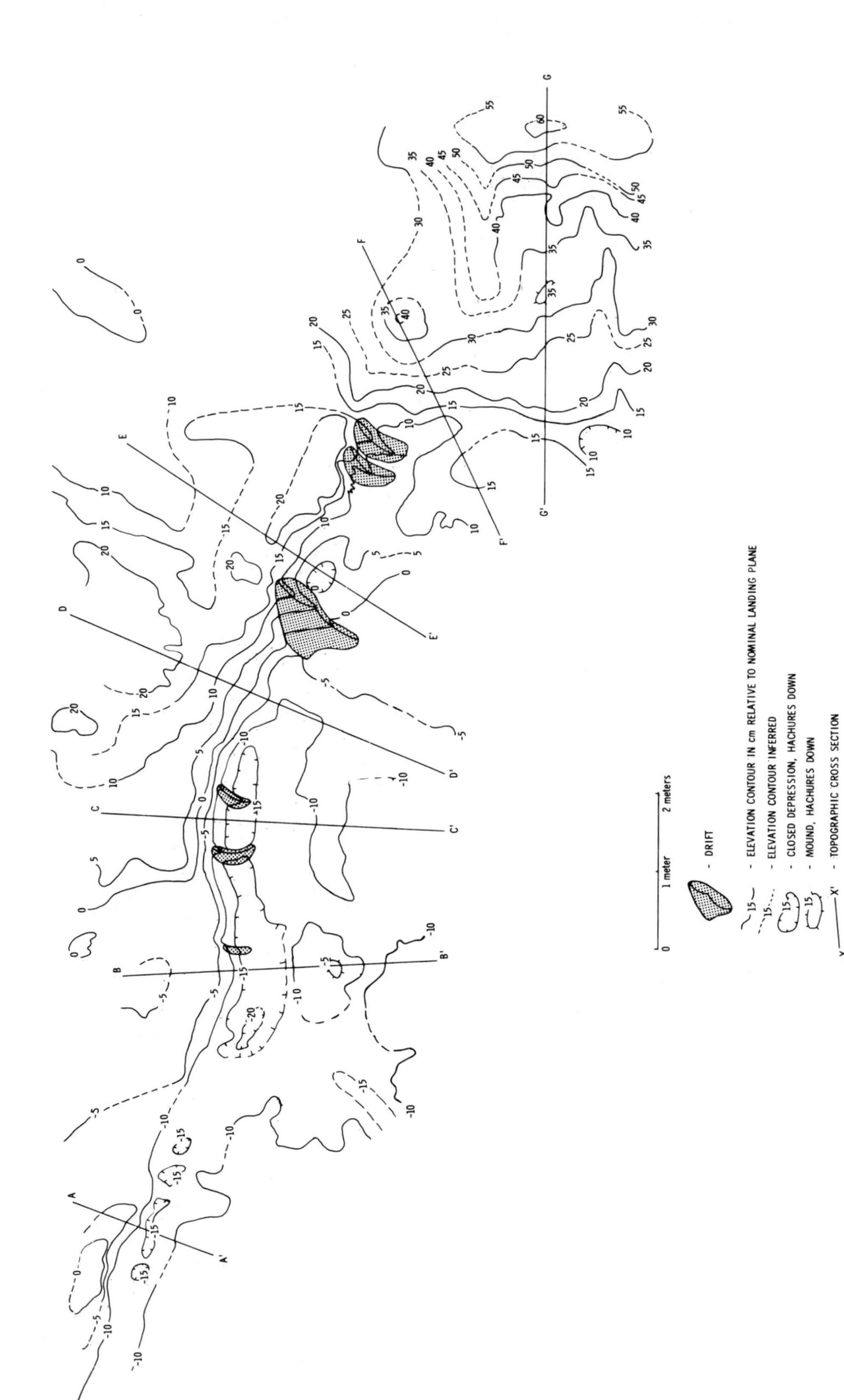

Fig. 6*a*. Topographic map of the trough shown in Figure 4 (derived from stereoscopic data [*Liebes and Schwartz*, 1977]).

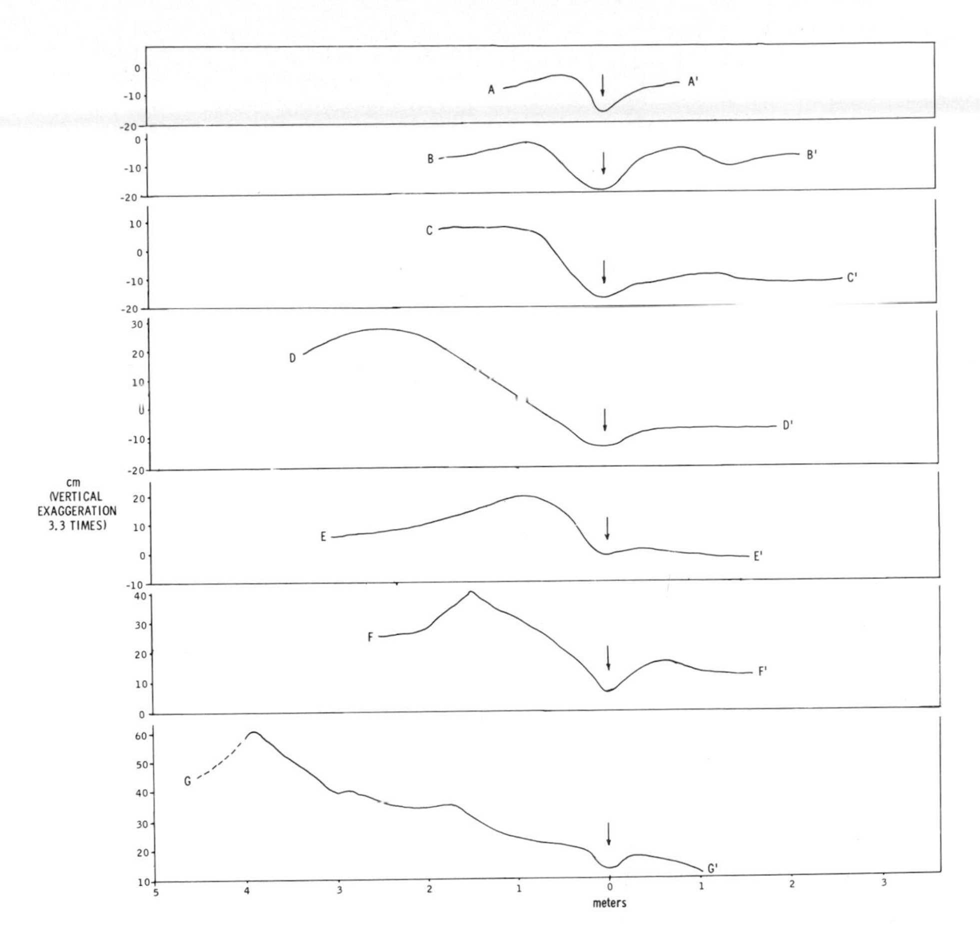

Fig. 6*b*. Cross sections of the trough shown in Figure 4 (derived from stereoscopic data [*Liebes and Schwartz*, 1977]).

zone was commonly bowed upward and fractured, indicating a strength greater than that of deeper, unconsolidated fine material (Figure 11).

Areas of polygonally fractured crust are observed adjacent to pebbly regions. Many of the small fragments have a slabby, angular appearance. The hypothesis that many of the pebbles are not 'rocks' but instead cemented fines is supported by the sampling results of the X ray fluorescence investigation. The sampling procedure was to obtain a scoop of mixed fines and pebbles, remove the fines by sieving, and then deliver the coarse fraction to the X ray instrument. Several times, no material remained for delivery after sieving. The apparent reason is that the pebbles comprise lightly cemented particles which are disaggregated by vibration [*Clark et al.*, 1976]. Those pebble fractions that have been analyzed contain 25–50% more sulfur than the fines. The elements Ca, Al, and Fe do not vary sympathetically with this increase, suggesting Mg or Na sulfate as the cement [*Clark et al.*, 1977]. Appar-

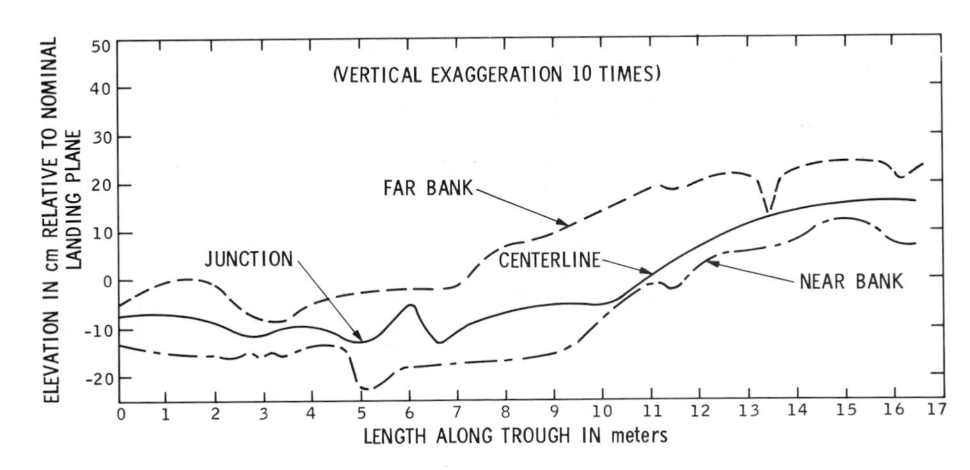

Fig. 6*c*. Longitudinal sections of the trough shown in Figure 4 (derived from stereoscopic data [*Liebes and Schwartz*, 1977]).

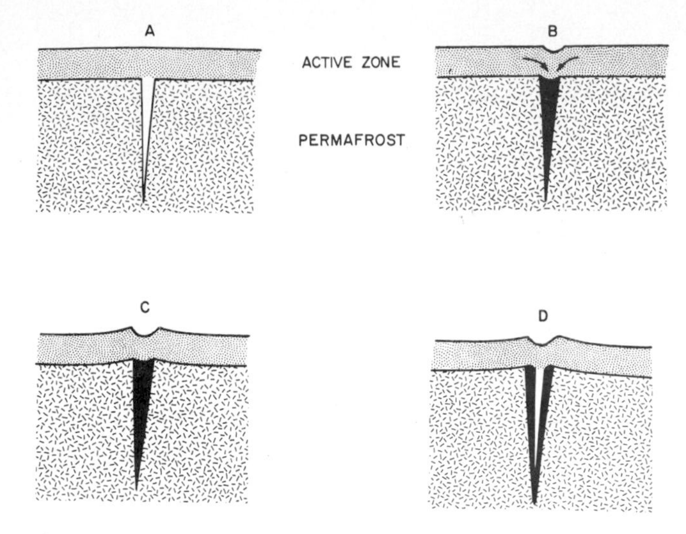

Fig. 7. Schematic evolution of an ice wedge [after *Lachenbruch*, 1962]. (*a*) Permafrost contracts during winter to form open fissure. (*b*) Spring meltwater from active zone fills crack and freezes. (*c*) Expansion of permafrost during summer causes upward bulge in vicinity of ice wedge. (*d*) Permafrost contracts during winter. Wedge opens, initiating another annual cycle of wedge growth.

ently, the duricrust forms by precipitation of a cementing agent due to evaporation of volatiles in the upper few centimeters of the regolith.

Viking Lander 1 and 2 Sites Compared

Since both landing sites were chosen, for reasons of safety, for their topographic blandness, it cannot be directly concluded that the small-scale features seen by the lander cameras are typical of the entire surface. Indeed, the view from the Viking orbiters suggests that there are many regions with more exotic small-scale landforms.

At first glance, both lander sites appear to be similar block-strewn landscapes. However, there are a number of differences in detail. The Viking Lander 2 site is generally flat. Bedrock is exposed at the first site but not at the second. Large drifts are observed at the first site but not at the second. Polygonal troughs are at the second site but not at the first. Viking Lander 1 rocks display a great deal of diversity in albedo, shape, and texture, whereas Viking Lander 2 rocks are almost all similar in albedo and vesicular texture. Individual boulders show more prominent evidence of wind faceting at the Viking Lander 1 site. However, the general impression of eolian stripping, or deflation, is much more striking at the Viking Lander 2 site. Blocks there rest on the ground, the situation that might have been expected had the surrounding fine-grained material been preferentially removed.

Fig. 8. Intersecting cracks in patterned ground formed in recent alluvium and outwash gravels, lower Victoria Valley, Antarctica. The troughs have been filled with windblown sand. The geologic hammer provides scale [from *Morris et al.*, 1972].

Fig. 9. The first picture taken at the Viking Lander 2 site, showing deeply pitted rocks. Some of the pits have polygonal outlines. The rock just at the top of the footpad is about 35 cm across. It appears more elongated than it actually is, because at large depression angles the azimuthal dimension of the scene is exaggerated. For further details, refer to the paper by *Huck et al.* [1977]. Clods of sediment, kicked up during landing maneuvers, are visible on the inner concave surface of the footpad. The camera event is 22A001. For the exact location of the picture, refer to the paper by *Mutch* [1977, Figure 1].

Fig. 10. Irregularly cemented material that occurs in a windblown, medium-grained, quartz sand deposit. The cemented zone forms dike-like masses with strike lengths of several kilometers. The surface outcrop is hard and cellular. Vertical sections reveal that the cemented zone decreases in hardness and vesicularity downward and, within 1 m, grades into unconsolidated sand. (Sample was supplied and described by A. O. Fuller.)

GEOLOGIC HISTORY

There are several competitive geologic histories that can be hypothesized for the Viking Lander 2 site, no one of which is uniquely persuasive. The various models are schematically shown in Figure 12.

A simple history specifies that the spacecraft is situated on a distal part of the Mie ejecta blanket. Pictures from orbit suggest that this is reasonable. Morphologically distinct tongues of ejecta appear to extend across the landing site (Figures 1 and 2). A bluff outlined on the horizon to the east of the lander may be the terminus of an ejecta lobe. The abundant blocks at the site may be the residue of a poorly sorted deposit emplaced by surface debris flow [*Carr et al.*, 1977] related to the formation of Mie. The finer materials in the debris flow may have been winnowed by wind to create a partly armored surface. Alternatively, the finer-grained sediment may have been selectively excluded by aerodynamic drag during ejecta emplacement [*Schultz*, 1977].

If all the boulders in the vicinity of the lander are impact emplaced, it is puzzling that they are so large, so evenly distributed, so petrologically similar, and so free of brecciation features. The terrain is remarkably flat, possessing none of the hummocky relief that is characteristically associated with ejecta deposits. This may be related to the winnowing of fine-grained material, a process that might also explain the subdued and 'buried' character of the ejecta material as viewed from orbit (Figure 1).

A second history specifies an episode of eolian deposition, followed by emplacement and erosion of thin but widespread basaltic lavas. If one were to remove the surface layer of boulders, the landscape would be a generally flat plain of fine-grained material. This distinctive geologic unit most probably is the same polar mantle previously postulated by *Soderblom et al.* [1973], the same deposit which was seen to subdue subjacent topography in Viking orbiter pictures acquired during site certification. Subsequent to deposition of the fine-grained material, low-viscosity lavas may have flowed across the surface. Evidence for volcanic activity is also contained in orbiter pictures. Several dome-shaped protuberances occur in the vicinity of the landing site (Figure 1). Originally a continuous stratum a few meters thick, the lavas may have been broken to yield the present boulder field. This fragmentation may have been favored by the initial scoriaceous character of the lava and probably proceeded along regularly spaced thermal contraction fractures. Erosion of underlying fine-grained material is indicated by the fact that some boulders are perched on pedestals of subjacent sediment. Where a protective boulder is missing, the fine-grained material has been excavated to a greater depth. The same sort of erosion might have enhanced basalt fragmentation. As supporting layers of sediment were removed, the overlying basalt might have been 'let down' and, in the process, broken up.

A generally analogous situation can be viewed here on earth. In Death Valley in southern California a rubble of basalt boulders covers fine-grained alluvium (Figure 13). Apparently, the boulders were transported downslope from a nearby extrusive flow. This last observation points up one of the major difficulties in the lava model. If a flow formerly extended across the Martian landscape, it is difficult to see how the bulk of the material could be preferentially and completely removed. Because the surface is so flat, one cannot invoke downslope dispersal, as is invoked in the Death Valley situation. Creation of in situ relief and slumping by erosion of underlying sediment, as described in the preceding paragraph, is an ad hoc hypothesis not observationally demonstrated.

A third history postulates extensive erosion of a poorly sorted sedimentary deposit. A bimodal size distribution can be attributed to some combination of interleaved eolian deposits, impact ejecta, and volcanic layers. Subsequent extensive deflation has created a lag deposit of large boulders on finer-grained sediment. Had this actually occurred, one might expect the boulders to be more petrologically diverse and to be more abundant at the surface, essentially acting as a continuous pavement.

A variant on the history just described specifies that a substantial polar ice cap once covered the region, encroaching on an eolian plain. Successive ejecta and/or volcanic deposits were emplaced on top of the ice. Following dissipation of the ice cap, the coarse-grained deposits remained as a residuum on the eolian plain. In this model, as in the previous model, one would expect a more substantial deposit of mantling coarse-grained material. In addition, there is no independent evidence for the former presence of a thick ice cap.

A fourth history involves deposition of eolian mantle, formation of a near-surface zone of cellularly cemented ferricrete, and subsequent deflation which leaves behind fragments of this 'rock' as isolated boulders.

Finally, the boulders might have been deposited by some massive sheetflood. One terrestrial analog is often cited. The scablands of eastern Washington were deeply scoured during Pleistocene time when a large lake to the west broke through its natural dam and flooded the adjacent terrain. Large boulders were carried along in this torrential flood [*Bretz*, 1969]. However, terrestrial sheetfloods typically have much less transportational capability [*Davis*, 1938]. Additionally, they almost always originate in neighboring highlands. The landscape at the Viking Lander 2 site is notable for the lack of relief.

Fig. 11. Fractured duricrust, near the center of the picture, that was subsequently penetrated during a trenching operation. The camera event is 22A007. For the exact location of the picture, refer to the paper by *Mutch* [1977, Figure 1].

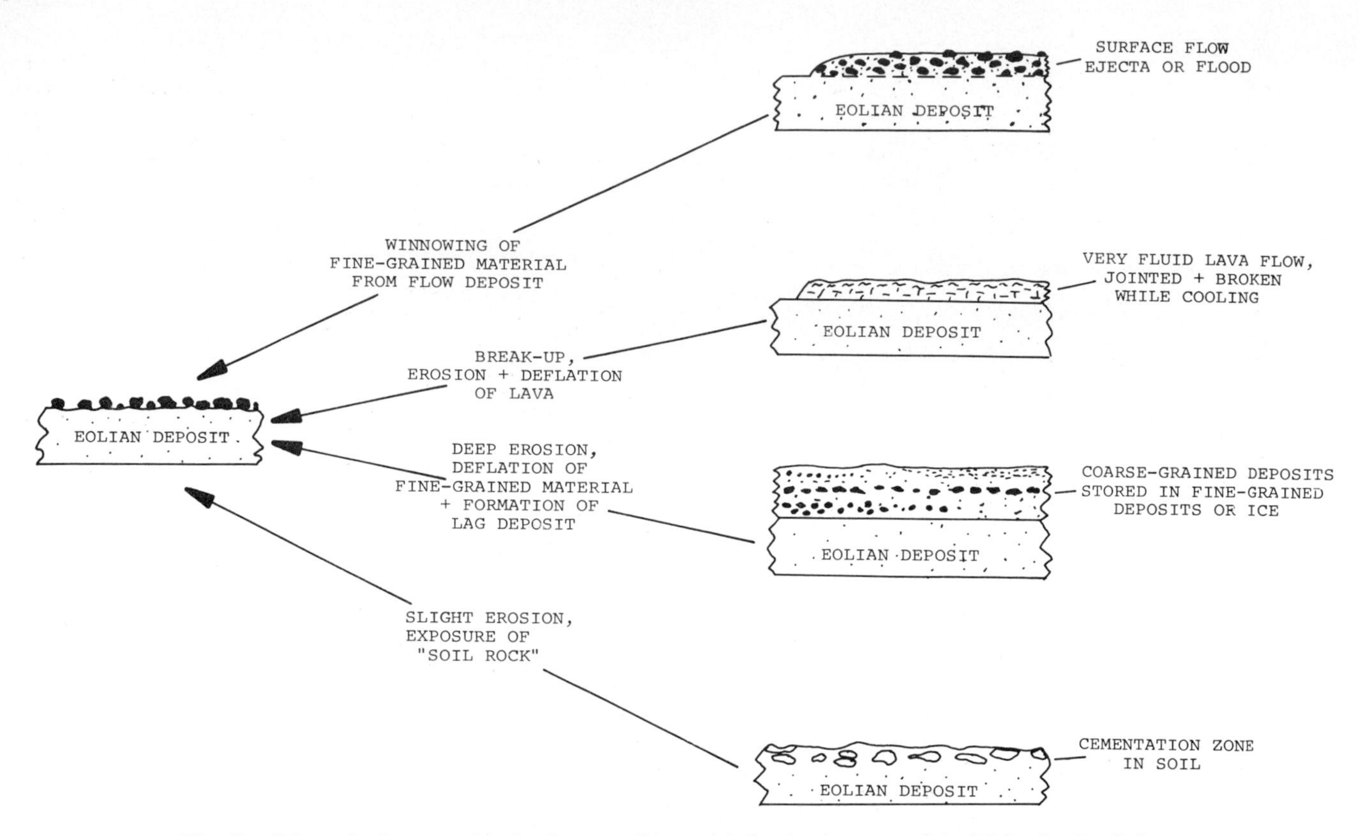

Fig. 12. Schematic diagrams showing four possible models for development of the Viking Lander 2 site.

After considering the several possible geologic histories we conclude that it is most likely that the lander is situated either on Mie ejecta or on eroded lava flows.

Acknowledgments. We are indebted to the many people within the Viking project who contributed to the acquisition and interpretation of the lander pictures. Our special thanks go to P. Avrin, S. H. Bosworth, C. E. Carlston, E. W. Dunham, P. L. Fox, S. U. Grenander, J. W. Head, F. O. Huck, K. L. Jones, R. A. Kahn, E. C. Levinthal, S. Liebes, Jr., B. K. Luchitta, D. Nummedal, W. R. Patterson, D. C. Pieri, J. B. Pollack, C. W. Rowland, C. Sagan, G. R. Taylor, R. B. Tucker, S. D. Wall, and M. R. Wolf, all of whom were closely involved with the lander imaging investigation. R. P. Sharp reviewed the manuscript and made valuable suggestions. This paper presents results of tasks carried out under the following NASA contracts: contract NAS1-9680 to the Department of Geological Sciences, Brown University; order L-9718 to the Branch of Astrogeologic Studies, U.S. Geological Survey; contract NAS1-11851 to Science Applications, Inc; and contract NAS1-13889 to the Department of Earth and Planetary Science, Washington University.

REFERENCES

Baird, A. K., P. Toulmin, B. C. Clark, H. Rose, Jr., K. Keil, R. P. Christian, and J. L. Gooding, Mineralogic and petrologic implications of Viking geochemical results from Mars: Interim report, *Science, 194,* 1288–1293, 1976.

Fig. 13. Rubble of basalt boulders on top of fine-grained alluvium along the east side of Death Valley. Apparently, the boulders were transported downslope from a nearby extrusive flow.

Binder, A. B., R. E. Arvidson, E. A. Guinness, K. L. Jones, E. C. Morris, T. A. Mutch, D. C. Pieri, and C. Sagan, The geology of the Viking Lander 1 site, *J. Geophys. Res., 82,* this issue, 1977.

Bretz, J. H., The Lake Missoula floods and the channeled scabland, *J. Geol., 77,* 505–543, 1969.

Carr, M. H., L. Crumpler, J. A. Cutts, R. Greeley, J. E. Guest, and H. Masursky, Martian craters and emplacement of ejecta by surface flow, *J. Geophys. Res., 82,* this issue, 1977.

Clark, B. C., A. K. Baird, H. J. Rose, P. Toulmin, K. Keil, A. J. Castro, W. C. Kelliher, C. D. Rowe, and P. H. Evans, Inorganic analyses of martian surface samples at the Viking landing sites, *Science, 194,* 1283–1288, 1976.

Clark, B. C., A. K. Baird, K. Keil, H. Rose, Jr., and P. Toulmin, Element concentrations in martian surface minerals, paper presented at Special Mars Session of 8th Lunar Science Conference, Johnson Space Center, Houston, Tex., March 15, 1977.

Davies, M. E., and D. W. G. Arthur, Martian surface coordinates, *J. Geophys. Res., 78,* 4355–4394, 1973.

Davis, W. M., Sheetfloods and stream floods, *Geol. Soc. Amer. Bull., 49,* 1337–1416, 1938.

Grim, R., *Applied Clay Mineralogy,* McGraw-Hill, New York, 1962.

Huck, F. O., D. J. Jobson, S. K. Park, S. D. Wall, R. E. Arvidson, W. R. Patterson, and W. B. Benton, Spectrophotometric and color estimates of the Viking lander sites, *J. Geophys. Res., 82,* this issue, 1977.

Lachenbruch, A. H., Mechanics of thermal contraction cracks and ice-wedge polygons in permafrost, *Geol. Soc. Amer. Spec. Pap., 70,* 66 pp., 1962.

Liebes, S., Jr., and A. A. Schwartz, Viking 1975 Mars lander interactive computerized video stereophotogrammetry, *J. Geophys. Res., 82,* this issue, 1977.

Malin, M. C., Salt weathering on Mars, *J. Geophys. Res., 79,* 3888–3894, 1974.

Masursky, H., and N. L. Crabill, Search for the Viking 2 landing site, *Science, 194,* 62–67, 1976.

Mayo, A. P., W. T. Blackshear, R. H. Tolson, W. H. Michael, Jr., G. M. Kelly, J. P. Brenkle, and T. Komarek, Lander locations, Mars physical ephemeris, and solar system parameters: Determination from Viking lander tracking data, *J. Geophys. Res., 82,* this issue, 1977.

Michael, W. H., Jr., A. P. May, W. T. Blackshear, R. H. Tolson, G. M. Kelly, J. P. Brenkle, D. L. Cain, G. Fjeldbo, J. N. Sweetnam, R. B. Goldstein, P. E. MacNeil, R. D. Reasenberg, I. I. Shapiro, T. I. S. Boak III, M. D. Grossi, and C. H. Tang, Mars dynamics, atmospheric and surface properties: Determination from Viking tracking data, *Science, 164,* 1337–1339, 1976.

Morris, E. C., T. A. Mutch, and H. E. Holt, Atlas of geologic features in the dry valleys of south Victoria Land, Antarctica: *Interagency Rep. Astrogeology 52,* 156 pp., U.S. Geol. Surv., Flagstaff, Ariz., 1972.

Mutch, T. A., The Viking lander imaging investigation: An introduction, *J. Geophys. Res., 82,* this issue, 1977.

Mutch, T. A., S. U. Grenander, K. L. Jones, W. Patterson, R. E. Arvidson, E. A. Guinness, P. Avrin, C. E. Carlston, A. B. Binder, C. Sagan, E. W. Dunham, P. L. Fox, D. C. Pieri, F. O. Huck, C. W. Rowland, G. R. Taylor, S. D. Wall, R. Kahn, E. C. Levinthal, S. Liebes, Jr., R. B. Tucker, E. C. Morris, J. B. Pollack, R. S. Saunders, and M. R. Wolf, The surface of Mars: The view from the Viking 2 lander, *Science, 194,* 1277–1283, 1976.

Neal, J., A Langer, and P. Kerr, Giant desiccation polygons of Great Basin playas, *Geol. Soc. Amer. Bull., 79,* 69–90, 1968.

Pullen, R. A., A morphological classification of lateritic ironstones and ferruginized rocks in northern Nigeria, *Nigerian J. Sci., 1,* 161–173, 1967.

Schultz, P. H., Impact ejecta in a martian atmosphere, paper presented at Special Mars Session of 8th Lunar Science Conference, Johnson Space Center, Houston, Tex., March 15, 1977.

Soderblom, L. A., R. J. Kreidler, and H. Masursky, The latitudinal distribution of a debris mantle on the Martian surface, *J. Geophys. Res., 78,* 4117–4122, 1973.

Wade, F. A., and J. N. DeWys, Permafrost features on the martian surface, *Icarus, 9,* 175–185, 1968.

Washburn, A. L., *Periglacial Processes and Environments,* 320 pp., St. Martin's, New York, 1973.

(Received April 13, 1977;
revised June 14, 1977;
accepted June 14, 1977.)

Lander Imaging as a Detector of Life on Mars

ELLIOTT C. LEVINTHAL

Department of Genetics, Stanford University, Stanford, California 94305

KENNETH L. JONES

Department of Geological Sciences, Brown University, Providence, Rhode Island 02912

PAUL FOX AND CARL SAGAN

Laboratory for Planetary Studies, Cornell University, Ithaca, New York 14853

Biological goals were among the important science objectives of the Viking lander camera. The camera performance characteristics relevant to these goals are discussed. They include the ability to observe (1) morphological detail, (2) color and reflectance spectra, and (3) motion and change. The scenes obtained by the cameras were scrutinized in many ways: monoscopically, stereoscopically, in color, and by computerized differencing of camera events. At the lander sites and during the times that observations were carried out on the surface of Mars, no evidence, direct or indirect, has been obtained for macroscopic biology on Mars. No obvious examples of geometric distortion that might have been motion induced have been observed. Using the repeated line scanning mode of the camera has revealed no changes or motion suggesting life. These negative results may be due to limitations in sampling, in camera design, or in our understanding of Martian biology, but they are certainly consistent with the hypothesis that macroscopic life is absent on Mars.

INTRODUCTION

The purpose of this paper is to describe the role of the Viking lander cameras in the search for life on Mars. This function of the camera is dependent on its ability to observe morphology, texture, color, or motion that would be suggestive of biological origin due either to the biological organisms themselves or to their spoor, artifacts, or fossil remains.

Organisms large enough for the cameras (or the unaided human eye) to see are called 'macrobes' and have been considered possible inhabitants of Mars [*Sagan and Lederberg*, 1976, and references therein]. From the beginning of the Viking mission design the search for macrobes has been among the principal objectives of the lander camera system [*Mutch et al.*, 1972]. Such macrobes might be detectable by their regular geometry, dynamic instability, or motion, even if they were very different from terrestrial life forms.

The ability of the camera to observe changes in surface features that might be interpreted as a sign of life depends not only on the camera system design but also on the mission profile and the opportunities for camera events that it generates. In this paper we wish to identify the limitations of the use of the cameras for this purpose, to discuss some of the methodology by which the search was carried out, and to describe the results to date.

RESOLUTION AND COVERAGE

Spatial resolution is the most important parameter for our search. The details of the camera design have been described elsewhere [*Huck et al.*, 1972]. The relevant issues of image quality are considered in a paper by *Huck and Wall* [1976]. Figure 7 of *Patterson et al.* [1977] shows the smallest linear dimension resolvable by the camera as a function of distance and focus step. High-resolution pixel sizes range from 1.0 mm at a distance of 1.5 m from the camera to 2 m at the horizon, 3 km away. It should be kept in mind that typically, 2–3 pixels across an object or feature are required to resolve its presence. Exactly how many are required depends on its shape and the signal-to-noise ratio, straight lines being the easiest to observe. The orbiter provided a much coarser resolution. At a periapsis altitude of 1500 km an orbiter image pixel was 38 m.

Figure 1 graphically illustrates the surface area inspected by the cameras, a smooth spherical surface being assumed and obscuration by the spacecraft, its hardware, and shadows being ignored. The total coverage for each lander, out to the horizon, is an area of 2.8×10^7 m^2, representing (2×10^{-7})% of the planet.

Safety considerations and therefore geological blandness were paramount considerations in selecting the Viking Lander 1 landing site and were important in selecting the local environment of Viking Lander 2. While we can conceive of biological models in which such sites are anticorrelated with macrobes which are abundant elsewhere, there is no well-developed hypothesis supporting this point of view. Were terrestrial analogy a useful guide, the coverage provided by the lander cameras would be considered a representative sample of the surface for the observations of extant macroscopic forms of life. On earth there are surprisingly few places (some deserts and ice-covered regions) at the interface of the atmosphere and the solid surface where a sample of this size would have failed to show some signs of macroscopic life. The ocean floors might provide very different circumstances. There is a specific model of Martian biology which proposes abundant life only in a limited number of microenvironments [*Lederberg and Sagan*, 1962], quite different from the Viking Lander 1 and 2 landing sites. If, however, one were to consider the probabilities of finding, on earth, fossil remains or ancient ruins of past civilizations, then the situation would be reversed, and less than (2×10^{-7})% of the surface would be much too small a sample to make any meaningful statement.

THE SEARCH FOR SIGNS OF LIFE

The search for signs of life is an exercise in pattern recognition. Its success depends on the skill and experience of the

Paper number 7S0501.

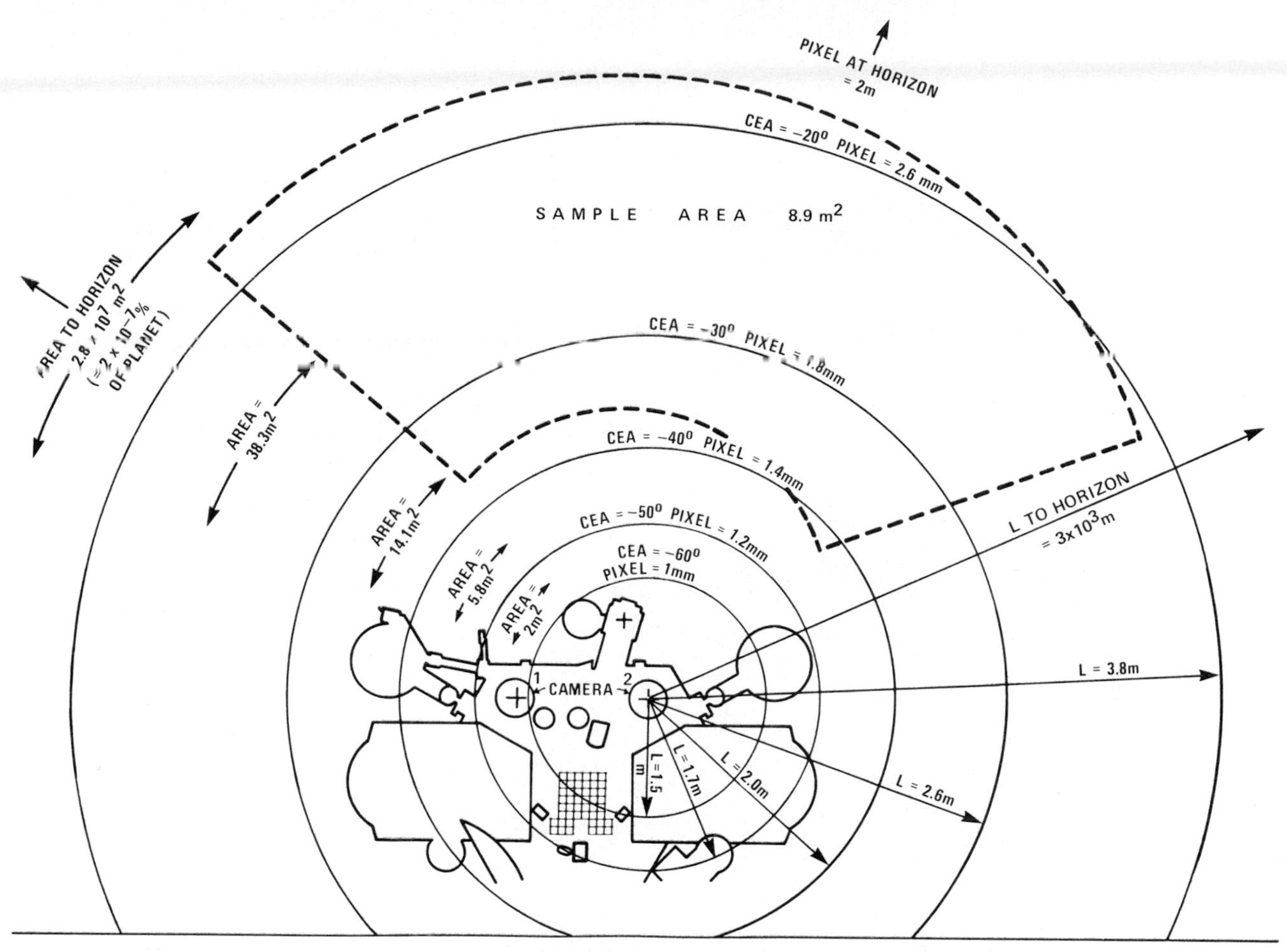

Fig. 1. Partial overhead view of the surface coverage provided by lander camera 2. *CEA* is the camera elevation angle; area, the area of annulus formed by a 10° change in *CEA*; pixel, the size of the picture element for the high-resolution (0.04°) diode at the indicated *CEA*; and *L*, the distance from the camera photogrammetric reference point to the surface at the indicated *CEA*. The sample area is that area accessible to the surface sampler arm.

observers and the degree of effort expended. It cannot be demonstrated to be complete.

Tests were run prior to landing to simulate the first 8 days of the mission. An artificial scene was created at a test facility at the Jet Propulsion Laboratory (JPL). The details of the scene were not revealed to the scientists who were to evaluate the received data. One of the camera team members with skill and experience in field observations immediately recognized in a camera image the fossil remains of a trilobite which was located about 2 m from the camera. Figure 2 shows the received image and illustrates the capability of making this kind of observation. Even if we had had no prior experience with trilobites, the bilateral symmetry of the fossil would have strongly suggested a biological origin.

The accessible scene at both landers has been completely imaged with each camera of both landers during the primary mission in high resolution at early morning and late afternoon sun angles. In addition, the data have also been acquired at noon for Viking Lander 2. The scene has been scrutinized carefully by experimenters from a variety of disciplines, on different science teams, and for a range of different purposes. There has been a systematic scrutiny as a result of the process of preparing mosaics of these camera events. Furthermore, the process of preparing the final team data record involved inspecting each camera event many times. This included not only the events that are the basis of the high-resolution mosaics but also many other events providing either coverage at other times of day or repeated coverage at approximately the same time of day.

Coverage obtained from both cameras provides stereo images of about one-half the forward and rear directions out to about 150 m. The arrangement of the cameras causes the stereo base to shrink to zero as the viewing direction moves to the vertical plane containing the cameras. In generating elevation contours and vertical profiles over the area with good stereo coverage, three-dimensional views of the scene were examined with great care.

The areas in front of both spacecraft that are accessible to the surface sampler cover 120° in azimuth and from about 2 to 4 m in range and are equal to 8.9 m². This area was examined with the greatest care by the largest group of observers. The Biology, Organic Chemistry, and Inorganic Chemistry teams all had to select sites for their soil samples. The Physical Properties team, in addition to their role in determining the character of the samples that might be obtained, searched this area and the remaining near fields for clues to soil properties and to the interaction of the spacecraft with the surface. Other engineering groups viewed the imaging data to determine the method and safety of sample acquisition and rock movements.

In addition to the above searches, some of which did not have biology as their primary objective, the authors and their colleagues have made a number of independent examinations of these data for patterns suggestive of life.

No morphological or other signs suggestive of biology were

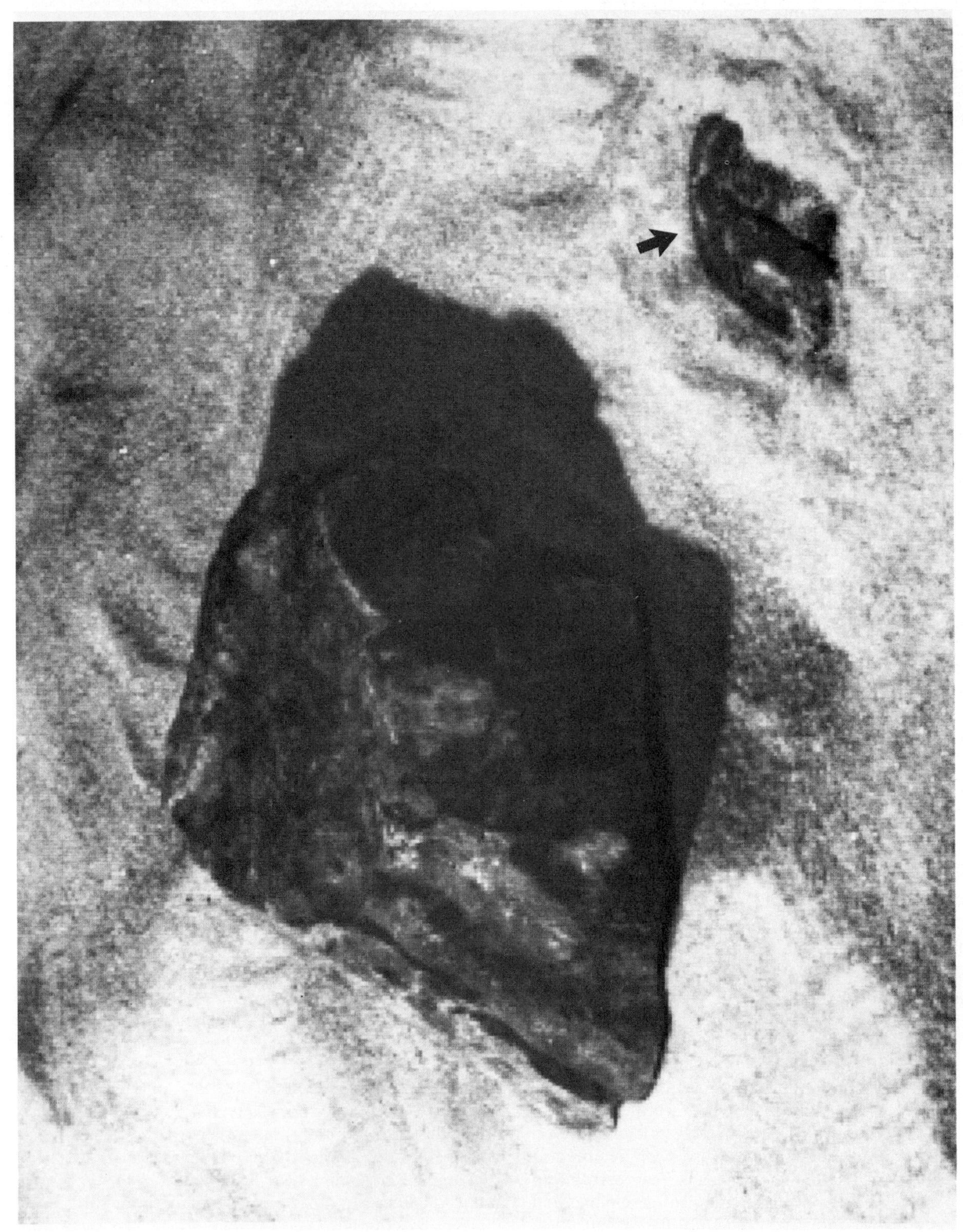

Fig. 2. Image of a trilobite (arrow) located in the sand in the lower right-hand corner. This is a portion of an image taken during demonstration test 4 (DT-4) of the Test and Training Program (C2R003/D01).

Fig. 3. The rock exhibiting the letter **B** (arrow) is the large rock near the middle of the left-hand margin of the picture, above the housing containing the surface sampler. At a camera elevation angle of −10° it is 7.5 m from the camera reference point; CE (camera event) label 11A030/004.

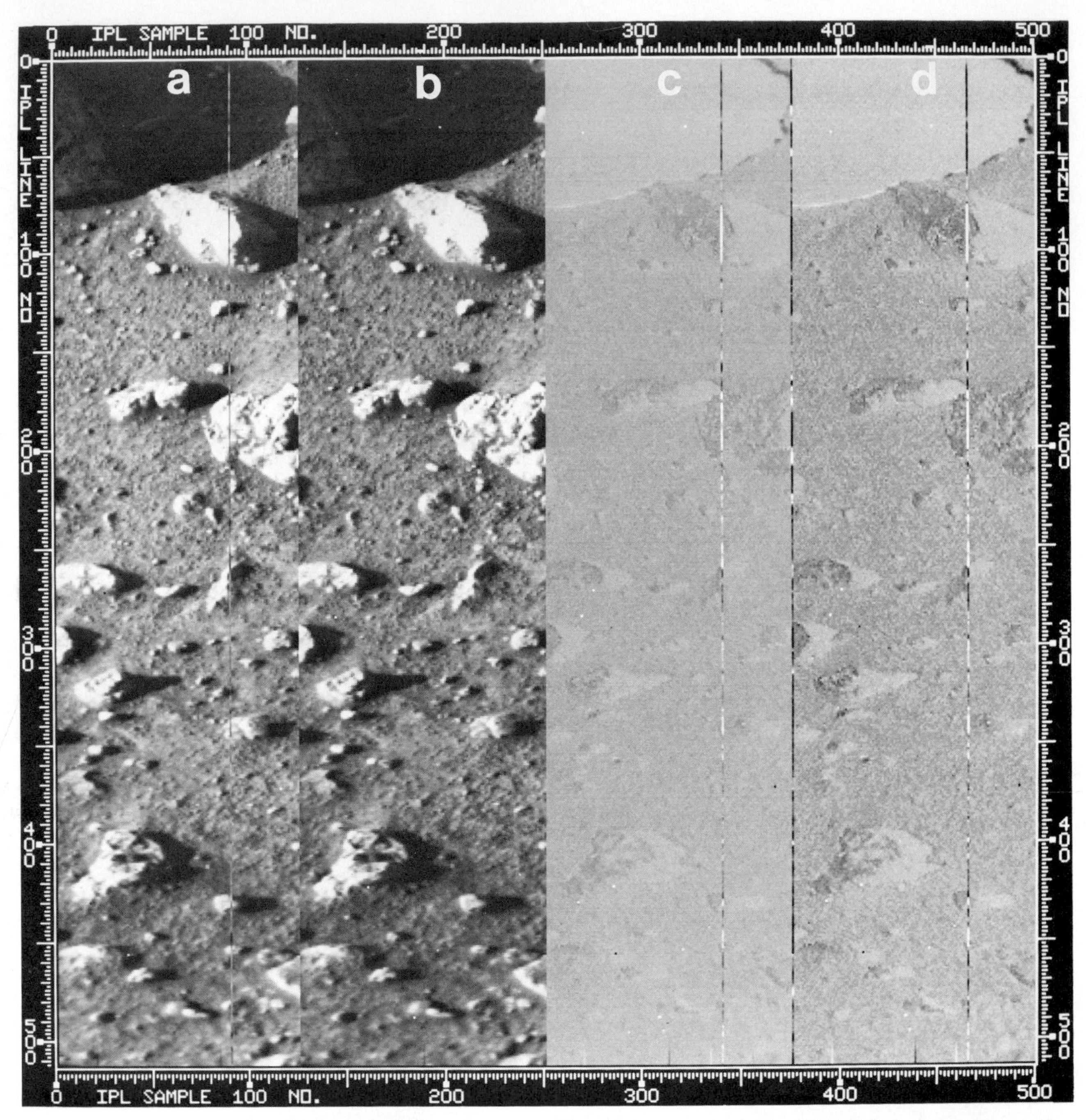

Fig. 4. Viking Lander 1 scene centered at $CEA = -50°$, from azimuth 112.5° to 117.5°, imaged on sols 16 and 21. (*a*) CE label 12A124/021. (*b*) CE label 12A103/016. (*c*) Part *b* minus part *a*. (*d*) Contrast-stretched difference. No change is observed. In part *c*, differences of a few digital numbers in the general background and larger changes on the facets of rocks are shown, indicating the limitation of the method.

detected. Clearly, such a statement refers to forms which reflect our present notions of life. With one exception, nothing was observed which suggests an artifact of intelligent life. The exception is the rock exhibiting the letter 'B' shown in Figure 3. 'Lettered' rocks are sometimes found in the desert sculpted by eolian and other erosion in a manner visually indistinguishable from that sculptured by a chisel. It is clear that a single letter by itself is not sufficiently improbable to force the hypothesis of intelligent origin. This is in addition to the unlikely universality of the Roman alphabet.

The same scenes have been examined by using visual color and infrared diodes. *Huck et al.* [1977] describe the problems of reducing these data and generating the 'true' color of the surface as it would be seen by an observer on Mars or as viewed through the earth's atmosphere. A noise spike occurring in the data in any of the diodes that may be used to generate the color of the scene will cause an anomaly in the color of a pixel. Clues to biology due to unusual color patches would have to include pixels from neighboring scan lines to avoid the correlated noise effects within a line. No such patches have been found. Small patches were selected in a variety of locations: on rocks, on the soil surface, and in the trenches dug by the sampler arm. From the multidiode data averaged over the areas, spectral reflectance curves were generated. These curves showed no unusual features that would distinguish the small areas from the remainder of the surface. No features have been observed which are unique attributes of photosynthetic pigments.

THE SEARCH FOR MOTION AND CHANGE

We have so far considered the search for objects of biological interest that are static features of the scene. Motion is an important signature of animal life. We now wish to discuss the ability of the cameras to discover objects that have appeared or disappeared from their field of view or are moving through it.

In the fast scan mode at 16,000 bps (bits per second) a scan line lasts 0.22 s, and in the slow scan mode at 250 bps it lasts 14.13 s. At a fast scan rate, which was used for most camera events in the primary mission, a 10° azimuth scan at high resolution thus consists of 250 lines over a time interval of 55 s.

Objects several scan lines or more in dimension which appeared or disappeared from the scene while being scanned, in times less than a scan line interval, would have a distinctive appearance. There would be straight edges or steps in precisely the direction of vertical scan on the left side if they were newly arrived and on the right side if they were just departing. No such objects were observed. Objects moving more slowly while being scanned by the camera would be compressed, stretched, or otherwise modified in shape depending on the direction of motion with respect to the azimuth scan direction. Such shapes would be difficult to distinguish within a matrix of various rock forms. In fact, such motion could obscure a symmetry that might be suggestive of biology.

To search for such objects, camera sequences were planned to look for changes in the scene over various intervals of time. For intervals of days or longer the ideal is to image a given area on many days of the mission at identical sun angles so that the observation of the motion of objects would not be made difficult because of photometric effects or the motion of shadows. To do this perfectly is obviously impossible without artificial lighting. To do this extensively in time and space was difficult because of the constraints of data transmission and storage, the timing of other activities in the sequence of events that evolved in mission planning, and the competition with other science goals.

Several camera event sequences were carefully planned to optimize this search for change. Several areas in the near field were imaged at different intervals at almost identical times of the day, yielding very similar sun angles.

All camera events that for any reason have areas in common with previous camera events have been compared. When the events are images taken by the same camera, pseudo stereo, effected by the fusing of the common areas, allows a good observer to readily detect changes and hence the movement, appearance, or disappearance of objects, except for the lighting difficulties mentioned above. This method succeeded in observing slight movement of the rock under the footpad of Viking Lander 2 and the displacement of surface material as a result of shroud ejection.

A complete listing of all events in the primary mission was scanned for candidate pairs for computer image differencing. Without any reference to scene content, only optimum lighting conditions being considered, for Viking Lander 1, three pairs at −50° elevation pointing angle and two pairs at −30° were chosen; for Viking Lander 2, one pair at −50° was chosen.

By using the interactive computer hardware and software facilities at JPL's Image Processing Laboratory the overlapping regions of these events were registered, differenced, and then contrast-stretched to enhance the changes. If no changes have occurred, the difference picture will be essentially featureless.

The next several figures illustrate the results and demonstrate the ability to observe small changes by using the 0.04° high-resolution diodes. Consider first Figure 4. The unstretched difference shows the characteristic easily identified effects of a shadow and a noisy line of data. The remainder of the image shows a difference of a few digital numbers, which is within the noise of the process. Figure 5 shows a pair of events covering a large area, 35° in azimuth, from which a sample had been acquired, taken at a 20-day interval. Again, no change is observed. Figure 6 shows the same area taken 20 days later. Here the difference does show a change, a characteristic adjacent dark and light patch, due to a slight slumping of material within the trench. Figure 7 demonstrates the ability to find objects that have appeared on the scene. The bright region in the upper left, causing a darkening of the difference image, is due to the spacecraft arm. At the lower edge, in the left-hand part of the image, we see the effect of the boom latch pin of the surface sampler, which fell to the surface on sol 5. The pin is 8.2 cm long and 0.6 cm in diameter, and it disturbed the surface about 2 cm from the axis of the pin.

The camera system is designed to provide an additional method of searching for variable features. The camera can operate without stepping in azimuth, a procedure resulting in rescanning the scene in elevation at a particular azimuth angle. There are three different conditions which can result in this rescan. A command bit can be set for a camera event, resulting in rescanning by the camera at the stop azimuth for a time stored in the data base of the lander's on-board computer. This time could be changed by uplink transmission. During the primary mission, six such camera events were executed on VL 1, and eight on VL 2. These consisted of all rescan lines, since the start azimuth equaled the stop azimuth. All were visual color triplets taken at the low scan rate and were used to carry out twilight studies to obtain information on the vertical distribution of scatterers in the Martian atmosphere [*Pollack et al.*, 1977].

If the event duration listed in the real-time imaging tables is longer than the time required to scan the commanded azimuth range, the extra time is consumed in rescanning at the final azimuth position. During the primary mission this type of rescanning was associated with 34 camera events on VL 1 and 40 camera events on VL 2. In 15 of these cases, only a few lines (<15) are acquired. These occurred at both high and low scan rates (real-time relay and real-time direct link) and sometimes consisted only of rescan lines. Most events were in the range of 25–100 rescan lines, eight having more than 100. In addition, rescan occurs owing to the assignment of extra time in recorded imagery to allow for tape recorder run-up and reversal. This process generally resulted in less than 12 rescanned lines. There was a total of 187 such occurrences on VL 1 and of 269 on VL 2 during the primary mission. Overall, 1568 rescan lines from VL 1 and 3633 rescan lines from VL 2 were searched for evidence of motion of objects large in comparison to a pixel. These rescan lines had a total duration of 194.46 min for VL 1 and 283.89 min for VL 2.

Figure 8 shows two direct link slow rate rescans of 50 lines and a duration of 708 s. The shadow of the meteorology boom moving through the azimuth position being scanned demonstrates the effect that would be created by moving objects. The wiggles in the shadow edge are due to the unevenness of the

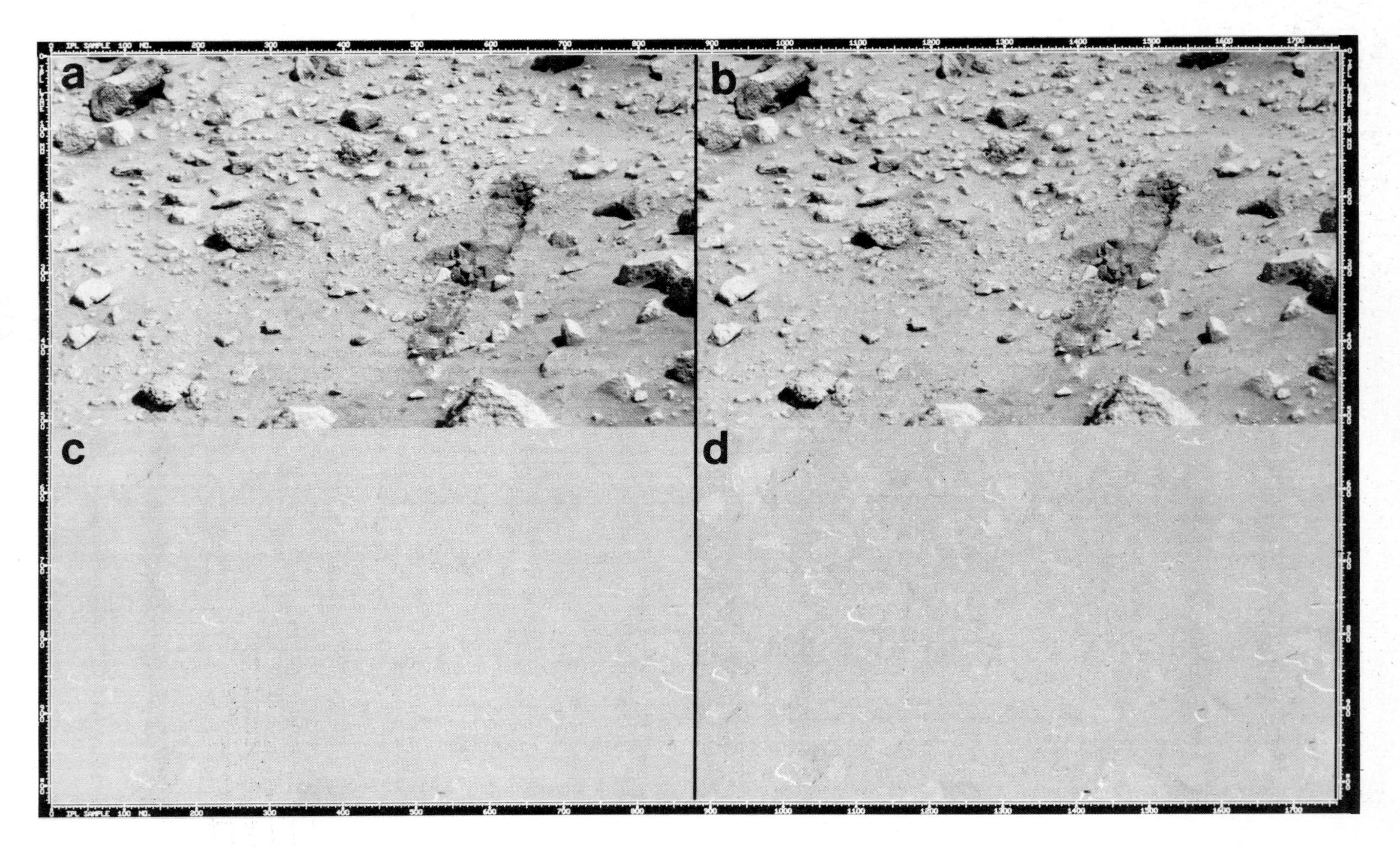

Fig. 5. Viking Lander 1 scene centered at $CEA = -30°$, from azimuth 85° to 120°, imaged on sols 53 and 73. (*a*) CE label 12B163/073. (*b*) CE label 12B138/053. (*c*) Part *b* minus part *a*. (*d*) Contrast-stretched difference. No change is observed.

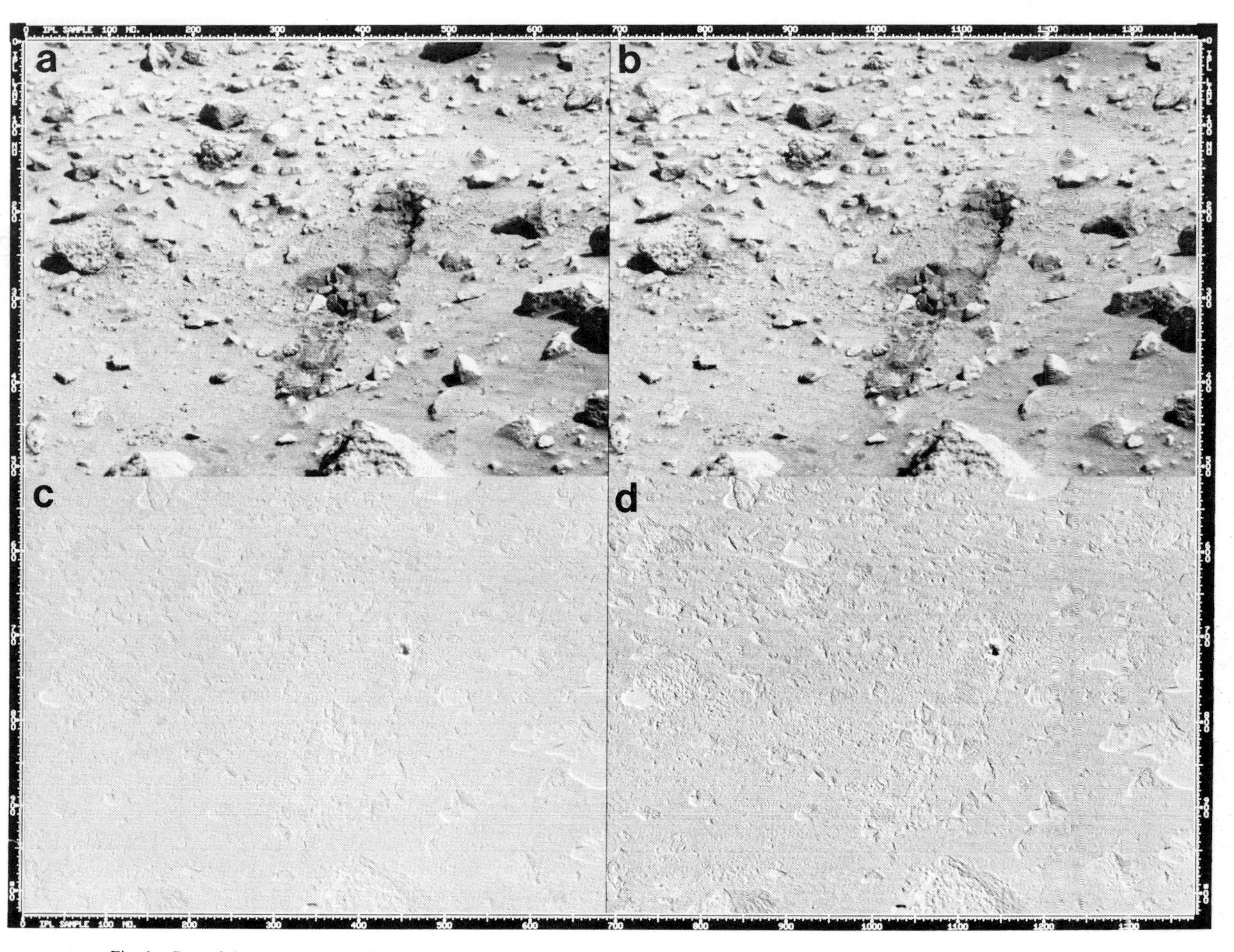

Fig. 6. Part of the same scene shown in Figure 5, from azimuth 92.5° to 120°, imaged on sols 73 and 93. (*a*) CE label 12B188/09[illegible]. (*b*) CE label 12B163/073. (*c*) Part *b* minus part *a*. (*d*) Contrast-stretched difference. Note the adjacent dark and light patches at line 700 and sample 450, caused by movement of the material in the trench.

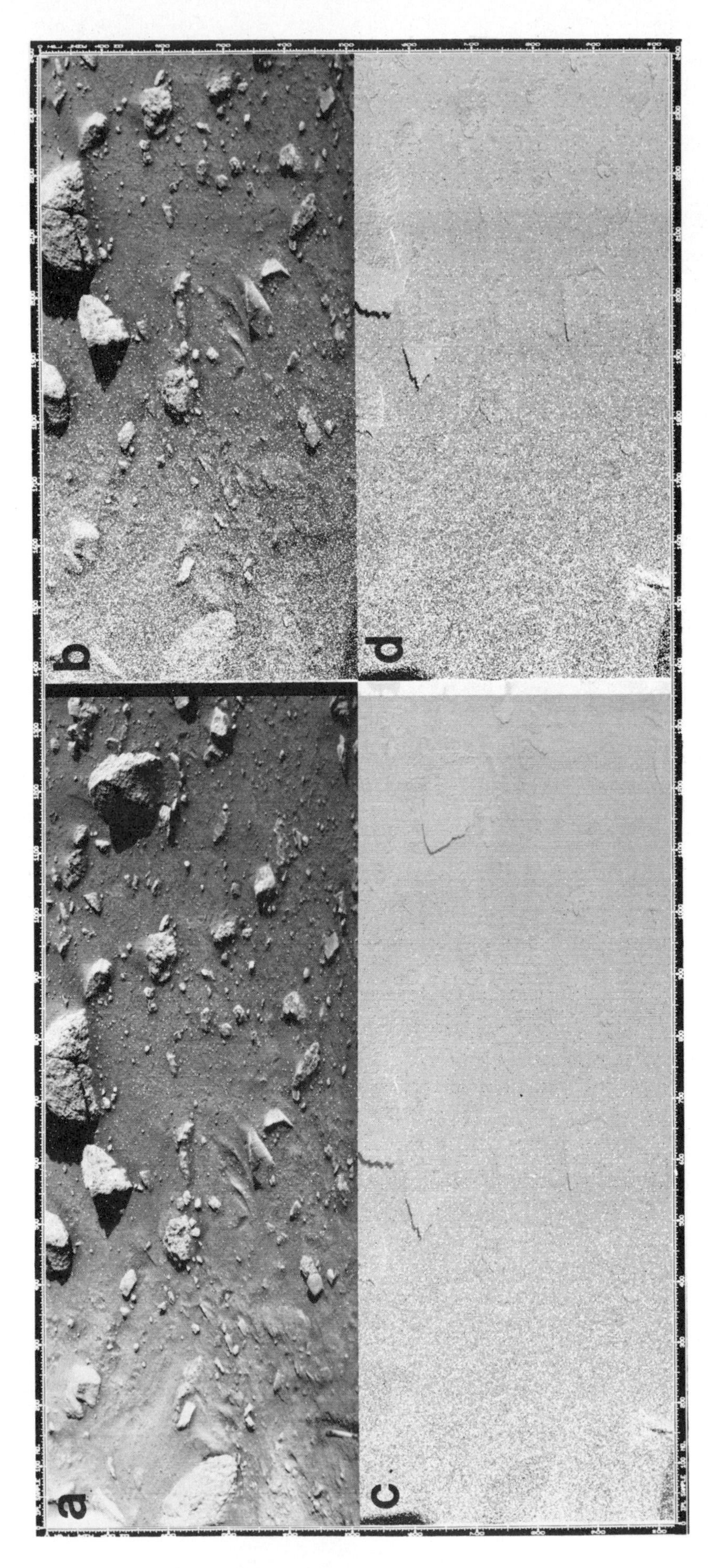

Fig. 7. Viking Lander 1 scene centered at *CEA* = −50° from azimuth 80° to 135°, imaged on sols 1 and 12. (*a*) CE label 12A081/012. (*b*) CE label 12A009/001. (*c*) Part *b* minus part *a*. (*d*) Contrast-stretched difference. Note the effect of the boom latch pin and the adjacent disturbed surface in the lower left-hand part of the image.

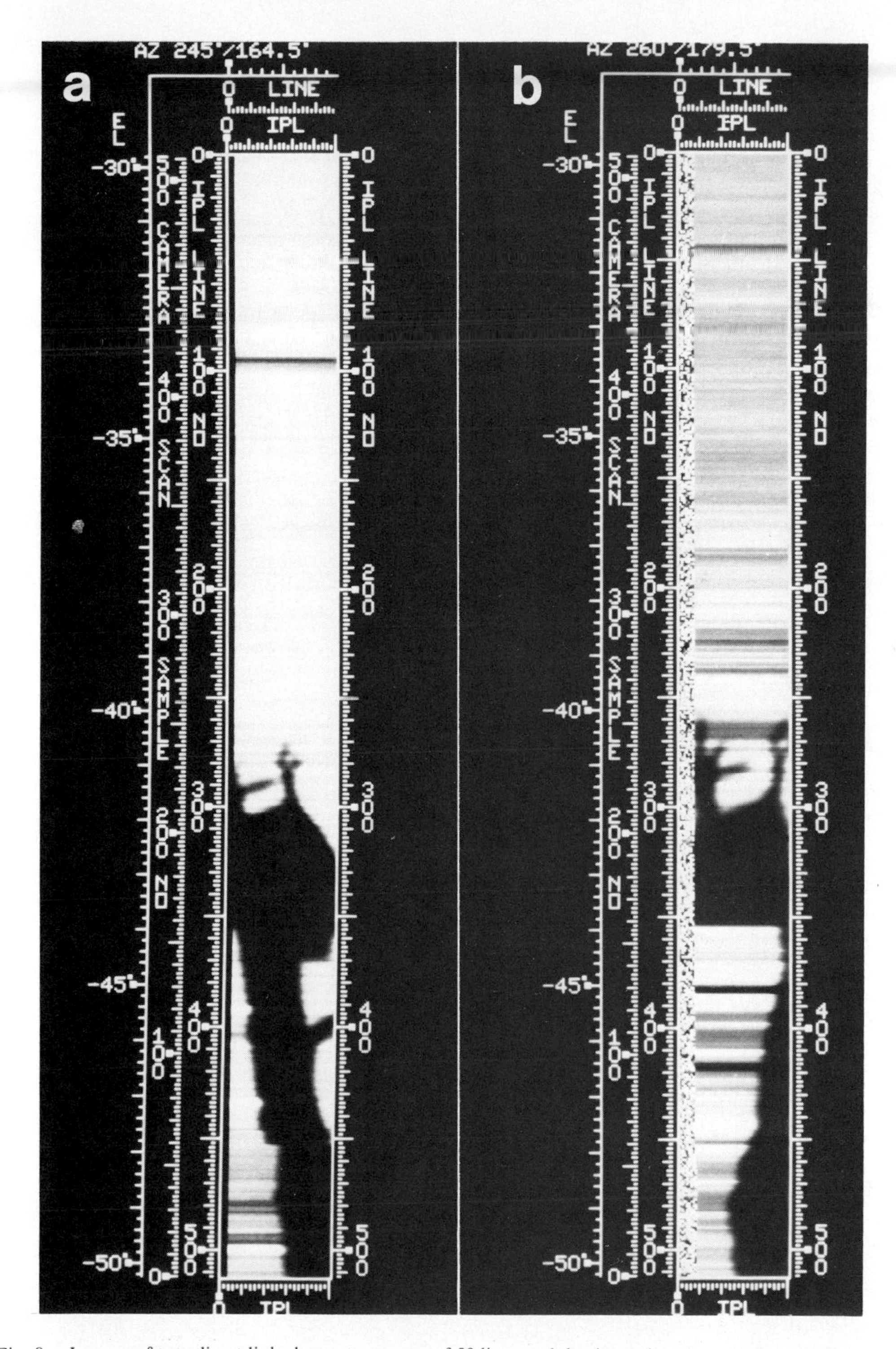

Fig. 8. Images of two direct link slow rate rescans of 50 lines each having a duration of 708 s at *CEA* = −40°. (*a*) CE label 21A078/011; azimuth = 245°, and local lander time is 1430:10. (*b*) CE label 21A085/012; azimuth = 260°, and local lander time is 1515:10. The shadow of the meteorology boom demonstrates the effect of moving objects.

surface over which the shadow is moving. The search of the rescanned lines for evidence of motion has found none other than that explained by shadows, as seen in Figure 8.

POSTCONJUNCTION OBSERVATIONS

As new images are being received in the extended mission, they continue to be compared with previous camera events covering the same area. Very significant changes in the lighting condition make this comparison difficult. A slumping of surface material has been noted in front of the large boulder which is on the scene observed by camera 1 on Viking Lander 1 in the northeast direction [*Jones,* 1977].

A nighttime search for light-emitting objects was carried out by using an image on VL 1 taken at 0155 local lander time on sol 269 by using the 0.12° aperture red diode and the highest gain and viewing a 140° azimuth range of the surface from elevation angles of −40° to −10°. No signal above noise was received.

CONCLUSION

The Viking lander camera system has found no evidence of macroscopic life on Mars. This may be because Mars harbors no contemporary macrobes, because our sample in space and time is too restricted, or because of limitations in the camera design. If macrobes are absent, the possible presence of microbes is in no way eliminated. We may be observing Mars relatively early in its biological evolution; for more than 80% of the history of life on earth, only microbes were to be found. There is yet no general theory of biology which states necessary conditions for the existence of life. Thus there are an indefinite number of possible biological models which could be the result either of our imagination or of real evolutionary processes in a large variety of conceivable environments. At the same time that we should be wary of rejecting unusual possibilities we should also be aware of the impossibility of disproving rigorously the existence of life. A model of biology can always be invoked which would have evaded detection by our instruments. For example, Martian photophobes could always be poised one scan line away, waiting for the reflected light from the nodding camera mirror to disappear.

Acknowledgments. The search for life on Mars has been stimulated in many ways and over many years by Joshua Lederberg of Stanford University. The search through the data was carried out in part by people other than the authors. They are too numerous to be acknowledged properly other than by references to the team activities given in the text. The efforts of William Benton of the Image Processing Laboratory at the Jet Propulsion Laboratory in producing Figures 3-7 are gratefully acknowledged. This paper represents one phase of the tasks carried out by the Genetics Department, Stanford University, under contract NAS1-9682; by the Department of Geological Sciences, Brown University, under contract NAS1-9680; and by the Laboratory for Planetary Studies, Cornell University, under contract NAS1-9683.

REFERENCES

Huck, F. O., and S. D. Wall, Image quality prediction: An aid to the Viking imaging investigation on Mars, *Appl. Opt., 15,* 1748-1766, 1976.

Huck, F. O., H. F. McCall, W. R. Patterson, and G. R. Taylor, The Viking Mars lander camera, *Space Sci. Instrum., 1,* 189-241, 1972.

Huck, F. O., D. J. Jobson, S. K. Park, S. D. Wall, R. E. Arvidson, W. R. Patterson, and W. B. Benton, Spectrophotometric and color estimates of the Viking lander sites, *J. Geophys. Res., 82,* this issue, 1977.

Jones, K. L., Geomorphological change on the surface of Mars, *Science,* in press, 1977.

Lederberg, J., and C. Sagan, Microenvironments for life on Mars, *Proc. Nat. Acad. Sci. U.S., 48,* 1473-1475, 1962.

Mutch, T. A., A. B. Binder, F. O. Huck, E. C. Levinthal, E. C. Morris, C. Sagan, and A. T. Young, Imaging experiment: The Viking lander, *Icarus, 16,* 92-110, 1972.

Patterson, W. R., III, F. O. Huck, S. D. Wall, and M. R. Wolf, Calibration and performance of the Viking lander cameras, *J. Geophys. Res., 82,* this issue, 1977.

Pollack, J. B., D. Colburn, R. Kahn, J. Hunter, W. Van Camp, C. E. Carlston, and M. R. Wolfe, Properties of aerosols in the Martian atmosphere as inferred from Viking lander imaging data, *J. Geophys. Res., 82,* this issue, 1977.

Sagan, C., and J. Lederberg, The prospects for life on Mars, a pre-Viking assessment, *Icarus, 28,* 291-300, 1976.

(Received May 23, 1977;
revised June 3, 1977;
accepted June 3, 1977.)

Properties of Aerosols in the Martian Atmosphere, as Inferred From Viking Lander Imaging Data

JAMES B. POLLACK,[1] DAVID COLBURN,[1] RALPH KAHN,[2] JUNE HUNTER,[3] WARREN VAN CAMP,[3] C. E. CARLSTON,[4] AND M. R. WOLF[5]

Observations of the Martian sky, Phobos, and the sun were taken with the Viking lander imaging cameras to obtain information on the properties of the atmospheric aerosols. Atmospheric optical depths were derived from the observations of the brightness of the celestial objects. Information on the absorption coefficient, mean size, and shape of the aerosols was derived from studies of the sky brightness. For this purpose we used a multiple-scattering computer code that employed a recently developed technique for treating scattering by nonspherical particles. By monitoring the brightness of the twilight sky we obtained information on the vertical distribution of the particles. Three types of aerosols are inferred to have been present over the landers during the summer and fall season in their hemisphere. A ground fog made of water ice particles was present throughout this period. It formed late at night during the summer season and dissipated during the morning. We infer that during the summer the frost point temperature was 195°K and the water vapor volume mixing ratio equaled about 1×10^{-4} near the ground at VL-2. Assuming that condensation occurs only on suspended soil particles, we estimate that the average particle radius of the fog was about 2 μm and that the fog's depth equaled approximately 0.4 km. A higher-level ice cloud was prominent only during the fall season, when it was a sporadic source of atmospheric opacity at VL-2. The formation of upper level water ice clouds during the summer may have been inhibited by dust heating of the atmosphere. Suspended soil particles were present throughout the period of observation. During the summer they constituted the only major source of opacity in the afternoon and most of the night. The cross-section weighted mean radius of these aerosols is about 0.4 μm. They have a nonspherical but equidimensional shape and rough surfaces. These soil particles have a scale height of about 10 km, which is comparable to the gas scale height, and they extend to an altitude of at least 30 km. The principal opaque mineral in these particles is magnetite, which constitutes 10% ± 5% by volume of this material. We propose that soil particles, as well as any associated water ice, are eliminated from the atmosphere, in part, by their acting as condensation sites for the growth of CO_2 ice particles in the winter polar regions. The resultant CO_2-H_2O-dust particle is much larger and therefore has a much higher fallout velocity than an uncoated dust or water ice particle.

Observations of the Martian sky and celestial objects have been made with the Viking lander imaging cameras to obtain estimates of the composition, size, vertical distribution, and optical depth of the aerosols above the two lander sites. Preliminary estimates of these quantities have been given by *Mutch et al.* [1976*a*, *b*, *c*]. In this paper we examine a much larger set of data than was used in the earlier papers and subject the measurements to a more thorough analysis. In the remainder of this introductory section we briefly review prior information about the aerosols in the Martian atmosphere and then indicate the manner in which the lander imaging experiment can add to this knowledge.

Aerosols made of soil material and water ice are known to be present in the Martian atmosphere, while the occurrence of carbon dioxide ice is suspected. We now briefly discuss what is currently known about each of these three aerosol species, beginning with the small suspended soil particles. Such particles are injected into the atmosphere during the time of great dust storms, which begin when Mars is close to its perihelion orbital position [*Leovy et al.*, 1973*a*]. The corresponding season on Mars is late spring in the southern hemisphere or, equivalently, L_s equals about 250°, where L_s is the aerocentric longitude of the sun relative to spring equinox in the northern hemisphere. Global dust storms begin in the southern hemisphere, usually totally envelope this hemisphere, and sometimes spread to the northern hemisphere as well.

Analysis of the infrared interferometer spectrometer and ultraviolet spectrometer observations conducted from the Mariner 9 orbiter have provided a determination of some of the properties of the suspended dust particles during the decaying phase of the 1971–1972 global dust storm. At its peak the optical depth was about 1.5, and the optical depth more or less monotonically declined to a value of about 0.2 over a 3-month time period [*Pang and Hord*, 1973; *Toon et al.*, 1977]. The differential distribution of particle sizes had a slope of −4 for particle radii between 1 and 10 μm, and surprisingly, the distribution remained approximately constant over the duration of the decay phase [*Toon et al.*, 1977]. This result has been interpreted as indicating that eddy mixing rather than Stokes-Cunningham fallout controls the elimination of dust particles from the atmosphere. The particles are made of either a mixture of acidic silicate materials or else a mixture of clay minerals and perhaps basalts [*Toon et al.*, 1977]. Chemical analyses of the soil at the Viking lander sites indicate that it is made either of iron rich clays or a mixture of basalts and carbonaceous chondritic material [*Baird et al.*, 1976]. In addition, the surface material contains 0.1–1% chemically bound water [*Biemann et al.*, 1976]. The smaller-sized suspended dust particles may be made of similar material. This conclusion may be consistent with the above cited results of *Toon et al.* [1977].

In addition to global dust storms, localized dust storms are known to occur occasionally throughout the Martian year [*Leovy et al.*, 1972; *Sagan et al.*, 1972]. These localized events can place modest quantities of dust in the atmosphere at times well removed from the time of the global storms. On the basis of the exceptional clarity of surface features seen on Mariner 9 pictures obtained during its extended mission, *Leovy et al.*

[1] Space Science Division, NASA Ames Research Center, Moffett Field, California 94035.
[2] Center for Earth and Planetary Physics, Harvard University, Cambridge, Massachusetts 02138.
[3] Informatics, Inc., Palo Alto, California 94305.
[4] Martin Marietta Corp., Denver, Colorado 80201.
[5] Jet Propulsion Laboratory, Pasadena, California 91103.

Paper number 7S0559.

[1973*b*] estimated that the optical depth was less than 0.04 for L_s values of 30–100, a time interval that slightly overlaps the beginning of the Viking mission.

Water ice clouds are known to form both in the polar regions during their fall and winter seasons and near topographically elevated places during summer [*Martin and McKinney*, 1974; *Leovy et al.*, 1973*b*]. The polar clouds are referred to as the 'polar hood.' It exhibits significant variability in its location on time periods of the order of days and even hours [*Martin and McKinney*, 1974]. On the average the northern polar hood has been observed to extend to latitudes as low as about 40°. Thus during fall and winter seasons in the northern hemisphere we might expect occasionally to observe the polar hood above the Lander 2 site (48°N) but not above the Lander 1 site (22°N).

During the summer, water ice clouds have been observed to form and intensify during the day above elevated regions, such as the Tharsis shield volcanos [*Leovy et al.*, 1973*b*]. Analysis of such clouds viewed by the infrared interferometer spectrometer experiment showed that the cross-section weighted mean radius of the ice particles was about 2 μm and the integrated vertical column density of water ice was about 0.5 precipitable microns (pr μm) [*Curran et al.*, 1973]. The Viking orbiter cameras have also seen what appears to be low lying fogs within a few depressed areas in the equatorial region [*Carr et al.*, 1976]. Detached high-altitude haze layers have been observed on both Mariner 9 and Viking orbiter photographs [*Leovy et al.* 1972; *Carr et al.*, 1976]. This structure is obvious only at certain seasons and latitudes. Near the beginning of the Viking mission the detached haze layer was prominent at southern latitudes, where its height and optical depth had values of about 40 km and 0.01, respectively [*Carr et al.*, 1976]. Conceivably, this haze layer could be formed from the condensation of CO_2 gas rather than H_2O vapor. Finally, general circulation calculations indicate that the air temperature in the inner portion of the winter polar cap regions (latitudes of > 60°) falls to the frost point of CO_2, and therefore CO_2 ice particles may form there [*Pollack et al.*, 1976].

With the above results in mind, let us consider the opportunities for studying atmospheric aerosols with the Viking lander cameras. By virtue of their geometry and photometric quality, observations with the lander cameras lead to a good determination of a number of important aerosol properties. The cameras have a linear response to incident radiation; their absolute calibration is well determined as a result of extensive ground-based tests as well as ones made on Mars [*Patterson*, 1977]; and accurate photometric comparisons can be made across a given picture, since the same diode views the whole scene. One of the diodes is a sun diode, which allows imaging of the sun. Such images permit an unambiguous determination of the optical depth of the atmosphere during the day, while images of Phobos obtained with other diodes provide comparable data during the night. By way of contrast, the derivation of optical depth from orbiter or ground-based observations of Mars is usually very model dependent simply because of the more unfavorable geometrical arrangement of observer, atmosphere, and light source. Also by virtue of their geometry the lander cameras can view the Martian sky close to the sun. Such observations provide information on the size of the aerosols. Also lander images of the sky are not contaminated by light originating from the surface, whereas such situations cannot generally be realized from other platforms.

The atmospheric observations discussed in this paper span the time period from mid-July 1976 to the end of February 1977. During this time period the season at the Viking 1 (22.5°N, 48.0°W) and the Viking 2 (47.9°N, 225.9°W) sites changed from early summer to midfall (L_s = 100–230). Below we first present values of the optical depth obtained from images of the sun and Phobos, then give estimates of particle size and shape found from sky brightness pictures, next discuss compositional information obtained from sky color measurements, and finally investigate the vertical distribution of the aerosols with the aid of photographs in which the sky brightness was monitored as a function of time before sunrise or after sunset. In each case we will discuss the manner in which the desired physical variable is extracted from the observations and then present the values obtained along with some preliminary interpretations. Following these sections we discuss our results in the more general context of their possible implications for Martian meteorology. In the last section of this paper our main conclusions are summarized.

OPTICAL DEPTH

Procedure

Observations of the sun and Phobos were obtained to determine the optical depth of the atmosphere during the day and night. Optical depth values τ were derived in a straightforward fashion from Beer's law, according to which the observed intensity of these objects I is related to τ by

$$I = I_0 \exp\left[-\tau/M(e)\right] \qquad (1)$$

where I_0 is the value that would have been obtained had $\tau = 0$, M is an air mass factor, and e is the elevation angle of the celestial object, as measured from the horizon for a hypothetically flat surface aligned perpendicular to the local gravity vector. Except for very small values of e, $M(e) = \sin e$. The value for τ can be found from (1) either by using the observed intensities from a pair of pictures taken close together in time but with the object at significantly different elevation angles or alternatively by knowing the value of I_0 and using the observed intensity of a single picture. Because of picture budget limitations a number of single sun pictures were obtained. We therefore averaged the I_0 values derived from pairs of pictures obtained with a given camera and diode on the same day at different elevation angles and used these average values to determine τ for each picture. Comparison of τ values found by the single-picture method for sun pictures taken close together in time indicated that on the average these values have an uncertainty of about ±5%.

The values of the sun's elevation angle needed for (1) were obtained from an ephemeris prepared by S. Synnott and T. Duxbury of the Jet Propulsion Laboratory. We adopted this approach instead of using the observed positions of the sun in the pictures because the latter procedure would have involved making corrections for the tilt of the lander with respect to the local surface and for the tilt of the local surface with respect to the local geoid. The chief uncertainty in the ephemeris values stems from an uncertainty in the location of the lander. This error translates into an error of a few tenths of a degree in the value of e, an error consistent with the difference between observed and predicted positions of the sun and Phobos in the pictures.

Images of the sun and Phobos were obtained with diodes whose instantaneous field of view was 0.12°. In obtaining these pictures the cameras were almost always stepped by 0.04° in azimuth and elevation so as to increase the number of discrete readings on the object and to increase the probability that one

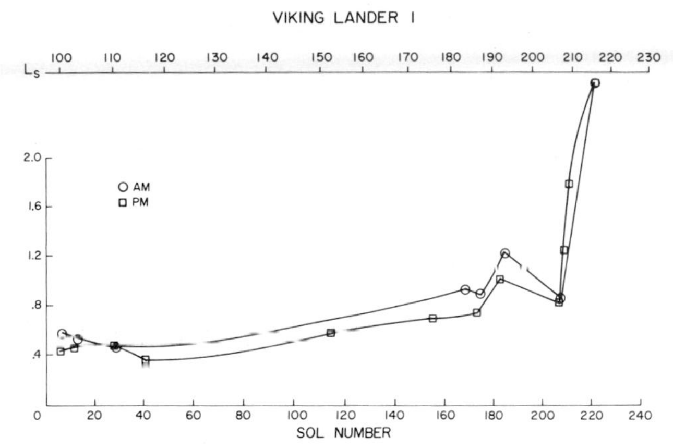

Fig. 1. Optical depth as a function of time at the Viking Lander 1 site. Separate curves are shown for results from (A.M.) morning sun pictures and (P.M.) afternoon sun pictures. The bottom horizontal axis shows time in terms of sols, the number of Martian solar days from the time of landing (sol 0). The top horizontal axis shows time in terms of the aerocentric longitude of the sun L_s, which provides a measure of the seasonal date on Mars. L_s had a value of zero at spring equinox in the northern hemisphere.

sample was taken when the object was almost on the optical axis of the camera. As is seen from the surface of Mars, the sun subtends an angle of about 0.3°, while Phobos typically measures about 0.1° across. In the case of the sun pictures we used the nine brightest intensity samples to derive the position of the optical axis in the image plane. In so doing, we took account of the camera smearing function and the known solar limb darkening [*Allen*, 1955]. Using this information and the derived position of the optical axis, we then found the intensity reading that we would have gotten for a sample exactly centered on the optical axis. Also a small correction was made to find this central intensity for a specified fixed size for the sun. The intensities so found had the effects of discrete sampling and varying solar size removed and therefore formed a common data set from which the optical depth could be found. In the case of Phobos a similar procedure was used. However, we employed the photometric function of Phobos's surface given by *Noland and Veverka* [1977] and converted the derived surface brightness to 0° phase angle by using their phase coefficient. Because the phase angle varies appreciably for pairs of Phobos pictures, the optical depths found from these pictures have a larger uncertainty than those derived from sun pictures.

The following procedure was adopted for modeling the smearing function of the camera. We assumed that a given point in object space mapped into a circle of uniform intensity in image space. The instantaneous contribution of that point to the observed sample value is therefore proportional to the product of the photometric function evaluated at that point and the area of overlap between the blur circle and the instantaneous field of view. This product was integrated over the object of interest and over the finite length of time of sampling, during which the image moves slightly and continuously in elevation [*Huck et al.*, 1973]. The size of the blur circle was found empirically by obtaining a least squares fit to the entire image of the sun.

The sun pictures were taken with a diode covered with the standard red filter, whose effective wavelength is about 0.67 μm [*Huck et al.*, 1977]. Most of the Phobos pictures were obtained with the blue diode so as to maximize the signal. Its effective wavelength is 0.50 μm. However, on one night, Phobos pictures were taken with all six color diodes. For each diode, pictures were obtained at several elevation angles so as to provide information on the wavelength dependence of the optical depth.

Results

Figures 1 and 2 show the variation of optical depth τ with time at the Lander 1 and 2 sites, respectively. The bottom horizontal axis gives time in units of sol number, the number of solar days on Mars from the time of landing. Sol 0 corresponds to the day of landing at each site. Since Viking 2 landed at its site 45 days after Viking 1 landed at its site, the zero point for the bottom horizontal axis in Figure 2 has been displaced by 45 sols. Consequently, common horizontal positions on the two figures correspond to the same seasonal data. The seasonal coordinate L_s is given along the upper horizontal axis of these figures. Two separate curves are given in each figure. One curve, labeled A.M., was derived from sun pictures taken during the morning, while the other curve, labeled P.M., was found from afternoon sun pictures. Typically, A.M. sun pictures were obtained 1–2 hours after sunrise, and P.M. sun pictures were otained 1–2 hours before sunset. When several A.M. or P.M. sun pictures were taken on the same day, we averaged together the optical depth values found from those pictures having adequate signal strength to obtain the value graphed in these figures. Individual data points are shown by circles and squares for the A.M. and P.M. optical depth values, respectively.

One striking result shown by Figures 1 and 2 is the high value of the optical depth at the two landing sites throughout the period covered by the observations. The lowest value of the optical depth was 0.18. During the early portions of the mission, τ had an average value of about 0.5 and 0.3 at the Lander 1 and 2 sites, respectively. At both sites, τ increased significantly at later times. These findings can be contrasted with the much lower values of optical depth derived in an indirect fashion from Mariner 9 orbital imaging data obtained during a seasonal period that overlaps the beginning of the lander mission, as was discussed in the introduction to this paper. Almost all of the optical depth is due to extinction by atmospheric aerosols. Using formulae given by *Hansen and Travis* [1974] and a mean surface pressure of 7.5 mbar at the two sites [*Hess et al.*, 1976*a*, *b*, *c*], we find that the molecular Rayleigh scattering optical depth is only about 1.5×10^{-3} at the effective wavelength of the sun diode.

The optical depth curves of Figures 1 and 2 exhibit three

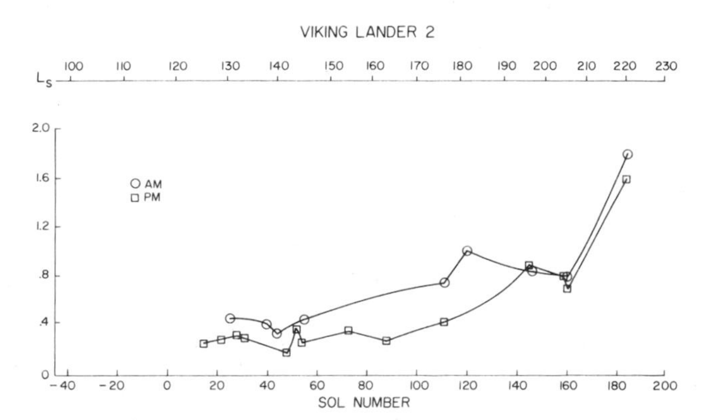

Fig. 2. Optical depth as a function of time at the Viking Lander 2 site. The A.M. and P.M. curves and the horizontal axes have the same meaning as they do in Figure 1.

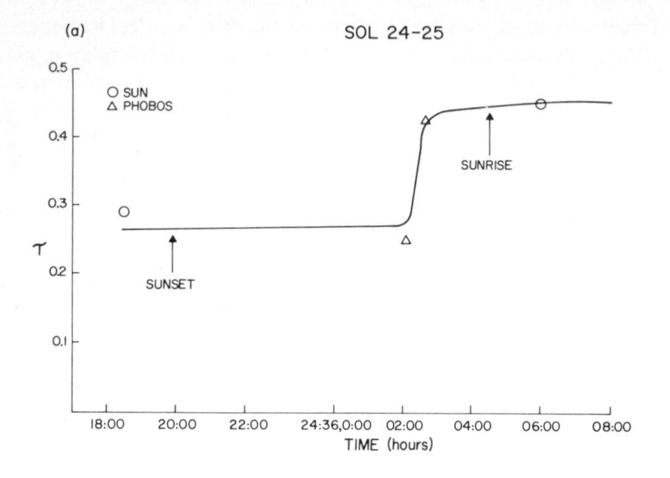

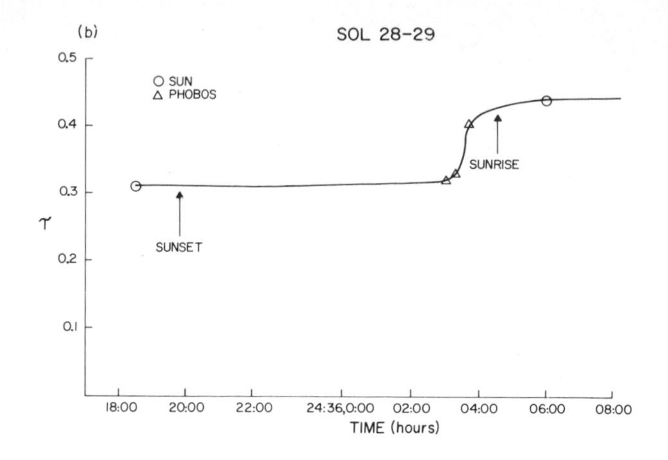

Fig. 3. (*a*) Variation of optical depth from the late afternoon of sol 24 to the early morning of sol 25 at the VL-2 site, as inferred from photographs of Phobos (triangles) and the sun (circles). The time coordinate gives hours in terrestrial units. (*b*) Similar data obtained on sols 28 and 29 at the VL-2 site.

different types of temporal changes: diurnal, random, and seasonal. Below we discuss each of these types of variations. The diurnal variation is characterized by the A.M. τ values lying systematically above the P.M. values.

Additional characterization of this diurnal difference has been achieved by taking photographs of Phobos or the sun at various times on the same day so as to determine when the optical depth changes occur. Figures 3*a* and 3*b* illustrate the temporal behavior of the optical depth over the portion of a day in which it is increasing. These diagrams were constructed from photographs of Phobos obtained late at night at the Lander 2 site on sols 25 and 29, respectively. In deriving optical depths from individual Phobos pictures on these nights we used an I_0 value obtained from Phobos pictures taken at an earlier time of the night on sol 48, when no optical depth changes should have occurred (see (1)). Besides the optical depth values given by the Phobos observations (triangles), Figures 3*a* and 3*b* also show optical depth values in the late afternoon and early morning (circles), which were obtained from the solar optical depth curves given in Figure 2. The vertical arrows along the horizontal axis show the times of sunrise and sunset. These observations refer to midsummer in the northern hemisphere (L_s = 110).

According to Figures 3*a* and 3*b* the increase in optical depth commences around 0200 hours local lander time, almost all of the optical depth increase occurring between this time and sunrise. The slightly different temporal behavior of the curves on sols 25 and 29 is most likely due to weather phenomena, i.e., fluctuations in the temperature and humidity characteristics of the air masses over the landing site.

The declining phase is illustrated in Figures 4*a* and 4*b*, which refer to a time period about 150 sols later than that for Figure 3. The season is now the start of fall (L_s = 190). Figures 4*a* and 4*b* were constructed from optical depth results obtained from sun pictures at the Lander 2 site on sols 160 and 184, respectively. We see that most of the decrease in optical depth takes place in the late morning and early afternoon. Whether the optical depth increases slightly or remains level in the middle of the afternoon cannot be established from Figure 4*b* because of the low signal to noise ratio characterizing the last two sun pictures.

We interpret the diurnal variations in optical depth shown in Figures 1–4 as being due to a ground fog that forms during the coldest portion of the day (the late night), survives during the early morning, and then dissipates by the middle of the afternoon. Such a fog has been predicted on theoretical grounds [*Flasar and Goody*, 1976; *Hess*, 1976].

In addition to consistent trends in the optical depth curves of Figures 1 and 2, there were also irregular variations. For example, between sol 48 and sol 52 at the second landing site,

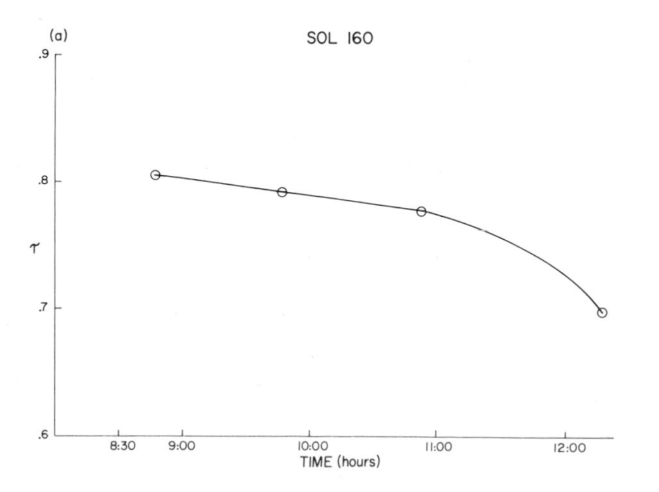

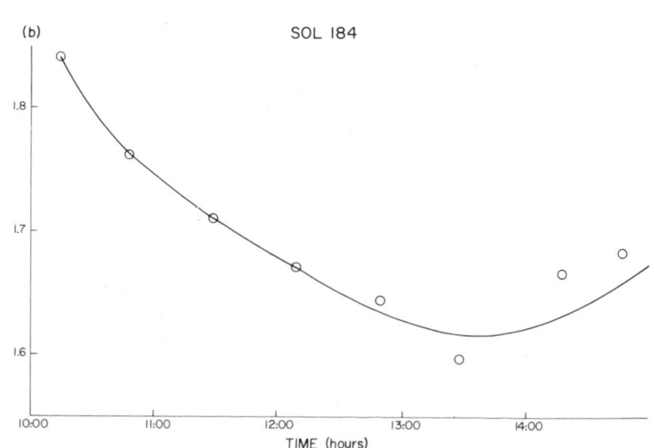

Fig. 4. (*a*) Variation of optical depth over part of the sunlight portion of sol 160 at the VL-2 site. (*b*) Similar information on sol 184, also at the VL-2 site.

the P.M. optical depth increased by about 0.2. Two sols later it had decreased by about 0.1. On each of these 3 sols, several sun pictures were taken, and the optical depths derived from the individual pictures agree quite well with one another. Therefore we believe these changes to be real. In a crude sense we interpret these changes to be due to weather, i.e., the passage of different air masses past the landing sites.

Figures 5*a* and 5*b* illustrate a very dramatic increase in the optical depth of the atmosphere that took place at the second lander site on sol 161 (L_s = 205) within an interval of just 2 hours. Figure 5*a* shows a picture of the surface obtained at a local lander time of 1100 hours, while Figure 5*b* shows a picture taken at 1310 hours on the same day (0000 hours is local midnight). The same gain and offsets and the same type of diode were used for both pictures. These photographs have been reproduced so as to illustrate approximately their true relative brightnesses. We see that the surface brightness was much lower in the later picture. This effect cannot be attributed to the photometric function of the surface: other lander images show the surface to backscatter incident illumination preferentially. Since the area covered by the later picture is much closer to the antisolar point than that in the earlier picture, the photometric function would produce an effect opposite to the one observed. This inference of a true darkening of the scene is also supported by the obvious presence of shadows in the earlier picture and a lack of notable shadows in the later one. Both the changes in average scene illumination and the visibility of shadows can be seen in a quantitative fashion from a comparison of the histograms of scene brightness given in the graphs next to the upper right-hand portions of the picture.

We interpret the changes seen in Figure 5 as being due to a

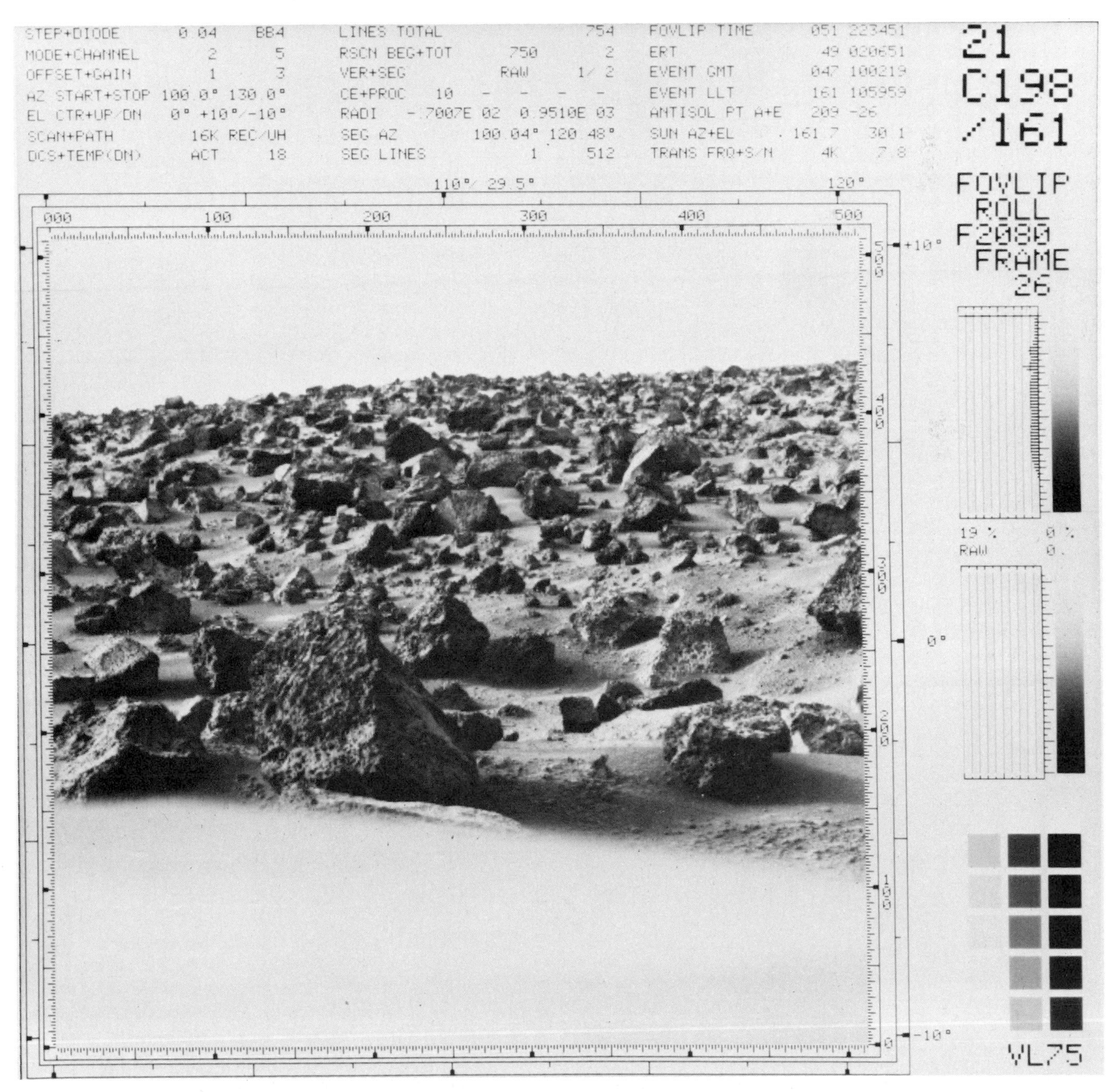

Fig. 5*a*. Photograph of the surface of Mars at the VL-2 site on sol 161 at 1100 hours local lander time.

large increase in the optical depth of the atmosphere over the 2-hour period separating the two pictures and further attribute this increase to the passage of the polar hood over the second lander. Such an increase in optical depth would both decrease the amount of light reaching the surface and lead to the result that a larger fraction of this light is skylight instead of the attentuated direct solar beam. The latter characteristic implies a more omnidirectional scene illuminance, which would reduce the contrast of shadows and lead to less variance in scene brightness. The latitude and season of these observations are consistent with the suggestion that the optical depth increase is due to the passage overhead of the polar hood, as is the time scale of the change [*Martin and McKinney*, 1974]. By way of contrast, no sudden increase in wind speed was noted by the meteorology experiment in this time period, so the occurrence of a local dust storm as the cause for the optical depth increase seems unlikely (J. Tillman, private communication, 1977).

Additional evidence for the occasional presence of the polar hood above the second lander at this season is given by a sequence of daily optical depth measurements initiated on sol 187. There were marked, erratic fluctuations in the optical depth from one day to the next, changes as large as ±1 occurring on a time scale of 1 day to several days. By way of contrast, no such large fluctuations were noted in a less frequent monitoring of the optical depth at the first landing site.

Fig. 5*b*. A photograph similar to that in Figure 5*a* obtained at 1300 hours on the same day. The two photographs (camera events 21C198 and 21C200) have been developed to display approximately their true relative brightnesses. The histogram in the upper right-hand corner of each picture shows the distribution of brightness in each picture. For the exact location of the pictures, refer to *Mutch* [1977, Figure 1].

The above observations are consistent with the known properties of the polar hood, which were discussed at the beginning of this paper.

The third type of temporal variability is a seasonal one. Toward the beginning of the mission, there was some tendency for the optical depth values given in Figures 1 and 2 to decrease, while at later times, there was a marked increase in optical depth. These trends occurred at similar times at the two landing sites and so may characterize the seasonal behavior of many equatorial and mid-latitude regions of the northern hemisphere.

Some insight into the cause for the increase in optical depth from summer to fall can be obtained from a correlation of lander and orbiter information. As is illustrated in Figure 1, a very dramatic increase in optical depth occurred at the first landing site between L_s = 205 and 215. Photographs obtained from the Viking orbiter show that a very violent dust storm began in the Thaumasia region of the southern hemisphere at L_s = 205. By L_s = 215 the entire southern hemisphere was covered with a thick dust cloud (G. Briggs, private communication, 1977). Consequently, the dramatic increase in optical depth observed at VL-1 in this time interval is most likely due to a spreading of the dust cloud into the northern hemisphere. In part, this dust storm may also be responsible for the increase seen over VL-2 at a similar time (see Figure 2). However, the presence of erratic fluctuations in optical depth at VL-2 noted at a slightly later time raises the possibility that the polar hood may have also contributed to the increase in optical depth at the VL-2 site shown in Figure 2.

More generally, dust storm activity in the southern hemisphere may be the dominant factor behind the seasonal increase in the optical depth observed during the fall at both landing sites. Orbiter observations show that local dust storms occurred in the southern hemisphere over this entire season (G. Briggs, private communication, 1977). These, however, did not become global in scale, as did the Thaumasia storm described earlier. Their cumulative effect could be a slow rise in the global dust content, consistent with the lander results.

Finally, we have obtained estimates of the wavelength dependence of the optical depth by observing Phobos with the various color and infrared diodes on sol 48 at the Lander 2 site. These results are shown by the triangles in Figure 6, where optical depth is given as a function of the effective wavelength of the blue, green, red, and IR1 (near infrared) diodes (the IR2 and IR3 diodes have an effective wavelength similar to that of the IR1 diode). Also shown in this figure is the optical depth value derived from sun pictures obtained in the late afternoon of the same day that the Phobos pictures were taken (circle). Its effective wavelength is the same as that of the red diode. We see that the optical depth is approximately independent of wavelength, as is illustrated by the fit of the horizontal line to the data points.

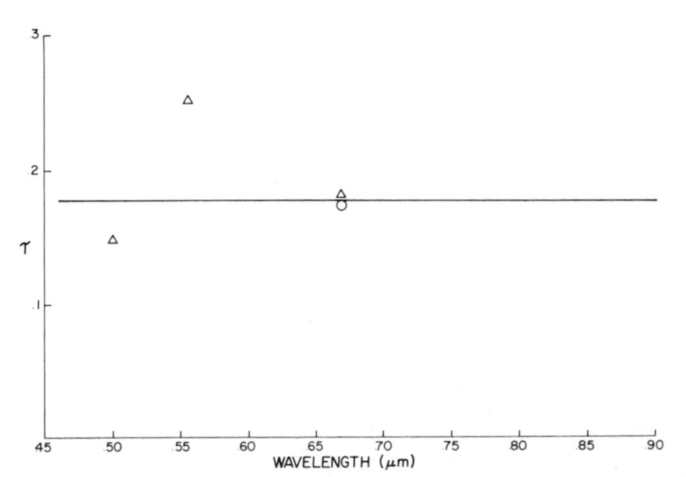

Fig. 6. Optical depth as a function of wavelength, as derived from observations of Phobos at VL-2 on sol 48 at 2300 hours (triangles). Also shown is the optical depth found from sun images taken 4½ hours earlier (circle). The horizontal line exhibits the fit of a wavelength independent optical depth to these results.

PARTICLE PROPERTIES

Procedure

We have used photographs of the sky brightness to obtain information on several particle properties. From the observed angular variation of sky brightness close to the sun we obtained estimates of the mean particle size. Similar data at larger distances from the sun provide some information on the shape of the particles. To carry out this analysis, we have used a computer program that has the following characteristics: the multiple-scattering problem was accurately solved, allowance for the possible nonsphericity of the particles was made in the derivation of their single-scattering characteristics, and the radiative interaction between the atmosphere and the ground was taken into account. Below, each of these items is discussed. We then describe how the desired particle properties were obtained from the observations.

Except possibly for the special geometry of the twilight situation, which is discussed in a later section, it is clear from the sizable values of the optical depth that a single-scattering description of the sky brightness is inadequate and that allowance should be made for multiple scattering. Given the single-scattering properties of the particles, as found in a manner to be described shortly, we solved the multiple-scattering problem for a plane parallel atmosphere in an accurate fashion by using a computer program based on the doubling principle [*Hansen*, 1969]. Aside from the single-scattering properties the only other input information needed to carry out this calculation was a specification of the optical depth, which was obtained from the results of the previous section. In performing this multiple-scattering computation we assumed that the single-scattering properties of the particles were invariant with altitude. Such an assumption may have limited validity during the morning, when ice particles may be located preferentially closer to the ground than suspended soil particles. Note, however, that the skylight at the bottom of the atmosphere does not depend on the altitude profile of the aerosols' number density.

To determine the single-scattering characteristics of the particles, we used a recently developed algorithm that allows for the possible nonspherical shape of the particles [J. B. Pollack and J. N. Cuzzi, manuscript in preparation, 1977; *Cuzzi and Pollack*, 1977]. This procedure was based upon an analysis of laboratory measurements of the scattering characteristics of irregularly shaped, randomly oriented particles. We briefly review the main features of this semiempirical model. When the size parameter α is less than some upper bound *ALFO*, irregularly shaped, randomly oriented particles were found to have approximately the same phase function for scattering as do spheres of equal volume and refractive index. Parameter α is the ratio of a particle's circumference, $2\pi r$, to the wavelength λ. Thus Mie scattering theory can be used for particles in this size regime. Typically, *ALFO* ranges from values of 3 to 8, larger values characterizing smoother particles.

The scattering characteristics of larger-sized particles, i.e., $\alpha > ALFO$, is divided into three components: a diffracted component, an externally reflected component, and an internally refracted and transmitted component. For randomly oriented particles the first two of these were equivalent to their counterparts for spheres. Physical optics theory was used to calculate the diffracted component, and Fresnel's reflection laws were employed to find the externally reflected part. Laboratory measurements suggested that the third component, $R(\theta)$, has the following simple behavior: $\ln (R) \sim \theta$, where θ is the scattering angle. The slope of this linear behavior was determined by a parameter *FTB* which was equal to $\int_0^{90} R\, d\theta / \int_{90}^{180} R\, d\theta$. Typically, *FTB* is about 2. Finally, the scattering portion of the interaction cross section, but not the absorption part, was enhanced by a factor *SAR* to allow for the larger surface to volume ratio of irregular particles as compared to spheres. Typically, *SAR* was about 1.3.

In our calculations we used nominal values of 6, 2, and 1.3 for *ALFO*, *FTB*, and *SAR*. To study the behavior of spherical particles, we simply set $ALFO = \infty$. For the real part of the index of refraction we used either 1.5, a value representative of soil material, or 1.31, the value of water ice at visible wavelengths. Usually, the first of these was used. The imaginary index was found by matching the absolute value of the sky brightness given by the calculations with the observed value.

Figure 7 shows a representative example of the fit of the semiempirical theory (solid line) to laboratory measurements (solid circles with error bars) for an ensemble of randomly oriented cubes having α values between 1.9 and 17.8. Also shown for comparison is the phase function found from Mie theory for an ensemble of equal volume spheres. The vertical coordinate is the phase function $p(\theta)$.

Using the above semiempirical theory for scattering by irregularly shaped particles and the doubling method, we obtained the scattering and transmission behavior of the atmosphere by itself. We then used the adding method to combine these results with the scattering characteristics of the surface to find the total sky brightness as viewed from the surface and the top of the atmosphere as well as the total amount of reflected ground light [*Lacis and Hansen*, 1974; *Pollack and Toon*, 1974].

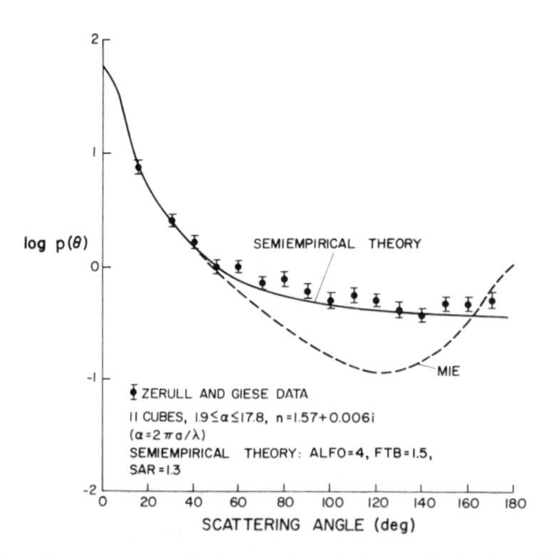

Fig. 7. Comparison of observed and theoretical phase functions for scattering p by a distribution of large and small cubes. The observed values are shown by filled circles with error bars, while the dashed and solid lines show the prediction of Mie theory and the semiempirical theory used in this paper, respectively. Data are from *Cuzzi and Pollack* [1977].

This procedure automatically yielded the fraction of the observed skylight due to light first reflected from the ground as well as the fraction of the observed ground light due to diffuse skylight.

We adapted the Hapke-Irvine photometric function to describe the scattering properties of the surface, in part because this law gave a good fit to the Mariner 9 observations of Phobos [*Noland and Veverka*, 1977]. This function is given by

$$I = [(2B_0 \cos i)/(\cos i + \cos e)]f(\alpha) \quad (2)$$

where I is the brightness; B_0 is the geometric albedo; and i, e, and α are the angle of incidence, angle of view, and phase angle, respectively. Furthermore, $f(\alpha = 0) = 1$, and $\log (f)$ can be approximated as having a linear dependence on α. With f in magnitude units the linear coefficient β has a value of about 0.02 magnitudes/deg for Phobos. We used this value as our initial value but found an improved estimate of 0.01 from the surface pictures.

By comparing calculated brightnesses with the observed brightnesses of the sky and ground we proceeded in an iterative fashion to determine the value of a number of the poorly known model parameters. These parameters include the imaginary index of refraction n_i, the cross-section weighted mean particle radius *RM*, the irregular particle scattering parameters *ALFO* and *FTB*, the surface geometric albedo B_0, and the surface phase angle coefficient β. To do this, we began with nominal choices of these parameters, adjusted n_i to match the absolute value of the sky brightness at an angular distance of about 50° from the sun, found *RM* by matching the variation of sky brightness close to the sun, determined *ALFO* and *FTB* from data on the angular variation of sky brightness at large distances from the sun, estimated β from the variation of ground brightness with angular distance from the antisolar point, and finally derived B_0 from the absolute value of the ground brightness. This cycle was then repeated until convergence was obtained. In all our calculations we used the particle size distribution of *Hansen and Hovenier* [1974], since one of the parameters of this distribution equals *RM*. The other parameter, b, which provides a measure of the width of the distribution, was set equal to 0.15, a value appropriate for a wide distribution. Sensitivity studies performed with single-scattering calculations indicated that while the sky brightness close to the sun depended sensitively on *RM*, it was very insensitive to the choice of b; i.e., in practice, only *RM* can be derived from the observations.

Here, as in subsequent sections, the raw digital numbers that make up a given picture were converted to actual brightness values by using ground-based calibration files [*Huck et al.*, 1975] together with a correction for the wavelength selective degradation of the diodes caused by the radioactive power source on the landers [*Patterson*, 1977]. In all cases, brightness was expressed as the ratio of the observed brightness to that of a perfectly reflecting Lambert surface, normally illuminated by sunlight at Mars's distance from the sun.

Results

We now present comparisons between observed and computed sky brightness values that illustrate the manner in which we estimated the cross-section weighted average particle radius *RM* and the shape factors *ALFO* and *FTB*. As we explained above, the various model parameters were derived in an iterative fashion. The comparisons given in this subsection show the results from the last iteration.

Figure 8 presents the sky brightness observed close to the

sun with the blue diode on sol 20 at VL-1 (Viking 1 landing site). Also shown in this figure are theoretical curves for several choices of *RM*, which have been normalized to agree with the observed value at an azimuth of 20°. In all cases, brightness is plotted as a function of azimuthal distance from the sun's position for a constant elevation angle of view of 32° in local horizon coordinates. In this coordinate system, elevation angles are measured from a horizon aligned with the local geoid. At the time of the picture the sun had an elevation angle of 25.5° in this coordinate system.

No comparisons were made in Figure 8 at azimuthal angles of less than 10° both because the picture is saturated at these angles and because of the possibility of contamination due to sunlight scattered within the camera. On the basis of the camera design and field tests, no such contamination should be present at azimuthal angles of greater than 10°. The field tests involved taking color pictures with a lander camera on a clear day in Boulder, Colorado. The sky color was a vivid blue at azimuthal angles of greater than 10° from the sun but was white at smaller azimuthal angles, with an abrupt boundary between these two color domains.

The comparison between theory and observation in Figure 8 implies that *RM* equals approximately 0.4 μm. This result is consistent with the size distribution found by *Toon et al.* [1977] from an analysis of data taken during the decay of the great dust storm of 1971–1972, when L_s was about 180° different from its value for the data shown in Figure 8. We can understand the behavior of the theoretical curves of Figure 8, hence our ability to infer *RM* as follows: For an *RM* value of 0.1, most of the particles are sufficiently small in comparison to the wavelength of observation (~0.5 μm), so their single-scattering phase function is close to the Rayleigh scattering one. As a result the computed sky brightness has only a slight angular dependence and is therefore incompatible with the observed angular variation. A contrasting case is provided by the curve

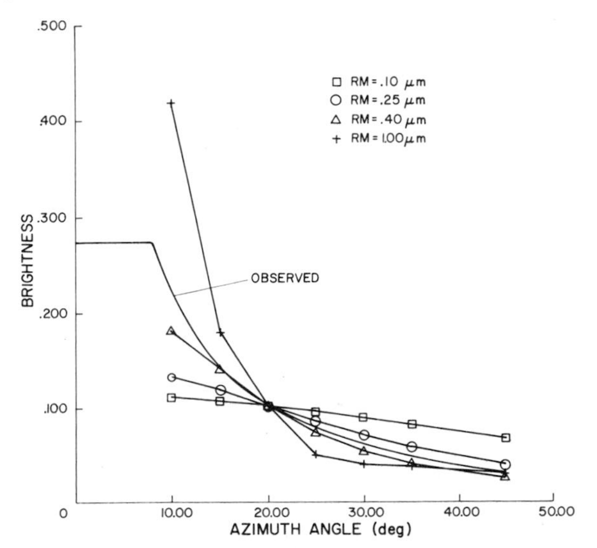

Fig. 8. Comparison between the observed angular variation of sky brightness and that of theoretical models having differing values of *RM* (in microns), the cross-section weighted average particle radius. The angular coordinate is azimuthal distance from the sun. The observed curve was derived from an image taken at VL-1 with the blue diode on sol 20 at 1708 hours (camera event 12A121). At that time the sun's elevation angle was approximately 25.5° in local horizon coordinates. The line scan shown in this figure refers to an elevation angle of view that is 6.5° higher than that of the sun. For the exact location of the picture, refer to *Mutch* [1977, Figure 1]. All theoretical curves have been normalized to agree with the observed value at an azimuth of 20°.

for *RM* = 1 μm. For the angles of interest its single-scattering phase function is dominated by its diffraction peak, which causes its computed sky brightness to increase too rapidly at azimuths less than 25° and to be too flat at larger angles. By decreasing *RM* by a factor of 2, the diffraction peak has its height diminished and its width increased, and this leads to the good agreement between the *RM* = 0.4 μm curve and the observed one.

In Figures 9*a* and 9*b* we examine the sky brightness at larger azimuthal distances from the sun than were considered in Figure 8 to obtain information on the shape of the particles. The observed curve was derived from the panorama picture taken with the survey diode on sol 0 at VL-1. The observed and computed curves refer to a constant elevation angle of 15° at a time when the solar elevation angle was 36.8°. The free parameters for the theory are now *ALFO* and *FTB*. As we discussed above, *ALFO* is the value of the ratio of particle circumference to wavelength separating the Mie regime of scattering from the irregular particle regime, while *FTB* is the ratio of the transmitted component of the phase function integrated over the forward hemisphere of scatter to its value for the backward hemisphere. In Figures 9*a* and 9*b* the theoretical curves have been normalized to agree with the observed value at an azimuth of 95°.

According to Figure 9*a*, values of *ALFO* of 6 or more lead to predicted angular variations that are grossly inconsistent with the observed variation, while smaller values lead to results in much better agreement with the observations. Altering the choice of *FTB* for the large *ALFO* cases causes little change in the computed brightness variations. Our ability to discriminate among various choices of *ALFO* can be attributed to the inability of Mie scattering particles to match the observations. For the wavelength of interest and the *RM* value determined earlier, most of the particles lie within the Mie regime for *ALFO* of 6 or larger. Thus we may conclude that the particles do not have a spherical shape. Furthermore, according to Figure 9*a*, *ALFO* equals approximately 3. The studies of J. B. Pollack and J. N. Cuzzi (manuscript in preparation, 1977) indicate that such a value of *ALFO* is characteristic of rough particles having sharply angular corners.

In Figure 9*b* we show a comparison between the observed sky brightness curve and the theoretical curves for several plausible choices of *FTB* and a fixed value of 3 for *ALFO*. This comparison implies that *FTB* values of 5 or more can be excluded. The best fit to the data is given by the theoretical curve having *FTB* = 2, although the fit for the *FTB* = 1 curve is only slightly worse. Again, referring to the work of J. B. Pollack and J. N. Cuzzi (manuscript in preparation, 1977), we interpret the derived value of *FTB* = 2 as indicating that the particles are approximately equidimensional. The physical reason for such a conclusion is the following: For equidimensional nonspherical objects, some of the light transmitted through the particle undergoes total internal reflection. Such rays enhance the amount of light scattered into the backward hemisphere. Much less light undergoes total internal reflection in the case of particles having one dimension much smaller than the others. Thus equidimensional objects have smaller values of *FTB* than do nonequidimensional objects.

PARTICLE COMPOSITION

Procedure

We used the same multiple-scattering program described in the previous section to obtain information on particle compo-

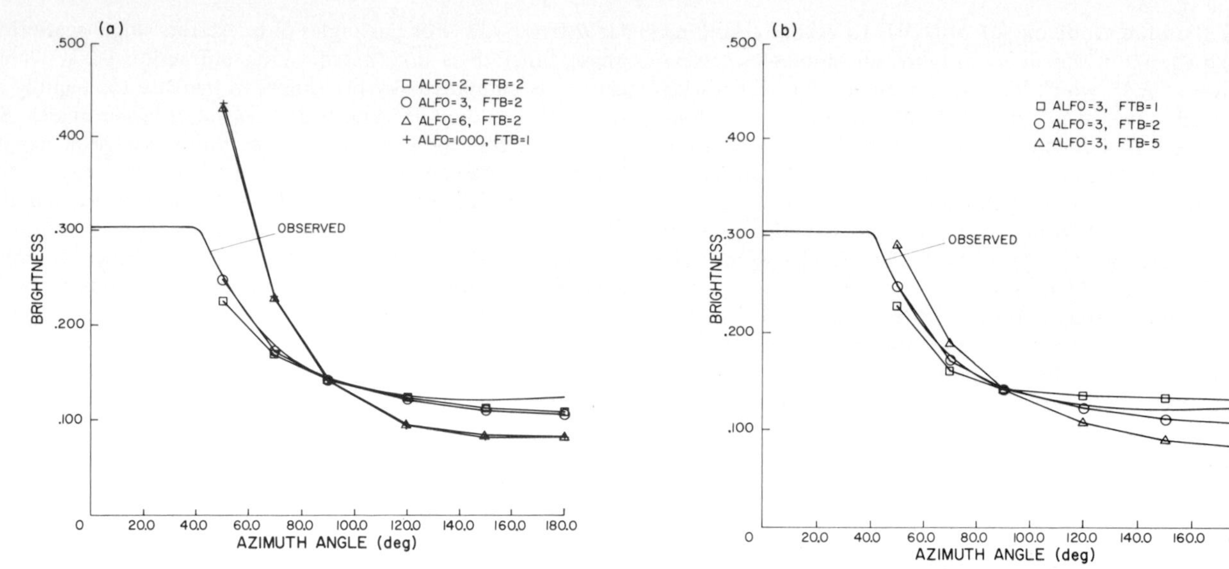

Fig. 9. (*a*) Comparison between the observed angular variation of sky brightness and that predicted by models having varying values of the parameter *ALFO*. The observed curve was obtained from the panorama picture taken at VL-1 on sol 0 at 1619 hours with the survey diode (camera event 12A002). At that time the sun's elevation angle was 36.8° in local horizon coordinates. The line scan shown in this figure refers to an elevation angle of view equal to 15°. Azimuth angles are measured from the azimuthal position of the sun at the time of the picture. For the exact location of the picture, refer to *Mutch* [1977, Figure 1]. All theoretical curves were computed with $RM = 0.4$ μm and $FTB = 2$. They have been normalized to agree with the observed value at an azimuth of 95°. (*b*) Comparison between the same observed curve shown in Figure 9*a* and the sky brightness values computed for aerosol models having varying values of *FTB*. All models have $RM = 0.4$ μm and $ALFO = 2$. The theoretical curves have been normalized to agree with the observed value at an azimuth of 95°.

sition from an analysis of the wavelength dependence of the sky brightness. For this purpose we used pictures taken close together in time with some or all of the six narrow band diodes. These six diodes are the blue, green, red, and three near-infrared channels (IR1, IR2, and IR3), which span the spectral region from about 0.4–1.1 μm [*Huck et al.*, 1977]. In performing this analysis we used values derived in the last section for most of the aerosol and ground properties. The only free parameters now are the values of the imaginary index of refraction of the aerosols n_i and the geometric albedo of the surface B_0. These are the two parameters that should vary the most with wavelength. They are determined in an iterative manner along the lines of the procedure used in the previous section. Note that the analysis carried out in the last section yielded n_i and B_0 only for the blue and survey diodes.

The analysis discussed above leads to estimates of n_i for each of the six narrow band channels. These values are inserted into a computer program analogous to the one described by *Park and Huck* [1976] to obtain a coarse resolution plot of n_i as a function of wavelength. In effect, this program removes the overlap in the wavelength coverage of the different channels. To obtain the desired information about the aerosols' composition, we compare the absolute value of n_i and its wavelength dependence with values that characterize plausible candidate materials.

In addition to the above analyses we also used color pictures to estimate the ratio of the red to blue brightness of the sky. Such data are useful in a relative sense. Diurnal and seasonal changes in sky color provide clues concerning the compositional changes that accompany optical depth variations.

Results

We obtained estimates of n_i and B_0 from an analysis of the absolute value of the sky and ground brightnesses found from images obtained on sol 39 at VL-1. Commencing at approximately 1311 hours, pictures were taken in quick succession with each of the narrow band diodes. At this time the sun's elevation angle was approximately 77.7°. The absolute brightness values of the sky were found by averaging a 20 × 20 array of samples centered at an elevation angle of 12° and an azimuth of 26° from the sun's azimuth, while a similar averaging was used at elevation and azimuthal angles of −7° and 54°, respectively, to find absolute brightness values for the surface.

Figure 10 illustrates the manner in which n_i was determined for the blue diode. The straight horizontal line shows the observed value of the sky brightness in the usual reflectance units. The curved line shows the theoretical dependence of the

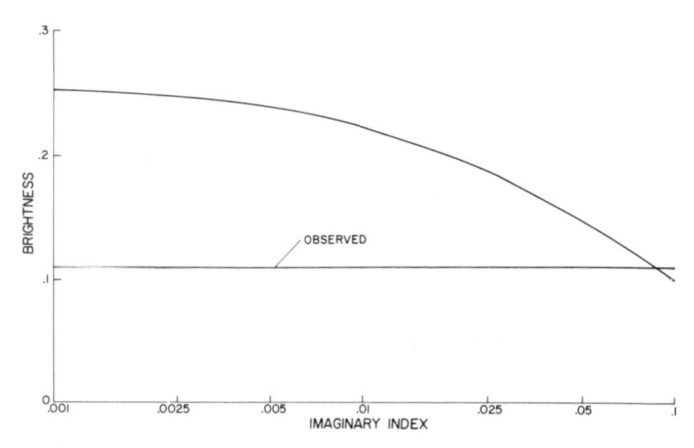

Fig. 10. Illustration of the procedure used to find the imaginary index of refraction n_i. The horizontal line shows the observed value of sky brightness, in reflectance units, found from a picture taken at VL-1 with the blue diode on sol 39 (camera event 12B069). The observed value refers to a position located at an elevation angle of 20° and an azimuthal distance of 26° from the sun. At the time of the picture, 1311 hours, the sun was at an elevation angle of 77.7°. For the exact location of the picture, refer to *Mutch* [1977, Figure 1]. The other curve of this figure shows the theoretical dependence of sky brightness at this position as a function of n_i. The inferred value of n_i is found from the intersection of the theoretical curve with the horizontal line.

sky brightness on n_i for a particular set of model parameters, including the final value of the surface brightness parameter B_0. The intersection of these two lines leads to an estimate of n_i. Note that the observed brightness lies far below the asymptotic limit of the theoretical curve for nonabsorbing particles. Thus it is possible to make a good estimate of n_i. As we mentioned above, the values of n_i and B_0 were determined in an iterative fashion by alternately matching the observed absolute values of the sky and surface brightnesses. Only two iteration cycles were needed.

Table 1 summarizes the values of n_i and B_0 derived for each of the six diodes found in the manner described above. In all cases, n_i is quite large. We also see that n_i first decreases with increasing wavelength from the blue to the red diode but then increases at still larger wavelengths. We note that the long-wavelength portion of the passbands of the IR1, IR2, and IR3 diodes extends to progressively longer wavelengths. This spectral behavior can be seen better in Figure 11, which shows n_i as a function of wavelength. This curve was derived by using the information of Table 1 in the wavelength deconvolution program described earlier in this section.

We now use the n_i results of Table 1 and Figure 11 to obtain information on the aerosols' composition. The absolute values of n_i are many orders of magnitude larger than those that characterize pure water ice at these wavelengths [*Irvine and Pollack*, 1968]. Thus the aerosols are not made solely of water ice. A logical alternative is that they are composed, at least in part, of soil material. The n_i information that we have obtained provides diagnostic data about the composition of the opaque portion of this material. Since the yellowish brown color of the Martian surface has been traditionally attributed to the influence of ferric oxide compounds, it seems only natural to propose that this same material is the determinant of the n_i values of the suspended soil particles. However, this suggestion encounters two serious problems. First, the absolute value of n_i of ferric oxide is too small. *Egan and Becker* [1969] found an n_i value of approximately 5.6×10^{-3} at 0.5 μm for a sample of limonite. As ferric oxide constituted 73% of this sample by weight, even aerosols made entirely of ferric oxide would have an n_i value substantially less than that of the Martian aerosols. In addition, n_i for this same sample exhibited a steady decline in value from 0.4 to 1.1 μm. This spectral behavior is inconsistent with the observed profile of Figure 11.

Both the above problems are eliminated when we consider the imaginary indices of magnetite. At 0.5 μm, n_i for magnetite equals approximately 0.65 [*Huffman and Stapp*, 1973]. Thus as seems reasonable, magnetite could be a minor constituent of the aerosols and still account for all their absorption. Furthermore, the shape of the n_i curve for magnetite is in remarkable agreement with the spectral curve shown in Figure 11. Magnetite's n_i has a local minimum at about 0.8 μm, with increasing values on either side of the minimum over the wavelength region of our data. Also n_i for magnetite increases more

TABLE 1. Estimates of the Imaginary Index of Refraction of the Atmospheric Aerosols n_i and the Geometric Albedo of the Surface B_0

Diode	n_i	B_0
Blue	0.086	0.061
Green	0.074	0.092
Red	0.041	0.213
IR1	0.035	0.239
IR2	0.054	0.201
IR3	0.063	0.190

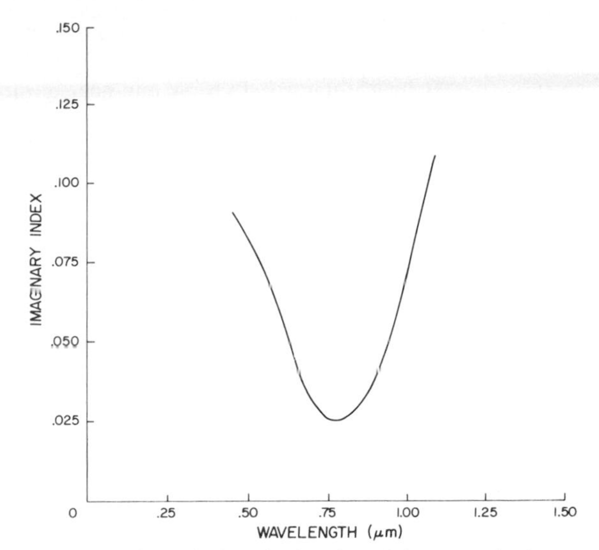

Fig. 11. Imaginary index of refraction of the atmospheric aerosols as a function of wavelength. This curve was generated with a wavelength deconvolution program described in the text. The input information for this program consisted of the n_i values found for each of the six narrow band filters. These results were derived in turn from pictures taken at VL-1 on sol 39 at 1311 hours.

steeply on the long-wavelength side of this minimum than it does on the short-wavelength side. We conclude that magnetite is the opaque material in the suspended soil material, which is responsible for the absorption of visible and near-infrared light by these particles.

From the ratio of the n_i values of the Martian aerosols to those of magnetite we can obtain an estimate of the volume mixing ratio of magnetite in these particles. We find this mixing ratio to be 10% ± 5%, where the error bars reflect the uncertainty in the absolute values of our derived imaginary indices. Magnetite's mass mixing ratio is probably somewhat higher than the above volume mixing ratio.

It is interesting to compare our compositional results with those obtained from studies of the great dust storm of 1971–1972. As we mentioned in the introduction, analysis of middle infrared spectral features centered near 10- and 20-μm wavelengths indicated that the dust was composed in large part of certain classes of silicate minerals. Our identification of magnetite in the suspended soil material above the lander sites is not inconsistent with these findings, since magnetite constitutes only a minor fraction of this material and since the silicate minerals of interest are much more transparent at visible wavelengths than is magnetite.

By modeling Mariner 9 ultraviolet observations of Mars at the time of the great dust storm of 1971–1972, *Pang and Ajello* [1977] and *Pang et al.* [1976] deduced the imaginary index and several other properties of the dust particles. Unfortunately, the results of our paper raise questions about the validity of several simplifying assumptions used by these authors: no allowance was made for the contribution from a backscattering surface, and the single-scattering phase function appropriate for spheres was used. Consequently, it is unclear how much in error their derived parameter values are. Nevertheless, it seems likely that the shape of their derived n_i curve in the 0.18- to 0.35-μm spectral region is at least qualitatively valid, since it depends chiefly on the spectral dependence of the observed absolute brightness. From a comparison of their n_i curve with plausible candidate materials, Pang and Ajello deduced that TiO_2 was the opaque mineral phase in the ultraviolet. Since

TiO_2 is extremely transparent between 0.45 and 1.1 μm, n_i being $\lesssim 10^{-5}$ [*Gray*, 1963], it cannot be responsible for the high values of n_i that we have obtained at visible and near-infrared wavelengths. Furthermore, n_i for magnetite has a maximum in the ultraviolet at a spectral position very close to that exhibited by the n_i curve of Pang and Ajello. Conceivably, magnetite and not TiO_2 was the dominant opaque mineral in the ultraviolet at the time of the great dust storm of 1971–1972.

We next consider the relationship between our compositional results for the atmospheric soil particles and those obtained for the surface from an analysis of analogous Viking lander data [*Huck et al.*, 1977]. For each situation the material consists of a mixture of minerals. The ability of a given mineral to impress an identifiable spectral signature on a set of observations depends strongly on both the wavelength range covered and the state of the material. In the case of observations of atmospheric aerosols at visible wavelengths, only the most opaque mineral will absorb a noticeable amount of light because of the very small mean size of the particles. In fact, magnetite, which ordinarily is totally black, i.e., completely absorbing, in the visible portion of the spectrum, is semitransparent in the atmospheric aerosols and so is able to impress its spectral signature on the color of the sky. As the particle size of the surface material is much larger than that of the suspended material, a mineral like magnetite would be totally absorbing in the surface material and hence have little influence on the color of the surface. Less opaque but sufficiently absorbing minerals, such as ferric compounds, would be expected to control the spectral characteristics of the surface. Thus there is no contradiction between inferring the presence of magnetite from multicolor observations of the sky and inferring the presence of some other iron compound from analogous studies of the surface. Indeed, both materials may be present in the two situations. This proposition is consistent with the results of Table 1, which indicate that the brightness of the sky decreases by a much larger amount from the IR1 to the IR3 channel than does the brightness of the surface.

Huguenin [1974] has suggested that iron-containing compounds on the surface of Mars experience a photostimulated oxidation that alters Fe^{2+} ions to Fe^{3+}. The end result of this process is a red-colored surface. It might appear that the presence of sizable amounts of magnetite, particularly in the tiny atmospheric aerosols, would be incompatible with such a surface oxidation process. However, Huguenin's process operates at a much slower rate for magnetite than it does for other ferrous compounds, so there is no necessary conflict between these two findings [*Huguenin*, 1974; R. L. Huguenin, private communication, 1977].

The above compositional results can be used to interpret some of the results obtained by other lander experiments. *Hargraves et al.* [1976] found a substantial amount of magnetic material at both landing sites. We suggest that magnetite was an important component of this magnetic material. *Clark et al.* [1976] found that iron constituted approximately 13% by weight of the fine material at both landing sites. If magnetite were as abundant in these samples as it is in the soil aerosols, much of the total iron of the Martian regolith would be in the form of magnetite.

We conclude this section by studying the temporal variations of the ratio of the red and blue sky brightnesses. The individual brightness values again are in reflectance units, so a ratio R/B of 1 would typify a sky that contained only water ice particles, while larger values would indicate the presence of soil material (see Table 1). The R and B values used to find R/B were found by averaging a number of sample values centered at an azimuth of 40° from the sun and an elevation angle of 20°. Figure 12 illustrates the diurnal variation of R/B obtained from color photographs taken on sol 132 at VL-2. The progressively increasing values of R/B during the morning can be attributed to the dissipation of the diurnal ground fog. That R/B is lower in the early morning than it is at later times implies that the material responsible for the larger optical depths at the earlier times is more neutral in color than the material present throughout the day. This deduction is consistent with the hypothesis that the diurnally varying aerosols are made of water ice.

Comparing R/B values in the afternoon, we find that little change in these ratios has occurred throughout the mission to date. For example, at VL-2, R/B had values of 1.81, 1.86, and 1.97 on sols 14, 132, and 173. Thus soil particles have been a major component of the atmospheric aerosols from the start of the nominal mission through the extended mission to date and have played a significant role in the seasonal increase in optical depth observed over both landing sites.

VERTICAL DISTRIBUTION

Procedure

We have monitored the sky brightness during the hour period preceding sunrise and that following sunset to obtain information on the vertical distribution of scatterers in the Martian atmosphere. In the discussion below we will refer to the sequence of events as they occur during evening twilight in order to simplify matters. Totally analogous statements will also apply to the morning twilight.

Figure 13*a* illustrates the geometry of the evening twilight. The sun has set at the lander site, point D. Point S is the closest place to the lander at which sunset is occurring. The camera is viewing the sky at an elevation angle e, measured from the local geoid, and at an azimuth assumed to be coincident with the solar azimuth for the purposes of this illustration. For this line of sight, the shadow edge occurs at point C, where its altitude above the surface equals z_1. Points along the line of sight closer to point D than point C are in darkness, while points further away than point C are being directly illuminated by the sun and thus make the dominant contribution to the observed sky brightness. Note that z_1 is the lowest altitude of any point along the line of sight DC that is being directly illuminated by the sun. The symbols R, O, θ, and δ in Figure

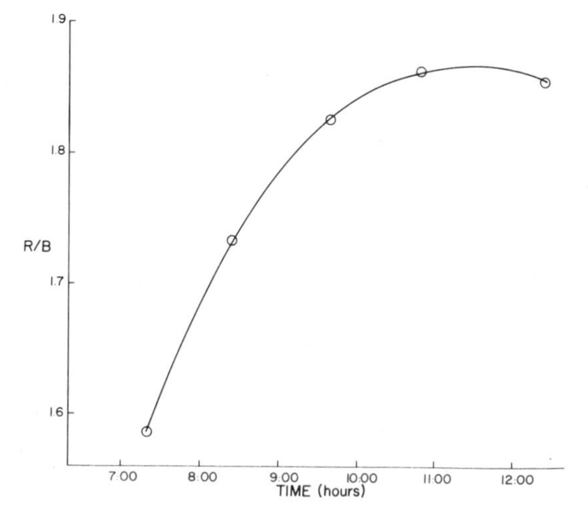

Fig. 12. Ratio of the red to blue sky brightnesses R/B, in reflectance units, as a function of time of day at VL-2 on sol 132.

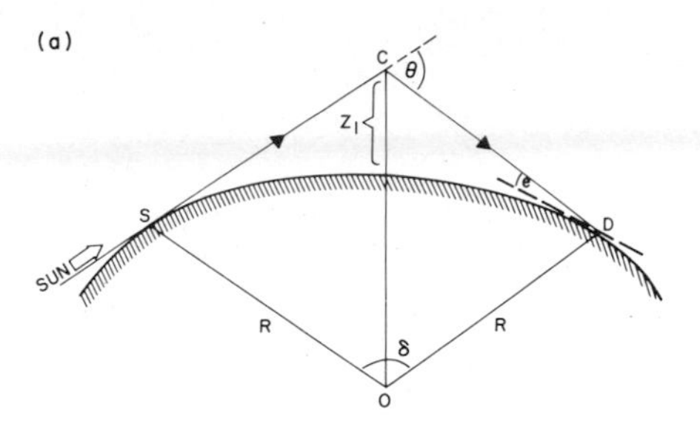

Fig. 13*a*. Geometry of the evening twilight. Symbols are explained in the text.

13*a* refer to the radius of Mars, the center of Mars, the scattering angle at point C, and the solar depression angle at the lander site, respectively.

At progressively longer times after sunset the shadow edge altitude z_1 progressively increases. This situation is illustrated in Figure 13*b* for sol 41 at VL-1. An evening twilight sequence was obtained at this time. The variation of z_1 with time (or alternatively with δ) is shown for two choices of camera elevation angle that are achievable with the lander camera. According to this figure, information on the vertical distribution of scatterers in the lowest 50–70 km of the atmosphere can be obtained by monitoring the sky brightness during the hour period after sunset.

To obtain the desired information, we operated the camera in the slow rescan color mode. In this case the camera acquires one vertical line of data every 14.4 s, 512 samples of data being obtained between $e = -20$ and $+40$. Because the camera is in the rescan mode, it does not step in azimuth between successive lines of elevation. By operating in the color mode the camera uses the blue diode for the first line of data, the green diode for the next line, and the red diode for the third line, this color cycling being continually repeated. For a typical twilight sequence we took three successive rescans, each lasting about 20 min. This was done partly to reset the gain of the camera and partly to keep the viewing azimuth almost precisely aligned with the solar direction. This strategy allowed us to view the brightest portion of the twilight sky and thus to obtain good signal to noise ratios at the higher altitudes.

We have constructed a computer program to simulate the single-scattering part of the twilight illumination. A sequence of solar rays, parallel to the shadow edge, are integrated through the atmosphere to produce source functions along the line of sight of the camera. A Mie scattering program is used to obtain cross sections and phase functions for the particles. For the small angles of scatter treated here (~20°), particle shape does not significantly affect the scattering behavior, and hence Mie theory provides an adequate approximation. The source functions are integrated along the line of sight to produce a predicted sky brightness at the camera's location. Account is taken of the spherical geometry of the atmosphere in this procedure.

Analysis of higher-order scattering for the twilight geometry indicates that it is not a dominant factor. Multiple scattering may contribute as much as 15–20% of the observed light for small z_1 and small θ. This contribution rapidly diminishes to values of less than 5% for $\theta \gtrsim 18°$ and $z_1 > 10$ km.

The basic input parameters used in the twilight analysis program include the optical constants, particle size distribution function, atmospheric optical depth, vertical distribution of the particle number density $n(z)$, and geometrical parameters, such as e. The real part of the index of refraction was obtained from the same a priori prescriptions used in the sky brightness analysis; the imaginary part of the refractive index and the size distribution were obtained from the results of this paper, as described in a previous section; and the optical depth was obtained from values derived from solar extinction photographs taken close to the time of the twilight rescan pictures.

We used several simple forms for the height dependence of the number density, such as exponential, Gaussian, and step function. For example, the exponential distribution is given by

$$n(z) = n_0 \exp(-z/H) \tag{3}$$

where n_0 is the number density at the ground, z is altitude, and H is the scale height of the aerosols. For a given value of H, n_0 is determined from the specified optical depth and size distribution function. Thus comparisons are made with the observations for several trial values of H to determine the value of this one free parameter. In making such comparisons we plot the sky brightness as a function of z_1 for a fixed value of θ, the scattering angle. We do this so as to minimize the sensitivity of the results to changes in the phase function, which depends strongly on the particle size distribution function. A further description of our treatment of the twilight observations will be given elsewhere (R. Kahn and J. B. Pollack, manuscript in preparation, 1977).

Results

Figures 14*a* and 14*b* show the sky brightness observed for an evening twilight picture sequence on sol 41 at VL-1. The observed values (diamonds) are plotted as a function of shadow height z_1 for a constant scattering angle of 20°. These data were obtained with the red diode. Very similar results were found with the blue and green diodes. The data shown in Figure 14*a* were derived from a picture taken during the first 25 min of twilight, while the data of Figure 14*b* were obtained from a picture commencing shortly after the end of the first picture and also lasting for 25 min.

Also shown in Figures 14*a* and 14*b* are theoretical curves of

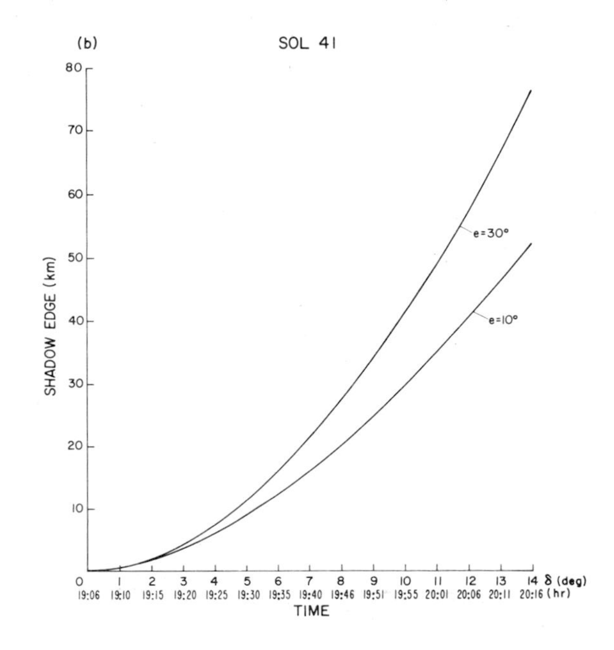

Fig. 13*b*. Shadow edge altitude as a function of local lander time and solar depression angle δ for an evening twilight on sol 41 at VL-1. Sunset occurs at 1906 hours. Two curves are shown corresponding to two choices of camera elevation angle e.

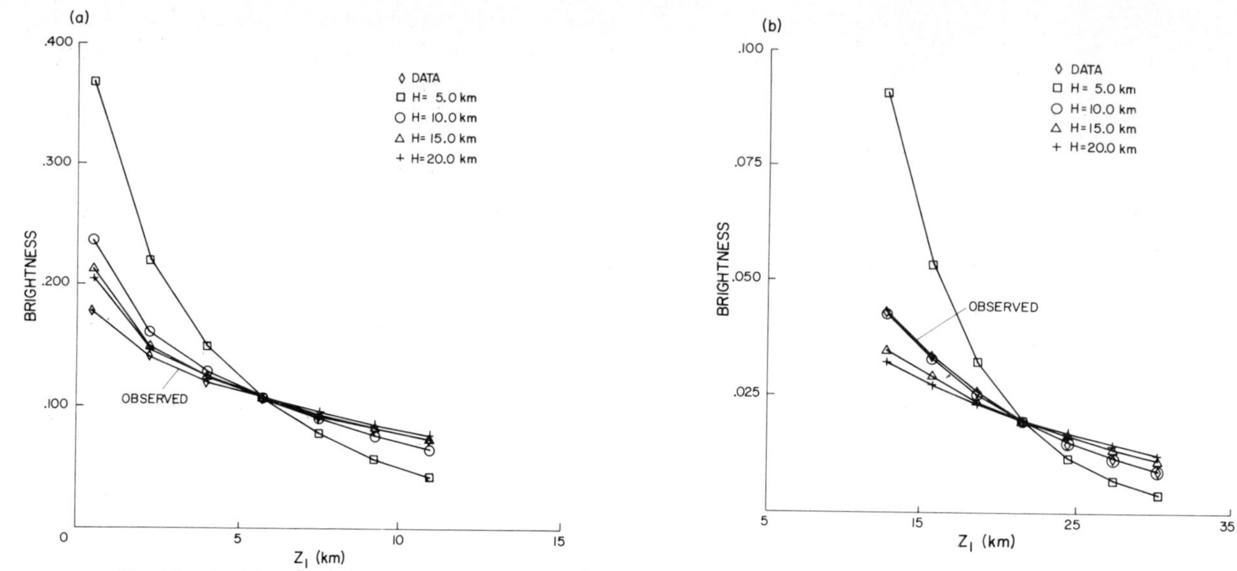

Fig. 14. (*a*) Sky brightness at twilight as a function of shadow edge height z_1 for a fixed angle of scattering θ of 20°. The observations (diamonds) were derived from a twilight rescan picture taken during the first 25 min of twilight on the evening of sol 41 at VL-1 (camera event 12B114). The observed values refer to results obtained with the red diode. The other curves in this figure show the results of theoretical calculations for models having an exponential distribution of aerosol number density, with varying choices of scale height H. Each theoretical curve has been normalized to agree with the observed value at z_1 = 6 km. (*b*) Same as Figure 14*a* except that the observations commence shortly after the end of the picture used in that figure and span the next 25 min of twilight (camera event 12B115). For the exact location of the two twilight pictures, refer to *Mutch* [1977, Figure 1]. Each theoretical curve has been normalized to agree with the observed value at z_1 = 21 km.

models having an exponential distribution function for the aerosol number density, as given by (3). The theoretical curves differ in the choice of the aerosol scale height H. Each of them have been normalized to agree with the observed brightness at a particular altitude: 6 km for Figure 14*a* and 21 km for Figure 14*b*.

Considering first Figure 14*b*, we see that the shape of the observed curve is matched almost exactly by the theoretical curve having H = 10 km. Noticeably poorer fits are given by the theoretical curves having both larger and smaller scale heights. The model with H = 10 km also approximately matches the data shown in Figure 14*a*, which cover a lower altitude region than those in Figure 14*b*. Deviations of this theoretical curve from the observed values at the smallest z_1 region of Figure 14*a* may be due in part to the increasing importance of multiple scattering at these altitudes, as is discussed above. While the data of Figure 14*a* provide a good discriminant for the smaller values of H, they do not do so for the larger values of H. We conclude that the aerosols have a scale height of about 10 km between the surface and an altitude of 30 km.

The results of the above comparison provide indirect information on the composition of the aerosols present near sunset for the summer season. Aerosols are present from the ground to quite high altitudes. They exhibit a smooth vertical distribution with a scale height of about 10 km, which is comparable to the gas scale height; i.e., to first approximation the aerosol mixing ratio is constant. Such a behavior is consistent with that expected for suspended soil particles but differs from a layered structure that might be expected for water ice clouds. We conclude that soil particles constitute almost all of the particles present in the afternoon during the summer season at the VL-1 site.

DISCUSSION

Our analysis of the Viking lander imaging data suggests that three types of aerosols were present over the landing sites: a ground fog, the polar hood, and suspended soil particles. Below we summarize the evidence for each of these components and then discuss some of their possible implications.

Evidence for the presence of a ground fog is given by the following data: Throughout the period of observations (L_s = 100–230) the optical depth is systematically larger in the morning than in the afternoon (see Figures 1 and 2). During the earlier portion of this period the diurnal optical depth increase commences in the late night, at a time when the air temperature close to the ground is near its minimum value. Almost all the optical depth increase occurs prior to sunrise. At a somewhat later seasonal date, dissipation occurs principally in the late morning (see Figures 3 and 4). Such a temporal behavior is crudely consistent with that expected for a fog on theoretical grounds [*Flaser and Goody*, 1976; *Hess*, 1976]. Furthermore, the diurnal variation in the red to blue brightness of the sky shows color changes consistent with this hypothesis (see Figure 12).

Accepting the above identification, let us try to characterize the ground fog further. According to Figure 3 the fog begins to form at approximately 0200 hours local lander time at VL-2, when L_s = 130. Measurements made from the Viking meteorology experiment indicate that the air temperature at 1.8 m above the surface equaled about 195°K at this time [*Hess et al.*, 1976*b*, *c*]. Since the fog is expected to begin forming close to the ground [*Flaser and Goody*, 1976; *Hess*, 1976], this value should be quite close to the frost point temperature. Furthermore, it seems reasonable to assume that condensation occurs when the relative humidity is not much above 100%, because of the presence of small dust particles that can serve as nucleation centers. Using the vapor pressure curve of water vapor above ice, the surface pressure at the VL-2 site [*Hess et al.*, 1976*c*], and the perfect gas law, we find that the partial pressure of water vapor P_{H_2O}, the density of water vapor ρ_{H_2O}, and the water vapor mixing ratio α_{H_2O} equaled approximately 7×10^{-4} mbar, 8×10^{-10} g/cm^3, and 10^{-4}, respectively, near the ground just prior to condensation.

We next obtain a crude estimate of the amount of water vapor in a vertical column above VL-2. According to the temperature profile determined during the descent of the second lander the frost point is not reached above the boundary layer until an altitude of about 30 km for a constant water vapor mixing ratio of 10^{-4} [*Seiff and Kirk*, 1977]. If we therefore assume the water vapor to be uniformly mixed at lower altitudes and employ the above estimate of ρ_{H_2O}, we find that there are approximately 8 pr μm in a vertical column. This value is in reasonable accord with numbers obtained from the orbiter water vapor experiment [*Farmer et al.*, 1977]. It should be noted that our value pertains to the amount of water vapor present late at night, at a time when some water vapor present during the day may have been absorbed onto the top layers of the soil [*Flaser and Goody*, 1976]. Also the orbiter value may be somewhat aliased owing to scattering by the atmospheric aerosols, whose optical depth is nontrivial.

We can estimate the depth of the ice fog by assuming that condensation takes place only on the suspended soil particles. To do this, we first estimate the number of soil particles n close to the ground. This number is related to the dust particles' optical depth τ_d, mean size $\bar{r}_d$, and scale height H_d by

$$\tau_d = 2\pi \bar{r}_d^2 n H_d \tag{4}$$

where a value of twice the geometric cross section has been used to allow for the diffraction component. According to our analysis, $\tau_d = 0.3$, $\bar{r}_d = 0.4$ μm, and $H_d = 10$ km. Hence $n \simeq 30$ particles/cm^3. If condensation occurs only on dust particles, the radius of the resulting ice-dust particles r_i is related to ρ_{H_2O} and the fractional amount of condensation f by

$$f\rho_{H_2O} = (4\pi/3)(r_i^3 - \bar{r}_d^3)n\rho_{ice} \tag{5}$$

where ρ_{ice} is the density of ice (0.92 g/cm^3). The temperature declines by about 4°K from 0200 to 0430 hours, when sunrise occurs and the minimum temperature is reached. Since the saturation pressure at 291°K is about half its value at 295°K, the frost point temperature, f approximately equals $\frac{1}{2}$. Using this value of f and values given earlier for ρ_{H_2O}, $\bar{r}_d$, n, and ρ_{ice}, we find that r_i is about 1.5 μm. Finally, we can use the observed enhancement in optical depth between morning and afternoon, $\Delta\tau$, to estimate the depth of the ice fog h. These two variables are related by

$$\Delta\tau = 2\pi r_i^2 nh \tag{6}$$

Setting $\Delta\tau = 0.15$ from Figure 3 and employing our estimates of r_i and n, we find that $h \simeq 0.4$ km.

We next consider the polar hood, which was occasionally present at VL-2 during the fall season. Evidence for its presence at VL-2 is given by large erratic fluctuations in optical depth that occurred on a time scale of a day as well as a dramatic darkening of the surface that happened in just a 2-hour time interval on sol 161 (see Figure 5). No such large random variations were observed at VL-1. These observations are consistent with earth-based studies of the seasonal, positional, and temporal characteristics of the polar hood [*Martin and McKinney*, 1974].

The polar hood is thought to be an ice condensation cloud, the ice phase being either CO_2, H_2O, or a mixture of these two materials [*Leovy et al.*, 1972]. Thus the lander data gave evidence of the presence of upper level ice clouds (in contrast to the ground fog) only during the fall season above VL-2. That water ice clouds are not more ubiquitous is at first surprising; theoretical calculations indicate that for almost all seasons and latitudes the temperature of the upper troposphere lies below the frost point of water vapor for an assumed uniform mixing ratio of 10^{-4} [*Pollack et al.*, 1976]. Such a mixing ratio approximately equals that inferred from our observations of the ground fog and also leads to column abundances of water vapor comparable to the average Martian value [*Farmer et al.*, 1977].

In our discussion above of the ice fog conditions during the summer we pointed out that in reality the frost point temperature above the boundary layer was not reached until a surprisingly high altitude of 30 km for a uniform water vapor mixing ratio of 10^{-4}. This conclusion was based on temperature profiles observed during the descent of the Viking landers. It differs from theoretical profiles for dust free atmospheres in being more subadiabatic in the troposphere than the theoretical ones. This difference could be attributed to the effect of the suspended soil particles, which absorb some of the incident sunlight and so stabilize the temperature structure [*Gierasch and Goody*, 1972]. Thus the presence of substantial quantities of suspended soil particles in the warmer regions of Mars may markedly inhibit the formation of water ice clouds.

Finally, we consider the suspended soil particles, which dominate the diurnally constant aerosol component during the summer season and make a significant contribution to this component in the fall. Evidence for the presence of this material is given by the absorption characteristics of the aerosols and their vertical distribution, as inferred from an analysis of images obtained during the summer season in the afternoon. We found that the aerosols' imaginary indices of refraction were quite substantial for all six diodes (see Table 1). Such values are consistent with the imaginary indices that might be expected from soil particles, but ice particles are much more transparent at visible wavelengths. At the times under discussion the aerosols were found to have a smooth vertical structure, whose scale height was comparable to the atmospheric scale height. Aerosols were detected up to altitudes of at least 30 km. Such a behavior is consistent with that of suspended soil particles but is incompatible with that of water ice particles, which should have a more layered structure.

We now discuss possible sources and sinks for the suspended soil particles. As we mentioned in the introduction, Mariner 9 observations of the decay phase of the 1971–1972 global dust storm showed that the optical depth declined from 1.5 to 0.2 during the first 3 months of observations. The seasonal date at the end of this period was about 200 days earlier than the seasonal date at the start of the Viking mission. Furthermore, the optical depth in the afternoon was typically 0.2–0.4 during the summer season at the Viking sites. Therefore we consider it highly unlikely that the soil particles present over the landing sites were remnants from the last global dust storm. It is more likely that they were generated more recently from a number of local dust storms. Evidence for the existence of such storms is given both by direct observations of them on Mariner 9 and Viking orbiter images [*Leovy et al.*, 1972, 1973*b*; G. Briggs, private communication, 1977] and by Mariner 9 photographs of numerous localized surface albedo changes [*Sagan et al.*, 1972].

It has been thought that the soil particles are eliminated by vertical eddy mixing, which caused them eventually to be deposited as a more or less uniform blanket across the surface of Mars. While we do not dispute that some elimination occurs in this manner, we wish to suggest a second mechanism that may also be important. We propose that some soil particles are removed from the atmosphere as a result of their serving as nucleation centers for the formation of carbon dioxide ice particles in the polar region of the winter hemisphere. As we show below, such particles are much larger than the soil particles themselves and are therefore more able to sediment gravitationally out of the atmosphere. Prior to the condensation of the carbon dioxide, water ice probably forms on the soil particles. But the increase in particle size from the water condensation is probably not great enough for the combined particle to fall rapidly out of the atmosphere. Thus the formation of carbon dioxide ice particles on existing atmospheric aerosols may serve to eliminate both water and dust from the atmosphere, both materials generally being present in such condensation centers.

To explore the above hypothesis, we first estimate the increase in the size of the dust particles due to ice formation and then relate these sizes to fallout velocities. Calculations given earlier in this section show that water vapor condensation on the soil particles typically results in growth to a particle radius of about 2 μm. While these calculations referred specifically to the ice particles in the ground fog, the size so found is a good approximation to a number of other situations involving water ice formation: The final radius given by (5) depends on the cube root of the amount of available water vapor; (5) can also be used to estimate the size of carbon dioxide particles that form on water ice–dust particles. We simply replace f and ρ_{H_2O} by their CO_2 counterparts and use earlier estimates of the other quantities appearing in this equation. According to the general circulation calculation of *Pollack et al.* [1976], atmospheric temperatures reach the frost point of CO_2 in the inner portion ($\gtrsim 60°$ latitude) of the seasonal cap during winter. The rate of condensation of CO_2 in this zone is about $\frac{1}{3}$ g/cm^2/d. If we assume that all the condensation occurs in the atmosphere and that the scale height of the condensed CO_2 is comparable to an atmospheric scale height, then the rate of CO_2 condensation is approximately 3×10^{-7} g/cm^3/d. Finally, if we assume that fresh air is brought into this region every 10 days [*Pollack et al.*, 1976], then $f\rho_{CO_2} \cong 3 \times 10^{-6}$ g/cm^3. Using this value in (5), we find that the radius of the resulting CO_2 ice particles equals about 25 μm. Thus the CO_2 ice particles are much larger than either the original soil particles or the combined water ice–soil particles.

We next estimate the fallout velocities of the various particles of interest. According to the calculations of *Arvidson* [1972] a CO_2 ice particle having the radius estimated above would have a terminal velocity of several tens of centimeters per second. Consequently, it would fall quickly to the ground. By way of contrast, the smaller-sized dust particles and water ice–dust particles have terminal velocities that are about 2 orders of magnitude smaller.

The above calculations show that CO_2 condensation can effectively remove dust and water ice particles from the atmosphere, and as a result, dust and water ice may be preferentially deposited in the winter polar regions. This inference is consistent with results from Viking orbiter photographs taken early in the mission, which show the atmosphere in the southern hemisphere to be much more transparent than the atmosphere in the other hemisphere [*Carr et al.*, 1976]. At this time, carbon dioxide was condensing in the southern polar regions. Our hypothesis also provides an explanation for why the seasonal CO_2 ice cap and the permanent H_2O ice cap have albedos significantly less than unity [*Briggs*, 1974; *Kieffer et al.*, 1976]: Both ice deposits contain dust particles intimately mixed with the ice because ice condensation in the atmosphere occurs on suspended dust particles. We also wish to point out that the clearing of global dust storms might be the result, in part, of the meridional transportation of dust particles to the winter polar region, where CO_2 condensation effectively leads to their removal from the atmosphere. This mechanism is consistent with the puzzling inference that the size distribution of the dust particles did not change during the decay phase of the global dust storm of 1971–1972, as was mentioned in the introduction. If CO_2 condensation is the chief agent by which dust particles are removed from the atmosphere, then the time scale for the decay of a global dust storm is determined by the time scale for atmospheric meridional motions. Such a time scale would be independent of the size of individual dust particles. Finally, our mechanism may have relevance for the formation of the polar laminae and debris mantle, as is discussed in detail by J. B. Pollack (manuscript in preparation, 1977).

CONCLUSIONS

In this section we summarize our major results. Three types of particles were present above the landing sites during the period from L_s = 100 to 230: a water ice ground fog, a higher-level ice cloud (the polar hood), and suspended soil particles. Below we describe the properties of each of these classes of aerosols as derived from the lander images. We also present some of the implications of these results, which were discussed in the preceding section.

A ground fog composed of water ice particles was present at both landing sites throughout the entire period of the observations. Observations taken near the beginning of the mission at VL-2 indicate that the fog begins to form at about 0200 hours and that most of its growth takes place prior to sunrise. Observations taken towards the end of this period at VL-2 imply that most of the dissipation of the fog takes place in the late morning. During the summer season we estimate that the frost point temperature is about 195°K and that the water vapor mixing ratio close to the ground at night equals about 1×10^{-4} prior to fog formation. Assuming that condensation occurs only on the suspended soil particles, we find that the average radius of the fog particles is about 2 μm and the depth of the fog is about 0.4 km.

A higher-level ice cloud is not prominent during the summer season, but it is occasionally present during the fall season above the VL-2 site. We speculate that the absorption of sunlight by the suspended soil particles warms the middle and upper parts of the troposphere and so inhibits water ice cloud formation in the summer. The opacity of the ice cloud above VL-2 shows marked temporal variations associated perhaps with weather disturbances.

Suspended soil particles were present throughout the mission. They constituted the dominant source of the nondiurnally varying component of the atmospheric opacity. These particles have a cross-section weighted average radius of 0.4 μm, are nonspherical but equidimensional in shape, and have rough surfaces. They are distributed in a smooth manner from the surface to altitudes in excess of 30 km. This distribution can be characterized by an exponential dependence on altitude, with a scale height of 10 km. This value is quite comparable to the gas scale height. The principal opaque mineral in the soil particles is magnetite, whose abundance by volume is 10% ± 5%. In addition to being eliminated from the atmosphere by eddy mixing processes in all regions of the planet, the soil particles may also be eliminated, in association with water ice, through their acting as condensation nuclei for CO_2 ice in the winter polar atmospheric region. This mechanism implies a preferential deposition of dust and water ice in the polar regions and may provide an explanation for the low albedo of the CO_2 and H_2O surface ice deposits as well as for the relative clarity of the winter hemisphere. In addition, it may play a role in the formation of the polar laminae and subpolar debris mantles.

Acknowledgments. We are very grateful to Richard Goody for suggesting to us that the lander cameras could be used to look for a ground fog and for encouraging the participation of one of us (R.K.), to many other members of the Lander Imaging Team and of the Jet Propulsion Laboratory for all their work in obtaining and processing the images used in this paper, and to Kenneth Bilsky for his help in the data reduction and analyses. The work of one of us (R.K.) was supported in part by NASA grant NGL-22-007-228.

REFERENCES

Allen, C. W., *Astrophysical Quantities*, p. 169, Athlone, London, 1955.

Arvidson, R. E., Aeolian processes on Mars: Erosive velocities, settling velocities, and yellow clouds, *Geol. Soc. Amer. Bull.*, *83*, 1503, 1972.

Baird, A. K., P. Toulmin III, B. C. Clark, H. J. Rose, Jr., K. Keil, R. P. Christian, and J. L. Gooding, Mineralogical and petrological implication of Viking geochemical results from Mars: Interim report, *Science*, *194*, 1288, 1976.

Biemann, K., J. Oro, P. Toulmin III, A. O. Nier, D. M. Anderson, P. G. Simmonds, D. Flory, A. V. Diaz, D. R. Rushneck, and J. A. Biller, Search for organic and volatile inorganic compounds in two surface samples from the Chryse Planitia region of Mars, *Science*, *194*, 72, 1976.

Briggs, G., The nature of the residual Martian polar caps, *Icarus*, *23*, 167, 1974.

Carr, M. H., H. Masursky, W. A. Baum, K. R. Blasius, G. A. Briggs, J. A. Cutts, T. Duxbury, R. Greeley, J. E. Guest, B. A. Smith, L. A. Soderblom, J. Veverka, and J. B. Wellman, Preliminary results from the Viking orbiter imaging experiment, *Science*, *193*, 766, 1976.

Clark, B. C., A. K. Baird, H. J. Rose, Jr., P. Toulmin III, K. Keil, A. J. Castro, W. C. Kelliher, C. D. Rowe, and P. H. Evans, Inorganic analyses of Martian surface samples at the Viking landing sites, *Science*, *194*, 1283, 1976.

Curran, R. J., B. J. Conrath, R. Hanel, V. G. Kunde, and J. C. Pearl, Mars: Mariner 9 spectroscopic evidence for H_2O ice clouds, *Science*, *182*, 381, 1973.

Cuzzi, J. N., and J. B. Pollack, Saturn's rings: Particle composition and size distribution as constrained by microwave observations, I, Radar observations, submitted to *Icarus*, 1977.

Egan, W. G., and J. F. Becker, Determination of the complex index of refraction of rocks and minerals, *Appl. Opt.*, *8*, 720, 1969.

Farmer, C. B., D. W. Davies, A. L. Holland, D. D. LaPorte, and P. E. Doms, Mars: Water vapor observations from the Viking orbiters, *J. Geophys. Res.*, *82*, this issue, 1977.

Flasar, F. M., and R. M. Goody, Diurnal behavior of water on Mars, *Planet. Space Sci.*, *24*, 161, 1976.

Gierasch, R., and R. M. Goody, The effect of dust on the temperature of the Martian atmosphere, *J. Atmos. Sci.*, *29*, 400, 1972.

Gray, D. E., *American Institute of Physics Handbook*, pp. 6–74, McGraw-Hill, New York, 1963.

Hansen, J. E., Radiative transfer by doubling very thin layers, *Astrophys. J.*, *155*, 565, 1969.

Hansen, J. E., and J. W. Hovenier, Interpretation of the polarization of Venus, *J. Atmos. Sci.*, *31*, 1137, 1974.

Hansen, J. E., and L. D. Travis, Light scattering in planetary atmospheres, *Space Sci. Rev.*, *16*, 527, 1974.

Hargraves, R. B., D. W. Collinson, R. E. Arvidson, and C. R. Spitzer, Viking magnetic properties investigation: Further results, *Science*, *194*, 1303, 1976.

Hess, S. L., The vertical distribution of water vapor in the atmosphere of Mars, *Icarus*, *28*, 269, 1976.

Hess, S. L., R. M. Henry, C. B. Leovy, J. A. Ryan, J. E. Tillman, T. E. Chamberlain, H. L. Cole, R. G. Dutton, G. C. Greene, W. E. Simon, and J. L. Mitchell, Preliminary meteorological results on Mars from the Viking 1 lander, *Science*, *193*, 788, 1976*a*.

Hess, S. L., R. M. Henry, C. B. Leovy, J. A. Ryan, J. E. Tillman, T. E. Chamberlain, H. L. Cole, R. G. Dutton, G. C. Greene, W. E. Simon, and J. L. Mitchell, Mars climatology from Viking 1 after 20 sols, *Science*, *194*, 78, 1976*b*.

Hess, S. L., R. M. Henry, C. B. Leovy, J. L. Mitchell, J. A. Ryan, and J. E. Tillman, Early meteorological results from the Viking 2 lander, *Science*, *194*, 1352, 1976*c*.

Huck, F. O., S. F. Katzberg, D. J. Jobson, and C. L. Fales, Jr., An analysis of the facsimile camera response to radiant point sources, *NASA Tech. Note*, *D 7389*, 1973.

Huck, F. O., E. E. Burcher, E. J. Taylor, and S. D. Wall, Radiometric performance of the Viking Mars lander cameras, *NASA Tech. Rep.*, *TMX-72692*, 1975.

Huck, F. O., R. E. Arvidson, D. J. Jobson, S. K. Park, W. R. Patterson, and S. D. Wall, Spectrophotometric and color estimates of the Viking lander sites, *J. Geophys. Res.*, *82*, this issue, 1977.

Huffman, D. K., and J. L. Stapp, Optical measurements on solids of possible interstellar importance, in *Interstellar Dust and Related Topics*, edited by M. Greenberg and H. C. van de Hulst, p. 297, D. Reidel, Hingham, Mass., 1973.

Huguenin, R. L., The formation of goethite and hydrated clay minerals on Mars, *J. Geophys. Res.*, *79*, 3895, 1974.

Irvine, W. M., and J. B. Pollack, Infrared optical properties of water and ice spheres, *Icarus*, *8*, 324, 1968.

Kieffer, H. H., S. C. Chase, Jr., T. Z. Martin, E. D. Miner, and F. D. Palluconi, Martian north pole summer temperatures: Dirty water ice, *Science*, *194*, 1341, 1976.

Lacis, A. A., and J. E. Hansen, A parameterization for the absorption of solar radiation in the earth's atmosphere, *J. Atmos. Sci.*, *31*, 118, 1974.

Leovy, C. B., G. A. Briggs, A. T. Young, B. A. Smith, J. B. Pollack, E. N. Shipley, and R. L. Wildey, The Martian atmosphere: Mariner 9 television experiment progress report, *Icarus*, *17*, 373, 1972.

Leovy, C. B., R. W. Zurek, and J. B. Pollack, Mechanisms for Mars dust storms, *J. Atmos. Sci.*, *30*, 749, 1973*a*.

Leovy, C. B., G. A. Briggs, and B. A. Smith, Mars atmosphere during the Mariner 9 extended mission: Television results, *J. Geophys. Res.*, *78*, 4252, 1973*b*.

Martin, L. J., and W. M. McKinney, North polar hood of Mars in 1969 (May 18 to July 25), I, Blue light, *Icarus*, *23*, 380, 1974.

Mutch, T. A., A. B. Binder, F. O. Huck, E. C. Levinthal, S. Liebes, Jr., E. C. Morris, W. R. Patterson, J. B. Pollack, C. Sagan, and G. R. Taylor, The surface of Mars: The view from the Viking 1 lander, *Science*, *193*, 791, 1976*a*.

Mutch, T. A., R. E. Arvidson, A. B. Binder, F. O. Huck, E. C. Levinthal, S. Liebes, Jr., E. C. Morris, D. Nummedal, J. B. Pollack, and C. Sagan, Fine particles on Mars: Observations with the Viking 1 lander cameras, *Science*, *194*, 87, 1976*b*.

Mutch, T. A., S. U. Grenander, K. L. Jones, W. Patterson, R. E. Arvidson, E. A. Guinness, P. Avrin, C. E. Carlston, A. B. Binder, C. Sagan, E. W. Dunham, P. L. Fox, D. C. Pieri, F. O. Huck, C. W. Rowland, G. R. Taylor, S. D. Wall, R. Kahn, E. C. Levinthal, S. Liebes, Jr., R. B. Tucker, E. C. Morris, J. B. Pollack, R. S. Saunders, and M. R. Wolf, The surface of Mars: The view from the Viking 2 lander, *Science*, *194*, 1277, 1976*c*.

Mutch, T. A., The Viking lander imaging investigation: An introduction, *J. Geophys. Res.*, *82*, this issue, 1977.

Noland, M., and J. Veverka, The photometric functions of Phobos and Deimos, III, Surface photometry of Phobos, *Icarus*, *30*, 212, 1977.

Pang, K., and J. M. Ajello, Complex refractive index of Martian dust: Wavelength dependence and composition, *Icarus*, *30*, 63, 1977.

Pang, K., and C. W. Hord, Mariner 9 ultraviolet spectrometer experiment: 1971 Mars dust storm, *Icarus*, *18*, 481, 1973.

Pang, K., J. M. Ajello, C. W. Hord, and W. G. Egan, Complex refractive index of Martian dust: Mariner 9 ultraviolet observations, *Icarus*, *27*, 55, 1976.

Park, S. K., and F. O. Huck, Spectral reflectance estimation technique using multispectral data from the Viking lander camera, *NASA Tech. Note*, *D 8292*, 1976.

Patterson, W. R., F. O. Huck, S. D. Wall, and M. R. Wolfe, Calibration and performance of the Viking lander cameras, *J. Geophys. Res.*, *82*, this issue, 1977.

Pollack, J. B., and O. B. Toon, A study of the effect of stratospheric aerosols produced by SST emissions on the albedo and climate of the earth, in *Proceedings of the Third Conference on the Climatic Impact Assessment Program*, p. 457, U.S. Department of Transportation, Washington, D. C., 1974.

Pollack, J. B., C. B. Leovy, Y. Mintz, and W. Van Camp, Winds on Mars during the Viking season: Predictions based on a general circulation model with topography, *Geophys. Res. Lett.*, *3*, 479, 1976.

Sagan, C., J. Veverka, P. Fox, R. Dubisch, J. Lederberg, E. Levinthal,

L. Quam, R. Tucker, J. B. Pollack, and B. A. Smith, Variable features on Mars: Preliminary Mariner 9 results, *Icarus*, *17*, 346, 1972.

Seiff, A., and D. B. Kirk, Structure of the atmosphere of Mars in summer at midlatitudes, *J. Geophys. Res.*, *82*, this issue, 1977.

Toon, O. B., J. B. Pollack, and C. Sagan, Physical properties of the particles composing the Martian dust storm of 1971–1972, *Icarus*, *30*, 663, 1977.

Zerull, R., and R. H. Giese, Microwave analog studies, in *Planets, Stars, and Nebulae; Studied With Photopolarimetry*, edited by T. Gehrells, University of Arizona Press, Tucson, 1975.

(Received April 12, 1977;
revised June 16, 1977;
accepted June 16, 1977.)

Surface Materials of the Viking Landing Sites

Henry J. Moore,[1] Robert E. Hutton,[2] Ronald F. Scott,[3] Cary R. Spitzer,[4]
and Richard W. Shorthill[5]

Martian surface materials viewed by the two Viking landers (VL-1 and VL-2) range from fine-grained nearly cohesionless soils to rocks. Footpad 2 of VL-1, which landed at 2.30 m/s, penetrated 16.5 cm into very fine grained dunelike drift material; footpad 3 rests on a rocky soil which it penetrated ≈3.6 cm. Further penetration by footpad 2 may have been arrested by a hard substrate. Penetration by footpad 3 is less than would be expected for a typical lunar regolith. During landing, retroengine exhausts eroded the surface and propelled grains and rocks which produced craters on impact with the surface. Trenches excavated in drift material by the sampler have steep walls with up to 6 cm of relief. Incipient failure of the walls and failures at the end of the trenches are compatible with a cohesion near 10–10^2 N/m^2. Trenching in rocky soil excavated clods and possibly rocks. In two of five samples, commanded sampler extensions were not achieved, a situation indicating that buried rocks or local areas with large cohesions (≥10 kN/m^2) or both are present. Footpad 2 of VL-2, which landed at a velocity between 1.95 and 2.34 m/s, is partly on a rock, and footpad 3 appears to have struck one; penetration and leg strokes are small. Retroengine exhausts produced more erosion than occurred for VL-1 owing to increased thrust levels just before touchdown. Deformations of the soil by sampler extensions range from doming of the surface without visible fracturing to doming accompanied by fracturing and the production of angular clods. Although rocks larger than 3.0 cm are abundant at VL-1 and VL-2, repeated attempts to collect rocks 0.2–1.2 cm across imbedded in soil indicate that rocks in this size range are scarce. There is no evidence that the surface sampler of VL-2, while it was pushing and nudging rocks ≈25 cm across, spalled, chipped, or fractured the rocks. Preliminary analyses of surface sampler motor currents (≈25 N force resolution) during normal sampling are consistent with cohesionless frictional soils ($\phi \approx 36°$) or weakly cohesive frictionless soils ($C < 2$ kN/m^2). The soil of Mars has both cohesion and friction.

Introduction

Viking lander 1 (VL-1) landed on Chryse Planitia on July 20, 1976, and was followed by Viking lander 2 (VL-2), which landed on Utopia Planitia, 6500 km away from VL-1, on September 3, 1976. Both landers successfully completed their primary missions on November 15, 1976, prior to solar conjunction. Extensive activities for VL-1 lasted 41 Martian days (sols) (the duration of 1 sol is about 24.66 hours) and continued at a reduced level for an additional 64 sols. Extensive activities for VL-2 lasted 60 sols. Both landers survived solar conjunction and are currently performing their extended missions [*Soffen*, 1976]. This report is concerned with the Physical Properties Investigation of the Martian surface materials based on lander activities during their primary missions. Preliminary results have been reported previously [*Shorthill et al.*, 1976*a*, *b*, *c*].

The objective of the Physical Properties Investigation is to further man's understanding of the Martian environment by determining the physical properties of the surface materials [*Shorthill et al.*, 1972] within the constraints defined in the Viking '75 Project Mission Definition [*National Aeronautics and Space Administration*, 1970, p. 32].

The primary objective of the mission is the exploration of Mars with an emphasis on the search for life. Three analytical instruments and experiments designed to meet the objective were carried to the surface of Mars: Biology, Molecular Analysis, and Inorganic Chemical Analysis. All of these experiments required the use of the Viking surface sampler to obtain samples on a priority basis. Thus physical properties of the Martian surface materials have been estimated insofar as possible from normal activities of the surface sampler while it was acquiring samples for these experiments and other spacecraft activities. Optimal use of the surface sampler for the Physical Properties Investigation awaits exhaustion of the capacities of the analytical experiments sometime in the extended mission.

[1] U.S. Geological Survey, Menlo Park, California 94025.
[2] Applied Mechanics Laboratory, TRW Systems Group, Redondo Beach, California 90278.
[3] Department of Engineering and Applied Sciences, California Institute of Technology, Pasadena, California 91125.
[4] NASA Langley Research Center, Hampton, Virginia 23665.
[5] Geospace Sciences Laboratory, University of Utah Research Institute, Salt Lake City, Utah 84108.

Paper number 7S0447.

General Description of the Landing Sites

Chryse Planitia

Panoramas taken by VL-1 are reminiscent of both terrestrial aeolian and lunar scenes. Large tracts of cross-bedded drifts of wind-eroded dunelike structures are superposed on rocky terrain [*Mutch et al.*, 1976*a*]. The terrain is similar in appearance to the immediate surroundings of the Surveyor 7 landing site of the moon, and both have similar size-frequency distributions of rocks and blocks [*Mutch et al.*, 1976*b*]. The surface in the immediate vicinity of VL-1 consists of an area underlain by very fine grained material (informally named Sandy Flats) and an area of rocks set in a matrix of finer-grained material (includes Rocky Flats) (Figure 1*a*). Finer-grained materials, which occupy 12–14% of the sample field, are probably the same material as those in the drifts and will be called drift material. Small grains and fragments propelled by engine exhausts during landing have produced chains of elongate rimmed to rimless craters by impact with the drift material. Elongation of the craters and alignment of the chains are parallel to radials from the three retroengines.

Footpad 2 penetrated drift material 16.5 cm, open fissures and both monoclinal and anticlinal flexures of the surface thus being produced near the footpad (Figure 2*b*). Penetration by footpad 2 was so large that it was completely covered by drift material. Drift material increases in thickness from the mapped boundary with the rocky area in the sample field toward footpad 2 (Figure 1*a*).

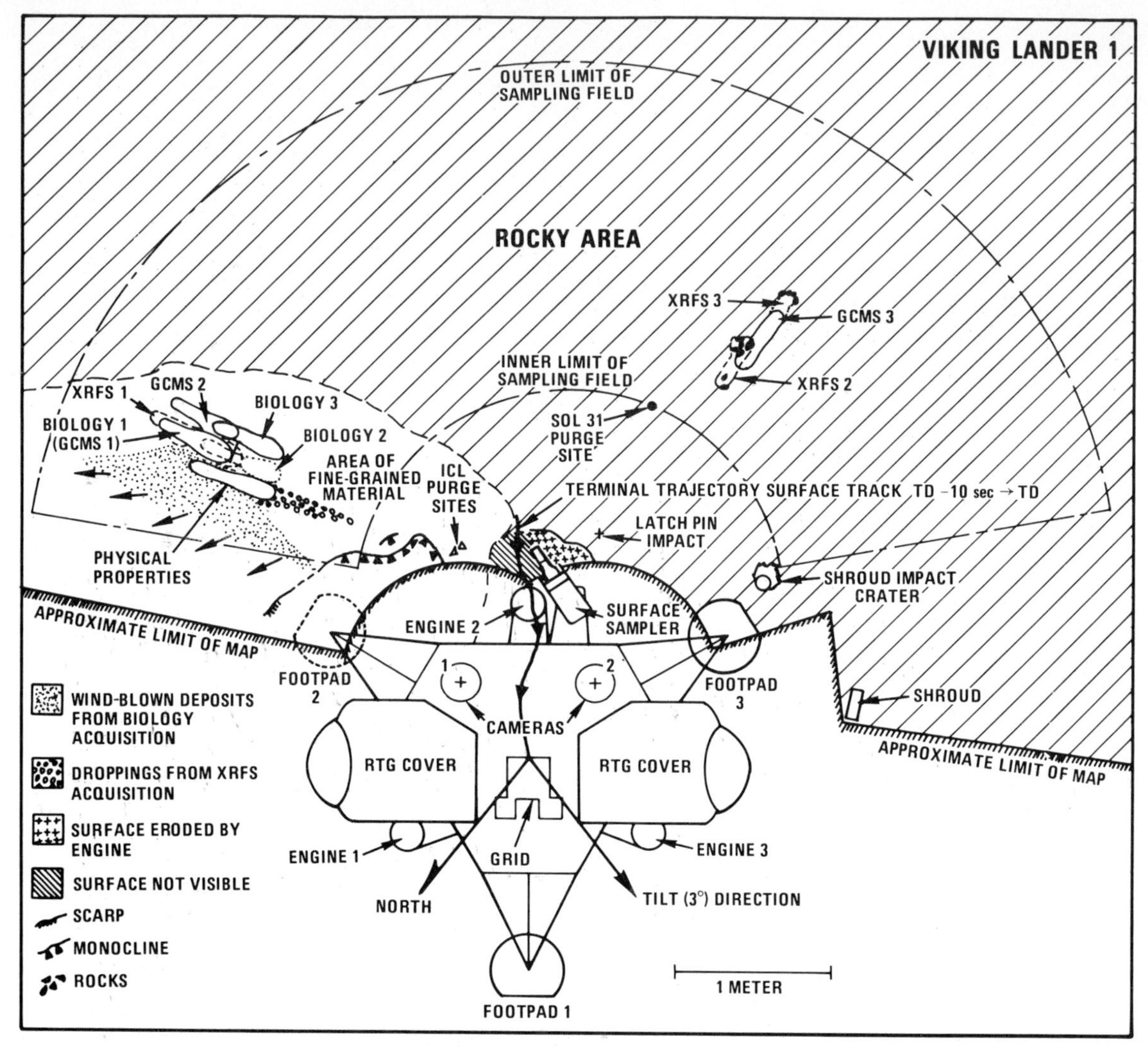

Fig. 1a

Fig. 1. Plan views of (*a*) VL-1 and (*b*) VL-2 showing the landed spacecraft and their orientations, locations of sample sites, selected rocks, and surface disturbances relevant to the Physical Properties Investigation, and the radioisotope thermal generator (RTG); surface sampler activities are summarized in Tables 4 (VL-1) and 5 (VL-2). Plan views are projected in the lander coordinate system.

Rocky materials occupy most of the sample field. Roughly 25% of the rocky area is exposed rocks, both on and partially beneath the surface. These rocks and exposures of rocks vary from a few centimeters to several tens of centimeters across within the sample field. Between the rocks, fine-grained material is commonly present as thick wind tails on the southern sides of rocks and as thinner deposits between the rocks. Wind tails are elongated in southerly directions. Small deflation hollows produced by wind erosion are common on the northern sides of rocks. Locally, loose fine-grained materials have been stripped away, and a residue of clods, small fragments, and tips of larger buried clods and rocks 0.5 cm or so across has thus been left. Footpad 3 rests on these rocky materials and, judging from footpad shadows assumed to be cast on a level surface, has penetrated the surface about 3.6 cm (Figure 2*a*). Near the spacecraft (Figure 1*a*), retroengine exhausts have eroded the surface, dislodged small rocks, produced fillets of debris on the sides of rocks facing retroengine 2, and propelled small grains and fragments that produced small craters upon impact with the surface. Near retroengine 2, fines have been stripped away to expose a horizontal planar surface underlain by more cohesive fractured material (Figure 2*a*).

Orbital pictures of the Chryse region taken by the Viking 1 orbiter and earth-based radar echoes from the region are compatible with the VL-1 panorama [*Tyler et al.*, 1976]. The pictures reveal that the uppermost Martian surface has been modified by the wind and impact cratering. Small craters a few hundred meters in diameter have excavated dark material by the ejection of debris and were subsequently modified by the wind which produced bright wind tails on the southwestern sides of the craters and dark northeast flanks by erosion. Radar reflection coefficients are larger than average lunar ones; surface roughness is comparable to that of the rougher lunar maria. Thus both aeolian features and rocky surfaces seen by VL-1 are consistent with the orbiter pictures and echoes from earth-based radars.

Utopia Planitia

In contrast with Chryse Planitia, deposits of cross-bedded drifts are virtually absent at Utopia Planitia, and the site is uniformly rocky. The rocky eroded appearance of the VL-2 site is consistent with orbital pictures of the site in particular and the region of Mars between 44°N and 48°N in general. The surface materials in the sample field (Figure 1*b*) can be described as blocks and fragments set in a matrix of finer-grained material. Typically, the rocks and blocks appear to be

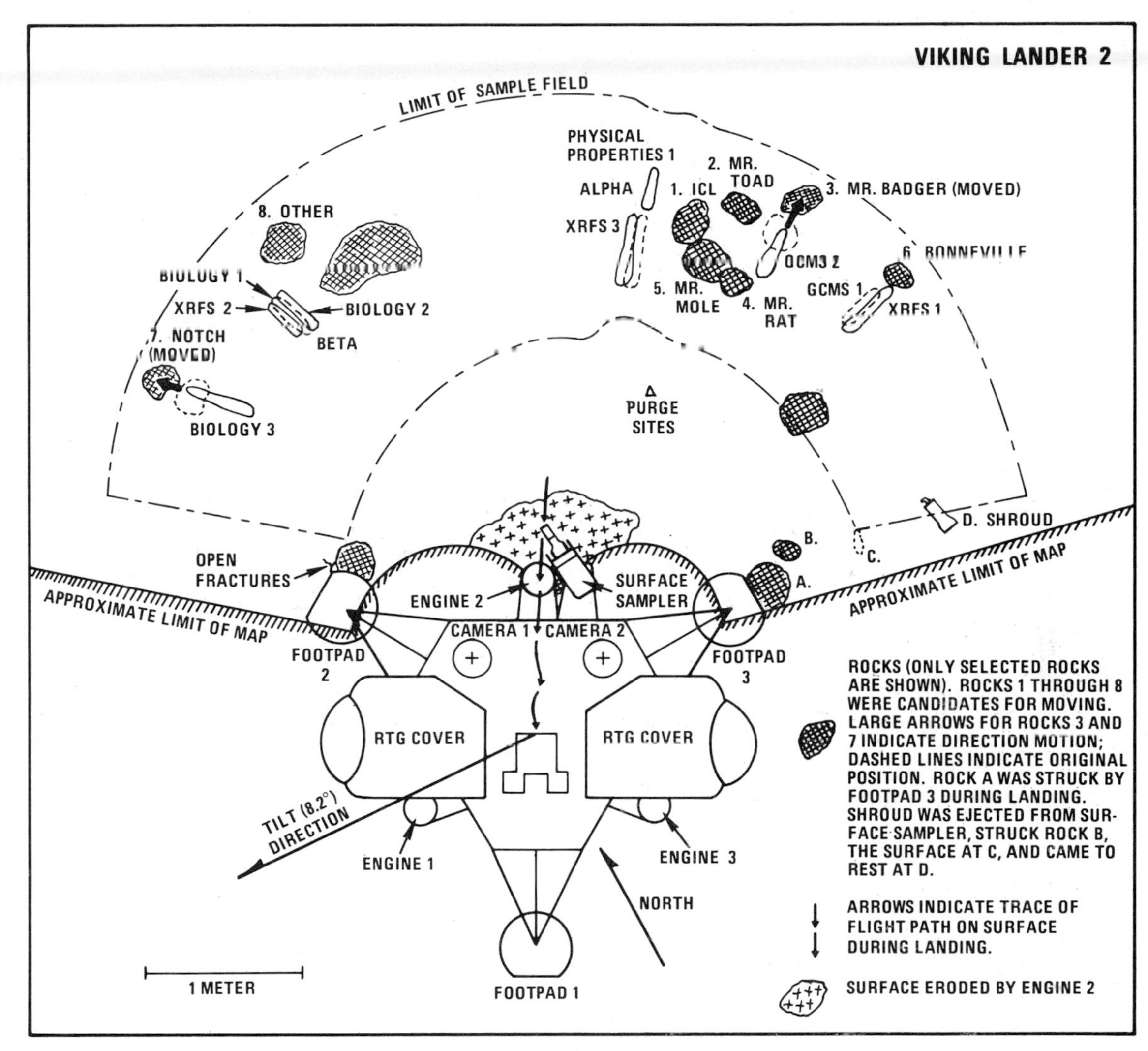

Fig. 1*b*

vesicular. Large rocks and blocks are approximately twice as abundant as those at the VL-1 site [*Mutch et al.,* 1976*b*], so abundant that they profoundly affected analysis of the landing data for physical properties and the subsequent sampling activities. Blocks within the sample field attain dimensions of 0.65 × 0.23 m, and 16–20% of the area is covered by rocks a few centimeters across and larger. Locally, beyond the sample field, small dunes in linear depressions or rilles have been pitted by rocks propelled by engine exhausts. Their form and orientation indicate a local wind direction near 320°. Finer-grained materials between the rocks are present as fillets and small drifts, but most surfaces have been stripped of fines to expose a weak platy crust, blocky fractured soil, and knobby surfaces of small clods and rocky fragments. Unlike the VL-1 site, where the sample field could readily be divided into two types of material, Utopia Planitia is more uniform. The sample field near footpad 3 and the ejected shroud is relatively rock free, however (Figure 1*b*). The footpads probably struck rocks on landing (footpad 1 is not visible) and have penetrated the surface a very small amount. A rock 22 cm across occupies part of the projected area beneath retroengine 2 (Figure 1*b*).

LANDING

Descent

Descent trajectories of both landers passed over the sample fields. Ten seconds before touchdown, VL-1 was 27 m above the surface traveling in the direction of leg 1 (in the landed configuration), and the body center was approximately 1.5 m uprange from its final location (Figure 1*a*). Thus the sample field was exposed to engine exhaust impingement to a greater degree than it would have been if the lander had approached from any other direction. One second before touchdown, VL-1 was descending at 2.44 m/s and moving about 0.15 m/s in a horizontal direction toward leg 1. A linear least squares fit to velocity increments measured by the inertial reference unit (IRU) during the last 2 s indicated a touchdown velocity of 2.30 m/s. Roll about the vertical axis was only 0.25°/s. Inflight tilt at touchdown required that leg 2 was about 1 cm lower and leg 1 was about 1.5 cm higher than leg 3. When these requirements were combined with local surface tilt and topography, leg 2 touched down first, followed by leg 3 and then leg 1.

At 10 s before touchdown, VL-2 was 26 m above the surface and about 1.7 m uprange of its final location. It too passed over the sample field moving in the direction of leg 1. VL-2 was descending at 2.44 m/s 1 s before touchdown. A linear least squares fit to velocity increments from the inertial reference unit during the last 2 s indicated a touchdown velocity of 2.34 m/s. Engine thrust levels increased during the last 0.4 s and probably reduced the velocity to 1.95 m/s. This is discussed in the following section. At touchdown the inflight tilt required that leg 1 was 3.5 cm lower and leg 2 was 7 cm higher than leg

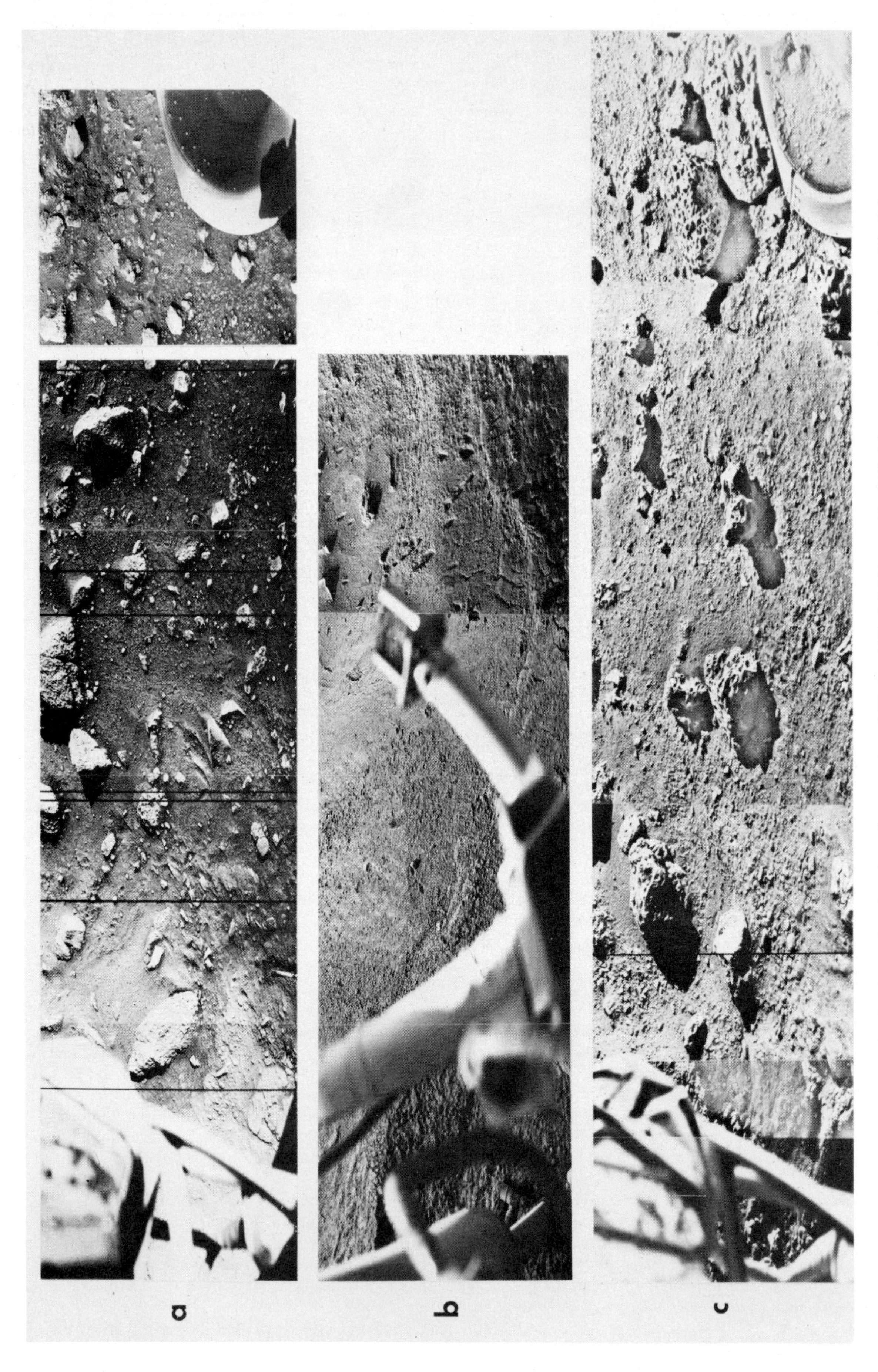

Fig. 2. Mosaic showing footpads and engine exhaust erosion of VL-1 and VL-2. (*a*) Footpad 3 of VL-1 has penetrated rocky material ≈3.6 cm. Some of the rocks were moved by engine exhausts; the crater produced by shroud impact is near footpad 3. Note the area stripped by engine exhausts to the left. (*b*) Footpad 2 of VL-1 has penetrated drift material and is buried. Note chains of small elongate craters produced by grains and fragments propelled by engine exhausts. Chains are alined along radials from engines. (*c*) Footpad 3 of VL-2; the footpad probably struck rock to the left of the footpad. The rock beyond the footpad was struck by the ejected shroud. Note the ridges, grooves, and fragments moved by engine exhaust gases to the left.

TABLE 1. Touchdown Parameters for Viking 1 and 2 Landers

Parameter	Viking 1	Viking 2
Touchdown velocity, m/s		
IRU	2.30	2.34
Engine thrusts	2.44	1.95
Latitude, °N	22.46	47.97
Longitude, °W	48.01	225.67
Leg 1 stroke, by stroke gauge, cm	7.0	2.5–3.2
Leg 2 stroke, cm		
By stroke gauge	3.2	7.6
By footpad travel	2.8	7.6
Leg 3 stroke, cm		
By stroke gauge	8.3	1.3
By footpad travel	8.3	1.3
Footpad 1 penetration (not visible)	...	...
Footpad 2 penetration, cm	16.5	2.5
Footpad 3 penetration, cm	3.6	0–0.3
Leg azimuth (east of north), deg	321.9	209.1
Tilt angle (relative to gravity vector), deg	3.0	8.2
Tilt azimuth (east of north), deg	285.2	277.7
Landed mass, kg	611	611

3. The local surface tilt indicated that leg 3 should have touched down first, followed by leg 1 and then leg 2.

Touchdown

A comparison of the leg strokes on the two landers indicates that the average leg stroke on VL-1 was larger than that on VL-2. The strokes of legs 2 and 3 determined from stroke gauges and leg travel for VL-2 are the same and imply that the load limiters at the secondary strut attachment did not yield, so energy dissipation was essentially due to irreversible crushing of the primary strut honeycomb tube core cartridge. This suggested that a larger fraction of the remaining energy must have been removed by the penetration of the footpads into the Martian surface on lander 2, but comparison of the footpad penetrations of the two spacecraft indicates that the opposite is true (touchdown parameters are given in Table 1).

The problem can be resolved by considering data on valve settings of the descent engines just prior to VL-2 touchdown. During the last 0.43 s before touchdown, thrust levels of engines 1, 2, and 3 increased by about 14, 81, and 91%, respectively [*Martin Marietta Corporation*, 1976]. The impulse associated with the total thrust level increase is about 1360 N s and produced an incremental velocity change of about 0.49 m/s. This makes the velocity of VL-2 at surface contact near 1.95 m/s and smaller than the nominal descent value of 2.44 m/s existing before the thrust suddenly increased. This smaller velocity is compatible with the smaller leg strokes on VL-2. Energy balance calculations indicate that the ratios of the energies absorbed by the primary strut stroking to the initial kinetic energies of VL-1 and VL-2 are 66 and 51%, respectively. More detailed energy balance calculations will be made at a later date.

Footpad-Surface Interactions

Footpad penetrations and leg strokes of VL-1 show that the two materials there have markedly differing mechanical properties. Footpad 3 of VL-1 only penetrated rocky material about 3.6 cm, whereas footpad 2 penetrated 16.5 cm and was buried in drift material (Figures 2*a* and 2*b*). Significantly different mechanical properties of the two materials are indicated because velocities of the two footpads at touchdown should be nearly the same, although footpad 2 may have touched first.

Footpad penetration cannot entirely resolve the mechanical properties of drift material because drift material is superposed on rocky material and thins to a feather edge at the mapped boundary of the two, so it is entirely possible that a hard substrate or buried rock caused penetration to cease. On the other hand, excavation of a trench 23 cm deep counterclockwise of previous trenches during the extended mission encountered no difficulty; a 15-cm rock was excavated, however. Footpad interactions with the surface by VL-2 are problematic. Footpad 3 of VL-2 penetrated 0.3 cm or less because of the reduced velocity, and it probably struck a rock at touchdown. Footpad 2 of VL-2 penetrated about 2.5 cm, judging from shadows assumed to be cast on a level planar surface, and it too may have struck a rock at touchdown.

Although results for VL-2 are difficult to interpret, footpad–surface material interactions by VL-1 provide valuable insight on the contrasting physical properties of the materials when the observations are combined with experimental data. The Viking footpad is an inverted frustum of a cone capped by a spherical segment, with an irregular skirt around the base (Figure 3). Calculations for VL-1 indicate that the total work done by plastic deformation of the honeycomb tube core, frictional sliding of the footpads, soil penetration, and the decaying engine thrusts during touchdown becomes equal to the kinetic energy at surface contact plus the change of potential energy from surface contact to the final rest position, when the effective coefficient of sliding friction is in the range of 0.3–0.5. If these coefficients and the force-stroke characteristics [*Martin Marietta Corporation*, 1973*a*] are used, forces on the footpads during touchdown are 6.8–7.6 kN, leg 1; 4.6–5.4 kN, leg 2; and 8.2–9.2 kN, leg 3. Comparison of these forces and the corresponding penetrations with results from essentially static tests of a $\frac{3}{8}$-scale footpad in three soil simulants (Table 2), appropriately scaled for the Viking lander on Mars (R. F. Scott, unpublished data, 1976), shows the contrasting difference in penetration resistance of drift material and rocky material (Figure 4). The force and penetration for footpad 3 of VL-1 lie to the upper left of the curves for the three soils and imply a stronger material. The force and penetration for footpad 2 of VL-1 lie to the lower right of the curves for the three soils and imply that the drift material is very weak and penetrable.

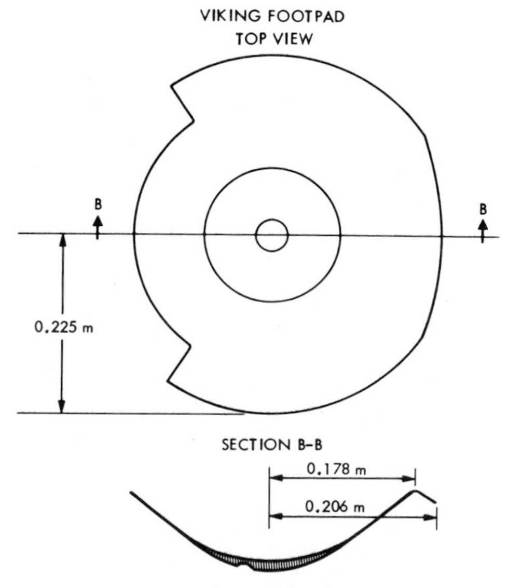

Fig. 3. Shape and dimensions of footpad of Viking lander.

TABLE 2. Mechanical Properties of Soil Simulants Used in Footpad Tests

Test	Simulant	Density, kg/m³	Cohesion,* kN/m²	Angle of Internal Friction,* deg	Grain Size
⅜ scale	lunar nominal	1100–1700	0.5–0.6	40–45	4–10⁴ μm, poorly sorted
⅜ scale	basalt dune sand	1150–1220	0–0.1	30–40	0.6-mm mean, well sorted
⅜ scale	white sand	1600–1660	0.06–0.08	30–35	0.21–0.30 mm
Full scale	lunar nominal	1440	0.5–0.6	35–40	4–10⁴ μm, poorly sorted
Full scale	lunar nominal	1630	0.8–1.0	35–40	4–10⁴ μm, poorly sorted
Full scale	basalt dune sand	1130–1280	...	37–52	0.6-mm mean, well sorted
Reduced scale	lunar nominal	1360			4–10⁴ μm, poorly sorted
		1600			4–10⁴ μm, poorly sorted
		1840			4–10⁴ μm, poorly sorted

*Blank spaces in these columns indicate that values were not actually measured but are similar to the values for the corresponding soil that appear in this table.

Dynamic tests of footpads were done because of the importance of several factors during dynamic loading such as (1) positive pore gas pressures resulting from compression of fine-grained materials during dynamic loading [*L. V. Clark*, 1971], (2) the geometry of the penetrating object [*Young*, 1967], and (3) the acceleration of gravity [*Pyrz*, 1969]. Full-scale tests of prototype skirted footpads were conducted in lunar nominal soil beds with two densities and basalt dune sand (Tables 2 and 3) at velocities of 2.44 m/s in 1 atm of air [*Martin Marietta Corporation*, 1971]. In one test, penetration into lunar nominal soil with a density of 1440 kg/m³ (run 14) [*Martin Marietta Corporation*, 1971] was an astonishing 26.5 cm, and leg stroke a mere 0.3 cm (Figure 5). This test result prompted an investigation of footpad penetration in both 5- and 1000-mbar ambient pressures with lunar nominal test beds of densities estimated to be 1360, 1600, and 1740 kg/m³ (Tables 2 and 3) [*L. V. Clark*, 1971]. These tests at reduced pressure showed that penetrations were functions of the ambient pressure and density of the test soil. Results from the low-pressure tests scaled to the Viking footpads and velocity by using equations developed for low-velocity impacts (see Figure 5 and *L. V. Clark and J. L. McCarty* [1963]) imply a penetration near 15 cm for a density of 1360 kg/m³ and 9 cm for a density of 1840 kg/m³ when the mass allotted to a single footpad is 200 kg (Figure 5). Penetrations calculated for run 14 adjusted for ambient pressure and the scaled low-pressure test data using low-density lunar nominal soil are in fair agreement with the actual penetration of footpad 2 on Mars, and this argues for a density of the drift material near 1300 kg/m³. Acceleration of gravity [*Pyrz*, 1969] (also see Figure 5) may have some effect on the penetration if the cohesion of the soil is low enough. The surface around footpad 2 is not level and flat but slopes about 20° away from the footpad toward the sample field. Such a slope would reduce penetration resistance [*Scott*, 1963]. Footpad 3 penetration is consistent with a weakly cohesive material with a density of 2300 kg/m³ according to Figure 5; however, the small penetration could be the result of a large cohesion and independent of density.

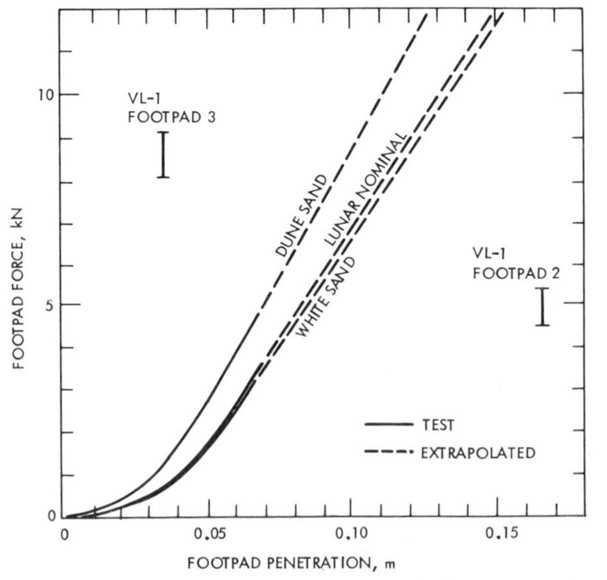

Fig. 4. Force-penetration curves for static loading test of ⅜-scale Viking footpad scaled to the Viking footpads and Mars gravity.

In any event, footpad analyses in comparison with laboratory studies show that there is a significant difference in mechanical properties of the drift and rocky materials. Drift material is somewhat like low-density lunar nominal test soil, whereas the rocky material is much stronger. As was noted previously, none of the analyses entirely resolve the problem of the drift material because the penetration by the footpad could have been arrested by a hard substrate or buried rock. Experiments are planned for the extended mission to learn more about the drift material.

Soil Erosion by Descent Engines

Exhausts from retroengines of both landers eroded the surface materials during landing. Evidence for erosion is found in the first few seconds of pictures taken by the landers 25 s after touchdown, debris found in unburied footpads, displaced rocks and fragments, chains of elongated craters alined along radials from the retroengines, shallow craters beneath individual en-

TABLE 3. Generalized Footpad Test Conditions

	Mars VL-1	Mars VL-2	Full-Scale Test	Reduced-Scale Tests
Maximum projected area of footpad, m²	0.135	0.135	0.151	0.092
Mass, kg	200	200	377	52.2
Velocity, m/s	2.30	1.95	2.44	3.48
Pressure, mbar	~8	~8	1000	5

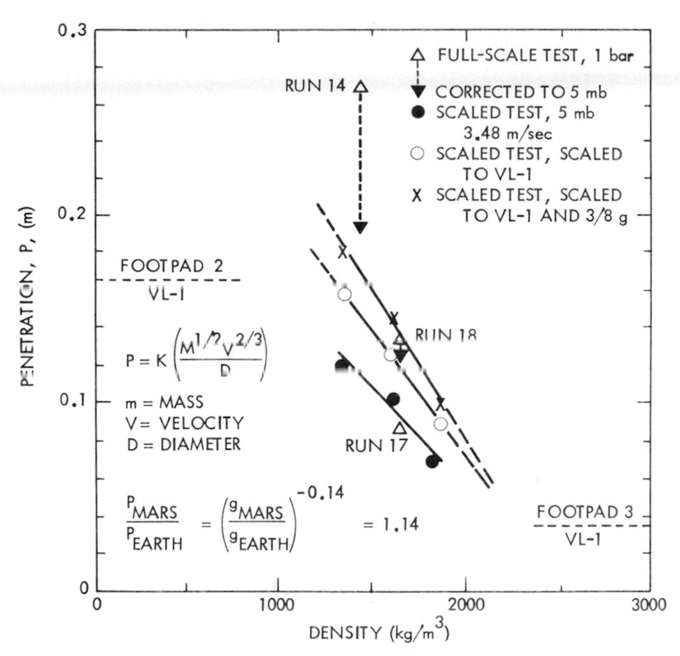

Fig. 5. Density-penetration curves for full-scale tests of Viking prototype footpads in 1 atm of air (open triangles), reduced-scale tests of footpads in a 5-mbar atmosphere (solid circles), tests of the reduced-scale tests scaled to the Viking lander (VL-1) (open circles), and tests of the reduced-scale tests scaled to the Viking lander and Mars gravity ($\frac{3}{8}g$) (crosses). Penetrations by footpads 2 and 3 of VL-1 are also plotted. The scaled test was adjusted to full scale by using the low-velocity impact penetration equation relating penetration P to mass M, velocity V, and diameter D [*L. V. Clark and J. L. McCarty*, 1963]. The equation for effect of gravity on penetration is from *Pyrz* [1969]. Adjustment of full-scale tests for reduced atmospheric pressure is from data of *L. V. Clark* [1971].

gine nozzles viewed through a boom-mounted mirror, and direct views of the area in front of the forward retroengines. The symmetry of erosion by retroengine 2 of VL-2 is more or less radial with a slightly larger extent of erosion toward footpad 3 because of spacecraft tilt (Figure 1*b*), whereas this symmetry was not produced by the VL-1 engine because of the contrasting properties of drift material at the left side of the engine (Figure 1*a*). Because of spacecraft tilt, erosion to the left of the engine should have been more extensive than it was to the right of the engine.

Vertical bands which appear in the first 40 s or so of the first pictures taken by each lander were produced by coarse particles falling out of a dust cloud generated by engine exhausts [*Shorthill et al.*, 1976*a*]. This is entirely consistent with site alteration tests with conditions similar to those of the actual landings [*Romine et al.*, 1973]. Two soils were used in separate tests: lunar nominal and basalt dune sand (see Table 2). For the lunar nominal test soil, blowing dust was observed as soon as the test engine was ignited at 12.2 m above the soil surface; the test area was obscured by blowing dust after engine shutdown, and several minutes were required for the chamber to clear. It seems likely, on the basis of the presence of fine grains on Mars, that the dust cloud caused by the landing would also last several minutes in view of the lower Mars gravity. In addition to the vertical bands in the first picture taken by VL-1, a shadow was cast in the field of view nearly 40 s after the first picture began. At the time of landing of VL-1 the sun was in the southwest, and winds were probably from the northeast, so a cloud produced by the engines of VL-1 would have passed between the lander and the sun.

The first pictures of Mars also viewed the area of footpad 3 at a time when the dust cloud had settled, and subsequently, pictures were obtained of the region of and around footpad 2. Debris that settled from the dust cloud is found in footpad 3 of both landers and footpad 2 of VL-2. Displaced rocks and fragments occur at distances from the lander that are compatible with the site alteration tests of 1971. In the tests, rocks located 0.6 m from the engine axis were moved by the exhaust gases. The largest of these had a mass of 74 g and was roughly equidimensional with a size of 3.0 cm. The pressure required to move this rock was about 0.7 kN/m². The rock traveled 1.65 m from the engine axis, and an initial pressure of about 0.35 kN/m² was required to keep it moving at this distance. On Mars a rock of the same dimensions and density would weigh 0.38 as much, so at a radial distance of 0.6 m, loose equidimensional rocks 7.0 cm across would be moved. At 1.65 m, rock fragments 3.5 cm across would be moved. Displacements of rocks on Mars are consistent with test results. About 1.6 m from the VL-1 engine near footpad 3, rocks 7 × 15 cm have moved a few tens of centimeters, and at VL-2, rocks 4 cm across that are 2 m from the engine have been moved. In one case near the first VL-2 sample site a 4-cm rock lies beside a rimless depression the same size as the rock. Apparently, engine exhausts lifted the rock from its socket.

As in the tests, fragments, grains, and clods propelled by the Viking lander engine exhausts produced small elongate craters. Site alteration tests suggest that velocities of the particles were several tens of meters per second. Morphologies of the craters vary. In general, craters large distances from the engines appear fresh, but those near the engines are partly filled by fines, a situation showing that some aerodynamic sorting occurred during dust cloud formation and the decay of engine thrusts. Near footpad 2 of VL-1, craters both near and at large distances from the engine appear to be fresh. Lack of filling of craters at large distances is probably the result of dispersion of fines over large areas. Although the fresh appearance of craters near footpad 2 may have partly resulted from the transport of fines to the southwest by wind, the impacts of grains and fragments have produced unusually deep craters in the relatively weak drift material, so more fines would be required to subdue them. It may also be that fewer fines were eroded to the left of engine 2 because the grain size of the drift material is so small that it is not readily eroded. In many cases the small grains and fragments that produced the craters lie in or near the craters. Small grains of quartz near 1 mm in diameter dropped into low-density (600 kg/m³) lunar nominal soil on earth at velocities of 3.5 m/s penetrate several diameters and are not visible after penetration. Larger velocities result in larger penetrations. In view of the presence of small fragments in and by the craters in drift material, it seems unlikely that drift material has a density substantially less than 1300 kg/m³ because their penetration has been small.

Erosion by the retroengine 2 exhaust gases of both landers produced small shallow craters beneath individual nozzles (viewed through a boom-mounted mirror) and scoured the surface in front of the engines (viewed directly), a crater with a rim of mixed fines and platy to equidimensional fragments of soil and rocks (Figures 2*a* and 2*c*) thus being formed. Fragments imbedded in the rim, which is near 0.5 m from the VL-1 engine axis, have intermediate diameters near 1 cm and indicate that erosion has occurred to depths of 1 or 2 cm. The surface to the left of the engine of VL-1 has not been stripped despite the tilt of the spacecraft, which would favor more stripping in this region. Evidence for extensive erosion is found on both sides of the VL-2 engine, but it is more extensive to the right front, presumably because of spacecraft tilt. Interpretation is somewhat clouded because a rock 22 cm across partly occupies the area beneath the engine. In contrast with VL-1 the rim of the erosional crater is not as clearly defined. At distances near 0.5–0.6 m, fragments of rock and soil exposed in the rim are roughly twice as large as those at VL-1, a result suggesting erosion to depths of 2–4 cm. Erosion by the retroengines on Mars is more extensive than that in the lunar nominal soil during site alteration tests and like that of the dune sand. In the lunar nominal soil of the site alteration tests the area below the descent engine was scrubbed, and some soil was removed to a depth of a few millimeters, but no crater was formed, and no signs of separate craters caused by the individual engine nozzles were apparent. Regions of lunar nominal soil prepared at lower density than that of the bulk of the test bed showed no difference in erosion. This indicates that the grain size of the soil is more important than the density in the erosion process. Erosion of the surface in the tests was 0.6 cm or less at distances of 0.5 m from the engine center line. For the coarse basalt dune sand, separate craters up to 4.3 cm deep beneath the individual nozzles were produced, and depths of erosion were near 0.6–1.5 cm at 0.5 m from the engine. Thus the response of the Martian surface to the retroengine exhausts is less than that of lunar material and like that of dune sand. Comparison between the relative amounts of erosion between the two landers is at best difficult because the surface did not respond in the exact same manner. The extent of fragments at VL-2 and their large size and concentrations beyond 0.5 m from the engine suggest that erosion of the surface by VL-2 was greater than it was by VL-1. Such a result could arise from a combination of factors related to landing conditions and the character of the surface. For a short interval of time just prior to thrust termination by VL-2 the increase in thrust subjected the surface to nearly a doubling of the plume impingement pressures and to the viscous shearing stresses exerted along the gas-surface interface. These higher stresses are consistent with a larger amount of erosion of material near retroengine 2 on VL-2 than that on VL-1. The grain size of the surface materials could also affect the amount of erosion. This appears to be the case at VL-1, where erosion of the fine drift material to the left of engine 2 appears to be substantially less than it is to the right, where coarser fragments and clods are present.

SURFACE SAMPLER ACTIVITIES

Information on the physical properties of the Martian surface materials during the primary mission has been gleaned from normal surface sampler activities during sample acquisitions for the analytical experiments: Biology, Molecular Analysis (gas chromatograph mass spectrometer, GCMS), and Inorganic Chemical Analysis (X ray fluorescence spectrometer, XRFS). One sol of the VL-1 mission and 1 sol and part of another sol of the VL-2 mission were allotted to the Physical and Magnetic Properties Investigations, for which sample acquisition procedures were the same as those for the analytical experiments. Surface sampler activities are generalized in Tables 4 and 5, and locations of trenches and rocks are shown in Figures 1*a* and 1*b*. More detailed data can be found in the summary of surface sampler activities [*L. V. Clark et al.*, 1977]. Typical sample acquisition sequences involve (1) positioning of the sampler boom to the desired azimuth, (2) extension of the collector head to the desired amount, (3) lowering of the collector head to the surface, (4) extension of the collector head into the soil about 0.16 m with the jaw open to acquire a sample, (5) retraction of the collector head with the jaw closed, and (6) elevation and delivery of the sample normally through 0.2-cm openings in the upper jaw of the collector head. After delivery the remaining coarse fraction is purged or dumped at a preselected position in front of the landers. Included in the normal surface sampler activities are the ejection of a protective shroud early in the mission and the extraction of the restraint (latch) pin of VL-1 on sol 5.

The collector head has a lower jaw 4.45 cm wide with a serrated tip [*Martin Marietta Corporation,* 1973*b*] (see also Figure 12); 10.2 cm from the serrated tip is a backhoe 6.1 cm wide and 6.45 cm high. An upper movable jaw is actuated by a solenoid capable of vibrating the upper jaw at two frequencies: 4.4 and 8.8 Hz. When the jaw is closed, the distance from the base of the lower fixed jaw to the top of the movable jaw is 5.5 cm. The upper surface of the movable jaw has 0.2-cm holes in it to separate the coarse fraction of samples from the fines during delivery of the samples. The collector head assembly is 24.3 cm long and is attached to a furlable boom. About 21 cm from the tip of the head the assembly pivots vertically through an arc of 10°. The collector head can be inverted about its longitudinal axis so that material can be sieved through the 0.2-cm holes in the upper jaw.

Surface sampler activities of the two landers were similar but differed in two important ways: (1) the period of intensive operations by VL-2 lasted for 58 sols, whereas that of VL-1 lasted for 41 sols, and (2) rocks were nudged and pushed, and samples were obtained from the newly exposed surfaces beneath the rocks by VL-2. Importantly, endeavors by both spacecraft to collect samples of small rocks and mineral grains in the 0.2- to 1.2-cm size range have not been successful. Samples analyzed in this size range by VL-1 have a composition consistent with clods of clayey soil [*Baird et al.*, 1976], and the lumps seen in VL-2 pictures are disaggregated to sizes less than 0.2 cm during vibration of the collector head.

VL-1

The first samples of Martian soil were acquired from drift material in Sandy Flats, because the profusion of rocks seen elsewhere was absent and fragments propelled by the engine exhausts made impact craters in the drift material there. Samples were subsequently collected from the rocky materials at Rocky Flats. Morphologies of the trenches from the two materials differed significantly. In drift material, surface-sampler-commanded azimuths were achieved in all cases, whereas they were not achieved on two occasions in the rocky material. Sample trenches in drift material (Figure 6*a*) typically had large depths, steep walls, domed and fractured surfaces around the far ends of the trenches produced by extension into the soil, lumpy floors, relatively large, uniform, but lumpy tailings piles produced by the backhoe during retraction, and highly reflective surfaces on disturbed material produced by tamping and rubbing of the surface sampler parts. In contrast, trenches in rocky material (Figure 6*b*) had gently sloping lumpy walls, and one trench disrupted the surface material, a chaotic array of fragments several centimeters across thus being produced.

The sample trenches provide quantitative data on the cohesion of the drift material (Figure 6*a* and Table 4, sols 8, 14, 36, 41, and 91). Lateral trench walls have arcuate outlines where slope failure has occurred and other outlines where the arcuate pattern outlines material that has not yet collapsed into the trench. Ends of two trenches have collapsed; this produces a steplike mass near the end of the trench and a trench that is longer than the commanded extension by several centimeters.

TABLE 4. Surface Sampler Activities Related to the Physical Properties Investigation, Viking Lander 1

Activity	Sol	Local Lander Time[1]	Surface Sampler Positions[2]			Remarks
			Azimuth, deg	Extension, inches	Elevation, deg	
Shroud ejection and retroengine 2 picture via boom mirror 2 (Figures 1*a* and 2*b*)	02	1028 : 02	255.4	6.0	40.1	Shroud ejected at 3.2 m/s, struck surface near footpad 3 at 2.6 m/s, producing a shallow crater 1 cm deep and 9 cm in diameter by displacement of rocks and ejection of fine debris, then ricocheted and came to rest about 1 m from crater (Figure 2*b*). Picture under retroengine 2 via boom mirror 2 showed a small rock and two shallow craters produced by engine exhausts.
Restraint (latch) pin ejection (Figure 2*b*)	05	1042 : 47	186.0	12.0	−7.2	Restraint (latch) pin (8.2 cm long, 0.6 cm in diameter, and 11.3 g) fell from 0.9–1.0 m to the surface, reaching a velocity of 2.6–2.7 m/s. Pin impacted on an end with small roller bearings, producing a small circular crater, and then fell over toward the spacecraft, producing an elongate crater. Very fine grained material was ejected from the craters to distances of 2.4 cm (Figure 2*b*).
Biology 1 sample (Sandy Flats, Figure 1*a*)	08	0705 : 10	104.5	90.2 96.7 90.2	25.0	Trenched by extending after surface contact and then retracting; trench about 45 cm long from rim to rim, 8.5 cm wide from rim to rim, and 4 cm deep; small lumps in and around trench are chiefly clods; material is very fine grained because materials rubbed and tamped by surface sample parts are smooth and reflective; small craters on left side of trench produced by wind-driven clods and debris falling from sampler and boom (Figure 1*a*).
Purge (Biology, Figure 1*a*)	08	0839 : 09	110.2	10.5	37.6	Purged material has grain size between 2.0 and 5.0 mm and may be small clods of surface material rather than rock or mineral grains; small impact pit produced by purged material (Figure 1*a*).
GCMS 1 sample (Sandy Flats, Figures 1*a* and 6*a*)	08	0903 : 32	104.5	90.2 96.7 90.2	25.6	Trenched by extending after contact and then retracting with identical commands as above. No pictures of the trench were obtained, engineering data indicate that amount of sample delivered was unusually small after two attempts and surface domed at far end of trench to 10 cm along trench axis and 7 cm in lateral directions from rim (Figure 6*b*).
	08	0931 : 56	104.5	90.2 96.7 90.2	25.6	
Purge (GCMS 1, Figure 1*a*)	08	0953 : 25	110.2	9.7	37.6	Available data suggest no purged material.
XRFS 1 sample (Sandy Flats, Figures 1*a* and 6*a*)	08	1045 : 55	104.5	94.1 100.7 94.1	24.3	Trenched by extending after surface contact then retracting; trench and tailings about 70 cm long, far end about 10 cm across, and trench about 4–6 cm deep. Surface domed to about 10 cm from rim; measurements and images indicate that wall at end of trench collapsed and incipient failure occurred on left side of trench (Figures 1*a* and 6*a*).
	08	1132 : 55	104.5	94.1 100.7 94.1	25.0	
Purges (XRFS 1, Figure 1*a*)	08	1110 : 18	110.2	11.5	37.6	Purged material has grain size between 2.0 and 5.0 mm and may be small clods of surface material rather than rock or mineral grains; small impact pits produced by purged material (Figure 1*a*).
	08	1157 : 18	110.2	11.5	37.6	
GCMS 2 sample (Sandy Flats, Figures 1*a* and 6*a*)	14	0635 : 42	107.7	90.2 96.7 90.2	25.6	Trenched by extending after surface contact then retracting; trench and tailings about 47 cm long, trench is 7 cm wide at far end and 2.3 cm deep, doming to about 7 cm from rim; septum between XRFS 1 trench and this trench was displaced toward XRFS trench during sampling. Sample was not analyzed by GCMS (Figures 1*a* and 6*a*).

TABLE 4. (Continued)

Activity	Sol	Local Lander Time[1]	Surface Sampler Positions[2]			Remarks
			Azimuth, deg	Extension, inches	Elevation, deg	
Purge (GCMS 2)	22	1227 : 49	87.5	9.9	37.6	Purged material on surface obscured by lander body.
GCMS 3 sample (Rocky Flats, Figures 1*a* and 6*b*)	31	1051 : 02	204.9	70.2 76.5 64.4	30.0	Trenched by extending after surface contact then retracting; trench and tailing about 42 cm long, far end about 7.6 cm across, and trench 3–5 cm deep (Figures 1*a* and 6*b*).
Purge (GCMS 3, Figure 1*a*)	31	1118 : 02	194.8	42.0	37.6	Purged material has grain size a few millimeters across and may be small clods; no impact pits (Figure 1*a*).
XRFS 2 sample (Rocky Flats, Figures 1*a* and 6*b*)	34	1019 : 21	203.6	62.0 67.5 56.2	33.2	Trenched by extending after surface contact and then retracting; trench and tailings about 33 cm long and 5–7 cm wide; disrupted region at far end is 19 cm across; first pass by surface sampler did not attain the commanded extension at 68.6 inches, an indication that forces near 210 N were applied to surface by surface sampler; second pass by surface sampler achieved the commanded extension (Figures 1*a* and 6*b*).
	34	1119 : 21		62.0 68.6 56.2	33.8	
Purges (XRFS 2, Figure 1*a*)	34	1041 : 45	194.8	42.0	37.6	Purged material a few millimeters across (Figure 1*a*).
	34	1141 : 45	194.8	42.0	37.6	
Biology 2 samples (Sandy Flats, Figures 1*a* and 6*a*)	36	1121 : 02	104.5	78.3 84.9 72.8	27.5	Trenched by extending after surface contact and then retracting; trench and tailings about 50 cm long, 6–7 cm wide, and 4–5 cm deep; trench transects tailings of previous sol 8 trenches (Figures 1*a* and 6*a*).
Purge (Biology 2, Figure 1*a*)	36	1322 : 40	194.8	42.0	37.6	Purged material a few millimeters across (Figure 1*a*).
XRFS 3 sample (Rocky Flats, Figure 1*a*)	40	1038 : 39	204.9	74.6 81.2 69.1	28.8	Trenched by extending after surface contact and then retracting; trench extends 8–9 cm beyond GCMS 3 trench along same azimuth. Clod and rock fragments 2–4 cm across. First pass achieved commanded extension, second pass did not achieve commanded extension, an indication that forces near 210 N were applied by surface sampler (Figure 1*a*).
	40	1218 : 39	204.9	74.6 80.9 69.1	28.8	
Picture of footpad 2 temperature sensor via boom mirror 1	40	1106 : 02	125.4	NA	35.1	Picture included part of temperature sensor housing; a subsequent picture taken during extended mission shows that temperature sensing elements are buried but entire housing is not.

Purge of fines (XRFS 3)	40	1042 : 29	204.9	69.1	7.9	Purge of fines cannot be recognized by using visual inspection of pictures.
	40	1222 : 29	204.9	69.1	7.9	
Purge of coarse material (XRFS 3, Figure 1*a*)	40	1114 : 18	189.8	42.0	37.6	Purge.
	40	1254 : 18	189.8	42.0	37.6	
Physical Properties 1 sample temperatures (Sandy Flats, Figures 1*a* and 6*a*)	41	1537 : 45	101.4	78.3 84.9 72.8	27.5	Trenched by extending after surface contact and then retracting; trench and tailings about 55 cm long, 6 cm wide, and 6 cm deep; trench is unusually narrow, and visual picture differencing shows that septum between this trench and the Biology 3 trench moved toward Physical Properties trench rather than toward the previously formed Biology 3 trench. Surface sampler collector head temperatures were 272°K at extension and 269°K in the surface material at 10.7 and 11 min after extension (Figures 1*a* and 6*a*).
Physical Properties magnification mirror image of front porch	41	1558 : 04	4.2	24.7	0.3	Front porch picture.
Purge of fines on lander grid (Physical Properties 1, Figure 7)	41	1603 : 42	4.2	24.7	3.5	Purge of fines on lander grid produced more or less conical mounds. Pictures taken with a sun elevation angle of 38° show that slopes of mound are shadowed, so angle of repose is >38° (Figure 7).
Purge of coarse material (Physical Properties 1, Figure 1*a*)	41	1617 : 41	189.8	42.0	37.6	Purge (Figure 1*a*).
Biology 3 sample (Sandy Flats, Figures 1*a* and 6*a*)	91	0711 : 02	107.7	78.3 84.9 72.8	28.8	Trenched by extending after surface contact and then retracting; trench and tailings about 50 cm long, 6 cm wide, and 4–5 cm deep; trench transects tailing of previous sol 14 GCMS 2 trench (Figures 1*a* and 6*a*).
Purge (Biology 3, Figure 1*a*)	91	0844 : 41	189.8	42.0	37.6	Purge (Figure 1*a*).

[1]Listed local lander times for the various activities correspond to (1) time of extension command for shroud ejection, (2) collector head open command just before extension, (3) extension commands for nudges and pushes, and (4) vibration commands for purges.

[2]Positions are from surface sampler potentiometer readouts; azimuths were measured from a line 80° counterclockwise from the (plus) Y_L direction (direction from camera 2 to camera 1). Azimuths should be reduced by 0.3° because of boom override due to lander tilt; extensions are reported here in inches to be consistent with engineering units used for the surface sampler systems and represent the increase in length of the boom from the stowed position (1 inch equals 2.56 cm); NA is not applicable; elevations are measured from the Z-Y plane (plane parallel to the upper surface of the lander body), passing through the surface sampler elevation axis; positive values are angles below the plane, and negative values are angles above the plane.

TABLE 5. Surface Sampler Activities Related to the Physical Properties Investigation, Viking Lander 2

Activity	Sol	Local Lander Time[1]	Surface Sampler Positions[2]			Remarks
			Azimuth, deg	Extension, inches	Elevation, deg	
Shroud ejection (see Figure 1*b*)	01	1052 : 02	255.4	10.2	38.9	Shroud ejected at 3.2 m/s, struck a rock near footpad 3 at 3.7 m/s, ricocheted from rock, impacting surface 0.6 m beyond rock, and came to rest 1.1 m beyond rock (Figure 1*b*); rock near footpad 3 moved a small amount as a result of the impact.
Biology 1 sample (Beta, see Figures 1*b* and 8*a*)	08	1610 : 20	124.7	85.2 91.7 79.6	23.1	Trenched by extending after surface contact and then retracting; trench and tailings about 40 cm long from rim to rim and 7.6 cm wide from rim to rim; sample delivered to Biology; coarse fraction purged to XRFS funnel, but no sample was received (Figures 1*b* and 8*a*). Small lumps in and around trench are probably chiefly clods.
GCMS 1 sample (Bonneville, see Figures 1*b*, 8*b*, and 9)	21	1010 :19	216.3	93.6 88.8 97.3 91.2	30.0	Trenched by retracting after surface contact, then extending, and finally by retracting; trench about 25 cm long, 5.6 cm wide at far tip, and 7.3 cm wide near tip. Collector head tunneled beneath crust, doming surface and crust near far tip. Platy fragment of crust near 7 cm across and 1 cm thick was moved by backhoe (Figures 1*a* and 8*b*).
Purge (GCMS 1, Figure 1*b*)	21	1053 : 46	190.4	39.9	36.3	Purged material greater than 2.0 mm; a fragment 2.7 cm across and additional finer grains or clods 2 mm and larger (Figure 2*b*).
Biology 2 sample (Beta, see Figure 1*b*)	28	1610 : 31	126.0	85.2 91.7 79.6	23.1	Trenched by extending after surface contact and then retracting. Trench and tailings about 40 cm long and 7.6 cm wide; sample delivered to Biology (Figure 1*b*).
Purge (Biology 2, Figure 1*b*)	28	1744 : 46	190.4	39.9	36.3	Purged material should be greater than 2.0 mm; no evidence for purged coarse particles (Figure 1*b*).
XRFS 1 sample (Bonneville, Figures 1*b*, 8*b*, and 9)	29	1339 : 56	217.5	93.0 99.4 90.9	29.4	Trenched by extending after surface contact and then retracting; first acquisition extension lifted rock at far end of trench about 0.4 cm; large platy fragments of crust in and around trench and fine-grained material in debris pile at tip (Figures 1*b* and 8*b*); trench and tailings 28 cm long, 7.7 cm wide at far end, and 9.4 cm wide at near end; note that material at far end has spread laterally to trench azimuth because of rock; rectangular fragment about 3 cm on an edge and 1 cm thick.
	30	1039 : 56	217.5	93.0 99.4 90.9	30.0	
Purge (XRFS 1, Figure 1*b*)	29	1405 : 45	190.4	39.9	36.3	Purged material greater than 2.0 mm; a large fragment has been displaced, and purged material 5 mm and smaller has been added (Figure 1*b*).
	30	1105 : 45	190.4	39.9	36.3	
Rock 1 nudge (initial computer load (ICL), see Figure 1*b*)	30	1129 : 34	186.4	75.4 78.6	33.2 30.6	Nudged by elevating after surface contact, then extending, and finally retracting; surface sampler did not achieve commanded extension, which was 83.1 inches; rock did not move (Figure 1*b*); force on rock was near 210 N (Figure 1*b*).
Rock 3 push (Mr. Badger, see Figures 1*b* and 10)	34	1039 : 40	201.1	84.4 96.5 82.0	30.6 30.0 . . .	Pushed by elevating after surface contact, then extending, and finally retracting; rock was pushed about 10 cm away from lander, rotated about 60° or 70° counterclockwise about vertical axis, and tilted near 45° about a horizontal axis. Collector head went beneath rock which leaned on collector head, deflecting it to the right; this forced the excavation of trench in front of rock (Figures 1*b* and 10).

Rock 3 push (Mr. Badger, see Figures 1*b* and 10)	37	1006 : 55	200.5	89.1 101.2 82.0	29.4, 28.1	Pushed by elevating after surface contact, then extending, and finally retracting; rock was pushed an additional 12 or 15 cm away from lander with little rotation. Commanded extensions compared with measurements of rock indicate that rock tilted backward and then forward to its present position (Figures 1*b* and 10).
GCMS 2 sample (under Mr. Badger, see Figures 1*b* and 10)	37	1614 : 48	201.1	93.0 84.1 87.0 93.6	28.8 . . . 20.5, 30.0 . . .	Trenched by retraction after surface contact to clear debris away, elevating, extending, and deelevating to touchdown, extending for sample acquisition, and then retracting. Trench about 30 cm long and 10 cm wide (Figures 1*b* and 10). Sample was delivered to GCMS from first acquisition, although surface sampler made two acquisitions.
	37	1651 : 50	201.1	93.0 84.1 87.0 93.6	29.4 . . . 20.5, 30.6 . . .	
Purge (GCMS 2, Figure 1*b*)	40	1601 : 01	192.9	39.9	36.3	Purged material should be greater than 2.0 mm; no evidence for purged material (Figure 1*b*).
Rock 6 nudge (Bonneville, see Figure 1*b*)	45	1010 : 28	217.5	99.1 103.0 98.0	25.6, 26.2	Nudged by elevating after surface contact and then extending; reversal of elevations was related to boom sag and overtravel; rock rotated about a nearly horizontal axis, and points on front surface of rock were displaced upward about 0.4 cm during nudge (Figure 1*b*); rock fell back again after collector head retraction, pushing materials from trench rim into trench.
Rock 7 nudge (Notch, see Figure 1*b*)	45	1022 : 07	105.8	84.1 87.1 60.2	23.1, 22.4	Nudged by elevating after surface contact and then extending; rock rotated in horizontal plane about an axis on right side of rock; left edge of rock was displaced about 3.8 cm (Figure 1*b*).
XRFS 2 sample (Beta, see Figure 1*b*)	46	1307 : 52	123.5	85.2 91.7 83.1	23.1	Final trench about 30 cm long and 8 cm wide (Figure 1*b*). Sampled to collect rocks, but little or no sample was received. Some fragments at surface may be rocks, but most are clods of soil.
	46	1347 : 52	123.5	85.2 91.7 83.1	23.7	
	47	1307 : 52	123.5	85.2 91.7 83.1	23.7	
	47	1347 : 52	123.5	85.2 91.7 83.1	23.7	
Purge (XRFS 2)	46	1316 : 34	123.5	83.1 40.2	−11.0 . . .	Purged material less than 2.0 mm; no evidence for purged material at trench site.
	46	1356 : 34	123.5	83.1 40.2	−11.0 . . .	

TABLE 5. (Continued)

Activity	Sol	Local Lander Time[1]	Surface Sampler Positions[2] Azimuth, deg	Extension, inches	Elevation, deg	Remarks
Purge (XRFS 2) (continued)	47	1316 : 34	123.5	83.1 40.2	−11.0 . . .	
	47	1356 : 34	123.5	83.1 40.2	−11.0 . . .	
Rock 7 push (Figure 1*b*)	51	0622 : 49	106.4	86.7 98.0 60.2	21.8, 21.8	Pushed by elevating after surface contact and then extending; rock rotated about 50° counterclockwise in horizontal plane and translated about 24 cm from original position by plowing and sliding away from surface sampler gimbal axis (Figure 1*b*).
Biology 3 sample (under Notch rock, see Figure 1*b*)	51	0655 : 35	107.1	93.6 78.1 88.0 94.6 82.0	20.5 . . . 15.5, 21.8	Trenched by retracting to clear away possible contaminating debris, then elevating, extending, and deelevating to surface contact, and finally extending for sample acquisition followed by retracting; backhoe trench about 7.6 cm wide, extension trench difficult to measure, trench and tailings about 46 cm long (Figure 1*b*). Sample was delivered to Biology.
Purge (Biology 3 Figure 1*b*)	51	0903 : 49	190.4	43.9	36.3	Purged material greater than 2.0 mm; some purged material, generally small grains (≤2.0 mm) (Figure 1*b*).
Physical Properties 1 sample, temperatures (Alpha, see Figures 1*b* and 8*b*)	56	1415 : 58 1433 : 12	180.9	88.0 95.2 88.0	28.8	Trenched by extending after surface contact and then retracting; trench about 28 cm long and 6 cm wide near tip; trench very shallow because of local slope and interference by excavated rock about 3 × 6 cm. Surface sampler temperatures were 272°K at extension (t = 0), 272°K (t = 4.7), 273°K (t = 8.7), 273°K (t = 12.7), and 271°K (t = 16.7).
Physical Properties magnification mirror image of front porch and footpad 2 picture via mirror 1	56	1441 : 58 1458 : 44 1501 : 18 1503 : 48	3.5 120.9 120.9 118.4	18.6 NA NA NA	1.0 28.1 33.2 33.2	Picture of front porch to study soil with magnification mirror. Three pictures intended to determine conditions of the footpad 2 temperature sensor did not include sensor.
Physical Properties retroengine 2 pictures via boom mirror 2	57	0649 : 37 0652 : 04 0654 : 31	255.4 251.6 247.8	NA NA NA	40.1 40.1 40.1	A series of three pictures showing craters produced by erosion by retroengine exhausts.
Purge (Physical Properties 1, Figure 1*b*)	57	0700 : 16	192.3	39.9	36.3	A few objects, 0.5 cm and smaller, purged from collector head (Figure 1*b*).
XRFS 3 sample (Alpha, see Figures 1*b* and 8*c*)	57	0808 : 29	180.3	75.2 81.7 63.3	33.2	Trenched by extending after surface contact and then retracting; sol 57 trench and tailings 56 cm long and 7.3 cm wide; largest clods 4.0 cm across; deformation producing clods extends 7.0 cm from rim in lateral directions. Sol 58 trench same size as sol 57 trench but overlaps it and extends lateral deformation only a small amount to left (Figure 1*a* and 8*c*).
	57	0853 : 29	180.3	75.2 81.7 63.3	33.2	
	58	0808 : 29	179.0	75.2 81.7 63.3	32.5	

	58	0853 : 29	179.0	75.2 81.7 63.3	32.5	
Purges (XRFS 3)	57	0812 : 40	180.3	63.3	−18.0	Material purged less than 2.0 mm.
	57	0857 : 40	180.3	63.3	−18.0	
	58	0812 : 40	179.0	63.3	−18.0	
	58	0857 : 40	179.0	63.3	−18.0	

[1]Listed local lander times for the various activities correspond to (1) time of extension command for shroud ejection, (2) collector head open command just before extension; (3) extension commands for nudges and pushes, and (4) vibration commands for purges.

[2]Positions are from surface sampler potentiometer readouts; azimuths were measured from a line 80° counterclockwise from the (plus) Y_L direction (direction from camera 2 to camera 1). Azimuths should be reduced by 0.6° because of boom override due to lander tilt; extensions are reported here in inches to be consistent with engineering units used for the surface sampler systems and represent the increase in length of the boom from the stowed position (1 inch equals 2.56 cm); NA is not applicable; elevations are measured from the Z-Y plane (plane parallel to the upper surface of the lander body), passing through the surface sampler elevation axis; positive values are angles below the plane, and negative values are angles above the plane.

These observations suggest that the trench slopes range from instability to marginal stability. The behavior of the ridge between the sol 36 and the sol 40 trench during sampling lends further evidence for marginal stability. In contrast with the other trenches for which the ridge between them was displaced toward the previously formed trench during the later insertion of the surface sampler, the ridge between the sol 41 and the sol 36 Biology trench was displaced away from the previously formed trench toward the later trench, presumably after retraction. The end result is an unusually narrow trench (Figure 6*a*). Slopes of the upper trench walls are commonly greater than 60°, and locally they are 80°, and trench depths are as large as 5–6 cm. Slope stability factors for soils possessing both friction and cohesion for slope angles greater than 60° and angles of internal friction between zero and ≈30° are [*Scott*, 1963]

$$4 < \rho g h/C \approx <15$$

For densities ρ in the range of 1000–1800 kg/m³, Mars gravity g, and trench depths h of 5–6 cm, the cohesion C of the drift material comes out to be in the range of 10–10² N/m².

Insertion of the surface sampler into the soil causes the surface around it to be deformed upward. Deformation of the surface around the trenches extends to one trench width (≈7.0 cm) in lateral directions and at least 10 cm along the trench azimuth from the far rim and represents failure by general shear which in turn implies internal friction.

The highly reflective surfaces produced by tamping and rubbing by the surface sampler show that drift material is fine grained and probably silt size or smaller. In preflight tests in sand, lunar nominal soil, and montmorillonite, only the finer-grained simulants became reflective when they were tamped and rubbed by the sampler. Lumps seen in and around the trench are weakly cohesive clods.

The X ray fluorescence spectrometer is capable of estimating the densities of the material in the analysis chamber [*B. C. Clark et al.*, 1976]. These authors find that the density of the delivered sample of drift material is 1100 ± 150 kg/m³. This density is consistent with the relatively low densities implied by the penetration by footpad 2. It is possible that in situ densities could be either slightly larger or smaller than the disturbed density of the delivered sample. Chemically, samples of drift material are best interpreted as weathering products of mafic igneous rock, possibly clay minerals [*Baird et al.*, 1976].

Trenches in rocky material do not lend themselves to slope stability analysis. The contrasting appearance of the trenches combined with the fact that the surface sampler failed to reach the commanded extensions on some occasions but not on others can be interpreted to result from materials with significantly variable cohesion which may be locally large. This is discussed later.

Sampling at Rocky Flats (Table 4, sols 34 and 40) for Inorganic Chemical Analysis was intended to collect rocks (Figure 6*b*). Initially, small knobs and lumps there were interpreted to be rocks. Analysis of the coarse fraction delivered for Inorganic Chemical Analysis revealed that the composition of the coarse fraction was similar to that of drift material, although the sulfur content was higher [*B. C. Clark et al.*, 1976]. The bulk density of the coarse fraction delivered was a mere 600 ± 100 kg/m³ and is consistent with soil clods having bulk densities of 1200 kg/m³ and a void volume between the soil clods of 50% [*B. C. Clark et al.*, 1976]. Thus it appears that the material that arrested footpad 3 may need only be cohesive and not dense, as is implied by the footpad analysis (see Figures 4 and 5).

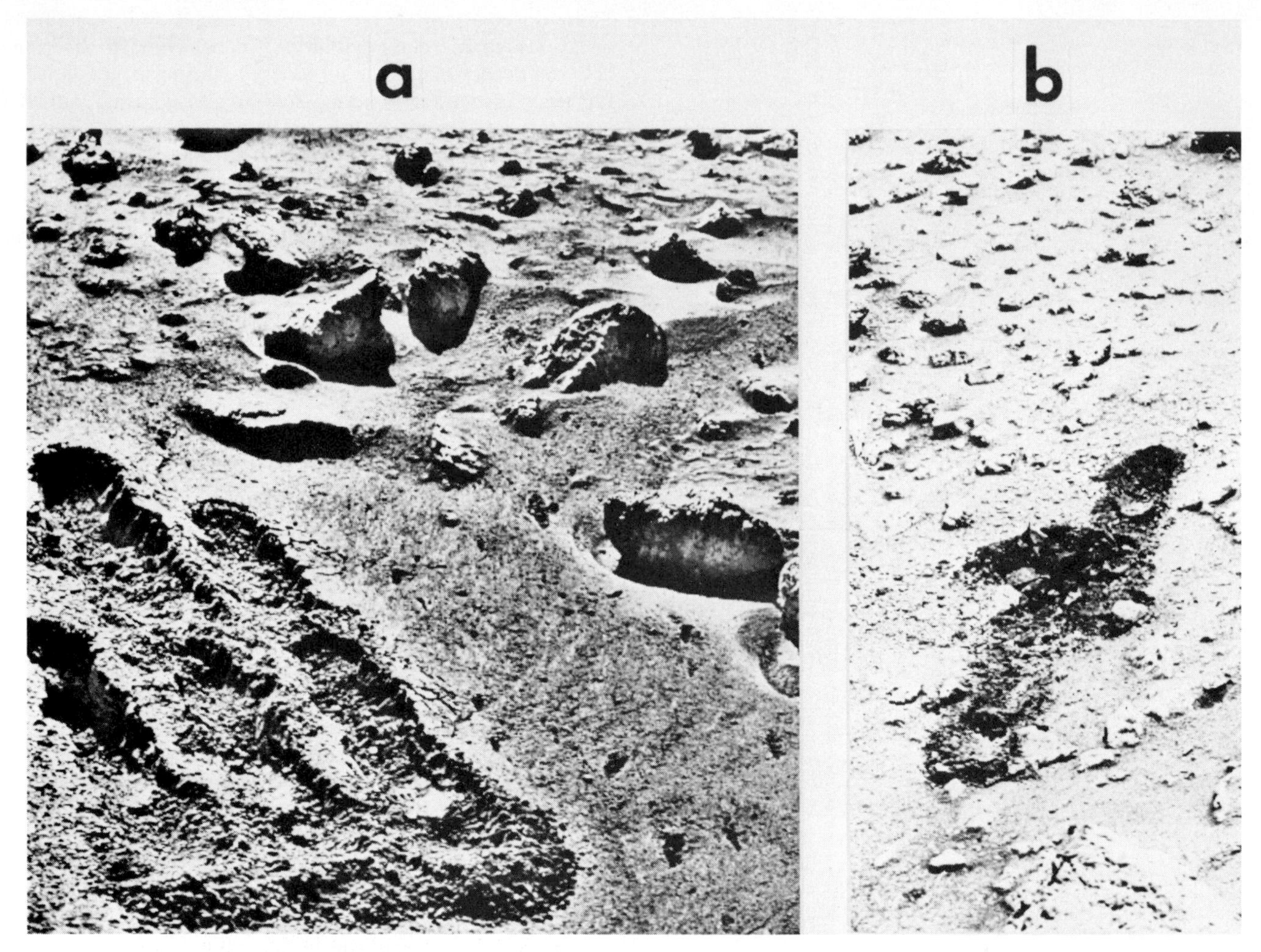

Fig. 6. Trenches excavated by VL-1 during the primary mission. (*a*) Trenches in drift material of Sandy Flats (see Figure 1*a*): the upper two trenches were excavated on sols 8 (upper left, XRFS 1) and 14 (upper right, GCMS 2). Actually, five acquisition strokes were made along the same azimuth on sol 8. The lower three trenches were excavated on sols 41 (lower left, Physical Properties), 36 (lower center, Biology 2), and 91 (lower right, Biology 3). Note that the left edge of the sol 8 trench at left center has slumped, the sol 14 trench to the right is unusually shallow, and the sol 41 trench at the lower left is unusually narrow; also note the steplike structure of the sol 41 trench where the end of the trench has collapsed. Local highly reflective surfaces in disturbed material were produced by tamping and rubbing of the surface sampler and show that the drift material is very fine grained. Dimensions of the trenches are given in Table 4. Note the deflation hollows in front of the rocks to the right and the wind tails behind the rocks in the background. (VL-1 camera 1, frame B180; sol 91, 0904:59 LLT; sun elevation, 47.2°; sun is to the right.) (*b*) Rocky material of Rocky Flats (see Figure 1*a*): the trench at right center was excavated on sol 31 (GCMS 2); the trench extending to the lower left was excavated on sol 34 (XRFS 2). Note the disrupted material with fragments about 4 cm across at the far end of the sol 34 trench. Dimensions of the trenches are given in Table 4. (VL-1 camera 1, frame B030; sol 34, 1255:00 LLT; sun elevation, 81.6°.)

After normal acquisitions, small amounts of coarse fragments 0.2–0.5 cm across were purged in the field of view. Although these coarse particles were initially thought to be rock fragments, the results of the Inorganic Chemical Analysis strongly suggest that they are clods with unusually large cohesion.

Drift materials were deposited on the grid of the upper lander body through the collector head sieve; thus a conical pile was produced, and debris was scattered (see Table 4, sol 41). Most of the grid was obscured, but with time, winds have removed some of the scattered debris. The angle of repose of the material is near 39° as shown by the dark shadow cast on the right side of the pile by the sun, which was at an elevation of 38.7° (Figure 7). This indicates that the angle of internal friction of drift material in a loose state [*Terzaghi,* 1943] is near 39°.

VL-2

Sampling operations by VL-2 (Table 5) were substantially affected by the profusion of rocks, and considerable effort was expended, first dodging and later pushing and nudging them. Three basic types of trenches (Figure 8) were produced during sampling activities by VL-2: (1) trenches with knobby raised rims, (2) trenches with rims of thin platy fragments mixed with finer debris, and (3) trenches that produced blocky fragments. In the first type the regular knobby raised rims are similar to trenches produced in coarse sand and rock fragments during tests on earth (Figure 8*a* and Table 5, sol 8). The rims composed of lumps a fraction of a centimeter across are uniform in appearance, and no large clods or cohesive blocks are present, so the lumps, while they are cohesive themselves, do not have much interlump cohesion. Teeth marks made during retraction of the collector head tip and backhoe are poorly preserved as they would be in sand, but no tamping or smoothing has occurred.

In the second kind of trench the presence of platy objects is striking (Figure 8*b* and Table 5, sols 21 and 29). Presample pictures suggest that the sampled area was a cohesive fractured material in which the small knobby fragments and fines seen elsewhere had been removed by erosion (Figure 9*a*). Prior to acquisition of the first sample at this site a picture of the backhoe during touchdown was obtained (Figure 9*b*). The contact switch stopped the sampler when the backhoe had penetrated the fractured material about 1 cm. The force required to cause this penetration is not great, about 10 N, and the penetration is about the same as the sampler achieves in a medium-dense fine-grained normal terrestrial soil like lunar nominal. Trenching during both extension and retraction shows that this is a thin weak crust. In the first trench formed, the surface sampler extended into the material, doming and fracturing the surface. After retraction, pictures revealed a cavity overlain by a thin arch of the domed surface at the far end. During retraction of the backhoe a plate of the crust was pulled backward (Figure 8*b*). Subsequent trenches left a residue of the platy crust intermixed with a pile of fines.

The third type of trench has thick angular clods of soil to distances of 7 cm or so from the edge of the trench (Figure 8*c* and Table 5, sol 57). Sizes of the upper surfaces of the clods are the same as those of undisturbed blocks of soil partly etched out by erosion in adjacent areas. The principal difference between this type of trench and the previous one is that the angular clods are much thicker than the plates, and they may extend well below the surface.

Although large rocks are abundant, attempts to collect small rocks 0.2–1.2 cm across from the Beta and Alpha sites (Figure 1*b*) have been unsuccessful. After eight attempts to collect 'rocks,' no sample was found to be present in the XRFS chamber (Table 5, sols 46, 47, 57, and 58). Examination of the purge site after a sample acquisition at Alpha (Table 5, sol 28) also gave no evidence for purged coarse material. Endeavors to build a rock pile during the extended mission by sifting samples with the collector head sieve from the same general area of the Beta site area have yielded two or three centimeter-size rocks after 11 tries. Thus it appears that the smaller lumps are chiefly weakly cohesive clods of material with grains finer than 0.2 cm etched out by weathering, wind, and possibly engine exhausts. Rocks are present, however, because one about 2.7 cm across was purged earlier in the mission (Figure 8*c* and Table 5, sol 31).

Rock Pushing, VL-2

Rock nudging and pushing were undertaken by VL-2 because samples collected from the newly exposed materials would have been shielded from the sun, which destroys biota and decomposes organic molecules. In view of the fact that the surface sampler was not designed to push rocks, the outcome was remarkably successful. Behavior of the rocks during nudging and pushing varied. Rock 1 (Figure 1*b*) did not move, perhaps because it is deeply buried. Rock 3 moved in a complicated way (Figure 10). During the first push (Table 5, sol 34), rock 3 probably tilted up, rotated counterclockwise (as viewed in Figure 1*b*), and translated about 6 cm (Figure 10*b*). The second push was accompanied by tilting and skidding as shown by the smooth appearing skid marks (Figure 10*c*). Rock 7 plowed and furrowed while it was rotating clockwise.

Fig. 7. Conical pile of drift material from Sandy Flats dumped on the lander grid during sequence on sol 41. Material is at angle of repose. (VL-1 camera 1, frame B107; sol 41, 1606:09 LLT; sun elevation, 38.7°; sun is to the left.)

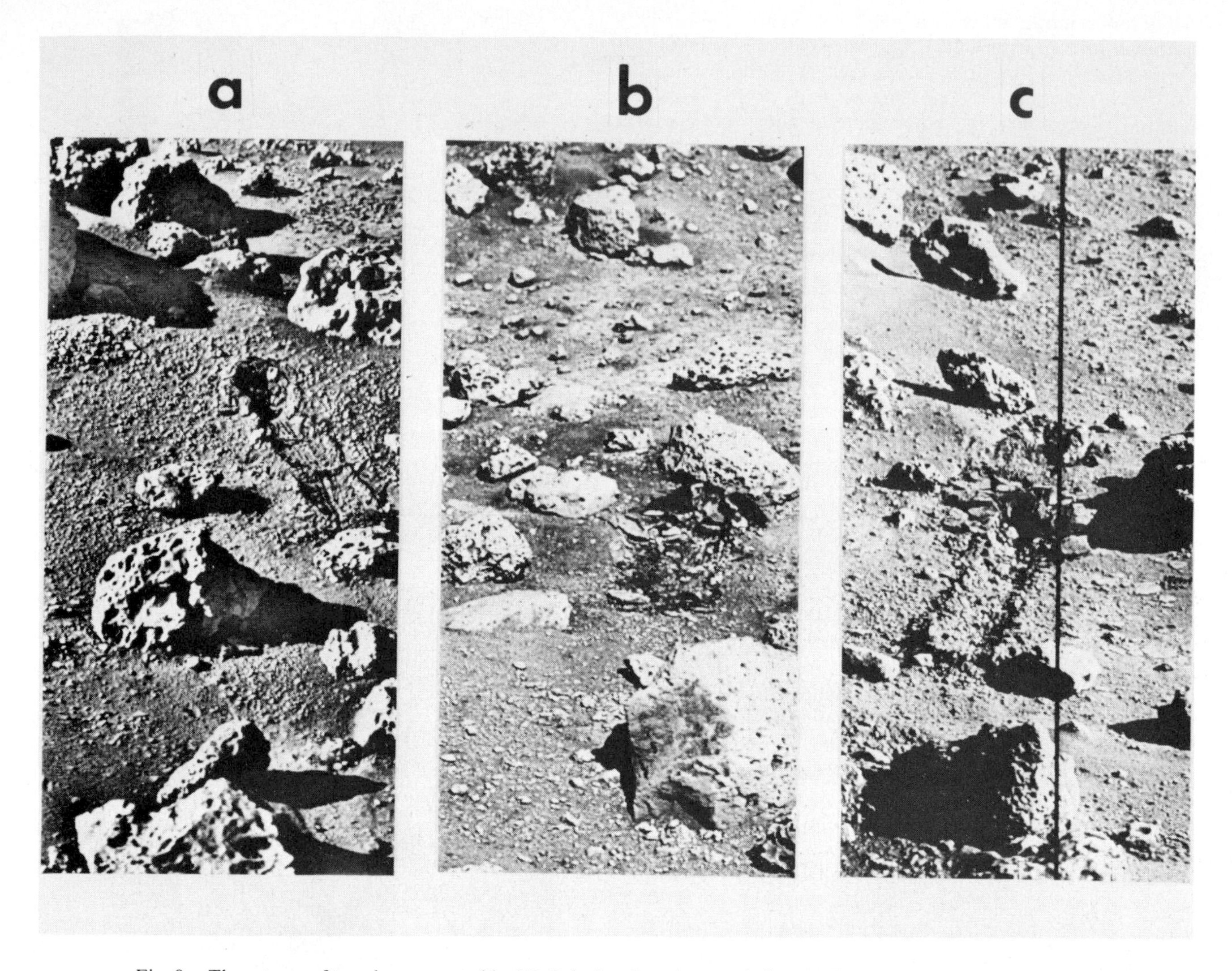

Fig. 8. Three types of trenches excavated by VL-2 during the primary mission. (*a*) Trench with knobby raised rim dug on sol 8 (Biology 1, VL-2 camera 2, frame A110; sol 15, 1659:59 LLT; sun elevation, 28.9°; sun is to the left). (*b*) Trenches with mixed fines and platy fragments dug on sols 21 (GCMS 2) and 29 (XRFS 2) (VL-2 camera 2, frame A242; sol 29, 1419:59 LLT; sun elevation 51.5°; sun is to the right). (*c*) Trench with blocky fragments dug on sol 57 (XRFS 3, VL-2 camera 2, frame C045; sol 57, 0940:59 LLT; sun elevation, 44.2°; sun is to the right). Dimensions of the trenches are given in Table 5. The small rock in the lower right corner of (*c*) was purged by the surface sampler.

Rock 6 merely tipped backward away from the lander. It is noteworthy that none of the rocks spalled or chipped noticeably while they were being nudged and pushed. Forces of about 200 N were exerted on rock 1 because commanded surface sampler extensions were not achieved when rock 1 failed to move. The teeth of the collector head should have exerted a sizable stress on the rock. The area of a tooth is about 1 mm^2, so the stress was of the order of 10^8 N/m^2. Thus a thick weak weathered rind is not present on that rock.

Samples Under Rocks, VL-2

Samples were acquired beneath rocks 3 and 6 (Figure 1*b*). Very rough estimates of the water evolved during heating were obtained by the Molecular Analysis experiment. The amount of water evolved by the soil from beneath rock 3 when it was heated from 50° to 200°C is much larger (0.15–1.1%) than that evolved from a sample exposed to the sun and heated in one step to 200°C (~0.002%) [*Biemann et al.,* 1976]. Heating of both samples from 200° to 350°C and then to 500°C evolved comparable amounts of water in each step. The Biology Pyrolytic Release results for a sample collected beneath rock 7 (Figure 1*b*) are also compatible with relatively large amounts of water [*Horowitz et al.,* 1976]. Larger amounts of water should be expected in samples of soil collected from beneath rocks than in samples of soil exposed to the sun. Field and laboratory studies on earth show that soil beneath rocks in a field of soil has detectably more adsorbed water at depths of 2.5–5.0 cm than soils exposed to the sun and atmosphere [*Jury and Bellantuoni,* 1976*a*, *b*]. The studies of these authors indicate that the net heat flow is toward the soil beneath the rocks, so that water vapor moves under the thermal gradient toward the area beneath the rocks, where it is cooler on the average than it is elsewhere at shallow depths of 2–5 cm. The rock cap inhibits evaporation. Additionally, ultraviolet radiation causes dehydration of exposed soils [*Huguenin,* 1976].

Surface Sampler Motor Currents

Viking surface sampler motor currents were sampled every 0.195 s with a resolution of 0.039 A (4 kbit/s) in the engineering data format (Format 5) as the surface sampler collector

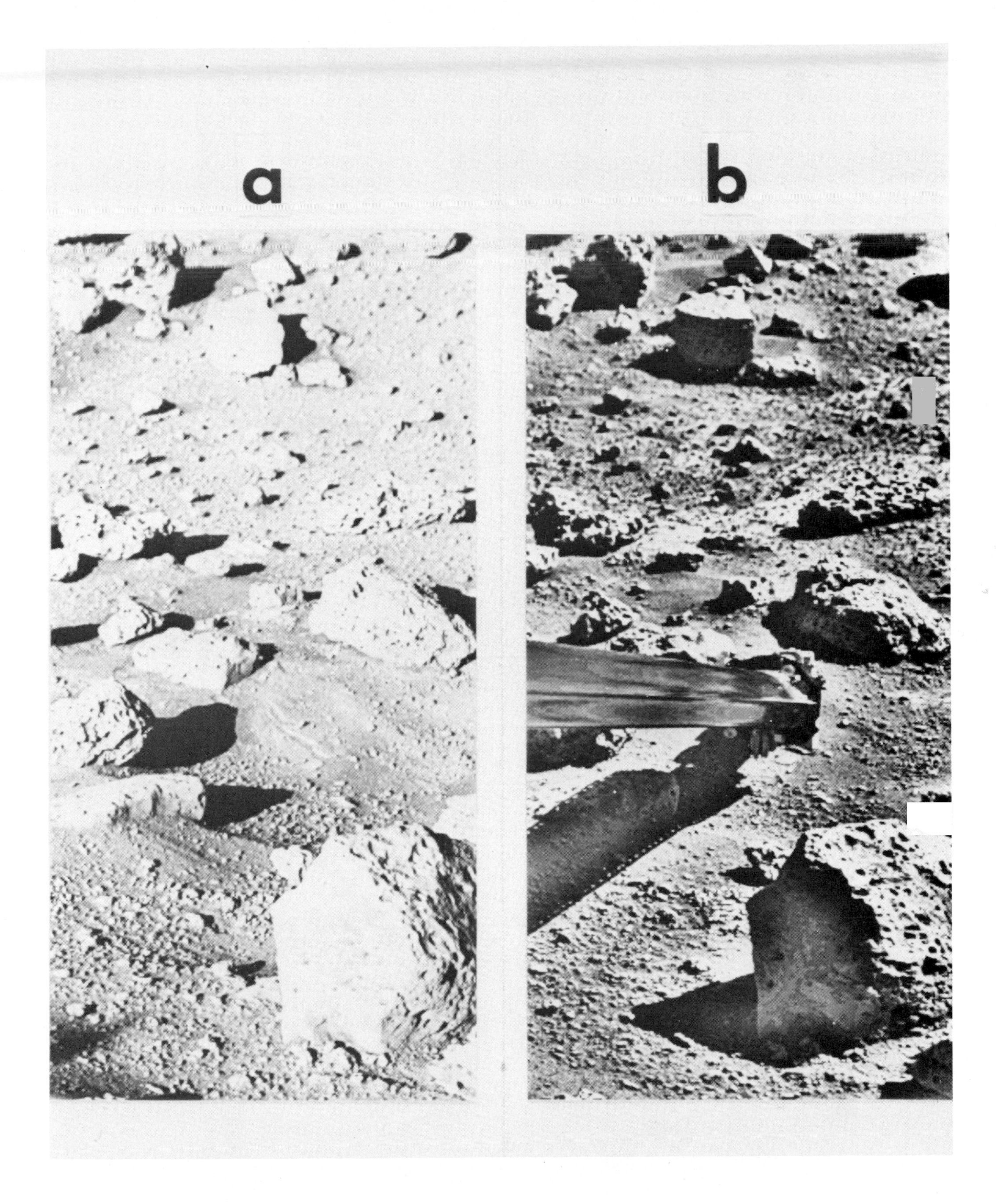

Fig. 9. (*a*) Bonneville Salt Flats prior to sample acquisitions on sols 21 and 29 (see Figure 8*b*, VL-2 camera 2, frame A005; sol 0, 1729:59 LLT; sun elevation, 25.6°; sun is to the left) and (*b*) backhoe touchdown picture on sol 21 (VL-2 camera 2, frame A154; sol 21, 1010:58 LLT; sun elevation, 52.2°; sun is to the right). Note that the surface in (*a*) has been locally stripped of fines to expose a smooth surface of material with open fractures; the backhoe in (*b*) has penetrated smooth surface material with open fractures about 1 cm; this shows that it is weak.

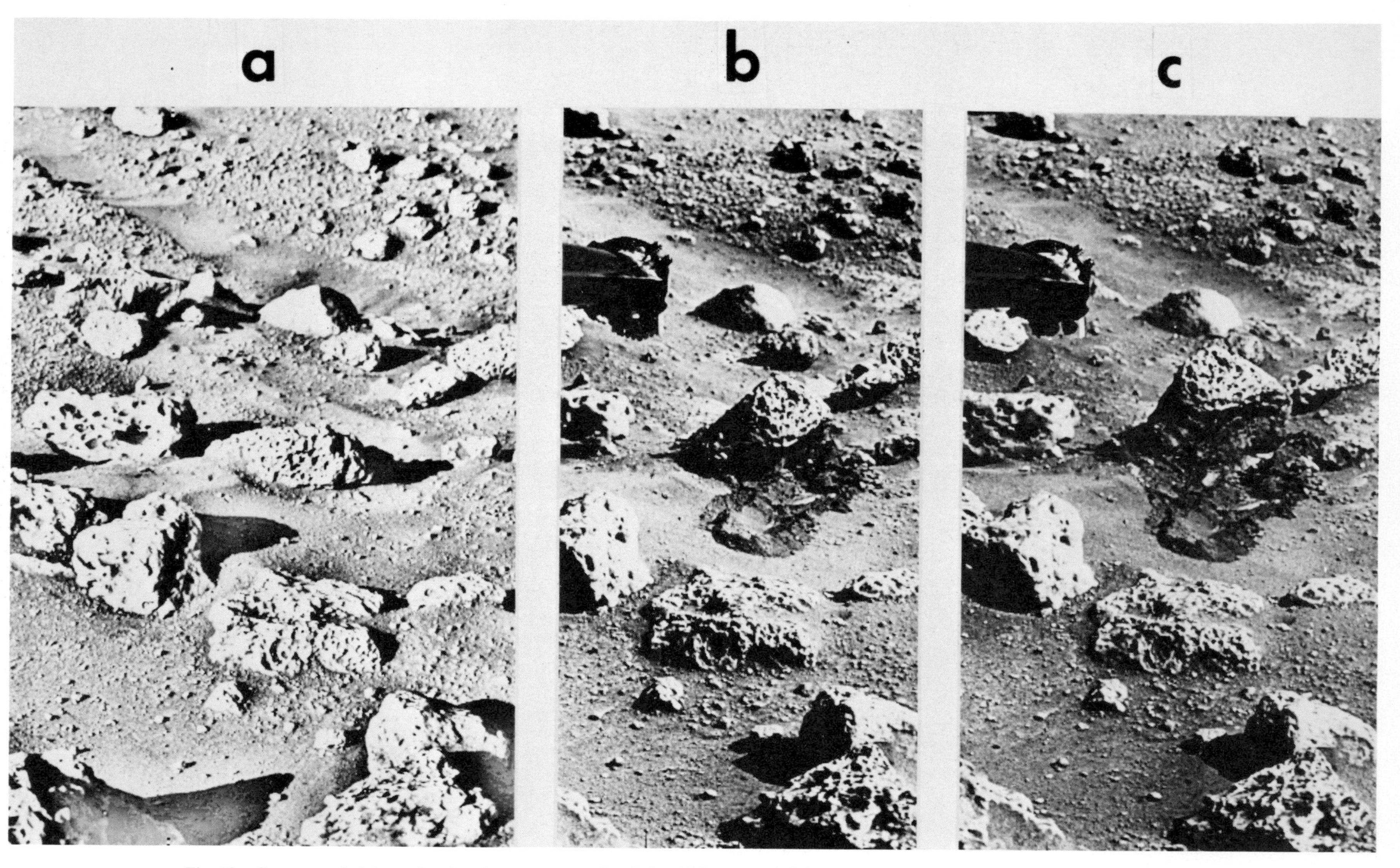

Fig. 10. Sequence of pictures showing the movement of rock 3, which was pushed by VL-2 to collect samples protected from ultraviolet radiation. (*a*) Rock prior to being pushed (VL-2 camera 2, frame A005; sol 0, 1729:59 LLT; sun elevation, 25.6°; sun is to the left). (*b*) Rock after push on sol 34 has rotated it 69° counterclockwise, tilted it about 40°, and translated it about 6.5–7.0 cm (VL-2 camera 2, frame B030; sol 34, 1048:10 LLT; sun elevation, 55.0°; sun is to the right). (*c*) Rock after push on sol 37 has rotated it 7° clockwise, tilted it 20° backward, and translated it 12–15 cm away from the lander (VL-2 camera 2, frame B046; sol 37, 1146:21 LLT; sun elevation, 58.6°; sun is to the right). The upper surface of the rock, the layer midway in the rock, and the base show that the rock tilts away from the spacecraft in (*a*). Note the 'water line' of soil adhering to the left side of the rock in (*b*); the large trench in (*b*) was inadvertently formed during retraction of the surface sample because the rock leaned on the surface sampler. Note the smooth surfaces produced by skidding on soil in (*c*). The sample of soil originally under the rock for Molecular Analysis (GCMS 2) was collected to the left of a long trench that was produced inadvertently.

head was pushed into the Martian surface materials during some sample acquisitions. After removal of cyclic currents and the base current, forces are related to motor currents and power by using calibration data from tests using flight type hardware [*Crouch,* 1976]. The resolution in force is about 25 N.

Despite this poor resolution, forces inferred from motor currents are consistent with soils having low cohesions. The surface sampler extends at a rate of about 0.025 m/s so that durations of motor currents may be correlated with extension or length of travel as the collector head plows through the surface material.

Forces on Mars exhibit four basic patterns (Figure 11): (1) an initial rise from zero to about 50 N followed by a rise to 75 N (Figures 11*a* and 11*b*), (2) an initial rise to 50 N followed by periodic rises to 75 N (Figures 11*c* and 11*d*), (3) a rapid rise to 75 N (Figures 11*e* and 11*f*), and (4) a steady force of 50 N (Figure 11*g*). The final peak force in Figure 11*c* was caused by an encounter with a rock. Tests were also conducted with the Science Test lander on earth to gain a clearer understanding of the motor current data. Forces for a sample acquisition into sand are similar to those of type 1 on Mars (Figure 11*h*). Forces of 50 N were measured without digging by extending in air (Figure 11*i*) and on a smooth metal surface (Figure 11*j*). Thus the force during excavation must account for the no-load condition of 50 N and is 25–50 N larger than that for the no-load condition (compare Figures 11*h* and 11*j*).

Dimensionless equations relating forces on small blades during plowing [*Luth and Wismer,* 1971; *Wismer and Luth,* 1972] may be applied to the results for the Science Test lander and Mars with the tacit assumption that soil-to-metal frictions are the same for the surface sampler and plow blades used in the tests to determine the constants in the dimensionless equations. In using the equations for cohesionless sand with an angle of internal friction of 35°–37° (see Figure 12), a plowing depth z of 5.6 cm, a plow or collector head width b of 4.4 cm, a plow height l of 7.6 cm, an angle of attack α of 80°, and a rate of 0.025 m/s (see Figure 12 for definitions), forces due to plowing would increase from zero to about 26 N as the collector head extended into cohesionless sand with a density of 1600 kg/m³ on earth. For Mars, in using the values above and an acceleration of gravity of 3.78 m/s², forces due to plowing would increase from zero to about 10 N. Thus the motor currents and inferred forces for the sols 46 and 47 sample acquisitions are compatible with cohesionless sand having an angle of internal friction near 35°–37°. Cohesion could also be present. Near the end of the sol 47 acquisition, forces rose above 75 N, whereas they did not during sol 46. Such a rise in force could be due to buried rocks or local clods with large cohesion. If the increase in force is due to cohesion, dimensionless equations for cohesive frictionless soil (see Figure 12) applied to Martian soils place an upper bound for cohesion near 2 kN/m².

Motor currents translated into forces for other trenches can also be interpreted within broad limits. Forces inferred for both the Molecular Analysis (GCMS 2, Figure 1*b*) sample acquisition from under a rock (rock 3, Figure 1*b*) and the VL-2 Physical Properties sample acquisition on sol 56 (Physical Properties 1, Figure 1*b*) exhibit an oscillatory pattern with values of 50 and 75 N. The collector head tended to skim the surface at shallow depth for both of these trenches because the local surface had relatively large tilts away from the spacecraft (Figures 11*c* and 11*d*). Uniform forces between 50 and 75 N were observed for the VL-1 Physical Properties sample acquisition on sol 41 (Physical Properties, Figure 1*a*), which are consistent with a uniform weakly cohesive surface material.

Similar results are inferred for the VL-2 sample acquisition for Inorganic Chemical Analysis (XRFS 1, Figure 1*b*). The large forces at the end of the current record (Figure 11*e*) are the result of interaction between the sampler, the soil, and rock 6 (Figure 1*b*), which was displaced upward 0.4 cm at the end of the extension stroke. Unusually low currents for the sample acquisition under rock 7 (Biology 3, Figure 1*b*) may be due to a very shallow trench, because stereoscopic pictures do not reveal the presence of a large deep trench.

Although motor currents are not available for the sample acquisitions for Molecular Analysis and Inorganic Chemical Analysis in the rocky area of VL-1 (GCMS 2, XRFS 1 and 2, Figure 1*a*), the commanded extensions were not achieved two out of five times. This means that the sampler motor clutched, and forces of at least 200 N were exerted on the surface materials there. Using the equations and dimensions above for a cohesive frictionless soil indicates that the cohesion could be ≥10 kN/m². Alternatively, buried rocks could be present. During the first extension for Inorganic Chemical Analysis in the rocky area of VL-1 (XRFS 2, Figure 1*a*) the surface sampler failed to reach its full extension, and the resulting trench (Figure 6*b*) had an area of disrupted clods and possibly small rocks several centimeters across. This soil behavior is consistent with a relatively large cohesion.

The surface sampler coupled with motor current measurements has not been fully utilized for Physical Properties Investigation at this time. Larger extensions into the Martian surface material with larger depression angles should produce larger forces and currents. This procedure is required to overcome the disadvantage of the poor resolution of 0.039 A or 25

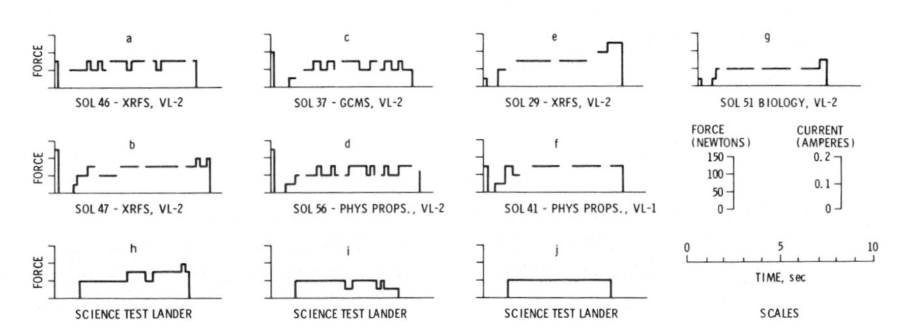

Fig. 11. (*a–g*) Surface sampler motor currents and forces for sample acquisitions on Mars and (*h–j*) Science Test lander on earth. Note the similarity in current records of (*a*), (*b*), and (*h*); (*c*) and (*d*); and (*e*) and (*f*). Four types of patterns mentioned in text are shown in (*a*) and (*b*), (*c*) and (*d*), (*e*) and (*f*), and (*g*). Generalized sampling sequences and conditions for Mars can be found in Tables 4 and 5 by using the sols for the appropriate lander.

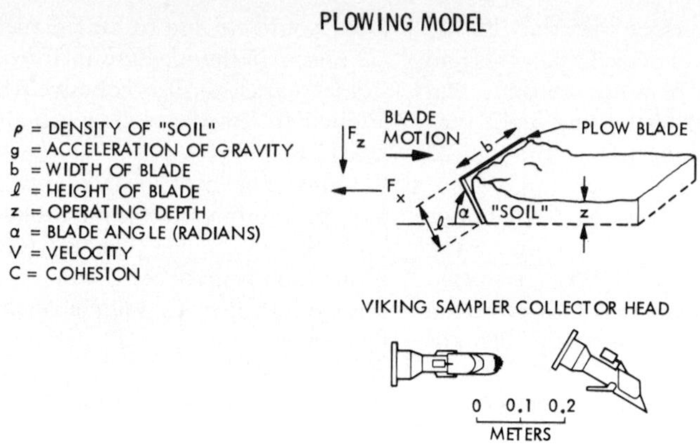

PLOWING MODEL

ρ = DENSITY OF "SOIL"
g = ACCELERATION OF GRAVITY
b = WIDTH OF BLADE
ℓ = HEIGHT OF BLADE
z = OPERATING DEPTH
α = BLADE ANGLE (RADIANS)
V = VELOCITY
C = COHESION

BLADE MOTION
PLOW BLADE
F_z
F_x
"SOIL"

VIKING SAMPLER COLLECTOR HEAD

0 0.1 0.2
METERS

PURE FRICTION

$$F_x = \rho g b z^{0.5} \ell^{1.5} \alpha^{1.73} \left\{\frac{z}{\ell \sin\alpha}\right\}^{0.77} \left\{1.05\left(\frac{z}{b}\right)^{1.1} + 1.26\frac{V^2}{g\ell} + 3.91\right\}$$

$$F_z = \rho g b z^{0.5} \ell^{1.5} \left\{0.193 - (\alpha - 0.714)^2\right\} \left\{\frac{z}{\ell \sin\alpha}\right\}^{0.777} \left\{1.31\left(\frac{z}{b}\right)^{0.966} + 1.43\frac{V^2}{g\ell} + 5.60\right\}$$

PURE COHESION

$$F_x = \rho g b z^{0.5} \ell^{1.5} \alpha^{1.15} \left\{\frac{z}{\ell \sin\alpha}\right\}^{1.21} \left\{\left(\frac{11.5C}{\rho g z}\right)^{1.21} \left(\frac{2V}{3b}\right)^{0.121} \left(0.055\left(\frac{z}{b}\right)^{0.78} + 0.065\right) + 0.64\frac{V^2}{g\ell}\right\}$$

$$F_z = \rho g b z^{0.5} \ell^{1.5} \left\{0.48 - (\alpha - 0.70)^3\right\} \left\{\frac{z}{\ell \sin\alpha}\right\} \left\{\left(\frac{11.5C}{\rho g z}\right)^{0.41} \left(\frac{2V}{3b}\right)^{0.041} \left(9.2\left(\frac{z}{b}\right)^{0.225} - 5.0\right) + 0.24\frac{V^2}{g\ell}\right\}$$

Fig. 12. Diagram illustrating plowing model, Viking surface sampler collector head, and dimensionless equations used to calculate expected forces on the collector head during sampling on Mars [*Luth and Wismer*, 1971; *Wismer and Luth*, 1972].

N of force. When this occurs in the extended mission, better estimates of cohesions and angles of internal friction of the surface materials will be possible.

Soil Adhesion

On sol 8 a picture taken of the collector head after it acquired a sample from a rather deep trench in Sandy Flats showed drift material adhering everywhere on the collector head. After this sample and a subsequent sample were delivered to Inorganic Chemical Analysis, and after the coarse fraction had been purged on sol 8, a second picture was acquired from the rather shallow trench on sol 14 (Figure 6*a*), and during retraction a malfunction occurred causing the collector head to stop. A picture of the collector head taken on sol 20 to discover the cause of the malfunction showed drift material adhering to the lower part of the collector head, but the upper part was clean, a result of the shallow excavation. The sample was delivered to Molecular Analysis on sol 22, and pictures of the collector head taken on sol 24 showed that it was again free of adhering drift material. Thus the adhering drift material had survived winds up to 15 m/s and boom oscillation during elevation from the surface, but vibrations of 8.8 and 4.4 Hz of the collector head upper jaw during sample delivery and purging removed the material. Natural boom frequencies for the extension (90.2 inches, or 229 cm) during elevation after sampling is 2.1 Hz. Tests simulating delivery and purge of the sol 14 sample on sol 22 using flight type hardware produced accelerations a of 118–529 m/s². For grains 0.01 cm across (diameter d) with a density ρ of 1500 kg/m³ the upper limit of adhesive stress σ, which strongly depends on the assumed grain size, is

$$\sigma = pda = 79 \quad \text{N/m}^2$$

Accelerations produced by the extended boom (90.2 inches, or 229 cm) during elevation after sampling are accompanied by 1.3- to 1.9-cm displacements at 2.1 Hz, so accelerations of 2.3 m/s² are indicated. Peak accelerations are 2.3 m/s² plus the acceleration due to gravity (3.8 m/s²) for a total of 6.1 m/s². Lower bound stresses are then 0.9 N/m². Thus the adhesion of the Martian soil is in the range of 0.9–79 N/m² if the grain size is in fact 0.01 cm.

Surface Temperature Measurements

The parachute phase [*Nier et al.*, 1976] ambient temperature sensors on VL-1 and VL-2 survived the shock of touchdown. The sensor is located on the inboard perimeter of footpad 2. Figures 13*a* and 13*b* show temperature plots of this sensor for both landers. The main leg support and secondary supports as well as other lander body parts cast shadows across the temperature sensor at different times during a sol. This fact resulted in a complicated insolation function for both landers. In the case of VL-1 footpad 2, which penetrated about 16 cm into the surface (see Figure 2*b*), however, it required a series of pictures taken by the boom-mounted mirror to determine that the temperature sensor was partially covered with surface material. This accounts for the elevated temperature curve compared with the predicted (H. H. Kieffer, personal communication, 1976) surface temperatures. In the case of VL-2, footpad 2 was not buried and only penetrated about 3 cm. It is unlikely that the sensor is in contact with the surface. Pictures of the temperature sensor have not yet been obtained on footpad 2 on VL-2 to verify this. If the shadow (see Figure 13*b*) were

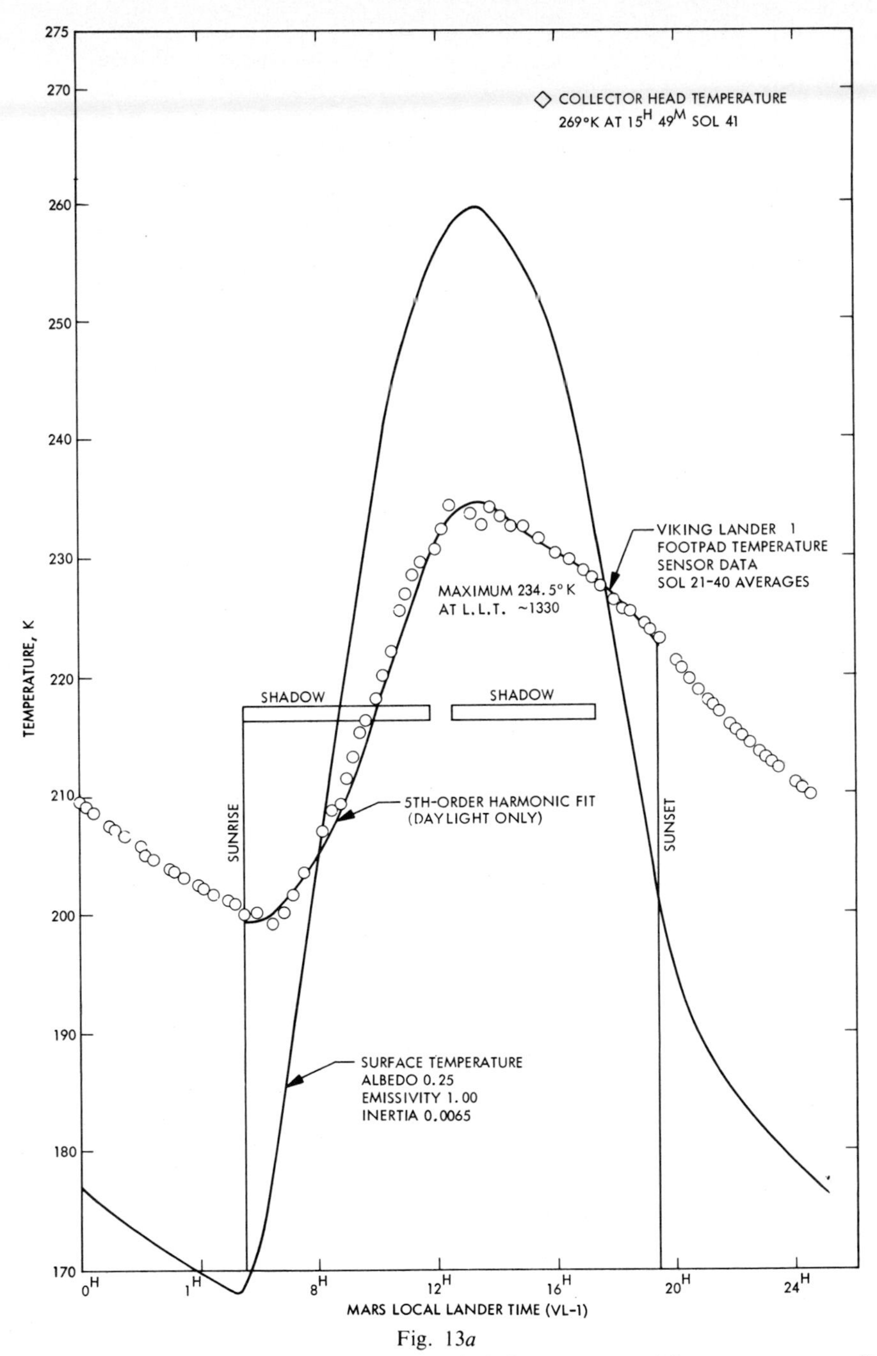

Fig. 13*a*

Fig. 13. Footpad 2 temperature sensor plots. Each open circle is an average of from one to ten readings. (*a*) VL-1 temperatures averaged over sol 21 through sol 40 for each 6-min interval. The point at 1530:00 LLT was obtained by the collector head after being in the soil for 10 min. (*b*) VL-2 temperatures averaged over sol 1 through sol 10 for each 6-min interval. The point at 1433:12 LLT was obtained by the collector head after being in the soil for 17 min. Solid curves are expected temperatures (H. H. Kieffer, personal communication, 1976). Times at which sensors were in shadow are shown by open bars.

removed, the temperature curve would probably peak slightly higher than 260°K, so it is unlikely that it is covered with surface material. Analysis of the temperature data from VL-1 and VL-2 footpad temperature sensors for the primary and extended missions will be reported elsewhere.

In addition to the footpad temperature sensor, there is a thermocouple on the bottom of the collector head intended to show that a soil sample did not reach a value in excess of 20°K above the maximum predicted surface temperature. A temperature reading was obtained each time the surface sampler boom or collector head was given a command. Upon the command to close the collector head jaws, readings were obtained 2 and 4 s later. The time constant of the thermocouple was of the order of 8 min, so the reading may be in doubt. Experiments were performed with the collector head buried several centimeters on VL-1 and just barely below the surface on VL-2 for 10 min or more. It is noteworthy that the collector head temperature of VL-2 was 272°K just before it was extended on the surface, next rose to 273°K after it was extended on the surface, and then declined to 270°K just after it was elevated from the surface. These values are plotted in Figure 13 for both landers. Since the thermocouple is physically bonded to the bottom of the collector head, conduction effects may be significant.

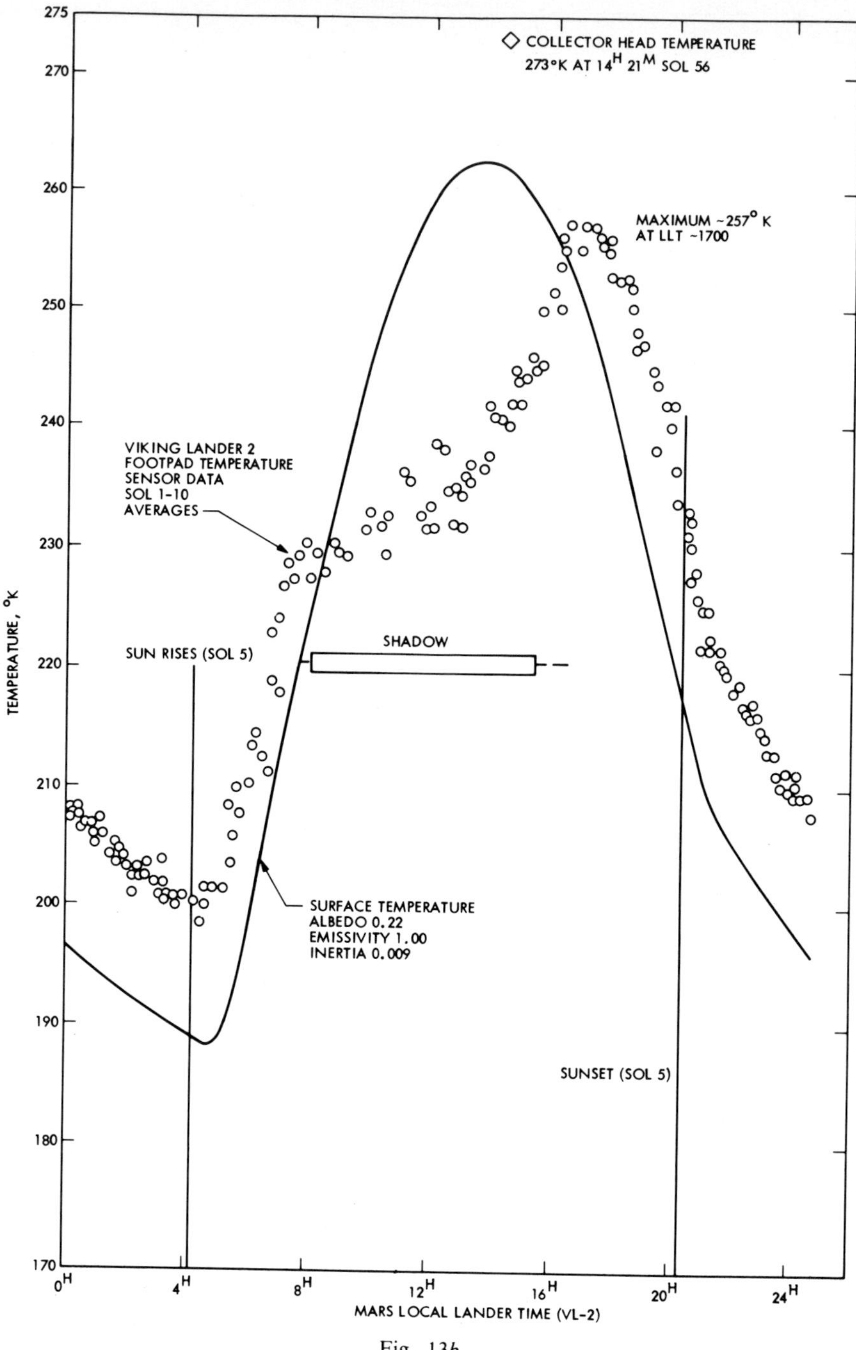

Fig. 13*b*

CONCLUDING REMARKS

Characteristics of terrestrial soils are commonly described in terms of bulk density, particle size, angle of internal friction, cohesion, and moisture content. Surface materials of Mars at the Viking sites have a broad spectrum of values for these characteristics. At VL-1, there is the relatively weak drift material, along with more cohesive materials of the rocky area, and rocks. A variety of materials are present at the VL-2 site also. Moisture content is low in comparison to terrestrial soils. Current estimates of some of the physical properties are given in Table 6 and discussed below.

Bulk Density

Bulk densities of the surface materials of Mars cover the range commonly found in natural terrestrial materials. Drift material delivered to Inorganic Chemical Analysis has a bulk density near 1100 kg/m^3 in the disturbed state. In situ densities could range from somewhat higher to somewhat lower than this by a small amount. Penetration by footpad 2, under the assumptions that results can be scaled by using low-velocity impact equations and that there is no hard substrate, lends some support to the notion that the in situ density is slightly larger than 1100 kg/m^3, say, near 1300–1400 kg/m^3. A similar conclusion is reached from the craters produced by grains and fragments propelled by engine exhausts. The coarse fraction of cohesive material from Rocky Flats has a disturbed density near 600 kg/m^3 in the XRFS chamber. If the sample in the chamber is 50% voids and 50% clods, the density of the clods is about 1200 kg/m^3. At least some of the fragments disrupted during the sol 29 acquisition are probably clods, and engine exhaust erosion exposed a fractured cohesive surface, so it

TABLE 6. Current Best Estimate of Soil Properties Deduced from Viking 1 and Viking 2 Data

Property	VL-1 Sandy Flats	VL-1 Rocky Flats	VL-2
Bulk density, kg/m^3			
Soil	1000–1600	1200–1600	1100–1480
Rock		2900	2600
Particle size, %			
>2 cm	0	25	20
Clods and fines	100	75	80
Cohesion of soil, N/m^2	$10–10^2$	$10–10^4$	$10–10^3$
Cohesion of rock, N/m^2	...	$>10^4$	$>10^4$
Angle of internal friction, deg	30–45	30–45	30–45
Penetration resistance, $N/m^2/m$	3×10^5	6×10^6	6×10^6
Adhesion, N/m^2	$1–10^2$	...	...
Coefficient of sliding friction	0.3–0.5	0.3–0.5	...

Particles as used here include clods as well as individual mineral and rock grains; thus the estimates of particle sizes must be considered approximate and preliminary. The frequency of rocks 10 cm and larger at VL-2 is twice as large as that at VL-1 (Figures 1 and 2). The estimated density of materials after delivery to the XRFS experiment is 1.1 ± 0.15 g/cm^3. The estimates may be revised at a future date.

seems likely that in situ bulk densities of this material are near 1200 kg/m^3. Since the Rocky Flats sample came from between larger rocks, the bulk density of the whole rocky area will be larger. For a material with 25% rocks with densities of 2900 kg/m^3 and 75% matrix with a density of 1200 kg/m^3 the bulk density of the entire rocky soil would be 1625 kg/m^3. For comparison, reflection coefficients of terrestrial radar imply a density near 2000 kg/m^3 for the Chryse region near the landing site [*Tyler et al.*, 1976]. Quantitative data for a more rigorous detailed comparison of lander and radar results are not yet available.

Densities of samples of fines determined by Inorganic Chemical Analysis by VL-2 are comparable to those of VL-1 (B. Clark, personal communication, 1977). Because about 20% of the sample field is covered by rocks, the bulk density of the entire material at VL-2 would be higher. If 20% of the volume of the material is rocks with a density of 2600 kg/m^3 (because they may be vesicular) and the remaining is matrix with a density of 1200 kg/m^3, the density of the whole would be 1480 kg/m^3.

The Inorganic Chemical Analysis indicates that the surface materials may be montmorillonite clays [*Baird et al.*, 1976]. If this is the case, individual mineral grains composing the fines at both landing sites should be comparable to those of montmorillonite clays on earth. These are typically 2500–2600 kg/m^3.

Particle Size

Again large variations are present at both sites. Rocks several meters across occur in the field of view of both landers. Rocks several centimeters in size are also found in the field of view at both landers. One rock, 2.7 cm across, was purged by VL-2. Many of the objects seen in the lander pictures a few millimeters across to several centimeters are clods as proved by the repeated attempts to collect rocks. In most cases, small clods and rocks cannot be separated visually from one another in the pictures. Local smoothing and tamping of disturbed material that produces highly reflective surfaces is consistent with granular material that is silt size and smaller.

Cohesion

Cohesion of the drift material of VL-1 determined from the relief of marginally stable trench walls is in the range of $10–10^2$ N/m^2. Failure of the surface sampler of VL-1 to reach its commanded extension at Rocky Flats combined with estimates of forces required for plowing suggests that cohesions there may exceed 10^4 N/m^2, but locally they are lower. A large cohesion and relatively large bulk density of the rocky material are consistent with the small penetration by footpad 3 of VL-1. Penetration by footpad 3 by all criteria is substantially smaller than would be expected for a lunar nominal soil with a cohesion of 10^3 N/m^2 and a density of 1600 kg/m^3. Thus it seems likely that the cohesion of the rocky area may be rather large. Soil between the rocks at VL-2 probably has a cohesion near 2×10^3 N/m^2, but it could be less at least locally. Support for a cohesion near 10^3 N/m^2 comes from the backhoe touchdown picture on sol 21 as well as surface sampler motor currents. The rocks at both sites clearly have very large cohesion, but measurement is not possible. Considerable strength for the rocks is implied by the attempt to push rock 1. Stresses exerted by the sampler collector head teeth must have exceeded $\sim 10^8$ N/m^2, but they were not large enough to chip or spall the rock.

Angle of Internal Friction

Careful analyses of deformations around trenches have not been made at this time, so the angles of internal friction that appear in Table 6 are based on rough estimates. Deformations and disruption of the soils extend considerable distances from a number of the trenches. Topographic evidence indicates that the surface was domed out to at least 10 cm from the rim of the sol 8 trench of VL-1 (Figure 6) and 7–8 cm in the lateral directions. Similar results for the sol 8 trench and other trenches show that the soils fail by general shear and have internal friction. The pile of drift material deposited on the grid of VL-1 (Figure 7) has slopes slightly larger than 39°, implying that an angle of internal friction in the loose state [*Terzaghi*, 1943] is near but somewhat larger than 39°.

Aeolian Transportability

Four observations on the physical properties of the Martian surface place constraints on aeolian processes: (1) the surface materials have cohesion, (2) engine exhausts transported surface materials, (3) material deposited on the body of VL-1 has been winnowed away by the wind, and (4) the profusion of rocks on the surface should affect the near-surface wind velocity gradients. Cohesion of the drift material of the VL-1 site and interrock fines at both sites indicate that the threshold wind velocity required to initiate grain motion will pass through a minimum just as terrestrial soils have a minimum [*Bagnold,* 1941]. Readily entrained very fine grained cohesionless soils [*Sagan and Bagnold,* 1975] are absent at the Viking landing sites and perhaps Mars. Additional products of cohesion are aggregates and clods of soils which require rather complicated aeolian transport models for the soils (see, for example, *Chepil and Woodruff* [1963]). Indeed, areas where surface fines at the VL-2 site have been stripped away to expose fractured and blocky soil units are similar to surfaces of underlying clay exposed during terrestrial tests [*Chepil and Woodruff,* 1963]. A final result of cohesion is the possibility that Martian dunes are made of aggregates and clods instead of mineral grains and rocky fragments.

The quantitative aspects of engine exhaust erosion have not yet been fully explored. It is clear, however, that the surface materials are erodible and can be transported. Soil deposited on the body of VL-1 has been eroded and winnowed away by the wind. Thus it is probable that the dunelike structures seen at both landing sites are in fact a result of aeolian transport.

Ubiquitous large blocks observed by both landers should substantially contribute to the stability of the rocky surfaces to wind erosion by altering the near-surface velocity gradient in a complicated way [*Chepil and Woodruff,* 1963]. Erosive effects of impacts of saltating grains along flat trajectories should be substantially reduced or even eliminated at times by the blocks. Such a process would tend to abrade the rocks and leave an indurated rock surface.

Volatiles

The large amounts of water (for Mars) evolved during heating as part of the Molecular Analysis experiment lend strong support to models requiring storage of water in the Martian regolith [*Fanale,* 1976; *Huguenin,* 1976]. Water evolved from the sample under rock 3 at the VL-2 site during heating from 50° to 200°C may represent adsorbed water. If this is the case and Mars is like the earth, adsorbed water may be present at larger depths, where it is cooler. Thus the Martian regolith may contain substantial amounts of chemically bound and adsorbed water.

Temperatures

Although they have not been fully analyzed, footpad 2 temperatures and collector head temperatures of the Martian soil lie well above the predicted curves. Footpad 2 temperature sensors of VL-1, which are immersed in soil from 0.2 to 1.5 cm, are 30°K higher just before sunrise than the predicted curves. Since temperatures are approximately corrected for spacecraft-related conduction, the difference is significant. A similar result is found for the unburied footpad 2 temperature sensor for VL-2, for which temperatures are 10°K higher than the predicted curve just before sunrise. Collector head temperatures are likewise higher than the predicted curves. For VL-2 the collector head temperatures are roughly 10°K higher than the predicted curve.

It is noteworthy that the collector head temperature of VL-2 reached 273°K or very close to the temperature of the triple point of water. Since surface pressures are substantially greater than 6 mbar [*Seiff and Kirk,* 1976], it is entirely possible that pressure-temperature conditions at the upper surface of Mars are in the stability field of liquid water for short periods of time and in local areas. This suggests that near-surface freeze-thaw cycles may exist. Perhaps more data on surface temperatures using the collector head temperature sensor can be obtained during the extended mission.

Acknowledgments. We acknowledge the continuing aid and support given to the Physical Properties Investigation Team by the Surface Sampler Team, L. V. Clark, D. S. Crouch, L. K. Schwab, K. Z. Bradford, and W. DeShazor. The Imaging Team kindly furnished the images used in this report. We thank R. B. Hargraves, D. W. Collinson, and E. C. Morris, who gave us able assistance throughout the mission. S. Liebes, Jr., provided the mensuration data for the rock pushes and nudges. We also thank I. M. Mack for her assistance as the Physical Properties intern during the month of August and P. Duffy for his assistance for the month of September. For typing the manuscript several times and for attending to many administrative matters we thank L. Crafton. We appreciate the support of R. Goldstein during sol 0 for VL-1 and VL-2 and throughout the mission. We also thank A. Castro and V. Gillespie for their additional support. The help of H. Zimmer and G. Neukum during the acquisition of data is appreciated. The trajectory data were kindly furnished by A. Fontana, F. W. Hopper, J. T. Findley, and J. W. Gerschultz. We acknowledge the help and assistance of S. Dwornik, particularly at the beginning of the project when the Physical Properties Investigation first started. We are especially appreciative of the assistance of P. Cates during the primary mission and particularly during the extended mission. Finally, the Physical Properties Team wishes to express its sincere 'thank you' to the project manager, Jim Martin; the mission director, Tom Young; and the project scientist, G. A. Soffen, for their untiring dedication to the goals of the mission, which maximized the science return. This work was supported by NASA contract NAS1-12705 to the Geospace Sciences Laboratory of the University of Utah Research Institute, NASA order L-9714 to the U.S. Geological Survey, and NASA contract NAS1-10534 to TRW Systems, Inc.

REFERENCES

Bagnold, R. A., *The Physics of Blown Sand and Desert Dunes,* 265 pp., Methuen, London, 1941.

Baird, A. K., P. Toulmin III, B. C. Clark, H. J. Rose, Jr., K. Keil, R. P. Christian, and J. L. Gooding, Mineralogic and petrologic implications of Viking geochemical results from Mars: Interim report, *Science, 194,* 1288, 1976.

Biemann, K., et al., Composition of the atmosphere and search for organic compounds at the Martian surface (abstract), *Eos Trans. AGU, 57,* 945, 1976.

Chepil, W. S., and N. P. Woodruff, The physics of wind erosion and its control, *Advan. Agron., 15,* 211, 1963.

Clark, B. C., A. K. Baird, H. J. Rose, Jr., P. Toulmin III, K. Keil, A. J. Castro, W. C. Kelliher, C. D. Rowe, and P. H. Evans, Inorganic analysis of Martian surface samples at the Viking landing sites, *Science, 194,* 1283, 1976.

Clark, L. V., Effect of ambient pressure on Viking Lander footpad penetration in nominal lunar soil, *Lett. 159,* Viking Proj. Office, NASA Langley Res. Center, Hampton, Va., Oct. 12, 1971.

Clark, L. V., and J. L. McCarty, The effect of vacuum on the penetration characteristics of projectiles into fine particles, *NASA Tech. Note, D-1519,* 1963.

Clark, L. V., D. S. Crouch, and R. D. Grossart, Viking '75 Project summary of primary mission surface sampler operations, *Doc. VFT-019,* 477 pp., Viking Flight Team, NASA Langley Res. Center, Hampton, Va., 1977.

Crouch, D. S., PTC surface sampler boom loading test with Format 5 and SSCA TM data, *Lett. SST-17870-DCS,* Martin Marietta Corporation, Littleton, Colo., June 25, 1976.

Fanale, F. P., Martian volatiles: Their degassing history and geochemical fate, *Icarus, 28,* 179, 1976.

Horowitz, N. H., G. L. Hobby, and J. S. Hubbard, The Viking carbon assimilation experiments: Interim report, *Science, 194,* 1321, 1976.

Huguenin, R. L., Mars: Chemical weathering as a massive volatile sink, *Icarus, 28,* 203, 1976.

Jury, W. A., and B. Bellantuoni, Heat and water movement under surface rocks in a field of soil, I, Thermal effects, *Soil Sci. Soc. Amer. J., 40,* 505, 1976*a*.

Jury, W. A., and B. Bellantuoni, Heat and water movement under surface rocks in a field of soil, II, Moisture effects, *Soil Sci. Soc. Amer. J., 40,* 509, 1976*b*.

Luth, H. J., and R. O. Wismer, Performance of plane soil cutting blades in sand, *Trans. ASAE, 14,* 255, 1971.

Martin Marietta Corporation, Footpad soil penetration tests, 1, Data summary, *Rep. VER-188,* Littleton, Colo., 1971.

Martin Marietta Corporation, Structures and mechanisms design data book, *Rep. VER-273,* Littleton, Colo., 1973*a*.

Martin Marietta Corporation, Surface sampler collector head assembly, *Drawing 837J-5500100,* Littleton, Colo., 1973*b*.

Martin Marietta Corporation, Entry data analysis for Viking landers 1 and 2, *Final Rep. TN 3770 218,* Littleton, Colo., 1976.

Mutch, T. A., R. E. Arvidson, A. B. Binder, F. O. Huck, E. C. Levinthal, S. Liebes, Jr., E. C. Morris, D. Nummedal, J. B. Pollack, and C. Sagan, Fine particles on Mars: Observations with the Viking 1 lander cameras, *Science, 194,* 87, 1976*a*.

Mutch, T. A., S. U. Grenander, K. L. Jones, W. Patterson, R. E. Arvidson, E. A. Guiness, P. Avrin, C. E. Carlston, A. B. Binder, C. Sagan, E. W. Dunham, P. L. Fox, D. C. Pieri, F. O. Huck, C. W. Rowland, G. R. Taylor, S. D. Wall, R. Kahn, E. C. Levinthal, S. Liebes, Jr., R. B. Tucker, E. C. Morris, J. B. Pollack, R. S. Saunders, and M. R. Wolf, The surface of Mars: The view from the Viking 2 lander, *Science, 194,* 1277, 1976*b*.

National Aeronautics and Space Administration, Viking '75 Project, Viking mission definition, *Rep. M 75-123-1 (RS-3703001), Append. D,* p. 35, Viking Proj. Office, NASA Langley Res. Center, Hampton, Va., 1970.

Nier, A. O., W. B. Hanson, A. Seiff, M. B. McElroy, N. W. Spencer, R. J. Duckett, T. C. D. Knight, and W. S. Cook, Composition and structure of the Martian atmosphere: Preliminary results from Viking 1, *Science, 193,* 786, 1976.

Pyrz, A. P., Gravity effects on low velocity penetration of a projectile into a cohesionless medium, *Rep. GSF/MC/69-6,* School of Engineering, Wright-Patterson Air Force Base, Ohio, 1969.

Romine, G. L., T. D. Reisert, and J. Gliozzi, Site alteration effects from rocket exhaust impingement during a simulated Viking Mars landing, *NASA Contract. Rep. CR-2252,* 1973.

Sagan, C., and R. A. Bagnold, Fluid transport on earth and aeolian transport on Mars, *Icarus, 26,* 209, 1975.

Scott, R. F., *Principles of Soil Mechanics,* Addison-Wesley, Reading, Mass., 1963.

Seiff, A., and D. B. Kirk, Structure of Mars' atmosphere up to 100 kilometers from the entry measurements of Viking 2, *Science, 194,* 1300, 1976.

Shorthill, R. W., R. E. Hutton, H. J. Moore, and R. F. Scott, Martian physical properties experiments: The Viking Mars lander, *Icarus, 16,* 217, 1972.

Shorthill, R. W., R. E. Hutton, H. J. Moore, R. F. Scott, and C. R. Spitzer, Physical properties of the Martian surface materials from the Viking 1 lander: Preliminary results, *Science, 193,* 805, 1976*a*.

Shorthill, R. W., H. J. Moore, R. F. Scott, R. E. Hutton, S. Liebes, Jr., and C. R. Spitzer, The 'soil' of Mars (Viking 1), *Science, 194,* 91, 1976*b*.

Shorthill, R. W., H. J. Moore, R. E. Hutton, R. F. Scott, and C. R. Spitzer, The environs of Viking 2 lander, *Science, 194,* 1309, 1976*c*.

Soffen, G. A., Scientific results of the Viking missions, *Science, 194,* 1274, 1976.

Terzaghi, K., *Theoretical Soil Mechanics,* John Wiley, New York, 1943.

Tyler, G. L., D. B. Campbell, G. S. Downs, R. R. Green, and H. J. Moore, Radar characteristics of the Viking 1 landing sites, *Science, 193,* 812, 1976.

Wismer, R. D., and H. J. Luth, Performance of plane cutting blades in clay, *Trans. ASAE, 15,* 211, 1972.

Young, C. W., The development of empirical equations for predicting depth of an earth-penetrating projectile, *Develop. Rep. SC-DR-67-60,* Sandia Corp., Albuquerque, N. Mex., 1967.

(Received March 31, 1977;
revised May 23, 1977;
accepted May 24, 1977.)

VOL. 82, NO. 28 JOURNAL OF GEOPHYSICAL RESEARCH SEPTEMBER 30, 1977

Seismology on Mars

DON L. ANDERSON,[1] W. F. MILLER,[1] G. V. LATHAM,[2] Y. NAKAMURA,[2] M. N. TOKSÖZ,[3] A. M. DAINTY,[3] F. K. DUENNEBIER,[4] A. R. LAZAREWICZ,[4] R. L. KOVACH,[5] AND T. C. D. KNIGHT[6]

A three-axis short-period seismometer has been operating on the surface of Mars in the Utopia Planitia region since September 4, 1976. During the first 5 months of operation, approximately 640 hours of high-quality data, uncontaminated by lander or wind noise, have been obtained. The detection threshold is estimated to be magnitude 3 to about 200 km and about 6.5 for the planet as a whole. No large events have been seen during this period, a result indicating that Mars is less seismically active than earth. Wind is the major source of noise during the day, although the noise level was at or below the sensitivity threshold of the seismometer for most of the night during the early part of the mission. Winds and therefore the seismic background started to intrude into the nighttime hours starting on sol 119 (a sol is a Martian day). The seismic background correlates well with wind velocity and is proportional to the square of the wind velocity, as is appropriate for turbulent flow. The seismic envelope power spectral density is proportional to frequency to the −0.66 to −0.90 power during windy periods. A possible local seismic event was detected on sol 80. No wind data were obtained at the time, so a wind disturbance cannot be ruled out. However, this event has some unusual characteristics and is similar to local events recorded on earth through a Viking seismometer system. If it is interpreted as a natural seismic event, it has a magnitude of 3 and a distance of 110 km. Preliminary interpretation of later arrivals in the signal suggest a crustal thickness of 15 km at the Utopia Planitia site which is within the range of crustal models derived from the gravity field. More events must be recorded before a firm interpretation can be made of seismicity or crustal structure. One firm conclusion is that the natural background noise on Mars is low and that the wind is the prime noise source. It will be possible to reduce this noise by a factor of 10^3 on future missions by removing the seismometer from the lander, operation of an extremely sensitive seismometer thus being possible on the surface.

INTRODUCTION

Because the primary emphasis on landed Viking science was on biology, organic chemistry, imagery, and meteorology, other areas such as surface chemistry, petrology, and geophysics were mostly relegated to future missions. The exceptions were the inorganic analysis experiment, the magnet experiment, and the seismometer. However, these studies were limited to reconnaissance measurements.

The ultimate goals of a seismic experiment are to determine the dynamics and internal structure of a planet. These are relevant to the composition and evolution of the interior. Before a comprehensive seismic experiment can be considered, however, it is necessary to establish the background noise level of the planet, the level, nature, and location of the seismicity, and the nature of the seismic signals. Estimates of some of these parameters have been obtained by the Viking seismometer. The ultimate goal, however, can only be accomplished with a network of sensitive seismometers deployed so as to minimize artificial sources of noise and wind-induced vibrations.

The Viking seismic experiment design was constrained by strict weight, power, and data allocations and perturbed by the conflicting demands of the other on-board experiments. The weight constraint precluded an ultrasensitive seismometer of the Apollo class or a broadband seismometer. The original desire to offload the seismometer was sacrificed because of the weight and complexity penalty of such an operation; thus an on-board location was dictated that immediately increased the noise level because of lander and wind activity (by at least 3 orders of magnitude). The data constraint required severe data compression using an on-board data processing capability and thus also a weight and power penalty. The most severe constraints on the Viking seismic experiment were limited data allocations and the on-board location of the seismometer.

These various constraints and trade-offs led to the design of a short-period three-component seismometer with on-board data compaction and triggering to optimize the data return [*Anderson et al.*, 1972]. The objectives were (1) to characterize the seismic noise environment at the landing sites, (2) to detect local events in the vicinity of the lander, and (3) to detect large events at teleseismic distances.

Under optimal conditions it would also be possible to determine the following: (1) the approximate distance of events by the separation of the various seismic phases, (2) the direction of events with a 180° ambiguity, (3) the attenuation and scattering properties of the crust to determine if the crust is lunar-like or earthlike in these characteristics (which are related to the volatile content), and (4) an estimate of crustal thickness if crustal and reflected phases can be identified. The Viking 1 seismometer failed to uncage, and no useful data were returned. The Viking 2 seismometer, emplaced on the surface of Mars in the Utopia Planitia region, 47.9°N, 225.9°W, successfully uncaged and has been operating nominally since that time [*Anderson et al.*, 1976].

STATE OF STRESS IN THE MARTIAN INTERIOR

The regional topography of the earth is in approximate isostatic balance, and regions of partial compensation are generally the most seismically active. Except for the nearside mascons the moon is also in equilibrium. Both bodies are seismically active, although the moon is many orders of magnitude less active than the earth. Large areas of Mars seem to

[1] Division of Geological and Planetary Sciences, California Institute of Technology, Pasadena, California 91125.
[2] University of Texas, Galveston, Texas 77550.
[3] Massachusetts Institute of Technology, Cambridge, Massachusetts 02139.
[4] University of Hawaii, Honolulu, Hawaii 96822.
[5] Stanford University, Palo Alto, California 94305.
[6] Martin Marietta Aerospace Corp., Denver, Colorado 80201.

Paper number 7S0408.

be only partially compensated [*Phillips and Saunders,* 1975] and could be seismically active. These are the younger areas which include the Tharsis plateau and the adjacent low areas of Chryse and Amazonis. Stresses implied by lack of compensation and dynamic stresses in the earth and the moon are in the 10- to 100-bar range. Stress drops in earthquakes are also in this range. Laboratory measurements on crustal rocks at modest pressure give short-term strengths of kilobars, but rocks creep at high temperatures at relatively low stress levels. Stresses in the crust of a planet therefore tend to decrease with time unless they are rejuvenated by plate tectonic motions, convective drag, or thermoelastic stresses caused by heating or cooling. *Phillips and Tiernan* [1977] argue that the Tharsis gravity anomaly represents a long-wavelength stress supported by the lithosphere or the asthenosphere for 10^8–10^9 years, implying stresses in the mantle of 100 bars or stresses in the lithosphere of a few kilobars. They favor at least partial dynamic support of Tharsis by mechanisms such as mantle convection or viscous and thermoelastic stresses associated with magmatic activity. The existence of large stresses in the Tharsis region, at least at the time of formation, is also implied by the pattern of lineaments, including apparent grabens and the great canyon, aligned generally radially with respect to the plateau summit [*Blasius and Cutts,* 1976]. The depth of compensation of the Martian topography is of the order of 150 km, and about 3 km of Tharsis is presently uncompensated [*Phillips and Saunders,* 1975]. It is likely that the uncompensated areas of Mars are seismically active, since the stresses are similar or greater than stresses in the earth and moon. This is true whether the planet is striving toward a state of isostatic equilibrium or whether the stresses are being rejuvenated. Most of the seismic activity on earth is associated with plate margins, although moderate to large (but infrequent) earthquakes also occur in plate interiors of both oceanic and continental plates. Plate motions on the moon and Mars are certainly less pronounced than on earth, so we might also expect Mars to be less seismically active than earth. On the other hand, the larger stresses required to support the topography and the gravity field imply the potential for seismic activity even if creep and flow occur. In the earth and in the laboratory, deformation proceeds by both slow aseismic processes (creep) and rapid seismic processes (earthquakes). The latter dominate at low temperatures such as exist in the lithosphere.

The variations of gravity and topography suggest that the seismicity of Mars is not uniform and that it is likely to be highest near Tharsis and its environs and similar young uncompensated constructs. The Viking 2 lander is approximately 110° from Chryse Planitia (VL-1 site) and 25° from Elysium Mons. Marsquakes at these locations would have to be quite large to be detected by the Viking 2 seismometer. In summary, we expect Mars to be a tectonically active planet, but the seismicity may be localized.

MODELS FOR THE INTERIOR OF MARS

Interpretation of Martian seismic data requires an a priori model of the planet's interior. Such a model must be calculated on the basis of available direct data, cosmochemical considerations, equations of state, and theoretical modeling of the planetary interior and its thermal evolution. Such efforts have been made by a number of investigators [*Urey,* 1952; *Anderson,* 1972; *Anderson and Kovach,* 1967; *Kovach and Anderson,* 1965; *Binder,* 1969; *Binder and Davis,* 1973; *Johnston et al.,* 1974; *Johnston and Toksöz,* 1977; *Ringwood and Clark,* 1971; *Toksöz and Johnston,* 1977]. Here we will briefly review the results most relevant to the seismic experiment. The only direct evidence concerning the internal structure of Mars prior to the Viking landings was the mean density, moment of inertia, topography, and gravity field. It is possible to calculate models of the Martian interior with plausible chemical models and temperature profiles that satisfy these few constraints. However, the process is highly nonunique.

The mean density of Mars, corrected for pressure, is less than that of earth, Venus, and Mercury but greater than that of the moon. This implies that Mars has a smaller total Fe-Ni content than do the other terrestrial planets and more than does the moon. Plausible models for Mars can be constructed which have solar or chondritic values for Fe [*Anderson,* 1972]. Mars is the only terrestrial planet, including the moon, which could have chondritic abundances of iron, and it is therefore likely to contain a larger proportion of volatiles such as water-, sulfur-, and potassium-bearing phases. The densities and compositions of the planets are consistent with the view that the volatile content increases with distance from the sun [*Anderson,* 1971]. This can be understood in terms of a primitive solar nebula in which the decreasing temperature away from the sun controlled the composition of the condensates which were available for incorporation into the various planets. In this view, Mars should be more volatile rich than the other terrestrial planets. Volatiles in this context include FeS, hydrous minerals, water, K_2O, and Na_2O. Low condensation temperatures also imply a higher oxidation state for the available iron; i.e., more Fe should be available as FeS, FeO, Fe_2O_3, and Fe_3O_4 than is the case for the primitive earth. The core of Mars therefore is probably richer in FeS, and the mantle is richer in FeO than is the case for the earth, where more free Fe is available. A large Fe-FeS core, an FeO-rich mantle, and a thick crust (Na_2O, K_2O, and hydrous minerals) are therefore expected. On the other hand, the relative proportions of CaO and Al_2O_3, components of crustal minerals, are expected to be less, and the total crustal thickness may be buffered by the availability of these constituents. However, SiO_2 and low-melting-point FeO-bearing compounds probably dominate the crustal composition.

With such broad chemical constraints, mean density ($\bar{\rho}$ = 3.96 g/cm^3), and moment of inertia factor (C/MR^2 = 0.365, *Reasenberg* [1977]) and under the assumption of a differentiated planet it is possible to trade off the size and density of the core and density of the mantle [*Anderson,* 1972; *Binder and Davis,* 1973; *Johnston et al.,* 1974; *Johnston and Toksöz,* 1977]. Most of these models favor an FeO enrichment of the Martian mantle relative to the mantle of the earth. *Anderson* [1972] concluded that Mars has a total iron content of about 25 wt %, which is significantly less than the iron content of earth, Mercury, or Venus but is close to the total iron content of ordinary and carbonaceous chondrites. The high zero-pressure density of the mantle suggests a relatively high FeO content in the silicates of the Martian mantle. The radius of the core can range from as small as one-third the radius of the planet for an iron core, or a core similar in composition to the earth's core, to more than half the radius of the planet if it is pure FeS. With chondritic abundances of Fe-FeS the size of the core would be about 45% of the planet's radius, or about 12% by mass. A small dense core would imply a high-temperature origin or early history because of the high melting temperature of Fe-Ni, while a larger light core, presumably rich in sulfur, would allow a cooler early history, since sulfur substantially reduces the melting temperature. A satisfactory model for the interior of Mars can be obtained by exposing ordinary chondrites to modest temperatures, melting the Fe-FeS, and removing most of it to the core [*Anderson,* 1972]. This would make the Martian core substantially more S rich than the terrestrial core and would favor the larger core radius. Alternatively, Mars could be a mixture of carbonaceous and ordinary chondrites (or hypersthene or bronzite chondrites) and satisfy the mean density and moment of inertia with partial removal of Fe-Ni-S from the silicate phase to the core.

The trade-offs between core size, composition, and mantle density are given by *Anderson* [1972] and *Johnston and Toksöz* [1977]. Under the assumption of an uncertainty of ±0.005 in the moment of inertia factor the core radius in these models could vary between about 1300 and 1800 km, and the mantle density, corrected to 0°C and 1-bar pressure, between ρ_m of 3.4 and 3.6 g/cm³. It is obvious that a direct determination of the size of the core by seismic means could constrain the overall composition of the planet.

The topography and gravity fields of Mars indicate that parts of Mars are grossly out of hydrostatic equilibrium and that the crust is highly variable in thickness. If variations in the gravity field are attributed to variations in crustal thickness, with a constant density ratio between crust and mantle, then reasonable values of the density contrast (<0.8 g/cm³) imply that the average crustal thickness is at least 30 km (W. M. Kaula, personal communication, 1976). This minimal bound is based on the assumption of zero crustal thickness in the Hellas basin (R. Phillips, personal communication, 1976). An impact large enough to excavate the Hellas basin would easily remove a 30-km-thick crustal layer (T. J. Ahrens, personal communication, 1976). This minimal average crustal thickness on Mars gives a crust/planet mass ratio that is more than 5 times the terrestrial value, indicating a well-differentiated planet.

The crust of the earth is enriched in CaO, Al_2O_3, K_2O, and Na_2O in comparison to the mantle. Ionic radii considerations and experimental petrological results suggest that the crust of any planet will be enriched in these constituents. A minimal average crustal thickness for a fully differentiated chondritic planet can be obtained by removing all of the CaO possible, with the available Al_2O_3, as anorthite to the surface. This gives a crustal thickness of about 100 km for Mars. Incomplete differentiation and retention of CaO and Al_2O_3 in the mantle will reduce this value, which is likely to be the absolute upper bound.

The average thickness of the crust of the earth is 15 km, which amounts to 0.4% of the mass of the earth. The crustal thickness is 5–10 km under oceans and 30–50 km under older continental shields. The thickness of the crust increases with age. The volume of the crust increases only slightly if presently active subducted regions are taken into account. The situation on the earth is complicated, since new crust is constantly being created at the midoceanic ridges and consumed at island arcs. It is probable that some of the crust is being recycled. If the present rate of crustal genesis was constant over the age of the earth and none of the crust was recycled, then 17% of the earth would be crustal material.

The moon apparently has a mean crustal thickness greater than that of the earth. If the average composition of the moon is similar to chondrites minus the Fe-Ni-S, then the crust could be as thick as an average of 62 km. This is about the thickness of the crust determined by seismic experiments on the frontside of the moon [*Toksöz et al.,* 1974] but less than the average thickness inferred from the gravity field [*Bills and Ferrari,* 1977*a*]. An alternative model based on a Ca-Al-rich moon [*Anderson,* 1973] can yield a much greater crustal thickness.

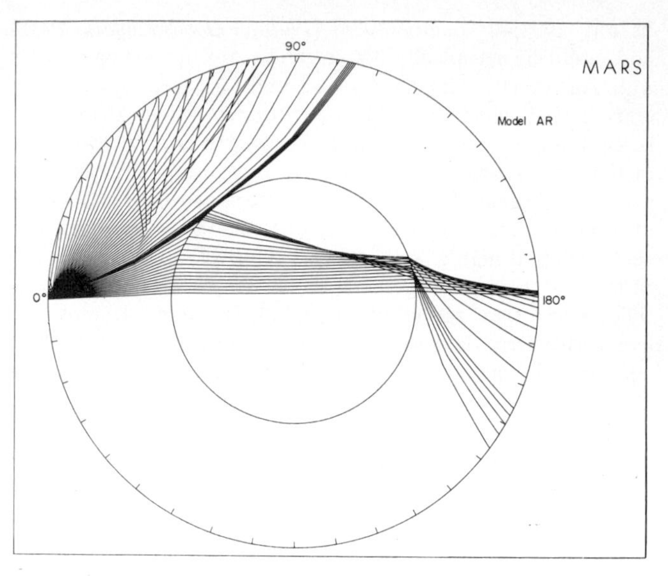

Fig. 1. Seismic ray paths for compressional waves through the interior of a theoretical Martian model that satisfies the mean density and moment of inertia. The 'mantle' is homogeneous in composition, the olivine-spinel transition causing a discontinuity at a 1100-km depth. Note the shadow zone at the surface caused by the decrease in velocity in the core. The core is assumed to be molten. Parameters for the model are shown in Figure 2.

The surface reflectance spectra of Phobos and Deimos and the low density of Phobos are also consistent with the fact that these bodies are similar to carbonaceous chondrites. Even if these are captured bodies, the suggestion is that Mars may have accreted from relatively volatile rich material. The state of water on Mars is unknown, but the evidence for running water at the surface early in its history and the inability of substantial amounts of water to escape from the atmosphere indicate that there should be free or bound water in the interior.

Atmospheric analyses suggest that Mars is a less outgassed planet than earth [*Owen and Biemann,* 1976], but the high ratio of ^{40}Ar relative to the other inert gases suggests that the Martian crust is enriched in ^{40}K in relation to the earth's crust.

The volatile content of a planet's interior along with the temperature of the interior are the two most important factors that control the attenuation of seismic waves. The attenuation properties of the Martian mantle may be similar to those of the earth's mantle with the possible exception of the low Q (high attenuation) zones of the earth's asthenosphere. Thus we would expect seismic waves to propagate through the Martian interior with similar or greater efficiency than they do on earth, particularly if the lithosphere is as thick as is implied by isostatic calculations [*Phillips and Saunders,* 1975]. Compressional wave ray paths are shown in Figure 1 for a typical Martian model based on an FeS core composition. Note the large shadow zone due to the low-velocity core. It is important to point out that these models are nonunique and actual seismic data are required to determine the internal structure and composition of Mars. The velocities and density models used to calculate the ray paths in Figure 1 are shown in Figure 2.

CRUSTAL STRUCTURE OF MARS

It is clear from the topography and gravity field of Mars that the crustal layer is highly nonuniform. Attempts have been

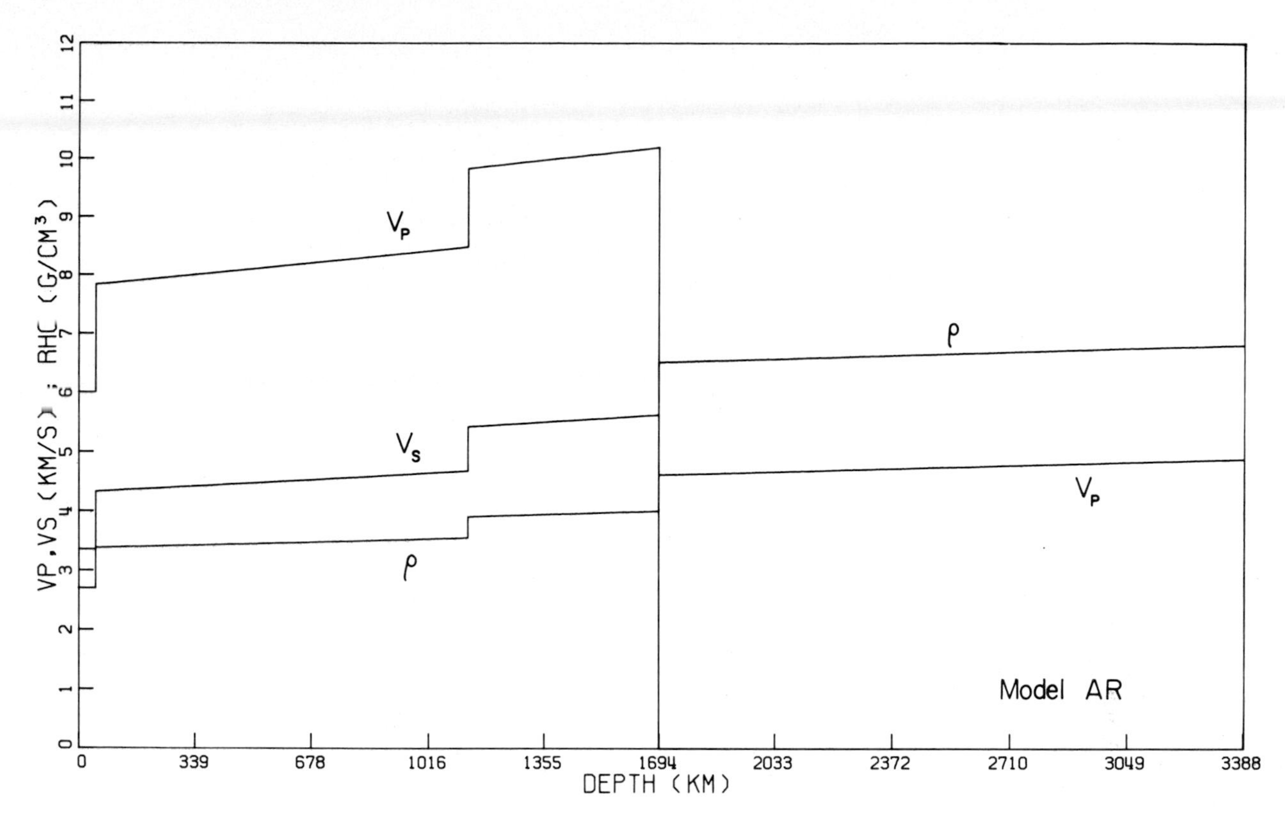

Fig. 2. Theoretical velocities and density in the interior of Mars. This model satisfies the known constraints of mean density and moment of inertia. The core size can be increased or decreased if the density of the core relative to that of the mantle is decreased or increased, respectively. Data used in this model are from *Anderson* [1972], *Okal and Anderson* [1977], and *Johnston et al.* [1974].

made to infer variations in crustal thickness [*Phillips et al.*, 1973; *Phillips and Tiernan*, 1977; *Bills and Ferrari*, 1977*b*], but the process is nonunique unless it can be tied to a direct measurement of the crustal thickness made by seismic methods. Interpretation of the Bouguer gravity field involves both the crustal thickness and the crust/mantle density ratio. Usually, the mean crustal thickness and the density contrast are assumed, and the Bouguer anomalies are interpreted in terms of deviations of the crustal thickness from the mean. Figure 3 gives a crustal thickness (isopach) map of Mars for an assumed mean crustal thickness of 40 km and a density contrast of 0.6 g/cm³ [*Bills and Ferrari*, 1977*a*]. Note that the thickest crust, 75 km, is under Tharsis at the head of Valles Marineris and the thinnest crust, 10 km, is under Hellas. A second crustal map for an assumed mean crustal thickness of 100 km is shown in Figure A-1 of the microfiche appendix.[1]

One way to bound the problem is to assume zero crustal thickness under the Hellas basin. If this large circular feature is of impact origin, deep excavation can be expected which would remove most, if not all, of the crustal material. For a density contrast of 0.3 g/cm³ the mean crustal thickness is about 65 km, and the thickness at the Viking 2 site would be 10–20 km. A higher density contrast would decrease the mean thickness and decrease the variation in crustal thickness. The maximum crustal thickness is 140 km for $\Delta\rho$ = 0.3 g/cm³ and 90 km for $\Delta\rho$ = 0.6 g/cm³. These calculations are due to R. Phillips (personal communication, 1976). He has demonstrated that the crust in the vicinity of the Viking 2 landing site is quite thin in comparison to the rest of the planet. If the crustal thickness could be determined at a single site and if the density contrast is constant, then we could immediately determine the average crustal thickness of the planet and, for example, the crustal thickness under the Tharsis plateau.

An average crustal thickness of even 30 km would indicate that Mars is an extremely well differentiated planet having a crust/planet mass ratio of over 5 times that for earth. The average thickness would be less if the density contrast were greater. There is evidence that the Martian mantle is denser than the terrestrial mantle [*Anderson*, 1972], but there is no direct information on the density of the crust. If ice, water, and hydrated minerals are abundant, then the crustal density may be quite low. On the other hand, a hydrous mantle could result in a well-differentiated planet because of the effect of water on melting temperatures and viscosity.

DESCRIPTION OF THE VIKING SEISMOMETER

The Viking seismometer includes sensors, amplifiers, filters, and electronics for automatic event detection, data compaction, and temporary data storage. The instrument package, shown in Figure 4, is 12 × 15 × 12 cm and weighs 2.2 kg. It is located on top of the lander's equipment bay near the attachment of leg 1 (Figure 5). The nominal power consumption is 3.5 W. The useful frequency range is 0.1–10 Hz with a minimum ground amplitude resolution of 2 nm at 3 Hz and 10 nm at 1 Hz.

Figure 6 compares the Viking seismometer displacement response in each of its operating modes with the displacement response of a typical station in the U.S. Geological Survey World-Wide Standard Seismograph Network (WWSSN). The

[1] Supplement is available with entire article on microfiche. Order from American Geophysical Union, 1909 K Street, N. W., Washington, D. C. Document J77-004; $1.00. Payment must accompany order.

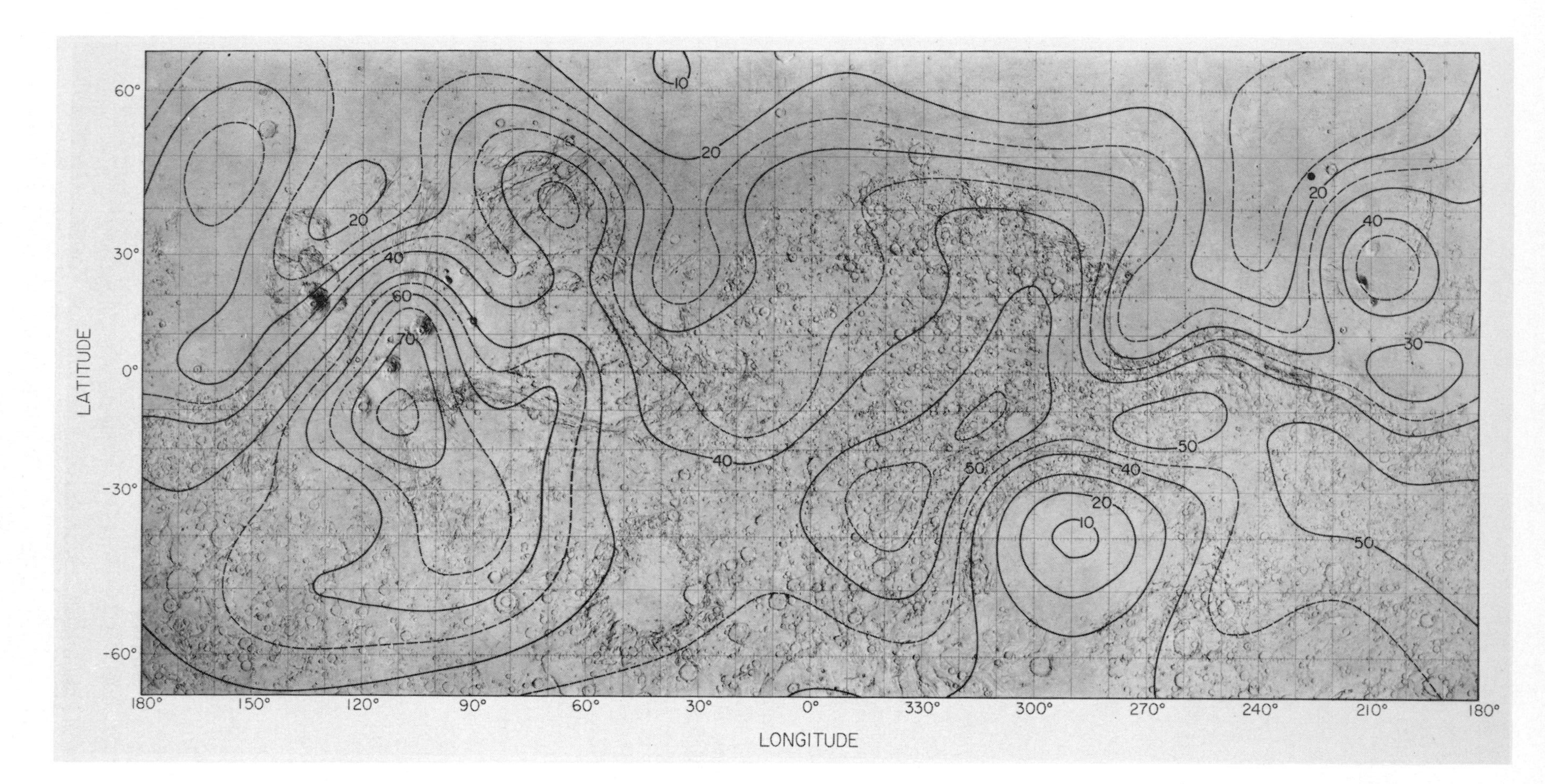

Fig. 3. Crustal thickness map for Mars based on the Bouguer gravity anomaly [*Bills and Ferrari*, 1977*b*] under the assumption of an average crustal thickness of 40 km and a crust-mantle density contrast of 0.6 g/cm³. The thinnest crust is in regions of large basins, and the thickest is in the highland and young volcanic regions. Isostatically uncompensated regions, such as the Tharsis plateau (10°N, 120°W), suggest that dynamic stresses and tectonic activity may be present. A second map for a mean crustal thickness of 100 km is presented in the microfiche appendix (Figure A-1).

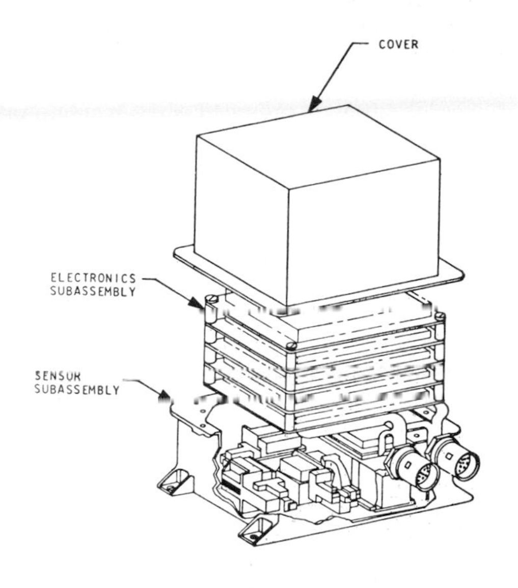

Fig. 4. Seismometer package. The dimensions of the unit, including sensors and electronics, are 12 × 15 × 12 cm with a weight of 2.2 kg. The nominal power consumption is 3.5 W.

peak magnification is 218,000 at 3 Hz when the returned data are plotted at a scale of 0.44 mm per digital unit of seismometer output. Figure 7 compares the acceleration sensitivity of the Viking seismometer with that of conventional accelerometers, the U.S. Geological Survey WWSSN, and the Apollo lunar instruments. The Viking seismometer has a maximum sensitivity equivalent to that obtainable at a relatively quiet site on earth.

The seismic sensors are three matched, orthogonally mounted (one vertical and two horizontal) inertial seismometers fitted with velocity transducers. One sensor is shown schematically in Figure 8. Each sensor occupies a volume measuring 7.5 × 3.8 × 3.8 cm. A 16-g mass-coil assembly is supported on twin booms by two Bendix Free-Flex elastic hinges such that the flat transducer coil is poised between the facing poles of two channel magnets arranged in series. Motion of the frame causes the transducer coil to move in the field of the magnets and generates a signal which is in proportion to the relative velocity of the coil with respect to that of the magnet. The undamped natural frequency of each instrument is 4 Hz, the coefficient of damping is 0.6, and the generator constant is 177 V/(m/s). The natural undamped frequency of the sensors was chosen such that the instrument would meet the constraints of weight and volume and to insure that the sensors would operate over the largest expected tilt of the lander (15°) without the use of any mechanical zeroing adjustments. With this design the horizontal units will tolerate up to 23° of tilt, and the vertical up to 35°.

The inertial mass of each sensor is individually caged to protect it from the shock and vibration encountered during launch, separation of the lander from the orbiter (the greatest shock peaking at 1200 *g*), and the landing. The caging is provided by spring-loaded plungers which hold the mass firmly against a stop seat. The plungers are secured by a palladium-aluminum fuse wire. Uncaging is accomplished by causing the fuse wire to fail nonexplosively when it is electrically heated. All three axes failed to uncage on the Viking 1 lander. Because the uncaging system was triply redundant except at the interface with the lander and because the seismometer functioned properly in all other aspects, it is believed that the failure was

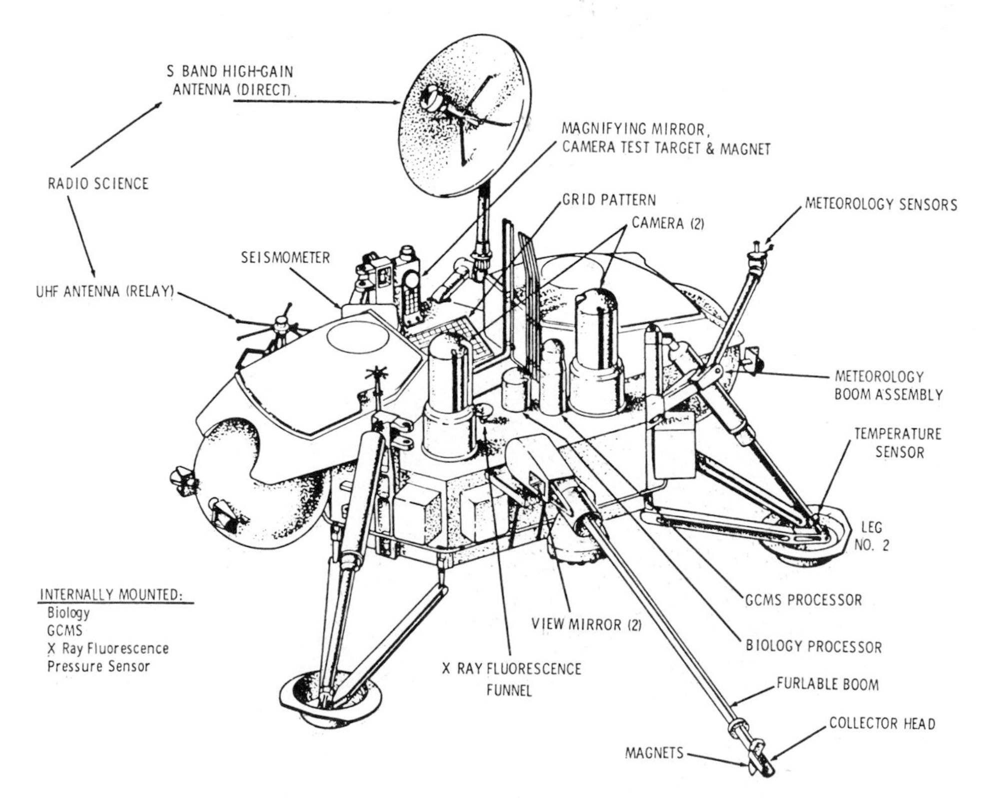

Fig. 5. The Viking lander showing the location of the seismic package and other instruments. The distance from footpad to footpad is about 2.5 m, and the lander mass is 605 kg.

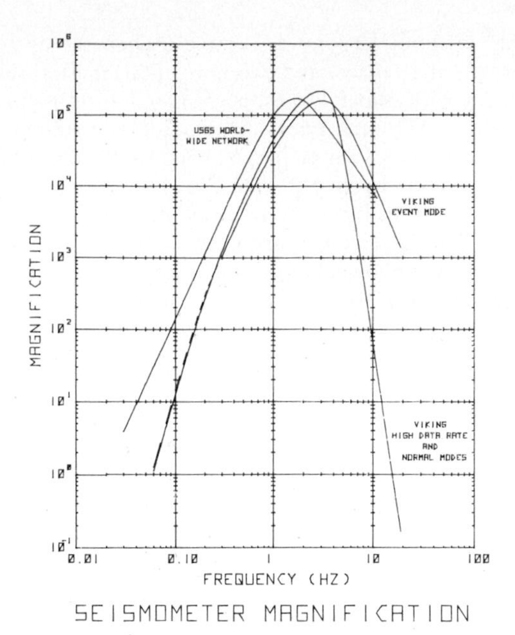

Fig. 6. Magnification of the Viking seismograph in each of the operating modes. The magnifications are based on the assumption that the digitized data are plotted at scales of 0.44, 0.51, and 0.76 mm/du for the high data rate, event, and normal modes, respectively. A typical U.S. Geological Survey short-period instrument used in the WWSSN is shown for comparison.

at the interface. The Viking 2 lander seismometer uncaged in a nominal fashion.

Each inertial sensor is equipped with a calibration mechanism by which the mass may be magnetically deflected approximately 4 μm. The deflection and release of the mass in each direction produce a pair of pulse doublets, shown in Figure 9, from which the operation and characteristics of the instrument may be ascertained and which may be used to determine the attitude of the seismometer package relative to the local vertical by measuring the asymmetries of the calibration doublet which are a function of the tilt of the instrument. This comparison of the Viking 2 seismometer calibration doublet with prelaunch calibration data showed the instrument attitude to be 9.5° ± 2.8° down at an azimuth of 278° ± 17° from north. This is in agreement with the lander's inertial reference system, which determined the tilt of the lander to be 8.2° down at an azimuth of 278° from north.

A block diagram of the instrument is shown in Figure 10. Each axis has an amplifier with a bandwidth of 0.2–1.2 Hz and an amplification, which is selectable upon earth command, from 6×10^3 to 4×10^5 in six increments. After amplification, prefiltering, and analog multiplexing, the seismic signals are converted to a 7-bit plus sign digital word at the rate of 121.21 samples per second by a dual-slope integrating analog-to-digital converter. The subsequent digital processing of the data includes filtering, averaging, compression, event detection, and buffer memory storage. These functions, as well as digital multiplexing, command decoding, timing, and control, are implemented in custom large-scale-integrated (LSI) circuitry.

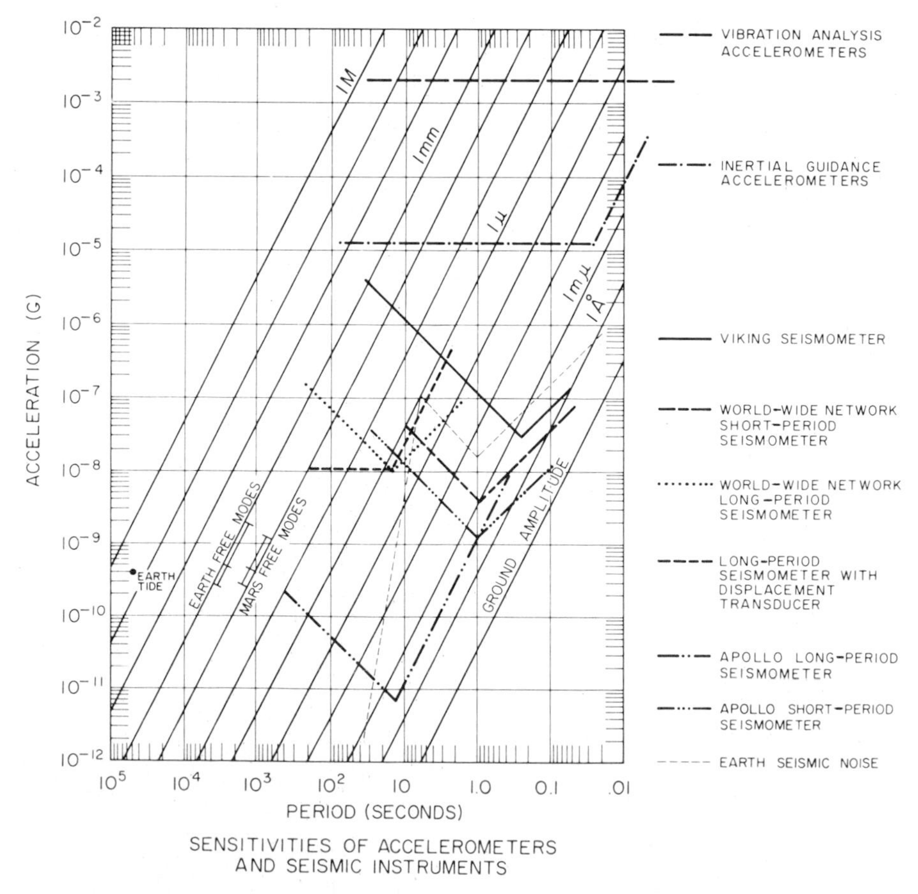

Fig. 7. Comparison of the acceleration sensitivity of the Viking seismometer with the sensitivities of conventional accelerometers, the U.S. Geological Survey WWSSN, and the Apollo lunar seismometers. The maximum sensitivity of the Viking seismometer is equivalent to that of a relatively quiet short-period seismometer on earth.

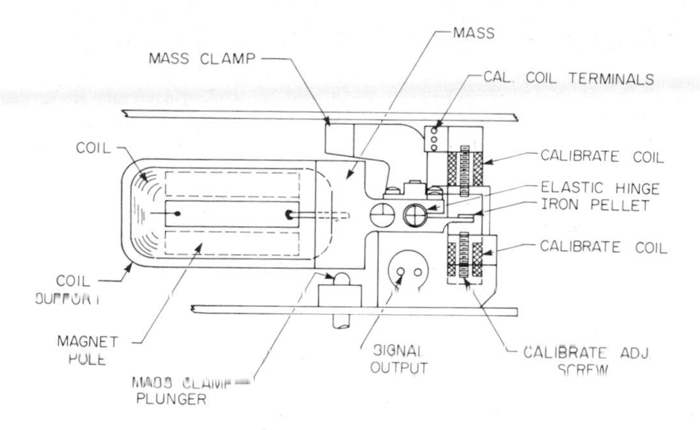

Fig. 8. The Viking seismic sensor. As the mass-coil assembly moves in relation to the frame, a signal is induced in the coil as it passes through the field of the magnets. The signal produced is proportional to the velocity of the relative motion.

The instrument may operate in any of three data processing modes.

High data rate mode. Each channel is digitally filtered, and the 7-bit plus sign word is temporarily stored in one of two 2048-bit recirculating memories to await servicing by the lander's data acquisition and processing unit (Dapu). The data rate is 20.2 samples/s/channel.

The filter is a digital implementation of a sixth-order maximally flat (Butterworth) low-pass filter with command-selectable cutoff frequencies of 0.5, 1.0, 2.0, and 4 Hz.

Normal mode. The normal mode is the lowest data rate mode, operating at 4.04 samples/min/channel. Its purpose is to investigate the average level and spectral content of the microseismic background. A form of 'comb filtering' is performed by using the digital low-pass filter in conjunction with the frequency response of the inertial sensor. The equivalent noise bandwidth at each filter cutoff is shown in Figure 11. The rectangular response represents an ideal band-pass filter with the same power and peak value as those of the total system frequency response.

After filtering, the absolute value of the data is then passed through a low-pass filter to obtain the 12.7-s running average of the microseismic level which is in turn sampled at the rate of one sample every 14.85 s on each of the three axes. The digital low-pass filter may be fixed or automatically stepped through each cutoff frequency at the rate of 2 min per step.

Event or triggered mode. This mode refers to the data compaction mode of operation, where the envelope of the seismic signal and the number of zero crossings rather than the signal itself are sampled at the moderate rate of 1.01 sample/s/axis. To produce the envelope, the absolute value of the seismic signal is smoothed by passing it through the digital filter operating at the 0.5-Hz cutoff frequency. Simultaneously, a running count of the positive axis crossings (a measure of the dominant frequency of the signal) is sampled at the same rate. This combination of sampling the envelope (7-bit word) and axis crossing (5-bit word) results in a 12.3 to 1 reduction in the data required to encode the original signal over the high data rate. The effects of smoothing the envelope, demonstrated by a computer synthesis of its operations, are shown in Figure 12. Recall that the resulting envelope is sampled once per second. Of particular note are the delay in the start of the signal and the rise time, nominally 1 s each. In addition, there are effects due to transients and incomplete smoothing at frequencies below 0.4 Hz.

The event mode may be initiated by earth command or automatically by enabling an event detector on any one or combination of the three axes. The purpose of the detector is to record seismic events which are transient in nature (Marsquakes or meteor impacts) at the higher data rate of the event mode while it is conserving data by monitoring the seismic

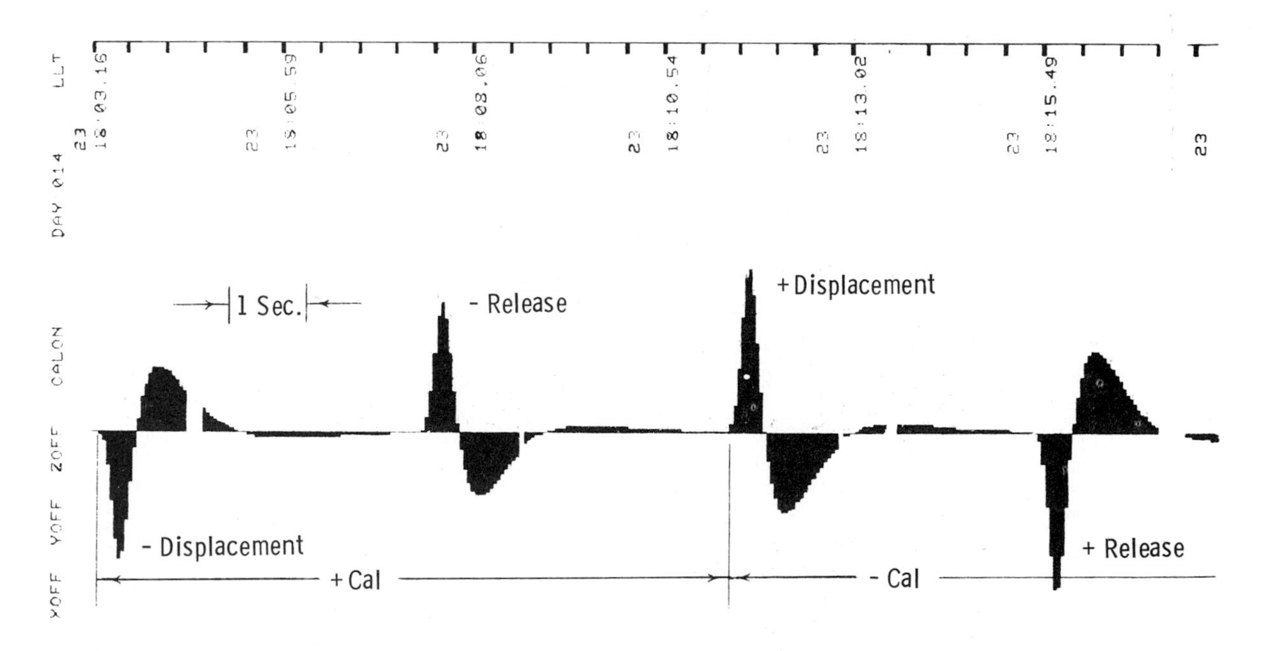

Fig. 9. Seismometer calibration sequence. The deflection and release of the mass by the calibration coils produce a pair of pulse doublets from which the operational characteristics and the attitude of the seismometer can be ascertained. Calibration sequences are run in the high data rate mode.

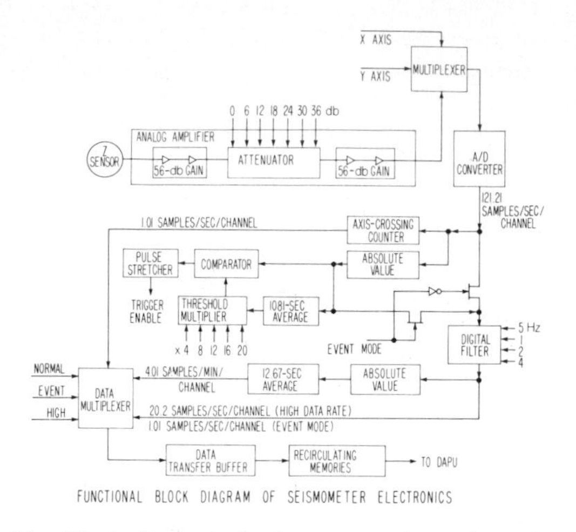

Fig. 10. Block diagram of seismometer electronics. The analog amplifier, shown for the Z axis only, also includes a low-pass filter to prevent aliasing.

background in the normal mode. The trigger level of the detector is a multiple of the long-term (1081 s) average microseismic level. The multiple is selected by ground command from values times 4, 8, 12, 16, or 20. When the signal level drops below the average background, the instrument will revert to the normal mode. To reduce the number of data taken from false triggering from lander thermal 'pops,' the event detector on time beyond the normal off time is controlled to be proportional to the time that the level of the event is above the trigger level. This time increase varies from a minimum of 2 s to a maximum of 1 min for the longest event.

This autotrigger mode was used during the conjunction period when the total number of Viking science data were limited to those which could be stored on the lander tape recorder during the 44 days of conjunction. This mode is also used in the automatic mission, which is designed to continue lander operations in the event that the ability to load new commands into the lander is lost. The instrument will produce 6170 bits/h in the normal (or background monitoring) mode, 1.47×10^5 bits/h in the event mode, and 1.77×10^6 bits/h when it is operating in the high data rate mode.

The instrument contains dual 2048-bit data buffer memories. Two are necessary so that data flow continues while the full buffer is serviced by the lander. The data are in turn transferred along with engineering and scientific data from other experiments to the lander's magnetic tape recorder. Normally, the data are transmitted by the lander to an orbiter for relay to earth; however, they can be transmitted directly to earth at a much lower data rate. The seismometer will transfer 69 buffers per day if it is operating wholly in the normal mode, 1716 buffers per day in the event mode, and 20,700 buffers per day in the high data rate mode. The number of data buffers that the seismometer transfers each day is controlled by the lander and is part of the overall mission planning. It varies from zero to about 3500 buffers per day but is nominally about 1000–1500 buffers per day. The tape recorder on the lander cannot physically hold more than about 18,000 buffers of seismic data.

Normally, the operating time of the instrument is divided among the three modes so as to maximize the amount of time in the event mode with a sampling of high data rate mode during the quietest part of the sol and for calibration. The normal mode is used only to fill in or when other lander activities would produce exceptional lander vibrations. When the data allocation for the seismometer is more than 1700 buffers, the excess data are recorded in the high data rate mode.

INSTRUMENT PERFORMANCE

Following touchdown the azimuths of the horizontal (Y and Z) component seismometers were determined with regard to the lander inertial reference system. The positive output from the Y component corresponds to lander motion toward the northwest (N31°W), and the positive output from the Z component corresponds to lander motion toward the southwest (S59°W). The polarity of the vertical (X) component follows the usual convention of positive output for upward displacement. The 'vertical' is 8.2° from the true vertical toward N82°W.

The initial calibration pulse amplitudes were in good agreement with predicted values. A gradual increase in calibration pulse amplitudes (18%) occurred during the first 140 sols, but the relative sensitivities remained closely matched throughout this period (see Figure A-2 of the microfiche appendix). The instrument responses were matched to within approximately 5% during prelaunch adjustment. The gradual change in pulse amplitudes is a consequence of the decreasing temperature of the lander and thus the lowering of the temperature (resistance) of the transducer and calibration coils. The difference in calibration pulse amplitudes among the three components is caused by the tilt of the instrument.

It was noted early in the experiment that signal amplitudes were consistently smallest on the Y component (10–20% less than those on X) and largest on the Z component (20–50% greater than those on X) even though the evidence from calibration data showed that the component sensitivities were well matched. In addition, the three components were nearly always in phase, with a high degree of coherence between channels during the brief intervals of high data rate operation in

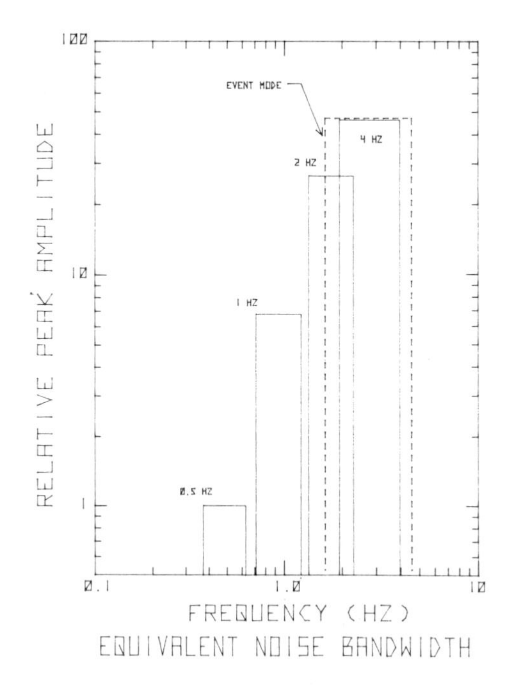

Fig. 11. Equivalent noise bandwidth at each filter setting and of the event mode. The rectangular response represents an ideal band-pass filter with the same area and peak value as the total system frequency response.

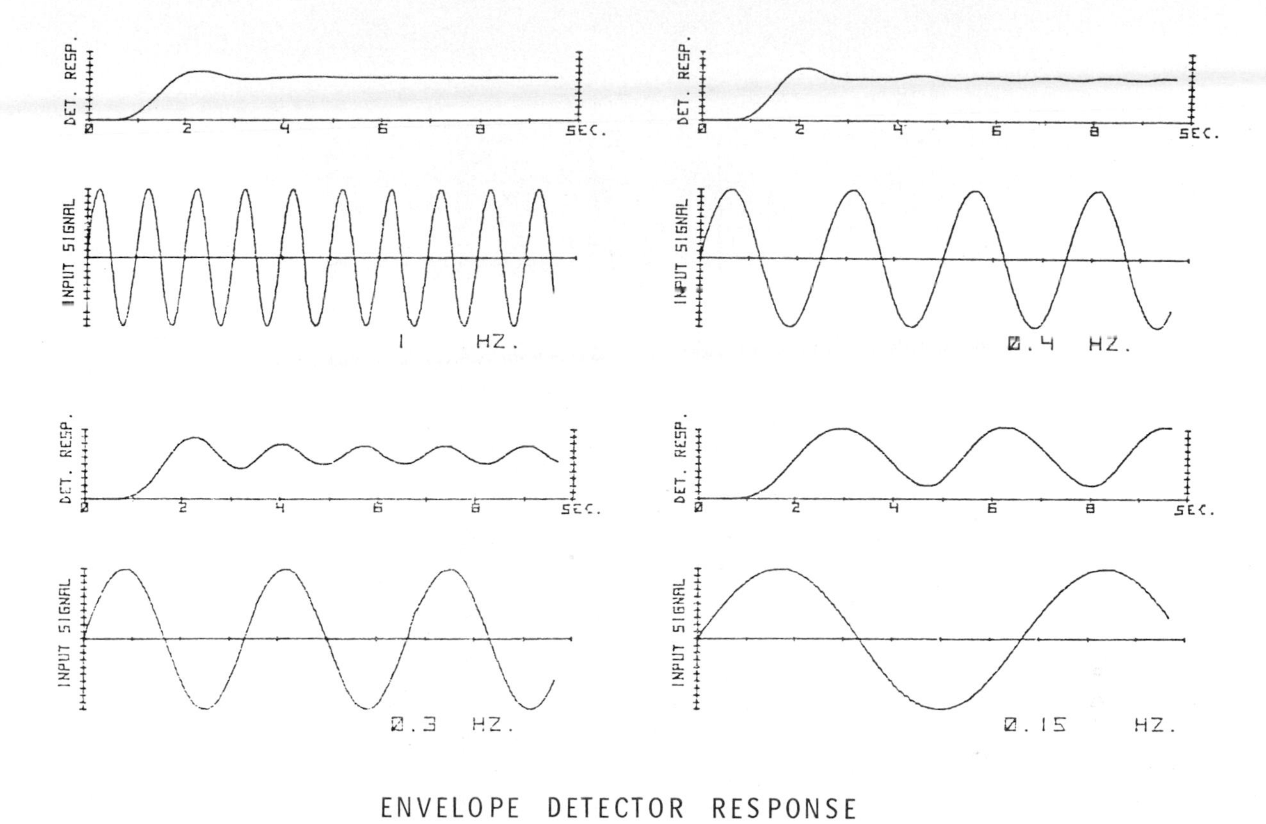

Fig. 12. Event mode envelope detector response. The effects of smoothing low-frequency rectified signals with a 0.5-Hz low-pass filter are shown. The resulting signal is then sampled once per second. The input signals are zero before time zero.

which phase comparisons can be made. This observation can only be explained by rectilinear motion of the lander in a preferred direction. In-phase correlation requires that the preferred direction of motion be up to the west and down to the east, i.e., along the direction of tilt of the lander. This direction is consistent with a rocking motion of the lander about the line connecting footpads 1 and 2 (see Figure A-3). The seismometer is located approximately 1 m above ground level. The remaining dimensions and angles are shown in Figure A-4. From this the expected amplitude ratios for a rocking motion about the line connecting footpads 1 and 2 are as follows:

$$A_x:A_y:A_z = 1:0.83:1.44$$

These values are in approximate agreement with those observed.

Postlanding imagery revealed that footpad 3 is balanced on a rock. Thus a rocking motion in the direction inferred from seismic data is a reasonable expectation.

A related observation is that wind-induced vibrations of the lander are highest for a given velocity for winds from the east. We would expect that the rocking motion described above would be most easily excited by easterly and westerly winds.

SUMMARY OF OPERATIONS

On earth, operations such as change of gain and filter setting are usually trivial. During the Viking mission, however, these and other operations (such as change in operating mode and trigger levels) had to be planned many days in advance of the actual execution of the command. This advance planning was necessary to avoid conflicts with other experiments on the lander, to insure the safety of the lander, and because commands could be sent to the lander generally only every other day. During the primary mission, covering the first 62 sols, planning of commands to be sent to the lander was divided into 6-day cycles, each cycle covering three command sequences (uplinks) to be sent to the lander. Planning for each cycle started 16 days before the first command uplink was to occur. As the time for a particular uplink approached, the sequence of commands became less flexible, and more justification was required to incorporate a change. With these restrictions each command could be tested before it was executed on the lander, thus insuring against disastrous errors. During the conjunction period when Mars was behind the sun and totally out of communication with the earth (sols 63–108), the lander operated in a reduced mode, executing commands uplinked before conjunction. During the extended mission (sol 109 to the end of the mission), planning cycles cover 2 weeks, and commands are sent to the lander once per week. During the primary mission, data were returned to earth (downlink) once per day, while downlinks during the extended mission average about twice per week.

The seismometer commands are stored in the lander computer together with the time to be spent in each command. When the end of the command sequence is reached, the sequence is restarted automatically. Uplinks need to be sent only to change a command or command sequence. Each command must specify the mode of operation, vertical gain, horizontal gain, filter frequency, trigger state, trigger threshold, calibration state, and time to be spent in this configuration. In addition, the number of buffers to be recorded and the number of commands to be used can be specified. Until sol 141, two tables of 12 commands each plus a secondary table of three commands could be specified. After sol 141, two tables of 34 entries each became available.

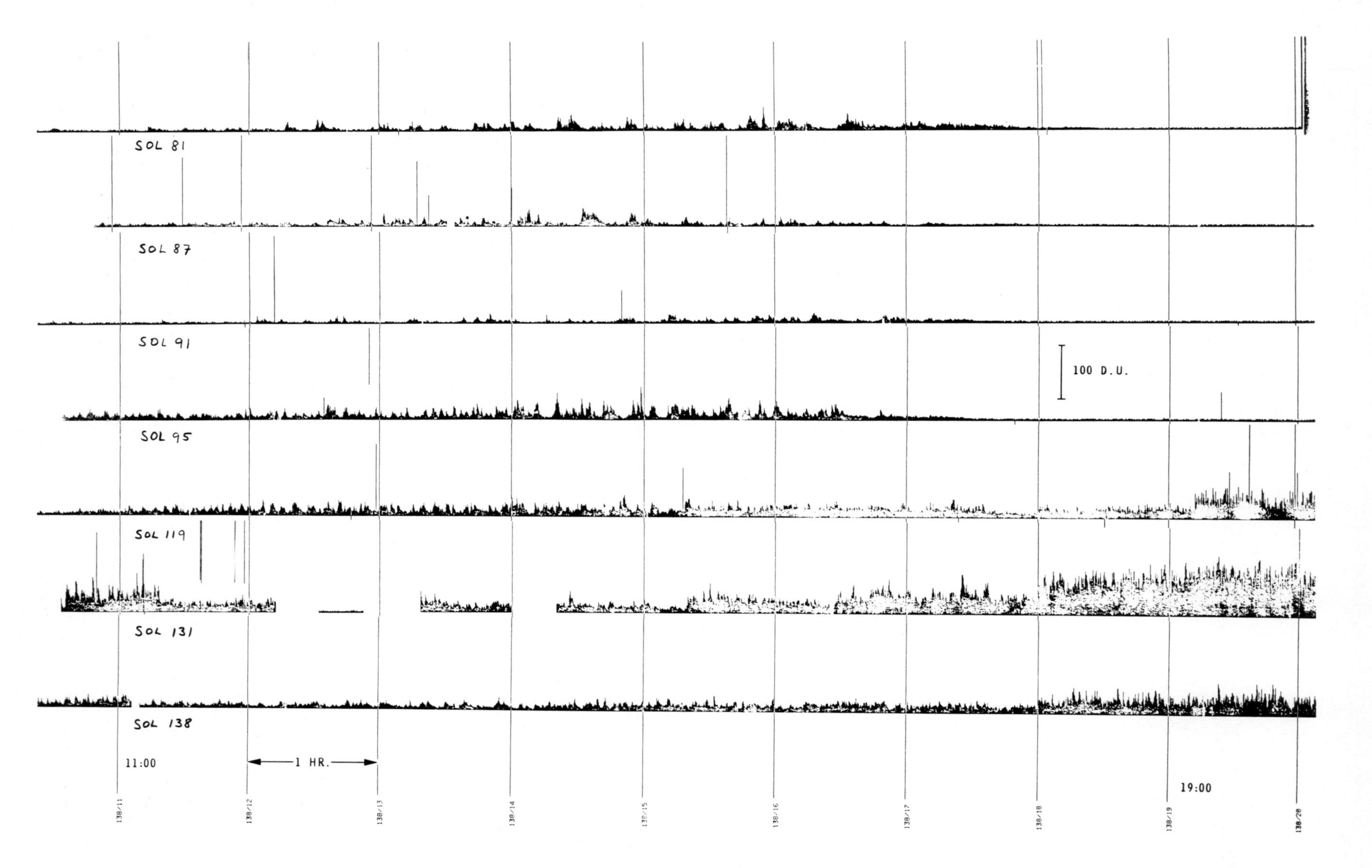

Fig. 13. Representative samples of seismic background from 1100 to 2000 LLT. Records from seven different sols are presented to show the change in wind-generated noise as the season at the landing site progresses from summer to fall. The first four traces are typical of activity in the early part of the mission and show the low noise that was typical after 1800 LLT. A change of conditions occurred near sol 119, when noise often built up during the night. These traces are for the Z axis in the normal mode.

TABLE 1. Examples of Time Lines

Period Beginning		Period End		
Sol	Time, LLT	Sol	Time, LLT	Mode, Other Comments
		Primary Mission, Sol 17 Downlink		
16	0845:48	16	0947:39	Relay link, no data collected
16	0947:39	16	1053:59	Event mode
16	1053:59	16	1422:34	Direct link, normal mode
16	1422:34	16	2300:31	Event mode
16	2300:31	16	2317:31	Normal mode, filter stepping
16	2317:31	16	2320:31	Calibration, 30-dB attenuation
16	2320:31	16	2424:20	High data rate mode, 4.0 filter
16	2424:20	17	0843:16	Event mode
		17	0843:16	Power off, 2187 buffer dumps
		Conjunction, Daily Sequence		
N	0400:23	N	1900:01	Normal mode, 4.0 filter
N	1900:01	$N + 1$	0300:01	Normal mode, 4.0 filter, X trigger on, threshold × 12
$N + 1$	0300:01	$N + 1$	0330:01	Event mode
$N + 1$	0330:01	$N + 1$	0330:17	Calibration, 30-dB attenuation
$N + 1$	0330:17	$N + 1$	0359:07	Event mode
		$N + 1$	0359:07	Power off, 152 buffer dumps (see text)
		Extended Mission, Sol 129		
129	0000:00	129	0200:02	Event mode
129	0200:02	129	0200:36	Calibration, 30-dB attenuation
129	0200:36	129	0700:01	Normal mode, 4.0 filter
129	0700:01	129	1800:01	Normal mode, 4.0 filter, 6-dB attenuation
129	1800:01	129	2012:31	Normal mode, 4.0 filter
129	2012:31	130	0000:00	Event mode, 500 buffer dumps
		Extended Mission, Sol 156		
156	0000:00	156	0159:48	Event mode
156	0159:48	156	0200:18	Calibration, 30-dB attenuation
156	0200:18	157	0000:00	Event mode, 1720 buffer dumps

Horizontal and vertical attenuations are the same and are 0 dB unless they are otherwise noted. No filter means that the low-pass filter was set to 0.5-Hz cutoff.

The planning procedures and structure for the lander operation have been described by *Lee* [1976]. Readers are referred to this paper for a more complete description of the uplink procedures. Although a large amount of time was spent by each team in these planning sessions, the necessity of the system cannot be denied.

Before touchdown the commands loaded into the tables were to survey the ambient seismic noise at various gain levels, with approximately 500 buffers of data allocated per sol. The data returned from this phase of the operation demonstrated that the seismometer could be operated at all times at maximum gain. After sol 9, high enough data rates were available to use the event mode most of the time. During conjunction, 152 buffers per sol were available to the seismometer. Only 1 hour of event mode per sol was scheduled with 8.5 hours of the remaining 23.5 hours of normal mode set aside for trigger-enabled operation. If the trigger was not activated, 141 buffers would be taken; thus 11 buffers were set aside for triggered event mode. In the extended mission a more flexible schedule is followed. On occasion, 1700 buffers are available, continuous event mode recording thus being permitted. At other times, fewer data, down to 500 buffers per sol, were available.

To date, the seismometer has been operated nearly continuously except for two periods: sols 103 and 104 and sols 141–146. One hundred and seventy-one sols of data have been collected, consisting of 87 sols of normal mode data, 75 sols of event mode data, and 63 hours of high data rate mode data (the balance is lost time due to playback periods of the lander tape recorder for relay to the orbiter). Some detailed examples of data collection are given in Table 1. Periods when the seismometer was inhibited by other experiments have been ignored except for inhibitions due to the relay link and direct link, since these occurred on a regular basis during the primary mission and were generally of short duration.

LANDER NOISES

Since the seismometer is located on top of the lander, it is important to understand all internal sources of noise, the response of the lander to wind, and the effect of the lander on external seismic signals. A catalog of these noise sources and representative signatures of each are given in the microfiche appendix to this report. Noises generated inside the lander are generally of short duration, infrequent, and distinctive in nature and occur at predictable times. There is therefore little likelihood that they would be confused with natural events, but they do complicate data processing, and it has taken considerable effort to identify and catalog them. The internal sources identified are tape drive, X ray, sample dump, antenna tracking, camera motions, operations of the soil sampler, and electrical transients from other experiments. The known vibrational frequencies of these activities are in agreement with the frequencies measured by the seismometer. These are generally greater than 4.7 Hz. When the soil sampler arm is in the extended configuration, frequencies of 2–4 Hz are observed, the dominant frequency depending on the extension. The Q of the arm is about 200. Wind-induced vibrations of the lander are also more pronounced when the arm is extended. Fortunately, the arm is usually stowed. Before sample delivery the collector head is usually vibrated at 8.8 or 4.4 Hz for a short period of time. These frequencies are clearly observed by the seismometer at the appropriate times.

There is no firm evidence that any high Q lander resonances

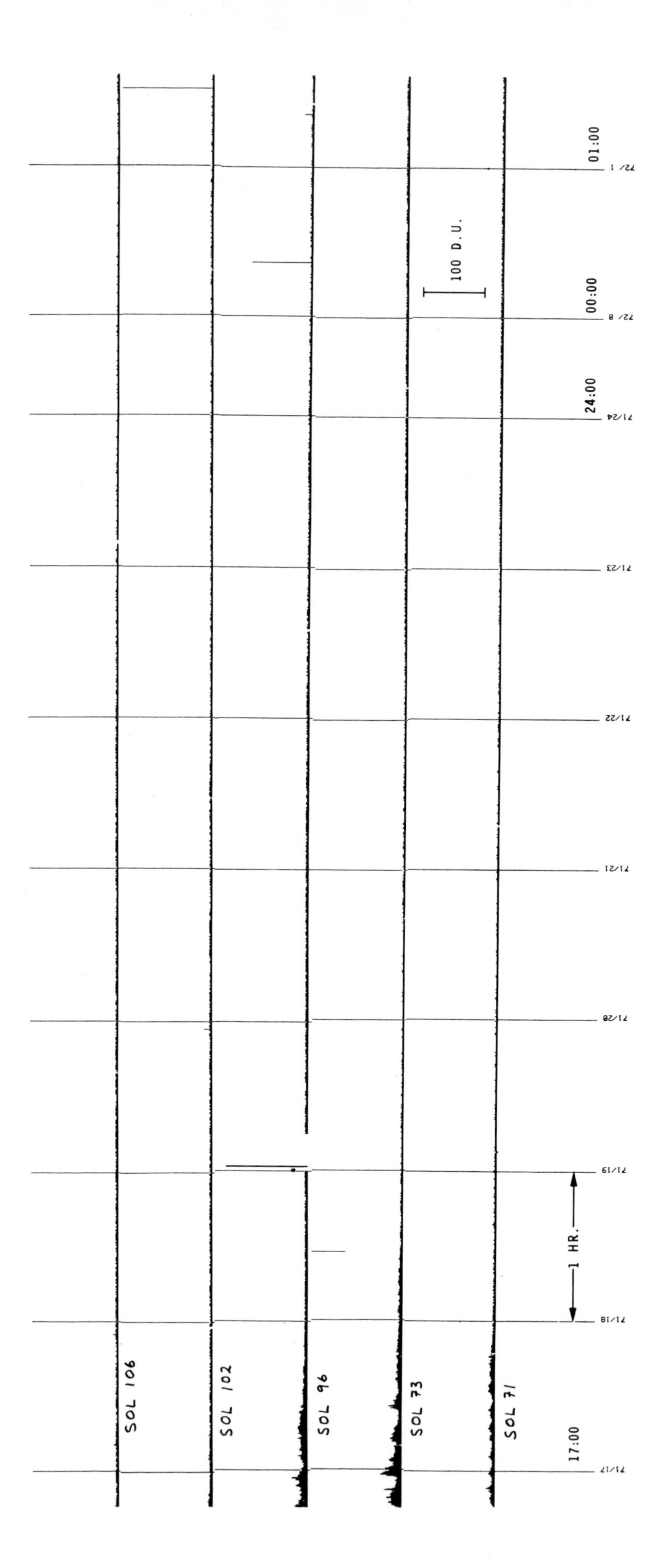

Fig. 14. Representative samples of seismic background noise from 1700 to 0100 LLT. These records show the decrease in noise level that was typical after sunset. Early in the mission, calm conditions usually lasted until sunrise. Beginning after sol 100, the wind noise continued well after sunset on some sols. These traces are for the Z axis in the normal mode.

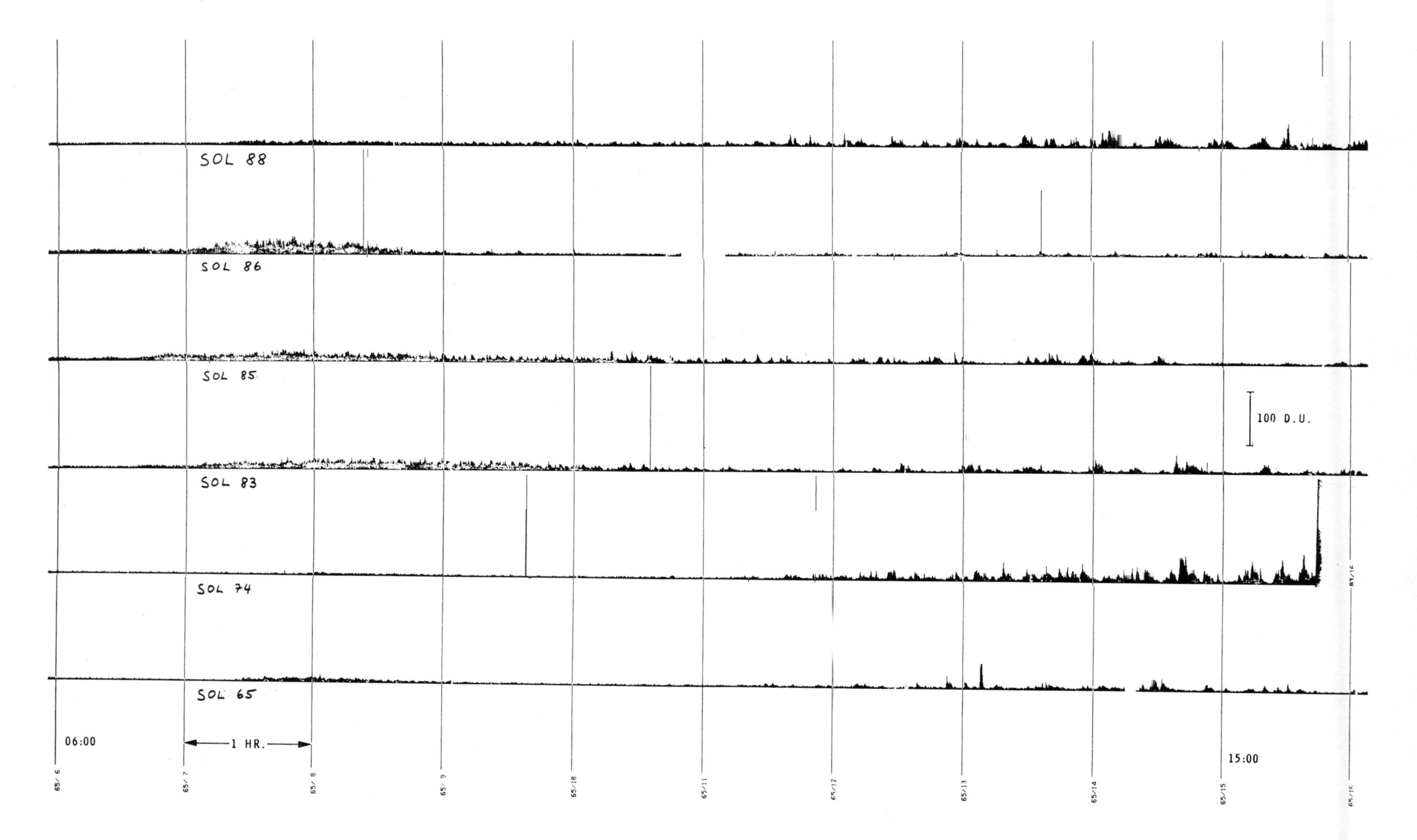

Fig. 15. Amplitude of seismic noise from 0600 to 1600 LLT. The gradual buildup in amplitude near 0600 LLT reflects wind changes after sunrise. Note the change in 'gustiness' of the wind from the morning to the afternoon. Sol 86 is particularly noisy in the morning, and sol 74 is particularly quiet. The reverse is the case for the afternoon. Sol 65 is unusually quiet all day. These traces are for the Z axis in the normal mode.

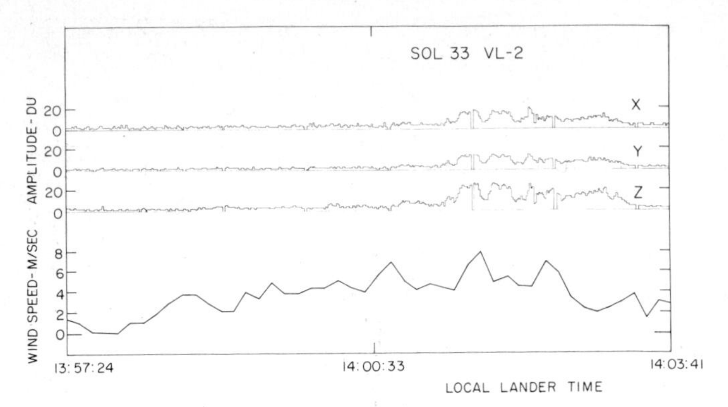

Fig. 16. Correlation of wind speed with seismic noise on sol 33. In this case, wind gusts of 7–8 m/s produced noise levels of 16–20 du. The wind detection threshold for seismic noise is about 3–4 m/s.

are excited by any lander activity. There is a suggestion of a resonance at 7.6 Hz, seen during the initial slew of the high-gain antennae and occasionally during a sampling arm operation, and also a resonance at frequencies of 14–15 Hz based on infrequent high-amplitude impulsive events of possible thermal origin. High Q lander resonances were designed to be outside the useful frequency band of the seismometer. There is also no evidence for high Q wind-induced lander resonances.

Wind-Generated Noise

The main source of external noise is the wind blowing on the lander. Generally speaking, the wind noise background repeats from day to day, but the background is quite variable when it is looked at in detail. Some days are very noisy (windy), and some days are very calm. Martian winds and hence the background noise level vary distinctly with the season. Prior to conjunction, during summer in the northern hemisphere (sols 0–60), the average wind speeds were observed to be somewhat higher in the morning than in the afternoon, but individual wind gusts had higher peak speeds in the afternoon [*Hess et al.*, 1977]. This is reflected in the seismic data (Figure 13). The highest seismic background typically occurred between 1300 and 1700 hours (Mars local time), when the seismic amplitudes ranged from 10 to 40 digital units (du). Noise bursts were typically 1–3 min in duration, separated by time intervals of 10–50 min. The seismic background level drops between individual bursts of wind; this drop indicates that there is little direct transfer of energy into the ground by the wind and that the principal effect on the seismometer is produced by wind-induced vibrations of the lander.

Typically, the quietest recording interval occurs from about 1800 hours local lander time (LLT) (2 hours before sunset) to about 0400 hours the following morning (Figure 14). During this time interval the wind speed decreases to less than 1–2 m/s, and the seismic noise background level falls to between 1 and 2 du. (In the event mode at maximum sensitivity, 1 du corresponds to 2 nm of instrument displacement at 3 Hz.) Around Martian sunrise the background noise level generally increases to about 2–3 du (Figure 15).

An example of the correlation between seismic noise activity in the event mode and a wind gust recorded on sol 33 is shown in Figure 16. In this case a 7- to 8-m/s wind gust produced a seismic signal of 16- to 20-du amplitude. A similar figure is shown in the microfiche appendix (Figure A-18) covering a different period of time. The threshold of wind detectability of the seismometer, as seen in the event mode at maximum sensitivity, appears to be about 3–4 m/s.

Figure 17 shows the high degree of correlation of the amplitude of the seismic signals recorded by the seismometer in the normal mode (Z component) with the wind speeds recorded by the meteorology experiment. The data shown in Figure 17 indicate that the seismic amplitude is proportional to the square of the wind speed, as is expected for turbulent flow. Detailed correlations are difficult, since the seismic data represent averages and the wind data are widely spaced point samples. A similar figure for X component data in the event mode is shown in Figure A-19. In the high data rate mode of operation, where no averaging is involved, seismic activity peaks and wind speed peaks are also reasonably well correlated. Figure A-20 shows the correlation between Z component amplitudes and wind speeds in the high data rate mode on sols 35 and 36. Data observed in the high data rate mode also show that the seismic amplitude increases as the square of the wind speed. Data plotted in Figure 17 also show the correlation of seismic noise for different wind directions. The data are in good agreement with the relation $A = 0.15V^2$, where A is the mean seismic amplitude in digital units (event mode) and V is the mean wind speed in meters per second. The scatter of data at low speeds suggests a threshold of about 3 m/s.

During the period of conjunction (sols 60–90) the background noise level was extremely low from 1800 hours each evening to 0700 hours the following morning, the background level increasing slightly starting between 0630 and 0700 hours. The afternoon gusty wind period began between 1018 and 1230 hours each day, usually close to local noon. Peak gusts usually occurred near 1430 and 1520 hours. Typically, 8–12 wind gusts were observed between 1200 and 1800 hours lasting 8–15 min and separated by intervals of roughly 40 min. The gusts produced seismic amplitudes ranging from 10 to 20 du in the normal mode.

Northwest winds typically occur in the afternoon from 1300 to 1800 hours. Northeast winds generally occur in the early morning hours, changing to southeast winds from 0700 to 0900 hours and to southwest winds from 0900 to 1200 hours [*Hess et al.*, 1977]. During the postconjunction period from sols 110 to 118 the winds were generally less than the seismic threshold (~3 m/s) from about 1800 hours to sometime be-

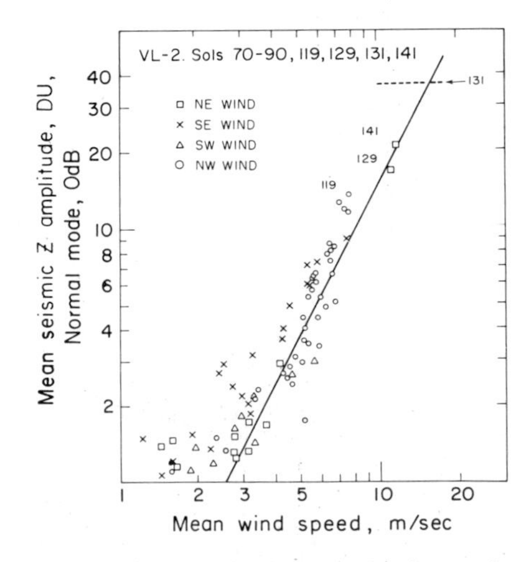

Fig. 17. Correlation between wind speed and normal mode seismic amplitude (Z component) for winds from different directions. The solid line corresponds to the slope expected if seismic amplitude is proportional to the square of the wind speed, as is appropriate for turbulent flow. The scatter of points below wind speeds of 3 m/s is due to noise from other sources than wind.

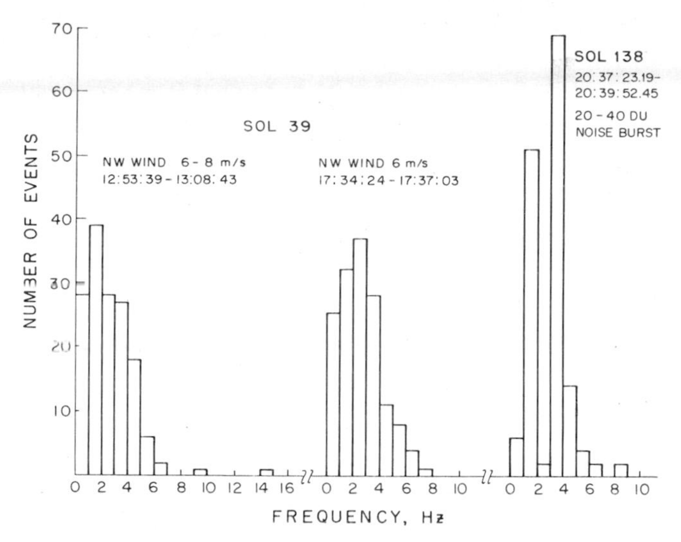

Fig. 18. Axis-crossing histograms for windy periods. The number of events (ordinate) is a measure of the amplitude at the corresponding frequency. If high Q lander resonances were excited by winds, these histograms would show sharp peaks at the resonant frequencies. Such peaks are not observed, and the shape of the histograms reflects the frequency response of the seismometer.

tween 0600 and 1030 hours LLT. Winds generally averaged from 2.5 to 4.9 m/s from 1000 to 1800 hours LLT, with maximum wind peaks from 6.4 to 11.5 m/s. The seismic data were characterized by quiet nights, relatively quiet mornings, and increased activity in the afternoon.

Unusually high seismic background was recorded on sol 119 beginning at 1100 hours LLT. The noise level was continuous and unlike the usual afternoon windy periods. Average wind speeds in this period were unusually high even after 1800 hours, when it is usually quiet. The wind speeds averaged greater than 7.5 m/s during the time interval from 1200 to 1800 hours, with peak gusts of 11.6, 13.2, and 15.1 m/s. Quite surprisingly, the wind speeds from 1800 hours to midnight ranged from 4.7 to 6.4 m/s for successive 1½-hour-long meteorology samples. Wind speed peaks of 11.4, 10.7, 9.5, and 8.9 m/s were observed throughout this normally quiet period.

Continuous, although lower-level, activity also occurred in the early morning hours on sols 118, 121, 128, 129, and 130. Exceptionally high activity was observed on sol 131 lasting until 0200 hours LLT on sol 132. The mean seismic Z amplitude was 37.1 du with a maximum of 56 du in the normal mode at 0 dB. Extrapolating the relation found earlier (Figure 18) suggests mean wind speeds of 16–18 m/s, a value higher than any previously reported. Unfortunately, because of a temporary malfunction, there were no meteorology data during this period. Imagery results show that at some time between sols 110 and 120 the optical depth apparently increased by an order of magnitude. Average temperatures were also observed to drop rapidly on sols 119 and 120, 128, and 131. All of these observations are suggestive of an unusual meteorological disturbance or the onset of a new meteorological regime due to the change of seasons.

Before landing, there was some question about whether the wind would excite high Q resonances in the lander. To examine this question, zero-crossing (frequency) histograms for different times of day and wind conditions were compiled. Figure 18 shows a zero-crossing histogram for a moderate northwesterly wind during sol 39 at two times during the afternoon windy period. The predominant frequency is between 2 and 3 Hz, close to the natural period of the seismometer system. Also shown are data from a noise burst on sol 138. Figures A-21 and A-22 show histograms compiled for a 15-min interval during an unusually windy period near Martian midnight on sol 133 and a similar 5-min interval on sol 139. As before, the main frequency appears to be concentrated between 2 and 3 Hz. There are no obvious lander resonances or preferred excitation frequencies. The shape of the histograms reflects the response of the instrument.

WIND STRESSES ON THE LANDER

Wind stresses on the lander are of the order of 0.002–0.2 dyn/cm², depending on wind velocity. Although the wind gusts themselves are of generally low frequency, large changes in amplitude occur of the order of seconds or less. The lander itself can introduce a high-frequency motion due to turbulent shedding at preferred frequencies. The latter are related to wind velocity, the characteristic dimensions of lander components, and the Strouhal number via

$$f = sV/h$$

where V is the wind velocity, h is the dimension of the lander or extended components, and s is the Strouhal number, which is about 0.2 for cylindrical or blunt obstacles. Taking 1 m as the characteristic dimension of the lander, the shedding frequency is 4 Hz for 20-m/s winds and 0.4 Hz for 2-m/s winds. Therefore vortex shedding frequencies are in the seismic band.

There is a slight indication from the seismic data that increased frequency correlates with increased amplitude (and wind velocity), but the passband of the instrument is too narrow to make a detailed correlation at this stage in the data processing.

The vibrations of the lander due to wind action are proportional to the square of the wind velocity V. Wind forces τ on the lander can be written

$$\tau \sim \rho A V^2$$

where A is the cross-sectional area of the lander and ρ is the density of the atmosphere. The seismometer threshold to winds is about 3 m/s. The location of the seismometer on the top of the lander magnifies the wind-induced strains, and the response of the lander to wind stresses can be written

$$\tau = M\epsilon = Md/l$$

where M is the effective modulus of the lander system, ϵ is the strain, d is the displacement, and l is the distance of the seismometer from the contact between the legs and the surface. In order to reduce wind noise therefore, an optimum seismic experiment would minimize A and l and maximize M. The cross-sectional area of the lander is about 10^4 cm². For an offloaded seismometer this could be reduced to 10^2 cm², an immediate decrease in wind stresses of a factor of 10^2 being given, raising the threshold wind speed to 30 m/s. The modulus M of a compact surface instrument could probably be increased by about an order of magnitude, and the height of the package could be decreased by an order of magnitude, an improvement of another 2 orders of magnitude thus resulting in the signal-to-noise ratio. In principle therefore a 10^4 improvement in seismometer sensitivity could be usefully employed for a Martian seismometer, and direct wind-induced noise would be no problem for winds as high as 300 m/s. Future missions to Mars could have seismometers of Apollo-like sensitivity. This is important not only to increase the range

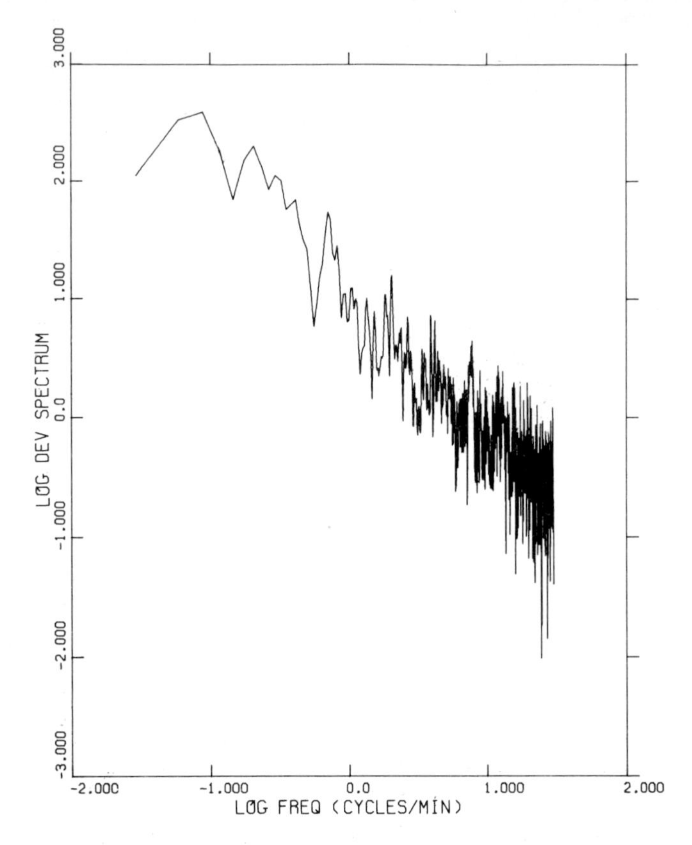

Fig. 19. Power spectrum of event mode amplitude envelope data for a 21½-min period starting at 1429:18 on sol 16. Two major wind gusts during this period provide the energy for this spectrum. The slope of the spectrum at frequencies higher than 1 cycle per minute is about −0.9.

of detectability but because the number of small events is undoubtedly much greater than the number of large events. Magnitude-frequency relationships indicate that for each order of magnitude increase in instrument sensitivity, there will be an order of magnitude increase in number of detected events. An optimal seismic network on Mars therefore could usefully employ magnitude zero events to define the seismicity and internal structure of the planet.

NOISE SPECTRA

The spectra of wind-induced seismic noise are not only useful for micrometeorological purposes but are potentially useful discriminants for identifying natural events. Most of the meteorological data have been taken at sampling intervals of 4 s or greater, although occasionally wind speed is sampled at 1.2-s intervals. Higher sampling is necessary to define the turbulent characteristics of the lower Martian atmosphere.

Figure 19 shows the power spectrum of the amplitude envelope for a 21.5-min section of event mode data (Z axis) starting at 1429:18 hours on sol 16. There are two major wind gusts in this period. At periods of less than 1 min the power spectrum has a mean slope of −0.9. Data at periods of less than 15 s may be aliased by high-frequency changes in the envelope and should be ignored.

Figure A-23 gives the power spectra of all three axes of the seismometer for a high-wind period of sol 131 starting at 2015:34 hours and extending for 55 min. The nearly monotonic decrease of power with increasing frequency with a slope slightly less than unity is typical of the wind noise spectra on Mars.

Table 2 summarizes the spectral slopes for 34 samples of wind noise. The slopes are substantially lower for the vertical component than they are for the horizontal. The slopes vary from 0.58 to 0.98 with a mean of 0.91 ± 0.10 on the Z axis.

TRANSIENT EVENTS

Many transient seismic signals have been detected by the Viking seismometer. However, as is described elsewhere in this paper, wind-induced vibrations of the lander greatly complicate the problem of establishing the sources of the observed signals. Winds were light during the midsummer to late summer period of operation at site 2; thus about half of each day was left relatively free of wind disturbances. With the approach of fall at site 2, however, winds began to increase to the point where wind disturbances occasionally persisted throughout the Martian day. In the absence of additional stations for intercomparison our only recourse is to search the records for signals of unusual character and to determine if these signals correlate with local-wind-generated or lander-generated events. We have seen that seismic signals do correlate quite closely with variations in wind speed and direction and that the patterns of these variations can be quite complex. Onsets of wind gusts can be abrupt, and isolated gusts can occur during otherwise quiet periods. Unfortunately, wind data are not obtained by the meteorology experiment on a continuous basis, so correlation with seismic data is not always possible. For example, during the conjunction period, wind data were only available for about 17% of the total time.

From the seismic data available through sol 144, eight signals were selected for further analysis prior to receipt of wind data. These are shown in Figures A-24 to A-29 and Figures 20 and 21. Selection was based primarily on their relatively impulsive onsets and relatively low axis-crossing counts. Following receipt of wind data, six of these events were found to correlate with wind gusts. Wind data were not obtained during the time intervals of the remaining two signals. The times, maximum winds, and average signal frequencies for each event are given in Table 3. Signal frequencies are average values for the vertical component over 40-s samples centered at the time of peak signal amplitudes. On the basis of these data, there appears to be some tendency toward increasing signal fre-

TABLE 2. Seismic Noise (Wind) Spectra, 55-Min Samples, Frequency Range 0.05–0.5 Hz, and PSD With Mean Removed

Axis	$-d \log PSD / d \log f$	σ	Samples
	Morning		
X	0.58	0.04	9
Y	0.68	0.08	9
Z	0.88	0.10	9
	Afternoon		
X	0.78	0.05	5
Y	1.00	0.14	5
Z	0.98	0.11	5
	Night		
X	0.73	0.06	3
Y	0.80	0.00	3
Z	0.90	0.00	3
	All Day		
X	0.66	0.10	17
Y	0.79	0.17	17
Z	0.91	0.10	17

PSD is power spectral density.

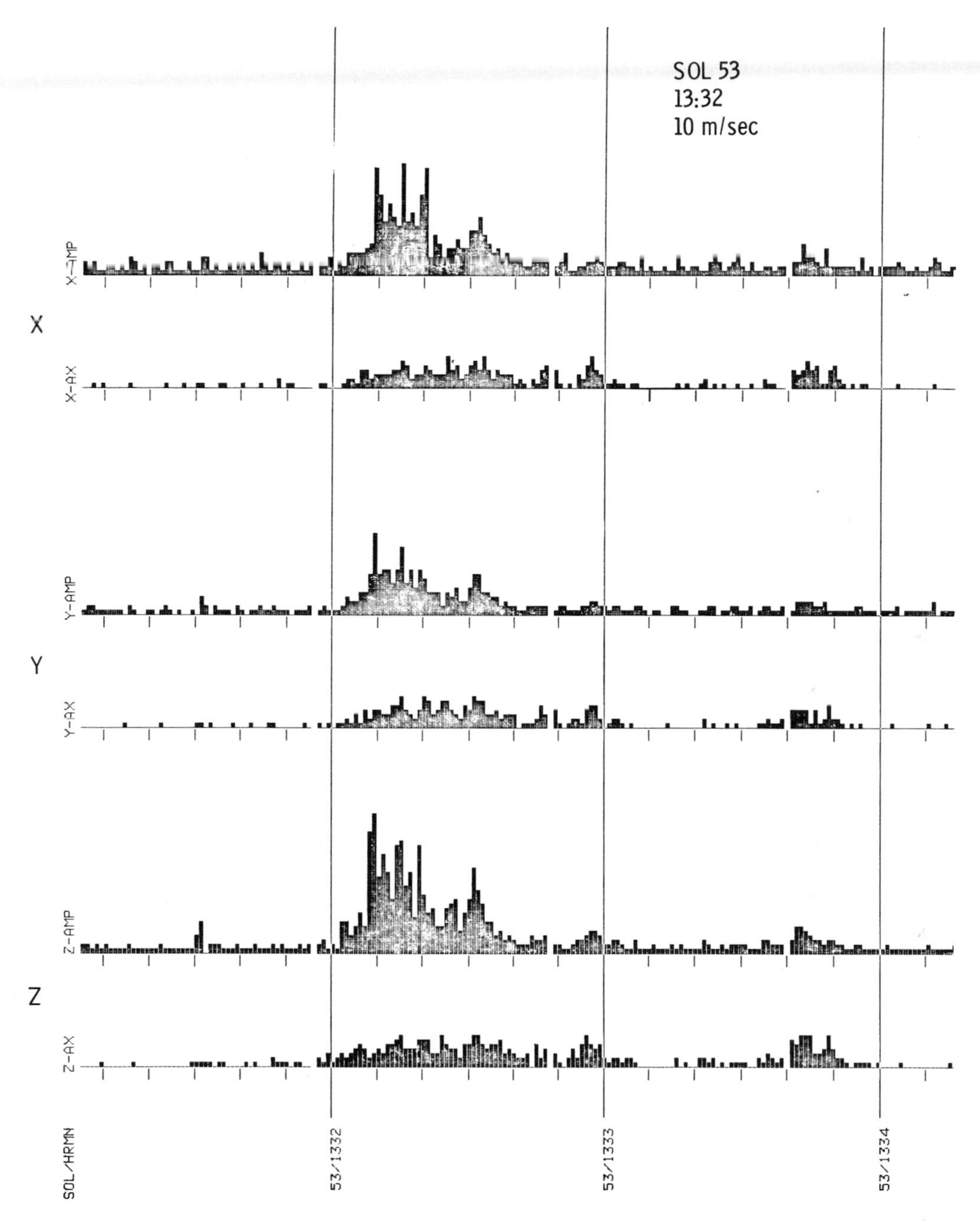

Fig. 20. Sol 53 1332 LLT event recorded in event mode. Although it was thought to be a possible Marsquake when it was first observed, this event was later correlated with a 10-m/s gust of wind.

quency with increasing wind speed. However, the frequency content of the sol 80 signal, for which there are no wind data, is unusually high in relation to the remainder of the data set. Other unusual characteristics of the sol 80 signal are worthy of note: (1) It is the only signal of the group that occurred during the normally quiet period of the early morning, (2) both amplitude and frequency variations occur at intervals of about 10 s, and (3) the amplitude variations are similar to those of two local events recorded by the Viking seismometers during operational tests in southern California. Regarding the first point an isolated wind gust was detected during the late evening hours of sol 82, so such gusts, though they are infrequent, can occur.

The 10-s periodicity in the amplitude variation of the sol 80 signal is clearly seen in the spectrum of the signal envelope in Figure 22. A pronounced spectral peak of 10 s is not seen in the spectra of any of the other candidate signals except that of the sol 42 signal. The signal character in the time domain is quite different in these two cases, however. The high-frequency character of the sol 80 and sol 42 signals is also distinctly different. Histograms giving the distribution of zero crossings for the sol 80 event and three wind events are given in Figure 23. The dominant frequency on the horizontal axes is 3–4 Hz in all cases, close to the natural period of the seismometer system. The same is true for the vertical component of the known wind events. The vertical (X) component for the sol 80 event has peaks at 2, 4, and 7 Hz. A peak at 6 or 7 Hz also shows up in the horizontal components of the wind gusts. The sol 80 event is distinctly bimodal in comparison to the other events.

The sol 80 signal is compared in Figure 24 with two terrestrial signals recorded with a test instrument operated at the

California Institute of Technology campus in Pasadena. The uppermost signal is an aftershock of the 1971 San Fernando earthquake, recorded at a range of 65 km, the middle signal was generated by 102 metric tons of chemical explosive detonated in a rock quarry at a distance of 69 km, and the bottom signal is the sol 80 event recorded on Mars.

Obviously, a strong case for seismic origin for any Martian signal cannot be made until a strong signal is obtained at a time when wind speeds are known to be low. However, on the basis of the similarities between the signals recorded from local shallow sources in southern California and the sol 80 signal, let us see if the amplitude-duration relations are consistent with the hypothesis that the sol 80 signal was generated by a Marsquake similar in crustal structure to that of the terrestrial examples. Interpreting the amplitude and frequency variations as P and S arrivals, as shown in Figure 25, the S-P time is approximately 13 s. This places the epicentral distance at about 110 km for a surface focus event if crustal velocities normal for the earth are assumed. The maximum amplitude of the shear wave arrival (peak 1, average of Y and Z components) is 188 nm at a frequency of about 5 Hz. The trace amplitude on a standard Wood-Anderson seismograph (T_0 of 0.8 s and magnification of 2800) would be 0.53 mm, yielding a Richter magnitude of 2.8 for the event.

The magnitude-duration relation determined for southern California is given by

$$M_t = -0.68 + 1.75 \log t + 0.0014\Delta$$

where t is the signal duration in seconds and Δ is the distance in kilometers. If t = 70 s and Δ = 110 km are used, this expression yields M_t = 2.7, in good agreement with the magnitude estimate from maximum signal amplitude. The duration

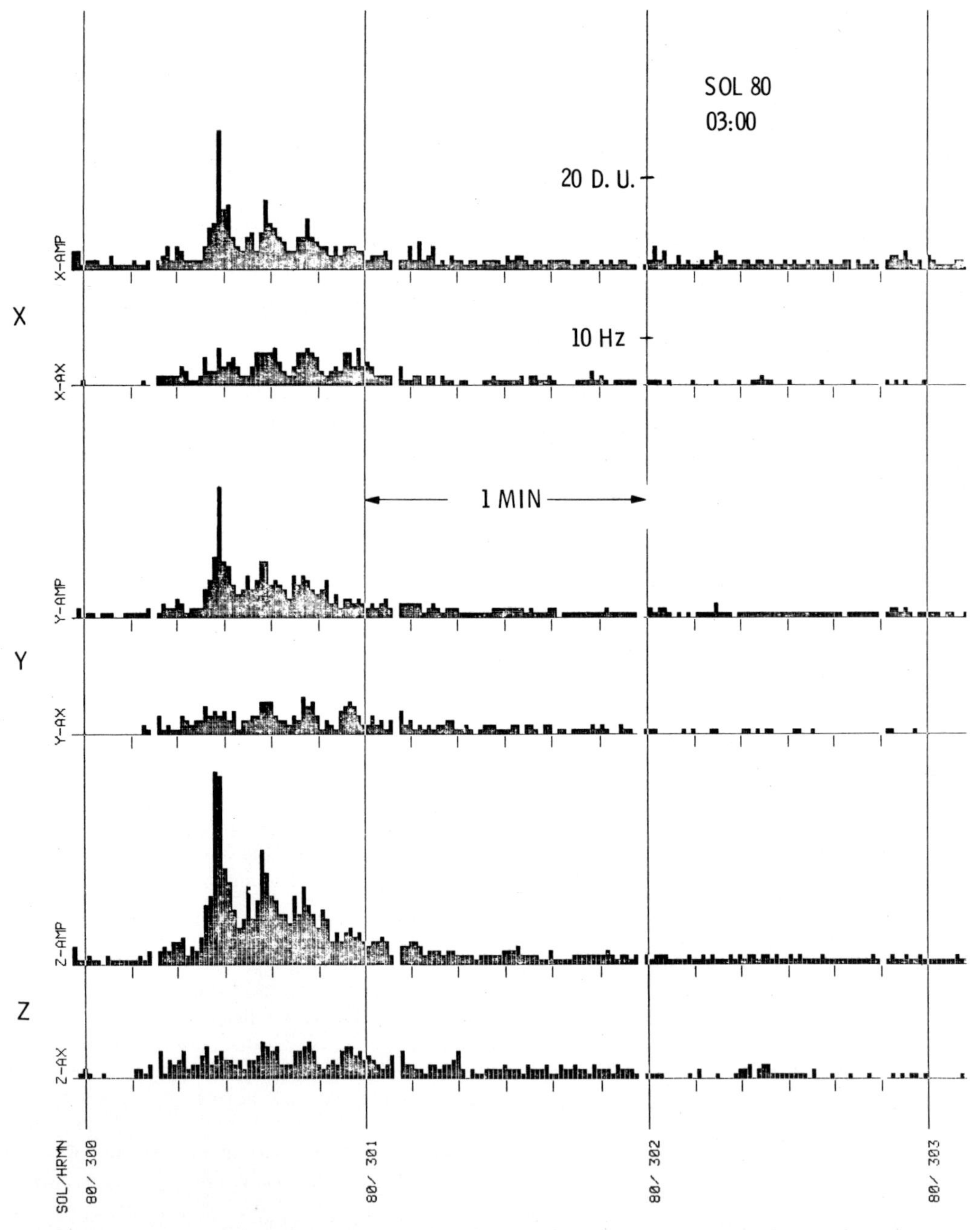

Fig. 21. Sol 80 0300 LLT event. This event is the most likely candidate for a Marsquake recorded to date. Because wind data were not recorded at the time of this event, it is possible that this event could also have been generated by the wind.

TABLE 3. Times, Average Frequencies, and Wind Speeds for Eight Seismic Signals

Sol	Local Lander Time, hours	Average Signal Frequency, Hz	Maximum Wind Speed, m/s
12	1650	3.48	no data
23	1515	3.60	12
42	1554	3.90	11
49	1345	3.73	12
49	1435	2.98	8
49	1548	3.50	10
53	1332	3.45	10
80	0300	4.80	no data

of the sol 80 event is therefore consistent with the amplitude for a Marsquake at a distance of 110 km in a crust having scattering and attenuation properties similar to those of the earth's crust.

There are at least two amplitude peaks (peaks 2 and 3) following the main peak at intervals of 10 s (peak 1 to peak 2) and 9 s (peak 2 to peak 3). One way to explain this pattern is by shear wave reflections in a surface layer. Many investigators have recognized such reflections in interpretation of seismograms and have used them in determination of crustal structure [e.g., *Gutenberg,* 1944]. If the two-way travel time for a shear wave traveling vertically in a surface layer is taken to equal 9 s and the layer shear wave velocity to equal 3.5 km/s, this interpretation gives a discontinuity at a depth of approximately 16 km beneath the region of landing site 2. The peak amplitude ratios and assumed Q of the layer can be used to calculate the reflection coefficient of the discontinuity. However, further speculation concerning the signal is of little value unless additional signals possessing similar characteristics are obtained.

The importance of positive identification of even a single event as being of internal (Marsquake) origin cannot be overstressed. Aside from possible structural interpretations of the type described above, much can be learned from signal characteristics alone. For example, if the sol 80 event is a Marsquake, the similarities in signal characteristics between it and the terrestrial cases would indicate that the source parameters and transmission characteristics of Mars and earth are similar. This would be in marked contrast to the lunar case, in which signals are prolonged, with little coherence between the three components of ground motion. This is ascribed primarily to the very high Q of the outer shell of the Moon and a sharp increase in velocity with depth in the superficial (regolith) zone. A moonquake with magnitude and range equal to those of the cases discussed above would have generated a seismic signal lasting several hours instead of 1 min. If similarity between Martian and terrestrial signals can be established, we can say immediately that a lunarlike regolith is unlikely on Mars and that absorption of seismic wave energy is much higher in the region of Mars traversed by the observed signal. The high Q in the case of the moon can be attributed to the absence of volatiles in the outer layers.

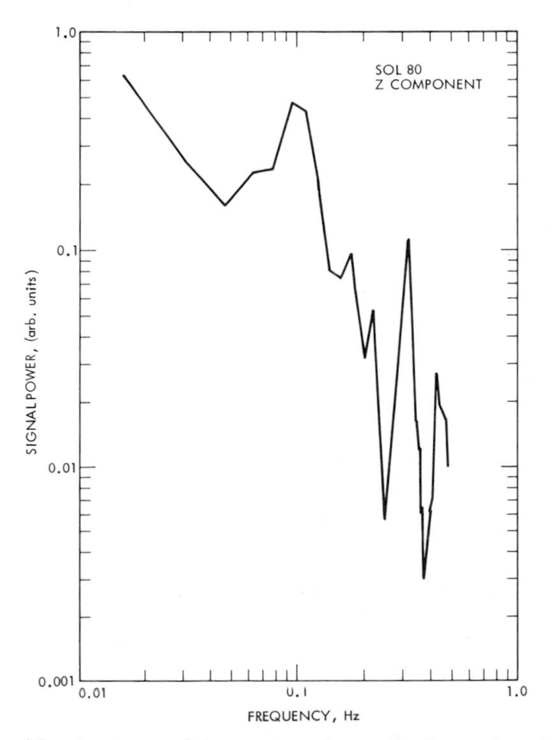

Fig. 22. Spectrum of the event mode amplitude envelope for the sol 80 event. The 10-s periodicity observed in the time domain is clearly visible in the spectrum.

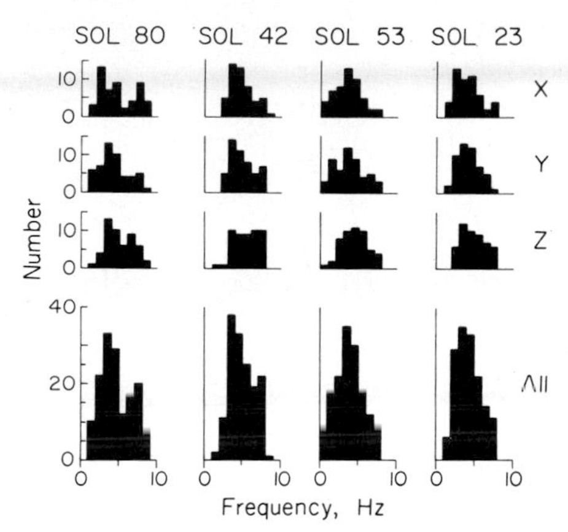

Fig. 23. Histogram of the sol 80 axis crossings compared with histograms from three other events. The sol 80 event is noticeably richer in high frequencies.

In summary, during the first 146 sols of operation of the Viking 2 seismometer, only one event has been detected that might be a local Marsquake. No events that can be interpreted as large teleseismic Marsquakes have been identified with certainty. These conclusions are based on visual inspection of the records. Only limited digital processing has been done to date. It should be pointed out that the seismometer has only been operating at highest gain and in the optimal modes (high data rate or event mode) during quiet periods, uncontaminated by lander activities or winds, for a cumulative total of about 30 days. This is too short a period of time to draw definitive conclusions about the seismicity of Mars. It should also be pointed out again that the seismicity of Mars is likely to be nonuniform both in time and space, as it is on the earth and the moon, and that the Viking 2 landing site is remote from the possibly most active regions of the planet inferred from photogeology and gravity information.

DETECTION CAPABILITY, SEISMICITY, AND METEOROID IMPACTS

The detection of Marsquakes by the Viking seismometer system depends on the size and distance of the event, the magnification (gain level) of the instrument, and the background noise level. The noise level is discussed extensively elsewhere in this report. Furthermore, the amplitude of the

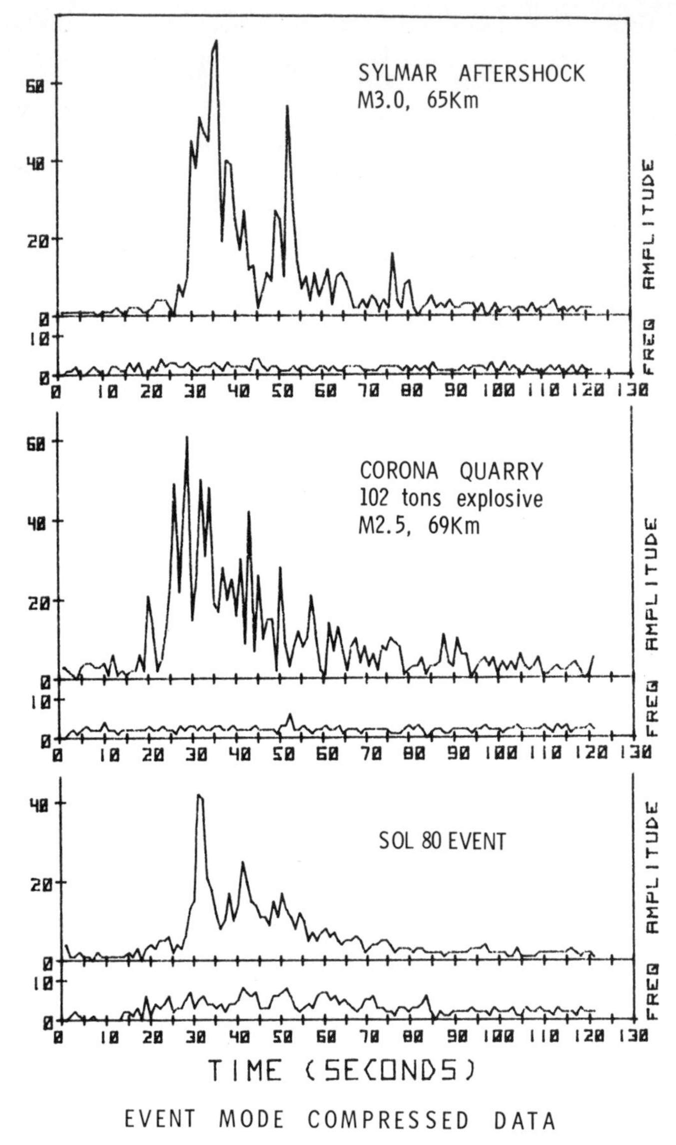

Fig. 24. The sol 80 event compared with similar events recorded in southern California by a Viking seismometer located at the California Institute of Technology. All events were recorded in the event mode.

ground motion produced at the Viking site from a given Marsquake depends on the velocity structure and the attenuation characteristics of the Martian interior. The internal structure models calculated on the basis of available constraints predict the presence of a core shadow zone starting at about 90°–120°. The seismic ray paths for one such Mars model are shown in Figure 1.

The event detection capability of the Viking seismic instrument has been determined by combining the actual test data obtained in the earth and the ray theoretical amplitudes of seismic body waves calculated for a theoretical model of Mars. An average mantle Q value of 2000 was assumed. The detection threshold as a function of distance is shown in Figure 26. This curve is based on the condition that an event can be identified if it produces a ground motion equivalent to 25 du or greater in the output seismic data. This is indeed a conservative case, since during wind-free periods the background noise rarely exceeds 3 du. Figure 26 shows that the Viking seismometer should detect Marsquakes of magnitude 3 at a distance of 200 km, magnitude 5 at 1000 km, and magnitude 6 over half the surface of the planet ($\Delta = 90°$). It is likely that an event of magnitude 6.5 or greater occurring anywhere on the surface of the planet can be detected by the Viking seismic instrument when the winds are relatively low.

To evaluate the level of seismic activity on Mars, it is necessary to analyze the available data, taking into account the operation modes of the instrument. The Marsquake detection capability of the Viking seismograph system has not been uniform throughout this period of observation because of the variability of wind interference, other lander activities, and limited data allocations. The detection capability is highest when the instrument is operated either in the high data rate or event mode, the level of wind-generated noise is less than 1–2 du, and there are no other interfering lander activities. Under these conditions, seismic signals of amplitudes greater than 25 du should easily be identified. These conditions were met for a total of 640 hours during the first 146 sols of operation, or about 18% of the total time of operation.

On the basis of these data the comparative seismicity of Mars relative to that of the earth and moon can be estimated. On the earth, there are about 45 earthquakes per year of magnitude 6.5 or greater. Most earthquakes are at shallow depths. If the average number of Marsquakes per unit area of the Martian surface were the same as that on the earth, with the detection capability of the Viking seismometer one would expect to detect 13 Marsquakes of magnitude 6.5 or greater per year, or approximately one per month. No such events have yet been recognized during the 640 hours of the high-sensitivity low-noise recording period of the Viking mission. From this limited sample it appears that Mars is seismically less active than the earth.

A more quantitative estimate of comparative seismicity of Mars relative to that of the earth and moon can be made by taking into account smaller events. If we assume that Marsquakes above a certain magnitude, like earthquakes, can be

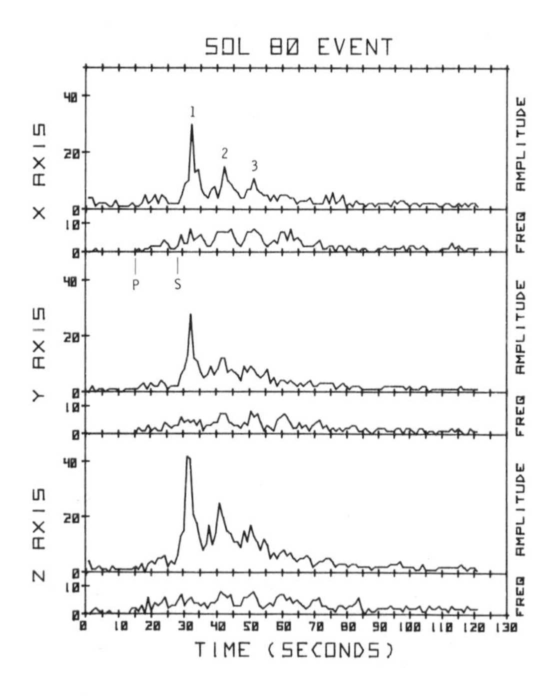

Fig. 25. The sol 80 event as recorded on all three axes. The P and S refer to the possible arrivals of compressional and shear waves. The peaks labeled 2 and 3 are possible reflections from the base of the crust.

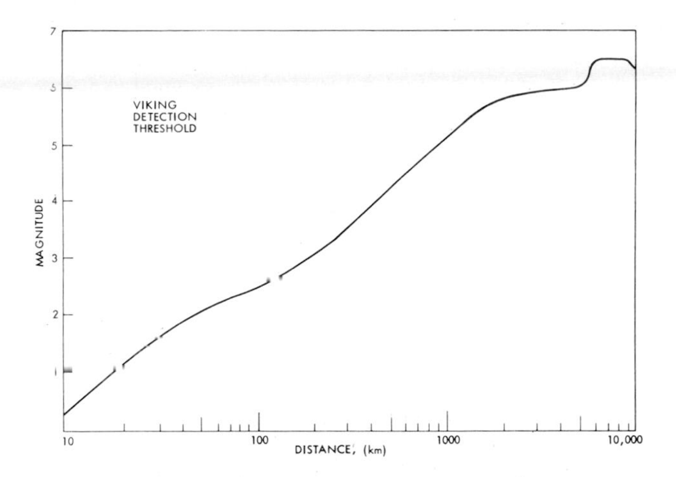

Fig. 26. Hypothetical Marsquake detection threshold values based on an average mantle Q of 2000 and a detection threshold amplitude of 25 du. It is likely that a Marsquake of magnitude 6.5 or larger occurring anywhere on Mars could be detected.

represented by a random Poisson process, the probability of observing n events over a period interval T is

$$P[n, T] = [(\alpha T)^n/n!]e^{-\alpha T}$$

and the probability of observing no events is

$$P[0, T] = e^{-\alpha T}$$

where α is the mean rate of occurrence of events. Substituting the above mentioned values into this equation, we obtain

$$P[0, 640] = 0.39$$

Therefore although it is more probable that Mars is less active than the earth, there still is a 39% probability that Mars can be as active as the earth if the landing site happens to be an aseismic region of Mars. If we consider events of magnitude 4 or greater and if we assume that Marsquakes are uniformly distributed in space (thus assuming the nonexistence of aseismic regions), the number of expected Marsquakes will be increased to about 40 per year at earth rate. Therefore the probability of not detecting any event during 640 hours of observation under ideal conditions becomes

$$P[0, 640] = 0.05$$

Thus with the data collected so far and with the assumptions stated, the seismicity of Mars per unit surface area is less than that of the earth with a probability of 0.95. The additional data that will be collected during the next 2 years clearly will place more definite limits on Martian seismicity. It is also probable, as we discussed earlier, that the seismicity of Mars is not uniformly distributed in space and that the Viking 2 site may be in a relatively aseismic part of the planet.

A comparison of Martian seismicity with lunar seismicity can be made by using the data obtained from the seismometers deployed on the lunar surface during the Apollo mission [*Latham et al.*, 1973]. The four seismic stations that are currently in operation are detecting moonquakes and meteoroid impact signals at a rate of about 2000 per year. The high rate of detection, however, is not due to the high seismicity of the moon but is due primarily to the high sensitivity of the instruments, the low background noise level, and the low attenuation characteristics of the lunar interior.

Correcting for the difference in instrument response, a Viking seismometer deployed at the Apollo 14 landing site would have detected three high-frequency transient events (possible shallow moonquakes at large distances [*Nakamura et al.*, 1974]) and one meteoroid impact during the 6 years of operation. Multiplying by a factor of 4 to account for the difference in the surface areas of the moon and Mars, we would thus expect to have two Marsquakes of detectable sizes annually if Martian seismicity were the same as lunar seismicity and if the Q of the Martian interior were as high as that of the lunar mantle, which is about 4000 [*Dainty et al.*, 1976, *Nakamura et al.*, 1976]. If we use statistics similar to those above, the chance that no Marsquakes are observed to date when an average 22 Marsquakes are expected per year (i.e., activity an order of magnitude higher than that on the moon) would be 20%. Thus for a Mars of low attenuation comparable to the Moon we can say with 80% confidence that the seismicity of Mars cannot be greater than an order of magnitude above the lunar seismicity.

With the presence of volatiles the Q of the Martian interior most likely is lower than that of the moon. If we assume a Q of 500, the detection of moonquake-sized events will be limited to within 500 km of a station. Then if we assume uniform areal distribution of Marsquakes, the expected number of detectable Marsquakes would be reduced to 1 in about 50 years if the Martian seismicity were the same as the lunar seismicity.

In summary, the Viking seismic observations to date are not sufficient to determine whether Mars is more active than the moon.

We next examine the question of the expected rate of detection of meteoroid impacts on Mars. Approximately 300 meteoroid impacts are detected on the lunar surface each year by the instruments of the Apollo Seismic Network [*Duennebier et al.*, 1976]. These fall in the mass range from about 0.5 kg to 1 metric ton. Most estimates of meteoroid flux in the vicinity of Mars are equal to or larger than those in lunar space, so we might reasonably ask, Where are the impact signals on Mars?

Three factors reduce the likelihood of impact detection on Mars in relation to the moon. First, according to the theoretical results of *Gault and Baldwin* [1970] the Martian atmosphere is surprisingly effective as a shield against meteoroid impacts. Their results indicate that the number of meteoroids of 10-kg mass reaching the surface would be only 10% of the number incident on the atmosphere, and the velocity of those that reach the surface would be reduced to several hundred meters per second. In fact, the impact velocity of objects as large as 1 metric ton will be significantly reduced. The predicted shielding from objects in the small to intermediate mass range is consistent with the angular appearance of the surface features of Mars compared with the much more subdued lunar morphology. A second factor bearing on relative rates of meteoroid detection is instrument sensitivity. The peak sensitivity of the Viking seismometer is about 70 times lower than that of the short-period Apollo seismometers. If the Apollo instruments had been operating at the sensitivity of the Viking seismometers, only one impact would have been recorded during the entire period of operation of the Apollo network (7 yr).

Finally, the influence of seismic wave absorption must be considered. The Q of the outer shell of the moon has been estimated to be about 6000 [*Latham et al.*, 1973]. Typical terrestrial values for shallow depths range from 200 to 300. This difference is ascribed primarily to the quantity of water contained in the medium: reduced water content resulting in higher Q. On this basis we might expect Q values in the outer shell of Mars to be much lower than those found in the moon but possibly higher than terrestrial values.

If we assume a Mars Q of 500 and a moon Q of 6000, the attenuation of a seismic wave at a range of 2000 km will be 40 times greater for Mars than for the moon.

If we take these three factors (atmospheric shielding, instrument sensitivity, and attenuation) together, the likelihood of detecting a meteoroid impact on Mars with the Viking seismometers during the planned lifetime of the experiment is extremely small.

IMPLICATIONS FOR FUTURE MISSIONS

The Viking experiment, even though it was highly curtailed by mission priorities and by the inability to uncage the first seismometer, has provided and will continue to provide valuable information for the planning of any future missions to Mars. First, it established that the seismic background noise on Mars due to winds and atmospheric pressure fluctuations is very low. The Viking seismometer, mounted on a relatively compliant spacecraft with a large surface area exposed to winds, could still operate at maximum sensitivity at least half of the time during the nominal mission, a situation that indicated that seismometers with much greater sensitivities can be operated on the planet. Emplaced by penetrators or deployed as small packages, seismometers more sensitive than the Viking instrument by a factor of at least 10^3 can operate on the planet without being affected by typical Martian winds.

Although the data on Martian seismicity are still preliminary, the indications are that Mars is probably less active than the earth. Greater sensitivity is a must for any future seismic instruments on the planet.

Another important consideration is the deployment of a network of instruments. With a well-placed network of five very sensitive Apollo seismographs it was possible to study the moon and determine its structure, even though most moonquakes are very small. A similar approach of deployment of a network of highly sensitive instruments is needed for Mars exploration. Since Mars is a larger planet than the moon and shows much greater geologic and tectonic diversity, both global and regional seismic networks are needed to understand Martian structure and tectonics. It is important that future missions deploy such networks.

Acknowledgments. The Viking seismometer was designed and tested by F. L. Lehner and W. Miller at the California Institute of Technology. Flight hardware was designed and fabricated by J. Lewko, D. Gibson, M. Van Dyke, T. Gaffield, and D. LaFeniere of Bendix Aerospace. The large-scale integration circuitry was fabricated by American Microsystems, Inc. Wyatt Underwood has been in charge of downlink operations. George Sutton and Ken Anderson were involved in various aspects of the experiment. We thank W. Kaula, R. Phillips, B. Bills, A. Ferrari, and E. Okal for useful discussions and for some of the calculations given in this paper. The cooperation of the Meteorology Team headed by S. Hess is gratefully acknowledged. This work was supported by NASA contract NAS1-9703. Contribution 2910 of the Division of Geological and Planetary Sciences, California Institute of Technology, Pasadena, California 91225.

REFERENCES

Anderson, D. L., Implications of the inhomogeneous planetary accretion hypothesis, *Comments Earth Sci. Geophys., 2,* 93–98, 1971.

Anderson, D. L., Internal constitution of Mars, *J. Geophys. Res., 77,* 789–795, 1972.

Anderson, D. L., The moon as a high temperature condensate, *Moon, 8,* 33–57, 1973.

Anderson, D. L., and R. L. Kovach, The composition of the terrestrial planets, *Earth Planet. Sci. Lett., 3,* 19–24, 1967.

Anderson, D. L., R. L. Kovach, G. Latham, F. Press, M. N. Toksöz, and G. Sutton, Seismic investigations: The Viking Mars lander, *Icarus, 16,* 205–216, 1972.

Anderson, D. L., F. K. Duennebier, G. V. Latham, M. N. Toksöz, R. L. Kovach, T. C. D. Knight, A. R. Lazarewicz, W. F. Miller, Y. Nakamura, and G. Sutton, The Viking seismic experiment, *Science, 194,* 1318–1321, 1976.

Bills, B. G., and A. J. Ferrari, Mars topography harmonics and geophysical implications, *J. Geophys. Res., 82,* in press, 1977*a*.

Bills, B. G., and A. J. Ferrari, A lunar density model consistent with topographic, gravitational, librational, and seismic data, *J. Geophys. Res., 82,* 1306–1314, 1977*b*.

Binder, A. B., Internal structure of Mars, *J. Geophys. Res., 74,* 3110–3117, 1969.

Binder, A. B., and D. R. Davis, Internal structure of Mars, *Phys. Earth Planet. Interiors, 7,* 477–485, 1973.

Blasius, K. R., and J. A. Cutts, Shield volcanism and lithospheric structure beneath the Tharsis plateau, Mars, *Proc. Lunar Sci. Conf. 7th, 3,* 3561–3574, 1976.

Dainty, A. M., M. N. Toksöz, and S. Stein, Seismic investigation of the lunar interior, *Proc. Lunar Sci. Conf. 7th,* 3057–3075, 1976.

Duennebier, F., Y. Nakamura, G. Latham, and H. Dorman, Meteoroid storms detected on the moon, *Science, 192,* 1000–1002, 1976.

Gault, D., and B. Baldwin, Impact cratering on Mars—Some effects of the atmosphere (abstract), *Eos Trans. AGU, 51,* 343, 1970.

Gutenberg, B., Reflected and minor phases in records of near-by earthquakes in southern California, *Bull. Seismol. Soc. Amer., 34,* 137–160, 1944.

Hess, S. L., R. M. Henry, C. B. Leovy, J. A. Ryan, and J. E. Tillman, Meteorological results from the surface of Mars: Viking 1 and 2, *J. Geophys. Res., 82,* this issue, 1977.

Johnston, D. H., and M. N. Toksöz, Internal structure and properties of Mars, *Icarus,* in press, 1977.

Johnston, D. H., R. T. McGetchin, and M. N. Toksöz, Thermal state and internal structure of Mars, *J. Geophys. Res., 79,* 3959–3971, 1974.

Kovach, R. L., and D. L. Anderson, The interiors of the terrestrial planets, *J. Geophys. Res., 70,* 2873–2882, 1965.

Latham, G. V., M. Ewing, F. Press, J. Dorman, Y. Nakamura, M. N. Toksöz, D. Lammlein, F. Duennebier, and A. M. Dainty, Passive seismic experiment, Apollo 17 Preliminary Science Report, *NASA Spec. Publ., 330,* 1–9, 1973.

Lee, B. G., Mission operations strategy for Viking, *Science, 194,* 59–62, 1976.

Nakamura, Y., J. Dorman, F. Duennebier, M. Ewing, D. Lammlein, and G. V. Latham, High frequency lunar teleseismic events, *Proc. Lunar Sci. Conf. 5th, 3,* 2883–2890, 1974.

Nakamura, Y., F. Duennebier, G. V. Latham, and J. Dorman, Structure of the lunar mantle, *J. Geophys. Res., 81,* 4818–4824, 1976.

Okal, E., and D. L. Anderson, Seismic models for Mars, submitted to *Icarus,* 1977.

Owen, T., and K. Biemann, Composition of the atmosphere at the surface of Mars: Detection of argon 36 and preliminary analysis, *Science, 193,* 801–803, 1976.

Phillips, R. J., and R. S. Saunders, The isostatic state of Martian topography, *J. Geophys. Res., 80,* 2893–2897, 1975.

Phillips, R. J., and M. Tiernan, Martian stress distributions: Arguments against the static support of Tharsis, submitted to *J. Geophys. Res.,* 1977.

Phillips, R. J., R. S. Saunders, and J. E. Conel, Mars: Crustal structure inferred from Bouguer gravity anomalies, *J. Geophys. Res., 78,* 4815–4820, 1973.

Reasenberg, R., The moment of inertia and isostasy of Mars, *J. Geophys. Res., 82,* 369–375, 1977.

Ringwood, A. E., and S. P. Clark, Internal constitution of Mars, *Nature, 234,* 89–92, 1971.

Toksöz, M. N., and D. H. Johnston, The evolution of the moon and the terrestrial planets, in *The Proceedings of the Soviet-American Conference on the Cosmochemistry of the Moon and Planets,* in press, 1977.

Toksöz, M. N., A. M. Dainty, S. C. Solomon, and K. R. Anderson, Structure of the moon, *Rev. Geophys. Space Phys., 12,* 539–567, 1974.

Urey, H. C., *The Planets, Their Origin and Development,* Yale University Press, New Haven, Conn., 1952.

(Received March 30, 1977;
revised May 9, 1977;
accepted May 9, 1977.)

The Viking Magnetic Properties Experiment: Primary Mission Results

R. B. HARGRAVES,[1] D. W. COLLINSON,[2] R. E. ARVIDSON,[3] AND C. R. SPITZER[4]

Three permanent magnet arrays were aboard each Viking lander: a strong array fixed on a photometric reference test chart (RTC) on top of the landers, and two other arrays, one strong and one weak, incorporated in the backhoe of the surface sampler. The RTC magnets on both landers have attracted magnetic particles from the dust cloud caused by the retro-rockets on landing and by dust raised in connection with surface sample acquisition and delivery. A considerable amount of magnetic particles has been attracted to both backhoe magnets from the surface material as a result of sample acquisition. We judge that the loose Martian surface material contains from 1 to 7% highly magnetic mineral. Preliminary spectrophotometric analysis shows the material adhering to both weak and strong magnets to be identical and indistinguishable from the normal Martian surface material exposed in trenches. The highly magnetic mineral could be present as discrete grains of iron, magnetite, or pyrrhotite, each with a red ferric iron oxide coating, or as particles of maghemite, γFe_2O_3, with or without other red oxide. Alternatively, the magnetic mineral could be uniformly distributed as a subsidiary component of composite clay or other silicate mineral particles which constitute all or part of the surface material sampled. Mechanical mixtures of the above alternatives are also possible. Although none of the possibilities can be unambiguously excluded, the simplest explanation which fits the available data is that the red pigment in or on all surface particles consists, in part at least, of γFe_2O_3 and that it is the principal source of the magnetic susceptibility of the surface material.

INTRODUCTION

This investigation [*Hargraves and Petersen,* 1972] was designed to detect magnetic particles and, if they are present in the Martian surface material around the Viking landing sites, to determine their composition and abundance. The experiment was simple, utilizing a series of permanent magnet arrays which were either inserted directly into the surface material or passively exposed to windblown particles. The magnets were periodically viewed with the lander imaging system, the resulting pictures being the primary data on which conclusions are based. Use of a $\times 4$ magnifying mirror was intended to aid in optical resolution of individual particles. A steel wire magnet-cleaning brush is being used during the Viking extended mission to allow the conduct of more controlled experiments. This paper constitutes a combination and elaboration of the published preliminary results from the first and second Viking missions [*Hargraves et al.,* 1976*a, b*].

MAGNET ARRAYS

The shape and dimensions of the samarium cobalt permanent magnets [*Hargraves and Petersen,* 1972] are illustrated in Figure 1 (left). The center and ring magnets are magnetized parallel to their axes but in opposite directions.

One of these annular arrays is mounted on the central photometric reference test chart (RTC) atop each lander [*Huck et al.,* 1975], the other two being incorporated in the backhoe of the surface samplers. These latter arrays (Figure 1 (right)) are fitted so that where the surface of one array is approximately 0.5 mm from the backhoe surface, the adjacent array is sunk 3 mm below the surface, providing two levels of attractive force. The weak array on one side of the backhoe is the strong array on the other and vice versa. The effective magnetic field and field gradient at the surface of a strong array (including the magnet on the photometric target) are approximately 2,500 G and 10,000 G/cm; for a weak array these values are 700 G and 3,000 G/cm, respectively. The magnetic attractive force provided by the strong and weak magnet arrays is in the approximate ratio of 12:1 for paramagnetic minerals where the force is proportional to $H \cdot dH/dz$. Ferromagnetic and ferrimagnetic particles may be near saturation in the above fields, and for such particles the ratio will be less. The minimum ratio in this case will be approximately 3:1 (i.e., the ratio of the field gradients) for small particles on the backhoe surface.

PREPARATORY STUDIES

Magnet Characteristics

In preparation for the Viking mission the magnetic force associated with each flight magnet and with the magnets used in laboratory testing was determined. This entailed the (separate) mapping of the magnetic field H and the field gradient dH/dz in planes normal to the surface of magnets, over a distance of 1 cm. As finally fabricated, all magnet arrays proved similar.

Sensitivity of Magnet Response to Magnetic Particles

Apart from the actual abundance of magnetic particles the main factors affecting the amount held on the backhoe magnets are (1) effective susceptibility χ_e or saturation magnetization J_s of the minerals, (2) grain size, (3) presence of composite grains, (4) reduced gravity of Mars, and (5) nature of the backhoe motion before extraction from the surface.

1. The important magnetic property is either the particle susceptibility χ_e in the fields near the backhoe surface or the saturation magnetization J_s. The particle susceptibility χ_e is independent of H in paramagnetic minerals but depends on H in ferromagnetic and ferrimagnetic minerals. The surface fields at the centers of strong and weak magnets are 2500 G and 750 G, respectively, and the former field is sufficient to virtually

[1] Department of Geological and Geophysical Sciences, Princeton University, Princeton, New Jersey 08540.

[2] Institute of Lunar and Planetary Sciences, School of Physics, The University, Newcastle upon Tyne, England.

[3] McDonnell Center for the Space Sciences, Department of Earth and Planetary Sciences, Washington University, St. Louis, Missouri 63160.

[4] Langley Research Center, National Aeronautics and Space Administration, Hampton, Virginia 23665.

Paper number 7S0506.

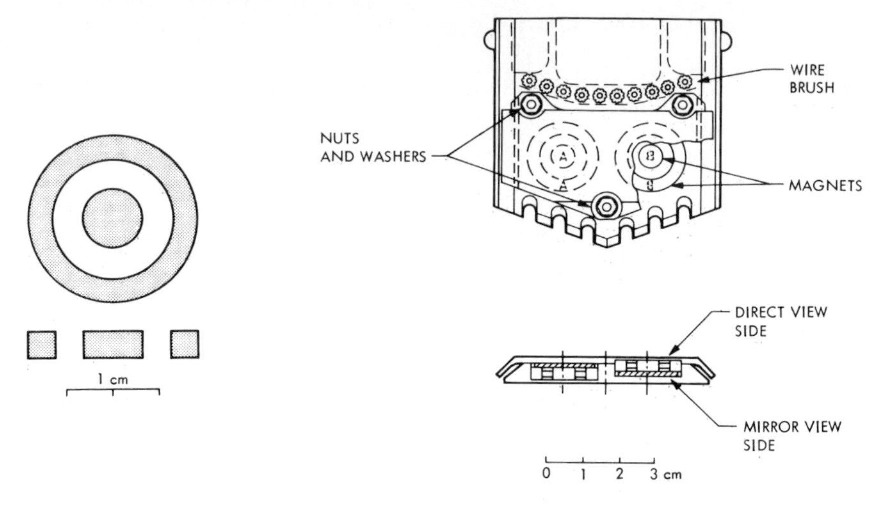

Fig. 1. (Left) Viking magnet array. When they are mounted, the magnets are completely covered by a magnesium metal plate, 0.5 mm thick over a strong array and 3.0 mm thick over a weak array. (Right) Surface sampler backhoe, showing the two magnet arrays. Although it normally stands perpendicular to the sampler axis, the backhoe is spring hinged so that it can fold back when the sampler advances into the surface.

saturate minerals such as magnetite, iron, and maghemite, and for these, J_s is the important property.

2. Because of the decrease in H and dH/dz with z (distance from magnet surface), there will be less force per unit mass on a larger grain than on a small one (say, 1.0 and 0.1 mm), and thus grain size will partially determine whether a given mineral is held. Because the weak magnet surface is 3 mm below the backhoe surface, this effect is not as marked as in the case of the strong magnet.

3. Composite grains will have an effective susceptibility or saturation magnetization depending upon the proportions of magnetic and nonmagnetic material and will be held or not according to the magnitude of these parameters and the particle size. A separating effect will occur, since grains with a higher proportion of magnetic material will be held (especially on the weak arrays) in preference to those that are more weakly magnetic.

4. The general effect of the reduced gravity (g) on Mars ($\approx 0.42g_{\text{earth}}$) will be an increase in the amount of material held and again will be more significant where a small amount of material is held. In the extreme case of a monolayer of grains, the grains will be held on Mars if their susceptibility is approximately half the lower limit for holding on earth, and for a given susceptibility, larger grains will be held on Mars than would be held on earth. The latter effect will be barely significant, though, since theoretically, the diametral ratio of grains on Mars and earth for the same gravitational force is given by $d_M/d_E = (1/0.42)^{1/3} \approx 1.3$. Because of decreasing H and dH/dz with increasing grain diameter, in practice the increase in d would be less than the above value.

5. Depending on the amount of magnetic material held on the magnets, the motion of the backhoe in the surface material during deployment can be important in determining how much material is held. During normal sampling acquisition, with an advancing motion of the scoop, the backhoe folds back, and there is a tendency for magnetic particles on the front to be scraped off. Alternatively, during the trenching or retraction mode a layer of soil in contact with the magnets moves with the backhoe, and thus there is not a continuous magnetic sampling as the backhoe moves through the soil. Our ignorance as to the precise behavior of the backhoe during a deployment of the surface sampler on Mars introduces significant uncertainty into the interpretation of the backhoe results obtained.

To collectively evaluate the first three of these factors, the response of a variety of pure materials and synthetic materials of varying susceptibility and particle size to manual insertions of a Viking-type magnet array was determined. The results of the laboratory investigation, adjusted arbitrarily to best match the response obtained during actual automated surface sampler operations with a test lander, are summarized in Figure 2.

Magnetic Particle Abundance Estimates

The increasing amount of magnetic particles attracted to a flight-type magnet array with repeated manual insertions of the backhoe into a variety of natural unconsolidated materials and some synthetic mixtures was measured and photographed. The composition and abundance of magnetic minerals in these terrestrial materials (see Table 1) were determined. With regard to the estimation of magnetic mineral abundance on Mars, manual laboratory simulations of backhoe deployment are, for reasons given above, only approximately related to the automated backhoe insertion on Mars.

MAGNET EXPERIMENTS ON MARS

All magnet images obtained from both Viking landers (VL-1 and VL-2) during the primary mission are listed in Tables 2 and 3 (see also *Hargraves et al.* [1976*a*, *b*]). The results are summarized and interpreted below.

Reference Test Chart Magnets

Survey mode images (lower resolution) of these magnets were received immediately after touchdown of both landers. While a faint bull's-eye pattern (signifying particles adhering to the magnet) could be seen more distinctly on VL-1, high-resolution images taken later confirmed their presence on both spacecraft. The amount of material adhering to the magnets increased during the first 30–40 days of the mission but thereafter seemed to remain relatively constant (Figure 3).

Three possible sources of these magnetic particles are (1) dust elevated into the atmosphere by the rocket exhausts on landing, (2) dust generated in connection with acquisition and

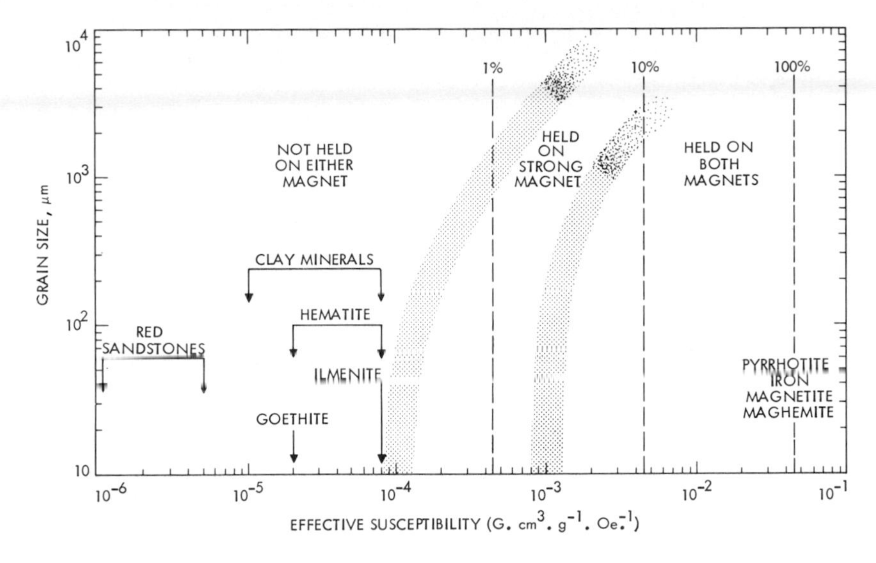

Fig. 2. Diagram illustrating the response to the backhoe magnets of the common magnetic minerals, according to their grain size and magnetic susceptibility. The diagram is based on extrapolation of the results of laboratory experiments with Viking backhoe magnets using natural and synthetic materials to the conditions which we consider to apply when the magnets are actually deployed during sampling activity on Mars. The vertical dashed lines refer to composite particles containing different percentages of any of the four strongly magnetic minerals. The broad stippled bands separate the three main levels of magnetic attraction between which there is no clear division.

delivery of samples to the major analytical instruments, and (3) dust particles normally suspended in the atmosphere.

The pink sky on Mars has been attributed to the presence of reddish dust particles of the order of 1 μm in diameter [*Mutch et al.,* 1976*a*] and in an amount equivalent to about 10 particles/cm³ at VL-1 (J. B. Pollack, personal communication, 1976). There was evidence from the atmosphere extinction coefficient at the two sites (0.35 and 0.25 at VL-1 and VL-2, respectively) that during the primary mission there were fewer atmospheric particles at the VL-2 site, if one assumes a similar size distribution [*Mutch et al.,* 1976*b*]. A 1-μm particle, even if of hematite, would be attracted to an RTC magnet. Weakly magnetic particles alone, however, would tend to be removed by stronger winds, and this has not been observed (see Figure 3 (top)).

If the atmosphere were the sole source of the magnetic particles on the RTC magnets, then assuming nominal wind velocities of 5 m/s and a 5% extraction efficiency, we calculate that a minimum of 15 days would be required to accumulate a discernible layer, 10 μm thick. The material causing the bull's-eye pattern present very soon after landing at both sites must have another source. Thus we conclude that the particles discernible on the RTC magnets of both landers prior to any surface sample activity must certainly have been attracted from the dust cloud raised into the atmosphere by the retro-rocket exhausts when both spacecraft landed [*Mutch et al.,* 1976*a, b*].

A significant increase in the amount of material on the RTC magnets is noticeable following acquisition of samples by the surface sampler on sol 8 and their delivery to the analytical instruments aboard the lander. The progressively increasing amounts discernible on subsequent images (see Figure 3) is most plausibly attributed to the continued surface sampling. The material sampled at both sites is known to contain a significant fraction of highly magnetic particles, and dust raised into the local atmosphere by the sampling activity is considered to be the principal source of the particles on the RTC magnets.

Comparison of images of the RTC magnets on the two landers at roughly equivalent stages of the mission (sol 31, VL-1, six acquisitions, and sol 33, VL-2, five acquisitions; see Figure 3) shows somewhat less material held on VL-2 than on VL-1. This could be due to variations in magnetic mineral

TABLE 1. Properties of Natural and Synthetic Soil Samples Used in Laboratory Testing

	J_s,* emu/g	Extractable Magnetic Component, wt %
Basalt ash/sand, Sunset Crater, Arizona		
Whole rock	0.8	32.8
−16, +25 mesh		70.4
−200, +325 mesh		100.0
White tephra, Settlement Layer, Iceland	0.2	12.7
Black tephra, Hekla, Iceland	0.7	7.0
White tephra, Hekla, Iceland	0.15	22.5
Green rhyolitic ash, Iceland	0.4	4.1
Basaltic beach sand, Braedermerkur-sandur, Iceland	0.45	23.2
Dust released from melting glacier, Braedermerkurjokull, Iceland	0.35	13.3
Rhyolitic ash/sand, Fenton Lake, New Mexico	0.4	3.7
Airfall tuff, Redonda, New Mexico	0.35	4.0
Booming dune sand, Nevada	0.05	3.5
Salt encrusted sand, north of Fallon, Nevada	0.12	5.3
Desert sand, Fallon, Nevada	0.4	21.6
Dune sand, Indio, California	0.65	2.0
Basaltic andesite sand, Red Mountain, Arizona	1.1	11.8
Red silt, Clinton, Oklahoma	0.01	0.3
Red sand, Route 11, Tennessee	0.04	0.4
Wadi sand, Sinai, Israel	1.4	7.8
Red sand, Umzumbi, South Africa	0.05	0.7
Synthetic		
Lunar nominal; crushed dacite	2.1	4.8
Pierre shale, crushed	0.02	0.06

*At 3000 G.

TABLE 2. Viking 1 Magnet Images Received

Sol	Image Reference Numbers*	RTC Magnet	Backhoe Magnets	Comments
0	12A002/000, F1001/24	Survey, shade		
3	11A018/003, F1005/13	Survey, shade		
5	11A032/005, F1007/5		HR, shade	Direct view before sampling.
7	12A054/007, F1012/19		HR, sun	Direct view before sampling.
8	11A060/008, F1012/42		HR, shade	Long-range direct view after fourth sample acquisition.
8	12A061/008, F1012/46		Survey, shade	Backhoe over inorganic analysis sample receiver after fifth sample acquisition; out of focus.
5	12A101/015, F1020/6	HR, sun		
17	11A105/017, F1023/2		Survey, shade	Long-range direct view after sixth sample acquisition.
19	12A110/019, F1026/33	HR, shade		
19	12A112/019, F1026/49		HR, shade	Direct view, slightly out of focus.
20	11A118/020, F1027/66,		HR, sun/shade	Direct view, strong magnet in sun.
24	12A140/024, F1031/23		HR, sun	Direct view after sample delivery.
26	12A145/026, F1033/2	Survey, shade		
27	12A160/027, F1034/68	Color, sun		First color image.
28	12A171/028, F1035/56	Color, IR, survey, sun		
31	12A247/031, F1038/65	Color, IR, sun		
31	12A242/031, F1038/41	HR, sun		
31	12A252/031, F1039/4		HR, sun	Direct view after seventh sample acquisition.
34	12B034/034, F1041/57	Color, IR survey, sun		
34	11B019/034, F1041/82 11B020/034, F1041/87		HR, sun	Direct view images after eighth and ninth sample acquisitions.
36	12B054/036, F1043/61		HR, shade	Direct view after eleventh sample acquisition.
38	12B067/038, F1045/36		HR, sun	Direct view, out of focus.
39	11B073/039, F1046/60	HR, shade		Out of focus.
40	11B088/040, F1047/67	Color, IR, survey, shade		

TABLE 2. (continued)

Sol	Image Reference Numbers*	RTC Magnet	Backhoe Magnets	Comments
40	11B092/040, F1047/8		Color, IR, sun	First color image of backhoe.
40	11B093/040, F1047/14		HR, sun	Direct view after thirteenth sample acquisition.
40	11B099/040, F1048/20	HR, sun		Out of focus.
41	12B104/041, F1049/7		HR, sun	Front of backhoe in ×4 mirror; after insertion into surface, before vibration.
41	12B105/041, F1049/12		Color	As above.
41	12B106/041, F1049/19		HR, sun	As above, after vibration.
41	12B107/041, F1049/24	Color, sun	Color, sun	Combined image of backhoe and RTC magnet.
41	12B112/041, F1049/68	HR, shade		Out of focus.
57	12B141/057, F1056/20	Color, shade		
76	12B166/076, F1063/12	Color, shade		

'Survey' indicates low-resolution camera mode; HR indicated high-resolution camera mode; 'direct view' is a view of the back surface of the backhoe; and IR means infrared image.

* The first number is the camera event number, and the second is the first-order Viking lander image processing (Fovlip) roll and frame number, the frame number of the best image being given.

content of the surface material, in grain size distribution, or in wind speed and direction.

As described below, the evidence from the backhoe magnets is that the magnetic particles in the surface at the two sites are comparable in amount and grain size, and the evidence favors contrasting wind conditions as the cause of the observed difference between the two RTC magnets.

Examination of the average of the first 20 sols of meteorological data obtained from each lander showed that the wind directions were variable at both sites [*Hess et al.*, 1976*a*, *b*; S. L. Hess, personal communication, 1976]. At the VL-1 site the winds blew from between SE and SW during the times of sampling activity, tending to transport dust raised into the atmosphere by this activity toward the RTC magnet. At VL-2, on the other hand, the prevailing winds during sampling periods blew from between SW and NW, tending to blow raised dust away from the RTC magnet. Thus the winds at the two sites are such as to favor particle capture by the RTC magnet on VL-1 over particle capture on VL-2. We attribute the contrast in the amounts adhering to the RTC magnets of the two landers to this cause.

Backhoe Magnet Arrays

The first good, high-resolution images of the VL-1 backhoe were obtained on sol 20, after the sixth sample acquisition, and the first good images of the VL-2 backhoe were obtained on sol 28, after the third sample acquisition. The sol 20 VL-1 image was taken after sample aquisition but before the vibration of the collector head, an operation which is performed in connection with delivery of each sample to any one of the analytical instruments. This image shows considerable material adhering to the magnets and elsewhere on the collector head. However, a sol 24 VL-1 image after vibration still shows a conspicuous concentration of magnetic particles on both weak and strong magnets (Figure 4*b*).

The sol 28 VL-2 image, after only three sample acquisitions and associated vibration, revealed a similar distinct bull's-eye concentration of magnetic particles on both magnets (Figure 5*a*).

In general, the amount adhering to the backhoe magnets appeared to increase slightly, with succeeding sample acquisitions, but evidently the magnets were already approaching saturation (i.e., the point at which they would hold as much of the available material as possible) when first observed (Figures 4 and 5). The subsequent changes noted [see *Hargraves et al.*, 1976*b*] could as well have been the result of variations in the sun angle when the images were taken, as of absolute changes in the amount adhering.

Larger individual particles, 2–3 mm across, were seen clinging to the stronger of the backhoe magnets after sol 34 on VL-1 (Figure 4*c*) and sol 56 on VL-2 (Figure 5*c*). These particles seem to survive subsequent sample acquisitions, indicating a significant effective susceptibility χ_e. Their adhesion to the strong magnet only, however, suggests that they are composite mixtures of highly magnetic plus nonmagnetic phases.

The lack of backhoe images following the earliest sample acquisitions on both landers hinders estimates of magnetic particle abundance. The data available from the primary mission, however, suggest that the magnets approach saturation early, doing so after only three sample acquisitions at VL-2.

TABLE 3. Viking 2 Magnet Images Received

Sol	Image Reference Numbers*	RTC Magnet	Backhoe Magnets	Comments
0	22A002/000, F2001/39	Survey, sun		
6	21A040/006, F2013/20	HR, sun		Before any sampling activity.
10	21A076/010, F2019/2		HR, shade	Out of focus, after first sample acquisition (beta site).
12	22A086/012, F2021/47		Survey, shade	Out of focus, after first sample acquisition (beta site).
12	21A081/012, F2021/21	Color, sun		First color image.
13	21A090/013, F2022/20	HR, sun		
16	21A120/016, F2025/10	Survey, shade		
18	21A129/018, F2026/3	HR, sun		Out of focus.
21	22A160/021, F2030/39	Color, IR, shade		
22	22A166/022, F2031/46	Color, sun		
23	22A181/023, F2034/7	Color, IR, survey, shade		
23	21A184/023, F2034/14	Color, IR, survey, shade		
24	21A196/024, F2035/12	Color, sun		
25	21A204/025, F2035/40	HR, sun		Out of focus.
25	21A212/025, F2035/102	Color, IR		
28	21A229/028, F2039/38		HR, sun	After a total of three sample acquisitions.
29	22A239/029, F2040/32		HR, sun	Weak magnet visible only.
29	21A241/029, F2040/40		HR, shade	
30	21A246/030, F2040/56		HR, sun	After a total of four sample acquisitions.
30	21A248/030, F2040/64		HR, sun	Long-range direct view; images taken during rock-moving activity.
30	21A249/030, F2040/68		HR, sun	As above.
30	21A250/030, F2040/72		HR, sun	As above.
33	22B020/033, F2043/19	HR, sun		
33	21B022/033, F2043/15		HR, sun	Magnets partially shadowed.
34	22B027/034, F2044/28	Color, sun		

TABLE 3. (continued)

Sol	Image Reference Numbers*	RTC Magnet	Backhoe Magnets	Comments
36	22B041/036, F2047/74		HR, sun	After a total of six insertions into surface.
36	22B045/036, F2047/84		HR, shade	
36	21B013/036, F2047/78	Color, sun		
42	22B101/42, F2053/7	Color, shade		
42	22B098/042, F2052/34	HR, sun		
43	21B104/043, F2054/5	Color, shade		
44	22B111/044, F2054/51	Color, IR, survey, sun		
45	21B121/045, F2055/24		HR, partly shaded	After a total of eight insertions into the surface.
46	22B129/046, F2056/37	Color, IR, survey, sun		
46	22B136/046. F2057/20		Color, IR, survey, sun	First color image of material on backhoe; total of nine insertions.
47	22B146/047, F2058/47		HR, sun	After a total of 11 insertions into the surface.
48	22B151/048, F2058/64	Color, IR, survey, sun		
49	22B180/049, F2059/51	Color, IR, survey, sun		
50	22B190/050, F2060/104	Color, IR, survey, sun		
51	21B203/051, F2061/4		HR, shade	After total of 13 insertions into the surface.
51	22B207/051, F2061/21	Color, IR, sun		
56	22C021/56, F2067/20		HR, sun	Front of backhoe in ×4 mirror.
56	22C022/056, F2067/24		HR, sun	Direct view, before insertion.
56	22C026/056, F2067/40		HR, sun	Direct view, after total of 14 insertions into surface; after vibration.
56	22C030/056, F2067/56		HR, sun	Front of backhoe in ×4 mirror.
57	22C037/057, F2067/83	Color, IR	Color, IR	Combined image of backhoe and RTC magnet.
57	22C039/057, F2067/94	HR, shade	HR, sun	As above.

'Survey' indicates low-resolution camera mode; HR indicates high-resolution camera mode; 'direct view' is a view of the back surface of the backhoe; and IR means infrared image.

* The first number is the camera event number, and the second is the Fovlip roll and frame number, the frame number of the best image being given.

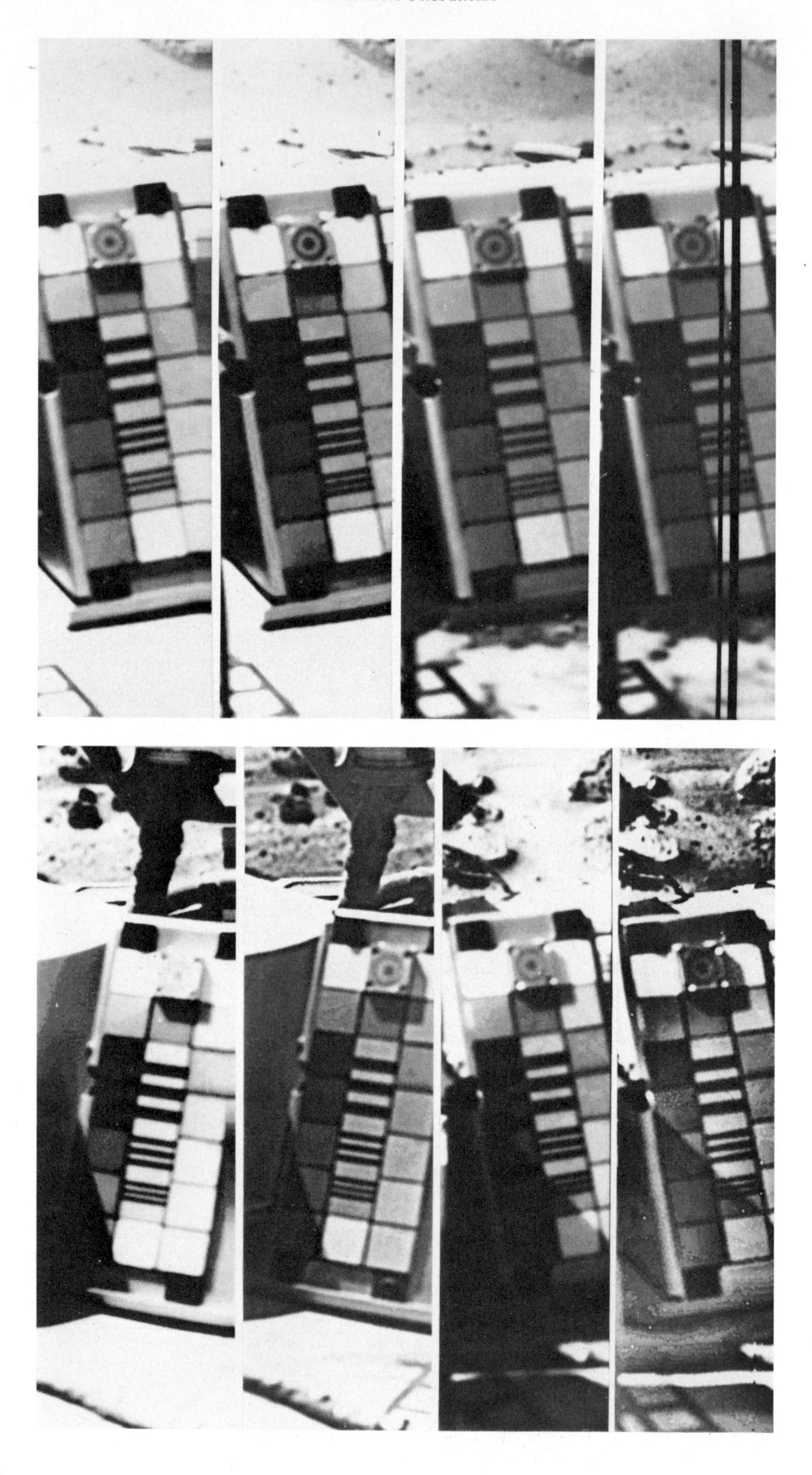

Fig. 3. Reference test chart magnet images (top) for VL-1 on (from left to right) sols 15, 31, 151, and 168 and (bottom) for VL-2 on (from left to right) sols 6, 13, 33, and 42. Note that the VL-1 series includes images taken during the extended mission which show no apparent change in the appearance of the magnet despite the long exposure.

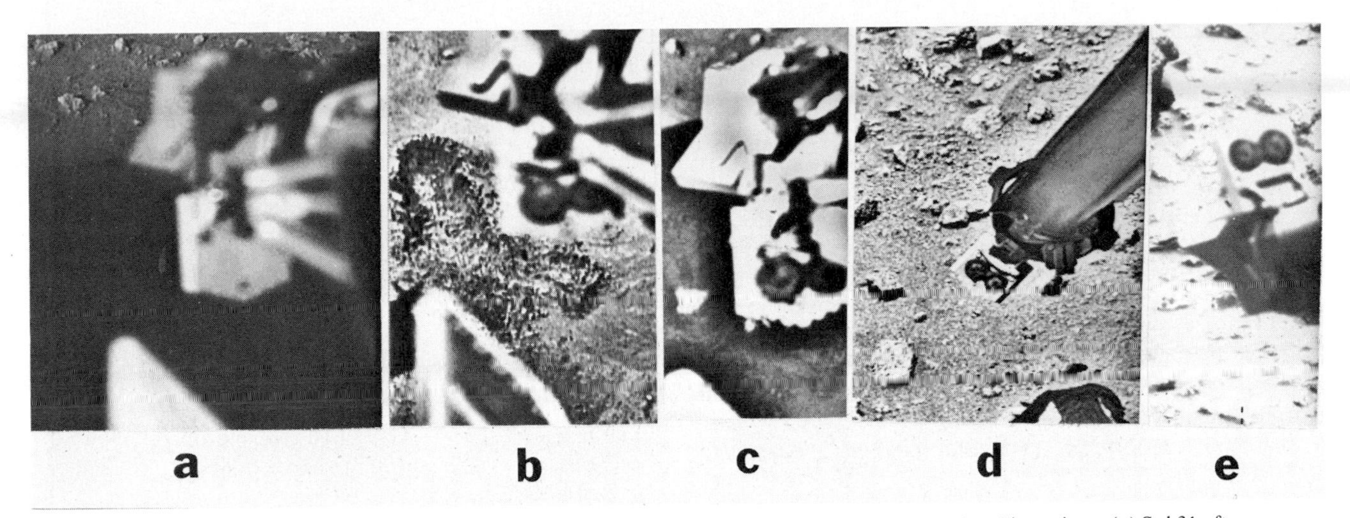

Fig. 4. VL-1 backhoe magnet images. (*a*) Sol 7 before any insertions in soil. (*b*) Sol 24 after 6 insertions. (*c*) Sol 31 after 7 insertions. (*d*) Sol 34 after 9 insertions. (*e*) Sol 40 after 13 insertions. In (*b*), (*c*), and (*d*) the weak magnet is to the right, and in (*e*) it is to the left.

This fact, coupled with the similarity in amount on both weak and strong magnets, provided the basis for our original [*Hargraves et al.*, 1976*a, b*] estimate that the surface material contains from 3 to 7% highly magnetic mineral. Our inclination, following consideration of the probable very fine grain size of the surface particles (<100 μm [*Shorthill et al.*, 1976]), is to reduce the lower limit to 1%. (Note that the inorganic chemical analysis results [*Clark et al.*, 1976] indicate the presence of ample Fe in the surface material to account for this magnetic component.) More refined abundance estimates must await the results of the extended mission experiments. The mode of occurrence of the magnetic mineral, whether as discrete single-phase or composite particles, and its composition are discussed further in a later section.

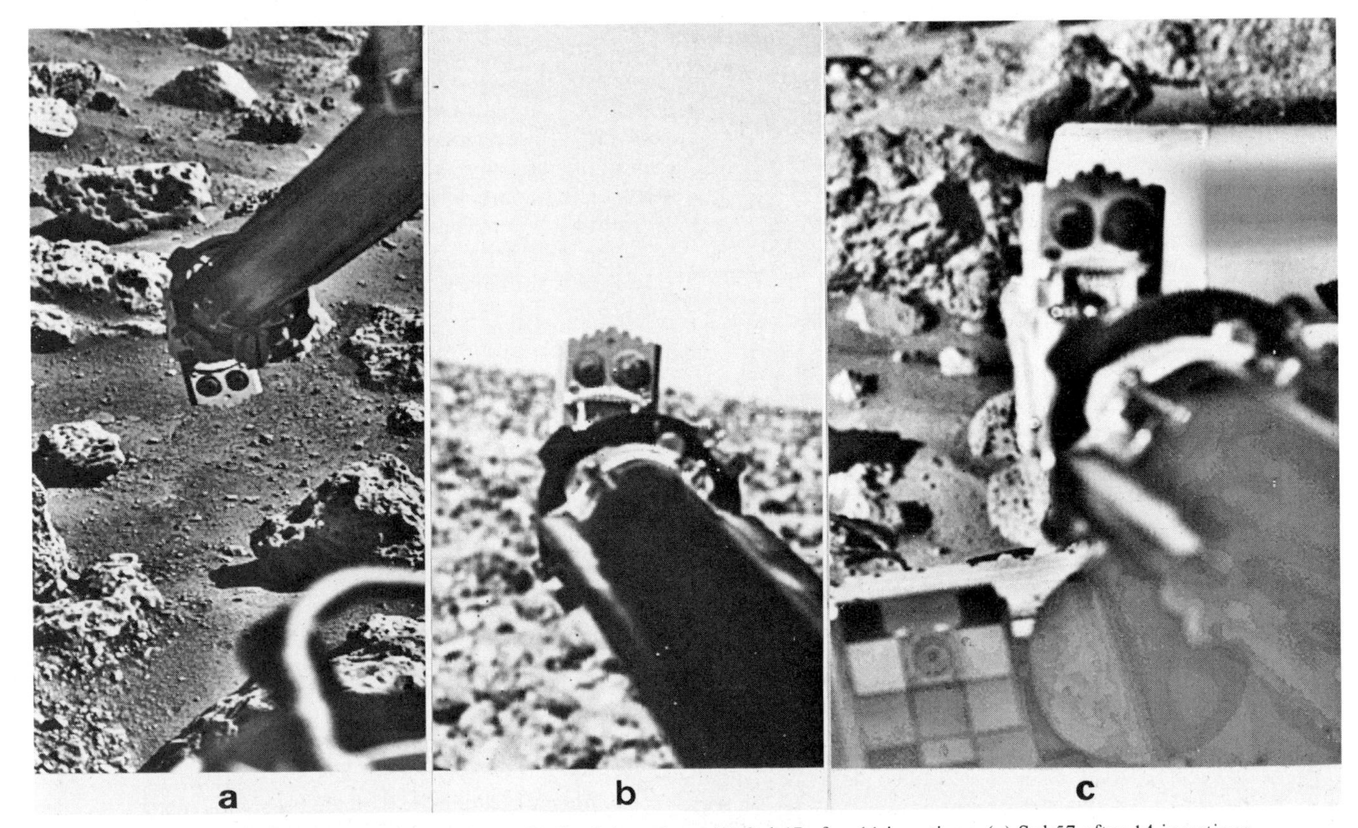

Fig. 5. VL-2 backhoe images. (*a*) Sol 28 after 3 insertions. (*b*) Sol 47 after 11 insertions. (*c*) Sol 57 after 14 insertions (RTC magnet also shown). In (*a*) the weak magnet is to the right, and in (*b*) and (*c*) it is to the left. It may be noted that whereas there are particles bridging the gap between the two magnets on VL-1 (Figure 4), these are absent on VL-2. This suggests that whereas the polarities of the two larger ring magnets are opposed on the VL-1 backhoe (as they should be), they are the same on VL-2. This inadvertent reversal presumably occurred before launch, during reassembly of the backhoe after sterilization.

Spectrophotometric Analysis

The lander cameras have six channels for use in defining spectral reflectivity in the wavelength range from 0.4 to 1.1 μm [*Huck et al.*, 1975]. Three of the channels have appreciable out-of-band response. Because of the nature of the channels it is difficult to assign to each an effective wavelength and then plot wavelength versus reflectance; this is especially true for two of the IR channels [*Kelly et al.*, 1975]. A preliminary comparison of the reflectivity of material clinging to the magnets can be derived by utilizing the blue, green, red, and IR 1 channels as discrete wavelengths [*Hargraves et al.*, 1976*b*]. Although treating the channels as having effective wavelengths degrades the reflectivity estimate, it does serve to delineate similarities and differences. Effective wavelengths for these channels are 0.500, 0.556, 0.669, and 0.867 μm, respectively [*Kelly et al.*, 1975].

Figure 6 shows estimated reflectance values for material on the backhoe magnet and for a portion of the trench imaged on sol 26. It is now known that the central RTC on VL-1 was partially covered with dust by the time that color and IR images of the magnet and chart were taken on sols 34 and 39 [*Bragg*, 1977]. Since the reflectance curve for material clinging to the RTC magnet reported by *Hargraves et al.* [1976*b*] was derived by calibration to a gray patch on the central test chart and since the chart was dirty, results reported for the reflectance curve of material on the RTC magnet are invalid. However, subsequent analysis of the material on the RTC magnet, with calibration to a clean gray patch on another RTC (number 1), demonstrates that the reflectance of material clinging to the RTC magnet on VL-1 is similar to the reflectance of material on the backhoe magnets. Note that these reflectivities are actually a product of normal albedo and the photometric function; i.e., lighting and viewing geometry effects on reflectivity have not been numerically taken into account.

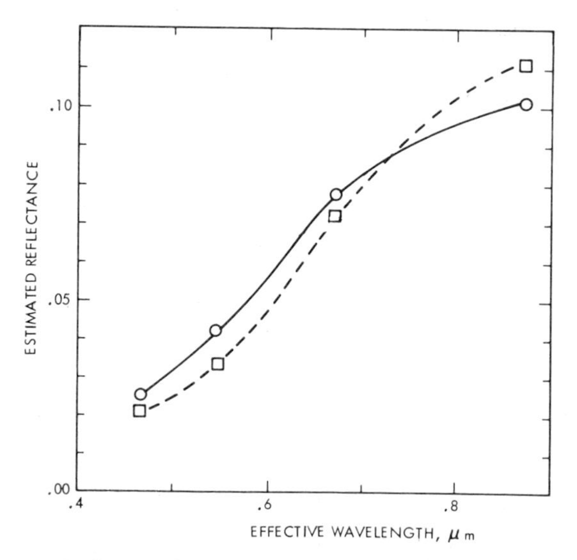

Fig. 6. Estimates of the spectral reflectance of soil exposed in part of the trench and of the material on the backhoe magnets. Estimates were derived by calibration to a reference test chart patch as described by *Hargraves et al.* [1976*b*]. The difference in reflectance between the weak and strong magnets is so small (2–3%) and erratic that the estimates are plotted as one point for each wavelength. Lighting and viewing geometries for the magnets and trench materials are similar, allowing direct comparison of the spectra. Errors induced for each estimate by variance in brightness over the field of interest and by uncertainties in test patch reflectance average about ±15%.

Nevertheless, some preliminary observations can be made. First, the material on the strong and weak backhoe magnets is spectrally indistinguishable to the accuracy that can be discerned by the technique of data reduction used. Second, the material on the backhoe magnets (and the RTC magnet) is similar to the material in the trench. As the particular region chosen for reflectivity of the trench has lighting and viewing angles similar to the backhoe images (thus reducing brightness differences due to lighting and viewing geometry), we feel that this conclusion is soundly based.

DISCUSSION

Purity of Magnetic Particles

An image of the VL-2 backhoe received on sol 47, after 11 insertions, was taken with a sun angle of about 15° relative to the backhoe surface. In this picture the extent of the shadow cast by the particles adhering to the strong magnet suggests a pile 2–4 mm high [*Hargraves et al.*, 1976*b*]. If this height reflects the maximum amount of Martian surface particles which can be attracted to the strong magnet, it is indicative of the effective susceptibility of the particles. By reference to Figure 2 this suggests that the particles are composite, containing between 1 and 10% of a highly magnetic phase.

A most conspicuous feature evident in even the earliest of the images is the similarity (approaching equality) in the amount of material adhering to the weak and the strong magnets (see Figures 4 and 5). In the laboratory this tendency was noted in tests with fine-grained material containing a significant magnetic component. If the magnetic phase is present as discrete grains, then a substantial number of nonmagnetic particles are entrained together with the magnetic fraction. This, however, is particularly true of the strong magnet, and a contrast between weak and strong magnets is still usually apparent. Vibration preferentially purges nonmagnetic material on the weak magnet, and with repeated insertions and vibrations the amount of material on the weak magnet increases and approaches that on the strong, but it is relatively enriched in the magnetic mineral.

If, on the other hand, the highly magnetic phase is present as a uniform subsidiary component of each individual particle (i.e., every particle has a significant effective susceptibility), then the particles will be held, or not, by either magnet, depending on the particle size. As may be seen from Figure 2, particles less than 10 μm, containing 5–10% magnetite or maghemite, will be held by both magnets. The strong magnet (and possibly the weak) could be saturated at the first insertion; but there may be a tendency for the amount on the weak magnet to increase as more of the smaller particles, or grains with slightly higher χ_e, are acquired with succeeding insertions. There should, however, be no difference in the mineralogy of the material adhering on weak and strong magnets, which is in keeping with the spectrophotometric evidence cited earlier. These considerations lead us to conclude that the bulk of the particles clinging to the magnets are in fact composite, each containing a small percentage of highly magnetic minerals.

Composition of Particles

The potential candidates for the highly magnetic mineral or minerals present in the surface material are metallic iron, Fe (or Ni-Fe), magnetite (Fe_3O_4, or titaniferous magnetite), maghemite (γFe_2O_3), and monoclinic pyrrhotite (approximately $Fe_{0.9}S$). Lower susceptibility minerals such as hematite (αFe_2O_3), ilmenite ($FeTiO_3$), and goethite and lepidocrocite

($\alpha FeO \cdot OH$ and $\gamma FeO \cdot OH$, the principal components of limonite) can be excluded, as they could not possibly adhere to the weak backhoe magnet in the amount observed (see Figures 4 and 5).

The only other direct evidence bearing on the composition of the particles comes from the spectrophotometric studies: for VL-1 these show the magnetic particles to be red and essentially indistinguishable from material exposed in trenches. Results are not yet available for VL-2. The reddish coloration of Mars is reasonably attributed to the presence of ferric iron oxides or hydroxides. The results of the inorganic chemical analysis experiments indicate, however, that if these phases are present as a coating on grains, then the coating must be very thin (<0.25 μm) or discontinuous [*Clark et al.*, 1976]. If the phases coloring the magnetic particles are those common on earth, hematite and goethite (or limonite), which are nonmagnetic, then they must coat a core which contains one or the other of the highly magnetic reduced iron phases or maghemite (γFe_2O_3).

Alternatively, the colored coating could contain or consist of maghemite. On earth, maghemite usually occurs in association with, and is subordinate to, hematite, with respect to which it is metastable. The synthetic forms of γFe_2O_3 are characteristically yellowish or reddish brown rather than red, but natural occurrences, particularly if admixed with αFe_2O_3, could well be red. We note that the color estimate of the Martian soil, computed from reflectance estimates derived from Lander camera multispectral data, is yellowish brown [*Huck et al.*, 1977].

The data available, however, simply do not permit an unambiguous identification of the highly magnetic mineral phase present in the surface material of Mars. The possibilities are discussed further in the next section.

SYNTHESIS

The surface material at both Viking landing sites is fine grained, reportedly less than 100 μm in average grain size [*Shorthill et al.*, 1976]. In particular, the particles adhering to the RTC magnet, which are small enough to have been transported some distance laterally in the Martian atmosphere, are probably 1 μm or smaller and certainly less than 10 μm in diameter. The reddish color of the Martian surface is consistent with the relatively high partial pressure of oxygen in the atmosphere ($pO_2 \sim 1 \times 10^{-6}$ atm [*Owen and Biemann*, 1976]), and ferric oxide is the stable form of iron. Despite the low temperatures on Mars it seems unlikely that small particles (<10 μm) of reduced iron minerals (metallic Fe, magnetite, or pyrrhotite) could escape complete oxidation. Furthermore, at equilibrium the vapor pressure of sulfur above magnetic pyrrhotite (approximately $Fe_{0.9}S$) is sufficiently high that, if present, sulfur should have been detected after the 500°C heating for organic analysis. Sulfur was not detected, but the rapidity of this heating cycle and the unlikelihood that equilibrium could be achieved cast doubt on the validity of this criterion.

If this argument, concerning the probability of small particles being completely oxidized, is valid, it implies that the highly magnetic mineral on Mars is likely to be maghemite. In addition to hematite, maghemite has been reported to result from the photostimulated oxidation of magnetite under Martian conditions [*Huguenin*, 1973]. On earth, maghemite is a minor but widely occurring mineral [*Deer et al.*, 1962]. It is seen in some oxidized basalts, in oceanic red clays, and in some red sandstones [*Stacey and Banerjee*, 1974, p. 31], as well as in lateritic rocks [*Frankel*, 1966]. It is clearly a low-temperature oxidation or weathering product.

Maghemite is widely believed to form only by a topotactic reaction involving low-temperature oxidation of preexisting magnetite [*Bernal et al.*, 1959], and evidence of such an origin has been frequently documented [*Stacey and Banerjee*, 1974]. If this is true, then the occurrence of maghemite on Mars would suggest that the surface material is the direct oxidation product of finely comminuted magnetite-bearing igneous rock or carbonaceous chondrite.

Lepidocrocite ($\gamma FeO \cdot OH$), containing ferric iron only, however, also transforms to maghemite upon dehydration [*Deer et al.*, 1962]; it is difficult to conceive of the production of ferrous-iron-bearing magnetite ($Fe^{2+}Fe_2^{3+}O_4$) as an intermediary in this reaction. *Giovanoli and Brütsch* [1975] document the direct formation of maghemite from lepidocrocite, without passing through magnetite at an intermediate stage. *MacKenzie and Rogers* [1977] report the early generation of maghemite by experimental dehydration of nontronite under air and argon atmospheres. These results appear to indicate that the magnetic mineral in the Martian surface material could in addition be the product of photostimulated dehydration of ferric-iron-bearing hydroxides or clays derived from preexisting silicates. Laboratory experiments are planned to test this mechanism.

Summary and Conclusions

The possibilities as to the nature of the magnetic particles detected on Mars are here summarized. Some or all could be (1) highly magnetic, unoxidized mineral grains (metallic Fe, magnetite, pyrrhotite) forming the core beneath a reddish coating of limonite or hematite; (2) grains composed of γFe_2O_3, with or without other iron oxides; (3) igneous rock or mineral particles which consist of an admixture of unweathered silicate mineral or minerals with a significant fraction of a highly magnetic phase, again with a reddish coating; (4) igneous rock or mineral particles, intrinsically nonmagnetic, but having a reddish coating containing γFe_2O_3; and (5) clay mineral particles which contain and/or are coated by Fe_2O_3, of which at least a substantial fraction is in the γFe_2O_3 form.

Highly magnetic reduced iron minerals, if present, could reflect meteoritic infall or be inherited from some preexisting igneous or metamorphic rocks. Such minerals are not stable under Martian conditions, however [*O'Connor*, 1968], and their presence would signify that weathering and oxidation have not been pervasive. The fine grain of the surface material makes this possibility less likely.

Alternatively, the surface material could be completely weathered, with maghemite present as individual particles and/or uniformly contained within, or in the coating on, each clay grain. The effective susceptibility of these latter particles would depend upon their size.

Earth-based spectral reflectance studies of Mars [*Adams and McCord*, 1969] are interpreted to signify that the dark areas are composed of relatively less oxidized basalt containing considerable FeO. The light areas, on the other hand, are interpreted as being more oxidized, low in FeO relative to ferric iron and rich in clay alteration products. From examination of the map entitled 'Martian Albedo Features and Topography' published by the Lowell Observatory [*Lowell Observatory Planetary Patrol Program*, 1973], it appears that the Viking landing sites are located in light rather than dark areas, and hence the fine-grained unconsolidated material should be dominated by weathering products and ferric iron.

All these arguments favor the interpretation that the magnetic mineral on Mars is predominantly maghemite. Unease with this view derives primarily from relative ignorance as to the details of its origin and distribution on earth.

Acknowledgments. Our participation in the Viking Project was supported in part by the NASA Langley Research Center. We acknowledge with thanks the assistance of the Surface Sampler Team and the Lander Imaging Team in obtaining the pictures of the magnet arrays on which our interpretations are based. We also thank Naoma Dorety, who performed most of the preflight magnet calibrations and tests. NASA contract NAS1-9705.

REFERENCES

Adams, J. B., and T. B. McCord, Mars: Interpretation of spectral reflectivity of light and dark regions, *J. Geophys. Res., 74,* 4851–4856, 1969.

Bernal, J. D., D. R. Dasgupta, and A. L. McKay, Oxides and hydroxides of iron and their structural interrelationships, *Clay Miner. Bull., 4,* 15–30, 1959.

Bragg, S. L., Characteristics of Martian soil at Chryse Planitia as inferred by reflectance properties, magnetic properties, and dust accumulation on Viking Lander 1, A.M. thesis, Washington Univ., St. Louis, Mo., 1977.

Clark, B. C., A. K. Baird, H. J. Rose, P. Toulmin III, K. Keil, A. J. Castro, W. C. Kelliher, C. D. Rowe, and P. H. Evans, Inorganic analyses of Martian surface samples at the Viking landing sites, *Science, 194,* 1283–1288, 1976.

Deer, W. A., R. A. Howie, and J. Zussman, *Rock Forming Minerals,* vol. 5, 371 pp., John Wiley, New York, 1962.

Frankel, J. J., Some mineralogical observations on Australian lateritic rocks, *Aust. J. Sci., 29,* 115–117, 1966.

Giovanoli, R., and R. Brütsch, Kinetics and mechanism of the dehydration of γFeO·OH, *Thermochim. Acta, 1,* 15–36, 1975.

Hargraves, R. B., and N. Petersen, Magnetic properties investigation: The Viking Mars lander, *Icarus, 16,* 223–227, 1972.

Hargraves, R. B., D. W. Collinson, and C. R. Spitzer, Viking magnetic properties investigation: Preliminary results, *Science, 194,* 84–86, 1976*a*.

Hargraves, R. B., D. W. Collinson, R. E. Arvidson, and C. R. Spitzer, Viking magnetic properties investigation: Further results, *Science, 194,* 1303–1309, 1976*b*.

Hess, S. L., R. M. Henry, C. B. Leovy, J. A. Ryan, J. E. Tillman, T. E. Chamberlain, H. L. Cole, R. G. Dutton, G. C. Greene, W. E. Simon, and J. L. Mitchell, Preliminary meteorological results on Mars from the Viking 1 lander, *Science, 193,* 788–791, 1976*a*.

Hess, S. L., R. M. Henry, C. B. Leovy, J. L. Mitchell, J. A. Ryan, and J. E. Tillman, Early meteorological results from the Viking 2 lander, *Science, 194,* 1352–1353, 1976*b*.

Huck, F. O., H. F. McCall, W. R. Pattersen, and G. R. Taylor, The Viking Mars lander camera, *Space Sci. Instrum., 1,* 189–241, 1975.

Huck, F. O., R. Arvidson, D. Jobson, S. Park, W. Patterson, and S. Wall, Spectrophotometric and color estimates of the Viking lander sites, *J. Geophys. Res., 82,* this issue, 1977.

Huguenin, R. L., Photostimulated oxidation of magnetite, 1, Kinetics and alteration phase identification, *J. Geophys. Res., 78,* 8481–8493, 1973.

Kelly, L., F. O. Huck, and R. E. Arvidson, Spectral response of the Viking lander camera: Preliminary evaluation, *NASA Tech. Memo., X-72778,* 27 pp., 1975.

Lowell Observatory Planetary Patrol Program, Martian albedo features and topography, 1:25,000,000, *Lowell Observ. Map Ser.,* Lowell, Mass., 1973.

MacKenzie, K. J. D., and D. E. Rogers, Thermal and Mossbauer studies of iron-containing hydrous silicates, I, Nontronite, *Thermochim. Acta, 18,* 177–196, 1977.

Mutch, T. A., R. E. Arvidson, A. B. Binder, F. O. Huck, E. C. Levinthal, S. Liebes, E. C. Morris, D. Nummedal, J. B. Pollack, and C. Sagan, Fine particles on Mars: Observations with the Viking 1 lander cameras, *Science, 194,* 87–91, 1976*a*.

Mutch, T. A., S. U. Grenander, K. L. Jones, W. Patterson, R. E. Arvidson, E. A. Guinness, P. Arvin, C. E. Carlston, A. B. Binder, C. Sagan, E. W. Dunham, P. L. Fox, D. C. Pierce, F. O. Huck, C. W. Rowland, G. R. Taylor, S. D. Wall, R. Kahn, E. C. Levinthal, S. Liebes, R. B. Tucker, E. C. Morris, J. B. Pollack, R. S. Saunders, and M. R. Wolf, The surface of Mars: The view from the Viking 2 lander, *Science, 194,* 1277–1283, 1976*b*.

O'Connor, J. T., Mineral stability at the Martian surface, *J. Geophys. Res., 73,* 5301–5311, 1968.

Owen, T., and K. Biemann, Composition of the atmosphere at the surface of Mars: Detection of argon 36 and preliminary analysis, *Science, 193,* 801–803, 1976.

Shorthill, R. W., H. J. Moore, R. F. Scott, R. E. Hutton, S. Liebes, and C. R. Spitzer, The soil of Mars (Viking 1), *Science, 194,* 91–97, 1976.

Stacey, F. D., and S. K. Banerjee, *The Physical Principles of Rock Magnetism,* 195 pp., Elsevier, New York, 1974.

(Received April 5, 1977;
revised June 6, 1977;
accepted June 6, 1977.)

VOL. 82, NO. 28 JOURNAL OF GEOPHYSICAL RESEARCH SEPTEMBER 30, 1977

Meteorological Results From the Surface of Mars: Viking 1 and 2

S. L. Hess,[1] R M. Henry,[2] C. B. Leovy,[3] J. A. Ryan,[4] and J. E. Tillman[5]

We deal here primarily with the surface meteorological data for both Viking landers during the nominal missions (44 sols for lander 1 and 61 sols for lander 2). The diurnal patterns of wind, temperature, and pressure were strongly similar from sol to sol, as was expected in the summer. The chief characteristics of the wind data are that winds were light (a few meters per second), with a complex hodograph at VL-1 dominated by counterclockwise turning of the wind and a simpler hodograph at VL-2 marked by clockwise turning of the wind. This repetitive pattern of wind has begun to break down at VL-2 with advancing season, and several episodes of protracted northeasterly winds have occurred. Some of these are associated with lower than normal temperatures. Examples are given of wind and temperature traces over short periods, illustrating the effects of convection, static stability, and lander interference. We present a theoretical argument based upon the horizontal scale dictated by heating of slopes and upon vertical mixing of momentum to explain the different sense of rotation of the wind vectors at the two sites. Analysis of the semidiurnal pressure oscillation suggests that absorption of solar radiation is an important thermal drive but that convective heat flux from the surface is also significant. The seasonal variation of pressure extending past the end of the nominal missions shows a decrease of pressure to a minimum at $L_s \approx 149°$ with a rapid rise thereafter. This is clearly due to condensation and sublimation of CO_2 on and from the southern polar cap.

Introduction

The meteorology experiment aboard the Viking Mars landers (VL-1 and VL-2) was designed to measure atmospheric temperature, wind speed, wind direction, and pressure. The scientific objectives have been described previously [*Hess et al.*, 1972]. In brief, the objectives are to obtain information about the local environment and meso-scale and planetary-scale systems and processes, to investigate boundary layer phenomena, and to obtain a better understanding of earth's atmosphere through comparison with the simpler Mars atmospheric behavior.

The instrumentation has been described by *Chamberlain et al.* [1976]. In brief, the temperature and wind sensors are mounted at the end of a boom which was deployed shortly after landing. Figure 1 shows the boom-mounted sensors, while Figure 2 shows the deployed boom on Mars (VL-1 site). In the deployed position the sensors are nominally 1.6 m above the ground, 0.7 m above the top of the lander body, and slightly more than 0.3 m horizontally outward from the closest portion of the lander body. The objective of the deployment was to place the sensors as far as practicable from lander effects. The pressure sensor is mounted underneath the lander body and is the same sensor used to obtain pressure profiles during the parachute phase of entry. It is of the variable reluctance, stressed diaphragm type and measures over the range 0–20 mbar with least digital step about 0.09 mbar. Its measurement accuracy is much better than 0.09 mbar.

The ambient temperature sensor consists of three Chromel-Constantan thermocouples wired in parallel. It is capable of measuring over the entire range of expected Martian temperatures with an accuracy of about ±1.5°C. Wind speed is measured by means of two hot film (platinum) sensors mounted 90° apart in the horizontal plane and maintained at a nominal overheat temperature of 100°C above ambient as measured by a reference temperature sensor. Wind speed accuracy is about ±10% over most of the range (~2–150 m s^{-1}) but degrades somewhat for very light winds. Lander interference, i.e., wind from over the lander, also degrades accuracy, but ground testing indicates that the degradation should not be much in excess of ±10% for the system as a whole. The wind speed sensors also measure wind direction but with a fourfold ambiguity. Selection of the proper quadrant is accomplished by a quadrant sensor, consisting of a heated cylindrical core surrounded by four thermocouple junctions at equal angles and distance about the core. The thermocouples sense the thermal wake from the heated center core. Overall accuracy in wind direction measurement is about ±10°. All electronics, except the bridge circuit for the ambient temperature sensor, are located within the lander body.

The instruments have performed well during the primary mission (from the Mars landing to the solar conjunction period beginning in early November 1976) except for two anomalies. The first concerned the ambient temperature sensor on VL-2 and was observed shortly before launch. The readings exhibited a temperature-dependent error, increasing as lander body temperature decreased. Also, the error gradually decreased after application of power to the electronics, approaching a steady state value after 30–40 min of operation. A series of checkouts was performed during the cruise phase, and it was concluded that the probable cause of the anomaly was a temperature-dependent resistance change somewhere within the electronics. Analysis of the landed data and comparison with the reference temperature sensors on VL-1 and VL-2 and the ambient sensor on VL-1 permitted a suitable correction to be applied. This required that power be on continuously to eliminate the drift. This was done from VL-2 sol 25 onward. (A sol is defined as 1 Martian day, and a VL-2 sol as the number of sols after VL-2 landing. In like manner a VL-1 sol is the number of sols after VL-1 landing.) Because of this anomaly the reference temperature sensor was utilized as the measure of VL-2 atmospheric temperature throughout most of the primary mission. It is much more susceptible to radiation and conduction errors, but corrections for these have been applied, and it is believed to have a residual error of less than ±4°C. The correction to the VL-2 ambient temperature sensor should bring its error to within ±3°C.

The second anomaly involved the VL-1 quadrant sensor. The heater for the sensor core exhibited intermittent behavior on VL-1 sol 45 and failed shortly thereafter. This precluded

[1] Florida State University, Tallahassee, Florida 32306.
[2] NASA Langley Research Center, Hampton, Virginia 23365.
[3] University of Washington, Seattle, Washington 98195.
[4] California State University, Fullerton, California 92631.
[5] University of Washington, Seattle, Washington 98195.

Paper number 7S0459.

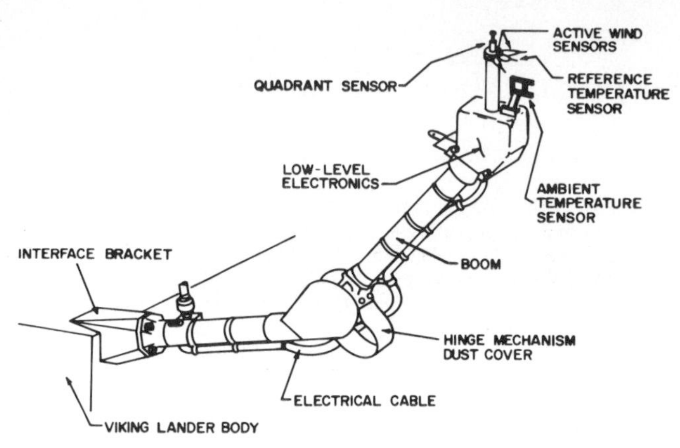

Fig. 1. Schematic diagram of the meteorology boom and sensors.

unambiguous determination of wind direction. Analysis of flight data has revealed that even without power being applied the core remained at a higher temperature than the surrounding thermocouples as a result of solar heating by day and at a lower temperature as a result of infrared radiative cooling by night. The temperature differences are sufficient to be detectable and software modifications to retrieve wind direction information subsequent to sol 45 are being implemented. Although the modifications appear to work well in the low wind speed regime of current operations, the question remains as to how well they will work under the high-wind conditions expected later in the mission.

The on-board lander software was designed to provide a high degree of flexibility in the manner in which meteorology data are taken. One can vary by ground command the rate at which samples are taken, the duration over which a given sample rate is maintained, the serial sequence of sampling rates and duration, the time between sampling sequences, the times and frequencies of sampling initiation, and the total amount of data taken per sol. Prior to landing, preplanned sequences were loaded into each lander. These consisted of modules with durations of about 9, 20, and 39 min spaced nominally 1½ hours apart throughout the sol. Intervals between individual samples were 4 and 8 s for the 9-min modules, 4 s for the 39-min modules, and 2 s for the 20-min modules. One complete sequence consisted of 18 modules. Sixteen of these were 9 min long and two, spaced about 12 hours apart, were of longer duration (one being 39 min long, the other 20 min long). The 18 periods were spaced so as to occupy somewhat more than a sol. As a result the pattern of sequences stepped ahead about 1½ hours each sol. This survey mode was adopted in order to define the diurnal cycle while moving long sequences through the sol, a study of fluctuation characteristics at all portions of the diurnal cycle thus being permitted.

Subsequently, extended periods of rapid sampling (about 1-s intervals) were included, for a time, to study boundary layer processes at as high a sampling rate as the system would allow. In addition, various alterations were made in sampling rates, durations, and measurement periods to determine an optimal pattern for sampling the Martian atmosphere. It became evident that the total science return would be maximized by taking samples nearly continuously but at rates slow enough to satisfy mission constraints on total meteorology data. This approach has been followed ever since, except when precluded by mission requirements.

Initially, pressure data were gathered in groups of four samples, 1 s apart, each group being at the start of a module of the other meteorological data. Thus about 18 groups of pressure data were gathered each sol. The pressures within each group of four were almost always the same. Consequently, to maximize the usefulness of the data, this was changed to a pattern in which one pressure sample was taken approximately every 17 min.

VL-1 landed on July 20, 1976, at 22.5°N, and VL-2 landed on September 3, 1976, at 48.0°N. These dates correspond to planetocentric longitudes of the sun, L_s, of 98° and 120°, respectively. Thus both landings were shortly after the beginning of northern hemisphere summer (L_s = 90°), and all of the nominal mission data reported here are for summer conditions.

Mission data are being sent to the archives of the National Space Science Data Center as quickly as is feasible. Ultimately, all will be available there to interested scientists.

Winds and Temperatures

At the VL-1 site, little secular change in wind behavior was noted during the first 44 sols on the surface. On sol 45 the quadrant heater, which serves to identify the quadrant from which the wind is blowing, began to exhibit erratic behavior and subsequently failed. The subsequent wind direction data can be recovered, as was noted previously, but as of this writing the data are not yet available. A composite wind hodograph for the first 44 sols is given in Figure 3. The composite was obtained by computing vector mean winds at approximately hourly intervals from individual wind vectors at the corresponding times on all the sols included.

The general behavior of the wind vector is a counterclockwise rotation with one complete rotation per sol. However, localized periods of clockwise rotation are evident. Nocturnal winds are light, generally less than 2 m s^{-1}, and change direction gradually from east to southwest from sunset (about 1910 local lander time (LLT)) to midnight. From midnight through dawn (about 0525 LLT) the wind is roughly from the southwest, the upslope direction at the site (Map M25M 3RMC, prepared by the U.S. Geological Survey for the Viking Program, 1976). This may thus represent a drainage wind. Since the lander is only about 1 km above the lowest point of the Chryse Basin, the earlier nocturnal winds may also represent drainage winds, the gradual shift from east to southwest being related to time after terminator passage, areas to the east cooling earlier.

Wind speed increases following sunrise, reaching a maximum of about 7 m s^{-1} at 1100 LLT, from south-southwest. Wind direction remains from the southwestern quadrant throughout the period. This is in contrast to sand drifts and lee deposits which indicate that the much stronger winds required to move granular material at the site are from the northeast [*Mutch et al.*, 1976*a*]. After 1100 LLT, wind speed declines until 1600 LLT, when nocturnal levels are reached. Concurrently, wind direction shifts through south to east-southeast, directions corresponding to the canyon mouth system associated with Valles Marineris. The mouths lie a few hundred kilometers from the VL-1 site, but the canyons are significant topographic features, in both depth and length, and hence could act as wind channels whose effects are felt at the VL-1 site. The data are also in reasonable accord with predictions from general circulation models [cf. *Pollack et al.*, 1976] extrapolated to the surface.

Detailed inspection of the VL-1 hodographs on a sol-by-sol basis reveals small variations between sols, a few moderately large directional changes lasting 1 hour or less between adja-

Fig. 2. The meteorology boom and sensors photographed on Mars by a lander camera (VL-1).

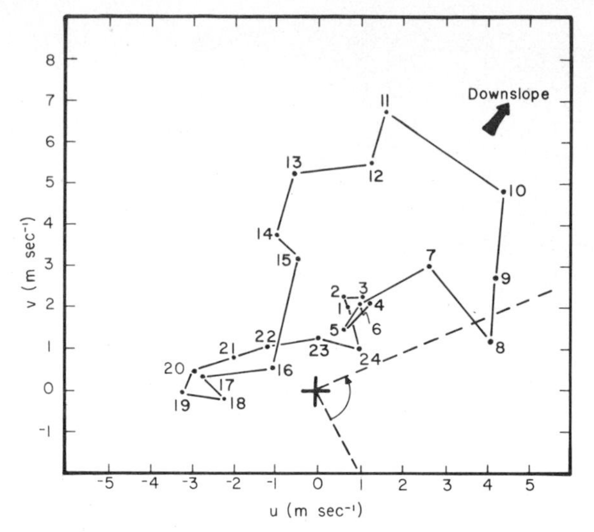

Fig. 3. Mean wind hodograph for the first 44 sols at VL-1. The west-to-east component of the wind is u, and the south-to-north component is v. Thus the wind blows from the origin (marked by a cross) to each plotted point. The number next to each point gives the mean LLT to the nearest hour. Each point represents the vector mean of all observed winds during the first 44 sols measured within 30 min of the indicated time. The sector of lander interference (wind from the lander to the sensors) is bounded by dashed lines connected by an arrow. The heavy black arrow indicates the approximate downslope direction. At VL-1 this ground slope is roughly 0.015.

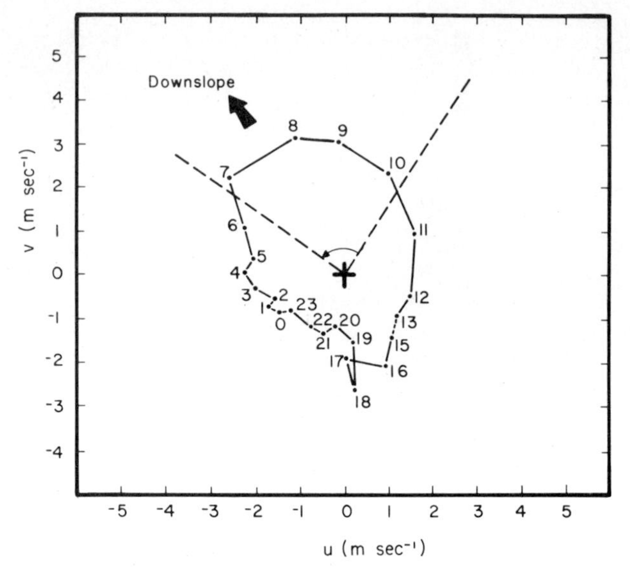

Fig. 4. Mean wind hodograph for the first 50 sols at VL-2 plotted as in Figure 3. A point for 1400 LLT is not shown, since data from only 5 sols are available then; the points for 1300 and 1500 LLT are connected by a dashed line. At VL-2 the ground slope is roughly 0.002.

cent sols, and two small secular trends over the primary mission. The largest short-duration directional change between adjacent sols at appreciable wind speed (>2 m s^{-1}) was an 80° shift which occurred on VL-1 at 1630 LLT each sol over a 3-sol period (sols 20–23). The shift was roughly from east-southeast on sol 20 to south-southeast on sol 21, returning to easterly by sol 23. Only three other changes of similar magnitude (>50°) were observed. No correlation between these shifts and time of day, particular sol, or any of the other measurables was evident.

Two gradual secular trends were observed for VL-1. First, during daylight hours the wind vector at any specific hour turned slowly clockwise from sol to sol, resulting in more southwesterly winds as the mission progressed. At night, however, no such trend was evident. Second, a slight gradual decline in wind speed at night occurred, and the maximum wind for a sol decreased by about 2.5 m s^{-1} during the period. The maximum mean wind typically occurred in the morning (the highest mean speed recorded on any sol being 9.5 m s^{-1}), so that the decline in speed was concurrent with the change in direction. It is premature to ascribe these small secular trends to any specific general or local seasonal changes in circulation.

The wind behavior at the VL-2 site also showed a generally uniform pattern, but significantly more sol-to-sol variation was evident, and several 'events' occurred of which one constituted a dramatic change in wind behavior. A composite hodograph for the first 50 VL-2 sols is given in Figure 4. Unfortunately, owing to the instrument problem on VL-1, the wind data presently available for the two landers do not overlap. The VL-2 composite is distinctly different from that for VL-1 in several ways. First, the wind vector rotates clockwise at the VL-2 site, completing one revolution each sol. Second, all wind directions are represented, whereas for VL-1 no winds from the west through north to east-northeast were observed. Third, the wind speeds are significantly less than at the VL-1 site, particularly during the daytime hours. (Sunrise at the VL-2 site at landing was 0420 LLT, while on sol 61 it was 0459 LLT. The corresponding sunset values are 2017 and 1920 LLT.) Finally, maximum winds are from the southeast versus south-southwest at VL-1. This is in contrast to sand deposits at the VL-2 site which indicate the much stronger winds required to move granular material fare from the northwest [*Mutch et al.*, 1976*b*]. Data at VL-2 during the primary mission were collected for 61 sols. The following discussion and the hodograph are limited to sols 0–50, since during sols 51–61 a dramatic change in the wind behavior was observed. This event is discussed later in this paper.

Nocturnal winds are light as at the VL-1 site and rotate from north to east as the night progresses. Peak wind occurs in midmorning followed by a decay to nocturnal values by late morning. Peak winds are from the southeast, the general direction of Hecates Tholus and Elysium Mons, roughly 1000 km distant. These are the dominant topographic features in the Utopia Planitia area with peaks more than 10 km above the VL-2 site and an apparently monotonic downward slope to that site. The topography at the VL-2 site is not as well known as that at the VL-1 site, but it is possible that the midmorning maximum winds represent drainage winds from the southeastern area.

As was noted above, significant departures from the composite picture, as well as trends, were observed. The magnitude, time of occurrence, and direction of maximum wind on a given sol varied significantly. The largest maximum wind during the first 50 sols was 7.1 m s^{-1}, occurring at 0700 LLT on sol 30 with wind from the east-southeast. The lowest maximum wind was 2.1 m s^{-1}, occurring in the time period 1500–1600 LLT on sol 19 with wind from the north. The largest and most frequent maximum winds on a sol-to-sol basis are from the east-southeast, but a secondary maximum, based on frequency of occurrence and magnitude, occurred with wind from the south-southwest. The complete directional range over which maximum wind occurred on individual sols was from east through south to north, but directions other than east-southeast and south-southwest were quite infrequent and represented instances when wind was light throughout the sol. Times of maximum wind varied from 0400 to 1730 LLT, but again the maximum occurred most frequently and was largest from 0700–0800 LLT. There is a suggestion of periodicity in the interdiurnal variation of maximum wind speed, but the evidence is inconclusive.

During the early morning hours, particularly from 0500 to 0800 LLT, wind speed gradually increased until approximately VL-2 sols 28–34, where it peaked at values 1.5–2 times those shortly after landing. No wind direction trend was associated with this. Subsequently, wind speed during the same time period decreased to previous values by sols 40–50 and thereafter increased to the event beginning on sol 52.

From 0900 to 1800 LLT the wind speed behavior was distinctly reversed from the early morning period, exhibiting pronounced minima in the period from sol 20 to sol 30 and immediately preceding the event, with pronounced maxima around sols 5–15 and around sols 35–45. No trends in wind direction were correlatable with these changes. These speed variations, particularly during the midafternoon hours, exhibited pronounced quasi-cyclic behavior with a peak-to-peak amplitude of about 2 m s^{-1} and a period of about 23 sols. From sunset through dawn, wind speed varied randomly through the sols except for a pronounced increase on about sol 50 leading up to the event.

Wind direction exhibited several trends through sol 50. From midnight through dawn the wind turned slowly counterclockwise from sol to sol, beginning about sols 20–25, the total turning by sol 50 amounting to 40°–50°. From 0700 through 1200 LLT the wind began a counterclockwise turning about sol 10, shifting about 50° by sol 25, at which point a clockwise turning began. This continued until about sol 50, reaching the original directions. The directional behavior during the afternoon hours was erratic but with some suggestion of a periodic behavior. Oscillation was most pronounced, by far, between 1400 and 1500 LLT, where wind would be from north through east for several sols and then shift to north through west, a total of five such shifts being discernable through sol 50. Wind

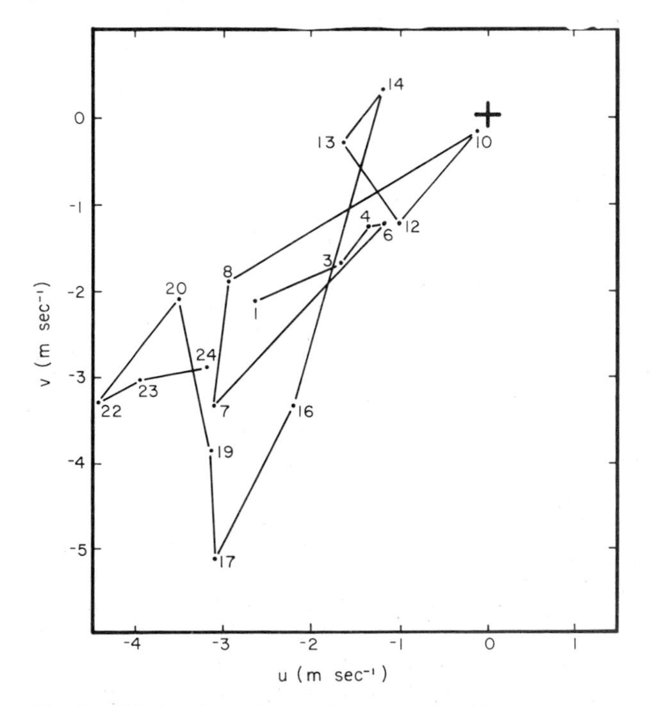

Fig. 5. Wind hodograph on sol 55 at VL-2, at the peak of the only major event during the primary mission. This is plotted as in Figure 3 except that only 1 sol is represented.

direction during the nocturnal hours was variable, no trends or patterns being evident.

A dramatic change in diurnal wind behavior began to be evident on sol 51 when wind during the late afternoon hours was from the north-northeast with a diurnal peak from this direction of 5.6 m s^{-1} at 1700 LLT. The same behavior was evident on sol 52. By sol 53 the wind was confined to the sector north through east to southeast, while throughout sols 54 and 55 the wind blew continuously from the northeast with a maximum of 6 m s^{-1}. Figure 5 shows the sol 55 behavior. On sol 56 the wind reverted to its previous behavior and continued in this 'normal' fashion through the remainder of the primary mission (VL-2 sol 61). This event was not accompanied by any definite temperature change. However, it was closely associated in time with increased fluctuation of pressure at both the VL-1 and the VL-2 site. The pressure decline noted throughout most of the primary mission ceased about this time, and it is not known whether or not the pressure fluctuations are directly associated with the wind event. However, it is quite likely that the event represents some type of north polar wave indicative of the approaching autumn and winter seasons. As of this writing (February 1977), several similar events have occurred subsequent to the end of the primary mission. Although these have not as yet been analyzed in detail, it appears that the north polar region is exerting an increasingly dominant influence upon the weather at the VL-2 site, not only with respect to stronger and more northerly winds but also with respect to declining temperatures. The repetitive pattern of northern summer appears to be changing.

As was previously reported [*Hess et al.*, 1976*a*, *b*], the diurnal temperature cycles at both landing sites have been strikingly repetitive from day to day. As is shown in Figure 6, this pattern has continued throughout the primary mission, and thus through most of the Martian summer, with little change except for very gradual seasonal cooling at the VL-2 site. Even the anomalous northeast winds of sols 54 and 55 produced little perturbation of the repetitive pattern, as can be seen by comparing the curves for sols 45 and 55 in Figure 6 (bottom panel).

Figure 7 shows the day-to-day variation of minimum, maximum, and mean ambient temperatures at the instrument height of 1.6 m. Interdiurnal variations are small at both landing sites. Very little seasonal change can be seen at VL-1, but a slow seasonal cooling with a small reduction in diurnal range is apparent at VL-2. Also shown in Figure 7 are the planetary surface brightness temperatures taken from *Kieffer* [1976]. The minimum air temperatures differ only slightly from the surface temperatures. As was expected [*Gierasch and Goody*, 1968], the maximum air temperatures are considerably lower than the planetary surface temperatures.

Variations of temperature, as well as of wind, become increasingly larger in the last few sols of the period, suggesting the approach of a season of greater variations.

TURBULENT CHARACTERISTICS

The planetary boundary layer and the surface layer of planets like the earth and Mars are turbulent, and their fundamentally important characteristics are the static stability and the vertical fluxes of heat and momentum. (For the purpose of this section we take the planetary boundary layer to mean the atmosphere from the surface to the first inversion, several kilometers in daytime, while the surface layer is the shallower layer where the heat and momentum fluxes do not vary significantly from the surface value.) To measure these fluxes directly

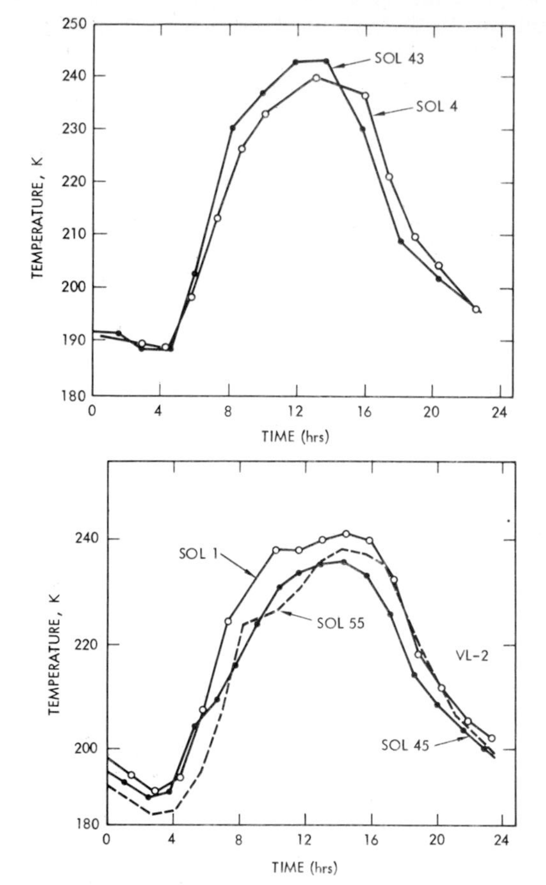

Fig. 6. Diurnal temperature variation at (top) VL-1 and (bottom) VL-2 for sols early and late in the primary Viking mission.

at heights of 1.6 m requires sampling rates of several tens of hertz and three-dimensional measurements.

Since neither three-dimensional measurements nor sampling rates greater than 1.0 Hz are available from Viking, the fluxes and stability can only be inferred indirectly by examining the time series, from statistical properties, or by other indirect means. To aid the reader in understanding the value and the limitations of these data in applications involving both mean and statistical analyses, we present several typical examples of the time series. Some comparisons of the statistics with terrestrial data are also presented.

Continuous time series have been obtained with sampling intervals ranging from 1.0 to 128.0 s and with the number of samples in a module ranging from 16 to 2000. However, modules having less than 64 samples are not generally useful for time series analysis. We have not completed analysis of the time series data because of the demands of the ongoing mission and because of the substantial effort required to validate these series properly for analysis. That work is in progress. To represent the large set of time series, we present a nighttime stable case, a daytime convective case, and an evening transitional case. As a further aid to understanding the data and its limitations a case with interference arising from flow over the lander is given.

Figure 8 is the nighttime stable case. The fluctuations exhibited are similar to the effects on earth of mechanical turbulence generated by roughness elements. The minimum scalar speed is 1.2 m/s (which is above the instrumental threshold), the maximum is 6.5 m/s, and the mean is 3.4 m/s. The wind direction varies between 190° and 225° with a mean of 204°. There is a slight clockwise turning of the wind from the beginning to the end of the record. Temperature data are from the thermocouple system (the prime source of temperature data) and from the wind reference temperature sensor, uncorrected for radiation and conduction errors. These errors can be as high as 8°C. It is therefore essential in using the output of this secondary source of temperatures to correct for radiation and conduction. We are routinely generating both corrected and uncorrected temperatures.

Figure 8 indicates that there is a modest amount of energy in the wind speed and direction spectra with periods of several minutes. The temperature traces are undistinguished except that the two traces diverge toward the end of the record.

Figure 9 is an example of late afternoon convection. The sequence ends about 2.8 hours before sunset. The major variations in wind speed have a period of about 20 min, with significantly less energy in shorter periods. This is due to thermal convection with horizontal scales of the order of several kilometers and is in distinct contrast with the nighttime stable case. In this example there is much better agreement between the two temperature sensors, despite the solar heating of the wind reference sensor, because the wind reference sensor has been corrected for radiation and conduction.

Figure 10 illustrates the transition from strong to decaying convection during late afternoon with a 2-hour module beginning 2.3 hours before sunset. The decay of the low-frequency convective input in the wind speed and direction trace with time is evident along with the almost monotonic decrease in temperature.

We have previously reported a high degree of uniformity of diurnal patterns of wind and temperature from sol to sol in the early part of the mission [*Hess et al.*, 1976*a*] and have taken advantage of that uniformity to calculate the composite values over a 20-sol period as a function of time during the sol. Table 1 compares the module means corresponding to Figures 8, 9, and 10 with the values at the same LLT from the 20-sol composite. The close agreement exemplifies the sol-to-sol uniformity that made it feasible to calculate meaningful composites.

Figure 11 illustrates the effect of lander interference. The wind azimuth trace shows that the wind blew from the lander toward the meteorology sensors near the end of the record. The region of such lander interference for VL-1 is a wind from 246°–332°, and that for VL-2 is a wind from 126°–212°. Lander interference will affect both temperature and wind, but its effects are more readily detected in the temperature traces than in the wind traces owing to the lander's thermal plume.

The entire lander body is a source of heat, but the major heated areas are the two radioisotope thermoelectric generators. They are capable of creating plumes several degrees Celsius above ambient as shown at the end of Figure 11. There are other physically small sources of heat such as the support structure of the meteorology sensor assembly (heated by the sun and by conduction) and the active wind sensors. These primarily affect the readings from the wind reference sensor, when it is downwind of these sources, since it is close to the local sources. The thermocouple sensor is much less affected, since the only structure close to it is the pedestal supporting the wind sensors. This sensor is located such that when the pedestal is upwind of the thermocouples, the wind is blowing from lander to sensor, so that both lander and pedestal interferences are in the same azimuth angle range. This difference in interference effects between the two temperature sensors is of

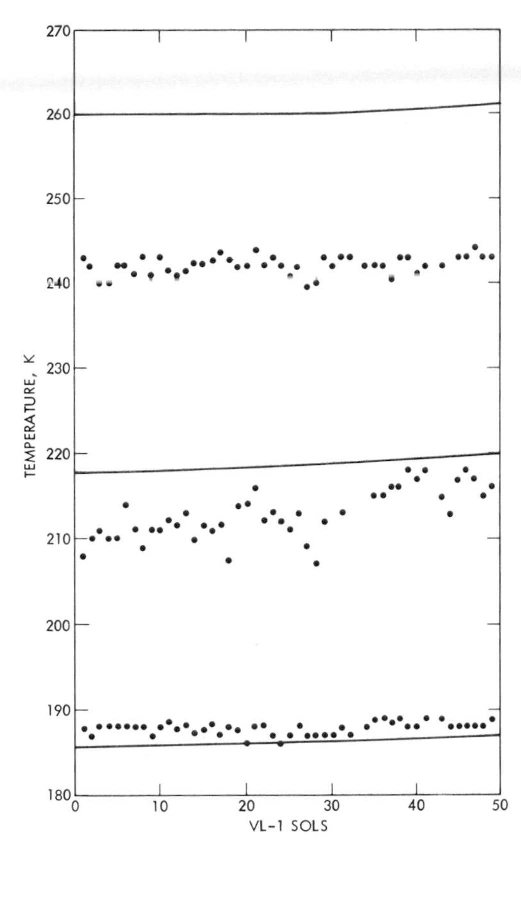

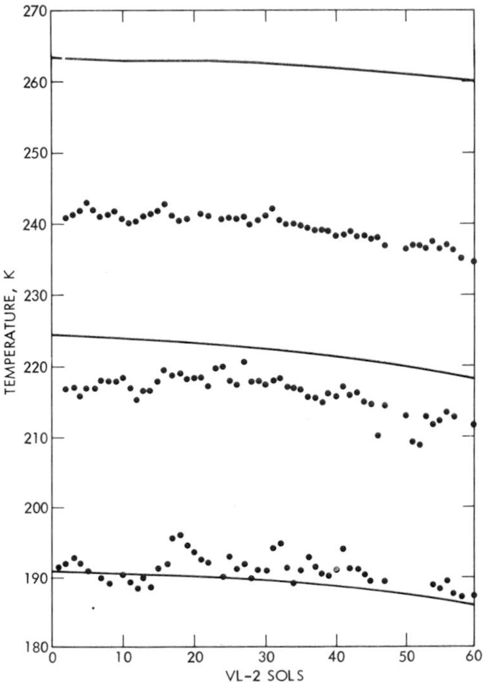

Fig. 7. Maximum, mean, and minimum temperatures for (top) VL-1 and (bottom) VL-2 for each sol of the primary Viking mission. On some sols, one or more of these quantities was in doubt because of winds from over the heated lander, and so no value is shown. Solid lines are maximum, mean, and minimum surface brightness temperatures from *Kieffer et al.* [1976].

particular importance for VL-2, where there is the electronic problem discussed earlier. The process of correcting the thermocouple values for this electronic problem involves comparison with the wind reference sensor readings. For this purpose it was necessary to exclude carefully from the comparison set those wind directions that would produce local as well as total lander effects on the wind reference sensor.

Figure 12 presents the diurnal variation of several statistical parameters for VL-1 sol 3 and the Massachusetts Institute of Technology Round Hill field station site at Buzzards Bay, Massachusetts. The Round Hill data include only data from the wind direction sector which has the most uniform upwind fetch. In the immediate few hundred meters upwind, Round Hill is probably as uniform (except for moisture) as the Viking sites, and it changes to a wooded terrain with a 20-m hill 1.0 km or more from the site [*Cramer et al.*, 1967].

Figure 12 (bottom) compares the gustiness ratio, the ratio of the standard deviation to the mean wind, for the two locations and for a similar range of wind speeds. The Round Hill standard deviations are computed after high-pass filtering at 0.0016 Hz, while the Mars values are computed around a 16-point mean. Since the 16-point mean corresponds to 2 min or less with the sampling intervals used on sol 3, the Mars data generally have smaller standard deviations, in convective conditions, than they would have if the longer periods were included. Note that the gustiness ratios at both sites are quite similar from 0200 to about 1000. After 0600, convection grows in depth, increasing the gustiness ratio until a maximum is reached at about 1400. The diurnal variation of the Round Hill data is smoother than that of the Mars data, since it is a composite of many weeks' data and a greater variety of synoptic situations. Also, the Martian diurnal cycle at VL-1 and VL-2 is probably dominated by thermally generated slope winds, including drainage winds, and by coupling to winds aloft [*Hess et al.*, 1976*a*, *b*] (see next section) at this season. Note that the Round Hill data are dashed during midday due to uncertain or missing data. This is mainly due to the fact that the good wind sector during summer is associated with good weather only during synoptic highs. Since the site is coastal, a sea breeze circulation often opposes the westerly flow during this portion of the day, thereby limiting the sample acquired. Experience at other sites indicates that the Round Hill gustiness is lower than at dry desert sites during low wind speed midday convective conditions.

Figure 12 (middle) compares the standard deviation of azimuth angle for the same two data sets. The lower Mars curve is obtained by computing a standard deviation around a 16-point record mean, while the upper curve is for the standard deviation around the module mean (4–8 records of 16 samples each). Including the lower frequencies essentially doubles the azimuth variations during the convective period. In general, the standard deviation increases with decreasing frequency, almost leveling off as the convective scale is passed, and again increasing as the diurnal variation is included.

Figure 12 (top) illustrates the diurnal variations of temperature variability after the record mean is removed. Note that it increases dramatically around 0600 with the onset of solar heating as the heat flux goes from a small downward value at night to a large upward value during the day. The diurnal variation is similar to that of the azimuth angle variation, and it is possible to compute the heat flux during the unstable portion of the day from 0700 to approximately 1600 [*Tillman*, 1972].

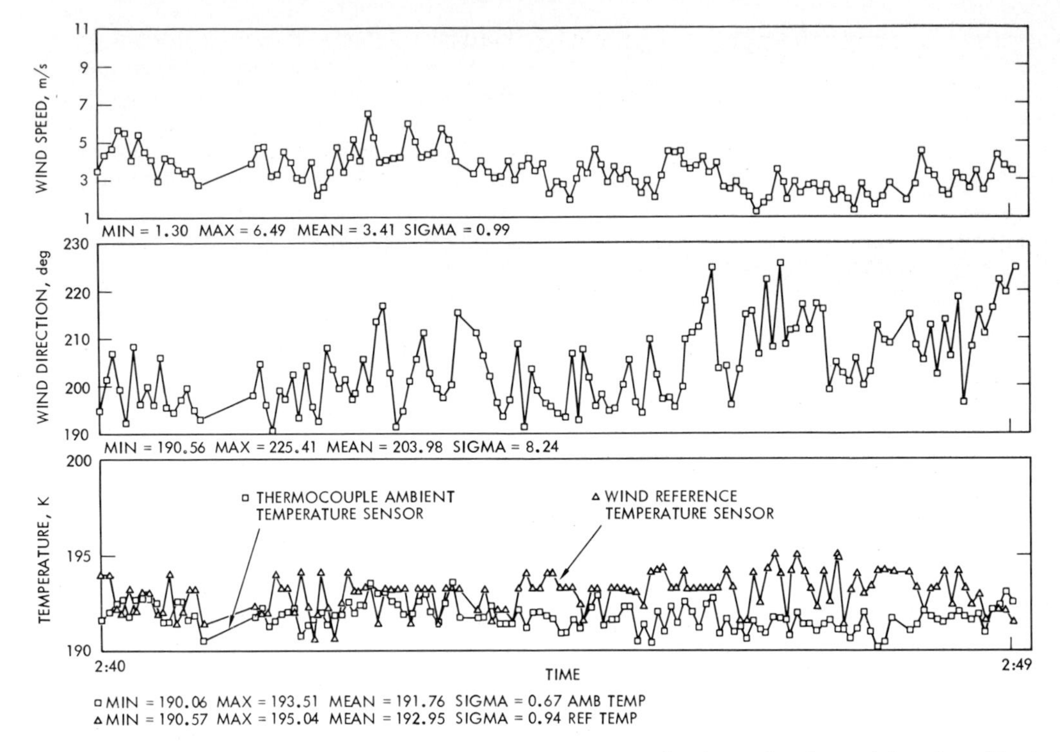

Fig. 8. Nighttime stable case from VL-1 sol 22. This series has a 4-s sampling interval and a total duration of 9 min. The wind reference sensor temperatures are uncorrected. The time scale is hours and minutes.

Interpretation of the Daily Variations of Wind and Pressure

The repeatable daily variations of wind and pressure can be attributed to a variety of diurnal and semidiurnal forcing factors. In general, the wind variations are dominated by processes of smaller scale than the pressure variations, since winds are influenced by the gradient of pressure. A small-amplitude pressure variation acting on a small scale can have more influence on local winds than a much larger amplitude pressure variation acting on a global scale. We believe that the observed pressure fluctuations are primarily planetary in scale for two reasons: (1) the phases of both diurnal and semidiurnal components of the pressure variation are similar at the two sites, and (2) the observed pressure variations would induce much stronger winds than those observed if they had the scale of local topographic variations, for example, the scale of the Chryse basin. Conversely, the winds at both sites seem to be associated primarily with local effects. The wind at both sites is approximately in the local upslope direction in the late after-

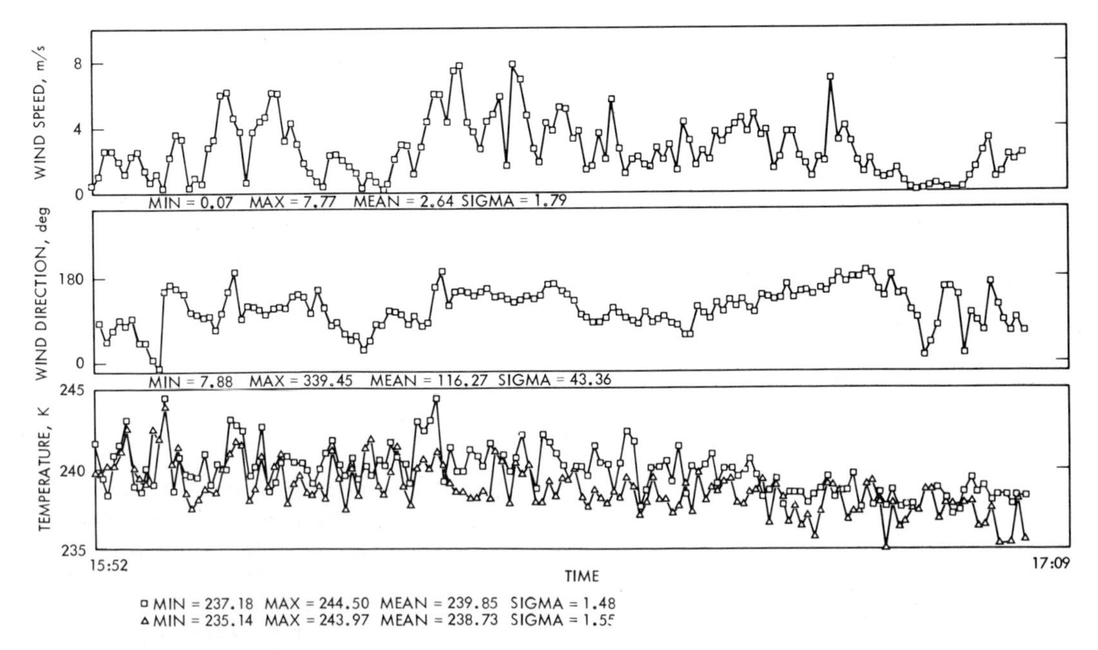

Fig. 9. Late afternoon unstable convective case from VL-1 sol 22. This series has a sampling interval of 32 s and a total duration of 76 min. The wind reference sensor temperatures are corrected. The time scale is in hours and minutes.

TABLE 1. Comparison of Module Mean Data Corresponding to Figures 9, 10, and 11 With the 20-Sol Mean Results at the Same LLT [*Hess et al.*, 1976*a*]

Source	0240 LLT Module Mean	0240 LLT 20-Sol Mean	0701 LLT Module Mean	0701 LLT 20-Sol Mean	1122 LLT Module Mean	1122 LLT 20-Sol Mean
Wind speed, m/s	3.4	2.5	4.0	3.3	6.7	6.8
Wind direction, deg	204	201	240	223	187	192
Temperature, °K	192	190	199	198	234	231

noon and is downslope in the early morning. This property, discussed in greater detail below, is expected for winds under local topographic influence [*Blumsack et al.*, 1973], but it is inconsistent with the variation expected for planetary scale tidal winds [*Zurek*, 1976]. The same separation between planetary scale diurnal pressure variations and local or regional scale daily wind variations occurs on earth [*Chapman and Lindzen*, 1970; *Wallace and Hartranft*, 1969]. Although this separation of planetary effects from local effects on the wind variation is likely to be a very crude approximation, we assume in the present analysis that the observed winds are controlled entirely by local factors while the observed pressure variations, particularly the semidiurnal variation, are entirely planetary in scale.

Wind Models

Two distinct processes have been suggested as drives for local wind oscillations over terrestrial land areas. *Blackadar* [1957] pointed out that the coupling between near-surface winds and those at higher levels varies diurnally as a result of the diurnal variation of the turbulent mixing rate. The influence of such a coupling is suggested by the mean VL-1 hodograph (Figure 3). The winds at the 1- to 2-km level during midafternoon were from the south-southeast at about 20 m/s, at least on the landing sol [*Seiff*, 1977]. The southerly component of the observed winds during midday suggests coupling to the upper-level wind at this time of day. The simplest analytic treatment of this effect is the Rayleigh friction model with a diurnally varying coupling coefficient, and this model will be used here.

The second forcing factor is the diurnally varying pressure gradient associated with the local slope [e.g., *Holton*, 1967]. This factor includes the drainage wind mechanism mentioned earlier. As the boundary layer is heated or cooled almost uniformly along surfaces parallel to the slope, a pressure variation is produced as a hydrostatic consequence of the resulting horizontal temperature variation. The magnitude of this effect is readily shown to be [e.g., *Blumsack et al.*, 1973]

$$\delta \mathbf{G} \approx (gh\delta\Gamma/\bar{T})\,\nabla z \tag{1}$$

where the left side represents the variation of the pressure gradient force induced by lapse rate variation $\delta\Gamma$ occurring through a boundary layer of depth h on a slope ∇z. Other quantities are the gravitational acceleration g and mean temperature $\bar{T}$.

With the Rayleigh friction assumption the wind $\mathbf{u}$ is described by the linearized equation of motion

$$\partial \mathbf{u}/\partial t + f\,\hat{\mathbf{k}} \times \mathbf{u} = \mathbf{G} - d(\mathbf{u} - \mathbf{u}_r) \tag{2}$$

where $\hat{\mathbf{k}}$ is the vertical unit vector, $\mathbf{u}_r$ is a 'relaxation wind' assumed to be approximately equal to the actual wind at the 1- or 2-km level, f is the Coriolis parameter, and d is the Rayleigh coupling coefficient. The quantities $\mathbf{u}$, $\mathbf{G}$, d, and $\mathbf{u}_r$ can be separated into daily mean and diurnal components, denoted by overbars and primes, respectively:

$$(\mathbf{u}, \mathbf{G}, d, \mathbf{u}_r) = (\bar{\mathbf{u}}, \bar{\mathbf{G}}, \bar{d}, \bar{\mathbf{u}}_r) + \mathrm{Re}\,[(\mathbf{u}', \mathbf{G}', d', \mathbf{u}_r')e^{i\omega t}]$$

This separation leads to the steady oscillation component of the solution for the complex amplitude of the diurnal wind oscillation:

$$\mathbf{u} = -\frac{[(i\omega + \bar{d}) - f\hat{\mathbf{k}} \times]}{[(i\omega + \bar{d})^2 + f^2]}\,[\mathbf{G}' - d'(\bar{\mathbf{u}} - \bar{\mathbf{u}}_r) + \bar{d}\mathbf{u}_r'] \tag{3}$$

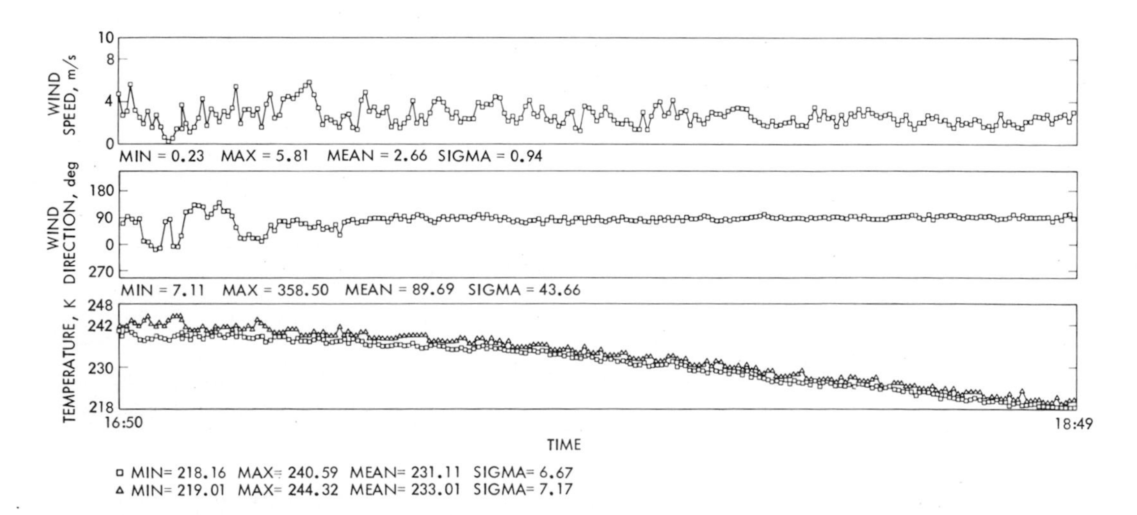

Fig. 10. Evening transitional case from VL-1 sol 30. This series has a sampling interval of 32 s and a total duration of 2 hours. The wind reference sensor temperatures are corrected. The time scale is in hours and minutes.

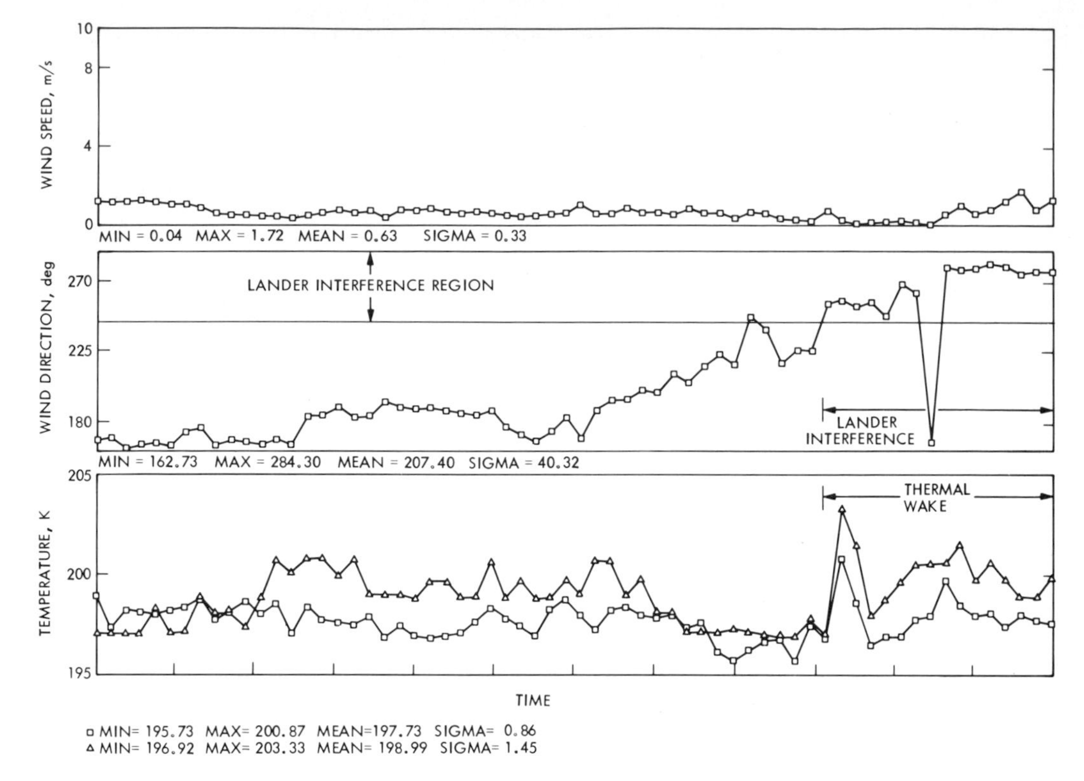

Fig. 11. A case of lander interference from VL-1 sol 29. This series has a sampling interval of 48 s and a duration of 51 min starting at 2310 LLT.

Equation (3) covers many possibilities depending on the magnitude and direction of $\mathbf{G}'$, the magnitude of $\bar{d}$ and d', and the magnitudes and directions of $\bar{\mathbf{u}}_f$ and $\mathbf{u}_f'$. To reduce the range of possible parameter variations to a manageable one, we neglect $\mathbf{u}_f'$ and assume that $\mathbf{G}'$ is given by (1), that is, $\mathbf{G}'$ is assumed to be directed only upslope or downslope. These simplifications are based on the assumption that planetary scale diurnal pressure and wind variations do not strongly affect the diurnal surface wind. Moreover, at the VL-1 site the direction of $\bar{\mathbf{u}}_f$ is taken to correspond to both the observed wind aloft and the slope-induced mean wind, i.e., it is parallel to the topographic contours with low elevations to the right of the flow. Finally, note that the nighttime surface winds are essentially decoupled from the winds aloft; a preliminary analysis combining VL-1 air temperatures and winds and ground temperature estimates from Viking orbiter 1 [*Kieffer*, 1976] suggests that the bulk Richardson number at night is very close to the critical one for the suppression of turbulent exchange across the boundary layer. Hence it is reasonable to assume that coupling vanishes at night. In terms of the Rayleigh friction model this implies that $\bar{d}$ and d' are comparable in magnitude, and we take them to be equal.

With these assumptions the solution hodographs depend on latitude, the relative phase of d' and $\mathbf{G}'$, and the ratio of the coupling factor to the slope-induced pressure gradient:

$$r \equiv |d'| \times |\bar{\mathbf{u}} - \bar{\mathbf{u}}_f| / |\mathbf{G}'|$$

We have explored the parameter ranges $0 \lesssim |\bar{d}| \lesssim 10^{-4}\ \mathrm{s}^{-1}$ and $0 \lesssim r \lesssim 4$ and two phase angles between d' and $\mathbf{G}'$: d' in phase with the maximum upslope pressure gradient (for example, if both occur in midafternoon) and d' leading the upslope pressure gradient by $\frac{1}{6}$ sol. The latter is thought to be somewhat more realistic and corresponds to maximum upslope pressure gradient in the midafternoon and maximum turbulent coupling in the late afternoon. Some sample hodograph results are shown in Figure 13. For the VL-1 site these hodographs have the following properties.

1. Although none of the calculated hodographs resemble the observed VL-1 hodograph in detail, the gross aspects of counterclockwise rotation and correct phase of the upslope wind are reproduced for some values of the parameters.

2. Counterclockwise hodograph rotation occurs only for rather large values of d' ($\gtrsim 0.5 \times 10^{-4}\ \mathrm{s}^{-1}$) and for the case with a $\frac{1}{6}$-sol time lag between the maximum of d' and the maximum upslope pressure gradient.

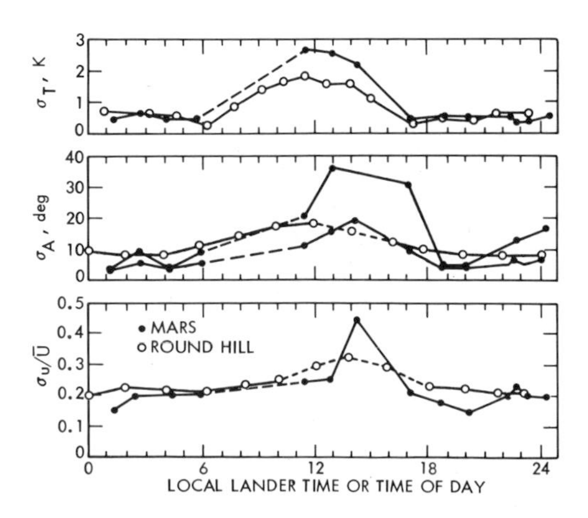

Fig. 12. Comparison of the diurnal variations on Mars (VL-1 sol 29) and Round Hill, Massachusetts: (bottom) gustiness ratio, (middle) standard deviation of wind angle, (top) standard deviation of temperature. In the top plots the terrestrial data have been multiplied by 4 and are derived from an earlier experiment.

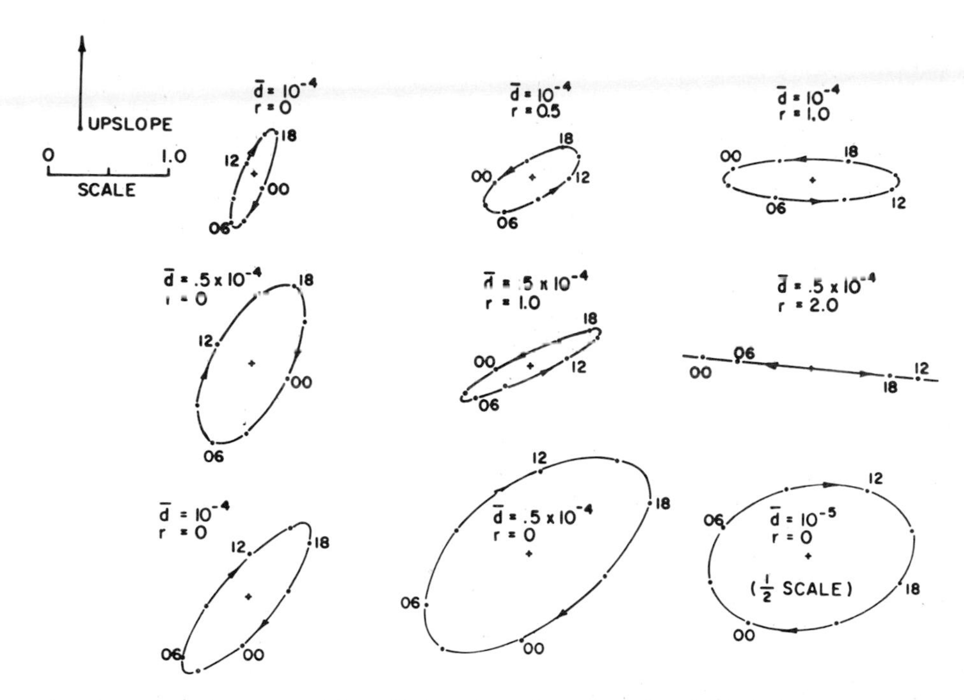

Fig. 13. Sample hodographs computed from (3). The upper six cases are for 23°N; the lower three are for 48°N. The four cases with nonzero coupling to winds aloft ($r \neq 0$) have a $\frac{1}{6}$-sol phase lag between the time of maximum coupling (1153 LLT) and the time of maximum upslope pressure gradient (1500 LLT). All values are nondimensionalized by dividing the amplitudes by $f^{-1}|\mathbf{G}|$ and are scaled as indicated except for the hodograph at lower right which is one half the scale of the others.

3. These cases also show very low wind amplitudes, partly as a result of the competition between $\mathbf{G}'$ and $d'(\bar{\mathbf{u}} - \bar{\mathbf{u}}_f)$, which gives out-of-phase forcing.

4. Values of r greater than about 1 yield increasingly large discrepancies between the model and observed times of maximum upslope winds.

5. If the three cases shown in Figure 13 having counterclockwise wind rotation are taken to be representative of a reasonable fit to the VL-1 data, one finds an amplitude of $\mathbf{G}'$ and a corresponding estimate of $h|\Gamma'|$ for the particular slope and latitude of VL-1 from (1). ($|\Gamma'|$ is the amplitude of the diurnal oscillation of lapse rate.) The result is $h|\Gamma'| \sim 2$–3°C. This corresponds, for example, to a lapse rate variation of 1–1.5°C/km through a depth of 2 km. For the VL-2 site we find the following.

1. The predicted phase of $\mathbf{u}'$ disagrees with the observed phase to an increasing degree as $|d'|$ decreases from 10^{-4} s^{-1}.

2. The shapes of the model and observed hodographs agree well when $r = 0$ and $|d'| \approx 0.5 \times 10^{-4}$ to 1×10^{-4} s^{-1}.

3. For these parameters the disagreement in phase between the observed and calculated hodographs is still about 45°; note, however, that the correct phase reference, the upslope direction, is quite uncertain for this site.

4. As $|d'|$ increases beyond 0.5×10^{-4} s^{-1}, the model hodographs become thinner than those observed, and there is no significant improvement in the phase discrepancy.

5. If $r = 0$, $|d'| = \bar{d} = 0.5 \times 10^{-4}$ s^{-1} are taken to be 'best fit' values for the VL-2 data, the corresponding value of $h|\Gamma'|$ which is consistent with the slope obtained from the Mars topographic map (Map M25M 3RMC, prepared by the U.S. Geological Survey for the Viking Program, 1976) is ~4–5°C. This value is in reasonable agreement with the value derived for the VL-1 site.

Although no firm conclusions can be drawn from the above analysis, several results are suggested. The rough qualitative agreement between model and observed winds for similar parameter values at both sites supports the hypothesis that the dominant factors are the local slope pressure gradient and coupling with winds aloft. These are the factors suggested by Blackadar and Holton for diurnal boundary layer wind variations, and they have been developed in some detail for Mars by Gierasch, Blumsack, and Wessel. At the VL-1 site both factors appear to be comparable in magnitude; at the VL-2 site there seems to be no need to invoke coupling to winds aloft. A large damping factor seems to be required, but its value may not be unreasonable for winds within 2 m of the surface. A damping time scale of the order of 4 hours for the surface winds is inferred. Finally, the two hodographs seem to be most consistent with a 'slope heating parameter' $h|\Gamma'| \approx 2$–5°C.

Pressure Models

The repeatable daily pressure variations for the early sols have been analyzed into diurnal and semidiurnal tidal components at both sites [*Hess et al.*, 1976*a*, *b*]. The peak-to-peak amplitudes of the diurnal and semidiurnal components were found to be 0.16 and 0.07 mbar at VL-1, and 0.03 and 0.03 mbar at VL-2. On Mars as on earth the diurnal component may experience substantial regional modulations as a result of interactions with the topography and other spatially variable surface factors. It is also sensitive to details of the vertical thermal structure of the atmosphere [*Zurek*, 1976]. The semidiurnal component of the daily pressure variation on earth shows very little longitudinal dependence and does not seem to be much affected by orography, continent-ocean differences, or vertical structure variations [*Chapman and Lindzen*, 1970], and similar behavior can be expected for the semidiurnal tide on Mars. The reason for this difference in behavior holds on both planets: the predominant semidiurnal tidal modes have a much larger vertical scale than those of the diurnal tide. Consequently, the semidiurnal tide is much less sensitive to the precise lower boundary condition and to irregularities in heating. The amplitude and phase of the semidiurnal tide should relate in a fairly direct way to the semidiurnal heating. Thus we adopt the following strategy: (1) use the observed properties of the semidiurnal tide at the VL-1 and VL-2 sites to infer some properties of the heating and (2) use these properties to constrain models of the diurnal tide. Only the first of these two steps is attempted here.

Following Chapman and Lindzen, we can express the semidiurnal surface pressure ratio $\delta p/\bar{p}$ in the form

$$\delta p/\bar{p} = \mathrm{Re}\ \{(i\gamma/\omega) \sum_n \theta_n(\phi) Y_n(0) e^{i\omega t}\} \tag{4}$$

where $\gamma = c_p/c_v$ is the ratio of specific heats, ω is the semidiurnal frequency, and $\theta_n(\phi)$ is the nth semidiurnal Hough function, or latitude-dependent modal solution of LaPlace's tidal equation [cf. *Chapman and Lindzen*, 1970]. $Y_n(x)$ is the solution of the vertical structure equation

$$\frac{d^2 Y_n}{dx^2} + \lambda_n^2 Y_n = \frac{(\gamma - 1) J_n e^{-x/2}}{\gamma^2 g h_n} \tag{5}$$

where x is the vertical coordinate divided by the local scale height H, g is the gravitational acceleration, and J_n is the heating per unit mass in the nth tidal mode. The parameter λ_n^2 is

$$\lambda_n^2 = \left[\frac{(\gamma - 1)H}{\gamma} + \frac{dH}{dx}\right] \Big/ h_n - \frac{1}{4}$$

where h_n, the 'equivalent depth,' is a separation parameter for the nth mode in LaPlace's tidal equation. We consider the class of heating models considered by *Zurek* [1976].

$$J_n = J_0 e^{-(\nu + i\alpha)x} \tag{6}$$

where J_0, ν, and α are constants. The solution to (5) is sought which satisfies the radiation condition $Y_n \rightarrow \exp(i\lambda_n x)$ as $x \rightarrow \infty$ and which also has vanishing vertical velocity at $x = 0$. The latter condition applied to Y_n takes the form

$$(dY_n/dx) + \beta_n Y_n = 0 \qquad x = 0 \tag{7}$$

where $\beta_n = (H/h_n - \frac{1}{2})$.

For the semidiurnal tide in the Mars atmosphere it is sufficient to treat λ_n as a constant. Then the required solution to (5) evaluated at $x = 0$ is

$$Y_n(0) = \{\beta_n[(\nu + \tfrac{1}{2}) + i(\alpha - \lambda_n)] \cdot [1 + i\lambda_n \beta_n^{-1}]\}^{-1} \left[\frac{(\gamma - 1)J_0}{\gamma^2 g h_n}\right] \tag{8}$$

The surface heating rate can be evaluated in terms of the net surface heat flux in the nth semidiurnal mode, F_0:

$$J_0 = \frac{gF_0}{p_s} \cdot \frac{[\alpha^2 + (\nu + 1)^2]}{[\nu + 1]} \tag{9}$$

where p_s is the mean surface pressure. We also assume that F_0 is a constant fraction A of the total solar flux available in the nth semidiurnal mode at every latitude. Fourier decomposition of the total solar flux then gives

$$F_0 = \frac{2AF_s}{\pi} \int_{-\pi/2}^{\pi/2} \left(\int_0^{t_s} \cos Z \cos 2t \, dt \right) \theta_n(\phi) \cos \phi \, d\phi \tag{10}$$

where Z is the solar zenith angle and F_s is the solar constant. The local time t (relative to local noon) and the time of sunset t_s are measured in radians, and the Hough function normalization has been used:

TABLE 2. Parameters Used in the Calculation of the Semidiurnal Tide

n	h_n, km	λ_n	β_n	θ_n(23°N)	θ_n(48°N)
2	5.56	0.1841	1.6221	0.796	0.185
4	1.49	0.9002	7.4188	0.548	0.6555
6	0.58	1.4400	16.8515	−0.944	1.0255

F_s = 520 W/m², $H(0)$ = 11.80 km, dH/dx = 0.87 km/scale height, P_s/g = 161.3 kg/m², declination of sun = +23°.

$$\int_{-\pi/2}^{\pi/2} \theta_n^2(\phi) \cos \phi \, d\phi = 1$$

Equations (4), (8), (9), and (10) combine to give $\delta p/\bar{p}$ in terms of the parameters A, ν, α, the solar constant, and the atmospheric mass (p_s/g), as well as the desired declination and latitude. The lowest even order Hough function mode is the dominant one in the summation of (4), but we have included the first three even modes in this calculation. The corresponding parameters are given in Table 2.

The resulting amplitudes and phases of the semidiurnal tide for various values of ν and α are shown in Figures 14*a* and 14*b*. Phases shown represent the phase advance relative to local noon. The amplitudes have been divided by the unknown fractional absorption A. Figure 15 shows the ratio of amplitudes at the VL-1 and VL-2 sites predicted by the model. The observed phases reported by *Hess et al.* [1976*a*, *b*] are recapitulated in Table 3. Comparing the observations in Table 3 with the model results, we conclude that reasonable agreement can be achieved with $A \sim 0.1$, $\alpha \sim 1$–2, and essentially any positive value of ν. The general agreement of the ratio of amplitudes at both sites with the model for most values of ν and α, as well as the phase consistency at the two sites lends credence to the model, although a slight phase advance of the pressure at VL-2 relative to VL-1 is predicted in all cases, while a slight phase retardation is actually observed.

The required value of A is quite large, a reflection of the fact that the observed semidiurnal pressure oscillation is itself large. It is unlikely that such a large solar heat input could be accomplished by convective transport from the surface alone. For example, at the latitude of the VL-1 site the maximum daily flux of heat into the atmosphere would have to be of the order of 50 W/m² to produce the observed semidiurnal tide. But the actual heat flux F can be estimated from the bulk transfer relation

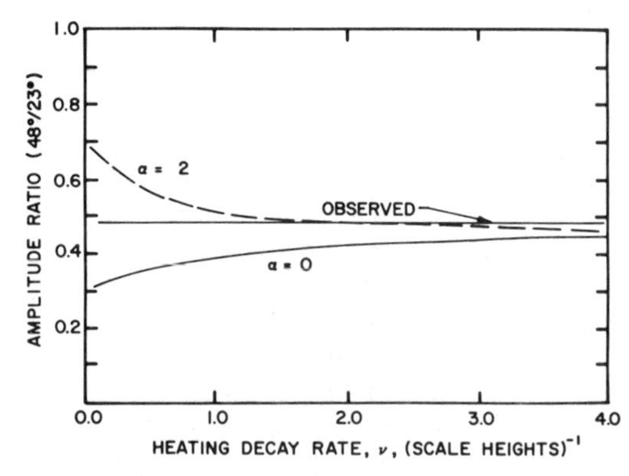

Fig. 14*a*. Dependence of the amplitude of the semidiurnal pressure ratio on ν, the inverse vertical scale of heating, for two values of α, the heating phase lag parameter. The solid lines are for $\alpha = 0$, and the dashed lines are for $\alpha = 2$. Values of $\delta p/\bar{p}$ shown are divided by the absorption parameter A.

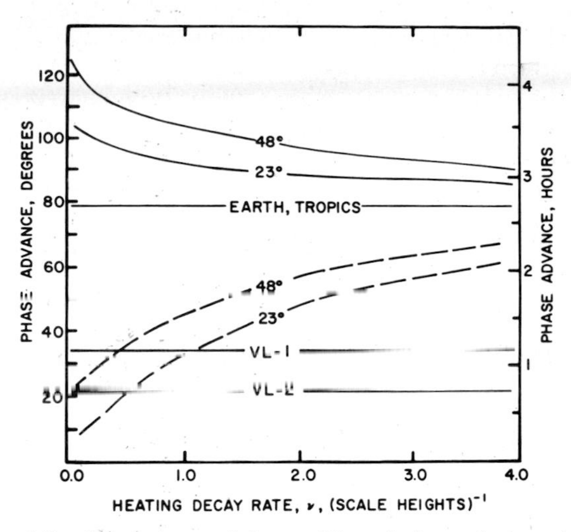

Fig. 14*b*. Dependence of the semidiurnal phase advance, relative to local noon, on ν for two values of α. The solid lines are for $\alpha = 0$, and the dashed lines are for $\alpha = 2$. The observed lags at VL-1 and VL-2 and in the earth's tropics are also shown.

$$F_c = \rho C_p C_D u \Delta T \tag{11}$$

where the transfer coefficient $C_D \approx [k_0/\ln(z/z_0)]^2$; k_0 is von Karman's constant, z is instrument height (1.6 m), and z_0 is roughness length (estimated to lie in the range 0.01–1 cm). Other parameters are the measured wind u and the temperature difference ΔT between the atmosphere at instrument height and the ground. The latter can be estimated with the aid of ground temperatures inferred from the Viking orbiter infrared experiment [*Kieffer et al.*, 1976]. Rough evaluations give F_c of about 10 W/m^2 at the VL-1 site. Although this may be underestimating the global average of F_c, it is very unlikely to be in error by as much as a factor of 3. Since convection alone does not seem adequate to provide the required heating, absorption of solar radiation must be playing an important role. It appears that absorption of solar radiation is providing at least half of the required heating. The absorbing agent is apparently atmospheric dust. The opacity due to dust has been monitored by the imaging systems on both landers, and the optical depths for normal incidence measured during this period were substantial, ranging from 0.2 to 0.4 [*Mutch et al.*, 1976*c*; *Pollack et al.*, 1977].

Even though radiation absorption must provide an important thermal drive, convective heating also appears to be playing a role. Radiation absorption alone should exhibit very little phase variation with height, with maximum heating near local noon. This would produce a semidiurnal pressure maximum very near to 0850 LLT (+90° phase shift; see Figure 14). Such a phase shift is observed for the semidiurnal tide on earth which is known to be driven primarily by radiation [cf. *Chapman and Lindzen*, 1970]. The observed Martian pressure phases indicate that the heating maximum occurs considerably later than noon at some levels in the atmosphere. Such a phase retardation of heating is expected at levels above the surface for convective heating. We conclude that both factors, convection and absorption of solar radiation, play significant roles in driving the global tides.

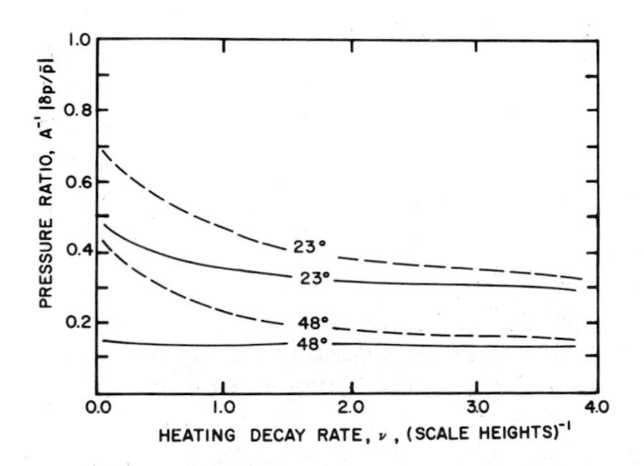

Fig. 15. Dependence of the ratio of the semidiurnal amplitudes at the VL-2 and VL-1 latitudes on ν and α. The solid lines are for $\alpha = 0$, and the dashed lines are for $\alpha = 2$.

TABLE 3. Observed Tidal Properties on Mars

	VL-1	VL-2	(VL-1)/(VL-2)
Amplitude ratio	0.0050	0.0022	0.44
Phase, hours	1.17	0.75	

The absorption fraction estimated from the semidiurnal tidal model can be used to estimate an effective diurnal temperature change for the equatorial atmosphere. For example, values $A = 0.1$, $\nu = 2$ imply an average daily temperature amplitude of 10°C through a layer 5 km deep. Although this variation is quite large, so far as we know it is not inconsistent with any other data. It is much larger than the temperature variations which were required to drive the boundary layer winds, but the boundary layer winds are likely to reflect primarily the convective contribution to heating. We note that the inferred values of $h|\Gamma'|$ are quite consistent with the convective heating estimate at the VL-1 site from (11). Moreover, the apparent requirement of a phase lag between the time of maximum thermal drive for these winds and the time of maximum vertical coupling is consistent with convective driving. The boundary layer winds are probably not very responsive to radiative heating through a deep layer.

SEASONAL VARIATION OF PRESSURE

We have previously reported [*Hess et al.*, 1976*a*, *b*] a secular decline of the sol mean pressure at the two lander sites. When initially detected at VL-1, this decline was at the rate of 0.0122 mbar per sol or 0.16% per sol, a rapid rate of the sort that is never seen on earth. Although we did not believe that this was due to a defective instrument, that possibility could not then be ruled out. When VL-2 landed, it too showed a long-term decrease in pressure, thus making it still more unlikely that an instrumental problem was the cause. We now have a long enough run of data to demonstrate that a minimum has been reached and that the pressure is rising. This makes it quite certain that we are dealing with a real physical process and not with instrumental anomalies.

The sol mean pressure results from both landers are given in Figure 16, extending well past the end of the nominal mission. The minima are quite broad, and it is impossible to specify the date of minimum with great accuracy. However, the data indicate that the time of minimum pressure is the same at both landers and is approximately VL-1 sol 100 (VL-2 sol 56), or L_s = 149°. The uncertainty in this seems to be ±10 sols or ±5° of L_s.

The obvious interpretation of these results is the loss of atmospheric mass through condensation of CO_2 on the winter

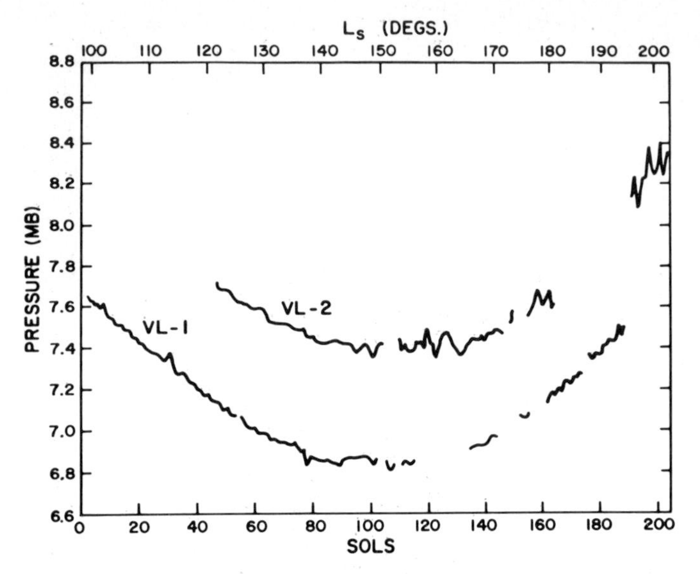

Fig. 16. Surface atmospheric pressures averaged over each sol for VL-1 and VL-2, plotted as a function of time (VL-1 sols) and L_s. Gaps are due to missing data or to data that have not yet been reduced.

(southern) cap and the subsequent gain of atmospheric mass through sublimation of CO_2 from that cap when the season progresses sufficiently. A seasonal variation of atmospheric pressure was anticipated before the Viking landers touched down [*Leighton and Murray*, 1966; *Briggs*, 1974; *Dzurisin and Ingersoll*, 1975; *Pollack et al.*, 1976]. As we have previously reported [*Hess et al.*, 1976*a*], the results of Pollack et al., based upon a numerical general circulation model for the Viking season, gave a nearly exact fit to the initial rate of decrease found at VL-1. Briggs' calculation resulted in a rate of change of pressure that was some 25% too slow. This discrepancy is understandable because Briggs' model constrained the edge of the polar cap to lie no further equatorward than 57°S, in agreement with the best data then available. Pollack et al. placed no such constraint on the cap but allowed it to grow as dictated by the model physics. Viking orbital pictures indicate that the southern cap does extend farther north than 57°S, so that the choice made in the general circulation model was probably better. The additional condensation of CO_2 involved in a cap extending equatorward of 57°S is probably responsible for the excellent agreement between the general circulation model and the observed rate of change of pressure during the early part of the VL-1 mission.

For the very practical reason that general circulation models consume large amounts of computer time, Pollack et al. did not calculate results as the season changed but obtained predictions for only one value of L_s. Thus we have not as yet had the capability to compare the full range of observed data to the most successful model. The portion of the seasonal curve given in Figure 14 can be compared to the results of Briggs, which do not include any sorption of CO_2 on the regolith, and to those of Dzurisin and Ingersoll, which do include plausible parameterized interactions with the regolith. Since sorption on the regolith would buffer the exchange of mass between atmosphere and cap, its effect would be to decrease the amplitude of the pressure curve and to shift the extrema to earlier times. Thus Dzurisin and Ingersoll predict a rate of change of pressure that is 70% slower than is observed. This discrepancy suggests that the CO_2 sorption mechanism is not significant on a seasonal time scale. However, the time of occurrence of minimum pressure from Figure 14 is appreciably closer to the time predicted when sorption is included (Dzurisin and Ingersoll) than to that predicted when sorption is omitted (Briggs). Thus amplitude considerations argue against the importance of CO_2 sorption, while phase considerations favor it.

It is likely that this contradiction may be another consequence of the artificial restriction of the edge of the cap to poleward of 57°S in the Briggs model. The condensate that actually deposits equatorward of that latitude would be the first to sublime and could be responsible for the early occurrence of the minimum without the intervention of a sorption mechanism. The Leighton-Murray model, which had no restrictions on the latitude of the cap edge, is in good agreement with our observations. This suggests that the above analysis is correct.

It should be mentioned that adsorption of CO_2 on the regolith is potentially significant for the important issue of climatic changes on Mars. For cyclic climate change to occur it is necessary to have a reservoir into and out of which large amounts of CO_2 (or some other gas) can pass. Prior to Viking it was supposed that the residual summer polar cap was condensed CO_2 and that it was the needed reservoir [*Ward et al.*, 1974]. Viking orbital infrared measurements have demonstrated that the residual northern polar cap is too warm to contain condensed CO_2 in equilibrium with the atmosphere [*Kieffer et al.*, 1976] and therefore cannot serve as a reservoir for atmospheric CO_2. The same conclusion is probably applicable to the southern cap. Therefore the idea that appreciable CO_2 can be stored by adsorption in the regolith [*Fanale and Cannon*, 1974] supplies the only other reasonable reservoir. If it ultimately develops that the seasonal pressure variation requires sorption of CO_2, that would be powerful support for the idea of the regolith as a storehouse for CO_2. If the reverse conclusion is drawn, it will weaken the idea somewhat, but not fatally, because the process could conceivably be insignificant on the scale of 1 year but significant on a scale of thousands or millions of years.

For the range of L_s represented in Figure 14 there is probably little or no contribution from the northern cap. At the beginning of the Viking missions that cap likely had no CO_2 left in it, and the contributions to atmospheric mass by H_2O are insignificant. For the latest dates represented in Figure 14, little if any CO_2 could have been condensed on the northern cap. Thus we probably have seen only the consequences for total atmospheric pressure of the southern cap. Subsequent data will reveal the full interplay of the atmosphere with both caps.

Solar Conjunction and Extended Mission

The two Viking landing sites were deliberately selected so that (among many other criteria) expected wind speeds would be low during landing and the primary mission. This has been found to be the case, and the wind behavior has been quite regular throughout northern summer, particularly at the near-equatorial VL-1 site. Temperatures have also been mild by Martian standards, and very little secular change was observed. In fact, the most striking aspect of Martian weather has been its remarkable regularity. The only notable departures have been the secular pressure decrease and the wind event at the VL-2 site.

The primary mission ended early in November 1976, with the beginning of the solar conjunction period. Meteorological data were taken throughout this period and stored on the onboard tape recorder for postconjunction transmission to earth.

Tape recorder capacity limited the amount of data to 16 wind and temperature samples each 1½ hours during the night, 32 each 1½ hours during the day, and a pressure sample every 17 min. This was sufficient to maintain a continuous record of the diurnal cycle and pressure trends at both sites. An increase in wind variability was observed at the VL-2 site with continued temperature decrease. However, conditions at the VL-1 site remained essentially the same as before. Pressure began a slow rise at both sites.

It is planned to operate the meteorology experiment for nearly 2 earth years in order to cover all four seasons on Mars. With the establishment of this extended mission we have been able to return to quasi-continuous sampling and expect to remain in this mode. Measurements will be coordinated with lander imaging and the orbital experiments. Northern autumn began in early January 1977, and perihelion will occur in late April 1977, northern winter beginning on May 30, 1977 (L_s = 270°). We expect to see some dramatic meteorological changes during this period. During northern autumn, and more certainly by the time of perihelion, high winds, possibly in excess of 50 m s^{-1}, should occur. These should be accompanied by blowing sand and dust, both locally and globally. Associated with these could be local and meso-scale vortex systems. Additionally, frontal systems are believed to occur [*Briggs and Leovy*, 1974]. Evidence for the occurrence of at least some of these phenomena comes from a long history of earth-based observations, from the Mariner and Viking orbiters, and from the Viking landers. For example, lee deposits of granular material are present at both landing sites, indicating deposition by strong northeast winds at VL-1 and northwest winds at VL-2. These directions are opposite to those of the maximum winds observed during the primary mission (southwest at VL-1 and southeast at VL-2). To date, in the extended mission we have already noted significant changes in wind behavior from the summer pattern. A series of events with moderate (10–15 m s^{-1}) winds from northerly directions has occurred at VL-2. Some events have been associated with temporary temperature decreases superimposed upon the general temperature decline. These events are suggestive of cold front passages. Temperature at VL-2 is decreasing at an accelerating rate with minima as low as 165°K now being observed. The seasonal effect at the VL-1 site, however, continues to be small. Details of these observations will be presented when data reduction and analysis are completed.

Continued large pressure variations are expected as the CO_2 polar caps form and sublimate. Monitoring of pressure throughout the Martian year should allow determination of CO_2 behavior, particularly with respect to the amounts deposited at, and cycled between, the polar caps. It is possible that CO_2 condensation may occur at the VL-2 site during northern winter, and temperature measurements will be important in the study of the condensation-sublimation behavior. In this regard we have found, during the primary mission at both VL-1 and VL-2 sites, a suggestion that the temperature sensors are detecting the H_2O frost point. This is indicated by a temporary decline in the nocturnal rate of decrease of temperature, occurring in the time period 0100–0300 LLT and at temperatures of about 191°K (VL-1) and 196°K (VL-2). Such an effect could be caused by the formation of a ground ice fog acting to decrease the rate of surface radiative loss to a small extent. Much more work is required to verify this; hence these data are not presented here. However, it is worth noting that observations by *Pollack et al.* [1977] indicate the possible presence of nocturnal ice fog. If we are indeed detecting the frost point, then such data combined with the observations of *Farmer et al.* [1977] can provide a basis for modeling the vertical distribution of water vapor and its seasonal behavior. Following temperature and wind trends will allow verification (and modification) of circulation and surface thermal models. Seasonal variations in the diurnal and semidiurnal pressure oscillations will allow us to monitor the atmospheric tidal responses and their thermal drives.

The most interesting meteorology lies ahead, and we expect that the meteorological sensors will survive throughout the extended mission and beyond.

Acknowledgments. The section of this paper on interpretation of the daily variations of wind and pressure has benefited from discussions with R. Zurek, J. M. Wallace, H. Tennekes, and S. Mullen. We are indebted to T. Chamberlain, H. Cole, R. Dutton, G. Greene, J. Mitchell, and W. Simon, without whose devoted efforts this work would not have been possible. This research was supported by the National Aeronautics and Space Administration under contract NASI 9693, NASI 11854, and NASI 9694.

REFERENCES

Blackadar, A. K., Boundary layer wind maxima and their significance for the growth of nocturnal inversions, *Bull. Amer. Meteorol. Soc., 38,* 283, 1957.

Blumsack, S. L., P. J. Gierasch, and W. R. Wessel, An analytical and numerical study of the Martian planetary boundary layer over slopes, *J. Atmos. Sci., 30,* 66, 1973.

Briggs, G. A., The nature of the residual Martian polar caps, *Icarus, 23,* 167, 1974.

Briggs, G. A., and C. B. Leovy, Mariner 9 observations of Mar's north polar hood, *Bull. Amer. Meteorol. Soc., 55,* 278, 1974.

Chamberlain, T. E., H. L. Cole, R. G. Dutton, G. C. Greene, and J. E. Tillman, Atmospheric measurements on Mars: The Viking meteorology experiment, *Bull. Amer. Meteorol. Soc., 57,* 1094, 1976.

Chapman, S., and R. S. Lindzen, *Atmospheric Tides*, Gordon and Breach, New York, 1970.

Cramer, H. E., F. A. Record, and J. E. Tillman, Investigation of low level turbulent structure for various roughnesses and gross meteorological circulations, final report, U.S. Army contract DA49-092-ARO-66, Dep. of Meteorol., Mass. Inst. of Technol., Cambridge, 1967. (Available from J. E. Tillman, Department of Atmospheric Sciences, University of Washington, Seattle, Washington 98105.)

Dzurisin, D., and A. P. Ingersoll, Seasonal buffering of atmospheric pressure on Mars, *Icarus, 26,* 437, 1975.

Fanale, F. P., and W. A. Cannon, Exchange of adsorbed H_2O and CO_2 between the regolith and atmosphere of Mars caused by changes in surface insolation, *J. Geophys. Res., 79,* 3397, 1974.

Farmer, C. B., D. W. Davies, A. L. Holland, D. D. LaPorte, P. E. Doms, Mars: Water vapor observations from the Viking orbiters, *J. Geophys. Res., 82,* this issue, 1977.

Gierasch, P. J., and R. M. Goody, A study of the thermal and dynamical structure of the Martian lower atmosphere, *Planet. Space Sci., 16,* 615, 1968.

Hess, S. L., R. M. Henry, J. Kuettner, C. B. Leovy, and J. A. Ryan, Meteorology experiments: The Viking Mars lander, *Icarus, 16,* 196, 1972.

Hess, S. L., R. M. Henry, C. B. Leovy, J. A. Ryan, J. E. Tillman, T. E. Chamberlain, H. L. Cole, R. G. Dutton, C. G. Greene, W. E. Simon, and J. L. Mitchell, Mars climatology from Viking after 20 sols, *Science, 194,* 78, 1976*a*.

Hess, S. L., R. M. Henry, C. B. Leovy, J. L. Mitchell, J. A. Ryan, and J. E. Tillman, Early meteorological results from the Viking 2 lander, *Science, 194,* 1352, 1976*b*.

Holton, J. R., The diurnal boundary layer wind oscillation above sloping terrain, *Tellus, 19,* 199, 1967.

Kieffer, H. H., Soil surface temperatures at the Viking landing sites, *Science, 194,* 1344, 1976.

Kieffer, H. H., S. C. Chase, T. Z. Martin, E. D. Miner, and F. D. Palluconi, Martian north pole summer temperatures: Dirty water ice, *Science, 194,* 1341, 1976.

Leighton, R. B., and B. C. Murray, Behavior of carbon dioxide and other volatiles on Mars, *Science, 153,* 136, 1966.

Mutch, T. A., A. B. Binder, F. O. Huck, E. C. Levinthal, S. Liebes, E. C. Morris, W. R. Patterson, J. B. Pollack, C. Sagan, and G. R. Taylor, The surface of Mars: The view from the Viking 1 lander, *Science, 193,* 791, 1976*a*.

Mutch, T. A., et al., The surface of Mars: The view from the Viking 2 lander, *Science, 194,* 1277, 1976*b*.

Mutch, T. A., R. E. Arvidson, A. B. Binder, F. O. Huck, E. C. Levinthal, S. Liebes, E. C. Morris, D. Nummedal, J. B. Pollack, and C. Sagan, Fine particles on Mars: Observations with the Viking 1 lander cameras, *Science, 194,* 87, 1976*c*.

Pollack, J. B., C. B. Leovy, Y. H. Mintz, and W. Van Camp, Winds on Mars during the Viking season: Predictions based on a general circulation model with topography, *Geophys. Res. Lett., 3,* 479, 1976.

Pollack, J. B., D. Colburn, R. Kahn, J. Hunter, W. Van Camp, C. E. Carlston, and M. R. Wolfe, Properties of aerosols in the Martian atmosphere as inferred from Viking lander imaging data, *J. Geophys. Res., 82,* this issue, 1977.

Seiff, A., D. B. Kirk, and R. Blanchard, The Viking atmospheric structure experiment: Preliminary results, paper presented at the annual meeting of the Division of Planetary Sciences, Amer. Astron. Soc., Honolulu, Hawaii, Jan. 1977.

Tillman, J. E., The indirect determination of stability, heat and momentum fluxes in the atmospheric boundary layer from scalar variables during dry unstable conditions, *J. Appl. Meteorol., 11,* 783, 1972.

Wallace, J. M., and F. R. Hartranft, Diurnal wind variations: Surface to 30 km, *Mon. Weather Rev., 96,* 446, 1969.

Ward, W. R., B. C. Murray, and M. C. Malin, Climatic variations on Mars, 2, Evolution of carbon dioxide atmosphere and polar caps, *J. Geophys. Res., 79,* 3387, 1974.

Zurek, R. W., Diurnal tide in the Martian atmosphere, *J. Atmos. Sci., 33,* 321, 1976.

(Received March 28, 1977;
revised May 26, 1977;
accepted May 26, 1977.)

Report of the Viking Inorganic Chemical Analysis Team: Introductory Statement

PRIESTLEY TOULMIN III,[1] A. K. BAIRD,[2] B. C. CLARK,[3] KLAUS KEIL,[4] AND H. J. ROSE, JR.[1]

The first of the three papers comprising this report [*Clark et al.*, 1977] deals with the elemental analysis of Martian samples collected during the primary Viking mission (July 20, 1976–November 15, 1976) and the early part of the extended Viking mission (November 16, 1976–March 31, 1977). Calibration, operation, and data reduction are therein described in sufficient detail to give the reader a basis for understanding the significance of the results reported here and previously [*Clark et al.*, 1976*a*, *b*; *Baird et al.*, 1976; *Toulmin et al.*, 1976].

The second paper of the present group [*Baird et al.*, 1977] describes the procedures by which sample sites were selected and samples obtained and discusses the microgeological environments of the samples analyzed. It also gives details of the laboratory studies carried out in conjunction with mission operations in order to quantify the spectra returned from Mars and to facilitate the geochemical and mineralogical interpretation of the results.

The third paper [*Toulmin et al.*, 1977] discusses the geochemical, mineralogical, and lithologic conclusions that we have been able to reach at this early stage of a continuing process of data reduction and interpretation. The extended Viking mission continues actively at this time, and for our experiment it is operationally even more complex and demanding than the primary mission. Operational matters thus continue to occupy most of the Inorganic Chemical Analysis Team's attention, and the purely interpretative phase of the investigation is still in a preliminary stage. Much of the interpretation to follow therefore is excerpted and recast from the previously published reports of preliminary and interim results. The truly 'final' report of our findings will be forthcoming only after extensive further work.

[1] U.S. Geological Survey, Reston, Virginia 22092.
[2] Department of Geology, Pomona College, Claremont, California 91711.
[3] Martin-Marietta Corporation, Denver, Colorado 80201.
[4] Department of Geology and Institute of Meteoritics, University of New Mexico, Albuquerque, New Mexico 87131.

REFERENCES

Baird, A. K., P. Toulmin III, B. C. Clark, H. J. Rose, Jr., K. Keil, R. P. Christian, and J. L. Gooding, Mineralogic and petrologic implications of Viking geochemical results from Mars: Interim results, *Science, 194,* 1288–1293, 1976.

Baird, A. K., A. J. Castro, B. C. Clark, P. Toulmin III, H. J. Rose, Jr., K. Keil, and J. L. Gooding, Viking X ray fluorescence experiment: Sampling strategies and laboratory simulations, *J. Geophys. Res., 82,* this issue, 1977.

Clark, B. C., A. K. Baird, H. J. Rose, Jr., P. Toulmin III, K. Keil, A. J. Castro, W. C. Kelliher, C. D. Rowe, and P. H. Evans, Inorganic analyses of Martian surface samples at the Viking landing sites, *Science, 194,* 1283–1288, 1976*a*.

Clark, B. C., P. Toulmin III, A. K. Baird, K. Keil, and H. J. Rose, Jr., Argon content of the Martian atmosphere at the Viking I landing site: Analysis by x-ray fluorescence spectroscopy, *Science, 193,* 804–805, 1976*b*.

Clark, B. C., III, A. K. Baird, H. J. Rose, Jr., P. Toulmin III, R. P. Christian, W. C. Kelliher, A. J. Castro, C. D. Rowe, K. Keil, and G. R. Huss, Viking X ray fluorescence experiment: Analytical methods and early results, *J. Geophys. Res., 82,* this issue, 1977.

Toulmin, P., III, B. C. Clark, A. K. Baird, K. Keil, and H. J. Rose, Jr., Preliminary results from the Viking x-ray fluorescence experiment: The first sample from Chryse Planitia, Mars, *Science, 194,* 81–84, 1976.

Toulmin, P., III, A. K. Baird, B. C. Clark, K. Keil, H. J. Rose, Jr., R. P. Christian, P. H. Evans, and W. C. Kelliher, Geochemical and mineralogical interpretation of the Viking inorganic chemical results, *J. Geophys. Res., 82,* this issue, 1977.

(Received May 2, 1977;
revised May 28, 1977;
accepted May 28, 1977.)

Paper number 7S0465.

The Viking X Ray Fluorescence Experiment: Analytical Methods and Early Results

BENTON C. CLARK III,[1] A. K. BAIRD,[2] HARRY J. ROSE, JR.,[3] PRIESTLEY TOULMIN III,[3] RALPH P. CHRISTIAN,[3] WARREN C. KELLIHER,[4] ANGELO J. CASTRO,[1] CATHERINE D. ROWE,[1] KLAUS KEIL,[5] AND GARY R. HUSS[5]

Ten samples of the Martian regolith have been analyzed by the Viking lander X ray fluorescence spectrometers. Because of high-stability electronics, inclusion of calibration targets, and special data encoding within the instruments the quality of the analyses performed on Mars is closely equivalent to that attainable with the same instruments operated in the laboratory. Determination of absolute elemental concentrations requires gain drift adjustments, subtraction of background components, and use of a mathematical response model with adjustable parameters set by prelaunch measurements on selected rock standards. Bulk fines at both Viking landing sites are quite similar in composition, implying that a chemically and mineralogically homogeneous regolith covers much of the surface of the planet. Important differences between samples include a higher sulfur content in what appear to be duricrust fragments than in fines and a lower iron content in fines taken from beneath large rocks than those taken from unprotected surface material. Further extensive reduction of these data will allow more precise and more accurate analytical numbers to be determined and thus a more comprehensive understanding of elemental trends between samples.

INTRODUCTION

The original science payload selections for the Viking mission heavily emphasized the exobiological aspects of this first exploratory journey to the surface of Mars. A proposal by two of us [*Baird and Clark*, 1969] for an X ray fluorescence geochemical analyzer was judged premature but led to 'proof of concept' studies supported by NASA Headquarters. As a result of a demonstration test (see *Clark and Baird* [1973*a*] for results) and growing interest from the planetary science community, both within and outside of NASA, in broadening the scientific base of Viking, the X ray fluorescence spectrometer (XRFS) was added to the payload in the spring of 1972. A science team designated the Inorganic Chemical Analysis Team (Icat), composed of five of us (Toulmin, Baird, Clark, Keil, and Rose), was selected to guide the development, operation, and data analysis of the XRFS experiment. Since it was added to the Viking mission 2 years later than the other experiments, the time available for instrument development was considerably shortened. Furthermore, spacecraft design was already firm, imposing serious constraints upon the XRFS instrument design, including size and configuration. For example, there was no opportunity to influence the design of the surface sampling system, to implement instrument deployment schemes, or to include sample processing. Likewise, multiple-sample handling had to be accommodated by retaining analyzed material within the volume envelope allocated for the instrument, and electronic multichannel analyzer circuitry could not be accommodated because of weight constraints. Nonetheless, the experiment as implemented has met the fundamental science objectives of detecting and quantifying some dozen major, minor, and trace elements in the Martian soil. In addition, it has analyzed what appear to be duricrust fragments, determined an upper limit for argon in the Martian atmosphere, measured the soil bulk density, established a new upper limit for thickness of iron oxide coatings on mineral grains, and demonstrated the highly adhesive nature of Martian soil.

Interim descriptions of instrument design and performance capabilities have been given previously [*Clark et al.*, 1977; *Clark and Baird*, 1973*b*; *Toulmin et al.*, 1973]. Portions of the following section include information given in interim descriptions, and some of the data given in the section on results have been published in a previous report [*Clark et al.*, 1976].

INSTRUMENT DESCRIPTION

X ray fluorescent emissions from a sample irradiated by two radioisotope sources are analyzed in the energy dispersive mode by four miniature proportional counter detectors. Characteristics of these components are summarized in Table 1. The instrument is shown in Figures 1, 2, and 3. The minimum X ray energy for which useful sensitivity is obtained is the 1.25-keV K_α emission from magnesium. Excitation energies are at 5.9, 22.2, and 87.7 keV by ^{55}Fe electron capture, ^{109}Cd electron capture, and ^{109}Cd nuclear gamma emissions, respectively. Thus fluorescent emissions from all elements in the periodic table between Mg and U are within the excitation and detection range of the instrument, provided they are present in amounts above their minimum detection limit (see Table 2 of *Clark and Baird* [1973*b*]) and are not interfered with by emissions from other elements.

Design. The XRFS is mounted within the thermally controlled compartment of the Viking lander body. Samples are delivered by the surface sampler scoop in several possible modes (see *Baird et al.* [1977] in this issue). A convex screen with 12.5 × 12.5 mm openings is mounted inside the XRFS inlet funnel to prevent the entrance of large rocks or clods which could cause clogging of the instrument. The sample is held in a 25 × 25 mm (cross section) sample cavity and is irradiated through two 20-mm diameter analyzing ports at right angles. The quantity of samples required to fill the cavity to the top of the ports is ~25 cm³. Analyses can be performed

[1] Planetary Sciences Laboratory, Martin Marietta Aerospace Corporation, Denver, Colorado 80201.
[2] Department of Geology, Seaver Laboratory, Pomona College, Claremont, California 91711.
[3] U.S. Geological Survey, Reston, Virgina 22092.
[4] NASA/Langley Research Center, Hampton, Virginia 23665.
[5] Department of Geology and Institute of Meteoritics, University of New Mexico, Albuquerque, New Mexico 87131.

Paper number 7S0448.

TABLE 1a. XRFS Proportional Counter (PC) Detectors

PC	Gas Composition	Window Material
1	20% Ne, 75% He, 5% CO_2	5-μm Al
2	10% Xe, 73% Ne, 10% He, 7% CO_2	25-μm Be
3	40% Xe, 47% Ne, 10% He, 3% CO_2	50-μm Be
4	40% Xe, 47% Ne, 10% He, 3% CO_2	25-μm Be

on less sample material, but there would be an increase in uncertainties in the deduced concentrations.

Selection of the thin film windows used in the ports required considerable study. The films had to be extremely thin to pass the low-energy Mg X ray, heat resistant, tough enough to withstand impact from rock fragments free falling into the cavity from the sampler, and radiation resistant because of the dosage of approximately 3×10^7 rads by the ^{55}Fe source during the exposure time of 2 years. Beryllium foil, the first choice, was ruled out because of its susceptibility to fracture from rock impact. Some candidate organic films were found to be incapable of maintaining integrity when subjected to heat sterilization sequences. Others became embrittled or completely disintegrated during high radiation dose tests with a ^{60}Co source. Polycarbonate film was found to be acceptable for the thinnest film requirement, although the commercially available material still had to be stretched by radiatively heating to the softening point under controlled tension to attain the thickness desired. Specially cast polyimide film was found acceptable for the port window exposed to radiation from the ^{109}Cd source. Both films were supported by high-purity metallic grids (Ni for the ^{55}Fe window and Al for the ^{109}Cd window).

Radioisotope sources were fabricated by Isotope Products Laboratories, Inc., Burbank, California, from accelerator grade material with specific activities of $\geq$500 Ci/g. After additional purification they were electroplated onto suitable backing foils, heat diffusion bonded, and sealed into a beryllium-windowed capsule by using metal sealing techniques only (welding and brazing). The radioisotope spot size was kept below 5 mm diameter. A 180-μm aluminum foil filter attenuated to an acceptable level fluorescence emission from copper in the brazing alloy of the ^{109}Cd source. Source holders and radiation collimators were carefully designed to protect the detectors from direct radiation and to avoid the use of any material that would produce unwanted fluorescent radiation. The ^{109}Cd holder is a composite assembly of tungsten alloy and high-purity silver with thicknesses calculated for optimum attenuation of the 87.7-keV gamma ray. This was necessary to prevent detectable increase of background counting rate in the ^{14}C radioassay detectors of the integrated biology instrument adjacent to the XRFS. The resultant design is such that leakage radiation is extremely small, and the assembled XRFS instrument can be handled safely without 'radiation worker' classification.

The miniature proportional counters (Reuter-Stokes, Inc., Cleveland, Ohio) are also of custom design for this instrument. Gas compositions, given in Table 1, were selected to minimize

TABLE 1b. XRFS Sample Cavity Windows

Window	Material	Grid
^{55}Fe	1.5-μm polycarbonate	>86% open area, pure Ni
^{109}Cd	2.5-μm polyimide	>81% open area, pure Al

TABLE 1c. XRFS Spectra Usage

Element	Proportional Counter	
	Primary	Secondary
Mg	1	
Al	1	2
Si	2	4
S	2	
Cl	2	1
Ar	2	
K	2	1
Ca	2	1, 4
Ti	2	1, 4
Fe	4	3
Zn	4	3
Ba	3	4
Traces (Rb, Sr, Y, Zr)	4	3
Bulk density	3	4
Sample fill level	4, 1, 3	2

escape peak interferences and to optimize spectral response for typical element abundances for geologic specimens. Entrance windows are of unsupported beryllium foil, with the exception of proportional counter (PC) 1, which employs a 5-μm pure Al foil on an Al supporting grid to transmit Mg and Al fluorescent X rays and to filter out the interfering Si fluorescence signal. The only method deemed reliable for sealing these thin windows to the counter bodies was epoxy bonding. All other counter parts were joined and sealed by either welding, brazing, or cold-welding. Early in development testing it was noticed that the prototype detectors exhibited an undesirable gain sensitivity to temperature. Literature and informal surveys revealed that our requirement for gain stability was beyond the state-of-the-art as it applied to spacecraft missions. Thermogravimetric and evolved gas analysis of the bonding epoxy showed that partial decomposition occurred during the 8-hour 150°C bakeout used for outgassing the counters. Numerous changes in manufacturing procedures were made to improve the gas purity of the sealed detectors. This included air cure followed by vacuum cure of the epoxy, outgassing the assembled counter at 140°C for 75 hours, two-cycle gas purification, and temperature soak to accelerate equilibration. These changes reduced the temperature sensitivity coefficient below the maximum acceptable value of 0.06% per degree Fahrenheit. During the final test of the first flight instrument, one PC detector suddenly failed to produce any output but 'revived' after a few hours with power off. Subsequent testing of the flight detectors showed that many exhibited similar behavior after a long period of operation, i.e., their output suddenly ceased and could be recovered only after a period of quiescence with high-voltage bias removed. Consultation with a number of workers in the ionization detector field revealed a similar case (L. W. Acton and R. E. Varney, unpublished data, 1974) in which it was demonstrated that the effect was due to gradual charging of thin epoxy surfaces on the inside of the counter until pinhole field emission was stimulated (the Malter effect [*Malter,* 1936]) and a continuous discharge was initiated. Intensive laboratory testing showed that it was the extremely rigorous outgassing procedure that cleaned up the epoxy coating to the degree necessary to initiate this phenomenon in our detectors. In a redesign the thin windows were mounted outside of the tube wall rather than inside, so that the epoxy bond could be applied under greater control and seepage of excess epoxy onto internal surfaces could be eliminated.

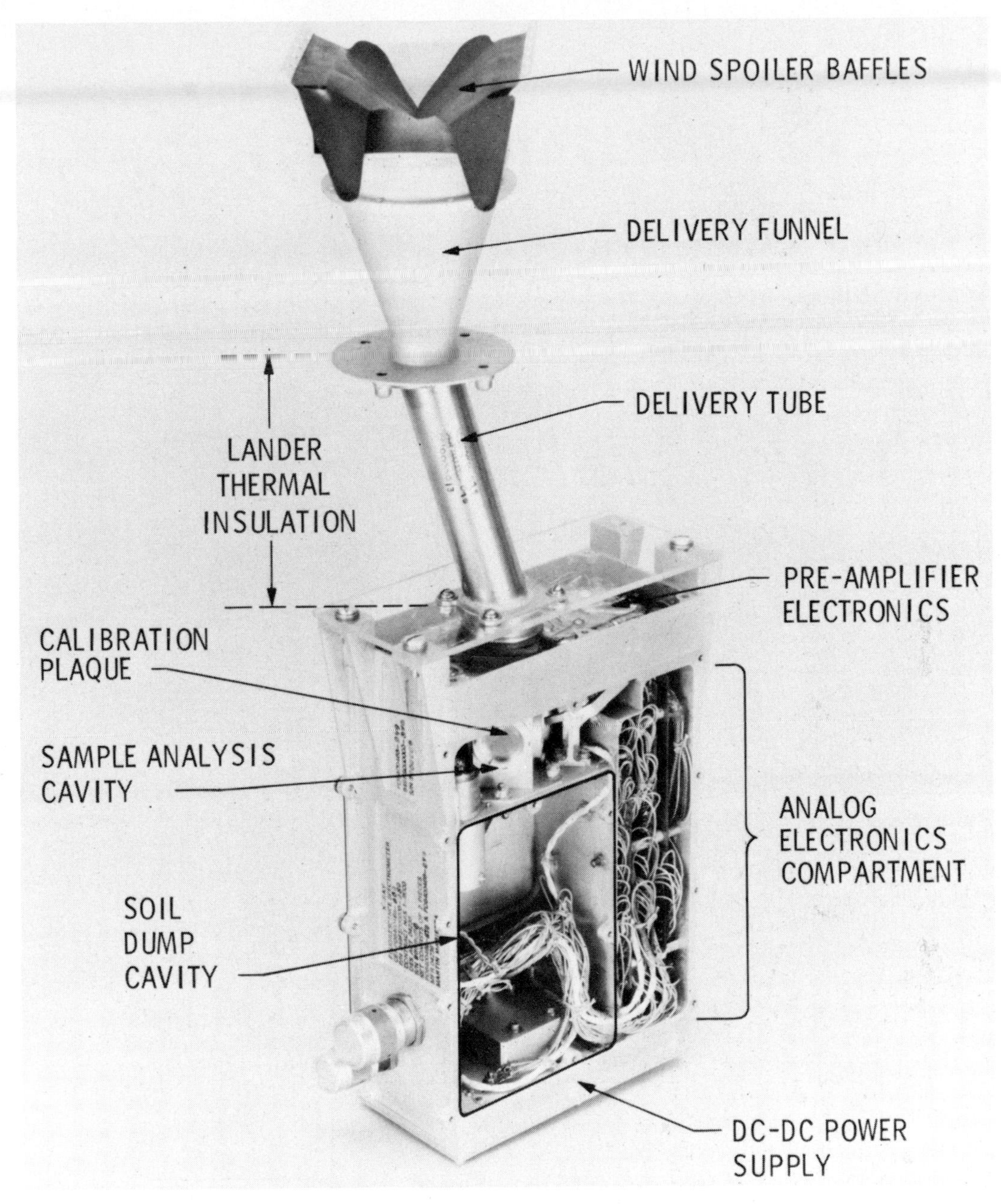

Fig. 1. Photograph of spare XRFS flight unit (FLT-1).

No detectors manufactured by this modified design have been observed to fail via this mechanism.

Considerable effort went into the design and test of XRFS electronics to maximize operational stability, particularly for the amplifier string and the high-voltage bias supply. The bias supply has an output commandable in 5-V increments from 715 to 1350 V, allowing adjustment of PC detector gain over a total range of 10,000:1. All four detectors are connected through signal isolation networks to the single high-voltage power supply.

Electronic data encoding. Data collected by the instrument must be stored until readout by the Viking guidance control and sequencing computer (GCSC). Eight channels of a 128-channel spectrum are scanned, the information is held until the data are read out by the GCSC (usually for less than 1 s), and the scan is resumed. Each readout is 312 bits of data (256 of data and 56 of identification) and is called a data frame. XRFS data frames are interspersed among all other lander science and engineering data. Retrieval is obviously a complicated process. A feature unique to the XRFS instrument is that the data are highly encoded to minimize the impact of data transmission and handling errors.

Counts accumulated during each measurement interval (commandable intervals are 7.7, 30.7, 61.4, 123, 246, and 492 s/channel) are stored in a 16-bit 'pseudorandom counter' that operates in the following manner [*Blizard,* 1974]: each time a count is received, it is shifted one bit to the right, and the leftmost bit is set equal to the modulo 2 sum of the previous contents of bits 4, 13, 15, and 16. To the observer the pattern appears random for successive counts, but in fact, a unique mapping function exists between each of the 65,535 possible states and the accumulated count. An important feature is that a single bit error results in an equivalent count much different than the true one, whereas for a binary counter, errors in bits

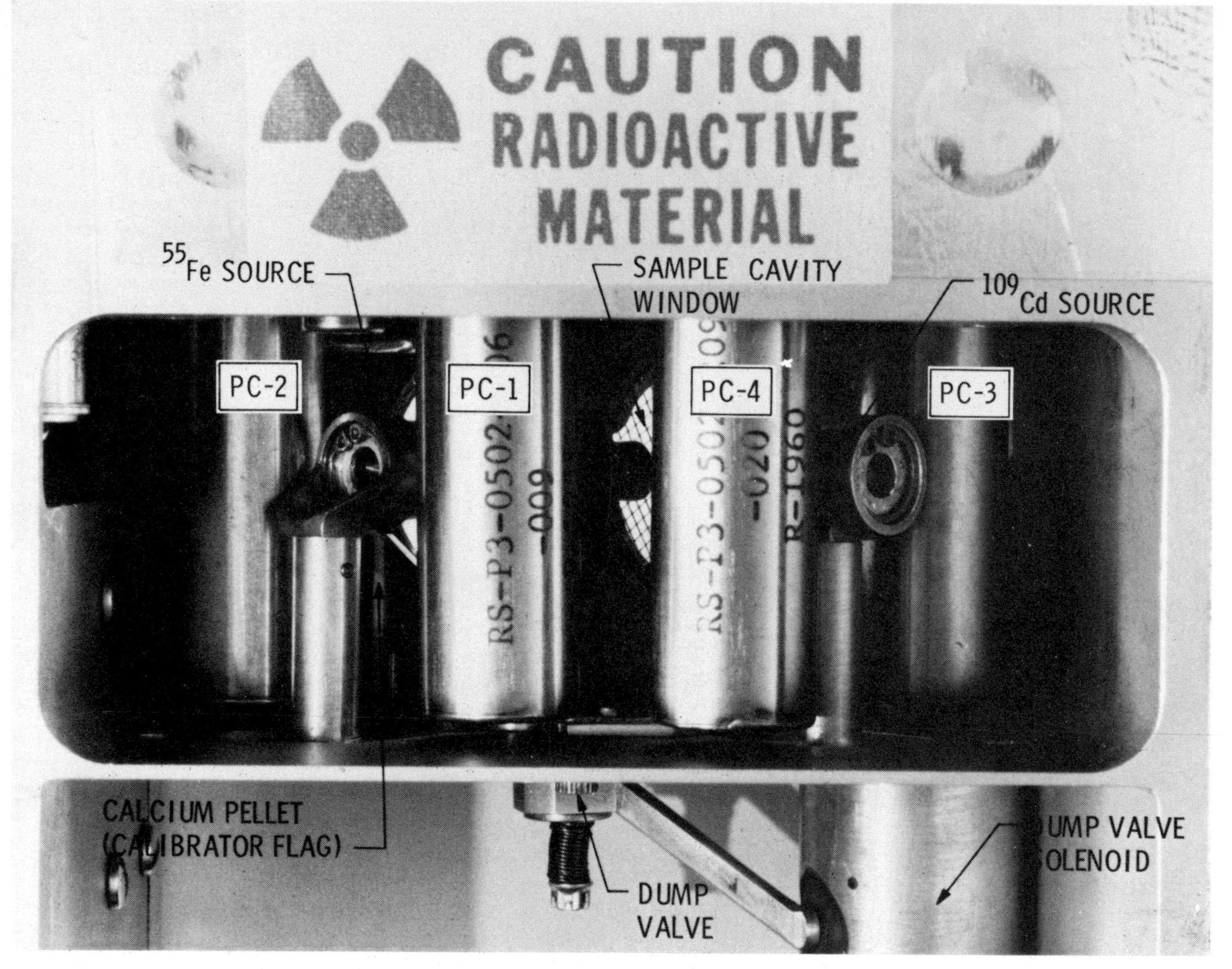

Fig. 2*a*

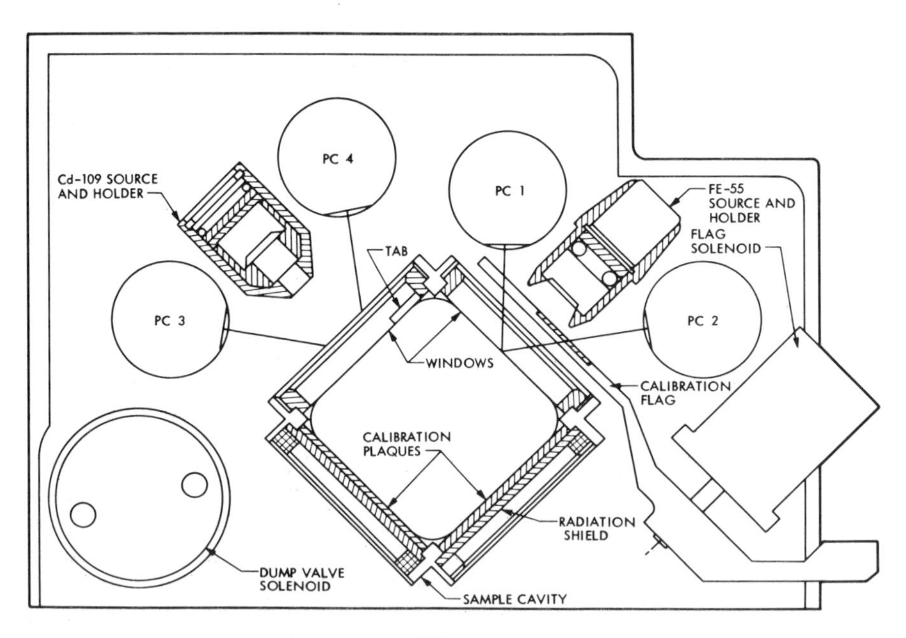

Fig. 2*b*

Fig. 2. Analysis section, indicating the radioisotope sources, proportional counter detectors, and sample cavity. (*a*) Photograph of analysis section. (*b*) Cross-sectional diagram of analysis section with the flag calibrator shown in the activated position.

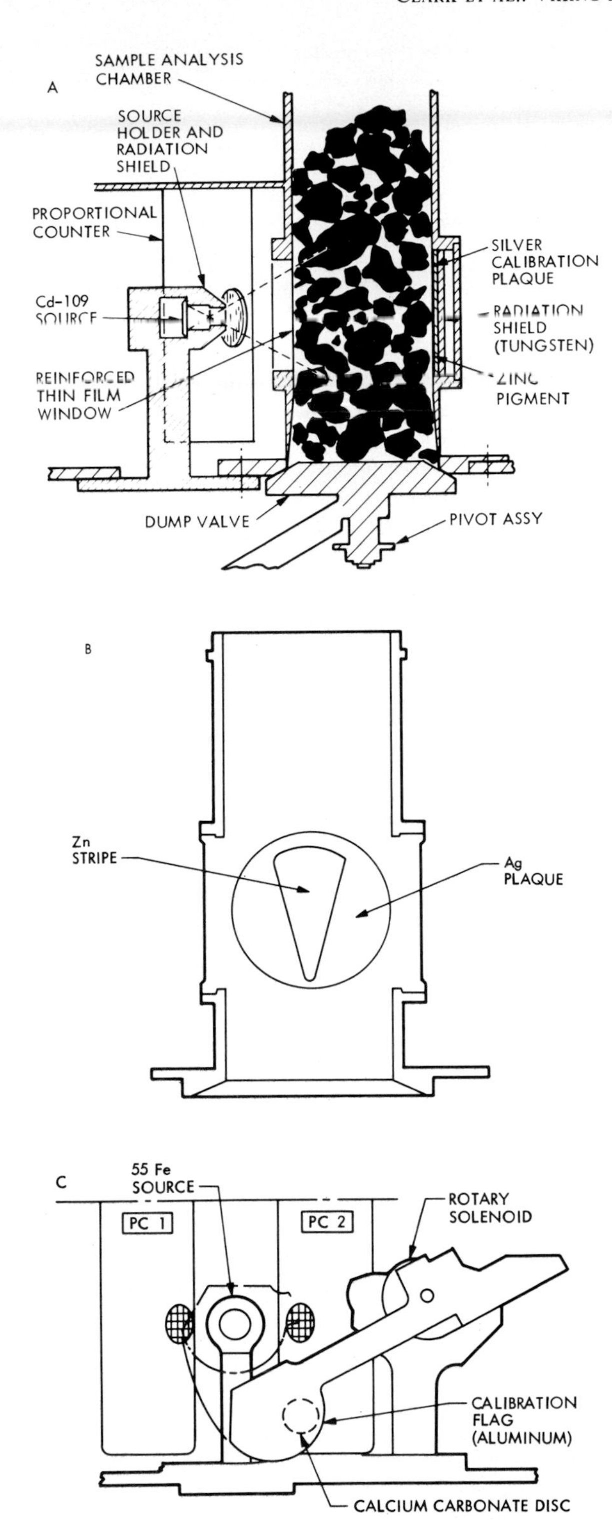

Fig. 3. Details of the sample cavity and calibration targets. (*a*) Cavity shown filled with lithic and crustal fragments in the 2- to 12-mm size range. (*b*) Plaque target for the ^{109}Cd source. (*c*) Flag target for the ^{55}Fe source.

of lesser significance produce subtle count errors that are difficult to recognize. To protect the data further, the 16-bit word is transmitted with a second image of itself formed by adding 32 counts. Also, the counter is purposely not reset between accumulations, so that the transmitted count represents the integral. Our data reduction program, ICAN (inorganic chemical analysis), includes an algorithm that decodes the pseudorandom word in a highly efficient manner by the sequence partial decode, table look up, final decode. Both images are decoded. If they differ by the 32 counts, they are considered error free, but if not, error-correcting subroutines are brought into play. These subroutines progress the first word forward 32 counts and the second one backward 32 counts and compares them with the received words. In this way, single- and double-bit errors can be recognized, as well as six-bit consecutive errors, which can result when the (32, 6) biorthogonal code used in orbiter-to-earth transmissions contains an uncorrectable six-bit error. Correction is automatically performed, and a message flag is set. Simulation tests have demonstrated that bit error rates as high as one bit in 10 can be tolerated, and experience with the Viking data stream shows that the impact of poor data links, when they occur, is principally to cause loss of data (because the master stripping programs misidentify data) rather than to affect data quality. Since the XRFS data are encoded within the instrument itself and not decoded until the final processing step (ICAN), the dozen or more pieces of equipment in the data chain on Mars, in space, and on earth cannot introduce unknown errors into the data. This error detection/correction scheme results in at least 99.99% of our data being a perfect replica of the data generated by the XRFS instrument, which means that all sources of error originate only within the instrument and not in data handling. The ICAN program also performs searches of previously received data and tags those data frames that are simply copies of old data. Because of the many data storage devices in the data loop, redundant data are a common occurrence. Identification of copies is greatly facilitated by transmitting the integral spectrum, which results in many fold fewer chance coincidences than would occur if differential spectra counts were transmitted.

Calibration targets. Rather than use electronic calibration, such as a dual-amplitude pulser, it was decided to employ an end-to-end calibrator, as it is the overall system response that must be determined. Calibration targets were built into the far walls of the sample analysis cavity (Figures 3*a* and 3*b*) so that with no sample present the elements in the targets would fluoresce. Because the geometry of the system is fixed and the X ray emission energies do not vary with the environmental parameters, the use of fluorescence targets provides an absolute method for establishing the energy scale, element response function, and relative source strength level for each detector-source combination. These targets are called 'calibration plaques.' Opposite the ^{55}Fe source is a plaque of aluminum, and opposite the ^{109}Cd source is a pure silver plaque with a recessed triangular stripe of ZnO-pigmented paint (Figure 3*b*). This stripe is oriented with its point downward. As material fills the cavity, the stripe becomes progressively obscured and the Zn peak detected in PC 3 and PC 4 diminishes to a negligible level. Note that the ratio of intensities of Ag/Zn will diminish as the chamber fills. Calibration spectra for a clean cavity are shown in Figures 4 and 5, with the Al, Ag, Zn, and backscatter peaks indicated. Also present in the PC 3 and PC 4 spectra is a peak due to ^{55}Fe scattered from the plaques. Indeed, if it were not for a pure aluminum tab added to the soil cavity window on the ^{109}Cd source side, visible in Figure 2*a*, a direct path would exist for entry of ^{55}Fe radiation into PC 3; this would produce an unacceptably high background counting rate.

An additional method of calibration has been provided for PC 1 and PC 2 to facilitate accurate determination of Mg/Al and K/Ca ratios. This consists of a separate mechanism, with

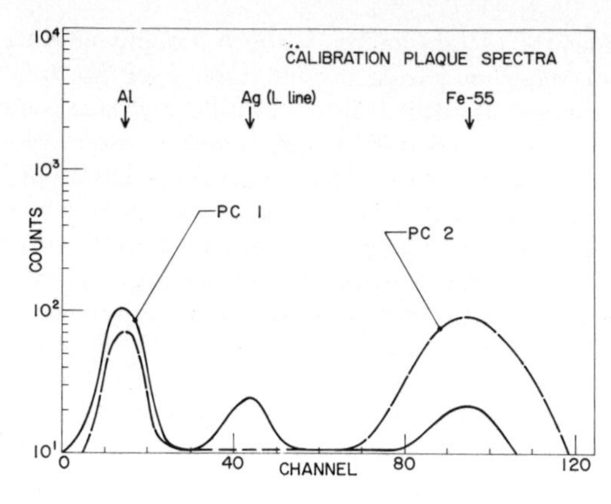

Fig. 4. Typical calibration plaque spectra for an empty and clean sample cavity. Because of the geometrical configuration, PC 1 can 'see' the Ag plaque, while PC 2 cannot.

Ca and Al targets that can be inserted between the ^{55}Fe source and the sample by energizing a rotary solenoid to which it is attached. The advantages of this 'flag' calibrator are that (1) it can be used even when soil fills the cavity, unlike the plaques which can be used only when the cavity is empty, (2) it remains clean at all times, and (3) it produces a relatively high counting rate, allowing the buildup of adequate statistics in a single spectral scan. The construction of this target is shown in Figure 3*c*.

PRELAUNCH CALIBRATIONS

A number of methods were devised to certify and calibrate properly the XRFS instruments subsequent to their fabrication and final acceptance tests. All flight units were manufactured to the same drawings and procedures, and all subassemblies were individually tested electronically and/or mechanically against predetermined specification limits.

Performance verification. Reliability considerations mandated that all Viking flight hardware be protected from exposure to dust or other particulate contamination prior to flight. This prevented the calibration of XRFS units with compositionally characterized soil and rock powders. Instead, 29 solid rock specimens were selected and shaped by sawing and grinding to fit snugly within the sample cavity. Four of these samples were used to verify instrument analytical capabilities. These 'performance standards' are listed in Table 2. The remaining 25 standards (given in Table 3) were used to characterize the response of each instrument to a wide range of rock compositions. In addition, pure element standards for Mg, Al, Si, S, K, Ca, Ti, Fe, Rb, Sr, and Zr were constructed by machining the pure element or by mixing an appropriate powder with epoxy in a special mold. The elemental concentration in the molded standards was adjusted to produce an overall counting rate of ~1000 counts/s in the detector under test, and in some cases, aluminum or nickel powder was added to reduce background from backscatter. Typical response functions for pure elements are shown in Figure 6. These functions were generated by a least squares fit of Gaussian curves to the peaks and by hand fitting over non-Gaussian regions. Peak resolutions are proportional in first approximation to the theoretical inverse square root of energy law. These data are used to extrapolate response functions for elements not directly measured.

Following final assembly, each unit was tested by taking spectra on each of the four standards and reducing the data to demonstrate that the analyses for 10 elements were within acceptable error limits. The unit was then tested for ster-

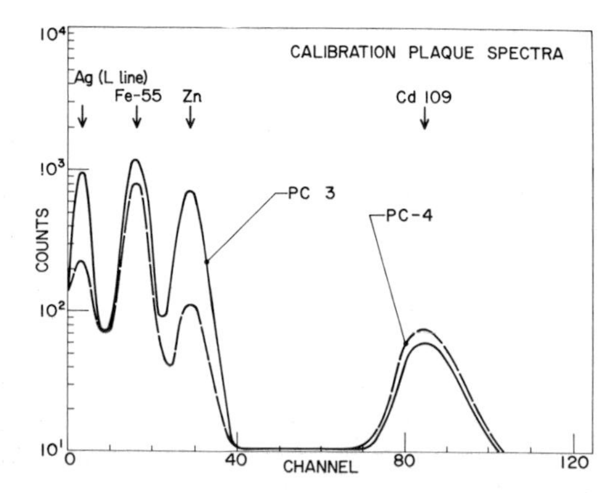

Fig. 5. Calibration plaque spectra for the ^{109}Cd detectors. The ^{55}Fe signal reaches these detectors when the sample cavity is empty by scattering off of the plaques and interior walls. A small Al tab, visible in Figure 2*a* and called out in Figure 2*b*, prevents a direct path from the ^{55}Fe source to PC 3 and also partially obscures the Zn stripe to PC 4, which accounts for the lower Zn peak for that detector.

TABLE 2. Performance Standards

Sample	Site Code	Description	Location
8007	BCR-1	basalt	Columbia River Plateau, Bridalvail Quad, Oregon
8010	AGV-1	andesite	Lake County, Oregon
8083	QLO-1	quartz latite	Lake County, Oregon
8084	RGM-1	rhyolitic obsidian	Glass Mountain, Siskiyou County, California

TABLE 3. Calibration Standards

Sample	Description	Location
908	taconite	Black River Falls, Jackson County, Wisconsin
6017	biotite quartz monzonite	Keller Peak, San Bernardino Mountains, California
6019	granodiorite	Mount Edna, San Jacinto Mountains, California
6021	granite pegmatite	Mount Rubidoux, Riverside, California
6027	norite	Perris, California
8060	pyroxene syenite	Empire, Colorado
8061	nepheline monzonite	Empire, Colorado
8062	biotite-hornblende granite	Eldorado, Colorado
8063	syenogabbro	Empire, Colorado
8064	biotite peridotite	Caribou, Colorado
8065	graywacke	Reston, Virginia
8066	shale	Knoxville, Tennessee
8067	rhyolite welded tuff	Mount Rodgers, Virginia
8068	latite	Table Mountain, Colorado
8069	arkosic sandstone	Connecticut
8070	spodumene	Helena Mine, South Dakota
8071	dolomitic shale	Green River, Colorado
8072	brick	Rockville, Maryland
8073	quartz	Brazil
8074	limestone	Solenhofen, Germany
8076	sandstone	Osage County, Oklahoma
8078	slightly sandy limestone	Stone County, Arkansas
8079	dolomite	Cape Girardeau County, Missouri
8080	shale	Lincoln County, Oklahoma
9001	basalt	Molokai, Hawaii

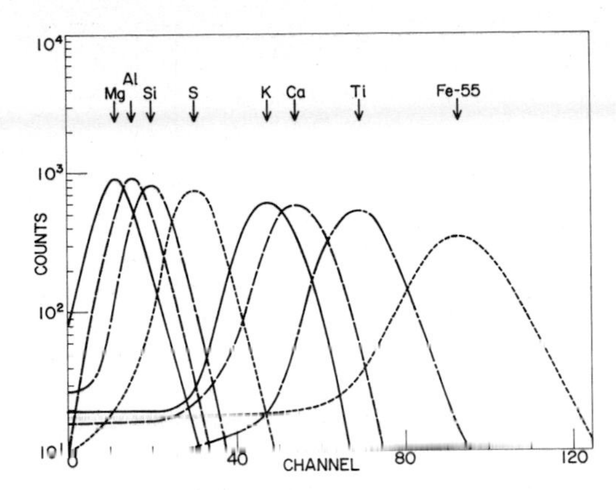

Fig. 6. Pure element response functions for PC 2. The appreciable overlap between adjacent elements mandates accurate gain knowledge. The ^{55}Fe backscatter peak provides an internal calibration for gain.

ilization compatibility by heating in a dry nitrogen atmosphere for 54 hours at 233°F. Low-level vibration tests, with both random and sweeping sinusoidal frequencies, verified the mechanical integrity of the unit. Thermal and pressure environments were simulated by four cycles of 0°F to +100°F to 0°F in an atmosphere of 5-torr CO_2, and the unit was tested during 4 hours at the high and low temperatures. During the last two cycles the temperature was varied in 12.5° and 25°F increments, and the units operated at each temperature level to establish the temperature coefficient of each detector. This also verified that neither corona nor arc discharge from the high-voltage power supply occurs in a low-pressure Marslike environment. Subsequently, the four standards were rerun, and the unit was required to pass performance specifications again. All of the above operations were accomplished with special radiation sources whose activities closely approximated the target source activities at the time of the landing on Mars (240 mCi for ^{55}Fe and 55 mCi for ^{109}Cd).

Characterization. Just prior to final instrument installation into the lander spacecraft the radiation sources were replaced with the freshly prepared flight level sources, and 225 spectra were taken to obtain final characterization of the response of the unit. This included the 25 solid rock standards of Table 3, as well as numerous calibration plaques and flags, and pure argon gas. All rock spectra were taken for 500 s with an external multichannel analyzer which bypassed the internal single-channel analyzer. Spectral signatures of the 25 standards are unambiguously unique in comparison to one another. Indeed, the discriminatory capability of the instrument is large: a conservative estimate is that at least 1.5×10^7 distinct elemental compositions should be distinguishable.

Tests. Numerous quasi-blind, blind, and double-blind tests have been performed during development of the instrument. An early breadboard model was evaluated with respect to four powdered rock samples supplied by NASA, and the results were discussed by *Clark and Baird* [1973*b*, 1974]. At the spacecraft system level, samples of Pikes Peak granite and Pierre shale and an unknown sample delivered by the Viking surface sampler were analyzed on a test lander during the 'science end-to-end test.' An interesting deviation in the apparent granite composition, which was in the form of pea-sized gravel, was later shown to be a result of a particularly large microcline fragment whose natural cleavage face covered a major fraction of the viewing port window. The unknown sample was supplied by NASA and was successfully identified as rhyolite from a suite of five preselected candidates of diverse compositions. Other tests, of an inadvertent nature, have served further to increase confidence in the analyses. For example, our sample 8066, Chattanooga shale, was found to contain relatively high quantities of sulfur, whereas the available conventional analysis showed only an unusually high ignition loss. Subsequent laboratory analysis confirmed the high sulfur (4.6%). Similarly, when Riverside nontronite was run, low but approximately equal amounts of strontium and yttrium were indicated; the ex post facto laboratory analysis gave 80 ppm Sr and 57 ppm Y. A field portable version of the instrument has successfully been used to identify this nontronite occurrence at Riverside, California, detect a high zirconium content in a Nevada rhyolite, identify sulfate salts in Utah lake samples, determine copper and zinc in aluminum alloys and chlorine in epoxies, find contamination residues on surface-passivated metals, detect inhomogeneities in the standard rocks and deviations from the composition of their correlative powders, etc. On the other hand, magnesium and aluminum are quite difficult to analyze (see below), and experience has shown that considerable errors can occur unless the most stringent attention is paid to the details of the analysis. For example, a currently unresolved problem is that the Al_2O_3 content of Riverside nontronite was estimated at a level that is low by 1.3 wt % in comparison with the concentration determined conventionally.

ANALYSIS METHODOLOGY

From the 25 standards, three have been selected for detailed calibration studies. Two of these, 8063 (syenogabbro) and 8064 (biotite peridotite), are, like the Martian soil, relatively high in iron and have a high Mg/Al ratio. The third, 8066 (Chattanooga shale), is of importance in quantifying sulfur because it is the only preflight standard that contains an appreciable amount of this element. The preflight spectra of these standards are compared to selected Mars spectra in Figures 7*a*–7*c*. None of these standards is a close match to the Mars spectra; similarly, of the spectra taken of several hundred different geological materials prior to and since landing, none matches Mars samples in all respects [*Baird et al.*, 1976; *Toulmin et al.*, 1977]. Furthermore, as of this writing it can be stated that the naturally occurring sample that most closely matches the first sample taken at the lander 1 site is the first sample taken on Mars at the lander 2 site, some 6500-km away.

Two independent methods of determining elemental concentrations from spectra are used in the analysis of the Viking XRFS data and are described below.

Transfer calibration. In the first technique the spectra are corrected by subtracting the background contributions (see below), and the pure element response functions are then employed to determine the relative number of X ray photons detected at each X ray fluorescence energy interval of the element and at the energy region of backscattered primary radiation. These relative contributions are determined by the iterative successive approximation approach whereby an initial distribution is assumed and then corrected at each iteration step by determining the ratios of the measured number of counts in each chosen energy interval to the calculated number of counts and modifying the distribution accordingly. By this procedure, only that portion of the pure element spectrum in the chosen energy interval actually contributes to the determination of concentration for that particular element. More sophisticated matrix formulation methods, described by

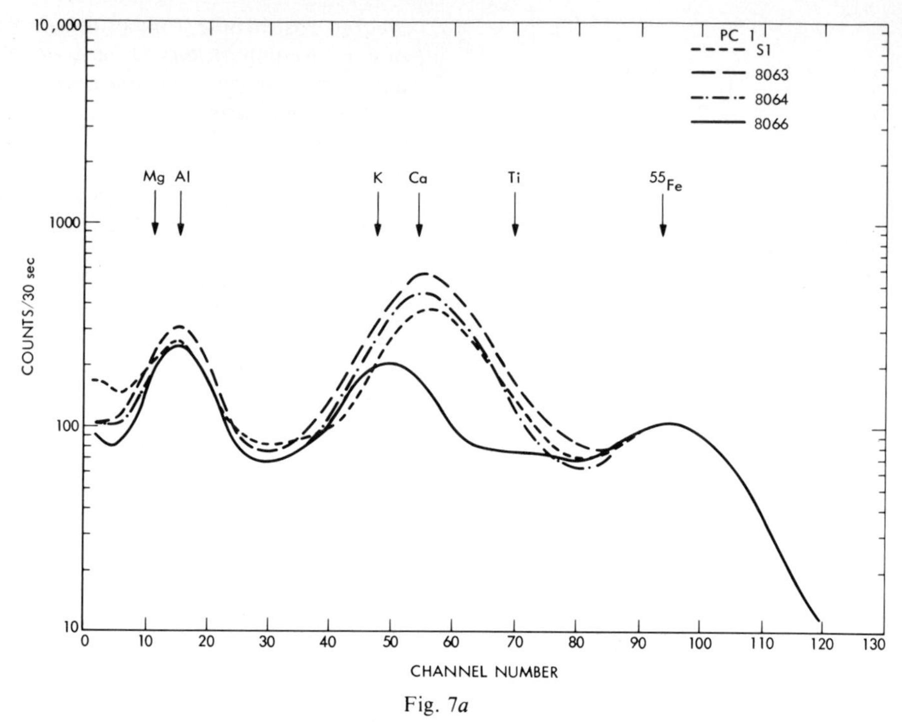

Fig. 7*a*

Fig. 7. Comparison of lander 1 spectra with prelaunch spectra of the three standards selected as the most useful in the derivation of concentration values for Martian samples. (*a*) PC 1 spectra and Mars sample S1. (*b*) PC 2 spectra and sample S5. (*c*) PC 4 spectra and sample U1.

Trombka and Schmadebeck [1968], for example, not only employ the entire response functions for component determination but also allow for iterative compensation for electronic gain and/or offset uncertainties. Application of such methods is planned for the final data analysis. From the relative detected photon distribution, element composition is calculated on the basis of a mathematical model of this instrument [*Clark,* 1974] that includes the following calculated factors: transmission of excitation photons (5.9 keV for ^{55}Fe and 22.2 keV for ^{109}Cd) through the thin film soil cavity windows (Table 1) into the sample; primary fluorescence; secondary fluorescence (fluorescence of element *j* by fluorescence radiation from element *i* (tertiary fluorescences are negligible)); sample absorption; scattering of excitation radiation by the sample; transmission of fluorescent and scattered photons through the soil cavity window and proportional counter entrance window; and absorption in counter gas. Overall matrix absorption and enhancement effects (second, third, and fourth factors

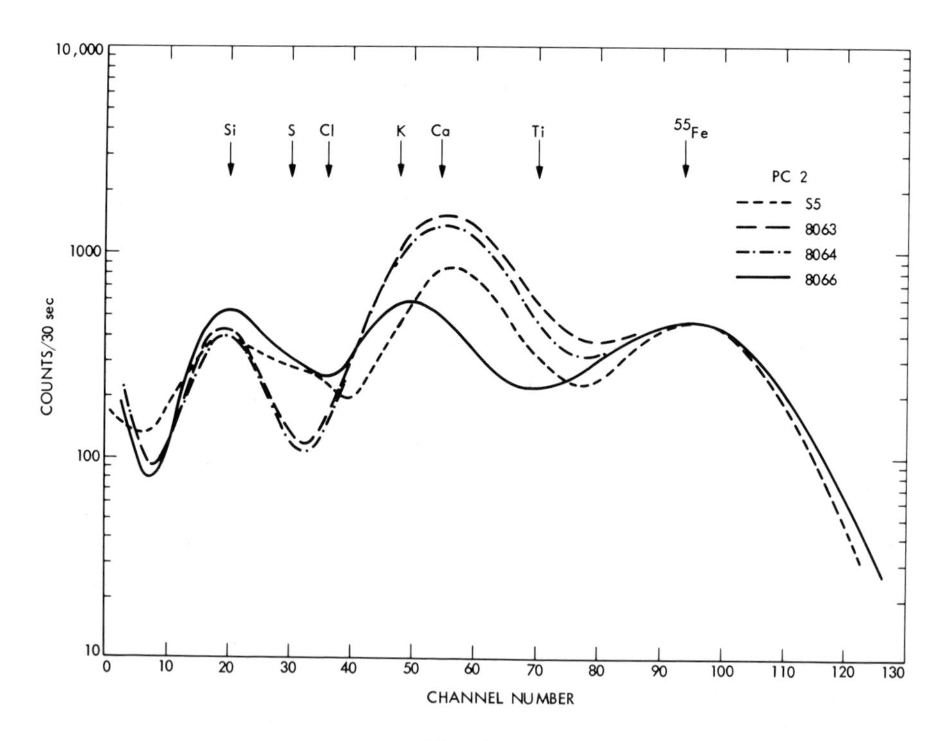

Fig. 7*b*

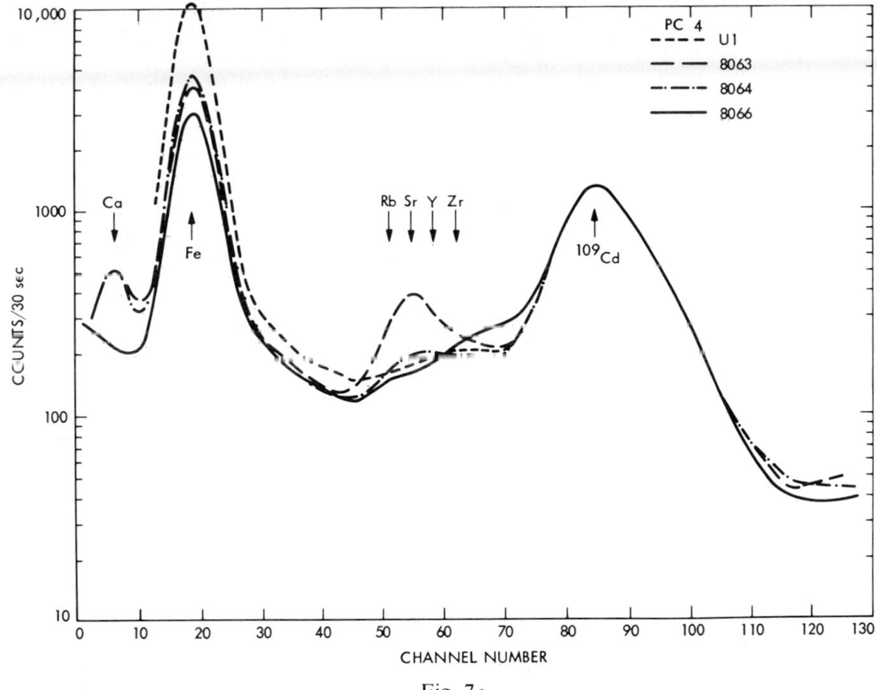

Fig. 7*c*

above) are calculated by the fundamental parameter method originally proposed by *Criss and Birks* [1968], which includes the secondary fluorescence formula of *Shiraiwa and Fujino* [1966]. Backscattered radiation is computed by the formula of *Clark* [1974]. Data base constants include the X ray mass absorption and scattering coefficients of *Veigele et al.* [1971], the X ray fluorescent yields from *Bambynek et al.* [1972], the window compositions and thicknesses and detector gas compositions (Table 1), and pressures as specified in the procurement contracts. Because of small uncertainties in all constants and allowances for manufacturing tolerances the actual algorithm used includes an adjustment parameter for each element and proportional counter called the *T* factor for that element-detector combination. As formulated, the *T* factor is a multiplier of the fluorescent yield. *T* factors are determined empirically for each individual instrument by employing the mathematical model as embodied in the computer program ICAN to calculate the composition of a known sample from its spectra. Each *T* factor is adjusted manually until the calculated composition is identical to the 'true' composition. Ideally, the *T* factors would be equal to 1.00; however, in practice, they range from 0.55 to 1.3. Thus the true composition of a sample could be calculated from its spectra by knowledge of the construction details of the instrument and the use of physical principles. In effect, no calibration is needed, and the calculation is independent of source strength, length of analysis, sample density, etc. This represents the most complete modeling of an X ray fluorescence analyzer to an accuracy of ±50%.

Ideally, the *T* factors would be independent of the type of standard employed to find them. Actually, variations are found. These are generally of second order, but because of this, both sample 8063- and sample 8064-generated *T* factors will be used in the final analyses. For the element sulfur a *T* factor determined by analysis of 8066 will be used, possibly supplemented by the ones found from extrapolation between the Si, K, and Ca *T* factors of 8063 and 8064. Uncertainties in the exact composition of the particular areas irradiated on the three standard rocks, which are known to be inhomogeneous, are problems. These uncertainties are currently being resolved by laboratory studies.

Sample simulation. The second technique for determination of the composition is the method of analogues. In this approach a series of samples are prepared to encompass a range of compositions bracketing that calculated by the technique outlined above. The samples are then analyzed by a flightlike instrument, and the spectra are compared to the spectra obtained on Mars. By making allowances for minor differences between the ground instrument and the Mars instrument, as determined by a comparison of prelaunch data and plaque and flag calibrations, the deviations in response are used to guide the construction of new analogues, and the process is repeated until a match is obtained that is well within statistical limits. Once a best match analogue has been achieved, new suites of samples in which each element is varied individually in a small range about its nominal value are analyzed to determine elemental concentration sensitivites. From this the absolute error of measurement for a sample of this matrix (mineralogic phases and particle size distributions) can be determined. However, the elemental composition cannot uniquely define the possible mineralogic assemblage [*Toulmin et al.*, 1977; *Baird et al.*, 1976], but other plausible alternatives must also be investigated. In addition to being used for direct spectral comparisons, the analogue is also used to augment the *T* factor method described above by estimating more accurate factors through comparison of the set derived from the analogue with the set(s) from the prelaunch standards. Preparation of analogue suites is underway [*Baird et al.*, 1977], and some preliminary results have been presented [*Baird et al.*, 1976].

DATA REDUCTION METHODOLOGY

The complexity in the reduction of XRFS data is chiefly due to the scanning mode of operation. To obtain adequate statistics, multiple fast scans are made over a period of several days, and the data are pooled only after each spectrum is individ-

TABLE 4. Accumulated Analysis Time

Sample	PC 1	PC 2	PC 3	PC 4
S1	3010	2030	340	1840
S2	370	250	31	280
S3	3780	2520	320	2835
S4	1350	1350	170	1520
S5	2580	1375	5750	2150
S6	3380	1550	8400	4400
U1	1840	800	660	1270
U2	1230	800	470	1510
U3	1905	1110	1010	2120
U4	4590	1290	4520	7860

The accumulated analysis time is the equivalent multichannel analysis time, in seconds per spectrum.

ually corrected for gain drift. If an electronic multichannel analyzer could have been included, the equivalent information would have required less than 10^{-2} the operating time. Stated another way, the data production and duration of XRFS operations on Mars have been sufficient for the analysis of hundreds of different samples. The equivalent multichannel analysis time for each sample is given in Table 4. A full discussion of the sequence of data processing, the corrections to be applied, and methods for estimating probable errors are given in Appendix 1.

INSTRUMENT PERFORMANCE ON MARS

At the time of this writing (March 31, 1977, sol 248 for Viking lander 1 and sol 203 for Lander 2), the units have accumulated over 5300 spectra and logged 1570 and 1167 operating hours, respectively, with no electronic or mechanical failures.

Since spacecraft constraints did not permit the incorporation of a multichannel analyzer, it was recognized long ago that the electronic circuitry had to be extremely stable so that none of the inherent scientific performance potential of this technique would be sacrificed. Control of stability was achieved by carefully designed circuits coupled with the aforementioned calibration aids capable of in situ operation on Mars. The measured temperature sensitivity of gain for three flight units over the range of 0°–100°F is shown in Figure 8.

Comparisons between postlanding flag and plaque spectra with the prelaunch spectra indicate a high degree of fidelity between performance on Mars and performance under laboratory conditions. However, studies of resolution and means of backscatter peaks indicate a significant change in operation characteristics of the PC 2 on lander 1 during the first few

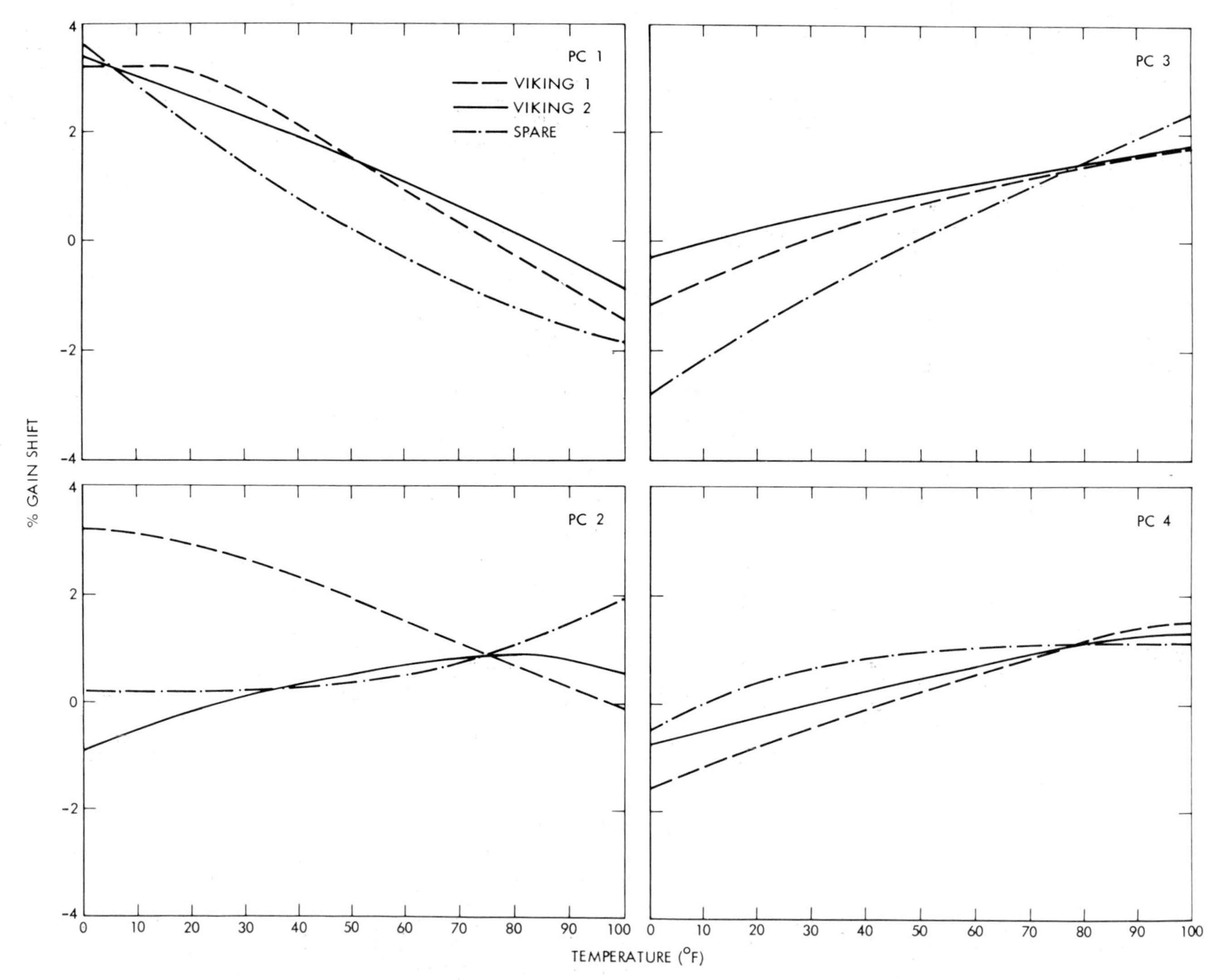

Fig. 8. Gain sensitivity as a function of temperature as measured on each flight unit during unit acceptance tests.

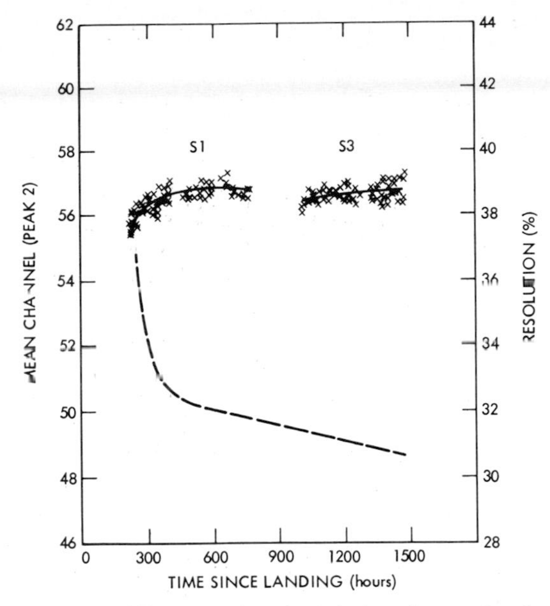

Fig. 9. Gain drift (crosses) and resolution changes for the PC 2 detector during the first 57 sols of mission 1. Resolution of this detector steadily improved during this time period. Gain stability is evaluated by plotting the mean channel of the Ca peak of samples S1 and S3. Most other detectors were more stable than this one.

weeks of the mission. As it is plotted in Figure 9, the gain of this detector increased and its resolution improved, possibly due to the release of a contaminant gas during the entry landing shock and thermal spike, with subsequent slow gettering of the contaminant. Other detectors exhibit stability of the same order or better.

Concern over gain stability had arisen during prelaunch check-out because of a gradual loss of ~3% gain by two of the proportional counters in the XRFS mounted on lander 2. This unit was removed shortly before lander sterilization and was replaced with the flight spare (FLT-4 unit). The removed instrument, FLT-1, was subjected to extensive tests, including heat sterilization, which showed unpredicted decreases in detector gains of up to 15%. Previous sterilization cycling of the PC detectors during manufacture and upon delivery and as part of unit acceptance tests showed negligible gain changes. Tests of residual PC spares verified the presence of a long-term (>6 months) aging effect, possibly due to migration of an absorbed gaseous component from regions in the epoxy not fully outgassed during fabrication into the volume of the bond line internal to the detector or a depolymerization product of the epoxy itself. Mass spectrometry of the fill gas of spare detectors revealed the presence of a gas contaminant with a mass of ~200 amu. No action was taken primarily because launch was imminent and secondarily because it was found that the gain drop was reversible with time. All proportional counters had acquired sensitivity to sterilization (see Figure 10) but recovered fully. Resolution performance was unaffected by sterilization; however, temperature sensitivity was altered radically. Tests with the FLT-1 unit have proved that this enhanced temperature sensitivity also disappears with time since sterilization (Figure 11). Calibration data from the two instruments landed on Mars also demonstrate that temperature response sensitivities have not been degraded permanently. However, some sensitivity remains, and it has been a continuing strategy to obtain most data during the most temperature stable periods, viz., at night. With the progression of the seasons on Mars, the diurnal temperature cycle has been

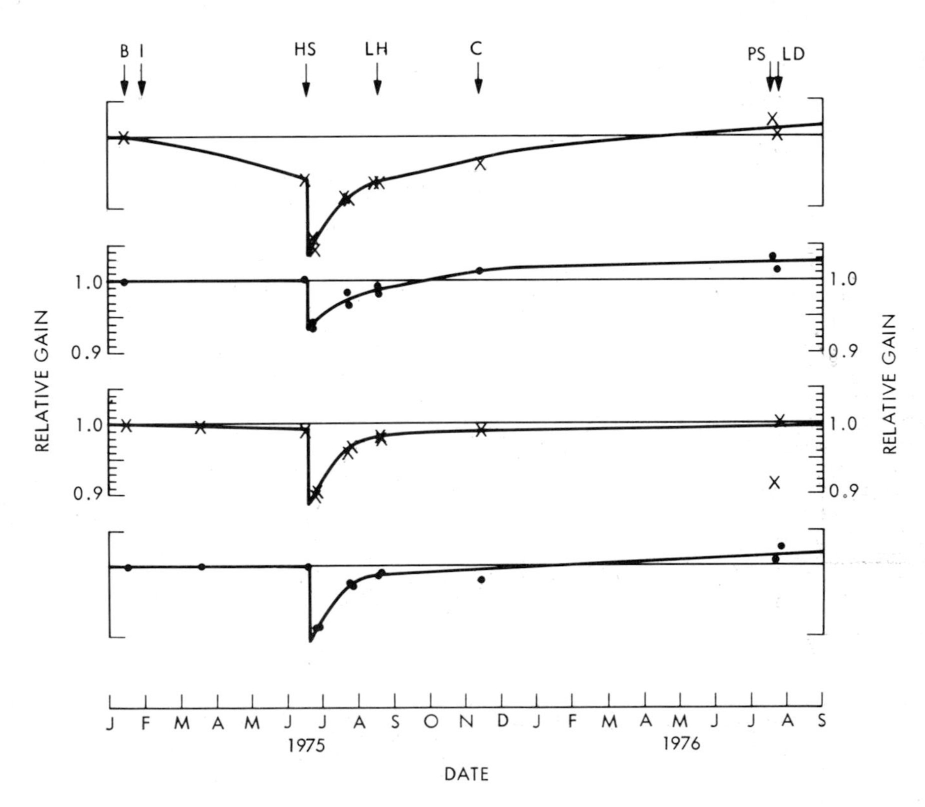

Fig. 10. Gains in the lander 1 XRFS unit detectors at times of bench tests (B), installation in the lander (I), before and after lander heat sterilization (HS), just prior to launch (LH), at cruise (C) and preseparation (PS) check-outs, and immediately after landing (LD). Recovery from the sterilization insult was apparently completed during the cruise phase of the mission.

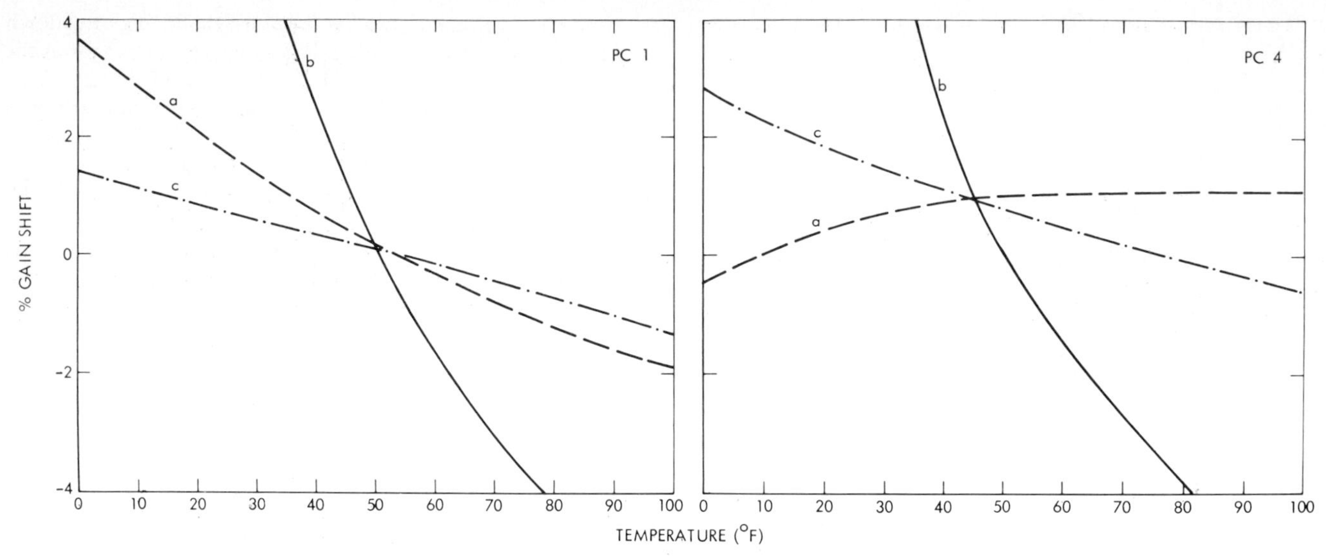

Fig. 11. Temperature sensitivity of the spare flight unit as measured in the laboratory at three different times: (*a*) during acceptance test, (*b*) after heat sterilization contemporaneous with lander 2 sterilization, and (*c*) 2 months prior to landing. Recovery of gain was accompanied by a return of gain temperature sensitivity to acceptable values.

modified to one of lower total excursion (Figure 12), and long-term spectral integrations will soon be possible at nearly constant temperature (±4°F), which will eliminate gain drift uncertainties due to temperature sensitivity. Also, because of recent reprograming of the spacecraft GCSC computer to allow more flexibility in commanding XRFS operations it has become possible to operate the instrument continuously for several days at a time, thereby allowing maximum high-voltage bias stability and amplifier gain stability.

RESULTS

A preliminary report [*Toulmin et al.*, 1976] described the first sample taken on Mars. Interim analytical numbers have also been given for the composition of samples at both landing sites [*Clark et al.*, 1976]. These values are given in Table 5 and plotted as histograms in Figure 13; the error bar for sulfur has been increased somewhat because of current uncertainties in the S content of the preflight standard. Refinement of the analyses is in progress but has not been completed because (1) a large fraction of the available man power and computer resources are currently allocated to the planning and execution of new sample acquisitions and analyses and because (2) more favorable instrument operating conditions are now possible, which will provide superior data from which to derive compositional information.

Some general conclusions (see *Clark et al.* [1976] for details) based upon the analyses made to date are given here.

1. From a compositional standpoint the fines (<2-mm size fraction) are remarkably similar at both lander sites in spite of a distance of 6500 km between the two sites (a separation of 178° in longitude and 25° in latitude).

2. The fines are composed predominantly of one or more silicate minerals (bulk SiO_2 = 45 ± 5 wt %) with a high content of iron (Fe_2O_3 equivalent to 19 ± 3%).

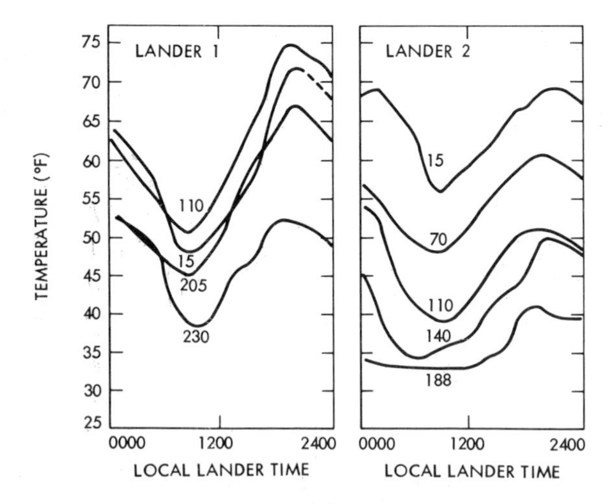

Fig. 12. Diurnal temperature variations within the XRFS unit (inside the lander thermally controlled compartment) for the sol after landing as numbered on each curve. As the seasons have progressed, the temperature excursions have decreased from 25°F to approximately 8°F, allowing more stable operation of the XRFS instrument.

TABLE 5. Elemental Concentrations in Martian Regolith Samples

	S1	S2	S3*	U1
Mg	5.0 ± 2.5		5.2	
Al	3.0 ± 0.9		2.9	
Si	20.9 ± 2.5	20.8	20.5	20.0
S	3.1 ± 0.5	3.8	3.8	2.6
Cl	0.7 ± 0.3	0.8	0.9	0.6
K	<0.25	<0.25	<0.25	<0.25
Ca	4.0 ± 0.8	3.8	4.0	3.6
Ti	0.51 ± 0.2	0.51	0.51	0.61
Fe	12.7 ± 2.0	12.6	13.1	14.2
L†	50.1 ± 4.3			
X‡	8.4 ± 7.8			
Rb	≤30 ppm			≤30 ppm
Sr	60 ± 30 ppm			100 ± 40 ppm
Y	70 ± 30 ppm			50 ± 30 ppm
Zr	≤30 ppm			30 ± 20 ppm

*All S3 concentrations have been adjusted by a preliminary correction factor of 0.9 to compensate for the effect of a partial fill on the analysis. The Fe is also affected by the contribution of ^{55}Fe to the peak. Trace elements analyze low.

†*L* is the sum of all elements not directly determined.

‡If the detected elements are all present as their common oxides (Cl excepted), then *X* is the sum of components not directly detected, including H_2O, Na_2O, CO_2, and NO_x.

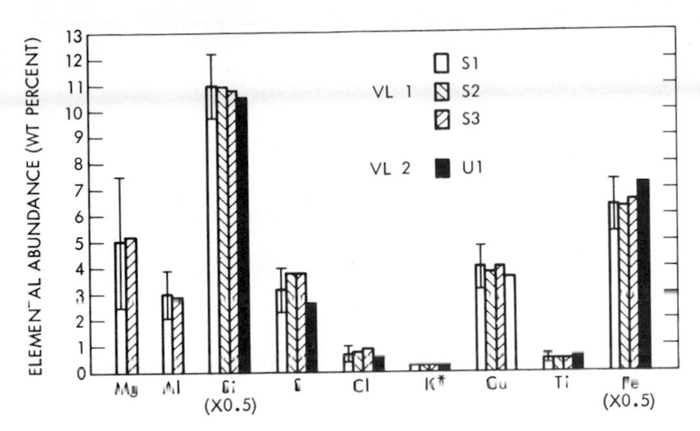

Fig. 13. Element concentrations for analyses completed to date. Error bars on the first section of each element apply to absolute concentration. The relative concentrations measured by the same instrument are considerably more accurate; for example, the sulfur contents of duricrust samples S2 and S3 are definitely higher than the sulfur in the fines sample, S1. Only the detection limit is shown for K.

3. Sulfur content is unexpectedly high, approximately 100 times higher than the average for the earth's crust. Samples that may contain fragments of a duricrust (hardpan) observed by the lander camera have higher sulfur contents than the loose fines and may be soil cemented by a magnesium sulfate salt comprising approximately 10% of the total sample.

4. Potassium content is less than 0.25%, at least 5 times lower than the average content for the earth's crust. If this content is also representative of the Mars average crustal potassium content, then the low content of the ^{40}Ar observed in the Martian atmosphere may reflect the dearth of source material (^{40}K) rather than a lower degree of planetary outgassing on Mars compared to the earth.

5. If an iron oxide coating exists over all mineral grains (as is proposed by some), then such a coating must be extremely thin (<0.25 μm) or highly discontinuous.

6. The bulk density of fines after delivery to the XRFS is 1.10 ± 0.15 g/cm³, implying a bulk soil porosity of $60 \pm 15\%$.

Detailed interpretation of the mineralogic and petrologic implications of these data regarding the geological evolution of the surface of Mars are presented elsewhere [*Baird et al.*, 1976; *Toulmin et al.*, 1977]. A total of 10 samples have been acquired to date. Descriptions of their locations and the manner in which they were delivered to the XRFS instruments are given by *Baird et al.* [1977]. Representative spectra from PC 2 for different samples on lander 1 are presented in Figure 14. Although gross compositions are remarkably similar, important differences have been detected directly from the spectra without the benefit of detailed data reduction. For example, the implied duricrust sample S5 taken on lander 1 at the 'Atlantic City' site contains higher sulfur than duricrust taken at 'Rocky Flats' (sample S3). Sulfur differences between implied duricrust and fines do not correlate well with iron and correlate negatively with calcium.

At lander 2, although duricrust is apparently more abundant, none has been acquired and analyzed successfully. However, a sample of fines (U2) taken from beneath 'Notch' rock contains approximately 10% less iron than fine material not protected from eolian modification, solar illumination, etc. (sample U1 from 'Bonneville' site and sample U3 collected from a previously dug trench). Furthermore, a sample taken beneath 'Badger' rock (sample U4) verified this trend; it contained 18% less iron than the uncovered fines. The PC 4 spectra, from which these differences were noted, are shown in Figure 15. Each of the spectra has been normalized to the backscatter peak of sample U1. Although the differences in Fe peak height appear subtle because of the logarithmic count

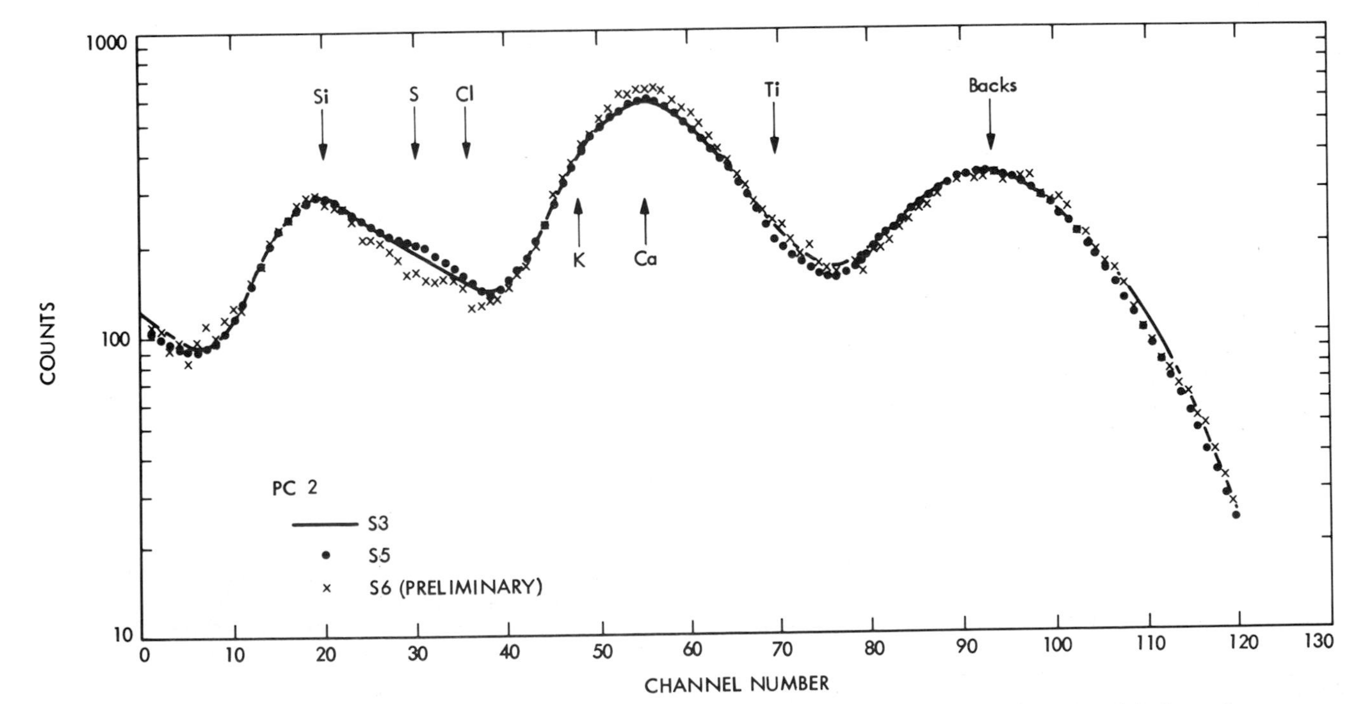

Fig. 14. Comparison of PC 2 spectra of samples taken at lander 1 site. S3 and S5 are both samples of duricrust, but sample S5 contains more sulfur. In contrast, the S6 (deep hole) sample of fines is significantly lower in sulfur and higher in calcium than crustal material.

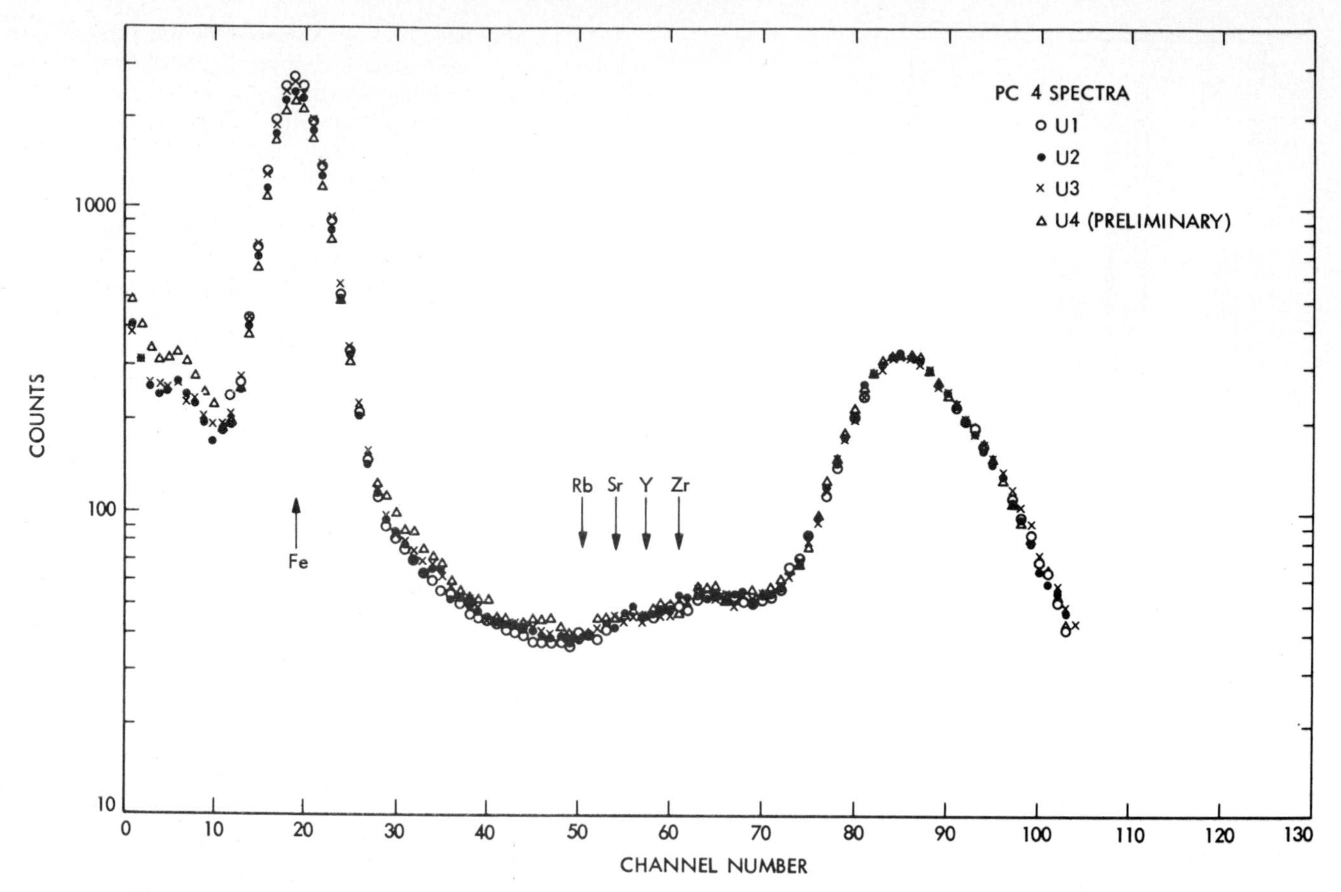

Fig. 15. Comparison of PC 4 spectra of samples taken at lander 2 site. The differences in Fe, although appearing small because of the logarithmic format, are significant and real (see text for discussion). Samples taken under rocks contain less Fe.

scale, they are far greater than the inherent limitations due to counting statistics, for which the standard deviations of the integrated peaks is less than $\pm 0.5\%$.

An important aspect of future analysis will be the determination of whether other elements vary sympathetically or antipathetically with these changes in iron content.

Appendix 1: Data Processing and Reduction

The sequence for data processing, the corrections to be applied, and methods for estimating probable errors are discussed in this appendix, and a schematic version of the sequence in which data reduction proceeds is given in Appendix 2.

First-order processing includes the decoding of the data frame identifiers and data words, error detection and correction, command list correlation and search, and identification of redundant data. From printouts of the decoded and tagged frames, spectra are constructed under interactive control of retrieval and splice routines. Upon completion of first-order processing, complete spectra are available from which instrument performance can be monitored and quick look estimates of sample composition can be made. Because these results are critical to uplink planning, this process always receives the highest priority and is performed as soon as the data become available.

A strategy for optimizing available counting time has been to use long counting times over portions of each spectrum for which statistical precision is the limiting error, e.g., the trace element region for PC 4, pooled with fast scans over the rest of the spectrum. Second-order processing includes the pooling of these data to make complete spectra normalized to the standard count period of 30.7 s/channel. Also included is verification by inspection of each spectrum and transfer of accepted spectra to the specialized files that contain all the data taken by a given PC detector for a given sample.

At the third-order level the mean channel and resolution are calculated for each significant peak in each spectrum. This includes Al, Ca, Ag, Zn, and backscatter peaks for calibration spectra, selected elements, such as Fe, K-Ca, and Mg-Al, and backscatter peaks for spectra from each sample. Resolution is calculated as $2.35\sigma/(\bar{X} + X_0)$, where $\bar{X}$ is the mean channel, X_0 is the instrument offset (in channels), and σ is the standard deviation of a least squares fit of a Gaussian curve to the peak taken between the limits specified by the ICAN program operator. The algorithm for calculating the mean (i.e., the first moment of the spectra counts) of each peak uses these same limits. A refinement for such calculations has been devised to solve two problems frequently encountered: (1) selection by the operator of channel limits that do not symmetrically bracket the peak and (2) peaks only a few channels in width, no symmetrical channels being available. In either case the calculated mean depends upon which limits are selected. The refinement is, once limits are specified, that the software subroutine automatically calculates additional means for both limits moved upscale by one, two, and three channels and downscale by the same amounts. For each of the seven means so generated a difference variable is assigned a value which is the mean channel minus the midpoint between the two limits. For perfectly symmetrical limits this difference variable is obviously zero. The algorithm therefore scans the difference

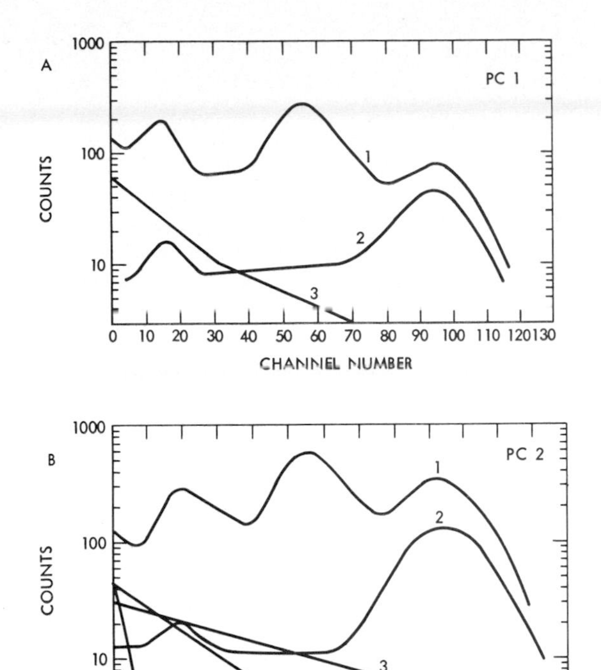

Fig. 16. Component contributions to observed spectra for (*a*) PC 1 and (*b*) PC 2. In each figure the components are (1) total spectrum, (2) stray radiation scattered from sample cavity structure, (3) RTG background radiation, (4) cross talk, and (5) count rate dependent noise.

value to locate the plus to minus transition and then performs a linear interpolation to find the zero-crossing point, which is then taken as the best estimator of the true mean. This algorithm has great utility in that it finds the mean even when poor limits are chosen, and it also detects whether a peak is indeed located within the interval selected and sets a message flag to the operator when no peak (no zero crossing) can be found.

Fourth-order processing comprises studies of instrument gain changes as a function of operating time and temperature. This includes plots of peak means, inspection for trend patterns, separation into segments (where appropriate) of data exhibiting the same trend, and multivariate regression analyses. For the analyses, linear and quadratic terms are allowed for both temperature and time after landing. No attempt has been made to model drifts associated with more complex factors, such as time from instrument turn on, duty cycle, etc. In the cases studied to date, corrections of gain by the regression equation result in a standard deviation of gain that is improved by up to a factor of 2 over the uncorrected data but is still higher than the component of standard deviation due to the statistical limitations of the data alone. Hence unmodeled factors remain, and it is for this reason that continuous operation of the instrument for several days at a time has been planned for later portions of the extended mission.

Under fifth-order processing, corrected and pooled spectra from the previous step are compared in the context of the total mission. Tests for differences in composition between samples and changes in background levels are made by subtracting spectra after adjusting them to the same scale by normalizing them in relation to the backscatter peak.

Sixth-order processing is relatively elaborate because of the numerous calibration and background corrections that must be made. Final derivation of the best energy scale for each spectrum is accomplished by study of the plaque and flag calibration spectra, as well as by the internal standardization method that uses the backscatter peak as reference and the offset/gain ratio R determined from prelaunch calibration data. In practice, this results in several sets of A_{cal}, the energy offset for channel zero, and B_{cal}, the energy increment per channel. Typically, $A_{cal} \sim 0.6$ keV for ^{55}Fe spectra, and $A_{cal} \sim 2.3$ keV for ^{109}Cd; and $B_{cal} \sim 57$ eV/channel for ^{55}Fe spectra, and $B_{cal} \sim 217$ eV/channel for ^{109}Cd. From pairs of these values, as generated by different in situ calibration techniques, different corrected spectra can be generated, resulting in several estimates of composition, some elements being affected much more than others. This spread in results is used to estimate the probable error of the estimate considered most reliable.

After adjustment of the spectra for gain and offset, sources of background must be removed. The most important background source is the leakage gamma radiation from the two radioisotope thermoelectric generators (RTG) used to provide primary electrical power to the landers. Each RTG contains quantities of ^{238}Pu, responsible for much of the gamma emission, accompanied by trace quantities of ^{236}Pu ($\sim$1 ppm), whose decay chain includes ^{208}Tl, a short half-life energetic gamma emitter that figures prominently in the total emission spectrum. Early in the XRFS design phase a number of sensitivity tests were performed [*Clark et al.*, 1973] using available RTG's and prototype PC detectors. No peaks were detected because the sensitive volume of the detectors is much too small for total absorption of the Compton electrons and photoelectrons produced in the counter gas and walls; rather, a continuum spectrum monotonically decreasing in energy is observed. This background is small in comparison with the sample and calibration plaque signals and therefore is difficult to determine in situ on Mars. Because it is only important at low energies, its interference is serious only for the peaks from Mg and Al. For these two elements the ratio of fluorescence signal to RTG background is approximately 1:1, whereas for Si it is >10:1, and for Ca and Fe it is >100:1. The relative contribution of the RTG radiation to the total level can be seen in Figures 16*a* and 16*b*, which plot the current estimate of RTG interference based on the sol 1 calibration data and the numerous ground-based interference tests. As the mission progresses, our ability to measure this background improves because radioisotope decay lowers the calibration plaque peak heights while the RTG level increases due to growth of the ^{208}Tl daughter product. Once improved data on the RTG level and spectral shape are obtained it will be possible to extrapolate backward in time by use of an independent monitor of the growth of the RTG gamma emissions. This monitor is the ^{14}C detector system of the pyrolytic release (PR) experiment in the integrated biology instrument. Data from this experiment have a low signal to background ratio, necessitating long controlled counts of the background alone. In Figure 17 are plotted the PR background data (G. L. Hobby and F. S. Brown, private communication, 1977) taken to date. In comparison with the RTG background at the beginning of the mission, sol 0, the RTG background on lander 1 had increased by approximately 11% by sol 200 and is projected to increase 22% by sol 400. In contrast, ^{55}Fe source levels at sols 200 and 400 are 86% and 74% of their strength at landing, respectively; ^{109}Cd strengths are 73% and 54%, respectively.

Additional sources of background are the natural decay of thorium in the Mg-Th alloy used for nearly all instrument

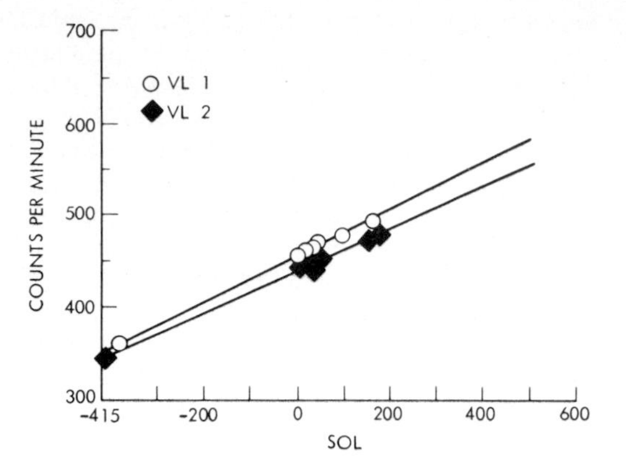

Fig. 17. Increase in RTG background radiation levels since time of installation on the Viking lander (VL) spacecraft as determined by background counts with the ^{14}C detectors in the biology instrument pyrolytic release experiment (after G. L. Hobby and F. S. Brown, unpublished data, 1977).

housings. This source is weaker than the RTG's by at least a factor of 10 [*Clark et al.*, 1973]. Likewise, the effects of cosmic rays and the natural radioactivity of the soil are negligible.

Another type of background correction is for stray radiation resulting from scattering of ^{55}Fe source radiation from the structure of the sample cavity. This correction does not apply to ^{109}Cd because it is tightly collimated so that no emitted radiation intersects the cavity. However, because of the RTG background the ^{55}Fe collimator was opened to maximize irradiation of the sample. As a measure to minimize this type of interference the sample cavity was plated with 3.7-μm Ni to absorb Al fluorescence emissions. Determination of the precise amount of this stray radiation has proved to be difficult because even with an empty cavity, spectral contributions occur from the circular plaques and the interior housing walls. After evaluating several methods the most satisfactory mode was found to be the comparison of the calculated spectrum for pure calcium carbonate, for which the backscatter peak is extremely low due to competitive absorption of ^{55}Fe by the high-calcium photoelectric cross section, the spectrum being measured from Solenhofen limestone. After normalizing to the calcium peak, the difference in the backscatter peak is ascribed to the stray component. These studies, combined with plaques-out measurements, show that 90% of the backscatter peak of the PC 1 calibration plaque spectrum is due to this stray component from the front surface of the soil cavity. This guarantees that residual soil on the plaque cannot significantly affect the mean channel for this peak as it can for the Al and Ag fluorescence channels. The corresponding value for PC 2 is 98%. The stray spectra plotted in Figure 16 each show a lower energy peak. For PC 1 it is due to fluorescence of the aluminum entrance window of the detector by the scattered photons. For PC 2 it is the escape peak from Xe in the counter gas. (Note that the ^{55}Fe-scattered radiation produces this escape peak; fluorescent emissions of interest all have energies below the Xe *L* absorption edges and do not produce Xe escape peaks.) Another type of background arises from electronic effects. These may be categorized as preamplifier noise, cross talk, count rate dependent noise, and external pick up. The first of these does not produce a significant number of counts in any of the 128 channels of interest, as is proven by special sequences run on both landers where the PC bias was reduced to 715 V, and essentially zero output was obtained. All preamplifier noise counts are contained within the ~11 channels offset from true zero pulse height. It is known, however, that the resolution of peaks is affected, in part, by preamplifier noise.

Cross talk is important only for PC 2. Laboratory measurements show it is primarily a coupling of the Fe peak signal from PC 3 and PC 4 into the PC 2 preamplifier. However, it is a minor contribution to the background (Figure 16*b*).

Power supply ripple arises from ac components of the high-voltage bias supply and the low-voltage dc-dc convertor. When the PC tubes are not processing signals (i.e., when sources are removed), this ripple is undetectable. However, during operation at higher than normal voltages a noise spectrum in the very low channels becomes apparent. An additional count rate dependent noise component occurs for PC 2. Although its cause is still under study, it is a calibratable component that can be subtracted once the total counting rate is determined (see Table 6).

External pick up is defined as counts occurring as a result of operation of equipment other than the XRFS. No such interference has occurred on lander 1 nor was it ever observed on any of the test landers. However, the data after touchdown of lander 2 showed a high noise component in low-energy channels (Figure 18). Correlation of these spurious counts to the lander event profile showed that the seismometer was in its high data rate mode at the time. Subsequent analysis of data, from a diagnostic 'quiet period' test performed on sol 27, showed that any activity by the spacecraft computer (GCSC) produced noise in the lower 24 channels of the XRFS. Circuit analysis combined with tests performed on flightlike XRFS units showed that (1) no single component failure could cause the extra counts to be different in each PC section, (2) increased power noise currents in the lander equipment plate could induce counts into the PC tube/preamplifier circuits, and (3) impedance changes in XRFS filter components would create very high count rates that are unlike the data obtained on lander 2.

It is concluded that the XRFS unit on lander 2 is performing normally. It is more likely that a drastic change in one of the isolation impedances of a lander equipment unit is causing an abnormally high current to flow in the equipment plate when the GCSC is powered up or powered down. This high noise current is being sensed by the XRFS preamplifier and is counted in the lower-energy channels of the spectra. The effect of these extra counts has been minimized by increasing the gain of the PC tube effectively to place significant peaks above the lower channels.

In the case of PC 1 this requires runs at two different gain settings because it is not possible to keep all peaks simultane-

TABLE 6. Detector Counting Rates

	Rate, counts/s			
	PC 1	PC 2	PC 3	PC 4
	Lander 1			
Plaque calibration (sol 1)	200	270	1170	600
Flag calibration (sol 25)	1950	5470		
Sample S1	430	1320	1010	1070
Sample S6	390	1140	700	660
	Lander 2			
Plaque calibration (sol 1)	260	310	960	500
Flag calibration (sol 61)	2480	6940		
Sample U1	500	1310	1040	980
Sample U4	450	1150	740	660

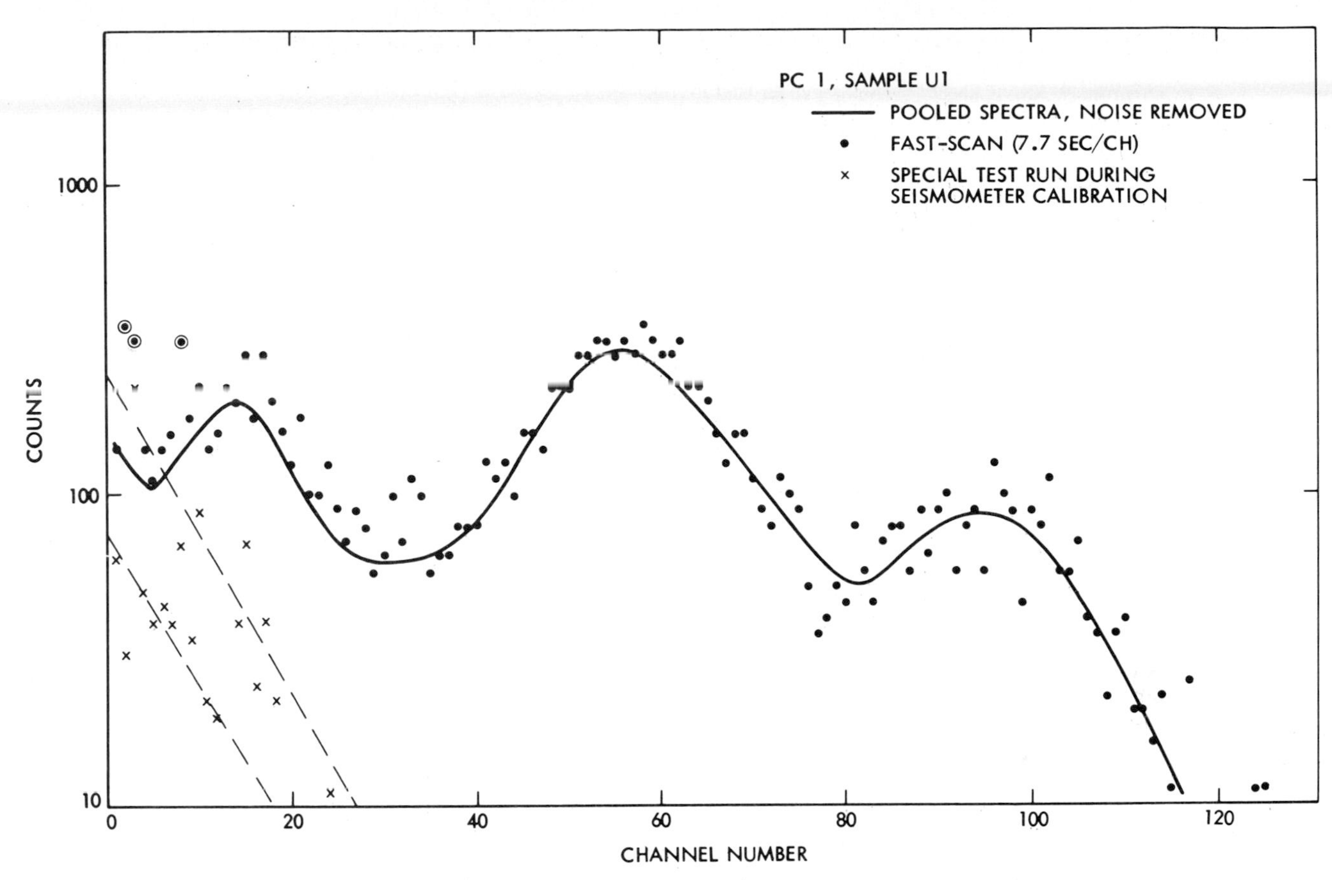

Fig. 18. Examples of low-channel electronic pick up noise experienced on lander 2. The crosses are noise counts taken with PC 1 bias lowered to the point that no signal should be present. This test was performed during a period when the seismometer instrument was particularly active in data transfer. The solid line is a noise free spectrum; the spectrum is a typical example. The three circled points are obviously affected by these noise bursts. The dashed lines indicate the two general trends of the apparently bimodal noise distribution.

ously on the remaining usable scale. Use of the flag calibrator at both gain settings provides the required correlation factors between scales.

After subtraction of the various sources of background from the gain-adjusted spectra a spectral set is formed which is composed of one corrected spectrum for each of the four PC tubes. Under seventh-order processing the element composition of the sample is calculated from the spectral set. Before this can be done, however, the T factor calibration data must be available; they are therefore considered part of this level of processing. Estimates of the absolute accuracy of the analyses are made by performing calculations using the different T factors determined for the three calibration rocks and by using different sets for which the gain parameters (A_{cal} and B_{cal}) and background corrections are varied over their range of probable errors. Estimates of instrument precision are made by analyses of successively taken spectra corrected in the same manner. For an experiment of this type, errors in precision are much less than errors in absolute concentrations of the elements.

Under eighth-order processing are included all the data handling necessary to compare laboratory analogue samples run in the flight spare unit with the spectra obtained on Mars.

APPENDIX 2: DATA PROCESSING HIERARCHY

First order — decode data frames and command word tags, print decoded data, correlate frames with commanded sequence, and assemble frames into complete spectra.

Second order — correct bad data points, missing frames, etc., certify and tabulate spectra, pool partial spectra, transfer spectra to specialized files, and verify.

Third order — calculate peak means and resolutions, tag time and temperature, and edit peak mean files.

Fourth order — plot and drift fit for gain and offset, find correction algorithms, construct composite corrected spectra, and splice lander 2 PC 1 spectra.

Fifth order — construct master spectral summary and compare for composition differences, background changes, etc.

Sixth order — derive best gain and offset factors, determine RTG, noise, cross talk, and other corrections, and construct spectral sets.

Seventh order — determine composition of three standard rocks, adjust T factors for three standard rocks, and run analyses for combinations: sets and T factors.

Eighth order — fabricate analogues, analyze chemically, characterize flightlike unit, compare analogue spectra with Mars spectra, and modify and rerun analogues.

Acknowledgments. Without the advice and early encouragement of those who perceived the potential importance of this experiment, development of the XRFS concept would not have proceeded to a state of readiness for the Viking mission. In chronological order, they are G. W. Wetherill, S. E. Dwornik, E. C. Morris, W. W. Bender, F. Cuttitta, G. A. Soffen, and G. J. Wasserburg. We are particularly indebted to S. E. Dwornik, NASA Chief of Planetology, for continuing financial support and to L. C. Acton and R. E. Varney of Lockheed Research Laboratories for sharing their unpublished data. The flawless operation of the two XRFS units on Mars is directly attributable to over 250 engineers and technicians of Martin Marietta Aerospace Corporation and its subcontractors, who contributed to the fabrication, development, and test of the instruments. The work was supported by NASA grants NAS 1-11855, NAS 1-11858, L-9717, and NAS 1-9000.

REFERENCES

Baird, A. K., and B. C. Clark, Geochemical investigation of the surface composition of Mars, report, 54 pp., NASA Headquarters, Washington, D. C., 1969.

Baird, A. K., P. Toulmin III, B. C. Clark, H. J. Rose, K. Keil, R. P. Christian, and J. L. Gooding, Mineralogic and petrologic implications of Viking geochemical results from Mars: Interim report, *Science*, *194*, 1288–1293, 1976.

Baird, A. K., A. J. Castro, B. C. Clark, P. Toulmin III, H. J. Rose, K. Keil, and J. L. Gooding, The Viking X ray fluorescence experiment: Sampling strategies and laboratory simulations, *J. Geophys. Res.*, *82*, this issue, 1977.

Bambynek W., B. Crasemann, R. W. Fink, H. U. Freund, H. Mark, C. D. Swift, R. E. Price, and P. V. Rao, X-ray fluorescence yields, auger, and Coster-Kronig transition probabilities, *Rev. Mod. Phys.*, *44*, 716–813, 1972.

Blizard, R. B., Error correction with minimum hardware, *New Technol. Disclosure Rep. 492,* Martin Marietta Corp., Denver, Colo., 1974.

Clark, B. C., X-ray cross-sections in design and analysis of non-dispersive systems, *Advan. X Ray Anal.*, *17*, 258–268, 1974.

Clark, B. C., and A. K. Baird, Ultraminiature X-ray fluorescence spectrometer for in situ geochemical analysis on Mars, *Earth Planet. Sci. Lett.*, *19*, 359–368, 1973*a*.

Clark, B. C., and A. K. Baird, Martian regolith X-ray analyzer: Test results of geochemical performance, *Geology*, *1*, 15–18, 1973*b*.

Clark, B. C., and A. K. Baird, Martian regolith X-ray analyzer: Reply, *Geology*, *2*, 23–24, 1974.

Clark, B. C., P. G. Kase, J. P. Martin, and J. G. Morse, Use of nuclear data in designing space science experiments, in *Nuclear Data in Science and Technology*, vol. 1, pp. 577–593, Int. At. Energy Agency, Vienna, 1973.

Clark, B. C., A. K. Baird, H. J. Rose, P. Toulmin, K. Keil, A. J. Castro, W. C. Kelliher, C. D. Rowe, and P. H. Evans, Inorganic analyses of Martian surface samples at the Viking landing sites, *Science*, *194*, 1283–1288, 1976.

Clark, B. C., A. K. Baird, P. Toulmin, H. J. Rose, and K. Keil, X-ray fluorescence geochemical analysis on the surface of Mars, in *Nuclear Methods in Minerals Exploration*, edited by J. G. Morse, Elsevier, New York, in press, 1977.

Criss, J. W., and L. S. Birks, Calculation methods for fluorescent X-ray spectrometry, *Anal. Chem.*, *40*, 1080–1086, 1968.

Malter, L., Thin film field emission, *Phys. Rev.*, *49*, 48–58, 1936.

Shiraiwa, T., and N. Fujino, Theoretical calculations of fluorescent X-ray intensities in fluorescent X-ray spectro-chemical analysis, *Jap. J. Appl. Phys.*, *5*, 886–897, 1966.

Toulmin, P., III, A. K. Baird, B. C. Clark, K. Keil, and H. J. Rose, Inorganic chemical investigation by X-ray fluorescence analysis: The Viking Mars lander, *Icarus*, *20*, 153–178, 1973.

Toulmin, P., III, B. C. Clark, A. K. Baird, K. Keil, and H. J. Rose, Preliminary results from the Viking X-ray fluorescence experiment: The first sample from Chryse Planitia, Mars, *Science*, *194*, 81–83, 1976.

Toulmin, P., III, A. K. Baird, B. C. Clark, K. Keil, H. J. Rose, R. P. Christian, and P. H. Evans, Geochemical and mineralogical interpretation of the Viking inorganic chemical results, *J. Geophys. Res.*, *82*, this issue, 1977.

Trombka, J. I., and R. L. Schmadebeck, A numerical least-square method for resolving complex pulse-height spectra, *NASA Spec. Rep. 3044*, 1968.

Veigele, W. J., E. Briggs, L. Bates, E. M. Henry, and B. Bracewell, X-ray cross-section compilation from 0.1 keV to 1 MeV, *Def. Nucl. Agency Rep. DNA-2433F, I, II,* 1st rev., Dep. of Defense, Washington, D. C., 1971.

(Received May 1, 1977;
revised May 23, 1977;
accepted May 24, 1977.)

VOL. 82, NO. 28 JOURNAL OF GEOPHYSICAL RESEARCH SEPTEMBER 30, 1977

The Viking X Ray Fluorescence Experiment: Sampling Strategies and Laboratory Simulations

A. K. BAIRD,[1] A. J. CASTRO,[2] B. C. CLARK,[2] P. TOULMIN III,[3] H. ROSE, JR.,[3] K. KEIL,[4] AND J. L. GOODING[4]

Ten samples of Mars regolith material (six on Viking Lander 1 and four on Viking Lander 2) have been delivered to the X ray fluorescence spectrometers as of March 31, 1977. An additional six samples at least are planned for acquisition in the remaining Extended Mission (to January 1979) for each lander. All samples acquired are Martian fines from the near surface (<6-cm depth) of the landing sites except the latest on Viking Lander 1, which is fine material from the bottom of a trench dug to a depth of 25 cm. Several attempts on each lander to acquire fresh rock material (in pebble sizes) for analysis have yielded only cemented surface crustal material (duricrust). Laboratory simulation and experimentation are required both for mission planning of sampling and for interpretation of data returned from Mars. This paper is concerned with the rationale for sample site selections, surface sampler operations, and the supportive laboratory studies needed to interpret X ray results from Mars.

INTRODUCTION

This paper describes earth-based activities in support of the Inorganic Chemical Analysis Team's Viking Mission operations on Mars. Included are (1) rationales for selecting sampling sites and methods of surface sampler operations, (2) plans for operations during the Extended Mission period, (3) laboratory studies to permit rapid assessments of instrument response on Mars and to generate spectral libraries from terrestrial materials, and (4) development and documentation of terrestrial analog materials (based upon the current level of accuracy of analytical results from Mars) to refine Mars chemical results further and to support various other laboratory tests (e.g., ultraviolet radiation and reflectance studies).

MARS SAMPLING STRATEGIES

Constraints and Supports

In order to evaluate properly the science data return from the X ray experiment on Mars as a function of samples actually delivered to the instruments an understanding of the overall operational constraints is needed. The constraints result from (1) the X ray instruments themselves, (2) the capability of the surface sampler to provide samples of different types, and (3) the mission objectives, priorities, and requirements. These are considered in turn.

1. The X ray fluorescence spectrometer (XRFS) instruments were designed to receive (via a funnel on top of the lander) untreated material acquired by the surface sampler. The material flows by gravity down a delivery tube (2.5 cm in diameter) through the lander insulation section and into the XRFS measurement cavity. Approximately 25 cm³ of material is required to fill this cavity to the top of the detection windows, an order of magnitude more material than is needed for either the gas chromatograph/mass spectrometer (GCMS) or biology instruments. The XRFS sample cavity (2.5 × 2.5 × 3.8 cm) is designed to accommodate not only fine material but also larger particles ('pebbles'). Because the round delivery tube could be bridged and/or blocked by particles of sizes larger than one-half the diameter of the tube, however, the top of the funnel is covered with a convex 1.2-cm-square wire mesh. To prepare the sample cavity for another sample, material in the cavity is dumped through a vibrating trapdoor located at the bottom of the sample cavity and into a dump chamber 600 cm³ in volume. This capacity, plus one additional cavity fill, is the maximum amount of material that the XRFS can analyze (Figure 1). These limitations on sample particle size and volume were dictated by engineering constraints of weight and volume which were exaggerated by the late decision to add the X ray experiment to the Viking program. A deployable X ray excitation/detection assembly to cover larger areas of the surface and to analyze large rocks would have been a highly desirable addition to the experiment. In addition

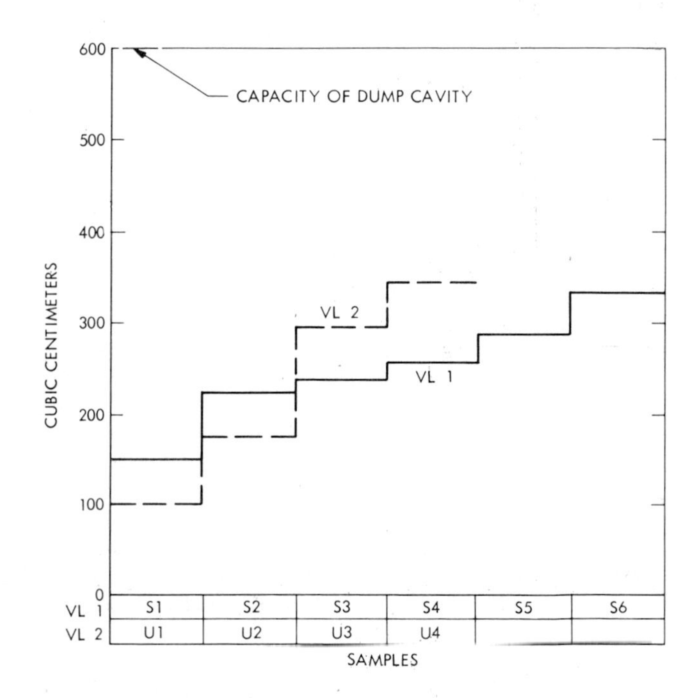

Fig. 1. Progressive filling of dump cavities in X ray fluorescence spectrometers on Viking Lander 1 and 2. S and U refer to samples, as discussed in text.

---

[1] Department of Geology, Pomona College, Claremont, California 91711.
[2] Planetary Sciences Laboratory, Martin Marietta Corporation, Denver, Colorado 80201.
[3] U.S. Geological Survey, Reston, Virginia 22092.
[4] Institute of Meteoritics and Department of Geology, University of New Mexico, Albuquerque, New Mexico 87131.

Paper number 7S0451.

to or in place of such an assembly, means to pretreat samples (e.g., crusher, grinder, fractionator) would have been valuable.

2. The surface sampler, consisting of a collector head assembly (Figure 2) on the end of a retractable boom, can reach an area of approximately 15 m^2 in front of the lander, from 1.5 to 3 m outboard (Figures 3 and 4). The collector head has a jaw opening of about 3.1 × 4.2 cm and an effective maximum capacity of approximately 100 cm^3. Material can be delivered to the XRFS in three modes: (1) directly from the collector head by vibrating the jaw, (2) by inverting the collector head and sieving the material through 2-mm openings in the 'roof' of the collector head, and (3) by sieving away from the lander, reinverting the collector head, and delivering, as is done in mode 1 above, the residue of particles greater than 2 mm in size. Therefore mode 1 should result in an XRFS cavity fill of all particles in the sampler below 12 mm in size, mode 2 in all particles below 2 mm in size, and mode 3 in all particles between 2 and 12 mm in size. For purposes of later discussion, mode 2 is referred to as a 'fines delivery,' and mode 3 as a 'rock delivery.' These modes have been used in both successful and unsuccessful attempts to acquire samples during the Primary Mission, and mode 1 is planned for use during the Extended Mission (see discussion below).

By means of a full-scale test model lander with flightlike sampler boom a variety of studies were conducted to test the capability to acquire and deliver various types of samples to the XRFS. Materials such as sand, sand-pebble mixtures, and synthetic lunar soils were used with varying ground slopes and elevations (with respect to the lander), thus requiring different boom extensions and angles of attack of the sampler into the surface. These tests demonstrated that successful acquisitions (more than 25 cm^3 of material) principally require accurate imaging and analysis of ground slopes in the sampling area. When the surface slopes away from the lander, the jaw tends to skim over the surface. Conversely, when the surface slopes toward the lander, the jaw bites excessively into the ground, and careful selection of boom extension is required to avoid possible collector head or boom damage. Slope and elevation information on Mars can be obtained from stereo imaging provided by the two lander cameras. For the safest and most reliable sampler operation, sampling is generally restricted to areas of the total reachable field exposed to the view of both cameras. Tests with synthetic lunar soil (simulating lunar size distributions from submicron to small pebble) proved that the Viking surface sampler is fully capable of providing an adequate sample to the XRFS in all three delivery modes, provided that the ground surface meets slope requirements and that the sampled material has enough cohesion to minimize flow out of the sampler during retraction of the boom. For example, the collector head is capable of retaining only 15–20 cm^3 of dry, coarse sand acquired at boom depression angles practical in Mars operation. The cohesion of the Martian regolith material is far greater than that of dry sand and has not proved to be a constraint in landed operation.

3. Viking mission objectives, priorities, and requirements placed severe constraints upon the operation of the surface sampler for acquisitions of samples for the X ray experiments during the early part of the Primary Mission. These constraints were the result of placing a higher priority on samples for biologic and organic analyses, considerations of sampler boom safety and longevity of operation, and actual anomalies during sampler operations. The last constraint caused minimal delays in planned operations (because of the dedicated efforts of the Surface Sampler Team) but served to reinforce concerns about the useful life of the sampler. In the latter parts of the Primary Mission and in the Extended Mission operations of both landers, however, use of the sampler for XRFS acquisitions was increased significantly. The total number of samples delivered (as of March 31, 1977) to XRFS instruments is 10, a number which makes the Viking X ray experiment a complete success by any measure. In addition, the surface sampler has been used for 20 other attempts to acquire rock material (see discussion below). For a detailed discussion of all Primary Mission surface sampler operations, see the report by *L. V. Clark et al.* [1977].

Because of the relatively low power requirements and relatively low data output of the XRFS instruments the mission placed few constraints on instrument operation time. This was especially fortunate in view of the long counting durations (up to 10 hours per analysis per day) required by the stepwise single-channel mode of operation of the XRFS [*B. C. Clark et al.*, 1977].

Geochemical Sampling Considerations of Landing Sites

The goal of the Inorganic Chemical Analysis Team is to characterize chemically as many 'different' samples of Martian materials as can possibly be acquired and delivered to the XRFS instruments. 'Different' refers to any characteristic of apparent physical nature, such as grain size, albedo, color, and degree of induration and also refers to position (surface material, material at depth, material shielded from ultraviolet radiation), which might influence chemical composition. Because of the constraints outlined above, our goal was met only partially in the Primary Mission; more than twice the Primary Mission activity has been (or is planned to be) done during the Extended Mission. Even with this activity, geochemically exciting sites in much of the area around each lander (Figures 3 and 4) are inaccessible to chemical analysis by Viking because they are beyond the reach of the sampler.

This section makes no attempt to reiterate the detailed descriptions of the local geology of the lander sites [see *Mutch et al.*, 1976*a*, *b*, *c*; *Shorthill et al.*, 1976*a*, *b*, *c*] but rather defines the practical sampling goals for the X ray experiments. At both landers, large blocks of generally igneous-appearing rocks are scattered on and partially buried in fine-grained red orange material, mostly in particle sizes below the resolution of lander imagery. Study of imagery suggests that a continuum of particle sizes exists from meter scale down to camera resolution limits (~2.5 mm). However, many of the smallest particles clearly resolved (about 1 cm to a few centimeters in size) have platelike shapes unlike those of the larger, more equidimensional blocks. These platy fragments are especially evident in areas that have been disturbed in spacecraft landing (Figure 5*b*) or by sampler operations and are particularly abundant at the Viking Lander 2 site. Uncertainty exists as to the presence of pebble-sized fresh rock material that might be delivered to the XRFS instruments. At this writing, all attempts to deliver pebbles have resulted either in no sample (VL 2), probably because they disaggregated during sieving in the mode 2 delivery sequence, or in delivery (VL 1) of samples chemically similar to fines and with bulk densities too low for fresh, consolidated rock. From the latter result [*B. C. Clark et al.*,

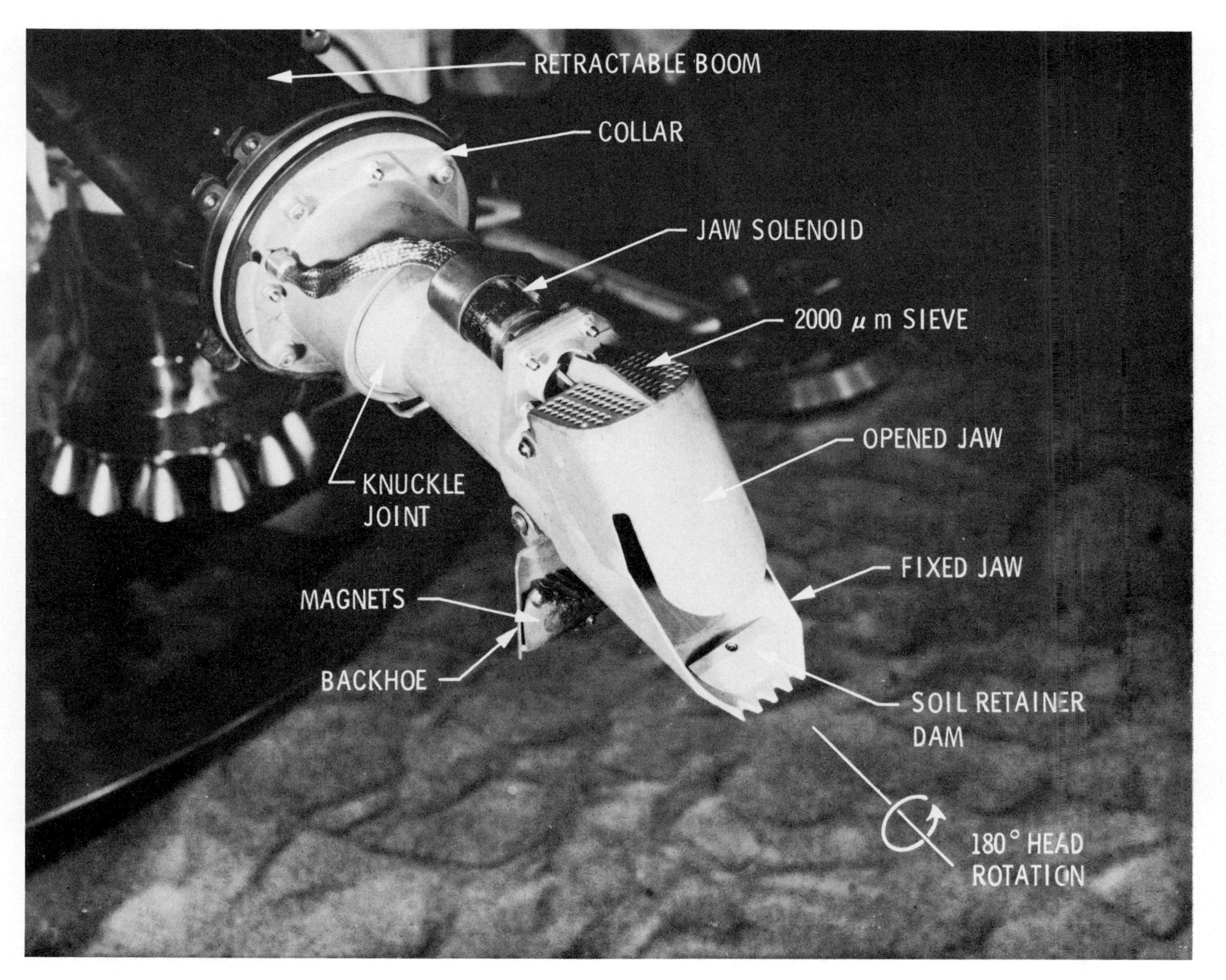

Fig. 2. Viking lander surface sampler head. Jaw solenoid operates at 4.4 or 8.8 Hz. Backhoe is shown in its extended position; in stowed position it folds backward toward knuckle joint.

Fig. 3. Mosaic of Viking Lander 1 sample field. Rectangles indicate locations of sample sites, keyed to figure numbers. Compare with Figure 7.

Fig. 4. Mosaic of Viking Lander 2 sample field. Rectangles indicate locations of sample sites, keyed to figure numbers. Compare with Figure 8.

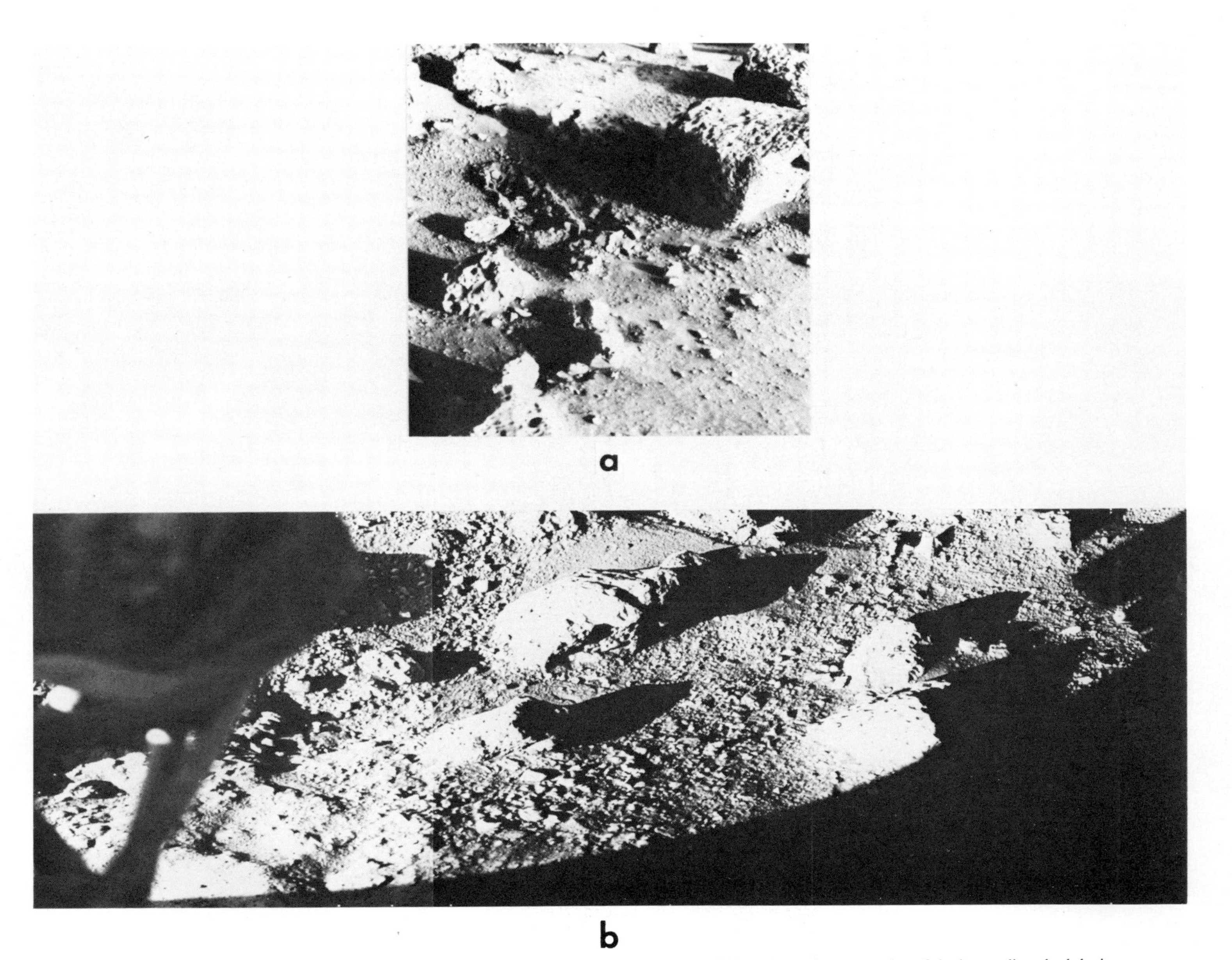

Fig. 5. (*a*) Pebble-sized material in partial halo pattern around rock at right side of view. For other examples of duricrust disturbed during sampler activities, see Figures 9–19. (*b*) Surface material disturbed during landing of VL 2. Pebble-sized material is interpreted as broken-up duricrust,

1976*b*] we conclude that the plate-shaped particles are cemented duricrust and not rock pebbles. The acquisition of rock material, however, remains a high-priority goal in the Extended Mission. In support of this priority, common purge sites have been established for each lander onto which the sampler dumps coarse particles following sieve operations after all sample acquisitions. These sites, in stereo view of the cameras, are monitored to determine if sufficient pebble-sized material can be accumulated for later reacquisition and delivery by the sampler to the XRFS instruments. At VL 2, where the most extensive attempts to build a pebble pile on the surface have occurred (12 separate purges), only three pebbles have been deposited to date. We are therefore forced to conclude that the two lander sites are grossly depleted in firm, fresh rock material in the millimeter to centimeter size range. A measure of the degree of depletion is provided by comparisons with rock acquisitions in simulated lunar regolith material containing 8% (by weight) of particles larger than 2 mm (and up to 1 cm in maximum size). In more than 100 tests of the surface sampler in this material using the rock acquisition mode a usable volume of pebbles for XRFS analysis was always obtained (J. Gliozzi, personal communication, 1977). Several explanations for this depletion of pebble-sized rocks on Mars have been offered: (1) The larger blocks were deposited by a process (impact ejection?, catastrophic flooding??) resulting in deposition of cobble and boulder sizes only, and the very fine material results from a separate process (aeolian?) incapable of delivering pebble sizes to the sites. (2) Weathering processes, subsequent to deposition of particles of all sizes, have completely altered particles in small pebble sizes, so that both their mechanical behavior during sieving and their chemical composition are indistinguishable from those of the very fine material. (3) Unusual aeolian processes have selectively removed pebble-sized rocks, leaving large blocks and very fine materials below the wind traction limit. Geologic evidence from the landing sites does not conclusively support or refute any of these explanations. However, the conspicuous dunelike structures at VL 1, apparently encroaching upon blocky rubble, show that aeolian processes have been active. That Martian winds are capable of selectively removing small pebble sizes has not been demonstrated and seems unlikely. In fact, it is difficult to understand an aeolian process that could remove pebble-sized particles of firm rock (specific gravity about 2.6–3.0) and leave pebble-sized particles of duricrust (specific gravity 1.?)

At VL 2, ringlike deposits of pebble-sized material around larger rocks (Figure 5*a*) could be weathering spalls from the large rocks. If this were their origin, one would expect most of the larger rocks to have distinct weathering crusts on their exteriors; high-resolution imagery does not support this hypothesis. Large rocks appear to be fresh, having clearly defined vesicles and other textural features. In addition, several rocks have been nudged and pushed by the surface sampler during underrock sample acquisition sequences, and their surfaces have not been damaged as one might expect if they were covered with a weathering product. (To test rock surface character further, we plan to use the sampler backhoe to scratch surfaces of rocks that are too large to be moved. This activity is proposed for late in the Extended Mission.)

From our sampling efforts to date, we favor the first explanation for the origins of regolith material at the lander sites, but the lack of pebble-sized material remains an enigma. To processes of probable impact ejecta deposition and aeolian deposition we believe a third process, duricrust formation, should be added. It is possible that surface crusts could be formed on Mars by a mechanism similar to that observed in

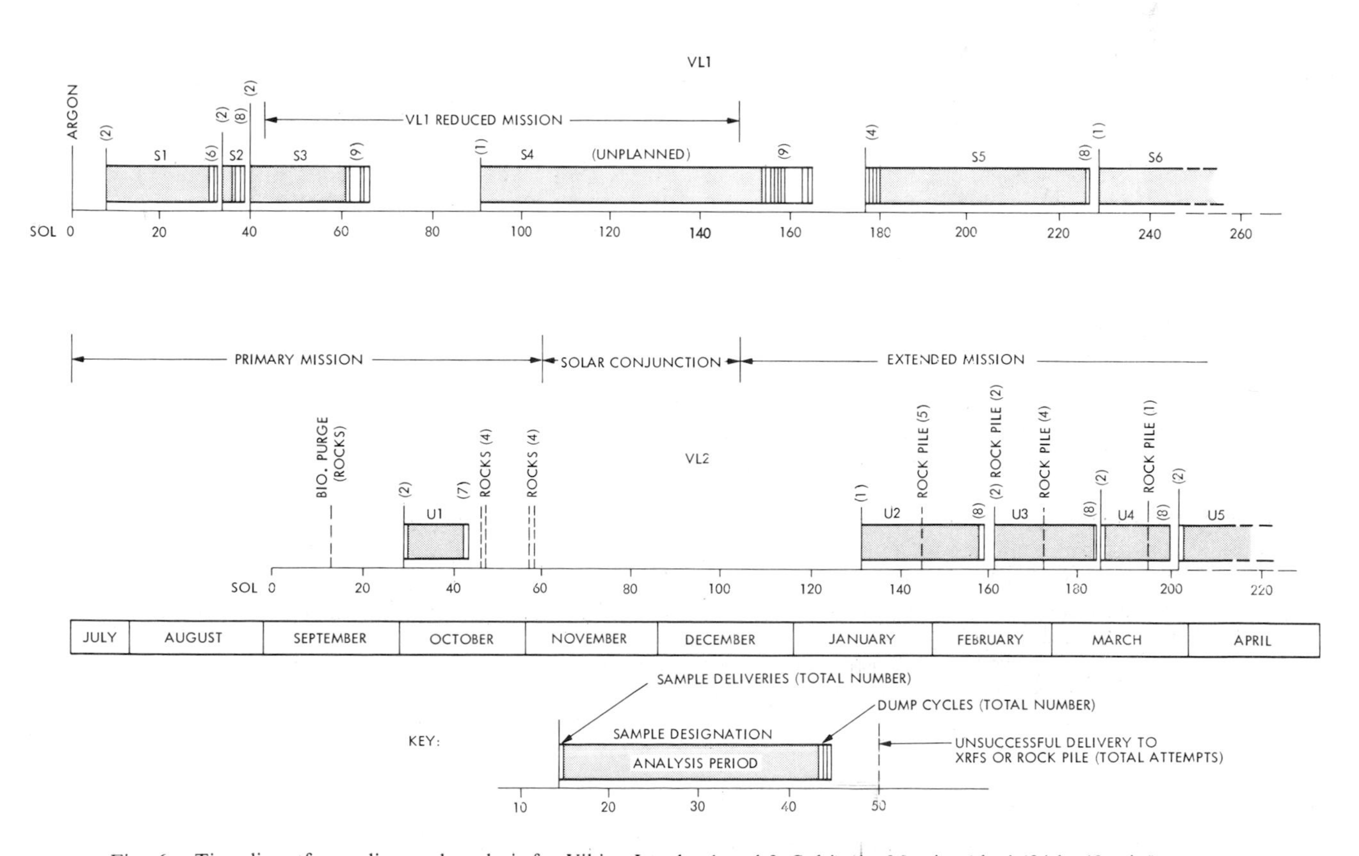

Fig. 6. Time line of sampling and analysis for Viking Lander 1 and 2. Sol is the Martian 'day' (24 h, 40 min).

TABLE 1. XRFS Sample Acquisitions

Sample	Sample Site	Sol	Type	Depth, cm
	Viking Lander 1			
S1	Sandy Flats	8	fines	4–6
S2	Rocky Flats	34	duricrust	0–4
S3	Rocky Flats	40	duricrust	2–5
S4	Sandy Flats	91	fines	0–4
S5	Atlantic City	177–180	duricrust	0–4
S6	Sandy Flats	229, 250	fines	23–25
	Viking Lander 2			
U1	Bonneville	29, 30	fines	0–4
U2	under Notch Rock	131	fines	0–4
U3	Spalling Valley	161	fines	4–6
U4	under Badger Rock	185, 186	fines	4–6

arid regions on earth, namely, upward migration of solutions and precipitation of cementing minerals in the top few centimeters of the surface. Preliminary results from our measurements [*B. C. Clark et al.*, 1976*b*, 1977] show that S and possibly Cl are enriched in the duricrust relative to subsurface fines. Further conclusions await refinements in data analysis and the acquisition and analysis of a sample skimmed from the surface only (scheduled for the Extended Mission).

Thus in view of the limited practical sampling options available, only very fine material, what we interpret as cemented very fine material, has been acquired and analyzed to date. In the sections that follow, we describe Primary Mission activities and those additional activities planned for the Extended Mission.

Sampling Rationale

Figure 6 shows the time line of sampling activities and periods during which XRFS analyses were carried out from touchdown of VL 1 through approximately March 31, 1977. Table 1 summarizes the sampled localities. These are keyed to planimetric maps of the lander sites (Figures 7 and 8). The Primary Mission for both landers ended with solar conjunction, on November 15, 1976, but most activities on VL 1 terminated at touchdown of VL 2; the remaining period during the Primary Mission for VL 1 is referred to as the Reduced Mission. The Extended Mission follows solar conjunction and, it is hoped, will operate until January 1979.

Appendix Tables A1 and A2 provide details on the locations of sample sites with respect to lander reference coordinates and lander camera control azimuths. Appendix Tables A3 and A4 give precise times and commands for each surface sampler activity as well as estimates of the amounts of material delivered to XRFS instruments in Viking Lander 1 and 2, respectively. These estimates must be based upon the degree of obscuration of calibration plaque spectra by the delivered sample, upon postacquisition images of sample sites, and upon known characteristics of the surface sampler, since the XRFS instruments cannot determine how much excess material was delivered. In the appendix tables and text to follow, sample site names coined during mission activities are used; they are also indicated on the maps of the lander sites (Figures 7 and 8).

Planning of sampling at both landers became interactive almost from the first moments after touchdown. Upon ground receipt of images of the preselected sites for first sampling (the preprogrammed activity, or initial computer load) it was quickly determined that safer and more desirable sites existed for initial sampling. Operating within the constraints described above, subsequent sampling became, to a large extent, a 'trial and error' process in which sampling goals were set by the Inorganic Chemical Analysis Team, approval was sought from mission managers, acquisitions were attempted, and the results were used to set the next sampling procedure. The missions, however, could not be perfectly interactive and adaptive to prior results because of required planning times for development of command sequences (see *Lee* [1976] for details). Therefore the sample acquisition attempts for XRFS were necessarily, in part, compromises brought about by insufficient knowledge at the time that planning decisions had to be made.

Viking Lander 1. The part of the landing site available for sampling consists of a small area of fine material in which footpad 2 is buried (dubbed 'Sandy Flats') and the remaining area of blocky rubble buried in and sitting on fine material of appearance similar to that at Sandy Flats (Figure 3). For sampler safety and maximum assurance of acquiring a sample the first samples for all analytical instruments were collected from Sandy Flats. Sample S1 for XRFS came from the headwall area of the freshly dug biology and GCMS trench. Two identical acquisitions were made 47 min apart in the late morning of sol 8. Figures 9*a* and 9*b* show the sample area before any sampler activity and after the biology, GCMS, and XRFS samples were acquired. A nominal total of 85 cm^3 was probably delivered in the fines mode with 60 s of sieving. The sample most likely consisted of material from 4- to 6-cm depth.

While subsequent sampling planned by the Inorganic Chemical Analysis Team for Viking Lander 1 (S2, S3, and S5 in the Extended Mission) was aimed at acquiring pebble-sized rock material, only duricrust was obtained (see discussion above and *B. C. Clark et al.* [1977]). These acquisitions were made by using detailed mapping studies of the pebble distributions (Figure 10) in the areas designated 'Rocky Flats' (Figure 11) and 'Atlantic City' (Figure 12), with delivery in the rock mode following sievings of 60 s (S2, sol 34), 90 s (S3, sol 40), and 120 s (S5, sols 177–180). The rationale for sieving was based upon tests in soil simulating lunar grain size distributions. Under terrestrial conditions it was found that 60-s sieve times effectively cleaned pebbles of adhering fine material which would otherwise absorb X ray emissions from the rocks and effectively yield an analysis only of the fines. As is reported elsewhere [*B. C. Clark et al.*, 1976*b*], S2 yielded spectra virtually identical in all elements except S (in first analysis) to those from S1 fines. Because of the extreme adhesiveness of Martian fines (demonstrated by the difficulties encountered in our attempts to dump S1 from the sample cavity [*B. C. Clark et al.*, 1976*b*]) we at first concluded that the S2 pebbles were still heavily coated because of insufficient sieving. On this basis the decision was made to dump this sample and reacquire pebbles with a longer sieve time before delivery. S3, using a 90-s sieve, was obtained 6 sols later in an extremely fast adaptive action for the Viking Mission. Speed was required because the forthcoming Reduced Mission would have precluded any further sampling (for XRFS) for 4 months. Unfortunately, the increased sieve time for S3 may have served only to reduce the amount of deliverable sample (by disaggregation), and we entered the Reduced Mission with a cavity partially (but usefully) filled with material which gave X ray spectra very similar to those obtained from S2. Extensive analysis of spectra of S3 and comparison with S1 [*B. C. Clark et al.*, 1976*b*] led us to conclude that no rock material had been acquired despite the fact that careful study of imagery showed about 50% of the

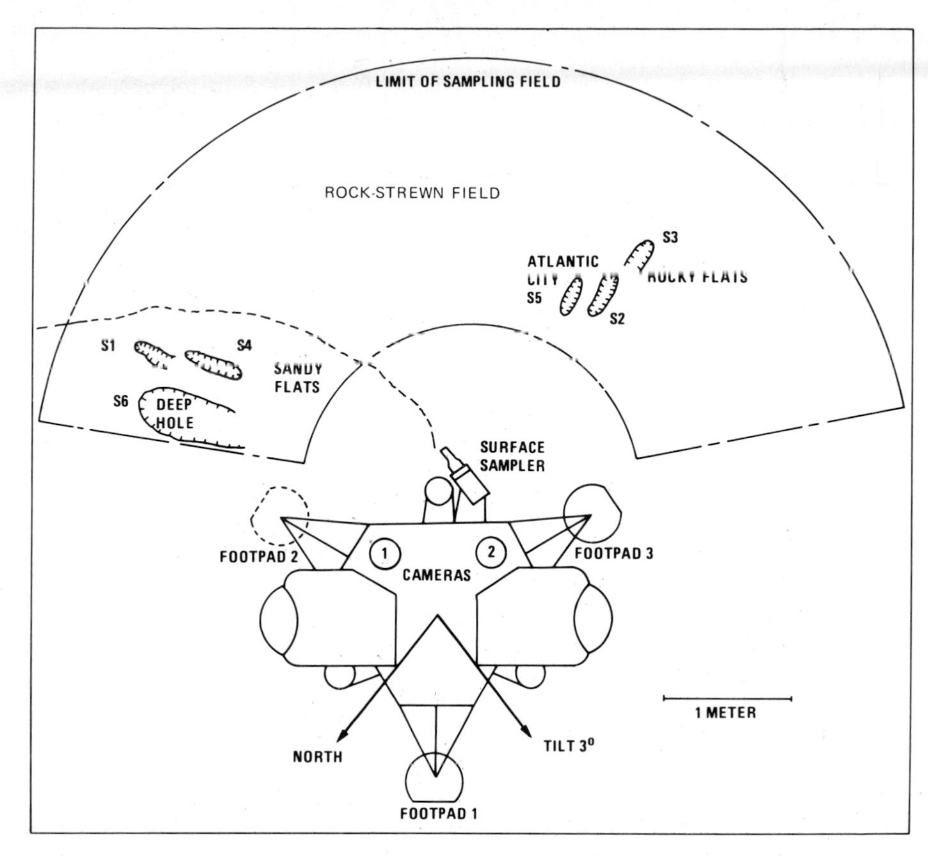

Fig. 7. Map of sampled sites, Viking Lander 1. Dashed line divides Sandy Flats from rock-strewn field.

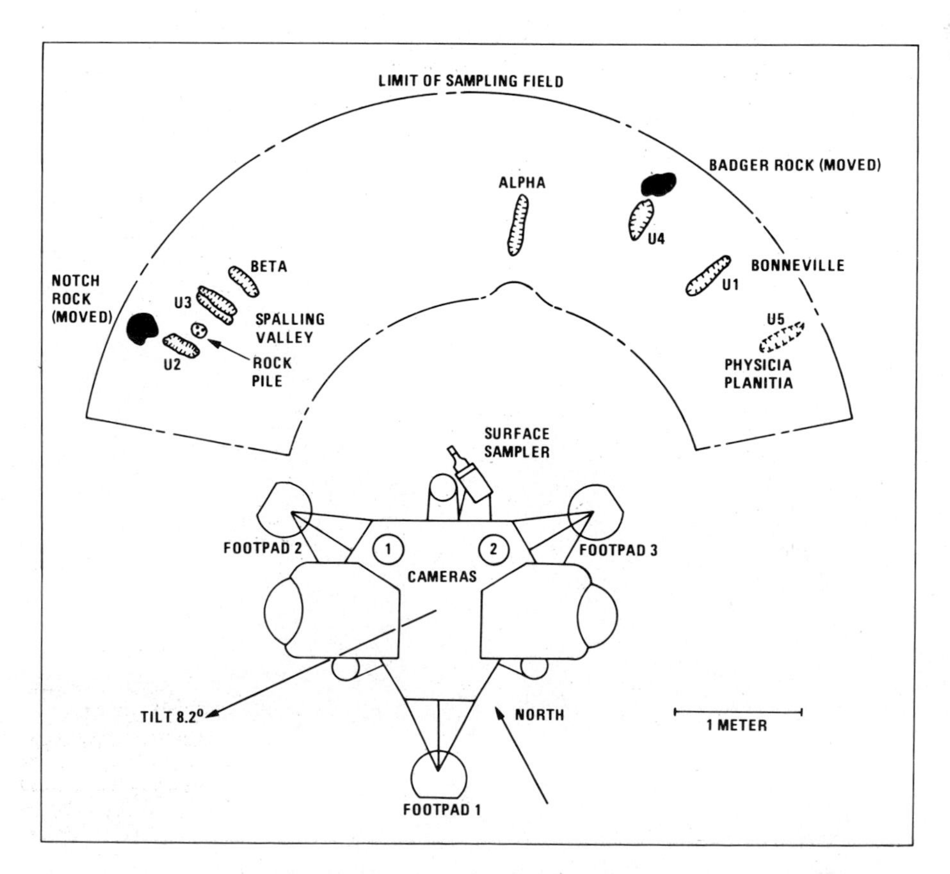

Fig. 8. Map of sampled sites, Viking Lander 2.

resolvable particles to be between 5 and 10 mm in size (Figure 10). Figures 11*a* and 11*b* show the Rocky Flats area before and after acquisitions for S2 and S3; the platy nature of the disturbed surface material is especially obvious in the post-acquisition images.

Continuing our quest for deliverable rock material, we assigned a high priority to acquisitions at the Atlantic City site (Figure 12) for the first activity following solar conjunction. The plan for this sampling activity was to increase the sieve time to the maximum permissible for the surface sampler (limited because of solenoid heating) in the hope of completely breaking up and sieving out duricrust material, thus leaving only rock material for delivery to the XRFS. Four acquisitions were made on successive sols (177–180), intervening analyses being made to monitor sample cavity fill (see Table A3). Preliminary results [*B. C. Clark et al.*, 1977] from this sample indicate that again only duricrust was obtained.

During the Reduced Mission on VL 1 the S3 duricrust sample was dumped on sols 61–67 because sufficient analyses had been performed and because we wished to determine if the XRFS cavity might possibly collect wind-blown material during the remaining Reduced Mission and solar conjunction period. However, on sol 91 the surface sampler acquired a sample for biology from the Sandy Flats area. During delivery (in sieve mode) to the biology sample processor screen (located 7.6 cm higher and within 5 cm of the XRFS funnel), approximately 15–20 cm^3 of material inadvertently entered the X ray funnel. The exact reason for the 'spill' into our funnel is

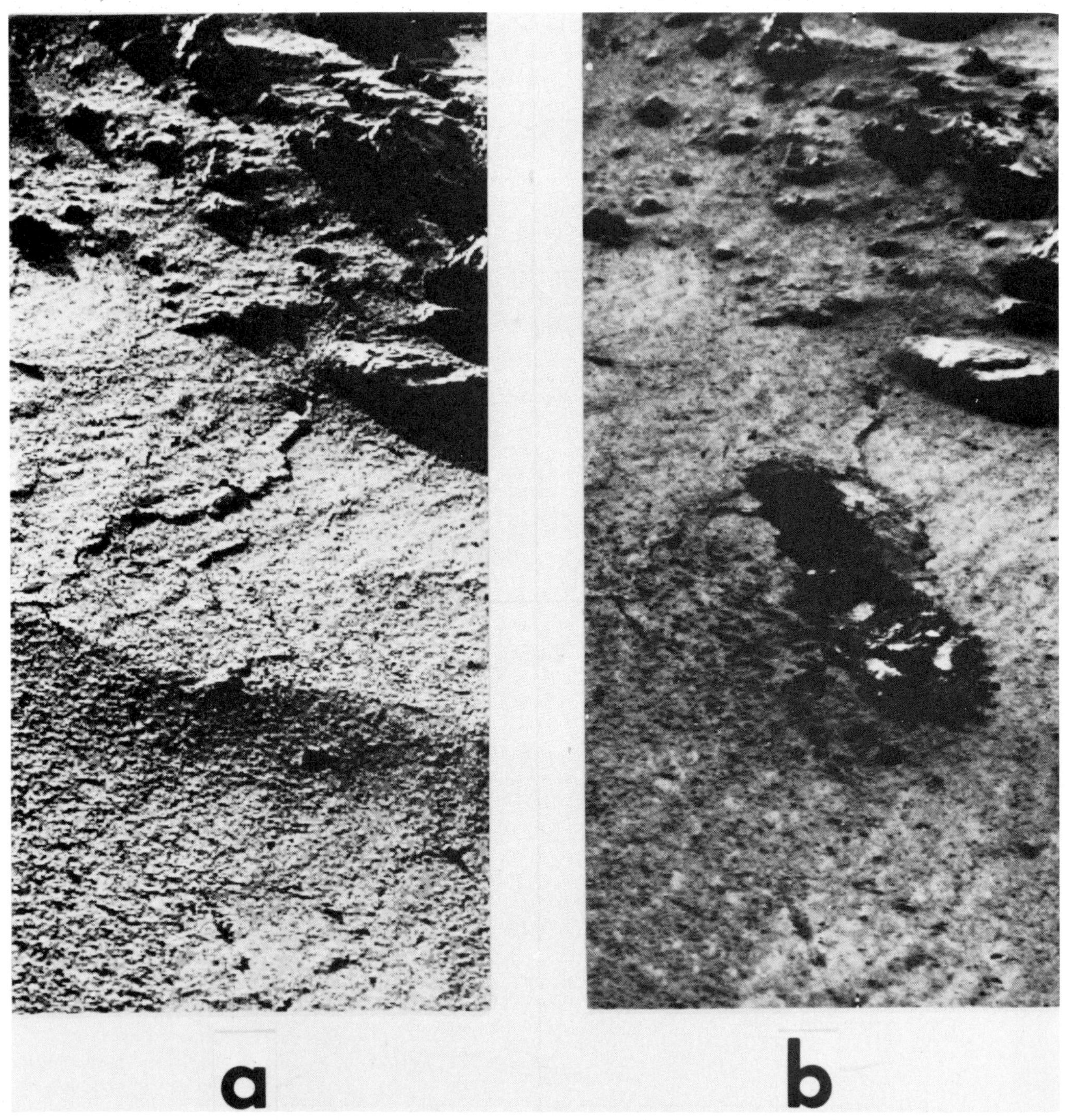

Fig. 9. Site of S1 sample, Viking Lander 1, at Sandy Flats. (*a*) Before sampling. (*b*) After XRFS acquisitions.

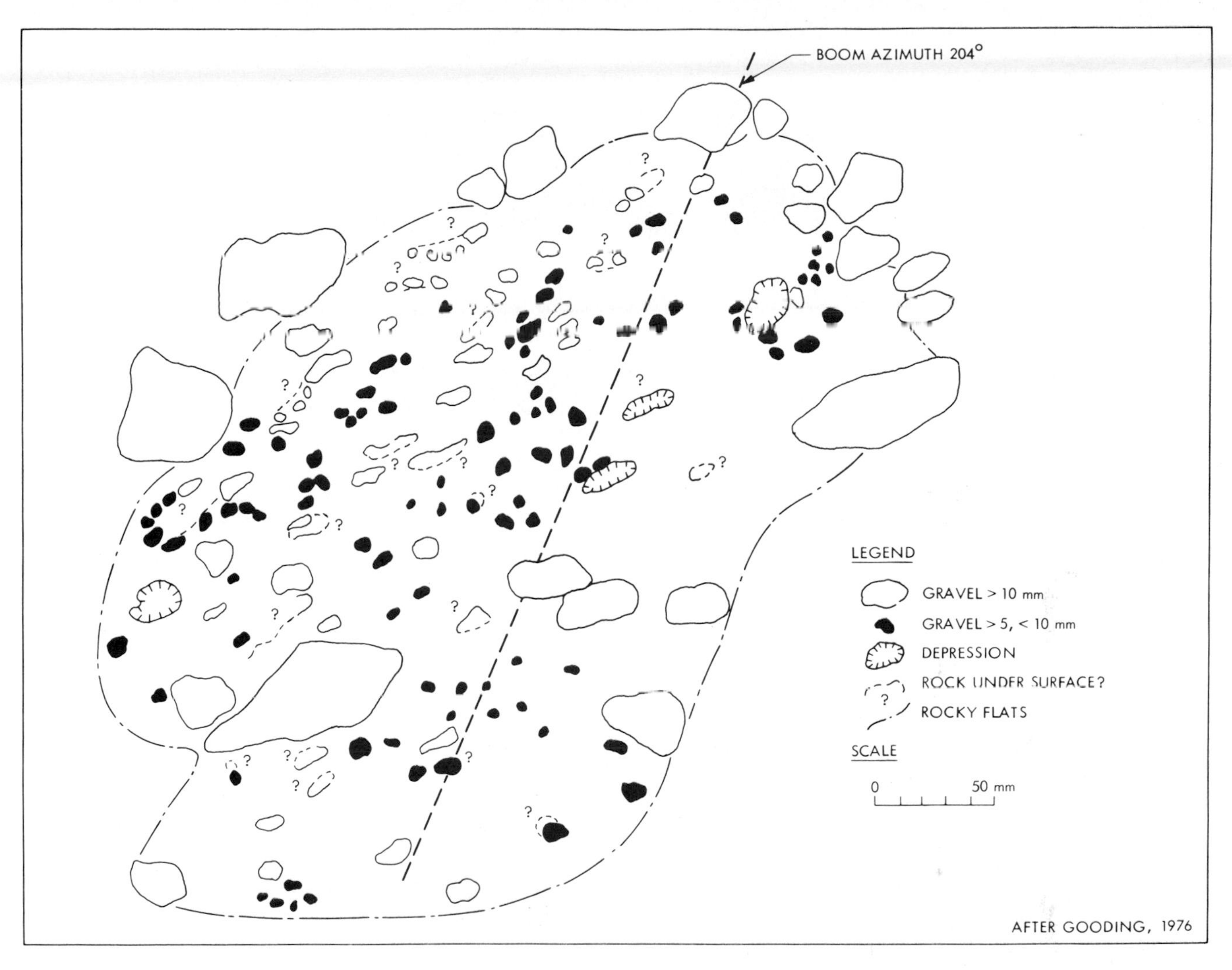

Fig. 10. Microgeologic map of Rocky Flats sample site, Viking Lander 1. Rock (pebble) acquisition attempts were made on each side of boom azimuth 204°. Compare with images in Figure 11.

unknown but might be due to release of the soil-laden backhoe during rotation of the collector head over the biology screen. This sample of fines (S4) was analyzed for 10 sols, through solar conjunction. Preliminary results [*B. C. Clark et al.*, 1977] indicate similarity in composition to sample S1 from Sandy Flats.

The latest (as of this writing) sampling activity for X ray, sample S6, is a fines mode delivery from the bottom of the 'Deep Hole' (Figure 13), dug during 4 sols of extensive backhoeing operations in Sandy Flats. S6, delivered on sol 229, came from a depth of about 25 cm below the Martian surface and will be especially valuable as a test for chemical differences between surface and subsurface regolith material. (First results show that the XRFS sample cavity was only about 80% filled in the single acquisition attempt; a second acquisition was successfully made on sol 250.) Images of the Deep Hole (Figure 13) show typical Martian fines throughout its depth and give no indication of rock material or that a floor to the drift deposit had been reached. The Deep Hole will likely be the deepest penetration of Martian regolith in the Viking Mission, though we intend to request further hole-digging attempts for both landers in the remaining Extended Mission.

Viking Lander 2. The immediate site available for sampling is similar in mesoscale geology to the Rocky Flats area of VL 1 but lacks the obvious rock-free drift material typical of Sandy Flats (Figure 4). Two major differences from the Rocky Flats area are that (1) the blocky rubble appears much more uniform in color and texture (as if it is mainly composed of a single rock type) and (2) the surface of the red orange fines, forming the matrix between blocks, apparently is cemented more distinctly into a duricrust and has prominent fractures similar in appearance to desiccation cracks or even small fumarolic vents in some terrestrial materials. Much of the duricrust is broken up into platelike fragments; these fragments are especially evident near footpad 2 (where they probably formed by fracture during spacecraft landing) and in 'halos' around some of the larger rocks (Figure 5).

The sampling rationale for VL 2 was strongly influenced by prior results from VL 1 on samples S1 and S2. At the time of landing of VL 2 we knew the general chemical nature and extreme adhesiveness of Martian fines, and we were perplexed by the failure of our initial attempts to acquire and analyze pebble-sized rock material. We placed rock acquisition as the first priority for VL 2 in order that the sample cavity would be uncontaminated by fine material which, on the basis of our experience with VL 1, might be difficult to dump from the cavity. However, our Science Team request for an early (sol 8) rock acquisition and delivery was denied by project manage-

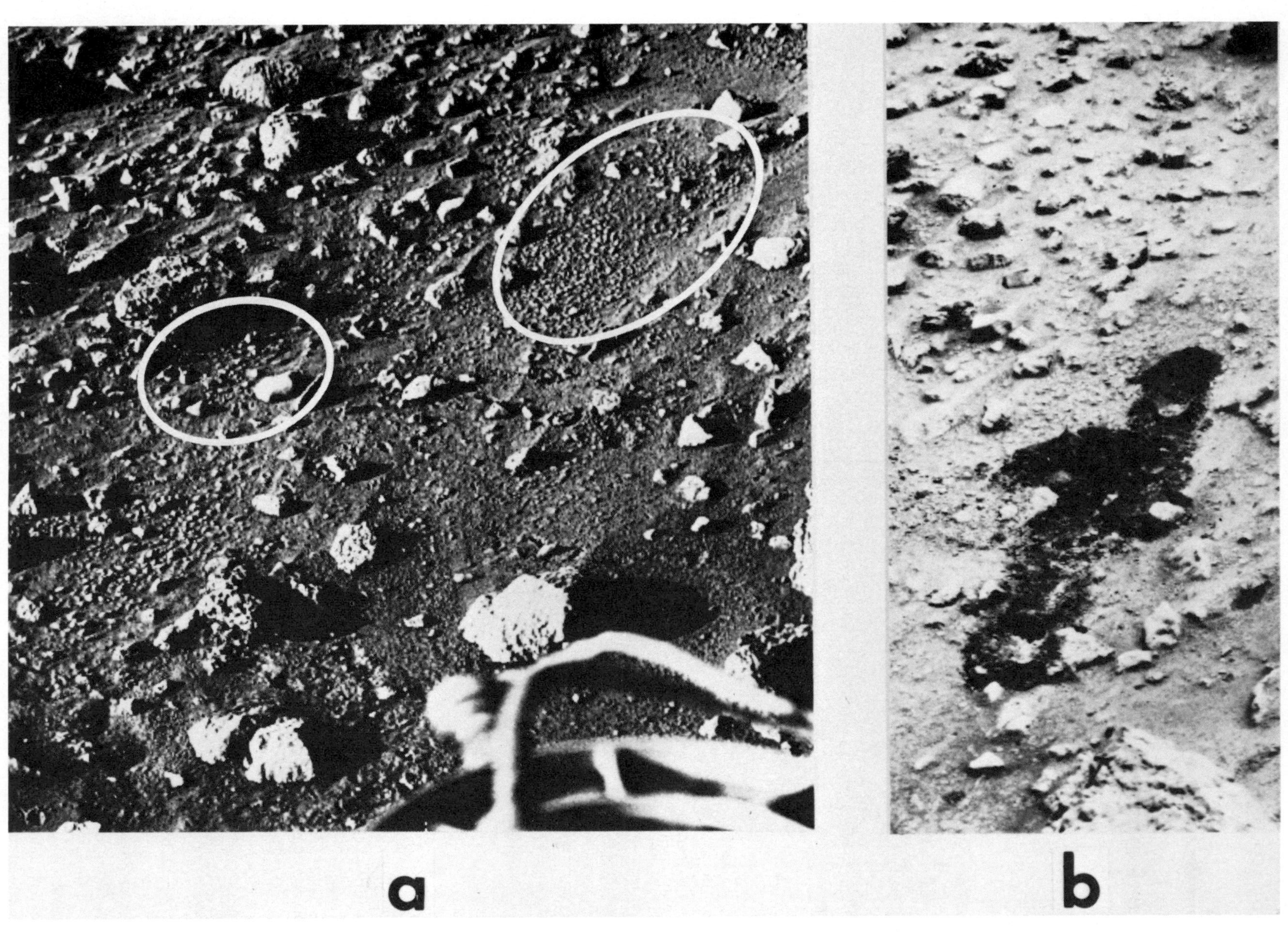

Fig. 11. Rocky Flats and Atlantic City sample sites, Viking Lander 1. (*a*) Preacquisition image. Rocky Flats is ovoid area in upper right part of view. Atlantic City is in front of large cobble in upper left. (*b*) Postacquisition image of Rocky Flats.

Fig. 12. Atlantic City, Viking Lander 1, postacquisition. Compare with Figure 11*a*.

ment on the grounds that it would endanger the sampler before the acquisitions of higher-priority samples for biology and GCMS. A compromise was reached in which the residue of material in the collector head, following an acquisition for biology on sol 8, was to be delivered (purged) to the XRFS funnel. A 'no-go' anomaly (collector head rotation switch malfunction) stopped sampler operations following the biology delivery, but the attempted delivery to XRFS was completed on sol 13. No sample was obtained, however.

The XRFS on VL 2 remained empty until sol 29 because of the priority restrictions on use of the sampler for XRFS acquisitions before other analytical instruments. In the interim, two samples of fines had been obtained for biology, and one for GCMS. At this stage of planning for VL 2, considerable spectral information had been obtained from S3 on VL 1, and it was obvious that true rock acquisitions at VL 2 might be very difficult to achieve. Accordingly, the decision was made to postpone such attempts until a fines acquisition with more assurance of success had been made. The area dubbed 'Bonneville' (Figure 14) was selected because the presence of distinct surface cracks suggested an apparently different type of sample material and because the sample would be in conjunction with a previous GCMS acquisition on sol 21 from the same area. As is discussed elsewhere [*Toulmin et al.*, 1977], potentially important supportive analytical data (e.g., H_2O and CO_2 contents) on the sampled material might be obtained from GCMS analyses. Two acquisitions in the fines delivery mode were made for XRFS on sols 29 and 30 with an intervening analysis period. This sample, U1, was analyzed through sol 41.

Resuming the attempts to acquire fresh rock material, we scheduled a series of eight separate acquisitions in two areas (termed 'Beta' and 'Alpha,' Figures 15 and 16, respectively) of the VL 2 site which were performed on sols 46–47 and 57–58. On each try, 90 s of sieving was done, followed by conventional rock mode deliveries. None of the attempts yielded measurable material even though images of the areas selected showed abundant particles of the desired size on the presample surfaces and in the postsample trenches. We conclude that the material acquired from the trenches must have contained a significant fraction of duricrust fragments but that these fragments (along with fines) were completely disaggregated and sieved out in the 90-s operation. The XRFS on VL 2 therefore remained empty through solar conjunction; no measurable amount of wind-blown material entered the sample cavity.

Following conjunction (and to date), four additional surface sampler operations have been performed to provide a variety of different materials to XRFS while the search for deliverable rock material has also been continued. In these activities a common surface purge site (Figure 17) has been established onto which the sampler deposits the residue material of all sieve operations. As was discussed above, few particles have so far been accumulated in the 'Rock Pile,' but this activity will continue in the remaining Extended Mission.

Sample U2 was obtained from the area (Figure 18) under the original position of 'Notch Rock' (moved during the Primary Mission) and was delivered to the XRFS in fines mode in a single acquisition on sol 131. Preceding and following acquisition of U3 on sol 161 an extensive series of surface sampler operations (see Table A4) in an area named 'Spalling Valley' (Figure 15) was performed in attempts both to dig a hole as deep as possible and to acquire and purge material to the Rock Pile. Digging was done in a sample acquisition mode and not by backhoe operations, as was done for the Deep Hole on VL 1, in order to try to accumulate pebbles. The strategy was to sample the hole site on sol 161 (sample U3) to obtain a 'quasi-deep' sample of partially subsurface fines, to continue digging on sol 172, and to resample the hole on sol 185 for a 'deep' sample. As was planned, the quasi-deep sample was obtained (probably from a depth of 4–6 cm), but further digging failed to deepen the hole significantly. Thus the attempt at a deep subsurface fines acquisition was changed to permit collection of a sample (U4) from under 'Badger Rock' (Figure 19). This was done in a double acquisition on sols 185

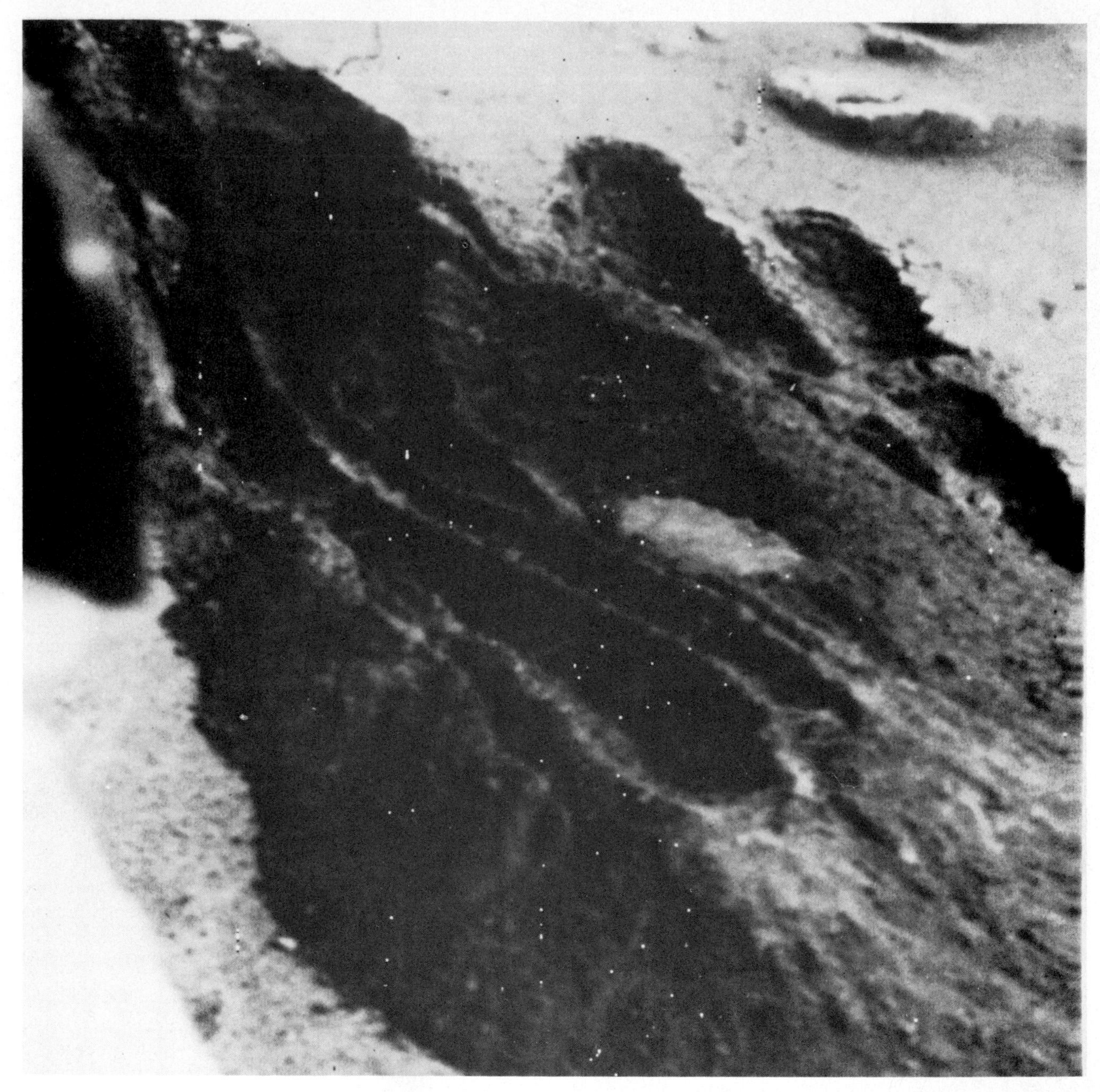

Fig. 13. Deep Hole site, Viking Lander 1, Sandy Flats. Compare with Figure 3.

and 186. On sol 195 the surface sampler stopped operating during a planned biology acquisition. The reason for the failure is unknown, but a decision was made to make no further attempts at sampler operations, at least until more favorable temperatures prevail. Viking Lander 2 is at 48°N latitude, and the onset of Martian winter will lower ambient temperatures below the mission operating limits (−65°C) for the surface sampler. Sample U4 will be retained in the XRFS instrument until sol 265; dump cycle commands will then be issued, and the sample cavity purged to give us the opportunity to collect airborne material from the global dust storms predicted to begin in May.

Additional Plans for Extended Mission

Sample U4 will be the last acquired from the surface of Mars at VL 2 until sometime in the fall of 1977. Before the sol 195 sampler failure we had planned to acquire and characterize a reasonably pure sample of the duricrust material from an area (Physicia Planitia) clockwise from Bonneville (see Figure 20). This sample (U5) would have been delivered on sol 203 but now will be postponed until fall 1977 as the first operation after Martian winter. Tests with the Science Test Lander have proven that previously used acquisition and delivery sequences must be modified in order to skim only the surface and to produce acceptable amounts of delivered material. For this operation the collector head will be knuckled (Figure 2) on the ground by commanding an additional 1-s downward movement of the boom after the sampler has reached the surface. The knuckling causes the jaw to assume an orientation nearly parallel to the ground surface and permits skimming during subsequent extension of the boom. Because the duricrust has apparently completely disaggregated in all prior sieve oper-

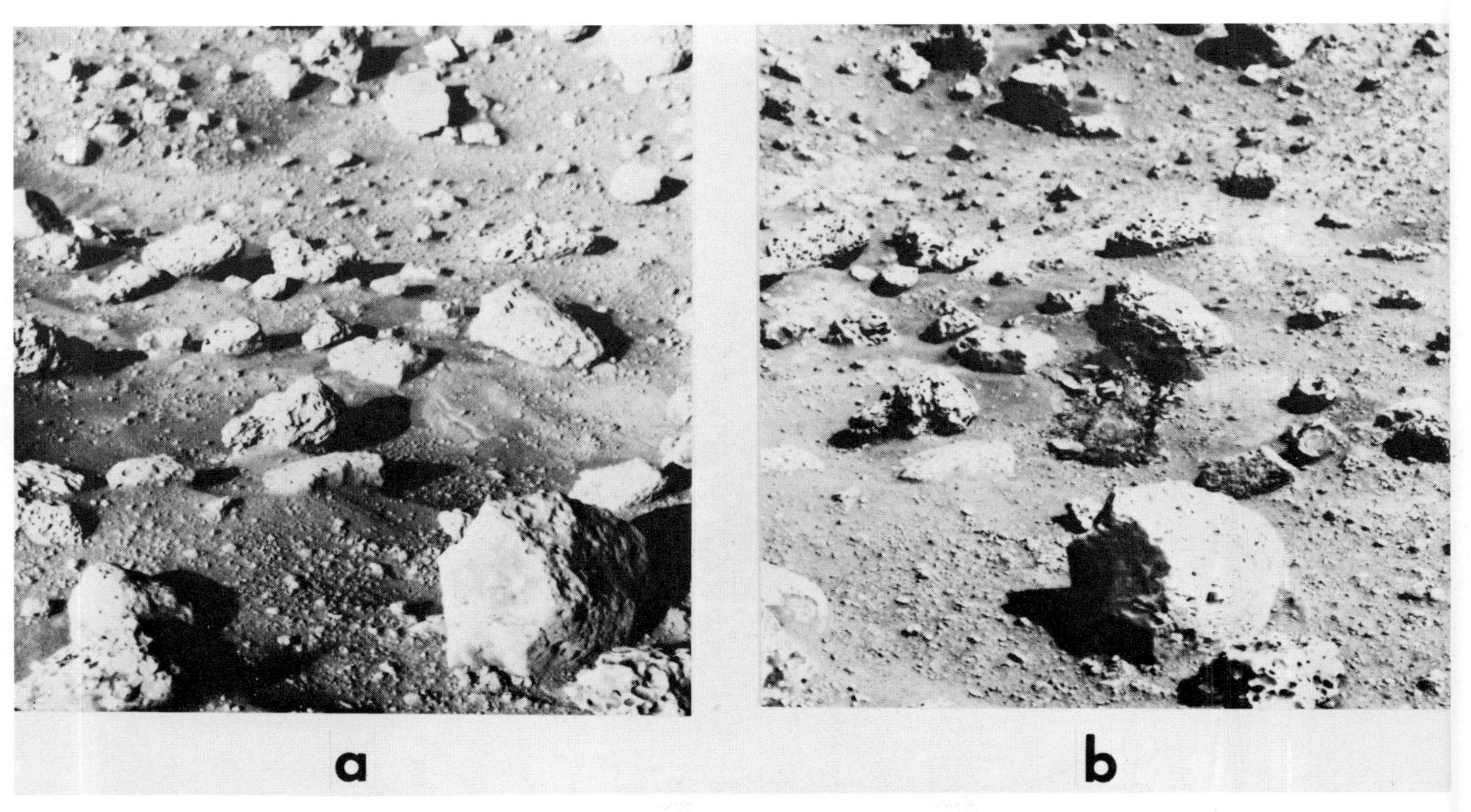

Fig. 14. Bonneville sample site, Viking Lander 2. (*a*) Preacquisition image. (*b*) Postacquisition image.

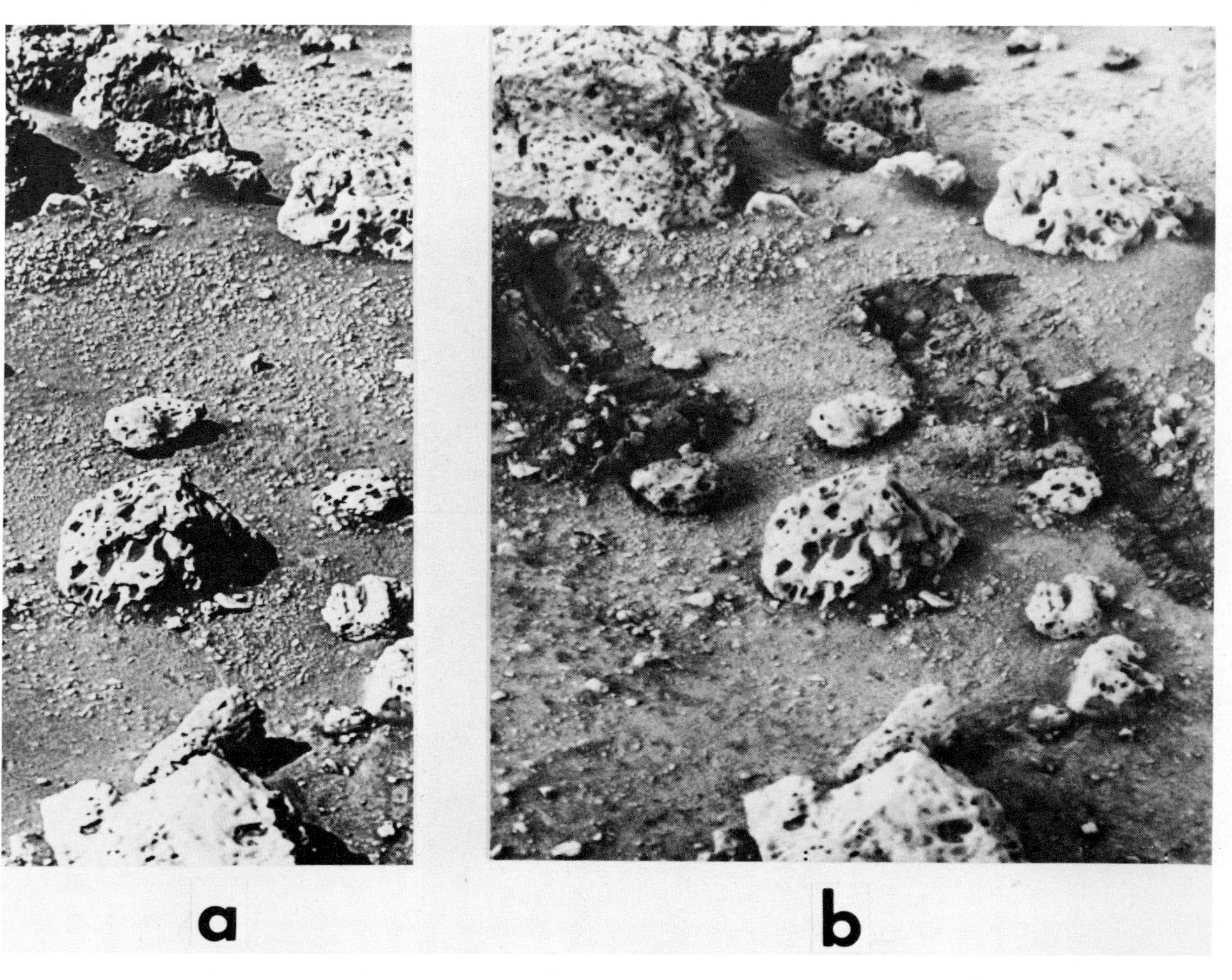

Fig. 15. Beta and Spalling Valley sample sites, Viking Lander 2. Trenches to right are Beta, and those to left are Spalling Valley. (*a*) Preacquisition. (*b*) Postacquisition.

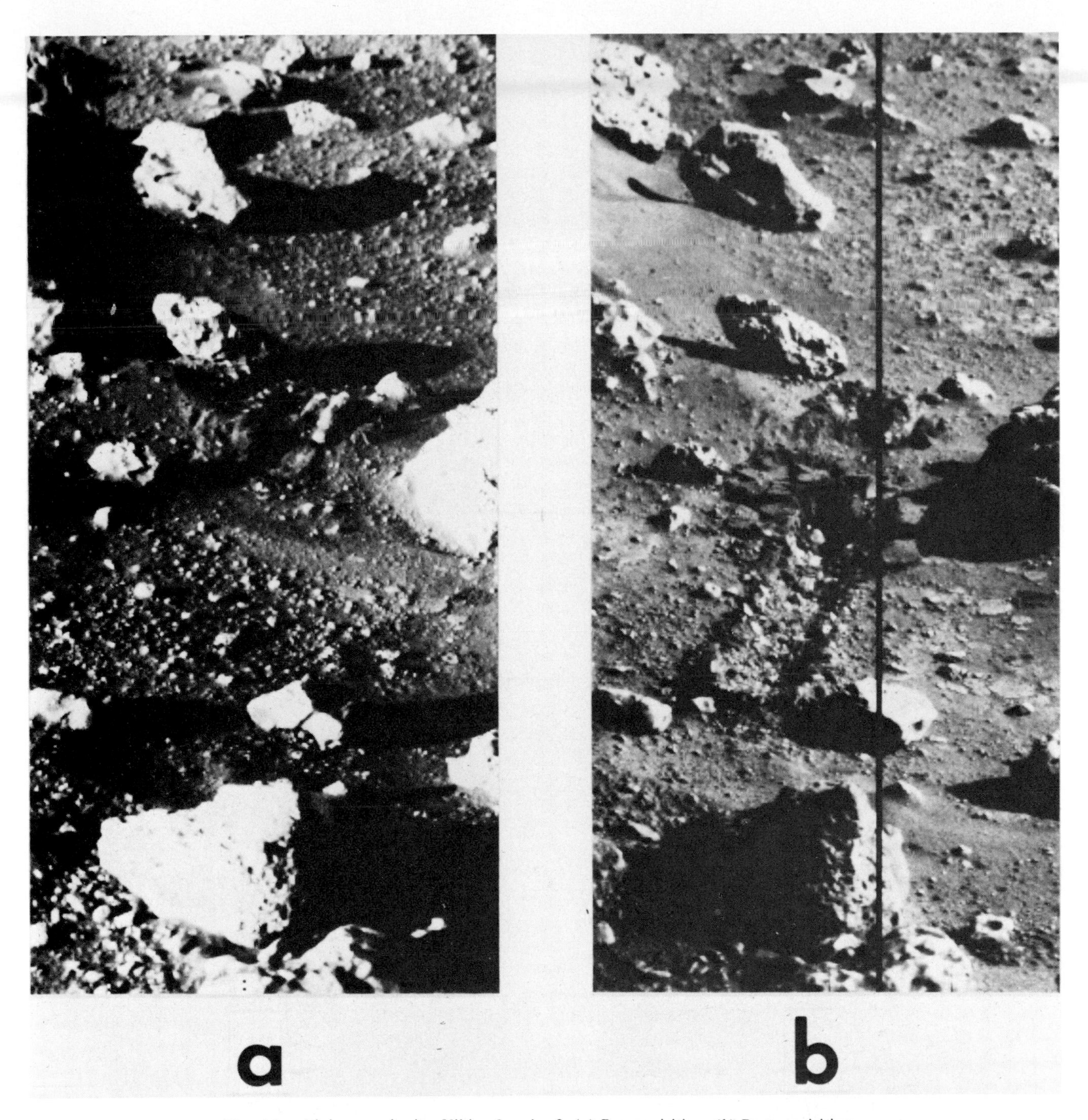

Fig. 16. Alpha sample site, Viking Lander 2. (*a*) Preacquisition. (*b*) Postacquisition.

ations, this skimmed surface sample will be delivered to the XRFS by direct purge from the collector head (mode 1) operation.

On the assumption that Viking Lander 2 activities can be resumed in the fall of 1977 a number of other operations would be desirable: (1) further attempts to acquire fresh rock material, (2) further attempts to penetrate the surface as deeply as possible, and (3) attempts to acquire surface materials of different albedos and/or colors.

Viking Lander 1, at 22.5°N latitude, will not experience thermal problems requiring cessation of sampler activities. Therefore we expect to continue operations until at least October 1, 1977 (unless global windstorms preclude operation). Desired activities will be similar to those listed for VL 2. In addition, we wish to test the surface properties of larger rocks (which appear to be more diverse in nature than those at the VL 2 site). We propose to use the sampler to scratch rock surfaces and possibly to produce fragments which could be acquired for analysis. Another test now being developed in simulation is to acquire and analyze a fines sample from an area previously depleted in magnetic material by repeated insertions of the backhoe magnets. Material from under rocks

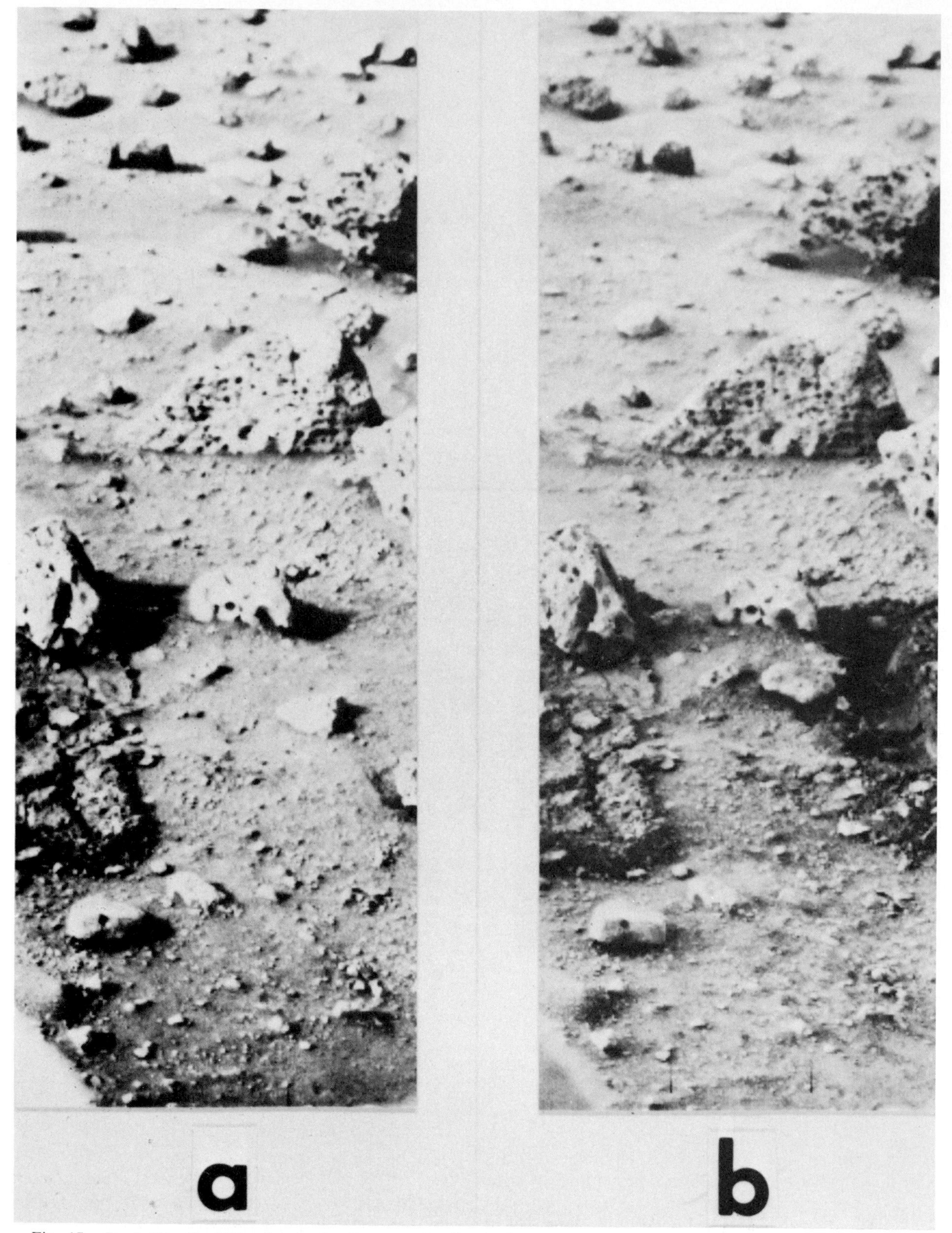

Fig. 17. Rock Pile site, Viking Lander 2, adjacent to Spalling Valley. (*a*) Sol 145, before piling attempt. (*b*) Sol 172, after 11 piling attempts. 'Rock Pile' is located midway between two trenches visible in part *b*.

(as obtained at the VL 2 site) should also be acquired on VL 1. On both landers, sufficient dump cavity capacity (Figure 1) remains to accommodate the planned samples.

LABORATORY STUDIES

In addition to required instrumental testing and certification by the builder (Martin Marietta Corporation) of the X ray fluorescence spectrometers the Science Team undertook a number of studies aimed at interpreting data returned from Mars. The premission focus of these studies was mainly in four areas: (1) understanding instrument response to calibration samples and to samples of varying particle sizes, (2) rapid recognition of incomplete sample cavity filling and incomplete sample cavity purging, (3) rapid recognition of classes of po-

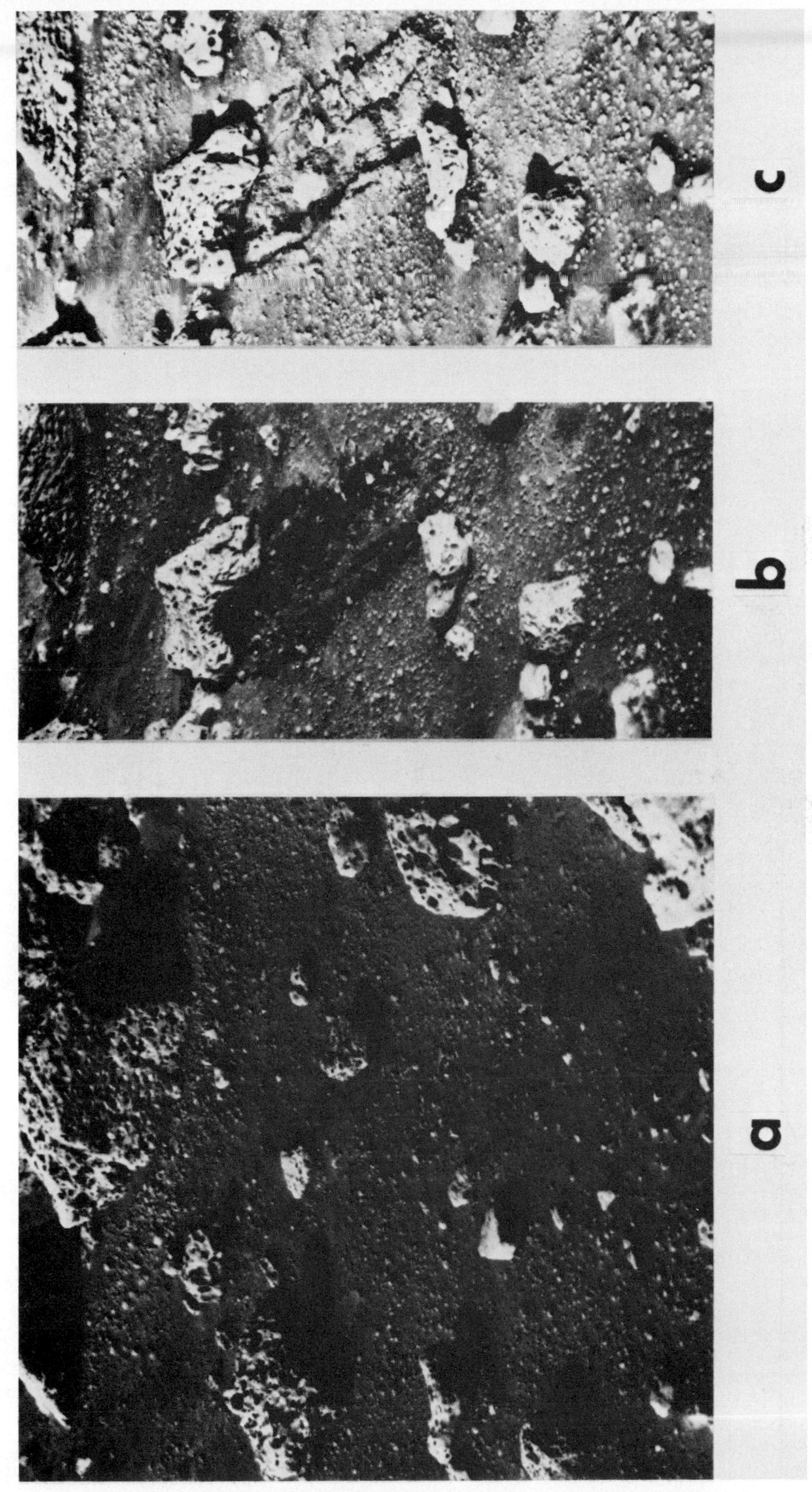

Fig. 18. Notch Rock sample site, Viking Lander 2. (*a*) Before rock move. (*b*) After rock push. (*c*) After acquisition.

Fig. 19. Badger Rock sample site, Viking Lander 2, after rock move and acquisition.

tential Martian materials from spectral signatures of terrestrial rocks and minerals and meteoritic materials (no lunar samples were available), and (4) proper assessments of the nominal operating characteristics of the XRFS (e.g., detector cross talk, high-voltage power supply ripple) which would affect later, more detailed interpretations of Mars results. Results from the second area of study were essential for quick response to day-to-day mission planning operations. The applications of results from other studies are more appropriately described in accompanying reports [see *B. C. Clark et al.*, 1977; *Toulmin et al.*, 1977] and in *B. C. Clark et al.* [1976*a*].

In support of these activities, laboratories at the Martin Marietta Corporation (Littleton, Colorado) and the U.S. Geological Survey (Reston, Virginia) were equipped with Science Team sensor assemblies (STSA) consisting of flightlike sample cavities, flight strength radioactive sources, and flightlike X ray detectors (Figure 21). The Martin Marietta facility was moved to the Jet Propulsion Laboratory (JPL) (Pasadena, California) in February 1976 and has been in continuous use throughout the Viking Mission. The laboratory at JPL also includes developmental models of the flight XRFS and a flight-qualified spare XRFS (FLT-1), all of which can be operated with electronics closely simulating Mars operation in a thermally and atmospherically controlled chamber. Only the gravitational conditions of Mars cannot be simulated. The heat-sterilized flight spare XRFS has been stored continuously in vacuum, except for brief periods during transfer to and from the test chamber. This instrument has been operated period-

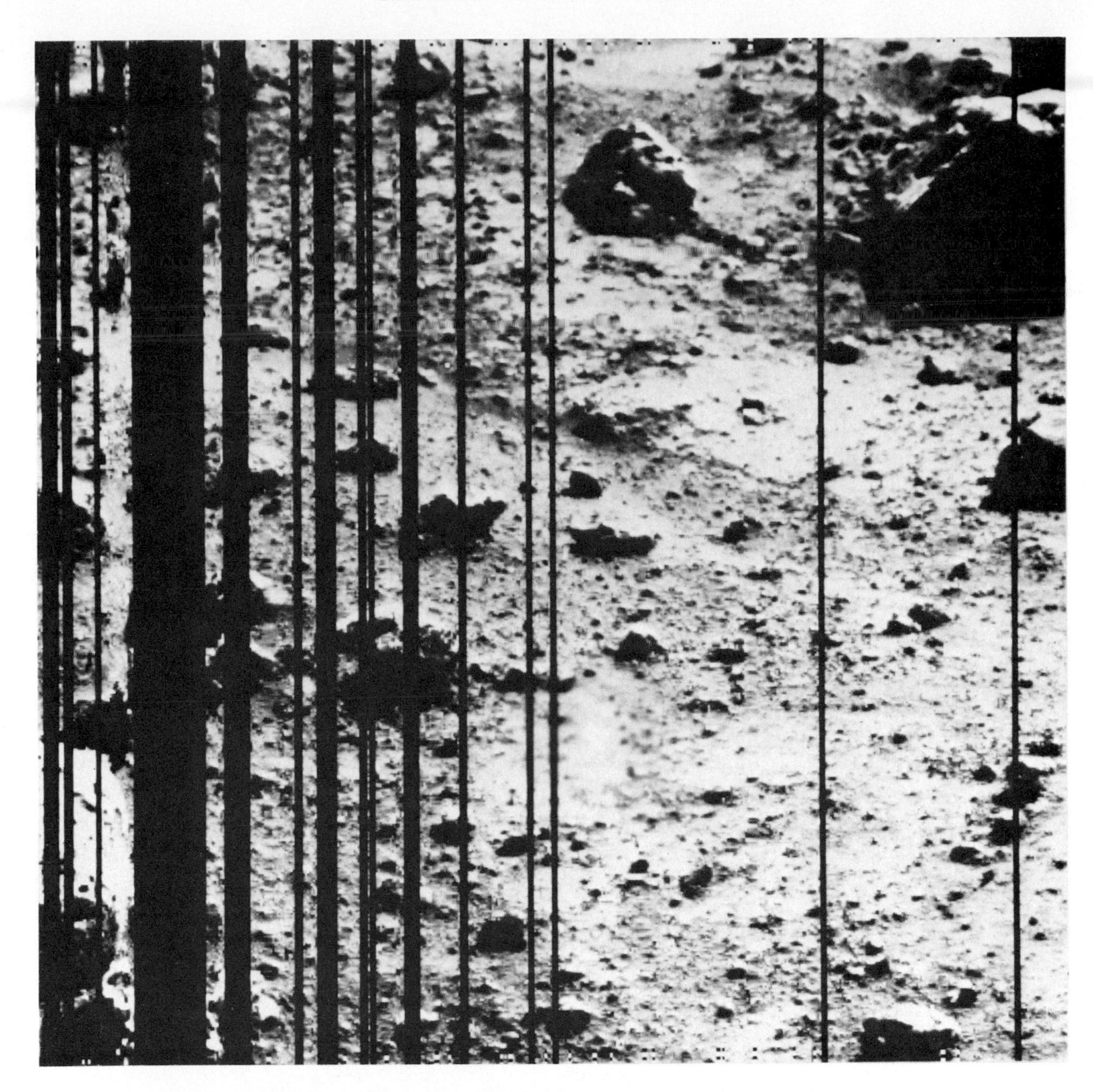

Fig. 20. Physicia Planitia skim sample site, Viking Lander 2. Vertical black bars are lost data.

ically in the plaque calibration mode and with rock standards to check ageing characteristics; no response changes have been observed during the period of landed operation on Mars.

During the Primary Mission and Extended Mission the laboratory emphases shifted to testing of special instrument operating modes (both to increase science data return and to provide work-arounds for low-energy noise problems on VL 2 [*B. C. Clark et al.*, 1977]) and to development of analog materials which simulate X ray spectra of returned data from Mars.

Special Operating Modes

By adjusting the high-voltage bias on the proportional counter detector tubes the effective gain of the XRFS is changed, and portions of a given spectrum can be expanded or contracted. Expansions were determined for operating the detectors on VL 2, so that all useful data from Mars fell above channel 20 (of the 128 channels) to avoid the low-energy noise pulses, whose origin(s) is (are) poorly understood [*B C. Clark et al.*, 1977]. Other tests, operating the detector tubes at reduced voltages and thus contracting the spectra, were aimed at utilizing the nuclear gamma emission (88 keV) from the ^{109}Cd radioactive source. As was reported earlier [*B. C. Clark et al.*, 1976*b*], this emission can be used to determine the bulk density and therefore the estimated porosity of the sample in the cavity.

A further application of the 88-keV emission is the potential for excitation of the K series X ray spectra of heavier elements

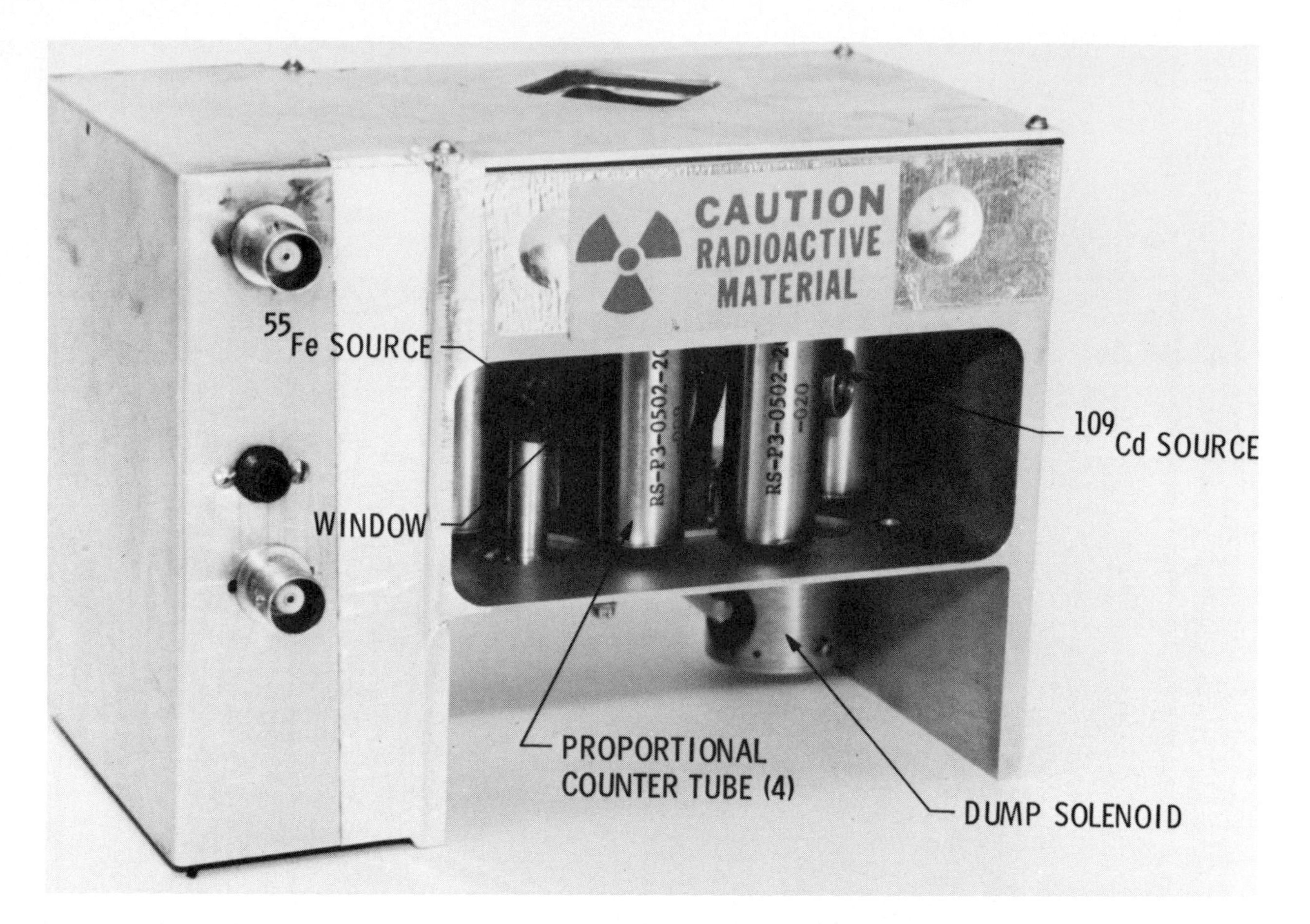

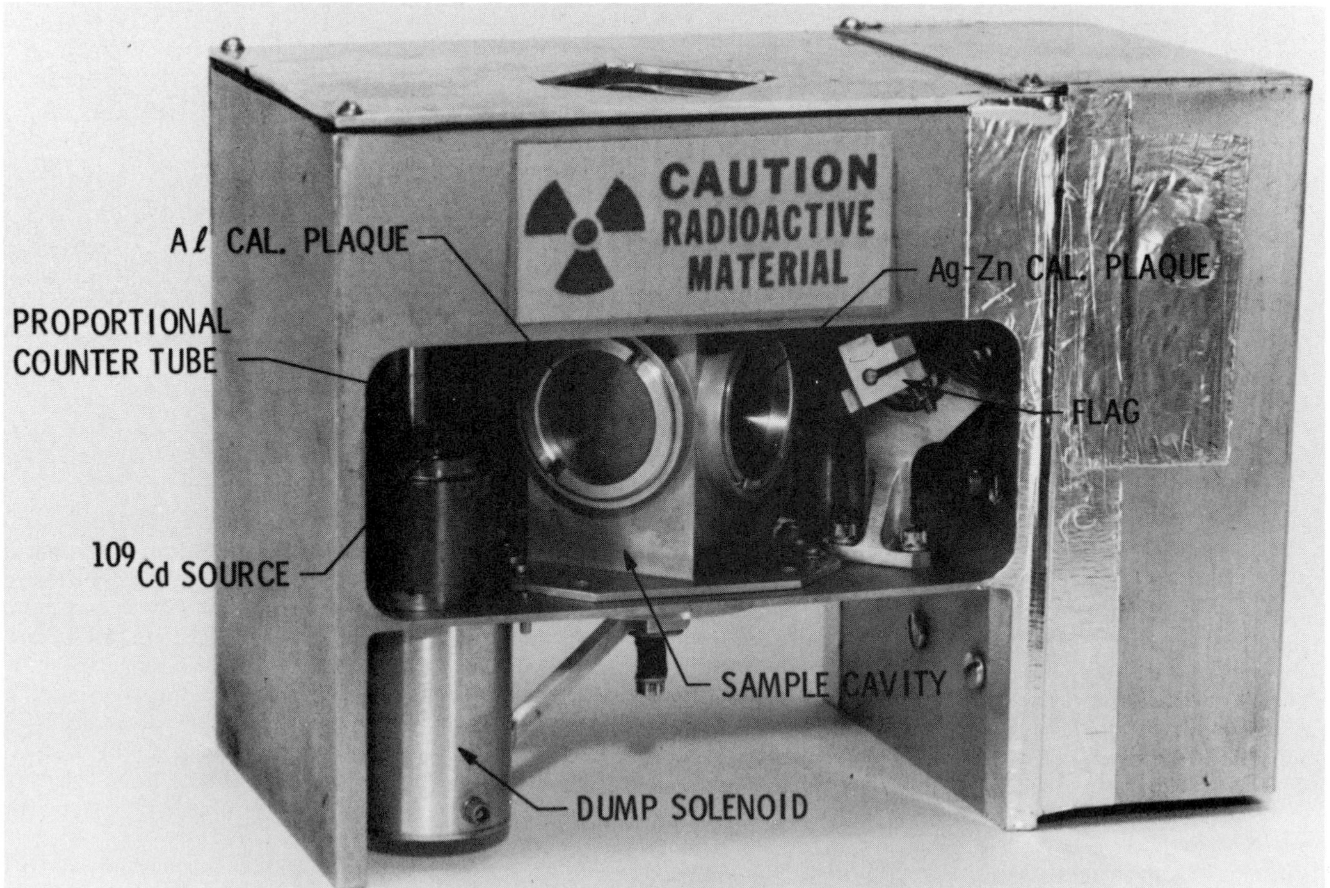

Fig. 21. Science Team sensor assembly (STSA) X ray fluorescence spectrometer. (Top) Front. (Bottom) Back.

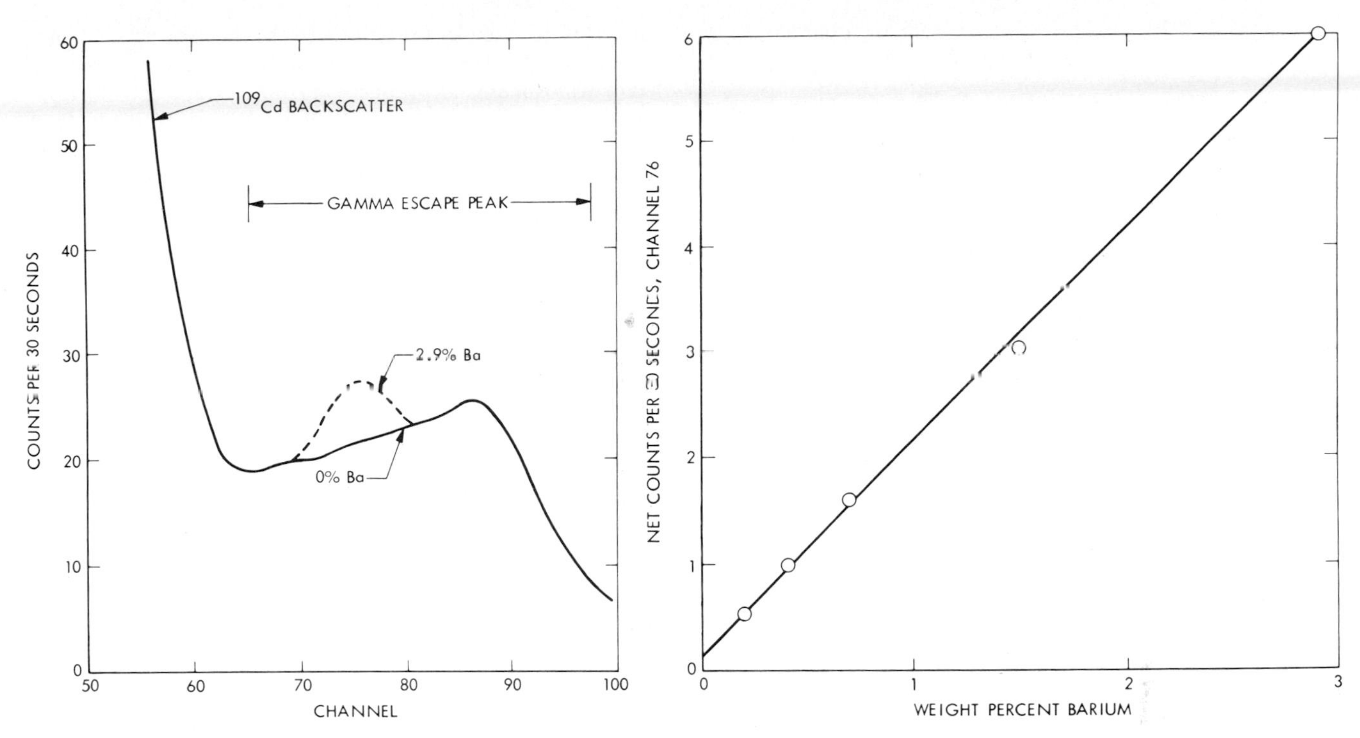

Fig. 22. Determination of Ba in montmorillonite-Fe_2O_3 mixtures. (Left) Portion of reduced voltage PC 3 spectrum over region of 88-keV gamma emission. (Right) Calibration curve for various barium concentrations. See text for discussion.

($Z \cong 47$–81) in the sample. In particular, the element barium is of special interest because its L series emissions are essentially identical to the K series emissions from titanium (4.5 keV). We have reported [*B. C. Clark et al.*, 1976*b*, 1977] that S1 on Mars contains 0.8% TiO_2; however, some portion of this might be due to Ba interference. Figure 22 shows the result of doping a sample (80% montmorillonite, 20% Fe_2O_3) with increasing amounts of Ba (as $BaSO_4$). Total counting time (multichannel analyzer) was 5000 s. From the calibration curve we estimate detection limits of about 0.1% Ba, providing the maximum

TABLE 2. Chemical and Mineralogical Compositions of Analog Material Compared to Mars Sample S1

Name	Riverside Nontronite (6039)	Wyoming Bentonite (7021)	First Biology Analog	Analog (6040)	Analog (6041)	Mars S1
			Oxides			
SiO_2	45.6	63.6	39.5	40.8	44.6	44.7
Al_2O_3	5.1	17.3	7.0	6.8	5.5	5.7
Fe_2O_3	25.0	2.7	21.7	15.5	18.6	18.2
MgO	2.2	1.5	4.3	5.1	5.4	8.3
CaO	6.4	1.4	7.0	7.4	6.0	5.6
Na_2O	0.06	1.7	0.5	0.4	0.4	
K_2O	0.05	0.57	0.17	0.15	0.12	<0.3
TiO_2	0.48	0.24	0.30	0.84	0.70	0.8
P_2O_5		0.10	0.02			
MnO	0.23	0.03	0.20	0.12	0.12	
SO_2	<0.10	0.30	5.45	7.0	8.4	7.7
Cl	<0.10	0	0	0.6	0.5	0.7
Total	85.52	89.44	86.14	84.71	90.34	91.9
			Minerals			
Nontronite	100	0	51.1	59.1	49.5	
Bentonite	0	100	25.5	21.7	18.2	
Kieserite	0	0	9.4	11.8	13.7	
Quartz	0	0	0	0	7.5	
Calcite	0	0	6.0	5.9	4.9	
Hematite	0	0	5.0	0	2.5	
Magnetite	0	0	3.0	0	2.4	
Halite	0	0	0	1.0	0.8	
Leucoxene	0	0	0	0.5	0.4	

Compositions are in percent by weight.

count period (492 s/channel) can be repeatedly used on a Mars sample. These data are now being taken.

Instrument Calibration Standards

Of the 25 XRFS rock standards used for instrument calibration, three (8063, 8064, and 8066) will be of particular importance in the detailed analysis of Mars returned data (see *B. C. Clark et al.* [1977] for discussion). It is especially important that we accurately know the chemical compositions of those exact parts of the cut-rock prisms which were viewed by the radioactive sources and X ray detectors of the XRFS instruments now on Mars and the flight spare XRFS now in our laboratory. Chemical analyses (U.S. Geological Survey, Reston, Virginia) have been made of extra pieces sawn from the rock standards but, of course, could not be made of exactly the same material that composes the actual faces of the prisms. Furthermore, we know that the rock prisms are not perfectly chemically homogeneous (as can be demonstrated by differing X ray spectra from each of the four faces of the prisms).

In order to determine prism face compositions we are using a flightlike sample cavity and flight strength sources but are substituting a high-resolution (157 eV full width at half maximum at Mn $K\alpha$) Si(Li) detector for the proportional counter tubes. The spectra from each face of each rock prism are compared with spectra from analyzed powders of scrap material from the prisms. Because the interelement (matrix) effects are close for the prism and powder of each standard, emissions for each element can be directly ratioed, following normalization to the backscattered source radiations.

Mars Chemical and Mineralogic Analog Samples

The preparation of analogs to simulate Mars results has two largely distinct goals. First, we need mixtures of terrestrial materials with the correct atom proportions to duplicate exactly the spectral response of samples in instruments on Mars, as measured in flightlike (and eventually flight spare) XRFS instruments on earth. These chemical analogs, whatever their mineralogic character, are required to refine the compositions assigned to Mars samples and to establish elemental detection limits and analytical accuracies in a matrix as chemically identical as possible to the Mars sample (see *B. C. Clark et al.* [1977] for discussion). First attempts at the preparation of these analogs have been made [*Baird et al.*, 1976], and further work is proceeding.

The second goal of analog preparation is to simulate not only Martian sample chemistry but also the mineralogical/petrological characteristics of the Martian samples; this is a task of far greater uncertainty and difficulty. In fact, any proof of the validity of such an analog must await another mission to Mars with far more elaborate analytical instruments for mineralogic studies (e.g., X ray diffractometer, differential thermogravimetric and gas effluent analyzer) or a mission to return to earth with samples of the Martian regolith. The state of our present understanding of the mineralogic/petrologic character of Mars samples, as deduced from the results of the XRFS experiments (and as augmented by results from the GCMS, biology, magnetic and physical properties, and lander imaging experiments), is the subject of another report in this issue [*Toulmin et al.*, 1977] and will not be repeated here.

The development of mineralogic analogs is not only important as an interpretive goal of Viking results, but the analogs (and their parent materials) are also needed for related laboratory experiments aimed at testing hypotheses of origins and causes of geochemical reactions on Mars. Important among these is the effect of solar ultraviolet radiation (combined with Mars atmosphere) to produce unusual surface chemical phenomena, as suggested by results from biology [*Horowitz et al.*, 1976; *Levin and Straat*, 1976; *Oyama et al.*, 1977]. We are also concerned about the potential role of UV radiation in the apparent mineralogic contradiction of coexisting products of a highly oxidizing environment (the red orange iron oxides) and possible products of a reducing environment (the montmorillonitic clay minerals proposed). Specific questions are as follows. (1) Under what conditions, similar to those existing now (or earlier) on Mars, could clay minerals be produced from mafic to ultramafic igneous rocks? (2) What happens to clay minerals under present Mars conditions? (3) Of the variety of possible parental iron-bearing compounds, what sequence of reactions best produces material consistent with imaging and magnetic properties results?

For the reasons discussed by *Toulmin et al.* [1977] we favor a mineralogic analog containing a large fraction of iron-bearing smectite clays, most of the remainder being a mix of a sulfur compound, a carbonate compound, and iron oxides. To produce such a mixture of terrestrial minerals that also has the appropriate chemical composition presents problems. Most

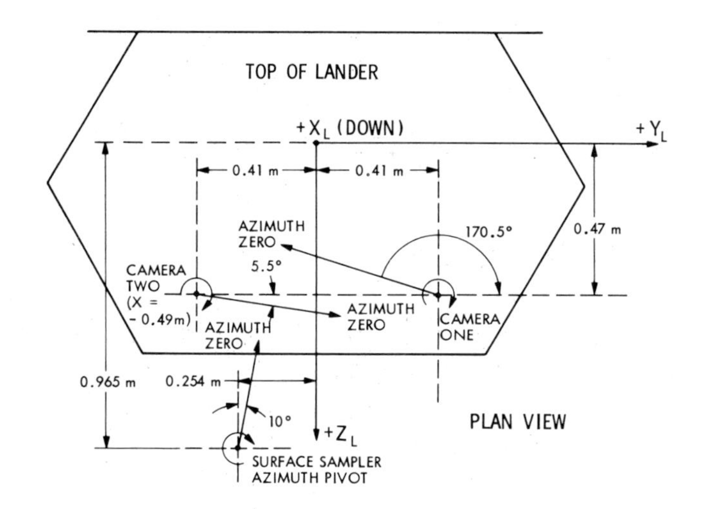

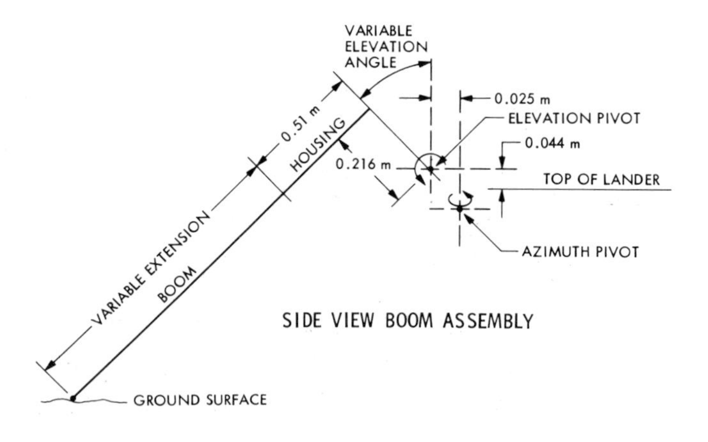

Fig. A1. Viking lander, camera, and surface sampler coordinate systems. Plan view shows relationships between cameras and sampler and their zero azimuth positions. Side view shows relationship between the variable boom extension and elevation angle and the positions of pivotal axes of the sampler.

TABLE A1. Summary of Sample Site Locations: Primary Mission

						Camera Viewing Angles,* deg			
			Lander Coordinates,* m			Camera 1		Camera 2	
Sol	Sample	Sample Site	X_L	Y_L	Z_L	Az.	Elev.	Az.	Elev.
				Viking Lander 1					
8	S1	Sandy Flats	+0.95	+2.25	+2.11	212.2	−30.3	26.1	−24.7
34	S2	Rocky Flats	+0.92	+1.30	+2.54	300.1	−27.6	107.8	−31.9
40	S3	Rocky Flats	+0.92	−1.53	+2.80	300.4	24.9	110.2	−28.6
91	S4	Sandy Flats	+1.05	−1.93	+2.11	217.7	−34.5	29.5	−28.2
				Viking Lander 2					
8		Beta	+0.73	+1.59	+2.77	233.4	−26.2	43.5	−22.2
29, 30	U1	Bonneville	+1.18	−2.20	+2.74			22.5	−29.9
46, 47		Beta	+0.80	+1.61	+2.73	232.6	−26.8	42.7	−23.1
57, 58		Alpha	+1.10	−0.64	+3.10	282.3	−29.2	89.5	−31.0

Az., azimuth; Elev., elevation.
*See Figure A1 for lander and camera coordinate systems.

terrestrial montmorillonitic clays contain less than 6% or more than 25% Fe_2O_3 [*Ross and Hendricks,* 1945]. Of the 20% Fe_2O_3 in Mars soils, 1–7% may be in magnetic oxides [*Hargraves et al.,* 1977], and an additional few percent in nonmagnetic red pigment; this may leave only about 10–13% in the proposed Mars clay. Accordingly, no single terrestrial clay can be used in analog preparation. However, we have found a clay (termed 'Riverside nontronite' from its locality at New City Quarry in Riverside, California), formed as an alteration product of a hedenbergite pyroxenite, which can compose 50% or more of an appropriate chemical analog.

The finest fraction of this material contains (preliminary estimates) 85% nontronite, 7–8% epidote, 3–4% hedenbergite, and 1–2% quartz. The bulk chemical composition is shown in Table 2 along with the composition of a bentonite (7021) which, with the nontronite, forms 60–80% of the mineralogic analogs so far developed. Other possible constituents of the chemical analog are kieserite ($MgSO_4 \cdot H_2O$), calcite ($CaCO_3$), hematite (Fe_2O_3), maghemite (Fe_2O_3), magnetite (Fe_3O_4), rutile (TiO_2), halite (NaCl), and quartz (SiO_2).

Table 2 also shows the chemical and mineralogical compositions of Martian analogs that we have produced to date in comparison with chemical values for sample S1 from Viking Lander 1. Until further refinements are made in our Mars chemical results, improvements in the mineralogic analogs will not be practical. We especially need to have better knowledge of the Mg and Al values for all samples, and we need to complete our analyses of data from the Deep Hole (S6), Atlantic City (S5), and underrock samples (U2 and U4) in order to establish elemental correlations with S and Fe, which exhibit the most prominent elemental variations in our results as we now understand them.

Unfortunately, the UV irradiation tests (and perhaps other tests yet to be proposed) cannot wait until our analytical results on Mars samples are completely refined, and these additional studies may provide important new information on the correct mineralogic model of the Mars regolith. Therefore a large volume of analog material is now being developed which will have the chemical composition of S1 and the approximate mineral composition of 6041 in Table 2.

TABLE A2. Summary of Sample Site Locations: Extended Mission (to April 1, 1977)

						Camera Viewing Angles,* deg			
			Lander Coordinates,* m			Camera 1		Camera 2	
Sol	Sample	Sample Site	X_L	Y_L	Z_L	Az.	Elev.	Az.	Elev.
				Viking Lander 1					
177–180	S5	Atlantic City	+0.98	−0.86	+2.72	290.0	−29.6	95.8	−32.5
229, 250	S6	Deep Hole	+1.01	+2.27	+1.77	205.3	−33.4	20.3	−26.6
				Viking Lander 2					
131	U2	Notch Rock	+0.75	+2.14	+2.12	214.2	−27.2	27.5	−22.1
161	U3	Spalling Valley	+0.80	+1.99	+2.63	224.3	−25.6	36.4	−21.7
185, 186	U4	Badger Rock	+1.13	−1.54	+3.10	297.1	−26.3	107.8	−29.5
131, 145 172, 185 186		Rock Pile	+0.77	+1.94	+2.27	220.2	−28.9	31.9	−23.5

Az., azimuth; Elev., elevation.
*See Figure A1 for lander and camera coordination systems.

TABLE A3. Surface Sampler Activities in Support of XRFS: Viking Lander 1

	Activity	Local Lander Time		Surface Sampler Coordinates*				Delivery/Purge Mode	Estimated Total Material Delivered,¶ cm^3		
		Sol	Time	Azimuth, deg	Elevation Surface Contact, deg	Extension, in. Start	End		Max	Min	Nom
Sample S1, Sandy Flats	Fines delivery								150	50	85
Acquisition 1		8	1046:17	104.5	−24.3	94.1	100.7	Sieve using 60 s of			
Delivery 1		8	1059:32	4.8	−8.5	11.5		HF vibration.			
Acquisition 2		8	1133:17	104.5	−25.0	94.1	100.7	Same			
Delivery 2		8	1146:32	4.8	−8.5	11.5					
Sample S2, Sandy Flats	Rock delivery								75	25	50
Acquisition 1		34	1019:43	203.6	−33.2	62.0	67.5†	Sieve using 60 s of			
Sieve 1		34	1024:11	194.8	−7.9	57.3		HF vibration.			
Delivery 1		34	1033:16	4.8	−10.4	8.4		Delivery using 60 s of LF vibration through jaw.			
Acquisition 2		34	1119:43	203.6	−33.8	62.0	68.6	Same			
Sieve 2		34	1124:11	194.8	−7.9	57.3					
Delivery 2		34	1133:16	4.8	−10.4	8.4					
Sample S3, Rocky Flats	Rock delivery								15	10	12
Acquisition 1		40	1039:01	204.9	−28.8	74.6	81.2	Sieve using 90 s of			
Sieve 1		40	1042:29	204.9	−7.9	69.1		HF vibration.			
Delivery 1		40	1051:22	4.8	−10.4	8.4		Delivery using 60 s of LF vibration through jaw.			
Acquisition 2		40	1219:01	204.9	−28.8	74.6	80.9‡	Same			
Sieve 2		40	1222:29	204.9	−7.9	69.1					
Delivery 2		40	1231:22	4.8	−10.4	8.4					
Sample S4, Sandy Flats	Spill of fines from biology acquisition								20	10	15
Acquisition		91	0711:24	107.7	−28.8	78.3	84.9	Delivery to biology			
Delivery to biology		91	0725:18	14.3	3.4	10.5		PDA using 45 s of HF vibration through sieve.			

Sample S5, Atlantic City	Rock delivery								30	15	20
Acquisition 1		177	1441:11	189.1	−35.1	62.0	67.0§	Sieve using 120 s of HF vibration.			
Sieve 1		177	1445:56	133.6	15.4	56.5		Delivery using 60 s of LF vibration through jaw.			
Delivery 1		177	1454:41	4.8	−10.4	8.4					
Acquisition 2		178	1451:11	189.1	−35.7	62.0	67.5§	Same			
Sieve 2		178	1455:56	133.6	15.4	56.5					
Delivery 2		178	1504:41	4.8	−10.4	8.4					
Acquisition 3		179	1041:11	189.1	−35.1	62.0	67.3§	Same			
Sieve 3		179	1045:56	133.6	15.4	56.5					
Delivery 3		179	1054:41	4.8	−10.4	8.4					
Acquisition 4		180	1041:11	189.1	−35.1	62.0	67.8§	Same			
Sieve 4		180	1045:56	133.6	15.4	56.5					
Delivery 4		180	1054:41	4.8	−10.4	8.4					
Sample 6, Sandy Flats	Deep Hole fines delivery								45	25	40
Acquisition 1		229	0817:32	97.6	−26.2‖	90.9	98.6	Delivery 1 using 120 s of HF vibration through sieve.			
Delivery 1		229	0829:21	4.8	−8.5	11.5		Purge 1 using 60 s of LF vibration through jaw.			
Purge 1		229	0840:33	203.6	−28.8	63.1					
Acquisition 2		250	1117:32	97.6	−26.2‖	90.9	99.6	Delivery 2 using 60 s of HF vibration through sieve.			
Delivery 2		250	1129:21	4.8	−8.5	11.5		Purge 2 using 60 s of LF vibration through jaw.			
Purge 2		250	1139:39	203.6	−28.8	63.1					
Sandy Flats	Deep Hole biology acquisition										
Acquisition		250	0817:34	97.6	−26.2‖	90.9	99.6				
Delivery		250	0834:14	14.3	3.4	10.5		Delivery to biology using 45 s of HF vibration through sieve.			
Purge		250	1010:27	203.6	−28.8	63.1		Purge using 60 s of LF vibration through jaw.			
Total delivery to XRFS									335	135	222

*See Figure A1 for sampler coordinate system.
†Commanded value was 68.6 in.
‡Commanded value was 81.2 in.
§Commanded value was 68.6 in.
‖Boom was preloaded to elevation of −28.8 prior to acquisition stroke.
¶Max, maximum; Min, minimum; Nom, nominal.

TABLE A4. Surface Sampler Activities in Support of XRFS: Viking Lander 2

	Activity	Local Lander Time		Surface Sampler Coordinates*				Delivery/Purge Mode	Estimated Total Material Delivered,‡ cm³		
		Sol	Time	Azimuth, deg	Elevation Surface Contact, deg	Extension, in. Start	Extension, in. End		Max	Min	Nom
Beta site	Rock delivery, purge of biology acquisition								8	0	4
Acquisition		8	1610:42	124.7	−23.1	85.2	91.7†	Delivery using 90 s of LF vibration through jaw.			
Delivery		13	1646:38	4.2	−9.2	8.4					
Sample U1, Bonneville	Fines delivery								100	50	75
Acquisition 1		29	1342:35	217.5	−29.4	93.0	99.4	Delivery using 90 s of HF vibration through sieve.			
Delivery 1		29	1354:04	4.2	−8.5	11.5					
Acquisition 2		30	1042:35	217.5	−30.0	93.0	99.4	Same			
Delivery 2		30	1054:04	4.2	−8.5	11.5					
Beta site	Rock delivery										
Acquisition 1		46	1308:14	123.5	−23.1	85.2	91.7	Sieve using 90 s of HF vibration.	2	0	1
Sieve 1		46	1316:34	123.5	+11.0	83.1					
Delivery 1		46	1330:50	4.2	−9.2	8.4		Delivery using 60 s of LF vibration through jaw.			
Acquisition 2		46	1348:14	123.5	−23.7	85.2	91.7	Same	2	0	1
Sieve 2		46	1356:34	123.5	+11.0	83.1					
Delivery 2		46	1410:50	4.2	−9.2	11.5					
Acquisition 3		47	1308:14	123.5	−23.7	85.2	91.7	Same	2	0	1
Sieve 3		47	1316:34	123.5	+11.0	83.1					
Delivery 3		47	1330:50	4.2	−9.2	11.5					
Acquisition 4		47	1348:14	123.5	−23.7	85.2	91.7	Same	2	0	1
Sieve 4		47	1356:34	123.5	+11.0	83.1					
Delivery 4		47	1410:50	-4.2	−9.2	11.5					
Alpha site	Rock delivery										
Acquisition 1		57	0808:51	180.3	−33.2	75.2	81.7	Sieve using 90 s of HF vibration.	2	0	1
Sieve 1		57	0812:40	180.3	+18.0	63.3					
Delivery 1		57	0821:38	4.2	−9.2	8.4		Delivery using 60 s of LF vibration through jaw.			
Acquisition 2		57	0853:51	180.3	−33.2	75.2	81.7	Same	2	0	1
Sieve 2		57	0857:40	180.3	+18.0	63.3					
Delivery 2		57	0906:38	4.2	−9.2	8.4					
Acquisition 3		58	0808:51	179.0	−32.5	75.2	81.7	Same	2	0	1
Sieve 3		58	0812:40	181.0	+18.0	63.3					
Delivery 3		58	0821:38	4.2	−9.2	8.4					
Acquisition 4		58	0853:51	179.0	−32.5	75.2	81.7	Same	2	0	1
Sieve 4		58	0857:40	181.0	+18.0	63.3					
Delivery 4		58	0906:38	4.2	−9.2	8.4					
Sample U2, Notch Rock	Fines delivery, purge to Rock Pile								75	25	50
Acquisition		131	1043:51	105.8	−21.2	88.0	94.6	Delivery using 90 s of HF vibration through sieve.			
Delivery		131	1053:01	4.2	−8.5	11.5					
Purge		131	1108:08	111.5	−19.9	79.9		Purge using 120 s of LF vibration through jaw.			

Beta site	Biology acquisition, purge to Rock Pile										
Acquisition		145	0910:52	126.0	−26.2	74.1	80.7	Delivery to biology using 45 s of HF vibration through sieve. Purge using 60 s of LF vibration through jaw.			
Delivery to biology		145	0924:38	13.6	+3.4	10.2					
Purge		145	1034:48	111.5	−19.9	79.9					
Spalling Valley	No delivery, purge to Rock Pile										
Acquisition 1		145	1339:09	116.5	−21.2	93.6	100.1	Sieve using 120 s of HF vibration. Purge using 60 s of LF vibration through jaw.			
Sieve 1		145	1347:57	238.4	+10.4	10.5					
Purge 1		145	1358:06	111.5	−19.9	79.9					
Acquisition 2		145	1439:21	115.3	−21.2	93.6	100.1	Same			
Sieve 2		145	1448:51	238.4	+10.4	10.5					
Purge 2		145	1458:16	111.5	−19.9	79.9					
Acquisition 3		145	1539:09	116.5	−21.2	93.6	100.1	Same			
Sieve 3		145	1547:57	238.4	+10.4	10.5					
Purge 3		145	1558:06	111.5	−19.9	79.9					
Acquisition 4		145	1639:21	115.3	−21.1	93.6	100.1	Same			
Sieve 4		145	1648:51	238.4	+10.4	10.5					
Purge 4		145	1658:16	111.5	−19.9	79.9					
Acquisition 5		172	1209:09	114.0	−20.5	93.6	100.1	Same			
Sieve 5		172	1217:57	238.4	+10.4	10.5					
Purge 5		172	1228:06	111.5	−19.9	79.9					
Acquisition 6		172	1249:09	114.0	−20.5	93.6	100.1	Same			
Sieve 6		172	1257:57	238.4	+10.4	10.5					
Purge 6		172	1308:06	111.5	−19.9	79.9					
Acquisition 7		172	1329:09	114.0	−21.2	93.6	100.1	Same			
Sieve 7		172	1337:57	238.4	+10.4	10.5					
Purge 7		172	1348:06	111.5	−19.9	79.9					
Acquisition 8		172	1409:09	114.0	−21.2	93.6	100.1	Same			
Sieve 8		172	1417:57	238.4	+10.4	10.5					
Purge 8		172	1428:06	111.5	−19.9	79.9					
Sample U3, Spalling Valley	Quasi-deep fines, delivery and purge to Rock Pile								120	40	80
Acquisition 1		161	1118:50	116.5	−21.2	93.6	100.1	Delivery using 90 s of HF vibration through sieve. Purge using 60 s of LF vibration through jaw.			
Delivery 1		161	1133:14	4.2	−8.5	11.5					
Purge 1		161	1142:24	111.5	−19.9	79.9					
Acquisition 2		161	1203:50	116.5	−21.2	93.6	100.1	Same			
Delivery 2		161	1218:14	4.2	−8.5	11.5					
Purge 2		161	1227:24	111.5	−19.9	79.9					
Sample U4, Badger Rock	Fines delivery, purge to Rock Pile								50	25	30
Acquisition 1		185	1410:06	201.1	−30.6	87.0	93.6	Delivery using 90 s of HF vibration through sieve. Purge using 60 s of LF vibration through jaw.			
Delivery 1		185	1425:52	4.2	−8.5	11.5					
Purge 1		185	1435:02	111.5	−19.9	79.9					
Acquisition 2		186	1310:06	201.1	−30.6	87.0	93.6	Same			
Delivery 2		186	1325:52	4.2	−8.5	11.5					
Purge 2		186	1335:02	111.5	−19.9	79.9					
Total delivery to XRFS									369	140	247

*See Figure A1 for sampler coordinate system.

†Commanded value was 91.7 in. but was not achieved.

‡Max, maximum; Min, minimum; Nom, nominal.

Acknowledgments. Without the dedicated help of the Surface Sampler Team (L. V. Clark, D. S. Crouch, W. R. DeShazor, R. D. Grossart, D. D. Pike, and L. K. Schwab) in the design, testing, and execution of sampling operations, our X ray experiment would have been seriously compromised. J. Gliozzi, J. F. Elzey, and K. Z. Bradford very ably coordinated sampling activities in the planning of mission operations. R. R. Moore was responsible for operating and maintaining laboratory equipment and for preparing instrument performance summaries. S. Kenley compiled summaries of surface sampler operations, sample site locations, and associated lander imagery. R. Weldon, as Science Intern, contributed greatly to preparation and analysis of analog materials. N. Brosnahan built laboratory apparatus and aided in the nondispersive X ray analysis of rock calibration standards. This work was supported by NASA (Viking program) grants NAS 1-11855, NAS 1-11858, L-9717, and NAS 1-9000.

REFERENCES

Baird, A. K., P. Toulmin III, B. C. Clark, H. J. Rose, Jr., K. Keil, R. P. Christian, and J. L. Gooding, Mineralogic and petrologic implications of Viking geochemical results from Mars: Interim report, *Science, 194,* 1288–1293, 1976.

Clark, B. C., P. Toulmin III, A. K. Baird, K. Keil, and H. J. Rose, Jr., Argon content of the Martian atmosphere at the Viking I landing site: Analysis by X-ray fluorescence spectroscopy, *Science, 193,* 804–805, 1976*a*.

Clark, B. C., A. K. Baird, H. J. Rose, Jr., P. Toulmin III, K. Keil, A. J. Castro, W. C. Kelliher, C. D. Rowe, and P. H. Evans, Inorganic analyses of Martian surface samples at the Viking landing sites, *Science, 194,* 1283–1288, 1976*b*.

Clark, B. C., A. K. Baird, H. J. Rose, Jr., P. Toulmin III, R. P. Christian, W. C. Kelliher, A. J. Castro, C. D. Rowe, K. Keil, and G. Huss, The Viking X-ray fluorescence experiment: Analytical methods and early results, *J. Geophys. Res., 82,* this issue, 1977.

Clark, L. V., D. S. Crouch, and R. D. Grossart, Summary of Primary Mission surface sampler operations, July 20–Nov. 2, 1976, *Viking Intern. Rep. VFT-019,* NASA, 1977.

Hargraves, R. B., D. W. Collinson, R. E. Arvidson, and C. R. Spitzer, The Viking magnetic properties experiment: Primary Mission results, *J. Geophys. Res., 82,* this issue, 1977.

Horowitz, N. H., G. L. Hobby, and J. S. Hubbard, The Viking carbon assimilation experiments: Interim report, *Science, 194,* 1321–1322, 1976.

Lee, B. G., Mission operations strategy for Viking, *Science, 194,* 59–62, 1976.

Levin, G. V., and P. A. Straat, Viking labeled release biology experiment: Interim results, *Science, 194,* 1322–1329, 1976.

Mutch, T. A., A. B. Binder, F. O. Huck, E. C. Levinthal, S. Liebes, Jr., E. C. Morris, W. R. Patterson, J. B. Pollack, C. Sagan, and G. R. Taylor, The surface of Mars: The view from the Viking 1 lander, *Science, 193,* 791–801, 1976*a*.

Mutch, T. A., R. E. Arvidson, A. B. Binder, F. O. Huck, E. C. Levinthal, S. Liebes, Jr., E. C. Morris, D. Nummedal, J. B. Pollack, and C. Sagan, Fine particles on Mars: Observations with the Viking 1 lander cameras, *Science, 194,* 87–91, 1976*b*.

Mutch, T. A., S. U. Grenander, K. L. Jones, W. Patterson, R. E. Arvidson, E. A. Guiness, P. Avrin, C. E. Carlston, A. B. Binder, C. Sagan, E. W. Dunham, P. L. Fox, D. C. Pieri, F. O. Huck, C. W. Rowland, G. R. Taylor, S. D. Wall, R. Kahn, E. C. Levinthal, S. Liebes, Jr., R. B. Tucker, E. C. Morris, J. B. Pollack, R. S. Saunders, and M. R. Wolf, The surface of Mars: The view from Viking 2 lander, *Science, 194,* 1277–1283, 1976*c*.

Oyama, V. I., B. J. Berdahl, and G. C. Carle, Preliminary findings of the Viking gas exchange experiment and a model for Martian surface chemistry, *Nature, 265,* 110–114, 1977.

Ross, C. S., and S. B. Hendricks, Minerals of the montmorillonite group: Their origin and relation to soils and clays, *U.S. Geol. Surv. Prof. Pap., 205-B,* 23–79, 1945.

Shorthill, R. W., R. E. Hutton, H. J. Moore II, R. F. Scott, and C. R. Spitzer, Physical properties of the Martian surface from the Viking 1 lander: Preliminary results, *Science, 193,* 805–809, 1976*a*.

Shorthill, R. W., H. J. Moore II, R. F. Scott, R. E. Hutton, S. Liebes, Jr., and C. R. Spitzer, The 'soil' of Mars (Viking 1), *Science, 194,* 91–97, 1976*b*.

Shorthill, R. W., H. J. Moore II, R. E. Hutton, R. F. Scott, and C. R. Spitzer, The environs of Viking 2 lander, *Science, 194,* 1309–1318, 1976*c*.

Toulmin, P., III, A. K. Baird, B. C. Clark, K. Keil, H. J. Rose, Jr., R. P. Christian, and P. H. Evans, Geochemical and mineralogical interpretation of the Viking inorganic chemical results, *J. Geophys. Res., 82,* this issue, 1977.

(Received May 1, 1977;
revised May 23, 1977;
accepted May 24, 1977.)

Geochemical and Mineralogical Interpretation of the Viking Inorganic Chemical Results

PRIESTLEY TOULMIN III,[1] A. K. BAIRD,[2] B. C. CLARK,[3] KLAUS KEIL,[4]
H. J. ROSE, JR.,[1] R. P. CHRISTIAN,[1] P. H. EVANS,[2] AND W. C. KELLIHER[5]

The elemental analyses whose basis is described in the preceding two papers represent the composition of samples of Martian fines; the only undetermined major constituents thought to be present are H_2O, CO_2, Na_2O, and possibly NO_x. The samples are principally silicate particles, with some admixture of oxide and probably carbonate minerals; the fines appear to have been indurated to a variable degree by a sulfate-rich intergranular cement. The overall elemental composition is dissimilar to any single known mineral or rock type and apparently represents a mixture of materials. Close chemical similarity among samples at each site, and between the two sites, indicates effective homogenization of the fines, presumably by planetary windstorms, and further suggests that the samples analyzed represent the fine, mobilizable materials over a large part of the planet's surface. Low trace element, alkali, and alumina contents suggest that the great preponderance of the materials in the mixture is of mafic derivation; highly differentiated, salic igneous rocks or their weathering products are insignificant components of the samples. Normative calculations, comparisons with reference libraries of analytical data, and mathematical mixture modeling have led to a qualitative mineralogical model in which the fines consist largely of iron-rich smectites (or their degradation products), carbonates, iron oxides, probably in part maghemite, and sulfate minerals concentrated in a surface duricrust. The original smectites may have formed by interaction of mafic magma and subsurface ice, and the sulfates (and carbonates?) may have been concentrated in the surface crust by subsurface leaching, upward transport, and evaporation of intergranular moisture films. Testing and refinement of this and competing models will accompany continuing acquisition of samples and data and refinement of the analyses, particularly with respect to the critical light elements Mg, Al, and Si.

INTRODUCTION

This paper presents the current state of geochemical, mineralogical, and petrologic interpretation of the data whose acquisition and refinement to date are described in the preceding two papers [*Clark et al.*, 1977; *Baird et al.*, 1977]. In order to present a comprehensive review in this collection of Viking results, we have necessarily included here some material that has been previously reported [*Toulmin et al.*, 1976; *Clark et al.*, 1976*a*, *b*; *Baird et al.*, 1976].

The analyses that have been reduced to numerical terms are tabulated (as oxides) for reference in Table 1. The close similarity among them is evident. Although we have not yet reduced the spectral data for the light elements Mg and Al in sample U1 to numerical abundances, inspection of the spectra indicates that their abundances do not differ greatly from those found in the two samples from Chryse Planitia.

The principal intersample difference demonstrated by the tabulated data is that the sulfur content of S2 and S3 (the 'pebble' samples from Rocky Flats) is higher than that in the fines. Qualitative examination of spectra from sample S5, an additional pebble sample taken at a different location ('Atlantic City') at the Chryse landing site, indicates an even higher sulfur content than that of S3, while S6, a sample of fines from a 23-cm-deep trench, has less sulfur (and more calcium) than S3, thus strengthening the association of high sulfur with pebble-type samples. As will be discussed in greater detail below, several lines of reasoning lead us to the interpretation that the pebbles we have acquired are actually fragments of a sulfate-cemented duricrust that appears to be widely distributed at both landing sites. At Utopia the spectra indicate significantly lower Fe contents in fines collected under rocks than in samples collected on the unprotected surface.

In view of the general similarities among the samples, S1 will be taken as the principal sample for discussion and interpretation of the general lithologic and mineralogic character of the Martian surface materials. Differences between it and other samples are believed to result from relatively minor variations in the proportions of phases present. Inasmuch as the reduction of analysis S1 is more advanced than the reductions of the other samples, we will first attempt to interpret the general features of its mineralogy and lithology and then proceed to a discussion of the variations represented by other samples.

GEOCHEMICAL INTERPRETATION

Constituents Not Determined Directly

The deficit in the reported analysis, i.e., the difference between 100% and the sum of constituents reported, must be attributed to (1) analytical error, (2) the presence of elements lighter than Mg (i.e., $Z \leq 11$), and (3) the sum of small amounts of heavier elements present in concentrations below their respective detection limits. Analytical errors are not likely to account for the low totals; the analyses have been preliminarily corrected for most known systematic errors, and random errors should not lead to systematic underestimation of constituents. Furthermore, the effects of many potential sources of systematic error are removed by normalization of fluorescent X ray intensities with respect to the intensity of the backscattered primary radiation. Finally, frequent calibration of instrument response by use of both calibration plaques and flags [*Toulmin et al.*, 1973; *Clark et al.*, 1977] and by long-term

[1] U.S. Geological Survey, Reston, Virginia 22092.
[2] Department of Geology, Pomona College, Claremont, California 91711.
[3] Martin-Marietta Corporation, Denver, Colorado 80201.
[4] Department of Geology and Institute of Meteoritics, University of New Mexico, Albuquerque, New Mexico 87131.
[5] NASA Langley Research Center, Hampton, Virginia 23665.

Paper number 7S0464.

repetitive analysis minimizes the probability of serious undetected systematic errors leading to general underestimation of elements. With regard to elements heavier than Na, the X ray fluorescence spectrometer (XRFS) is particularly sensitive to the major rock-forming elements and to the geochemically important elements Rb, Sr, Zr, and Y, but other elements with relatively high limits of detection [*Clark et al.*, 1976*a*] might in principle contribute to the analysis deficit. On general geochemical or cosmochemical grounds, however, most of these elements would be expected to be present at such low levels that they could not contribute significantly to the analytical deficit. Possible exceptions are Mn and Ni, each of which has a detection limit of 2% at the present stage of data analysis. More refined analysis may permit closer tolerances on these two elements, although their detection will depend greatly on resolving their emission intensities from the much larger emission of iron.

The principal constituents of the deficit seem most likely to be substances inherently not detectable by the Viking XRFS, the most important of which are H_2O, CO_2, Na (as Na_2O), and NO_x. Some very light elements or their compounds, e.g., Li_2O, B_2O_3, NH_3, and F, are also possible but would not be expected with as high a probability as the others. Furthermore, several of the possible candidates generally occur in readily volatilizable form and should have been detected by the gas chromatograph–mass spectrometer (GCMS). We will therefore limit our principal discussion of deficit constituents to H_2O, CO_2, Na_2O, and NO_x.

Both CO_2 and H_2O are evolved from the heated samples analyzed by the Viking GCMS [*Biemann et al.*, 1977]. The semiquantitative results obtained indicate that a probable H_2O content of several tenths of a percent to several percent of the sample can be driven off by heating at 500°C. Earth-based observations of infrared reflection spectra have been interpreted to show that the surface materials contain of the order of 1% 'bound' H_2O [*Houck et al.*, 1973], and some data of the Viking Mars Atmospheric Water Detection Experiment suggest the possibility of diurnal cycling of H_2O between the surface and the lower atmosphere [*Farmer et al.*, 1976*a*]. Even if the diurnal cycling is only local, or entirely within the atmosphere (i.e., between vapor and suspended crystalline particles), seasonal transport between a buried permafrost and the atmosphere seems highly likely [*Farmer et al.*, 1976*b*; C. B. Farmer, private communication, 1977]. Such H_2O would presumably be rather loosely adsorbed on the surfaces of mineral grains, but hydrated and hydroxylated minerals that would decompose wholly or in part in the GCMS heating may well exist in the Martian surface materials. We therefore believe it reasonable to suppose that H_2O may account for a significant fraction of the mass deficit in our analyses.

Simple thermodynamic calculations show that carbonates of the alkaline-earth elements (e.g., calcite, dolomite, magnesite) are stable in relation to their common pyrogenic silicates (e.g., pyroxene, plagioclase) at the temperatures and CO_2 pressures existing at the Martian surface [*Toulmin et al.*, 1973, p. 172; *O'Connor*, 1968; *Gooding*, 1977].

The rates of direct carbonation reactions between solid and gas are generally regarded as impracticably slow for effective weathering, and in the past the presence of liquid water to promote reaction has been regarded as essential. In recent years, however, some studies [*Huguenin*, 1974; *Booth and Kieffer*, 1977] have suggested that the direct reactions may proceed at significant rates in the presence of ultraviolet radiation fluxes of strengths comparable to those believed to characterize the surface environment of Mars [*Huguenin*, 1974, p. 3896]. Given a reaction mechanism and a thermodynamic potential favoring the development of carbonate minerals, we infer that over geologic time, some carbonate minerals should have formed as weathering products at or near the surface of the planet. If our samples contain such carbonates, their CO_2 would contribute to the analytical deficit. CO_2 sufficient to account for all the CaO as calcite, for example, would amount to 4–4.5% of the samples. The CO_2 detected by the GCMS is thought to be degassed in part from surface sites and perhaps in part from partial decomposition of carbonate minerals (equilibrium vapor pressures for such common alkaline-earth carbonates as calcite, dolomite, and magnesite range from 10^{-4} to 10 atm at 500°C, the highest temperature attained by the GCMS ovens (data are from *Robie and Waldbaum* [1968]; see also *Stern and Weise* [1969])).

The role of Na_2O, the third major candidate to explain the deficit, is more difficult to assess, for we have little external information to guide us. On grounds of geochemical coherence the very low content of other alkali metals (K, Rb) may be consistent with a low value for sodium as well.

It has been suggested [*Yung et al.*, 1977] that nitrate minerals in the surface materials may account for a significant part of the planet's nitrogen. Our data cannot exclude the possibility of such materials, some of which may be sufficiently stable thermally not to reveal their presence by evolving nitrogen oxides on heating in the GCMS (though nitrogen oxides are copiously evolved from soils collected in the Dry Valleys of Antarctica when heated in a simulated GCMS [*D. M. Anderson*, 1977]). The only nitrate minerals of more than accidental occurrence are salts of Na and K, which for the reasons mentioned above seem unlikely as major constituents. Certain ammonium compounds, especially silicates, would probably not be detected by either the XRFS or the GCMS.

In summary, the indirect evidence available points to H_2O and CO_2 as the principal components of the deficit; Na_2O or nitrates cannot be excluded but are not as well supported by other evidence.

One last point should be made with respect to the deficit: It is based upon the convention of calculating the analysis in oxide form. In an iron- and sulfur-rich sample such as this one, variations in the assumed oxidation state could greatly increase the deficit.

Oxidation State of Sulfur and Nature of Pebbles

The question of the oxidation state of sulfur requires some discussion. The element occurs in minerals principally in the valence states −2 and −1 (sulfides), 0 (native sulfur), and +6 (sulfates). Only the highest oxidation state is stable in environments equilibrated with the terrestrial atmosphere, and most of the sulfur in surficial weathering products (except in unusually reducing environments such as euxinic muds) on earth is present as sulfate minerals. There are several lines of evidence leading to the inference that the surface environments of both Chryse and Utopia Planitiae are highly oxidizing. The pervasive red color is generally regarded as resulting from the abundance of ferric oxides or hydroxides derived by oxidation of ferrous minerals in the primary rocks [*Salisbury and Hunt*, 1969; *Binder and Jones*, 1972; *Adams and McCord*, 1969]. The results of two of the Viking biology experiments indicate that elemental oxygen is released upon moistening the soil and that the soil is capable of rapidly oxidizing the carbon of organic nutrients with which it comes into contact [*Klein et al.*, 1976;

Levin and Straat, 1976]. The data of the magnetic properties experiment show that the Martian surface materials contain a few percent of a rather highly magnetic mineral whose color is more suggestive of maghemite (γ-Fe_2O_3) than of less highly oxidized, strongly magnetic minerals such as magnetite (Fe_3O_4), pyrrhotite ($Fe_{1-x}S$), or native iron [*Hargraves et al.,* 1976]. In such a highly oxidizing environment we would expect to find sulfur in sulfates rather than in more reduced forms.

Most mineral sulfates are stable against thermal decomposition to temperatures above 500°C and therefore would not be expected to evolve detectable amounts of sulfur-containing volatile species in the GCMS experiments [*Stern and Weise,* 1966; *Knopf and Staude,* 1955; *Biemann et al.,* 1976, 1977]. Sulfide minerals vary more widely in their thermal stability and equilibrium partial vapor pressures of sulfur-containing species [*Barton and Skinner,* 1977]; specifically, equilibrium values of pS_2 for the iron sulfides commonly found in rocks may range from $\sim 10^{-15}$ for troilite (FeS) to $\sim 10^{-5}$ for the assemblage pyrite (FeS_2)-pyrrhotite ($Fe_{1-x}S$) at 500°C [*Toulmin and Barton,* 1964]. In the presence of 1 atm H_2, as in some GCMS experiments, large pressures of H_2S would be generated at equilibrium, and although the brief heating period (30 s) would probably not permit attainment of equilibrium, the absence of detectable sulfur-containing volatiles in the GCMS experiments tends to support the inference of sulfates.

As has been remarked previously, sulfur (and possibly chlorine) [*Baird et al.,* 1977] is significantly more abundant in the samples of pebble-sized fragments lying on the surface than in the fine-grained material more generally distributed at the Chryse landing site (improving the precision of our Cl determinations, which bear importantly on the nature of the pebbles, is of high priority in our planned future refinement of the data). The intensity of the backscattered 88-keV gamma ray emitted by the ^{109}Cd source indicates that the pebble-sized fragments have about the same density as the fines, about 1.1 g cm^{-3} [*Clark et al.,* 1976*a*].

After our early tentative interpretation of the fines as probable weathering products, we concentrated on the acquisition of pebbles in the hope that they would prove to be small fragments of the apparently volcanic rocks that are so obviously and abundantly strewn about the Martian landscape at both landing sites. In view of the close similarity in composition between the pebbles and the fines, the high sulfur content of the pebbles (which is not typical of primary igneous rocks), and the low density of the pebbles, it appears much more likely that the pebbles acquired are aggregates of fine material indurated to a sufficient degree to withstand the mechanical agitation of the sieving process in the rock-mode sample delivery [*Baird et al.,* 1977]. The Martian surface at both landing sites shows areas of an apparent crustified layer one to a few centimeters thick. The most reasonable interpretation of our results to date is that the patches of pebbles on the surface are fragments of this crust, which we would interpret as a caliche-like duricrust, probably cemented by sulfate and possibly chloride minerals. The cement may have formed by the leaching of soluble ions from rock particles into a thin intergranular film of moisture and its ultimate deposition close to the surface when the moist intergranular film evaporated into the atmosphere. Considerable evidence now exists [*Farmer et al.,* 1976*a*, *b*; C. B. Farmer, personal communication, 1977] for seasonal and perhaps diurnal variations in the H_2O vapor content of the lowermost layers of the Martian atmosphere, and the data are consistent with cycling between the atmosphere and the surface, where the H_2O presumably resides as thin intergranular adsorbed films. If these films become thick enough to behave as a bulk water phase, even locally and temporarily, they should be capable of dissolving soluble materials in the manner described. Indeed, the chemical reactivity of adsorbed films too thin to behave as a true liquid phase may be sufficient to permit the leaching process, though the surfaceward migration may require bulk movement of a solution phase.

Additional support for our interpretation of sulfate as the cementing medium for the pebbles of duricrust results from qualitative and semiquantitative examination of the data we have so far received in the extended phase of the Viking mission. As was noted previously, S5, a pebble sample taken about 30 cm from S2, shows even higher sulfur than the two previous pebble samples at Rocky Flats (S2 and S3), while fines at Chryse (S1 and S6) show significantly less sulfur than any of the indurated samples, confirming our previously inferred association of high sulfur content with induration of surface materials.

Comparison of Elemental Composition With Various Materials

A direct comparison of the major abundances and ratios in the Martian samples with major classes of terrestrial and lunar igneous rocks (Figures 1 and 2) shows that despite some apparent similarities the Martian samples are distinct from these populations and do not fall on the trends defined by them. Relationships involving Al_2O_3 are especially anomalous, reflecting the low Al content relative to other silica-saturating bases that characterizes the Martian material. Aluminum plays an important 'fluxing' role in the partial fusion of ultramafic compositions, such as those believed to characterize the mantles of the terrestrial planets; it is concentrated in the low-melting fraction and is expressed mineralogically by the abundance of feldspars in the generally basaltic rocks that dominate the direct products of partial fusion of such materials. Only igneous rocks representing much more primitive compositions (i.e., more nearly complete fusion of the planetary mantle) are as low in Al_2O_3 relative to the other bases, and in these the SiO_2 content is similarly reduced so as to produce silica-undersaturated minerals like olivines. The Martian samples, by contrast, are quartz normative and do not correspond to the ultramafic compositions generally believed to characterize planetary interiors. Thus the major element composition of the Martian samples does not seem to correspond to any recognized igneous rock type.

One group of minerals characterized by high iron and generally low aluminum and bases is the iron-rich smectite clays, such as nontronite and lembergite. We will return to a discussion of this topic in a later section.

Special note should be taken of the work of *Hunt et al.* [1973] and *Logan et al.* [1975], who interpreted certain features in the infrared spectra of the Martian dust cloud observed during the Mariner 9 mission [*Hanel et al.,* 1972] as indicative of montmorillonitic clay particles as a major component of the dust. This result is, of course, in general accord with our model for the surface fines, as is the suggestion that the data are consistent with the presence of sulfates [*Logan et al.,* 1975, p. 141]. It should be pointed out, however, that the same authors state that major amounts of hematite or carbonates cannot be present in the dust, a conclusion not in such good agreement with our model. Furthermore, alternative interpretations of the infrared spectra are possible—*Aronson and Emslie* [1975], for example, point out that the data are compatible with other plausible materials, including tectosilicates as well as true

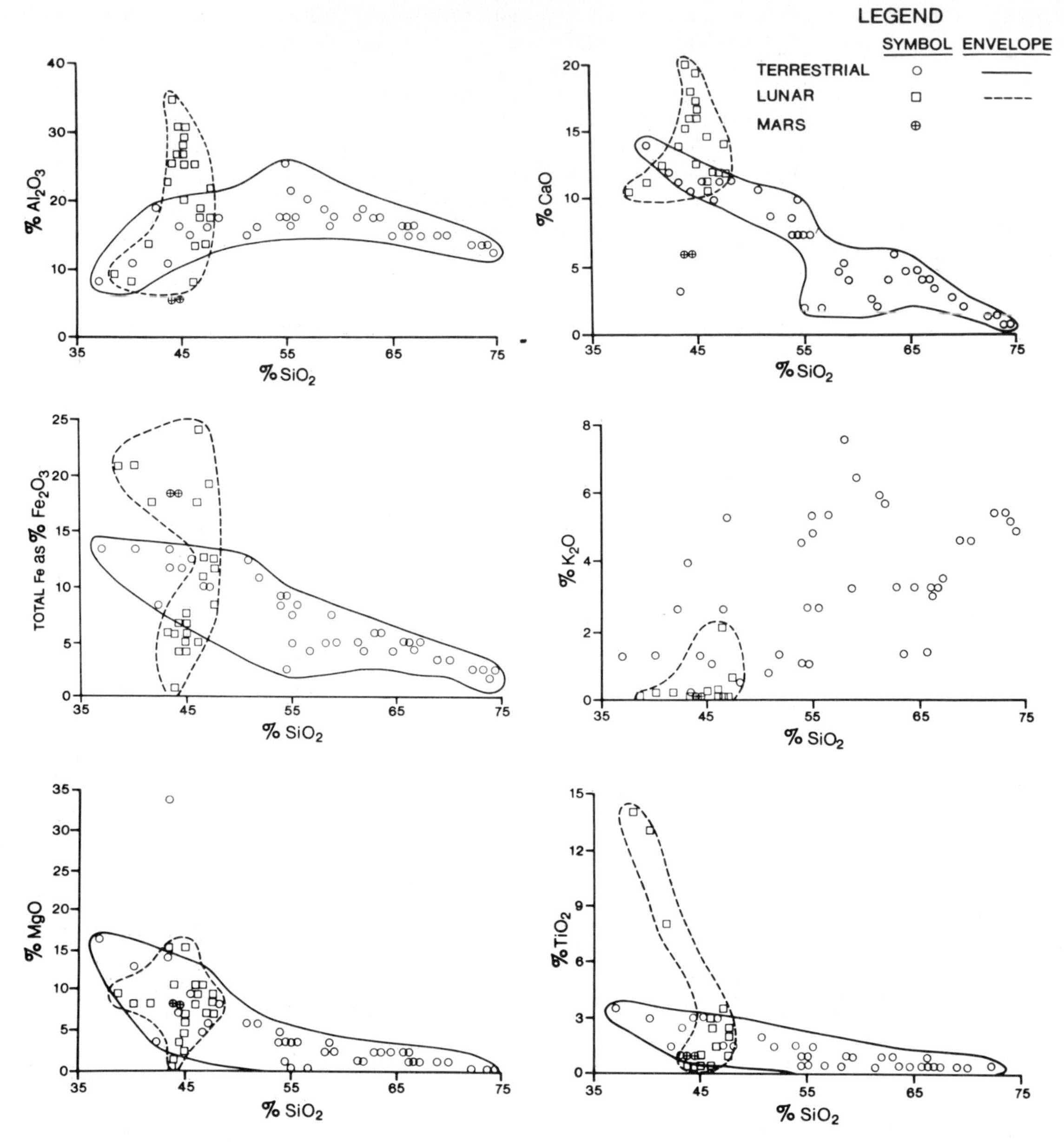

Fig. 1. Harker variation diagrams for terrestrial and lunar igneous rocks and Martian fines. Lunar rocks are basaltic and anorthositic rocks and breccias from the Apollo 11, 12, 14, 15, 16, and 17 missions [*Rose et al.*, 1973, 1975; *Taylor*, 1975]. Terrestrial igneous rocks are averages of common rock types [*Wedepohl*, 1969]. Fields enclosing the vast majority of the two groups have been delineated subjectively.

micas, not all of which, however, are consistent with our elemental data on the surface fines.

Similarly, comparison with meteorites, using the classes and compositions of *Keil* [1969], reveals a clear distinction between the Martian samples and the trends defined by meteorite compositions. Figure 2 illustrates several of these points. The Martian samples fall on the linear CaO-Al_2O_3 trends defined by meteorites or by some lunar samples, clearly distinct from the much more highly differentiated population of terrestrial igneous rocks (Figure 2*a*). In these coordinates the Martian samples fall within the field dominated by meteorites. But when igneous differentiation is emphasized, as in the plot of CaO/Al_2O_3 versus MgO/SiO_2, the Martian samples are seen to be quite sharply distinguished from the meteorites.

The trace elements Rb, Sr, and Zr, to which the Viking XRFS is particularly sensitive, are characterized by low abundances in the Martian fines, while Y occurs at levels nearer the middle of its range in common terrestrial and lunar rocks (Table 1). On earth, such low values are typical of primitive, undifferentiated igneous rocks such as peridotites. Some lunar basalts are also very low in these elements. High values, especially of Rb, are typical of more highly differentiated rocks, such as those of the granitic family, and the generally low trace element content of the Martian fines tends to confirm our interpretation that they are derived from a relatively primitive, mafic source terrane and that they do not contain an appreciable contribution of highly differentiated (e.g., granitic) material.

Development of a Mineralogic/Lithologic Model

The bulk chemical composition of the surface materials, while of considerable intrinsic interest, achieves maximum scientific value when interpreted in terms of mineralogy and genetic processes. Such an interpretation is necessarily in-

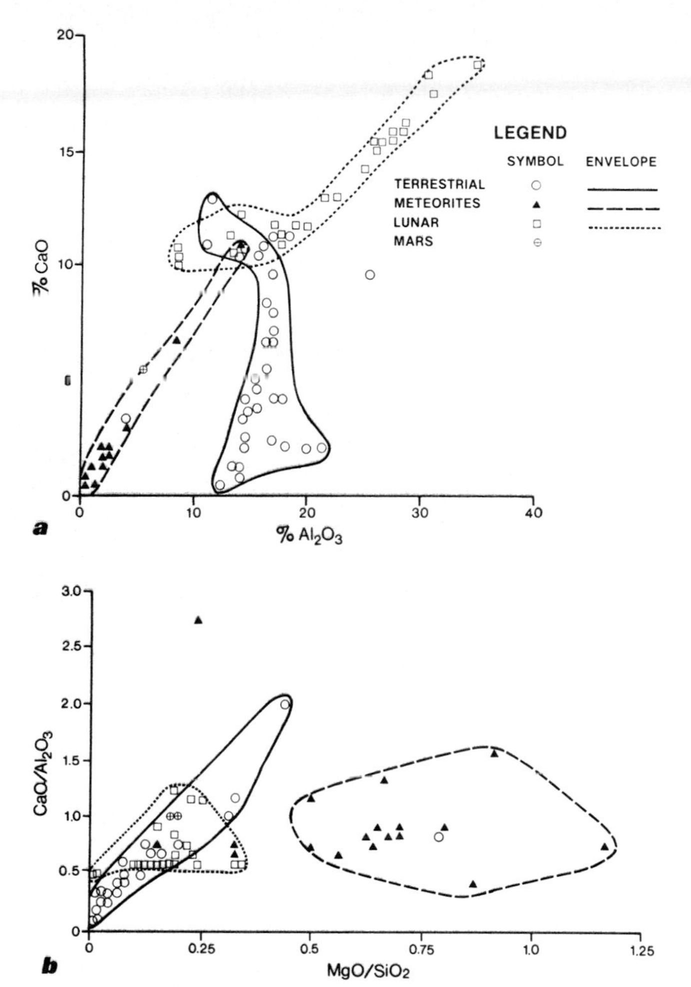

Fig. 2. Ca-Al relationships among Martian fines, meteorites and lunar and terrestrial igneous rocks. Meteorite compositions are from *Keil* [1969]; other data are as in Figure 1. (*a*) The coherent CaO/Al_2O_3 ratio of lunar, meteoritic, and Martian materials as contrasted with the effects of extensive igneous differentiation on terrestrial igneous rocks is emphasized. (*b*) The systematic trend of CaO/Al_2O_3 with igneous differentiation (measured by MgO/SiO_2) in terrestrial igneous rocks and the clear separation of meteoritic compositions from lunar and terrestrial igneous rocks are shown. Note also that the Martian fines, which seem closely related to the main field of meteorite compositions in (*a*), are clearly separated from them in (*b*).

ductive; in the absence of specific mineralogical data we strive for a model that is consistent with the existing chemical data and plausible in the light of the processes suggested by our general knowledge of the geologic environment of the landing sites. Our approach to this problem, and the results, are essentially unchanged from those described by *Baird et al.* [1976]. We will therefore recapitulate those arguments here with some minor additions and modifications.

Computer Searches for Matches

The Martian chemical analyses were compared with a group of 1000-entry, computerized data files containing compositions of a wide variety of analyzed reference materials. Terrestrial and lunar rocks, minerals, meteorites, soils, and other materials were included. The searches yielded rank-ordered matches of file entries to the Martian analyses, the closeness of match required being adjustable by setting both the limits for each element and the number of individual element matches required. As might be expected for a complex mixture of phases, most matches were selected either because of good fit in all but one or two elements, which typically agreed very poorly with the reference, or because of specific peculiarities of the Martian analyses (e.g., high S, low trace elements, or low Al/Si), even though general agreement may have been poor.

The best fits include mafic and ultramafic rocks, amphiboles, some lunar rocks, a few analyses of meteorites (especially carbonaceous chondrites), and some Fe-rich soils and clays. Fe-rich montmorillonites are generally good matches for most elements except S, suggesting that such minerals may be a major component of the mixture of materials analyzed. Table 2 shows an example of a typical search (in which, however, sulfur was not used as a criterion) and illustrates the rather large divergences in individual elements that may occur.

Normative Calculations

We have previously presented norms calculated from the S1 analysis in three different ways [*Baird et al.*, 1976]. One calculation neglected sulfur entirely so as to represent the silicate fraction only. A second calculation arbitrarily assumed all sulfur to be present in $MgSO_4$, all Ca to be in $CaCO_3$, and a ferric/ferrous ratio of 9:1. This calculation, in effect, partially adjusted the norm for assumed modal kieserite, calcite, and magnetite. A third norm was calculated to represent an unoxidized precursor material in which all sulfur was present as sulfide (FeS), and the ferric/ferrous ratio was arbitrarily set at 1:9. The results of the second and third calculations are shown in Table 3, along with 'strict' CIPW norms of the S1 analysis as reported in Table 1, calculated on the assumptions that the deficit includes both enough Na to balance Cl and enough Na to balance both Cl and SO_3. The most notable feature of all the norms is the large amount of free quartz, in apparent contrast with the rather low SiO_2 content of the material. This is the result of (1) the high assumed oxidation state of the iron (in all but the precursor case), which effectively removes it from normative silicates, (2) the low alkali and alkaline-earth elements, and most important (3) the low value for Al_2O_3, which is required for the formation of aluminosilicates such as feldspars, in which the ratio of silica to bases is higher than in nonaluminous normative silicates. This chemical peculiarity, as noted previously, is one of the most distinctive characteristics of the analyses and may be interpreted as indicating the presence of Al-poor, Fe-rich smectite clays. In any event, the norms calculated under any reasonable set of assumptions do

TABLE 1. Compositions of Martian Samples

	S1	S2	S3	U1	Estimated Absolute Error
SiO_2, wt %	44.7	44.5	43.9	42.8	5.3
Al_2O_3, wt %	5.7	n.y.a.	5.5	n.y.a.	1.7
Fe_2O_3, wt %	18.2	18.0	18.7	20.3	2.9
MgO, wt %	8.3	n.y.a.	8.6	n.y.a.	4.1
CaO, wt %	5.6	5.3	5.6	5.0	1.1
K_2O, wt %	<0.3	<0.3	<0.3	<0.3	...
TiO_2, wt %	0.9	0.9	0.9	1.0	0.3
SO_3, wt %	7.7	9.5	9.5	6.5	1.2
Cl, wt %	0.7	0.8	0.9	0.6	0.3
Sum	91.8	n.y.a.	93.6	n.y.a.	...
Rb, ppm	≤30			≤30	
Sr, ppm	60 ± 30			100 ± 40	
Y, ppm	70 ± 30			50 ± 30	
Zr, ppm	≤30			30 ± 20	

The abbreviation n.y.a. means not yet available.

TABLE 2. Comparison of Library Analyses Matched to Martian Analysis SI

	SI	L141	L1503	L2128	L5011	L6030	L143	L152	L846	L847	L1401
SiO_2	44.7	51.4	48.7	48.2	49.6	47.0	44.9	48.5	53.9*	53.8*	47.9
Al_2O_3	5.7	3.9	9.3	9.4	8.7	13.5*	11.2*	6.6	8.0	3.1	12.0*
Fe_2O_3	18.1	20.5	20.7	15.0	18.4	16.4	18.0	14.6*	22.5*	23.8*	17.8
MgO	8.3	11.2	9.5	17.5*	16.1*	7.8	10.4	14.8*	5.6	6.9	8.4
CaO	5.6	10.2*	10.8*	7.0	6.5	7.0	12.1*	5.6	3.7	6.6	12.0*
K_2O	0.1	0.1	0.0	0.4	0.3	3.5	0.8	2.2	0.0	0.0	0.1
TiO_2	0.9	0.7	1.5	0.8	0.1	1.6	1.5	1.3	0.6	0.2	1.5
(SO_3)	(7.8)	...	(0.1)	...	(0.8)	...	...	...	...	...	...

Identification of library analyses is as follows: L141, actinolite, New Zealand [*Deer et al.*, 1963, p. 253]; L1503, lunar rock 15076 (Elbow Crater; mare basalt) [*Apollo 15 Preliminary Examination Team*, 1972]; L2128, olivine diabase, Palisade Sill, New Jersey [*Turner and Verhoogen*, 1960, p. 215]; L5011, howardites, average [*Keil*, 1969]; L6030, biotite hornblendite, Rattlesnake Mountain, San Bernardino County, California (A. K. Baird, unpublished data, 1976); L143, hornblende, Idaho [*Deer et al.*, 1963, p. 279]; L152, magnesiokataphorite, Montana [*Deer et al.*, 1963, p. 360]; L846, nontronite, Westerwald, [*Koester*, 1960]; L847, nontronite, Westerwald [*Koester*, 1960]; and L1410, lunar rock 14053 (basalt) [*Lunar Sample Preliminary Examination Team*, 1971].

*Outside search criteria.

not resemble a plausible igneous rock, and this is in agreement with the results of our direct comparison of bulk compositions in an earlier section of this paper.

Computer-Generated Mixtures

Compositions of mineralogic and petrologic components can be mixed mathematically to yield close fits to the chemical composition of the Mars samples. For example, Table 4 [from *Baird et al.*, 1976] shows the progressive improvement in fit, beginning with three idealized end members of the montmorillonite family of clay minerals (montmorillonite, $Mg_{0.3}Al_{1.70}Si_4O_{10}(OH)_2 \cdot Ca_{0.15}$; saponite, $Mg_3Al_{0.5}Si_{3.5}O_{10}(OH)_2 \cdot Ca_{0.25}$; and nontronite, $Fe_2Al_{0.5}Si_{3.5}O_{10}(OH)_2 \cdot Ca_{0.25}$) and adding kieserite ($MgSO_4 \cdot H_2O$), calcite ($CaCO_3$), and rutile ($TiO_2$). Other mixtures can be constructed that account well for the bulk composition of the Martian samples: mixtures of basaltic compositions, sulfates, and smectites somewhat different in composition from those used in the calculation of Table 4, for example, or even a mixture of approximately equal parts of carbonaceous chondrite and tholeiitic basalt compositions. We believe that the significant point to be noticed is that all these possible mixtures are dominated by mineral assemblages characteristic of mafic igneous rocks or their weathering or alteration products.

Probable Mineralogical Constitution of the Martian Fines

Although it is obvious that no definitive affirmative conclusions as to the mineralogy of the Martian surface materials can be drawn from the data at hand, a number of constraints can be established and plausible models constructed. The close compositional similarity of the samples, both locally at each landing site and at the two widely separated sites, coupled with the visual evidence for fine grain size and the activity of aeolian processes, argues strongly for effective homogenization of material over large areas, presumably of planetary or at least hemispheric dimensions. The inference is that the samples analyzed are in fact reasonably representative samples of the fine-grained surface materials over a large part of the planet. This does not necessarily imply that the source region(s) for the aeolian deposits we have sampled is (are) homogeneous—only that the samples more or less adequately represent the average composition of the surface fines. If this inference is correct, the compositional characteristics of the sample take on added significance.

As has been noted, the overall composition of the sample is suggestive of mafic rocks and, more specifically, their weathering/alteration products, rather than of more highly differentiated salic igneous rocks. The very low K_2O content and low Al_2O_3, in particular, tend to preclude any large contribution of intermediate or granitic terranes to the materials analyzed. In a terrestrial environment, weathering of intermediate and granitic rocks typically leads to the separation of water-soluble constituents (e.g., especially, the alkalies and alkaline earths) from a suite of resistates enriched in Al_2O_3 and SiO_2, and under some circumstances, Fe_2O_3. Although the apparent lack

TABLE 3. Normative Calculations Based on Analysis S1

	Norm 1	Norm 2	Norm 3	Norm 4
QZ (SiO_2)	39	21	26	23
FSP	1	20	18	16
OR ($KAlSi_3O_8$)	*100*	*5*	*4*	*4*
AN ($CaAl_2Si_2O_8$)		*95*	*96*	*96*
PX	12	47	28	25
WO ($CaSiO_3$)		*14*		
EN ($MgSiO_3$)	*100*	*53*		
FS ($FeSiO_3$)		*33*		
DI ($Ca(Fe, Mg)Si_2O_6$)			*31*	*31*
HY ($(Fe, Mg)SiO_2$)			*69*	*69*
HT (Fe_2O_3)	15		17	18
MT (Fe_3O_4)	3	2		
IL ($FeTiO_3$)	2	2		
TI ($CaTiSiO_5$)			2	2
COR (Al_2O_3)	6			
KST ($MgSO_4$)	13			
CCT ($CaCO_3$)	9			
HLT (NaCl)			1	1
THN (Na_2SO_4)				14
PY (FeS_2)			7	
TR (FeS)		8		

Norm 1 is a modified norm assuming all SO_3 as $MgSO_4$, all CaO as $CaCO_3$, and 10% of Fe as FeO; norm 2 is a modified norm of hypothetical precursor composition assuming all S as FeS and 90% of Fe as FeO; norm 3 is a strict CIPW norm assuming sufficient Na to combine with Cl as HLT; and norm 4 is a strict CIPW norm assuming sufficient Na to combine with Cl as HLT and with SO_3 as THN. FSP = OR + AN, and PX = WO + EN + FS = DI + HY. Relative proportions of 'feldspar' and 'pyroxene' components are shown in italics. All values are in weight percent.

TABLE 4. Chemical Compositions of Computer-Modeled Mixtures Compared to S1

Item	Composition, wt %			
	Mixture 1	Mixture 2	Mixture 3	S1
Oxide				
SiO_2	55.1	46.0	43.6	44.7
Al_2O_3	8.3	8.0	6.9	5.7
Fe_2O_3	19.5	19.0	18.4	18.2
MgO	10.1	9.6	9.0	8.3
CaO	2.4	2.0	5.6	5.6
K_2O	0.0	0.0	0.0	0.1
TiO_2	0.0	0.0	0.9	0.8
SO_3	0.0	9.4	7.3	7.7
Mineral				
Nontronite	51	52	47	
Montmorillonite	19	21	17	
Saponite	30	13	15	
Kieserite		16	13	
Calcite			7	
Rutile			1	

Data are from *Baird et al.* [1976].

of long periods of stable, humid climate in the history of Mars would not favor this terrestrial style of surficial weathering, one might expect that the weathering products of granitic or intermediate igneous rocks on Mars, if they show chemical variation from the source material, would show some tendency to separate soluble from insoluble constituents. Yet, as we have noted, the Mars fines seem to be characterized by low values of both alkalies and alumina. In relation to the average composition of igneous rocks in the near-surface portion of the earth's crust [*Wedepohl*, 1969], the Martian samples are depleted in K_2O by a factor of about 25 and in Al_2O_3 by about $2\frac{1}{2}$. SiO_2 is only about two thirds of the terrestrial value, while total iron oxides are enriched by a factor of 4, MgO by about $3\frac{3}{4}$, and CaO by about $1\frac{1}{2}$. By this yardstick as well as the others that have been discussed, the Martian samples suggest derivation from mafic igneous rocks, with no evidence for admixture of highly differentiated salic rocks. If our samples do in fact represent a sample of a major portion of the planet's surface, then we must conclude that salic differentiates are not abundant over large areas. Even 10% of granitic rocks would have yielded K_2O values several times the upper limit set by our data.

This inferred low abundance of potassium in the near-surface portion of at least large areas of the planet has implications for the history of Mars' atmosphere and the degassing of the solid planet. *Owen and Biemann* [1976] reported that the atmosphere of Mars contains about 100 times less nonradiogenic ^{36}Ar than that of the earth, scaled on the basis of mass, while the ratio of radiogenic to nonradiogenic argon, $^{40}Ar/^{36}Ar$, is 10 times higher than on earth. One explanation that they offer requires near-surface concentration of K on Mars and concomitantly greater production and outward migration of the radiogenic isotope ^{40}Ar. The low K content of our samples does not encourage belief in that model. One possible alternative is that Mars is inherently poor in volatile elements, depleted more in primordial Ar than in K (in relation to earth), which would be reflected in low values of total Ar, $^{36}Ar/^{40}Ar$, and K on the planet today. Unique comparison of Martian and terrestrial outgassing histories cannot be made on the basis of these data alone.

The ratio Fe/Fe + Mg is in general a useful index of differentiation in igneous rocks, especially mafic ones. Taken at face value, the high value of this ratio (~52 at. %) might suggest a rather high degree of differentiation. We refrain from such an interpretation because (1) the large uncertainty in the value for Mg at this time makes any such interpretation dangerous (the range in the ratio permitted by our present stated uncertainties is 38–72%), (2) the high degree of oxidation of the iron and the possible presence of a sulfate intergranular cement suggest the action of surficial processes that may also have altered the amounts of both Fe and Mg in the surface fines, and (3) this interpretation of the Fe/Fe + Mg ratio rests on the assumption that one is dealing with a mafic igneous rock, but as we have pointed out, the bulk composition of the Martian fines does not correspond to such a rock.

The bulk chemical composition of a geologic material results from the operation of processes that, in general, transform preexisting phase aggregates (or 'mineral assemblages' in a broad sense) into new ones, either by destroying old phases and producing new ones or by re-sorting the minerals so as to modify their relative proportions. In either case the properties of the minerals influence the final result. It is important to recognize, however, that further mineralogical transformations may occur after the bulk composition at a given scale of sampling has been 'set' (i.e., after the material has become a 'closed system') and that for this reason the ultimate mineral assemblage may differ from that which set the bulk composition of the unit. For example, our speculation that Fe-rich smectite such as nontronite is an important constituent of the Martian fines is based on bulk chemistry and thus refers to the stage in the evolution of the material when its elemental composition was fixed; the present mineralogical constitution of the material may be quite different and may, for example, represent the products of destruction, under present surface conditions, of an earlier assemblage reflecting the conditions under which the bulk composition was defined.

Virtually all the interlayer H_2O can be removed from terrestrial nontronite, for example, in room temperature dry atmospheres, and it seems likely that a similar mineral at the Martian surface would retain only its 'hydroxyl water.' One may speculate that even more fundamental changes may result under the influence of intense ultraviolet radiation and low vapor pressures, such as dehydroxylation and the attendant crystal-structural rearrangements leading to the production of free iron oxides. In this regard, it is interesting to note that the first product of structural breakdown in the thermal dehydroxylation of nontronite under oxidizing or inert atmospheres is maghemite (γ-Fe_2O_3) [*MacKenzie and Rogers*, 1977], the ferromagnetic form of ferric oxide that for a variety of reasons is thought to be the magnetic material adhering to the Viking magnet arrays [*Hargraves et al.*, 1977]. Maghemite formed by thermal decomposition of nontronite is converted to the more stable oxide hematite (and magnetite), only at much higher temperatures in the process (800°–1000°C); it is plausible that dehydroxylation under Mars ambient conditions might be arrested at a stage where maghemite persisted, probably as tiny crystallites finely intergrown with residual clay and other dissociation products, such as free SiO_2 in amorphous or crystalline form or calcite.

A considerable literature derived from earth-based observation and experiment concerns the mineralogical nature of the long-hypothesized iron oxides on the surface of Mars. Summaries, and references to earlier literature, are given by *Binder and Jones* [1972]. The two main issues have been the hydration state (hematite (Fe_2O_3) versus goethite (FeOOH) or one of its polymorphs (lepidocrocite, akaganêite)) and whether the 'limonite' was a dominating component of the surface materials or only a surface stain on rocks or rock fragments. The data of the present experiment are of little help with regard to the first question but are directly relevant to the second. It is clear that silicate minerals, rather than iron oxides/hydroxides, must predominate in both unconsolidated and indurated fines. Furthermore, as was discussed by *Clark et al.* [1976*a*], our data are not consistent with a continuous iron oxide coating more than $\frac{1}{4}$ μm thick on the silicate grains. If such a coating existed, its absorption of primary and fluorescent radiation would cause the algorithm we use to calculate elemental abundances to overestimate the light elements (Mg, Al, Si) so greatly that oxide sums much higher than 100% would result. Thus we must conclude that the iron oxides present in the primarily silicate fines do not form continuous coatings more than $\frac{1}{4}$ μm thick on the silicate grains but must instead be present as discrete grains and/or inclusions within silicate grains.

Among the several good matches of possible mixtures with the Martian regolith samples is also one that is constituted roughly of equal proportions of average tholeiitic basalt and type I carbonaceous chondrites. Although tholeiitic basalt may well be a rock type on the surface of Mars (e.g., as indicated by the gigantic shield volcanoes such as Olympus Mons), the occurrence of type I carbonaceous chondrite material in such great abundances in the Martian regolith is unlikely. It is unlikely that Mars itself is a source of great amounts of material having the composition of type I carbonaceous chondrites: There is no evidence for indigenous type I carbonaceous chondrite material on the terrestrial moon, and Mars' apparently greater degree of differentiation and melting would most likely have destroyed, by melting and differentiation, any primitive, ancient material of such composition. Furthermore, influx of significant quantities of type I carbonaceous chondrites from either meteoritic or cometary sources is unlikely in view of the large quantities of material required and the high erosion, weathering, and mixing rates on the planet (see also *Toulmin et al.* [1973]). Thus we conclude that the apparent good match of about equal proportions of tholeiitic basalt and type I carbonaceous chondrites to Mars soil is probably fortuitous and does not necessarily imply occurrence of large amounts of primitive, undifferentiated material. Rather, this mixture may match the composition of the Mars fines simply because both represent mixtures of iron-rich silicates, sulfates, iron oxides, etc., of mafic affinity.

Possible Origin of Fe-Rich Smectites

If iron-rich clays such as nontronite are major constituents (past or present) of the Martian fines, the question arises whether a plausible mechanism for their formation exists. On earth, these minerals commonly form as weathering or alteration products of basaltic rocks [*Grim*, 1974; *Ross and Hendricks*, 1945] and also form both by direct precipitation and by devitrification of volcanic glass in some deep-sea environments [*Bischoff*, 1972; *Bonatti*, 1965]. The geologic arguments against widespread, long-lived Martian seas [*Masursky et al.*, 1977] seem compelling, so we will concentrate on possible mechanisms related to weathering or alteration of mafic rocks. *Huguenin* [1974] has proposed that under the stimulus of ultraviolet radiation, both hydration and oxidation of silicate minerals should proceed at the Martian surface by direct reaction with the atmosphere. Clays with high contents of ferric iron are not unreasonable products of such reactions and may indeed be so produced, but as far as we are aware, no experimental or observational confirmation of their actual production by this mechanism exists. In addition to weathering processes capable of proceeding under present Martian conditions, however, endogene processes may also yield smectites. Hydrothermal alteration associated with volcanic activity, for example, is a possibility. On earth, smectites are produced abundantly in hydrothermal environments and also characterize some products of related processes. One that seems particularly consistent with the Martian environment is the production of palagonite tuffs by subglacial volcanic activity, well documented in Iceland [e.g., *Noe-Nygaard*, 1940; *Walker and Blake*, 1966], where basaltic eruptions under large glaciers have produced both catastrophic floods (jökulhlaups) and large masses of smectite-rich altered volcanic glass, known as palagonite. The possible analogy with the Martian case is striking: large areas of chaotic terrain at the heads of major canyons have been interpreted as the result of subsidence caused by melting of permafrost in response to local magmatic activity. If violent explosive interaction between iron-rich basaltic magma and subterranean ice has been widespread on Mars, a ready source for large quantities of finely divided, clay-rich material may be at hand. The hydration of basaltic glass to palagonite apparently occurs essentially simultaneously with the chilling of the magma in contact with water [*Bonatti*, 1965, esp., pp. 261–262]; devitrification and smectite formation may occur later, but some smectite is probably formed during cooling of the glass [*Bonatti*, 1965, p. 266]. Thus at least the early stages of smectite formation on Mars could have occurred independently of the surface 'weathering' environment. Several lines of argument based on the theory of solar nebula condensation [e.g., *Grossman and Larimer*, 1974] and overall planetary physical properties [e.g., *Lewis*, 1972; *D. L. Anderson*, 1972] have led others to suggest that Mars' mantle should be richer in oxidized iron than that of the earth, which would lead to the enrichment of FeO in Martian basalts relative to terrestrial ones; thus we might expect palagonites derived from them to be especially rich in iron-rich smectites.

SUMMARY

To recapitulate, we believe that the following are the most important inferences to be drawn at the present time from the data available on the composition of the Martian surface materials.

1. The materials are dominantly fine silicate particles admixed with or including, rather than simply coated by, iron oxide particles.
2. Both major element and trace element abundances in all samples so far analyzed are indicative of mafic source rocks rather than more highly differentiated, salic materials.
3. The surface fines are nearly identical in composition at the two widely separated Viking landing sites, despite visual evidence for some lithologic diversity at the 100-m scale, implying that some agency, presumably aeolian processes, has thoroughly homogenized them on a planetary scale. This in turn implies that the samples analyzed are representative of the average composition of available fines over a large portion of the planet. This does not imply that the source region is homogeneous, only that the samples fairly represent what has been derived from it.
4. The most plausible model for the mineralogical constitution of the fine-grained surface materials at the two landing sites is a fine-grained mixture dominated by iron-rich smectites or their degradation products, with ferric oxides, probably

including maghemite, minor amounts of carbonates such as calcite, but not less stable phases such as magnesite or siderite, and sulfate minerals locally cementing the fines into a continuous or fragmented duricrust. The latter may have been formed by a process of intergranular leaching, surfaceward migration, and evaporative precipitation related to diurnal (or longer term) exchange of H_2O between the atmosphere and the solid surface of the planet. Much of the fine-grained surface material may be derived ultimately from palagonitelike products of encounters between mafic magmas and subterranean ice deposits.

Acknowledgments. Many of the interpretations presented here had their origins in discussions, formal and informal, during mission operations in which many of our Viking scientific colleagues participated. John Hower (National Science Foundation) and James Bischoff (U.S. Geological Survey) made special trips to Pasadena to share with us their expert knowledge of clay minerals. In addition, we are grateful to Jane Hammarstrom, Cristina Zen, James Lindsay, Paul Hearn, Floyd Brown, James Jensen, and Richard Larson of the U.S. Geological Survey for their assistance in various aspects of the laboratory studies supporting our interpretive efforts. Finally, we thank the exceptional group of students who served with our team as Viking Science Interns: K. Brau, Harvard University; K. Molenaar, Gustavus Adolphus College; B. Sawhill, Stanford University; R. Wall, U.S. Naval Academy; and R. Weldon, Pomona College. The work was supported in part by NASA (Viking Program) grants NAS 1-11855, NAS 1-11858, L-9717, and NAS 1-9000.

REFERENCES

Adams, J. B., and T. B. McCord, Mars: Interpretation of spectral reflectivity of light and dark regions, *J. Geophys. Res.*, *74*, 4851–4856, 1969.

Anderson, D. L., Internal constitution of Mars, *J. Geophys. Res.*, *77*, 789–795, 1972.

Anderson, D. M., Analysis of volatile constituents of the Martian regolith by the Viking GCMS (abstract), in *International Colloquium on Planetary Geology 1975*, University of Rome, in press, 1977.

Apollo 15 Preliminary Examination Team, The Apollo 15 lunar samples: A preliminary description, *Science*, *175*, 363–375, 1972.

Aronson, J. R., and A. G. Emslie, Composition of the Martian dust as derived by infrared spectroscopy from Mariner 9, *J. Geophys. Res.*, *80*, 4925–4931, 1975.

Baird, A. K., P. Toulmin III, B. C. Clark, H. J. Rose, Jr., K. Keil, R. P. Christian, and J. L. Gooding, Mineralogic and petrologic implications of Viking geochemical results from Mars: Interim results, *Science*, *194*, 1288–1293, 1976.

Baird, A. K., A. J. Castro, B. C. Clark, P. Toulmin III, H. J. Rose, Jr., K. Keil, and J. L. Gooding, Sampling strategies and ground support for inorganic chemical analyses of Mars regolith, *J. Geophys. Res.*, *82*, this issue, 1977.

Barton, P. B., Jr., and B. J. Skinner, Sulfide mineral stabilities, in *Geochemistry of Hydrothermal Ore Deposits*, 2nd ed., edited by H. L. Barnes, John Wiley, New York, 1977.

Biemann, K., J. Oro, P. Toulmin III, L. E. Orgel, A. O. C. Nier, D. M. Anderson, P. G. Simmonds, D. Flory, A. V. Diaz, D. R. Rushneck, and J. A. Biller, Search for organic and volatile inorganic compounds in two surface samples from the Chryse Planitia region of Mars, *Science*, *194*, 72–76, 1976.

Biemann, K., J. Oro, P. Toulmin III, L. E. Orgel, A. O. Nier, D. M. Anderson, P. G. Simmonds, D. Flory, A. V. Diaz, D. R. Rushneck, J. E. Biller, and A. L. Lafleur, The search for organic substances and inorganic volatile compounds in the surface of Mars, *J. Geophys. Res.*, *82*, this issue, 1977.

Binder, A. B., and J. C. Jones, Spectrophotometric studies of the photometric function, composition, and distribution of the surface materials of Mars, *J. Geophys. Res.*, *77*, 3005–3020, 1972.

Bischoff, J. L., A ferroan nontronite from the Red Sea geothermal system, *Clays Clay Miner.*, *20*, 217–223, 1972.

Bonatti, E., Palagonite, hyaloclastites, and alteration of volcanic glass in the ocean, *Bull. Volcanol.*, *28*, 257–269, 1965.

Booth, M. C., and H. H. Kieffer, Carbonate formation in Mars-like environments, *J. Geophys. Res.*, *82*, in press, 1977.

Clark, B. C., III, A. K. Baird, H. J. Rose, Jr., P. Toulmin III, K. Keil, A. J. Castro, W. C. Kelliher, C. D. Rowe, and P. H. Evans, Inorganic analyses of Martian surface samples at the Viking landing sites, *Science*, *194*, 1283–1288, 1976*a*.

Clark, B. C., III, P. Toulmin III, A. K. Baird, K. Keil, and H. J. Rose, Jr., Argon content of the Martian atmosphere at the Viking I landing site: Analysis by x-ray fluorescence spectroscopy, *Science*, *193*, 804–805, 1976*b*.

Clark, B. C., III, A. K. Baird, H. J. Rose, Jr., P. Toulmin III, R. P. Christian, W. C. Kelliher, A. J. Castro, C. D. Rowe, K. Keil, and G. R. Huss, The Viking X ray fluorescence experiment: Analytical methods and early results, *J. Geophys. Res.*, *82*, this issue, 1977.

Deer, W. A., R. A. Howie, and J. Zussman, *Rock-Forming Minerals*, vol. 2, 379 pp., John Wiley, New York, 1963.

Farmer, C. B., D. W. Davies, and D. D. La Porte, Mars: Northern summer ice cap—Water vapor observations from Viking 2, *Science*, *194*, 1339–1341, 1976*a*.

Farmer, C. B., D. W. Davies, and D. D. La Porte, Viking Mars atmospheric water vapor mapping experiment—Preliminary report of results, *Science*, *193*, 776–780, 1976*b*.

Gooding, J. L., Chemical weathering on Mars: Thermodynamic stabilities of primary minerals (and their weathering products) from mafic igneous rocks, *Icarus*, in press, 1977.

Grim, R. E., *Clay Mineralogy*, 2nd ed., 596 pp., McGraw-Hill, New York, 1974.

Grossman, L., and J. W. Larimer, Early chemical history of the solar system, *Rev. Geophys. Space Phys.*, *12*, 71–101, 1974.

Hanel, R., B. Conrath, W. Hovis, V. Kunde, P. Lohman, W. Maguire, J. Pearl, J. Pirraglia, J. Prabhakara, B. Schlachman, G. Levin, P. Straat, and T. Burke, Investigation of the Martian environment by infrared spectroscopy on Mariner 9, *Icarus*, *17*, 423–442, 1972.

Hargraves, R. B., D. W. Collinson, R. E. Arvidson, and C. R. Spitzer, Viking magnetic properties investigation: Further results, *Science*, *194*, 1303–1309, 1976.

Hargraves, R. B., D. W. Collinson, R. E. Arvidson, and C. R. Spitzer, The Viking magnetic properties experiment: Primary mission results, *J. Geophys. Res.*, *82*, this issue, 1977.

Houck, J. R., J. B. Pollack, C. Sagan, D. Schaak, and J. A. Dekker, Jr., High-altitude infrared spectroscopic evidence for bound water on Mars, *Icarus*, *18*, 470–480, 1973.

Huguenin, R. L., The formation of goethite and hydrated clay minerals on Mars, *J. Geophys. Res.*, *79*, 3895–3905, 1974.

Hunt, G. R., L. M. Logan, and J. W. Salisbury, Mars: Components of infrared spectra and the composition of the dust cloud, *Icarus*, *18*, 459–469, 1973.

Keil, K., Meteorite composition, in *Handbook of Geochemistry*, vol. I, edited by K. H. Wedepohl, chap. 4, pp. 78–115, Springer, New York, 1969.

Klein, H. P., N. H. Horowitz, G. V. Levin, V. I. Oyama, J. Lederberg, A. Rich, J. S. Hubbard, G. L. Hobby, P. A. Straat, B. M. Berdahl, G. C. Carle, F. S. Brown, and R. D. Johnson, The Viking biological investigation: Preliminary results, *Science*, *194*, 99–105, 1976.

Knopf, H. J., and H. Staude, Untersuchungen ueber das Gleichgewicht $MgSO_4 = MgO + SO_3$, *Z. Phys. Chem.*, *204*, 265–275, 1955.

Koester, H. M., Nontronit und Picotit aus dem Basalt des Oelberges bei Hundsangen, Westerwald, *Beitr. Mineral. Petrogr.*, *7*, 71–75, 1960.

Levin, G. V., and P. A. Straat, Viking labeled release biology experiment: Interim results, *Science*, *194*, 1322–1329, 1976.

Lewis, J. S., Metal/silicate fractionation in the solar system, *Earth Planet. Sci. Lett.*, *15*, 286–290, 1972.

Logan, L. M., G. R. Hunt, and J. W. Salisbury, The use of mid-infrared spectroscopy in remote sensing of space targets, in *Infrared and Raman Spectra of Lunar and Terrestrial Minerals*, edited by C. Karr, Jr., pp. 117–142, Academic, New York, 1975.

Lunar Sample Preliminary Examination Team, Preliminary examination of lunar samples from Apollo 14, *Science*, *173*, 681–693, 1971.

MacKenzie, K. J. B., and D. E. Rogers, Thermal and Mossbauer studies of iron-containing hydrous silicates, I, Nontronite, *Thermochim. Acta*, *18*, 177–196, 1977.

Masursky, H., J. M. Boyce, A. L. Dial, G. G. Schaber, and M. E. Strobell, Classification and time of formation of Martian channels based on Viking data, *J. Geophys. Res.*, *82*, this issue, 1977.

Noe-Nygaard, A., Sub-glacial volcanic activity in ancient and recent times, 1, Studies in the palagonite-system of Iceland, in *Folia Geographica Danica*, vol. I, no. 2, 65 pp., Royal Danish Geographical Society, Copenhagen, 1940.

O'Connor, J. T., Mineral stability at the Martian surface, *J. Geophys. Res.*, *73*, 5301–5311, 1968.
Owen, T., and K. Biemann, Composition of the atmosphere at the surface of Mars: Detection of argon-36 and preliminary analysis, *Science*, *193*, 801–803, 1976.
Owen, T., K. Biemann, D. R. Rushneck, J. E. Biller, D. W. Howarth, A. L. Lafleur, The composition of the atmosphere at the surface of Mars, *J. Geophys. Res.*, *82*, this issue, 1977.
Robie, R. A., and D. R. Waldbaum, Thermodynamic properties of minerals and related substances at 298.15 K (25.0 C) and one atmosphere (1.013 bars) pressure and at higher temperatures, *U.S. Geol. Surv. Bull.*, *1259*, 256 pp., 1968.
Rose, H. J., Jr., M. K. Carron, R. P. Christian, F. Cuttitta, E. J. Dwornik, and D. T. Ligon, Jr., Elemental analyses of some Apollo 16 samples, *Lunar Sci. Conf. 4th Abstr.*, 631–633, 1973.
Rose, H. J., Jr., P. A. Baedecker, S. Berman, R. P. Christian, E. J. Dwornik, R. B. Finkelman, and M. M. Schnepfe, Chemical composition of rocks and soils returned by the Apollo 15, 16, and 17 missions, *Proc. Lunar Sci. Conf. 6th*, 1363–1373, 1975.
Ross, C. A., and S. B. Hendricks, Minerals of the montmorillonite group, their origin and relation to soils and clays, *U.S. Geol. Surv. Prof. Pap.*, *205-B*, 23–79, 1945.
Salisbury, J. W., and G. R. Hunt, Compositional implications of the spectral behaviour of the Martian surface, *Nature*, *222*, 132–136, 1969.
Stern, K. H., and E. L. Weise, High temperature properties and decomposition of inorganic salts, 1, Sulfates, *Nat. Stand. Ref. Data Ser. Nat. Bur. Stand.*, *7*, 38 pp., 1966.
Stern, K. H., and E. L. Weise, High temperature properties and decomposition of inorganic salts, 2, Carbonates, *Nat. Stand. Ref. Data Ser. Nat. Bur. Stand.*, *30*, 27 pp., 1969.
Taylor, S. R., *Lunar Science: A Post-Apollo View*, 372 pp., Pergamon, New York, 1975.
Toulmin, P., III, and P. B. Barton, Jr., A thermodynamic study of pyrite and pyrrhotite, *Geochim. Cosmochim. Acta*, *28*, 641–671, 1964.
Toulmin, P., III, A. K. Baird, B. C. Clark, K. Keil, and H. J. Rose, Jr., Inorganic chemical investigation by x-ray fluorescence analysis: The Viking Mars lander, *Icarus*, *20*, 153–178, 1973.
Toulmin, P., III, B. C. Clark, A. K. Baird, K. Keil, and H. J. Rose, Jr., Preliminary results from the Viking x-ray fluorescence experiment: The first sample from Chryse Planitia, Mars, *Science*, *194*, 81–84, 1976.
Turner, F. J., and J. Verhoogen, *Igneous and Metamorphic Petrology*, 2nd ed., 694 pp., McGraw-Hill, New York, 1960.
Walker, G. P. L., and D. H. Blake, The formation of a palagonite breccia mass beneath a valley glacier in Iceland, *Quart. J. Geol. Soc. London*, *122*, 45–61, 1966.
Wedepohl, K. H., Composition and abundance of common igneous rocks, in *Handbook of Geochemistry*, vol. I, edited by K. H. Wedepohl, chap. 7, pp. 227–249, Springer, New York, 1969.
Yung, Y. L., D. F. Strobell, T. Y. Kong, and M. B. McElroy, Photochemistry of nitrogen in the Martian atmosphere, *Icarus*, *30*, 26–41, 1977.

(Received May 2, 1977;
revised May 28, 1977;
accepted May 28, 1977.)

The Composition of the Atmosphere at the Surface of Mars

TOBIAS OWEN,[1] K. BIEMANN,[2] D. R. RUSHNECK,[3] J. E. BILLER,[2]
D. W. HOWARTH,[4] AND A. L. LAFLEUR[2]

We have confirmed the discovery of N_2 and ^{40}Ar by the Entry Science Team, and we have also detected Ne, Kr, Xe, and the primordial isotopes of Ar. The noble gases exhibit an abundance pattern similar to that found in the terrestrial atmosphere and the primordial component of meteoritic gases. Xenon appears to be underabundant in comparison to the meteoritic ratio, as it is on earth. The isotopic ratios $^{15}N/^{14}N$, $^{40}Ar/^{36}Ar$, and $^{129}Xe/^{132}Xe$ are distinctly different from the terrestrial values, implying different evolutionary histories for volatiles on the two planets. The noble gas abundances indicate that at least 10 times the present atmospheric amount of N_2 and 20 times the CO_2 abundance were released by the planet during geologic time; the outgassing of a large amount of water must also have taken place. There is thus an explanation for the high surface pressure and abundance of water required at some early epoch to cut the dendritic channels observed on the Martian surface.

This report is a summary of the current status of our investigations of the composition of the Martian atmosphere. We have been using the mass spectrometers that function as the analytical components of the molecular analysis experiments on the two Viking landers. The basic concepts underlying the experiment and its instrumentation have been discussed by *Anderson et al.* [1972]; a detailed description of the flight instruments is currently being prepared for publication (D. R. Rushneck, private communication, 1977).

The first accounts of our analyses of the Martian atmosphere have appeared in three previous papers [*Owen and Biemann*, 1976; *Biemann et al.*, 1976*a*; *Owen et al.*, 1976]. Since these publications, we have obtained additional data from the two landers, and we have refined some of our earlier conclusions. This process is continuing, special emphasis being placed on the acquisition of supplementary calibrations using the Engineering Breadboard (EBB) version of the instrument and the spare flight-configured model. The instruments on the spacecraft were turned off on March 13, 1977 (Viking Lander 1), and April 5, 1977 (Viking Lander 2), to avoid potential hazards to other experiments from problems that developed with the high-voltage supplies.

In order to discuss the atmospheric data we must briefly review the manner in which they were acquired. Although the molecular analysis instrument was designed primarily for the detection of organic compounds in the gas chromatographic mode [*Biemann*, 1974], the mass spectrometer's high sensitivity (dynamic range, 6–7 orders of magnitude), high mass range (m/e, 12–200), and resolution (1:200 at m/e = 200; better at lower mass) were used to advantage in determining the composition of the atmosphere, particularly its minor constituents. The penalty that one pays for resolution and sensitivity is a certain loss of accuracy, mainly because the residual background in the instrument becomes more significant and the long-term reproducibility of the fragmentation pattern is lowered.

Prior to the Viking missions the detection of even traces of N_2 in the Martian atmosphere was deemed to be extremely important because of its relevance both for the history of volatiles on Mars and for possible Martian biology. Previous data suggested that it must be a minor component or could be almost completely absent [*Barth et al.*, 1969; *Dalgarno and McElroy*, 1970; *McElroy*, 1972]. One of the major problems in a mass spectrometric determination of N_2 in the Martian environment is the interference of CO^+ (from CO or CO_2) with the ion current of N_2^+ at m/e = 28. For this reason the gas reservoir of the instrument is coupled via separate valves to two cavities, one containing Ag_2O and LiOH for the oxidation of CO to CO_2 and the absorption of all CO_2 and the other containing $Mg(ClO_4)_2$ for the removal of the resulting water (Figure 1).

We thus have three distinct options for analyzing the Martian atmosphere: (1) we can admit the atmosphere directly to the mass spectrometer (an unaltered or unfiltered sample), (2) we can use the chemical scrubbers to reduce CO and CO_2, and (3) we can repeat the scrubbing procedure, progressively admitting fresh samples to the gas reservoir and thereby building up the partial pressure of the trace gases.

During the fourth and fifth days after the landing of Viking Lander 1 (July 20, 1976) a total of six atmospheric analyses were performed at approximately 6-hour intervals. In the first four of these analyses, CO and CO_2 were removed; in the last two analyses, samples of unaltered atmosphere were used. During the third analysis the spectrometer shut down temporarily, leaving us with a total of five sets of mass spectral scans. Analysis of these spectra confirmed the presence of nitrogen and argon in the Martian atmosphere as reported by *Nier et al.* [1976*a*] and led to the discovery of ^{36}Ar [*Owen and Biemann*, 1976]. The isotope ratio $^{36}Ar/^{40}Ar$ was found to be 3.34×10^{-4}, approximately one-tenth the terrestrial value (3.43×10^{-3}), while the total abundances of ^{36}Ar and ^{40}Ar in the Martian atmosphere relative to the planet's mass were 0.0075 and 0.08, respectively, of the comparable terrestrial values.

In these first analyses we also detected molecular oxygen and (of course!) carbon dioxide, previously known from ground-based observations. We observed considerable scatter in the oxygen measurements (a factor of ± 2), which we have attributed to instrumental causes. No correlation with external effects (time of day, operation of other instruments, etc.) was noted, nor have we detected a seasonal variation in the abundance of any atmospheric component during the lifetime of the mission.

[1] Department of Earth and Space Sciences, State University of New York, Stony Brook, New York 11794.

[2] Department of Chemistry, Massachusetts Institute of Technology, Cambridge, Massachusetts 02139.

[3] Interface, Inc., Fort Collins, Colorado 80522.

[4] Guidance and Control Systems Division, Litton Industries, Woodland Hills, California 91364.

Paper number 7S0515.

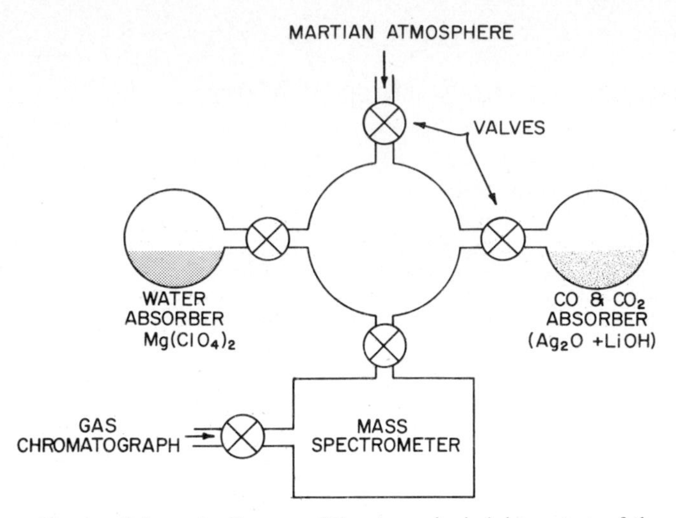

Fig. 1. Schematic diagram of the atmospheric inlet system of the molecular analysis experiment. The valves can be cycled independently. (Diagram is not to scale.)

Variable amounts of water vapor, 0.07% carbon monoxide, and up to 0.03 ppm of ozone (also variable) have also been detected by ground-based or spacecraft observations [*Barth,* 1974; *Owen,* 1974*a*; *Young and Young,* 1977]. The mass spectrometric data on H_2O are rather meaningless because of adsorption on the relatively large surface area of the interior of the atmospheric inlet system and of the ion source (which is not heated in the atmospheric analysis mode). Low concentrations (less than a few percent) of CO cannot be detected with our system because of the interference by the large amount of CO_2 in analyses of the unmodified atmospheric sample and the concomitant removal of CO along with CO_2 when the sample was exposed to Ag_2O in the other type of analysis. The abundance of ozone is below our detection limit.

Subsequent analysis of enriched samples on Viking Lander 1 (VL 1) led to the establishment of ratios for the abundant isotopes, including confirmation of the report by *Nier et al.* [1976*b*] that the abundance of the naturally occurring heavy isotope of nitrogen (^{15}N) is enhanced in the Martian atmosphere by a factor of approximately 1.7 over the terrestrial value [*Biemann et al.,* 1976*a*]. Other isotope ratios measured during this sequence ($^{13}C/^{12}C$, $^{18}O/^{16}O$, and $^{36}Ar/^{38}Ar$) appear to exhibit terrestrial values within the errors of measurement ($\pm 10\%$).

The successful deployment of Viking Lander 2 (VL 2) on Mars (September 3, 1976) afforded us the desired opportunity to increase the sensitivity of our experiment by further enrichment of atmospheric samples. We had established from tests of the two instruments carried out during the cruise from earth to Mars that the background in the mass spectrometer on VL 2 was much lower than that in the instrument on the first lander, making it superior for the detection of trace amounts of atmospheric gases. It was also clear from experience gained with the operation of VL 1 that the best way to maintain this low instrumental background would be to perform the atmospheric analyses prior to any analyses of the Martian soil. A delay in obtaining the first soil sample for our instrument with VL 2 provided the opportunity to design an optimized enrichment sequence for detecting trace gases in the atmosphere.

Using chemical scrubbing sequences of 5 and 10 cycles, we obtained enrichments of 4 and 6.3 times the yield from a single cycle. Mass spectra of the $\times 6.3$ enriched sample showed an indication of krypton, but the identification was not conclusive. After evaluating the performance of the instrument we changed the internal timing of the sequence and obtained a ninefold enrichment with 15 cycles. This sample was analyzed 16 times by the mass spectrometer with the electron multiplier gain increased by a factor of 5.3 over its nominal value. These spectra gave clear evidence of the presence of krypton. The analysis was subsequently repeated after an additional 15 cycles with a multiplier gain of 28 times nominal.

The results are shown in Figure 2, which indicates the appearance of the averaged mass spectrum in the vicinity of the krypton and xenon isotopes. The characteristic isotopic pattern of krypton is clearly evident; the broad peak at $m/e = 80$ is an artifact caused by the high partial pressure of argon in the instrument. In the case of xenon, ^{129}Xe is much more abundant, in relation to the other xenon isotopes, in the Martian atmosphere than in the terrestrial atmosphere. The absence of high molecular weight organic compounds in the soil [*Biemann et al.,* 1976*b*], the absence of spectroscopically active high molecular weight compounds in the atmosphere [*Barth,* 1974; *Owen,* 1974*a*], and the clean instrumental background in this mass range make us quite confident that the peaks shown in this region of Figure 2 are chiefly caused by xenon on Mars.

It is not yet possible to compute accurate abundances for

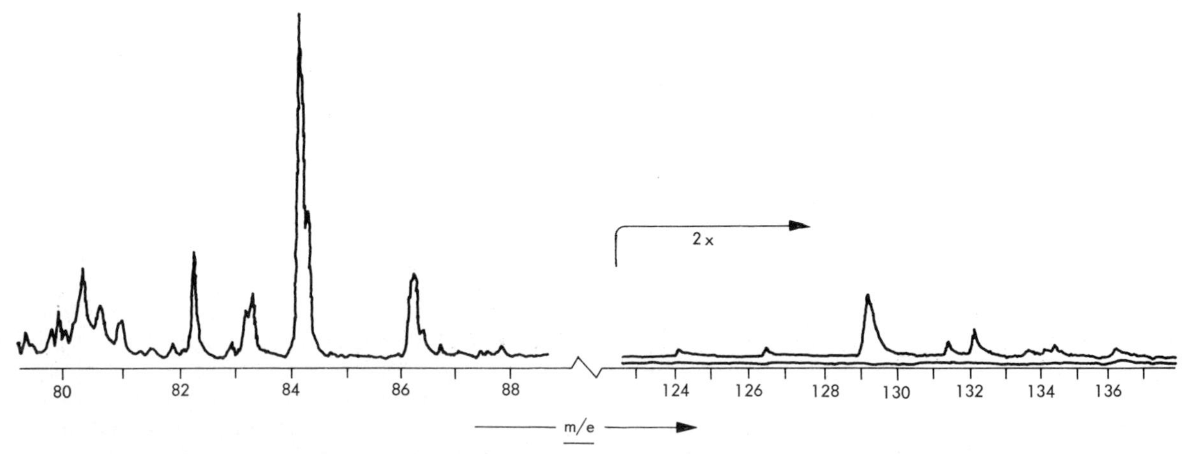

Fig. 2. Mass spectra of enriched samples of the Martian atmosphere in the region of krypton and xenon. The spectra are averages of nine scans, the lower lines are averages of three background scans. The vertical scale is linear; it has been increased by a factor of 2 for xenon.

these two gases or even to give precise values for the ratios of their isotopes. At this low level of detection the instrumental response is noisy, nonlinear, and distorted by memory effects (degassing from the pump). Thus the small peaks at m/e = 124 and 126, while appearing to be anomalously high for xenon, are probably caused by a very weak hydrocarbon background from the ion pump, not by Martian gases. There is, however, no question that ^{129}Xe is much more abundant than ^{132}Xe and ^{131}Xe rather than almost equal, as is the case on earth. To make further progress, the mass spectrometer must be recalibrated in a manner that duplicates experimental conditions on Mars. Such tests are now being conducted with the third flight unit that was built for this mission as a spare instrument.

This completed the studies of the atmosphere carried out prior to solar conjunction (November 15, 1976). The prime objective for postconjunction work was a search for neon. The difficulty posed by this gas is that each of the abundant isotopes is overlapped in the mass spectrum by doubly ionized species that are much more abundant than neon on Mars: $^{20}Ne^+$ by $^{40}Ar^{++}$ and $^{22}Ne^+$ by $^{44}CO_2^{++}$. By using a lower ionizing energy (~40 eV versus 70 eV nominal) we could reduce the contribution from the doubly ionized species, but we had already found that this was inadequate to reveal the presence of the neon isotopes with our standard procedures. There was no way in which we could affect the argon concentration, but we felt that we might be able to reduce the residual CO_2 abundance by exposing the enriched sample to the LiOH capsule during the entire solar conjunction period (~2 months). This indeed proved to be the case. When we analyzed the mixture at low ionizing energy on sol 118, the residual concentration of CO_2 was found to be reduced from 0.33 to 0.015% of its abundance in the free atmosphere.

At this level we found that $^{44}CO_2^{++}$ contributed less than one third of the intensity of the observed peak at m/e = 22 at low ionizing energy. This evaluation was made by using preflight calibration data for the CO_2 cracking pattern that we confirmed by subsequent analyses of an unaltered sample of the Martian atmosphere (sol 191). The resulting abundance for ^{22}Ne in the Martian atmosphere is 0.25 ppm. To determine the total neon abundance, we have to make an assumption for the value of $^{20}Ne/^{22}Ne$ on Mars. Both a 'planetary' value of 8.2 ± 0.4 and a 'solar' value of 12.5 ± 0.4 have been identified in meteorites (see the work of *Heymann* [1971] for a review); in the earth's atmosphere the value is 9.8, and the value at the surface of the sun is probably close to 13 [e.g., *Geiss et al.*, 1970]. In view of this potential range of values we adopt $^{20}Ne/^{22}Ne$ = 10 ± 3 for Mars. Including the other uncertainties in this determination, we find a neon abundance of $2.5^{+3.5}_{-1.5}$ ppm for the Martian atmosphere.

TABLE 1. Composition of the Lower Atmosphere

Gas	Proportion
Carbon dioxide (CO_2)	95.32%
Nitrogen (N_2)*	2.7%
Argon (Ar)*	1.6%
Oxygen (O_2)	0.13%
Carbon monoxide (CO)	0.07%
Water vapor (H_2O)	0.03%†
Neon (Ne)*	2.5 ppm
Krypton (Kr)*	0.3 ppm
Xenon (Xe)	0.08 ppm
Ozone (O_3)	0.03 ppm†

* Discovered by Viking experiments.
† Variable.

TABLE 2. Isotope Ratios in Atmospheric Gases

Ratio	Earth	Mars
$^{12}C/^{13}C$	89	90
$^{16}O/^{18}O$	499	500
$^{14}N/^{15}N$	277	165
$^{40}Ar/^{36}Ar$	292	3000
$^{129}Xe/^{132}Xe$	0.97	2.5

Uncertainties in the Mars values are presently ±10% except for Ar and Xe (see text).

A summary of our current best estimate for the composition of the atmosphere near the surface of Mars is given in Table 1; isotope ratios are summarized in Table 2. The uncertainties in abundances not specifically discussed in the text are of the order of ±20% of the values. We expect to be able to improve this precision after we have completed the additional calibrations that are currently under way. A detailed interpretation of the results must obviously await these improvements, but a few general comments can be made at this time.

In our previous papers we have used the relative abundances of the noble gases in the terrestrial and Martian atmospheres as a guide to the total volatile inventories on both planets [*Owen and Biemann*, 1976; *Owen et al.*, 1976]. The discovery of neon lends additional weight to this approach, since it increases the similarity between the noble gas abundance pattern on Mars and that found in the earth's atmosphere and in the primordial meteoritic component (Figure 3). The difference in the relative abundances of ^{36}Ar and ^{40}Ar on the two planets (Table 2) may be interpreted in one of several ways:

1. Mars has a volatile inventory identical to that of the earth, but either (1) an early catastrophic event removed over 90% of its atmosphere, or (2) Mars never degassed as much as the earth.

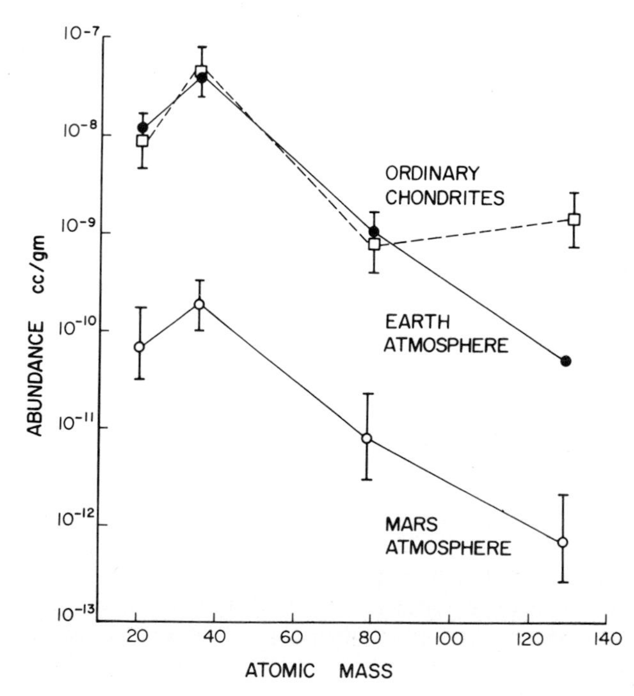

Fig. 3. Abundances of noble gases in ordinary chondrites and in the atmospheres of the earth and Mars. Error bars are only approximate, since systematic errors remain to be assessed. Meteoritic abundances are from *Signer* [1964]. Abundances are in cubic centimeters per gram (at STP) of planet (or meteorite).

2. Mars is deficient in the most volatile elements in comparison to the earth, and degassing is less complete.

The choice among these alternatives is not obvious. In our previous papers we favored part 2 of alternative 1, but in collaboration with E. Anders we are developing a strong case for alternative 2. Catastrophic loss seems unlikely, as will be shown below.

The end result of either of these approaches is the same: they lead to the prediction that at least 10 times more nitrogen and 20 times more carbon dioxide were outgassed on Mars than the atmosphere now contains. The equivalent of a layer of water about 10 m deep over the entire planet would also have been produced. (The basic argument involves a comparison of the Martian atmosphere with the earth's volatile inventory, using the noble gases as a guide. It goes back to the work of *Rubey* [1951, 1955], *Brown* [1952], and *Holland* [1964]; it has been developed for Mars by *Owen* [1966, 1974*a*, *b*, 1976], *Fanale* [1971, 1976], and *Levine* [1976] (and work cited in these papers); its first application to Viking data was in our three previous papers. Subsequent treatments by *Rasool and Le Sergeant* [1977] and by *Anders and Owen* [1977] lead to similar conclusions regarding the estimates of total outgassing.) This prediction for nitrogen agrees with the minimum value deduced from the enhancement of $^{15}N/^{14}N$ by *McElroy et al.* [1976], and the water abundance agrees with the 10-m layer required by the 'normalcy' of $^{16}O/^{18}O$ (M. B. McElroy, private communication, 1977). This agreement may be fortuitous; preferential degassing of water could produce a 40-m layer, and some of the degassed water may be bound in rocks and thus not exchangeable with the atmospheric CO_2. In any event, even the smaller value for the outgassed water abundance is adequate to explain the carving of the fluvial features observed on the planet's surface [*Milton*, 1973] at a time when the ancient atmosphere exhibited the higher surface pressure implied by the missing CO_2. We must emphasize, however, that there is still no adequate model for climatic change on Mars that permits the planet's surface to be warm enough for this to occur. A water vapor controlled greenhouse is an attractive possibility (R. D. Cess, private communication, 1977).

Examining Figure 3 again, we note that xenon is clearly underabundant in the earth's atmosphere in comparison with the meteoritic pattern, and the same deviation may hold for Mars. The present uncertainties in the xenon and krypton abundances are too large to permit a rigorous comparison, but this ambiguity should be resolved by the calibration work currently in progress. It is an important point, since the xenon deficiency on earth has been attributed to the preferential trapping of xenon in shales and other sedimentary material after it was outgassed [*Canalas et al.*, 1968; *Fanale and Cannon*, 1972]. One is thus led to the tentative conclusion that similar processes have been active on Mars, perhaps in association with the epochs of fluvial erosion that have left their imprint on the planet's surface. An alternative (or supplementary) suggestion is that some of the xenon could be adsorbed in the regolith (F. P. Fanale et al., unpublished manuscript, 1977).

At this stage of our investigation of Martian xenon isotopes we can only be sure of the enhancement of ^{129}Xe. It is generally agreed that ^{129}Xe anomalies in meteoritic and terrestrial gas samples result from the production of this isotope by decay of extinct ^{129}I [*Reynolds*, 1960, 1963]. We find that the ratio $^{129}Xe/^{132}Xe$ is 2.5^{+2}_{-1}; the terrestrial atmospheric value is 0.97, and in the carbonaceous and ordinary chondrites, values as high as 4.5 and 9.6 have been reported [*Signer*, 1964; *Mazor et al.*, 1970; *Pepin*, 1964]. There are several possible interpretations of this anomaly. Catastrophic loss of an early atmosphere at just the right number of ^{129}I half-lives could produce the observed abundance of ^{129}Xe, but such an event would have to 'shut off' very abruptly. A gradual decrease in the intensity of a T-Tauri solar wind (for example) would seem to require that a point would eventually be reached at which the lighter noble gases would be preferentially swept away from the upper atmosphere. This in turn would lead to a fractionation from the pattern observed in the terrestrial atmosphere and the meteorites which is apparently not the case (Figure 3). But a quantitative model should be developed before this alternative is completely abandoned.

A more satisfactory explanation may be found by postulating a difference in the manner in which a volatile-rich veneer was accumulated by the earth and Mars. This is one aspect of a model that attempts to account for the volatile abundances on all of the inner planets in terms of late-accreting veneers with a composition resembling that of the C-3V meteorites [*Anders and Owen*, 1977].

There is clearly much more to be done in the way of improving the precision of the existing data and refining their interpretation, but the following points seem well established:

1. The present atmosphere on Mars represents only a small fraction of the total amount of volatiles outgassed by the planet. The potential for a high surface pressure (>100 mbar) and abundant water at some past epoch is therefore available.

2. The noble gases in the Martian atmosphere exhibit a relative abundance pattern similar to that in the earth's atmosphere and (except for Xe) to that in the primordial component of the meteorites. The existence of a 'planetary component' is thus proven, supporting the arguments of those who favor a fractionation of noble gases prior to the formation of the planets.

3. Despite this basic similarity between the volatile inventories on earth and Mars the histories of the two atmospheres, perhaps even the preformation histories, have been very different. The isotopic ratios of nitrogen, argon, and xenon provide ample data for this assertion.

Acknowledgments. This work was supported by NASA research contracts NAS 1-10493 and NAS 1-9684. We thank E. Anders, B. Clark, O. A. Schaeffer, and P. Toulmin III for helpful discussions.

REFERENCES

Anders, E., and T. Owen, Origin and abundance of volatiles on the earth and Mars, *Science*, in press, 1977.

Anderson, D. M., K. Biemann, L. E. Orgel, J. Oro, T. Owen, G. P. Shulman, P. Toulmin III, and H. C. Urey, Mass spectrometric analysis of organic compounds, water, and volatile constituents in the atmosphere and surface of Mars, *Icarus*, *16*, 111, 1972.

Barth, C. A., The atmosphere of Mars, *Annu. Rev. Earth Planet. Sci.*, *2*, 333–368, 1974.

Barth, C. A., W. G. Fastie, C. W. Hord, J. B. Pearce, K. K. Kelley, A. I. Stewart, G. E. Thomas, G. P. Anderson, and O. F. Raper, Mariner 6: Ultraviolet spectrum of Mars' upper atmosphere, *Science*, *165*, 1004, 1969.

Biemann, K., Test results on the Viking gas chromatograph-mass spectrometer experiment, *Origins Life*, *5*, 417, 1974.

Biemann, K., T. Owen, D. R. Rushneck, A. L. Lafleur, and D. W. Howarth, The atmosphere of Mars near the surface: Isotope ratios and upper limits on noble gases, *Science*, *194*, 76–78, 1976*a*.

Biemann, K., J. Oro, P. Toulmin III, L. E. Orgel, A. O. Nier, D. M. Anderson, P. G. Simmonds, D. Flory, A. V. Diaz, D. R. Rushneck, and J. E. Biller, Search for organic and volatile inorganic compounds in two surface samples from the Chryse Planitia region of Mars, *Science*, *194*, 72–76, 1976*b*.

Brown, H., Rare gases and the formation of the earth's atmosphere, in *The Atmospheres of the Earth and Planets*, edited by G. P. Kuiper, chap. IX, University of Chicago Press, Chicago, Ill., 1952.

Canalas, R. A., E. C. Alexander, Jr., and O. K. Manuel, Terrestrial abundance of noble gases, *J. Geophys. Res.*, *73*, 3331–3334, 1968.

Dalgarno, A., and M. B. McElroy, Mars: Is nitrogen present?, *Science*, *170*, 168, 1970.

Fanale, F. P., History of Martian volatiles: Implications for organic synthesis, *Icarus*, *15*, 279–303, 1971.

Fanale, F. P., Martian volatiles: Their degassing history and geochemical fate, *Icarus*, *28*, 179–202, 1976.

Fanale, F. P., and W. A. Cannon, Origin of planetary primordial rare gas: The possible role of adsorption, *Geochim. Cosmochim. Acta*, *36*, 319–328, 1972.

Geiss, J., P. Eberhardt, F. Bühler, and J. Meister, Apollo 11 and 12 solar wind composition experiments: Fluxes of He and Ne isotopes, *J. Geophys. Res.*, *75*, 5972, 1970.

Heymann, D., The inert gases, in *Handbook of Elemental Abundances in Meteorites*, edited by B. Mason, pp. 29–66, Gordon and Breach, New York, 1971.

Holland, H. D., On the chemical evolution of the terrestrial and cytherean atmospheres, in *The Origin and Evolution of Atmospheres and Oceans*, edited by P. J. Brancazio and A. G. W. Cameron, pp. 86–101, John Wiley, New York, 1964.

Levine, J. S., A new estimate of volatile outgassing on Mars, *Icarus*, *28*, 165–169, 1976.

Mazor, E., D. Heymann, and E. Anders, Noble gases in carbonaceous chondrites, *Geochim. Cosmochim. Acta*, *34*, 781–824, 1970.

McElroy, M. B., Mars: An evolving atmosphere, *Science*, *175*, 443, 1972.

McElroy, M. B., Y. L. Yung, and A. O. Nier, Isotopic composition of nitrogen: Implications for the past history of Mars' atmosphere, *Science*, *194*, 70–72, 1976.

Milton, D. J., Water and processes of degradation in the Martian landscape, *J. Geophys. Res.*, *78*, 4037–4048, 1973.

Nier, A. O., W. B. Hanson, A. Seiff, M. B. McElroy, N. W. Spencer, R. J. Duckett, T. C. D. Knight, and W. S. Cook, Composition and structure of the Martian atmosphere: Preliminary results from Viking 1, *Science*, *193*, 786–788, 1976*a*.

Nier, A. O., M. B. McElroy, and Y. L. Yung, Isotopic composition of the Martian atmosphere, *Science*, *194*, 68–70, 1976*b*.

Owen, T., The composition and surface pressure of the Martian atmosphere: Results from the 1965 opposition, *Astrophys. J.*, *146*, 257–270, 1966.

Owen, T., What else is present in the Martian atmosphere?, *Comments Astrophys. Space Phys.*, *5*, 175, 1974*a*.

Owen, T., Martian climate: An empirical test of possible gross variations, *Science*, *183*, 763, 1974*b*.

Owen, T., Volatile inventories on Mars, *Icarus*, *28*, 171–177, 1976.

Owen, T., and K. Biemann, Composition of the atmosphere at the surface of Mars: Detection of argon-36 and preliminary analysis, *Science*, *193*, 801–803, 1976.

Owen, T., K. Biemann, D. R. Rushneck, J. E. Biller, D. W. Howarth, and A. L. Lafleur, The atmosphere of Mars: Detection of krypton and xenon, *Science*, *194*, 1293–1295, 1976.

Pepin, R. O., Isotopic analyses of xenon, in *The Origin and Evolution of Atmospheres and Oceans*, edited by P. J. Brancazio and A. G. W. Cameron, pp. 191–234, John Wiley, New York, 1964.

Rasool, S. I., and L. Le Sergeant, Implications of the Viking results for volatile outgassing from earth and Mars, *Nature*, *266*, 822–823, 1977.

Reynolds, J. H., Determination of the age of the elements, *Phys. Rev. Lett.*, *4*, 8–10, 1960.

Reynolds, J. H., Xenology, *J. Geophys. Res.*, *68*, 2939, 1963.

Rubey, W. W., Geologic history of seawater, *Geol. Soc. Amer. Bull.*, *62*, 1111, 1951.

Rubey, W. W., Development of the hydrosphere and atmosphere, with special reference to the probable composition of the early atmosphere, *Geol. Soc. Amer. Spec. Pap.*, *62*, 631–650, 1955.

Signer, P., Primordial rare gases in meteorites, in *The Origin and Evolution of Atmospheres and Oceans*, edited by P. J. Brancazio and A. G. W. Cameron, pp. 183–190, John Wiley, New York, 1964.

Young, L. D. G., and A. T. Young, Interpretation of high-resolution spectra of Mars, IV, New calculations of the CO abundance, *Icarus*, *30*, 75–79, 1977.

(Received April 22, 1977;
revised June 8, 1977;
accepted June 8, 1977.)

The Search for Organic Substances and Inorganic Volatile Compounds in the Surface of Mars

K. BIEMANN,[1] J. ORO,[2] P. TOULMIN III,[3] L. E. ORGEL,[4] A. O. NIER,[5] D. M. ANDERSON,[6] P. G. SIMMONDS,[7] D. FLORY,[8] A. V. DIAZ,[9] D. R. RUSHNECK,[10] J. E. BILLER,[1] AND A. L. LAFLEUR[1]

A total of four Martian samples, one surface and one subsurface sample at each of the two Viking landing sites, Chryse Planitia and Utopia Planitia, have been analyzed for organic compounds by a gas chromatograph–mass spectrometer. In none of these experiments could organic material of Martian origin be detected at detection limits generally of the order of parts per billion and for a few substances closer to parts per million. The evolution of water and carbon dioxide, but not of other inorganic gases, was observed upon heating the sample to temperatures of up to 500°C. The absence of organic compounds seems to preclude their production on the planet at rates that exceed the rate of their destruction. It also makes it unlikely that living systems that behave in a manner similar to terrestrial biota exist, at least at the two Viking landing sites.

1. INTRODUCTION

One of the major goals of the Viking mission was to find out whether or not organic compounds exist on the surface of the planet Mars and, if they do exist, to determine their structures and measure their abundances. This seemed important because we hoped that the nature of Martian organic molecules would provide a sensitive indicator of the chemical and physical environment in which they were formed. Furthermore, we hoped that the details of their structures would indicate which of many possible biotic and abiotic syntheses are occurring on Mars. The relatively simple compounds expected from the photochemical reaction of the components of the atmosphere [*Hubbard et al.*, 1973] differ greatly, for example, from the organic compounds found in carbonaceous chondrites [*Oro*, 1972; *Nagy*, 1975] and which in turn differ greatly from the complex and highly ordered structurally specific substances produced by living cells. Furthermore, since much is known about the degradation of organic compounds under the influence of high temperature, pressure, irradiation, etc. [*Miller et al.*, 1976], the absence of organic compounds above a certain limit of detection might eliminate certain sets of conditions that otherwise could be postulated to exist or to have existed at the surface.

To achieve our goal, a sensitive technique of high structural specificity and broad applicability was required. A gas chromatograph coupled to a mass spectrometer was chosen [*Anderson et al.*, 1972] in order to combine the sensitivity and the structural specificity of electron impact mass spectrometry with the separation power of gas chromatography. An obvious approach would have been to mimic terrestrial laboratory procedures by digesting the surface material to be analyzed by wet chemical methods, followed by solvent extraction and possibly chemical separation. However, the automation and the miniaturization of such a system, which has to be of the utmost reliability, was clearly beyond the technical and economic resources available for the experiment and was certainly far outside the weight and power allocations. In the task at hand, where nothing at all is known about the organic chemistry of the Martian surface, a general or group identification is a major advance, should one encounter a very complex mixture. This is in contrast to terrestrial investigations, where much more specific information is required and can be achieved.

For these reasons, thermal volatilization (without or with thermal degradation) of the organic compounds from the surface material was selected as the simplest and most reliable approach. It was expected that the nature of the thermal degradation products (pyrolyzate) would lead to the identification of the set of parent compounds originally present in the sample.

The original plans [*Anderson et al.*, 1972] also included a sample oven which was directly monitored by the mass spectrometer, thus making it possible to detect more complex and less volatile substances which would not pass through the gas chromatograph and associated interfaces. However, this part was eliminated during the design and test phase to simplify the final instrument package [*Biemann*, 1974]. At the same time the number of sample ovens connected to the gas chromatograph was reduced from eight to three. Since each oven could utilize only one soil sample, the flexibility of the experiment was considerably reduced. In addition, during interplanetary cruise it was found that one oven in each instrument was not operable. Fortunately, the excellent data transmission quality and quantity throughout the mission and the seemingly uniform surface composition made this reduction in the number of samples less damaging than might have been anticipated. The loss of the direct input oven which could be heated slowly and continuously was, in retrospect, more regrettable because it could have been used to obtain more information concerning the mineralogy of the surface material.

2. INSTRUMENTATION

The instrument has, in principle, been described previously [*Anderson et al.*, 1972; *Biemann*, 1974], but the final flight hardware differed in some of the parameters. It is therefore

[1] Department of Chemistry, Massachusetts Institute of Technology, Cambridge, Massachusetts 02139.
[2] Department of Biophysical Sciences, University of Houston, Houston, Texas 77004.
[3] U.S. Geological Survey, Reston, Virginia 22092.
[4] Salk Institute for Biological Studies, San Diego, California 92102.
[5] School of Physics and Astronomy, University of Minnesota, Minneapolis, Minnesota 55455.
[6] Division of Polar Programs, National Science Foundation, Washington, D. C. 20550.
[7] Organic Geochemistry Unit, School of Chemistry, University of Bristol, Bristol, England.
[8] Spectrix Corporation, Houston, Texas 77054.
[9] NASA Langley Research Center, Hampton, Virginia 23665.
[10] Interface, Inc., Fort Collins, Colorado 80522.

Paper number 7S0556.

necessary to describe here briefly the final version and its mode of operation. Figure 1 illustrates the major features of the instrument. It consists of the sample ovens with their associated valves, the gas chromatograph, an effluent divider which protects the mass spectrometer from excessive gas pressure, the carrier gas separator, the mass spectrometer, and the associated electronics and logic system.

The sample ovens (ceramic, 2-mm inside diameter, 19 mm long) are mounted in a circular holder that can be positioned in a preset sequence upon a command from the instrument's internal logic system, but this sequence in turn can be modified (within limits) by ground command if the experimental results call for a change.

There are two positions to which any of the ovens can be moved in any sequence. The 'load position' is directly under the sampling system, which delivers about 1–2 cm³ of surface material that after having been ground is passed through a 0.3-mm sieve. A mechanical poker pushes the material through a funnel into the oven. This operation is timed in such a manner that the filling of the oven is complete with any of the terrestrial test soils (including finely ground basalt, commonly referred to as 'lunar nominal'). However, there is no sensor measuring the final level or completeness of the filling operation. Thus one has to assume that the oven is filled to capacity, i.e., approximately 60 mm³ of surface material is being analyzed. The oven is then moved by rotation of the circular holder to the 'analysis position,' where the lines leading to valves V1 and V3 are clamped onto both ends of the oven. A gas-tight seal is achieved by pressing the circular knife edges, into which the gas lines terminate, into gold rings attached to each end of the oven. In the normal use of an oven this seal is established only once. It had been shown in a series of tests that the seal can be reestablished more than a dozen times without leaking because the precision of the positioning mechanism is such that the knife edge always hits the groove made in the first sealing operation.

Each oven can be heated to 50°, 200°, 350°, or 500°C in 1–8 s (depending on the temperature selected) and held there until a total of 30 s has elapsed. Valves V1 and V3 are opened for 30 s prior to the heating of the oven and are closed immediately after the heating period. Any volatile material that emerges from the sample is swept onto the gas chromatographic column with about 2–3 ml of ^{13}C-labeled carbon dioxide (99+% isotopic purity, Mound Laboratories). The use of CO_2 rather than H_2, which is the carrier gas for the gas chromatograph, avoids catalytic or thermally induced reduction of the organic material possibly present in the sample. It causes certain complications, however, which will be discussed later in connection with the operation of the so-called effluent divider.

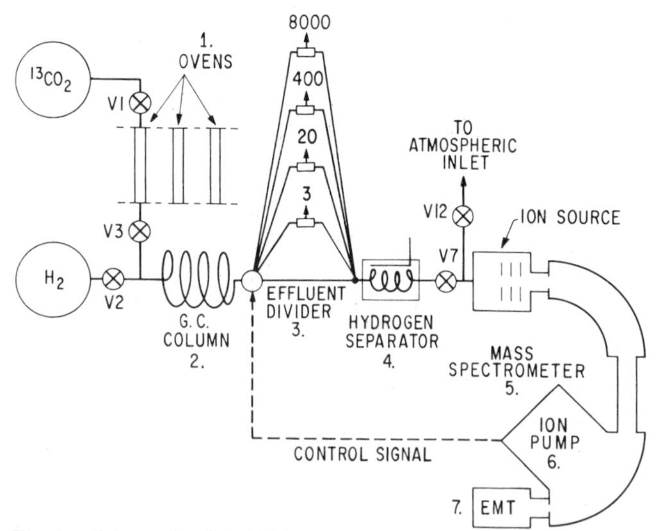

Fig. 1. Schematic of the Viking gas chromatograph–mass spectrometer.

At this point it should be noted that during the 50 min preceding the heating of the sample the entire instrument is turned on in the following sequence: the 'thermal zone' (which contains the effluent divider with valves V4, V4A, V5, and V6; the hydrogen separator; and valves V7 and V12) and the ion source of the mass spectrometer are heated; then the tubing leading to the oven and to the column are heated; finally, the mass spectrometer is energized. The recording of data begins just prior to the first opening of valves V1 and V3. Valve V7 opens 1 min later; thus six scans of the mass spectrometer background are obtained just prior to the analysis.

Simultaneously with the closing of valves V1 and V3, valve V2, which connects the hydrogen tank (initial pressure of ~750 psi (52 bars)) to the gas chromatograph, is opened, thus starting the gas chromatographic phase of the experiment.

The gas chromatographic column is filled with a liquid-modified organic adsorbent consisting of 60- to 80-mesh Tenax-GC (2, 6-diphenyl-p-phenylene oxide) coated with polymetaphenoxylene. This specific packing was developed to maximize the separation of water and CO_2 from organic compounds, to transmit efficiently most compound classes at the low nanogram level, to have exceptional thermal stability, and to have mechanical strength compatible with the rigors of space flight [*Novotny et al.*, 1975]. The column consists of a stainless steel tube of 0.76-mm inside diameter 2 m in length. Initially, the temperature of the gas chromatographic column is held at 50° for 12 min, followed by a linear increase to 200° over a period of 18 min, and then held at this temperature for 18, 36, or 54 min. The holding period can be selected by commands previously sent from earth. This permits control of the amount of data produced to avoid filling valuable tape recorder space with data not containing useful information. There were a number of additional commands which allowed one to manipulate the volume of data recorded during each experiment, but they were rarely used during the mission.

The restrictor at the hydrogen tank was adjusted to give a flow rate of about 2.5 ml/min for the first experiment. It slowly decreased in each additional run because of decreasing tank pressure. The amount of hydrogen carried along was such that one could expect to perform a large number of experiments (prelaunch and postlaunch tests and experiments on the Martian surface) before the flow rate falls below 1.5 ml/min, which was still deemed to be an acceptable value.

The interface between the gas chromatograph and the mass spectrometer includes a network of valves and restrictors (the 'effluent divider'). These are designed to prevent the entry into the mass spectrometer of large quantities of gases which would raise the pressure above the capacity of the small ion pump, which would then cease to function. Such an event would permanently disable the spectrometer and thus the entire instrument, including its capability to perform atmospheric analyses. To protect itself from such a catastrophe, the ion pump controls the flow rate of material into the spectrometer by causing the effluent divider to vent to the Martian atmosphere such a fraction of the gas chromatographic effluent that the remainder which enters the spectrometer can be pumped efficiently. This is accomplished by a control loop in which the magnitude of the ion pump current, in combination with its rate of rise, controls the valves between the exit of the gas chromatographic column and the hydrogen separator. Four such valves are employed, each of which contains a restrictor. The restrictors are of such a conductance that they produce nominal split ratios (parts entering the mass spectrometer to parts vented) of 1:3, 1:20, 1:400, and 1:8000. Naturally, this causes a decrease of total system sensitivity by the same ratio, but the higher split ratios are reached only when very large amounts of material emerge from the column. It was expected that this would only be water or carbon dioxide from decomposing minerals and the $^{13}CO_2$, all of which appear at the very beginning of the gas chromatogram. If the ion pump current exceeds a predetermined value, valve V7 (the one just ahead of the mass spectrometer) closes, affording final protection. The reversal of these steps is controlled both by the ion pump current falling below a certain level and by the time elapsed since the last action to avoid valve 'chatter' (rapid changing of the split ratio back and forth). There are two modes of operation that can be selected by ground command, each involving different sets of lapse times. One is termed the 'hydrous' mode and requires 45 s before the effluent divider switches from 1:8000 to 1:400, 2 min for the next step (1:400 to 1:20), and 15 min each for the third and fourth steps (1:20 to 1:3 to 1:0). Although the gas chromatographic column was designed [*Novotny et al.*, 1975] to elute water early as a sharp peak, it is not reasonable to expect the pump current to decrease from maximum to 1:20 of its value in less than 45 s and to 1:400 in less than 2 min 45 s, etc. Since this large overload was expected only from water expelled upon heating the sample, the term hydrous was adopted for this operating mode. For situations where it is reasonable to assume that negligible quantities of water are produced (i.e., if it turned out that the Martian surface material is free of thermally labile water or hydrates), another mode, termed 'anhydrous,' is available upon command; in that case the maximum waiting period is 15 s between each step. If the pressure falls below a predetermined level, the waiting period becomes 10 s in either mode.

The hydrogen separator [*Dencker et al.*, 1972] consists of a 60-cm-long silver-palladium tube of 0.15-mm inside diameter surrounded by a sodium hydroxide–water eutectic which melts at about 160° and is operated at 220°C. The Ag-Pd tube forms the anode, and another larger tube placed within the electrolyte but opening to the atmosphere forms the cathode of an electrolytic cell ($\Delta V = 0.7V$). The separator removes more than 99.99995% of the hydrogen carrier gas and thus prolongs the life of the ion pump and keeps the ion pump current well below the threshold at which the effluent divider network starts to operate. In the organic analysis mode the ion source of the mass spectrometer is held at 225°C, the filament is operated at 70 eV, and the accelerating voltage is scanned from 2350 V ($\approx m/e$ 11.5) to 125 V ($\approx m/e$ 215) every 10.24 s during the entire experiment. The electron multiplier is set to a gain of about 10^3 but specifically to such a value that the ion currents for CO_2 in the atmospheric mode are equal for both the Viking lander 1 (VL-1) and the Viking lander 2 (VL-2) instruments, which have inlet leaks of slightly different conductance. There are two higher gains that can be selected by ground command, but these were not used during the mission except for the atmospheric analysis [*Owen et al.*, 1977].

During each scan, 3840 samples of the output of the electron multiplier amplifier were taken after conversion to a log value and were encoded to 9 bit. The original plans for handling the data had called for peak selection and mass assignment at that point and perhaps even further compression of the data. However, when an on-board tape recorder with high capacity became available and the predicted data rate for transmission from lander to orbiter and from orbiter to earth increased to 16 kbit/s and 4 kbit/s, respectively, the transmission of all raw data (about 17 Mbit per analysis) became possible. Thus all the data processing to mass spectra and chromatograms, mass scale calibration, noise spike recognition, averaging of data, etc., was done on earth, either at the Jet Propulsion Laboratory (JPL) or at the Massachusetts Institute of Technology (MIT) by using algorithms previously developed in the latter laboratory [*Hertz et al.*, 1971; *Biller and Biemann*, 1974; *Biller et al.*, 1977]. The final data were recorded on 16-mm microfilm and displayed on a film reader for inspection and interpretation. For the immediate first-order processing of the data during the mission these programs were adapted by R. Williams for use on the JPL computer system.

The performance of the instrument will be demonstrated by the results obtained with a laboratory version. It corresponds almost exactly to the flight instrument with the exception of the associated test equipment and the capability of reusing a single sample oven which is filled manually. A sample of Antarctic soil (sample 638 [*Cameron et al.*, 1970]) was heated to 200° and 500°C and was analyzed as described above. The gas chromatogram (as total ion current plot) is shown in Figure 2. The interpretation of the mass spectral data led to the identification of a large number of compounds, the more abundant of which are marked in the figure and listed in the following tabulation.

Code in Figure 2	Compound
AR-1	benzene
AR-2	toluene
AR-3	xylene (benzene-C_2)
AR-5	benzene-C_3
AR-9	naphthalene
AR-10	methyl naphthalene
AR-11	biphenyl
O-1	furan
O-2	acetone
O-4	methyl vinyl ketone
O-5	methyl furan
O-6	furane-C_2
O-7	benzofuran
O-8	phenol
O-10	dibenzofuran
N-1	acetonitrile
N-2	propionitrile
N-3	pyridine
N-4	benzonitrile
N-5	toluonitrile
S-1	thiophene
S-2	methylthiophene
S-3	thiophene-C_2
S-4	benzthiophene
S-5	sulfur dioxide
HC-1	cyclohexene

Code 'AR' stands for aromatic hydrocarbons; 'O,' for oxygen-containing compounds; 'N,' for nitrogen-containing compounds; 'S,' for sulfur-containing compounds; and 'HC,' for aliphatic hydrocarbons. The compounds detected range in size from acetonitrile to dibenzofuran. Although the latter does not give rise to a discernible peak in the gas chromatogram, a well-developed peak is displayed in the mass chromatogram of the m/e 168, the molecular ion of dibenzofuran. As a measure of sensitivity the quantities represented by the peaks representing benzonitrile and benzofuran were calculated and found to correspond to 150 and 43 ppb, respectively. In fact, acetoni-

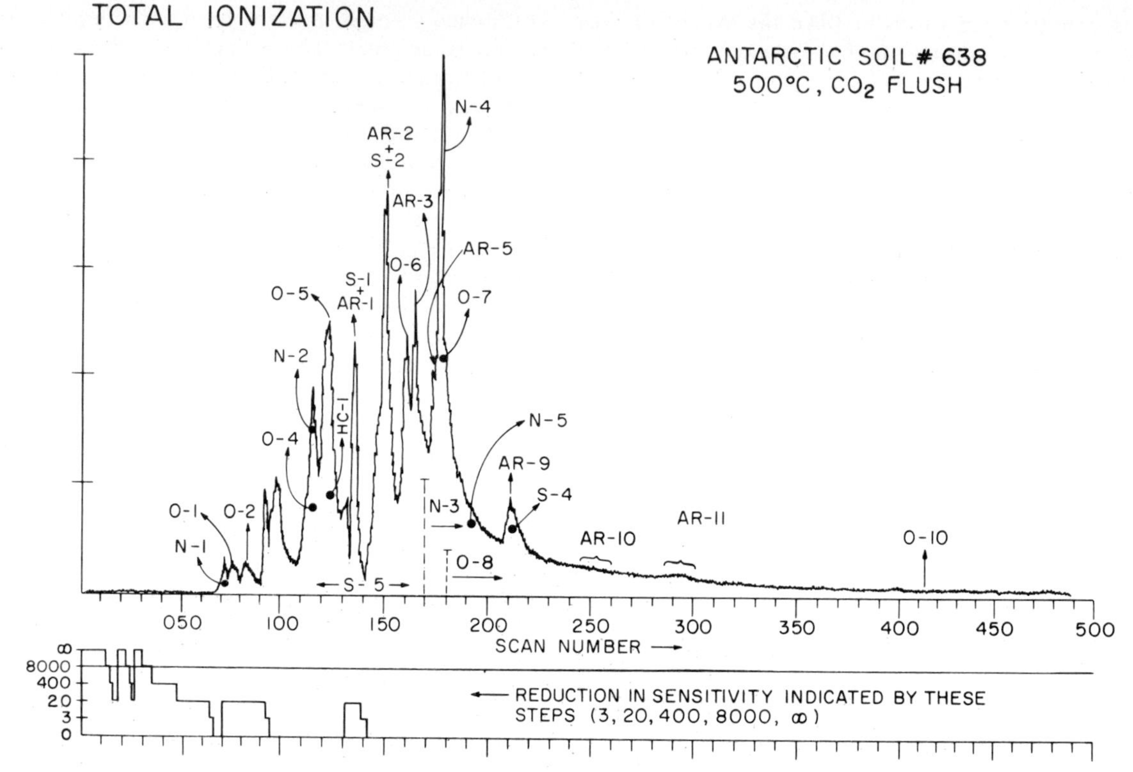

Fig. 2. Gas chromatogram (plot of the sum of intensities above m/e 47) obtained with a sample of Antarctic soil. The identifying codes for the various components are listed in the tabulation in section 2 of the text. Components N-3 (pyridine) and O-8 (phenol) appear (as judged from the mass chromatograms of their molecular ions) as tailing peaks with a sharp front represented by the vertical dashed lines at scans 170 and 180, respectively. The abscissa represents scan numbers of the consecutive mass spectral scans of 10.24 s each; each tick mark thus corresponds to 102.4 s of elapsed time of the gas chromatogram. The ordinate is linear, with the largest peak plotted to full scale.

trile and benzene are major components that cause the effluent divider to switch into the 20:1 state. Compounds eluting before acetonitrile are obscured by the large water peak which causes the venting of most or all of the material emerging from the gas chromatograph.

TABLE 1. Acquisition Sites and Analysis Conditions for the Four Martian Samples

Identification Number*	Date of Analysis	Oven Temperature, °C	Mode	Oven Purge Gas	Time Column Held at 200°, min	Oven
VL-1 Sample 1 (Subsurface), Acquired on Sol 8						
10015	sol 17	200	hydrous	$^{13}CO_2$	18	1
10018	sol 23	500	anhydrous	$^{13}CO_2$	36	1
VL-1 Sample 2 (Surface), Acquired on Sol 31						
10023	sol 32	350	hydrous	$^{13}CO_2$	54	2
10024	sol 37	500	hydrous	$^{13}CO_2$	54	2
10025	sol 43	500	hydrous	$^{13}CO_2$	36	2
VL-2 Sample 1 (Bonneville Duracrust), Acquired on Sol 21						
10032	sol 24	200	hydrous	H_2	36	2
10033	sol 26	350	hydrous	H_2	36	2
10034	sol 35	500	hydrous	H_2	36	2
10035	sol 37	500	hydrous	$^{13}CO_2$	36	2
VL-2 Sample 2 (Under Badger Rock), Acquired on Sol 37						
10036	sol 41	50	hydrous	H_2	36	3
10037	sol 43	200	hydrous	H_2	36	3
10038	sol 45	350	hydrous	H_2	36	3
10039	sol 47	500	hydrous	H_2	36	3
10041	sol 61	500	hydrous	$^{13}CO_2$	36	3

*This number is a data processing code listed here to facilitate correlation of data in this table with the figures in the text.

3. RESULTS

Altogether, four samples of Martian surface and subsurface materials were analyzed for organic compounds and inorganic volatiles. Two samples were obtained at each landing site, Chryse Planitia (VL-1) and Utopia Planitia (VL-2). The results of the former have been summarized previously [*Biemann et al.*, 1976] and will be discussed here in more detail. The VL-2 data, which gave essentially the same results, have not been described before and thus will require more discussion.

Table 1 lists all samples, including dates and conditions, that were analyzed during the prime mission (before conjunction) at both landing sites. Two major changes were made for the experiments at the second site. First, the $^{13}CO_2$ purge mode was replaced in the first few analyses of each sample by the 'hydrogen expansion mode,' in which the opening of valve V1 was replaced by the opening of valve V2 just before the heating of the sample. In this way the effect which the large amount of carbon dioxide has on the effluent divider (which remains in the 1:8000 mode for a relatively long time and even closes valve V7) is eliminated because the oven becomes filled with hydrogen at relatively high pressure (about 0.5 atm), which during the heating of the sample expands back into the carrier gas stream. In addition, this mode replaces the gas in all the void areas between valves V1 and V3 by hydrogen, thus allowing more reliable detection of CO_2 which may evolve from the sample in subsequent heating cycles.

The second change that is evident from Table 1 is the fact that more consecutive analyses were performed on the samples from Utopia than from Chryse. Confidence in the communication system had substantially increased since the beginning of the first mission. New data could therefore be transmitted in place of the automatic retransmission of each set of data (as was done previously to avoid accidental loss of any of the irretrievable information).

The results are summarized in Figure 3, in which the gas chromatograms of all VL-1 experiments are plotted at the same scale. These chromatograms are shown as the summed intensity of all ions above m/e 47 (to eliminate the effect of H_2O and CO_2) in each scan versus scan number along the abscissa. The gas chromatogram shown at the top is the blank experiment performed during cruise (oven 2 at 500°C).

As is discussed in the preliminary report [*Biemann et al.*, 1976], the sharp peak in the very first gas chromatogram (identification number 10015, 'Sandy Flats' site, 200°) returned to earth from Mars was identified as methyl chloride. This identification was based on the mass spectrum recorded at that point and the fact that the mass chromatograms of all ions characteristic for CH_3Cl showed the same sharp maximum. As an example the mass chromatogram of m/e 50 ($CH_3{}^{35}Cl$) is shown in Figure 4. The calculation of the abundance of this compound is based on the integrated intensity of this ion, corrected for the effluent divider state. The sensitivity of the mass spectrometer for this or any compound can be determined from the signal obtained from the m/e 44 ion during the analysis of the Martian atmosphere [*Owen and Biemann*, 1976] combined with the known atmospheric pressure, 7.65 mbar at that time [*Hess et al.*, 1976], the conductance of the leak in the atmospheric inlet system, and the relative intensities of m/e 44 for CO_2 and of any chosen peak in the compound of interest. These relative intensities can be taken from the *American Petroleum Institute* [1955] collection of mass spectral data. This method had been chosen because it would have been impossible to calibrate the flight instrument with many compounds without severely contaminating it. For this reason, and owing to the uncertainty in the amount of surface material contained in the oven (see the discussion in section 2), all values for abundances may deviate from the actual values by a factor of 2 or even more, but this is quite sufficient for the purpose at hand. Needless to say, the relative abundance ratios could be determined more accurately by reconstituting the appropriate mixture and analyzing it on the spare flight instrument, but this has not been necessary because no complex mixtures of organic compounds were encountered.

On the basis of such a calculation the amount of methyl chloride represented by the peak centered around scan 27 in the first gas chromatogram from Mars (identification number 10015) was found to represent about 15 ppb with respect to the sample. Comparisons of all the gas chromatograms from the VL-1 samples (Figure 3) indicates that this is the only clearly identifiable gas chromatographic peak. All other changes in signal correspond to changes in effluent divider status (plotted in the top part of each chromatogram) and thus to changes in the amount of material which slowly and continuously elutes from the gas chromatographic column (column bleed) and enters the mass spectrometer. This should be noted particularly for the pattern of the 500° analysis of sample 1 (third from the top in Figure 3), which has the appearance of a series of peaks. This experiment was run in the anhydrous mode (see the discussion in section 2), but the unexpectedly large amount of water evolved at 500° kept the residual ion pump current at such a level as to cause the effluent divider to cycle between the 0:1 state (no venting) and the 20:1 split rate. Since the gas chromatogram is plotted as the sum of all ions above m/e 47, these fluctuations are due not to the ion current of H_2O but to the above-mentioned traces of material continuously eluted from the column and inner surfaces of the interfaces which are attenuated by the valve switching. For this reason all further experiments were conducted in the hydrous mode.

From the sharpness of the peak for methyl chloride and its height, which corresponds to only 15 ppb, it is evident that there is no other compound observable that is of this magnitude or larger, except possibly for those regions of the gas chromatograms where the effluent divider is in a state where a large portion of the material is vented. This occurs mostly at the very beginning of the chromatogram, when $^{13}CO_2$ and H_2O emerge. The only other identifiable compounds were fluorocarbons of the freon-E type (polyperfluoropropylene oxides), which can be detected on the basis of the mass chromatograms of their characteristic ions at m/e 69, 97, 101, 119, 147, etc. The small peaks in the region from scan 50 to scan 150 in the cruise test (top of Figure 3) are due to a series of oligomers of this type. A typical spectrum is shown in Figure 5. It corresponds well with that of any of these oligomers, which all give very similar spectra in this mass region. The ions at m/e 47 (CFO), 97 (C_2F_3O), and 147 (C_3F_5O) are characteristic for these oxygen-containing fluorocarbons; so are ions at m/e 51 (CF_2H) and 101 (C_2F_4H) because the oligomers contain one hydrogen atom per unit.

Traces of freon-E and methyl chloride had been detected in previous tests on earth. There is no doubt that all the freon-E is of terrestrial origin. The low level, a few tens of parts per billion, did not interfere with the experiment; in fact, it provided a welcome calibration of the mass scale of the instrument. The methyl chloride, or part of it, could conceivably be indigenous to Mars. However, if it were, one would expect that other related compounds like ethyl chloride or methyl bromide would also be formed, but none were detected. The abundance ratio of m/e 50 to n/e 52 was about 3:1, corresponding to the terrestrial isotope ratio of chlorine; but this does not necessarily confirm its origin, because there is no reason to predict a different ratio for Martian chlorine. Considering all these facts, we tend to believe that all the methyl chloride is from terrestrial sources (chlorinated solvents or from adsorbed traces of methanol and HCl).

Using the abundance calculation for the methyl chloride as an example, one can then proceed to search for the presence of any compound by inspection of the mass chromatogram of a characteristic ion in the region of the determined or the predicted retention time (the retention behavior of a sufficient number of compounds had been determined on the laboratory version of the instrument or on a gas chromatograph fitted with a flight column). Such searches were carried out for quite a number of compounds, particularly those encountered in the Antarctic soil discussed in section 2 and in a sample of Murchison meteorite analyzed on the laboratory version of the Viking gas chromatograph–mass spectrometer. For none of these compounds could a peak be observed in the mass chromatograms of the VL-1 experiments, thus demonstrating their absence below the detection limit of the instrument. This detection limit was determined by calculating the amount of material that would correspond to the background signal integrated over 10 scans in the region of the appropriate retention time and assuming that that amount, when present in addition to the background signal, would have produced an observable peak in the mass chromatogram. Some of these detection limits for a few typical compounds are listed in Table

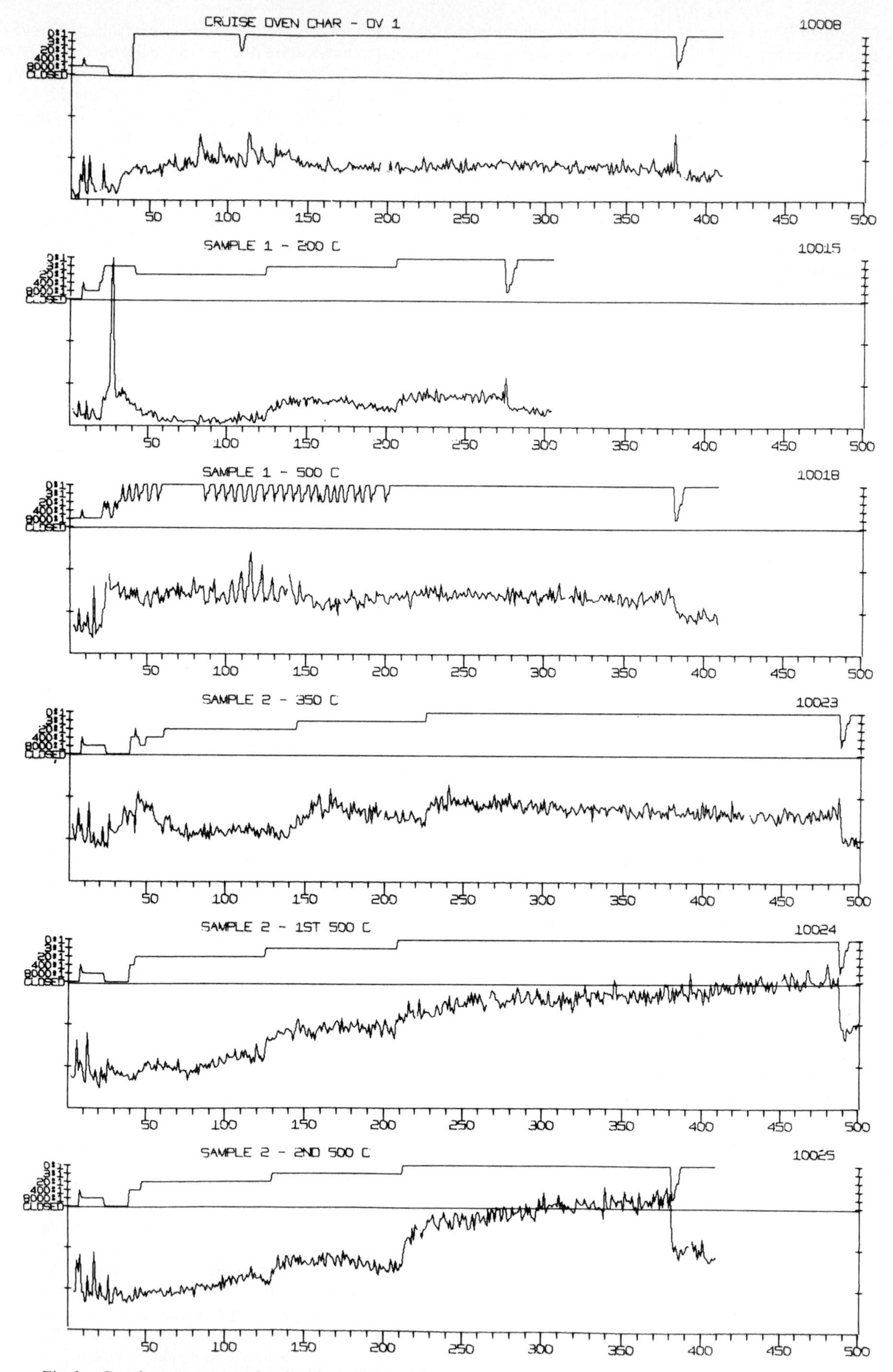

Fig. 3. Gas chromatograms of cruise blank and five VL-1 experiments, plotted as sums of all ions above m/e 47. All are plotted at the same scale for comparison. The upper quarter of each plot indicates the effluent divider status during each experiment (also in Figures 7 and 8).

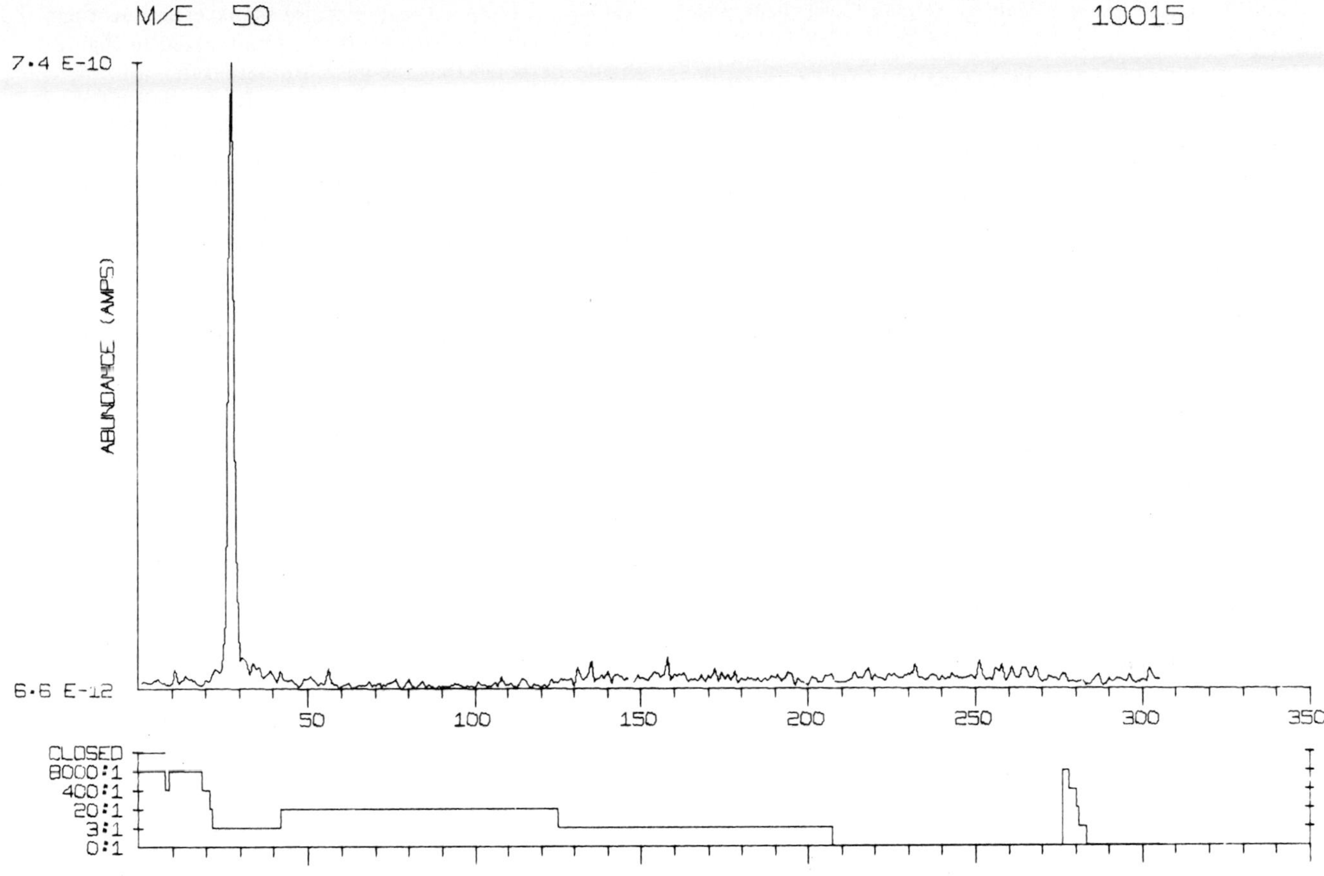

Fig. 4. Mass chromatogram of *m/e* 50 from the 200° experiment of the first VL-1 sample. In this figure and in Figures 6 and 9 the effluent divider status is plotted underneath the graph (the horizontal line indicates the 1:8000 level, and the step function represents the effluent divider status).

2. The mass chromatograms of two typical examples are shown in Figure 6, where *m/e* 58 is chosen as being representative of acetone and *m/e* 128 of naphthalene. These compounds are expected to emerge in the regions around scans 90 (acetone) and 200 (naphthalene). It will be noted that there are no maxima in these regions, and the signal is around 1×10^{-11} A for *m/e* 58 and about 2×10^{-13} A for *m/e* 128. Furthermore, the split ratio of the effluent divider is 20:1 where acetone would emerge, while it is 3:1 in the naphthalene region. Therefore the detection limits quoted in Table 2 for these two compounds differ by about 2 orders of magnitude.

As was the case on the instrument on Viking lander 1, the one on Viking lander 2 had also been shown probably to contain one inoperable oven and thus to limit the experiments to the collection and analysis of two rather than three samples. While the mass spectrometer on this instrument displayed a much lower background than that on VL-1 and was thus particularly suitable for the analysis of the Martian atmosphere at a trace level, those subsystems involved in the soil analysis, namely, the sample ovens and the tubing and valving prior to and possibly after the gas chromatographic column, were less clean. For this reason the interpretation of the soil

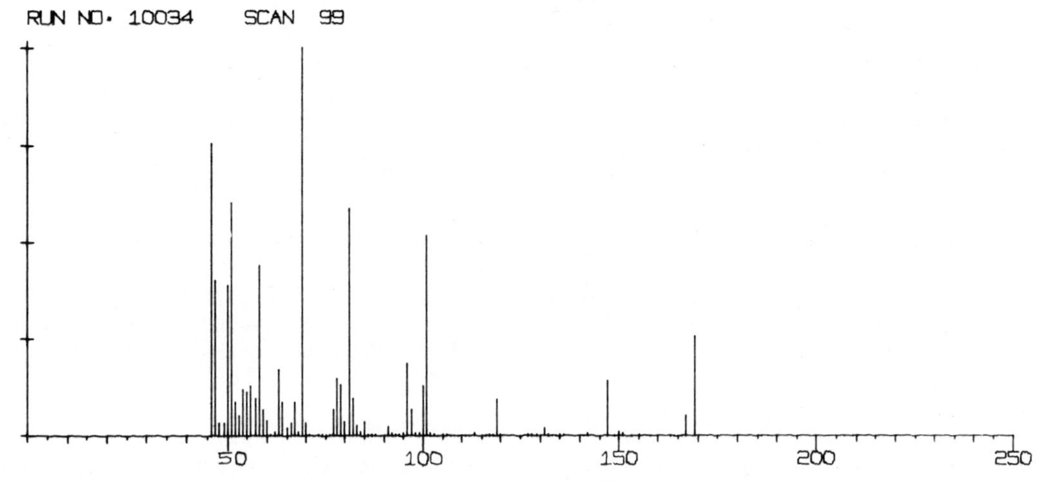

Fig. 5. Typical mass spectrum of the freon-E contaminants. Peaks in the region below *m/e* 46 have been deleted to avoid distortion of the plot due to contributions of H_2O and CO_2.

TABLE 2. Upper Limits of Selected Organic Compounds Which Would Be Detected If Present in the Martian Soil Samples

Compound	Range of Detection Limits, parts per 10^9	
	Lander 1 (Chryse Planitia)	Lander 2 (Utopia Planitia)
Simple Molecules		
Methanol	*	<300–3000†
Ethanol	*	<9–90
Formaldehyde	*	<1200–12,000‡
Ethane	*	<1200–12,000‡
Propane	*	<3–30
Aliphatic Hydrocarbons		
Butane	<1–10	<3–30
Hexane	<1–10	<0.5–5
Octane	<1–10	<0.15–1.5
Aromatic Hydrocarbons		
Benzene	<0.5–5	<8–80
Toluene	<0.5–5	<3–30
Naphthalene	<0.05–0.5	<0.0015–0.015
Oxygen-Containing Compounds		
Acetone	<10–50	<250–2500
Furan	<0.1–1	<0.05–0.5
Methylfuran	<0.2–2	<0.15–1.5
Nitrogen-Containing Compounds		
Acetonitrile	<1–10	<0.5–5
Benzonitrile	<0.2–2	<0.015–0.15
Sulfur-Containing Compounds		
Thiophene	<0.1–0.5	<0.015–0.15
Methylthiophene	<0.1–0.5	<0.015–0.15

*The limit of detection estimated for these compounds is of the order of tens of parts per million for most of these experiments because the eluting $^{13}CO_2$ keeps the effluent divider in the 8000:1 state or even closes valve V7.

†Effluent divider in 400:1 split ratio.

‡Effluent divider in 8000:1 split ratio.

analysis data requires the subtraction in a qualitative and a quantitative sense of the background data acquired during a complete blank run performed during cruise. This blank was obtained on oven 2 and therefore holds strictly for that oven, and the use of these background data with the data obtained with oven 3 is based on the assumption that the contamination level of the two ovens would be similar.

In the region of Utopia Planitia, two samples were analyzed for volatile materials. The first was collected from the area named Bonneville and was thought to represent duracrust-type material; the second was obtained from underneath a rock (Badger). These were the two regions in the accessible area from which a sample could be obtained and which were believed to be possibly different from the region of the exposed surface area [*Shorthill et al.*, 1976]. Exposed regolith had been analyzed on VL-1 as well as by the biology experiment on VL-2, which indicated that it may not substantially differ from the VL-1 material. For this reason it was decided to reserve the two samples which could be analyzed by the gas chromatograph–mass spectrometer for the most different materials within the VL-2 region.

In an effort to lower the detection limit for the most volatile components that could be expected the experiments on VL-2 were run in the so-called hydrogen expansion mode with the exception of the last experiment on each sample, which was performed in the normal mode (see Table 1). The gas chromatograms obtained in the cruise blank experiment and the first four sample experiments are summarized in Figure 7. It should be noted that the gas chromatogram from the blank experiment (top) has been corrected for a small difference in the behavior of the gas chromatographs between cruise and surface experiments. Inspection of the column temperature engineering data revealed that it stayed rather constant during the cruise test but always increased slowly to about 70° in the sample runs on the surface of Mars. This speeds up the elution of acetone and at the same time broadens its gas chromatographic peak. Simulation of this column temperature shift before plotting the cruise data resulted in a gas chromatogram which is practically identical with that of the 200° Bonneville experiment (identification number 10032). Thus there is no difference in the nature of the materials eluted in these two experiments except that there is only half as much in run 10032. Table 3 represents a semiquantitative evaluation of the major components observed in the experiments run on the VL-2 instrument. (Rather than labeling all the peaks in all chromatograms, which would greatly confuse the figures, only the middle peak in Figure 7, i.e., Bonneville, 350°, is labeled.) It is obvious that with the exception of methylene chloride all other compounds are identical with those detected in the cruise experiment. Their quantities vary while always remaining in the same order of magnitude as those in the blank experiment. This is to be expected because of the difference in the temperature regime to which each sample was heated and the fact that some of the impurities deplete (like the more volatile acetone) or are lower at the lowest temperatures and higher at higher temperatures.

The only still puzzling peak is the one always eluting around scan 110 and labeled 'methyl fluorosiloxane.' Its mass spectrum, which was equally prominent in the cruise experiment and was observed to a much lesser extent in the VL-1 data also, is dominated by a peak of *m/e* 81 and also shows, among others, an ion of *m/e* 96. A group of peaks at *m/e* 207–209 also maximizes in intensity at that same point. While the spectrum from *m/e* 96 on down is very similar to that of dimethyldifluoro-silane, $(CH_3)_2SiF_2$, this compound is ruled out by the gas chromatographic behavior (it elutes much earlier). The best interpretation of the spectra, at present, also involves the peak at *m/e* 207, which is typical for polymeric dimethyl siloxanes (common silicones). Thus a compound like $F[Si(CH_2)_2\text{-}O]_nSi(CH_3)_2F$, which may arise from a reaction of a silicone with HF or another fluoride, could give rise to the observed mass spectrum by a fluorine transfer, accompanied by cyclization of the dimethyl siloxane portion after ionization in the mass spectrometer. (The mass spectra of dimethyl silicones have been discussed by *Biemann* [1962].) This interpretation is still to be verified in laboratory simulations.

The second and last sample analyzed from the Utopia Planitia region was collected from underneath so-called 'Badger Rocks.' Great care was taken to assure that the collected material was entirely from the area which the rock protected from ultraviolet radiation. In an effort to determine the amount of carbon dioxide that is released from the sample upon heating and to differentiate it from that which was merely due to the occluded atmosphere sealed into the oven with a sample, the first experiment was run without appreciable heating of the sample. This is accomplished in the 50° mode, where the sample is heated to that temperature or the ambient temperature of the sample oven, whichever is higher. Thus the sample from underneath the rock was analyzed five times, namely, at 50°, 200°, 350°, and 500° in the hydrogen

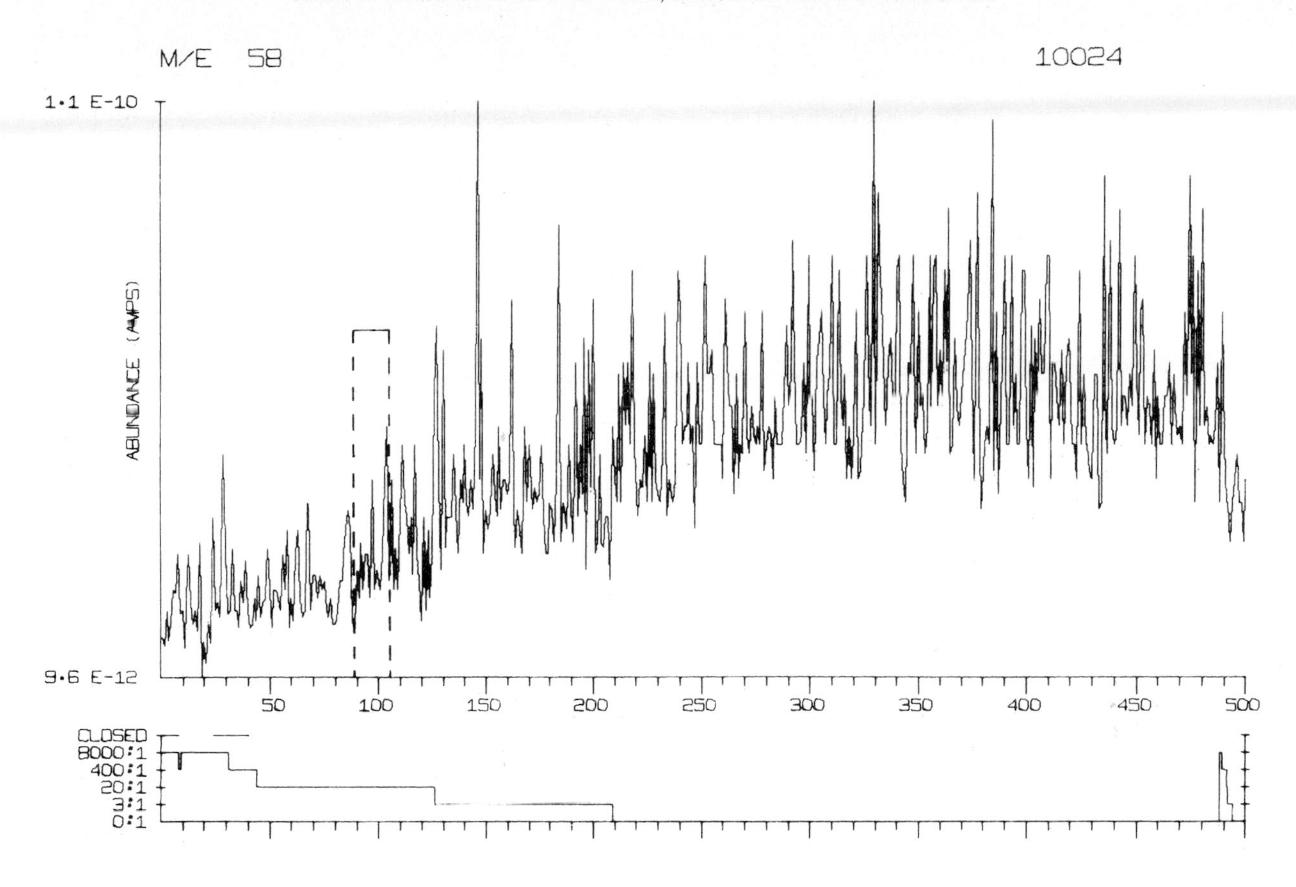

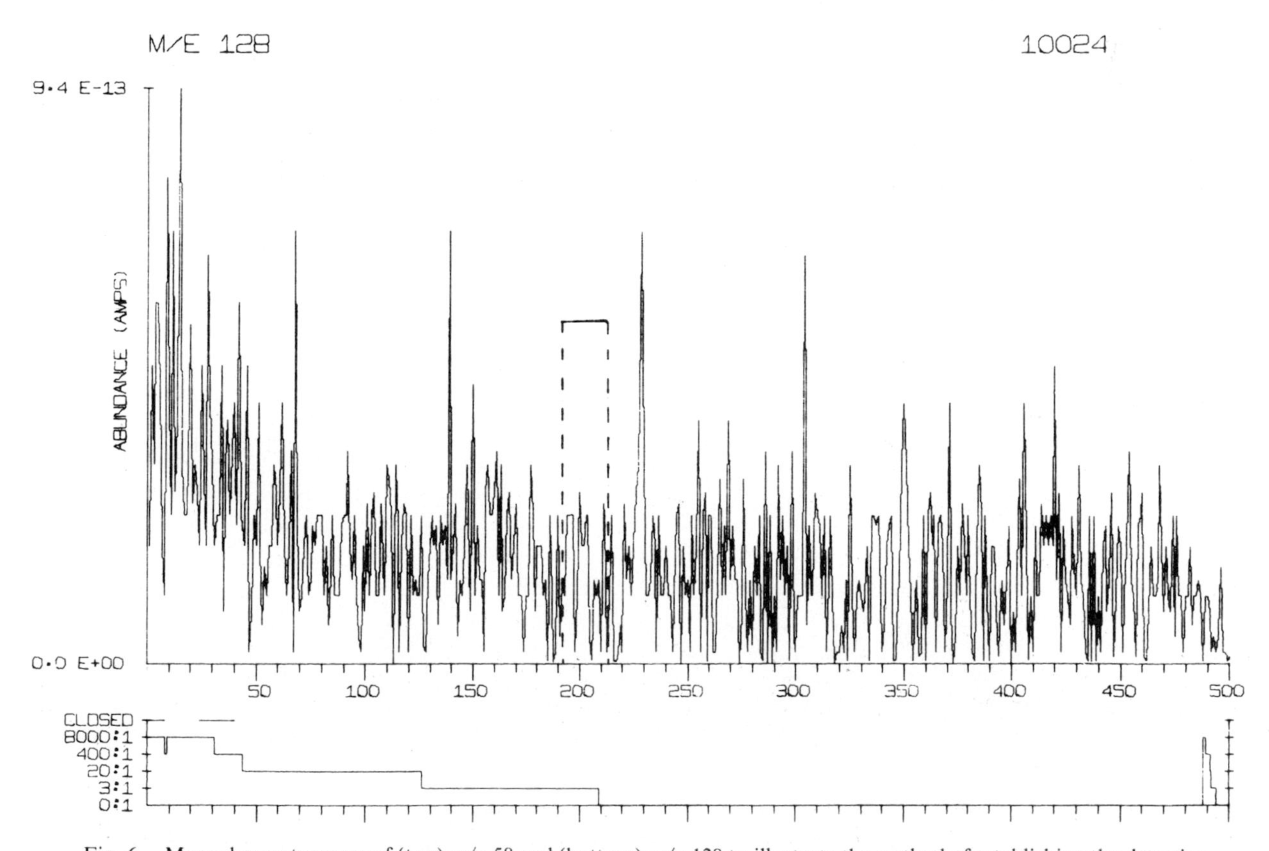

Fig. 6. Mass chromatograms of (top) m/e 58 and (bottom) m/e 128 to illustrate the method of establishing the detection limits listed in Table 2. The dashed lines bracket the retention times of (top) acetone and (bottom) naphthalene.

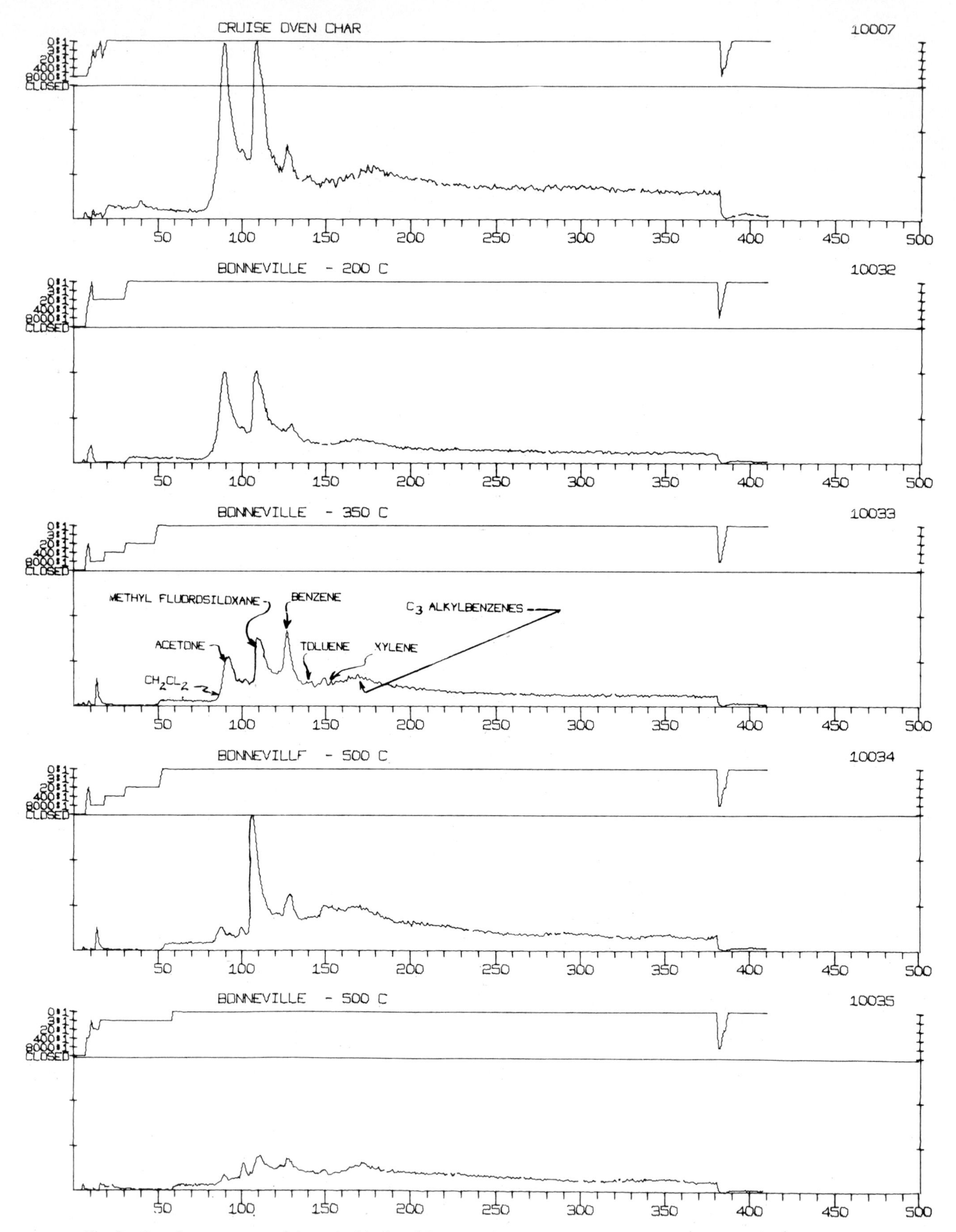

Fig. 7. Gas chromatograms of the cruise blank and four experiments with the first VL-2 sample (Bonneville). For details, see the Figure 3 legend.

TABLE 3. Terrestrial Contaminants Identified in VL-2 Samples

Sample	Temperature, °C	Mode	Methylene Chloride (89)	Acetone (92)	Freons (40, 105, 130)	Methyl Fluoro-Siloxane (110)	Benzene (129)	Toluene (140)	Xylene (150)	C_3 Alkyl Benzene (162)
Blank (oven 2)	500	CO_2	ND	120–240	10–20	60–120	4–8	1–2	0.6–1.4	0.6–1.0
Bonneville	200	H_2	ND	60–120	6–10	100–200	1–2	2–3	0.3–0.5	0.1–0.3
(oven 2)	350	H_2	6–14	40–70	10–20	70–130	3–6	2–3	0.3–0.5	0.09–0.16
	500	H_2	6–14	10–20	10–20	160–320	3–6	0.8–1.6	0.4–0.8	0.1–0.2
	500	CO_2	2–6	1–2	10–20	35–70	2–4	0.6–1.4	1–3	0.04–0.08
Under Badger	50	H_2	ND	ND	ND	20–40	0.2–0.4	0.1–0.3	ND	ND
Rock	200	H_2	0.04–0.08	200–400	4–8	40–85	0.6–1.4	0.4–0.8	0.3–0.5	0.8–1.6
(oven 3)	350	H_2	10–20	30–60	2–4	30–55	0.6–1.4	0.3–0.5	ND	0.04–0.08
	500	H_2	<1	<5	0.04–0.08	140–280	1–2	0.04–0.08	ND	ND
	500	CO_2	20–40	5–10	5–10	50–90	0.75–1.75	1–1.5	0.1–0.2	ND

Values are in parts per billion, 100-mg samples being assumed.
Numbers in parentheses indicate the approximate scan numbers where components elute.
ND, not detected.

expansion mode and again at 500° in the CO_2 purge mode. Figure 8 again summarizes the five chromatograms obtained in this series of experiments headed by the plot of the cruise experiment (which is, of course, the same experiment as that shown at the top of Figure 7). It is clear that very little is evolved in the 50° experiment and that a rather large amount of acetone (see Table 3) is produced in the 200° experiment. This would imply that there was more residual acetone in oven 3 than there was in oven 2.

It should be noted, however, that in the chromatograms obtained from the subrock sample, as in all the previous experiments, no new materials were observed in addition to those detected in the blank run performed during cruise. Therefore none of the gas chromatographic peaks observed are deemed to be due to compounds indigenous to the Martian soil sample. As was the case with the Bonneville sample, traces of methylene chloride were produced from the second sample as well, but since this was a common laboratory solvent which had been used in the cleaning of the oven–gas chromatograph subsystem, it was considered a terrestrial contaminant although it had not been detected at those levels in the blank run.

There is one peak which appears to be new in the 500° hydrogen expansion experiment (second from the bottom, Figures 7 and 8). It appears at scan 130, and an inspection of the mass spectrum corresponding to this gas chromatographic peak suggests that it is most likely due to gas chromatographic column bleed that is pushed through the column by the water which evolves upon heating of the sample to 500°. Because the Tenax column transmits water very quickly [*Novotny et al.*, 1975], this peak emerges only after the water peak trails off (it should be noted that all of these gas chromatograms are generated by summing the ion intensities above mass 47 and therefore do not show the contributions of water which otherwise would completely dominate the gas chromatograms).

The appearance of the last gas chromatogram (500° CO_2 purge) is somewhat unusual because it shows a blank region from scan 100 to scan 135. A similar effect is noted, but less clearly, in the 500° hydrogen expansion experiment between scan 70 and scan 106. As is noted in the effluent divider status plot at the top of each gas chromatogram, this is a region where the effluent divider is in the 1:20 mode. A detailed assessment of the mass spectral intensities before and after the switch from 1:3 to 1:20 and vice versa indicates that the 1:20 divide ratio is not true for the VL-2 instrument. It is most likely due to blockage of the corresponding restrictor. Blockage of this restrictor would result in an effluent divide mode that corresponds to the next step, in this case, 1:400. Thus in the two regions mentioned the sensitivity of the instrument is decreased by a factor of 400 instead of 20, leading to the apparent blocking of this region.

The level of these contaminants, while clearly recognizable, is still within or not much above the instrument specifications, which called for no more than 1 ppm total organic contaminants. While this value may seem high, one has to keep in mind how difficult it is to manufacture such a complex instrument and still keep organic materials completely out. Because of the greater than expected sensitivity of the instrument, they are very noticeable but would not severely hamper the detection of other substances not present in the cruise blank. The major disadvantage of the presence of these impurities is that they greatly increase the detection limits of these compounds in the Martian soil, as is evident from Table 2. However, their presence demonstrates that both the gas chromatograph and the mass spectrometer operated reliably and reproducibly and that compounds of this type are not readily oxidized by the Martian soil under the conditions of the experiment. In this connection it should be noted that the relatively sharp gas chromatographic peaks demonstrate that the impurities evolve during the heating of the oven and must be there, rather than elsewhere in the system.

In summary, one can state that the two samples analyzed in the Utopia Planitia region do not contain any detectable amounts of organics. Furthermore, the hydrogen expansion mode (in which most of these experiments were conducted) did not reveal any of the small organic molecules that could have escaped detection in the VL-1 experiments.

4. DISCUSSION

Mechanisms for the Production and Destruction of Organic Molecules on Mars

The data outlined in section 3 demonstrate that if organic materials are present in the samples analyzed, they must be there at extremely low levels. This is an important finding as such. At present, it is impossible to provide a unique, detailed interpretation that would lead to a single set of circumstances responsible for the apparent absence of organic substances. It is only possible to outline a number of processes that could lead to the accumulation of organic molecules on Mars and a second set of processes that could destroy these organic molecules. We know very little about the rates at which the various

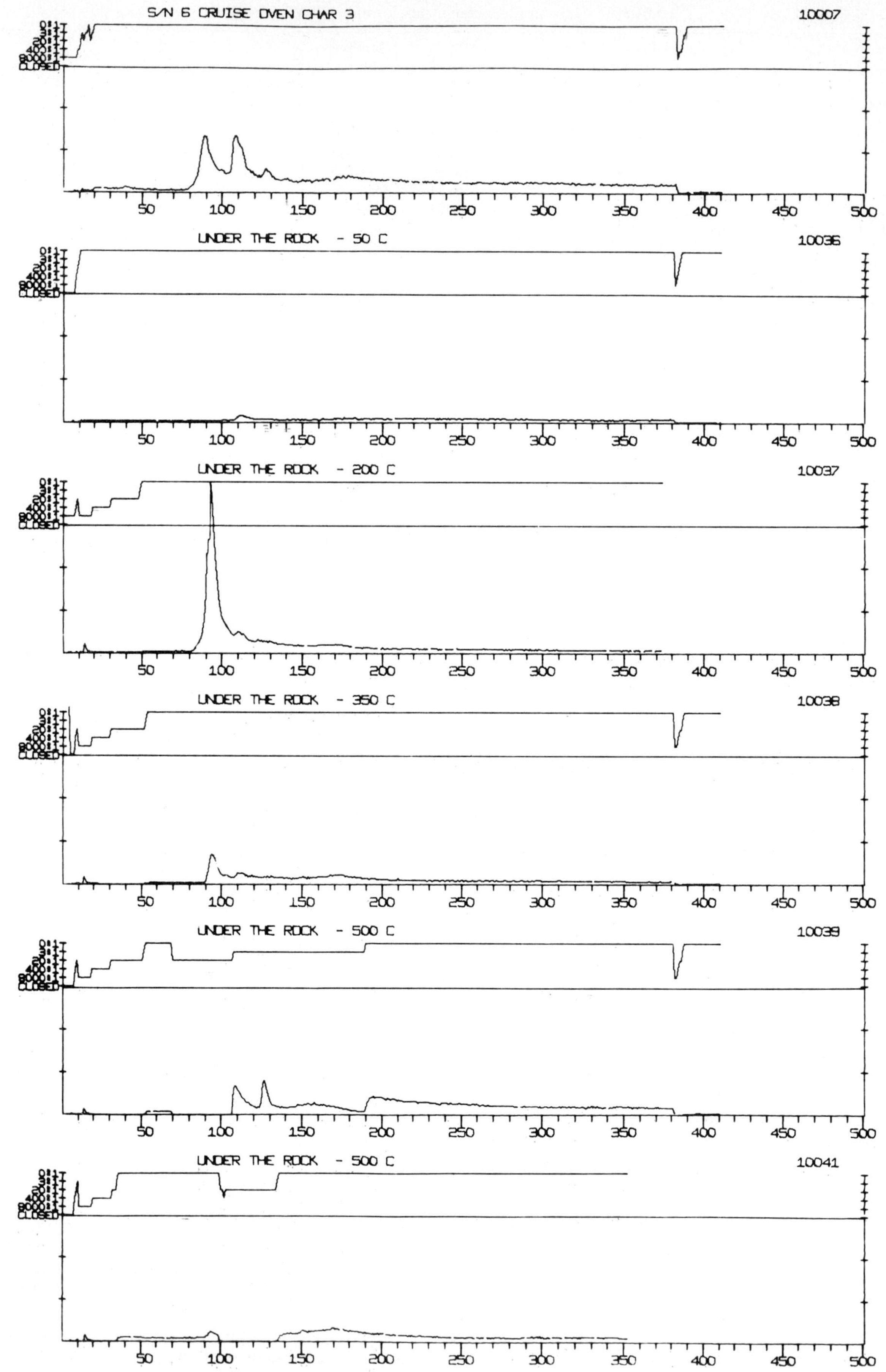

Fig. 8. Gas chromatograms of the cruise blank and five experiments with the second VL-2 sample (from under Badger Rock). For details, see the Figure 3 legend.

processes are presently occurring or about the rates at which they occurred in the past. Consequently, the conclusions drawn must be largely qualitative.

While one could postulate that neither now nor in the past has there existed a process that leads to the formation of organic compounds at the surface or in the atmosphere of the planet and that none of the organic compounds possibly present in the material from which the planet accreted has survived, one must admit that organic molecules present in meteoritic material must be arriving at the Martian surface. As on the moon [*Anders et al.*, 1973; *Morgan*, 1976], type 1 carbonaceous chondrites and micrometeorites of similar composition would be expected to make up a significant fraction of the meteoritic material reaching the surface of Mars. On the moon this material is more or less uniformly distributed over the entire lunar surface and comprises about 1.1% of well-exposed lunar maria soils at three Apollo sites [*Morgan*, 1976]. It is also known that type 1 carbonaceous chondrites contain on the average 3–5% carbon in the form of organic matter [*Mason*, 1971; *Wasson*, 1974]. Thus if the Martian surface contained unmodified meteoritic material in substantial amounts, the organic material contained therein should be detectable by the Viking gas chromatograph–mass spectrometer.

One approach to the estimation of the meteoritic input again utilizes data from the moon. The average micrometeoritic and carbonaceous chondrite input at the three Apollo sites is equivalent to a layer of meteoritic debris from type 1 carbonaceous chondrites about 5 cm deep. This material has been diluted into a regolith about 4.6 m deep, on the average. The input of meteorites on Mars has been estimated by some authors to be only twice as great as that on the moon [*Soderblom et al.*, 1974] and the depth of the regolith to be as much as 2 km [*Fanale*, 1976]. If these rather extreme estimates are used, the meteoritic material would be much more 'dilute' in the Martian regolith than on the moon and would be equivalent to a 0.005% contribution of material from type 1 carbonaceous chondrites. Typical organic compounds found in meteorites, for example, naphthalene, are detected in type 1 carbonaceous chondrites at levels of about 1 ppm by using a laboratory version of the Viking gas chromatograph–mass spectrometer. Thus if the above estimates are correct, they should not be detected in the Martian surface material, since the anticipated abundance would be about 0.05 ppb for naphthalene, which is below our detection limit (see Table 2).

However, other authors have made very different estimates, both of the rate of meteoritic input on Mars and of the depth of the regolith. If one supposes that the regolith is only 100 m deep on the basis of arguments involving the adsorption of xenon onto the regolith (T. Owen, private communication, 1976) and accepts a value for the meteoritic input that is 25 times higher than that on the moon [*Anders and Arnold*, 1965], the expected abundance for naphthalene would be as high as 12.5 ppb, a value well above the detection limit of this experiment (Table 2).

A second, much less likely, exogenous source of organic material on Mars is the solar wind. Here the situation is quite unlike that on the moon because carbon-containing ions in the solar wind would be unlikely to penetrate the Martian atmosphere. Furthermore, any of those reaching the surface would be extensively diluted into the regolith when the surface is disturbed by winds, etc. Thus one can safely exclude the solar wind as a contributor to organic material deposited at the surface of Mars.

There is one process known that must be producing organic molecules on the surface of the planet at the present time, and that is the photochemical reduction of carbon monoxide (and possibly also carbon dioxide) at the interface between solid particles and the atmosphere. Recent studies by *Hubbard et al.* [1973] have established that formic acid, glycolic acid, formaldehyde, acetaldehyde, etc., are produced when a number of solids, including Vycor and volcanic ash, are irradiated under a simulated Martian atmosphere with ultraviolet light corresponding to the solar spectrum incident on Mars. The organics are derived from carbon monoxide rather than carbon dioxide.

The biology experiments on the two Viking spacecraft have given perplexing results. However, the pyrolytic release experiment suggests that the surface material on Mars is more active than the samples studied by *Hubbard et al.* [1973] in bringing about the photochemical fixation of CO when irradiated at relatively long wavelengths. On the other hand, very recent terrestrial experiments have shown that darkly colored iron-containing materials are less effective than the lightly colored samples previously studied when irradiated at shorter wavelengths (J. S. Hubbard, private communication, 1976). Thus it is hard to estimate whether the total rate of production on Mars is greater or less than that originally suggested on the basis of these experiments.

The steady state concentration of primary products found in the terrestrial experiments, when the sample was thoroughly mixed throughout the irradiation, was not very much below the detection limits of the Viking gas chromatograph–mass spectrometer [*Biemann et al.*, 1976, footnote 18]. However, when the samples were left undisturbed during the irradiation, the steady state yields were much lower. The Martian samples which were analyzed for their organic content had not been disturbed for a substantial time before their acquisition. Thus it is not necessary to postulate special destructive mechanisms on Mars to explain the failure to detect in these samples organic molecules that could have been produced by the above mechanism because the amounts would be too low.

A number of energy sources are effective in producing organic molecules when they act on reducing gas mixtures [*Miller and Orgel*, 1974]. These processes could not be taking place to any great extent on the bulk of the Martian surface at the present time, since the only reducing form of carbon that has been detected in the atmosphere, carbon monoxide, is present at very low levels. However, they might occur at special sites, for example, where volcanic gases are vented. Also, they may have been important at an earlier stage in the history of the planet if the Martian atmosphere was ever more reducing.

Finally, organic molecules could be formed by biological processes if there were living organisms on Mars. Vast quantities of CO_2 are fixed on the earth by photosynthetic mechanisms. There is no overriding reason why similar processes, or possibly nonphotochemical fixation, should not occur on Mars. However, it should be emphasized that at the present time there is no persuasive evidence for the existence of organisms on Mars.

Prior to the Viking mission it was anticipated that the lifetime of organic molecules at the surface of Mars, although limited by destructive photochemical processes, would still be long enough to permit some compounds to accumulate. The situation is changed by the discovery that Martian surface material has the capacity to oxidize organic compounds. Under these circumstances is is not possible to estimate the half-life of typical organics in the environment from which the Viking samples were obtained.

Short-wavelength ultraviolet radiation would destroy most organic compounds sooner or later. In the presence of oxygen, hydrogen peroxide, certain metal oxides, and many other oxidizing agents this radiation would remove organic compounds very much faster. Some oxidizing agents which might be present on Mars, for example, hydrogen peroxide [*McElroy and Kong*, 1976], would be destructive even in the dark. It is clear, then, that a number of different destructive mechanisms are operating on Mars. One knows little about the detailed mechanisms of oxidation and nothing about the rates of destruction for different classes of organic molecules under Martian conditions.

As was mentioned in section 3, it is unlikely that the failure to detect organics is due to their destruction in the ovens of the gas chromatograph–mass spectrometer by the oxidizing components of the Martian surface material because the terrestrial contaminants that were detected survived, thus ruling out the total destruction of all organic compounds. Furthermore, it must be recalled that the material in the surface which generates oxygen must be present in very small amounts. The oxygen (about 1 μmol) evolved in the gas exchange experiment [*Klein et al.*, 1976] is equivalent to only about 6 μg of reduced carbon per gram of surface material. The efficiency of oxidation is unlikely to be very high, and thus the total amount of organic carbon that could be oxidized must be far below the parts per million level. Finally, the amount of water present in the sample as mineral hydrate would be ample to convert the oxidizing agent, whatever it is, to O_2. Molecular oxygen would hardly have much effect on the organic material under the pyrolytic conditions of the experiment (30 s at temperatures of up to 500°C), particularly at such a low abundance.

Given all the uncertainties discussed above, it does not seem to be useful to elaborate specific models in great detail. The fact remains that the samples, when heated, did not evolve organic compounds at detectable levels. One cannot be certain that organic materials are transported to or synthesized at the Martian surface in quantities large enough to be detected by this instrument. It is reasonable to assume that if they were, they would be photooxidized so quickly that the steady state concentrations would still remain below the detection limit as discussed earlier.

Intuitively, it appears unlikely that organic compounds will be found in substantial amounts in the red iron-containing material that covers much of the Martian surface. It would be interesting to examine subsurface material taken from as great a depth as possible at the Viking landing sites or elsewhere. Material taken from very different sites, for example, the polar regions, where organic substances may have been cold trapped, would be of much greater interest. Nothing that has been learned about Mars at this point rules out the possibility that organic compounds have concentrated at favored sites on the surface of the planet.

Implications for Biology

The Viking gas chromatograph–mass spectrometer was not designed to search for life on Mars. Consequently, the demonstration that very little, if any, organic material is present does not exclude the existence of living organisms in the samples analyzed and certainly does not rule out the possibility of a rich biota out of range of the Viking landing sites. However, at the very least the results of this experiment demonstrate the absence of a widely distributed Martian biota which, like that on earth, has left its mark almost everywhere.

In an extensive study of soils collected in the Antarctic and ranging from practically sterile to highly populated with microorganisms, *Cameron et al.* [1970] have shown that there is on an average at least 10^4 times more organic carbon in these soils than there is in the bacteria present in them. Thus the organic material that was detected in the Antarctic soil sample discussed at the close of section 2 must be due to the organic debris derived from a long period of biological activity rather than from the microorganisms themselves present in the soil sample (this particular soil sample discussed in section 2 is believed to have contained about 100 bacteria and few algae per gram).

In view of the low detection limits for organic compounds (see Table 2) it is difficult to maintain the possibility that living organisms based on the familiar carbon-hydrogen-nitrogen-oxygen chemistry are present in appreciable numbers in the samples analyzed, unless one postulates that they differ in one important respect from terrestrial organisms: they must be much more efficient in scavenging organic carbon. It is hard to know whether this is a severe requirement, since one knows so little about the potential of living organisms to deal with a limitation of this kind. Alternatively one could postulate that organisms are present and that the absence of organic carbon is due to the unusual oxidizing properties of the soil rather than the peculiarity of the organisms. We think this to be unlikely.

In summary, the results of the organic analysis experiment do not rule out completely the possibility that there are living organisms in the samples analyzed, but they should not give encouragement to those who hope to find life on Mars.

Mineralogical Interpretations

A subsidiary goal of these experiments is to derive mineralogical constraints on the constitution of the surface materials by detection and measurement of inorganic volatiles evolved from the samples on heating and comparison of the results with the thermal decomposition behavior of appropriate mineral phases [*Anderson et al.*, 1972]. Inorganic volatiles of particular interest in this regard include H_2O, CO_2, sulfur-containing species, and oxides of nitrogen. Of these, only the first two were detected at levels regarded as possibly of mineralogical significance (see Table 4).

The results of other investigations [*Toulmin et al.*, 1977] have led to mineralogical models for the Martian surface materials incorporating hydrous silicates such as iron-rich montmorillonitic clays; hydrates and hydrated oxides, chiefly of iron; sulfates, possibly hydrated; and carbonates. The high level of sulfur reported by the inorganic chemical investigation [*Clark et al.*, 1977] implies the abundance of minerals containing major amounts of sulfur; both reduced (sulfide) and oxidized (sulfate) forms of this element occur in such minerals on earth. Rather wide variations exist in the thermal stability of mineral sulfides and sulfates; information with respect to the evolution of sulfur species from the Martian surface material on heating should therefore be relevant to the interpretation of the mineralogical occurrence of the element.

Several factors severely complicate the determination of the amount of water evolved upon heating Martian soils. The principal obstacle is the operation of the effluent divider network because the sample flow dynamics resulting when it changes state rapidly make quantitative determinations very difficult. Under these conditions it is impossible to determine the ratio of sample introduced into the mass spectrometer to sample vented by the effluent divider. It is only when the effluent divider is in a given split ratio for the complete duration of a gas chromatographic peak that a reliable determination can be made. Water emerges quite early (in approximately 100 s) as a relatively sharp peak at a time when the effluent divider is allowing greater and greater quantities of sample to flow into the mass spectrometer. This often results in the closing of valve V7 for a significant length of time (up to 44.5 s), and when this is the case, the mass spectrometer does not generate useful data for a large fraction of the time during which the water peak emerges from the gas chromatograph.

To obtain more reliable values for the amount of water evolved in a particular experiment on Mars, we carried out simulations on the earth-based laboratory instrument. Although it is impossible to duplicate exactly test results from such complex instruments involving highly complex modes of operation, a reasonable simulation of the data can be achieved by injection of known quantities of water into the laboratory instrument while key parameters such as effluent divider status, ion pump current, and resolved ion current as a function of time are recorded. A series of tests were carried out to narrow the range down to a 'best' value. The results shown in Figure 9 were obtained by the injection of 0.2 mg of water into the laboratory instrument. The shapes of the two curves (mass chromatograms of m/e 17, which was used because m/e 18 would be off scale in one of them) are nearly identical, as is the effluent divider behavior. Other parameters which were recorded, such as the ion pump current, also give credence to the assumption that the two sets of data correspond to nearly identical quantities of water. The ion intensity obtained was much higher in the Martian data than in the laboratory result mainly because the sensitivity of the laboratory instrument had degraded through constant use. The problems of duplicating the Martian results were further complicated by the fact that one of the flow restrictors (20 1) in the lander 2 gas chromatograph–mass spectrometer effluent divider was found to have been blocked almost completely. This condition was simulated as closely as possible in the laboratory tests.

The results of the simulation tests lead us to the conclusion that the amount of water evolved in the experiment on sol 43 (identification number 10037) is equivalent to approximately 0.2 mg (which in the very worst case could be off by a factor of 5). This corresponds to 0.2% by weight based on a 100-mg soil sample. Values estimated for the other experiments are listed in Table 4.

Inspection of the data in Table 4 shows that most samples lost little H_2O at 200°, but all evolved significant and generally similar amounts at 350° and 500°. Samples from areas of surface crustification (Bonneville, Rocky Flats) seem to evolve decreasing amounts of H_2O on successive heatings. This is not so apparent for other samples. However, it must be noted that the VL-1 H_2O data are much more ambiguous because the use of $^{13}CO_2$ as a purge gas further complicates the behavior of the effluent divider.

The most distinctive difference among the samples was the increased evolution of both H_2O and CO_2 at 200° from the sample taken under Badger Rock. Considerably more H_2O and CO_2 was driven off at this temperature than from other comparable samples; results at 350° and 500° were not distinctively different from other samples. The CO_2 values listed in Table 4 are still very uncertain at the present state of data analysis.

The CO_2 evolved upon heating the samples presumably is partly desorbed from mineral surfaces and partly derived by partial decomposition of carbonate minerals. Table 5 shows approximate values of partial pressure of CO_2 in equilibrium

TABLE 4. Water and Carbon Dioxide

Sample	Temperature, °C	Mode	Water, wt %	CO_2, ppm
		*VL-1**		
Oven 1 (cruise)	500	$^{13}CO_2$	<0.1	
Sandy Flats	200	$^{13}CO_2$	<0.1	
	500	$^{13}CO_2$	0.1–1.0	
Rocky Flats	350	$^{13}CO_2$	0.1–1.0	
	500	$^{13}CO_2$	0.1–1.0	
	500	$^{13}CO_2$	0.1–1.0	
		VL-2		
Bonneville	200	H_2	0.05	<50
	350	H_2	0.3	50–500
	500	H_2	1.0	50–500
	500	$^{13}CO_2$	0.25	
Under Badger Rock	50	H_2	<0.01	<50
	200	H_2	0.2	50–500
	350	H_2	0.3	40–400
	500	H_2	0.8	70–700
	500	$^{13}CO_2$	0.6	

*Values for water are much less precise for VL-1 than for VL-2 because of experimental conditions ($^{13}CO_2$ mode).

with some common carbonate minerals at the temperature to which the samples are heated, calculated from the data of *Robie and Waldbaum* [1968]. It is hardly likely that gas-solid equilibrium is attained or even closely approached in the short time the samples are heated (30 s), but the equilibrium values presumably are upper limits, and their relative values give some indication of the pattern of thermal evolution of CO_2 to be expected. Although one can make only approximate estimates of the partial pressure of CO_2 in the oven from the abundance data, it would seem that thermal decomposition of calcite alone is inadequate to explain the amount of CO_2 evolved from the Martian samples. The relative importance of adsorbed CO_2 and CO_2 evolved from more labile carbonates is very difficult to assess without more specific kinetic data.

Interpretation of the data on the evolution of H_2O in terms of presumed mineral phases is more difficult both because of the greater variety of possible sources and because of the uncertainties attached to the analytical results. Other than adsorbed H_2O on mineral grain surfaces, the sources for evolved H_2O suggested by the most widely proposed mineralogical models include clays, goethite, and hydrated sulfates. Thermal studies of iron-rich montmorillonitic clays generally support the conclusion that most dehydroxylation occurs above 200° and may take place progressively over a temperature range of several hundred degrees [*MacKenzie and Rogers*, 1977]. Thermal dehydration of a nontronite from Riverside, California, whose major element composition is appropriate for a major constituent of the Martian surface samples showed a maximum dehydroxylation rate at about 350° in an inert gas atmosphere. Goethite varies widely in its dehydration

TABLE 5. Equilibrium Vapor Pressure P_{CO_2} Over Some Carbonate Minerals, Calculated From Data of *Robie and Waldbaum* [1968]

	P_{CO_2}, atm		
	200°C	350°C	500°C
Calcite ($CaCO_3$)	5×10^{-12}	2×10^{-7}	1.3×10^{-4}
Dolomite ($CaMg(CO_3)_2$)	9×10^{-6}	2×10^{-2}	2.4
Magnesite ($MgCO_3$)	1.4×10^{-4}	0.2	10

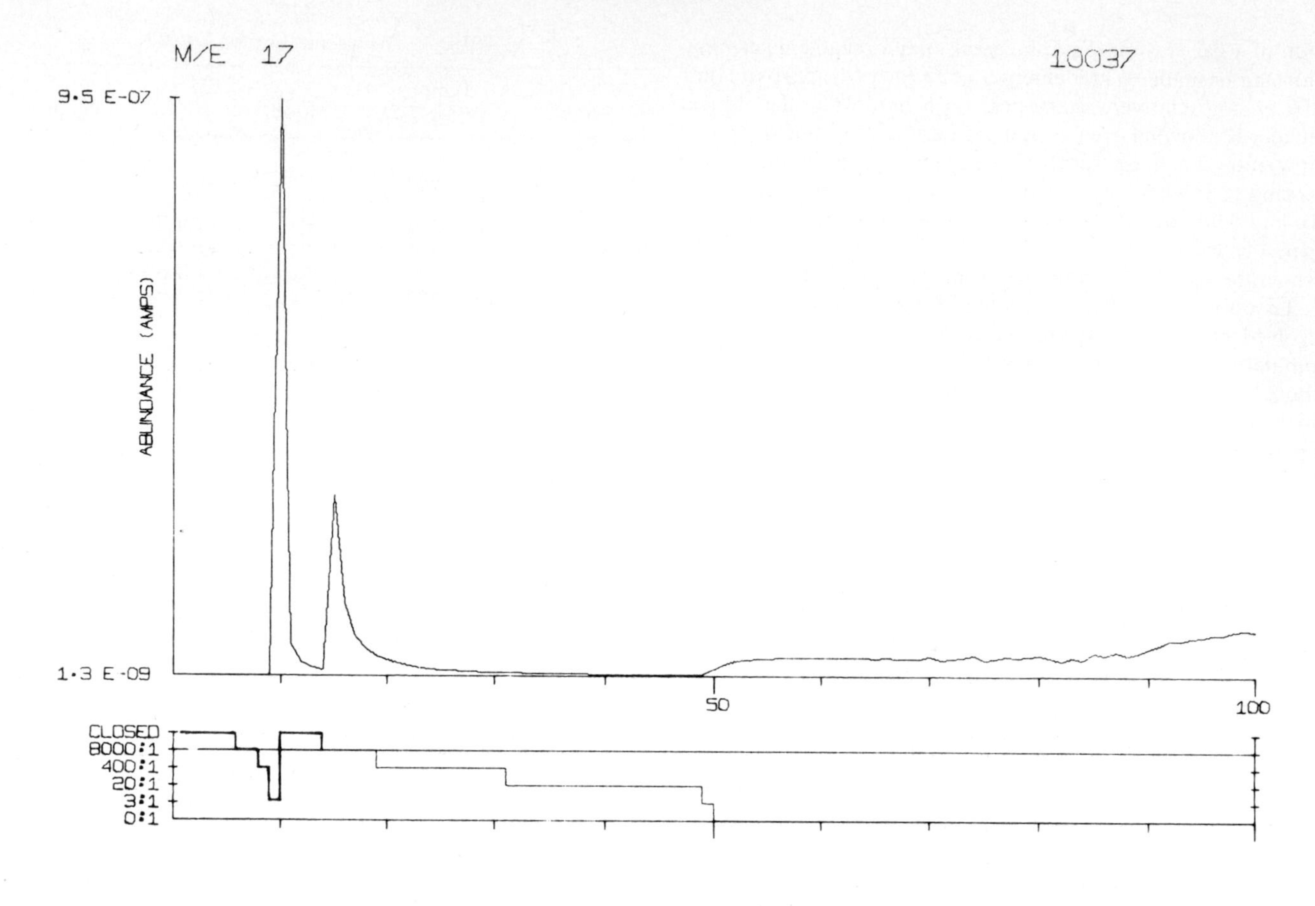

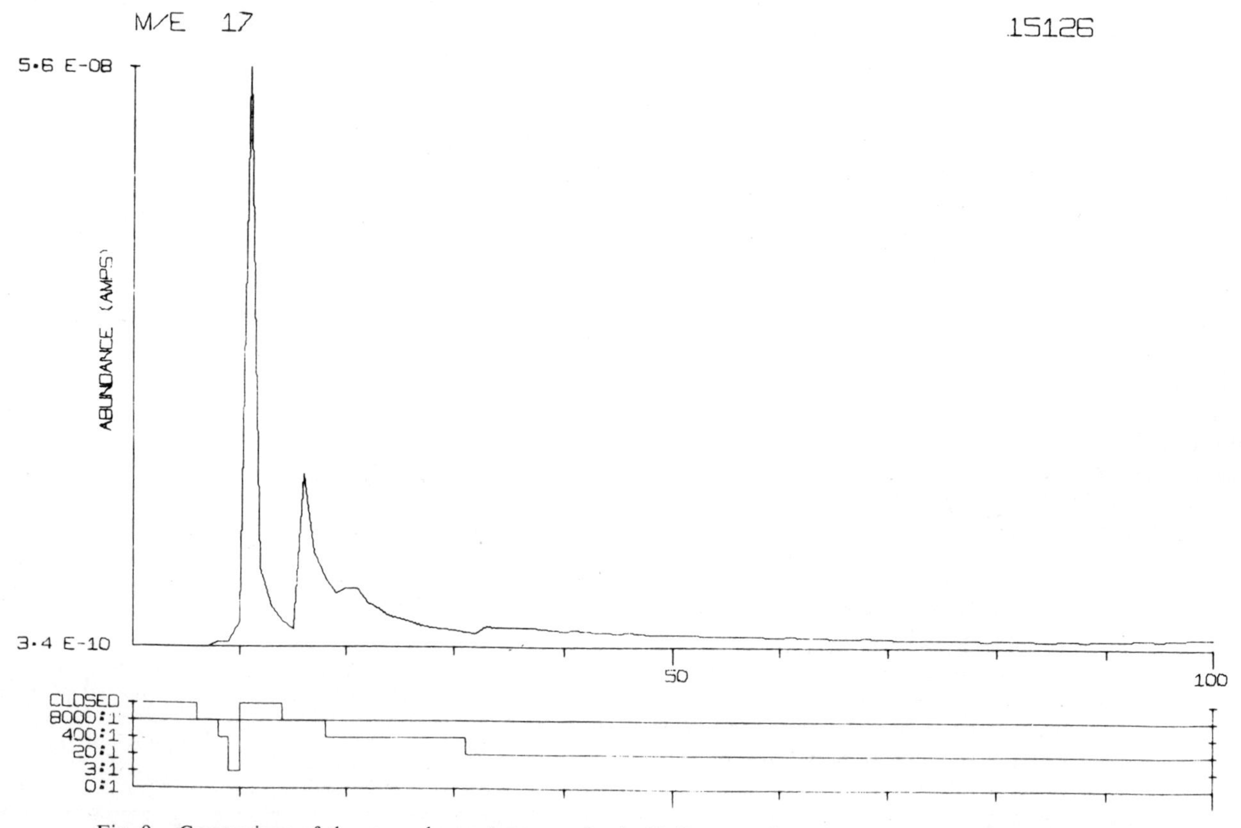

Fig. 9. Comparison of the mass chromatograms of m/e 17 (fragment ion of H_2O) and the effluent divider behavior between the 200° experiment on the second VL-2 sample and an injection of 0.2 mg of H_2O into the laboratory version of the flight instrument.

TABLE 6. Equilibrium Vapor Pressures in Inert Atmospheres (P_{S_2}) and in 1 atm H_2 (P_{H_2S}) Over Iron Sulfide Minerals, From Data of *Toulmin and Barton* [1964] and *Barton and Skinner* [1977]

	Equilibrium Vapor Pressure, atm		
	200°C	350°C	500°C
		Troilite (FeS)	
P_{S_2}	2×10^{-28}	2×10^{-20}	2×10^{-15}
P_{H_2S}	8×10^{-14}	2×10^{-10}	2×10^{-8}
		Pyrite (FeS$_2$)	
P_{S_2}	5×10^{-18}	4×10^{-10}	3×10^{-5}
P_{H_2S}	2×10^{-3}	3	370

rate and temperature range; crystallinity and particle size are especially important parameters. A well-crystallized specimen, ground to a fine powder, dehydrated mostly over the range 290°–410°, with the maximum rate (0.9–1.4% min^{-1}) at 360°–380° (heating rate 10° min^{-1}). Both these materials are being studied in a simulated flight instrument at MIT in order to elucidate their behavior under Vikinglike conditions. Similar experiments on hydrated sulfates are contemplated.

The absence of any indication of sulfur-containing species in our data is of interest in view of the rather high content of that element in the surface materials, as was shown by the inorganic analysis experiment. This result is consistent with the supposition that the sulfur occurs as a thermally stable sulfate, with a low vapor pressure (at least with respect to sulfur-containing species) below 500°. Table 6 shows the partial pressure of S_2 and H_2S (the latter in the presence of 1 atm H_2) in equilibrium with two common iron sulfides at the temperatures of interest [*Toulmin and Barton*, 1964; *Barton and Skinner*, 1977]. Common terrestrial pyrrhotites are solid solutions, $Fe_{1-x}S$, and have equilibrium sulfur pressures intermediate between pyrite (FeS_2) and troilite (FeS). Although the prima facie implication of these data is that the presence of pyrite and probably most pyrrhotites is inconsistent with the lack of detectable sulfur species in the evolved gases, caution is indicated until the appropriate simulations on the laboratory instrument have been carried out. Because of the danger to the laboratory instrument (copious amounts of sulfur or hydrogen sulfide in the gas stream poisons the H_2 separator) this experiment must be delayed until all other contemplated experiments have been completed.

Acknowledgments. We are indebted to all of the many individuals whose efforts and talents have contributed to this work. We thank specifically H. C. Urey, G. A. Shulman, and R. A. Hites for their contributions at the earlier phases of the Viking activities; our fellow team member T. Owen for his continuous interest and stimulation; D. Howarth, R. Rampacek, and C. Haag for their expert assistance before and during the mission; R. Williams and E. Ruiz through whose dedication and efforts the necessary software was developed at JPL and MIT; J. Lavoie, who helped in running the many tests on the laboratory instrument at MIT; and the Viking Surface Sampler Team for their successful efforts to acquire the samples needed for the experiments. Above all, the authors are indebted to those individuals at JPL who were involved in the design of the entire instrument, particularly C. E. Griffin, who was instrumental in the design of the mass spectrometer, and to Litton Industries, Perkin Elmer Aerospace Division, and Beckman Instruments, who constructed the flight hardware. Finally, our gratitude goes to the entire Viking Flight Team, whose dedication and expertise made possible the successful conduct of our experiments. This work was supported by NASA research contract NAS 1-9684 and many other related ones. P.G.S. is a research fellow at the University of Bristol.

REFERENCES

American Petroleum Institute, *Catalog of Mass Spectral Data, Res. Proj. 44*, New York, 1955.

Anders, E., and J. Arnold, Age of craters on Mars, *Science, 149*, 1494–1496, 1965.

Anders, E., R. Ganapathy, U. Krahenbuhl, and J. W. Morgan, Meteoritic material on the moon, *Moon, 8*, 3–24, 1973.

Anderson, D. M., K. Biemann, L. E. Orgel, J. Oro, T. Owen, G. P. Shulman, P. Toulmin III, and H. C. Urey, Mass spectrometric analysis of organic compounds, water, and volitile constituents in the atmosphere and surface of Mars: The Viking Mars lander, *Icarus, 16*, 111–138, 1972.

Barton, P. B., Jr., and B. J. Skinner, Sulfide mineral stabilities, in *Geochemistry of Hydrothermal Ore Deposits*, 2nd ed., edited by H. L. Barnes, John Wiley, New York, in press, 1977.

Biemann, K., *Mass Spectrometry: Organic Chemical Applications*, pp. 171–172, McGraw-Hill, New York, 1962.

Biemann, K., Test results on the Viking gas chromatograph–mass spectrometer experiment, *Origins Life, 5*, 417–430, 1974.

Biemann, K., J. Oro, P. Toulmin III, L. E. Orgel, A. O. Nier, D. M. Anderson, P. B. Simmonds, D. Flory, A. V. Diaz, D. R. Rushneck, and J. E. Biller, Search for organic and volatile inorganic compounds in two surface samples from the Chryse Planitia region of Mars, *Science, 194*, 72–76, 1976.

Biller, J. E., and K. Biemann, Reconstructed mass spectra, a novel approach for the utilization of gas chromatograph–mass spectrometer data, *Anal. Lett., 7*(7), 515–528, 1974.

Biller, J. E., W. C. Herlihy, and K. Biemann, Identification of the components of complex mixtures by GCMS, in *Computer Assisted Structure Elucidation*, edited by D. H. Smith, American Chemical Society, Washington, D. C., in press, 1977.

Cameron, R. E., J. King, and C. N. David, *Soil Sci., 109*, 110–120, 1970.

Clark, B. C., III, A. K. Baird, H. J. Rose, Jr., P. Toulmin III, R. P. Christian, W. C. Kelliher, C. D. Castro, K. Keil, and G. R. Huss, The Viking X ray fluorescence experiment: Analytical methods and early results, *J. Geophys. Res., 82*, this issue, 1977.

Dencker, W. D., D. R. Rushneck, and G. R. Shoemaker, Electrochemical cell as a gas chromatograph–mass spectrometer interface, *Anal. Chem., 44*, 1753–1758, 1972.

Fanale, E. P., Martian volatiles: Their degassing history and geochemical fate, *Icarus, 28*, 179–202, 1976.

Hertz, H. S., R. A. Hites, and K. Biemann, Identification of mass spectra by computer-searching a file of known spectra, *Anal. Chem., 43*, 681–691, 1971.

Hess, S. L., R. M. Henry, C. B. Leovy, J. A. Ryan, J. E. Tillman, T. E. Chamberlain, H. L. Cole, R. G. Dutton, G. C. Greene, W. E. Simon, and J. L. Mitchell, Preliminary meteorological results on Mars from the Viking I lander, *Science, 193*, 788–791, 1976.

Hubbard, J. S., J. Hardy, G. E. Voecks, and E. E. Golub, Photocatalytic synthesis of organic compounds from CO and water: Involvement of surfaces in the formation and stabilization of products, *J. Mol. Evol., 2*, 149–166, 1973.

Klein, H. P., G. V. Levin, V. I. Oyama, J. Lederberg, A. Rich, J. S. Hubbard, G. L. Hobby, P. A. Straat, B. J. Berdahl, G. C. Carle, F. S. Brown, and R. D. Johnson, The Viking biological investigation: Preliminary results, *Science, 194*, 99–105, 1976.

MacKenzie, K. J. D., and D. E. Rogers, Thermal and Mossbauer studies of iron-containing hydrous silicates, I, Nontronite, *Thermochim. Acta, 18*, 177–196, 1977.

Mason, B. (Ed.), *Handbook of Elemental Abundances in Meteorites*, Gordon and Breach, New York, 1971.

McElroy, M. B., and T. Y. Kong, Oxidation of the Martian surface: Constraints due to chemical processes in the atmosphere, *Geophys. Res. Lett., 3*, 569–572, 1976.

Miller, S. L., and L. E. Orgel, The formation of the solar system, in *The Origins of Life on the Earth*, edited by W. D. McElroy and C. P. Swanson, pp. 5–15, Prentice-Hall, Englewood Cliffs, N. J., 1974.

Miller, S. L., H. C. Urey, and J. Oro, Origin of organic compounds on the primitive earth and in meteorites, *J. Mol. Evol., 9*, 59–72, 1976.

Morgan, J. W., Chemical fractionation in the solar system, in *Proceedings of the International Conference on Modern Trends in Activation Analysis*, Munich, Germany, 1976.

Nagy, B., *Carbonaceous Meteorites*, Elsevier, New York, 1975.

Novotny, M., J. M. Hayes, F. Bruner, and P. G. Simmonds, Gas chromatographic column for the Viking 1975 molecular analysis experiment, *Science, 189,* 215–216, 1975.

Oro, J., Extraterrestrial organic analysis, *Space Life Sci., 3,* 507–550, 1972.

Owen, T., and K. Biemann, Composition of the atmosphere at the surface of Mars: Detection of argon-36 and preliminary analysis, *Science, 193,* 801–803, 1976.

Owen, T., K. Biemann, D. R. Rushneck, J. E. Biller, D. W. Howarth, and A. L. Lafleur, The Atmosphere of Mars: Detection of krypton and xenon, *Science, 194,* 1293–1295, 1976.

Robie, R. A., and D. R. Waldbaum, Thermodynamic properties of minerals and related substances at 298.15 K (25.0 C) and one atmosphere (1.013 bars) pressure and at higher temperatures, *U.S. Geol. Surv. Bull., 1259,* 256, 1968.

Shorthill, R. W., H. J. Moore II, R. E. Hutton, R. F. Scott, and C. R. Spitzer, The environs of Viking 2 lander, *Science, 194,* 1309–1318, 1976.

Soderblom, L. A., C. D. Condit, R. A. West, B. M. Herman, and T. J. Kreidler, Martian planet-wide crater distribution implications for geologic history and surface processes, *Icarus, 22,* 239–263, 1974.

Toulmin, P., III, and P. B. Barton, Jr., A thermodynamic study of pyrite and pyrrhortite, *Geochim. Cosmochim. Acta, 28,* 641–671, 1964.

Toulmin, P., III, A. K. Baird, B. C. Clark, K. Keil, H. J. Rose, Jr., R. P. Christian, P. H. Evans, and W. C. Kelliher, Geochemical and mineralogical results, *J. Geophys. Res., 82,* this issue, 1977.

Wasson, J. T., *Meteorites: Classification and Properties,* Springer, New York, 1974.

(Received June 1, 1977;
revised June 15, 1977;
accepted June 15, 1977.)

Viking on Mars: The Carbon Assimilation Experiments

N. H. HOROWITZ AND G. L. HOBBY

Division of Biology, California Institute of Technology, Pasadena, California 91125

JERRY S. HUBBARD

School of Biology, Georgia Institute of Technology, Atlanta, Georgia 30332

A fixation of atmospheric carbon, presumably into organic form, occurs in Martian surface material under conditions approximating the actual Martian ones. The reaction showed the following characteristics: The amount of carbon fixed is small by terrestrial standards; highest yields were observed in the light, but some dark activity was also detected; and heating the surface material to 90°C for nearly 2 hours had no effect on the reaction, but heating to 175°C for 3 hours reduced it by nearly 90%. New data from Mars do not support an earlier suggestion that the reaction is inhibited by traces of water. There is evidence of considerable heterogeneity among different samples, but different aliquots from the same sample are remarkably uniform in their carbon-fixing capacity. In view of its thermostability it is unlikely that the reaction is biological.

INTRODUCTION

With respect to its surface environment, Mars is more earth-like than any other extraterrestrial body of the solar system. Of particular importance from a biological point of view is the fact that Mars possesses both an atmosphere of light elements and a temperature regime that is compatible with the existence of organic matter and even of some terrestrial life forms. These characteristics make Mars the most plausible habitat of extraterrestrial life in the solar system, but at the same time it is clear that they represent necessary, not sufficient, conditions for life. In particular, the absence of liquid water on the Martian surface excludes the possibility of terrestrial types of organisms on Mars. All known terrestrial species have high and, within narrow limits, apparently irreducible requirements for water. On this ground alone it has been evident for some time that Martian life could not be built on a terrestrial model [*Horowitz et al.*, 1972*a*; *Horowitz*, 1976*a*, *b*].

The carbon assimilation experiment (also known as the pyrolytic release experiment) is designed to perform a biological test on Mars under Martian conditions of temperature, pressure, water activity, and atmospheric composition. Although these conditions depart from our geocentric sense of what constitutes a proper biological environment, it is obvious that Martian life, if any, is adapted to Martian conditions, not terrestrial ones. By preserving these conditions as far as possible we maximize the chance of survival of indigenous life and make reasonably sure that whatever events we may detect are not artifactual. In actual practice, our experiment has operated on Mars under a reasonable approximation of local conditions except for temperature, which, owing to heat sources within the spacecraft, has been consistently higher than the ground temperature.

Preliminary accounts of the data returned from Mars have been published [*Klein et al.*, 1976; *Horowitz et al.*, 1976]. In this paper we summarize all of the data, including the results of two new experiments on Mars, and we conclude that they are unlikely to have a biological explanation.

EXPERIMENT DESCRIPTION

The carbon assimilation experiment is designed to detect the synthesis of organic matter in Martian surface material from atmospheric CO or CO_2 or both. The experiment assumes that Martian life would be based on carbon and that this carbon would necessarily cycle through the atmosphere. The rationale for these assumptions has been presented [*Horowitz*, 1976*a*]. Descriptions of the experiment, including the results of tests carried out on a variety of terrestrial soils, have also been published [*Hubbard et al.*, 1970; *Horowitz et al.*, 1972*b*; *Hubbard*, 1976].

In brief, 0.25 cm^3 of Martian surface material ('soil') is enclosed in one of the 4-cm^3 test chambers of the instrument under Martian atmosphere at the ambient pressure. The chamber is closed by a window that transmits wavelengths longer than 320 nm originating in a 6-W high-pressure xenon lamp mounted above the chamber. The radiant energy entering the chamber, integrated between 335 and 1000 nm, is about 8 mW cm^{-2}, or approximately 0.2 times the energy in the same wavelength interval at the Martian subsolar point. Although solar wavelengths as short as 200 nm reach the surface of Mars, we have excluded wavelengths shorter than 320 nm from the test chamber in order to prevent a nonbiological, photocatalyzed synthesis of organic compounds from CO and adsorbed water vapor that we find to occur on silicate and other mineral surfaces irradiated with ultraviolet shorter than 300 nm [*Hubbard et al.*, 1971, 1973, 1975]. Since these short wavelengths are generally destructive to organic matter except under the special conditions of the synthesis, we assume that Martian life, if it exists, must also avoid this spectral region; its deletion from the radiation entering the test chamber therefore does not constitute a significant change in the biological environment. The use of the lamp during any experiment is optional and is commandable from earth. Also commandable is the injection of approximately 80 μg of water vapor (or integral multiples thereof) into the test chamber at the start of an experiment.

The atmosphere in the test chamber is labeled by the injection of 20 μl of a mixture of $^{14}CO_2$ and ^{14}CO (92:8 by volume, total radioactivity 22 μCi) at the start of an experiment. The resulting pressure increase is 2.2 mbar over ambient, which was 7.6 mbar initially at both landing sites. The Martian atmosphere contains about 95% CO_2 and 0.1% CO. The injection of radioactive gases increases the partial pressures of CO_2 and CO by 28% and 23-fold, respectively.

The sample is incubated for 120 hours at temperatures which have ranged from 8°C to 26°C in the experiments

Paper number 7S0503.

conducted to date. The lamp is then turned off, and the chamber is brought to 120°C while the radioactive atmosphere is vented. The chamber is next heated to 635°C to pyrolyze organic matter in the sample. The volatile products, together with a large amount of $^{14}CO_2$ and ^{14}CO desorbed from the soil grains and walls of the chamber, are swept by a stream of He into a column packed with a mixture of 25% cupric oxide and 75% Chromosorb-P (a form of diatomaceous earth). The column, which operates at 120°C, retains organic molecules larger than methane but allows all but a small fraction of the CO_2 and CO to pass into a radiation counter, where their radioactivity is measured. This count is referred to as peak 1.

The column temperature is then raised to 640°C, the high temperature causing the release of organic compounds and their oxidation to CO_2 by the CuO in the column. The radioactivity of this gas is peak 2; it represents organic matter synthesized from ^{14}CO or $^{14}CO_2$ during the incubation. Peak 2 also contains the small fraction of CO and CO_2 which failed to elute with peak 1, presumably because of the presence of some high-affinity sites in the column. The radioactivity of this fraction, referred to as peak 2(0), must be subtracted from peak 2 in order to estimate the amount of C fixed in organic matter. The size of peak 2(0) is known from laboratory tests carried out with heat-sterilized soils or with no soils in the test chamber by using flight-configured columns. These measurements show that for values of peak 1 less than 7×10^5 dpm (disintegrations per minute), peak 2(0) is only weakly dependent on peak 1, i.e., that it is almost constant. It can be estimated from the regression line peak $2(0) = 28.8 + 2.84 \times 10^{-5}$ peak 1, peak heights being expressed in disintegrations per minute. The standard deviation of peak 2(0) is 27 dpm. The regression line is plotted in Figure 1.

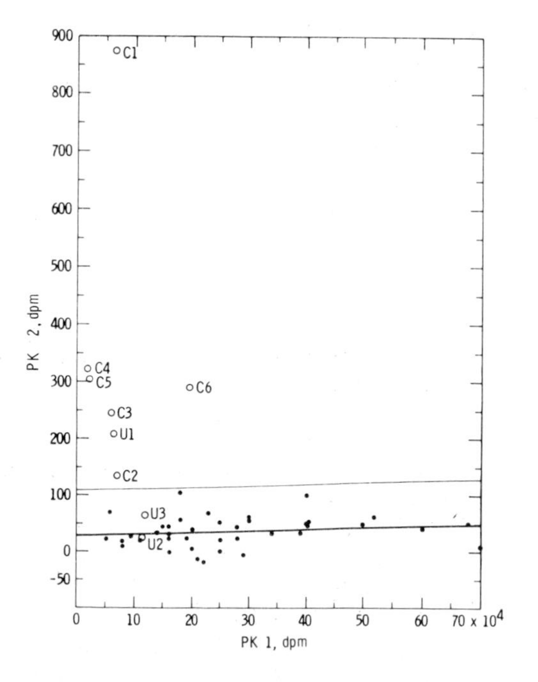

Fig. 1. ^{14}C fixation results from Mars (open circles) compared with laboratory data obtained with heat-sterilized soils or no soils (solid circles). The laboratory data are fitted with the regression line peak $2(0) = 28.8 + 2.84 \times 10^{-5}$ peak 1. The light line marks the 3σ level. The high point at peak $1 = 41 \times 10^4$ dpm was not used in computing the regression line. C is Chryse; U, Utopia; and dpm, disintegrations per minute.

RESULTS

Ten experiments were performed on Mars, six at the Chryse site (numbered chronologically C1–C6) and four at the Utopia site (U1–U4). Of these, U4 is not usable owing to an apparent valve failure which allowed peak 2 to escape before it could be counted. Attempts to correct or circumvent the failure proved ineffectual. A summary and a statistical analysis of all usable results are shown in Figure 1 and Table 1. Descriptions of all experiments but C5 and C6 have been given in previous reports and will not be repeated here.

Experiment C5. C5 was performed with a surface sample that had been acquired before the solar conjunction and stored in the hopper of the soil distribution assembly for 69 sols before C5 was initiated (a sol is a Martian day, equal to 24.65 hours). Since the three test chambers of the Viking Lander 1 (VL 1) instrument had already been used, it was necessary to run the experiment in the 'soil-on-soil' mode, as was also the case with C4. The purpose of C5 was to determine the effect of water vapor injection, followed by evaporation of the water at an elevated temperature, on the soil activity. Since the test chamber now contained two soil aliquots, two injections of water vapor were made. After 4 hours the chamber was vented, and the temperature was brought to 120°C for 1 or 2 min before it dropped to about 90°C, where it was held for approximately 112 min. The temperature was then lowered to 17° ± 1°C for the rest of the experiment. Three 20-μl injections of radioactive gas were made to compensate in part for the reduced pressure in the VL 1 gas bottle, which, after 160 sols of use and normal leakage, should by now have dropped to about 10% of its initial value. The remainder of the experiment proceeded normally.

The results of C5 are hardly distinguishable from those of C4. The unusually low first peaks of both experiments are explained by the depletion of radioactive gas mentioned above. The second peaks of C4 and C5 are statistically identical, an important result in view of the different thermal histories of the two samples. This finding will be considered further in the discussion section below.

Experiment C6. This experiment tested the effect of water vapor injection without heating or evaporation. It repeated U2, the result of which had suggested that water vapor inhibits the fixation reaction [*Horowitz et al.,* 1976]. It differed from U2 and C5 in that the order of addition of water vapor and radioactive gases was reversed, the latter being added first in C6 in consideration of their low pressure in the gas bottle. The change is probably not significant. The surface sample used in C6 was an aliquot of the same material that had been used in C4 and C5 (see Table 1). This material had by now been stored in the soil hopper for 139 sols at temperatures from 5°C to 24°C. Two injections of water vapor and six injections of radioactive gas were made; incubation was performed with the lamp on.

The high value of peak 1 (Figure 1 and Table 1) confirms that water vapor injection occurred. (In C5 the water was evaporated before the addition of radioactive gas, and a high first peak was not obtained or expected.) Of great interest is peak 2, which is seen not to differ significantly from peak 2 of C4 or of C5. We have to conclude, despite earlier indications to the contrary, that water vapor does not inhibit the fixation reaction. It should be noted here that two anomalous data segments, identifiable by their non-Poisson statistics, were edited from the C6 data. Similar anomalies had been encountered in the C2 data [*Klein et al.,* 1976]. Despite these episodes we do not doubt the validity of the results.

TABLE 1. Carbon Fixation Statistics

Experiment	Sample	Conditions	Incubation Temperature, °C	Disintegrations per Minute ± Standard Error				
				Peak 1	Peak 2	Peak 2(0)	Peak 2 − Peak 2(0)	P
C1	Sandy Flats 1, fresh	light, dry	17 ± 1	67,464 ± 536	873 ± 10	30.7 ± 27	842 ± 29	~0
C2	Sandy Flats 1, stored 19 sols	light, dry, 175°C heat treatment	15 ± 1	69,536 ± 545	136 ± 12	30.8 ± 27	105 ± 29.5	2×10^{-4}
C3	Sandy Flats 2, fresh	light, dry	13–26	61,027 ± 527	245 ± 8.9	30.5 ± 27	214.5 ± 28	10^{-14}
C4	Sandy Flats 3, fresh	light, dry	16 ± 2	18,545 ± 381	318 ± 15	29.3 ± 27	289 ± 31	6×10^{-21}
C5	Sandy Flats 3, stored 69 sols	light, H_2O, 90°C heat treatment	17 ± 1	20,295 ± 395	304 ± 11	29.4 ± 27	275 ± 29	10^{-21}
C6	Sandy Flats 3, stored 139 sols	light, H_2O	15 ± 2.5	193,803 ± 864	289 ± 15	34.3 ± 27	255 ± 31	10^{-16}
U1	Beta 1, fresh	dark, dry	15 ± 3	64,845 ± 527	209 ± 15	30.6 ± 27	178 ± 31	5×10^{-9}
U2	Beta 2, fresh	light, H_2O	18 ± 1.5	113,845 ± 690	25 ± 8.4	32.0 ± 27	−7 ± 28	0.5
U3	Under Notch Rock, fresh	dark, dry	10 ± 2	118,309 ± 400	68 ± 23	32.2 ± 27	36 ± 35	0.15

Samples are identified by sampling location, acquisition number, and age at the start of the experiment. Conditions indicate whether the lamp was on or off, whether water vapor was injected, and whether the sample was heated before starting incubation. P is the conventional one-tailed probability that a positive deviation from peak 2(0) as large as or larger than that found would be obtained by chance.

DISCUSSION

The data show that a fixation of atmospheric carbon occurs in the surface material of Mars under conditions approximating the Martian ones. The highest activity was seen in experiment C1, where approximately 10 pmol of CO (or 30 pmol of CO_2, which has a lower specific activity in the gas space) was fixed. This activity is quite small by terrestrial standards, but it is significant. Indeed, when it is recognized that the only identified nonbiological organic synthesis, the surface-catalyzed photoreaction referred to earlier, had been eliminated in designing the experiment, the results are startling. Nevertheless, a biological interpretation of the results is unlikely in view of the thermostability of the reaction. Thus the second peaks of experiments C4, C5, and C6 are not significantly different from one another, although the thermal histories of the samples are very different. The three samples were aliquots from the same scoopful of surface material (Table 1). In C4 the material was tested fresh from the surface. In C5 it had been stored for 69 sols at temperatures between 10°C and 24°C before the experiment was initiated, and it was heated for nearly 2 hours at 90°C before incubation began. In C6 it had been in storage for 139 sols at temperatures from 5°C to 24°C before the start of the experiment. Given these histories and the observation that the temperature of the ground, the habitat of any presumed organisms, does not reach 0°C at any time of the year at either landing site [*Kieffer*, 1976], it is difficult to reconcile the constancy of the responses with a biological origin.

Further evidence is found in experiment C2, in which a sample from the same dig that had supplied the C1 sample was heated at 175°C for 3 hours before starting the incubation. The activity was reduced considerably but not to the level of sterile soils; the peak 2 count is 3.5σ above the mean of such soils. It appears that the agent responsible for the reaction is somewhat heat labile but not as labile as we expect living organisms to be.

In an earlier report we drew the inference from the results of experiments U2 and U3 that the fixation reaction is inhibited by traces of water [*Horowitz et al.*, 1976]. Unless the two landing sites are fundamentally different in their surface chemistries (an assumption that is opposed by all the available evidence), we are forced by the outcome of the C6 experiment to conclude that the results of U2 and U3 are related not to the water content of the samples but more probably to sample heterogeneity. That significant heterogeneity exists can be seen in experiments C1, C3, and C4. These experiments were run in the same way with different samples collected at the Sandy Flats site. The peak 2 counts all differ significantly from one another, and they vary over a 3.5-fold range. These results are in striking contrast to the uniformity seen in C4, C5, and C6, the samples for which were all aliquots from the same dig.

The Utopia samples are known to be heterogeneous with respect to at least one relevant characteristic: their irradiation histories. Sample U3 was obtained from under Notch Rock and presumably had not been exposed to solar radiation in recent times. If the plausible assumption is made that only recently irradiated surface material can fix atmospheric carbon, then the difference between experiments U1 and U2 could reflect variations in the quantity of such material transferred to the incubation chambers; a similar explanation is possible for the other differences noted above.

If organic matter is being synthesized on Mars, it does not accumulate above the sensitivity threshold of the GCMS (gas chromatograph mass spectrometer), the Viking organic analysis instrument [*Biemann et al.*, 1976]. The amount of carbon fixed even in experiment C1 is well below the detection limit of this sensitive instrument. A low-level, steady state synthesis like that described by *Hubbard et al.* [1971, 1973, 1975] would be compatible with the observations, providing it could operate under the conditions described in this paper. We are investigating the possibility that organic compounds synthesized by the Hubbard reaction on Mars analog surface materials can exchange with ^{14}CO or $^{14}CO_2$ under the conditions of the carbon assimilation experiment [*Hobby*, 1977].

Other mechanisms that have been suggested to explain the results described here include carry-over of particulate matter from the incubation chamber to the column [*Huguenin*, 1977], incorporation of ^{14}CO into carbon suboxide polymer preformed on the Martian surface [*Oyama*, 1977], and reduction of ^{14}CO by H_2O_2 in the surface material [*Hobby*, 1977]. It remains to be seen whether any of the proposed mechanisms can account for the intriguing observations.

Acknowledgments. We owe a debt of gratitude to all those connected with the Viking mission, whose efforts helped make the mission the historic achievement that it has been. We acknowledge in particular the important contributions made by Ron Gilje, Fred Brown, and Steve Loer to the successful operation of our instrument during the active phase of the mission, and we thank Paul K. Cartier III and William Ashley for their invaluable technical assistance in the laboratory phases of the investigation. This work was supported by NASA contracts NAS1-12311 (to N.H.H.) and NAS1-13422 (to J.S.H.) and NASA grants NGR-05-002-308 (to N.H.H.) and NSG-7069 (to J.S.H.).

REFERENCES

Biemann, K., J. Oro, P. Toulmin III, L. E. Orgel, A. O. Nier, D. M. Anderson, P. G. Simmonds, D. Flory, A. V. Diaz, D. R. Rushneck, and J. A. Biller, Search for organic and volatile inorganic compounds in two surface samples from the Chryse Planitia region of Mars, *Science, 194,* 72–76, 1976.

Hobby, G. L., The pyrolytic release experiment, paper presented at Annual Meeting, Amer. Ass. Advan. Sci., Denver, Colo., Feb. 22, 1977.

Horowitz, N. H., The search for life in the solar system, *Accounts Chem. Res., 9,* 1–7, 1976*a*.

Horowitz, N. H., Life in extreme environments: Biological water requirements, in *Chemical Evolution of the Giant Planets,* edited by C. Ponnamperuma, pp. 121–128, Academic, New York, 1976*b*.

Horowitz, N. H., R. E. Cameron, and J. S. Hubbard, Microbiology of the dry valleys of Antarctica, *Science, 176,* 242–245, 1972*a*.

Horowitz, N. H., J. S. Hubbard, and G. L. Hobby, The carbon assimilation experiment: The Viking Mars lander, *Icarus, 16,* 147–152, 1972*b*.

Horowitz, N. H., G. L. Hobby, and J. S. Hubbard, The Viking carbon assimilation experiments: Interim report, *Science, 194,* 1321–1322, 1976.

Hubbard, J. S., The pyrolytic release experiment: Measurement of carbon assimilation, *Origins Life, 7,* 281–292, 1976.

Hubbard, J. S., G. L. Hobby, N. H. Horowitz, P. J. Geiger, and F. A. Morelli, Measurement of $^{14}CO_2$ assimilation in soils: An experiment for the biological exploration of Mars, *Appl. Microbiol., 19,* 32–38, 1970.

Hubbard, J. S., J. P. Hardy, and N. H. Horowitz, Photocatalytic production of organic compounds from CO and H_2O in a simulated Martian atmosphere, *Proc. Nat. Acad. Sci. U.S., 68,* 574–578, 1971.

Hubbard, J. S., J. P. Hardy, G. E. Voecks, and E. E. Golub, Photocatalytic synthesis of organic compounds from CO and water: Involvement of surfaces in the formation and stabilization of products, *J. Mol. Evol., 2,* 149–166, 1973.

Hubbard, J. S., G. E. Voecks, G. L. Hobby, J. P. Ferris, E. A. Williams, and D. E. Nicodem, Ultraviolet-gas phase and photocatalytic synthesis from CO and NH_3, *J. Mol. Evol., 5,* 223–241, 1975.

Huguenin, R. L., Photochemical weathering and the Viking biology experiments on Mars, paper presented at Annual Meeting, Amer. Astron. Soc., Div. Planet. Sci., Honolulu, Hawaii, Jan. 20, 1977.

Kieffer, H. H., Soil and surface temperatures at the Viking landing sites, *Science, 194,* 1344–1346, 1976.

Klein, H. P., N. H. Horowitz, G. V. Levin, V. I. Oyama, J. Lederberg, A. Rich, J. S. Hubbard, G. L. Hobby, P. A. Straat, B. J. Berdahl, G. C. Carle, F. S. Brown, and R. D. Johnson, The Viking biological investigation: Preliminary results, *Science, 194,* 99–105, 1976.

Oyama, V. I., The chemical potential demonstrated in the gas exchange experiments and its meaning and confirmation of a model of Martian chemistry in the biology experiments, paper presented at Annual Meeting, Amer. Ass. Advan. Sci., Denver, Colo., Feb. 22, 1977.

(Received April 1, 1977;
revised June 6, 1977;
accepted June 6, 1977.)

VOL. 82, NO. 28 JOURNAL OF GEOPHYSICAL RESEARCH SEPTEMBER 30, 1977

Recent Results From the Viking Labeled Release Experiment on Mars

GILBERT V. LEVIN AND PATRICIA ANN STRAAT

Biospherics Incorporated, Rockville, Maryland 20852

Additional results have recently been obtained from the Labeled Release (LR) life detection experiment on Mars. On Viking Lander 2 an experiment using surface material obtained from under a rock shows a response essentially identical with those obtained from other surface samples. Further, after a second nutrient injection, there is an initial drop in the level of radioactive gas present in the test cell, followed by a slow, gradual evolution of radioactivity over the long incubation period. This gas evolution, which proceeds at a rate declining with time, attains a final level similar to that seen just prior to second injection. Following completion of this experimental cycle a fresh Martian surface sample was heat sterilized for 3 hours at 46°C prior to nutrient injection. The resultant evolution of radioactive gas is substantially reduced in agreement with results reported previously. These results are consistent with a biological response and also greatly narrow the number of possible chemical reactants. The current status of the Labeled Release experiment on Mars is summarized.

The Labeled Release (LR) life detection experiment seeks detection of heterotrophic metabolism by monitoring radioactive gas evolution following the addition of a radioactive nutrient containing seven ^{14}C-labeled organic substrates to surface material. Extensive terrestrial tests in flightlike equipment have previously established the characteristics of typical LR biological responses and their abolition following heat treatment of the organism-containing soil for 3 hours at 160°C. The nutrient utilized in the test, as well as the scientific concepts and instrumentation of the experiment, has been described elsewhere [*Klein et al.*, 1976; *Levin*, 1972; *Levin and Straat*, 1976*a*, *b*].

LR results obtained on Mars prior to conjunction [*Levin and Straat*, 1976*b*] showed rapid evolution of radioactive counts upon addition of the radioactive nutrient to a fresh surface sample. At both landing sites the responses were quite similar, and the magnitude of the evolved counts was consistent with utilization of only one of the labeled carbon atom positions available in the nutrient. Although radioactivity was continuously evolved, even over a 60-sol (1 sol equals 24.6 hours) incubation period, no evidence was seen of exponential growth. In one cycle, which utilized a surface sample obtained from under a rock, the initial LR response was essentially identical in kinetics and magnitude with those of other active samples; this indicates that if a chemical reaction is responsible for the LR results, its activity does not depend upon recent ultraviolet activation. Further, the 'active' agent(s) in the Mars sample is stable to 18°C but completely inactivated by 3 hours of heating at 160°C and substantially reduced following similar pretreatment at 50°C. This behavior is consistent with a biological response, although it has been postulated [*Klein et al.*, 1976; *Levin and Straat*, 1976*b*] that a limited number of nonbiological reactants could produce such results.

Since our last data summary [*Levin and Straat*, 1976*b*] prior to conjunction, two additional cycles have recently been completed for the Labeled Release experiment on Mars. The results, summarized below, provide further information on the gas kinetics following a second injection of nutrient and on the effect of 'cold sterilization' of the Mars surface material.

VIKING LANDER 2, CYCLE 3

The third LR cycle on Viking Lander 2 (VL-2) was initiated prior to conjunction by using a fresh sample obtained from under Notch rock. The data obtained over 7 sols following the first injection have been previously presented along with interpretations of the results [*Levin and Straat*, 1976*b*]. After these 7 sols a second aliquot of nutrient was then added to the sample, and incubation was continued for the ensuing 80 sols, all data being accumulated on the lander tape recorder for playback to earth after conjunction. This extensive incubation

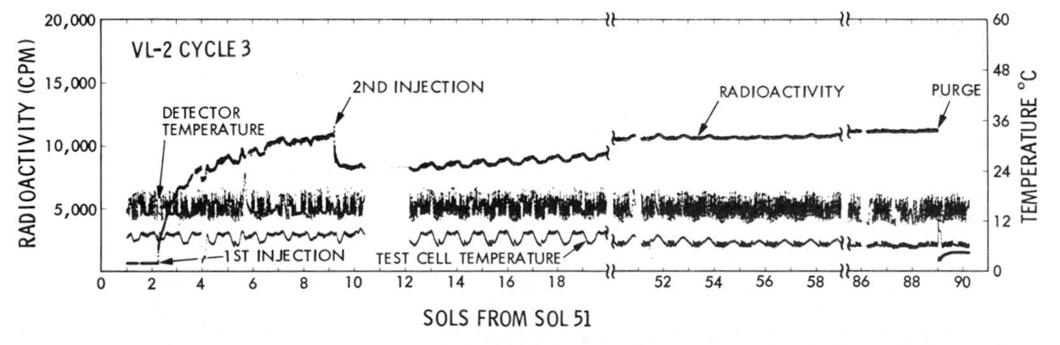

Fig. 1. Plot of LR data from third sample analysis on VL-2. An active sequence was used on a fresh sample that was acquired from surface material exposed by pushing aside a rock. Radioactivity was measured at 16-min intervals throughout the cycle except for the first 2 hours after each injection, when readings were taken every 4 min. Radioactivity data include a background count of 659 cpm prior to the onset of the cycle. Data obtained in the single-channel counting mode between sols 53 and 60 have been corrected to the dual-channel mode of operation for comparison with the remainder of the cycle and with data from previous cycles. Data from sols 61 and 62 were lost. Detector and test cell temperatures were measured every 16 min.

Paper number 7S0469.

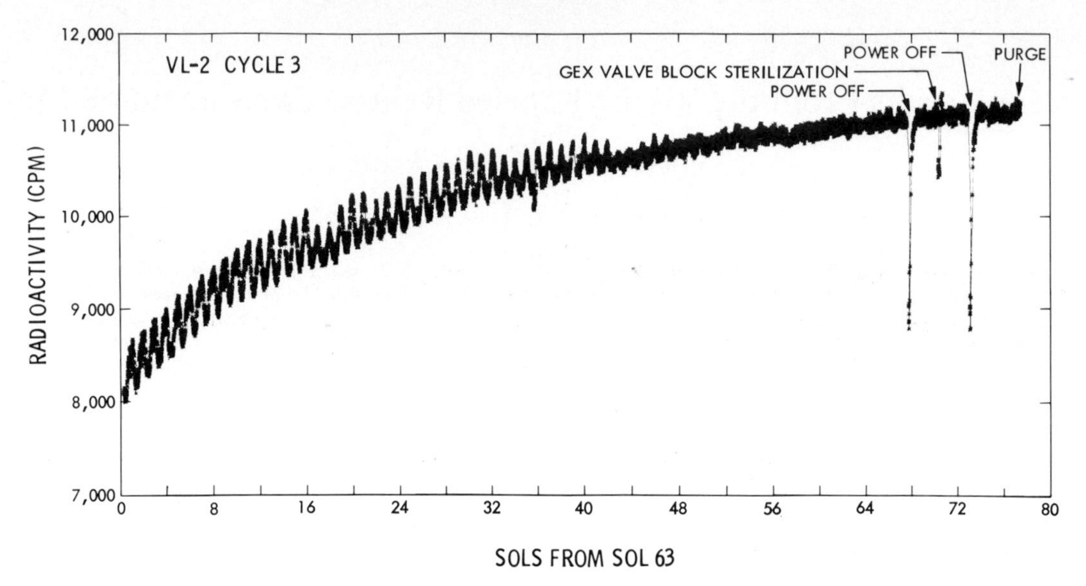

Fig. 2. Plot of LR data following second injection to third sample analyzed on VL-2. Data after second injection on sol 60 were expanded from data presented in Figure 1. The pronounced drops in radioactivity occurring between sols 66 and 74 correspond to sharp drops in test cell temperature resulting from brief periods in which the power was turned off or during which the valve block in the Gas Exchange (GEX) experiment was sterilized. Throughout the indicated period, radioactivity was measured at 16-min intervals.

period provided an opportunity to examine in detail the kinetics of gas evolution following the second injection.

The complete results for VL-2, cycle 3, are shown in Figure 1 on a scale identical with those used [*Levin and Straat,* 1976*b*] for all previous cycles. As is shown, immediately upon the second injection a spike of radioactivity was observed, followed by a rapid disappearance of approximately 35% of the volume of radioactive gas previously evolved. This percentage drop is essentially the same as that seen following second injections for all active cycles thus far conducted (G. V. Levin and P. A. Straat, manuscript in preparation, 1977). This decrease may represent solubility of the gas in the added aqueous nutrient with no further production of gas, perhaps aided by *p*H changes from the additional soil wetting.

After the immediate decrease following the second injection, a slow, gradual evolution of radioactivity is observed. In VL-1, cycle 3, this increase appeared linear over the 16-sol period following the second injection [*Levin and Straat,* 1976*b*], although the exact kinetics are difficult to determine because of the strong temperature interference. (The amount of radio-

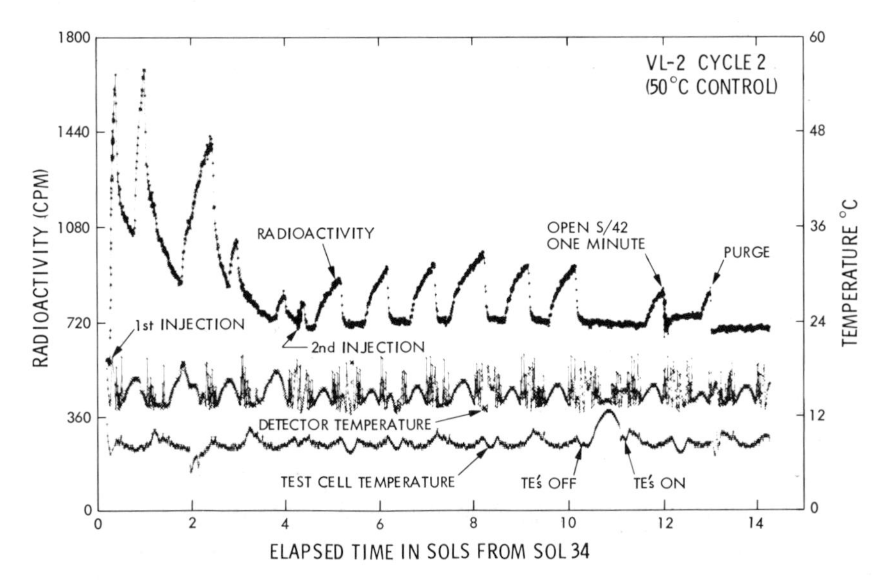

Fig. 3. Expanded plot of LR data from second sample analysis on VL-2. A control sequence was used in which a fresh surface sample was heat sterilized for 3 hours at approximately 51.5°C just prior to nutrient injection. Radioactivity was measured at 16-min intervals throughout the cycle except for the first 2 hours after each nutrient injection, when readings were taken every 4 min. Radioactivity data include a background count of 590 cpm prior to sterilization of the sample. Toward the end of the cycle the thermoelectric coolers (TE's) were turned off for 1 sol to determine their contribution to the diurnal fluctuation in radioactive counts. The test cell was then pressurized with low-pressure helium through valve S/42. Because the counts immediately dropped as radioactive gas was pushed from the detector and then returned to essentially the original value, the possibility of a leak was eliminated. Following purge the level of radioactivity was lower than was previously observed in the cycle. This confirms the absence of a leak during the cycle. Detector and test cell temperatures were measured every 16 min.

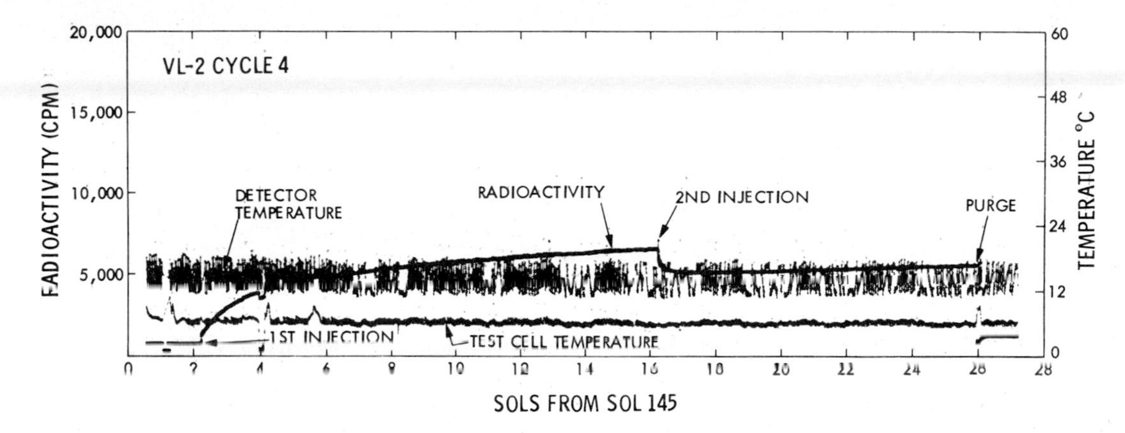

Fig. 4. Plot of LR data from fourth sample analysis on VL-2. A control sequence was used in which a fresh surface sample was heat sterilized for 3 hours at approximately 46°C about 26 hours prior to nutrient injection. Radioactivity was measured at 16-min intervals throughout the cycle except for the first 2 hours after each nutrient injection, when readings were taken every 4 min. Radioactivity data include a background count of 754 cpm prior to sterilization of the sample. Detector and test cell temperatures were measured every 16 min.

active gas in the detector fluctuates with test cell temperature.) From the long incubation period following the second injection in VL-2, cycle 3, however, it is apparent that the evolution is not linear but continues at a declining rate with incubation time (Figure 2). In this cycle it should be noted that as the external ambient temperatures at the lander site decrease with the approach of the Martian winter, the diurnal test cell temperature fluctuations become less pronounced and gradually disappear. Correspondingly, the radioactivity fluctuations become less pronounced. The radioactivity level attained just prior to purge approaches plateau and is approximately equal to that seen immediately prior to the second injection. At this point it is not yet certain whether this evolution represents a desorption of the absorbed or dissolved gas or is caused by additional oxidation of the radioactive substrates provided in the nutrient.

VIKING LANDER 2, CYCLE 4

In VL-2, cycle 2, a freshly acquired surface sample was heated for 3 hours at approximately 50°C and allowed to cool prior to the injection of nutrient. The purpose of this cold sterilization was to differentiate further between possible biological or chemical reagents responsible for the LR data. Since the 50°C temperature is likely to have an adverse effect on any Mars organisms, a subsequent reduction in gas evolution would be consistent with a biological response. In any event it would narrow the range of possible chemical reactants to those stable between 18°C (the highest storage temperature to which samples were exposed prior to nutrient injection) and 50°C.

In VL-2, cycle 2, the response was significantly attenuated by pretreatment at 50°C. However, the peculiar kinetics obtained (Figure 3) suggested a possible instrument anomaly.

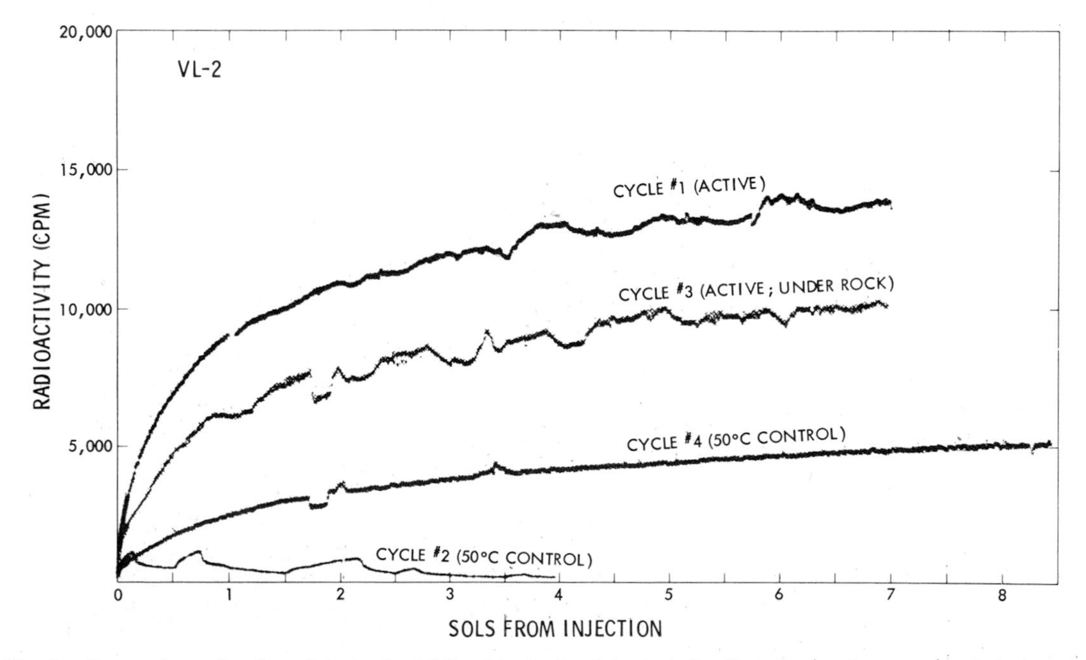

Fig. 5. Comparison of radioactivity evolved following the first injection of radioactive nutrient to each analysis cycle of VL-2. A fresh surface sample was used for each cycle. The sample used in cycle 3 was obtained from under a rock. Cycles 1 and 3 were active sequences, whereas cycles 2 and 4 were control sequences in which the samples were heated for 3 hours at approximately 51.5°C and 46°C, respectively, prior to nutrient injection. All data have been corrected for background counts observed prior to injection.

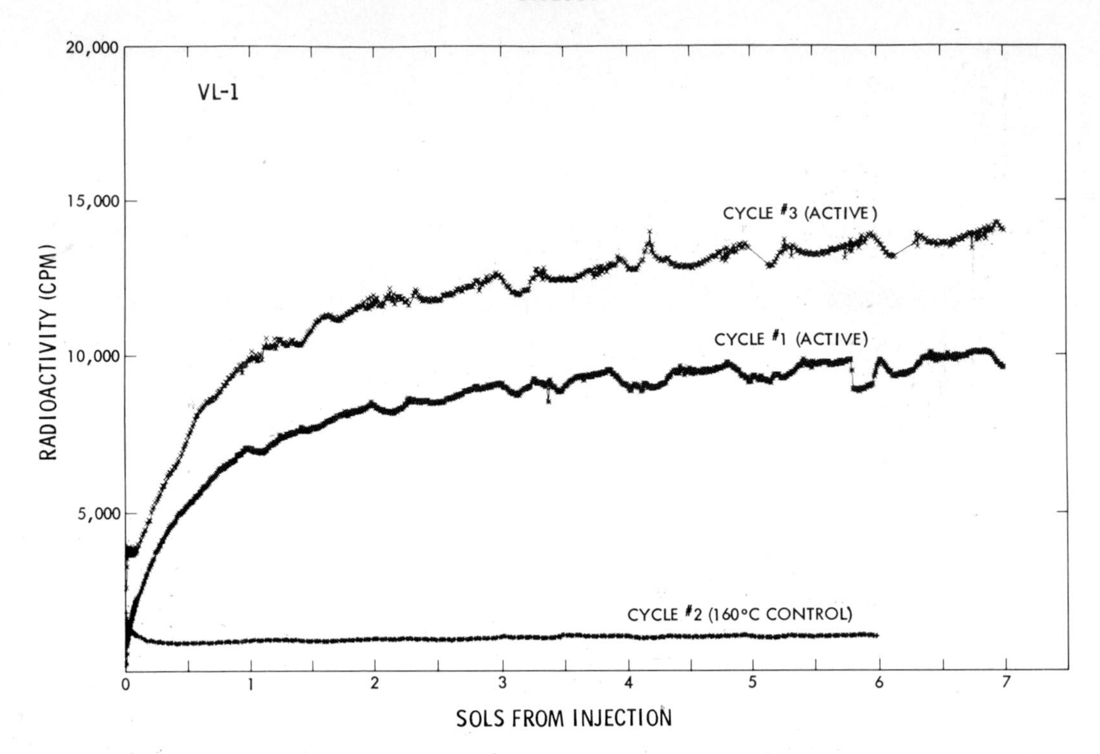

Fig. 6. Comparison of radioactivity evolved following the first injection of radioactive nutrient to each analysis cycle of VL-1. A fresh surface sample was used for the active sequences of cycles 1 and 3. For cycle 2 a stored portion of the same sample used for cycle 1 was heated for 3 hours at 160°C prior to nutrient injection. All data have been corrected for background counts observed prior to nutrient injection.

Diagnostic sequences conducted remotely by the engineering team on the instrument toward the end of that cycle showed no indications of a hardware malfunction. The count fluctuations are somewhat related to the thermoelectric coolers (TE's), since turning them off for a 19-hour period starting on sol 44 resulted in a disappearance of the unusual counting phenomenon. Since electronic cross talk between the TE's and the detectors does not appear to be the cause (the problem did not exist after purge), the phenomenon may result from some temperature-dependent gas movement between the sample in

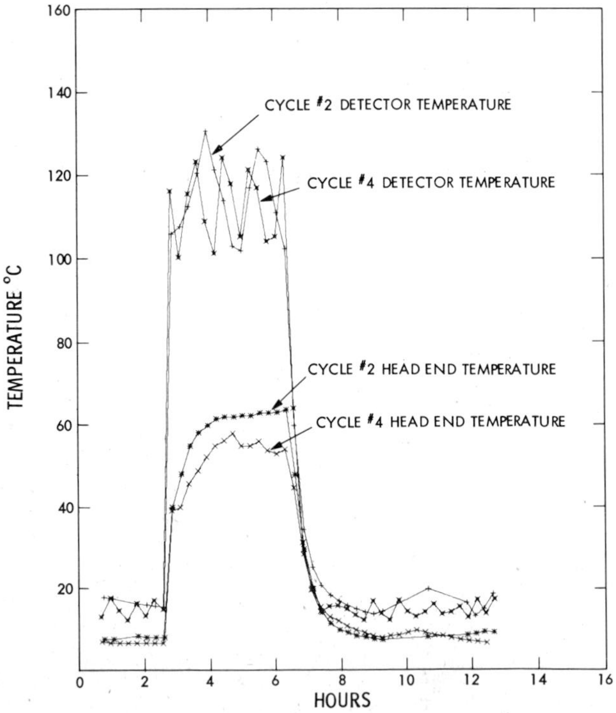

Fig. 7. Temperatures attained during heat sterilization of cycles 2 and 4 on VL-2. During sample sterilization, temperatures were measured every 16 min at both the head end of the test cell and the detector, as is indicated. The LR heaters were turned on for a total of 3.75 hours to provide the sample with approximately 3 hours of heating at the desired sterilization temperature.

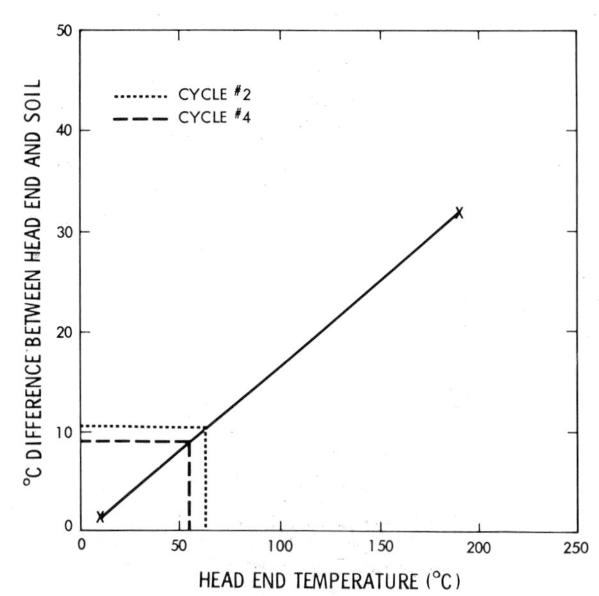

Fig. 8. Estimation of sample temperatures attained during cold sterilization on VL-2. Because the LR heaters are located in the head end of the LR test cell, sample temperatures attained are less than those recorded at the head end. From known temperature differences x between sample and head end a straight line relationship has been assumed. From this relationship the sample during sterilization in cycle 2 is calculated to be 10.5°C below the average head end temperature, or 51.5°C. Similarly, the cycle 4 sample reached a sterilization temperature of 46°C. The sample sterilization temperatures are considered correct within 2°C.

the test cell and the detector chambers. Although these patterns are not understood, the diagnostics indicate that the LR instrument was functioning properly.

Because of the importance of the cold sterilization for understanding the nature of the Mars reactant in the LR response, the experiment was repeated in VL-2, cycle 4, with another fresh surface sample. The results of this cycle are given in Figure 4, and the magnitude of the evolved radioactivity is compared to that from each of the other three VL-2 cycles in Figure 5. For further comparisons, evolved radioactivity from each VL-1 cycle is shown in Figure 6. As can be seen (Figures 4–6), the kinetics resulting from the VL-2, cycle 4, cold sterilization resemble those seen in all active cycles. However, the magnitude of the response following the first injection is again significantly attenuated. This validates the previous [*Levin and Straat,* 1976*b*] conclusion from VL-2, cycle 2, that pretreatment at 50°C significantly destroys the Mars reactant. That the 50°C pretreatment does not affect gas adsorption or solution following the second injection is seen by the immediate 35% drop following that injection. Although the subsequent gradual rise in radioactivity is smaller in magnitude than it is seen to be in active cycles, it represents a similar percentage rise (G. V. Levin and P. A. Straat, manuscript in preparation, 1977).

The experimental sequences of cycles 2 and 4 were identical with two exceptions. In cycle 4, nutrient injection was performed approximately 26 hours after soil sterilization, similar to the sequence used in the 160°C control (VL-1, cycle 2) and in all earth-based flightlike instruments. In VL-2, cycle 2, however, injection was performed within 3 hours after sterilization, and although the head end of the test cell had returned to ambient temperatures, it is possible that the surface sample contained therein was still somewhat warmer than ambient. In addition, the average temperature of the soil for cycle 2 sterilization was 51.5°C versus 46°C in cycle 4 (see the appendix). However, these factors do not seem to explain the observed kinetic differences between cycles 2 and 4. The cycle 4 result is quite believable when the kinetics of gas evolution are compared to those from active cycles (Figure 1, for example). On this basis the validity of the cycle 2 kinetic data might be questioned, even though the engineering tests discussed above failed to detect any instrumental malfunctions.

The results obtained from the cold sterilization experiment are extremely important in helping to differentiate a biological from a nonbiological reaction. Further, cycle 4 results establish that the reactant is severely degraded by exposure to 46°C, although it is known to be stable for at least 2 sols at 18°C [*Levin and Straat,* 1976*b*]. This behavior both resembles a biological response and greatly narrows the range of possible chemical reactants.

As of this writing then, the LR results are entirely consistent with a possible biological interpretation. However, on separating fact from speculation the following points have been established:

1. A remarkably uniform production of gas issues from the Labeled Release nutrient when it is placed on Martian soil obtained at both lander sites.

2. After the reaction approaches completion, the addition of more nutrient results in a net loss of radioactive gas from the instrument headspace.

3. Direct exposure of the surface sample to visible or ultraviolet light prior to placing it in the test cell is not a prerequisite for a positive LR response.

4. The reactant in the Mars soil is completely inhibited by heating the soil to 160°C and largely inhibited by heating to 46°C. In contrast, exposure to 18°C for 2 sols does not appear to inhibit the reaction.

Additional LR experiments are now in progress at both Mars sites and may shed further light on the nature of the reactive agent in the soil. Our thrust in the laboratory is to examine the narrowed range of chemical candidates in an attempt either to repeat the flight data and thus show that the reaction could be chemical or to eliminate all candidate nonbiological reactants. These two efforts will be pressed along with a detailed analysis of our Viking results and those of other Viking experiments in an effort to answer the tantalizing question posed by the remarkable results of the LR experiment on Mars.

APPENDIX

The soil temperatures attained during each cold sterilization are calculated from the head end temperature readings shown in Figure 7. The Mars surface sample, being contained at the bottom of the test cell, cannot reach temperatures as high as those reached by the head end, where the heaters are located. Because the instrument was not intended to perform a cold sterilization, no 'library' data are available to relate head end temperature to soil temperature in this range. However, it is known [*Glenn,* 1974] that at 10°C, soil temperature is 1°–2°C lower than the head end temperature, whereas for this particular lander the soil temperature remains 32°C lower than the head end temperature during normal sterilization, in which the head end attains 191°C. Under the assumption of a straight line relationship between these two points (Figure 8), temperatures attained by the soil during each cold sterilization can be estimated. Thus the soil in cycle 2 reached a temperature 10.5°C below the head end temperature, or 51.5°C. Similarly, the cycle 4 sample reached 46°C. While the absolute value of each estimate is probably correct to within only a few degrees, the difference between 46°C and 51.5°C is probably accurate to within 1.0°C. No significant difference occurred in the temperature reached by the detectors in both cycles.

Acknowledgments. The authors thank Cynthia Ann Waldman for excellent technical assistance in the reduction of flight data. The many helpful contributions of various members of the Viking Biology Flight Team are also gratefully acknowledged. We again thank all those persons cited in publications over the past 15 years for their efforts in the development of the Labeled Release experiment. Finally, we wish also to express appreciation to the National Aeronautics and Space Administration for continued support over the years. NASA contract NAS1-9690.

REFERENCES

Glenn, E., Module thermal test report, *Rep. 7138,* TRW Syst. Group, Redondo Beach, Calif., Oct. 18, 1974.

Klein, H. P., N. H. Horowitz, G. V. Levin, V. I. Oyama, J. Lederberg, A. Rich, J. S. Hubbard, G. L. Hobby, P. A. Straat, B. J. Berdahl, G. C. Carle, F. S. Brown, and R. D. Johnson, The Viking biological investigation: Preliminary results, *Science, 194,* 99–105, 1976.

Levin, G. V., Detection of metabolically produced labeled gas: The Viking Mars lander, *Icarus, 16,* 153–166, 1972.

Levin, G. V., and P. A. Straat, Labeled release—An experiment in radiorespirometry, *Origins of Life, 7,* 293–311, 1976*a*.

Levin, G. V., and P. A. Straat, Viking labeled release biology experiment: Interim results, *Science, 194,* 1322–1329, 1976*b*.

(Received March 30, 1977;
revised May 31, 1977;
accepted June 1, 1977.)

The Viking Gas Exchange Experiment Results From Chryse and Utopia Surface Samples

VANCE I. OYAMA AND BONNIE J. BERDAHL

NASA Ames Research Center, Moffett Field, California 94035

Immediate gas changes occurred when untreated Martian surface samples were humidified and/or wet by an aqueous nutrient medium in the Viking lander gas exchange experiment. The evolutions of N_2, CO_2, and Ar are mainly associated with soil surface desorption caused by water vapor, while O_2 evolution is primarily associated with decomposition of superoxides inferred to be present on Mars. On recharges with fresh nutrient and test gas, only CO_2 was given off, and its rate of evolution decreased with each recharge. This CO_2 evolution is thought to come from the oxidation of organics present in the nutrient by γ Fe_2O_3 in the surface samples. Atmospheric analyses were also performed at both sites. The mean atmospheric composition from four analyses is N_2, 2.3%; O_2, $\leq$0.15%; Ar, 1.5%; and CO_2, 96.2%.

INTRODUCTION

The gas exchange experiment (GEX), one of three experiments comprising the Viking lander biology instrument, periodically samples the headspace gases above a Martian surface sample incubating under dry, humid, or wet conditions and analyzes the gases with a gas chromatograph (GC). The GEX was designed to distinguish between gas changes arising from microbial metabolism and those arising from purely chemical reactions or physical phenomena, such as sorption and desorption, by recycling the soil sample. A chemical or a physical reaction would be reduced or eliminated in subsequent cycles, whereas a biological system would perpetuate itself. The gas changes from the former would be reduced or disappear; from the latter they would continue or increase. The recycling is accomplished by flushing the test chamber of gas and nutrient, adding fresh nutrient and test gas, and continuing incubation. A more detailed account of the experimental design and operation can be found elsewhere [*Merek and Oyama,* 1970; *Oyama,* 1972; *Oyama et al.,* 1976; *Klein et al.,* 1976*a*].

Preliminary GEX Viking data have been reported [*Klein et al.,* 1976*b*] and interpreted [*Oyama et al.,* 1977]. In this report all the data received up to March 29, 1977 (sol 234 on Viking lander 1 (VL-1) and sol 184 on Viking lander 2 (VL-2)), are summarized.

METHODS AND MATERIALS

Five surface samples were tested by the GEX experiment: two from the VL-1 Chryse site, Sandy Flats area, and three from the VL-2 Utopia site. The first and third samples from Utopia were from the Beta area, while the second was from under Notched Rock. The first sample from Chryse was tested as received; the second was heated at 145°C for 3.5 hours before testing. The first and second samples from Utopia were tested as received, while the third sample was heated at 145°C for 3.5 hours before testing. Table 1 summarizes the tests performed by GEX with the Martian surface samples.

A number of operational sequences were used repeatedly, and their general features and functions are given below.

Initialization sequence. A 1-cm^3 soil sample was placed in the 8.7-cm^3 test cell, and the cell was sealed. After adding the test gas (91.65% He, 5.51% Kr, and 2.84% CO_2), which brought the cell pressure to approximately 200 mbar, we had three options available: (1) add no moisture, (2) add enough nutrient to humidify the soil (approximately 0.5 cm^3), and (3) add enough nutrient to wet the soil (approximately 2 cm^3). The nutrient medium is a complex mixture of organic compounds and inorganic salts [*Oyama et al.,* 1976].

Recharge sequence. At the end of an incubation cycle the test cell was flushed of test gas and nutrient by flowing He through the cell and out the drain valve. (Approximately 0.5 cm^3 of nutrient remained in the cell sump, and 0.35 cm^3 remained in each cubic centimeter of soil.) The test gas and fresh nutrient were then added.

Termination sequence. This sequence was followed prior to receiving another soil sample because the GEX experiment has only one test cell. The cell was flushed of test gas and nutrient as it was in the recharge sequence except that the drain valve was left open, with no He flow, for 3 or more sols to dry the soil. With the drain valve open the inside of the test cell was open to the Martian environment.

Heat treatment sequence. A fresh surface sample was added to the test cell, which had been terminated as described above. The cell was sealed and heated to 145°C for 3.5 hours with one standard cubic centimeter per minute of He flowing through the test cell and out the drain valve. The test cell was allowed to cool, the drain valve was closed, and the He flow stopped.

The GEX GC system can measure H_2, Ne, N_2, O_2, Ar or CO, NO, CH_4, Kr, CO_2, N_2O, and H_2S. Argon is not separated from CO by the GC column employed [*Oyama et al.,* 1976], but we assume that the Ar/CO peaks measured were Ar unless stated otherwise, since the Ar abundance is estimated to be at least an order of magnitude greater than CO in the Martian atmosphere [*McElroy et al.,* 1976], and the CO contribution at the atmospheric Ar concentration is not measurable by our GC.

RESULTS

Chryse Sandy Flats Sample

Figure 1 shows the gas changes which occurred in the test cell headspace throughout four cycles of incubation of the first Sandy Flats sample. Trapped Martian atmospheric gases and the added test gas were present in the test cell before humidification occurred in cycle 1; the quantities of these gases calculated to be in the test cell at the outset of humidification were 61, 4, 40, and 3600 nmol for N_2, O_2, Ar, and CO_2, respectively. Comparing these values with those obtained from the sol 9 analysis, one can see that the concentrations of all the

Paper number 7S0508.

TABLE 1. Summary of GEX Tests

Surface Sample	Soil Test Events	Duration of Incubation, sols	Number of Analyses	Amounts of Nutrient Injected, cm³	Cell Temperature, °C	
					Mean	Range
Chryse, Sandy Flats	first cycle, humid	7	5	0.56	10.3	(8.3–11.6)
	first cycle, wet	13	6	1.8	11.7	(8.3–24.4)
	second cycle, wet	39	14	2.3	11.2	(8.5–24.1)
	third cycle, wet	35	8	1.7	11.1	(8.8–14.7)
	fourth cycle, wet	103	13	1.7	10.4	(8.5–15.0)
Chryse, Sandy Flats (in hopper for 139 sols), heated	first cycle, dry	1	1		9.3	(8.0–10.8)
	first cycle, humid	0.1	1	0.49	8.6	(8.3–9.1)
Utopia, Beta	first cycle, dry	1	3		11.6	(10.9–12.4)
	first cycle, humid	7	5	0.59	11.1	(10.4–11.6)
	first cycle, wet	19	8	2.0	12.2	(10.9–21.3)
	second cycle, wet	12	5	2.2	11.5	(10.4–13.4)
Utopia, under Notched Rock	first cycle, dry	2	3		10.5	(8.8–11.1)
	first cycle, wet	78	14	1.8	9.2	(8.3–18.2)
	second cycle, wet	16	5	1.8	9.1	(8.3–13.4)
	third cycle, wet	22	6	1.8	9.0	(8.3–11.1)
Utopia, Beta (in hopper for 32 sols), heated	first cycle, dry	1	1		8.8	(8.6–8.8)
	first cycle, humid	4	2	0.53	8.7	(8.6–8.8)

gases, without exception, were greater than expected upon humidification of the soil sample. The greatest change in concentration was observed with O_2, from an expected 4 to 520 nmol. By sol 11, virtually all gas production had ceased.

From laboratory studies we have concluded that early, rapid gas changes induced by humidification are associated only with physical and chemical phenomena and are therefore unreliable as indices of biological activity. Further, we had never seen native terrestrial soils evolve O_2 in the dark. We postulated, therefore, that the extremely fast production of gases was indicative of desorption processes or unusually rapid chemical reactions, but we were not certain whether these physical or chemical processes were caused by bringing cold Martian soil to a warmer environment or by the humidification. Since the soil after acquisition had been exposed to temperatures of 9°C for at least 0.28 hour [*Oyama et al.*, 1977], it is conceivable that most of the thermal degassing had already occurred prior to the soil being dumped into the test cell. Therefore it seemed more likely that at the relatively steady test cell temperature (8°–11°C), humidification rather than thermal desorption caused the gaseous evolutions.

The increase in O_2 production ceased between sols 11 and 15, and after injection of additional nutrient on sol 16 the concentration of O_2 began to decrease. We attribute the uptake of O_2 to the ascorbate ion in the nutrient, whose capacity to react with O_2 is well established [*Steinman*, 1942]. Thermal stability tests of our nutrient medium had shown that the ascorbate ion was not destroyed during the instrument and spacecraft sterilization regimes. We estimated that the ascorbate ion in the 2.36 cm³ of nutrient injected in the first cycle had an O_2 absorbing capacity of 350 nmol, and we assumed that exactly this amount of O_2 disappeared.

In contrast to the O_2 changes, the N_2 and Ne kept slowly increasing. The extremely slow equilibration of Ne from the nutrient solution in the test cell was a reproducible phenomenon that had been demonstrated in tests using the GEX laboratory flightlike instrument. Nitrogen had been shown to be desorbed very slowly from sterile terrestrial soils in the same way it appeared to be desorbed from the Martian soils. Nitrogen probably penetrates deeper into micropores in the soil structure and is most difficultly dislodged by water vapor.

The overall wet mode data show that only CO_2 evolution persisted beyond the first cycle. In general, all of the CO_2 curves were similar, except for the first wet cycle. Here a decrease lasting for 1 sol occurred after addition of more nutrient. We consider that this decrease was caused by readsorption of CO_2 into an alkaline Martian sample. Following this there was an accelerated increase which lasted for a few sols followed by a slower log linear increase which persisted to the end of the cycle. The apparent acceleration in CO_2 evolution, we feel, was due to the equilibration of CO_2 with the soil carbonate system. (There was an aberrant point on sol 39 in which the CO_2 increased sharply. This sharp increase coincided with increased cell temperatures of about 10°C and was probably due to a decreased solubility of CO_2.) The first few gas analyses following a recharge were made before the bulk of the carbonate reserve in the wet soil had equilibrated with the atmosphere in the headspace. The relatively large swing observed for CO_2 equilibration in the beginning of the cycle was in contrast to the slow log linear increases in the headspace CO_2 for the rest of the cycle. The persistence of the CO_2 log linear increase over the four wet cycles suggested a very slow oxidation of the organics in the nutrient by an oxidant in the Martian soil. This oxidant was not very soluble because it was not washed out by the three recharges of aqueous nutrient. The log linear CO_2 production would be expected of a diffusion-limited system in which the movement of organics in the liquid phase was the rate-limiting step.

The Chryse sample was incubated under humid conditions for 6 sols followed by a wet incubation period of 190 sols, interrupted only for nutrient recharges, for a total of 196 sols (~201 days). We had observed with Antarctic soils having low microbial populations that up to 200 days of incubation were required to elicit a biological gas response from them. Thus we felt it necessary to incubate a Martian surface sample for a long period of time to allow any microorganisms present in the Martian soil to manifest themselves.

Utopia Beta Sample

Figure 2 shows the gas changes which occurred in the headspace of the test cell over two incubation cycles with the Beta sample. The nominal sequence of events was changed in the

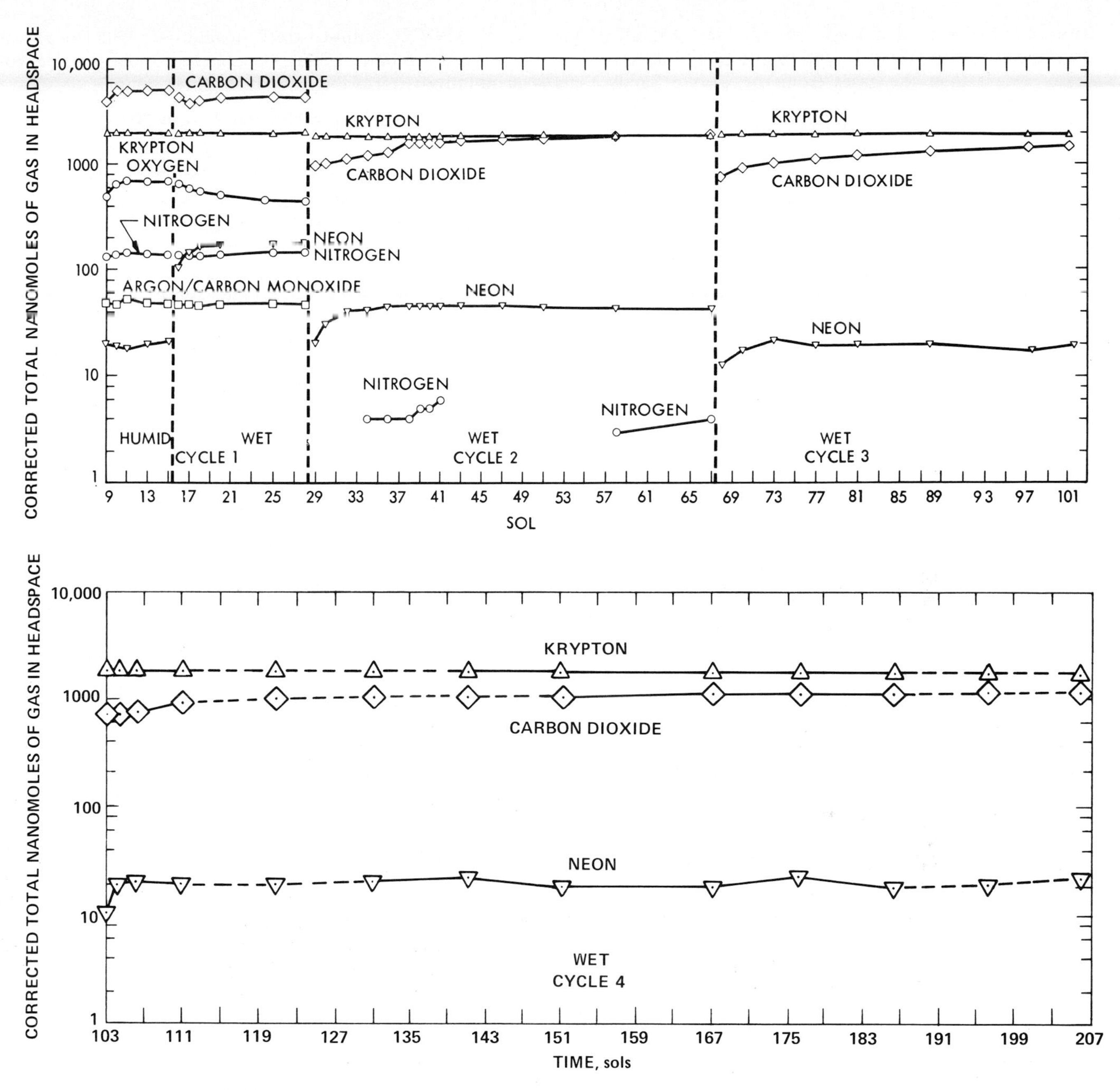

Fig. 1. VL-1, Chryse Sandy Flats sample, gas changes in GEX headspace. The corrected total nanomoles were calculated as follows. The GC detector data were sampled at 1-s intervals, digitized, and fitted to a skewed Gaussian distribution from which peak heights were obtained. The gas in the headspace was obtained from the ratio of the sample loop volume to the total headspace volume. The gas composition was corrected for cumulative sampling losses by referencing absolute changes in the krypton values for successive samples. Corrections were made for pressure sensitivity in this flight instrument caused by a partial restriction in the gas sampling system which prevented total evacuation of the sample loop to ambient pressure before filling (three times) from the test cell. The volume for krypton was corrected for pressure as follows: nanomoles of Kr = $37.77P_c \exp - 0.118V_p \exp 1.016$, where P_c is the test cell pressure in millibars and V_p is the peak height in volts. The value for each gas was corrected by the ratio of the term $37.77P_c \exp - 0.118$ to the similar Kr value from a pressure insensitive instrument. P_c was determined from an inventory of all sources of pressure such as the enclosed atmosphere, the pressure of the initialization gas injected, and any additional injection of helium required to increase cell pressure. The points lying within the dashed lines in cycle 4 were meant only to indicate when analyses were performed and had no quantitative significance except for removal of headspace gas during sampling. These data were lost in transmission.

VL-2 GEX tests to test the thermal hypothesis of the immediate gas release when the Martian surface sample was placed in the test cell and humidified. This involved exercising an option in the initialization sequence such that three gas analyses were performed with the dry soil in the test cell before humidification. No significant changes in the headspace gases were noted. Within 2.8 hours after humidification, however, CO_2, Ar, N_2, and O_2 were already being released. The calculated contributions of gases to the headspace from the Martian atmosphere and test gas at the start of the humid cycle were 61, 4, 40, and 3500 nmol for N_2, O_2, Ar, and CO_2, respectively. By comparing these calculated values with the measured values from the first humid analysis one can see that all of the headspace gases, without exception, increased but to a much lesser extent than the gases from the Chryse Sandy Flats sample. The O_2 production ceased by sol 12, and the total O_2 evolved in the headspace was ~110 nmol, only 0.15 as much as that from the Chryse sample. After the soil was wet on sol 17, the O_2 steadily decreased so that by sol 25, unlike the Chryse Sandy Flats sample, the O_2 could not be detected. We estimated that the O_2 absorbing capacity of the ascorbate ion in the nutrient medium added was equivalent to 390 nmol of O_2.

The N_2 and Ne slowly increased, as occurred with the Sandy Flats sample. The steady increase in CO_2 within each cycle, but with an overall decline in CO_2 evolution from cycle to cycle, paralleled the results obtained with the Chryse Sandy Flats sample. There was a drop in the CO_2 concentration immediately after the soil was wet which was consistent with an alkaline soil, as was also observed with the Sandy Flats sample.

Utopia Under Notched Rock Sample

The data from the under Notched Rock sample (Figure 3) show that like the Beta sample, no significant changes in the gas composition occurred with the dry soil. The calculated contributions of gases to the headspace from the test gas and the Martian atmosphere at the start of the wet cycle were 56, 4, 37, and 3100 nmol of N_2, O_2, Ar, and CO_2, respectively. The O_2 produced during the wet phase in cycle 1 had only a very short existence in the test cell. Within 4 sols all of the O_2 had disappeared. Based upon the amount of nutrient delivered to the test cell, we estimated that the medium's ascorbate ion capacity to scavenge O_2 was equivalent to 270 nmol of O_2. The decreased amount of O_2 produced correlated with a reduction in N_2. The CO_2 output agreed qualitatively with the results obtained from both the Chryse and the Utopia samples, but quantitatively it was the least of the three samples.

SUMMARY OF HUMIDIFICATION DATA

Table 2 summarizes the gas output data from the three surface samples during humidification. The under Notched Rock sample data were included, even though the sample was wet by a single injection of 2.2 cm^3 of nutrient. The range of O_2 estimated for the Utopia Notched Rock sample was 70–270 nmol. The range limits were based upon the largest amount of O_2 observed and the O_2 scavenging capacity of the ascorbate ion in the medium. Since virtually all O_2 had disappeared between sols 57 and 59, the O_2 output was probably much closer to the 70- than to the 270-nmol limit. In the case of the Beta and Chryse Sandy Flats samples the O_2 yields exceeded the O_2 scavenging capacity of the ascorbate ion, and more accurate O_2 outputs could be calculated.

If the assumption was made that soil material from all sites had about the same surface area per unit volume, calculations could be made of the relative nanomoles of each gas desorbed with respect to the CO_2 desorbed. Absolute values were not calculable because surface area estimates and the amount of actual water molecules adsorbed were not exactly known. Table 2 shows the predicted desorption of Ar, N_2, and O_2 in nanomoles for each of the three samples upon humidification and the actual values found. For all three samples the Ar and N_2 values found were less than the predicted values, whereas the O_2 values were above those predicted: 790/4.4, 190/3.4, and 70–270/2.7 for VL-1 Sandy Flats, VL-2 Beta, and VL-2 under Notched Rock, respectively. We ascribe the differences in the Ar and N_2 found in the three surface samples to the

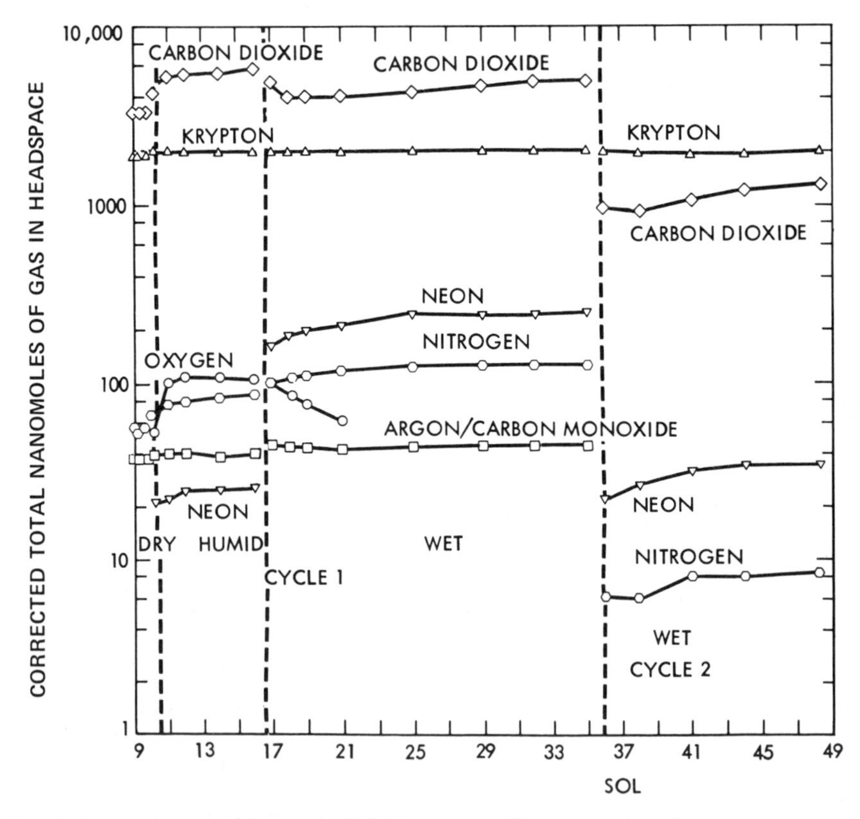

Fig. 2. VL-2, Utopia Beta sample, gas changes in GEX headspace. The corrected total nanomoles were calculated in the same manner as indicated in the legend for Figure 1 except that no pressure sensitive corrections were included.

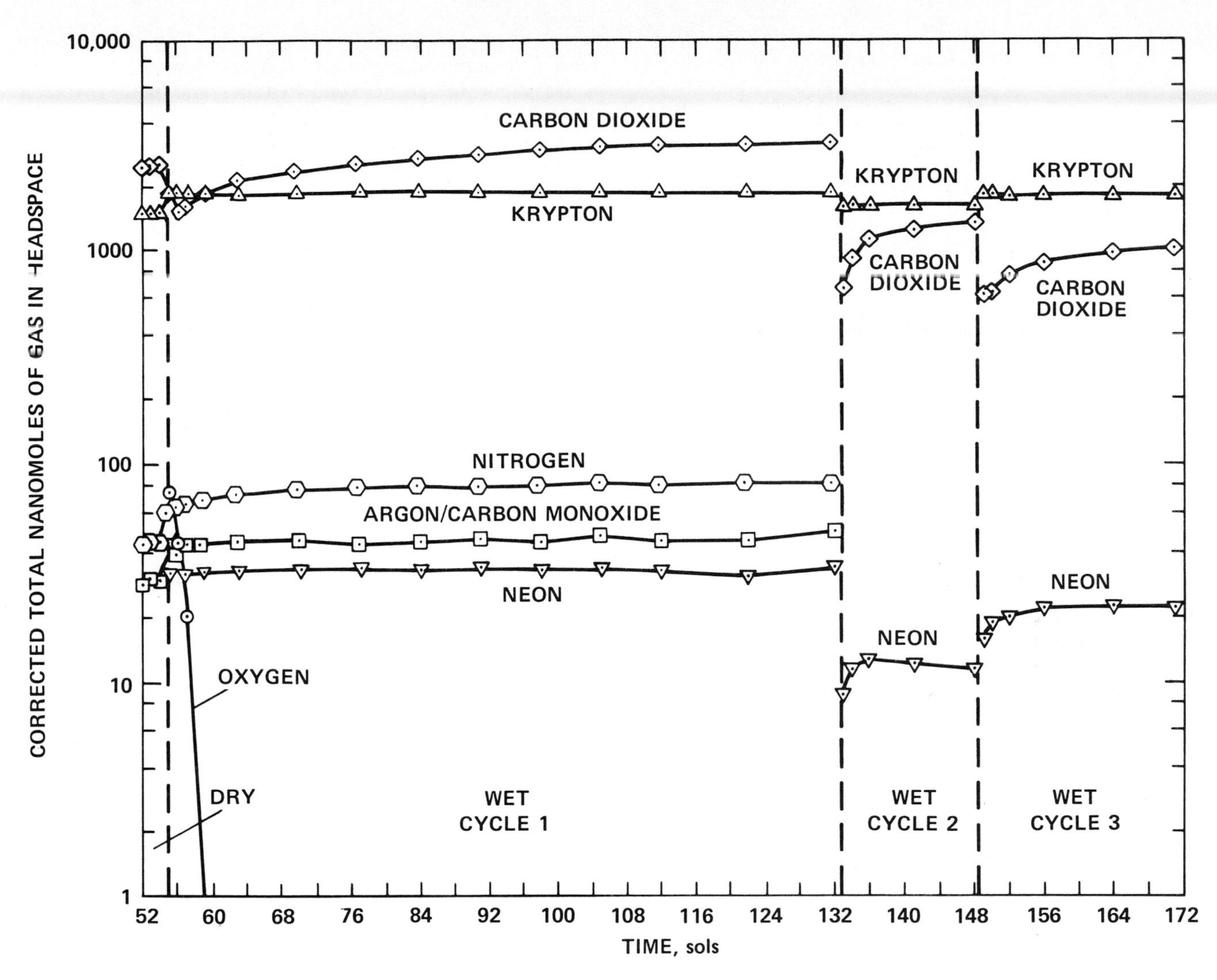

Fig. 3. VL-2, Utopia under Notched Rock sample, gas changes in GEX headspace. See legend for Figure 2.

TABLE 2. Predicted Nanomoles of Gas Desorbed by Humidification, Based on Atmospheric Abundance and Relative Heats of Adsorption, Normalized to CO_2

		Ar	N_2	O_2	CO_2
Heat of adsorption* E_1 (−183°C silica gel), cal/mol		2096	2127	2215	5172†
Gas mol % (M), GC measurement		1.5	2.3	0.1	96.2
VL-1, Sandy Flats (N), nmol	predicted	62	96	4.4‡	9800
	found	13	83	790	9800
VL-2, Beta (N), nmol	predicted	49	76	3.4	7750
	found	4	30	190	7750
VL-2, under Notched Rock (N), nmol	predicted	39	60	2.7	6110
	found	6	13	70–270	6110

The net nanomoles of gas found were estimated in the manner given in Table 3 of *Oyama et al.* [1977]. The O_2 values were further corrected by adding the O_2 absorbing capacity of the ascorbate ion in the nutrient added; it is assumed that at the end of the humid cycles for the Chryse and Beta samples all of the ascorbate ion was oxidized. For the under Notched Rock sample the range of O_2 was bounded by the largest observed O_2 and the total O_2 absorbing capacity of the ascorbate ion added, since the O_2 disappeared entirely.

**Brunauer et al.* [1938].

†−78°C silica gel.

‡Example: $N_{O_2} = (N_{CO_2} \times M_{O_2} \times E_{1,O_2})/(M_{CO_2} \times E_{1,CO_2}) = (9800 \times 0.1 \times 2215)/(96.2 \times 5172) = 4.4$ nmol.

TABLE 3. Corrected Total Gas Composition in the GEX Cell Headspace, VL-1, Heated Sandy Flats Sample

Gas	Corrected Total Gas Composition, nmol	
	Sol 233*	Sol 234†
N_2	0	42
O_2	0	250
CO_2	160	230
Ar	0	· 0
Kr	1400	1800

The contribution of gases to the test cell from the added test gas at the start of incubation on sol 233 was 940 nmol of CO_2 and 1800 nmol of Kr. The corrected total nanomoles were calculated as described in the Figure 1 legend.

*Analysis with dry soil.

†Analysis after humidification.

difference in adsorbed water vapor; i.e., the greater the difference between the predicted and the found values, the greater the degree of prior water binding. Thus we concluded that the under Notched Rock sample was the wettest, followed by a less wet Beta sample and finally by a very dry VL-1 Sandy Flats sample.

The O_2 values found were in substantial excess of the predicted desorption values. The evolution of O_2 on humidification of the Martian surface samples was clearly a chemical reaction involving one or more reactive species such as ozonides, superoxides, and peroxides. On the basis of our model of Martian morphology the most plausible of these was superoxide [*Oyama et al.*, 1977] (E. V. Ballou, unpublished data, 1977). Among the candidate superoxides likely to be present on Mars were potassium, calcium, and sodium superoxide, all of which were known to react very quickly in the presence of water vapor [*Cohen and Margrave*, 1957; *Vol'nov and Shatunina*, 1957].

RESULTS OF THE SOIL HEATING EXPERIMENTS

The first soil heating experiment had two purposes: (1) to demonstrate that superoxide could be identified by its thermal stability and (2) to establish the nonbiological origin of the O_2 evolution. A sample from the VL-1 site, where the soil was driest, was the most appropriate to test because any water released during the heating of wetter samples could attack the superoxides. It is known that potassium and calcium superoxides are thermally stable; KO_2 begins to decompose at 425°C [*Sokovnin*, 1963] and $Ca(O_2)_2$ at 290°C, liberating superoxide oxygen [*Vol'nov et al.*, 1956]. Thus if soil was relatively dry, heating the soil should release only a small amount of water vapor, which could be removed by the purging helium. Any O_2 released by later humidification would be a measure of the superoxide remaining. Oxygen was observed 2.5 hours after humidification (Table 3). The amount was about 48% of that given off with the original Sandy Flats sample for the same time interval after humidification, thus demonstrating that the original soil sample had been adequately dried prior to the addition of the second sample and verifying at the same time the requirement that a superoxide should survive the 145°C heating. The spontaneous evolution of O_2 on humidification of the heated soil samples ruled out a biological origin for this gas, since heating at 145°C for 3.5 hours suffices to destroy or reduce significantly the population of soil organisms.

The second soil heating experiment, using a VL-2 Beta sample, was performed like the first but for an additional reason. Based upon the known conversion of superoxides by water vapor to the peroxide hydrates and peroxyhydrates [*Makarov and Grigor'eva*, 1955*a*] and partial reversal of these to superoxides by heating [*Makarov and Grigor'eva*, 1955*a*, *b*], we tested the hypothesis that at the Utopia site most of the superoxide had been spent by the atmospheric water vapor, some of it having been converted to the peroxide hydrate and peroxyhydrate forms. After heating, therefore, a test with water vapor could demonstrate that superoxides were reformed by an increase in the O_2 concentration. The data in Table 4, however, show that virtually no O_2 was evolved after humidification.

We would have concluded from these data that no additional superoxide was generated by heating the soil if we had not observed the Martian atmospheric constituents, N_2 and Ar, in the test cell prior to humidification (Table 4). The presence of these gases could only mean that the drain valve was closed or that the drain line was partially blocked during the soil heating at 145°C. Having the exhaust partially or totally blocked would allow water, desorbed from the soil at 145°C, to remain in the test chamber and destroy any superoxide present. The O_2 resulting from the destruction of superoxides and peroxides could oxidize CO. If CO were present, O_2 would be lost and no conclusion about superoxides in this surface sample could be made.

The possibility that CO was generated during the soil heating is suggested by the high concentration of Ar with respect to N_2. If the 36 nmol of N_2 (Table 4) came solely from the Martian atmosphere (Table 5), then 23 nmol of Ar would have been expected from the atmosphere. Thus the 110 nmol of Ar actually measured provided an excess of ~87 nmol of CO, in Ar equivalents, presumably generated from the heat treatment of the soil. From these rationalizations the absence of O_2 before or after humidification cannot be used to deny the presence of superoxides in the sample.

The source of the CO could be the thermal degradation of dried organics from the nutrient medium which had wet the earlier surface samples or from other unidentified carbon sources in the Martian surface material. The excess amount of CO_2, ~1300 nmol (Table 4), observed in the test cell prior to soil humidification, could have resulted from the thermal degradation of organics as well.

TABLE 4. Corrected Total Gas Composition in the GEX Cell Headspace, VL-2, Heated Beta Sample

Gas	Corrected Total Gas Composition, nmol		
	Sol 179*	Sol 180†	Sol 184
N_2	36	36	47
O_2	0	7	0
CO_2	4700	2400	520
Ar	110	110	120
Kr	1400	1400	1400

The contributions of gases to the test cell from the added test gas and the Martian atmosphere (assuming that the atmospheric gases were not flushed from the cell) at the start of incubation on sol 179 were (in nanomoles) N_2, 62; Ar, 41; CO_2, 3400; and Kr, 1400. The corrected total nanomoles were calculated as described in the Figure 1 legend except that no pressure sensitive corrections were included.

*Analysis with dry soil.

†First analysis after humidification

TABLE 5. GEX Atmospheric Analyses

Gas	Composition, %			
	VL-1, 1 (Sol 63)	VL-1, 2 (Sol 94)	VL-2, 1 (Sol 7)	VL-2, 2 (Sol 8)
N_2	2.3	2.2	2.3	2.2
O_2	$\leq$0.15	$\leq$0.15	$\leq$0.15	$\leq$0.15
Ar	1.6	1.5	1.5	1.6
CO_2	96.1	96.3	96.2	96.2

ATMOSPHERIC ANALYSES

Actual estimates of the Martian atmospheric contributions to the headspace gases in the GEX test cell were made by sampling the atmosphere. The composition of the Martian atmosphere as measured by the GEX at the VL-1 and VL-2 sites is given in Table 5. The precision of the measurements for each gas was determined from the data bit ratios and was estimated to be $\pm$5% for CO_2, $\pm$12.5% for N_2, and $\pm$20% for Ar. The concentrations of the gases were within the ranges estimated for each of the atmospheric constituents by the Viking Molecular Analysis Team [*Owen and Biemann,* 1976]. Our data were also in close agreement with atmospheric values extrapolated to the surface [*McElroy et al.*, 1976]. The mixing ratios given by these authors were 2.4×10^{-2}, 1.5×10^{-2}, and 1.6×10^{-3} for N_2, Ar, and O_2, respectively.

DISCUSSION

Superoxide, Peroxide, and Hydrogen Peroxide

The spontaneous evolution of O_2 when the Martian surface samples were humidified and the decreasing capacity to produce O_2 relative to the increasing native water content of the samples from Chryse Sandy Flats, Utopia Beta, and under Notched Rock suggested that the superoxides were modified by the water vapor available at these sites.

At incubation temperatures in the test cell the reaction of alkaline earth or alkali metal superoxides with water is known to produce O_2 and alkaline earth or alkali metal hydroxides [*Cohen and Margrave,* 1957; *Vol'nov and Shatunina,* 1957]. On the other hand, at cold temperatures, water vapor acts on the superoxides to produce the alkaline earth and alkali metal peroxide hydrates [*Mel'nikov and Firsova,* 1961], and we could expect this to occur on a cold Mars. Now, upon wetting at test cell temperatures the alkaline earth and alkali metal peroxides react stoichiometrically with liquid water to produce hydrogen peroxide [*Gilles and Margrave,* 1956]. Thus the final nature of the oxidant determines how effective it will be as a source of O_2 or as an oxidant for the oxidation of organics and is offered as an explanation for the O_2 evolution of GEX and the rapid CO_2 evolution generated by the labeled release experiment (LR) [*Klein et al.,* 1976*b*].

The key to the explanation is γ Fe_2O_3. It has been shown [*Krause and Gawrychowa,* 1937; *Krause,* 1939; *Krause and Wisniewski,* 1953; *Krause,* 1958; *Krause and Binkowna,* 1959; *Krause and Lezuchowska,* 1959] that γ FeOOH, amorphous iron hydroxides, and Cu^{++} are catalysts for the oxidation by hydrogen peroxide of a variety of organics. We have determined (V.I. Oyama, unpublished data, 1977) that γ Fe_2O_3 can catalyze the oxidation of organics present in the LR medium by hydrogen peroxide to CO_2, simulating the sharply rising portion of the LR curves, and that γ Fe_2O_3 can also oxidize organics directly at rates closely simulating the slower log linear production of CO_2 demonstrated by both GEX and LR.

Minor Gaseous Phenomena

Since we did not observe any peaks at the retention times corresponding to H_2, NO, CH_4, and N_2O, we have assumed that these were not present to the extent measurable by the GC, namely, at partial pressures of about 0.55, 0.019, 0.037, and 0.066 mbar, respectively, in the headspace of the test chamber. The inability to measure hydrogen gas does not completely rule out the presence of an H_2-producing source such as metallic iron but does permit estimation of an upper limit for metallic iron in Martian samples. We had reported that the presence of metallic iron in lunar fines was measurable by H_2 evolution provided that O_2 had been consumed or eliminated [*Oyama et al.*, 1971, 1972]. On the basis of the detectable limit for H_2 the upper limit of metallic iron in the Chryse sample would be 0.0066%. This value differs from the value of 0.003% reported earlier [*Oyama et al.*, 1977] because we neglected to include O_2 in our computations.

The absence of methane contraindicates, but does not rule out entirely, the presence of carbides or other reducing systems. The absence of NO and N_2O reduces the likelihood that biological denitrification was responsible for the N_2 evolution, since these products are associated with mixed biological populations. Their absence does not rule out altogether the presence of nitrite in the Martian surface material that could generate N_2 by way of the Van Slyke reaction with the α amino acids in the nutrient medium [*Oyama et al.,* 1977]. However, the amount of N_2 evolved is fully accommodated by that amount allowable by desorption (Table 2).

The highly oxidizing nature of the Martian surface that includes superoxides, peroxides, and γ Fe_2O_3 implies the atmospheric formation of highly active species that are formed by solar wind or flare, UV, and cosmic radiation. The decomposition of the most abundant atmospheric species, CO_2, to active fragments must be of prime importance in the ultimate nature of the surface of Mars.

SUMMARY

The response of Martian surface soil samples to water vapor resulting in O_2 output is ascribed to superoxides in the Martian surface material. The desorption of most of the CO_2, Ar, and N_2 from the Martian surface samples is inversely related to previous water exposure of the soil surface. The reactions of superoxide and water vapor at low temperature in the Martian environment could account for the presence of peroxides that form hydrogen peroxide on contacting liquid water. The residual superoxide suggests a highly active atmosphere that is likely to give rise to a whole spectrum of carbon-oxygen combinations. The slow log linear production of CO_2 is ascribed to the direct oxidation of organics by γ Fe_2O_3, which has been demonstrated to be responsible for the catalytic oxidation of organics by H_2O_2. All the gas changes observed can most easily be explained or demonstrated by plausible chemical reactions that require no biological processes.

Acknowledgments. The authors thank Harold P. Klein and Melvin P. Silverman for helpful suggestions for the text; Glenn C. Carle for his participation in the early flight team work; Fritz Woeller and Marjorie Lehwalt for their assistance on the γ Fe_2O_3 studies; Mary Ann Rowland in Test Standard Module Work; Fred Brown, Donald L. DeVincenzi, and John Billingham for encouragement; and the management and participants of NASA Langley, Martin Marietta, and especially TRW Systems, who made it all happen.

REFERENCES

Brunauer, S., P. H. Emmett, and E. Teller, Adsorption of gases in multimolecular layers, *J. Amer. Chem. Soc., 60,* 309–319, 1938.

Cohen, S. H., and J. L. Margrave, Reaction rates in the analytical determination of some inorganic peroxides and superoxides, *Anal. Chem.*, *29*, 1462–1463, 1957.

Gilles, P. W., and J. L. Margrave, The heats of formation of Na_2O_2, NaO_2, and KO_2, *J. Phys. Chem.*, *60*, 1333–1334, 1956.

Klein, H. P., J. Lederberg, A. Rich, N. H. Horowitz, V. I. Oyama, and G. V. Levin, The Viking mission search for life on Mars, *Nature*, *262*, 24–27, 1976*a*.

Klein, H. P., N. H. Horowitz, G. V. Levin, V. I. Oyama, J. Lederberg, A. Rich, J. S. Hubbard, G. L. Hobby, P. A. Stratt, B. J. Berdahl, G. C. Carle, F. S. Brown, and R. D. Johnson, The Viking biological investigation: Preliminary results, *Science*, *194*, 99–105, 1976*b*.

Krause, A., Active iron and copper compounds as inorganic catalysts in reactions of peroxidative oxidation, in the light of the theories of oxygen and hydrogen activation, *Rocz. Chem.*, *19*, 481–486, 1939.

Krause, A., Chemistry of the ferromagnetic iron oxides, *Przem. Chem.*, *37*, 637–639, 1958.

Krause, A., and A. Binkowna, The catalytic properties of natural magnetite, *Rocz. Chem.*, *33*, 819–821, 1959.

Krause, A., and M. Gawrychowa, Amorphous and crystallized oxide hydrates and oxides, XXXI, 'Peroxidase' properties of amorphous ferric hydroxide, The catalytic oxidation of formic acid by hydrogen peroxide, *Chem. Ber.*, *70B*, 439–443, 1937.

Krause, A., and J. Lezuchowska, The catalytic decomposition of hydrogen peroxide and the peroxidatic oxidation of formic acid on amorphous iron (III) hydroxide as a function of the particle size of the latter, Note on the mechanism of these reactions, *Z. Anorg. Allg. Chem.*, *301*, 294–300, 1959.

Krause, A., and J. Wisniewski, Activation of γFeOOH by amphoteric metal hydroxides and abnormal behavior of some complex catalysts of this type, *Rocz. Chem.*, *27*, 232–241, 1953.

Makarov, S. Z., and N. K. Grigor'eva, System with concentrated H_2O_2, VII, Isotherm 0° of the system $NaOH$-H_2O-H_2O_2, *Izv. Akad. Nauk SSSR Otd. Khim. Nauk*, 17–20, 1955*a*.

Makarov, S. Z., and N. K. Grigor'eva, Systems with concentrated hydrogen peroxide, VIII, Thermal characteristics and stability of the solid phases for the system $NaOH$-H_2O_2-H_2O, *Izv. Akad. Nauk SSSR Otd. Khim. Nauk*, 203–215, 1955*b*.

McElroy, M. B., T. Y. Kong, Y. L. Yung, and A. O. Nier, Composition and structure of the Martian upper atmosphere: Analysis of results from Viking, *Science*, *194*, 1295–1298, 1976.

Mel'nikov, A. K., and T. P. Firsova, Reaction of sodium superoxide with water vapor at low temperature, *Zh. Neorg. Khim.*, *6*, 169–176, 1961.

Merek, E. L., and V. I. Oyama, Integration of experiments for the detection of biological activity in extraterrestrial exploration, in *Life Sciences and Space Research*, vol. VIII, edited by W. Vishniac, pp. 108–115, North-Holland, Amsterdam, 1970.

Owen, T., and K. Biemann, Composition of the atmosphere at the surface of Mars: Detection of argon-36 and preliminary analysis, *Science*, *193*, 801–803, 1976.

Oyama, V. I., The gas exchange experiment for life detection: The Viking Mars lander, *Icarus*, *16*, 167–184, 1972.

Oyama, V. I., E. L. Merek, M. P. Silverman, and C. W. Boylen, Search for viable organisms in lunar samples: Further biological studies in Apollo 11 core, Apollo 12 bulk and Apollo 12 core samples, in *Proceedings of the Second Lunar Science Conference*, vol. 2, edited by A. A. Levinson, pp. 1931–1937, MIT Press, Cambridge, Mass., 1971.

Oyama, V. I., B. J. Berdahl, C. W. Boylen, and E. L. Merek, Search for viable organisms in lunar samples: Gas changes over Apollo 14 fines wet by aqueous media, paper presented at the Third Lunar Science Conference, Lunar Sci. Inst., Houston, Tex., Jan. 10, 1972.

Oyama, V. I., B. J. Berdahl, G. C. Carle, M. E. Lehwalt, and H. S. Ginoza, The search for life on Mars: Viking 1976, Gas changes as indicators of biological activity, *Origins Life*, *7*, 313–333, 1976.

Oyama, V. I., B. J. Berdahl, and G. C. Carle, Preliminary findings of the Viking gas exchange experiments and a model for Martian surface chemistry, *Nature*, *265*, 110–114, 1977.

Sokovnin, E. I., Thermal stability of potassium ozonide, *Izv. Akad. Nauk SSSR Otd. Khim. Nauk*, 181–182, 1963.

Steinman, H. G., On the mechanism of the ascorbic acid–ascorbic acid oxidase reaction, The hydrogen peroxide question, *J. Amer. Chem. Soc.*, *64*, 1212–1219, 1942.

Vol'nov, I. I., and A. N. Shatunina, Reaction capacity of the superoxides of the alkaline earth metals toward water and carbon dioxide, *Zh. Neorg. Khim.*, *2*, 1474–1478, 1957.

Vol'nov, I. I., V. N. Chamova, V. P. Sergeeva, and E. I. Latysheva, Synthesis of the superoxides of alkaline earth metals, I, The reaction of $CaO_2 \cdot 8H_2O$ with perhydrol at 100°, *Zh. Neorg. Khim.*, *1*, 1937–1942, 1956.

(Received April 4, 1977;
revised June 8, 1977;
accepted June 8, 1977.)

The Viking Biological Investigation: General Aspects

H. P. KLEIN

NASA Ames Research Center, Moffett Field, California 94035

The Viking biological investigation has tested four different hypotheses regarding the possible nature of Martian organisms. While significant results were obtained for each of these, tests of three of the hypotheses appear to indicate the absence of biology in the samples used, while the fourth is consistent with a biological interpretation. The original assumptions for each experiment and the experimental procedures that were utilized to test these assumptions are reviewed.

INTRODUCTION

Before the two Vikings reached Mars, speculation about the prospects for life on that planet ranged from extreme pessimism to optimism. At that time, when the biology experiments were actually selected for final inclusion on Viking, our information about Mars was far from complete. Under these circumstances it is not surprising that different ideas emerged concerning what Martian organisms might be like and what procedures or techniques would best elicit evidence of their metabolism. As a result of these diverse ideas it was decided to incorporate several different biological experiments into the Viking payload in order to test a number of different, and sometimes conflicting, assumptions about the characteristics of Martian organisms [*Young,* 1976]. Indeed, had it been possible to fly additional biological experiments and thereby extend the number of different ideas and techniques that could be tested, this would have been a valuable thing to do. However, the biological portion of the payload was constrained to approximately 0.03 m^3 of volume and 15 kg of weight [*Klein,* 1976].

By supplying four different sets of environmental conditions within which to conduct incubations (Table 1), the three experiments that were finally selected for the Viking biology instrument actually tested four different hypotheses. One of these assumed that active Martian metabolism was limited by the availability of water. Another assumed that biological activity would best be seen under conditions approximating those on Mars. The remaining two tested for heterotrophic metabolism, one by using a very dilute aqueous solution of simple organic compounds, the other by utilizing a concentrated mixture of many organic compounds. All of these have now been tested, and the results from these are the subjects of the ensuing papers.

As of this writing the two Viking landers have been operating on the surface of Mars for 8½ and 7 months, respectively. All during this time the Viking biology instrument has performed exceptionally well, with few instances of instrument anomalies, and those that have occurred are sufficiently well understood so that they do not interfere with the interpretations of the data. Accordingly, for the papers that follow, one can safely assume that all of the information is based on statistically significant data with strong 'signal-to-noise' ratios and that we are not engaged here in describing or explaining artifacts produced by the instrumentation.

Initially, in the so-called 'nominal' mission we expected to perform a total of 13 or 14 separate experiments on the two spacecraft. With no a priori knowledge of the local Martian surface environments our overall initial strategy was essentially based on the concept that there might be local heterogeneity in the surface of Mars in the vicinity of the spacecraft. We planned therefore to perform the biological experiments in a 'survey' mode, testing sample after sample by using the same experimental sequences until any one of the experiments yielded a 'presumptive' positive result. (For a discussion of the criteria that were to be used to arrive at this judgment, see *Hubbard* [1976], *Levin and Straat* [1976*a*], and *Oyama et al.* [1976].) Once such a presumptive positive result was observed, our strategy called for repeating that particular experiment by using a heat-'sterilized' sample. This procedure, of using heated 'controls' to confirm biological processes, is deeply rooted in biological tradition. More important, it was based on hundreds of ground-based tests of the Viking concepts using terrestrial soil samples (from desert soils to rich garden soils), lunar samples, and pure cultures of microorganisms, the results of which enabled us accurately to determine the presence or absence of living organisms.

In point of fact, our intended nominal strategy was quickly discarded as the very first sample was analyzed. Two of the three experiments yielded presumptive positive results, while the third produced evidence for oxidizing surface material at that site [*Klein et al.,* 1976]. This combination of findings, together with the demonstrated lack of organic compounds [*Biemann et al.,* 1976], required substantial changes to the original experimental strategies and resulted in major departures from experiment sequences that we had anticipated using in these experiments. In all cases every element of flexibility inherent in the instruments was called into play in an effort to use the equipment both as biological and as chemical laboratories in order to discriminate between the two mechanisms that might be responsible for the presumptive positive results.

By now, considerable progress has been made in resolving the issues raised by the first set of analyses. To achieve this, we have exceeded the number of experiments initially planned, having carried out 26 experiments. A few additional experiments still remain to be performed, but on the basis of all the information now available, it is likely that the results of one of the two experiments that yielded presumptive positives, the Pyrolytic Release experiment [*Horowitz et al.,* 1976], are nonbiological in origin. The Labeled Release experiment, which also has consistently yielded presumptive positive biological results [*Levin and Straat,* 1976*b*], remains ambiguous.

In arriving at any final judgment on the fundamental question, Is there life on Mars?, we must carefully consider not only the actual experimental results that have been obtained but also the context within which these data were obtained. We must examine the assumptions on which each of the experimental techniques was based, the conditions under which the

Paper number 7S0439.

TABLE 1. Environmental Parameters in the Viking Biology Investigation

Experiment	Nutrients Added	Water Added	Illumination
Pyrolytic Release	none*	none	light and dark
	none	trace§	light and dark
Gas Exchange	none	moist‖	dark
Labeled Release	dilute solution of simple organic compounds†	moist¶	dark
Gas Exchange	concentrated solution of organic and inorganic compounds‡	wet**	dark

*A mixture of CO_2 and CO was introduced into the incubation chamber.

†See *Levin and Straat* [1976*b*].

‡See *Oyama* [1972].

§Approximately 80 μg of H_2O was injected into an incubation chamber (2.6-cc volume) containing a 0.25-cc sample.

‖Approximately 0.5 cc of nutrient solution was added below the 1-cc sample in the chamber (8.7-cc volume).

¶Approximately 0.115 cc of nutrient solution was added to the 0.5-cc sample in the chamber (3.25-cc volume).

**Approximately 2.5 cc of nutrient solution was added to the 1-cc sample in the chamber (8.7-cc volume).

experiments were actually carried out, and the data themselves.

THE GAS EXCHANGE EXPERIMENT (HUMID NONNUTRIENT MODE)

As is indicated in Table 1, the Gas Exchange (GEX) experiment tested two different concepts of Martian biology. In one mode the fundamental assumption was that the sole limiting factor to growth of Martian organisms is water. Here it was assumed that nutrients, perhaps in the form of simple organic compounds formed photochemically as is described by *Hubbard et al.* [1971], already are present in the Martian surface and that organisms would be dormant in the dry Martian environment until enough moisture became available to stimulate the dormant organisms into metabolic activity, which was to be measured by analyzing the atmosphere above the incubating system with a gas chromatograph system.

Now let us examine some of the experimental conditions under which this assumption was actually tested. First, the Martian samples were incubated in the presence of Martian atmosphere to which additional carbon dioxide, krypton, and helium were added in order to bring the total pressure to approximately 200 mbar to facilitate subsequent gas sampling [*Edelson et al.*, 1975]. After introduction of approximately 0.5 cc of nutrient solution into the incubation cell, under conditions in which the nutrient does not come into contact with the samples, the atmosphere rapidly becomes saturated with water at the incubation temperatures of 8–15°C. After an incubation period for this phase of the experiment of approximately 7 days the experiment was terminated.

In this experiment, which was performed twice, once at each of the landing sites, the findings for both were essentially the same. While physical (desorption of some gases) and chemical (generation of oxygen) phenomena were noted, there was nothing in the data to suggest the presence of metabolic activity on the basis of the criteria that had been developed for this experiment, and therefore the results of this experiment can be said to be negative with regard to biology.

In terms of interpreting the results obtained, if we regard both the assumptions and the experimental conditions to be valid, we must conclude that the samples that were assayed did not contain metabolizing organisms. However, the original assumption may be incorrect in that some source of energy may be a requirement to stimulate metabolic activity of organisms on Mars. In this case a negative result would not preclude the existence of 'life' in the samples tested. Alternatively, it is possible that one or more of the experimental conditions employed during these tests prevented the accumulation of biological signals. For example, in this experiment, as in all of the biological experiments, incubation temperatures ran some tens of degrees warmer than ambient surface temperatures at the two sites [*Kieffer*, 1976]. Another issue is whether a total incubation period of 7 days was sufficient to demonstrate metabolic activity in view of tests with Antarctic soils that required months of incubation to produce presumptive positive results [*Oyama et al.*, 1976]. Other potential sources of possible inhibition of metabolism include the high pressure and alteration of the incubation atmosphere, as indicated above.

THE GAS EXCHANGE EXPERIMENT (WET NUTRIENT MODE)

In running the GEX experiment, in the presence of added organic compounds and inorganic salts the fundamental assumption made was that a significant fraction of the Martian 'biota' is composed of heterotrophic organisms [*Oyama et al.*, 1976]. Therefore the addition of organic compounds was deemed necessary to elicit metabolic response. Furthermore, it was assumed that this response would be expressed only in an aqueous environment, and finally, the presence of a large number of different organic and inorganic compounds [*Oyama*, 1972] was assumed not to be inhibitory to the expression of this metabolism.

This experiment was performed three times, for periods of 200 (Viking 1), 31 (Viking 2), and 116 (Viking 2) sols. However, Martian atmosphere was present for only a portion of these incubation periods, 13, 19, and 78 sols, respectively. For the remainder of the time the atmosphere consisted of carbon dioxide, krypton, and helium. Once again the total atmospheric pressure was approximately 200 mbar, and incubation temperatures were in the 8–15°C range.

While some gas changes were noted in the three trials of this experiment [*Oyama*, 1977], none of these fit the criteria for biological activity. Consequently, on the basis of the original assumptions for this experiment and provided that the conditions under which they were tested were adequate, we can conclude that no viable organisms were present in the samples. On the other hand, a negative finding for this experiment does not rule out the possible presence of autotrophs (i.e., chemosynthetic organisms) in the samples. In addition, the corollary assumptions concerning the nutrient mixture used, as well as the high water activity under which these tests were conducted, must also be assessed in arriving at the biological significance of the data from this experiment. Finally, even if the original assumptions are correct, some of the experimental conditions (temperature, pressure, and 'artifical' atmosphere) may have precluded positive biological findings in this experiment.

THE PYROLYTIC RELEASE EXPERIMENT

The fundamental assumption for this experiment stems from considerations of the characteristics of the planet Mars.

Since both carbon dioxide and carbon monoxide were known to be present in the atmosphere of Mars, it was assumed that organisms in the local ecology of that planet would have developed the capacity to assimilate one or both of these gases [*Horowitz et al.,* 1972] and convert these to organic matter. A basic tenet of this experiment is that metabolic activity would best be demonstrated under conditions approximating the ambient conditions on Mars as closely as possible.

The actual Pyrolytic Release (PR) experiments were conducted under conditions which, in many aspects, did approximate those on Mars, but in some they did not. Incubations were carried out either in the light or in the dark for 5-day periods. In those cases in which illumination was used, wavelengths below about 320 nm were filtered out in order to avoid false positives [*Hubbard,* 1976]. The incubation temperatures again were in the 10–18°C range.

As has been reported [*Horowitz et al.,* 1976], initial results of the PR experiment, as well as later attempts to repeat the original conditions, resulted in weak but significant presumptive positives. However, later experiments designed to elucidate the mechanisms yielding these results appear to rule out a biological explanation for these results.

As with the GEX experiments, it is possible that in the case of the PR experiment the basic assumptions are not correct. The photochemical synthesis of simple organic compounds has not been ruled out on the surface of Mars. If in the steady state this process were to supply organic matter to the surface, the need for autotrophic fixation in the Martian ecology may be obviated, and only heterotrophs may be present. In this case a negative result in this experiment would not preclude the existence of life in the samples tested.

In examining the experimental conditions under which the PR tests were conducted, the duration of the incubation periods, the incubation temperatures, and the fact that short-wavelength solar radiation was unavailable to the surface samples all raise questions regarding the adequacy of these experiments in providing the requisite conditions for measuring this type of metabolism even if Mars should have indigenous autotrophic organisms.

THE LABELED RELEASE EXPERIMENT

This experiment [*Levin,* 1972] makes the assumption that heterotrophic organisms are present on Mars and that these organisms would be capable of decomposing one or more simple organic compounds of the type reported to be produced from the so-called 'primitive reducing atmosphere' in laboratory simulations [*Miller and Urey,* 1959] and including some that have been found in carbonaceous chondrites [*Kvenvolden et al.,* 1970]. For this experiment, even more than for the others, the inclusion of a heated control was deemed absolutely necessary in order to avoid the possibility of obtaining false positive results. The conditions under which samples that were not heat sterilized were incubated, now four times, for periods of about 13, 52, and 90 days, were Marslike with the exception that incubation was carried out at around 10°C; and of course, there was the addition of a small volume of water containing the dilute solution of organic substrates [*Levin,* 1972]. As with the GEX experiment, the incubation cell was somewhat pressurized (to approximately 60 mbar) during incubation. Heat sterilization was carried out on three samples, first at approximately 160°C, and later at 50°C and 44°C.

What has been observed in all of the analyses performed on nonsterilized samples is an active initial decomposition of the nutrient, the reaction leveling off, in each case, at a time when about 95% of the added radioactivity still remained in the surface material. Prior heating of the Martian samples had substantial effects on this process, 160°C for 3 hours completely abolishing the reaction [*Levin and Straat,* 1976*b*].

TABLE 2. Comparison of Data From the LR and GEX Experiments

Sample	Oxygen Released (GEX)*	Carbon Dioxide Produced (LR)*
Viking 1 (surface)	770	~30
Viking 2 (surface)	194	~30
Viking 2 (subrock)	70	~30

*Nanomoles per 1-cc sample.

On the basis of all of the experiments performed to date, the Labeled Release (LR) experiment, unlike the other biological experiments, yielded data which met the criteria originally developed for a positive. On this basis alone the conclusion would have to be drawn that metabolizing organisms were indeed present in all samples tested. Can we believe such a conclusion? Clearly, we must be wary of this in the face of information indicating that all of the samples tested yielded oxygen in the GEX experiment upon introduction of water. The evidence for strongly oxidizing chemicals in these samples is quite convincing. However, it should also be recognized that the two phenomena are not directly related (Table 2). From the relatively constant amount of decomposition that took place in all cases in the LR experiment, it would appear that all of the Mars samples contained excess oxidants. Thus it is reasonable to assume either that the factor limiting the LR reaction in each case was the depletion of some constituent in the mixture of substrates supplied in the nutrient or that the samples analyzed in these experiments contained at least two kinds of oxidants.

While a nonbiological (i.e., a chemical 'oxidant') theory may well explain the LR data, it does not seem likely that the ambiguity in interpreting this experiment will be resolved on Mars by the remaining Viking experiments.

SUMMARY

We have tested a number of different concepts about the nature of hypothetical Martian organisms over the course of the past few months. We have seen significant and quite reproducible results in each of our experiments. For each experiment, except for the LR experiment, we must conclude that there were no organisms present within the limits of detectability for these experiments and that all of the observed reactions for these were the result of nonbiological phenomena. Otherwise, we must question the fundamental assumptions made for these experiments and the conditions under which they were actually carried out. In the case of the LR experiment, from the very beginning of operations on Mars we recognized the possibility that the striking changes seen during incubation could be the result of nonbiological processes, but our attempts to discriminate between biological and nonbiological mechanisms by manipulating sterilization temperatures and the length of incubation have not resolved the issue.

Finally, we must not overlook the fact, in assessing the probabilities of life on Mars, that all of our experiments were conducted under conditions that deviated to varying extents from ambient Martian conditions, and while we have accumulated data, these and their underlying mechanisms may all be coincidental and not directly relevant to the issue of life on that planet.

REFERENCES

Biemann, K., J. Oro, P. Toulmin III, L. E. Orgel, A. O. Nier, D. M. Anderson, and P. G. Simmonds, Search for organic and volatile inorganic compounds in two surface samples from the Chryse Planitia region of Mars, *Science, 194,* 72–76, 1976.

Edelson, H. E., F. S. Brown, O. W. Clausen, A. J. Cole, J. T. Cragin, R. J. Day, C. H. Debenham, R. E. Fortney, R. I. Gilje, D. W. Harvey, F. A. Jackson, J. A. Katherler, J. L. Kropp, S. J. Loer, J. L. Logan, Jr., O. D. Minnick, E. M. Noneman, W. D. Potter, G. T. Rosiak, and J. S. Shapiro, The Viking lander biology instrument, *Rep. 21020-6003-RU-00,* TRW Syst. Group, Redondo Beach, Calif., 1975.

Horowitz, N. H., J. S. Hubbard, and G. L. Hobby, The carbon-assimilation experiment: The Viking Mars lander, *Icarus, 16,* 147–152, 1972.

Horowitz, N. H., G. L. Hobby, and J. S. Hubbard, The Viking carbon-assimilation experiments: Interim report, *Science, 194,* 1321–1322, 1976.

Hubbard, J. S., The pyrolytic release experiment: Measurement of carbon-assimilation, *Origins of Life, 7,* 281–292, 1976.

Hubbard, J. S., J. P. Hardy, and N. H. Horowitz, Photocatalytic production of organic compounds from CO and H_2O in a simulated Martian atmosphere, *Proc. Nat. Acad. Sci. U.S., 68,* 574–578, 1971.

Kieffer, H., Soil and surface temperatures at the Viking landing sites, *Science, 194,* 1344–1346, 1976.

Klein, H. P., General constraints on the Viking biology investigation, *Origins of Life, 7,* 273–279, 1976.

Klein, H. P., N. H. Horowitz, G. V. Levin, V. I. Oyama, J. Lederberg, A. Rich, J. S. Hubbard, G. L. Hobby, P. A. Straat, B. J. Berdahl, G. C. Carle, F. S. Brown, and R. D. Johnson, The Viking biological investigation: Preliminary results, *Science, 194,* 99–105, 1976.

Kvenvolden, K., J. Lawless, K. Pering, E. Peterson, J. Flores, C. A. Ponnamperuma, I. R. Kaplan, and C. Moore, Evidence for extraterrestrial amino-acids and hydrocarbons in the Murchison meteorite, *Nature, 228,* 923–926, 1970.

Levin, G. V., Detection of metabolically produced labeled gas: The Viking Mars lander, *Icarus, 16,* 153–166, 1972.

Levin, G. V., and P. A. Straat, Labeled release—An experiment in radiorespirometry, *Origins of Life, 7,* 293–311, 1976*a*.

Levin, G. V., and P. A. Straat, Viking labeled release biology experiment: Interim results, *Science, 194,* 1322–1329, 1976*b*.

Miller, S. L., and H. C. Urey, Organic compound synthesis on the primitive earth, *Science, 130,* 245–251, 1959.

Oyama, V. I., The gas exchange experiment for life detection: The Viking Mars lander, *Icarus, 16,* 167–184, 1972.

Oyama, V. I., Preliminary findings of the Viking gas exchange experiment in a model for Martian surface chemistry, *Nature, 265*(5590), 110–114, 1977.

Oyama, V. I., B. J. Berdahl, G. C. Carle, M. E. Lehwalt, and H. S. Ginoza, The search for life on Mars: Viking 1976 gas changes as indicators of biological activity, *Origins of Life, 7,* 313–333, 1976.

Young, R. S., The origin and evolution of the Viking mission to Mars, *Origins of Life, 7,* 271–272, 1976.

(Received April 1, 1977;
revised May 20, 1977;
accepted May 20, 1977.)